U0943905

国家科学技术学术著作出版基金资助出版

中国木本植物种子

SEEDS OF WOODY PLANTS IN CHINA

国家林业局国有林场和林木种苗工作总站　主编

中　国　林　业　出　版　社

图书在版编目（CIP）数据

中国木本植物种子/国家林业局国有林场和林木种苗工作总站主编．
—北京：中国林业出版社，2000.8
ISBN 7-5038-2576-6

Ⅰ．中…　Ⅱ．国…　Ⅲ．木本植物-种子-中国　Ⅳ．S722

中国版本图书馆 CIP 数据核字（2000）第 28134 号

出版　中国林业出版社（100009　北京市西城区刘海胡同 7 号）
http：/www. naturalbook. com
E-mail：cfphz@public. bta. net. cn
Tel：66184477

发行　新华书店北京发行所
印刷　中国科学院印刷厂
版次　2001 年 2 月第 1 版
印次　2003 年 3 月第 2 次
开本　787mm×1092mm　1/16
印张　72.5　彩页 8 面
字数　1760 千字
印数　2001～5000 册

定价　180.00 元

《中国木本植物种子》编辑委员会

撰稿供稿人

（共76人，按编委会随机编号顺序排列；正文各个表1“供稿”栏下使用的便是这套编号）

编号	姓　名	通信地址及其邮政编码
101	周陛勋	
102	陈运大	515040 广东省汕头市园林管理处
103	罗丽芬	150040 哈尔滨市东北林业大学种苗教研室
104	邹　铨	150040 哈尔滨市东北林业大学种苗教研室
105	王彩荣	150040 哈尔滨市黑龙江省林业厅林木种子站
106	林淑云	150040 哈尔滨市黑龙江省林业厅林木种子站
107	周德本	150040 哈尔滨市黑龙江省科学院自然资源研究所植物室
108	穆福忠	150080 哈尔滨市东北林业大学哈尔滨试验林场
109	王向宏	150040 哈尔滨市黑龙江省科学院自然资源研究所
201	李荣晨	010019 呼和浩特市内蒙古林学院林学系

202 赵美华 010019 呼和浩特市内蒙古林学院教务处
203 周君英 830052 乌鲁木齐市新疆农业大学林学院
204 陈开秀 830052 乌鲁木齐市新疆农业大学林学院
205 郭志中 730020 兰州市甘肃省林业科学研究所
301 陶章安 100091 北京市中国林业科学研究院林业研究所种子室
302 钱耀明 100091 北京市中国林业科学研究院林业研究所种子室
303 于淑兰 100091 北京市中国林业科学研究院林业研究所种子室
305 王木林 100091 北京市中国林业科学研究院森林生态环境与保护研究所
306 李建文 100091 北京市中国林业科学研究院林业研究所植物室
307 傅紫芰 100091 北京市中国林业科学研究院林业研究所育种室
308 高秀岩 100093 北京市农林科学院林业果树研究所
309 宋朝枢 100091 北京市中国林业科学研究院森林生态环境与保护研究所
310 张毅萍 100091 北京市中国林业科学研究院林业研究所经济林室
311 邓明全 100091 北京市中国林业科学研究院林业研究所经济林室
401 宋廷茂 100083 北京市北京林业大学造林教研室
402 印佩文 100083 北京市北京林业大学造林教研室
403 刘长江 100093 北京市香山中国科学院植物研究所北京植物园
404 郭春景 150040 哈尔滨市黑龙江省科学院自然资源研究所
501 赵德铭 610082 成都市四川省林木种苗站
502 阙再凫 625014 四川省雅安市四川农业大学林艺学院
504 郑行生 650215 昆明市云南省林业厅种苗站
505 向性明 610082 成都市四川省林业科学研究院生态室
506 郭玉明 556000 贵州省黔东南自治州林业科学研究所
507 向国中 634000 重庆市万州区林业局种子站
508 杨绍安 614003 四川省乐山市乌龙坝管理处
509 罗祖筠 550001 贵阳市贵州省林业厅
510 陆宝珍 556000 贵州省黔东南苗族侗族自治州林业科学研究所
511 王道植 550025 贵阳市贵州农学院林学系
512 徐玉蓉 550025 贵阳市贵州农学院林学系
601 王宏志 530001 南宁市广西壮族自治区林业科学研究院
602 韦增健 530001 南宁市广西壮族自治区林业科学研究院
603 张声燕 530001 南宁市广西壮族自治区林业科学研究院

中心

1006　高捍东　210037 南京市南京林业大学国家林业局南方林木种子检验中心

1007　沈永宝　210037 南京市南京林业大学国家林业局南方林木种子检验中心

1008　陈智建　210037 南京市南京林业大学国家林业局南方林木种子检验中心

绘图人员

(共 19 人，按汉语拼音顺序排列)

姓　名	通信地址及其邮政编码	
曹雅范	150040	哈尔滨市黑龙江省科学院自然资源研究所
陈荣道	210014	南京市江苏省中国科学院植物研究所
黄光郁	528306	广东省顺德市桂川镇容兴燃具公司
黄应钦	530001	南宁市广西壮族自治区林业科学研究院
胡冬梅	100083	北京市北京林业大学 116 号信箱
梁　鸣	150040	哈尔滨市黑龙江省自然资源研究所
林　平	100083	北京市北京林业大学 116 号信箱
马　平	100020	呼和浩特市内蒙古大学
马信祥	650223	昆明市中国科学院西双版纳热带植物园昆明分部
孟　玲	100091	北京市中国林业科学研究院森林生态环境与保护研究所植物标本室
史渭清	210014	南京市江苏省中国科学院植物研究所
田恒德	210037	南京市南京林业大学森林昆虫教研组
童军平	310013	杭州市杭州植物园
吴新安	830001	乌鲁木齐市北京北路 11 号蝶王公司
许芝源	150040	哈尔滨市东北林业大学植物教研室
杨向晨	610081	成都市四川省林业勘察设计院
姚桂英	316000	浙江省舟山市浙江海洋学院校报编辑部
叶可佳	625014	四川省雅安市四川农学院
张世经	210037	南京市南京林业大学树木学教研组

责任编辑　李德林

英文翻译　周谟智

封面设计　黄华强

序

国家林业局国有林场和林木种苗工作总站主持编纂的大型图书《中国木本植物种子》，历时十余年，已告完成。它的问世，值得庆贺。

纵览全书，有诸多特点值得称道。

特点之一是鸿篇巨制，前所未有。全书共收木本植物492属1 222种以及54个种以下分类等级，已占我国木本植物总数的1/7。我国含木本种类的植物有180余科，收入本书的达132科，占2/3。尤其是裸子植物部分，除了我国不产的百岁兰科以外，国产的裸子植物11个科和引进的2个科，已全数囊括，收集得极为完整。以木本植物种子为内容的专著，篇幅如此之大，树种如此之多，前所未见。

特点之二是撰写队伍，精干强大。本书主编单位遴选的76位专家学者遍布全国，长期从事木本植物种子的生产、科研或教学，熟悉所在地区树木种子的习性，许多树种本来就是他们研究关注的对象，掌握了不少第一手资料，写来自然驾轻就熟。对于接触不多的树种，他们更是兢兢业业，严格要求，务必亲自观察、采集、描述、研究而后成文。由他们执笔撰稿，不仅能总结前人，尤其是总结20世纪50年代以来林木种子生产和科研的巨大成就，更融进了他们自己潜心研究的大量成果。书中许多材料都是首次发表，不少材料填补了文献空白。说本书反映了20世纪后期我国木本植物种子生产和科研的最新水平，当不为过。

特点之三是图文并茂，相得益彰。种子的外形和结构，种子的萌发和幼苗的初期形态，是树种早期鉴认的重要依据，可惜有些分类学的专著往往只注重成年树体的枝叶花果，较少涉及种子，很难满足种苗生产和科研的需要。本书配有插图900多幅，多数是对实物的精心描绘，准确细致，不仅可供读者看图识物，区别异同，具有重要的实用价值，也使本书的文字更为简明精炼。

特点之四是言必有据，“述而不作”。本书许多内容来自供稿人和撰稿人自己的观察研究，真实可靠。引用他人文献时本书又能做到忠实客观，不误导读者；书末并附所引文献的目录，供有兴趣的读者追踪查阅，体现出严肃认真、向读者负责的精神。

特点之五是精心统编，反复锤炼。全书篇幅如此之大，撰稿人数如此之多，但能具有基本统一的编写体例，具有相当一致的行文风格，能够在原稿的基础上精益求精，校正一些难以避免的差错，统稿诸君严谨细腻，认真执着，功不可没。

写书，也常常难免遗憾：稿成之后发现疏漏而又无法弥补。本书也有这种情况。例如，引用的文献数量尚嫌太少；有些比较重要的被子植物值得写而未写，比如粘木科、金虎尾科、茶茱萸科和清风藤科等科中的一些属和种，又如大戟科中的五月茶、肥牛树、血桐等属种。这当然是从严要求之言。本书的编纂诸公因此殷切地期望祖国统一，两岸学者能够合作修编；期望新世纪的木本植物种子科技工作者能够薪火相传，把这本书一代一代地补充修订下去。词意恳切，人们听到的是一代学人对事业的拳拳之心。

展望新的世纪，林业在生态环境建设中的主体作用必将更加突出。在这新旧世纪交替之时，行业主管部门主持编写这样一部巨著，把它看成是种苗事业中一项重要的基本建设，极具战略眼光。为本书供稿撰稿和统稿的专家学者，多数人年已古稀，仍然以事业为重，齐心协力，发挥余热，为国家的林业建设作出了极有意义的贡献。披览全书，至感欣慰。是为序。

国家林业局副局长 李育材

2000年3月12日

Preface

Seeds of Woody Plants in China, a major reference work compiled under the auspices of the National Service Center for State-Owned Forest Farms and Forest Seed and Seedling Affairs of the Forestry Ministry, has, after more than ten years of arduous work, been finally completed. The publication of the book is an event deserving of hearty congratulations.

This book has a number of characteristic features worthy of special commendation. First and foremost is its unique size, which is of an enormousness never before attempted. The book concerns itself with 492 genera, 1, 222 species and 54 taxa under species of woody plants, amounting to 1/7 of the country's total. Of the more than 180 families of plants found in China that contain woody species, 132, or 2/3 of the total, have been treated of in the book. Even more outstanding is the part dealing with *Gymnospermae*. With the exception of *Welwitschiaceae*, which is not found in China, the 11 families of indigenous *Gymnospermae* and two families introduced from abroad have all been given a place in the book, which points to its very high degree of exhaustiveness. That a book discussing seeds of woody plants should cover so much ground, including in it such a large number of tree species is indeed unparalleled.

A second feature characterizing the book is the very great capability of its writers and compilers. The 76 scholars and specialists chosen for the task come from all over the country, from among those who have for long periods of time been engaged in the production, research or teaching of woody plant seeds. For this reason, they are thoroughly familiar with the habits and characteristics of woody plant seeds found in their area, many species of trees being in fact their objects of research. It is only natural that, possessed as they are of first-hand knowledge, they should be able to go about the writing and compilation with great ease. When they were confronted with species of trees they were not sufficiently acquainted with, they would collect specimens, observe and study them with meticulous care and try to describe them as accurately as possible; and only then would they put their findings down in black and white. As writers of the book, they have not only helped to summarize the enormous achievements of their predecessors, especially those made since the 1950s, in the production and research of woody plant seeds, but have also blended into their writings a rich store of findings derived from their own researches. Many of the facts and data included in the book have been made public for the first time, and many more have served to fill up voids in the literature of forestry. It would be no exaggeration to describe the book as representing the latest achievements in China's production and research of woody plant seeds in the latter half of the 20th century.

Yet another feature characterizing the book is the excellence in both text and illustration, which are mutually complementary. The form and structure of seeds and the form of seeds in germination and seedlings in the early stages constitute important bases for the early identification of tree species. It is regrettable, however, that in earlier works on dendrotaxonomy, importance was often attached only to the branches, leaves, flowers and fruit of mature trees, but seldom if ever to their seeds, so that they fell far short of meeting the needs of the production and research of seeds and seedlings. The present book is illustrated with more than 900 plates, most of which are true-to-life representations executed with great care and accuracy. These plates will be of important practical value: they will not only enable the reader to identify the objects in question and compare their similarities and differences, but also to better comprehend the text.

A fourth characteristic feature is that the writers adhere to the principle that everything they say must be on good grounds, making sure that none of their statements are garbled, nor are any of the quotations taken out of context. Much of the information contained in the book has been provided by the writers and suppliers of articles themselves, who derived it from their own observations or researches, and so is based on true experience and therefore reliable. Whenever other authors were quoted, the compilers made sure that the quotations were unbiased and faithful to the original source so that the reader may have no fear of being misled. For the benefit of readers who wish to trace a quotation to its source, a bibliography is appended at the back of the book listing the books, magazines or articles from which the quotations are taken. This again attests to the serious and responsible attitude of the compilers toward their readers.

Last but not least is the painstaking editing that has been done to make the book as nearly perfect as possible. For a book of such size and with so many people participating in its preparation, it is necessary not only to unify the style and manner of writing, but also to improve on the original manuscript and remove unavoidable errors. That is exactly what has been done to the book. The general editors should be accorded the fullest credit for bringing the book to its present form.

Regrets sometimes bother writers of books. Very often omissions and oversights are discovered only after the book has already been written so that nothing can be done about them. This, too, has happened to the present volume. Quotations from relevant documents, for instance, are felt to be too few and far between. Furthermore, some of the more important *Angiospermae*, such as certain genera and species of *Ixonanthaceae*, *Malpighiaceae*, *Icacinaceae* and *Sabiaceae* as well as members of the *Euphobiaceae* family including bignay chimalaurel (*Actidesma bunins*), Chinese cephalomappa (*Cephalomappa sinensis*) and common macaranga (*Macaranga tanasius*) -items such as these that deserve a place in the book have unfortunately been excluded. Things of this kind are of course what ought to be done when the compilers set strict demands on themselves. It is the ardent wish of the compilers that the unification of the Motherland would be realized at an earliest possible date, so that scholars from both sides of the Taiwan Strait may be able to pool their expertise. It is also their wish that scientists and workers in the field of woody plant seeds of the new Century would continuously try and improve on the present work and pass it on from one generation to the next. Coming from the bottoms of their hearts, these words bespeak the sincere and fond sentiments that the compilers cherish about their lifelong undertakings.

As we look ahead into the new Century, we feel certain that forestry will play an even greater role in the construction of ecological environment. The compilation of a monumental work such as this at the juncture at which the Old Century slides into the New, is indeed a most important strategic move, which goes to show the foresight of the organizers, since this is a major item of capital construction in the field of seeds and seedlings. Most of the scholars who have supplied articles for the book, participated in its compilation and/or the general editing are already in their seventies. Choosing to devote their declining years to this most significant undertaking, they threw themselves into the work with concerted efforts and have made a major contribution to the country's forestry construction. As I browse through the pages, my mind fills with gratitude and admiration.

Li Yucai

March 12, 2000

前　　言

一

《中国木本植物种子》一书，由国家林业局国有林场和林木种苗工作总站主持编写。编辑委员会成立于 1986 年 8 月。参加撰稿和提供资料的有全国高等林业院校、林业和植物研究院所、林木种苗生产和管理部门的专家学者共 76 人。

本书原则上以属为单位编写。一属只有一种，或者虽有多种但资料不足的，改以种为单位编写。全书有独立成篇的文章 492 篇，亦即共收木本植物 492 属，分隶于 132 科。收入树种计 1 222 种 3 亚种 45 变种 4 变型 2 品种，合为 1 276 种（含种以下分类等级），其中包括少量引种成功的外来树种。本书的资料未按科归纳，但按科的系统排列。裸子植物采用郑万钧（1978）的系统，被子植物采用哈钦松（J. Hutchinson）（1959）修订的系统，并都略有增补。科以下的属和属以下的种，均按学名的拉丁字母顺序排列。一个属有若干个种而由不同的人提供资料的，在该文表 1 的“供稿”栏下以编号的方式列出供稿人（编号的含义请见本书撰稿供稿人名单）。表 1 未设“供稿”栏的，表示该文的资料全由撰稿人自己收集。

本书树种的分类及其名称以《中国植物志》或《中国树木志》为准，也吸收近人的研究成果，例如金松属独立成科，又如从罗汉松属中分出竹柏属并建立竹柏科。除必要者外，本书对树木的中文名和拉丁学名一般不列异名、别名。

本书有插图 921 幅。其中 435 篇文章各有插图两幅，一示种子外形及其纵切面，一示种子萌发及幼苗初期形态；有两篇文章绘插图 3 幅，45 篇文章有插图各 1 幅，仅 10 篇文章无图。本书插图大部分根据实物描绘，凡引自公开发行的出版物的，都在图下注明了出处。本书谨向原书作者以及所引原图的作者表示感谢。

二

本书主编单位的设想是，通过编委会和广大撰稿供稿人的努力，在我国历史上首次为木本植物种子编出一部专著，反映我国木本植物种子资源丰富的实际状况，反映 20 世纪 50 年代以来木本植物种子生产和科研的巨大成就，体现 20 世纪最新水平，具有先进性、科学性和实用性；面向种子生产和管理的基层技术人员，也供从事种子科学与技术、遗传育种、生物工程、种子生产、苗木培育、飞机播种、园林建设、环境保护、植物利用、药物栽培、林产工业的专业人员以及农林和生物院校的师生参考。

本书以各论的形式分属或种描述花、果、种子的形态特征和解剖结构，分析它们开花结实、采收、调制、加工、贮藏、催芽和播种的特点。生产上需要而在其它书籍中难以查寻的技术资料，例如花期、果实成熟期、种子散落期、成熟果实的形状大小和颜色、出种率、千粒重以及最适宜的发芽条件等，都列出了详细数据。只要可能，物候期还细分出初期、盛期和末期，并注明观察地点和年份。以属为单位编写的，各类数据都以表格的方式归纳，查找十分方便。

我国拥有大量具有潜在价值的木本植物。它们的开花结实和种子习性人们知之不多，是这些树种开发利用不够的原因之一。《中国植物红皮书》（第一册）共列珍稀濒危木本植物 330 多种，其中收入本书的就有 128 种。这些树种或地处偏远，或开花结实稀少，本书的供稿人和撰稿人对它们观察、采集、研究而后成文，克服了不少困难，本书也因此在这方面填补了大量空白。

本书篇幅巨大，未能着重于理论上的探索阐发，但对问题的描述既要求浅显明白，通俗易懂，又要求保持学术上的严谨性。例如本书对蜡梅果实的描述是，“发育长大的坛状果托蒴果状，通常称为果实，成熟时半木质化，先端撕裂，内含瘦果 1～11 枚，通称种子。”又如对胡颓子属果实的描述是，“瘦果为肉质化宿存的萼筒所包围，呈核果状……果核椭圆形或纺锤形。核壳为宿存萼筒去除外面肉质层后留下的骨质层……（采收调制后）所得果核即为播种材料，通称种子。”这种写法在书中随处可见，具有植物学基础知识的人读来可以温故知新，一般的读者从中也能得到启发。着墨不多而使不同的读者群都能开卷有益，是本书编辑的宗旨之一。果实的类别常有见仁见智，各家说法不一的情况，本书一般采用专治该科该属的学者的意见。

编委会希望本书既是木本植物种子生产和科研的总结，又要验之以供稿撰稿人自身的实践，力求资料数据翔实可靠。有些树种，植株极其稀少，种子极其难采，有关数据只好根据少量样本的一次测定推算。尽管数据来之不易，填补了文献空缺，但文中仍必如实交代观测的依据，不让读者产生误解。本书的记载有时与前人不尽一致。例如，A. B. 伦德勒在《有花植物分类学》（中译本第二册）中说，紫金牛种子萌发时“子叶一直留在种子内”。但本书紫金牛属撰稿人的实际观察结果是，这个种属于出土萌发型。本书没有深入探讨这种差异的原因，而是让两种记述并列共存。这样做，既是实事求是的科学态度，想必又有利于读者参照阅读。

本书希望收集并反映 20 世纪 50 年代以来，特别是 80 年代以来生产和科研的最新成果。对某些文献的某些方面存有疑义时，编委会坚持严谨求实的精神，想方设法弄清事实，核实数据，既忠实于读者，又对原作者负责。对同一个问题，不同的文献常有不同的看法。例如杉木常常形成涩粒，有的学者把它归因于自交效应，有的学者则认为应当从受精和胚胎发育期间的气象条件中寻找原因。本书如实介绍各种观点，“述而不作”，希望做到忠实客观，不限于一家之言，书末并附有参考文献，供有兴趣的读者追踪查阅，扩大信息量，开拓思路。

为了统一全书的编写体例，编委会事先分发了经过反复研究的编写大纲，既便于风格不同的撰稿人构思行文，又使篇幅巨大的全书尽可能具有统一的框架，便于读者阅读。稿成之后，主编单位又请包括树木分类学家在内的几位编委为全书统稿，核对树种中名和拉丁学名，统一形态术语，查阅并补充文献资料，核实数据，修补插图，校正差错，润饰行文，力求本书以尽可能高的质量呈献给广大读者。

三

不同的学科用不同的视角认识种子。在植物学教科书中，种子是高等植物特有的繁殖器官。遗传学家认为种子是基因的载体，世代之间的桥梁。植物利用工作者手中的种子是获取淀粉、蛋白质、油脂、药物、香料的来源。农林园艺工作人员采收调制种子，为的是取得播种材料。生态学的一个基本观点是生物和环境之间相互联系相互依存。从这个角度观察，种子又是地球生态系统中维系生态平衡的重要环节。世世代代依靠种子繁殖更新的森林和灌丛，不仅为野生动物包括鸟类提供隐蔽的栖息之地，种类繁多的果实和种子还是它们赖以生存的食物之源。反过来，野生动物和鸟类也以不同的方式传播种子，帮助木本植物更新繁衍，扩大种群范围。

我国南方有一种野生雀鸟，名叫白鹇。秋季解剖它们的嗉囊和胃，明显可见的食物颗粒主要是种子，而且主要是木本植物种子。按质量、数量和出现频率计算的重要程度顺序排列，这些种子是：苦槠、檵木、软条七蔷薇、樟树、枳椇、乌药、黄肉楠属的一种，以及光叶山矾、油桐和猕猴桃属的一种（彭长根、楚国忠等，1994）。它们绝大多数不是目前人工造林选用的树种。

一次调查研究不可能反映森林生态系统的全部关系。但是已可看到，从生态环境建设的长远目标出发，人们必须用更加广阔的视野来看待种子，规划林业生产活动，不能单纯考虑人类自身的生存和发展。我国地域辽阔，野生动物种类繁多。除了留鸟以外，冬去春来，大地上还有许多候鸟来往迁徙。行程万里，它们也要补充大量食物。因此，更新造林时怎样有意识地保存天然植被，保护生物的多样性；为人工造林和景观建设选择树种时怎样充分展现我国木本植物资源丰富的优势，更好地发挥林业在生态环境建设中的主体作用，单单从这个角度出发，也还有许多课题需要研究。本书汇集了 1 200 多种木本植物种子的资料，希望能在国家生态环境建设的宏伟工程中起到应有的作用。书稿初成之日，正值长江突发洪水，黄河频频断流，母亲河的生态性灾难引发举国深思之时。林业在国土保安中的重要性从来没有显得像今天这样突出。“让河山妆成锦绣，把国土绘成丹青”，美好的前景需要几代人为之奋斗。人类已跨入新千年，我们谨寄厚望于新世纪的林业工作者。

四

本书收列树种之多，可以说前无古人。但我国的木本植物多达8 000多种，我们的能力和学识有限，只能迈出这小小的一步。祖国的宝岛台湾树种资源丰富，这本书的编写，台湾学者未能参加，台湾树种的情况书中反映太少，十分遗憾。我们期盼祖国早日统一，两岸学人能够携手合作，让木本植物种子领域中的这本力作一代一代地增补修订下去，不久的后来人便可看到树种更多、资料更全、水平更高、同《中国木本植物种子》这个名称真正当之无愧的传世之作。

编写本书的建议发出后，各地从事木本植物种子工作的专家学者纷纷响应，或提供资料，或惠赐书稿，好多还是他们从未发表的第一手材料。为了符合本书的体例要求，有的专家还不厌其烦地修改书稿。他们的支持为本书的出版创造了最坚实的基础，本书的主编单位谨向他们表示衷心的感谢。本书能够出版，全国有关高等农林院校、林业和植物研究机构、有关省（自治区、直辖市）林木种苗管理部门，都给予了帮助和支持，特别是广西壮族自治区林业厅种苗站、南京林业大学和国家林业局南方林木种子检验中心，为本书的编写提供了极为方便的条件，我们谨向他们表示深深的谢意。书中涉及病虫害的问题，承南京林业大学钱范俊教授惠予审订；杉木、松属和落叶松属关于种源变异问题的文字，承南京林业大学王章荣教授提供资料并惠予审校；本书序和前言的汉译英，承北京外国语大学周谟智教授执笔，译文优美流畅；本书从组稿到最后统编，吴琼美编委一直给予热情耐心的帮助，做了大量工作；吴婷婷、李淑娴和关亚丽三位女士，认真细腻，完成了全书的文字录入。我们谨向他们表示真挚的感谢之情。

最早倡议编写这本书的是周陛勋教授，为了这本书，他精心筹划，多方呼吁，体现了老一辈科学家忠于事业的一片赤诚。不幸的是他过早地离开了人间。此书今天终于问世，当是对这位老人最好的纪念。

先后参加本书统稿的，以汉语拼音字母为序，有：陈幼生、黄鹏成、王宏志、王棋、赵德铭、周德本、周建铭，最后总其成的是陈幼生和黄鹏成。

限于我们的水平，加上有些事当初考虑不周，事后极难弥补，书中错误和疏漏在所难免。衷心地盼望海内外专家读者不吝赐教。

编　者

2000年1月

Foreword

1

Seeds of Woody Plants in China has been compiled under the auspices of the National Service Center for State-Owned Forest Farms and Woody Plant Seed and Seedling Affairs, of the Forestry Ministry. The Compilation Committee was set up in August 1986. Participating in the preparation of the articles or providing material and data were 76 scholars and specialists from forestry colleges and universities, forestry and botany research institutes and woody plant seed and seedling production and administrative departments all over the country.

In principle, the book was compiled with genus as the basic unit. Where the genus consisted of only one species or where there was a lack of data though the genus in question consisted of more than one species, then the unit used would be species rather than genus. The book is made up of 492 mutually independent essays, which means a total of 492 genera of woody plants, belonging respectively to 132 families, are treated of. With reference to tree species, 1,222 species, 3 subspecies, 45 varieties, 4 forms and 2 cultivars have been dealt with, totaling 1,276 different species (including taxa under species); and this includes a small number of exotic tree species successfully introduced from abroad. In the book, the materials and data have not been sorted out according to families, but have been arranged in the family system. *Gymnospermae* have been arranged according to the Cheng Wan-chun system (1978), and *Angiospermae* to the revised J. Hutchinson system (1959), though both have been slightly supplemented. The scientific names of the genera under each family and those of the species under each genus are arranged in alphabetical order. In the case of a genus consisting of several species for which the material and data have been supplied by different persons, different numbers may be found under the heading Data Suppliers in Table 1 appended at the end of the article in question. (The reader is referred to the list of names of the writers and suppliers of articles appended at the back of the book for the meaning of the said numbers). Absence of a Data Suppliers column from Table 1 implies that the writer him- or herself has collected the material and data used in that article.

The tree plants are classified and named in the book with *Sylva Sinica* and *Flora Reipublicae Popularis Sinicae* as the standard, though occasional changes have been made according to recent researches. For example, the genus *Sciadopitys* (Japanese umbrella pine) has now been separated from *Taxodiaceae* and established as an independent family to be known as *Sciadopitysaceae*; and *Podocarpus nagi* (Thunb.) Zoll. et Mor. is now regarded as a new genus named *Nageia* and established as an independent family to be known as *Nageiaceae*. Unless absolutely necessary, other names and synonyms are not used for trees in this book apart from their Chinese and Latin scientific names.

The book is illustrated by 921 plates. Of all of the articles, 435 are illustrated each by two plates, with one showing the outward form and the vertical section of the seed under discussion and the other showing the seed in germination and the form of the seedling in an early stage. Two of the articles are illustrated each by three plates, and 45 each by one. Only 10 articles are not illustrated in any way. The majority of the plates are true-to-life drawings of the actual objects. Where a plate has been taken from an open publication, a note may be found below the plate giving its source. Plates of this latter type are taken mainly from the following four sources: *Collection of Illustrative Plates on Seeds and Seedlings of Major Tree Species* (Ma Dapu et al, 1978), *Seed and Nursery Technology of Tropical and Subtropical Tree Species* (Wang Hongzhi, 1985), *Illustrated Manual on Seedlings of Major Tree Species in Northeast China* (Huang Puhua et al, 1992) and *Seedling Morphology of Major Tree Species in China* (Zhang Ruohui et al, 1993). The compilers wish to publicly acknowledge their sincere thanks to the compilers of the four plate collections mentioned above and the artists who did the original drawings.

2

As envisaged by the Forestry Ministry department supervising the compilation affairs, the project was to produce a book on woody plant seeds, the very first of its kind ever written in this country, through the concerted efforts of the Compilation Committee and the mass of scholars and specialists who wrote articles or provided materials and data. The book was to make known to the public the country's rich resources in woody plant seeds, reflect the enormous achievements that we have made since the 1950s in the production of and research on seeds of woody plants and embody the latest developments of the twentieth century. It was to be of the most advanced level and of the greatest possible scientific and practical value. And it was to be geared not only to the needs of rank-and-file technical personnel in tree-seed production and administration departments, but also to those of the professionals working in departments concerning the science and technology of tree seeds, genetics and breeding, biotechnology, seedling cultivation, air seeding, landscape architecture, environmental protection, plant utilization, herb cultivation and forest industry, as well as to those of the faculty and students of colleges of agronomy, forestry and biology. All of these aims, we are happy to say, have been basically achieved.

As mentioned earlier on, the book is made up of 492 separate articles, which describe, on the basis of the genus or species, the morphological characteristics and structural anatomy of flowers, fruit and seeds, and analyze the special features they exhibit while flowering, bearing, seed collection, seed conditioning, seed processing, seed storage, sprouting or sowing takes place. Much of the technical data frequently needed in production but usually not available in ordinary manuals may often be found in the present volume. This refers to such items as anthesis, fruit maturation period, seed dispersal period, the form, size and color of ripe fruit, seed percentage, 1000-seed weight, the most favorable conditions for germination and the like. For items such as these, detailed data are listed in the book. Wherever possible, the phenological period is subdivided into the initial, the peak and the final phase, with the time and place for observation indicated. Where the facts are presented with genus the basic unit, all the various kinds of relevant data are tabulated, which lends itself to easy access.

Our country has an immense wealth of woody plants that have great potential value, but little is known about what happens to them during the flowering and bearing periods, nor is much known about the characteristics of their seeds, which is one of the major reasons why they have not been fully developed and utilized. *The Red Book on Plants in China* (Vol. 1) lists more than 330 precious and rare species of woody plants that are endangered, of which 128 have been included in this book. Because they are found only in remote areas or because their flowers and fruits are much too scarce, the specialists who described them or supplied material and data about them were able to observe them, collected samples of them, conducted researches and wrote down their findings about them only after surmounting one obstacle after another. It is owing to their work that we have been able to fill up a considerable void in this area.

We have not been able to lay special emphasis in the book on theoretical exploration and elucidation, but we do require that our discussion of issues and statement of facts be not only plain and easy to understand, but also rigorous and precise from an academic point of view. Take for example what is said in the book about the fruit of wintersweet (*Chimonanthus praecox*): "A fully developed and grown-up urceolar fruit receptacle is capsular and it is usually called fruit. It is semi-lignified when ripe, with its tip lacerate, and it contains 1～11 achenes, which are generally referred to as its seeds. "And the fruit of *Elaeagnus* is described in these terms: "An achene is surrounded by fleshy and persistent calyx tube, drupaceous in shape. The stone is oval or shaped like spindle. The mesocarp is the osseous coat that is left when the fleshy stratum has been removed from the persistent calyx tube....... The stone that results (from collection and processing) will be seeding material, and this will generally be referred to as a seed. " The book is permeated throughout with this kind of statements, which would be readily understood by people with an elementary grounding in botany, who would feel that they are gaining new knowledge while reviewing old; the general reader would also be enlightened. One of the aims that the compilers have been trying to achieve has been, with succinct descriptions like those quoted above, to enable readers from different walks of life to benefit from what they read every time they open the book. It frequently happens that experts differ as to the categorization of a particular fruit. What we have done is to abide by the views of those who specialize in the family or genus in question.

The Compilation Committee hopes that the book will be a general summary of the production of and researches on seeds of woody plants that is supported by the field experience of the writers and suppliers of the articles. For this reason, great importance is attached to the fullness, accuracy and reliability of the materials and data used. Certain species of trees, for example, are extremely scarce so that their seeds are very hard to come by; and the relevant data could only be calculated and determined once for all on the basis of a very limited number of samples. Although the data had not been obtained easily and a void in the literature had been filled as a result, we had nevertheless to give a true account of what our observations had been based on, so as to cause no misunderstanding on the part of the reader. The facts presented in this book do not always accord with those presented by earlier writers. A. B. Rendle, for instance, said in *The Classification of Flowering Plants* (Vol. II of the Chinese translation) that while seeds of Japanese ardisia (*Ardisia japonica*) are in germination, "the cotyledons remain inside the seeds all the time" . However, according to the observations of the writer whose article about *Ardisia* has been included in the book, this particular species belongs to the epigeal type. We do not propose to probe into the cause that gave rise to the difference; instead, we think fit to allow both descriptions to be included in the book, which is in keeping with the motto "Be practical and realistic" and which will necessarily be of help to readers who wish to compare and form their own judgment.

It is the desire of the compilers to collect documents that reflect the most outstanding achievements since the 1950s, since the 1980s especially, in the production of and researches on woody plant seeds throughout the country. Whenever doubts arose about certain aspects of certain documents, the Compilation Committee would take a strict and realistic approach to the problem and try every means to set the facts straight and check the data so as to be not only faithful to the reader but responsible to the original author as well. It often happens that different documents interpret the same phenomenon in totally different ways. Take for example Chinese fir (*Cunninghamia lanceolata*), which often form seeds with tannin-like substances. Some scholars attribute this to the effect of selfing, while others are of the opinion that the cause is to be found in the climatic conditions during fertilization and during the embryonic development period. The book makes it a rule to faithfully record all different points of view, which is not unlike what Confucius said of himself, "I just expound the teachings of earlier sages, but produce nothing myself". This is done in order that we may remain objective and not limit the readers to the views of one particular school. A bibliography is appended at the end of the book for the benefit of those who wish to trace a point of view to its source, broaden their store of information or open up new trains of thought.

In order that the different writers may use the same style of writing, an outline listing the guidelines of the style to be followed in the preparation of the articles was drawn up by the Compilation Committee and, after repeated deliberation and revision but before the compilation officially started, distributed to the article writers. This was done not only for the convenience of the writers, but also, as far as possible, to give the gigantic book a uniform framework so that readers may find it easy to use. When all of the articles were handed in, a number of Compilation Committee members, one of whom was a dendrologist, were invited to take charge of the general editing of the book, checking the Chinese and Latin names of the tree species dealt with, unifying the form terminology, looking up relevant documents and supplementing them, checking the data, revising and adding illustrative plates, removing errors and polishing the writing: doing all of these with a view to presenting to the reading public a work of the highest possible quality.

3

Different disciplines regard seeds each from a different angle of view. In textbooks on botany, seeds are the reproductive organs peculiar to higher flora. Geneticists regard seeds as the carriers of genes that serve as a bridge between one generation and the next. In the hands of workers in the utilization of plants, seeds are sources from which to obtain such useful substances as starch, protein, oil, medicine and perfume. Farmers and gardeners collect and condition seeds for the purpose of obtaining seeding material. One of the basic viewpoints of ecology is that the living beings and their environment are interrelated and mutually dependent. Viewed from this angle, seeds constitute an important link in the Earth's ecosystem that serves to maintain ecological equilibrium. Generations of forests and coppices that depend on seeds for regeneration and propagation not only provide wildlife, including birds, with habitats, but the huge variety of fruits and seeds available in the forests and coppices provide them with sources of food, on which they depend for their very existence. Conversely, wild animals and birds scatter seeds in all sorts of ways, which helps the regeneration and propagation of woody plants and enlarges the range of their population.

In the south of our country, there exists a wild bird named silver pheasant (*Lophura nycthemera*). When they are dissected in autumn, granules of food, mainly woody plant seeds, are clearly visible in their craw and stomach. Arranged in the order of importance determined on the basis of calculations carried out according to weight, quantity and frequency of appearance, the seeds are evergreen chinkapin (*Castanopsis lanceolata*), Chinese loropetalum (*Loropetalum chinense*), Henry rose (*Rosa henryi*), camphor trees (*Cinnamomum camphora*), raisin trees (*Hovenia acerba*), combined spice-bush (*Lindera aggregata*), *Actinodaphne* sp., smooth leaf, sweet leaf (*Symplocos lamcifolia*), tung oil trees (*Vernicia fordii*) and *Actinidia* sp. (Peng Changgen, Chu Guozhong et al. 1944). The majority of these are not seeds chosen for current afforestation projects.

It is impossible for just one single fact-finding field trip to reveal all of the relationships in the forest ecosystem. But it can already by seen that, in the perspective of the construction of ecological environment as a long-term objective, we must take a still broader view of the question of woody plant seeds and draw up an overall plan for forest production and management rather than consider solely the existence and development of man alone. Our country has a vast territory and rich resources of fauna of a great many varieties. Apart from resident birds, there are vast numbers of migratory birds that travel from one part of the country to another as winter changes to spring and then back to where they come from when the weather there gets warm enough. Flying over long distances, they need great amounts of food to sustain them. For this reason, we would naturally be concerned with questions such as these: what conscious efforts should be made while afforestation and reforestation are carried out so as to preserve indigenous vegetation and protect biodiversity; and in our choice of tree species for afforestation, reforestation and landscape construction projects, what should be done to fully display our country's superiority in its resources of woody plants and give play to the major role of forestry in the construction of ecological environment. There would be many issues that deserve our closest attention even if we proceed from these considerations only. We have included in this book relevant materials and data about a total of 1, 200 tree species, which we hope will play their proper role in our country's gigantic ecological environment construction projects. At the time when the book's first draft was completed, a huge flood broke out along the Yangtze River, and the Yellow River frequently ran dry on its lower reaches. The ecological disasters that struck the Mother Rivers gave the nation much food for thought. The importance of forestry in the country's land preservation projects has never appeared as prominent as it does today. "Clothe our rivers and mountains in brocade and fill our vast land with splendors." To turn these aspirations into reality, the efforts of several generations will be needed. As mankind cross the threshold of the new Millennium, we place high hopes on forestry workers of the new Century.

4

The large number of tree species included in this book may indeed be said to be unprecedented. Since woody plant species found in China are more than 8,000 in number, we have only moved a very small step forward, limited as we are in both capability and knowledge. China's treasure island Taiwan is very rich in woody tree resources, which is hardly reflected in our book; and we regard it as a great pity that Taiwan scholars were not able to participate in the compilation of this book. It is our earnest hope that our Motherland will be unified at the earliest date possible so that scholars from both sides of the Taiwan Strait may work hand in hand and make it possible for this book to be continuously supplemented and revised by future generations. Should this really come true, a book dealing with still more species of trees, supported by a still more complete store of data and of a still higher level of scholarship, a book that is really worthy of its title Seeds of Woody Plants in China, would be available to our successors. Years ago, when the proposal for the writing of the book was first dispatched, it was greeted with enthusiastic response from scholars and specialists working in woody plant seeds all over the country. They either provided materials and data or wrote and sent us articles at the request of the Compilation Committee, much of which was first-hand material that had never been previously published. Some specialists took pains to revise their articles in order to conform to our stylistic requirements. Their support helped to lay the foundation on which the publication of the book depended. The Forestry Ministry department supervising the book's compilation extends to them its heart-felt thanks. Cordial thanks are due also to all of the forestry and agriculture colleges concerned, forestry and botanical research institutes, woody plant seed and seedling administrative departments of the provinces and autonomous regions concerned, all of whom afforded us help and support. Cordial and sincere thanks are due especially to the Nanjing Forestry University and the Forestry Ministry Southern Tree Seed Inspection Center, who made things convenient for the book's compilation. The earliest person that suggested compiling a book such as this was Professor Zhou Bishun, who did careful planning and appealed to various quarters for support, which testified to the devotion of an old-generation scientist to his career. It is unfortunate that he departed this life much too early. The publication of the book will be the best possible way to commemorate him.

Scholars who participated in the general editing at one time or another are, arranged in order of the *Pin Yin* alphabet, Chen Yousheng. Huang Pengcheng, Wang Hongzhi, Wang Qi, Zhao Deming, Zhou Deben and Zhou Jianming. Those to brought the work to completion are Chen Yousheng and Huang Pengcheng.

Limited by our level of expertise and owing to the lack of careful planning and the fact that remedial measures are practically impossible, the book must necessarily contain errors and oversights. Advice and comments from experts both at home and abroad are cordially invited.

The Editors

January, 2000

目　录

被子植物 ANGIOSPERMAE

双子叶植物 DICOTYLEDONEAE

单子叶植物 MONOCOTYLEDONEAE

裸子植物 GYMNOSPERMAE

苏 铁 属
Cycas L.

（苏铁科 Cycadaceae）

生长习性、分布和用途 本属约 85 种，我国 24 种，引栽 4 种（王定跃，2000），本文描述 3 种。常绿木本植物，不分枝，高 2～4.5 m，直径 20～40cm。不耐严寒，生长缓慢。产于华南至西南热带、亚热带地区及东南亚。多为庭园种植和盆栽，供观赏。干髓含淀粉，可食。嫩叶及外种皮亦可食。大孢子叶和种子供药用。这 3 个种的名称、分布和用途等见表 1，其中攀枝花苏铁已收入《中国植物红皮书》，被列为濒危种。

表 1 苏铁属树种的名称、生长、分布和用途

中 名	学 名	树 高 (m)	分 布	用 途	供 稿
攀枝花苏铁	*C. panzhihuaensis* L. Zhou et S. Y. Yang	4	川西南	观赏、药用	501
苏铁	*C. revoluta* Thunb.	4.5	闽、粤、台。华南、西南露地栽培，华中、华北室内盆栽。日本、菲律宾、印度尼西亚	观赏、药用	902
台东苏铁	*C. taitungensis* Shen，Hill，Tsou et Chen	3～5	台。粤、闽栽培	观赏、药用	902

开花结实 10 年生开花结实，有大小年现象，间隔期 1～5 年。球花单性，异株。雄球花（小孢子叶球）单生于茎顶，直立，长卵形或圆柱形，长 30～70cm，径 8～15cm。小孢子叶扁平，楔形，螺旋状排列，长 2.5～6cm，宽 1.5～2.5cm，下面着生多数单室的花药（小孢子囊）。花药 3～5 个聚生，纵裂，无花丝。小孢子萌发时产生两个有纤毛的游动精子。大孢子叶扁平，长 9～20cm，通常密被茸毛，不形成雌球花。大孢子叶上部通常羽状分裂，下部狭窄成柄状，两侧着生（1）2～8（14）个胚珠。花期随地区及树龄不同而异，一般还随纬度以及海拔高度的增加而推迟。同一地区，雄花较雌花约早 10 天开放（表 2）。

表 2 苏铁属树种开花结实物候期

树种	观察地点	观察年份	开花期						果实成熟期		种子脱落
			始期		盛期		末期		始期	盛期	
			雌	雄	雌	雄	雌	雄			
攀枝花苏铁	攀枝花市	1981～1983	4月上旬	3月中旬	5月上旬	4月中旬	5月下旬	5月中旬	8月中旬	9月上旬	10月上旬
	福建南平（向阳）	1987～1988	6月上旬	5月下旬	6月中旬	6月上旬	6月下旬	6月中旬	10月上旬	10月中旬	11月中旬
	福建南平（背阴）	1987	6月上旬	6月上旬	6月中旬	6月中旬	6月下旬	6月下旬	—	—	—
苏铁	福州鼓山	1987	6月中旬	6月上旬	6月下旬	6月中旬	7月上旬	6月下旬	10月下旬	11月上旬	12月上旬
	福州市区	1987	6月上旬	5月下旬	6月中旬	6月上旬	6月下旬	6月中旬	—	—	—
	福建永安	1988	6月上旬	5月下旬	6月中旬	6月上旬	6月下旬	6月中旬	—	—	—
台东苏铁	福建永安	1988	6月中旬	—	6月下旬	—	7月上旬	—	—	—	—

零星栽培的苏铁要获得种子，须进行人工授粉。在福建，苏铁于5月中旬开始出现花芽，6月上中旬开花，6月5～15日为花粉成熟期，从下向上逐渐成熟。成熟时花粉囊开裂，应及时在每天中午或傍晚收集花粉。6月10～20日大孢子叶张开，此时为最佳授粉期，可授期5～8天，应每隔1～2天授粉1次。接受花粉后大孢子叶重新闭合起来，可以见到胚珠由金黄色变为浅绿色，并逐渐转变为绿色；约经20天，又由绿色转变为橘红色并迅速生长。到8月中旬，种子迅速增大，逐渐成熟。种子核果状。外种皮肉质，橘红色，中种皮木质，常具2（3）棱，内种皮膜质。种子卵圆形、近圆形或倒卵圆形，长3～5cm，径2～3.5cm，常有密生的灰褐色茸毛，后渐脱落。胚乳丰富。子叶2枚，长3～6mm。胚长1～2.1cm，基部有较长的螺旋状胚柄。苏铁种子的形态见表3、图1。

表 3 苏铁属树种种子的形态特征

树种	形状	颜色		长（cm）	径（cm）	种子附属物
		未熟时	成熟时			
攀枝花苏铁	近球形或近圆形	黄绿色	红色或橘红色	2.5～3.0	2.5	外种皮具薄纸质、分离而易碎的外层
苏铁	倒卵圆形，稍扁，顶端凹	橘红色	深橘红色或玫瑰红色	3～5	2～3.5	基部密被灰黄色绒毛，后渐脱落
台东苏铁	卵圆形或长圆形，稍扁	橘红色	红褐色	3～4.5	1.5～3	基部密被灰红褐色绒毛，后渐脱落

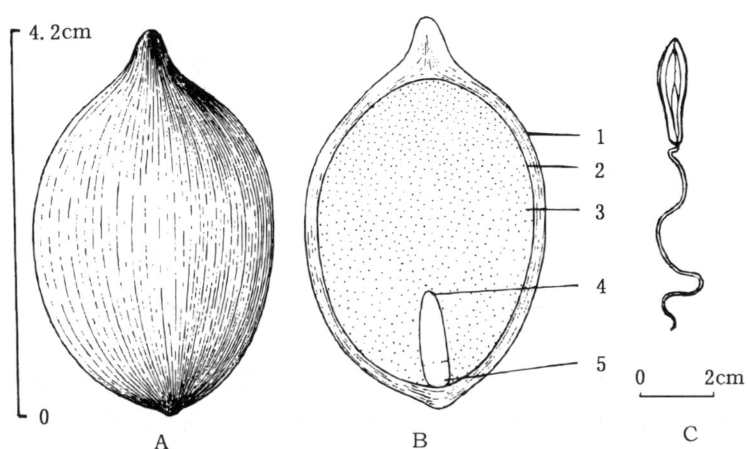

图 1　苏铁种子除去外种皮后的外形（A）及其纵切面（B）和胚（C）
1. 中种皮　2. 内种皮　3. 胚乳　4. 子叶　5. 胚根
（A、B. 黄应钦绘，C. 田恒德仿《中国主要树木幼苗形态》）

种子的采收调制和贮藏　种子成熟盛期采摘。采回的种子用手工剥去或浸水 4～6 天后除去外种皮，即为播种材料。也可不去皮而直接沙藏，种子含水量应保持在 44%左右，忌失水，运输或贮藏时需混湿沙。混沙贮藏于室内阴凉处，贮藏期可达 1 年。苏铁属树种种子的质量等见表 4。

表 4　苏铁属树种除去外种皮后种子的得率和净度、质量

树　种	除去外种皮的种子得率（%）	净度（%）	粒重（g）	每千克纯净种子粒数（粒）
攀枝花苏铁	70	99	6.3～7.8	130～160
苏铁	62	97～100	12.5～13.5	74～80
台东苏铁	65	96～100	12.6～13.7	70～80

发芽和播种　种子有休眠习性。苏铁和台东苏铁的种子去掉外种皮后，可用 100μg/g 的赤霉素处理 12 小时并湿沙层积于 0～5℃的低温下 1～2 个月，再行播种，播后 3 个月左右发芽，发芽整齐。卢小根（1988）的经验是，苏铁种子休眠期长，采收后应贮藏 1 年再行春播。

苏铁属树种的种子发芽时日均温宜在 20～30℃。未经处理的种子，播后要经过 7 个月才能发芽。检验时可用四唑法测定生活力。苏铁属种子的发芽测定条件及发芽能力见表 5。

表 5　苏铁属树种的发芽测定条件及发芽能力

树　种	预处理	发芽测定条件		发芽率（%）			生活力（%）		场圃发芽率（%）
		基质	温度（℃）	计算天数	一般数值	变动范围	一般数值	变动范围	
攀枝花苏铁	—	—	—	—	—	—	35	—	—
苏铁	去外种皮并用赤霉素处理 12 小时后低温层积	湿沙或蛭石	25	120	70	40～80	78	70～85	65
台东苏铁	去外种皮后低温层积	湿沙或蛭石	25	150	40	15～60	50	20～65	35

春季用经过层积的种子点播，覆土 3～5cm。留土萌发。2 子叶包于丰富的胚乳和种壳内。无上胚轴。由胚芽发出两枚羽状裂叶。2 叶之间为 2 片幼叶，白色，密被淡褐色毛。主根粗大，直伸。侧根近平展。主侧根均白色（图 2）。生产中多用茎上萌发的不定芽移栽，成活率可达 100％。

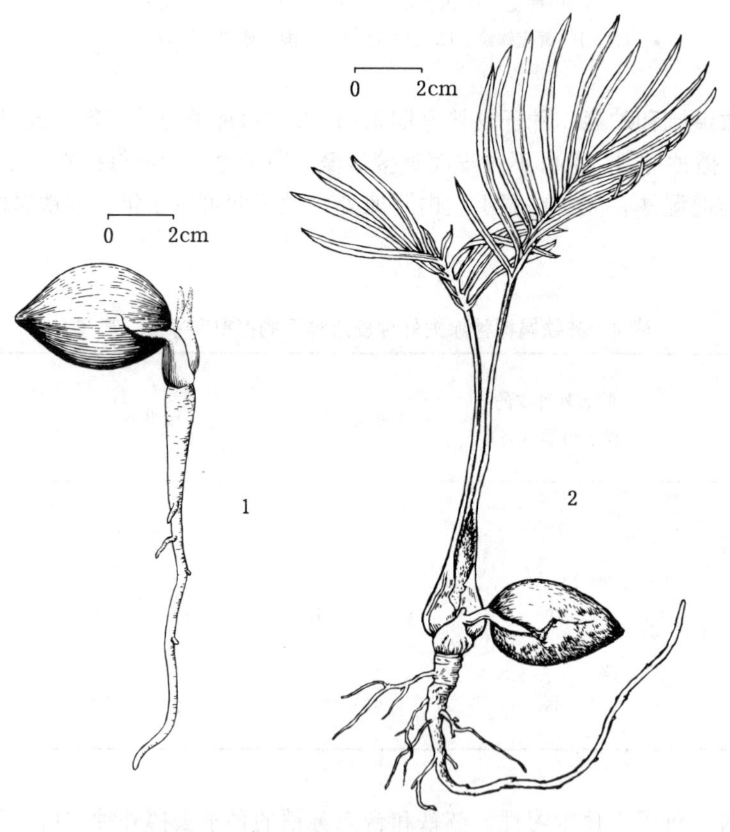

图 2　苏铁种子的萌发和幼苗的生长情况

1. 地下萌发，初生叶初露　2. 幼苗形成

（1. 黄应钦绘；2. 田恒德仿《中国主要树木幼苗形态》）

（郑郁善）

银　杏

Ginkgo biloba L.

（银杏科　Ginkgoaceae）

生长习性、分布和用途　银杏属是我国特有的一个单型属。中生代第四纪冰川后几近灭绝，目前仅我国天目山散生有野生的银杏植株，但数量稀少，在《中国植物红皮书》中被列为稀有种。人工栽培范围很广：北起辽宁，南抵广东、广西，东自台湾，西至川、滇、黔，不少地方还形成了银杏的集中产区，选育出了许多地方品种。世界各国都有引种栽培。树形伟岸，叶形独特，秋色悦目，是著名的观赏树种。寿命长，各地名胜古迹常有数百年至上千年的参天古树。对空气污染有很强的抗性。木材坚实致密，是珍贵用材。种子营养丰富，是传统的保健食品并可入药。叶和外种皮含银杏黄酮和银杏类脂，有较高的药用价值。

开花结实　实生树 20 年左右开始开花结实，结实能力可历数百年不衰。雌雄异株。有的地方雄株过少因而结实不良。邓荫伟和黄连桂（1986）比较过促进银杏结实的几种措施，效果最好的是人工授粉。科学工作者正在探索利用植物的药效反应来鉴定银杏植株的性别〔参见白志华（1986）译文〕。

球花单生于短枝顶端的叶腋。雄球花柔荑花序状。雌球花具长梗，先端分 2 叉，叉端具 1 珠座，每个珠座 1 胚珠。花期在 3 月下旬～4 月中旬。种子 10 月成熟。1984 年安徽歙县和陕西西安两地的开花始期分别为 4 月 28 日和 4 月 15 日，盛期分别为 5 月 3 日和 4 月 17 日，末期分别为 5 月 8 日和 4 月 22 日。在歙县，种子成熟始于 10 月 5 日，盛期在 10 月中旬，10 月 23 日种子开始脱落。

种子核果状，椭圆形、倒卵圆形或近球形，长 2.5～3.5cm，径约 2cm，未熟时青绿色，熟时黄色至橙黄色，外被白粉。外种皮肉质，有臭味。中种皮骨质，白色。内种皮膜质，黄褐色。胚乳丰富。去外种皮的种子椭圆形，有 2 或 3 棱，长 1.5～2.8cm（图 1）。

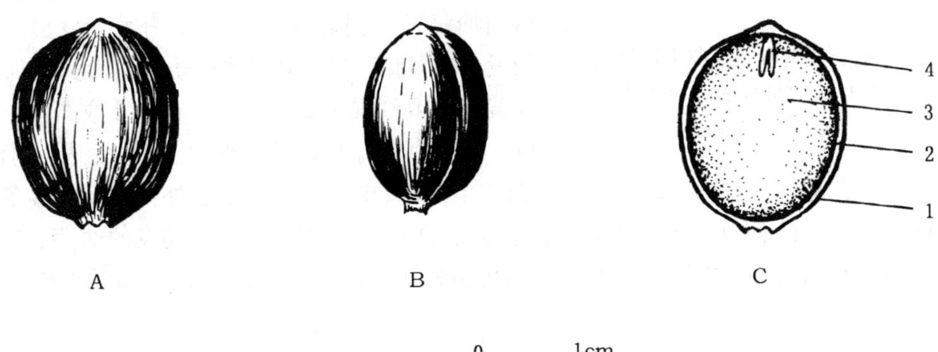

图 1　银杏种子（已去肉质外种皮）的正面（A）、侧面（B）及其纵切面（C）

1. 骨质中种皮　2. 膜质内种皮　3. 胚乳　4. 胚

（孟玲绘）

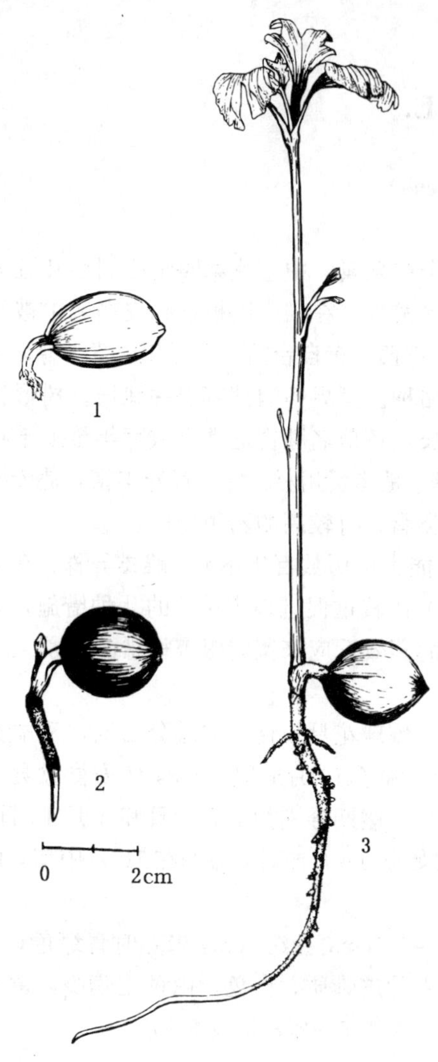

图 2 银杏种子萌发和幼苗生长情况
1. 幼根延伸 2. 子叶柄伸长，幼芽生长
3. 幼苗形成（孟玲绘）

种子的采收调制和贮藏 充分成熟后上树采摘，也可以在自然脱落后或人工击落后在地面拾集。采回堆在阴湿处或加水沤于缸内，让肉质外种皮腐烂。数日后淘净外种皮，置阴凉处风干。这种去掉外种皮仅存中种皮的部分即为播种材料，通称种子。出籽率 25%～30%。千粒重约 2 000～3 200g，每千克有纯净种子 400 粒左右。种子不耐干燥，应混沙湿藏并在低温层积状态下解除休眠。

播种前的处理 银杏播种前需要层积处理。梁玉堂和龙庄如等（1987）观察到，在 1～10℃室内层积 4 个月后，银杏的胚显著伸长。他们认为层积实际上是为银杏胚的形态后熟提供条件。中国林业科学研究院林业研究所认为银杏在低温下层积 2 个月即可。在他们的试验中，0～5℃下层积 1 个月的银杏发芽率为 11%，层积 2 个月和 3 个月的发芽率分别为 92%和 93%。银杏在层积和发芽过程中常有种粒腐坏。一般的推测是，至少有一部分是因为没有受精，虽然胚乳发育起来，但其中并没有胚。层积后播种前清除坏粒，可以使苗木匀齐。

发芽和播种 据中国林业科学研究院林业研究所试验，经过层积的银杏，室内发芽测定的最佳条件是 30℃/20℃的昼夜变温，每天并在高温时段给予光照。

冬播，或早春播种经过层积处理的种子。播种沟距 25～30cm，深约 4cm，每隔 10cm 点播 1 粒。覆土后稍加镇压。春播的约一个半月开始萌发出土。留土萌发。有多胚现象。上胚轴和幼茎初为黄绿色，渐转为黄褐色或红褐色。有初生不育叶 2 或 3 枚，偶为 4 枚，以后为发育叶。下胚轴短。主根发达，侧根较短（图 2)。当年苗高 25～30cm，每 666m^2 产苗约 1 万株。也有将银杏不去外种皮而直接冬播，但无法清除其中无胚的种粒。李家玉和齐之尧等人（1993）对层积过程中萌发的幼芽施行截根，并继续培育数日后移栽，据说同对照的 1 年生苗相比，地下部分的干重可以增加 29%，苗高可以增加 24%。

银杏也可以扦插繁殖和嫁接繁殖。齐之尧和李家玉（1991）报道过银杏的嫩枝单芽扦插技术。

（钱耀明）

贝　壳　杉

Agathis dammara（Lamb.）Rich.

（南洋杉科　Araucariaceae）

生长习性、分布和用途　贝壳杉属约20余种，我国引入3种，本文描述1种。常绿乔木，在原产地高达45m，胸径4m以上。喜温暖湿润气候，适生于肥沃、深厚、排水良好的土壤。生长迅速。原产马来西亚、印度尼西亚、菲律宾和大洋洲，我国台湾、福州、厦门、广州等地有栽培。树形优美，供庭园观赏，木材供建筑用，玳玛（dammar）树脂在工业和医药上有广泛用途。

开花结实　15～20年生开始结实。正常结实在50年生以上。结实有大小年现象，丰年间隔期1～3年。球花单性，通常雌雄同株。雄球花单生于叶腋，具短柄，圆柱形，长5～7.5cm。雄蕊多数，密集，螺旋状着生。雌球花单生于枝顶，卵状椭圆形或近球形，由多数螺旋状着生的苞鳞组成。苞鳞顶端宽圆，两侧扁平，基部一侧有时两侧各斜生1胚珠，胚珠与苞鳞离生。在福州3～4月开花，翌年7月下旬球果开始成熟，8月上旬为成熟盛期，球果鳞片开始开裂。球果近球形或宽卵形，长6.5～12cm，径6～10cm。苞鳞顶端圆钝或具短突尖，向上反曲。种子1，着生于基部的一侧。每球果约有种子370粒。种子倒卵圆形，略扁，褐色，长0.8～1cm，宽0.5～0.6cm。种子的一侧具膜质翅，种翅与苞鳞离生，另一侧具一小突起，很少发育成翅。有胚乳，子叶2（图1）。

球果的采收调制和种子贮藏　球果由青绿色转为黄绿色时上树采摘球果，摊晾1周或曝晒2～3天，使种鳞开裂，种子脱出。搓除种翅并筛净杂质即得纯净种子。根据1份样品测定，千粒重85g，每千克种子约11 700粒。贮藏的种子含水量宜在10%以下，用塑料袋密封，在0～5℃的低温下存放。

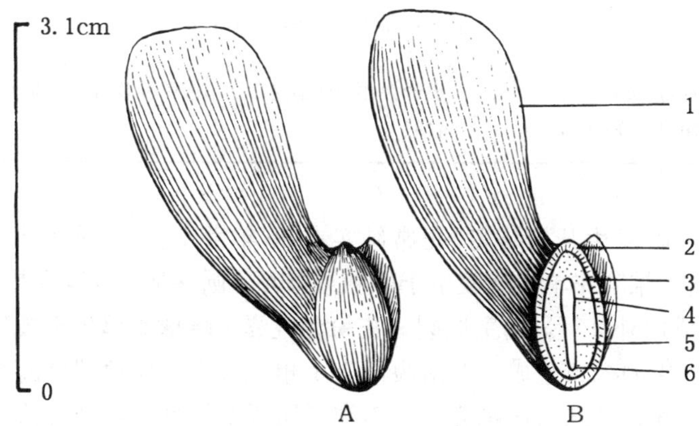

图1　贝壳杉种子外形（A）及其纵切面（B）

1. 翅　2. 种皮　3. 胚乳　4. 子叶　5. 胚轴　6. 胚根

（黄应钦绘）

发芽和播种 种子无休眠习性。播种前用冷水或始温50℃的水浸种24小时。用沙盘播种，温度控制在25℃左右，播后第5天开始发芽，第15天发芽结束，发芽率可达90%。生产上多随采随播，每平方米播种10～12g，播后覆盖一层细沙，以不见种子为度，上面再盖一层草。经10～15天出土萌发，30天以后长出初生叶。2年生苗出圃。可以扦插繁殖。用3～4年生幼树的枝条扦插，成活率达60%。

（李玉科）

南洋杉属

Araucaria Juss.

（南洋杉科 Araucariaceae）

生长习性、分布和用途 本属约15种，我国引入6种，栽培于华南及台湾，本文描述3种。常绿乔木。适生于温暖、湿润的气候和土壤肥厚、排水良好的环境。幼年稍耐荫。生长迅速。原产南美洲、大洋洲及太平洋岛屿。闽、粤、琼、桂、台等地有栽培。树形高大，树姿优美，供庭园观赏。木材供建筑、家具等用。这3个树种的名称、生长、分布和用途见表1。

表1 南洋杉属树种的名称、生长、分布和用途

中 名	学 名	树 高（m）	分 布	用 途
大叶南洋杉（毕氏南洋杉）	*A. bidwillii* Hook.	50	原产大洋洲沿海地区。华南引种栽培长势良好，能开花结实，北部地区盆栽	材用、观赏
南洋杉（肯氏南洋杉）	*A. cunninghamii* Sweet	70	原产大洋洲东南沿海地区。华南引种栽培的生长快，能开花结实，北部地区盆栽	材用、观赏
异叶南洋杉（诺福克南洋杉）	*A. heterophylla* (Salisb.) Franco	50	原产大洋洲诺福克岛。华南引种栽培，北部地区盆栽	材用、观赏

开花结实 14～16年生开始结实，正常结实在50年生以上。结实有大小年现象。球花单性，雌雄同株或异株。雄球花单生或簇生于叶腋或枝顶，圆柱形。雄蕊多数，下缘悬垂着4～20丝状花药，常排成内外2行，药室纵裂，花粉无气囊。雌球花单生于枝顶，椭圆形或近球形，由多数螺旋状着生的苞鳞组成。苞鳞腹面有1相互合生、仅顶端分离的舌状珠鳞。每珠鳞基部有1个倒生胚珠，胚珠与珠鳞合生。南洋杉属两个树种开花结实的物候期见表2。

球果椭圆形或近球形，成熟时苞鳞脱落。苞鳞上部边缘具锐利的横脊，中央有三角状或尾状尖头，向外反曲或向上弯曲。发育苞鳞具1粒种子。种子有胚乳，子叶（南洋杉）4枚。种子形态见图1，球果和种子的形状、大小见表3。

表 2　南洋杉属树种的开花结实物候期

树　种	观察地点	观察年份	开花期			果实成熟期		种子散落		种子丰年间隔期
			始期	盛期	末期	始期	盛期	始期	盛期	
大叶南洋杉	福州	1981～1985	9月下旬	10月下旬	11月下旬	翌年8月上旬	8月中旬	8月下旬	9月上旬	5～6年
南洋杉	福州	1986～1988	9月中旬	10月中旬	11月下旬	翌年7月下旬	8月上旬	8月中旬	8月下旬	1～2年

表 3　南洋杉属树种的球果、种子形状和大小

树　种	成熟球果			种　子		
	形　状	长（cm）	径（cm）	形　状	长（cm）	径（cm）
大叶南洋杉	球形	16～20	15	圆锥形	3.7～5.7	2.1～3.0
南洋杉	椭圆形	6～10	4～7	长椭圆楔形	2.0～2.2	0.8～1.0
异叶南洋杉	椭圆形	7～12	6～10	长椭圆楔形	—	—

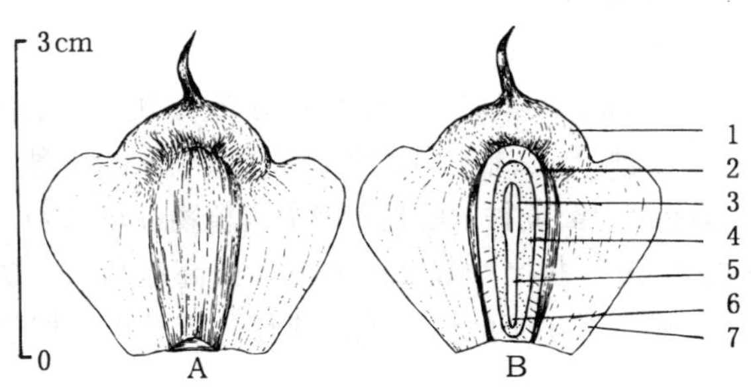

图 1　南洋杉种子的外形（A）及其纵切面（B）
1. 苞鳞　2. 种皮　3. 胚乳　4. 子叶　5. 胚轴　6. 胚根　7. 翅
（黄应钦绘）

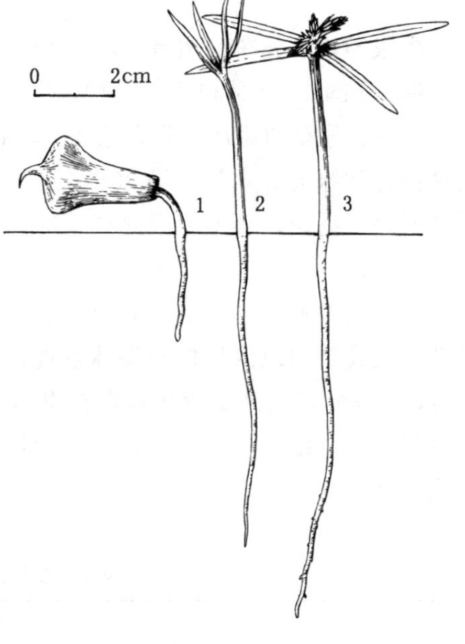

图 2　南洋杉种子的萌发和幼苗生长情况
1. 下胚轴延伸　2. 子叶出土　3. 初生叶出现
（黄应钦绘）

表 4　南洋杉属树种每千克纯净种子的粒数

树　种	一　般	变动范围	样品份数
大叶南洋杉	80	70～90	4
南洋杉	4 400	3 300～6 600	4
异叶南洋杉	570	550～620	4

球果的采收调制和种子贮藏　球果由青绿色变为黄绿色或褐色时上树采摘。采回的球果置于室内摊放阴干，苞鳞开裂后剔除空瘪粒，收集饱满的种子，去除鳞脊和种翅，即可贮藏

或播种。这3个树种种子的质量见表4，资料来自《美国木本植物种子手册》。

在常温条件下种子容易丧失发芽力，应在采收后1个月内播种。如果不能及时播种，应混湿沙贮藏，在3℃的低温条件下可以贮藏4～6年。从低温贮藏条件中取出的种子应立即播种。

发芽和播种 种子无休眠习性。1988年在福州用沙盘播种南洋杉，当时室温20～30℃，播后第4天开始发芽，第15天发芽结束，发芽率28%。南洋杉可以撒播，每平方米播种150g，播后盖1层细沙，以不见种子为度，再盖1层草。10～15天后种子开始萌发。翌年春季进行移植，再培育1年出圃。南洋杉还可扦插育苗，但侧枝扦插的苗难形成主干，顶枝扦插才可形成主干。扦插成活率50%～90%。南洋杉、异叶南洋杉属于出土萌发（图2）。大叶南洋杉属于留土萌发，且发芽方式比较奇特：子叶包藏在种壳内，仅子叶柄伸出连接主根；主根上端膨大成纺锤状；此后子叶柄逐渐腐烂，再由“纺锤”顶端发生幼茎。

（李玉科）

冷 杉 属
Abies Mill.

（松科 Pinaceae）

生长习性、分布和用途 本属约50种，主产欧、亚和北美，少数种分布到中美及北非高山地带。我国有22种和几个变种，并引入栽培6种。本文描述12种1变种及引入的1种，它们的名称、分布和用途已列入表1。其中长苞冷杉为渐危种，已收入《中国植物红皮书》。这些种都是常绿大乔木，胸径30～60cm，有的达1m，树高一般20～40m，最高达60m。多数树种耐寒、耐荫，喜温凉湿润或干冷的高山气候，适生于酸性山地灰化棕色森林土、棕色森林土和山地草甸棕色森林土。

我国的冷杉属树种主要分布在东北、华北、西北、西南以及湖北、浙江和台湾的高海拔地带，组成高山地区独特的森林景观，在涵养水源、保护河川上游水土资源方面起着重要的作用，是各分布区高山造林和森林更新的重要树种。冷杉属木材结构细，材质软，易加工，是用材和木材纤维工业的重要原料。冷杉和臭冷杉皮层分泌胶液，可以提制冷杉胶，作光学仪器的粘合剂。

表1 冷杉属树种的名称、分布和用途

中 名	学 名	分 布	用 途	供 稿
苍山冷杉	*A. delavayi* Franch.	滇西北、藏东南。海拔2 600～4 000m	材用、保土	504
黄果冷杉	*A. ernestii* Rehd.	川西北、鄂西、藏东。海拔2 600～3 600m	材用、保土	508

（续）

中　名	学　名	分　布	用　途	供　稿
云南黄果冷杉	*A. ernestii* Rehd. var. *salouenensis* (Borderes-Rey et Gaussen) Cheng et L. K. Fu	滇西北、藏东南。海拔 2 500～3 200m	材用、保土	504
冷杉	*A. fabri* (Mast.) Craib	川西、川西南。海拔 2 000～2 400m	材用、采脂、保土	508
岷江冷杉	*A. faxoniana* Rehd. et Wils.	川西北、甘中、甘南。海拔 2 800～4 000m	材用、保土	508
中甸冷杉	*A. ferreana* Borderes-Rey et Gaussen	滇西北、川西南。海拔 3 000～3 800m	材用、栲胶	504
日本冷杉	*A. firma* Sieb. et Zucc.	原产日本。我国辽、鲁、苏、浙、赣、台有栽培。海拔 400～1 000m	材用、观赏	704
川滇冷杉	*A. forrestii* C. C. Rogers	滇西北、川西南、藏东南。海拔2 800～4 000m	材用、保土	504
长苞冷杉	*A. georgei* Orr	滇西北、川西南。海拔 3 400～4 300m	材用、保土	504
杉松（沙松）	*A. holophylla* Maxim.	辽、吉。朝鲜半岛、俄罗斯远东地区。海拔 500～1 200m	材用、香料	107
臭冷杉（臭松）	*A. nephrolepis* (Trautv.) Maxim.	辽、吉、黑、晋、冀。朝鲜半岛、俄罗斯远东地区。海拔 500～2 700m	材用、采脂、栲胶、保土	102
紫果冷杉	*A. recurvata* Mast.	川西北、甘南。海拔 2 300～3 600m	材用、栲胶、保土	508
西伯利亚冷杉	*A. sibirica* Ledeb.	新。蒙古、西伯利亚。海拔 1 400～2 350m	材用、栲胶、保土	204
鳞皮冷杉	*A. squamata* Mast.	藏东、川西北、青东南。海拔 2 800～4 400m	材用、保土	508

开花结实　正常开始结实年龄多在 25～40 年，但也因树种、林分组成及生长环境而有不同（表 2）。吉林市用作行道树的杉松 15 年左右已开始结实，哈尔滨栽植的杉松到 20 年生还未见结实。结实间隔期多为 3～5 年，庐山引种的日本冷杉为 2～3 年。

表 2　冷杉属树种成年时的树高、开始结实年龄和结实间隔期

树　种	成年时树高（m）	开始结实年龄	结实间隔期（年）
黄果冷杉	60	30～40 (60)①	3～5
云南黄果冷杉	60	20～40	4～5
冷杉	30～40	40～50 (80)①	4～5

（续）

树　种	成年时树高 (m)	开始结实年龄	结实间隔期 (年)
岷江冷杉	20～30	30～40 (70)①	3～5
中甸冷杉	20	25～40	4～5
日本冷杉②	23	25～30	2～3
川滇冷杉	20	—	—
长苞冷杉	30	25～40	4～5
杉松	30	40	3～4
臭冷杉	30	30	3～4
紫果冷杉	20～40	20～30 (60)①	3～5
西伯利亚冷杉	35	15～20 (40～50)③	3～5
鳞皮冷杉	30～40	30～40 (70)①	3～5

① 括弧内的年龄指林木，括弧外的是林缘木、孤立木

② 江西庐山植物园资料

③ 括弧外的数字系指散生树

雌雄同株。球花单性，着生于树冠上部去年生的叶腋，雄球花在树冠的着生部位比雌球花低。雄球花着生于小枝下面及两侧的叶腋；雌球花单个或小群生于小枝上面的叶腋。雄球花幼时呈长椭圆形或矩圆形，后成穗状圆柱形，下垂，有梗。雄蕊多数，螺旋状着生。花药2枚，黄色或紫红色，花粉有气囊。雌球花呈短圆柱形，直立，有短梗或近无梗，具多数螺旋状着生的珠鳞与苞鳞，苞鳞大于珠鳞。每片珠鳞有2枚胚珠。

开花期多在4月中下旬～5月上中旬，有时也会迟至6月上旬。球果成熟期在9月中下旬～10月（表3）。

表3　冷杉属树种的开花结实物候

树　种	观察地点	开花期			果实成熟期		种子散落期		
		始　期	盛　期	末　期	始　期	盛　期	始　期	盛　期	末　期
苍山冷杉	云南大理	4月下旬	5月上旬	5月中旬	10月中旬	10月下旬	—	11月	—
黄果冷杉	四川马尔康	4月中旬	4月下旬	5月上旬	9月中旬	9月下旬	10月初	10月上旬	10月中旬
云南黄果冷杉	云南丽江	4月下旬	5月上旬	5月中旬	10月中旬	10月下旬	—	11月	—
冷杉	四川峨眉	4月下旬	～	5月上旬	9月下旬	10月上旬	10月中旬	～	10月下旬
岷江冷杉	四川马尔康	4月下旬	5月上旬	5月中旬	9月下旬	9月末	10月上旬	10月中旬	10月下旬
中甸冷杉	云南中甸	4月下旬	5月上旬	5月中旬	10月中旬	10月下旬	—	11月	—
日本冷杉	江西庐山	4月	～	5月	10月上旬	10月中旬	10月下旬	11月上旬	—
川滇冷杉	云南丽江	4月下旬	5月上旬	5月中旬	10月中旬	10月下旬	—	11月	—

（续）

树　种	观察地点	开　花　期			果实成熟期		种子散落期		
		始　期	盛　期	末　期	始　期	盛　期	始　期	盛　期	末　期
长苞冷杉	云南丽江	4月下旬	5月上旬	5月中旬	10月中旬	10月下旬	—	11月	—
杉松	黑龙江镜泊湖	4月下旬	～	5月	9月下旬	10月上旬	10月	～	11月
臭冷杉	黑龙江植物园	4月下旬	～	5月	9月下旬	10月上旬	10月	～	11月
紫果冷杉	四川马尔康	4月中旬	4月下旬	5月上旬	9月中旬	9月下旬	10月初	10月上旬	10月中旬
西伯利亚冷杉	新疆	5月下旬	～	6月上旬	9月下旬	10月上旬	10月	～	11月
鳞皮冷杉	四川马尔康	4月中旬	4月下旬	5月上旬	9月中旬	9月下旬	10月初	10月上旬	10月中旬

球果未成熟时绿色、淡绿色、紫色或紫黑色，成熟时呈褐色、黄褐色、紫褐色或深紫黑色。种鳞木质，排列紧密。苞鳞不露出的有黄果冷杉、云南黄果冷杉、杉松、紫果冷杉和西伯利亚冷杉，其它树种的苞鳞露出或微露出。球果表面常有树脂分泌。只有球果中部的每片种鳞才有两枚发育完好的种子。据观察，西伯利亚冷杉一个中等大小的正常球果，上端和下部各有15枚鳞片着生的都是空粒和发育不良的种子。球果成熟后10～11月种鳞和种子一齐散落，树枝上留下钉状中轴。种子上部具宽大的膜质长翅，可借风力传播。种皮颇软，且具树脂囊。种翅下端边缘包卷种子，不易分离。

据少量样品观察，每个球果所含种子数如下：黄果冷杉166～310粒，岷江冷杉214～427粒，日本冷杉247粒（其中饱满种子194粒），紫果冷杉200～407粒，鳞皮冷杉144～316粒。胚乳和胚白色。子叶3～12枚，以4～8为多。黄果冷杉种子的形态和纵切面见图1。冷杉属球果和种子的颜色、形状等性状见表4。

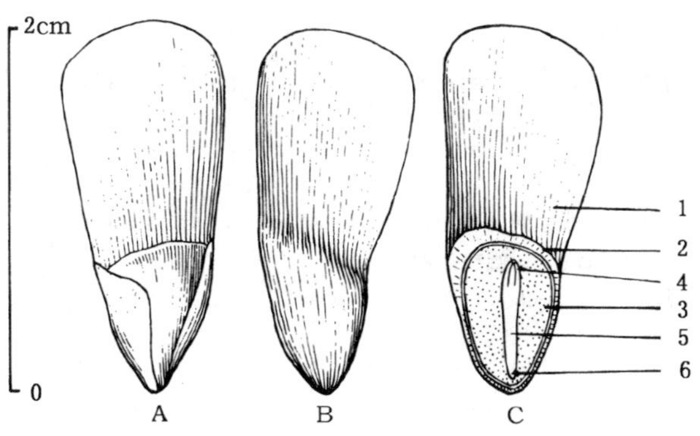

图1　黄果冷杉种子的腹面（A）、背面（B）及其纵切面（C）

1. 翅　2. 种皮　3. 胚乳　4. 子叶　5. 胚轴　6. 胚根

（黄应钦、孟玲绘）

表 4　冷杉属树种的球果和种子的颜色、形状及其大小

树　种	未成熟球果颜色	成熟球果				成熟种子		
		颜　色	形　状	大 小 (cm) 长	径	色　泽	形　状	长 (mm)
苍山冷杉	紫色	黑色，被白粉	圆柱形或卵状圆柱形	6～11	3～4	淡紫褐色	三角状卵形	7～10，翅较种子短
黄果冷杉	绿色或淡黄绿色或淡褐绿色	淡褐黄色或淡褐色	卵状圆柱形或圆柱形	5～10	2.8～3.5	褐黄色	斜三角形	7～9，连翅长 15～27
云南黄果冷杉	淡黄绿色或褐绿色	褐黄色或淡褐色	长卵状圆柱形	10～15	4～6	褐色	斜三角形	7～13
冷杉	紫绿色	暗黑色或浅蓝黑色，微被白粉	卵状圆柱形或短圆柱形	3.5～11	3～4.5	黑褐色	长椭圆形	6～10，翅与种子近等长
岷江冷杉	紫黑色	紫黑色或蓝黑色，有白粉	卵状椭圆形或圆柱形	4～10	3～4	黄褐色	倒三角状卵形	6～8，与翅近等长
中甸冷杉	紫色	紫黑色或蓝黑色，微被白粉	短圆柱形或卵状圆柱形	5～7	3～4	淡紫褐色	斜三角形	7～10，连翅长 14～18
川滇冷杉	紫色	黑褐色或褐紫色	卵状圆柱形	7～12	3.5～6	褐红色	斜三角形	约10，翅宽大
长苞冷杉	紫色	黑色	卵状圆柱形	7～11	4～5.5	褐色	长椭圆形	10～12，翅长约 7
杉松	绿色	淡黄褐色或淡褐色	圆柱形	6～14	3.5～4	淡褐色或紫褐色	倒三角形	8～12，翅宽大，比种子长
臭冷杉	淡绿色	紫褐色或紫黑色	卵状圆柱形或圆柱形	4.5～9.5	2～3	淡褐色	倒卵状三角形	5～7，翅略长于种子或近等长
紫果冷杉	紫色或深紫色	紫褐色，淡蓝褐色或暗黑色	椭圆状卵形或圆柱状卵形	4～8	3～4	褐黄色	倒卵状斜方形	5～8，连翅长 11～13
西伯利亚冷杉	绿黄色	褐色或浅褐色	圆柱形或长椭圆形	4.5～9.5	2～4	淡褐色	倒卵状三角形	4～6，翅长为种子的 1～2 倍
鳞皮冷杉	淡黑色	黑色或紫黑色	短圆柱形或长卵圆形	5～8	2.5～3.5	黄褐色	扁长卵形	5～7，翅与种子近等长

冷杉种子产量不高，最突出的现象是空粒多。据检测，冷杉这个树种的空粒率有时在75%以上。西伯利亚冷杉的空粒率平均为60%，个别的达80%～90%。杉松空粒亦多。这种现象可能是球花花期不遇或授粉不足造成的。不利的气象条件也会降低种子产量。花期多雨会阻碍花粉粒飞散。春季的严寒会使部分球花甚至全部球花中途败育。庐山引种的日本冷杉1980年本应是结实丰年，但4月中旬受到强寒流袭击，该年因而未能结实。

松鼠会盗食相当数量的种子。某些昆虫以冷杉属的球果或种子为蛀食对象，它们大量发生时会严重破坏一年的收成。据观察，危害西伯利亚冷杉球果和种子的主要害虫有冷杉大痣小蜂（*Megastigmus strobilobius*）、冷杉梢斑螟（球果螟）（*Dioryctria abietella*）和冷杉球果卷叶蛾（*Evetria margarotana*）。个别单株的球果受害率可以高达80%。

球果的采收调制和种子贮藏 冷杉属从球果成熟到种子开始飞散，通常为时只有一个月，且种粒不大，散落后很难收集。因此球果出现成熟特征时即需采集。冷杉属树种多半生长在高山地区，降雪早，交通不便，要及早安排人力上树或者从伐倒木上人工采摘尚未开裂的球果。冷杉属树体高大，枝条较脆，树梢容易断裂，上树采集球果要特别注意安全。

经验表明，将新采收的冷杉属球果在调制脱粒之前先存放一段时间，种子的发芽能力会有所提高。因此生产上通常是将采回的球果在通风干燥的室内堆放几周，使种子经历1个后熟阶段。不过，新采收的球果含水量高，堆放不能过厚，以免发热生霉。冷杉属球果大都利用日光曝晒脱粒。摊晒时以不超过两个球果的厚度为宜，并应经常翻动，使种鳞干燥均匀。夜晚或遇雨雪时要用篷布或塑料薄膜遮盖球果。种鳞和种子裂散后用簸扬或过筛的办法清除种鳞、苞鳞、中轴和其它夹杂物。通常是手工搓揉去翅。冷杉属种翅的基部边缘紧贴着卷包种子，难以剥离，种皮又薄又脆，要防止去翅操作过度而伤及种子。冷杉属的出种率和种子质量等数据见表5。

表5 冷杉属树种的球果出种率和种子质量

树 种	出种率（%）	净 度（%）	千粒重（g）	每千克纯净种子粒数（万粒）
苍山冷杉	29	75～85	19～29	3.4～5.3
黄果冷杉	11～15	80～95	11.5～14.0	7.1～8.7
云南黄果冷杉	20.5	90～98	19～30	3.3～5.3
冷杉	11～15	55～75	10.2～14.0	7.1～9.8
岷江冷杉	11～17	80～90	7.8～14.8	6.7～12.8
中甸冷杉	17.6	80～92	15.6～20.0	5.0～6.4
日本冷杉	6～8	—	51.5～57.8	1.7～1.9
川滇冷杉	7.2	60～80	14～18	5.5～7.1
长苞冷杉	20	80～90	17.6～23.0	4.3～5.7
杉松	2～3	65～70	45～54	1.8～2.2
臭冷杉	5	80～90	8.0～8.3	12.0～12.5
紫果冷杉	11～15	80～95	12.9～14.1	7.1～7.8
西伯利亚冷杉	2～3	80～90	7.5～9.5	10.5～13.3
鳞皮冷杉	11～17	80～90	7.5～12.8	7.8～13.3

冷杉属种子的贮藏寿命研究得还不多。一般认为贮藏的适宜含水量为8%～10%，并以密

封低温贮藏效果最好。曾经将冷杉这个树种的种子在0～5℃条件下密封贮藏过2年，种子生活力无明显变化。不密封的种子在不控温的库房贮藏1年，生活力便显著下降。

发芽前的处理 一般认为冷杉属的种子具有轻度的生理性休眠，播前常需低温层积一段时间。西伯利亚冷杉需要低温层积2～3个月，播种前2周取出种子，再在20℃左右的室温下催芽。杉松种子则常在12月下旬在室外坑内按1与2之比混雪埋藏，播前约1个月取出拌以湿沙，在5～20℃下催芽，经常翻动并保持湿润。催芽的种子一般都是等到有1/3的种粒裂口时播种。西伯利亚冷杉的休眠也有可能是胚萌发需要的氧气受到限制引起的。据研究，西伯利亚冷杉低温层积2个月的种子发芽率为10%，未经层积的种子剥去种皮后7天的发芽率为30%，离体胚培养两天的发芽率就超过了95%（肖东玉和陈开秀，未刊稿）。西南高山地区的黄果冷杉、冷杉、岷江冷杉、紫果冷杉和鳞皮冷杉，被认为没有明显的休眠习性，播前可以用温水浸种催芽，但为了使发芽迅速整齐，也常用1～5℃的低温层积半个月。

发芽测定 测定前可用始温45℃水浸种24小时，或者还用0.5%高锰酸钾液浸种30分钟，用清水洗净后再用温水浸种。用得较多的发芽基质是纱布或滤纸，有的地方习惯使用泡沫塑料垫。

冷杉属种子发芽需要的温度比较低。据试验，岷江冷杉、冷杉、黄果冷杉、紫果冷杉和鳞皮冷杉发芽的适宜温度为15～20℃；西伯利亚冷杉为10～15℃；苍山冷杉、云南黄果冷杉、中甸冷杉、川滇冷杉、长苞冷杉等的适宜发芽温度为13～15℃。发芽测定中一般建议给予光照。

冷杉属种子的发芽率不高，一般为10%～25%，不少样品甚至还不到10%。庐山植物园引种的日本冷杉，发芽率通常只有20%～30%，从50年生的大树上采集的种子，发芽率曾经高达60%。冷杉属常常有许多不饱满的种粒，净种时如果能清除这样的种粒，或者某一年饱满种粒的比例特别高，也会得到较高的发芽率。辽宁省林木种子站测定的1份杉松种子，发芽率就高达70%。表6汇总了冷杉属部分树种的发芽测定条件及发芽数据的实例。

表6 冷杉属树种发芽能力的实例

树种	预处理	基质	温度（℃）	发芽势（%）			发芽率（%）		
				计算天数	一般数值	变动范围	计算天数	一般数值	变动范围
苍山冷杉	0.5%高锰酸钾液浸种30分钟，清水冲洗	纱布	15	10	27	15～35	14	34	20～45
黄果冷杉	始温45℃水浸种24小时	泡沫塑料垫	20	7	9	5～12	19	14	6～20
云南黄果冷杉	0.5%高锰酸钾液浸种30分钟，清水冲洗	纱布	15	10	18	15～30	14	14	20～40
冷杉	始温45℃水浸种24小时	泡沫塑料垫	20	7	7	5～10	19	13	10～34
岷江冷杉	始温45℃水浸种24小时	泡沫塑料垫	20	7	7	2～9	19	18	5～25
中甸冷杉	0.5%高锰酸钾液浸种30分钟，清水冲洗	滤纸	15	10	26	20～30	14	32	28～45
川滇冷杉	0.5%高锰酸钾液浸种30分钟，清水冲洗	滤纸	15	10	10	5～15	14	17	10～30

（续）

树　种	预处理	基　质	温　度（℃）	发芽势（%）			发芽率（%）		
				计算天数	一般数值	变动范围	计算天数	一般数值	变动范围
长苞冷杉	0.5%高锰酸钾液浸种30分钟，清水冲洗	滤纸	15	10	12	10～25	14	17	15～35
紫果冷杉	始温45℃水浸种24小时	泡沫塑料垫	20	7	8	3～11	19	12	6～16
鳞皮冷杉	始温45℃水浸种24小时	泡沫塑料垫	20	7	6	3～10	19	19	7～25

播种　播种前通常需要层积催芽。无明显休眠习性的树种可不作催芽处理。春播，西南高海拔山区常在4月，东北、西北地区多在5月。条播或撒播。条播幅宽5～10cm，条距10～15cm。播种沟深1cm。条播时每666m^{2}播种30～40kg；撒播时每666m^{2}播种40～60kg。经过催芽的种子播后约20天大部分种子发芽出土。未经催芽的种子播后约40天左右少量发芽出土。发芽出土以后分批揭除盖草。川西北地区在揭除盖草后还需搭设荫棚。小苗阶段透光度要小，可以是25%～35%，2年生苗的透光度可以是75%～90%。3年生苗不再遮荫。冷杉属幼苗生长缓慢，常需培育4～6年才能出圃，塑料大棚育苗可以缩短为2～3年。用于园林绿化的多用近10年生经过移植培育的大苗。

出土萌发，子叶通常4～6枚，条形，扁平，深绿色，中脉在上面隆起并有两条白色气孔带，在下面平，无气孔带。初生叶螺旋状互生。1年生苗侧根稀疏。黄果冷杉种子的萌发和幼苗生长情况见图2。

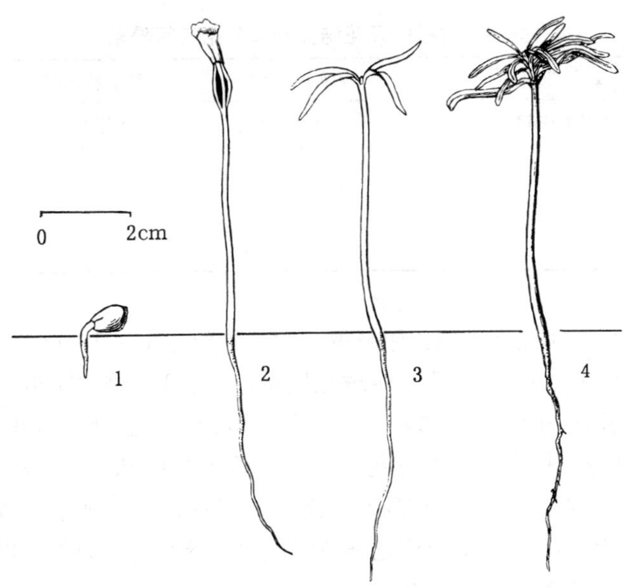

图2　黄果冷杉种子萌发及幼苗生长情况

1. 幼根伸出，种壳出土　2. 子叶初现　3. 子叶展开　4. 幼苗形成

（1、2、3. 为黄果冷杉，孟玲、黄应钦绘；4. 为川滇冷杉，田恒德仿《主要树木种苗图谱》）

（赵德铭）

银　　杉

Cathaya argyrophylla Chun et Kuang

（松科　Pinaceae）

生长习性、分布和用途　银杉，古老的残遗植物，本世纪 50 年代发现，是我国特有的松科单型属植物，在《中国植物红皮书》中列为稀有种。常绿乔木。贵州道真仡佬族苗族自治县沙河林区的银杉，107 年生，胸径 40cm，树高近 19m。喜温暖湿润或冬无严寒、夏无酷热的温凉湿润环境。适生于山地黄壤或山地黄棕壤。分布在湘桂接壤的越城岭和川黔连界的大娄山两大区及其相邻山区，海拔 800～1 840m，大都生长在岩石裸露的山脊或悬崖上。

开花结实　雌雄同株。球花单生，但常 2～3 个邻近生长而成假轮生。雄球花着生于 2～4 年生或更老的枝条上部，也有生于枝条中部的，直立，未开放前长椭圆状卵形，长约 2cm，开放时呈穗状，长 5～6cm。雄球花初开时淡黄色，以后黄褐色。花粉粒两侧有气囊。雌球花生于新枝下部或基部叶腋，卵圆形或长椭圆状卵圆形，长 8～10mm。苞鳞紫红色，10～23 枚，一般 14～16 枚。珠鳞淡黄色，胚珠 2 枚，稀 1 枚，球花基部珠鳞无胚珠。

开始开花结实年龄有待观察。目前开花结实的银杉树龄都在百年以上。

各分布区的开花结实物候期大体接近，惟结实间隔期因树木生长状况、土壤性质和气候条件不同而有差异（表 1）。

表 1　银杉的开花结实物候和结实间隔期

产　地	开花期	果实成熟期	种子散落期	结实间隔期（年）
湖南界福山	雄球花:4 月上中旬～5 月上旬 雌球花：4 月下旬～5 月	开花翌年 9 月下旬～10 月上旬	开花翌年 10 月中旬	2～3
四川金佛山	5 月初～5 月底	开花翌年 10 月上旬	开花翌年 10 月中下旬	4～6

据彭德纯等（1984）对湖南界福山银杉的观察，雄球花 4 月上中旬出现，随即迅速膨大；4 月下旬～5 月上旬撒粉，历时仅几天；雌球花比雄球花晚开约 2 周；叶芽鞘膜脱落抽出新梢后两三天，雌球花开放，一般在上午 8 时左右，遇阴雨天或低温则推迟开放；1 个雌球花可开放 5～7 天，林分中开花末期出现的雌球花，开放时间可达 12 天；单株雌球花花期约 1 个月，林分内雌球花花期延续约 1 个半月。雌球花数多于雄球花，其比例有 7∶1 的，只有少数树木雄球花多于雌球花，为 2∶1。由于雄球花开放时间早于雌球花，且撒粉时间短，雌球花受粉机遇少，不利的天气条件和病虫灾害更会影响授粉，所以银杉自然座果率一般只有 15%～20%。雌球花授粉后苞鳞闭合。苞鳞变成青绿色时球果停止生长。翌年 4 月中下旬球果迅速膨大，5 月份基本完成径向生长，9 月下旬～10 月上旬球果成熟，10 月中旬种子脱落。

四川金佛山的银杉 5 月开花，翌年 10 月上旬球果成熟。银杉种子飞散后，球果宿存数年不落。

球果未熟时青绿色，成熟时黄褐色或暗褐色，长 1.9～5cm，径 1～2cm，卵圆形、长卵圆形或长椭圆形，种鳞背面密被微透明的短柔毛。种子长 5～8mm，褐色或黑褐色，有不规则的浅色斑纹，长椭圆形、长倒卵形或斜倒卵形，上部具膜质翅，长 1～1.5cm。种翅下端边缘不包卷种子，仅与种子向轴面贴生，易与种子分离。胚乳和胚呈乳白色。子叶 3 枚（少数 4～5 枚）。

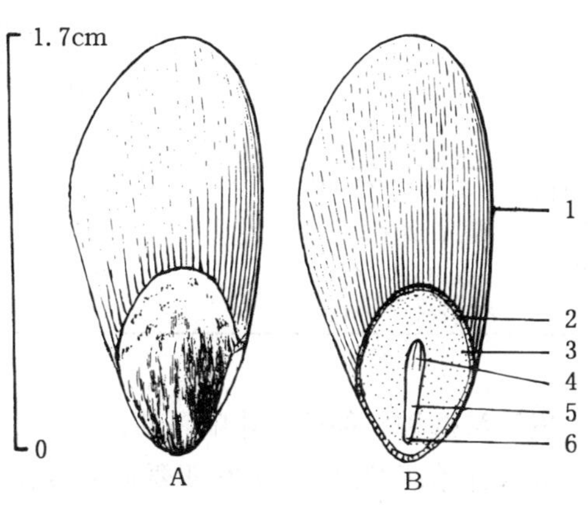

图 1　银杉种子的外形（A）及其纵切面（B）

1. 翅　2. 种皮　3. 胚乳　4. 子叶　5. 胚轴　6. 胚根

（黄应钦、田恒德绘）

银杉种子的形态及纵切面见图 1。银杉球果和种子颜色、形状和大小列入表 2。

球果的采收调制和种子贮藏　银杉结实丰年有明显的间隔期，应当趁大年采收。9～10 月间球果呈黄褐色或暗褐色时即可上树或在树下用采种工具采集球果。银杉多生长在悬崖峭壁或岩石裸露的山脊上，上树摘取球果要注意安全。

表 2　银杉球果和种子的颜色、形状和大小

分布区	成熟球果				成熟种子		
	颜　色	形　状	大　小（cm）		色　泽	形　状	长（mm）
			长	径			
贵州紫云万峰山自然保护区	暗褐色	卵形、长卵形或扁卵形	2.1～5.0	1.0～2.1	黑褐色	斜倒卵形或长倒卵形，略扁	6.0～8.0
广西龙胜花坪自然保护区	黄褐色	卵形或长卵形	1.9～4.1	1.3～2.0	黑褐色	长椭圆形	6.5～7.0
四川南川金佛山自然保护区	黄褐色	卵形	3.0～5.0	1.5～2.0	黑褐色	斜倒卵形，略扁	5.0～6.0
贵州道真沙河银杉自然保护区	暗褐色	卵形或长卵形	2.0～5.0	1.0～2.0	黑褐色或褐色	斜倒卵形，略扁	6.0

球果采用自然干燥法脱粒，去翅，风筛去杂后即得纯净种子。调制所得的饱满种子约占 50%。

银杉种子的贮藏通常采用“先干后湿法”，即先将种子干藏 2～4 周，再混湿沙层积，种与沙之比为 1∶10。层积期间应保持湿润状态，但也不应过湿，以防种子霉烂。层积 2 个月后可以取出播种。种子不宜在常温下干藏。据彭德纯等人的干藏试验：用纸袋干藏 52 天，发芽率为 3.4%，127 天后发芽率为 0.4%，142 天则全部丧失发芽能力。

出种率和种子质量的数据列入表 3。

表 3　银杉球果出种率和种子质量

分布区	出种率 (%)	千粒重 (g)	每千克纯净种子数 (万粒)
贵州紫云万峰山自然保护区	3.1	15.0～22.0	4.5～6.6
广西龙胜花坪自然保护区	2.1	16.0	6.2
四川南川金佛山自然保护区	3.2	13.6	7.3
贵州道真沙河银杉自然保护区	4.0	17.0	5.8

发芽和播种　种子有休眠习性。10 月下旬～11 月上旬在田间播种未经处理的种子，播后约 200 天开始萌发出土，萌发期要延续四五十天，且部分种子要到下一年才能发芽。本文作者在实验室按常规检查发芽能力时，虽一再延长测定时间亦难以达到预期结果；又在室外播种，4 个月后仍未见有种子萌发。据王必农等（1980）试验，经过“先干后湿”贮藏的种子于早春 1 月播种，用 50μg/g 的吲哚丁酸溶液浸种 16 小时，再用 25℃温水浸种 3 小时，捞出阴干后播种，播后约 30 天，幼苗相继出土。

银杉结实量少，又有结实间隔期，种子获取不易，生产上常采用容器育苗节约用种。容器多为塑料袋、纸袋、折叠式蜂窝纸杯或瓦盆等。容器内装填营养土（用富含腐殖质的酸性土或加堆肥的壤土，经整细后加 3%的过磷酸钙和菌根土充分搅拌而成），播种前还可用 0.5%高锰酸钾液浸种 30 分钟灭菌，清水冲洗阴干后播入容器，每个容器 2～3 粒，覆土 0.5cm 并盖草。

出土萌发。子叶 3 枚，少数 4～5 枚，线形，革质，初为浅绿色，后转为深绿色，下面有两条白色气孔带。子叶出土后 15～18 天生出初生叶，初生叶较短，条形。下胚轴长 3～5cm，上胚轴近于无。幼茎初时淡绿色微带紫红色，后转为绿褐色。根黄褐色，早期极不发达，1 年生苗仅近根颈处有少量须根。苗期生长缓慢，1 年生苗高约 6cm。银杉种子的萌发及幼苗生长

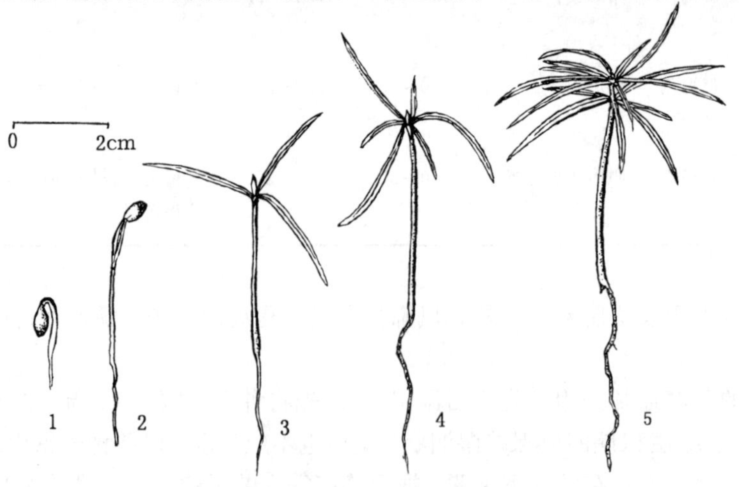

图 2　银杉种子的萌发及幼苗生长情况

1. 下胚轴延伸　2. 子叶初现　3. 子叶展开　4. 初生叶出现　5. 幼苗形成

（田恒德仿《热带亚热带主要树种采种育苗技术》）

情况见图 2。出圃时间根据培育要求而定。

银杉可以嫁接繁殖，嵌接或切接。砧木可以选用与银杉亲和力较强的湿地松。穗条采用健壮树上有饱满顶芽的 1～2 年生粗壮主侧枝，长 10cm。3 月下旬至 4 月上旬嫁接成活率最高。

（赵德铭）

雪　松　属
Cedrus Trew

（松科　Pinaceae）

生长习性、分布和用途　本属 4 种，分布于非洲北部、小亚细亚、阿富汗、尼泊尔和印度。我国引种雪松始于 20 世纪初，目前旅顺、大连以南，特别是长江中下游各大城市已普遍栽培作为庭园观赏树、风景树和行道树，南京市已用于造林。其它 3 个种我国引种栽培较少。本文以描述我国栽培面广的雪松为主，也汇集了其它 3 个种的一些资料。它们的名称、生长、分布和引种情况列入表 1。

雪松属的树种均为常绿乔木。喜温和凉润气候，抗寒性强，较喜光，幼树稍耐庇荫。酸性土、微碱性土均能适应，在粘重黄壤及干旱瘠薄土壤上也能生长，但不耐水涝。浅根性，抗风力弱。幼树对氟化氢及二硫化碳等气体相当敏感，在工矿区常受害。木材坚硬，抗腐性强，具芳香，蒸馏可得芳香油，是珍贵用材。雪松树干高耸，主枝先斜出，后平展，小枝下垂，冠形既伟岸挺拔，又飘逸秀雅，是世界著名的庭园观赏树种。

表 1　雪松属树种的名称、生长、分布和引种情况

中　名	学　名	树高 (m)	胸径 (m)	分　布	引种情况
北非雪松（大西洋雪松）	*C. atlantica* Manetti	40	3.0	地中海南部沿岸的阿特拉斯山区、阿尔及利亚、摩洛哥	美国、英国、澳大利亚、新西兰和南非有栽培。我国有少量引种
塞浦路斯雪松（短叶雪松）	*C. brevifolia* (Hook f.) Henry	12	0.7	塞浦路斯	南京有少量引种
雪　松（喜马拉雅雪松）	*C. deodara* (Roxb.) G. Don	80	4.3	阿富汗、尼泊尔、印度	旅顺、大连以南各大城市广泛栽培
黎巴嫩雪松	*C. libani* Rich.	25～40	3.0	黎巴嫩、叙利亚向南到小亚细亚	俄罗斯、英国、澳大利亚、南非和美国有栽培。上海、庐山曾引种

雪松属树种在树形和叶色等方面存在着种内变异。例如原苏联的 A. N. 考高斯尼柯夫曾将雪松划分为金叶雪松、银叶雪松等 7 个自然类型。日本学者也在雪松这个种中划分出雪白雪松、变色雪松、轮生雪松。我国南京地区也初步将雪松分出厚叶雪松、垂枝雪松、翘枝雪

松等几个类型。

南京地区栽培的雪松生长迅速，适应性强：单株孤立木胸径的年平均生长量可达1.5cm，20多年生的小片人工林的胸径年平均生长量也在1.0cm以上，年高生长量可达50～80cm。

开花结实 雪松寿命长，在原产地可达700～800年。生殖能力到来得较迟。南京地区栽培的雪松20～30年生开始开花结实。昆明的雪松15年生开始开花，18年生开始结实。大量结实始于30～40年。结实有大小年现象，每2～3年有1次丰年。据在南京观察，近95%的雪松植株为雌雄异株，大约只有5%的植株为雌雄同株。雄球花直立，多着生在树冠中下部，通常在6月份出现，6月底开始膨大，8～9月迅速增大，长3～5cm，直径0.8～1.5cm，近黄色，10月下旬～11月上旬开始撒粉。雄球花在开放时急骤伸长到5～7cm，径2cm。雌球花出现在8～9月，呈长卵形，多数着生于树冠上部或顶端，直立，10月底～11月初成熟，呈浅绿色，微有白粉，此时长仅1～2cm，径0.4～0.5cm。受粉闭合后呈淡紫色。花粉粒大而重，虽具气囊，但飘浮能力较差，雌花难以接受花粉。此外，雪松雌雄异熟现象十分明显。南京地区雪松的雄球花往往比雌球花早熟10～20天，所以结实很少，空瘪粒多，需要辅以人工授粉才能获得饱满的种子。

雪松球果在开花的翌年秋季成熟。成熟球果大，长7～12cm，径5～9cm，近卵圆形至椭圆状卵圆形，种鳞多数，排列紧密，木质。种鳞倒三角形，顶端宽平，两侧边缘薄，有不规则细锯齿，背面密被锈色毛。每片种鳞着生种子2枚。种子上部有倒三角形翅。种翅褐色，宽大，膜质，比种子长。种子连翅长2.2～3.7cm，种子长1.4～1.6cm，宽0.6～0.7cm，蜡黄色至灰白色，外种皮薄，革质，易破损。内种皮膜质，有韧性。种子富含油脂，易分泌，粘性强。胚乳丰富，黄白色，具芳香。胚粗壮，黄色，有时绿色。雪松种子的外形和解剖结构见图1。

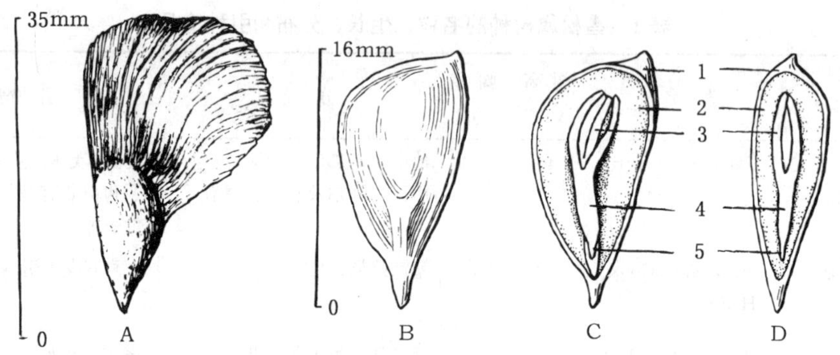

图1 雪松带翅种子外形（A）、去翅种子外形（B）及其纵切正面（C）和纵切侧面（D）

1. 种皮 2. 胚乳 3. 子叶 4. 胚轴 5. 胚根

（A. 田恒德仿《主要树木种苗图谱》；B、C和D. 张世经绘）

球果的采收调制和种子贮藏 球果成熟后，种鳞与种子一同脱落，仅中轴暂时不落。因此当球果颜色由浅绿色微被白粉转向浅棕色且鳞片微裂时即应采摘球果。晒干，使球果自然开裂，抖出带翅种子。轻揉去翅，簸扬剔出种翅和杂质，筛去瘪粒，即得纯净种子。雪松种子种皮较薄，易受损伤，调制收藏和运输中要避免挤压。雪松种子寿命不长，一般是在普通干藏后翌年春播。如果种皮完好无损，且能将种子含水量降至10%以下，在1～3℃温度下密封干藏，生活力可以保持3～6年。雪松种子呈不规则的长三角形，浅黄色。净度可达90%以上。千粒重85～120g，每千克种子8 000～12 000粒。发芽率为60%～80%，近年从印度进口的雪松种子中曾发现种子小蜂（*Eurytoma* sp.），这种小蜂国内尚无记载。雪松属树种的球果和种子性状见表2。

表2　雪松属树种的球果和种子性状

树　种	球果成熟期	球　果		种　子	
		长（cm）	宽（cm）	千粒重（g）	每千克粒数
北非雪松[①]	9～10月	5.1～7	4.4	55～130	7 500～18 000
塞浦路斯雪松[①]	9～11月	7.1	4.1	—	—
雪　　松	10～11月	7～12	5～9	85～120	8 300～12 000
黎巴嫩雪松[①]	8～10月	8.3～10.2	3.8～6.4	35～180	5 500～26 900

①　引自《美国木本植物种子手册》

发芽和播种　雪松种子无休眠习性。国际种子检验协会（ISTA，1995）规程规定的发芽测定条件是，种子在3～5℃湿润条件下冷冻21天后置于吸水纸上，在20℃恒温或白天30℃、夜间20℃的变温条件下测定21天。发芽过程无需光照。美国和加拿大官方种子检验人员协会（AOSA，1965）的规定是20℃的恒温，在吸水纸上用经过预先冷冻的种子进行测定。根据我国国家林业局南方林木种子检验中心的经验，未经处理的雪松种子置床后7天开始发芽，发芽过程35天，发芽高峰不明显；而经过预冷处理的种子，置床后3天即开始发芽，25天发芽结束，发芽高峰出现在第5～20天。雪松属种子发芽测定情况见表3。

表3　雪松属种子发芽测定情况

树　种	发芽势		发芽率	
	累计（%）	天数	平均（%）	变动范围（%）
北非雪松[①]	27	20	53	9～74
雪　　松	50	20	64	49～88
黎巴嫩雪松[①]	23	20	46	13～92

①　引自《美国木本植物种子》1974年版

春播或冬播。一般多用春播。条播，每米长播种行上播8～15粒，每666平方米约播3～4kg。或集中催芽后将芽苗移栽于圃地或育苗容器中。干藏种子可在播种前浸种1～2天，其间换水1～2次。雪松属种子为出土萌发型，子叶6～13。雪松子叶9～10（13），线状锥形，粉绿色，上面两侧各有6～7条白粉气孔线。上胚轴短而不明显，幼茎淡黄绿色，被白粉。初生叶螺旋状互生，针形。下胚轴圆柱形，粗壮，长5～6.2cm，径2.5mm（图2），初呈淡绿色，后变淡红褐色。主根长。幼苗出土初期稍耐荫。1年生苗高15～20cm。国内多用换床大苗定植。雪松还可扦插繁殖。

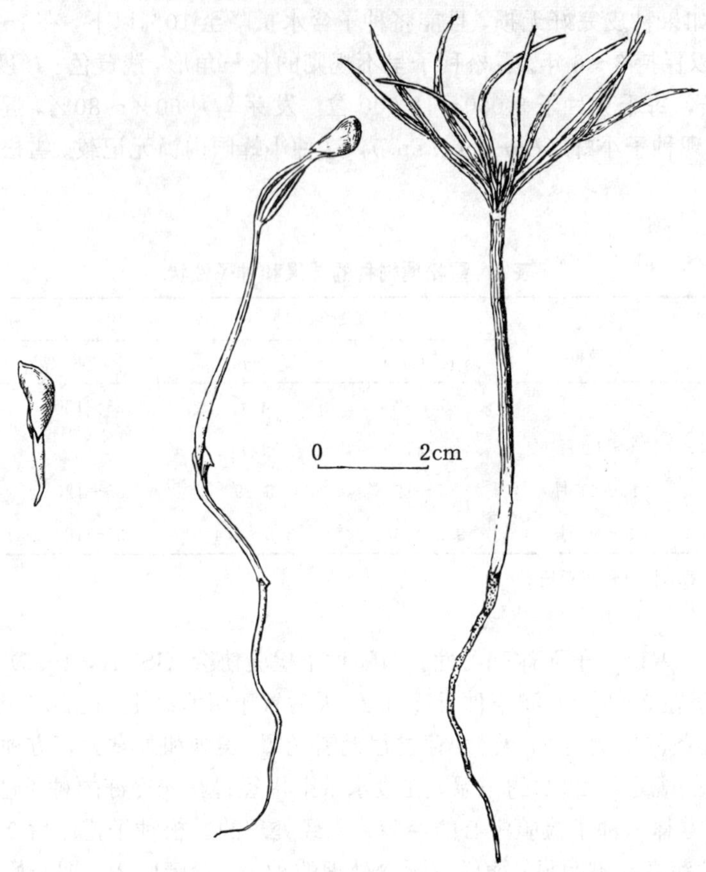

图 2 雪松种子萌发后第 1、5、12 天的幼苗生长情况
（张世经、田恒德绘）

（吴琼美）

油 杉 属
Keteleeria Carr.

（松科 Pinaceae）

生长习性、分布和用途 油杉属约 12 种和 1 变种，我国有 10 种和 1 变种，本文描述 8 种 1 变种。常绿乔木，高达 20m 以上，胸径 80cm 以上。苗期较耐庇荫，成龄树喜光。生长较迅速，主干通直圆满。多数种喜温暖湿润、干湿季分明的气候。较能耐干旱贫瘠土壤，多数种适生于酸性土，少数种分布钙质土。抗逆性较强。目前多属野生，近年已推广人工造林。木材硬度适中，不变形，较耐腐，为优质用材。种子富含油脂。树形美观。它们的名称、生长、分布及用途见表 1，其中黄枝油杉和油杉为渐危种，柔毛油杉为稀有种，都已列入《中国植物红皮书》。

表 1　油杉属树种的名称、生长、分布和用途

中　名	学　名	树高 (m)	胸径 (cm)	分　布	用　途
黄枝油杉	*K. calcarea* Cheng et L. K. Fu	30	180	桂、黔	材用、观赏、油脂
江南油杉	*K. cyclolepis* Flous	30	160	滇、黔、桂、粤、湘、赣、浙	材用、观赏
铁坚油杉	*K. davidiana* (Bertr.) Beissn.	50	250	秦岭以南的甘、陕、川、湘、鄂、桂、黔	材用、观赏
青岩油杉	*K. davidiana* (Bertr.) Beissn. var. *chien-peii* (Flous) Cheng et L. K. Fu	20	80	黔、桂	材用、观赏
云南油杉	*K. evelyniana* Mast.	40	100	滇、黔、桂	材用
台湾油杉	*K. formosana* Hayata	35	250	台	材用
油杉	*K. fortunei* (Murr.) Carr.	30	100	浙、闽、粤、桂	材用、观赏、油脂
矩鳞油杉	*K. oblonga* Cheng et L. K. Fu	30	150	桂	材用、观赏
柔毛油杉	*K. pubescens* Cheng et L. K. Fu	30	160	桂、滇、湘	材用、造纸

开花结实　10～15 年生开始结实，25 年以上为正常结实期。大小年间隔期一般为 1 年。球花单性，雌雄同株。雄球花 4～8，簇生于侧枝顶端，间或生于叶腋。雄蕊多数，花药 2 枚，花粉粒有气囊。雌球花单生侧枝顶端，直立，由多数螺旋状排列的珠鳞与苞鳞组成。珠鳞生于苞鳞之上，基部合生。每苞鳞胚珠 2，花后珠鳞增大成种鳞。1979 年对广西西北部隆林县分布的江南油杉、青岩油杉、云南油杉、油杉、矩鳞油杉、柔毛油杉等 6 个种的天然林进行了物候观察，1978～1984 年对南宁市郊人工栽培的江南油杉、矩鳞油杉进行了 6 年的物候观察，各个种的开花结实期在同一地区基本接近。2 月上旬至中旬，雄球花开始成簇出现，黄绿色略带紫色，着生于头年生的枝顶。2 月中旬雄球花伸长为圆柱状，长约 1cm，紫色略带黄色。2 月下旬少数雄球花开放，花粉黄色，3 月上中旬为盛花期，下旬为末花期。雌球花 2 月中旬出现，3 月上旬至中旬开放。果实当年 10 月下旬开始成熟，11 月为成熟盛期。近年发现云南油松的球果常受蛀心昆虫严重危害，有时受害株几占 95%以上，球果受害率在 60%以上，武春生等人（1987）鉴定这种昆虫为一新属新种。

球果圆柱状，直立。种鳞木质，宿存。苞鳞短于种鳞。成熟时球果由粉绿色转为青色，充分成熟时为褐色。11 月中下旬，种鳞陆续开裂。每种鳞含种子 2 粒，开裂后带翅的种子随风飘落。种子三角状长椭圆形，剖面近似菱形，各个种的种翅的形状和大小有差异。翅厚膜质，有光泽。子叶 2～4，胚乳丰富，含油质。成熟球果和种子的形态见表 2 和图 1。

表 2　油杉属树种球果和种子的形态特征

树种	球果		种鳞开裂特征	种子形态						
	长(cm)	径(cm)		种子带翅长(mm)	种翅			种子		
					形态	长(mm)	宽(mm)	长(mm)	宽(mm)	厚(mm)
黄枝油杉	11～14	4～5.5	边缘外曲或不曲而先端微内曲。背面露出部分密生短毛，无白粉	24～30	中下部较宽	18～22	10～12	13～15	5～7	4～5
江南油杉	7～18	3.5～6	边缘微内曲，稀外曲。背面露出部分无毛或近无毛	27～29	中上部较宽	22～24	10～12	14～15	4～6	4～5
铁坚油杉	8～21	3.5～6	边缘外曲。背面露出部分无毛或疏生短毛	25～28	先端反曲，下部较宽	20～23	8～10	13～15	4～6	4～5
青岩油杉	8～15	3.5～6	边缘外曲	28～33	先端反曲，下部较宽	20～26	10～11	14～16	4～5	4～5
云南油杉	9～22	4～6.5	上部边缘向外反曲。背部有毛或近无毛	29～39	先端微曲，下部较宽	25～32	10～13	12～17	5～7	4～5
台湾油杉	5～15	4～4.5	上部边缘向外反曲	25～28	中部较宽	20～24	10～11	11～14	5～6	4～5
油杉	8～18	5～6.5	上部边缘内曲背面露出部分无毛	24～27	上部较宽	18～20	13～15	13～15	5～6	4～5
矩鳞油杉	15～22	4.5～6.5	边缘不反曲或先端微内曲。背部无毛	24～26	中下部较宽，近矩形	18～22	10～12	14～16	5～7	4～5
柔毛油杉	7～11	3～3.5	边缘微向外反曲。背面露出部分密生短毛	25～27	中上部较宽	18～23	10～12	11～14	4～5	4～5

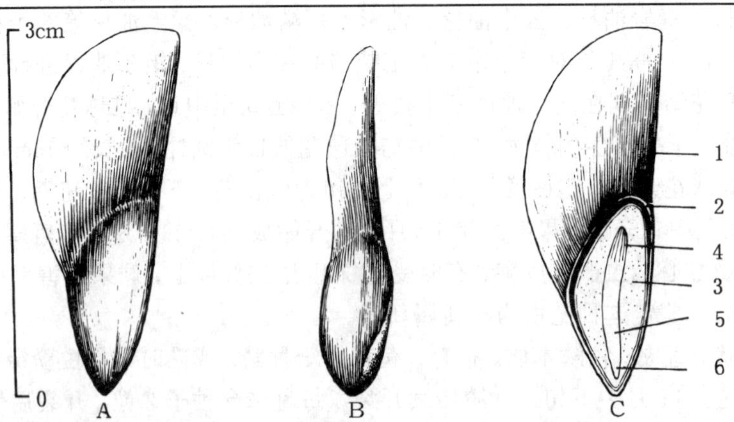

图 1　黄枝油杉种子的正面外形（A）、侧面外形（B）及其纵切面（C）

1. 种翅　2. 种皮　3. 胚乳　4. 子叶　5. 胚轴　6. 胚根

（黄应钦绘）

球果的采收调制和种子贮藏　11 月上旬为主要采果期。果实成熟的主要标志是球果外表附着的白粉已部分脱落；种皮呈黄褐色，表面平展有光泽；胚乳发育饱满，呈固体状态，可以压出油脂。用高枝剪或钩刀采下球果。球果采回后堆放 3～5 日，有利于部分晚熟种子充分成熟，再摊放于弱光下晒 1～3 日，俟种鳞变为褐色时转至室内通风处摊放，任其自然开裂。种鳞全部开裂后，将球果翻动数次并稍加敲打，种子即可脱出。筛出种子，轻揉去翅后扬去种翅和瘪粒，即得纯净种子。鲜果出种率、种子的净度和质量见表 3。

表 3　油杉属树种球果出种率和种子的净度、质量

树　种	按质量计算的出种率（%）①	净　度（%）	千粒重（g）	每千克纯净种子粒数
黄枝油杉	2～6	88～98	142	7 000
江南油杉	3～7	80～96	120	8 300
铁坚油杉	2～7	85～95	128	7 800
青岩油杉	2～6	86～96	132	7 600
云南油杉	3～7	85～95	138	7 200
台湾油杉	2～6	85～95	122	8 200
油　杉	3～6	86～98	127	7 900
矩鳞油杉	4～8	90～98	138	7 200
柔毛油杉	2～6	85～95	108	9 300

①　不含瘪粒种子

油杉属种子富含油脂，忌日晒和高温，脱粒后应在室内通风处摊晾 2～4 日。当种子的含水量下降到 12%左右时，方可装袋或装入竹筐内，贮存于室内凉爽干燥处。用直观方法鉴别种子优良的尺度是，外表有光泽，无皱纹，无油脂外溢，种粒饱满。种子发芽力的保存期短，在常温条件下，从当年 11 月起，约可保存 5 个月。4 月初以后，华南地区日均温多在 19 ℃以上，发芽能力即开始下降，4 月中旬可降至 40%～50%，5 月中旬降至 5%以下。在 5～8 ℃的低温条件下贮存，发芽能力的保存期约可延长 5 个月，但 9 月份以后，发芽率也明显下降至 30%以下。一般认为，影响发芽能力保存的主要原因是，种子富含油脂，当日均温上升到 15 ℃以上或存放时间过久，便有油脂外溢，表明已经丧失发芽能力。

发芽和播种　种子无休眠习性。发芽时要求日均温在 20℃以上。广西林业科学研究所 1987 年 3 月下旬、5 月中旬和 1988 年 12 月先后在室外沙床和室内温箱，对黄枝油杉、江南油杉、矩鳞油杉和油杉作过发芽测定，结果见表 4。

表 4　油杉属树种的发芽能力及其测定条件①

树　种	测定温度（℃）		发芽势（%）		发芽率（%）	
	发芽箱	室外沙床	计算天数	一般数值	计算天数	一般数值
黄枝油杉	—	24～26	13	35	33	52
江南油杉	—	24～26	—	不明显	35	66
油　　杉	28	—	—	不明显	31	50
矩鳞油杉	—	24～26	14	40	35	69

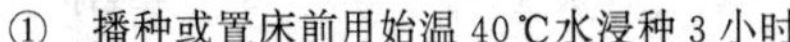
①　播种或置床前用始温 40℃水浸种 3 小时

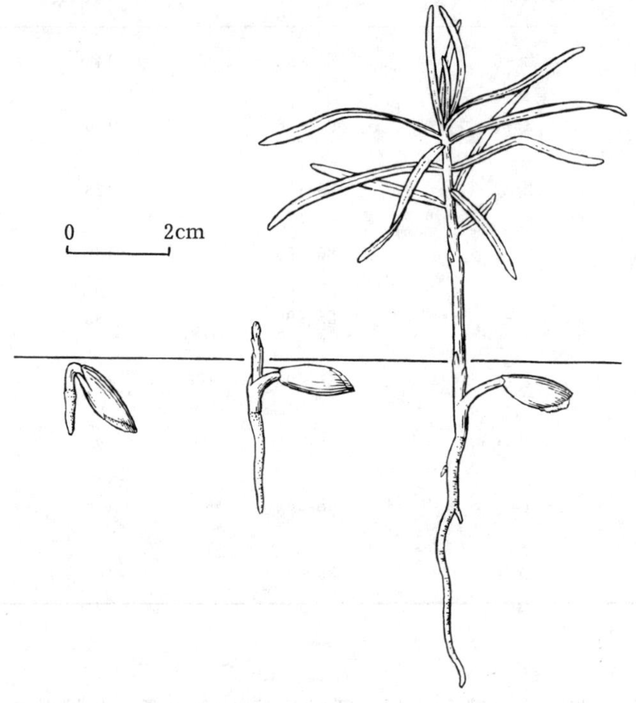

图 2　矩鳞油杉种子萌发后第 3、10、30 天的幼苗生长情况

（黄应钦仿《热带亚热带主要树种采种育苗技术》）

留土萌发。播种后 9～12 天，胚根开始萌出，胚芽随后生出。子叶 3～5 枚，针状，存留于种壳内的胚乳中。上胚轴长 0.2～0.3cm，下胚轴近于无。种子萌发后 10～12 天幼苗出土，同时发出初生叶，茎基部具初生不育叶 7～16 枚（图 2）。

3 月上中旬为适宜播种期，播种前用 0.5%的高锰酸钾溶液浸种 30 分钟，取出洗净后用 40℃温水浸种至自然冷却，可提早 2～3 天发芽。条播。每 666m² 播种 12～15kg，覆土约 1cm。1 年生苗可以出圃。油杉属树种采用裸根苗造林，早期生长缓慢，近年多改用容器育苗。早春将种子保温催芽，点播至容器内，培育 3～4 个月后可以出圃定植。

（韦增健）

落叶松属
Larix Mill.

（松科　Pinaceae）

生长习性、分布和用途　本属约18种，分布于北半球温带高山与寒温带、寒带地区。我国有10种1变种，引种栽培2种。本文介绍7种（表1）。落叶乔木。树高25～50m，胸径80～100cm。喜光，耐寒，速生，适应性强。主产东北、华北、西北和西南地区（见表1）。据傅辉恩、刘建勋（1987）报道，华北落叶松在祁连山北坡引种，初期生长比当地惟一的建群种青海云杉迅速。本属木材坚韧细致，纹理通直，耐水湿，抗腐朽，供建筑、桥梁、造船、矿柱、电杆、木纤维等用，为分布区重要树种。

表1　落叶松属树种的名称、生长和分布情况

中　名	学　名	树高(m)	胸径(cm)	分　布	供　稿
落叶松（兴安落叶松）	*L. gmelini* Rupr.	35	90	大、小兴安岭。俄罗斯	103
日本落叶松	*L. kaempferi* (Lamb.) Carr.	30	100	原产日本。东北、华北、西北、西南引入栽培	105
四川红杉	*L. mastersiana* Rehd. et Wils.	25	80	川西北	508
黄花落叶松（长白落叶松）	*L. olgensis* Henry	30	100	东北。朝鲜半岛、俄罗斯远东	106
红　杉	*L. potaninii* Batal.	50	100	川西、甘南	508
华北落叶松	*L. principis-rupprechtii* Mayr	30	100	晋、冀。陕、甘、宁、青、内蒙古栽培	401
西伯利亚落叶松	*L. sibirica* Ledeb.	40	80	新。俄罗斯、蒙古	204

开花结实　开始结实年龄一般为10～30年，多在15年左右。西伯利亚落叶松的大量结实在70～100年，每株可产种子3～4kg，150年以后结实能力逐渐减弱，每株产种量减到2.5kg。丰年结实间隔期3～5年。大兴安岭林区落叶松的天然林，在1950～1971年的21年间出现过7次丰年，即1950、1955、1958、1961、1963和1971年，其中1961年和1963年产量略低，人称小丰年。刘益康、王冰（1993）报道了小兴安岭发现的5种落叶松球果花蝇（*Lasiomma* spp.），在球果歉年或平年，受它们危害的球果几达100%。据他们研究，每年5月中旬球果种蝇开始羽化，成虫必须寻找蜜源补充营养。此时用诱集液毒杀，可使球果被害率由对照区的74.0%下降到5.5%。在辽宁东部，落叶松球果卷叶蛾（*Petrova perangustana*）危害的球果可达40%，刘振陆和王洪魁（1985）报道过它的被害状和防治方法。分布在黑龙江、内蒙古和山西等地的落叶松种子广肩小蜂（*Eurytoma laricis*）能随种子调运作远距离传播，是国内森林植物检疫对象。

球花单性，雌雄同株。雌球花和雄球花均单生于短枝顶端，无规则地着生在2～5年生小枝的各个侧面，先叶或与针叶同步开放。雄球花黄色，球形或长圆形，花粉无气囊。雌花球较小，常具短柄，直立，红色或紫红色。花芽分化于前一年的6～7月，翌年4～5月开花，当年秋季种子成熟。王景章和王有才（1985）连续两年对辽宁省清源县日本落叶松种子园64个无性系的花期做过观察，并分析过它们雌花和雄花同步的程度。落叶松属成熟的球果黄褐色或紫褐色，每片种鳞基部着生2粒种子。种子成熟时种鳞开裂，种子散落，主要由风传播。空球果在树上宿存一段时间。7种落叶松开花结实习性见表2。

表2 落叶松属树种开花结实习性

树种	观察地点	开花结实年龄（年）	丰年间隔期（年）	花期	果实成熟期	球果成熟特征	种子散落期
落叶松	黑龙江	15～20	3～5	5月上旬～5月下旬	8月下旬～9月中旬	黄绿或暗紫色	9月上旬～10月中旬
日本落叶松	辽宁	15	3	4月上旬～4月下旬	9月上旬～10月上旬	黄褐色	10月下旬～11月中旬
四川红杉	四川小金	20～30	3～5	4月上旬～4月下旬	9月下旬～10月上旬	淡褐色	10月上旬～10月中旬
黄花落叶松	吉林	15	2～4	4月下旬～5月上旬	8月下旬～9月下旬	褐色或淡黄色	9月中旬～10月下旬
红杉	四川小金	20～30	3～5	4月上旬～4月下旬	9月下旬～10月上旬	紫褐色或灰褐色	10月上旬～10月中旬
华北落叶松	河北围场	8～10	2～5	5月上旬～5月下旬	8月下旬～9月下旬	黄褐色	9月下旬～10月下旬
西伯利亚落叶松	新疆	13～15	3～5	5月中旬～5月下旬	9～10月	褐色或淡褐黄色	9～10月

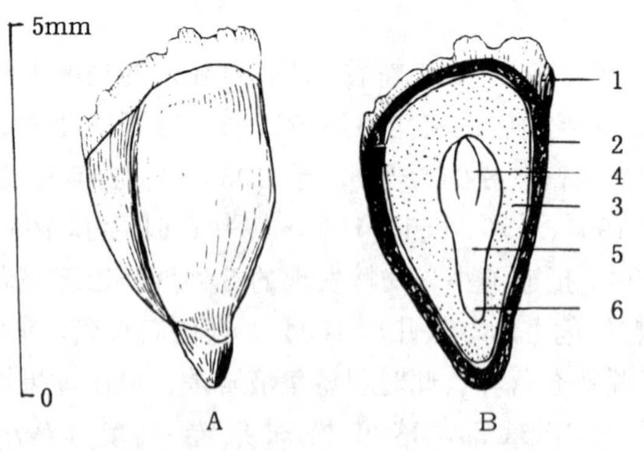

图1 华北落叶松种子外形（A）及其纵切面（B）
1. 种翅（部分） 2. 种皮 3. 胚乳 4. 子叶 5. 胚轴 6. 胚根
（胡冬梅、林平绘）

种子几为三角形或斜倒卵状椭圆形，有膜质翅，不易脱落。外种皮硬，淡褐色或灰白色，有光泽。内种皮膜质，灰栗褐色。具胚乳（图1）。落叶松属各个种的种子外形十分相似。人们正在寻求种子识别的新途径（周学权，1989）。

在探讨山西华北落叶松种群变异规律时，李文荣和齐力旺等人（1992）研究过球果的长、宽和种子千粒重的变异范围。落叶松属成熟球果和种子的大小与特征见表3。

表 3　落叶松属树种成熟球果及种子的大小和特征

树　种	成熟球果（cm）		成熟种子（mm）		种子特征
	长	径	长	带翅长	
落叶松	1.2～3.0	1.0～2.0	3～4	10	灰白色，斜卵圆形
日本落叶松	2.0～3.5	1.8～2.8	3～4	11～14	灰白色，倒卵圆形
四川红杉	2.5～4.0	1.5～2.0	2～3	7～9	灰白色，斜倒卵圆形
黄花落叶松	1.5～4.6	1～2	3～4	9	淡黄白色，斜倒卵圆形
红杉	3～5	1.5～2.5	3～4	7～10	灰白色，斜倒卵圆形
华北落叶松	2～4	2.5	3～5	10～12	灰白色，斜倒卵圆形
西伯利亚落叶松	2～4	1.5～3.0	4～5	10～15	灰白色，斜倒卵圆形

红杉有一个自然类型，特点是果型较大，种鳞较厚。根据雌球花和球果的颜色，西伯利亚落叶松可以区分出 3 种类型：紫红果型，果大，长 3.5～4.8cm，径 1.6～2.4cm，种子千粒重 8～9g；绿果型，果小，长 3～4cm，径 1.5～1.7cm，种子千粒重 6.5～7.5g；绿果型的花期迟于紫红果型，可能是由于这个原因球果虫害较轻。另外还有一个中间类型，但性状不甚稳定，一般是球果上部紫色，下部绿色，果型居以上两种类型之间，长 3.6～4.8cm，径1.6～1.8cm，种子千粒重 7～8g。日本落叶松的成熟球果黄褐色，在我国引种栽培的人工林中出现了球果性状各异的几个类型，生长也比较迅速：红果型，幼果呈红紫色；杂色果型，种鳞边缘绿色，中间红紫色，或基部绿色而上部红紫色；大果型，幼果绿色或淡黄绿色，成熟时紫褐色，球果比较大。西伯利亚落叶松单个球果重 1～3.7g，平均 2.4g，其中有种子 21～74 粒，平均 57 粒。

影响种子产量的主要因素有温度、光照、水分和病虫危害等。种子丰收两年后，若 6～7 月间气温高，降水少，日照持续时间长，则花芽分化顺利，翌年有可能种子丰收。花期冻害和落叶松球果花蝇（*Lasiomme* spp.）危害是我国东北地区几种落叶松种子减产的重要原因。落叶松球果花蝇于开花时羽化为成虫，在球果基部种鳞间产卵，经 8～10 天孵化为幼虫，绕果轴自下而上危害种子。幼虫期约 1 个月左右，6～7 月间离开球果落地化蛹。在羽化成虫产卵前喷药毒杀，效果最佳。危害西伯利亚落叶松的主要害虫有落叶松球果花蝇（*Lasiomma lanicicola*）、落叶松球果瘿蚊（*Resseliella sibirica*）和冷杉梢斑螟（球果螟蛾）（*Dioryctria abietella*）。刘振陆和王洪魁（1985）报道过落叶松球果卷蛾（*Retinia perangustana*）的防治方法。

地理变异　落叶松属许多树种存在着明显的地理变异（徐化成，1990；马常耕，1990）。例如，黄花落叶松的种源试验表明，生长性状在种源间呈现海拔梯度渐变为主，纬向渐变为辅的基本变异模式。低海拔、等效纬度偏低的种源比高海拔、高等效纬度的种源生长快。落叶松的生长变异是经向变异为主、纬向变异为辅的经纬双重连续渐变模式。这种变异反映了各种源对气候生态因子的干—湿、冷—暖渐变模式的适应。华北落叶松和日本落叶松生长性状的变异则无明显的地理规律性，可能是由于随机遗传漂变的缘故。种源试验表明，日本落叶松在抗鞘翅蛾危害和豪猪危害，落叶松在抗溃疡病，黄花落叶松在抗旱、抗病，以及华北落叶松在造林成活率方面，种源间都有很大的差异。

根据种源试验结果，并参考分布区内气候、植被的特点，我国于1986年颁布了落叶松、黄花落叶松和华北落叶松的种子区划。落叶松区划为大兴安岭北部种子区、大兴安岭东南部种子区（包括甘河亚区、库都尔亚区、阿尔山亚区）和小兴安岭种子区。黄花落叶松区划为牡丹江种子区和长白山种子区（包括北长白山亚区、南长白山亚区）。华北落叶松区划为北部区（包括冀北和蒙古高原山地）、中部区（含山西中北部和东部的吕梁山、太行山山地）和南部区（以黄土高原为主体的太岳山、吕梁山、陇山、六盘山、小陇山山地）。原则上各地只能使用本种子亚区或种子区的种子。至于尚未划分种子区的树种，用种原则是首先使用本地优良种源种子，也可以使用与本地区气候、土壤条件类似的种源所生产的种子。张万雄和刘君（1992）研究过兴安落叶松优良采种林分的划分方法。

球果的采收调制 球果达到正常大小，由绿色转为黄褐色或紫褐色且种鳞微裂，这些特征标志种子成熟。落叶松属球果成熟后种鳞容易开裂，种粒轻小有翅，短期内便极易飞散。落叶松从球果出现成熟颜色到种子飞散，历时只有15～20天，应当不失时机地立即组织采种。可从伐倒木上采摘球果，或上树用钩、镰截断小果枝于地面上收集。为保护母树，有的地方要求截断的果枝直径不得超过1.5cm，总枝数不得超过树冠小枝的1/3。不过，伤断枝条多少都会影响以后的产量，甚至延长丰年间隔期。

采回的球果摊放在阳光直射的晒场或通风良好的室内自然干燥。有条件的地方可以采用人工加热干燥。自然干燥时，球果摊放厚度不要超过7～10cm，每天翻动2～3次，经3～5天种鳞开裂，种子全部脱出。人工加热干燥球果，温度不宜超过40～45℃，烘烤时间为7～9小时。

球果开裂后，通过震荡机或筛选，将种子分离出来，并用去翅机或将种子装入袋中搓擦去翅，最后经风扬筛选清除杂质后再充分晾晒。调制后的种子，要求净度为95%以上，含水量在10%以下。我国目前7种落叶松球果出种率和有关种子质量的数据见表4。

表4 落叶松属树种球果出种率和有关种子质量的数据

树　种	出种率（%）	净　度（%）		千粒重（g）		每千克纯净种子粒数（万粒）	
		一　般	变动范围	一　般	变动范围	一　般	变动范围
落叶松	1.5～3.0	90	80～95	4.0	2.3～5.1	25.0	19.6～43.5
日本落叶松	3.0～4.0	94	90～96	3.6	2.8～4.6	28.0	21.3～35.7
四川红杉	2.0～5.0	70	67～87	2.3	2.2～2.4	43.5	41.6～45.0
黄花落叶松	4.0～6.0	96	94～98	3.8	3.1～4.6	26.3	22.0～32.3
红　杉	2.0～5.0	86	75～97	2.9	2.4～3.1	35.0	32.4～42.4
华北落叶松	3.0～4.0	95	90～98	5.2	4.5～7.5	19.2	13.0～22.0
西伯利亚落叶松	2.0～5.0	96	90～98	7.0	3.7～9.6	14.3	10.3～27.0

落叶松属种子中空瘪粒较多，优良度一般只有50%～60%，特别是歉年时这个现象更为严重。

种子贮藏 影响落叶松种子寿命的主要因素是种子含水量和贮藏库的温度。将种子含水量控制在 8%～9%密封贮藏，普通库可贮藏 2～3 年，机械冷库可贮藏 3～4 年。吉林省对黄花落叶松种子的研究结果表明，在±5℃的机械冷库中用铁桶密封贮藏，生命力能保持 10 年以上。据李锦文和崔红（1988）研究，在 0～5℃的冷库中贮藏 39 个月的华北落叶松，只有含水量为 5%～7.5%的种子发芽率基本上没有变化，含水量为 9%～10.8%的种子发芽率下降了 10 个百分点，含水量为 17%～18%的发芽率则已降为零。

发芽测定　落叶松属的种子发芽并不困难，即使有休眠，程度也比较轻，不过很多地方习惯于在置床前用温水浸种 1 昼夜。这个属的种子中空粒较多，不易同饱满种子分离开来，因此发芽率不高。多数树种发芽的适宜温度在 20～30℃（表 5），例如据肖东玉、陈开秀研究（未刊稿），西伯利亚落叶松在 10～15℃、20℃、25℃和 30℃下 20 天的发芽率分别为 17%、71%、83%和 85%。不过通常认为变温有利于发芽。据于淑兰和孙秀琴（1993）研究，在1～5℃温度下层积 14 天的落叶松种子，室内测定发芽能力的最佳条件是 25℃/20℃的昼夜变温，每天光照 8 小时。马常耕和王建华（1994）研究过落叶松属 4 个种的种子发芽时对水分胁迫的反应。

表 5　落叶松属树种种子发芽能力及测定条件

树　种	基　质	发芽温度（℃）		发芽率（%）		
		有光时	无光时	计算天数	一般数值	变动范围
落 叶 松	滤纸	28	20	13	40	25～72
日本落叶松	滤纸	25	20	21	55	30～80
四川红杉	泡沫塑料	25	20	17	37	23～51
黄花落叶松	滤纸	30	25	15	55	40～79
红　　杉	泡沫塑料	25	20	17	34	23～55
华北落叶松	滤纸	30	20	11	50	30～83
西伯利亚落叶松	滤纸	30	25	20	57	30～68

播种　春播或秋播。10 月上中旬播种的西伯利亚落叶松，翌年 5 月上旬即可发芽出土。为了使出苗迅速整齐，生产上常在入冬前将供春播的落叶松、日本落叶松、黄花落叶松、华北落叶松和西伯利亚落叶松种子，在室外混雪埋藏，或者是在春播前的 2～3 周在 1～5℃下用湿沙层积，当 5cm 处地温稳定升至 8～10℃时取出播种。经过 15～20 天层积的黄花落叶松大约已有 1/3 的种粒裂嘴。经过雪藏，落叶松播后约半个月，日本落叶松播后 6～15 天出苗即可结束。对未经雪藏也未经层积的种子，有的场圃是在春播前作浸种催芽处理，希望能够加速萌发，但雪藏、低温层积和浸种催芽三者比较，以雪藏后播种效果最好，出苗迅速，幼苗抗性强。常培英和刘曼玲（1989）特别推荐利用雪藏二次干湿交替的方法处理华北落叶松种子。四川红杉和红杉的种子播前无需特殊处理，通常是在地面温度回升到 5℃以上的 4 月上中旬播种，播后 20～40 天开始出苗。

条播或撒播。每平方米床面的播种量在落叶松和日本落叶松为 15～17g，黄花落叶松为 9～15 g。四川红杉和红杉每平方米播 32～42g，西伯利亚落叶松每平方米播 50～75g。宋廷茂和白兆瑞等人（1985）研究过华北落叶松的播种量，认为每 666m^2 以播种 6.5kg，产苗 43.5

万株为宜。落叶松属种子的覆土厚度 0.5 cm 左右，覆土材料可用 1∶1 过筛河沙与苗圃土的混合土。播后用苇帘覆盖，也可全光育苗。出土萌发。子叶 4～7 (8)。萌发时下胚轴延伸，将种壳带出土面，5～10 天后种壳脱落，子叶展开。华北落叶松幼苗萌发形态见图 2。1～3 年生苗出圃定植。

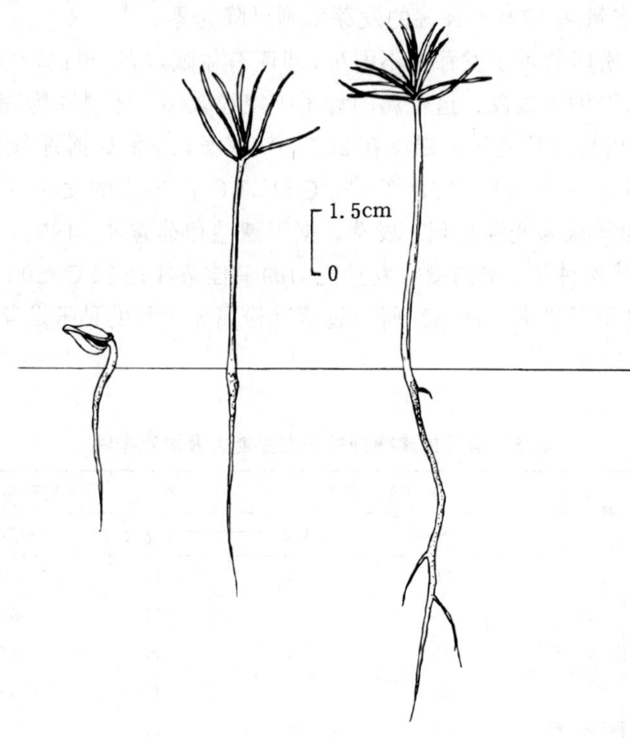

图 2　华北落叶松萌发后第 1、7、22 天的幼苗生长情况
(胡冬梅、林平绘)

落叶松属的扦插繁殖已经在几个树种中取得成功。在全光喷雾的条件下，选用 1.5～2 年生实生苗当年生半木质化枝条，采用生长激素处理，于 6～7 月间进行嫩枝扦插，插后 1～1.5 月即可生根。华北落叶松扦插生根率可达 96%～100%；日本落叶松生根率可达 98%；落叶松生根率可达 93%。

(宋廷茂)

云　杉　属
Picea Dietr.

(松科　Pinaceae)

生长习性、分布和用途　本属约 40 种，分布于北半球。我国有 20 种，5 变种，引入栽培 6 种。本文描述 12 种和 3 变种。常绿大乔木，胸径 25～100cm，树高 20～50m，有的高达 60m。

浅根性树种。适生于土层深厚、排水良好的酸性或微酸性的山地棕壤、山地暗棕壤、山地灰棕壤、山地褐色土，栗钙土或淡栗钙土也能生长。

我国云杉属分布地域广阔。东北的大小兴安岭与长白山，华北山地，华中的鄂西、豫西山地，西南的巫山、大巴山、邛崃山、沙鲁里山、岷山、玉龙雪山、横断山脉纵谷区和西藏境内的喜马拉雅山，西北的秦岭、阴山、祁连山、昆仑山、阿尔泰山、天山，以及台湾省中央山脉均有分布，分布的经度跨越40余度，纬度跨越约30度，以在西南地区分布的树种为多。

云杉属材质优良，纹理直、结构细致，轻软有弹性，易加工，是制作乐器的优良用材，还可用作建筑、桥梁、车辆、航空器材、电杆、体育器具、家具、胶合板、包装箱和造纸原料。树皮含单宁，可提取栲胶。

云杉属树种分布在高山地区，组成这些地区重要的水源涵养林。许多树种已经用于造林，是所在地区造林和迹地更新的主要树种。有些种还用于庭园观赏。云杉属树种的名称、分布和用途已汇入表1。

表1　云杉属树种的名称、分布和用途

中　名	学　名	分　布	用　途	供　稿
云杉	*P. asperata* Mast.	川西、甘东、甘南、陕南。海拔2 400～3 600m	材用、栲胶、保土	508
川西云杉	*P. balfouriana* Rehd. et Wils.	川西、川西北、青南、藏东南。海拔3 000～4 400m	材用、保土	508
黄果云杉	*P. balfouriana* Rehd. et Wils. var. *hirtelle* (Rehd. et Wils.) Cheng	川西北、滇西北、藏。海拔3 000～4 000m	材用、保土	504
油麦吊云杉	*P. complanata* Mast.	川西、川西南、滇西北、藏东及藏东南。海拔2 000～3 200m	材用、保土	504
青海云杉	*P. crassifolia* Kom.	甘、青、宁、内蒙古。海拔1 600～3 000m	材用、保土	201
鱼鳞云杉	*P. jezoensis* Carr. var. *microsperma* (Lindl.) Cheng et L. K. Fu	辽、吉、黑。朝鲜半岛、西伯利亚。海拔300～1 000m	材用	102
红皮云杉	*P. koraiensis* Nakai	辽、吉、黑、内蒙古。朝鲜半岛、西伯利亚。海拔300～1 600m	材用、栲胶、观赏	102
丽江云杉	*P. likiangensis* (Franch.) Pritz.	川西南、川西北、滇西北。海拔2 300～3 800m	材用、保土	504
林芝云杉	*P. linzhiensis* (Cheng et L. K. Fu) Cheng et L. K. Fu	川西北、滇西北、藏东南。海拔2 700～3 900m	材用、保土	504
西伯利亚云杉	*P. obovata* Ledeb.	新。蒙古、西西伯利亚。海拔1 200～1 800m	材用	204
紫果云杉	*P. purpurea* Mast.	川西北、甘南、青东。海拔2 000～3 800m	材用、保土	508

（续）

中 名	学 名	分 布	用 途	供 稿
鳞皮云杉	*P. retroflexa* Mast.	川西北、青东南。海拔 3 000～3 800m	材用、保土	508
雪岭云杉	*P. schrenkiana* Fisch. et Mey.	新。西西伯利亚。海拔 1 200～3 000m	材用	204
天山云杉	*P. schrenkiana* Fisch. et Mey. var. *tianshanica* (Rupr.) Cheng et S. H. Fu	新。西西伯利亚。海拔 1 200～3 500m	材用、栲胶、保土	204
青 杆	*P. wilsonii* Mast.	晋、冀、甘、陕南、鄂西、川北、青东、内蒙古。海拔 1 400～2 800m	材用、保土	508

开花结实 开始结实的年龄因树种以及生长环境等不同而异，人工林稍早于天然林。结实间隔期也因树种及分布区域而有差异（表 2）。

表 2 云杉属树种的树高、开始结实年龄和结实习性

树 种	成年时树高（m）	开始结实年龄[①]（年）	结实间隔期（年）
云杉	45	30～40（80）	2～4
川西云杉	40～45	20～30（60）	2～3
黄果云杉	40～45	20～30（60）	2～3
油麦吊云杉	30	20～30	4
青海云杉	30	30（40～60）	4
鱼鳞云杉	30～40	30～35	4～5
红皮云杉	35	30～35	3～5
丽江云杉	50	20～30（30～40）	3～4
林芝云杉	40	20～30（30～40）	3～4
西伯利亚云杉	25～35	35	3～4
紫果云杉	40～50	20～30（60）	2～3
鳞皮云杉	45	30～40（80）	2～4
雪岭云杉	40	30～35	4～5
天山云杉	30～60	30～35	4～5
青 杆	50	20～30（60）	2～4

① 括弧外的数字系孤立木和林缘木结实年龄，括弧内数字系林木结实年龄

云杉属球花单生，雌雄同株。雄球花单生在去年生枝的叶腋处，稀单生枝顶，椭圆形或圆球形，黄色、紫色或深红色。雄蕊多数，呈螺旋状排列，花药 2，花粉粒具气囊。雌球花单生

枝顶，圆柱形，绿色或紫红色，由多数螺旋状排列的珠鳞组成，珠鳞腹面基部有胚珠 2 枚，背面有苞鳞，极小。

开花期因树种及分布环境不同而有差异。分布在川西的云杉花期在 4 月，川西云杉、黄果云杉在 4 月中旬～5 月上旬。纬度偏北的地区春季温度回升较晚，那里的云杉属树种的花期普遍迟于纬度偏南的地区，例如，分布在黑龙江和新疆的云杉属树种的开花期，比西南地区的要迟 1～2 个月左右。

雌球花完全张开接受花粉的时间仅几天。接受花粉后珠鳞闭合，几天内便受精。幼果形成后下垂。多数树种球果于当年 9～10 月成熟，少数树种球果于当年 8 月中下旬成熟（表 3）。

表 3　云杉属树种的开花结实物候

树　种	观　察 地　点	开花期			果实成熟期		种子散落期		
		始　期	盛　期	末　期	始　期	盛　期	始　期	盛　期	末　期
云杉	四川马尔康	4 月上旬	4 月中旬	4 月下旬	9 月上旬	9 月中旬	9 月下旬	10 月中旬	11 月
川西云杉	四川马尔康	4 月中旬	4 月下旬	5 月上旬	9 月下旬	10 月上旬	10 月上旬	10 月中旬	11 月
黄果云杉	四川马尔康	4 月中旬	4 月下旬	5 月上旬	9 月下旬	10 月上旬	10 月上旬	10 月中旬	11 月
	云南迪庆	4 月中旬	4 月下旬	5 月上旬	10 月上旬	10 月中旬	10 月下旬	～	11 月
油麦吊云杉	云南迪庆	4 月中旬	4 月下旬	5 月上旬	10 月上旬	10 月中旬	10 月下旬	～	11 月
青海云杉	青海、内蒙古		5 月		9 月			10 月	
	甘肃		4～5 月		9～10 月		9 月下旬	～	11 月
鱼鳞云杉	黑龙江 植物园		5～6 月		9～10 月上旬		10	～	11 月
红皮云杉	黑龙江 植物园		5～6 月		9～10 月上旬		10	～	11 月
丽江云杉	云南丽江	4 月中旬	4 月下旬	5 月上旬	10 月上旬	10 月中旬	10 月下旬	～	11 月
林芝云杉	云南丽江	4 月中旬	4 月下旬	5 月上旬	10 月上旬	10 月中旬	10 月下旬	～	11 月
西伯利亚云杉	新疆阜康		5～6 月		9 月上旬	9 月中旬	9 月下旬	～	10 月
紫果云杉	四川马尔康	4 月中旬	4 月下旬	5 月上旬	9 月下旬	10 月上旬	10 月上旬	10 月中旬	11 月
鳞皮云杉	四川马尔康	4 月上旬	4 月中旬	4 月下旬	9 月上旬	9 月中旬	9 月下旬	10 月中旬	11 月
雪岭云杉	新疆阜康	5 月下旬	6 月上旬	6 月下旬	9～10 月			10～11 月	
天山云杉	新疆乌鲁 木齐南山	6 月初	6 月上旬	6 月中旬	8 月中旬	9 月上旬		9 月中旬～10 月	
	新疆阜康	5 月下旬	6 月上旬	6 月下旬	8 月下旬	9 月中旬	9 月中旬	9 月下旬	10 月
青　杄	四川马尔康	4 月上旬	4 月中旬	4 月下旬	9 月中旬	9 月下旬	10 月上旬	10 月中旬	11 月

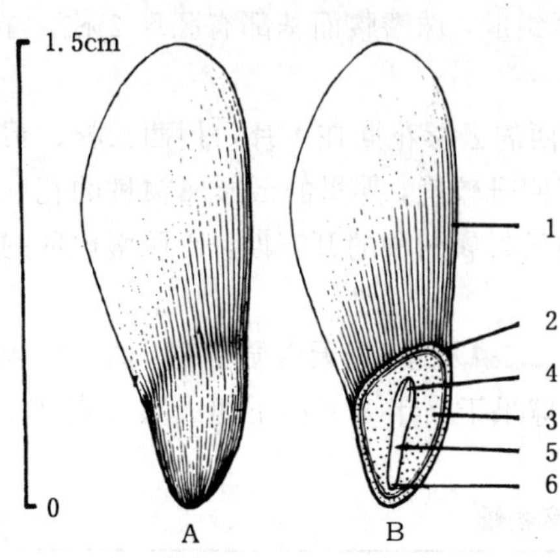

图 1 云杉种子的外形（A）及其纵切面（B）
1. 翅 2. 种皮 3. 胚乳 4. 子叶 5. 胚轴 6. 胚根（孟玲、黄应钦绘）

云杉属的成熟球果呈黄色、黄褐色、淡褐色、栗褐色、褐色、黑褐色、红紫色、紫红色或紫黑色。有几个种，例如川西云杉、丽江云杉和紫果云杉，近熟球果与未成熟球果几乎同色。球果卵状长圆形至圆柱形。树种之间球果的大小有明显差异。果型小的有紫果云杉（长 2.5～5.5cm）、鱼鳞云杉（长 4～6cm），果型大的有鳞皮云杉（长 6～13cm）、雪岭云杉和天山云杉（长 8～10cm）、油麦吊云杉（长 7～15cm）。云杉、青海云杉、红皮云杉、西伯利亚云杉、鳞皮云杉、雪岭云杉、天山云杉和青杆，球果种鳞较厚，排列多紧密；丽江云杉、林芝云杉、川西云杉、黄果云杉和紫果云杉的球果种鳞较薄。

发育种鳞腹面有种子 2 枚。种子具胚乳。子叶 4～9（15）枚。种子长 2.5～6mm，倒卵圆形或卵圆形，上端有膜质翅。图 1 绘出的是云杉种子的外形及其纵切面。云杉属各树种球果和种子的形态、颜色与大小见表 4。云杉属的种子 9～11 月散落，借风力传播。

表 4 云杉属树种球果和种子的颜色、形态与大小

树种	未成熟球果颜色	成熟球果				种子		
		颜色	形态	大小（cm）		色泽	形状	长(mm)
				长	径			
云杉	绿色	淡褐色或栗褐色	圆柱形或圆柱状长圆形	5～6	2.3～3.5	黑褐色	倒卵圆形	约 4
川西云杉	紫红色或黑紫色	与未熟球果近同色	卵圆形或圆柱形	4～9	2.5～4.5	深褐色	近卵圆形	约 4
黄果云杉	绿黄色或黄色	褐黄色	圆柱形或卵状圆柱形	5～9	2～4	褐色或深褐色	卵圆形	2.5～5
油麦吊云杉	紫黑色或黑紫色	黑褐色	卵状圆柱形	7～15	3～5	褐色	卵圆形	2.5～5
青海云杉	种鳞背面部分绿色，上部边缘紫红色	褐色	圆柱形或矩圆状圆柱形	7～11	2～2.5	淡褐色	斜倒卵圆形	约 4
鱼鳞云杉	绿色	淡黄褐色或褐色	长圆状圆柱形或长卵圆形	4～6	2～3.2	黑褐色	卵圆形	3～3.5
红皮云杉	绿色	黄褐色或褐色	卵状圆柱形或长卵状圆柱形	5～8	2.5～3.5	淡黑褐色	倒卵圆形	约 4
丽江云杉	红褐色或紫黑色	褐色或近于未成熟时颜色	卵状长圆形或圆柱形	7～12	3～5	棕色	近卵圆形	3～5

（续）

树　种	未成熟球果颜色	成熟球果				种　子		
		颜　色	形　态	大　小（cm）		色　泽	形　状	长(mm)
				长	径			
林芝云杉	紫红色	深褐色	短卵柱状或卵状长圆形	7～10	2.5～4	褐色	卵圆形	2.5～4
西伯利亚云杉	黄绿色并常带紫色	褐色	卵状圆柱形或圆柱状矩圆形	5～11	2～3	暗褐色	倒三角状卵圆形	3～6
紫果云杉	紫黑色或淡紫红色	与未成熟时近于同色	圆柱状长卵形或椭圆形	2.5～5.5	1.7～3	深褐色	近卵圆形	3.5
鳞皮云杉	种鳞背面绿色，上部边缘紫红色	褐色或淡褐色	圆柱形或柱状椭圆形	6～13	2.5～4	黑褐色	斜卵圆形	约4
雪岭云杉	绿色	褐色	椭圆状圆柱形或圆柱形	8～10	2.5～3.5	浅褐色或棕褐色	斜卵圆形	3～4
天山云杉	紫红色或暗红色	褐色	椭圆状圆柱形或圆柱形	8～10	2.5～3.5	浅褐色或棕褐色	斜卵圆形	3～6
青　杆	绿色	黄绿色、黄褐色或淡褐色	卵圆状圆柱形或圆柱状卵圆形	6～13	2.5～4	淡黑褐色	倒卵圆形	3～4

球果和种子常遭病虫危害，影响结实，影响产量。病害为云杉球果锈病，病原菌是盖痂锈菌属的杉李盖痂锈菌（*Thekopsora areolata*）和云杉盖痂锈菌（*Th. sparosa*）；还有1个病原菌是鹿蹄草金锈菌（*Chrysomyxa pyrolae*）。在新疆天山林区，天山云杉感染鹿蹄草金锈菌的球果数量有时占到10%，受害球果难以正常发育，种鳞极小，球果中的种子粒数显著减少，据测定，发芽率只有0.9%～12.2%。害虫有两种：云杉球果小卷蛾（*Pseudotomoides strobilellus*）和云杉梢斑螟（*Dioryctria schuetzeella*）。

球果的采收调制和种子贮藏　过去多半是在一般林分或采伐后留存的母树上采集云杉属球果。80年代以来，有些地方建立了母树林，在一般林分中采种也能注意到选择优良母树。云杉属结实有明显的间隔期（请见表1），应当有周密计划并组织种子产量预测预报工作。

云杉属有的树种种子成熟后很快便飘飞散落，例如红皮云杉和鱼鳞云杉，从球果开始成熟到种子飞落一般只有15天，因此需要严格掌握采种期。可以根据颜色判断云杉属球果是否成熟。有的树种成熟球果同未熟球果颜色差异不大（参见表4），就要看种鳞是否变硬，是否微有开裂。当前，生产上多数是人工上树采集球果，也有的是在采种季节配合采伐从伐倒木上采摘。云杉属树体高大，上树采集要注意安全。

采回的球果摊在晒场上曝晒，使种鳞收缩开裂，种子脱落。球果摊晒的厚度以不超过6cm为宜。摊晒时须经常翻动球果，使种鳞均匀张开并使种子从张开的种鳞上脱出。每天扫集脱落的带翅种子。夜晚或有雨雪时须用覆盖物遮盖球果。

也可以使用干燥室或球果烘干机干燥云杉属的球果，温度以30～40℃为宜，超过45℃会降低种子质量。烘干机如有转动装置，种子便会从滚动的网孔自动落下。有些干燥室没有转动装置，就要在干燥程序结束后迅速取出开裂的球果，立即震动，抖出带翅的种子。如果球果含水量高，干燥时温度宜逐渐升高，初始温度过高也会降低种子质量。新疆有的地方曾经利用太阳能作能源调制过天山云杉球果（陈开秀等，1988）。他们的做法是将球果预干后放在有网眼的盛果盘内，置于干燥室的干燥架上。干燥架装有滚动装置，可以摇动，使带翅种子落入下面的盛种器中。

人工搓揉使种翅与种子分离，再筛除去种翅及夹杂物即得纯净种子。云杉属球果的出种率和种子质量数据见表 5。

表 5　云杉属树种的出种率、种子净度与质量

树　种	出种率（%）	净度（%）	千粒重（g）	每千克纯净种子粒数(万粒)
云杉	2.1～3.5	80	4.3～5.1	19.6～23.3
川西云杉	3.0～3.8	80～90	3.1～3.8	26.3～32.3
黄果云杉	2.5～3.1	80	3.3～3.9	25.6～30.3
油麦吊云杉	3.4	80～90	4.5～5	20.0～22.2
青海云杉	2.0～3.0	80	4.3～4.9	20.4～23.2
鱼鳞云杉	3.0～4.0	80～90	2～2.4	41.6～50.0
红皮云杉	2.9～3.3	80～90	5～6	16.6～20.0
丽江云杉	3.0	80	4.8～5.2	19.2～20.8
林芝云杉	2.9	80	2.9～4.3	23.2～34.5
西伯利亚云杉	—	—	6.3	15.8
紫果云杉	2.3～3.0	80～90	2.2～3.9	25.5～45.4
鳞皮云杉	2.1～3.5	85	4～5.8	17.2～25
天山云杉	3.8	85	6～10	10.0～16.6
	6.4	90	—	—
青　杆	2.5	90	5.1	19.6

云杉属中的云杉已经制订了种子区划，作为国家标准发布。根据云杉的天然分布和林木的生长发育状况划分为 3 个种子区和 4 个亚区，即甘南区、川西北区和陇南秦西区。甘南区又

区分为北部亚区和南部亚区。川西北区区分为西部亚区和东部亚区。该标准对云杉种子的使用和调拨做了规定：以使用本区种子为主，也规定了邻区和相邻亚区的调用范围及其海拔高度。

云杉属的其它树种也各有其特性及地域性，在调用种子上也要考虑树种与环境的关系，不宜盲目大量调拨和引进。

云杉属各树种的种子对贮藏条件的要求和反应大致相似，它们都适于干藏。干藏种子的适宜含水量为7%～9%或再低一些。容器可选用塑料袋、桶、罐、麻袋或布袋。越冬后即将播种的种子可以在干燥通风的室内存放。在普通室内存放的天山云杉种子，3年后发芽率仅降低了10%～12%。必须贮藏较长时间的种子则应密封后在低温处保存。在0～5℃冷库内密封贮藏4年的云杉种子，生活力没有显著下降。

发芽前的处理　云杉属的种子没有明显的休眠习性。为了促进萌发，发芽测定之前或播种之前也可以用始温45℃左右的水浸种1～2天，或者用0.3%～0.5%的高锰酸钾浸渍30分钟。有的地方是温水浸种之后再用高锰酸钾浸泡。高锰酸钾浸泡过的种子用清水冲洗后播种或置床作发芽测定。云杉属的种子虽无明显的休眠，有些生产单位还是倾向于将种子低温层积一段时间，使发芽迅速，出苗整齐。据认为，长期干藏过的种子更需要这样做。常用的层积基质是湿沙，冬季雪多的地方更习惯于用雪。层积的温度为1～5℃，时间一般是1～2个月。有的地方还在播种之前的1～2周将层积过的种子移到大约是20～25℃的温暖处催芽，等到有1/3左右的种粒裂嘴时取出播种。

发芽测定　经过预处理的种子放入衬有纱布或泡沫塑料垫的发芽盒内，在光照发芽箱或恒温箱中发芽。发芽能力及其测定条件见表6。

表6　云杉属树种种子发芽能力及其测定条件

树　种	预处理	基　质	温　度（℃）		发芽势（%）			发芽率（%）		
			有光时	无光时	计算天数	一般数值	变动范围	计算天数	一般数值	变动范围
云杉	始温45℃水浸种24小时	泡沫塑料垫	—	25	5	28	16～52	15	62	46～77
川西云杉	始温45℃水浸种24小时	泡沫塑料垫	—	25	5	26	14～37	15	46	32～67
黄果云杉	0.5%高锰酸钾液浸种30分钟清水冲洗	纱布	20	20	6	45	33～53	11	53	41～61
油麦吊云杉	0.5%高锰酸钾液浸种30分钟清水冲洗	纱布	20	20	6	41	34～48	11	43	37～52
青海云杉	始温45℃水浸种24小时	纱布	25	25	5	40	—	15	85	80～90

（续）

树种	预处理	基质	温度（℃）		发芽势（%）			发芽率（%）		
			有光时	无光时	计算天数	一般数值	变动范围	计算天数	一般数值	变动范围
丽江云杉	0.5%高锰酸钾液浸种30分钟清水冲洗	纱布	20	20	6	40	30～45	11	57	34～60
林芝云杉	0.5%高锰酸钾液浸种30分钟清水冲洗	纱布	20	20	6	35	20～45	11	37	25～50
西伯利亚云杉	始温40℃水浸种24小时，0.3%高锰酸钾液浸种30分钟清水冲洗	纱布	10～15	10～15	—	—	—	20	27	—
			20	20	—	—	—	20	74	—
			25	25	—	—	—	20	70	—
紫果云杉	始温45℃水浸种24小时	泡沫塑料垫	—	25	5	22	7～35	15	45	26～62
鳞皮云杉	始温45℃水浸种24小时	泡沫塑料垫	—	25	5	31	20～54	15	63	61～67
天山云杉	始温40℃水浸种24小时，0.3%高锰酸钾液浸种30分钟清水冲洗	纱布	25	25	—	—	—	20	66	最高95
青杄	始温45℃水浸种24小时	泡沫塑料垫	—	25	5	64	—	15	76	—

播种 一般采用春播。播种期因纬度、海拔高度不同而有先后。川西北海拔2 000～2 600m的地带，播种期在4月上旬，海拔2 600～3 200m的地带则在4月下旬。纬度偏北的地方，如东北的黑龙江、西北的新疆等地于5月播种。

条播或撒播。条幅5～10cm，条距12～15cm，播种沟深1cm。条播时每平方米播种50～60g。撒播时每平方米播种75～120g。播后用土覆盖，厚1cm，轻压后浇水、盖草。播后20天左右种子发芽出土。种子发芽出土后分批揭除盖草。川西北地区于盖草揭除后还需搭设荫棚给幼苗遮荫，避免日灼。

高寒地区生长期短，1年生苗高仅5～6cm，出圃时间需要4～5年，有的还需6年才能出圃。塑料大棚可以为幼苗生长提供良好的温度、水分和光照条件，一般3年即可出圃。

出土萌发。子叶线状锥形，(4) 5～9 (11) 枚，长10～20mm，横切面呈三角形，底边略呈弧形，两侧边有2至多条白色气孔线，边缘常具疏齿毛，浓绿色，有气孔线者带粉绿色。上胚轴极短，一般直伸或斜上生出1至数个初生叶即结顶，成圆球形或圆卵形的芽越冬，或当年不生初生叶就结顶。初生叶锥形或线形，边缘常疏生细齿毛，粉绿色，叶枕淡白绿色。下胚轴长圆柱形，长2.5～5.1cm，淡绿色至黄绿色、褐色，具多条纵脊。主根细长，直伸，侧根疏生，向下较短而细，淡褐色至褐色。图2显示的是天山云杉种子的萌发及幼苗生长情况。

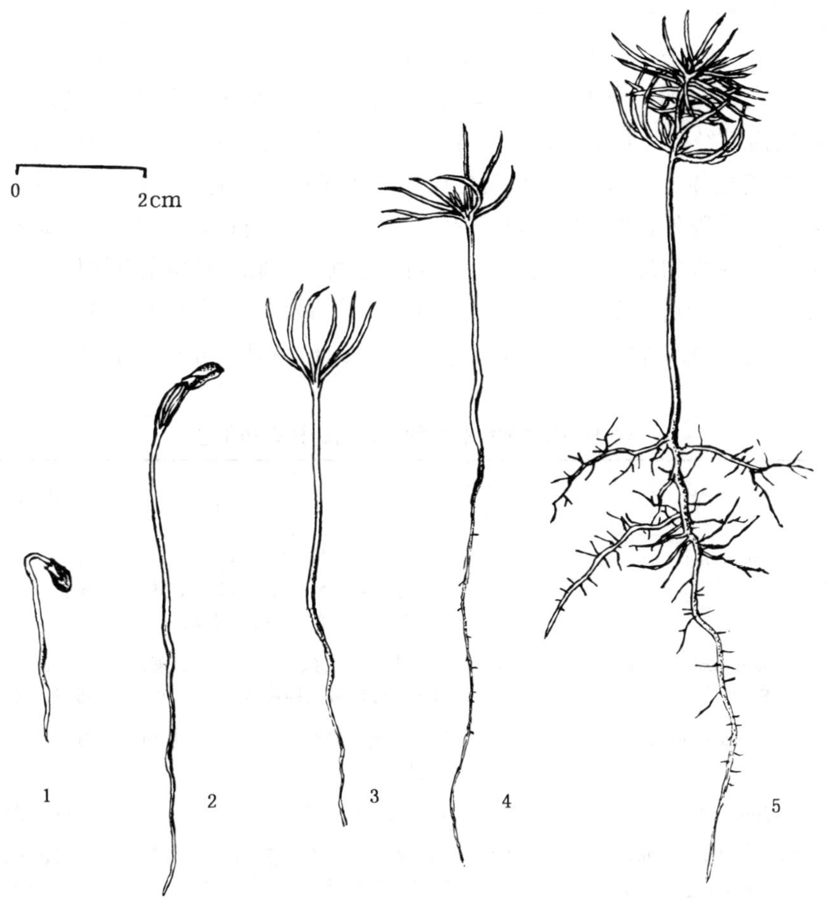

图 2　天山云杉种子萌发及幼苗生长情况

1. 下胚轴延伸　2、3. 子叶出土　4. 初生叶初露　5. 2 年生苗

（田恒德仿《主要树木种苗图谱》）

云杉还可以扦插繁殖。母株年龄愈小，扦插生根率愈高。一般生根率约 70%，成活率可达 90%。

（赵德铭）

松　　属

Pinus L.

（松科　Pinaceae）

生长习性、分布和用途　松属是林业生产中最重要的一个属，近百个种，还有许多变种，广泛分布于北半球的各种气候带。在亚洲，松属树种从西伯利亚向南一直分布到印度尼西亚的苏门答腊。我国有 23 种 10 变种，分布几乎遍及全国，是许多地方森林的重要组成树种。有些树种，例如马尾松、油松和华山松，自然分布区极其辽阔。有些树种的自然分布区则比较狭窄，例如长白松仅见于长白山地区，台湾五针松（*P. morrisonicola* Hayata）只在台湾中央山脉海拔 300～2 300m 的山脊上散生或与其它树种混生。樟子松天然分布地域原来也比较狭窄，但是现在向西已经引种到了新疆的伊犁地区，经度向西扩展了 37°；向南引种到了甘肃天水地区，纬度向南延伸了 13°，成了我国松属中栽培区扩大得最为显著的一个树种。

除了原有的树种以外，我国还引进了 30 多个外来松属树种，其中不少已取得成功，例如原产美国东南部和南部的湿地松、火炬松、长叶松、刚松，原产加勒比海地区的加勒比松以及原产日本的黑松。这些树种在不同程度上丰富了我国森林树种的组成。本文描述我国原产的 16 种 3 变种以及引进的 7 种。它们的名称、成年时的树高以及分布和主要用途已列入表 1。

表 1　松属树种的名称、树高、分布和用途

中　名	学　名	树　高 (m)	分　布	用　途	供　稿
华山松	*P. armandii* Franch.	35	藏、滇、桂、黔、湘、鄂、川、青、甘、宁、陕、豫、晋。内蒙古、冀、鲁、苏、浙、皖、赣引种栽培	材用、食用、栲胶	303
白皮松	*P. bungeana* Zucc. ex Endl.	30	甘、陕、晋、豫、鄂、川。辽、冀、鲁至长江流域引种栽培	材用、食用、药用、观赏	303
加勒比松	*P. caribaea* Morelet	45①	原产加勒比海地区。闽、粤、桂引种栽培	材用	（综合）
高山松	*P. densata* Mast.	30	青、川、藏、滇	材用、采脂	508
赤松	*P. densiflora* Sieb. et Zucc.	30	黑、吉、辽、鲁、苏。日本、朝鲜半岛、俄罗斯	材用、采脂	101
长白松	*P. sylvestriformis* (Takenounchi) T. Wang ex Cheng	30	吉。黑有引种	材用、采脂 栲胶、观赏	107 307
萌芽松（短叶松）	*P. echinata* Mill.	40①	原产北美。福州、南京引种	材用、采脂	901
湿地松	*P. elliottii* Engelm.	30①	原产美国东南部。长江以南广为引种	材用、采脂	（综合）
海南五针松（葵花松）	*P. fenzeliana* Hand.-Mazz.	30	琼、桂、湘、川、黔。引种栽培的有滇、鄂、浙、赣等	材用、食用、采脂	509
巴山松	*P. henryi* Mast.	40	鄂、川、陕	材用	507
思茅松	*P. kesiya* Royle ex Gord. var. *langbianensis* (A. Chev.) Gaussen	30	滇。越南、缅、印度、老挝也有。四川引种	材用、采脂	504
红松	*P. koraiensis* Sieb. et Zucc.	40	黑、吉。俄罗斯、朝鲜半岛、日本也有。辽宁引种	材用、采脂	103
华南五针松	*P. kwantungensis* Chun ex Tsiang	30 (40)	湘、黔、桂、粤、琼	材用、采脂	（综合）

（续）

中　名	学　名	树　高 (m)	分　布	用　途	供　稿
南亚松	*P. latteri* Mason	25	琼、桂、粤。东南亚也有	材用、采脂	601
马尾松	*P. massoniana* Lamb.	40	豫、川、陕、鄂、皖、苏、浙、赣、湘、黔、桂、粤、闽、台	材用、采脂	701
长叶松	*P. palustris* Mill.	45	原产美国东南部和南部。江南各城市少量引种	材用、采脂	（综合）
偃松	*P. pumila* (Pall.) Regel	1～6	吉、黑、内蒙古。俄罗斯、朝鲜半岛、日本也有	保土、观赏、食用	103
刚松	*P. rigida* Mill.	25	原产美国东部。旅顺、大连、青岛、南京、杭州少量引种	材用	（综合）
西伯利亚红松	*P. sibirica* (Loud.) Mayr	35	新。俄罗斯、蒙古也有	材用、采脂	204
樟子松	*P. sylvestris* L. var. *mongolica* Litv.	30	黑、内蒙古。吉、辽、陕、甘、新引种栽培	材用、固沙	104
油松	*P. tabulaeformis* Carr.	30	内蒙古、宁、甘、青、川、陕、豫、晋、冀、鲁、辽、吉	材用、采脂	303
火炬松	*P. taeda* L.	30①	原产美国东南部。淮河以南广泛引种	材用	（综合）
黄山松	*P. taiwanensis* Hayata	30	皖、浙、赣、鄂、湘、桂、滇、黔、豫、台	材用、采脂	801
黑松	*P. thunbergii* Parl.	30	原产日本、朝鲜半岛。辽、鲁、苏、浙、皖、豫、陕、鄂、湘引种栽培	材用、海岸防沙	301
毛枝五针松	*P. wangii* Hu et Cheng	20	滇	材用、采脂	615
云南松	*P. yunnanensis* Franch.	35	滇、桂、黔、川、藏	材用、采脂	504

①　指在原产地

松属都是常绿树种。除了偃松以及云南松的一个变种地盘松〔*P. yunnanensis* Fr. var. *pygmaea* (Hsueh) Hsueh〕为灌木或小乔木外，我国松属树种都是大乔木，在立地条件适宜的地方成年时树高可达 30～40m 甚至更高，树龄可以高达一百多年至几百年。

松属是世界上木材生产的重要资源。不少松树，例如马尾松、华山松、云南松、红松，还是造纸工业的重要原料。我国许多松树，例如马尾松、云南松、南亚松、油松、赤松、樟子松以及引进的湿地松等，可以采割松脂，是重要的工业原料。松树种子富含油脂，是鸟类和某些野生动物的重要食物。特别是红松、华山松、海南五针松和偃松，种子颗粒大，种仁含油量高达 60%左右，是风味独特的干果食品。松属许多树种能够忍受瘠薄干燥的生境，例如油松、巴山松、高山松、云南松、思茅松、南亚松、马尾松和黄山松，是荒山荒地造林的重要先锋树种，在迅速恢复森林植被、改善生态环境方面具有重要作用。

地理变异　林业生产实践的一条重要经验教训是造林用种必须来自适宜的种源。70 年代后期以来，我国林业科技人员对若干重要造林树种开展了种源研究工作。以下引述的是这项课题以及某些学者对松属几个树种研究的初步结果。

华山松 研究表明，华山松种子的形状、大小、质量、颜色、胚轴长度以及子叶数目、子叶长度、初生叶颜色、生长期、越冬能力和幼树生长等性状，都呈现出清晰的同纬度相平行的单向渐变的地理变异模式。云贵高原起源的种子略呈圆形，粒大而重，种壳颜色较深，发芽较快；北方种源则发芽所需天数较多，所成的幼苗生长较慢，但可耐－15℃以下的低温。云贵高原的种源引种到黄河流域，由于无法越冬会全部失败，引入长江中游山区初期表现为生长迅速，但在冬季严寒的年份也可能遭受轻微的冻害。北方种源引入南方则表现为生长不良，逐渐被其它树种排斥。马常耕和张云跃等（1990）也主张，在调运种子时必须将抗寒性放在首位而不能只考虑生长力。

红松 红松是分布在东北长白山、老爷岭、张广才岭、完达山和小兴安岭地区的珍贵用材树种，自然分布区南北间跨越的纬度超过 9°。已经发现种子的千粒重由南往北有减轻的趋势。辽宁省引种栽培的红松人工林，种源来自吉林的生长优于来自黑龙江的种源。

马尾松 马尾松的水平分布跨越经度约 20°，纬度约 12°。种源研究的多点试验表明，马尾松的地理种源之间明显存在着纬向的渐变群变异，这种变异同产地的纬度呈线性相关。低纬度种源生长期长，地上部分生长量大，抽梢次数多，开始开花结实较晚 ，封顶迟，适应性较差。根据地理种源的变异模式并参考植被区划，已将马尾松的分布区划分成 3 带 9 区 22 个亚区。生产用种调拨的基本原则是，优先选用本亚区和相邻的南面或西面的优良种源，并应有选择地引进南面或西面的适生优良种源，但对引进跨度过大的种源应当注意耐寒性和成活率等问题。种源研究还倾向于推测，四川盆地和广西盆地的马尾松有可能是第四世纪冰川以后保存下来的两个原始类型，中亚热带和北亚热带广大地区的马尾松则由这两个原始类型向外扩散进化而来。

樟子松 樟子松被认为是欧洲赤松（*P. sylvestris* L.）的一个变种，是我国大兴安岭西坡、北坡、东坡和呼伦贝尔草原沙地主要的天然成材树种。近 30 年来，樟子松已被广泛引种。引种地区从东到西包括了中温带湿润森林、森林草原、干旱草原、半荒漠草原直到荒漠，向南则引种到了暖温带的南缘。在多种多样的气候条件下樟子松生长发育基本正常，在不少地方的生长甚至超过了原产地，也超过了当地引种的其它一些树种。引种到科尔沁沙地南缘章古台的樟子松已能正常生产种子。樟子松的各个种源在北京低海拔地区表现极差，但随着海拔增高而有改善的趋势（徐化成等，1991）。樟子松能在自然分布范围以外表现良好，可能是它的原种欧洲赤松分布广，能适应多种生境的遗传特性的反映。樟子松的引种范围虽然广阔，但种源研究还有待开展。目前比较一致的看法是，东北、华北、西北干旱的草原和沙地发展樟子松最好采用呼伦贝尔草原的种源；东北山地则以采用大兴安岭或小兴安岭的种源为宜。

油松 油松分布范围之广可能稍次于马尾松。油松分布区的走向基本上呈东北—西南向，且东部的南北范围较宽 ，而在西北地区则呈明显的不连续性。油松的地理变异表明，种子的大小和质量，以及在一定程度上球果的大小均与纬度呈正相关，而与降水量和温度呈负相关，即气候越是干燥寒冷，球果越大，种子也越大越重。这种相关关系在分布区的中部和东部表现得最为明显。在种源试验中，从造林成活率、越冬保存率以及幼林生长来看，总的趋势似乎都是当地种源表现较好。因此目前的结论是，油松的大规模生产性造林仍应采取就近用种的方针。李长喜和徐化成（1989）研究过油松同巴山松的形态差异和地理分界。

云南松　云南松是我国西南地区分布范围相当大的松属树种，主要分布在中亚热带，向北跨入暖温带，向南伸入南亚热带。为时不长的试验表明，苗木高生长的趋势是，南部、西部水热条件较好的种源优于北部、东部水热条件较差的种源。有几个种源，例如马关、双江、潞西、保山，在云南省境内四个栽培点上的高生长都处于领先地位，只是目前还不能在这一方面作出明确的结论。云南松已被鉴定的变种中有一种地盘松呈灌木状。此外，人们还发现云南松的整个分布范围内存在着一种叫做扭松的新变型，纹理扭曲的特性受遗传控制。云南松人工林中地盘松和扭松的比例有逐渐增加的迹象，林业工作者已经在采种和种子收购等环节上采取措施扭转这种趋势。

开花和结实　据秦国峰和汪名昌（1991）观察，有一片马尾松人工林在定植的第2年便有20%的植株开出雌球花，到6年生时，开出雌球花的植株达到87.9%；定植的第3年有1.4%的植株出现雄球花，到6年生时，出现雄球花的植株达到49.8%；定植的第3年开始结出球果，到6年生时，结出球果的植株占全林植株总数的81.4%。他们还注意到，6年生马尾松人工林所结出的种子，无论是千粒重还是发芽率，都同成林没有明显差异。他们因此认为6年生的马尾松人工林已经进入正常结实。油松和思茅松大约也是在5～7年生时开始结实。据王九龄（1979）观察，昆明地区的云南松天然林多在8～11年开始结实，最早的为6年，最迟的为16年。华山松、高山松、赤松、海南五针松、巴山松和偃松一般是在10年生左右或稍迟，西伯利亚红松则要到40～60年生时才有形成雌雄球花的能力。我国松属中已知开始结实最迟的是红松。天然林中红松开始结实的年龄大约是80～140年。一般认为这同天然更新的幼树长期处于光照不足的林冠之下有关。据刘继国（1963）报道，辽宁草河口地区的红松人工林，7年生就有个别植株开花并在第2年收到种子；9年生时林内已有4.5%的植株结实，19年生时结实株数的百分数上升到了34.3%；33年生时上升到79%；35年生时结实的植株已占全林的91%；而在70、80和140年生的红松天然林中，结实株数的比例分别是0%、40%和80%。李财德（1983）在小北湖天然林中的观察是，80～140年生的红松天然林虽已开始结实，但结实量仍不多，此后逐渐增加，到180～210年生时才进入结实盛期。李永多等人（1981）报道过红花尔基天然更新的樟子松结实能力同年龄的关系。据他们观察，个别的樟子松幼树7～10年生就能开花结实，15年生后普遍结实但数量不多，30年生以后进入结实盛期且结实量随年龄而增加，到110年生时结实量未见下降。

松属结实盛期延续的时间随树种而不同，不过这方面的文献记载还不多。据观察，华山松约在30～80年，马尾松约在15～50年，西伯利亚红松则在80～140年，而且有的植株可以延续到200～300年。在条件优越的地方400～450年生的西伯利亚红松仍有相当旺盛的结实能力。李财德（1983）报道说，红松的结实能力可以延续到370年。李永多等人（1981）报道，有一株樟子松树龄已达287年，1974年时结出球果8 706个。

松属结实丰年的间隔期随树种而不同，间隔1～2年的有思茅松、黑松；间隔期为2～3年的有巴山松、马尾松、黄山松和云南松；华山松、油松、樟子松大约每2～4年出现一次结实丰年；西伯利亚红松则相隔3～5年才会有一次丰盛的结实。一般认为海南五针松没有明显的结实大小年现象，但据贵州习水地区观察，这个树种每4～5年会有一次丰富的结实。地处长白山的露水河林业局1958～1976年对一片红松林作过连续19年的观察记载：19年中丰年一共出现过7次，即1958、1961、1964、1967、1970、1973和1976年，大体上是每隔3年1次，不过2个丰年之间有时是连续两年属于中等产量。刘继国（1963）报道过辽宁草河口地区红松人工林1949～1961（林龄21～33）共13年的结实量，其中1954、1958和1961年为丰年，结实量相当于平均结实量5～8倍；丰年之前的1年则均为大歉年。宛志沪（1986）研究过湿地松结实量与气候因子的关系。

松属都是雌雄同株的树种，间或也会出现性比例失常的个体，因而有的个体雄球花极少，表现得接近于雌株，有的则雌球花极少而接近于雄株。王九龄（1979）报道说，昆明地区云南松林内偏雄性的植株数约占2%～3%，有的散生植株一直到22年生仍未着生一个球果。孔繁彬（1994）在湖北宜都市（原枝城市）马尾松天然林中的调查结果是，偏雌性的植株占12.7%，偏雄性的植株占11.2%。

松属的雌球花和雄球花都着生在当年的新梢上，着生的部位相对固定：雌球花主要出现在树冠上部，有短梗，单生或2～4枚着生于新枝近顶端。据秦国峰和汪名昌（1991）调查，大部分马尾松球果着生在2级和3级侧枝上。雄球花多数出现在树冠中下部，无梗，螺旋状簇生于新梢下部，密集在1.0～2.6cm（海南五针松），或4～7cm（赤松），或5～10cm（白皮松）的范围内。

松属球花从芽中出现的时间是春季或春末夏初，雄球花略早于雌球花。多数松树的雄球花撒粉时呈黄色或红黄色如红松，或淡红褐色如黄山松。撒粉完毕后雄球花干枯并逐渐脱落。吴天林和朱德俊等（1993）在浙江富阳研究过马尾松57个种源的雄球花数量。他们发现，雄球花数量多的主要是北部种源的植株，南部种源多数表现为雄球花数量少。雷振民和美存焕等（1993）报道过陕西洛南11～12年生油松雌雄球花和球果的物候进程。刘继国（1963）观察过辽宁草河口地区红松开花结实的物候进程。王晓茹和沈熙环（1987）在辽宁兴城连续3年研究过油松无性系种子园的花粉飞散规律。他们发现，油松雄球花发育到成熟散粉，要求开花当年≥10℃的有效积温达到110～120℃，且散粉的高峰期只有2～3天。据他们测定，这片油松种子园在11年生时每公顷产生花粉9.5kg。黄启强（1988）在一片已经进入结实盛期的马尾松初级种子园中也作过测定，每公顷的花粉产量是26.8kg；只是各无性系的花粉产量相差很大：在所观察的28个无性系中，有3个无性系的花粉量占花粉总量的48.2%，有16个无性系的花粉量在花粉总量中仅占15.7%。张卓文和许大明等人（1994）在浙江余杭研究过湿地松球花的特性和种子园雌雄配子的比例。

松属雌球花呈直立状，椭圆形或卵圆形。珠鳞螺旋状排列，多数，例如海南五针松每个雌球花有（56）60～70（88）枚珠鳞。每枚珠鳞的腹面基部具2个倒生胚珠。传粉前珠鳞闭合使雌球花呈绿色或略带红色，传粉时肥厚肉质状的珠鳞张开因而雌球花呈紫色或紫红色。接受花粉后珠鳞由下而上逐渐闭合，紫色消失。闭合的幼小球果在开花当年几乎不发育，越冬以后，即开花次年的4～5月（如高山松）、5～6月（如红松）、6～7月（如西伯利亚红松）开始受精。此时的球果已因新梢的生长而呈侧生或下垂状态。受精以后球果体积迅速增大。成熟时球果由绿色逐渐向黄色、褐色或深褐色转变，有的转为黄绿色或灰绿色，如红松；有的则转为紫色或紫褐色，如偃松；有的还密被白粉，如海南五针松。我国松属树种的球果多在冬季或冬末春初成熟。多数松树成熟的球果逐渐开裂，带翅的种子随风飞散。此时已是授粉后的第2个冬季或花芽分化后的第3个冬季。有性生殖周期经历的时间如此漫长，影响种子产量和质量的因素也就相对较多，是松属树种大小年现象比较明显的重要原因。

松属树种的球果呈卵圆形或圆锥状卵圆形，有短梗，或近于无梗如黄山松和萌芽松。如果把我国松属树种的球果粗略地区分为大、中、小 3 种类型，则赤松、巴山松、偃松、樟子松、黄山松的球果可以归入小型球果这一类，长度一般在 2.5～5.0cm；偃松球果略为大一些，但长度一般也不超过 6cm。果型中等的树种较多，例如白皮松、高山松、思茅松、马尾松、西伯利亚红松、长白松和油松，它们的球果长度多数在 4～7cm，其中油松可以长到 9cm。可以视为大型球果的有华山松（长 10～20cm，径 5～ 8cm）和红松（长 9～16cm 甚至 20cm，径 5～10cm）。海南五针松和云南松的球果虽然不及华山松和红松那样硕大，但仍然明显地超过属于中等类型的那些球果。

松属多数树种的球果成熟后因含水量的下降而在树上自然开裂，种子便陆续随风散落。在树上自然开裂后种子散落的空球果，通常在几个月之内就会凋落，但思茅松和油松等树种的空球果常常在树上宿存几年，黄山松有的球果可以宿存 7 年之久。有的树种球果脱落后会在枝条上留有果痕。宿存的球果或果痕可以用来追溯历年结实量的波动情况。红松、偃松和西伯利亚红松这几个树种的球果成熟后在树上不开裂或仅有微裂，种子不散出，自然脱落的是带有种子的整个球果。

我国松属树种的开始结实年龄、结实盛期、丰年间隔期、花期、球果成熟期、球果成熟特征以及种子散落期等资料都已汇入表 2。

表 2　松属树种的开花结实习性

树　种	开始结实年龄（年）	结实盛期的开始年龄（年）	丰年间隔期（年）	开　花　期	球果翌年成熟期及球果成熟特征	种子散落期
华山松	10～15	20～25	2～3	4～5 月。1987～1988 年在北京观察，雄球花 5 月上旬～5 月中旬，雌球花 5 月中旬～5 月下旬	9～10 月。球果由绿色转为黄褐色，种鳞微裂，白粉增多，种子呈黑褐色。球果圆锥状长卵圆形，长 10～20cm，径 5～8cm	10 月
白皮松	8～15	20	—	4～5 月。1987～1988 北京观察，雄球花 4 月下旬～5 月中旬，雌球花 5 月上旬～5 月中旬	10～11 月。球果由淡绿色转为黄绿色或黄褐色，卵圆形或圆锥状卵圆形，长 5～7cm，径 4～6cm	
加勒比松	10～12	20	—	2～3 月	8～9 月，球果由青绿色转呈黄褐色至浅褐色，种鳞微裂	—
高山松	10～15	—	—	4～5 月	10 月下旬。栗褐色至红褐色，种鳞微裂。球果卵圆形，长 5～6.5cm，径 4～5cm	11 月上旬
赤松	15 以上	—	—	4～5 月	9～10 月。淡黄褐色或暗黄褐色，种鳞尚未开裂。球果长 3～5.5cm，径 2.5～4.5cm	—

（续）

树　种	开始结实年龄（年）	结实盛期的开始年龄（年）	丰年间隔期（年）	开　花　期	球果翌年成熟期及球果成熟特征	种子散落期
长白松	人工林 9～11	—	—	5～6月	9月中旬～10月中旬。球果由灰绿色变为淡褐色。果长卵圆形或三角状卵圆形。长4～5cm，径3～4.5cm	11月
萌芽松	8	—	3～5	福州地区3月下旬～4月上旬	11月上旬。由绿色变为浅褐色或暗褐色。球果卵形或近圆锥形，长4～8cm，径2.5～3m，近无梗，常宿存	11月中旬
湿地松	7～10	15	3	2～4月。在广州，雄球花2月上旬～3月中旬，雌球花2月中旬～3月中旬。在武汉，雄球花3月下旬～4月下旬，雌球花4月上旬～4月下旬	10月。球果从绿色转为褐色。种鳞微裂。果柄脆，容易采摘	—
海南五针松	10～12	15	4～5	海南岛2～3月；贵阳、习水5月。1988年贵阳观察，雄球花4月28日始花，5月3日进入盛期，5月9日～5月12日凋落；雌球花5月1日始花，5月6日进入盛期；5月16日授粉期结束	海南2～3月。1988年贵阳观察，9月上旬开始成熟，9月下旬～10月中旬为成熟盛期。球果由绿色转为灰绿褐色，密被白粉，有光泽。长卵圆形或卵状椭圆形，长6～12cm，径3.5～6cm，柄长0.6～2.5cm。球果宿存至翌年6月	10月下旬
巴山松	11～13	—	2～3	始期3月中旬，盛期4月上旬，末期4月中旬	10月下旬～11月下旬开始成熟，盛期在11月中旬。球果由绿色转为栗褐色。圆卵形或圆锥状卵形，长2.5～5cm，径2.5～5cm	2～3月
思茅松	5	15	1～2	1985～1987年云南思茅观察，始期在2月上旬，2月中旬进入盛期，3月上旬为末期	11月上中旬开始成熟，盛期在12月。球果由青绿色转为黄褐色或黑褐色。卵圆形，长5～6.5cm，径约3.5cm；宿存多年不落	3月
红松	天然林 80～140 人工林 15～20	天然林 180年以后，人工林 30年以后	2～3	1963～1965年小兴安岭凉水林场观察，始花期在6月中旬并迅速进入盛期，6月下旬为末期	9月下旬开始成熟，10月初进入盛期。球果由绿色转为黄绿色、深绿色、灰绿色。卵状圆锥形，长9～16(20)cm，径5～10cm	10月至翌春整个球果脱落

（续）

树　种	开始结实年龄（年）	结实盛期的开始年龄（年）	丰年间隔期（年）	开　花　期	球果翌年成熟期及球果成熟特征	种子散落期
华南五针松	—	—	—	4～5月	10～11月。球果熟时淡红褐色。圆柱形或圆柱状卵形，长4～14（17），径3～6（7）cm，柄长0.7～2cm	—
南亚松	12	20	1	粤、琼3～4月。1983～1988年在广西南宁观察，3月上旬始花，盛期在3月中旬，末期在3月下旬	粤、琼9～10月。广西南宁9月中旬开始盛熟，9月下旬～10月初为盛熟期。球果卵形，长约10.5cm，径2.5～4cm	—
马尾松	5～6年	15	2～3	3～4月。湖北宜都市观察，1991年雄球花4月4日～5月4日，雌球花4月4日～5月2日；1993年雄球花3月25日～4月26日，雌球花3月20日～4月28日	10～11月。成熟时由青色变为黄褐色或栗褐色。球果卵圆形或圆锥状卵形，长4～7cm，径2.5～4cm	12月～翌年2月
长叶松	20	—	5～7	—	—	—
偃松	15	—	—	6～7月	9～10月。由绿色或紫色转为淡紫褐色。球果长3～6cm，径2.5～3.5cm	不散落
刚松	3～5	—	3～5	4～5月	10月	—
西伯利亚红松	40～60	80	3～5	6～7月	9～10月。球果由绿色转为紫色至浅灰色或褐色。果长5～8cm，径3～5.5cm	—
樟子松	人工林15年左右	25	2～4	5～6月。1963～1984年哈尔滨观察，始期平均在5月22日，盛期在5月30日。1975年红花尔基观察，雄球花始于6月14日，雌球花始于6月16日	9～10月。成熟球果淡褐色至灰绿褐色。长3～6cm，径2～3cm。球果成熟后并不迅速开裂脱落，10月～4月均可采收	4月
油松	6～10	25	2～4	4月下旬～5月上旬	9月下旬～10月中旬。球果由深绿色变为灰褐色，部分球果先端鳞片出现黄褐色，失去光泽，微裂，常宿存数年不落。长4～9cm，径与长略等	10月中旬以后

（续）

树 种	开始结实年龄（年）	结实盛期的开始年龄（年）	丰年间隔期（年）	开 花 期	球果翌年成熟期及球果成熟特征	种子散落期
火炬松	8～15	20	—	—	10～11月。球果由绿色转呈黄褐色	—
黄山松	7～8	15	2～3	4～5月	10月中旬～11月中旬。球果由绿色转为栗褐色或暗褐色。近无梗。长3～5cm，径3～4cm。常宿存树上2～7年	—
黑松	10～15	20	1～2	4月下旬～5月上旬	10月中旬～11月中旬。球果由深绿色变成黄褐色，种鳞微裂。果长4～6cm，径3～4cm	11月上旬以后
毛枝五针松	—	30	3～4	3～4月	10月上旬～11月下旬。球果熟时淡黄褐或褐色，圆柱状长卵圆形，长4.5～9cm，径2～4.5cm，柄长1.5～2cm	—
云南松	8～11	25	2～3	1965～1967年昆明观察，始期在3月上旬，3月中旬进入盛期，3月下旬进入末期	云南11～12月。四川2～3月。球果由青绿色转为黄褐色或深灰褐色。果长5～11cm，径3.5～7cm	2月上旬～3月中旬

蛀食松属球果的害虫很多。常见的情况是同一种害虫能够危害几种松树，同一种松树常常受到几种害虫的侵袭。例如松实小卷蛾（*Petrova cristata*）蛀食的对象既有马尾松、黄山松，也有引进的黑松、晚松、火炬松、湿地松和长叶松；危害马尾松球果的除了松实小卷蛾以外，还有油松球果小卷蛾（*Gravitarmata margarotana*）、松小梢斑螟（*Dioryctria pryeri*）、微红梢斑螟（*Dioryctria rubella*）等。赵锦年和陈胜等人（1989）在浙江淳安县姥山林场15年生马尾松林中采摘了1 677个球果，总的虫害率为64.9%。受害球果中的种子品质明显下降。据赵锦年和陈胜等人（1988）测定，湿地松健康球果和受害球果种子的千粒重分别是42.4g和29.3g，相应的发芽率分别是45%和12.8%。陈胜（1993）对火炬松球果受害的过程和后果也有详细的报道。赵锦年和陈胜等人（1991，1993）报道过马尾松种子园松实小卷蛾和油松球果小卷蛾的危害特点及其防治方法。

成熟种子和种翅的大小、形状、颜色的种间变异很大（表3）。我国松属中种粒最大的是红松（长12～17mm，宽8～12mm）、华山松（长10～15mm，宽6～10mm）和海南五针松（长12～15mm，宽6～7mm）。西伯利亚红松和偃松的种粒也较大。马尾松、樟子松、黑松、赤松和云南松的种子通常较小，长度约为4～6mm。松属种子多数具翅，是外种皮的延伸物，在种子发育的早期就已经出现。种子成熟时种翅干缩成膜质或纸质。多数松树种翅的基部有关节，容易同种子分离。我国松属种子不具翅的有红松、偃松和西伯利亚红松，它们的球果成熟后不开裂或仅微微开裂，种子没有飞散传播的功能。华山松种子也无翅或仅在两侧和顶端具棱脊，有时少数种粒有极短的木质翅。

松树种子的解剖结构十分相似，都有1层木质化的外种皮，通常较薄，但也有厚而坚硬的，例如红松。里面的内种皮是一层干缩的薄膜，紧贴在富含营养物质的雌配子体（习称胚乳）上。胚乳中央有1枚已经分化出子叶、胚轴和胚根的胚。借助于射线摄影技术有时可以发现胚腔中有2～3枚胚，是为多胚现象，不过通常都是其中1个胚发育得比较大。有时其中的1个或几个胚还未发育成形，称为"点状胚"。胚乳是否充满种腔以及胚在胚腔中的相对大小，是肉眼判断松树种粒发育是否完好以及是否充分成熟的重要标志，通过射线摄影很容易作出这种判断。没有受精的胚珠也能形成"种粒"，但其中没有胚乳和胚，成为空粒种子，当然没有发芽成苗的能力。王丽娟和衣俊鹏（1990）调查过吉林省净月潭樟子松种子园中35个无性系，发现败育种子的百分率达到73.5%。林分密度过大，花粉浓度过低，种子园中各无性系的花期不遇，以及开花授粉时阴雨天气多，都会使空粒的比例上升，需要在种子调制时认真清除。图1绘出的是红松、华山松、油松和马尾松四种松树种子的外形及其纵切面。

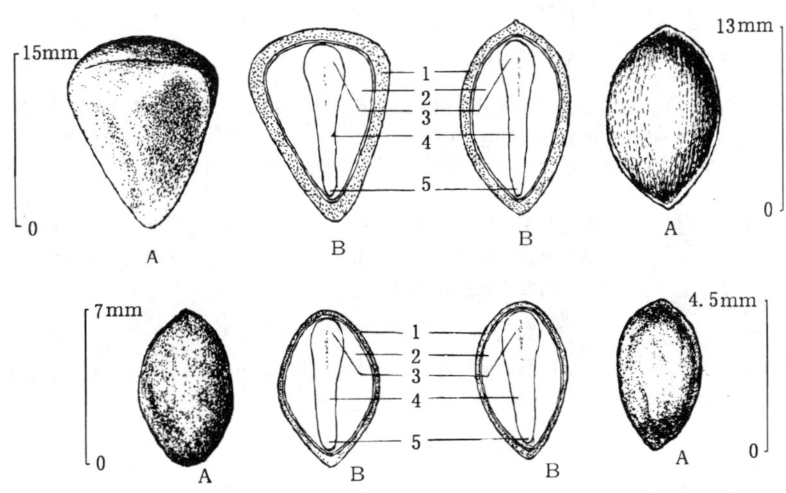

图1　松属种子的外形（A）及其纵切面（B）

红松（上左）、华山松（上右）、油松（下左）和马尾松（下右）

1. 种皮　2. 雌配子体（胚乳）　3. 子叶　4. 胚轴　5. 胚根

（田恒德绘）

表3　松属树种种子的形态特征

树　种	种　子			种　翅	
	长　度 (mm)	宽　度 (mm)	形状及颜色	长　度 (mm)	颜色及其它
华山松	10～15	6～10	扁卵形或倒卵形。黄褐、暗褐或黑色	无翅或两侧及顶端具棱脊，罕具极短的木质翅	—

（续）

树 种	种子			种翅	
	长 度 (mm)	宽 度 (mm)	形状及颜色	长 度 (mm)	颜色及其它
白皮松	10	5～6	近倒卵圆形。灰褐色	4～6	赤褐色。基部有关节，易与种子分离
加勒比松	5.7～6.5	3	窄长卵圆形，具棱脊。灰色或浅棕色，具斑点	20～25	棕色，膜质，原变种种翅基部通常不易脱落。但其它两个地理变种的关节松，容易脱落
高山松	4～8	3.5～4.5	椭圆状卵圆形，微扁。浅灰褐色或具黑褐色斑点	9～18	灰褐色或淡紫色。基部有关节，易与种子分离
赤松	4～7	—	倒卵状椭圆形或卵圆形	10～12	基部有关节，易与种子分离
长白松	4	2.4	长卵圆形或三角状卵圆形。黑色外种皮斑驳脱落或大部分保留	—	淡褐色，有少数褐色条纹。基部有关节，易与种子分离
萌芽松	4～6	2～3	卵形。黑褐色	15～20	基部同种皮融合，难与种子分离
湿地松	5～7	2.5～3.5	卵圆状，略呈三角形，稍具棱脊。淡灰色、灰色至黑色，具斑点	8～35	淡灰色、灰色、褐色至红褐色。易与种子分离
海南五针松	12～15	6～7	倒卵状椭圆形，先端圆，基部楔形。背面灰黑色，无光泽，腹面浅褐色并具栗褐色斑块或斑点	2～4（8）	纸质，易与种子分离；但基部厚纸质，常残留种子上
巴山松	6～8	—	椭圆状卵圆形。褐色	12～14	紫色，基部有关节，易与种子分离
思茅松	5～6.5	—	椭圆形。黑褐色	12～15	基部有关节，易与种子分离
红松	12～17	8～12	倒三角状卵形，微扁。棕褐色，基部有深褐色种脐	无翅	—
华南五针松	8～12	—	倒卵形或椭圆状倒卵形	8～12	—
南亚松	5～8	4～5	椭圆状卵圆形，微扁，顶端略尖。浅褐色至褐色	7～14，宽 6～7	—
马尾松	4～6	2.5～3.0	卵圆形。灰褐色或黄褐色，杂有不规则的褐色条纹	15～20	淡黄色，有褐色平行条纹。基部有关节，易与种子分离
长叶松	12	—	—	4	种翅基部与种子紧密结合，不易分离
偃松	7～10	5～7	三角状倒卵形，微扁。暗褐色	无翅	—

（续）

树　种	种　　子			种　翅	
	长　度（mm）	宽　度（mm）	形状及颜色	长　度（mm）	颜色及其它
刚松	6	—	长卵圆形	20	基部与种子接合紧密，常不易分离
西伯利亚红松	7～12	5～10	倒卵形或近三棱锥形，表面光滑。黄褐色至深褐色	无翅	—
樟子松	4.5～5.5	—	长卵圆形或倒卵圆形。黑褐色至黄褐色	6～10	基部有关节，易与种子分离
油松	6.5～9.5	3.5～5.0	卵圆形或长卵圆形。黑褐色或灰褐色	10～20	有褐色条纹。基部有关节，易与种子分离
火炬松	6	—	卵圆形或近菱形，微有棱脊。深褐色，有黑色斑点	25	基部有关节，易与种子分离
黄山松	4～6	2.0～3.5	椭圆形至倒卵状椭圆形，表面具不规则斑纹。红褐色	7～12	浅褐色，基部有关节，易与种子分离
黑松	5～7	4	倒卵状椭圆形。灰褐色，表面有深色条纹	10～12	灰褐色，具深色条纹。基部有关节，易与种子分离
毛枝五针松	7～12	5～6	椭圆状卵形。淡褐色	16	与种子结合而生
云南松	3.5～6.0	—	卵圆形或倒卵形，微扁。深灰褐色	12～15	基部有关节，易与种子分离

球果的采收　生产上多半是根据球果的颜色判断成熟期。松属球果的成熟特征已列入表2。有的地方除了球果以外，还要观察种子的某些性状，例如华山松要看种皮是否已经转成黑褐色，红松要看胚的大小是否已经发育到了胚腔长度的2/3。生产上常常是在大约有10%的球果出现成熟特征之时开始采种。通常认为略欠成熟的种子在采下来的球果中还可以继续成熟，不过这方面还没有看到确切的研究报告。四川省西昌地区习惯于在1月份采收云南松。但李滨等人（1989）的研究表明，这个地区2月份以前的采种应当属于掠青早采，表现为种子发芽能力弱，即使是低温贮藏，活力也会显著下降。该项研究认为，这个地区云南松的适宜采种期是2月上旬～3月上旬，此时球果的含水量已从12月份的40%左右下降到13%～18%，少数球果微裂但尚无种子散落。这个时候采集不仅种子质量高，而且由于球果含水量低，因而调制时间短，工效高。王培蒂和秦国峰（1994）对马尾松的研究表明，9月下旬采收的种子千粒重8.55g，发芽率只有22%，发芽测定中发霉的种子则多达78%；11月上旬和中旬采收的，千粒重为10.10g，发芽率上升到80%，发芽测定中发霉的种子仅为2%。张海廷和黄泽昌等（1990）报道过红松采种期和脱粒时间对种子成熟度的影响。有些国家是用测定新鲜球果比重

的方法来判断松属球果的成熟期，我国还未见采用。

我国多数是上树直接或辅以采种钩、采种刀采摘球果，或者是在地面用带有长柄的钩刀钩摘。有的树种例如萌芽松，果柄坚韧，通常都要上树采摘。红松和西伯利亚红松是带有成熟种子的整个球果自然坠落，可以在地面拾集，但多半还是要上树敲落球果再从地面收集。偃松也有这种情况，不过偃松树形比较矮小，可以在地上采摘。范忠仁（1985）测定过樟子松球果同果柄的结合力，发现即使是结合力最小的春季 3 月，这个数值也高达 2 400g，不可能用震动法采种。

松属采种时，树上既有已经发育成熟的球果，也有还未受精发育的小球果，应当避免伤断枝条而破坏下一年的收成。在有采伐任务的地区可以调整采伐季节，从伐倒木上采收球果。这种方法可以免除爬树这道费力而且极不安全的工序，但一经伐倒，就不容易从表型性状上选择母树。红松和西伯利亚红松种粒较大，是松鼠之类啮齿动物喜爱的食物，从它们的贮藏洞穴中有时可以收集到不少种子，缺点是难以保证种子的成熟程度，当然更难判断母树的质量。

球果调制 我国松属球果的脱粒多数仍然沿用传统的手工作业，不过用球果干燥房烘干脱粒的方法也在逐渐推广。特别是像樟子松这样处于高寒地区的树种，球果采集以后便进入气温极低的冬季，为了加速取得种子，人工加热干燥已经成了脱粒的主要手段。据范亚奇（1983）试验，樟子松在 26～52℃范围内的 7 种温度下烘干 48 小时，以 40℃时出种最多，发芽率也最高；如果烘前用 45～50℃温水浸泡 1 小时然后预干，再经受烘干，可以多出种子。陆康年等人（1982）研究过烘烤温度同樟子松球果开裂的关系；他们还发现，虫害球果难于开裂，其中受害的部分没有种子，其它部分的种子质量也不高，应在烘干前剔除。烘烤后仍未开裂的樟子松球果可以在 25～30℃的水中浸 5～20 分钟后再放回干燥室烘至开裂。如果仍有少量球果没有开裂，可用同样的方法再处理一次。用这种办法处理，一批樟子松球果一般 5～8 天便可脱粒完毕。从文献报道上看，采用过球果干燥脱粒的还有高山松、马尾松、长白松、油松和黑松。据认为，黑松球果的干燥温度以 40℃为好。徐明柏（1986）报道过马尾松球果干燥设备的一次生产性试验，所用的温度为 50～60℃。我国的人工干燥房多数是用脱粒以后的球果作为热源，准确地控制温度便成为保证种子质量的关键所在。如果球果不是静止地平摊在网架上，而是装进一种能在干燥室内缓慢转动的网式滚筒，使脱出的种子随即通过滚筒的网眼并经过一个漏斗落入下面的常温室，这样做对于保证种子质量效果会更好一些。

华山松、白皮松、赤松、萌芽松、思茅松、西伯利亚红松和油松的球果开裂并不困难，一般是用翻晒使之自然干燥的方法脱粒。摊放的厚度以不超过两三个球果为宜，否则表层和下层球果之间干燥失水速度的差异太大，翻动也很困难。

海南五针松、巴山松、马尾松和黄山松的球果开裂比较困难，一般认为是由于富含松脂，使得种鳞相互粘结的缘故，也有人对这种解释表示怀疑，不过都缺乏确切的研究材料来证实各自的推测。陆康年等（1982）讨论过樟子松球果组织结构同开裂的关系。马尾松的传统脱粒方法是，将球果聚集成堆，浇水（或草木灰水，或 2%的石灰水沉清液），盖草，大约堆沤半个月，球果变成黑褐色时移出翻晒 7～10 天便可脱粒。广东省的高州、信宜地区使用一种球果刨，切掉马尾松球果的两端并削去部分“果皮”然后摊晒，2～3 天就能全部脱粒。红松和偃松的球果比较容易干燥，球果采下后用木棒敲击 ，或者摊放几天后敲击，种鳞便可散裂脱出种子。

松属种子用人工搓揉的方法去翅，但搓揉不可过度，以免使种子受到损伤。特别是萌芽松、长叶松和刚松，它们种翅的基部同种皮结合在一起，要把种翅的基部完全清除掉而不伤及种子，并非易事。去翅所造成的损伤一般不会立即危及种子质量，但这样的种子容易感受潮气，孳生霉菌。

采用风扬之类的办法清除空粒和夹杂物以后，种子最好再摊晒一段时间，让含水量下降到适于包装、运输和贮藏的程度。陈运大（1981）研究过红松种粒大小同苗木质量的关系，他建议红松种子应在分级后播种。

松属树种球果调制的方法已列入表 4。出种率以及种子质量等数据已汇入表 5。

表 4　松属树种球果的调制方法

树　种	球果调制方法
华山松	采收后堆放 5～7 天再摊晒 3～4 天，待种鳞大部张开，敲打脱粒。使用球果干燥房脱粒时，温度不应超过 55℃
白皮松	在通风良好处摊开晾晒，勤加翻动，种鳞开裂后轻加敲打即可脱粒
高山松	摊晒，也可用球果干燥房烘干脱粒
赤松	烘干，或晒干后敲打脱粒
长白松	在球果干燥房内烘干脱粒
萌芽松	新采球果在室内堆放一周后移出翻晒脱粒
湿地松	摊晒 2～3 天后翻动脱粒。人工干燥脱粒时温度不宜超过 45℃
海南五针松	采回后在通风干燥处堆放并浇水一次，5～7 天后移出翻晒，3～5 天即可脱粒
巴山松	浇以 5%石灰水沉清液后堆沤 10 天左右，球果变为黑褐色时移出翻晒。脱出的种子在弱光下晒干或阴干
思茅松	自然干燥脱粒
红松	敲打或晾晒数日后敲打脱粒
华南五针松	采回堆放 3～5 天，摊开曝晒，并经常翻动敲打脱粒
马尾松	在阴湿处浇水或 2%的石灰水沉清液，盖草堆沤，约半个月球果变为黑褐色后移出翻晒 4～8 天脱粒。用刀削去部分“果皮”并切去球果两端后摊晒，2～3 天可以脱粒。也可用球果干燥房在不超过 45℃的温度下加热脱粒
长叶松	在 38℃的球果干燥房内烘干 48 小时
偃松	摊晒并敲击脱粒
西伯利亚红松	摊晒或室内晾干球果，约 1 个月可以脱出种子
樟子松	在人工干燥房用 25℃预热一昼夜，再升温至 45℃，但不得超过 55℃，脱出的种子应随即离开高温，剩余尚未开裂的球果可浸水后再烘至开裂
油松	摊晒翻动，5～7 天球果可以开裂脱粒
火炬松	摊晒 7～10 天后翻动脱粒
黄山松	与马尾松基本相同。堆沤后摊晒约 10 天球果开裂，翻动脱粒
黑松	连续摊晒 3～5 天即可开裂脱粒。人工加热干燥温度以 40℃为宜，不能超过 55℃
毛枝五针松	球果采后摊开曝晒，种鳞开裂，种子即可脱出
云南松	摊晒 5～6 天部分球果开始脱粒，可以收集种子。未开裂的球果浇水堆沤一昼夜再摊晒 2～3 天，又可以收集一批种子。如此反复处理，约 1 个月可以脱粒完毕

表 5 松属树种的出种率和种子质量等数据

树 种	每千克球果数（个）	每个球果的种粒数	出种率（%）	净 度（%）	千粒重（g）		每千克纯净种子粒数（万粒）	
					一 般	变动范围	一 般	变动范围
华山松	4～10	120～150	7～12	80～99	—	230～290	—	0.34～0.44
白皮松	—	—	5～8	90～99	—	140～165	—	0.60～0.72
加勒比松	—	—	6	—	14	12～19	7.1	5.2～8.3
高山松	—	—	2	80～90	16	11～26	6.2	3.8～9.1
赤松	—	—	2	85～95	8.7	7～12	11.5	8.3～14.3
长白松	—	—	4～5	90	6.6	—	15	—
萌芽松	—	—	2～3	94～99	10.8	9～14	9.3	7.0～11
湿地松	6～24	—	2～8	95～99	32	25～40	3.0	2.5～4.0
海南五针松	—	50～80	8～10（15）	95	140 206	110～150 172～215	0.7 0.48	0.6～0.9 0.52～0.62
巴山松	—	—	2～2.5	85～95	19	18～20	5.2	5.0～5.5
思茅松	—	—	1.5～3	88～97	16	13～21	6.2	4.7～7.7
红松	4～6	90～120	13～18	96～99	500	350～600	0.20	0.16～0.28
马尾松	30～85	—	2～4	80～95	11	8.5～13.6	9.0	7.4～12
长叶松	—	—	0.8～1.2	—	90	70～170	1.0	0.6～1.5
偃松	—	—	—	93～96	300	260～360	0.33	0.28～0.38
刚松	—	—	1	—	7.4	5.5～10.6	13.5	9.4～18
西伯利亚红松	10	75～100	23～25	>90	250	210～280	0.4	0.35～0.47
樟子松	100～150	25～40	1～2	85～98	8.0	5～10	12.5	10～20
油松	30～80	40～50	3～5.5	90～94	40	32～50	2.5	2.0～3.1
火炬松	25～35	—	1.5～2.5	95～99	28	22～32	3.5	3.1～4.5
黄山松	—	—	2.5～3.0	71～95	12	10～16	8.0	6～10
黑松	45～65	—	2.5～3.1	60～97	—	16～20	—	5.0～6.3
毛枝五针松	32～38	35～47	8～12	60～70	63	55～70	1.6	1.5～1.7
云南松	18～20	—	1.5～2.8	85～96	14	11～18	7.1	5.5～9.0

种子贮藏　生产上需要得最多的是越冬的短期贮藏。为了以丰补歉的储备性贮藏为时一般不会超过 1～2 个结实周期。只要贮藏得法，松属种子能在这种要求的时间里保持较高的发芽能力。含水量 6%～10%的松属种子在 0～5℃的条件下一般都能安全贮藏 2～3 年。在吉林，1963 年开始在 5～－5℃冷库中贮藏含水量为 7.6%的红松种子，20 年后生活力仅下降了 8%（刘澄和宿仁敬，1983）。罗丽芬、王彩荣和林淑云（1983）也报道过不同温度的库房中红松种子的贮藏效果。不同的树种对同样的贮藏条件反应不会完全相同。曾经在南京的室温条件下同样用布袋贮藏过黑松和思茅松，11 个月后用四唑测定，黑松生活力从 72.6%下降到了 26.7%，思茅松则从 81%仅下降到 72.9%（严圭，1982）。王培蒂（1987）将来自五个产地、发芽率为 80%、含水量为 9.0%～10.4%的马尾松种子放在 2～5℃的温度下用塑料袋密封贮藏，4 年后测定，发芽率平均为 72.2%。杨国华（1988）也研究过马尾松种子的贮藏条件。于淑兰（1988）研究过油松种子的贮藏问题。钟淑英等人（1986）发现，在最高室温不超过25℃、含水量不超过 11%的条件下，云南松种子的平均寿命为 6 年，思茅松种子的平均寿命为 4 年。董鸿运和班青等（1987）观察过马尾松种子在人工老化过程中的某些生理生化变化。王东馥等（1985）报道过贮藏前后油松种子活力和内源激素的变化。表 6 列出了我国松属树种常用的贮藏条件。

表 6　松属树种种子常用的贮藏条件

树　种	种子贮藏条件
华山松	0～5℃，种子含水量 7%～8%，硬质容器或聚丙烯编织袋盛装，可安全贮藏 2 年
白皮松	0～5℃，种子含水量 7%～8%，用硬质容器或聚丙烯编织袋盛装，可安全贮藏 2 年
高山松	0～5℃，种子含水量 8%～9%。越冬贮藏可用麻袋包装。用塑料袋密封包装，在凉爽、干燥、温度低而变幅小的库房中可以贮藏 3～4 年
长白松	含水量 9%～10%的种子在常温下密封于瓶中，可以贮藏 1～2 年
萌芽松	越冬贮藏可用麻袋、木箱盛装，置于干燥通风处，种子含水量不应超过 10%。贮藏 1 年以上的，种子含水量不应超过 8%，用聚乙烯薄膜袋密封包装，置于 0～5℃冷库
海南五针松	在普通条件下不耐长期贮藏。据贵州习水地区观察，在通风荫凉处普通贮藏 1 年零 5 个月后发芽率仅 20%
巴山松	5℃，种子含水量低于 10%，密封，可安全贮藏 3 年
红松	含水量 7%～8%的种子在 0～5℃下铁桶密封贮存 6 年，生活力可以不低于 94%
华南五针松	室内普通干藏
马尾松	含水量 10%的种子在 15℃下可以安全贮藏 1 年。如用常温贮藏 1 年，含水量不应超过 7%～8%
偃松	含水量 8%～9%的种子在低温库中密封贮藏，生活力可以保持 4～5 年；普通库袋藏可以保持 1 年

（续）

树 种	种子贮藏条件
西伯利亚红松	含水量13%～16%的袋装种子在低温干燥的室内贮藏，生活力可保持2～3年
樟子松	含水量9%～11%的种子，在0～5℃、相对湿度50%～60%的条件下可以贮藏5～6年
油松	采用低温密封干藏，发芽能力可以保持2～3年
黄山松	含水量不高于10%的种子在低温下密封，可以安全贮藏2年。一份含水量8.2%的种子在0～5℃下干藏1年后发芽率仍有93%
黑松	室内普通干藏，生活力可以保持1年
毛枝五针松	在通风干燥处干藏
云南松	在通风干燥处普通干藏，可以保存2～3年

种子含水量和库房温度是松属种子能否安全贮藏的两个最重要的因素。具备低温贮藏手段的地方不能因而放松对种子含水量的要求 。如果没有低温库房当然就更应严格控制种子的含水量。据在南京地区试验，含水量为10%的马尾松种子在15℃下可以安全贮藏1年；如果要求在常温下贮藏1年，则含水量不能超过7%～8%（严圭，1982）。据王培蒂和秦国峰（1994）试验，含水量9.4%、发芽率为87%的马尾松种子在2～4℃下密封贮藏10年，发芽率仅下降两个百分点；在室温下开放贮藏的，则3年后全部死亡。在李锦文和崔红（1988）的研究中，在5%～10.5%含水量范围内油松种子的各个样品，在0～5℃下密封贮藏39个月，发芽能力没有显著差异；但是在没有控温设备的普通库内密封存放，则是含水量越低，贮藏效果越好：含水量为11%～11.5%的种子，39个月后发芽率已降为1.5%。

发芽前的处理 松属的许多树种，例如高山松、赤松、海南五针松、马尾松、黄山松、黑松和云南松，它们的种子没有休眠习性，春播前种子不经任何处理也能顺利发芽生长。有些树种的种子则不经处理便难于发芽，而且在同一个树种中休眠深度往往还因种源（如华山松），甚至因种批以及是否经过贮藏而有不同的表现。据陶章安（1956）试验，发芽率为7.3%的白皮松种子层积132天后发芽率上升到64%。不过，据于淑兰（1983）报道，在干燥冷藏的过程中，白皮松种子的休眠深度便逐渐减弱，冷藏1年后就能正常发芽。董丽芬（1987）发现，白皮松的种皮阻碍透气透水并含有发芽抑制物质。她的试验表明，流水冲淋或浸种且经常换水可以明显地提高白皮松的发芽率。佘祥威和周光明等（1987）和汪企明（1986）研究过日本五针松（*Pinus parviflora* Sieb. et Zucc.）种子的休眠和催芽。还有些树种，例如樟子松和油松的种子，没有明显休眠习性，但是播前层积曾经收到过很好的效果。周佑勋和段小平（1993）报道，华南五针松新采的种子有生理休眠习性，外种皮含有发芽抑制物质，种皮透性也差，施加外源赤霉素对解除休眠没有明显效果；但5℃下层积60天能够有效地解除这个树种的休眠，发芽率可以从层积前的13%提高到74%。据周佑勋（1987）报道，低温（5℃）下层积30天或赤霉素浸种对火炬松种子的萌发有显著的促进作用。不过她发现，不同种源的火炬松种子对低温层积的反应不同，要求不同的层积期。

徐化成和唐季林（1989）研究过不同生态型的油松种子对低温层积的反应。他们发现，冬季比较温暖湿润的南部的油松种子，在3℃下层积37天后发芽率明显提高。他们认为这是南部型油松在自然选择中形成的防御性对策。李晓洁和徐化成（1989）对白皮松的研究也发现，种子的休眠习性随生态型而不同：暖温带生态型的种粒较重，有休眠习性，无论是在0～5℃下层积40天，还是在0～5℃下层积20天后转入25℃继续层积20天，或者是在25℃下层积40天，发芽率都显著高于对照；中亚热带和北亚热带生态型的白皮松种子则基本上没有休眠。李晓洁（1991）研究过自然分布区内6个种源的华山松种子，发现这个树种种子的休眠存在着显著的地理变异：温带草原地带的种子大而重，休眠程度弱或无休眠；暖温带落叶阔叶林地带和东部亚热带常绿阔叶林地带的种子则小而轻，休眠性较强。不仅如此，她还发现，在含油率、脂肪酸和氨基酸的组成上，这个树种也存在着随植被带不同而变异的情况。无论休眠深浅，华山松种子在0～5℃下层积20天都可以促进发芽。

对松属种子休眠讨论得最多的是红松（例如王文章，1978；王文章等，1980和1986；张良诚等，1981；谭志一，1983），休眠情况同红松类似的可能还有西伯利亚红松。陈永盛等人发现，剥去坚硬外壳的红松种子能够顺利发芽。他们因此认为红松发芽的主要障碍是外种皮阻碍了氧气进入种子、外种皮对发芽有机械阻力并含有发芽抑制物质（陈永盛、谷凤坤，1994）。曾经试用过不少方法来解除红松种子的休眠，不过生产中常用的还是层积，通常是层积240～270天，到春季4月间取出播种。金万昌和李海臣等人（1989）认为红松室外窖内层积开始的时间最好是8月下旬～9月上旬，可以避开夏季高温积水造成的霉烂。松属树种常用的层积方法或其它发芽前的预处理措施见表7。

表7　松属树种种子常用的层积条件或其他发芽预处理措施

树　种	常用的层积条件或其他发芽预处理措施
华山松	有的种源无休眠习性。有休眠习性的种子曾在0～5℃下层积20天，再用20℃层积7天，得到过较好的效果。有的地方播种前层积7～10天后，每天用50～60℃水浇喷两次，以加速催芽
白皮松	亚热带生态型的种子基本上无休眠习性。暖温带生态型的种子休眠深度在冷藏过程中逐渐减弱，冷藏1年即可正常发芽。新鲜种子需在1～5℃下层积30～40天。用100～200μg/g赤霉素浸种48小时后，在光照条件下用20～25℃变温发芽也收到过好的效果
高山松	无休眠习性
赤松	无休眠习性。有的地方水浸24小时后混沙置10～20℃条件下层积15～20天
长白松	无休眠习性，但催芽后发芽迅速整齐。一般是播前10～15天在10～20℃下催芽，如用15～25℃，5～6天即可裂嘴播种
萌芽松	有休眠的种批可以在1～5℃下层积15天，干藏过的种子建议层积15～60天
湿地松	原产地建议贮藏过的种子在1～5℃下层积15～60天。我国常用的措施有：①始温50～60℃水浸种1昼夜后置竹箩中，每天淋水并翻动，直至裂嘴后播种。②1～5℃下湿沙层积20～30天

（续）

树　种	常用的层积条件或其他发芽预处理措施
海南五针松	无休眠习性。有的地方播前用始温 40～50℃水浸种 1～2 昼夜
巴山松	无休眠习性。播前可用始温 40℃水浸种 1 昼夜
思茅松	无休眠习性
红松	在地下水位低、排水良好的地方挖坑，深约 1m，宽度和长度取决于种子数量。坑底铺放约 10cm 厚的卵石，石上铺约 15cm 厚的粗沙。按 1∶2～3 的容积比将种子与含水量约 60％的湿沙均匀混拌后倒入坑内，厚约 0.4～0.5m，并覆以 20cm 厚的粗沙，盖上草帘，最后用土堆成屋脊形封顶。坑内均匀竖放通气孔。坑的四周开沟排水。埋藏的时间一般选在 7 月下旬，有人建议在暑热之后的 8 月下旬至 9 月上旬，让种子利用较高的温度完成形态后熟再转入低温解除生理休眠
南亚松	无休眠习性
马尾松	无休眠习性
长叶松	贮藏过的种子建议在 1～5℃下层积 30 天
偃松	1～5℃层积 120 天
刚松	贮藏过的种子建议在 1～5℃下层积 30 天
西伯利亚红松	1～5℃下层积 100～150 天后在 20～25℃下催芽。有的场圃用始温 50～60℃水浸种，每天置换室温水，5～7 天后拌沙在 1～5℃下层积 3～5 个月，播前 2 周转入 20～25℃下催芽，待有 1/3 的种粒裂嘴时播种，但可能仍有少量种子翌年才发芽出土
樟子松	无明显休眠习性。但生产上采用以下方法收到过较好的效果：①雪藏 4 个月；②10～20℃下层积15～20 天。据认为雪藏过的种子萌发的幼苗抗性较强
油松	无明显休眠习性，始温 45℃水浸种 24 小时，但产地冬季湿润温暖的种子在 1～5℃下层积 40 天发芽率明显提高。有的地方用始温 45～60℃水浸种 1 昼夜后转入 15～20℃下催芽
火炬松	原产地建议 1～5℃下层积 30～60 天。我国采用过的方法有：①始温 50～60℃水浸种 1 昼夜，在常温下保湿并加翻动，至部分种粒裂嘴后播种。②1～5℃下湿沙层积 1 个月，但不同的种源要求不同的层积期
黄山松	无明显的休眠习性
黑松	无明显休眠习性。有的地方用始温 40～50℃水浸种 1 昼夜，在常温下保湿并翻动，4～5 天后播种
云南松	无明显休眠习性

敖复和高永丽等人（1987，1988）对油松种子做过空气成分的调控试验。他们将经过催芽，已经有 1/3 的种粒开始裂嘴的油松种子含水量回降到 13％左右，在 18～20℃下装入氧气分压为 5％～7％，二氧化碳分压为 6％～8％的气调袋中等候适宜的撒播造林天气。据他们报道，在甘肃地区撒播这种经过调控的油松种子，有缩短出苗期、提高出苗率的效果。据姚显明和刘天斌（1991）报道，油松种子飞播前用 HL 粉剂拌种，对某些鼠类有很好的驱避效果，而且不影响种子发芽。

发芽测定　表 8 列举的是我国松属种子发芽测定的条件及其结果的某些实例，其中有的树种只给出了用常规方法通常可以得到的发芽势和发芽率。从表 8 可以看出，施加某种预处理常常能使某些松树种子的发芽能力得到充分表现。松属中有些树种的种子为了充分发芽还明显地需要光照。就目前所知，这样的树种有白皮松、油松、黄山松、黑松和云南松。对发芽测定时需要光照的树种，通常是每 24 小时内给以 8 小时的光照。测定的温度条件通常是 25℃的恒温或 20～30℃的变温。在发芽测定中采用变温时每昼夜的高温时间维持 8 小时，如果同时需要光照，就在这 8 个小时中加光，其余 16 小时则维持黑暗和低温的条件。据徐化成和唐季林（1989）研究，油松种子发芽的最适条件是 20℃或 25℃恒温、有光、水分供应充足；30～20℃的昼夜变温对发芽并无促进作用；30℃特别是 35℃、无光、低于－2 000hPa 的水势，均对油松种子发芽不利。浙江和福建的种子检验工作人员发现，黄山松的种子发芽需要的是 15℃这样的低温，较高的温度反而明显地抑制发芽。在比较了各种测定条件之后，李晓洁（1990）认为黑松发芽测定前无需浸种或仅用室温水浸种 24 小时；她建议的其它条件是，发芽基质为两层滤纸，温度为 25℃恒温。她认为，对于油松种子发芽，每天光照 8 小时同无光照的没有明显差异。据周佑勋（1987）研究，层积后火炬松种子最适宜的发芽温度是30～20℃的昼夜变温，而且要在每天 1 000lx 光照 8 小时的条件下才能充分萌发。周佑勋和段小平（1993）对解除休眠后的华南五针松种子建议的最适发芽条件是，30～20℃的昼夜变温加光照，其次是有光情况下的 25℃恒温。

表 8　松属树种种子发芽测定条件及其结果举例

树　种	预处理	基　质	温　度（℃）		发芽势		发芽率	
			有光时	无光时	天　数	数　值（%）	天　数	数　值（%）
华山松	（1979 年山西种源）始温 45℃水浸种 24 小时	沙	25	20	14	38	30	55
	（1987 年四川凉山种源）1～5℃层积 14 天后 20℃层积 7 天	沙	25	20	8	41	23	82
白皮松	1～5℃层积 30～40 天	沙	25	20	28	68	32	85
	始温 45℃水浸 48～72 小时，再用 100～200 μg/g 赤霉素浸 48 小时	沙	25	20	42	67	50	83
	始温 45℃水浸 48～72 小时	沙	25	25	50	51	70	71
	始温 45℃水浸 48～72 小时	沙	—	25	50	0	70	0
	始温 45℃水浸 48～72 小时	沙	—	20～25	—	—	36	16

（续）

树 种	预处理	基 质	温 度（℃）		发芽势		发芽率	
			有光时	无光时	天 数	数 值（%）	天 数	数 值（%）
高山松	温水浸种 24 小时	滤纸	—	25	7	33(15～43)	17	59(31～87)
赤松	—	滤纸	—	25	—	—	21	65～90
萌芽松	—	滤纸	—	25	7	—	28	66
湿地松	1～5℃层积 20 天	滤纸	28	28	14	50～60	21	85～95
巴山松	始温 40℃水浸 24 小时	滤纸	—	25	—	35	—	40～65
思茅松	温水浸种	滤纸	25	—	7	45～70	21	75～90
红松	（具深休眠。常用染色法测定种子品质，生活力一般为 75%～95%）							
南亚松	—	沙	（日均温 20）		12	50	15	65
马尾松	—	滤纸	—	25	10	30～50	20	60～90
长叶松	—	沙或土加蛭石	22（每天光照 16 小时）	22	10	90	30	95
偃松	（具中度休眠。常用感官鉴定法测定优良度）	—	—	—	—	—	—	—
刚松	—	吸水纸	30	20	18	60	45	70
西伯利亚红松	（具深休眠。曾用离胚法测定，生活力可达 70%）	—	—	—	—	—	—	—
油松	始温 70℃水浸种 24 小时	滤纸	—	20～25	7	31	20	74
	始温 40℃水浸种 24 小时	滤纸	—	20～25	11	33	22	67
	始温 20℃水浸种 24 小时	滤纸	—	20～25	7	37	24	73
	始温 45℃水浸种 24 小时	滤纸	13～30	—	4	75	10	95
	始温 45℃水浸种 24 小时	滤纸	—	13～30	4	44	10	64
	无处理	滤纸	—	20～25	9	30	23	67

（续）

树　种	预处理	基　质	温　度（℃）		发芽势		发芽率	
			有光时	无光时	天　数	数　值（%）	天　数	数　值（%）
火炬松	1～5℃层积 30 天	滤纸	30	20	14	50～60	28	85～95
黄山松	—	石英砂	—	15	16～21	28～49	34	85～89
		石英砂	15	15	18	47	25～37	89～92
黑松	始温 45℃水浸种 24 小时	滤纸	20～25	—	8	26	23	70
		滤纸	—	30	5	23	23	51
云南松	温水浸种	滤纸	25	—	7	40～60	21	65～90

红松、偃松和西伯利亚红松的种子休眠较深，通常是用四唑或靛蓝染色的方法测定生活力。陈永盛等特别推荐红松的离胚发芽测定，据他们报道，红松离胚发芽初期由白色变为黄色的胚所占的比例同去皮种子 25 天的发芽率，以及同层积 9 个月后的场圃发芽率颇为一致（陈永盛和谷凤坤，1994）。据陈开秀试验，在 20～30 ℃下层积过的西伯利亚红松种子在近 3 个月内无一粒发芽；剥去种皮后在温箱中培养 2 个月也发芽甚少且多腐烂；但离体胚的发芽率可达 70%。研究表明，利用软 x 射线摄影技术、软 x 射线衬比摄影技术以及 IDX 技术，可以相当准确地估测松属几个常用树种种子的发芽能力（陈幼生、吴琼美和陈智建等，1992，1993a，1993b，1994；高捍东，1998a，1998b；沈永宝，1996）。李淑娴等（1996）研究过测定湿地松种子活力的途径。喻方圆等（1996）探索过马尾松种子的发芽标准。松属中有几个树种种子的外观相近，难以区分。高捍东和沈永宝等（1995）研究过利用过氧化物同工酶鉴定松属种子的真实性。种子收购现场为了在更短的时间内估测种子品质，常常采用切开法来测定松属种子的优良度，只要经验丰富，测定的结果还是相当可靠的。

播种　松属树种是我国造林的主要树种，各地播种育苗经验都比较丰富。一般是春季播种。为了利用自然条件解除种子休眠，或者为了争取春季气温回暖之时种子能及时萌发出土，有些场圃对偃松、油松等树种也实行过晚秋播种。萌芽松和马尾松等树种也有冬播的，但多在中亚热带或其以南地区，这样的冬播其实就是春播。春播前，即使是没有休眠习性的树种，有的地方也用温水浸种。浸种时间取决于种子吸胀的难易程度，有的只要 1～2 天，如海南五针松、巴山松，有的要 5～7 天，如华山松、油松和黑松。提高水的初始温度有时可以缩短浸种时间。如果其它条件都配合得好，许多场圃宁可将种子处理到有 1/3 左右的种粒裂嘴露白之时再播种。有些树种种子成本较高，例如近年来引进的湿地松、火炬松以及优良杂交组合得到的少量种子，生产上还普遍采用芽苗移栽的办法提高成苗率。

松属树种多已改用条播，但有些场圃仍然沿用撒播以提高单位面积的产苗量。松属育苗的技术要点可以从许多专著和小册子中查到。

松属树种都是出土萌发（图 2）。子叶顶带着种壳出土，极易遭受鸟害。子叶数、子叶长以及下胚轴的长度随树种不同而有比较固定的数目（表 9）。

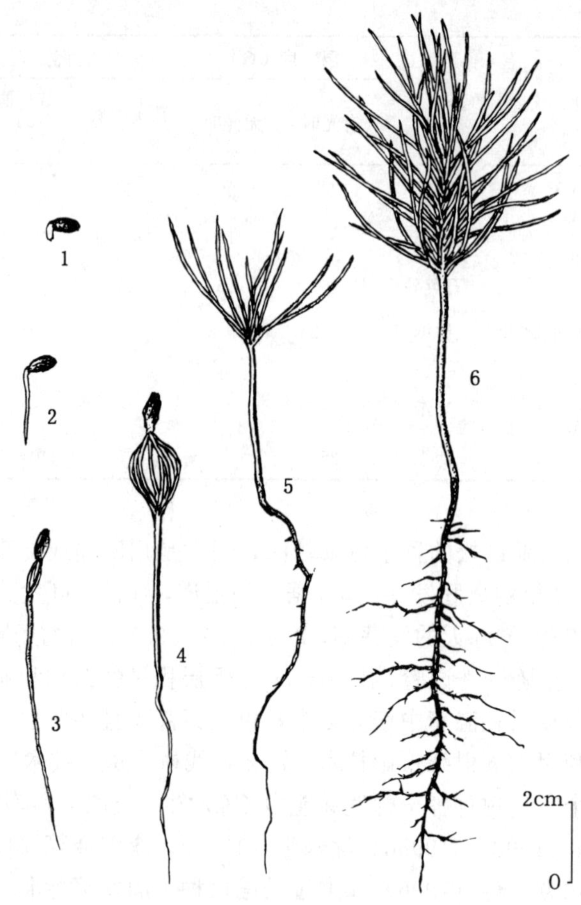

图 2 马尾松种子的萌发和幼苗生长情况
1. 胚根初生 2. 下胚轴延伸 3、4. 子叶初露 5. 初生叶出现 6. 幼苗形成
(田恒德仿《主要树木种苗图谱》)

表 9 松属树种幼苗的子叶数、子叶长度和下胚轴长度①

树 种	子叶数	子叶长（mm）②	下胚轴长（mm）
华山松	（8）10～15	40～66	45～60
白皮松	（9）11（13）	31～38	25～48
高山松	（5）7～8	23～26	28～38
赤松	（5）6（7）	（22）	—
萌芽松	（5）6～7	（25）	—
湿地松	（6）7～9（10）	35～42	（35）40～55（65）
海南五针松	（8）10～11（14）	—	—
思茅松	6～7（9）	30～35	25～38
红松	（10）11～16	38～50	40～65

（续）

树　种	子叶数	子叶长（mm）②	下胚轴长（mm）
马尾松	（5）6～8（9）	14～35	37～40
刚松	（4）5～6（7）	（20～25）	—
西伯利亚红松	10～15	（20～30）	—
樟子松	（4）5～6（7）	18～24	—
油松	（8）9～10（12）	50～60	55～78
火炬松	6～7（9）	20～35	35～48（55）
黄山松	6～8（9）	21～40	35～55
黑松	（6）7～8（10）	30～35	50～60
毛枝五针松	8～9（10）	50	55～75
云南松	（4）6～8（9）	28～38	26～40

① 黄鹏成供稿

② 子叶长度加括弧者，是测定时子叶仍在生长，相应的下胚轴长度未记录

下胚轴的颜色因树种而有区别，例如华山松呈绿色而微带紫色，白皮松下部白色而上部深绿色，马尾松为黄绿色至褐绿色，油松下部颜色淡而上部褐黄绿色，湿地松呈淡黄褐色，云南松为褐绿色，黑松为红色或紫红色。这些特征可以作为鉴定树种的辅助依据。

中亚热带以南的地区培育马尾松还盛行所谓的百日苗。容器育苗在松属中使用得相当普遍。有些松树的嫁接已经成功。从湿地松实生幼龄植株上截取穗条，甚至摘取针叶束进行扦插也有过报道。

（陈幼生）

金　钱　松

Pseudolarix kaempferi（Lindl.）Gord.

（松科　Pinaceae）

生长习性、分布和用途　金钱松属仅有本文介绍的1种，我国特产，《中国植物红皮书》列为稀有种。落叶乔木，高可达40m，胸径1.5m。喜温暖湿润的气候和土层深厚肥沃、排水良好的酸性土和中性土。能耐短时－20℃低温，较耐水湿，不耐干旱瘠薄，不适应盐碱地和积水低洼地。深根性，抗风性强。人工林生长较天然林快。分布于浙、赣、湘、皖南、苏南、闽北、鄂西、川东等地。木材用于建筑、家具，树皮可提取栲胶并提制土槿皮酊。树姿优美，秋季叶色金黄，为著名的庭园观赏树种。

开花结实　15～18年生开始开花结实，25年生以后进入正常结实期。大小年明显，结实间隔期3～5年。花单性，雌雄同株。雄球花数个簇生于短枝顶端，黄色，具细短梗，有多数螺旋状排列的雄蕊，每雄蕊具2花药。雌球花单生短枝顶端，有短梗，具多数螺旋状排列的珠鳞和苞鳞。苞鳞大于珠鳞。每珠鳞有2胚珠，倒生。发育的种鳞有2枚种子。花期4～5月。

据浙江安吉观察，初花期4月中旬，末花期5月上旬。球果当年10～11月上旬成熟。10月底起种鳞和种子同时开始散落。球果卵圆形，长6.0～7.5cm，径4～5cm。种鳞木质，扁平。种子黄白色，呈不规则卵形，有宽大的种翅，几与种鳞等长。种子长5～7mm，宽3～4mm，胚乳白色，胚细小，绿色，胚根稍尖（图1）。

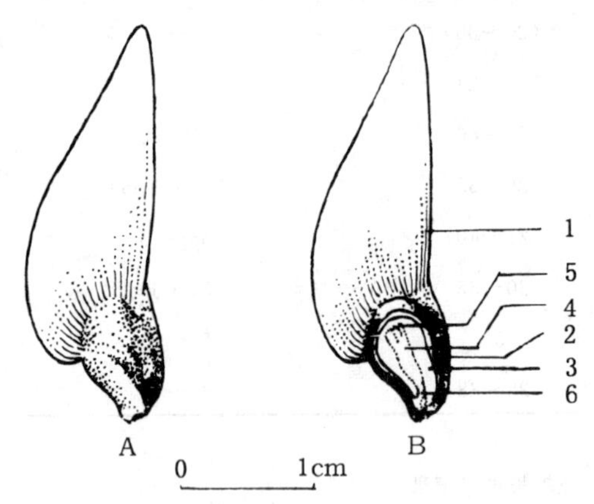

图1 金钱松种子的外形（A）及其纵切面（B）
1. 种翅 2. 种皮 3. 胚乳 4. 胚轴 5. 子叶
6. 胚根（童军平绘）

球果的采收调制和种子贮藏 有研究表明，从100年生树上采集的金钱松种子发育出的幼苗，其高、径和根系生长皆优于较小年龄（18～50）的母树和更大年龄（130）的母树。采种不能过早。据许绍远（1990）报道，9月上旬采得的种子，场圃发芽率仅56%，1年生苗高仅9.7cm；达到形态成熟时的11月上旬左右采得的种子，这两个指标分别是93.5%和12.4cm。过迟则稍有震动种子便与种鳞同时脱落，难以收集。针叶开始脱落，种鳞转为褐黄色表示成熟，应及时采集。人工摘取球果。球果采回后堆放干燥通风处，待种鳞开裂，抖出种子，再揉去种翅，筛或风选后得纯净种子。丰年球果出种率高，可达16%～20%，种子千粒重40～50g，每千克有种子2.0万～2.5万粒，发芽率达75%以上。歉年球果出种率10%以下，千粒重30g以下，发芽率不足30%。

曾经认为金钱松种子不宜曝晒。翁尧富（1991）在研究中发现，金钱松种子不耐贮藏的主要原因恰恰是种子含水量太高，往往高达22%～25%；而用日晒或温箱把种子含水量降至7%～8%以下的种子，可以贮至翌年春季；把种子含水量降至5%～6%，在0～5℃条件下密封贮藏1年，种子发芽率还相当于贮藏前的85%以上。蔡克孝和孙鸿有（1988）在比较了各种温度和种子含水量的效应后认为，金钱松种子贮藏的最佳条件是－15～－18℃，含水量4%～6%。他们还发现，入库种子的品质对贮藏效果影响很大：1980年和1982年所采种子的发芽率分别是48.7%和79%，在上述贮藏条件下，1980年的种子1年后发芽率已降为6%，1982年的种子贮藏3年后仍在30%以上。

发芽和播种 种子有休眠习性。未经预处理的种子发芽期长，发芽不整齐。用0.2%～0.4%硝酸钾或硝酸铵溶液浸种，发芽期可以明显缩短，发芽整齐：在25℃下培育，16天发芽30%～40%，25天的发芽率可达50%～85%。过氧化氢和赤霉素对发芽无明显作用。

鉴定金钱松种子优良度的标准是：种粒黄亮饱满，胚绿色，胚乳白色为优良种子。

3月播种。条播，条距20cm。每平方米播种30～38g。播后20～25天发芽出土，30～40天出土基本结束。幼苗期生长缓慢，一般用2年生苗出圃造林。

出土萌发。子叶4～6，线形，常微弯，柔软。初生叶线形，微弯如镰状。种子萌发和幼苗生长情况见图2。

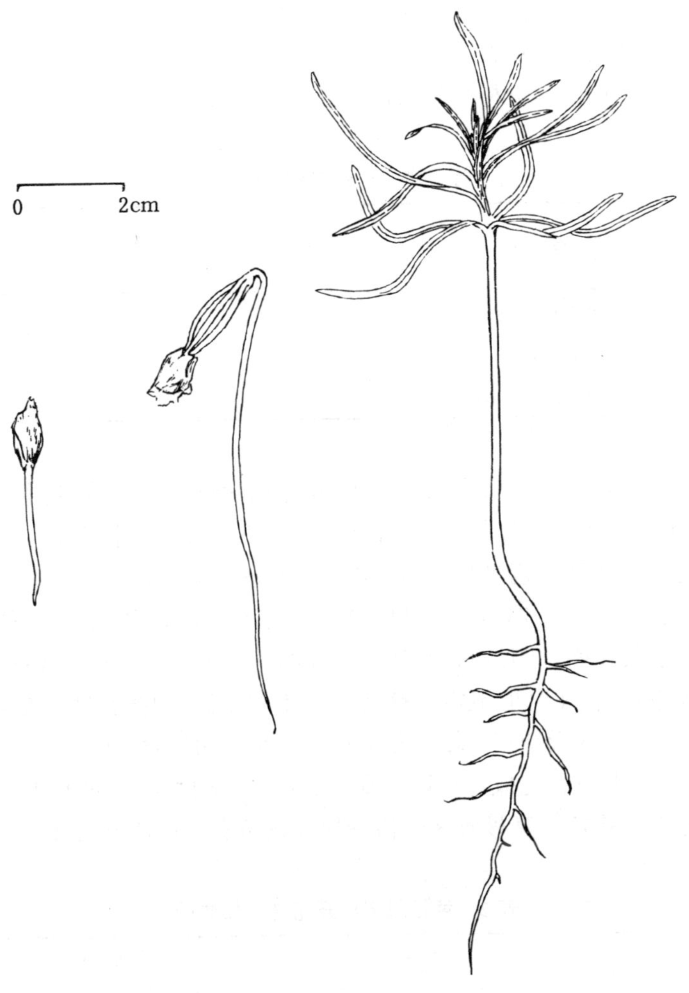

图 2　金钱松种子萌发后第 3、10、45 天幼苗生长情况
（童军平绘，右图仿《主要树木种苗图谱》）

金钱松可以嫁接或扦插繁殖。嫁接方法以削芽接与贴枝接效果较好（钱莲芳等，1990）。

（翁尧富）

黄　杉　属
Pseudotsuga Carr.

（松科　Pinaceae）

生长习性、分布和用途　本属约 8 种，我国 5 种。从北美引进的两个种，北美黄杉（*P. menziesii*）和北美大果黄杉（*P. macrocarpa*）在庐山植物园生长不良。本文描述的 3 种都是我国的特有树种。常绿乔木。高达 40～50m，胸径达 1～2m。喜温暖湿润气候和土层深厚、排水良好的酸性土壤。材质坚韧，纹理细致，有弹性，耐久用，是建筑、桥梁、车辆、枕木、家具及木纤维工业原料的优质用材。黄杉属这3个树种的名称、生长和分布见表1。在《中国植物红皮书》中，华东黄杉和黄杉列为渐危种，台湾黄杉列为濒危种。

表1　黄杉属树种的名称、生长和分布

中　名	学　名	树　高 (m)	胸　径 (m)	分　布
华东黄杉	*P. gaussenii* Flous	40	1	皖、赣、浙、闽、川
黄　杉	*P. sinensis* Dode	50	1.5	川、滇、黔、鄂
台湾黄杉	*P. wilsoniana* Hayata	50	2	台

开花结实　20～25年生开始结实，30～40年生以后进入正常结实。结实有大小年，丰年间隔期2～4年。雌雄同株。球花单生。雄球花腋生，圆柱状。雌球花着生在侧枝顶端，下垂，卵圆形，由多数螺旋状排列的苞鳞与珠鳞组成。苞鳞显著，先端有3裂。珠鳞小，生于苞鳞基部，腹面有2枚胚珠。据1979年在杭州观察，华东黄杉雌雄球花2月下旬出现，3月上旬雌球花明显伸长，呈紫红色。3月中旬雌球花中部的鳞片张开，此时为开花初期。3月下旬鳞片全部张开，进入盛花期。4月上中旬授粉结束，鳞片闭合，为终花期。雄球花在雌球花初开3～5天后撒粉，又经3～5天结束。雌球花受粉后球果开始发育。6月下旬～7月上旬，球果生长停止。8月中下旬胚发育完成，可以见到橙黄色的胚轴和幼小的子叶。10月中旬种子成熟，球果由淡绿色变为褐色。黄杉属3个树种的开花结实物候期见表2。

表2　黄杉属树种开花结实物候期

树　种	观　察 地　点	开花期		果实成熟期		种子 散落期
		雄　花	雌　花	始　期	盛　期	
华东黄杉	江西德兴	3月上旬～中旬	4月上旬～中旬	9月下旬	10月上旬～中旬	10月下旬
	江西三青山	3月中旬～下旬	4月中旬～下旬	9月中旬	9月下旬～10月上旬	10月中旬～11月上旬
	浙江杭州	3月上旬	3月下旬～4月上旬	10月上旬～中旬	10月中旬～下旬	10月下旬～11月上旬
黄　杉	江西庐山	4月中旬～下旬	5月上旬～中旬	10月中旬	10月下旬	10月下旬～11月上旬
台湾黄杉	台湾中央山脉	4月下旬	5月上旬～中旬	10月中旬	10月下旬	10月下旬～11月上旬

黄杉属球果卵圆形，下垂，有柄。种鳞木质，坚硬，蚌壳状，宿存。发育种鳞具有2粒种子。种子三角状卵形，稍扁，上端有翅。黄杉属3个树种的球果和种子特征见表3。黄杉的种子形态见图1。

表 3　黄杉属树种的球果和种子的形态特征

树　种	成熟球果				成熟种子		
	形　状	长（cm）	径（cm）	颜　色	形　状	长（mm）	色　泽
华东黄杉	圆锥状卵形或卵圆形	3.5～5.5	2～3	褐色被白粉	三角状卵形，上部有褐毛	8～10	深褐色
黄　杉	卵圆形、椭圆状卵形	4.5～8	3.9～4.5	褐色有短毛	三角状卵圆形，上部有短褐毛	约 9	褐色，背面白色
台湾黄杉	卵圆形、椭圆状圆形	4～6	3～9	深褐色无毛	三角状卵圆形，无毛	约 6	深褐色，光亮

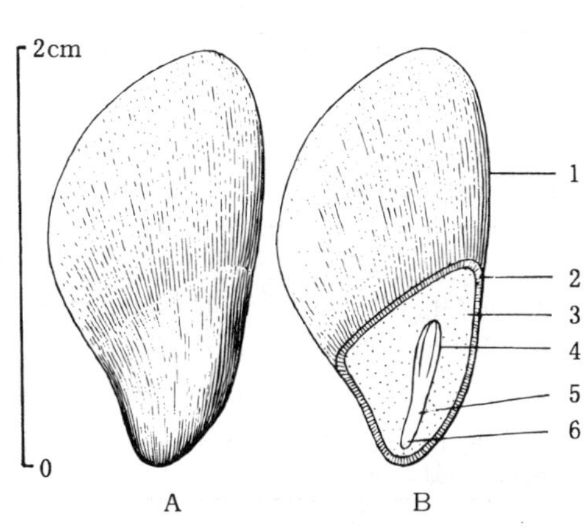

图 1　黄杉种子的外形（A）及其纵切面（B）
1. 翅　2. 种皮　3. 胚乳　4. 子叶　5. 胚轴　6. 胚根
（黄应钦绘）

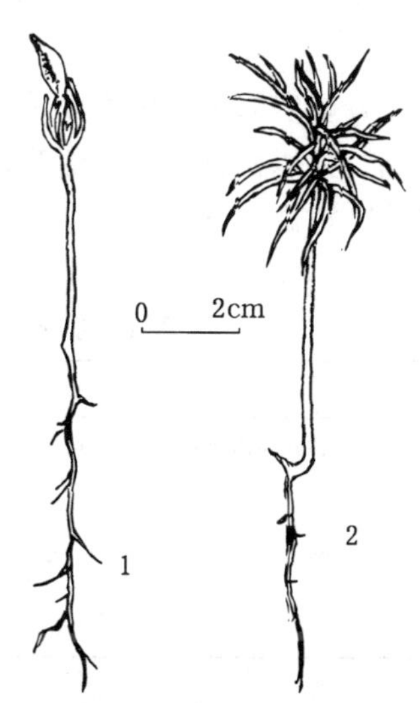

图 2　台湾黄杉的幼苗生长情况
1. 出土的种壳行将脱落，子叶露出　2. 初生叶螺旋状互生
（田恒德仿《中国主要树木幼苗形态》）

球果的采收调制和种子贮藏　球果成熟后种鳞很快开裂，脱出种子，应在球果由淡绿色转为褐色，球果尚未开裂时及时采摘。摊晒 5～6 天种鳞即可裂开，种子脱出。人工搓去种翅，经风选过筛后即得纯净种子。华东黄杉球果出种率低，且空粒种子的比例很高，常常占到 40%～50%。据分析，黄杉属这几个树种种子产量低而且质量不高的可能的原因是花粉飞散能力差，自花授粉难以形成饱满种子。杭州地区人工栽培的华东黄杉 1977 年结出的饱满种子仅占 3%，千粒重约 50g。同年人工辅助授粉后，饱满种子的比例提高到了 10%～ 35%（姜国武等，1985）。种子在 5℃左右的干燥环境中可以贮藏 1 年。

发芽和播种　适于发芽测定的温度为 15～20℃。发芽率通常只有 10%～20%。条播或撒播。3 月中下旬播种，播后 20～30 天开始出土。出土萌发。台湾黄杉子叶 7～9，轮生，针形，横切面三角形。初生叶螺旋状互生，线形。下胚轴长 3～4cm，绿褐色或紫红色。根颈白色。主根明显，侧根稀疏，茶色。台湾黄杉的幼苗生长情况见图 2。

（李　华）

铁 杉 属

Tsuga Carr.

(松科 Pinaceae)

生长习性、分布和用途 本属约16种，分布于亚洲东部和北美洲。我国有7种1变种。本文描述2种。常绿大乔木，胸径0.3～1.5m，有的达2.7m，树高20～40m，有的高50m。耐荫性强，寿命长，在天然林中生长缓慢。喜温凉湿润的气候。适生于棕色森林土地带及黄棕壤山地。

材质优良，坚实耐用，可供建筑、车船、家具、木纤维原料等用；树皮可提取栲胶；树干可割取树脂，提炼松香和松节油等，是分布区高山地带造林和森林更新的主要树种。本文两个种的名称、树高、分布和用途见表1。

表1 铁杉属树种的名称、树高、分布和用途

中 名	学 名	树高（m）	分 布	用 途	供 稿
铁杉	*T. chinensis*（Franch.）Pritz.	20～30（50）	黔、川、甘、陕、豫、鄂。海拔1 200～3 300m	材用、栲胶、保土	501
云南铁杉	*T. dumosa*（D. Don）Eichl.	20～40	藏、滇、川。海拔1 700～3 500m。印度、不丹、尼泊尔、缅甸	材用、割胶、采脂、保土	504

开花结实 30～40年生开始开花结实，150年以上的大树仍能保持旺盛的结实能力。丰年间隔期3～4年。

球花单性，雌雄同株。雄球花单生于叶腋，由多数雄蕊组成，有短梗。每雄蕊有2花药，药隔节状。本文所写的两个树种花粉无明显气囊。雌球花单生于侧枝顶端，直立，紫蓝色或

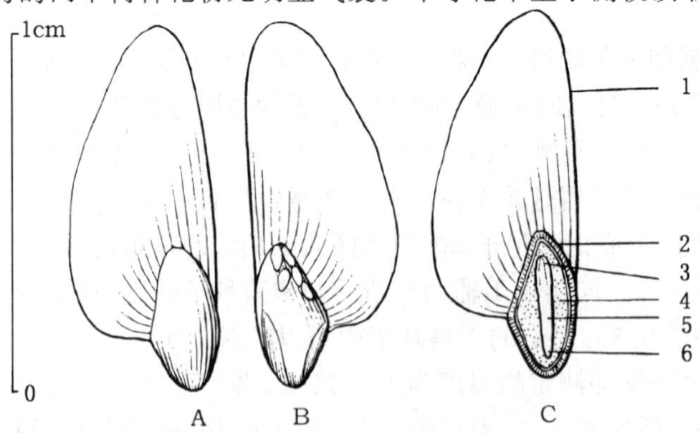

图1 云南铁杉种子外形（A）和腹面（具油点）（B）及其纵切面（C）

1. 翅 2. 种皮 3. 胚乳 4. 子叶 5. 胚轴 6. 胚根

（黄应钦绘）

浅紫红色，具多数螺旋状着生的珠鳞和苞鳞。珠鳞圆形，覆瓦状，约与苞鳞等长或较大，基部具胚珠 2 枚。球果小，下垂。中部种鳞腹面有 2 粒种子，种子上部有膜质翅。苞鳞短小，不露出。铁杉球果有短柄，中部种鳞近五角状卵形或近圆形，先端微内曲。云南铁杉球果无柄，中部种鳞卵状矩圆形或倒卵状矩圆形，上部边缘薄，微向外反曲。

胚乳和胚为乳白色。子叶 3～6 枚。云南铁杉种子形态见图 1。

花期 4～5 月，果实 10～11 月成熟。11～12 月种子散落（表 2）。球果未熟时绿色、淡绿色或浅蓝色，成熟时黄褐色或淡褐色。种子浅褐色至深褐色或黑褐色，三角状扁卵圆形或长卵圆形，腹面有油点。球果和种子的形态特征见表 3。

表 2 铁杉属树种的开花结实习性

树　种	开始结实年龄（年）	丰年间隔期（年）	开花期	果实成熟期	种子散落期
铁杉	30～40	3～4	4 月	10 月	11 月
云南铁杉	30～40	3～4	4～5 月	10～11 月	11～12 月

表 3 铁杉属树种球果和种子的形态特征

树　种	未成熟果颜色	成熟球果				种　子		
		颜　色	形　状	长（cm）	径（cm）	色　泽	形　状	长（mm）
铁杉	绿色或浅蓝色	黄褐色	长卵圆形或卵圆形	1.5～2.5	1.2～1.6	浅褐色或黑褐色	三角状扁卵圆形	连翅 7～9
云南铁杉	绿色或淡绿色	淡褐色或黄褐色	卵圆形或长卵圆形	1.5～3	1～2	浅褐色或深褐色	卵圆形或长卵圆形	连翅 8～12

球果的采收调制和种子贮藏　应在球果呈黄褐色或淡褐色而种鳞尚未开裂时从树上采摘，或从球果成熟期的伐倒木上采摘。采摘的球果摊晒脱粒，或置于球果干燥设备中烘干脱粒。人工去翅后筛去杂质即得纯净种子。

种子贮藏适宜的含水量为 8%左右。干藏。只需越冬的种子用布袋或塑料袋盛装，置库房内或通风、干燥、凉爽的地方存放。如需贮藏较长时间，应当密封置 0～5℃的干燥环境中。

球果出种率和种子质量、数量见表 4。

表 4 铁杉属树种的球果出种率、种子质量和数量

树　种	出种率（%）	净　度（%）	千粒重（g）	每千克纯净种子数（万粒）
铁杉	—	85	5.2	19.2
云南铁杉	4.4	85～95	2.0～3.0	33.3～50.0

发芽和播种 发芽测定的种子用始温 45℃水浸种 24 小时，用 0.5%高锰酸钾溶液浸 30 分钟，取出用清水冲洗后，置铺有纱布或滤纸的发芽盒内，放入恒温箱内发芽。种子发芽能力及其测定条件见表 5。

表 5 铁杉属树种的发芽能力及其测定条件

树种	预处理	基质	温度（℃）	发芽势（%）		发芽率（%）	
				计算天数	一般数值	计算天数	一般数值
铁杉	始温 45℃温水浸种 24 小时	纱布	25	3	22	18	48
云南铁杉	0.5%高锰酸钾溶液浸 30 分钟后清水冲洗	滤纸	20	10	23	14	30

春播。播种期 3 月下旬～4 月。播种前种子一般不处理。撒播。每平方米播种 50～60g，播后用细土覆盖，厚 1cm，轻压后洒水，再用草类覆盖。播后半月左右开始发芽。出土萌发。当年苗高约 6cm。在四川西部海拔 2 000m 以上山区育苗需 4～5 年出圃。采用塑料大棚育苗出圃时间可以缩短到 3 年。

图 2 绘出的是南方铁杉 *T. tchekiangensis* Flous 种子萌发和幼苗的生长情况。

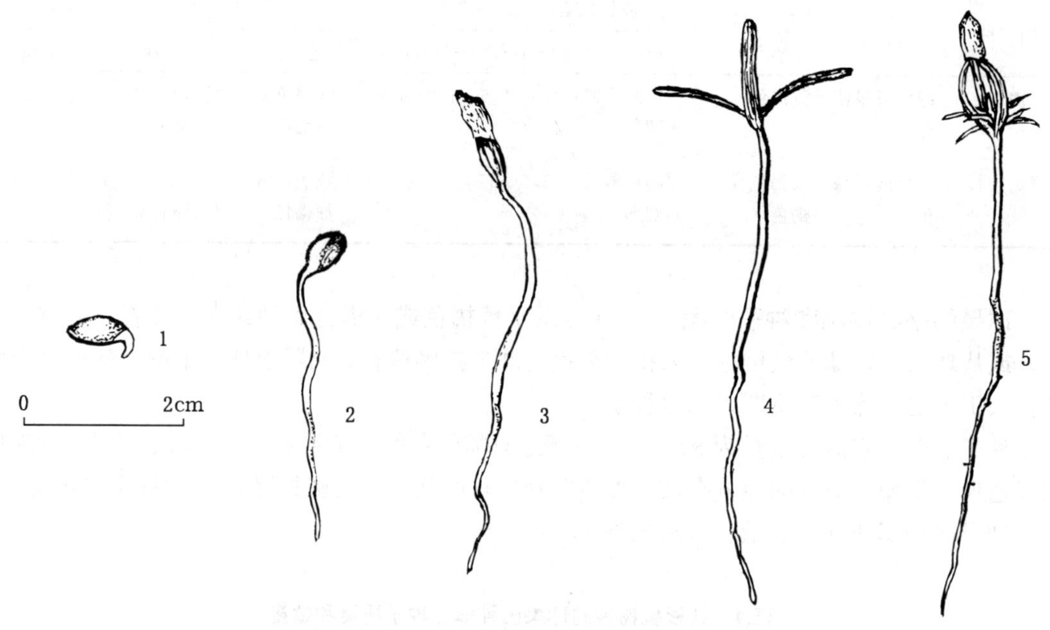

图 2 南方铁杉种子萌发和幼苗生长情况

1. 胚根初萌 2. 下胚轴延伸 3. 子叶初露 4. 子叶展开 5. 初生叶出现

（黄鹏成供稿，田恒德临）

（赵德铭）

金 松

Sciadopitys verticillata (Thunb.) Sieb. et Zucc.

(金松科 Sciadopityaceae)

生长习性、分布和用途 金松属仅有本文描述的1种，又名日本金松，为日本山区特产的孑遗植物。常绿乔木，在原产地树高达40m，胸径3m。喜光，生长较慢，喜温暖湿润环境。产于日本四国、九州至本州中部，北纬32°05′～37°37′，东经131°～140°，垂直分布在海拔100～1 500m。在海拔500～1 000m的山地常呈纯林。我国青岛、上海、杭州、南京、庐山、武汉等地引种栽培。金松是世界著名庭园观赏树种之一。木材耐水湿，供建筑、桥桩、造船等用。

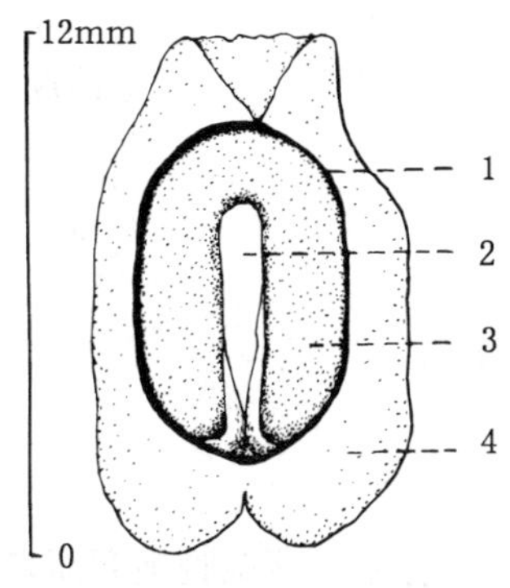

图1 金松种子纵切面
1. 种皮 2. 胚 3. 胚乳 4. 种翅
(张世经仿《日本の树木种子》)

开花结实 球花单性，雌雄同株。雄球花在秋季形成，15～30朵簇生于植株中上部侧枝顶，第2年3～5月开放，椭圆形，长约2cm，径约1.3cm。雄蕊多数，宽矩圆形，花药2室，纵裂。雌球花单生于植株上部侧枝顶。珠鳞圆扇形，长1.2cm，宽9mm，腹面基部约有9个椭圆球形、带浅橙色的胚珠，横向排成一列。苞鳞与珠鳞结合而生，仅先端分离。4～5月授粉，至10月受粉雌球花长3.5～4cm，径2.8～3cm，越冬后至翌年7月初才完成受精，生长至10月球果成熟。球果圆筒状椭圆形，有短梗，长6～8（10）cm，径4.5～5.5cm，种鳞宽2.8～3.6cm，背面绿色，先端宽圆，边缘外翻，腹面密被细毛，发育种鳞具6～9粒种子。种子矩圆形或椭圆形，扁平，周围具窄翅，连翅长（8）10～13mm，宽（6）7～9mm，黄褐色。日本本州中部结实周期常为3年。

球果的采收调制和种子贮藏 球果种鳞由绿色变为褐绿色时即可采集。采回的球果摊晒并勤加翻动，使种鳞开裂脱出带翅种子，搓揉去翅，筛选去杂，即得纯净种子。据日本长野县测定，平均每个球果有202粒（5.13g）种子，其中饱满者106粒，占52.4％。另一资料报道，每个球果得310粒种子，饱满者仅62粒，占20％，千克鲜球果有种子1 600～3 000粒（浅川澄彦等，1981）。种子形态见图1。据日本文献记载，种子千粒重(15)21～27(40) g，每千克37 000～47 000（多者达65 900）粒。风干后含水量约10％或更低的种子，在5℃低温下密封贮藏，发芽能力有可能至少保存2年。1973年吴中伦教授去日本考察林业，带回一批金松种子，分赠南京林业大学一些种子，当时实测20g种子为758粒（其中空瘪粒320粒，占42.2％）。

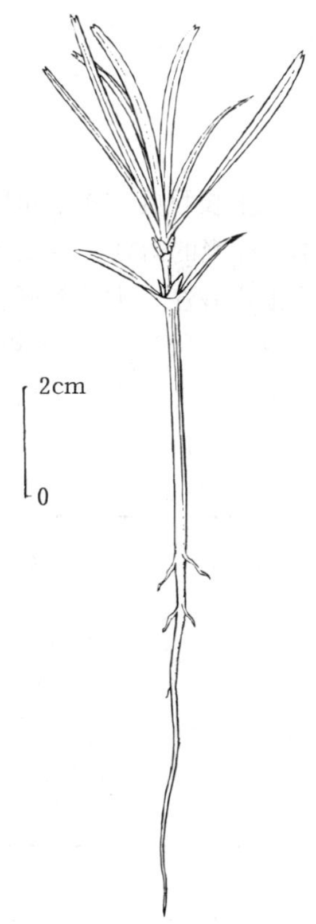

图2 金松2年生幼苗
(张世经仿柳田由藏)

据此推算的千粒重为26.4g，每千克约37 000粒。

播种 播种前需层积处理。未经处理的种子春播后发芽不整齐，部分当年发芽，大部分要到第2年和第3年才发芽。有人建议在17～20℃下层积100天，也有人建议在0～10℃下层积90天。一般是在20～25℃条件下层积2个月，再在1～4℃下层积3个月。据日本资料记载，经过层积处理的种子在室内30（8小时，光照）～20℃（16小时，黑暗）的变温条件下，50天开始发芽，80天发芽率可达90%（浅川澄彦等，1981）。原产地曾将种子浸泡于0.02%～0.1%的硝酸银溶液中24小时，据说也有促进发芽的效果。我国现有种子不多，有的地方播种前曾作如下处理：将30cm宽的棉布消毒浸水后在棉布一端平铺一层种子，折叠一层后再平铺一层种子，待全折叠起来后装入聚乙烯袋，置于－3～3℃的低温环境中，两三个月后取出播种。目前我国多用盆播，处理过的种子播后一个月可以发芽。出土萌发。子叶2，同真叶一样，也是合生叶。幼苗形态见图2。生长缓慢，当年常仅生2子叶即封顶越冬。芽鳞被红褐色毛。庐山植物园已成功采用扦插繁殖。

（黄鹏成）

柳 杉 属

Cryptomeria D. Don

（杉科 Taxodiaceae）

生长习性、分布和用途 本文介绍本属仅有的两个种。常绿高大乔木。喜温暖湿润而夏季较凉爽的海洋性或山区气候。适生于山地黄棕壤、红黄壤、黄壤，以在土层深厚、疏松湿润的壤质酸性土生长较好。柳杉分布于华东、华中、西南和华南诸地。日本柳杉原产日本，江西庐山引种已有60余年。这两个种材质轻软，纹理直，可作家具；有优美的树形，是营建风景林和园林观赏的优良树种。本属树种的名称、生长、分布和用途见表1。

表1 柳杉属树种的名称、生长、分布和用途

中 名	学 名	树 高 (m)	胸 径 (m)	分 布	用 途
柳杉	*C. fortunei* Hooibrenk ex Otto et Dietr.	40	2～3	华东、华中、华南、川、滇、黔	材用、观赏
日本柳杉	*C. japonica* (L. f.) D. Don	30～40	—	原产日本。鲁、豫、皖、苏、浙、赣、湘、鄂、川引种栽培	材用、香料

开花结实 柳杉10年生左右开始开花结实，正常结实年龄在15年生以后，间隔期1～2年。日本柳杉在原产地20～25年开始结实，结实周期有的为2～3年，有的为3～4年或4～5年（浅川澄彦等，1980）。庐山引种栽培的日本柳杉8年生开始开花结实，丰年间隔期为1年（居翔汉和徐正法，1986）。球花单性，雌雄同株。雄球花长圆形，无梗，单生于小枝上部

的叶腋，多数密集成穗状，花药 3～6。雌球花近球形，单生枝顶，无梗。珠鳞螺旋状排列，胚珠 2～ 5，苞鳞与珠鳞合生，仅先端分离。钱莲芳（1988）报道过柳杉花芽分化的进程。

开花期因产地气候环境不同而有差异，果实于开花当年 10～11 月成熟，大部分产区种子于 11 月散落（表 2）。

表 2　柳杉属树种的开花结实物候

树　种	观察地点	观察年份	开花期	果实成熟期	种子散落期
柳杉	浙江杭州	1977～1980	2 月下旬～3 月上旬	10 月中下旬	11 月上中旬
日本柳杉	浙江杭州	1990	2 月下旬～3 月中旬	10 月中旬	11 月上旬
	日本高知	—	4 月中旬	10 月下旬～11 月上旬	12 月（11 月～翌年 1 月）

种鳞约 20～30，宿存，木质，盾形。发育的种鳞具 2～5 粒种子。种子呈不规则的扁椭圆形或扁三角状椭圆形，边缘具窄翅。子叶 2～3（图 1）。柳杉属球果和种子的形态特征见表 3。

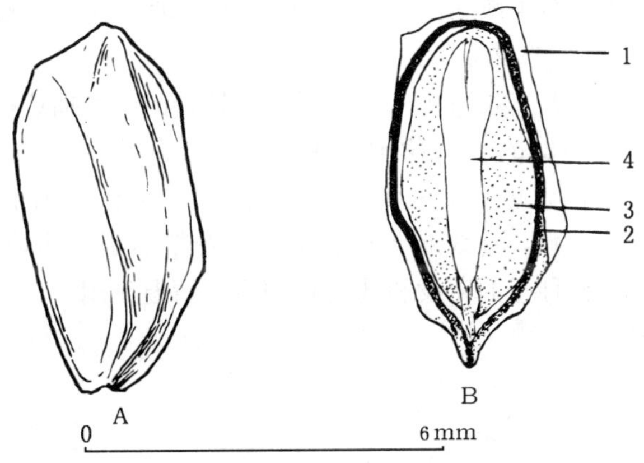

图 1　日本柳杉种子外形（A）及其纵切面（B）

1. 种翅　2. 种皮　3. 胚乳　4. 胚

（A. 童军平绘，B. 田恒德仿《日本の树木种子》）

表 3　柳杉属球果和种子的形态特征

树　种	球果 形　状	球果 大小（cm）	球果 颜色	种子 形　状	种子 大小（mm）	种子 色　泽
柳杉	近球形。种鳞 20 片左右，上部具 4～5 短三角形裂齿，齿长 2～4mm	径 1.2～2.0	黄褐色	不规则的扁三角状椭圆形，边缘有窄翅	长 4～6.5 径 2～3.5	深褐色
日本柳杉	近球形。种鳞 20～30 片，上部 4～5 深裂，裂齿长 6～7mm	径 1.5～2.5	黄褐色	扁椭圆形或不规则多角形，边缘具窄翅	长 6～7 径 2～3	深褐色

球果的采收调制和种子贮藏　选择生长良好、无病虫害的中龄林木作母树。当球果变为黄褐色时即应采摘。可以用竹竿击落球果，或用高枝剪采摘果枝。采后摊晒3～5天，种鳞开裂后抖出种子。柳杉属树种球果的出种率和种子质量等数据见表4。

表4　柳杉属树种的出种率和种子质量

树　种	出种率（%）	净　度（%）	千粒重（g）		每千克纯净种子粒数（万粒）	
			一　般	变动范围	一般	变动范围
柳杉	7～9	95～99	3.5	3.2～4.2	28.5	23.8～31.3
日本柳杉	8.0～8.1	95～99	4.0	3.3～4.5	25	22.2～30.3

脱粒后未加晾晒的柳杉种子含水量可以高到15%～17%。作者的实际经验是，批量贮藏的柳杉种子，含水量降到12%以下才能在冬季常温下安全贮藏到翌年春播。含水量高于12%的柳杉种子，在0～5℃条件下密封贮藏1年，发芽率明显降低。

发芽和播种　种子无明显休眠现象。涩粒含量高，一般占30%～60%。实验室发芽测定前有用温水浸种的，但通常不作预处理。发芽基质多为滤纸或纱布。测定温度25℃。14天的发芽势为10%～25%，25天的发芽率为20%～45%。

春播。播种期一般在3月，也有1月下旬或2月播种的。条播或撒播。每666m^{2}条播通常为5～7kg，撒播为10～12kg，播后覆土盖草。柳杉1年生苗高约20cm，日本柳杉10～20cm。有用1年生苗造林的，也有用留床或移植后的2年生苗栽植的。

出土萌发，子叶3（稀4），宽线形。初生叶第1～3轮3叶轮生或近轮生，第2～4轮后螺旋状互生，窄线形。柳杉种子的萌发和幼苗初期生长情况见图2。

柳杉还可扦插繁殖。

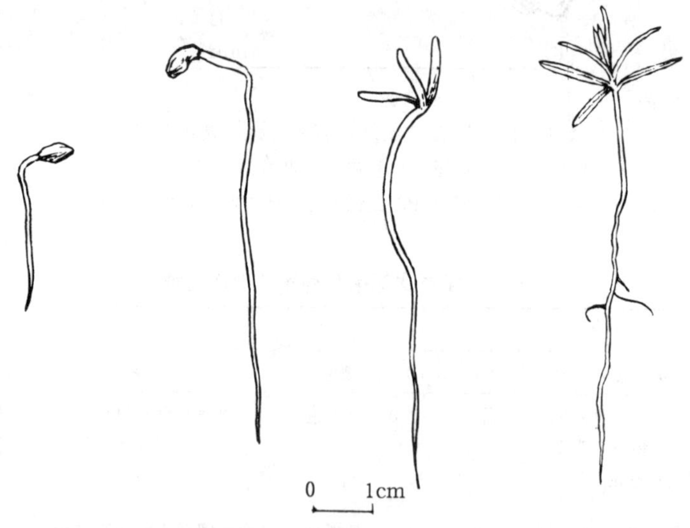

图2　柳杉种子萌发后第3、8、15、35天幼苗生长情况

（童军平绘）

（翁尧富）

杉　木　属

Cunninghamia R. Br.

（杉科　Taxodiaceae）

生长习性、分布和用途　本属只有本文描述的台湾杉木和杉木这两个种。常绿乔木，胸径可达2.5～3.0m。杉木树高可达30m，台湾杉木树高可达50m。适生于肥厚湿润的土壤，忌瘠薄干燥，也忌排水不良。萌芽力强，可以萌芽更新、扦插繁殖和插条造林。杉木材质轻韧，强度适中，为建筑、家具、车船、器具等优良用材，是我国南方重要的造林树种。杉木属树种的名称、分布和用途见表1。

表1　杉木属树种的名称、分布和用途

中　名	学　名	分　布	用　途
台湾杉木	*C. konishii* Hayata	台湾	材用
杉　　木	*C. lanceolata* (Lamb.) Hook.	秦岭、长江以南及台湾山区。越南	材用

开花结实　杉木开始开花结实年龄因种源、林分起源和母树生长条件而不同，早的3～5年，一般6～10年。边缘产区开始结实比中心产区早。实生杉木人工林的结实盛期在15年生以后，种子丰年间隔期1～4年。从平均单株球果数和种子千粒重两个方面衡量，中国科学院林业土壤研究所（1973）认为杉木实生林的种子品质以25年生以后为好。

球花单性，雌雄同株，偶有偏雄性的单株。雄球花簇生于侧枝顶部，呈扁圆球状，外被盾状鳞片，花药3。雌球花1～3朵顶生，卵球形，苞鳞与珠鳞的下部合生，螺旋状排列。苞鳞大，有锯齿。珠鳞小，3浅裂，胚珠3，倒生。雌球花多生于树冠中上部，雄球花多生于树冠中下部。

杉木的球花芽孕育于当年枝梢。表2列出了杉木属两个树种在几个点的物候期。温远光和刘世荣（1994）认为，包括雄球花始花期在内的上半年的杉木物候期，有由南向北推迟的趋势；包括球果成熟期在内的下半年的杉木物候期，则有由北向南推迟的趋势。用不同种源的无性系材料营建种子园时，应当考虑无性系之间花期是否匹配。

表2　杉木属树种的开花结实物候

树　种	地　点	开花期		球果成熟期		采种期
		雌球花	雄球花	始　期	盛　期	
台湾杉木	阿里山	—	—	—	—	10月中旬以前
	台湾大学实验林和社营林区	—	—	—	—	10月下旬～11月上旬
	天平山	—	—	—	—	11月中旬
	竹　东	—	—	—	—	11月上旬

（续）

树　种	地　点	开　花　期		球果成熟期		采种期
		雌球花	雄球花	始　期	盛　期	
杉 木	福建洋口	2月中旬～3月中旬	2月中旬～3月中旬	10月下旬	11月上中旬	10月下旬～11月中旬
	贵州天柱	3月上旬～下旬	3月中下旬	9月中旬	10月中旬	10月下旬～12月上旬
	安徽休宁	3月中旬～4月上旬	3月中旬～4月上旬	10月上旬	10月下旬	10月上旬～11月下旬
	湖北武汉	3月中下旬	3月中下旬	—	—	—

据福建南平地区观察，杉木雄球花芽在6月下旬开始分化，7月形成苞片，8月上旬出现雄球花原基，10月中下旬初步形成雄球花和花药，次年1月下旬形成花粉，2月下旬花粉成熟。同是南平地区，雌球花的分化则较迟，9月上中旬出现苞鳞突起，11月上中旬雌球花雏形初现，12月中下旬苞鳞基部出现珠鳞和胚珠突起，翌年2月中旬胚珠发育完成，2月下旬～3月上旬雌球花苞鳞张开，胚珠裸露，接受花粉。

杉木小孢子叶背面呈盾形，有纵向肋条并具蜡层。撒粉期如遇雨天，小孢子叶会以背面朝外，整个小孢子叶球便为排列有序的不透水层所包被，保护花粉。但如降水强度过大，持续时间过长，雨水也会进入小孢子叶腹面，甚至使花粉粒霉变死亡（张卓文、林平，1990），导致当年种子减产。

据俞新妥（1983）在福建观察，同一植株的授粉期为10～15天，接受花粉后约1周雌球花闭合；5～6月间花粉管中的精子细胞进入颈卵器，与卵细胞结合形成合子；9月间胚胎发育完成；11月种子成熟，12月散落。根据在浙江姥山林场的观测，迟健和邵蓓蓓（1988）报道说，杉木球果体积和质量的增长主要是在8月底以前，种子体积和质量的迅速增长期是6～8月。他们认为应当顺应这种节律安排杉木种子园的抚育管理和施肥。史忠礼和许月明等（1990）报道，7～8月，杉木胚的长度增长7.4倍，核酸总量增加2倍，蛋白质和类脂物质的含量也迅速增加。

杉木种子有许多空粒和涩粒（王华缄，1956；俞新妥，1960；叶培忠和陈岳武，1964，1981；蒋恕，1980；杜宏彬，1980；何福基，1985；余象煜等，1989；符梅忠和傅远志等，1989；林思祖和叶建春，1990）。叶培忠、陈岳武和蒋恕等人（1981）对44个无性系55个样株连续3年观察的结果是，平均而言，空粒占26%，涩粒占36%。这意味着杉木无效种子的比例平均高达六成。空粒中仅有残遗的珠心薄膜。涩粒中只有含单宁物质的败育胚乳和残存的胚柄下连接的败育胚。涩粒种子在外形、色泽、大小、质量上同正常种子几乎没有区别。王华缄（1956）提出过清除涩粒的方法。由于同自由授粉和人工异交相比，自交组合的材料中涩粒的比例高出40%～100%，因此叶培忠、陈岳武和蒋恕（1981）推测，在杉木无性系种子园中，自交效应是决定涩粒数量的主要因素，是自交导致了胚胎败育，使胚乳物质产生了变异。陈益泰和何贵平（1989）的试验也倾向于这种推测。马常耕（1989）对杉木分布区内70个种子园的混合种子作过分析，发现涩粒的比例在年份间和种子园间都存在变异。他认为，同千粒重相比，涩粒的多少受地域性生态环境和结实当年具体条件制约的程度更强。他支持迟健（1988）的看法：应当从受精和胚胎发育期间气象因子的不适性来认识杉木胚的败育，而不把自交看成是杉木形成涩粒的主导因素。

杉木种子园之间种子产量差异很大。面积为 5.57hm^2 的浙江淳安县姥山杉木种子园，1985～1989 年连续 5 年平均每公顷年产种子 66.1kg（王嫩良和方炳炎，1991）。有些种子园，特别是杉木中心产区的种子园则产量较低。迟健（1988，1992）的看法是，中心产区的生态条件能充分满足杉木营养生长的需求，但对开花结实不一定有利。他建议把杉木种子园从中心产区移往撒粉期雨量不过多、不常遭冻害的一般产区，否则就应当选择立地条件稍差而阳光充足之处建园，以适当抑制营养生长。马常耕（1989）认为，杉木分布区北带东部的浙、皖、赣邻境地区和南部湘、桂、黔毗邻的地区更适合于营建杉木种子园。

促进结实是杉木种子园工作者关心的问题。迟建和胡德治等人（1993）在浙江、广东共 6 个杉木种子园研究过施肥的效应。他们报道说，最佳施肥配方可以使种子增产 30%～50%。王赵民和吴隆高等人（1993）报道，叶面喷施 100mg/kg 或 150mg/kg 的 GA_3 能提高杉木的出种率、千粒重和发芽率，减少涩粒。李锦清和韩宁林（1991）对浙江低丘红壤上开始结实的 1.5 代杉木种子园喷施 MG 粉，使种子增产 75%。王赵民和张健忠等（1994）报道过疏伐去劣对杉木初级种子园的增产作用。

科技工作者早就在寻求预测杉木种子产量的方法（中国科学院四川分院林研所园林绿化室，1960）。近年来在杉木种子园中研究过的预测途径有可见半面树冠法、开花结实信息段法和球果切开法（陈幼生和方升佐，1990；喻方圆和陈幼生，1992 和 1996；喻方圆和肖石海等，1992；陈幼生和喻方圆，1993；沈永宝和陈幼生等，1996；莫钊志和喻方圆等，1996）。

杉木球果近球形或卵圆形。根据苞鳞的松紧程度，杉木球果可以分成紧包、松张和反卷三个类型。类型不同，种子的成熟期和散落期也有差异。反卷型球果的种子成熟早，散落也早。紧包型球果成熟迟，种子散落也迟。松张型球果的成熟期介于两者之间。为了便于球果的采收调制，营建种子园和母树林时应当考虑种内的这种物候差异。林平和张卓文等人（1990）还发现，每个球果的种子粒数以及其中的饱满粒数、涩粒数和空粒数，在不同类型的球果之间有显著的差异。

杉木属的成熟苞鳞革质，扁平，有细锯齿，宿存。发育的种鳞小，较种子短。种鳞腹面基部有种子 3 粒，偶有 4 粒、5 粒、6 粒的（林平和张卓文等，1990）。台湾杉木种子黑褐色。杉木种子棕褐色，扁平，两侧边缘有窄翅。有胚乳。成熟时胚略呈粉红色。子叶两枚。杉木属球果和种子的大小见表 3。种子的外形及其内部结构见图 1。

表 3　杉木属树种成熟球果和种子的大小

树　种	球　果		种　子	
	长（mm）	径（mm）	长（mm）	宽（mm）
台湾杉木	24～32	19～23	4～4.8	2.8～3.5
杉木	25～50	30～40	7～8	5

已经发现几种害虫影响杉木种子产量。例如，杉木球果麦蛾（*Dichomeris bimeculatus* Lin

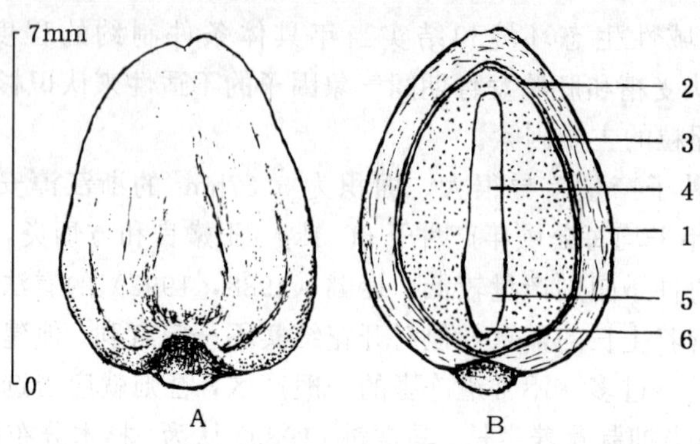

图 1 杉木种子的外形（A）及其纵切面（B）
1. 翅 2. 种皮 3. 胚乳 4. 子叶 5. 胚轴 6. 胚根
（黄应钦、田恒德绘）

et Qian）和杉木球果织蛾（*Macrobathra flavidus* Qian et Liu，sp. n.）直接危害球果和种子。杉梢小卷蛾（*Polychrosis cunninghamiacola* Liu et Pai）和杉木扁长蝽（*Sinorsillus piliferus* Usinger）则既能危害嫩梢和花芽，又能危害球果。由于它们的侵袭，1987 和 1988 年福建洋口林场的杉木种子园几乎绝收。

种子区 杉木是我国主要造林树种，分布广，变异大。林业科技工作者对杉木生态特性、地理种群状况及其分布规律作过不少研究（例如吴中伦，1955；阳含熙等，1958；叶培忠等，1964；盛炜彤、王岚等，1981；南方十四省区杉木栽培科研协作组，1981）。80 年代组织的全国杉木种源试验研究表明，杉木在生长、生物量、物候、结实、寒害等性状方面存在着显著或极显著的种源变异，多数性状同产地的纬度有密切的线性相关，呈现出以纬度为主的渐变。产地的温度和湿度是杉木种源变异的主导因子。由于分布区内复杂多样的气候生境条件的长期选择，杉木已经形成对生态因子要求各异，形态特征有不同程度变异的地理群体（全国杉木种源试验协作组，1994）。根据这种地理变异和各地特点，国家标准（GB 88.22.2-88）将杉木分布区划分为 10 个种子区。其中的第Ⅷ区，即南岭山地种子区，包括桂北、黔东南、湘西南、赣南和闽北是杉木的中心产区，也是杉木生产力最高、适应性较广的优良种源区。这个区中以广西融水、贵州锦屏为代表的南岭西部融江、清水江和都柳江流域的种源表现最为突出。中心产区的种子在一般产区和边缘产区育苗造林，生长量都超过当地种源；边缘产区的种子种植在一般产区或中心产区，生长量均小于当地种源。

球果的采收 表 2 已经列出了杉木属 2 个树种在几个地点的采种期。生产上是将多数球果种鳞（苞鳞）先端渐呈黄褐色作为杉木采种期的标志。迟健和邵蓓蓓（1988）建议除了球果颜色以外，还应当考虑球果的含水量和比容。他们认为适宜于采收的杉木球果含水量应在 65%左右，比容应小于 0.95。方升佐和许献文等人（1989）在安徽休宁县山斗种子园发现，不同采收期杉木球果的比容，同所得种子生成优良幼苗的比例相关显著，可以作为衡量杉木种子成熟程度的指标。他们认为，种子园中有 70%的球果比容下降到 0.93 左右时即可采收。

除了上树用简易的钩竿钩拉枝条采摘球果以外，杉木还可以用震落法直接采收种子。因为无需采摘并运输球果，也无需翻晒脱粒，震落法比采摘球果可以显著降低成本，而且种子成熟度高，质量好。震动的方法可以是敲击着生球果的小枝，也可以用一种叉钩摇动树冠（林夏馨，1982），但不能敲击树干造成损伤。生产上希望研制出轻便的震动采种机具，减轻劳动强度。震落法要求在干燥无风的天气进行，要求同一林分的种子成熟期和散落期一致，一般要多次震动才能采收干净。

收购球果时常常根据单位质量的球果数划分等级。颜松河（1954）报道过的标准是，一级、二级和三级每500g球果数分别为25、35和45个。

加工调制　采回的球果应及时摊晒脱粒。摊晒的厚度不宜超过3～4个球果，并应经常翻动，只要采收期正确，球果含水量不高，5～6个晴天后种子即可脱出。用干燥房脱粒时，球果应事先摊晾预干。烘干室的温度应逐步升高，最高时不得超过50℃。用固定干燥床烘烤，干燥温度取50℃时，在36小时以内干燥时间对发芽能力几乎没有影响；在55℃下经历4小时，杉木种子的发芽率便下降到只有6%（南京林产工业学院球果干燥科研组，1979）。烘烤期中最好翻动一两次。脱出的种子要迅速离开烘干室。风扬筛选后的种子净度可以达到90%～95%。进入贮藏的种子含水量应控制在10%以下。

据全国杉木种源试验协作组（1994）报道，杉木球果的大小在分布区内无一定的变异形式。出种率的变化幅度为1%～5.6%，总平均为3.58%，其中黔东南、桂北、湘、鄂、赣、浙等地在3.5%以上，陕南、桂南、豫、粤、川、皖等地在3.5%以下。千粒重同产地的纬度相关不密切，呈不连续变异：北带（陕西南郑）为6.61g，中带（川、湘、黔交界处）一般高于7g，中带南部或南带（广西融水等）仅5.56～5.94g。马常耕（1989）认为，大致上以南岭山地、武夷山和黄山为界，这条线的东南地区杉木种子千粒重要比此线西北为重，平均值分别为7.4g和6.7g。

林平和张卓文等人（1990）在浙江姥山林场观察，杉木每个球果有种子59.8～95.5粒，其中饱满种子14.3～43.5粒，涩粒27.8～46.6粒，空粒5.2～41.2粒。喻方圆和萧石海等（1992）报道，福建省两个杉木种子园全果种子总数在23～180粒范围内变动，平均约80粒；全果饱满种子数在0～55粒范围内变动，平均约22粒。杉木属两个树种的球果质量、出种率和种子质量的数据见表4。

表4　杉木属树种的球果质量、出种率和种子质量

树　种	每千克的球果个数	每个球果的种粒数	出种率（%）	净　度（%）	千粒重（g）	每千克种子粒数（万粒）	每升种子质量（g）
台湾杉木	—	—	4.9①	85.7②	3.2②	31.2②	31
杉木							
福建大田	80	80～120	4.3	95	7.2	14	—
贵州天柱	130	50～150	2.7～6.4	95	5.2～7.9	12.7～19.2	—
安徽休宁	80	105	4.5～5.0	—	9.3	10.8	—
综合资料③	—	80～120	3.0～5.0	85～95	5.9～9.7	10.3～16.9	34

① 按容积计算，据《台湾木本植物志》

② 一份样品的测定结果

③ 1976年11个省23份样品的综合数据

种子贮藏　杉木种皮疏松，容易感受潮气，丧失发芽能力。采后供翌春播种的越冬贮藏虽然时间不长，但是杉木分布区，特别是分布区的南带，冬春时节空气仍然比较潮湿，气温也比较高，对种子贮藏十分不利。因此，如果没有冷藏设备，杉木分布区的南部应当更加重视降低种子含水量：脱出的种子必须再行摊晒，充分干燥后再用防潮的容器包装，存放于干燥的环境之中。福建省林木种苗总站的研究（未刊稿）认为，在10℃的温度下贮藏半年的杉木种子，含水量不能高于9.8%。用一般方法贮藏的杉木种子，经过一个夏季发芽能力便显著下降甚至完全丧失。他们认为，如果要求贮藏后发芽能力不低于原来发芽能力的90%，则贮藏1年的种子含水量不能高于9.2%，且贮藏环境的温度不能高于5℃；如果在同样的要求下希望在5℃的环境中贮藏2年，则杉木种子的含水量不能超过7.0%。

发芽 杉木属的种子没有休眠习性。杉木发芽对温度的反应是，在15～30℃范围内发芽率无显著差异，发芽进程随温度升高而加速，至30℃时又渐趋缓慢。国家标准《林木种子检验规程》规定杉木发芽测定的温度为25℃。杉木种子置床后4～6天开始萌发，9～17天发芽结束，室内发芽率多数为20%～40%，少数种批可以达到60%～70%。台湾杉木发芽率往往低于20%。本文作者对1份台湾杉木种子样品测得的发芽率为23%。

俞新妥（1983）发现，500～700mg/kg的硫酸铜溶液浸种24小时的杉木种子，同对照相比，发芽率提高了30%～34%，发芽时间也能缩短。他还认为，用磷酸浸种后的杉木种子不仅发芽率提高，幼苗早期生长也好。李晓储和黄利彬（1990）报道过，0.05%高锰酸钾浸种4～12小时，对杉木种子发芽有促进作用。

杉木种子也可以用四唑或靛蓝检验生活力。陈幼生和吴琼美等（1993）研究过利用x射线衬比法测定杉木种子的发芽能力。曾广文和傅远志等（1989）观察过吸胀期间杉木种子亚细胞结构的动态过程。他们推测，杉木种子吸胀期间胚细胞内淀粉粒形成的速度，有可能成为种子活力和苗期生长势的早期鉴定指标。

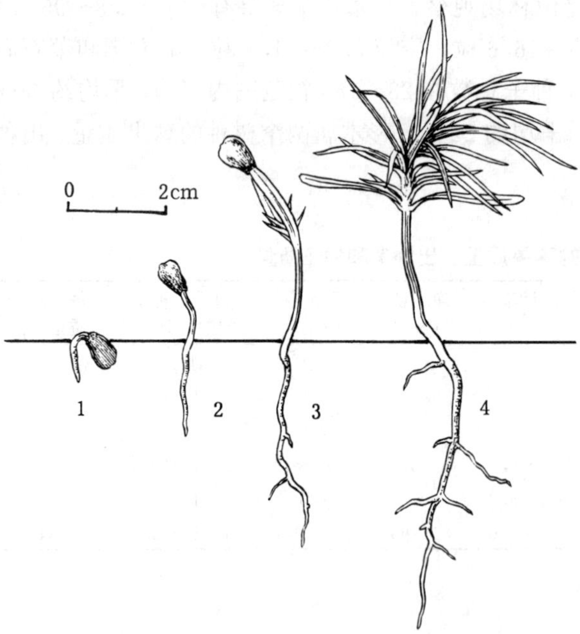

播种 春季适时早播可以使杉木幼苗及早出土，增强抗病抗旱能力。一般是在早春平均气温稳定在10℃以上时播种。低于8℃时播种，幼苗出土时间长，容易烂种。

条播或撒播。杉木条播时，每666m^{2}用种3～5kg，撒播时每666m^{2}用种4～8kg。覆土厚约0.5cm，盖草。播后15～20天开始出土时分批揭草。大多数幼苗出土，苗高约2cm，下胚轴呈红色时揭去全部覆草。台湾杉木曾在0.8m^{2}面积上撒播0.02L（0.02L台湾杉木种子重约0.62g，约200粒——本文作者注）。

0 2cm
1 2 3 4

图2 杉木种子的萌发和幼苗生长情况
1、2. 胚轴延伸 3. 初生叶出现 4. 幼苗形成
（黄应钦仿《主要树木种苗图谱》）

杉木种子出土萌发，萌发时胚轴伸长，将种壳带出土面。出土后15～20天种壳脱落，子叶张开。台湾杉木

子叶长 9～10mm，宽 2mm。杉木子叶长 12～18mm，宽 2.5～3mm。初生叶 2，对生，以后螺旋状互生。胚轴淡红褐色。有主根。侧根稀疏，水平伸展（图 2)。1 年生苗出圃。

杉木可以无性繁殖。

（李玉科、陈幼生）

水　　松

Glyptostrobus pensilis (Staunt.) Koch

（杉科　Taxodiaceae）

生长习性、分布和用途　水松属只有本文描述的 1 种，为我国特有的孑遗树种，现存植株零散分布，被《中国植物红皮书》列为稀有植物。半常绿乔木，高 8～20m。喜光，要求温暖湿润的气候。深根性，耐水湿，产粤、桂、闽、赣、川、滇等地，长江流域一些城市有零星栽培。木材轻软细密，供建筑、板料、造船、水闸板等用。根部木质松软，可作瓶塞和救生圈。树姿优美，供庭园观赏。根系发达，可固堤护岸。

开花结实　5～6 年生开始开花结实，10 年生后进入正常结实期。结实无大小年现象。球花单性，雌雄同株，单生于具鳞叶的小枝顶端。雄球花椭圆形或圆球形，雄蕊 15～20 枚，螺旋状着生，花药通常 5～7。雌球花近球形或卵状椭圆形，由 20～22 枚螺旋状着生的珠鳞与苞鳞组成。珠鳞小，中部珠鳞内有胚珠 2 枚。苞鳞大，与珠鳞近合生。据 1987～1988 年在福建南平观察，1 月下旬出现球花，2 月上旬始花，中旬为盛花期，下旬为末花期。球果 11 月中旬开始成熟，12 月上旬为成熟盛期，12 月中旬为成熟末期。成熟后种鳞开裂，种子散落。球果中部发育的种鳞有 2 粒种子，稀 1 粒。每球果有种子 6～12 粒，大量种鳞无种子。球果倒卵形，未成熟时绿色，成熟时黄褐色，长 1.6～2.5cm，径 1～1.5cm。种子椭圆形，微扁，褐色 ，长 5～8mm，径 3～4mm，基部有向下的长翅。翅褐色，矩圆状披针形，长 3～7mm。种

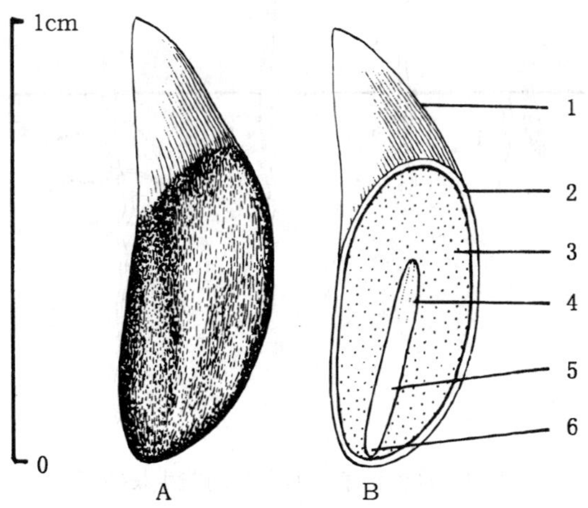

图 1　水松种子的外形（A）及其纵切面（B）
1. 翅　2. 种皮　3. 胚乳　4. 子叶　5. 胚轴　6. 胚根
（黄应钦绘）

子有胚乳，子叶3～5枚（图1）。

球果的采收调制和种子贮藏　11月下旬或12月上旬少量球果开始开裂时，上树用采种刀采割果枝。采回的球果摊放在室内通风处，厚5～10cm，7～10天后移至阳光下曝晒3～5天，球果开裂，搓揉去翅并去除杂质，即得纯净种子。种子含水量宜在8%以下。球果出种率为7.5%。种子净度一般为87%，变动范围80%～95%。千粒重13g，变动范围11～15g。每千克种子7.7万粒，变动范围6.7万～9.1万粒。晒干后的种子在0～5℃的低温下密封贮藏1年，发芽率未见明显降低。短期贮藏的种子可以装入麻袋或木箱中，存放在干燥通风处。

发芽和播种　种子无休眠习性。在25℃恒温、无光的湿润条件下35天发芽结束，发芽率55%～75%。未发芽粒多为空粒和腐坏粒。场圃出苗率约为30%～55%。春播。条播时每平方米播种约15g，覆土厚度0.8～1cm，播后盖草。播后25天左右开始萌发，40天左右大量萌发。出土萌发。子叶（3）4～5，线状针形，横切面三角形，长12～18mm，径1.2mm。初生叶为3～5叶轮生，与子叶同数交叉，向上的叶近轮生或螺旋状互生。下胚轴圆柱形，地下部白色，上部绿色渐变灰褐色。主根浅短，侧根近水平伸展。种子萌发和幼苗生长情况见图2。供园林绿化的多用2～3年生移植苗出圃。水松可以扦插繁殖。

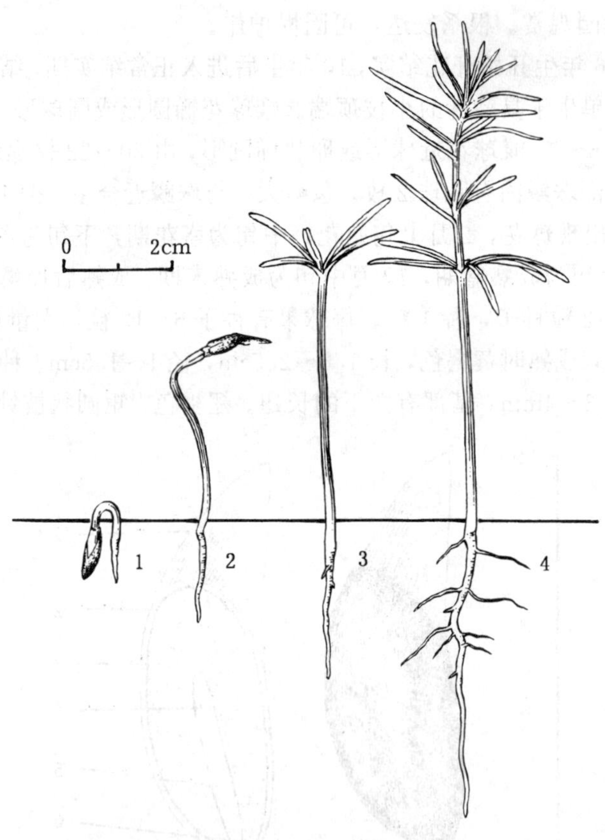

图2　水松种子的萌发和幼苗生长情况

1. 胚轴延伸　2. 子叶初露　3. 子叶展开　4. 初生叶轮生至互生

（黄应钦绘，2～4. 仿《主要树木种苗图谱》）

（郑郁善）

水　　杉

Metasequoia glyptostroboides Hu et Cheng

（杉科　Taxodiaceae）

生长习性、分布和用途　水杉属在中生代白垩纪及新生代约有 10 种，广泛分布于北半球。第四纪冰期后仅存本文描述的这一个种，20 世纪 40 年代在湖北省利川县的谋道溪附近发现，被称为“活化石”，在《中国植物红皮书》中列为稀有种。

落叶大乔木。高 35～44m，胸径 2.5m。喜光，速生。对气候适应性强，在年均温 12～20℃，年降水量在 1 000mm 以上的地方生长良好。冬季能耐－47℃低温。对土壤要求不严，酸性山地黄壤、黄褐土、钙质土均可生长。不耐水涝，不耐干旱瘠薄。在土壤湿润、深厚、肥沃的沙壤土上生长迅速。天然植株集中分布在湖北利川市的小河、四川石柱县和湖南龙山县等地。现在已在东起鲁、苏、浙、台，西至川、滇，北自辽南、北京、延安，南达粤、桂的范围内广为栽培。在秦岭、淮河以南，南岭以北的广大地区生长快、长势好。亚、非、欧、美有 50 多个国家和地区都有引种栽培。

龙跃华等（1984）在湖北潜江的广华寺进行变异种群调查时发现丛枝水杉变种（*M. glyptostroboides* Hu et Cheng var. *caespitosa* Y. H. Long et Y. Wu），无明显主干，呈丛生大灌木状，23 年生时树高不超过 8m。美国加利福尼亚的普拉塞维尔森林遗传研究所栽培的 6 株水杉，有一半呈灌木状，据推测是从孤立木上采集了自花授粉种子的结果。

水杉生长迅速，树形伟岸，经霜呈现秋色，普遍用于观赏和庭园环境栽培。木材轻软，易于加工，可作用材。木纤维强度较大，可用于造纸。

开花结实　水杉幼年期长，而且先出现雌球花，25～40 年后才结出雄球花，40～60 年开始大量结实，百年结实不衰。浙江余杭县长乐林场从已结实的母树上采集穗条，接在 2～3 年生的杉木上，次年即有 16%～42%的植株开花结实，3 年平均单株结果 53 个。

球花单性，雌雄同株。雄球花单生叶腋或枝顶，排成总状花序或圆锥状花序，总长 21～24（31）cm。雌球花单生于枝顶或近枝顶，珠鳞 20～28，交叉对生，中部珠鳞具 5～9 枚胚珠。在湖北武汉市 3 月开花并完成授粉，4～5 月球果迅速增大，6～7 月球果大小趋于定型，10 月以后球果逐渐成熟。表 1 汇集了江苏南京和安徽歙县若干年的开花结实物候记录。

表 1　水杉开花结实物候期

观察地点	观察年份	开花期			果实成熟期	种子散落	
		始　期	盛　期	末　期		始　期	末　期
安徽歙县	1964	3 月中旬	3 月中旬	4 月上旬	10 月下旬	—	—
安徽歙县	1978～1980	2 月中旬～3 月上旬	2 月下旬～3 月中旬	3 月上旬～下旬	10 月上旬～中旬	11 月上旬～中旬	11 月下旬
江苏南京	1964、1965	3 月中旬	3 月下旬	3 月下旬	—	12 月中旬	—

球果近球形，长1.3～1.9（2.5）cm，径1.2～1.7（2.5）cm。发育种鳞具5～9粒种子。种子扁平，长椭圆形，长3.8～4.0mm，宽1.5～1.8mm，褐色，表面皱缩或具肋条状突起。种子先端渐尖，顶部钝圆，基部楔形。种脐带有短梗，黑褐色。种子两侧由新月形薄翅包裹。翅淡褐色或黄色，上宽下窄，呈波状反曲。带翅种子圆形或倒卵形，顶端一侧微凹，边缘褐色，长约6.0mm，宽5.0mm，淡黄色。种子表面具树脂囊，树脂粘稠，黄褐色。有胚乳，子叶2，偶为3。种子形态见图1。

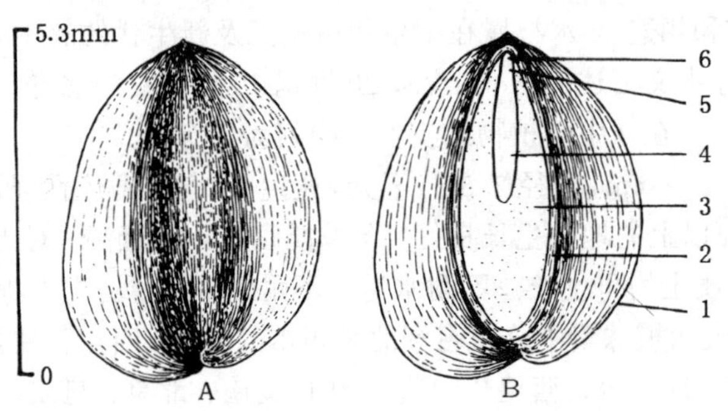

图1 水杉种子的外形（A）及其纵切面（B）
1. 翅 2. 种皮 3. 胚乳 4. 子叶 5. 胚轴 6. 胚根
（黄应钦绘）

球果的采收调制和种子贮藏 球果霜降后成熟。成熟时种鳞黄褐色，微裂。种子轻小具翅，容易散失，应及时采摘。采回的球果薄摊曝晒数日，待种鳞张裂时轻击球果，筛取种子。种子无需去翅。出种率为6%～8%。净度85%以上。据湖北利川水杉原产地对11株不同年龄母树上采得的101个球果分析，单果鲜重为3.2（2.0～6.4）g，每个球果平均有种子69.3（44.0～92.0）粒。通常每千克有鲜果462～560个，每个球果有种子80～114粒。种子千粒重1.75～2.28g，每千克种子44万～58万粒。贮运的种子含水量宜在8%～10%。种子晒干后在1～4℃下密封贮藏效果很好。在一般干藏条件下，种子生活力虽可保持两年，但发芽能力显著下降。

发芽和播种 种子无休眠习性。发芽适宜温度为18～25℃。光有促进发芽的作用。水杉常有胚珠发育不正常，珠心早期发育中止，雌配子体和胚发育不良，因此空瘪粒经常多达80%～90%。在25℃下发芽15天，发芽率常仅8（5～15）%。

长江中下游的播种期在3月中下旬。条播或撒播。条播的条距20cm，条幅3cm。播种前有不浸种即行播种的，也可在播种前浸种5～6小时后拌细土2～3倍，每平方米播种量为1.5～2.3g。随播随筛一层细土，以不见种子为度。不浸种的覆土后轻压、洒水、盖草，浸种的覆土轻压后盖草。播后10～14天发芽。出土萌发。出土3～5天后种壳脱落，露出子叶。子叶2（稀为3），质柔软，条形或略呈镰状，先端钝。上胚轴短，幼茎略呈四棱形。下胚轴圆柱形。主根、侧根发达，较纤细。水杉发芽情况见图2。每平方米可产1年生播种苗40～50株，一二级苗占85%。吴侠中和胡素梅（1993）报道过水杉温床播种而后芽苗移栽的技术，2年生苗平均高1.74m。在他们实施这项技术的两年中，每千克种子得苗数分别是44 620株和15 920株。

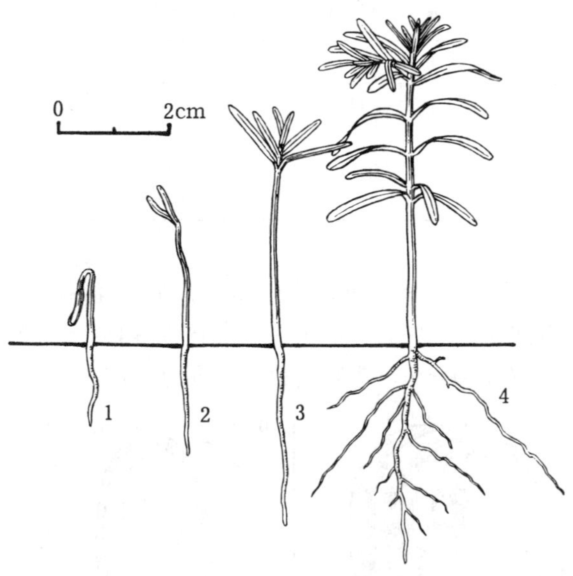

图 2　水杉种子萌发及幼苗生长情况

1、2. 子叶出土　3. 初生叶对生　4. 幼苗形成

（黄应钦仿《主要树木种苗图谱》）

水杉属于扦插极易生根的树种，生产上常用扦插繁殖，但应从 1 年生实生苗或 1 年生实生苗的截干苗上采集插穗。四旁绿化多用 3～4 年生移植苗。

（杨国华）

北美红杉

Sequoia sempervirens（Lamb.）Endl.

（杉科　Taxodiaceae）

生长习性、分布和用途　北美红杉属仅有本文描述的 1 种。常绿大乔木，在原产地树高 60～90m，最高达 120m。喜冬季多雨、夏季多雾的气候和排水良好、深厚肥沃的酸性至中性土壤。世界著名速生珍贵树种。寿命长，最老的树龄约 2 200 年。产美国太平洋沿岸长 700km，宽约 60km，海拔 500～1 000m 的狭窄地带。我国上海、杭州、舟山、南京、鸡公山、庐山、昆明等地引种栽培，以舟山、昆明等地生长最好。在浙江年高生长量为 60～130cm，最高达 209cm。昆明 14 年生树高 20m，胸径 30cm。心材红色，质坚，易加工，保存期长，为世界名材，供作建筑、胶合板、纸浆原料、桩柱等用。树形雄伟壮丽，是优良庭园树种。

开花结实　原产地 5～15 年后开始开花结实，盛期在 60 年生以后。据昆明树木园观测，7 年生实生树便开始开花结实。球花单性，小，雌雄同株，但着生于不同的枝条。雄球花单生枝顶或叶腋，雄蕊多数，螺旋状排列。雌球花生于短枝顶端，珠鳞 15～20，每珠鳞有胚珠 3～7，直立，呈新月状排列。开花结实物候因产地不同而有差异（见表 1）。

表1 北美红杉的开花结实物候

地 点	雌球花	雄球花	球果成熟期
原产地	11月～翌年3月		9月
日本本州	2月中下旬		10月中旬
杭 州	2月下旬～3月中旬	2月下旬～3月中旬	10月上中旬
昆 明	9月中旬	9月中下旬	翌年9月中旬

球果长椭圆形或近球形，长1.3～3cm，径1.3～2.2cm，淡红褐色，下垂。种鳞15～18，木质，菱状盾形，背部有凹槽，中央有一小尖头。发育种鳞具2～5种子。种子扁圆形或不规则三角形，长3～6mm，宽3～4mm，两侧具棕色窄翅（见图1）。

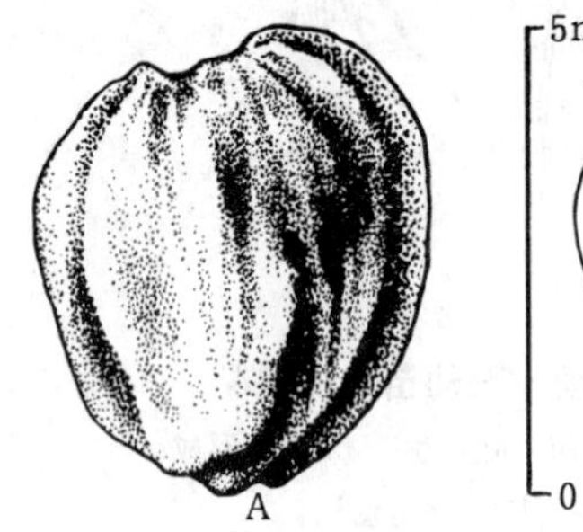

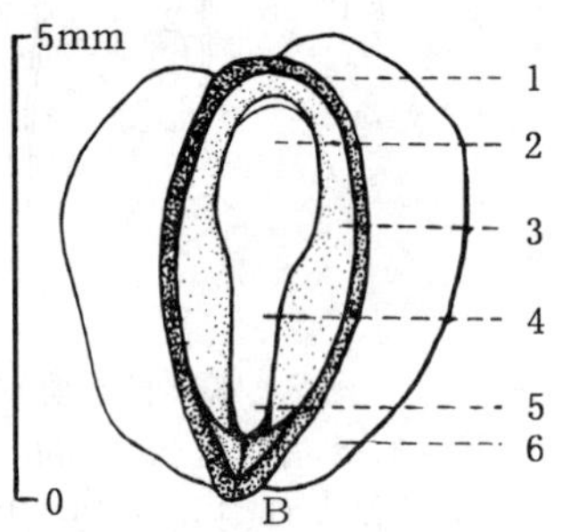

图1 北美红杉种子外形（A）及其纵切面（B）
1. 种皮 2. 子叶 3. 胚乳 4. 胚轴 5. 胚根 6. 种翅
（张世经绘）

球果的采收调制和种子贮藏 球果种鳞由绿色变为黄绿色或种鳞微微张开时，即应采集。采回的球果置于室内通风干燥处7～10（14）天，待种鳞开裂，用木耙翻动球果，脱出种子。搓掉种翅，经风筛去杂后即得纯净种子。据云南林业科学研究所测定，每球果有种子36～68粒，平均56粒；鲜果出种率为14.4%；种子的净度95%；千粒重3.4g，每千克有纯净种子29.4（25～33）万粒。北美红杉种子中空粒较多。原产地曾对成熟林自然散落的北美红杉种子作过为期5年的观察，发现饱满种子只有2.5%～12.4%。

据原产地的经验，在－2～5℃下密封贮藏1年的北美红杉种子，发芽能力保持得很好，但从冷库取出后便迅速降低；含水量为6%～10%的种子在5℃下密封贮藏3年，发芽率为14%，16年后完全丧失发芽能力。对北美红杉种子，我国尚无成熟的贮藏经验，一般是在室内晾干后用布袋干藏，贮藏期约半年。

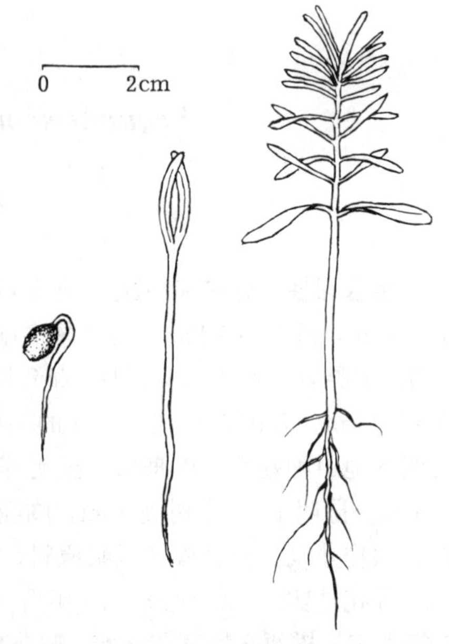

图2 北美红杉种子萌发后1、7、25天的幼苗生长情况（田恒德绘）

发芽和播种 种子无休眠习性，发芽前不需预处理。发芽测定用20℃恒温或30（8小时）～20℃（16小时）的变温。1989年3月上旬，云南林业科学研究所将14年生北美红杉母树所产种子播于室外沙床，当时日平均温度为15～20℃，约20天发芽，播后50天发芽率14%。出土萌发，子叶2（图2）。在昆明，1年生苗可以出圃，如果用于庭园绿化，则需培育2～3年生苗。杭州植物园还成功地用过扦插、分蘖、嫁接繁殖法。邢章美（1988）在秋末冬初用1年生半木质化枝条扦插，成活率达90%。

（黄鹏成、张茂钦）

巨　杉

Sequoiadendron giganteum **（Lindl.）Buchholz**

（杉科　Taxodiaceae）

生长习性、分布和用途　巨杉属仅有本文描述的1种，常绿大乔木，在原产地高达100m，胸径10m。耐荫，喜肥沃疏松的酸性土壤，也可在石灰性土壤上生长。抗寒性强，可耐－24℃的低温，能耐干旱。10～15年生以前生长较慢，以后渐快。产美国加尼福利亚州中部内华达山脉西坡海拔1 350～2 250m的山地。寿命长，有2 000～3 000年的大树。我国上海、南京、杭州、南宁等地引种栽培，但不少因不适应夏季高温或梅雨气候而死亡。木材较轻软，纹理直，材质较北美红杉脆，较难加工，与土壤接触表现出能耐久用，林中倒木数百年不腐朽。供建筑、桩柱等用，并供观赏。

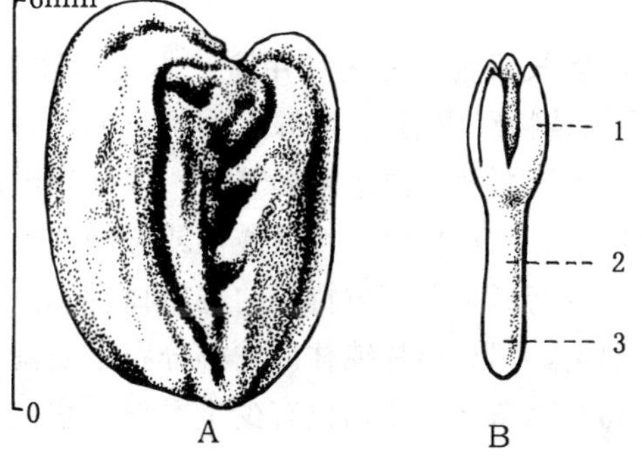

图1　巨杉种子外形（A）及从中取出的胚（B）
1. 子叶　2. 胚轴　3. 胚根
（张世经绘，B. 仿《美国木本植物种子手册》）

开花结实　在原产地20年生后开始开花结实，盛期可至200多年。雌雄同株。雄球花单生短枝顶，雄蕊多数，螺旋状排列。雌球花顶生，珠鳞25～40，每珠鳞有胚珠3～12，排成2行。球花于2～3月开放，4～5月授粉，待8月受精后球果迅速增大，直至翌年9～10月成熟。球果和其中的许多种子可宿存树上多年。球果下垂，椭圆形，长5～8.9cm，径4～5.5cm，暗红褐色。种鳞木质盾形，背部有凹槽，发育种鳞具3～9种子。种子长椭圆形，长3～6mm，两侧具宽翅，连翅宽2.5～4.5mm，淡褐色。胚具4（3～5）子叶（图1）。

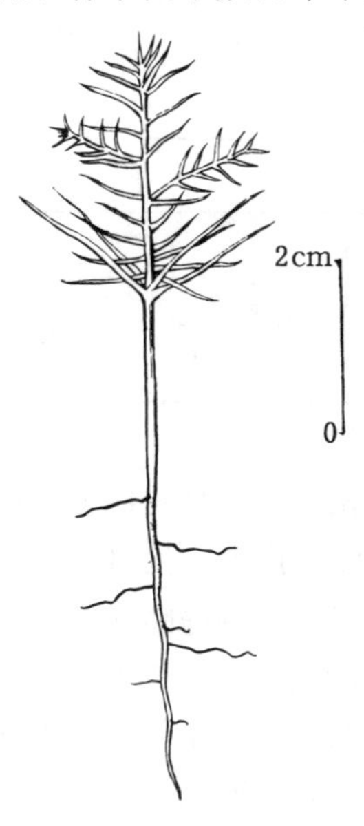

图2　巨杉幼苗
（张世经仿柳田由藏）

球果的采收调制和种子贮藏　球果由青绿色变黄绿色时即可采集。球果采回后置室内通风处或在日光下晾晒，种鳞开裂后抖出种子，除去种翅杂质即得纯净种子。据《美国木本植物种子手册》记载，每球果平均有种子230粒，球果出种率1.6%，每千克有纯净种子178 000（136 000～220 000）粒，折合千粒重5.6（4.5～7.3）g。净度 81%。饱满种子约占 41%。在 0～5℃的环境中干藏的种子，生命力可以保持较长的时间。

发芽和播种 发芽的适宜温度为 15～20℃的恒温，20～30℃的昼夜变温效果也很好。发芽温度不能低于 5℃，也不能高于 30℃。春播。南京林业大学树木园曾得到国际交换的巨杉种子 27 粒，盆播后出苗 3 株，发芽率 11%。出土萌发。子叶 3～5，以 4 为多，三角状锥形，上面两侧粉绿色，下面茶褐色带红色，密被白粉气孔线。初生叶与子叶同数，近轮生，再上的叶近对生至互生。10～15 叶的叶腋生出嫩枝，上生较短而宽的叶（图 2）。下胚轴胭脂红色，最为醒目。

（黄鹏成）

台湾杉属

Taiwania Hayata

（杉科 Taxodiaceae）

生长习性、分布和用途 本文描述该属仅有的 2 种，它们都是我国的稀有种，已列入《中国植物红皮书》。常绿大乔木，胸径可达 2～3m，树高可达 60～75m。浅根性。喜夏无酷热，冬无严寒的温暖湿润气候，或气候温凉，夏秋多雨的环境。适生于山地红壤、黄壤、红黄壤或棕色森林土，在微酸性或微碱性土壤上均能正常生长。

台湾杉是台湾省特产的稀有树种，在中央山脉海拔 1 800～2 600m 处，与红桧或台湾扁柏等混生，间有小片纯林。秃杉分布于云南西部怒江上游、澜沧江流域、贵州东南部、湖北西部和福建东部。缅甸也有少量秃杉。它们的名称、分布和用途见表 1。

表 1 台湾杉属树种的名称、分布和用途

中 名	学 名	分 布	用 途	供 稿
台湾杉	*T. cryptomerioides* Hayata	台。闽、川等地引种栽培	材用、保土	501
秃杉	*T. flousiana* Gaussen	滇、黔、闽、鄂。苏、浙、川等地引种栽培。缅甸	材用、保土、观赏	506，510

木材轻软，纹理直，结构细，易加工，可供建筑、桥梁、船舶、家具、板材及造纸原料等用。台湾杉还以其对海港蚀材虫的抵抗力特强，为台湾省最优良的海岸岸壁防浪材。

台湾杉和秃杉均为分布区及适生区的主要造林树种。秃杉树姿优美，可供庭园观赏。

开花结实 雌雄同株。雄球花 1～9 个，簇生于小枝顶端。台湾杉雄蕊 10～15 枚，秃杉雄蕊 30 枚以上（刘业经，1981）。每个雄蕊有花药 2～4 个，药隔鳞片状。雌球花单生于小枝顶端，直立，椭圆形，有覆瓦状排列的珠鳞多枚，每珠鳞有胚珠 2 枚，苞鳞退化。

开花结实年龄较迟，40～50 年生，有的要在 60 年生以上才开花结实。但结实期较长，盛果期可以历经一二百年。结实间隔期 2～3 年。

开花期3～4月。台湾杉10～11月，秃杉10月中下旬～12月上旬球果成熟。球果未熟时绿色，成熟时黄褐色或黄绿褐色，长1.5～2.5cm，径0.6～0.9cm，球形、卵形、椭圆形或短圆柱形。种鳞倒心形。台湾杉种鳞较少，12～20片或15～21片；秃杉种鳞较多，27～36片或21～39片。这也是台湾杉与秃杉的区别之一。种鳞扁平，革质，无种阜，发育的种鳞具2枚种子。种子表面平滑，扁平，黄褐色，长圆状卵形、长椭圆形或长卵形，两侧边缘有窄翅。翅的上下两端有缺口。种子连翅长4～7mm，宽3～4.5mm。子叶2枚。秃杉种子的外形和纵切面见图1。

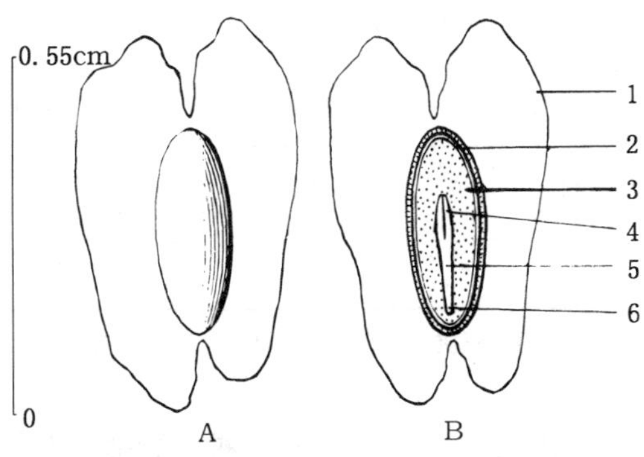

图1　秃杉种子的外形（A）及其纵切面（B）
1. 翅　2. 种皮　3. 胚乳　4. 子叶　5. 胚轴　6. 胚根
（黄应钦绘）

球果的采收调制和种子贮藏　应在球果呈黄绿褐色，种鳞未张开时将球果采下，置通风干燥处摊晾，在种鳞开裂的过程中经常翻动球果，种子即可脱出。有些地方在采集秃杉种子时，为保护母树，在种鳞已裂而种子未飞散时，上树用软兜套住球果并抖动，使种子落入兜内，每株进行3～4次可基本收完种子（潘盛荣，1987）。

据胡宝信（1993）报道，秃杉种子只有在－10℃以下的低温中贮藏才能较好地保持发芽能力。在他的试验中，在较高的温度（≥10℃）下贮藏1年以及在室温下贮藏7个月的秃杉种子，发芽率已大幅度下降。不过他没有确切说明用于试验的种子含水量。一般的看法仍然是，台湾杉属的种子适于在充分干燥后用普通的低温保存。

台湾杉属树种种子的质量等数据见表2。

表2　台湾杉属树种种子净度、质量和数量

树　种	产　地	净度（%）	千粒重（g）		每千克纯净种子数（万粒）	
			一　般	变动范围	一　般	变动范围
台湾杉	台湾	98	0.80	0.77～0.91	125.0	110～130
秃杉	云南	—	—	1.2～1.6	65.3	63～83
	贵州	97	1.55	—	64.5	—

发芽测定 通常用始温45℃水浸种24小时，或用0.5%高锰酸钾液浸种30分钟，取出冲洗后作发芽测定。也有不作浸种处理的。常用的发芽基质为滤纸或纱布，在25℃的恒温下测定。据郭玉明等的试验，秃杉种子在25℃或30℃恒温条件下，种子发芽分别为8～9天或6～7天。采种后立即作发芽测定，发芽率较高，一般在50%以上。表3所示是经调运及越冬贮藏后的发芽测定结果，种子发芽能力有所降低。

表3 台湾杉属树种种子发芽能力及其测定条件①

树 种	发芽势（%）		发芽率（%）	
	计算天数	一 般	计算天数	一 般
台湾杉	11	14	22	22
秃杉	15	20	21	26

① 测定前用始温45℃水浸种24小时；基质为纱布；测定温度为恒温25℃

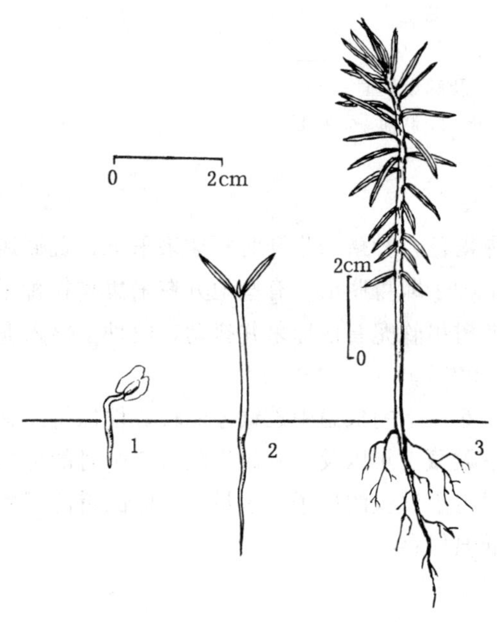

图2 秃杉种子萌发和幼苗生长情况
1. 下胚轴延伸，种壳出土 2. 子叶展开 3. 幼苗形成
（叶可佳、黄应钦绘）

秃杉种子萌发和幼苗生长情况如图2。

播种 春播。播种期约在1月底～2月。条播，条距与条幅根据圃地情况及需要而定，沟深0.5cm。每平方米苗床播种2～3g（杨思平，1993）。播后用细土覆盖，轻压后洒水，盖草。据观察，秃杉种子播种后，积温在250～330℃条件下，18天即发芽出土，20余天为发芽出土高峰期，40天左右发芽出土基本结束。在种子发芽出土期间分批揭除盖草，发芽出土基本结束前将盖草全部揭除，并搭设荫棚，为小苗遮荫，荫棚遮荫度以50%较好。

出土萌发，子叶2，线形。每克种子可产一级苗40株左右（杨思平，1993）。1年生苗高一般20cm左右，最高可达32cm，地径约4mm，可达7mm。

秃杉可以扦插繁殖，秋季扦插成活率高达96.5%（徐光田，1990）。另据王挺良（1995）报道，秃杉一年四季都可扦插，成活率均在80%以上。

（赵德铭）

落羽杉属
Taxodium Rich.

（杉科　Taxodiaceae）

生长习性、分布和用途　落羽杉属只有本文描述的 3 种。落叶、半常绿或常绿乔木。寿命长。速生。喜光。落羽杉、池杉喜温暖、湿润生境，耐水湿性强。墨西哥落羽杉喜温暖气候，不耐寒，较耐干旱。落羽杉、池杉喜深厚疏松湿润的酸性土壤，墨西哥落羽杉能适应碱性土壤。原产北美及墨西哥，本世纪初引入我国，已成为江南水网地区造林树种。其分布和原产地生长情况见表 1。木材纹理直，耐腐，是建筑、电杆、造船和家具用材。

表 1　落羽杉属树种的名称、分布和原产地的生长情况

中　名	学　名	习　性	树高（m）	胸径（cm）	寿命（年）	分　布
池杉	*T. ascendens* Brongn.	落叶	25	—	—	原产美国东南部及墨西哥湾沿海地带。长江流域引种栽培
落羽杉	*T. distichum*（L.）Rich.	落叶	50	3	3 000	原产美国东南部、墨西哥湾沿海地带及亚拉巴马河与密西西比河沿岸。鄂、湘、苏、浙、粤、桂引种栽培
墨西哥落羽杉	*T. mucronatum* Tenore	半常绿或常绿	50	4	4 000	原产墨西哥东部、美国西南部和危地马拉。苏、湘引种栽培

开花结实　落羽杉 7～9 年生开始结实，池杉 2～3 年生实生苗即能出现雌球花。10～12 年后每年都能结出种子，丰年间隔期 3～5 年。雌雄同株。雄球花在枝端排成总状或圆锥花序状，有多数或少数雄蕊，每雄蕊 4～9 花药，药隔明显。雌球花单生枝顶，珠鳞与苞鳞螺旋状排列，每个株鳞具 2 枚胚珠，苞鳞与珠鳞几合生。

在杭州，池杉 7～8 月花芽分化，次年 3 月中旬花芽膨大，3 月下旬～4 月下旬开花。雌雄球花花期一致，气温高则花期早而短。受粉后雌球花逐渐发育，6～7 月球果不再膨大，9～10 月果色渐渐变黄，种子成熟。池杉和落羽杉的开花结实物候期见表 2。

球果球形或卵圆形，径 2.2～3.6cm，熟时由青绿色变为黄褐色。每球果含种子 9～18（2～21）粒。发育种鳞具种子 2 粒。种子呈不规则三角形，红褐色或黄褐色。种皮厚，具角，具疣。种皮边缘突出，三棱脊。有胚乳。子叶 4～9 枚。球果和种子特征见表 3。落羽杉种子的外形和剖面结构见图 1。

表 2　落羽杉属树种的开花结实物候期

树　种	观察地点	观察年份	现蕾期	开花期			果实成熟期	
				始　期	盛　期	末　期	始　期	盛　期
池杉	杭州	1972	—	3 月 18～23 日	3 月 25～28 日	3 月 28 日～4 月 14 日	9 月中旬	10 月中上旬
	杭州	1978	3 月 20 日	3 月 28 日	4 月 1 日	4 月 3 日	—	—
落羽杉	杭州	1978	3 月 10 日	3 月 16 日	3 月 22 日	3 月 26 日	—	10 月

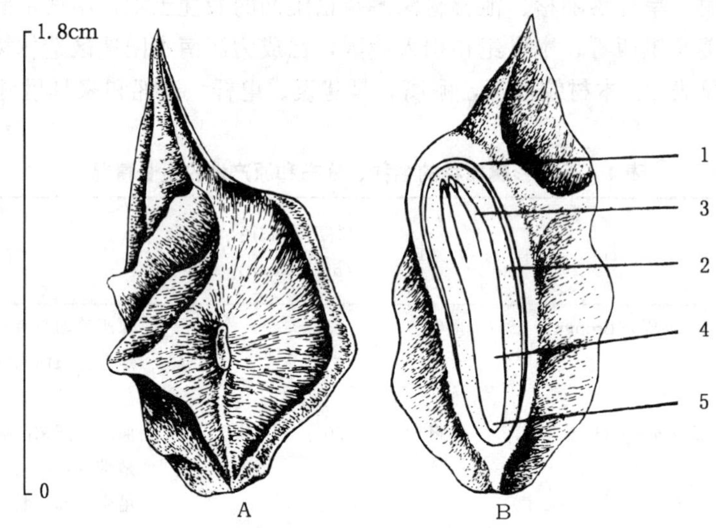

图 1　落羽杉种子的外形（A）及其纵切面（B）

1. 种皮　2. 子叶　3. 胚乳　4. 胚轴　5. 胚根

（黄应钦仿《美国木本植物种子手册》）

表 3　落羽杉属树种球果和种子的形态特征

树　种	球　果		种　子		
	形　状	大小（cm）	形　状	大小（cm）	色　泽
池　杉	圆球形或钝圆形	长 2～4，径 1.8～3.0	不规则三角形，微偏，边缘有锐棱脊	长 1.3～1.8 径 0.5～1.1	红褐色
落羽杉	球形或卵圆形	1.2～2.5	不规则三角形，有锐棱	1.2～1.8	褐色
墨西哥落羽杉	卵圆形	—	—	—	—

球果的采收调制和种子贮藏　球果变为褐色，种鳞变干但未开裂时采摘。摊开阴干或日晒，用连枷敲击球果，使之开裂。种鳞碎片很难同种子分离，因此常将种子与种鳞碎片一起播种。出种率和有关种子质量的数据见表 4。将带壳种子放麻袋中，在 0～5℃下干藏，或置通风干燥处。湿沙层积有提早发芽的作用。种子寿命短，一般贮藏不超过 1 年。

表 4　落羽杉属树种的出种率和种子的质量

树　种	出种率（%）	千粒重（g）		每千克净种子数		每　果种子数
		一　般	变　幅	一　般	变　幅	
池　杉	10～15（去壳） 30～40（带壳）	80	35～128	12 300	8 000～28 500	9～18 （2～21）
落羽杉	50（带壳）	90	54～178	11 470	5 600～18 400	18～30

发芽和播种　种子具明显的内休眠。王成霖和邵蓓蓓（1987）在 0～5℃下将池杉种子分别层积 0 周、2 周、4 周和 8 周后，得到的发芽率分别为 10%、28%、45%和 49%。邵蓓蓓（1989）报道说，在 0～5℃下层积 14 天的池杉种子，在 30℃/20℃的温度条件下发芽率为 95%；将层积时间延长到 28 天、42 天或 60 天后在同样温度条件下发芽，发芽率没有显著变化。她还认为，池杉的萌发障碍同种壳无关，未经层积的种子能否顺利萌发，更多地是取决于发芽温度：在 20℃、25℃和 30℃的恒温下发芽率分别为 0%、10%和 40%；如果改用 30℃/20℃的变温，发芽率可以提高到 80%。播种前，生产上常将干藏过的池杉种子用始温 45℃水浸种 4～5 天，每天换水一次，也有一定的促进发芽作用。落羽杉属种子的发芽测定条件和结果见表 5。

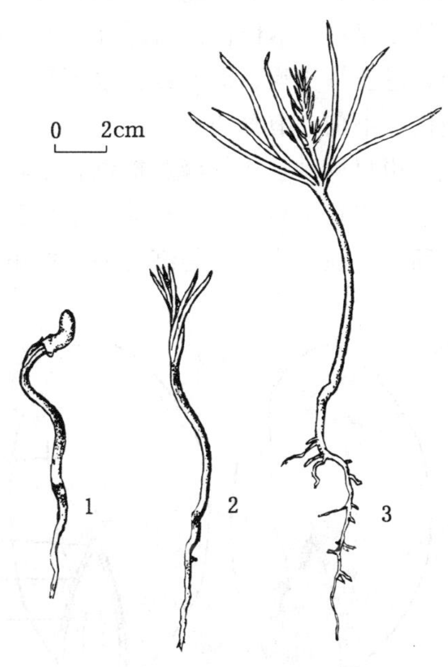

图 2　池杉种子的萌发及幼苗生长情况
1. 子叶出土　2. 子叶初展　3. 初生叶互生
（田恒德仿《主要树木种苗图谱》）

表 5　落羽杉属树种发芽测定条件和发芽能力

树　种	层　积		发芽温度（℃）		发芽率	
	温度（℃）	天数	有光时	无光时	天数	%
池　杉	0～5	30	30	20	30	40～60
落羽杉	0～5	30	30	20	15	30～60

播种　每 666m² 播种 8～12kg，带壳种子播 20～30kg。覆土厚 0.6～1.3cm。出土萌发。子叶 4～9 枚。发芽情况见图 2。每 666m² 产苗 2.5 万～3.0 万株。1 年出苗高 80～100cm。也可用 1～2 年生实生苗的枝、干扦插繁殖。

（杨国华）

翠　柏（大鳞肖楠）
Calocedrus macrolepis Kurz

（柏科　Cupressaceae）

生长习性、分布和用途　翠柏属有2种，我国和北美各产1种。本文描述我国大陆地区所产的这1个种。本种在台湾还有1个变种。常绿乔木。高达35m，胸径1.2m。喜温暖湿润的环境，忌土壤干燥、瘠薄或排水不良。分布于滇中、滇南，也间断分布于黔西、桂西和琼中。越南、缅甸亦有分布。

优良用材树种，边材淡黄褐色，心材黄褐色，纹理直，结构细，有香气，耐久用，供建筑、桥梁、家具之用。种子榨油，也可药用。鳞叶苍翠碧绿，是理想的观赏树种。翠柏资源已逐渐稀少，在《中国植物红皮书》中被列为渐危种。

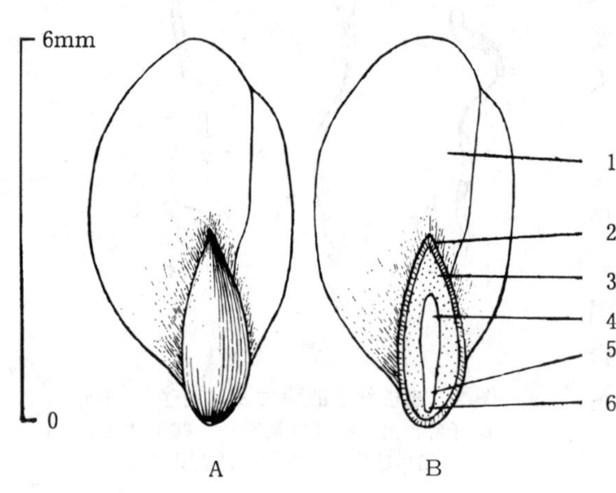

图1　翠柏种子的外形（A）及其纵切面（B）
1. 翅　2. 种皮　3. 胚乳　4. 子叶　5. 胚轴　6. 胚根
（黄应钦绘）

开花结实　10～15年生开始结实，15～25年生后正常结实。种子大小年明显，丰年间隔期1～2年。雌雄同株。球花单生于枝顶。雄球花椭圆形或卵圆形，黄色。雄蕊6～8对，花药2～5。雌球花有3对交叉对生珠鳞，珠鳞基部着生2枚胚珠。3月开花，10月球果成熟。球果成熟前被有白粉，成熟时红褐色，矩圆形、椭圆形或长卵状圆柱形，长1.5～2.2cm。种鳞木质，扁平，下面1对形小，上面1对结合而生，均无种子，仅中间1对种鳞各具2粒种子。种子近卵圆形或椭圆形，微扁，暗褐色，长约6mm。种子上部具1长1短的膜质翅。长翅长卵形或长圆形，连同种子几与中部种鳞等长。短翅线状长圆形。有胚乳。胚直立，子叶2枚。种子形态见图1。

球果的采收调制和种子贮藏　秋季果熟时采种。球果在室内后熟1周左右，再摊晒2～3天，种鳞裂开，种子脱出，晾干后干藏。出种率2%～4%。

发芽和播种　春播。撒播时每平方米约播38～50g。种子发芽容易。3月中下旬播种，20天左右即可发芽，但空粒多，发芽率约20%。出土萌发。子叶2枚，线形，长5～7mm，宽1.5～1.8mm。初生叶线状刺形，长4～7mm，宽约1mm，背面无白粉，初为交叉对生，后4叶轮生。

（李　华）

扁　柏　属
Chamaecyparis Spach

（柏科　Cupressaceae）

生长习性、分布和用途　扁柏属有 6 种，产北美、日本和中国。我国有 1 种 1 变种，引入 5 种和若干栽培变种。本文描述 5 种 1 变种，见表 1。另北美产黄扁柏 *Ch. nootkatensis* (D. Don) Spach，在台湾引栽。常绿乔木，寿命长。台湾阿里山巨木——红桧高 53m，地径 10.9m，树龄约 3 000 年，惜于 1960 年枯死。现在保存的以在台湾新竹的桧山巨木为最大，高 51m，地径 8.2m，树龄约 3 000 年。红桧是我国特有树种，在《中国植物红皮书》中被列为稀有种。浅根性。喜温暖湿润或冬季湿、夏季干、温度变化不大的气候环境。适生于肥沃湿润的沙质土壤，也能耐干旱瘠薄之地。木材芳香，有光泽，供建筑、桥梁、造船、家具等用。树姿优美，可供观赏。

表 1　扁柏属树种的生长和分布

中　名	学　名	树　高 (m)	胸　径 (m)	分　布
红　桧	*Ch. formosensis* Matsum.	57	6.5	台湾特有，产中央山脉、阿里山和北插天山等地
美国扁柏	*Ch. lawsoniana* (A. Murr.) Parl.	73	3.6	原产美国。庐山、南京、杭州等地引种，庭园观赏
日本扁柏	*Ch. obtusa* (Sieb. et Zucc.) Endl.	40	1.5	原产日本。青岛、南京、庐山、鸡公山、广州、台湾等地引种。在庐山造林生长旺盛
台湾扁柏	*Ch. obtusa* (Sieb. et Zucc.) Endl. var. *formosana* (Hayata) Rehd.	40	3	台湾特有，产中央山脉北部及中部
日本花柏	*Ch. pisifera* (Sieb. et Zucc.) Endl.	50	1	原产日本。青岛、南京、庐山、杭州和台湾引种作庭园树
美国尖叶扁柏	*Ch. thyoides* (L.) B. S. P.	27	1	原产美国。庐山、南京、杭州等地引种作庭园树

开花结实　8～13 年开始结实，15～20 年后正常结实。实生植株 20～25 年生后进入盛果期。有大小年，丰年间隔期 1～3 年。据居翔汉和徐正法（1986）记载，庐山引种的日本扁柏 13 年生开始开花结实，结实丰年间隔期约 2～3 年。

雌雄同株。球花小，单生于小枝顶端。雄球花长，椭圆形或卵圆形，具 3～4 对交叉对生的雄蕊，各具 3～5 个花药。雌球花圆球形或卵圆形，带绿色，有 3～6 对交叉对生的珠鳞。胚珠 1～5 枚，直立，生于珠鳞内侧。在庐山，日本扁柏和日本花柏的花粉在 2 月下旬～3 月上旬形成，3月中旬～下旬花粉成熟。雌球花在3月中下旬～4月上旬珠鳞张开。接受花粉后10～15天，小球果便不断生长发育，7～8月球果停止生长，9～10月球果由绿色变为褐色。10月下旬～11月上旬种子成熟，11月中下旬种子散落。扁柏属树种开花结实物候见表2。

表2 扁柏属树种的开花结实习性

树 种	结实年龄	丰年间隔期（年）	开花期		果实成熟		种子散落期
			雌 花	雄 花	时 间	颜 色	
红 桧	15～25	3～5	4月下旬	4月中旬	10～11月	褐色	11月
美国扁柏	5～20	3～5	4月上旬	3月下旬	9～10月	绿黄色至红褐色，有白粉	9月～翌年5月
日本扁柏	13～25	1～4	4月上旬	3月下旬	9～10月	红褐色	10月～翌年4月
台湾扁柏	15～25	1～4	4月中旬	4月上旬	9～10月	红褐色	10月～翌年4月
日本花柏	15～25	2～4	4月上旬	3月下旬	9～10月	暗褐色	9～12月
美国尖叶扁柏	3～20	1～2	4月上旬	3月下旬	10～11月	红褐色，有白粉	10月～翌年3月

球果圆球形或稀矩圆形。种鳞3～6对，木质，盾形，顶部中间有小尖头。发育的种鳞内有1～5（通常3）粒种子。种子卵圆形或圆形，微扁，有棱角，两侧具窄翅。子叶2。

扁柏属树种球果和种子的形态特征见表3和图1。

表3 扁柏属树种球果和种子的形态特征

树 种	成熟球果				成熟种子			
	形 状	长（mm）	径（mm）	种鳞对数	形 状	长（mm）	宽（mm）	色 泽
红 桧	矩圆形、矩圆状卵形	10～12	6～9	5～6	倒卵圆形	2～3	1.8～2	红褐色，微有光泽
美国扁柏	圆球形	10～12	8～9	4	卵圆形	2～3	1.6～2	深褐色、有光泽
日本扁柏	圆球形	—	8～10	4～5	近圆形	2.6～4	2～3	褐色、有光泽
台湾扁柏	圆球形	—	10～11	4～5	扁，倒卵圆形	3～3.5	2～3	红褐色、微有光泽
日本花柏	圆球形	—	6～8	5～6	三角状卵圆形	2～3	1.8～2	淡褐色
美国尖叶扁柏	圆球形	—	6	3～6	扁，卵圆形	2～2.5	1.6～2	褐色

球果的采收调制和种子贮藏 球果呈现成熟颜色且种鳞微裂时，应及时采摘。本文描述

的这几个种都有明显的结实周期，小年时种子较少，最好趁结实丰年安排采种。球果可以用 43℃以下的温度烘干，或摊晒 3～5 天，种鳞开裂后经翻动震荡即可脱出种子。美国尖叶扁柏脱粒较难，可以先将球果干燥，浸水过夜后再行干燥，这样做脱粒比较容易。扁柏属种子易受伤害，净种时不应去翅。扁柏属树种的出种率和种子质量见表 4。

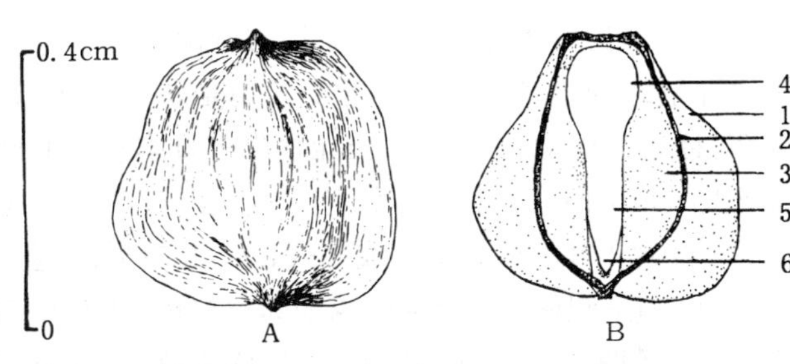

图 1　日本扁柏种子的外形（A）及其纵切面（B）
1. 翅　2. 种皮　3. 胚乳　4. 子叶　5. 胚轴　6. 胚根
（A. 黄应钦绘　B. 田恒德仿《日本の树木种子》）

表 4　扁柏属树种的出种率和种子质量

树　种	出种率（%）	净　度（%）	千粒重（g）	每千克种子粒数（万粒）
美国扁柏	10	80～90	2.2～2.5	40～45.5
日本扁柏	8～11	80～90	2.0～2.4	42～50
日本花柏	4～5	80～85	0.7～0.9	110～142
美国尖叶扁柏	4～5	80～85	1.5～2.0	50～67

扁柏属这几个树种引入我国后还没有做过严格的种子贮藏试验，看来最好在 0～5℃条件下密封干藏，适于贮藏的含水量应在 10%以下。美国扁柏在原产地的一项试验表明，含水量为 8%的种子用 0℃和 -17.8℃两种温度贮藏，效果同样好：初始生活力为 56%，7 年后降至 43%，但在室温下贮藏的则完全丧失了生命力。散落在林下地被物中的美国尖叶扁柏种子经历了至少两年仍然保持着生命力。原产地对日本扁柏和日本花柏贮藏寿命的研究表明，这两个树种的种子，只要含水量不超过 10%，也适于低温密封干藏。

发芽和播种　在原产地，美国扁柏和美国尖叶扁柏种子的发芽率不高。据认为，一个原因是种子本身质量低，另一个原因是胚在某种程度上有休眠习性。据报道，有 1 份未经层积、优良度为 48%的美国扁柏种子，在 30℃/20℃的昼夜变温和每天 8 小时光照的条件下，不到 28 天便全部发芽。英国苗圃播种的美国扁柏，则一直是以播前层积的效果最好。从文献资料看，美国尖叶扁柏的发芽能力较高：有 1 份未经层积的种子，在上述发芽条件下 60 天的发芽率为 84%。但是也曾报道过，秋播的美国尖叶扁柏大约有一半具有生命力的种子迟到播种的次年才发芽（Schopmeyer，1974）。我国庐山植物园引种栽培的美国尖叶扁柏植株已经进入结实年龄，但所产种子空粒多，发芽率还没有超过 10%。

日本扁柏种子在原产地的一次试验中，水浸 24 小时并层积 0 天、3 天、6 天、30 天后，在 21℃恒温下的发芽率都在 95%以上。庐山植物园引种的日本扁柏和日本花柏，所结种子未经处理，在 15～20℃的温度下 20～25 天开始发芽，发芽率可达 30%～40%，其中日本花柏空粒多，发芽率较低。

3月中下旬～4月上旬播种。条播时行距13～18cm。日本扁柏和美国尖叶扁柏撒播时每平方米播种量约30～35g，苗木密度大约是每平方米320～540株。覆土厚度为种子厚度的3～4倍。播后30～40天发芽。出土萌发。日本扁柏的幼苗形态如下：子叶2枚，线形，顶端平圆或微凹，上面有气孔线，绿色，微被白粉；初生叶刺形，扁平；第1对初生叶与子叶交叉对生，以后4叶轮生，交叉成8列，4～5轮后幼茎开始分枝；分枝上的叶由刺形逐渐变为鳞叶；下胚轴细圆柱形，由黄绿色变为黄褐色；1年生苗高8～14cm，2～3年出圃。日本扁柏的幼苗形态见图2。扁柏属树种可以扦插繁殖。在庐山植物园，扦插成活率可达80%～95%。

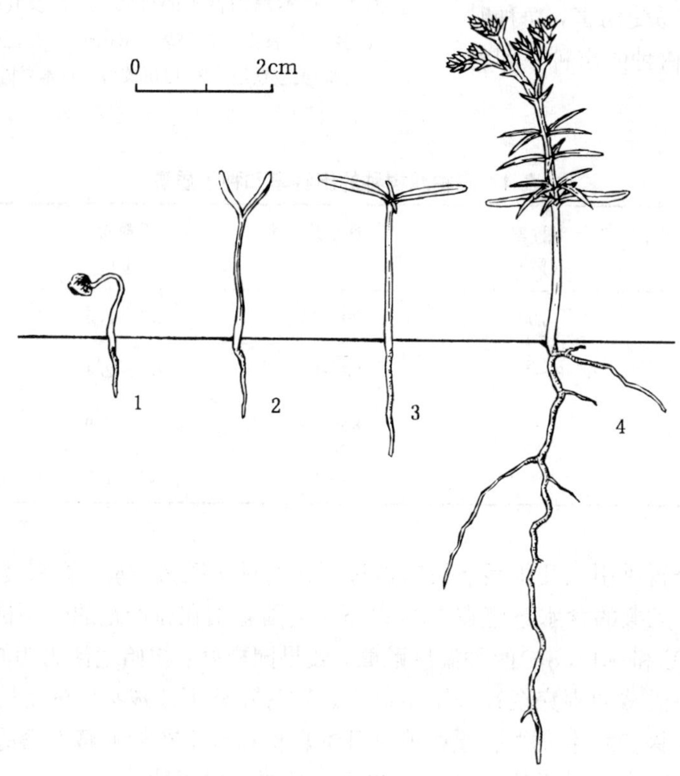

图2 日本扁柏种子萌发及幼苗生长情况

1. 下胚轴延伸，种壳出土 2、3. 子叶展开 4. 初生叶轮生并分枝出现鳞叶

（黄应钦仿《主要树木种苗图谱》）

（李 华）

柏木属

Cupressus L.

（柏科 Cupressaceae）

生长习性、分布和用途 本属约20种，产北美、东亚、喜马拉雅山区和地中海等温暖地带。我国产5种，引入栽培数种。本文描述我国产的4种和引入栽培的4种。其中岷江柏木

为渐危种，已列入《中国植物红皮书》。常绿乔木，胸径 20～40cm，有的达 200cm，树高 20～35m。喜光，能耐侧方庇荫。

我国柏木属中以柏木分布最广，水平分布南北有 900km，东西约有 1 700km：东起浙、闽，西至川、黔，南达粤、桂北部和滇东南，北抵陕、甘南部，但以川、黔和鄂西等钙质土地区为多。岷江柏木、干香柏和西藏柏木分布较窄。岷江柏木主产于川西和甘南。干香柏主要分布在滇西北至滇东南以及毗邻的川、黔部分地方。西藏柏木产于藏东南，邻近的不丹和尼泊尔也有分布（见表 1）。柏木属树种在中性、微酸性和钙质土上均能生长，以在湿润、深厚、富含钙质的土壤上生长最快。能耐干旱瘠薄。加利福尼亚柏木能生长在贫瘠的沙地；岷江柏木能生长在干燥的山地阳坡；干香柏、墨西哥柏木和西藏柏木能耐瘠薄的山地；柏木在浅薄的钙质紫色土和岩石缝隙的石灰土上也能正常生长，它构成的纯林是亚热带钙质土上典型的指示植被。

木材结构细，纹理直，坚韧耐腐，是建筑、桥梁、船舶、家具、木模及细木工等优良用材。种子可榨油。球果、根、枝、叶可入药。根、干、枝、叶可提芳香油。

表 1　柏木属树种的名称、分布和用途

中　名	学　名	分　布	用　途	供　稿
绿干柏	*C. arizonica* Greene	原产美国和墨西哥。苏、赣、川等引种栽培	材用、观赏	501
岷江柏木	*C. chengiana* S. Y. Hu	川、甘	材用、保土、观赏	508
干香柏	*C. duclouxiana* Hickel	滇、川、黔、桂	材用	504
柏木	*C. funebris* Endl.	陕、甘、川、黔、滇、桂、湘、鄂、皖、浙、赣、闽、粤	材用、药用、香料	502
加利福尼亚柏木	*C. goveniana* Gord.	原产美国加利福尼亚州。苏、滇等引种栽培	材用、保土、观赏	614
墨西哥柏木	*C. lusitanica* Mill.	原产墨西哥、危地马拉、萨尔瓦多、洪都拉斯。苏、浙、皖、赣、湘、鄂、川、滇、黔等引种栽培	材用、观赏	614
大果柏木	*C. macrocarpa* Hartw	原产美国加利福尼亚州。苏、浙、湘、赣、滇引种栽培	材用、保土	614
西藏柏木	*C. torulosa* D. Don	藏东南。印度、不丹、尼泊尔。川、滇等引种栽培	材用、保土、观赏	614

柏木属耐干旱瘠薄，对土壤适应性广，是分布区和适生区的主要造林树种。树形美观，是重要的观赏树。四川北部古驿道两旁人工栽植的柏木绵延 200km 以上，据查定，有的树龄已有 1 700 多年（王继贵，1991）。

开花结实　雌雄同株。球花单生枝顶。雄球花长椭圆形或卵圆形，黄色。雄蕊多数，每雄蕊有花药 2～6，药隔明显，鳞片状。雌球花近球形，具 4～8 对珠鳞，中部珠鳞具 5 至多数排成一行至数行的胚珠。球果翌年成熟，球形或近球形。种鳞木质，盾形，中部种鳞发育，各具 5 至多数种子。柏木每球果中有 30～50 粒种子。墨西哥柏木每球果中有 56～68 粒种子。种子长圆形或长圆状倒卵形，稍扁，有或无棱脊，两侧具窄翅。子叶 2～5 枚。柏木种子的形态见图 1。

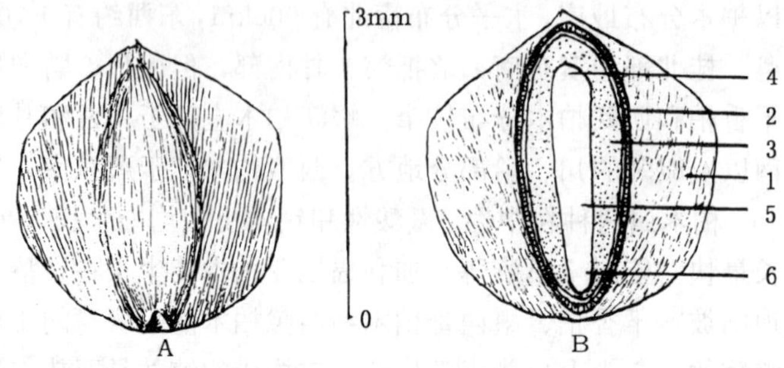

图 1 柏木种子外形（A）及其纵切面（B）

1. 翅 2. 种皮 3. 胚乳 4. 子叶 5. 胚轴 6. 胚根

（田恒德绘）

开花结实年龄因种类和生境因子等而有不同。岷江柏木 15～20 年生开始开花结实；柏木则在 10 年生左右，15 年生以后结实逐渐增多，20～30（40）年生以后进入盛果期。人工栽植的柏木也有在 3～4 年生时开花结实的。据在昆明树木园观察，西藏柏木和墨西哥柏木 4～5 年生开始开花结实，大果柏木和加利福尼亚柏木在 5～6 年生开花结实。

开花结实的物候期见表 2。

表 2 柏木属树种的开花结实物候

树 种	观察地点和年份	花 期	球果成熟期	种子散落期
岷江柏木	四川汶川 1983～1987	4 月	翌年 10 月初～10 月中旬	10 月底～12 月上旬
干香柏	云南昆明 1984～1987	12 月下旬～2 月下旬	翌年 9～10 月	12 月～第 3 年 4 月
柏 木	四川北碚 1975～1976	3 月中旬～4 月初	翌年 8 月中旬～下旬	—
	四川宜宾 1975～1976	1 月底～2 月下旬	翌年 8 月上旬～中旬	8 月下旬
	四川广安 1977～1978	2 月中旬～3 月上旬	翌年 7～8 月	—
	四川梓潼 1974～1977	2～3 月	翌年 8 月中旬～下旬	9 月中旬
	湖南常德 1985～1986	3 月中旬～3 月下旬	翌年 10 月	11 月中旬
	广东广州 1985～1986	2 月下旬～3 月中旬	翌年 11 月	12 月中旬

（续）

树　种	观察地点和年份	花　期	球果成熟期	种子散落期
加利福尼亚柏木	云南昆明 1987～1989	12 月下旬～2 月	翌年 5 月下旬～6 月中旬	—
墨西哥柏木	江苏南京 —	3 月中旬～3 月下旬	翌年 4 月下旬～5 月	—
	云南昆明 1987～1989	12 月下旬～2 月	翌年 5 月底～6 月中旬	6 月中旬～7 月
大果柏木	云南昆明 1987～1989	12 月下旬～2 月	翌年 5 月底～6 月中旬	6 月下旬～7 月
西藏柏木	云南昆明 1987～1989	12 月下旬～2 月	翌年 5 月底～6 月上旬	6 月中旬～下旬

球果未熟时绿色、青绿色或灰绿色，成熟时绿褐色、深褐色、紫褐色或暗褐色，球形、近球形或椭圆形，长 0.8～3cm。种子长 3～6mm，黄褐色、棕褐色、紫褐色或浅褐色，近卵圆形、椭圆形、矩圆形，两侧有窄翅，有的种有棱脊（见表 3）。

表 3　柏木属树种球果和种子的形态特征

树　种	未熟球果颜色	成熟球果				成熟种子		
		颜　色	形　状	大　小（cm）		色　泽	形　状	长（mm）
				长	径			
绿干柏	青绿色	暗紫褐色	宽椭圆状球形，顶部具显著锐尖头	1.5～3	1.8～2	棕褐色	倒卵圆形	5～6
岷江柏木	绿色	绿褐色	近球形或稍长，顶部平，呈不规则的扁四边形或五边形	1.5～2.5	—	黄褐色或棕褐色	近卵圆形	3～5
干香柏	青绿色	深褐色或紫褐色，微被白粉	球形，顶部五角形或近方形，具不规则向四周放射的皱纹，中央平或稍凹，有短尖	1.6～3	—	黄褐色或深褐色	近卵形，微扁，有棱脊	3～4
柏木	青绿色	黄褐色或暗褐色	球形，顶端为不规则的五边形或方形，中央有凸尖	0.8～1.5	—	黄褐色或深褐色	近圆形，微扁，有棱脊	2.5～3

（续）

树种	未熟球果颜色	成熟球果				成熟种子		
		颜色	形状	大小（cm）		色泽	形状	长（mm）
				长	径			
加利福尼亚柏木	绿色	深褐色	球形或椭圆形	1.8～2.5	1.6～2	棕褐色	不规则扁圆形，顶部有一尖头	3～4
墨西哥柏木	青绿色	褐色，微被白粉	球形，顶部有尖头	1.2～1.5	—	黄褐色或深褐色	近卵形，有棱脊	3～5
大果柏木	灰绿色	深褐色	椭圆形	1.5～2	1.8～2	红棕色	不规则扁圆形	3～5
西藏柏木	绿色	紫褐色或深褐色	球形，顶部五边形，有放射状条纹，中央具短尖头或无	1.8～2	—	棕褐色	椭圆形或不规则扁圆形	5～6

球果的采收调制和种子贮藏 当球果呈现成熟颜色，种鳞硬化且微裂时即可采集。生产上有两种采集方法：一是摇枝击果，即当球果呈现成熟颜色，种鳞已裂而种子又未散落时，选择晴朗天气，铺放收种布，震动树干或树枝，或用竹竿轻击球果，让种子落于布上。另一种是上树用采种刀、叉、钩等工具将球果采下。上树采摘时要注意枝上有两种球果，一是确已成熟的2年生球果，为采摘对象；一是仍呈青绿色外被白粉未成熟的1年生小球果，应当留等翌年成熟时采摘。

采回的球果在日光下摊晒，勤翻动，数日后种子从开裂的种鳞中脱出，去杂后即得纯净种子。

原西南林业试验场在重庆对柏木采种期与种子质量的关系做过研究。结果表明，从7月中旬～9月下旬每旬采集1次，以9月上旬采集脱粒的种子发芽率最高，比7月中旬采集的高35%，比8月上旬采集的高26%，比8月下旬采集的高18%（蒋临轩，1964）。

柏木属树种的球果出种率、千粒重和每千克种子粒数，因种类不同而有差异。岷江柏木的出种率可以低到5.5%，西藏柏木可以高达24%。柏木的千粒重一般只有3.3g，绿干柏则常为8.4g（见表4）。罗成荣（1989）按直径将柏木球果分为大（径1.6～1.4cm）、中（径1.39～1.2cm）、小（径1.19～1.0cm）3级并分别脱粒，据测定，所得种子的千粒重分别为4.1g、3.35g和2.67g。

表4 柏木属树种球果的出种率和种子质量

树种	出种率（%）	净度（%）	千粒重（g）	每千克纯净种子数（万粒）
绿干柏	—	90～98	6.8～10.8	9.3～14.7
岷江柏木	5.5～7.4	80～95	4.5～4.8	20.8～22.2
干香柏	6	70～90	4.0～5.4	18.5～25.0

（续）

树　种	出种率（%）	净　度（%）	千粒重（g）	每千克纯净种子数（万粒）
柏木	9～11（14）	80～95	3.0～3.9	25.6～33.3
加利福尼亚柏木	16.3～20	80	4.7～5.4	18.5～21.2
墨西哥柏木	11.7～13.6	90～99	4.5～5.5	18.1～22.2
大果柏木	10.7～19	—	4.3～5.8	17.2～23.2
西藏柏木	19.5～24	99	7.3～9.5	10.5～13.7

柏木属种子的耐贮性较强。贮藏的适宜含水量为8%～10%。越冬贮藏可以在普通库房或干燥凉爽处袋装或散装。散装的可用苇箦围成囤，囤的中央竖设通气孔。贮藏时间在2年以上的宜低温（－5℃）密封冷藏。据试验（阙再旵，未刊稿），来自四川江油的柏木种子，贮藏前含水量为11.7%，发芽率为56.3%，在0～5℃下用双层塑料薄膜密封存放两年半后，发芽率为40%～47%。

柏木丽松叶蜂（*Augomonoctenus smithi* Xiao et Wu）幼虫是蛀食柏木幼果的单食性昆虫。幼虫在4月中旬～6月上旬蛀食球果，1头幼虫1代可以危害5～12个球果。幼果受害后枯黄脱落，种子颗粒无收。

发芽测定　习惯上用始温45℃的水浸种一昼夜后置滤纸、纱布、泡沫塑料或石英沙上，放入恒温箱或光照发芽箱测定发芽能力。也有不经温水浸种直接放入温箱的。王成霖（1986）比较过不同温度下柏木的绝对发芽率，发现15℃和20℃下的发芽能力最高，只是15℃时发芽历程略长；25℃时发芽能力显著下降，30℃时便无一发芽。同一份研究还表明，测定前浸种与否，测定中是否给予光照，对柏木种子的发芽均无显著影响。种子发芽能力的数据列入表5。

表5　柏木属树种种子发芽能力和测定条件

树　种	基　质	温　度（℃）	发芽势（%）			发芽率（%）		
			计算天数	一般数值	变动范围	计算天数	一般数值	变动范围
绿干柏	纱布	25	12	19	17～23	24	38	37～39
岷江柏木	纱布	25	7	42	34～51	19	59	56～65
干香柏	滤纸	25	10	15	10～20	20	25	20～40
柏　木	纱布	20	15	18	16～22	30	34	30～42
加利福尼亚柏木	—	—	—	—	—	50	5	—
墨西哥柏木	纱布	25	14	15	12～20	24	25	16～29
大果柏木	—	—	—	—	—	40	15	—
西藏柏木	纱布	25	—	—	—	24	24	—

注　始温45℃水浸种24小时后置床

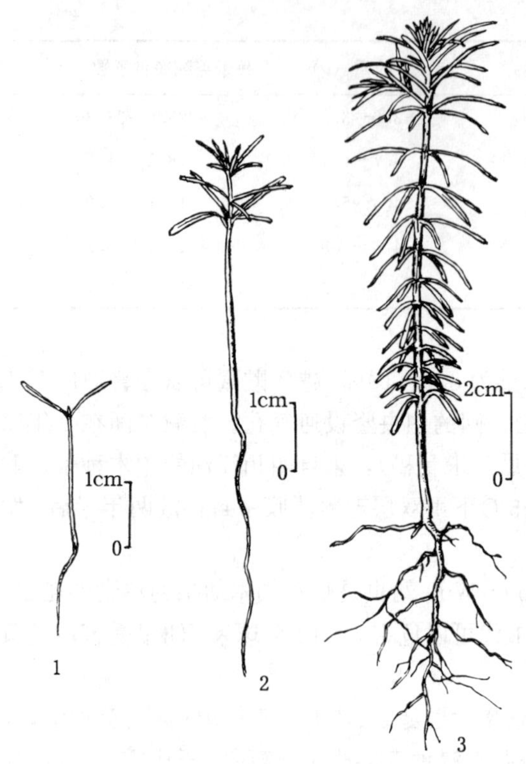

图 2 柏木幼苗的生长情况
1. 子叶展现 2. 初生叶轮生 3. 幼苗形成
（田恒德仿《主要树木种苗图谱》）

播种前的预处理 柏木属中，至少柏木、岷江柏木和绿干柏的种子并无明显的休眠习性，但播种前生产上也常对它们做浸种催芽。常用的方法有 2 种：一种是用始温约 40℃水浸种一昼夜后播种，例如岷江柏木、加利福尼亚柏木和绿干柏。第 2 种方法是像这样浸种之后每天再用温水淋浇两三次，大约半个月后半数种子裂嘴露白时取出播种，例如柏木、大果柏木和西藏柏木。发芽比较困难的是墨西哥柏木。未经处理的墨西哥柏木种子播后不仅出苗缓慢而不整齐，出苗率也低。云南省林业科学研究所用上述第 2 种方法处理的墨西哥柏木，出苗率由 5%提高到了 19%，出苗时间由 2 个月缩短为 1 个月。他们采用的另一方法是，将始温 40℃水浸泡 1 昼夜的种子置搪瓷盆中，厚度不超过 5cm，上覆塑料薄膜并再盖 1 层玻璃，曝晒 1 天(中午盆内温度可达 38～40℃)后混以 3 倍容积的湿沙再回置盆中，仍旧覆以薄膜并盖玻璃，再晒两三天，到大约 10%的种粒露白时取出播种。播后约 5 天出苗，10 天即可出齐。采用这种方法处理墨西哥柏木，从浸种到出苗终结历时仅 25～30 天，出苗率可以高达 30%（张茂钦，未刊稿）。

播种 春播，也可以秋播。秋播在采种结束后的 9 月下旬～10 月上旬进行。云南省因干湿季节分明，夏秋正是雨季，常在 7～8 月播种。南方其他省（自治区）春播在 2 月下旬～3 月中旬。岷江柏木的海拔较高，春播期约在 4 月中下旬～5 月上旬。条播或撒播。条播的条距 20～25cm，条幅 5～10cm，每 666 平方米播种量：柏木为 4～5kg，岷江柏木、干香柏、加利福尼亚柏木、墨西哥柏木和大果柏木为 10～15kg，绿干柏、西藏柏木为 20kg。撒播时每 666 平方米柏木播种 5～8kg。播后用混有草木灰的细土覆盖种子，以不见种子为度，轻压后盖草、洒水。秋播的，播种 1 周后发芽出土，冬季前苗高 5～6cm 或 8～12cm，翌年秋后苗高 30～45cm 或 60cm 以上，可出圃栽植。春播的种子一般需 3 周左右发芽出土。除岷江柏木外，其他种 1 年生苗高 20～40cm，或 60～70cm（墨西哥柏木）。大苗、壮苗可出圃造林，小苗留床或移植，于次年出圃造林。岷江柏木播于高寒地带，生长期短，1 年生幼苗仅几厘米，需培育 3～4 年方可出圃造林。如果用塑料大棚育苗，培育期可以缩短为 2～3 年。

出土萌发，子叶条形。岷江柏木、干香柏、柏木子叶 2；大果柏木子叶（2）3（4）；加利福尼亚柏木、墨西哥柏木 3 或 4；绿干柏 4（5）。柏木子叶长 8～13mm，宽 1.3～1.7mm，先端钝圆，上面中脉微隆起，有不明显的白粉气孔点，绿色；下面平，无气孔点，淡黄绿色。子叶夏后枯黄。上胚轴不明显，幼茎有棱角。初生叶扁平，线状披针形，较柔软，第 1 对与子叶交叉对生，以后 4 叶轮生，交叉成 8 列。节间长 3～6mm。叶上面平或微隆起，两面有白粉气孔线，粉绿色。生长 12～18 轮后分枝，小枝上叶仍为线状披针形。下胚轴圆柱形，长1.7～2.2cm，径约 0.9mm，初绿色后变红褐色。具主根，侧根近平展，褐色。柏木幼苗生长情况见图 2。干香柏子叶长 10～16mm，宽 2mm，初生叶第 5～6 轮后分枝。

柏木属树种还可以扦插繁殖。例如柏木既可春插、秋插，也可夏插。墨西哥柏木扦插生根率也很高（彭金贵等，1984a 和 1984b）。

（赵德铭）

福建柏（建柏）

Fokienia hodginsii（Dunn）Henry et Thomas

（柏科　Cupressaceae）

生长习性、分布和用途　福建柏属仅有本文描述的 1 种。常绿乔木，高达 25m，胸径 80cm。较耐荫。适生于温暖湿润的气候环境。浅根性，侧根发达，穿透能力强。稍耐干燥瘠薄。产于闽、粤、桂、黔、滇及浙、赣、湘、川南部。越南北部也有分布。材质轻软，纹理直，结构细，为家具、雕刻、铅笔杆的良好用材。树形美观，供庭园栽培。天然林中现存数量不多，且繁殖更新能力较弱，被《中国植物红皮书》列为渐危种。

开花结实　8～10 年生开始结实，15 年生以后进入正常结实期。结实有大小年现象，丰年间隔期为 1～4 年。球花单性，雌雄同株，单生于枝顶。雄球花有 6～8 枚雄蕊，交叉对生，每雄蕊有 2～4 个花药。雌球花有 6～8 对交叉对生的珠鳞，每珠鳞的基部有 2 枚胚珠。据福建永泰大湖林场 1983～1987 年观察，10 月中旬始花，11 月上旬盛花，11 月中旬末花，翌年 9 月下旬～10 月中旬球果成熟，11 月上旬种子散落。球果呈矩圆形，径 1.7～2.5cm。种鳞 6～8 对，木质。发育的种鳞有种子 2 枚，每个球果有种子 12～18 粒。球果成熟时褐色。种子黄褐色，椭圆形或卵形，长约 4mm，上部具 2 枚不等大的薄翅，小翅有时不明显。种子有胚乳，子叶 2 枚（图 1）。

球果采收和调制　福建地区多在 10 月上旬上树采摘球果。球果摊晒之前应将枝叶剔除，以免晒干后鳞叶碎片混在种子中难以分离，影响净度。球果摊晒 3～5 天后种鳞开裂，种子脱出，揉碎种翅，筛去杂质，将纯净种子晒干至含水量 8%以下即可贮藏。100kg 球果出种 3.5～4kg。净度85%～95%，一般 90%。千粒重 4.4～7.4g，一般 6g。每千克种子 13.5 万～22.7 万粒，一般 16.6 万粒。

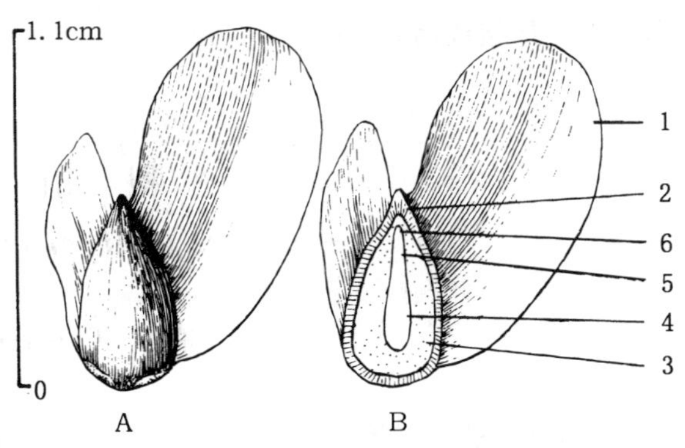

图 1　福建柏种子外形（A）及其纵切面（B）
1. 翅　2. 种皮　3. 胚乳　4. 子叶　5. 胚轴　6. 胚根
（黄应钦绘）

种子贮藏　福建柏种子的发芽率本来就不高，发芽能力的保存期又短，即

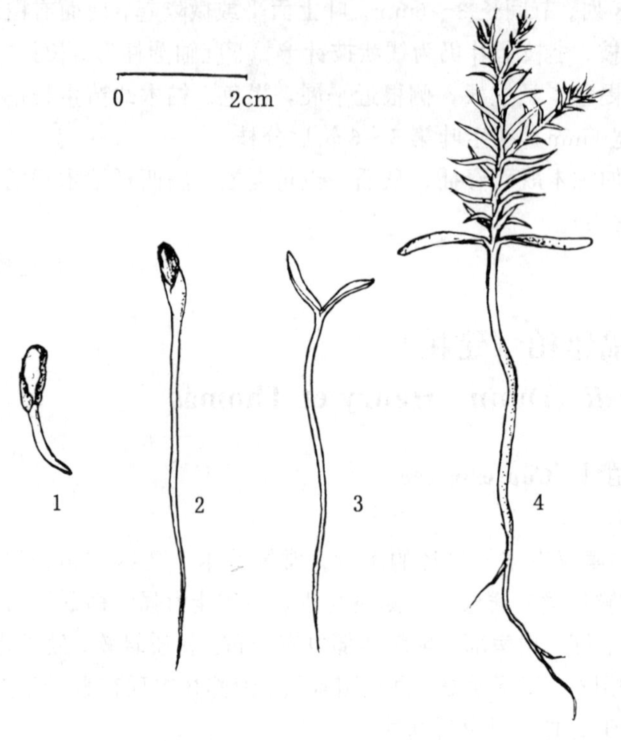

图 2 福建柏种子的萌发和幼苗生长情况
1. 幼根延伸 2. 下胚轴延伸，种壳带子叶出土 3. 子叶展开
4. 幼苗形成（黄鹏成供稿，田恒德绘）

使是秋季采收的新鲜种子，存放到春季播种时发芽率往往便已明显下降。据福建省林木种苗总站研究，如果要求在 6 个月的贮藏期内福建柏种子的存活率不低于 90%，则种子含水量不能高于 5%，贮藏温度不能高于 0℃。

发芽和播种 种子无休眠习性。发芽时无需光照，温度宜在 25℃左右。置床后第 6～8 天开始发芽，经 25～27 天发芽结束。发芽能力随母树年龄和结实的年份不同而异。10 年生以下的母树发芽率只有 10%左右，一般发芽率 20%～40%，结实大年发芽率可达 50%～60%，结实小年发芽率在 20%以下。种子收购、调拨时可用四唑测定种子生活力。

春播。通常用撒播，每平方米播种 10g，覆土厚度以不见种子为度，上面再盖草。播后 30 天左右子叶出土。子叶条形，长 15～22mm，宽 2.8～3.0mm，上面具多条白粉气孔线，粉绿色，下面黄绿色。上胚轴不明显。初生叶扁平刺形，第一对叶长 3～4mm，与子叶交叉对生，以后 4 叶轮生，交叉，叶长 9～12mm，宽 1.2mm，上面黄绿色，下面有白粉带。(3) 4～6 (8) 轮初生叶后出现分枝。分枝上的叶交叉对生，两侧叶鳞状刺形，背腹叶鳞形。下胚轴圆柱形，暗绿色渐变为淡红褐色。主根细长，侧根稀疏，褐色（图 2)。1～2 年生苗出圃。可以无性繁殖，扦插成活率 90%，嫁接成活率 80%。

（李玉科）

刺 柏 属
Juniperus L.

（柏科 Cupressaceae）

生长习性、分布和用途 本属有 10 余种，我国 3 种，引入栽培 1 种，本文描述 2 种。常绿乔木。耐干旱瘠薄，酸性土、中性土、钙质土均能生长。杜松极喜光，喜生于向阳湿润的沙质山坡，耐旱、耐寒，须根发达，适应性强。这两个种的名称、生长、分布和用途见表 1。

表 1　刺柏属树种的名称、生长、分布和用途

中　名	学　名	树高（m）	分　布	用　途	供稿
刺柏	*J. formosana* Hayata	12	西北东部、西南、华中、华东、晋、粤、台	材用、观赏、保土	303
杜松	*J. rigida* Sieb. et Zucc.	10	东北、华北、西北东部。朝鲜半岛、日本	材用、药用、观赏	201

开花结实　杜松在 15～20 年生开始结实，25～30 年生以后进入结实盛期。雌雄异株或同株，球花单生叶腋。雄球花具 5 对雄蕊。雌球花具 3 枚轮生的珠鳞，胚珠 3，生于珠鳞之间。球果肉质，浆果状，2～3 年成熟，近球形。种鳞 3，合生，苞鳞与种鳞结合而生，仅顶端尖头分离。熟时球果不张开或仅顶端微张开。种子 3 或 2，稀 1，卵圆形，有棱脊及树脂槽。杜松种子形态见图 1。

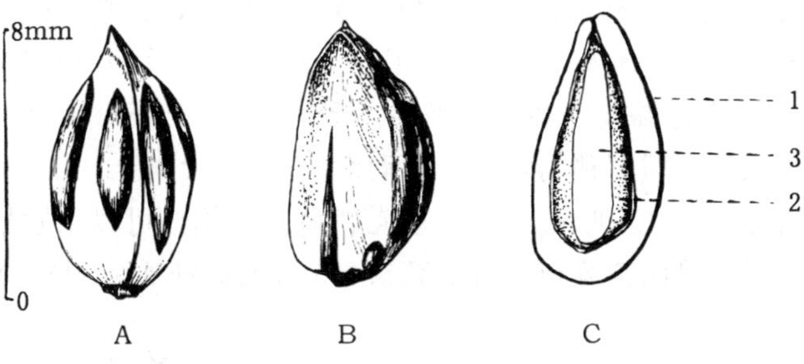

图 1　杜松种子外形背面（A）、腹面（B）及其纵切面（C）
1. 种皮　2. 胚乳　3. 胚（孟玲绘）

据内蒙古呼和浩特市观察，杜松始花期为 5 月中旬，末期为 5 月下旬；果实成熟始于翌年 10 月上旬，成熟末期在 10 月中下旬；种子 10 月中旬～11 月下旬脱落。刺柏属树种球果和种子的特征见表 2。

表 2　刺柏属树种球果和种子特征

树　种	成熟球果				成熟种子	
	形　状	长（mm）	径（mm）	色　泽	形　状	长（mm）
刺柏	近球形或宽卵圆形	7～10	6～9	淡红色或淡红褐色，被白粉或白粉逐渐脱落	半圆形，具 3～4 棱脊，基部有 3～4 树脂槽	5～6
杜松	球形	7～9	6～8	淡褐黑色或蓝黑色，被白粉	近卵圆形，先端尖，有 4 条钝棱	6

球果的采收调制和种子贮藏　球果成熟后可从树上采摘球果，经碾压，或水沤后反复冲洗，去除果肉及杂质，晾干，即可得到纯净种子。种子含水量控制在 9%～10%。最好是密封贮藏，温度不高于 5℃可贮藏 3 年。这 2 个树种的出种率、净度和质量见表 3。

表 3　刺柏属树种的出种率、净度和质量

树　种	出种率（%）	净度（%）	千粒重（g）	每千克纯净种子粒数(万粒)
刺柏	45	—	16	6.2
杜松	28	80～90	8～22	4.5～12.5

发芽和播种　刺柏和杜松种子种皮厚而坚硬，具树脂，吸水困难，影响种子的萌发，有隔年发芽的习性，需混沙埋藏催芽。曹泽猷（1993）认为杜松种子难以萌发的原因是含有萌发抑制物质，且采种时胚尚未充分发育。在他的试验中，混以湿沙的杜松种子在室内 20℃条件下处理 90 天，转室外在 0～5℃下处理 30 天，再转室内处理 60 天，共 180 天，经过这样处理的种子发芽率达到 95%；如果每天 8：00～23：00 给以 20～25℃，其余时间给以 5～8℃，处理 120 天后发芽率为 98%；在室外背阴处层积 200 天的种子，发芽率只有 65%～80%。常用四唑染色法测定种子生活力。具体方法是，种子充分吸水膨胀后，沿纵轴切割种子，以不伤胚为度，四唑浓度为 1%，染色温度为 30℃。条播，行距为 15～20cm。每平方米播种 75～100g，覆土 1～2cm，盖草。经过层积催芽的种子春播后约 20～30 天出土。出土萌发。要注意防止鸟害。杜松幼苗生长状况见图 2。通常 3～4 年生苗出圃。刺柏亦可以用圆柏、侧柏为砧木进行嫁接。杜松也可采用露地遮荫覆膜苗床扦插育苗，5 月上旬扦插，70～80 天生根。

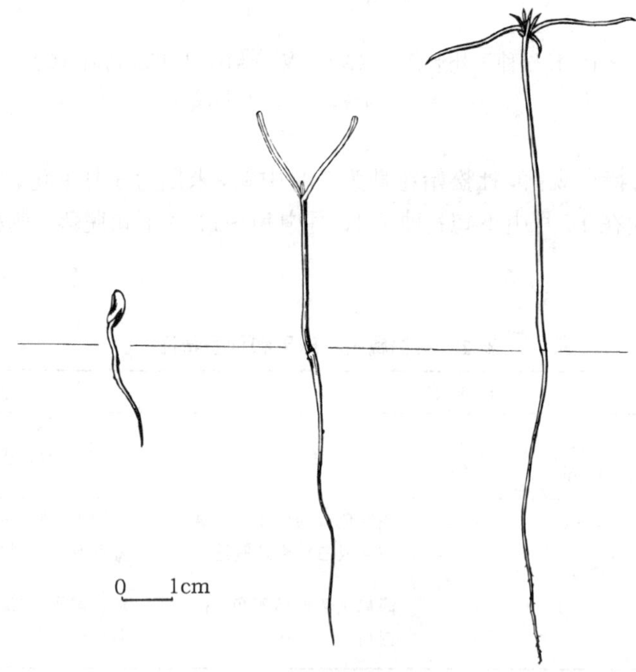

图 2　杜松种子发芽后第 8、18 和 30 天幼苗生长情况

（孟玲绘）

（于淑兰）

侧　　柏

Platycladus orientalis（L.）Franco

（柏科　Cupressaceae）

生长习性、分布和用途　侧柏属仅 1 种，常绿乔木，高 20m 以上，胸径达 2.4m。产于东北南部、华北、西北东部、西南、中南、华东，自然分布极广，尤以淮河以北、黄河流域为多，栽培几遍全国。我国许多古刹圣地多有千年以上古柏。极耐干旱和贫瘠，有一定的抗盐性，能生长在含盐量 0.2%的轻盐土上。多用于黄土高原、石灰岩山地、沙荒和轻盐碱土造林。侧柏材质细，纹理斜而匀，易加工，耐腐力强，油漆和粘胶性能好，供建筑、造船、家具、农具、雕刻用。种子、根、枝叶、树皮均可入药。种子含油率约 20%左右，出油率 18%左右，在医药和香料上有较高的经济价值。树形美观，多用于庭园绿化。

开花结实　侧柏没有明显的结实周期。球花单性，雌雄同株，球花单生枝顶。雄球花具 6 对雄蕊，各有花药 2～4。雌球花具 4 对珠鳞，仅中部的 2 对珠鳞各有 1～2 胚珠。3～4 月开花，9～10 月球果成熟。1984 年全国 15 个地点的开花结实物候期如表 1。

表 1　1984 年侧柏的开花结实物候期

观察地点	开花期			球果成熟期		种子散落期
	始　期	盛　期	末　期	始　期	末　期	
哈尔滨	5 月 12 日	5 月 13 日	5 月 20 日	—	—	—
新疆石河子	4 月 19 日	—	4 月 30 日	9 月 26 日	—	10 月 22 日
呼和浩特	—	5 月 8 日	5 月 16 日	—	—	—
北京	4 月 4 日	4 月 9 日	4 月 15 日	—	—	—
秦皇岛	4 月 26 日	5 月 2 日	5 月 10 日	9 月 21 日	9 月 30 日	10 月 6 日
洛阳	3 月 8 日	3 月 9 日	3 月 11 日	9 月 5 日	—	10 月 10 日
西安	3 月 10 日	3 月 12 日	4 月 2 日	9 月 26 日	—	10 月 6 日
江苏盐城	3 月 16 日	—	—	—	—	—
上海	3 月 17 日	—	—	10 月 21 日	—	—
合肥	3 月 13 日	3 月 19 日	3 月 22 日	10 月 27 日	11 月 3 日	11 月 7 日
武汉	4 月 15 日	4 月 25 日	5 月 3 日	9 月 1 日	9 月 15 日	9 月 30 日
长沙	3 月 17 日	—	—	—	—	—
贵阳	3 月 14 日	3 月 17 日	3 月 20 日	—	—	—
桂林	3 月 14 日	3 月 16 日	3 月 19 日	10 月 29 日	—	—
厦门	2 月 15 日	2 月 18 日	2 月 28 日	9 月 20 日	10 月 5 日	12 月 20 日

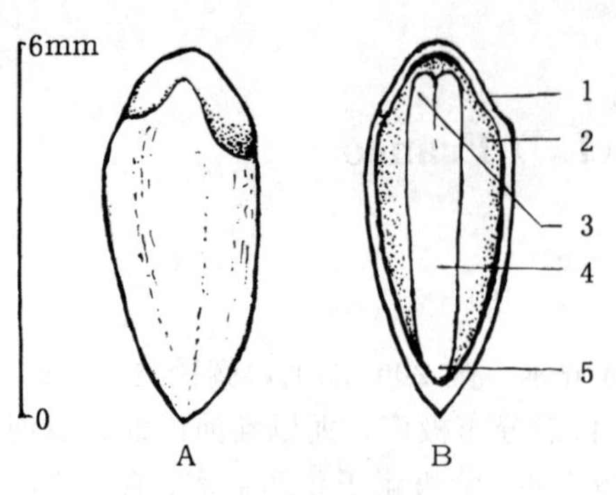

图1 侧柏种子外形（A）及其纵切面（B）
1. 种皮 2. 胚乳 3. 子叶 4. 胚轴 5. 胚根
（孟玲绘）

球果卵圆形，长1.5～2.5cm，成熟前肉质，蓝绿色，外被白粉，成熟时木质，黄褐色。球果当年成熟，种鳞4对，中部2对各具1～2枚种子。种子椭圆状卵圆形，长6～8mm，宽3～3.5mm，基部稍偏斜而尖，褐色，无翅或有棱脊（图1）。

球果的采收调制和种子贮藏 在种鳞尚未开裂时采集球果，曝晒5～6天球果开裂后，筛取种子。空粒较多，播种前水选剔除。鲜果出种率8%～10%；净度85%～95%；千粒重一般22g。陈晓阳和陈振炳等（1992）对7个省、13个县13个林分单株的种子做过分析，发现千粒重在种源间的变异幅度为18.3～28.3g。每千克纯净种子一般4.5万粒，变动范围3.5万～5.5万粒。含水量9%～10%的种子在0℃条件下可以安全贮藏2年，在普通库中可以贮藏1年。由于侧柏每年结实都较多，生产上多数都是越冬贮藏（于淑兰，1988）。

发芽和播种 侧柏种子无休眠习性。1987年在北京采集的种子，净度为94%，含水量8.8%，千粒重20.96g，不加任何处理，在恒温25℃，不加光照的条件下，22天的发芽率为75%。北方地区播种前常用始温30～40℃水浸种12小时，捞出置于篮筐内，放在背风向阳的地方，每天用清水淘洗一次，并经常翻动，当种子有一半裂嘴时取出播种。春播，条播或撒播。条播行距20～25cm，每平方米床面播种3g，覆土厚度1cm，15天左右开始发芽出土，20～25天为出土盛期，场圃发芽率在70%左右。出土萌发。发芽情况见图2。

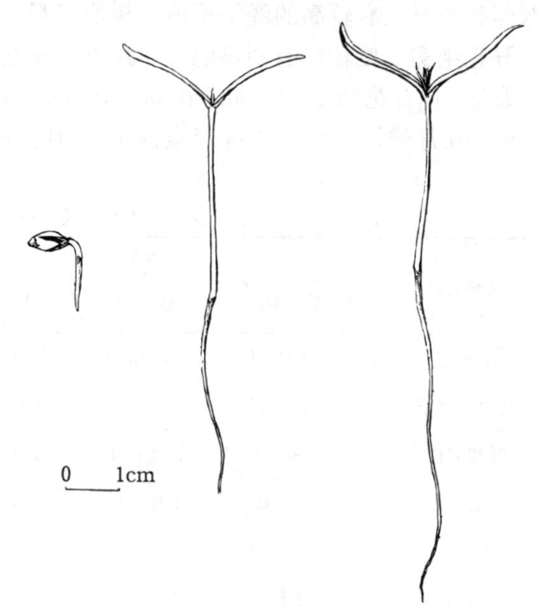

图2 侧柏种子发芽出土后第2、8、11天幼苗的生长情况
（孟玲绘）

（钱耀明）

圆 柏 属
Sabina Mill.

（柏科 Cupressaceae）

生长习性、分布和用途 圆柏属约50种，分布在北半球，北自北极圈，南至热带高山。

我国有 17 种，多数产于西北和西南高山，另引入栽培 2 种。本文描述 4 种（包括引入 1 种）。常绿乔木、灌木或匍匐灌木。耐寒、耐干旱瘠薄，叉子圆柏更耐沙压。圆柏对土壤要求不严，适宜于湿润肥沃的沙质壤土或壤土，酸性或微碱性土（pH 值 8）都能生长。引入的北美圆柏也较耐寒，冬季遇－14（徐州）～－22.8℃（北京）低温亦不受冻害。北美圆柏抗旱力强。在干旱比较严重的徐州石质山地用实生苗造林，不少树种死亡，北美圆柏的成活率能达 90％。北美圆柏也较耐盐碱，在土壤含盐量 0.3％的盆栽条件下苗木生长正常。据细胞质壁分离和四唑试验，其耐盐力可达 0.6％。江苏射阳土壤含盐碱较重，pH 值 9.22，这个树种的幼树长势仍然旺盛。北美圆柏是本属中生长快、适应性较广的一个值得推广的树种。又因材质细腻，均匀一致，耐切削，北美圆柏被认为是最优良的铅笔杆材，在铅笔工业上占有重要地位，所以我国又习称铅笔柏。圆柏也是一种优良的用材和园林观赏树种，在我国有悠久的栽培历史，各地都有数百年至千余年的古树名木。例如江苏苏州光福司徒庙有“清、奇、古、怪”4 株古圆柏。河南济源县济渎庙有“古汉柏”，树高 30m，胸径 2.29m，树龄已约 2 000 年。河北临漳县“邺都古柏”，树高 19m，胸径 1.43m，据传曹操曾在此拴马，也有 1 800 多年的历史。这 4 个树种的名称、生长、分布和用途见表 1。

表 1　圆柏属树种的名称、生长、分布和用途

中　名	学　名	树高(m)	胸 径(cm)	分　布	用　途	供　稿
昆仑方枝柏	*S. centrasiatica* Kom.	12～14(18)	35～60	新疆西昆仑山北坡。塔吉克斯坦	材用	204
圆柏(桧)	*S. chinensis* (L.) Ant.	20(30)	350	华北、西北、西南、中南、东南。辽、藏栽培。朝鲜半岛、日本	材用、油料、观赏、药用	1002
北美圆柏(红柏、铅笔柏)	*S. virginiana* (L.) Ant.	原产地 30	原产地 300～400	原产加拿大南部、美国、墨西哥北部。17 世纪引入欧洲，20 世纪初引入我国，现京、冀、鲁、豫、苏、皖、赣、湘栽培	铅笔杆、观赏	1002
叉子圆柏(臭柏)	*S. vulgaris* Ant.	匍匐灌木 0.5～1	—	内蒙古、陕、宁、甘、新、青。南欧、中亚	固沙、绿篱、饲料	201

开花结实　本属树种的开始开花结实年龄以叉子圆柏较早，5～6 年生即可开花结实，北美圆柏为 10～12 年生，圆柏和昆仑方枝柏约 20 年生。它们分别在 7～8 年生、15～20 年生和 20 年生以后正常结实。昆仑方枝柏每年都能结实，2～3 年出现一次丰年。球花单性，雌雄异株，偶有同株。球花单生短枝顶端。雄球花卵圆或矩圆形，黄色。雄蕊 5～7 对，交叉对生，各具 2～4 花药。雌球花珠鳞 6～8，3～4 对交叉对生或 3 枚轮生。胚珠 1～6 着生于珠鳞腹面基部。珠鳞渐变为肉质，并联合成很小而不开裂的球果。球果呈浆果状，通常 2 年（圆柏、昆仑方枝柏）或当年（北美圆柏）或第 3 年（叉子圆柏）成熟。成熟的球果蓝 黑色至红褐色，通常被有白色蜡霜。圆柏属 4 个树种的开花结实物候见表 2。球果及种子形态见表 3。

表 2 圆柏属树种的开花结实物候

树 种	观察地点	开花期 始 期	开花期 盛 期	开花期 末 期	球果成熟期	球果脱落期
昆仑方枝柏	新 疆	—	4～5 月	—	翌年 10 月	—
圆 柏 (1963 年)	内蒙古呼和浩特	5 月 5 日	—	—	—	—
	北京	雄 3 月 25 日	3 月 31 日	—	—	—
		雌 3 月 31 日	4 月 10 日	—	—	翌年 9 月
	陕西武功	2 月 27 日	3 月 1 日	3 月 22 日	—	—
	江苏南京	3 月 16 日	3 月 18 日	3 月 23 日	翌年 4 月 28 日	第 3 年 1～2 月
	安徽芜湖	雄 2 月 15 日	2 月 21 日	—	—	—
		雌 2 月 23 日	3 月 1 日	—	翌年 3～6 月	6 月 22 日
	浙江杭州	2 月 23 日	2 月 27 日	3 月 20 日	—	—
	湖北武昌	2 月 25 日	3 月 1 日	3 月 7 日	翌年 3 月 21 日	—
	云南昆明	3 月 3 日	3 月 5 日	3 月 8 日	—	—
北美圆柏	江苏南京	—	3 月中旬	—	当年 11 月中下旬	—
叉子圆柏	内蒙古	—	4～5 月	—	第 3 年 10 月中下旬	—

表 3 圆柏属树种球果及种子形态

树 种	成熟球果 形 态	成熟球果 颜 色	成熟球果 大 小 (mm)	每球果种子粒数	成熟种子 形 态	成熟种子 颜 色	成熟种子 大 小 (mm)
昆仑方枝柏	卵圆形或纺锤形	褐黄色或淡褐色，外被蜡质	长 9～13 径 7～9	1	扁卵圆形，先端较尖，基部圆，种皮有多条纵沟纹	淡黄褐色或黄褐色	长 6～9 宽 5～7 厚 4.5～5
圆 柏	近圆球形	暗褐色或紫褐色，被白粉或白粉脱落	径 6～8 (11)	(1) 2～4 (5)	卵圆形或不规则圆锥形，大小不等，表面凸凹不平，具长短不等纵棱及树脂槽	黄褐色至栗褐色，有光泽	长 2.8～4.6 宽 3.2～5
北美圆柏	圆球形或卵圆形	蓝绿色，被白粉	长 5～6 径 4～5	1～2 (3、4)	卵圆形，先端尖锐，基圆钝，表面有瘤状突起及树脂槽	黄色，先端黑褐色，基部乳黄色，有光泽	长 2～4.2 宽 2～2.8
叉子圆柏	倒三角状球形或叉状球形	未熟时蓝绿色，熟时紫蓝色或紫褐色，被白粉	长 5～8 径 5～9	(1) 2～3 (4、5)	卵圆形，微扁，先端钝或尖，具纵脊或树脂槽	黄褐色	长 4～5 宽 3～4

圆柏属树种的空粒种子往往较多。据江苏植物研究所观察，南京地区 20 年生的北美圆柏种子，1974 年饱满粒占 37%，1975 年只有 10%。但同为 1975 年，山东泰安 17 年生的北美圆柏的种子中，饱满粒达 43%（王名金、刘克辉等，1990）。种子无翅。种皮坚硬骨质。具胚乳。本文的这几个种，胚具 2 子叶。圆柏种子外形及其纵切面见图 1。

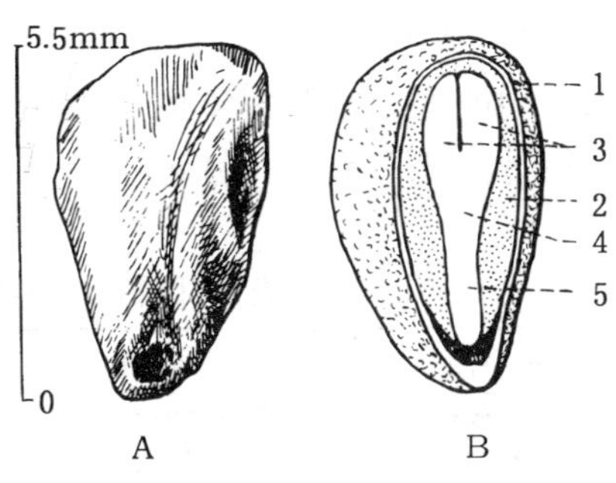

图 1　圆柏种子外形(A)及其纵切面(B)
1. 种皮　2. 胚乳　3. 子叶　4. 胚轴　5. 胚根
（张世经绘）

球果的采收调制和种子贮藏　本文的 4 个种中，只有北美圆柏的球果在开花当年成熟，其余 3 个种的球果多为 2 年（稀 3 年）成熟，采集时不要将当年的青绿色球果误采下来。可以击落后在地面收集，也可剪取果枝。种鳞在树上软化后易为鸟雀啄食，应及时采收。球果采收后可用石磙碾压，或用热水或草木灰水浸沤数日，再用木棒捣烂球果，搓揉，淘去种鳞和瘪粒等杂质，即得纯净种子。种子应摊开晾干，使含水量降至 10%以下。球果出种率和种子质量等数据见表 4。

表 4　圆柏属树种的球果出种率、净度和种子质量

树　种	产　地	出种率（%）	净　度（%）	千粒重（g）		每千克种子数（万粒）	
				一般数值	变动范围	一般数值	变动范围
昆仑方枝柏	新疆叶城	12～20	90	27	26～28	3.7	3.5～3.8
圆　柏	江苏南京	20～25	90	17	13～30	5.9	3.3～7.7
北美圆柏	原产地	14～18	—	10.4	8.3～12	9.6	8.2～12.1
	江苏南京	20～26	—	10	8～12	10.0	8.3～12.5
叉子圆柏	内蒙古	51	85	28	25～30	3.6	3.3～4.0
	陕西榆林	—	—	19.8	16～25	5.0	4.0～6.3
	甘肃高台	20～21	—	14.2	12～17	7.0	5.8～8.3

种子质量变幅较大。北美圆柏在原产地的千粒重有记载为近 26g。种子有后熟作用，如不随采随播，应将种子放在低温下混沙层积。圆柏属树种结实间隔期明显，常 2～3 年才出现一次丰年。为了弥补歉年的需要，应有一定数量的种子放入冷库干藏。某些种子易受线虫危害，可放入小包花椒预防。

发芽和播种　圆柏属树种的种子属于深休眠，发芽较困维，播后常需 2～3 年才能萌发。例如昆仑方枝柏当年新采的种子秋播，来年不发芽，即使到第 3 年也只有少数种子能够发芽。种子具不透水的种皮和休眠的胚，胚和胚乳均含脂溶性发芽抑制物质，外源植物激素和暖层积均不能打破休眠，必须经受低温层积才能解除休眠。种皮附有油脂或蜡质，阻碍吸水，低温层积前先行处理，可收到较好的效果。但郑振鸿（1994）研究后认为，铅笔柏种皮透水性良好，未经层积的种子，无论是否破开种皮，发芽率都很低，表明铅笔柏种子的休眠并不明显地取决于种皮，因此用改善种皮透性的药液，例如硫酸、乙醚、乙醇等加以处理，均不能解除休眠。他认为最有效的措施是层积，且层积的温度应当是 5～8℃，在这种温度下层积 90 天已足；高于或低于这个温度都会引发二次休眠。徐本美和闪崇辉等（1989）也发现，圆柏种子浸水 40 小时便已达到饱和状态，意味着这个树种的种皮并不妨碍吸水。不同树种层积的时间不尽相同，现将有关材料综合汇入表 5。

表 5 圆柏属树种种子层积前预处理、层积时间和发芽率

树 种	层积前预处理	5℃时的层积天数	发芽率（%）
昆仑方枝柏	小苏打液浸泡或 1%肥皂水浸泡	300	40～60
圆 柏	5%福尔马林液浸泡 3 分钟	100①	40
	未作处理	120～180	40～50
北美圆柏	带浆果状种鳞	90	10
	去除种鳞	90	30
	用 1%柠檬酸浸种 4 天	90	92
	用水浸种 4 天	90	72
	不处理	90	22
叉子圆柏	草木灰加 3 倍始温 55℃水浸种 24 小时	150	50～72
	0.2%的硝酸钾处理	150	75
	10μg/g 对苯二酚处理	150	67.9

① 层积 45 天时取出种子，稍干燥后浸润，再层积，层积期可以从 100 天缩短为 90 天（陈嵘，1933）

播种多用条播。秋播，或用经过层积处理的种子春播。行距 15～25cm，覆土厚 1～2cm。盖草。每平方米约播昆仑方枝柏 38g，圆柏 38～75g，北美圆柏 25～50g，叉子圆柏 50～80g。播后 15～25 天可出土。子叶 2，出土萌发。圆柏种子萌发和幼苗生长情况见图 2。

圆柏类皆可扦插繁殖。

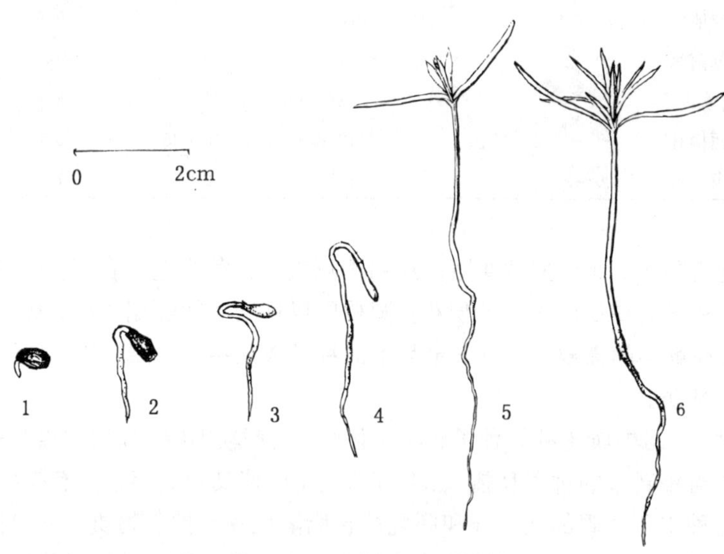

图 2 圆柏种子的萌发和幼苗生长情况

1. 胚根初露 2、3. 胚轴延伸 4. 子叶出土 5. 初生叶轮生 6. 幼苗形成

（张世经临《主要树木种苗图谱》）

（黄鹏成）

崖　柏　属
Thuja L.

（柏科　Cupressaceae）

生长习性、分布和用途　崖柏属约 6 种，产北美和东亚。我国 2 种，引入栽培 3 种。本文描述我国产的 1 种和引入的 2 种。常绿乔木。耐荫。喜凉爽湿润的气候和富含腐殖质的山地黄壤和棕壤，在瘠薄山地或含石灰质的湿润地区也能生长。崖柏属这 3 个树种的名称、生长、分布和用途见表 1。野生的朝鲜崖柏仅星散分布于我国长白山西南侧的狭小范围内，在《中国植物红皮书》中被列为渐危种。

表 1　崖柏属树种的名称、生长、分布和用途

中　名	学　名	树　高 (cm)	胸　径 (cm)	分　布	用　途
朝鲜崖柏（长白侧柏）	*T. koraiensis* Nakai	10	75	吉林。朝鲜半岛	材用、香料、观赏
北美香柏	*T. occidentalis* L.	20①	100①	原产北美。我国华东、华中、西南各地引种	材用、香料、药用、观赏
日本香柏（大叶香柏）	*T. standishii* (Gord.) Carr.	25～30①	60①	原产日本。我国华东、华中、西南各地引种	材用、香料、观赏

①　系原产地生长情况

开花结实　10～12 年生开始开花，20～30 年生后正常结实。北美香柏和日本香柏在土壤瘠薄之地开花结实较早。结实大小年明显，丰年间隔期 2～4 年。雌雄同株。球花单生枝顶。雄球花球形或卵圆形，淡褐色或黄色，具多数雄蕊，花药 4。雌球花长圆形或长卵形，有 3～5 对珠鳞，仅下部 2～3 对珠鳞各有 1～2 枚直生胚珠。这 3 个树种的开花结实物候见表 2。球果矩圆形或长卵形，长 0.8～1.3cm，径 0.6～1.0cm。种鳞薄，近革质，扁平，淡褐色，顶端有钩状突起，发育种鳞各有 1～2 粒种子。种子扁平，椭圆形，两侧有翅。具胚乳。子叶 2 枚。球果和种子的形态特征见表 3、图 1。

表 2　崖柏属树种的开花结实物候

树　种	观察地点	开花期		球果成熟期		种子散落期
		雄　花	雌　花	始　期	盛　期	
朝鲜崖柏	吉林延吉	5 月中旬	5 月中旬	9 月下旬	10 月上旬	10 月下旬～11 月
北美香柏	江西庐山	4 月上旬～中旬	4 月中旬～下旬	9 月下旬	10 月上旬	10 月中旬～11 月下旬

（续）

树　种	观察地点	开花期		球果成熟期		种　子散落期
		雄　花	雌　花	始　期	盛　期	
	江西九江	3 月上旬～4 月上旬	3 月上旬～4 月上旬	10 月下旬	10 月下旬	11 月下旬～12 月
日本香柏	江西九江	3 月下旬	4 月上旬	10 月上旬	10 月中旬	11 月下旬
	山东青岛	3 月中下旬	4 月上中旬	10 月中旬	10 月下旬	11 月下旬～12 月

表 3　崖柏属树种球果和种子的形态特征

树　种	成熟球果				成熟种子			
	形　状	长（cm）	径（cm）	颜　色	形　状	长（mm）	宽（mm）	颜　色
朝鲜崖柏	椭圆状球形	9～10	6～8	深褐色	椭圆形	4	1.5	淡褐色
北美香柏	长椭圆形或长卵形。幼时直立，熟时向下弯垂	8～13	5～6	淡褐色或黄褐色	椭圆形、长圆形	4～7	2.5～4	淡褐色
日本香柏	卵圆形或长圆形	8～10	4～5	暗褐色	椭圆形或圆形	5～6	2	暗褐色

球果采收调制和种子贮藏　将采集的成熟球果摊晒 2～4 天，种鳞开裂后脱出种子。无需去翅，清除杂质后阴干，即可在低温下干藏。出种率 2%～3%。千粒重 1.1～1.3g，每千克有纯净种子 77 万～95 万粒。

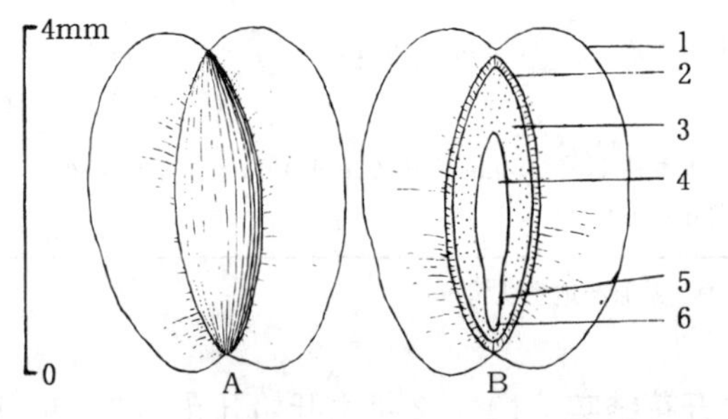

图 1　朝鲜崖柏种子的外形（A）及其纵切面（B）
1. 翅　2. 种皮　3. 胚乳　4. 子叶　5. 胚轴　6. 胚根
（黄应钦绘）

发芽和播种　《国际种子检验规程》规定北美香柏种子的发芽测定温度是每天给予 20～30℃的变温，并在 30℃的 8 个小时中加光照。据庐山植物园（朱国芳，未刊稿）试验，在20～25℃的恒温下，10 天开始发芽，20～25 天发芽结束，发芽率 8%～10%。出土萌发。子叶 2 枚，线形或披针形。北美香柏的子叶长 0.8～1cm，灰绿色；初生叶交叉对生，线形或披针形，长约 1cm；顶端钝尖或锐尖，被白粉（图 2）。出苗 2 个月后，长出鳞片叶。1 年生苗高6～12cm。引种栽培在庐山植物园的北美香柏和日本香柏结实很少，多用扦插繁殖，成活率可达 90%以上。

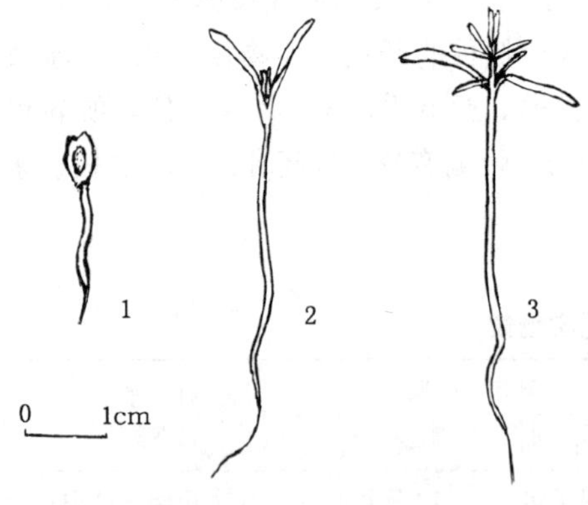

图 2　北美香柏种子萌发和幼苗生长情况
1. 胚轴延伸　2. 子叶出土　3. 初生叶对生
（田恒德绘）

（李　华）

罗汉柏（蜈蚣柏）

Thujopsis dolabrata（L. f.）Sieb. et Zucc.

（柏科　Cupressaceae）

生长习性、分布和用途　罗汉柏属仅本文描述的1种，为日本特有。常绿乔木，在原产地高15～30m，胸径30～100cm。耐荫性强，喜深厚湿润的山地黄棕壤，忌干燥瘠薄及排水不良。京、鲁、苏、浙、赣、闽、鄂、湘、粤、黔、滇引种栽培。据调查，罗汉柏在浙江四明山林场海拔800m以上的山区、贵州省林业科学研究所、湖南省南岳树木园和江西省庐山植物园均生长良好，在广州则生长不良。

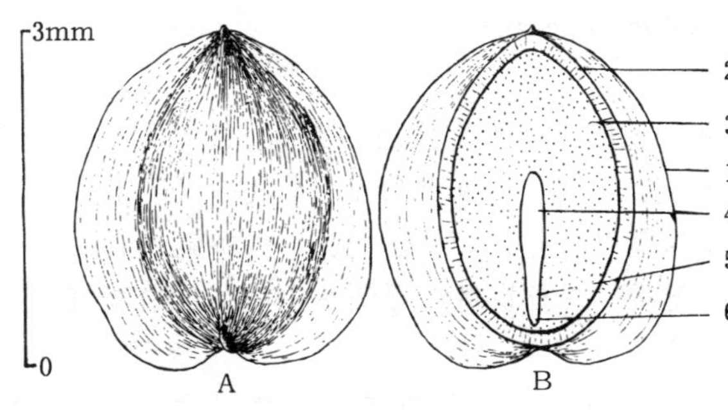

图1　罗汉柏种子的外形（A）及其纵剖面（B）
1. 翅　2. 种皮　3. 胚乳　4. 子叶　5. 胚轴　6. 胚根
（黄应钦绘）

罗汉柏树形优美，鳞叶绿白相间，可供庭园观赏。木材轻软，有芳香，易加工，可作枕木、桥梁用材。

开花结实　15～20年生开始结实，25～30年生后正常结实，结实有大小年，丰年间隔期1～3年。据庐山观察，花期在3月下旬～4月，球果当年9～10月成熟。可能是由于春季雨水过多，庐山的罗汉柏结实率低，种子空粒多。

雌雄同株。球花单生于短枝顶端。雄球花长椭圆形，绿黄色，6～8对雄蕊交叉对生。雌球花扁球形或卵圆形，淡黄绿色，具3～4对珠鳞，仅中间两对各有胚珠3～5枚。球果近圆球形，长1.2～1.5cm，熟时暗褐色，有白粉。种鳞3～4对，木质，扁平，在顶端的下方有一短尖头，中间两对发育种鳞各有3～5粒种子。种子近圆形，在原产地长4～4.5mm，引种到庐山的一般长2～3mm，淡褐色，有光泽，两侧有窄翅。子叶2枚。有胚乳（图1）。

球果的采收调制和种子贮藏　球果秋季成熟，可在种鳞微裂时采集。采回的球果摊晒2～3天，种鳞裂开后筛出种子，阴干。出种率4%，千粒重约4g。每千克纯净种子20万～25万粒。干藏。

发芽和播种　3～4月播种。在气温15～20℃时播后20～25天发芽，发芽率约

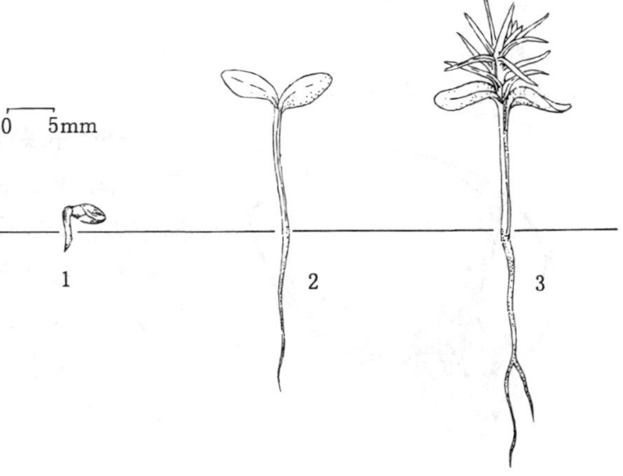

图2　罗汉柏种子萌发及幼苗生长情况
1. 种壳出土，下胚轴初露　2. 子叶展开
3. 初生叶四叶轮生并出现分枝（梁鸣绘）

20%。出土萌发。子叶 2 枚，线形，长 8～12mm，顶端平或微凹，上面有气孔线，绿色，下面黄绿色具白粉。幼茎淡黄绿色。初生叶扁平刺形，第 1 对叶长 4～6mm，与子叶交叉对生，以后 4 叶轮生，再后形成分枝。分枝上叶片鳞状刺形，逐渐变短变宽成鳞叶。鳞叶质地较厚，上端较钝，上面深绿色，叶背具有较宽的白色气孔带。下胚轴圆柱形，长 2.5～3.0cm，黄绿色转黄褐色。主根发达，侧根疏生，平展，黄褐色。罗汉柏种子的萌发和幼苗生长情况见图 2。一般用扦插繁殖，成活率 90%以上。

（李　华）

陆均松（泪柏）

Dacrydium pierrei Hickel

（罗汉松科　Podocarpaceae）

生长习性、分布和用途　陆均松属约 20 种，我国只有本文描述的 1 种。常绿大乔木。树高 35m，胸径 200cm 以上。幼龄期稍耐荫，成龄树喜光。分布热带地区，我国仅产于海南岛中部及南部。印度、菲律宾、泰国、柬埔寨和越南也有分布。材质优良，为桥梁、船舶、车辆、家具及建筑等用材。树形优美，可供庭园观赏。森林资源已逐渐稀少，在《中国植物红皮书》中被列为渐危种。

开花结实　实生苗栽植后 10～15 年开始结实，正常结实期约在 20 年生以后。结实有大小年现象，间隔期一般为 1 年。扦插苗植后 3～4 年便开花结实。球花单性，雌雄异株，稀同株。雄球花着生于近枝顶的叶腋，穗状，圆柱形，长 0.5～1.1cm，花药紧挤，2 室，药隔盾状，花粉有 3 气囊。雌球花无梗，单生或排成穗状，基部有少数苞片，最上一枚苞片发育成套被，内有一倒生胚珠。胚珠受精后直立。据在海南岛观测，花期在 3～5 月，种子 10 月中旬开始成熟，11 月为成熟盛期。种子坚果状，卵圆形，栗褐色，长 4～6mm，径约 3mm，先端钝尖，横生于较薄而干的杯状假种皮中。假种皮熟时红色或红褐色，无梗。种子的形态见图 1。种子具胚乳，子叶 2，出土。

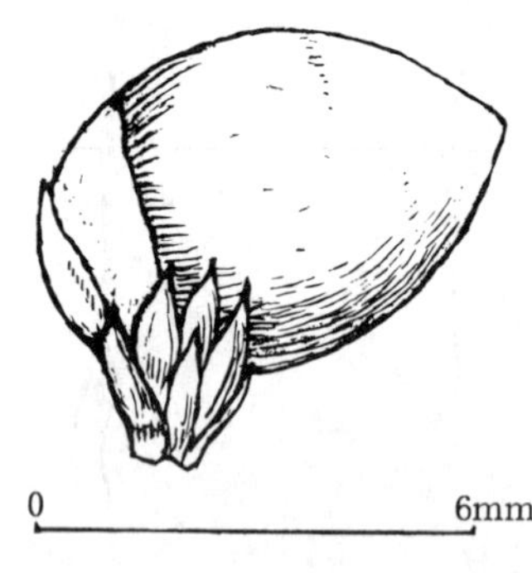

图 1　陆均松横生于枝顶的种子
（田恒德绘）

种子的采收调制和贮藏　种子成熟时用高枝剪剪落果枝，或在伐倒木上采摘，也可于成熟盛期用竹竿敲打或摇动树枝震落，然后在地面捡拾。采集的种子稍阴干后即可播种或贮藏，忌日晒或裸露存放。种子的净度可达 90%以上，千粒重 17～20g，每千克纯净种子 5 万～5.8 万粒。无鼠害地区可随采随播。贮藏越冬的种子宜混湿沙，贮藏至翌年 2～3 月播种。运输时宜用竹筐包装，每筐不超过 20kg，抵达目的地后立即混湿沙贮藏，贮藏期在半年以内。

发芽和播种　种子无明显休眠习性。发芽时日均温需在 20℃以上。播种前种子用始温40～50℃水浸种 24 小时。播后 20～30 天开始发芽。场圃出苗率可达 80%左右。喜与菌根菌共生，育苗时应当用松林下的土壤接种。条播。每平方米播种 4～5g。11～12 月播种时由于气温较低，发芽时间会延迟 2～4 个月之久。苗期生长缓慢，需搭棚遮荫。1 年生苗高仅 10～15cm。陆均松可以扦插繁殖，方法是取 1～2 年生健壮幼嫩枝条作插穗，长约 25cm，扦插于沙床上，注意遮荫保湿，发根后移至圃地培育，成苗率可达 50%以上。

（符史深）

罗汉松属

Podocarpus L'Her. ex Persoon

（罗汉松科　Podocarpaceae）

生长习性、分布和用途　本属约 100 余种，我国有 9 种和 3 变种。本文描述的 3 种 1 变种均为常绿乔木，高达 20m 以上（变种为常绿小乔木）。较耐庇荫，喜湿润温暖气候。适生于土壤肥沃深厚的酸性土。木材细致均匀，硬度适中，为优质材。它们的名称、生长、分布和用途见表 1，其中鸡毛松被《中国植物红皮书》列为渐危种。

表 1　罗汉松属树种的名称、生长、分布和用途

中　名	学　名	树高(m)	胸径(cm)	分　布	用　途
鸡毛松	*P. imbricatus* Blume	40	180	琼、粤、桂、滇。越南、菲律宾、印度尼西亚	材用、水源、油脂
罗汉松	*P. macrophyllus* (Thunb.) D. Don	30	60	东南、中南、西南。日本	特用材、油脂、观赏
狭叶罗汉松	*P. macrophyllus* (Thunb.) D. Don var. *angustifolius* Blume	6～10	10	长江以南庭院栽培	雕刻、观赏
百日青（竹叶松）	*P. neriifolius* D. Don	30	50	东南、中南、西南。东南亚各国	体育用材、雕刻、油脂、风景

开花结实　10～12 年生开始结实，正常结实期在 15～20 年生以后。大小年间隔期为 1 年，但不甚明显。花单性，雌雄异株。雄球花穗状，腋生，雄蕊多数，每雄蕊具 2 花药，花粉有气囊。雌球花单生叶腋或苞腋，稀顶生。基部数枚苞片的腋内无胚珠，顶端一枚苞片发育成囊状套被，内有胚珠 1 枚。花后套被增厚成肉质，俗称假种皮。种子当年成熟，成熟 6～10 天后陆续落地。脱落的持续时期在鸡毛松为 20～30 天，罗汉松、狭叶罗汉松和百日青为 30～45 天。广西及海南等地观察的开花结实物候期见表 2。

表 2 罗汉松属树种的开花结实物候期

树 种	观察地点和年份	开花期			种子成熟期		种子脱落期
		始 期	盛 期	末 期	始 期	盛 期	
鸡毛松	海南 1984	3 月下旬	4 月上旬	4 月中旬	9 月下旬	10 月中旬	10 月下旬～11 月中旬
罗汉松	南宁 1978～1984	4 月上旬	4 月中旬	5 月上旬	7 月上旬	7 月下旬	7 月下旬～8 月中旬
狭叶罗汉松	南宁 1983～1987	4 月下旬	4 月底～5 月初	5 月中旬	8 月下旬	9 月中旬	9 月上旬～10 月中旬
百日青	海南 1983～1984	5 月下旬	6 月中旬	7 月上旬	9 月上旬	10 月上旬	10～11 月

种子核果状，近球形，种子 1 粒，全部为肉质套被所包。鸡毛松种子生于小枝顶端，熟时肉质套被红色，苞片发育成红色肉质种托，较种子小，无梗。罗汉松、狭叶罗汉松、百日青种子着生于叶腋，不生胚珠的苞片和轴发育成红色或紫红色肉质种托，较种子大，外种皮膜质。种子具长梗，有胚乳，子叶 2。成熟时套被和种子的特征见表 3、图 1。

表 3 罗汉松属树种种子的形态特征

树 种	种托颜色	套被（假种皮）颜色	种子形状		种子大小（mm）	
			带套被	去套被	带套被种托长	去套被种托长
鸡毛松	红色	红色，有光泽	近球形	卵圆球形	8～10	长 5～6
罗汉松	红色或紫色	青色至紫黑色	椭圆形	卵状球形，先端圆	21～26	长 9～12，径 8～10
狭叶罗汉松	红色	青色	椭圆形	卵状球形	15～20	径 7～9
百日青	橙红色	紫红色	椭圆形	近球形	20～25	径 10～15

种子的采收调制和贮藏 将母树下清理平整并铺塑料薄膜，用竹竿敲打或上树摇动树枝，震落种子后收集，也可俟种子自然成熟脱落后在地面拾集。植株不高的母树，可在地面或上树选择成熟种子采摘。采回应将肉质种托及时脱去，或置水中搓洗，去掉种托和套被，以免霉坏。洗净的种子摊放晾干后即为播种材料。种子忌脱水，不宜在阳光下曝晒，亦忌堆沤。含水量应保持在 35%以上。用于随采随播的种子，也可带套被直接播种。种子贮藏或运输时均需混以湿沙。室内裸露存放的种子，20～30 天便失去发芽能力。混沙贮藏的种子，贮藏期一般为半年左右。秋后采集的种子可贮藏至次年春播。

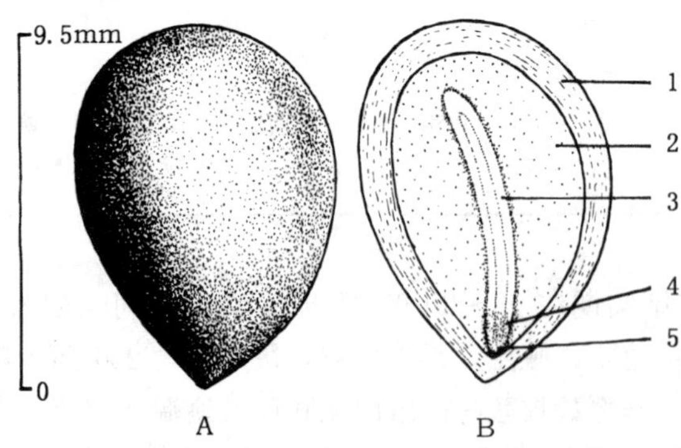

图 1 罗汉松种子外形（A）及其纵切面（B）
1. 种皮 2. 胚乳 3. 子叶 4. 胚轴 5. 胚根
（黄应钦绘）

9月以前采集的种子宜随采随播。除去种托、套被的出种率、种子净度、质量等见表4。

表4　罗汉松属树种去除种托、套被的出种率、净度和质量

树　种	出种率（%）	净度（%）	千粒重（g）	每千克纯净种子粒数（粒）
鸡毛松	50～60	97～100	65～75	13 000～15 400
罗汉松	25～35	97～100	400～600	1 700～2 500
狭叶罗汉松	28～35	97～100	380～420	2 300～2 600
百日青	30～37	97～100	600～700	1 400～1 700

发芽和播种　种子无休眠习性。发芽时日均温宜在20℃以上。气温高，种子的发芽进程较快。1978～1987年广西林业科学研究所利用室外沙床对罗汉松和狭叶罗汉松进行过发芽测定，结果见表5。

表5　罗汉松属树种的发芽能力及其测定条件（室外沙床）

树　种	温度（℃）	发芽势（%）		发芽率（%）	
		计算天数	一般数值	计算天数	一般数值
罗汉松	22～27	7	54	17	90
狭叶罗汉松	22～28	6	72	14	85

出土萌发。随采随播的种子，播后2～5天胚根萌发，14～17天子叶带壳出土，20～22天发出初生叶。罗汉松种子的发芽情况和幼苗生长见图2。

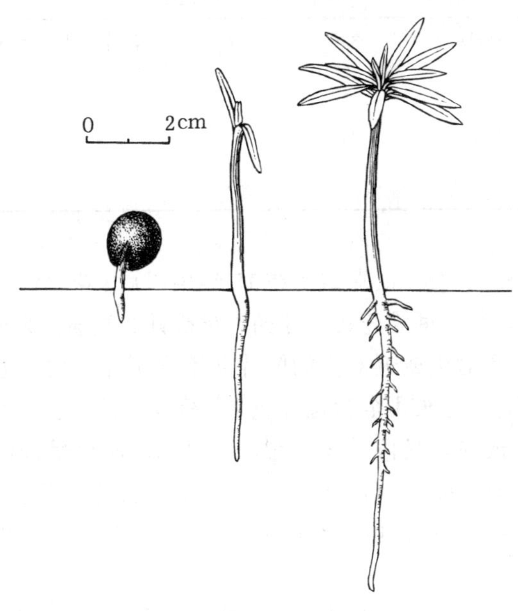

图2　罗汉松种子萌发后第3、20、35天幼苗的生长情况

（黄应钦仿《热带亚热带主要树种采种育苗技术》）

条播。覆土厚为种子直径的2～3倍。每平方米的播种量，鸡毛松为10～12g，罗汉松和狭叶罗汉松38～50g，百日青60～75g。早春播种时也可将种子密播，并加盖塑料薄膜保温催芽，待种粒萌发后将芽苗分批移至大田。一般培育1～2年生苗出圃。用于庭园栽培的常需培育3～5年。可以扦插育苗。

（王宏志）

竹柏属
Nageia Gaertner

（竹柏科 Nageiaceae）

生长习性、分布和用途 竹柏科竹柏属是从罗汉松科罗汉松属分出的新科新属（傅德志，1992）。本属6种，分布于东亚、南亚至菲律宾及新几内亚等南太平洋岛屿。我国4种，本文描述2种。常绿乔木，高20～30m。叶无中脉，具多条平行脉，耐荫或较耐荫，适于暖热潮湿气候和深厚疏松、多腐殖质的沙壤或轻粘土。在阴坡比在阳坡生长快5～6倍。材质轻软，细致均匀，不翘不裂，优质用材。种子含油量30%，优质食用油或供工业用。它们的名称、生长、分布和用途见表1。长叶竹柏在《中国植物红皮书》中列为渐危种。

表1 竹柏属树种的名称、生长、分布和用途

中 名	学 名	树 高 (m)	胸 径 (cm)	分 布	用 途
长叶竹柏	*N. fleuryi* (Hickel) De Laub.	30	70	滇、桂、粤。越南、柬埔寨	材用、乐器、油脂、观赏
竹柏	*N. nagi* (Thunb.) Kuntze	25	80	东南、华南、川。日本	材用、乐器油脂、观赏

开花结实 10年生开始结实，正常结实在15年生以后。花单性，雌雄异株。雄球花穗状，1至数个生于叶腋，雄蕊多数，每雄蕊具2花药，花粉具2气囊。雌球花生于叶腋，其轴端具不发育的顶芽，轴上有数个对生或近对生的鳞片状不育苞片。轴的上部具2个对生的可育苞片，每苞片着生一倒生胚珠，胚珠几全为苞片包被。苞片下部伸长，远轴面近基部与下方1枚苞片部分合生。种子当年成熟，具假种皮。雌球花之轴木质化呈种柄状，稀膨大成绿色肉质种托[例如肉托竹柏 *N. wallichiana* (Presl) Kuntze]，其上具宿存苞片或脱落的痕迹。种子成熟后10天左右开始陆续脱落，脱落的持续时间可达4～5个月。广州、南宁等地观察的开花结实物候期见表2。

种子核果状，圆球形，外为肉质假种皮，外种皮骨质，内种皮膜质，具胚乳，子叶2。成熟时假种皮和种子的特征见表3、图1。

表 2　竹柏属树种的开花结实物候期

树　种	观察地点和年份	开花期			种子成熟期		种子脱落期
		始　期	盛　期	末　期	始　期	盛　期	
长叶竹柏	广州 —	雄球花： 4 月中旬	4 月下旬	5 月上旬	10 月下旬	11 月中旬	12 月～翌年 4 月
		雌球花： 4 月下旬	4 月下旬	—			
竹柏	南宁 1978～1987	4 月下旬	5 月上旬	5 月中旬	11 月上旬	11 月下旬	12 月上旬～翌年 4 月
	海南（尖峰岭）	雌球花： 2 月上旬	—	3 月下旬	8 月上旬	9 月中旬	—

表 3　竹柏属树种假种皮和种子的形态特征

树　种	未熟时颜色	成熟时颜色		形　状		大　小（mm）	
		假种皮	外种皮	带假种皮	去假种皮	带假种皮	去假种皮
长叶竹柏	青绿色	黄绿色至蓝色	深褐色	球形，无种托	球形，基部长	径 17～22，梗长 20	径 15～19
竹　柏	粉绿色	暗紫色	黄褐色	球形，无种托	球形，基部尖	径 11～15 梗长 7～13	径 8～12

种子的采收调制和贮藏　将母树下地面清理平整，铺张布幕，敲打或摇动树枝，震落后收集，或待自然脱落后拾集。采后可堆放数日，待假种皮充分软化后用水加粗沙搓擦，去除假种皮即得纯净种子。种子最忌曝晒，如曝晒 3 天即完全丧失发芽能力。含水量应保持在 30%以上。种子即使混有湿沙，也不宜久藏。贮藏期超过 4 个月则发芽率大为降低。用于随采随播的种子也可带假种皮直接播种。出种率、种子净度、质量等见表 4。

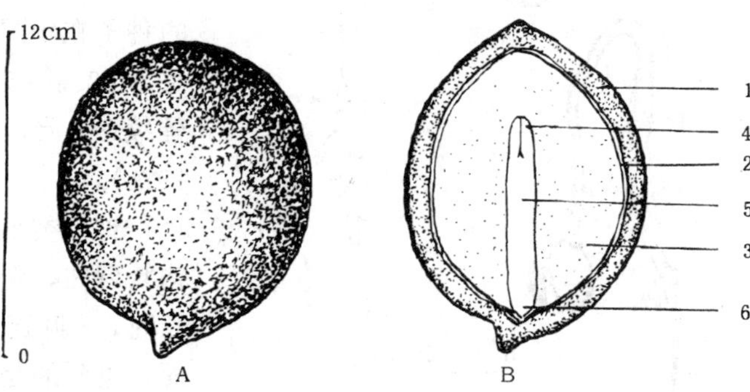

图 1　竹柏带假种皮种子外形（A）及其纵切面（B）
1. 假种皮　2. 种皮　3. 胚乳　4. 子叶　5. 胚轴　6. 胚根
（田恒德仿《日本の树木种子》）

表 4　竹柏属树种去除假种皮的出种率和质量

树　种	产　地	出种率（%）	净度（%）	千粒重（g）	每千克纯净种子数
长叶竹柏	广州	70～80	96～100	1 800～2 300	430～560
竹柏	南宁	70～80	96～100	300～500	2 000～3 300
	湖南	50	—	407～560	1 800～2 450
	海南万宁	55	96.5	300	3 300
	万宁（不去除假种皮）		98	624	1 600

发芽和播种　种子无休眠习性，发芽时日均温宜在 20℃以上。1978～1987 年，广西林业科学研究所在室外沙床作过发芽测定，结果如表 5。

表 5　竹柏属树种的发芽能力及其测定条件（室外沙床）

树　种	温　度（℃）	发芽势（%）		发芽率（%）	
		计算天数	一般数值	计算天数	一般数值
长叶竹柏	16～20	16	56	40	90
竹柏	20～28	9	76	30	90

中国林业科学研究院热带林业研究所在海南尖峰岭对万宁产的竹柏种子做过观察：除去假种皮的种子在场圃播种后 17 天发芽，发芽过程延续 74 天，发芽率 85%～95%；不除假种皮的种子在场圃播种后 27 天发芽，延续 64 天，发芽率 37%～50%。前者在室内播种，10 天发芽，延续 31 天，发芽率 92%～95.5%（陈荷美，1978）。

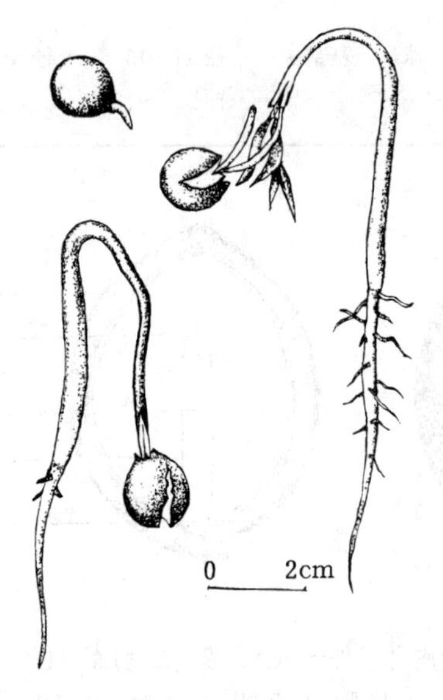

图 2　竹柏种子萌发后第 2、15、25 天的幼苗生长情况
（黄鹏成供稿，田恒德绘）

竹柏类出土萌发，下胚轴伸长，弓形。子叶钩状弯曲，藏于种皮及胚乳内。当下胚轴继续伸长时，子叶连同胚乳和种皮脱落，很少能带出土面，见到的是对生的初生叶（图 2）。

开沟点播。粒距 10cm，覆土厚 3cm。每平方米播种 35～50g。2 年生苗出圃。亦可扦插育苗。

（黄鹏成、王宏志）

三　尖　杉

Cephalotaxus fortunei Hook. f.

(三尖杉科　Cephalotaxaceae)

生长习性、分布和用途　三尖杉属有 8 种，我国 6 种，本文描述三尖杉这 1 种。常绿乔木，高达 20m，胸径 40cm。喜温凉湿润气候和肥沃排水良好的沙质壤土，在险峻的石灰岩上亦能生长。多散生于溪边、阴坡、山谷、山洼等针阔叶混交林中。分布区北起秦岭、大别山，东及浙、闽，南至华南北部，西至甘南和川、滇。苏、鲁栽培。木材坚实，供细木工用。种子可食并可入药。枝、叶、根和种子可提取三尖杉酯碱，对白血病、肺癌及淋巴肉瘤等有一定疗效。种仁含油 52%，可榨油，供制皂、油漆、蜡和硬化油。枝叶繁茂，种子成熟时红艳夺目，挂满枝头，为优良观赏树种。

开花结实　4～5 年生开始开花结籽。球花单性，雌雄异株。雄球花 8～10 聚生成头状花序状，单生叶腋。每一雄球花有 6～10 枚雄蕊，各具 3 花药。雌球花具长梗，生于小枝基部或近枝顶苞片的腋部。花梗上部的花轴上具数对交叉对生的苞片，每一苞片的腋部有两枚直立胚珠。胚珠生于瓶状珠托之上。花期 4 月。3～8 枚胚珠发育成种子。种子于翌年 8～10 月成熟。种子核果状，全包于由珠托发育成的肉质假种皮中，椭圆状卵形或近圆球形，长约2.5～3cm。假种皮成熟时紫色或紫红色。去除假种皮后种子卵状椭圆形，长 2.2～2.5cm。外种皮坚硬骨质，内种皮薄膜质。具胚乳，子叶 2（图 1）。

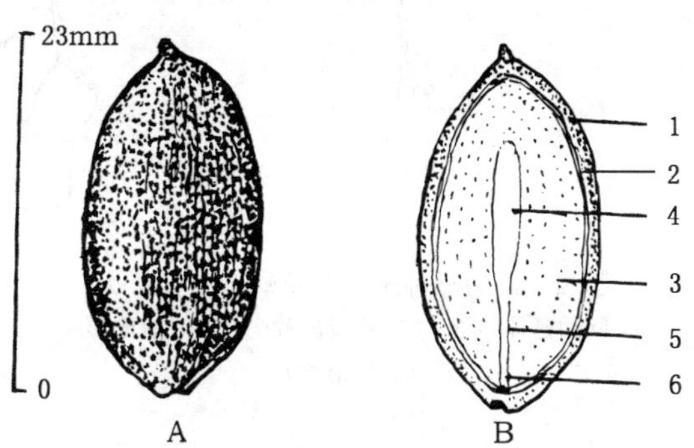

图 1　三尖杉种子外形（A）及其纵切面（B）

1. 外种皮　2. 内种皮　3. 胚乳　4. 子叶　5. 胚轴　6. 胚根

（田恒德绘）

种子的采收调制和贮藏　选取 20～50 年生母树，当假种皮由青绿色变紫色或紫红色时采摘或敲落后收集，装入箩筐，置流水中搓去假种皮，洗净，阴干，混湿沙贮藏。出种率 45%。种子千粒重 1 170g（武夷山），变动在 1 050～1 250g。每千克有纯净种子 860（800～950）粒。

发芽和播种　种子有休眠习性。1961 年冬季南京林学院树木园收到采于赣南的种子，1962 年 2 月 28 日播种，到 1963 年 4 月中旬种子萌发。福建来舟林业试验场于 1986 年冬在省内武夷山采得种子，调制后湿沙层积 1 年，1988 年春季条播，当年出苗较整齐。

条播，种子宜横放，每平方米播种 50～75g，覆土 3～4cm，盖草。出苗后揭草并立即搭设荫棚。当年苗高 10～15cm。2～3 年生苗方可出圃。

出土萌发。子叶 2，窄条形，长 2.5～4cm，宽 2～3mm，中脉在上面凹下，凹槽中有窄的白粉带。初生叶 1～3 对，近对生，披针状条形，长仅 4～15mm。以后叶螺旋状互生，镰状条形，长 3.5～7cm（图 2）。

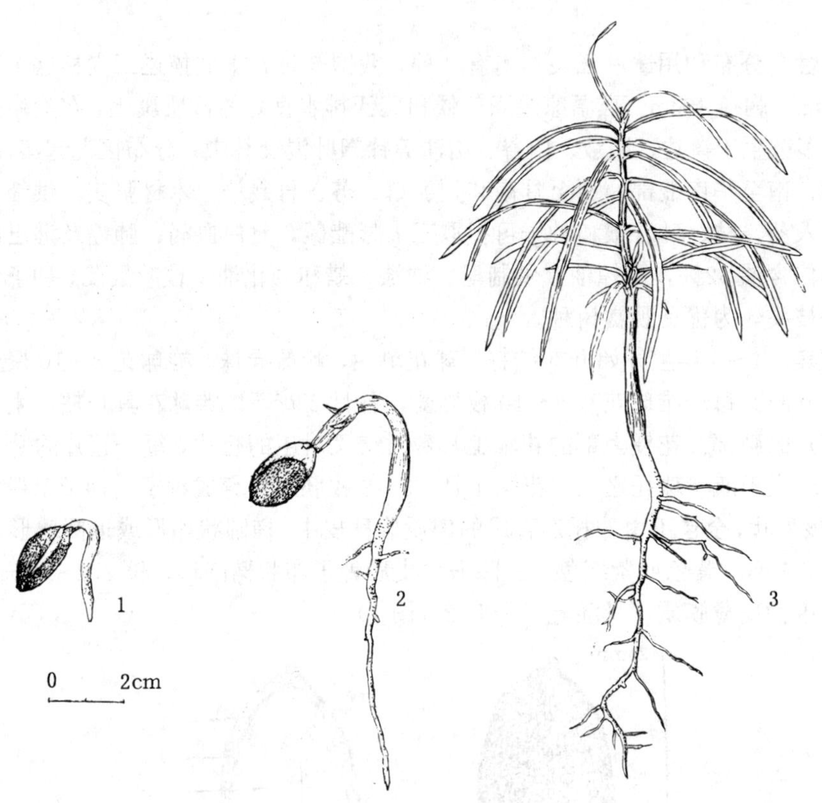

图 2　三尖杉种子萌发及幼苗生长情况

1. 胚轴延伸　2. 子叶将脱出种壳　3. 幼苗形成

（田恒德绘）

（黄鹏成）

云南穗花杉

Amentotaxus yunnanensis Li

（红豆杉科　Taxaceae）

生长习性、分布和用途　穗花杉属约 3 种，均产我国。本文描述的云南穗花杉这个种产云南东南部和贵州西南部，在越南北部也有分布。常绿乔木，高 15m，胸径 25cm。深根性。生长较慢。喜温暖湿润气候。材质优良，树形美观，种子可榨油。分布狭窄而资源又日渐枯竭，在《中国植物红皮书》中被列为濒危种。

开花结实　开始开花结实的年龄不明，30 年生以后正常结实。结实有大小年现象，间隔期 2～3 年。球花单性，雌雄异株或同株。雄球花多数，对生，组成穗状球花序，4～6 穗集生于近枝顶的苞腋，长 10～15cm。雄蕊多数，盾形，每雄蕊花药 4～8，辐射排列，药室纵裂。雌球花单生于新枝上的苞腋或叶腋，花梗长。胚珠 1 个，直立，着生于花轴顶端的苞腋，为一漏斗状珠托所托，基部有 6～10 对交叉对生的苞片。花后珠托发育成囊状假种皮。4 月开花，雌球花受粉并不即时受精，翌年 4～5 月种子成熟。种子核果状，椭圆形，长 1.8～3.2cm，径0.8～1.4cm。成熟时假种皮红紫色，微被白粉。种梗长约 1.5cm。胚乳丰富。胚短小，子叶 2（图 1）。

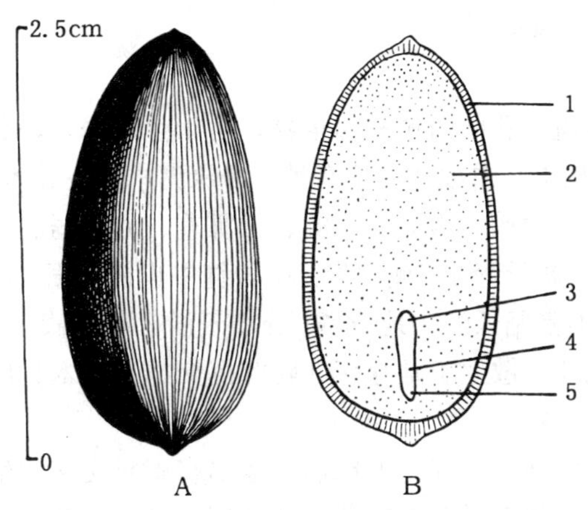

图 1　云南穗花杉种子的外形(A)及其纵切面(B)
1. 种皮　2. 胚乳　3. 子叶　4. 胚轴　5. 胚根
（黄应钦绘）

种子的采收调制和贮藏　种子成熟后易受虫、鸟、鼠害，应及时采集。可以用竹竿敲打或摇动树枝，震落后在地面拾集，也可上树采摘。采回后剥去鲜红色肉质假种皮即为播种材料。按质量计算，去除假种皮的出种率为 88%～95%。种子净度 85%～95%。种子千粒重约 2 200（1 000～2 500）g，每千克有种子约 450（400～1 000）粒。种子忌失水，不宜曝晒或干藏。短途运输前不宜去掉假种皮，最好是抵达目的地后再行调制。运输时间在 1 周以上的还需混以湿沙或锯屑。贮藏时也应混拌湿沙或置 5～8℃的低温处贮藏。

发芽和播种　种子有深休眠习性。据云南林业科学研究所试验，1987 年 9 月 10 日采集的种子，剥去假种皮并经沙藏，1988 年 1 月 9 日在室外荫棚下苗床条播，种子当年不发芽，也不腐烂。直至 1989 年 4 月 22 日开始萌发，种子在苗床休眠时间长达 16 个月。发芽过程长达 70 天，且无明显的发芽盛期。出苗率约 55%。

出土萌发。胚根萌发后迅速向下伸展，长 3.5～7cm，径 0.4cm，并形成侧根，长 1～2cm，径 0.1cm。紫红色胚轴成弓形出土，子叶带种壳留土或出土。经过 15～30 天，子叶脱出种壳，初生叶展出。子叶 2，长 5.5cm。初生叶 4，长 2.4cm。发芽情况见图 2。目前尚无大田育苗经验，多用催芽后点播。一般培育 2 年出圃。苗期需遮荫。

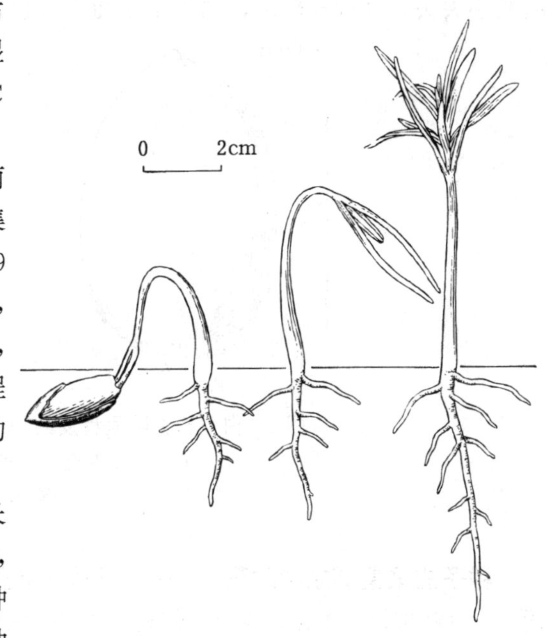

图 2　云南穗花杉种子萌发后第 15、22、32 天的幼苗生长情况
（黄应钦绘）

（李达孝）

白 豆 杉

Pseudotaxus chienii (Cheng) Cheng

（红豆杉科 Taxaceae）

生长习性、分布和用途 白豆杉是第三纪残遗于我国的单种属植物，由于生境恶化，分布区逐渐萎缩，资源日益枯竭，在《中国植物红皮书》中被列为稀有种。常绿灌木或小乔木，高 4～7m。分布星散，个体稀少，幼年生长缓慢。耐荫，喜温凉阴湿、云雾重、光照弱的环境。也耐一定的干旱和较强的光照，但生长萎缩，树干弯曲。侧根发达，岩缝中也能扎根，但形成丛生灌木。喜酸性（pH4.2～5.0）而有机质丰富肥沃的山地黄壤。分布于浙、赣、湘、粤、桂等亚热带海拔 740～1 500m 的陡坡深谷或悬岩上，散生于常绿落叶阔叶混交林下。木材纹理均匀，结构细致，可用于美工和细木工。

开花结实 球花单性，雌雄异株。球花单生叶腋。雄球花近球形，基部有 4 对交叉对生的苞片。雄蕊盾形，6～12，交叉对生，各有 4～6 辐射排列的花药，花丝短。雌球花基部有 7 对交叉对生的苞片，花轴顶端的苞腋有 1 直立胚珠着生于圆盘状的珠托上。受精后珠托发育成杯状、肉质、白色的假种皮。由于受孕率低，所以雌株结实常不稳定。花期 3 月下旬～5 月，种子 9 月下旬～10 月上旬成熟。种子卵圆形，长 5～8mm，径 4～5mm，上部微扁，顶端有突起的小尖头。外种皮较硬，内种皮薄而带红色，胚乳丰富，子叶 2（图 1）。

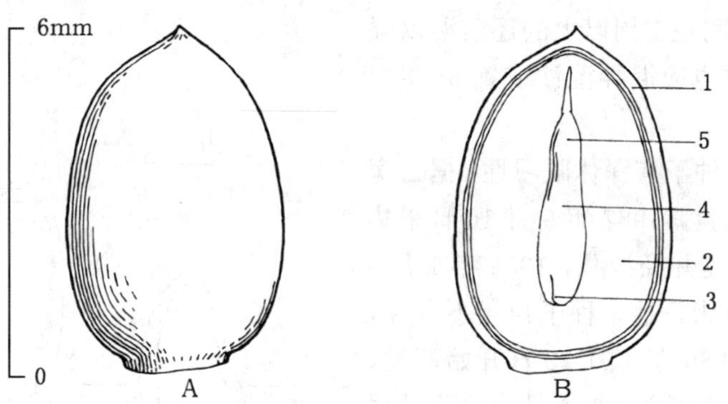

图 1 白豆杉去除假种皮的种子外形（A）及其纵切面（B）

1. 种皮 2. 胚乳 3. 子叶 4. 胚轴 5. 胚根

（童军平绘）

种子的采集调制和贮藏 9～10 月待杯状假种皮呈肉质、白色而有光泽时，即可及时采摘。采后略堆沤、搓揉，再洗去假种皮并去除瘪粒和杂质，摊晾于通风处阴干，不宜曝晒。曾经对 1996 年和 1997 年采自浙江龙泉和衢县的少量种子做过测定，出种率为 50%～56%，千粒重 38～48g，每千克纯净种子 20 800～26 300 粒（史晓华，未刊稿）。供翌春播种用的种子要混湿沙贮藏。

发芽和播种 种子有休眠习性，需隔年发芽。据观察，成熟的种子发育完全，种皮有良好的透水性，但种子吸胀后仍长期不能萌发，即使剥去种皮也不见萌发。1996 年 10 月本文作者将采自浙江凤阳山的白豆杉种子用浓硫酸处理 30 分钟，同年 12 月 11 日将经过酸蚀的种子用拌有赤霉素溶液的湿沙在杭州室外条件下层积，赤霉素溶液分为 250μg/g、500μg/g 和 1 000μg/g3种，1997 年 2 月 17 日结束层积，但盆播后 10 个月仍未见发芽迹象（史晓华，未刊稿）。这似乎表明，白豆杉种子的休眠可能与胚的生理后熟有关，需要在湿润和较低的温度下经历较长的时间才能解除休眠。

种子随采随播或低温层积后翌春播种。种子数量不多，以盆播为宜，播后覆土 0.8～1.0cm，盖草，幼苗出土后应及时揭草并遮荫。出土萌发。亦可扦插繁殖，成活率在 85%以上。

（黄鹏成、史晓华）

红豆杉属

Taxus L.

（红豆杉科　Taxaceae）

生长习性、分布和用途　本属约 11 种，我国有 4 种 1 变种，本文描述 1 种 1 变种。常绿乔木，高 20m，胸径 40～100cm。喜湿润凉爽的气候环境，适生于湿润疏松、排水良好的土壤。木材细密，坚韧耐用。它们的名称、生长、分布及用途见表 1。

表 1　红豆杉属树种的名称、生长、分布和用途

中　名	学　名	树　高 (m)	胸　径 (cm)	分　布	用　途	供　稿
南方红豆杉	*T. chinensis*(Pilger)Rehd. var. *mairei* (Lemee et Levl.)Cheng et L. K. Fu	15～20	100	陕、甘、西南、中南、华东。苏、沪有栽培	观赏、材用	1 003
东北红豆杉	*T. cuspidata* Sieb. et Zucc.	20	40～100	东北。朝鲜半岛、俄罗斯、日本	材用、油脂	107

开花结实　10 余年生开始开花结实，正常结实期在 20 年生以后。结实有大小年现象。球花单性，雌雄异株。球花单生叶腋。雄球花球形，有梗。雄蕊 6～14，盾状。花药 4～9，辐射排列。雌球花近无梗，有一顶生胚珠，基部托以圆盘状珠托，受精后珠托发育成肉质、杯状、红色的假种皮。据在南京对南方红豆杉观察，雌雄球花同期开放，2 月中下旬～3 月上旬为授粉期，10 月下旬种子成熟。另据黑龙江观察，东北红豆杉 5～6 月开花，9～10 月种子成熟，11 月种子脱落。种子坚果状，成熟时肉质假种皮红色。南方红豆杉种子宽倒卵形，上宽，下端微尖，略扁，长约 7～8mm，宽约 5mm，外种皮骨质，浅褐色，表面光滑，种脐椭圆形。内

种皮厚膜质。胚乳含油。胚小，柱形，位于中央，长约1.5～2mm。东北红豆杉的种子卵圆形，稍扁，有2～3条棱线，长约6mm，径5mm，紫红色，上部具3～4钝脊，顶端有小钝尖头。种脐三角形或四方形。有胚乳（图1）胚长2～3mm。

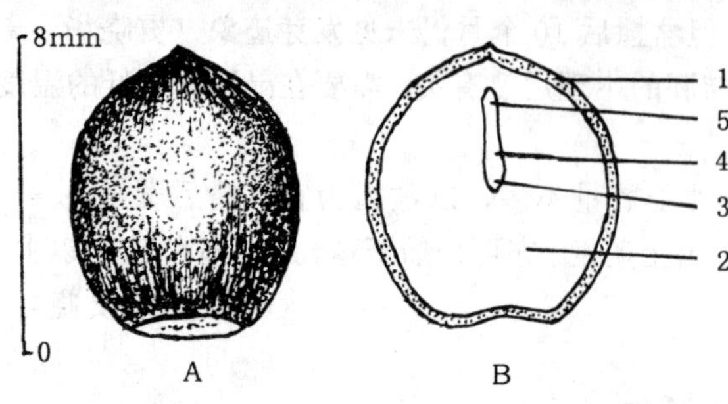

图1 南方红豆杉种子外形（A）及其纵切面（B）
1. 种皮 2. 胚乳 3. 子叶 4. 胚轴 5. 胚根
（田恒德绘）

种子的采收调制和贮藏 种子成熟后易遭鸟食，应及时采收。当假种皮呈红色时，摇动树枝或用竹竿轻轻敲击，在地面捡拾，忌剪断树枝，以免影响次年结实。采后浸水3～5小时，搓擦并淘去假种皮，晾干即得纯净种子。出种率为80%～95%。南方红豆杉种子千粒重约80g，每千克纯净种子约12 500粒。东北红豆杉千粒重约45g，每千克纯净种子约22 000粒。调制后的种子不宜裸露存放，贮藏或运输时需混拌湿沙。混沙贮藏的时间一般为8个月至一年半。

发芽和播种 种子有深休眠习性，播种前需要层积处理。秋季采得的东北红豆杉种子，调制后应立即混沙层积，翌春取出种子，再用冷热交替等措施催芽处理，待大约1/3的种子开始萌动时即可播种。哈尔滨地区适宜的播种期为4月下旬～5月上旬，发芽率可达60%左右。据谭一凡（1991）观察，南方红豆杉种子达到形态成熟时，多数胚尚未发育完全。湖南南岳树木园将11月采收经过调制的南方红豆杉种子混沙后在地窖内层积14个月，至第3年2月取出播种，据说效果较好。江苏省植物研究所曾将1983年11月采收调制的南方红豆杉种子混沙装入塑料袋，埋置室外树荫下，至1985年2月25日开始萌发，3月中旬为发芽盛期，4月1日发芽终止，发芽率为81.8%，出苗率为75%～80%。出土萌发。子叶2。见图2。

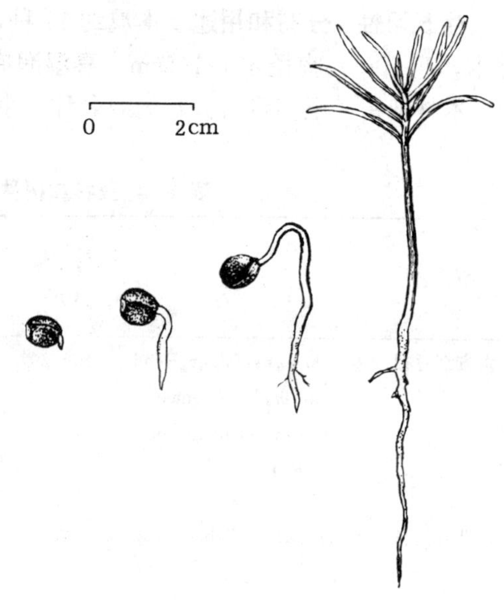

图2 南方红豆杉种子萌发后第1、5、10、40天的幼苗生长情况
（田恒德仿《主要树木种苗图谱》）

条播。经过层积的南方红豆杉种子每平方米苗床播种8～12g，东北红豆杉每平方米床面播种约5～8g。苗期宜遮荫，一般育2～3年生苗出圃。可用扦插育苗，春季和夏季为适宜扦插期。

（王宏志）

榧 树 属

Torreya Arn.

（红豆杉科　Taxaceae）

生长习性、分布和用途　本属有7种，我国产4种，引入栽培1种，本文描述1种和1栽培品种。其中长叶榧树被《中国植物红皮书》列为渐危种。常绿乔木。喜温暖湿润环境。适生于排水良好、深厚肥沃的沙壤土或石灰质风化土，一般在酸性或微酸性土壤上生长。

木材结构细致，耐水湿，抗腐性强，是建筑和家具优良用材。香榧是著名的干果，营养丰富，具有特殊香味。这两个树种都可作庭园观赏树种。其名称、生长、分布和用途见表1。

表1　榧树属树种的名称、生长、分布及用途

中　名	学　名	树　高 (m)	胸　径 (cm)	分　布	用　途	供　稿
香榧	*T. grandis* Fort. cv. Merrillii	10～15	100	原种产皖南、苏南、浙、闽西北、赣、湘南、黔东北。品种产浙江诸暨和东阳	食用、材用	803
长叶榧树	*T. jackii* Chun	8～12	20～30	浙、闽	材用	903

开花结实　5～15年生开始结实，20～40年生进入正常结实年龄。结实大小年不明显。花单性，雌雄异株，稀同株。雄球花单生于叶腋，椭圆形或圆柱形，具短梗，雄蕊4～8轮，每轮4枚，花药4（稀3），向外一边排列。雌球花无梗，成对生于叶腋。胚珠1，直立，生于漏斗状珠托上。花期3月中旬～5月上旬。种子翌年秋季成熟，核果状，全包在肉质假种皮中。假种皮被白粉，含柠檬醛和芳香脂。两个树种的结实情况见表2（其中香榧，据马正山等，1983）。开花结实物候见表3。香榧雌雄异株而开花结实较迟，因此，植株性别的早期鉴定成为人们关注的问题。黄少甫和王雅琴等（1990）通过根尖或茎尖染色体的观察和核型分析，建立了香榧雌雄性别的核型模式。苏梦云（1987）企图利用生理生化指标鉴定香榧的性别，但未能取得理想结果。

表2　榧树属树种结实情况

树　种	观察地点	开始结实年龄（年）		开始进入正常结实的年龄（年）		经济寿命（年）
		实生苗	嫁接苗	实生苗	嫁接苗	
香榧	浙江诸暨	20	8～10	30～40	20	300～400
长叶榧树	福建武夷山	5～10	—	20～40	—	—

表3　榧树属树种的开花结实物候

树　种	观察地点	花　期	种子成熟期	种子散落期
香榧	浙江诸暨	4月中旬～5月上旬	翌年8月下旬～9月上旬	9月中旬
长叶榧树	浙江杭州	3月中旬～4月上旬	翌年10月中旬～下旬	11月上旬

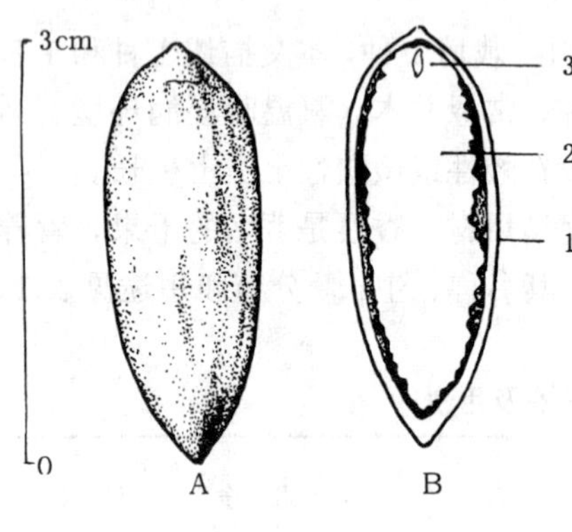

图1　香榧种子的外观和纵剖面
A. 种子外观　B. 纵剖面
1. 种壳　2. 胚乳　3. 胚
（童军平绘）

种壳木质，表面有纵向浅槽，基部尖或钝圆，顶端有小凸尖头。种脐椭圆形，两侧有2个小孔眼，膜质。香榧胚乳微内皱，长叶榧树胚乳向内深皱。胚乳肥大，胚小，皆为白色（图1）。种子的形态特征见表4。

种子的采集调制和贮藏　假种皮变为黄褐色或紫黑色，且大部分开裂，少数种子开始脱落时即应采集。可上树采摘，但不要伤及新枝和未成熟种子。也可以刈除树下杂草，任其自行落下收集。采回后堆沤，剥除假种皮，用清水冲洗后阴干，即得纯净种子。种子不耐干藏，需混湿沙层积。在层积期内应注意保持沙的湿度，不能干燥，越冬后于翌春播种。贮藏1年或1年以上的，需少量装于透气保湿的容器中，置1～5℃冷库内。这两个树种的出种率、种子净度和质量等数据见表5。

表4　榧树属树种种子的形态特征

树　种	未熟时颜色	成熟种子			去假种皮后的种子		
		形　状	大小（cm）	颜　色	形　状	大　小（cm）	颜色
香榧	青绿色	倒卵圆形或宽长圆形	长2.5～4.2 径1.5～2.0	黄绿色	长圆状倒卵形或圆柱形，种脐较小	长2.0～3.5 径1.2～1.9	淡棕褐色
长叶榧树	绿色	倒卵圆形	长2.0～3.0 径0.7～1.0	紫黑色	倒卵圆形，种脐较大	长1.6～2.2 径1.2～1.7	淡咖啡色

表5　榧树属树种的出种率和种子净度、质量的数据

树　种	出种率（%）	净　度（%）	千粒重（g）		每千克纯净种子粒数	
			一　般	变动范围	一　般	变动范围
香榧	23～25	95～99	2 500	1 880～3 000	400	330～530
长叶榧树	28～32	88～99	1 700	1 500～2 200	590	450～670

发芽和播种　种子休眠期长。这是因为种子成熟时，胚的分化还很不完善，需要1个后熟期。解除休眠的措施通常是在10～20℃条件下湿沙层积2个月。在层积的后期，当胚根长达0.5～1.5cm时拣出播种。条播，条距25cm，沟深1～1.5cm。播后浅覆土，盖草并保持湿润。冬季或早春播种，一般在4月下旬出苗，5月中下旬齐苗。这2个树种的播种期、播种量

和出苗率见表 6。幼苗期不耐高温，应遮荫。一般为 2 年出圃。

表 6　榧树属树种的播种期、播种量和出苗率

树　种	层积处理	播种期	播种量 (g/m^2)	出苗率 (%)	1 年生苗高 (cm)	每 $666m^2$ 的产苗量（万株）
香榧	10～20℃层积 2 个月	12 月下旬～翌年 3 月上旬	100～125	80～90	15～25	1.2
长叶榧树	室内常温层积 2 个月	11 月下旬	300	35～40	10	2.0

两个树种皆为留土萌发。子叶包于坚硬的种壳和胚乳内，子叶柄长 5～8mm。初生不育叶 10～20 片，三角状鳞形。发育叶线形，微弯成镰状，先端渐尖具尖头，上面平，无明显中脉，下面中脉两侧微凹，有气孔带。长叶榧树的幼苗生长情况见图 2。

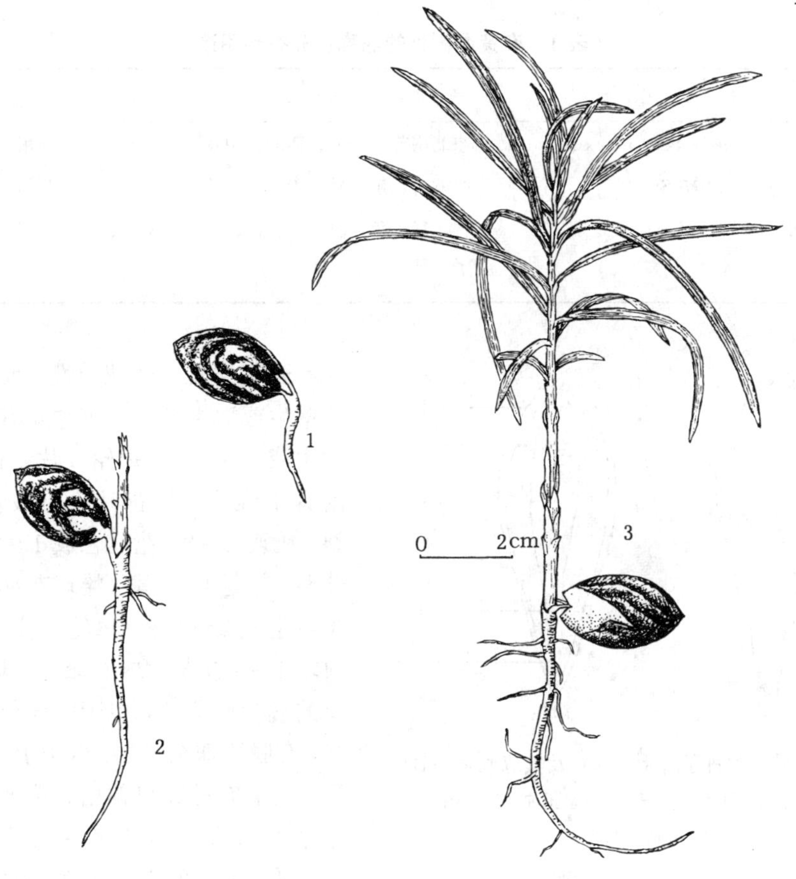

图 2　长叶榧树种子萌发后幼苗的生长情况

1. 子叶留土萌发　2. 示初生不育叶　3. 幼苗形成

（黄鹏成供稿，田恒德绘）

两个树种都能用无性繁殖。香榧嫁接成活率高。长叶榧树扦插繁殖成活率较高。

（翁尧富）

麻 黄 属

Ephedra Tourn ex L.

（**麻黄科 Ephedraceae**）

生长习性、分布和用途 本属约 40 种，广布于亚洲、美洲、欧洲东南部及非洲北部等干旱、荒漠地区。我国有 12 种 4 变种，除长江下游及珠江流域外，其它各地均有分布，以西北各地及云南、四川高山地带种类较多。灌木、亚灌木或草本状灌木。茎直立或匍匐，多分枝，小枝绿色，有节，对生或轮生。高度不一，最高可达 150cm，矮者仅有几厘米。本文只描述沙漠地带常见的 3 种，其名称、分布和用途见表 1。

表 1 麻黄属树种的名称、分布和用途

中 名	学 名	分 布	用 途
中麻黄	*E. intermedia* Schrenk ex Mey.	东北南部、华北、西北。中亚	药用、固沙、薪材
膜果麻黄	*E. przewalskii* Stapf	内蒙古、宁、甘、青、新。蒙古	药用、固沙、薪材
草麻黄（麻黄）	*E. sinica* Stapf	吉、辽、鲁、冀、晋、豫、陕、内蒙古、宁、甘	药用、固沙、薪材

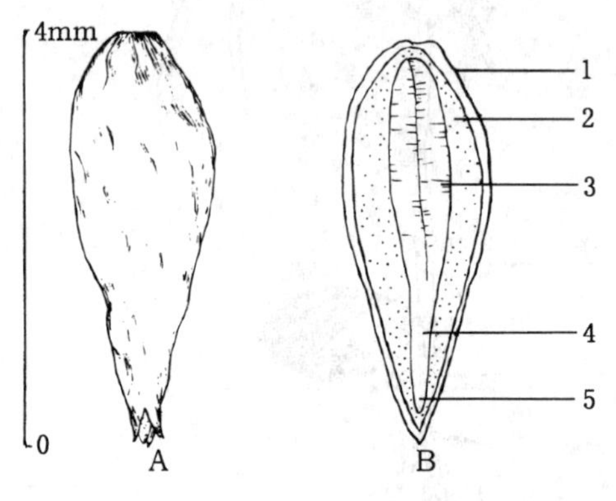

图 1 膜果麻黄种子外形（A）及其纵切面（B）
1. 种皮 2. 胚乳 3. 子叶 4. 胚轴 5. 胚根
（姚桂英绘）

开花结实 球花单性，雌雄异株。雄球花单生或数个丛生。雄球花有 2～8 对交叉对生或 2～8 轮（每轮 3 片）苞片，每苞片内具 1 雄花。雄花具膜质假花被，先端 2 裂。雄蕊 2～8，花丝合成 1～2 束或于先端分离，花药 1～3 室。雌球花亦具 2～8 对交叉对生或 2～8 轮（每轮 3 片）苞片，仅在顶端 1～3 枚苞片内有雌花。雌花具顶端开口的囊状假花被，包于胚珠外。胚珠 1，直立，有膜质珠被 1 层。珠被上部延长成珠被管，由假花被管口伸出，珠被管直或弯曲；花序上部的 4～6 枚苞片于种子成熟时变为肉质，呈红色或橘红色，膜果麻黄则为干膜质，淡褐色，成了外面另一层被覆物。假花被发育为革质假种皮。种子 1～3 粒，当年成熟。胚乳丰富，子叶 2 枚，出土萌发。目前麻黄属植物均为野生，人为破坏严重影响麻黄的自然繁衍。麻黄属 3 个树种开花结实习性见表 2。种子形态如图 1。

表 2　麻黄属树种开花结实习性

树　种	开始结实年龄（年）	丰年间隔期（年）	开花期	果　实成熟期	种子成熟特征	种　子散落期
中麻黄	3～5	1	5～6 月	7～8 月	雌球花成熟时肉质，红色，椭圆形、矩圆状卵圆形。种子藏在苞片内，2～3 粒，呈卵形或长卵形，长 5～6mm，径约 3mm	8～9 月
膜果麻黄	3	1	5～6 月	6～7 月	雌球花成熟时苞片增大成干膜质，淡褐色。种子（2）3 粒，包于膜质苞片内，长卵形，长约 4mm，径约 2～2.5mm，顶端尖，暗褐红色	7～8 月
草麻黄	—	—	4～5 月	6～7 月	雌球花成熟时肉质，红色，矩圆状卵形或近圆形。种子包于苞片内，通常 2 粒，不外露，或与苞片等长，三角状卵圆形或宽卵圆形，长 5～6mm，径 2.5～3.5mm，黑红色或灰褐色	7～8 月

种子的采集　麻黄属的种子成熟期在种内较为一致，苞片呈现成熟颜色时应及时采摘。也可连同枝条一起收割，运到晒场集中采摘，剩下的枝条可供药用。麻黄属的萌生能力很强，采割时只要不伤根系，对植株的生长影响不大。

种子的调制和贮藏　采摘下的浆果状假果应当及时放到水里揉搓，滤去肉质苞片，留下纯净种子。经过充分晾晒，当种子含水量在 8%以下时即可入库贮存。库房应通风、干燥，阴凉。在年气温变幅－29～38℃，年平均相对湿度 47%的甘肃民勤地区，经过上述处理的种子，贮存 4 年之后仍有 20%的发芽率。麻黄属 3 个树种的出种率及种子质量见表 3。

表 3　麻黄属树种的出种率及种子质量

树　种	出种率（%）	净　度（%）	千粒重（g）	每千克纯净种子粒数（万粒）
中麻黄	15	40～55	3.5～3.9	25.6～28.6
膜果麻黄	30	30～70	2.1～2.5	40～47.6
草麻黄	25	50～70	7.1～7.4	13.5～14.1

表 4　麻黄属树种种子发芽能力

树　种	发芽势（%）			发芽率（%）		
	计算天数	一般数值	变动范围	计算天数	一般数值	变动范围
中麻黄	3	28	15～40	6	30	15～45
膜果麻黄	3	66	60～70	6	76	68～80
草麻黄	3	52	33～70	6	58	34～82

注　测定条件：25℃恒温

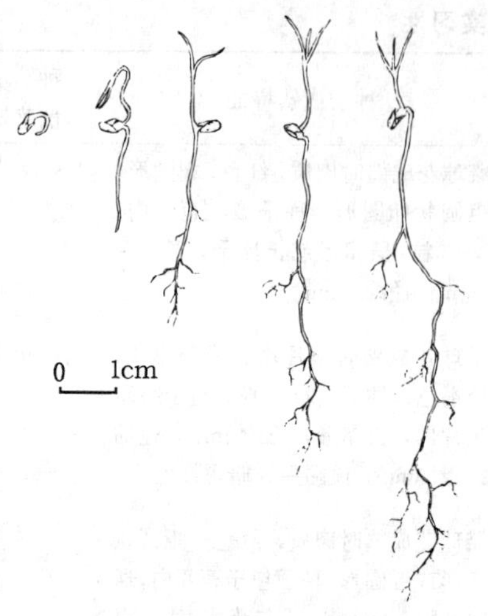

图 2 膜果麻黄种子萌发后第 2、4、7、15 天及以后的幼苗生长情况
（姚桂英绘）

发芽前的处理及发芽测定 麻黄属树种的种子没有休眠习性，无需采取特殊的促进发芽措施。由于种子主要采自野生母树，受自然条件影响较大，空粒及不饱满粒较多，有时高达 50%以上，所以发芽能力一般较低，需要采用水选法清除空粒种子。发芽能力及其测定条件见表 4。

播种 播种期在内蒙古、甘肃以 4 月中下旬为宜，每平方米播种 2.5～3.8g。单行式条播，行距20～25cm。覆土不能厚，一般 1～2cm，膜果麻黄不能超过 1cm。播后用地膜覆盖。4 月中旬播种，4 月下旬即可出苗。发芽时子叶出土，初生叶似棒状。当年出圃，苗高 15～25cm。幼苗形态如图 2。

（郭志中）

买 麻 藤

Gnetum montanum Markgr.

（买麻藤科 Gnetaceae）

生长习性、分布和用途 属于裸子植物的买麻藤科只有买麻藤属 1 属，约 30 余种，我国有 7 种，本文描述 1 种。常绿木质藤本，长达 12m，茎较纤细，节间膨大。喜肥沃湿润的酸性土壤，多生于热带常绿阔叶林中。分布于广东、广西、云南、海南。越南、老挝、泰国、缅甸和印度也有分布。茎皮纤维质地坚韧，可织麻袋、鱼网等，种子可炒食或榨油。

开花结实 6～8 年生始开花结实，正常结实期在 15 年以后。花单性，雌雄异株。雄球花穗单生或数穗组成顶生及腋生聚伞花序。雄球花穗圆柱形，具 13～17 轮环状总苞，各轮环状总苞排列紧密，不露花穗轴。每轮总苞有雄花25～45，排成两行。雄花具杯状肉质假花被，雄蕊 2，稀 1，花丝合生，药 1 室。雌球花穗单生或数穗组成聚伞状圆锥花序，每轮总苞有雌花 5～8，胚珠具两层珠被，并紧包于囊状假花被之中。内珠被的顶端延伸成珠被管，从假花被顶端开口伸出。外珠被分化为肉质外层和骨质内层，肉质外层与假花被合生并发育成假种皮。在广西南宁和那坡观察，花期 7 月，9 月上旬种子开始成熟，10 月上旬为成熟盛期。种子核果状，具橙红色肉质假种皮，长椭圆形，长 1.5～2cm，径 1～12cm。种皮灰白色，具纵槽

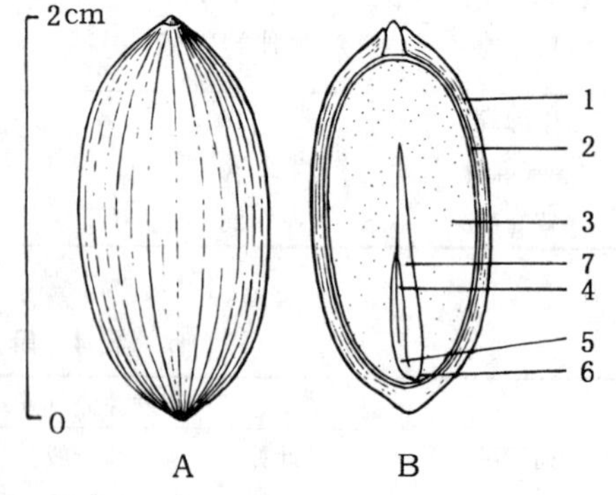

图 1 买麻藤种子的外形(A)及其纵切面(B)
1. 假种皮 2. 种皮 3. 胚乳 4. 子叶
5. 胚轴 6. 胚根 7. 吸器
（黄应钦绘）

纹。胚乳丰富，肉质。子叶 2（图 1）。

种子的采收调制和贮藏　直接采摘或敲打落地后在地面捡拾。采后堆放 1～2 日，待充分软熟后装入筐内置水中搓去假种皮，晾干即得种子。按鲜种质量计算的出种率约为 56%，种子的净度可达 95%以上。新鲜种子含水量 25%～30%，种子千粒重约 4 200g，变动范围3 500～5 000g。每千克有纯净种子约 240 粒，变动范围 200～300 粒。种子忌失水，不能日晒或裸露存放。1～2 天的短期运输可暂不去除假种皮，运到目的地后再行调制。长途运输则需调制后稍加晾干，混以湿沙包装运输。贮藏时需混湿沙。贮藏期为半年以内。

发芽和播种　种子有短期休眠习性。发芽时温度宜在 20℃以上。1990 年 10 月 26 日，广西林业科学研究所对当年新采的种子用发芽箱进行发芽测定，基质为湿沙，箱内温度经常保持 28℃，置床后 76 天于 1991 年 1 月 14 日开始发芽，盛期不明显，至 2 月 8 日发芽终止。发芽率为 90%。出土萌发，萌发时子叶被长的下胚轴带出土面。下胚轴基部一侧的突起物（吸器）则留在种子中继续从胚乳摄取营养。图 2 为短柄垂子买麻藤 *G. pendulum* C. Y. Cheng f. *intermedium* C. Y. Cheng 的种子和幼苗。

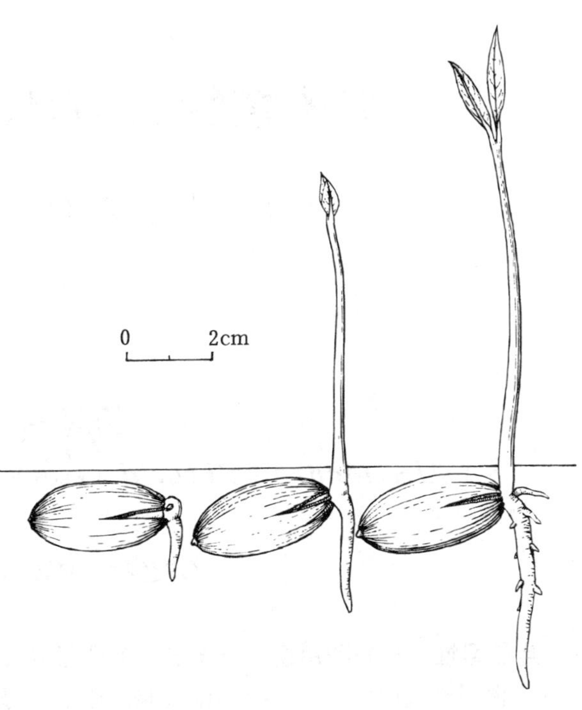

图 2　短柄垂子买麻藤种子发芽后第 5、7、10 天的幼苗生长情况
（黄应钦绘）

目前尚无批量育苗经验。在试验性育苗中，一般是将种子播于沙盘内催芽，子叶出现时再移至容器培育。苗期需稍加遮荫。幼苗生长至 30cm 左右时可出圃定植。

（曾　玲）

被子植物 ANGIOSPERMAE

双子叶植物 DICOTYLEDONEAE

长蕊木兰

Alcimandra cathcartii（Hook. f. et Thoms.）Dandy

（木兰科　Magnoliaceae）

生长习性、分布和用途　长蕊木兰属只有本文描述的 1 种。常绿乔木，高达 25m，胸径 85cm。深根性。耐荫。喜温凉潮湿气候，适生于海拔 1 800～2 300m 的亚热带高山地区，分布于藏、滇。印度、不丹、缅甸和越南也有。材质优良，花香而美，已濒临灭绝，被《中国植物红皮书》列为濒危种。

开花结实　10～15 年生开始开花结实，30 年生以后为正常结实期，结实有大小年现象，但不甚明显。结实量较大。花两性，单生枝顶。花被片 9，3 轮，纯白色，有透明油点。雄蕊长约 3.5cm，花药伸长，内向开裂，药隔伸长成舌状。雌蕊群具柄，圆柱形，长约 2cm。心皮多数，离生，胚珠 2～5。据 1984 年在云南腾冲观察，4 月中旬为始花期，4 月下旬为盛花期，5 月中旬为末花期，10 月中旬果实开始成熟，10 月下旬进入成熟盛期。小气候环境不同，果实成熟期可以相差一个月。聚合果长4～8.5 cm，成熟时黄色。蓇葖扁球形，径 7～9mm，有白色皮孔，内有种子 1～4 粒。外种皮外层黄色，肉质，内层黑色，骨质。种子扁圆形，长 8.5mm，宽 6.5mm，厚 4mm。胚小。胚乳丰富（图 1），富油质。

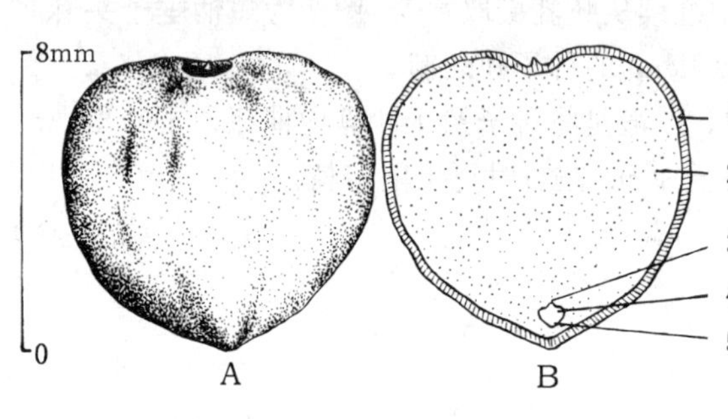

图 1　已去肉质外层的长蕊木兰种子外形（A）及其纵切面（B）

1. 种皮　2. 胚乳　3. 胚芽　4. 胚轴　5. 胚根

（黄应钦绘）

果实的采收调制和种子贮藏　果实成熟后易受鼠类等危害，应及时采集，可在果实成熟盛期用采种钩刀截断果梗或上树采摘聚合果。采回后放在室内摊晾，忌堆积，以免发热霉坏。当蓇葖微裂，能见黄色外种皮时，剥开蓇葖，取出种子，置水中搓擦漂淘，除去种皮外层的肉质，即为播种材料。种子净度 85%～

95%。千粒重100～160g，每千克有纯净种子6 200～10 000粒。种子含油脂，忌失水，不能曝晒，也不能干藏，运输或贮藏时均需混拌湿沙。混沙贮藏时间一般为半年左右。

发芽和播种　种子有休眠习性。混沙湿藏能够起到层积的作用。发芽时日平均气温宜在16℃以上。1989年云南林业科学研究所在室外荫棚下的苗床上进行过一次发芽测定：3月12日用湿沙层积过的种子播种，播后35天左右开始发芽，至发芽盛期的20天内有50%的种子萌发。整个发芽过程约90天，最终发芽率55%。出土萌发。胚根萌出后约15天，下胚轴成弓状将种粒拉出土面，子叶随后脱出种皮，10～15天初生叶展出，发芽情况见图2。

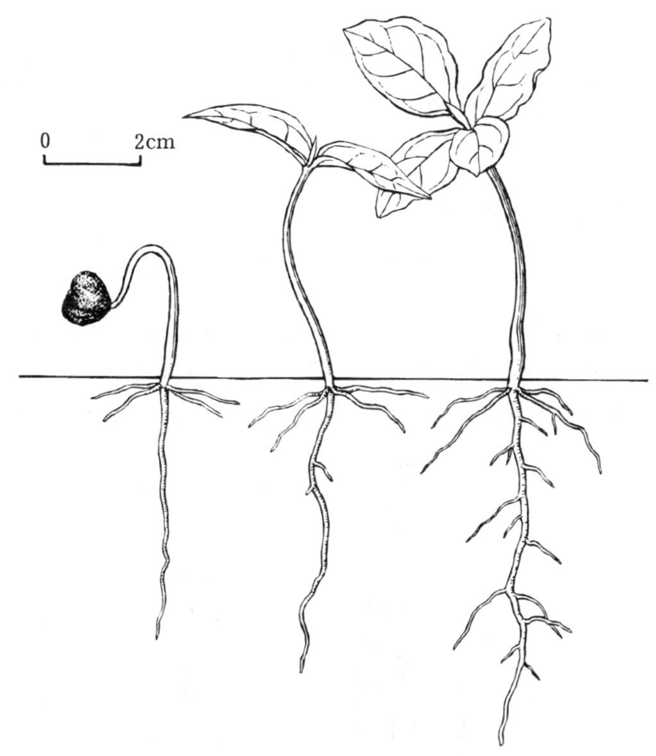

图2　长蕊木兰种子萌发后第15、28、40天的幼苗生长情况
（黄应钦绘）

条播。每平方米苗床播种20～25g，覆土约2cm。1年生苗可以出圃。也可用扦插繁殖。

（李达孝）

鹅掌楸属

Liriodendron L.

（木兰科　Magnoliaceae）

生长习性、分布和用途　本属仅有本文描述的2个种，为新生代以后残存的古老树种，一产我国，一产北美并在30年代引入我国，生长良好。南京林业大学已在60年代培育出这2个种的杂交种，有明显的杂交优势。落叶大乔木，速生，寿命长，喜湿润温暖避风环境。不耐干旱及水渍。在深厚肥沃的酸性土上生长良好。天然大树常遭砍伐，更新能力很弱，在《中国植物红皮书》中列为稀有种。鹅掌楸有东部和西部两大分布亚区。通过两大亚区种群表现的比较，贺善安和郝日明等（1996）探讨过导致鹅掌楸天然植株稀少的生态因素。叶形独特，花冠美观，树干通直，少有病虫，是优良的庭园绿化树种和行道树种，作为用材树种造林也很有前途。它们的名称、成年树高、分布和用途见表1。

表 1　鹅掌楸属树种的名称、成年树高、分布和用途

中　名	学名	树高（m）	分　布	用　途
鹅掌楸	*L. chinense* Sarg.	40	陕、川、黔、鄂、皖、赣、浙、闽。越南	材用、观赏、绿化
北美鹅掌楸	*L. tulipifera* L.	60	北美东南部。20 世纪 30 年代引入我国，已在长江中下游各省及云南栽培	材用、观赏、绿化

开花结实　15～20 年生开始正常结实，大小年不明显。花两性。花大，杯状，直径 4～6cm，单生枝顶，与叶同放或后叶开放。花被片 9，3 片 1 轮，近相等。雄蕊多数，分离，螺旋状排列于隆起的花托下部。花药条形，2 室，内向或侧向开裂。离心皮雌蕊多数，螺旋状排列于花托上部，最下部的不育。胚珠 2。聚合果纺锤形，长约 7～9cm，含 150 左右的翅状小坚果。小坚果顶端延伸成翅，木质，成熟时自聚合果的中轴脱落，中轴宿存。坚果内含种子 1～2 枚。种皮薄而干，附着于内果皮，胚乳丰富（图 1）。据在南京地区多年观察，花期为 4 月下旬 ～5 月下旬，当年 10 月果实成熟。

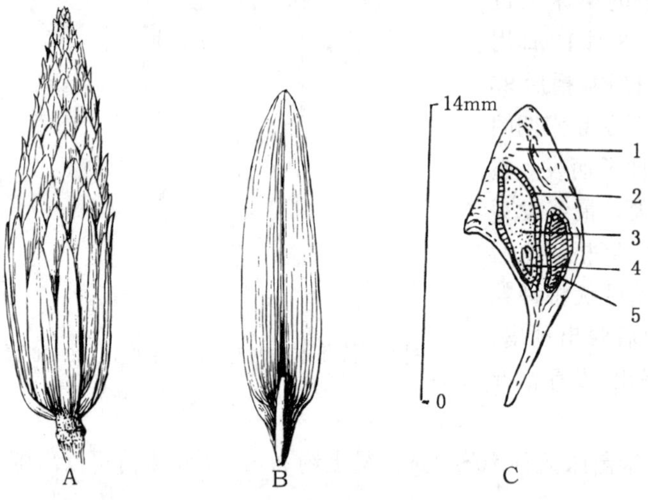

图 1　鹅掌楸的聚合果（A）和具翅坚果（B 放大）与北美鹅掌楸坚果的纵切面（C）
1. 果皮　2. 种皮　3. 胚乳　4. 胚　5. 败育种子
（史渭清绘，A、B. 仿《主要树木种苗图谱》，C. 仿《美国木本植物种子手册》）

鹅掌楸的有性生殖过程存在着严重障碍，能育的饱满种子极少。樊汝汶和叶建国等（1992）研究鹅掌楸的胚胎发育后报道，在南京自然条件下，饱满种子的比例不足 1%；控制授粉后 9 个聚合果中的饱满种子平均也只有 17.7%。樊汝汶和尹增芳等（1990）观察过鹅掌楸花芽分化的细胞学形态变化，认为结籽率低的原因是单花花期不遇、花粉对胚珠的比值低、可授粉期短。周坚和樊汝汶（1994）发现鹅掌楸的花粉管在花柱中遇到不亲和性反应。秦慧贞和李碧媛（1996）则从胚囊的败育分析了鹅掌楸的生殖障碍，并发现东部亚区种群胚囊败育的程度和比例高于西部亚区。目前提高饱满种子比例的措施是人工辅助授粉。

果实的采收调制和种子贮藏　聚合果成熟后，带翅的小坚果自中轴脱落，随风飘散，应在坚果已呈褐色而尚未飘散前及时采摘。一般是树冠上部 2/3 处的坚果比较饱满。剪摘下的聚合果先在室内摊放 7～20 天，或摊晒 2～3 天，聚合果即自行散开，扬去杂质后收入布袋干藏。生产上就以这种带翅的坚果作为播种材料，常不去翅，通称种子。千粒重 22～35g。干燥的种子在低温（2～4℃）下密闭贮藏，生命力可保存 3 年。翌年播种的种子可在采后低温层积处理促进发芽，效果明显。

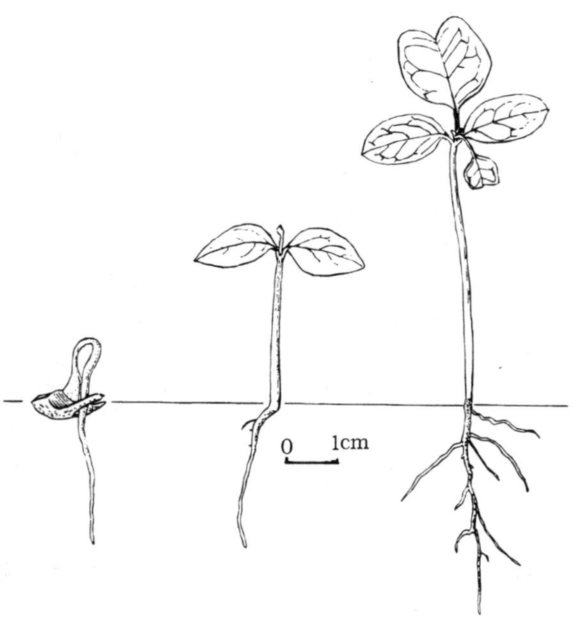

图 2　鹅掌楸萌发后第 1、10、30 天的幼苗生长情况
（史渭清仿《主要树木种苗图谱》）

发芽和播种　种子有休眠。播种前用 2℃恒温或 2～12℃变温处理 90 天左右可以解除休眠。经过层积处理的种子在春季 3 月上旬播种，20～30 天出苗。也可在秋季采后即播。苗圃育苗可用撒播或条播。条播行距为 20～30cm，覆土 0.6cm。每平方米播种 25～40g。苗期有根腐病（*Cylindrocladum soparium*），应注意防治。出土萌发。初生叶全缘，先端有凹缺，第 4 叶以上具正常叶形（图 2）。可以扦插繁殖，但需运用各种技术提高扦插成活率。目前只能由人工杂交制取杂种马褂木种子，工效不高且发芽仍低。采用嫁接可以长期利用杂种优势。嫁接成活率可达 90%（赵书喜，1991）。

（何泽瑛）

木　兰　属
Magnolia L.

（木兰科　Magnoliaceae）

生长习性、分布和用途　木兰属约 90 种，我国有 30 余种，本文描述 14 种和 1 亚种（表 1）。在《中国植物红皮书》中，天目木兰、黄山木兰、厚朴、凹叶厚朴和小花木兰列为渐危种，大叶木兰和宝华木兰列为濒危种。本文的这些种适生于肥沃湿润而排水良好的微酸性或中性土壤，不耐水淹。厚朴、凹叶厚朴、紫玉兰和玉兰等树种萌蘖力较强。分布于中国、日本、马来群岛、中美洲和北美洲。木兰属许多树种因花色美丽，叶色浓绿，果实红艳而成为著名的观赏树种，大都用于观赏栽培。滇藏木兰早年引入欧、美，被誉为珍贵的园艺树种。有些树种的皮、根、花入药，是名贵的中药材。有些树种因生长快，材质好而作为建筑、家具的优良用材。有些树种则因花和叶可提取香料，是香料工业的重要原料。木兰属植物种子含油丰富，也是有待开发的木本油料植物资源。本文所列的树种中，荷花玉兰原产美洲，日本辛夷原产日本，二乔木兰是19世纪初在法国用玉兰和紫玉兰育成的杂交种。这3个种传入我国后，均用于观赏栽培。

表1 木兰属树种的名称、分布和用途

中 名	学 名	分 布	用 途	供 稿
天目木兰	*M. amoena* Cheng	浙、苏、皖、闽	观赏、香料、材用	702
滇藏木兰	*M. campbellii* Hook. f. et Thoms.	藏、滇。不丹、尼泊尔、印度、缅甸	观赏、材用、药用、香料	615
黄山木兰	*M. cylindrica* Wils.	皖、浙、赣、闽	观赏、药用、材用	702
山玉兰	*M. delavayi* Franch.	滇、川、黔	观赏、材用、药用	614，902
玉 兰	*M. denudata* Desr.	皖、浙、赣、湘、粤。北京及黄河流域以南至西南各地普遍栽培	观赏、药用、材用、香料、油脂	702
荷花玉兰	*M. grandiflora* L.	原产美洲东南部。我国长江流域以南各地栽培	观赏、材用、香料、药用	702
大叶玉兰	*M. henryi* Dunn	滇、藏。缅甸、泰国、老挝	材用	615
日本辛夷	*M. kobus* DC.	原产日本和朝鲜半岛南部。青岛、南京、杭州栽培	观赏、材用	702
紫玉兰	*M. liliflora* Desr.	鄂、川、滇。长江流域及鲁、黔、桂栽培	药用、观赏、香料	702
馨香玉兰	*M. odoratissima* Law et Zhou	滇、粤	材用、香料、观赏	615
厚 朴	*M. officinalis* Rehd. et Wils.	甘、陕、川、黔、桂、鄂、湘、赣、浙、皖	药用、观赏、材用	702
凹叶厚朴	*M. officinalis* Rehd. et Wils. ssp. *biloba* (Rehd. et Wils.) Law	闽、浙、皖、赣、湘、桂、粤	药用、观赏、油脂、材用	702
天女花	*M. sieboldii* K. Koch	辽、吉、鲁、皖、赣、浙、桂。日本、朝鲜半岛	观赏、药用、香料、材用	107
二乔木兰	*M. soulangeana* Soul. -Bod.	我国各地栽培	观赏	702
宝华玉兰	*M. zenii* Cheng	江苏宝华山。长江流域以南各省引种	观赏	902

开花结实 开始开花结实年龄因树种、母树起源和生长条件不同而异，早的3～6年生，一般8～12年生便可开花结实，正常结实年龄往往在15年生以后。起源于嫁接苗的母树，结实比实生苗的早。陶金川、宗世贤等人（1988）报道，1980年播种在苗圃的天女花，1982年春季便已开花。紫玉兰花朵繁多而结实甚少。木兰属树种的结实一般无明显的丰年间隔期，但若花期气温较低且又阴雨连绵，也会严重影响种子产量。木兰属不少树种寿命长，江苏云台山有800年以上的玉兰大树，仍能连年开花。木兰属树种的开花结实年龄和丰年间隔期见表2。

花两性，大，单生枝顶。花色各异，多具芳香。花被片9～21，每轮3～5片，近相等，有时外轮花被片小，带绿色，呈萼片状。雄蕊多数，花丝扁平。雌蕊群和雄蕊群相连接。心皮多数，且分离，螺旋状着生于一伸长之花托上。胚珠2。花期一般在2～5月。聚合果8～10月成熟。木兰属树种的开花结实物候期见表3。

表 2　木兰属树种成年时的树高、开始结实年龄和丰年间隔期

树　种	成年时树高（m）	开始结实年龄（年）	开始正常结实年龄（年）	丰年间隔期
天目木兰	12～15	10～15	—	不甚明显
滇藏木兰	30	10	20～35	不甚明显
黄山木兰	8～10（20）	8～10	—	0
山玉兰	13～15	3～5	7	不甚明显
玉　兰	20	8～12	—	0
荷花玉兰	30	8～12	—	0
大叶玉兰	20	8～10	20～30	不甚明显
日本辛夷	20	嫁接苗 4～6	—	0
紫玉兰	3～5	6～8，嫁接苗 3	—	—
馨香玉兰	6	5	15～30	不明
厚　朴	20	8～15	—	1
凹叶厚朴	15	11～15	—	不甚明显
天女花	10	5	—	不明
二乔木兰	6～10	10～12，嫁接苗 3～4	—	0
宝华玉兰	7	5～8	10	不甚明显

表 3　木兰属树种的开花结实物候

树　种	观察地点和年份	开花期	果实成熟期	种子开始散落期
天目木兰	浙江杭州			
	1977～1980	2 月下旬～3 月中旬	9 月	—
	1989	—	8 月下旬	9 月上旬
滇藏木兰	云南昆明			
	—	3 月上旬～4 月上旬	10 月	—
黄山木兰	浙江杭州			
	—	3 月中下旬	9 月上旬	9 月中旬
山 玉 兰	云南昆明			
	1986～1988	5～7 月	9 月下旬或 10 月上旬	10 月中旬
玉　兰	浙江杭州			
	1977～1980	3 月中旬～4 月上旬	9 月	—
	1989	—	—	9 月中旬
	陕西西安			
	1977	3 月中旬～4 月上旬	9 月中旬	—
	云南昆明			
	1975	1 月上旬～2 月中旬	8 月中旬	—
荷花玉兰	浙江杭州			
	1977～1980	5 月下旬～6 月下旬	9～10 月	—
	1989	—	9 月中旬	10 月上旬
大叶玉兰	湖南常德			
	1974	5 月上旬～6 月上旬	9 月中下旬	9 月下旬
	—	4 月～5 月上旬	8～9 月	—

（续）

树　种	观察地点 和年份	开花期	果实成熟期 散落期	种子开始
日本辛夷	浙江杭州			
	1978	2月下旬～4月上旬	9月中下旬	10月上旬
紫玉兰	浙江杭州			
	1982～1983	3月下旬～4月下旬	9月上中旬	9月下旬
	安徽霍山			
	1967	4月上旬～5月上旬	9月中旬	9月下旬
馨香玉兰	云南昆明			
	—	盛花期5月	9月上旬～10月上旬	—
厚朴	广西桂林			
	—	4月中旬～5月中旬	10月下旬	—
凹叶厚朴	浙江杭州			
	1977～1980	4月中旬～5月上旬	9～10月	—
	1989	—	9月中旬	9月下旬
	安徽歙县			
	1978	5月上中旬	9月下旬	—
天女花	吉林集安			
	—	5月下旬～6月中旬	9月中旬	10月
	浙江杭州			
	—	5月中下旬	9月上旬	9月中旬
二乔木兰	浙江杭州			
	—	3月上旬～4月上旬	9月上旬	9月中旬
宝华玉兰	—	4月中旬～5月下旬	8月中旬～9月下旬	9月下旬

果实由多数蓇葖果聚合而成。每个蓇葖含种子1～2粒，沿背缝线开裂。外种皮外层红色，肉质，含油分，内层坚硬。内种皮薄，膜质。胚乳丰富，胚较小，深埋于胚乳之中（图1）。珠柄有细丝（螺纹导管）与胎座相连，种子靠这根富有弹性的细丝悬挂在开裂的蓇葖上一段时间，有招引鸟类之效。果实和种子的形态特征详见表4。

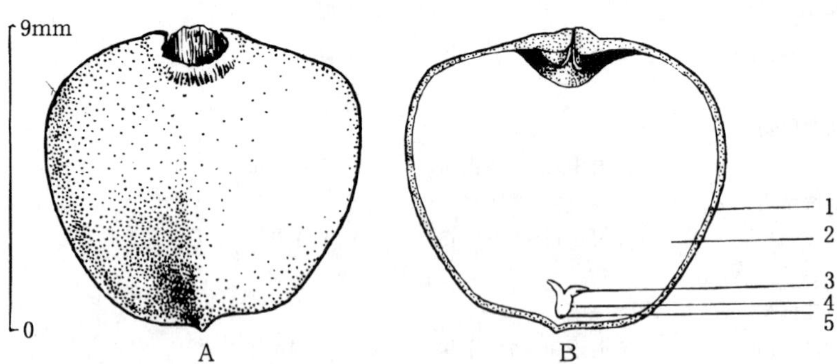

图1　除去外种皮肉质外层的玉兰种子外形（A）及其纵切面（B）

1. 外种皮的内层　2. 胚乳　3. 子叶　4. 胚轴　5. 胚根

（童军平绘）

表 4　木兰属树种果实和种子的形态特征

树　种	未熟果颜色	成熟果实			种子(已去外种皮肉质外层)		
		形　状	大小(cm)	颜　色	形　状	大小(cm)	颜　色
天目木兰	绿色	圆柱形,常弯曲。蓇葖扁球形,有瘤状小突起	长 7.0～11.0,径 2.0～2.5	褐红色	宽倒卵形或宽三角状倒卵形。宿存花柱略突出	长 0.6～0.75 宽 0.6～0.85	黑色或棕褐色
滇藏木兰	绿色	圆柱形。蓇葖紧贴	长 11.0～20.0	褐色,有光泽	宽倒卵形	长 0.9～1.0 宽 0.8	黑色
黄山木兰	绿色	圆柱形。蓇葖平滑	长 7.0～11.0 径 3.0～4.0	暗紫红色	三角状倒卵形或三角状矩圆形	长 0.65～0.90 宽 0.75～1.0	黑褐色或棕褐色
山玉兰	绿色	卵状长圆形。蓇葖被细柔毛,先端喙外弯	长 8.0～15.0 径 3.0～6.0	褐色	三角状倒卵形或三角扁圆形,表面粗糙	长 0.7～1.0 宽 0.6～0.9	黑褐色或棕褐色
玉兰	绿色	圆柱形	长 10.0～18 径 3.5～5.0	暗紫红色或红褐色	宽倒卵形或倒卵形,微扁	长 0.7～1.2 宽 0.8～1.0	黑色或棕黑色
荷花玉兰	黄绿色	圆柱状长圆形或卵形,密被褐色或灰黄色绒毛。蓇葖先端具长喙	长 6.0～10.0 径 4.0～6.0	淡紫红色	椭圆状倒卵形或矩圆状倒卵形,侧扁	长 0.7～1.1 宽 0.5～0.7	乳黄色
大叶玉兰	绿色	卵状椭圆形。蓇葖先端具短尖的喙	长 12.0～16.0 径 3.0～5.0	—	长椭圆形	长 1.0 宽 0.7	黑色
日本辛夷	绿色	圆柱形,常弯曲。蓇葖具白色皮孔	长 6.0～11.0 径 2.0～2.5	暗红色	矩圆形或宽倒卵形,宿存花柱突出,中间有一小孔	长 0.65～0.85 宽 0.8～1.0	棕褐色或棕黑色,有光泽
紫玉兰	黄绿色	圆柱形,果皮粗糙	长 4.0～10.0	淡褐色	半球形或宽倒卵形,背面隆起	长 0.6～0.9 宽 0.6～0.85	棕褐色
馨香玉兰	绿色	长椭圆形。蓇葖具短喙	长 3.8～7.0 径 2.0～2.5	黄绿色	横椭圆形	长 0.6 宽 1.5	黑色或淡黄色
厚　朴	青绿色	圆柱状卵形,基部宽圆。蓇葖具 2～3mm 的喙	长 9.0～15.0 径约 6.0	棕褐色	三角状倒卵形,表面粗糙	长约 1.0	黑色或黑褐色
凹叶厚朴	绿色	圆柱状卵形	长 8.0～11.0 径约 4.0	褐红色	形状不规则,多为倒卵形或半球状倒卵形,表面粗糙	长 0.8～1.0 宽 0.6～0.8	棕黑色或黑褐色
天女花	绿色	卵形或长圆形	长 5.0～7.0	红色	宽倒卵形或三角状倒卵形	长 0.45～0.6 宽 0.55～0.8	黑色
二乔木兰	绿色	圆柱形,常弯曲,果皮糙,具白色皮孔	长 7.0～12.0 径 3.0～4.0	暗紫红色	宽倒卵形或倒卵形,微扁	长 0.65～1.0 宽 0.7～1.2	棕黑色或黑色
宝华玉兰	绿色	圆柱形。蓇葖近圆形,有疣点状突起	长 5.0～7.0	绿褐色	倒卵形或矩圆形	长 0.7～0.9 宽 0.6～0.8	棕褐色或褐黑色

果实的采收调制和种子贮藏　当果实呈现成熟颜色，并有少数蓇葖开裂时，就可及时从树上采摘，也可以在树下用高枝剪剪取或用竹竿击落果实。采集后摊晒 2～3 天，待蓇葖全部开裂，取出种子。也可将果实摊于室内干燥，待蓇葖开裂，脱出种子后用清水或 0.3%～0.5%

的碱水浸泡 2～3 天（史晓华，1982），搓洗并漂净外种皮的肉质外层，晾干，即得播种材料，通称种子。种子宜摊晾阴干，切忌曝晒，并应经常翻动，使种子均匀干燥。木兰属树种的出籽率和有关种子质量的数据详见表 5。

表 5 木兰属树种的出籽率和有关种子质量的数据

树 种	出籽率（%）	净度（%）	千粒重（g）		每千克纯净种子粒数（粒）	
			一般	变动范围	一般	变动范围
天目木兰	7～11	86～96	110	90～140	9 100	7 100～11 000
滇藏木兰	20～50	85～95	200	150～230	5 000	4 300～6 600
黄山木兰	4～10	90～98	150	130～172	6 600	5 800～7 700
山玉兰	2～9	90～98	145	135～160	6 900	6 200～7 400
玉兰	9～11	95～98	150	130～167	6 600	6 000～7 700
荷花玉兰	2～3	94～97	73	66～86	13 700	11 600～15 100
大叶玉兰	12～16	70～85	150	130～160	6 600	6 200～7 700
日本辛夷	8～9	94～98	170	165～185	5 900	5 400～6 000
紫玉兰	10～12	90～98	120	106～130	8 300	7 700～9 400
馨香玉兰	29	75～85	160	—	6 200	—
厚 朴	5	98	310	—	3 200	—
凹叶厚朴	3～4	90～98	170	150～210	5 900	4 700～6 700
天女花	—	80	64	—	15 600	—
二乔木兰	5～8	95～98	150	130～168	6 500	5 900～7 700
宝华玉兰	5	94～98	—	136～142	—	7 000～7 300

短期贮藏的种子可用塑料袋密封后放于阴凉的室内。玉兰种子这样贮藏至翌春，很少丧失生活力，但生产上多用混沙湿藏。史晓华（1986）认为，贮藏用的木兰属种子要适度干燥，含水量以 12%～14%为宜；沙子要清洗、曝晒，使用前再用 0.2%高锰酸钾溶液消毒；沙子的湿度以手握紧时沙能成团，放松时能慢慢散开为度，太湿种子容易腐烂，太干种子易丧失生活力；贮藏期间 1 周喷水 1 次，半月换沙 1 次。

木兰属种子既可除去外种皮的肉质外层后贮藏，也可带着干燥的肉质外层贮藏。据《美国木本植物种子手册》介绍，无论用哪种方法，只要使用密封容器，并置于 0～5℃温度中，木兰属的种子就可贮存几年而很少丧失生活力。种子能耐短时间干燥运输，但运到目的地后要立即用湿沙层积。长途运输时要与保湿材料混装，并适当通气。

发芽前的处理和发芽测定 表 1 所列的木兰属树种，除山玉兰种子无明显的休眠现象外，一般都有休眠习性，需经 2～5 个月的层积处理破除休眠。孙昌高和徐秀瑛等人（1987）描述过玉兰胚在层积处理期间的形态变化。当然，秋播本身便是天然的层积处理。据史晓华（1987）报道，玉兰种子用赤霉素（GA）处理有打破休眠的效果。尤其是层积前用 GA 处理，能缩短层积时间，加快萌发速度，比单纯层积的处理效果好。

经过适当层积处理的木兰属种子，在常温下置于常用的发芽基质（沙）中就能顺利萌发，发芽能力因树种而异。发芽测定时，除少数树种要求变温外，多数为 20℃或 25℃恒温（表 6）。玉兰种子萌发对光照不敏感（史晓华，1987）。

表 6　木兰属树种种子的发芽能力及其测定条件[①]

树　种	预 处 理[②]	温度（℃）		发芽势（%）			发芽率（%）		
		昼	夜	计算天数	一般数值	变动范围	计算天数	一般数值	变动范围
天目木兰	室温层积 84 天	20	20	10	47	38～52	20	82	80～85
黄山木兰	室温层积 113 天	25	25	6	35	30～40	25	80	70～86
山玉兰	未处理	25	25	—	—	—	15	75	68～83
玉　兰	室温层积 56 天	25	25	11	43	38～44	25	88	85～94
	GA100μg/g 浸种 24 小时后室温层积 14 天	25	25	12	45	40～46	25	84	82～86
荷花玉兰	室温层积 120 天	20	20	10	50	42～60	20	85	80～96
日本辛夷	室温层积 56 天	30	15	18	61	56～64	25	90	80～96
紫玉兰[③]	室温层积 120 天	25	25	11	64	—	17	92	—
凹叶厚朴	室温层积 70 天	30	20	13	30	22～38	25	93	90～98
天女花[③]	室温层积 120 天	30	20	11	38	—	17	95	—
二乔木兰	室温层积 113 天	20	20	9	38	21～52	20	85	74～95
宝华玉兰	未处理	25	25	12	17	13～20	23	25	21～30

① 发芽基质为沙，测定时均无光照

② 室温层积的温度变动在 3～15℃

③ 一份样品测定的数据

播种　木兰属种子秋播或春播均可，多数采用越冬层积后春播，特别是鼠害严重的地区。条播，行距 20～30cm，覆土 2～3cm，并盖草。层积后春播的一般 35～50 天萌发。播种量因树种而异（表 7）。

表 7　木兰属树种育苗技术措施

树　种	预 处 理[①]	播种期	播种量（g/m^2）	出苗期	出苗率（%）	当年苗高（cm）
天目木兰	室温层积 150 天	3 月中旬	18～25	4 月中旬	40～70	40～60
滇藏木兰	层积处理[②]	2 月下旬	—	5 月上旬	35～50	5～9
黄山木兰	室温层积 150 天	3 月中旬	25～38[⑥]	4 月下旬	50～80	40～70
山玉兰	室温层积 150 天	3 月上旬	25～40	4 月下旬	60～70	—
玉兰	室温层积 150 天	3 月上旬	38～50	4 月中旬	75～90	30～50
荷花玉兰	室温层积 135 天	3 月上旬	18～25	4 月下旬	47～65	7～10
大叶玉兰	未处理[③]	8 月	—	11 月中旬	50～60	27
日本辛夷	室温层积 140 天	3 月中旬	38～50	5 月上旬	46～85	40～60
紫玉兰	室温层积 130 天	3 月下旬	—	4 月下旬	—	40～55
馨香玉兰	层积处理[④]	1 月中旬	—	3 月下旬	43～50	—
厚　朴	层积处理	2 月上旬至 3 月上旬	30～38	—	50～80	35
凹叶厚朴	室温层积 140 天	3 月下旬	38～50	4 月下旬	45～86	30～40
天女花	层积处理[⑤]	5 月上旬	25	5 月下旬	—	20～30
二乔木兰	室温层积 150 天	3 月下旬	38～50	4 月下旬	70～80	25～50
宝华玉兰	层积处理[④]	3 月上旬	25～38	4 月上旬	13～18	—

① 室温层积的温度变动在 3～15℃

② 种子于 1986 年 10 月 17 日在云南云龙采集后，湿沙层积至播种

③ 随采随播，在云南昆明室外沙盘中测试

④ 种子调制后湿沙层积至播种

⑤ 种子调制后在 6～20℃室温下层积至 4 月，后转入 18～30℃下层积至播种

⑥ 据周家骏、高林（1985）

出土萌发。子叶2枚，出土后10～15天展现初生叶。玉兰子叶为椭圆形或卵状椭圆形，全缘，微呈波状，柄及子叶下面中脉基部有红色斑点。初生叶互生，椭圆形，全缘，基部楔形，叶脉在上面凹下，背面凸出。下胚轴长4.2～6.5cm，紫红色。主根伸出后在根颈下轮生4条侧根（图2）。木兰属树种除播种繁殖外，还可行嫁接、压条、扦插繁殖。萌蘖力强的树种还可分蘖繁殖和萌芽更新。

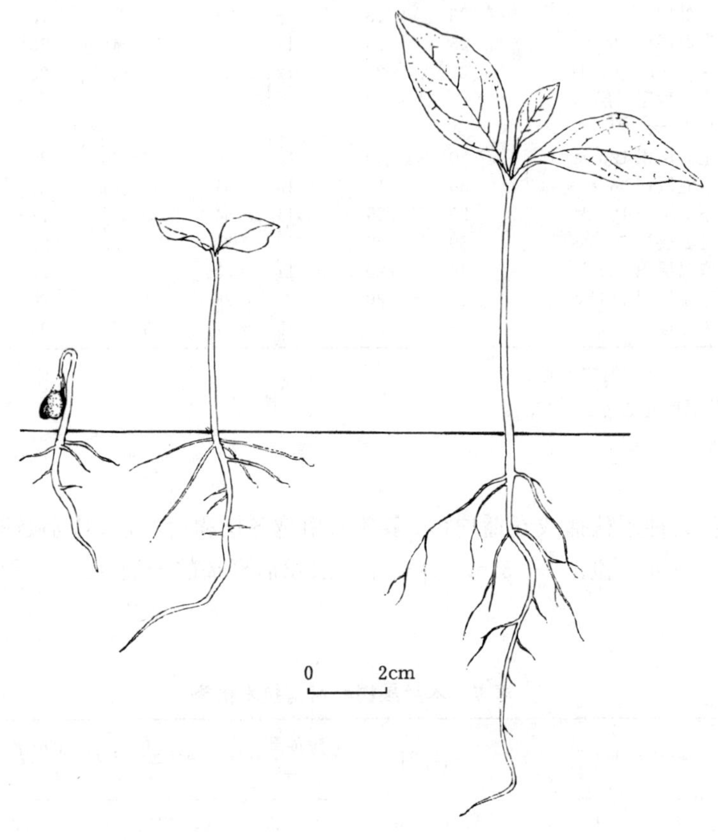

图2 玉兰种子萌发后4、10、25天幼苗的生长情况

（童军平绘）

（史晓华）

木 莲 属
Manglietia Blume

（木兰科 Magnoliaceae）

生长习性、分布和用途 木莲属约35种，我国产20余种，本文描述13种和1引进种。其中香木莲、大果木莲、红花木莲和大叶木莲已列入《中国植物红皮书》，红花木莲为渐危种，其他3种为濒危种。常绿乔木，有的树种树高可达40m，胸径1.1m。喜温暖湿润气候，忌干燥、瘠薄与排水不良。树体高大，树姿雄伟。花大，着生枝顶，形如莲花，香气浓郁，是重要

的香料及园林绿化观赏树种。木材通直，材质优良，易加工，不变形，少开裂，耐腐，为建筑、家具优良用材。木莲属树种的名称、生长、分布及用途见表1。

表1 木莲属树种的名称、成年树高、分布和用途

中名	学名	树高(m)	胸径(m)	分布	用途	供稿
香木莲	*M. aromatica* Dandy	25	0.5	滇、桂。越南	材用、观赏、香料	615
桂南木莲	*M. chingii* Dandy	20	0.6	桂、粤、湘	材用、观赏、香料、药用	703、615
川滇木莲	*M. duclouxii* Finet et Gagnep.	15	0.35	川、滇	材用、观赏、香料	615
木莲	*M. fordiana* Oliv.	40	1.1	皖、浙、赣、闽、粤、桂、滇、黔、湘	材用、观赏、香料、药用	703、615
滇桂木莲	*M. forrestii* W. W. Smith ex Dandy	25	0.5	滇、桂	材用、观赏	616
灰木莲	*M. glauca* Blume	30	1.0	原产越南和印度尼西亚。我国粤、桂、琼引入栽培	材用、观赏、药用	601
大果木莲	*M. grandis* Hu et Cheng	12	0.4	滇	材用、观赏、香料	615
海南木莲(绿楠)	*M. hainanensis* Dandy	30	1.0	琼	材用、观赏	601
中缅木莲	*M. hookeri* Cubitt. et Smith	25	0.5	滇、黔。缅甸	材用、观赏、香料、环保	615
红花木莲	*M. insignis* (Wall.) Blume	30	0.4	藏、滇、桂、黔、湘。印度、缅甸、越南、尼泊尔	材用、观赏、香料	罗仲春
马关木莲	*M. maguanica* Chang et B. L. Chen	18	0.6	滇	材用、观赏、香料	615
大叶木莲	*M. megaphylla* Hu et Cheng	25	0.7	滇、桂	材用、观赏、香料	615
卵果木莲	*M. ovoidea* Chang et B. L. Chen	20	0.4	滇	材用、观赏、香料	615
乳源木莲	*M. yuyunensis* Law	20	0.7	粤、湘、赣、皖、浙	材用、观赏、香料、药用	702

开花结实 5～15年生开始结实，20～30年生以后进入正常结实，有大小年现象，丰年间隔期通常2～3年。花两性，单生枝顶，芳香。花被9（6～13），排成3轮或2至数轮。雄蕊多数，花药条形，2室，内向纵裂。雌蕊群无柄，心皮多数，螺旋状排列，分离。每心皮具胚珠4或更多。胚珠倒生，具2层珠被。聚合果紧密，卵形或长椭圆状卵形。蓇葖木质，沿背缝及腹缝2瓣裂，或腹缝开裂，通常顶端具喙，宿存。果熟时由绿色转为淡红色至紫色，种子1至多枚。外种皮的外层肉质，红色，内层骨质，黑褐色。种子充分成熟时由胚柄发育的螺旋导管悬挂下垂，招引鸟类啄食。木莲属树种的开花结实物候期见表2。

表 2 木莲属树种的开花结实物候

树 种	观察地点 观察年份	花芽 出现期	开花期			果实成熟期		
			始 期	盛 期	末 期	始 期	盛 期	末 期
香 木 莲	云南马关 1988	3 月上旬	—	4 月下旬	5 月上旬	9 月上旬	9 月中旬	10 月上旬
桂南木莲	湖南新宁 1987	4 月中旬	5 月初	5 月上旬	5 月中旬	—	9 月初	—
	云南麻栗坡 1988	4 月上旬		4 月下旬	5 月上旬	9 月上旬	9 月下旬	10 月上旬
川滇木莲	云南	3 月	—	5 月上旬	5 月下旬	9 月	—	10 月
木莲	湖南新宁 1987	5 月上旬	5 月中旬	5 月下旬	—	—	9 月上旬	—
	云南滕冲 1986～1988	—	—	5 月	—	10 月上旬	10 月中旬	10 月下旬
滇桂木莲	云南景洪普文 1987～1989	4 月上旬	6 月	～	8 月	9 月	10 月	11 月
灰木莲	广西南宁 1981～1984	3 月中旬	4 月上旬	4 月中旬	4 月下旬	8 月下旬	—	9 月上旬
大果木莲	云南 1988	—	—	4 月	—	9 月下旬	9 月中旬	10 月上旬
海南木莲	广西南宁 1981～1984	4 月中旬	4 月下旬	5 月上旬	5 月中旬	8 月下旬	—	9 月上旬
中缅木莲	云南	3 月上旬	—	4 月中旬	5 月中旬	—	11 月	—
红花木莲	湖南新宁 1987	4 月下旬	5 月初	5 月中旬	5 月底	—	8 月下旬	—
马关木莲	云南马关 1988	—	—	3～4 月	—	9 月上旬	9 月下旬	10 月上旬
大叶木莲	云南	3 月上旬	—	4 月中旬	5 月上旬	—	10 月	—
卵果木莲	云南马关 1988			4 月		9 月上旬	9 月中旬	10 月上旬
乳源木莲	浙江杭州	—	5 月中旬	5 月下旬	6 月上旬	9 月中旬	—	10 月上旬

果实的采收调制和种子贮藏 蓇葖果成熟时间较一致。果实采回后应及时薄摊于通风处，待果开裂，取出种子，置水中浸泡 1～2 天使外种皮的肉质外层软化。搓去肉质层，淘洗干净，晾干 2～3 天，即得播种材料，通称种子。木莲属树种果实和种子的形态特征见图 1 和表 3。种子品质数据见表 4。种子生命力难以长久保存，应混沙湿藏，注意通气、保湿。

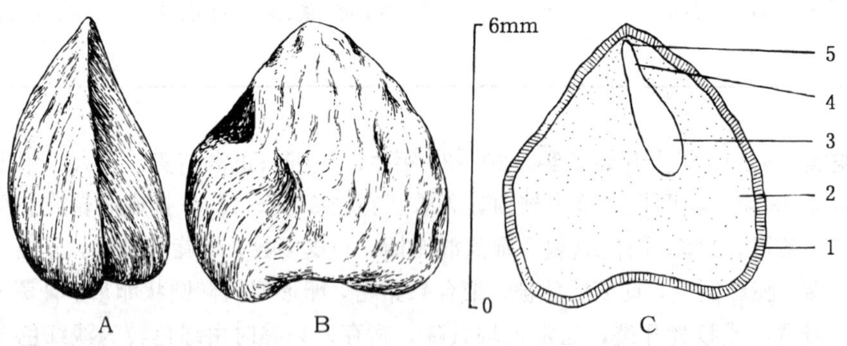

图 1 灰木莲种子除去肉质外层后的侧面（A）、正面（B）及其纵切面（C）

1. 外种皮的内层 2. 胚乳 3. 子叶 4. 胚轴 5. 胚根

（黄应钦绘）

表 3　木莲属树种果实和种子形态特征

树　种	聚合果			蓇葖果			种　子	
	形　状	长(cm)	色　泽	形　状	长(cm)	色　泽	形　状	长(mm)
香木莲	近球形或卵状	径 5～8	紫红	厚木质，沿腹缝和背缝开裂	—	红	心脏形	9
桂南木莲	卵形	4～5	红	具疣状凸起，先端具短喙	—	棕褐	心脏形	8
川滇木莲	卵状椭圆形	3.5	鲜红	—	—	红褐或鲜红	扁圆形	5
木莲	卵形	7～10	红	露出面有点状凸起先端具 1mm 的短喙	—	棕褐	心脏形	8
滇桂木莲	卵形或卵状球形	4～6	鲜红	密生瘤状凸起，具短喙，背腹缝同时开裂	—	红褐或鲜红	—	—
灰木莲	圆柱形	—	黄褐	露出面具 1 纵沟，先端具外弯的喙	—	黄褐或黄绿	心形、桃仁形	5～7.5
大果木莲	长圆状卵形	16～20	鲜红	先端尖，微内曲，沿背缝及腹缝开裂	3～4	鲜红	卵圆形	10
海南木莲	卵形或椭圆状卵形	5～6 径 3～4	暗褐	露出面有疣状凸起先端无喙	2	淡黄或红褐	稍扁，具 3 棱	7～8
中缅木莲	卵圆形或近球形	7～11.5	紫红	露出面菱形，具短喙平滑，无瘤状凸起，背缝开裂	—	紫红	椭圆形、圆锥形	11
红花木莲	卵状长圆形	5～10	紫红	具明显瘤状突起，先端具短喙，背缝全裂	—	紫红	心脏形	8
马关木莲	卵状圆筒形	7.5～11	紫红	—	—	紫红	椭圆形	8
大叶木莲	卵球形或长圆状卵形	6.5～11	紫红	先端尖，稍向外弯，沿背缝及腹缝开裂	2.5～3	紫红	三角形、卵圆形	7
卵果木莲	近球形	5.7～7.1	红	—	—	红	近心形	7～9
乳源木莲	卵形	2.5～3.5 径 2～4	未熟时淡黄色，熟后褐色	先端具短喙，背缝开裂	—	棕红	心形	—

表 4　木莲属树种出籽率和有关种子质量的数据

树　种	出籽率(%)	净　度(%)	千粒重（g）		每千克纯净种子粒数	
			一般	变动范围	一般	变动范围
香木莲	2.4～6.5	85～95	185	130～240	5 400	4 100～7 700
桂南木莲	10～15	75～85	73	70～85	13 000	11 000～14 000
川滇木莲	7～12	80～90	32	25～40	31 000	25 000～40 000
木　莲	11～14.5	75～88	54	50～60	18 400	16 700～20 000
滇桂木莲	4～6	—	56	—	17 800	—
灰木莲	23	97～99	50	40～60	20 000	16 000～25 000
大果木莲	3.2～7	85～95	200	—	5 000	—
海南木莲	23	97～99	50	45～65	20 000	15 000～22 000
中缅木莲	9.5～21	85～90	186	130～230	5 300	4 300～7 700
红花木莲	—	39～49	30	23～40	33 000	25 000～44 300
马关木莲	10～14	80～90	100	90～110	10 000	9 000～11 000
大叶木莲	4～23	85～95	45	40～50	22 000	20 000～25 000
卵果木莲	8～14	75～85	90	80～100	11 000	10 000～12 500
乳源木莲	4～6	93～95	40	32～50	25 000	20 000～31 000

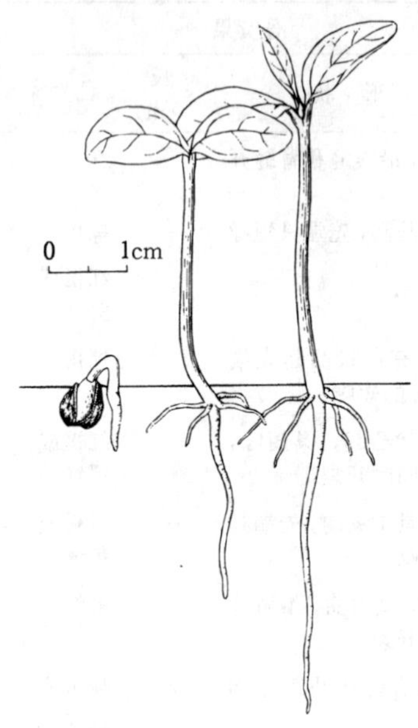

图 2 灰木莲种子萌发后第 3、9、26 天幼苗的生长情况
(黄应钦绘)

发芽与播种 海南木莲、滇桂木莲的种子发芽期长，且不整齐，播前需要层积处理。木莲属种子的发芽能力并不太低，问题是难以长期保存，最好是随采随播或混沙层积后春播。发芽率一般在 35%～70%（表 5）。播种前种子用 0.5%高锰酸钾浸种 30 分钟，清水冲洗后晾干。据王成霖和邵蓓蓓（1987）试验，在 0～5℃下层积 0 周、10 周、14 周和 19 周的乳源木莲种子，层积后的发芽率分别为 1%、24%、40%和 67%。木莲属种子出土萌发，1 年生苗高 15～20cm，萌芽力强，可萌芽更新，也可用嫁接繁殖。灰木莲种子的萌发及幼苗生长情况见图 2。

表 5 木莲属树种的发芽能力

树种	测定方法	发芽势（%）		发芽率（%）	
		天数	数值	天数	变动范围
香木莲	室外发芽	15	34	35	46～66
桂南木莲	室外发芽	5	15	10	21～33
川滇木莲	室外发芽	—	—	10	40～50
木莲	室外发芽	15	54	30	54～65
滇桂木莲	沙盘催芽 17 天后室外发芽	—	—	60	40～60
灰木莲	室外发芽	14	26	33	40～50
大果木莲	室外发芽	15	52	20	48～73
海南木莲	室外发芽	18	28	60	45
中缅木莲	室外发芽	—	—	22	45～55
红花木莲	室内 27℃，无光	—	—	30	46
马关木莲	室外发芽	15	51	30	50～65
大叶木莲	室外发芽	—	—	25	45～55
卵果木莲	室外发芽	15	20	30	28～35
乳源木莲	变温层积，室外发芽	—	—	—	80～85

（廖舫林）

华　盖　木

Manglietiastrum sinicum Law

（木兰科　Magnoliaceae）

生长习性、分布和用途　华盖木属只有本文描述的这1个种。常绿大乔木，树高达40m，胸径120cm。深根性，速生。喜温暖湿润气候。分布于滇东南海拔1 300～1 500m的山沟常绿阔叶林中。材质优良，花芳香，树形美观。现存大树极少，已被《中国植物红皮书》列为稀有种。

开花结实　12年生左右开始开花结实，30年生以后进入正常结实期。在天然林内结实量很少，甚至不结实，大小年现象较明显，间隔期3～5年。花两性，白色，单生枝顶。花被9片，3轮。雄蕊约65，药室内向开裂，药隔伸出成长尖头。心皮13～16，离生，具雌蕊群柄，果时长约1cm，径约1.3cm。每心皮有胚珠3～5枚。据1985～1988年在自然分布区观察，5月中旬开始开花，5月下旬为开花盛期，6月上旬为开花末期，10月上旬果实开始成熟，10月中旬为果实成熟盛期，10月下旬进入果实成熟末期。

聚合果成熟时绿色，稍带红晕，干时暗褐色，卵形、长圆状卵形或倒卵形，长3.3～8.5cm，径3.0～6.5cm。蓇葖厚木质，窄长圆状椭圆形或倒卵状椭圆形，长2.5～4cm，径1.5～2.5cm，沿腹缝线全裂且顶端2浅裂，背面具粗皮孔。每个蓇葖具种子1～3（4）粒。外种皮外层红色，肉质，内层黑色。种子横椭圆形，腹孔凹入，中有凸点，背棱微凸，长7～12mm，宽10～13mm，厚2～3mm，有胚乳，胚小（图1）。

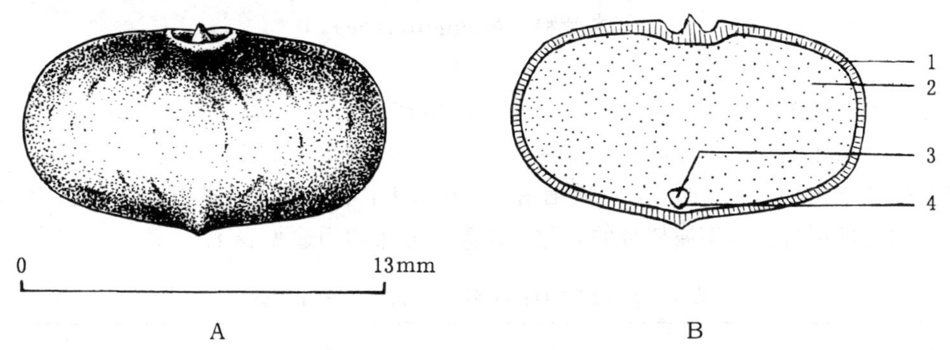

图1　除净外种皮肉质外层后华盖木种子外形（A）及其纵切面（B）

1. 外种皮内层　2. 胚乳　3. 胚芽　4. 胚根

（黄应钦绘）

果实的采收调制和种子贮藏　果实成熟后有鼠、鸟盗食，应及时采集。用采种刀将果柄截断或用竹竿打落后在地面拾集。采回的聚合果堆放于室内阴凉处，俟其充分成熟后再摊开，蓇葖微裂时剥取种子，搓洗，淘去外种皮的外层即为播种材料。每千克有聚合果30～33个。每果含种子4～18粒，平均约9粒。按鲜果质量计算的出籽率为3%～7%。种子净度可达95%。千粒重约100g，变动范围57～150g。每千克有纯净种子约10 000粒，变动范围为6 600～

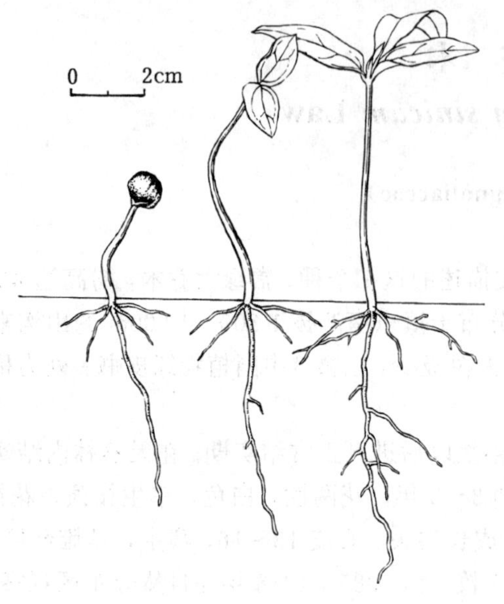

图2　华盖木种子萌发后第6、10、25天幼苗的生长情况
（黄应钦绘）

16 000粒。种子含油脂，不能日晒，也不能干藏。运输或贮藏时均需混以湿沙。混沙贮藏兼有层积催芽的作用，为期半年左右。

发芽和播种　种子有休眠习性。发芽时日均温宜在18℃以上。1989年2月下旬，云南林业科学研究院曾用头年10月采集并经层积催芽的种子在室外荫棚苗床做发芽测定，播种后50天左右，即4月中旬开始发芽。从播种之日起算，80天的发芽率为69%，在发芽盛期的15天中发芽百分数约60%。出土萌发。胚根萌发后主根迅速向下伸展并形成4条侧根，上胚轴带种壳出土，子叶随即伸出并展开，再过10～15天初生叶出现（图2）。

目前尚无批量育苗经验。1年生苗可以出圃。也可扦插育苗。

（李达孝）

含　笑　属
Michelia L.

（木兰科　Magnoliaceae）

生长习性、分布和用途　含笑属约60种，我国产40余种，本文描述24种。常绿乔木或灌木。适生于温暖湿润气候下的酸性土壤，忌干燥瘠薄和排水不良。花芳香，是香料工业和庭园观赏的重要树种。木材通直，材质细密，软硬适中，有香气，耐腐朽，是建筑、家具及胶合板工业的优良用材。含笑属树种的名称、树高、分布和用途见表1。

表1　含笑属树种的名称、树高、分布和用途

中　名	学　名	树高(m)	胸径(cm)	分　布	用　途	供稿
铜色含笑	*M. aenea* Dandy	25	70～80	滇	材用、香料	615
石灰含笑	*M. calcicola* C. Y. Wu	20	50	滇、桂	材用、香料	615
黄　兰	*M. champaca* L.	30～40	100	藏、滇。印度、缅甸、越南。闽、台、粤、琼、桂引种	观赏、香料、材用、药用	614
乐昌含笑	*M. chapensis* Dandy	15～30	100	赣、湘、桂、粤	材用	703
麻栗坡含笑	*M. chartacea* B. L. Chen et S. C. Yang	30	60～98	滇	材用、香料	615
西畴含笑	*M. coriacea* Chang et B. L. Chen	10	20	滇	材用、香料	615

（续）

中　名	学　名	树高（m）	胸径（cm）	分　布	用　途	供　稿
南亚含笑	*M. doltsopa* Buch. -Ham. ex DC.	30	40	藏、滇	材用	615
含　笑	*M. figo* (Lour.) Spreng.	5	—	长江流域以南	观赏、香料、药用	702
多花含笑	*M. floribunda* Finet et Gagnep.	20	40	滇、川。缅甸	材用、香料	615
金叶含笑	*M. foveolata* Merr. ex Dandy	30	80	湘、赣、粤、琼、桂、黔、滇	材用	703
福建含笑	*M. fujianensis* Q. F. Zheng	25	100	闽。浙江引种	材用、绿化	902
金花含笑	*M. ingrata* B. L. Chen et S. C. Yang	16	35	滇	材用、香料	615
壮丽含笑	*M. lacei* W. W. Smith	15	18	滇	材用	615
长蕊含笑	*M. longistaminata* Law	15	50	粤	材用	601
醉香含笑（火力楠）	*M. macclurei* Dandy	30	100	桂、粤、琼。越南	材用、观赏	601
黄心夜合	*M. martinii* (Lévl.) Dandy	20	65	豫、鄂、湘、川、滇、黔	香料、材用	615
深山含笑	*M. maudiae* Dunn	20	45	浙、闽、湘、桂、黔	材用、观赏	703
小果含笑	*M. microcarpa* B. L. Chen et S. C. Yang	30	95	滇	材用、香料	615
畴阳含笑	*M. nitida* B. L. Chen	20	40	滇	材用、香料	615
马关含笑	*M. opipara* Chang et B. L. Chen	20	70	滇	材用、香料	615
阔瓣含笑	*M. platypetala* Hand. -Mazz.	20	50	黔、鄂、湘、桂、粤	材用	703
毛果含笑	*M. sphaerantha* C. Y. Wu et Y. W. Law	18～25	80	滇	材用、香料、观赏	615
绒叶含笑	*M. velutina* DC.	30	230	藏、滇	材用	615
云南含笑（皮袋香）	*M. yannanensis* Franch. ex Finet et Gagnep.	2～4	—	滇	香料、观赏	614

开花结实　灌木类 2～3 年开花，乔木类 5～10 年开花。结实有大小年现象，间隔期一般为 1～2 年。花两性，单生叶腋，芳香。花被 6～21，3 或 6 片 1 轮。雄蕊多数，药室侧向或近侧向开裂。雌蕊群有柄，心皮多数或少数，分离。胚珠 2 至数枚。开花结实的物候期见表 2。

表 2　含笑属树种的开花结实物候

树　种	观察地点和年份	开花			果实成熟		果实脱落	
		始　期	盛　期	末　期	始　期	盛　期	始　期	末　期
铜色含笑	—	—	3～4 月	—	9 月下旬	10 月上旬	—	—
石灰含笑	云南麻栗坡 1987～1988	—	4 月中旬	4 月下旬	8 月下旬	9 月上旬	—	—
黄　兰	云南景洪 1979～1980	—	4 月中旬	6 月下旬	8 月上旬	8 月中下旬	—	—
乐昌含笑	湖南澧陵 1988～1989	3 月上旬	3 月下旬	4 月上旬	8 月下旬	—	9 月下旬	10 月上旬
麻栗坡含笑	—	—	3 月中旬	4 月	—	10 月	—	—

（续）

树种	观察地点和年份	开花			果实成熟		果实脱落	
		始期	盛期	末期	始期	盛期	始期	末期
西畴含笑	云南西畴 1987～1988	—	4月	—	9月上旬	9月中旬	—	—
南亚含笑	云南腾冲 1983～1988	—	4～5月	—	10月上旬	10月中旬	—	—
含笑	浙江杭州 1978～1980	4月上中旬	4月下旬	5月中旬	9月上旬	9月中旬	—	—
多花含笑	云南腾冲 1986、1988	—	4月	—	10月上旬	10月中旬	—	—
金叶含笑	湖南新宁 1987	3月上旬	4月上旬	4月下旬	—	9月中旬	—	9月下旬
	浙江杭州 1987	4月中旬	4月下旬	5月上旬	9月上旬	9月中旬	10月上旬	
福建含笑	福建三明 1982～1987	1月中旬	1月下旬～2月上旬	2月中下旬	10月下旬	—	11月中旬	12月上旬
金花含笑	云南马关 —	—	3～4月	—	9月中旬	9月下旬	—	—
壮丽含笑	—	—	1月下旬	2月下旬	—	9月	—	—
长蕊含笑	广西南宁 1987～1988	4月中旬	4月下旬	5月上旬	10月中旬	10月下旬	11月上旬	11月下旬
醉香含笑	广西南宁 1981～1984	1月上旬	2月中旬	3月上旬	11月上旬	11月中旬	11月中旬	12月中旬
黄心夜合	云南马关 1988	—	2～3月	—	9月上旬	9月中旬	—	—
深山含笑	湖南新宁 1986～1987	1月下旬	2月	3月中旬	—	10月上旬	10月中旬	10月下旬
小果含笑	云南麻栗坡 1988	—	4～5月	—	9月上旬	9月中旬	—	—
畴阳含笑	云南西畴 1987～1988	—	2～3月	—	9月上旬	9月中旬	—	—
马关含笑	—	3月上旬	3月下旬	4月上旬	—	9～10月	—	—
阔瓣含笑	湖南新宁 1986～1987	1月上旬	3月中旬	4月中旬	—	9月下旬	9月下旬	10月上旬
毛果含笑	—	—	4月上旬	5月上旬	—	10月	—	—
绒叶含笑	云南腾冲 1983、1986、1988	—	5月	—	10月上旬	10月中旬	—	—
云南含笑	云南昆明 1986～1988	2月	3月	4月上中旬	8月初	—	8月下旬	—

聚合果，通常部分蓇葖不发育。蓇葖背缝线开裂或背腹2瓣裂，宿存。果熟时呈紫红色或淡黄色。每个蓇葖有种子2至数粒。外种皮外层红色，肉质松软，内层黑色或黑褐色，木质。胚乳丰富，富含油脂，胚细小。含笑属果实和种子的形态见表3、图1。

表3 含笑属树种果实和种子形态特征

树种	聚合果长(cm)	蓇葖果			种子			
		形状	长(cm)	颜色	形状	长(mm)	外种皮外层	外种皮内层
铜色含笑	8～12	卵圆形	0.5	黄褐色	扁状卵圆形	7	红色	黑色
石灰含笑	9～13	—	—	—	心脏形	7	红色	黑褐色
黄兰	7～10	倒卵状长圆形，有皱纹	1～1.5	淡黄色	近扁圆形	—	红色，有皱纹	黑色
乐昌含笑	约10	长圆形或卵形，扁而稍斜，先端具短尖，皮孔稀疏且不显著	1.2～1.5	紫红色	卵形或长圆状卵形	约10	红色	黑色
麻栗坡含笑	10～28	卵球形或椭圆形，具皮孔，具短喙	1.3～1.5	淡黄色	扁圆形或扁椭圆形	9.5	红色	黑色
西畴含笑	4～5	卵形	1.9～2.1	—	椭圆形	9～18	红色	黑色
南亚含笑	2.5～10	近倒卵形，先端具凸尖喙	1.5	淡黄色	扁圆形	6	红色	黑色

(续)

树　种	聚合果长(cm)	蓇葖果			种　子			
		形　状	长(cm)	颜　色	形　状	长(mm)	外种皮外层	外种皮内层
含　笑	5～7.5	扁卵圆形或扁球形，先端喙短尖，有白色皮孔	—	黄绿色	—	5～9	红色	黑色
多花含笑	2～7扭曲	圆球形或长圆形，先端微尖，有白色皮孔	0.6～1.5	黄红色	心脏形	6	红色	黑色
金叶含笑	7～20	长圆状椭圆形	1～2.5	淡红色	扁卵形	8～10	红色	黑色
福建含笑	—	倒卵形	1.5～2	绿褐色	扁卵形	10～15	红色	黑色
金花含笑	5～15	卵圆形	1.4	黄褐色	椭圆形	5	红色	黑色
壮丽含笑	—	—	—	黄绿色	宽卵形或三角状卵形	10	红色	黑色
长蕊含笑	8～12	宽倒卵形或长圆形或近球形，两侧微扁，先端有喙，下弯	1.5～2.1	黄褐色或黄绿色	长椭圆形或扁形	7～12	红色	黑色
醉香含笑	3～7	长圆形或倒卵状长圆形或倒卵形，基部宽，疏生白色皮孔	1～3	淡黄色或红褐色，星状斑点	扁卵形或三角形或倒卵状椭圆形	8～10	红色	黑色
黄心夜合	6.5～18扭曲	倒卵形或长圆状卵形，先端有短喙，具白色皮孔	—	淡黄色	长椭圆形	12	红色	黑色
深山含笑	10～12	长圆形或倒卵形或卵形，先端圆钝或具短尖头	1.5～2	紫红色	斜卵形，稍扁	10	红色或红棕色	黑褐色
小果含笑	3～6	卵球形	—	灰白色	—	6	红色	黑色
畴阳含笑	3～5	卵形或倒卵形	2.1～3.5	淡黄色	宽椭圆形	5	红色	黑色
马关含笑	11～14	卵形，具短喙及皮孔	1.5～2	—	长椭圆形	11	红色	黑色
阔瓣含笑	5～15	长圆形或卵形，无柄，先端圆，有时偏上部一侧有短尖，皮孔灰白色	1.5～2	紫红色	三角状扁卵形或三角状椭圆形	5～8	淡红色	黑色
毛果含笑	10～40	卵形或椭圆形	—	—	圆形微扁	9	红色	黑色
绒叶含笑	5～13	倒卵形，先端圆钝或短尖，蓇葖疏离，具明显皮孔，下端收缩成柄	0.5～0.8	淡黄色	心脏形	6	橙黄色	黑色
云南含笑	—	扁球形，先端具短尖，有毛	—	绿褐色	—	—	—	—

果实的采集调制和种子贮藏　果实呈现成熟颜色并有少数蓇葖开裂时，应及时采集。采回的聚合果薄摊于荫凉通风处，让其自然干燥开裂。取出的种子用0.5%的碱水浸泡1～1.5

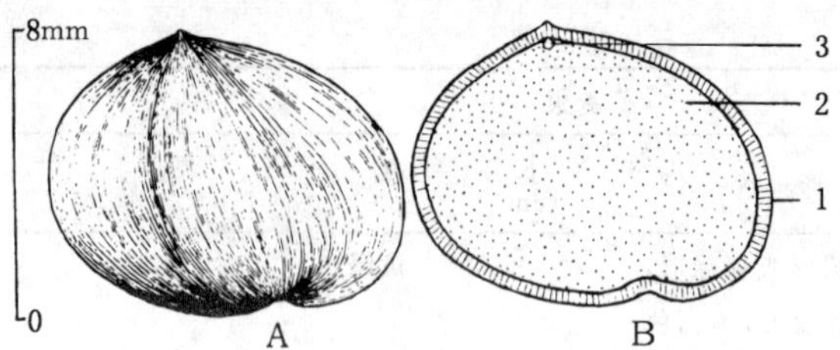

图 1 醉香含笑除去肉质外层后的种子外形（A）及其纵切面（B）
1. 外种皮内层 2. 胚乳 3. 胚
（黄应钦绘）

天，或清水浸泡 3～4 天，待种皮的肉质外层软化后，置水中搓擦并冲洗干净，即得仅由外种皮内层包被的播种材料，通称种子。调制后稍加阴干的含笑种子含水量为 23%～35%，一般都是混沙湿藏越冬，供来年春播。湿藏容器中央部位需插放有助于通气的设施。贮藏期间应经常检查，如果沙子干燥，应及时适量加水，重新翻拌均匀。鲜果的出籽率和种子质量数据见表 4。

表 4 含笑属树种的出籽率和有关种子质量的数据

树 种	每千克聚合果的个数	每个聚合果中的蓇葖数	出籽率（%）	净 度（%）	千粒重（g）	每千克纯净种子数（粒）	
						一般	变动范围
铜色含笑	60～120	28～56	20～30	85～95	120～150	7 400	6 500～8 300
石灰含笑	30～35	20～80	11～15	75～85	110～130	8 200	7 700～9 100
黄兰	—	—	5～8	85～95	65～85	13 000	12 000～15 500
乐昌含笑	—	—	27	—	260～300	3 500	3 300～3 900
麻栗坡含笑	53～98	22～100	30～34	80～90	134～158	7 000	6 200～7 500
西畴含笑	200～300	—	5～12	60～75	50～100	16 000	10 000～20 000
南亚含笑	90～120	1～21	10～19	75～85	80～100	11 000	10 000～13 000
含笑	—	—	12～18	95～99	76～96	11 660	10 500～13 200
多花含笑	160～200	7～17	19～27	85～95	90～110	10 000	9 000～11 000
金叶含笑	—	—	5.3	98	55～70	16 200	15 300～18 200
福建含笑	—	—	6.5	98	80～88	12 000	11 000～13 000
金花含笑	90～105	2～36	8	70～80	80～105	11 000	9 500～13 000
壮丽含笑	98～140	48～75	15～25	88～95	130～165	6 800	6 000～7 800
长蕊含笑	—	—	3～4	80～90	110～180	6 800	5 500～9 200
醉香含笑	—	—	3～5	90～95	110～170	7 400	5 800～9 100
黄心夜合	70～100	1～28	15～18	75～85	140～160	6 600	6 200～7 200
深山含笑	—	—	6.3	96～99	70～80	11 500	12 000～14 300
小果含笑	350～400	—	20～24	80～90	135～160	6 800	6 200～7 500
畴阳含笑	55～77	1～6	5～6	85～95	100～150	8 000	6 600～10 000
马关含笑	56～89	8～55	40～50	85～95	150～200	5 800	5 000～6 800
阔瓣含笑	—	—	3.4	93～99	38～60	21 000	16 000～26 400
毛果含笑	36～37	60～120	17～36	85～95	80～100	10 800	10 000～12 500
绒叶含笑	127～180	4～23	15～18	75～85	80～150	9 500	6 800～12 800
云南含笑	—	—	12.5	90～99	140～175	6 400	5 700～7 200

发芽　从现有的经验看，含笑属有的树种种子没有明显的休眠习性，例如黄兰、福建含笑和云南含笑，播前不加预处理便能在适宜的条件下顺利萌发（表5）。有些树种的种子则具有短期休眠习性，播前必须经过湿沙层积。而且即使如此，不少树种仍然发芽迟缓，田间的发芽过程需要很长时间才能结束（表6，据李达孝未刊稿）。

表5　无明显休眠习性的含笑属树种的发芽情况①

树　种	置床至开始萌发的天数	置床至发芽盛期的天数	发芽势		发芽率	
			计算天数	数值（%）	计算天数	数值（%）
黄　兰②	20	23～26	27	45	50	65～75
	22	25～28	27	45	50	65～75
福建含笑③	—	—	—	—	21	53④
云南含笑③	—	—	12	31～37	25	42～54

① 播前种子未经预处理，发芽基质为沙

② 随采随播，田间观测

③ 室内测定。发芽条件：25℃恒温，未加光照

④ 未发芽粒中腐烂粒占14%，空粒占33%

表6　含笑属树种有休眠习性的种子室温层积后的田间发芽情况

树　种	层积天数	播种期	播种至开始萌发的天数（天）	播种至发芽结束的天数（天）	发芽率（%）
铜色含笑	110	2月下旬	73	95	55～65
石灰含笑	140	2月中旬	75	135	45～54
麻栗坡含笑	130	3月中旬	35	53	55～65
南亚含笑	110	2月下旬	73	90	25～40
多花含笑	110	2月下旬	69	100	39～55
金花含笑	110	2月中旬	72	87	48～55
壮丽含笑	150	2月下旬	30	71	50～65
黄心夜合	140	2月下旬	74	90	46～73
小果含笑	90	1月初	61	130	35～66
畴阳含笑	140	2月下旬	92	112	60～87
马关含笑	130	3月中旬	34	52	55～65
绒叶含笑	110	2月下旬	74	104	23～31

含笑和阔瓣含笑的种子也有明显的休眠习性。据史晓华（1983～1987，未刊稿）在杭州试验，没有经过层积处理的含笑种子不能发芽，室温层积12周或100μg/gGA_3浸种24小时再室温层积8周，效果十分显著，而且都优于低温层积（表7）。据周佑勋和胡春姿（1990）观察，阔瓣含笑胚的形态在层积前后变化不大，且完整种子和去皮种子的吸水速度并无明显差异，只是层积前和层积120天后，每小时每克去皮干种子的吸氧量比完整种子高。她们因此推测阔瓣含笑种子发芽迟缓的原因是种皮透气性差，也可能是缺乏内源性萌发促进物质。据她们试验，室温层积80天即可有效地解除阔瓣含笑种子的休眠，发芽率可达75%。层积时间延长到120天并无提高发芽率的作用。这个树种低温（5℃）层积的效果不及室温层积。她们的研究

还表明，用300μg/g的GA_3浸种48小时，可以取代层积而得到66%的发芽率，将GA_3的浓度提高到1 800μg/g，发芽率可达76.7%。

表7 含笑种子不同预处理的效果

预 处 理	层积时间（周）	恒温25℃下20天的发芽率（%）
清水浸种24小时后室温（2～17.5℃）层积	12	82～90
$GA_3$100μg/g浸种24小时后室温（2～17.5℃）层积	8	70～84
清水浸种24小时后低温（3～5℃）层积	12	50～70
对照	0	0

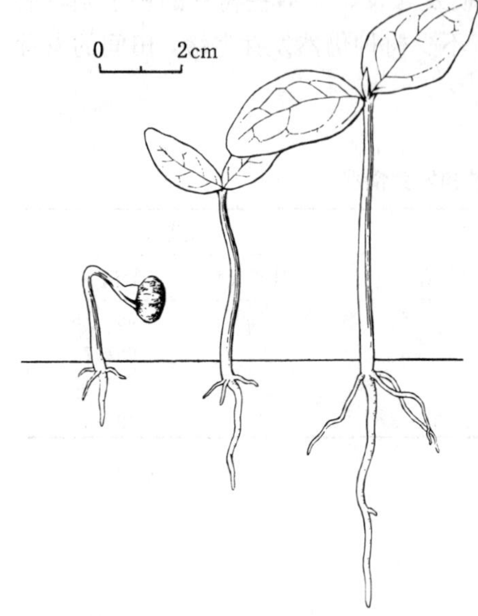

图2 醉香含笑种子萌发后第5、10、32天的幼苗生长情况
（黄应钦仿《热带亚热带主要树种采种育苗技术》）

经过层积的阔瓣含笑种子发芽的最佳条件是恒温30℃的黑暗环境（周佑勋和胡春姿，1990）。目前只知道含笑和福建含笑发芽时对光不敏感。

播种 含笑属树种可以随采随播或冬播，也可以室温层积后春播。条播。条距20～30cm，条幅5～7cm，播种沟深2～3cm。每平方米用种量福建含笑为10～13g，长蕊含笑和醉香含笑为13～15g，黄兰为30～35g，云南含笑为20～30g。播后覆细土并盖草或松针。有的地方是集中播种，待幼苗长到一定大小时移至育苗容器继续培育。出土萌发。子叶2。醉香含笑子叶斜方形或椭圆形，出土后继续生长至长4cm，宽2.2cm；上胚轴长约1.5mm，初生叶互生；下胚轴长5～8cm，近子叶处渐细，上部暗绿色，有紫褐色斑点，下部渐粗，近根颈处白色；根白色。醉香含笑种子的萌发和幼苗初期形态见图2。含笑属有些树种也可用扦插、压条、嫁接繁殖。

（廖舫林）

拟单性木兰
Parakmeria Hu et Cheng

（木兰科 Magnoliaceae）

生长习性、分布和用途 本属约6种，我国5种，本文描述2种。常绿乔木，树高30m，胸径达140 cm。深根性。速生。喜温凉湿润气候。较耐寒，抗风力差。云南拟单性木兰的残存大树已不多见，被《中国植物红皮书》列为濒危种。这两个树种的名称、生长、分布和用途见表1。

表 1　拟单性木兰属树种的名称、生长、分布和用途

中　名	学　名	树高 (m)	胸径 (cm)	分　布	用　途
光叶拟单性木兰	*P. nitida* (W. W. Smith) Law	30	140	滇、桂	材用、水源、观赏
云南拟单性木兰	*P. yunnanensis* Hu	30	150	滇、桂、黔	材用、水源、观赏

开花结实　10～15 年生开始开花结实，30 年生以后进入正常结实期。结实丰年间隔期 2～3年，但不甚明显。花杂性异株，单生枝顶。雄花中的雄蕊 10～30（60）枚，花丝短，花药条形，内向开裂。两性花中的雄蕊群与雄花同，雌蕊约 10～20 枚，具雌蕊群柄。心皮发育时全部互相愈合。花芳香。光叶拟单性木兰心皮绿色，柱头深红色，雄花未见。云南拟单性木兰雄花及两性花异株。在云南西畴等地观察的物候期见表 2。

表 2　拟单性木兰属树种的开花结实物候期

树　种	观察地点和年份	开花期			果熟期		种子散落期
		始　期	盛　期	末　期	始　期	盛　期	
光叶拟单性木兰	—	3 月	—	5 月	10 月上旬	10 月中旬	10 月下旬～11 月
云南拟单性木兰	云南西畴 1988	4 月中旬	4 月下旬	5 月上旬	9 月上旬	9 月下旬	10～11 月

聚合果，成熟时由青色转变为黄绿色。光叶拟单性木兰每心皮有种子 1～3，云南拟单性木兰每心皮有种子 1～2。木质蓇葖成熟沿背缝及顶端开裂后，种子垂悬于丝状而有弹性的珠柄上。外种皮的外层红色，肉质，内层黑色。胚小，胚乳丰富，富含油质。果实及种子的形态见表 3、图 1。

表 3　拟单性木兰属树种果实和种子的形态特征

树　种	果　实			种　子	
	形　状	大小（cm）	颜　色	形　状	大小（mm）
光叶拟单性木兰	聚合果卵状长圆形或椭圆状卵形。蓇葖椭圆形	长 3～7，径 1.5～2.5	黄绿色	长椭圆形	高约 7，宽约 15，厚约 4
云南拟单性木兰	聚合果长圆状卵形。蓇葖菱形	长 4.5～9，径 2.5～3.5	黄绿色	扁椭圆形	高约 8，宽约 10，厚约 4

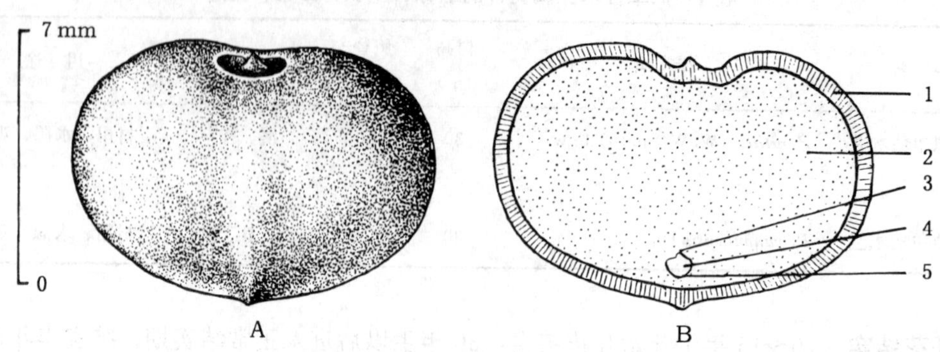

图 1 除净肉质外层的云南拟单性木兰种子外形（A）及其纵切面（B）

1. 外种皮内层 2. 胚乳 3. 胚芽 4. 胚轴 5. 胚根

（黄应钦绘）

果实的采收调制和种子贮藏 成熟的种子有鼠、鸟盗食，应及时采收。用采种钩刀将聚合果钩落，在地面捡拾。采回的果实可堆放于室内 3～5 天，充分成熟后摊开晾干。当蓇葖微裂，可见红色种子时，剥开蓇葖取出种子，装入布袋或竹筐，置水中搓擦，洗净外种皮的外层，同时漂去上浮的瘪粒，即得纯净的播种材料，通称种子。种子的净度、质量等数据见表 4。

表 4 拟单性木兰属树种的出籽率和种子净度、质量

树 种	出籽率（%）	净 度（%）	千粒重（g）		每千克纯净种子粒数（粒）	
			一 般	变动范围	一 般	变动范围
光叶拟单性木兰	8	80～95	225	210～240	4 400	4 000～4 700
云南拟单性木兰	15	85～95	200	150～270	5 000	3 700～6 700

种子忌失水，不能日晒，运输和贮藏均需混以湿沙。混沙湿藏还兼有解除休眠的作用。

发芽和播种 种子有休眠习性。云南林业科学研究院曾将 1988 年采集的种子混沙层积，至 1989 年 2 月中旬在室外沙床播种，5 月上旬开始发芽。从开始萌发起算的 90 天，光叶拟单性木兰发芽率为 40%，无明显的发芽盛期；从开始萌发起算的 15 天内，云南拟单性木兰的发芽数为 40%，到发芽结束的 86 天内发芽率为 56%。

出土萌发。主根向下伸延，4 条侧根轮生。下胚轴成弓形出土，将子叶拉出土面，10～15 天出现初生叶。发芽及幼苗生长情况见图 2。目前尚无大田育苗经验，多为集中播种，待萌发后移到苗床或育苗容器中继续培育。1 年生苗可以出圃，但用于庭院观赏的大苗需培育 3 年。

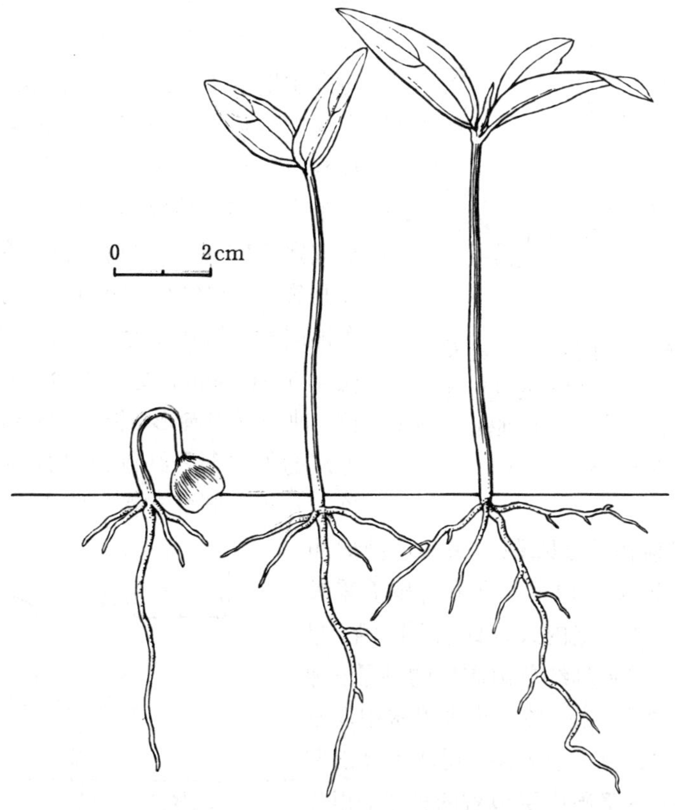

图 2　云南拟单性木兰种子萌发后第 7、10、15 天幼苗的生长情况
（黄应钦绘）

（李达孝、张茂钦）

合果木（山桂花）
Paramichelia baillonii（Pierre）Hu

（木兰科　Magnoliaceae）

生长习性、分布和用途　合果木属约 3 种，我国约 2 种，本文描述 1 种。常绿至半落叶乔木，高达 35m，胸径 100～150cm。喜光。能耐－3℃的短期低温。在砖红壤、红色石灰山土、冲积土上均能成材。分布于云南南部和西南部。广西、广东有引种。印度、缅甸、泰国、越南也有分布。材质坚韧，生长迅速，树干通直，水源涵养效益高。成年大树已日渐稀少，被《中国植物红皮书》列为渐危种。

开花结实　12～14 年生开始结实，18～35 年生结实量增多。大小年现象较为明显，丰年间隔期 1 年。花两性，单生于叶腋，黄白色，芳香。花被片 12～18，每轮 6 片，披针形，覆瓦状排列。雄蕊多数，长 6～7mm。花药侧向或近侧向开裂，药隔伸出成尖头。雌蕊群具柄，

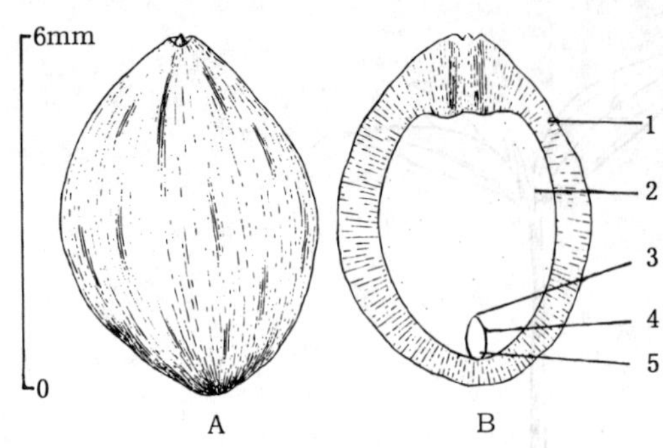

图 1　除净外种皮肉质外层后合果木种子的外形(A)及其纵切面(B)
1. 外种皮内层　2. 胚乳　3. 胚芽　4. 胚轴　5. 胚根
(黄应钦绘)

花柱红色，心皮多数，胚珠 2～6。心皮完全合生或部分合生，但结果时完全合生，形成近圆筒形的肉质聚合果，有凸起的皮孔，成熟后呈不规则小块状脱落。1988 年在云南景洪普文林场附近观察，3～5 月开花，9 月仍有少数花开放；7 月底个别果实开始成熟，8 月中旬～9 月上旬为果实成熟盛期。果实由粉绿色转变为黄绿色时有部分开裂，现出红色的种子，即示成熟。成熟后果实连同种子陆续落地。聚合果长 8.5～11.5cm，径 4.5～6.0cm。每一聚合果内有种子 85～160 粒。种子近扁圆形，长 8～9mm，宽 7～8mm。外种皮的外层部分红色，松软，肉质；内层骨质，黑色。胚小，胚乳丰富（图 1)。

果实的采收调制和种子贮藏　果实成熟时可在树下拾集落地的果实和种子，或用高枝剪采摘聚合果。采回的果实摊于室内或置日光下曝晒，果壳开裂后除去杂质，得到具红色肉质层的种子。随即拌以草木灰或适量洗衣粉，装入袋内置水中反复搓揉，淘净外种皮的红色肉质外层，用清水漂洗并稍加晾干，即为外种皮坚硬内层包被的种子，用作播种材料。按鲜果质量计算的出籽率为 4%～6%。种子的净度为 90%～95%，千粒重约 60g，每千克纯净种子约有 1.7 万粒。种子的含水量宜保持在 20%左右，忌失水，不宜日晒和干藏。短期运输可将果实用麻袋或竹箩包装，运抵后再行调制；也可在调制后混少量湿锯木屑或用青苔包装。贮藏时应混湿沙，贮藏期为半年以内。

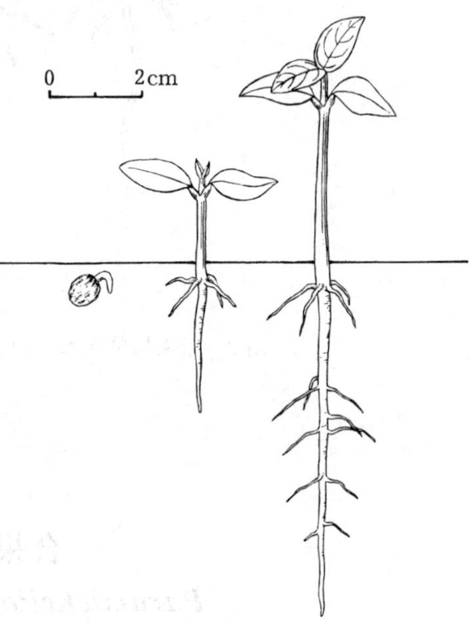

图 2　合果木种子萌发后第 3、7、12 天幼苗的生长情况
(黄应钦、马信祥绘)

发芽和播种　种子无明显休眠现象。发芽时的日均温宜在 20～30℃。为使种子发芽整齐，缩短出苗时间，播前可将种子混沙催芽，待部分种粒萌发时取出播种。条播。每平方米苗床约播 50g。据云南景洪普文林场的经验，8 月下旬用当年新采种子播种，播后 20 天开始发芽，30～35 天为发芽盛期；9 月中旬播种后 23 天开始发芽，35～42 天为发芽盛期。发芽全程约 30～40 天。从播种之日起算，38 天的发芽率为 35%，其中前 10 天为发芽盛期。出土萌发。子叶出土后 7～9 天发出初生叶。主根明显，根颈以下 4 条侧根轮生（图 2)。1 年生苗可以出圃。

（龙素珍）

观光木（香花木）

Tsoongiodendron odorum Chun

（木兰科　Magnoliaceae）

生长习性、分布和用途　观光木属为我国特产，且只有本文描述的一种。常绿乔木，高达25m，胸径达1m。速生。能耐短期－3℃左右的低温。喜湿润的酸性土。产滇、黔、桂、湘、赣、闽、粤、琼。越南亦产。材质轻软，结构细，宜作细木工、乐器、胶合板。树冠浓密，可供庭园观赏，涵养水源的效益也高。分布范围虽广，但多零星分散，又因繁殖能力较弱，大树已不多见，在《中国植物红皮书》中被列为稀有种。

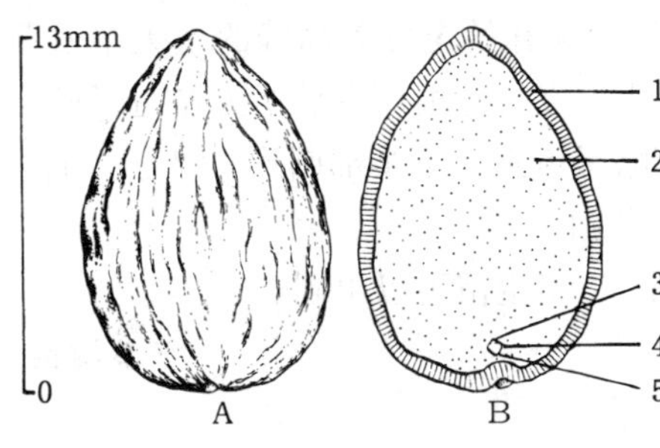

图1　除净外种皮肉质外层后观光木种子的外形（A）及其纵切面（B）
1. 外种皮内层　2. 胚乳　3. 胚芽　4. 胚轴　5. 胚根
（黄应钦绘）

开花结实　8～10年生开始结实。20年生后进入正常结实期。结实丰年间隔期为1年，较明显。花两性，淡黄白色，单生于叶腋。雄蕊多数，花丝短，圆柱状。花药条形，侧向开裂，药隔突出成尖头状。具柄的雌蕊群包于雄蕊群内。心皮少数，部分相互连合且基部与中轴愈合，受精后全部合生。胚珠12～16，成2列叠生。据1978～1984年在广西南宁观察，4月下旬始花，5月上旬为盛花期，5月中旬为末花期；9月下旬果实开始成熟，10月上旬为果熟盛期。果为聚合果，外表弯拱起伏，未成熟时近肉质而浅绿色；成熟时木质，黄绿色略带紫红色，干后深棕色，有明显的黄色皮孔；长椭圆形或卵形，长8～12cm，径5～10cm。成熟后小果裂成2果瓣，每蓇葖有种子4～6粒。外种皮外层肉质红色，内层脆壳质，淡黄色至黄色，表面有明显皱纹。蓇葖开裂后种子悬垂于丝状珠柄上，扁卵形、近矩形、椭圆形或三角状倒卵形或形状不规则，长11～17mm，径6～10mm，厚6～9mm。胚小，胚乳丰富（图1），含油质。

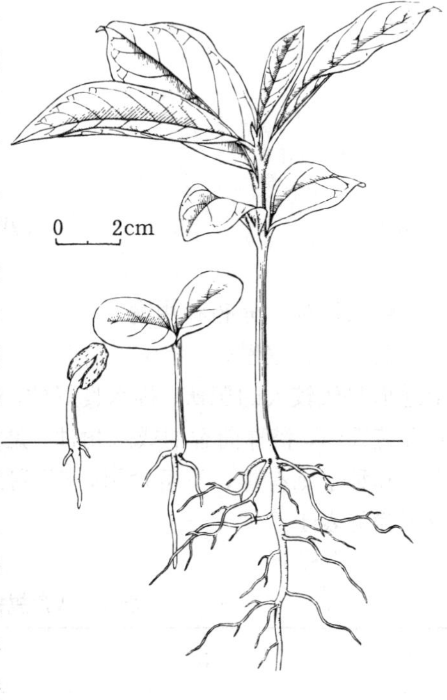

图2　观光木种子萌发后第5、8、40天幼苗的生长情况
（黄应钦仿《热带亚热带主要树种采种育苗技术》）

果实的采收调制和种子贮藏　果实接近开裂或部分开裂时，上树采摘或用采种刀钩下聚合果。采回后摊放于室内，待果实全部开裂种子外露时轻加敲打，种

子即可脱出。脱出的种子装入布袋或竹筐内，置水中搓洗，淘去红色的种皮外层，即得播种材料，通称种子。鲜果出籽率约为4%。种子的净度可以接近100%。千粒重变动在350～480g，一般为380g。每千克纯净种子粒数在2 000～2 900粒，一般为2 600粒。种子的大小和质量变异较大。据黄鹏成和林金姬（未刊稿）在福建来舟林业试验场测定，除净外种皮肉质外层后的观光木种子千粒重为210g，每千克有纯净种子约4 760粒。调制出来的种子可稍加晾干。种子含水量降到20%还未见影响发芽能力，但忌过分失水，不能日晒。曾见在干燥的室内裸露存放1周以上，种子发芽率便显著下降。短途运输的可以在运抵目的地后再行调制。长途运输或贮藏，均需洗净外种皮的肉质外层后及时混以湿沙，以免引起霉烂。贮藏期一般为3～6个月。

发芽和播种 种子无明显的休眠现象。发芽时的日均温宜在18℃以上。1979年广西林业科学研究所曾在室外沙床进行过一次发芽测定：1月5日播种，3月15日气温18℃时开始萌发，3月16～22日为发芽盛期，4月10日发芽终止。从发芽开始至发芽高峰的7天中发芽45%；从播种至发芽终了共106天的发芽率为60%。出土萌发。胚根萌发后7天左右子叶出土，11天左右发出初生叶。初生叶互生。下胚轴上半部略呈方形，下半部圆形，绿色。主根发达。根颈处4侧根轮生，波状平展或斜展（图2），淡黄白色。

条播。每平方米床面可播60～80g，覆土1.5～2cm。1年生苗可以出圃。

（韦增健）

八 角 属

Illicium L.

（八角科 Illiciaceae）

生长习性、分布和用途 本属约50种，我国约30种，本文描述2种。常绿乔木或小乔木。苗期生长较慢，1年生以后转快。喜湿润温暖、年降水1 200mm以上的气候环境。适生于地形起伏较大的肥沃、排水良好的酸性土壤。八角是我国重要的经济树种，果皮、种子、树叶含芳香油，俗称茴香油或八角油，是化妆品、调味品的重要原料，又可入药。披针叶茴香的果实和种子则有毒，不可食用，但可提取香料或入药。这两个树种的名称、生长情况、分布及用途见表1。

表1 八角属树种的名称、生长、分布和用途

中名	学名	树高 (m)	胸径 (cm)	分布	用途	供稿
披针叶茴香（莽草）	*I. lanceolatum* A. C. Smith	3～10	—	苏、皖、浙、赣、闽	香料，有毒可入药，农药	702
八角	*I. verum* Hook. f.	20～30	30	桂、滇、粤、闽有栽培	香料、药用、调味、材用、水源	602

开花结实 10～12年生开始结实。八角20～80年生为结实盛期。丰年间隔期为1年，但

不甚显著。影响结实的主要原因是丰产年养分消耗过多，收果后补施综合肥料，可减轻大小年现象。花两性，单生或2～3朵簇生于头年生嫩枝的叶腋或近顶端。花被7～21，粉红至深红色或黄色。披针叶茴香雄蕊6～11，心皮10～13。八角雄蕊11～20，心皮通常为8。花丝舌状或近圆柱状，药室内侧向纵裂。花柱短钻形。子房1室，胚珠1。据浙江杭州1977～1980年对披针叶茴香的观察，4月下旬为始花期，5月上旬为盛花期，5月下旬为末花期，10月果实成熟。不少文献认为八角每年开花结实两次，第1次为3～4月开花，当年9～10月果实成熟（称大造果或正造果）；另一次在7～8月开花，翌年3～4月果熟（称小造果或春造果）。据黄卓民等人研究，在广西的宁明、德保等地，八角每年7月上旬～中旬为初花期，7月下旬～8月下旬为盛花期，9～11月为末花期，其余各月均有花零散开放；在南宁市郊，7月中旬～下旬为初花期，8月上旬～9月上旬为盛花期，9～11月为末花期，3～4月有花零散开放；在玉林和藤县，7月下旬～8月上旬为初花期，8月中旬～9月中旬为盛花期，9月下旬～11月为末花期，3～4月有花零散开放；在广西较北的桂林和海拔较高的金秀等地，8月上旬～中旬为初花期，8月下旬～9月下旬为盛花期，10～11月为末花期，无零散花期。他们认为，3～4月成熟的小造果系上年8月中旬以前初花期所结的果；10月份成熟的大造果系上年8月中旬盛花期所结的果；3～4月份的零散花成果率很低甚至不能坐果；因此，无论是3～4月的小造果还是10月的大造果，均系上一年8月前后初花期和盛花期所结的果。育苗宜用大造果的种子。聚合蓇葖单轮排列，未熟时直立，熟时各个蓇葖向外伸长成放射状。腹缝线开裂，每蓇葖内有种子1粒。胚乳丰富，含油脂，胚较小（表2、图1）。

表2　八角属树种果实和种子的形态特征

树　种	未熟果颜色	成熟果实			成熟种子		
		形　状	大小（mm）	颜　色	形　状	大小（mm）	色　泽
披针叶茴香	青绿色	蓇葖数10～13，大小不整齐，先端有细长而弯曲的尖头	径25～40	黄褐色	广椭圆形	长6～7.5 宽4～5.5	淡褐色，有光泽
八角	绿色	蓇葖数(7)8(9)，大小整齐，先端钝或钝尖	径32～40 高7～8	黄褐至红褐色	扁椭圆形	长8～10 宽6～8 厚4～5	黄褐色

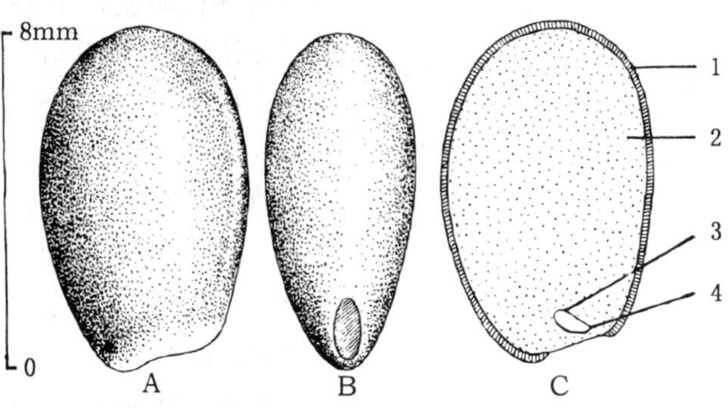

图1　八角种子的侧面（A）、腹面（B）及其纵切面（C）
1. 种皮　2. 胚乳　3. 子叶　4. 胚根
（黄应钦绘）

果实的采收调制和种子贮藏　非育苗用的八角，当果实呈黄绿色或黄色时即可采收。用于育苗的披针叶茴香和八角，采果需推迟10～15天，待果实呈深黄褐色时方可采收。上树采摘或用高枝剪采收。采回的果实在室内摊晾几天后蓇葖开裂，即可翻动脱粒，或用竹签挑出种粒。这个方法所得的种子品质较好。也可当树上蓇葖开裂后上树摇动枝条，在地面收捡震落物，筛去果皮和杂质，得到种子，但品质略差。

八角属胚乳富含油脂，种子又忌脱水，发芽能力的保存期较短。10月初以前处理出的八角种子宜随采随播。10月上旬末以后处理出的八角种子和披针叶茴香的种子，因气温转低，播种后发芽不齐，宜混湿沙贮藏，或将种子裹上黄泥浆，阴干后装入瓦缸内贮藏。采用这种方法，种子的贮藏期约为8个月，次年3月中旬即应播种，超过5月份以后，发芽能力将明显下降。常温条件下种子不能贮藏越夏。不宜用裸露的种子直接贮藏。3～5天内的短期运输，可将种子装入木箱或麻袋，运输途中需防雨防晒，运抵后应及时混沙贮藏。八角每千克有鲜果约160个，果粒大的每千克有果80～90个，果形小的每千克约250个。这两个树种的出种率和种子的净度、质量等见表3。

表3　八角属树种果实出种率和种子净度、质量

树　种	出种率（%）	净　度（%）	千粒重（g）		每千克纯净种子粒数（粒）	
			一　般	变动范围	一　般	变动范围
针叶茴香	3～13	90～99	70	50～110	14 300	9 000～20 000
八角	18～24	95～100	120	110～130	8 300	7 700～9 000

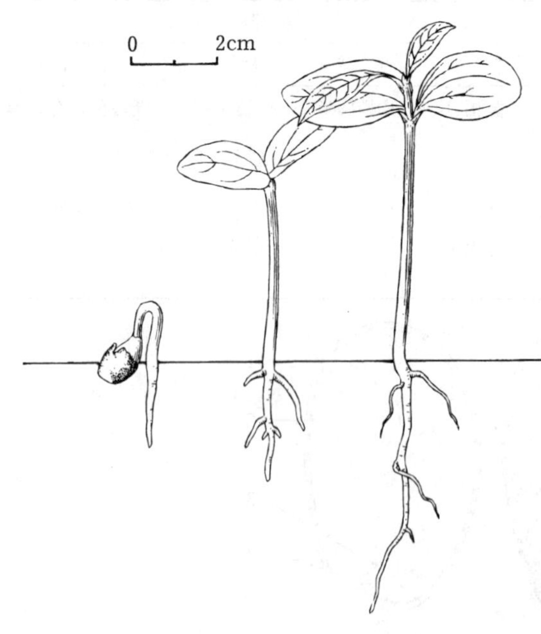

图2　八角种子萌发后第10、18、45天幼苗的生长情况
（黄应钦仿《主要树木种苗图谱》）

发芽和播种　种子无休眠习性。萌发时的日均温在18℃以上。新鲜种子发芽率可达80%，一般为70%左右，发芽高峰不明显。据杭州植物园试验，经过层积的披针叶茴香种子，60天的发芽率为68%。广西林业科学研究所1982年3月中旬用沙藏越冬的八角种子在室外沙床播种，播后10～15天日平均温度达18℃时胚根萌发，第18天进入发芽盛期；从发芽开始至发芽高峰的28天中发芽百分数为52%，从播种之日起算，68天的发芽率为68%～75%。出土萌发。胚根萌发后10天左右下胚轴拱出地面，再10天后子叶出土并展平，再20天后发出初生叶。下胚轴上部圆而略扁，淡绿色，近根颈处白色。具主根，根颈处有2～3根侧根，淡褐黄色（图2）。

条播。苗圃宜用新开荒地。新鲜种子可不作任何处理直接播种。经过沙藏的种子用清水浸泡4～5小时。每平方米播种12～20g。1年生苗可以出圃。

（韦增健）

五　味　子

Schisandra chinensis（Turcz.）Baill.

（五味子科　Schisandraceae）

生长习性、分布和用途　五味子属有 25 种，我国 19 种，本文描述 1 种。落叶藤本，长达 8m。稍耐荫，喜湿润环境，耐寒，喜湿润肥沃、排水良好的土壤。常生于沟谷溪旁及次生阔叶林或针阔叶混交林中，缠绕它树而生。分布于东北、华北及豫、鲁。朝鲜半岛、日本也有分布。果为著名中药，又可酿果酒，制饮料。茎皮可作调料。可作庭园垂直绿化材料供观赏。

开花结实　人工栽培 3 年生，野生 3～4 年生开始开花结实，4～10 年生为结实盛期。结实大小年现象明显，丰年间隔期一般为 1～2 年。花单性，异株。有细长花柄，雌花柄长，雄花柄短，稍下垂。花单生或 2～4（偶见 5～7）簇生于叶腋，乳白色，偶见粉红色，稍有香气。花被片6～9，雄蕊 5。雌蕊群长圆状椭圆形，长 2～4mm，心皮 17～40。子房卵形或卵状椭圆形，柱头鸡冠状。果时花托伸长，成熟心皮排列成穗状的聚合浆果。长 1.5～8.5cm。小浆果近球形，深红色，径 0.6～1.1cm，内有 1～2 粒种子，含两粒种子者多数仅 1 粒成熟。种子肾形有光泽，长 4～6mm，宽 3～5mm；淡橘黄色，种脐明显凹入呈 U 形，种皮透水性不强。胚小，长仅0.2～0.5mm，胚乳丰富（图 1）。开花结实物候见表 1。

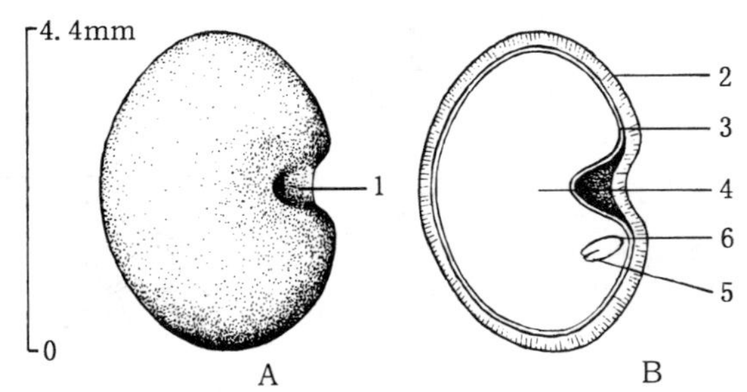

图 1　五味子种子外形（A）及其纵切面（B）
1. 种脐　2. 外种皮　3. 内种皮　4. 胚乳　5. 子叶　6. 胚根
（梁鸣、黄应钦绘）

表 1　五味子开花结实物候

观察年限 地点	开花期			果熟期			果实散落期
	始　期	盛　期	末　期	始　期	盛　期	末　期	
1988 黑龙江逊克县	6 月上旬	6 月中旬	6 月下旬	8 月下旬	9 月上旬	9 月下中旬	9 月下旬～10 月上旬
1991 黑龙江五常县	5 月下旬	5 月下旬	6 月上旬	8 月下旬	9 月上旬	10 月上旬	9 月下旬～10 月上旬
黑龙江江山骄林场	5 月下旬	6 月上旬	6 月中旬	8 月中旬	9 月上中旬	9 月下旬	—

果实的采收调制和种子贮藏　聚合浆果呈红色并稍变软后，从藤上摘取果穗。堆沤，捣碎，用水漂洗出果皮与果肉，稍阴干后去杂，即可得纯净种子，或提取果汁后留种。出种率为 2.8%～8.0%，净度 90%～98%，千粒重 21～25g，出种率与千粒重由北向南递增。人工

栽培植株的种子千粒重约 30g，每千克 3.3 万～4.7 万粒，晾晒至含水量 10%～11%时贮藏。低温（0～5℃）干藏可保存 1～2 年。在种子库密封低温（0～5℃）贮藏，可保存 3 年以上，但发芽率明显降低。

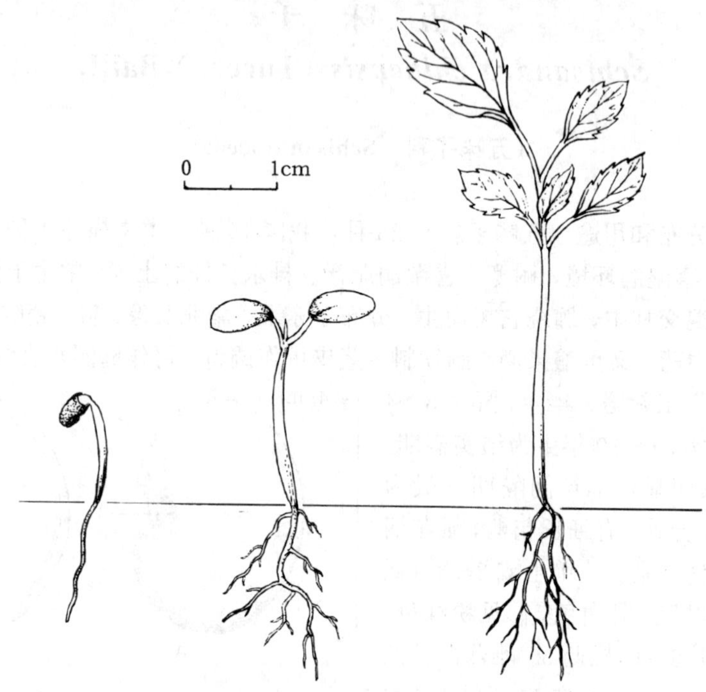

图 2　五味子种子发芽后第 4、17、30 天幼苗生长情况

（曹雅范绘）

发芽和播种　胚小且发育不全，需要经过形态后熟。常自立等人（1990）报道，温水浸种 3～5 昼夜后层积至翌春解冻，可以解除形态休眠。据范长生（未刊稿）介绍，10℃以下为裂口温度，适于发芽的温度为 13～17℃，超过 20℃则不利于发芽。黑龙江省林副特产研究所中草药研究室（1988）报道，只有温度在 5～15℃范围内，五味子才能完成其胚的分化，在形态上表现出生长而转入发芽。在他们的研究中，在 13～14℃的恒温中层积 55 天，只有少数种粒能够发芽；层积 70～80 天时发芽率为 15%；在 10～12℃下层积 60～70 天后转入 5℃下继续层积，可以取得满意的结果。秋播也能起到层积的作用。据刘德旺（1990）报道，当年采收的种子用浓硫酸浸泡 7 分钟，清水冲洗干净后用 500μg/g 赤霉素溶液处理 12 小时，以沙为基质，在 23℃的温箱中培养 15 天，发芽率可达 85%。

周德本等人（1991）重复过上述方法，没有取得预期效果。周德本等人（1991）的结果是，层积时变温优于恒温：在 110 天的层积期中，无论是先用 3～5℃后转 13～19℃，还是先用 13～19℃后转 3～5℃，发芽裂口的种子都可达到 90%以上，且发芽集中，出苗早。

苗床条播。每平方米播种 25g。播后 20～30 天出苗，子叶出土萌发（图 2）。幼苗需要遮荫。种子播前处理得好，当年平均苗高可达 20cm，最高可达 60cm，可以出圃。种子播前处理得不好，当年苗高只有 5～8cm，继续培养 1 年才能出圃。

（周德本、王彩荣）

领　春　木

Euptelea pleiosperma Hook. f. et Thoms.

（领春木科　Eupteleaceae）

生长习性、分布和用途　领春木属有2种，我国只产本文描述的1种，是典型的东亚植物区系成分的特征种，古老的残遗植物，在《中国植物红皮书》中列为稀有种。落叶小乔木，高5～15m，胸径可达28cm。主根发达，侧根较少。适生于土层深厚，富含有机质的沙壤土和壤土，大都生于山谷、溪边杂木林中，多系零星散生。分布于冀、晋、豫、陕、甘、西南、华中和浙、皖。印度、缅甸也有分布。木材淡黄色，可作家具、农具。树皮可作栲胶原料。树姿优美，先花后叶，花果簇生而红色，可供观赏。

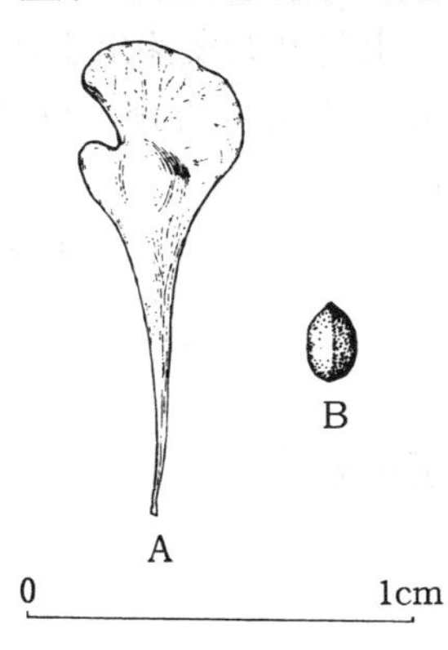

图1　领春木翅果（A）和种子（B）的外形
（童军平绘）

开花结实　花两性，6～12朵簇生叶腋。无花被。雄蕊6～14，花药红色，较花丝长，药隔顶端延长成附属物。心皮6～12，离生，子房歪斜，有长柄，1室，每室有1～3（—4）枚倒生胚珠。花期4～5月，7～10月果实陆续成熟。结实较多，挂果时间长。1991年在浙江西天目山观察，12月中旬树上还挂有许多翅果。

果实为聚合翅果。小翅果为不规则倒卵圆形，边缘具膜质翅，先端圆，一边凹缺，下端渐细成明显的柄。小翅果长7～13mm，宽3～5mm。每果有种子1～3（4）粒，多数1～2粒。种子卵圆形或卵形，两边厚薄不均，薄的一边有一条明显的棱脊。种子紫黑色或黑色，有光泽，长1.5～2.0mm，宽0.8～1.3mm。胚小，胚乳丰富，含油分。果实和种子的外形见图1。

果实的采收调制和种子贮藏　果实由绿色转为棕褐色或褐色时上树采集，或用竹竿击落至采种布上收集。果实采回后，清除树枝和树叶等杂物，在通风处摊晾阴干，不宜曝晒。清除果翅果皮后可以取得纯净种子。纯净种子千粒重0.7～1.2g，每千克纯净种子83.3万～143万粒。不过领春木果实较小，去除果皮果翅比较费工，所以供翌春播种用的种子多用布袋或塑料袋带果皮干藏。

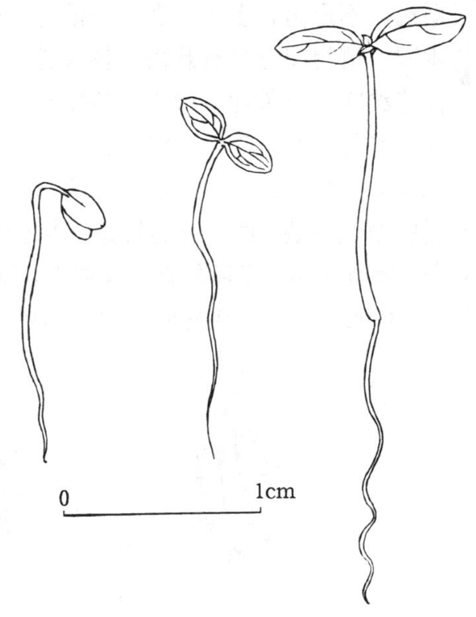

图2　领春木种子萌发后第1、3、30天幼苗的生长情况
（童军平绘）

发芽测定　领春木种子萌发需要光照。据万才淦（1988）报道，在25℃、每天光照（900～1 100Lx）8小时的条件下，20天的发芽率为60%，无光照的只有3%；冷层积（3～6℃）30天的种子，在同样的温度和光照条件下发芽率为73%，无光照的只有12%；赤霉素（GA_3）可以部分地取代领春木种子对光照的需求：100～500μg/gGA_3浸种28小时再冷层积30天的种子，在无光条件下发芽率可以达到54%～64%。

播种　春播。多用条播。南岳树木园对种子的播前处理是用清水浸种24小时后，薄摊在室内光照充足的地方并保持湿润，经常翻动，使种子均匀受光。据认为，这样处理7～10天再行播种，能明显提高发芽率（庾府诗，1984）。3月上旬播种，4月上旬开始发芽出土，5月中旬结束发芽。出土萌发，子叶2枚。出土子叶淡黄绿色，长约2mm。出土后约20～26天出现初生叶。种子发芽情况见图2。1年生苗高60～65cm，可以出圃。

（史晓华）

连香树

Cercidiphyllum japonicum Sieb. et Zucc.

（连香树科　Cercidiphyllaceae）

生长习性、分布和用途　连香树科仅1属1种1变种。本文描述1种，为东亚孑遗植物之一，稀有种，已列入《中国植物红皮书》。在我国星散分布于黔、川、甘、陕、晋、豫、鄂、湘、赣、浙、皖等地，以鄂西及川中两地为多。日本也有分布。落叶大乔木，高可达40m。喜温暖湿润，不耐荫，多生长在沟谷溪边。幼时生长迅速，寿命长，材质优良，树形美观。各地树木园、植物园已多有栽培，供观赏、研究。

开花结实　花单性，腋生，雌雄异株，无花被。苞片在花期呈红色。雄花具短梗，雄蕊15～20。雌花具梗。离心皮雌蕊2～6，每一心皮含胚珠数个。聚合蓇葖果2～6，荚果状，长8～18mm，径2～3mm，暗紫褐色，微弯，先端渐细，有宿存花柱，每果含种子10～15粒。成熟时沿腹缝线开裂。种子一端具翅，连翅长5～7mm。种子扁椭圆形，长2～3mm。种皮膜质，棕褐色。胚乳丰富，子叶较薄，长卵形，下胚轴与胚根部圆柱形，约与子叶等长（图1）。4月中旬～5月上旬开花，花先叶开放或与叶同放，果熟期8～10月。

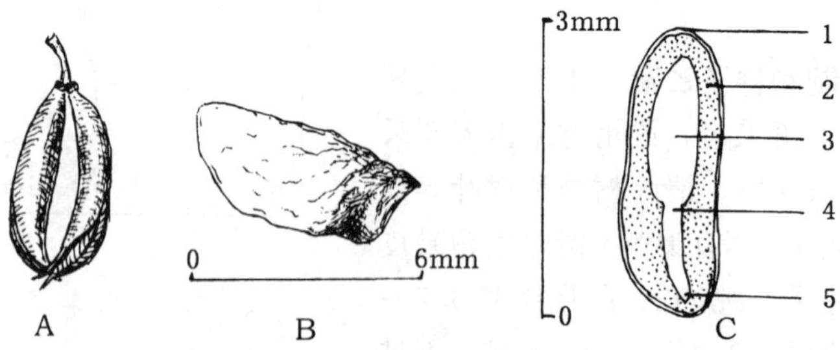

图1　连香树蓇葖果外形（A）、具翅种子外形（B）及种子纵切面（C）

1. 种皮　2. 胚乳　3. 子叶　4. 胚轴　5. 胚根

（史渭清绘）

果实的采收调制和种子贮藏　连香树树体高大，采种比较困难。现有母树又零星分布，结实量少。为避免有限的资源散失，应在9～10月初，蓇葖果出现开裂征兆时及时采摘果穗，摊晾阴干，抖出带翅种子后轻揉去翅，筛去杂质即得纯净种子。也可以带果干藏后在播前再调制脱粒。种子适于干藏，但不宜久藏。种子千粒重0.75～0.9g，每克种子有1 100～1 300粒。净度可达90%以上。

发芽和播种　种子无休眠习性，室内发芽测定20天可以结束。春播。据南岳树木园报道，1980年3月11日播种，4月11日～5月14日发芽。场圃发芽率可达65%。连香树种子资源不多，幼苗出土时生长细弱，常受大雨袭击，可先播在盆内，待苗高4～6cm，真叶出现后按株行距10cm×15cm移栽。幼苗期需遮荫防雨。子叶出土萌发，初生叶对生（图2），全株呈亮绿色，幼茎、叶柄、叶缘和叶脉浅红色。连香树萌蘖力强，生长旺，可扦插繁殖或压条繁殖。

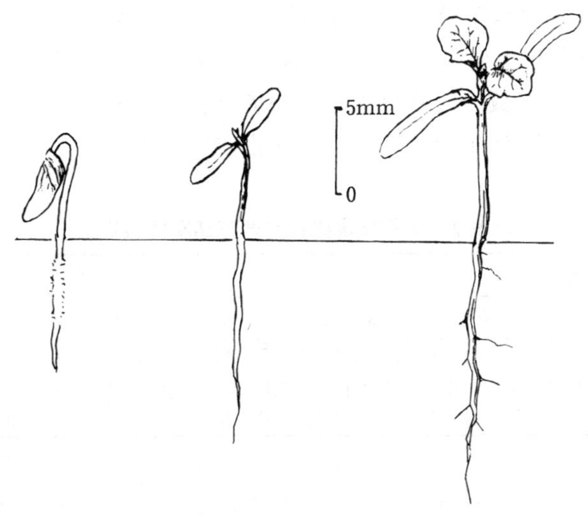

图2　连香树种子萌发后第3、10、20天的幼苗生长情况

（史渭清绘）

（何泽瑛）

番荔枝属

Annona L.

（番荔枝科　Annonaceae）

生长习性、分布和用途　本属约120种，产非洲和美洲的热带地区，我国引入7种，本文描述3种。小乔木至乔木，高4～10m，径10～20cm。较速生。不耐霜冻，也不耐较长期的5℃低温。喜肥沃湿润的酸性土。这3个树种的名称、树高，在我国的栽培地区及用途见表1。其中牛心果、红毛榴莲为常绿树种，番荔枝为落叶树种。

表 1 番荔枝属树种的名称、生长、分布和用途

中 名	学 名	树高 (m)	胸径 (cm)	我国栽培地区	用 途
牛心果（圆滑番荔枝）	*A. glabra* L.	10	20	粤、琼、桂、滇、浙、台	观赏、纤维、材作瓶塞
红毛榴莲（刺果番荔枝）	*A. muricata* L.	5～8	8～12	粤、琼、桂、滇、台	热带名果，食用、观赏、纤维、紫胶虫寄主
番荔枝	*A. squamosa* L.	4～6	8～12	粤、琼、桂、滇、闽、浙、台	热带名果，食用、纤维、药用、紫胶虫寄主

开花结实 4～6 年生始开花结实，结实后第 3 年所产的种子即可用于播种。结实丰年间隔期为 1 年。影响结实的主要原因是养分供应不足，在我国栽培则还有冬季气温过低这个因素。如果冬无严寒而且加强培肥管理，也可连年丰产。花两性，单生或 2～4 朵集生、顶生或与叶对生。萼片 3，镊合状排列。花瓣 6，2 轮，或内轮 3 花瓣退化成鳞片状。雄蕊多数，药隔膨大，顶端截形。心皮多数，通常合生。每心皮有胚珠 1 枚，基生，直立。据广西南宁及海南观察，花蕾于 4 月中下旬出现，开花结实的物候期见表 2。

表 2 番荔枝属树种的开花结实物候期

树 种	观察地点	观察年份	开花期			果实成熟期		果实脱落期
			始 期	盛 期	末 期	始 期	盛 期	
牛心果	南宁	1987～1988	5 月上旬	5 月中下旬	6 月上旬	7 月上旬	8 月中旬	7 月中旬～9 月上旬
红毛榴莲	南宁	1987～1988	5 月中旬	5 月下旬	7 月上旬	7 月上旬	8 月下旬	7 月中旬～9 月中旬
番荔枝	海南	—	5 月上旬	5 月中下旬	7 月中旬	7 月下旬	11 月下旬	7～11 月

花期不甚整齐，除上表所列的主要开花结果期外，尚有零星不定期的开花结实现象；同一树上常有成熟果与花共存。聚合浆果，球形、心形或圆锥形，肉质，成熟后不开裂。每果有种子多粒。胚乳丰富，皱折状，胚微小。果实及种子形态见表 3、图 1。

表 3 番荔枝属树种果实和种子的形态特征

树 种	未熟果颜色	果实 形 状	果实 大小 (cm)	果实 色 泽	种子 形 状	种子 大小 (mm)	种子 色 泽
牛心果	浅绿至青绿色	椭圆形或卵圆形，平滑	长 5.5～9 径 4～7.8	淡黄色	矩圆形	长 13～16 宽 8～10 厚 4.5～6.5	浅黄色
红毛榴莲	青带浅黄色	阔卵圆形或近球形，幼时有下弯的刺，渐脱落后残存小突体	长 9～11 (35) 径 8～9 (15)	淡黄色	矩圆形或扁椭圆形	长 12～16 宽 9～12 厚 5.5～8	黄褐色
番荔枝	青绿色	球形或心状圆锥形	径 3～8	黄绿色，被白粉	扁椭圆形，两端尖	长 10～16 宽 6～8 厚 4～5.5	紫褐色

果实的采收调制和种子贮藏　采后即供食用的果实，应在果实充分成熟软化后用手采摘。需要贮存、运输的食用果实，应在果皮由青色转变为黄色但尚未软化时采下，用竹筐装运，每筐不宜超过25kg，以免堆沤发热。红毛榴莲和番荔枝的果实软化食用后所得的种子，用清水洗净并漂去空粒，即为播种材料。牛心果的风味较差，一般不作食用，可在提取果汁后，或任其自然软熟后，置水中浸泡搓洗，漂去果肉残渣，得出种子。种子含油脂，不宜脱水，忌日晒，含水量宜保持在20%以上。9月以前采集的种子宜随采随播，不需贮藏。10月以后处理出的种子宜混沙贮藏越冬，贮藏期以不超过8个月为宜。种子的净度、质量等见表4。

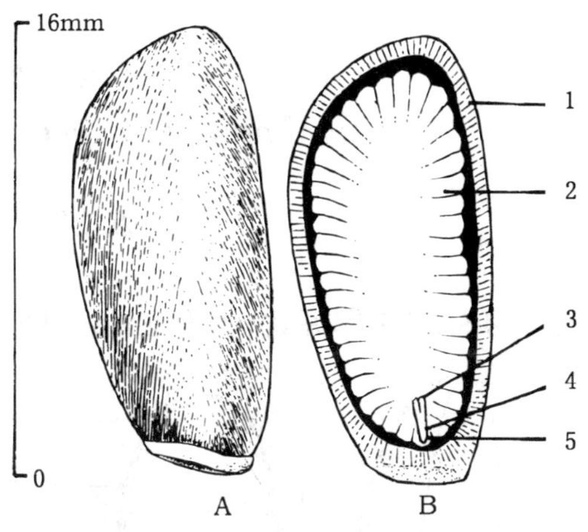

图1　番荔枝种子的外形(A)及其纵切面(B)
1. 种皮　2. 胚乳　3. 子叶　4. 胚轴　5. 胚根
(黄应钦绘)

表4　番荔枝属树种果实的出种率和种子净度、质量

树　种	出种率 (%)	净　度 (%)	千粒重 (g)	每千克纯净种子粒数 (粒)
牛心果	35～48	95～99	230～280	3 570～4 350
红毛榴莲	25～30	95～99	250～350	2 860～4 000
番荔枝	25～35	95～99	150～250	4 000～6 670

发芽和播种　番荔枝可在13℃左右的低温条件下发芽，其余两个种发芽时日平均温度需在20℃以上。采用鲜种或经混沙湿藏的种子，播种前均可不作任何处理。广西林业科学研究所曾对这3个种用沙床作过发芽测定：经过沙藏的番荔枝种子1978年10月20日播种，播后74天即1979年1月2日开始发芽，2月1日发芽终止，发芽率为52%。红毛榴莲和牛心果使用的都是新鲜种子，分别于1987年7月20日和9月11日播种，红毛榴莲播后24天开始发芽，8天中发芽52%，从开始萌发到发芽结束的42天中发芽率为82%。牛心果播后36天开始发芽，从开始萌发到发芽结束的58天中发芽率为50%，未见明显的发芽盛期。

出土萌发。红毛榴莲胚根萌发后5天子叶出土，9天发出初生叶。牛心果和番荔枝种子萌发后约9天子叶出土，11天发出初生叶。

条播。牛心果、红毛榴莲每平方米播种40～50g，番荔枝每平方米播种30～40g。1年生苗可以出圃。

（韦增健）

鹰　爪　花

Artabotrys hexapetalus（L. f.）Bhandari

（番荔枝科　Annonaceae）

生长习性、分布和用途　鹰爪花属约100种，我国有4种，本文描述1种。常绿攀援灌木，高达4m。耐庇荫。萌生力强。喜高温高湿的气候环境。忌干旱，较耐水湿，适生于肥沃湿润的酸性土，钙质土亦能生长。分布于浙、台、闽、赣、粤、琼、桂、滇，多为栽培，少数野生。印度、斯里兰卡和东南亚也有栽培或野生。枝叶翠绿，花极芳香，可提制香料，可作花篱、花墙、花架，为常见的观赏花木。

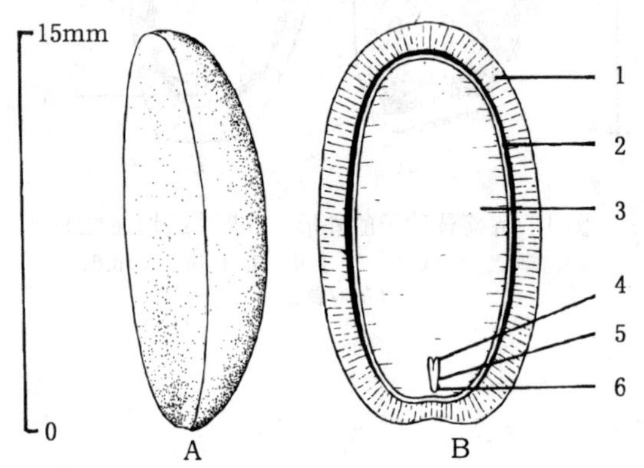

图1　鹰爪花种子的外形（A）及其纵切面（B）
1. 外种皮　2. 内种皮　3. 胚乳　4. 子叶
5. 胚轴　6. 胚根
（黄应钦绘）

开花结实　实生植株4～6年生开始开花结实，大小年现象不明显。花两性，单生或两朵共生于钩状的总花梗上，淡绿或淡黄色。萼片3。花瓣6，2轮，长圆状披针形，长3～4.5cm。雄蕊多数，长圆形，药隔三角形，无毛。心皮4至多数，长圆形。柱头条状长椭圆形。胚珠2，基生。据在广西南宁观察，5月上旬为始花期，6月上中旬为盛花期，下旬为末花期。有些栽培品种的花期可延至8月下旬。果实成熟期为12月～翌年2月。小浆果卵圆形，离生，肉质，聚生于坚硬的果托上，顶端尖，未成熟时绿色，成熟后为黄色，长2.5～4cm，径1.5～2.5cm。每果有种子两粒。种子龟背状椭圆形，深褐色，长12～17mm，宽8～12mm，厚6～7.5mm，胚乳丰富，胚小（图1）。

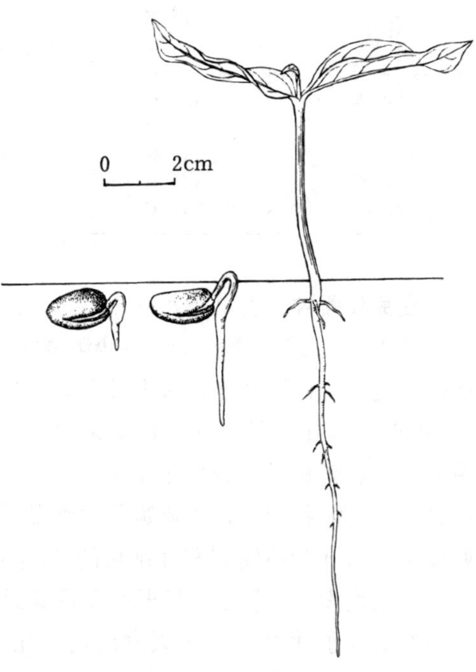

图2　鹰爪花种子萌发后第5、13、25天幼苗的生长情况
（黄应钦绘）

果实的采收调制和种子贮藏　采集的果实堆沤5～7日，俟果肉果皮软熟后置水中搓擦，淘去果皮等杂质，洗净种子即可。按鲜果质量计算的出种率为15%～20%。种子含水量约为20%，忌失水，不能日晒或干藏。种子的净度为92%～98%。千粒重为610（560～660）g。每千克纯净种子为1 600

(1 500～1 800) 粒，运输时可以果实的状态包装，运抵后再行调制。宜随采随播种，一般不作贮藏。半个月内的短期贮存也可以暂不调制而带果肉存放。已调制的种子贮藏时需混湿沙，贮藏期一般 3～5 个月。

发芽和播种　种子有浅休眠现象。发芽时日均温宜在 20℃以上。1989 年 1 月 7 日，广西林业科学研究所用刚采集的新鲜种子在室外沙床播种，播后至 5 月 3 日开始萌发，发芽盛期不明显，5 月 24 日发芽终止，发芽率为 50%。出土萌发。胚根萌发后约 7 天下胚轴拱出地面，13～15 天子叶带壳出土，约 20 天子叶展开，约 25 天发出初生叶（图 2)。

条播。每平方米播种 100～150g，覆土 1.5～2cm。生产中常用扦插育苗。1 年生苗出圃。

（曾　玲）

假鹰爪属
Desmos Lour.

（番荔枝科　Annonaceae）

生长习性、分布和用途　本属约 30 种，我国 4 种，本文描述 2 种。常绿直立灌木或攀援灌木，高 0.7～3m。极耐荫，空旷地亦能生长。喜湿润肥沃的酸性土。这两个种的名称、生长情况、分布及用途见表 1。

表 1　假鹰爪属树种的名称、生长、分布和用途

中　名	学　名	生长习性	分　布	用　途	供　稿
假鹰爪（酒饼叶）	*D. chinensis* Lour.	直立，上部蔓延，高 0.7～1m	粤、桂、滇、黔。印度和东南亚	花芳香，果美观，根叶入药	601
毛叶假鹰爪	*D. dumosus* (Roxb.) Saff.	攀援状，高 2～3m	滇、黔、桂。印度和东南亚	花极香，茎皮纤维作绳索	619

开花结实　4～6 年生开始结实，10 年后进入正常结实期，大小年现象不明显。花两性，单生叶腋或与叶对生或互生，淡黄色。花萼裂片 3，镊合状排列。花瓣 6，2 轮，镊合状排列，外轮较大。雄蕊多数，药室线形，外向，药隔顶端近圆形或截形。心皮多数，分离，柱头 2 裂，胚珠 1～8，倒生。1987～1988 年在广西南宁观察，假鹰爪 5～8 月陆续开花，其中主花期 2 次，一次在 5 月上旬，果实成熟于 7 月中旬；一次在 7 月下旬～8 月下旬，果实成熟于 11 月中旬至翌年 1 月下旬。1985～1987 年在云南勐仑观察，毛叶假鹰爪 5 月下旬始花，6 月为盛花期，7 月中旬为末花期，11 月中旬果实开始成熟，12 月中旬为果熟盛期。小浆果

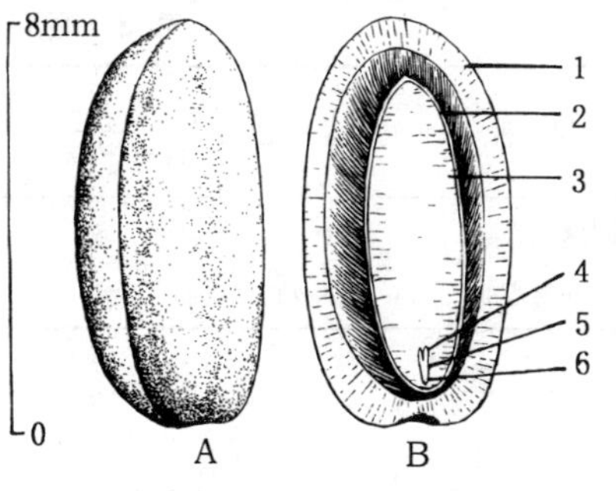

图 1　假鹰爪种子的外形(A)及其纵切面(B)
1. 外种皮　2. 内种皮　3. 胚乳
4. 子叶　5. 胚轴　6. 胚根
（黄应钦绘）

(成熟心皮)多数，长1.8～8cm，伸长而在种子间缢缩成念珠状。每果有种子1～8粒，多数为3～5粒。种子有胚乳，子叶薄。果实及种子形态见表2、图1。

表2 假鹰爪属树种果实和种子的形态特征

树 种	果实			种子		
	形 状	大 小(cm)	颜 色	形 状	大 小(mm)	颜 色
假鹰爪	念珠状	长1.8～8 径0.5～0.7	黄转红色	卵形或椭圆形	长5～8 径4～6	浅褐色
毛叶假鹰爪	念珠状	长2～5 径0.6～0.8	紫黑色	卵形或椭圆形	长6～10 径5～7	浅褐色

果实的采收调制和种子贮藏 果实呈现成熟颜色时用手采摘。采后在室内堆沤3～5日，果肉软熟后装入竹筐或布袋，置水中反复搓揉，淘去果皮等杂质即得种子。种子忌失水，不能日晒，含水量应保持在30%以上。运输时可以果实的状态装运，运抵后再行调制。贮藏时需混湿沙，贮藏期一般为半年左右。种子的质量、净度等见表3。

表3 假鹰爪属树种的出种率和种子的净度、质量

树 种	出种率(%)	净 度(%)	千粒重(g)		每千克纯净种子粒数	
			一 般	变动范围	一 般	变动范围
假鹰爪	50～60	95～99	128	100～150	7 800	6 600～10 000
毛叶假鹰爪	—	90～98	270	—	3 700	—

发芽和播种 种子无明显的休眠现象。可以随采随播。1987年，广西林业科学研究所和西双版纳热带植物园用新采种子，在室外沙床上分别对假鹰爪和毛叶假鹰爪进行过发芽测定，结果见表4。

表4 假鹰爪属树种的发芽能力及其测定条件①

树 种	气 温(℃)	发芽势(%)		发芽率(%)		
		计算天数	一般数值	计算天数	一般数值	变动范围
假鹰爪	29～35	8	70	29	83	80～92
毛叶假鹰爪	15～23	20	65	40	80	75～90

① 室外沙床

假鹰爪留土萌发，发芽后约5天上胚轴出土，具初生不育叶3～5，鳞形，9天左右发出初生叶。毛叶假鹰爪出土萌发，子叶出土后6～10天展出初生叶。

条播或点播。假鹰爪每平方米播种20～25g。毛叶假鹰爪每平方米播种60～75g。1年生苗出圃。也可将半年生苗移至花盆内培育盆景。

(王宏志)

哥　纳　香

Goniothalamus chinensis Merr. et Chun

（番荔枝科　Annonaceae）

生长习性、分布和用途　哥纳香属约50种，我国10种，本文描述1种。常绿灌木，高3m。喜酸性土。较耐荫，常见于湿润的林中和林缘。耐寒力低。分布于海南和广西十万大山。为水源林下层林木，果可食。

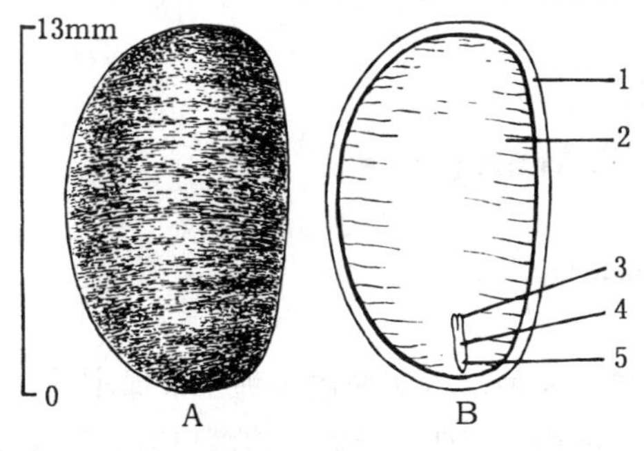

图1　哥纳香种子的外形(A)及其纵切面(B)
1. 种皮　2. 胚乳　3. 子叶　4. 胚轴　5. 胚根
(黄应钦绘)

开花结实　3～6年生开始开花结实，正常结实期在8年生以后。结实无大小年现象。花两性，黄绿色，1～2朵生于叶腋。花梗长约1cm，基部有小苞片数枚。萼片3，镊合状排列。花瓣6，2轮。外轮花瓣窄披针形，长2.2～3cm，宽0.7cm。内轮花瓣卵圆形，长1.2cm，宽1.0cm。雄蕊多数，条状长圆形，长约2mm。心皮多数，长圆形。每心皮有胚珠2。花柱与心皮等长，柱头2深裂。据广西上思1987年观察，花始期在6月中旬，盛期在7月上旬，末期在8月下旬。果实成熟始于9月初，盛期在9月下旬，11月初为果熟末期。聚合浆果。小果8～20，分离，椭圆形，长1～1.8cm，径0.8cm。每小果有种子1颗。种子椭圆形，一面略平，表面光滑，褐色至灰褐色，长1～1.3cm，径6～7mm。胚小，直立。胚乳丰富（图1）。

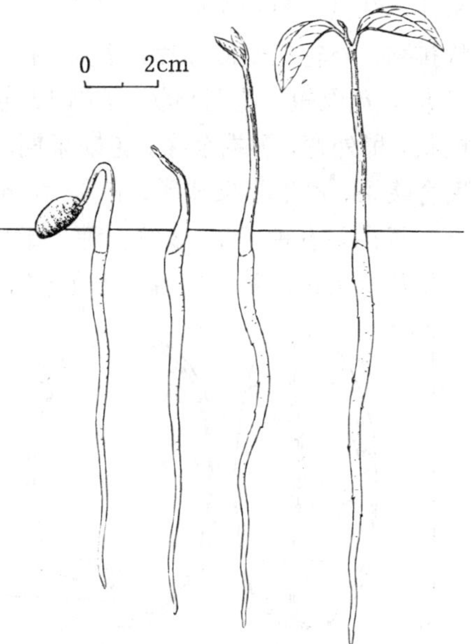

图2　哥纳香种子萌发后第12、14、22、30天幼苗的生长情况
(黄应钦绘)

果实的采收调制和种子贮藏　果实易受鸟兽危害，在成熟盛期用手或枝剪采摘。采回的果实在室内堆放1～3天，软熟后食用或装入布袋，加入少量细沙轻轻搓揉，用水淘出种子。种子净度可达100%。千粒重320（280～350）g。每千克有纯净种子3 100（2 900～3 600）粒。种子忌失水，不宜曝晒或裸露贮存。运输及贮藏时均需混入湿沙。混沙贮藏期约为1年。

发芽和播种　种子有休眠现象。当年采收的种子宜混沙层积到翌年春末夏初播种。1987年11月10

日，广西林业科学研究所在室外沙床作发芽试验：播种时日均温约26℃，1988年5月6日开始发芽，5月10日为发芽盛期，5月15日发芽终止。发芽盛期4天中的发芽80%，从播种到发芽终止，发芽率为90%。出土萌发。胚根萌出后12天下胚轴出土，子叶随即脱落，10天后初生叶出现。主根单一，粗壮（图2）。

点播。种子沙藏至翌年4月中旬播种，每平方米播种50～60g，覆土1.5cm。夏日宜遮荫，2年生苗出圃。

（曾 玲）

野 独 活

Miliusa chunii W. T. Wang

（番荔枝科 Annonaceae）

生长习性、分布和用途 野独活属约30种，我国有3种，本文描述1种。常绿小乔木，高约5m，胸径10cm。适生于酸性土壤。分布于广东、海南、广西、云南。越南也有。为水源涵养树种，木材可作农具。

开花结实 5～7年生开始开花结实，正常结实期在10年生以后。结实大小年不显著。花两性，红色，单生于叶腋，径1.3～1.6cm。萼片3，卵形，长约2mm。花瓣6，2轮，镊合状排列。外轮花瓣似萼片，略大；内轮花瓣卵圆形，长达18mm，宽8～12mm。雄蕊倒卵形，多数，花丝短，花药卵状，药室毗连，药隔顶端尖。心皮多数，弯月形，胚珠2～3。柱头圆柱状，稍外弯，顶端全缘。花期果期均较长：花期4～7月，果期可以从7月延续到翌年春季。聚合浆果，离生小果球形，径7～10mm。小果梗细，长1～2cm，集生于长4～7.5cm的果柄上。每小果有种子1～3粒。种子半球形或扁圆形，黄褐色，厚5～8mm，长8～9mm，扁侧面具环状条纹。种皮较薄，胚乳丰富，胚小，位于一端（图1）。

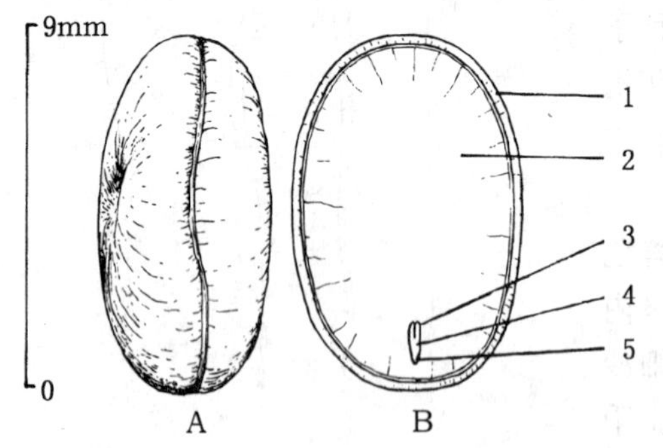

图1 野独活种子的侧面外形(A)及其横切面(B)
1. 种皮 2. 胚乳 3. 子叶 4. 胚轴 5. 胚根
（黄应钦绘）

果实的采收调制和种子贮藏 果实成熟时用手或采种刀具截断总果柄。采回的果实在室内堆放1～2日，果实充分成熟后装入筐内，置水中搓擦，淘去果皮等杂质即得纯净种子。按鲜果质量计算的出种率约32%。调制后稍晾干的种子含水量约为25%。种子净度可达95%。千粒重375（350～400）g，每千克有纯净种子2 700（2 500～2 900）粒。种子忌失水，不宜裸露贮存，运输时宜混以湿沙。夏季或春季采集的种子可以不经贮藏而随采随播。秋冬季采集的种子可混湿沙贮藏。贮藏期一般为3～5个月。

发芽和播种 种子无休眠习性。发芽时日均温宜在20℃以上。1990年10月26日广西林业科学研究所在发芽箱内用当年新采种子作过发芽测定：温度经常保持28℃，基质为湿沙，播种后29天开始发芽，至11月29日发芽终止。

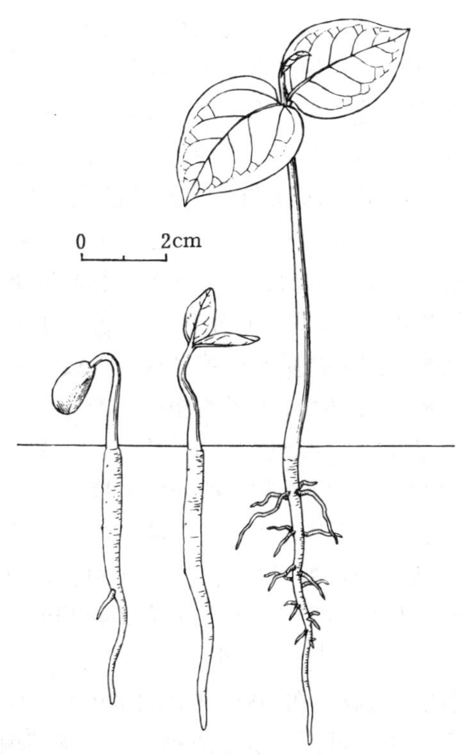

图 2　野独活种子萌发后第 10、27、50 天的幼苗生长情况
（黄应钦绘）

开始萌发后第 2 天便进入发芽盛期，以 3 天计算，发芽势为 50%。从播种之日起算，34 天的发芽率高达 90%。出土萌发。室外沙床播种后 29～35 天种子开始萌发，过 10 天左右子叶带壳出土，出土后约 17 天，种壳脱落，子叶展现，再过 1 个月左右初生叶发生，发芽情况见图 2。

撒播或条播。撒播时每平方米播种 80～120g，条播时每平方米播种 40～60g。苗期需遮荫。1 年生苗可以出圃定植。也可以密集播种后将萌出的芽苗移至容器内继续培育。这样的容器苗 3～4 个月可以出圃定植。

（曾　玲）

银　钩　花

Mitrephora thorelii Pierre

（番荔枝科　Annonaceae）

生长习性、分布和用途　银钩花属约 40 种，我国 3 种，本文描述 1 种。常绿乔木，高达 25m，胸径 50cm。能耐轻霜。喜肥沃疏松的酸性沙壤土，肥力中等以上的粘性土亦能生长良好。早年生长较慢，10 年生以后转快。分布于海南和广西。越南、老挝、柬埔寨和泰国亦产。材质坚实，宜供建筑、家具之用。花芳香，可提制香料。

开花结实　8～10 年生开始开花结实，正常结实在 20 年生以后。结实丰年的间隔期通常为 1 年。花两性，单生或总状花序腋生或与叶对生。花淡黄色，径 1～1.5cm。花序梗、花梗、花萼、花瓣均密被锈色柔毛。萼片 3。花瓣 6，2 轮，外轮大于内轮。雄蕊多数，楔形，药隔盘状。心皮多数，被毛，离生，长圆形，花柱圆柱状，每心皮有胚珠 8～10 枚，2 排。据 1987 年在海南观测，3 月上旬当年嫩枝抽出时，花芽相随出现，3 月下旬花始开，4 月上旬为盛花期，5 月上旬开花结束；6 月下旬果实开始成熟，7 月为果实成熟盛期，8～9 月仍有少量果熟。

果实成熟时由青色转变为红色，有明显环纹，密被褐色绒毛，成熟后不开裂。充分成熟的果实在树上存留4～7日后自行脱落。多心皮浆果离生，小果卵形或近球形，长1.6～2cm，径1.4～1.6cm。每果有种子8～10粒，2排列于果壁内。种子半月形，一侧平，另一侧隆起，表面粗糙，暗褐色，长10～14mm，宽6.5～8.5mm，厚3～5mm。胚乳丰富，胚小（图1）。

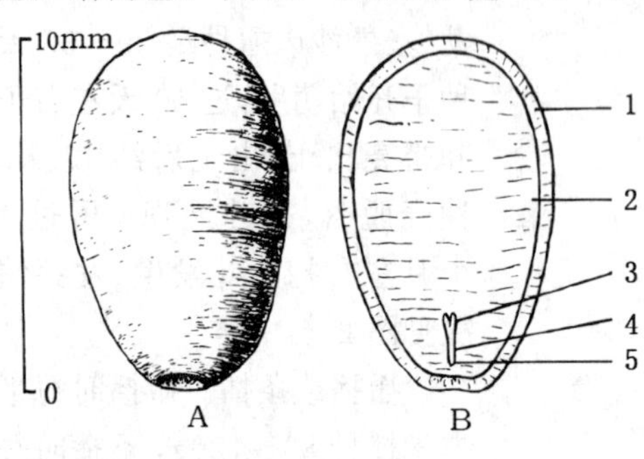

图1　银钩花种子的外形(A)及其纵切面(B)
1.种皮　2.胚乳　3.子叶　4.胚轴　5.胚根
（黄应钦绘）

果实的采收调制和种子贮藏　果实呈现红色时用钩刀带总果梗采下，也可用竹竿敲打或摇动震落。果形较大而色红，在地面容易捡拾。果实采回后堆沤2～3天，充分软化后盛入竹筐置水中反复搓洗，淘去杂质即得纯净种子。按鲜果质量计算的出种率约40%。种子的净度一般为94%～98%。千粒重320（300～350）g。每千克有纯净种子3 100（2 800～3 300）粒。种子含水量20%～25%，不宜失水，忌日晒，亦不耐久藏。洗净的种子置室内通风处晾干后即可播种。运输时可拌湿锯木屑或湿沙，用竹筐或木箱装运。贮藏需混拌湿沙。混沙贮藏的时间不宜超过10个月。

发芽与播种　种子有休眠现象。发芽时的日均温需在20℃以上。1980年3月1日（日均温约17℃），广西林业科学研究所用头年8月采集经过沙藏越冬的种子作过发芽测定：播后68天，至5月7日（日均温26℃）开始萌发，5月13～25日为发芽盛期，28日发芽终止。从开始萌发至发芽高峰的18天中，发芽65%，从播种至发芽终了的89天中，发芽率为76%。留土萌发，或子叶稍出土即脱落。胚根萌发后约7天幼苗出土，10天左右发出初生叶。

条播。每平方米播种40～60g，覆土1.5～2cm。培育7～9个月后次年春季可以出圃。

（韦增健）

细　基　丸

Polyalthia cerasoides (Roxb.) Benth. et Hook. f. ex Bedd.

（番荔枝科　Annonaceae）

生长习性、分布和用途　暗罗属约120种，我国有17种，本文描述细基丸这1种。3月间新叶发出前有约10天的落叶期。乔木，高20m，胸径40cm。分布于琼、粤、桂、滇。东南亚地区亦产。木材宜作室内一般用材。树皮含单宁，纤维坚韧可用于编织。花可提制芳香油。

开花结实　10年生左右开始结实，正常结实期在20年生以后。结实丰年的间隔期为1年，但大小年的差异不甚明显。花两性，单生于当年生嫩枝的叶腋。萼片3。花瓣6，2轮，镊合状排列，绿色，内外轮近等长。雄蕊多数，楔形，药室外向，药隔截形。心皮多数，离生，柱头长圆形，顶全缘，被柔毛。每心皮有胚珠1颗。据海南尖峰岭1986年观测，3月下旬～4月上旬花蕾形成，5月上旬为始花期，中旬至下旬为盛花期，6月上旬为末花期。7月上旬

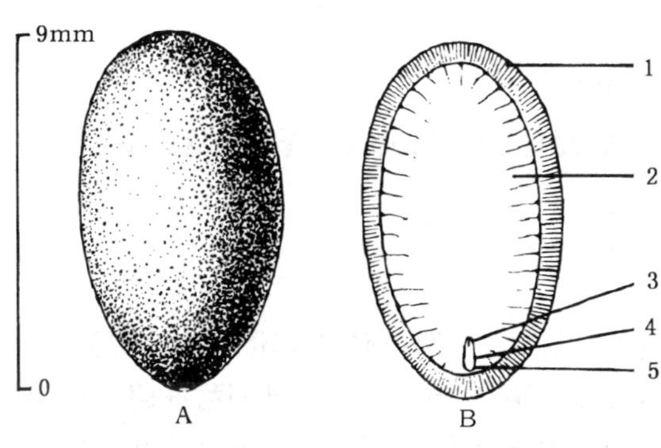

图 1　细基丸种子的外形（A）及其纵切面（B）
1. 种皮　2. 胚乳　3. 子叶　4. 胚轴　5. 胚根
（黄应钦绘）

果实开始成熟，中旬为盛熟期，下旬末及 8 月上中旬尚有少数果实成熟。另据广西南宁 1987 年观测，6 月中旬始花，7 月中下旬花期结束，8 月中旬果实开始成熟，9 月中旬～下旬尚有果实陆续成熟，花期果期较海南南部迟 1～2个月。多心皮浆果离生，小果集生于总果柄上，成熟时由青色转红色，干后黑色，球形或卵圆形，长 8～10mm，径 4～7mm，成熟后在树上悬挂 1～3 天后自行脱落或遭鸟食。种子椭圆形，间或圆形，棕褐色，长 7～9mm，径 3.5～6mm。种壳薄革质，内壁具纤维。有胚乳，皱折状，胚小（图 1）。

果实的采收调制和种子贮藏　果实大熟时清除地面的灌木杂草，铺上塑料薄膜，用竹竿敲打或上树摇动震落，就地收集。采回的果实不宜摊放或日晒，以免果皮干缩，难以脱粒。果实堆沤 2～4天，俟果皮全部软熟，置竹筐内浸入水中搓洗，淘净果皮杂质即得纯净种子。按鲜果质量计算的出种率为 25%～35%，净度约为 97%。种子千粒重 113（100～140）g。每千克有纯净种子 8 850（7 000～10 000）粒。种子的含水量为 25%～30%，忌失水，不宜日晒。洗净晾干的种子即可播种。常温条件下，裸露的种子存放 10 天以上发芽率便明显降低。贮藏或运输时均需混以湿沙。混沙贮藏的种子可延迟至次年 3 月播种。

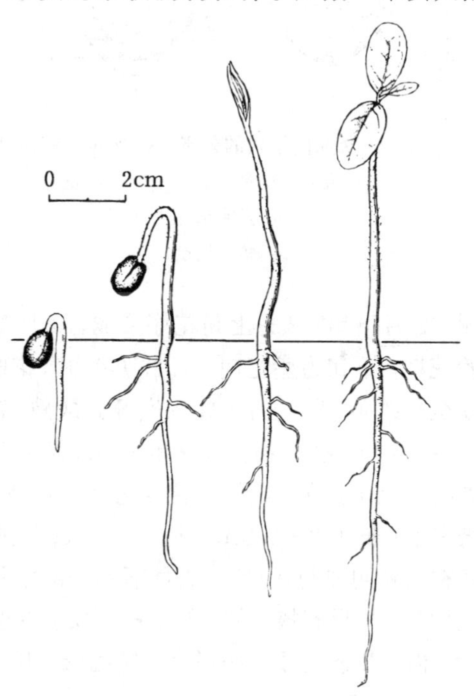

图 2　细基丸种子萌发后第 3、5、8、10 天的幼苗生长情况
（黄应钦绘）

发芽和播种　种子无休眠习性，发芽较整齐。广西林业科学研究所用 1987 年 7 月 20 日采集，经调制并保湿运输的种子，至 8 月 10 日播种（此时日均温 29℃）。播种后 18 天开始萌发，萌发后第 3 天进入发芽盛期。从发芽开始至发芽高峰的 5 天中，发芽百分数为 40%。从播种之日起算，29 天的发芽率为 51%。出土萌发。种子萌动后 3 天下胚轴拱出土面，5～6 天子叶出土并脱壳，8 天左右发出初生叶（图 2）。

条播。播前将种子浸泡 2～3 小时，捞出后稍晾干即可播种。每平方米播种 25～40g，覆土约 1cm。1 年生苗出圃。

（韦增健）

山潺琼楠

Beilschmiedia appendiculata（Allen）S. Lee et Y. T. Wei

（樟科 Lauraceae）

生长习性、分布和用途 琼楠属有200余种，我国约35种，本文描述1种。常绿乔木，高25m，胸径40～50cm。早期耐庇荫，成年树多处于林冠上层。喜高温高湿气候和土层深厚肥沃的微酸性至酸性土壤。产于海南。东南亚热带地区亦有。广西南宁有栽培。材质优，适用于建筑及各种家具器具。种子含油脂。

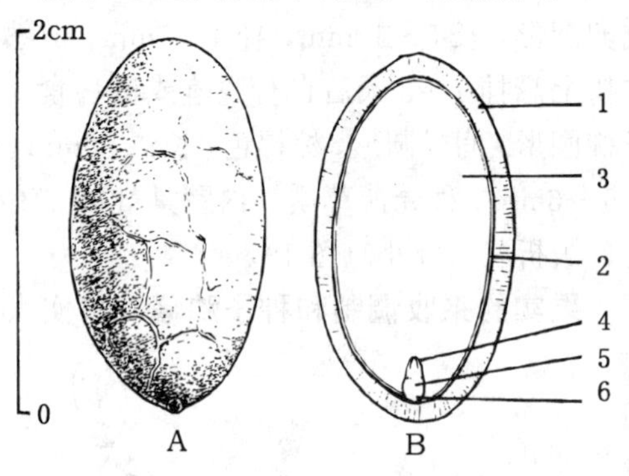

图1 山潺琼楠果核的外形(A)及其纵切面(B)
1. 内果皮 2. 种皮 3. 子叶 4. 胚芽
5. 胚轴 6. 胚根
（黄应钦绘）

开花结实 10年生左右开始开花结实，正常结实期在20年生以后。结实丰年间隔期1年，较明显。两性花组成的圆锥花序在头年小枝上腋生，长1～2cm。花黄色。花被筒短，裂片6，稀8。能育雄蕊6枚，有时8枚，3轮，花药2室，第1、2轮向内，第3轮外向，退化雄蕊3（4）位于最内轮。子房上位，无柄，1室，1胚珠，花柱顶生。据1987～1988年在南宁观测，2月下旬～3月上旬花序形成，3月中旬为始花期，下旬为盛花期，4月中旬为末花期，花期全程约40天。7月上旬果实开始成熟，7月中旬～下旬为成熟盛期。浆果状核果，椭圆形，未熟时青色带黄色，成熟时转变为紫黑色或黑色，被白粉；长1.8～2.8cm，径1～1.4cm，成熟后在树上约可维持10天，7月下旬～8月中旬陆续脱落。每果有核1粒。外果皮膜质。中果皮浆状。内果皮脆壳质，深褐色，椭圆形，长1.5～2.4 cm，径0.9～1.2cm。无胚乳。子叶厚，肉质（图1）。

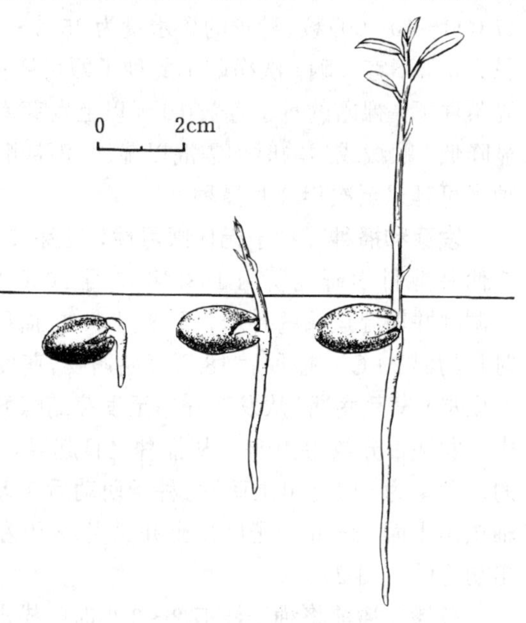

图2 山潺琼楠种子萌发后第2、5、12天幼苗的生长情况
（黄应钦绘）

果实的采收调制和种子贮藏 果实常为雀鸟取食，且熟后逐渐脱落，应在盛熟时及时用竹竿敲打或摇动树枝，震落后在地面收捡。采得的果实不宜日晒，亦忌堆沤过久发热腐坏。可在室内堆放1～2天，待中果皮充分软熟时装入竹筐，置水中反复搓洗，淘尽果皮及浆质。调制时不得损伤内果皮。彻底淘洗后所得的果核用作

播种材料，通称种子。按鲜果质量计算的出籽率约为50％。种子净度一般为97％。种子的含水量约30％～40％。千粒重约1030g。每千克有纯净种子约970粒。调制出来的种子不能日晒，忌失水，也不耐久藏。运输时用竹筐包装，每筐不宜超过20kg，并需混以湿锯木屑。也可在采后以果实的状态运输，运抵后再进行调制。宜随采随播。种子裸露的贮藏期不宜超过7天。混湿沙贮藏也不宜超过两个月。

发芽和播种　种子无休眠习性。发芽时日均温宜在20℃以上。1987年7月20日（日平均气温约30℃），广西林业科学研究所用当年新采种子在室外沙床作发芽测定，播后第14天开始萌发，第15天便进入发芽盛期，从开始萌发至萌发高峰的8天中，发芽百分数为75％。至8月21日发芽终止。从播种到发芽终止的33天中发芽率为90％。场圃播种的出苗率一般为80％～85％。留土萌发。种子萌动后4～5天上胚轴出土，具初生不育叶2～4枚。8～10天出现初生叶（图2）。

条播。随采随播的种子播前可不作任何处理。经过存放或沙藏的种子，播种前需要用清水浸种2～3小时。每平方米播种100～150g，覆土1.5～2cm。苗木培育1年半出圃。

（韦增健）

樟　属

Cinnamomum Trew.

（樟科　Lauraceae）

生长习性、分布和用途　樟属约250种，我国约46种。本文描述的12个种都是常绿乔木或大乔木，高10～34m，胸径20～300cm。生长速度较快，萌芽性强。寿命长。喜温暖湿润气候，适生于肥沃湿润的酸性土至中性土。多数种的种子、木材及枝叶含芳香油。有些种可提炼樟脑，有些种的皮为珍贵药材。材质优良，枝叶浓密，经济效益及涵养水源的效益均很高，为我国亚热带至热带地区重要的树种。这12个树种的名称、生长、分布及用途见表1。其中沉水樟在《中国植物红皮书》中被列为渐危种。

开花结实　少花桂3～4年生即开始结实，斯里兰卡肉桂5～6年生开始结实，沉水樟13～15年生开始结实，其余9个种约在8～10年生开始结实。正常结实期多在20年生以后，樟树和黄樟等大乔木100年以上的老树仍可结实不衰。结实大小年间隔期一般为1年，但不甚明显。天然林每年结实均很少，人工栽培及生长于宅旁、林缘等处的结实量多。花两性，稀杂性。圆锥花序着生于当年生嫩枝的叶腋或近顶生。花被筒短，杯状或钟状，花后脱落或下半部残留，稀宿存。阴香、樟树的花冠为绿色或黄绿色，其它树种多为淡黄色或黄白色。发育雄蕊通常9枚，第1、2轮无腺体，第3轮近基部有2腺体。花药4室，稀第3轮为2室，第1、2轮内向，第3轮外向。退化雄蕊3。子房1室，胚珠1枚，倒生，悬垂。花柱细，与子房等长。柱头头状或盘状，具3齿裂。这12个树种花果的物候期见表2。

表 1　樟属树种的名称、生长、分布和用途

中　名	学　名	树　高 (m)	胸　径 (cm)	分　布	用　途	供　稿
钝叶桂	*C. bejolghota* (Buch. -Ham.) Sweet	5～25	10～30	滇、桂、粤、琼。印度、老挝、越南	材用、油脂	619
猴樟	*C. bodinieri* Lévl.	25	90	黔、川、鄂、湘、滇	材用、香料、油脂	620
阴香	*C. burmanii* (C. G. et Th. Nees) Bl.	20～25	80	粤、琼、桂、滇、闽。印度、缅甸、越南、印度尼西亚、菲律宾	香料、材用、药用、水源、观赏、油脂	601
樟树	*C. camphora* (L.) Presl.	30	100～300	台、闽、赣、粤、桂、湘、鄂、滇、浙	材用、樟脑、香料、观赏、养蚕	601
肉桂	*C. cassia* Presl.	20	20～40	桂、粤、琼、台、滇、闽、赣、湘	药用、香料、材用	601
坚叶樟	*C. chartophyllum* H. W. Li	20	30	滇	水源、用材、油脂	619
云南樟	*C. glanduliferum* (Wall.) Nees	20	30	滇、黔、川、藏。印度、尼泊尔、缅甸、马来西亚	樟脑、油脂、材用、水源	619
沉水樟	*C. micranthum* (Hay.) Hay.	34	180	浙、台、闽、赣、湘、粤、桂。越南	材用、水源、香料、油脂	701
少花桂	*C. pauciflorum* Nees	14	15	湘、鄂、川、滇、黔、粤	香料、材用、药用、水源	619
黄樟	*C. porrectum* (Roxb.) Kosterm.	20～25	40～50	粤、琼、桂、闽、赣、黔、滇。巴基斯坦、印度、马来西亚、印度尼西亚	樟脑、香料、材用、养蚕	902
细毛樟	*C. tenuipilum* Kosterm.	25	30	滇	材用、油脂、水源	619
斯里兰卡肉桂	*C. zeylanicum* Bl.	10～15	30	印度、斯里兰卡、马来西亚。琼、粤、桂、闽、台、滇引种	药用、香料、材用	610

表 2　樟属树种的开花结实物候期

树　种	观察年份	观察地点	开花期			果实成熟期		果实脱落期
			始　期	盛　期	末　期	始　期	盛　期	
钝叶桂	1987～1988	勐腊	3 月上旬	3 月中旬	4 月上旬	7 月上旬	7 月下旬	8 月
猴樟	—	贵阳	5 月中旬	5 月下旬	6 月上旬	9 月下旬	10 月上旬	10 月中旬
阴香	1978～1984	南宁	3 月下旬	4 月上旬	4 月下旬	11 月上旬	12 月中下旬	12 月中旬～翌年 1 月下旬
樟树	1978～1984	南宁	3 月下旬	4 月上旬	4 月下旬	10 月中旬	10 月下旬	11 月上中旬
肉桂	1978～1984	南宁	6 月上旬	6 月中旬	6 月下旬	翌年 3 月下旬	4 月上旬	4 月上旬～下旬
坚叶樟	1987	勐腊	7 月中旬	8 月下旬	10 月下旬	9 月上旬	10 月中旬	10 月中旬～12 月中旬

（续）

树　种	观察年份	观察地点	开花期			果实成熟期		果实脱落期
			始　期	盛　期	末　期	始　期	盛　期	
云南樟	1987	勐腊	12 月下旬	翌年 1 月中旬	2 月中旬	9 月下旬	10 月中旬	11 月上旬～12 月中旬
沉水樟	—	江西	7 月上旬	7 月中下旬	8 月上旬～下旬	翌年 11 月上旬	12 月上中旬	12 月下旬
少花桂	1987	勐腊	3 月中旬	4 月下旬	6 月上旬	7 月下旬	8 月中旬	8 月中旬～9 月中旬
黄樟	1987	福建南靖	3 月中旬	4 月上中旬	5 月中旬	10 月上旬	11 月	10 月中旬～11 月中旬
细毛樟	1987	勐腊	2 月中旬	3 月上旬	4 月上旬	7 月中旬	8 月上旬	7 月下旬～8 月下旬
斯里兰卡肉桂	1987	海南儋县	2 月上旬	3 月中旬	3 月下旬	7 月下旬	8 月上旬	8 月中旬～9 月上旬

浆果状核果椭圆形或球形。果托杯状、盘状或钟状。每果有种子 1 粒。猴樟、樟树、坚叶樟、云南樟、沉水樟、黄樟、细毛樟等结果时花被裂片脱落。钝叶桂、阴香、肉桂、少花桂和斯里兰卡肉桂则果时花被裂片宿存，或上部脱落而下半部残存于花被筒上。果实成熟后在树上存留 1 周至半个月后自行脱落。每果有核 1 粒。种皮薄，无胚乳。子叶厚，肉质（表 3、图 1）。

表 3　樟属树种果实及种子的形态特征

树　种	未熟果颜色	成熟果实			成熟种子（核）		
		形　状	大小（mm）	颜　色	形　状	大小（mm）	颜　色
钝叶桂	青绿色	椭圆形	长 3～6 径 9～12	紫黑色，有光泽	椭圆形或圆矩形	长 10～12 径 6～8	深褐色
猴樟	绿色	球形	径 7～8	紫黑色	—	—	—
阴香	浅绿色	长椭圆形或矩圆形	长 8～12 径 6～8	紫黑色	长椭圆形，顶端钝尖	长 7～10 径 5～7	暗褐色
樟树	青色	近球形或卵形	径 6～8	紫黑色	近球形	径 6～7.5	褐色有光泽
肉桂	青色	椭圆形	长 10～13 径 7～10	紫黑色	椭圆形	长 8～11 径 6～8	黑褐色
坚叶樟	黄绿色	椭圆形，顶端小尖	长 10～15 径 7～9	紫黑色	球形	径 5～6	深褐色
云南樟	黄绿色	球形	径 10～12	紫黑色	球形	径 5～7	深褐色
沉水樟	青黄色	椭圆形或扁球形	长 15～25 径 10～20	紫红至紫黑色	卵圆形，具缝棱	长 10～14 径 7～9	棕褐色
少花桂	青绿色	椭圆形或卵形	长约 11 径约 5	褐色至紫黑色	椭圆形或卵形	长约 9 径 4～5	褐色
黄樟	绿色	近球形	径 6～8	紫黑色或黑色	近球形	径 5～6	暗褐色
细毛樟	绿色	近球形	径约 15	红紫色	近球形	径 5～7	深褐色
斯里兰卡肉桂	青色	椭圆形	长 12～15 径 7～8	紫或黑色	椭圆形	长 10～11 径 5～6	灰褐色

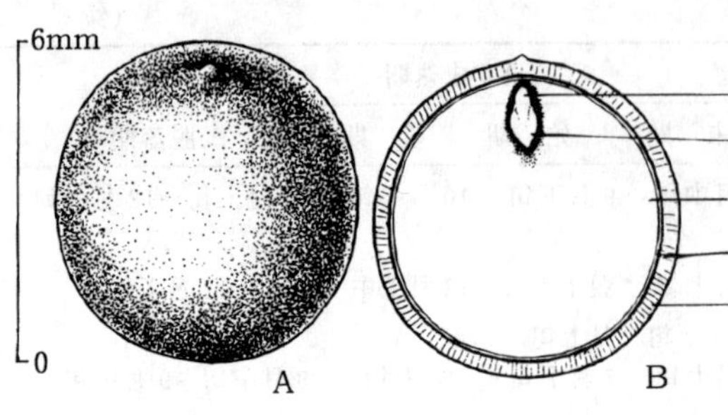

图 1 樟树果核的外形（A）及其纵切面（B）
1. 内果皮 2. 种皮 3. 子叶 4. 胚芽 5. 胚根
（黄应钦绘）

果实的采收调制和种子贮藏 阴香、樟树、肉桂、云南樟、细毛樟和斯里兰卡肉桂的果实成熟期较整齐，可在成熟盛期上树采摘，或清理树下杂草灌木，摇动树枝，用铺在地面的薄膜或布单承接震落的果实。坚果樟和少花桂同一株树上的果实成熟期不一，需要分批采集。采得的果实在室内堆放3～4天，充分软熟后装入竹筐或布袋内，置水中反复搓擦，淘去果皮等杂质以及漂浮的瘪粒。所得果核用作播种材料，通称种子。种子忌失水，不能日晒。已调制的种子在室内裸露存放一周以上发芽率便明显下降。新调制的种子含水量一般在30%左右。稍加晾干而不影响发芽能力的含水量宜在20%以上。运输时需混以湿沙或湿锯木屑。秋冬季成熟的种子适于湿沙贮藏越冬，贮藏期为3～6个月。肉桂、少花桂的种子宜随采随播，调制后不作贮藏。这12个树种的出籽率及种子的质量数据见表4。

表 4 樟属树种果实出籽率和种子的净度、质量

树 种	出籽率（%）	净度（%）	千粒重（g）	每千克纯净种子粒数（粒）
钝叶桂	20	70～90	270～300	3 300～3 700
猴樟	33	—	100～125	8 000～10 000
阴香	37	96～100	83～123	8 000～1 200
樟树	32	97～100	80～150	6 700～12 500
肉桂	43	96～100	280～380	2 600～3 600
坚叶樟	38	96～100	310～340	2 940～3 230
云南樟	32	95～98	360～390	2 560～2 780
沉水樟	20～30	—	1 500	670
少花桂	35	97～99	160～200	5 000～6 300
黄樟	30	90～100	80～150	6 700～12 500
细毛樟	38	96～99	360～380	2 600～2 800
斯里兰卡肉桂	36	96～99	340～400	2 500～2 900

发芽和播种 樟属种子无明显休眠现象。发芽时日均温宜在20℃以上。低于20℃或高于15℃，有些树种的种子也可发芽，但发芽进程比较缓慢。表5汇集的是1987年在广西南宁、云南勐腊、海南儋县（现名儋州市）、贵州贵阳和江西等地对樟属12个树种所作的一次发芽测试情况。

留土萌发。播种后4～5天胚根萌发，7～10天上胚轴出土，幼茎由绿色渐变为紫黑色，具初生不育叶4～6枚。出土后2～4天展出初生叶。主根发达，基部有4条侧根轮生（图2）。

表 5　樟属树种的发芽能力及其测定条件

树　种	测试地点	室外测定环境温度（℃）	发芽势%		发芽率（%）	
			计算天数	一般数值	计算天数	一般数值
钝叶桂	勐腊	土床，24～26	15	60	32	70
猴樟	贵阳	土床，—	—	—	—	70
阴香	南宁	沙床，24～26	—	—①	30	66
樟树	南宁	沙床，20～22	—	—①	32	65
肉桂	南宁	沙床，23～27	—	—①	28	70
坚叶樟	勐腊	沙床，16～25	16	65	—	80
云南樟	勐腊	沙床，16～25	15	60	—	75
沉水樟	江西	土床，—	—	—	—	40
少花桂	勐腊	沙床，16～25	16	60	—	80
黄樟	南宁	沙床，25～27	14	60	30	76
细毛樟	勐腊	沙床，16～25	16	65	35	80
斯里兰卡肉桂	儋县	沙床，28	14	75	30	80

①　没有明显的发芽高峰

条播。每平方米沉水樟播种 150～200g，钝叶桂、肉桂、坚叶樟、云南樟、细毛樟、斯里兰卡肉桂播种 75～85g，少花桂播种 50～60g，阴香、樟树、黄樟播种 25～35g。覆土 1.5～2.5cm。据谢善高和莫金莲(1993)报道，发芽率 50%的肉桂种子，每平方米苗床播 0.25kg，幼苗出现两片初生发育叶并有侧根时间苗，1 年生苗高可达 30cm，每平方米可得合格苗 100 株；间出的苗移入容器继续培育，1 年生可以高达 25cm。黄樟在福建有用扦插繁殖的，肉桂和斯里兰卡肉桂在广西有用高压繁殖的，其余未见有无性繁殖。1～2 年生苗出圃。

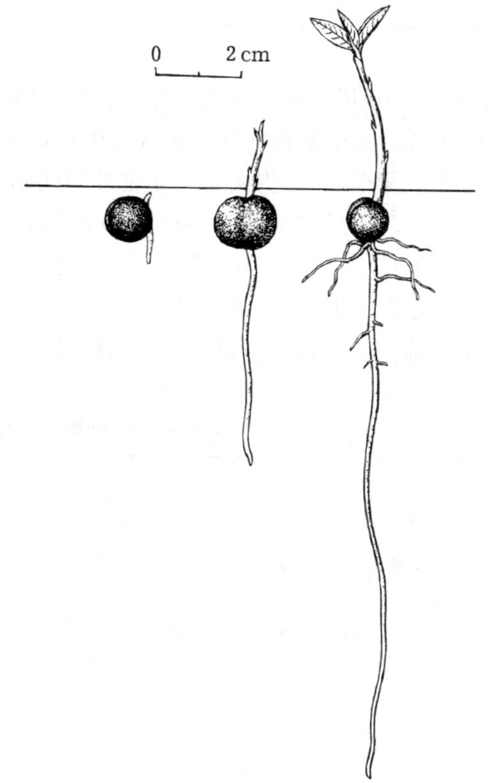

图 2　樟树种子萌发后 4、10、15 天幼苗的生长情况

（黄应钦绘）

（王宏志）

厚壳桂属
Cryptocarya R. Br.

（樟科 Lauraceae）

生长习性、分布和用途 本属约 200～250 种，我国约 19 种，本文描述 3 种。常绿乔木，高 25～28m，胸径 40～70cm。生长较慢。喜湿润、温暖、凉爽的气候。能耐短期－2℃左右的低温，但忌严寒。适生于疏松湿润的酸性土壤。它们的名称、生长、分布及用途见表 1。

表 1 厚壳桂属种树种生长情况、分布和用途

中 名	学 名	树高（m）	胸径（cm）	分 布	用 途	供 稿
厚壳桂	*C. chinensis* (Hance) Hemsl.	25	60	川、桂、粤、琼、闽、台	材用、水源	602
黄果厚壳桂	*C. concinna* Hance	25	40	粤、桂、琼、赣、台。越南也有	材用、水源	602
云南厚壳桂	*C. yunnanensis* H. W. Li	28	40～70	滇	材用、水源、油脂	618

开花结实 约 10 年生开始结实，正常结实在 20 年生以后。结实大小年间隔期为 1 年。花两性，组成腋生或近顶生圆锥花序。花被筒宿存，花后顶端收缩，花被裂片 6，早落。发育雄蕊 9 枚，药 2 室，第 1、2 轮内向，第 3 轮外向或侧外向。子房无柄，为花被筒所包被。1 室 1 胚珠，倒生，悬垂。据广西南宁 1986～1988 年观察，黄果厚壳桂每年开花结实 2 次：第 1 次为小造果，3 月下旬始花，4 月上旬盛花，4 月下旬为末花期，10 月中旬果始熟，10 月下旬果盛熟；第 2 次为大造果，7 月中旬始花，7 月下旬盛花，8 月下旬为末花期，12 月中旬～翌年 1 月上旬果实成熟并脱落。厚壳桂的花期和果实成熟期比黄果厚壳桂约早半个月。在海南，

表 2 厚壳桂属树种果实和种子的形态特征

树 种	成熟果实			成熟种子		
	形 状	大小（mm）	色 泽	形 状	大小（mm）	色 泽
厚壳桂	球形或扁圆形，具纵棱 12～15 条	长 7～9 径 9～12	紫黑色	扁圆形	高 5～7 径 7～10	褐色
黄果厚壳桂	长椭圆形，有纵棱 12 条，但有时不明显	长 15～24 径 8～11	黄褐色至紫黑色或蓝黑色	椭圆形	高 13～20 径 6～8	黄褐至褐色
云南厚壳桂	卵球形，有不明显纵棱 12 条	长 13～17 径 10～12	黑紫色	椭圆形	高 10～15 径 8～10	粉红色

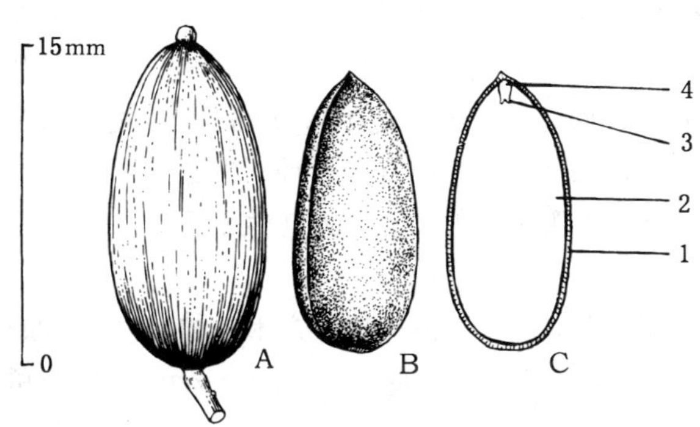

图 1　黄果厚壳桂果实（A）和种子外形（B）及其纵切面（C）
1. 种皮　2. 子叶　3. 胚芽　4. 胚根
（黄应钦绘）

厚壳桂第 1 次开花在 1～2 月，6～7 月果熟；第 2 次在 7～8 月开花，12 月果熟。同样是在海南，黄果厚壳桂第 1 次开花在 2～3 月，7～8月果熟；第 2 次在 8～9 月开花，12 月～翌年 1 月果熟。据云南勐仑 1988～1989 年观察，云南厚壳桂 3 月上旬为始花期，中旬为盛花期，下旬为末花期，7 月上旬果实开始成熟，中下旬为成熟盛期。果为核果状，全部包藏于肉质或硬化增大的花被筒内，顶端有一小开口。每果有种子 1 粒。种子无胚乳，子叶厚，胚小（表 2、图 1）。

图 2　黄果厚壳桂种子萌发后第 12、16 天幼苗的生长情况
（黄应钦绘）

果实的采收调制和种子贮藏　成熟盛期前清理林下杂草灌木，垫上布单或薄膜，用竹竿敲打或摇动树枝，震落果实。采回的果实可不经调制，直接用作播种材料；也可以堆沤 1～2 天，装入竹筐置水中搓擦花被筒及果皮，淘去杂质，取出下沉的种子。种子忌失水，不能日晒。新鲜种子的含水量约为 40%，稍加晾干而不影响种子发芽的含水量约为 30%。不经调制的果实宜于就地随采随播。短途运输时可以直接包装果实，每筐不宜超过 20kg，运抵后再行调制。调制后如需贮藏，应混拌湿沙，贮藏期为 3～4 个月。种子的净度、质量等数据见表 3。

表 3　厚壳桂属树种果实出种率和种子净度、质量

树　种	出种率（%）	净度（%）	千粒重（g）	每千克纯净种子粒数（粒）
厚壳桂	50～60	90～95	700～900	1 100～1 400
黄果厚壳桂	50～60	90～95	700～900	1 100～1 400
云南厚壳桂	60～65	90～95	680～720	1 400～1 500

发芽和播种　黄果厚壳桂有轻度休眠习性。云南厚壳桂的种子发芽并不困难。目前还缺乏厚壳桂的发芽测定资料。发芽时要求的日均温，黄果厚壳桂为 15～18℃，云南厚壳桂在

20℃以上。1987年10月23日，广西林业科学研究所在南宁室外沙床上用当年新采的黄果厚壳桂种子播种，当时日平均气温为25℃，播后80天至翌年1月13日，日平均气温13℃时开始发芽，2月10～24日为发芽盛期，2月29日发芽终止，得到的发芽率为85%。云南西双版纳热带植物园1988年7月在室外播种云南厚壳桂，播后20天即有少量种子发芽，27天的发芽率为74%。留土萌发。黄果厚壳桂胚根萌发后约6天上胚轴出土，无初生不育叶，11天左右初生叶出现（图2）。

条播。苗圃地宜选择阴坡。如果是调制后的种子，这3个树种每平方米播种120～200g；如果是用带花被筒的果实作为播种材料，每平方米可播250～380g。覆土1.5～2cm。1年生苗可以出圃。

（韦增健）

山胡椒属
Lindera Thunb.

（樟科 Lauraceae）

生长习性、分布和用途 本属约100种，我国约有50种。本文描述的3种都是常绿小乔木或乔木。适生于酸性土，喜温暖湿润气候。它们的名称、生长、分布及用途见表1。

表1 山胡椒属树种的名称、生长、分布和用途

中名	学名	树高 (m)	胸径 (cm)	分布	用途	供稿
乌药	*L. aggregata* (Sims) Kosterm.	5	10	皖、浙、闽、赣、鄂、湘、粤、桂、台。越南、菲律宾	药用、农药、香料、油料	602
香叶树	*L. communis* Hemsl.	13	36	陕、甘、湘、鄂、赣、浙、闽、台、粤、桂、黔、滇、川。越南	材用、油料、药用、香料	602
黑壳楠	*L. megaphylla* Hemsl.	25	60	陕、甘、川、滇、黔、鄂、湘、赣、皖、浙、闽、桂、粤	材用、油料、香料	902

开花结实 乌药和香叶树7～8年生开始结实，黑壳楠约10年生开始结实，正常结实期在15年生以后。结实大小年现象不明显。花单性，雌雄异株，伞形花序。乌药的花序集生于叶腋，香叶树花序单生或成对生于叶腋，黑壳楠的花序常成对生于叶腋。花被片6。雄花能育雄蕊9（偶有12），3轮（偶4轮），花药2室，全部内向。雌花通常具退化雄蕊9。子房上位，1室，1胚珠。它们的物候期见表2。

表 2　山胡椒属树种的开花结实物候

树　种	观察地点	观察年份	开花期 始　期	开花期 盛　期	开花期 末　期	果实成熟期 始　期	果实成熟期 盛　期	果实脱落期
乌 药	广西 南宁	1987～1988	4 月中旬	5 月上旬	6 月中旬	8 月中下旬	9 月中旬	9 月下旬～10 月上旬
香叶树	广西南宁	1984～1988	3 月上旬	3 月中旬	3 月下旬	9 月下旬	10 月上旬	10 月上旬～11 月上旬
黑壳楠	福建南平	—	3 月下旬	4 月上旬	4 月下旬	9 月中旬	10 月上旬	10 月下旬～11 月中旬

浆果状核果，每果有核 1 粒。外果皮薄膜质，中果皮肉质，内果皮壳质。子叶肥大充实，无胚乳。这 3 个树种果核的形态特征见表 3。图 1 是香叶树果核的外形和纵切面。果实成熟后在树上可存留 1 周至半月。

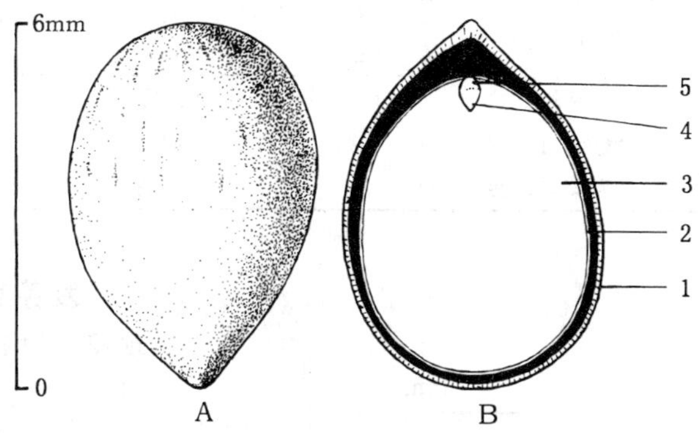

图 1　香叶树果核的外形（A）及其纵切面（B）

1. 内果皮　2. 种皮　3. 子叶　4. 胚芽　5. 胚根

（黄应钦绘）

表 3　山胡椒属树种果实和果核的形态特征

树　种	未熟果颜色	成熟果实 形　状	成熟果实 大小（mm）	成熟果实 颜　色	成熟果核 形　状	成熟果核 大小（mm）	成熟果核 颜　色
乌药	青黄色	卵形或球形	长 6～9 径 5～7	紫红色或黑色	卵形	长 4～6 径 3～5	褐色
香叶树	青绿色	卵形近球形	长 8～12 径 6～9	红色	卵形	长 5～6 径 4～5	浅褐色
黑壳楠	绿色	椭圆形或卵形	长约 18 径约 13	黑色	卵形	长约 14 径约 10	浅褐色

果实的采收调制和种子贮藏　成熟盛期时摇动树枝或用竹竿敲打，在地面收集震落的果实。采回的果实可以堆沤 2～3 天或浸水 3～4 天。果肉充分软熟后装入竹筐或布袋，置水中搓擦，淘去皮肉杂质，再拌草木灰以脱去油脂，一天后再行搓洗 ，所得果核用作播种材料，通称种子。种子忌日晒，不宜过度失水。经过调制的种子可摊放室内阴干 2～3 天。当种子的含水量降到 30%左右时即可贮藏。采得的果实也可直接用竹筐包装运输，每筐以不超过 25kg 为

宜，运抵后再行调制。已经调制好的种子可用麻袋等包装运输。贮藏时需混以湿沙，贮藏期一般为半年左右。种子的净度、质量等见表4。

表4　山胡椒属树种的出籽率及净度和质量

树种	出籽率（%）	净度（%）	千粒重（g）	每千克种子粒数（粒）
乌药	30～35	92～98	45～55	18 000～22 000
香叶树	30	92～98	50～60	16 000～20 000
黑壳楠	25～30	95～99	270～390	2 500～3 700

表5　山胡椒属树种的发芽能力及其测定条件

树种	测定条件		发芽势（%）		发芽率（%）	
	基质	温度（℃）	计算天数	一般数值	计算天数	一般数值
乌药	滤纸	发芽箱，28	—	不明显	95	50
香叶树	细沙	室外，13～20	—	不明显	110	63
黑壳楠	滤纸	发芽箱，28	21	85～93	90	88～95

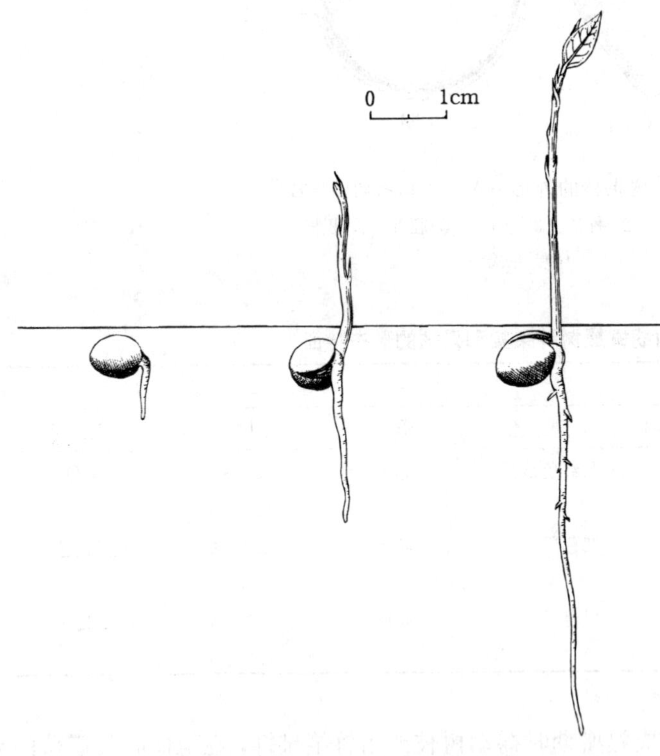

图2　香叶树种子萌发后第3、11、21天幼苗的生长情况
（黄应钦绘）

发芽和播种　种子有休眠现象。日均温15℃上下时种子可以发芽，但发芽进程较慢。25～28℃的条件下发芽较快，从发芽开始至终结的全程为35～45天。1988年在广西南宁和福建南平进行过发芽测试，结果见表5。

留土萌发。胚根萌发后，乌药约11天胚芽出土，有初生不育叶7枚。胚芽出土后10天左右发出初生叶；香叶树约6天胚芽出土，具初生不育叶4～6枚，出土后21天左右发出初生叶；主根明显，侧根稀疏（图2）。

条播。种子以湿沙贮藏越冬至3月份播种较好。乌药和香叶树每平方米播种7～12g；黑壳楠每平方米约播30g。覆土3～5cm，1年生苗出圃。

（韦增健）

木姜子属
Litsea Lam.

(樟科　Lauraceae)

生长习性、分布和用途　本属约200种，我国约有72种18变种和3变型。本文描述的4个种为常绿至落叶乔木，分布于热带至亚热带地区，苗期忌霜冻。适生于肥沃、疏松而湿润的酸性土壤，在干旱贫瘠或粘重土上生长不良。它们的名称、生长、分布和用途见表1，其中天目木姜子和五桠果叶木姜子资源稀少，更新困难，在《中国植物红皮书》中被分别列为濒危种和渐危种。山鸡椒（俗称山苍子）果实蒸馏所得的油中富含柠檬醛，用于香料工业。

表1　木姜子属树种的名称、生长、分布和用途

中　名	学　名	树　高 (m)	胸径 (cm)	分　布	用　途	供　稿
天目木姜子	*L. auriculata* Chien et Cheng	10～20	40～60	浙、皖、豫、赣	材用、药用、观赏	1002
山鸡椒（山苍子，木姜子）	*L. cubeba* (Lour.) Pers.	10	15	长江以南。东南亚各国	香料、药用、油脂、材用	903
五桠果叶木姜子	*L. dilleniifolia* P. Y. Pai et P. H. Huang	20～26	28～30	滇西南	材用、香料	619
滇南木姜子	*L. garrettii* Gamble	4～12	11～19	滇、藏。泰国、缅甸	材用、香料	619

开花结实　山鸡椒3～4年生开始结实，6年生以后进入正常结实期，立地条件好的地方盛果期可以延续到20年以上。吴美春和施拱生等（1988）认为，以采果为目的营造山鸡椒林，雌雄比例以10∶1为宜。木姜子属的其余3个种5～8年生开始结实，10年生以后进入正常结实。大小年现象比较明显，丰年间隔期为1年。花单性，雌雄异株。天目木姜子伞形花序簇生，山鸡椒伞形花序单生或簇生，总梗细长。五桠果叶木姜子和滇南木姜子由伞形花序组成总状花序，腋生。苞片4～6，交互对生，开花时宿存。花被裂片6。雄花具发育雄蕊9或12。花药4室，内向，瓣裂。雌花子房上位，1室1胚珠。花柱显著，柱头盾状。在云南西双版纳勐仑及福建南平观察的物候期见表2。浆果状核果，每果有核1粒。子叶2，半球形，无胚乳（表3、图1）。

表2　木姜子属树种的开花结实物候期

树　种	观察地点和年份	开花			果实成熟		种子散落期
		始　期	盛　期	末　期	始　期	盛　期	
天目木姜子	浙江	3月	4月中旬	—	9月上旬	9月中旬	9月下旬
山鸡椒	福建南平1987	12月中旬	翌年1月上中旬	2月上旬	8月下旬	9月上旬	10月中旬～11月中旬

（续）

树　种	观察地点和年份	开花			果实成熟		种子散落期
		始　期	盛　期	末　期	始　期	盛　期	
	浙江龙泉	雄花 2 月下旬	3 月上中旬	3 月下旬	—	—	—
	—	雌花 3 月上旬	3 月中旬	3 月下旬		8～9 月	—
五桠果叶木姜子	云南勐仑 1987～1988	4 月上旬	4 月中旬	4 月下旬	7 月中旬	7 月下旬	7 月下旬～8 月上旬
滇南木姜子	云南勐仑 1986～1988	4 月中旬	4 月下旬	5 月上旬	7 月中旬	7 月下旬	7 月下旬～8 月上旬

表 3　木姜子属树种果实和种子（核）的形态特征

树　种	未熟果颜色	成熟果实			成熟种子(核)		
		形　状	大　小(mm)	颜　色	形　状	大　小(mm)	颜　色
天目木姜子	青色	卵圆形或椭圆形	长 13～17	紫黑色	椭圆形	径 8～9	褐色
山鸡椒	绿色	球形，有一圈微棱，有光泽	径 4～5	黑色	圆形	径 3～4	黄褐色
五桠果叶木姜子	绿色	扁球形，有光泽，先端凹	高 5～20 径 22～28	紫红色	扁球形	厚 10～15 径 12～20	黑褐色
滇南木姜子	黄绿色	椭圆形或矩圆形，有光泽	长 12～15 径 8～10	紫黑色	椭圆形或矩圆形	长 9～13 径 5～7	褐色

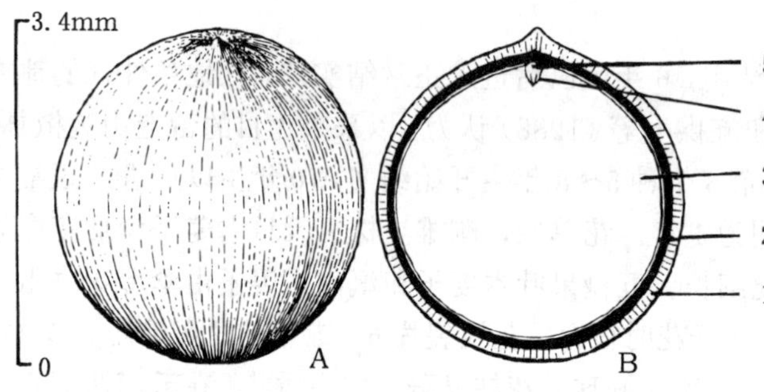

图 1　山鸡椒果核的外形(A)及其纵切面(B)
1. 内果皮　2. 种皮　3. 子叶　4. 胚芽　5. 胚根
(黄应钦绘)

果实的采收调制和种子贮藏　用于蒸油的果实需在成熟前采集。用于育苗的应在成熟盛期自然散落之前及时采收，形态表现是果皮转呈紫黑色。用采种刀截断果梗，或用竹竿敲打树枝后在树下拾集。采得的果实在室内堆沤 2～5 天，果肉充分软熟后在水中搓擦，淘去皮肉杂质和漂浮的瘪粒。所得果核即为播种材料，通称种子。种子忌失水，不宜日晒。山鸡椒核壳包有一层蜡膜，对种子起保护作用，但阻碍通气透水，用始温 55℃水浸泡 15～20 分钟即可脱去。木姜子属种子调制后稍加晾干的种子含水量应在 35%以上。也可以在采集后随即用竹筐或木箱包装运输，到达目的地后再行调制。已经调制的种子，运输时需混以湿沙或湿润的锯木屑。山鸡椒和天目木姜子通常都是混湿沙贮藏越冬，次年春播。据试验，袋装悬挂于室内、直播入露天花盆以及室内混沙湿藏的天目木姜子，发芽率分别为 2.4%、57.0%和 98.4%(汪五星，1986)。五桠果叶木姜子和滇南木姜子则宜随采随播，从调制至播种以不超过 15～20 天为宜。种子的净度、质量等数据见表 4。

表 4　木姜子属树种出籽率和种子的净度、质量

树　种	出籽率(%)	净度(%)	千粒重(g)	每千克种子(核)粒数(粒)
天目木姜子	46	95	450～540	1 850～2 200
山鸡椒	18～26	90～97	30～43	23 000～33 000
五桠果叶木姜子	36	97～100	2 350	430
滇南木姜子	45	94～98	220～340	2 900～4 500

表 5　木姜子属两个树种的发芽能力（西双版纳，室外沙盆）

树　种	发芽势（%）		发芽率（%）	
	计算天数	一般数值	计算天数	一般数值
五桠果叶木姜子	15	70	60	95
滇南木姜子	13	72	40	82

发芽和播种　山鸡椒种子有短暂的休眠现象，天目木姜子、五桠果叶木姜子和滇南木姜子的种子无休眠习性。发芽时的日平均气温宜在 20℃以上。1985 和 1988 两年的 7 月下旬，在西双版纳热带植物园的室外沙盆中对两种木姜子进行过发芽测定，当时气温为 25～ 30℃，测定结果见表 5。

天目木姜子在福建南平引种，发芽率在 95%以上。留土萌发。初生不育叶 3～5 片。出苗后 3～5 天展出初生叶(图 2)。

点播或条播。每平方米山鸡椒播种 4～7g，天目木姜子约 25g，五桠果叶木姜子约 300g，滇南木姜子约 45g。覆土并盖草。天目木姜子从播种到幼苗出齐约 30～40 天。木姜子属也可在幼苗具 2～4 片初生叶时按既定的株行距移植。半年至 1 年生苗出圃。天目木姜子 1 年苗高约 65cm，每平方米可产 50 株。

（程必强）

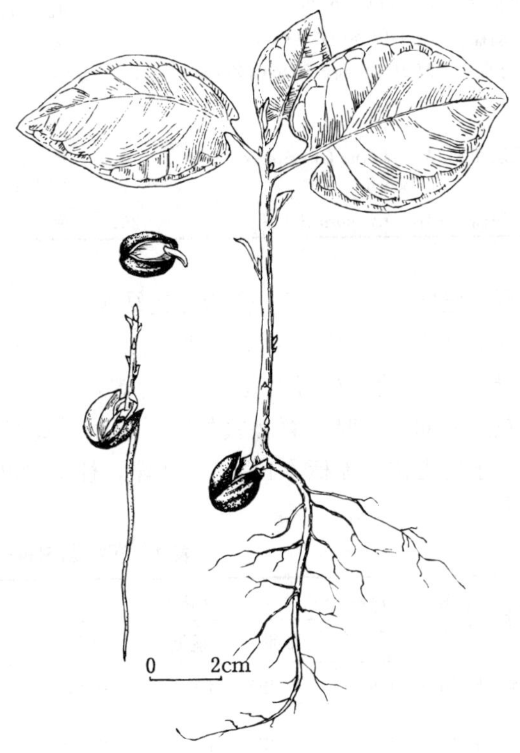

图 2　天目木姜子种子萌发后第 2、12、25 天幼苗生长情况

（黄鹏成供稿、田恒德抄）

润　楠　属
Machilus Nees

（樟科　Lauraceae）

生长习性、分布和用途　本属约100种，我国70种，本文描述7种。常绿。耐荫。深根性。喜温暖湿润气候，适生于土层深厚肥沃的土壤。分布于亚洲热带和亚热带地区。木材用于建筑、高级家具和细木工。树皮和叶研粉，可制熏香。枝叶和果可提取芳香油。这7个树种的名称、生长、分布和用途见表1。

表1　润楠属树种的名称、生长、分布和用途

中　名	学　名	树高（m）	分　布	用　途	供　稿
薄叶润楠	*M. leptophylla* Hand. -Mazz.	28	苏、皖、浙、闽、赣、湘、粤、桂、黔	材用、观赏	902
建润楠	*M. oreophila* Hance	8	闽、粤、湘、桂、黔	防风、护岸	902
刨花润楠	*M. pauhoi* Kanehira	20	皖、浙、闽、赣、湘、粤、桂	材用、工业原料	902
柳叶润楠	*M. salicina* Hance	20	粤、琼、桂、黔、滇	材用	619
红润楠	*M. thunbergii* Sieb. et Zucc.	20	鲁、苏、皖、浙、赣、闽、台、湘、粤、桂。日本、朝鲜半岛	材用、香料工业原料	902
绒毛润楠	*M. velutina* Champ. ex Benth.	18	粤、琼、桂、湘、赣、闽、浙。中南半岛	材用	902
黄枝润楠	*M. versicolora* S. Lee	20	闽	材用	902

开花结实　4～10年生开始开花结实，7～20年生进入正常结实年龄。结实有大小年现象，丰年间隔期为1～2年。花两性，圆锥花序。花被筒短，花被裂片6，2轮，近相等或外轮较小。发育雄蕊9，3轮。花药4室，外2轮雄蕊无腺体，花药内向；第3轮雄蕊近基部具2有柄腺体，花药外向，有时下面2室外向，上面2室侧向。第4轮退化雄蕊形小，先端箭头形，有短柄。子房上位，无柄，1室，1胚珠，柱头盘状或头状。开花结实的物候期见表2。

表2　润楠属树种开花结实物候

树　种	观察地点	观察年份	开花期			果实成熟			果实脱落		
			始期	盛期	末期	始期	盛期	末期	始期	盛期	末期
薄叶润楠	福建南平	1987～1988	3月中旬	4月中旬	4月下旬	8月中旬	8月下旬	9月上旬	9月上旬	9月中旬	9月下旬
建润楠	福建南平	1988	3月中旬	4月上旬	4月中旬	6月上旬	7月	8月上旬	7月中旬	8月上旬	8月下旬
刨花润楠	福建南平	1987～1988	5月上旬	5月中旬	5月下旬	11月上旬	11月下旬	12月上旬	12月上旬	12月中旬	12月下旬
柳叶润楠	云南西双版纳	（第1次）	5月下旬	6月中旬	7月中旬	8月下旬	9月上旬～10月中旬	11月下旬	—	—	—

（续）

树　种	观察地点	观察年份	开花期			果实成熟			果实脱落		
			始期	盛期	末期	始期	盛期	末期	始期	盛期	末期
柳叶润楠	云南西双版纳	（第2次）	11月上旬	11月中旬	12月中旬		12月上旬～翌年3月		—	—	—
红润楠	福建南平	1988	2月中旬	3月中旬	3月下旬	7月上旬	7月中旬	8月上旬	7月下旬	8月上旬	8月中旬
绒毛润楠	福建南平	1987～1989	10月下旬	11月	12月中旬	翌年3月上旬	3月上旬	3月中旬	3月中旬	3月下旬	4月上旬
黄枝润楠	福建南平	1988～1989	4月上旬	4月中旬	4月下旬	6月中旬	6月下旬	7月中旬	7月上旬	7月中旬	7月下旬

浆果状核果，球形或扁球形或少有椭圆形。花被裂片宿存，常反曲。果柄不增粗或微增粗。外果皮薄膜质，中果皮肉质，内果皮骨质。种子无胚乳。子叶2枚，厚肉质，乳白色（图1）。这7个树种成熟果实的形状、颜色和大小见表3。

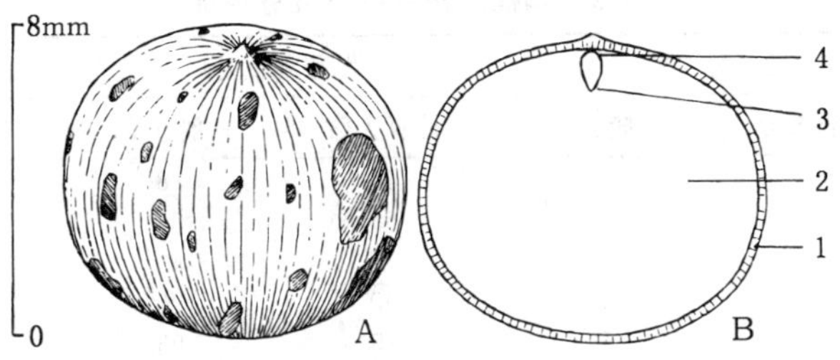

图1　柳叶润楠果核的外形（A）及其纵切面（B）

1. 内果皮　2. 子叶　3. 胚芽　4. 胚根

（黄应钦绘）

表3　润楠属树种果实的形状、颜色和大小

树　种	形　状	颜　色	直　径（mm）
薄叶润楠	球形	紫黑色	10
建润楠	球形	紫黑色，果梗有小柔毛	8～10
刨花润楠	球形	黑色	10
柳叶润楠	球形稍扁	紫黑色或褐色，果梗红色	7～10
红润楠	扁球形	黑紫色，果梗鲜红色	8～10
绒毛润楠	球形	蓝紫色，果梗红色	4
黄枝润楠	扁球形	紫黑色	7～10

果实的采收调制和种子贮藏　即使是同一个树种，润楠属各个植株果实的成熟期也不完全一致。应在果实出现成熟颜色时上树采摘。充分成熟的果实也可以振动树枝，摇落后在地上拾捡。采回的果实应及时调制。将果实放在加有少量苏打的水中浸3～5天，使果皮果肉软化。反复揉搓并用水淘洗，除去皮肉等杂质以及漂浮的空瘪粒，即得果核，用作播种材料，通称种

子。种子忌失水，切忌曝晒或裸露陈放，以免影响发芽能力。出籽率和种子质量等数据见表 4。

表 4　润楠属树种鲜果出籽率和种子质量

树　种	出籽率（%）	净度（%）	千粒重（g）		每千克种子粒数（粒）	
			一般数值	变动范围	一般数值	变动范围
薄叶润楠	90	94～98	350	315～395	2 860	2 530～3 170
建润楠	92	95～99	330	290～360	3 030	2 780～3 450
刨花润楠	93	96～99	340	305～365	2 940	2 740～3 280
柳叶润楠	95～98	97	318	—	3 140	—
红润楠	92	95～99	335	305～360	2 980	2 780～3 280
绒毛润楠	89	95～99	325	290～350	3 080	2 860～3 450
黄枝润楠	90	96～99	320	290～345	3 120	2 900～3 450

种子含水量应保持在 30%～40%，低于 30%会影响种子的生命力。春夏季成熟的种子宜随采随播，秋末或冬季成熟的种子，可混拌湿沙置于 5～10℃的低温下贮藏越冬，贮藏期可达 10 个月。

表 5　润楠属树种种子发芽情况

树　　种	发芽势（%）		发芽率（%）		场圃发芽率（%）	
	计算天数	一般数值	计算天数	一般数值	计算天数	一般数值
薄叶润楠	40	55	60	88	86	60
建润楠	45	53	65	84	85	57
刨花润楠	50	55	75	87	90	58
柳叶润楠	40	52	60	90	—	—
红润楠	45	58	70	93	85	65
绒毛润楠	40	60	65	80	85	62
黄枝润楠	40	50	60	78	86	55

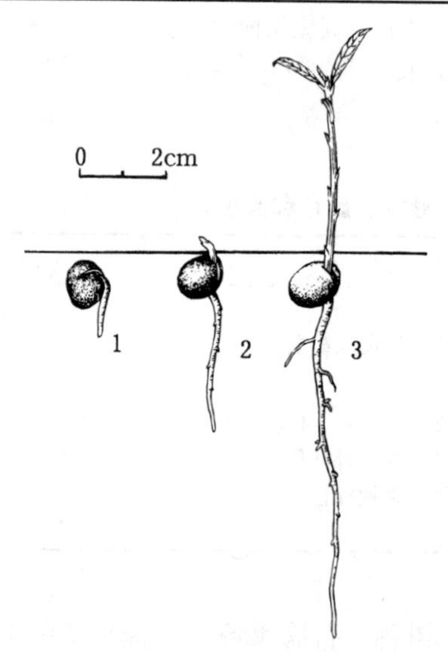

图 2　柳叶润楠种子的萌发和幼苗生长情况

1. 胚根伸出　2. 幼苗初露　3. 初生不育叶互生，发育叶互生

（黄应钦绘）

发芽和播种　种子发芽期长。福建林学院和西双版纳热带植物园曾在室内用沙或蛭石作基质，在气温 20～25℃的条件下作过发芽测定，发芽情况见表 5。

播种　每平方米播种 50～75g。播种沟深 2～5cm，间距 20～25cm。覆土 1～2cm 并盖草。播后两个月左右开始萌发，再经 15 天左右幼苗出土，再约 8 天长出叶片，2～3 个月出土结束。留土萌发。初生不育叶互生。柳叶润楠种子的萌发过程见图 2。1～2 年出苗出圃。也可扦插育苗。

（郑郁善）

滇新樟

Neocinnamomum caudatum（Nees）Merr.

（樟科　Lauraceae）

生长习性、分布和用途　新樟属约 7 种，我国有 5 种，本文描述滇新樟这 1 种。常绿乔木，高 5～20m，胸径 25～40cm。生长较快。喜光。能耐－4℃左右的短期低温。在肥沃湿润的酸性土或石灰土上均能生长。产于滇、桂的热带和亚热带地区。印度、尼泊尔、缅甸、越南亦有分布。木材中等，叶可以提制樟脑并入药，果可制取芳香油。

开花结实　4～6 年生开始结实，10 年生以后进入正常结实期。大小年现象明显，丰年间隔期一般为 1 年。花两性。团伞花序通常具 5～6 朵花，组成圆锥花序，长 6～10cm。花小，黄色。花被裂片 6。发育雄蕊 9，药 4 室，第 1、2 轮上 2 室小，内向，下 2 室大，内向或侧内向；第 3 轮下 2 室外向，上 2 室几与下 2 室横排成一列，侧向。子房梨形，1 室 1 胚珠，倒生，悬垂。花柱短，柱头盘状。1986～1988 年在云南西双版纳勐仑观察，8 月中旬花芽始现，着生于当年新枝叶腋或顶生；9 月上旬～10 月上旬花序形成，花蕾膨大；10 月下旬～11 月中旬花盛开，12 月中旬开花结束，幼果形成；翌年 2 月中旬果实开始成熟，3 月中旬为果实成熟盛期。浆果状核果，椭圆形或球形。果托大而浅，肉质增厚，高脚杯状。宿存花被片略增大，直伸或开展。成熟时外果皮由黄绿色转变为紫黑色，有光泽。果长 1.4～1.8cm，径 1～1.4cm，成熟后 5～7 日脱落。每果有核 1 粒，椭圆形或卵圆形，长 1.1～1.6cm，径 0.6～0.9cm。内果皮褐色。种仁黄色。无胚乳，子叶厚，肉质（图 1）。

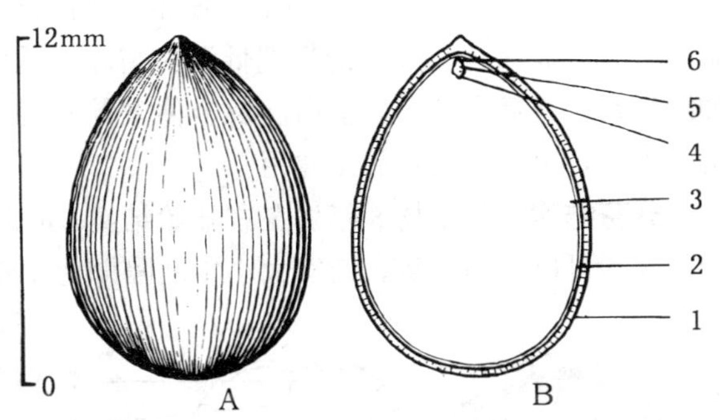

图 1　滇新樟果核的外形（A）及其纵切面（B）

1. 内果皮　2. 种皮　3. 子叶　4. 胚芽　5. 胚轴　6. 胚根

（黄应钦绘）

果实的采收调制和种子贮藏　果实成熟盛期时上树采摘，或摇动树枝震落后在地面拾集。采得的果实当天剥去果皮，或装入竹箩置水中反复搓擦，淘去果皮杂质及漂浮在水面上的瘪粒，所得果核用作播种材料，通称种子。按鲜果质量计算的出籽率为 35%～40%。种子的净度为 90%～98%。千粒重 440～460g。每千克有纯净种子 2 100～2 300 粒。调制后稍加晾干的种子含水量为 23%～25%。忌失水，不宜日晒。短期运输时可用木箱等包装，并混以湿润的木屑或沙。贮藏时也需混以湿沙，但最好随采随播，不作贮藏。

发芽和播种　种子没有明显的休眠现象。调制出来的种子应及时播种。发芽时的日均温宜在 20℃以上。1988 年 3 月 9 日，西双版纳热带植物园在室外沙盆中用新鲜种子作过发芽测定：4 月 11 日开始发芽，从发芽开始至发芽高峰的 20 天中发芽百分数为 35%。从播种之日起算，70 天的发芽率为 80%～90%。留土萌发。具初生不育叶 1～3 片，出土后约 5～8 天展出初生叶。

点播或条播。每平方米播种 100～130g。覆土 1～2cm。幼苗具 2～4 片初生叶时移植。1 年生苗出圃。

（程必强）

鳄　梨

Persea americana Mill.

（樟科　Lauraceae）

生长习性、分布和用途　鳄梨属约 50 种，本文描述我国引入的 1 种。原产热带美洲，我国引种于 1920 年前后，华南地区、台湾、福建、四川南部和云南西双版纳都有栽培。常绿乔木，高 10～20m，胸径 20～30cm。速生。喜高温高湿。耐寒力因品种而异，最耐寒的墨西哥品系可耐－6.7～－3.7℃的极端低温和霜冻，适生于肥沃湿润的沙壤土，忌积水。果肉含多种维生素，并富含脂肪、蛋白质及多种微量元素，为世界名果。种子含油，有香气，可以食用，入药，也可供化妆品工业用。

开花结实　实生树 8 年生左右开始结实，正常结实期在 12 年生以后。无性繁殖所成的植株第 3 年可以开始结实。结实大小年间隔期为 1 年。在水肥条件较好的地方结实大小年现象不明显。花两性。聚伞状圆锥花序近顶生。花淡绿色而带黄色，密被黄褐色柔毛。花被裂片 6，2 轮，花后增厚。发育雄蕊 9，3 轮。花药 4 室，第 1、2 轮药室内向，第 3 轮药室外向。子房卵形，1 室 1 胚珠。花柱纤细，柱头盘状。1978～1987 年在海南及广西南宁观测，鳄梨花芽潜伏越冬，3 月中旬为始花期，3 月下旬～4 月上旬为盛花期，5 月上旬仍有少数花开；8 月下旬果实开始成熟，9 月为果实成熟盛期。肉质核果，梨形，有时球形或卵形，成熟时由青色转为黄绿色或红棕色。果实的大小因品种和栽培条件而不同，差异较大。经过改良的栽培品种，果实长 8～18cm，径 6～13cm，每果重约 0.5kg。有果核 1 枚，梨形或卵形。外果皮膜质，中果皮浆状肉质，内果皮膜质，有明显网状脉。果核长 4.5～8cm，径 3.5～6cm，无胚乳。常有多胚现象。子叶肉质，充满种子（图 1）。

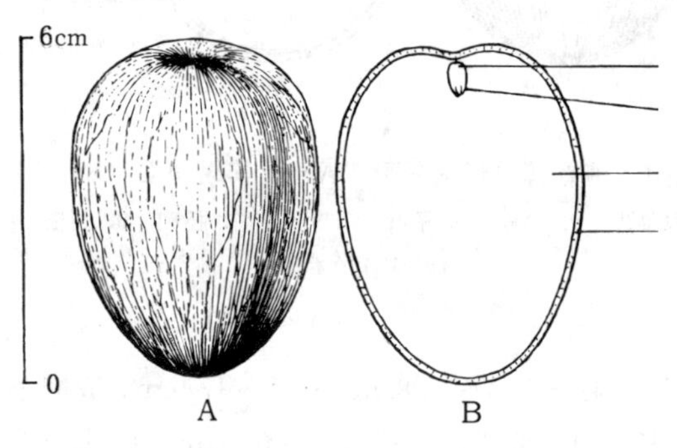

图 1　鳄梨果核的外形（A）及其纵切面（B）
1. 内果皮和种皮　2. 子叶　3. 胚芽　4. 胚根
（黄应钦绘）

果实的采收调制和种子贮藏　在果熟季节分批选采已经成熟的果实。上树采摘，或用钩刀等采种器具钩落果实，并在树下离地面拉上布单承接，以免坠地破伤果皮。采下的果实可存放 1～3 天，果肉软熟后可供食用并取得果核，或将食品加工后取得的果核洗净，即得播种材料，通称种子。按鲜果质量计算的出籽率约为 14％，净度可达 92％～99％。每个果核重约 70g，变动在 40～80g。每千克有纯净种子 14（12～25）粒。种子含水量 45％～60％，忌日晒，不宜失水。裸露种子存放期不宜超过 1 周。海南等热带地区可以随采随播，无需贮藏。我国大陆南部地区宜混沙湿藏越冬，贮藏期为 6～8 个月。

发芽和播种　种子有短暂的休眠现象。发芽时日均温需在 20℃以上。1980 年广西林业科学研究所在南宁室外沙床做过发芽测定，使用的是湿沙贮藏 3～4 个月的种子，分别在采种当年的 12 月和次年 1 月播种。4 月下旬开始萌发，5 月中下旬进入发芽盛期，6 月中旬发芽结束。发芽全程近 50 天，无明显的发芽高峰。发芽率为 80%。留土萌发。每粒种子可出苗 2～3 株。胚根萌发后约 20 天上胚轴出土，具不育叶多枚，出土后 11 天展出初生叶（图 2）。

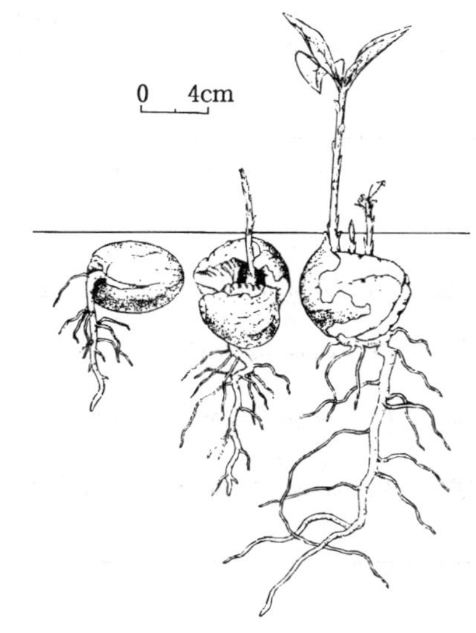

图 2　鳄梨种子萌发后第 10、20、32 天幼苗的生长情况
（黄应钦仿《热带亚热带主要树种采种育苗技术》）

生产上一般是将种子密播于苗床，待胚根萌发后将芽苗移栽到圃地继续培育，也可在苗床上直接点播。覆土 2～3cm。播后需防治鼠害。苗圃或沙床均忌积水。用于庭院绿化的一般是 1 年生实生苗。以采果为目的时多用实生苗作砧木，嫁接成苗后出圃定植。

（韦增健）

楠　属
Phoebe Nees

（樟科　Lauraceae）

生长习性、分布和用途　本属约 94 种，分布于亚洲、美洲热带和亚热带。我国约 34 种，广布于长江流域以南地区。本文描述 5 种（表 1），其中闽楠、浙江楠和桢楠已列入《中国植物红皮书》，属于渐危种。常绿乔木。耐荫。喜温暖湿润气候。适生于土层深厚、肥沃而排水良好的中性或微酸性壤土。在平原、沟谷、溪边、阴坡下部及河边台地的阴湿环境中生长良好。木材坚实，结构细致，不易变形和开裂，为建筑、家具、船板等优良用材，尤其是桢楠和闽楠，素以材质优良而闻名于世。

开花结实　湘楠 4 年生开始开花结实。闽楠 12 年生开始结实，正常结实年龄在 20 年生以后，丰年间隔期 3 年。浙江楠、紫楠与桢楠未见明显的结实间隔期。花两性。聚伞状圆锥花序，腋生于新枝上，长 3～15cm。花小，长约 4～5mm，花被裂片 6，卵形，2 轮，直立，宿存。花的构造与润楠属近似。子房 1 室。花期 4～6 月，果实成熟期 10～11 月，散落期11～12 月。这 5 个树种开花结实的物候期见表 2。

表 1 楠属树种的名称、生长、分布和用途

中 名	学 名	树高 (m)	胸径 (cm)	分 布	用 途	供 稿
闽楠	*Ph. bournei* (Hemsl.) Yang	40	150	赣、闽、粤、湘、鄂、浙南、桂西北及桂、东北、黔东南及黔东北、豫南及豫西南	材用	903
浙江楠	*Ph. chekiangensis* C. B. Shang	20	50	浙、闽北、赣东	材用、观赏	702
湘楠	*Ph. hunanensis* Hand. -Mazz.	8	8	陕南、甘南、赣西南、鄂、湘、黔东	材用	620
紫楠	*Ph. sheareri* (Hemsl.)Gamble	15	50	苏、浙、闽、皖、湘、黔、粤北	材用、香料、药用、观赏、油料	702
桢楠	*Ph. zhennan* S. Lee et F. N. Wei	30	60	川西、黔西北、鄂西、湘西北	材用、观赏	501

表 2 楠属树种的开花结实物候

树种	观察地点	花 期	果实成熟期	种子散落期
闽楠	福建三明	4 月中旬～5 月上旬	10～11 月	—
	福建北部	—	11 月中下旬	12 月中下旬
浙江楠	浙江杭州	5 月～6 月上旬	10 月下旬～11 月中旬	11 月下旬～12 月中旬
湘楠	贵州贵阳	5 月	10 月	11 月
紫楠	浙江杭州	5 月中旬～6 月上旬	10 月中旬～11 月上旬	11 月中旬～12 月中旬
桢楠	四川成都	5 月中旬～6 月上旬	10 月下旬～11 月中旬	11 月上旬～12 月

浆果状核果，卵形或椭圆形，花被裂片宿存，包被果实基部，内含种子 1 粒。内果皮薄，脆壳质，与薄膜质种皮常合而难分。无胚乳，子叶厚，肉质。图 1 绘出的是紫楠种子的外观及其解剖构造。5 个树种果实和果核的形态特征见表 3。

表 3 楠属树种果实和果核的形态特征

树 种	未成熟果实颜色	成熟果实			成熟果核			
		形 状	大小(cm)	颜 色	形 状	大小(cm)	颜色	胚数
闽楠	青绿色	椭圆形或长椭圆形	长 1.1～1.5 径 0.6～0.8	蓝黑色	椭圆形或长椭圆形，两端尖	长 0.9～1.4 径 0.4～0.6	黄褐色	单胚，稀 2～3 胚
浙江楠	绿色	椭圆状卵形，外略被白粉	长 1.2～1.5 径 0.6～0.9	乳黑色	椭圆状卵形，顶端微弯，腹面略平，背面隆起	长 1.0～1.3 径 0.5～0.7	黄褐色	多胚，子叶不等大
湘楠	绿色	卵形	长 1.0～1.2 径 0.7～0.8	黑色	卵形	—	—	—
紫楠	绿色	卵形	长 0.8～1.2 径 0.6～0.8	黑色	卵形，种皮有不规则黑斑	长 0.7～1.1 径 0.5～0.7	黄褐色	单胚，两侧对称
桢楠	青绿色	椭圆形或矩圆形	长 1.2～1.5 径 0.6～0.8	蓝黑色	椭圆形或矩圆形	长 1.0～1.2 径 0.4～0.6	灰黄褐色	单胚或 2～3 胚

果实的采收调制和种子贮藏　果皮变为蓝黑色或黑色时即达成熟，应及时采集。可用高枝剪、采种刀等工具剪采，也可用竹竿击落于地面收集。采回的浆果倒入木桶或缸中，水浸至果肉软化，捣碎或搓去果肉，用清水冲洗干净，将调制出的果核置通风处阴干。果核用作播种材料，通称种子。楠属种皮（包括内果皮和种皮）薄，容易失水开裂因而丧失发芽能力，所以阴干后宜随即播种，或混湿沙层积越冬。据试验（史晓华，未刊稿），紫楠种子在3～5℃低温条件下，用塑料袋密封可保存约1年。5个树种的果实出籽率、种子净度和有关质量的数据见表4。

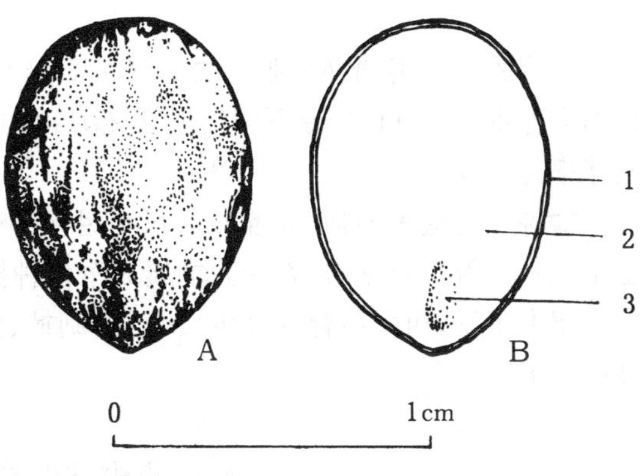

图1　紫楠果核外形（A）及其纵切面（B）
1. 内果皮及种皮　2. 子叶　3. 胚
（童军平绘）

表4　楠属树种的出籽率和有关种子质量的数据

树　种	出籽率（%）	净　度（%）	千粒重（g）	每千克纯净种子粒数（粒）
闽楠	40～50	90～99	250～345	2 900～4 000
浙江楠	50～60	90～98	240～290	3 400～4 200
湘楠	64	90～97	290	3 400
紫楠	50～60	90～98	220～270	3 700～4 500
桢楠	50～60	99	300～360	2 750～3 300

发芽前的处理　楠属种子的休眠与种皮有关，但树种之间不尽相同。据研究，紫楠种子种皮透水性和透气性均差（史晓华和田丽浩，1988），浙江楠种子种皮透水性能虽好，但透气性差。层积处理能解除休眠，尤其是变温层积，能缩短层积时间，效果较好（史晓华和史忠礼，1990）（表5）。

表5　楠属树种种子的层积处理及其发芽测定结果①

树　种	层积处理条件	层积天数（天）	发芽势（%）			发芽率（%）		
			计算天数	一般数值	变动范围	计算天数	一般数值	变动范围
浙江楠	室温（1.5～11.5℃）层积	21	25	48	42～54	40	82	76～86
	15～25℃变温层积	21	13	32	26～40	40	90	86～96
	对照	0	—	—	—	—	15	—
紫楠	室温（1～17℃）层积	63	22	38	32～42	30	70	64～82
	15～25℃变温层积	49	10	31	28～32	30	89	86～96
	对照	0	—	—	—	—	0	—

①　基质为沙，测定温度为25℃

发芽测定　闽楠种子萌发的适宜温度为17～20℃,发芽率为75%～89%。浙江楠和紫楠宜剥去种皮,在30℃恒温下测定发芽能力。有1份长途运输的桢楠种子,在恒温25℃条件下的发芽率为64%。

播种　可随采随播,或混沙层积至翌春播种。条播,行距15～20cm,条宽6～10cm。播后覆土1～2cm,盖草或锯屑、谷壳。楠属5个树种播种育苗的某些技术参数见表6。

留土萌发,初生不育叶互生(图2)。闽楠、浙江楠、桢楠种子有多胚现象。闽楠和紫楠亦可扦插繁殖。

表6　楠属树种播种育苗的某些技术参数

树　种	播种期	每平方米播种量(g)	出苗期	出苗率(%)	1年生苗高(cm)	出圃年龄
闽楠	1～2月	50～60	4月上旬	54～95	30～40	1
	3月上旬	50～60	5月上旬	54～95	30～40	1
浙江楠	3月下旬	60～75	5月下旬	62～74	15～25	2
湘楠	3月上旬	38	4月上中旬	86	—	—
紫楠	3月下旬	60～75	5月下旬	64～80	15～25	2
桢楠	3月上旬	50～60	3月下旬	85	15～25	2

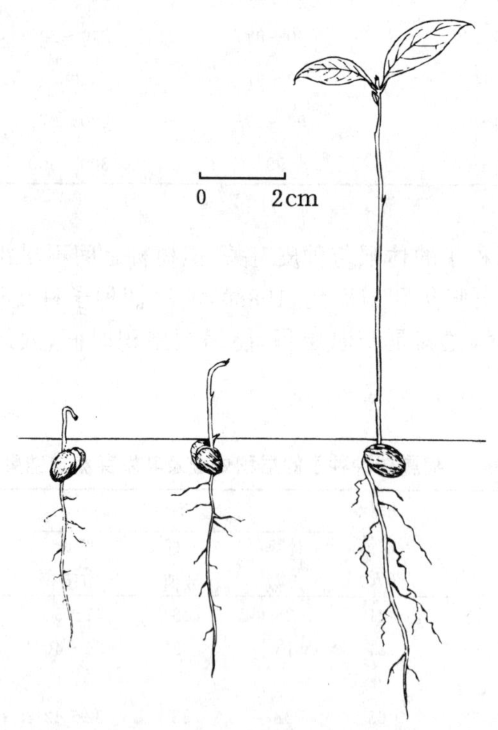

图2　紫楠种子萌发后第1、6、36天幼苗的生长情况

(童军平绘)

(史晓华)

檫　木

Sassafras tzumu（Hemsl.）Hemsl.

（樟科　Lauraceae）

生长习性、分布和用途　檫木属 3 种，分布于东亚和北美。我国 2 种，檫木产长江以南，台湾檫木产于台湾。本文描述檫木这 1 个种。落叶乔木，高可达 35m，胸径 2.5m。根深，适生于土层深厚肥沃、排水良好的酸性土壤。不耐旱，忌水湿。常散生，也与杉木、毛竹等星散混生，人工纯林生长不良。分布在苏、浙、闽、川、滇、黔、湘、鄂、桂、赣及皖西北、粤北。材质优良，纹理美观，具香气，可用于造船、建筑及家具。种子含油率 20%，可制油漆。根和树皮可入药。树形挺拔，秋叶红艳，是良好的观赏树。

开花结实　6～8 年生开始开花结实，正常结实在 10 年生以后，大小年现象不明显。花杂性异株。总状花序顶生。花被裂片 6。花药 4 室，雄花具能育雄蕊 9，退化雄蕊 3。雌花具退化雄蕊 12，子房卵形，1 室 1 胚珠。花柱细，柱头盘状。花期长，一般在 1 月中旬～3 月上旬先叶开花，黄色。果实成熟期在 6 月底～8 月中旬。浙江新昌在 7 月下旬～8 月中旬成熟。浆果状核果近球形，径 0.5～0.7cm。果托盘状。果实由红转黑色或蓝黑色，外被白蜡状粉末，大部分果托由青变红时即表示成熟。每果含果核 1 粒。果核近球形，径 4～5mm，内果皮稍坚硬，内有膜质种皮，无胚乳，胚白色。子叶肥大，半球形，胚芽和胚根细小，呈点状（图 1）。

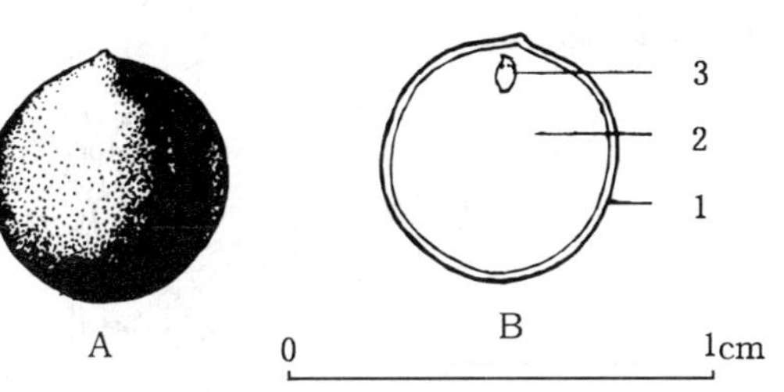

图 1　檫木种子外形（A）及其纵切面（B）
1. 内果皮及种皮　2. 子叶　3. 胚轴
（童军平绘）

果实的采收调制和种子贮藏　成熟果实占 60%～70%时用采种刀连同果穗采下或用杆击落。采种要及时，成熟一批采一批。过迟果实会遭受鸟类啄食或失水干缩。核果肉质，且核壳外被蜡层，因此调制工作包括去皮和脱蜡两大内容。先将采得的果实盛入筐箩，用清水漂去空瘪粒，并在筐边反复搓擦，洗掉果皮果肉，滤出包有蜡层的果核。脱蜡时常用 5%～10%草木灰或 30%～50%的泥沙与种子拌和，反复搓揉，直至核壳乌黑发亮，白蜡完全脱掉为止。所得脱除蜡层的果核即为播种材料，通称种子。鲜果出籽率 25%～28%。种子千粒重 58～75g，每千克有种子 1.3 万～1.7 万粒。

果实盛夏成熟，调制后应及时贮藏。过去贮藏的成功率不高。看来檫木种子的贮藏，一方面要最大限度地保持生命力，一方面还要为解除休眠创造条件。林宏彬等（1982）根据贮藏时间的长短采取不同方法，获得了较好的效果。

1. 越冬贮藏（贮藏后翌春播种）　多藏于山洞或防空洞或地窖内。贮藏介质为黄心土（用直径 4mm 的筛孔过筛）或用清水沙。筛后的黄心土含水量为 30%～40%，清水沙以手捏成团，触之散开为度。贮藏时先在洞窖底部铺 5cm 的潮湿介质，上面堆放种子与介质以 1∶3

的比例拌匀的混合物，堆顶盖 4～6cm 介质，堆高不超过 50cm。种子含水量控制在 25%～32%（下同）。贮藏初期气温较高，每隔 10 天翻堆 1 次，并注意加水，拣去霉烂粒。随着气温逐渐降低，翻堆间隔时间可相应延长。11 月份以后可不再翻堆。用黄心土作介质的效果优于清水沙。贮藏结束时种子发芽率为 30%～60%，无明显的发芽盛期，仍有部分种子休眠。

2. 隔年贮藏　种子与黄心土以 1∶4～5 的比例拌和，用缸、甏盛装。选择室内阴凉处挖一洞穴，把容器的 1/3 埋入地下。种子离容器口 35cm 左右处覆土。如覆土太薄，翌年 4～6 月间上部种子会陆续发芽。整个贮藏期间应注意加水保湿。贮藏结束时，种子发芽率可达 90% 以上。

3. 2 年贮藏（到第 3 个春天播种）：种子与黄心土以 1∶1 的比例拌和，装入甏中或塑料袋中，贮藏在 0℃条件下（低于 0℃种子会受冻害）。贮藏结束后，种子发芽率为 40%～50%，部分种粒不能萌发，有人认为仍属休眠。

檫木种子长途运输的技术关键是通气和保湿。运输以冬末初春为宜。最好采用能够通气的容器，并用潮湿的苔藓等将种子分层包装。运输中途应经常检查是否发热，并浇水保湿。运到后立即湿藏。

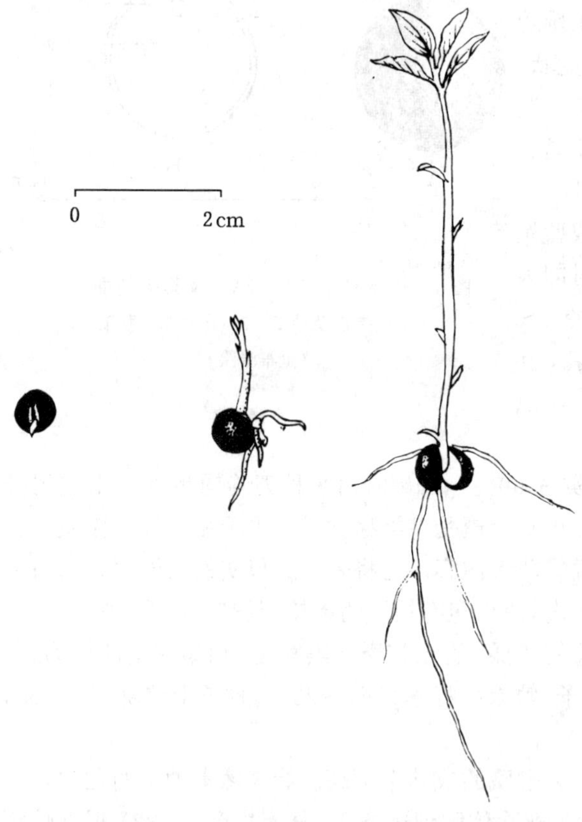

图 2　檫木种子萌发后第 1、7、16 天幼苗的生长情况
（童军平绘）

种子质量检验　对于解除了休眠的檫木种子，常用的发芽条件是 25℃恒温，测定期 28 天。但是檫木种子个体间休眠的深度不一，有的种粒播后 20 个月才萌发。生产中通常是检查优良度或生活力。

优良度测定时，首先用手捏，能捏破的为低劣种子。捏不破的再用解剖法观察：优良种子的剖面表现是子叶新鲜饱满，剖面中间部位颜色较深，带有油光，靠近核壳一圈的子叶呈淡白色或淡绿色。子叶剖面中部和边缘颜色一致，呈暗黑色，无油光，或壳肉分离的，为无生命种子。

优良度测定还可采用刺孔滴水法：用刀片沿种脊垂直切去种子的 1/3，切口要求平整。在留下的种子切面上，用细针刺 2～3 个小孔（直径 0.5mm，分布均匀，以不刺破种皮为度），将种子置于垫有湿滤纸的培养皿中，随即滴上清水，5 分钟左右，观察水的颜色变化。生活力高的种子，剖面上有白色乳浊液并使种子剖面模糊不清。无生活力的种子没有白色乳浊液产生，能清晰地看清剖面。只产生少量白色乳浊液的种子生活力已衰退，需进一步观察胚芽、胚轴、胚根是否完好。

播种　一般为春播，每平方米播种6～10g。平均气温18℃左右时开始出土。据浙江临海县观察，2月15日播种的出土期为74天，4月8日播种的出土期为22天。1年生苗高达1m左右，根径1cm左右，可以出圃。每666m^{2}产苗约1万株。

留土萌发。子叶2，萌发时仍包于核壳内，下胚轴不明显。具初生不育叶2～6，互生。檫木种子的萌发和幼苗初期生长情况见图2。

（翁尧富）

风吹楠属
Horsfieldia Willd.

（肉豆蔻科　Myristicaceae）

生长习性、分布和用途　本属约90余种，我国有5种，本文描述3种。常绿乔木。速生。不耐干旱，常生于土壤肥沃湿润的沟谷密林或山坡密林环境。能耐0℃左右的短期极端低温，但苗期忌霜冻。种子富含以十四碳为主要成分的油脂，是重要的工业油料。这3个种的名称、生长、分布和用途见表1。其中琴叶风吹楠和滇南风吹楠是我国热带季节性雨林中的特有种，分布区狭小，数量特别是雌株数量稀少，在《中国植物红皮书》中被分别列为渐危种和濒危种。

表1　风吹楠属树种的名称、生长、分布和用途

中　名	学　名	树高(m)	胸径(cm)	分　布	用　途
风吹楠	*H. glabra*（Bl.）Warb.	10～15	20～40	滇、琼、桂。越南、缅甸、印度及安达曼群岛	材用，种子含油29%～33%
琴叶风吹楠	*H. pandurifolia* Hu	15～24	30～45	滇南、滇西南	材用，种子含油57%
滇南风吹楠	*H. tetratepala* C. Y. Wu	15～25	35～40	滇南	材用，种子含油34%

开花结实　风吹楠栽培后3年开始结实，5年后进入正常结实期，结实大小年现象不明显，每年主花期有8月和12月两次。琴叶风吹楠和滇南风吹楠10～12年生开始结实，15年生以后为正常结实期，大约每1～2年有一个结实较多的年份。花单性异株或同株（琴叶风吹楠）。花小，无花瓣，花被3（2～4）裂，黄色。雄花序常为复合圆锥状；分枝疏散，有时聚合成团。雄蕊10～12，花丝合生成雄蕊柱，花药2室，外向纵裂，由背部贴生，通常包被雄蕊柱。雌花子房上位，1室，有近基生的倒生胚珠1枚。柱头无柄，微合生。在云南西双版纳观察的物候期见表2。

果卵形。琴叶风吹楠、滇南风吹楠的果皮近木质。风吹楠果皮肉质，较厚，外表光滑。果实成熟时常开裂为2果瓣。种子具肉质鲜红色或橙红色的假种皮，种皮外层壳质状肉质，中层常木质，内层薄膜质，常透入胚乳内。胚乳嚼烂状，含固体油。子叶基部合生，近贝壳状，位于种子中部的一侧。果实和种子的形态见表3、图1。

表 2 风吹楠属树种的物候期

树 种	观察年份	开花期			果实成熟期		果实脱落
		始 期	盛 期	末 期	始 期	盛 期	
风吹楠	1988～1989	8 月上旬	8 月中旬	8 月下旬	12 月	翌年 1～2 月	1～2 月
		12 月上旬	12 月中～1 月中	2 月下旬	4 月上旬	4 月中下旬	4 月下旬～5 月中旬
琴叶风吹楠	1985～1988	5 月中旬	5 月下旬	7 月中旬	翌年 3 月中旬	4 月中旬	4～5 月上旬
滇南风吹楠	1985～1988	6 月上中旬	6 月中下旬	8 月中下旬	翌年 2 月中旬	3～4 月	4～5 月上旬

注 观察地点云南勐腊

表 3 风吹楠属树种果实和种子的形态特征

树 种	未熟果颜色	成熟果实			成熟种子		
		形 状	大小(cm)	颜 色	形 状	大小(cm)	颜 色
风吹楠	绿色	卵圆形至椭圆形，先端具短喙	长 3～3.5(4) 径 1.5～2.5	橙黄色	卵形，珠孔在中部以下，至基部有显著的宽线形疤痕	长 2.4～2.7 径 1.4～1.6	灰褐色，具脉纹
琴叶风吹楠	深绿色	卵状椭圆形，先端锐尖，基部不偏斜	长 3～4.5 径 2～2.5	黄褐色	卵球形至卵球状椭圆形，顶端具明显的突尖，基部偏生卵形疤痕	长 2.5～3.2 径 1.6～1.8	灰白色，光滑，具脉纹和斑块
滇南风吹楠	绿色光滑	椭圆形，先端钝圆，基部偏斜，花被片宿存	长 4.5～5 径 2.8～3.5	橙黄色	卵状椭圆形，两端钝，平滑，珠孔下陷，中央有小突起	长 3.5～4 径 2.0	淡黄褐色，具疏脉纹

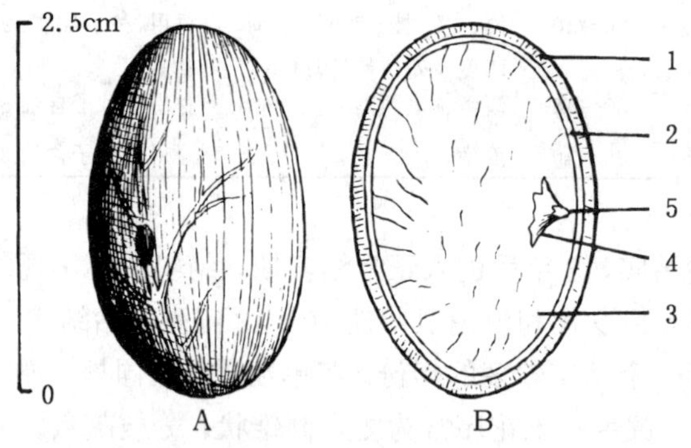

图 1 风吹楠种子去除假种皮后的外形(A)及其纵切面(B)
1. 外种皮 2. 内种皮 3. 胚乳 4. 子叶 5. 胚根
(黄应钦、马信祥绘)

果实的采收调制和种子贮藏

果实进入成熟盛期而果皮尚未裂开时，上树采摘或敲打落地后捡拾。种子容易丧失发芽力，采集的果实应及时调制，忌堆沤。可以摊放于通风荫凉处，果皮开裂后取出种子，并及时剥去假种皮，即得纯净种子。种子忌失水，不能日晒或干藏。种子的含水量需保持在40%以上。短期运输时需混湿润锯木屑。贮藏时需混湿沙并存放于

表 4 风吹楠属树种的出种率和种子净度、质量

树 种	出种率(%)	净度(%)	千粒重(g)		每千克纯净种子粒数(粒)	
			一 般	变动范围	一 般	变动范围
风吹楠	20～22	90～98	2 750	2 700～2 800	360	350～370
琴叶风吹楠	18～20	90～98	4 500	4 000～5 900	220	170～250
滇南风吹楠	24～30	90～98	6 000	5 500～7 500	170	130～180

荫凉处，贮藏期为3～5个月。这3个树种种子的净度和质量等数据见表4。

发芽和播种　种子无休眠习性或有短期休眠现象。发芽时日均温宜在23～25℃。1989年西双版纳热带植物园在室外沙盆中对风吹楠种子进行过发芽测定：1月下旬播种的，播后50天开始萌发；3月中旬播种的，播后18天开始出土；没有明显的发芽盛期，发芽率约70%。留土萌发。具初生不育叶4～5枚，18天左右初生叶展现（图2）。

条播。每平方米播种风吹楠370～600g，琴叶风吹楠750～1 200g，滇南风吹楠1 000～1 400g。覆土1.5cm左右。1年生苗出圃。

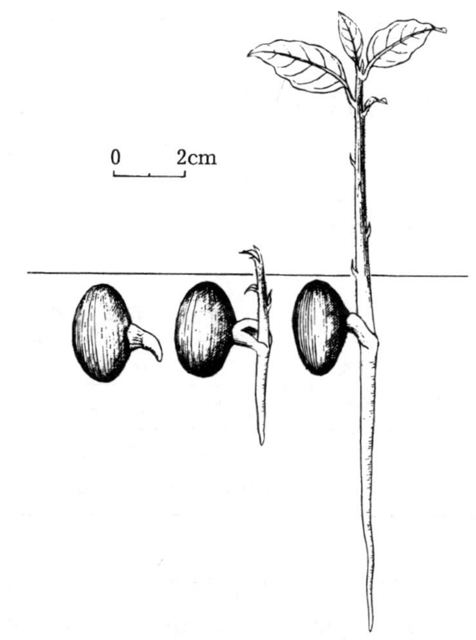

图2　风吹楠种子萌发后第3、8、28天幼苗的生长情况

（黄应钦、马信祥绘）

（马信祥）

五桠果属
Dillenia L.

（五桠果科　Dilleniaceae）

生长习性、分布和用途　本属约60种，我国有4种，本文描述2种。常绿乔木，高达25～30m，胸径60～80cm。适生于肥沃、湿润、疏松的酸性土壤，颇耐水渍。抗寒力低，持续半个月5℃左右的低温便全株严重受冻。苗期忌霜冻。分布于热带亚洲。我国分布于滇、桂、琼。木材轻软，花大而美丽，果可食。这2个树种的名称、生长、分布及用途见表1。

表1　五桠果属树种的名称、生长、分布和用途

中　名	学　名	树　高（m）	胸　径（cm）	分　布	用　途
五桠果（第伦桃）	*D. indica* L.	30	60～80	滇、桂。印度、印度尼西亚、中南半岛	工艺材、果可食、观赏、种子含油
毛五桠果（大花第伦桃）	*D. turbinata* Finet et Gagnep.	25	60	滇、桂、琼。越南	材用、观赏、栲胶、果可食

开花结实　约10年生开始开花结实，正常结实期在20年生以后。结实大小年现象较为

明显，间隔期1年。偶尔一遇的冬季严寒会加重大小年现象。花两性。萼片5，覆瓦状排列，宿存，厚革质或硬肉质。花瓣5，早落。雄蕊多数，离生，2轮，内轮较长，花药顶孔开裂。心皮离生于隆起的花托上，花柱条形，胚珠多数。五桠果花单生于近枝顶的叶腋，白色，颇肥大，径12～20cm，花瓣长7～9cm。心皮16～20。毛五桠果的总状花序顶生，花序柄长3～5cm，花3～5朵，黄色或淡红色，花径10～12cm，花瓣长5～7cm。心皮8～9。在南宁和海南观测的物候期见表2。

表2 五桠果属树种的开花结实物候期

树种	观察地点	观察年份	开花期			果实成熟期		果实脱落期
			始期	盛期	末期	始期	盛期	
五桠果	南宁	1987～1988	6月中旬	6月下旬	7月上旬	8月下旬	9月上旬	9月上旬～9月下旬
毛五桠果	海南	—	4月下旬	5月上中旬	5月下旬	6月下旬	7月上中旬	7月中旬～8月上旬

聚合浆果，球形，常为宿存的肥厚萼片所包裹，成熟后在树上存留10～15天即逐渐脱落。每果的种子数在五桠果为100～150粒，毛五桠果约为100粒。五桠果的种子有假种皮，毛五桠果的种子无假种皮。胚乳丰富，肉质，胚微小。这两个树种的果实及种子形态见表3。五桠果的种子外形及其内部结构见图1。

表3 五桠果属树种果实和种子的形态特征

树种	未熟果颜色	成熟果实			种子		
		形状	大小（cm）	色泽	形状	大小（mm）	色泽
五桠果	绿至黄绿色	近球形	径8～12（15）	暗红色	近肾形或扁椭圆形，表面具毛，有假种皮	长5～6.5 宽3～4.5 厚1.5～2	棕褐色
毛五桠果	绿至黄绿色	近球形	径4～5	暗红色	扁倒卵形，无毛，无假种皮	长4～5 宽3～4 厚1.2～1.8	深褐色

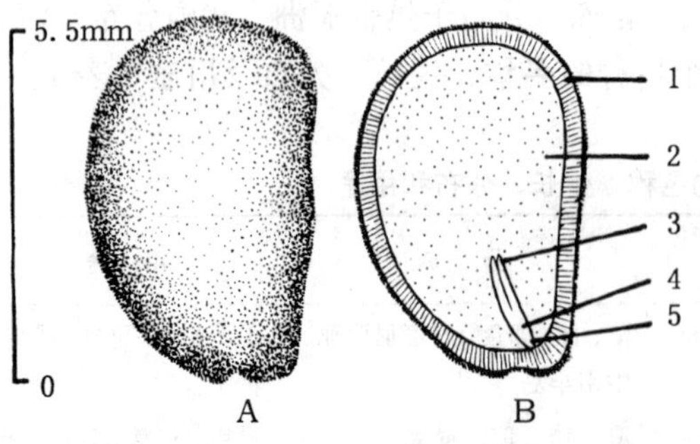

图1 五桠果种子的外形（A）及其纵切面（B）
1.种皮 2.胚乳 3.子叶 4.胚轴 5.胚根
（黄应钦绘）

果实的采收调制和种子贮藏 果实由黄绿色转为淡红色时即可用采种钩刀钩断果梗，或用采种叉将果叉落，在地面捡拾。采回的果实先堆沤3～5天，果肉充分软熟后掰开果实取出种子，用布袋或纱布包裹，置水中洗去皮肉和种子上的粘稠汁液，即得纯净种子。种子的净度、质量等见表4。

种子含水量应保持在20%左右，不宜失水，忌日晒。经过处理的纯净种子，稍晾干后即可播种或贮运。种子宜随采随

表 4　五桠果属树种的出种率和种子净度、质量

树　种	出种率（%）	净度（%）	千粒重（g）	每千克纯净种子粒数（万粒）
五桠果	1	90～95	18～22	4.5～5.6
毛五桠果	1	90～95	15～19	5.2～6.7

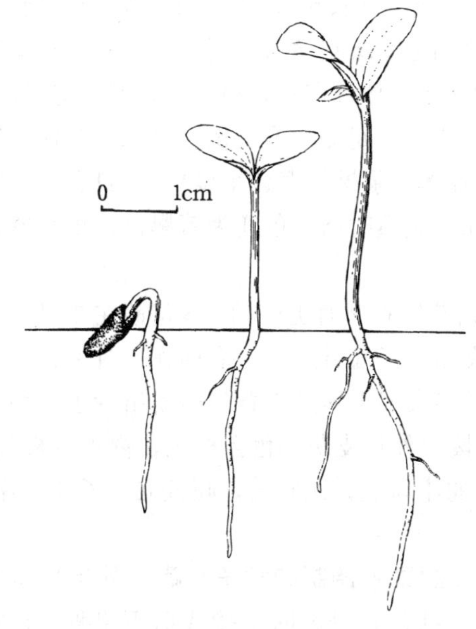

图 2　五桠果种子萌发后第 5、8、26 天幼苗的生长情况
（黄应钦绘）

播，一般不宜贮藏越冬。运输时需混以湿沙或锯屑，并用紧密布袋包装。如需贮藏，应混拌湿沙。贮藏期一般为半年左右。经过沙藏的种子比新鲜种子的发芽率会降低 10%～30%。裸露存放的种子，10 天左右发芽率便开始明显下降，半个月左右会全部失去发芽能力。

发芽和播种　种子无休眠习性或有短暂休眠期。发芽时的日均温需在 18℃以上。1987 年 7 月下旬和 9 月上旬，广西林业科学研究所在南宁用当年新采的种子于室外沙床播种，分别对两个树种做发芽测定：播种后约 50 天种子开始萌发，发芽时的日均温为 25～32℃，无明显的发芽盛期，在 68 天中五桠果的发芽率为 64%，毛五桠果的发芽率为 65%。出土萌发。胚根萌发后 7～8 天子叶出土，再过 16～18 天初生叶出现（图 2）。

撒播。每平方米播种 5～8g，播后稀薄覆土，上盖塑料薄膜低棚以防雨水冲击。用喷雾器淋水。也可以将种子播入沙盆。播种后约 1 个月左右将芽苗移至容器或大田继续培育。容器苗 3 个月可以出圃，大田苗 1 年生出圃。

（韦增健）

马　　桑

Coriaria nepalensis Wall.

（马桑科　Coriariaceae）

生长习性、分布和用途　马桑属 15 种，我国 3 种，本文描述 1 种。落叶小乔木或灌木。树高 1.5～2.5m，最高达 6m，胸径可达 12cm。喜光。耐干旱瘠薄土壤，萌芽力极强，是荒山荒地常见的先锋树种。喜偏钙质的紫色土，在该类土上生长的马桑，根的结瘤多，固氮量高，生长量大。

分布于甘、陕、川、藏、滇、黔、桂、湘、鄂、豫。多生于海拔 300～1 200m 的山地和丘陵灌丛中，在云南丽江玉龙山可以分布到海拔 2 850m。印度、尼泊尔也有分布。

保持水土的优良树种，能提供薪柴。种子含油率较高，可榨油，供制作涂料。茎皮、根皮及叶含鞣质，可提取栲胶。寄生在马桑植株上的桑寄生和毛顺桑寄生的干燥叶可入药。全株含马桑碱，有毒，可作土农药。幼嫩枝叶压青可作肥料。

开花结实　3～4 年生即可开花结实。无大年小年现象。花单性或杂性。总状花序 1～5 枚生于 2 年生的枝条上，腋生。花小，红色，径约 2～3mm。雄花序先叶开放，长 1.5～2.5cm，多花密集，序轴被腺状微柔毛。雄蕊 10，花丝线形，开花时伸长至 3～3.5mm。花药长圆形，具细小瘤状物，花隔伸出，花药基部短尾状。不育雌蕊存在。雌花序与叶同出，长 5～8cm，紫色。苞片和小苞片卵形，膜质，半透明，内凹。花梗无毛。萼片卵形，边缘半透明。花瓣小，肉质，龙骨状。心皮 5，耳形，侧向压扁，分离。子房上位，每心皮有 1 倒生胚珠。花柱长约 1mm，具小瘤体，柱头上部外弯，紫红色，具多数小瘤体。浆果状聚合小瘦果 5，红色，为增大的肉质花瓣所包被。果球形，有光泽，径 3～ 6mm，成熟时肉红色变为紫黑色。每小瘦果有种子 1 枚。

本文作者 1987～1988 年在四川雅安观察，马桑始花期在 2 月上中旬，盛花期在 2 月中下旬，末期在 3 月上中旬。雄花先于雌花开放。果实成熟始于 6 月上旬，盛期在 6 月中旬，末期在 6 月下旬。每果穗着果 3～20 个。成熟末期的果实大多干缩呈 5 背裂而露出种子。种子呈卵状椭圆形，微扁，背面具 1 棱，两侧略现 1～2 棱。外种皮灰棕色至棕色，内种皮褐色。种子长约 2mm，径约 1mm，表面具无数排列不整齐的小斑点。无胚乳，胚直立。子叶 2 片，较肥大，胚根短，均呈乳白色。

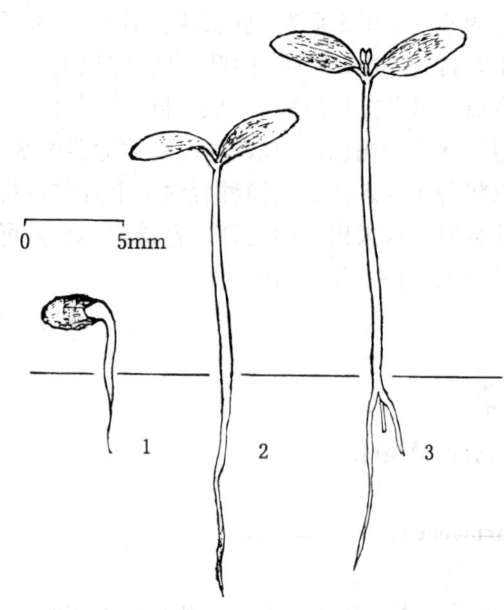

图 1　马桑种子的萌发和幼苗生长情况
1. 胚轴延伸　2. 子叶展开　3. 初生叶初露
（田恒德绘）

果实的采收调制和种子贮藏　外果皮呈紫黑色或紫红色且变软时采摘或剪下果穗，在室内摊晾阴干，数天后果背裂开，搓揉并筛去杂质即得纯净种子。鲜果出种率 15%～16%，干果出种率 47%～58%。净度 94%～96%。千粒重 0.90～1.45g。每克纯净种子有 680～1 100 粒。

风干的种子用布袋盛装，置室内通风处存放，供当年秋季播种。如需翌年春播，应当用塑料袋密封，在 0～5℃条件下贮藏。

发芽和播种　种子无休眠习性。有的地方发芽测定前用始温 45℃水浸种 24 小时，也有的并不浸种，置纱布上用 20～30℃的变温发芽。有一批种子置床 4 天后萌动，14 天的发芽百分数为 24%，45 天的发芽率为 45%～47%。

可以随采随播，也可于翌年 3 月上旬春播。每平方米播种 1.8～2.5g。覆土厚 0.2～0.3cm 后盖草。播种后 7 天发芽出土，再经 20 天发芽基本结束。

出土萌发。子叶 2，长卵形，略呈肉质，全缘，先端平圆。初生叶交叉对生，卵圆形。幼茎 4 棱形，红褐色。马桑幼苗生长情况见图 1。

1 年生苗呈多枝丛生状，枝条长度悬殊，一般长 15～25cm，最长达 45cm。1 年生苗可以出圃定植。

生产上常用直播造林，随采随播，或春播。也有用插条或埋条等方法繁殖的。

（阙再旦）

木　瓜　属

Chaenomeles Lindl.

（蔷薇科　Rosaceae）

生长习性、分布和用途　本属约 5 种，分布于亚洲东部。本文描述 2 种，它们的名称、树高、分布和用途见表 1。落叶灌木或小乔木。生长缓慢。喜温暖。要求土壤排水良好，不耐盐碱。春季花色红艳，入秋果实芳香，是观花观果的优良树种。皱皮木瓜可编成花篱。果可泡制药酒。木瓜种子含油，供食用或工业用。

表 1　木瓜属树种的名称、成年时树高、分布和用途

中　名	学　名	树高(m)	分　布	用　途	供　稿
木瓜	*C. sinensis* (Thouin) Koehne	5～10	陕、晋、豫、鲁，长江中下游及粤、桂	观赏、药用、食用	1006
皱皮木瓜（贴梗海棠）	*C. speciosa* (Sweet) Nakai	2～4	晋、豫、陕、甘、川、滇、黔、桂、粤、湘。冀、鲁至长江中游多栽培。缅甸	观赏、药用	614、306、307

开花结实　木瓜大约 10 年生开始开花结实。皱皮木瓜开始开花结实的年龄较早。木瓜和皱皮木瓜的结实大小年都不明显。这两个树种的开花结实习性、花期和果实成熟期见表 2。1986～1988 年在昆明观察，皱皮木瓜在 3 月形成花蕾，3 月中下旬至 4 月中旬盛开，末花期在 4 月下旬～5 月上旬。

表 2　木瓜属树种的开花结实习性

树　种	开始结实年龄（年）	花期（月）	果实成熟期（月）
木瓜	10	4～5	10～11
皱皮木瓜	6～7	3～5	9～10

花两性，红色。木瓜花单生于叶腋，皱皮木瓜 3～5 朵簇生于 2 年生短果枝上。萼片 5，花瓣 5。雄蕊多数。花柱 5，基部连合。子房下位，5 室，每室有胚珠多数，排成 2 列。梨果大，果皮光滑，质硬，萼片脱落，花柱常宿存，芳香。种子长三角形或圆锥状卵圆形，棕色，无胚乳。皱皮木瓜种子顶端有凹陷的种脐。这两个树种果实和种子的形态特征见表 3。木瓜果实的外形和种子的纵剖见图 1。

表 3　木瓜属树种的开花结实习性

树　种	观察地点	未熟果颜色	成熟果实			成熟种子		
			形　状	大　小 (cm)	颜　色	形　状	大　小 (mm)	色　泽
木瓜	南京	青绿色	长椭圆形，有短梗	长 10～15，径 6～8	暗黄色	长三角形，扁平	长 8.5，宽 5，厚 1.5	棕褐色
皱皮木瓜	昆明	绿色	长卵圆形，果梗短或近于无	长 7.5～9，径 6～7	黄色或黄绿色	圆锥状卵圆形	长 8～11，径 5～7	紫红褐色
	山东	—	球形或卵形	径 (2) 3～5	黄色或黄绿色	扁卵形	长 7，宽 6 厚 4	紫红褐色

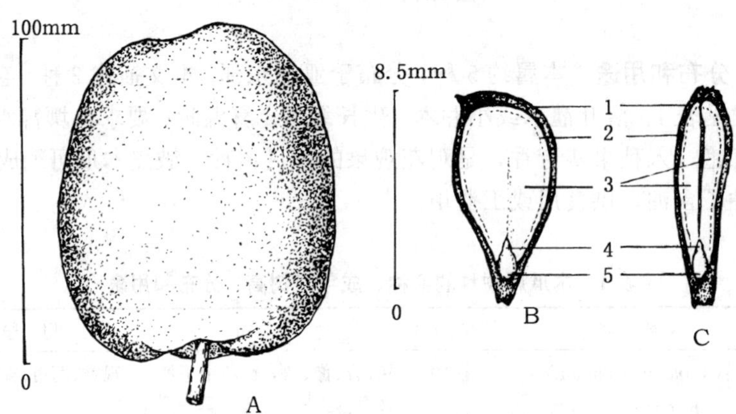

图 1　木瓜果实的外形 (A) 以及种子的纵切正面 (B) 和纵切侧面 (C)

1. 外种皮　2. 内种皮　3. 子叶　4. 胚芽　5. 胚根

(田恒德绘)

果实的采收调制和种子贮藏　在果实变为黄色且有香气逸出时采摘。采回后放置数日，至果实松软时剖开，取出种子淘洗干净，晾干。皱皮木瓜每个果实可得饱满种子 38～75 粒。木瓜每果饱满种子数约 170 粒，种子中的空瘪粒不多。1989 年南京采集的一份木瓜种子样品中空瘪粒在 5%以下。这两个树种的净度可以接近 100%。它们的出种率和种子质量数据已列入表 4。目前还没有这两个树种种子贮藏寿命的研究报道。这两个树种可以随采随播或湿沙层积后来年春播。

表 4　木瓜属树种的出种率和种子质量数据

树　种	产　地	出种率 (%)	千粒重 (g)	每千克纯净种子粒数 (万粒)
木瓜	南京	6.5	50～58	1.7～2.0
皱皮木瓜	昆明	4.0	110～130	0.76～0.9
	山东	—	38～46	2.1～2.6

发芽和播种　这两个树种的种子都需要低温层积的条件打破休眠。1989 年林业部南方林

木种子检验中心将1份木瓜种子在2～5℃下层积2个月，在25℃恒温、每天光照8小时条件下置于脱脂棉上发芽，28天的发芽率达82%。层积至3个月时，有的种子在低温下也能萌发。在上述发芽条件下，未经层积的种子28天的发芽率仅为4%。

皱皮木瓜的种子需要在低温条件下层积50～90天。云南省林业科学研究所曾将皱皮木瓜种子层积至翌年3月初，用清水浸泡1～2天，取出稍晾干后播种，播后20天出苗，播后35天发芽70%，播后50天的发芽率为88%。

可以随采随播，或低温层积至翌春播种。条播，行距30cm。木瓜每平方米播种8～11g，皱皮木瓜20～25g。出土萌发。子叶2，椭圆形，先端平圆或微凹。初生叶互生，托叶大，边缘有细锯齿，齿端有腺（图2）。皱皮木瓜当年生苗高40～60cm，可以出圃。木瓜和皱皮木瓜可以扦插繁殖，皱皮木瓜还可用分株、压条法繁殖。木瓜还可以用楸子（海棠果，*Malus prunifolia* Borkh.）作砧木嫁接繁殖。

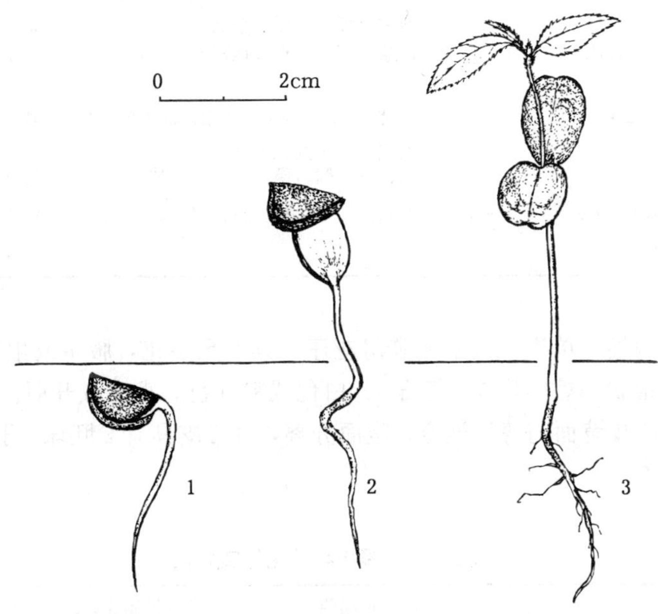

图2　木瓜种子的萌发和幼苗生长情况

1. 胚根延伸　2. 子叶带壳出土　3. 初生叶互生

（田恒德绘）

（高捍东）

栒　子　属

Cotoneaster B. Ehrh.

（蔷薇科　Rosaceae）

生长习性、分布和用途　本属90余种，分布于亚洲（日本除外）、欧洲、北非温带地区。

我国50余种，主产我国西部及西南部。灌木，稀小乔木。本文描述8种，其名称、树高、分布情况和用途见表1。喜光树种。可作绿篱和护岸树，也供观赏。木材坚韧，可做手杖、器具柄或工艺品。

表1 栒子属树种的名称、生长、分布和用途

中 名	学 名	树高（m）	分 布	用 途	供 稿
灰栒子	*C. acutifolius* Turcz.	1.5～2	华北、西北东南部、西南。蒙古	药用，绿化	201
西南栒子	*C. franchetii* Bois	1～3	川、滇、黔。南京引种栽培，生长良好。泰国	观赏	1003
平枝栒子	*C. horizontalis* Decne.	—	湘、鄂、陕、甘、川、滇、黔。尼泊尔	观赏	310
黑果栒子	*C. melanocarpus* Lodd.	1～2	黑、吉、内蒙古、冀、甘、新。蒙古、西伯利亚、西亚、东欧	种子含油，观赏	201
蒙古栒子	*C. mongolicus* Pojark.	1～2	内蒙古浑善达克沙地西部。蒙古东部	观赏	201
水栒子	*C. multiflorus* Bge.	2～4	东北、华北、西北、西南。高加索、西伯利亚、中亚、西亚	观赏，保土	310
华中栒子	*C. silvestrii* Pomp.	1～2	苏、皖、赣、豫、鄂、川、甘	观赏	1003
准噶尔栒子	*C. soongoricus* (Reg. et Herd.) Popov	1～2.5	内蒙古、宁、甘、新、川、藏	观赏，保土	201

开花结实 花两性，单生，聚伞或伞房花序，具2至多花，腋生或生于短枝顶端。萼筒钟状、筒状或倒圆锥状，短萼片5。花瓣5，白色或粉红色，直立或开展。雄蕊20（5～25）。花柱2～5，离生，心皮背面与萼筒连合，腹面分离，每心皮具有2胚珠。子房下位或半下位。开花结实物候见表2。

表2 栒子属树种开花结实物候

树 种	开始结实年龄（年）	观察地点	开花期		果实成熟期		果实散落期
			始 期	末 期	始 期	盛 期	
灰栒子	—	内蒙古呼和浩特	5月	6月	9月	10月	10月至次年
西南栒子	—	南京	5月上旬	6月下旬	9月	10月中下旬	熟后不易脱落，1月底2月初仍可采到果实
平枝栒子	3～5（正常结实7以后）	北京香山	4月下旬	5月中旬	9月中旬	9月下旬	12月以后
黑果栒子	8～10	内蒙古呼和浩特	6月	7月	8月	9月	10月以后
蒙古栒子	8～10	内蒙古呼和浩特	6月	7月	8月	9月	10月以后
水栒子	3～5（7～15）	内蒙古呼和浩特	6月	7月	9月下旬	10月中下旬	11月下旬以后
		北京香山	5月上旬	5月中旬	7月下旬	8月中旬	8月下旬以后
华中栒子	—	南京	5月上旬	5月下旬	9月上旬	9月下旬	—
准噶尔栒子	—	内蒙古乌拉山	5月	6月	9月	10月	10月以后

栒子属果实为梨果，红色、褐红色或紫黑色。萼片宿存，内有1～5（6～7）骨质小核，包被种子。种子扁平，子叶平凸。果实和种子形态特征见表3。图1绘出了平枝栒子果核的外形及其纵剖面。

表3　栒子属树种果实和种子形态特征

树　种	未熟果颜色	成熟果实			种子		
		形　状	大　小（cm）	色　泽	形　状	大　小（mm）	色　泽
灰栒子	紫色	倒卵形稀椭圆形	径7～9，常有2小核，稀3	紫黑色	上大下小，呈倒三角形，背略隆起	长5，径6	褐色
西南栒子	绿色	梨状卵球形	长7～9，径6～7，含3～5粒小核	红色	倒三角形，背略隆起，腹面平凹	长5	棕灰色，表面具浅纹，腹面纵向纹
平枝栒子	绿色	近球形	长4～6，径4～6	红色	扁圆	长3～4 宽2.5～3	褐色
黑果栒子	紫色	近球形	径7～9，含2～3（4）小核	蓝黑色或黑色，被蜡粉	倒三角形	长7，上宽4，下宽1.5～2	棕褐色
蒙古栒子	紫色	近球形	径6.5～7，长7～7.5	黑色	椭圆形，背隆起腹面平凹	长3.2～4.2，径3	浅褐色
水栒子	暗红色	球形	长7～8，径7～8，有小核（坚果）1	红色	扁椭圆形	长3～3.5 径2.5～3	黄色
华中栒子	绿色	球形	径8～10	鲜红带粉色	缸状，有2小核，通常合成1个核	—	灰褐色，有光泽
准噶尔栒子	绿色	卵形或椭圆形	长7～10，有1～2小核	红色或暗红色	—	—	—

果实的采收调制和种子贮藏　果实成熟后变成红色、褐红色或黑紫色，陆续脱落，或延续到次年脱落，常被鸟类啄食或被虫蛀，应及时采集。摘下的果实浸泡水中，去掉果皮及果肉，除去浮粒及瘪粒，用水漂洗干净，所得纯净小核即为播种材料，通称种子。种子阴干后装入容器，贮藏在一般条件下即可。种子的净度及质量见表4。

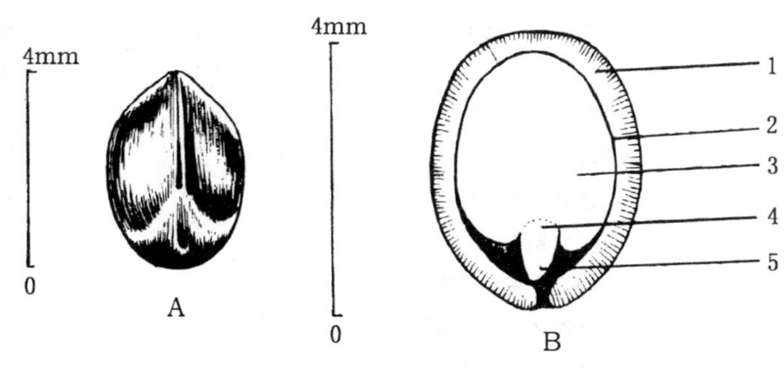

图1　平枝栒子果核外形（A）及其纵切面（B）

1. 核壳　2. 种皮　3. 子叶　4. 胚轴　5. 胚根

（孟玲绘）

表 4　栒子属树种种子净度及质量

树　种	出籽率（%）	净　度（%）	千粒重（g）	每千克纯净种子粒数（万粒）
灰栒子	25～35	80～85	23	4.4
西南栒子	—	—	23	4.4
平枝栒子	2～4	80～90	12	8.3
黑果栒子	31～35	90～98	22～24	4.2～4.5
蒙古栒子	30	90～98	17～19	5.2～5.9
水栒子	6～10	85～90	3.6～4.0	25～28
华中栒子	16～17	—	77	1.3

发芽和播种　栒子属核壳坚硬，发芽困难，为深休眠。只有低温层积才可解除休眠，但是必须在核壳的不透性得到改变之后，低温才能起作用。栒子属树种层积处理的条件及其效果见表 5。

表 5　栒子属树种的层积条件、层积期及其效果

树　种	层积条件	层积期	层积后的发芽率（%）
灰栒子	浸水后 1～3℃	3～4 个月	未萌动（仍处于休眠状态）
	1～3℃	3～4 个月	未萌动（仍处于休眠状态）
西南栒子	混湿沙后用塑料袋在下列条件下层积		
	室温	2 月初～5 月 6 日	27（约 65%为活种子，但仍处于休眠状态）
	室外	2 月初～4 月 26 日	26
	5～10℃	2 月初～4 月 24 日 （层积尚未结束）	67
平枝栒子	冬季低温混沙层积	3～4 个月	84
黑果栒子	1～3℃	3～4 个月	仍处于休眠状态
	0℃以下	3～4 个月	仍处于休眠状态
	室温（20～22℃）	3～4 个月	未萌动
蒙古栒子	1～3℃	3～4 个月	未萌动
	0℃以下	3～4 个月	未萌动
	室温	3～4 个月	未萌动
	浓硫酸浸 2 小时后 1～3℃层积	4～5 个月	未萌动
	冷冻及冷藏后用浓硫酸处理 2～3 小时再置床	101 天	未萌动
水栒子	采回调制后室外混沙层积	210 天	未萌动
		570 天	少部分发芽

栒子属树种可秋播或经过层积后春播。出土萌发，子叶一般 5～7 天即可出土。播种深度 2cm 左右，每平方米的播种量取决于种子质量，大约为 0.5～18g。幼苗形态如图 2。

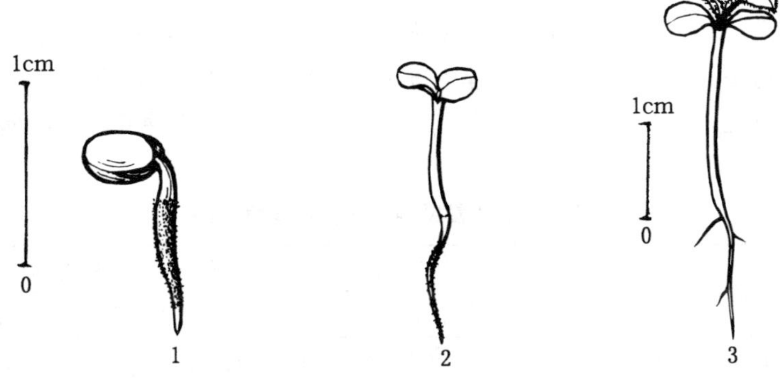

图 2　平枝栒子种子萌发和幼苗生长情况
1. 胚轴延伸　2. 子叶展开　3. 初生叶出现
（孟玲绘）

（李荣晨）

山　楂　属
Crataegus L.

（蔷薇科　Rosaceae）

生长习性、分布和用途　本属 1 000 余种，广布于北温带，主产北美，我国约 17 种，本文描述 4 种。落叶小乔木，成熟时树高 4～8m，胸径 15～25cm。喜光，耐严寒和干旱，适生各类土壤，以通气良好的沙壤土最好。根系发达，寿命较长，萌蘖力也较强。本文描述的几种均可作山里红等优良品种砧木，其名称、分布和用途见表 1。

表 1　山楂属树种名称、分布和用途

中　名	学　名	分　布	用　途
阿尔泰山楂	*C. altaica*（Loud.）Lange	新疆中部至北部。俄罗斯伏尔加河下游、西伯利亚	食用、观赏、药用、材用
山楂	*C. pinnatifida* Bge.	东北、华北、华东，豫、陕。朝鲜半岛、西伯利亚	食用、观赏、药用、材用
辽宁山楂	*C. sanguinea* Pall.	东北，内蒙古、冀、新。蒙古、西伯利亚	食用、观赏、药用、材用
准噶尔山楂	*C. songorica* K. Koch	新疆天山、准噶尔西部山地、伊犁地区。哈萨克斯坦、阿富汗、伊朗	食用、观赏、药用、材用

开花结实　5～6 年生开始开花结实，结实旺盛年龄在 10～30 年，每年结实。两性花。伞房花序（阿尔泰山楂为复伞房花序）由 10 余朵密集小花组成，径 4～6cm。单花直径约 0.8～

1.5cm。萼筒钟状，萼片5，三角状卵形，披针形，萼片宿存。花瓣5，倒卵形或圆形，长6～8mm，白色。雄蕊15～20，比花瓣略短。花柱2～5，心皮1～5，大部分与花托合生，仅先端和腹面分离。子房下位或半下位，1～5室，每室2胚珠，其中1个常不发育。5～6月开花，昆虫传粉。果柄长1.2～4.7cm。果实成熟后自然脱落。这4个树种的开花结实物候见表2。

表2　山楂属树种的开花结实物候期

树　种	观察地点	观察年份（年）	开花期			果实成熟期		果实开始散落期
			始　期	盛　期	末　期	始　期	末　期	
阿尔泰山楂	乌鲁木齐	1990	5月上旬	5月中旬	5月下旬	7月下旬	9月上旬	9月上旬
山楂	北京	—	5月中旬	6月上旬	6月中旬	9月中旬	10月中旬	10月中旬
辽宁山楂	石河子	1989～1990	4月下旬～5月上旬	5月中旬	5月下旬	7月中旬	8月中旬	8月中旬
准噶尔山楂	石河子	1989～1990	5月上旬	5月中旬	5月下旬	8月下旬	9月中旬	9月中下旬

梨果，近球形，肉质心皮熟时为内果皮，骨质，硬化成1～5个小核，每核含1粒种子。种子无胚乳。果实和种子的形态特征见表3。图1绘出的是准噶尔山楂种子的纵切面。

表3　山楂属树种果实和种子的形态特征

树　种	未熟果颜色	成熟果实			种子（果核）		
		形　状	果径（cm）	颜　色	形　状	大　小（mm）	色　泽
阿尔泰山楂	青绿	球形	0.8～1.0	金黄色，果肉粉红色	近肾形，两侧有凹痕	长6～7 宽3～4 厚3～4	淡黄色
山楂	青色	近球形	1～1.5	红、橙红，有白色、褐绿色皮孔点	扁卵圆形、肾形	长3～4 宽2～3 厚2～3	黄白色
辽宁山楂	青绿	近球形	0.9～1.1	艳红色	不规则肾形，两侧有凹痕	长4～6 宽3～4 厚3～4	黄色
准噶尔山楂	青绿	球形	1.2～1.6	深红、黑，有淡色皮孔点，果肉黄色	半球形，平滑，有3条浅裂	长7～8 宽6～7 厚3～4	黄褐色

影响结实的主要因素是光照和水肥条件。春、夏开花时的低温、大风和干旱对传粉、受精、果实发育均有不良影响。山楂核壳厚，空粒多，特别是准噶尔山楂，空粒达70%以上，可能是未受粉或自花受粉不育的结果。空瘪粒种子在外形上与健全种子无明显差异。

果实的采收调制和种子贮藏　8～10月果实呈黄色或红色而变软时，可用高枝剪剪取果枝，也可手摘或用木棒敲落后在地面收集。果熟前掉落的果实中多为空粒和瘪粒种子。采集的果实可结合加工果汁、果糕获取种子，或切开果实取出种子，还可用钻孔器取出种子而保持果实完整。如不食用，也可堆沤数日后装入容器，用木棒等物捣烂，漂淘冲去果皮、果肉等杂物，将沉入水中的小核洗净，摊晾、阴干，即为播种材料，通称种子。种子含水量降低到10%时即可装入麻袋、布袋，在低温、通风的室内可以贮存1～2年。这4个树种的鲜果出籽率、种子净度和质量见表4。

表4　山楂属树种出籽率和种子净度、质量

树　种	出籽率（%）	净　度（%）	千粒重（g）		每千克纯种粒数（万粒）	
			一　般	变动范围	一　般	变动范围
阿尔泰山楂	14～16	90～98	25	22～28	4.0	3.5～4.5
山楂	14～19	—	86	62～110	1.2	0.9～1.6
辽宁山楂	16～18	90～98	30	27～33	3.3	3～3.7
准噶尔山楂	18～20	90～98	108	98～118	0.9	0.8～1.0

发芽和播种　种子有明显的休眠习性，多数长达10个月以上。主要是生理性休眠，而且骨质的核壳坚硬，透性差，也严重影响萌发。曾经用始温60℃热水浸泡山楂种子，自然冷却后每天换水1次共处理7天，剥去种皮，以胚在发芽箱内培养20天，竟无一粒发芽。用四唑测定生活力为34%的1份阿尔泰山楂种子，按1∶3的比例与湿沙混匀，湿度保持60%，置于不同温度条件下层积处理。结果是，在0～7℃的低温下层积90天后，置于15～20℃发芽箱内，7天发芽32%，9天发芽39%；但用7～15℃的温度层积90天，发芽率仅10%，用15～25℃的温度层积120天，发芽率仅2%。钱万杰（1984）研究过山楂种子的休眠和萌发生理。

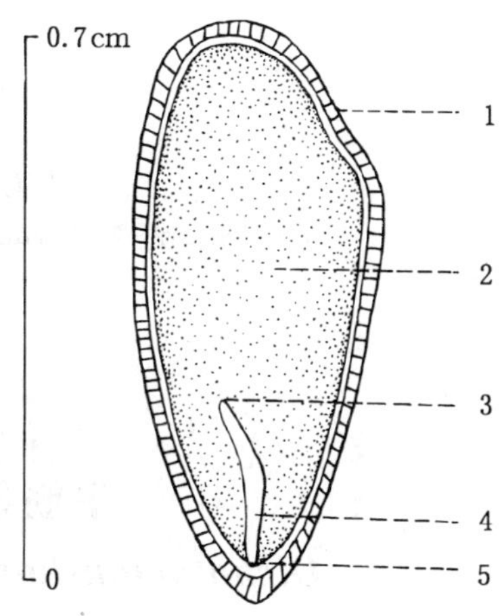

图1　准噶尔山楂种子(去核壳)的纵切面
1. 种皮　2. 子叶　3. 胚芽　4. 胚轴　5. 胚根
（吴新安绘）

完整的山楂种子在13℃的水中24小时吸水便可达到饱和状态。徐本美和闪崇辉等人(1984)据此推断，充分吸胀的山楂种子不能萌发，原因是核壳限制了氧气供应，因此建议在层积之前用下列方法之一对种子作“裂口”处理：用5%的NaOH溶液浸种4小时，裂口率可达80%，而后干湿处理反复10次；用浓硫酸浸种25分钟，而后浸种、日晒，反复3次；直接干湿处理，即浸种后每天摊晒3～4小时，以晒干表面为度，晒干后再行浸种，反复10余次。她们报道说，经过以上处理后的种子再行室外埋藏，在北京地区经历一个冬季（4个月）即可满足种子对低温的要求，春播后当年出苗率可达44.6%。新疆地区曾经采用成熟的种子秋播，次年4月14日开始出土，4月17日大量出土，且出苗整齐。

育苗时采用条播，行距30～40cm，播深2～3cm，每平方米播种量38～50g。培育2年后移植成大苗。出土萌发。准噶尔山楂幼苗形态如图2。

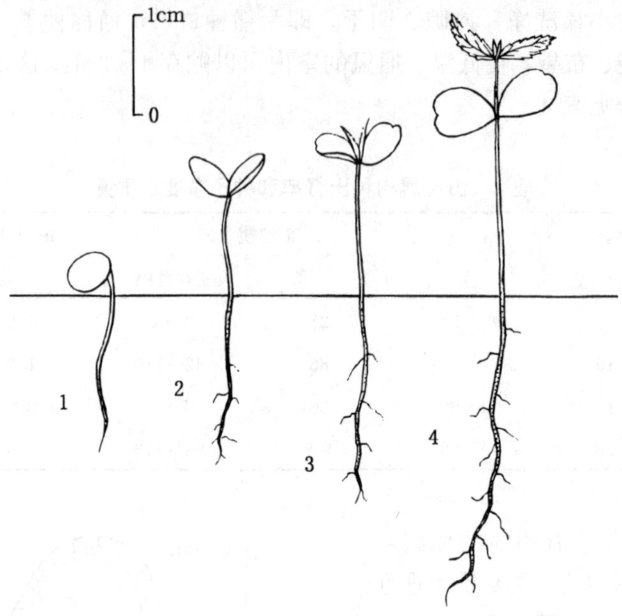

图 2 准噶尔山楂幼苗生长情况
1. 下胚轴延伸 2. 子叶出土 3. 初生叶出现 4. 初生叶近对生
（吴新安绘）

（陈开秀）

牛筋条（白牛筋）
Dichotomanthus tristaniaecarpa Kurz

（蔷薇科 Rosaceae）

生长习性、分布和用途 牛筋条属只有本文描述的 1 种。常绿小乔木或灌木，树高 2～5m。萌芽力强。可耐－7℃的低温。适生于山地红壤、黄红壤等酸性土壤。分布于滇、川。材坚韧，富弹性，宜做手杖、工具柄等。实生苗可以用作矮化苹果的砧木。

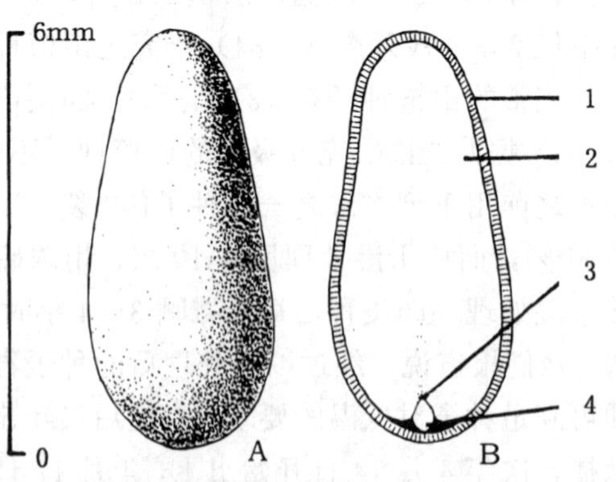

图 1 牛筋条果实的外形（A）及其纵切面（B）
1. 果皮及种皮 2. 子叶 3. 胚芽 4. 胚根
（黄应钦绘）

开花结实 4～5 年生开始开花结实。花两性，多花密集成顶生复伞房花序。萼片 5。花瓣 5，白色。雄蕊 15～20。心皮 1，花柱侧生，柱头头状。子房上位，1 室，2 胚珠并生。据云南观测，3 月初形成花蕾，4 月为盛花期，5 月上中旬为末花期；6～9 月幼果发育，10～11 月果实成熟，12

月果实脱落。果期心皮干燥，革质，成小核状，长圆柱状，顶端稍具短柔毛，长5～7mm，褐色至黑褐色，突出于肉质红色的杯状萼筒之中。每果常发育成熟1粒种子，少数2粒。子叶平凸，无胚乳或有少量胚乳（图1）。

果实的采收调制和种子贮藏　果实成熟时手工采摘。采回后用清水浸泡1～2天，搓擦脱去萼筒，所得果实即为播种材料，通称种子。按鲜果质量计算的出籽率为60%～70%。千粒重22～25g，每千克有纯净种子4万～4.5万粒。种子洗净后在阳光下曝晒，至含水量低于12%时即可装入布袋或其它容器，置室内干燥处贮藏。贮藏时间4～5个月。

发芽和播种　种子无休眠习性或有短暂休眠现象。3月中旬播种。播前不作处理，或用清水浸泡24小时。播后30天出苗。发芽盛期的15天中发芽百分数为32%；从播种之日起算的60天内发芽率为55%。出土萌发。子叶出土后8天展现初生叶（图2）。

条播。播种沟深宽各1～1.5cm，播种沟距10～12cm。每平方米苗床播种4～6g。1年生苗出圃栽植或作嫁接用砧木。

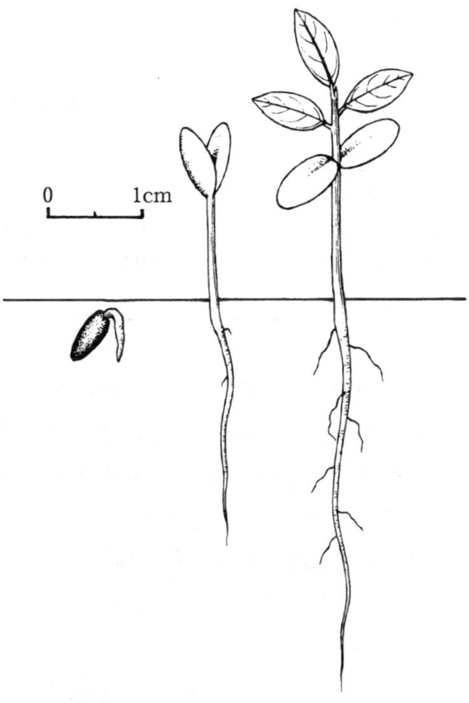

图2　牛筋条种子萌发后第3、8、20天幼苗的生长情况
（黄应钦绘）

（张茂钦）

枇　杷

Eriobotrya japonica（Thunb.）Lindl.

（蔷薇科　Rosaceae）

生长习性、分布和用途　枇杷属约30种，我国13种，本文描述1种。常绿乔木，高10～15m。喜温暖湿润气候，不耐严寒。适生于肥沃排水良好的酸性或微碱性土壤。产陕、甘、豫、皖、苏、浙、闽、台、赣、湘、鄂、川、黔、滇、桂、粤、琼。日本、印度、越南、缅甸、泰国、印度尼西亚也有栽培。类型和品种很多。果可食，叶、花、果、种仁及根供药用，主治咳嗽。木材可制木梳、手杖、农具柄及工艺品等。

开花结实　3～5年生开始结实。大小年现象不甚明显。花两性，圆锥花序顶生。每花序有花70～100朵。花萼、花瓣各5枚。雄蕊20枚。子房下位，5室，每室2胚珠。花柱5。花期和果期因地区、品种不同而异，在花期内温度11～14℃时开花最多，10℃以下则花期延长。果实为由子房、萼片及花托发育而成的梨果。一般是纬度低，果实成熟期早。各地枇杷的开花结实物候期见表 1。

表 1 枇杷开花结实物候期

地 区	初花期	盛花期	末花期	果实成熟期
福建莆田	10 月下旬～11 月上旬	11 月下旬～12 月中旬	12 月上旬～1 月上旬	4 月上旬～5 月中旬
浙江塘栖	10 月下旬～11 月中旬	11 月中旬～11 月下旬	12 月中旬～1 月中旬	5 月上旬～6 月中旬
江苏洞庭山	10 月下旬～11 月中旬	11 月上旬～12 月下旬	12 月上旬～1 月中旬	6 月上中旬
湖南沅江	—	10 月下旬～12 月上旬	—	5 月上旬～6 月中旬

果球形或长圆形，有种子 1～5 粒。未成熟果青绿色，熟时黄色或橘黄色，被锈色绒毛，后脱落，径 2～5cm。种子球形或扁球形，子叶 2 枚，无胚乳（图 1）。

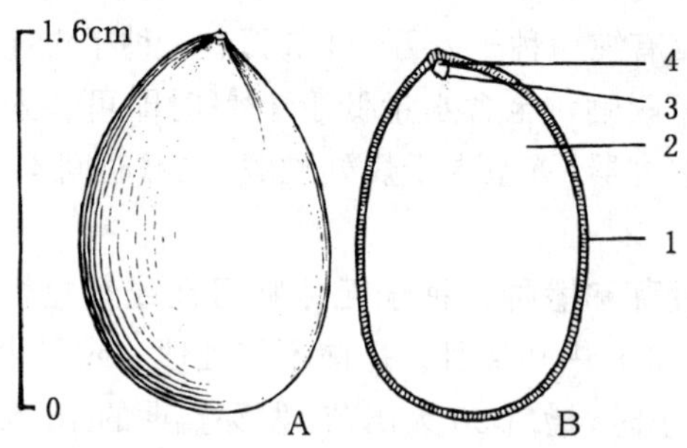

图 1 枇杷种子的外形（A）及其纵切面（B）
1. 种皮 2. 子叶 3. 胚芽 4. 胚根
（黄应钦绘）

果实的采收调制和种子贮藏 果皮颜色变黄，果肉组织软化时即可采收。同一树上果实的成熟期也不一致，需分期分批采收。剥开果实取出种子，用水淘净杂质，捞起晾干即可。每百千克鲜果可处理出种子 20～30kg，种子含水量约 45%，净度可达 100%。千粒重 1 900g，变幅为 1 570～2 390g，每千克种子 530 粒，变幅为 420～640 粒。贮藏时需注意保湿，可将 1 份种子与 2～3 份湿沙混合，放置室内荫凉处。以这种方式贮藏 6 个月，发芽率可以保持 70%～80%。

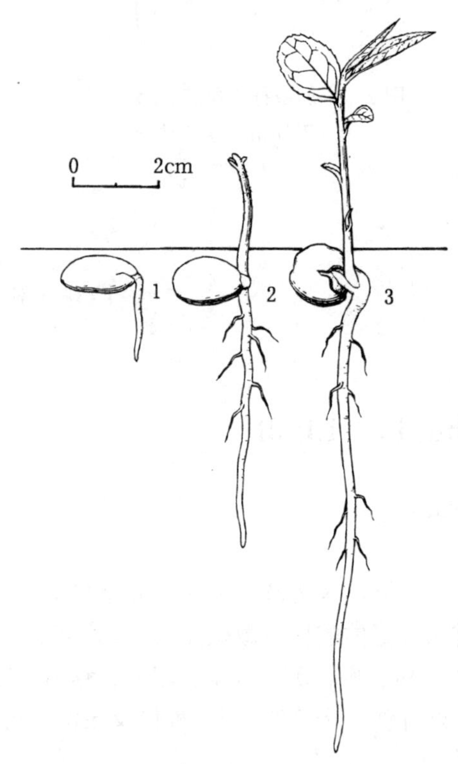

图 2 枇杷种子的萌发和幼苗生长情况
1. 幼根延伸 2. 胚芽出土 3. 幼苗形成
（黄应钦绘）

发芽和播种 种子休眠现象不明显。发芽测定时用沙床，在室温 20～30℃时播后 12 天开始萌发，20 天进入发芽盛期，33 天发芽结束，发芽率可达 95%。生产上可在 4～5 月随采随播，每平方米播种 100～130g，覆土不宜过厚。留土萌发。幼苗具初生不育叶 2～3 枚。出土 10 天左右展出初生叶 1 片，以后逐渐增生新叶。幼叶和茎密被褐色绒毛，主根发达，侧根细（图 2）。1 年生苗出圃，嫁接苗 2 年生出圃。

（李玉科）

白鹃梅属
Exochorda Lindl.

（蔷薇科　Rosaceae）

生长习性、分布和用途　本属4种，产亚洲；我国3种，本文描述2种。落叶灌木，树高2～5m，单叶互生，全缘或中部以上有锯齿。喜光，耐旱，耐寒，耐瘠薄，也稍能耐荫。喜深厚肥沃湿润的土壤，不耐涝。春季白花满树，是优美的观赏树种，北京、青岛等地多有引栽。根皮、枝皮入药。这2个树种的名称、树高、分布及用途见表1。

表1　白鹃梅属树种的名称、树高、分布及用途

中　名	学　名	树　高（m）	分　布	用　途	供　稿
白鹃梅	*E. racemosa* (Lindl.) Rehd.	3～5	豫、苏、浙、赣。北京、青岛栽培	观赏，药用	403
齿叶白鹃梅	*E. serratifolia* S. Moore	2	辽、冀。朝鲜	观赏	305、308

开花结实　一般5年生开始开花结实，正常结实年龄在8年生以后，大小年现象不明显。花两性。4～12朵花组成顶生总状花序。花直径2～4cm。萼筒钟状，萼片5。花瓣5。雄蕊15～25。心皮5，合生，花柱分离，子房上位。蒴果黄褐色，倒圆锥形，具5脊，径1～1.5cm，5室，沿背腹两线开裂，每室有1～2粒种子。种子薄片状，半圆形，红褐色，边缘围绕薄翅；长7～9mm，宽4.5～5.5mm，厚0.7～0.8mm；表面密生黄褐色细茸毛。种脐稍向直边一侧歪斜。胚直立，子叶片状，大而扁，胚根短，被子叶包围，无可见胚乳。本文两个树种果实种子区别不明显。花期和果期见表2。

表2　白鹃梅属树种开花结实物候

树　种	观察地点和年份	开花期			果实成熟期		种子脱落期
		初　期	盛　期	末　期	初　期	盛　期	
白鹃梅	北京香山 1986～1988	4月上旬	4月下旬	5月上旬	9月上旬	9月中旬	9月中旬
齿叶白鹃梅	北京香山 1984、1987、1988	4月中旬	4月下旬	5月上旬	10月上旬	10月中旬	10月下旬
	哈尔滨 1963～1980	5月下旬	—	6月中旬	—	—	—

种子的采收调制和种子贮藏　果实呈现黄褐色并有少量开裂时即可采收。采回放在通风干燥处晾干，待蒴果全部开裂时抖动果壳，脱出种子，少数不能脱出的，用手掰开。因果壳

和种子比重相近，难以风选分开，需要人工拣出果壳杂质，取得净种。种子装入纸袋或容器，密封贮藏在低温干燥处。曾经观察到在10～24℃的地下室内，贮藏4年的白鹃梅种子仍有60%的发芽率。这2个树种的出种率、净度、质量见表3。

表3 白鹃梅属的出种率、种子净度、质量

树种	出种率（%）	净度（%）	千粒重（g）	每千克纯净种子粒数（万粒）
白鹃梅	3～5	—	12	8.3
齿叶白鹃梅	3～5	70～75	11	9.1

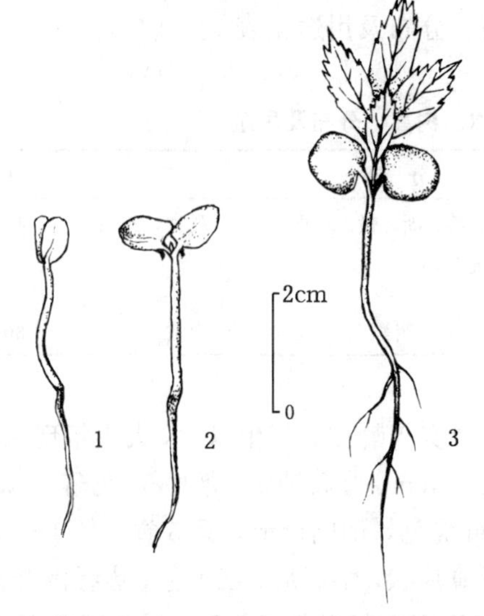

图1 白鹃梅幼苗生长情况

1. 子叶出土 2. 子叶展开 3. 幼苗形成

（田恒德仿黄鹏成供稿及《中国东北主要树木幼苗图说》）

发芽处理及播种 一般在1月份，先用始温40℃水浸种一昼夜后，用凉水浸泡3天，再混2～3倍的湿沙，置于15～25℃温度下层积1个月，再转到0～5℃的低温下1个月左右。春季温度升高时，种子要经常倒翻，加水保持湿度，当有1/3的种子裂嘴时即可播种。在1988年的一次测定中，齿叶白鹃梅发芽率为71%。每平方米播种10～20g，覆土1cm左右，保持床面湿润，4～6天即可出土，1周长出初生叶。当年苗高5～10cm，2～3年生苗可出圃栽植。幼苗形态见图1。白鹃梅也可以扦插繁殖。

（王木林、高秀岩）

苹 果 属

Malus Mill.

（蔷薇科 Rosaceae）

生长习性、分布和用途 本属约35种，我国产20余种，本文描述3种。落叶，稀半常绿乔木或灌木。喜光，较耐寒，抗旱，不耐水湿，适生于排水良好的沙壤土及深厚肥沃的土壤。寿命较长，抗逆性较强，新疆野苹果寿命可达150年。分布于北温带地区，在我国以东北、华北和西北地区较为普遍。是北温带的重要水果，品种甚多，有些种类为优良的果树砧木，生长较快，产量高，病虫害少，与苹果嫁接亲合力较强。花为蜜源，美丽芳香，花繁满树，亦为观赏树种。果实食用或加工成果品，也可酿酒。木材供农具、家具、细木工用。这3个树种的名称、生长和分布情况见表1。

表 1　苹果属树种的名称、生长和分布

中　名	学　名	树高(m)	胸径(cm)	分　布	供　稿
山荆子（山定子，山丁子）	*M. baccata* Borkh.	5～14	5～25	东北、华北、西北东部。蒙古、俄罗斯、朝鲜半岛、日本	309
新疆野苹果	*M. sieversii* (Ledeb.) Roem.	5～18	10～80	新疆天山西部和准噶尔西部山地。中亚	204
海棠花	*M. spectabilis* (Ait.) Borkh.	4～8	5～20	华北、华中至西南地区，各地常有栽培	309

开花结实　4～8 年生开始开花结实，正常结实在 10 年生之后，盛果期中的新疆野苹果单株结实量可达 90kg。结实大小年的间隔期为 1～3 年。花两性。伞形或近伞形花序由 3～6 朵花组成，着生于果枝顶端，花梗较粗，密被白色茸毛。萼筒钟状。花瓣倒卵形或卵圆形，白色或粉红色，直径 2～5cm。雄蕊 15～25，花丝长短不等。花柱 5，稀 4，基部连合。子房下位，5（4）室，每室 2 胚珠。花期 4～5 月，果实成熟期 8～9 月。因树种、品种及分布区不同，开花期和果熟期也不一致（表 2）。危害新疆野苹果的害虫有苹果蠹蛾（苹果皮小卷蛾）(*Laspeyresia pomonella* L.）和苹长尾小蜂（*Syntomapsis druparum* Boh.），被害的种子有时达 90%或更多。

表 2　苹果属树种的开花结实物候期

树　种	观察地点	观察年份	开花期			果实成熟期	
			始　期	盛　期	末　期	始　期	盛　期
山荆子	北京	1989	4 月下旬	5 月下旬	6 月上旬	8 月下旬	10 月上旬
新疆野苹果	新疆天山	1986	5 月上旬	5 月中旬	5 月下旬	8 月	9 月
海棠花	北京	1988	4 月中旬	5 月上旬	5 月下旬	9 月中旬	10 月上中旬

梨果肉质，无石细胞或微有石细胞，内果皮软骨质。新疆野苹果每果有 1～8 粒种子。据调查，平均而言，新疆野苹果林缘木树冠上部每果有种子 4.4 粒，树冠下部 5.7 粒，林内树冠中部每果仅有种子 3.2 粒。种子褐色，无胚乳。子叶平凸，卵圆形（表 3、图 1)。

表 3　苹果属树种的果实和种子特征

树　种	成熟果实			种子		
	形　状	大小（cm）	色　泽	形　状	大小（mm）	色　泽
山荆子	近球形	0.8～1.0	红色或黄色带红晕	扁卵圆形	4×3×2	灰褐色
新疆野苹果	球形、扁球形	2.5～6.0，单果重 6～25g	黄白色或黄白色带红晕	楔形，长卵形	7×4×2	暗灰色、灰褐色
海棠花	近球形	2～2.5	黄色带红晕	三角状卵圆形	4×3×2	深褐色

果实的采收调制和种子贮藏　果实成熟时采摘或击落后收集。采收后不能堆积，以免果肉发酵，产生高温，降低种子发芽能力。宜放置阴凉通风处，使其充分成熟，用木棒捣烂，压碎果肉，在水中漂洗，去杂，将所得种子摊晾，经常翻动，使其充分干燥，要求种子含水量在10%以内。再进行筛选，使种子净度在95%以上。装袋室内贮存。新疆野苹果曾经用玻璃瓶密封，在气温为10℃以上的环境中保存3年，仍有92%的发芽率。在通风干燥的室内存放的新疆野苹果的种子，两三年后发芽率仍在50%以上。据试验，在相对湿度50%～70%，气温为16～20℃的室内贮存的新疆野苹果，1年后发芽率仅下降1%～6%；在10℃左右的温度下用装有氯化钙的干燥器密封贮存时，3年后仍有92%的发芽率。种子的净度和质量等数据见表4。

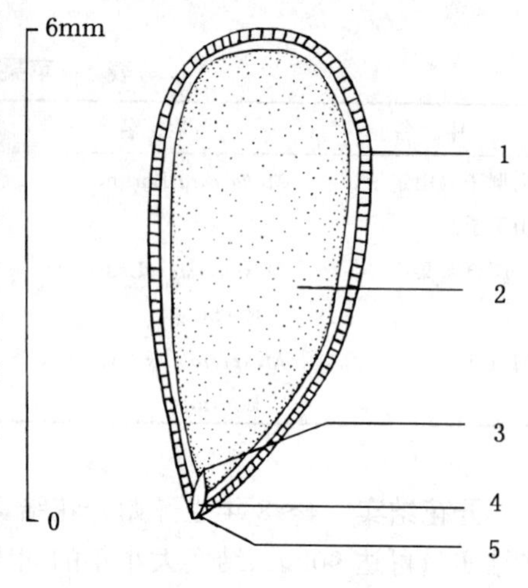

图1　新疆野苹果种子纵切面
1. 种皮　2. 子叶　3. 胚芽　4. 胚轴　5. 胚根
（吴新安绘）

表4　苹果属鲜果出种率和种子净度、质量

树　种	出种率（%）	净度（%）	千粒重（g）	每千克纯净种子粒数（万粒）
山荆子	2.5～3	90～99	5.6～7.6	13.2～17.8
新疆野苹果	1～2	90～99	19～29	3.4～5.3
海棠花	1～2	90～99	20～23	4.4～5.0

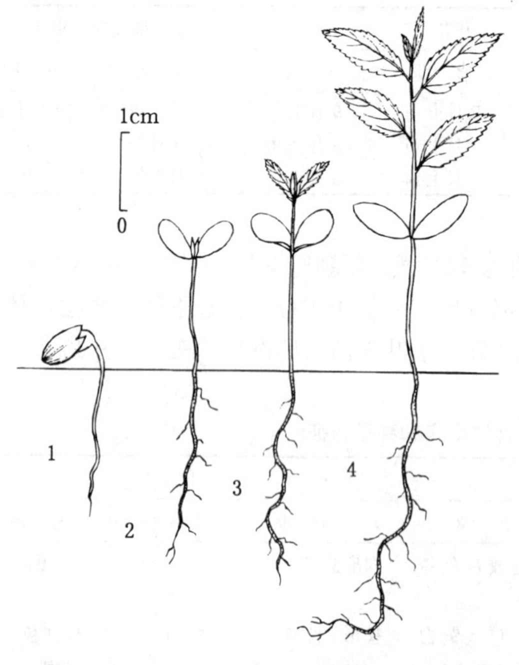

图2　新疆野苹果种子萌发及幼苗生长情况
1. 下胚轴延伸　2. 子叶出土　3. 初生叶出现　4. 幼苗形成
（吴新安绘）

发芽和播种　种子的休眠期达3～4个月，不经过处理的种子春播后当年不发芽。秋季采收的种子用5～0℃的温度层积处理，翌年春季气温升至15～20℃时取出播种。这样的种子播后10～20天幼芽带壳出土，10天左右种壳脱落，出现2片子叶。出土后要注意防治鸟害。条播，播种量每平方米5～6g，覆土厚度1～2cm。秋播效果好，发芽整齐。秋播后要冬灌1次。曾经用以下几种措施处理过新疆野苹果的完整种子、去皮种子和去子叶的胚，但都未能发芽。这些措施是：用18～25℃的温度催芽22天；在18～20℃下层积35天；用Cris培养基培养22天；用GA和6-BA处理24小时。因此推测新疆野苹果的种子具有生理休眠习性，建议2月上中旬在2～8℃的温度下层积，春播前 1 周移到 15～18℃的室内催芽，待有 1/3 的种粒裂口吐白时取出播种。伊犁地区果农的做法是延迟采种，到冬季再剥取种子而后在 12 月或 1 月犁开雪层，在冻土上播种新疆野苹果。苹果属种子出土萌发，当年苗高 20～50cm。种子发芽及幼苗形态见图 2。

（宋朝枢）

石　楠　属
Photinia Lindl.

（蔷薇科　Rosaceae）

生长习性、分布和用途　本属 60 多种，分布于东亚、南亚和北美。我国约 40 种，产西南到中部地区。本文描述 4 种，见表 1。多生于 1 000m 以下的山谷灌丛疏林中。喜温暖湿润，耐瘠薄，稍耐荫，多用于庭园栽培。

表 1　石楠属树种的名称、生长习性、分布和用途

中　名	学　名	生长习性	树高(m)	分　布	用　途	供　稿
中华石楠	*Ph. beauverdiana* Schneid.	落叶小乔木	10	陕、豫和长江中下游以南及西南各地	绿化、材用	902
椤木石楠	*Ph. davidsoniae* Rehd. et Wils.	常绿乔木或灌木	6～15	陕南、西南、中南及华东。越南、缅甸、泰国	绿篱	1003
桃叶石楠	*Ph. prunifolia* (Hook. et Arn.) Lindl.	常绿乔木	10～20	西南、中南。日本、越南	绿化、材用	902
石楠	*Ph. serrulata* Lindl.	常绿小乔木	6～12	陕南、甘南、豫、粤、桂、长江中下游及西南各地	药用、绿化、材用、油用	1003

开花结实　花两性，多数，成顶生的伞形、伞房或复伞房花序。萼筒杯状、钟状或筒状，短萼片 5。花瓣 5，白色，开展。雄蕊约 20。心皮（2）3～5，花柱离生或基部合生。子房半下位，2～5 室，每室 2 胚珠。果实为 2～5 室的小梨果，球形或长球形。果肉由花托形成，果皮木质，成熟后不裂，顶端或上部与萼筒分离，有宿存萼片。每果含种子 1～4 粒不等，种子呈不规则的卵圆形或椭圆形。外种皮较薄，光滑，褐色或棕褐色。内种皮膜质。胚乳近于无，子叶肥大，胚根短（图 1）。4～5 月开花，9～10 月果熟。开始结实树龄为 5～8 年或 10～15 年（表 2）。一般每年均能结实，但有的年份结实量较少。

表 2　石楠属树种开花结实物候期和成熟果实的颜色

树　种	开始结实年龄（年）	开花期	果熟期	果实采收期	成熟果实颜色
中华石楠	5～8	5 月	8 月	8 月下旬	紫红色
椤木石楠	10～15	5～6 月	9～10 月	11 月中下旬	黑褐色
桃叶石楠	8～30	4 月上旬～5 月上旬	11 月中旬～12 月下旬	11～12 月	红褐色
石楠	10～20	4～5 月	10～11 月	11～12 月	红色-红褐色

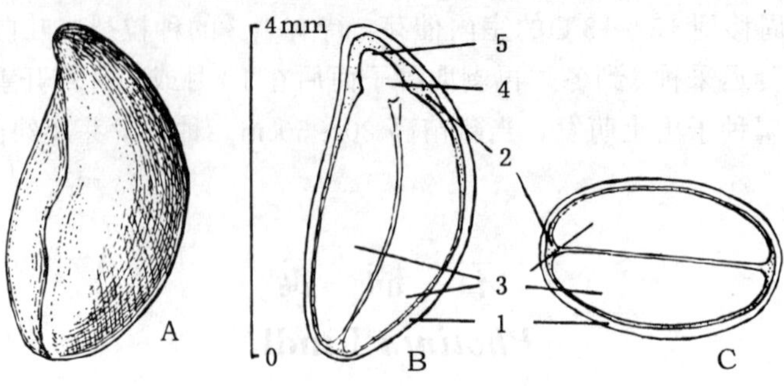

图 1　石楠种子的外形（A）及其纵
切面（B）和横切面（C）
1. 种皮　2. 胚乳　3. 子叶　4. 胚轴　5. 胚根
（史渭清绘）

果实的采收调制和种子贮藏　当果实呈现成熟颜色，少量果实开始脱落时即可采摘。采后摊晾，避免堆放过久而引起果肉发酵，影响种子的活力。取种前浸泡 2 天，搓揉淘洗，去除果肉等杂物及空粒，取沉水的种子晾干备用。种皮较薄，切忌搓揉过重而损伤种子。出种率 70%以上。成品种子净度可达 92%以上。椤木石楠果皮木质坚硬，难以去除，也可以整粒果实作为播种材料，习惯上仍称种子。供春播用的种子常温干藏即可。含水量 10% 以下的种子在 2～5℃的温度下密封存放，可以较长期地保持发芽能力。石楠种子极易腐烂。果实和种子形态特征及质量数据见表 3。

表 3　石楠属树种果实和种子特征及种子的质量数据

树　种	果　实			种　子					
	形　状	长 (mm)	径 (mm)	形　状	长 (mm)	径 (mm)	颜　色	千粒重 (g)	每千克粒数 (万粒)
中华石楠	卵形	7～8	5～6	长卵形	4～5	3～4	红色	2.8～3.8	26.3～35.7
椤木石楠	近球形	7～10	7～10	尖卵形	4～5	2～3	深褐色	5.4	18.5
桃叶石楠	椭圆形	7～9	3～4	不规则椭圆形	3～4	2～3	红色	3.5～4.5	22.2～28.6
石楠	近球形	5～6	5～6	不规则卵形	4～5	2～3	棕红色	6.5～7.0	14.3～15.4

发芽和播种　本文所述 4 个种的种子没有明显的休眠习性。福建林学院曾在 25℃的恒温条件下测定过中华石楠和桃叶石楠的发芽能力，基质为滤纸，未加光照，两个树种的种子样品都能在 25 天结束发芽，发芽率分别为 78%和 82%，相应的场圃发芽率分别为

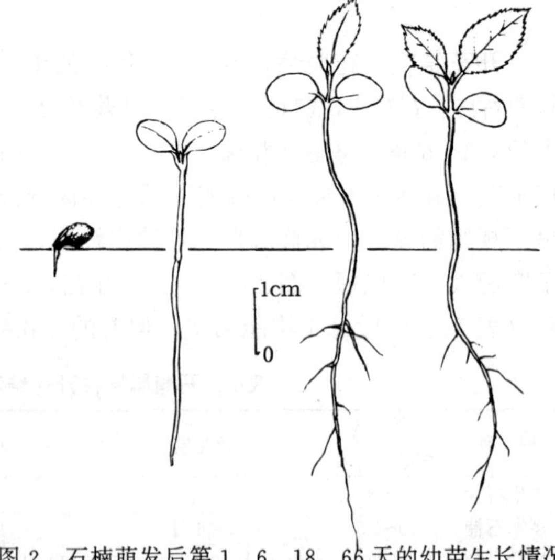

图 2　石楠萌发后第 1、6、18、66 天的幼苗生长情况
（史渭清绘）

65%和 70%。中华石楠和石楠净度高而质量好的种子，发芽率也可达 90%以上。早春 2 月中下旬播种，温暖地区可以在冬季播种。条播，每平方米床面播 2.5～5g。覆土约 1cm。盖草。出土萌发。初生叶互生（图 2）。石楠下胚轴绿色，基部淡黄色。石楠属幼苗较耐寒，也无需遮荫。1 年生苗可以出圃，用于庭园点缀的则常用大苗。可以扦插繁殖。

（何泽瑛）

风　箱　果

Physocarpus amurensis（Maxim.）Maxim.

（蔷薇科　Rosaceae）

生长习性、分布和用途　风箱果属约 20 种，主产北美，我国仅有本文描述的 1 种。落叶灌木，高达 3m，冠大，花密，是优良的观赏灌木。喜光，耐寒，耐旱，耐瘠薄，常丛生于黑龙江和河北部分地区海拔 800m 的山地。俄罗斯远东及朝鲜半岛北部亦产。近年我国各地均有栽培。观花灌木，可作防护林和水土保持林下木，花园、假山处可单株点缀，也是蜜源植物。

开花结实　3～5 年生开始开花结实。花两性。伞形总状花序，顶生，径 3～4cm，花梗长 1～1.8cm。花径 0.8～1.3cm，白色。雄蕊 20～30，花药紫色。雌蕊 2～4，外披星状毛，基部合生。子房上位，1 室。1984 年、1987 年、1988 年 3 年在北京香山地区观察，初花期 5 月中旬，盛花期 5 月下旬，末花期 6 月上旬。据哈尔滨多年观察，开花始期最早为 5 月 28 日，末期最晚在 6 月 22 日，有二次开花现象。蓇葖果膨大，卵形，成熟时沿背腹两线开裂，内含光亮黄色种子 2～5 枚。种子具胚乳。北京香山地区 7 月下旬～8 月中旬果实开始成熟，成熟盛期在 8 月上中旬。哈尔滨初熟期在 6 月末，成熟末期在 9 月中旬。成熟后 20～30 天脱落。

果实的采收调制和种子贮藏　蓇葖果有少量开裂时即可采收。一般用手摘下，放在阴凉通风处晾干，用手搓揉或小棍敲打使种子脱落，筛选扬净杂物，得纯净种子。装入纸袋内普通干藏即可。种子形态见图 1。

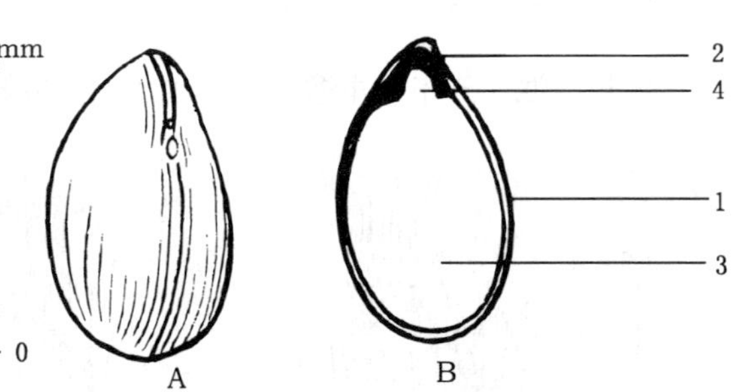

图 1　风箱果种子外形（A）及其纵切面（B）
1. 种皮　2. 胚乳　3. 子叶　4. 胚根
（孟玲绘）

发芽和播种　先用始温 40℃水浸种 1 昼夜，捞出后装入布袋或包以纱布，放在温暖处催芽，每天用温水冲洗。10 天后将种子排放于发芽皿内湿滤纸上，在 18～20℃室温下 5～6 天开始发芽，10～15 天发芽结束，发芽率可达 80%。浸种后也可混沙催芽，见有部分种子发芽时播种。千粒重 1g，每克约 1 000 粒。每平方米播种量 2g，产苗 500 株。种粒小，覆土要薄而均匀，约 0.5cm 即可。保持床面湿润，4～5 天即可出土，1 周左右长出初生叶。出土萌发（图 2）。

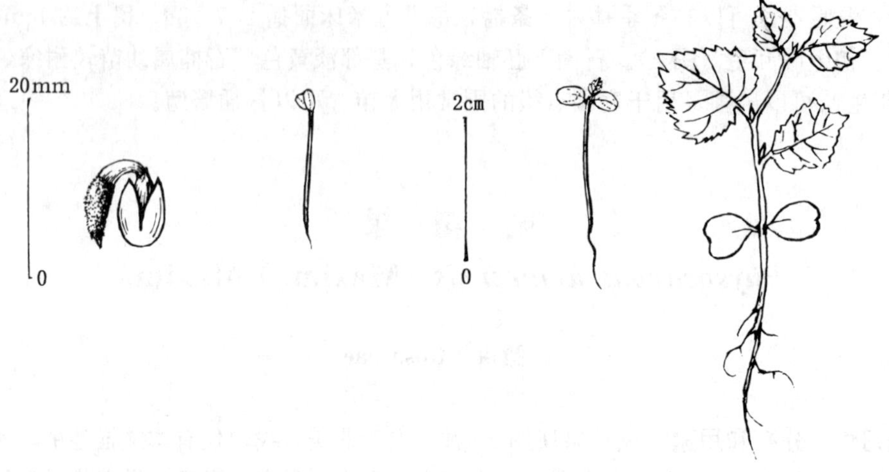

图 2　风箱果种子萌发后第 2、7、45、60 天的幼苗生长情况

（孟玲绘）

（王木林、高秀岩）

金露梅（金老梅）

Potentilla fruticosa L.

（蔷薇科　Rosaceae）

生长习性、分布和用途　委陵菜属约 300 种，我国约 85 种，其中木本 3 种，本文描述 1 种。落叶灌木，高可达 1.5m，多分枝。喜光，耐寒，对土壤要求不严，但喜湿润。广泛生于北温带山区，亚洲、欧洲、北美均产。我国分布于东北、华北、西北至西南各地。叶可代茶，花可入药，又为饲料，骆驼爱吃，藏民广泛用为建筑材料。枝叶密茂，黄花鲜艳，为较好的庭园观赏灌木。

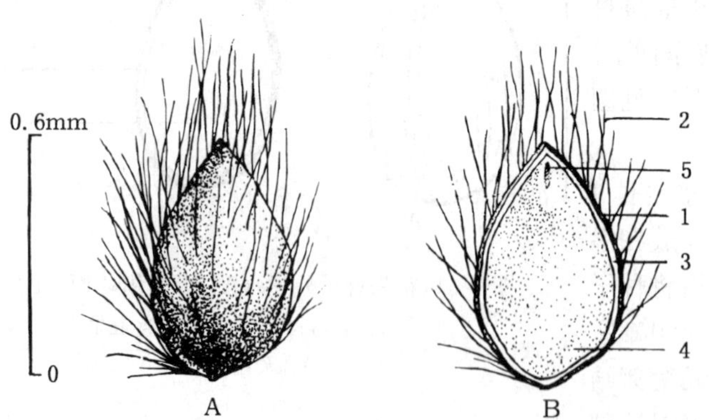

图 1　金露梅瘦果外形（A）及其纵切面（B）

1. 果皮和种皮　2. 长柔毛　3. 胚乳　4. 子叶　5. 胚根

（梁鸣绘）

开花结实　土肥条件好的地方个别当年生苗木便能开花结实，2 年生普遍开花结

实，5年生以后为正常结实年龄。花两性，单生或数朵成聚伞花序顶生。萼筒下凹，多呈半球形，萼片5，镊合状排列。副萼片5，与萼片互生。花瓣5，黄色，径2～3cm，后叶开放，数量多。雄蕊10～30枚。子房上位，心皮多数，离生。每心皮具1颗下垂胚珠。瘦果10枚左右，着生在干燥凸起的花托上，花萼宿存。瘦果甚小，长卵圆形，棕褐色，密生长柔毛，长1mm左右，径0.5mm左右。种皮膜质，种子有少许胚乳。据黑龙江省森林植物园1963～1980年观察，开花始期为6月上旬，末期在10月上旬，开花日数约120天。果实初熟期在7月中旬，末熟期在10月中旬。果实翌春落尽。

果实的采收调制和种子贮藏　树上采摘，晾干后搓揉，风选所得的纯净瘦果即为播种材料，通称种子。瘦果的外形及其纵剖面见图1。出种率8%～12%，净度25%～50%，千粒重约0.2g，每克有种子4 760～5 260粒。一般低温下袋装可贮存半年。

发芽和播种　条播或撒播都应选在无风天，最好是播后2～3天无暴雨天气。宜密播，微量覆土，每平方米播种量5g左右。播后15～20天齐苗，5～7天伸出第1片初生叶。出土萌发（图2）。1年生苗高30～60cm即可出圃。每平方米产苗量可达200～500株。

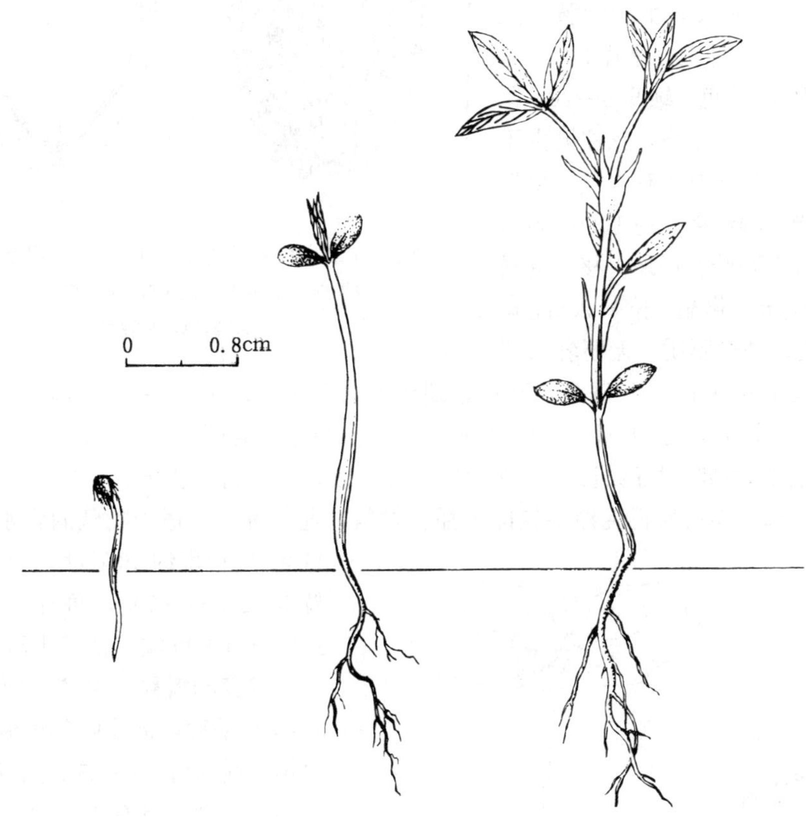

图2　金露梅种子萌发出土后第2、7、19天幼苗生长情况

（梁鸣绘）

（周德本、陈运大）

东北扁核木

Prinsepia sinensis（Oliv.）Kom.

（蔷薇科　Rosaceae）

生长习性、分布和用途　扁核木属有5种，我国4种，本文描述1种。落叶灌木，高达2～3m，分枝多，枝有刺。喜光，不耐庇荫，耐干燥瘠薄。产于东北。朝鲜半岛、俄罗斯远东地区也有分布。茎干可作手杖及各种柄材。花果均有香气。果色红艳，可食。种仁入药，能清热明目。观赏树种。

开花结实　3～5年生开始开花结实，人工栽植的扁核木开花结实后易为红蜘蛛及几种食叶害虫危害，应注意防治。花两性，1～4朵簇生于叶腋，径1.2～1.8cm。萼筒钟状，萼片5，三角状卵形。花瓣5，黄色。雄蕊8～20，花丝与花药等长。子房1室，花柱侧生于子房基部。花后叶开放，有香气。核果球形至卵圆形，鲜红色，径1.5～2cm，果肉多浆汁，有香味。每果1核，黄褐色，坚硬，卵球形，微扁，长1.2cm，宽1cm，有雕纹。种皮膜质。无胚乳（图1）。据哈尔滨1963～1980年观察，开花始期至末期为5月9日～5月29日，开花日数约15天；果初熟至末熟为7月18日～8月16日；初熟至脱落间隔期16～20天。

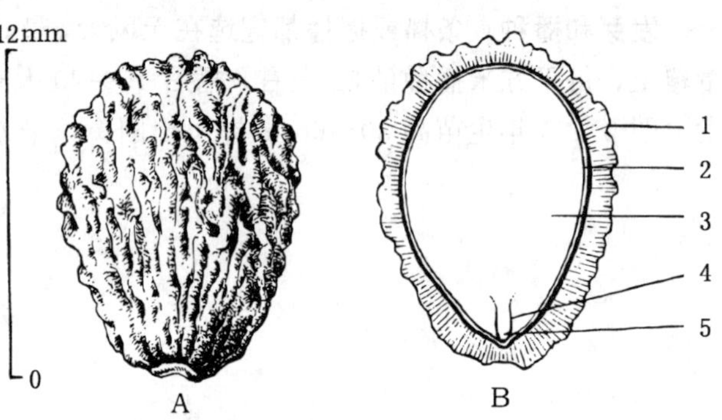

图1　东北扁核木果核外形（A）及其纵切面（B）

1. 内果皮　2. 种皮　3. 子叶　4. 胚轴　5. 胚根

（梁鸣、黄应钦绘）

果实的采收调制和种子贮藏　果熟后应及时采收，随即将果实堆放，促使果肉软熟后放水中搓洗，水选去杂后所得纯净果核即为播种材料，通称种子。适于贮藏的含水量为9%～11%。出籽率约40%，净度95%以上，千粒重208～350g，每千克纯净种子2 800～4 800粒。低温干藏。

发芽和播种　果核坚硬，不易发芽。8～9月采种后需作隔年埋藏处理。或在翌年春播前置15～25℃温度下催芽，保持种子湿度，待有1/3的种粒裂嘴后播种。发芽率约65%。床面点播，每平方米床面播种150g，覆土2～3cm。出土萌发（图2）。2年生苗出圃。

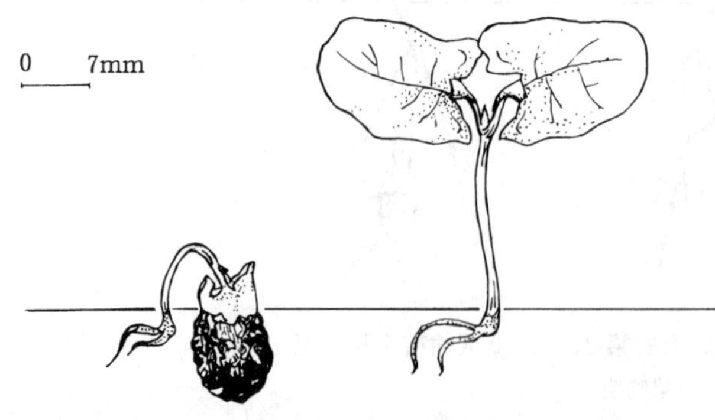

图2　东北扁核木种子萌发后第11、23天幼苗的生长情况

（梁鸣绘）

（周德本、邹铨）

李　属
Prunus L.

（蔷薇科　Rosaceae）

生长习性、分布和用途　本属约 280 种，我国约 90 余种。本文描述 12 种和 1 变种，是果树中较抗旱、耐寒、适应性强、寿命长、结果早、栽培管理较易的树种。特别适生于山区平川地、丘陵、平原等各种沙质或粘质壤土，也耐一定程度的盐碱，在深厚肥沃的土壤上生长良好。栽培历史悠久，如杏树有 2 500 多年，桃树达 4 000 年以上。李属树种在我国广泛栽培，以华北、西北、华东地区为主（见表 1），出现了许多栽培品种。李属多产优良水果，又多为花木供观赏。种子多为优良油料原料，还可入药，如杏仁。不少树种是嫁接果树的砧木。

表 1　李属种树种的名称、生长和分布

中　名	学　名	树高(m)	胸径(cm)	分　布	供　稿
巴旦杏（扁桃）	*P. amypdalus* Batsch	4～10	10～20	原产中亚。新疆喀什栽培悠久，西北，鲁栽培	203
杏	*P. armeniaca* L.	5～8（12）	30～50	西北、华北，东北、西南、华东	1003
山杏	*P. armeniaca* L. var. *anus* Maxim.	5～10	20～40	东北南部、华北、西北，鲁、苏。日本、朝鲜半岛	309
山桃	*P. davidiana* (Carr.) Franch.	6～10	20～50	黄河流域，川、滇	309
欧李	*P. humilis* Bunge	1	—	东北、华北、鲁、豫	309
斑叶稠李（山桃稠李）	*P. maackii* Rupr.	6～16	30～40	东北。朝鲜半岛、俄罗斯	107
梅	*P. mume* Sieb. et Zucc.	15	—	原产西南。长江流域以南及苏、豫栽培最广。日本、朝鲜半岛	1003
稠李	*P. padus* L.	10～13	20～50	东北、华北、西北东部。朝鲜半岛、日本、欧洲	103
桃	*P. persica* (L.) Batsch	4～8	20～50	原产我国。各地广泛栽培，世界各地引种	1003
樱桃	*P. pseudocerasus* Lindl.	6～10	15～40	西北东部、华北南部、东北南部、苏、浙、赣、川	1004
李	*P. salicina* Lindl.	6～12	15～40	西北东部、华北、东北南部、华东、中南，台。世界各地栽植	309
毛樱桃	*P. tomentosa* Thunb.	2～3	—	东北、华北、西北、西南，苏、皖、鄂	107
大叶桂樱	*P. zippenliana* Miq.	20～25	25～40	西北东部、西南、中南、东南，台。越南、日本	604

开花结实　3～5 年生开始开花结实，正常结实年龄多在 6～8 年生之后。结实有大小年现

象，间隔期1～2年。王白坡和戴文圣等（1990）报道，樱桃栽植后第1年有74%的植株开花，到第3年有87%的植株结实，平均每株产果7.7kg。李属树种花先叶开放或与叶同时开放，花期短，开花早，多在2月下旬～4月中旬。有明显的落花落果现象。王白坡和仰新民等人（1990）在浙江萧山观察过樱桃的落花落果。开花结实的物候期见表2。

表2 李属树种的开花结实物候期

树 种	观察地点	观察年份	开花期			果实成熟期		果实脱落期
			始 期	盛 期	末 期	始 期	盛 期	
巴旦杏	新疆	1988	3月下旬	4月上旬	4月上旬	8月	9月	9月
杏	北京	1989	4月上旬	4月中旬	4月下旬	6月中旬	7月中旬	7月中旬
山杏	北京	1988	4月上旬	4月下旬	5月上旬	6月下旬	7月下旬	7月下旬
山桃	北京	1988	3月下旬	4月上旬	4月中旬	7月	8月	8月
欧李	北京	1989	4月中旬	4月下旬	5月上旬	7月上旬	7月下旬	8月上旬
斑叶稠李	哈尔滨	1986	5月上旬	5月中旬	5月下旬	6月下旬	9月上旬	9月中旬
梅	浙江	1987	2月下旬	3月上旬	3月中旬	4月下旬	5月	5月
稠李	哈尔滨	1986	5月中旬	5月下旬	6月上旬	7月中旬	8月下旬	9月上旬
桃	北京	1988	3月下旬	4月中旬	4月下旬	7月	8月	8月
樱桃	南京	1984	3月11日	3月14～17日	3月22日	5月上旬	5月6～13日	5月下旬
李	北京	1988	4月下旬	5月上旬	5月下旬	7月上旬	8月上旬	8月中旬
毛樱桃	哈尔滨	1986	5月上旬	5月中旬	5月下旬	6月下旬	7月中旬	7月下旬
大叶桂樱	广西南宁	1988～1989	10月20日	11月6日	11月中旬末	3月中旬	3月下旬	—

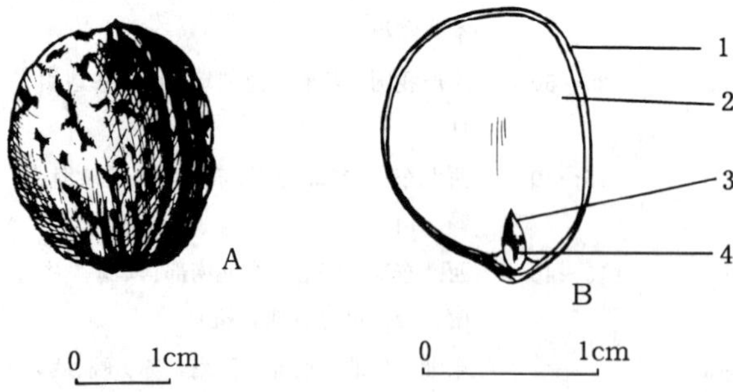

图1 山桃果核外形（A）及其种子纵切面（B）
1. 种皮 2. 子叶 3. 胚芽 4. 胚根
（孟玲绘）

花两性，单生，簇生或伞形、伞房状或短总状花序，或长总状花序。萼片5。花瓣5，白色、粉红色或红色。雄蕊多数。心皮1，子房上位，胚珠2，垂悬。花柱顶生，柱头稍扩大。核果肉质，稀成干燥无汁。核平滑，有沟纹或孔穴，内含1粒种子。子叶肉质，背部隆起。某些种无胚乳，某些种有胚乳。山桃果核的形态和解剖结构见图1和表3。

表3　李属树种果实和种子的形态特征

树　种	未熟果颜色	成熟果实			成熟种子		
		形　状	大小（cm）	颜　色	形　状	大小（cm）	色　泽
巴旦杏	淡绿色	椭圆形、长圆卵形	长3.2～5.5 径2.0～2.5	黄、黄绿或黄红色	椭圆形	2.5～3.0	黄白色
杏	黄绿色	球形	2.5以上	白、黄、至红黄色	扁圆形	1.2～1.8	淡黄色
山杏	黄绿色	近球形	2～2.5	橙黄色	扁心脏形	0.8～1.2× 0.6～1.0× 0.6～1.0	黄褐色
山桃	绿白色	近球形	2.5～4.0	淡黄绿色	扁球形	1.2～1.8× 0.6～1.4	暗褐色
欧李	绿色	卵状球形	1.5～1.8	红色或紫红色	卵状球形	0.4～0.6× 0.3～0.5	黄白色
斑叶稠李	绿色	卵圆形	0.5～0.7	黑色	矩椭圆形	0.4～0.6× 0.3～0.4	淡黄色
梅	绿白色	近球形	2～3	黄色或绿白色	狭楔形	—	—
稠李	绿色	近球形	0.6～1.0	黑色至紫红色	倒卵形	0.6～0.8× 0.4～0.5	淡黄色
桃	绿白色	卵形、椭圆形或扁圆形	3～7（12）	绿白、橙黄，具红晕	椭圆形、近圆形	2.5～3.0	红褐色
樱桃	绿色	近球形	1.0～1.5	朱红色或深红色	矩椭圆形	0.9～1.0× 0.5～0.7	黄色
李	绿色	球形或卵状球形	2～2.5	绿黄或紫红色	卵形	1～1.5× 0.7～1.2× 0.5～0.8	黄白色
毛樱桃	绿色	球形或椭圆形	0.8～1.2	红色或红白色	球形或椭圆形	0.6～0.8× 0.5～0.7	黄白色
大叶桂樱	绿色	卵状椭圆形	1.3～1.8× 1.0～1.3	紫红色	椭圆形	1.2～1.7× 0.7～1.0	浅褐色

果实的采收调制和种子贮藏　应当采收成熟的果实，其种仁发育充实，发芽率高。杏、桃等果实成熟的标志是果肉变软，果肉与果核离开。通常都是收集食用或食品加工后的果核作为播种材料，通称种子。肉薄的种类或品种，采得的果实堆积一处，待软化后搓揉，洗去果肉，所得的果核放置通风干燥处摊开晾干即为播种材料，通称种子。干藏，种子含水量要求在10%以下，装袋贮藏。如贮藏1年以上，宜放置在5℃以下的低温环境。果核具厚硬壳，播前可用湿沙层积处理，时间约100天以上。徐本美（1988）认为樱桃种子不适于干藏。据她报道，用牛皮纸盛装放在5～9℃冰箱中干藏的樱桃种子（含水量不明），4个月后发芽率为28%，10个月后便全部丧失生命力。各树种的种子净度、质量等数据见表4。

表 4　李属树种的出籽率和种子的净度、质量

树　种	出籽率（%）	净度（%）	千粒重（g）	每千克纯净种子粒数
巴旦杏	—	70～96	4 760	210
杏	10～40	—	3 000～4 000	260～340
山杏	25～35	95～98	950～1 250	800～1 100
山桃	15～25	94～98	1 200～1 380	750～850
欧李	15	93～97	350～450	2 200～2 800
斑叶稠李	12	85～95	14～16	62 500～71 400
梅	—	—	2 000～3 000	300～500
稠李	40	98～100	46～60	16 700～21 700
桃	20	90～96	1 640～6 250	160～600
樱桃	15～20	95～98	180～240	4 100～5 600
李	15～20	94～98	800～1 200	800～1 200
毛樱桃	11	—	62～70	14 500～16 200
大叶桂樱	48	98～100	500～650	1 540～2 000

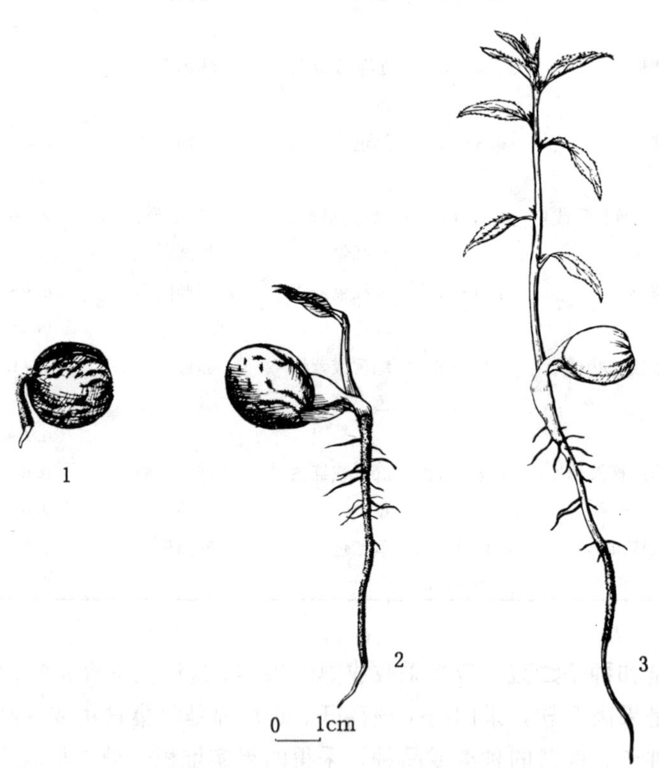

图 2　山桃种子萌发及幼苗生长情况

1. 胚根伸长　2. 初生叶出土　3. 幼苗形成

（孟玲绘）

发芽和播种　种子有休眠习性。秋播可以使种子在自然条件下通过休眠，来年出苗早而整齐。春播种子要进行湿沙层积，待种子裂嘴时播种。据徐本美（1988）报道，在 5～9 ℃下层积 20 天的樱桃种子，发芽率为 44%，层积 20 天并用 500μg/gGA_3 处理的发芽率为 92%。条播，覆土厚度为种子的 3～5 倍。播后地温稳定在 15℃以上时，10～20 天可以发芽，许多种是留土萌发的，例如巴旦杏、桃、梅和大叶桂樱，还有些种是出土萌发的，例如李、稠李、毛樱桃，还有些种是地面萌发的，例如山桃、杏、山杏。山桃的发芽及幼苗生长情况见图 2。

（宋朝枢）

臀　形　果

Pygeum topengii Merr.

（蔷薇科　Rosaceae）

生长习性、分布和用途　臀形果属约 40 余种，我国约 6 种。本文描述的这个种为常绿乔木，高达 25m，胸径 50cm。喜温暖湿润凉爽气候，不耐严寒。适生于肥沃湿润的酸性土。分布于闽、琼、粤、桂、滇、黔。材质优，宜作家具及工艺品。涵养水源作用强。种子可榨油。

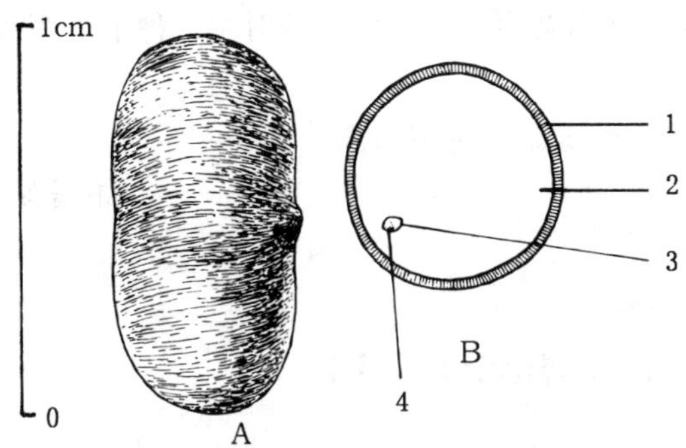

图 1　臀形果果核侧立外形（A）及其纵切面（B）
1. 内果皮和种皮　2. 子叶　3. 胚芽　4. 胚根
（黄应钦绘）

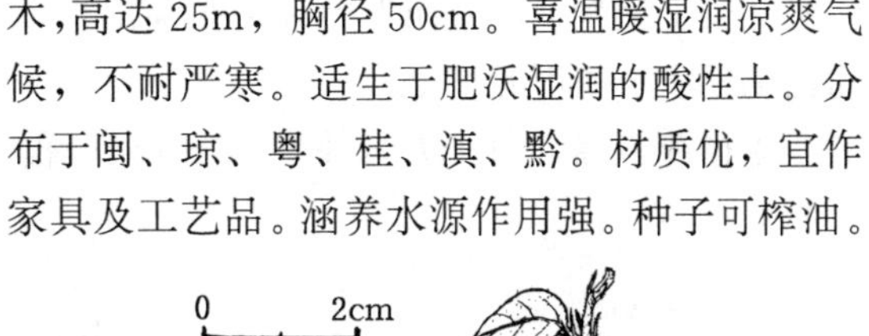

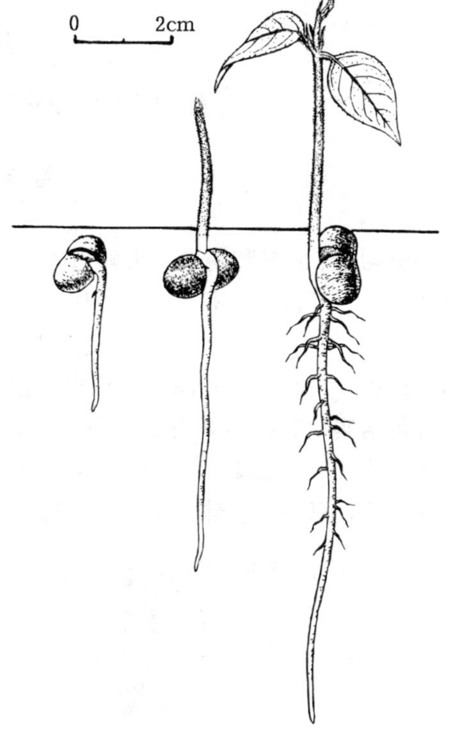

图 2　臀形果种子萌发后第 4、8、20 天幼苗的生长情况
（黄应钦绘）

开花结实　约 10 年生开始开花结实，正常结实期在 20 年生以后。结实大小年间隔期 1 年，但不甚明显。杂性异株。花白色或黄色。总状花序单生或簇生于当年或头年生的叶腋。梗及萼密被褐色柔毛。萼筒倒锥圆形。花被片 10～12，长 1～2mm。萼片与花瓣各 5～6，有时不易区分。雄蕊约 20。子房上位，由单心皮组成，有下垂胚珠 2 枚，花柱顶生。据在广西南宁观测，6 月上旬为始花期，6 月中旬为盛花期，7 月中下旬为末花期；翌年 3 月上旬果实开始成熟，3 月下旬为果熟盛期。在海南 2～3 月果熟。核果，未成熟时青色，熟时黄褐色，臀形，高 7～12mm，宽 8～13mm，厚 6～12mm，3 月中旬～4 月下旬脱落。每果有核一粒。果核臀形或横向长圆形，

浅褐色至褐色，高 6～10mm，宽 5～8mm，厚 4～8mm。无胚乳，子叶肉质（图 1）。

果实的采收调制和种子贮藏 果实进入成熟盛期时上树采摘，或摇动树枝震落后在地面捡拾。采得的果实在室内堆放 1～2 日，软熟后浸水搓洗，清除皮肉等杂质，所得果核用作播种材料，通称种子。生产上也有不加调制，直接用果实播种的。果实出籽率约 50%。调制所得的果核含水量 20%～30%。千粒重约 325g，变动在 250～400g。每千克有纯净种子(核)3 000 粒，变动在 2 500～4 000 粒。种子不宜脱水。短途运输时可以以果实的状态用竹篓或布袋装运，运抵后再行调制或直接播种。果实成熟期正是适宜播种的季节，一般不作贮藏。如需贮藏，应混拌湿沙，贮藏期约半年。

发芽和播种 种子无休眠习性。发芽时日均温需在 20℃以上。1988 年 3 月 15 日，广西林业科学研究所用当年新采种子在广西南宁室外沙床做过发芽测定：播前种子未作任何处理，播后 25 天即 4 月 8 日（日均温 20℃）开始发芽，9～17 日为发芽盛期，25 日发芽终止。从开始发芽之日至发芽高峰的 9 天中，发芽百分数为 48%，从播种之日起算的 42 天中，发芽率为 62%。留土萌发。胚根萌发后 8 天胚芽出土，15 天展出初生叶。初生叶近对生。种子发芽及幼苗初期生长情况见图 2。

条播或点播。每平方米播种 38～75g。覆土约 2cm。1 年生苗出圃。

（韦增健）

火　　棘

Pyracantha fortuneana（**Maxim.**）**Li**

（**蔷薇科　Rosaceae**）

生长习性、分布和用途 火棘属 10 种，我国产 7 种，本文描述 1 种。常绿灌木，高达 3m，具枝刺，枝叶茂密，花密集，果橘红或深红色。喜温暖湿润气候和深厚土壤。主产西北南部、西南、华南至华东。花果密集，颜色鲜艳，北京及以南地区多栽培供观赏，亦可作绿篱。果含淀粉和糖类，可食用、酿酒及作饲料。根、叶及种子入药，根皮可提制栲胶。

开花结实 3～4 年生开始开花结实。花两性。复伞房花序，径 3～4cm，花径约 1cm 左右。萼筒短，萼片 5。花瓣 5，白色，雄蕊 20。心皮 5，在腹面离生，在背面约 1/2 与萼筒相连。子房半下位，每室 2 胚珠。梨果，近球形，萼片宿存，径约 5mm，深红色或橘红色，内含 5

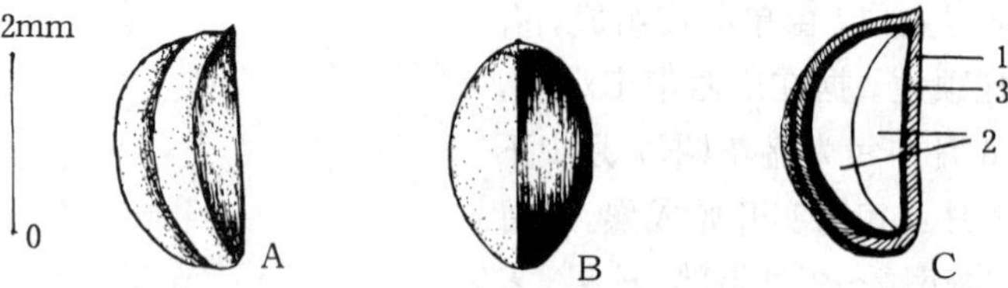

图 1　火棘果核外形侧面（A）和正面（B）及其纵切面（C）

1. 内果皮（核壳）　2. 两种子　3. 种皮

（孟玲绘）

小核。花期 3～5 月，果熟期 9～11 月。

种子的采收调制和种子贮藏　9～11 月间，果实由绿色变为深红色或橘红色时剪下果穗或摘下果实，拣出小枝，碾碎果肉，洗净，晾干，风选后所得纯净果核即为播种材料，通称种子。种子近紫色或紫黑色，有光泽，偏斜三角状，长 1.5～2mm，径约 1mm（图 1）。

出籽率 52%左右。种子千粒重 2.1g，每千克 48 万粒左右。种子可以装袋干藏。

发芽和播种　火棘核壳骨质，坚硬，透水性差，需要层积处理才能发芽。秋末混湿沙层积 3～4 个月，翌春发芽率可达 70%左右。

火棘种子细小，出土能力较弱，覆土要薄而均匀，厚度在 0.5cm 左右。充分层积的种子春播后 14 天左右幼苗出齐。出土萌发。种子萌发情况见图 2。

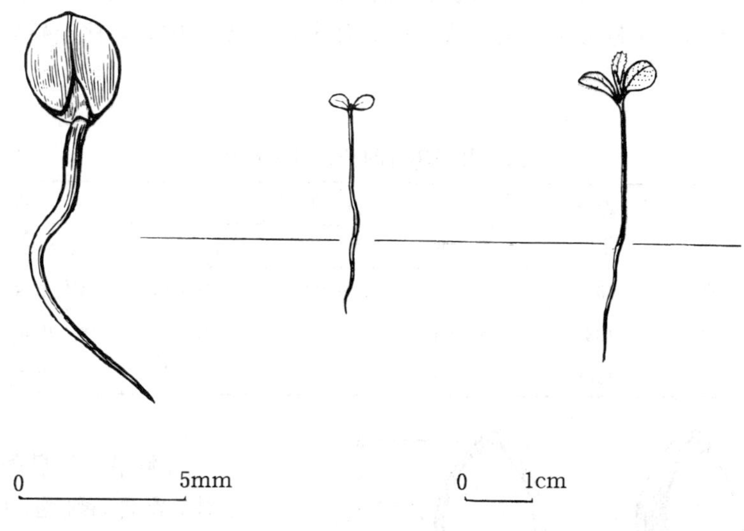

图 2　火棘种子萌发出土后第 2、6、27 天幼苗生长情况

（孟玲绘）

（王木林、高秀岩）

梨　属

Pyrus L.

（蔷薇科　Rosaceae）

生长习性、分布和用途　本属约 26 种，我国 15 种，本文描述 3 种。落叶，稀半常绿，乔木或灌木。喜光。适应性强，耐寒，抗旱，亦较耐低温和盐碱，适生于湿润深厚的沙壤土。根系发达，萌蘖力强，寿命较长，生长较慢。主要分布在长江流域以北。温带地区的重要水果，有些种类是梨树的主要砧木。花为蜜源植物。木材红褐色，材质坚硬，供制高级器具、印刷木板、细木工等用。这 3 个树种的名称、树高和分布情况见表 1。

表 1　梨属树种名称、生长和分布

中　名	学　名	树高（m）	胸径（cm）	分　布	供　稿
杜梨（棠梨）	*P. betulaefolia* Bge.	10	30	东北南部、黄河流域至长江流域	309
豆梨（糖梨）	*P. calleryana* Decne.	6～12	20	黄河中下游、长江中下游至华南	309
秋子梨	*P. ussuriensis* Maxim.	10～15	30	东北、华北、西北南部	107

开花结实　5～8 年生开始开花结实，正常结实在 8 年生以后。结实大小年间隔期为 1 年，但不甚明显。花两性。伞形总状花序由 5～15 朵花组成。花先叶开放，花梗长 1.5～5cm。花瓣卵圆形，白色，在花刚开时微现粉红色。雄蕊 20。花柱 2～5，离生。子房 2～5 室，每室 2 胚珠。花期 4～6 月，果期 8～10 月（表 2）。9 月中旬～10 月中旬，秋子梨果实成熟，陆续脱落。

表 2　梨属树种的开花结实物候

树　种	观察地点	观察年份	开花期			果实成熟期	
			始　期	盛　期	末　期	始　期	盛　期
杜梨	北京	1988	4 月中旬	5 月上旬	5 月下旬	8 月下旬	9 月中旬
豆梨	山西	1988	4 月中旬	5 月上旬	5 月下旬	8 月中旬	9 月上旬
秋子梨	哈尔滨	1986	5 月上旬	5 月中旬	6 月上旬	8 月下旬	10 月中旬

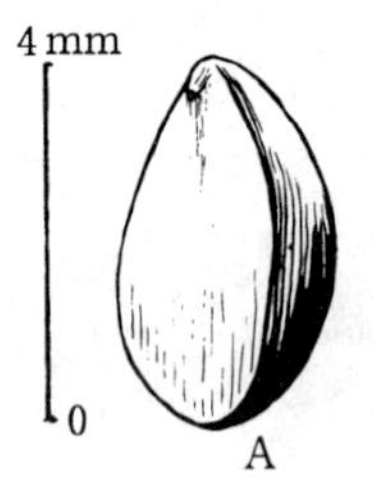

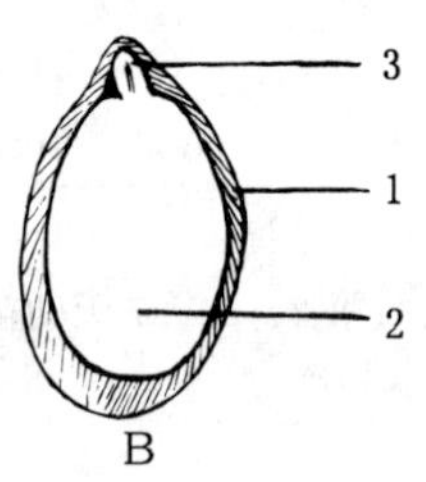

图 1　杜梨种子外形（A）及其纵切面（B）
1. 种皮　2. 子叶　3. 胚根
（孟玲绘）

梨果，具显著皮孔，果肉多汁，富含石细胞。内果皮软骨质，每果有种子数粒，黑色或黑褐色。种子无胚乳，子叶平凸。果实和种子的形态特征见图 1 和表 3。

表 3　梨属树种果实及种子的形态特征

树　种	成熟果实			成熟种子		
	形　状	大小（cm）	颜　色	形　状	大小（mm）	颜　色
杜梨	近球形	0.5～1.0	赭褐色	扁三角状卵形	3.5～4.5 ×3～3.5 ×1.5～2.0	暗褐色
豆梨	近球形	约 1.0	褐色	三角状卵圆形	3～4 ×2.5～3 ×1.5～2.0	棕褐色
秋子梨	近球形	2～6	褐色	三角状卵圆形	4～6 ×3～3.5 ×2～3.0	黑色或黑褐色

果实的采收调制和种子贮藏　当果实成熟时，摘取果实或击落后收集。秋子梨成熟后自然脱落，可在地面收集落果。采回的果实堆集捣碎，用水洗净取得种子，或切开果实取种。洗净的种子置通风干燥处摊放晾干，装袋贮存。曾经用含水量在10%以下的种子在0～10℃的低温密封的条件下保存，2年后仍有90%的发芽率。种子的净度、质量等数据见表4。

表4　梨属树种的鲜果出种率和种子净度

树　种	出种率（%）	净度（%）	千粒重（g）	每千克纯种粒数（万粒）
杜梨	10	80～90	27～29	3.4～3.7
豆梨	6	85～95	18～22	4.5～5.5
秋子梨	1	60～80	42～48	2.1～2.4

发芽和播种　种子的休眠程度不深，秋季播种，覆土3cm，翌春可以大部出苗。春播的种子要在0～10℃的低温条件下层积处理30～40天，发芽率可达80%以上。发芽时的日均温需在15℃以上。垄播或条播，播种量每平方米50～75g，播种深度1～1.5cm。当年苗高30～40cm，可出圃栽植或用作嫁接砧木。子叶出土萌发。杜梨种子萌发和幼苗生长情况见图2。

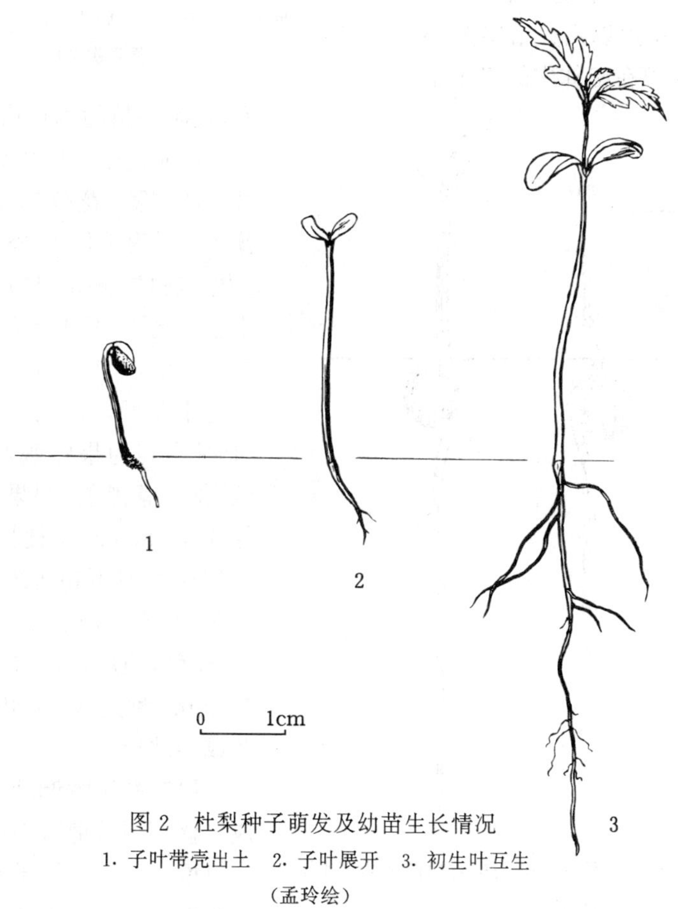

图2　杜梨种子萌发及幼苗生长情况

1. 子叶带壳出土　2. 子叶展开　3. 初生叶互生

（孟玲绘）

（宋朝枢）

石斑木（车轮梅、春花木）

Rhaphiolepis indica（L.）Lindl.

（蔷薇科　Rosaceae）

生长习性、分布和用途　石斑木属约15种，我国7种。本文描述的这1个种为常绿小乔木，高4～6m，胸径约10cm。适生于排水良好的酸性土壤。分布于皖、浙、赣、闽、台、湘、黔、滇、桂、粤。日本、越南亦有。果可食。根入药，医治跌打损伤。材质坚韧致密，为雕刻、工艺良材。

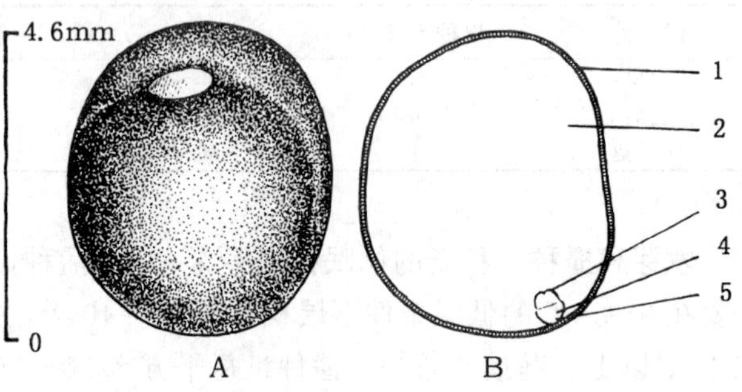

图1　石斑木种子的外形（A）及其纵切面（B）
1. 种皮　2. 子叶　3. 胚芽　4. 胚轴　5. 胚根
（黄应钦绘）

开花结实　8～10年生开始结实，正常结实期在15年生以后。结实大小年现象不明显。生长在密林中的多不结实，庭院种植的每年均能正常结实。花两性。圆锥花序或总状花序顶生于1年生枝。萼5裂。花瓣5，白色或淡红色。雄蕊15。子房下位，2室，每室有直立胚珠2枚。花柱2～3。1978～1984年在广西南宁观测，2月中旬花序形成，2月下旬为始花期，3月上中旬为盛花期，3月下旬为末花期，10月下旬果实开始成熟，11月中旬为果熟盛期。果实未熟时青色，熟后紫黑色。梨果，核果状，近球形，径6.5～8.5mm，梗长5～10mm，11月下旬～12月下旬陆续脱落。每果有种子1粒，少有2粒。种子近球形或半球形，棕褐色，径5.5～7.5mm，厚4.5～5.5mm。种皮薄。无胚乳。子叶肥厚，平凸或半球形。

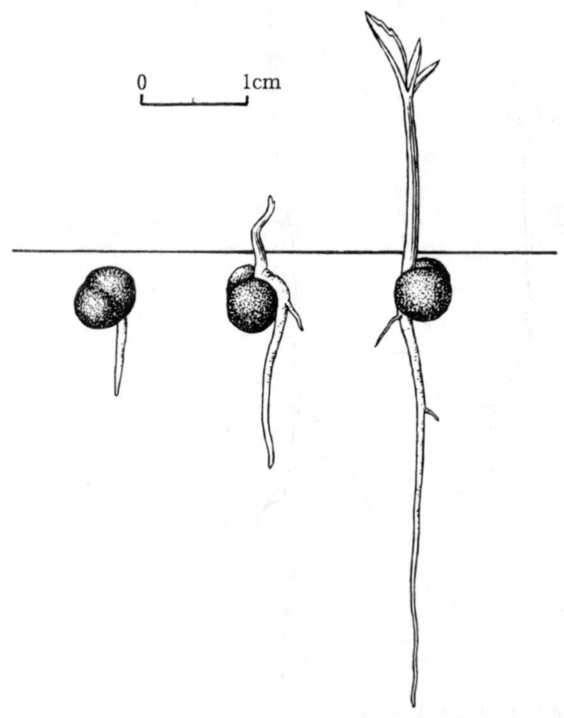

图2　石斑木种子萌发后第3、6、15天幼苗的生长情况
（黄应钦绘）

果实的采收调制和种子贮藏　果实成熟盛期用枝剪带果梗剪下。采得的果实可堆沤3～5天。果实充分软熟后置水中搓擦漂洗，淘去皮肉等杂质，即得种子。按质量计算的出种率约55%。净度

92% ～98%。千粒重 110（90～130）g。每千克有纯净种子 9 000（7 700～11 000）粒。调制所得种子的含水量为 25%～35%，忌失水，不宜日晒。运输时可连带果梗装入竹筐或布袋内，运抵后再行调制。贮藏时需混沙，贮藏期 4～6 个月。

发芽和播种　种子有休眠现象。发芽时日平均气温宜在 15℃以上。1987 年 10 月 27 日，广西林业科学研究所在室外沙床用当年新采种子做过发芽测定：播后 58 天即 12 月 23 日开始发芽，25～29 日为发芽盛期，至 1988 年 1 月 6 日发芽终止。从发芽开始至发芽高峰的 7 天中，发芽百分数为 45%。从播种至发芽终止的 72 天中发芽率为 53%。留土萌发。胚根萌发后约 3 天上胚轴出土，13 天初生叶发生。主根明显，侧根稀少（图 2）。

条播。播前种子可不作处理。每平方米播种 12～20g，覆土 0.5cm。1 年生苗出圃。

（韦增健）

蔷　薇　属
Rosa L.

（蔷薇科　Rosaceae）

生长习性、分布和用途　本属约 200 种，广布于北半球寒温带至亚热带地区。我国 82 种，本文介绍 5 种。落叶或常绿灌木，茎直立、蔓延或攀援，常有刺，稀无刺。这 5 个种的名称、生长习性、分布和用途见表 1。

表 1　蔷薇属树种的名称、生长习性、分布和用途

中　名	学　名	生长习性	分　布	用　途	供　稿
山刺玫（刺玫蔷薇）	*R. davurica* Pall.	落叶灌木，高约 1.5m，花单瓣，粉红色	东北、华北。朝鲜半岛、蒙古、东西伯利亚	花可提芳香油；果含 Vc 达 4.3%～7.2%，制玫瑰酱；果、根药用；观赏	108
贵州缫丝花	*R. kweichowensis* Yu et Ku	常绿或半常绿，攀援状灌木，高 1～1.5m。花单瓣，白色	黔、桂	果含多种营养元素	602、1002
金樱子	*R. laevigata* Michx.	常绿攀援灌木，高达 5m，花单瓣，白色	华东、中南、西南、陕	根、叶、果药用，果可制饮料	1003
缫丝花	*R. roxburghii* Tratt.	落叶或半常绿灌木，高 1～2.5m，花重瓣至半重瓣，淡红至粉红色	东南、华中、西南，甘、陕均有野生或栽培	果入药并制饮料，花供观赏	1003
黄刺玫	*R. xanthina* Lindl.	落叶灌木，高达 3m，花重瓣至半重瓣，黄色	东北、华北，陕、甘、青	珍贵观赏树种，花可提芳香油	107

开花结实　2～4 年生开始结实，5 年后正常结实。山刺玫大小年现象不明显，贵州缫丝花大小年间隔期为 1 年，黄刺玫很少结实。花两性，整齐，单生或成伞房状稀复伞房状或圆锥花序。萼筒（花托）球形、壶形至杯形，颈部缢缩，萼片 5，稀 4。花瓣 5，稀 4，有时为重瓣，白色、黄色、粉红至红色。雄蕊多数，着生花盘周围。离生心皮多数，稀少数，着生在萼筒内。成熟时聚合瘦果包藏在肉质或粉质萼筒（花托）内，特称蔷薇果。果熟时花托红色、黄色或近黑色。这 5 个树种的开花结实物候期见表 2，果实和种子（瘦果）的形态特征见表 3。山刺玫的瘦果外形及其纵切面见图 1。

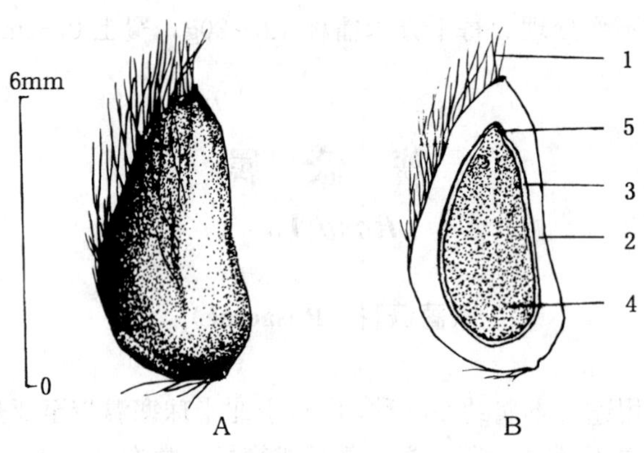

图 1　山刺玫瘦果的外形（A）及其纵切面（B）

1. 果毛　2. 果皮　3. 种皮　4. 子叶　5. 胚根

（梁鸣绘）

表 2　蔷薇属树种的开花结实物候

树　种	观察年限地点	开花期			果实成熟期			成熟特征	果实脱落期
		始　期	盛　期	末　期	始　期	盛　期	末　期		
山刺玫	1963～1980 哈尔滨	6 月上旬	～	7 月上旬	7 月中旬	～	9 月下旬	果变红色	初熟至脱落间隔 30～40 天，或部分不脱落
贵州缫丝花	1987 广西河池	3 月下旬～4 月上旬	4 月中旬	4 月下旬	8 月中旬	8 月下旬～9 月上旬	～	青带红色，干后黑褐色	熟后 30～60 天脱落
金樱子	～	4 月下旬	～	5 月中下旬	9 月下旬	～	11 月中旬	果变红色	熟后经久不落
缫丝花	～	5 月中旬	～	6 月中旬	8 月中旬	～	9 月上旬	果由绿变黄	—
黄刺玫	1963～1980 哈尔滨	5 月下旬	～	7 月上旬	7 月下旬	～	8 月中旬	果变红黄色	初熟至脱落间隔 15～20 天

表 3　蔷薇属树种果实和种子的形态特征

树　种	成熟果实（cm）			果实含瘦果粒数	瘦果大小（mm）		
	形　状	长	径		形　状	长	宽
山刺玫	球形或卵形	—	1～1.5	多粒	不规则三角状长卵形，淡褐色	4～6	2～3
贵州缫丝花	壶形密被尖刺	1.6～2.2	1.4～2.2（带刺径）	15～70	卵形或椭圆形，黄褐色	4～5	2～4.5
金樱子	长倒卵形	2～4	—	多粒	不规则三角状长卵形，淡黄至黄褐色	5.3～6.0	3.8～4.0
缫丝花	扁球形密被尖刺	—	2.5～3.5	20～95	不规则三角形	—	—
黄刺玫	球形或倒卵圆形	0.8～1.2	0.7～1.0	多粒	扁卵形，黄褐色至赤褐色	2～5	1～1.5

果实的采收调制和种子贮藏　果实呈现成熟时特有的红色、红黄色、黄色时即可采收。有的果实密被尖刺，需戴防护手套或用刀剪剪取。采后堆放使之变软，捣烂，淘净、晾干所得的瘦果即为播种材料，通称种子。也可从制果酒、果酱、果汁所余果渣中获得种子。种子低温袋装可贮藏 1 年。其出籽率、净度及种子质量见表 4。

表 4　蔷薇属树种的出籽率及种子质量

树　种	出籽率（%）	净度（%）	千粒重（g）		每千克纯净种子粒数（万粒）	
			一　般	变动范围	一　般	变动范围
山刺玫	26～29	90～99	7.2	6.4～11.3	14	8.8～15.6
贵州缫丝花	10	92～98	15	10～20	6.6	5.0～10
金樱子	—	—	—	17～18.7	—	5.3～5.9
缫丝花	—	—	24	—	4.2	—
黄刺玫	24～28	90～98	12	10～25	8.3	4.0～10

发芽和播种　蔷薇属种子（瘦果）一般外壳坚硬且有休眠习性。徐本美和张治明等（1993）研究过包括缫丝花在内蔷薇属 7 个树种种子的萌发和休眠。她们认为多数属于综合性休眠，即既有外壳造成的强迫性休眠，也有胚的生理性休眠，但不同的种休眠深浅不一；在北京地区，浅休眠的可以在春季早播；多数只需一个冬季的低温层积便可解除休眠；深休眠的可以在采收后立即酸蚀处理再行层积，或酸蚀后经过干湿交替处理再行层积。经过酸蚀和干湿交替，达到裂口程度的缫丝花种子，5℃下层积 3 个月后发芽率可以达到 62.5%。贵州缫丝花种子无明显休眠习性，在 20℃左右温度下 30～40 天即可发芽。据 1987 年 9 月 21 日在广西南宁对贵州缫丝花所作的发芽试验，播后 36 天（10 月 26 日）开始发芽，至 12 月 5 日发芽终止（盛期不明显），发芽率为 51%。出土萌发。子叶出土后 5 天发出初生叶。山刺玫种子萌发情况见图 2。蔷薇属种子一般都可秋播，翌春出苗率可达 50%～80%。苗床条播，每平方米床面播种 10～15g。山刺玫与黄刺玫当年苗高 50～80cm，1～2 年生出圃。蔷薇属树种还可通过分株、压条及扦插繁殖。

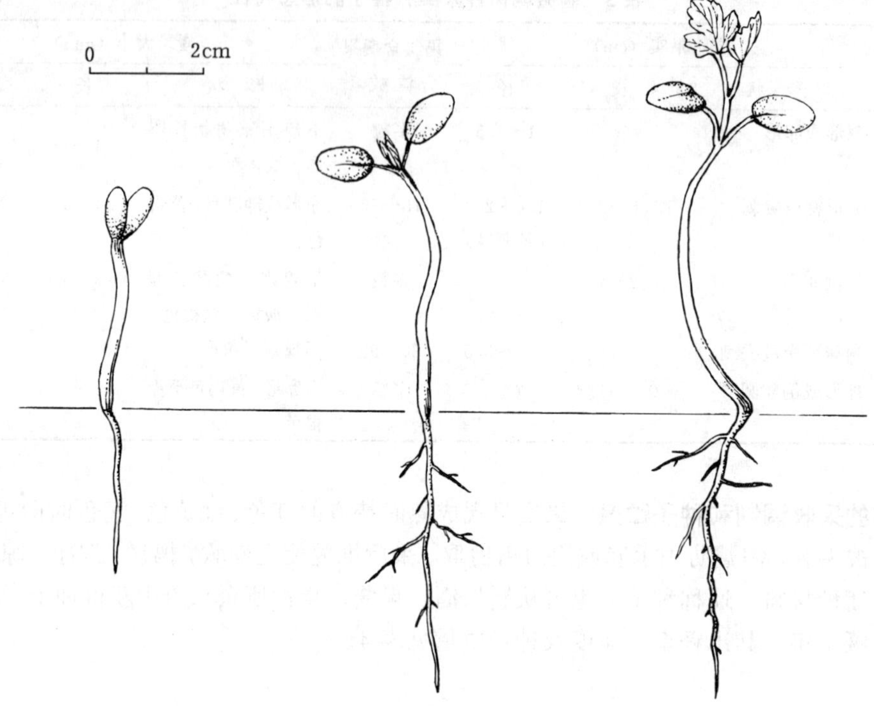

图 2　山刺玫种子萌发后第 4、9、16 天幼苗生长情况

（曹雅范绘）

（周德本）

悬钩子属

Rubus L.

（蔷薇科　Rosaceae）

生长习性、分布和用途　本属约 700 种，我国 194 种，绝大多数为木本，本文介绍 3 种。越南悬钩子为攀援状常绿灌木，高 3～5m，能耐轻霜及短期 0℃左右低温，忌冰雪。牛迭肚与库页悬钩子为落叶灌木，高 0.6～2（3）m，耐干旱瘠薄。它们的名称、分布和用途见表 1。

表 1　悬钩子属树种的名称、分布和用途

中　名	学　名	分　布	用　途	供　稿
越南悬钩子（蛇泡筋）	*R. cochinchinensis* Tratt.	桂、粤、滇。越南	可食、作果汁。根药用	602
牛迭肚（托盘、马林果）	*R. crataegifolius* Bge.	东北、华北。朝鲜半岛、俄罗斯远东、日本	可食，作果汁、果酱、果酒。果、根药用	107
库页悬钩子（毛叶悬钩子）	*R. matsumuranus* Lévl. et Van.	东北、华北、西北。俄罗斯，欧洲，朝鲜半岛，日本	可食，作果汁、果酱、果酒、入药	107

开花结实　2～3 年生开始开花结实，5 年生以后正常结实。取代老枝的萌生条或修剪后发生的新枝又可正常结实。花两性，稀单性异株，组成聚伞状圆锥花序、总状花序、伞房花序或数朵簇生及单生。花萼 5（稀 3～7）裂，宿存。花瓣 5，稀缺，白色或红色（本文 3 种皆白色）。雄蕊多数，着生花萼上部。心皮多数（稀少数），分离，着生于球形或圆锥形花托上，子房 1 室，每室 2 胚珠。果实为由小核果集生于花托而成的聚合果，多浆或干燥。小核果的核称为小核，半月形或不规则形，有皱纹或麻点。种子无胚乳。越南悬钩子果球形，红黑色，径 0.6～1.4cm；种子浅红色，长 2～2.2mm，径 1～1.5mm。牛迭肚果近球形，暗红色，有光泽，果径 1cm 左右；种子长 2～2.3mm，径 1～1.5mm（见图 1）。库页悬钩子果卵球形，红色，具绒毛，径 1～1.5cm；种子长 2～2.4mm，径 1～1.5mm。越南悬钩子果实成熟后可在树上存留 3～4 天。牛迭肚和库页悬钩子的果实初熟至脱落间隔 20～25 天。这 3 个种的开花结实物候见表 2。

表 2　悬钩子属树种的开花结实物候

树　种	观察年限 地点	开花期			果实成熟期			成熟 特征
		始　期	盛　期	末　期	始　期	盛　期	末　期	
越南悬钩子	1987 南宁	5 月上旬	5 月下旬	6 月中下旬	6 月下旬	7 月中旬～8 月中旬	7 月下旬～9 月中旬	果幼时红色，熟时变黑色
牛迭肚	1962～1980 哈尔滨	6 月 11 日	～	7 月 19 日	7 月 17 日	～	9 月 15 日	果为暗红色，果汁粉红色，种子淡褐黄色
库页悬钩子	1962～1980 哈尔滨	6 月 4 日	～	7 月 7 日	7 月 7 日	～	8 月 5 日	果实由鲜红转深红，种子暗黄色

果实的采收调制和种子贮藏　果实在 7～9 月份成熟，应随熟随采，用手摘取。果实采回后经搓洗，或堆沤 2～3 天后挤出果汁后搓洗，用水漂出果皮和果肉，晾干去杂，所得小核即为播种材料，通称种子。生产上正在开发利用果实资源，可从提取果汁后的果渣中获得种子。

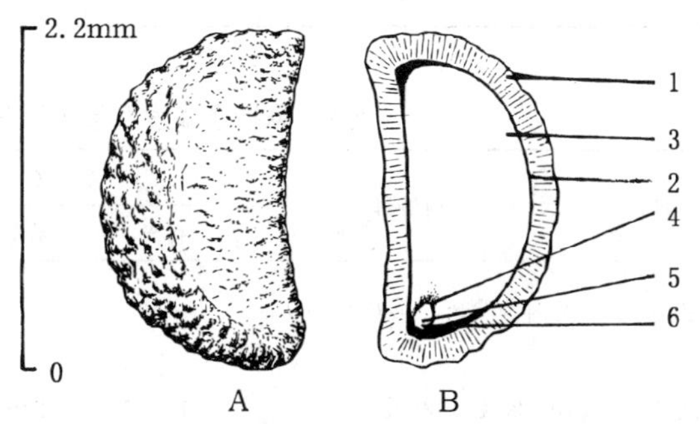

图 1　越南悬钩子小核外形（A）及其纵切面（B）
1. 内果皮　2. 种皮　3. 子叶　4. 胚芽　5. 胚轴　6. 胚根
（梁鸣绘）

越南悬钩子调制后忌失水，不宜日晒，运输应保持湿润，贮藏应在低温处混以湿润细沙，贮藏期为半年。牛迭肚和库页悬钩子低温密封可贮藏 1 年。这 3 个种的出籽率及种子质量见表 3。南京中山植物园从苏、皖、浙、闽、湘、赣、黔引进悬钩子属植物近 50 个种。顾姻、王传水等人（1996）在 1993～1995 年对其中 30 个种的开花结实物候、果重、果的高和径、小果数、以及果实的氨基酸、维生素和矿物质含量做过测定，并对涉及到植物育种的 2 个种品质作了评价。

表 3　悬钩子属树种的出籽率及种子质量

树　种	出籽率（%）	净度（%）	千粒重（g）	每克纯净种子粒数
越南悬钩子	10	86～96	0.8～1.2	830～1250
牛迭肚	2.4	85～95	1.5～1.8	550～660
库页悬钩子	7.6	85～95	1.6～1.85	540～630

发芽和播种　悬钩子属的种子均有休眠现象，催芽处理后播种才能达到苗齐苗壮。表 4 汇编了这 3 个树种种子催芽处理的实例及其效果。表 5 汇编了这 3 个树种发芽能力测定的实例。

表 4　悬钩子属种子催芽处理条件及效果的实例

树　种	时间和地点	催芽的条件	催芽天数	效　果
越南悬钩子	1987 年 8 月～11 月 南宁	滤纸，室内自然光，20℃左右	90	20%～35%发芽
牛迭肚	1975 年 2 月 哈尔滨	种子水浸 2 天，混沙，10～25℃	80～100	30%裂嘴，播后 50%～60%出苗
库页悬钩子	1975 年 2 月 哈尔滨	种子水浸 2 天，混沙，10～25℃	80～100	35%裂嘴，播后 40%～50%出苗

表 5　悬钩子属发芽能力及其测定条件的实例

树　种	预处理	温　度（℃）		发芽势（%）		发芽率（%）	
		有光时	无光时	计算天数	一般数值	计算天数	一般数值
越南悬钩子	室内自然光，20℃，90 天	20	—	11	27	15	35
牛迭肚	10～25℃，80 天	15～25	10～15	15	30	20	45
库页悬钩子	10～20℃，80 天	15～25	10～15	15	42	20	49

出土萌发。越南悬钩子发芽后约 3 天子叶出土，14 天左右发出初生叶。牛迭肚与库页悬钩子在哈尔滨应于 4 月下旬～5 月上旬播种，可采用苗床条播，每平方米播种量 5g，覆土厚 0.5～1.0cm。1 年生苗高 15～30cm，1～2 年生苗出圃（图 2）。

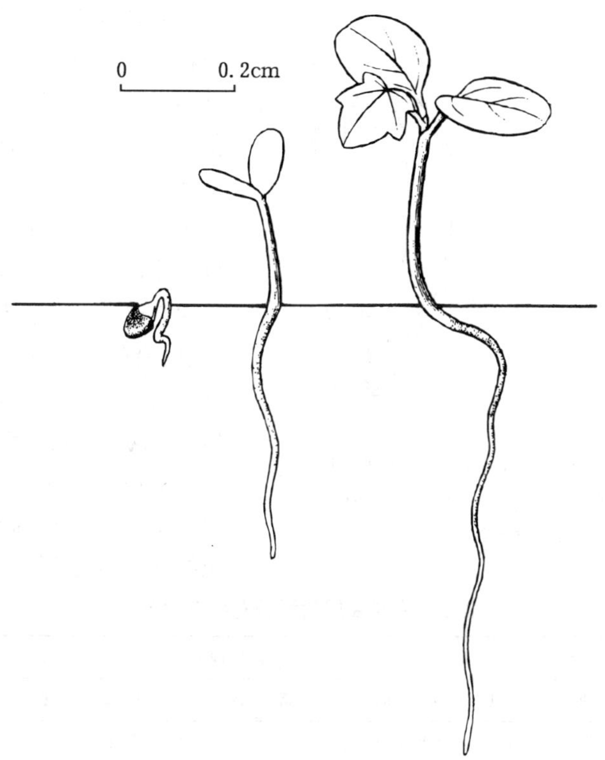

图 2　越南悬钩子种子萌发后第 3、5、15 天幼苗的生长情况
（梁鸣绘）

（周德本、韦增健）

珍珠梅属
Sorbaria（Ser.）A. Br. ex Aschers.

（蔷薇科 Rosaceae）

生长习性、分布和用途　本属约有 9 种，分布于亚洲。我国约 4 种，本文描述 2 种。落叶灌木。奇数羽状复叶，互生，小叶对生。叶形秀丽，花序大，花期长，供绿化观赏。抗污染能力强。这 2 个树种的名称、生长习性、分布及用途见表 1。

表 1　珍珠梅属树种名称、生长习性和分布及用途

中　名	学　名	树高（m）	分　布	用　途	供　稿
华北珍珠梅	*S. kirilowii*（Regel）Maxim.	3	西北东南部、华北、豫、鲁	观赏	403
珍珠梅	*S. sorbifolia*（L.）A. Br.	2	东北、内蒙古。朝鲜半岛、日本、蒙古、俄罗斯	观赏、枝果药用	404

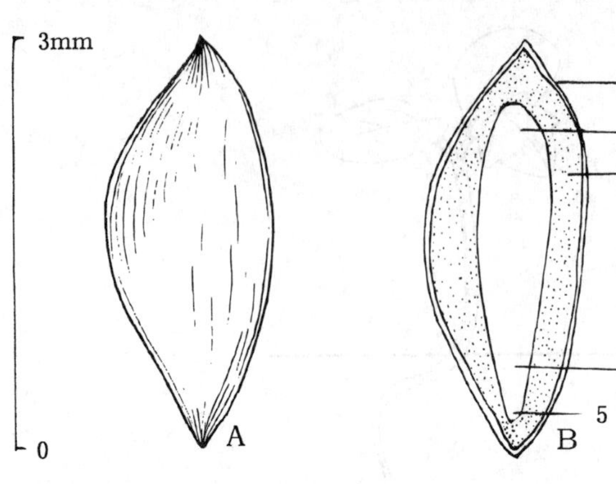

图 1　珍珠梅种子外形（A）及其纵切面（B）
1. 种皮　2. 胚乳　3. 子叶　4. 胚轴　5. 胚根
（胡冬梅、林平绘）

开花结实　一般 5 年生以后便进入开花结实盛期，花果量大，每年结实，大小年现象不明显。花两性。大型密集圆锥花序顶生。花小，萼片 5。花瓣 5，白色，覆瓦状排列。雄蕊20～50。心皮 5，基部合生，与萼片对生。子房上位，每室具多数倒生胚珠。萼片宿存，反折，稀开展。果为 5 蓇葖，长圆柱形，长约 3mm，黑褐色，沿腹缝线开裂。华北珍珠梅每个蓇葖有种子12～24粒，少至 6 粒。种子长条形，黄褐色，表面带数条纵肋条，长 2～3mm，径 1～2mm。有胚乳，胚直立，线型（见图 1）。珍珠梅属这 2 个树种的开花结实物候已汇入表 2。

表 2　珍珠梅属树种开花结实物候

树　种	观察年份和地点	开花期			果实成熟期			种子散落期		
		始　期	盛　期	末　期	始　期	盛　期	末　期	始　期	盛　期	末　期
华北珍珠梅	1986～1980 北京地区	6 月上旬	6 月下旬～7 月中旬	8 月下旬	8 月下旬	9 月上旬	9 月中旬	9 月上旬	9 月中旬	10 月中旬
珍珠梅	1963～1980 哈尔滨	7 月上旬	～	8 月下旬	9 月中下旬	～	10 月上旬	9 月下旬	～	10 月下旬

果实的采收调制和种子贮藏　果实成熟时由深绿色渐变为深褐色。应在蓇葖开始开裂，种子散落前剪取果穗。如果实已经开裂，捋取果穗时不得振落种子。采回摊于通风处晾干，待果实全部开裂后将果穗倒置，轻轻振动即可脱出种子。或将果穗装入袋内揉搓，但勿用力过度而折碎种子。用相应孔径的筛除去杂质，即可得纯净种子。果实出种率、净度、种子质量见表 3。干燥种子装入袋或罐内，在冷凉处干藏。在北京地区一般室内条件下，纸袋内贮存 9 个月的种子，只有 10%的发芽率。

表 3　珍珠梅属树种果实出种率、净度及种子质量

树　种	果实出种率（%）	净度（%）	千粒重（g）	每克纯净种子粒数
华北珍珠梅	10	60～70	0.25～0.34	2 900～4 000
珍珠梅	13	65～80	0.9	1 100

发芽与播种　这 2 种树种种粒细小，发芽测定可以用发芽皿内的滤纸作床。在 25 ℃的恒温下，第 3～4 天开始发芽。育苗时宜先行室内盆播，出苗后移栽露地。也可以在苗圃条播，每平方米苗床播种 5g 左右。覆土厚不超过 0.5cm，保持床面湿润，播后 10～15 天出苗。过密的幼

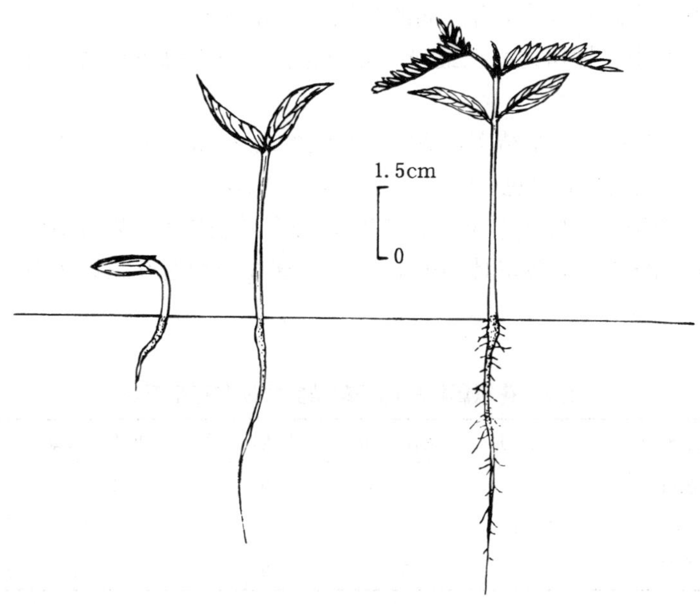

图 2　华北珍珠梅萌发出土后第 1、6、18 天的幼苗生长情况
（胡冬梅、林平绘）

苗可以间苗移植。出土萌发(见图 2)。当年苗高 20～30cm。生产上多用分蘖或扦插方法繁殖。

（刘长江）

花　楸　属

*Sorbus*L.

（蔷薇科　Rosaceae）

生长习性分布和用途　本属约 80 种，分布于北温带。我国有 50 余种，主产西南、西北及东北地区。本文描述 3 种，它们的木材可作家具和工具。花序大，入冬果实颜色鲜艳，有观赏价值。果实富含维生素，可以利用，也可为鸟雀提供食物。这 3 个种的名称、树高、分布及用途见表 1。

表 1　花楸属树种的名称、树高、分布和用途

中　名	学　名	树高（m）	分　布	用　途	供　稿
黄山花楸	*S. amabilis* Cheng ex Yu	10	浙、皖	材用、观赏、食用、砧木	1003
花楸树	*S. pohuashanensis* (Hance) Hedl.	8	东北、华北、甘、鲁。西伯利亚、朝鲜半岛	材用、食用、药用、观赏	105
天山花楸	*S. tianschanica* Rupr.	4～6	新、青、甘。阿富汗、土耳其	材用、食用、药用、观赏	203、204

花楸属为落叶乔木或小乔木，喜湿润的酸性或微酸性土壤，常生于山坡或山谷杂木林中，较耐荫，能耐寒，其中花楸树耐寒性最强。天山花楸生长于海拔较高的山林中，但不耐盐碱和干旱。

开花结实 5～15 年生开始结实，每年结实较多或 2～3 年有一次丰年。密集型复伞房花序顶生，大而明显。花两性，白色。萼片、花瓣各 5。雄蕊 15～25。这 3 个种的心皮 3～5，大部分与花托连合。子房半下位或下位，2～5 室，每室具胚珠 2。梨果小型，近球形，萼片宿存。内果皮软骨质，2～5 室，各室具种子 1～2 枚。这 3 个种的开花结实物候及花果形态见表 2。

表 2 花楸属树种的开花结实物候和花果形态

树 种	观察地点	花 期（月）	花径（mm）	果熟期（月）	果径（mm）	成熟果实的颜色
黄山花楸	安徽黄山	5	7～8	9～10	6～7	红色
花楸树	哈尔滨	5～6	6～8	9～10	6～8	红色或橘红色
天山花楸	新疆天山	5～6	15～18	9～10	10～12	鲜红色至紫红色

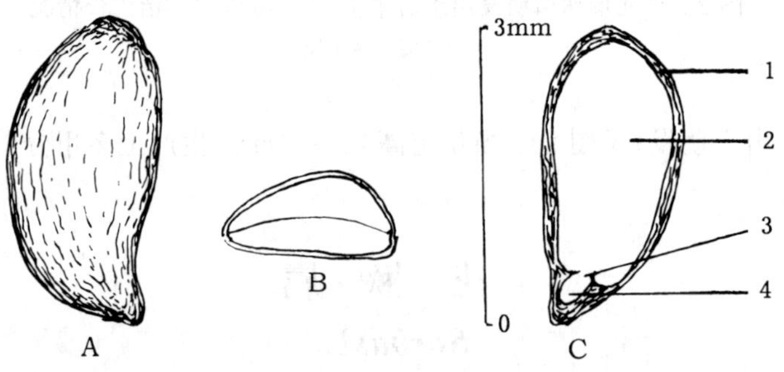

图 1 黄山花楸种子外形（A）及其横切面（B）和纵切面（C）

1. 种皮 2. 子叶 3. 胚轴 4. 胚根

（史渭清绘）

种子长卵形略扁，背拱腹平，棕褐色至近黑色；胚根端较尖而歪，种皮厚膜质，表面基本平滑，放大观察则可看到纵向细纹；胚乳不明显；胚大，紧贴种皮。子叶长椭圆形，不明显的胚芽藏在两片子叶间。图 1 绘出的是黄山花楸的种子外形及剖面结构。天山花楸种子有花楸长尾小蜂（*Syntomapsis ancupariae* Rodz.）危害。

果实的采收调制和种子贮藏 成熟果实可以在树上存留 1 个月以上，经霜雪后渐次脱落。果熟后应及时采摘，以免鸟雀啄食。果实的成熟度如果不高，果肉较硬，难以脱粒取种，宜堆放室内，待果肉变软后搓揉淘洗，清除果皮、果肉等杂物，漂去杂质及空瘪种子，将下沉的饱满种子洗净收集晾干。调制过程宜轻揉细淘，以免伤及种皮。若能精心调制，净度可达 95%以上。这 3 个种的出种率和种子的有关数据见表 3。

表 3　花楸属树种的出种率及种子质量

树　种	出种率（%）	种子长度（mm）	种子千粒重（g）	每克种子粒数
黄山花楸	—	3～3.5	2.5～2.9	340～400
花楸树	1.4～2	4～4.5	2～2.4	420～500
天山花楸	1～3	4～4.5	3.7～4.0	250～270

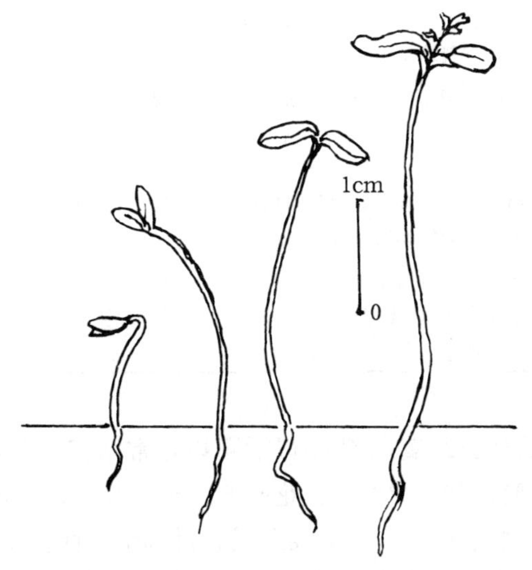

图 2　美洲花楸〔*S. americana*（March.）DC.〕种子萌发后 1、3、7、24 天的幼苗
（引自《美国木本植物种子手册》供参考）
（史渭清临）

花楸树种子含水量降到 9%～10%，在 0～±5℃下用密闭容器贮藏，种子生命力至少可以保持 3 年。天山花楸的种子在干燥通风的室内一般干藏，2～3 年仍有一定的发芽能力。供翌年春播的种子可用室温干藏，但最好是在室外层积。室外低温层积具有催芽作用，春季播种后发芽迅速而且整齐。

发芽和播种　花楸属种子具休眠习性。春播用种需经 60 天以上的 0～5℃或 5～10℃的低温层积催芽处理，当种子有少数裂嘴时即可播种。也可将种子混沙堆积室外，在冬季自然变温条件下处理 2～3 个月。经解除休眠处理的种子场圃发芽率可达 60%～80%，且发芽整齐。也可在种子采收调制后秋播。室内发芽测定的种子事先不经层积，很难发芽，可用四唑或靛蓝测定生活力。

苗圃播种时通常混沙浅播，覆土 1cm，也可用湿沙或锯木屑覆盖。出土萌发（图 2）。幼苗较耐寒，也抗病虫害。2 年生苗可以出圃。

（何泽瑛）

绣线菊属

Spiraea L.

（蔷薇科　Rosaceae）

生长习性、分布和用途　本属 100 余种，我国约 50 种，本文描述 7 种 1 变种。落叶灌木，高 1～3m，多数种类具耐寒、耐旱、适应性强的特点，但在湿润肥沃的土壤上生长更好。分布在北半球温带至亚热带山区。其名称、树高、分布和用途见表 1。

表 1 绣线菊属树种的名称、树高、分布和用途

中 名	学 名	树 高（m）	分 布	用 途	供稿
麻叶绣线菊	*S. cantoniensis* Lour.	1.5	东南、华南。华北南部、华东、豫、陕、川栽培。日本	观赏，入药	1003
中华绣线菊	*S. chinensis* Maxim.	1.5～3	西北南部、华北、东南、中南至西南	观赏，入药	1003
毛花绣线菊	*S. dasyantha* Bge.	2～3	华北、辽、苏、赣、鄂	观赏，入药	403
粉花绣线菊	*S. japonica* L. f.	1～1.5	原产日本、朝鲜半岛。我国各地均有栽培	观赏，入药	305、308
单瓣李叶绣线菊	*S. prunifolia* Sieb. et Zucc. var. *simpliciflora* Nakai	2	华中、华东各地庭园常见栽培	观赏	1003
绣线菊	*S. salicifolia* L.	1.5～2	东北、内蒙古。欧洲东南部及蒙古、朝鲜半岛、日本	观赏，入药	305、308
毛果绣线菊	*S. trichocarpa* Nakai	1.5～2	辽、内蒙古。朝鲜半岛	观赏，蜜源	305、308
三裂绣线菊	*S. trilobata* L.	1.5～2	东北南部、华北、西北东部、皖。西伯利亚	观赏，栲胶	305、308

开花结实 本属树种一般 2～3 年生开始开花结实，4 年生以后可以正常结实。花两性，稀杂性。伞形、复伞形、伞房或圆锥花序。萼筒钟状，萼片 5。花瓣 5。雄蕊 15～60。心皮 5，离生。蓇葖果 5，常沿腹缝线开裂，内具数粒细小种子。种子线形或长椭圆形。种皮膜质，胚乳少或无。花序及花的形态见表 2。果实与种子形态特征见表 3。花、果实的物候期见表 4。

表 2 绣线菊属树种的花序及花的形态

树 种	花 序	具花数	总 梗	花梗长（mm）	花 色	花 径（mm）	雄蕊数
麻叶绣线菊	伞形	多数（16～25 朵）	有	8～14	白色	5～7	20～28
中华绣线菊	伞形	多数（16～25 朵）	有	5～10	白色微绿	3～4	22～25
毛花绣线菊	伞形	多数（10～20 朵）	有	6～10	白色	4～8	20～22
粉花绣线菊	复伞房	多数，密集	有	4～6	粉红	4～7	25～30
单瓣李叶绣线菊	伞形总状	3～6 朵	无	6～10	白色	10～12	20
绣线菊	矩圆形或金字塔形圆锥花序	多数密集	有	4～7	粉红	5～7	50
毛果绣线菊	复伞房	多花	—	5～9	白色	5～7	18～20
三裂绣线菊	伞形	多数（15～30 朵）	有	8～13	白色	6～8	18～20

表 3 绣线菊属树种果实与种子的形态特征

树 种	蓇葖果状态	开裂方式	果 长（mm）	果 壳	种子形状	种子长（mm）	胚 形
麻叶绣线菊	开张，无毛，花柱顶生，渐尖，倾斜开展	沿腹缝线上端 1/3～1/2 开裂，裂口窄而短，种子不易掉落	2.5～3.0	较硬，壳质，褐色	不规则梭形，有棱，褐色	1.5～2.0	长圆形，长 1.5～1.7mm

（续）

树　种	蓇葖果状态	开裂方式	果　长 (mm)	果　壳	种子形状	种子长 (mm)	胚　形
中华绣线菊	开张，被短柔毛，花柱顶生，直立或稍倾斜	沿腹缝线上端 1/3～1/2 开裂，裂口窄而短，种子不易掉落	3.0～3.7	较硬，壳质，褐色	梭形，褐色	1.8～2.0	长圆形，长 1.5～1.7mm
单瓣李叶绣线菊	开张，腹缝线上具短柔毛，花柱顶生于背部	沿腹缝线从上到下充分开裂，裂口大，呈心形，种子易掉落	3.2～4.0	较薄，纸质，棕色	椭圆形，两端膜质种皮延长成翅状，浅褐色或棕色	2.0～2.2	椭圆形，长 1mm

表 4　绣线菊属树种的开花结实物候

树　种	观察地点	开花期			果熟期		种子散落期
		初　期	盛　期	末　期	始　期	盛　期	
麻叶绣线菊	南京	4 月下旬	4 月底	5 月初	—	7 月上中旬	9 月底
中华绣线菊	南京	5 月下旬	6 月上旬	7 月下旬	—	9 月上旬	9 月底
毛花绣线菊	北京香山地区	5 月	～	6 月	—	10 月	10 月底
粉花绣线菊	北京香山地区	6 月初	6 月中上旬	6 月中下旬	9 月上旬	9 月下旬	10 月上旬
单瓣李叶绣线菊	南京	2 月中旬	2 月下旬	4 月上旬	—	5 月底	6 月初
绣线菊	北京香山地区	6 月上旬	6 月中旬	6 月底	8 月下旬	8 月底	9 月中旬
毛果绣线菊	北京香山地区	5 月上旬	5 月中旬	5 月下旬	8 月下旬	9 月中旬	9 月底
三裂绣线菊	北京香山地区	4 月底	5 月初	5 月上旬	6 月中旬	6 月下旬	6 月底

果实的采收调制和种子贮藏　蓇葖果呈黄褐色并有少量开裂时即可采种。剪下整个果穗，于阴凉通风处风干，待蓇葖果沿腹缝线全部开裂，抖动果穗即可脱粒，去掉杂质得到净种。绣线菊种子不能过分干燥，脱粒后即可装入布袋或纸袋，放在阴凉干燥处普通干藏。如长期存放，应密封在干燥器皿内低温（4℃左右）贮藏；隔年播种，发芽率仍能达 70%～80%。出种率、种子净度和种子质量见表 5。

表 5　绣线菊属树种出种率、净度和种子质量

树　种	出种率（%）	净度（%）	千粒重（g）	每克纯净种子粒数（万粒）
麻叶绣线菊	—	—	<0.1	>1
中华绣线菊	—	—	<0.1	>1
毛花绣线菊	5～10	60～70	0.026～0.029	3.4～3.8
粉花绣线菊	4～6	80～90	0.14～0.18	0.5～0.7
单瓣李叶绣线菊	—	—	0.05～0.07	1.4～2
绣线菊	4～5	70～80	0.16～0.20	0.5～0.6
毛果绣线菊	4～6	80～90	0.18～0.21	0.4～0.6
三裂绣线菊	3～5	70～80	0.04～0.06	1.6～2.5

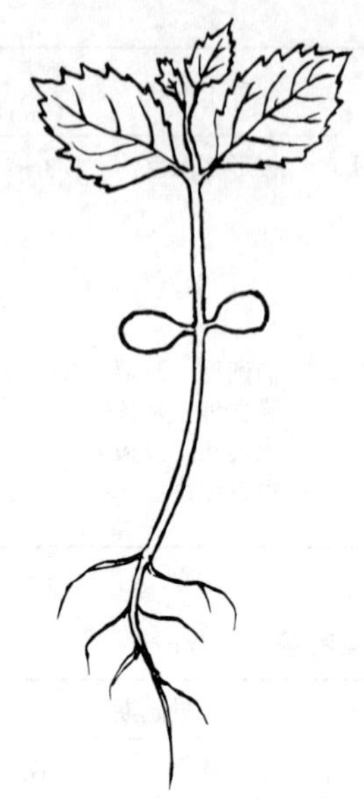

图 1　石蚕叶绣线菊的幼苗
（田恒德仿《中国东北主要树木幼苗图说》）

发芽和播种　本属各树种种子均无休眠习性。曾经在下述条件下测定过发芽能力：种子用 30℃温水浸泡一昼夜，取出放在发芽皿内湿润的滤纸上，在 18～20℃和室内自然散射光条件下，8～10 天开始发芽，约 10 天结束，发芽率可达 80%～95%。

种子在初夏成熟的种，如单瓣李叶绣线菊，可以在采后即播。其它一些种在南方地区秋季播种也能很快发芽，以幼苗状态越冬。北京地区 4 月初开始播种。条播，覆土以不见种子为宜，并覆地膜保湿增温。7～10天即可出土。出土萌发，7 天后可长出初生叶。每平方米播种 5g。图 1 绘出的是石蚕叶绣线菊（*S. chamaedryfolia* L.）的幼苗。

（王木林、高秀岩）

红　果　树

Stranvaesia davidiana Decne.

（蔷薇科　Rosaceae）

生长习性、分布和用途　红果树属约 5 种，我国有 4 种，本文描述 1 种。常绿灌木或小乔木，成年时树高 2～10m。喜温暖湿润气候。适生于土层深厚的山地黄棕壤。产滇、桂、川、黔、赣、陕、甘。越南北部也有分布。叶亮绿。果橘红色或深红色，挂于枝上经久不落，供绿化及观赏。

开花结实　5～7 年生开始开花结实，10 年后进入结实盛期。无大小年现象。花两性。复伞房花序顶生，或枝梢上端有 1～2 个腋生。腋生花序的成果率低。据本文作者 1988～ 1989 年在四川蒙顶山观察，花期在 5～6 月。花多，萼筒钟状，萼片 5。花瓣 5，白色。雄蕊约 20。花柱 5，连合至中部以上；子房半下位，基部与萼筒合生，上部离生，5 室，每室有胚珠 2 枚。每果穗有果 9～60 个，多数为 25～35 个，但每个腋生的果穗有果仅 3～4 个。梨果，未熟时绿色，10 月中旬果实进入生理成熟阶段，11 月中下旬形态成熟，呈橘红色或深红色。心皮成熟时变为革质。梨果长 5.3～6mm，径 7～8mm，每室有发育种子 2 粒，每果含种子 10 粒。成熟种子长卵形，两面微扁平，近似橘瓣，长 3.5～4mm，径 2mm 左右。种皮软骨质，紫褐色至暗紫褐色。空粒种子背腹面均凹入。胚和子叶均呈乳白色，无胚乳。果实宿存期较长，至翌年 2 月尚有 80%以上的梨果挂于枝上。

红果树种子的外形及纵剖面见图 1。

果实的采收调制和种子贮藏　外果皮呈橘红色时即可采集。采回的鲜果用5%的石灰水浸泡2～3天后，在清水中搓揉，漂去果皮等杂质，将种子摊晾阴干，装入布袋或瓦钵内干藏。

据本文作者测定，鲜果出种率8%～12%，净度70%～80%，千粒重1.7～1.9g，每克纯净种子500～600粒。

发芽和播种　可以用四唑染色检验种子生活力。有1份种子，用始温45℃水浸种24小时，基质用纱布，在恒温25℃下6天发芽势为12%～18%；15天发芽率为27%～29%。

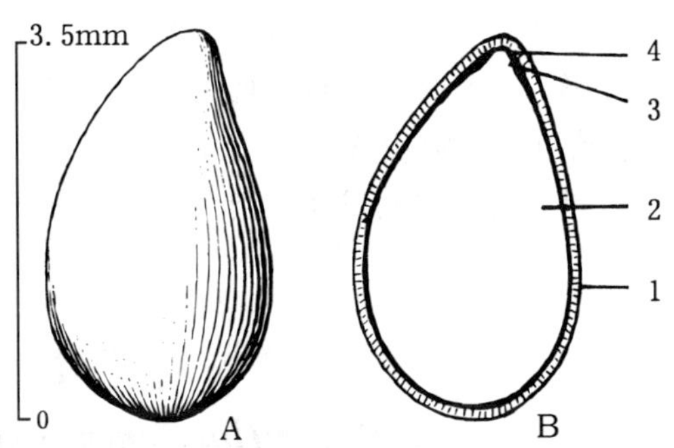

图1　红果树种子的外形（A）及其纵切面（B）
1. 种皮　2. 子叶　3. 胚轴　4. 胚根
（黄应钦绘）

春播。播期2～3月。条播，条距15～20cm，条幅3～5cm，播种沟深1cm。播前用始温45℃水浸种8小时，捞出拌细沙或草木灰混播。每平方米播种量2.5～4g。播后覆细土，以不见种子为度，轻压后盖草。播后约15天开始发芽出土，再约经3周发芽基本结束。幼苗基本形成时可以间苗。

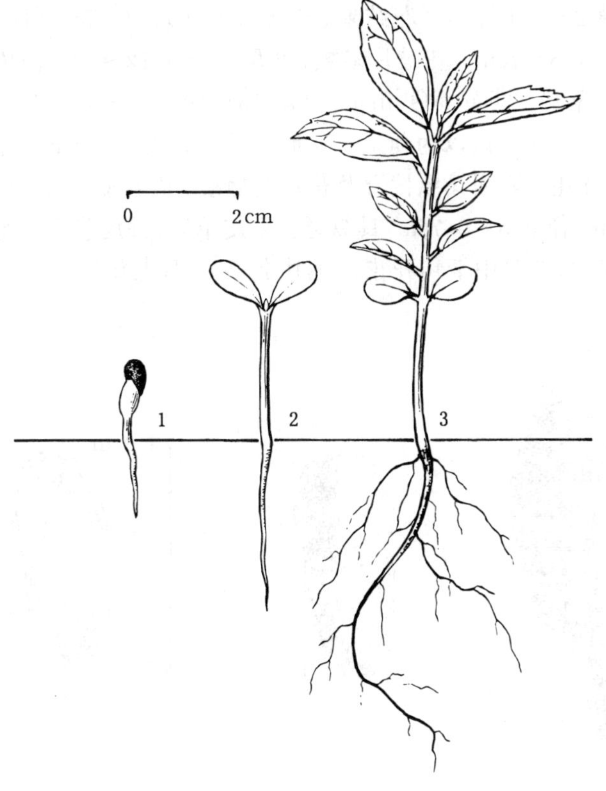

图2　红果树种子萌发及幼苗生长情况
1. 子叶初现　2. 子叶展开　3. 幼苗形成
（叶可佳、田恒德绘）

出土萌发。子叶 2，椭圆形。初生叶互生，卵形，随着幼苗的生长所形成的叶片逐渐增大，呈长卵形。叶片多为全缘，稀具 1～4 锯齿。叶背多呈淡紫绿色。茎紫红色，密被白色柔毛。2 年生苗可以出圃栽植。

红果树种子的萌发及幼苗生长情况见图 2。

（阙再忠）

夏 蜡 梅

Calycanthus chinensis Cheng et S. Y. Chang

（蜡梅科 Calycanthaceae）

生长习性、分布和用途 夏蜡梅属约 4 种，我国仅产本文描述的 1 种。落叶灌木，高 1～3m，适生于温暖湿润的沟谷林下，耐荫。分布区域窄，首先发现于浙江临安、天台一带的山林中，近年各地陆续引种，被《中国植物红皮书》列为渐危种。在杭州、庐山、南京的植物园中生长健壮，开花结实正常，适应性较强。可用于庭园绿化，供观赏。

开花结实 定植 2～3 年开始开花结果，未见大小年现象。花两性，单生于当年生小枝顶端。花无香气。花径 4.5～7cm。花被片多数，2 型，外片 12～14，白色，淡紫色边晕；内片 9～12，较厚，红褐、紫褐、淡黄等色相间。雄蕊 16～19。心皮 11～12，离生，每个心皮含胚珠 2。花托发育成果托，通常误以为蒴果，梨形或钟形，长 3～5cm，近顶端微收缩，成熟时果托先端撕裂，外被柔毛，表面具纵棱脊及苞片的疤痕，内含瘦果 3～11 枚。瘦果暗褐色，矩圆形，长 1.2～1.6cm，径 0.5～0.7cm，具纵脊。果皮革质，种皮膜质。无胚乳。近倒三角形的肥大子叶席卷于种皮内。5 月中下旬开花，9 月中旬～10 月上旬果熟。果实和种子形态见图 1。

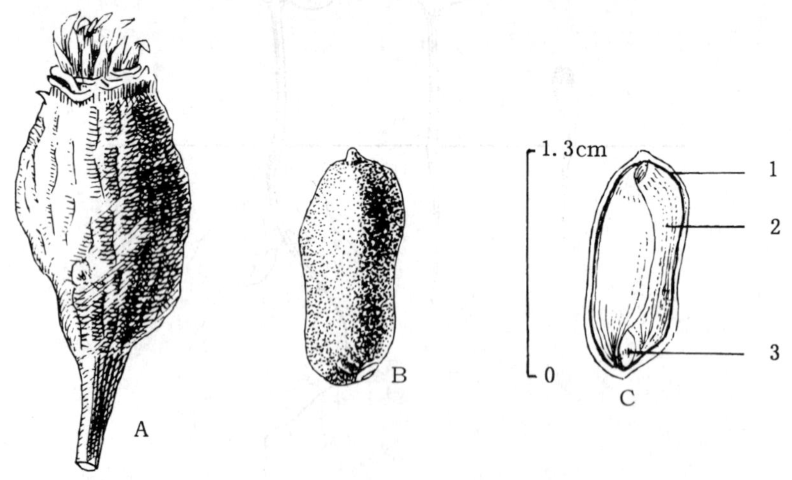

图 1 夏蜡梅坛状果托的外形（A）瘦果的外形（B）及剥除部分果皮和种皮，示胚（C）

1. 果皮及种皮 2. 子叶 3. 胚根

（陈荣道绘）

果实的采收调制和种子贮藏　9～10 月间采摘果实状的果托，置室内阴干。一般将果托干藏到播种前脱粒，得到的瘦果即为播种材料，通称种子。净度可达 98%以上。种子千粒重 230～260g，每千克含种子 3 800～4 400 粒。这个种目前还处在引种试验阶段，植株数量不多，还没有大量采种的条件。

发芽和播种　种子无休眠习性，适于春播。据浙江临安市林业科学研究所的经验，3 月初播种，4 月中旬开始出土，5 月中旬齐苗。当年苗高可达 17cm，最高达 50m。换床培育 1 年后苗高 1m 以上。出土萌发（图 2）。植株有很强的萌蘖能力。

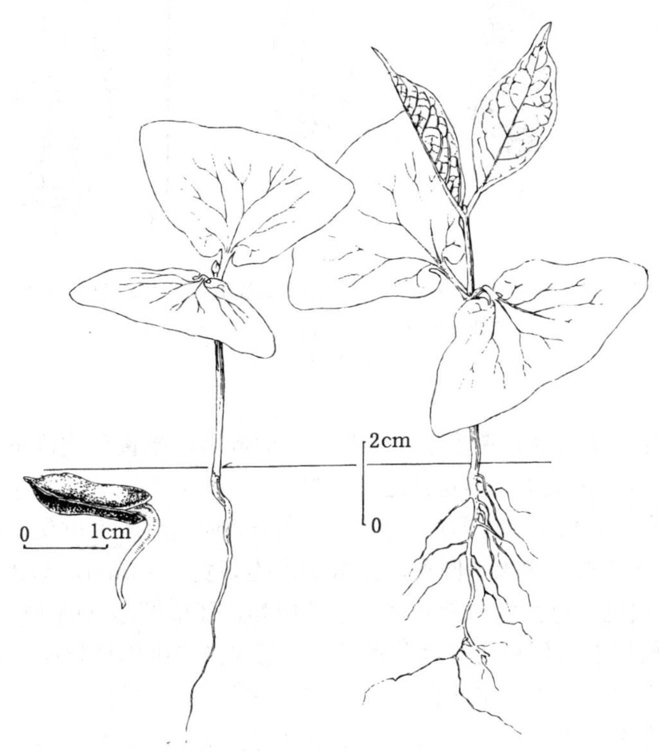

图 2　夏蜡梅种子萌发后第 2、10、50 天的幼苗生长情况

（陈荣道绘）

（何泽瑛）

蜡　　梅

Chimonanthus praecox（L.）Link

（蜡梅科　Calycanthaceae）

生长习性、分布和用途　本属共 6 种，我国特产。本文描述其中价值最高、栽培最广的 1 种。落叶灌木，高可达 4m。原产我国中部地区，秦岭南坡及鄂西山区有野生分布，目前自

北京以南已广泛栽培，寿命可达百年。经长期选择培育，品种繁多，是我国冬季名花。耐寒、耐旱、喜光，适生于排水良好、肥沃湿润的沙质壤土，忌水湿，积水1周以上即可致死。花供观赏，并可提取芳香油。根、茎、花均可入药。

开花结实 3～5年生始花，6～8年生开始结实。有大小年但未发现明显规律，有的年份

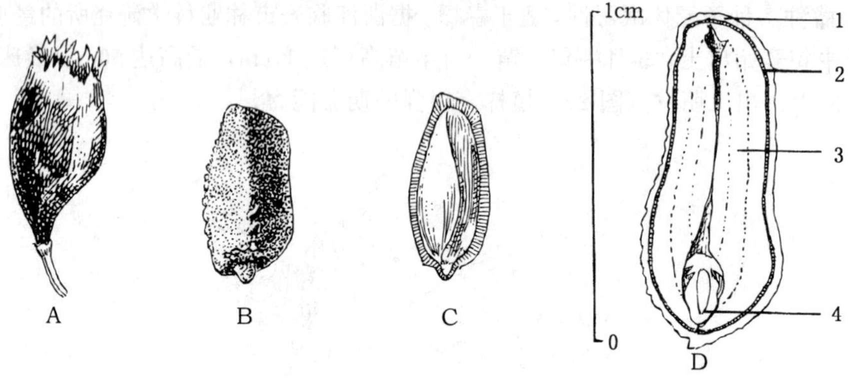

图1 蜡梅坛状果托的外形（A）、瘦果外形（B）、剥除部分果皮种皮示胚（C）及其纵切面（D）
1. 果皮 2. 种皮 3. 席卷的子叶 4. 胚根
（陈荣道绘）

某些植株甚至不开花。花两性，单生于1年生枝条的叶腋，蜡黄色或浅黄色，芳香，径2～2.5cm。花被片14～20，着生于壶形花托之外。雄蕊5～7。心皮7～14，分离。发育长大的坛状果托蒴果状，通常称为果实，长卵形，长1.5～4cm，成熟时半木质化，先端撕裂，内含瘦果1～11枚，多为2～7枚。瘦果矩圆形，长1cm左右，宽3～4mm，厚2～3mm。果皮硬革质，深褐色或浅灰褐色，有光泽。无胚乳，子叶旋卷，胚根短小（图1）。

开花结实的物候期因品种和栽植地点而不同。根据南京中山植物园1981～1982年对41株蜡梅的观察，大多数品种的初花期在12月中旬～翌年1月中旬；1月下旬为盛花期，3月中旬终花。单朵花的开放时间为15～25天。枝条上的各花蕾开放时间不一致，因而同一花枝的着花时间较长，观赏价值随之提高。开花后期有落花现象。6月下旬果熟。

果实的采收调制和种子贮藏 蒴果状果托由绿色转黄绿色，瘦果外皮由白变褐表示成熟。果实成熟后短期不落，6～8月均可采摘。室内阴干后搓揉取出瘦果或剥取瘦果即为播种材料，通称种子，漂除瘪粒后即可干藏待播。净度可达98%。普通干藏的种子寿命可以保持1年。鲜果的出籽率为25%～45%，干果为65%。种子饱满率约90%。千粒重180～260g，每千克含种子3 800～5 600粒。

发芽和播种 种子无休眠习性，遇有适宜条件随时可以萌发，7～10天内发芽率可达85%～90%。过分干燥会导致种子出现强迫休眠，种皮变硬，吸水困难，使发芽时间延长，也降低发芽率。因此蜡梅种子以随采随播为好。幼苗健壮耐寒，在南京可以露地越冬。母树下常有天然下种苗。也可在3月底春播。如果是干藏的种子，春播前宜用温水浸泡12小时。播后40天左右出苗。子叶大，出土萌发。出苗过程见图2。

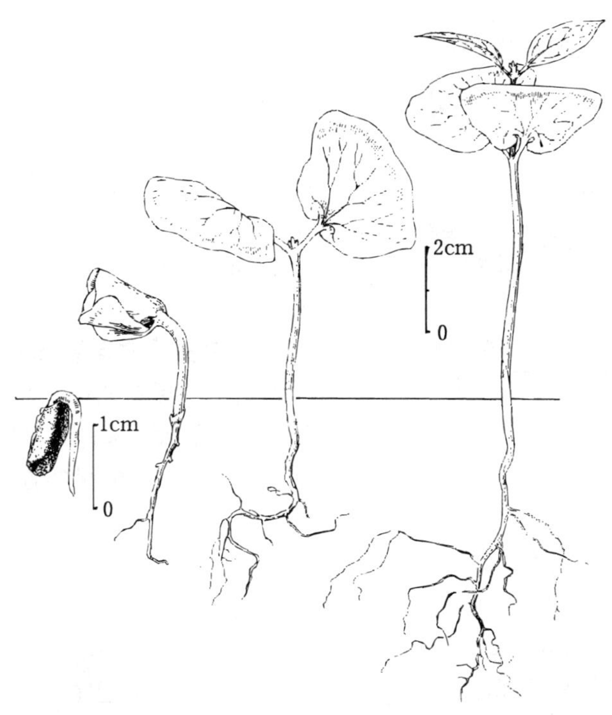

图 2　蜡梅种子萌发后第 1、3、8、30 天的幼苗生长情况
（陈荣道绘）

（何泽瑛）

顶果木（梣叶豆）

Acrocarpus fraxinifolius Wight ex Arn.

（苏木科　Caesalpiniaceae）

生长习性、分布和用途　顶果木属有 2 种，产热带亚洲，我国只有本文描述的 1 种。落叶大乔木，高达 50m，胸径达 1.2m。具板根。喜高温高湿的静风山谷地。对土壤水肥条件的要求较高，肥力中等以下的土壤不能长成大材。石灰岩山地及酸性土壤均能适应。产滇西南和滇东南、黔西、桂西。老挝、泰国、缅甸、印度、斯里兰卡和印度尼西亚亦产。木材宜作纤维原料和板材。花艳丽，供观赏。树干通直，材质好，常遭砍伐，破坏严重，在《中国植物红皮书》中列为稀有种。

开花结实　约 10 年生开始开花结实，正常结实年龄在 20 年生以后。结实大小年间隔期为 1 年。花两性。总状花序，腋生。萼片 5，近相等。花瓣 5，绯红色。雄蕊 5，花丝长，远伸出花冠之外，花药背着。子房上位，1 室，具长柄，胚珠多数，花柱短，内弯，柱头小，顶生。据广西宁明观测，3 月上旬花芽与叶芽同时抽出，3 月下旬为始花期，4 月上旬为盛花期，中下旬为末花期；6 月中旬果实开始成熟，6 月下旬～7 月中旬为果熟盛期，有的地方果熟期

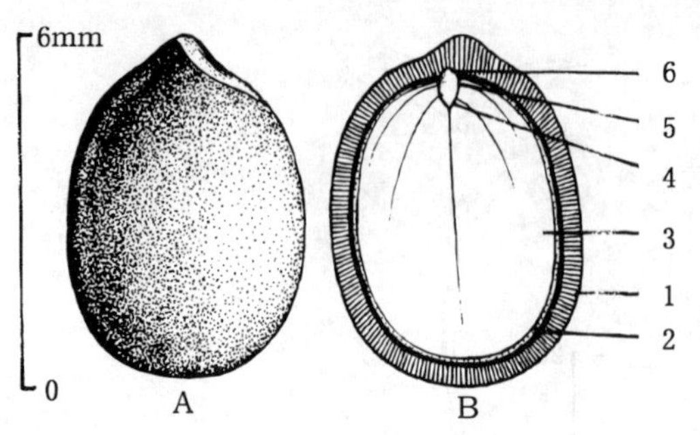

图 1 顶果木种子外形（A）及其纵切面（B）
1. 种皮 2. 胚乳 3. 子叶 4. 胚芽 5. 胚轴 6. 胚根
（黄应钦绘）

可延迟至 8 月下旬。荚果，带状，长 10～20cm，宽 1.9～2.3 cm，腹缝具窄翅，子房柄长约 1.5cm。萼筒碗状，木质宿存。果黑褐色，不开裂，可在树上宿存半年左右。每果有种子 8～20粒。种子扁倒卵形至扁椭圆形，浅褐色至黑褐色，有光泽，长 5.0～7.2mm，宽 4.2～5.2mm，厚 1.5～ 1.8mm。子叶大而扁平，有胚乳（图 1）。

果实的采收调制和种子贮藏 果荚成熟后在树上的保存期长，从 7 月至次年 2 月均可采收。果穗大，采种时可用钩刀带果穗采下后摘取果荚。采回的果实置日光下曝晒，充分干燥后用手搓揉或用木棒敲打，经过筛选后即得种子。荚果出种率为25%～35%，种子的净度可达 99%。种子的含水量在 10%以下。千粒重 35g，变动范围为 26～36g，每千克有纯净种子 28 500 粒，变动范围为 2.7 万～3.8 万粒。种子可用袋或坛罐贮藏，置室内干燥处。常温条件下，这样贮藏 3 年以后发芽力仍可保持 80%左右。

发芽和播种 种子无生理性休眠。发芽时要求日均温在 20℃以上。种皮外层有胶质，不易吸水，播前用硫酸拌种 10 分钟后清水洗净，或用沸水烫种至自然冷却，使胶质膨胀，洗净再浸泡 1 昼夜，即可播种。经过预处理的种子发芽迅速整齐。1981 年 5 月下旬，日均温 27℃时，广西林业科学研究所在南宁室内对顶果木种子进行过发芽测定。种子用酸蚀法处理后播于发芽器内，光照为室内自然光，播后 1 天即开始萌动，2 天便进入发芽盛期，3 天发芽终止。播种后 2 天，发芽百分数已达 82%，3 天时发芽率为 85%。出土萌发。发芽后第 3 天子叶出土，6～7 天发出初生叶（图 2）。

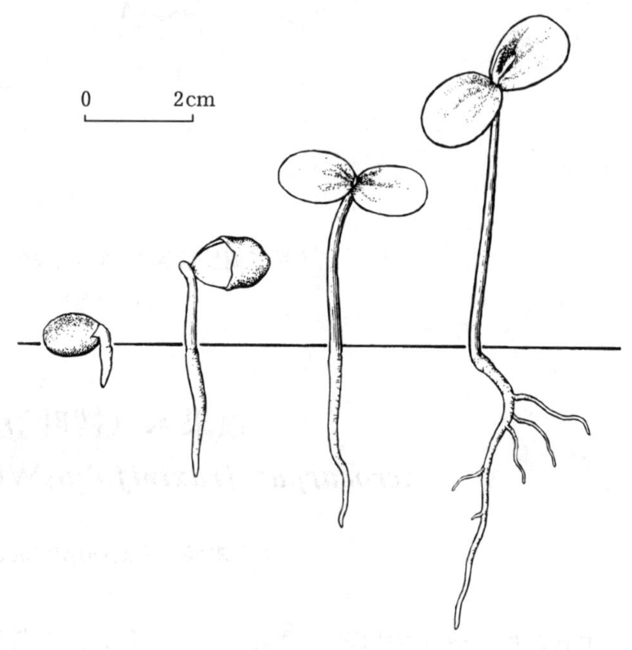

图 2 顶果木种子萌发后第 2、3、5、7 天幼苗的生长情况
（黄应钦仿《热带亚热带主要树种采种育苗技术》）

条播。每平方米播种 2.5～4g，覆土 1.5～2cm。半年生～1 年生苗出圃。也可将种子催芽后点播在容器内，培育 3 个月左右出圃。

（韦增健）

羊蹄甲属

Bauhinia L.

（苏木科　Caesalpiniaceae）

生长习性、分布和用途　本属约570种，我国约40种，本文描述5种。分布于热带至南亚热带地区，少数种可分布至中亚热带。较耐水湿，喜肥沃湿润土壤，石灰岩山土或酸性土均适宜生长。这5个种的名称、生长习性、分布和用途见表1。

表1　羊蹄甲属树种生长情况、分布和用途

中　名	学　名	习　性	分　布	用　途
白花羊蹄甲	*B. acuminata* L.	小乔木，高2～4m	粤、琼、桂、滇、台。南亚、东南亚	观赏、纤维
金叶羊蹄甲（火索藤）	*B. aurea* Lévl.	具卷须藤本，长20～30m	滇、川、黔、桂	遮荫、纤维、观赏、药用
羊蹄藤	*B. kerrii* Gagnep.	具卷须藤本，长20m	粤、桂、琼。越南、泰国	遮荫、纤维、观赏
黄花羊蹄甲	*B. tomentosa* L.	灌木，高约3m	产印度。粤、桂有栽培	观赏、纤维、药用
洋紫荆（羊蹄甲）	*B. variegata* L.	乔木，高10～15m	粤、桂、滇、台、琼。东南亚及南亚	观赏、纤维、材用、鞣料、饲料、根皮药用

开花结实　开花结实期较早，白花羊蹄甲1～2年生即可开花结实，3年生以后便进入正常结实期。洋紫荆和黄花羊蹄甲2～3年生开始开花结实，正常结实期在5年生以后。金叶羊蹄甲和羊蹄藤8年生开始开花结实，正常结实期在12年生以后。白花羊蹄甲、洋紫荆和黄花羊蹄甲结实无大小年现象，正常天气条件下每年均能正常结实。金叶羊蹄甲和羊蹄藤结实大

表2　羊蹄甲属树种的开花结实物候期①

树　种	观察地点	观察年份	开花期			果实成熟期		果实开裂和脱落
			始　期	盛　期	末　期	始　期	盛　期	
白花羊蹄甲	南宁	1986～1988	5月上旬	5月中旬	6月下旬	8月下旬	9月上旬	10月脱落
金叶羊蹄甲	南宁	1987～1988	6月上旬	6月中旬	7月中旬	10月中旬	10月下旬	11月中旬～12月下旬开裂
羊蹄藤	南宁	1987～1988	4月中旬	5月上旬	6月上旬	8月中旬	8月下旬	9月开裂
黄花羊蹄甲	云南	1987～1988	8月上旬	8月中旬	9月上旬	11月上旬	11月下旬	12月开裂
洋紫荆	南宁	1986～1987	2月中旬	2月下旬	3月上旬	4月中旬	4月下旬	5月开裂

①　表中所列为主花期。除主花期外，白花羊蹄甲还会在7月下旬～8月下旬和3月2次开花，分别在10月中旬～10月下旬和4月果熟。其它4个树种也有多季开花现象

小年间隔期为1年。白花羊蹄甲为总状花序伞房状腋生。金叶羊蹄甲为伞房花序顶生。羊蹄藤和洋紫荆为总状花序顶生，少数着生于枝梢的叶腋。黄花羊蹄甲花单生或2～3簇生于叶腋。花两性。萼佛焰苞状或开花时分裂为5萼片。花瓣5，近等长。能育雄蕊10、5或3，花丝长短不等，分离，花药背着。子房具柄，胚珠2至多数。花期发育快，从花芽出现到开花一般只需10～15天的时间，由下至上次第开放。下部的果荚已经形成，顶部还在陆续出现花蕾。花期长，不甚整齐。这5个树种开花结实的物候期见表2。

荚果，木质或革质，成熟后在树上可保持2个月不脱落。洋紫荆的果荚成熟后约1个月左右开裂，其余4个树种不开裂。每个果荚有种子数粒至数十粒。种子具胚乳。洋紫荆种子的外形及其解剖结构见图1。这5个树种果实和种子的形态特征见表3。

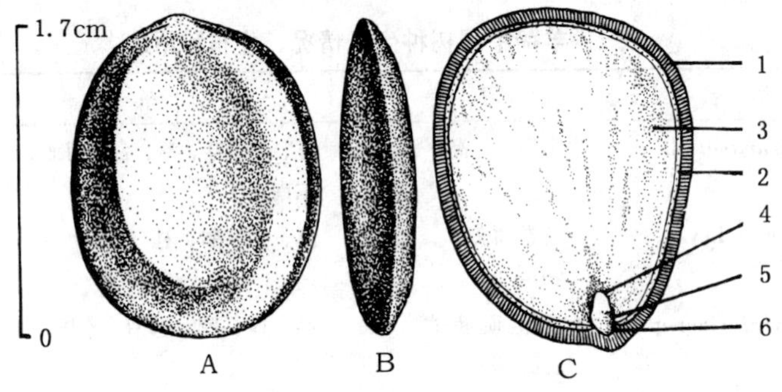

图1　洋紫荆种子外形正面（A）、侧面（B）及其纵切面（C）

1. 种皮　2. 胚乳　3. 子叶　4. 胚芽　5. 胚轴　6. 胚根

（黄应钦绘）

表3　羊蹄甲属树种果实和种子的形态特征

树种	未熟果颜色	果实			种子		
		形状	大小（cm）	颜色	形状	大小（mm）	颜色
白花羊蹄甲	绿色	带状，微镰形	长6～8 宽1.2～1.5	黄褐色	扁椭圆形	长7～8.5 宽5～6.5 厚2～3	黄褐色
金叶羊蹄甲	黄绿色	倒披针形或带形	长10～25 宽3～6	灰黑褐色，被锈色绒毛	扁椭圆形	长18～26 宽15～20 厚3.5～5	黄色或黄褐色
羊蹄藤	黄绿色	倒披针状长圆形	长15～25 宽3.5～6	黄褐色，密被红棕色锈毛	扁椭圆形	长8～10 宽4.5～8.5 厚1.5～2	褐黄色或黑褐色
黄花羊蹄甲	青绿色	扁平，近刀形	长10～14 宽1.3～1.8	黑褐色	扁椭圆形	长7～9.5 宽5.5～7.5 厚2～2.5	黄色或黄褐色
洋紫荆	青绿色	近刀形，扁平，上部渐阔	长13～25 宽1.7～2.1	黑褐色	扁平，近圆形或椭圆形	长12～17 宽10～12.5 厚2～2.5	棕褐色或暗褐色

果实的采收调制和种子贮藏　白花羊蹄甲的果荚可用手就地采摘，其余 4 个树种，一般是在全果穗的果荚已大部分成熟时，用采种刀或高枝剪剪下果穗，除去果穗上部种子发育不饱满的果荚，其余果荚置日光下曝晒，大部分果荚便会失水开裂。少数仍不开裂的果荚可用木棒敲打，使果荚破裂，种子脱出，筛选除去杂物，即得种子。白花羊蹄甲、洋紫荆和黄花羊蹄甲的种子可以在日光下晒干，待含水量降至 10%左右，用袋或坛贮藏在常温条件下，它们的发芽能力可以保持 2 年左右。金叶羊蹄甲和羊蹄藤的种子则不宜曝晒，应在室内通风处摊晾，使含水量降至 18%左右后混干沙贮藏在常温条件下，它们的发芽能力可以保持 1 年左右。这 5 个树种的种子净度和质量见表 4。

表 4　羊蹄甲属树种果实出种率和种子净度、质量

树　种	出种率（%）	净度（%）	千粒重（g）	每千克纯净种子粒数
白花羊蹄甲	15～25	96～100	85～90	11 100～11 800
金叶羊蹄甲	25～30	96～100	800～1 000	1 000～1 250
羊蹄藤	40～50	98～100	65～70	14 300～15 400
黄花羊蹄甲	6～10	95～99	55～80	12 500～18 200
洋紫荆	15～25	98～100	200～260	3 850～5 000

发芽和播种　种子无休眠习性。当种子吸足水分，日均温在 20℃以上时，很快便能发芽。华南地区 4～9 月的气温高，种子在这段时间成熟的宜随采随播。10～11 月成熟的则可以贮藏至次年春播。随采随播的种子，播前用冷水或始温 50℃水浸种。经过贮藏的种子，播前需用 80～90℃水浸泡。1979 年 5 月和 1987 年 9 月，广西林业科学研究所在室内对这 5 个树种进行过发芽测定。播种时日均温 25～28℃，光照为室内自然光，播前种子做过浸泡处理。测定结果见表 5。出土萌发。播种后 3～5 天胚根萌动，6～7 天子叶出土，10～11 天发出初生叶。本属中有的种类是留土萌发，如龙须藤 *B. championii* Benth.。洋紫荆种子的萌发和幼苗生长情况见图 2。

表 5　羊蹄甲属树种的发芽能力及其测定条件

树　种	室内温度（℃）	发芽势（%）		发芽率（%）	
		计算天数	一般数值	计算天数	一般数值
白花羊蹄甲	25～26	4	43	12	60
金叶羊蹄甲	27～28	4	58	11	70
羊蹄藤	27～28	4	58	11	72
黄花羊蹄甲	25～26	4	41	12	61
洋紫荆	27～28	4	42	11	72

条播。视需要而在播前对种子进行预处理。每平方米的播种量，金叶羊蹄甲为 75～120g，洋紫荆 25～38g，其余 3 个树种 12～15g。覆土 1.5～2cm。一般 1 年生苗出圃。金叶羊蹄甲和羊蹄藤用容器育苗，3 个月出圃。洋紫荆用于庭院或街道绿化，需培育 2 年生苗出圃。

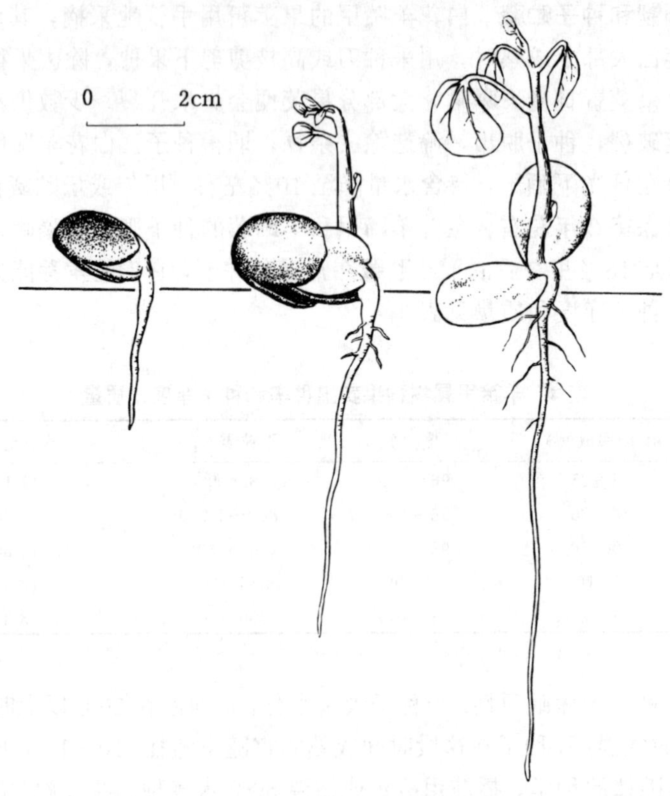

图 2 洋紫荆种子萌发后第 5、7、15 天幼苗的生长情况

(黄应钦仿《热带亚热带主要树种采种育苗技术》) (韦增健)

苏 木 属

Caesalpinia L.

(苏木科 Caesalpiniaceae)

生长习性、分布和用途 本属约 100 种，我国约 20 种，引入 5 种，本文描述 4 种。树干有刺。多数种分布于南亚热带至热带地区，少数种可分布至中亚热带地区。对水肥条件的要求不甚苛刻，肥力中等的石灰岩山地或酸性土均能生长。这 4 个树种的名称、生长习性、分布及用途见表 1。

表 1 苏木属树种的名称、生长、分布和用途

中 名	学 名	习 性	分 布	用 途	供 稿
云实	*C. decapetala* (Roth) Alston	落叶攀援灌木	秦岭淮河以南。亚洲热带和温带	刺篱，观赏，鞣料，油脂，药用	1003

（续）

中　名	学　名	习　性	分　布	用　途	供　稿
喙荚云实（南蛇簕）	*C. minax* Hance	有刺藤本，长达4～5m	桂、粤、琼、滇、黔、川南。越南	刺篱，根、叶可清热解毒，种子补脾去热毒	603
金凤花（洋金凤）	*C. pulcherrima* (L.)Sw.	落叶灌木或小乔木，高2～3m，少数达5m	西印度群岛。我国滇、桂、粤、琼、台引种	观赏，全株可作强性通经药，根退热	603
苏木	*C. sappan* L.	常绿小乔木，高10m，胸径12cm	粤、琼、桂、滇、黔、台。中南半岛，印度	刺篱，材制小提琴并提取珍贵红色染料，药用	603

开花结实　金凤花1～2年生即能开花结实，4年以后进入正常结实期。喙荚云实、云实和苏木3～4年生开始开花结实，6年生以后进入正常结实期。水肥条件较好的地方无大小年现象，连年结实繁多。花两性。喙荚云实和苏木为圆锥花序顶生，云实为总状花序顶生，金凤花为伞房状花序，顶生或腋生。萼片5，下方一片较大。花瓣5，最上方一片较小，稍带红色。其余的花瓣在喙荚云实为白色，云实和苏木为黄色，金凤花为黄色或橙色。雄蕊10，2轮，分离。花丝基部加粗，被毛，花药背着。子房1室，无柄或具短柄，有胚珠2至多颗。花柱圆柱形，柱头截平或凹入。广西南宁及江苏南京观察的开花结实物候期见表2。

表2　苏木属树种的开花结实物候期

树　种	观察地点	观察年份	开花期			果实成熟期		果实开裂期
			始　期	盛　期	末　期	始　期	盛　期	
云实	南京	—	4月下旬	～	5月上旬	7月下旬	8月中旬	8月中旬～下旬
喙荚云实	南宁	1986～1988	3月中旬	4月上旬	4月下旬	8月中旬	9月上旬	9月中旬～10月下旬
金凤花	南宁	1986～1988	3月中旬	5～9月	11月上旬	6月中旬	—	成熟1个月后开裂
苏木	南宁	1978～1984	5月上旬	5月下旬	7月上旬	9月中旬	9月下旬	不开裂，不脱落

荚果。每个荚果内云实和金凤花有种子5～12粒，其余2种为3～7粒。种子无胚乳。子叶肥厚。果实及种子的形态见表3、图1。

表3　苏木属树种果实和种子的形态特征

树　种	未熟果颜色	果　实			种　子		
		形　状	大小（cm）	颜　色	形　状	大小（mm）	色　泽
云实	青绿	椭圆形，稍扁	长6～12 宽2.3～3	棕褐色	椭圆形	长13 宽5.5	黑色
喙荚云实	青绿	扁椭圆形，密被针刺	长8～13 宽4～6	棕色	长椭圆形	长16～21 径8.5～11	黑色

（续）

树 种	未熟果颜色	果实			种子		
		形 状	大小（cm）	颜 色	形 状	大小（mm）	色 泽
金凤花	绿	带状，狭而薄，有长喙	长 5～9 宽 1.5～1.8	黑褐色	扁矩圆形或三角状圆形	长 7.5～9.5 宽 5.5～7.5 厚 2.5～4	棕褐色或暗褐色
苏木	黄绿	近棱形，压扁，基部稍狭，顶喙尖	长 6～8.5 宽 3～4	红棕色	扁椭圆形	长 17～20 宽 7～12.5 厚 4.5～8	黄褐至暗褐色，有光泽

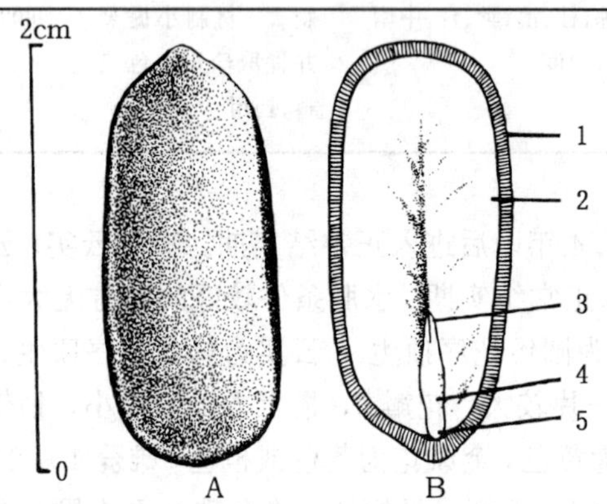

图 1 苏木种子外形（A）及其纵切面（B）
1. 种皮 2. 子叶 3. 胚芽 4. 胚轴 5. 胚根
（黄应钦绘）

果实的采收调制和种子贮藏 云实等植株带刺，宜用高枝剪或钩刀采下果穗。苏木每年果熟 2 次，以 9 月成熟的果实较佳。金凤花从 6～12 月可随熟随采。采得的果荚置日光下曝晒，用木棒敲打，开裂后除去果壳杂物，即得纯净种子。金凤花和苏木的种子可以干藏，但也忌烈日久晒，以免过分失水降低发芽能力，种子含水量宜在 12%～16%。喙荚云实和云实的种子可以在烈日下晒干，种子含水量可以降到 10%左右。种子的净度和质量等数据见表 4。

表 4 苏木属树种果实的出种率和种子净度、质量

树 种	出种率（%）	净度（%）	千粒重（g）	每千克纯净种子粒数
云实	—	98	200～280	3 600～5 000
喙荚云实	50～60	99	1 150～1 350	740～870
金凤花	25～35	98	125～140	7 140～8 000
苏木	25～35	98	600～850	1 170～1 700

苏木的种子易受虫蛀，处理出的种子宜用始温约 50℃水浸泡 2～3 昼夜，使幼虫闷死，捞出后稍晾干即可播种或混沙贮藏。沙藏的种子，发芽力可保持 6～8 个月，贮藏期过长，发芽能力将逐渐下降。在室内常温条件下贮藏，金凤花种子的发芽力约可保持 1 年。喙荚云实的种子耐久藏，晒干后贮藏于室内，在常温条件下发芽能力可以保持 2～4 年。

发芽和播种 金凤花、喙荚云实和苏木的种子无生理性休眠。据南京中山植物园研究，云实种子有的无休眠，有的有休眠。休眠的主要原因是种皮不透水以及种脐吸水机能受阻。据广西林业科学研究所对云实所作的发芽试验，用始温 80℃水浸种，24 小时后播种，发芽很快，未见休眠现象。金凤花和苏木的种子播前用冷水或始温 45℃水浸种半天，喙荚云实需用始温 80℃水或沸水浸种，并在冷却过程中继续浸泡 1 昼夜。这 4 个树种发芽时日均温需在 20℃左右。1982 和 1987 年，广西林业科学研究所对 3 个树种进行了发芽测定：金凤花 9 月 16 日播种，19 日开始发芽，22 日发芽终止。喙荚云实 9 月 2 日播种，8 日开始发芽，14 日发芽终止。

苏木 5 月 4 日播种，10 日开始发芽，18 日发芽终止。金凤花出土萌发。云实、喙荚云实和苏木留土萌发。胚根萌发后约 3 天子叶或上胚轴出土，第 5～8 天初生叶展出，发芽情况见表 5、图 2。

表 5　苏木属树种的发芽能力及其测定条件

树　种	预处理	发芽测定			发芽势（%）		发芽率（%）	
		基　质	温度（℃）		计算	一般	计算	一般
			室　内	室　外	天数	数值	天数	数值
喙荚云实	始温 80℃水浸种	沙	—	29	3	72	12	95
金凤花	始温 45℃水浸种	纸	28	—	2	40	6	55
苏木	冷水浸种	沙	—	27	4	34	14	57

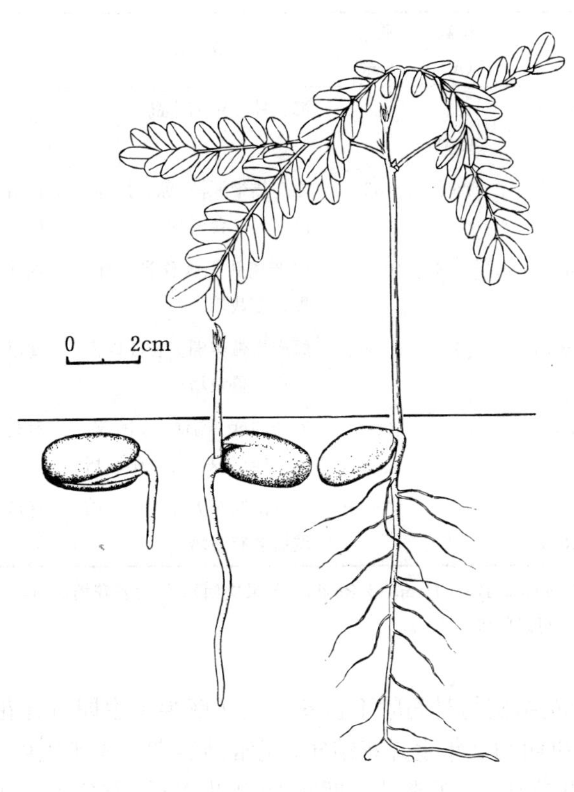

图 2　苏木种子萌发后第 2、4、15 天幼苗的生长情况

（黄应钦仿《热带亚热带主要树种采种育苗技术》）

条播。每平方米播种量分别为：喙荚云实和苏木 100～120g，金凤花 15～20g。覆土1.5～2.5cm。一般培育半年或 1 年生苗出圃。

（张声燕）

铁刀木（决明）属
Cassia L.

（苏木科 Caesalpiniaceae）

生长习性、分布和用途 本属约 600 种，我国约 25 种（其中引入的约有 10 种），本文描述 6 种。适生于高温高湿的气候环境，多年生木本的适生区年均温为 21℃。少数亚灌木可引至纬度较高的地区作为 1 年生植物栽培。适生于肥力中等以上的石灰岩山地或酸性土。有些种花艳丽，花期长，供观赏。有些种的干叶入药作缓泻剂。它们的名称、生长、分布及用途见表 1。

表 1 铁刀木属树种的名称、生长、分布和用途

中 名	学 名	树高 (m)	胸径 (cm)	分 布	用 途	供 稿
神黄豆（粉花山扁豆）①	*C. agnes* (de Wit) Brenan	15 (30)	25	滇、桂。中南半岛	材用、观赏、种子入药	602
腊肠树	*C. fistula* L.	20	35	原产热带亚洲。琼、粤、桂、滇、闽、台引种	材用、烟草香剂、食用、观赏	602
粉叶决明	*C. glauca* Lam.	5～8	10～15	原产南亚至大洋洲。桂、粤、琼栽培	观赏、材用	602
望江南	*C. occidentalis* L.	2	—	原产热带美洲。长江以南及京、鲁栽培	食用、健胃、治蛇伤	602
铁刀木	*C. siamea* Lam.	20	40～50	原产印度、缅甸、泰国等。滇、桂、琼、粤、闽栽培	薪柴、材用、栲胶、绿肥	617
黄槐	*C. suffruticosa* Koen. ex Roth	6～10	10～20	原产南亚至大洋洲。华南城镇多有栽培	材用、观赏、叶为缓泻剂	602

① 本种学名曾经误为 *C. nodosa* Buch-Ham. ex Roxb.（节果决明），但后者花瓣深黄色而非粉红色，且荚果较短（长 30～45cm），产于夏威夷群岛，可以区别

开花结实 开始开花结实的时间颇不整齐。望江南半年生即可开花结实，正常结实期为 1～3年生；黄槐和粉叶决明约 2 年生开花结实，正常结实期在 4 年生以后；腊肠树、神黄豆和铁刀木 6～7 年才开始开花结实，正常结实期在 12 年生以后。结实无大小年现象。影响结实的主要因素是土壤肥力，种植在肥沃地或采用人工施肥，可以连年花繁果茂。霜冻和低温也影响开花结实。花两性。腋生总状花序或呈伞房状，或为顶生圆锥花序。花多为黄色，但神黄豆为粉红色。萼 5 深裂。花瓣 5。雄蕊 5～10，常不相等，或有退化雄蕊。花药背着或基着，顶孔开裂。子房有柄或无柄，胚珠多数，花柱内弯，柱头小。广西和云南观察的开花结实物候见表 2。

表 2 铁刀木属树种的开花结实物候期①

树 种	观察地点和年份	开花期			果实成熟期		果实开裂期
		始 期	盛 期	末 期	始 期	盛 期	
神黄豆	南宁 1978～1984	5 月上旬	5 月中旬	7 月上旬	翌年 3 月下旬	4 月中旬	4 月下旬～5 月下旬落果
腊肠树	南宁 1978～1988	5 月中旬	5 月下旬	7 月下旬	12 月中旬	翌年 1 月上旬	1 月中旬～2 月下旬落果
粉叶决明	南宁 1987～1988	4 月下旬	5 月上旬	5 月中旬	5 月中旬	5 月下旬	5 月下旬～6 月中旬开裂
望江南	南宁 1987～1988	4 月下旬	5 月中旬	9 月上旬	6 月下旬	10 月下旬	8 月上旬～翌年 2 月开裂
铁刀木	景洪 —	7 月	8～9 月	12 月	翌年 1 月	3 月	2～3 月开裂
黄槐	南宁 1987～1988	5 月上旬	5 月中旬	6 月中旬	7 月上旬	8 月中下旬	9 月上旬～9 月下旬开裂

① 表中所列为主花期。除主花期外，黄槐和粉叶决明每年还有 1 次开花。黄槐的第 2 次开花在 8 月上旬～12 月上旬，11 月下旬～12 月果熟。粉叶决明第 2 次开花在 9 月中旬～12 月上旬，11 月中旬～12 月果熟

荚果，成熟时由青色转为紫褐色或黑褐色，扁平带状或圆柱形。腊肠树与神黄豆的果荚不开裂，其余 4 种成熟后果荚开裂，种子散落。种子有胚乳，子叶扁平。荚果及种子特征见表 3、图 1。

表 3 铁刀木属树种果实和种子的形态特征

树 种	未熟果颜色	果 实			种 子		
		形 状	大小（cm）	颜 色	形 状	大小（mm）	颜 色
神黄豆	青绿色	圆柱形	长 50～68 径 1.7～1.9	黑褐色	扁椭圆形，外层种瓤包裹成圆饼形	长 6.5～7.5 宽 5～6.5 厚 3～4.5	黄褐至棕褐色
腊肠树	绿色	圆柱形具纵纹	长 30～72 径 2～2.5	黑褐色	扁圆形，外层种瓤包裹成圆饼形	长 6～8 宽 5～7	棕褐色
粉叶决明	绿色	带状	长 15～20 宽 1.2～1.8	黄褐至灰褐色	扁椭圆形，顶端稍尖	长 5～8 宽 3～4 厚 1.3～1.8	黄褐至暗褐色
望江南	绿色	带状，微弯	长 10～13 宽 0.8～1	黄褐色，被疏毛	扁圆形	宽 5～10 厚 3～7	黄褐色
铁刀木	绿色	扁平，带形，边缘增厚	长 15～30 宽 1.4～1.6	紫褐色，被绒毛	扁平，近圆形	长 7～11 宽 4.5～7	深褐色，有光泽
黄槐	绿色	带状，顶喙尖	长 8～10 宽 1.1～1.3	黄褐至灰褐色	扁椭圆形，顶端稍尖	长 5.5～8 宽 3～4 厚 1.5～2	黄褐至暗褐色

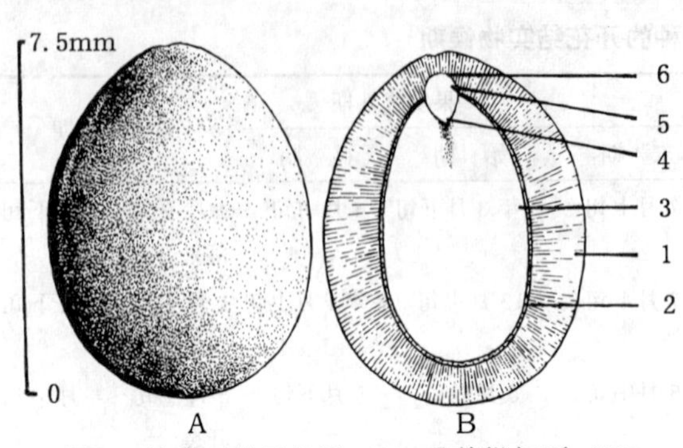

图 1　神黄豆种子外形（A）及其纵切面（B）
1. 种皮　2. 胚乳　3. 子叶　4. 胚芽　5. 胚轴　6. 胚根
（黄应钦绘）

果实的采收调制和种子贮藏　荚果成熟后应及时用采种钩刀、高枝剪剪取或直接摘下果穗或果荚，置日光下曝晒。望江南、铁刀木、黄槐和粉叶决明晒干后荚壳开裂，种子脱出，除去荚瓣，筛去杂质即得种子。腊肠树和神黄豆的果荚不开裂，需充分晒干并用木棒敲碎果荚才能得出带横隔的种子，剥去横隔方得纯净种子。运输和贮藏前种子应充分晒干，使含水量降到 10%以下。这 6 个树种种子的出种率、净度和质量见表 4。

表 4　铁刀木属树种果实的出种率和种子净度、质量

树　种	出种率（%）	净度（%）	千粒重（g）	每千克纯净种子粒数
神黄豆	23～25	98	120～185	5 400～8 300
腊肠树	20～25	98	110～140	7 000～9 000
粉叶决明	10～15	98	19～21	47 000～53 000
望江南	20～25	98	40～60	17 000～25 000
铁刀木	15～20	98	25～30	33 000～40 000
黄槐	10～15	98	17～19	53 000～59 000

运输可用麻袋包装。贮藏时装入麻袋或瓦缸，存放在室内通风干燥处。常温条件下，室内裸露贮藏的种子发芽能力可保持 1～2 年，混干沙贮藏或晒干后密封贮藏的种子，发芽能力可保持 2～3 年。

发芽和播种　种子无生理性休眠。望江南的种子在日平均气温 17～20℃时即可发芽，其余 5 种发芽时日平均气温需在 20℃以上。播种前需用始温 60～80℃的热水浸种，在冷却的过程中浸泡 1 昼夜。神黄豆和腊肠树的种子用硫酸浸渍 10 分钟，洗净后浸泡 1 昼夜，可以收到 80℃热水浸种同样的效果。广西林业科学研究所 1982 年和 1987 年分别对腊肠树、神黄豆、铁刀木、黄槐和粉叶决明进行过发芽测定：经过预处理的种子播后第 4 天开始萌发。出土萌发。发芽后约 3 天子叶出土，再过 4～7 天发出初生叶。发芽情况见表 5、图 2。

表 5　铁刀木属树种的发芽能力及其测定条件

树　种	预处理	基　质	温度（℃）		发芽势（%）		发芽率（%）	
			室　内	室　外	计算天数	一般数值	计算天数	一般数值
神黄豆	80℃水浸种	沙	—	26～28	8	52	16	76
腊肠树	80℃水浸种	沙	—	27～28	7	50	15	75
粉叶决明	80℃水浸种	纸	28	—	4	65	12	85
望江南	60℃水浸种	圃地	—	24～26	3	65	10	80
铁刀木	70℃水浸种	圃地	—	26～28	8	40	20	60
黄槐	60℃水浸种	纸	28	—	发芽盛期不明显		11	52

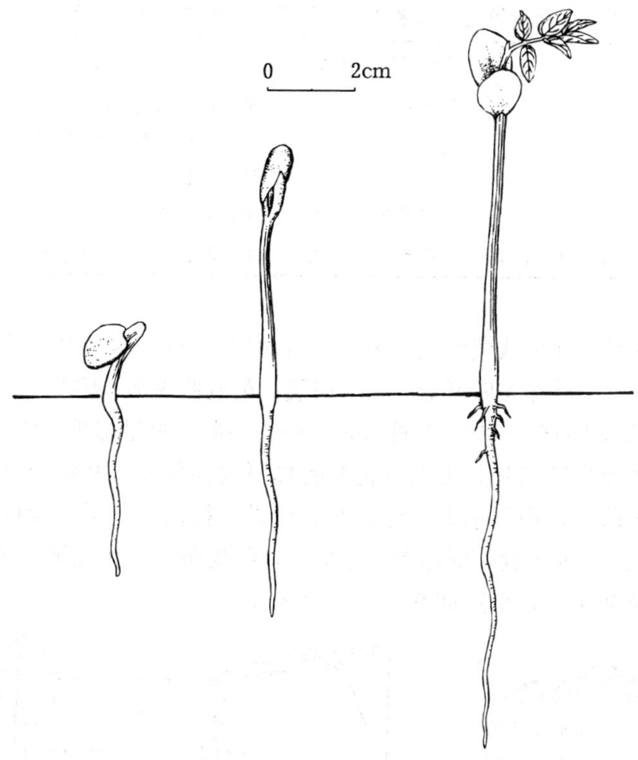

图 2　神黄豆种子萌发后第 3、4、7 天幼苗生长情况
（黄应钦仿《热带亚热带主要树种采种育苗技术》）

条播。每平方米播种量：腊肠树和神黄豆为 17～20g；铁刀木、黄槐和粉叶决明 5～ 8g；望江南 7～10g。腊肠树和神黄豆多用于庭院绿化，需培育 3～4 年生苗。黄槐和粉叶决明育 1～2年生苗，铁刀木育半年至 1 年生苗。望江南一般播种后留圃作叶菜经营，不再起苗栽植。

（韦增健）

紫　荆　属

Cercis L.

（苏木科　Caesalpiniaceae）

生长习性、分布和用途　本属约 11 种，分布于北美、东亚和南欧。我国有 6 种，本文描述 3 种，见表 1。落叶小乔木或丛生灌木，后者具不分枝且无顶芽的细长枝条。适生于排水良好的微酸性土壤。这 3 个树种春初先叶开花，花簇繁茂，嫣红夺目，是常见的观赏树。紫荆对氯气有一定的抗性，滞留尘埃的能力也强。紫荆和黄山紫荆的皮、根均可入药。

表 1　紫荆属树种的名称、生长习性、树高、分布与用途

中　名	学　名	生长习性	树高 (m)	分　布	用　途
紫荆	*C. chinensis* Bge.	丛生灌丛	2～4	黄河流域以南，西北至陕、甘、新，西至川、黔、藏、滇，南至粤、桂均有栽培	观赏、药用、蜜源
黄山紫荆	*C. chingii* Chun	丛生灌木	6	皖、浙、粤	观赏、药用、蜜源
巨紫荆	*C. gigantea* Cheng et Keng f.	乔木	20	浙、皖、豫、鄂、湘、粤、黔	材用、观赏、蜜源

开花结实　花两性，红色或紫红色，稍呈左右对称，5～14 朵簇生于老干上，或成总状花序长在短侧枝及主干上，早春先叶开放。萼 5 齿裂。花冠假蝶形。雄蕊 10，分离。单雌蕊，子房上位，1 室，具短柄，胚珠 2～10。花柱线形，柱头头状。荚果扁平，长 3～14cm，具柄。果瓣上边直下边弯，表面具细网脉，尖端有花柱的加厚残存部分。紫荆和巨紫荆果瓣腹缝具窄翅，开裂后果瓣不扭曲。黄山紫荆果无翅，开裂后果瓣扭曲。荚果内含种子若干粒。这 3 个树种荚果的形态见表 2。种子扁，倒卵形，棕褐色，种皮坚硬，胚乳量中等，子叶大，胚根较短（图 1，表 3）。花期 4 月，果熟期 9～11 月（表 4）。

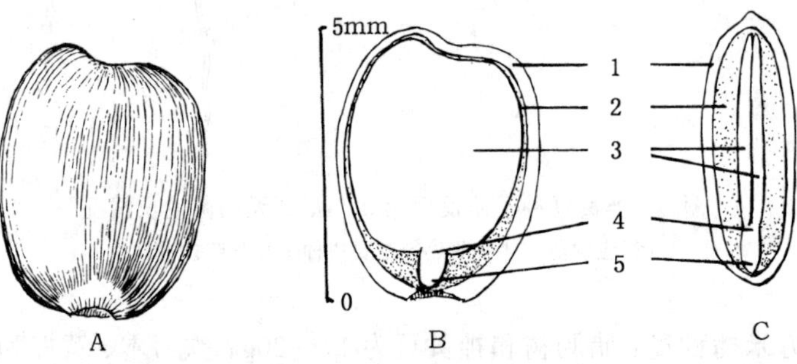

图 1　紫荆的种子外形（A）及其正面纵切（B）和侧面纵切（C）

1. 种皮　2. 胚乳　3. 子叶　4. 胚轴　5. 胚根

（史渭清绘）

表 2　紫荆属树种荚果的形态特征

树　种	荚　长 (cm)	荚　宽 (cm)	颜　色	果皮质地	腹缝线的翅	先端形状	表面网纹	每荚含种子粒数
紫荆	3～10	1.3～1.5	黄褐色	纸质	具窄翅	短尖	明显	7～9
黄山紫荆	6～8	1～1.1	棕色	厚革质	无翅	长喙状	无	3～8
巨紫荆	6～14	1.5～2	深紫红色	厚纸质	具翅	渐尖	明显	6～9

表 3　紫荆属树种种子的特征

树　种	形　状	长 (mm)	千粒重 (g)	每千克种子数（万粒）
紫荆	扁椭圆形，偏斜有凸尖	5～6	24	4.1
黄山紫荆	扁圆形	6～8	32	3.1
巨紫荆	扁倒三角圆形，顶平	5～6	20	5

表 4　紫荆属树种的开花结实物候

树　种	花　色	花梗长（cm）	簇生花朵数	开花期（月）	果熟期（月）
紫荆	红色，萼紫红色	0.6～1.5	5～8	4～5	9～10
黄山紫荆	淡紫红色	<1	（6）8～10	4	10
巨紫荆	淡紫红色，萼暗红色	1.2～2	7～14	4	10～11

果实的采收调制和种子贮藏　紫荆属荚果成熟后挂在树上，入冬前还不裂不落，采收时间较长。摘下荚果摊晒，充分干燥后击打脱粒，剔除空瘪粒和杂质。短期存放的种子可以在室内干藏。用玻璃或金属容器在 3～5℃条件下贮存，发芽能力可以保持 2～3 年。或湿沙层积后供翌年春播。紫荆属荚果出种率 15%～25%。净度可达 90%以上。种子质量的数据已列入表 3。紫荆属种子虫害严重。特别是紫荆种子虫害率极高，严重影响种子质量。

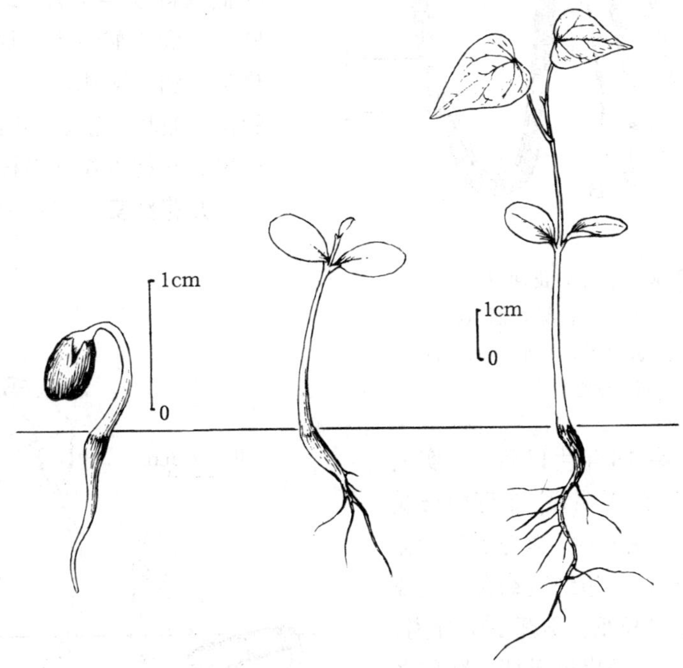

图 2　紫荆种子萌发后第 2、31、45 天的幼苗生长情况

（史渭清绘）

发芽与播种　紫荆属种子有休眠习性，除了种皮坚韧，透性不良而使吸胀受阻以外，可能还由于胚本身需要低温解除生理休眠。南京中山植物园曾用湿沙在室外自然温度条件下进行过这 3 个树种的层积试验。黄山紫荆于 12 月中旬层积，翌年 3 月中旬开始萌发，13 天内发芽率高达 80%。紫荆和巨紫荆可能是因为虫害粒或空瘪粒较多，层积效果不佳，发芽率分别为 24%和 31%。据文献介绍，用浓硫酸浸种 25～60 分钟，或用热水浸泡过夜，可以改善种皮的透性，再经 3～5℃低温层积 5～12 周，可以解除休眠，促进萌发。不过，每批种子的种皮“硬度”不同，休眠深度不同，必须通过试验，分别种批确定硫酸或热水的温度及其处理的时间。发芽的适宜温度为 20℃左右。出土萌发（图 2）。紫荆初生叶无毛，巨紫荆初生叶背面基部有毛。可以扦插繁殖。

（何泽瑛）

凤　凰　木

Delonix regia (Boj.) Raf.

（苏木科　Caesalpiniaceae）

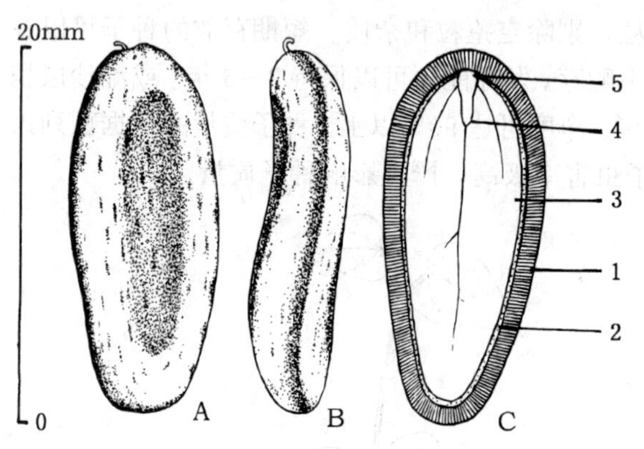

图 1　凤凰木种子外形正面（A）、侧面（B）及其纵切面（C）
1. 种皮　2. 胚乳　3. 子叶　4. 胚芽　5. 胚根
（黄应钦绘）

生长习性、分布和用途　本属 3 种，本文描述我国引入的 1 种。落叶乔木，高达 20 余米，胸径达 1m。耐寒力较差，忌霜冻，月平均气温低于 12℃，或 6℃左右的低温持续半个月，即有严重冻害。喜肥沃湿润的酸性土。原产马达加斯加等热带非洲。我国琼、粤、桂、闽、台、滇、黔南有栽培。花大而艳丽，供观赏或作行道树。木材可作火柴杆，也可造纸。

开花结实　5～6 年生开始开花结实，正常结实年龄在 15 年生以后。结实大小年间隔期为 1 年，但不明显。冬春嫩枝受冻则影响开花结实。花两性。伞房状总状花序。萼片 5，内面红色，边缘黄绿色。花瓣 5，匙形，鲜红色，具长爪。雄蕊 10，分离，红色花丝长 3～6cm。子房近无柄，胚珠多数，花柱长 3～4cm，柱头小，截形。据广西南宁 1987～1988 年观察，4 月上旬花芽与叶芽同时抽出，着生于当年生嫩枝的叶腋，4 月下旬～5 月上旬初花始开，5 月中旬～6 月中下旬为盛花期，6 月下旬为末花期，7～8 月尚有个别植株零星开花；11 月中旬果实开始成熟，11 月下旬～12 月上旬为成熟盛期。荚果长带状，成熟时由青色转为黑褐色，长 30～60cm，宽 4.5～5cm。成熟后果荚不开裂，在树上存留到翌年 3 月前后才陆续脱落。每荚有种子 10～20 粒。

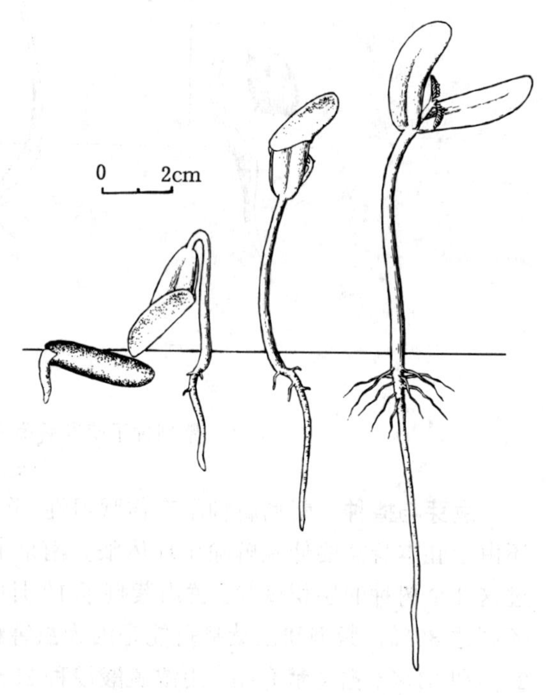

图 2　凤凰木种子萌发后第 1、3、4、6 天幼苗的生长情况
（黄应钦仿《热带亚热带主要树种采种育苗技术》）

种子椭圆状矩圆形，略扁，稍扭曲；酱紫色，有光泽，两侧各有1乳白色环状花纹；长13～25mm，宽6～8mm，厚3.5～5mm；具少量胚乳（见图1）。

果实的采收调制和种子贮藏　结合冬春季修整树形，用高枝剪或采种钩刀采下果荚，或将整个枝条锯下后摘取果实。采回的果实置日光下曝晒1至数天，用木棒敲打使果荚裂开，除去杂质后即得种子。荚果出种率约20%～25%。种子净度可达97%～99%。千粒重一般为400g，变动范围为350～750g，每千克有纯净种子2 500粒，变动范围为1 300～2 860粒。种子宜晒干使含水量降到10%以下。贮藏时可用袋装或装入一般容器，置室内干爽处存放。常温条件下发芽能力可保持1～2年。

发芽和播种　种子无生理性休眠。发芽时日平均气温需在20℃以上。1982年6月下旬日平均气温28℃时，广西林业科学研究所在室外沙床作过发芽测定：播前用80℃热水浸种，在冷却过程中浸泡1昼夜，播后第3天开始发芽，第5天进入发芽盛期，第18天发芽终止；从发芽之日起算的10天中发芽百分数为45%。从发芽之日起算的16天中发芽率为60%。出土萌发。胚根萌发后第3天子叶出土，第6天长出初生叶（见图2）。

条播。每平方米播种50～75g，覆土2～2.5cm。1年生苗出圃。

（张声燕）

格　　木

Erythrophleum fordii Oliv.

（苏木科 Caesalpiniaceae）

生长习性、分布和用途　格木属15种，我国只有本文描述的1种。常绿大乔木，高25m，胸径达1m。适生于年平均气温21℃以上的南亚热带至北热带的湿润型气候区，喜较肥沃湿润的酸性土壤。广西南部为主产区，浙、闽、粤、黔、台亦产。越南有分布。木材深褐色，坚实耐腐，有铁木之称，为珍贵用材。枝叶浓密，涵养水源的能力强。叶可作饲料及肥料。格木的天然植株已不多见，在《中国植物红皮书》中列为渐危种。

开花结实　约15年生开始开花结实，正常结实年龄在25年生以后，结实大小年间隔期为1年。花两性。总状花序或数序再排列为圆锥花序，腋生于当年嫩枝。萼钟状，5齿裂。花瓣5，白至淡黄色。雄蕊10，分离，长为花瓣的2倍，花药纵裂。子房长圆形，具柄。花柱短，柱头圆。胚珠10～12。据南宁1978～1984年观测，4月中旬～下旬从当年生嫩枝抽出花芽，5月上旬伸长，形成花序；5月中旬为始花期，5月中旬～下旬初为盛花期，下旬末花期结束；6月上旬～中旬幼果形成，10月中旬果实开始成熟，下旬为果实成熟盛期。荚果，成熟时由青色转变为黑褐色；木质，长圆形，扁；长10～18cm，宽3.8～4.3cm。果实成熟时不立即脱落，熟后约30天在树上开裂。果荚开裂后，种子间因有胶着物粘连而成串脱落。果瓣存留至翌年2月前后脱落。每荚有种子5～12粒。种子扁椭圆形或矩圆形，黑色，长13～19mm，宽12～16mm，厚5～8mm。有胚乳，子叶扁平（图1）。

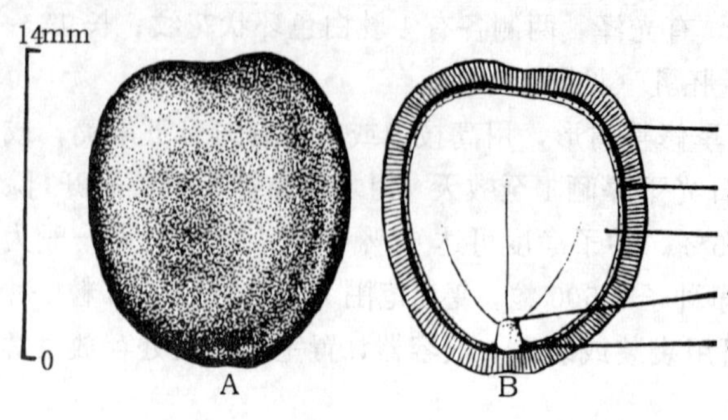

图 1　格木种子外形（A）及其纵切面（B）

1. 种皮　2. 胚乳　3. 子叶　4. 胚芽　5. 胚根

（黄应钦绘）

果实的采收调制和种子贮藏　用钩刀将果荚钩落，置日光下曝晒，开裂后得出种子。格木的生物学特性之一是，常常数粒种子胶连成串，从开裂的果荚中脱落出来，而且落地的种子不受虫蛀，又无鼠鸟危害，所以最省工的方法是在地面拾集成熟自落的种子。当然也可以将正在开裂的果实击落。采得的种子通常要置烈日下晒干，使含水量降到10%左右。种子的净度可达 95%～99%。千粒重 890～1 100g，每千克有纯净种子 900～1 150 粒。种子极耐贮藏，常温条件下干藏，发芽能力可以保持 4 年不至于显著下降。

发芽和播种　种子无生理性休眠。发芽时日均温宜在 20℃以上。但种皮具胶质，妨碍吸水，未经处理的种子播后发芽极不整齐，少数种子可能迟至第 3 年方始发芽。预处理的方法最好是用硫酸浸渍 30 分钟，胶质膨胀后用清水彻底洗净，再用水浸泡 1 昼夜。经过这样酸蚀的种子播后很快发芽。也可以用沸水烫种，冷却后洗净再浸泡 1 昼夜。经过沸水浸烫的种子播后发芽也较整齐，但比硫酸处理的迟发芽 7～10 天。1982 年 4 月中旬，广西林业科学研究所在室外沙床上对硫酸处理过的种子做过发芽测定。当时日平均气温 20℃，播后第 5 天开始萌发，萌发后第 2～3 天进入发芽盛期，到第 11 天发芽率为 84%。出土萌发。种子萌发后 4 天子叶出土，10～15 天长出对生的初生叶（见图 2）。

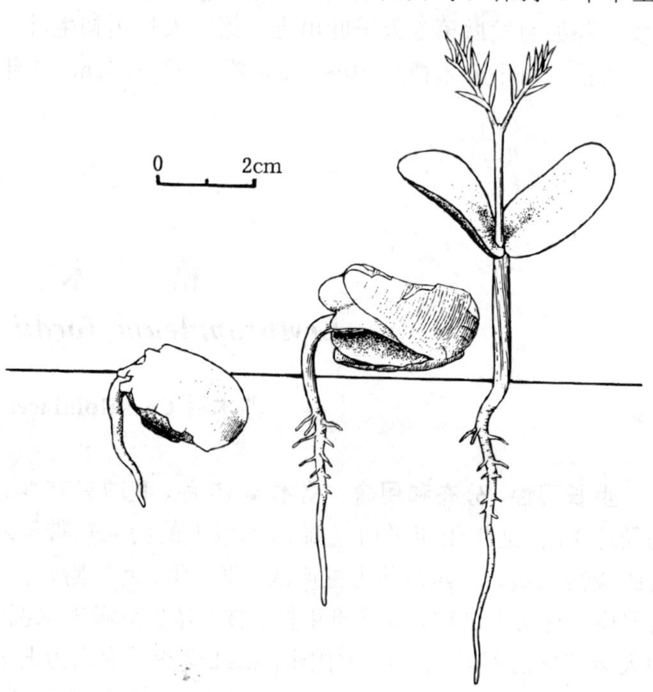

图 2　格木种子萌发后第 2、4、15 天幼苗的生长情况

（黄应钦仿《热带亚热带主要树种采种育苗技术》）

条播。每平方米约播 100g，覆土 2～3cm。一般培育 1 年生苗，到 3 月下旬～4 月气温稳定回升后出圃。适宜于与松、杉等混交造林，或营造水源林。

（张声燕）

皂　荚　属

Gleditsia L.

（苏木科　Caesalpiniaceae）

生长习性、分布和用途　本属约15种，我国产9种，又引入1种，本文描述4种。落叶乔木。喜光，不耐庇荫。喜深厚湿润土壤，但耐干旱，在石灰岩山地、石质山坡、中性、微酸性及轻度盐碱土壤上均能生长。有些种类能在干旱瘠薄的山坡上生长，但生长不良。这4个树种的名称、树高、分布和用途见表1。

表1　皂荚属树种的名称、树高、分布和用途

中　名	学　名	树高（m）	分　布	用　途	供　稿
日本皂荚	*G. japonica* Miq.	15（25）	吉、辽、冀。南京栽培。日本、朝鲜半岛	用材，观赏	305、308
山皂荚	*G. melanacantha* Tang et Wang	14	华东、晋	观赏	305、308
野皂荚	*G. microphylla* Gordon ex Y. T. Lee	4	晋、冀、鲁、皖、豫、陕	用材，四旁绿化，绿篱	305、308
皂荚	*G. sinensis* Lam.	30	晋、冀、陕以南至华南、华东、华中、西南	用材，观赏，果代肥皂，果、种子入药	302

开花结实　8～15年生开始结实，25年生以后进入盛果期，盛果期可持续50～80年或更长。花杂性或单性异株。总状花序。萼筒钟状，3～5裂。花瓣3～5。雄蕊6～10，离生。花药2室，丁字着生，纵裂。子房上位，1室，胚珠1至多数，边缘胎座，花柱短。荚果带状，不裂或迟裂。幼果绿色，成熟时变为黑褐色或紫红色，可存留越冬，遇风逐渐脱落。种子1至多粒。种皮革质，坚硬，有光泽，有角质胚乳。这4个树种的开花结实物候见表2。果实、种子形态见表3。图1。

表2　皂荚属树种开花结实物候

树　种	观察地点	开花期			果实成熟期	
		初　期	盛　期	末　期	初　期	盛　期
日本皂荚	北京	5月上旬	5月中下旬	6月上旬	9月中旬	10月
山皂荚	—	5月	～	6月	9月上旬	10月
野皂荚	—	5月	～	6月	9月中旬	10月
皂荚	北京	5月中旬	5月下旬	6月上旬	9月中下旬	10月

表 3　皂荚属果实和种子的形态

树种	果实形状	果实长（cm）	果实宽（cm）	每果种粒数	种子形状	种子长（cm）	种子颜色
日本皂荚	带状扁平，具短尖头，直或微弯，子房柄长约2cm，暗红褐色	15～25	2～3.5	多数	长圆形，卵状椭圆形，稍扁	0.9～1	黑褐或栗褐色
山皂荚	镰形，常扭曲，有泡状突起，红褐色，微被白粉，子房柄较短	20～35	3～4.5	多数	长圆形，扁	0.9～1.1	绿褐色
野皂荚	斜椭圆形，红褐色，具喙，薄革质，子房柄长1.2cm以上	3～6	1.5～2	1～3	长圆形，扁	0.8～1	—
皂荚	带形，弯或直，饱满，黑褐或紫红色，木质，子房柄长7mm以上	5～35	3～4	多数	长圆形，扁灰绿色	1.0～1.3	亮棕色或灰绿色

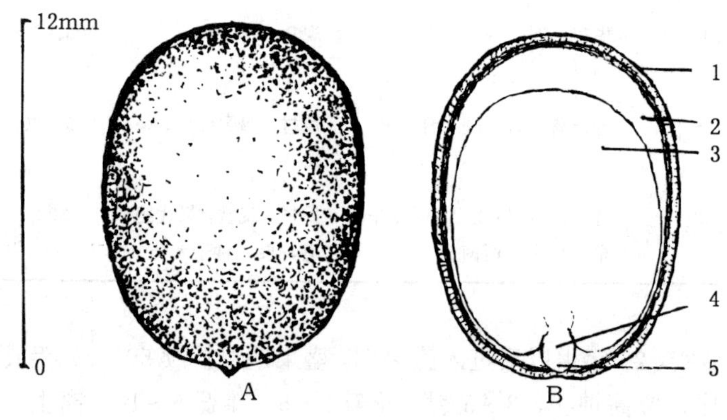

图 1　皂荚种子外形（A）及其纵切面（B）
1. 种皮　2. 胚乳　3. 子叶　4. 胚轴　5. 胚根
（孟玲、田恒德绘）

果实的采收调制和种子贮藏　荚果由绿变为黑褐色或紫红色、红褐色时表示开始成熟，剪下或打下荚果，摊开曝晒，砸碎荚果，除去果皮及杂质，即得纯净种子。曾经从鲜重 1 350g，晒干后重约 1 000g的野皂荚荚果中脱出种子 248g，以鲜果质量计算的出种率约为 18%。这 3 个树种的出种率、种子净度及质量见表 4。

表 4　皂荚属的出种率、种子净度和质量

树种	出种率（%）	种子净度（%）	千粒重（g）	每千克纯净种子粒数
日本皂荚	5.6～10	52～72	190	5 200
野皂荚	18.4	88	130	7 700
皂荚	20～25	—	450～600	1 600～2 200

发芽和播种　皂荚属种子的种皮革质，较厚，坚硬致密，透水透气性差，播前需经催芽处理。生产上常用的催芽方法见表 5。

表 5　皂荚属树种常用的催芽方法及发芽能力

树　种	催芽方法	催芽时间	一般发芽率（%）
日本皂荚	①秋末冬初浸种，充分吸水后混沙低温层积	4～5个月	28
	②种子充分吸水后秋播		
	③浸泡种子，种皮破裂时播种		
野皂荚	①秋末冬初浸种，充分吸水后混沙低温层积	4～5个月	—
	②种子充分吸水后秋播		
	③浸泡种子，种皮充分吸水破裂后即播	1个多月	—
皂荚	①播前浸种，每5～7天换水1次，种皮开裂即播	1～2个月	
	②用始温60～90℃水浸种，在18～25℃室内覆盖催芽，每天用40℃水冲淋1～2次，待20%～30%种子开裂即播		80
	③浓硫酸浸种0.5～1小时，用清水冲洗后层积	3～5天	
	④秋末浸种，充分吸水后层积	4～5个月	

条播，覆土2～4cm。日本皂荚每平方米播种量15～25g，野皂荚10～18g，皂荚50～100g。出土萌发。幼苗破土能力较差，应注意及时耙松表土。种子萌发和幼苗生长情况见图2。

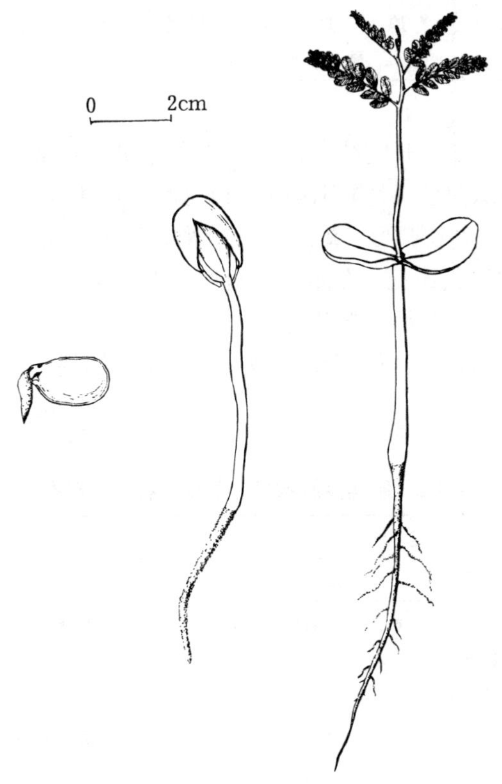

图2　皂荚种子发芽出土后第1、9、16天幼苗生长情况

（孟玲绘）

（王木林、高秀岩）

肥皂荚属
Gymnocladus L.

（苏木科 Caesalpiniaceae）

生长习性、分布和用途 肥皂荚属5种（北美1种，东亚4种），我国3种，又引入北美1种。本文介绍肥皂荚和美国肥皂荚这2个种。落叶大乔木，高25～30m，胸径可达1m以上。喜温暖湿润气候和肥沃土壤。肥皂荚亦可在石灰岩低山及微盐碱地生长。木材坚重耐久，可作家具、农具。荚果肥厚，富含皂素，为优良制皂原料。美国肥皂荚种子炒食，可代咖啡，被称为“肯塔基咖啡树”。肥皂荚果药用，种子可榨油作油漆等工业用油，种仁可食。两种肥皂荚的名称、分布和用途见表1。

表1 肥皂荚属树种名称、分布和用途

中 名	学 名	分 布	用 途
肥 皂 荚	*G. chinensis* Baill.	豫、皖、苏、浙、闽、赣、鄂、湘、粤、桂、黔、川。	材用、观赏、制皂、药用
美国肥皂荚	*G. dioicus* K. Koch	原产加拿大东南部和美国东北部至中部。青岛、泰安、南京、杭州栽培	材用、观赏、种子代咖啡、药用

开花结实 7～10年生开始开花结实，15～20年生进入正常结实年龄。花雌雄异株、杂性异株或同株。圆锥花序绿白色（美国肥皂荚）或总状花序淡紫色（肥皂荚），顶生。花整齐，萼筒状4～5裂。花瓣4～5，长圆形。雄蕊10，分离，长短相间，着生于花萼筒口，花丝粗，花药背着。雌蕊在雄花中退化或无。在雌花和两性花中子房上位，1室，侧膜胎座具2～8胚珠。花柱直，稍粗而扁，柱头偏斜。肥皂荚4～5月花叶同放；美国肥皂荚6月先叶后花。果成熟 期9～10月。荚果肥厚，肉质，干后坚实近木质，2瓣裂。这2个树种的果实和种子形态特征见表2。

表2 肥皂荚属树种的荚果和种子的形态

树 种	荚果				种子		
	形 状	大小（cm）	颜 色	种子数	形 状	大小（cm）	颜 色
肥皂荚	长圆形	长7～12（14） 宽3～4 厚约1.5	暗褐色	2～4	近球形	径约2	黑色
美国肥皂荚	长圆状镰刀形	长10～25 宽3.5～5 厚1～2	暗褐色或红褐色	4～8	扁圆形或卵状球形	径1.5～2	暗褐色

种皮极度坚硬，骨质。具胚乳，子叶肥厚。美国肥皂荚种子的外形和内部结构见图1。

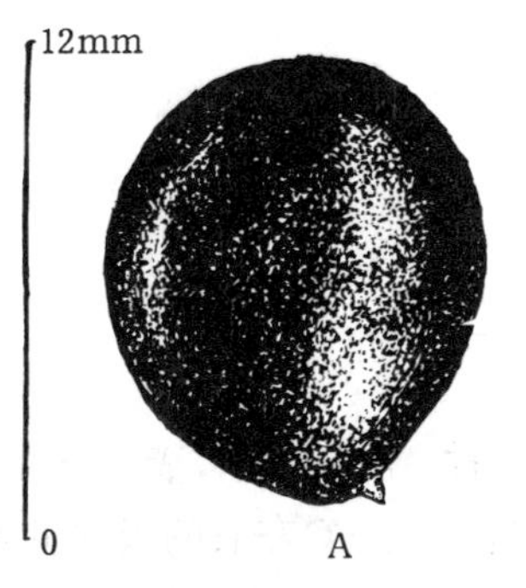

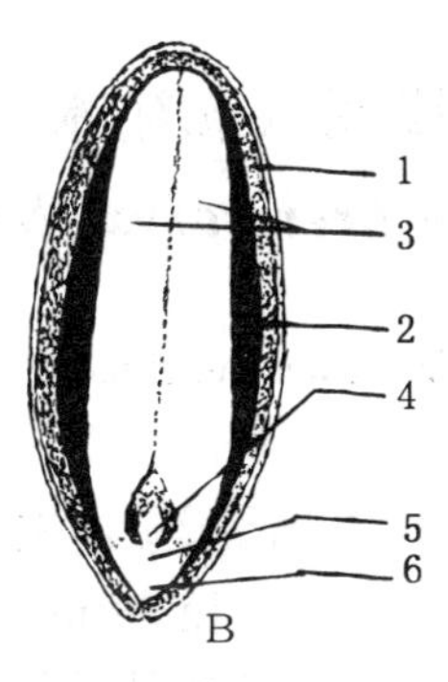

图 1　美国肥皂荚种子的外形（A）及其纵切面（B）
1. 种皮　2. 胚乳　3. 子叶　4. 胚芽　5. 胚轴　6. 胚根
（田恒德绘，B 仿《美国木本植物种子手册》）

果实的采收调制和种子贮藏　美国肥皂荚的荚果成熟后可持续到翌年早春才脱落。当 10 月下旬果实成熟时即可用竹竿敲落。两种荚果采收后摊开曝晒，干裂后取出种子或浸水软化后取出种子晾干，干藏。种子寿命长，在自然状态下发芽能力可以保存数年之久。美国肥皂荚雌雄异株，在自然状态下雄株比例较大，种源较少。荚果、种子有时受皂角食心虫 *Cryptophlabia ombrodalta*（Lower）、皂角豆象 *Bruchidius dorsalis* Fabricius 危害，应烧毁有虫的荚果和种子。肥皂荚种子千粒重约 1 810g，每千克约 550 粒（陈嵘，1933）。美国肥皂荚种子千粒重 1 520～2 270g，每千克 440～660 粒，平均 505 粒（据《美国木本植物种子手册》）。净度均可高达 95%～100%。

发芽和播种　2 个树种的种皮坚硬，不易透水，发芽较慢。用温水浸种或浓硫酸短时间浸泡，可使种子提早发芽。南京林学院树木园 1953 年 3 月 18 日播的美国肥皂荚种子，未经处理，至 4 月 23 日开始萌发，前后经历 36 天。据记载，美国曾用浓硫酸处理 2 小时，20 天的发芽百分数为 48%，25 天为 82%，30 天后为 86%。出苗率 60%～75%。南京林学院树木园 1960 年从湖南收到少量肥皂荚种子，1961 年 1 月中旬播种，2 月上旬萌发。两种皆为留土萌发。肥皂荚幼苗具 1 或 2 初生不育叶，互生，以后为一回羽状复叶，小叶 10～20 片（图 2）。美国肥皂荚幼苗无初生不育叶，初生叶为 4 对小叶的羽状复叶，互生。

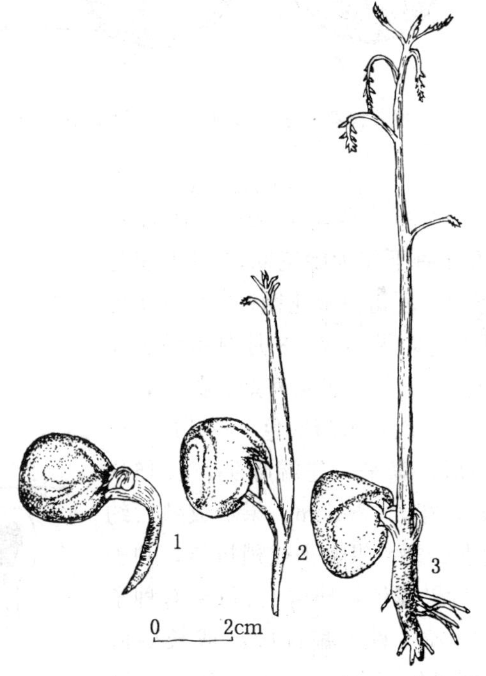

图 2　肥皂荚种子的萌发和幼苗生长情况
1. 子叶不出土　2. 上胚轴伸长
3. 具初生不育叶和发育叶（根部略）
（黄鹏成供稿，田恒德临）

单行条播，每 10 平方米用种 2～3kg。覆土厚 2～3cm。盖草。1 年生苗可以出圃。

（黄鹏成）

短萼仪花（麻轧木）

Lysidice brevicalyx Wei

（苏木科 Caesilpiniaceae）

生长习性、分布和用途 仪花属 2 种，我国皆产。本文描述 1 种。常绿乔木，高 25m，胸径 80cm。能耐轻霜及 0℃左右的极端低温。喜湿润肥沃的石灰岩山地或酸性土壤。分布于桂、粤、滇、黔。越南亦产。材质优，较珍贵。根、茎、叶有小毒，可止血止痛。种子可食。花艳丽，供观赏。

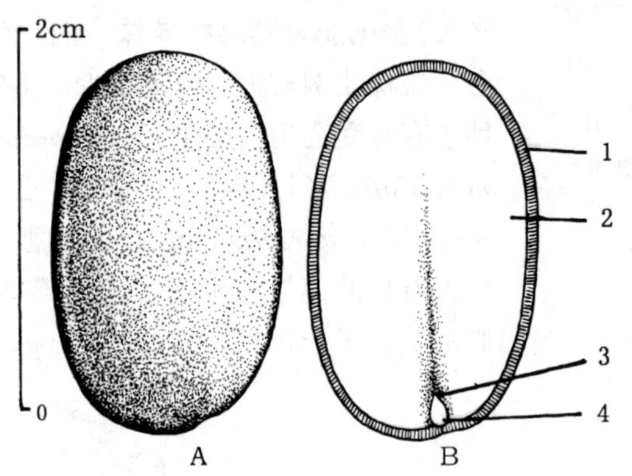

图 1 短萼仪花种子外形（A）及其纵切面（B）
1. 种皮 2. 子叶 3. 胚芽 4. 胚根
（黄应钦绘）

开花结实 12 年生左右开始开花结实，正常结实期在 20 年生以后。结实大小年间隔期为 1 年，但不甚明显。花两性。圆锥花序顶生。花萼管比萼裂片短，顶部 4 裂片，花后反曲。花瓣 3，具长爪，紫堇色，前面 2 片退化。苞片白带绿色。发育雄蕊 2，花药背着，退化雄蕊 5～6（8）。子房具短柄，胚珠 9～14。据广西南宁 1978～1984 年观察，4 月下旬为始花期，5 月上旬为盛花期，下旬为末花期；5 月下旬至 6 月上旬幼果形成；8 月中旬果实开始成熟，下旬为果熟盛期。荚果，成熟时由青色转为暗褐色，厚革质，长倒卵形，扁平，先端喙尖，长15～26cm，宽 3.5～5cm。果实成熟后约 1 周即开裂。果瓣平或稍扭转，种子脱落后果瓣仍悬挂树上。每果有种子 5～12 粒。种子扁椭圆形或长矩圆形，栗褐色有光泽，长 15～28mm，宽 12～18mm，厚 3.3～5.6mm。无胚乳，子叶肥大（图 1）。

果实的采收调制和种子贮藏 果实成熟尚未开裂前用钩刀采下果荚，置日光下曝晒，开裂后捡去果荚，即得种子。荚果的出种率 25%～30%。种子净度可达 98%～100%，

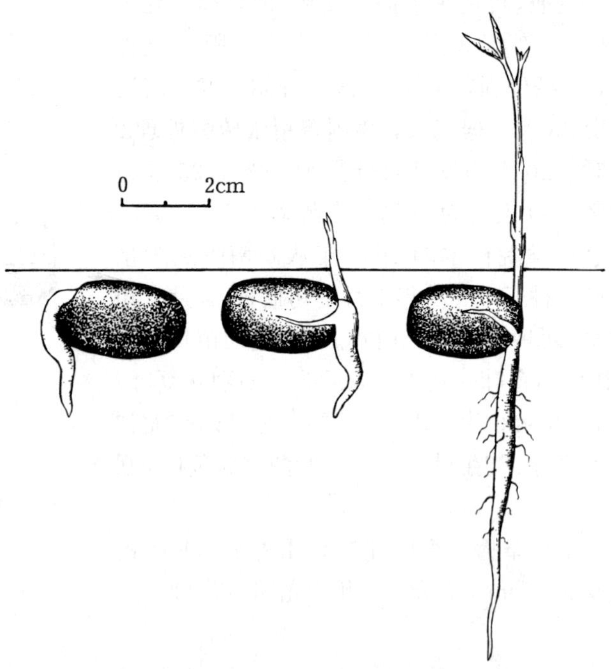

图 2 短萼仪花种子萌发后第 3、6、15 天幼苗的生长情况
（黄应钦仿《热带亚热带主要树种采种育苗技术》）

千粒重1 000（800～1 200）g，每千克有纯净种子 1 000（800～1 200）粒。种子可以晾干，但忌在烈日下曝晒。种子的含水量可以降到 14%左右。贮藏时需混以干润沙。常温条件下室内混沙贮藏的种子，发芽能力可以保持 1 年半至 2 年。

发芽和播种　种子无生理性休眠，但种皮透性较差，阻碍吸水萌发，播前需要处理。发芽时日均温需在 20℃上下。1987 年 9 月下旬，广西林业科学研究所用当年新采种子在室外沙床做发芽测定。播前用硫酸浸渍 6 分钟，彻底洗净后用水浸泡半天。播种时日均温 27℃。播后 9 天种子开始萌发，至 10 月 14 日发芽终止。从发芽之日起至发芽高峰的 9 天中，发芽百分数为 60%；从播种之日起算的 24 天中，发芽率为 74%。留土萌发。胚根萌发后 6～7 天上胚轴出土，具初生不育叶 4～6 片，13～15 天初生发育叶长出（图 2）。

条播。每平方米播种 100～125g，覆土 2～2.5cm。1 年生苗出圃。

（张声燕）

扁　轴　木

Parkinsonia aculeata L.

（苏木科　Caesalpiniaceae）

生长习性、分布和用途　扁轴木属约 4 种，产热带美洲和南非，本文描述引入的 1 种。常绿小乔木，高达 8m，胸径 10cm，有刺。生长速度中等。喜肥力中等以上的酸性土壤。原产墨西哥。我国琼、粤、桂等地栽培。形态奇特，供观赏并可作绿篱。树皮及叶入药，为滋补剂。

开花结实　约 5 年生开始开花结实，正常结实年龄在 8 年生以后。结实大小年间隔期为 1 年，但不甚明显。花两性。总状花序具长柄，腋生。萼筒短，裂片 5，膜质。花瓣 5，黄色，上面一片较宽，具爪。雄蕊 10，分离，丁字着生，具花盘。子房具柄，胚珠多数。据南宁1986～1988 年观察，花芽于 3 月下旬形成，花期长，从 5～10 月陆续开花；较明显的花期有 2 次，第 1 次开花在 5 月上旬～6 月下旬；6 月下旬果实开始成熟，7 月上旬为盛果期；第 2 次开花在 9 月下旬～10 月下旬，10～11 月为盛果期。荚果，成熟时由绿色转变为褐色，念珠状，长 5～15cm，径 0.8～1cm，成熟后 1 个月内不开裂。每果有种子 2～6 粒。种子长椭圆形，有光泽，黑褐色，长 10～11mm，宽 4～5mm，厚 2～3mm。有胚乳，子叶扁平（图 1）。

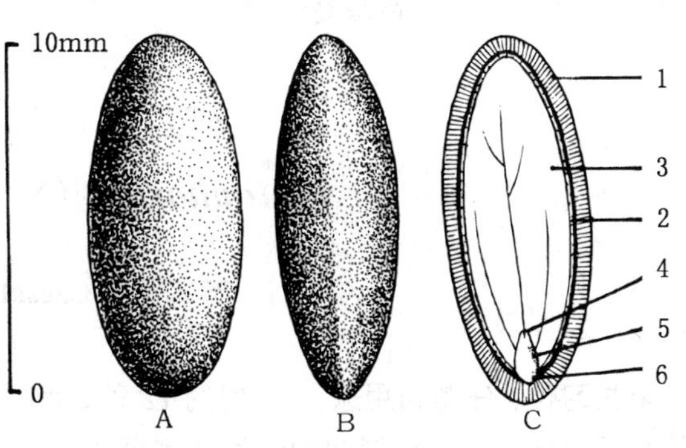

图 1　扁轴木种子外形正面（A）、侧面（B）及其纵切面（C）
1. 种皮　2. 胚乳　3. 子叶　4. 胚芽　5. 胚轴　6. 胚根
（黄应钦绘）

果实的采收调制和种子贮藏　果实成熟期长，7～12 月均可采集。树干树枝带刺，难于用手采摘，宜用高枝剪或钩刀将果穗采下，置日光下曝晒，果荚即开裂，种子脱出。鲜果出种率为 38%～45%，种子净度可达 97%～99%。千粒重约 84g，每千克纯净种子为 11 900 粒。种子可以晒干，使含水量降到 10%以下。贮藏时用袋或坛罐，置室内通风处保存。常温条件下 1 年以内发芽能力不致明显降低。贮藏于 5℃左右的低温条件下，发芽能力可保持 2 年以上。

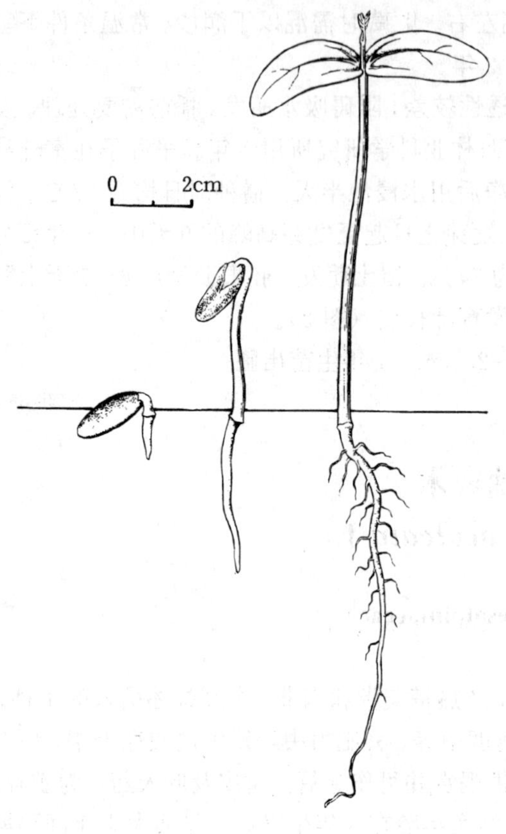

图 2　扁轴木种子萌发后第 1、2、6 天幼苗的生长情况
（黄应钦绘）

发芽和播种　种子发芽时日均温需在 20℃以上。如果温度适宜，发芽快而整齐。1987 年 9 月中旬，广西林业科学研究所在室内做过发芽测定。播前用始温 55℃水浸种，在冷却过程中浸泡 1 昼夜，发芽时日均温 27～28℃，光照为室内自然光。置床后第 3 天胚根萌出，4 天的发芽百分数为 40%～45%，9 天的发芽率为 60%。贮藏半年以上的种子播前需用始温 80℃水浸种。出土萌发。发芽后约 2 天子叶出土，6 天发出初生叶（见图 2）。

条播。每平方米播种 10～12g，覆土 1.2～2cm。目前种子较少，可将种子处理后密播，当子叶出土初生叶初露时，将芽苗移至容器或场圃。半年或 1 年生苗出圃。用于庭院绿化的需培育 2～3 年生苗。

（张声燕）

双翼豆（盾柱木）属
Peltophorum（Vogel）Benth.

（苏木科　Caesalpiniaceae）

生长习性、分布和用途　本属约 12 种，我国 1 种，引入 2 种，本文描述 2 种。落叶乔木。幼龄时忌霜冻。喜肥沃湿润的酸性沙壤至壤土。它们的名称、生长、分布和用途见表 1。

表 1　双翼豆属树种的名称、生长、分布和用途

中　名	学　名	树高（m）	胸径（cm）	分　布	用　途
盾柱木	*P. pterocarpum* (DC.) Backer	25	60	东南亚各国。琼、粤、桂栽培	材用、绿化、皮可制黄色染料
银珠（双翼豆）	*P. tonkinense* (Pierre) Gagnep.	30	80	琼、桂栽培。越南	材用、水源、绿化

开花结实　约10年生开始开花结实，正常结实年龄在15年生以后。结实大小年间隔期为1年。花两性，黄色，有香味。银珠为总状花序，盾柱木为圆锥化序，腋生或顶生。萼片5，花瓣5，均为覆瓦状排列。雄蕊10，分离，花丝稍伸出花冠外，基部密被粗毛。花药长圆形，背着。子房无柄，被毛。花柱长，柱头大，盾状、头状或盘状。胚珠3～4。在海南观测的开花结实物候期见表2。

表2　双翼豆属树种的开花结实物候期（海南）

树　种	开花期			果实成熟期		果实开裂和脱落期
	始　期	盛　期	末　期	始　期	盛　期	
盾柱木	5月下旬	6月上旬	6月下旬	8月下旬	10月中旬	熟后不开裂，悬挂树上
银珠	3月上旬	3月中旬	3月下旬	6月中旬	7月中旬	熟后不开裂，悬挂树上至12月

荚果，披针状长圆形，扁平，沿背腹两缝线具窄翅，不开裂。盾柱木每荚内种子多为3粒，银珠每荚内有种子2～4粒。无胚乳，子叶扁平。果实及种子的形态特征见表3、图1。

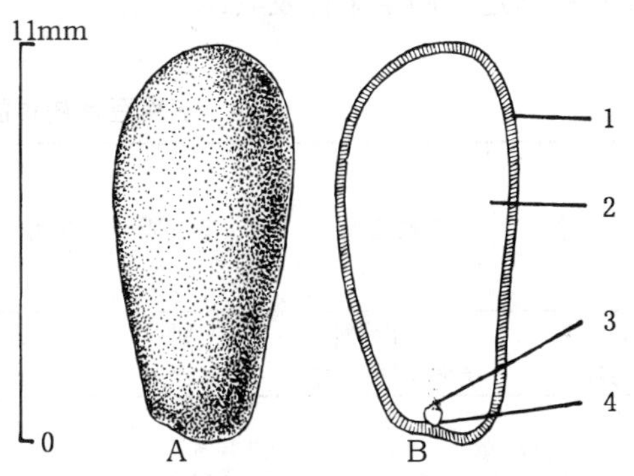

图1　盾柱木种子外形（A）及其纵切面（B）
1. 种皮　2. 子叶　3. 胚芽　4. 胚根
（黄应钦绘）

果实的采收调制和种子贮藏　果实成熟后在树上的存留期长，采种时期因此也较长，但一般仍应在成熟后1个月内采集。用竹竿敲打或用采种刀带果枝钩落后在地面拾集。果荚难开裂，采集后置烈日下曝晒，使果荚及翅干脆，然后用木棒反复敲打或用木臼冲捣，使果荚破碎，脱出种子。筛扬除去破碎荚壳等杂物即得纯净种子。但调制时不得破伤种皮。出种率和净度、质量等见表4。

贮藏前种子需晒干，使含水量在10%以下。常温条件下，室内裸露贮藏的种子，发芽能力可以保持2年。贮藏于5～8℃的低温条件下，发芽能力可以保持4年以上。

表3　双翼豆属树种果实和种子的形态特征

树　种	未熟果	成熟果实			种　子		
	颜　色	形　状	大小（cm）	颜　色	形　状	大小（mm）	颜　色
盾柱木	黄绿色	扁平，长椭圆形或椭圆形，有纵纹	长7～11 宽1.5～2 翅宽0.3～0.5	紫红色	扁椭圆形	长11～13 宽4～5 厚1.5～2	浅褐色
银珠	黄绿色	扁平，纺锤形，中部无条纹	长8～13 宽2～3 翅宽0.5～0.7	黄褐至红褐色	扁椭圆状倒卵形	长13～16.5 宽5～7 厚1.8～2	浅黄褐色

表 4　双翼豆属树种果实的出种率和种子净度、质量

树　种	出种率（%）	净　度（%）	千粒重（g）		每千克纯净种子粒数（粒）	
			一　般	变动范围	一　般	变动范围
盾柱木	15～25	98	76	73～80	13 000	12 500～13 700
银珠	15～25	98	100	80～140	10 000	7 100～12 500

发芽和播种　种子无生理性休眠，但种皮妨碍吸水，播前需作处理。随采随播的种子需用 70℃热水浸种至自然冷却；晒干的种子或经过贮藏的种子需用硫酸拌种或沸水浸种。未经处理的种子，播后发芽极不整齐，有些种子可延迟数月至第 2 年方陆续发芽。发芽时气温需在 18℃以上。1982 年和 1987 年 8 月，广西林业科学研究所对这两个树种前 1 年采得的种子做过发芽测定，测定条件及其结果见表 5。

表 5　双翼豆属树种的发芽能力及其测定条件

树　种	预处理	基　质	温度（%）	发芽势（%）		发芽率（%）	
				计算天数	一般数值	计算天数	一般数值
盾柱木	沸水浸种	纸	28	5	66	12	85
银珠	硫酸拌种 15 分钟	纸	28	5	60	13	82

出土萌发。播后 4～5 天开始萌发，胚根萌发后约 2 天子叶出土，5～6 天长出初生叶（见图 2）。

条播。每平方米播种 10～13g，覆土 1.5～2cm。苗期忌霜冻。培育半年至 1 年生苗出圃。

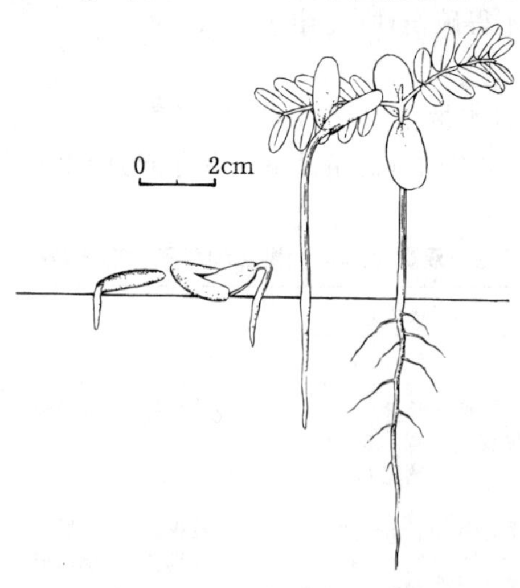

图 2　盾柱木种子萌发后第 1、2、3、7 天幼苗生长情况

（黄应钦仿《热带亚热带主要树种采种育苗技术》）

（张声燕）

无　忧　花

Saraca asoca（Roxb.）De Wilde

（苏木科　Caesalpiniaceae）

生长习性、分布和用途　无忧花属约20种，我国3种，本文描述1种。常绿乔木，高20m，胸径80cm。能耐轻霜。喜肥沃湿润的石灰岩山地或酸性土。分布于广西西南部及云南南部。越南、印度等亚洲热带地区亦产。花色橙黄或绯红，花时如火焰，供观赏。木材供建筑用。树皮及根治风湿及子宫下坠。

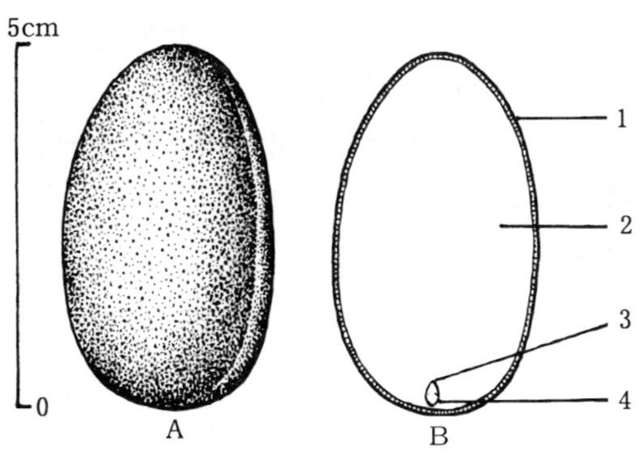

图1　无忧花种子外形（A）及其纵切面（B）
1. 种皮　2. 子叶　3. 胚芽　4. 胚根
（黄应钦绘）

开花结实　10年生左右开始开花结实，正常结实期在20年生以后。结实有大小年现象，间隔期一般为1年。伞房花序花密生，着生于前1年生的枝梢或叶腋。花两性，无瓣。小苞片呈花瓣状。花萼管状，裂片4，花瓣状，橙黄色至绯红色。雄蕊8，分离。子房具柄，胚珠6～8。据广西南部观察，3月中旬为始花期，3月下旬～4月中旬为盛花期，5月上旬为末花期；8月中旬果实开始成熟，下旬至9月中旬为果熟盛期。荚果，木质，扁平，带状长圆形，成熟时由青色转为黑褐色；长(10)15～30(45)cm，宽6～8cm；表面有光泽，有网状纹隆起；成熟后约10天开裂，种子脱落。每果有种子5～8粒。种子扁平，肾形、倒卵形或长椭圆形，长3.2～5.5cm，宽2～2.9cm。无胚乳，子叶较肥大（图1）。

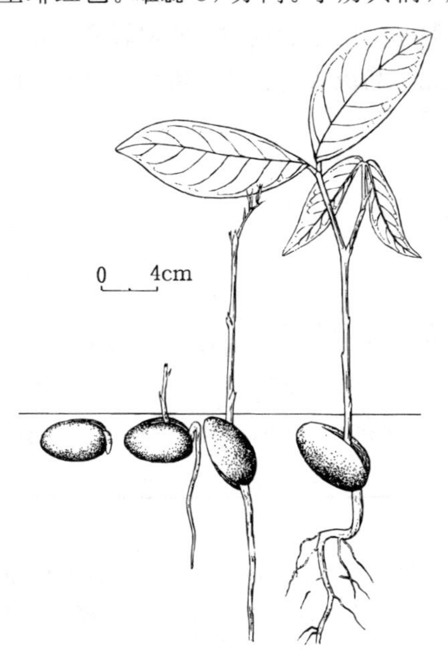

图2　无忧花种子萌发后第1、10、20、50天的幼苗生长情况
（黄应钦仿《热带亚热带主要树种采种育苗技术》）

果实的采收调制和种子贮藏　果荚尚未开裂时用采种钩刀采下，摊放于室内通风处，晾干后荚壳自行开裂，捡去果荚即得种子。荚果出种率为50%～60%。种子的净度可以高达100%。千粒重6 850g，变动在6 000～7 500g，每千克纯净种子约150粒。种子含水量35%～50%，忌失水，不能日晒，贮藏或运输均需混以湿沙。裸露存放的种子，半月以后发芽率明显降低。混湿沙贮藏的种子，发芽能力的保存

期也在 6 个月以内。

发芽和播种 种子无休眠习性，但少数种粒可能有休眠现象。发芽时日均温需在 20℃以上，一般宜随采随播。沙藏至次年春暖后播种，发芽率有所下降。1981 年 8 月 24 日日均温 28℃时，广西林业科学研究所用当年新采种子在室外沙床播种，播后 22 天（9 月 15 日）开始发芽，发芽不甚整齐，至 9 月 25 日方进入盛期；10 月以后，可能因气温转低或因种子休眠，少数种子迟至 1982 年 3 月下旬发芽。从发芽之日至发芽高峰期，20 天的发芽百分数约为 60%。从播种至发芽终止，210 天的发芽率为 80%。留土萌发。种子发芽后约 8 天上胚轴出土，16～20 天初生叶出现（图 2）。

条播。新鲜种子播前不需作任何处理。混沙贮藏过的种子，播前用冷水浸泡半天。每平方米播种 620～880g，覆土 2cm。1 年或半年生苗出圃。用于庭院栽植或作行道树，需培育 2～3 年生苗。

（王宏志）

油 楠 属

Sindora Miq.

（苏木科 Caesalpiniaceae）

生长习性、分布和用途 本属约 20 种，我国产 1 种并引入 2 种，本文描述 2 种。常绿大乔木，分枝低，主干多不通直。耐干旱，喜高温而干湿季节明显的气候，能耐轻霜。喜肥沃湿润的酸性土。材质优，宜作雕刻、收音机壳、钢琴外壳及高级胶合板等。树脂可点灯。种柄可供雕刻。它们的名称、生长、分布及用途见表 1。

表 1 油楠属树种的名称、生长、分布和用途

中 名	学 名	树高 (m)	胸径 (cm)	分 布	用 途	供 稿
油楠	*S. glabra* Merr. ex de Wit	30	120	海南。越南	材用、树脂、绿化、水源、雕刻	613
东京油楠	*S. tonkinensis* A. Cheval. ex K. et S. S. Larsen	25	80	越南。广西引种	材用、绿化、雕刻	601

开花结实 约 13 年生开始开花结实，正常结实期在 25 年生以后。结实大小年间隔期为 1 年，但不甚明显。花两性。圆锥花序顶生，或腋生于上部。萼管短，裂片 4，镊合状排列，外被软刺毛。花瓣 1，无爪。雄蕊 10，9 枚短而偏斜合生。花药背着，纵裂，另 1 枚无花药。子房具短柄，胚珠 2～7 颗，花柱长，拳卷。在广西南宁及海南乐东观察的开花结实物候期见表 2。

荚果，扁平。油楠果具刺，每果有种子 1（2）粒，种柄与种子等长。东京油楠果无刺，每

表 2　油楠属树种的开花结实物候期

树　种	观察地点	观察年份	开花期			果实成熟期		种子脱落期
			始　期	盛　期	末　期	始　期	盛　期	
油楠	乐东	1988	4月下旬	5月中下旬	6月上旬	8月下旬	9～10月	10～11月
东京油楠	南宁	1978	6月上旬	6月中旬	6月下旬	8月中旬	8月下旬	9～10月

果有种子1～5粒，种柄略小于种子。成熟后约1周，果荚开裂，种子脱落。种子基部有黄色肉质种柄。种皮黑色，坚硬。无胚乳，子叶肥大。果荚及种子形态见表3、图1。

表 3　油楠属树种果实和种子的形态特征

树　种	果实			种子		
	形　状	大小（cm）	颜　色	形　状	大小（mm）	色　泽
油楠	扁圆或椭圆，喙尖	长4～6 宽4～5	黑褐色	半圆或近圆卵形	长13～20	黑色，有光泽，种柄黄色
东京油楠	扁长椭圆，喙尖	长5～12 宽3.8～6	黑褐色	矩圆或扁椭圆形	长11～20 宽11～15 厚7～10	黑色，有光泽，种柄黄色

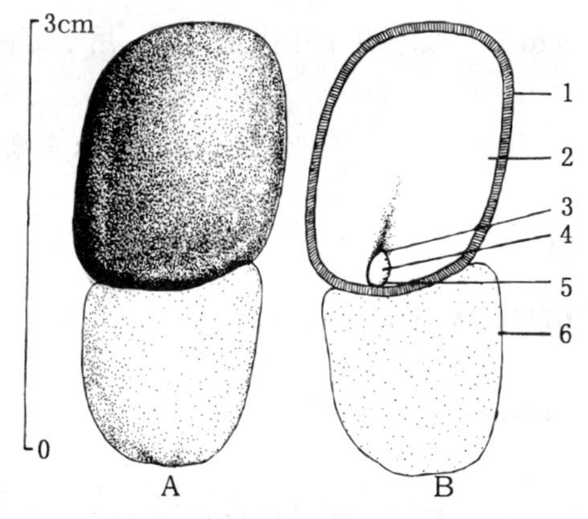

图1　东京油楠种子外形（A）及其纵切面（B）
1. 种皮　2. 子叶　3. 胚芽　4. 胚轴
5. 胚根　6. 种柄
（黄应钦绘）

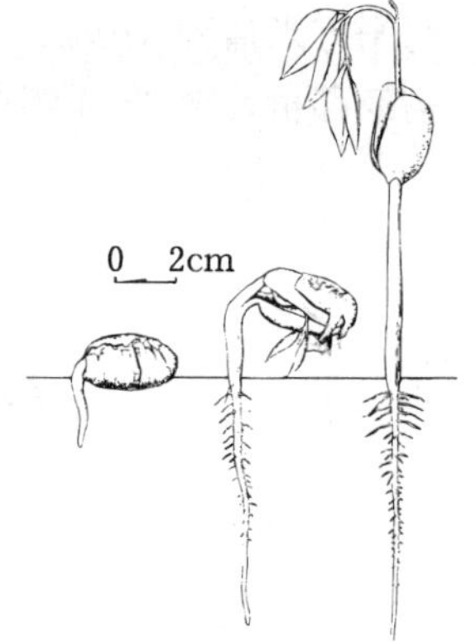

图2　东京油楠播种后第11、28、35天幼苗的生长情况
（黄应钦仿《热带亚热带主要树种采种育苗技术》）

果实的采收调制和种子贮藏　果荚开裂前及时用钩刀带荚果采下，采下的果荚可置日光下曝晒使之开裂，或摊放于室内，任其晾干后自行开裂。除去果荚即得带有种柄的种子。种柄与种子胶着牢固，不易分离，运输或贮藏时一般都带有种柄。种子需晒干，含水量宜在12%以下。种子可以摊放于室内干燥处或装入瓦缸内贮藏。常温条件下，室内贮藏的种子发芽能力可以保持2年。种子的质量等见表4。

发芽和播种　种子无生理性休眠，但种皮致密，妨碍吸水。发芽时日均温需在 20℃以上，播前需用始温 80～90℃水浸种。未经处理的种子，发芽进程可能延续半年以上。用热水烫种并在冷却过程中继续浸泡 1～2 昼夜以后，将种子置竹筐内用木棒冲捣，使种柄与种子分离并洗去种柄，以免播种后影响发芽或引起霉病。1987 年 9 月，广西林业科学研究所和中国林业科学研究院热带林业试验站，分别对油楠和东京油楠用始温 80～90℃水浸种处理后进行过发芽测定，情况见表 5。

表 4　油楠属树种的出种率和种子净度、质量（带种柄）

树　种	出种率（%）	千粒重（g）		每千克纯净种子粒数	
		一　般	变动范围	一　般	变动范围
油楠	35	3 090	2 500～3 500	320	280～400
东京油楠	25～30	2 550	2 000～3 000	390	330～500

表 5　油楠属树种的发芽能力及其测定条件（室外）

树　种	基　质	温度（℃）	发芽势（%）		发芽率（%）	
			计算天数	一般数值	计算天数	一般数值
油楠	场圃	28～30	10	63.5	40	80
东京油楠	沙	26～28	11	53	35	75

出土萌发。播种后约 11 天胚根开始萌发，再过 16～17 天子叶出土，同时出现初生叶（图 2）。

条播。每平方米播种 370～500g，覆土 2～2.5cm。一般培育 1 年生苗出圃。用于庭院绿化植树的，需经移植，育 3～4 年苗出圃。

（王宏志）

酸　豆

Tamarindus indica L.

（苏木科　Caesalpiniaceae）

生长习性、分布和用途　酸豆属只有 1 种。原产非洲中部，现已广泛栽培于全球热带亚热带地区。我国粤、桂、滇、台、闽、琼以及川南有栽培。常绿大乔木，高 25m，胸径 120cm。抗风性能强。对土壤要求不严格，在深厚肥沃的粘性土或海滨沙地都可正常生长发育。木材红色，坚硬，适于制作各种家具、农具。果肉味酸，含酒石酸、钙、维生素 B_1、烟碱酸等，营养丰富，可生食，调以糖浆可作清凉饮料，或制成果酱。树形美观，宜作风景树。酸豆也是紫胶虫寄主树。

开花结实　10 年生左右开始开花结果，正常结实期在 15 年生以后。每年结实量均较多，无大小年现象。花两性。总状花序顶生或腋生于上部。萼筒陀螺形，裂片 4。花瓣前方 2 片退化呈鳞片状或刚毛状，仅后方 3 片发育，淡黄色，杂以紫红色条纹。雄蕊 1 束，仅 3 枚发育，

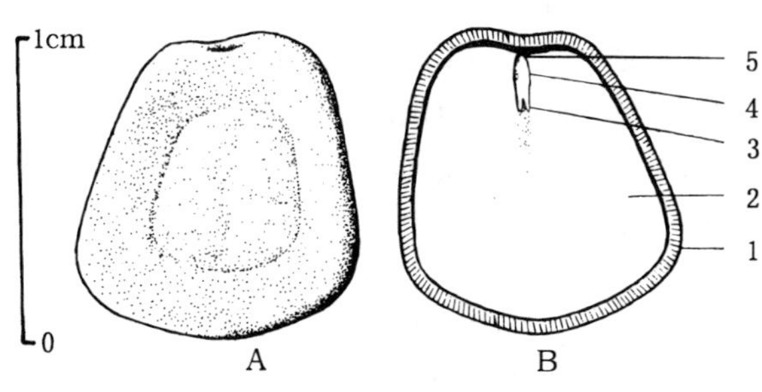

图 1　酸豆种子外形（A）及其纵切面（B）
1. 种皮　2. 子叶　3. 胚芽　4. 胚轴　5. 胚根
（黄光郁、黄应钦绘）

花药背着。具花盘。子房具柄，胚珠多数。据海南儋州观测，4 月中旬为始花期，5～6月为盛花期，7 月下旬为末花期，有时花期可延至 10 月；果熟期为翌年1～4 月，主要成熟期在2～3月。荚果，肉质，圆柱形，略弯，常不规则缢缩，长5～15cm，径1.5～2cm。果熟时黄锈色，被秕糠状鳞片。外果皮脆壳质，中果皮厚纤维肉质，内果皮薄膜质。果实成熟后不开裂，种子间有隔膜。种子深褐色，斜卵圆形或斜长方形，扁，无胚乳，径约1～1.5cm（图 1）。

果实的采收调制和种子贮藏　果实呈浅咖啡色或褐色时，用高枝剪或采种刀将荚果采下，放置室内 3～5 天，待荚果充分后熟，用细沙反复摩擦，除净果胶，再用清水搓擦冲洗，即得播种材料。鲜果荚的出种率约 40%左右。种子净度可以达到 98%。种子千粒重约 900g，每千克有纯净种子约 1 120 粒。调制出来的种子含水量约 17%。种子忌失水，不宜日晒或裸露陈放。运输时宜用布袋或木箱包装，每袋（箱）不宜超过 20kg，运抵后即行播种。贮藏的种子需混以湿沙，贮藏期一般为 3 个月以内。

发芽和播种　种子无休眠现象，宜随采随播。1989 年 4 月 21 日，华南热带作物研究所在海南儋县室外沙床用新鲜种子做过发芽测定：播后 7 天开始发芽，5 月 11 日发芽结束，21 天的发芽率为 66%。出土萌发。发芽后 8 天展出初生叶。种子萌发和幼苗生长情况见图 2。

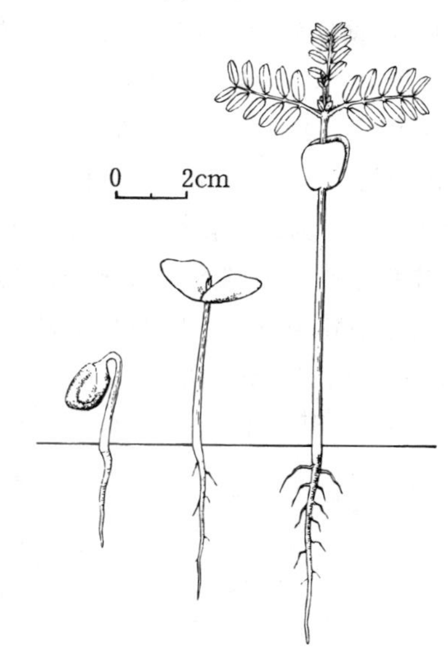

图 2　酸豆种子萌发后第 2、4、17 天的幼苗生长情况
（黄光郁、黄应钦绘）

条播。每平方米播种 150～200g，覆土 1.5～2cm。1 年生苗出圃。用作风景树的苗木需培育 2 年出圃。

（丁慎言）

翅荚木（任豆）

Zenia insignis Chun

（苏木科　**Caesalpiniaceae**）

生长习性、分布和用途　翅荚木属为我国特产，只有本文描述的 1 种。落叶或半落叶大乔木，高达 30m，胸径达 1m。喜肥沃湿润深厚的土壤，自然分布区主要为石灰岩山地，酸性土亦能生长。能耐 −6℃左右低温及冰雪。原产我国北热带至中亚热带南部的滇、黔、桂、粤、湘等地，现已引种至中亚热带的广大地区。木材纤维长而韧，为优质浆粕及胶合板原料；是紫胶虫寄主，叶可喂牲畜，萌芽力强，有“砍头树”之称，宜营造纸浆、薪柴，饲料，肥料兼用林。翅荚木大树日渐稀少，亟需保护，在《中国植物红皮书》中被列为稀有种。

开花结实 7 年生开始开花结实，正常结实年龄在 15 年生以后，结实有大小年现象，间隔期为 1 年。花两性，圆锥花序顶生。萼片 5。花瓣 5，最上片稍大，覆瓦状排列。发育雄蕊 4 (5)，花药基着，花丝短。花盘小，深波状分裂。子房具短柄，胚珠 6～9。花柱钻状，稍弯曲，柱头小。据广西南宁 1978～1984 年的物候观测，4 月上旬，当年新枝顶端形成疏松的圆锥状复伞花序，4 月下旬为始花期，下旬末～5 月上旬初为盛花期，上旬末花期结束；7 月上旬果始熟，中旬为成熟盛期。荚果，膜质，压扁，阔披针形或近矩圆形，腹缝一侧具宽翅，中偏上侧有主脉条纹，两侧网纹明显。成熟时荚果由绿色转变为红棕色，干后棕褐色，近革质，长 9～12 (15) cm，宽 3～4cm，翅宽 5～10mm。成熟后果荚不开裂，可在树上维持 3 个月左右。每荚内有种子 5～10 粒，种子扁圆形或椭圆形，黄褐色或茶褐色，有光泽，长 5～7mm，宽 4～6mm，厚 1.5～2mm，有胚乳，子叶扁平。形态见图 1。

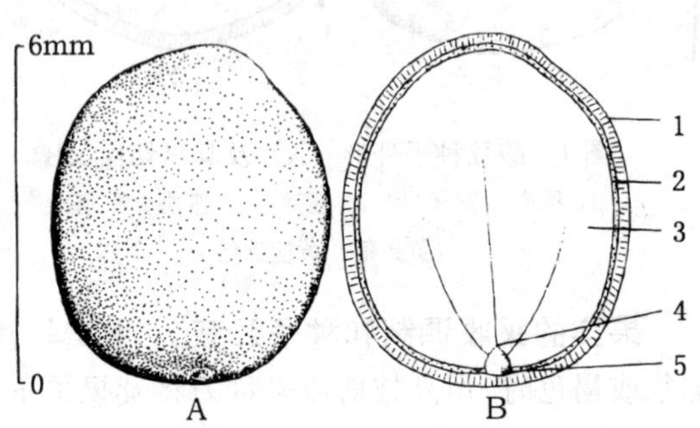

图 1 翅荚木种子外形 (A) 及其纵切面 (B)
1. 种皮 2. 胚乳 3. 子叶 4. 胚芽 5. 胚根
(黄应钦绘)

果实的采收调制和种子贮藏 7 月下旬～9 月间，果荚尚未飘落前，用采种刀采下果穗，摘取果荚，置日光下曝晒，充分干燥后用木棒敲打并搓揉，使膜质果荚破碎，脱出种子，经筛选簸扬，除去破碎的果荚和其它杂质，即得纯净种子。出种率 15%～25%，净度可达98%～99%。千粒重 46～50g，每千克有纯净种子 20 000～22 000 粒。种子需晒干，贮藏的种子含水量宜在 10%以下，装于布袋或坛缸均可。据苏冬梅 (1991) 报道，翅荚木种子室温条件下贮藏 1 年，发芽率为 66%，贮藏 2 年时发芽率为 61%，贮藏 3 年后发芽率仍有 52%。由此推测，在 5～ 8℃下低温贮藏，发芽能力可以保持 3 年以上。

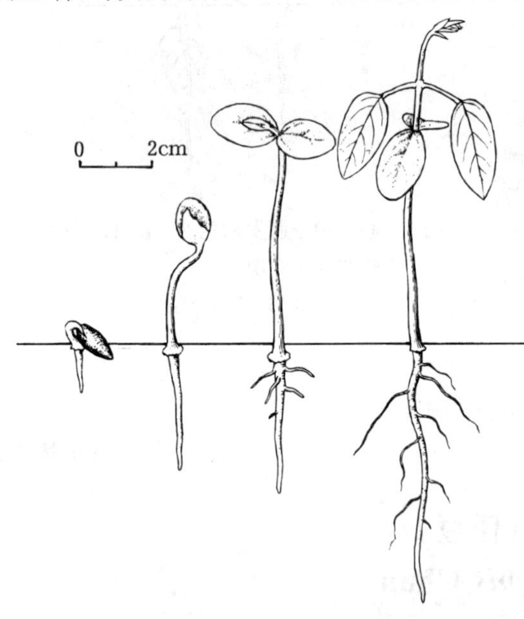

图 2 翅荚木种子萌发后第 3、8、16、25 天幼苗的生长情况
(黄应钦仿《热带亚热带主要树种采种育苗技术》)

发芽和播种 种子无休眠习性。发芽时日均温宜在 18℃以上。播前用始温 60～70 ℃

水浸种，可加速种子发芽。苏冬梅（1992）的试验表明，在38℃和96%的相对湿度环境中处理35小时后，翅荚木种子的室内发芽率可以从对照的6%提高到47%，相应的田间出苗率也高于对照。1982年3月中旬，日均温约19℃，广西林业科学研究所在室外沙床播种作过发芽测定：播前用始温60℃水浸种，冷却后继续浸泡1昼夜，播后4天开始发芽，6天中发芽百分数为65%，12天的发芽率为83%。出土萌发。胚根萌发后8天子叶出土，15～20天长出初生叶。翅荚木种子的萌发和幼苗初期生长情况见图2。

条播。每平方米播种5～8g，覆土1.5～2cm。培育半年或1年生苗出圃。近年生产中多将催芽萌动的种子点播至容器内育苗3个月出圃。湿润肥沃的林地也可直播。

（韦增健）

金合欢属

Acacia Willd.

（含羞草科　Mimosaceae）

生长习性、分布和用途　本属约900多种，我国约10种，又引入40多种，其中基本成功的约20种。本文描述12种。常绿稀落叶乔木、灌木或藤本，高4～30m。速生。广布于热带至南亚热带地区，颇耐干热。多数种对土壤肥力的要求不苛，适生于酸性或石灰岩土壤。有根瘤菌。珍珠相思入冬形成球形花蕾，银色，可供观赏。为热带、亚热带地区用于保持水土，防风固沙及栲胶、浆粕、用材、薪柴、绿肥的重要造林树种。广东湛江地区引种的厚荚相思8年生时树高12m，平均胸径12.8cm（潘志刚、游应添，1994）。金合欢属的名称、生长、分布及用途见表1。

表1　金合欢属树种的名称、生长、分布和用途

中　名	学　名	树　高 (m)	胸　径 (cm)	分　布	用　途	供　稿
耳叶相思	*A. auriculiformis* A. Cunn. ex Benth.	30	60	大洋洲。琼、闽、粤、桂栽培	材用、水源、防风、蜜源、薪柴	602
儿茶	*A. catechu* (L.) Willd.	15	20	印度、非洲。滇、桂、粤、琼、浙、台栽培	鞣料、药用、珍贵用材、紫胶虫寄主、薪柴	602
台湾相思	*A. confusa* Merr	20	60	闽、台、琼、粤、桂、滇、川、浙栽培。菲律宾、印度尼西亚	材用、鞣料、水源、薪柴	602
厚荚相思	*A. crassicarpa* A. Cunn. ex Benth.	20	40～60	澳大利亚。琼、桂、粤、闽栽培	材用、纸浆、薪柴、水源	602
肯氏相思	*A. cunninghamii* Willd.	6～8	8～12	澳大利亚。琼、粤、桂栽培	薪柴、纸浆、保土、观赏	602
银荆树	*A. dealbata* Link.	15	20	澳大利亚。滇、黔、川、桂、台栽培	鞣料、材用、观赏、改土、绿化、绿肥	801

（续）

中 名	学 名	树 高（m）	胸 径（cm）	分 布	用 途	供 稿
金合欢	*A. farnesiana* (L.) Willd.	9	15	热带美洲。台、闽、粤、桂、滇、川栽培	材用、树胶、鞣料、薪柴、绿篱	602
灰金合欢	*A. glauca* (L.) Moench	4～5	—	西印度群岛。闽、粤、桂、琼、川栽培	紫胶虫寄主、薪柴、绿肥	602
绢毛相思	*A. holosericea* A. Cunn. ex G. Don	6～8	10～15	澳大利亚。桂、粤栽培	绿化、材用、薪柴、保土	602
马占相思	*A. mangium* De Willd.	30	40～50	澳大利亚。琼、粤、桂栽培	绿化、材用、纸浆、薪柴、改土、鞣料	602
黑荆树	*A. mearnsii* De Willd.	10～18	25	澳大利亚。粤、桂、滇、川、台、闽、赣、浙栽培	鞣料、观赏、材用、纸浆、薪柴、绿肥	801
珍珠相思	*A. podalyriifolia* A. Cunn. ex G. Don	6～8	10	澳大利亚。粤、桂、闽栽培	观赏、材用、薪柴	602

开花结实 金合欢属树种的开花结实期较早，灰金合欢等灌木类树种一般1年生即开始结实；小乔木类的肯氏相思、金合欢、绢毛相思、珍珠相思的始花期一般为2～3年生，正常结实期在4年生以后；乔木类的耳叶相思、马占相思、银荆树、黑荆树的始花期一般为3～5年生，正常结实期在6年生以后；乔木类的厚荚相思、儿茶、台湾相思的始花期一般为5～7年生，正常结实年龄在10年生以后。1988年利用巴布亚新几内亚的几个种源在广东湛江营造的厚荚相思母树林，4年生时有48%～76%的植株开花；2.4hm^2的母树林在1991年收获种子13kg（潘志刚、游应添，1994）。结实无大小年现象，影响结实的重要因素是越冬期的低温。遇特寒年份，枝梢受冻害，开花结实量骤减。正常气温条件下每年结实均较多。

花两性，3～5基数。头状或穗状花序腋生。耳叶相思、儿茶、厚荚相思、肯氏相思、金合欢、绢毛相思、马占相思等为穗状花序；台湾相思等为头状花序；银荆树、黑荆树、珍珠相思等为头状花序复组成总状或圆锥花序式排列。灰金合欢和儿茶的花白色至淡黄色，台湾相思花瓣淡绿色，其余9种的花为黄色至深黄色。萼钟状或漏斗状，齿裂。花瓣分离或于基部合生。雄蕊多数，花丝分离，或于基部稍连合。子房无柄或具长柄，胚珠多数，花柱丝状，柱头小。黑荆树、银荆树分布较广，各地的物候期有差异。据广西南宁、云南昆明和浙江等地观察，这12个树种的开花结实物候期见表2。

表2 金合欢属树种的开花结实物候期

树 种	观察地点	观察年份	开花期			果实成熟期		种子脱落期
			始 期	盛 期	末 期	始 期	盛 期	
耳叶相思	南宁	1978～1984	7月下旬	8月上中旬	10月下旬	翌年4月下旬	5月上中旬	5月下旬开裂，露出种子，1个月内种子脱落
儿茶	南宁	1978～1984	7月上旬	7月下旬	8月下旬	翌年2月下旬	3月中下旬	荚果不开裂，4月上旬～5月中旬落果

（续）

树　种	观察地点	观察年份	开花期			果实成熟期		种子脱落期
			始　期	盛　期	末　期	始　期	盛　期	
台湾相思	南宁	1987～1988	4月上旬	4月中旬	5月下旬	7月下旬	8月上旬	8月中旬至8月下旬开裂，种子脱落
厚荚相思	南宁	1987～1988	10月中旬	11月上旬	12月下旬	翌年4月下旬	5月上中旬	5月中旬～6月上旬开裂，种子脱落
肯氏相思	南宁	1986～1988	10月中旬	11月下旬	翌年1月下旬	4月下旬	5月中下旬	成熟后即开裂，半个月左右种子脱落
银荆树	浙江	—	2月	～	3月	5月	6月	熟后开裂，种子脱落
	云南	—	12月上旬	翌年1月	3月	5月	6月	
金合欢	—	—	3月	4～5月	6月	7月	8～10月	10～12月
灰金合欢	南宁	1986～1988	5月中旬	6月中旬	11月下旬	11月下旬	12月下旬	翌年1月中旬～6月中旬落果
绢毛相思	南宁	1986～1988	12月中旬	12月下旬	翌年1月下旬	4月下旬	5月中下旬	熟后开裂，1个月内种子脱落
马占相思	南宁	1986～1988	9月下旬	10月下旬	11月中旬	翌年5月中旬	5月下旬	5月下旬～6月下旬开裂
黑荆树	南宁	1978～1984	4月上旬	4月中旬	4月下旬	7月下旬	8月中旬	熟后即开裂，种子脱落
	浙江	—	11月	11月	翌年6月	—	6月中下旬	
珍珠相思	南宁	1987～1988	1月下旬	2月中旬	3月上旬	5月中旬	5月下旬	5月下旬～6月上旬开裂

荚果，长圆形或条形，多扁平，稀圆筒形。每果有种子3～10粒，稀1或20多粒。种子无胚乳。果实及种子形态见表3、图1。

表3　金合欢属树种果实和种子的形态特征

树　种	未熟果颜色	成熟果实			成熟种子		
		形　状	大小（cm）	颜色	形　状	大小（mm）	色　泽
耳叶相思	青色	带状，螺旋状扭曲成环状	长8～15 宽约1	黑褐色至灰黑色	扁椭圆形，光滑，种脐附丝状体	长4.5～5.5 宽3.5～4.5 厚1.8～2.2	深褐色，有光泽，丝淡黄色
儿茶	赭色	薄带状，条形	长7～9 宽1.4～1.7	紫褐色	扁圆形或扁椭圆形	长5.5～10.5 宽5～9.5 厚1.3～1.6	棕褐或黄褐色
台湾相思	绿色	条形，压扁，微缢缩	长8～11 宽0.8～1.1	褐色或棕褐色	扁椭圆形，两侧有明显的椭圆环状纹，光滑	长4.9～6.2 宽3.5～4.2 厚1.8～2.2	褐色至深褐色

（续）

树　种	未熟果颜色	成熟果实			成熟种子		
		形　状	大小（cm）	颜色	形　状	大小（mm）	色　泽
厚荚相思	青色	条形，不扭曲	长 5～9 宽 2～3	褐色	扁椭圆形，丝状体折叠于种子一端	长 3.5～5 宽 2.5～3 厚 1～1.5	黑色，有光泽，丝浅黄色
肯氏相思	绿色	条形，微扁，扭曲，荚室间缢缩，呈浅波状	长 5～10 宽 0.4～0.5	黄褐至棕褐色	扁椭圆形，光滑，丝状体折叠于种子一端	长 4.2～5.3 宽 2～2.5 厚 1.4～1.6	黑褐色或棕褐色，有光泽，丝黄色
银荆树	绿色	带状扁平	长 6～9 宽约 1	紫褐色，有光泽	扁椭圆形，种脐位于种子一端	长 3.5～5 宽 3～4.5 厚 1～1.5	黑褐色，有光泽
金合欢	绿色	圆柱形，两侧有纵槽纹	长 4～11 径 1～1.3	棕褐色或黑褐色	扁椭圆形、卵形或不规则形，光滑	长 6～9.5 宽 4.5～7 厚 3～4.5	棕褐色
灰金合欢	青绿	带状，荚室微隆起	长 4～9 宽 1.2～1.4	红褐色	扁椭圆形，两侧各具一环状痕	长 3～4 宽 2.5～3.5 厚 2.3～2.5	棕褐色
绢毛相思	绿色	条形，微扁，荚室隆起，螺旋状扭曲成环	长 3.5～4.7 宽 0.3～0.5	黑褐色	扁椭圆形，种脐附丝状体	长 3～3.3 宽 2～2.3 厚 1.3～1.5	黑色，有光泽，丝金黄色
马占相思	绿色	条形，压扁，卷曲，荚室间缢缩	长 7～14 宽约 0.3	黑褐色	扁椭圆形，光滑，种脐附丝状体	长 3～4 宽 2.3～2.7 厚 1.4～1.7	黑色，有光泽，丝金黄色
黑荆树	绿色	条形，荚室间缢缩	长 3～9 径 0.5～0.8	暗褐色	扁倒卵形、扁椭圆形，种脐位于种子的一端	长 4～5.5 宽 3～4.5 厚 0.5～0.8	黑褐色，有光泽
珍珠相思	绿色	长椭圆形或弯形，压扁，荚室稍隆起，下端凹尖或尾尖	长 6～9.5 宽 1.4～2.2	暗褐至棕褐色	扁椭圆形，光滑，丝状体折叠于种子一端	长 5.5～6 宽 3.5～4.2 厚 1.5～3	黑色，有光泽，丝浅黄褐色

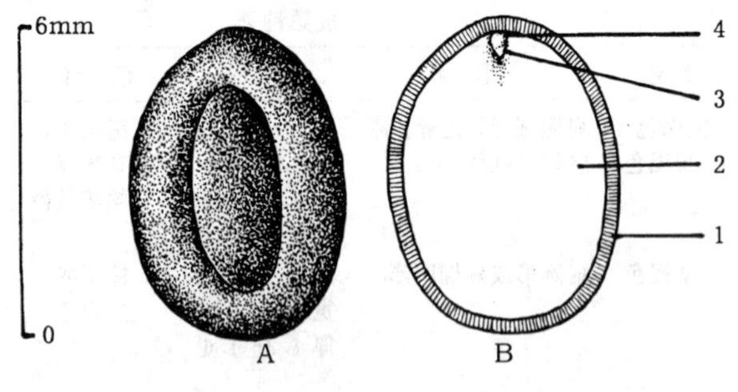

图 1　台湾相思种子外形（A）及其纵切面（B）

1. 种皮　2. 子叶　3. 胚芽　4. 胚根

（黄应钦绘）

果实采收　由于地理、遗传等因素，金合欢属同一树种各植株间的生长速度、干形、抗寒力等差异较大，采种时应注意选择优良母树。对一些抗寒力较低的种类，如耳叶相思、马占相思等，以及浙江等地的黑荆树，还应选择抗寒能力较强的植株作采种母树。金合欢属的荚果成熟后多数开裂，其中耳叶相思、肯氏相思、珍珠相思、马占相思和台湾相思等，成熟期整齐，成熟后开裂也较快，台湾相思的种子 2～3 天内便可全部散落，果熟期应密切观察，及时采收，以免过期种子落地，难以收集。儿茶、金合欢的荚果成熟后不开裂，在树上的存留期也较长，但存留过久，种子易遭虫蛀，也应及时采收。采种时可用高枝剪或采种刀带果穗采下。

表 4　金合欢属树种的出种率、种子净度和质量

树　种	出种率（%）	净度（%）	千 粒 重（g）	每千克纯净种子粒数（万粒）
耳叶相思	3～5	97～99	20～23	4.3～5.0
儿茶	16	97～99	30～55	1.8～3.4
台湾相思	15～20	98～100	25～30	3.3～4.0
厚荚相思	18	97～99	20～25	4.0～5.0
肯氏相思	17	96～98	10～11	9.0～10
银荆树	19	98～100	14～17	5.9～7.2
金合欢	56	98～100	120～130	0.75～0.85
灰金合欢	55	97～99	14～15.5	6.6～7.2
绢毛相思	19	96～98	9.5～17.5	5.7～10.5
马占相思	4～6	96～98	9～12	8.3～11
黑荆树	17	98～100	9～11	9～11
珍珠相思	17	96～98	27～30	3.3～3.7

果实的采收调制和种子贮藏　采集的果荚置日光下曝晒，开裂后敲打或搓揉，脱出种子。儿茶、金合欢的果荚不开裂，晒干后可用木棒敲打或反复搓揉，种子方可脱出。儿茶种子忌曝晒，不宜过分失水，含水量宜保持在15%左右。本文描述的其余11个种的种子，可在日光下充分曝晒，含水量应降到10%以下贮藏。常温条件下，儿茶种子的裸露贮藏期一般为4个月以内，超过4个月发芽率便开始下降；超过半年，发芽率将降至30%以下；如果混以干润沙贮藏，发芽能力可保持1年左右。其余11个种的种子可以干藏，装袋或装于坛、缸置室内荫凉干爽处，常温条件下，发芽能力可以保持2年；在3～8℃的低温条件下贮藏，发芽能力可保持3年以上。在贮藏中，有的树种的种脐易受虫蛀，会影响发芽。春、夏、秋3季发虫率较高，应经常注意检查翻动，及时取出曝晒，搓去虫卵。金合欢属12个树种的种子净度、质量数据见表4。

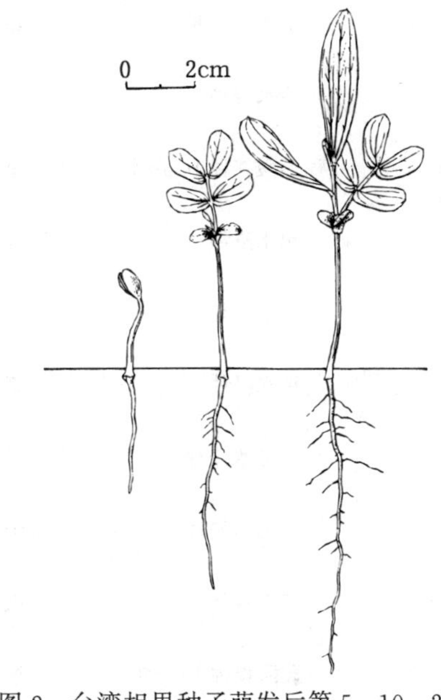

图 2　台湾相思种子萌发后第5、10、25天幼苗的生长情况
（黄应钦仿《主要树木种苗图谱》）

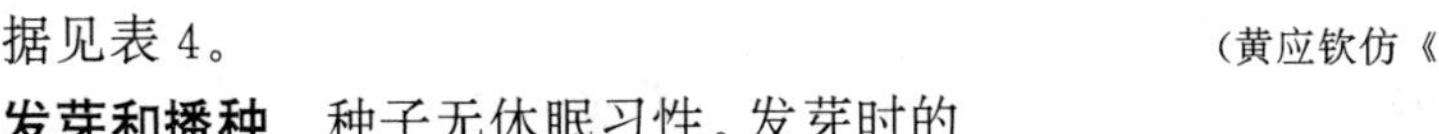
发芽和播种　种子无休眠习性。发芽时的日均温宜在20℃以上。儿茶种子的成熟期较早，3月下旬调制出种子后，4月上旬即宜播种。其余11个种的种子成熟期均在4～8月，其中以4、5、6月成熟的种子最多，适值高温季节，

均适宜随采随播，也可以贮藏至次年或第3年春播。播前除儿茶等少数树种可以用冷水浸种外，其余11个种的种子均须用热水、沸水浸种或用硫酸浸蚀。一般在幼龄母树上所采的种子或新采收未经晒干的种子宜用始温80℃热水浸种；母树年龄较大的或晒干后经过贮藏的种子，宜用沸水浸种或浓硫酸浸蚀。1981～1987年，广西林业科学研究所先后对金合欢属这12个树种在室内用发芽皿进行3次发芽测定，结果见表5。

出土萌发。台湾相思种子发芽后2～5天子叶出土，5～10天发出初生叶，生长情况见图2。

金合欢属树种育苗时多采用经过预处理的种子，先在沙床或苗床内密播，俟子叶出土、初生叶尚未发生或刚发生时，移芽苗至容器内培育3个月出圃造林。圃地育苗多用条播，每666m^{2}播种量为：金合欢、台湾相思8～10kg，儿茶、厚荚相思、珍珠相思3～5kg，耳叶相思、银荆树、灰金合欢、肯氏相思、黑荆树、马占相思0.6～1kg；绢毛相思0.4～0.6kg。覆土1～1.5cm。半年或1年生苗出圃。

表5 金合欢属树种的发芽能力及其测定条件

树种	预处理①	基质	气温（℃）		发芽势（%）		发芽率（%）		
			室内	室外	计算天数	一般数值	计算天数	一般数值	变动范围
耳叶相思	80℃水浸种	纸	27	—	6	62	15	85	80～90
儿茶	冷水浸种	沙	—	22～24	5	48	16	55	50～80
台湾相思	硫酸拌种10分钟	纸	27	—	6	52	15	71	60～90
厚荚相思	沸水浸种	沙	—	22～24	5	56	15	70	65～90
肯氏相思	沸水浸种	纸	28	—	4	45	13	62	55～90
银荆树	沸水浸种	纸	22～24	—	4	55	14	72	65～90
金合欢	80℃水浸种	纸	28	—	6	42	15	55	50～80
灰金合欢	80℃水浸种	纸	28	—	—	不明显	9	50	45～70
绢毛相思	沸水浸种	纸	28	—	4	62	10	81	70～90
马占相思	硫酸拌种10分钟	纸	28	—	6	61	14	76	70～90
黑荆树	80℃水浸种	纸	20～24	—	6	56	14	72	65～90
珍珠相思	100℃水浸种	纸	28	—	5	58	12	71	60～90

① 水温均指浸种时的初始温度

（韦增健）

海红豆（孔雀豆）

Adenanthera pavonina L. var. *microsperma* (Teijsm. et Binnend.) Nielsen

（含羞草科　Mimosaceae）

生长习性、分布和用途　海红豆属约10种，我国只有本文描述的1变种。落叶乔木，高30m，胸径达60cm。喜肥沃湿润疏松的酸性土。能耐0℃的极端最低气温，但忌严重霜冻。分布于闽、台、琼、粤、桂、黔、滇。东南亚亦有分布。心材坚实，供造船和制枪托等用，并可作红色染料。种子鲜红色，可作装饰品。全株有毒。

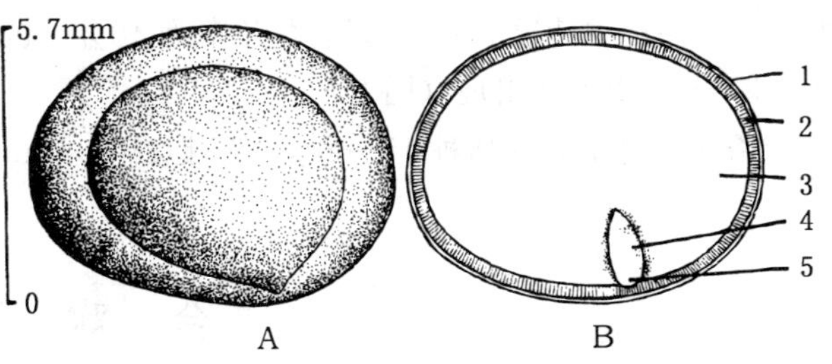

图1　海红豆种子外形（A）及其纵切面（B）
1. 外种皮　2. 内种皮　3. 子叶　4. 胚轴　5. 胚根
（黄应钦绘）

开花结实　约8年生开始开花结实，正常结实期在15年生以后。结实大小年间隔期为1年，但不甚明显。影响结实的主要原因是水肥条件，在良好的立地上，每年都能正常结实。花两性，腋生。穗状的总状花序，或排列为顶生的圆锥状花序，长12～16cm。花萼钟状具5短齿。萼及花梗被金黄色柔毛。花瓣5，白色或淡黄色。雄蕊10，分离，花药卵形，顶端具1脱落性腺体。子房具短柄，胚珠多数。据广西南宁1978～1986年的观测，5月中旬为始花期，下旬为盛花期，6月上旬为末花期；9月上旬果实开始成熟，9月中下旬为果熟盛期。荚果，成熟时由青转变为黄褐色至褐色，带状扁平，盘旋；长9～23cm，宽1～1.5cm，厚约4mm。9月下旬～11月果荚陆续开裂，不脱落，开裂后果瓣扭曲，种子外露。10月上旬至翌年3月中旬种子逐渐脱落。种子宽卵形至扁椭圆形，鲜红色有光泽，长6.5～7.5mm，宽5.7～6.8mm，厚4～5mm，无胚乳（图1）。

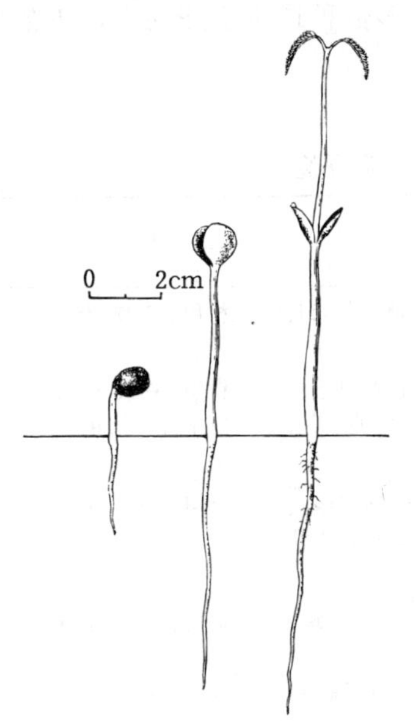

图2　海红豆种子萌发后第3、4、6天幼苗的生长情况
（黄应钦仿《热带亚热带主要树种采种育苗技术》）

果实的采收调制和种子贮藏　9月下旬～10月上旬，当果实成熟且部分果实逐渐开裂时，用高枝剪或采种钩刀带果穗采下。如果一部分采回的果荚尚呈青色或黄色，可在室内堆放3～5日，俟果荚全部转为黄褐色时置日光下曝晒，开裂翻出种子后用木棒敲打，脱出种子，清除果瓣并筛去果梗等杂质，即

得种子。果实出种率为23%～25%。种子的净度常为96%～100%，含水量约10%。千粒重90～132g，每千克有纯净种子7 600～11 000粒。种子可用袋、箱、坛、缸等容器干藏，置室内避光干爽处。常温条件下贮藏期约为1年，1年以后发芽能力逐渐下降。

发芽和播种 种子无生理性休眠，但种皮坚厚致密，妨碍吸水，播前需经处理。发芽时日均温需在20℃上下。1981年11月上旬和1982年3月中旬，广西林业科学研究所先后用硫酸浸种、始温80～90℃热水烫种和不加处理等3种方法在室外沙床播种，播种时日均温18～20℃。用浓硫酸浸种30分钟的，播后第2天开始发芽，3～4天便进入发芽盛期，第8天发芽结束；从播种之日起算，到发芽盛期的4天，发芽百分数为78%；以10天计，发芽率为90%。用热水烫种的播后3～4天开始发芽，发芽率为85%～90%。不作处理的种子播后约1周开始发芽，发芽过程持续3～5个月，发芽率为40%～60%。出土萌发。胚根萌发后约3天子叶出土，5～6天初生叶出现（图2）。

条播。每平方米播种15～20g，覆土2～2.5cm。1年生苗出圃。

（韦增健）

合 欢 属
Albizia Durazz.

（含羞草科 Mimosaceae）

生长习性、分布和用途 本属约150种，我国有14种，又引入2种。本文描述9种。多为落叶乔木，常为荒山造林先锋树种。少数种能耐水湿。多数种适生于酸性土，少数种可分布至石灰岩山地。它们的名称、生长、分布及用途见表1。

表1 合欢属树种的名称、生长、分布和用途

中 名	学 名	树高 (m)	胸径 (cm)	分 布	用 途	供 稿
楹树	*A. chinensis* (Osb.) Merr.	25	40	粤、琼、桂、湘、黔、滇、藏、闽、浙。印度、越南、缅甸	材用、栲胶、纸浆	603
南洋楹	*A. falcataria* (L.) Fosberg	45	100	印度尼西亚。热带非洲和亚洲广为引种。我国闽、台、粤、琼、桂、滇引种	材用、栲胶、放养白木耳、绿化	603
合欢	*A. julibrissin* Durrazz.	16	20	黄河以南各地。朝鲜半岛、日本、越南、泰国、缅甸、印度、伊朗及东非	材用、药用、栲胶	702
阔荚合欢（大叶合欢）	*A. lebbeck* (L.) Benth.	20	40	热带非洲、亚洲。台、闽、粤、琼、桂有栽培	材用、饲料、虫胶	603
光叶合欢	*A. lucidior* (Steudel) I. Nielsen	30	60	滇、桂、台。印度、缅甸、越南、泰国	材用、绿化	603
山合欢（山槐）	*A. macrophylla* (Bge.) P. C. Huang	15	30	黄河流域以南，西至陕、甘、川、滇，南至桂、粤、闽、台。朝鲜半岛、日本	材用、栲胶、纤维	603，1002

（续）

中　名	学　名	树高（m）	胸径（cm）	分　布	用　途	供　稿
香合欢（黑格）	*A. odoratissima* (L. f.) Benth.	20	30	粤、桂、黔、滇、川、浙。印度、缅甸、越南、马来西亚	材用、雕刻、栲胶、虫胶	603
红荚合欢（白格）	*A. procera* (Roxb.) Benth.	30	100	粤、桂、滇。印度、缅甸、老挝、越南、马来西亚、菲律宾	材用、造纸、栲胶	603
白花合欢（光腺合欢）	*A. speciosa* (Jacq.) Benth.	20	35	粤、桂、湘、黔、川、滇。印度、缅甸、越南	材用、虫胶	1002

开花结实　多在6～8年生开始结实，正常结实期在15年生以后。结实大小年间隔期一般为1年，但不甚明显。花两性。头状或圆柱状穗状花序，腋生或生于枝顶。萼钟状或漏斗状，具5齿或5浅裂。花瓣中部以下合生成漏斗状，上部具5裂片。雄蕊多数，花丝细长，基部稍连合，淡红或白色。子房有胚珠多数。广西南宁及浙江杭州等地观察的开花结实物候期见表2。

表2　合欢属树种的开花结实物候期

树　种	观察地点	观察年份	开花期			果实成熟期		果实脱落期
			始期	盛期	末期	始期	盛期	
楹树	南宁	1978～1984	4月中旬	4月下旬	5月中旬	11月上旬	11月下旬	翌年1月上旬～2月上旬
南洋楹	南宁	1978～1984	4月下旬	5月中旬	6月中旬	7月下旬	8月中旬	8月下旬～9月中旬荚果开裂或脱落
合欢	杭州	1963～1980	6月上旬	6月上中旬	7月上中旬	10月中旬	10月下旬	成熟后1～2个月
阔荚合欢	南宁	1987～1988	5月下旬	6月上旬	6月中下旬	12月下旬	翌年1月下旬	2月下旬～4月上旬
光叶合欢	广西桂平	1977	8月中旬	8月下旬	9月	翌年3月下旬	4月上旬	4月下旬荚果开裂
山合欢	南宁	1978	5月下旬	6月中旬	7月上旬	9月中旬	9月下旬	11～12月
香合欢	南宁	1978～1984	6月中旬	6月下旬	8月中旬	翌年1月上旬	2月上旬	3月中旬～4月中旬荚果开裂
红荚合欢	南宁	1978～1984	9月中旬	9月下旬	10月上旬	翌年3月中旬	4月上旬	4月上旬～下旬荚果开裂
白花合欢	湖南道县	1978	5月下旬	6月中旬	7月上旬	8月	10月	荚果不易脱落

荚果，薄带状，革质，扁平。种子圆形或卵形或长卵形，扁平。种皮厚，两侧常具马蹄形痕。成熟后种柄丝状。无胚乳。荚果及种子形态见表3、图1。

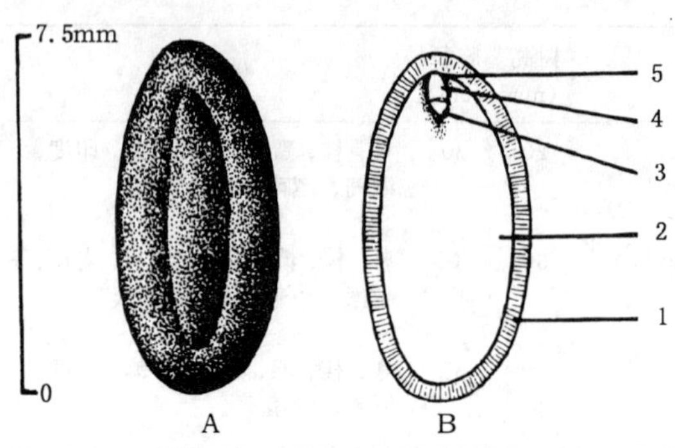

图1 南洋楹种子外形（A）及其纵切面（B）

1. 种皮 2. 子叶 3. 胚芽 4. 胚轴 5. 胚根

（黄应钦绘）

表3 合欢属树种果实和种子的形态特征

树 种	未熟果颜色	果实			种子		
		形 状	大小（cm）	色 泽	形 状	大小（mm）	色 泽
楹树	绿色	扁长条状	长8～15 宽1.8～2.3	黄褐色或灰褐色	扁椭圆形	长4～5.3 宽3～4 厚1.5～2	浅褐色
南洋楹	绿色	条形，边缘较厚	长10～13 宽1.4～2	黑褐色	扁椭圆形，光滑	长5～9 宽2～3 厚1.5～2	黄褐色
合欢	绿色	带状，扁平	长8～17 宽1.2～2.5	黄褐色	椭圆形，扁平	长6～9 宽3～4.5	淡褐色
阔荚合欢	青色	阔带状，室具隆起外缘	长15～32 宽2～5.5	黄褐色	扁广椭圆形	长7～11 宽6.5～9 厚2～2.5	黄褐色
光叶合欢	绿色	长线形，扁平，两端尖	长8～30 宽2～3.4	黑褐色，表面平滑有光泽	扁圆形	长7～9	褐色
山合欢	绿色	带状扁平	长7～18 宽1.5～3	黄褐色	扁椭圆形或长卵形	长6～9 宽2.5～4	茶褐色
香合欢	绿色	长带形	长7～24 宽2～4	褐色	扁椭圆形，光滑	长6～8 宽4.5～6.3 厚1.5～2	黄褐色
红荚合欢	绿色	长条形，两端尖	长10～17 宽1.2～3.5	赭红色	扁椭圆形	长6～7.5 宽4.5～6 厚1.8～2.3	黄褐色
白花合欢	绿色	长圆形	长9～16 宽2.6～3.8	褐或黄褐色，有光泽	矩圆形或卵形	长8～9 宽4～7	黄褐色

果实的采收调制和种子贮藏 荚果成熟后较难脱落，宜用高枝剪或采种钩刀采集，不宜用竹竿敲打。采回的荚果摊放于日光下曝晒，果荚开裂或脆干后搓揉或用木棒敲打，扬去果

荚等杂质，筛出种子。种子的净度和质量等见表 4。光叶合欢的种子忌脱水，不宜日晒或裸露存放，含水量应保持在 20%以上，调制后宜随即播种或混沙贮藏，发芽能力保存期一般仅有 3 个月。阔荚合欢的种子可稍加干燥，但忌日晒，含水量宜保持在 15%左右。其余 7 个树种的种子可以晒干，含水量保持在 10%以下，袋藏于室内干爽通风处，贮藏期可达 1 年以上。

表 4　合欢属树种的出种率和种子净度、质量

树　种	出种率（%）	净　度（%）	千粒重（g）		每千克纯净种子粒数（万粒）	
			一　般	变动范围	一　般	变动范围
楹树	18	97	25	20～30	4	3.3～5
南洋楹	18.5	97	22	18～26	4.5	3.8～5.6
合欢	20～40	85	33	27～40	3	2.5～3.7
阔荚合欢	14	98	157	120～200	0.63	0.5～0.8
光叶合欢	17	97	178	150～200	0.56	0.5～0.7
山合欢	14～16	97	30	28～33	3.3	3～3.5
香合欢	18	90	35	30～40	2.9	2.5～3.3
红荚合欢	27～33	88	47	40～55	2.1	1.8～2.5
白花合欢	15	90	65	60～70	1.5	1.4～1.7

发芽和播种　种子无生理性休眠，但有时需要破除种皮对吸水的障碍。发芽时日均温宜在 20℃以上。随采随播的种子可用始温 40℃水浸种 1 昼夜。尚未脱水干燥的新鲜种子也可不经浸泡即行播种。经过干藏的种子一般需用始温 80℃水浸种。1980～1988 年，广西林业科学研究所和杭州植物园先后对 8 种合欢进行过发芽测定，情况见表 5。经过预处理的种子发芽迅速整齐，播种后 2～4 天胚根开始萌发，6～7 天子叶出土，8～ 11 天初生叶出现。经过干藏的种子如不作预处理，不仅无明显的发芽盛期，发芽率也明显降低，发芽过程可能延长至 2 个月以上。出土萌发。种子发芽及幼苗生长情况见图 2。

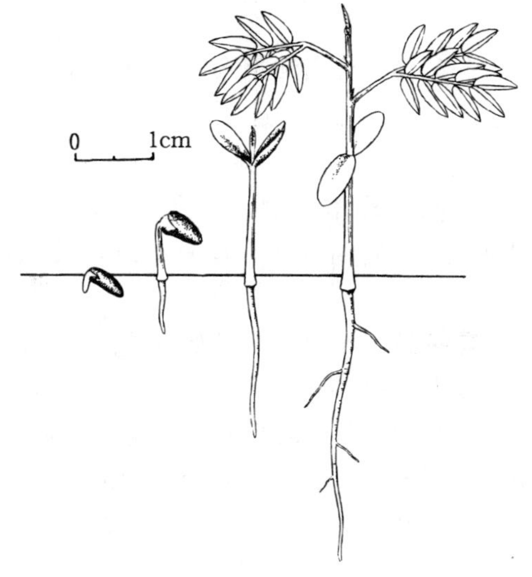

图 2　南洋楹种子萌发后第 1、3、6、15 天幼苗的生长情况

（黄应钦仿《热带亚热带主要树种采种育苗技术》）

表 5　合欢属树种的发芽能力及其测定条件（室内）

树　种	预处理（水温均指初始温度）	测定时室温（℃）	发芽率（%）		发芽率（%）	
			计算天数	一般数值	计算天数	一般数值
楹树	80℃水浸种	22	6	60	13	74
南洋楹	沸水浸种	24	4	84	11	86
合欢	70～85℃水浸种	30	—	—	7～9	96
阔荚合欢	80℃水浸种	24	5	55	12	70
光叶合欢	新鲜种子，未处理	24	5	70	12	85
香合欢	80℃水浸种	24	6	84	13	90
红荚合欢	80℃水浸种	24	6	70	13	80
白花合欢	80℃水浸种	24	5	65	12	80

条播。每平方米播种量为：楹树、南洋楹 3.7～6.2g；合欢、山合欢为 5～8.5g；香合欢、红荚合欢 7～10g；阔荚合欢、光叶合欢、白花合欢 12～30g。也可将种子密播，发芽后移芽至容器内培育，或待幼苗稍显木质化时移至容器或大田继续培育。容器苗 3 个月出圃，大田苗一般 1 年生出圃。

（张声燕）

朱缨花（美蕊花）

Calliandra haematocephala Hassk.

（含羞草科　Mimosaceae）

生长习性、分布和用途　朱缨花属约 200 种，我国引入 3 种，本文描述 1 种。常绿小乔木或呈灌木状，高 3～4m。喜高温高湿，苗期忌霜冻。适生于土壤较肥沃湿润的酸性土。原产南美。琼、粤、台、桂引种栽培。花可观赏，皮具纤维，枝供胶虫寄生。

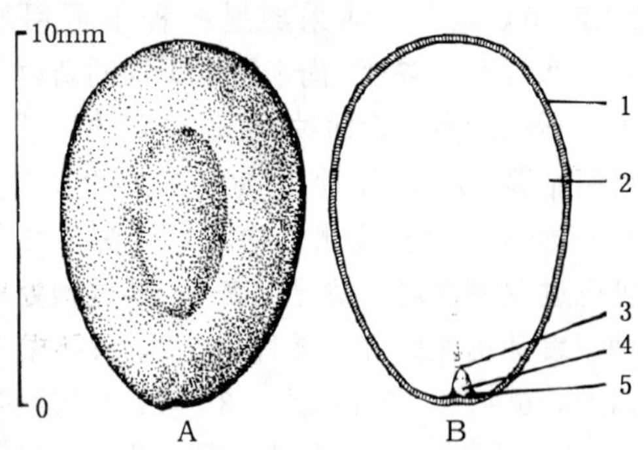

图 1　朱缨花种子外形（A）及其纵切面（B）
1. 种皮　2. 子叶　3. 胚芽　4. 胚轴　5. 胚根
（黄应钦绘）

开花结实　约 2 年生开始开花结实，正常结实期在 4 年生以后，结实无大小年现象。影响结实的主要因素是水肥条件以及早春气温。立地条件较差的地方结实不多。早春长期低温会使当年减产。花杂性。头状花序腋生。萼钟状，具齿，绿色。花冠管长 3.5～5mm，淡紫红色，顶端 5 裂片反折。雄蕊多数，突露于花冠之外。雄蕊管长约 6mm，白色，上部离生的花丝长约 2cm，深红色，花药常具腺毛。子房无柄，胚珠多数，花柱丝状，柱头头状。据广西南宁 1987～1988 年观察，花果期均很长。每年主花期为 2 次：第 1 次始花期 5 月上旬，盛花期 5 月中旬～6 月，末花期 7 月上旬，8～9 月果实成熟；第 2 次始花期 10 月中旬，盛花期11月，末花期12月上旬，11月中旬～12月下旬果实成熟。荚果条状倒披针形，未成熟时绿色，成熟时暗褐色，长8～12cm，宽10～15mm，厚3～4mm。果荚成熟开裂后自顶端向基部翻卷。种子扁椭圆形，黑褐色，长8～10mm，宽5～6.5mm，厚1.5～2mm。无胚乳。种子的形态见图1。

果实的采收调制和种子贮藏　果实盛熟时选摘饱满形大的荚果。采回的果荚置日光下曝晒，开裂后用手搓揉或用木棒敲打，脱粒后扬去荚瓣即得种子。鲜果的出种率为25%～40%。种子可以晒干贮藏，含水量宜保持10%以下。种子净度可达98%。千粒重约52g，每千克纯净种子数约1.9万粒。用布袋或瓦缸等容器贮藏。常温条件下，室内贮藏期为1年，4～8℃的低温条件下可达4年。

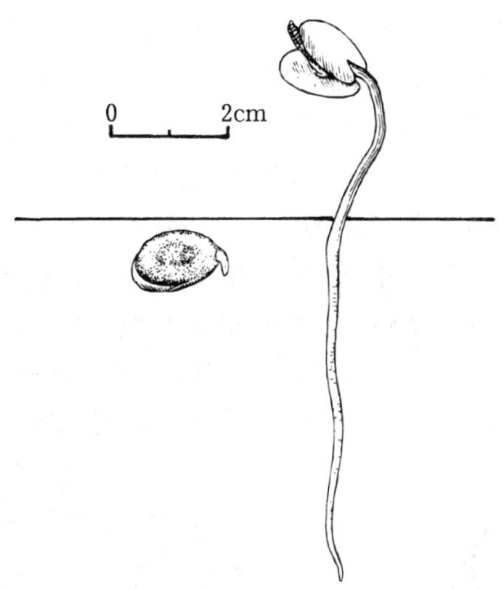

图2　朱缨花种子萌发后第1、5天幼苗的生长情况
（黄应钦绘）

发芽和播种　种子无生理性休眠，但种皮可能妨碍吸水。发芽时日均温需在20℃以上。1987年10月下旬，广西林业科学研究所用当年新采种子在室内用发芽皿作过发芽测定，播前用80℃热水浸种并在自然冷却过程中继续浸泡1昼夜。基质为滤纸，光照为室内自然光，温度约20℃。置床后第2天开始发芽，发芽盛期不明显，14天发芽率为62%。出土萌发。胚根萌发后4天子叶出土，5天子叶展开，初生叶出现（见图2）。

植株少，种子少，多用花盆等容器育苗。半年至1年生苗出栽。

（韦增健）

大棋子豆

Cylindrokelupha eberhardtii（Nielsen）T. L. Wu

（含羞草科　Mimosaceae）

生长习性、分布和用途　棋子豆属13种，我国10种，本文描述1种。高16～20m，胸径50～60cm。忌严重霜冻。喜肥沃湿润的酸性土壤。分布于广西西南部。越南北部亦产。优质用材，果大供观赏。落叶层厚，涵养水源效果好。

开花结实　10年生左右开始开花结实，正常结实期在15年生以后。结实大小年间隔期为1年，但在水肥条件好的地方丰歉年差异不明显。花两性。头状花序再组成圆锥状复花序，花序具花3～5。花萼杯状，裂齿5，三角形。花冠漏斗状，长11～12mm，裂片5，长3～3.5mm。雄蕊多数，基部连合成短管，与花冠管等长。子房无毛，长约2.5cm，子房柄长约4mm。花柱线形，柱头点状。据广西南宁1978～1984年观察，3月上旬～中旬花序形成，着生于头年生枝的叶腋或叶痕处，4月中旬初为始花期，下旬为盛花期，5月上旬为末花期；9月上旬～

中旬果实开始成熟，中下旬为果熟盛期。荚果特大，圆柱形，两端渐尖，成熟时由浅绿色转变为黄绿色以至棕褐色，长10～30cm，径4.5～6.5cm。果荚近木质。果实成熟后在树上可以维持10～15天不脱落，亦不开裂，半月以后果荚才陆续开裂，种子脱落。也有一部分果荚从基部脱落，落地后再开裂。每果有种子1～10粒。种子黑褐色，形态颇不一致。每荚内多粒种子的，居中部的种子为扁平圆形，状如较大型的象棋子，径4～5.5cm，厚2.5～4cm；居两端的种子为圆锥状，径与高均约4～6cm。荚果仅有1粒种子的，种子为扁椭圆形，长5～6.5cm，宽5～6cm，厚4～5cm。种皮薄，胚根位于子叶腹面，无胚乳（图1）。

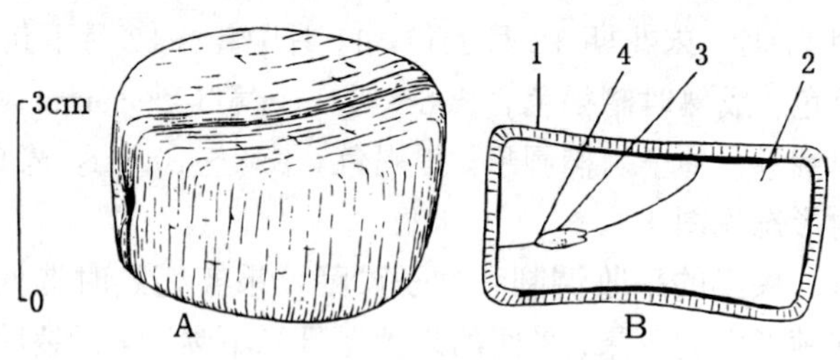

图1 大棋子豆种子外形（A）及其纵切面（B）
1. 种皮 2. 子叶 3. 胚轴 4. 胚根
（黄应钦绘）

果实的采收调制和种子贮藏 用采种钩刀从果柄处钩落，也可在树上摇动震落。果体大而沉重，脱落时可能伤人，应注意安全。采回的荚果置日光下曝晒1～2日或摊于室内晾干，俟开裂后剥去果荚，即得种子，鲜果出种率约65%。种子净度达100%。含水量35%～45%，忌日晒，不能脱水。种子千粒重35～95kg，每千克有种子10～28粒。9月下旬以前调制出的种子可以随采随播，10月中旬以后调制出的种子，可用湿沙贮藏至次年3月春暖后播种。种子混沙的贮藏期约为8个月。

发芽和播种 种子休眠现象不明显。发芽时日均温需在20℃以上。在适宜的气温下，发芽全程为半个月。如果秋播后遇到低温，种子会延迟至次年春暖后发芽。1987年9月上旬，广西林业科学研究所用新采的种子在室外做过发芽测定（播种时日均温28℃），播后10天开始萌发，发芽后第2天进入萌发盛期，第15天发芽终止；从发芽之日至高峰期的6天发芽百分数为71%。从播种之日起算，25天发芽率为78%。留土萌发。胚根萌发后第4～5天后幼芽出土，14～15天发生初生叶（图2）。

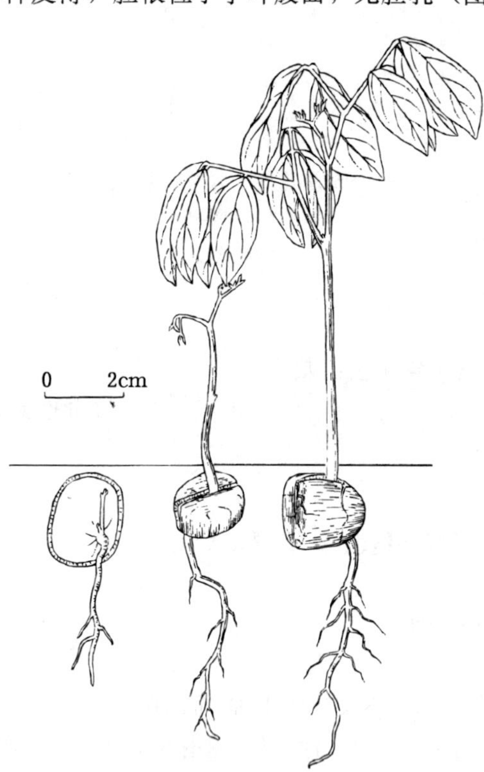

图2 大棋子豆种子萌发后第4、15、30天幼苗的生长情况
（黄应钦仿《热带亚热带主要树种采种育苗技术》）

条播。每平方米约播50～70粒，覆土2.5～3cm。1年生苗出圃。

（韦增健）

象耳豆属
Enterolobium Mart.

（含羞草科　Mimosaceae）

生长习性、分布和用途　本属约11种，我国只有本文描述的2种，原产南美，我国南亚热带地区引种栽培。落叶乔木，高20m，胸径40～50cm。对水肥的要求较高，喜肥沃疏松的酸性土。它们的名称、生长、分布及用途见表1。

表1　象耳豆属树种的名称、生长、分布和用途

中　名	学　名	树高(m)	胸径(cm)	分　布	用　途
青皮象耳豆	*E. contortisiliquum* (Vell.) Morong	20	40	热带美洲。琼、粤、桂、滇、闽、赣、浙栽培	胶合板、纸浆、栲胶、饲料、皂素、观赏
红皮象耳豆	*E. cyclocarpum* (Jacq.) Griseb.	20	50	热带美洲。琼、粤、滇、桂栽培	胶合板、纸浆、栲胶、胶虫寄主、皂素、观赏

开花结实　8～9年生开始开花结实，正常结实期在15年生以后。结实大小年现象不明显。影响结实的主要因素为寒害及水肥条件。立地条件好的庭园、路旁，正常年份结实均较多。特寒年份嫩枝受冻，次年不能结实。花两性，头状花序单生、簇生或排成总状花序，着生于当年生嫩枝的叶腋。花无柄，5出数。花萼钟状，具5短齿。花瓣合生至中部成漏斗状，绿白色。雄蕊多数，基部合生成管状。子房无柄，胚珠多数。在广西南宁及海南观察的开花结实物候期见表2。

表2　象耳豆属树种的开花结实物候期

树种	观察地点	观察年份	开花期			果实成熟期		果实脱落期
			始期	盛期	末期	始期	盛期	
青皮象耳豆	南宁	1987～1988	4月中旬	4月下旬	5月中旬	8月下旬	9月中旬	荚果不开裂，9月下旬～11月下旬自然脱落
红皮象耳豆	海南	1988	4月中旬	4月下旬	5月上旬	8月上旬	8月下旬	荚果不开裂，9月上旬～10月下旬自然脱落

荚果卷曲或内弯成肾形或耳状，扁平，成熟时由青色转为黑褐色或红褐色，成熟后不开裂，存留于枝上约1个月脱落。外果皮硬革质，中果皮海绵质，种子间有隔膜。青皮象耳豆每果有种子10～20粒，红皮象耳豆每果有种子12～14粒。无胚乳。果实及种子的形态见表3、图1。

果实的采收调制和种子贮藏　果实成熟落地后易遭虫蛀，应及时采收。用采种钩刀或高

表 3 象耳豆属树种果实和种子的形态特征

树 种	未熟果颜色	果实			种子		
		形 状	大小（cm）	色泽	形 状	大小（mm）	色泽
青皮象耳豆	黄绿色	近半圆形，稍弯曲，耳状	长 5～18 宽 4～6	赤褐色至黑褐色	扁卵状椭圆形，两侧各有 1 环纹	长 9～13 宽 5～8 厚 4～6	栗褐色，椭圆环纹暗黑色
红皮象耳豆	青色	近半圆形，稍弯曲，耳状	长 12～16 宽 5～7	暗红色至黑褐色	扁椭圆形，两侧各有 1 环纹，明显	长 10～17 宽 7～11 厚 4～7	红棕色，椭圆环纹黄色，内黑褐色

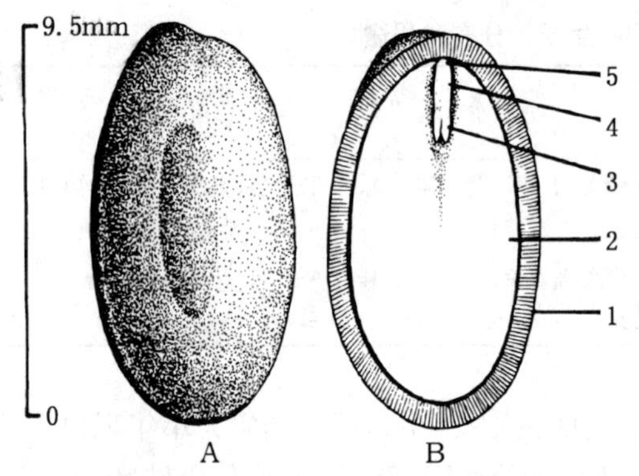

图 1 青皮象耳豆种子外形（A）及其纵切面（B）
1. 种皮 2. 子叶 3. 胚芽 4. 胚轴 5. 胚根
（黄应钦绘）

枝剪带果穗采下，摘取荚果，置日光下曝晒。干燥后用木棒敲碎果荚，使种子脱出并筛出种子。亦可先将荚果浸水，待果皮沤软后搓洗，取出种子晾干。破碎的果荚及隔膜等可作洗涤剂。种子的净度、质量等见表 4。

种子可以晾干或置弱光下晒干，但不宜置烈日下久晒。含水量宜保持在 10%～12%。晒干后的种子可贮藏于布袋或坛、缸等容器内，置室内干爽处。如发现有虫蛀，应在晒干后拌少量杀虫粉，或将种子置始温 50℃水中浸泡 3～4 日，捞出晒干后再行贮藏。

表 4 象耳豆属树种的出种率和种子净度、质量

树 种	出种率（%）	净度（%）	千粒重（g）		每千克纯净种子粒数	
			一 般	变动范围	一 般	变动范围
青皮象耳豆	25～30	98	285	220～350	3 500	2 800～4 500
红皮象耳豆	25～30	98	426	350～550	2 350	1 800～2 860

发芽和播种 种子无生理性休眠，但种皮致密，妨碍吸水萌发。播种前要用硫酸拌种10～15 分钟或用沸水浸种，发芽方可快速整齐。经过预处理的种子播后第 2～3 天开始萌发。发芽时的日均温需在 18℃以上。1987 年广西林业科学研究所用硫酸拌种 10 分钟，洗净后在室内进行过发芽测定，情况见表 5。

出土萌发。发芽后 2 天子叶出土。青皮象耳豆出土后 1 天即展出初生叶，红皮象耳豆出土后 2～3 天展出初生叶（图 2）。

表 5　象耳豆属树种的发芽能力及其测定条件

树　种	基　质	预处理	温度（℃）	发芽势（%）		发芽率（%）	
				计算天数	一般数值	计算天数	一般数值
青皮象耳豆	纸	硫酸浸渍 10 分钟	26～28	3	40	8	52
红皮象耳豆	纸	硫酸浸渍 10 分钟	26～28	3	50	8	54

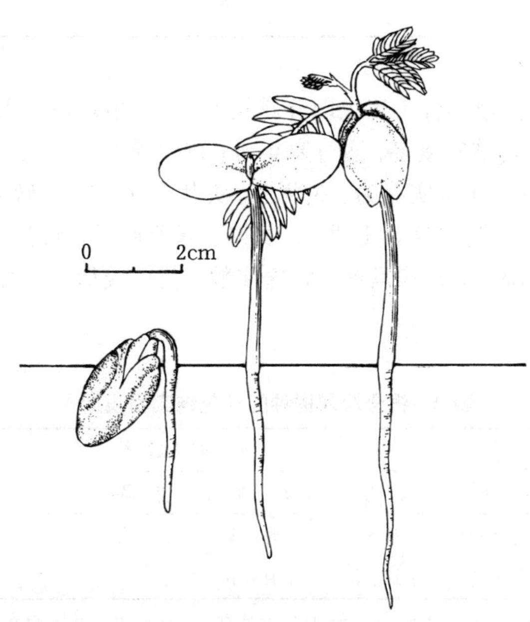

图 2　红皮象耳豆种子萌发后第 2、3、7 天幼苗的生长情况
（黄应钦仿《热带亚热带主要树种采种育苗技术》）

条播或容器育苗。圃地每平方米约播 75～100g，覆土 2～2.5cm。1 年生苗出圃。容器育苗时每容器点播已萌芽的种子 1 粒，育 3 个月苗出圃。

（韦增健）

银合欢属
Leucaena Benth.

（含羞草科　**Mimosaceae**）

生长习性、分布和用途　本属约 25 种，我国引入本文描述的 1 种及其 1 栽培品种。常绿。喜土质较肥沃的石灰岩山区钙质土，酸性土亦能生长。忌严重霜冻。原产热带美洲，世界上热带地区多有引种。我国热带至南亚热带地区广有栽培。为重要能源、饲料、肥料树种。它们的名称、生长、分布及用途等见表 1。

表 1 银合欢属树种的名称、生长、分布和用途

中 名	学 名	树 高 (m)	胸 径 (cm)	分 布	用 途
银合欢	*L. leucocephala* (Lam.) De Wit	8	10	热带美洲。粤、琼、桂、滇、川、闽、台栽培	保土、薪柴、肥料、饲料、食用、观赏
新银合欢	*L. leucocephala* (Lam.) De Wit cv. Salvador	20	30	中美洲。琼、粤、桂、闽栽培	材用、薪柴、肥料、饲料、观赏

开花结实 开花结实期颇早，1 年生即开花结实，2 年生以后进入正常结实。结实无大小年现象，正常情况下每年均结实繁多。遇特寒年份，枝梢受冻害，次年开花结实期稍推迟，但萌发的新枝仍能具有正常开花结实能力。花两性。头状花序，1～3 腋生于当年生嫩枝的上部，或数花序簇生于枝梢顶部。花白色，无梗，5 基数。萼管钟形，短齿裂。花瓣分离。雄蕊 10，分离，突出，花药顶端无腺体。子房具柄，胚珠多数，花柱线形。广西南宁观察的开花结实物候期见表 2。

表 2 银合欢属树种的开花结实物候期①

树 种	开花期			果实成熟期		果实开裂期
	始 期	盛 期	末 期	始 期	盛 期	
银合欢	4 月中旬	5 月上旬	5 月下旬	7 月中旬	8 月上旬	8 月上旬～9 月上旬
新银合欢	4 月中旬	5 月上旬	5 月下旬	7 月中旬	8 月上旬	8 月上旬～9 月上旬

① 本表所列为主花期。除主花期以外，3～11 月均有花开，6 月至翌年 2 月均有果实成熟（广西南宁，1978～1984）

荚果薄带状，革质，成熟后 5 天左右微开裂，但此时种子不脱落，7～10 天以后种子才逐渐脱落。种子脱落后果荚仍留在树上。每荚有种子 6～30 粒。种子无胚乳。果荚及种子的形态见表 3、图 1。

表 3 银合欢属树种果实和种子的形态特征

树 种	未熟果颜色	成熟果实			种 子		
		形 状	大小（cm）	色 泽	形 状	大小（mm）	色 泽
银合欢	青绿色	带状，基部狭，荚室隆起	长 9～18 宽 1.5～2.0	黄褐色	扁椭圆形，两面具椭圆形环纹	长 6～7.5 宽 3.5～4.5 厚 1.5～1.7	棕褐色，有光泽
新银合欢	青绿色	带状，基部狭，荚室隆起	长 18～24 宽 1.6～2.5	褐色或紫褐色	扁椭圆形，两面具椭圆形环纹	长 7～10 宽 4～6 厚 1.5～2	棕褐色，有光泽

果实的采收调制和种子贮藏 当果荚已转呈革质即将开裂或微有开裂时，用枝剪或钩刀

带果穗采下。采得的荚果置日光下曝晒1～2天即全部开裂，稍加振动，种子极易脱出。除去果荚即得种子。出种率和种子的净度、质量等见表4。

表4　银合欢属树种的出种率和种子净度、质量

树　种	出种率（%）	净度（%）	千粒重（g）	每千克纯净种子粒数（粒）
银合欢	50～60	97	30～35	28 600～33 300
新银合欢	46～56	97	50～60	16 700～20 000

种子可以干藏。晒干种子的含水量约为10%。袋藏或用坛罐等容器贮藏于室内。常温条件下种子发芽能力可以保持2年。贮藏于3～10℃的低温条件下，发芽能力可以保持3年以上。

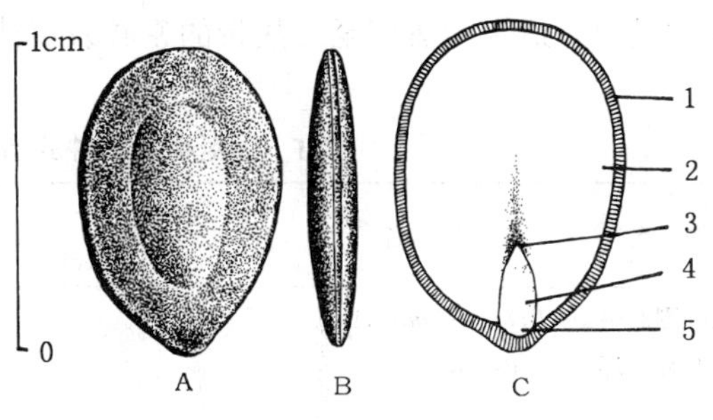

图1　新银合欢种子外形正面（A）、侧面（B）及其纵切面（C）

1. 种皮　2. 子叶　3. 胚芽　4. 胚轴　5. 胚根

（黄应钦绘）

发芽和播种　种子无生理性休眠，但种皮致密，妨碍吸水萌发。播前需用沸水浸种，或用硫酸拌种10分钟后彻底清洗，再用冷水浸泡1昼夜。未经处理的种子播后发芽极不整齐。发芽时的日均温需在18℃以上。1987年10月上旬，广西林业科学研究所在室外沙床做过发芽测定，播前用沸水浸种，播时日均温26℃。播后4～5天胚根萌发，从发芽开始至发芽终了，全程仅4～5天。以发芽高峰的前3天计，发芽百分数为60%。从播种之日起算以10天计，发芽率为84%。出土萌发。胚根萌发后约2天子叶出土，再过4～5天发出初生叶（图2）。

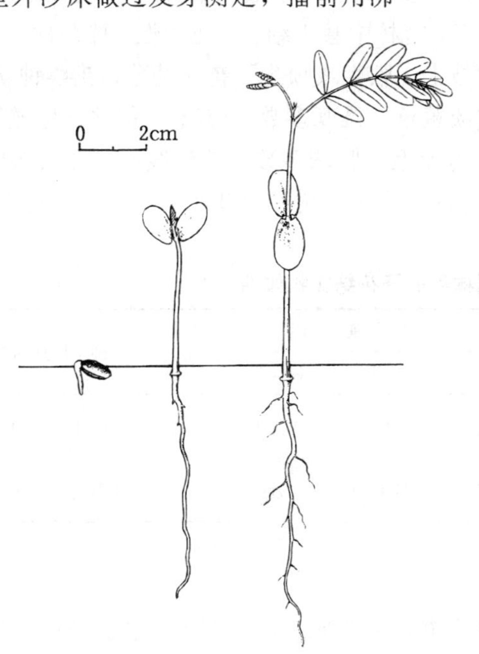

图2　新银合欢种子萌发后第2、6、12天幼苗的生长情况

（黄应钦仿《热带亚热带主要树种采种育苗技术》）

目前多用容器育苗。种子经处理催芽萌动后点播至容器内，每容器播种2粒。圃地育苗多用条播，每平方米播种6～8g，覆土1～1.5cm。半年或1年生苗出圃。这2个种多与果树间作，或用于采割嫩枝叶作饲料或绿肥。在立地条件较好的造林地可以用经过催芽的种子直播造林。

（韦增健）

围涎树（猴耳环）属
Pithecellobium Mart.

（含羞草科 Mimosaceae）

生长习性、分布和用途 本属约120种，我国约5种，又引入2种，本文描述2种。常绿乔木，高约18m，胸径40～50cm。能耐－5℃左右的极端低温及较轻霜雪。喜质地疏松肥沃的酸性土壤。适生于热带至亚热带的密林及溪边等处。它们的名称、生长、分布及用途见表1。

表1 围涎树属树种的名称、生长、分布和用途

中名	学名	树高（m）	胸径（cm）	分布	用途
亮叶围涎树（亮叶猴耳环）	*P. bigeminum*（L.）Benth.	17	40	台、浙、赣、闽、粤、琼、桂、滇、黔、湘、川。印度、越南	材用、枝叶入药，果有毒
围涎树（猴耳环）	*P. clypearia*（Jack）Benth.	18	45	台、浙、闽、湘、粤、琼、桂、滇。越南、缅甸	材用、栲胶、药用、胶虫寄主

开花结实 6～7年生开始开花结实，正常结实期在15年生以后。结实大小年间隔期一般为1年。人工栽培或溪边生长的母树结实较多，密林中甚少结实。花两性，稀杂性。头状花序组成圆锥状或总状排列，着生于当年生的嫩枝上，腋生或顶生。花5基数，花萼钟状，齿裂。花冠漏斗状，花瓣在中部以下合生，白色或淡黄色。雄蕊多数，突出，花丝合生成管，药顶端无腺体。围涎树子房被毛，亮叶围涎树子房无毛。胚珠多数。花柱线形，柱头头状。据广西南宁对人工栽培的10～14年生幼龄母树观察，这两个树种的开花结实物候期见表2。

表2 围涎树属树种的开花结实物候期

树种	开花期			果实成熟期		种子脱落期
	始期	盛期	末期	始期	盛期	
亮叶围涎树	5月下旬	6月上旬	6月下旬	9月中旬	9月下旬	9月下旬～10月中旬
围涎树	3月上旬	3月中旬	4月上旬	5月上旬	5月中旬	5月中旬～6月上旬

注 广西南宁，1987～1988

荚果，带状旋卷成圆环，边缘在种子间缢缩。果实成熟后1周内开裂。围涎树每果有种子4～10粒，亮叶围涎树每果有种子6～13粒。种子悬挂于细长的丝状种柄上，7～10天左右即脱落。种皮薄，无胚乳。果实及种子形态见表3、图1。

表 3　围涎树属树种果实和种子的形态特征

树　种	果　实			种　子		
	形　状	大小（cm）	色　泽	形　状	大小（mm）	色　泽
亮叶围涎树	带形，边缘在种子间缢缩，旋卷呈环状	长 8～15 宽 2.2～3.0	棕红色或红褐色，无毛	椭圆形	长 15～18 径 12～13.5	紫黑色，表面被白粉
围涎树	带形，边缘在种子间缢缩，旋卷呈环状	长 11～13 宽 1.1～1.5	棕红色，被短硬毛	椭圆形，光滑	长 13～17 径 8.5～12.5	紫黑色，表面被灰色粉

果实的采收调制和种子贮藏　当果荚由青色转呈褐色而尚未开裂时应及时采收。采回的果荚如尚有部分为青色，可堆放于室内 3～4 天，俟种子充分成熟后置弱光下晒干，或摊放于室内通风处晾干。果荚自行开裂后用手搓揉或用木棒敲打，种子即可脱落。清除果荚等杂质即得种子。种子的净度和质量数据见表 4。

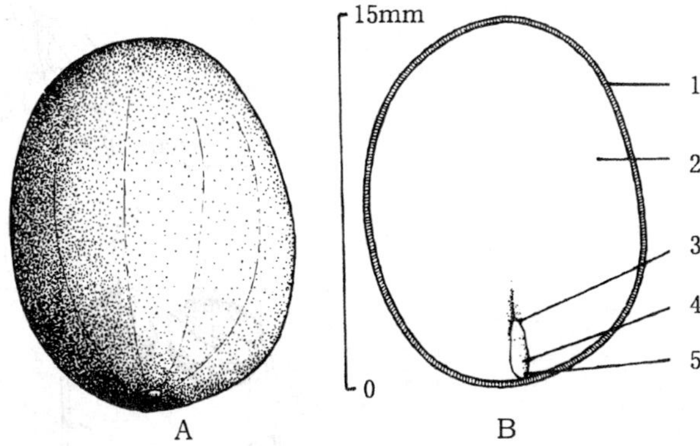

图 1　亮叶围涎树种子外形（A）及其纵切面（B）
1. 种皮　2. 子叶　3. 胚芽　4. 胚轴　5. 胚根
（黄应钦绘）

种子可以轻微脱水，但不宜过分干燥，忌日光曝晒，适宜的含水量为25%～40%。运输时可用袋或箱包装，混以湿润锯木屑。贮藏需混湿沙，贮藏期约为 7 个月左右。9～10月成熟的种子，裸露贮存 1～2 个月便会失去发芽能力。即使混沙贮藏，次年 5 月以后气温升高，发芽能力也会明显下降。一般以贮藏至次年 3～4 月为宜。

表 4　围涎树属树种的出种率和种子净度、质量

树　种	出种率（%）	净度（%）	千粒重（g）		每千克纯净种子粒数（粒）	
			一　般	变动范围	一　般	变动范围
亮叶围涎树	60～75	99	1 500	1 250～1 750	670	570～800
围涎树	50～60	99	970	800～1 200	1 030	830～1 250

发芽和播种　亮叶围涎树的种子有轻度休眠，围涎树的种子休眠现象不明显。可以随采随播，也可以沙藏越冬，次年 3 月中旬～3 月下旬播种。少数种子在 14℃这样低的气温条件下也可以发芽，但发芽极不整齐。多数种子发芽时的日均温宜在 18℃以上。播种前不需作预处理。1981 年 10 月和 1987 年 10 月，广西林业科学研究所对亮叶围涎树和围涎树进行过发芽测定，结果见表 5。

表 5 围涎树属树种的发芽能力及其测定条件①

树 种	温度（℃）	发芽势（%）		发芽率（%）	
		计算天数	一般数值	计算天数	一般数值
亮叶围涎树	14～26	—	不明显	97	82
围涎树	18～24	10	67	28	90

① 室外沙床

留土萌发。胚根萌发后约 2 天上胚轴出土，8 天展出初生叶，有时第 1～第 3 片叶上的小叶发育不完整（图 2）。

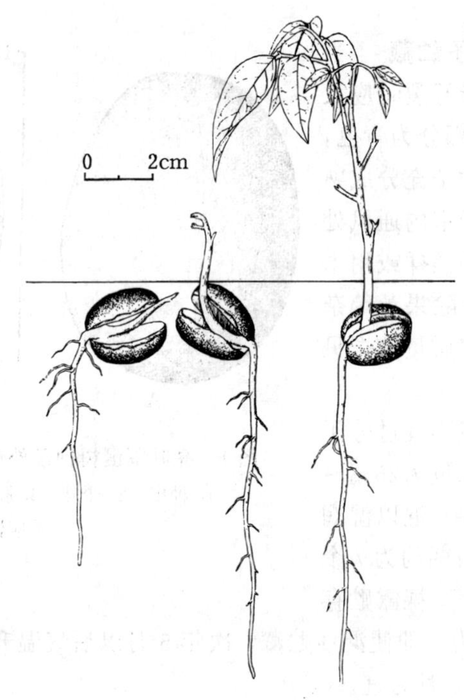

图 2 亮叶围涎树种子萌发后第 4、8、25 天幼苗的生长情况
（黄应钦仿《热带亚热带主要树种采种育苗技术》）

条播。每平方米的播种量：围涎树 120～150g，亮叶围涎树 170～200g，覆土 2～3cm。1 年生苗出圃。

（韦增健）

雨树（雨豆树）
Samanea saman Merr.

（含羞草科 Mimosaceae）

生长习性、分布和用途 雨树属 20 种，本文描述我国引入的 1 种。落叶乔木，高达 24m，胸径 80～100cm，主干短。耐修剪。较耐干旱。抗风力强。一般立地条件能生长成材。原产热带美洲。琼、台、粤、闽、滇引种栽培。木材供作家具、雕刻等，叶及果作饲料。果肉富含糖，可制酒精。树姿优美，供观赏。

开花结实　5～8 年生开始结实，10 年以后进入正常结实期，大小年现象不明显。花两性。头状花序腋生，总花梗长 10～12.5cm。花 5 基数。花萼管状。花冠漏斗状，长约 1.7cm，粉红色。雄蕊 20，基部连合，深红色。子房无柄，花柱线形，胚珠多数。据海南乐东尖峰岭的观察，4 月上旬花蕾始现，着生于当年生枝叶腋；4 月中旬为始花期，4 月下旬为盛花期，5 月下旬为末花期；6 月上旬幼果形成；翌年 3 月上旬果实开始成熟，3 月中旬为果熟盛期，果实成熟后陆续脱落。荚果直，长 15～20cm，宽 1.5～2.5cm，稍厚或略呈圆柱形，边缘厚，肉质，成熟时变成近木质，不裂，种子间有隔膜。果实无柄。每荚有种子 10～25 粒。成熟时外果皮由青绿色转变为深棕色。内果皮骨质，褐色。种子长椭圆形，基部稍凹，长约 11mm，径约 5mm。种皮薄，无胚乳。种子形态见图 1。

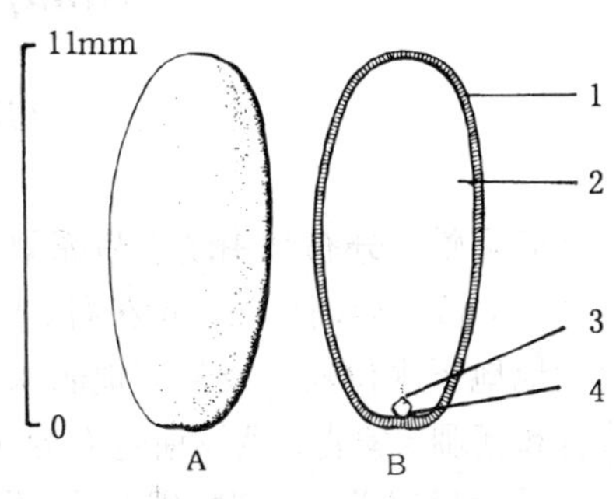

图 1　雨树种子外形（A）及其纵切面（B）
1. 种皮　2. 子叶　3. 胚芽　4. 胚根
（黄应钦绘）

果实的采收调制和种子贮藏　果实成熟时在树下拾集或用竹竿打落后收集。采得的果实置烈日下晒干，用铁锤或木棒敲打开裂后再用小刀挖出种子，置水中浸泡冲洗，即得纯净种子。晾干或短时间晒干即可，此时含水量约在 10%左右。鲜果的出种率约 18%。净度可达 95%。千粒重约 220g，每千克纯净种子约 4 500 粒。贮藏可用袋或一般容器。常温条件下室内贮藏期为 8～10 个月。

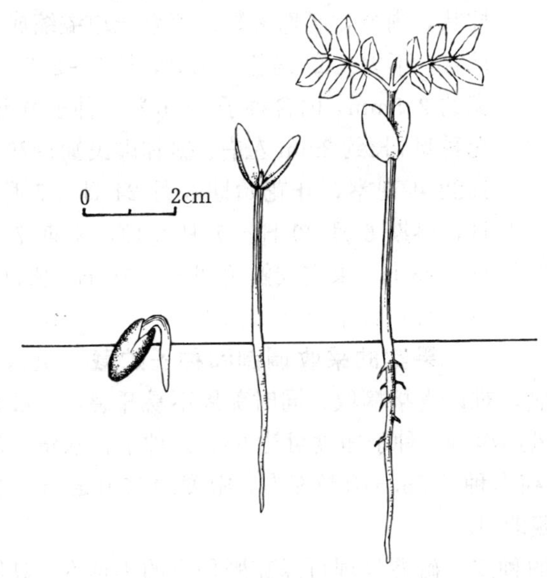

图 2　雨树种子萌发后第 2、4、8 天幼苗的生长情况
（黄应钦绘）

发芽和播种　种子无生理性休眠。发芽时日均温宜在 20℃以上，中国林业科学研究院海南热带林业试验站在场圃进行过发芽测定，播前用始温 80℃水浸种，播后 2 天萌发。从发芽之日起算，5 天的发芽百分数约为 57%；从播种之日起算，19 天的发芽率为 91%。出土萌发。胚根萌发后约 4 天子叶出土，5～6 天子叶开展并发出初生叶（图 2）。

条播。每平方米播 120～200g。播前需作预处理。也可将经过催芽的种子点播于容器内，每容器点播 1 粒。容器苗 80～90 天出圃。大田育苗半年～1 年出圃。

（符史深）

骆 驼 刺

Alhagi sparsifolia Shap.

（蝶形花科 Fabaceae）

生长习性、分布和用途 骆驼刺属有 5 种，我国 3 种，本文描述 1 种。多刺落叶半灌木，高 0.3～0.6m，最高可达 1m 左右。枝叶稠密，沙埋后容易产生不定根和不定芽。生于低平盐渍化沙地地下水位较高或常有地表水供给的各类土壤上。分布于甘肃河西走廊（东起民勤）、内蒙古和新疆。蒙古、俄罗斯也有分布。良好的饲料及蜜源植物，种子可药用。在吐鲁番盆地夏秋两季大风时节，针刺扎破叶片，伤口处泌出糖液，风吹日晒，浓缩成小块状结晶，称刺糖，是一种珍贵药材。

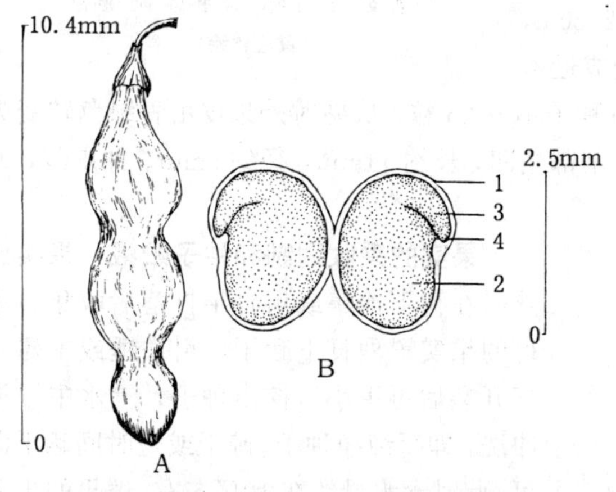

图 1 骆驼刺荚果外形（A）及其种子纵切面（B）
1. 种皮 2. 子叶 3. 胚轴 4. 胚根
（姚桂英绘）

开花结实 1～2 年生即可开花结实，2 年以后正常结实，间隔期不明显。花两性，总状花序腋生，总轴先端针刺状。花萼钟形，5 齿裂。花冠蝶形，红色，各瓣近相等。雄蕊两体（9+1）。子房上位，1 室，近无柄，长柱状，内弯，胚珠多数。荚果一边缢缩成念珠状，不裂，红褐色，无毛，长 1～2.5cm，宽约 2.5mm，内含种子 1～6 粒。种子肾形，无种阜，长约 3mm 左右。据甘肃民勤1979～1982 年观察，开花始期 6 月 24 日～7 月 3 日，盛期 6 月 30 日～7 月 6 日，末期 7 月 18～26 日。果实成熟始期 8 月 19 日，盛期 9 月 9 日。种子形态如图 1。

果实的采收调制和种子贮藏 荚果 9 月份成熟后可在植株上存留几个月。茎多针刺，难以直接摘取，同时荚果不易开裂，一般是用镰刀收割地上部分，晒干后打碾，筛选或风选净种。种子净度可达 95%。种子含水量一般在 7%以下，千粒重 3.60～3.85g，每千克有纯净种子 26.5 万粒左右，密度约 754kg/m^3。把气干种子装入布袋放到干燥、通风的地方贮藏即可。

发芽和播种 骆驼刺种子有 1 层不透水的种皮，硫酸处理可以克服种皮的不透性。具体方法是先用热水浸泡 1 昼夜，用孔径 3mm 的筛子选出已经吸水膨胀的种子，将剩余硬粒捞出晒干后，浸入比重 1.84、浓度 98%的硫酸 30～35 分钟。从酸中取出的种子必须用流动的凉水冲洗 5～10 分钟。据试验，测定前用热水浸种 24 小时的，在白天 18℃，夜间 10℃的温度条件下，15 天的发芽率只有 8%。用浓硫酸浸泡 5、10、15、35 分钟的，在相同条件和相同的时间内，发芽率分别为 40%、58%、78%和 88%。播种时间在春季，条播或撒播。每平方米播种 2.5g。覆土厚度不能超过 2cm，在墒情好的情况下深度以 1cm 左右最佳。子叶两枚，出土萌发。幼苗形态如图 2。骆驼刺可用根蘖繁殖。

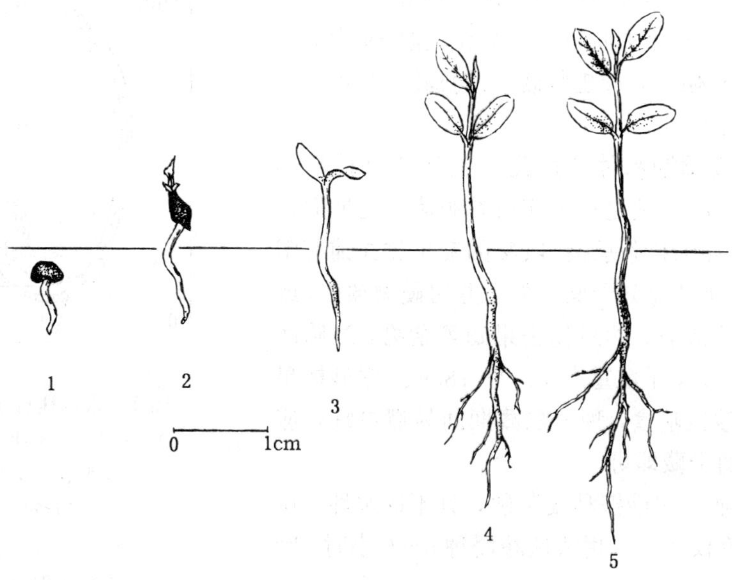

图 2　骆驼刺幼苗生长情况

1. 种子萌发　2、3. 子叶出土　4. 初生叶出现　5. 幼苗形成

（姚桂英绘）

（郭志中）

银　砂　槐

Ammodendron argenteum（Pall.）Jakovl.

（蝶形花科　Fabaceae）

生长习性、分布和用途　银砂槐属有 6 种，我国产 1 种。落叶灌木，高 0.5～2.0m。喜光，耐干旱，耐风蚀沙埋。产新疆霍城的塔合尔莫乎尔沙漠。俄罗斯西伯利亚、哈萨克斯坦的沙漠地区也有分布。生于固定或半固定沙丘、沙坡地和平坦沙地上。吐鲁番已引种成功。

银砂槐为浅根性树种，水平根发达，交错盘结，集中分布在 30～50cm 沙层内，长可达 10m 以上，垂直根深入沙层 2m 余。我国珍贵的固沙植物。枝叶牲畜不喜食，可作为荒漠牧场“草库仑”的生篱。中亚地区群众取其根作染料，花可为蜜源。

开花结实　3～4 年生开始结实，正常结实年龄在 6 年生以后，结实量大，每年结实。花两性。总状花序，长 2～10cm，宽 1.5～2.0cm，具花 5～30 朵，着生于幼枝顶部，偶有侧生。花梗长 4～10mm。萼钟形，具 5 齿，长约 3mm，被白绒毛。花冠蝶形，紫色，少有黄色，长 5～7mm。雄蕊分离，10 枚。子房上位，1 室。花期 5 月。荚果未成熟时暗绿色，成熟时浅黄色，矩圆状或菱状披针形，直或扭曲，长 1.2～2.0cm，宽 5～8 mm，边缘有翅，无毛，不开裂。果熟始期在 7 月上旬，盛期 7 月中旬，脱落始期在 7 月中旬，盛期 7 月中下旬，末期 7 月下旬。引种到吐鲁番的银砂槐物候期约提前1个月。每果常含1粒种子。种子卵状肾形，光滑，呈淡黄色，长4～6mm，宽3～4mm，种皮坚硬，无胚乳，两片子叶白色。种子形态如图1。

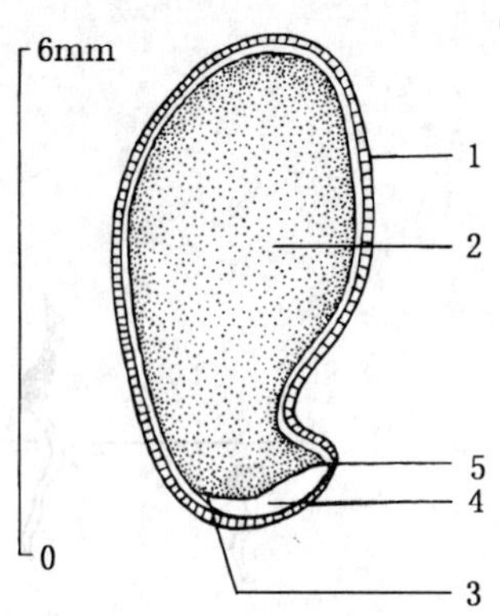

图1　银砂槐种子纵切面
1. 种皮　2. 子叶　3. 胚芽
4. 胚轴　5. 胚根
（吴新安绘）

果实的采收调制和种子贮藏　银砂槐果枝具小刺，难以直接采收，一般是待果实成熟脱落后在地面扫集，未脱落的可摇动枝条承接。收集的荚果放在铺有塑料布的地上，待果荚充分干燥后敲击并用硬物揉搓，或放在袋内敲击并揉搓，然后扬去果翅等杂物，过筛净种。出种率约50%，千粒重在13.9～18.6g。银砂槐果实脱落后易遭沙鼠啮食，种子贮藏期间易遭虫蛀，应注意防治。普通干藏即可。

发芽和播种　银砂槐种皮坚硬，具不透水性。层积3个月萌发率仅36%。用浓硫酸浸种5～6小时，吸胀率可达89%。据试验，用浓硫酸处理后再用始温90℃水浸烫，效果比单用浓硫酸好，吸胀率达90%以上，发芽率达80%。吸胀的种子盆播，用锯末做基质，气温25～30℃时播后4天出苗，7天可出齐。光照充足，气温30℃左右时成苗率可达80%。子叶出土萌发。幼苗形态如图2。

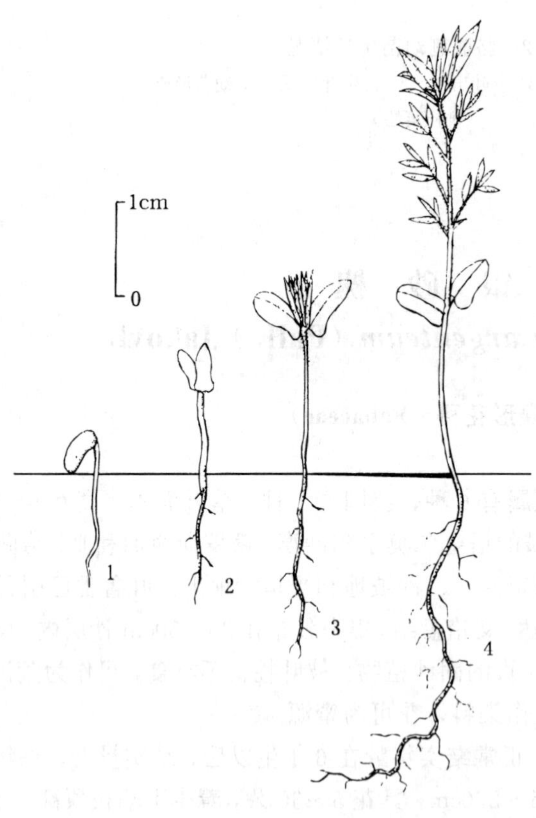

图2　银砂槐幼苗生长情况
1. 下胚轴延伸　2. 子叶出土　3. 初生叶出现　4. 幼苗形成
（吴新安绘）

经浓硫酸处理的种子，圃地应选在盐渍化轻的沙土地上，沙粒含量在70%以上，沙粒粒径必须大于0.065mm，以利通透性，否则出苗极少或不出苗。条播或垅播均可。银砂槐幼苗怕涝，以垅播为好。覆土厚度2cm左右，当年生苗高平均57cm，最高达85cm，可出圃，但不宜裸根。

（周君英）

沙冬青属

Ammopiptanthus Cheng f.

（蝶形花科　Fabaceae）

生长习性、分布和用途　本属仅有本文描述的 2 种，我国均产。常绿灌木或小灌木，分枝多。喜光，耐旱，抗沙压，能在流沙边缘生长。产于内蒙古、甘、宁、新。蒙古南部和中亚地区也有分布。生于固定沙地、沙质石质山坡。可做绿篱并供观赏。枝叶含黄花木素，牲畜不食，但可煎汤熏洗冻疮。是优良的固沙树种。张涛（1988）曾经从叶、茎和幼根的解剖结构特征论证了沙冬青耐旱、抗风和抗日灼的机理，并报道沙冬青具根瘤。这两个种都是第三纪古亚热带常绿阔叶林的残遗植物，分布面积日趋缩小，如不加保护，有濒临灭绝的危险，都已收入《中国植物红皮书》中。其名称、树高、分布和用途见表 1。

表 1　沙冬青属树种的名称、生长、分布和用途

中　名	学　名	树高（m）	分　布	用　途
沙冬青	*A. mongolicus* (Maxim. ex Kom.) Cheng f.	1～2	甘、宁、内蒙古。蒙古	固沙、药用
新疆沙冬青（矮沙冬青）	*A. nanus* (Popov.) Cheng f.	0.4～0.7	新疆喀什地区。吉尔吉斯斯坦	绿篱、观赏

开花结实　4 年生开始结实，5 年生进入正常结实，结实有大小年现象，但不甚明显。花两性。总状花序顶生。萼钟状，上部 2 齿合生，下部 3 齿，齿短三角形。花冠蝶形，黄色，旗瓣和翼瓣等长，龙骨瓣下部分离。雄蕊 10，分离。子房上位，边缘胎座，子房具柄。这两个种的物候期如表 2。

表 2　沙冬青属树种开花结实物候

树　种	观察地点	开花期	果实成熟期	种子脱落期
沙冬青	甘肃河西走廊	4 月中旬～5 月下旬	5～7 月上旬	6～7 月
	内蒙古乌海市	4～5 月	7～8 月	8 月
新疆沙冬青	新疆	5～6 月	6～7 月	7～8 月

荚果，具喙，扁平，矩圆形，黄褐色。沙冬青荚果无毛，新疆沙冬青荚果被稀疏短柔毛。子房柄（果颈）长 6～12mm。荚果裂开后种子脱落。每果有种子 2～10 粒。种子肾形或圆肾形，黄绿色，种阜小。果实和种子形态特征见表 3。种子的外形和解剖结构见图 1。

表 3 沙冬青属树种果实和种子的特征

树种	成熟果实			种子		
	形状	大小（cm）	颜色	形状	大小（mm）	颜色
沙冬青	条状矩圆形，扁	长 5～8，宽 1.5～2	黄褐色，无毛	圆肾形	径 6～7	黄绿色
新疆沙冬青	矩圆形，扁，微膨胀，具皱纹	长 3～5，宽 1.2～1.5	黄褐色，疏被短柔毛	圆肾形	径 5～6	黄绿色

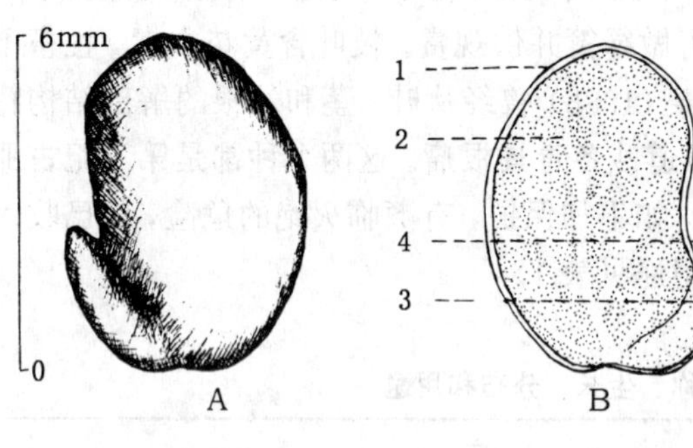

图 1 沙冬青种子外形（A）及其纵切面（B）
1. 种皮 2. 子叶 3. 胚轴 4. 胚根
（姚桂英绘）

果实的采收调制和种子贮藏 应从10～15 年生母树上采种。从树上摘取，或摇晃树干使果实落在承接单上，用木棒敲打荚果即可取出种子。风选净种。沙冬青果实出种率约 60%，种子净度 95%左右，千粒重 40～46g，每千克种子 2.2 万～2.5 万粒，发芽率 85%～95%。种子耐贮藏。密封于陶瓷罐中 15 年零 7 个月的种子含水量为 5.17%，发芽率仍达 68%，干藏于布袋中的种子，发芽率可保持5～6 年。

发芽和播种 用始温 50～60℃的水浸种，种子吸胀后播种。硬粒种子要反复用温水浸种，直至全部吸胀为止。每平方米播种量为 8～12g。低床条播，行距 25～30cm. 覆土厚度 2～3cm，5～6 天即可发芽出土。出土萌发，有 2 枚肥厚的子叶。幼苗形态如图 2。

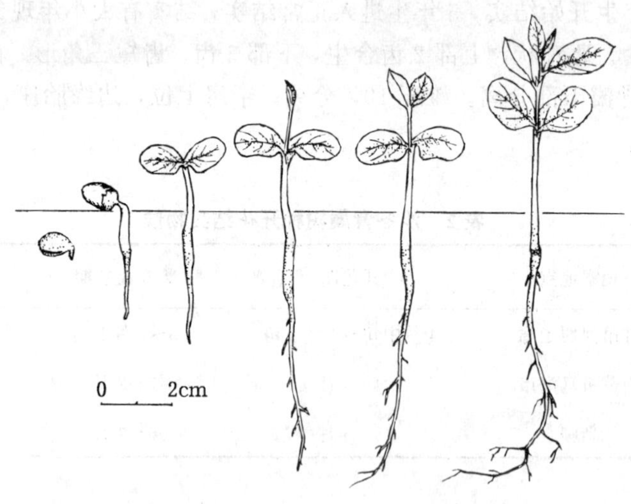

图 2 沙冬青种子萌发后第 1、8、13、22、35、41 天的幼苗生长情况
（姚桂英绘）

（李荣晨）

紫　穗　槐

Amorpha fruticosa L.

（蝶形花科　Fabaceae）

生长习性、分布和用途　紫穗槐属约 25 种，分布北美和墨西哥，本文介绍我国引入的 1 种。落叶灌木，高 1～4m。该种原产北美，我国东北南部、华北、华东、华中、西南及西北东部已广为栽培，以华北平原生长最好。适应性很强，耐旱、耐涝，耐瘠薄、耐盐碱，也耐干冷、高温气候。在极端低温－30℃，冻土深达 1.2m 的地区，枝条虽被冻枯，仍能从根际萌发新株。在宁夏腾格里沙漠东缘，地面温度有时高达 70℃，也能生长。对沙层含水量约 2.7%，干沙厚达 30cm，或水淹 45 天的环境，均可适应。叶可作饲料，也是优质绿肥。枝条可编织生产和生活用具。种子可榨油，为制造油漆、润滑油的原料。花为蜜源。具根瘤菌，能改良土壤，且根蘖性强，是荒山、沙荒、盐碱地的保土固沙树种。

开花结实　2 年生即可开花结实。花两性，总状花序顶生、直立。萼钟形，5 齿裂，有油腺点。花冠仅有旗瓣，心形，蓝紫色。果穗长 10～15cm，一个果穗有 100～150 枚荚果，荚果长7～9mm，宽 2mm，短镰状，密被瘤状油腺点。花期一般 5～6 月。据在北京妙峰山观察（侯惠宗，1956），初花期 4 月 27 日，末花期 5 月 27 日，开花平均 30 天。果期 9～10 月。荚果棕褐色时成熟，不开裂，内含种子 1 枚，少为 2 枚。种子椭圆形或长肾形，长 3～4mm，宽 1～1.5mm，光滑，黄绿色（见图 1）。

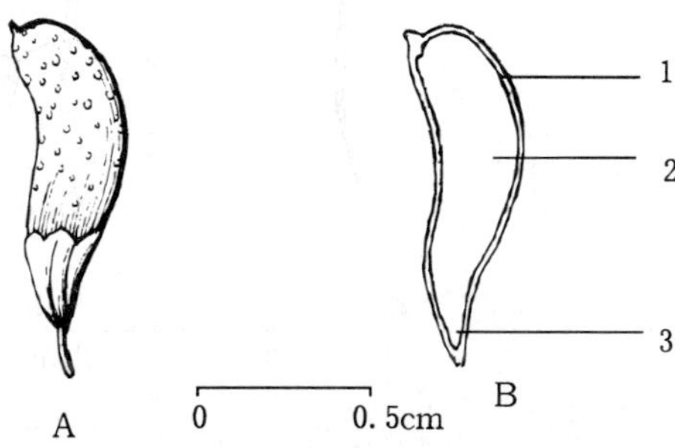

图 1　紫穗槐荚果外形（A）及其种子纵切面（B）
1. 果皮及种皮　2. 子叶　3. 胚根
（孟玲绘）

果实的采收调制和种子贮藏　荚果成熟后宿存枝上不自然脱落，可用手捋取荚果或采摘果穗，在阳光下摊晒 5～6 天，每天翻动几次，干后用风车风选去杂。荚果即为播种材料。据中国林业科学研究院试验，荚果气干含水量 9.3%时，在地下室布袋贮藏，2 年后发芽率由 72%降至 67%；地下室玻璃瓶贮藏发芽率达 81%。荚果千粒重变动范围为 9.2～12.4g，每千克约有 8～ 10.8 万粒。

发芽和播种　果皮含有蜡质，种子具有不透水种皮，吸水困难。播种前用 70℃热水浸种，水冷后继续浸种 1～2 天，捞出后每天洒温水 1～2 次。另外可用碾子碾破荚壳或用 6% 的草木灰水浸种 6～7 小时，也可以提高发芽率。据室内试验，去掉果皮置床，2 天后即开始发芽，发芽势 29%，发芽率 85%；如果不去果皮，7 天开始发芽，发芽势 15%，发芽率 41%。另据试验，同黑暗恒温相比，光照变温发芽可以提高发芽率和发芽势，最高的发芽势可以高出 23%，发芽率可以高出 21%。

种子常受紫穗槐豆象危害，一般被害率达 15%～20%，最高达 85%，对种子产量、质量影响很大，调运时要严格检疫。对种子进行熏蒸处理（在常温条件下，每立方米用溴甲烷或硫酰氟 30～35g 熏蒸 3 天），或在播种前用始温 80℃水浸种 6 小时，可以杀死幼虫。

春播。黄河流域以南地区在 2 月下旬至 3 月上中旬播种；黄河流域以北地区在土壤解冻后播种，北京地区以 3 月底播种为宜。条播，行距 20cm，播幅宽 8cm，播种沟深 5cm，覆土厚 1～1.5cm，每平方米播种 8～10g，播种后 5～7 天出苗。出土萌发。子叶 2（见图 2）。1 年生苗高约 1m，地径 0.6～1.0cm。在盐碱地上播种育苗保苗率低，可用径粗 1～1.5cm 的 1 年生条作插穗，扦插育苗。扦插前，先将插穗浸水 2～3 天，可提早发芽生根。

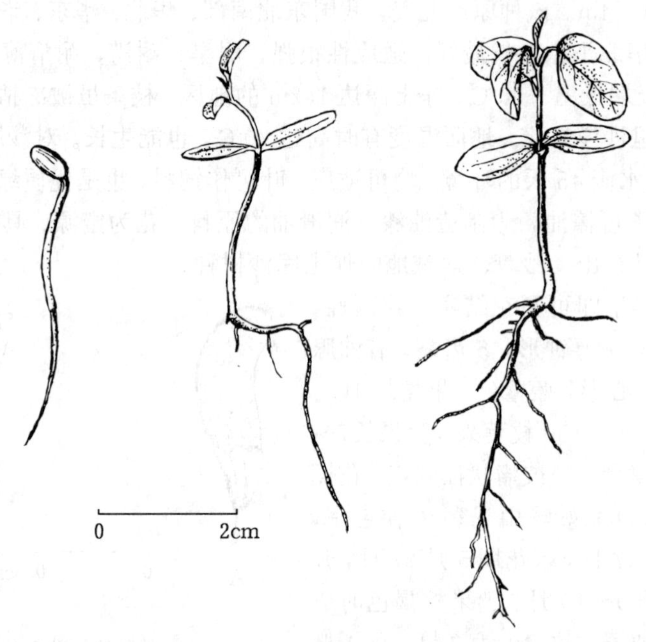

图 2　紫穗槐种子萌发后第 3、8、12 天幼苗生长情况

（田恒德临《主要树木种苗图谱》）

（陶章安）

木　豆

Cajanus cajan（L.）Millsp.

（蝶形花科　Fabaceae）

生长习性、分布和用途　木豆属 2 种，产热带非洲和亚洲，本文描述 1 种。常绿或半落叶灌木，高达 3m，速生。多分枝，主干不甚明显。喜肥但较耐干旱。忌霜雪。浙、闽、台、湘、粤、桂、琼、川、滇、黔等地有栽培。种子可加工制作食品和榨油，干枝可供胶虫寄生，叶作饲料，根清热止痛，茎皮作纤维用。印度作为主要粮食作物。

开花结实　1 年生即开花结实，1～2 年生为结实盛期，3～4 年生结实减少，无大小年现象。花两性。总状花序腋生。萼裂片披针形，具5齿。花冠黄色或带紫色条纹，旗瓣和翼瓣有耳。雄蕊10，2体（9+1）。子房近无柄，胚珠多数。据广西南宁1987～1989年的观察，11月下旬为始花期，12月下旬为盛花期，翌年2月中旬为末花期；3月中旬果实开始成熟，4月上旬为成熟盛期。荚果条形，先端渐尖，基部具宿存萼，长4～7cm，径0.6～1cm，种子间有深缢斜槽纹，先端有长喙，2瓣裂。未成熟时绿色，成熟后黄褐色。4月中下旬～5月上旬果荚开裂，种子脱落。每果有种子3～5粒。种子扁圆形或扁椭圆形，灰褐色或红褐色，长5～6mm，宽5～6.5mm，厚4～5mm，种皮薄，无胚乳（图1）。

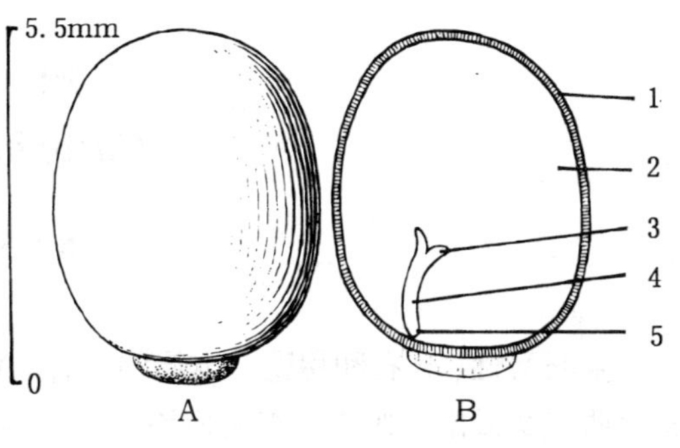

图1　木豆种子外形（A）及其纵切面（B）

1. 种皮　2. 子叶　3. 胚芽　4. 胚轴　5. 胚根

（黄应钦绘）

果实的采收调制和种子贮藏　果实成熟后分期采摘。采得的果实置日光下曝晒。果荚开裂后，用手搓揉或用木棒敲打，脱出种子。筛去或扬去果荚等杂质即得纯净种子。果实的出种率为30%～50%。净度可达98%。晒干的种子含水量在10%以下。千粒重约87g，每千克纯净种子约1.1（1～1.3）万粒。种子久藏易受虫蛀。生产上多为采种的当年播种，一般不作长期贮藏。贮藏一般用瓦缸。为防止虫蛀，贮藏时应拌以少量草木灰或石灰。贮藏期为1年。

发芽和播种　种子无休眠习性。发芽时日均温需在18℃以上。1987年7月12日广西林业科学研究所在室内发芽箱进行过发芽测定，播前用冷水浸种，基质为滤纸，发芽箱内保持恒温28℃。播种当天即开始发芽。从置床时起的一昼夜内，发芽百分数为52%，以3天计算，发芽率为72%。留土萌发。胚根萌发后约经2天幼苗出土，再过2天初生叶出现（图2）。

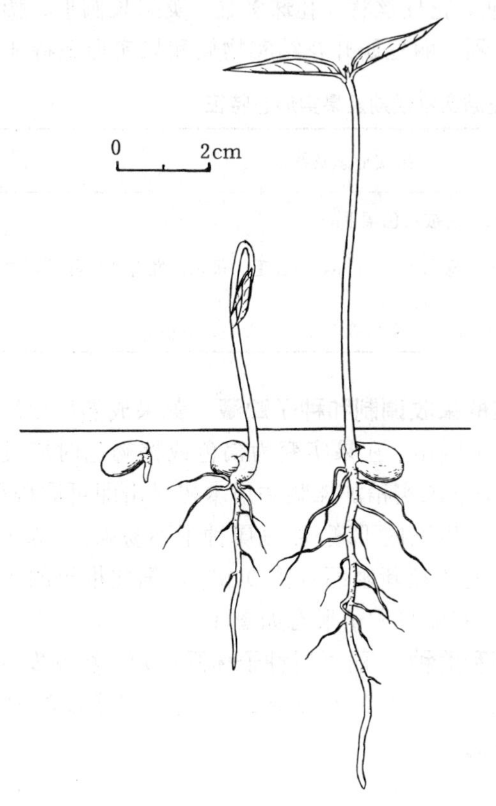

图2　木豆种子萌发后第1、3、5天幼苗的生长情况

（黄应钦绘）

种子浸泡后密播于圃地，发芽后15～30天移至容器或造林地栽植。也可将种子点播在容器或预先整理好的造林地，每容器或每穴点播2～4粒，覆土0.5cm。容器苗3个月出圃。

（韦增健）

丽 豆 属

Calophaca Fisch.

（蝶形花科 Fabaceae）

生长习性、分布和用途 本属约 7 种，我国 4 种，灌木或半灌木，株高 0.2～2m。分布于我国西北及云南、西藏。本文描述 3 种，其名称、树高、分布及用途见表 1。

表 1 丽豆属树种的名称、树高、分布及用途

中 名	学 名	树高 (m)	分 布	用 途
塔城丽豆（中国丽豆）	*C. chinensis* Boriss.	0.3	新疆塔城托里山区	饲料，纤维
新疆丽豆	*C. hovenii* Schrenk	1	新疆塔城	饲料，纤维
丽豆	*C. sinica* Rehd.	1～2	山西、内蒙古	观赏，饲料

开花结实 花两性，4 至多朵排成总状花序，腋生。萼筒倾斜，萼齿 5。花冠蝶形，黄色，旗瓣直立。雄蕊 10，两体 (9+1)。子房无柄，花柱丝状，胚珠多数。荚果长圆形，膨大，被毛，顶端有宿存花柱，两瓣裂，种子间无横隔。丽豆属开花结实物候和果实形态特征见表 2。

表 2 丽豆属树种开花结实物候期及果实形态特征

树 种	花果期	果实形态及特征
塔城丽豆	7～8 月	荚果矩圆形，长 1.5～1.8cm，密被白色柔毛
新疆丽豆	5～6 月	荚果长圆筒形，长约 2～4cm，宽 0.7～1.4cm，被腺毛。每果只发育 2 枚种子，种子卵圆形，棕色，长 0.4～0.5cm，宽 0.3cm
丽豆	5～6 月	荚果矩圆形，长约 3cm，宽约 8mm，密生腺毛和白色柔毛，有宿存花柱

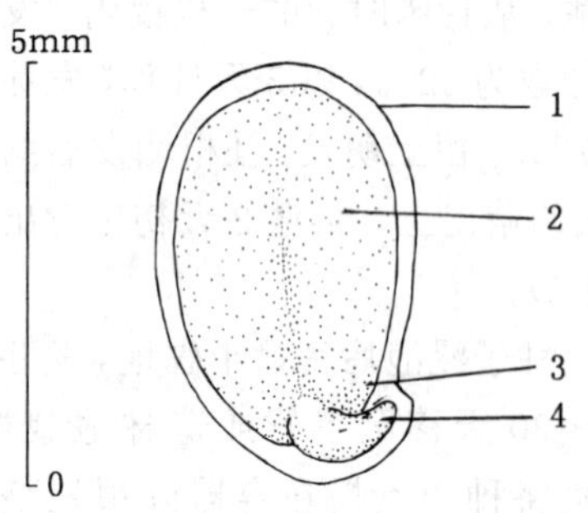

图 1 新疆丽豆种子纵切面

1. 种皮 2. 子叶 3. 胚轴 4. 胚根

（姚桂英绘）

果实的采收调制和种子贮藏 荚果成熟后由缝线自动开裂，种子脱出。在果实变为褐色或棕褐色时应及时从树上采摘。采下果实稍加摊晒并用木棒敲击即可取得种子。纯净种子晒干后室内干藏。丽豆属种子不易采到。为了撰写本文而采得的 4 粒新疆丽豆重 0.05g，据此推算的千粒重为 12.5g。新疆丽豆种子形态如图 1。

发芽和播种 丽豆属种子休眠期短，容易发芽。发芽前用始温 40～50℃的水浸泡 24 小时后置床，发芽较快，发芽率也较高。

（周君英）

锦鸡儿属
Caragana Fabr.

（蝶形花科　Fabaceae）

生长习性、分布和用途　本属约百种，分布东亚、中亚、东欧。我国约60种，本文描写12种和1变种。落叶灌木，稀为小乔木。树高1～3m，最高可达7m（树锦鸡儿）。喜光，深根性，耐干旱瘠薄，有根瘤菌，抗性强，是保持水土、改良土壤和防风固沙的优良树种。花为蜜源。锦鸡儿属树种的名称、分布及用途见表1。

表1　锦鸡儿属树种的名称、分布和用途

中　名	学　名	分　布	用　途	供　稿
树锦鸡儿	*C. arborescens* (Amm.) Lam.	东北、华北、西北。西伯利亚、蒙古、哈萨克斯坦	固沙，绿化	201
短叶锦鸡儿	*C. brevifolia* Kom.	甘南、青东、川西、藏	固沙	201
川西锦鸡儿	*C. erinacea* Kom.	青、甘、川、滇	保土，固沙，观花	201
金雀锦鸡儿（黄刺条）	*C. frutex* (L.) Koch.	新疆、内蒙古。蒙古、俄罗斯	保土，绿化荒山	201
中间锦鸡儿	*C. intermedia* Kuang et H. C. Fu	内蒙古、陕北、宁	保土，固沙	201
柠条锦鸡儿（毛条）	*C. korshinskii* Kom.	内蒙古、陕、宁、甘。蒙古	固沙，保土，饲料，绿肥，纤维，蜜源，薪柴，固沙	201
小叶锦鸡儿（柠条）	*C. microphylla* Lam.	东北西部、华北、西北。西伯利亚、蒙古、日本	保土，绿肥，薪柴，观赏	201
灰色小叶锦鸡儿	*C. microphylla* Lam. var. *cinerea* Kom.	内蒙古锡林郭勒盟	固沙，绿肥	201
甘蒙锦鸡儿	*C. opulens* Kom.	晋、陕、甘、宁、青、川、藏	固沙	201
猫耳锦鸡儿	*C. roborovskyi* Kom.	内蒙古、宁、甘、新、青	保土，固沙	201
狭叶锦鸡儿	*C. stenophylla* Pojark.	内蒙古、晋北、陕北、宁、甘西北。蒙古、西伯利亚	饲料，保土，固沙，纤维	201
川青锦鸡儿	*C. tibetica* Kom.	内蒙古、宁、甘、青、川、藏	保土，固沙	201
松东锦鸡儿	*C. ussuriensis* (Rgl.) Pojark.	黑龙江饶河；哈尔滨有栽培。俄罗斯远东、日本	水保，庭园绿化	107

开花结实　2～4年生开始开花结实，5～6年生进入结实盛期。花两性，单生，稀2～3朵成小伞形花序，花梗具关节。花萼筒状或钟状，5齿裂，近相等或上面2枚较小。花冠蝶形，黄色，稀淡紫色和浅红色，或旗瓣带橘红色、土黄色。旗瓣与翼瓣具长爪。雄蕊2体（9+

1)。单雌蕊，子房上位，1室，无柄，稀具柄，花柱直或稍内弯，胚珠多数。荚果条形，成熟时圆柱状，先端尖，2瓣裂。种子偏斜椭圆形或近球形，无种阜。锦鸡儿属的开花结实物候见表2，果实和种子的形态特征见表3。图1绘出的是柠条锦鸡儿种子的外形及其剖面结构。

表2　锦鸡儿属树种的开花结实物候

树　种	观察地点	开花期		果实成熟期		种子散落期	
		始期	末期	始期	盛期	始期	末期
树锦鸡儿	甘肃	5月	6月	8月	9月	8月	9月
短叶锦鸡儿	甘肃	6月	7月	8月	9月	8月	9月
川西锦鸡儿	甘肃	6月	—	7月	8月	7月	8月
金雀锦鸡儿	内蒙古东部地区	5月	6月	7月	—	7月	—
中间锦鸡儿	内蒙古伊克昭盟	5月	6月	6月上旬	7月中旬	6月	7月
柠条锦鸡儿	内蒙古伊克昭盟	5月上中旬	6月上旬	6月下旬	7月上中旬	6月下旬	7月上中旬
	宁夏中部沙坡头	4月下旬	5月上旬	5月中旬	6月中旬	5月中旬	6月中旬
小叶锦鸡儿	内蒙古伊克昭盟	5月	6月	6月下旬	7月下旬	7月	8月
灰色小叶锦鸡儿	内蒙古	5月	6月	6月	7月	7月	8月
甘蒙锦鸡儿	甘肃民勤，内蒙古巴彦淖尔盟，伊克昭盟	5月	6月	6月	7月	6月	7月
猫耳锦鸡儿	内蒙古阿拉善盟	5月	—	6月	7月	6月	7月
狭叶锦鸡儿	内蒙古伊克昭盟	4月	6月	7月	8月	7月	8月
川青锦鸡儿	甘肃西部地区	5月	7月	7月	8月	7月	8月
松东锦鸡儿	黑龙江哈尔滨市	5月中旬	6月中旬	6月中旬	7月中旬	6月	7月

表3　锦鸡儿属树种果实和种子的特征

树　种	成熟果实			种　子		
	形　状	大小（cm）	色　泽	形　状	大小（mm）	色　泽
树锦鸡儿	圆筒形	长3.5～6 宽0.3～0.5	棕色	圆形或椭圆形	长3～4 宽2.5～3	棕红色
短叶锦鸡儿	圆筒形	长1～2.5 宽0.2～0.4	黑褐色	椭圆形	长3～4 宽2	淡褐绿色
川西锦鸡儿	披针形或条形	长1.5～1.8 宽0.3～0.5	棕色	椭圆形	长3 宽2.2	棕绿色
金雀锦鸡儿	圆筒形	长2～3 宽0.3～0.4	棕色	椭圆形或卵圆形	长0.4 宽0.25～0.35	黑色
中间锦鸡儿	条状披针形或矩圆状披针形	长2.5～3.5 宽0.5～0.6	棕色	肾形	长6～9 宽4	黄绿或黄褐色，有褐色斑纹

（续）

树　种	成熟果实			种　子		
	形　状	大小（cm）	色　泽	形　状	大小（mm）	色　泽
柠条锦鸡儿	扁披针形或长条形	长 1.5～3.5 宽 0.6～0.7	棕褐色	肾形或扁长圆形	长 7.6 宽 4	淡褐色或黄褐色
小叶锦鸡儿	圆筒形稍扁	长 4～5 宽 0.5～0.6	深红褐色	长椭圆形	长 5～7 宽 3～3.5	黄绿色或米黄色
灰色小叶锦鸡儿	条形、扁	长 4～5	棕褐色	扁椭圆形或长肾形	长 6～8 宽 3.5～4.5	棕褐色或具紫斑
甘蒙锦鸡儿	圆筒形	长 2.5～4 宽 0.4～0.5	紫褐色	椭圆形	长 2.5～3.8 宽 2.0	棕绿色
猫耳锦鸡儿	圆柱形	长 2.5～3 先端长尖	棕褐色	椭圆形	长 4，宽 2	棕褐或褐绿色
狭叶锦鸡儿	圆柱形	长 2～3 宽 0.2～0.3	棕褐色	椭圆形	长 3.5～4.5 宽 2.0	绿色带黑条纹
川青锦鸡儿	椭圆形	长 0.7～0.8 宽 0.4	棕褐色	椭圆形	长 3～3.5 宽 2.5	棕褐色
松东锦鸡儿	扁条形，先端渐尖	长 3.3～3.5	黄褐色	—	—	浅黄褐带绿色

果实和种子常有虫害，常见的是柠条种子小蜂（*Eurytoma neocaraganae*）、柠条豆象（*Kytorhinus immixtus*）、柠条种子螟（*Epiepischnia keredjella*）和柠条种子鞘蛾（*Coleophora* sp.）。

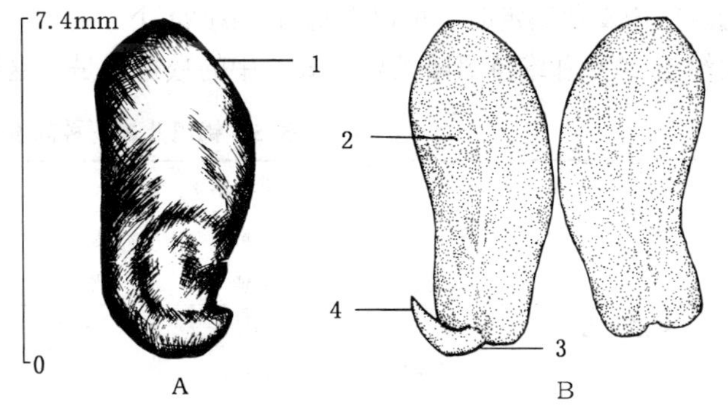

图 1　柠条锦鸡儿种子外形（A）及其纵切面（去种皮）（B）
1. 种皮　2. 子叶　3. 胚轴　4. 胚根
（姚桂英绘）

果实的采收调制和种子贮藏　盛果期时选生长旺盛、无病虫害感染的母树采种。荚果由绿色变坚硬而呈现黄棕色，枝上部荚果里的种子呈米黄色时采集。从荚果成熟到开裂的时间很短，单株2～4天，片林 3～7 天，因此要随熟随采。采集的荚果经摊晒、敲打或碾压，通过风选，除去果壳和夹杂物，即得纯净种子。锦鸡儿属种子的净度和质量见表 4。

种子贮藏前要用药剂熏蒸消灭害虫。不同树种的种子贮藏寿命不完全一致。小叶锦鸡儿采种当年，种子的发芽率 90%左右，存放 3 年后，种子变成暗灰色，开始离皮，发芽率下降到 30%左右，4 年后则完全丧失发芽能力。柠条锦鸡儿种子的生活力很强，当年发芽率高达 99%，发芽势 77%；在保存好的情况下，2 年后的发芽率仍有 93%，发芽势 67%；6 年之后发芽率降到 30%。

表 4　锦鸡儿属树种种子的净度及质量

树　种	净度（%）	千粒重（g）	每千克纯净种子粒数（万粒）
树锦鸡儿	90～98	25～28	3.6～4.0
短叶锦鸡儿	85	12.5	8.0
川西锦鸡儿	80～90	11.4～12.8	7.8～8.8
金雀锦鸡儿	90～98	27.8	3.6
中间锦鸡儿	85～95	50～60	1.6～2.0
柠条锦鸡儿	90～98	50～60	1.6～2.0
小叶锦鸡儿	70～80	35～37	2.7～2.9
灰色小叶锦鸡儿	90～95	24.6	4.0
甘蒙锦鸡儿	90～95	18.2	5.5
猫耳锦鸡儿	85～95	15～18	5.5～6.7
狭叶锦鸡儿	80～90	11.5	8.7
川青锦鸡儿	90～95	11～13	7.7～9.0
松东锦鸡儿	50～80	15.6～17.6	5.7～6.4

发芽和播种　锦鸡儿属种子休眠期短，不需特殊处理，用始温 50～60℃的温水浸泡 24 小时后，除去浮在上面的杂质及瘪粒，种子吸水膨胀后即可播种。可以用靛蓝或四唑测定生活力。室内发芽测定时，可以在每昼夜的 24 个小时中，16 小时给予 30℃并加光照，其余 8 小时给予 20℃并维持黑暗条件。表 5 中的数据就是在这种条件下测得的。

表 5　锦鸡儿属树种的发芽能力

树　种	发芽势（%）			发芽率（%）		
	计算天数	一般数值	变动范围	计算天数	一般数值	变动范围
树锦鸡儿	3	6	5～6	14	32	14～40
	25～40	45～75	—	—	94	55～100
	21	80	—	—	—	—
	21	45	—	—	—	—
川西锦鸡儿	4	59	—	14	86	85～95
金雀锦鸡儿	6	46	—	8	51	50～70
中间锦鸡儿	3	70	—	5	85	70～90
柠条锦鸡儿	2	90	85～95	3	95	95～98
小叶锦鸡儿	3	—	50～60	7	80	80～90
灰色小叶锦鸡儿	3	42	—	10	57	50～70
甘蒙锦鸡儿	2	41	—	16	79	70～85
猫耳锦鸡儿	2	36	—	16	69	60～80
狭叶锦鸡儿	—	—	—	7	60	—
川青锦鸡儿	2	20	—	7	28	—

春播。条播，行距 25cm，覆土厚 2～3cm。因树种而不同，每平方米播种量变动在 5～ 25g 之间。播后 5～7 天出土。子叶两枚，出土萌发。柠条锦鸡儿幼苗形态如图 2。

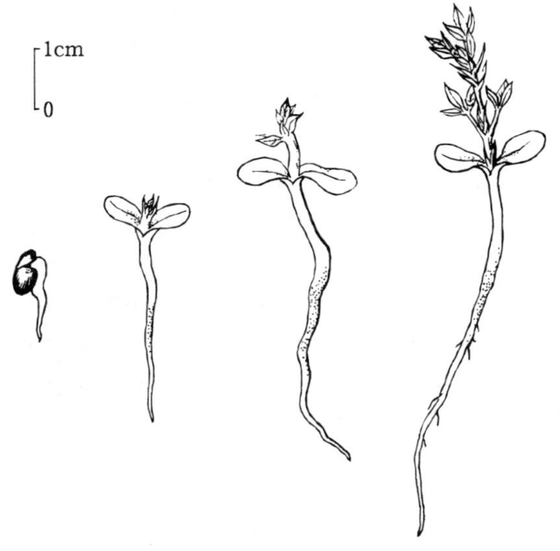

图 2　柠条锦鸡儿种子萌发后第 7、12、19、37 天幼苗生长情况

（姚桂英绘）

（李荣晨）

舞　　草

Codariocalyx motorius（Houtt.）Ohashi

（蝶形花科　Fabaceae）

生长习性、分布和用途　舞草属 2 种，我国均产，本文描述 1 种。灌木，高达 1.8m。适生于比较肥沃湿润的酸性土壤。分布于闽、赣、台、粤、桂、黔、川、滇。印度、越南、菲律宾亦产。全株药用，舒筋活血，夜间叶能转动，供观赏。

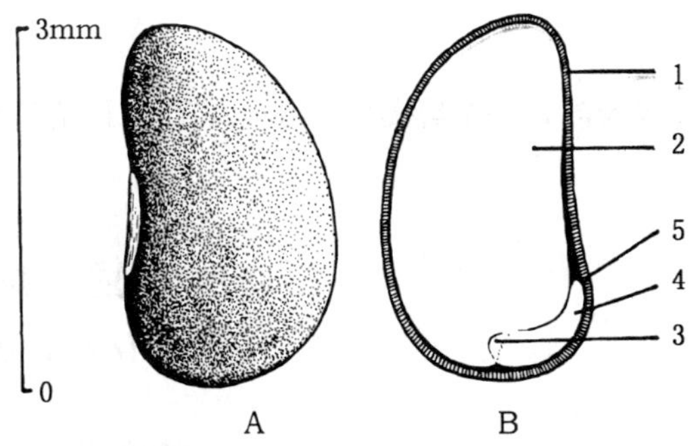

图 1　舞草种子外形（A）及其纵切面（B）

1. 种皮　2. 子叶　3. 胚芽　4. 胚轴　5. 胚根

（黄应钦绘）

开花结实　约 2～3 年生开始结实，正常结实期为 3～5 年生，结实无大小年现象。花两性，簇生，集成总状或圆锥花序，腋生或顶生于枝条上部。花序长约 20（24）cm。萼钟状，5 齿。花冠蝶形，紫红色，旗瓣近宽圆形，翼瓣基部卵形，龙骨瓣有矩状附属物。雄蕊 2 体（9+1）。子房上位，1 室，胚珠多数，花柱无毛。据广西南宁 1987～1988

年的物候观测，6 月下旬为始花期，7 月中旬为盛花期，8 月上旬为末花期；9 月下旬果实开始成熟，10 月中旬为果熟盛期，10 月下旬～12 月上旬背缝开裂，种子脱落。荚果，长椭圆形至线形，压扁，腹缝直，背缝于种子间稍缢缩，未成熟时绿色，成熟时黄褐色，疏被柔毛，长 1.5～4cm，宽 4～7mm。每果荚有种子 2～7 粒。种子有假种皮及种阜，扁椭圆形或肾形，褐色，种脐白色，长 2.5～3.5mm，宽 2～2.4mm，厚约 1mm。种皮薄，无胚乳，胚根内曲而位于子叶边缘（图 1）。

果实的采收调制和种子贮藏　果实成熟盛期采摘荚果，置日光下曝晒或在室内通风处摊晾。果荚开裂后用手搓揉，脱出种子。扬去果荚及杂质，即得纯净种子。荚果的出种率约 40%。晒干后种子含水量宜在 10%以下。种子的净度可达 97%。千粒重约 4.5g，每千克纯净种子约 22 万粒。少量种子短期贮藏可用布袋包装，悬挂于室内干燥阴凉处。贮藏期超过半年以上宜用坛、罐等瓦器贮藏，并拌以少量草木灰或石灰，以防虫蛀，贮藏期一般为 1 年。

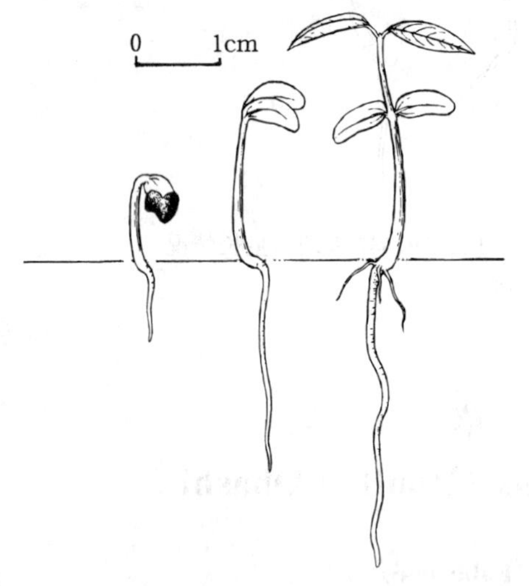

图 2　舞草种子萌发后第 2、3、5 天幼苗的生长情况
（黄应钦绘）

发芽和播种　种子无休眠习性。发芽时日均温宜在 20℃以上。1988 年 7 月上旬，广西林业科学研究所用头年 10 月所采的种子在室内作过发芽测定。基质为滤纸，光照为室内自然光，播前用始温 50℃水浸种 12 小时，播种时日均温 29℃。播后第 2 天开始发芽，第 3 天进入发芽盛期，第 23 天发芽终止。从开始发芽至发芽高峰，6 天的发芽百分数为 30%。从播种之日起，25 天发芽率为 55%。出土萌发。种子萌发后 2 天子叶出土，再过 1 天发出初生叶（图 2）。

点播或条播。大田每平方米育苗用种约 2.5g。一般培育 1 年生苗出圃。如用容器育苗，可将经过预处理的种子点播在容器内，每个容器约 3 粒，培育百日苗出圃。

（韦增健）

猪屎豆

Crotalaria pallida Ait.

（蝶形花科　Fabaceae）

生长习性、分布和用途　猪屎豆属（野百合属）约 350 种，我国约产 60 种，本文描述 1 种。灌木或半灌木，高约 1m。适生于弃耕地、田野、溪畔的疏松湿润土壤，贫瘠地及荒山未见生长。能耐 0℃左右低温，但忌霜冻。在热带地区可以成为多年生灌木或亚灌木，亚热带地区一般用作 1 年生绿肥栽培。分布于闽、台、琼、粤、桂、滇。非洲、美洲和亚洲热带也有分布。嫩茎及叶作绿肥和饲料，皮作纤维，种子有毒，适量可作强壮剂。

开花结实　7～9 个月即开花结实，可以正常结实 1～3 年。结实无大小年现象。花两性。总状花序顶生，长 15～30cm。花萼 5 裂。花冠蝶形，黄色。雄蕊 10，花丝基部合生，近轴面分离，花药 2 型，基着药与背着药互生。子房有胚珠多颗，花柱被毛。开花结实期较长。据广西南宁 1986～1988 年的观察，正常年份多数植株在 10 月上旬开始开花，10 月下旬～11 月为盛花期，12 月中下旬为末花期；11 月上旬早期开花的果实成熟，11 月下旬～12 月中旬为果熟盛期；冬季气温高的年份，结实期可延迟至 12 月下旬。花蕾随新梢伸展而陆续出现并开放，花与果常共存于 1 枝。荚果圆柱状，先端喙尖，果瓣膨胀，腹部有 1 槽纹，成熟时由青色转为淡黄色至棕褐色，长 3～5 cm，径约 6mm。每果有种子 20～30 粒。种子黑褐色，椭圆形，无种阜，长2.5～3.2mm，宽 2～2.5mm，厚 1.2～1.5mm。无胚乳，胚弯曲，胚根尖（图 1）。

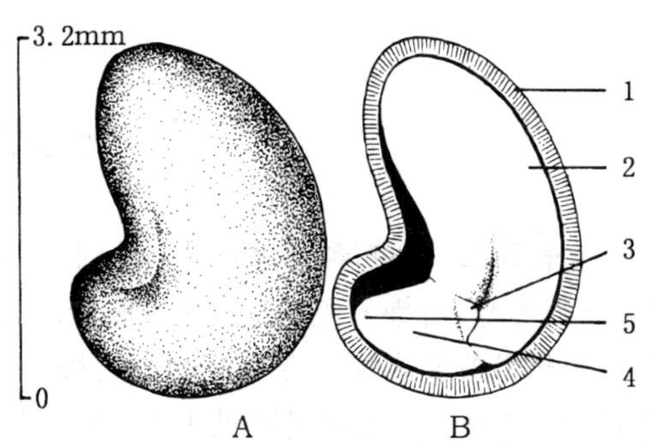

图 1　猪屎豆种子外形（A）及其纵切面（B）
1. 种皮　2. 子叶　3. 胚芽　4. 胚轴　5. 胚根
（黄应钦绘）

果实的采收调制和种子贮藏　荚果呈现淡黄色，种子呈黄褐色时应及时采收。成熟荚果在树上存留过久，种子会有虫蛀。采回的果实置日光下曝晒，果瓣开裂后用木棒敲打并用手搓揉，清除杂质即得纯净种子。出种率约 27%。净度可达 97%。千粒重约为 6.6g，每千克有纯净种子 15 万粒。种子可晒干贮藏。干种子的含水量约 10%。贮藏宜用坛罐等容器，置于室内干爽处。常温条件下室内裸露贮藏的种子，发芽能力的保持期一般为 1 年，密封贮藏的种子发芽能力约可保持 2 年。

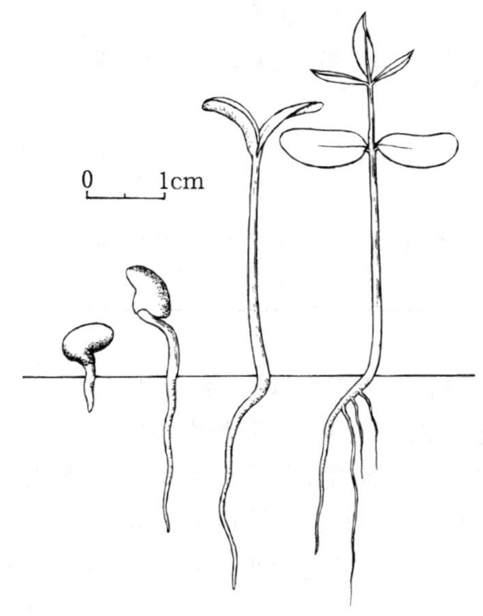

图 2　猪屎豆种子萌发后第 2、4、10、16 天幼苗的生长情况
（黄应钦绘）

发芽和播种　种子无休眠习性。发芽时日均温需在 20℃以上。1987 年广西林业科学研究所在发芽箱内作过发芽测定。置床前用 60℃热水浸种并在自然冷却过程中浸泡 1 昼夜，基质为滤纸，箱内温度保持在 28℃。播后 1 天开始发芽，第 2 天至第 5 天为发芽盛期。从播种之日起算，6 天的发芽百分数为 55%，14 天发芽率为 65%。出土萌发。胚根萌发后 4 天子叶伸出，再过 12 天初生叶出现（图 2）。

直播或用容器育苗。种子预处理后撒播或点播于林地，或与果树等混种。容器苗一般 2 个月左右出圃定植，多种植作围篱。

（韦增健）

黄 檀 属

Dalbergia L. f.

（蝶形花科 Fabaceae）

生长习性、分布和用途 本属约120种，我国约产25种，本文描述7种。落叶或常绿乔木，少数为藤本。萌芽力强。喜疏松肥沃的酸性土或钙质土，粘重或干旱瘠薄的土壤生长不良。广布于热带及亚热带。有些种的心材极珍贵，嫩枝为胶虫寄主，固土改土效果好。这7个种的名称、生长、分布和用途见表1。其中降香黄檀在《中国植物红皮书》中列为濒危种，亟需保护。

表1 黄檀属树种的名称、生长、分布和用途

中 名	学 名	树 高 (m)	胸 径 (cm)	分 布	用 途	供 稿
南岭黄檀	*D. balansae* Prain	15～20	30～40	东南、中南、西南	胶虫寄主、材用、纤维	601
黑黄檀	*D. fusca* Pierre	25	50～80	滇。越南、老挝、缅甸	材用、胶虫寄主	616
藤黄檀	*D. hancei* Benth.	藤本	—	华东、中南、川、黔	纤维、栲胶、根茎药用	601
黄檀	*D. hupeana* Hance	20	40	华东、中南、西南，陕	优质用材，胶虫寄主	601
钝叶黄檀（牛肋巴）	*D. obtusifolia* Parin	20～27	25～30	滇。桂、粤、闽、湘、黔、川栽培。缅甸、老挝	胶虫寄主、绿化、薪柴	601
降香黄檀（花梨）	*D. odorifera* T. Chen	20	80	琼。粤、桂、闽栽培	珍贵用材，心材制佛香，可提降香油	601
思茅黄檀（秧青）	*D. szemaoensis* Prain	20	80	滇。桂、粤、闽、黔、川栽培	优质胶虫寄主、材用、薪柴	601

开花结实 7～10年生开始开花结实，正常结实期在18年生以后。结实大小年间隔期一般为1年，但不甚明显。黄檀属的多数种可以作为胶虫的寄主树。放养胶虫会影响结实，用作采种母树的应禁放胶虫。花两性。降香黄檀为复聚伞花序，腋生，其余6种均为圆锥花序，顶生或腋生。黄檀、钝叶黄檀和降香黄檀花冠为淡黄色，黑黄檀、南岭黄檀、藤黄檀的花冠为白色或白绿色，思茅黄檀的花冠为紫色。萼钟状，萼齿5裂。花冠蝶形，花瓣皆具柄。雄蕊9，单体或10，两体（9+1、5+5）。花药小，药室顶裂。子房上位，1室，有柄，具少数胚珠。花柱短，内弯，柱头小，顶生。据广西南宁及云南景洪普文林场和思茅观察，开花结实的物候期如表2。

表 2　黄檀属树种的开花结实物候期

树　种	观察地点	观察年份	开花期			果实成熟期		果实脱落期
			始　期	盛　期	末　期	始　期	盛　期	
南岭黄檀	南宁	1987～1988	5 月下旬	6 月上中旬	6 月下旬	10 月下旬	11 月下旬	11 月下旬～12 月下旬
黑黄檀	普文	1987～1989	4 月上旬	4 月中旬	5 月中旬	10 月下旬	11 月	12 月～翌年 2 月
藤黄檀	南宁	1987～1988	5 月上旬	5 月中旬	5 月下旬	10 月上中旬	11 月下旬	12 月上旬～12 月下旬
黄檀	南宁	1987～1988	5 月下旬	6 月上中旬	6 月下旬	10 月上旬	10 月中下旬	11 月中旬～12 月中旬
钝叶黄檀	南宁	1987～1988	3 月上旬	3 月下旬	4 月上旬	7 月上旬	7 月中旬	7 月中旬～8 月上旬
	思茅	—	2 月上旬	3 月上中旬	3 月下旬	4 月下旬	5 月上旬	—
降香黄檀	南宁	1987～1988	5 月上中旬	5 月下旬	6 月上旬	11 月上旬	11 月中旬	12 月上旬～12 月下旬
思茅黄檀	南宁	1987～1988	4 月下旬	5 月上中旬	5 月下旬	11 月中下旬	12 月上旬	12 月～翌年 1 月

荚果，短带状，长圆形或长椭圆形，扁而薄。果瓣在种子处常具网脉，成熟后不开裂。果实脱落期较长。每果有种子 1～2（4）粒。种子肾形，扁平。无胚乳，子叶厚，胚根内曲于子叶边缘。荚果及种子的形态见表 3、图 1。

表 3　黄檀属树种果实和种子的形态特征

树　种	未熟果颜色	果　实			种　子		
		形　状	大小（cm）	色　泽	形　状	大小（mm）	色　泽
南岭黄檀	青色稍带黄	带状，荚室隆起	长 5～9.2 宽 2～2.5	黄褐色	扁平，肾形或椭圆形	长 6.5～11 宽 4～6	棕褐色或棕红色
黑黄檀	青色	舌状狭带形，两端尖	长 4～10 宽 1～1.5	浅褐色	椭圆形	长 7～10 宽 5～7	棕色
藤黄檀	黄绿色	矩圆形，荚室突起	长 3～9 宽 1.2～1.6	暗褐色	扁椭圆形	长 7～10 宽 4～6	深褐色
黄檀	绿色，略黄	长圆形或带状	长 3～9 宽 0.8～1.8	黄褐色	肾形或扁椭圆形	长 6.5～12 宽 4～6.5	褐色
钝叶黄檀	浅赭色	长椭圆形，具明显脉网	长 4～8 宽 1.3～1.6	暗褐色	肾形，扁平，薄	长 8～12 宽 4～7.5	黄褐色至棕褐色
降香黄檀	青色	舌状长椭圆形，具明显细纹	长 4.5～8 宽 1.5～2.3	褐色	肾形，扁平	长 9～11 宽 5～6.5	褐色，光滑
思茅黄檀	青色稍带黄	长圆形或带状	长 3～8 宽 1.8～2.5	黄褐色	扁椭圆形	长 6～10 宽 4～6	褐色

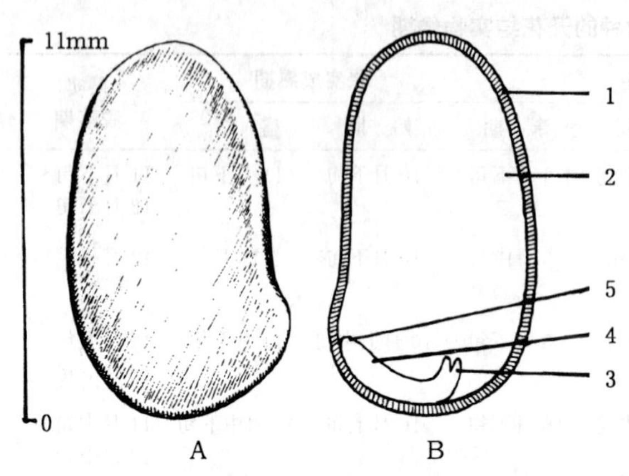

图 1　黄檀种子外形（A）及其纵切面（B）
1. 种皮　2. 子叶　3. 胚芽　4. 胚轴　5. 胚根
（黄应钦绘）

果实的采收调制和种子贮藏　黄檀属采种的时期虽较长，但成熟的果实存留在树上或落地的时间过久，雨后种子易受虫蛀，仍应在果实大熟期及时采集。用高枝剪或采种钩刀带总柄采下，也可上树采摘。采得的果荚不宜久堆，应及时曝晒。黄檀属多数种的果荚难开裂，较难脱出纯净种子。一般是将果荚晒至干脆，用木棒敲打或用手工剥开果荚。种子的质量等数据见表 4。

种子可以干藏，但不宜过分干燥，忌烈日曝晒，应在弱光下晒干或晾干，含水量保持在 10%上下，干藏于袋内或密封于坛罐内。袋藏的种子发芽能力可保持 6～8 个月。密封贮藏的种子发芽能力可保持 1 年。由于果荚极难裂开，种子不易脱出，生产上常将荚果晒干后，不加处理，袋装贮藏过冬，至翌年 3 月下旬取出，直接以荚果作为播种材料，也通称种子。为了方便播种并减少运输、贮藏的体积，可在荚果曝晒至果瓣脆干时，用木棒敲打或将果荚搓碎，扬去碎细的果瓣，得出裹有荚壳的种子用于播种。

表 4　黄檀属树种的出种率和种子净度、质量

树种	出种率（%）	净度（%）	千粒重（g）		每千克纯净种子粒数（万粒）	
			一般	变动范围	一般	变动范围
南岭黄檀	10～13	97	42	30～50	2.4	2～3.3
黑黄檀	26～28	96	49	43～54	2	1.8～2.4
藤黄檀	12～18	97	42	35～50	2.4	2～3
黄檀	10～15	97	40	35～50	2.5	2～3
钝叶黄檀	35～38	95	28	25～33	3.6	3～4
降香黄檀	11～15	96	70	55～85	1.4	1.2～1.8
思茅黄檀	12～18	97	38	30～45	2.6	2.2～3.3

发芽和播种　种子无休眠习性。发芽时日均温宜在 20℃以上。1980 年、1982 年和 1987 年，广西林业科学研究所对 6 种黄檀进行过发芽测定。播种前均用冷水浸种 1 昼夜，测定结果见表 5。

1988 年 5 月 24 日，云南景洪普文林场在室外沙床对黑黄檀也做过发芽测定：播后 5 天开始发芽，7 月 3 日发芽终止，发芽盛期不明显，发芽率为 50%。黄檀属种子为出土萌发。胚根萌发后 2～3 天子叶出土，5～6 天初生叶出现（图 2）。

表 5　黄檀属树种的发芽能力及其测定条件

树种	基质	温度（℃）		发芽势（%）		发芽率（%）	
		发芽箱	室外	计算天数	一般天数	计算天数	一般数值
南岭黄檀	纸	28	—	3	46	10	58
藤黄檀	纸	28	—	3	47	10	60
黄檀	沙	—	22	9	42	20	64
钝叶黄檀	纸	28	—	2	51	9	55
降香黄檀	沙	—	22	6	38	19	63
思茅黄檀	沙	—	22	5	44	11	62

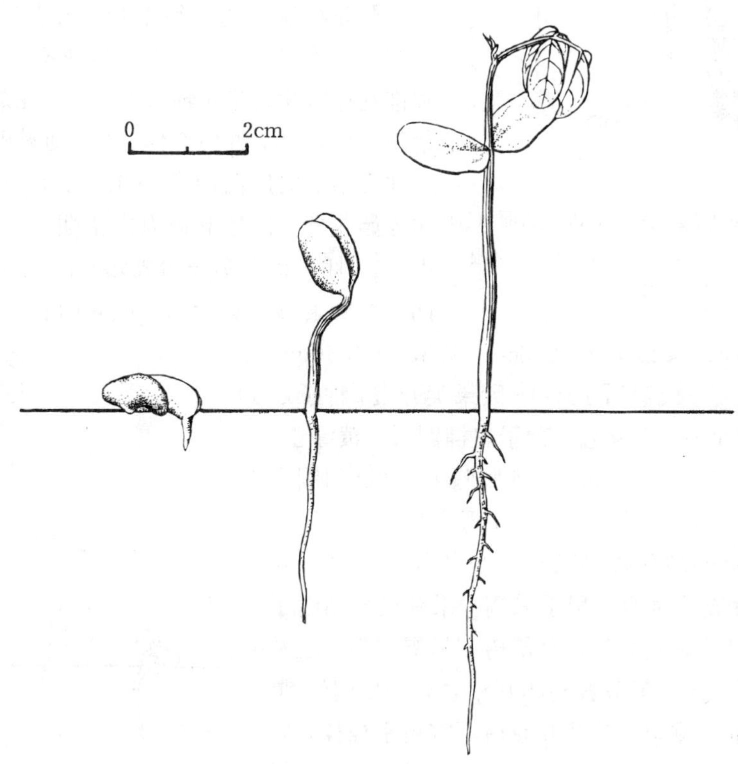

图 2　钝叶黄檀种子萌发后第 2、3、6 天幼苗的生长情况
（黄应钦仿《热带亚热带主要树种采种育苗技术》）

条播。每平方米播种量：降香黄檀 12～15g，钝叶黄檀 5～8g；其余 5 种 7～10g。如带荚播种，需相应增加播种量。降香黄檀一般需育 2 年生苗出圃，其余 6 种育 1 年生苗出圃。

南岭黄檀、钝叶黄檀、降香黄檀和思茅黄檀亦可扦插繁殖。南岭黄檀还可分根繁殖。

（王宏志）

假 地 豆

Desmodium heterocarpon（L.）DC.

（蝶形花科　Fabaceae）

生长习性、分布和用途　山蚂蝗属约 250 种，我国约产 50 种，本文描述 1 种。灌木，高 2～3m。喜生于水肥条件较好的疏残林下，土壤多为酸性，稍耐水湿，贫瘠地生长不良。能耐 －6℃的极端低温。分布在华东、中南和西南。印度、缅甸、泰国、日本亦产。全株供药用，治跌打骨折、蛇伤。茎皮作纤维原料。

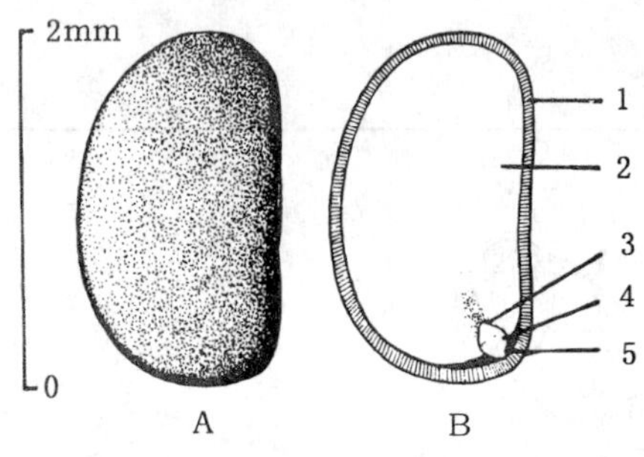

图 1　假地豆种子外形（A）及其纵切面（B）
1. 种皮　2. 子叶　3. 胚芽　4. 胚轴　5. 胚根
（黄应钦绘）

开花结实　约 3 年生开始开花结实，正常结实期在 6 年生以后。结实大小年现象不明显。花两性。圆锥花序腋生。花序轴被长柔毛。花成对着生，花冠紫色。雄蕊 10，近似 2 体，有 1 雄蕊略分开。子房有胚珠多数。据广西南宁 1987～1988 年观测，9 月中旬为始花期，9 月下旬为盛花期，10 月中旬花期结束；10 月下旬果实开始成熟，11 月上旬为果实盛熟期。荚果条形，具荚节，成熟时由黄绿色转为棕褐色，被灰白色钩毛。果长 1.2～2.6cm，宽 3.2～4.3mm，腹缝直，背缝波状。成熟后 10～15 天果荚开裂，种子逐渐脱落。每果有种子 1～9 粒。种子扁椭圆形，黄绿色，长 1.6～2mm，宽 1.2～1.6mm，种脐侧生，种脐周围有环形假种皮。无胚乳，胚根短。种子形态见图 1。

果实的采收调制和种子贮藏　当部分果荚呈棕褐色，种子已发育成熟时即可用手采摘。果荚极易粘附于人畜或野兽身上，造成失落，存留树上易遭损失，应及时采摘。采回的荚果，荚节容易相互粘着，结成 1 团，曝晒时可用齿耙推开摊匀。荚节开裂后不宜用手搓揉，以免种子粘附在果瓣上，难以分出，一般是用小竹竿轻轻敲打，使种子脱出，然后捡出粘结成团的果瓣。清除杂质后即得纯净种子。荚果的出种率约 28%。千粒重约 1.37（1.2～1.5）g，每千克有纯净种子 73（67～83）万粒。种子晒干供运输或贮藏。贮藏的种子用袋或坛罐包装，存放于室内通风避光处。常温条件下，贮藏期可达 1 年以上。

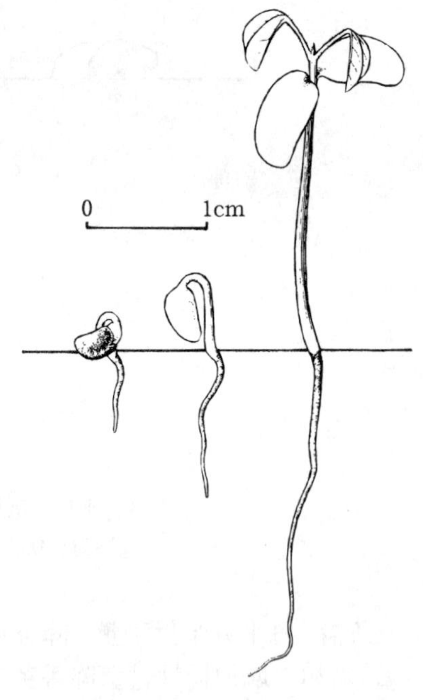

图 2　假地豆种子萌发后第 4、5、10 天幼苗的生长情况
（黄应钦绘）

发芽和播种　种子无生理性休眠。种皮可能妨碍吸水萌发，稍加处理，播后 3 天左右即可发芽。发芽时日

均温宜在18℃上下。1987年11月中旬，广西林业科学研究所在室内发芽箱内用当年新采的种子作过发芽测定，箱内温度保持18～20℃，利用自然光照。播前用始温80℃水浸种，自然冷却后再浸泡1昼夜。播后3天胚根萌发，再过2天进入发芽盛期，第16天发芽终止。从发芽开始至发芽高峰期的6天中，发芽百分数为35%。从播种之日起算的19天中发芽率为65%。出土萌发。胚根萌发后约5天子叶出土，9天左右初生叶出现（图2）。

条播。整地要求细致，开沟宜浅于3cm，沟底刮平。播前用始温80～100℃水浸种，每平方米播种量约为0.7～1.2g，覆土2～3mm。培育1年生苗出圃。

（韦增健）

千　斤　拔

Flemingia macrophylla（Willd.）Merr.

（蝶形花科　Fabaceae）

生长习性、分布和用途　千斤拔属约40种，我国19种，本文描述1种。亚灌木，高达3m。多生于田野空旷地，林中少见。根系较深，能耐干旱，亦耐水湿。喜肥沃疏松土壤，贫瘠地生长不良。广布于皖、赣、闽、台、湘、粤、桂、黔、川、滇、藏等地。亚洲热带地区也有分布。全株供药用，治风湿骨痛，又可作兽药，治猪脚软、牛骨折、牛胃炎等。

开花结实　约2年生开始开花结实，3年生以后可以正常结实，结实无大小年现象。花两性。总状花序腋生，花多而密。花冠淡红色，长约1cm。雄蕊2体（9+1）。子房近无柄，胚珠2。据南宁1987～1988年观察，9月下旬为始花期，10月上旬为盛花期，10月中旬为末花期；10月末果实开始成熟，11月上旬为果熟盛期。荚果成熟时由黄绿色转为黄褐色，被短毛，椭圆形，膨胀，顶端喙状，长1.1～1.6cm，径5.2～7.5mm。果荚成熟后15～20天开裂，裂后种子一般不脱落。每荚果有种子2粒。种子扁圆形或扁椭圆形，黑色有光泽，无种阜，径3～3.5mm，高2～2.8mm。无胚乳，胚根短（图1）。

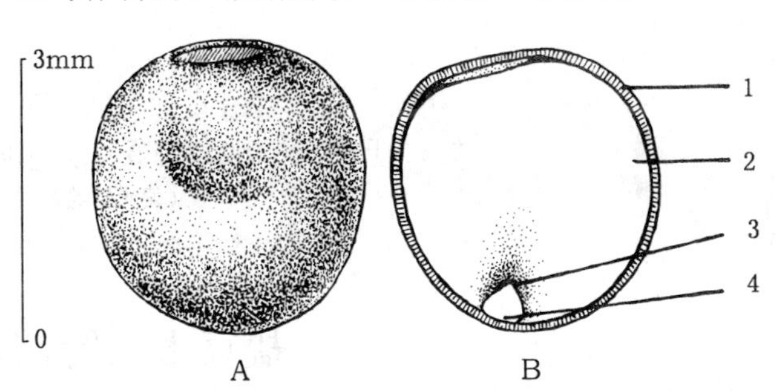

图1　千斤拔种子外形（A）及其纵切面（B）
1. 种皮　2. 子叶　3. 胚芽　4. 胚根
（黄应钦绘）

果实的采收调制和种子贮藏　荚果大部分成熟而尚未开裂或初裂时用手采摘。采得的果实置日光下曝晒。果瓣全部开裂后收集成堆，下垫塑料薄膜，先用木棒敲打，后用手搓揉，脱出种子。筛去果壳及杂质即得纯净种子。果实的出种率约18%。净度96%～99%。含水量约10%。千粒重16（14～18）g，每千克纯净种子为6.2（5.5～7.1）万粒。种子可以干藏，用布袋或坛罐等容器盛装，藏于室内避光干燥处。常温条件下贮藏期可达1～2年。

发芽和播种　种子无休眠习性。发芽时日均温宜在20℃以上。1987年11月，广西林业科学研究所用内垫滤纸的发芽皿在室内作发芽测定，种子为当年所采的新鲜种子，光照为室内自然光，温度在20℃上下。11月上旬末置床，中旬中期胚根开始萌发，发芽盛期不明显，12月中旬发芽终止。以36日计，发芽率为54%。留土萌发。胚根萌发后约4天上胚轴伸出，再过5天左右初生叶出现（图2）。

3月中下旬为适宜的播种期。播种前宜将种子浸水1昼夜，捞出稍干后即可播种。条播。每平方米播种量约3.5～5g，播后覆土1～5mm。1年生苗出圃。

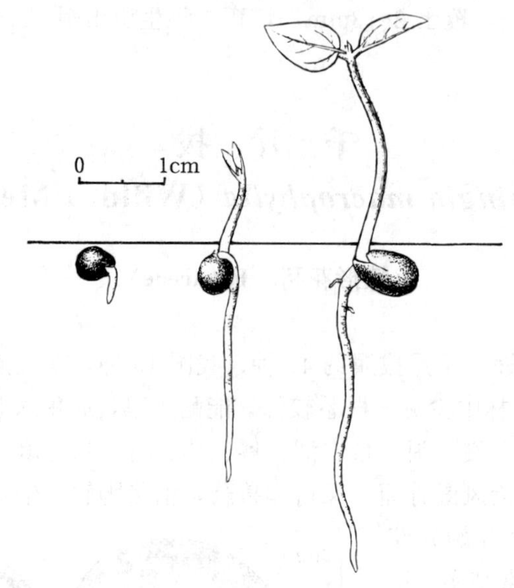

图2　千斤拔种子萌发后第1、5、10天幼苗的生长情况

（黄应钦绘）

（韦增健）

盐豆木（铃铛刺）

Halimodendron halodendron（Pall.）Schneid.

（蝶形花科　Fabaceae）

生长习性、分布和用途　盐豆木属仅1种，产于新疆、内蒙古巴丹吉林沙漠和甘肃河西走廊沙地。蒙古、哈萨克斯坦等亦产。落叶灌木，高达1～3m。喜光、耐寒，耐旱，耐盐碱。生于荒漠盐化沙土和河流沿岸。可用于固沙及改良盐碱土。可做绿篱，花艳丽，供观赏。

开花结实　3～4年生开始开花结实，4年就进入结实盛期。花两性，2～5朵簇生于老枝节上或组成腋生的总状花序。花冠蝶形，淡紫色或紫色。萼杯状，5齿裂，短，上部2枚稍合生，花瓣近等长，旗瓣近圆形。雄蕊2体（9+1），上部分离。子房具柄，内有多数胚珠。花柱内弯。花期5～7月，果期6～8月。荚果，先端突尖，矩圆状卵形，革质，膨胀，两缝线下凹；长1.8～2.5cm，宽8～12mm，光滑，黄褐色，仅基部有毛。果颈（子房柄）长为萼筒的2倍，长7～9mm。种子约7粒，肾形，淡褐色，长约4mm，宽2～3mm。种子的形态如图 1。

果实的采收调制和种子贮藏　荚果变为黄褐色即已成熟，应及时从树上摘取。采回后摊在通风干燥处晾晒，并敲击果实使种子脱出，除去果皮及夹杂物得到纯净种子。净度 70%～80%，出种率约 6%，千粒重约 10g，每千克种子 10 余万粒。种子耐贮藏，最适于贮藏的种子含水量为 7%以下。贮藏时将种子装入布袋或麻袋置于通风干燥处。种子易遭虫害，空粒较多。可在风选或筛选后再用 1%的食盐溶液水选，捞去漂浮种子，取下沉种子用清水洗净。播前用始温 60～70℃的水浸种半小时，再加凉水降温，浸种 24 小时，可起到催芽和灭虫作用。

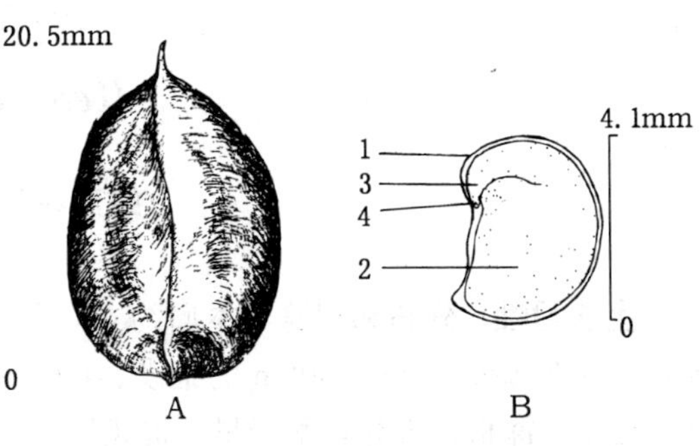

图 1　盐豆木荚果外形（A）及其种子的纵切面（B）
1. 种皮　2. 子叶　3. 胚轴　4. 胚根
（姚桂英绘）

发芽和播种　浸种 24 小时后取出其中膨胀种子播种，未膨胀的继续用温水浸种，直至膨胀。也可以把种子置于 0～5℃的条件下，混 3 倍湿沙（或锯木屑、珍珠岩）层积 10～15 天，种子萌动后取出播种，或直接置于 20～25℃的室温条件下催芽，露白即可播种。

发芽测定　温水浸种 24 小时后置于发芽皿内，用滤纸及脱脂棉作基质，放在发芽箱内发芽，在昼温 30℃，夜温 20℃的变温下，曾对一份样品测定了 12～14 天，测得发芽率为 39%。

种子出土萌发，子叶 2 枚，卵形。条播，行距 20～25cm。每平方米播种 12～15g。覆土厚度 1～1.5cm。1 年生苗可以出圃。幼苗形态如图 2。

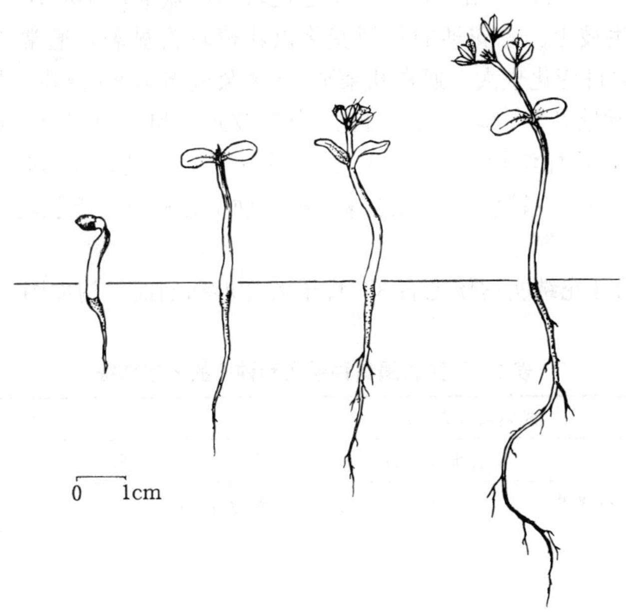

图 2　盐豆木播种后第 8、24、37、53 天幼苗生长情况
（姚桂英绘）

（李荣晨）

岩黄蓍属

Hedysarum L.

（蝶形花科 Fabaceae）

生长习性、分布和用途 本属约150余种，我国25种，产于北方，本文描述2种。落叶灌木，生于海拔400～1 000m的荒漠、半荒漠和干旱草原地带的沙地、戈壁，是优良的固沙树种。速生，深根，喜光，耐干旱，抗风蚀，耐严寒酷热，喜适度沙埋，耐流沙环境。花期长，为优良蜜源树种。纤维长，可制麻。其名称、分布和用途见表1。

表1 岩黄蓍属树种的名称、树高、分布和用途

中 名	学 名	树 高 (m)	分 布	用 途
羊柴（塔落岩黄蓍）	*H. laeve* Maxim.	1.5	内蒙古、甘。蒙古、西伯利亚	固沙，饲料，燃料
花棒（细枝岩黄蓍）	*H. scoparium* Fisch. et Mey.	7	内蒙古、宁、青、甘、新。陕、晋栽培。蒙古、哈萨克斯坦	固沙，蜜源，制麻，种子榨油可食用，燃料，饲料

开花结实 2～3年生开始结实，3～4年生进入结实盛期，有大小年现象。花两性。腋生总状花序生于当年生枝上。总花梗的长短变化以花棒较为显著，通常5～15cm，长者可达40cm。花序上小花数目变化也大，通常几朵至10余朵或多达20余朵。花萼钟形，具近等形5齿。花冠蝶形，通常紫红色、淡黄色或白色，翼瓣较旗瓣和龙骨瓣短。雄蕊2体（9+1）。子房上位，1室，无柄，花柱丝状。荚果，有1～6个荚节，不裂，成熟时由荚节间断落。荚节扁平或两面突起，有肋纹或针刺，有时边缘具齿。种子无种阜，无胚乳。果实和种子特征见表2。

岩黄蓍属树种的开花结实物候见表3。种子的外形和剖面结构如图1。

表2 岩黄蓍属树种果实和种子的形态特征

树 种	成熟果（荚节）			种 子		
	形 状	大小（mm）	颜 色	形 状	大小（mm）	颜 色
羊柴	椭圆形，扁，2～3荚节	长6 宽4	淡黄色，有皱纹，无毛	卵形	长4 宽2	褐色
花棒	球形，2～4荚节	径6	灰白色，并密被灰白色毡毛	卵圆形	3～4	褐色

表 3　岩黄耆属树种的开花结实物候

树种	观察地点和年份	开花期			果实成熟期		荚节散落	
		始期	盛期	末期	始期	末期	始期	盛期
羊柴	甘肃民勤 1974～1981	5月下旬～6月上中旬	7月下旬～8月中下旬	—	9月	10月	9月	10月
	宁夏中卫沙坡头 1966～1980	5月中旬	7月下旬	—	9月下旬	—	9月下旬	—
花棒	甘肃民勤 1974～1981	6月上旬～6月下旬	8月上旬～8月中下旬	9月中旬～10月上中旬	8月	10月	8月	10月
	宁夏中卫沙坡头 1966～1980	8月上旬	9月上旬	9月下旬	10月中旬	10月下旬	10月中旬	10月下旬

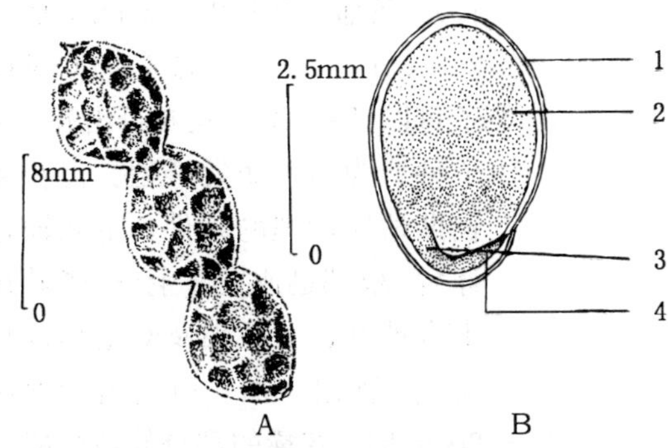

图 1　花棒荚果外形（A）及其种子纵剖面（B）
1. 种皮　2. 子叶　3. 胚轴　4. 胚根
（马平绘）

果实的采收调制和种子贮藏　当荚果由绿色转为灰白色或黄白色，少数荚节断落时应及时采收。选 5 年以上的壮龄母树为采种树。采集的果实要及时摊晒，去杂后得到断裂的荚节即为播种材料，通称种子。这两个种都适于在通风干燥处贮藏。花棒含水量 7.3%，发芽率 94%的种子，保存在一般通风干燥的条件下，至第 5 年发芽率还可达 80%以上。于卓和孙祥（1993）用四唑测定过室温下存放 1 年的 3 个来源的羊柴，生活力都接近或达到 100%，也未发现硬粒。常见的种子净度和质量见表 4。

表 4　岩黄耆属种子的净度和质量

树种	净度（%）	千粒重（g）	含水量（%）		每千克种子粒数（万粒）
			湿基	干基	
羊柴	90～95	13～17	7.5	8.1	5.9～7.7
花棒	90～95	20～40	7.2	7.8	2.5～5

发芽前的处理　种子休眠期短。生产上常将 1 份果实与 3 份含水量 60%的湿沙（或湿锯末）混匀后置 18～20℃条件下催芽 5～7 天，以促进发芽。在 1～3℃条件下层积 10～15 天后取出播种，则发芽快，发芽率高而整齐。

发芽测定　这两个树种的发芽测定条件及其结果见表 5。据于卓和孙祥（1993）报道，3 个来源的羊柴带有果荚时发芽率分别为 40.7%、40.3%和 49.6%，去掉果荚后的发芽率分别是 62.7%、67.0%和 76,1%；如果在种脐或子叶处刺破种皮，则发芽率分别提高到 99.2%、99.5%和 100%。

表 5　岩黄蓍属树种的发芽能力及其测定条件①

树　种	基　质	温度（℃）		发芽势（%）			发芽率（%）		
		有　光	无　光	计算天数	一般数值	变动范围	计算天数	一般数值	变动范围
羊柴	滤纸、脱脂棉	30	20	3～5	57	55～65	5～7	80	80～90
花棒	滤纸	30	20	3	60	60～70	5～7	85	80～90

①　测定前用始温 40～50℃水浸种 24 小时

播种　可以先将节荚用始温 40～50℃的水浸泡 24 小时，捞出沥干直接播种。每平方米的播种量，花棒是 15～20g，羊柴是 8～10g。条播，行距 25～30cm，播深 3～4cm。5cm 内土温 6～8℃时种子可以发芽，7 天左右可出齐。两个种皆为出土萌发类型。但花棒长期适应沙埋的条件，下胚轴延伸后，根部迅速垂直生长，子叶在土内展开，上胚轴及幼茎具锥状尖端，顶出土面。所以当沙埋达 20cm 时，花棒的部分幼苗尚能出土，而使子叶形同留土萌发。花棒幼苗形态见图 2。

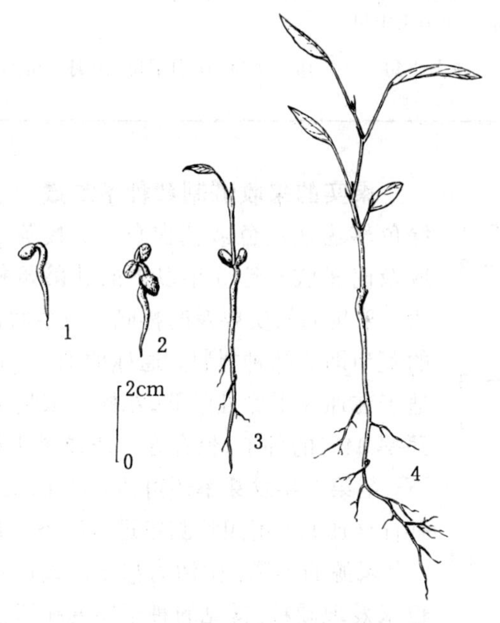

图 2　花棒幼苗生长情况

1. 胚根、胚轴延伸　2. 种皮脱落，子叶在土下展开
3. 上胚轴伸长，初生叶出现　4. 幼苗形成

（马平绘）

（李荣晨）

胡　枝　子

Lespedeza bicolor Turcz.

（蝶形花科　Fabaceae）

生长习性、分布和用途　胡枝子属约 60 种，我国约 20 种。本文仅描述胡枝子 1 种。落叶灌木，高达 3m，多分枝。喜光，耐干旱瘠薄。根系发达，萌芽性强，速生，耐割，耐踏。分布东北、华北、西北、豫、鄂、皖、鲁、浙。朝鲜半岛 、日本、俄罗斯西伯利亚和远东地区也有分布。有根瘤菌，有保土、改土作用，是三北防护林主要灌木树种之一。枝叶可作绿肥和饲料。叶、茎入药，有退热之功效。茎皮可制纤维，枝条可编筐篓。

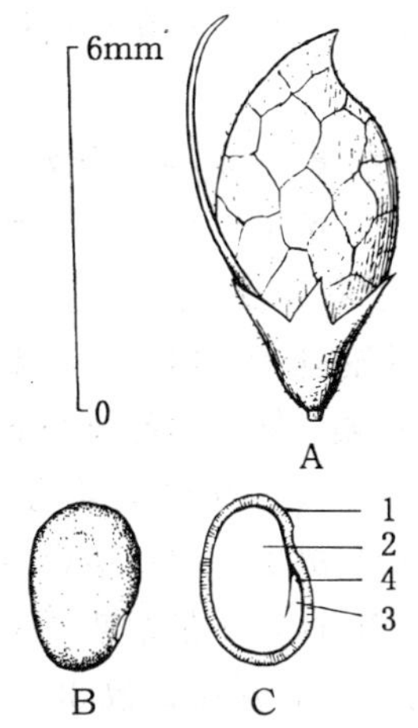

图 1　胡枝子荚果外形（A）、种子外形（B）及其纵切面（C）
1. 种皮　2. 子叶　3. 胚轴　4. 胚根
（梁鸣、黄应钦绘）

开花结实　2～3 年生开始开花结实。花两性。总状花序腋生，比复叶长，全部成为顶生圆锥花序。花梗无关节，花萼杯状，4 齿裂。花冠蝶形，紫色，旗瓣与龙骨瓣等长，龙骨瓣基部有长爪。雄蕊 10（9+1），2 体。子房上位，1 室 1 胚珠。花期 7～9 月，7 月上旬为始期，8 月中旬为盛期，9 月中旬为末期。9 月中旬果实开始成熟，10 月上旬为果实成熟盛期。据 1963～1980 年在哈尔滨观察，果实初熟至脱落间隔期 30～40 天。荚果斜倒卵形，扁平，先端针刺状突起，基部斜削，末端钝圆，长 6～7mm，宽 3～4mm；两面微凸，表面具网脉，密被柔毛。荚果不开裂，具 1 粒种子。种子长 2.8～3.0mm，宽 1.8～2.3mm，卵形或椭圆形；两端钝圆，表面有光泽，紫褐色，具黄绿色或棕色斑纹。无胚乳，种脐侧生，圆形（图 1）。

果实的采集调制和种子贮藏　荚果呈黄褐色时从树上摘取。晾干后搓掉果柄，清除杂物，所得不开裂的饱满荚果即为播种材料，通称种子。出种率 40%～50%。生产上要求净度达到 93%～95%。种子千粒重 9～10g，每千克有 10 万～12 万粒。含水量 9%～10%时入库贮藏，在 0～5℃低温干燥条件下可保存 3 年以上。因每年都结实，一般干藏即可。

发芽和播种　种皮坚硬不透水。用始温 80℃水浸种 24 小时，置光照发芽箱中，发芽率 70%～80%。可直播，播前先用始温 60℃水浸种 1 天，以后每天换 1 次清水，3～4 天后播种。10～20 天出苗。常用垄播，每平方米播种量 1～1.5g，覆土 0.5～1cm。出土萌发（图 2）。1 年生苗出圃造林。胡枝子可在落叶后萌动前分根繁殖，还可扦插育苗。

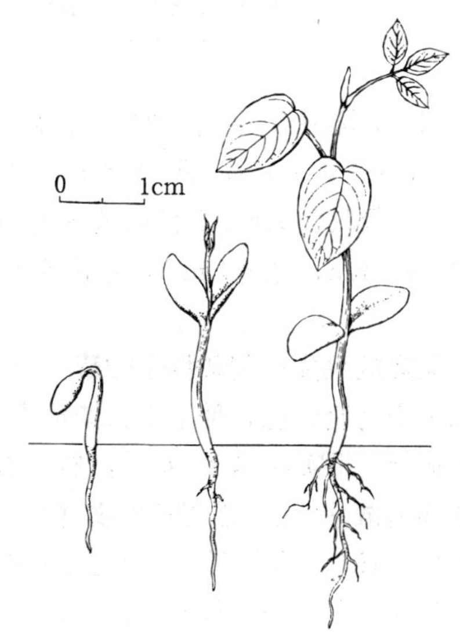

图 2　胡枝子种子发芽后第 3、6、18 天幼苗生长情况
（曹雅范、黄应钦绘）

（周德本、王彩荣）

怀槐（黄色木）

Maackia amurensis Rupr. et Maxim.

（蝶形花科　Fabaceae）

生长习性、分布和用途　马鞍树属约 12 种，我国 7 种，本文描述 1 种。落叶乔木，高达 25m，胸径 60cm。产于东北、华北东部。俄罗斯远东地区和朝鲜半岛、日本亦有分布。木材供建筑、器具、细木工、薪炭用；树皮作染料并可入药；怀槐也是蜜源植物和观赏树种。

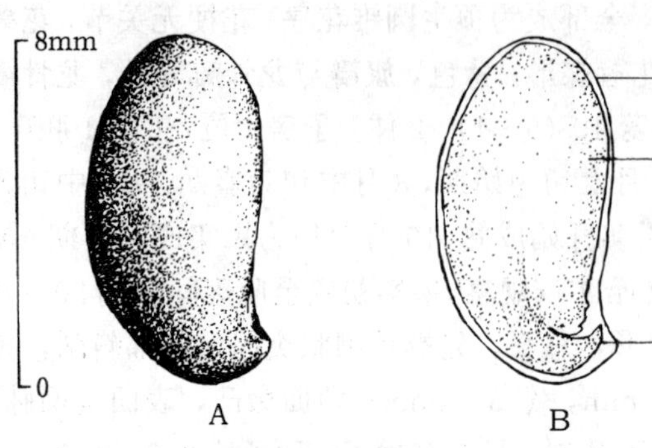

图 1 怀槐种子外形（A）及其纵切面（B）
1. 种皮 2. 子叶 3. 胚根
（梁鸣绘）

开花结实 约 8～10 年生开始结实，正常结实年龄在 15 年生以后。花两性。顶生总状花序或复总状花序，长 9～15cm。萼钟形，5 齿裂。花冠蝶形，白色，旗瓣倒卵圆形，长 7mm，翼瓣、龙骨瓣各长 8mm。雄蕊 10，花丝基部合生。子房密被柔毛。据黑龙江省森林植物园1963～1980 年观察，花蕾出现期为 5 月 26 日，开花始期至末期为 6 月 28 日～8 月 4 日，开花日数 30 天，后叶开放；果实初熟至末熟为 8 月 27 日～10 月 17 日，初熟至脱落间隔期为 20～27 天。荚果扁平，椭圆形至长椭圆形，暗褐色，长3～5cm，宽约 1cm，腹缝有翅。种子 1～6 枚，肾状长圆形，褐色或黄褐色，长 8（6～9）mm，宽约 4 mm，无胚乳（图 1）。

果实的采集调制和种子贮藏 荚果由绿色变为暗褐色，种子变为淡黄褐色即为成熟。果实成熟 20 天后自然开裂，种子逐渐散落，应注意及时采收。树上采摘荚果，晾干后自然开裂，或经过敲打，筛去果荚即可得纯净种子。出种率 17%左右，种子净度 50%～80%，千粒重 40～50g，每千克纯净种子 2 万～2.5 万粒。种子含水量 11%～12%以下，密封低温可贮存 4～5 年。

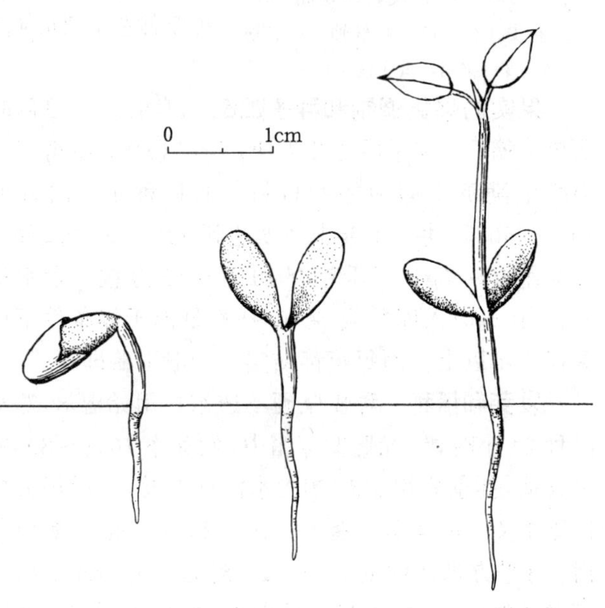

图 2 怀槐种子萌发后第 4、5、14 天的幼苗生长情况
（曹雅范、黄应钦绘）

发芽和播种 哈尔滨市 4 月下旬～5 月上旬播种，播后 10～20 天出土。1988 年 11 月曾在发芽皿内的滤纸上播放 100 粒种子，在 10～20℃温度下，10 天内发芽百分数为 60%，最终发芽率 85%。育苗时可采用垄播或苗床条播，每平方米播种量不超过 10g，覆土厚 1.5cm 左右。出土萌发，7～13 天子叶出土，15 天后初生叶展开（图 2）。当年生苗高50～70cm，可以出圃。

（周陛勋、周德本）

常春油麻藤

Mucuna sempervirens Hemsl.

（蝶形花科　Fabaceae）

生长习性、分布和用途　油麻藤属约160种，我国约25种，本文描述1种。常绿大藤本，长达30m，径30cm。耐荫，喜温暖湿润气候。多生于石灰岩地区，但微酸性和中性土壤亦能适应。抗干旱、耐水湿。分布于浙、闽、赣、鄂、川、滇、黔。日本也有分布。茎、根供药用。茎皮纤维可用于编织和造纸。种子可食，还可榨油，从中提取的“左旋多巴”能治震颤麻痹症。本种也可用作棚荫和垂直绿化材料。

开花结实　结实的大小年现象不明显。花两性，多数，成总状花序而着生于老茎上。花序下垂，长25～30cm。花萼钟形，5齿，上面2齿合生。花冠蝶形，深紫色，长约6.5cm。雄蕊两体（9+1）。花药2型：一种长而直立，一种短而横生。子房无柄，花柱无毛。花期4～5月，果熟期9～10月。据姜桦等1977～1980年在浙江杭州观察（未刊稿），花期在4月10日～5月12日，果熟期在10月。荚果成熟时果皮由绿色转为棕褐色或黑褐色，密被锈黄色刺毛。果木质，扁平条形，长30～70cm，宽2.5～3.0cm，种子间缢缩，表面无皱褶，不会自裂。每果有种子6～17粒，通常为8～14粒。种子扁矩圆形，两面凹陷，大而坚硬，棕黑色，沿种脐两面种皮为棕色，略有光泽，长2.3～3.0cm，宽1.8～2.2cm，厚0.9～1.1cm。种脐灰色或淡褐灰色，约占种子表面的3/4。无胚乳，子叶肥大。种子的形态特征见图1。

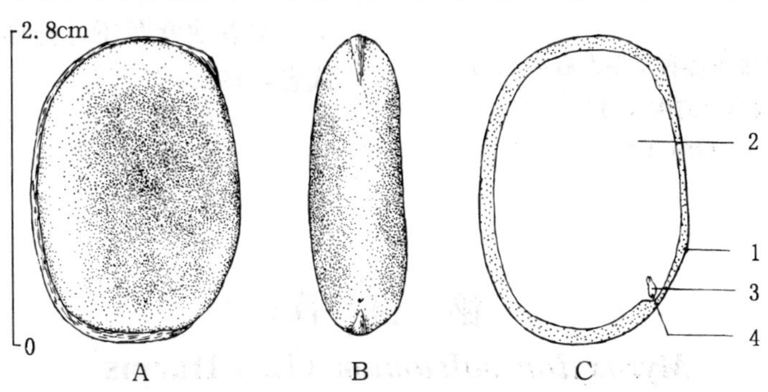

图1　常春油麻藤种子的正面（A）、侧面（B）及其纵切面（C）

1. 种皮　2. 子叶　3. 胚轴　4. 胚根

（童军平绘）

果实的采收调制和种子贮藏　果实成熟时，可站在被常春油麻藤缠绕的树上采摘果穗，或在树下用高枝剪剪取果穗，有时大风之后从地上也可以收集到一些果荚。采后摊晒数天，分节剪断荚果，剥出种子。操作时要戴手套，防止刺毛接触皮肤，引起奇痒。出种率为48%～60%。净度可达94%～98%。千粒重3 600～4 400g。每千克纯净种子230～280粒。调制后，风干种子的含水量为18%～20%。供翌春播种的种子，可用布袋、塑料袋或麻袋盛装后干藏。长期贮藏的应在0～5℃低温环境中密封保存。

发芽前的处理和发芽测定　常春油麻藤种皮透性不良，发芽困难。适当地伤蚀或软化种

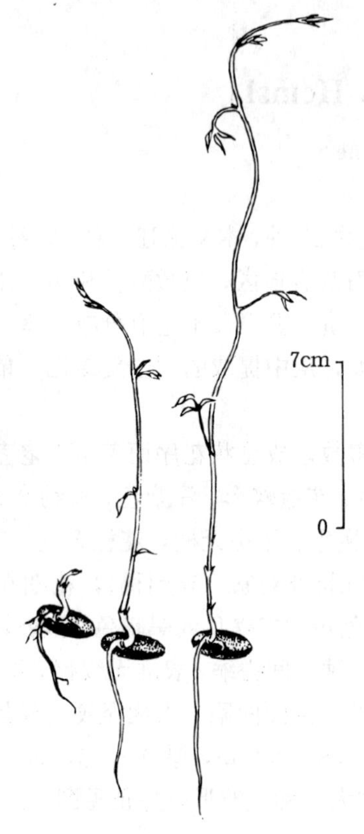

图 2 常春油麻藤种子萌发后第 2、10、20 天幼苗的生长情况
（童军平绘）

皮能促进种子萌发。据作者试验，效果较好的方法有：(1)始温 80℃水浸种 72 小时后，湿沙层积 5 周；(2)浓硫酸浸种 20 分钟，彻底洗净后再水浸 72 小时。在 1988 年的田间播种试验中，同未经处理的种子相比，上述两种处理的发芽率分别提高了 36%和 23%。

伤蚀过的种子发芽测定时，可用沙、石英沙和滤纸等作基质，温度可为 30℃。发芽测定不需要光照。

播种 多采用春播。2～3 月播种。条播或点播。条距 20cm，条幅 8cm，沟深 5cm。每平方米播种 75～100g。播后覆土约 4cm，盖草。1988 年 4 月 2 日，在浙江杭州苗圃地用处理过的种子播种，播后 28 天发芽出土，30 天为发芽盛期，发芽全程为 33 天。从萌发出土之日起算，3 天的发芽百分数为 33%～39%，最终发芽率为 69%～90%。留土萌发。发芽和幼苗生长情况见图 2。

常春油麻藤还可扦插或压条繁殖（周武忠，1986）。

（史晓华）

秘 鲁 香

Myroxylon balsamum（L.）Harms.

（蝶形花科 Fabaceae）

生长习性、分布和用途 南美槐属我国仅引入本文描述的 1 种。常绿大乔木，高达 30 m。原产南美洲，分布于巴西、阿根廷、秘鲁、哥伦比亚、委内瑞拉等地。我国 60 年代初期引入海南试种，70 年代已开花结果。耐旱。在年均温 20℃，极端最低温不低于 0℃，年降水量 1 100～1 700mm，相对湿度 75%～80%的气候环境中可以正常生长发育。适生于深厚肥沃的沙壤土。抗风力较差。树干割伤后渗出棕黑色的树脂，经处理后即得肉桂酸苄脂（Benzyl cinnamate），供药用，内服可杀菌祛痰，外敷有止血、杀菌等功效，亦为名贵香料。

开花结实 8～10 年生开始开花结实。正常结实期在 15 年生以后，大小年现象不明显。花两性。总状花序腋生。花冠白色。萼片及花瓣 5 枚。雄蕊 10。柱头 1。据在海南儋州观察，每年 5 月中旬花序形成，6 月中旬为盛花期，6 月底～7 月初花谢，荚果翌年 2～5 月成熟。荚果翅状长弯形。每果有种子 1 粒，位于果之一端。种子近肾形，无胚乳，胚根小（图 1）。

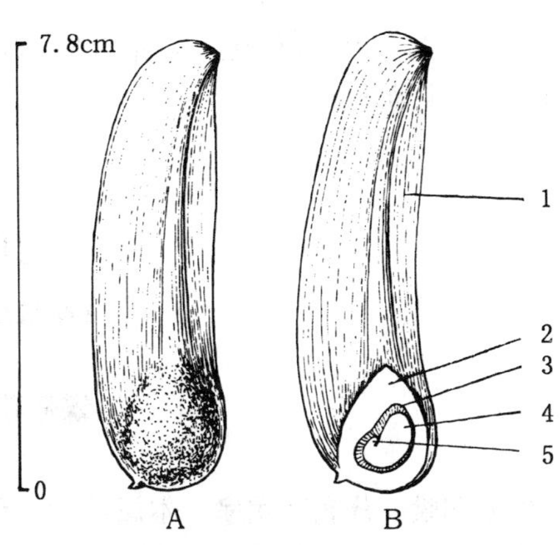

图 1　秘鲁香荚果外形（A）及其纵切面（B）

1. 果翅　2. 果皮　3. 种皮　4. 子叶　5. 胚根

（黄光郁绘）

果实的采收调制和种子贮藏　外果皮由青转变为黄色标志着果实成熟，此时即可采收。摇动树枝将果实震落，在树下拾集，晾干即可作为播种材料。也可剥开果荚取出种子。鲜果的出种率约 20%。荚果千粒重 760g，每千克有果荚 1 500（1 400～1 600）粒。干荚果中的种子含水量约 9%。生产中都是以果实的状态运输或贮藏。贮藏可用袋或罐、瓶等容器盛装，贮藏期约为 10 个月。在室内常温条件下，存放 14 个月的种子发芽率仅为 26%，存放 15 个月便完全失去发芽能力。

发芽和播种　种子无休眠习性。发芽时日均温宜在 22℃以上，宜随采随播。1987 年 7 月 18 日，华南热带作物研究所用当年采集的种子在室外沙床作发芽测定，播种时日均温 30℃左右。播后 25 天（8 月 13 日）开始发芽，9 月 2 日发芽结束，没有明显的发芽盛期，发芽率为 52%。留土萌发。发芽 4～5 天上胚轴出土，6 天左右初生叶展现（图 2）。

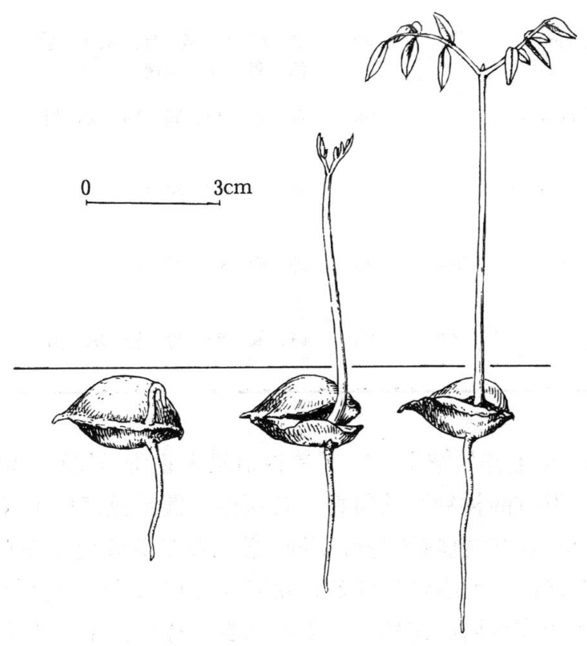

图 2　秘鲁香种子萌发后第 2、5、7 天幼苗的生长情况

（黄光郁绘）

目前尚无大田育苗经验。常用点播，每平方米播带荚的种子约100g。种子平放，覆土1cm左右。1年生苗出圃。

（丁慎言）

红豆树属

Ormosia Jacks.

（蝶形花科 Fabaceae）

生长习性、分布和用途 本属约120种，我国30余种，本文描述6种。常绿乔木。适生于肥沃湿润的土壤。喜高温高湿气候，不耐寒冷。木材花纹美丽，材质坚硬优良，为高级家具及美术工艺用材。这6个树种的名称、生长、分布及用途见表1。其中红豆树（又名花梨木）为我国特有，是本属中经济价值较高的珍贵用材树种，树冠浓荫覆地，又是优良的庭园树种，目前成年树日益稀少，已被《中国植物红皮书》列为渐危种。

表1 红豆树属树种的名称、生长、分布及用途

中 名	学 名	树 高 (m)	胸 径 (cm)	分 布	用 途	供 稿
肥荚红豆	*O. fordiana* Oliv.	15	100	粤、桂、滇。越南	材用、食用	601
花榈木	*O. henryi* Prain	13	20	湘、皖、赣、闽、浙、鄂、川、黔、滇、粤、桂。越南	材用、药用、观赏	903
红豆树（花梨木）	*O. hosiei* Hemsl. et Wils.	30	100	苏、浙、皖、闽、鄂、陕、川	材用、观赏	903
海南红豆	*O. pinnata* (Lour.) Merr.	25	60	粤、桂、琼。越南	材用	601
软荚红豆（红豆、相思子）	*O. semicastrata* Hance	10～15	40	琼、闽、赣、粤、桂	材用	601
木荚红豆	*O. xylocarpa* Chun ex L. Chen	20	80	赣、闽、湘、粤、桂、琼、黔	材用	903

开花结实 10～15年生开始结实，20年生以后进入正常结实期。结实有大小年现象，丰年间隔期一般为1年，但有的树种不甚明显。花两性。圆锥花序顶生或腋生。萼5齿裂，被柔毛。旗瓣宽卵圆形。花冠的颜色随树种而不同：肥荚红豆紫红色；花榈木黄白色或绿色，边缘淡紫色；红豆树和海南红豆黄白色带红色；软荚红豆和木荚红豆白色。雄蕊10，分离，稀5（软荚红豆）。子房无柄或具柄，花柱长，末端拳卷，柱头侧生。各个树种的胚珠数也有差异：肥荚红豆和海南红豆4；软荚红豆2；红豆树5～6；木荚红豆7～9。这6个树种的开花结实物候期见表2。荚果。成熟果皮革质或木质，开裂。种皮红色。无胚乳，子叶肥厚。红豆树属果实和种子的形态特征见表3、图1。

表2　红豆树属树种的开花结实物候

树　种	开始结实年龄	观察地点	观察年份	开　花			果实成熟		种子散落期
				始　期	盛　期	末　期	始　期	盛　期	
肥荚红豆	10	广西南宁	1981～1983	6月中旬	6月下旬	7月上旬	9月上旬	9月中旬	9月下旬
花榈木	15	福建	1987	6月中旬	6月下旬	7月上旬	10月下旬	11月上旬	11月下旬
红豆树	20	福建	1987	3月下旬	4月中旬	—	10月上旬	10月中旬	11月中旬
海南红豆	10	海南乐东	1978～1987	7月下旬～8月上旬	8月上中旬	8月下旬	11月上旬	11月中下旬	12月下旬
软荚红豆	10	海南	1987	5月中旬	5月下旬	6月上旬	9月中旬	9月下旬	10月中旬
木荚红豆	15	福建	1987	7月上旬	7月中旬	—	10月中旬	10月下旬	11月中旬

表3　红豆树属树种果实和种子的形态特征

树　种	果　实					种　子			
	形　状	大小（cm）	果　皮	颜　色	种子数	形　状	大小（mm）	种脐长（mm）	颜　色
肥荚红豆	椭圆形或倒卵圆形，扁平，顶端喙尖	长4～12.5 宽4.7～6.7	木质	淡黄色，干时褐色	1～4	椭圆形	长25～32	1～3	鲜红色
花榈木	圆形扁平，顶端喙状	长7～11 宽2～3	厚革质	黄褐色，干时紫黑色	2～7	椭圆形	长8～15	2.5～3	红色
红豆树	卵圆形，扁平，顶端喙状	长3.5～4.7	厚革质	干时褐色	1～2	近扁圆形	径10～17	7～8	深红色，有光泽
海南红豆	卵形或圆柱形，有隔膜，稍缢缩	长3～8 宽2～2.5	厚木质	深黄色至黄红色，干时黑褐色	1～4	椭圆形	长15～20 宽7～10	2	鲜红色
软荚红豆	圆形或扁圆形	长1.4～2 宽1.5～2	薄革质	黄褐色至黑褐色，有光泽	1	扁圆形	径约9	1～3	鲜红色，有光泽
木荚红豆	倒卵形或长椭圆形，扁平，先端具短喙	长5～7 宽3～3.5	厚木质	淡黄色至黄褐色	1～5	椭圆形或卵状椭圆形	长9.5～12 宽4.5～7	2.5～3	红色，有光泽

果实的采收调制和种子贮藏　部分果实成熟开裂时便应上树用采种刀割断果柄，或用竹竿敲落果荚。但肥荚红豆的子叶肥大脆嫩，忌直接敲打。采得的荚果置于室内通风处摊晾，荚果开裂后用木棒敲打或用手剥取，经过筛选即得纯净种子。据刘文明和宋学之（1990）研究，海南红豆的成熟指标不是果荚或种皮的颜色，只有子叶中叶绿素的含量趋于零，即子叶由青色变成黄色，才标志着种子完全成熟，而且只有充分成熟的种子才具有高的品质。在他们的试验中，青色、黄青色和黄色的子叶每克鲜重含有的叶绿素毫克数分别为0.100、0.016和0.008，发芽率均为100%，但最终的成活率差异很大，分别是56.3%、97.5%和100%。为了取得完全成熟的种子，他们的办法是将接近成熟的荚果装入薄膜袋，在25℃的温度下每天

给以 4 000lx 的光照 9 小时，约两周后，果荚由淡黄色变为橙黄色，子叶由黄青色转为黄色时取出种子。

这个属的种子忌失水，不能曝晒，表皮阴干即可贮藏。肥荚红豆和海南红豆的种子贮藏时应保持较高的含水量，贮藏和运输时均需混以湿沙。刘文明和宋学之（1990）认为，海南红豆种子含水量在 50.5%～53.0%时才能较好地保持生命力，含水量如果降到 45% 以下就会全部死亡。其余 4 个种的种子适于贮藏的含水量为 20 %～28%，贮藏和运输时也需混沙。贮藏期一般为 1 年。运输中用麻袋或竹篓包装，肥荚红豆和海南红豆每件不宜超过 25kg。红豆树属这 6 个树种的出种率和种子净度、质量等数据见表 4。

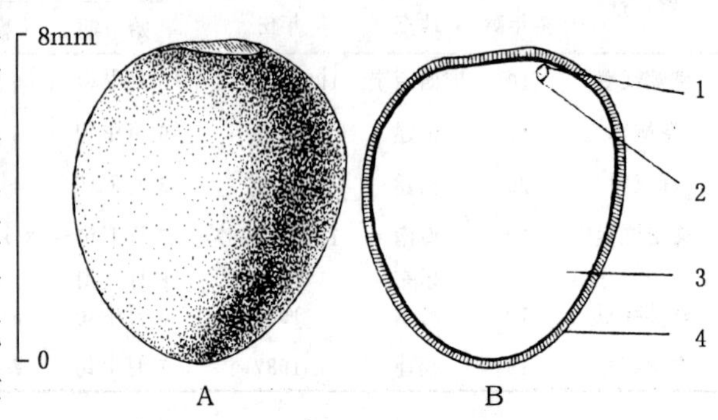

图 1　花榈木种子的外形（A）及其纵切面（B）

1. 种皮　2. 子叶　3. 胚芽　4. 胚根

（黄应钦绘）

表 4　红豆树属树种鲜果出种率和种子净度、质量

树　种	出种率（%）	净度（%）	千粒重（g）		每千克纯净种子粒数（粒）	
			一　般	变动范围	一　般	变动范围
肥荚红豆	30～40	98～100	6 600	6 000～7 000	150	140～170
花榈木	18	70～90	250	200～300	4 000	3 300～5 000
红豆树	35	70～80	1 000	900～1 170	1 000	850～1 100
海南红豆	8～9	98～100	950	800～1 100	1 050	900～1 250
软荚红豆	10～15	98～100	220	—	4 500	—
木荚红豆	10～15	72～92	210	180～250	4 760	4 000～5 600

发芽和播种　肥荚红豆种子有休眠习性，且个体间有差异。其余 5 个种无生理休眠现象，但常因种皮不易吸水而被迫休眠。解除休眠的常用方法是湿沙层积贮藏，或用始温 50～80℃水烫种，并在自然冷却过程中继续浸种 10～24 小时。取已经膨胀的种子用于播种，剩下的硬粒再用上述方法反复处理，直到硬粒全部解除为止。新鲜种子或经过层积的种子，播前可不作任何处理。

点播或条播。通常春季播种。适于海南红豆萌发和幼苗生长的条件是土壤 pH 值在 3.8～6.6，温度是 29～33℃。红豆树属的播种量见表 5。

留土萌发。肥荚红豆和红豆树有初生不育叶，花榈木和软荚红豆无初生不育叶。花榈木种子的萌发和幼苗初期生长情况见图 2。1 年生苗高 30cm 以上，地径 0.6cm 以上可以出圃。

表 5　红豆树属播种量

树　种	发芽率（%）		播种量 (g/m²)	树　种	发芽率（%）		播种量 (g/m²)
	一般	变动范围			一般	变动范围	
肥荚红豆	90	85～96	380～500	海南红豆	90	—	85～100
花榈木	82	79～86	25～32	软荚红豆	60	55～75	38～50
红豆树	82	78～85	85～100	木荚红豆	90	85～93	20～25

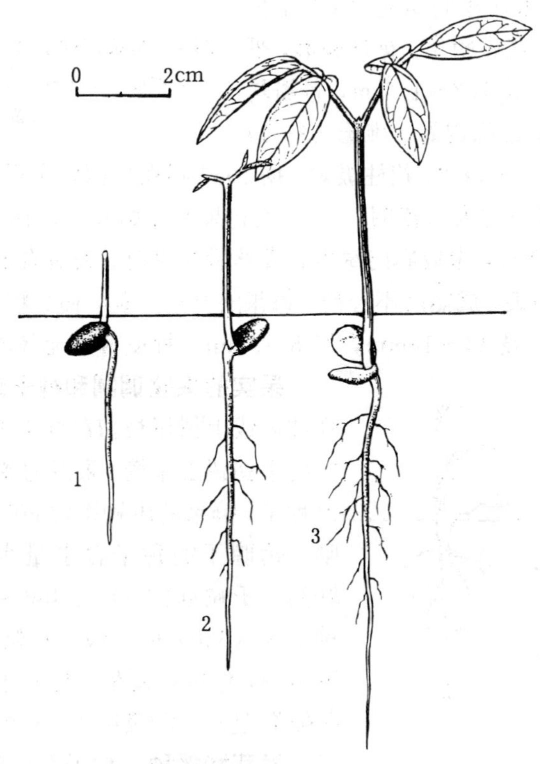

图 2　花榈木种子的萌发和幼苗生长情况

1. 上胚轴出土　2. 初生叶近对生，初现　3. 初生叶为 3 小叶复叶

（黄应钦仿《热带亚热带主要树种采种育苗技术》）

（李玉蕾）

水　黄　皮

Pongamia pinnata（L.）Merr.

（蝶形花科　Fabaceae）

生长习性、分布和用途　水黄皮属只有本文描述的 1 种。常绿乔木，高 8～15m，主干不明显，胸径约 20cm。能耐短期 5℃低温及 0℃左右的极端低温，不耐冰雪。喜肥沃湿润的轻度盐土及酸性土，较耐水湿，干旱地生长不良。分布于粤、桂、琼、台。印度、日本、马来西亚、新几内亚、澳大利亚及波利尼西亚群岛、马斯卡林群岛也有分布。材质松软，易受虫蛀，经处理后可作家具等用。叶作水田肥料和饲料。种子含油30%，可治癣疥、脓疮、风湿。全株可作催吐剂。茎枝可毒鱼。抗风力强，可作海岸防护树种。

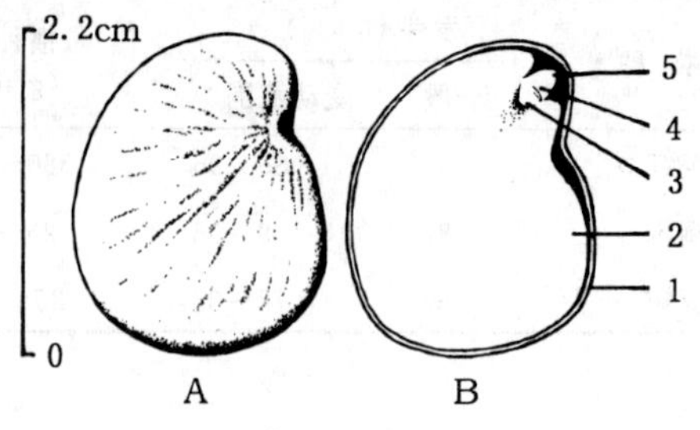

图1　水黄皮种子外形（A）及其纵切面（B）

1. 种皮　2. 子叶　3. 胚芽　4. 胚轴　5. 胚根

（黄应钦绘）

开花结实　8～10年生开始结实，正常结实期在20年生以后。结实大小年间隔期为1年，但不甚明显。花两性。总状花序腋生。花萼钟形，平截。花冠白色或淡红色，长1.2～1.6cm，各瓣均具爪，旗瓣背面被丝毛，边缘内卷。雄蕊（9+1）2体。子房被锈色短柔毛，胚珠2，花柱极短。据1987年在广西南宁观察，5月上旬为始花期，5月中旬为盛花期，5月下旬为末花期；10月上旬果实开始成熟，10月中旬为果熟盛期。荚果厚革质或木质，斜长圆形，未成熟时绿色，成熟后淡褐色，表面有不明显的疣凸。果长4～6cm，宽1.5～3cm，两端尖，成熟后不开裂。每果有种子1粒，稀2粒。种子近肾形，具皱纹，黄褐色，长17～24mm，宽14～18mm，厚5～9mm。种皮薄，无胚乳，子叶稍弯（图1）。

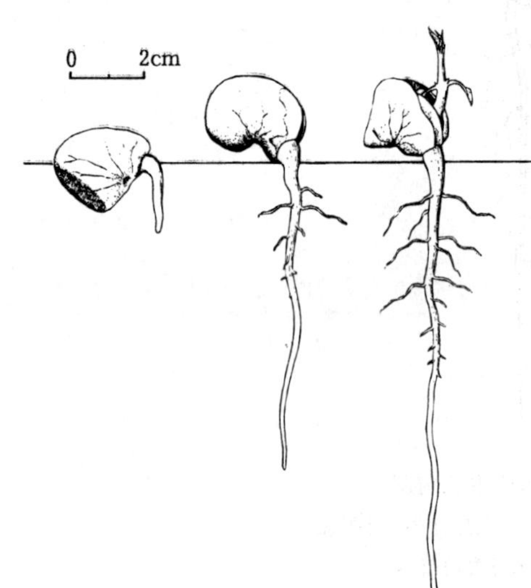

图2　水黄皮种子萌发后第2、4、5天幼苗的生长情况

（黄应钦绘）

果实的采收调制和种子贮藏　果实成熟盛期用高枝剪或上树用枝剪截断果梗，在地面捡拾，也可直接用手在树上采摘。采得的荚果摊放于室内，及时剥出种子。鲜果的出种率约60%。种子不宜脱水，忌日晒，稍晾干的种子含水量为20%～40%。净度达99%。千粒重1 500（1 400～1 600）g，每千克纯净种子为（620）670（720）粒。可以以果实的状态包装运输，也可在调制后将种子混以少量湿沙运输。贮藏需混湿沙，贮藏期4～6个月。

发芽和播种　种子无休眠习性。发芽时日均温宜在20℃以上。1987年广西林业科学研究所用新采种子在室外沙床作过发芽测定：10月4日播种，9日开始发芽，14～15日为发芽盛期，18日发芽终止。从发芽开始至发芽高峰，7天的发芽百分数为88%；从播种之日起算，14天的发芽率为93%。出土地面萌发。下胚轴短，胚根萌发后5天子叶仍包于种皮内，上胚轴出土展出1对生的初生叶。种子萌发及幼苗生长情况见图2。

点播。每平方米播种110～150g。随采随播或翌年春播均可。覆土1.5～2.0cm。1年生苗出圃。也可扦插育苗。

（王宏志）

紫　檀　属
Pterocarpus Jacq.

（蝶形花科　Fabaceae）

生长习性、分布和用途　本属约30种，分布于热带。我国引入栽培4种，本文描述其中2种。半常绿或落叶乔木。喜暖热气候。对土壤要求不甚严格，在较瘠薄的沙质壤土和干旱的荒地上均能生长成材。材质坚重，为著名珍贵用材，用于制作高级家具。它们的名称、生长、分布及用途见表1。

开花结实　紫檀7～8年生开始结实，檀香紫檀5～6年生开始结实。紫檀的正常结实期在15年生以后，檀香紫檀在10年生以后。这两个树种结实大小年现象均不明显。花两性。紫檀为圆锥花序，顶生或腋生于上部。花黄色，芳香。花梗与萼等长。花瓣边缘皱折。雄蕊10，单体，花药背着。檀香紫檀为总状花序。花黄色。花梗较萼短，萼5齿裂。子房有胚珠2～6。据云南景洪县普文林场和华南热带作物研究所观察，这2个树种的开花结实物候如表2。

表1　紫檀属树种的名称、生长、分布和用途

中　名	学　名	树　高(m)	胸　径(cm)	分　布	用　途	供　稿
紫檀	*P. indicus* Willd.	30	150	印度尼西亚、马来西亚、菲律宾。台、闽、粤、琼、桂、滇栽培	珍贵用材，树脂药用	617
檀香紫檀	*P. santalinus* L. f.	20	50	印度、泰国、马来西亚、越南。台、琼、粤栽培	珍贵用材，树脂药用	610

表2　紫檀属树种开花结实的物候期

树种	观察地点和年份	开花			果实成熟	
		始　期	盛　期	末　期	始　期	盛　期
紫檀	云南景洪 1988	5月中旬	6月中下旬	7月上旬	10月下旬～11月上旬	12月上中旬
檀香紫檀	海南那大 —	—	10～11月	—	—	5月中下旬

荚果扁平，近圆形，不开裂，周围具宽翅。紫檀每果有种子1～3粒，檀香紫檀每果有种子1～2粒。种子无胚乳，子叶厚。果实及种子的形态特征见表3、图1。

表3　紫檀属树种果实和种子的形态特征

树　种	果实			种子		
	形　状	大小（cm）	色　泽	形　状	大小（mm）	色　泽
紫檀	圆形，具翅	径4～6	深褐色	月牙形	长10～17 宽3～5	紫红色
檀香紫檀	圆形，具翅	径3～5	黄褐色	月牙形	长12～20 宽5～8	红棕色

果实的采收调制和种子贮藏　果实成熟盛期用高枝剪采集或打落后在地面拾集。由于种子极难脱粒，生产中常将翅状荚果置阳光下晒干作为播种材料，也通称种子。为了加速种子发芽，也可从荚果中剥出种子。紫檀荚果的千粒重为 600（500～700）g，每千克有荚果（1 400）1 660（2 000）粒。晒干的荚果含水量为 8%～11%。如剥出种子，檀香紫檀鲜果的出种率约 12.5%，种子的净度可达 99%，千粒重为 205(180～250)g，每千克有纯净种子（4 000）4 880（5 600)粒。种子含水量约为 10%。这两个树种的种子都适于干藏，用布袋盛装置室内干燥通风处。贮藏期以不超过 3 个月为宜。

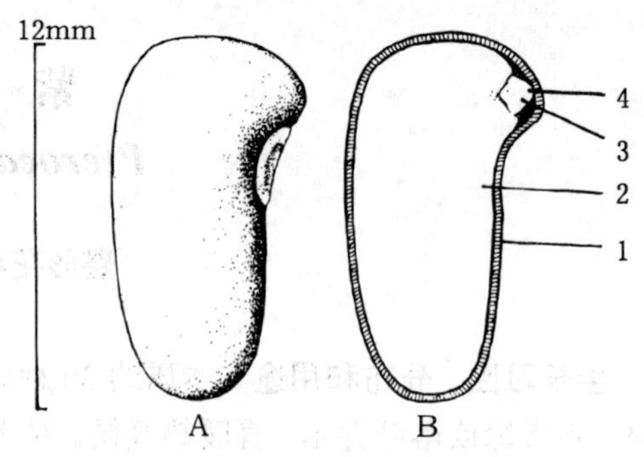

图 1　紫檀种子外形（A）及其纵切面（B）
1. 种皮　2. 子叶　3. 胚轴　4. 胚根
（黄应钦绘）

发芽和播种　种子无休眠习性。发芽时日均温要求在 18℃以上。1972 年 5 月上旬云南景洪县普文林场、1989 年 5 月华南热作研究所分别对紫檀和檀香紫檀作过发芽测定，播前用冷水浸种 8～12 小时。紫檀播后 6 天开始发芽，檀香紫檀播后 8 天开始发芽（表 4）。

出土萌发。紫檀的子叶出土后 6～7 天发出初生叶。檀香紫檀的子叶出土后约 8 天发出初生叶。紫檀种子的萌发和幼苗生长情况见图 2。

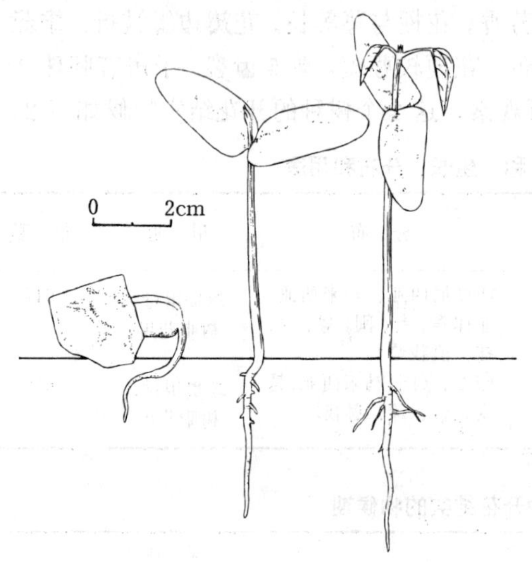

图 2　紫檀种子萌发后第 2、5、8 天幼苗的生长情况
（黄应钦仿《热带亚热带主要树种采种育苗技术》）

条播或点播。播种前用始温 50～60℃水浸荚果，并在冷却的过程中继续浸泡 1～2 天，捞出即可播种。紫檀每平方米播荚果 60～75g。檀香紫檀每平方米播 75～125g。覆土约 1cm。培育 1 年生苗出圃。

表 4　紫檀属树种的发芽能力及其测定条件（室外沙床）

树　种	温度（℃）	发芽势（%）		发芽率（%）	
		计算天数	一般数值	计算天数	一般数值
紫檀	22～26	15	40	30	70
檀香紫檀	22～26	16	50	35	60

（丁慎言）

葛　藤

Pueraria lobata（Willd.）Ohwi

（蝶形花科　Fabaceae）

生长习性、分布和用途　葛属约15种，我国12种，本文描述1种。藤本，块根肥厚。植株被黄色硬毛。除新疆、西藏外，分布几遍全国。朝鲜半岛、日本也有分布。喜光、耐干旱瘠薄，陡壁、荒谷及多石砾河滩地也能生长，萌芽力强。鲜根含淀粉约20%，可制葛粉，供食用或制作糊料及酿酒。种子含油约15%，可榨油。叶可作饲料。茎皮纤维作纺织原料。枝叶繁茂重叠，交错穿插，能形成厚的地被物；根系发达，有根瘤，用于改良土壤，保持水土。根及花均可入药。

开花结实　花两性。总状花序，腋生或集生成圆锥状，小花多而密。萼钟形，长0.8～1.0cm，蝶形花冠长1～1.2cm，紫红色，旗瓣近圆形，基部具黄色胼胝体，具短爪，翼瓣及龙骨瓣具耳。雄蕊10。子房有胚珠多粒。花期8～9月。1988年在北京植物园观察，初花期8月5日，盛花期8月29日，果期9～10月。荚果条形、扁平，薄革质，长5～10cm，宽约9mm，密被黄色硬毛。种子多数，椭圆形或肾形，黄褐色或黄绿色，长3～4.2mm，直径2～3mm（图1）。

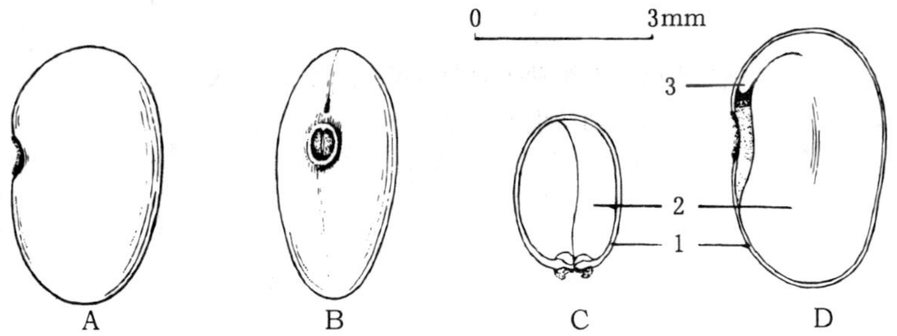

图1　葛藤种子外形侧面（A）和腹面（B）及其横切面（C）和纵切面（D）

1. 种皮　2. 子叶　3. 胚根

（孟玲绘）

果实的采收调制和种子贮藏　9月开始种子逐渐成熟，果荚颜色由绿色变为赤褐色，果皮干硬。应在荚果未开裂前剪下果枝，摊开晾晒，碾压或用脱粒机脱粒，扬去果皮和夹杂物后将种子阴干。出种率约为14%，气干种子含水量6%～6.5%。种子可干藏。种子的净度和质量见表1。

表1　两个产地葛藤种子的净度和质量

产　地	采　期	净　度（%）	千粒重（g）	每千克种子粒数
北京九龙山	1987年10月	74	15.3	65 400
赞皇	1987年10月	80	11.9	84 000

发芽和播种 孙秀琴和田树霞（1990）认为，光照变温（20～30℃）比黑暗变温（20～25℃）更有利于葛藤种子发芽。但发芽率通常不高，室内测定时常仅20%左右。种子中有硬粒，有时多达36%。除硬粒外，采种母株分散，成熟期差异大，采种时混采了未成熟种子也是发芽率低的重要原因。播种育苗时种子用80℃热水浸1分钟后再用室温水浸种2日，其中硬粒再用热水或温水浸种，可以在一定程度上消除硬粒。4月下旬播种，覆土厚0.6～0.8cm，播种后14天出苗。出土萌发。出苗后15天初生叶展开。图2是室内营养钵内的幼苗，播种后4～7天子叶展开，9天初生叶展现。

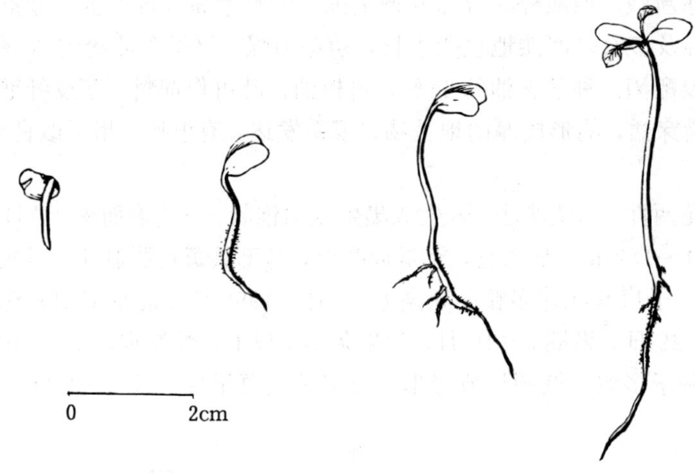

图2 葛藤种子萌发后第2、4、7、9天幼苗生长情况

（孟玲绘）

（陶章安）

刺　槐

Robinia pseudoacacia L.

（蝶形花科 Fabaceae）

生长习性、分布和用途 刺槐属约20种，产北美及墨西哥，我国引入3种，本文描述1种。落叶乔木，高达25m，胸径1m。萌蘖性强，根系发达，具根瘤，有一定的抗旱、抗烟、耐盐碱能力。在石灰性土壤上生长良好，酸性土、中性土上也能生长。在积水或地下水位过高的地方常引起烂根、枯梢以致死亡。原产北美东部，20世纪初从欧洲引入，从宁夏、内蒙古、河北、辽宁一线往南至长江流域广为栽培，以黄河中下游、辽东半岛及淮河流域生长最好。木材用于矿柱、桩材或建筑、家具。刺槐也是优良的薪炭树种和蜜源植物，是华北和西北地区水土保持、防风固沙、改良土壤和四旁绿化的重要树种。在我国栽培的还有无刺槐、伞刺槐和红花刺槐3个变型。刺槐种子虫害主要有刺槐荚螟（*Etiella zinckenella* Freitschke），刺槐种子麦蛾（*Sitotroga* sp.）和刺槐种子小蜂（*Bruchophagus philorobiniae* Liao）。

开花结实　3～5 年生开始开花结实，10 年生以后大量结实。花期 4～5 月，果期 9～10 月。花两性。总状花序下垂，从当年生叶腋长出。萼钟状，5 齿裂，稍 2 唇形。花冠白色。蝶形，各瓣具爪。雄蕊 9+1，2 体。单雌蕊，子房具柄，胚珠多数。荚果矩圆状条形，扁平，棕褐色，长 4～10cm，宽 1～1.5cm，沿腹缝有窄翅。种子 3～10，矩圆状肾形，长约 5mm，宽 3mm，厚 1.5mm，黑色或暗棕色，有淡色斑纹。无胚乳。种子形态和解剖结构见图 1。

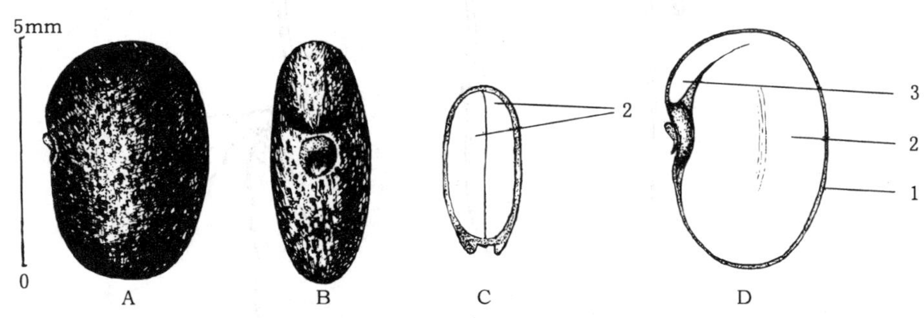

图 1　刺槐种子外形侧面（A）、腹面（B）及其横切面（C）和纵切面（D）
1. 种皮　2. 子叶　3. 胚根
（孟玲绘）

果实的采收调制和种子贮藏　荚果成熟后干硬，长期宿存枝上，易遭虫蛀，应及时采集。采后摊晒，碾压或用脱粒机脱粒，用风车扬去果皮、秕粒和夹杂物即得纯净种子。但碾压不可过重，以免造成机械损伤。种子净度应不低于 90%。出种率 20%（15%～33%）。千粒重 17.3～24.8g，每千克有种子 4.0 万～5.8 万粒。发芽率 60%～80%。含水量 6%～8%的种子在不高于 15℃的环境中干藏，生命力可保存 4 年。在夏季温度不高于 30℃的环境中贮存 2 年，发芽率未见明显降低。

发芽和播种　刺槐种皮坚硬，栅状表皮细胞排列紧密，胞壁木质化，种皮外覆有厚角质层，不易透水，发芽缓慢而不整齐。用机械伤擦、浓硫酸浸蚀及热水烫种，适当地伤蚀种皮能促进种子萌发。据试验，浓硫酸浸种 20 分钟的效果（发芽率 82%）优于始温 80℃的水浸种 24 小时（发芽率 64%）。种子预处理后，可在黑暗和恒温 25℃条件下进行发芽测定。刺槐种子也可用四唑染色法快速测定生活力。四唑溶液浓度为 1%，染色温度为 30℃。吴国源（1987）发现利用红墨水也可以快速测定刺槐种子的生活力。胡云楚和周培疆等人（1994）利用微热量计测定过刺槐种子的萌发热谱。

春季晚霜终止后播种。刺槐种子之间"硬粒"表现的程度很不一致，生产上常用热水反复浸烫，逐次取出吸胀的种粒播种。条播，行距 25～30cm。每平方米播种 10～12g，覆土厚0.6～0.8cm。播后约 8～10 天发芽出土。出土萌发，子叶 2，肥厚肉质，矩圆形。初生第 1 叶仅具 1 个小叶，第 2 甚至第 3 叶为 3 小叶复叶，以后为 5 小叶羽状复叶。刺槐种子的萌发和幼苗初期生长见图 2。

刺槐容易无性繁殖，根插、枝插、根蘖和嫁接均可。

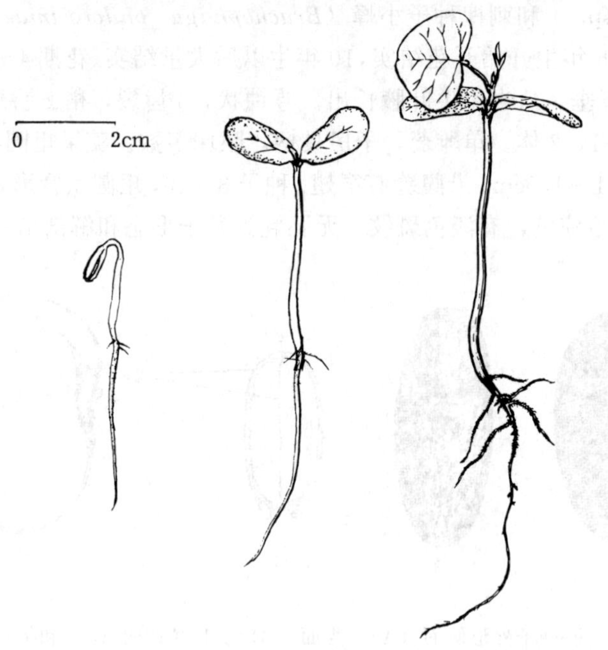

图 2 刺槐种子发芽后第 4、6、9 天幼苗的生长状况
（孟玲仿《主要树木种苗图谱》）

（于淑兰）

田 菁

Sesbania cannabina (Retz.) Pers.

（蝶形花科 Fabaceae）

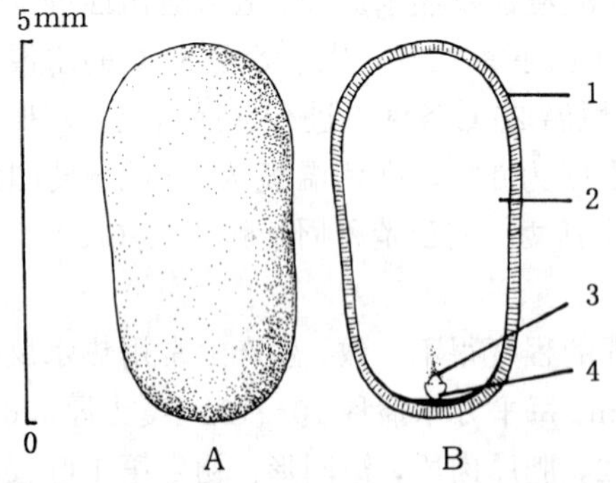

图 1 田菁种子外形（A）及其纵切面（B）
1. 种皮 2. 子叶 3. 胚芽 4. 胚根
（黄应钦绘）

生长习性、分布和用途 田菁属约 70 种，我国产 5 种，本文描述 1 种。灌木，高 1.5～3m。生长迅速，1 年生高可达 2m 以上。多分布于旷野，也可人工栽培作为绿肥。适生于低湿地或农田，酸性土和盐碱土均宜。山坡及干旱贫瘠地生长不良。忌严重霜冻。寿命较短。琼、粤、桂、滇、台、闽等地寿命可达 3～4 年，长江流域一带多作 1 年生栽培。茎叶为优质绿肥并可作饲料。皮纤维拉力强，可作麻袋及绳索。

开花结实 7～8 个月生便开始开花结实，正常结实期为 1～2 年生。结实无大小年现象。水肥条件充足的地方，无强寒天气的年份均可正常结实。花两性。总状花序腋生，每花序具花 2～6。花冠淡黄色，翼瓣和龙骨瓣基部有耳。子房具柄，胚珠多数。据南宁 1987～1988 年观测，9 月上旬为始花期，9 月中旬为盛花期，10 月上旬为末花期；11 月中旬果实开始成熟，11 月下旬为果熟盛期。荚果，细棍状条形，稍弯，顶端喙尖，成熟时由青色转为黄褐色，长 15～23cm，径 2～4mm。成熟后约半个月果实陆续开裂，种子散落。种子扁椭圆形，黄褐色，长 3.5～5 mm，宽 2.5～3mm，无胚乳（图 1）。

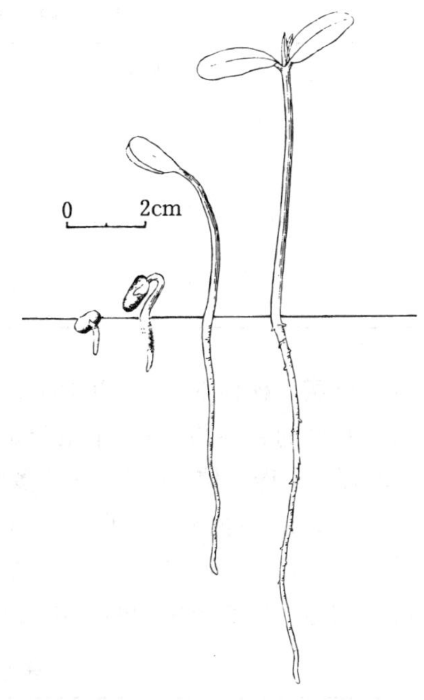

图 2　田菁种子萌发后第 1、2、6、9 天幼苗的生长情况

（黄应钦绘）

果实的采收调制和种子贮藏　少部分荚果转变为黄褐色，大部分果实开始由青转黄时，即可及时采收。如果目的是栽培 1 年生绿肥，可将枝叶全部收割，摊放在晒场上 1～2 日晒干，敲打出种子后将秸秆还田。多年生的母株一般用手摘采果实，晒干后敲打或用手搓揉脱出种子，清除果瓣及其它杂物，即得纯净种子。荚果的出种率为 25%～40%，净度可达 99%，种子可以晒干贮藏，晒干的千粒重约 16.8g，每千克有纯净种子 6 万粒。用袋或坛罐等容器盛装置室内避光干爽处存放。常温条件下贮藏期可达 1 年。

发芽和播种　种子无休眠习性。发芽时日均温需在 18℃以上。1987 年 11 月上旬，广西林业科学研究所用当年新采种子在室内用发芽皿作过发芽测定：基质为滤纸，室内的温度为 22℃，光照为自然光，播前用冷水浸种 20 小时；播后 1 天即开始发芽，并随即进入盛期，2 天后发芽终止，发芽率为 87%。出土萌发。胚根萌发后约 2 天子叶伸出，再过 7 天初生叶出现（图 2）。

生产中多在果园或田间直播作为绿肥。3 月中下旬撒播。播前用冷水浸种 12 小时或用始温 40～50℃水浸种 4 小时。每 666m^2 直播用种量为 2～3kg。

（王宏志）

槐　　属

Sophora L.

（蝶形花科　Fabaceae）

生长习性、分布和用途　槐属 50 余种，主产东亚、北美。我国 16 种，本文描述 2 种。槐树为落叶乔木，在酸性土、中性土、石灰性土及轻盐碱土上均能正常生长；深根，抗风力强；对二氧化硫、氯化氢及烟尘等抗性较强。苦参为落叶小灌木，耐干旱瘠薄，生于沙地、河岸石砾地、干草原、溪边、向阳山坡草丛中。这 2 个树种的名称、生长、分布和用途见表 1。

表 1 槐属树种的名称、生长、分布和用途

中 名	学 名	树高（m）	分 布	用 途	供 稿
槐树	*S. japonica* L.	25	东北南部，华北、陕、甘南部，西至川、滇，南至粤、桂普遍栽培。日本、朝鲜半岛	城市绿化、材用	303
苦参	*S. flavescens* Ait.	0.5～2	北至黑、内蒙古，东到台，西至青，西南至川、黔、滇等地均产。朝鲜半岛、日本、西伯利亚	绿化、水土保持、根入药、种子榨油	107

开花结实 槐树 30 年生以上生长健壮的树木出种率高，种仁饱满。结实大小年现象不明显。人工栽培的苦参约 3～4 年生开始结实。花两性。花序顶生。萼宽钟状。花冠蝶形。雄蕊 10。单雌蕊，子房上位，1 室，胚珠多数。荚果念珠状。槐树为圆锥花序，花冠黄白色，长1～1.5cm。果长 2.5～8cm。种子 1～6 个，肾形，长 6～9mm，宽 3～6mm，深棕色（图 1）。苦参为总状花序，花冠黄白色、黄色或粉红色，长约 1.5cm；果长 5～11cm；种子长圆形，长约 6.5mm，褐色。这 2 个树种的开花结实物候见表 2。槐树荚果肉质，成熟后经冬不落。

表 2 槐属的开花结实物候

树 种	观察地点	观察年份	开花 始 期	开花 盛 期	开花 末 期	果实成熟 始 期	果实成熟 末 期
槐树	北京[①]	1953	7 月中旬	～	7 月下旬	9 月	10 月
		1987	6 月下旬	7 月中旬	8 月上旬	—	—
苦参	哈尔滨	1968～1986	6 月下旬	～	7 月下旬	9 月上旬	9 月下旬

① 中国林业科学研究院

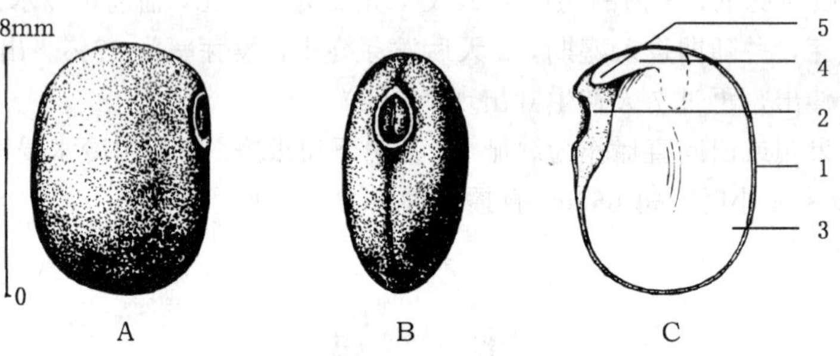

图 1 槐树种子外形侧面（A）、腹面（B）及其纵切面（C）
1. 种皮 2. 种脐 3. 子叶 4. 胚轴 5. 胚根
（孟玲绘）

果实的采收调制和种子贮藏 槐树的肉质荚果成熟时转为暗绿色或淡黄色，失水皱缩后略呈半透明状，悬挂树上经冬不落，也不开裂，整个冬季均可采收。采摘后用水浸泡 6～8 天，搓去果肉，洗净晾干，得出净种。在不高于 5℃，含水量为 8%～10%时，槐树种子可安全贮藏 3 年。贮藏环境夏季温度不高于 30℃，种子含水量为 8%～10%时，可安全贮藏 2 年。苦参荚果变为深褐色，种子呈黄色即为成熟。成熟后荚果中部开裂，种子外露。树上采摘，晾干后搓擦或敲打，筛选后即可得纯净种子。种子在低温干燥条件下可贮藏 1～2 年。种子的净度、

质量等见表 3。

表 3　槐属树种的出种率、净度和质量

树　种	出种率（%）	净　度（%）	千粒重（g）	每千克纯净种子粒数（万粒）
槐树	20	90 以上	120～150	0.7～0.8
苦参	34	60～80	50	2.0

发芽和播种　槐树种皮具有紧密结合的栅栏层细胞，透水性差，发芽前需将种子用始温 80℃水浸泡 24 小时。1986 年中国林业科学研究院测定过槐树种子发芽对光照条件的反应。种子来自北京，置床前用始温 80℃水浸种 24 小时。发芽基质为蛭石，12 天中每天给予 8 小时光照的一组，发芽率为 68%，不给光照的发芽率为 66%，没有显著差异。

槐树种子也可用四唑染色法测定生活力。测定时四唑溶液浓度 1%，温度 30～35℃，染色时间约 2～3 小时。

槐树可以春播，播前用始温 80℃热水浸种，或混沙层积约 1 个月。条播，行距 20～25cm，覆土厚 2～3cm，每平方米播种 25～38g。15～20 天发芽。出土萌发。子叶 2，肥厚肉质，椭圆形。初生叶互生，奇数羽状复叶，小叶对生（图 2）。1 年生苗木高 60～100cm，可出圃。槐树亦可分蘖繁殖或萌芽更新。

在哈尔滨，苦参可在 4 月下旬～5 月上旬播种。垄播或床面条播。床面条播时每平方米约播 5g，覆土厚 1cm 左右。播后 10～20 天出苗，出苗率为 70%～90%。当年苗高 20～30cm，可以出圃。

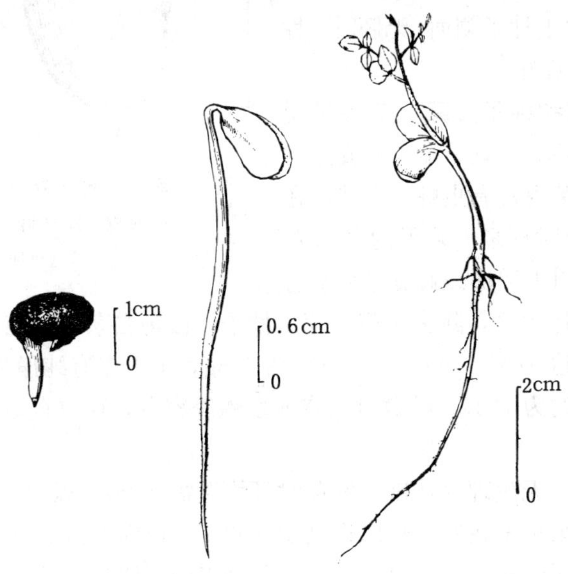

图 2　槐树种子发芽第 4、7、14 天的幼苗生长情况

（孟玲绘）

（于淑兰）

葫芦茶

Tadehagi triquetrum（L.）Ohashi

（蝶形花科 Fabaceae）

生长习性、分布和用途 葫芦茶属有 3 种，我国只有本文描述的 1 种。常绿灌木，高达 2m。萌芽力极强。多生于田野、林缘及灌丛。适生于土质比较疏松湿润的酸性土。能耐轻霜及短期 0℃左右极端低温，但忌冰雪。产闽、粤、琼、桂、滇、黔。印度、缅甸、越南、菲律宾亦产。全株供药用，治肝炎等多种炎症。可作凉茶。浸出液灭蝇蛆、孑孓。改良土壤和药用价值都高。

开花结实 约 2 年生开始开花结实，正常结实期在 3 年生以后，结实无大小年现象。花两性。总状花序，每节具花 2～3，顶生或生于上部的叶腋。萼 4 齿裂。花冠紫红色，花瓣脉纹显著，翼瓣与龙骨瓣近等长，基部有距。雄蕊 2 体（9＋1）。子房无柄，有胚珠数颗。据广西南宁 1986～1988 年观察，7 月中旬为始花期，8 月上中旬为盛花期，9 月上旬为末花期；早期开花所结的果实 11 月上旬开始成熟，中旬陆续有果实成熟，下旬为果实成熟盛期。荚果未成熟时为青色或赭红色，成熟时为棕褐色，狭长圆形，压扁，腹缝线直，背缝线稍缢缩。果长 2～4.3cm，宽 4～5.6mm，有荚节 5～8，顶端喙尖，表面被柔毛，成熟后约 1 个月开裂。种子扁椭圆形，青黄色或棕褐色，长 2～2.6mm，宽 1.5～2mm，种脐生于种子侧面。无胚乳。种子的外形和内部结构见图 1。

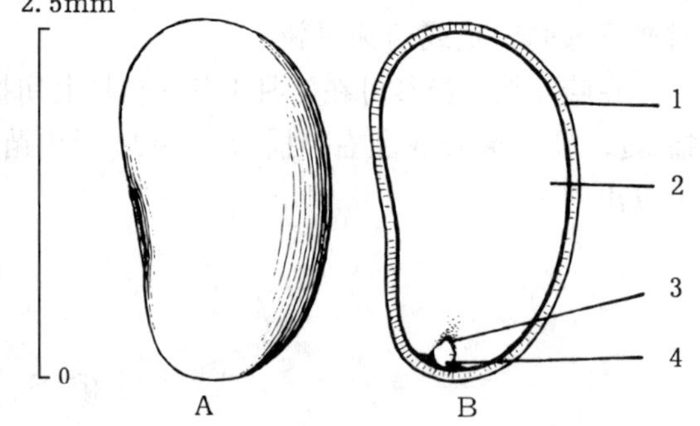

图 1 葫芦茶种子外形（A）及其纵切面（B）
1. 种皮 2. 子叶 3. 胚芽 4. 胚根
（黄应钦绘）

果实的采收调制和种子贮藏 荚果呈现赭红色时种子尚未成熟，不宜采摘，但成熟后久不采收，种子存留荚内常有虫蛀。应掌握在荚果初转棕褐色时及时摘取。采得的荚果置日光下曝晒，开裂后用手搓揉。果瓣与种子易粘连，可用小孔筛边搓边筛，除去果瓣，筛出种子，簸扬去杂即得纯净种子。荚果出种率约 40%。种子的净度可达 97%～ 99%。千粒重约 2.5g，每千克有纯净种子约 40 万粒。种子可以晒干贮藏，含水量约为 10%。贮藏时用袋或坛罐等容器，存放在室内干燥荫凉处。常温条件下贮藏期约 1 年。

发芽和播种 种子无生理性休眠。发芽时日均温需在 20℃以上。1987 年 12 月广西林业科学研究所用当年新采种子在发芽箱内作过发芽测试。箱内保持 28℃，基质为滤纸，播前用始温 60℃水浸种 1 昼夜。播后 2 天开始发芽，发芽盛期不明显，至第 18 天发芽终止。从播种之日起算，以 18 天计，发芽率为 40%。出土萌发。胚根萌发后约经 6 天子叶伸出，15 天初生叶发生（图 2）。

条播。播前种子用热水浸烫，每平方米约播 1g，覆土 5mm。半年或 1 年生苗出圃。也可将经过预处理的种子稍加晾干后，直播于地边或菜园四周作围篱。

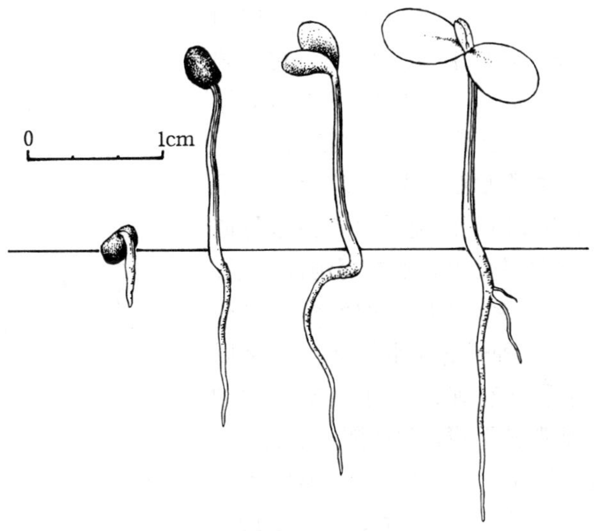

图 2　葫芦茶种子萌发后第 2、6、10、15 天幼苗的生长情况
（黄应钦绘）

（王宏志）

短萼灰叶（山毛豆）

Tephrosia candida DC.

（蝶形花科　Fabaceae）

生长习性、分布和用途　灰叶属 400 余种，我国 4 种，本文描述 1 种。常绿灌木，高达 3m。较耐干旱，在肥力中等以上的酸性土上能生长良好。能耐 0℃左右的极端低温，忌霜冻。产于琼、粤、桂及滇南。马来西亚亦产。为优良绿肥及檀香树的早期混交树种，也可作围篱并供观赏。枝干可培养木耳。

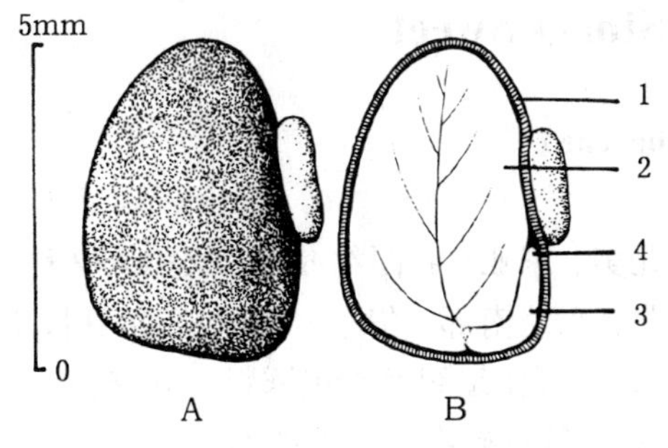

图 1　短萼灰叶种子外形（A）及其纵切面（B）
1. 种皮　2. 子叶　3. 胚轴　4. 胚根
（黄应钦绘）

开花结实　1 年生即可开花结实，正常结实期为 2～6 年生，结实无大小年现象。影响结实的主要原因是低温，如果冬春严寒，就会结实不饱满以至枝叶受冻不能结实。花两性。总状花序顶生，长 15～20cm。花序轴、花梗、萼均密被锈色丝毛。萼 5 齿裂。花冠白色或淡红色，旗瓣近圆形。雄蕊 10（9+1），2 体。子房无柄，有胚珠多颗，花柱内弯，柱头头状。据广西南宁 1987～1988 年观察，10 月上旬为始花期，10 月下旬为盛花期，12 月下旬为末花期；翌年 1 月下旬果实开始成熟，2 月中旬为果熟盛期。荚果条

状，下端喙尖，未成熟时绿色，成熟后黄褐色，被棕色毛，长 8～11cm，宽 7～8.5mm。2 月下旬至 3 月中旬荚果开裂，种子散落。每果有种子 10～15 粒。种子扁椭圆形，褐色，长 4.5～5.5mm，宽 3.2～3.6mm，厚约 2mm。无胚乳，胚根内曲于子叶边缘（图 1）。

果实的采收调制和种子贮藏　果熟盛期采摘果穗。采回的荚果置日光下曝晒，开裂后用木棒敲打或用手搓揉，脱出种子。筛除果荚等杂物即得纯净种子。已脱粒的种子可在烈日下晒干。果实出种率约 30%～35%。种子的净度达 97%，种子的含水量宜在 10%以下。千粒重约 24g，每千克有纯净种子（3.6）4.2（5）万粒。种子可以短期裸露干藏，贮藏期一般为 3～5 个月。种子易受虫蛀，贮藏中应注意翻晒。常温条件下，6 个月以后发芽率明显下降，1 年后发芽率仅 10%左右。

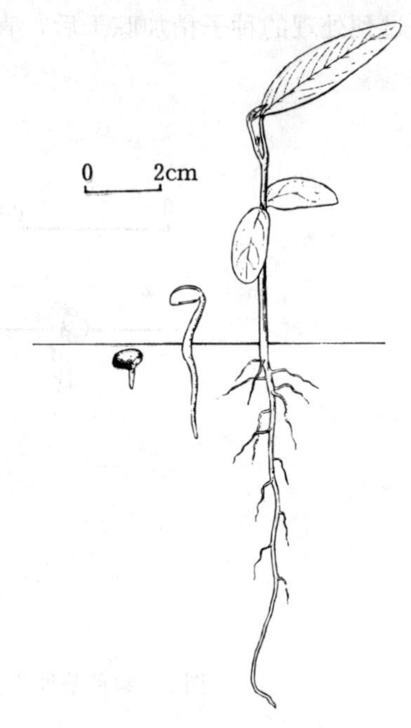

图 2　短萼灰叶种子萌发后第 1、3、7 天幼苗的生长情况

（黄应钦绘）

发芽和播种　种子无休眠习性。发芽时日均温宜在 20℃以上。1988 年 5 月 16 日，广西林业科学研究所用当年新采种子在室外沙床作过发芽测定，播种前曾用冷水浸种 8 小时，播时日均温 27℃左右。播后第 2 天开始发芽，并随即进入发芽盛期，27 日发芽终止。从播种之日起算，5 天的发芽百分数为 60%。12 天发芽率为 80%。出土萌发。胚根萌发后 2 天子叶带壳出土，出土后 2 天子叶成 2 片展开，5 天初生叶出现。种子萌发和幼苗生长情况见图 2。

直播或用容器苗种植作绿肥。每穴或每个容器点播种子 3～4 粒。播前浸种 6～ 8 小时，覆土 2cm。容器苗一般 2～3 个月出圃。

（王宏志）

紫　　藤

Wisteria sinensis（Sims）Sweet

（蝶形花科　Fabaceae）

生长习性、分布和用途　紫藤属约 9 种，分布北美、东亚。我国约 5 种，又引入 2 种，本文描述紫藤这 1 种。落叶大藤本，产西北东部、华北、东北南部、华东、华中及粤、川。花含芳香油，可食。花序和种子可供药用。树皮纤维柔韧，可作编织及纺织原料。叶密花浓，枝条粗壮，缠绕力强，为庭园绿化良好的棚架树种，对二氧化硫有较强的抗性。

开花结实　北京地区初花期 4 月 20 日～5 月 5 日，末花期 5 月 10～20 日。果实 9 月 25 日～10 月 10 日初熟，末熟期在 10 月底。1984 年底 5 个地区紫藤花果期物候观察的结果见表 1。

表 1　1984 年五个地区紫藤的开花结实物候期

观察地点	开花			种子成熟	
	始 期	盛 期	末 期	始 期	末 期
北京	4 月 30 日	5 月 7 日	5 月 15 日	—	—
西安	4 月 15 日	4 月 18 日	5 月 8 日	11 月 3 日	—
合肥	4 月 18 日	4 月 27 日	5 月 8 日	—	—
杭州	4 月 22 日	4 月 28 日	5 月 3 日	10 月 5 日	11 月 25 日
桂林	4 月 6 日	4 月 9 日	4 月 16 日	—	—

花两性，组成侧生、下垂的总状花序，长 15～30cm。花梗长 1～2cm。花梗、花序总轴及萼均被白色短柔毛。萼钟状。花冠蝶形，紫色或深紫色，芳香，长达 2.5cm，旗瓣内面近基部有 2 胼胝体状附属物。雄蕊 10（9＋1），2 体。子房具短柄，胚珠多数。荚果扁平，长条形，长 10～20cm，顶端有喙，未熟时淡绿色，熟时密生黄色绒毛，木质，常在种子间微缢缩，呈念珠状，开裂甚迟。种子 1～5，扁，近圆形，径约 1.5cm，褐色（图 1）。

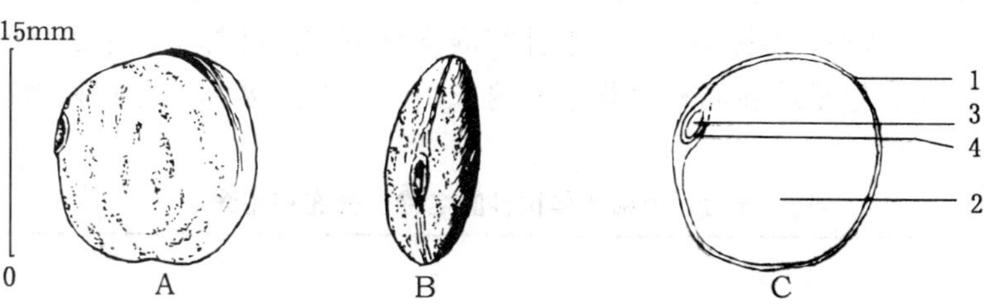

图 1　紫藤种子外形（A、B）及其纵切面（C）

1. 种皮　2. 子叶　3. 胚轴　4. 胚根

（孟玲绘）

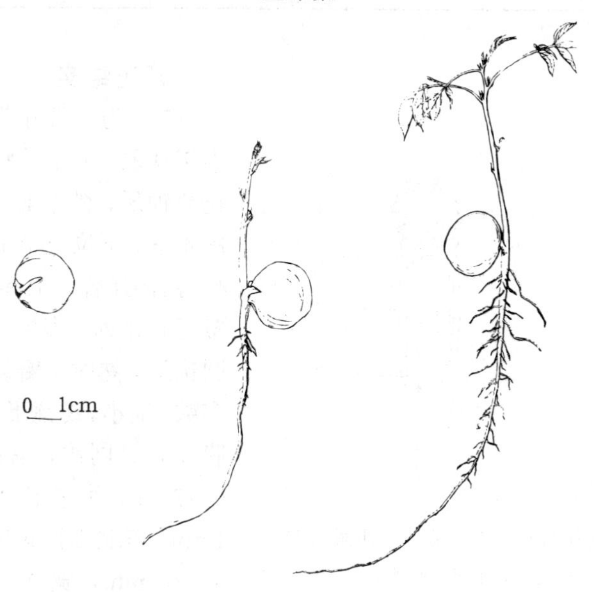

图 2　紫藤种子播后第 9、11、14 天幼苗生长情况

（孟玲绘）

果实的采收调制和种子贮藏 果实成熟采回后晒干，使荚果开裂。种子袋装悬挂于通风处贮藏。种子千粒重 450～550g，每千克纯净种子 1 800～2 200 粒。

发芽和播种 曾经在温度 20～25℃，每天光照 8 小时的条件下，将千粒重 530g 的种子浸种 12 小时后进行常规发芽测定，发芽率为 79%。留土萌发（图 2）。生产中一般采用嫁接、压条或分蘖方法繁殖。

（钱耀明）

山梅花属

Philadelphus L.

（山梅花科 Philadelphaceae）

生长习性、分布和用途 本属约 75 种，我国约 18 种，本文描述 2 种。落叶灌木，稀常绿，枝具白色实心髓。京山梅花高约 2～3m，小枝红褐色，生于低山山坡阔叶林或灌丛中。东北山梅花高约 2～2.5 m，小枝褐色，生于针阔混交林下或阔叶林下。这 2 个种适应性强，耐寒，耐旱，稍耐荫，很早就在园林绿化中普遍应用。它们的名称、分布和用途见表 1。

表 1 山梅花属树种的名称、分布和用途

中 名	学 名	分 布	用 途
京山梅花（太平花）	*Ph. pekinensis* Rupr.	辽南、华北、川。朝鲜半岛	观赏
东北山梅花	*Ph. schrenkii* Rupr.	东北。朝鲜半岛、俄罗斯远东、日本	观赏

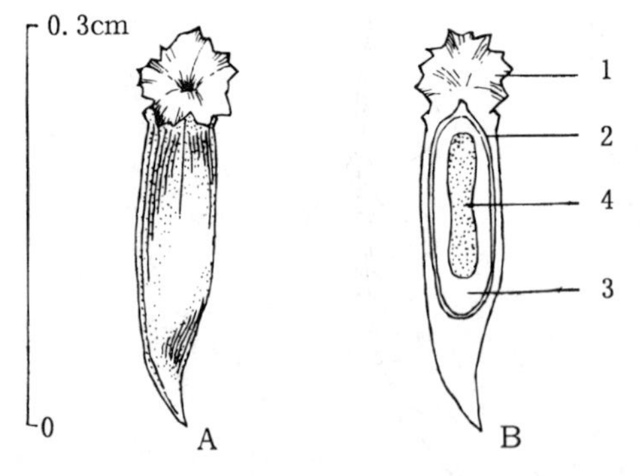

图 1 东北山梅花种子外形（A）及其纵切面（B）
1. 种翅 2. 种皮 3. 胚乳 4. 胚
（梁鸣绘）

开花结实 3 年生以上开花结实，5～15 年为正常开花结实年龄，无明显大小年现象。顶生总状花序。花两性，白色。花萼钟形，裂片 4。花瓣 4。雄蕊多数。子房 4 室，下位或半下位，中轴胎座。花柱 4，基部连合，上部分离。胚珠多数，呈覆瓦状排列，多层，下垂。蒴果，球状倒圆锥形，熟时 4 瓣裂，萼裂片宿存。种子多数，细小，纺锤形。种皮褐色或淡黑色，膜质，具网纹，有胚乳。京山梅花果径 5～7mm，种子长 2～3mm，宽 0.5～1mm。东北山梅花果径 6～9mm，种子长 3～3.5mm，宽 0.6～1mm（图 1）。这 2 个种的开花结实物候见表 2。

表 2　山梅花属树种的开花结实物候

树　种	观察年限 地　点	开　花		果实成熟		种子散落期
		始　期	末　期	始　期	末　期	
京山梅花	1963～1980 哈尔滨	6月9日	7月21日	9月20日	10月18日	初熟至脱落间隔期20～27天
东北山梅花	1963～1980 哈尔滨	5月18日	6月20日	9月20日	10月14日	初熟至脱落间隔期15～20天

果实的采收调制和种子贮藏　蒴果成熟时呈黄褐色，熟后15～20天即裂为4瓣，种子散落，故应准确掌握果实成熟期，及时从树上采摘。采后阴干，揉搓后去杂即得纯净种子。种子细小，含水量低于9%，低温（±5℃）下密封贮藏。它们的出种率和种子质量见表3。

表 3　山梅花属树种的出种率及种子质量

树　种	出种率（%）	净度（%）	千粒重（g）	每千克纯净种子粒数（万粒）
京山梅花	2.7～3.0	80～90	0.18～0.21	4 800～5 600
东北山梅花	2.4～2.8	80～90	0.23～0.28	3 600～4 400

发芽和播种　山梅花属种子细小，在苗床撒播或条播，应选无风无雨天气，微量覆土（厚约0.3cm），播后镇压。每平方米床面播种量3g。播后20天内保持床面潮润，10～20天出苗，当年苗高20～30cm，2年生可出圃。子叶出土萌发（图2）。

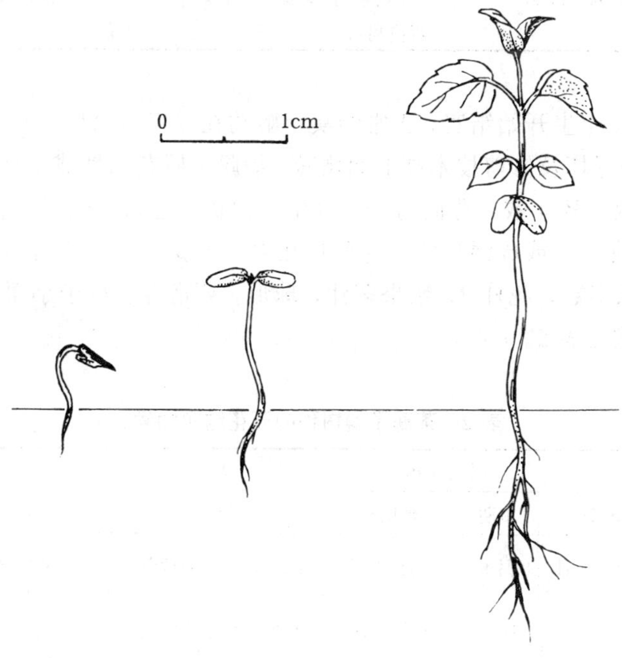

图 2　东北山梅花种子发芽后第2、4、17天幼苗的生长情况

（梁鸣绘）

（周德本、林淑云）

茶藨子属

Ribes L.

（醋栗科 Grossulariaceae）

生长习性、分布和用途 本属约150种，我国约50种，本文介绍3种。灌木，幼枝有刺或无刺。长白茶藨高1.5～2m，生于山坡阔叶林、林缘或灌丛中，岩石裸露山地也有生长。东北茶藨高1～2m，较耐荫，常生于针阔混交林或次生阔叶林下。兴安茶藨高1～1.5（2）m，分枝多，生于兴安落叶松林下或林缘，喜光耐寒，喜肥沃土。这3个树种的名称、分布和用途见表1。

表1 茶藨子属树种的名称、分布和用途

中 名	学 名	分 布	用 途	供 稿
长白茶藨	*R. komarovii* A. Pojark.	东北。朝鲜半岛、西伯利亚	果酒、果酱、观赏	109
东北茶藨	*R. mandshuricum* (Maxim.) Kom.	东北、华北、西北各省。朝鲜半岛、西伯利亚	果酒、果酱、观赏，果可入药	108
兴安茶藨	*R. pauciflorum* Turcz.	黑、内蒙古。朝鲜半岛、蒙古、西伯利亚	果汁、果酱、果酒、果糖、可生食	109

开花结实 2～4年生开始结实，正常结实年龄约在5年生以后。人工栽培的结实量随水肥条件、防寒越冬的好坏及修剪技术高低而增减。茶藨子属花两性或单性异株。总状花序，稀簇生或单生，花5数，稀4数。萼筒与子房合生，钟状、管状或碟形，裂片4～5，直立或开展。花瓣4～5，通常小，或成鳞片状，与萼片互生。雄蕊4～5，与花瓣互生。子房下位，1室，侧膜胎座，胚珠多数，花柱2。浆果多汁，球形，萼宿存。种子小而多，有胚乳。这3个树种的开花结实物候见表2。

表2 茶藨子属树种的开花结实物候①

树 种	开 花		果实成熟		成 熟	种子散
	始期	末期	始期	末期	特 征	落 期
长白茶藨	5月1日	5月22日	8月6日	9月23日	果红色，种子褐黄色	初熟至脱落间隔期15～20天
东北茶藨	5月18日	5月26日	7月16日	9月4日	果红色，种子淡红色	初熟至脱落间隔期15～20天
兴安茶藨	4月25日	6月12日	6月30日	9月14日	果暗紫褐色，种子红褐色	初熟至脱落间隔期17～29天

① 哈尔滨，1963～1980

长白茶藨果熟时红色，径7～8mm，种子长约2mm。东北茶藨果熟时红色，径7～9mm，种子约8～19粒，种皮光滑。兴安茶藨果未熟时红色至褐色，熟后暗紫黑色，径9～14mm，散生黄色腺点。长白茶藨种子形态见图1。

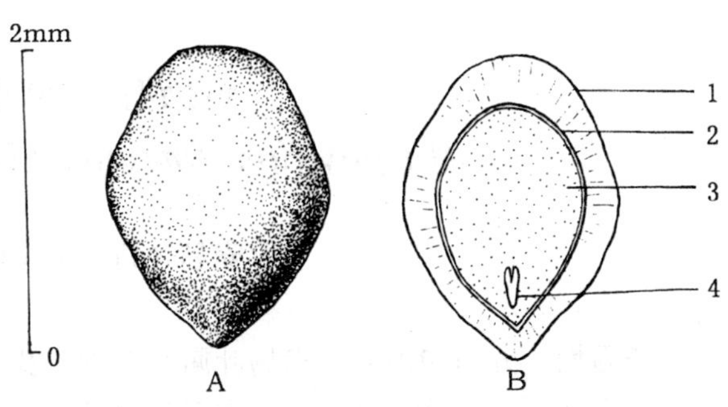

图1　长白茶藨种子外形（A）及其纵切面（B）
1. 外种皮　2. 内种皮　3. 胚乳　4. 胚
（梁鸣、黄应钦绘）

果实的采收调制和种子贮藏　盛果期时随熟随采，用手采摘或用手拍打并以容器承接。果实采收后可置容器中，待其腐熟变软后捣碎，经多次漂洗去掉果肉与果皮，阴干后去杂，即可得纯净种子。大量种子应从提取果汁后过滤所剩的果渣中获得。种子低温密封可贮藏1年。出种率、净度及种子质量见表3。

表3　茶藨子属树种的出种率及种子质量

树　种	出种率（%）	净度（%）	千粒重（g）	每千克纯净种子粒数（万粒）
长白茶藨	10～15	87～92	3.0～3.9	23～33
东北茶藨	10～15	86～91	3.1～4.3	23～32
兴安茶藨	2～5	86～97	1.3～1.8	55～77

发芽和播种　种子催芽处理的关键是需要有1个月左右0～5℃的低温层积阶段。催芽可于播种前2个月进行，将种子水浸1～2天，混沙后置于10～25℃温度下1个月左右，后转0～5℃温度下10～20天，可使大部分种粒发芽；亦可水浸2～3天，混沙后5～10℃温度下20～30天。苗床条播，播种量每平方米5g，覆土0.5～1cm，保持床面湿润，当年苗高10～25cm，1年或2年生苗出圃。子叶出土萌发，见图2。

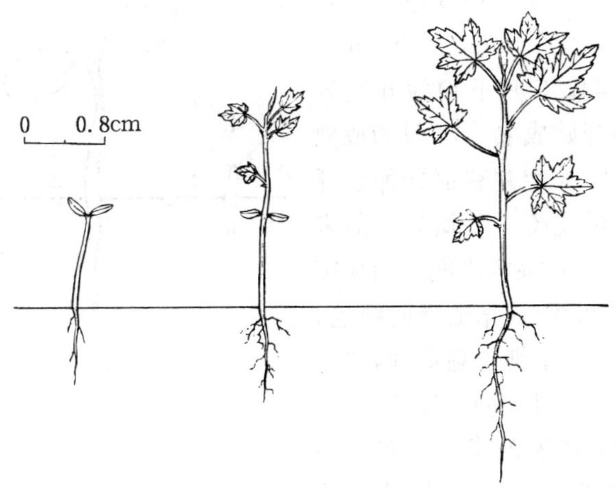

图2　长白茶藨种子萌发后第12、21、36天的幼苗生长情况
（许芝源、黄应钦绘）

（周德本）

赤杨叶（拟赤杨）

Alniphyllum fortunei（Hemsl.）Makino

（安息香科 Styracaceae）

生长习性、分布和用途 赤杨叶属有5种，我国全产，本文描述1种。落叶乔木，高达20m，胸径70cm。喜光，适生于土层深厚肥沃湿润的山地。分布于长江以南各地，南至海南岛。印度、越南、缅甸也有。木材纹理通直、疏松、轻软，可作包装箱、家具、农具、火柴杆、胶合板、纸浆等。

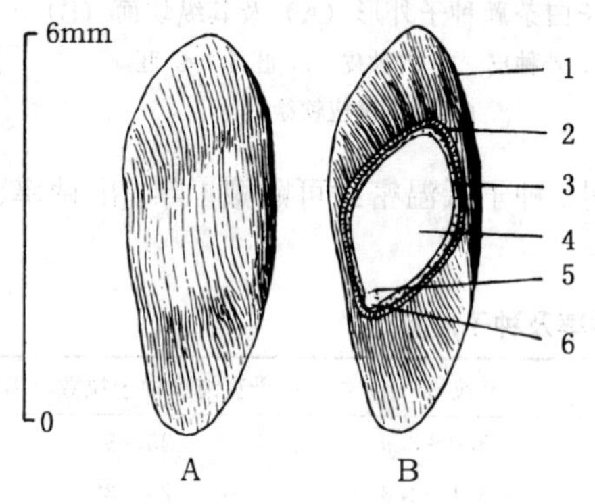

图1 赤杨叶种子的外形（A）及其纵切面（B）
1. 种翅 2. 种皮 3. 胚乳 4. 子叶
5. 胚芽 6. 胚根
（黄应钦绘）

开花结实 6～8年生开始开花结实，10～15年后进入正常结实年龄。结实有大小年现象，但不甚明显。花两性。圆锥花序顶生，长8～15cm，或总状花序腋生。花序轴、花梗及萼均被灰黄色短绒毛。萼筒浅碟形，萼具5齿，萼齿三角状披针形。花冠钟形5裂，裂片覆瓦状，白色或微红色，裂片长圆形，长1.5～2cm，两面密被柔毛。雄蕊10，5长5短。花药卵形，基部心形，内向。花丝宽扁，下部连成短筒，基部贴生于花冠。子房近上位，5室，每室胚珠8～10。胚珠直立或斜举，2列生于中轴上。花柱丝状，柱头5浅裂。子房密被黄褐色绒毛。1986～1988年在福建南平观察，3月上旬出现花芽，4月下旬花始开，5月上旬～中旬为开花盛期，5月下旬为开花末期；果实于10月中旬开始成熟，11月为成熟盛期，12月上旬为成熟末期并有部分种子开始散落，12月中旬大量脱落，下旬为脱落末期。在江西6月开花，9月果实成熟。蒴果窄长圆形，木质，长1.4～2.0cm；未成熟时绿色，成熟时褐色，室背5裂。种子细长扁薄，两端有翅，褐色，长6～9mm，有胚乳，胚直(图1)。

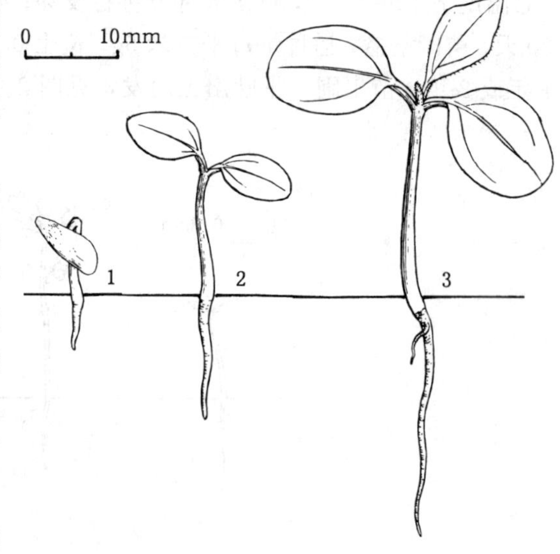

图2 赤杨叶种子的萌发和幼苗生长情况
1. 下胚轴延伸，种壳出土 2. 子叶展开 3. 初生叶互生
（黄应钦绘）

果实的采收调制和种子贮藏 果实由绿色变为黄褐色至深棕色且尚未开裂时采摘。采回后在室内摊放阴干，10天左右果实可以开裂，或用圆木轻轻滚压，脱出种子，筛除杂质，并晒干。生产上常不去翅。出种率为7%～

10%。种子净度一般为60%，变动范围50%～75%。千粒重0.6g，变动范围0.55～0.66g。每克纯净种子约1 700粒，变动范围1 500～1 800粒。种子含水量降低到9%以下后用塑料袋密封，在0～5℃的低温下贮藏1年对种子生命力无明显影响。短期贮藏可用布袋、木桶等盛装，置干燥通风处。运输时可用布袋包装，注意勿受潮湿。

发芽和播种　种子无明显休眠习性，但发芽率通常较低。在25℃恒温的无光处发芽，20天的发芽率为7%～13%，有时也可高达30%。场圃出苗率5%～11%。春播。播种后30天开始发芽出土，40天大量出苗，50天左右基本出齐。撒播时每平方米约播5g，条播时每平方米播2.5～4g。覆土以不见种子为度。出土萌发。初生叶卵形。种子萌发和幼苗生长情况见图2。

苗期需搭棚遮荫。幼苗基本形成时可以间苗移栽。1年生苗高70cm，地径0.6cm，可以出圃。每666m^2可产合格苗1万～2万株。

（郑郁善）

银　钟　花

Halesia macgregorii Chun

（安息香科　Styracaceae）

生长习性、分布和用途　银钟花属是我国与北美间断分布的属，约5种，我国只产本文描述的1种，为我国特有，在《中国植物红皮书》中列为稀有种。落叶乔木，高25m，胸径30cm。速生。根系发达，有抗风、耐旱、耐寒等特性。喜湿润而光照较充足的生境。分布在浙、闽、赣、湘、粤、桂，多生于海拔500～1 800m的山谷、山坡。材质轻软，纹理细，具有工艺价值，可用于建筑和家具。花色洁白，有清香，果形奇特，树叶入秋呈红色，为优美的观赏树种。

开花结实　花两性，2～7朵排成短缩的总状花序，生于2年生小枝的叶腋。萼筒倒圆锥形。花冠白色，宽钟状，裂片4，长约9～12mm。雄蕊8，4长4短相间，伸出花冠外面。子房下位，4室，每室有胚珠4枚。花柱钻状，较雄蕊长。核果椭圆形或倒卵形，有时近于圆形，长2.5～4cm，径2～3.5cm，红褐色，具4宽翅，近翅边有一凸起的条纹，顶端有尖头状宿存花柱。内果皮木质，坚硬。每果有种子1～4粒，多数只有1粒，长椭圆形。种子有胚乳，胚直立。花期3～4月，果熟期9～10月。银钟花果实的形态特征见图1。

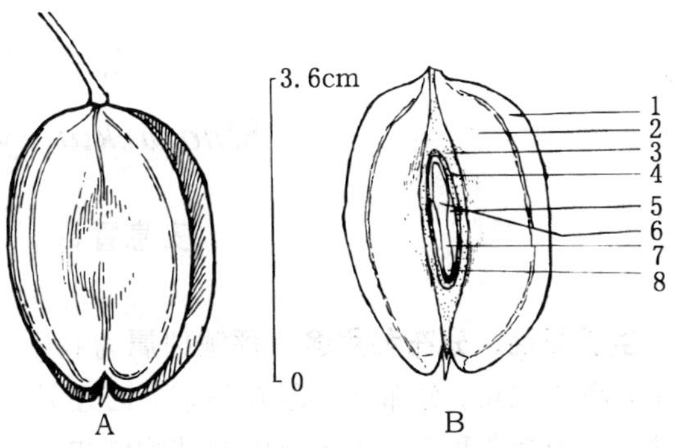

图1　银钟花果实的外形（A）及其纵剖面（B）
1.果翅　2.外果皮　3.内果皮　4.种皮
5.胚乳　6.子叶　7.胚轴　8.胚根
（童军平绘）

果实的采收调制和种子贮藏　果实由绿色转为黄褐色或红褐色时从树上采摘，或击落后在地上收集。采后将果实置通风处摊晾。为了减少体积并便于操作，可在干燥后搓去果翅。有些地方常不去翅。通常无需取出果核。去翅或不经去翅的核果即为播种材料，通称种子。用风选或筛选法清除杂物。净度95%～99%。核果去翅后千粒重为125～220g，每千克有去翅核果4 600～8 000粒（谢早荣，1986；史晓华，未刊稿）。供翌春播种的种子可混湿沙贮藏（谢早荣，1986；庾府诗，1984）。

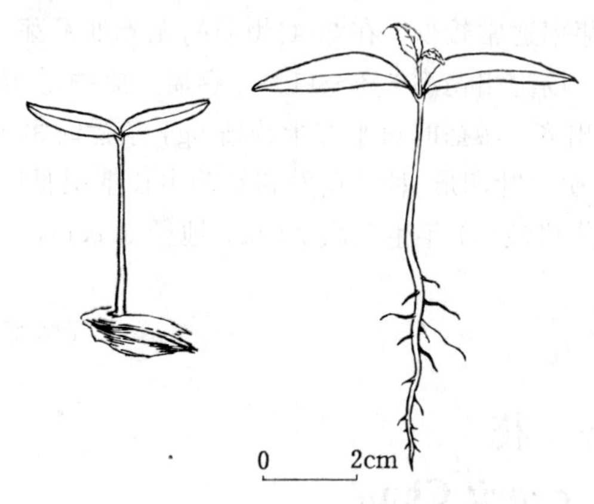

图2　银钟花种子萌发后第16、32天幼苗的生长情况
（童军平绘）

发芽和播种　种子有休眠现象，需隔年发芽。随采随播或层积后春播。湖南南岳树木园层积后春播的银钟花，翌年3月下旬～4月上旬萌发出土，4月中旬结束发芽，场圃发芽率为55%～75%（谢早荣，1986）。浙江杭州植物园1989年10月采集的种子，11月23日置室外层积至次年3月8日播种，一直到1991年3月27日才发芽出土，4月10日发芽结束；从萌发之日起算，15天的发芽率为60%（史晓华，未刊稿）。出土萌发，子叶2枚，出土子叶黄绿色，卵状披针形，全缘，长3.5～4cm，宽4～5mm，先端渐尖，基部宽楔形，主脉在叶下隆起。出土后17～22天展现初生叶，互生，有尖锐细锯齿。下胚轴白色。侧根稀疏，白色。银钟花种子的萌发和幼苗生长情况见图2。

条播，行距25cm，播种沟深8cm，施入3cm的土杂肥，再垫土2cm。播后覆土3cm，盖草。部分核果每果出苗2～3株，可以分株移植。1年生苗高约70～105cm，可以出圃。亦可扦插繁殖。

（史晓华）

秤　锤　树

Sinojackia xylocarpa Hu

（安息香科　Styracaceae）

生长习性、分布和用途　秤锤树属为我国特有，共4种。本文描述秤锤树这一种。落叶乔木，高6～7m。分布区局限在南京，已近灭绝，仅南京、杭州、上海、武汉等地植物园有少量栽培，为濒危种，已列入《中国植物红皮书》。春夏之交花白似雪，入秋果实繁多形如细丝下悬秤锤，随风摆动，颇为奇特，有很高的观赏价值。

开花结实　栽植后8～10年开始开花。花两性，白色，具长约3cm的梗，常下垂。3～5朵集成开展的、形似总状的聚伞花序，腋生。萼筒倒圆锥形。花冠钟形，裂片长圆形。雄蕊10～14。子房半下位，2～4室，每室含胚珠6～8枚。核果木质，下垂，不开裂，卵形；顶部约2/3处留有球状萼檐的残迹，顶部有子房上部发育成的稍钝或凸尖的圆锥形喙；连喙长1.5～2.5cm；棕褐色，表面有白色斑纹和浅棕色皮孔。中果皮木栓质。内果皮木质，似枣核，表面有高低不平的不规则纵棱。外果皮与中果皮粘结，不易分离。核内含细柱形种子 1 枚，有时 2 枚，长约 1.3cm。种皮厚膜质，具少量胚乳。花期 4 月中旬～5 月上旬。果熟期 9 月，果实脱落期 10～12 月。大小年现象不明显。果实外形和果实的解剖结构见图 1。

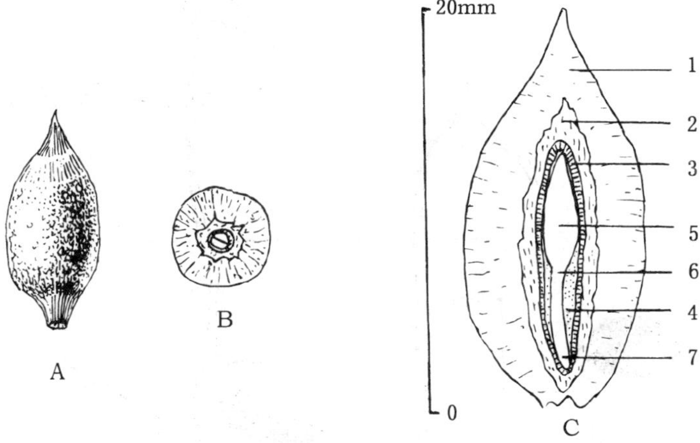

图 1　秤锤树的果实外形（A）及其横切面（B）和纵切面（C）
1. 外果皮和中果皮　2. 内果皮　3. 种皮　4. 胚乳　5. 子叶　6. 胚轴　7. 胚根
（史渭清绘）

果实的采收调制和种子贮藏　果实无明显的成熟标志，既不变色也不变软，可在果实脱落 以前的秋冬采摘。果实具很厚的木栓质，且含单宁等物质，鸟兽不食，又经久不烂，所以也可待果实落地后在树下拾集。果实不开裂，果皮又难以剥离，只能以果实作为播种材料，通称种子。果实比重轻，无论饱满或空瘪都浮水不沉而难以分离。果实中发育饱满的种粒常仅占 10%～20%。千粒重 380～430g，每千克有果实约 2 300～2 600 粒。由于具有特殊的保护组织，秤锤树的果实既不容易腐烂也不易失水干燥，干藏、沙藏或水浸均无损于发芽能力。一般的室内贮藏，发芽能力可保存 2～3 年。

发芽和播种　可能是由于果皮的障碍，发芽困难。庄茂长（1990）认为秤锤树有胚休眠习性。有的文献推荐过越冬层积和钳去果实两端的方法破除休眠，看来效果不大。南京中山植物园曾在 1988 年 3 月 7 日将种子浸泡水中，10 个月后捞出继续保湿贮藏，到 1989 年底果实仍完好无损。万才淦和王诗云（1990）曾经采用过室内层积、剥去外果皮、剥去外果皮并剪去内果皮两端、提前采收、60℃和 80℃干热处理以及沸水浸泡等措施，都未能打破休眠。据他们报道，1984 年 10 月开始层积的种子，直到 1987 年春季才有 19%发芽 。秤锤树结实率很高，但由于饱满粒少，休眠深，场圃发芽率很低。建议将果实放在室外层积，每年春季选取已经萌动者移于苗圃培育。也可于 3 月间利用母树下腐殖土中的“自出苗”移栽培育。还可以采用嫩枝扦插繁殖。

秤锤树种子为出土萌发型。初生叶互生。下胚轴淡绿色。主根明显，侧根发达，有根毛，白色。幼苗健壮，常见幼苗根部还带有尚未烂尽的残果。秤锤树的萌发和幼苗初期生长情况见图 2。

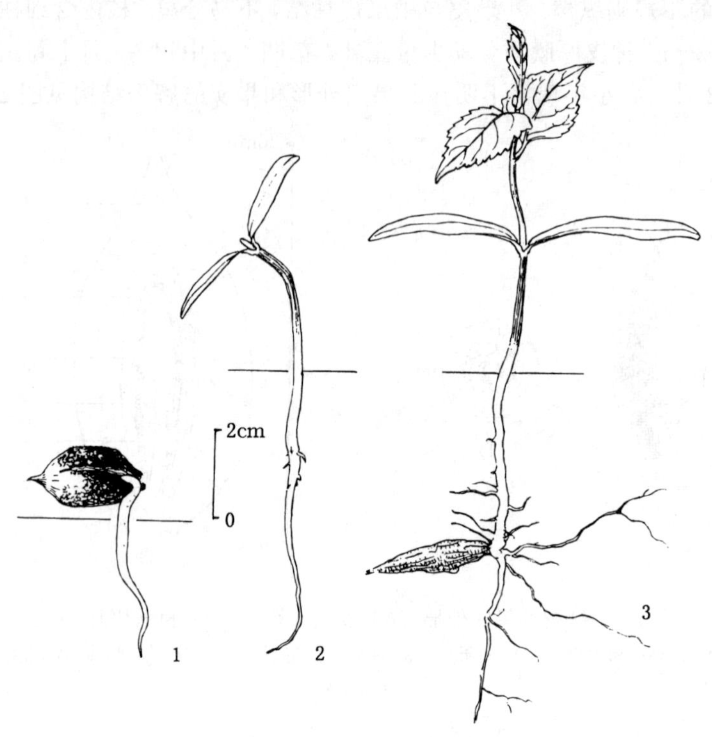

图 2 秤锤树果实萌发和幼苗生长情况
1. 下胚轴延伸 2. 子叶出土 3. 初生叶互生
（史渭清绘）

（何泽瑛）

安息香（野茉莉）属
Styrax L.

（安息香科 Styracaceae）

生长习性、分布和用途 本属约 130 种，我国约产 30 种，本文描述 6 种。适生于湿润凉爽、冬无严寒的气候环境。喜肥沃疏松的酸性土。树脂为著名芳香原料和名贵药材。它们的名称、生长、分布及用途见表 1。

开花结实 中华安息香和栓叶安息香 7～8 年生开始结实。其余 4 个种 3～5 年生开始结实。结实无大小年现象。生于海拔 300～1 200m 地段的母树结实量稍少，但寿命较长。引种至海拔 100m 左右的低丘后结实特多，但寿命短，植株难成大材。花两性。总状花序或圆锥花序。萼杯状或钟形。花冠多为 5 裂，花瓣白色至浅黄色。雄蕊多为 10 枚，花丝基部连合，贴生于花冠筒上，稀离生。花药长圆形，黄色，内向。子房上位，上部 1 室，下部 3 室，每室有胚珠 1～4。花柱钻状，柱头 3 浅裂或头状。广西南宁、海南和南京等地观察到的这几个种的开花结实物候期列入表 2。

表 1　安息香属树种的名称、生长、分布和用途

中　名	学　名	树　高 (m)	胸　径 (cm)	分　布	用　途	供　稿
中华安息香	*S. chinensis* Hu et S. Y. Liang	10～20	20～34	桂、滇	材用、树脂、药用、油料	601
垂珠花	*S. dasyanthus* Perk.	6～20	10～24	鲁、豫、皖、苏、浙、闽、赣、湘、鄂、川、黔、桂、滇	材用、药用、油料	1003
青山安息香①	*S. macrothyrsus* Perk.	20～25	20～40	桂	材用、树脂、药用、油料	601
齿叶安息香	*S. serrulatus* Roxb.	12	25	藏、滇、桂、粤、琼。越南、印度、缅甸	材用、油料	601
栓叶安息香（红皮）	*S. suberifolius* Hook. et Arn.	20	80	长江以南各地。越南	材用、树脂、油料、药用	601
越南安息香（白花树）	*S. tonkinensis* (Pierre) Craib ex Hartw.	30	70	滇、黔、桂、粤、湘、赣、闽。越南	树脂、材用、药用、油料	601

①　《中国植物志》60（2）卷将青山安息香作为越南安息香的异名。本文因它们具有不同特征，仍暂作 2 个种对待。——编者

表 2　安息香属树种的开花结实物候期

树种	观察年份	观察地点	开花			果实成熟		种子散落期
			始　期	盛　期	末　期	始　期	盛　期	
中华安息香	1978～1984	南宁	4 月上旬	4 月中下旬	5 月中旬	8 月中旬	8 月下旬末	9 月上旬～下旬
垂珠花	—	南京	5 月中旬	～	5 月下旬	—	9 月下旬	—
青山安息香	1987	海南	4 月中旬	4 月下旬	5 月上旬	7 月下旬	8 月上中旬	8 月下旬～9 月
齿叶安息香	1987	海南	4 月上旬	4 月中旬	4 月下旬	7 月下旬	8 月中旬	8 月下旬～9 月
栓叶安息香	1979～1984	南宁	4 月上旬	4 月下旬	5 月中旬	10 月上旬	10 月中旬	10 月中旬～11 月中旬
越南安息香	1987～1988	南宁	4 月上旬	4 月中旬	4 月下旬	10 月上旬	10 月中旬	11 月上旬～翌年 2 月中旬

核果，萼宿存，具星状绒毛，不开裂或不规则 3 瓣裂。外果皮肉质，干燥，内果皮木质较坚硬。每果有果核 1 粒。垂珠花、栓叶安息香和齿叶安息香稀有 2 粒。种脐大。胚乳丰富，肉质或近角质。胚直立。子叶卵形，较薄。果实形态见表 3、图 1。

表 3　安息香属树种果实和种子的形态特征

树　种	未熟果颜　色	果　实			种子（果核）		
		形　状	大小（cm）	色　泽	形　状	大小（mm）	色　泽
中华安息香	青绿色	阔卵形	高 2.3～2.8 径 2.6～3.3	黄绿色	近圆形，有凸凹纹各 3 条	高 15～21 径 16～25	棕褐色，有光泽
垂珠花	绿色	卵形或球形	高 0.9～1.3 径 0.5～0.8	黄绿色	球形或半球形，具纵沟 5 条	—	有光泽
青山安息香	绿色	椭圆形	高 2.7～3.2 径 1.5～2	黄褐色	椭圆形，有 6 条纵纹，具麻点	高 16～22 径 7～11	棕褐至黑褐色
齿叶安息香	青绿色	椭圆形	高 0.8～1.5 径 0.6～1.3	灰褐色	椭圆形	高 6～9 径 4～5	深褐色
栓叶安息香	暗绿色	球形	径 1～1.8	黄褐色	球形，有 3 条棱线突起	径 7～9.5	暗褐色
越南安息香	青绿色	近圆形或倒卵形	高 1～1.3 径 0.8～1.1	灰褐色	椭圆形	高 6～9 径 5～7	棕褐色，粗糙

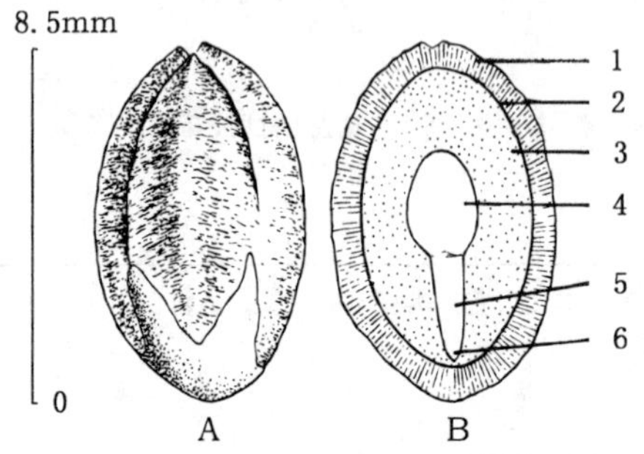

图 1　越南安息香果核外形（A）及其纵切面（B）
1. 内果皮　2. 种皮　3. 胚乳　4. 子叶　5. 胚轴　6. 胚根
（黄应钦绘）

果实的采收调制和种子贮藏　果实成熟盛期用竹竿敲打，或摇动树枝震落后在地面捡拾。采回的果实置日光下曝晒或在室内通风处摊晾。果壳开裂后用手搓揉，扬去果壳等杂质，所得果核用作播种材料，通称种子。种子富含油脂，已去果壳的种子不能曝晒或长期裸露存放。鲜种的含水量为 33%～37%，不影响种子发芽的最低含水量在 20%左右。也可带果壳运输，运抵后再行调制。如需贮藏越冬，贮藏时需混以干沙，贮藏期一般在 8 个月以内。出种率和种子的净度、质量等见表 4。

发芽和播种　种子有休眠习性。发芽时日均温宜在 15℃以上。不宜随采随播。8 月以前采集的种子宜沙藏至翌年 2 月上旬播种。9～10 月采集的种子宜沙藏至翌年 3 月下旬播种。1980 年 10 月和 1987 年 8～9 月，广西林业科学研究所对中华安息香、青山安息香和越南安息香 3 个树种的种子，在室外沙床作过发芽测定。播后约半年，至翌年 2 月中旬和 4 月上旬才开始发芽，见表 5。

表 4　安息香属树种果实出籽率和种子净度、质量

树　种	出籽率（%）	净度（%）	千粒重（g）	每千克种子粒数
中华安息香	55～60	99	4090	240
垂珠花	—	—	130	7 700
青山安息香	37	99	720	1 380
齿叶安息香	30	99	90	11 000
栓叶安息香	40	99	360	2 780
越南安息香	35	99	180	5 560

表 5　安息香属树种的种子发芽能力

树种	发芽率（%）	
	计算天数	一般数值
中华安息香	210	62
青山安息香	260	73
越南安息香	240	35

出土萌发。胚根萌发后 15～25 天子叶出土，再过 5～10 天发出初生叶。越南安息香种子的萌发和幼苗生长情况见图 2。

条播。每平方米中华安息香播种 450～620g，青山安息香播种 70～100g，栓叶安息香播种 38～50g，垂珠花、齿叶安息香和越南安香播种 15～25g。培育 1 年生苗出圃。

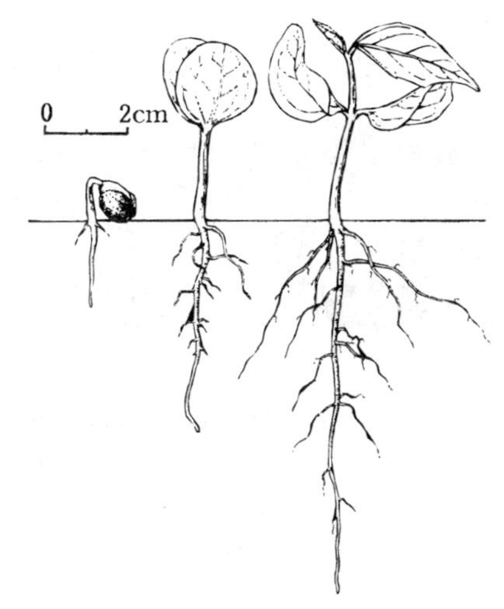

图 2　越南安息香种子萌发后第 15、20、25 天幼苗的生长情况
（黄应钦绘）

（王宏志）

山　矾　属

Symplocos Jacq.

（山矾科　Symplocaceae）

生长习性、分布和用途　本属约 300 种，我国约 80 种，本文描述 3 种。较耐干热及水湿，疏林荒坡及密林中肥沃地均能生长。它们的名称、生长、分布及用途见表 1。

表 1　山矾属树种的名称、生长、分布和用途

中　名	学　名	树高（m）	胸径（cm）	分　布	用　途	供　稿
华山矾	*S. chinensis* (Lour.) Druce	0.6～1	—	鲁、苏、皖、浙、闽、台、赣、湘、粤、桂、滇、黔、川、陕	根叶治跌打、烧伤，鲜叶治蛇伤，种子含油	601
光叶山矾	*S. lancifolia* Sieb. et Zucc.	4～6	10	东南、中南、西南。日本	和肝健脾、止血生肌、材用	601
白檀	*S. paniculata* (Thunb.) Miq.	2～5	—	东北、华北、华东、中南、西南、陕。朝鲜半岛、日本、印度	药用、油料、材用、观赏	1003

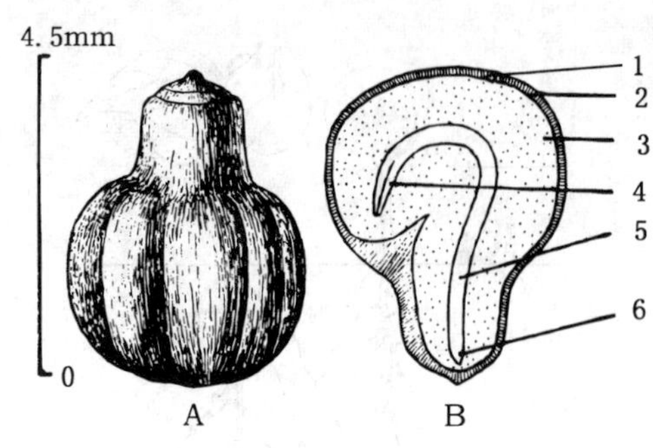

图1　华山矾果核外形（A）及其纵切面（B）
1. 内果皮　2. 种皮　3. 胚乳　4. 子叶
5. 胚轴　6. 胚根
（黄应钦绘）

开花结实　华山矾约3～5年生开始开花结实，正常结实期在8年生以后；光叶山矾8～10年生开始结实，正常结实期在15年生以后。结实无大小年现象。花两性。花萼5裂，常宿存。花冠5，深裂，几达基部。子房下位或半下位。光叶山矾穗状花序，长1～4cm，腋生；花冠淡黄色，雄蕊约25，子房3室。华山矾和白檀圆锥花序，长4～7cm，顶生；花冠白色，芳香。雄蕊40～60。子房2室，每室2～4胚珠。这3个种的开花结实物候期见表2。

核果，球形或椭圆形，顶端具宿存花萼裂片。外果皮肉质浆状，中果皮薄。内果皮木质或骨质，光滑或具棱，2～3室，每室1种子。种皮膜质。胚长形，弯曲。胚乳丰富，油质。子叶较短。果实及种子的形态见表3、图1。

表2　山矾属树种的开花结实物候期

树　种	观察地点和年份	开　花			果实成熟		果实脱落期
		始　期	盛　期	末　期	始　期	盛　期	
华山矾	南宁 1987～1988	4月中旬	5月上中旬	5月下旬 （7月上旬）	8月下旬	9月中旬	9月中旬～ 11月下旬
光叶山矾	南宁 1978～1984	6月上旬	6月中旬	7月中旬	11月中旬	12月上旬	12月中旬～ 翌年1月中旬
白檀	南京 —	4月中旬	5月上旬	—	7月	9月	—

表3　山矾属树种果实和果核的形态特征

树　种	未熟果颜色	果　实			果　核		
		形　状	大小（mm）	颜　色	形　状	大小（mm）	色　泽
华山矾	黄色	椭圆形	长7～9 径6～7.5	紫黑色	乳头状倒卵形	长4.5～6 径3.5～5.5	深褐色
光叶山矾	绿色	球形或卵圆形	长6.5～9 径5～7	紫黑色	乳头状倒卵形	长4～5 径3～4	深褐色
白檀	黄绿色	卵球形稍压扁	长5～8 径4～7	蓝黑色	歪梨状，具纵向不规则凹纹	长约5	灰褐色

果实的采收调制和种子贮藏　果实成熟盛期用手或高枝剪采摘。采得的果实可堆沤3～5日。果皮软熟后装入布袋置水中反复搓揉。淘去果皮等杂质，所得果核即为播种材料，通称种子。种子的含水量宜保持在30%左右。忌失水，不宜日晒或干藏。需要时可以果实的状态包装运输，运抵后再行调制。也可不作调制而带果皮播种。调制出来的种子，贮藏时需混以湿沙，贮藏期在半年以内。出籽率和种子的净度、质量等见表4。

表 4　山矾属树种果实的出籽率和种子净度、质量

树　种	出籽率（%）	净度（%）	千粒重（g）		每千克种子粒数（万粒）	
			一　般	变动范围	一　般	变动范围
华山矾	38	96	45	35～50	2.2	2～2.9
光叶山矾	28	96	33	25～40	3	2.5～4
白檀	—	—	35	30～38	2.9	2.6～3.3

发芽和播种　种子有休眠习性。发芽时日均温宜在20℃以上。1987年9月中旬和1988年1月中旬，广西林业科学研究所用当年新采华山矾和光叶山矾种子在发芽箱作过发芽测定，置床前种子未作处理。播后68天方始发芽，结果见表5。

表 5　山矾属树种的发芽能力及其测定条件

树　种	基　质	温度（℃）	发芽势（%）	发芽率（%）	
				计算天数	一般数值
华山矾	纸	28	不明显	112	30
光叶山矾	纸	28	不明显	110	28

出土萌发。胚根萌发后约11天子叶带壳出土。早期生长缓慢，出土后约56天种壳方脱落，子叶展现，再过4天左右发出初生叶（图2）。

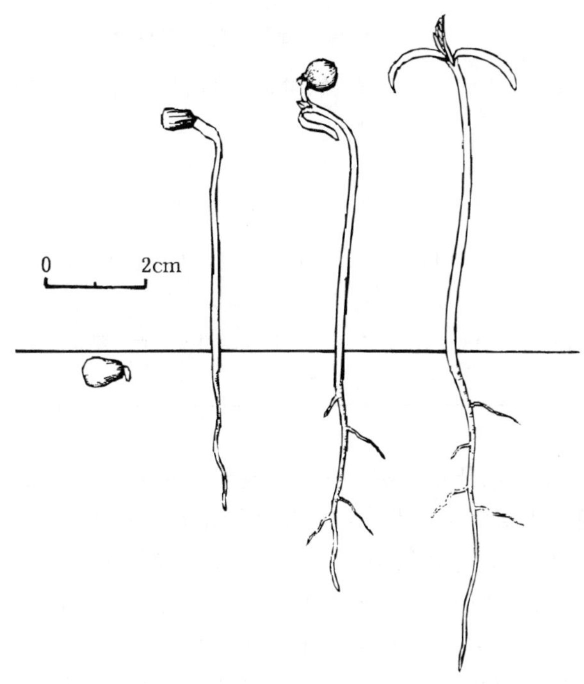

图 2　华山矾种子萌发后第3、56、60、63天幼苗的生长情况

（黄应钦绘）

撒播或条播。目前尚无大田育苗经验。每平方米可播种12g，覆土宜浅。1年生苗出圃。

（王宏志）

梾木属

Cornus L.

（山茱萸科 Cornaceae）

生长习性、分布和用途 梾木属33种，我国约20种，本文描述5种。落叶乔木或灌木。对土壤要求不严，中性、酸性或微碱性土壤上均能生长，在湿润深厚肥沃的土壤上生长尤为旺盛。

木材细致均匀，纹理直，坚硬，易干燥，车镟性能好，可供建筑、家具、玩具、雕刻、农具及制胶合板等用。果肉及种子含油量高，可供食用、药用，并作轻工业原料。树皮可提制栲胶。叶作饲料及绿肥。花是蜜源。是荒山造林、水土保持和园林绿化树种。它们的名称、分布和用途见表1。

表1 梾木属树种的名称、树高、分布和用途

中 名	学 名	树 高 (m)	分 布	用 途	供 稿
红瑞木	*C. alba* L.	3	东北、华北、陕、甘、青、苏、赣。浙江有栽培。朝鲜半岛、西伯利亚	油料、观赏	106
灯台树	*C. controversa* Hemsl.	20	辽、吉、陕、甘、华北、华中、华南、西南	材用、油料、栲胶、观赏	502
梾木	*C. macrophylla* Wall.	20	华东、华中、西南、晋、鲁、陕、甘、台。日本、印度、巴基斯坦、尼泊尔	油料、材用、药用、蜜源	701
毛梾	*C. walteri* Wanger.	15 (30)	辽、华北、西北、华东、华中、西南	油料、药用、栲胶、饲料	501
光皮梾木	*C. wilsoniana* Wanger.	18 (25)	陕、甘、川、黔、中南、闽、浙	材用、油料、饲料、观赏	502

开花结实 3～6年生开始开花结实。乔木类树种结实量大，结实期长。花两性，小，白色、黄白色或淡黄白色。伞房状聚伞花序或圆锥状聚伞花序顶生，下无总苞。萼小，4齿裂。花瓣4，镊合状排列。雄蕊4，花药长圆形。花盘垫状。子房下位，倒卵圆形或近倒卵圆形，2室，每室有胚珠1枚。花柱圆柱状或短圆柱状，柱头头状。花期因树种而不同，有的在4月下旬～5月，有的在5月下旬～7月。核果当年成熟（表2），未成熟时绿色，成熟时颜色因树种而不同，球形或倒卵圆形（表3）。外果皮薄，中果皮肉质，内果皮骨质。果核圆球形或卵圆形，稍扁，具沟槽，灰白色。仅灯台树果核顶端有近于方形的深孔。核内有种子1～2粒。胚小，乳白色，包于丰富的胚乳中（图1）。

表 2　梾木属树种的开花结实物候

树　种	观察地点	开　花			果实成熟	
		始　期	盛　期	末　期	始　期	盛　期
红瑞木	哈尔滨	5 月下旬	6 月中旬	7 月上旬	7 月上旬	8 月
灯台树	峨眉		5 月		9 月中旬	10 月
梾木	江西	4 月下旬	～	5 月	10 月	11 月
毛梾	峨眉		5 月		9 月中旬	10 月
光皮梾木	都江堰	4 月下旬	～	5 月	9 月下旬	10 月

表 3　梾木属树种核果和果核的颜色、形状与大小

树　种	核　果			果　核		
	颜　色	形　状	大小（mm）	色　泽	形　状	大小（mm）
红瑞木	乳白色或蓝白色	斜卵圆形，两头尖，微扁	5～8	浅灰色	长圆形，微扁	约 4
灯台树	紫红色至蓝黑色	球形	6～7	灰白色	球形，稍扁，具沟槽 8～9，顶端有一个近方形的小孔	约 5
梾木	紫红色至青黑色	近球形	4.5～6	灰白色	扁球形，左右各有 1 条浅沟，两侧有 6 条不明显的脉纹	3～4
毛梾	黑色	球形	6～7（8）	灰白色	扁球形至球形，略有不明显肋纹	约 5
光皮梾木	紫黑色至黑色	球形	6～7	灰白色	球形，肋纹不明显	4～4.5

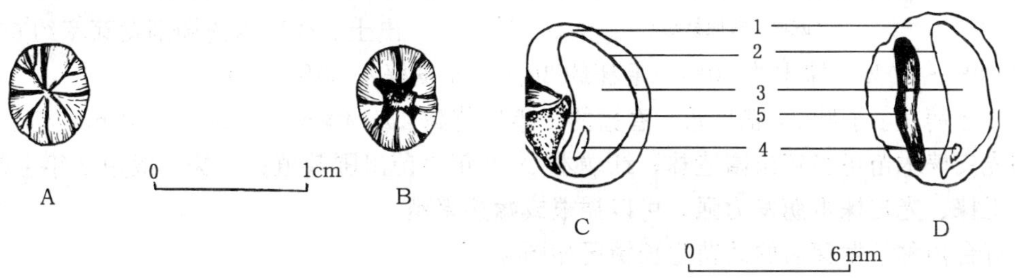

图 1　灯台树果核外形的底面（A）和端面（B）及其纵切面（C）和横切面（D）

1. 内果皮　2. 种皮　3. 胚乳　4. 胚　5. 胚珠未发育的子室

（孟玲绘）

果实的采收调制和种子贮藏　当核果呈成熟颜色（表 3）时应及时剪摘果枝，或用竹竿击落果穗后收集。采集的核果堆沤几天，待果皮变软后在水中搓揉，漂去果皮和果肉等杂质，所得的果核即为播种材料，通称种子。种子应混沙湿藏。沙的湿度为饱和含水量的 60%。种子与湿沙的容积比为 1∶2。堆积的厚度不宜超过 30cm。每隔 10 天检查翻动一次，酌情加水，使保持一定湿度，忌失水干燥。核果出籽率及种子（果核）的质量和数量见表 4。

表 4　梾木属树种核果的出籽率的果核质量与数量

树　种	出籽率（%）	净度（%）	千粒重（g）	每千克纯净果核数（万粒）
红瑞木	12	90～96	20～24	4.2～5.0
灯台树	30～40	95～98	62～120	0.8～1.6
光皮梾木	12	85～95	23.3～29.5	3.4～4.3

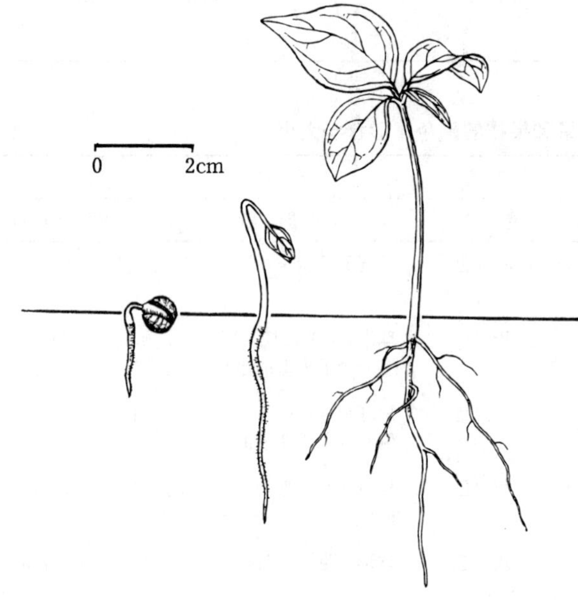

图 2　灯台树种子萌发后第 14、18、72 天的幼苗生长情况
（孟玲、黄应钦绘）

发芽前的处理　种子有休眠习性，可以在秋季随采随播或湿沙层积后春播。春播的需作催芽处理：春播前 1 个月将果核浸泡 1 天，混沙后置于 10～20℃温度下，保持 60%的湿度，20 余天取出播种。也可以将果核用始温 50℃水浸泡 36 小时，取出放入竹筐中，用纱布覆盖，每天用温水淋洒 1～2 次，待有部分种子裂嘴时播种。

播种　春播较普遍，也有秋播的。秋播在 10 月下旬左右。层积过的种子春播期约在 2 月中旬～3 月上中旬。条播，条距 20～30cm，条幅 5～10cm，播种沟深 1.5cm。每平方米播种 20～38g。播后覆细土 1.5cm 并盖草。秋播的翌年 2 月上中旬发芽出土。春播的 30～50 天后发芽出土。红瑞木的场圃发芽率约 60%，灯台树 40%～50%，梾木约 80%，毛梾约 30%，光皮梾木 50%～60%。

出土萌发。子叶 2，椭圆形。春播的 1 年生苗高 20～40cm，或 80～ 100cm 。灯台树、梾木和光皮梾木苗可当年出圃造林；红瑞木 1～2 年生苗出圃栽植；毛梾一般用 2 年生苗造林。

毛梾、光皮梾木萌发力强，可以插根或嫁接繁殖。

灯台树种子发芽后的幼苗生长情况如图 2。

（赵德铭）

香港四照花

Dendrobenthamia hongkongensis（Hemsl.）Hutch.

（山茱萸科　Cornaceae）

生长习性、分布和用途　四照花属约 20 种，我国约产 19 种，本文描述 1 种。常绿乔木，高达 18m，胸径约 40cm。喜空气湿润的生态环境，能耐短期－8℃左右的低温。适生于肥沃湿润的疏松土壤，荒山不宜。产浙、闽、赣、湘、粤、桂、黔、滇、川。木材结构致密，可作建筑、家具等。树冠浓绿，宜作庭院观赏树。果可食，或酿酒。

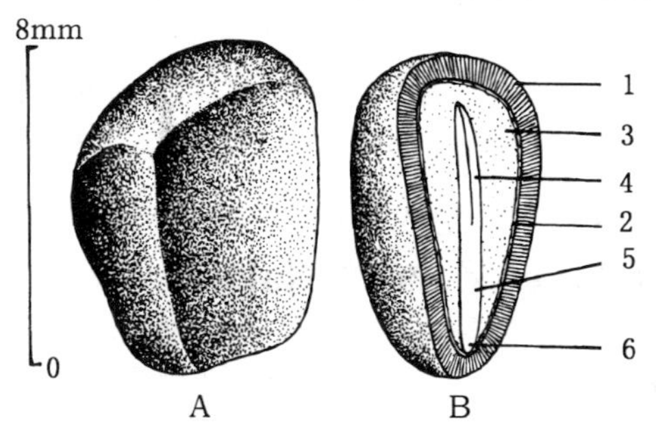

图1　香港四照花果核外形（A）及其纵切面（B）
1. 内果皮　2. 种皮　3. 胚乳
4. 子叶　5. 胚轴　6. 胚根
（黄应钦绘）

开花结实　约10年生开始开花结实，正常结实期在20年生以后，结实大小年现象不甚明显。花两性，小型，多数集合成1圆球状的头状花序，总梗长4～10cm。每花序具花50～70，花序径约1cm。总苞苞片4，白色，花瓣状，宽椭圆形或倒卵状宽椭圆形，长2.8～4cm。萼齿不明显。花瓣4，长椭圆形。雄蕊4。花盘8浅裂。子房下位，2室，每室有倒垂胚珠1颗。花柱圆柱形，粗，柱头平截。各花的子房完全合生于发育的花托中。据广西南宁1987～1988年观察，花芽于3月下旬抽出，4月中旬～下旬头状花序形成，腋生或顶生；5月上旬为始花期，5月中下旬为盛花期，6月上旬为终花期；10月下旬果实开始成熟，11月中旬为果实成熟盛期。核果长圆形，多数藏于花托发育而愈合的球状果穗中，聚花果径2.2～3cm。果穗梗长4～10cm。果未熟时黄绿色，成熟时黄色至红色，成熟后约10天于11月上旬～12月上旬逐渐散落。每果有核1～2粒。果核扁卵形、椭圆形或不规则形，淡红褐色，长5～12mm，宽5～8mm，厚4～6mm。胚直立，胚乳丰富（图1）。

果实的采收调制和种子贮藏　果实成熟盛期用采种刀或高枝剪截断果梗，采下头状果穗集中堆沤。肉质果皮软熟后装入竹筐置水中搓擦淘洗，漂去果皮等杂质。所得果核用作播种材料，通称种子。鲜果的出籽率约11%。净度可达98%。千粒重155（140～170）g，每千克种子约(5 800)6 500(7 200)粒。种子含水量20%～35%，忌日晒，亦不宜裸露存放。1周内的短途运输，可用竹筐包装果实运输，抵达后再行调制。未经调制的果实不宜贮藏，以免霉坏影响种子的发芽能力。调制出来的果核通常用湿沙层积贮藏，贮藏期为4～6个月。

发芽和播种　种子休眠现象不明显。发芽时的日均温宜在17℃以上。1987年11月18日，广西林业科学研究所在室外沙床播种，对当年新采的种子作发芽测定。播后1个月（12月18日，日均温16～18℃）开始发芽，21日进入发芽盛期，30日发芽终止。从发芽至发芽高峰，以7天计，发芽百分数为50%。从播种之日起算，43天发芽率为70%。出土萌发。胚根萌发后约7天子叶出土，再过13天初生叶展现。种子萌发和幼苗生长情况见图2。

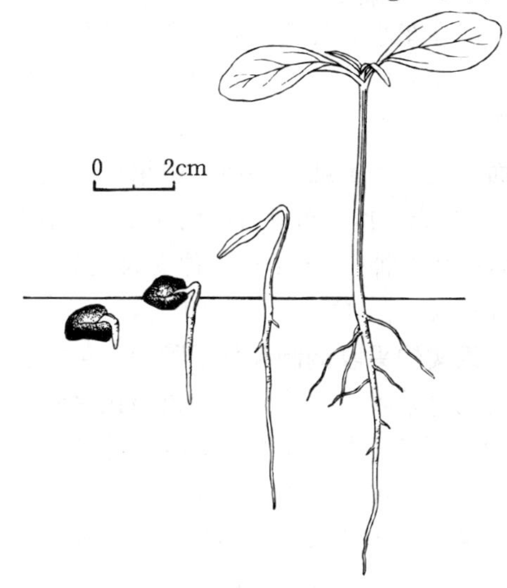

图2　香港四照花种子萌发后第3、7、10、20天幼苗的生长情况
（黄应钦绘）

种子发芽所需温度不高，宜随采随播。也可短期贮藏至翌年 2 月中下旬播种，不宜迟过 3 月下旬。条播。每平方米播种 20～25g，覆土 1.5～2cm。1 年生苗出圃。用于庭院绿化的，一般需培育 2～3 年生苗。

（曾 玲）

山 茱 萸

Macrocarpium officinale (Sieb. et Zucc.) Nakai

（山茱萸科 Cornaceae）

生长习性、分布和用途 山茱萸属共 4 种，我国有 2 种，本文描述的这 1 种为我国特有的名贵药用树种。落叶乔木，高 4～10m。主产浙、皖。苏、鲁、豫、晋、赣、湘、鄂、川等地多有栽培。生于山沟、溪边、坡地，喜肥沃湿润土壤，在干燥瘠薄地生长不良。果肉为中药材，称萸肉或山茱肉。周兆祥、陈经梧等（1988）分析过萸肉的化学成分。

开花结实 生长缓慢，8～10 年开始结实，15～20 年进入盛果期，结实期可持续百年以上，经济寿命长。结实大小年现象十分明显，结实率低。黎章矩和曹燕如（1986）报道过浙江淳安和临安两地山茱萸 1968～1983 年产量的波动情况。他们认为，花芽分化与果实生长在时间上同步，树体营养在二者之间分配上的矛盾是山茱萸结实出现大小年现象的主要原因之一。4 月中旬～5 月中旬，山茱萸落花落果现象较重，可能同花期的天气、树体营养状况以及自花授粉有关。据黎章矩试验，喷施 5～10μg/gIBA 或 10～20μg/gNAA 等药剂，有保花保果的功效。山茱萸在陕西周至县有 200 多年的栽培历史，但长期处于半野生状态，产量低。通过修剪、施肥，树龄 36 年的单株可年产萸肉 6.6kg，比对照提高 1 倍（刘培华和王小纪，1991）。

两性花小，黄色。伞形花序着生于枝侧粗壮的总花梗上，具花 15～35 朵。总苞片 4，黄绿色，花后脱落。萼 4 齿裂。花瓣 4。花盘垫状。子房 2 室，各具 1 胚珠。花期 3～4 月，花期不一。果熟期 8～10 月，11 月后成熟脱落。核果，椭圆形，长约 1.2～2cm，径 0.6～0.7cm，表面光亮，色鲜红。果肉（中果皮）浅红色，厚，浆质。内果皮坚硬，骨质，浅棕色，厚约 1.5～2mm，其中布满直径约 1mm 大小不等的油室，内含浅黄色树脂状液体，干藏后变为松香状固体。种仁 1～2 枚。种皮为无明显膜质结构的乳白色物质。胚乳丰富，胚长 6～10mm，子叶狭长（图 1）。

果实的采收调制和种子贮藏 果实转呈红色且果肉变软即为成熟标志，从 9～11 月果落之前均可采收。去除果肉的果核比水轻，洗净晾干即为播种材料，通称种子。忌曝晒。有空粒种子，但通常仅在 10%以下。干藏种子半年后几乎未失重，种仁也新鲜如初，但因不易发芽，通常用湿沙层积。千粒重 140～300g，每千克有种子（果核）3 300～7 200 粒。

发芽前的处理 山茱萸种子具深休眠习性。孙昌高、潘晓飙等人（1988）1985 年 11 月将新采收的种子在室内层积，1986 年春仅个别种粒发芽，1986 年 11 月将种子移到室外，到 1987 年 3 月才大量萌发，发芽率 82%。他们还报道说浓硫酸和过氧化氢处理无效。管康林（1990）推测山茱萸深休眠有两方面的原因，一是内果皮（核壳）含有单宁类物质阻隔氧气进入，二是胚有生理后熟现象。他将种子播在覆有弓形薄膜的沙床中，使种子先在 13 ～27℃的自然变温

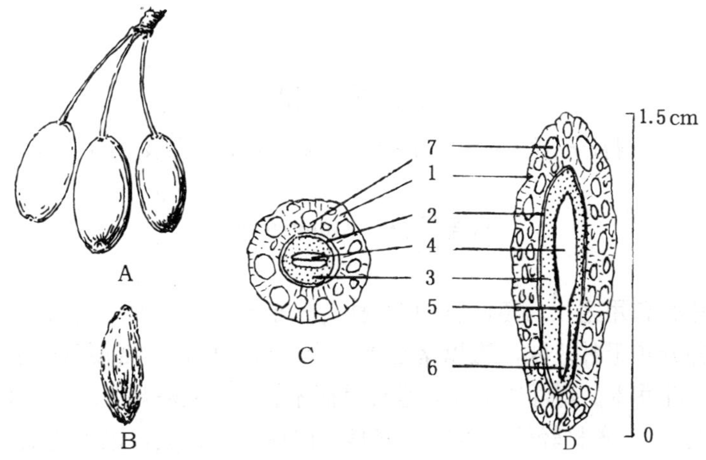

图 1　山茱萸的果实外形（A）、果核外形（B）及果核横切面（C）和纵切面（D）
1. 内果皮　2. 种皮　3. 胚乳　4. 子叶　5. 下胚轴　6. 胚根　7. 油室
（史渭清绘）

中经受 3 个月的暖层积，再在冬季接受 5～12℃的较低温度。据他报道，经过这样的处理，到春天气温回升时发芽率可达 70%～90%。

根据种子休眠的特性，生产上曾经有过以下一些育苗方法：(1) 采用刚转色而尚未变软的中等成熟果实，去果肉后及时层积。9 月中下旬移至阳畦播种越冬，翌春 3～ 4 月出苗，出苗率为 40%～60%。(2) 成熟果实采下后用始温 50℃水浸泡 40 分钟，除去果肉后略微晾干即播。(3) 将种子（果核）与草木灰和人粪混合放土坑内，并围以圈肥发酵，次年 2～3 月播种，5 月可出苗。(4) 11 月采果，条播。播种沟深 16cm，宽 26～30cm，垫圈肥 10cm，盖土，播种，覆以草木灰，以掩没种子为度。耙平，畦面铺干草厚约 13cm，点火烧尽。次年 4～5 月发芽出土，出苗率 90%以上。

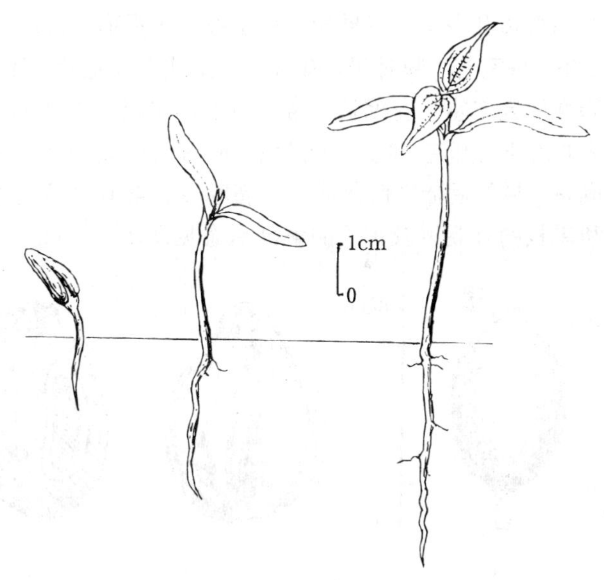

图 2　山茱萸种子萌发后第 10、25、39 天的幼苗生长情况
（史渭清绘）

播种　春季条播。经过先暖层积后冷层积的种子，每平方米约播 25～ 38g，田间出苗率约 50%～60%（管康林，1990）。子叶出土，初生叶对生（图 2）。下胚轴暗紫色，近根颈处白色。

（何泽瑛）

八　角　枫

Alangium chinense（Lour.）Harms

（八角枫科　Alangiaceae）

生长习性、分布和用途　八角枫属约 30 种，分布于亚洲、大洋洲和非洲。我国约 9 种。本文描写的这 1 种分布于甘、陕、豫以南至闽、台、粤、桂，西抵西藏南部。东南亚和非洲东部也有分布。落叶乔木，高可达 15m，但常呈灌木状。小枝略呈“之”字形曲折。喜光。根皮入药，称“白龙须”。木材轻软，可制作家具、门窗、胶合板，也可造纸。

开花结实　花两性。二歧聚伞花序腋生，总梗长 5～15mm，花梗长 5～15mm，有花 7～30（50）朵。萼齿 6～8。花瓣 6～8，条形，长 1～1.5cm，宽 1.0mm，镊合状排列成管状，开花后上部分离反卷，初为白色，后变乳黄色。雄蕊 6～8。花盘近球形。子房下位，2 室。柱头头状，2～4 裂。核果卵圆形，长 6～12mm，径 5～8mm，幼时青色，成熟时黑色，顶端有宿存的萼齿和花盘。每果含果核 1 枚，卵圆形，暗灰色，长 5～7mm，径 4～6mm。单室或少见 2 室，每室有 1 粒种子，有时 1 室空或多少退化。种子扁圆形，具有大型直立胚，子叶肥大，油脂性胚乳丰富。花期长，常见有的幼果已形成，同一花序上还有正在开放的花。南京地区 6 月中旬开始开花，6 月下旬为盛期；8 月中旬果实开始成熟，8 月下旬为盛果期，9 月中旬全部成熟。果熟后陆续脱落或为鸟雀啄食，常有鸟播现象。结实有大小年，间隔期 3～4 年。果实和果核的外观以及种子的解剖形态见图 1。

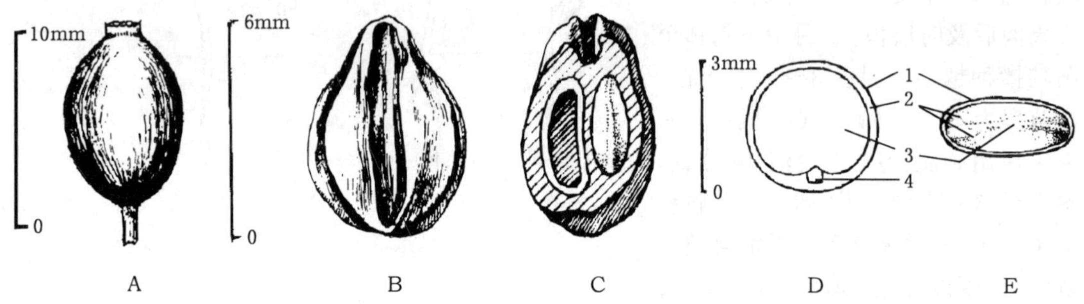

图 1　八角枫核果外形（A）、果核外形（B）及其纵切面，两室，仅一室有种子（C）和种子的纵切面（D）、横切面（E）

1. 种皮　2. 胚乳　3. 子叶　4. 胚根

（A. 田恒德仿《中国树木志》，B～E. 张世经绘）

果实的采收调制和种子贮藏　果实成熟期极不一致，又易被鸟雀啄食，应在盛果期分批采收。南京地区可在 8 月下旬～9 月上旬采摘已呈黑色的成熟果实，也可截取果穗后选摘成熟果实供调制。充分成熟的果实收集后即可揉搓淘洗，清除果皮果肉等杂物，晾干，所得果核即为播种材料，通称种子，净度可达97%。鲜果出籽率为31%～38%。千粒重约75g，每千克有种子约1.3万粒。种子宜低温密封干藏，或低温层积供翌年春播。干藏种子的含水量应降低到8%左右。

发芽和播种　种子有休眠习性。层积90天的种子，在25℃条件下第5天开始发芽，23天发芽率可达80%。未发芽种子多为空粒。育苗时宜将种子层积3个月后春播，也可在秋季播种。条播，深3cm。出土萌发。子叶2，卵形或椭圆形，长1.7～2.2cm。先端钝圆。初生叶互生，卵形，先端渐尖，基部心形，边缘有缺齿。主侧根发达（图2）。

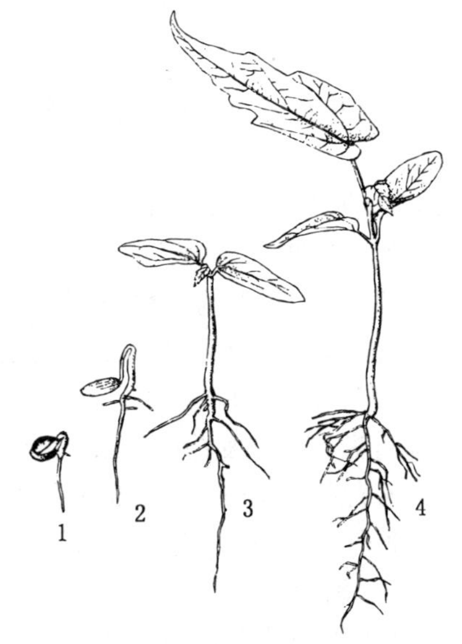

图2　八角枫的种子萌发和幼苗生长情况

1. 下胚轴延伸　2. 子叶出土　3. 子叶展开　4. 初生叶互生

（黄鹏成供稿，田恒德绘）

（沈永宝）

喜　树

Camptotheca acuminata Decne.

（蓝果树科　Nyssaceae）

生长习性、分布和用途　喜树属为单型属，只有本文描述的1种，我国特产。落叶乔木，高达25m。生长迅速，主干端直，喜温暖湿润气候。深根性，喜肥沃湿润土壤，不耐贫瘠，但较耐水湿。多生于平地。萌芽力强。病虫害少。分布于川、鄂、湘、赣、滇、黔、粤、桂、浙，豫、皖、苏、闽。木材轻软，结构细，但较脆，不耐腐，一般用于造纸和制作火柴杆。喜树的果实等部位含有生物碱，有抗癌作用，但亦有副作用。喜树生长快，树干端直，是四旁绿化的常用树种。

开花结实　花单性，雌雄同株。球形头状花序，雌花序顶生，雄花序腋生。花小，苞片3，舟状。花萼5齿裂，杯状。花瓣5，淡绿色。雄蕊10，排成两轮，外轮较长，内轮较短。子房下位，1室，有1胚珠。花期5～7月，果期9～11月。果实未熟时淡绿色，熟时黄褐色。坚果呈翅果状，窄矩圆形，不具梗，两侧不对称，顶端截形，有窄翅，花柱宿存。外果皮革质，具2～3条纵棱脊和众多纵纹。果长2.5～3.0cm。每果含种子1粒。种子细长，橄榄形。胚白色，长度约占胚腔的60%，有胚乳（图1）。多数果实集合成球形果穗，成熟后逐渐散落。

果实的采收调制和种子贮藏　果实成熟于秋末冬初，落叶后球形果穗长期悬着树上，散落期长；果虽有翅，但飘不远，可在散落盛期在树下扫集。翅果采回后在通风干燥处摊晾阴干，筛选去杂后用布袋或麻袋盛装干藏，含水量约10%～12%。翅果即播种材料，通称种子，净度可达100%。千粒重39(34～45)g。每千克有种子(果) 2万～3万粒。精选的种子千粒重可达67g。

发芽前处理　喜树种子有休眠习性。据宋纯清(1981)报道,从喜树果实中能分离出脱落酸。采用低温湿沙层积、变温发芽或变温加光照,以及去果皮等措施都能促进发芽。浸种时水的始温以45～60℃为宜,其中以低温层积3～4周最为适用,效果也最好。据周佑勋 (1989) 报道,喜树休眠的主要原因是果皮和种皮透气性差并存在发芽抑制物质,用低于5℃的温度层积10天即可见到促进发芽的效果,她认为最好是层积40天。王成霖和邵蓓蓓 (1987) 在0～5℃的温度下将喜树种子分别层积0周、1周、2周和3周后,得到的发芽率分别为2%、43%、68%和82%。邵蓓蓓(1989)将未经层积的喜树种子放在30℃下培养30天,得到的发芽率仅为3%;在0～5℃下层积30天后用同样的温度培养,8天的发芽率为83.5%。

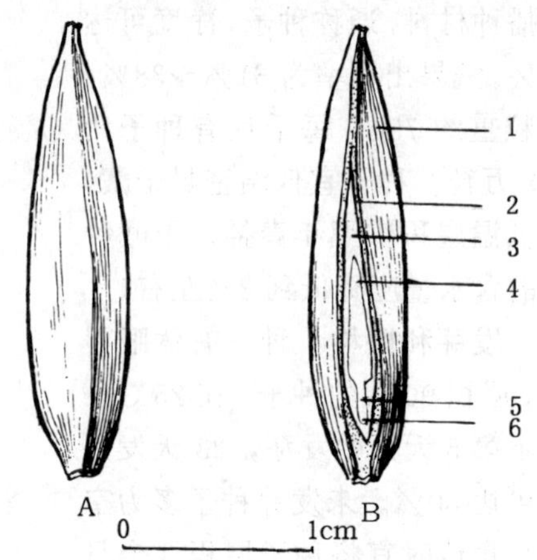

图1　喜树果实外形 (A) 及其纵切面 (B)
1. 果皮　2. 种皮　3. 胚乳　4. 子叶
5. 胚轴　6. 胚根
(童军平绘)

发芽测定　发芽温度以20～30℃的昼夜变温或30℃的恒温较好,光照不仅能提高发芽势,也能提高发芽率。发芽测定条件及发芽能力见表1。

表1　喜树种子发芽测定条件及发芽能力 (基质为石英沙)

预　处　理	每天光照时数	温度(℃)	发芽势(%)			发芽率(%)		
			计算天数	一般数值	变动范围	计算天数	一般数值	变动范围
剥去果皮	0	30	5	41	16～60	17	82	80～84
0～5℃层积4周	0	30	6	76	74～78	8	84	82～86
始温45℃水浸种24小时	8	20～30	9	48	42～50	14	80	76～82

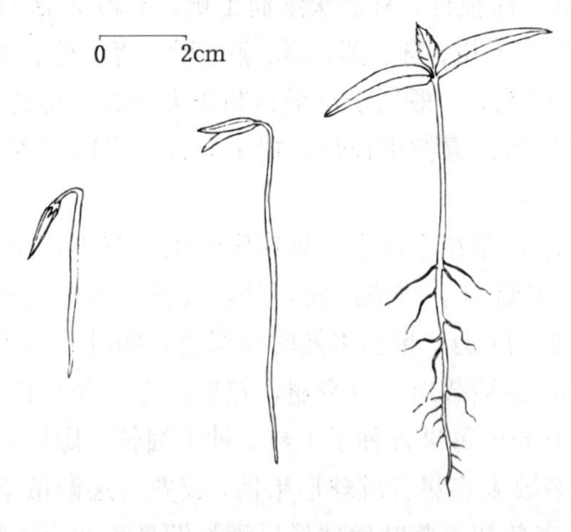

图2　喜树种子萌发后第1、6、22天幼苗的生长情况
(童军平绘)

播种　春季播种。层积越冬的种子播前无需预处理。条播,条距25～30cm,播种沟深2～3cm,每666m^{2}播种3～4kg。播后覆土约1cm,盖草。3月份播种的,约20～30天开始发芽出土。出苗率75%～80%。

出土萌发。子叶2枚。上胚轴短,长7～11mm,下胚轴圆柱状,长3.5～4.6cm。出土后约20天初生叶出现,互生;主根向下骤细,侧根发达。1年生苗高约60～80cm,可以出圃。行道树及庭园绿化常用2～3年生的移植苗。喜树种子的萌发和幼苗初期生长情况见图2。

(王成霖)

蓝果树（紫树）

Nyssa sinensis Oliv.

（蓝果树科　Nyssaceae）

生长习性、分布和用途　蓝果树属有10余种，我国产7种，本文描述1种。落叶乔木，高达30m，胸径100cm。喜湿润凉爽气候，忌干热，能耐－10℃左右的短期低温。适生于肥沃湿润的酸性土。干旱瘠薄地早期虽能生长，但4～5年生后即呈现衰退。产粤、桂、滇、黔、川、湘、鄂、赣、闽、浙、苏、皖。木材结构致密，但不耐腐，可作室内用具。落叶层厚，有改良土壤涵养水源的功效。

开花结实　7～8年生开始开花结实，正常结实期在20年生以后，结实大小年间隔期为1年。花单性，雌雄异株。雄花为伞形或总状花序，总梗长3～5cm，密被柔毛；花托盘状；花萼裂片5，细小；花瓣5，窄长圆形，早落；雄蕊5～10，生于花盘外缘。雌花近无柄，组成头状花序；花瓣鳞片状，长约1.5mm；子房下位，与花托合生，1室稀2室，每室具倒生胚珠1，花柱反曲。据广西南宁1978～1984年观察，3月中旬花芽与叶芽同时展现，雌花芽着生于当年新生枝的上部叶腋。4月上旬为始花期，中旬为盛花期，下旬为末花期；9月下旬果实开始成熟，10月上旬～中旬为果熟盛期。核果矩圆状椭圆形或长倒卵形，微扁，未成熟时绿色略带紫色，接近成熟时为紫色稍带青色，成熟时为深蓝色至蓝黑色，干时为紫褐色。果长1～1.5cm，径0.6～1cm。常3～4个果着生于头状果穗上呈簇生状。果实成熟后在树上可存留10天左右，10月中旬～11月上旬果实脱落。果核微扁，椭圆形，淡黄色至灰褐色。核坚硬，有5～10条纵向沟纹，长6～12mm，径4～7mm，每核有发育种子1（稀2室各有1种子发育）。种皮薄，具胚乳，胚直伸（图1）。

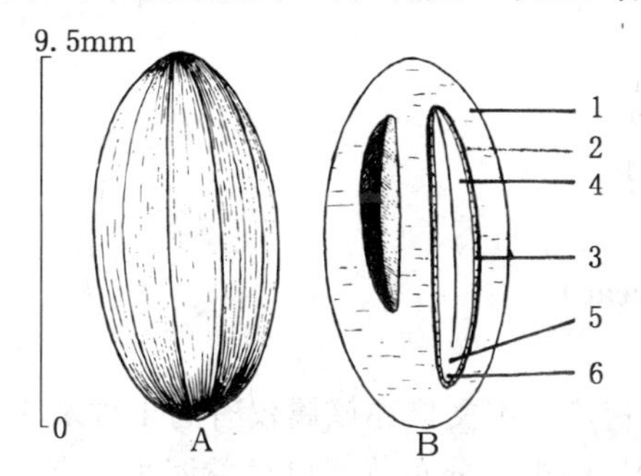

图1　蓝果树果核外形（A）及其纵切面（B）
1. 内果皮　2. 种皮　3. 胚乳　4. 子叶
5. 胚轴　6. 胚根
（黄应钦绘）

果实的采收调制和种子贮藏　成熟的果实易被鸟食，应及时采收。通常是在果实成熟盛期清理树下灌丛，用竹竿敲打或摇动树枝，使果实震落后在地面收集。落地过久，外果皮已呈干缩状的果实不宜选用。采得的果实可堆沤3～5日，果皮软熟后装入筐内置水中搓洗，淘去腐烂的外果皮和中果皮。所得果核用作播种材料，通称种子。鲜果出籽率20%～38%。净度95%～98%。含水量15%～20%。千粒重155g，变动在125～240g，每千克有核6 500粒，变动在4 200～8 000粒。种子忌失水，不宜日晒。3～5天的短期运输，果实采集后可暂不调制而直接包装运输，运抵后再行调制。未经调制的果实不能贮藏，以免霉烂影响发芽。已经调制出来的种子，稍晾干后即混湿沙贮藏。贮藏期一般为3～5个月。在华南地区，蓝果树的贮藏不宜超过翌年4月中旬。

发芽和播种　发芽时的气温要求较低，日均温15℃以上即可正常发芽。1979年1月上旬，

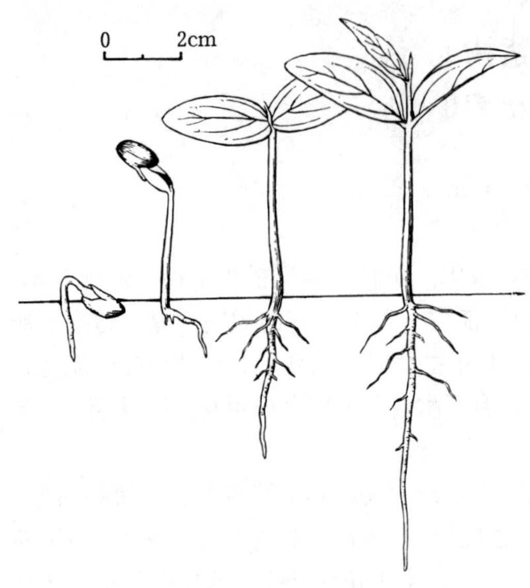

图 2 蓝果树种子萌发后第 4、6、16、25 天的幼苗生长情况

（黄应钦仿《热带亚热带主要树种采种育苗技术》）

广西林业科学研究所对头年 10 月采集，调制后经过沙藏的种子，在室外沙床作过发芽测定。播种后 35 天即 2 月上旬末，日均温约 16℃时开始发芽，2 月 17～26 日为发芽盛期，3 月中旬发芽终止；从开始发芽至发芽高峰日的 15 天，发芽百分数为 40%；从播种之日起算的 70 天，发芽率为 80%。据邵蓓蓓（1989）报道，未经层积的蓝果树种子在 30℃/20℃的温度条件下培养 60 天，发芽率只有 2%；如果先在室温下层积 120 天，再在同样的温度下培养，14 天发芽率便可达到 83%。

蓝果树出土萌发。胚根萌发后约 5 天子叶出土，再过 10～15 天初生叶展现。种子萌发和幼苗生长情况见图 2。

经过沙藏的种子播前可不作处理。条播，每平方米播种 12～20g，覆土约 2cm。1 年生苗出圃。

（王宏志）

珙 桐 属

Davidia Baill.

（珙桐科 Davidiaceae）

生长习性、分布和用途 珙桐科只有 1 属，我国特产。本文描述该属仅有的 1 种及 1 变种，它们是第三纪古热带植物区系的孑遗种。目前数量较少，分布范围也日益缩小，有被其它阔叶树更替的危险。在《中国植物红皮书》中，它们被列为稀有植物，需要加强保护。落叶乔木，胸径 30～50cm，最大逾 1m，树高 15～20m，可达 28m。喜降水较多、相对湿度大、云雾多的温凉气候。适生于微酸性的山地黄壤或黄棕壤。稍耐水湿。浅根性。

木材有光泽，纹理直或斜，结构均匀轻软，干燥后不翘裂，可供雕刻及美术品等用。花序下面两片白色大苞片形如飞鸽，国外通称中国鸽子树，欧美许多国家曾相继引种，为著名观赏树种。珙桐属树种的名称、分布和用途见表 1。

表 1 珙桐属树种的名称、分布和用途

中 名	学 名	分 布	用 途
珙桐	*D. involucrata* Baill.	陕南、鄂西、湘西北、黔、川、滇东北	材用、观赏
光叶珙桐	*D. involucrata* var. *vilmoriniana* (Dode) Wanger.	鄂西、湘西北、黔、川、滇西北，与珙桐同产地者常与之混生	材用、观赏

开花结实　萌生植株约 9 年生、实生植株约 12 年生才有结实能力。正常结实在 15～20 年生以后。有大小年现象，结实间隔期 1～2 年。

花杂性，两性花与雄花同株。多数雄花与一个雌花或两性花组成近于球形的头状花序，着生于嫩枝顶端，基部具 2～3 枚初为淡绿后转乳白色叶状的大苞片，苞片矩圆形或卵形，长7～15cm，宽 3～5cm，花后脱落。雄花无花萼和花瓣，有雄蕊 1～7。雌花或两性花子房下位，6～10 室，每室有胚珠 1 枚，常仅 3～5 室发育。

开花结实的物候期因生长地域而不同。花期多在 4 月中旬～5 月。同一地区的花期还因海拔高度而不同，例如峨眉山报国寺 5 月 1 日开花，万年寺 5 月 10 日开花，九老洞则在 5 月 15 日开花（四川省林业科学研究所谢瀫提供）。果实成熟期在 10～11 月中旬。成熟的果实悬吊在枝上至翌春 2 月末还有未落的（表 2）。

表 2　珙桐的开花结实物候

观察地点	观察年份	开花			果实成熟		果实脱落	
		始期	盛期	末期	始期	盛期	始期	末期
四川宝兴	1987	4 月下旬	5 月初	5 月上旬	10 月上旬	11 月中旬	—	—
四川峨边	1987	5 月初	5 月中旬	5 月下旬	11 月初	11 月上旬	—	—
四川荥经	1988～1989	4 月上旬	4 月中旬	4 月下旬	10 月上旬	11 月中旬	翌年 2 月下旬	—

核果，未熟时青绿色，成熟时褐色或紫绿色，密被锈褐色皮孔，长卵圆形或椭圆形，单生。据陶金川和宗世贤等人（1986）对采自四川雷波 118 个果实的测定，果长 2.40～4.16cm，平均 3.45cm，宽 1.40～2.78cm，平均 1.73cm；每果重 2.19～9.55g，平均 4.85g。外果皮薄；中果皮肉质；内果皮骨质，具 7～8 条沟槽。3～5 室。果核长 2.4～4cm，径 1.2～2.8cm。每个果核具能育种子 1～5 粒，多为 1～2 粒。种子条状，具 2 棱，有一面扁平，两端锐尖。种皮淡黄色。胚乳薄，胚乳和胚均为白色。子叶长椭圆形。珙桐果核的形态及其纵切面见图 1。

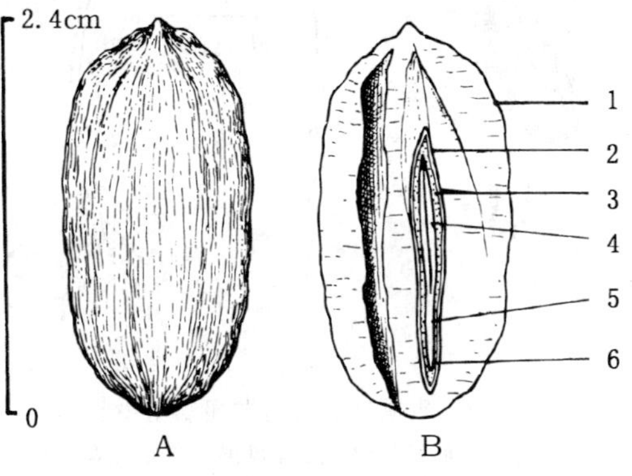

图 1　珙桐果核的外形（A）及其纵切面（B）
1. 内果皮　2. 种皮　3. 胚乳　4. 子叶
5. 胚轴　6. 胚根
（叶可佳、黄应钦绘）

果实的采收调制和种子贮藏　外果皮呈黄褐色或紫褐色，皮孔呈黄褐色时即可设法击落而从地面拾捡。采集的果实堆沤 5～7 天，待中果皮变软时捣破果肉，用水漂去果皮与果肉等杂质后所得的纯净果核即为播种材料，通称种子。内果皮极其坚硬，难以击破果核从中取出真种子。据本文作者 1987～1988 年两年测定，鲜果出籽率为 27%～29%，果核千粒重一般为 4 500g，变动范围为 3 400～5 500g。每千克有纯净果核 220 粒，变动范围为 180～290 粒。

调制出的果核应混沙湿藏，沙的含水量应始终保持在5%左右。贮藏期间应常检查，发现果核发霉时应取出清洗后再贮藏，翌春播种。

发芽前的处理　珙桐种子具休眠习性。即使采后随即秋播，也只有在第3年才能大批发芽出土。秋采层积后春播的，大部分种子仍然要到翌年才能发芽出土。本文作者曾经用过几种方法试图打破休眠，例如，切去果尖后分别在室内室外层积；在0℃下冷冻1～3月后播种或再层积；用浓硫酸浸蚀果核等，都未收到效果。无论是否夹破核壳，浸水8天的珙桐都能吸收同样多的水分。陈坤荣和方文德等人（1990）因此认为珙桐种子的休眠并非由于内果皮透性不良。他们发现，新采收的珙桐，胚的发育尚不完全，胚芽尚未分化，只有在层积过程中胚根才缓慢伸长；只有当胚完成其形态发育时，种子才能进入萌发状态。陶金川和宗世贤等（1986）则认为，有50%的珙桐果实中的种子无胚，因而不能发芽。除了形态后熟以外，陈坤荣和方文德等（1990）还从干种子和内果皮的提取液中得到酸性抑制物质，主要是酚类物质和ABA。用层积1年的种子播种，在适宜条件下，可能有部分种子萌发，大部分则不萌发。可见种子个体间休眠程度也有差异。层积1年后尚未打破休眠的种子，用50μg/g6-苄基嘌呤处理有促进胚根萌发，解除休眠的效果（李卓然等，1989）。刘家德和路永祺（1981）、杨业勤（1982）也作过珙桐种子的育苗试验。

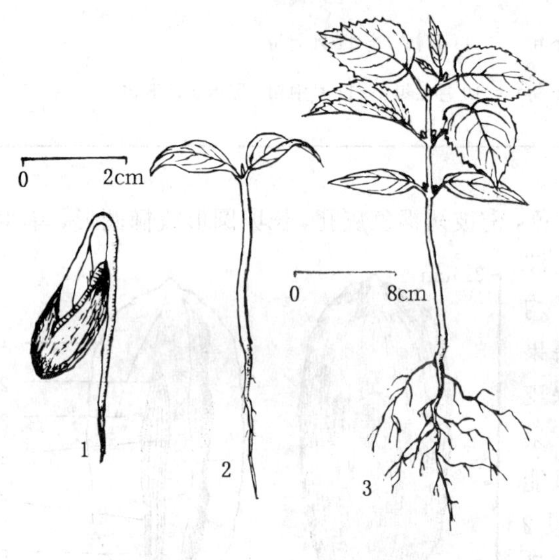

图2　珙桐种子的萌发和幼苗生长情况

1. 下胚轴延伸　2. 子叶展开　3. 幼苗形成

（叶可佳、黄应钦、田恒德绘）

播种　秋采即播或翌年春播，或层积1年后取已经裂缝、胚根伸出或即将伸出的果核春播。这样做可以得到较高的出苗率。春播以1月为宜。点播，株行距10～15cm，覆土3～4cm，轻压后覆以塑料薄膜。播种床切忌干燥，也忌积水。秋采即播（冬播）的可以播入露天湿润沙床，并经常洒水保持沙床湿润，经130天开始发芽。露天湿沙层积后春播的，少数种粒播后55天开始发芽（刘家德等，1981）。珙桐发芽出土一般在3月上旬至4月上旬。

出土萌发。子叶2片，长卵形，长约4.6cm，宽1.5cm，全缘，有羽状脉。上胚轴长约1.8cm，绿色。第1～2片初生叶对生，第3片起互生，阔卵形，先端锐尖，边缘具三角形而锐尖的粗锯齿（图2），叶表和叶背疏生白色柔毛。一个果核常能长出几株幼苗。当幼苗的第一对初生叶普遍形成，有的已开始长出第3～4片初生叶时，可以选择阴雨天对小苗分株移植，并搭棚遮荫。1年生苗高30～40cm，2年生苗高可达1～1.7m。一般用2年生苗出圃。珙桐还可采用嫩枝扦插繁殖。

（阚再旦）

五　加　属

Acanthopanax Miq.

（五加科　Araliaceae）

生长习性、分布和用途　五加属约 35 种，我国约 20 种，本文描述 3 种。灌木，直立或蔓生，稀为乔木。枝有刺，稀无刺。掌状复叶，有小叶 3～5，托叶不存在或不明显。五加高 2～5m，有时蔓生状。刺五加通常高 1～3m，有时可达 5m。白簕高 1～7m，常蔓生状。这 3 个树种的名称、分布和用途见表 1。

表 1　五加属树种的名称、分布和用途

中　名	学　名	分　布	用　途	供　稿
五加	*A. gracilistylus* W. W. Smith	甘南、晋南，西南至川中、滇西北、苏、浙、东南沿海	根皮药用	1 003
刺五加	*A. senticosus* (Rupr. et Maxim.) Harms	东北、华北。朝鲜半岛和西伯利亚也有	药用，观赏	107
白簕	*A. trifoliatus* (L.) Merr.	湘、鄂、赣、皖南、苏南、浙、闽、台、粤、桂、黔、滇、川	根叶药用	1 003

开花结实　人工栽植的刺五加 4～5 年生开始开花结实。全光下植株结实量显著较多。天然的刺五加绝大多数靠窜根萌生新植株，实生苗少见。花两性，稀单性异株。伞形花序或头状花序常组成复伞形花序或圆锥状花序。花瓣 5，稀 4。雄蕊 5，花丝细长。子房 5～2 室。花柱 5～2，宿存。浆果状核果，球形或扁球形，有 5～2 棱，具薄肉质的外皮及硬壳质的内皮。每果有核 2～5，扁平。这 3 个树种的开花结实物候见表 2。

表 2　五加属树种的开花结实物候

树　种	观察年份和地点	花期			果实成熟		成熟特征	果实散落期
		始　期	盛　期	末　期	始　期	末　期		
五加	—	4 月	～	8 月	6 月	11 月	果变软变黑	成熟期不一致
刺五加	1972～1978 伊春丰林	7 月上旬	7 月中旬	7 月下旬	8 月中旬	9 月下旬	果由绿变黑	11 月中旬落尽
	1960～1980 哈尔滨	—	—	—	8 月中旬	10 月上旬	果实变为黑色，种子淡黄色	11 月中旬落尽
白簕	—	8 月	～	11 月	9 月	12 月	—	成熟期不一致

刺五加果球形或卵球形，熟时紫黑色，直径 5～8mm，长 6～10mm，有 5 棱，内有 5 个分核。分核半圆形，薄而扁，长 5～9mm，宽 2～4mm，厚 1mm 左右；胚乳丰富；胚小，肉眼不可见（图 1）。五加与白簕果近球形，侧扁，内有果核 2，核内含种子 1 枚。五加果径约 5～6mm，果核扁，果瓣状。白簕果径约 5mm，胚小，长约 0.6mm。

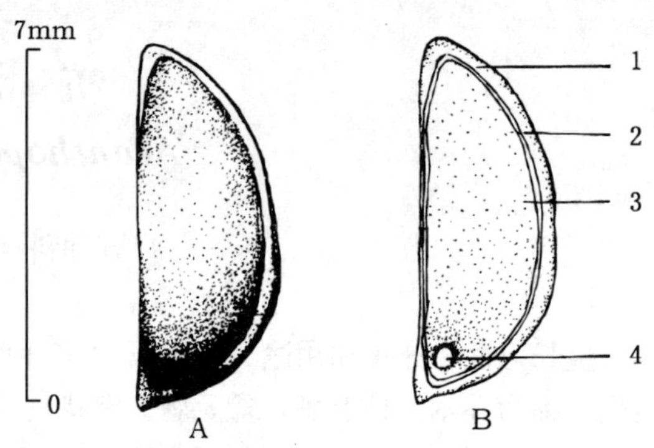

图 1 刺五加果核外形（A）及其纵切面（B）
1. 内果皮 2. 种皮 3. 胚乳 4. 胚
（梁鸣绘）

果实的采收调制和种子贮藏 果实变为黑色或紫黑色即为成熟。刺五加 9 月下旬大量脱落，落地后很快被虫蛀或被动物所食，应及时采收。五加与白簕种子成熟期不一致，应随熟随采。采后堆沤，腐熟变软后捣烂，用水搓洗，晾干去杂，所得纯净果核即为播种材料，通称种子。出籽率、净度及种子质量见表 3。

表 3 五加属树种的出籽率及种子质量

树 种	出籽率（%）	净度（%）	千粒重（g）	每千克纯净种子粒数（万粒）
五 加	—	—	8.0～9.0	11～12.5
刺五加	25～35	55～65	6.0～6.6	15～16.7
白簕	—	—	10	10

刺五加种子混沙贮藏可以保存 1.5 年，也可以低温密封贮藏。

发芽和播种 这 3 个种的种子均有休眠习性，属于胚生理后熟型，采后应立即混沙层积：水浸 3～4 天，混沙后在 10～20℃下层积 4 个月，再转入 5～0℃条件下继续层积 2 个月便可陆续发芽。刺五加在哈尔滨地区 4 月下旬播种。条播，每平方米播 100g 左右，播深 1～1.5cm。出土萌发。播后 15～20 天子叶出土，又 17～20 天放出第 1 片初生叶。刺五加的萌发和幼苗初期生长情况见图 2。幼苗出土后需遮荫，遮荫量 30%～40%，8 月中旬撤除。1 年生苗高 5～20cm。2 年生苗高 50cm 左右即可出圃。

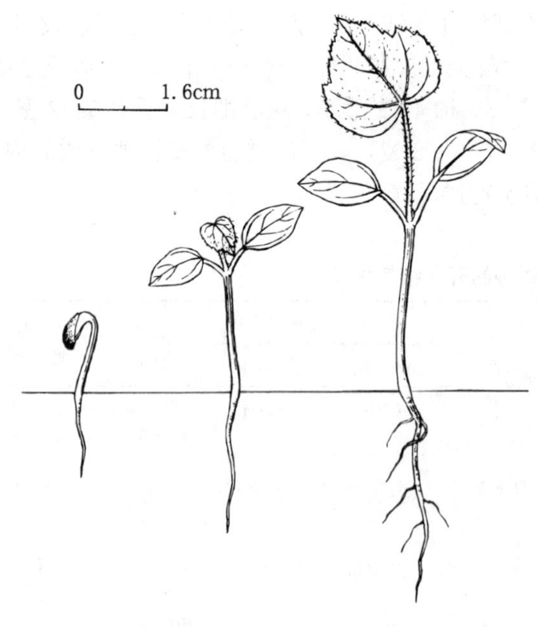

图 2 刺五加发芽后第 13、23、34 天的幼苗生长情况
（梁鸣、黄应钦绘）

（周德本）

楤　木　属
Aralia L.

（五加科　Araliaceae）

生长习性、分布和用途　本属约 40 种，我国约 28 种，本文描述 2 种。落叶小乔木、灌木或多年生草本，通常有刺，稀无刺。叶大，一至数回羽状复叶。楤木为灌木或小乔木，高 2～5（8）m，胸径达 10～15cm。辽东楤木为小乔木或灌木状，高 1.5～4（6）m，胸径可达 6～9cm。这 2 个种的名称、分布及用途见表 1。

表 1　楤木属树种的名称、分布和用途

中　名	学　名	分　布	用　途	供　稿
楤 木	*A. chinensis* L.	华北、华东、中南、西南	种子榨油，根皮药用	1003
辽东楤木（龙牙楤木）	*A. elata*（Miq.）Seem.	东北、冀、鲁。朝鲜半岛、俄罗斯远东、日本	嫩叶可食，著名山野菜；种子榨油，根皮药用，观赏	107

开花结实　辽东楤木 5～7 年开花结实。1988 年在黑龙江伊春见过被砍后两年的辽东楤木萌条开花。未见明显的大小年现象。花杂性。伞形花序再组成圆锥花序，顶生，长 25～60cm 以上。花萼边缘有 5 齿。花瓣 5，绿白色或黄白色。雄蕊 5，花丝细长。子房 5 室，花柱 5。1988 年在黑龙江观察，辽东楤木的初花期在虎林县为 7 月 25 日，尚志县为 7 月 30 日，伊春市带岭为 8 月 7 日。这一年这 3 个地点的盛果期分别为 9 月 14 日、9 月 16 日和 8 月 21 日。两个树种的开花结实物候期见表 2。

表 2　楤木属树种开花结实物候

树　种	观察时间 地　点	花期			果期			散落期
		始　期	盛　期	末　期	始　期	盛　期	末期	
楤木	—	6 月	～	7 月	8 月	～	10 月	—
辽东楤木	1988 年黑龙江	7 月下旬～8 月上旬	8 月上旬	8 月下旬	9 月上旬	9 月中旬	9 月下旬	至 11 月落尽

浆果状核果，成熟时黑色。楤木果扁球形，径 3～4mm，果肉浆状；果核 5 枚，核壳（内果皮）骨质，侧扁，长约 2.2mm，宽 1mm，每核内含种子 1 粒；种皮薄膜质，紧贴胚乳；胚小于 0.2mm。辽东楤木果扁球形，具 5 棱，径 3～5mm；果核侧扁，长 3mm 左右，宽 1.5mm 左右。辽东楤木果核形态见图 1。

果实的采收调制和种子贮藏　果实变为黑色即为成熟。辽东楤木至 11 月初落尽，应注意

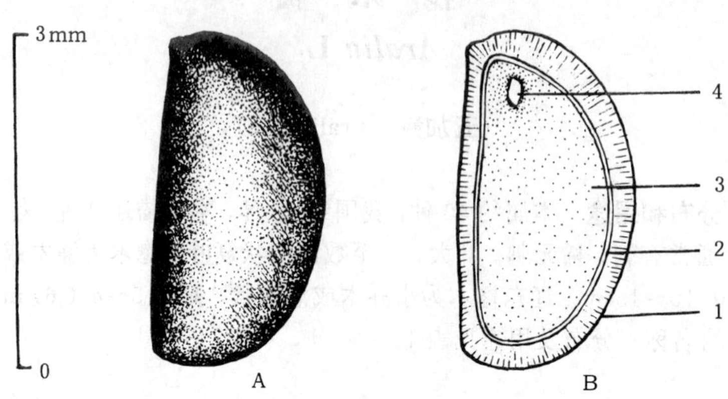

图 1 辽东楤木果核外形（A）及其纵切面（B）

1. 内果皮 2. 种皮 3. 胚乳 4. 胚

（梁鸣绘）

采收期。采后堆放，使果实变软，搓洗后用水漂出果皮与果肉，阴干后去杂。所得纯净果核即为播种材料，通称种子。出籽率及种子质量见表 3。

表 3 楤木属树种出籽率及种子质量

树 种	出籽率（%）	净 度（%）	千粒重（g）	每千克纯净种子数（万粒）
楤木	—	—	5.3	19
辽东楤木	12～16	80～90	0.8～1.5	66.7～125

辽东楤木种子含水量 8%～10%，密封瓶中常温下存放 1 年，发芽率无明显变化。

发芽和播种 两种树种均属于胚生理后熟型。南京地区秋季采得的楤木种子，胚长小于 0.2mm，尚未分化成熟；2 月中旬将种子放入湿沙袋，5 月中旬剥开观察，胚长虽略有生长，长约 0.75cm，大约相当于种子长度的 1/3，子叶端已见裂开，但整个胚仍呈圆柱形，直至 12 月底仍未见萌发。一般认为这两个树种采后应立即混沙层积处理。方法为将种子浸水2～3天，混沙后置于 10～20℃温度下，3 个月后转入 0～5℃温度下 2 个月。经过这样处理，辽东楤木的种子可以发芽，干籽则发芽迟缓。比较稳妥的办法是采用隔年埋藏法。

辽东楤木在哈尔滨市 4 月下旬至 5 月初播种。苗床条播时每平方米播种量 5g，播深 0.8cm 左右。幼苗期需适当庇荫，播后 20 天内保持床面湿润，否则易受日灼危害。当年苗高 30～50cm。哈尔滨人工栽植的辽东楤木，由于脱离了森林环境，地上部分几乎年年冻死，次年又萌生出 1m 多高的枝条。子叶出土萌发（图 2）。

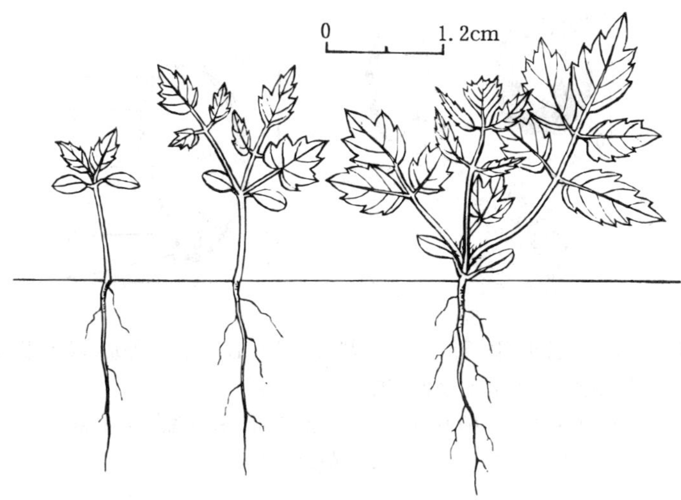

图 2　辽东楤木种子发芽后第 15、31、49 天幼苗的生长情况
（许芝源绘）

（周德本）

八角金盘

Fatsia japonica（Thunb.）Decne. et Planch.

（五加科　Araliaceae）

生长习性、分布和用途　八角金盘属共 2 种，我国 1 种，为台湾特产。本文描述的 1 种原产日本，我国各地已广为引种栽培。常绿丛生灌木，高 3～5m，较耐荫，在湿润背风处生长良好。长江流域以南可在庭园露地栽培，华北地区多在温室盆栽，供观赏。

开花结实　花两性或杂性。顶生圆锥花序，花序长 30～40cm。花黄白色，柄长 1～1.5cm，花萼具 5～6 齿。花瓣 5，镊合状排列。雄蕊 5。花柱 5，分离。子房下位，5 室，每室 1 胚珠。花期 10～11 月，翌年 4 月果实成熟。果实近球形，直径约 7～8mm，熟后呈黑色，肉质。每果含饱满种子 2～5 枚。种子扁卵形，长 6mm，宽 3mm，厚 1.5mm。外种皮深灰绿色或近黑色，角质膜状。内种皮薄，透明，膜质。胚乳丰富，略硬。采收时胚已分化完全，但长度仅为种子长度的 1/3。果实及种子剖面结构见图 1。

果实的采收调制和种子贮藏　本种为秋花春实，冬季幼果如不受冻害，均可成熟。一般 4 月下旬有一半以上果实转黑时即可采摘。5 月上旬果实脱落。采收后，趁果肉未干搓洗取种。出种率约 6.5%。种子千粒重约 18g，每千克含 5.5 万粒种子。多随采随播。如需贮藏，可将果实晾干后带果肉贮藏。

发芽和播种　种子无休眠习性。采收调制后的种子，在常温湿润条件下 10～15 天即可大量发芽，发芽迅速整齐。据在南京室内测定，11 天内发芽率可达 99%。4 月下旬播种，5 月中旬出苗。出土萌发。幼苗早期生长缓慢，出苗约 1 个月后才出现初生叶。初生叶不呈掌状分裂（图 2）。母树下常可挖掘天然萌生的幼苗，移栽成活良好。

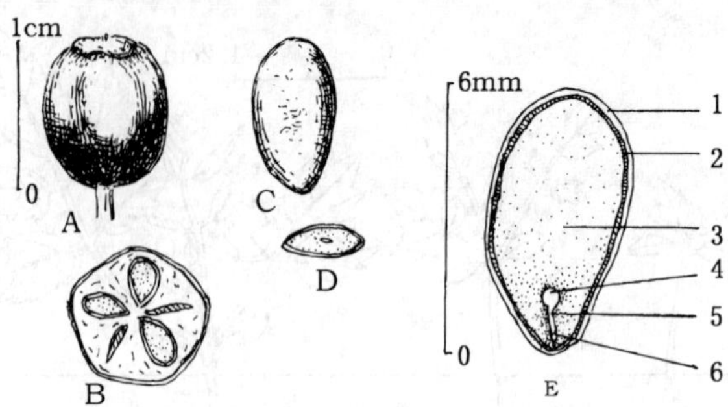

图 1 八角金盘果实外形（A）及其横切面（B），种子外形（C）及其横切面（D）和纵切面（E）

1. 外种皮 2. 内种皮 3. 胚乳 4. 子叶 5. 胚轴 6. 胚根

（史渭清绘）

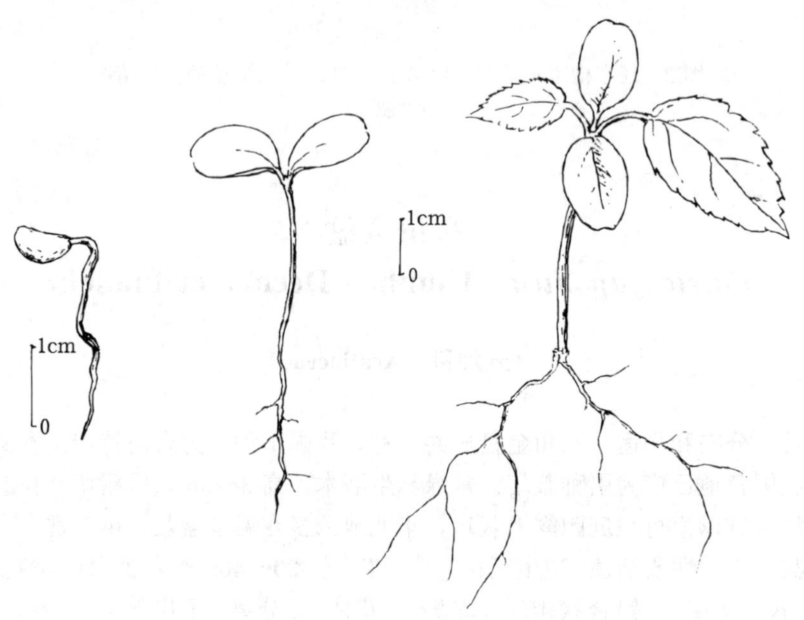

图 2 八角金盘种子萌发后第 6、9、52 天的幼苗生长情况

（史渭清绘）

（何泽瑛）

幌伞枫

Heteropanax fragrans（Roxb.）Seem.

（五加科 Araliaceae）

生长习性、分布和用途 幌伞枫属约 5 种，我国皆产，本文描述其中 1 种。常绿大乔木，高达 20～30m，胸径 70cm。分布于滇、桂、琼、粤。印度、孟加拉国、缅甸、印度尼西亚亦产。根皮药用。羽叶巨大，奇特，为优美的观赏树种。木材可作家具等用。

开花结实　12～18 年生开始开花结实，30 年生后进入结实盛期，大小年间隔期为 1～2 年。花杂性，腋生花序常为雄花，顶生花序为两性花。伞形花序组成总状或圆锥状，长(40) 60～80cm，密被锈色星状毛，后渐脱落。苞片及小苞片宿存。花梗无关节。萼筒 5 小齿。花瓣 5。雄蕊 5。子房下位，2 室，每室 1 胚珠。花柱 2，分离。据在广西南宁 1987～1988 年观察，现蕾期在 8 月上旬，始花期在 9 月下旬，盛花期在 11 月上旬，末花期在 11 月下旬。幼果期时果序梗逐渐增粗，星状毛脱落，长 80～110cm。果实需越冬成熟。如遇气温在 3℃上下，或 8～12℃的阴雨天持续 1 周以上，落果严重。果熟期始于翌年 3 月上旬，盛期在 3 月下旬，成熟较早的果实在 4 月中旬基本上全部脱落。除个别果穗外，5 月下旬果实全部脱落。果实成熟时由青紫色转为紫黑色，每果穗可收鲜果 4～6kg。核果状浆果，圆形略扁，高 7～11mm，径7～12mm，每果有种子 2。种子长卵形，一面稍平，表面有褐红色花纹，长 6～9mm，径5～6mm。有胚乳，嚼烂状。幌伞枫种子的外形和解剖构造见图 1。

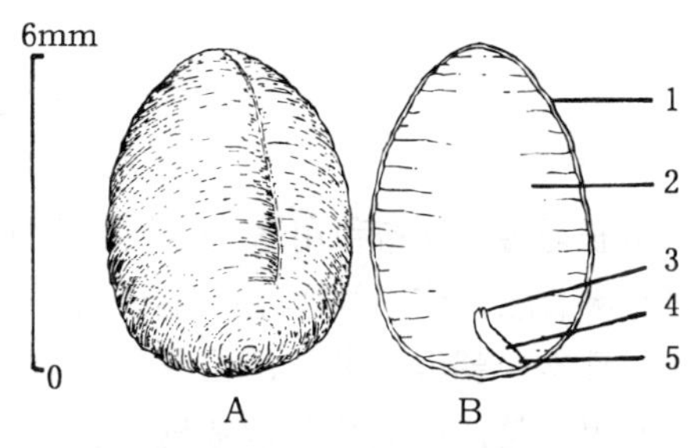

图 1　幌伞枫种子外形（A）及其纵切面（B）
1. 种皮　2. 胚乳　3. 子叶　4. 胚轴　5. 胚根
（黄应钦绘）

果实的采收调制和种子贮藏　果实盛熟时选择成熟较整齐的果穗，用采种刀截断总果梗。采得的果实在室内堆沤 10～15 天后置水中搓擦，淘净，得到种子。鲜果的出种率约 60%，种子净度约 97%。千粒重 95g，变动在 75～110g，每千克有纯净种子 1 万粒，变动在 0.9 万～1.3 万粒。种子忌失水，不宜日晒。短途运输以直接运输果实为宜。也可用布袋盛装调制好的种子，每袋最好不超过 2kg。长途运输和贮藏则需混以湿沙，贮藏期一般在 3 个月以内。

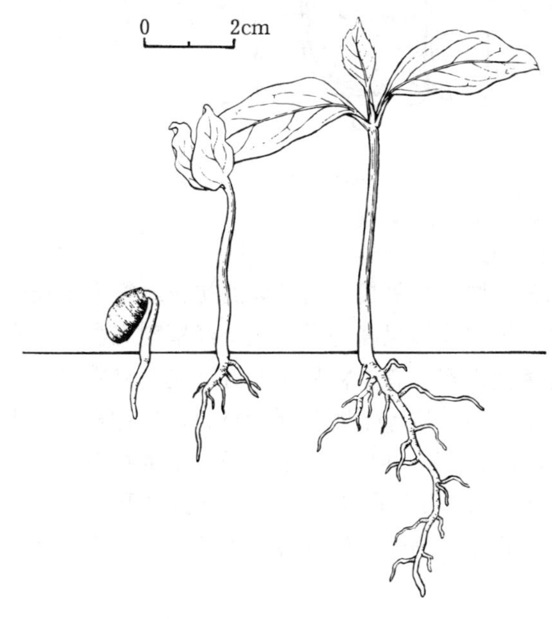

图 2　幌伞枫种子萌发后第 10、15、23 天幼苗的生长情况
（黄应钦绘）

发芽和播种　种子无休眠习性，可随采随播。广西林业科学研究所 1987 年 5 月 8 日曾在室内作过发芽测定，当时气温在 27℃左右，基质为细沙。置床 10 天胚根萌发，20 天子叶带壳出土，22 天以后初生叶展现。发芽高峰期为 5 月 26 日。从发芽之日起算的 8 天中，发芽百分数为 46%；从播种之日起算，以 24 天计，发芽率为 60%。出土萌发。幼苗生长情况见图 2。

条播。每平方米播 12～20g，覆土约 1cm。1 年生苗木可以出圃，2～3 年生大苗供庭院绿化植树。

（曾　玲）

刺 楸

Kalopanax pictus（Thunb.）Nakai

（五加科 Araliaceae）

生长习性、分布和用途 本属仅刺楸1种，落叶乔木，高10～30m，胸径50～100cm。在湿润肥沃的酸性或中性土壤上生长迅速。我国北起辽东，南至长江流域及粤、桂、川、滇均有分布。日本、朝鲜半岛和西伯利亚也有。优质用材树种，用途广泛。皮、枝可以入药，种子油工业用，是产区的重要造林树种。

开花结实 花两性，白色或淡黄色。花序大，由直径约1.5cm的伞形花序再组成圆锥花序或复伞形花序，顶生，直径20～30cm。花序梗细，长2～6cm。萼5齿裂。花瓣5。雄蕊5，花丝细长。子房下位，2室。花柱合生为柱状，柱头2叉。浆果状核果，呈两侧压扁的球形，径4～5mm，成熟后紫黑色，内含果核1对。果核长3～4mm，具3棱。内果皮纤维木质，极坚韧，棕灰色，表面凹凸不平。种皮膜质。胚乳丰富，油质。果熟采收时，发育早期的胚位于种子内侧下端，长0.1～0.9mm或不足0.1mm（图1）。花期7～8月，果熟期（9）10～11月。

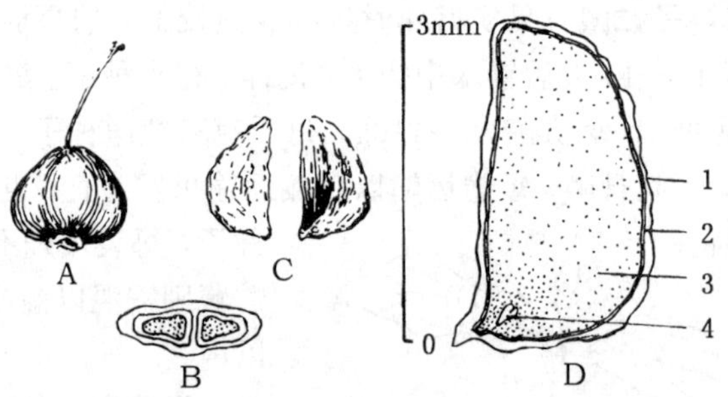

图1 刺楸浆果状核果的外形（A）及其横切面（B）和二果核（C）及果核纵切面（D）

1. 内果皮 2. 种皮 3. 胚乳 4. 胚

（史渭清绘）

果实的采收调制和种子贮藏 果实大部分变黑即可采摘，也可等到初霜后整个果枝带果落地后在树下拾集。因开花结实先后不一致，果实成熟程度差异较大。经解剖观察，只有采下前完全呈黑色的果实含有饱满的种子，且多数果实只有1粒种子发育。选摘老熟果实搓揉漂洗，去除果皮杂物及上浮的空瘪果核，将下沉的果核晾干即为播种材料，通称种子。也可将果实直接风干用作播种材料。新鲜种子含水量20%～29%，气干种子含水量在10%左右。纯净种子（果核）的千粒重6.3～7.0g，每千克约14万～16万粒。风干果实的千粒重约12g，每千克约8万粒。湿沙层积可以使种子内部发育不致停顿，供翌春播种使用。有报道认为低温密封贮藏可以长期保存发芽能力。

发芽和播种 刺楸在我国南北方均有分布。据余清珠和吴越（1986）报道，采自陕西勉县的种子有休眠习性，但休眠程度不深。湿沙层积的时间如果太长，种子在12℃的温度下也能较快地发芽。他们因此建议最好是在春播前1个月开始层积。东北地区所产的刺楸种子则有深休眠现象，春播前不加层积，便要等到冬季接受低温解除休眠，到再一个春季出苗。当然，也可能有部分种粒根本没有受精因而不能发芽。无论是当年新采收的种子，还是保存 3 年的刺楸种子，浸种 48 小时都能达到饱和含水量。许绍惠和韩忠环等人（1991）因此认为东北地区刺楸种子难发芽的原因不是吸水困难，而是种子含有发芽抑制物质并有形态后熟现象，并且认为消除这种生理休眠和形态休眠的途径是层积。他们在未经层积的气干种子中检出了 ABA。如果以此时的 ABA 含量为 100，则层积 64 天后的含量降至 32.5%，层积 4 个月后就未检出 ABA。据他们观察，新采收的种子大部分为心形胚，少数还处于鱼雷胚阶段。层积 30 天的胚依然很小，但子叶已开始生长，胚长 0.2～0.3mm。层积 60 天的胚长已达 0.3～0.6mm。层积 90 天，胚已达种子长度的 1/2。到 110 天时胚的长度相当于种子长度的 3/4，已经完成形态后熟。他们认为东北地区刺楸种子解除休眠的最佳层积条件是，在 15℃下层积 80 天后转入 5℃下再层积 40 天（许绍惠和韩忠环等，1991）。通过实验观察，黄玉国（1986）得到了与此大体相同的结论。如果把 15～20℃的层积称作暖层积，把 0～5℃的层积称为冷层积，黄玉国（1986）证明了无论是 6 个月的冷层积，还是 6 个月的暖层积，还是先冷层积 3 个月再暖层积 3 个月，解除休眠的效果都不如先暖层积 3 个月之后再冷层积 3 个月。黄玉国（1987）还发现，酸性磷酸脂酶同刺楸的后熟过程有比较紧密的联系。

据余清珠和吴越（1986）报道，经过层积的刺楸种子室内发芽测定时，光照的有无影响不大，在 18～24℃的温度下发芽率可达 86%。

播种育苗时，建议在秋末选采充分成熟的果实，采后随即播种，或在春季播种经过层积处理的种子。这样的种子春播后很快便能发芽，发芽率可以高达 95%（周德本，1986）。

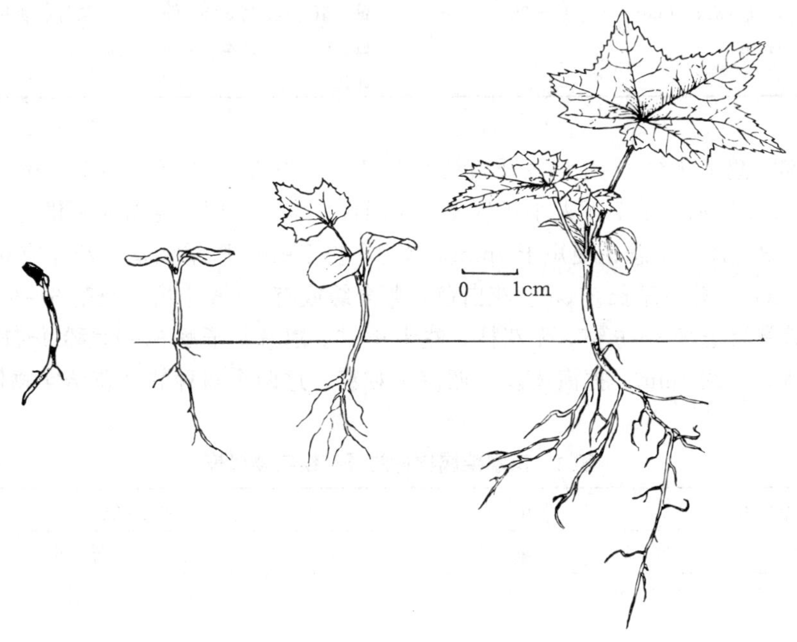

图 2　刺楸的种子萌发后第 2、4、37、60 天幼苗的生长情况

（田恒德、史渭清仿《主要树木种苗图谱》）

出土萌发。上胚轴不发达。初生叶互生但因幼茎抽生迟缓而呈簇生状。下胚轴圆柱形，上部紫色，下部白色。具主根，侧根多而细长，白色（图 2）。1 年生苗可高达 30cm，常用 2 年生苗出圃。土壤条件较好的母树下常有天然下种幼苗，也可采集培育后出栽。根部具萌芽力，生产上也常用分根法繁殖。

（何泽瑛）

鹅掌柴属

Schefflera J. R. et G. Forst.

（五加科 Araliaceae）

生长习性、分布和用途 本属约 400 种，我国约产 38 种，本文描述 2 种。常绿乔木或灌木（有时为藤本）。喜湿热气候，不耐严寒，适生于湿润的山冲、溪旁的酸性土。它们的名称、生长、分布及用途见表 1。

表 1 鹅掌柴属树种的名称、生长、分布和用途

中 名	学 名	树 高 (m)	胸 径 (cm)	分 布	用 途
鹅掌藤（七叶莲）	*S. arboricola* Hayata	3～4（藤本状）	—	琼、台、桂	药用、止痛、活血、强筋骨
鹅掌柴（鸭脚木）	*S. octophylla*（Lour.）Harms	15～20	30～40	藏、滇、桂、粤、湘、赣、闽、台、浙。日本、越南、印度	火柴杆、食用菌、冬季蜜源、治感冒

开花结实 鹅掌藤约 4～5 年生开始结实，鹅掌柴 6～8 年生开始结实，正常结实期分别在 8 年和 12 年生以后。结实无大小年现象。花两性。伞形花序，复组成圆锥花序状。鹅掌藤伞形花序有花 3～10 朵，总花梗短于 4mm，总花序长 20cm。鹅掌柴伞形花序有花 10～15 朵，总花梗长 1～2cm，总花序长 30cm。花白色，萼全缘或有 5～6 小齿。花瓣 5～6。雄蕊 5～6。子房下位。鹅掌藤子房 5～6 室，无花柱，柱头 5～6。鹅掌柴子房 5～7，稀 9～10，花柱合生成粗短的柱状，长约 1mm。据海南及广西南宁观察，这两个树种的开花结实物候期见表 2。

表 2 鹅掌柴属树种的开花结实物候期

树 种	观察年份和地点	开 花			果实成熟		果实脱落
		始 期	盛 期	末 期	始 期	盛 期	
鹅掌藤	海南	8 月下旬	9 月	10 月上旬	11 月中旬	12 月上旬	12 月上旬～下旬
鹅掌柴	1978～1984 南宁	11 月上旬	11 月中下旬	12 月中旬	翌年 3 月下旬	4 月上旬	4 月上旬～下旬

浆果，球形或卵形。鹅掌藤每果有种子 6 粒。鹅掌柴每果有种子 6 粒，稀 4～8 粒。果实成熟后约 1 周自行脱落。种子近半圆形，胚乳丰富。果实及种子形态见表 3、图 1。

表 3　鹅掌柴属树种的果实和种子形态特征

树　种	果　实			种　子		
	形　状	大　小（mm）	颜　色	形　状	大　小（mm）	颜　色
鹅掌藤	卵球形，柱头略膨大	径 4～6	紫黑色	镰形或半月形	长 3～4 宽 1.5～2 厚约 1	白色
鹅掌柴	球形，花柱突出	径 5～7	紫黑色	梳形或半月形	长 3～4 宽 1.5～2 厚 1～1.5	浅褐色

果实的采收调制和种子贮藏　果实成熟盛期用高枝剪或采种刀截断总果梗，摘取成熟果实。也可直接在树上用手采摘或用竹竿将果实敲落后在地面捡拾。采集的果实可在室内堆放 2～3 天。果皮软熟后装入布袋置水中搓擦，淘去果皮等杂质即得纯净种子。种子忌失水，不宜日晒或裸露存放。不影响种子发芽能力的含水量应在 20%左右。运输或贮藏时需混以湿沙，贮藏期一般在半年以内。出种率和种子的净度、质量等见表 4。

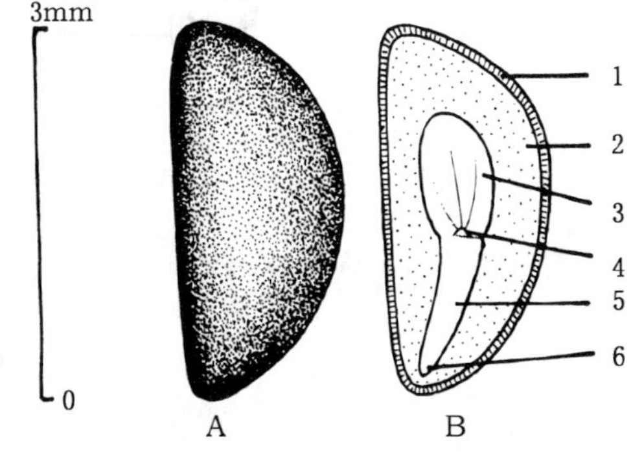

图 1　鹅掌藤种子外形（A）及其纵切面（B）
1. 种皮　2. 胚乳　3. 子叶　4. 胚芽　5. 胚轴　6. 胚根
（黄应钦绘）

表 4　鹅掌柴属树种的出种率和种子的净度、质量

树　种	出种率（%）	净　度（%）	千粒重（g）	每千克纯净种子粒数（万粒）
鹅掌藤	22～25	90～97	2～4	25～50
鹅掌柴	17～20	85～95	4～8	13～25

发芽和播种　种子无休眠习性。发芽时要求日均温在 20℃以上。1987 年 12 月和 1989 年 3 月下旬末，广西林业科学研究所分别在室内发芽箱和室外沙床上对当年新采且未作预处理的种子作过发芽测定，发芽情况见表 5。

表 5　鹅掌柴属树种的发芽能力及其测定条件

树　种	基　质	测定温度（℃）		发芽势（%）		发芽率（%）	
		发芽箱	室　外	计算天数	一般数值	计算天数	一般数值
鹅掌藤	滤纸	28	—	5	70	14	87
鹅掌柴	沙	—	18～24	6	48	15	64

出土萌发。胚根萌发后 6～8 天子叶出土，再过 20～30 天发出初生叶。鹅掌藤种子萌发和

幼苗生长情况见图 2。

撒播。经过沙藏催芽的种子再浸种 2～3 小时，捞出晾干后均匀撒播。每平方米播种 3～5g，初生叶发出后约半个月移至容器或大田培育。容器苗 60～90 天出圃。大田苗一般培育 1 年出圃。

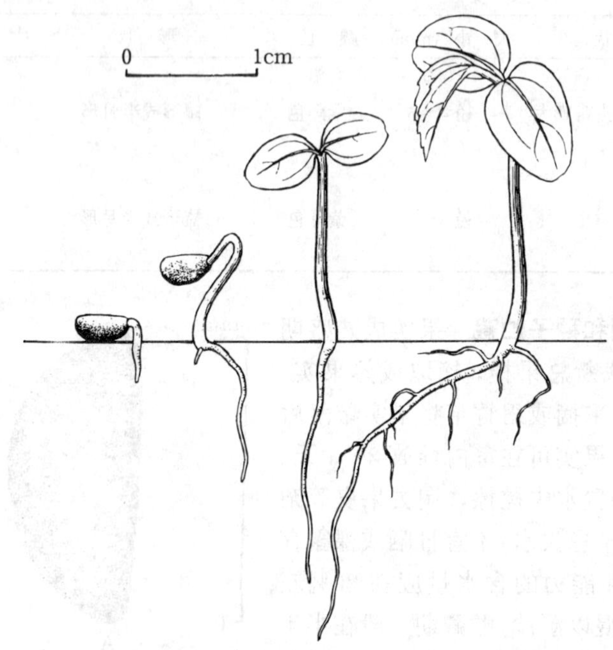

图 2 鹅掌藤种子萌发后第 6、10、15、30 天幼苗的生长情况

（黄应钦绘）

（王宏志）

通 脱 木

Tetrapanax papyrifer（Hook.）K. Koch

（五加科 Araliaceae）

生长习性、分布和用途 通脱木属仅有 1 种，我国特产。无刺灌木或小乔木，高 1～3.5m，地下有匍匐茎。喜光，生于土壤肥厚的向阳处。分布广，陕、川、黔、滇、桂、粤、湘、鄂、皖、赣、闽、台均有。长江流域以南常栽培供观赏。南京露地栽培冬季地上部分常枯萎，翌春 4～5 月间再度萌发。茎髓为著名中药及工艺品原料“通草”。

开花结实 花两性。多数排成伞形花序，径约 1～2cm，并再组成顶生、宽广的大型圆锥花序，分枝多，长达 50cm 以上。花浅黄白色，花瓣 4（5）。雄蕊 4（5）。子房下位，2 室。花柱 2，丝状，分离。浆果状核果近球形，径约 3.5～4mm，熟时紫黑色，内含半球形果核一对。果核长约 3mm，宽 2～2.5mm，厚 1.5～2mm。内果皮灰白色，软木质，较厚，约 0.2mm，表面有纵向不规则凹穴与沟纹。种皮膜质，棕褐色。种仁倒卵形，充满胚乳。采收时处于发育早期的椭圆形胚长仅 0.3～0.5mm，位于下端胚乳内（图 1）。花期 10～12 月。果实成熟于翌年 1～2 月。

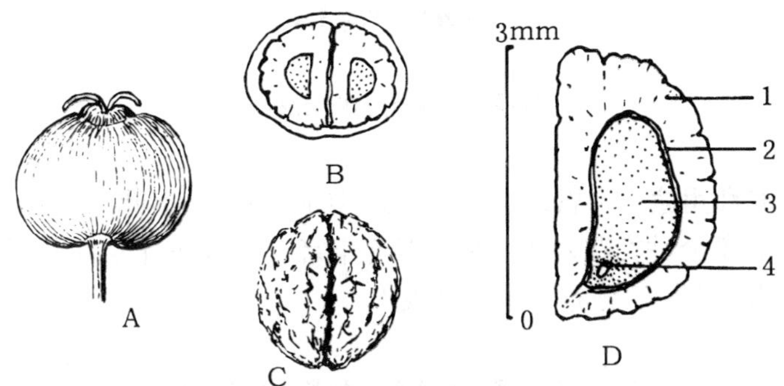

图 1　通脱木浆果状核果外形（A）及其横切面（B）和果核外形（C）及其纵切面（D）
1. 内果皮　2. 种皮　3. 胚乳　4. 胚
（史渭清绘）

果实的采收调制和种子贮藏　果穗极大，果实成熟先后不一。可在大部分果实成熟变黑时剪下果枝，选摘充分成熟的果实搓揉淘洗，去除果肉等杂质后，将灰白色果核晾干收藏作为播种材料，通称种子。不宜曝晒。种子质地轻，饱满种子和空瘪种子均上浮水面，不易分离，应认真选果以保证种子质量。种子萌发前有形态后熟过程，以湿沙层积为好，不过短期干藏也无妨。种子千粒重约 2.5g，每千克含种子约 40 万粒。种子在贮藏中不易腐烂，但空瘪粒和种胚败育的种粒容易受到霉菌侵染。

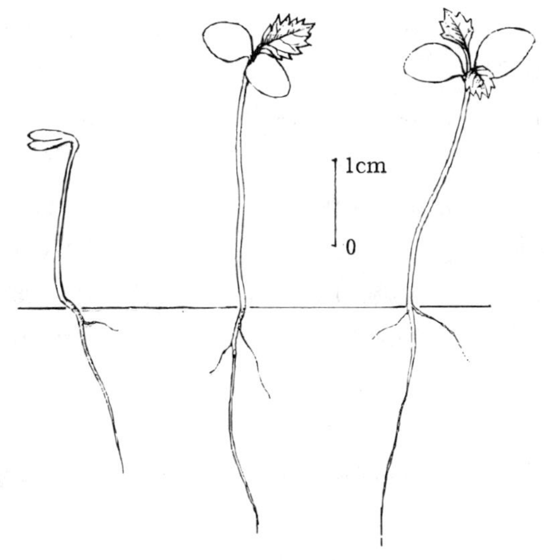

图 2　通脱木种子萌发后第 2、10、18 天的幼苗生长情况
（史渭清绘）

发芽和播种　种子有深休眠习性。果实成熟采收时胚还处在发育早期，分化不完全。本文作者曾将 1988 年 1 月采收取得的种子在 2 月 15 日用湿沙层积，置于室外自然温度条件下观察。结果是，当年 6 月 8 日～8 月 8 日有一部分种子萌发，发芽率为 33%；翌年 3 月 15 日～4 月 6 日又有 28%的种子萌发。两年合计发芽率为 62%。湿沙层积于室内的种子则未见萌发，原因还难以解释。解除休眠以后的发芽最适温度也尚难确定，可能以10～30℃范围之内的变温为好。

秋播或层积后翌春 3～4 月播种。出土萌发。初期子叶披针形，出土展开后逐渐长大为宽卵形（图 2）。通常用根蘖繁殖。

（何泽瑛）

忍 冬 属

Lonicera L.

（忍冬科 Caprifoliaceae）

生长习性、分布和用途 本属约 200 种，我国约 100 种，落叶或常绿，直立或缠绕灌木。本属多为观赏树种，部分种类花入药，果实营养丰富，可食并可酿酒或制作饮料。本文描述 8 种 1 变种，其名称、生长、分布和用途见表 1。

表 1 忍冬属树种名称、生长、分布和用途

中 名	学 名	树高 (m)	分 布	用 途	供 稿
蓝靛果	*L. caerulea* L. var. *edulis* (Turcz.) Rgl.	1.5	东北、华北、西北、川、滇。欧、亚、美 3 洲北部	果可食、饮料、观赏	108
金花忍冬（黄花忍冬）	*L. chrysantha* Turcz.	4.0	东北、华北、西北、川、鄂、皖。朝鲜半岛、日本、西伯利亚	观赏	107
葱皮忍冬（秦岭忍冬）	*L. ferdinandii* Franch.	3.0	西北、华北、豫、川，东北有栽培。朝鲜半岛	观赏	107
忍冬（金银花）	*L. japonica* Thunb.	5.0	辽、华北、华东、华中、西南。朝鲜半岛、日本	药用、香精、观赏	107
金银忍冬（金银木）	*L. maackii* (Rupr.) Maxim.	3 (5)	东北、华北、华东、西南、陕、甘。朝鲜半岛、俄罗斯远东、日本	花可入药、观赏	108
紫花忍冬（紫枝忍冬）	*L. maximowiczii* (Rupr.) Rgl.	3.0	东北东部。朝鲜半岛、俄罗斯远东	观赏	107
早花忍冬	*L. praeflorens* Batalin	2.0	东北。朝鲜半岛、俄罗斯远东	观赏	107
长白忍冬	*L. ruprechtiana* Rgl.	3 (5)	东北、冀。朝鲜半岛、俄罗斯远东、日本	观赏	108
华北忍冬（藏花忍冬）	*L. tatarinowii* Maxim.	2.0	东北、华北、陕、甘、川、鄂。朝鲜半岛	观赏	107

开花结实 2～3 年生开始开花结实，正常结实在 5 年生以后。丰年间隔期不太明显，蓝靛果，早花忍冬一般为 1 年，余者为 0～1 年。

花两性，常成对生于总梗上，腋生，或花无梗而轮生于枝端。花左右对称。萼 5 齿裂，花冠管状、漏斗状、钟状或三角形。雄蕊 5，花药内向。子房下位，2～3 室，稀 5 室，每室胚珠多数，下垂。花柱头状，长于雄蕊。浆果，红色、蓝黑色或黑色，相邻两果离生或结合，具多数种子。种子卵圆形，光滑或粗糙，有胚乳（图 1）。忍冬属树种的花期日数及花色、花形等开花习性见表 2。开花结实物候见表 3。除了早花忍冬的花先叶开放或与叶同时开放以外，本文描述的其余几个种的花都是后叶开放。

表 2　忍冬属树种花冠长度、开花日数及花色等形态特征

树　种	花冠长（cm）	开花日数	花　色	芳香味	花　形	花　量
蓝靛果	1.0～1.3	20	浅黄色至黄色	微香	似唇形	多
金花忍冬	0.7～2.0	10	黄白色变金黄色	无	二唇形	多
葱皮忍冬	1.5～2.0	15	鲜黄色	无	二唇形	多
忍冬	2.0～6.0	—	白色变为黄色	清香	二唇形	极多
金银忍冬	2.0	15	银白色变为淡黄色	微香	二唇形	极多
紫花忍冬	1.0	10	紫红色	无	似唇形	多
早花忍冬	1.0～1.2	7	淡紫色	微香	较整齐	较少
长白忍冬	1.5～1.8	10	白色变为淡黄色	无	似唇形	多
华北忍冬	0.8～1.0	13	深紫色	无	二唇形	多

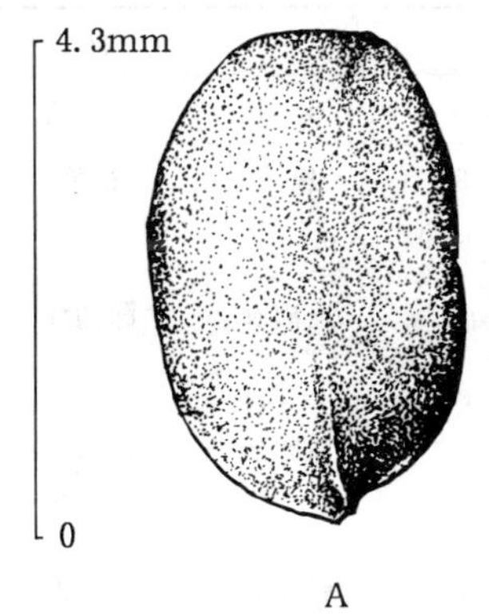

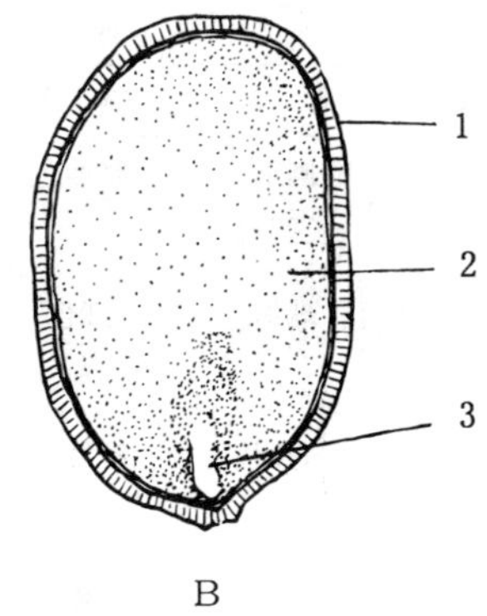

图 1　金银忍冬种子外形（A）及其纵切面（B）

1. 种皮　2. 胚乳　3. 胚

（梁鸣绘）

表 3　1963～1980 年哈尔滨地区忍冬属树种开花结实物候

树　种	花　期		果　期		果实初熟至果实脱落间隔期（天）
	始　期	末　期	始　期	末　期	
蓝靛果	5 月 4 日	5 月 26 日	6 月 3 日	7 月 14 日	18～24
金花忍冬	5 月 19 日	6 月 3 日	6 月 25 日	10 月 21 日	25～40
葱皮忍冬	6 月 2 日	6 月 30 日	9 月 14 日	10 月 21 日	15～26
忍冬	4 月	6 月	10 月	11 月	—

（续）

树　种	花　期		果　期		种子初熟至果实脱落间隔期（天）
	始　期	末　期	始　期	末　期	
金银忍冬	5月29日	6月23日	8月20日	11月4日	70天以上
紫花忍冬	5月26日	6月14日	7月4日	8月15日	25～30
早花忍冬	4月30日	5月15日	5月30日	6月16日	6～10
长白忍冬	5月20日	6月7日	6月23日	8月30日	15～20
华北忍冬	5月18日	6月5日	7月7日	9月13日	20～26

果实的采收调制和种子贮藏　果实呈现成熟颜色（表4）并达到软化即为成熟。应根据初熟至脱落间隔期的长短安排采收。采后可堆放数日或立即捣碎或搓揉，漂去果皮果肉与杂质，晾干后即得纯净种子。种子含水量应低于8%～10%。密封瓶中常温下可贮2年。果实和种子形态特征见表4。出种率及种子质量见表5。

表4　忍冬属树种果实和种子形态特征

树　种	未熟果颜　色	成熟果实			种　子		
		形　状	颜　色	大小（cm）	形　状	颜　色	大小（mm）
蓝靛果	绿色	椭圆形或长圆形	暗蓝色	1.0～1.8（径0.5～0.8）	扁圆	褐色	长2～3 宽1～1.5
金花忍冬	—	球形	红色	0.5～0.6	扁椭圆	深橘红	长3～4 宽2～3
葱皮忍冬	—	椭圆状	鲜红色	1.2	扁卵形	黄稍带淡紫色	长5～7 宽3～5
忍冬	—	球形	蓝黑色	0.6～0.7	扁椭圆	灰褐色	长3～4 宽2～3
金银忍冬	莹绿色	圆球形	鲜红色	0.5～0.9	扁圆形	淡黄色	长4～5 宽2～3
紫花忍冬	—	卵形	鲜红转深红色	—	扁椭圆	淡黄色	长4～5 宽2～3
早花忍冬	—	—	红色	—	椭圆形	淡黄色	长3～4 宽1.5～2
长白忍冬	绿色	圆球形	橘红色	0.4～0.8	扁圆	淡黄褐色	长2～3 宽1.5～2
华北忍冬	—	球形	鲜红色	0.5～0.6	扁圆	土黄色	长4～5 宽2～3

表 5　忍冬属树种出种率及种子质量

树　种	出种率（%）	净度（%）	千粒重（g）	每千克纯净种子粒数（万粒）
蓝靛果	2.2	80～90	0.7～1.0	100～140
金花忍冬	10	85～95	2.7～3.0	33～37
葱皮忍冬	2.8	85～95	17.9	5.6
金银忍冬	8.3	80～90	5.2～5.5	18.2～19.2
紫花忍冬	4.0	75～85	5.0～6.5	15～20
早花忍冬	2.8	75～85	2.0～2.4	42～50
长白忍冬	8	75～85	2.7～2.9	34～37
华白忍冬	4	80～90	5.4～6.1	16～18

发芽和播种　忍冬属种子无明显休眠习性，但播前必须进行催芽处理才能达到苗齐苗壮。一般将种子水浸 2～3 天，每天换清水 1 次，混沙置于 10～25℃温度下，保持湿度，勤翻动，10～60 天可发芽播种。播种后大体上是蓝靛果和早花忍冬最先发芽，次为忍冬、金银忍冬、金花忍冬、紫花忍冬、长白忍冬和华北忍冬，最迟为葱皮忍冬，需 60～70 天发芽。苗床条播。哈尔滨地区应于 5 月初播种，每平方米播种 0.5～1g，因种而异，播深 0.5～1.0cm。出土萌发（图 2）。1 年生苗高 20～50cm，当年可出圃，园林绿化中应培养大苗栽植。忍冬主要采用扦插、压条和分株方法繁殖，其它种亦可采用分株繁殖。

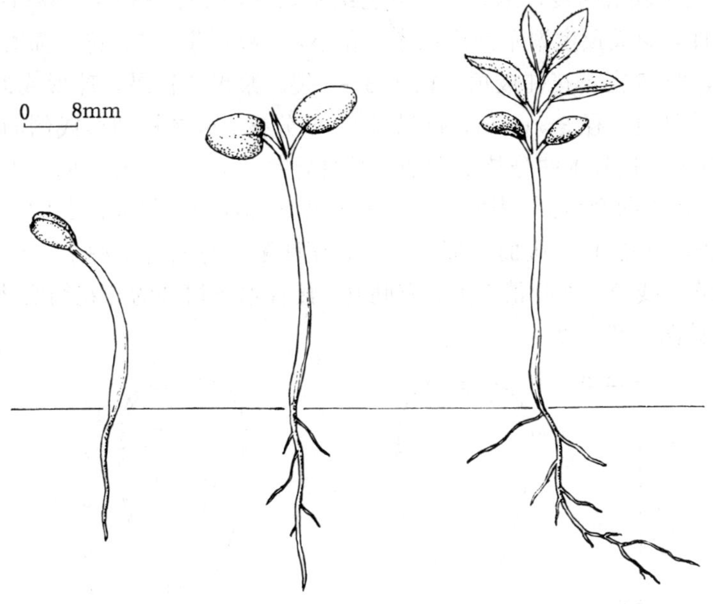

图 2　金银忍冬种子萌发后第 6、12、25 天幼苗的生长情况

（曹雅范绘）

（周德本）

接骨木属

Sambucus L.

（忍冬科　Caprifoliaceae）

生长习性、分布和用途　本属约 20 种，我国 5～6 种。落叶灌木或小乔木，稀为多年生草本。多为观赏树种。嫩枝入药，可治跌打损伤。本文描述 2 种 1 变种，其名称、生长习性和分布见表 1。

表 1　接骨木属树种的名称、生长习性和分布

中　名	学　名	生长习性	分　布	供　稿
朝鲜接骨木	*S. coreana* (Nakai) Kom. et Alis.	灌木，高 5m	东北。朝鲜半岛、俄罗斯远东	404
接骨木	*S. williamsii* Hance	灌木或小乔木，高 6～8m	东北、华北、华东、中南、西南至川、滇，西北至陕、甘。朝鲜半岛、俄罗斯远东、日本	403
毛接骨木	*S. williamsii* var. *miquelii* (Nakai ex Kom. et Alis.) Y. C. Tang	灌木，高 5～6m	东北、内蒙古。朝鲜半岛、日本	404

开花结实　人工栽培的接骨木 3 年生以上结实，7 年生以后进入结实盛期，结实大小年现象不明显。花两性，组成顶生聚伞圆锥花序。花小，5 数，萼齿 5。花冠辐射状，黄白色。雄蕊 5。子房下位，3～5 室，花柱极短，柱头 3～5 裂。浆果状核果，球形或近球形，紫黑色、蓝紫色、红色或紫红色，径 3～6mm，每果具 2～5 分核。分核椭圆形或倒卵形，有皱纹，软骨质，长 2.5～5mm。接骨木的分核上端圆，下端渐尖，背面中央有纵向圆形隆起，腹面中央有纵向脊状隆起，形成两个斜面，表面有明显的瘤状凸起或连成横向凸起的短棱，凸起间形成带角的孔眼或短沟。种脐位于腹面基部，白色，稍凹陷。每核含 1 种子。种子常靠鸟类啄食果实而传播。胚直，线型，子叶稍膨大，有胚乳。接骨木属树种的开花结实物候期见表 2。果实和果核的形态见图 1 和表 3。

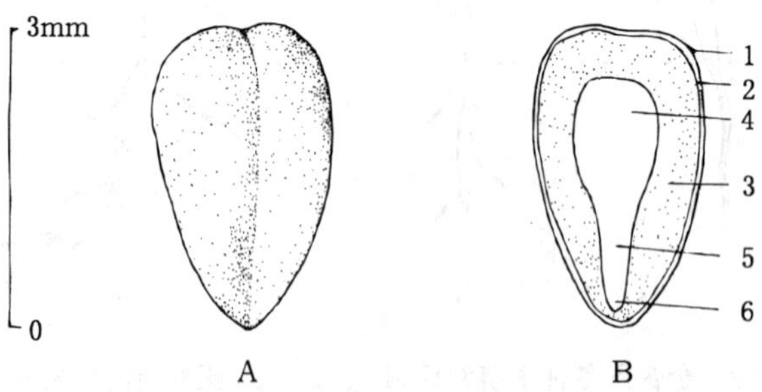

图 1　接骨木果核外形（A）及其纵切面（B）

1. 内果皮　2. 种皮　3. 胚乳　4. 子叶　5. 胚轴　6. 胚根

（胡冬梅、林平绘）

表 2　接骨木属树种的开花结实物候

树　种	观察年份和地点	开花 始期	开花 末期	果实成熟 始期	果实成熟 盛期	果实成熟 末期	果实散落期
朝鲜接骨木	1963～1980 哈尔滨	5月上中旬	6月中旬	6月下旬	～	7月中旬	初熟至脱落间隔20～30天
接骨木	1963～1980 哈尔滨	5月中下旬	6月上旬	6月下旬	～	8月上旬	—
	1986～1988 北京地区	4月中旬	4月下旬	7月中旬	7月下旬	8月中旬	8月上旬～8月下旬
毛接骨木	1963～1980 哈尔滨	5月中旬	6月上中旬	6月下旬	～	7月中旬	初熟至脱落间隔15～25天

表 3　接骨木属树种果实和果核的形态特征

树　种	成熟果实 形　状	成熟果实 大小（mm）	成熟果实 颜　色	果核 形　状	果核 大小（mm）	果核 颜　色
朝鲜接骨木	近球形	径3～6	红色	长圆形，有皱纹	长3～5 宽2～3	淡黄色
接骨木	球形	径3～5	紫黑色	倒卵形或椭圆形，有皱纹	长2.5～4 宽1.7～2	黄色带棕色
毛接骨木	球形	径3～6	红色或暗红色	倒卵形，有皱纹	长3～5 宽2～3	黄棕色

果实的采收调制和种子贮藏　果实由绿色变为红色、暗红色、紫黑色，果肉由硬变为多汁状即为成熟。应在果实充分成熟后和鸟类啄食前采收。剪取整个果穗或用手将果实摘下。堆放数日，待果肉腐熟变软时装入袋内或桶内，捣烂果肉，揉搓，漂洗，除去果肉、秕籽及杂质，得到的纯净果核即为播种材料，通称种子。其出籽率，净度及种子质量见表4。

表 4　接骨木属树种的出籽率、净度及质量

树　种	出籽率（%）	净度（%）	千粒重（g）	每千克纯净种子粒数（万粒）
朝鲜接骨木	11	90～99	1.8～2.6	38～56
接骨木	10	95～98	1.2～1.6	62～83
毛接骨木	11.5	90～99	1.7～2.7	37～59

曾经将纯净种子装入袋内，放在3～5℃条件下干藏1年，发芽率未见明显降低。

发芽和播种　接骨木属的果核坚硬，胚又具休眠习性，适于秋播或层积越冬后翌年春播。层积前可将果核水浸3～5天，再混湿沙在10～25℃下层积50～70天，即可播种。或在20～30℃的温度下层积30～60天，再在3～5℃的低温下层积90～150天。测定发芽能力时可用20～30℃的昼夜变温，也可以用四唑测定生活力。

条播，$10m^2$ 的苗床约播100g。播深1cm。覆盖薄层木屑以保持地面湿润。出土萌发（见

图 2)。当年苗高 20～30cm。未经任何处理的种子秋播，翌年春季也可得到良好发芽。也可用扦插、分根方法繁殖。

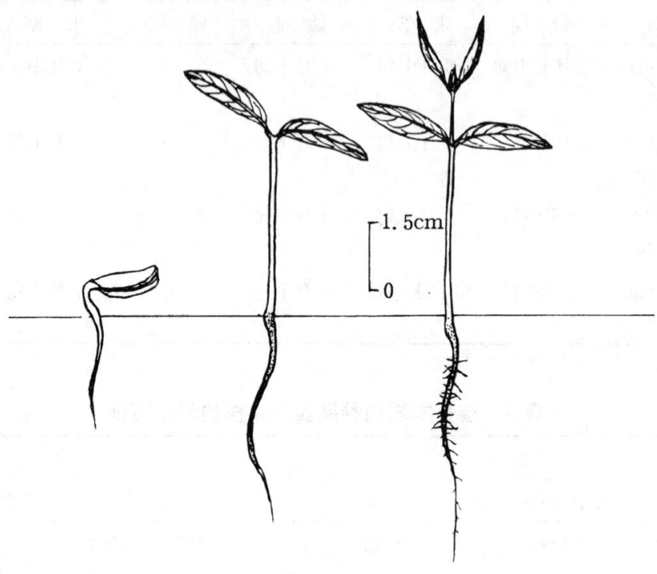

图 2 接骨木种子萌发出土后第 1、5、14 天的幼苗生长情况
(胡冬梅、林平绘)

(刘长江)

荚 蒾 属
Viburnum L.

(忍冬科 Caprifoliaceae)

生长习性、分布和用途 本属约 200 种，分布于东亚和北美。我国约有 80 余种，本文描述 6 种 1 变种 1 变型，它们的名称、分布和用途见表 1。

表 1 荚蒾属树种的名称、分布和用途

中 名	学 名	分 布	用 途	供 稿
珊瑚树	*V. awabuki* K. Koch	浙江普陀舟山及台湾。苏、皖、赣、鄂有栽培。日本、朝鲜半岛	观赏、绿篱、材用	1003
修枝荚蒾	*V. burejaeticum* Regel et Herd.	东北及华北。朝鲜半岛和俄罗斯远东地区	观赏、油脂、蜜源	105
水红木	*V. cylindricum* Buch. -Ham. ex D. Don	滇、川、黔、桂、粤、湘、鄂、甘	药用、油用、栲胶	1003
荚蒾	*V. dilatatum* Thunb.	陕、豫、冀，长江以南，东至台，西南至滇。华东最常见	观赏、药用、油用	1003

（续）

中　名	学　名	分　布	用　途	供　稿
琼花	*V. macrocephalum* Fort. f. *keteleeri* (Carr.) Nichols.	苏南、皖西、赣西北、湘南及鄂西	观赏、药用	1003
鸡树条荚蒾	*V. sargentii* Koehne	东北、华北、陕、甘、川、鄂、皖、浙。朝鲜半岛、日本及俄罗斯远东地区	观赏、油用、药用	105
天目琼花	*V. sargentii* var. *calvescens* Rehd.	除青海以外的黄河流域、东北、长江中下游各省。日本、朝鲜半岛	观赏	1003
茶荚蒾	*V. setigerum* Hance	陕南、长江流域以南、台湾	观赏、药用、果酿酒	1003

以上各种除珊瑚树为常绿灌木外，其余都系落叶灌木或小乔木，树高 2～5m。

开花结实　开始结实的年龄有的种为 2～3 年，有的种为 8～10 年，大小年不明显。两性花较小，白色，集成顶生的圆锥花序或伞形花序式的聚伞花序。有些种类花序边缘的花呈放射状，不孕，有的栽培品种全部花不孕。萼有 5 微齿。花冠轮形或钟状，5 裂。雄蕊 5。子房下位，1 室，胚珠 1 至多粒，花柱极短，柱头 3 浅裂。核果，成熟时由绿色转为鲜红、紫红或蓝黑色。5～6 月开花，9～10 月果熟（表 2）。大多数种的果实成熟后短期不落。

表 2　荚蒾属树种的开花结实物候期

树　种	观察地点和年份	开　花			果实成熟		
		始　期	盛　期	末　期	始　期	盛　期	末　期
珊瑚树	南京 1979，1988	5 月下旬	6 月中旬	6 月下旬	8 月下旬	9 月上旬	9 月中旬
修枝荚蒾①	哈尔滨 1964～1987	5 月 20 日	～	6 月 11 日	9 月 4 日	～	10 月 29 日
水红木	—	—	5 月	—	—	10～11 月	—
荚蒾	南京 1964，1988	5 月上旬	5 月中旬	6 月上旬	9 月上旬	10 月中旬	11 月上旬
琼花	南京 1978，1988	4 月中旬	4 月中下旬	5 月中旬	9 月上旬	10 月下旬	11 月
鸡树条荚蒾①	哈尔滨 1964～1987	5 月 28 日	～	6 月 23 日	9 月 4 日	～	10 月 18 日
天目琼花	南京 1985，1988	5 月上旬	5 月中旬	6 月上旬	9 月中旬	10 月	11 月中旬
茶荚蒾	南京 1978，1988	4 月上旬	4 月中下旬	5 月上旬	9 月下旬	10 月中下旬	11 月上旬

①　始期、末期系指观察年份内最早、最晚的日期

果实的形状、颜色和果核的形态，特别是种皮表面的沟槽等常为分类的重要特征（表 3、图 1）。果核多呈压扁状。内果皮木质，坚韧，黄色至灰褐色，同种皮不易分离。种皮膜质。胚乳丰富，具韧性。果实初熟时，发育早期的小胚位于胚乳尖端内（图 2），随着成熟逐渐长大。

表 3 荚蒾属树种果实和果核的形态特征

树 种	果实				果核					
	形 状	长 (mm)	径 (mm)	颜 色	形 状	长 (mm)	宽 (mm)	厚 (mm)	颜 色	表面特征
珊瑚树	长卵形	7～8	4	红色	长倒卵形	6～8	3～3.5	2.5～3	灰白色	具 1 深腹沟
修枝荚蒾	椭圆形	9～10	4	蓝黑色	长圆形	8	3	—	黄色	背面 2 浅槽，腹面 3 浅槽
荚蒾	卵圆形	7～9	7～8	鲜红色	扁椭圆形	6.5～7	4～4.5	2～2.5	浅红色	表面较粗糙
琼花	椭圆形	5～11	8～9	蓝黑色	扁长圆形	8	5	1.7	灰褐色	背面 2 浅槽，腹面 3 浅槽
鸡树条荚蒾	近球形	8～10	8～10	深红色	扁圆形	6	6	—	淡红色	无沟槽
天目琼花	近球形	1～1.2	1	橙红色	圆片状	8～10	8～8.5	1	浅棕色	无沟槽
茶荚蒾	卵圆形	9～11	7～8	红色	扁长卵形	6～7	3～4	2～3	浅棕褐色	背光，腹有凹槽

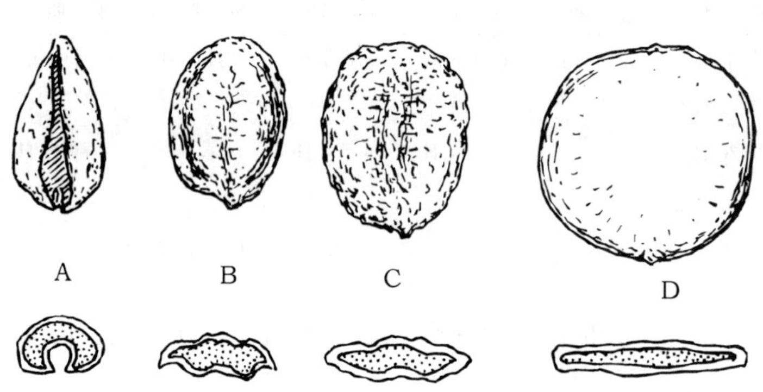

图 1 荚蒾属 4 种果核的外形（上）及其横切面（下）
A. 珊瑚树 B. 荚蒾 C. 琼花 D. 天目琼花
（史渭清绘）

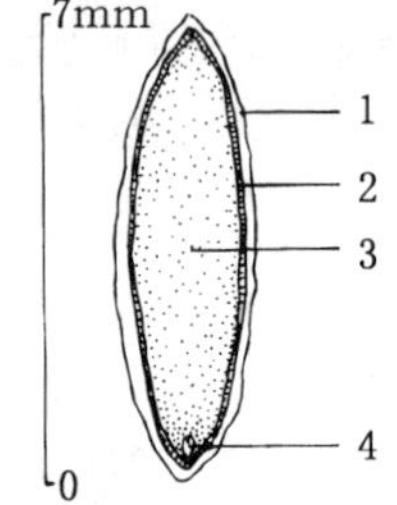

图 2 荚蒾果核侧面纵切
1. 内果皮 2. 种皮 3. 胚乳 4. 胚
（史渭清绘）

果实的采收调制和种子贮藏 果实变色、变软表示充分成熟，应在尚未脱落时采收。采集后摊放数天，促使果皮果肉继续软化，但忌久堆发热。搓揉，取出果核，淘净杂质，晾干即为播种材料，通称种子。可以干藏，或混湿沙层积。一般认为淘洗干净的种子比带果肉干藏的发芽率高。据说北美洲荚蒾属的气干种子在 1～4℃下用密闭容器贮藏，生命力可以保持 10 年。种子质量的数据见表 4。出籽率约 10%，种子净度可达 93%～96%。

表 4　荚蒾属树种的种子（果核）质量

树　种	千粒重（g）	每千克粒数（万粒）
珊瑚树	23～25	4.0～4.4
修枝荚蒾	32	3.2
荚蒾	26～30	3.3～3.9
琼花	52～60	1.6～2.0
鸡树条荚蒾	40	2.5
天目琼花	48～51	1.9～2.1
茶荚蒾	22～34	2.9～4.5

发芽和播种　荚蒾属多数种子因胚休眠或种皮的不透性障碍而不易发芽。修枝荚蒾及鸡树条荚蒾产于北方，种子需要先经 15～20℃的暖层积约 1 个月，入冬再转移到室外冷层积，待来春自然解冻时播种，发芽率约为 50%～60%。据万才淦和张炳坤（1994）报道，水红木的胚并无休眠习性，发芽的障碍在于核壳。在他们的初步研究中，不去核壳发芽率为零，去核壳的 16 天发芽率可达 68%，而且室内发芽时似乎要求 30℃/20℃变温。周德本（1986）的经验是，修枝荚蒾的种子在 20～35℃下层积 70～100 天即可发芽。对本文所述的荚蒾、琼花、天目琼花和茶荚蒾，曾经在南京地区做过一组层积试验，结果见表 5。

表 5　荚蒾属 4 个树种室外层积发芽试验结果

树　种	种子采收日期（年-月-日）	开始层积日期（年-月-日）	萌　发				供试种子粒数
			开　始（年-月-日）	结　束（年-月-日）	历时天数（天）	发芽率（%）	
荚蒾	1987-11	1988-02-06	1988-10-10	1988-10-26	17	96.6	505
琼花	1988-10-25	1988-12-16	1989-09-07	1989-10-10	34	96.9	650
天目琼花	1988-10-07	1988-12-16	1989-09-07	1989-10-10	34	78.7	465
茶荚蒾	1987-10-11	88-02-06	1988-10-18	1988-11-16	30	91.0	80

为了解除休眠，荚蒾属种子一般应尽早秋播，或在自然温度条件下层积。根据表 5 所述的初步试验来看，为减少田间管理的麻烦，种子可层积到正常发芽期来临之前播种下地。出土萌发。荚蒾种子萌发和幼苗生长情况见图 3。庭园绿化栽培时多用 2～3 年生或年龄更大的移植苗带土栽植。有些树种，如珊瑚树，可用扦插繁殖。有些种还可以压条繁殖。

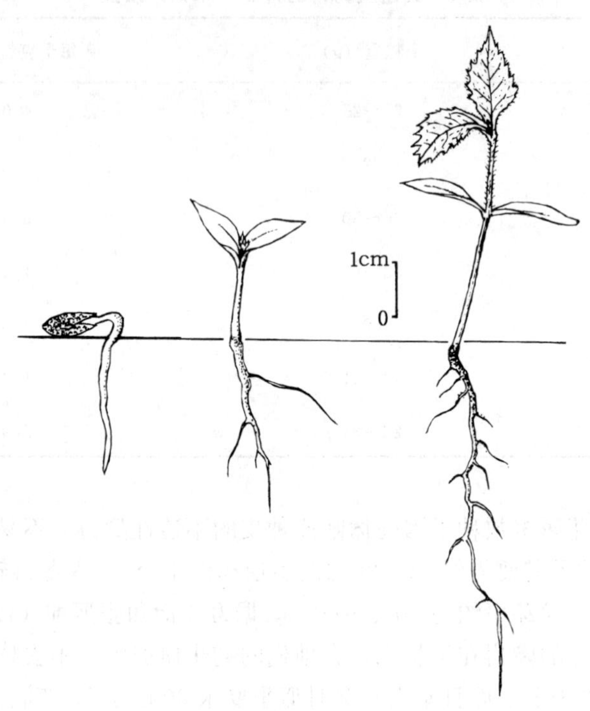

图 3 荚蒾种子萌发后第 2、8、20 天的幼苗生长情况
（史渭清绘）

（何泽瑛）

锦带花属
Weigela Thunb.

（忍冬科 Caprifoliaceae）

生长习性、分布和用途 本属约 12 种，我国 4 种，本文描述 2 种。落叶灌木，高 3～5m。生于湿润谷中。喜光，半阴处也能生长。耐寒，耐旱，喜富含腐殖质、排水良好的土壤，低洼积水则易烂根。产于东亚（表 1）。欧美许多国家有引种。锦带花属花朵多，颜色艳丽，花期长，为优良花灌木。

表 1 锦带花属树种的名称、树高、分布和用途

中 名	学 名	树高（m）	分 布	用 途
海仙花	*W. coraeensis* Thunb.	5	华东。广州、北京、庐山、武汉有栽培。日本	观赏
锦带花	*W. florida* (Bge.) A. DC.	3	东北、华北、陕、豫、苏。各地多栽培。俄罗斯远东、朝鲜半岛、日本	观赏

开花结实　开始开花结实年龄因繁殖方法而有不同：实生苗5年，扦插苗2～3年。结实大小年现象不明显。两性花集成聚伞花序，生于短枝顶及叶腋。海仙花初开时为淡玫瑰色或乳黄色，后变为深玫瑰紫色；锦带花为玫瑰红色，里面稍淡。花较大，长2.5～4cm，径约3.5cm。萼5裂。花冠漏斗状钟形，5裂，不整齐或近整齐，花冠筒长于裂片。雄蕊5。子房下位，2室，胚珠多数。花柱细长，柱头头状。蒴果柱状，长1.5～2.5cm，顶端具喙，室间开裂成2瓣，中间有一中轴。种子多数，被果瓣包住，并不立即散出。种子细小，胚小，呈线型，胚乳丰富（图1）。本文这两树种的开花结实物候期见表2。这两个种果实和种子形态及颜色相近，但大小和附属物不同，见表3。

表2　锦带花属的开花结实物候

树种	观察地点和年份	开花			果实成熟		种子散落		
		始期	盛期	末期	始期	盛期	始期	盛期	末期
海仙花	北京 1986～1988	5月上旬	5～6月	9月中旬	10月中旬	10月下旬	10月下旬	11月上旬	11月上旬
锦带花	北京 1986～1988	4月下旬	5～6月	9月上旬	10月上旬	10月中旬	10月中旬	10月下旬	11月上旬
	哈尔滨 1963～1980	5月中下旬	～	6月中下旬	8月末	10月上中旬	—	—	—

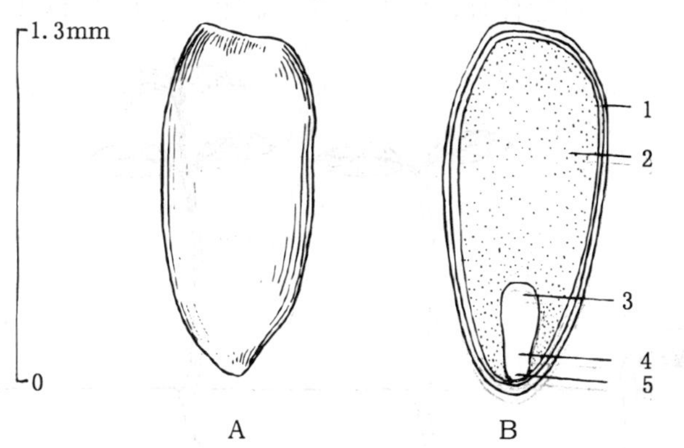

图1　锦带花种子外形（A）及其纵切面（B）
1. 种皮　2. 胚乳　3. 子叶　4. 胚轴　5. 胚根
（胡冬梅、林平绘）

果实的采收调制和种子贮藏　成熟时果皮由绿色渐变为黑褐色。应在蒴果稍见开裂时采摘。如果种子已经进入散落初期，采种时需注意不要振落种子。采下的蒴果摊平晾干，待果瓣充分开裂，种子自然脱出。或放入袋中轻轻揉搓，折断果瓣，脱出种子。用细筛除去果皮等杂物，即可得到纯净种子（表4）。种子晾干后装入袋内，存放于冷凉处即可。在北京地区室内贮藏，寿命可保持1年左右；在地下室（最低10℃，最高24℃）贮藏3年后全部丧失发芽能力。

表 3 锦带花属的果实和种子形态

树种	果实				种子		
	形状	颜色	长（cm）	每果种子粒数	形状	颜色	大小（mm）
海仙花	柱状	黑褐色	1.5～2	10～25	椭圆状，条形；有翅	黄褐色	长 0.6～0.7 宽 0.1～0.2
锦带花	柱状	黑褐色	1.5～2.5	15～20	圆柱状，两端斜截；无翅	浅褐色	长 1～1.4 宽 0.6～0.8

表 4 锦带花属的出种率和有关种子质量的数据

树种	出种率（%）	千粒重（g）	每克纯净种子粒数
海仙花	1～2	0.21～0.28	3 500～4 800
锦带花	1～2	0.29～0.32	3 100～3 500

发芽和播种 发芽测定时将种子排放在发芽皿内的滤纸床上，在恒温 25℃下第 5 天开始发芽，发芽全程约 12 天。出土萌发（图 2）。春播。先水浸种子 1 日，搀细土或细沙撒播。10m² 床面用种子 25g 左右。覆土厚 0.3cm，并覆盖稻草或塑料薄膜保持土壤湿润。播后 10 天即可出苗，过密时可以间苗。少量播种时，可采用室内盆播。还可以用扦插、分株方法繁殖。

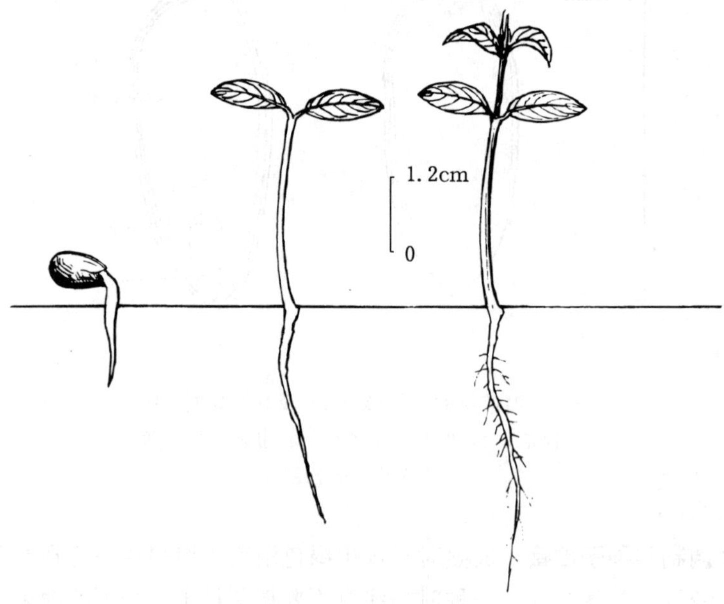

图 2 锦带花萌发出土后第 1、5、15 天的幼苗生长情况

（胡冬梅、林平绘）

（刘长江）

水　青　树

Tetracentron sinense Oliv.

（水青树科　Tetracentraceae）

生长习性、分布和用途　水青树属只有本文描述的1种，古老的孑遗植物，在《中国植物红皮书》中被列为稀有种。落叶大乔木，高达40m，胸径1～1.5m。根深。适生于土层深厚、疏松潮湿而排水良好的酸性或中性山地黄壤或黄棕壤。萌生力强。分布于豫、陕、甘、川、藏、滇、黔、鄂、湘。缅甸、越南、尼泊尔也有分布。材质轻软，纹理通直，干缩小，不翘裂，可做家具以及胶合板和造纸原料。树姿美观，可供观赏或作行道树。

开花结实　水青树结实较晚。在陕西宁西林区，大量结实的多系胸径20cm以上的大树（庚府诗，1984）。花两性。穗状花序细长，生于短枝顶端，长约10～15cm。花多数，4朵成一簇，无梗，淡黄色或黄绿色，直径1～2mm。苞片小，萼片4，无花瓣。雄蕊4。心皮4，沿腹缝线合生。花柱4。子房上位，4室。每室具4～10胚珠，通常4，生于腹缝线上。花期6～7月，果期8～10月。在陕西宁西林区，7月底为开花末期，10月初果实成熟。蒴果，倒卵状圆柱形或圆柱形，4深裂，宿存花柱位于果实基部，下弯。果长3～5mm，褐色或棕褐色，室背开裂。种子小，长2～3mm，黄褐色，条状，长椭圆形，有棱脊。胚小。胚乳丰富，含油质。果实和种子的外形见图1。

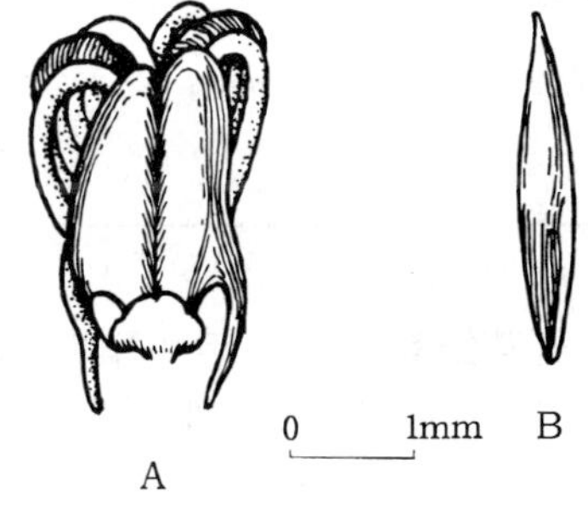

图1　水青树果实（A）和种子外形（B）
（童军平绘，A仿《中国珍稀濒危植物》）

果实的采收调制和种子贮藏　蒴果由绿色转为黄褐色时即可采摘果穗，摊晾2～3天后放在干燥通风的室内干藏。翌春播种前轻搓果穗，筛去杂质即得纯净种子。种子千粒重0.100～0.145g，每克有纯净种子6 900～10 000粒。

发芽测定　据任祝三等（1990）报道，光照处理能促进水青树种子萌发，其中光照强度为1 500lx，每天光照12小时，效果最佳。万才淦（1986）的研究表明，水青树种子对光的反应随温度而异：25℃时暗发芽良好，不需光；如果温度较高（白天24～33℃，8小时；晚上28～30℃，16小时），或较低（10～20℃），则光对发芽有促进作用。赤霉素（GA_3）可以代替光照而促进水青树种子萌发：种子用200μg/g、400μg/g和600μg/gGA_3浸泡3天，在25℃无光的环境中发芽率分别为91.2%、91.6%和90%，但GA_3的浓度过高会降低发芽率（任祝三等，1990）。

水青树种子最适宜的萌发温度为25℃（万才淦，1986），30℃时发芽率明显下降（任祝三等，1990）。

播种　春播。播前种子用0.15%高锰酸钾溶液处理3～5分钟，洗净后再用始温25℃水浸泡24小时，捞出装入纱布袋中催芽；每天早晚各洒温水1次，翻动数次，保持种子均匀湿润；经10～12天，种子开始萌动时即可播种。

条播。种子用草木灰拌匀后均匀撒入播种沟内，筛土覆盖至看不见种子为度。播后及时搭上弓形棚架，上覆塑料薄膜保温保湿。出土萌发。据陕西宁西林业局林科所（1985）报道，4月11日播种，5月9日出土，6月中旬出现初生叶。幼苗生长缓慢，1年生苗高约1.5～12cm。

（史晓华）

蕈树属

Altingia Noronha

（金缕梅科 Hamamelidaceae）

生长习性、分布和用途 本属约12种，我国8种，本文描述2种。常绿大乔木。它们的名称、生长、分布及用途见表1。

表1 蕈树属树种的名称、生长、分布和用途

中 名	学 名	树 高(m)	胸 径(cm)	分 布	用 途	供 稿
蕈树（阿丁枫）	*A. chinensis* (Champ.)Oliv.	20～25	50～80	浙、闽、赣、湘、粤、琼、桂、黔、滇。越南	材用、育香菇、制蕈香油、药用	601
细青皮	*A. excelsa* Noronha	30～40	80～200	滇、藏。印度、不丹、缅甸、马来西亚、印度尼西亚	材用、芳香树脂、嫩叶可食	616

开花结实 6～9年生开始开花结实，正常结实期在20年生以后。大小年间隔期一般为1年，但不甚明显。花单性，雌雄同株。雄花无花被，雌花无花瓣。花序被柔毛，顶生或腋生于上部。雄花短穗状花序（蕈树）或头状花序（细青皮），常多个再组成圆锥花序或总状花序。雄蕊多数，花丝极短，近于无柄。花药倒卵圆形，先端平截，2室，纵裂。雌花头状花序，每序具花14～26。萼管与子房合生，无萼齿或萼齿乳点状。子房近下位，2室，胚珠多数。花柱2，长3～4mm，先端向外弯曲。据广西南宁及云南南岛河观察，开花结实的物候期见表2。

表2 蕈树属树种的开花结实物候期

树 种	观察地点	观察年份	开 花			果实成熟		种子落脱
			始 期	盛 期	末 期	始 期	盛 期	
蕈树	南宁	1978～1986	3月下旬	4月上中旬	4月下旬	10月下旬	11月中旬	11月下旬～12月上旬
细青皮	南岛河	1987～1988	3月中旬	3月下旬	6月中旬	12月	翌年1～2月	2～3月

蒴果藏于头状果穗内，表面粗糙，未成熟时黄绿色或青色，成熟时棕褐色或浅黄色，果室间开裂为2瓣，每瓣2浅裂。每果有发育的种子1～2粒，位于胎座基部。种子样式不一，

有角质种皮、薄层胚乳及直胚，子叶大而扁平。果穗及种子形态见表3，种子的外形和解剖结构见图1。

表3　蕈树属树种果实和种子的形态特征

树　种	未熟果颜色	果穗			种子		
		形　状	大小（cm）	色　泽	形　状	大小（mm）	色　泽
蕈树	黄绿色	球形	1.7～2.6	褐色	龟背状扁椭圆形	长5～6.5 宽3～4 厚约1～1.5	褐至黄褐色，一侧有光泽
细青皮	青色	球形	1.5～2	浅黄色	扁三角形或扁椭圆形，具翅	长约5 厚约1	浅黄至褐色

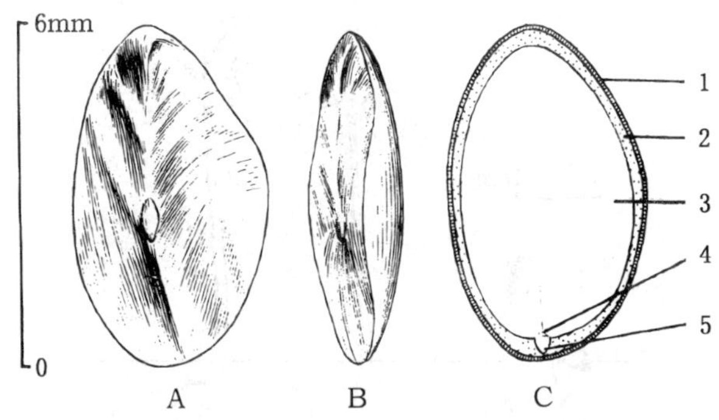

图1　蕈树种子外形正面（A）、侧面（B）及其纵切面（C）

1.种皮　2.胚乳　3.子叶　4.胚芽　5.胚根

（黄应钦绘）

果实的采收调制和种子贮藏　果实成熟而尚未开裂时，用高枝剪或钩刀截断果穗梗，在地面收集，或上树用手采摘。采得的果穗可以堆放3～5天，以利于少数果实后熟，然后摊放于室内通风处或置日光下曝晒。果实开裂后稍加敲打，使种子散落。捡去已脱粒的果穗，再用细筛除去杂质瘪粒，即得纯净种子。已脱粒的种子不宜在烈日下久晒，只能在弱光下晒干或晾干。蕈树种子的含水量约为11%，细青皮约为10%。蕈树种子可以贮藏越冬，翌年春播。细青皮宜随采随播。这2个树种的出种率和种子的净度、质量等见表4。

表4　蕈树属树种出种率和种子净度、质量

树　种	出种率（%）	净度（%）	千粒重（g）	每千克纯净种子粒数（万粒）
蕈树	0.7～1	93	7～11	9～14
细青皮	1	70	1.2～1.8	55～83

发芽和播种　种子无休眠现象。发芽时日均温需在18℃以上。1988年，广西林业科学研究所对当年新采的蕈树种子，在室内发芽箱内作过发芽测定。同年3月，云南普文苗圃在室外苗床上对当年新采并经精选的细青皮种子进行过发芽测定。置床前不作处理，测定结果见

表 5。

表 5 蕈树属树种的发芽能力及其测定条件

树 种	基 质	测定温度（℃）		发芽势（%）		发芽率（%）	
		发芽箱	室 外	计算天数	一般数值	计算天数	一般数值
蕈树	纸	28	—	5	35	13	60
细青皮	沙土	—	18～22	8	45	28	78

出土萌发。在发芽箱内，蕈树胚根萌发后约 6 天，子叶带壳伸出，11 天左右子叶展开，再过 22 天初生叶始现（图 2）。在苗圃，细青皮子叶出土后约 30 天展出初生叶。

撒播。种子密播于播种盘或经过细致平整的苗床。蕈树每平方米约播 10g。初生叶 4～6 片且幼苗半木质化时，移至大田培育。早期需搭棚遮荫并防暴雨冲击。1 年生苗出圃。

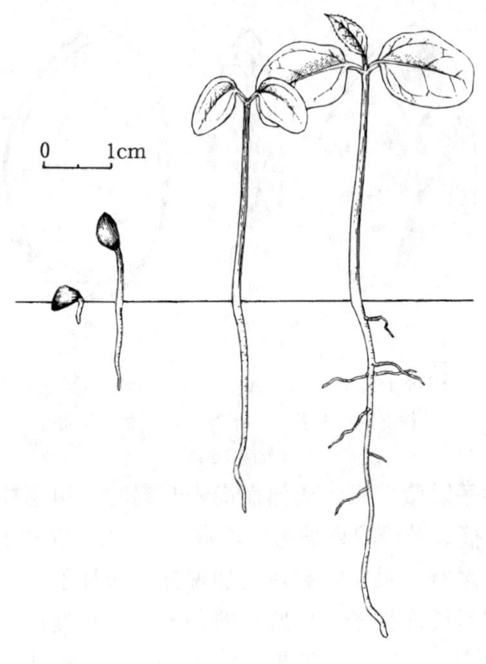

图 2 蕈树种子萌发后第 5、7、10、32 天幼苗生长情况

（黄应钦绘）

（王宏志）

瑞木（大果蜡瓣花）

Corylopsis multiflora Hance

（金缕梅科 Hamamelidaceae）

生长习性、分布和用途 蜡瓣花属约 29 种，我国有 20 种，本文描述 1 种。落叶或半常绿

小乔木，高达 5m。适生于中亚热带至北热带 550～1 200m 的山地，对土壤水肥条件要求中等。分布于鄂、湘、闽、台、粤、琼、桂、黔、滇。心材边材区别不明显，坚重，纹理斜，干燥易裂，用于细木工，常栽培供观赏。

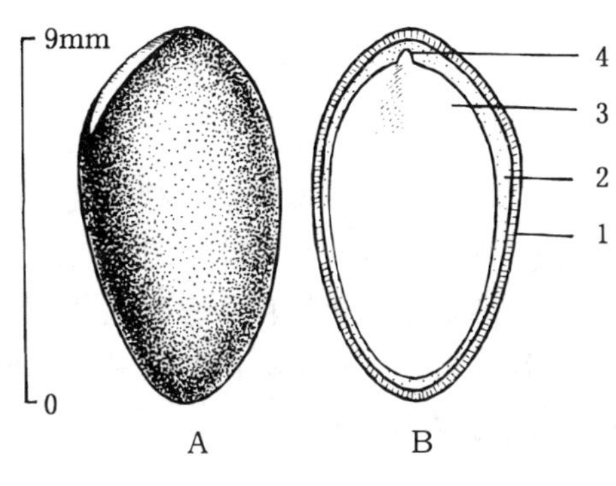

图 1　瑞木种子外形（A）及其纵切面（B）
1. 种皮　2. 胚乳　3. 子叶　4. 胚根
（黄应钦绘）

开花结实　6～8 年生开始开花结实，正常结实期在 15 年生以后。结实大小年现象较明显，间隔期一般为 1 年。花两性，先叶开放。总状花序下垂，长 2～4cm，基部具 1～5 叶。总苞片卵形，长 1.5～2cm，被灰白色柔毛。萼筒与子房下部合生。萼齿 5，卵形，花后脱落。花瓣 5，倒披针形，长 4～5mm。雄蕊 5，长 6～7mm，花药 2 室，纵裂。退化雄蕊 5，不裂，与雄蕊互生。子房 2 室，半下位，无毛，每室 1 胚珠，垂生，花柱 2。据广西野外调查资料，花期 4～5 月，9 月下旬果始熟，10 月上中旬为果实成熟盛期。蒴果，木质，卵圆形，长 1.2～2cm，径 1.0～1.4cm，无毛，室间及室背 4 裂。未熟果浅绿色，成熟后浅黄色至紫黑色。11 月中旬后果实开裂，种子散落。每果有种子 1～2 粒。种子长卵状圆锥形，棕褐色，有光泽，长 9～12mm，径 5～7mm，基部有宽大的种脐。种皮骨质。胚乳薄肉质。胚直，子叶扁平，长圆形（图 1）。

果实的采收调制和种子贮藏　果实成熟盛期用采种刀或用手采摘。采回的果实摊开曝晒，开裂后取出种子，除去杂质即得纯净种子。鲜果出种率约 26%。种子净度一般为 90%。千粒重 126（120～130）g，每千克有纯净种子 7 700～8 300 粒。可随采随播或密封贮藏，贮藏期一般为 5 个月。

发芽和播种　种子有浅休眠，发芽时气温要求在 15℃以上。1989 年 10 月 4 日，广西林业科学研究所用当年新采的种子在室外沙床作发芽测定，播种后 60 天（12 月 4 日）方始发芽，发芽盛期不明显，翌年 1 月 25 日发芽终止。自播种之日起算，以 114 天计，发芽率为 50%。出土萌发。胚根萌发后约 15～20 天子叶出土，再过 10 天左右长出初生叶（图 2）。

撒播或催芽移植。每平方米播种 12～20g，覆土 1cm 左右。苗期需要遮荫。由于种子发芽不整齐，最好是采用先催芽后移植的方法，将种子密播于沙床，待发芽后 1 个月左右将幼苗移至圃地，培育 1 年生苗出圃。

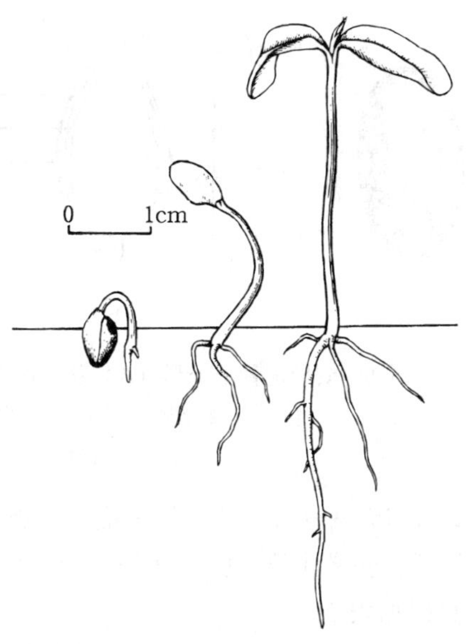

图 2　瑞木种子萌发后第 10、16、30 天幼苗的生长情况
（黄应钦绘）

（曾　玲）

蚊母树属

Distylium Sieb. et Zucc.

（金缕梅科　Hamamelidaceae）

生长习性、分布和用途　本属共18种，分布于东亚，我国有12种，本文描述列入表1的2种。常绿灌木或小乔木。蚊母树高16～25m，但栽培修剪后常为灌木状。耐阴湿，适应性强，多生于丘陵地带阔叶林中，分布于西南至东南地区，为常见的栽培绿化树种。根据江苏省植物研究所的观察，蚊母树对大气中的二氧化硫和氯气有抗性，适合于在工矿区栽培。蚊母树叶面常有“矢日本扁蚜”的虫瘿，影响观赏价值。杨梅叶蚊母树不长虫瘿。

表1　蚊母树属树种的名称、分布及用途

中　名	学　名	分　布	用　途
杨梅叶蚊母树	*D. myricoides* Hemsl.	粤、桂、黔、川、湘、赣、闽、浙、皖	绿化、鞣质、药用
蚊母树	*D. racemosum* Sieb. et Zucc.	粤、湘、闽、浙、台、琼。朝鲜半岛、日本	绿化、材用

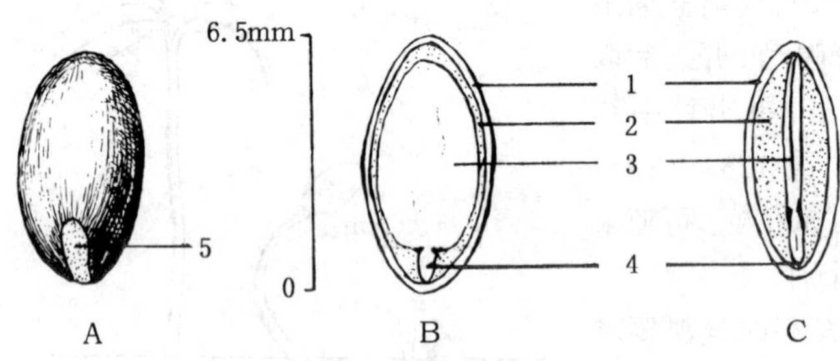

图1　蚊母树种子外形（A）及其两个纵切面（B、C）
1. 种皮　2. 胚乳　3. 子叶　4. 胚根　5. 种脐
（史渭清绘）

开花结实　花单性或杂性，同株或同序。穗状或总状花序腋生，长1～3cm。萼筒极短，萼齿2～6，大小不等。无花瓣。雄蕊2～8。子房上位，2室，每室有胚珠1。花柱2，长6～8mm，柱头锐尖。蒴果木质，卵圆形，长1～1.3cm，被黄褐色星状绒毛；室背及室间裂开，上半部2瓣裂，每瓣再2浅裂，先端尖，基部无宿存萼筒；2室，每室含种子1。种子卵圆形或长卵形，深褐色，有光泽，种脐白色；种皮角质，脆，内种皮膜质。胚乳丰富，胚大而扁，子叶卵圆形（图1）。江苏南部观察的花果物候及果实和种子的数据见表2。

表2　蚊母树属树种花果物候及果实、种子数据

树　种	花　期	果熟期	种子散落期	蒴果 形　状	蒴果 长　度（cm）	种子 长　度（mm）	种子 千粒重（g）	种子 每千克粒数（万粒）
杨梅叶蚊母树	3月下旬～5月上旬	8～10月上旬	10月下旬～11月	卵圆形	1～1.2	4～5	28	3.6
蚊母树	3月下旬～5月上旬	8～10月上旬	10月下旬～11月	卵形	1～1.3	6～7	30	3.3

表 3　蚊母树属树种发芽试验结果（南京，播前室外层积越冬 90～110 天）

树种	萌发日期	6 天的发芽百分数（%）	最终发芽率（%）	幼苗形态区分点
杨梅叶蚊母树	4.6～4.20	82	92	初生叶无红边
蚊母树	4.6～4.20	52	61	初生叶有紫红色边晕

果实的采收调制和种子贮藏　当蒴果内种子已经发育充实，种皮呈深褐色，蒴果尚未开裂或仅少数开始开裂时，将果枝剪下，放在室内摊晾后熟 2～3 天，晒果取种。晒时需用网筛罩住，以免果实炸裂，种子弹失。漂除空粒，晾干，一般干藏即可。

发芽和播种　种子有休眠习性，秋播或层积越冬可以解除休眠。据江苏省植物研究所试验，种子采收后混沙在室外自然变温条件下层积，翌年 3 月播种，4 月萌发，发芽良好，详见表 3 。出土萌发，幼苗形态见图 2。

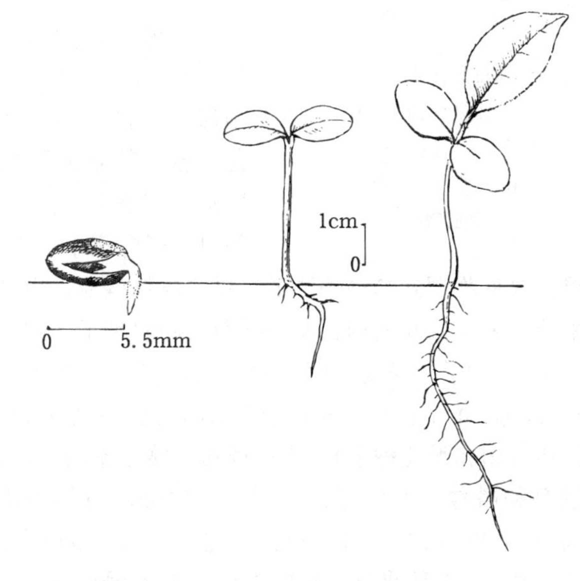

图 2　蚊母树萌发后第 2、20、40 天的幼苗生长情况
（史渭清绘）

（何泽瑛）

马蹄荷（白克木）

Exbucklandia populnea（R. Br.）R. W. Brown

（金缕梅科　Hamamelidaceae）

生长习性、分布和用途　马蹄荷属 4 种，我国产 3 种，本文描述 1 种。常绿乔木，高达 20m，胸径 1m。适生于中亚热带南部至热带的山地和丘陵地区。对土壤水肥条件的要求中等，分布于藏、滇、黔、桂。印度、尼泊尔、不丹、越南亦有分布。材质较优，用于家具、建筑、雕刻。树冠浓密，树形美观，可供观赏。落叶层厚，有涵养水源的功效。树叶含水量高，树皮耐火能力强，为优良防火树种。

开花结实　8～10 年生开始开花结实，15 年生以后进入正常结实期。结实有大小年现象，但不甚明显，间隔期一般为 1 年。花两性或单性同株。花 8～12 朵组成头状花序，单生或再组成圆锥花序状，腋生。花序梗被柔毛。萼齿鳞片状。花瓣 2～5，条形，长 2～3mm，或无花瓣。雄蕊 10～14，花丝细，花药基着，纵裂。子房被褐色柔毛，半下位，藏于肉质头状花序

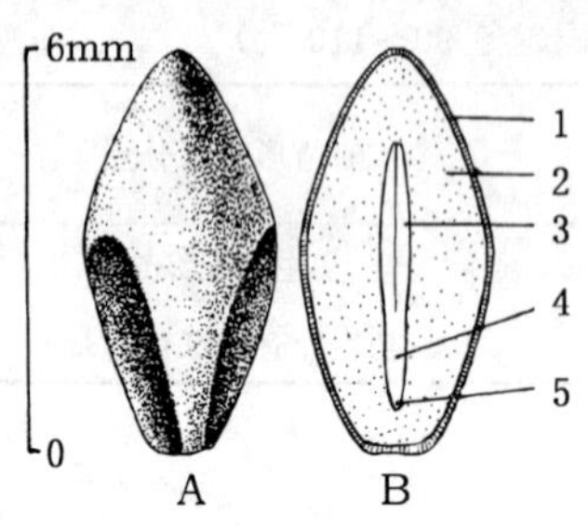

图 1 马蹄荷种子外形（A）及其纵切面（B）
1. 种皮 2. 胚乳 3. 子叶 4. 胚轴 5. 胚根
（黄应钦绘）

内，2 室，每室有胚珠 6～8 个。花柱 2，柱头尖细。据广西南宁 1978～1985 年观察，马蹄荷 1 年 2 次开花结实。第 1 次，2 月下旬始花，3 月上旬为盛花期，3 月下旬为末花期，10 月上旬果始熟，10 月中下旬为果熟盛期。第 2 次，9 月中下旬始花，10 月下旬为末花期，翌年 1 月下旬至 2 月上旬果始熟，2 月中下旬为果熟盛期。

蒴果，椭圆形，平滑，未成熟时青绿色，成熟时黄褐或棕褐色，长 8～10mm，径 5～6mm。果成熟后 1 个月内，上半部 2 片裂开，种子散落。每果有种子 8～12 粒，一般仅有位于胎座基部的数粒发育正常。发育种子棱形、梭形或不规则形，棕褐色，长 5～7mm，宽 2.5～3.5mm，种皮薄，具胚乳，子叶扁平（图 1）。具翅种子常为空粒，无发育能力。

果实的采收调制和种子贮藏 果实成熟盛期，用采种钩刀钩取或上树采摘。采回摊开曝晒，蒴果开裂后筛取种子。鲜果出种率一般约 8%～10%，种子的净度一般为 85%～95%。千粒重 7.5g，变动范围为6.5～8.5g，每千克有纯净种子 13.3 万粒，变动在 11.8 万～15.4 万粒（包括有翅和无翅种子）。用袋或坛干藏于室内干燥通风处。

发芽和播种 种子无休眠习性。发芽时日均温需在 15℃以上。1983 年 1 月 29 日，广西林业科学研究所在室外沙床作过发芽测定。种子为新采种子，播种前曾用始温 30～40℃水浸种 30～ 60 分钟。2 月 6 日开始发芽，2 月 16 日为发芽盛期，2 月 26 日发芽终止。从发芽开始期至发芽高峰期，10 天的发芽百分数为 22%。从置床之日起算，28 天的发芽率 25%。发芽率低的原因，可能是由于供测种子为春季成熟，且混有一半以上的无发芽能力的有翅种子。秋季成熟的种子还缺少测定材料。出土萌发。胚根萌发后 6 天子叶出土，再过 30 天长出初生叶（图 2）。

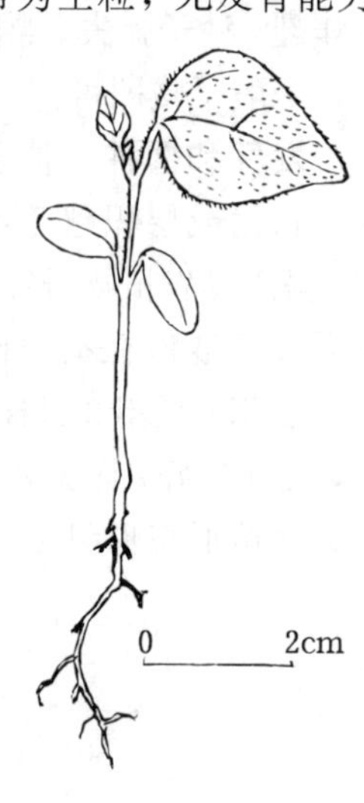

图 2 马蹄荷的幼苗
（田恒德绘）

为了清除无发芽能力的空粒（有翅），留下饱满种子（无翅）用于播种，可在播前用水或其它适宜的溶液对播种材料试行浮洗。撒播。每平方米播种 5～8g。苗期需遮荫，培育 1 年生苗出圃。亦可用容器培育 3～4 个月的小苗上山造林。

（王宏志）

牛 鼻 栓

Fortunearia sinensis Rehd. et Wils.

（金缕梅科 Hamamelidaceae）

生长习性、分布和用途 牛鼻栓属仅本文描述的 1 种，我国特产。落叶小乔木或灌木，高 5～9m，生于低山丘陵地区的山坡杂木林中。分布于陕、豫、川、鄂、赣、皖、苏、浙等地。

材质坚硬，常用来制作牛鼻栓，枝叶可入药。

开花结实　花单性或杂性。两性花总状花序顶生，萼筒倒锥形，萼齿5，花后脱落。花瓣5，退化为针状或披针形，细小。雄蕊5，花丝极短。子房半下位，2室，每室有胚珠1。花柱2，分离，线形，反卷。雄花为柔荑花序，具退化雌蕊。蒴果木质，卵圆形，灰褐色，长约1.5 cm，表面密布白色皮孔，沿室间2裂，每片又具2浅裂，果瓣尖，果梗长5～10mm。宿存萼筒与蒴果合生。种子长卵形，长约1cm，径约0.5cm，种皮黑色，有光泽，壳硬，质脆，种脐马蹄形，白色。胚乳丰富，胚大，子叶近圆形，宽，两侧边各向反向卷折（图1）。据江苏南部的物候记载，花期3月下旬～4月上旬，果熟期7～8月下旬。种子在蒴果开裂时弹出。

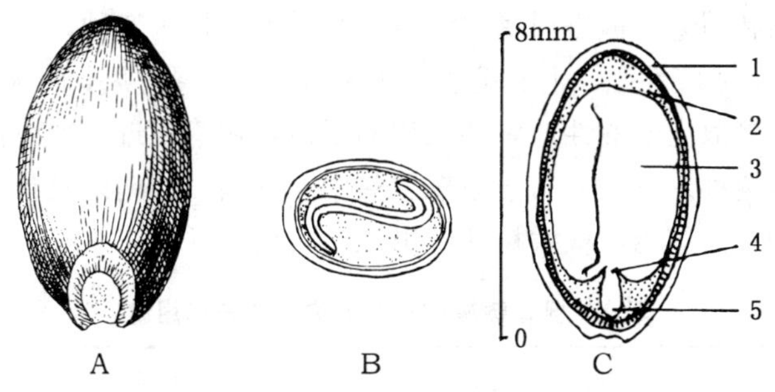

图1　牛鼻栓的种子外形（A）及其横切面（B）和纵切面（C）

1. 种皮　2. 胚乳　3. 子叶　4. 胚轴　5. 胚根

（史渭清绘）

果实的采收调制和种子贮藏　果实成熟时间不一。果实如过分成熟，种子即自行弹失。因此应选已出现裂缝而尚未开裂的成熟果实分批采摘。采回的蒴果置室内阴干，促使一部分接近成熟的种子后熟，随后晾晒，并用网筛罩盖，以免种子弹失。漂洗，清除瘪籽。普通干藏。种子千粒重约110g，每千克含种子9 000粒左右。

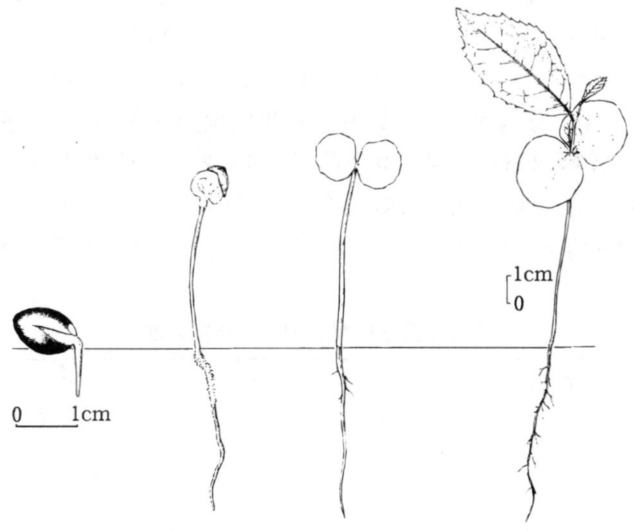

图2　牛鼻栓种子萌发后第2、18、20、45天的幼苗生长情况

（史渭清绘）

发芽和播种 种壳坚硬，种子可能有休眠习性，播种前需要层积处理约3个月。春播，4月初开始萌发，发芽期约20天，发芽率40%左右。出土萌发，子叶近圆形，生长较快(图2)。

（何泽瑛）

枫香树属

Liquidambar L.

（**金缕梅科 Hamamelidaceae**）

生长习性、分布和用途 枫香树属共5种，我国产2种1变种，本文描述2种。落叶乔木。枫香树主干高耸，适应性强，生长迅速，性喜阳光，根深抗风，耐火烧，萌发力强，常为次生林的优势树种，多散生于酸性或中性土壤的平地和村落附近。缺萼枫香抗寒性强，常为常绿阔叶林中的优势树种。它们分布于华东、华中、华南、西南以及台湾等地区。木材可供建筑、家具、农具和食品箱用。枫香树秋叶彤红，可供观赏（表1）。

表1 枫香树属树种的名称、分布和用途

中 名	学 名	树 高(m)	分 布	用 途
缺萼枫香	*L. acalycina* Chang	25	苏、浙、皖、鄂、赣、湘、粤、桂、黔、川	材用
枫香树	*L. formosana* Hance	30	华东、中南、西南、台。越南、老挝、朝鲜半岛	材用、药用、绿化

开花结实 春季开花。花单性，雌雄同株，无花瓣。雄花多数，聚合为短穗状花序，再组成总状花序。每一雄头状花序具苞片4，萼片缺如，雄蕊多且密集。雌花序圆头状，由多数雌花组成，萼筒与子房合生，子房半下位，藏于头状花序轴内，2室，胚珠多数，花柱2，柱头线形。花序柄长3～6cm。枫香树每一雌花序有花24～43朵，萼齿针形，宿存。缺萼枫香每一雌花序有花15～26朵，萼齿无，或为鳞片状。果实成熟期10月。果穗木质球形，径2.5～4cm，由多数蒴果组成，蒴果室间2裂。每个蒴果仅1～2枚可孕性种子。种子黑褐色扁平，顶端及两侧具翅。其余多为空瘪粒，细小，多角形，黄白色，无翅。枫香属这两个树种的开花结实习性见表2，它们种子的形状和大小见表3。胚乳薄，胚直立，胚根和子叶大体等长(图1)。

表2 枫香树属树种开花结实习性

树 种	花期 始 期	花期 盛 期	花期 末 期	果实成熟期	成熟特征	种子开始散落期
缺萼枫香	—	—	—	10月	果序为黄褐色	—
枫香树	3月中旬～4月上旬	3月下旬～4月上旬	3月末～4月上旬	10月上旬	果序由绿色转为黄绿色，种子褐色	11月中旬

表 3　枫香树属树种可孕性种子的形状和大小（含翅，空瘪粒无翅）

树　种	形　状	颜　色	长（mm）	宽（mm）
缺萼枫香	扁平，长椭圆形或长卵形，种皮具皱纹	褐色	6.0	2.0
枫香树	扁平，长椭圆形或长卵形	黑褐色	8～10	2.2

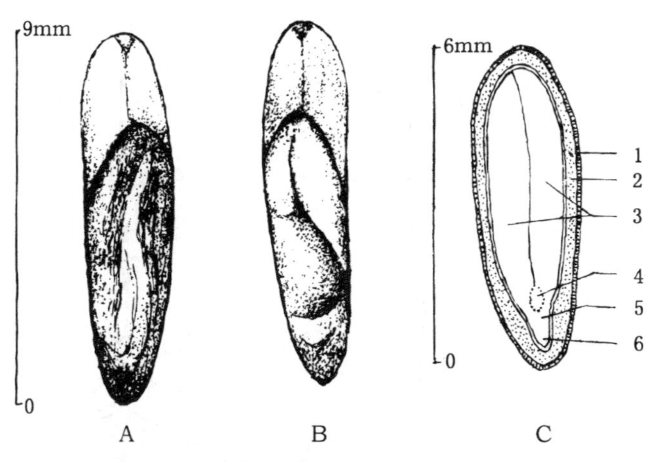

图 1　枫香树种子的外形（A、B）及其纵剖面（C）

1. 种皮　2. 胚乳　3. 子叶　4. 胚芽　5. 胚轴　6. 胚根

（田恒德绘）

果实的采收调制和种子贮藏　选择生长良好，无病虫害的母树，在果实充分成熟时，用竹竿敲击果枝，收集震落的种子。通常是上树采摘绿色已渐褪去的果球，曝晒 2～4 天，球中的蒴果开裂后用棍棒敲击脱粒，筛去果壳、果梗和无翅种子等杂质，搓揉去翅，即得纯净种子。出种率 0.3%～1.0%。千粒重等数据见表 4。种子干藏越冬。

表 4　枫香树属种子的出种率和种子质量

树　种	出种率（%）	千粒重（g）	每千克纯净种子粒数（万粒）
缺萼枫香	0.3	4.1～4.3	23～24
枫香树	0.5～1.0	4.3～6.4	15～23

发芽测定　种子发芽前无需处理。据王成霖（未刊稿）观察，变温加光照对枫香树的发芽势有抑制作用；对缺萼枫香的发芽势有促进作用，且随着每天光照时间的延长，发芽势逐步提高。枫香树种子的发芽温度以 30℃为好，发芽率在 70%以上。缺萼枫香种子在 25℃和 30℃的恒温下，或者在 20～30℃和 15～30℃的变温下，发芽率都在 85%以上，无显著差异（表 5）。

据本文作者观察，含水量为 8%的枫香树种子在 25℃下用石英沙床发芽，到第 2 天吸水量为种子干重的 67%并开始发芽，到第 3 天吸水量为 74%，发芽率达 92%，发芽已基本结束。

播种　播种期在 2 月中旬～3 月下旬。条播。播种沟深约 2cm，每平方米播枫香树种子

2.5～4g，缺萼枫香 2～5g。播后筛细土覆盖，以不见种子为度，稍加镇压。盖草，出苗约一半时揭草。枫香树出苗率约 50%，缺萼枫香出苗率约 40%。播种早的发芽期长，出苗率高，迟播的发芽期短，出苗率低。子叶 2，出土萌发（图 2）。1 年生苗出圃。

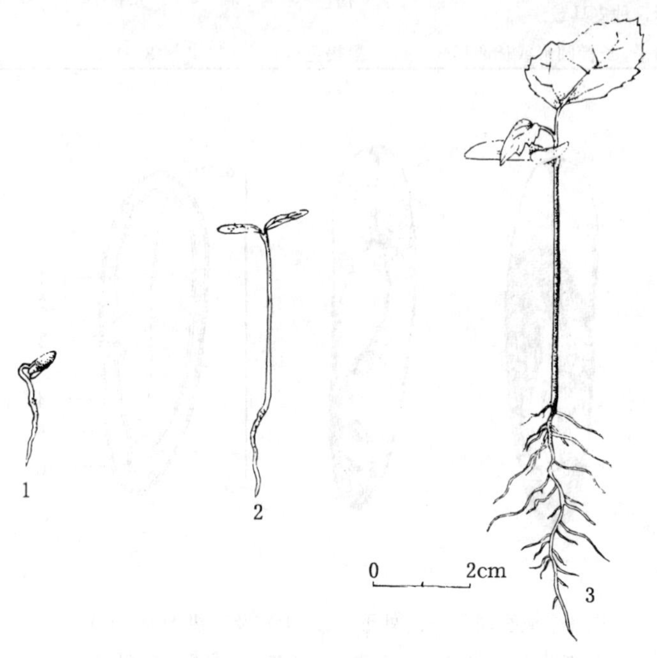

图 2　枫香树种子萌发后幼苗的生长情况

1. 下胚轴延伸　2. 子叶出土展开　3. 初生叶互生

（童军平仿《主要树木种苗图谱》）

表 5　枫香树属种子发芽能力及其测定条件①

树　种	测定条件		发芽势（%）			发芽率（%）		
	光照时数	温度（℃）	计算天数	一般数值	变动范围	计算天数	一般数值	变动范围
缺萼枫香	—	30	7～9	60	50～70	13～15	93	91～94
	12	20～30	11	72	64～76	14	93	90～98
枫香树	—	30	6～7	66	40～98	8～10	83	71～100
	8	20～30	7	40	26～50	11	84	76～88

①　种子经过精选，发芽基质为石英砂

（王成霖）

檵　木

Loropetalum chinense（R. Br.）Oliv.

（金缕梅科　Hamamelidaceae）

生长习性、分布和用途　檵木属有 4 种，分布于东亚的亚热带地区。我国有 3 种，本文只描述檵木1种。落叶或半落叶小乔木，高可达10m，常呈灌木状，见于江南低山丘陵的林缘灌丛中，喜光，多生于空旷山地，喜酸性土。分布于鲁东及长江中下游以南至华南、西南各地。全株均可作药用，春日白花繁盛，可供观赏。

开花结实　花两性，3～8朵簇生，花梗短。萼筒倒锥形、杯状，与子房合生。萼齿4，卵形，花后脱落。花瓣4，白色，狭带形，长1～2cm。雄蕊4。子房半下位，2室，每室有胚珠1。根据南京中山植物园1978和1979年的观察，4月上旬～5月上旬开花，10月上旬～10月下旬果实成熟。蒴果木质，卵球形，长7～8mm，径6～7mm，2瓣裂，每瓣2浅裂。种子卵形，长4～5mm，黑色，有光泽，脐部白色。种皮角质，脆。胚乳丰富。胚较窄长，子叶长椭圆形。种子的形态见图1。

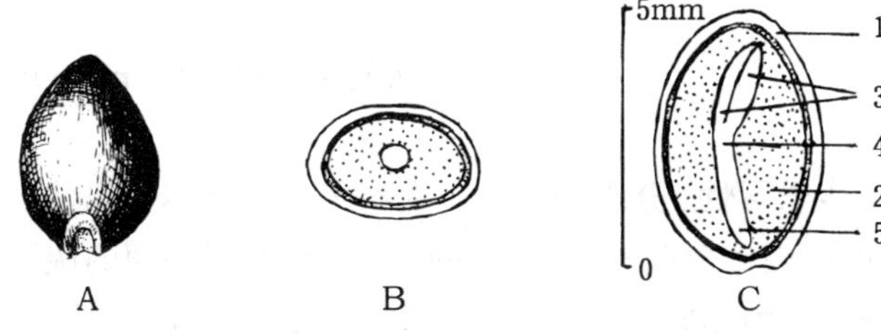

图1　檵木种子外形（A）及其横切面（B）和纵切面（C）
1. 种皮　2. 胚乳　3. 子叶　4. 胚轴　5. 胚根
（史渭清绘）

果实的采收调制和种子贮藏　10月蒴果转呈黑色尚未开裂时即可采摘。蒴果开裂时能将种子弹出，晾晒蒴果时宜用网筛覆盖其上，以免种子弹失。脱粒后用水漂除空粒。净度可达98%以上。种子可以普通干藏，也可层积越冬后春播。千粒重约20g，每千克约含种子4.8万粒。

发芽和播种　种子有休眠习性，可用秋播或层积越冬。混湿沙层积90天后春播，播后4月初萌发，发芽期15～20天。发芽率40%。子叶出土。萌发过程见图2。

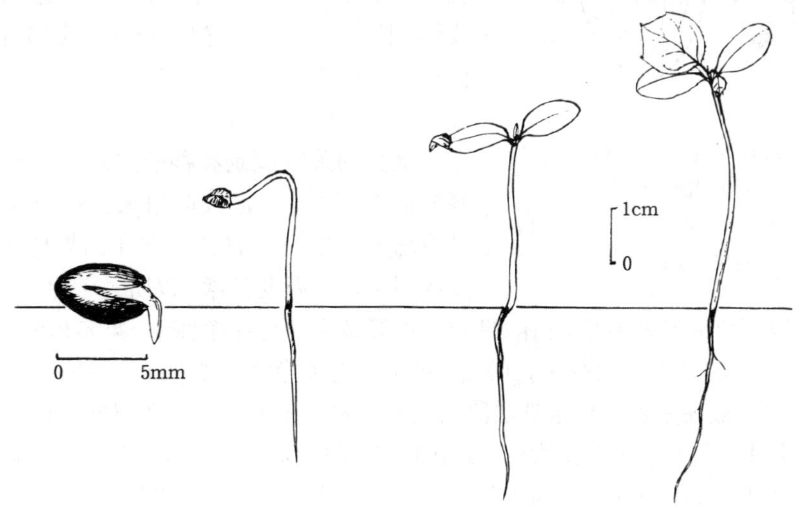

图2　檵木萌发后第1、10、20、45天的幼苗生长情况
（史渭清绘）

（何泽瑛）

壳菜果（米老排）

Mytilaria laosensis Lec.

（金缕梅科　Hamamelidaceae）

生长习性、分布和用途　壳菜果属只有本文描述的1种。常绿乔木，高达30m，胸径80～100cm。能耐短期－4℃左右的低温。喜肥沃湿润的疏松土壤，酸性沙壤至轻黏土上生长正常，石灰岩山地不宜，干旱贫瘠地生长不良。分布于滇、桂、粤南部。老挝、越南亦产。材质较优，不受虫蛀，是优良的速生用材，用于建筑、雕刻。落叶层厚，为重要的水源涵养树种。

开花结实　约10年生开始开花结实，正常结实期在20年生以后。结实丰年间隔期一般为1年，施肥可缩小大小年差异。花两性。肉穗状花序顶生或腋生。花多数，紧密排列在花序轴上。萼筒与子房连合，藏于肉质花序轴内。萼片5～6。花瓣5，带状舌形，白色。雄蕊10～13，着生于环状萼筒的内缘，花丝极短，花药4室藏于药隔内。子房下位，2室，每室胚珠6。花柱2，柱短，柱头有乳状突起。据广西南宁1978～1984年的物候观察，3月下旬花芽从顶梢或上部叶腋抽出，4月中旬花始开，4月下旬花盛开，花期不齐；5月上旬以后幼果陆续形成，但花期至6月中旬结束；10月中旬果实开始成熟，11月上旬为果熟盛期。

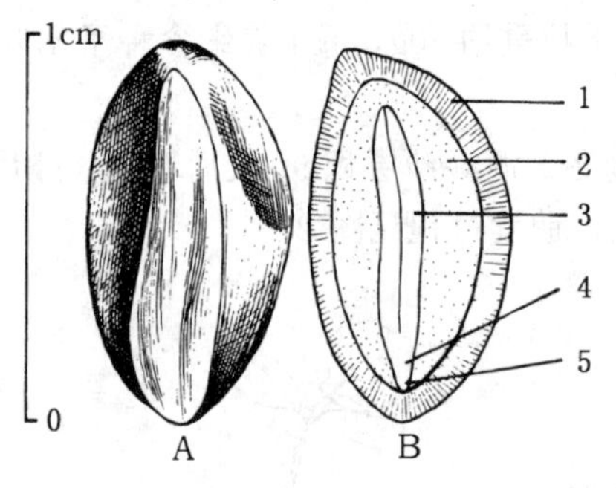

图1　壳菜果种子外形（A）及其纵切面（B）
1. 种皮　2. 胚乳　3. 子叶　4. 胚轴　5. 胚根
（黄应钦绘）

蒴果椭圆状卵形，长1.5～2cm，径约1.5cm。外果皮稍带肉质，松脆易碎。内果皮木质或软骨质，较外果皮薄。未熟时黄绿色，成熟时浅黄色或棕褐色。成熟后约10天，11月中旬至下旬，果实上半部2片裂，每片再2浅裂，种子弹出。种子近纺锤状或长三棱形，长1～1.4cm，宽5～6mm，腹部有明显带状白色种脐。种皮角质，黑褐色，有光泽。胚乳肉质，胚直立，位于中央（图1）。

果实的采收调制和种子贮藏　当果实进入成熟盛期尚未开裂时，应及时用采种钩刀或高枝剪截断果穗总梗，注意少伤树枝。采集的果穗在室内通风处摊放阴干，四周设屏障，以防种子弹失。大约摊晾1周时间，种子即可全部弹跳出来，清除果壳果轴等杂质即得纯净种子。果实出种率3%～5%，净度可达98%。千粒重160（146～180）g，每千克有纯净种子5 500～7 000粒。种子中的空粒较多，生产上一般在脱粒后用水选法除去空粒。种子不宜在烈日下曝晒，但可以晾干或置弱光下晒干，将种子的含水量降至大约12%即可运输贮藏。短期贮藏用袋缸等容器。2个月以上的贮藏应混以湿沙，贮藏期一般为半年以内。10月采集的种子宜于翌年3月中下旬播种，不宜贮藏至5月。

发芽和播种　种子休眠现象不明显。贮藏越冬的种子播前不作催芽处理，播后18～20天可以发芽。种子发芽时日均温需在20℃以上。1981年3月下旬和1987年3月下旬，广西林业科学研究所在室外沙床作发芽测定，所用种子一份经过水选，一份未经水选。经过水选的种子从开始发芽至发芽高峰之日，20天的发芽80%；从置床之日起算，45天的发芽率达95%。未经水选的种子则发芽盛期不明显，发芽率为42%。出土萌发。胚根萌发后5～6天子叶出土，再过12～15天初生叶展出。初生叶3浅裂，盾状着生。壳菜果种子的萌发和幼苗生长情况见图2。

为了促使种子快速发芽，生产中常在播前用始温50℃水浸种1昼夜，捞出后装入竹箩，裹以湿稻草，置于20～30℃的温暖处催芽。经过这样处理，大约10天可以萌动，然后播至圃地。每平方米播种约20g。条播，覆土约1cm。1年生苗出圃。也可将已催芽的种子点播于容器内，培育百日苗出圃。

（王宏志）

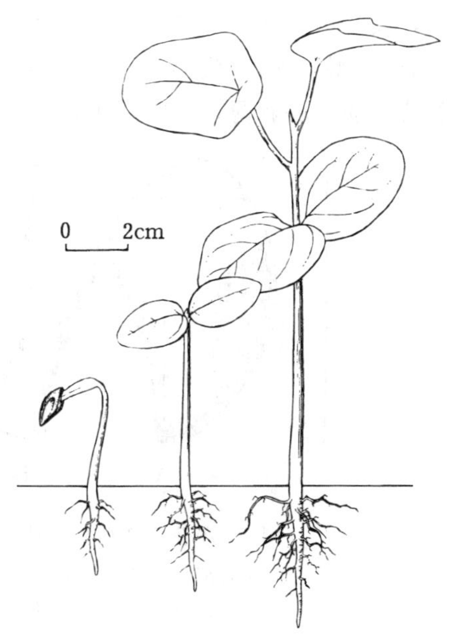

图2　壳菜果种子萌发后第5、20、30天幼苗的生长情况

（黄应钦仿《热带亚热带主要树种采种育苗技术》，有修正）

小花红花荷（红苞木）
Rhodoleia parvipetala Tong

（金缕梅科　Hamamelidaceae）

生长习性、分布和用途　红花荷属约9种，我国产6种，本文描述1种。常绿乔木，高达20m，胸径30cm，生长速度中等。天然下种能力强，天然更新良好。对土壤要求稍高，常生于酸性低山地，在干旱瘠薄山地或低丘地不能正常生长。分布于滇、黔、桂、粤。越南亦产。木材红褐或黄褐色，有光泽，结构细致，耐腐，易加工，供优良家具及建筑等用。花玫瑰红色，早春开放，且树形美观，可用于庭院观赏或作行道树。叶药用，可止血，治刀伤。

开花结实　8～10年生开始结实，18年生以后进入正常结实期，结实大小年现象不明显。花两性。头状花序腋生，花序长2～2.5cm，花序梗长1～1.5cm；总苞片5～7，卵圆形，大小不等，长7～10mm，被褐色柔毛。萼筒极短，先端平截，包围在子房基部。花瓣4～6，红色，匙形，长15～18mm。雄蕊6～8，与花瓣等长。花柱2，线形，与雄蕊等长，先端尖。子房半下位，2室，每室胚珠12～18，2列着生。据广西南宁1978～1983年的物候观察，1月中旬为始花期，2月上旬为盛花期，3月下旬为末花期；10月中旬果实开始成熟，10月下旬果盛熟。头状果序径2～3.5cm，有蒴果5。果未熟时深绿色或灰青色，成熟时黄绿色或褐色。

蒴果卵圆形，长 8～13mm，径 14～18mm。果皮薄木质，熟后约 20 天上部室间、室背 4 裂，种子散落。种子多数，扁平，棱形或不规则形，浅褐色，长 4～5.5mm，宽 3～3.5mm，厚0.5～0.7mm。种皮角质。胚乳薄，子叶扁平，长圆形（图 1）。

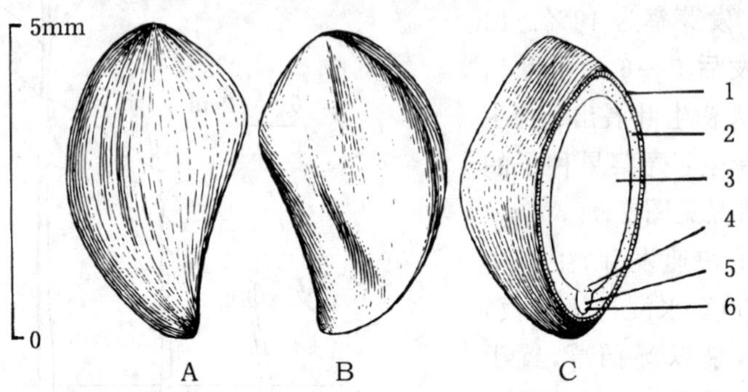

图 1　小花红花荷种子外形（A、B）及其纵切面（C）
1. 种皮　2. 胚乳　3. 子叶　4. 胚芽　5. 胚轴　6. 胚根
（黄应钦绘）

果实的采收调制和种子贮藏　果实成熟盛期用采种刀钩截果穗，或上树用手采摘。采回的果穗摊晒至果壳开裂，筛取种子。鲜果出种率约 3%。净度约 70%，千粒重 2.5～2.9g，每千克有纯净种子约 34 万～40 万粒。含水量约为 10%。种子不能曝晒，宜晾干。干藏，贮藏期一般为半年。

发芽和播种　种子无休眠习性。发芽时日均温要求 15℃以上。1983 年 1 月 29 日，广西林业科学研究所将前一年 10 月所采的种子经水浸处理后，在室外沙床作发芽测定。播后 9 天（2 月 6 日）开始发芽，发芽盛期不明显，2 月 21 日发芽终止。自播种之日起算，24 天的发芽率仅 10%。发芽率低的原因可能是采种母树尚未进入正常结实年龄。出土萌发。胚根萌发后 5 天子叶出土，再过 30 天发出初生叶。

撒播。每平方米播种量 2～3g。也可以催芽密播，苗木半木质化时移植至大田继续培育。苗期需遮荫，一般培育 1 年生苗出圃。如用于庭院绿化，需培育 2～3 年生苗。

（王宏志）

悬铃木属

Platanus L.

（悬铃木科　Platanaceae）

生长习性、分布和用途　本属是悬铃木科仅有的 1 个属，约 10 种，分布于北美至中美墨西哥、欧洲东南部、亚洲西南部至印度。本文描述我国引入的 3 个种，它们的名称、成年时树高、分布和用途见表 1。其中我国栽培最多的二球悬铃木是三球悬铃木和一球悬铃木在英国育成的杂交种。三球悬铃木在陕西户县有大树，胸径达 3m，相传为晋朝时引入（陈嵘，1933），今已不存在。这 3 个种都是落叶大乔木，在原产地高达 30～50m。喜光，不耐蔽荫，生长迅速，喜温湿气候，在肥沃排水良好的微酸性或中性土壤上生长良好，忌积水。抗空气污染能力较强。树形高大，枝叶繁茂，耐修剪，易整形，树皮白色，是优良的行道树。据观察，扦插苗根系浅，易遭风害，实生苗根系发达，抗风力强。

表1　悬铃木属树种的名称、生长、分布和用途

中　名	学　名	树　高 (m)	分　布	用　途
二球悬铃木	*P. hispanica* Muench.	35	英国育成，广植世界各地。我国黄河及长江中下游一带广泛引种栽培	庭园、行道树、材用
一球悬铃木	*P. occidentalis* L.	50	原产北美。苏、鲁、豫、陕、冀、粤等有栽培。	庭园、行道树、材用
三球悬铃木	*P. orientalis* L.	30	原产欧洲东南部、亚洲西部。我国陕、晋、冀、鲁、豫、苏、川、新等有栽培	庭园、行道树、材用

开花结实　实生树木4～6年生开始开花结实，以后逐年增多，直至树木衰老。一球悬铃木每隔1～2年有一个丰年，二球悬铃木和三球悬铃木无明显的丰歉年。开花期4月或4～5月。果熟期9～10月或10～11月。据1963年在南京地区对二球悬铃木的观察，4月3日花蕾始现，9日花初开，14日花盛开，18日花期结束，10月下旬果实开始成熟。又据1988年在南京观察，一球悬铃木的花期比二球悬铃木花期迟20天左右，果实成熟期则大体一致。花单性，雌雄同株，头状花序。雄花序无苞片，雌花序有苞片。雌花暗红色，一般沿小枝着生。雄花浅绿色，着生在枝顶。悬铃木属3个树种的果序和花柱情况见表2。

表2　悬铃木属3个树种的果穗数和花柱长度

树　种	每个总柄上球形果穗的数目（个）	果穗直径（cm）	花柱长度（mm）
二球悬铃木	2（1～3）	2.5	2～3
一球悬铃木	1（稀2）	3～4	极短
三球悬铃木	3（2～6）	2～2.5	3～4

果为聚花果，其球状果穗由600～1 400枚小坚果组成。成熟后球状果穗悬挂枝上，经冬不落。小坚果栗褐色，狭长倒圆锥形，基部围有褐黄色长毛，顶端花柱宿存。每个坚果含种子1枚，线形，胚乳层薄，胚直伸（图1）。

果实采收、调制和贮藏　球形果穗变成褐色即可采收，以落叶之后进行最为方便。由于果穗及其坚果要到来年春季新叶展现之后才逐渐散落，所以整个冬季都有充裕的时间采收。可用高枝剪直接剪下果穗或击落果穗，也可在春季从地面收集散落的果穗或坚果。采回的果穗适当摊晒干燥后在干燥通风的室内贮藏。翌春播种前取出搓揉或用小木棒敲打，取得的小坚果即为播种材料，通称种子。搓揉或敲打过程中会产生刺激性的粉尘，操作人员应有防护设

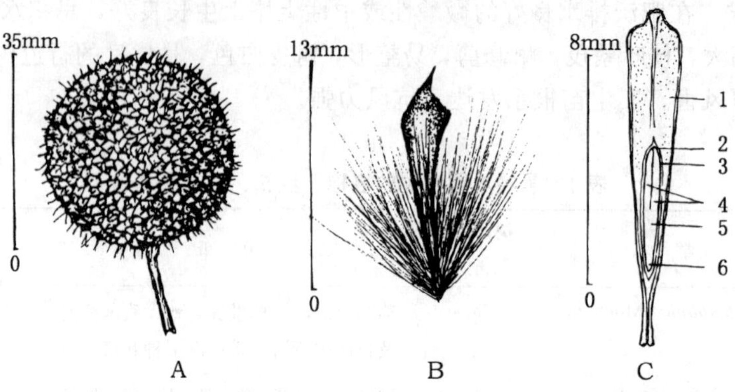

图 1　二球悬铃木的球状果穗（A）、坚果外形（B）及其纵切面（C）
1. 果皮　2. 种皮　3. 胚乳　4. 子叶　5. 胚轴　6. 胚根
（田恒德绘）

备。坚果所附的绒毛不妨碍种子发芽，可以不必清除。种子发芽能力较难保存，且结实量大而用种量少，一般都是采后翌春播种，不作长期贮藏。据美国报道，如果贮藏时间超过 1 年，应使种子含水量干燥到 10%～15%，并在 0℃左右的低温下密封贮藏。单性结实现象很普遍，有报道说二球悬铃木的空粒可达 60%～90%。一球悬铃木种子千粒重一般 2.3g，变动范围1.7～3.4g，每千克种子一般 43 万粒，变动范围 29 万～59 万粒。二球悬铃木种子千粒重一般 3.8g，变动范围 1.4～6.2g，每千克种子一般 26 万粒，变动范围 16 万～70 万粒。

发芽和播种　据南京市农林局测定，二球悬铃木种子的发芽以在 15～25℃温度范围内比较适宜，其中以 20℃为最好。在 25～35℃温度范围内，发芽率随温度增高而降低，超过 35℃，种子几乎不能发芽。据孙秀琴和田玉霞（1990）报道，二球悬铃木发芽测定时，25℃/30℃的昼夜变温优于 30℃/20℃昼夜变温；且这个种的种子能在黑暗中发芽，但有条件时仍以每昼夜给予 8 小时光照为好。

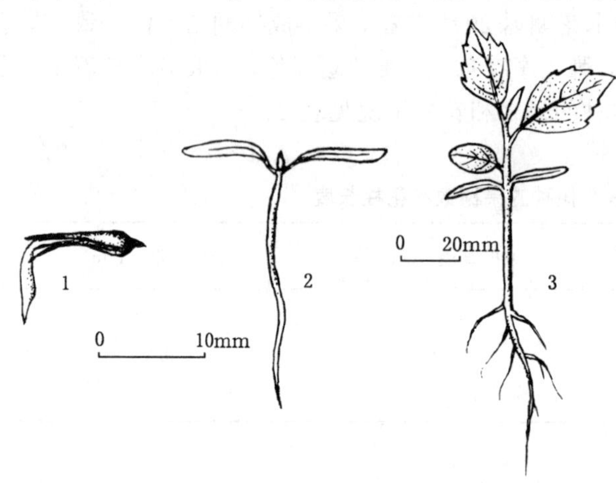

图 2　二球悬铃木种子的萌发和幼苗生长情况
1. 胚根伸出　2. 子叶展开　3. 初生叶互生
（田恒德绘）

二球悬铃木种子的发芽率通常只有 20%左右。如果空粒更多，发芽率会更低。一球悬铃木的发芽率有高达 80%的报道，但有些种批的发芽率也仅为 20%～35%。据报道，三球悬铃木的发芽率一般为 40%。悬铃木属这 3 个种的种子播种前无需处理。不过据南京市的经验，二球悬铃木播种前如果用 5℃左右的低温层积 20～30 天，发芽效果更好。如果每平方米希望得苗 30 株，种子发芽率为 20%，每平方米约需播种 3.5g。播前床面充分浇水湿透，播后覆盖细土或锯木屑 0.5cm。播后 3～5 天即可发芽。出土萌发。初生叶不呈掌状分裂（图 2）。1 年生实生苗高可达 1m，可以出圃供庭园栽植。行道树则常用4～5 年生的大苗。悬铃木属的树种很容易扦插繁殖。

（杨余康）

黄　杨　属

Buxus L.

（黄杨科　Buxaceae）

生长习性、分布和用途　本属约 70 种，分布于亚洲、欧洲、热带非洲以及中美洲的古巴、牙买加等处。我国约 17～20 种。本文描述 2 种，其中 1 种为引入栽培。常绿灌木或小乔木。喜温暖湿润气候。黄杨较耐荫，锦熟黄杨较喜光。对土壤要求不严，喜石灰岩山地，忌积水涝洼地，萌芽力强，耐修剪。枝叶茂密、四季常青，能抗烟尘，供园林观赏和工矿区绿化。材质坚硬致密，可作木雕等美工制品。全株可入药。两个树种的名称、树高和分布见表 1。

表 1　黄杨属两个树种的名称、树高和分布情况

中　名	学　名	树高（m）	分　布
锦熟黄杨	*B. sempervirens* L.	6～9	原产西亚、北非、南欧。京、鲁、豫引种栽培
黄　杨	*B. sinica* (Rehd. et Wils.) Cheng ex M. Cheng	1～6（8）	陕、甘、川、黔、桂、粤、赣、鄂、皖、苏、浙。鲁、冀、晋、豫广泛栽培

开花结实　5～8 年开始结实，12～15 年为正常结实期，大小年现象不明显。花单性，雌雄同序。头状花序或丛生叶腋，雌花 1 朵生花序顶端，雄花数至 10 朵生花序下方或四周。花小，黄绿色，无花瓣。雄花萼片 4，2 轮，雄蕊 4，与萼片对生，有长为萼片一半的退化雌蕊（锦熟黄杨）或有与萼片几等长的退化雌蕊（黄杨）。雌花萼片 6，2 轮，子房上位，3 室，花柱 3，分裂。花柱粗扁，柱头倒心形，下延至花柱近中部，宿存，子房每室 2 胚珠。3～4 月开花，6～7 月果实成熟。蒴果卵圆形或近球形，长 6～8（10）mm，径 8～10mm，成熟时由绿色变为黄褐色，沿室背裂为 3 片，宿存花柱角状，长 2～3mm，外果皮和内果皮分离。每室 2 种子，种子长圆形或三角状椭圆形，有三侧面，长 5～5.5mm，种皮黑色，有光泽。胚乳肉质，胚位于中央直立（图 1）。

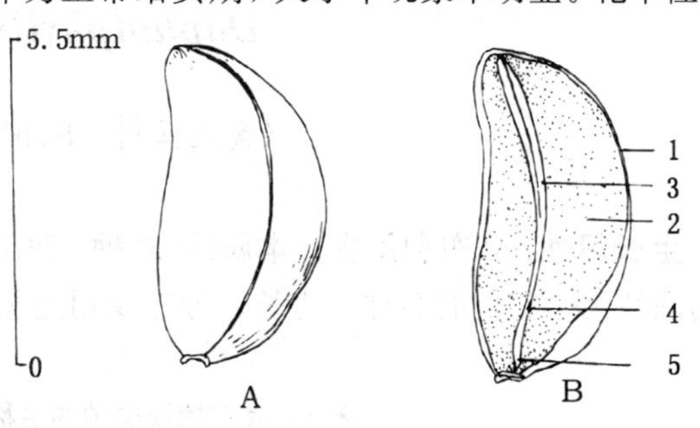

图 1　锦熟黄杨种子外形（A）及其纵切面（B）
1. 种皮　2. 胚乳　3. 子叶　4. 胚轴　5. 胚根
（胡冬梅、林平绘）

果实采收、调制和种子贮藏　6～7 月蒴果由青变黄褐色时即需及时采摘，稍迟则蒴果开裂，种子弹出，易散失。采后置阴凉处，使其开裂，晾干，脱粒，去杂，净种，即得播种材料。种子忌干燥失水，混湿沙贮藏。2种黄杨种子净度和质量见表2。

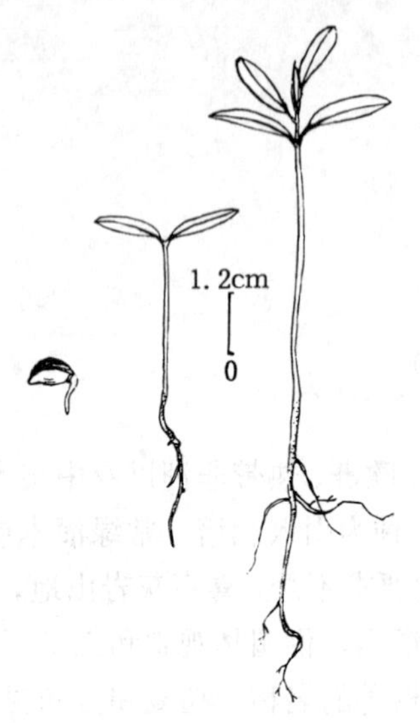

图2 锦熟黄杨发芽出土后第1、6、20天的幼苗生长情况

（胡冬梅、林平绘）

表2 黄杨属两个树种种子的净度和质量

树种	净度（%）	千粒重（g）	每千克纯净种子数（万粒）
锦熟黄杨	95～98	17～19	5.2～5.8
黄杨	95～98	25～28	3.5～4.0

发芽和播种 种子有休眠现象。发芽时的日均温宜在20℃左右。锦熟黄杨的发芽率一般为60%，变动范围为50%～70%。最好进行秋播，一般于10月中旬采用低床条播或撒播。为了防止冬天受冻，11月中旬结冻前盖草防寒，至翌年3月上旬转暖后去掉防寒物。为促使种子提早发芽出土，可在播种床上设高40cm左右的弓形塑料棚，以增高床内温度和湿度，出苗后即可撤去塑料薄膜，每平方米床面留苗80～100株。亦可扦插繁殖，四季均可进行。种子萌发情况见图2。

（印佩文、黄鹏成）

虎皮楠（交让木）属
Daphniphyllum Blume

（虎皮楠科 Daphniphyllaceae）

生长习性、分布和用途 本属约30种，我国约产13种，本文描述2种。常绿。喜疏松肥沃的酸性土。它们的名称、生长、分布及用途见表1。

表1 虎皮楠属树种的名称、生长、分布和用途

中名	学名	树高（m）	胸径（cm）	分布	用途
牛耳枫	*D. calycinum* Benth.	4～8	10～15	滇、桂、粤、湘、赣。越南、日本	种子榨油，根清热散瘀
虎皮楠	*D. oldhamii* (Hemsl.) Rosenth.	15	30	台、闽、浙、赣、鄂、湘、粤、桂、黔、川、陕。朝鲜半岛、日本	材用、雕刻

开花结实　牛耳枫5～6年生开始结实，正常结实期在10年生以后。虎皮楠约10年生开始结实，20年生以后进入正常结实期。结实大小年现象不明显。花单性，雌雄异株。总状花序，腋生。无花瓣。雄花：花萼小，4～6裂或盘状3～4浅裂；雄蕊7～12，花丝短，花药2室，纵裂，无退化雌蕊。雌花：子房上位，2室，每室有悬垂的倒生胚珠2，花柱短或近无，柱头2裂，卷曲。据在广西南宁1982～1988年观测，牛耳枫1年中主花期2次，虎皮楠为1次，雌雄花的花期同时。它们的开花结实物候期见表2。

表2　虎皮楠属树种的开花结实物候期（广西南宁，1982～1987）

树　种	开　花			果实成熟		果实脱落
	始　期	盛　期	末　期	始　期	盛　期	
牛耳枫	5月下旬	6月中旬	7月上旬	8月上旬	8月中旬	8月中旬～8月下旬
	8月上旬	8月中旬	9月上旬	10月上旬	10月中旬	10月下旬～11月
虎皮楠	4月上旬初	4月上旬末	4月中旬	10月下旬	11月上旬	11月中旬～12月上旬

浆果状核果，花柱宿存，椭圆形或球形。外果皮薄膜质，被瘤点，中果皮肉质，内果皮骨质。每果具1种子。种皮膜质，胚乳丰富，胚小，顶生。果实及种子形态见表3、图1。

表3　虎皮楠属树种果实和种子的形态特征

树种	未熟果颜色	果实			种子		
		形　状	大小（mm）	色　泽	形　状	大小（mm）	色　泽
牛耳枫	青绿色	卵圆形，小瘤状突起，基部具宿萼	7～10	黑色，被白粉	圆球形	径5～7	暗褐色
虎皮楠	黄绿色	椭圆形，具不明显瘤点，无宿萼	长8～14 径5～8	暗红色至黑色	椭圆形	长5～9 径3～5	深褐色，有星状点

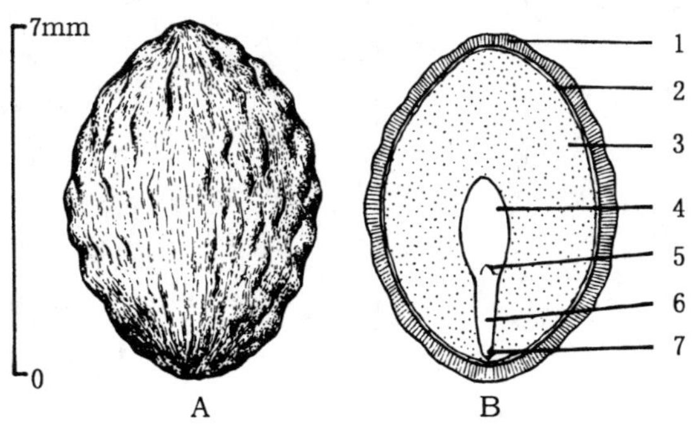

图1　虎皮楠果核外形（A）及其纵切面（B）

1. 内果皮　2. 种皮　3. 胚乳　4. 子叶　5. 胚芽　6. 胚轴　7. 胚根

（黄应钦绘）

果实采收调制和种子贮藏 牛耳枫的树形较矮，果实成熟后可用手采摘。采收虎皮楠的果实之前先清理地面，用竹竿敲打或摇动树枝震落后在地面捡拾，也可上树采摘。采得的果实堆沤 2～3 日，待果皮充分软熟后置水中搓擦，淘去果皮等杂质，所得果核用作播种材料，通称种子。这两个树种的种子都忌失水，不能日晒，含水量宜保持在 20%～25%。运输或贮藏均应混以湿沙，贮藏期一般为半年以内。它们的出籽率和种子的净度、质量等见表 4。

表 4 虎皮楠属树种的出籽率和种子净度、质量

树 种	出籽率(%)	净度(%)	千粒重(g)		每千克种子粒数	
			一 般	变动范围	一 般	变动范围
牛耳枫	60	97	104	100～110	9 600	9 000～10 000
虎皮楠	36	97	83	75～90	12 000	10 000～13 000

表 5 虎皮楠属树种的发芽能力及其测定条件①

树 种	温度（℃）	发芽势（%）		发芽率（%）	
		计算天数	一般数值	计算天数	一般数值
牛耳枫	15～18	15	50	36	62
虎皮楠	15～16	20	76	38	86

注 ①室外沙床

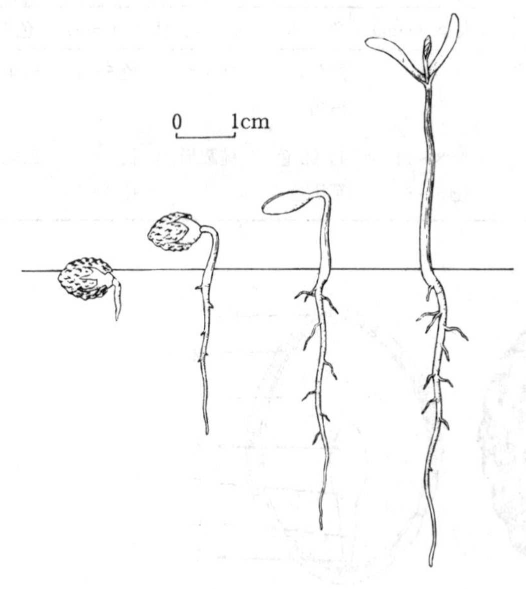

图 2 虎皮楠种子萌发后第 5、10、16、20 天幼苗生长情况

（黄应钦绘）

发芽和播种 种子无明显休眠现象。发芽时要求的气温不高，日均温仅需在 15℃以上。1987 年 9 月和 11 月，广西林业科学研究所用在室外沙床上用当年新采种子对牛耳枫和虎皮楠作过发芽测定，结果见表 5。

出土萌发。虎皮楠胚根萌发后约 10 天子叶带核壳出土，16～18 天壳脱落，子叶外露，20 天左右初生叶出现。种子发芽及幼苗生长见图 2。

条播或点播。每平方米牛耳枫播种20～25g，虎皮楠播种 12～15g，覆土 1.5cm。育 1 年生苗出圃。

（王宏志）

杨　属
Populus L.

（杨柳科　Salicaceae）

生长习性、分布和用途　杨属约100余种，除2种产于赤道非洲外，均产于欧、亚和北美的北温带地区。我国60余种（包括6天然杂种和引入栽培的4种），还有很多变种、变型和引进的栽培品种，分布于北纬25°～53°34′，东经80°～134°的范围内。本文描述21种。落叶中乔木至大乔木，耐寒，喜光，速生。木材白色，轻软，细致，可供建筑、板材、火柴杆、纤维造纸。叶可做牛、羊饲料。芽脂、花序、树皮可以药用。本属多为用材林、防护林、水土保持林或四旁绿化的优良速生树种。

杨属共分5组，我国均产。各组形态主要特征见表1。

表1　杨属各组主要形态特征①

组与亚组	芽	叶　片	叶　柄	雄　花	雌　花	果
白杨组						
白杨亚组	小，被绒毛	短枝叶较小，常为椭圆形，稀圆形。长枝与萌枝叶分裂，下面密被绒毛	短而圆	柔荑花序长8～10cm，雄蕊6～10	柔荑花序长5～11cm，柱头2～4裂	蒴果，2瓣裂
山杨亚组	小，无毛，有光泽，稀有柔毛，或芽鳞边缘有毛	短枝叶常圆形或卵圆形。长枝与萌枝叶不分裂，不为永久性毛	长而扁	柔荑花序长8～10cm，雄蕊5～20	柔荑花序长10～12cm，柱头2裂	蒴果，2瓣裂
大叶杨组	富粘脂，有光泽	大形、基部心脏形	圆	雄蕊6～40	柱头2～3裂	蒴果，2～3瓣裂
青杨组	常较大，有粘性，具芳香味	卵形，卵圆形，或椭圆形，长大于宽，基部常圆形或楔形，下面常苍白色	圆	雄蕊8～60	柱头2～4裂	蒴果，2～4（5）瓣裂
黑杨组	大型、有粘脂，具光泽	正三角形、三角形或菱形。叶缘具半透明边	扁	雄蕊12～40（60）	柱头2～4裂	蒴果，2～4瓣裂
胡杨组	有毛，稀疏毛或无毛	变异大，大部常具齿牙，长短枝叶异形	圆或先端稍扁	雄蕊15～35，柔荑花序大而红	柱头3～6裂，红色	蒴果，3瓣裂

① 引自FAO出版的《杨树》（1958年），有修正

杨属各种的名称、分布和用途见表2。

表 2 杨属树种的名称、分布及用途

中 名	学 名	树 高 (m)	分 布	用 途	供 稿
白杨组	Sect. *Populus*				
白杨亚组	Subsect. *Albidae* Dode				
银白杨	*P. alba* L.	15～30	新(额尔齐斯河)。东北南部、华北、西北栽培。北非、欧洲、西亚	建筑，造纸，树皮可提取栲胶，绿化，沙荒造林	201
银灰杨	*P. canescens* (Ait.) Smith	20	新(额尔齐斯河)。甘、黑栽培。欧洲、高加索、小亚细亚、哈萨克斯坦	建筑，造纸，树皮可提取栲胶，绿化，沙荒造林	203
毛白杨	*P. tomentosa* Carr.	30	东北南部、华北、华东、西北东部（以黄河中下游为中心）	用材，造纸，花序入药。庭园绿化，行道树	201
山杨亚组	Subsect. *Trepidae* Dode				
响叶杨	*P. adenopoda* Maxim.	15～30	陕、豫、华东、华中至西南	用材，器具，造纸，叶为饲料	1001
山杨	*P. davidiana* Dode	25	东北、华北、西北、华中至西南高原地带。朝鲜半岛、俄罗斯远东、日本	建筑，树皮入药或提取栲胶，造纸、水土保持	201、107
河北杨	*P. hopeiensis* Hu et Chow	30	华北、西北，河北山区习见	建筑，农具，水保，用材，行道树	201
欧洲山杨	*P. tremula* L.	10～20	新(阿尔泰山西南坡至天山北坡及伊犁山区)。西伯利亚、欧洲、高加索、北非	纸浆，纤维，建筑，树皮可提取栲胶	203
大叶杨组	Sect. *Leucoides* Spach				
大叶杨	*P. lasiocarpa* Oliv.	20	陕、鄂、川、黔、滇。川东、鄂西为多	家具，板料	201
青杨组	Sect. *Tacamahaca* Spach				
青杨	*P. cathayana* Rehd.	30	东北南部、西北、华北，川	家具，建筑，绿化，防护	201
二白杨	*P. gansuensis* C. Wang et H. L. Yang	20	甘(河西走廊)。内蒙古等地引种	防护林	201
香杨	*P. koreana* Rehd.	30	东北。朝鲜半岛、俄罗斯远东	建筑，造纸，火柴杆	107
苦杨	*P. laurifolia* Ledeb.	10～25	新(阿尔泰和塔城地区)，北疆多栽培。哈萨克斯坦、蒙古、西伯利亚	用材，环保，燃料，造纸，饲料	203
小青杨	*P. pseudosimonii* Kitag	20	东北，华北，西北东南部，川	用材、防护	201、107
冬瓜杨	*P. purdomii* Rehd.	30	冀、豫、陕、甘、川、鄂	建筑、造纸	201

（续）

中　名	学　名	树　高（m）	分　布	用　途	供　稿
小叶杨	*P. simonii* Carr.	20	东北、华北、华中、西北、西南	用材、造纸、防护	303
密叶杨	*P. talassica* Kom.	20～25	新(天山中部至西部)。哈萨克斯坦	用材，涵养水源	203
大青杨	*P. ussuriensis* Kom.	30	东北东部山地。朝鲜半岛、俄罗斯远东	建筑，舟船，造纸，火柴杆	103
滇杨	*P. yunnanensis* Dode.	20	滇、黔、川	火柴杆，行道树	615
黑杨组	Sect. *Aigeiros* Duby				
黑杨	*P. nigra* L.	30	新（额尔齐斯河和乌伦古河流域）。中亚、西亚至欧洲	材用，树皮可提取栲胶，做黄色染料，芽药用	203、107
胡杨组	Sect. *Turanga* Bge.				
胡杨	*P. euphratica* Oliv.	12～20	新、青、甘、内蒙古西部。蒙古、中亚、西亚、南亚	建筑，造纸。“胡杨碱”是优良生物碱，食用、工业用。造林	201
灰胡杨	*P. pruinosa* Schrenk	10（20）	新(准噶尔盆地至塔里木盆地)。中亚	防风，固沙	203

表 3　杨属树种的开花结实物候

树　种	开始结实年龄	观察地点	花　期	果实成熟期	种子散落期
银白杨	10～15	河北	3～4 月	4 月	4～5 月
		内蒙古西部	4～6 月	4～6 月	4～6 月
		新疆	3～4 月	5 月中旬	5 月
		北京	3～4 月	4 月上旬	4 月
		山东	4～5 月	5 月	5 月
		黑龙江	5 月	6 月上旬	6 月
银灰杨	6～7	新疆玛纳斯林场	4 月上旬	4 月上旬～5 月下旬	4 月中下旬
毛白杨	10（20 以上为结实盛期）	北京	3 月下旬～4 月下旬	4～5 月	4～5 月
		山西太谷	2 月下旬～3 月中旬	4 月	4 月
		陕西	2 月下旬～3 月上旬	4 月	4 月
		河南洛阳	2 月中旬～3 月中旬	4 月上旬～4 月下旬	4 月
		河北	3 月	5 月	5 月

（续）

树　种	开始结实年龄	观察地点	花　期	果实成熟期	种子散落期
		山东泰安	3月上旬～3月下旬	5月	5月
		山东淄博	3月中旬～3月下旬	5月	5月
响叶杨	5～8	南京	2～3月	4月中下旬	4月中下旬
山杨	8	北京	3月下旬	4～5月	4～5月
		黑龙江森林植物园	4月下旬～5月上旬	5月14日～6月10日	初熟至脱落间隔4～6天
		内蒙古西部	4～5月	5～6月	5～6月
河北杨	—	河北	4月	5～6月	5～6月
		山西	4月	5～6月	5～6月
		内蒙古西部	4～5月	5～6月	5～6月
欧洲山杨	8～10	新疆阿尔泰，额尔齐斯河流域	4月上旬～5月下旬	5月中旬	5月中旬
大叶杨	—	湖北	4～5月	5～6月	5～6月
		陕西南部	5月	6月	6月
青杨	—	内蒙古呼和浩特市	4月	5～6月	5～6月
		黑龙江	4月下旬～5月	7月中旬	7月
		陕西关中	3～4月	4月中下旬	4月中下旬
		青海西宁	3～4月	6月中旬	6月中旬
二白杨	5～7	甘肃武威市	3月下旬～4月上旬	5月下旬	5月下旬
香杨	8～15	内蒙古呼和浩特市	4月下旬	6月	6月
		黑龙江森林植物园	—	6月5日～6月18日	初熟至脱落间隔10天
苦杨	7～8	新疆玛纳斯林场	4～5月	6月	6月
小青杨	8～15	内蒙古呼和浩特市	4月下旬	6月	6月
		黑龙江森林植物园	4月下旬～5月上旬	6月上中旬	初熟至脱落间隔3～5天
		河北	3～4月	4～5月	4～5月

（续）

树　种	开始结实年龄	观察地点	花　期	果实成熟期	种子散落期
冬瓜杨	—	陕西秦岭林区（随海拔高度增高而延迟）	4～5 月	6 月上旬～7 月下旬	6 月上旬～7 月下旬
小叶杨	5～6	北京地区	3～5 月（初花 3 月 28 日，盛花期 3 月 29 日，末期 4 月 6 日）	4～6 月（初期 5 月 7 日，盛期 5 月 8 日，末期 5 月 14 日）	4～6 月
		内蒙古呼和浩特市	4 月下旬	5 月下旬～6 月上旬	5～6 月
		内蒙古奈曼旗	4～5 月	5 月下旬～6 月上旬	5 月下旬～6 月上旬
		陕西关中	3～4 月	4 月中下旬	4～5 月
		河南	3～4 月	4 月下旬	4～5 月
		陕西北部	4～5 月	5 月中下旬	5 月
		宁夏	4～5 月	5 月中下旬	5 月
		吉林	5 月	5 月下旬～6 月上旬	5～6 月
		辽宁南部	4～5 月	5 月下旬	5～6 月
		黑龙江森林植物园	4 月	5 月 13 日～6 月 13 日	初熟至脱落间隔 4～8 天
密叶杨	—	新疆乌鲁木齐市南山河谷	4～5 月	6～7 月	6～7 月
大青杨	8～9	黑龙江森林植物园	4 月下旬～5 月上旬	6 月上旬～6 月中下旬	初熟至脱落间隔 4～7 天
		吉林长白山林区	5 月上旬	6 月中下旬	6 月中下旬
滇杨	8～10（15～25 年为正常结实期）	云南昆明	3 月～4 月上旬	6 月	6 月
黑杨	5	新疆阿尔泰额尔齐斯河流域	4 月上旬	5 月下旬	5 月下旬
		黑龙江森林植物园	5 月上旬	6 月 3 日～6 月 18 日	6 月上旬初熟至脱落间隔 3～5 天
胡杨	6～7	内蒙古巴彦淖尔盟	4 月下旬～5 月上旬	7 月中旬～8 月上旬	7～8 月
		内蒙古阿拉善盟	4 月下旬～5 月上旬	7 月中旬～8 月上旬	7～8 月
		甘肃民勤植物园	4 月～5 月上旬	7 月上旬～8 月上中旬	7～8 月
		新疆	5 月上旬	7 月上中旬～8 月下旬	7～8 月
灰胡杨	8～15	新疆塔里木盆地	5 月	7～8 月	7～8 月

开花结实 杨树一般6～10年开花结实。杨属的多数树种没有明显的结实丰歉现象，但胡杨每隔1年，欧洲山杨每隔4～5年出现一次结实丰年。杨属各树种的开花结实物候期见表3。花单性，雌雄异株，极稀两性或雌雄同株、同穗。柔荑花序下垂。花无花被，基部具杯状花盘，着生于撕裂或条裂的苞腋内。苞片膜质，早落。花常先叶开放，雄花穗稍早于雌花穗。雄蕊5至多数，花药暗红色，2室，纵裂，花丝短，离生。杨属为风媒传粉，花粉量大，花粉粒小。100个雄花穗可得50～100mL花粉。雌蕊由2～4（5）心皮构成，子房1室，侧膜胎座，具多数倒生胚珠。花柱短，柱头2～4裂。蒴果，2～4裂，极稀5裂。果未熟时绿色，熟时黄绿色。每个果有种子1～55粒，有的多达1 000粒以上。种子微小，种皮薄，胚直立，无胚乳或有少量胚乳。种子基部围有白色絮状长毛。李文钿和朱彤（1988，1989）研究过胡杨的花粉和胚囊，报道过胡杨的受精作用和胚胎发育过程。姚秀玲和杨得基（1993）研究过甘肃河西地区小叶杨采种时间和种子品质的关系。杨属各树种果实和种子的形态特征见表4。图1绘出了密叶杨种子的解剖构造。

表4 杨属树种果实和种子的形态特征

树种	成熟果实			种子		
	形状	大小（mm）	颜色	形状	大小（mm）	颜色
银白杨	细圆锥形	长5。2瓣裂，无毛	黄绿色	—	—	—
银灰杨	细长卵形	长3～4。2瓣裂。每穗120～160个蒴果，每果5粒种子	黄绿色	—	—	—
毛白杨	圆锥形或长卵形	长约4～5。有短柄，2瓣裂，内含种子2～3粒	黄绿色	长圆卵形	—	微黄色
响叶杨	卵状长椭圆形	长4～6，稀2～3。2瓣裂。每穗80～170个蒴果	黄绿色	匙状	长（1.5）2.5～3（3.5），径1～1.5（2）	浅黄色
山杨	卵状圆锥形或椭圆状纺锤形	长约5。2瓣裂，有短柄	浅黄绿色	长卵圆形	—	浅黄色
河北杨	长卵形	2瓣裂，有短柄	—	—	—	—
欧洲山杨	细圆锥形近无柄，无毛	2瓣裂。每穗150个蒴果	黄绿色	—	—	—
大叶杨	卵形	长10～17。密被绒毛，3瓣裂	—	棒状	长3～3.5	暗褐色
青杨	卵圆形	长6～9。3～4瓣裂，稀2瓣裂	黄绿色	长椭圆形	长2～2.5，径1.0	灰褐色
二白杨	长卵形	长4～5。2瓣裂，果柄长0.5	黄褐色	长椭圆形	长（2.2）2.5～3，径1.0	灰白带浅褐色

（续）

树　种	成熟果实			种　子		
	形　状	大小（mm）	颜　色	形　状	大小（mm）	颜　色
香杨	卵圆形	无柄，无毛，2瓣裂	浅黄绿色	—	—	浅黄白色
苦杨	卵圆形	长5～6。2～3瓣裂。每穗30～40个蒴果，每果种子23～37粒	淡褐色偏灰白	长椭圆形	长2.5，径1.5	淡褐灰色
小青杨	长椭圆形	长5。2～3瓣裂。每果5～7粒种子	黄绿色带褐色	长椭圆形	长2.0，径1.0	浅黄色
冬瓜杨	球状卵形	长约7。（2）3～4瓣裂	黄色	—	径0.1～0.2	—
小叶杨	卵形	2（3）瓣裂。无毛	黄褐色	卵圆形	长1.8～2.0，径1.0	浅黄色
密叶杨	卵圆形	长5～8。3瓣裂。每穗25～40个蒴果，每果约30粒种子	浅黄色	棍棒状	长2.6 径1.2	浅褐灰白
大青杨	—	长7。无毛，近无柄，3～4瓣裂	黄绿色	—	—	白黄色
滇杨	圆锥形	果穗长27cm，果长10～12，径5～6。每穗有20～40个蒴果，每果有种子40～50粒	棕褐色	长卵形	长1.8 径0.8	浅褐色
黑杨	球形或卵圆形	长5～8，径3～5，2瓣裂。每穗约26个蒴果，每果11～15粒种子	黄白绿色	长卵圆形	长2.5 径1.5	乳白色至灰白色
胡杨	长卵圆形或卵圆形	长7～10（10～12），径4～5。2～3瓣裂，无毛	黄绿色	椭圆形	长1～1.2 径0.5～0.7	黄褐色
灰胡杨	长卵圆形	长5（8）～10，径2～3。3瓣裂	灰黄色	长圆形	长0.9～1.0 径0.2～0.3	灰褐色

果实的采收调制和种子贮藏　杨属树种的果实成熟后迅速开裂，种粒轻小且带有絮毛，天气干燥时会在短时间里飘飞散失，应在果皮由青变黄，个别蒴果开始开裂时立即采集树上成熟的果穗。采回的果穗放在通风良好的室内薄薄摊开，多数蒴果失水开裂，经抽打、筛选，得到纯净种子。如果制种量大，可用杨树蒴果调制机。有时从路边或水面上也可以收集到不少飞落聚积的种子。如果需种量很少，还可以在临近成熟时将果枝剪回水培，以取得种子。杨属果实出种率约2%～6%（少数7%～14%）。据陈耀华和阎万祥（1991）报道，北京地区毛白杨的种子败育现象严重，从1984、1989和1990年3年采取的2 088个果穗中，只得到273粒种子。新脱粒的种子含水量24%～40%，要进行干燥，干燥后的含水量要求达到5%～6%。有晒干、阴干和人工干燥3种方法。杨属种子的出种率、净度及质量见表5。

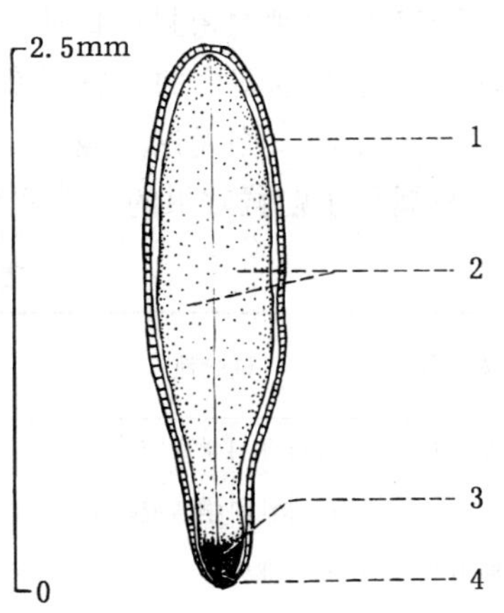

图1　密叶杨种子纵切面

1. 种皮　2. 子叶　3. 胚轴　4. 胚根

（吴新安绘）

表 5　杨属树种的出种率、种子净度及质量

树　种	出种率（%）	净　度（%）	千粒重（g）	每克纯净种子粒数
银白杨	2～6	80～90	0.54	1 800
银灰杨	—	85～92	0.34	2 900
毛白杨	2～6	80～90	0.55	1 800
响叶杨	3～4	85～95	1～1.5	600～1 000
山杨	5	80～90	0.36～0.42	2 300～2 800
河北杨	—	90	0.32	3 100
欧洲山杨	4～6	86～90	0.14	7 000
青杨	2～4	88～98	0.2	5 000
二白杨	4～6	85～95	0.88	1 100
香杨	6～14	85～95	0.74～0.83	1 200～1 400
苦杨	5	80～90	0.36～0.42	2 300～2 800
小青杨	5～6	80～90	0.3～0.4	2 500～3 400
小叶杨	2～3	85～95	0.3～0.5	2 000～3 400
密叶杨	4～7	85～95	0.5	2 000
大青杨	3～6	85～92	0.4～0.7	1 400～2 500
滇杨	9.6～12.4	88～95	0.38～0.46	2 100～2 600
黑杨	4～6	80～90	0.46～0.55	1 800～2 100
胡杨	2～4	85～95	0.06～0.20	5 000～16 700

杨树种粒小，寿命短，自然条件下生命力的保存期约为 2 周到 1 个月。贮藏前应将种子含水量降低到 6%以下，并装入放有氯化钙或硅胶等吸湿剂的玻璃瓶中，种子不要与吸湿剂直接接触，瓶口用石蜡密封，在 0～5℃的冰箱中贮藏。据姚秀玲和杨得基（1993）报道，1980 年用装有氯化钙的密封广口瓶贮藏的小叶杨种子，到 1981 年 11 月检查时仍有 89% 的发芽率；存放在普通室内的则发芽率已降至 29%。据赵君以（1958）测定，小叶杨的种子如果含水量超过 10%，则无论贮藏于一般库房还是－1℃的冷库，300 天后都全部失去发芽能力。表 6 是杨属种子贮藏效果的一些实例。

表 6　几种杨树种子贮藏效果的实例

树　种	试验地点	含水量（%）	贮藏条件	贮藏时间	发芽率（%）
小叶杨	中国林业科学研究院	6	0～5℃	半年	由 95%降至 91%
	中国林业科学研究院	4～6	0～5℃	2 年	—
响叶杨	南京林业大学	4	0～5℃密封干藏	2 年	由 98%降至 90%
胡杨	新疆生物土壤沙漠所	6	密封瓶放在坎儿井	9 个月	由 94%降至 75%
	内蒙古巴盟林科所	6.2	密封后窖底浅坑贮藏	290 天	由 77%降至 66%
	内蒙古巴盟林科所	6.5	密封后窖底浅坑贮藏	290 天	由 77%降至 33%
大青杨	黑龙江	5～6	0～5℃密封干藏	—	有较高发芽率

据报道，日本山杨 *P. sieboldiana* Miq.（*P. tremula* var. *villosa* Franch. et Sav.）在低于0℃的温度下密封贮藏6.5年还有81%的发芽率，据认为是生活力保持得最长的杨树，同样条件下，小叶杨在3年3个月以后生活力已降至26%。

发芽和播种　杨树种子无休眠习性，极易发芽。发芽测定时常用垫有脱脂棉的滤纸做发芽基质，发芽床要保持湿润。适于发芽的温度为20～30℃的变温或25℃的恒温。发芽率一般在80%～95%，发芽势50%～90%。杨属种子常见的发芽能力见表7。

表7　杨属树种的发芽能力

树　种	发芽势（%）			发芽率（%）		
	计算天数	一　般	变动范围	计算天数	一　般	变动范围
银白杨	—	—	—	2～3	96	—
银灰杨	1	90	—	2～3	98	90～98
响叶杨	—	—	—	1	—	85～95
山杨	—	—	—	—	80	65～93
河北杨	2	58	—	5	95	95～100
欧洲山杨	1	81	—	2～3	87	80～90
青杨	2～3	80	50～85	6～7	78	65～90
苦杨	1	80	75～80	2～3	98	90～98
小青杨	3	80	70～80	5～6	90	85～95
冬瓜杨	—	—	—	1	95	—
小叶杨	3	80	75～90	6	85	85～95
密叶杨	1	84	82～90	2～3	98	88～98
大青杨	2～3	60	50～75	6～7	78	65～90
滇杨（室外测定）	7～8	50	—	15	80	76～83
黑杨	1	80	73～90	2～3	98	87～98
胡杨	1	62	55～71	2	90	60～95
灰胡杨	1	84	75～85	2～3	90	87～93

杨树一般在播前开沟并灌足底水后条播，播后不覆土，但必须保持床面湿润。新鲜种子通常几个小时就开始发芽，12小时下胚轴开始伸出种壳。吉林敦化县寒葱岭林场是将大青杨的蒴果放在沟内，不覆土，曝晒2天，果裂后浇水，3～4天即可出齐。杨树播种量依种粒大小而异，一般每666m^2播0.25～1kg。银灰杨、胡杨、山杨种子较小，可减至0.4～0.6kg。王锡林和田乃祥等人（1985）的经验是，发芽率在50%或60%以上的胡杨，每666m^2播种350g，发芽率在40%左右的胡杨每666m^2播种500g。出土萌发，子叶两枚（图2）。生产实践中，杨树优良品种多采用扦插繁殖。

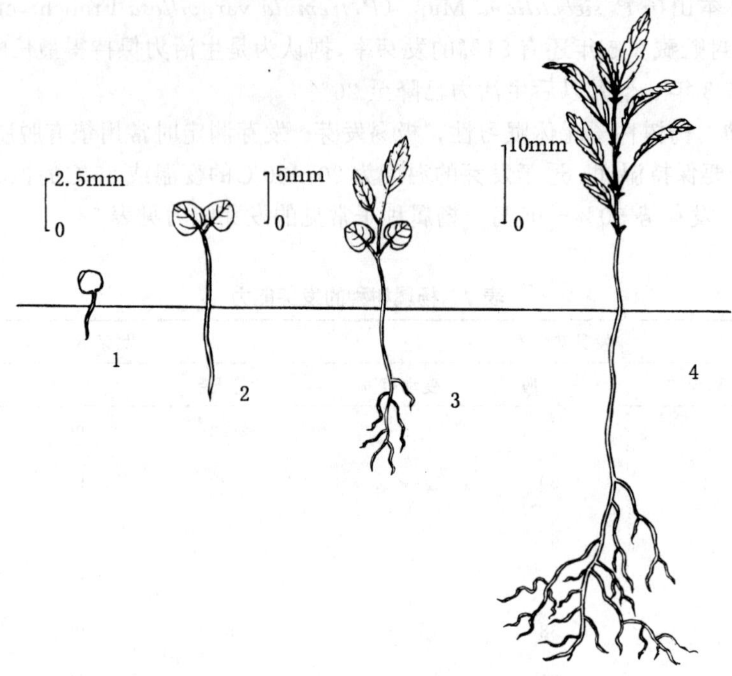

图 2 密叶杨的种子萌发和幼苗生长情况

1. 下胚轴延伸 2. 子叶出土 3. 初生叶互生 4. 幼苗形成

（吴新安绘）

（李荣晨）

柳 属

Salix L.

（杨柳科 Salicaceae）

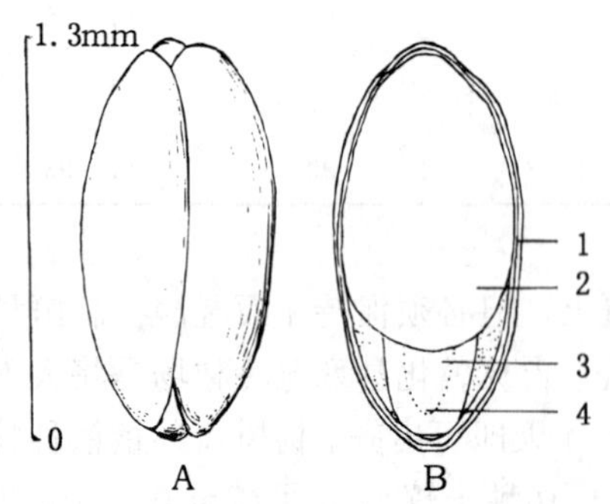

图 1 垂柳种子外形（A）及其纵切面（B）

1. 种皮 2. 子叶 3. 下胚轴 4. 胚根

（A. 胡冬梅、林平绘，B. 田恒德绘）

生长习性、分布和用途 本属约 520 种，主产北半球温带地区，寒带次之。我国 257 种，主产东北、西北、西南山区。本文描述 4 种。乔木或灌木。枝圆柱形，髓心近圆形。无顶芽，侧芽通常紧贴枝上，芽鳞单一。速生，喜光，喜湿润土壤。本属多为保水、固堤、固沙及“四旁”绿化树种。其名称、生长、分布和用途见表 1。

开花结实 花单性，雌雄异株。柔荑花序直立或斜展，侧生于去年生枝上。先叶或与叶同时开放。花无花被，着生于苞片与花序轴间，苞片全缘，具 2（腹生和背生）或 1（腹生）腺体。雄花：雄蕊 2，花丝分离（旱柳、垂柳）或合成单体（簸箕柳、北沙柳）。雌花：雌蕊由 2 心皮合成，1 室，子房无毛（旱柳、垂柳）或被毛（簸箕柳、北沙柳）。蒴果 2 瓣裂，种子多数，小，暗褐色，种皮薄，胚直伸，无胚乳。自珠柄（种柄）上长出许多长绢毛包被种子，俗称柳絮，能使种子借风或水迅速传播。种子的形态特征见图 1。旱柳、垂柳正常结实年龄在 15～30 年生。簸箕柳、北沙柳2～3 年生开始开花结实，4～20 年生大量结实，大小年现象不明显。它们的开花结实物候期见表 2。

表 1　柳属树种的名称、生长、分布和用途

中　文	学　名	树　高 (m)	胸　径 (cm)	分　布	用　途
垂 柳	*S. babylonica* L.	12～18	42	长江流域和黄河流域，其它各地栽培	材用，编织，栲胶，园林绿化
旱 柳	*S. matsudana* Koidz.	20	80	东北、华北平原、西北黄土高原西至甘、青，南至淮河流域及苏、浙。朝鲜半岛、俄罗斯远东、日本	材用，编织，固沙保土，四旁绿化，蜜源，干叶作饲料
北沙柳	*S. psammophila* C. Wang et Ch. Y. Yang	2～3	—	内蒙古、晋、陕、宁	防风，固沙。编织
簸箕柳	*S. suchowensis* Cheng	3～4	—	浙、苏、鲁、豫东、淮河中下游冲积平原，多栽培	固沙，护堤，编织

表 2　柳属树种的开花结实物候期

树　种	观察年份	观察地点	开花期			果实成熟期	种子散落期	
			始　期	盛　期	末　期		始　期	末　期
垂柳	1963～1984	各省物候观测点	3 月 5 日～4 月 25 日	3 月 10 日～4 月 25 日	3 月 15 日～4 月 30 日	3 月 25 日～5 月 20 日	4 月 10 日～5 月 25 日	4 月 20 日～5 月 31 日
	1969～1975	哈尔滨	4 月 29 日～5 月 12 日	5 月 4 日～5 月 14 日	5 月 10 日～5 月 16 日	5 月 20 日～6 月 2 日	5 月 31 日	6 月 2 日
旱柳	1963～1984	各省物候观测点	3 月 10 日～4 月 30 日	3 月 10 日～5 月 10 日	3 月 20 日～5 月 20 日	4 月 15 日～5 月 20 日	4 月 15 日～5 月 31 日	4 月 20 日～5 月 31 日
	1969～1975	哈 尔 滨	5 月 2 日～5 月 12 日	5 月 5 日～5 月 14 日	5 月 7 日～5 月 19 日	5 月 28 日～6 月 5 日	5 月 30 日～6 月 7 日	5 月 31 日～6 月 9 日
北沙柳	1985～1986	内蒙古伊克昭盟	4 月中旬	4 月下旬	4 月下旬～5 月上旬	5 月中旬～5 月下旬	5 月下旬	5 月下旬
簸箕柳	—	徐州	—	3 月	—	4～5 月	5 月	—

果实采收、调制和贮藏 柳属的蒴果由绿色变为黄色或黄绿色时标志种子成熟，应立即剪取果枝。迟则蒴果开裂，2～3天种子便连同种絮飞散。采集后在通风的室内摊开晾干，在室温条件下1～2天内蒴果即大部开裂，搓揉或用柳条抽打使种子脱出。去杂，净种，即为播种材料。垂柳、旱柳亦可收集在地面或水面聚集的柳絮，调制后立即播种，出苗效果亦很好。柳属的种子如在室温下贮藏，10天之后生活力便迅速下降；必须在充分干燥后采用密封容器冷藏，贮藏时安全含水量为5%～6%。种子的净度、质量等数据见表3。

表3 柳属树种种子的净度和质量

树 种	出种率（%）	净 度（%）	千粒重（g）	每克纯净种子粒数
垂 柳	5～8	85～95	0.09～0.11	9 000～11 000
	3～4（吉林市）	—	0.43～0.57（吉林市）	1 750～2 300
旱 柳	2～4	85～95	0.15～0.17	5 900～6 700
	3～4（哈尔滨）	—	0.28（哈尔滨）	3 570
北沙柳	2～3	85～95	0.11～0.13	7 700～9 100
簸箕柳	2～4	85～95	0.12～0.14	7 100～8 300

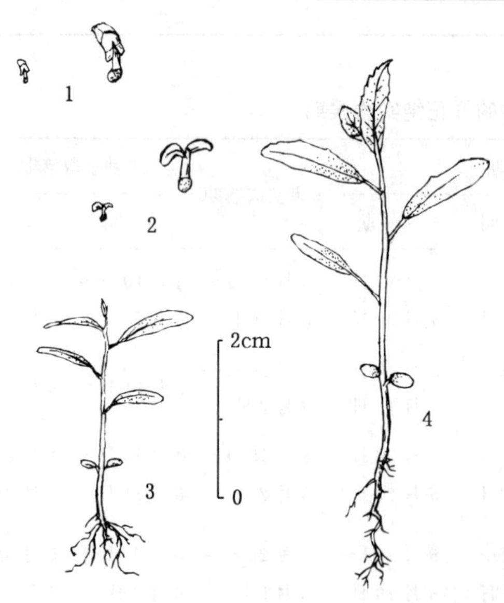

图2 垂柳种子萌发和幼苗生长情况
1. 胚根伸出，出现环圈根毛（放大） 2. 子叶展现（放大）
3. 初生叶互生 4. 幼苗形成
（田恒德仿《主要树木种苗图谱》）

发芽和播种 种子无休眠现象，一般多随采随调制随播种。室内发芽需光照，发芽温度20～25℃。通常12～24小时发芽，2～3天结束，发芽率一般为80%，变动范围为70%～85%。发芽率随种子老化而迅速下降。出土萌发，子叶2片。垂柳子叶椭圆形，两端圆，黄绿色，长3～3.5mm；上胚轴长7～9mm，幼茎淡黄绿色；初生叶互生，长椭圆形，第1叶或第1、2叶全缘，第2、第3叶以上，叶缘疏生锯齿；种子初萌时，根颈处增粗成圆环并被长毛，后渐不明显；主根发育弱，侧根发达，淡褐色（图2）。干旱地区采用低床育苗，要求土壤细碎，床面平整。播前灌足底水。条播或撒播，每平方米播种量0.8～1.2g，细沙覆盖，使种子似露非露。在幼苗出齐之前苗床必须保持湿润。在适宜的温湿度条件下，播后5～8小时即出现子叶，而后扎根，注意防止蚂蚁危害。2～3天苗木出齐。每平方米留苗15～100株，当年苗高0.3～1m。柳属容易无性繁殖，为保持原种优良特性，生产上广泛采用扦插繁殖。

（印佩文）

杨　梅

Myrica rubra Sieb. et Zucc.

（杨梅科　Myricaceae）

生长习性、分布和用途　杨梅属约50种，我国4种，本文描述1种。常绿乔木，高达15～20m，胸径60cm，寿命较长。耐荫。浅根性。根部有菌根菌共生，能耐干旱瘠薄，适生于酸性土壤。萌芽力强。分布于长江以南各省、自治区和台湾。朝鲜半岛、日本和菲律宾也有分布。杨梅有许多不同的类型和品种。果可生食，也可酿酒或制成蜜饯；树皮可提取栲胶；根皮药用；种仁含油脂，可食；木材可作家具及雕刻；抗火性强，可作防火林带树种。

开花结实　10年生开始结实。15年生以后为正常结实。结实有大小年现象，间隔期1年。花单性，无花被。雌雄异株或偶有同株，柔荑花序。雄花序着生于去年生叶腋或新枝基部。雌雄同序者基部为雄花。雌花序常生于叶腋。雄花具2～4小苞片，雄蕊4～6。雌花具2～4小苞片。雌蕊由2心皮合成，1室，1胚珠。花柱短，柱头2裂，羽状张开，鲜红色。花期因地区、品种而异。在福建莆田，雄花于2月中旬～3月上旬开放，雌花于2月下旬～3月中旬开放，盛花期均在3月上旬；5月下旬果实开始成熟，6月上旬为成熟盛期。核果球形，直径约2cm，有的品种可达3cm。每果有果核1粒。外果皮肉质，有密集的乳头状突起，成熟时深红色或紫色，有的品种为白色。果核卵形而稍扁，长1.2cm，宽0.8cm，厚0.75cm。内果皮坚硬，骨质，外附纤维。种子无胚乳，子叶2，胚直伸（图1），含油脂达40%。

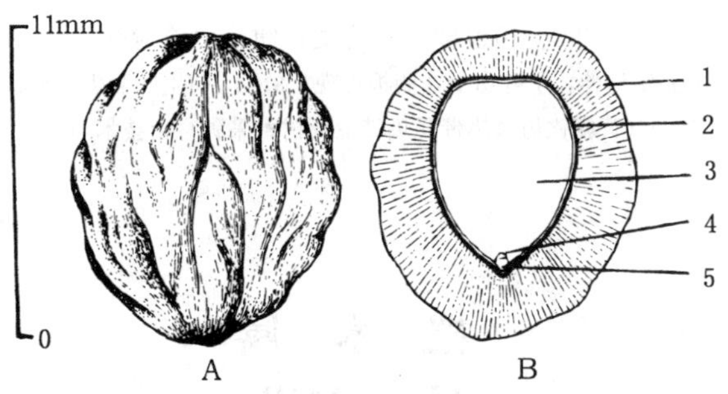

图1　杨梅果核外形（A）及其纵切面（B）

1.内果皮　2.种皮　3.子叶　4.胚芽　5.胚根

（黄应钦绘）

果实的采收调制和种子贮藏　选择20～30年生健壮的实生母树，在果实成熟盛期采摘。食去果肉或堆沤3～5天，待果肉腐烂，再浸入水中淘洗，清除果肉、杂质及干瘪粒，所得果核即为播种材料，通称种子。种子不宜曝晒，置于通风处稍晾干，含水量在25%左右时即可贮藏。每百千克鲜果可出种子7～12kg，净度为90%～98%。实生树所得种子千粒重560g，变幅为500～625g；栽培品种的千粒重可达1 000g，变幅为770～1 250g。实生树每千克有种子1 800粒，变幅为1 600～2 000粒；栽培品种每千克有种子1 000粒，变幅800～1 300粒。种子忌脱水，应混湿沙贮藏。贮藏期间要定期翻动并防鼠害。

发芽和播种　种子休眠期约1年，湿沙层积越冬可以解除休眠。发芽能力因种子来源不同而异。据华南农学院编著的《果树栽培学各论》记载，实生母树的种子发芽率约45%，变幅为30%～60%；栽培品种的种子发芽率约20%，变幅为10%～30%。春季播种，条播。每平方米播种100～140g，覆土1～1.5cm。层积过的种子播后20～30天发芽。出土萌发。上胚轴黄绿色，后转绿色。下胚轴肥大。子叶倒卵状，近似匙形，肉质。子叶出土2～3天初生叶出现，螺旋状互生，羽状裂，以后全缘。茎叶均被绒毛。主根发达（图2）。1～2年生苗可以出圃或供嫁接，嫁接后第2年出圃。防火林带多用实生苗，果园及庭院植树多用嫁接苗。

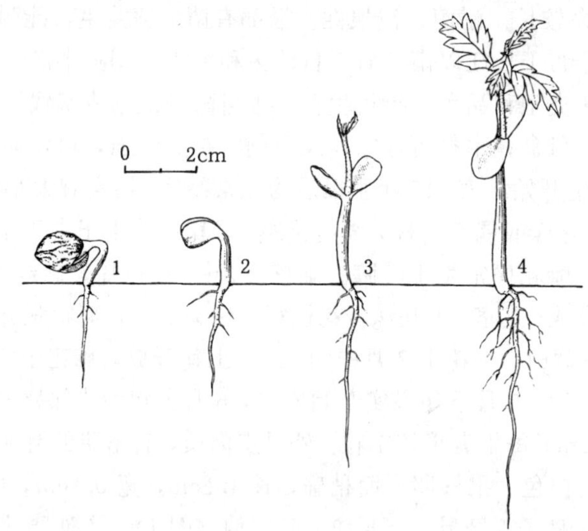

图2　杨梅种子的萌发和幼苗生长情况

1、2. 胚轴延伸，子叶出土，侧根出现　3、4. 初生叶互生，羽状分裂

（黄应钦仿《热带亚热带主要树种采种育苗技术》）

（李玉科）

桤木属

Alnus Mill.

（桦木科　Betulaceae）

生长习性、分布和用途　本属有40余种，我国产11种，本文描述5种（表1）。落叶乔木，生长迅速，主干圆满通直。多数种喜生于阳光充足，土壤肥沃的水边和山谷。根系发达，具根瘤菌，能改善土壤，护岸保土，是水湿地带和河岸护堤的重要造林树种。分布于东北、华

北、华东、中南和西南等地。桤木原产西南地区，自 60 年代起，已经陆续引种到江苏等 7 个省。材质软，耐水湿，是水土设施、坑木和矿柱的优良用材。树皮和果穗含鞣料，可提取栲胶。鲜叶含氮量达 2.7%，在混交林中能促进柏木和樟树的生长。

表 1　桤木属树种的名称、分布和用途

中　名	学　名	树　高（m）	分　布	用　途
桤　木	*A. cremastogyne* Burkill	40	甘、陕、川、黔。苏、浙、皖、湘、鄂、赣、粤有栽培	材用、薪柴、防风、保土、固沙、环保、绿肥、鞣料
赤　杨	*A. japonica* (Thunb.) Stued.	20	辽、吉、冀、鲁、皖、苏、台。日本、朝鲜半岛	材用、薪柴、鞣料
旱冬瓜（日本桤木）	*A. nepalensis* D. Don	20	滇、桂、黔、川、藏。印度、锡金、不丹、尼泊尔、缅甸	材用、绿肥、鞣料
辽东桤木（尼泊尔桤木）	*A. sibirica* Fisch. ex Turcz.	20	辽、吉、黑、鲁。西伯利亚、朝鲜半岛、日本	材用、染料、鞣料、蜜源
江南桤木	*A. trabeculosa* Hand. -Mazz.	20	豫、苏、浙、皖、鄂、湘、赣、闽、粤。日本	材用、绿肥

开花结实　桤木 3～6 年生开始结实，正常结实在 8 年生以后，大小年不明显。春花秋实，但旱冬瓜是秋季开花，翌年冬季果实成熟（表 2）。花单性，雌雄同株，柔荑花序，风媒。雄花序生于上 1 年生枝条顶端，圆柱形下垂。雄花每 3 朵生于 1 苞鳞内，花被 4 枚，雄蕊多为 4 枚，与花被对生。雌花序短，单生，或聚成总状或圆锥状，秋季出自叶腋或生于短枝。雌花每 2 朵生于 1 苞鳞内，无花被，子房 2 室，每室具 1 枚胚珠。果穗球果状，长椭圆形，长 10～20（35）mm，直径 10～15mm。果苞木质，鳞片状，宿存，由 3 枚苞片、2 枚小苞愈合而成，顶端具 5 枚浅裂片。每个果苞内具 2 枚小坚果。坚果轻小，近卵圆形，扁平，两面具棱纹，先端平截，呈深浅不一的褐色，边缘有翅，翅宽窄不一，纸质或膜质（图 1）。无胚乳。子叶较大，占据种腔的大部分，胚轴和胚根较小。种子成熟时随风飞散，传播的距离可远达 500m，有的树种种子也可随水流传播。

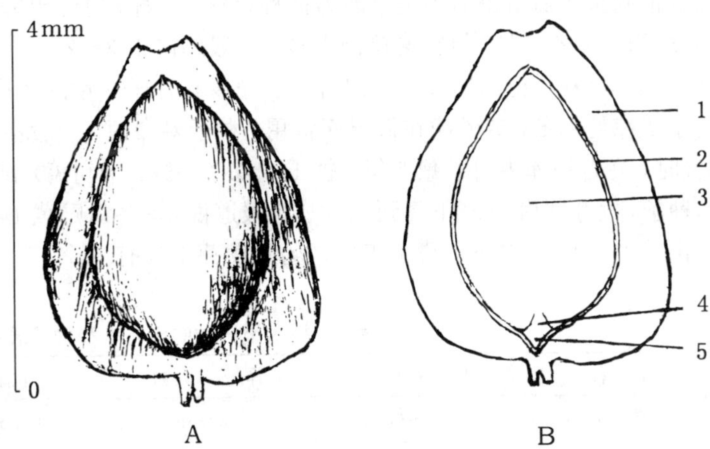

图 1　江南桤木小坚果的外形（A）及其纵切面（B）
1. 果皮（及翅）　2. 种皮　3. 子叶　4. 胚轴　5. 胚根
（童军平、田恒德绘）

表 2 桤木属树种开花结实习性

树 种	花 期	果实成熟期	成熟特征	种子散落期	种子(小坚果)特征	
					形 状	颜 色
桤 木	2～3月上旬	11月中旬～1月上旬	果穗由青绿转为暗褐色,种仁呈白色蜡质状	11月下旬～翌年2月	三角状卵圆形,长约3mm。翅膜质,基部延伸	黄褐色
赤 杨	2～3月	9～10月	—	—	卵圆形,长3～4mm,宽2～3.5mm。翅厚纸质	棕褐色
旱冬瓜	9～10月	翌年11月下旬～12月中旬	果穗由绿转黄褐色	1月下旬开始	矩圆形,长约2mm,翅膜质	黄褐色
辽东桤木	5月	8～9月	—	—	卵圆形,长约3mm,翅厚纸质	棕褐色
江南桤木	2～3月	9～10月	果穗转为黄褐色	—	宽卵形,长3～4mm,宽2～2.5mm,翅纸质	深褐色

果实的采收调制和种子贮藏 果苞微裂时，选无风的晴天，在树下铺采收布，轻击树枝，或震动树干，收集震落的小坚果。这样采得的种子质量较好。也可爬树，在立木上或在伐倒木上采集成熟的果穗，摊晾在通风处，数周后种子脱出，或将果穗摊晒1～3天脱粒。风选或筛选清除杂质后所得的小坚果即为播种材料，通称种子。出籽率约3%～9%。经过精细清选，净度可达90%左右，但空瘪粒仍很多，一般约占1/3～2/3。千粒重0.25～2.2g。每克有种子450～4 000粒。桤木属种子的出籽率、净度和种子质量的数据见表3。生产上常用布袋盛装，置于阴凉处越冬，或贮藏在低温库备用。种子寿命短，一般不耐贮藏。桤木种子含水量4%～6%时，在普通库内可贮藏2年。据王成霖（1987，未刊稿）试验，在0～5℃下贮藏1年的桤木种子，发芽率由55%下降到45%。有报道说，旱冬瓜贮藏1年，发芽率由40%下降到15%；江南桤木在0～5℃下贮藏1年后，场圃发芽率只有1%。

表 3 桤木属树种的出籽率和种子质量

树 种	出籽率（%）	净 度（%）	千粒重（g）	每克纯净种子粒数
桤 木	2.7～6.5	75～89	0.3～2.2	450～3 300
赤 杨	—	—	1.1～2.2	450～900
旱冬瓜	3～5	—	0.25～0.3	3 300～4 000
辽东桤木	—	—	0.9	1 100
江南桤木	7～9	—	0.25～1.7	580～4 000

发芽前处理和发芽测定 本属多数树种的种子无休眠习性。光照能促进旱冬瓜种子的发芽，提高发芽率，缩短发芽期。

江南桤木有的种批种子含脱落酸，有轻度休眠现象，低温层积1周即可解除休眠；也可采用流水浸种、赤霉素溶液浸种或发芽时采用变温（20～30℃）加光照等方法促进发芽。江南桤木种子在石沙、滤纸或纱布等不同的发芽基质上，发芽结果基本相同。

表 4　桤木属种子的发芽能力及其测定条件

树　种	预处理	发芽测定条件			发芽势(%)		发芽率(%)	
		基质	光照时数	温度(℃)	计算天数	数　值	计算天数	数　值
桤　木	—	滤纸	—	20	7～8	38	11～12	(30)51～61(70)
赤　杨	—	滤纸	—	30	6	24	8	26
	—	滤纸	—	20/30	6	17	7	24
旱冬瓜	—	滤纸	—	30	14～15	21	24～25	30～45
	—	滤纸	—	20/30	13	29～35	19～20	50～56
	—	滤纸	8	20/30	10	24	—	—
辽东桤木	—	滤纸	—	30	4	12～17	5	18～22
	—	滤纸	—	20/30	5	13	6	18～20
江南桤木	—	石英沙	8	20/30	7	18	9	33
	流水浸种 7 天	石英沙	—	30	5	11	13	28
	低温层积 8 天	石英沙	—	30	6	25	7	28
	—	石英沙	—	30	6	43	11	70

在 30℃下，桤木种子吸水率达 134%，旱冬瓜种子吸水率达 148%时开始发芽；吸水率分别达 177%和 255%时，发芽率分别为 42%和 54%，此时饱满种子的发芽基本结束。在 20～30℃的变温下，江南桤木种子吸水率达 47%时开始发芽，达 69%时绝对发芽率为 92%（表 4）。

播种　春秋两季都可播种，但多数采用春播。以撒播为主，也可条播。播种沟深 1.5～3cm，行距 20～30cm。每平方米约播桤木 4～5g，旱冬瓜 2.5g，江南桤木 6～8g。播后稍镇压，筛盖腐殖质土或土沙混合物，以不见种子为度，也可不盖土。床面盖草。幼苗基本形成时可以间苗。1 年生苗可以出圃。

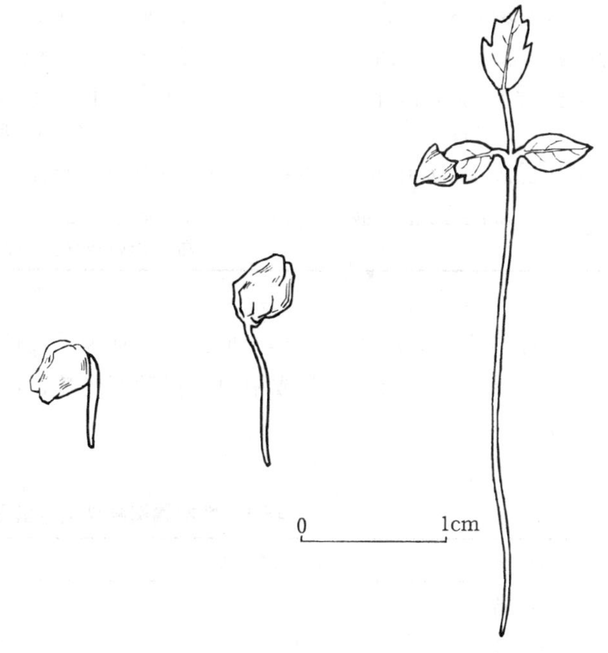

图 2　江南桤木种子萌发后 1、3、21 天幼苗的生长情况
（童军平绘）

出土萌发。桤木子叶多为椭圆形，薄，淡绿色，下面略带黄色，叶脉不明显。初生叶倒卵形，基部楔形，互生。绿色，下面淡绿色，叶缘上部有锯齿。上胚轴和下胚轴都呈细圆柱形，淡绿色而带红色，被短毛（图 2）。

（王成霖）

桦 木 属

Betula L.

（桦木科 Betulaceae）

生长习性、分布和用途 本属约100种，我国约30种，本文描述7种。桦木属为乔木或灌木，主要分布北温带，少数种类分布至北极圈，或向南至亚热带高山地区。本属树种多喜光，耐寒，耐干旱瘠薄，其名称、生长分布及用途见表1。

表1 桦木属树种的名称、生长、分布和用途

中名	学名	树高（m）	分布	用途	供稿
红桦	*B. albo-sinenses* Burk.	30	湘、鄂、陕、晋、豫、冀、滇、川、青、甘、宁	材用	508
西桦	*B. alnoides* Hamilt.	30	川、滇、桂、琼。尼泊尔	建筑、家具、胶合板等	617
硕桦	*B. costata* Tuartv.	30	东北、冀、鲁。西伯利亚	胶合板、枕木、车辆	106
黑桦	*B. dahurica* Pall.	20	东北、华北。朝鲜半岛、蒙古	材用、防护	106
光皮桦	*B. luminifera* H. Winkl.	25	滇、黔、川、甘南、陕、豫、皖、浙、闽、鄂、湘、桂、粤北	建筑、家具、造纸、绿化	502
垂枝桦	*B. pendula* Roth.	20～30	新。欧洲、西西伯利亚	材用、桦树汁饮料、绿化、防护	204
白桦	*B. platyphylla* Suk.	20	东北、华北、西北、西南。朝鲜半岛、东西伯利亚、蒙古、日本	材用、火柴杆。桦皮油入药。桦树汁饮料。绿化	106

开花结实 本属树种每年大量结实，常为采伐迹地或火烧迹地天然更新的先锋树种，繁生成次生林，以后渐为耐荫或稍耐荫树种所代替，或形成混交林。其结实年龄及丰年间隔期见表2。

表2 桦木属树种结实年龄及丰年间隔期

树种	开始结实年龄	正常结实年龄	丰年间隔期（年）
红桦	15～20	—	—
西桦	12～14	18～35	0
硕桦	15～20	20～40	0～1
黑桦	15～20	20～30	1～2
光皮桦	3～4	10～30	0
垂枝桦	10～15	20～30	2～3
白桦	15～20	20～30	1～2

花单性，雌雄同株。雄柔荑花序生于上1年枝顶或侧生，秋季形成，裸露越冬。每3朵

雄花生于苞腋内，花萼 4，膜质，雄蕊 2。雌柔荑花序在春季发育于芽鳞内，单生或 2～5 生于短枝顶部，直立或下垂，每苞腋内具 3 朵雌花和 2 小苞片，无花被。子房裸露，2 室，每室具倒生胚珠 1。花柱 2，分离。果穗穗状。果苞革质，由 3 枚苞片愈合而成，鳞片状，3 裂，每果苞具 3 个小坚果。坚果扁平，两侧具膜质翅，柱头宿存，成熟时与果苞同落。种子单生，具膜质种皮，无胚乳，胚直伸，子叶扁平（图 1）。开花结实物候及果实和种子（小坚果）的形态特征见表 3、表 4。

表 3　桦木属树种开花结实物候

树　种	观察年限 地　点	花　期			果　期			果　实 成熟特征	果实 散落期
		始　期	盛　期	末　期	始　期	盛　期	末　期		
红　桦	—	4 月	～	5 月	9 月	—	9 月	果苞变黄褐色	—
西　桦	1988 年 云南景洪	1 月中下旬	2 月上旬	2 月下旬	3 月上旬	3 月中旬	—	果由绿变褐黄色	—
硕　桦	—	5 月初	5 月中	5 月末	9 月初	～	9 月末	—	—
黑　桦	1963～1980 哈尔滨	4 月下旬	5 月上旬	5 月上旬	8 月上旬	～	8 月下旬	果苞深黄色，果淡黄褐色	初熟至脱落间隔期为 16～20 天
光皮桦	1987～1988 四川雅安	3 月中旬	3 月下旬	～	4 月下旬	～	5 月上旬	果由绿变为黄褐色	4 月下旬至 5 月上旬
垂枝桦	— 新疆	4 月上中旬	～	5 月上旬	7 月中旬	～	7 月下旬	果由青绿变浅黄色	开始脱落至大量脱落间隔 10～15 天
白　桦	1963～1980 哈尔滨	5 月上旬	5 月上旬	5 月上旬	7 月上旬	～	8 月中旬	果苞褐绿，果浅黄色	初熟至脱落间隔 20～25 天

表 4　桦木属树种果穗和种子（小坚果）的形态特征

树　种	果穗形状	果　穗（cm）		小坚果形状	小坚果大小（mm）		果翅长宽（mm）
		长	径		长	宽	
红　桦	圆柱形，单生或 2～4 排成总状	3.0～6.0	0.3～0.7	卵圆形	2～3	1.5～2.0	宽为果的 1/2
西桦	长圆柱形，3～5 排成总状	5～10（12）	0.4～0.6	倒卵形	1.5～2	—	翅长 2.0，宽为果的 2 倍
硕桦	短圆柱形，单生	1.5～2.2	1.0	倒卵形	2.5	1.5	与果等宽或较窄
黑桦	矩圆状圆柱形，单生直立或微下垂	2.0～3.0	1.0	宽椭圆形 稀倒卵形	2.5～3.0	2.0～2.5	宽为果的 1/2
光皮桦	长圆柱形，单生（每果穗含 400～700 粒小坚果）	3.5～14	0.6～1.0	倒卵形	2.5	1.5	宽为果的 1～2 倍
垂枝桦	矩圆状圆柱形，单生（每果穗含 100～425 粒小坚果）	2～4（5）	0.8～1.0（1.5）	长倒卵形	2～3	1～2	较长于果，宽为果的 2 倍
白桦	圆柱形，单生或斜展	2.2～3.0（4.5）	0.8～1.0	窄矩圆形、宽椭圆形或倒卵形	2.0	1.5	翅长 2.5，与果等宽或稍宽（宽1.5～1.6）

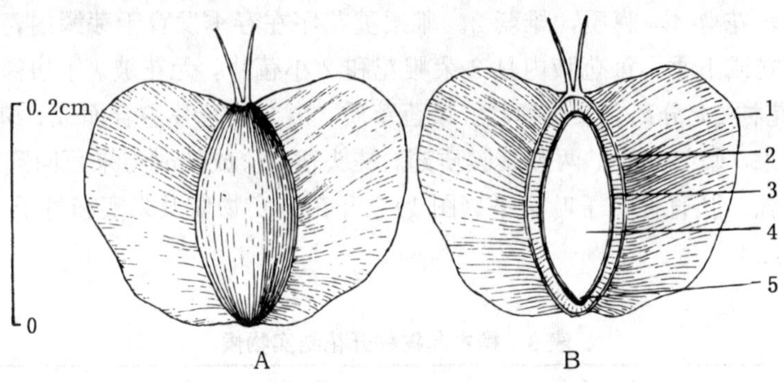

图 1　白桦坚果的外形（A）及其纵切面（B）

1. 果翅　2. 果皮　3. 种皮　4. 子叶　5. 胚根

（梁鸣，黄应钦绘）

果实的采收调制和种子贮藏　果苞由绿色变为黄褐色或褐黄色即可采收果穗。桦木属树种果实成熟期短，脱落也快，应及时采收。采收后阴干，每天翻动 1～2 次，一般经 3～5 天开始脱落。轻揉果穗，用簸箕轻簸或风选净种去杂，所得小坚果即为播种材料，通称种子。

桦木属种子贮藏时的含水量均应小于 10%，短期贮藏于通风干燥处。在种子库中用麻袋、布袋、木箱或铁桶密封，可贮藏较长时期。据有关单位试验，硕桦、黑桦、白桦种子含水量小于 10%，用麻袋或塑料桶装只能贮 1 年，用铁桶密封于相对湿度 45%～65%及±5℃条件下可贮 2～3 年；红桦种子含水量 5.5%～6.0%，放于一般种子库可贮 2 年；垂枝桦种子含水量 7%～8%，装麻袋、布袋、木箱可贮 1 年，用玻璃瓶或罐在种子库中可贮 2 年；西桦种子含水量 8%～12%，用布袋包装，在干燥通风避光处可贮 5 个月；光皮桦可用塑料袋密封，低温贮藏。这 7 个树种的出籽率及种子质量见表 5。

表 5　桦木属树种出籽率及种子（小坚果）质量

树　种	出籽率（%）	净度（%）	千粒重（g）		每克纯净种子粒数	
			一　般	变动范围	一　般	变动范围
红　桦	10～15	44～96	0.338	0.226～0.427	2 960	2 340～3 760
西　桦	12～14	45～55	0.07	0.06～0.08	14 290	12 500～16 700
硕　桦	8	35～45	0.7	0.6～0.9	1 430	1 110～1 670
黑　桦	10	45～55	0.9	0.7～1.0	1 110	1 000～1 430
光皮桦	—	70～78	0.22	0.21～0.23	4 550	4 500～4 800
垂枝桦	20～30	20～90	0.265	0.2～0.3	3 770	3 300～5 000
白　桦	5～20	30～40	0.3	0.27～0.4	3 330	2 500～3 700

发芽和播种　桦木属种子发芽快，达到发芽高峰期早，种子无明显休眠习性。据周佑勋（1984）研究，光皮桦是典型的需光萌发的种子，在黑暗中几乎不能萌发。但光皮桦对光非常敏感，吸水后一次性光照 20～24 小时，或每天连续光照 4 小时，便可满足对光的需求而充分

萌发。她的研究还表明，在满足光照的条件下，适于光皮桦发芽的温度是25℃恒温或30～20℃的昼夜变温。

可随采随播，一般应春播。播前10～20天用始温30℃水浸种24小时，捞出控干后混以2倍体积的河沙，在15～20℃下保持湿润，经常翻动，大约10%的种粒开始萌动即可播种。条播或撒播，播后20天内保持床面湿润。桦木属树种播种育苗的某些技术参数见表6。子叶出土萌发（图2）。

表6　桦木属树种播种育苗技术参数

树　种	播种量 (g/m²)	覆土深 (cm)	室内发芽测定数据			场圃发芽率 (%)	当年苗高 (cm)	出圃年龄
			温度（℃）	发芽势（%）	发芽率（%）			
红桦	4～6	0.3～0.5	25	17～33	15～49	—	—	2
西桦	0.2～0.3	不覆土	20以上	25～35（开始发芽后4天）	40（开始发芽后17天）	—	—	—
硕桦	5～6	0.5～0.7	—	—	—	30～60	15～30	1～2
黑桦	5～8	0.5～1.0	—	—	—	30～70	20～30	1～2
光皮桦	2.5～4	0.3～0.5	25；昼30夜20	55～58	59～62	—	—	—
垂枝桦	4～6	—	30	45	45	—	—	2～3
白桦	4～5	0.3～0.5	—	—	—	20～60	15～30	1～2

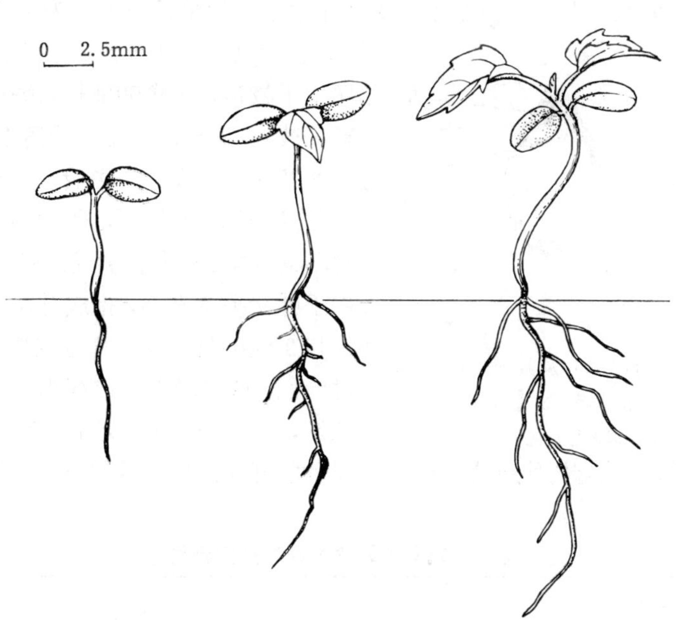

图2　白桦种子（小坚果）发芽后第7、12、21天幼苗的生长情况
（曹雅范绘）

（周德本）

鹅耳枥属

Carpinus L.

（榛科　Corylaceae）

生长习性、分布和用途　本属40余种，我国约30种，本文描述2种。落叶乔木，生于山坡或灌木丛中。喜湿润的中性或微酸性土壤，贫瘠的石质山坡也能生长。有些种类为喜钙树种，常生于石灰岩土地。木材坚韧，纹理美观，可作家具、农具。鹅耳枥种子可榨油，供食用及工业用。树姿秀丽，可供观赏。这两个树种的名称、树高、分布和用途见表1。普陀鹅耳枥目前仅浙江普陀山存有1株，在《中国植物红皮书》被列为濒危种。

表1　鹅耳枥属树种的名称、树高、分布和用途

中　名	学　名	树高(m)	分　布	用　途	供　稿
普陀鹅耳枥	*C. putoensis* Cheng	15	特产浙江普陀山	用材	702
鹅　耳　枥	*C. turczaninowii* Hance	15	辽、冀、晋、豫、陕、甘、鲁、苏、皖	用材，种子榨油，皮、叶制栲胶	305、308

开花结实　花单性同株，柔荑花序。雄花序生于1年生长枝上，为芽鳞包被。雄花无花被，每苞片具3～13雄蕊，药室分离，顶端有毛。雌花序生于枝顶，花序轴细长，每苞片具2雌花。雌花有花被，萼6～10齿裂，子房下位，不完全2室，每室2胚珠，其中之一败育；花柱短，柱头2，线形。花先叶开放或花叶同放。雌花被包在苞片内，开放时不易观察到。小坚果近椭圆形或宽卵形，具喙，着生于叶状果苞基部，排成总状果穗。果苞为1枚苞片和2枚小苞片在发育过程中愈合而成。小坚果幼时绿色，成熟时褐色或淡褐色。普陀鹅耳枥果宽卵圆形，长5～6mm，无毛。鹅耳枥果宽卵形，长3～4mm，疏生长柔毛或有树脂腺点。这2个树种的开花结实习性见表2。种子无胚乳。鹅耳枥种子形态见图1。

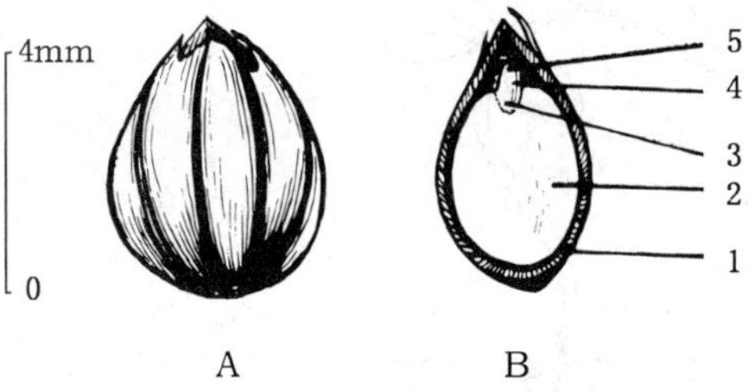

4mm
0
A　B
5
4
3
2
1

图1　鹅耳枥小坚果外形（A）及其纵切面（B）
1. 果皮和种皮　2. 子叶　3. 胚芽
4. 胚轴　5. 胚根（孟玲绘）

表2　鹅耳枥属树种开花结实物候

树　种	观察地点	花期		果熟期	果实散落期
		雌　花	雄　花		
普陀鹅耳枥	浙江	4月中旬	4月上旬	10月上中旬	成熟易落
鹅　耳　枥	北京	4月6～20日（雌花稍晚于雄花）		9月28日～10月10日	10～12月

果实的采收调制和种子贮藏　本属树种（特别是普陀鹅耳枥）的雌花晚于雄花，自然结实差，饱满种子所占比例少。发芽率亦低。小坚果由绿色变为褐色或淡褐色时应及时采摘果穗，晾晒后揉去果苞，除去杂质，得到的小坚果即为播种材料，通称种子。秋季播种或层积到翌年春播。普陀鹅耳枥种子不宜干藏，调制后如不播种，应层积至翌春播。鹅耳枥属 2 个树种的出籽率、千粒重等见表 3。

表 3　鹅耳枥属树种的出籽率和种子质量

树　种	出籽率（%）	饱满种子百分率（%）	千粒重（g）	每千克纯净种子粒数（万粒）
普陀鹅耳枥	—	8	18～20	5～5.5
鹅　耳　枥	40	90	8.5	12

发芽和播种　鹅耳枥属种子有较深的休眠习性，需要秋播在低温条件下越冬或层积处理解除休眠。如果鹅耳枥用低温层积催芽效果不好，可再用高温（18～25℃）进行催芽，提高发芽率。不过这两个树种的发芽率都很低，通常不到 5%。据高维孝（1984）报道，普陀山林场 1982 年秋播的普陀鹅耳枥，出苗率为 2.2%，层积至翌年春播的出苗率也只有 5.3%。

秋播在 10～11 月，春播在 3 月中旬至 4 月。条播或撒播，覆土要细薄，厚度约 0.5cm，出土期要防鸟害。春播后 15～25 天出齐。出土萌发。1 年生苗高 25～50cm。鹅耳枥的种子萌发和幼苗初期生长情况见图 2。普陀鹅耳枥可以扦插繁殖。

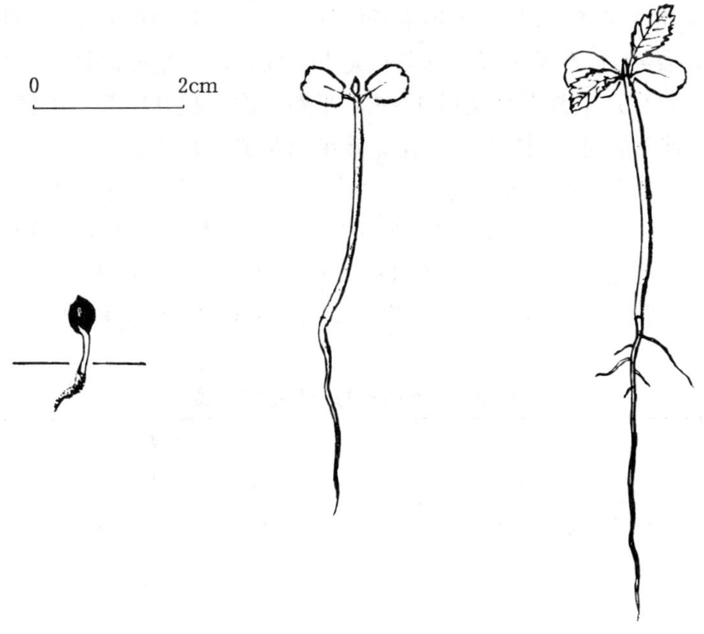

图 2　鹅耳枥种子萌发出土后第 1、6、13 天幼苗生长情况

（孟玲绘）

（王木林、高秀岩）

榛 属

Corylus L.

（榛科 Corylaceae）

生长习性、分布和用途 本属约20种，我国8种，本文介绍3种。落叶灌木或小乔木。喜光，耐寒。喜深厚肥沃、排水良好的土壤，亦能耐干旱瘠薄。其名称、树高、分布和用途见表1。

表1 榛属树种的名称、生长、分布和用途

中 名	学 名	树高(m)	分 布	用 途	供 稿
榛 子	*C. heterophylla* Fisch.	2(7)	东北、华北，陕。朝鲜半岛、蒙古、西伯利亚、日本	食用，榨油，入药(止咳)	107,108
毛 榛	*C. mandshurica* Maxim. ex Rupr.	3～4(6.5)	东北、华北、鲁、陕、甘。俄罗斯远东、朝鲜半岛、日本	食用，榨油	109
滇 榛	*C. yunnanensis* (Franch.) A. Camus	1.5～4 (7)	滇、川	食用、油料，树皮、果苞、叶可提制栲胶	614

开花结实 实生榛子和毛榛3～4年生开始结实，天然榛丛萌条第2年开始结实，3～4年达到高峰，5年以后结实量开始下降。结实有大小年现象，一般间隔期为1年。人工栽植的榛子3年生开始结实，生产上常在结实后4～6年平茬1次，促其结实。滇榛6～7年生开始结实。花单性同株，雄柔荑花序圆柱形，下垂。雄花无花被，雄蕊4～8。花丝短，花药2室，药室分离，顶端被毛。雌花序为头状，每苞鳞内具2雌花。雌花具花被，顶端4～8小齿。子房下位，2室，每室具1倒生胚珠。花柱2，柱头钻状。坚果簇生或单生，球形或卵圆形，全部或大部为钟状或管状总苞所包被，总苞上部裂片针刺状、叶状或管状。种子1，无胚乳，胚直伸，子叶肉质肥厚，发芽时留在壳内。开花结实物候见表2。果苞和种子的形态特征见表3。

表2 榛属树种的开花结实物候

树 种	观察年限 地 点	开 花			果实成熟			果实散落期
		始 期	盛 期	末 期	始 期	盛 期	末 期	
榛子	1963～1980 哈尔滨	4月上旬	4月中旬	4月中旬	7月上旬	～	8月下旬	初熟至脱落间隔12～15天
	1988 黑龙江逊克县	—	—	—	—	8月中旬	—	
毛榛	1963～1980 黑龙江尚志县	5月上旬	5月中旬	5月下旬	8月下旬	～	9月中旬	初熟至脱落间隔10～18天
	1988 黑龙江逊克县	—	—	—	—	8月下旬	—	

（续）

树　种	观察年限 地　点	开　花			果实成熟			果实散落期
		始　期	盛　期	末　期	始　期	盛　期	末　期	
滇榛	1988～1989	（雄花）	12月上旬～2月上旬	2月中旬～下旬	9月上旬	～	9月中旬	—
	昆　明	（雌花）	1月～2月中旬	2月下旬～3月上旬				

表3　榛属树种果苞和种子（坚果）成熟特征

树　种	坚果着生特征	果　苞			坚　果		
		形　状	大小(cm)	总苞颜色	形　状	大小(cm)	颜　色
榛子	坚果单生或2～6(多者达8～12)个簇生	果苞钟状，较果长1倍，密被短柔毛，间有散生红褐色刺毛状腺体	3.0	褐色	球形	1.5	淡褐色，被细绒毛
毛榛	坚果单生或2～5(6)个簇生	果苞在坚果上部缢缩，成长管状，密被黄褐色刚毛及短毛，杂有腺头	1.5～3 (5)	褐色	球形，顶端具小尖头	长1.1～1.7 径1.0～1.5	黄褐色，被白色细毛，顶端密被长硬毛
滇榛	坚果单生或2～3个簇生	果苞钟状，顶端缺裂，密被黄色绒毛及腺头毛	长2.2～3.0 径1.5～1.8	黄褐色	微扁，球形	长1.6～2.0 径1.4～1.6	密被灰色细绒毛

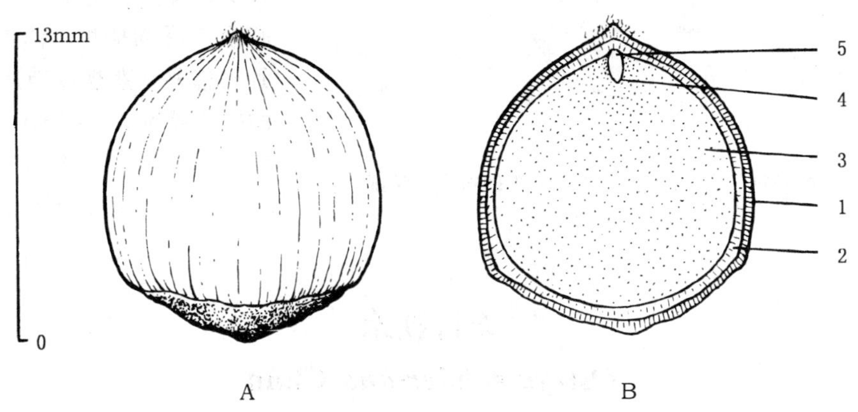

图1　榛子坚果外形（A）及其纵切面（B）

1. 果皮　2. 种皮　3. 子叶　4. 胚芽　5. 胚根

（梁鸣、黄应钦绘）

果实的采收调制和种子贮藏　果实由绿变黄，总苞与果实易脱落时采收，否则坚果落地，很快被虫蛀或被动物所食。采收摊晒可以脱去总苞，风选去杂后得到的坚果即为播种材料，通称种子。含水量低于13%并置于低温通风处贮藏1年，发芽率无显著降低。出籽率和种子质量见表4。

表 4 榛属出籽率和种子质量

树 种	出籽率（%）	净 度（%）	千粒重（g）	每千克纯净种子粒数（粒）
榛 子	30～40	97～99	1 000～1 280	790～1 000
毛 榛	15～25	98～99	650～970	1 030～1 550
滇 榛	57～60	95	850～1 000	1 000～1 180

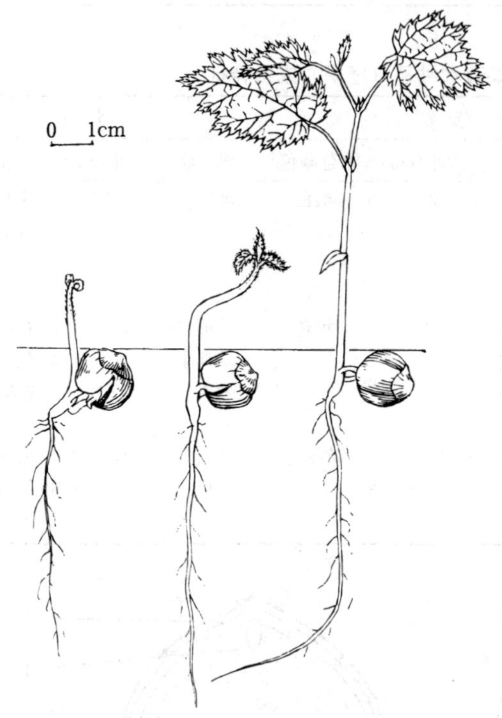

图 2 榛子种子发芽后第 3、5、19 天幼苗生长情况
（许芝源绘）

发芽和播种 有休眠习性。解除休眠的关键是要有 2 个月左右的低温阶段（0～5℃），不经催芽播后当年不出苗。可于播前 4 个月将种子水浸 2～3 天，混沙后置于冷室任其冻结再逐渐化解，天气转暖后勤加翻动，保持湿度，待部分种粒裂嘴后即可播种。亦可于播前 3 个月将种子水浸 4～5 天，混沙后置于 15～25℃温度下 1 个月，再转入 0～5℃下 2 个月，播前再日晒 2～3 小时，可增加裂口率。榛子和毛榛在哈尔滨 4 月底至 5 月初播种，滇榛在云南 3 月初播种。垄播或苗床点播，每 666m^2 播种 45～100kg，覆土厚 2.5～3.5cm，1 年生苗高 30～50cm。留土萌发（图 2）。

（周德本）

天目铁木

Ostrya rehderiana Chun

（榛科 Corylaceae）

生长习性、分布和用途 铁木属 8 种，我国 5 种，本文介绍 1 种。落叶乔木，高达 21m，胸径 80cm。野生于浙江西天目山海拔 250m 的山麓。由于人为破坏，现仅存几株，其中 1 株估计树龄已 300 多年，为濒危种，已列入《中国植物红皮书》。繁殖能力弱，天然下种的幼苗极少，亟需加强保护，扩大繁殖。木材材质致密，有光泽。

开花结实 花单性，雌雄同株，雄柔荑花序裸露越冬。每苞鳞具 1 雄花。雄花无花被。雄蕊 9～12，花药 2 室，分别着生于顶端 2 裂的花丝上。雌花序总状，生于新枝顶，每苞鳞内具

2 雌花。每一雌花具 1 苞片与 2 小苞片，基部囊状，上部 2 裂，绿色；花被 4～5 裂，与子房贴生。柱头 2，细长，淡红色。花柱短，深红色。子房下位，2 室，每室有 1 胚珠下垂。张若蕙和张金谈等人（1988）研究过天目铁木花的发育、结构和花粉形态。开花期 4 月，果熟期 9～10 月。坚果，由 2 小苞片发育而成的囊状、膜质、具网纹、呈倒卵状窄椭圆形或长椭圆形的果苞所包被。果穗总状，每果穗有坚果 5～26 枚。坚果披针形，长 9～10mm，宽 2～3mm，微扁，下部淡红褐色，上部淡褐色，微有光泽，两面各有 4～6 条微突起纵脉纹（见图 1）。

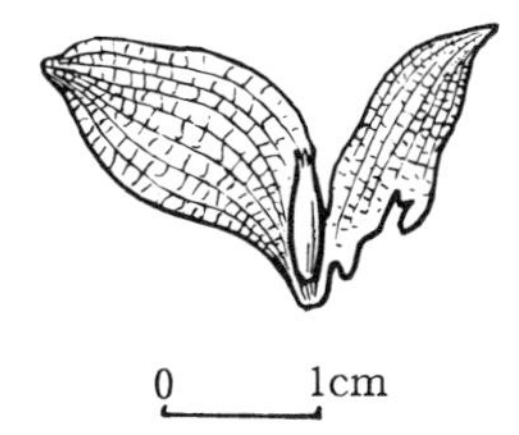

图 1　天目铁木囊状果苞破开露出坚果
（张世经绘）

果实的采收调制和种子贮藏　果苞由绿变黄时即可剪取果穗。果穗出果率 76.5%（带果苞），出籽率 80%（去果苞）。带苞坚果千粒重 14.7g，去苞坚果千粒重 11.9g，每千克去苞坚果 8.4 万粒。调制所得的坚果即为播种材料，通称种子。种子空瘪粒多，饱满粒仅为 20%～30%。采得的坚果需要混沙湿藏，但也难以长久保存。据浙江林学院试验，干藏或湿沙贮藏 1 年后，大部分种子会丧失生命力。

发芽和播种　种子（坚果）外具囊状果苞，果皮坚硬、透水性差，主要通过果苞网脉吸水。据管康林等（1988）试验，种子需经过 100～120 天亚高温—低温—亚低温层积才能通过生理后熟，破除休眠。采种后于 11 月作的室内发芽率为 16.2%；贮藏至翌年 3 月，室内发芽率降至 6.2%。多次测定的发芽率最高的也只有 22.7%。可在 10 月采种后剪破果苞，用 150μg/g 赤霉素浸泡 36 小时，随即播种，也可以将经过越冬层积的种子于 3 月上旬播于圃地。搭棚遮荫防雨。4 月发芽出土。幼苗长至 30cm 左右高时，逐步拆除荫棚。1 年生苗高 40～50cm。出土萌发，子叶 2，长椭圆形，长 1.3～1.5cm，宽 3～4mm，先端圆，基部箭形。初生叶互生，卵形或长卵形，边缘具锯齿或缺刻。种子的萌发和早期幼苗生长情况见图 2。也可利用幼树的幼枝扦插繁殖。

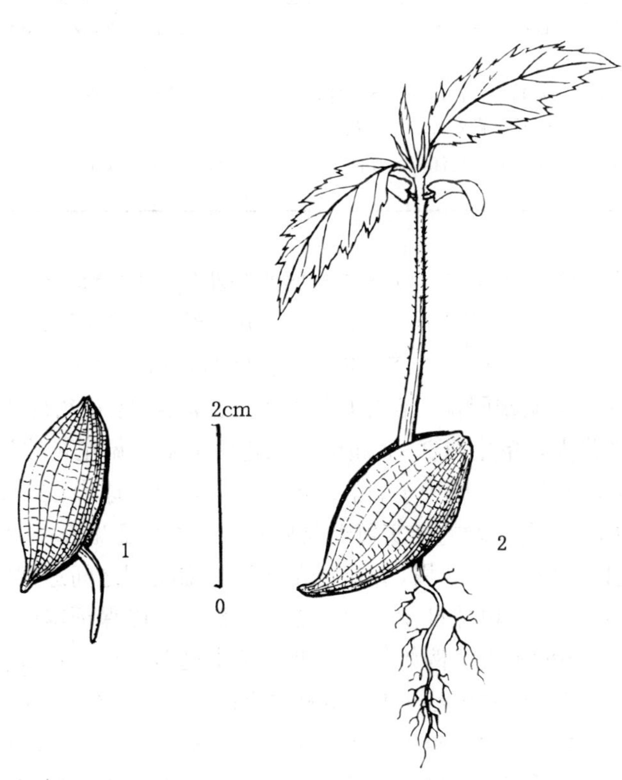

图 2　天目铁木种子萌发情况
1. 胚根伸出　2. 初生叶互生
（张世经仿《浙江林学院学报》）

（黄鹏成）

栗 属

Castanea Mill.

（壳斗科 Fagaceae）

生长习性、分布和用途 栗属约 12 种，分布于北半球的温带、亚热带地区；本文描述主产我国的 3 种（表 1）。其中板栗为世界上栗属最重要的 4 个经济树种之一，为我国特产干果，以风味甘糯，种皮易于剥离著称于世，栽培历史已有 2 600～3 000 年，地方品种有 300 个以上。1853 年、1903 年，特别是 1920 年大量引种到美国，已在美国东部广为栽培，世界各地也多有引种。其他 2 种可以作为嫁接板栗的砧木。板栗和锥栗又是优良用材树种。栗属为适应性强的广域树种，喜光、深根性、较耐旱，多生长于低山丘陵地，在深厚肥沃、排水良好的中性和微酸性土壤上生长最好。

表 1 栗属树种的名称、生长习性、分布和用途

中 名	学 名	生长习性	树高（m）	分 布	用 途	供 稿
锥栗	*C. henryi* Rehd. et Wils.	乔木	30	长江流域以南各省	干果、材用、砧木、鞣质	1003
板栗	*C. mollissima* Bl.	乔木	15～20	北起辽、吉，南至粤、桂，除新、青外各地均有栽培	干果、材用、鞣质	310
茅栗	*C. seguinii* Dode	灌木或小乔木	6～15	豫、陕及长江流域以南	砧木、薪炭、鞣质	1003

开花结实 1～2 年生的茅栗就能开始开花结实，板栗和锥栗的开花结实则晚得多。不过如果是嫁接的植株，锥栗也能在 4～5 年生、板栗能在 3～4 年生时开始结实。茅栗每年都能有较多的花和果。锥栗和板栗一般要到 15 年才进入结实盛期。在良好的栽培条件下，上百年生的板栗优良单株还能旺盛结实。花单性，雌雄同株。雄柔荑花序直立、腋生。雄花花被 6 裂，雄蕊 10～20。雌花 1～3（7），生于总苞内，单生或 2～3 簇生于雄花序下部。雌花花被片 6，子房下位，6 室，每室有胚珠 2，一般只 1 个发育成种子。坚果褐色，球形或扁球形，1～3 个聚生于一有刺、4 裂俗称刺苞的总苞内。栗属树种的结实习性及果实、种子特征见表 2。栗属的单性和两性柔荑花序上雄花的盛花期相隔约 2～3 周，雌花则在这两个雄花期之间达到盛花状态。雌花柱头露出总苞便进入可授期。许多单株自体不育，需异株授粉才能保证丰产。植株的树势衰弱，或者传粉受精期内外界环境不良，例如气温过低，雨水过多过少，也都会影响板栗的结实，结果是总苞虽然发育起来，其中并无坚果，俗称“空篷”或“空苞”，个别情况下空篷所占的比例可以高达 90%。任立中（1988）对板栗的空篷率作过研究。白仲奎（1988）报道过降低空篷率的技术措施。朱长进（1988）曾经利用生长调节剂促进板栗的开花和结实。

锥栗花期在 5 月中下旬，11 月上中旬果实成熟。板栗分布区域广阔，地区之间物候期差异较大。长江流域花期为 4～5 月，燕山地区则在 6 月中下旬；果熟期也随地区而不同，一般南方为 8～9 月，北方为 9～10 月。栗属的壳斗密被针刺，成熟后在树上自然开裂，内含坚果 1～3（5）个（表 2）。坚果外皮厚革质，较硬，浅褐、深褐至酱黑色，略具光泽，果脐部分灰黄色，略显粗糙；内层为薄而软的种皮，具茸毛。无胚乳，肥厚的子叶 2 片相贴甚紧，内含大量淀粉。胚根部较小，夹在子叶基部之间（图 1）。

表 2　栗属树种的结实年龄、结果习性及果实、种子特征

树　种	开始结实年龄	刺球苞（壳斗）			种子（坚果）		
		果枝上球苞数（个）	带刺直径（cm）	内含坚果数（个）	形　状	直　径（cm）	千粒重（g）
锥栗	8～10	1～5	2.5～3.5	1	圆锥形	1.5～2	2 000～2 850
板栗	5～7	1～3	4～6.5	2～3（5）	半球或扁球形	1.5～3	7 000～29 000
茅栗	1～2	3～7	3～5	3（5，稀 7）	扁球形	1～1.5	1 100～1 600

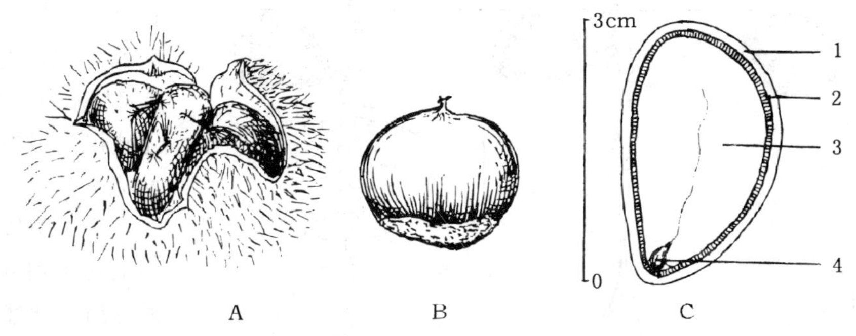

图 1　板栗成熟时开裂的总苞，内含 3 坚果（A）、坚果外形（B）及其纵切面（C）
1. 果皮　2. 种皮　3. 子叶　4. 胚根
（陈荣道绘）

果实的采收调制与种子贮藏　栗属的坚果就是播种材料，通称种子，属于顽拗型，不耐 0℃以下的低温，也不耐干燥，含水量不能低于 30%，极易腐坏，又容易遭受病、虫、鼠、兽危害，需要严格掌握采收调制和贮藏的各个环节。茅栗采种育苗的经验不多。锥栗和板栗应在优良类型中选择优良单株作为采种母树。最好是充分成熟、壳斗开裂、坚果自然脱落后在地面拾集，挑选其中粒大饱满、无病无虫的作为播种材料。也可以用细竹竿轻击已经成熟即将开裂的刺苞，在地面拾集刺苞。由于成熟期不一致，这样的拾集往往要进行几次。尚未裂开的刺苞放在阴凉通风处摊晾 2～7 天以促进后熟。堆高不超过 50cm；上面用稻草覆盖，每隔 1～2 天洒水一次，堆放 7～10 天后，扒开脱粒。剔除有虫眼、病斑及粒小未熟的劣质果，晾干坚果表面的潮气后混沙湿藏。坚果的净度可达 98%以上。坚果的千粒重随地区和品种而有很大差异，一般情况下的变化幅度参见表 2。坚果含水量及湿沙的含水量应分别控制在40%～50%和 15%～20%。我国北方产区通常用坑藏、窖藏或室内堆藏。

生产上贮藏栗属种子都是为了越冬后春播，目前还没有长期贮藏的成功经验。即便是几个月的短期贮藏，如果不能保证适度的潮润、通气和低温条件，还会出现种粒腐坏的现象。如果贮藏之前种粒未经仔细挑选，混入了已经感染病虫害、成熟程度不高或有机械损伤的种粒，腐坏情况就会更加严重。王晓峰（1993）综述过板栗在贮藏运输过程中大量腐坏的原因和文献报道过的保鲜方法。1987 年 10 月 19 日南京中山植物园对本文描述的 3 个树种做了一次混沙湿藏试验：至翌年 3 月 15 日止，锥栗腐坏率为 22.3%，已经发芽的为 6.2%，剩下的优良种粒为 71.5%；板栗由于未经优选，已经发芽的为 0.7%，腐坏率为 43%，优良种粒只有 56.3%；试验所用茅栗的质量较差，贮藏期中发芽粒为 1.5%，腐坏粒高达 58%，贮藏结束时的优良种粒仅为 40.5%。不过，经过这样贮藏的种子发芽情况良好，以板栗为例，在 16～21℃的温度下，28 天发芽率就达到 92%。俞志林和许仲明（1987）利用民间俗称盐壶的陶质容器贮藏板栗，贮藏 118 天后播种，场圃发芽率达到 62.7%。陕西省柞水县核桃板栗研究所（1985）报道过利用当地的石窑洞贮藏食用板栗的做法。

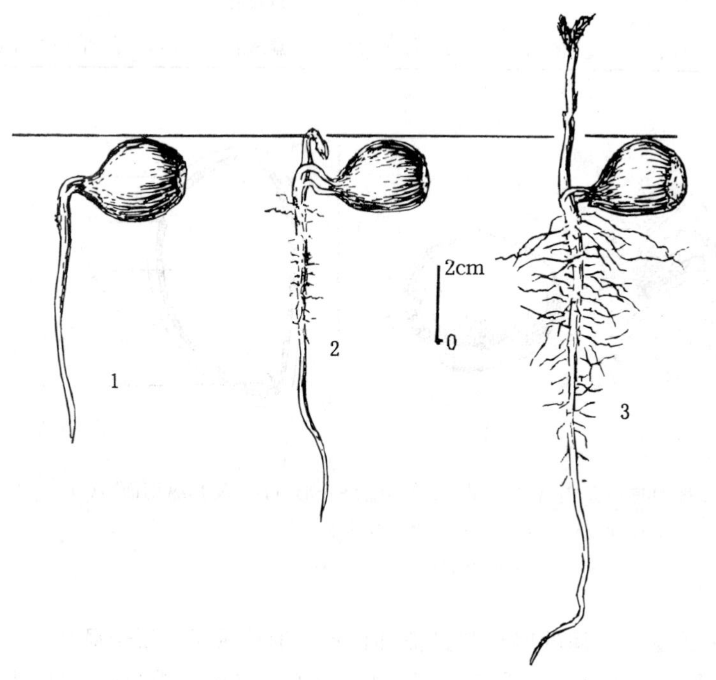

图 2　板栗种子萌发和幼苗生长情况
1. 幼根延伸　2. 胚芽伸出、子叶留土　3. 发育叶生长
（陈荣道绘）

发芽和播种　栗属种子通常有 3～4 个月甚至更长的休眠期，种粒又大，通常是用切开法鉴定优良度。国家标准《林木种子检验方法》还建议采用“胚方”测定发芽能力。

为了解除休眠，生产上常采用秋播或湿藏后春播。秋播时间在 10 月下旬～11 月上旬。解除了休眠的种子，发芽所要求的温度差异不大，大致是气温 15～20℃，土温 10～12℃。因此南方春播可在 3 月上中旬，北方可在 3 月下旬至 4 月上旬。条播时行距 15～20cm，在播种沟内每隔 8cm 平放 1 粒种子，覆土 3～5cm，盖草 2.5～10cm。锥栗每 666m^2 播 100～120kg；板栗在南方约播 100～125kg，北方约 75～100kg。春季种子萌发时分批揭除盖草。播种后应防治啮齿动物危害。播后 20 天左右苗木基本出齐，每 666m^2 出苗 1 万株左右，成苗率一般在 80%以上。留土萌发，有初生不育叶（图 2）。板栗 1 年生苗高40～60cm，可以出圃。为了促进根系发达，有的地方将 1 年生苗“平茬”后再培育 1 年出圃。杨镇和王志彦（ 1990）利用 400μg/g 的 NAA 处理，做过板栗的嫩枝扦插试验。

（何泽瑛）

栲（锥）属
Castanopsis Spach

（壳斗科　Fagaceae）

生长习性、分布和用途　本属约 130 种，我国 70 余种，本文描述 9 种，均为常绿乔木。生长快，萌芽力强，具菌根菌。幼龄期较耐荫。喜湿热气候，分布于亚洲热带和亚热带地区。木材坚硬，用于建筑、家具、造船、桥梁，为优质材，小材可作培养食用菌的原料。壳斗、枝皮可提取栲胶。种子含淀粉，多数可食。有些树种防火性能强，可作防火林带。涵养水源及改良土壤效益高。这 9 个种的名称、生长、分布及用途见表 1。其中青钩栲更新困难，数量极少，已被《中国植物红皮书》列为稀有种。

表 1　栲属树种名称、生长、分布和用途

中　名	学　名	树高(m)	胸径(cm)	分　布	用　途	供　稿
小红栲(米槠)	*C. carlesii* (Hemsl.) Hay.	20	80	江南各地	材用、栲胶、食用	903
桂林栲	*C. chinensis* Hance	20	60	粤、琼、桂、黔、滇	材用、栲胶、食用	612
罗浮栲	*C. fabri* Hance	20	45	江南各地。越南、老挝	材用、栲胶	612
栲树	*C. fargesii* Franch.	30	60	江南各地	材用、栲胶、食用	903
黧蒴栲	*C. fissa*(Champ.) Rehd. et Wils.	20	60	桂、粤、赣、闽、湘、黔、滇	材用、食用、防火	903
刺栲(红锥)	*C. hystrix* A. DC.	30	150	闽、湘、桂、粤、黔、琼、滇、藏。越南、缅甸、印度、老挝	材用、栲胶、食用、防火、涵养水源	612
青钩栲(格氏栲、吊皮锥)	*C. kawakamii* Hay.	40	150	台、闽、赣、桂、粤	材用、食用、栲胶、观赏	903
鹿角栲	*C. lamontii* Hance	25	100	闽、赣、湘、黔、桂、粤、滇	材用、水源涵养	903
苦槠	*C. sclerophylla* (Lindl.) Schott.	15	50	长江中下游以南各地	材用、食用、防火、涵养水源	903

开花结实　10～15 年生开始开花结实，正常结实期为 20～50 年生。结实大小年间隔期为 1～3 年。花单性，雌雄同株。雄花为柔荑花序直立。雄花花被 5～6 裂，雄蕊 10～12，花药近球形，退化雌蕊小，密被毛。雌花单生或 2～5 生于总苞内，花被 5～6 裂。子房下位，3 室，每室有胚珠 2 枚。花柱 3。花期 2～5 月。这 9 个树种中，只有黧蒴栲和苦槠的果实当年成熟，其余 7 个种的果实都是开花的翌年成熟。坚果。壳斗球形、卵形或椭圆形，稀杯状；开裂，稀不开裂；全包坚果，稀包一部分。壳斗外壁密生或疏生刺状或肋状苞片，稀为鳞片状。坚果 1～3 (7)，仅基部或至中部与壳斗内壁连生，果脐圆形或椭圆形，果顶喙尖。无胚乳，子叶平凸。除苦槠的胚着生于果的基部外，其它 8 种的胚均着生于顶部。栲属树种开花结实习性

见表 2。壳斗和坚果的形态见表 3 及图 1。

表 2 栲属树种开花结实习性

树 种	观察年份和地点	结实年龄（年）			开 花			果实成熟		果实散落期
		开 始	正 常	丰年间隔期（年）	始 期	盛 期	末 期	始 期	盛 期	
小红栲	1988 福建三明	20	25～50	2～3	3 月中旬	3 月中下旬	4 月上旬	翌年 11 月上旬	11 月中下旬	12 月上旬
桂林栲	1988 广西南宁	10～15	30～50	2	4 月下旬	5 月上旬	5 月中旬	翌年 10 月上旬	10 月下旬	11 月上旬
罗浮栲	1979～1981 广西南宁	15	—	2	3 月下旬	4 月上旬	4 月下旬	翌年 11 月上旬	11 月中旬	11 月下旬
栲树	1988 福建三明	20	25～50	2	4 月下旬	5 月上中旬	5 月下旬	翌年 9 月下旬	10 月中旬	11 月中旬
罴蒴栲	1988 福建三明	15	20～50	1	3 月中旬	3 月下旬	4 月中旬	10 月下旬	11 月上旬	11 月下旬
刺栲	1988 广西南宁	10～15	30～50	2	3 月上旬	3 月中下旬	4 月上旬	翌年 11 月上旬	11 月中旬	11 月下旬
青钩栲	1988 福建三明	15	30～90	3～5	3 月上旬	3 月中下旬	4 月上旬	翌年 10 月下旬	11 月中旬	12 月上旬
鹿角栲	1988 福建三明	20	23～50	2～3	2 月上旬	2 月中～3 月上旬	3 月中旬	翌年 10 月下旬	11 月中旬	12 月上旬
苦槠	1988 福建三明	20	23～50	2	4 月下旬	5 月上中旬	5 月下旬	10 月上旬	10 月下旬	11 月中旬

表 3 栲属树种壳斗和坚果的形态特征

树 种	壳 斗	坚 果
小红栲	径 0.9～1.5cm，开裂，被疣状突起或极短的尖刺。具 1 果	近球形或长圆锥形，径 0.8～1.3cm。果脐小，径 0.4～0.8cm
桂林栲	径 2～4cm，密被锐刺，内壁被棕色长绒毛，开裂。具 1 果	圆锥形，无毛，长 1.2～1.6cm，径 1～1.3cm。果脐与果底部同大
罗浮栲	径 2～3cm，被疏刺，开裂，具 3 果	圆锥形或三角状圆锥形，1～2 面平，径 0.8～1.2cm。果脐大于果底部
栲 树	径 1.5～3cm，被疏刺，刺长 0.8～1.5cm，多条合生至中部以上或刺束，刺束被短柔毛，开裂。具 1 果	圆锥形，长 1.0～1.5cm，径 0.8～1.2cm。果脐较大
罴蒴栲	被三角形或肋状凸起的鳞片，排列成 4～6 圆环，球形或椭圆形，无柄，开裂。具 1 果	宽卵形或圆锥形，无毛，长 1.3～1.8cm，径 1～1.6cm，果脐 0.4～0.6cm，子叶皱褶
刺栲	密被锐刺，径 2.5～4cm，壁厚 0.3～0.4cm，内壁无毛，开裂，具 1 果	宽圆锥形，几无毛，长 1.0～1.5cm，径 0.8～1.5cm。果脐占果底面积 1/3～1/4
青钩栲	径 6～8cm，密被锐刺，刺长 2～3cm，常合成刺束，有多次分枝。壁厚 0.3～0.4cm，开裂。具 1 果	扁圆锥形，密被黄棕色绒毛，长 1.2～1.5cm，径 1.7～2.0cm。果脐小于果底部

（续）

树　种	壳　斗	坚　果
鹿角栲	被长短粗细差异大的刺，多条连生成刺束。径4～6cm，近球形，壁厚0.3～0.7cm，不开裂。具3果	三角状圆锥形或圆锥形，径1.5～2.5cm，密被绒毛。果脐大于果底部
苦槠	被三角形或肋状凸起的鳞片，仅基部有时连成圆环。卵圆形，或扁球形，有柄，开裂。具1果	球形，有绒毛。果脐大于果底部，径1～1.4cm。子叶平凸

果实的采收调制和贮藏　果实成熟前清理地被物，待壳斗开裂，坚果散落，或上树用竹竿敲落果实，在地面及时收集。壳斗未开裂的果实应置通风处摊晾，待开裂后得出坚果。鹿角栲的壳斗不开裂，可剥开壳斗取出坚果。坚果即为播种材料，通称种子。栲属树种的出籽率和种子质量见表4。种子经水选阴干，拌上杀虫剂即可贮藏。种子的安全含水量高，需混沙湿藏，但贮藏期不宜超过半年。种子装在通气保湿的塑料袋内，置于1～5℃冷库中，可贮藏1年。据程必强和马信祥（1987）报道，产自云南省勐仑地区的刺栲，湿沙贮藏3～5个月后发芽率为38%～98%，未用湿沙的贮藏2～3个月后发芽率为10%～54%。椰子果实中棕色纤维的粉碎物俗称椰糠。宋学之和刘文明（1992）认为椰糠质量轻，疏松通气而又保水，是贮藏刺栲最适宜的介质。他们建议的最佳贮藏条件是种子含水量36%～37%，椰糠的含水量25%～27%，温度15～20℃并经常翻动。

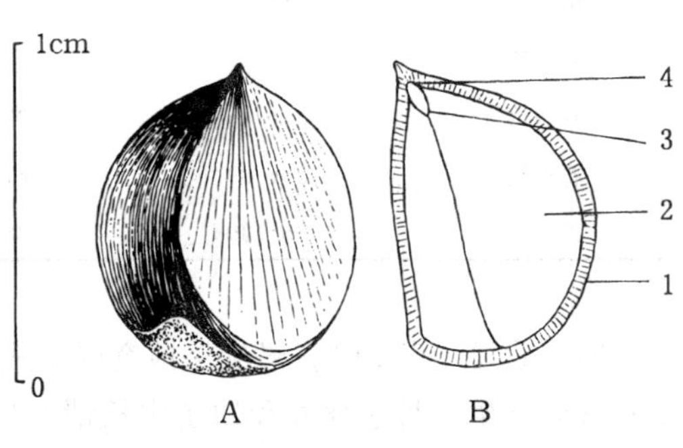

图1　罗浮栲坚果的外形（A）及其纵切面（B）
1. 果皮及种皮　2. 子叶　3. 胚芽　4. 胚根
（黄应钦绘）

表4　栲属树种的出籽率和种子（坚果）质量

树　种	出籽率（%）	净　度（%）	千粒重（g）		每千克粒数（粒）	
			一　般	变动范围	一　般	变动范围
小红栲	70～80	82～95	360	290～440	2 800	2 300～3 450
桂林栲	30～50	98～100	1 600	1 200～2 000	630	500～830
罗浮栲	30～50	98～100	350	300～400	2 850	2 500～3 300
栲　树	—	87～98	600	500～700	1 660	1 400～2 000
黧蒴栲	67～82	90～99	1 500	1 400～2 000	670	500～710
刺　栲	30～50	98～100	800	650～950	1 250	1 050～1 540
青钩栲	—	92～99	2 230	2 100～2 900	450	340～480
鹿角栲	22～29	92～99	2 400	2 200～2 600	420	380～450
苦　槠	—	92～99	1 200	1 000～1 400	830	710～1 000

发芽和播种　种子无明显休眠现象，发芽时日均温宜在15～28℃。用混沙湿藏越冬的种子在春3月播种。一般不随采随播，以减少鼠害。栲属树种的种子播种品质及播种量见表5。

表 5 栲属树种的种子播种品质及播种量

树 种	含水量（%）	优良度（%）	发芽率（%）	每平方米的播种量（g）
小红栲	35～40	80～90	72～89	80～100
桂林栲	—	—	70～80	200～240
罗浮栲	—	—	80～95	40～50
栲 树	35～40	88～94	78～85	100～140
黧蒴栲	41～46	90～96	70～80	190～230
刺 栲	—	—	60～85	130～180
青钩栲	34～46	90～95	80～90	500～620
鹿角栲	35～39	67～84	50～70	620～750
苦 槠	35～39	88～95	67～84	180～250

留土萌发。萌发时胚根从果顶突破果皮伸出，但苦槠的胚根从果脐伸出。下胚轴粗短，初生不育叶互生。种子萌发和幼苗生长情况见图 2。1～2 年生苗出圃。栲属中有些种还可扦插育苗。

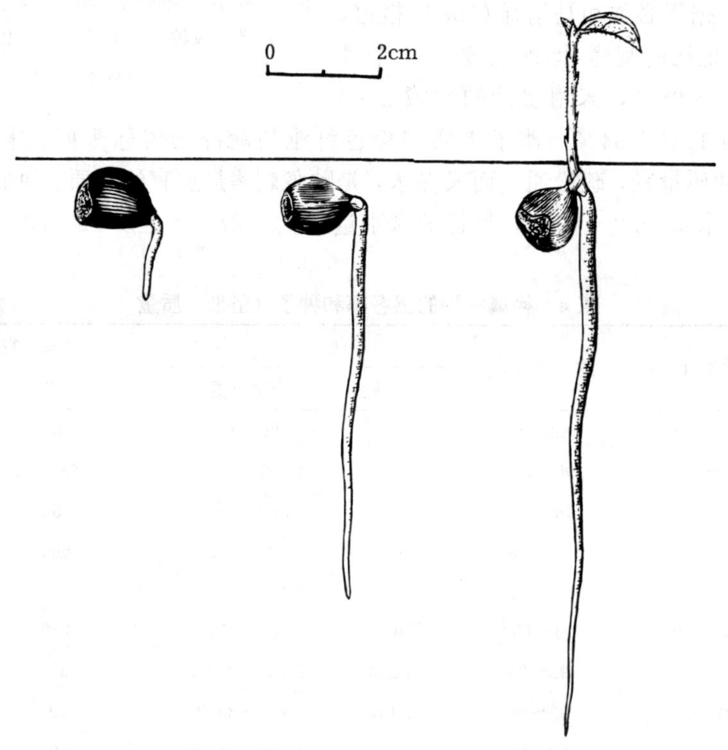

图 2 罗浮栲种子萌发后第 6、20、28 天幼苗生长情况

（黄应钦绘）

（李玉蕾）

青 冈 属
Cyclobalanopsis Oerst.

（壳斗科 Fagaceae）

生长习性、分布和用途 本属约 150 种，我国约 70 种，本文描述 10 种。常绿乔木，高 15～30m。适生于肥沃湿润的土壤，少数树种能耐干旱贫瘠。分布于我国秦岭及淮河流域以南各地，以及亚洲东南部地区。材质坚重，耐腐，可用于桩柱、车船、桥梁、工具柄、木制机械、刨架、枕木及运动器械等；树皮及壳斗可提取栲胶；种子含淀粉，可作饲料并可酿酒。这 10 个树种的名称、生长、分布和用途见表 1。

表 1 青冈属树种的名称、生长、分布和用途

中 名	学 名	树高（m）	分 布	用 途	供 稿
福建青冈	*C. chungii*（Metc.）Chun	15	赣、湘、闽、粤	材用	902
突脉青冈	*C. elevaticostata* Q. F. Zheng	20	闽	材用	902
饭甑青冈	*C. fleuryi*（Hick. et A. Camus）Chun	25	赣、闽、粤、桂、黔、滇。越南	材用、食用、栲胶	603
青冈	*C. glauca*（Thunb.）Oerst.	20	青、甘、陕、豫以南，苏、浙、闽、台向西达藏，南至粤、桂、滇。朝鲜半岛、日本、印度	材用、食用、栲胶	902
细叶青冈	*C. gracilis*（Rehd. et Wils.）Cheng et T. Hong	10	苏、皖、浙、闽、赣、鄂、湘、粤、桂、黔、川、甘	材用	902
大叶青冈	*C. jenseniana*（Hand. -Mazz.）Cheng et T. Hong	30	赣、浙、闽、鄂、湘、粤、桂、黔、滇	材用、食用、栲胶	902
小叶青冈	*C. myrsinaefolia*（Blume）Oerst.	20	北起陕、豫，东抵闽、台，西至川、黔，南至粤、桂、滇。越南、老挝、日本	材用、食用、栲胶	902
云山青冈	*C. nubium*（Hand. -Mazz.）Chun	25	苏、皖、浙、赣、闽、台、鄂、湘、粤、桂、黔、川。日本	材用、食用、栲胶	902
毛果青冈	*C. pachyloma*（Seem.）Schott.	20	台、闽、赣、粤、桂、黔	材用	902
褐叶青冈	*C. stewardiana*（A. Camus）Y. C. Hsu et H. W. Jen	12	皖、浙、闽、赣、鄂、湘、粤、桂、黔、川	材用	902

开花结实 10～15 年生开始结实，15 年生以后进入正常结实期。结实有大小年现象，丰年间隔期为 1～2 年。花单性，雌雄同株。雄花为柔荑花序，下垂，多簇生于新梢基部，花被 5～6 深裂，雄蕊 6（4～12）枚，花丝短，退化雌蕊细小或无。雌花序穗状，顶生，直立。雌花单生于总苞内，有时具细小退化雄蕊，花被 5～6 裂，子房下位，常 3 室，每室有胚珠 2 颗。花柱 3，柱头较短，侧生带状或顶生头状。本文这 10 个树种中，大叶青冈和褐叶青冈的果实在开花的次年成熟，其余 8 个种的果实都在开花当年成熟。它们的开花结实习性见表 2。

表 2 青冈属树种的开花结实习性

树种	开始开花结实年龄（年）	开始正常结实年龄（年）	丰年间隔期（年）	观察地点	观察年份	开花			果实成熟			种子散落		
						始期	盛期	末期	始期	盛期	末期	始期	盛期	末期
福建青冈	10～15	15	1～2	福建沙县	1987～1988	4 月上旬	4 月下旬	5 月上旬	11 月上旬	11 月中旬	12 月上旬	12 月上旬	12 月中旬	1 月上旬
突脉青冈	15～20	20	1～2	福建宁德	1987～1988	4 月上旬	5 月上旬	5 月中旬	11 月上旬	11 月中旬	11 月下旬	11 月下旬	12 月上旬	12 月中旬
饭甑青冈	8	20	1	广西南宁	1986～1988	4 月上旬	4 月中旬	4 月下旬	11 月上旬	11 月中旬	—	12 月上旬	12 月中旬	—
青冈	10～15	15	1	福建三明	—	3 月中旬	4 月上旬	4 月下旬	10 月中旬	11 月上旬	11 月中旬	11 月中旬	11 月下旬	12 月上旬
细叶青冈	约 12	15	1	福建南平	1987～1988	4 月下旬	5 月中旬	6 月中旬	10 月下旬	11 月上旬	11 月中旬	11 月中旬	11 月下旬	12 月上旬
大叶青冈	12～16	20	1	福建南平	1987～1988	4 月下旬	5 月中旬	6 月中旬	翌年 10 月中旬	11 月上旬	11 月中旬	11 月中旬	11 月下旬	12 月上旬
小叶青冈	11～15	15	1	福建邵武	1986～1987	6 月上旬	6 月中旬	6 月下旬	10 月上旬	10 月中旬	10 月下旬	10 月下旬	11 月上旬	11 月中旬
毛果青冈	约 15	20	2	福建南平	1987～1988	3 月上旬	3 月中旬	3 月下旬	10 月下旬	11 月上旬	11 月中旬	11 月中旬	11 月下旬	12 月上旬
云山青冈	约 15	20	1	福建三明	1986～1988	4 月中旬	5 月上旬	5 月下旬	10 月中旬	10 月下旬	11 月上旬	11 月上旬	11 月中旬	11 月下旬
褐叶青冈	约 10	15	1	福建长汀	1987～1988	7 月上旬	7 月中旬	7 月下旬	翌年 10 月上旬	10 月下旬	11 月上旬	11 月上旬	11 月中旬	11 月下旬

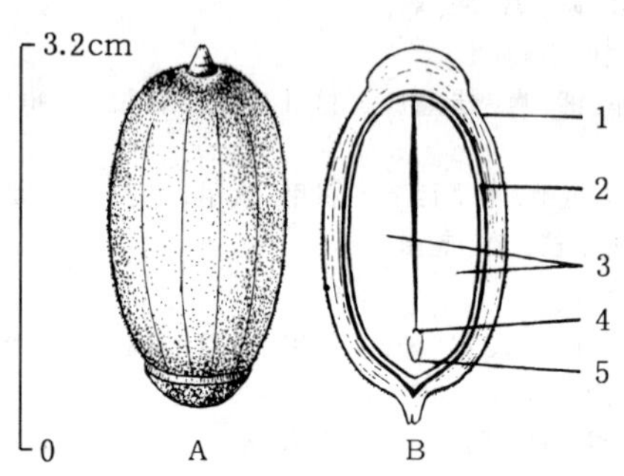

图 1 毛果青冈坚果的外形（A）及其纵切面（B）

1. 果皮 2. 种皮 3. 子叶 4. 胚芽 5. 胚根（黄应钦绘）

坚果。壳斗杯状、碟状、钟形、稀全包，鳞片愈合成同心环带，环带全缘或具齿裂。每壳斗有 1 坚果。成熟的坚果褐色，顶部有柱座。不育胚珠在种子顶部外侧。无胚乳。肉质子叶 2 枚，肥大。子叶和胚常为乳白色、浅黄色或米黄色（图 1）。

果实害虫主要是象甲（*Hylobius pales*），在幼果发育期间，成虫产卵于幼果内，孵化成幼虫后以果实为食。应在幼果形成期喷药，毒杀成虫。青冈属果实还易受鼠害。青冈属壳斗和果实的形态、色泽、大小和附属物的特征见表 3。

果实采收和调制 果实大多数在 10～11 月成熟并很快脱落，应在果实由绿色变为褐色或栗褐色时，上树用竹竿敲落果实，在树下铺采种布收集；或用采种刀截取果穗，也可待果实脱落后在地面及时收集。果实采回后切忌堆积或曝晒，应在通风阴凉处摊晾 5～10 天，坚果即从壳斗中脱出，用作播种材料，通称种子。毛果青冈、饭甑青冈壳斗较厚，包被较多而紧，阴干后还需要人工去除壳斗。果实的出籽率和种子质量等数据见表 4。

表 3 青冈属树种壳斗和果实的形态、色泽、大小和附属物

树种	壳斗					坚果				
	形状	色泽	高(cm)	径(cm)	附属物	形状	色泽	高(cm)	径(cm)	附属物
福建青冈	盘状；具 6～7 环带，下部两环有齿裂，余全缘	灰褐色	0.5	1.9	被灰褐色绒毛	扁球形，顶端平圆	灰褐色	1.5	1.7	微被细绒毛
突脉青冈	浅杯状，包果 1/3	深褐色	0.6～0.8	1.2～1.7	被灰黄色短绒毛	椭圆形或卵状椭圆形	深褐色	1.3～2.3	1.1～1.6	近顶端有灰黄色微柔毛，果脐微凸
饭甑青冈	钟形或近圆柱形，包果约 2/3；具 10～13 环带，近全缘	黄棕色	3～4	2.5～4	内外壁被黄棕色毡绒毛	柱状长椭圆形	黄棕色	3～4.5	2～2.5	密被黄色绒毛
青冈	碗形，包果 1/3～1/2；具 5～8 环带，全缘或有细缺刻	褐色	0.6～0.8	1～1.6	被薄毛	卵形或椭圆形	棕褐色	1～2.2	0.9～1.5	无毛
细叶青冈	碗形，包果 1/3～1/2；具 6～9 环带，常有裂齿	灰黄色	0.6～0.8	0.9～1.3	被灰黄色绒毛	椭圆形	浅褐色	1.5～2	0.9～1.2	顶端被毛
大叶青冈	杯形，包果 1/3～1/2；具 6～9 环带，有裂齿	灰黄色	0.8～1	1.3	无毛	长卵形或倒卵形	灰褐色	1.7～2.2	1.3～1.5	无毛。果脐突起
小叶青冈	杯形，薄，包果 1/3～1/2；具 6～9 环带，全缘	灰褐色	0.5～0.8	1.0～1.5	被灰白色柔毛	卵形或椭圆形，顶端圆	浅褐色	1.4～2.5	0.7～1	无毛
云山青冈	环形，包果约 1/3；具 5～7 环带，下部 2～3 环有裂齿，余近全缘	灰褐色	0.5～1	0.9～1.5	被灰褐色绒毛	倒卵形或长椭圆状倒卵形	灰褐色	1.3～2.4	0.8～1.4	柱座基部有几条环纹
毛果青冈	钟形，包果 1/2～1/3；具 7～8 环带，全缘	黄褐色	1.5～2	2～2.2	密生黄褐色绒毛	长椭圆形或倒卵形	黄褐色	2～3.3	1.2～1.7	幼时密被黄色绒毛，后渐脱落
褐叶青冈	杯形，包果约 1/2；具 6～9 环带，环带常与壳斗壁分离，边缘具粗锯齿	灰褐色	0.6～0.8	1.1～1.7	内壁被灰褐色绒毛，外壁被灰白色绒毛后渐脱落	宽卵形	浅褐色	0.8～1.5	1～1.5	无毛

表 4 青冈属树种的出籽率和种子（坚果）质量

树种	出籽率(%)	净度(%)	千粒重(g)		每千克种子粒数（粒）		容重(g/L)
			一般	变动范围	一般	变动范围	
福建青冈	87	93～98	1 430	1 340～1 520	700	660～750	655
突脉青冈	87	91～97	2 190	1 860～2 380	460	420～540	690
饭甑青冈	45	98～100	7 500	6 500～9 000	130	110～160	—
青冈	89	93～98	1 590	1 500～1 680	630	590～670	650
细叶青冈	86	94～98	1 470	1 380～1 550	680	640～720	635
大叶青冈	88	94～99	1 020	930～1 220	980	820～1 080	645
小叶青冈	85	89～98	1 050	970～1 140	950	880～1 030	654
云山青冈	88	94～99	1 640	1 530～1 750	610	570～650	660
毛果青冈	62	94～99	4 450	3 580～5 140	220	190～280	600
褐叶青冈	87	94～98	840	750～930	1 190	1 070～1 330	650

种子贮藏　适于贮藏的种子含水量为25%～30%，低于25%会影响种子生命力。贮藏时需用湿沙和种子混合，堆放在阴凉的室内，或装在敞开的木箱里，经常翻动，注意通气。防治蛀虫危害可将种子置密闭室内或密闭容器内，在温度25℃下，每立方米用二硫化碳30mL熏蒸处理20小时。生产中多将种子装入箩筐置流水中浸泡5～7天，稍晾干后混沙贮藏。如贮藏时间较长需将种子混沙后置于5～10℃的低温环境，贮藏期可达6个月至1年。短途运输的种子可用麻袋包装，到达目的地后及时散热沙藏。

发芽前的处理和发芽测定　种子无明显休眠习性。为使发芽整齐，发芽前1～2个月或采种后，将种子与湿沙混合，堆在通风透气良好的室内，厚度20cm左右，经常翻动，保持湿润。据福建林学院等单位试验，基质用沙或蛭石，在15～25℃的气温条件下测得的各项发芽能力指标见表5。

表5　青冈属树种的发芽能力及其测定条件

树　种	温　度（℃）	发芽势（%）		发芽率（%）		场圃发芽率（%）	
		计算天数	一般数值	计算天数	一般数值	计算天数	一般数值
福建青冈	17～22	18	82	25	89	40	84
突脉青冈	18～23	18	78	25	85	40	80
饭甑青冈	15～20	17	75	21	88	—	—
青　冈	18～25	15	68	20	85	65	77
细叶青冈	约22	15	90	20	98	45	97
大叶青冈	约25	17	81	20	88	60	85
小叶青冈	约22	20	85	25	95	80	90
云山青冈	约25	10	86	15	93	40	90
毛果青冈	约25	15	80	20	86	75	83
褐叶青冈	15～20	8	42	15	50	70	38

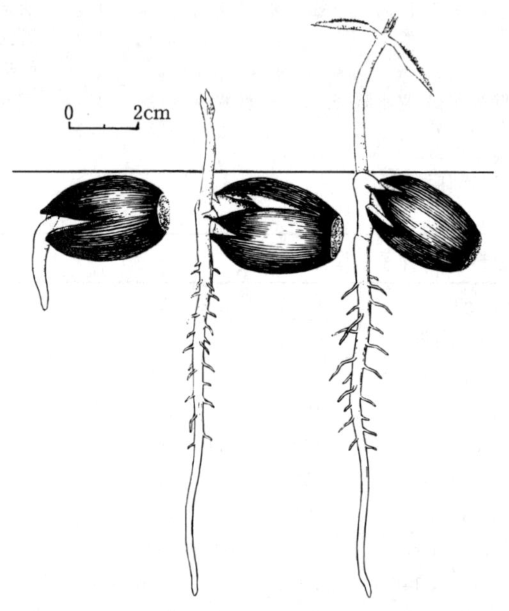

图2　毛果青冈种子萌发后第5、15、22天的幼苗生长情况
（黄应钦仿《热带亚热带主要树种采种育苗技术》）

播种　条播。经过混沙湿藏的种子每666m²的播种量为：青冈、细叶青冈、福建青冈和云山青冈约75kg，突脉青冈90kg，饭甑青冈200kg，毛果青冈150kg。条播的播幅为20～30cm，沟深3～8cm，覆土厚度2cm左右。播后1～2个月开始发芽出土。子叶留土萌发。青冈属有的树种有初生不育叶，如云山青冈，有的树种没有初生不育叶，例如青冈和毛青冈。幼苗出土后10天左右发出初生发育叶。初生发育叶2～4片，对生，如青冈、云山青冈和小叶青冈，有的互生。有的树种幼时密被绒毛。图2绘出的是毛果青冈种子萌发和幼苗的生长情况。少数树种如突脉青冈和青冈，1年生苗高可达40～50cm，地径5～6mm，可以

出圃。大多数树种 1 年生苗高 10～30cm，需要培育 2 年出圃。幼苗主根发达，生产上常在初夏进行切根，促使侧根生长。每 666m² 产苗 1 万～2 万株。青冈属的树种还可以扦插育苗。

（郑郁善）

水青冈属

Fagus L.

（壳斗科　Fagaceae）

生长习性、分布和用途　水青冈属约 11 种，我国 6 种，本文描述 3 种。落叶乔木，适生于土壤肥沃、湿润、多雾的高海拔环境。分布于秦岭、淮河以南，青藏高原以东。木材富韧性，稍耐腐，是坑木、车船、工具柄、农具、胶合板等的良好用材；种子榨油，可食用或作工业原料。这 3 个树种的名称、分布和用途见表 1。

表 1　水青冈属树种名称、分布和用途

中　名	学　名	成年时树高（m）	分　布	用　途
米心水青冈	*F. engleriana* Seem.	23	川、黔、滇、陕、豫、鄂、皖、闽	材用
水　青　冈	*F. longipetiolata* Seem.	25	陕、川、滇、黔、桂、粤、湘、鄂、皖、赣、浙、闽	材用、食用
亮叶水青冈	*F. lucida* Rehd. et Wils.	25	湘、鄂、皖、赣、闽、川、黔、桂	材用、食用

开花结实　6～10 年生开始开花结实，15 年后进入正常结实期。有大小年现象，丰年间隔期为 1～2 年。花单性，雌雄同株。雄花为总梗细长下垂的头状花序，有花 7～11（13）朵；近总花梗顶部有膜质线形或披针形的苞片 2～5 片；花被 4～7 裂，钟状；雄蕊 6～12，有退化雌蕊。雌花每 2 朵生于总苞内，稀 1 朵或 3 朵；花被 5～6 裂，细小；子房下位，3 室，每室有顶生胚珠 2 枚；花柱 3，基部合生。这 3 个树种的开花结实物候期见表 2。

表 2　水青冈属树种的开花结实物候

树　种	观察地点	观察年份	开花			果实成熟		果实脱落		
			始期	盛期	末期	始期	盛期	始期	盛期	末期
米心水青冈	福建邵武	1988	4 月上旬	4 月中旬	4 月下旬	8 月中旬	8 月下旬	9 月上旬	9 月中旬	9 月下旬
水青冈	福建将乐	1987	4 月中旬	5 月上旬	5 月下旬	8 月下旬	9 月中旬	9 月下旬	10 月中旬	10 月下旬
亮叶水青冈	福建邵武	1988	3 月下旬	4 月中旬	4 月下旬	8 月下旬	8 月下旬～9 月上旬	9 月上旬	9 月中旬	10 月上旬

坚果三角状卵形，有 3 棱脊，黄褐色。壳斗常 4 裂，小苞片为短针刺形、窄匙形、线形、钻形或瘤状突起。米心水青冈和水青冈每个壳斗具 2 个坚果，稀 1 个或 3 个。亮叶水青冈每

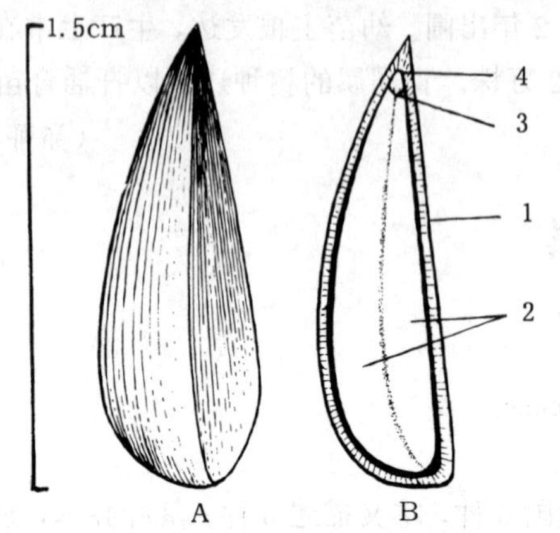

图1 水青冈坚果的外形（A）及其纵切面（B）
1. 果皮及种皮 2. 子叶 3. 胚芽 4. 胚根
（黄应钦绘）

壳斗具1～2个坚果。种子无胚乳。子叶2枚，折扇状，乳白色。坚果的大小见表3，水青冈坚果的外形及其剖面结构见图1。

主要虫害为象甲（*Hylobius pales*）之类的昆虫蛀食果实。可能是因为分布在高海拔多雾地区因而授粉不良，这几个树种种子空瘪粒多。

果实采收调制和贮藏 壳斗由绿色或浅褐色变为褐色，部分壳斗顶部开裂，少量坚果开始脱落时即可采收。可以上树采摘，也可以待坚果自然脱落后在地面收集。采回的果实切忌堆积，应及时摊放在阴凉处晾干，5～8天后壳斗开裂，脱出坚果，用作播种材料，通称种子。水青冈属树种的出籽率和种子质量见表4。

表3 水青冈属树种果实的大小

树 种	果实（带壳斗）		坚 果		
	长（cm）	径（cm）	长（cm）	宽（cm）	厚（cm）
米心水青冈	1.2～1.8	0.6～0.9	0.8～1.3	0.4～0.6	0.2～0.4
水 青 冈	1.8～3.0	0.8～1.2	1.3～2.0	0.6～1.0	0.3～0.7
亮叶水青冈	0.8～1.2	0.4～0.6	0.6～0.8	0.2～0.4	0.2～0.3

表4 水青冈属树种的出籽率和种子（坚果）质量

树 种	出籽率（%）	净 度（%）	千粒重（g）	每千克种子粒数（粒）	容 重（g/L）
米心水青冈	21	91～98	125～138	7 000～8 000	280～298
水 青 冈	11	90～98	75～88	11 000～13 500	268～282
亮叶水青冈	18	90～97	110～130	7 600～9 000	270～285

种子含水量低于18%会影响生命力，必须混沙湿藏在室内阴凉处或地窖内，供翌年春播。

发芽和播种 种子无休眠习性。发芽时适宜气温为20～25℃。种子的发芽率低。福建林学院测定的发芽情况见表5。

表5 水青冈属树种的发芽能力及其测定条件

树 种	温 度（℃）	发芽势（%）		发芽率（%）		场圃发芽率（%）	
		计算天数	一般数值	计算天数	一般数值	计算天数	一般数值
米心水青冈	20～25	15	2	25	3	60	2
水 青 冈	20～23	15	4	25	6	60	4
亮叶水青冈	23～25	15	6	25	10	60	8

冬播或春播　条播，播种沟间距 25～30cm，深 5cm。每 666m^2 播种量：水青冈约 50kg，其余 2 种约 80kg。覆土 1cm 并盖草。子叶出土萌发。初生叶对生。主根发达，侧根细密（图 2）。1～2 年生苗出圃。

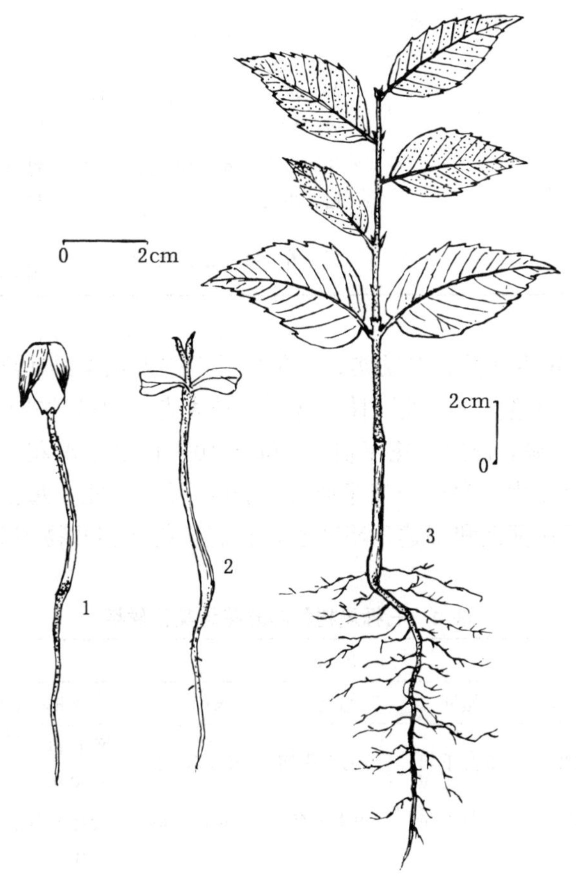

图 2　亮叶水青冈种子的萌发和幼苗生长情况

1. 子叶出土　2、3. 初生叶对生，以后互生

（黄鹏成供稿，田恒德绘）

（郑郁善）

石　栎　属

Lithocarpus Blume

（壳斗科　Fagaceae）

生长习性、分布和用途　本属约 250 种，我国约产 90 种，本文描述 3 种。常绿乔木，高 15～20m，胸径 20～50cm。萌芽性强。有较强的抗逆性，对土壤肥力要求不苛刻，适生性较广。为热带亚热带常绿阔叶林或针阔混交造林的重要树种。材质优，坚重耐腐，为多种工业

用材。它们的名称、生长、分布和用途等见表1。

表1　石栎属树种的名称、生长、分布和用途

中　名	学　名	树高(m)	胸径(cm)	分　布	用　途	供　稿
石栎(椆木)	*L. glaber* (Thunb.) Nakai	15	20	苏、皖、浙、闽、赣、鄂、湘、粤、桂。日本	材用、淀粉、油脂、鞣料、食用菌材	601
多穗石栎(甜茶)	*L. polystachyus* (DC.) Rehd.	20	50	长江以南各省。印度、缅甸、老挝、越南	材用、鞣料、食用菌材、饮料	902
滑皮石栎	*L. skanianus* (Dunn.) Rehd.	20	30	闽、赣、湘、粤、桂	材用、鞣料、淀粉、食用菌材	902

开花结实　8～12年生开始开花结实，15年生以后进入正常结实期。结实有大小年现象，间隔期为1年，但不甚明显。花单性同株，常同序排成直立的柔荑花序，单被花，花被4～6裂。雄花3～7朵聚成一簇，密生于花序轴上。雄蕊10～12枚。雌花3～7朵聚成一簇，生于雄花序之下的花序轴上或生于另外一花序轴上。子房下位，3室，每室胚珠2。花柱3，柱头顶生。据广西南宁、福建邵武和福建三明观察，它们开花结实的物候期见表2。

表2　石栎属树种的开花结实物候期

树　种	观察地点	观察年份	开花			果实成熟		果实脱落
			始期	盛期	末期	始期	盛期	
石栎	南宁	1978～1984	8月下旬	9月上中旬	9月下旬	翌年11月下旬	12月上旬	12月下旬
多穗石栎	三明	1986～1988	9月上旬	9月下旬	10月中旬	翌年9月中旬	10月上旬	10月下旬～11月上旬
滑皮石栎	邵武	1987～1988	9月中旬	10月上旬	10月下旬	翌年9月上旬	9月下旬	9月中旬～10月下旬

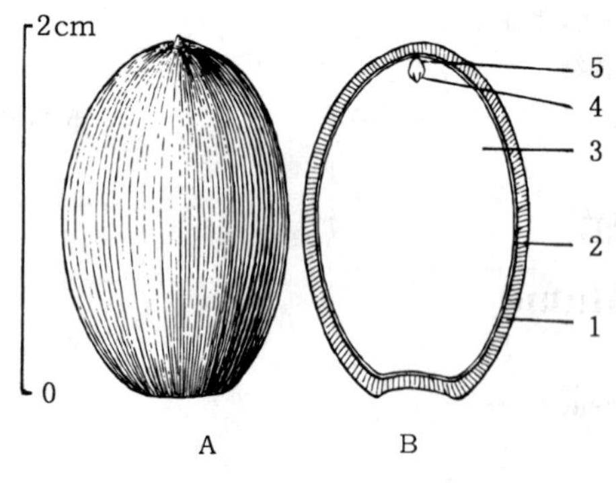

图1　石栎坚果外形（A）及其纵切面（B）
1. 果皮　2. 种皮　3. 子叶　4. 胚芽　5. 胚根
（黄应钦绘）

坚果。石栎果穗长8～10cm，轴细有短绒毛；壳斗杯状，近无柄，包围坚果基部。多穗石栎的果实3～5个簇生于果穗轴上，果穗长10～22cm；壳斗浅碟状，近无柄，包裹坚果基部。滑皮石栎果穗长10～13cm，壳斗近球形或扁球形，包裹坚果的3/4，有时全包。无胚乳，胚倒生，子叶大而充满种子。坚果的形态见表3、图1。

果实的采收调制和贮藏　果实由绿色转变为棕褐色，少量果实开始脱落时即可采收。在树下铺垫布单或塑料薄膜，用竹竿敲打或摇动树枝，将果实震落后收集。采得的石栎及多穗石栎，用手搓擦

表 3　石栎属树种坚果的形态特征

树　种	未熟时颜色	形　状	大小（cm）	色　泽
石　栎	青绿色	卵形或倒卵形，果脐内陷	长 1.4～2.5　径 1～1.5	栗褐色，被白粉
多穗石栎	绿色	卵形或近球形，果脐深陷	长 0.8～1.6　径 1.2～1.8	栗色，常有白粉
滑皮石栎	青绿色	扁球形或宽圆锥形，果脐内陷	长 1.1～1.7　径 1.4～1.8	栗褐色

表 4　石栎属树种的出籽率和种子的净度、质量

树　种	出籽率（%）	净度（%）	千粒重（g）		每千克纯净种子粒数	
			一　般	变动范围	一　般	变动范围
石　栎	80～86	99	2 000	1 500～2 400	500	420～670
多穗石栎	60～75	99	1 500	1 350～1 650	670	600～740
滑皮石栎	50～70	99	2 100	1 900～2 250	470	400～530

或抖落壳斗，即得坚果；滑皮石栎可置日光下稍晒，待壳斗开裂后脱出坚果。坚果即为播种材料，通称种子。已脱去壳斗的种子不宜置日光下曝晒，通常是摊放于通风阴凉处 2～4 天，即可供运输或贮藏。为防止虫蛀，可将种子置流水中浸 4～5 天，或混以少量杀虫粉剂。贮藏时需混沙，但应防止过湿，贮藏期一般为半年。这 3 个树种的出籽率和净度（脱去壳斗前后质量之比）、质量等见表 4。

发芽和播种　种子有短期休眠现象，休眠的原因可能是由于果壳坚硬而透水性不良。破伤果皮，或用浓硫酸处理 20 分钟，彻底洗净后再浸泡 2 天，有可能解除休眠。发芽时日均温宜在 20℃以上。种子的发芽能力见表 5。

留土萌发。胚根萌发后约 10 天胚芽出土，各个种具不育叶 5～6、9～13 或 8～ 10，再过 8 天左右展出初生叶。生长情况见图 2。

条播。播种期以 3 月中旬～4 月上旬，气温稳定回升后为宜。石栎和滑皮石栎每平方米播种 200～250g。多穗石栎每平方米播种 250～300g。苗期生长较慢，一般需培育 2 年方宜出圃造林。生产中常不育苗，而是在林地土壤湿润时，将已经催芽萌动的种子点播至已整地的穴内，每穴播种1～2 粒。

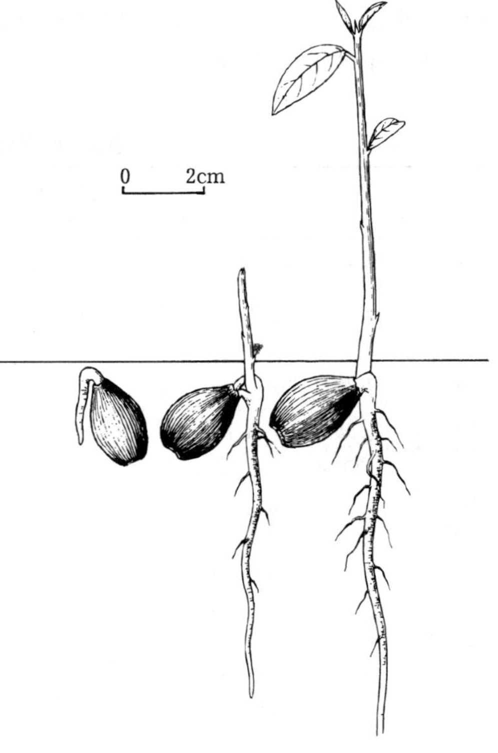

图 2　石栎种子萌发后第 4、10、20 天幼苗的生长情况

（黄应钦仿《热带亚热带主要树种采种育苗技术》）

表5 石栎属树种的发芽能力及其测定条件

树 种	预处理	基 质	室外温度（℃）	发芽率（%）	
				计算天数	一般数值
石 栎	层积催芽	湿沙	18～25	70	70
多穗石栎	层积催芽	圃地	20～28	120	55
滑皮石栎	层积催芽	圃地	20～28	70	60

（王宏志）

栎 属

Quercus L.

（壳斗科 Fagaceae）

生长习性、分布和用途 栎属约300种，分布于亚洲、非洲、欧洲和美洲。我国近60种，本文介绍10种1变种，包括引入1种。常绿或落叶乔木，稀灌木。喜光，耐旱，对土壤要求不严。主根发达，适应范围广，萌芽力强，广泛分布于我国南北各地（表1）。材质坚硬，强度大，耐腐，可供多种用途。树皮及壳斗富含鞣质，可提取栲胶。有的树皮可作软木，有的树叶喂养柞蚕。种子可酿酒或做饲料，也是许多野生动物和鸟类的食物。

表1 栎属树种名称、生长和分布

中 名	学 名	树 高 (m)	胸 径 (cm)	分 布	供 稿
岩 栎	*Q. acrodonta* Seem	12	—	甘、陕、豫、鄂、川、贵、滇。南宁引种	602
麻 栎	*Q. acutissima* Carr.	30	100	辽、冀、晋、陕、甘南至西南、华南。朝鲜半岛、日本、印度	401
槲 栎	*Q. aliena* Blume	25	100	辽、冀、晋、陕、甘南至华南、西南。朝鲜半岛、日本	401
锐齿槲栎	*Q. aliena* var. *acuteserrata* Maxim.	30	100	辽、冀、晋、陕、甘、西南、华中、华东。朝鲜半岛、日本	401
波罗栎（槲树）	*Q. dentata* Thunb.	25	100	东北、华北、西北东部、西南、华中、华东。蒙古、日本	401
辽东栎	*Q. liaotungensis* Koidz.	15	—	东北、华北、西北、川	401
蒙古栎（柞树）	*Q. mongolica* Fisch.	30	—	东北、华北。俄罗斯、朝鲜半岛、日本	401
尖叶栎	*Q. oxyphylla* (Wils.) Hand. -Mazz.	20	—	西北东部、西南东部、中南、皖、浙、闽	902
乌冈栎	*Q. phillyraeoides* A. Gray	10	—	西北东部、西南东部、华中、华南、东南。日本	902
夏 栎	*Q. robur* L.	40	90	原产欧洲。新疆有引种	204
栓皮栎	*Q. variabilis* Blume	30	100	辽、冀、晋、陕、甘南至西南、华南、台。朝鲜半岛、日本	401

开花结实　15～30年开始结实。丰年间隔期1～6年。花单性，同株。单被花。花芽分化于前年秋季新梢的叶腋。雄花柔荑花序，簇生，下垂。花被杯状，4～7裂，雄蕊6（4～12）。雌花序着生于结果枝的叶腋，穗状，直立。雌花单生于总苞内，花被5～6深裂。子房下位，3（2～5）室，每室具2胚珠。花柱与子室同数，短或伸长，柱头侧生带状或顶生头状。坚果（橡实）单生于1总苞（通称壳斗）内。壳斗杯状、碗状、碟状，半球形或近钟形，具鳞片，包被坚果的1/5～3/4。多为4～5月开花，在叶前开放或与叶同时开放。当年或翌年9～10月果实成熟，随后脱落。栎属开花结实物候见表2。

表2　栎属树种开花结实物候

树　种	观察地点	开始结实年　龄	丰年间隔期(年)	开花期	果熟期	种子散落期
岩　栎	广西南宁	8	1～2	5月上旬～5月下旬	10月上旬～11月上旬	10月下旬～11月下旬
麻　栎	北京植物园	10	1～2	4月下旬～5月上旬	翌年9月中旬～10月上旬	9月下旬～10月中旬
槲　栎	北京植物园	15	不明显	4月下旬～5月上旬	9月上旬～9月下旬	9月中旬～10月上旬
锐齿槲栎	北京植物园	15	不明显	4月中旬～4月下旬	9月中旬～10月上旬	9月下旬～10月中旬
波罗栎	北　京	16	2～4	4月中旬～4月下旬	9月上旬～9月下旬	9月中旬～10月上旬
辽东栎	北京植物园	20	3～5	4月下旬～5月上旬	9月上旬～9月下旬	9月上旬～10月上旬
蒙古栎	北京颐和园	20	3～5	4月下旬～5月中旬	9月上旬～9月下旬	9月上旬～10月上旬
尖叶栎	福建邵武	8～12	1～3	5月中旬～6月下旬	翌年9月中旬～10月下旬	10月下旬～11月下旬
乌冈栎	福建邵武	8～12	1～3	3月中旬～4月中旬	9月下旬～10月下旬	10月下旬～11月中旬
夏　栎	新疆塔城	20	2～4	5月上旬～5月中旬	9月上旬～9月下旬	9月上旬～9月下旬
栓皮栎	山西夏县	15	4～6	4月下旬～5月上旬	翌年8月下旬～9月中旬	9月上旬～9月下旬

栎属果实成熟前为绿色，成熟时变为黄褐色或栗褐色。坚果卵形、椭圆形或近圆形，顶部短尖，壳斗包被的基部有一突起的圆形果脐。栎属果实成熟时的颜色、形状和大小见表3。

表3　栎属果实的颜色、形状、大小和壳斗形状

树　种	颜　色	形　状	径(cm)	长(cm)	壳　斗	
					形　状	包被果实比例
岩　栎	棕褐色	长椭圆形	0.5～0.8	0.8～1.0	杯状	约1/2
麻　栎	栗褐色	卵形或椭圆形	1.5～2.1	1.6～2.5	杯状	约1/2
槲　栎	深栗褐色	椭圆状卵形	1.2～1.8	1.7～2.5	杯状	约1/2
锐齿槲栎	栗褐色	长卵形或卵形	1.0～1.4	1.5～2.2	杯状	约1/3
波罗栎	黄褐色	卵圆形或宽卵形	1.2～1.7	1.5～2.3	杯状	1/2～2/3
辽东栎	深褐色	卵形或卵状椭圆形	1.0～1.4	1.4～1.8	浅碗状	约1/3
蒙古栎	黄褐色	卵形或长卵形	1.2～1.8	1.8～2.3	杯状	1/3～1/2
尖叶栎	褐色	长椭圆形或卵形	1.0～1.4	2.0～2.5	杯状	约1/2
乌冈栎	褐色	长椭圆形	0.8～1.0	1.5～1.8	杯状	1/2～2/3
夏　栎	棕褐色	长椭圆形	1.0～1.5	2.0～3.5	钟形	1/5
栓皮栎	黄褐色	近球形或短柱状球形	1.5～2.1	1.5～2.0	杯状	约2/3

果实具 1 粒种子,极稀有 2 粒种子。不育胚珠通常在种子基部外侧。种皮膜质。子叶两片,肥大,肉质。无胚乳(图 1)。

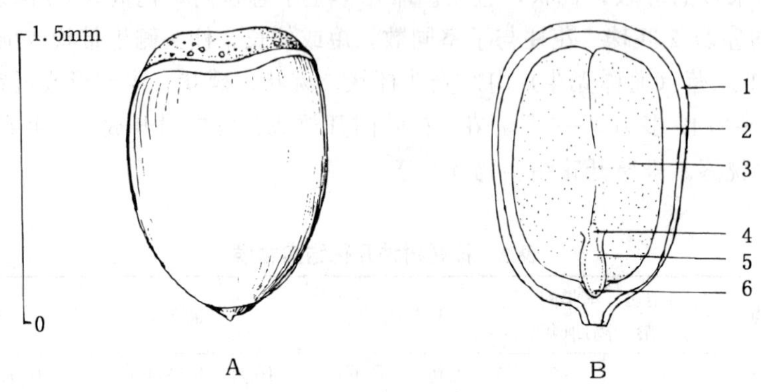

图 1　锐齿槲栎的坚果外形(A)及其纵切面(B)
1. 果皮　2. 种皮　3. 子叶　4. 上胚轴　5. 下胚轴　6. 胚根
(胡冬梅、林平绘)

栎属树种往往在 7～8 月间出现大量落果,严重影响橡实的收获量。产生落果原因主要是营养不足。据山西省林业科学研究所 1962 年对栓皮栎的调查,落果率达 37%以上。

果实的采集调制和贮藏　选择同一种子区或与造林地区的气候、土壤相近似的优良林分和母树采集果实。9～10 月橡实生长定型,颜色由绿变黄褐色或栗褐色,果皮光亮,标志果实成熟,继而自然脱落,应及时采集。一般初期脱落的多系发育不健全或遭受虫害的果实,大小不均,品质较差,数量也少。中期脱落的果实饱满,质量大,数量较多,品质最好。故采种应在脱落盛期及时从地面上拾取,或在树下铺设塑料布收集。橡实落地后会被野兽啮食,所以要及时采集。采收后要清除壳斗、小枝、叶片等夹杂物 ,捡去有缺陷的橡实(动物啃伤的、昆虫蛀食的、霉菌感染的、果皮开裂的,等等)。坚果即为播种材料,通称种子。橡实富含淀粉,自然含水量很高,怕干、怕热、怕冻,又无休眠期,落地后很快发芽。采集的果实若不立即播种,必须及时摊放在通风干燥处阴干。摊放厚度 10cm 左右,经常翻动,使含水量保持在 30%～60%,以防发芽、变质和干裂。

栎属果实象鼻虫(*Curculio* spp.)危害极为严重,于幼果期间产卵于果皮之下,橡实成熟后孵化为幼虫,啮食橡仁。山西省林业科学研究所曾剖开栓皮栎检查,有虫卵的橡实占 76%,每粒橡实内有虫卵 1～3 粒,最多的达 47 粒。在 22～28℃的温度下堆藏或袋藏 1 个月,危害率达 52.8%。用 50℃的热水浸泡 20 分钟可以灭杀潜藏的幼虫,并漂去浮起的空瘪橡实。热水要严格控制温度,因为水温超过 52℃就可把橡实烫死。经浸泡处理的种子,要及时摊开散热并阴干,以防发热霉烂。栎属树种的净度和有关种子质量见表 4。

橡实宜随采随播。若采后翌年播种,可在 0～3℃的冷湿条件下沙藏越冬,但不宜超过 6 个月。沙藏期间要经常检查,防止干燥失水、发芽、霉烂、低温受冻和动物危害等。吴振多和王昌杰(1986)介绍过用草袋包装麻栎种子的做法和效果。橡实不能密封贮藏。

发芽和播种　栎属种子不经任何处理便能发芽,但发芽时间较长(表 5)。低温层积处理后一般 2～3 周即可发芽出土，发芽率一般在 65%以上。秋播或春播。每 666m² 播种量 100～250kg。垅作或苗床育苗，条播，株行距 10cm×20cm。将橡实横放在播种沟沟底，以利胚根下扎，幼芽出土。覆土厚度 2～5cm。子叶留土萌发。根系发达，在胚芽未出土前初生根已长出侧根，胚芽出土后 1～2 周发育叶展开(图 2)。当年生苗高 20～40cm，每 666m² 产苗 2 万～3 万株。

表 4 栎属树种的净度和有关种子质量

树 种	净度(%)	千粒重(g)		每千克纯净种子粒数(个)	
		一 般	变动范围	一 般	变动范围
岩 栎	98～100	2 400	2 000～2 800	420	360～500
麻 栎	95～100	4 000	2 690～4 550	250	220～370
槲 栎	98～100	3 970	2 970～4 600	250	220～340
锐齿槲栎	98～100	3 030	2 030～3 700	330	270～490
波 罗 栎	98～100	2 850	2 530～4 500	350	220～400
辽 东 栎	98～100	2 900	1 900～3 400	340	290～520
蒙 古 栎	98～100	3 060	2 060～3 500	330	290～490
尖 叶 栎	94～99	1 820	1 670～2 000	550	500～600
乌 冈 栎	96～99	1 370	1 280～1 470	730	680～780
夏 栎	98～100	3 790	3 400～4 700	260	210～300
栓 皮 栎	95～100	4 250	3 000～5 000	240	200～330

表 5 栎属树种未经处理的种子发芽出土情况

树 种	发芽测定条件		发芽天数	出土和展叶持续天数		出苗率(%)
	基 质	温度(℃)		出 土	真叶展现	
岩 栎	沙	15～25	45	51	59	88
麻 栎	蛭石	10～20	46	96	113	80
槲 栎	蛭石	10～20	—	—	—	89
锐齿槲栎	蛭石	10～20	44	96	113	74
波 罗 栎	蛭石	10～20	9	32	42	86
辽 东 栎	蛭石	15～25	15	45	68	65
蒙 古 栎	蛭石	15～25	15	40	65	70
尖 叶 栎	沙	25	35	55	68	82
乌 冈 栎	沙	25	35	55	68	80
夏 栎	锯屑和土壤	20 左右	—	20	30	80
栓 皮 栎	蛭石	10～20	15	35	50	80

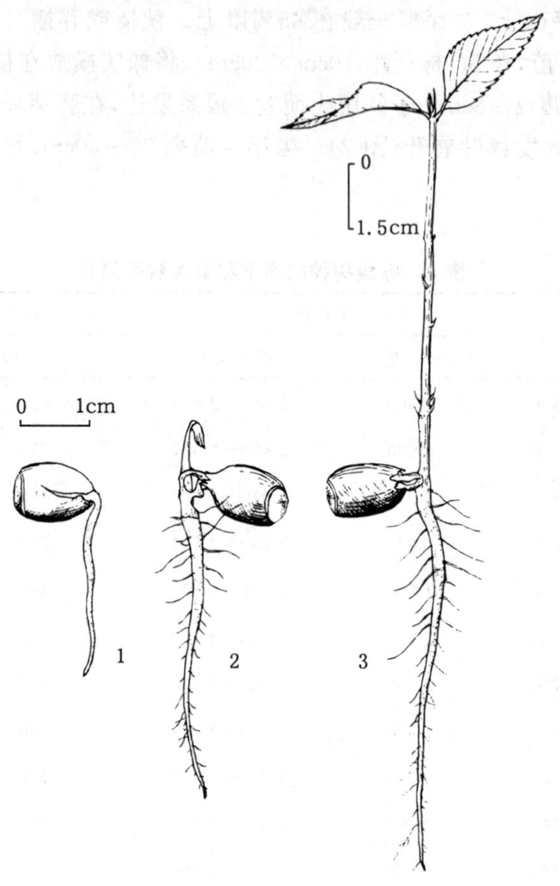

图 2 锐齿槲栎种子萌发及幼苗生长情况

1. 胚根延伸 2. 胚芽钩状出土 3. 具初生不育叶，发育叶对生

（胡冬梅、林平绘）

（宋廷茂）

喙 核 桃

Annamocarya sinensis（Dode）Leroy

（胡桃科 Juglandaceae）

生长习性、分布和用途 喙核桃属只有本文描述的 1 种。落叶乔木，高达 38m，胸径 1.2m。深根性。喜温热，耐水湿，也耐干热。喜深厚肥沃黄棕壤或赤红壤。分布于滇、桂、黔。越南亦产。优质用材，种仁可食，果皮含单宁。分布区狭窄，植株稀少，天然更新困难，《中国植物红皮书》列为稀有种。

开花结实 10 年生左右开始开花结实，20 年生以后为正常结实期。结实大小年间隔期为 1 年，但不甚明显。花单性，雌雄同株。无花被。雄柔荑花序 5～9 簇生于总梗上，着生于当年枝叶腋，序长 13～19cm，总梗长 2.5～4.3cm。雄花具 1 苞片及 2 小苞片，雄蕊 5～15。雌穗状花序具 3～5 花，顶生，直立。雌花苞片及小苞片同形，愈合而形成一顶端具 6～9 尖裂的壶状总苞，贴生于子房，花后随子房增大。子房下位，2 心皮组成。花柱膨大，近球形，柱头 2 裂，裂片半圆柱形，内面具 1 浅槽。据云南林业科学研究所观察，4～5 月开花，形成幼果，9 月上旬为果熟始期，9 月中旬为果熟盛期，10 月中旬为果熟末期。假核果（核果状坚果），近球形或卵状椭圆形，顶端尖，高 6～8cm，径 4～6cm。外皮即壶状总苞厚约 5～9mm，干后木质，黄褐色，密被灰黄色皮孔，4～9 瓣裂。果球形或卵球形，径 2.9～5cm，高 4.1～6.0cm，两侧略扁，具喙尖。果脐圆形、斜方形或一字形突起，长 7～18mm，宽 8～16mm。果壳厚 4mm，硬，骨质，久后自行破裂。无胚乳，胚有肉质而 4 裂起皱或为叶状的子叶。坚果的形态见图 1。

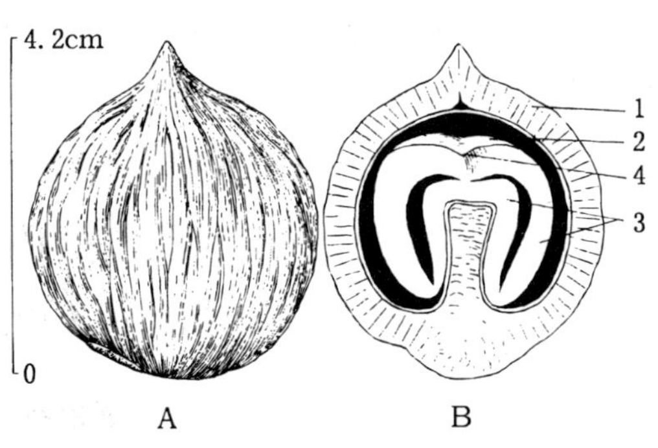

图 1　喙核桃坚果外形（A）及其纵切面（B）
1. 果皮　2. 种皮　3. 子叶　4. 胚根（黄应钦绘）

果实的采收调制和种子贮藏　果实成熟后易遭鼠类盗食，应及时采集。果实盛熟时上树采摘或敲打落地捡拾。采集的果实堆放 3～7 天，然后剥去外皮，洗净坚果，用作播种材料，通称种子。忌烈日曝晒，宜置弱光下晒干或晾干。当果皮外表变为淡黄褐色时，即可装袋贮存于荫凉通风处。育苗用的种子可用“水选法”进行精选，选取大半浸入水中，小半露出水面的种子。全沉水的是病变的坏种，大半浮于水面的是不饱满种子，均不宜选用。鲜果的出籽率约 45%。净度可达 95%以上。千粒重为 12 000（9 000～13 000）g，每千克约 80（75～110）粒。短期贮存可用布袋或瓦缸等容器，1 月以上的贮藏需混干沙。贮藏期一般为半年以内。

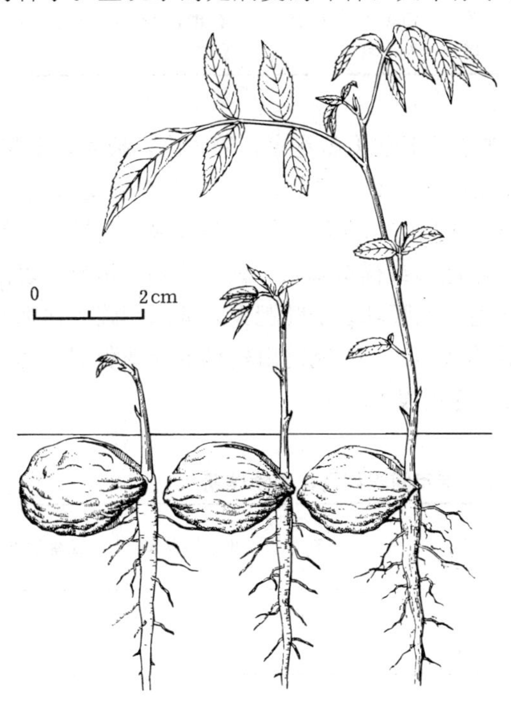

图 2　喙核桃种子萌发后第 20、30、40 天幼苗的生长情况
（李达孝、黄应钦绘）

发芽和播种　种子约有中度休眠现象。发芽时日均温宜在 18℃以上。1989 年 2 月 21 日，云南林业科学研究所在室外荫棚苗床进行过测定：播后 60 天种子发芽，发芽出土时日间最高温 23.3℃，夜间最低温 13.5℃；从开始发芽至发芽高峰之日，13 天的发芽百分数为 46%；从播种之日起算，80 天的发芽率为 75%。留土萌发，胚根萌发后，主根迅速向下伸展，并产生侧根。15 天左右，当主根约长 6cm，侧根约长 2cm 时，上胚轴出土，具初生不育叶 2～3 枚。出土 10 天左右展出初生叶，为奇数羽状复叶，小叶有锯齿。喙核桃的萌发和幼苗生长情况见图 2。

条播或点播。每平方米播种 1 000～1 200g。种子放于播种沟内，缝合线垂直地面，覆土厚度约 10cm。1 年生苗出圃。

（李达孝）

山核桃属

Carya Nutt.

（胡桃科 Juglandaceae）

生长习性、分布和用途 本属约 17 种，主产美洲东部、亚洲东南部，我国有 4 种。本文描述我国的 1 种以及早年引入并广为栽培的 1 种。它们的名称、树高、分布、栽培历史和用途见表 1。落叶大乔木。根深，速生，寿命长，材质好，果仁营养丰富，含油率高，是优良的干果、木本油料和用材树种。喜温暖湿润，薄壳山核桃能耐水湿，山核桃不耐涝。这 2 个种都有不少栽培品种。

表 1 山核桃属树种的名称、树高、分布、栽培历史及用途

中名	学名	树高（m）	分布	栽培历史	用途	供稿
山核桃	*C. cathayensis* Sarg.	30	浙西北、皖西南	我国特有，栽培已近 200 年	材，油，干果	801
薄壳山核桃	*C. illinoensis* K. Koch	34～43	原产北美。苏、浙、闽、赣、川等省已广为栽培	1900 年左右引入我国	材，油，干果，行道树	1003

开花结实 花单性同株，与叶同时开放。雄花成下垂柔荑花序，常 3 出，腋生，无花被，雄蕊 3～10。雌花 2～10 朵集生枝顶成短穗状，无花被；子房单生，下位，由 2 心皮组成 1 室，1 胚珠；无花柱，柱头盘状，2 浅裂；子房为一个 4 裂的总苞所包围。假核果（核果状坚果）。外皮具 4 棱，干后革质，4 瓣裂。果核圆球形、卵球形或椭圆球形。核壳木质，坚硬，表面基本平滑，浅褐色，基部 4 室，顶部（胚根端）2 室。无胚乳。子叶肥大，基部相连，顶部 2 裂（图 1）。结实大小年十分明显。黎章矩、钱莲芳等（1993）研究过山核桃的保花保果技术。开花结实年龄、结实大小年的间隔期以及花期果期等见表 2。

表 2 山核桃属树种的结实年龄、丰年间隔期及花果物候期

树种	结实年龄（年）		丰年间隔期（年）	花期	果熟期	果实脱落期	观察地点
	开始	盛果期					
山核桃	8～10	20 以后	1～3	4～5 月	9～10 月	10～11 月	浙江临安
薄壳山核桃	12～15	20 以后	1～2	3～5 月	10～11 月	10～12 月	南京

果实的采收调制和种子贮藏 当果实由绿变褐，外皮离核并开裂时即表示已充分成熟，此时可以摇动树身或敲击树枝使果实脱落，在地上拾集。未裂果实在室内保鲜堆放数天，促使

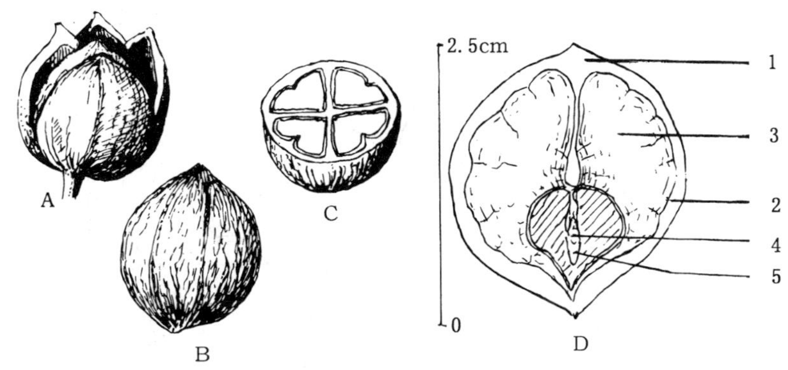

图1　山核桃带外皮的坚果（A）和坚果外形（B）及其横切面（C）和纵切面（D）

1. 果皮　2. 种皮　3. 子叶　4. 胚轴　5. 胚根

（何泽瑛绘）

后熟裂开，取出坚果，水选剔除空瘪粒，晾干至核壳发白。坚果即为播种材料，通称种子，净度可以达到98%以上。种子的形状、大小和质量随品种而不同，植株之间也有差异，表3列出的是一般的变化幅度。如果要存放3～5年，应当用密闭容器在4℃左右、相对湿度90%的条件下贮存。普通干藏8个月以上即失去发芽能力。春播用种最好是层积越冬。

表3　山核桃属树种种子（坚果）的形状、大小与质量

树　种	形　状	长（cm）	径（cm）	出籽率（%）	千粒重（g）	每千克粒数
山核桃	近圆球形	2～2.5	2～2.5	25～33	3 300～4 500	220～300
薄壳山核桃	长椭圆形或卵球形	2～4.8	1.4～2.8	50～75	2 800～8 200	120～360

发芽和播种　胚有休眠习性，1～4℃低温层积30～150天可以解除休眠。经层积的种子，发芽要求的温度为20～30℃的昼夜变温，发芽持续期30～90天。亚热带林业科学研究所的试验表明，山核桃的最适发芽温度为20～25℃。35℃以上及15℃以下都难发芽。这两个树种的发芽条件及发芽率见表4。

秋播或层积后春播。种子播前用冷水浸泡2～3天。条播，行距20～30cm，每平方米播放种子15～30粒，覆土厚3～5cm，盖草，出苗前揭除。播后约30～50天开始出苗。留土萌发。出苗率约50%～80%。山核桃1年生苗高40～50cm，薄壳山核桃1年生苗高30～40cm，可以出圃。主根极发达（图2）。也可用分根苗、根蘖苗繁殖，生产上多用嫁接繁殖。

表4　山核桃与薄壳山核桃发芽测定条件及其结果

树　种	低温层积天数	发芽测定条件			发芽率（%）
		基　质	温度（℃）	持续天数	
山核桃	54	—	20～25	25	75～90
薄壳山核桃	30～90	沙或土壤	20～30	45～60	50～85

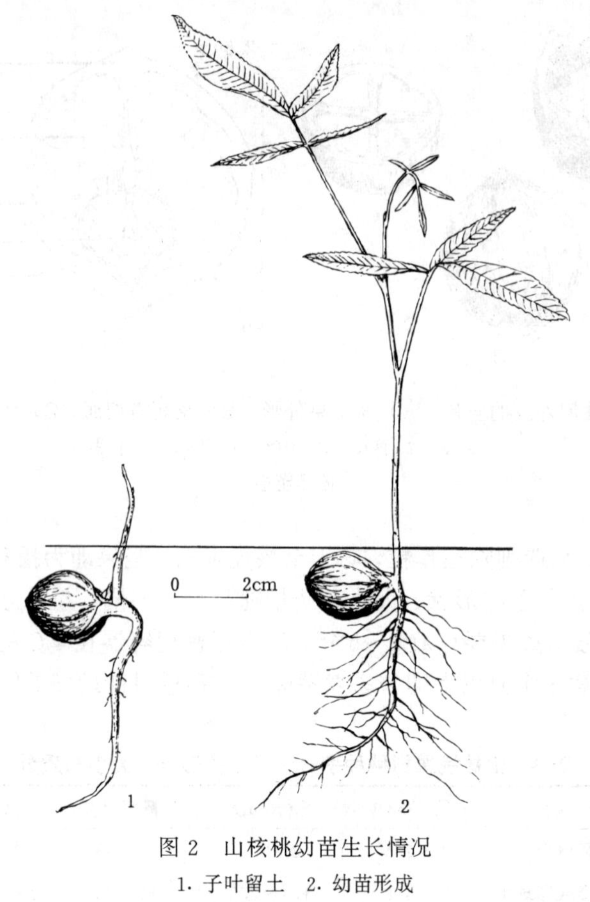

图 2 山核桃幼苗生长情况

1. 子叶留土 2. 幼苗形成

（史渭清绘）

（何泽瑛）

青 钱 柳

Cyclocarya paliurus（Batal.）Iljinsk.

（胡桃科 Juglandaceae）

生长习性、分布和用途 青钱柳属只有本文描述的 1 种。落叶乔木。高可达 39m，胸径 82cm。喜光。生长快。萌芽力强。要求深厚肥沃土地，稍耐干旱。抗病虫害。我国特有，福建省列为三级保护植物。分布在皖、苏、浙、赣、闽、台、粤、桂、湘、鄂、黔、滇、川、陕。木材轻软，有光泽，心材不明显，纹理交错，结构略细，加工容易，胶粘性和油漆性能良好，为家具良材。树皮可提制栲胶，也可用作纤维原料。

开花结实 20 年以后进入开花结实盛期。花单性，雌雄同株。雄柔荑花序长 7～18cm，3（稀 2～4）条集生在短总梗上，短总梗着生于去年生枝叶腋。雄花具 2 小苞片及 2 花被片，雄蕊（12）20～30。雌柔荑花序单独顶生，雌花具 2 小苞片及 4 花被片，子房下位，1 室，花柱短，柱头 2 裂，淡绿色。果穗下垂，长 15～25cm。坚果扁球形，径 3～7mm，有革质圆盘状翅，形如铜钱，径 2.5～6cm，顶端有 4 枚宿存花被和花柱。花期 5～6 月，果熟期 8～9 月。

果实的采收调制和种子贮藏　果实由青色转黄色时摘下果穗，在通风室内阴干后再摊晒 3～4 天。果翅晒至松脆时搓碎果翅，扬净，所得坚果即为播种材料，通称种子。出籽率 75％。千粒重 200g，每千克有种子 5 000 粒。混沙湿藏。

发芽和播种　种子有深休眠习性。胡少华等人（1987）推测，休眠的主要原因是果壳骨质，妨碍水分和氧气的吸收。他们将种子沙藏至第 3 年春播，场圃发芽率可达 80％以上。条播，条距 30～40cm。每平方米播种 18～25g，覆土厚 1cm，盖草。出土萌发。春播后约 25 天幼苗出土。子叶 2，掌状 4 裂，裂片先端微凹或再 2 浅裂。初生叶对生或互生，第 1～3（4）叶为单叶，柄红色；第 4 或第 5 片叶为 3～5 小叶复叶。下胚轴近子叶处绿色，中下部淡紫红色。主根发达，侧根疏生，黄白色。青钱柳果实的外形及幼苗生长情况见图 1。每 $666m^2$ 产苗约 1 万株。1 年生苗高 80～140cm，可以出圃。

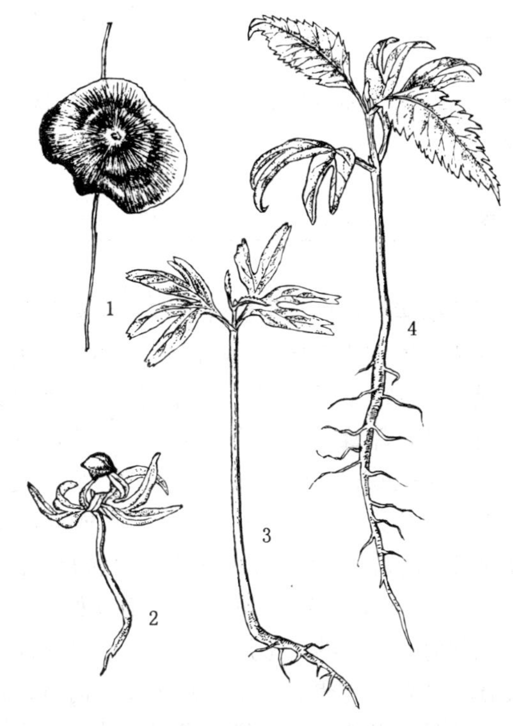

图 1　青钱柳翅果的外形及幼苗生长情况

1. 翅果　2. 子叶初现　3、4. 第 1～3 片初生叶为单叶

（黄鹏成供稿，田恒德临）

（杨国华）

核　桃　属
Juglans L.

（胡桃科　Juglandaceae）

生长习性、分布和用途　本属约 18 种，我国 4 种 2 变种。本文描述 3 种。落叶乔木，高 10～25（35）m，胸径可达 1.0～3.0m 以上。喜光。耐干冷，能耐－25～－40℃极端低温。不耐湿热。喜深厚肥沃、疏松湿润的微酸至弱碱性沙壤土，忌干燥、瘠薄、粘重和地下水位高及排水不良的土壤。种仁富含脂肪、蛋白质，可食用、药用。树皮、果皮可提取栲胶和染料，果实核壳可制活性炭；木材力学性质好，为军工、建筑、家具等的优良用材。本文 3 个种的名称、分布和用途见表 1。南齐（公元 6 世纪）时的《名医别录》说：核桃“出羌胡，汉时张骞使西域，始得种还，植之秦中，渐及东土。”据认为，此处所谓羌胡，指的是现在的青海、甘肃和新疆一带（李璠，1984）。

我国核桃一向采用实生繁殖，以致种质杂乱，品质良莠不齐。20 世纪 50 年代以来，科技人员在良种选育方面已取得很大进展。

表 1　核桃属树种的名称、分布和用途

中　名	学　名	分　布	用　途	供　稿
核桃楸	*J. mandshurica* Maxim.	东北、华北，陕、甘。朝鲜半岛	材用、食用	107
核桃	*J. regia* L.	辽南、华北、西北、西南、华中、华东。欧洲、西亚、中亚	食用、药用、材用	310
漾濞核桃	*J. sigillata* Dode	西南	食用、材用	504

开花结实　核桃实生树有早实、晚实两个类群，开始开花和结实的年龄因类群习性、栽培条件和繁殖方法而异。早实核桃 2～3 年，晚实核桃 8～10 年开始开花结实。漾濞核桃中的泡核桃，实生树 10 年左右开花结实。核桃楸 15～20 年开花结实。结实大小年间隔期约 2～3 年。花单性，雌雄同株。雄花芽为纯花芽，单生或簇生于去年枝条的叶腋，萌发后抽生柔荑花序，长 5～20cm，具 1 苞片，2 小苞片，3 花被片，皆贴生于花托上，雄蕊 6～30 枚。雌花芽为混合花芽，着生于结果母枝的顶端或近顶端数节成穗状花序，有 1～3 或 4～10 雌花，雌花花被片 4，高出于由苞片和 2 小苞片愈合成的一壶状总苞之上。子房下位，2 心皮组成 1 室，1 胚珠。花柱短，柱头 2，羽状，内面具柱头面。果为核果状坚果，外皮由总苞和花被发育而成，未熟时肉质，熟时不规则开裂，果皮硬骨质，即核壳，有刻纹及纵脊。种子 1，无胚乳，种皮薄。核仁中有隔膜，子叶常 4 裂，肉质，富含油脂（图 1）。

各树种的开花结实物候、果实及种子的形态特征如表 2、表 3。

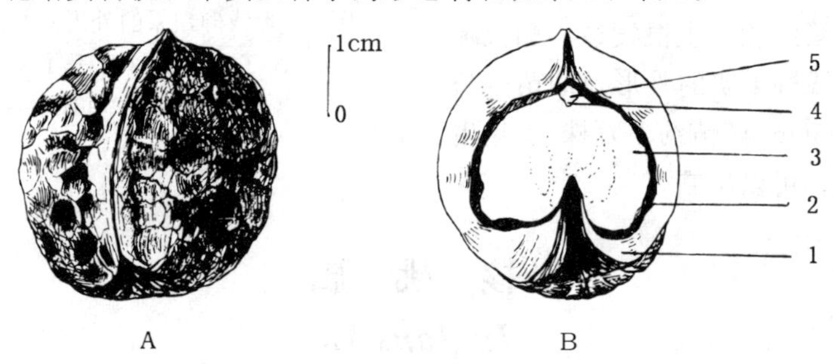

图 1　核桃种子（坚果）的外形（A）及纵切面（B）

1. 果皮　2. 种皮　3. 子叶　4. 胚轴　5. 胚根

（孟玲绘）

表 2　核桃属树种开花结实物候

树　种	观察地点	观察年份	花　期		果实成熟		果实散落期
			雌　花	雄　花	始　期	盛　期	
核桃楸	黑龙江森林植物园	1968～1986	5 月上旬～5 月下旬	5 月下旬～6 月上旬	8 月下旬	9 月下旬	10 月上旬
核桃	河北昌黎	1956，1962	4 月下旬～5 月上旬	4 月下旬～5 月上旬	9 月上旬	9 月中旬	9 月中下旬（采收）
	陕西商洛	—	4 月中旬～5 月上旬	4 月中旬～5 月上旬	9 月上旬	9 月中旬	—
漾濞核桃	云南昆明	1981	3 月中旬～4 月上旬	3 月上旬～3 月下旬	9 月上旬	9 月下旬	—
	云南漾濞	1980～1983	3 月下旬～4 月上旬	3 月中旬～3 月下旬	9 月上中旬	9 月中下旬	—

表 3　核桃属树种带外皮果实和去外皮果实（种子）的形态特征

树　种	带外皮果实				坚　果（种子）		
	形　状	径（cm）	颜　色		形　状	径（cm）	颜　色
			未熟时	成熟时			
核桃楸	近球形、卵形或椭圆形	3.0～7.5	棕绿色	绿褐色	长卵形或长椭圆形，先端锐尖，具 8 条纵脊	2.5～5.0	褐色
核桃	近球形、方圆形、椭圆形、倒卵形、广圆锥形	4.0～8.0	青绿色	黄绿色	圆形、方圆形、椭圆形或棱形，具 2 纵棱及浅刻纹	2.7～7.1	浅黄～黄褐色
漾濞核桃	近球形、倒卵圆形	3.1～6.0	青绿色	黄绿色	圆形、扁圆形、倒卵圆形，有 2 纵棱、深凹坑	3.1～4.8	棕黄～褐色

果实的采收调制和种子贮藏　当外皮由绿色变黄色（或绿褐色），全树果实有 1/3 以上外皮绽裂，部分果实能被敲落时即可采收。可以用竹竿击打枝条震落果实。采收时未脱皮的果实可在室内堆积 3～5 天后人工脱皮。张建国（1990）报道过喷洒乙烯利和萘乙酸混合液促进核桃外皮开裂的效果。脱皮后的坚果即为播种材料，通称种子。

核桃在调制、贮藏或播种后常有烂种现象，李近雨（1994）认为主要原因是采种过早以及调制时为脱去青皮而堆沤过久。他发现，8 月底采收的青绿色果实，发芽率为 63%，烂种率为 37%；9 月中旬采收的黄绿色果实，发芽率为 97%，烂种率为 3%；而且，早期采收的果实为脱青皮而堆沤 14 天后，烂种率会高达 46%。他认为，调制出来的种子最好是随即放在阳光下晾晒 7～10 天，使之充分干燥，便于干藏；如果需要湿藏，应先浸种 3～6 天，使种仁含水量达到 20%～25%，当地温降至 10℃以下时混沙埋藏。核桃属果实的出籽率、种子净度和种子质量见表 4。

表 4　核桃属树种果实出籽率和种子净度、质量

树　种	出籽率（%）	净　度（%）		千粒重（g）		每千克纯净种子粒数（粒）	
		一　般	变动范围	一　般	变动范围	一　般	变动范围
核　桃　楸	33.2	98	—	8 000	7 000～9 900	125	100～143
核　　　桃	27.2	93	85～98	13 020	8 000～19 680	77	51～125
漾濞铁核桃	—	—	—	12 000	5 300～20 000	83	50～188
漾濞泡核桃	—	90	—	13 000	7 700～20 000	77	50～130

发芽和播种　核桃坚硬的核壳（即果皮）并不妨碍种仁吸水。据李近雨（1994）观察，浸种 1 天，种仁的含水量就由气干状态的 3.8%上升到 13.3%，浸种 8 天能达到 38.1%。但核壳对发芽具有机械约束作用。室内测定发芽能力时，锯开或钳破核壳的尖端，第 5 天便开始发芽，发芽率达 83%的样品在第 8 天便可结束发芽；在 20℃、25℃和 28℃3 种发芽测定温度

中，以28℃为最好（李近雨，1994）。为了克服胚根萌发的机械阻力，山西汾阳一带核桃育苗的传统做法是，将浸水吸胀的种子捞出摊晒，裂口后播种。同未晒裂的相对照，晒至裂口的种子开始出苗早，出苗天数少，出苗率也高。点播，种子缝合线（或称棱脊）与地面垂直摆放。核桃每 666m² 播 90～140kg，覆土 3～5cm；漾濞核桃每 666m² 播 150～200kg，覆土 5～10cm；核桃楸每 666m² 播 75～100kg，覆土 8～10cm。留土萌发，见图 2。

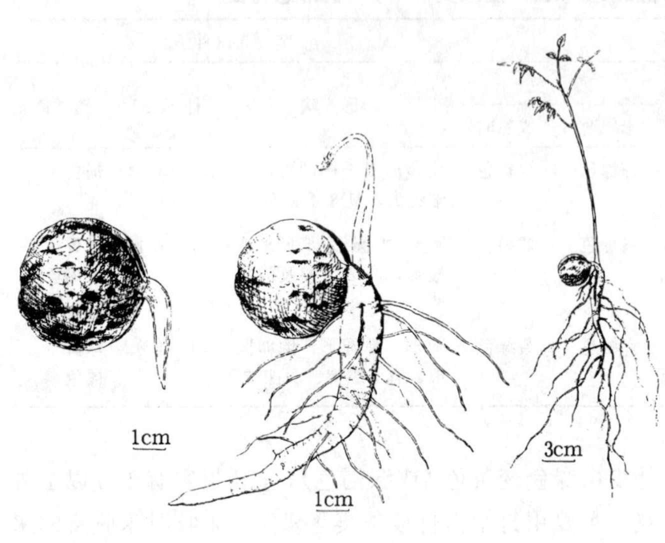

图 2　核桃种子萌发后第 12、30、42 天的幼苗生长情况
（孟玲绘）

生产中，以结实为目的繁殖多用嫁接方法。

（张毅萍）

化　香　树

Platycarya strobilacea Sieb. et Zucc.

（胡桃科　Juglandaceae）

生长习性、分布和用途　化香树属 2 种，本文描述 1 种。落叶乔木，高达 20m。喜光，耐干旱瘠薄，生于向阳山坡，为荒山荒地先锋树种，在酸性土、钙质土上均能生长。分布于甘、陕、豫、鲁等地及以南地区，尤以长江下游低山丘陵的次生林中习见。朝鲜半岛、日本也有分布。富含鞣质及纤维，为工业原料。萌芽力强，可以萌芽更新。

开花结实　花单性，雌雄同株。无花被。柔荑花序，长 5～10cm，直立。雄花序 3～15，伞房状集生枝顶，雌花序单生或 2～3 集生，有时位于中央顶端的常为两性花序，其下部为长 1～3cm 的雌花序，上端为雄花序。雄花具雄蕊 8～10，雌花生于披针形苞片腋部，小苞片与子房合生，果熟时发育成坚果的窄翅。子房下位，1 室。无花柱，柱头 2 裂，内面具柱头面。果穗球果状，卵状椭圆形至矩圆形，长 2.5～5cm，暗褐色，苞片木质，披针形，先端刺尖，果苞内含坚果 1 枚。坚果倒卵状矩圆形，扁平，两侧具狭翅，内含心形种子 1 枚。种皮黄褐色，膜质。无胚乳，大而薄的子叶掌状 4 裂，2 片子叶分向两边褶合在种子内（图 1）。据在南京观察，每年 5 月中旬～6 月底开花，8 月下旬～10 月初果实成熟。10 月中旬种子散落。

果实的采收调制和种子贮藏　10 月初果实从绿色转褐色时即可采收。摘回的果穗摊晒 2～3 天，或摊晾阴干 4～5 天后翻动敲击，脱出的坚果即为播种材料，通称种子。清除杂质后净度可达 90%。干燥后的种子不易腐坏，普通室内干藏即可。种子饱满率 30%～40%，空粒难以剔除。种子（坚果）千粒重 7～9g，每千克含种子 11 万～14 万粒。

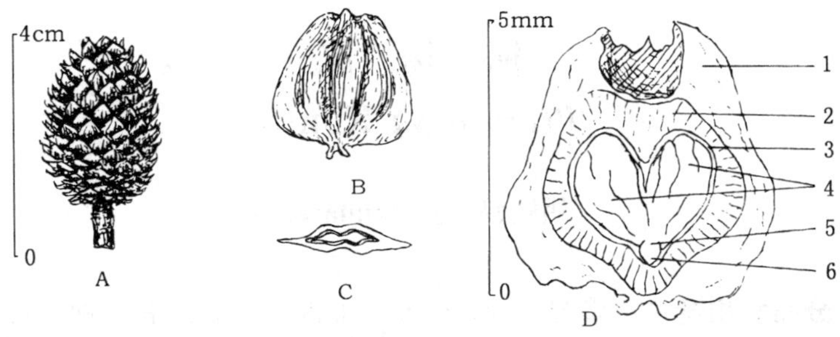

图 1　化香树果穗外形（A）与坚果外形（B）及其横切面（C）和纵切面（D）

1. 果翅　2. 果皮　3. 种皮　4. 子叶　5. 胚轴　6. 胚根（史渭清绘）

发芽和播种　种子有休眠习性，低温或自然变温层积 30～90 天可以解除。发芽率 30%左右。秋播，也可经层积后春播。4 月中旬～5 月初萌发。条播，行距 15～20cm，每平方米床面约播 25g，覆土厚 0.5～1cm，覆草。子叶出土萌发，初生叶为单叶互生，近于对生（图 2），第 4、5 叶以上为 5 小叶羽状复叶，以后小叶数渐多。1 年生苗高约 30cm，春季移植后用 2 年生苗出圃。

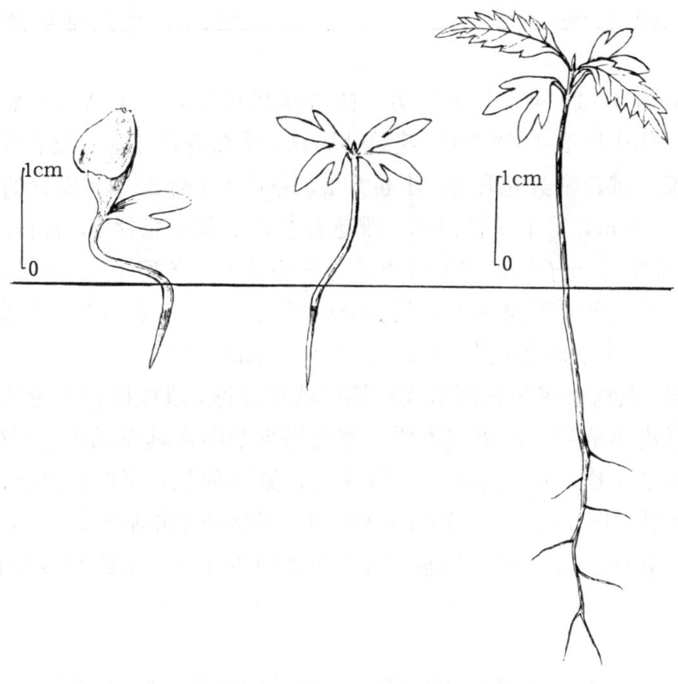

图 2　化香树种子萌发后第 2、10、37 天的幼苗生长情况

（史渭清绘）

（何泽瑛）

枫 杨 属
Pterocarya Kunth

（胡桃科　Juglandaceae）

生长习性、分布和用途　本属约 9 种，分布于北温带，我国有 7 种，其中 5 种为我国特有。本文描述 2 种，见表 1。落叶乔木，高达 30m，生溪涧河滩、阴湿山坡地。喜光，耐湿，速生，适应性较强。本文描述的这两个种属于本属的裸芽组，形态特征、生长习性及用途等基本近似。但枫杨在我国南北均产，并广泛栽植利用，为本属中最常见的种。滇桂枫杨的分布区较小。

表 1　枫杨属树种的名称、分布和用途

中　名	学　名	分　布	用　途	供　稿
枫　杨	*P. stenoptera* C. DC.	陕、晋、豫、鲁以南至华南，东抵台，西达藏。辽、冀有栽培。朝鲜半岛	材用、行道树、固堤、纤维、工业油、农药、鞣质	1003
滇桂枫杨	*P. tonkiensis* (Franch.) Dode	滇南、桂西。越南、老挝	材用、行道树、固堤、纤维、工业油、农药、鞣质	616

开花结实　枫杨 5～10 年生开始结实，15 年生以后进入正常结实年龄。滇桂枫杨 5 或 6 年生开始结实，10 年生以后正常结实。花单性同株，柔荑花序下垂。雄花序长5～10cm，单生于去年生枝的叶腋。雄花常具苞片 1，小苞片 2，并具 1（稀 2～3）发育的花被片，雄蕊 5～12。雌花序长 10～15cm，生于新枝顶端。雌花有苞片 1 和小苞片 2，各自离生，贴生于子房下部；花被片 4，亦贴生于子房，在子房顶端与子房分离。子房下位，2 心皮位于正中线上或位于两侧，内具 2 不完全隔膜而在子房底部分成不完全 4 室。花柱短，柱头 2 裂，裂片羽状。果穗长 13～45cm。坚果近球形或椭圆形，长约 6～7mm，柱头宿存，两侧具由小苞片发育形成的 2 片斜展果翅。枫杨果翅为长圆形或长椭圆状披针形，滇桂枫杨为舌状，翅长 1.5～2cm，翅脉近平行。外果皮薄革质。内果皮木质，在内果皮壁内常具充满疏松薄壁细胞的空隙，坑洼不平。果内含种子 1 枚，种皮厚膜质，浅棕色，紧贴种胚，子叶顶部 2 深裂，萌发过程中发展成 4 裂。无胚乳（图 1）。开花结实的物候期、开始结实年龄和每年结实量的多少均有单株差异，丰年间隔期为 2～3 年或不明显。这 2 个树种的开花结实物候以及种子的某些数据已列入表 2。

表 2　枫杨属树种的开花结实物候和种子的有关数据

树　种	花　期	果熟期	果实散落期	千粒重（g）	每千克种子数（万粒）	观察地点
枫　杨	4～5 月	9～10 月	9～11 月	80～100	1～1.2	江苏南京
滇桂枫杨	3 月中下旬	6～7 月	10 月	50～80	1.2～2	云南普文

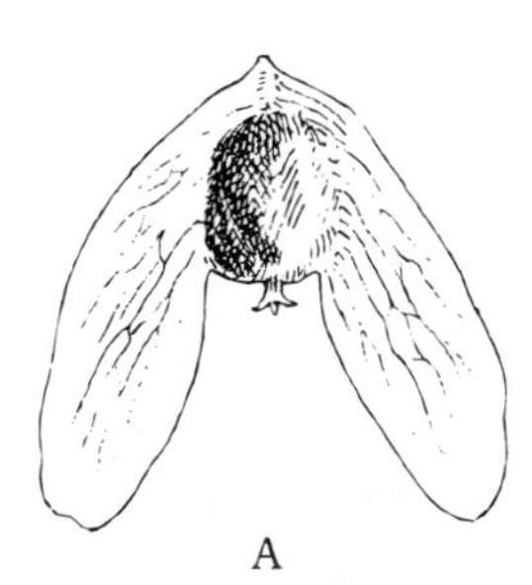

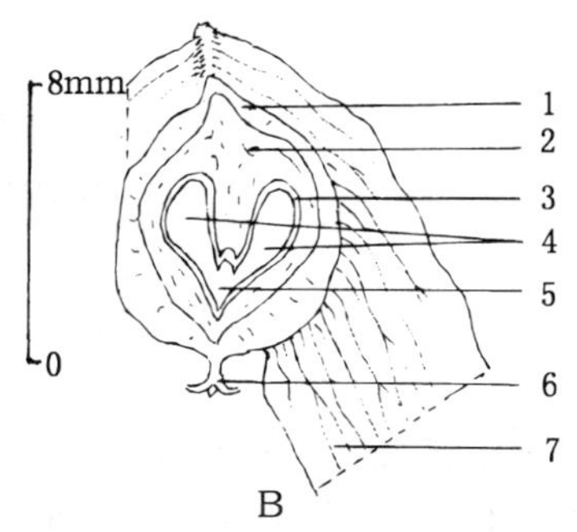

图 1 枫杨坚果外形（A）及其纵切面（B）

1. 外果皮 2. 内果皮 3. 种皮 4. 子叶 5. 胚根 6. 宿存柱头 7. 果翅（陈荣道绘）

果实的采收调制和种子贮藏 枫杨结实量大，树上常挂满果穗。成熟时果翅变为黄褐色或褐色，可以采摘果穗，或击落后在地上扫集果实，或待果实充分成熟自然脱落后在地面收集。埋在枯枝落叶下的果实不但不会腐坏，还有利于保湿，可以起到自然低温层积的作用。采集后无需特别调制处理，可以搓揉去翅，也可以不去翅。生产上一般就用带翅的坚果作为播种材料，通称种子。净度可达 98%以上。随采随播或混湿沙层积。种子不易腐烂，也不遭鸟雀鼠兽危害。贮藏期间不宜干燥失水。

发芽及播种 种子有休眠习性，需低温层积解除休眠，可以趁秋冬播种，春季即能发芽。本文作者在南京将 1987 年秋采到的枫杨立即混入湿沙层积，一直放在±5℃的冰箱内，到 4 月下旬种子便大量发芽，发芽率可达 80%～90%。而室温下干藏的则几乎不能萌发。又据王成霖和邵蓓蓓(1987)报道，0～5℃下层积 5 周后枫杨种子的发芽率为 77%，未层积的则无一发芽。云南林业科学研究所 1988 年 8 月 6 日在云南普文室外沙床上播种新采收的滇桂枫杨，8 月 16 日开始发芽，9 月 29 日发芽结束，发芽率为 30%～60%；沙藏至翌春 3 月初播种者，30 天开始发芽，30 天的发芽率为 20%～25%。条播，行距为 20～30cm，覆土 1.5～2cm，每平方米播种量约 20～30g。子叶出土时 4 裂。枫杨的初生叶为单叶互生，或第 1、2 叶对生或近对生，第 6～8 叶以上为 3～6 小叶的羽状复叶，以后小叶数逐渐增加（图 2）。1 年生苗高 1m 以上，可以出圃。

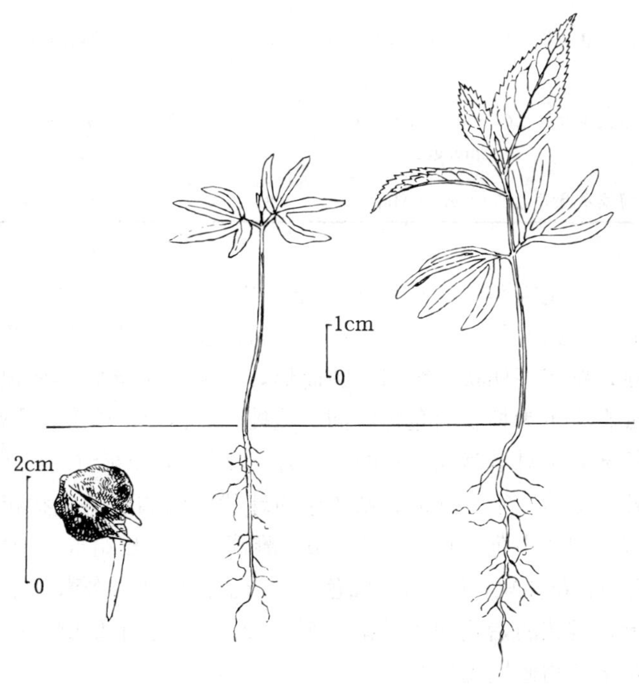

图 2 枫杨种子萌发后第 1、4、30 天的幼苗生长情况
（陈荣道绘）

（何泽瑛）

木麻黄属

Casuarina Adans.

（木麻黄科 Casuarinaceae）

生长习性、分布和用途 本属约 65 种，主产大洋洲。我国引入约 10 种，栽培面积较大的有 3 种，本文描述 5 种。常绿乔木，高 20～30m，胸径 35～70cm，速生。喜光。萌芽性强。抗风，耐沙埋，耐海潮浸蚀。能耐轻霜及短时间 0℃左右低温。耐盐碱，酸性土壤亦能生长，喜肥沃疏松的沙壤土。根具固氮菌，为热带、亚热带沿海重要防风树种。这 5 个树种的名称、生长、分布及用途等见表 1。

表 1 木麻黄属树种的名称、生长情况、分布和用途

中 名	学 名	树高（m）	胸径（cm）	分 布	用 途
鸡冠木麻黄	*C. cristata* Miq.	25	30	澳大利亚。粤、桂栽培	观赏、防护、绿化、材用
细枝木麻黄	*C. cunninghamiana* Miq.	25	40	澳大利亚。粤、琼、桂、闽、台栽培	防护、材用、栲胶、改土、绿化
木 麻 黄	*C. equisetifolia* L.	30	70	澳大利亚。粤、琼、桂、闽、浙、台栽培	防护、材用、栲胶、改土、绿化
粗枝木麻黄	*C. glauca* Sieb. ex Spreng.	25	30	澳大利亚。粤、琼、桂、闽、台栽培	防护、材用、绿化、观赏
海滨木麻黄	*C. littoralis* Miq.	20	30	澳大利亚。粤、琼、桂栽培	防护、材用、绿化

开花结实 5～6 年生始开花结实，正常结实期在 10 年生以后。除特寒年份嫩枝受冻会影响当年结实外，一般年份均能正常结实，无大小年现象。花单性，雌雄同株，稀异株。无花梗，雄花序穗状，纤细，圆筒形，着生于木贼状小枝顶端。雄花轮生于花序轴上，开放前隐藏在合生为杯状的苞片腋间；花被片 1 或 2，早落，顶端常呈帽状或 2 片合抱，覆盖着花药；基部有 1 对早落或宿存的小苞片；雄蕊 1，开花时花丝伸长将花被片推开，使花药伸出杯状苞外；花药 2 室，纵裂。雌花序为球状或椭圆状的头状花序，顶生于短的侧枝上。雌花生于 1 苞片和 2 小苞片腋间，无花被；雌蕊由 2 心皮组成，子房上位，初 2 室，后退化而为 1 室，胚珠 2；花柱短，柱头 2，红色，线形。据广西、海南观察，雄花序比雌花序大约早出现 15～20 天，雄花的始花期比雌花早 2～3 天，末花期又比雌花迟 2～3 天。海南和广西南宁记载的开花结实物候期见表 2。

果穗球形或椭圆形，球果状，每果穗有小坚果 30～90 粒。小坚果扁平，上部具膜质薄翅，密集纵列于果穗上；初包藏于 2 闭合的小苞片内，成熟时小苞片硬化，木质，展开露出小坚果。种子单 1，种皮膜质，无胚乳，胚直，具 1 对大而扁平的子叶，胚根短，形态见表 3、图 1。

表 2　木麻黄属树种的开花结实物候期

树　种	观察地点和年份	开　花			果实成熟		果实脱落
		始　期	盛　期	末　期	始　期	盛　期	
鸡冠木麻黄	南宁 1986～1988	7月中旬	9月上旬	10月下旬	翌年5月下旬	6月中旬	6月中旬～8月中旬
细枝木麻黄	海南 —	4月上旬	4月中旬	5月上旬	9月中旬	9月下旬	10月中旬
木　麻　黄	南宁 1978～1984	4月中旬	4月下旬	5月中旬	9月下旬	10月上旬	10月中旬～11月中旬
粗枝木麻黄	南　宁 1986～1988	4月下旬	5月中旬	5月下旬	10月上旬	10月下旬	11月上旬～翌年2月中旬
海滨木麻黄	南　宁 1986～1988	4月上旬	4月下旬	5月上旬	8月下旬	9月上中旬	9月下旬～10月

表 3　木麻黄属树种果序和小坚果的形态特征

树　种	未熟果颜　色	果　穗			小　坚　果		
		形　状	大小（cm）	颜　色	形　状	大小（mm）	颜　色
鸡冠木麻黄	黄绿色	椭圆形	长 2～3.2 径 1.6～2.3	黄褐色	带翅长椭圆形，翅膜质	带翅长 7～9 宽 3～4.5	浅褐色
细枝木麻黄	黄绿色	椭圆形或近球形	长 0.7～1.2 径 0.4～0.6	黄褐色	带翅长椭圆形，翅膜质	带翅长 3～5 宽 1.2～2	浅褐色
木　麻　黄	黄绿色	矩圆形	高 0.9～1.4（2.5） 径 1～1.3（1.5）	黄褐色	带翅长椭圆形，翅膜质	带翅长 4～7 宽 2～3.5	灰褐色至棕褐色
粗枝木麻黄	黄绿色	广椭圆形或近球形	长 1.2～1.6（2） 径 1～1.3（1.5）	黄褐色	带翅长椭圆形，翅膜质	带翅长 4～6 宽 1～2.2	淡灰褐色
海滨木麻黄	黄绿色	矩圆形	长 0.9～1.4 径 1～1.5	黄褐色	带翅长椭圆形，翅膜质	带翅长 4～4.8 宽 7～2.2	灰褐色

果实的采收　木麻黄属果穗未熟时黄绿色，熟时黄褐色。果熟后10～20天木质苞片开裂，小坚果飘飞散失。应掌握在果穗已转变为黄褐色但尚未开裂时采集。成熟度稍有差异的果实虽不显著影响种子的品质，但仍不宜过早采集。果穗柄短，又着生在当年生枝条的基部（头年生木质枝顶部），不应连枝采割，否则既损伤母树，又影响下一年的结实。对于优良母树，采种时更应注意保护枝条。近年出现一种特制的采种刀或齿耙，用来叉落果实，不伤树枝。采回的果实置水泥晒场或垫有薄膜的无风地曝晒，不宜在室内阴干，以免霉变。曝晒过程中注意翻动。苞片开裂，脱出的小坚果即为播种材料，通称种子。鲜果出籽率及种子的净度、质量等见表4。

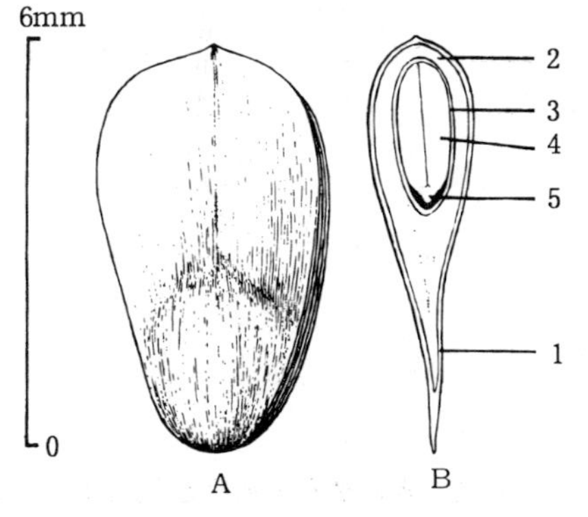

图1　木麻黄小坚果外形（A）及其纵切面（B）
1. 果翅　2. 果皮　3. 种皮　4. 子叶　5. 胚根
（黄应钦绘）

表 4 木麻黄属树种出籽率和种子（小坚果）质量

树 种	出籽率（%）	净度（%）	带翅千粒重（g）	每克种粒数
鸡冠木麻黄	5～6	85	3～5	200～340
细枝木麻黄	2～4	85	0.4～1.0	1 000～2 500
木 麻 黄	2～4	85	0.8～1.2	800～1 300
粗枝木麻黄	2～4	85	0.8～1.2	800～1 300
海滨木麻黄	2～4	85	1.0～1.4	700～1 000

木麻黄种子的千粒重因产地不同而差异较大。一般热量较高的地区，种子较重。据广东及福建测定，木麻黄种子的千粒重分别为 1.14～1.8g 和 1.0～2.5g，与表 4 所列数据稍有不同。

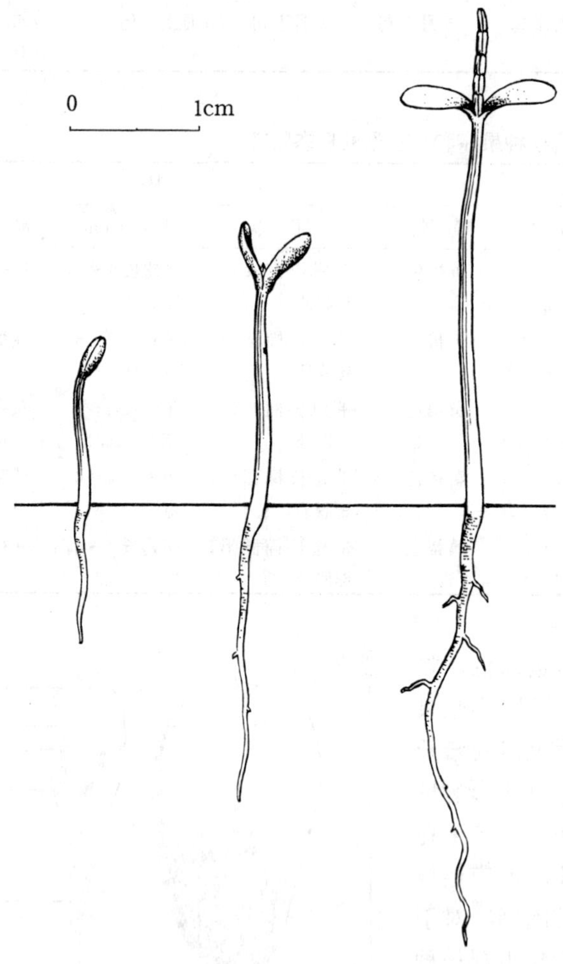

图 2 木麻黄种子萌发后第 3、8、20 天幼苗生长情况
（黄应钦仿《主要树木种苗图谱》）

加工调制和种子贮藏 种子的调制方法比较简单。清除果穗并经筛选或风选，除去杂质晒干后即可带翅运输或贮藏。运输宜用紧密布袋包装，装运过程中应防雨防潮。贮藏的种子含水量应在 10%以下，用布袋包装，悬挂于室内干燥、荫凉的通风处，常温条件下，发芽能力可以保持半年左右。如果密封贮藏于荫凉处的坛罐内，发芽能力可保持 1 年左右。贮藏于 0℃左右的低温条件下，发芽能力可保持 2 年以上。一般不宜用塑料薄膜袋贮藏。常温条件下贮藏种子，应定期检查，发现有虫害及返潮现象，应及时置日光下翻晒，清除虫卵。

发芽和播种 种子无休眠习性。发芽所需的温度不高，一般在 14℃以上即可以发芽。播种前用 0.1%的高锰酸钾或 1%的硫酸铜溶液浸种消毒，可促使种子提早 1～2天发芽。1983 年 3～5 月，广西林业科学研究所在室内用培养皿对 5 种木麻黄进行过发芽测定。测定前种子未作任何处理，置床后 4～6天开始萌动，发芽情况见表 5。出土萌发。胚根萌发后 2～3 天子叶出土，再过 14～18 天发出初生叶（图 2）。

表 5　木麻黄属树种的发芽能力及其测定条件

树　种	室内温度（℃）	发芽势（%）		发芽率（%）	
		计算天数	一般数值	计算天数	一般数值
鸡冠木麻黄	20～28	—	不明显	18	30
细枝木麻黄	18～26	6	25	16	31
木　麻　黄	18～26	6	26	16	30
粗枝木麻黄	14～24	6	20	18	35
海滨木麻黄	20～28	—	不明显	18	30

撒播。细致整地，播前淋足水，鸡冠木麻黄每平方米播种 30g，其余 4 种每平方米播种约 4g，稍加覆土。苗木基部转木质化后移至容器或苗床继续培育。容器苗 3 个月出圃，大田育苗一般育 1 年生苗出圃。也可将种子预处理后直接点播至容器，每容器点播种 6～10 粒，发芽后间苗，约 100 天出圃定植。海防林多用 1～2 年生大苗。截干造林可增强幼林抗风能力。

（王宏志）

糙　叶　树

Aphananthe aspera（Thunb.）Planch.

（榆科　Ulmaceae）

生长习性、分布和用途　糙叶树属有 8 种，产东亚和澳大利亚，我国 2 种，本文描述 1 种。落叶乔木，高达 20m，分布于华东、华南、西南及晋、陕，主产长江流域及其以南地区。材质坚硬，皮为纤维原料，叶面粗糙可摩擦金属、骨角等物，并可制农药。

开花结实　花单性同株。雄花排成稠密的聚伞花序，生于新枝基部叶腋，花萼 4～5 深裂，裂片钝头，凹陷。雄蕊 4～5 枚，与萼裂片对生。雌花单生于新枝上部叶腋或偶混生于雄花序中；萼裂片较狭；子房上位，无柄，花柱 2 深裂，内侧具柱头面。据在南京观察，花期在 4 月上旬～4 月下旬，果熟期在 9 月中旬～10 月上旬。核果近球形或卵球形，长 8～11mm，径8～10mm，有短柄，花萼花柱宿存。外果皮蓝黑色或黑色，被平伏硬毛，中果皮浆质，浅灰褐色。内果皮质地坚脆。果核卵球形略歪，长约 7～8mm，灰褐色，表面布满深褐色乳突状短棘刺，手感粗糙，有一条白色棱脊，为萌发时开裂处。种皮厚膜质，浅棕色。无胚乳。子叶狭窄，顺向内卷，外片较大，长约 18mm，包藏内片，内片较小，宽约 2～3mm，胚轴、胚根卷在外面。整个胚质地肥厚，硬肉质，子叶出土后质地仍不变。果实外形和剖面情况见图 1。

果实的采收调制和种子贮藏　果熟后陆续脱落，也可在果实呈黑色时敲击树枝，震落果实后拾集。果实采回后搓洗，除去外皮，漂除杂质和空粒，所得果核即为播种材料，通称种子。净度可达 98%以上。供翌春播种的可随即混湿沙层积，也可晾干后用普通干藏法保存至来年使用。纯净种子中饱满粒约占 85%。千粒重 112g，每千克约 8 900 粒种子。

发芽和播种　种子有休眠习性，可用低温（5～10℃）层积解除。1988 年 12 月中旬，本文作者将经过水选的种子用湿沙层积，第 2 年春天发芽相当整齐：4 月 12 日开始萌发；4 月

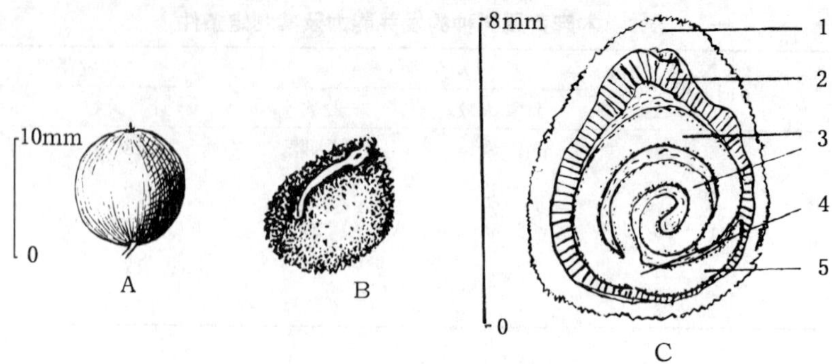

图 1　糙叶树核果外形（A）和果核外形（B）及其纵切面（C）

1. 内果皮　2. 种皮　3. 子叶　4. 胚轴　5. 胚根

（史渭清绘）

20 日发芽达到高峰，5 月 15 日发芽结束，发芽率达到 100%。这个初步试验表明，只要处理得当，糙叶树的饱满种粒都能萌发。生产实践中可在采收后随即秋播，或层积越冬后翌春播种。幼苗健壮，生长较快。1 个月的苗高可达 17cm。出土萌发。第 1、2 片初生叶对生，以后互生（图 2）。

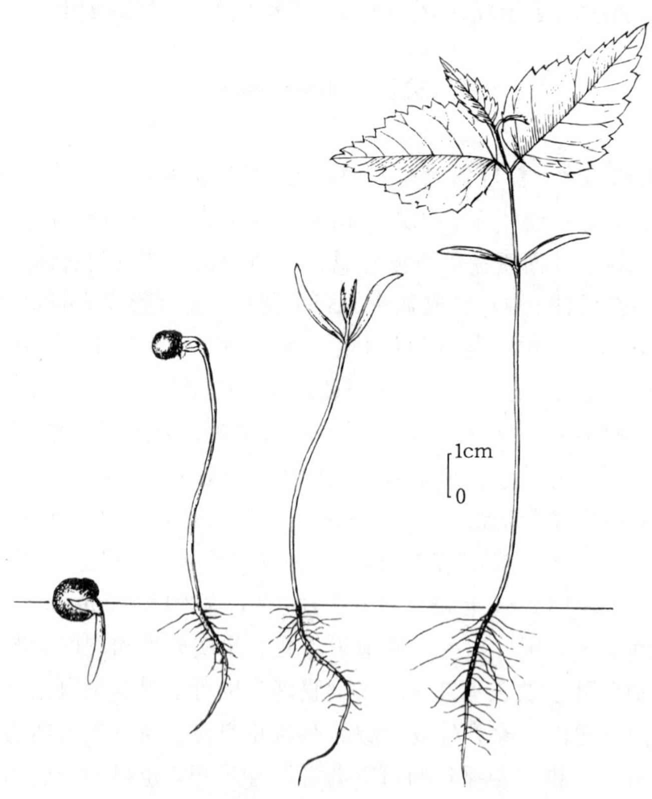

图 2　糙叶树种子萌发后第 1、5、14、30 天的幼苗生长情况

（史渭清绘）

（何泽瑛）

朴　属
Celtis L.

（榆科　Ulmaceae）

生长习性、分布和用途　本属约 70 种，分布于北温带和热带地区，我国有 20 余种，乔木或灌木。本文描述 5 种乔木，其中假玉桂常绿，其余 4 种落叶。耐干旱瘠薄，酸性土、中性土、钙质土上均可生长，但不耐盐碱。生长较快，有些种易生虫瘿。这 5 个种的名称、树高、分布和用途见表 1。

表 1　朴属树种的名称、树高、分布和用途

中　名	学　名	树高（m）	分　布	用　途	供　稿
小叶朴（黑弹树）	*C. bungeana* Bl.	20	华东、华中、华北、西南、西北及辽宁。朝鲜半岛	材用、薪炭、风景树	307
假玉桂	*C. cinnamomea* Lindl. et Planch.	5～12	藏南、滇、黔、桂、粤、琼、闽。印度、缅甸、越南、印度尼西亚及马来西亚	材用、油料	605
珊瑚朴	*C. julianae* Schneid.	25	豫、陕以南及长江流域，南至黔、粤、闽	材用、造纸、风景树	1008
昆明朴	*C. kunmingensis* Cheng et Hong	20	滇、川西南	材用、薪炭	614
朴树	*C. sinensis* Pers.	20	鲁、豫、长江中下游及滇、粤、桂、台。朝鲜半岛、日本、越南和老挝	材用、造纸	1003

开花结实　成年树每年多少都能结实，目前还没有这几个树种结实丰年间隔期的观察资料。花杂性同株。花与叶同时开放。雄花簇生于新枝下部叶腋。两性花或雌花单生或双生，也有的成聚伞花序生于新枝上部叶腋。花小，单被花，萼 4～5 裂，雄蕊 4～5。子房上位，1 室，胚珠 1。花柱 2 裂。核果近球形，外果皮的颜色随树种而异。中果皮肉质，味略甜，常因此而被鸟类传播。内果皮骨质，平滑或有凸网纹。胚弯曲，子叶旋卷，宽阔，胚乳不明显（图 1）。这 5 个树种的开花结实物候以及核果成熟的标志等见表 2。

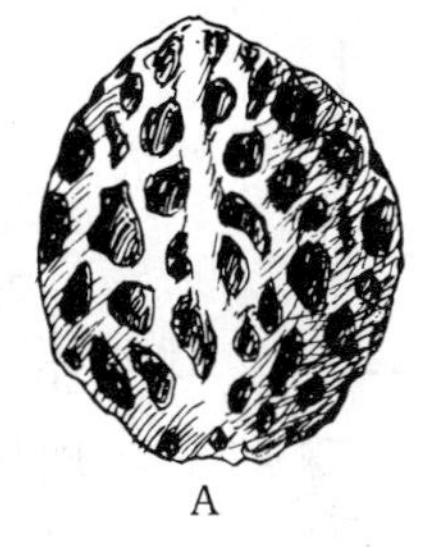

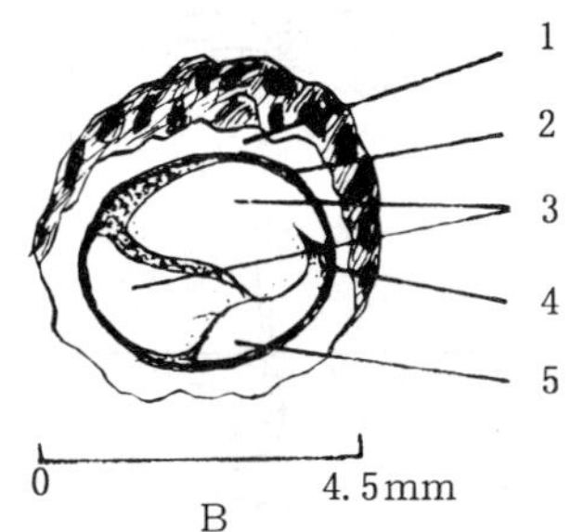

图 1　朴树果核的外形（A）及其纵剖面（B）
1. 内果皮　2. 种皮　3. 子叶　4. 胚轴　5. 胚根（史渭清绘）

果实的采收调制和种子贮藏　果实成熟时容易从树上摇落或击落，也可任其自然脱落后从地上扫集。采回的果实堆放或浸泡 1～3 天后搓擦去除果肉，漂除杂质后晾干。通常都以清除果肉后的果核作为播种材料，通称种子。净度可达 98%。表 3 所列的种子形态特征和种子质量都是指果核而言。生产上一般是普通干藏，或者是混沙层积后春播，还没有长期贮藏的经验。

表 2　朴属树种开花结实习性

树　种	开始结实年龄	物候观察地点	花　期	果熟期	成熟果实颜色	果　径 (mm)	果实散落期
小叶朴	—	北京	4～6月	9月中旬～10月	紫黑色	4～7	11月
假玉桂	5～8	海南	2月上旬～2月下旬	7月中下旬	橙红色或红色	5～7	7月下旬
珊瑚朴	—	南京	3月上旬～4月中旬	9～10月	橙红色	10～15	10月下旬
昆明朴	10	昆明	3月上旬～3月下旬	10～11月	黄褐色、黑褐色	6～8	11月下旬～12月
朴树	15	南京	5月	9～10月	红褐色	4～5	10月下旬

表 3　朴属树种的出籽率、种子形态特征及质量

树　种	出籽率 (%)	种子(果核)形态特征			千粒重 (g)	每千克种子粒数 (万粒)
		直径 (mm)	颜　色	表面纹饰		
小叶朴	—	4.4～5.5	灰-灰黄色	叶脉网纹状，2肋	73～87	1.2～1.4
假玉桂	40～50	4～5	乳白-浅褐	有网纹，具4肋	40～50	2.0～2.3
珊瑚朴	50～60	3～4.5	玉白色	网纹凹穴不明显	50～60	1.6～2.0
昆明朴	80～85	4～5	黄褐色	具4肋，先端突尖	130～150	0.6～0.8
朴　树	40～55	4～5	玉白色	网肋凹穴明显	30～40	2.4～3.3

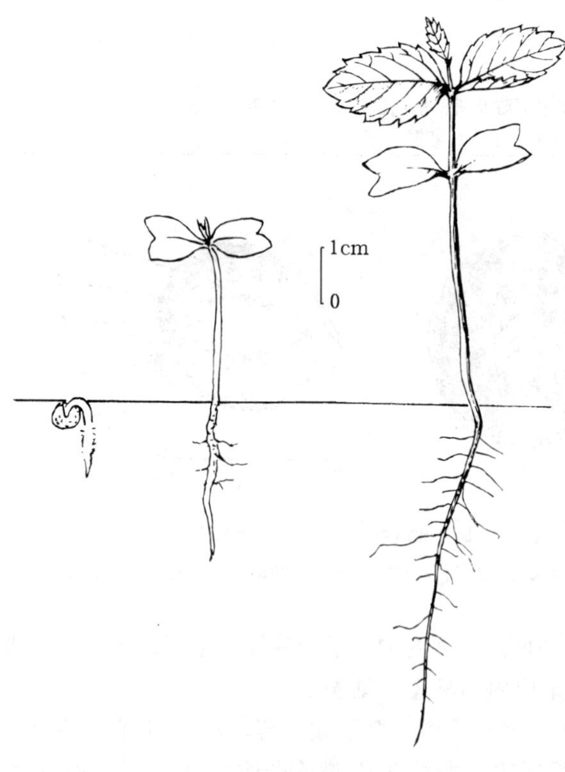

图 2　朴树种子萌发后第 1、4、35 天的幼苗生长情况
（史渭清绘）

发芽和播种　本文所描述的 5 个种中，假玉桂和昆明朴的种子没有休眠习性。假玉桂的果实夏季成熟，采集脱粒以后可以随即播种。1988 年，广西林业科学研究所曾以湿沙为基质对假玉桂做过一次试验：播后 13 天开始发芽，1 个月内发芽结束，不过发芽率仅约 40%。发芽率低的原因还不清楚。据云南省林业科学研究所观察，昆明朴春播后 30～40 天开始出苗，出苗期长约 20～30 天，出苗率可达 85%。小叶朴、珊瑚朴和朴树的种子看来都有休眠习性，可以冬播或层积后春播，不过休眠程度都不太深。据中国林业科学研究院林业研究所的一次观察，2 月下旬开始层积的小叶朴，3 月下旬便有部分种子开裂，即将萌发。珊瑚朴和朴树的种子需要层积约两个月。林业部南方林木种子检验中心 1988 年的一次试验表明，层积 2 个月的珊瑚朴在 25℃条件下 20 天可以结束发芽，发芽率达 85%。江苏省植物研究所 1987 年 2 月中旬层积的朴树种子，4 月中旬开始萌发，5 月下旬

发芽结束，发芽率可以达到 80%。

育苗常用条播，行距 20～25cm。覆土约 1cm，播后盖草。朴属 5 个树种每平方米的播种量一般是：小叶朴 10g，假玉桂约 6g，珊瑚朴约 8g，昆明朴约 22g，朴树约 5g。本文描述的 5 个种都是出土萌发。朴树子叶出土后 1 周左右出现 2 片对生的初生叶，以后真叶互生（图 2）。小叶朴当年苗高 1.2～1.5m，可以出圃；如果是供庭园绿化，可以移植 1～2 次，用 4～5 年生的大苗出圃定植。

（何泽瑛）

刺　　榆

Hemiptelea davidii（Hance）Planch.

（榆科　Ulmaceae）

生长习性、分布和用途　刺榆属仅有本文描述的 1 种，落叶小乔木，或呈灌木状，小枝坚硬成棘刺状。喜光，适应性强，耐干旱瘠薄，能生于固定沙丘上。产于东北南部、华北、西北、华中、华东。朝鲜半岛也有分布。哈尔滨引种栽培，长势良好。木材供建筑、农具、器具用，纤维可代麻，也是优良的固沙、保持水土和荒山造林树种，又可作绿篱。

开花结实　人工栽培的刺榆约 10 年后结实。杂性同株。花有短梗，单生或 2～4 朵簇生于当年生枝下部叶腋，与叶同放。单被花，花被杯状，4～5 浅裂。雄蕊 4～5，与花被裂片对生。子房侧向压扁，1 室，具 1 倒生胚珠。花柱 2 裂，内面为柱头。花期 5 月。果为小坚果，偏斜，黄绿色，斜卵圆形，两侧扁，长 5～7mm，基部有宿存花被，顶部有鸡冠状狭翅。果柄长 2～4mm，无毛。子叶宽展，胚乳少量（图 1）。据黑龙江省 1975～1980 年观测，果实初熟至末熟为 8 月 31 日～10 月 8 日，初熟至脱落间隔期为 12～18 天。

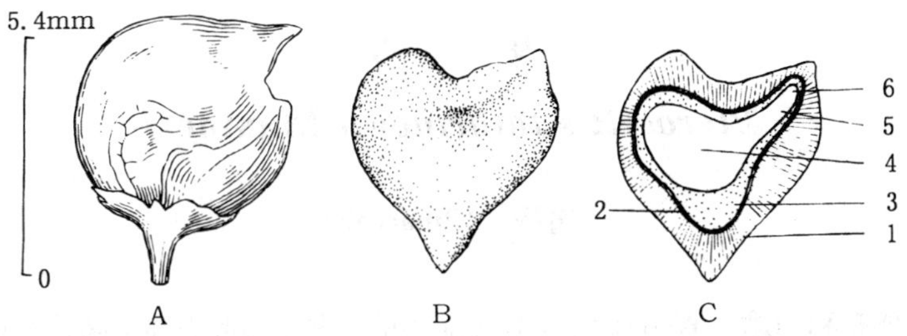

图 1　刺榆带翅坚果外形（A）、去翅坚果外形（B）及其纵切面（C）
1. 果皮　2. 种皮　3. 胚乳　4. 子叶　5. 胚轴　6. 胚根（梁鸣绘）

果实的采收调制和种子贮藏　果实呈黄绿色即标志成熟，用梯子攀摘或收集落果，去掉果柄及杂质，晾干后所得纯净果实即为播种材料，通称种子。出籽率 80%～90%，净度 90%。千粒重 1.4～1.8g，每千克有播种材料 55.5 万～71.4 万粒。种子在常温（5～20℃）下贮存半年，发芽率无显著降低。低温（0～5℃）散放可贮藏 1.5 年。

发芽和播种　春播，哈尔滨应于 5 月初播种。垄播时每 $666m^2$ 播种 2～3kg，覆土厚 0.5cm

左右，7～10天可出苗，出苗率70%～80%。出土萌发（图2）。1年生苗高30～50cm，可以出圃。

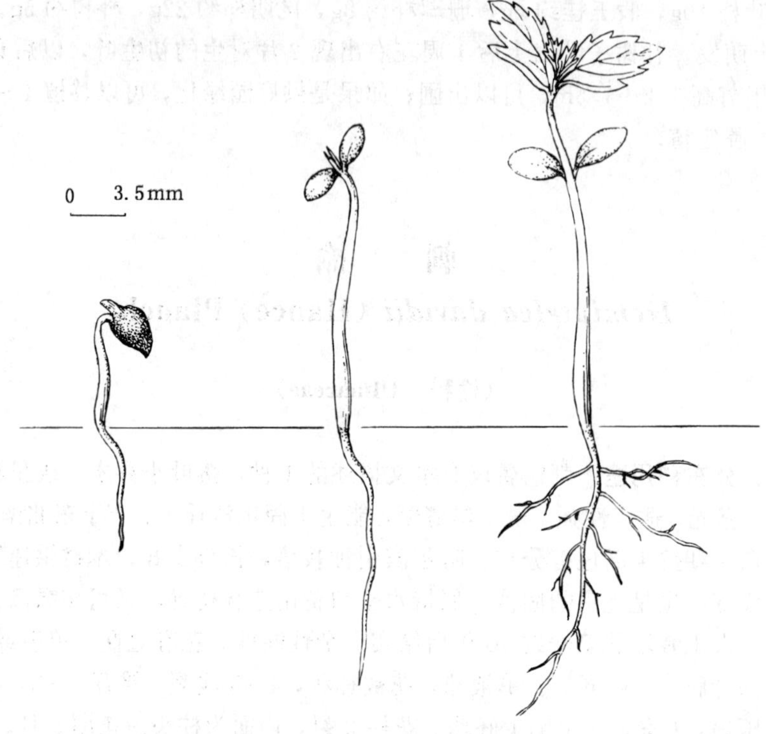

图2　刺榆种子发芽后第5、7、21天幼苗的生长情况

（曹雅范绘）

（周德本）

青　檀

Pteroceltis tatarinowii Maxim.

（榆科　Ulmaceae）

生长习性、分布和用途　青檀属仅有本文描述的1种。落叶乔木，高达20m，胸径1.7m，老树树干通常凹凸不圆。我国特产，为稀有种，已列入《中国植物红皮书》。耐干旱瘠薄。喜钙，在石灰岩山地生长良好，酸性的花岗岩山地和溪边河滩地亦能生长。萌芽性强。根系发达，在悬崖峭壁或石隙裂缝处也能散生。寿命长。零星分布于青、甘、陕及华北、华东、华中和西南。材质坚硬，富韧性，纹理直，结构细，供车辆、家具、细木工、绘图板等用。茎皮纤维绵韧易剥，是制作书法绘画用的"宣纸"的原料。

开花结实　花单性，雌雄同株。单被花，无花瓣，4月与叶同放。雄花簇生，雄蕊5，与花萼裂片对生，花药顶端有长毛。雌花单生叶腋，花萼4裂，裂片披针形。子房上位，微扁，由2心皮合成1室。花柱2裂，羽状，柱头面密被紫色粗毛。果8～9（9～10）月成熟。果横径 12～16mm，果梗长 1.5～2cm。坚果两侧具宽翅，翅薄木质。翅顶端凹缺，花柱宿存于缺口中央。两侧翅一大一略小，上窄下宽，形似折扇，具皱褶，翅缘波状，灰黄色。果核近球形（图1），种子无胚乳。

图 1　青檀的具翅坚果（A）及果核（B）
（张世经绘）

果实的采收调制和种子贮藏　15 年生左右开始结实。采种应选择 20～40 年健壮母树。当坚果由青变深黄色时及时采集，过熟则易飞散。采得的坚果薄摊在通风处阴干，除去果翅即得纯净果核，用作播种材料，通称种子。也有不去翅而直接用带翅坚果作为播种材料的。坚果出籽率为 76%。果核卵圆形，径 4～5mm，千粒重 21～28g，每千克 35 000～48 000 粒。果核和不经调制的带翅坚果可以干藏。

发芽和播种　春季播种。播前可用冷水浸种 2 天，换水 1 次，捞出稍阴干后即可播种。条播，行距 25～30cm，每平方米播种量 3～5g，播后覆土以不见种子为度，盖草。经 3～4 周发芽。出土萌发。子叶 2，长方状倒盾形，平展，先端 2 裂呈倒箭形，基部圆楔形，初生叶互生（图 2）。1 年生苗高 50～60cm，可以出圃造林。

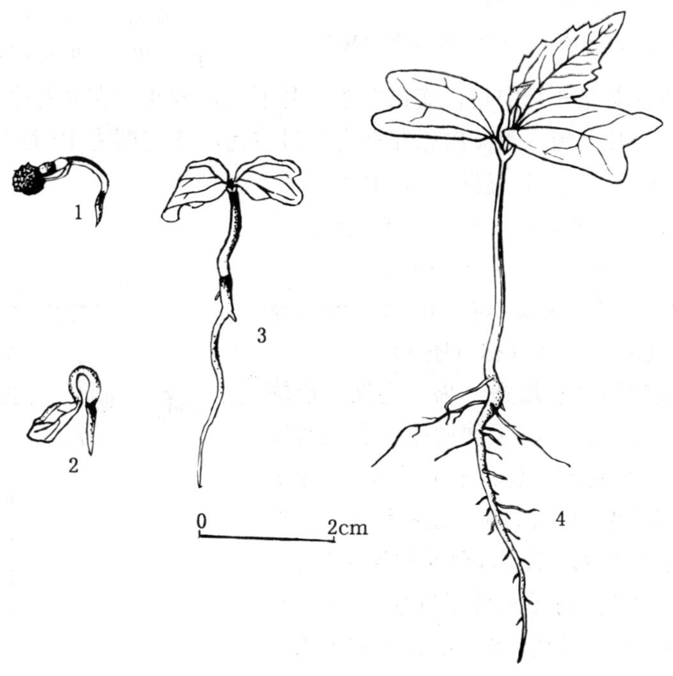

图 2　青檀种子萌发及幼苗生长情况
1～2. 子叶脱出　3. 子叶展开　4. 初生叶互生
（张世经仿《主要树木种苗图谱》）

（黄鹏成）

山 黄 麻

Trema tomentosa（Roxb.）Hara

（榆科 Ulmaceae）

生长习性、分布和用途 山黄麻属约50种，我国6种，本文描述1种。常绿小乔木，高5～8m，胸径10cm。适生于土质疏松的酸性土。产于藏、滇、黔、桂、湘、粤、琼、闽、台。日本、南亚、澳大利亚亦产。为低丘荒山重要的造林先锋树种，叶作绿肥，树皮为优质纤维原料，又含鞣质，可提制栲胶。种子油可用于制皂并作润滑油。

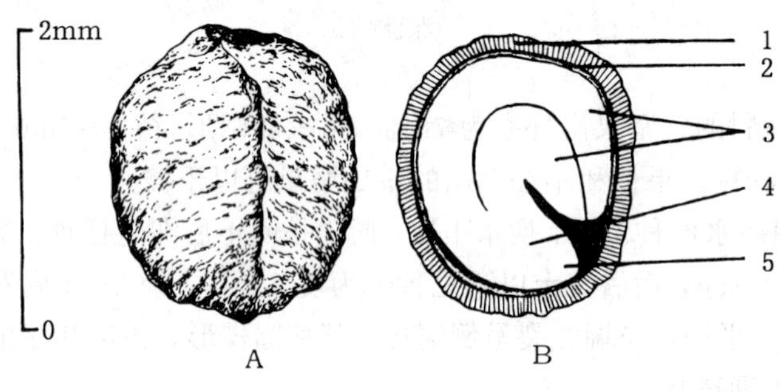

图1 山黄麻果核外形（A）及其纵切面（B）
1. 内果皮 2. 种皮 3. 子叶 4. 胚轴 5. 胚根（黄应钦绘）

开花结实 2～3年生开始开花结实，4～8年生后进入正常结实。结实无大小年现象。花单性同株。聚伞花序成对腋生，短而稠密，被柔毛，长约2cm。花小，单被花，无花瓣，萼裂片5。雄花长约1mm。雄蕊5，与萼裂片对生，蕾期直立。雌花子房上位，无柄，1室1胚珠，下垂，柱头2，被毛。开花结实期很长。据广西南宁1987～1988年观测，主要花果期有2次：第1次始花期在6月中旬，盛花期在7月下旬，末花期在8月中下旬，果实成熟始期8月下旬，盛期9月上旬；第2次始花期为9月中旬，盛花期为10月上旬，末花期10月下旬，果实成熟始期11月上旬，盛期11月中旬。果实成熟后10～15天脱落。核果，具宿萼，未熟时青黄色，熟后紫黑色，卵球形，长3mm，径2.4～3mm。果核扁圆形，褐色，表面具皱纹，宽1.8～2.2mm，厚1.3～1.5mm。无胚乳。子叶弯摺（图1）。

果实的采收调制和种子贮藏 果熟盛期，选摘发育饱满的果实，或用钩刀将果穗钩落，在地面捡拾。采得的果实装入紧密的布袋置水中搓洗，淘净果肉等杂质，所得果核用作播种材料，通称种子。鲜果按质量计算的出籽率为30%左右，净度约为95%。千粒重（2.5）2.8（3.2）g，每千克种子粒数在30万～40万。种子忌失水，不能日晒。调制所得的种子置室内稍晾干即可。运输时需混以等量湿沙，用布袋包装。贮藏期如超过半月，宜混以2倍以上的湿沙，装入坛罐，贮藏期一般为半年以内。

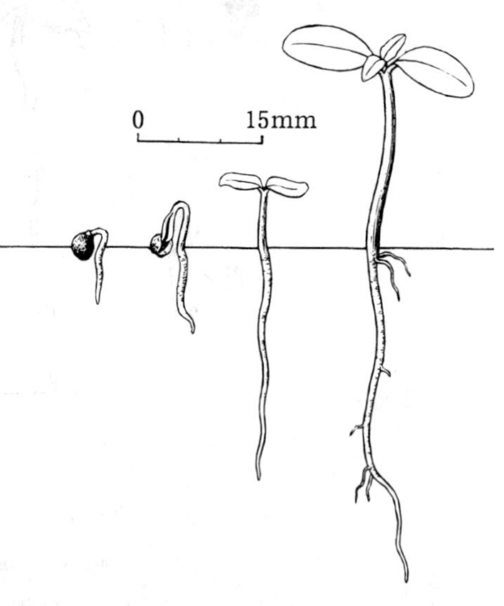

图2 山黄麻种子萌发后第3、4、6、8天幼苗的生长情况（黄应钦绘）

发芽和播种 种子无休眠习性。发芽时的日均温需在 20℃以上。1987 年 8 月下旬末，广西林业科学研究所用当年采集的新鲜种子在室内作过发芽测定：播前种子未作任何处理，基质为滤纸，光照为室内自然光，当时室内日均温为 24～26℃。播后 27 天至 9 月下旬开始发芽，盛期不明显，至 11 月 16 日发芽终止。供试种子空粒多，未经筛选，发芽率仅为 10%。出土萌发。播种后约 27 天胚根开始萌发，萌发后约 3 天子叶出土，再过 4 天初生叶展现。山黄麻种子的萌发和幼苗初期生长情况见图 2。

撒播。整地要求极细致，播前淋足水，播后可薄层覆盖火烧土或不覆土，盖草保湿即可。每平方米播种 2～4g。也可用容器育苗，每容器播种约 30 粒。容器苗 3～4 个月即可出圃。圃地培育 1 年生苗出圃。

（王宏志）

榆　　属

Ulmus L.

（榆科　Ulmaceae）

生长习性、分布和用途　本属约 40 种，分布于欧洲、亚洲和美洲。我国产 21 种 4 变种，本文描述 10 种 1 变种。落叶乔木稀半常绿。材质好，是广大平原地区“四旁”绿化、用材林、防护林和盐碱地造林的主要树种。其名称、生长、分布和用途见表 1。

表 1　榆属树种的名称、生长、分布和用途

中　名	学　名	树高(m)	分　布	供　稿
多脉榆（栗叶榆）	*U. castaneifolia* Hemsl.	20	东南、中南，黔、川。南京有栽培	1008
琅琊榆	*U. chenmoui* Cheng	15	皖、苏。南京等地栽培	1003
春　榆	*U. davidiana* Planch. var. *japonica* (Rehd.) Nakai	30	东北、华北、西北、华中、华东。俄罗斯远东、朝鲜半岛、蒙古、日本	107
长序榆	*U. elongata* L. K. Fu et C. S. Ding	25～30	浙、闽、皖、赣星散分布	1002
醉翁榆（毛榆）	*U. gaussenii* Cheng	20	皖。南京等地栽培	1003
脱皮榆	*U. lamellosa* C. Wang et S. L. Chang ex L. K. Fu	8	近年新发现，分布范围窄，内蒙古(包头)、京(房山)、冀(遵化)、晋(中条山)	201
欧洲榆（新疆大叶榆）	*U. leavis* Pall.	35	原产新疆及欧洲。甘、陕、青、内蒙古、冀、鲁、东北已引种栽培	303
大果榆	*U. macrocarpa* Hance	20	东北和华北	402
榔　榆	*U. parvifolia* Jacq.	20	华北、华东、中南、贵、川、藏。朝鲜半岛、日本	1002
白　榆（榆树、家榆）	*U. pumila* L.	25	东北、华北、西北、华东。朝鲜半岛、俄罗斯、日本	303
红果榆（川榆、蓉榆）	*U. szechuanica* Fang	20	浙北、苏南、皖南、赣、川中。山东栽培	1008

白榆分布广，种内存在着明显的地理变异。种源试验（马常耕，1993）表明，白榆的许多性状都呈随纬度和海拔变化的单向渐变模式。水、热是白榆发生地理变异的主导因素。南部种源生长快，干形通直，树冠小，主根发达，侧根多，叶片大，生长期长，耐旱耐寒性差。北部种源生长慢，干形差，根系发达，叶片小，耐旱耐寒能力强。根据种源试验的结果，已为白榆区划出了3个种子区（GB8822.1～8822.13～88）。

开花结实 春榆、欧洲榆和大果榆大体上在8～10年生开始开花结实。欧洲榆10年生以后，大果榆、榔榆和白榆15年生以后进入正常结实。花两性，小，无花瓣，排成腋生的总状花序或总状聚伞花序。通常是春季在叶前开放。有少数树种在夏末秋初开花。萼钟形，4～9裂。雄蕊与萼裂片同数。子房上位，2心皮合成1室，具1倒生胚珠。花柱短，2裂，内面为柱头，无柄或具柄。榆属树种的开花结实物候期见表2。果实为一扁平、圆形或卵形或窄椭圆形的翅果，周围有膜质翅，具宿萼，顶端常有缺口，成熟时借风散落。榆属果实的形态特征见表3。大多数树种每隔2或3年丰收1次。胚直立，无胚乳，子叶扁平或微凹，见图1。

表2 榆属树种的开花结实物候

树 种	观察地点	观察年份	开 花			果实成熟		果实散落
			始 期	盛 期	末 期	始 期	盛 期	
多脉榆	南京	—	3月中旬	3月中下旬	—	4月中旬	4月下旬	4月下旬
琅琊榆	—	—	—	3月中旬	—	—	4月下旬	4月下旬
春 榆	—	—		4～5月		5～6月		—
	哈尔滨	1978～1986	—	—	—	5月下旬	6月上旬	6月中旬
长序榆	福建南平	—	—	3月	—	—	4月上旬	4月中旬
醉翁榆	—	—	—	3月	—	—	4月下旬	5月上旬
脱皮榆	包头	—	—	4月下旬	—	5月上旬	5月下旬	6月上旬
欧洲榆	—	—		4月上旬～4月中旬		5月上旬	5月中旬	5月中旬
大果榆	—	—		4～5月		5～6月		5～6月
	北京	—	3月下旬	4月上旬	4月上旬	4月下旬	5月中旬	5月中旬
榔榆	—	—		8～9月		10月		10月
白榆	北京	1987		3月中旬～3月下旬		4月上旬～4月中旬		4月中旬
	黑龙江嫩江	—	—	4月下旬	—	5月下旬		5月下旬
	新疆乌鲁木齐	—	—	4月中旬	—	5月上旬		5月上旬
	山东金乡	—	—	3月下旬	—	4月下旬		4月下旬
	浙江海宁	—	—	3月下旬	—	4月下旬		4月下旬
红果榆	南京	—	—	3月中旬	3月下旬	4月下旬		5月上旬

表 3 榆属树种结实年龄和果实形态特征

树 种	果 实				种 子[1]			
	形 状	长(cm)	宽(cm)	颜 色	形 状	长(cm)	宽(cm)	颜 色
多脉榆	卵形或椭圆状倒卵形	1.8～3.3	1.2～1.5	—	—	1.0～1.2	0.4～0.6	黄褐色
瑯琊榆	窄至宽倒卵形	1.5～2.5	1～1.7	—	近扁圆形	0.4	0.3	浅棕褐色
春 榆	倒卵形	1.2～1.4	0.8～1.0	黄白色	—	—	—	棕色
长序榆	狭长	2.0～2.5	中部宽 0.3～0.4	黄棕色	椭圆形两端尖	—	—	—
醉翁榆	广卵形或近圆形	2～2.5	—	绿黄色	倒卵形或近圆形	0.8	0.5	浅棕色
脱皮榆	矩圆形或近圆形	2～3.0	1.5～2.4	黄色	椭圆形	—	—	—
欧洲榆	广椭圆形	1.2～1.6	—	—	扁圆形	—	—	黄白色
大果榆	近卵形	2.5～3.5	2.2～2.7	暗黄褐色	—	1.1	0.7	—
榔 榆	椭圆形或卵形	0.8～1.1	0.7～0.8	黄色或黄褐色	心形	—	—	—
白 榆	近圆形或倒卵圆形	1～1.5	—	黄白色	—	—	—	—
红果榆	倒卵形	1.1～1.6	0.9～1.4	—	—	0.25～0.45	0.2～0.3	紫红色或褐色

① 仅具形态分类意义

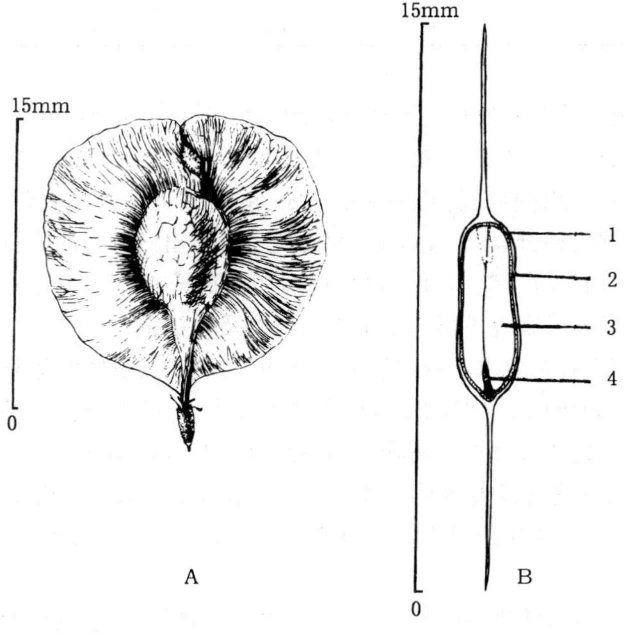

图 1 白榆果实外形(A)及其纵切面(B)

1. 果皮 2. 种皮 3. 子叶 4. 胚轴(孟玲绘)

果实的采收调制和贮藏　榆属树种果实成熟后很快脱落，要及时采收。可以等翅果自然成熟脱落后扫集，也可以选无风之日从枝条上击落后收集。为了避免飘飞散失，可以上树捋取翅果。采后置于通风处晾干几天，用风选或筛选法清除杂物。榆属的翅果通常就是播种材料，通称种子。榆属种子以较低含水量密封贮藏为好。含水量为 7%～8%，发芽率在 65%以上的白榆种子，在温度不高于 5℃的环境中可以贮藏 2 年；贮藏温度不高于 15℃时可以贮藏 1 年。种子(翅果)的净度、质量见表 4。

表 4　榆属种子(翅果)的净度和质量

树　种	净度(%)	千粒重(g)		每千克纯净种子粒数(万粒)	
		一　般	变动范围	一　般	变动范围
多脉榆	98	6.0	5.5～6.5	16.7	15.0～18.0
瑯琊榆	95 以上	34 (鲜重)	—	2.9	—
春　榆	80～90	—	5.0～7.0	—	14.3～20.0
长序榆	—	5.6	—	17.9	—
醉翁榆	—	31.8	—	3.1	—
脱皮榆	90～97	29.6	—	3.5	—
欧洲榆	—	10.0	—	10.0	—
大果榆	90～98	9.2	8.3～10.5	10.8	9.5～12.0
榔　榆	—	10.4	8.3～12.5	9.6	8.0～12.0
白　榆	70～88	7.8	5.9～8.9	12.5～14.0	11.2～16.9
红果榆	96	5.0	4.5～5.5	20.0	18～22

发芽和播种　瑯琊榆和榔榆种子有休眠习性。除了这两个树种以外，本文所述的榆属树种均易发芽。据邵蓓蓓(1987、1989)报道，在 20℃温度下培养 2 周后，剥除果皮和不剥除果皮的榔榆发芽率均为 3.3%，剥除种皮的发芽率则为 69%。她因此认为榔榆萌发的主要障碍来自种皮，同时胚也有浅休眠习性。榔榆一经吸胀，种皮便会分泌出一种透明无色的粘滑胶状液体。这种液体对油菜种子的萌发有抑制作用。据她试验，榔榆种子在 0～5℃下层积 30～40 天后移到发芽环境，便可完全解除休眠且不再有粘液渗出，12 天的发芽率可达 92.5%。

树种不同对发芽条件的要求不同，见表 5。室内利用变温条件发芽时，可以在白天 8 小时的高温时段给以光照，夜晚 16 小时的低温时段则在无光条件下进行。发芽基质可用沙、蛭石或多层纱布。通常 10 天之内发芽达到高峰；经过层积的种子，在 10～30 天内结束发芽。解除休眠后的榔榆种子最适宜的发芽温度是 30℃/15℃或 30℃/20℃的昼夜变温，在这样的温度条件下发芽持续 12～14 天；如果采用 20℃的恒温，发芽持续时间便会延长到 24 天；发芽无需光照(邵蓓蓓，1987)。

出土萌发。子叶 2 片，微肉质(图 2)。种子在春季成熟的榆属树种，通常都是采后即播；秋季成熟或者需要层积的，通常是来年春季播种。条播，行距 20～25cm。榆属多数树种每平方米播种量约 7～10g。覆土厚 0.5～1cm。播种到发芽结束之前，苗床都应保持湿润。一般 10 天左右可发芽出土。每 666m² 可以产苗 2 万～3 万株。1～2 年生苗可出圃。据张纪卯和陈巧女等(1993)试验，长序榆 1 年生苗高 85cm，地径 0.5cm，最佳密度应是每 666m² 有苗 3 万～4 万株。

榆属树种可通过扦插、嫁接进行无性繁殖。

表 5　榆属树种发芽测定条件及其结果

树　种	温度(℃)	天数(天)	发芽率(%)
多脉榆	25	25	2～3
醉翁榆	—	9	54
欧洲榆	20	30	65～80
大果榆	20	14	80(70～90)
榔　榆	20/30	—	55
白　榆	20～25	14	65～90
红果榆	25	20	53

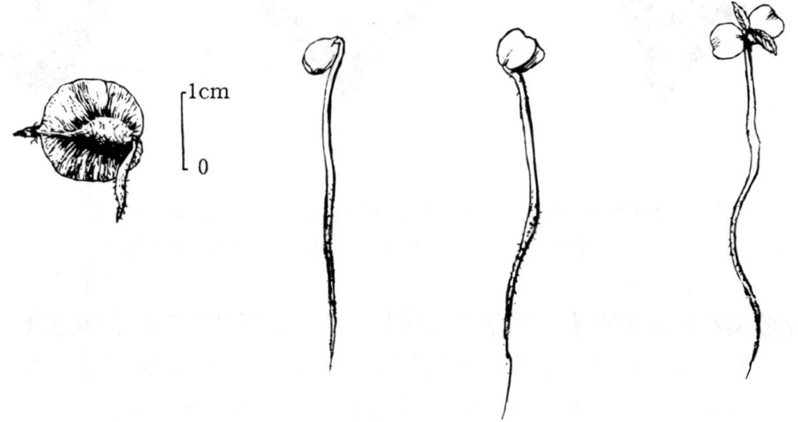

图 2　白榆种子萌发后第 2、4、6、13 天的幼苗生长情况
(孟玲绘)

(于淑兰)

榉树(大叶榉)

Zelkova schneideriana Hand.-Mazz.

(榆科　Ulmaceae)

生长习性、分布和用途　榉树属 5 种，产亚洲西部和中部，我国 3 种，本文描述 1 种。落叶乔木，高达 25m，胸径 1m，寿命长。分布于淮河流域和秦岭以南至粤、桂、滇、藏。在微酸性、中性、石灰质土及轻度盐碱土上均能生长。深根性，抗风力强。幼苗生长稍慢，6～7

年后生长转迅速，生长能力可以持续 70～80 年。榉树材质优良，树皮可造纸，是长江中下游的重要造林树种。

开花结实　10～15 年生开始结实，20 年生以后进入结实盛期，百年以上的大树仍可结实。结实丰年间隔期为 1～3 年。据成都市 1964 年观察，榉树 3 月 31 日花初开，4 月 1 日花盛开，4 日花期结束，9 月 22 日果实成熟。

花单性，稀杂性同株。雄花簇生于新枝下部的叶腋，雌花或两性花单生或 2～3 簇生于新枝上部叶腋。无花瓣。萼片 4～5，宿存。雄蕊 4～5。子房几无柄，1 室 1 胚珠，花柱 2，偏生。坚果，呈有棱角的不规则扁球形，上部歪斜，直径 2.5～4mm，成熟时灰黑色，借风力散落或与叶同时脱落，有时落叶后仍宿存枝上。胚弯曲，子叶阔，旋卷，无胚乳（图 1）。

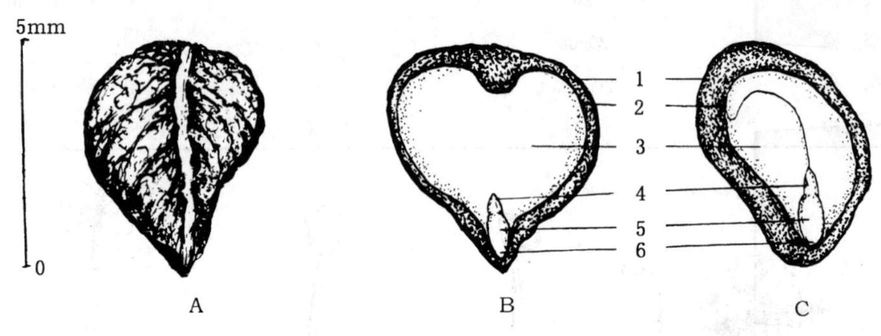

图 1　榉树的坚果外形（A）及其纵切正面（B）和纵切侧面（C）
1. 果皮　2. 种皮　3. 子叶　4. 胚芽　5. 胚轴　6. 胚根（田恒德绘）

果实的采收调制和种子贮藏　叶色转为枯黄，果实由青转为暗绿色或灰黑色时，沿小枝捋取坚果，或截取小枝后摊晒，搓揉，使坚果脱落，筛去小枝和树叶等杂质便可得到纯净的坚果，即为播种材料，通称种子。榉树果实多生于枝条上端，树身高大而种粒较小，采集比较困难，也可待坚果自然散落后在地面扫集，或于无风天摇撼树枝，震落后收集。榉树种子（坚果，下同）千粒重一般为 13g，变动范围为 12.5～15.5g；每千克纯净种子约为 7.7 万粒，变动范围为 6.4 万～8.0 万粒；气干种子含水量 10%～13%。供翌年春播的种子可用布袋悬挂在干燥通风处干藏越冬，也可以混沙层积越冬。

发芽和播种　榉树种子空粒较多，大年所采种子发芽率可达 50%～70%，小年种子发芽率往往只有 20%～30%。榉树种子有轻度休眠习性，可以随采随播，或在 1～5℃低温下层积 40 天后春播。王成霖和邵蓓蓓（1987）在 0～5℃的温度下将榉树种子分别层积 0 周、1 周、3 周和 5 周后，得到的发芽率分别为 28%、63%、67%和 68%。条播，行距 20cm，每平方米苗床播种量视种子质量而定，一般为 9～18g，播后覆土 0.5cm，盖草。播种后直到种壳出土应注意防止鸟害。经层积处理的种子播种后 3～5 天开始发芽。

出土萌发。子叶先端凹缺。初生叶 4，交叉对生，节间极短而呈轮生状，生长暂停后从上面的一对叶腋抽生 2 枝或一侧的叶腋抽生 1 枝，枝上叶互生，两行排列（图 2）。当年苗高约 50～80cm，可以出圃造林。

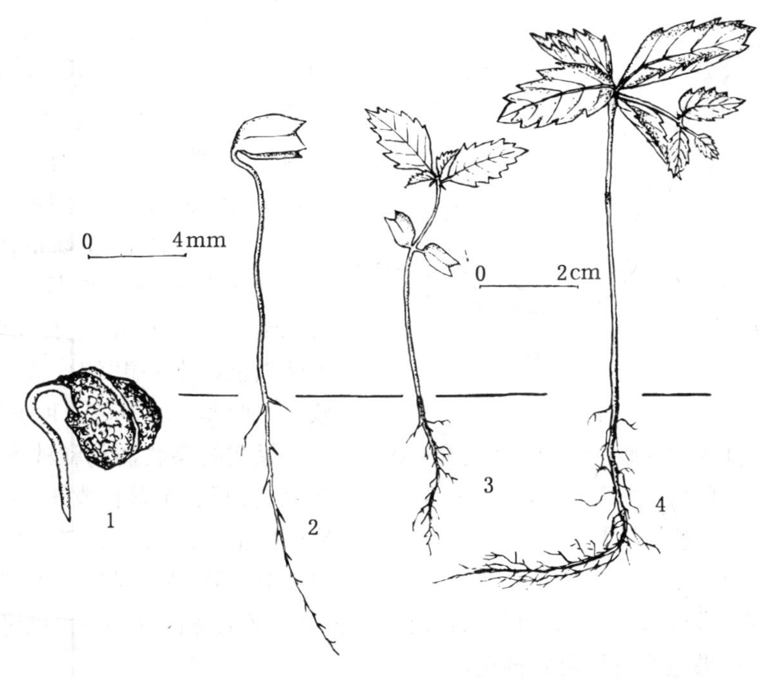

图 2　榉树种子的萌发和幼苗生长情况
1. 胚根伸出　2. 子叶脱出　3. 初生叶交叉对生　4. 分枝后真叶互生
（田恒德绘，3、4. 仿《主要树木种苗图谱》）

（杨余康）

见血封喉（箭毒木）

Antiaris toxicaria（Pers.）Lesch.

（桑科　Moraceae）

生长习性、分布和用途　见血封喉属约 5 种，我国只有本文描述的 1 种。常绿大乔木，高达 40m，胸径 65～80cm。能耐 2℃左右的极端低温。适生于肥沃湿润的酸性或微酸性土壤，亦可在石灰性土上生长。产滇、桂、粤、琼。印度、中南半岛、印度尼西亚亦产。树液有剧毒，含见血封喉甙，有强心、加速心律、增加心血输出量的作用，医药上有研究价值，曾用于制毒箭猎兽。木材可用于家具、建筑，是组成热带季雨林和雨林的主要树种之一。随着森林的破坏，植株逐渐减少，《中国植物红皮书》列为国家稀有种，应加强保护，扩大种植。

开花结实　10～15 年生开始结实，20 年生以后进入正常结实期。结实有大小年现象，大小年间隔期为 1 年。花单性同株。雄花密集于叶腋，生于一肉质、盘状、有短柄的花序托上。雄花萼片 4，稀 3，匙形。雄蕊 4 或 3，内藏，无退化雌蕊。雌花单生于一带鳞片的梨形总苞内，无花被。子房与总苞合生，1 室，胚珠自室顶悬垂。花柱 2 裂，裂片钻形，反曲。据云南

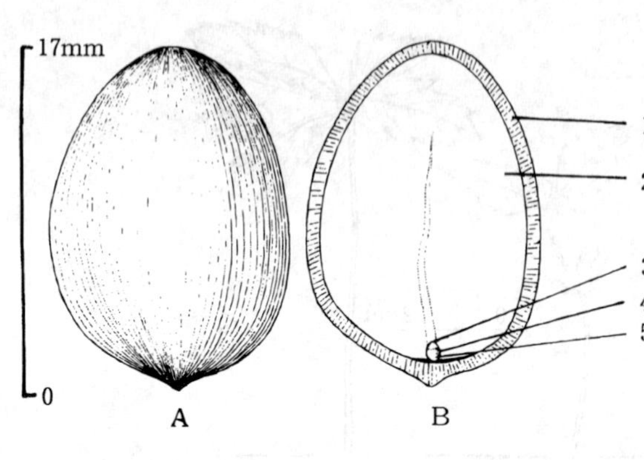

图 1　见血封喉种子外形（A）及其纵切面（B）
1. 种皮　2. 子叶　3. 胚芽　4. 胚轴　5. 胚根
（黄应钦绘）

西双版纳勐仑 1985～1988 年观察，2 月初肉质花托始现，3 月上旬为始花期，3 月中旬为盛花期，3 月下旬为末花期，4 月上中旬幼果形成，4 月下旬为果熟始期，5 月上旬为果熟盛期。肉质果卵形，果皮与总苞粘合，成熟时由绿色转变为鲜红色至紫红色，长 1.7～2cm，径 1～1.2cm，成熟后 3～4 天脱落。每果有种子 1 粒，卵形，长 15～18mm，径 8～10mm。种皮坚硬，灰白色，种脐褐色，另一端突尖。无胚乳。有双胚现象。子叶厚，乳白色。种子形态见图 1。

果实的采收调制和种子贮藏　果实成熟盛期，用竹竿敲打或摇动树枝，震落后在地面捡拾。采回的果实集中堆放一处，使果肉逐渐变软。果肉与种皮紧紧粘合，调制较难。但果肉内常有果蝇，3～4 天可将果肉吃净，露出干净的灰白色种子。也可于果实成熟落地后 3～4 天，待果蝇吃去果肉后再在地面收集，浸水 1 昼夜，搓洗后得出种子。鲜果的出种率为 60%～70%，种子净度可达 98%以上。千粒重约 1 500g，每千克有纯净种子 600～700 粒。稍晾干的种子含水量约 42%～45%。种子忌失水，不宜日晒。短期运输可用麻袋或木箱包装。在室内通风处可存放 2～4 周，混湿沙后约可贮藏半年。在低温（7～9℃）混沙条件下可贮藏 1 年左右。

发芽和播种　种子无明显的休眠现象，发芽所需日均温为 25℃左右。1988 年 5 月在云南西双版纳勐仑室外沙盆中用新采种子作过发芽测定：播后 25 天出土，39 天发芽结束，没有明显的发芽盛期，发芽率为 90%。留土萌发。初生不育叶线形，6～9 枚。种子的萌发和幼苗初期生长情况见图 2。

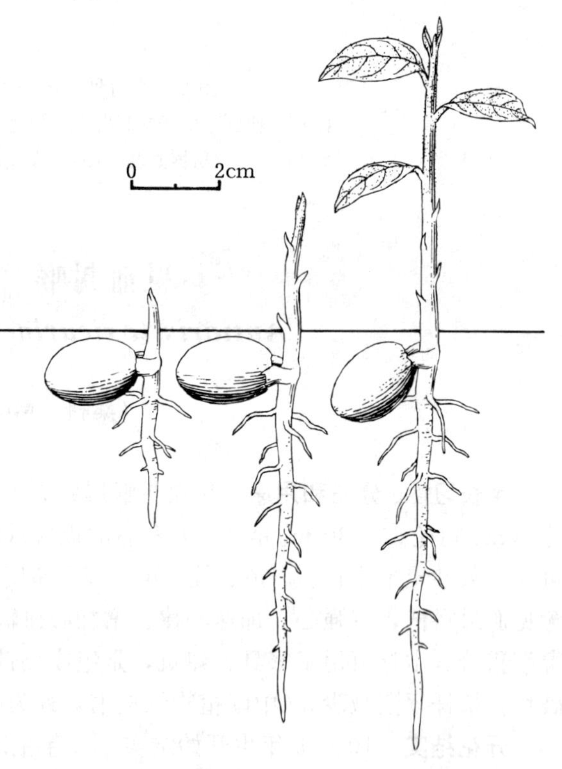

图 2　见血封喉种子萌发后第 5、10、25 天幼苗的生长情况（黄应钦绘）

点播。播前用冷水浸种 24 小时，捞出后稍晾干即可播种。每平方米播种 120～150g，覆土 1～2cm。1 年生苗可以出圃。

（马信祥）

桂　木　属

Artocarpus J. R. et G. Forst.

（桑科　Moraceae）

生长习性、分布和用途　本属约60余种，我国10种，本文描述4种。常绿乔木，高20～25m，胸径60～100cm。具乳汁。苗期及嫩枝忌霜冻，成龄枝能耐0℃左右的极端低温。喜土层深厚肥沃湿润的酸性土。它们的名称、生长、分布及用途见表1。

表1　桂木属树种的名称、生长、分布和用途

中　名	学　名	树高(m)	胸径(cm)	分　布	用　途
波罗蜜（木波罗）	*A. heterophyllus* Lam.	20	100	印度至马来西亚。粤、琼、桂、滇、台栽培	食用、材用、黄色染料、纤维、绿化
桂　木（大叶胭脂）	*A. lingnanensis* Merr.	25	80	粤、琼、桂、滇。越南，泰国	材用、食用、绿化
小叶胭脂（二色波罗蜜）	*A. styracifolius* Pierre	25	80	粤、琼、桂、黔、滇。越南	材用、食用、绿化
越南胭脂	*A. tonkinensis* A. Chev. ex Gagnep.	25	80	滇、黔、桂、粤。中南半岛	材用、食用、绿化

开花结实　波罗蜜4～8年生开始结实；小叶胭脂12～14年生开始结实，其余2种8～10年生开始结实。正常结实年龄在20年生以后，结实大小年的间隔期一般为1年。波罗蜜不耐低温，在我国大陆南部，如果冬春低温，则不能结实。花单性同株，生于一肉质的花序轴上。波罗蜜的花序着生于老茎、主枝以至树根上。桂木、小叶胭脂和越南胭脂的花序通常单个腋生。雄花序长圆形，柔荑状。雄花花萼2～4裂，雄蕊1，花丝在花蕾中直立，开花时伸出花被外，花药2室。雌花序球形。雌花花萼筒状，顶端3～4裂，下部埋于花序轴内。子房1室，胚珠倒生。花柱突出，顶生或侧生。广西南宁1978～1982年观察的开花结实物候期见表2。

表2　桂木属树种的开花结实物候（广西南宁，1978～1982年）

树　种	开花			果实成熟		果实脱落期
	始　期	盛　期	末　期	始　期	盛　期	
波罗蜜	2月上旬	2月中下旬	3月上旬	7月下旬	8月上中旬	8月下旬～9月下旬
桂　木	4月中旬	4月下旬	5月上旬	6月中旬	7月中旬	7月上旬～9月中旬
小叶胭脂	5月下旬	6月中旬	7月中旬	10月下旬	11月上旬	11月上旬～中旬
越南胭脂	4月中旬	5月上旬	5月中旬	6月下旬	7月上旬	7月上旬～9月中旬

波罗蜜人工栽培的历史悠久，已育出许多品种，其中的早熟品种可早在6月成熟，晚熟品种可迟至9～10月成熟。在海南，波罗蜜的成熟期比南宁早1个月，晚熟的果实又可迟至11月。瘦果，果皮膜质或薄革质，藏于肉质的花萼和花序轴所组成的聚花果内。萼齿成刺状

或瘤状。每个聚花果波罗蜜有瘦果数十粒至百多粒，桂木、小叶胭脂和越南胭脂有瘦果数粒至十数粒。种子无胚乳。这 4 个树种果实和种子的形态见表 3、图 1。

表 3 桂木属树种聚花果和种子(瘦果)的形态特征

树种	未熟果颜色	成熟聚花果			种子（瘦果）		
		形状	颜色	大小(cm)	形状	颜色	大小(mm)
波罗蜜	青绿色	椭圆形、圆形或不规则圆形	黄绿色	长 25～60 径 20～50	椭圆形或卵形	白色或浅灰白色	长 26～32 径 13～24
桂木	黄绿色	近圆形或扁圆形	黄色或粉红色	高 1.8～2.5 径 2.5～3	多棱状不规则形	白色	长 7～10 宽 6.5～10
小叶胭脂	黄绿色	圆球形	浅黄色	径 1.2～1.5	不规则形或椭圆形	白色或浅黄色	长 4～6 宽 3～6
越南胭脂	黄绿色	近圆形或阔圆形	浅黄色	高 4.5～6 径 4～5	扁椭圆形或倒卵形	浅褐色或淡黄色	长 8～11 宽 5.5～9

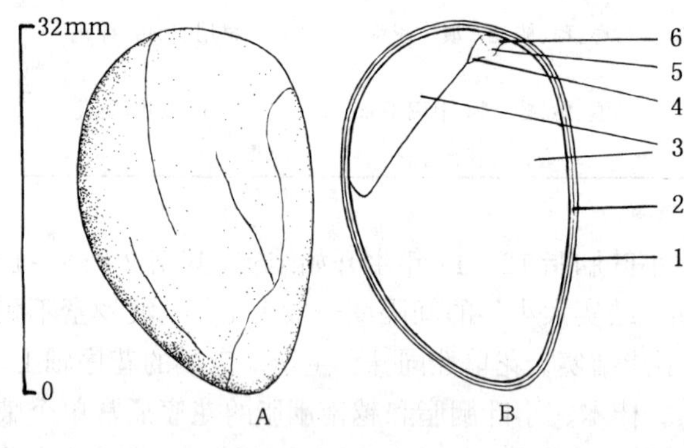

图 1 波罗蜜瘦果外形（A）及其纵切面（B）
1. 果皮 2. 种皮 3. 子叶 4. 胚芽 5. 胚轴 6. 胚根
（黄应钦绘）

果实的采收调制和种子贮藏 果实成熟盛期上树采摘。桂木、小叶胭脂和越南胭脂果形小，可以摇动树枝震落后在地面捡拾。采回的波罗蜜需堆放数天至十数天，软熟后供食用，其余 3 种堆放 1～2 日即可食用。它们的瘦果可以直接用作播种材料，也通称种子。也可将瘦果置水中搓洗，并淘去果肉等杂质，洗净后得出种子。种子不能失水，忌日晒，亦忌裸露存放。新处理种子的含水量一般为 30%～40%，不影响种子发芽的最低含水量为 20% 左右。宜随采随播。运输或贮藏均需混以湿沙，贮藏期为半年左右。这 4 个树种的出籽率、种子的净度和质量等数据见表 4。

表 4 桂木属树种鲜果的出籽率和种子的净度、质量

树种	出籽率（%）	净度（%）	千粒重（g）	每千克种子粒数
波罗蜜	10～20	98	5 500～7 000	140～180
桂木	5～6	97	250～300	3 300～4 000
小叶胭脂	4～6	97	80～120	8 000～13 000
越南胭脂	5～6	97	240～280	3 600～4 200

发芽和播种 波罗蜜、越南胭脂的种子无休眠习性，桂木和小叶胭脂的种子则有中等程度的休眠。1982 年和 1987 年，广西林业科学研究所分别对波罗蜜、桂木和越南胭脂作过发芽

测定，情况见表 5。

表 5　桂木属树种的发芽能力及其测定条件（广西南宁，室外沙床）

树　种	置床时间	室外温度（℃）	发芽势（%）		发芽率（%）	
			计算天数	一般数值	计算天数	一般数值
波罗蜜	7 月 29 日	27～35	8	60	14	80
桂　木	9 月 1 日	20～30	—	不明显	86	66
越南胭脂	6 月 30 日	27～35	—	不明显	34	63

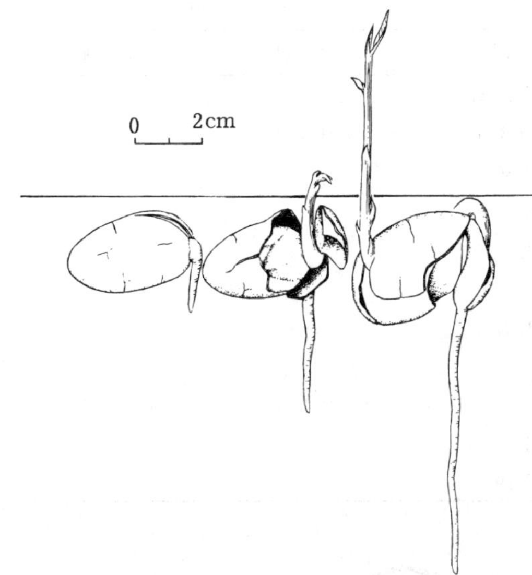

图 2　波罗蜜种子萌发后第 2、6、15 天幼苗生长情况
（黄应钦仿《热带亚热带主要树种采种育苗技术》）

留土萌发。波罗蜜的子叶 2 片，1 片特大，约为胚的 3/4～4/5，另 1 片小，约为胚的1/4～1/5，肉质。胚根萌发后 5～6 天上胚轴出土，具不育初生叶 5～7，11～12 天发出初生叶。其余 3 种胚根萌发后 11～14 天上胚轴出土，亦具不育叶，13～17 天发出初生叶。种子发芽及幼苗生长情况见图 2。

条播。波罗蜜每平方米播种 380～500g。桂木和越南胭脂每平方米播种 38～50g。小叶胭脂每平方米播种 12～15g。覆土 1～2cm。1～2 年生苗可以出圃。

（王宏志）

构　　树

Broussonetia papyrifera（L.）Vent.

（桑科　Moraceae）

生长习性、分布和用途　构属约 4 种，分布于东亚，我国 3 种，本文描述 1 种。落叶乔木，高达 16m，胸径可达 60cm。产黄河流域以南地区。越南、印度、日本亦产。适应性强，既耐干冷及湿热气候，又耐干旱瘠薄，也能生长于河边低湿地。多生于石灰岩山地，但酸性和中性土壤也能生长，常有鸟播现象。速生。萌芽性强。小枝、树皮纤维坚韧，韧皮纤维色泽洁白，含纤维素 36.4%，纤维长 5.5～11mm，是优良的造纸原料。木材黄白至淡黄褐色，纹理斜，轻软，易干燥，不耐久，供制器具及包装材料、薪炭材用。果可食用或入药。树液能治癣。叶可做饲料。抗烟性强，可做工矿区绿化树种。成熟果实肉质多汁，落地腐烂时有异味，容易招惹苍蝇，污染环境。

开花结实 花单性，雌雄异株。雄花组成柔荑花序，长6～8cm，下垂。雄花花萼4裂，雄蕊4，花丝在花蕾中内曲。雌花组成头状花序，径约2cm，花萼筒状不裂或顶部3～4小齿。子房内藏，具柄，花柱细长，侧生，柱头2，一长一短。聚花果球形，鲜果橙红色，径1.5～2.5cm，由多数外被宿存的肉质花萼及伸长的肉质子房柄构成的小核果组成。小核果扁球形，径1～1.6mm，红褐色，表面有小瘤点。内果皮革质。4～5月开花，7～9月果实成熟。1984年构树在全国8个地区的开花结实物候见表1。构树小核果的外形及其解剖构造见图1。

表1 构树的开花结实物候期（1984年）

观察地点	开花			种子成熟	
	始期	盛期	末期	始期	末期
洛阳	4月22日	4月24日	4月27日	—	—
西安	4月19日	4月23日	5月3日	7月30日	—
合肥	4月23日	4月27日	4月30日	—	—
南昌	4月3日	4月10日	4月19日	—	—
长沙	4月28日	5月2日	5月5日	—	—
贵阳	雄花：				
	4月5日	4月28日	5月6日	—	—
	雌花：				
	4月20日	4月23日	—	8月13日	8月16日
桂林	3月13日	4月15日	4月24日	8月1日	—
勐腊	2月20日	2月27日	3月19日	—	—

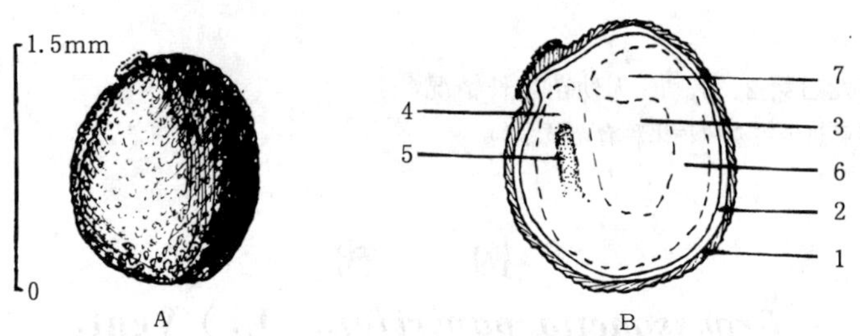

图1 构树果核的外形（A）及其纵切面（B）
1. 内果皮 2. 种皮 3. 胚乳 4. 子叶 5. 胚芽 6. 胚轴 7. 胚根
（孟玲绘）

果实的采收调制和种子贮藏 果实应随熟随采。采得的聚花果置桶中搓洗，除去果肉及腐烂物。所得果核即为播种材料，通称种子。在室内阴干后即可播种或沙藏。鲜果出种率6.6%；净度80%～100%。千粒重2～3g，每千克约含种子33万～50万粒。

发芽和播种 曾经利用一份采自北京的种子试测其发芽条件。种子的净度为100%，含水量为9.8%。发芽前用始温45℃水浸种24小时。试验表明，最适宜的发芽条件为30℃恒温加光照，持续15天，发芽率为81.5%。育苗时可以撒播，薄覆细土，盖草。经常浇水，约15～20天发芽出土。出土萌发（图2）。当幼苗形成4片真叶时，最好带土分床移植，移植前适当剪叶，移植距离 30～35cm。苗高 70～100cm 时可出圃。

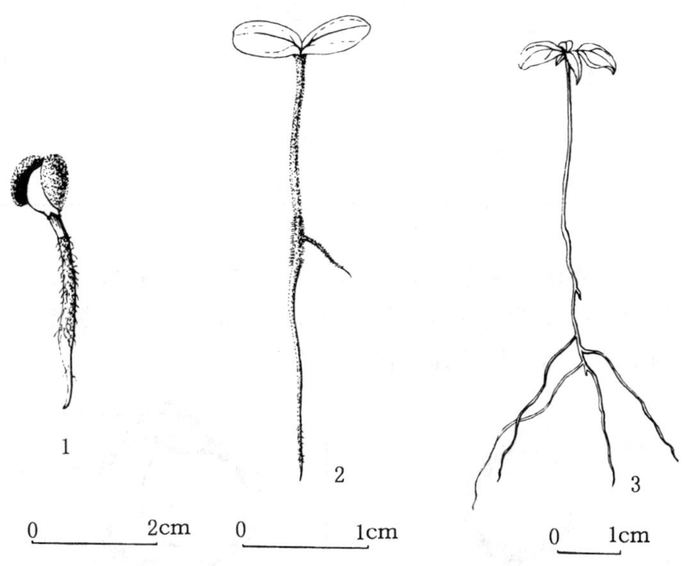

图 2　构树种子萌发和幼苗生长情况

1. 幼根伸出　2. 子叶展开　3. 初生叶对生（孟玲绘）

（钱耀明）

柘（葨芝）属
Cudrania Trec.

（桑科　Moraceae）

生长习性、分布和用途　本属约 10 种，落叶或常绿，乔木或灌木，分布于东亚至大洋洲，我国 4 种，本文描述 2 种（表 1）。喜光，喜温暖，但适应性强，耐干旱瘠薄。柘树喜钙。木材质地坚硬细致，心材黄色，可做染料，并可作家具、建筑和细木工材料。构棘可作绿篱。树皮纤维拉力强，可供纺织或制绳、造纸等。果可食。根皮可入药。柘树叶可以代替桑叶饲蚕。

表 1　柘属树种的名称、生长和分布情况

中　名	学　名	树高（m）	胸径（cm）	分　布	供　稿
构棘（葨芝）	*C. cochinchinensis* (Lour.) Kudo et Masam.	落叶直立或攀援状灌木，长 3～4	—	长江以南各省(自治区)。亚洲、大洋洲	604
柘树	*C. tricuspidata* (Carr.) Bur.	落叶小乔木，10	20	华北、华东、中南、西南。朝鲜半岛、日本	302

开花结实　6～8 年生开始开花结实，正常结实期在 10 年生以后。花期 5～6 月，果期9～10 月（表 2）。

表 2　柘属树种的开花结实物候期

树　种	观察地点	观察年份	开　花			果实成熟	
			始　期	盛　期	末　期	始　期	盛　期
构棘	南宁	1987	5月下旬	6月上旬	7月中旬	8月下旬	10月中旬
柘树	浙江	1989	5月中旬	5月下旬	6月中旬	9月下旬	10月中旬

花单性，雌雄异株。雌雄花均为头状花序，腋生。雄花萼片 4，基部具 2～4 个苞片，雄蕊 4。花丝直立。雌花萼片 4，紧包子房，柱头丝状。聚花果球形。瘦果包藏于肉质的苞片和花萼内，形成一肉质的头状体。种子具胚乳，见图 1、表 3。

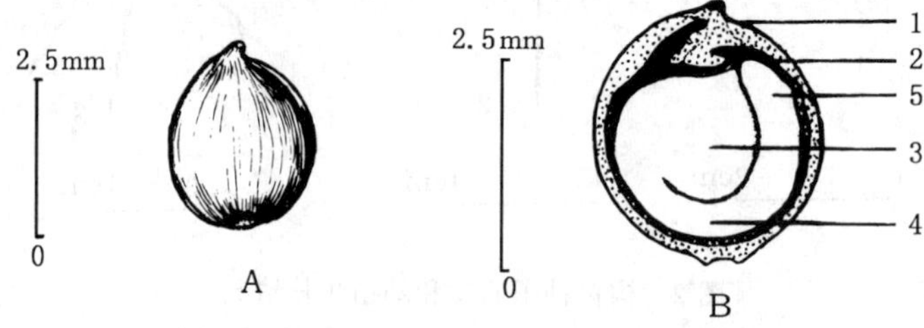

图 1　柘树瘦果外形（A）及其纵切面（B）

1. 果皮和种皮　2. 胚乳　3. 子叶　4. 胚轴　5. 胚根（孟玲绘）

表 3　柘属聚花果和种子（瘦果）的形态特征

树　种	未熟果颜　色	聚花果			种子（瘦果）		
		形　状	大小（cm）	颜　色	形　状	大小（mm）	颜　色
构棘	青绿色	矩圆形或不规则圆形	直径 2～5	鲜红	心脏形或扁卵形	长 4～5，宽 3～4，厚 2～3	灰白色
柘树	绿色	近球形	直径 2～3	橘黄或红色	扁卵形或肾形	长 2～3，宽 1.5～2.5，厚 1.5～2.5	乳白色

果实的采收调制和种子贮藏　可在秋冬果熟自动脱落时扫集或捡拾。作绿篱的构棘可结合修枝将果枝剪下采摘。将果实放水中或堆沤 2～3 天，搓擦漂淘，清除果皮等杂质，洗净捞出阴干，所得瘦果即为播种材料，通称种子。构棘果实也可在食用或果品加工后取得种子。调制所得的构棘种子含水量约为 20%，贮藏需混湿沙，贮藏期 3～6 个月。

表 4　柘属的出籽率和种子净度、质量

树　种	出籽率（%）	净度（%）	千粒重（g）	每克纯净种子粒数
构棘	6	95～98	8～11	90～130
柘树	3	85～95	2.4～3.0	330～420

发芽和播种　构棘种子无休眠习性，发芽时日均温宜在 20℃以上。1987 年曾经在南宁市室内条件下测定过新采构棘种子的发芽状况。发芽基质为滤纸，光照为室内自然光。8 月 29 日置床，9 月 10 日开始发芽，9 月 28 日发芽终止，发芽率 35%～50%，没有明显的萌发盛期。1988 年在北京室内自然光下测定过新采柘树种子的发芽能力：10 月 20 日置床，11 月 30 日开始发芽，盛期不明显，1 月 30 日发芽终止，发芽日均温为 18～24℃，发芽率为 30%～40%。出土萌发。初生叶对生，以后互生。发芽及幼苗生长情况见图 2。育苗时可将种子密播于沙盘或经过细致整地的苗床上，苗木转木质化时移至大田培育。每平方米播种 3～4g，覆土 0.2cm。

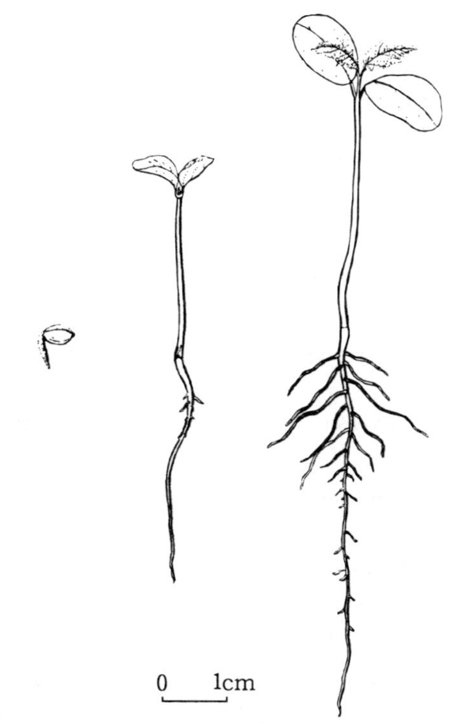

图 2　柘树种子萌发出土后第 1、10、19 天幼苗的生长情况
（孟玲绘）

（钱耀明）

榕　属

Ficus L.

（桑科　Moraceae）

生长习性、分布和用途　本属约 1 000 种，我国约 120 种，本文描述 10 种 1 亚种。常绿或落叶小乔木至大乔木，或灌木或藤状灌木。有乳汁。根系极发达，多数种具气根。耐湿热，耐水湿，喜湿润肥沃土壤。多数种能适应石灰质土和酸性土，干旱贫瘠土生长不良。树冠广阔，绿荫蔽日，多为华南重要四旁绿化树种。这些树种的名称、生长、分布及用途见表 1。

表 1　榕属树种的名称、生长、分布和用途

中　名	学　名	树高(m)	胸径(cm)	分　布	用　途	供　稿
高山榕	*F. altissima* Bl.	30	200	琼、粤、桂、滇。印度、斯里兰卡、越南、马来西亚	绿荫、风景、纤维	603
垂叶榕	*F. benjamina* L.	7～30	150	琼、粤、桂、滇、黔。印度至越南	绿荫、风景、纤维	603
无花果	*F. carica* L.	6～12	15～20	原产地中海和西南亚。我国各地栽培	食用、制酒、药用、纤维、油料	1003

（续）

中 名	学 名	树高(m)	胸径(cm)	分 布	用 途	供 稿
印度榕（印度胶树）	*F. elastica* Roxb. ex Hornem.	30	100～150	滇。印度、缅甸、马来西亚。桂、粤、琼、台、闽栽培	观赏、绿化，橡胶原料	603
水同木（空管榕）	*F. fistulosa* Reinw. ex Bl.	4～10	10～20	滇、桂、琼、粤、台。越南	花托可食、胶虫寄主	603
对叶榕	*F. hispida* L. f.	3～8	10～15	粤、琼、桂、黔、滇。越南、马来西亚、印度、缅甸、大洋洲	护堤、纤维、药用、食用菌	603
榕树	*F. microcarpa* L. f.	20～30	80～200	滇、黔、桂、粤、琼、台、浙。印度、马来西亚、缅甸	绿化、纤维、药用	603
薜荔	*F. pumila* L.	20～30（攀援状）		华东、华南、西南。日本、印度	药用、食用、橡胶	1003
菩提树（思维树）	*F. religiosa* L.	10～30	60～130	滇、粤。桂、琼、闽栽培。中南半岛	观赏、鞣质、硬性树胶、药用	603
斜叶榕	*F. tinctoria* Forst. f. ssp. *gibbosa* (Bl.) Corner	5～20	20～80	滇、黔、桂、粤、琼、闽、台。越南、缅甸、印度、斯里兰卡、马来西亚、菲律宾、大洋洲	绿化、纤维	603
笔管榕（黄葛树）	*F. virens* Ait.	5～20	80～200	黔、川、滇、桂、粤、琼、闽、赣、浙、台。印度、大洋洲	绿化、护岸、护路、叶解毒杀虫	603

开花结实 无花果、水同木和对叶榕等落叶或半落叶小乔木，3～6 年生开始开花结实，正常结实期在 15 年生以后。高山榕、垂叶榕、印度榕和榕树等常绿大乔木，10～14 年生开始开花结实，正常结实期在 20 年生以后，有的可以延续至 100 多年。多数树种结实大小年间隔期为 1 年，但不甚明显。无花果、对叶榕等每年结实均较多。花单性，雌雄同株或异株。花生于球形、卵形、梨形等肉质花序托的内壁上。垂叶榕、印度榕、对叶榕、榕树、菩提树和笔管榕的雌雄花同株同序，即雄花、瘿花和雌花混生在同一植株的同一花序托内。高山榕、无花果、斜叶榕、薜荔、水同木的雌雄花异株异序，即雄花和瘿花同生在一花序托内，雌花生于另一植株花序托内。高山榕、垂叶榕、印度榕、无花果、斜叶榕、榕树、菩提树、薜荔、笔管榕等花序成对生或单生于叶腋，对叶榕成对生于叶腋或簇生于树干或无叶的枝上，水同木簇生于由老干发出的瘤状短枝上。雄花被裂片 2～6，雄蕊 1～2 枚，花丝在蕾中直立。雌花被 2～6，有时不完全或缺。子房直或偏斜，有下垂的胚珠 1。花柱偏生。瘿花与雌花形状相似，子房内被一种膜翅目昆虫的蛹所盘踞，有助于雌雄花异序间的传粉。榕属树种开花结实的物候期因所在地区气候不同而差异较大。无花果是栽培种，花期差异更大。广西南宁 1987～1988 年观察的物候期见表 2，其中水同木每年开花 3 次，对叶榕和榕树每年开花 2 次。

瘦果，果皮骨质，包藏于花序托内。每花序托内有瘦果数十至数百粒，每果有种子 1 粒，多数为瘪粒。花序托外形近似果，也有称为隐花果或榕果的。种子颗粒极小，有胚乳，胚直立（表 3、图 1）。

表 2　榕属树种的开花结实物候期（广西南宁）

树　种	观察年份	开花			果实成熟		果实脱落期
		始　期	盛　期	末　期	始　期	盛　期	
高山榕	1987～1988	4 月下旬	5 月中旬	5 月下旬	9 月下旬	10 月上旬	10 月中旬～下旬
垂叶榕	1987～1988	4 月下旬	5 月中旬	6 月上旬	9 月上旬	10 月中旬	9 月下旬～翌年 1 月中旬
无花果	1988	3 月下旬	5～8 月	9 月下旬	6 月中旬	9 月中旬～12 月中旬	7 月中旬～翌年 2 月上旬
印度榕	1988	5 月下旬	6 月上中旬	6 月下旬	10 月中旬	10 月下旬	11 月上旬～下旬末
水同木	1987～1988	4 月中旬 6 月中旬 9 月上旬	5 月中旬 7 月上旬 9 月中旬	5 月中旬 7 月下旬 9 月下旬	5 月下旬 8 月上旬 10 月上旬	6 月上旬 8 月下旬 10 月下旬	6 月中旬～11 月中旬
对叶榕	1987～1988	3 月下旬 7 月下旬	4 月上旬 8 月中旬	4 月下旬 9 月上旬	6 月下旬 9 月中旬	7 月中旬 10 月上旬	7 月下旬～11 月中旬
榕　树	1987～1988	5 月上旬	5 月中旬	5 月下旬	8 月中旬	10 月中旬	9 月上旬～12 月上旬
		12 月下旬	1 月上旬	1 月中旬	4 月下旬	5 月下旬	5 月上旬～6 月下旬
薜　荔	—	4 月	5 月	6 月	8 月	9～10 月	11 月
菩提树	1987	4 月下旬	5 月中旬	5 月下旬	10 月下旬	11 月中旬	11～12 月
斜叶榕	1987～1988	6 月上旬	6 月下旬	7 月上旬	11 月中旬	11 月下旬	12 月中旬～翌年 1 月
笔管榕	—	5 月	6 月上中旬	7 月	9 月下旬	10 月中旬	11～12 月

表 3　榕属树种花序托和种子（瘦果）**的形态特征**

树　种	花序托未成熟时颜色	成熟花序托			种子　（瘦果）		
		形　状	大小（cm）	颜　色	形　状	大小（mm）	颜　色
高山榕	青绿色	卵形	高 1.8～2.5 径 1.2～1.6	深红或带黄色	卵形或圆锥形	长 1～1.2 径 0.6～0.8	浅褐色
垂叶榕	浅黄绿色	卵形或梨形	高 1～1.6 径 0.8～1.5	深黄至黄褐色	卵形、圆形或椭圆形	长 1～1.3 径 0.7～0.8	褐　色
无花果	青　色	倒圆锥形，顶略平或微凹	高 2～5 径 3～5.5	红色间杂黑褐色	卵形	长 0.9～1.3 径 0.6～0.7	黄褐色
印度榕	绿色	椭圆形，顶部浑圆	高 1～1.5 径 0.7～1	紫黑色	近卵形，先端尖	长 0.9～1.1 径 0.5～0.7	白色
水同木	青色	球形	径 2.6～4	紫红或暗红色	卵形	—	浅黄色
对叶榕	青绿色	球形	径 2.5～3.5	紫红或暗红色	椭圆形	长 1.4～1.6 径 0.8～1.2	浅褐色
榕树	淡红色	倒卵形或近球形	高 0.9～1.1 径 0.8～1	浅黄或浅紫褐色	卵形或椭圆形	长 0.6～1.2 径 0.5～0.8	浅褐色
薜荔	青色	倒卵形或梨形	高 4～6 径 3～4	黄红色	弯月形	长 2.7～3.1 径 1.2～1.4	深褐色
菩提树	青色	扁球形	高 0.7～1 径 0.8～1.2	黄红色	弯形或半月形	长 1.5～1.8 径 0.6～0.8	淡黄色
斜叶榕	黄绿色	扁球形或梨形	径 0.6～0.8	黄或橙红色	卵形或肾形	长 1.2～1.6 径约 0.7	黄褐色
笔管榕	浅黄色	近球形	径 0.5～0.7	浅红或紫红色	卵形或椭圆形	长 1～1.3 径 0.5～0.8	褐色

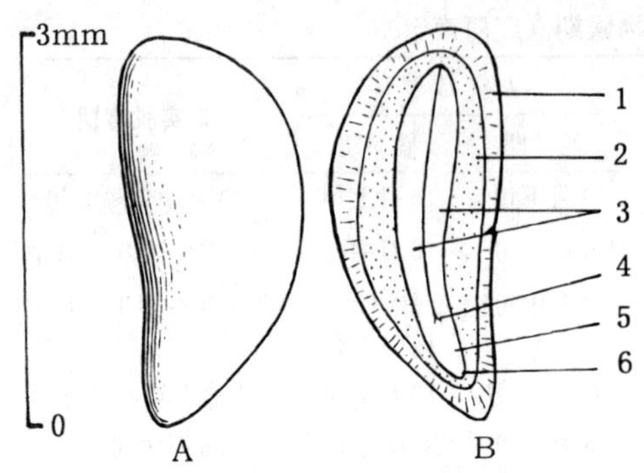

图 1 薜荔瘦果外形（A）及其纵切面（B）
1. 果皮和种皮 2. 胚乳 3. 子叶 4. 胚芽 5. 胚轴 6. 胚根（何泽瑛、黄应钦绘）

果实的采收调制和种子贮藏 榕属树种果实的成熟期长，当花序托的色泽呈现成熟特征时用手采摘，或用竹竿在树上敲打，落地后在地面捡拾。采回的花序托可堆沤数日，充分软熟后装入紧密布袋，置水中搓揉，反复搓擦漂淘，除去花序托的肉渣皮屑和空瘪粒等杂质，以在水中下沉的瘦果作为播种材料，通称种子。花序托的出籽率为2%～4%。净度和质量等数据见表 4。

榕属种子忌日晒，不宜失水。洗净的种子在室内稍加晾干即可供播种。运输或贮藏均需用湿滤纸包裹后，再装入塑料袋内，贮藏期一般为 1 个月左右。

表 4 榕属树种的种子净度、质量

树 种	净度（%）	千粒重（g）	每克种子粒数（粒）
高山榕	50～70	0.13	7 700
垂叶榕	50～70	0.38	2 600
无花果	75～85	0.37	2 700
印度榕	50～60	0.3	3 300
水同木	60～70	1.1	900
对叶榕	60～80	0.45	2 200
榕 树	50～70	0.25	4 000
薜 荔	—	0.66	1 500
菩提树	50～60	0.45	2 200
斜叶榕	50～70	0.4	2 500
笔管榕	50～65	0.24	4 200

发芽和播种 种子无休眠习性。发芽时日均温需在 20℃以上。种子的颗粒小，播种育苗的技术要求高，生产上多用扦插育苗。但实生苗的树形好，寿命长，播种育苗在生产上仍有应用。不同年份的榕属种子发芽能力差异很大。有些树种成龄期的种子，某一年发芽率很低，另一年发芽率又较高，还需进一步研究。1987 年，广西林业科学研究所对本文所描述的 11 种榕树，在室内作过发芽测定，结果有 7 种不发芽，发芽的只有无花果、斜叶榕、榕树和菩提树 4 种，测定情况见表 5。

种子无休眠习性，播后 10 天左右开始发芽。出土萌发。胚根萌发后 2～4 天子叶伸出。无花果在子叶伸出后 15 天左右发出初生叶；斜叶榕、榕树和菩提树在子叶伸出后 21～32 天才发出初生叶。无花

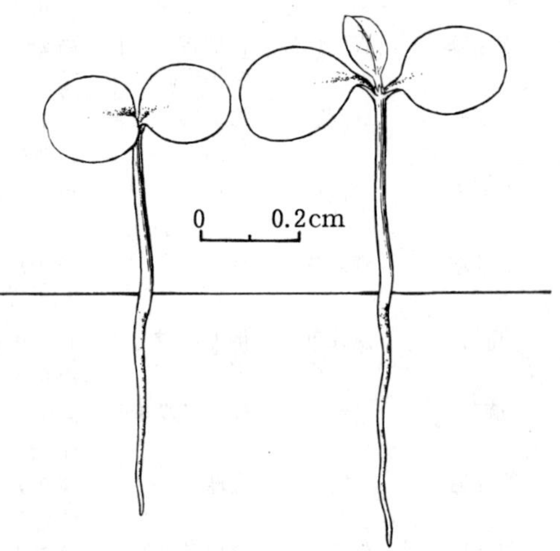

图 2 无花果种子萌发后第 6、15 天幼苗的生长情况
（黄应钦绘）

果种子的萌发和幼苗初期生长情况见图 2。

表 5　榕属树种的发芽能力及其测定条件（广西南宁，基质均为滤纸）

树　种	温度（℃）		发芽势（%）		发芽率（%）	
	发芽箱	室　外	计算天数	一般数值	计算天数	一般数值
无花果	—	25～29	8	64	28	80
榕　树	28	—	—	不明显	28	8
菩提树	28	—	8	17	26	23
斜叶榕	—	26～30	—	不明显	36	50

撒播。将种子密播于沙盘内，每平方米无花果播 3～5g，本文所述的其余 10 种因发芽率低，可播 5～10g。沙盘中的幼苗转木质化时移至圃地继续培育。一般 1 年生苗出圃，如用于庭院绿化，需培育 2～3 年，苗高 1.5m 以上时出圃。生产中多用扦插或压条育苗，成活率在 60%以上。

（张声燕）

桑　属

Morus L.

（桑科　Moraceae）

生长习性、分布和用途　桑属约有 12 种，产北半球温带和亚热带，我国有 9 种，本文描述两种（表 1）。落叶小乔木或灌木。耐寒，耐干旱瘠薄。在微酸性土、中性土、钙质土以及含盐量在 0.2%以下的盐碱土上都能生长，但以肥沃、排水良好的中性土壤为宜。桑在我国栽培历史悠久，公元前 1562～前 1066 的殷商时代就有栽桑的记载。公元前 4～5 世纪，我国栽桑养蚕的技术便传入南亚、中亚和欧洲。桑已在欧洲和北美自然归化。我国桑的栽培品种极多，著名的有珠江流域的广东荆桑，太湖流域的湖桑，四川盆地的嘉定桑，黄河下游的鲁桑以及新疆的白桑等，叶大，嫩而多汁，产量高。但蒙桑不能饲蚕。桑和蒙桑的木材可供雕刻和制作家具，茎皮可造纸，果可生食或酿酒，根、皮、茎、叶、果均可入药。

表 1　桑属树种的名称、树高、分布和用途

中　名	学　名	树高（m）	分　布	用　途	供　稿
桑	*M. alba* L.	15	新疆、内蒙古、哈尔滨以南至川、滇、黔、粤、台	饲蚕、材用、造纸、药用、酿造	1004
蒙　桑	*M. mongolica* Schneid.	3～8	内蒙古、辽、冀、鲁、晋、豫、鄂、湘、川、滇。朝鲜半岛	材用、造纸、药用、酿造	201

开花结实　4～7 年开始结实，没有明显的大小年现象。这 2 个树种的物候期因地区不同

而有差异。同是1975年，桑在桂林的花期为4月上中旬，在西安为4月中下旬，大约相差10天。蒙桑果实的成熟期在山东常为5～6月，在内蒙古则多为6月中下旬（表2）。

表2 桑属树种的开花结实物候期

树 种	观察地点和年份		开花			果实成熟		
			始 期	盛 期	末 期	成熟期	脱落始期	脱落末期
桑	桂林	1975	4月7日	4月11日	4月18日	5月5日	5月10日	5月30日
	西安	1975	4月16日	4月20日	4月28日	5月15日	—	—
蒙桑	内蒙古	—	5月中旬	～	5月下旬	6月中旬	6月下旬	7月中旬

花单性，同株或异株。柔荑花序。雄花萼片4；雄蕊4，芽时内弯；有退化雌蕊。雌花萼片4，果时增大而肉质，子房上位，1室。桑树雄花序长1.5～3.5cm，为雌花序的2倍，密生细毛；雌蕊无花柱，或花柱甚短，柱头2裂。蒙桑雄花序长3cm，穗状腋生，雌蕊花柱长，柱头2裂。聚花果（桑椹）卵形或卵圆形至圆柱形。每一桑椹有瘦果30～50枚，外被宿存的肉质苞片和花萼，初为绿色，以后逐渐变成红色，成熟时呈紫黑色，少数品种的桑椹成熟时为淡红色或玉白色。瘦果外果皮多少为肉质。内果皮硬壳质，黄褐色或浅褐色，扁卵形，表面平滑，微具棱。种皮膜质，胚乳丰富。子叶2枚，弯曲。图1绘出的是桑的聚花果和去掉肉质外果皮后的瘦果。桑属这两个树种的聚花果和清除外果皮后瘦果的大小及其成熟时的颜色见表3。

表3 桑属树种的聚花果和清除外果皮后瘦果的大小及成熟时的颜色

树 种	聚花果			瘦果的大小（mm）		内果皮颜色
	长（cm）	形 状	成熟颜色	长	径	
桑	1.3～2.5	卵圆形	紫黑色，少数淡红色、玉白色	2～2.5	1.5	淡黄、黄褐色
蒙桑	1～2.5	卵圆形或圆柱形	紫黑色	—	—	浅黄色

果实的采收调制和种子贮藏 同一植株上桑椹的成熟期不一致，采种时应分批采集成熟的桑椹。直接摘取或摇落地面后拾集。采回的成熟桑椹需及时淘洗取种。如不能立即调制，应把桑椹摊放在阴凉处，防止发热变质影响种子质量。调制时，将成熟的桑椹放在容器内揉搓，用水反复漂去果肉等杂质和空瘪粒，捞出沉在水底的饱满的清除了外果皮的瘦果，即为播种材料，通称种子。桑属种子忌曝晒，应薄摊在通风处阴干。阴干的种子可供夏播。如果供来年播种，应密封贮藏。生产上通常的做法是：用一只腹大口小的坛，坛底放生石灰，上铺几层草纸，把含水量已经阴干到5%～7%的种子装入布袋，置于草纸上，迅速加盖，封口后放在荫凉处。通常坛内约1/4放生石灰，1/2放种子，留出1/4的空间以利种子呼吸。用这种方法贮藏的桑籽，发芽能力可以保持3年。桑属2个树种的出籽率、种子净度和种子质量数据见表4。

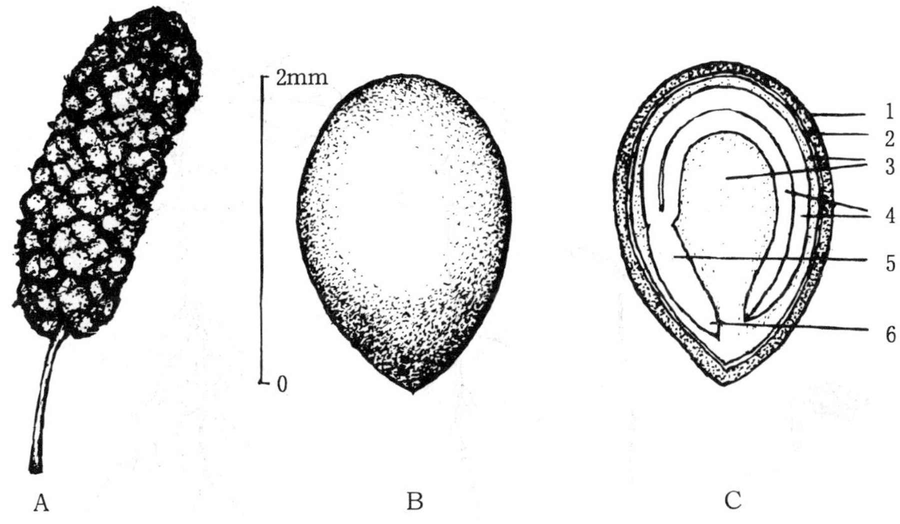

图 1 桑的聚花果外形（A）、去掉肉质外果皮后的瘦果外形（B）及其纵切面（C）
1. 内果皮 2. 种皮 3. 胚乳 4. 子叶 5. 胚轴 6. 胚根
（田恒德绘）

表 4 桑属树种的出籽率、种子净度和种子质量数据

树 种	桑椹出籽率（%）	净度（%）	千粒重（g）	每克纯净种子粒数
桑	2～4	85～95	1.5～1.7	580～680
蒙桑	2～3	85～95	1.2～1.4	700～840

发芽和播种 桑属种子在 25～35℃范围内的恒温下能正常发芽，其中以 28～32℃的发芽率最高。发芽测定时每昼夜如能提供 8 小时光照会更有利于发芽。温度在 15℃以下和 39℃以上均不能发芽，在 20℃和 37℃条件下发芽率也低。当年采集的新鲜种子发芽率可达 90%以上。如果贮藏得法，普通干藏到翌年春季发芽率也可以达到 90%。

桑属种子可以随采随播，夏播或秋播，也可以贮藏后翌年春播。夏播、秋播不必催芽。春播的具体时间随地区而异。以桑为例，珠江流域约在 3 月下旬，长江流域在 4 月上中旬，黄河流域在 4 月下旬～5 月上旬，东北地区在 5 月中下旬。春播前 5～6 天用始温 45℃的清水浸泡 24 小时后混拌 2～4 倍的湿沙，放在温暖之处催芽，经常翻动并保持湿润，待有 20%～30%的种子裂嘴时即可播种。一般采用条播，行距 20～30cm，每平方米播种 1.2～2.5g。覆细土 0.5cm，然后盖草。出土萌发。播后 5～10 天可发芽出土。第 1、2 片初生叶对生，以后互生（图 2）。

桑 1 年生苗高可过 80～120cm，可以出圃或作砧木。桑属树种还可以用嫁接和压条法繁殖。王林（1993）利用过生根粉对桑树施行硬枝扦插，成活率达 80%以上，当年生苗高 1.5m 以上。

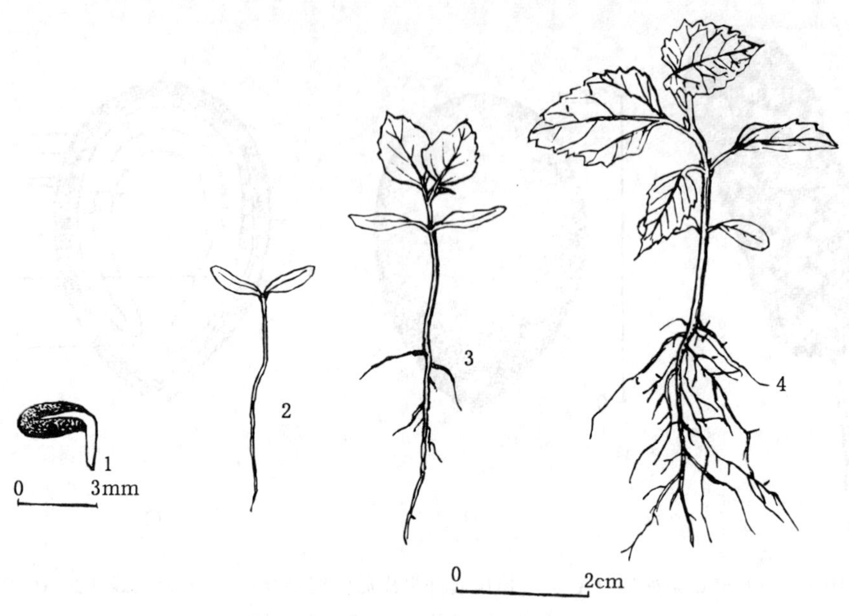

图 2 桑的萌发和幼苗生长情况

1. 胚根伸出 2. 子叶展开 3. 第 1～2 片初生叶对生 4. 互生叶出现

（田恒德仿《主要树木种苗图谱》）

（杨余康）

鹊 肾 树

Streblus asper Lour.

（桑科 Moraceae）

生长习性、分布和用途 鹊肾树属约 22 种，我国产 6 种，本文描述 1 种。大灌木或小乔木，高 2～10m，胸径 5～40cm，具乳汁。热带树种，产琼、粤、桂、滇。印度、尼泊尔、斯里兰卡、缅甸、马来西亚、泰国亦产。木材为小径材或短料材，可供制作家具。果可食。纤维可织袋。树脂入药。

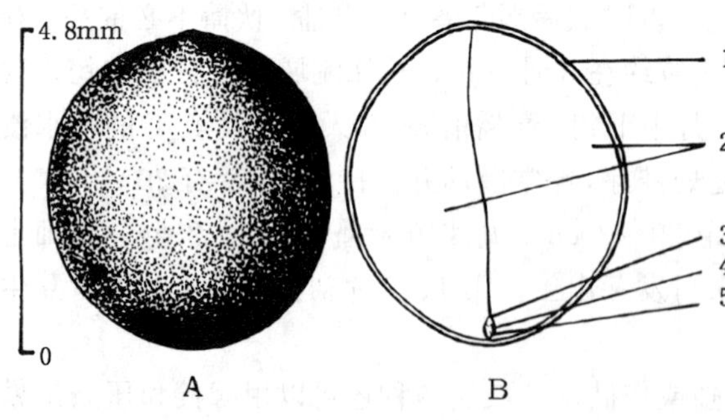

图 1 鹊肾树种子（瘦果）外形（A）及其纵切面（B）

1. 果皮和种皮 2. 子叶 3. 胚芽 4. 胚轴 5. 胚根

（黄应钦绘）

开花结实 5～8 年生开始结实，10 年生以后为正常结实期，大小年现象不明显。花单性，雌雄异株，少有同株。雄花序为小头状花序，直径 5～7mm，近无柄，单生或 2～3 序聚生。雄花萼片 4，雄蕊 4，花丝在蕾中内弯，退化雌蕊顶部扩大。雌花具柄，单生或 2～4 朵簇生于叶腋，有时单朵生于雄花序中央；苞片 3，萼片 4；子房上位，球形，胚珠下垂；花柱细长，中部以上2分枝，宿存。据海南尖峰岭1988年观察，每年开花结实2次：第1次3～4月开花，6月下旬～7月上旬果实大量成熟；第2次8～9月开花，次年3～4月果熟。果肉质，为花后增大的萼片所包围。内为瘦果，果皮与种皮难分。种子1粒，球形，径4～6mm。种皮膜质。无胚乳。胚球形，子叶肉质。鹊肾树瘦果的外观及其解剖结构见图1。

果实的采收调制和种子贮藏　果实成熟期中在树下拾集，或用竹竿敲打果枝震落后收集。果实可在食后得出其中瘦果，或沤烂后在水中搓擦，除去皮肉杂质，所得瘦果即为播种材料，通称种子。鲜果出籽率约18%，净度可达98%。千粒重约100g，每千克约有纯净种子10 000粒。调制后稍加晾干的种子含水量为18%～20%，忌失水，不宜日晒。短期运输可用湿椰糠或湿锯屑拌和种子装运，到达后即播种。宜随采随播，短期贮藏需混湿沙，贮藏期一般为1个月以内。

发芽和播种　种子无休眠习性。发芽时日均温需在22℃以上。1988年8月24日，中国林业科学研究院热带林业试验站用新鲜种子作过发芽测定，置床后5～6天开始萌发。从开始发芽至发芽高峰，10天的发芽百分数为50%～55%；从播种至发芽终结，21天的发芽率为70%～75%。出土萌发。子叶出土后3～4天长出初生叶（见图2）。

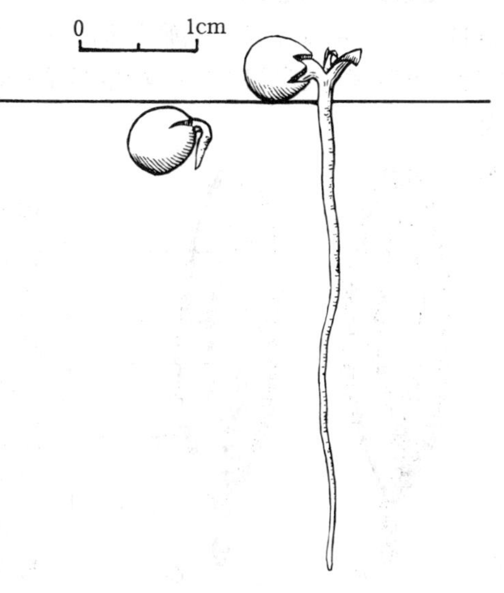

图2　鹊肾树种子萌发后第2、6天幼苗的生长情况
（黄应钦绘）

条播。每平方米播10～12g。发芽约1个月后间苗。培育半年生或1年生苗出圃。

（符史深）

杜　仲

Eucommia ulmoides Oliv.

（杜仲科　Eucommiaceae）

生长习性、分布和用途　世界上只此1种，我国特产，属稀有种，已列入《中国植物红皮书》。落叶乔木，树高可达20m，胸径可达40cm。喜光。深根性。喜温暖湿润气候。对土壤适应性强，在酸性土、微碱性土及钙质土上均能生长，以在土层深厚、湿润肥沃、质地疏松、排水良好的微酸性的黄壤、红壤上生长良好，土壤过于瘠薄，酸度过大对生长不利。

分布范围较广。秦岭、黄河以南，五岭以北，东至华东，西至云贵高原均有生长。主产贵州、四川、陕南、鄂西和湘西。引种成功的地方有北京、青岛、大连、福建的南平和三明等。1896年引入欧洲，1899年引入日本，1906年引入俄国。美、法、德、朝等国也曾引种繁殖。

杜仲树皮入药，果实富含杜仲胶，入药的干燥树皮亦称杜仲，是用途广泛的贵重药材。果实、树皮、树叶含杜仲胶，是良好的绝缘材料。

杜仲是剥割树皮入药的经济树种。生产上通常根据树皮特性划分类型，常见的类型有粗皮、光皮以及被认为是这两者之间自然杂交产生的中间类型。解剖研究表明，光皮杜仲内皮比粗皮杜仲内皮重22.3%；厚度也超过13.9%。药用价值也是光皮杜仲优于粗皮杜仲。光皮杜仲被认为是杜仲树种中的优良类型。

开花结实　据周政贤（1958）观察，杜仲8年生左右开始开花结实，20～30年生进入结实盛期，30年生后结实量逐渐下降，50年生后结实甚少，百年生大树虽仍能结实，但几乎全为空粒。林缘木和孤立树6～8年生开始开花结实。已被剥皮的雌株仍能结实。结实间隔期一般为1年。

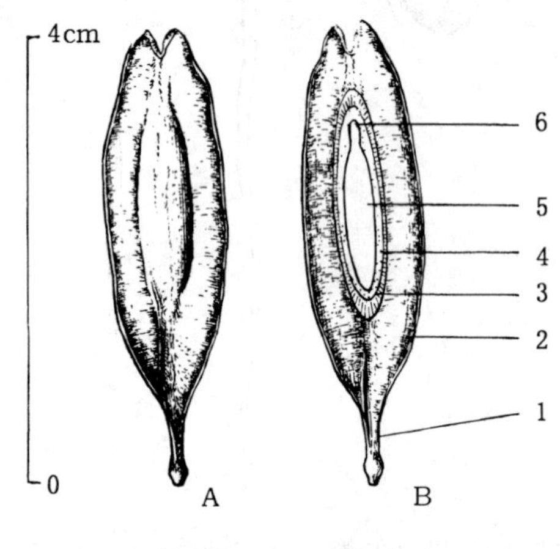

图1　杜仲带翅坚果的外形（A）及其纵切面（B）

1. 果柄　2. 果翅和果皮　3. 种皮　4. 胚乳　5. 子叶　6. 胚根

（黄应钦绘）

花单性，雌雄异株。花无花被，先花后叶或与叶同时开放，单生于小枝基部。雄花疏散，具苞片，密集成头状花序，生于短梗上，由（3）6～10个雄蕊组成。花药条形，花丝极短。雌花单生于每个苞腋内。雌蕊由2个合生心皮组成。子房扁，长椭圆形，1室，有胚珠2枚。3～4月开花，单个雄花和单个雌花的花期各在7天左右，但雄花先于雌花约3～4天。果5月中下旬生长定型。果熟期10～11月。带翅坚果未熟时青绿色，成熟时呈栗褐色、棕褐色或黄褐色，卵状长方形或长椭圆形，不开裂，有光泽，薄革质而有柄，扁而薄，长3～4cm，宽1～1.5cm，顶端2裂，中央微隆起，基部楔形。坚果内含种子1粒，间有2粒。种子垂生于室的顶端，扁条形，长1.4cm，宽0.2～0.3cm，两端圆。种皮膜质。种脊生于背面。胚乳丰富，白色。胚直立于中央，与胚乳近等长，上端具一多少呈扁形的幼根。子叶扁平，肉质，长圆状条形，较胚根长。杜仲坚果的外形及其纵切面见图1。

影响种子产量和质量的主要因子是温度、降水、光照和剥皮。开花期若气温下降或降水过多会影响授粉。果实发育期如遇干旱则种子不饱满。据调查，光照充分的孤立木单株产果量为8.5kg，林缘木和散生树单株产量为0.86～2.3kg，光照少的林内树木仅0.25kg。结实母树树皮被剥割，营养运输遭破坏，虽也结实，但种子发育不饱满，生活力弱，不能作种用。

果实的采集调制和种子贮藏　果实成熟后虽不立即脱落，但仍应适时采集。若采收过早，胚的发育不健全，果实含水量也高，采后容易发热霉变，丧失生活力。采集过晚，果实易风吹飘散，难以收集，寒冷地区还易遭受霜冻危害。一般是在霜降以后树叶大部分脱落，果皮呈栗褐色、棕褐色或黄褐色时，选无风或微风之日，用竹竿轻敲或摇动树枝，使坚果脱落在铺于母树附近的收种布上。采集的坚果筛选去杂后置通风荫凉处晾干，切勿烘烤曝晒。坚果即为播种材料，通称种子。坚果净度一般在90%以上，千粒重为58～130g，每千克带翅坚果7 600～17 500粒。贮藏时含水量应在10%以下。

在室温下普通干藏的杜仲种子很难长期保持生命力。供翌年育苗的种子如不冬播，应当混沙层积越冬，既是贮藏，也为萌发创造条件。如果需要较长时间地保持发芽能力，必须在低温下密封干藏。林坚和郑光华等人（1990）对采自北京的种子利用离体胚测定，在4℃条件下密封干藏半年后的发芽率为99%，室内普通干藏半年的则仅为14.7%。

发芽测定　未经处理的干燥种子置床后发芽率很低，置床前浸种也不能显著提高发芽率。据观察，杜仲种子吸水并不困难，因此有人推测影响发芽的可能是胚乳限制了吸胀后的气体交换。林坚和郑光华等（1989）则认为，杜仲果皮并不限制水分进入，果皮含有发芽抑制物质，但不是影响萌发的主要因素。他们的看法是果皮富含胶质，紧紧地束缚了种子，使胚根难以穿透萌发。

据王成霖等人（1986）研究，同样是在20℃条件下测定的绝对发芽率，干种子置床的为28.5%；室温水浸种24小时后置床以及浸种后在胚根一端划破种皮的均为32%；改用500μg/g赤霉素浸种24小时的为62%；室温水浸种24小时后在胚根一端切去部分种皮和胚乳的效果最好，达95%。又据同一份试验，杜仲的发芽测定温度以20℃为宜：同样是在胚根一端切去部分种皮和胚乳，在20℃、25℃、30℃和35℃条件下的绝对发芽率及其计算天数分别是95%（8天）、78%（16天）、45%（21天）和2%（15天）。在他们的试验中，发芽基质是石英沙，测定时未加光照。林坚和郑光华等（1989）也发现杜仲的发芽对温度极为敏感。在所用的5℃、10℃、18℃、20℃和25℃5种温度条件下，他们得到的发芽率分别为0%、72%、98%、74%和5%。吴楚材（1990）从层积过两个月的果实中取出种子，在10℃、15℃、20℃、25℃和30℃的5种恒温下发芽，经历10天得到的发芽率分别为57.8%、67.6%、73.9%、67.1%和31.4%。李碧涛（1991）将始温45℃的水浸种4天的果实切去胚根一端的胚乳约1mm（但未伤及胚根），12天的发芽率达82%，未切去胚乳的则无一发芽。徐本美和白克智（1995）认为杜仲是浅休眠型种子，需要适度的冷层积：在2～5℃下层积1、2、3个月的种子，以层积2个月的出苗率最高；未经冷层积的种子萌发后会形成少量的矮化苗，需要经历冬天的自然低温后才能转化为正常苗。

播种前的处理　春季播种的种子最好在5℃左右的室温下混以10倍的湿沙层积30～35天。大约1个月种粒便可充分吸胀，部分种子的胚根即将萌出时取出播种，场圃发芽率可以达到80%。未经层积的种子可以在播前浸种3～5天，但播种期必须提早到冬季。不过这样做实际上已是将种子在圃地里层积，起作用的已经不是浸种这项措施。

播种　冬播或春播。冬播期在采种当年的12月。一般多为春播。经过湿沙层积催芽的在2月～3月中旬日均温稳定在10℃以上时播种。宽幅条播。条距与条幅各为20cm，沟深1.5～2cm。宽1m的苗床，每一条幅播种50～60粒，均匀地铺于播幅内，覆细土1.5cm左右，盖草。每666m^{2}播种5～8kg。12月播种的，播后约50天种子发芽出土。混沙催芽的播后约20天发芽出土。

出土萌发。子叶2，条状披针形，长23～36mm，宽3.5～6.2mm，先端尖或钝，基部渐狭与柄界限不明，羽状脉。上胚轴长6～15mm，幼茎淡绿色，渐呈暗紫色，被毛。初生叶第1、第2片对生，以后互生。叶卵状椭圆形，先端长尖，基部楔形或心形，边缘具不整齐锯齿，被毛。下胚轴圆柱形，长3.5～6（10）cm，径约1.5mm，向根颈处较粗，径2mm；上部绿色，中部淡紫色，下部白色。具主根，侧根斜下伸展。1年生苗高50～70cm，有的达1m，可出圃栽植。

杜仲种子的萌发和幼苗生长情况见图 2。

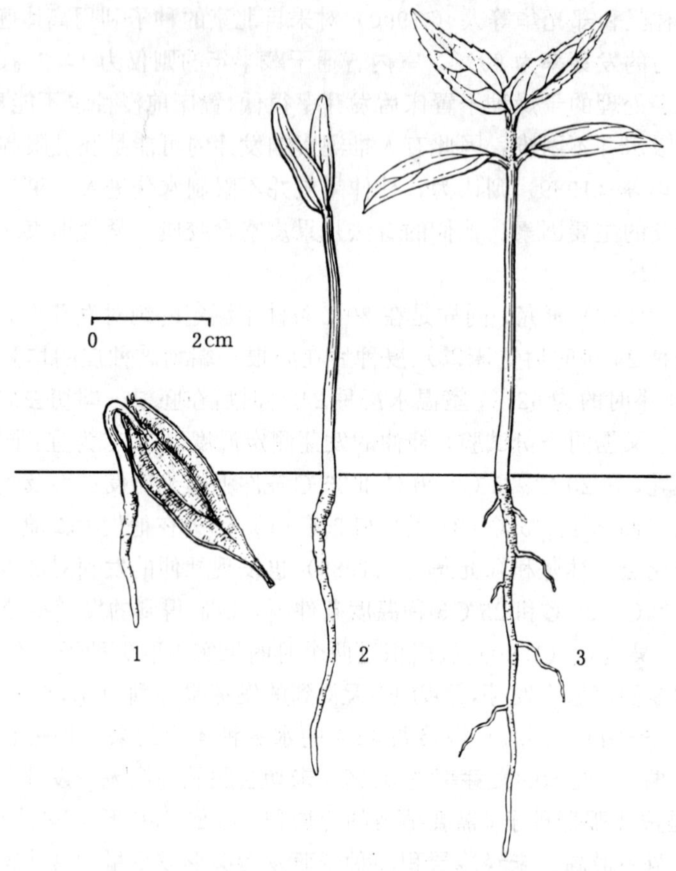

图 2　杜仲种子的萌发和幼苗生长情况

1. 下胚轴延伸　2. 子叶初展　3. 第 1、2 片初生叶对生

（黄应钦仿《主要树木种苗图谱》）

（徐玉蓉）

红木（胭脂木）

Bixa orellana L.

（红木科　Bixaceae）

生长习性、分布和用途　红木属约 4 种，产热带美洲，我国引入 2 种，本文描述 1 种。常绿灌木至小乔木，高 4～6m，多分枝，主干不明显。生长快，1 年生可高达 70cm。适生于肥沃湿润的酸性土壤。不耐轻霜或较长期的 4～5℃低温，南宁、广州露地栽培，寒冷年份常受冻害。西非、印度、缅甸等热带地区广泛栽培，已乡土化。我国滇、桂、粤、琼、闽有引种。肉质外种皮含红色素和黄色素，可作食品染料。树皮为优质纤维。种子为收敛退热剂。花果供观赏。

开花结实　2～3年生即可开花结实，4年生以后为正常结实年龄。无大小年现象，每年结实均较多。花两性，圆锥花序顶生。萼片5，花瓣5。雄蕊多数，花药顶孔开裂。子房上位，1室，或由于侧膜胎座突入中部而成假数室，胚珠多颗。据广西南宁1986～1988年观测，10月上旬花序形成，着生于当年生枝梢，10月下旬为始花期，11月上旬为盛花期，下旬至12月上旬为末花期；次年3月上旬至中旬果实开始成熟，3月中下旬为果实成熟盛期，5月中旬脱落。蒴果，密被软刺，成熟时由青色转变为绿红色以至红棕色，心脏形或扁圆形，高1.9～4.2cm，宽2.4～4.5cm，厚1.6～3.2cm，成熟后2瓣开裂，每果有种子16～24粒。种子外露，不脱落。外种皮红色肉质，内种皮木质，暗红色至赭色，圆锥形或近三角形，一侧有下凹槽纹。径3～4mm，高4～5mm。有胚乳，子叶弯褶。种子形态见图1。

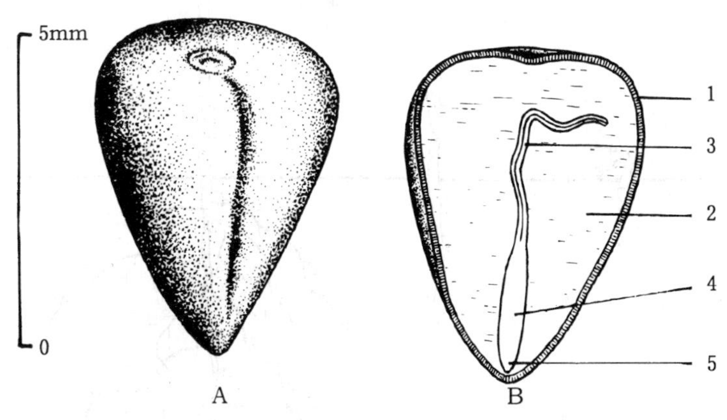

图1　红木种子（去外种皮）外形（A）及其纵切面（B）

1. 内种皮　2. 胚乳　3. 子叶　4. 胚轴　5. 胚根（黄应钦绘）

果实的采收调制和种子贮藏　每果穗着生果实数十个，下部的果实先熟，上部果实中的种子发育多不饱满，应选采果穗中下部形体较大的果实。采时用枝剪将整个果穗采下，选留较大的果实后，其余的用于提取红色素。选作种子用的果实摊放于室内通风处2～3天，果壳成2瓣开裂，手持两果瓣相对剥打，即可脱出种子。脱出的种粒浸水，搓去肉质红色外种皮，得到由内种皮包被的部分用作播种材料，通称种子。鲜果的出籽率为10%～15%。净度可达98%。千粒重一般为26（22～30）g，每千克有纯净种子38 000（30 000～45 000）粒。种子的含水量约为15%，可以适当晾干，但忌日晒。运输宜以果实的状态装运，运抵后再行调制。种子需混沙贮藏，贮藏期一般不宜超过2个半月。

发芽和播种　种子无休眠习性。1981年7月20日，日均温30℃时，广西林业科学研究所在室外沙床作过发芽测定，播后第5天开始发芽。出土萌发。发芽后第4天子叶出土，第9～10天展出初生叶。发芽的进程较快，但发芽率较低。以8天计，发芽势为25%，以18天计，发芽率为30%。红木种子的萌发和幼苗生长情况见图2。

种子宜随采随播，适宜的播种期为5～8月，9月以后气温转低，不宜播种。条播。每平方米约播5～8g，生产上一般是在湿沙床内密播，俟子叶发出后将芽苗移至容器或圃地培育，3个月左右出圃。

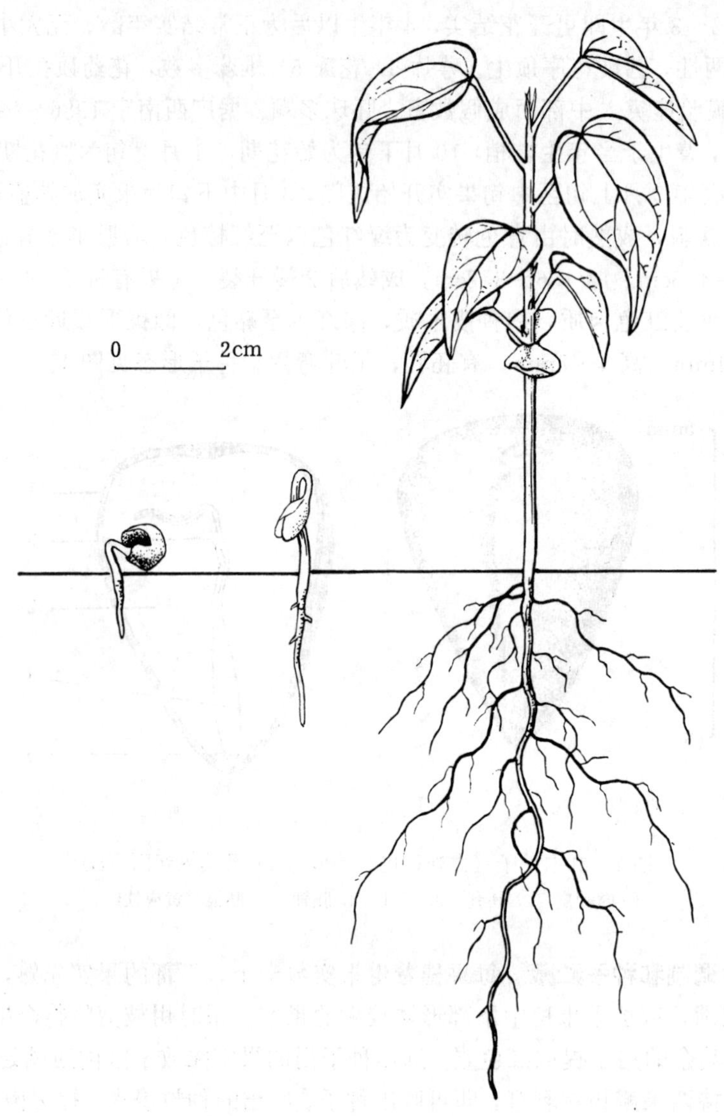

图 2　红木种子萌发后第 4、6、25 天幼苗的生长情况
（黄应钦仿《热带亚热带主要树种采种育苗技术》）

（张声燕）

半　日　花

***Helianthemum polifolium* DC.**

（半日花科　Cistaceae）

生长习性、分布和用途　半日花属约 100 种，我国仅有本文描述的 1 种，产于内蒙古（乌海市、鄂托克旗、阿拉善左旗）、甘肃（永昌北山）和新疆（准噶尔盆地、伊宁）。哈萨克斯坦也有分布。矮小灌木，高 5～15cm，多分枝，稍呈垫状。强旱生植物，生于草原化荒漠区的石质和砾石质山坡。李玉俊和朱勇（1991）测定过这个种的抗旱生理指标。在《中国植物红皮书》中列为稀有种。地上部分含红色物质，可作红色染料。

开花结实　3～4 年生开始结实，5～6 年生便能正常结实。花两性，辐射对称。花单生枝顶，径 1.4cm。萼片 5，黄绿色，不等大，外面的 2 个较小，条形；内面的 3 个卵形，背面有 3 条纵肋。花瓣 5，黄色，倒卵形。雄蕊多数。子房上位，1 室或不完全的 3 室，花柱丝状，柱头头状。据内蒙古伊克昭盟鄂托克旗 1990 年的物候观察，开花始期为 7 月，盛期在 8 月上旬，末期在 9 月上中旬；种子散落期在 9 月上中旬。蒴果，卵圆形，黄色，3 瓣裂，长约 5～7mm，径约 5mm，被短柔毛，室间开裂，具几粒或多粒种子。种子阔卵形，先端较尖，粒小，褐色，长约 2～3.5mm，径约 1.5～2.5mm。种子有胚乳，较薄一层紧贴于种皮与子叶之间。种子形态如图 1。

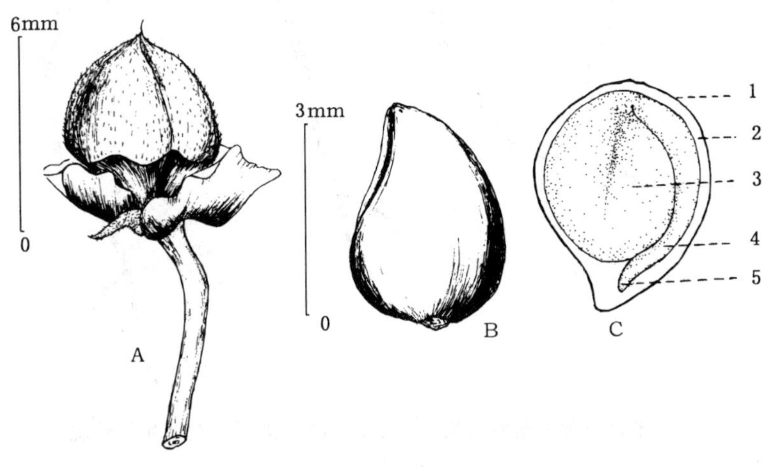

图 1　半日花果实（A）、种子外形（B）及其纵切面（C）

1. 种皮　2. 胚乳　3. 子叶　4. 胚轴　5. 胚根（姚桂英绘）

果实的采集调制和种子贮藏　半日花的结实率很低，种子很难采集。果实成熟后种子很快脱落，难以从植株上采种，常常只能从地面收集。收集的果实放在室内易于观察的地方取种。种粒小，瘪粒多而饱满粒少，蒴果又是室间开裂，所以取种时只能轻轻敲击或用镊子取出，否则极易散失。收集到的果实夹杂物多，净度很低，但取得的种子净度较高，可达85%～95%。100 粒种子重 0.18g，据此推算的千粒重为 1.8g，每克纯净种子约 500 粒。半日花资源稀少，种子又容易丧失发芽力，采得的种子最好密封于 1～5℃的条件下干藏。

发芽和播种　半日花的种子小，种皮薄，无休眠习性，极易发芽。1990 年 9 月采集到的种子于当年 9 月下旬～10 月进行发芽测定（由于种子少，选取较饱满种子 50 粒分 5 组，每组 10 粒）。发芽基质为脱脂棉上放滤纸，发芽床用钟罩式发芽器，置于室温 18～22℃的条件下发芽。发芽前第 1 组用冷水浸泡 24 小时，其余 4 组用始温 45℃的水浸泡 24 小时。发芽天数 12～21 天，其平均发芽率为 78%，变动范围 50%～100%；2～5 天的平均发芽百分数为 30%，变动范围为 20%～40%。从发芽结果来看，浸种的水温对种子发芽关系不大，关键在于发芽床要经常保持湿润，且室温不能太低。

播种要用沙质土壤，撒播。播前将苗床灌足底水，保持土壤湿润，播后覆极薄一层细沙。出土萌发。播后 3～4 天子叶即可出土，幼苗形态如图 2。

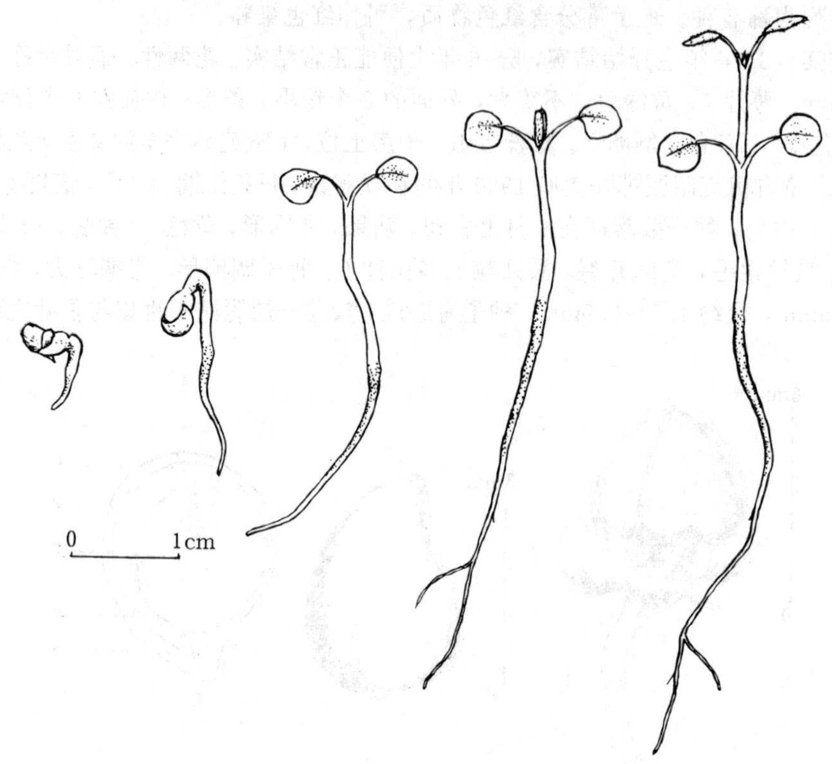

图 2　半日花种子萌发后第 2、3、5、14、26 天幼苗的生长情况

（姚桂英绘）

（赵美华）

大叶刺篱木（卢甘果）

Flacourtia rukam Zoll. et Mor.

（大风子科　Flacourtiaceae）

生长习性、分布和用途　刺篱木属约 15 种，我国 5 种，本文描述 1 种。半落叶乔木，高 15m。树干基部具长刺。耐旱。土壤以沙质壤土为好。深根性，抗风性能强。分布于滇、桂、粤、琼。越南、印度也有。果肉酸甜多汁，可作果酱、蜜饯，营养丰富。木材坚硬耐腐，可作小型农具。

开花结实　8～10 年生开始开花结实，正常结实期在 15 年生以后，大小年现象不明显。花单性同株。总状花序腋生。花无花瓣。花盘肉质，圆形，由与花萼对生的腺体组成。萼片 4～5 枚，卵形，基部稍连合。雄花：雄蕊约 25 枚，花丝丝状，插生在肉质的花盘内。雌花：花盘肉质 8 裂；子房上位，圆球形，不完全 6 室，每个侧膜胎座上有 2 胚珠；花柱 6，分离，柱头 2 浅裂。据海南儋县观察，花期 7～8 月，果实成熟期在 9 月下旬至 10 月上旬。浆果，圆形至倒卵形，直径约 1.4cm，顶端有宿存花柱。每果有种子约 12 粒。种子卵状微扁或形态多样，长 5～7mm，宽 3～5mm。种皮软骨质。胚乳肉质，子叶宽广，形态见图 1。

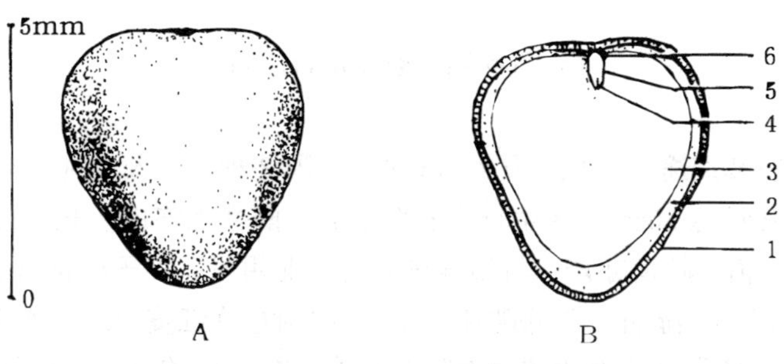

图 1　大叶刺篱木种子外形（A）及其纵切面（B）

1. 种皮　2. 胚乳　3. 子叶　4. 胚芽　5. 胚轴　6. 胚根

（黄光郁绘）

果实的采收调制和种子贮藏　果实由青绿色变为暗红色时，即可用高枝剪或长柄钩刀割取果穗。作果酱用的宜在未成熟时采集；供播种用的必须充分成熟，在果实转变为黑红色时逐个采摘，或用竹竿敲打落地后收捡。采得的果实放置 1～2 天，充分软化后装入布袋，置水中搓洗，淘去果肉等杂质，稍加晾干即为纯净种子。鲜果的出种率约 4.7%，种子的净度可达 98%。种子的千粒重在 80g 左右，每千克有纯净种子 12 000（10 000～14 000）粒。调制后稍晾干的种子含水量约为 14%。种子忌失水，不宜曝晒。运输时应以果实的状态装运，运抵后再行调制。贮藏宜混以细沙，贮藏期在 5 个月以内较宜。

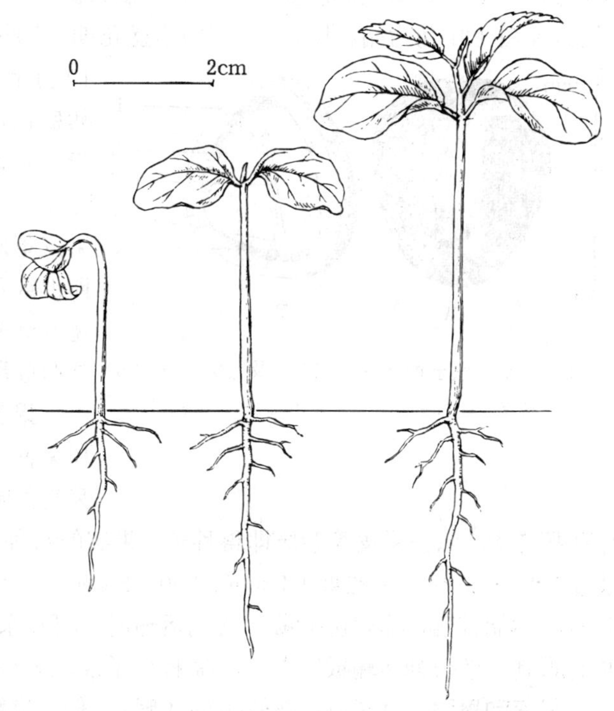

图 2　大叶刺篱木种子萌发后第 3、13、38 天幼苗的生长情况

（黄光郁绘）

发芽和播种　种子有休眠。1988 年 9 月 27 日，华南热带作物研究所在室外沙床用新采种子作发芽测定，播后历时 5 个月，至翌年 2 月 18 日才开始发芽，4 月 14 日发芽结束，没有明显的发芽盛期。发芽率为 40%。出土萌发。子叶出土后 32 天展出初生叶。幼苗初期的生长情况见图 2。

条播。每平方米播种 12～20g，覆土 3～5mm。1 年生苗可以出圃。

（丁慎言）

海南大风子（龙角）

Hydnocarpus hainanensis（Merr.）Sleum.

（大风子科 Flacourtiaceae）

生长习性、分布和用途 大风子属约45种，我国3种，引入2种，本文描述1种。常绿乔木，高达15m，胸径达50cm。多生长在石灰质土，酸性土亦能生长。忌霜冻及0℃以下低温。分布于琼、桂、滇。越南亦产。材质坚硬，为工业用材，种子油富含大风子酸和晁横酸，可治麻疯病、牛皮癣、风湿病。资源破坏严重，《中国植物红皮书》中列为渐危种。

开花结实 10～13年生开始开花结实，正常结实期在20年生以后。结实大小年间隔不明显，每年结实均较少。花单性，雌雄异株。总状花序腋生。雄花序具15～20花，萼片及花瓣各4，雄蕊10～12，花丝基部粗壮。雌花被与雄花相似，略大，退化雄蕊15。子房上位，卵状椭圆形，1室，有侧膜胎座5个，胚珠多数。花柱缺，柱头3。据广西南宁1987年的物候观测，4月上中旬为始花期，4月下旬为盛花期，5月中旬为末花期，9月下旬果实开始成熟，10月上旬为成熟盛期。另据海南资料，在海南夏初盛花，8～9月果实成熟。浆果圆球形，先端尖，未熟时青色至褐色，熟后灰褐色，被黑褐色绒毛，径4～7.5cm。10月下旬～11月中旬果实脱落。果皮革质。每果有种子15～30粒。种子椭圆形、卵形或不规则形，表面有条纹，长1.4～2.1cm，宽0.9～1.5cm，厚0.7～1.3cm。种壳坚硬。具油质胚乳，子叶扁平。种子形态见图1。

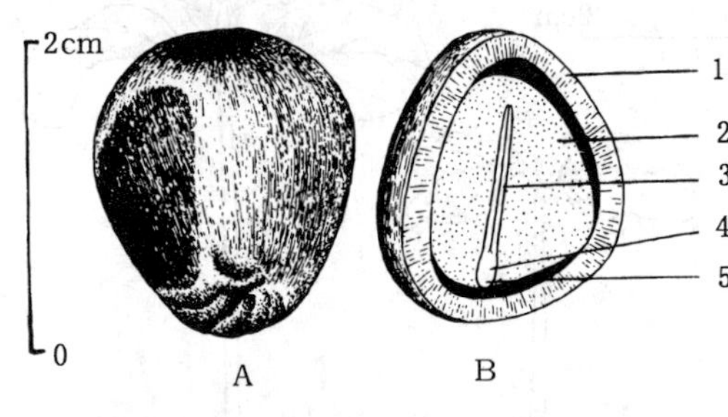

图1 海南大风子种子外形（A）及其纵切面（B）
1.种皮 2.胚乳 3.子叶 4.胚轴 5.胚根
（黄应钦绘）

果实的采收调制和种子贮藏 果熟盛期上树采摘，或用采种刀将果实钩落后在地面捡拾。采得的果实堆沤2～3天，软熟后装入筐内，置水中搓擦漂淘，除去果皮等杂质即得种子。果实的出种率约为17%。种子的净度可达98%，含水量30%～40%。千粒重1 550（1 300～2 000）g，每千克有纯净种子500～770粒。种子忌失水，不能日晒，需保湿贮藏。短期运输时，可在采集后以果实的状态装运，到达目的地后再行调制。较长期运输或贮藏，均需将种子混以湿沙，贮藏期一般为6～8个月。

发芽和播种 种子有中等程度的休眠，一般不宜随采随播。发芽时日均温需在20℃以上。1987年9月26日，广西林业科学研究所在室外沙床上用新采种子作过发芽测定：播前种子未作任何处理，播后1个月内日均温为20～27℃，但不见发芽，迟至1988年4月26日（日均温26℃）方始发芽，发芽盛期不明显，5月11日发芽终止。发芽率为50%。出土萌发。胚根萌发后17天子叶出土，再过9天左右初生叶展出。种子萌发及幼苗初期生长情况见图2。

种子经层积催芽后播种。条播或点播。每平方米播种150～200g，覆土1.5～2.0cm。幼苗期需搭棚遮荫。培育1～2年生苗出圃。

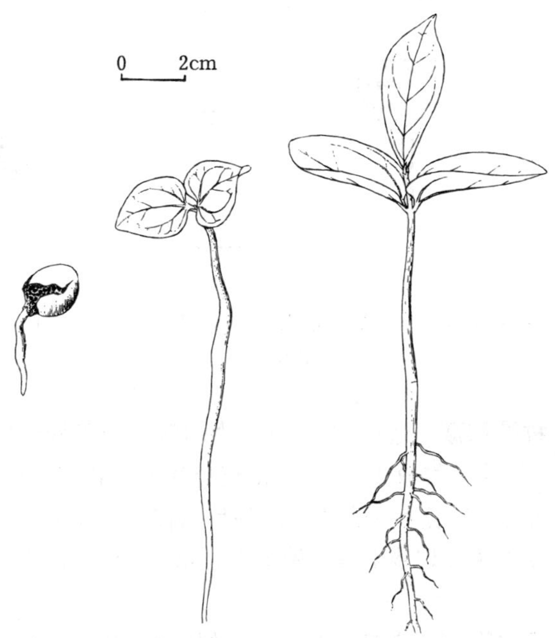

图 2　海南大风子种子萌发后第 18、26、40 天幼苗的生长情况

（黄应钦绘）

（王宏志）

山　桐　子

Idesia polycarpa Maxim.

（大风子科　Flacourtiaceae）

生长习性、分布和用途　山桐子属有 2 种 1 变种，我国全产，本文描述 1 种。落叶乔木，高 15m。喜温暖湿润气候和肥沃疏松、富含腐殖质且排水良好的酸性黄壤，在土层较厚的石灰岩山地的疏林中和林缘也能生长。分布于晋、豫、鄂、皖、浙、台、闽、赣、湘、粤、桂、黔、滇、川。朝鲜半岛、日本也有分布。木材黄褐色或黄白色，纹理直，结构细，质轻软，可作箱板、火柴杆、牙签，也可造纸。茎皮纤维可作绳索和编织麻袋。果肉含油 4.4%，种子含油 23.4%，为半干性油，可代桐油，亦可制肥皂或作润滑油。树形美观，雌株果穗大而下垂，秋天红果累累，鲜艳夺目，宜作行道树或供园林配植。

开花结实　10 年生开始开花结实，正常开花结实在 15 年生以后。花单性，雌雄异株。花淡黄色，无花瓣，成大型顶生圆锥花序。萼片 5（3～6），被短柔毛。雄花：雄蕊多数着生于小花盘上，花丝丝状，花药椭圆形，纵裂，具一小退化雌蕊。雌花：具一近圆形子房，基部围以多数短小退化雄蕊；子房上位，侧膜胎座 5（3～6）；胚珠多数；花柱常 5，开展，柱头倒卵形。花期 5～6 月，果熟期 9～10 月。果穗下垂，长 10～25cm。浆果圆球形，径 6～9mm，红色或橙黄色。种子多数（每果有 10～50 粒），细小，椭圆形或近卵形，长 1.5～2mm，径

1～1.5mm，两端渐尖，黑褐色。种皮坚硬壳质。胚乳丰富，肉质。胚直生，子叶圆叶状(图1)。

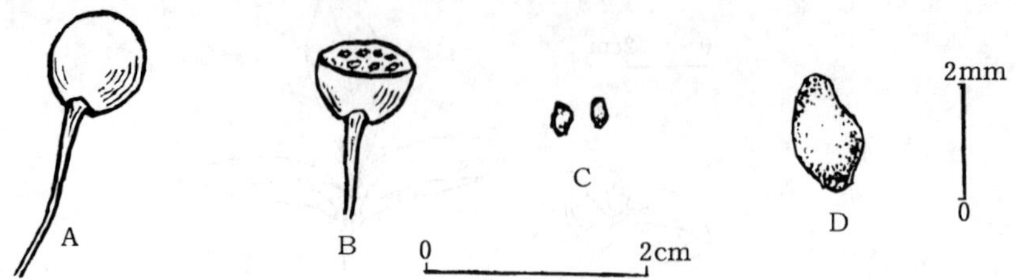

图1 山桐子浆果外形（A）及其横切面（B）和种子外形（C）及其放大（D）
（田恒德绘）

果实的采收调制和种子贮藏 选取15～30年生的雌株为采种母树。果实变为深红色时即可采收。剪下果穗，捋下浆果，堆放1～2天，待充分软熟后置水中搓洗，淘去果皮果肉等杂质。淘洗净种时用细孔容器，以免种子漏失。净种后再浸入草木灰水中1～2小时，擦去种皮外的蜡质，晾干，混湿沙贮藏或袋装干藏。种子千粒重2～3g，每千克有纯净种子34万～50万粒。

发芽和播种 笔者用采自福建来舟林业试验场树木标本园的种子做过发芽试验，发芽率15%～31%。育苗时宜用条播，条距约30cm，播种沟深3cm，播前用始温30～40℃水浸种24小时，拌细土条播。覆土厚1.5～2cm，上盖稻草。每平方米约播2～4g。1年生苗高40～60cm，可以出圃。

出土萌发。子叶2，矩圆形，初生叶对生，第3叶以上互生。种子萌发和幼苗初期生长情况见图2。

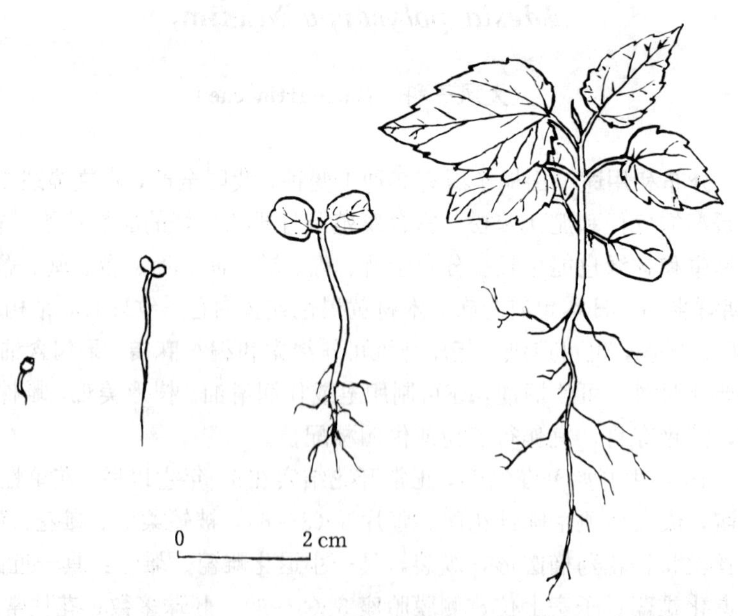

图2 山桐子种子萌发后第1、3、10、45天幼苗的生长情况
（黄鹏成供稿，田恒德临）

（黄鹏成）

栀子皮（伊桐）

Itoa orientalis Hemsl.

（大风子科　Flacourtiaceae）

生长习性、分布和用途　栀子皮属 2 种，我国只有本文描述的 1 种。常绿乔木，高 20m，胸径 30～40cm。喜肥沃湿润土壤。分布于滇、川、黔、桂、琼。越南亦产。材质较坚实，用于家具、建筑。

开花结实　约 10 年生开始开花结实，正常结实期在 20 年生以后。结实大小年间隔期一般为 1 年。花单性，雌雄异株。雄花成顶生直立的圆锥花序，雌花通常单独顶生。萼 3（4）深裂，裂片三角状，无花瓣。雄蕊多数。子房上位，生于侧膜胎座上，胚珠多数。据广西南宁 1987～1989 年的观测，5 月上旬为始花期，5 月中旬为盛花期，6 月上旬为末花期；10 月下旬果实开始成熟，11 月上中旬为成熟盛期，11 月中旬～12 月中旬果实开裂，种子飘落。蒴果，卵形至纺锤形，未熟时绿色，熟后黑褐色，长 4～6（9）cm，径 2～4（6）cm，多为 6 瓣裂，少有 4 裂，各裂瓣沿胎座自基向上至中部又 2 裂。外果皮革质，被橙黄色毡状毛。内果皮木质。每果有种子 600～800 粒。种子扁，四周带不规则形膜质翅，三角形至梯形，长 1.5～2cm，种子外向一侧下凹。种皮革质。具胚乳。子叶小，胚根位于 1 侧（图 1）。

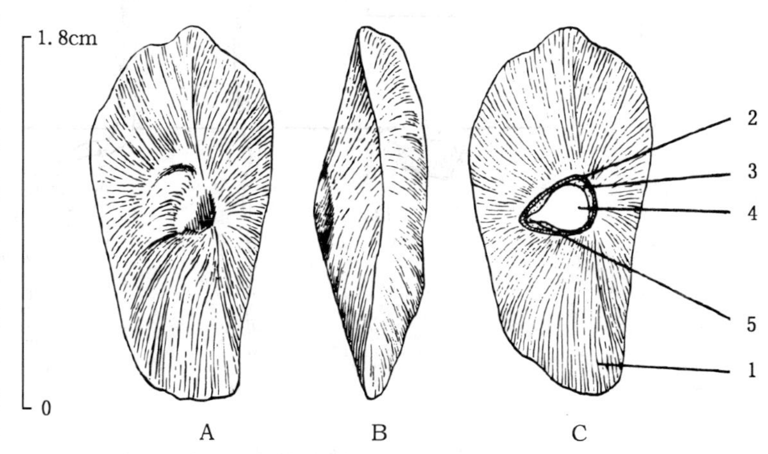

图 1　栀子皮种子外形正面（A）、侧面（B）及其纵切面（C）
1. 种翅　2. 种皮　3. 胚乳　4. 子叶　5. 胚根
（黄应钦绘）

果实的采收调制和种子贮藏　果实由绿色转呈灰褐色时用枝剪截断果梗，采得的果实堆放于室内，下垫塑料膜。果实大部分开裂后敲打翻动，清除果壳等杂质即得种子。鲜果出种率约为 20%。种子的净度为 60%～80%。每千粒带翅种子一般为 9.5g，每千克带翅种子 10 万～11 万粒。种子可以晾干，但忌烈日曝晒。可用瓶或瓦坛贮藏，置室内阴凉、干燥、避光处，贮藏期在 4 个月以内。

发芽和播种　种子无休眠习性，发芽时日均温宜在 18℃以上。1991 年 1 月 19 日，广西林业科学研究所用头年秋末所采种子，在发芽箱内作过发芽测定：基质为滤纸，箱内温度保持 28℃，置床 5 天（1 月 23 日）开始发芽，28 日为发芽盛期，31 日发芽终止；从开始发芽至发芽高峰，6 天的发芽百分数为 25%；从播种至发芽终了，13 天的发芽率为 30%。出土萌发。播种后 4 天胚根萌出，11 天子叶出土，23 天左右第 1 对初生叶展现。种子的萌发和幼苗初期生长情况见图 2。

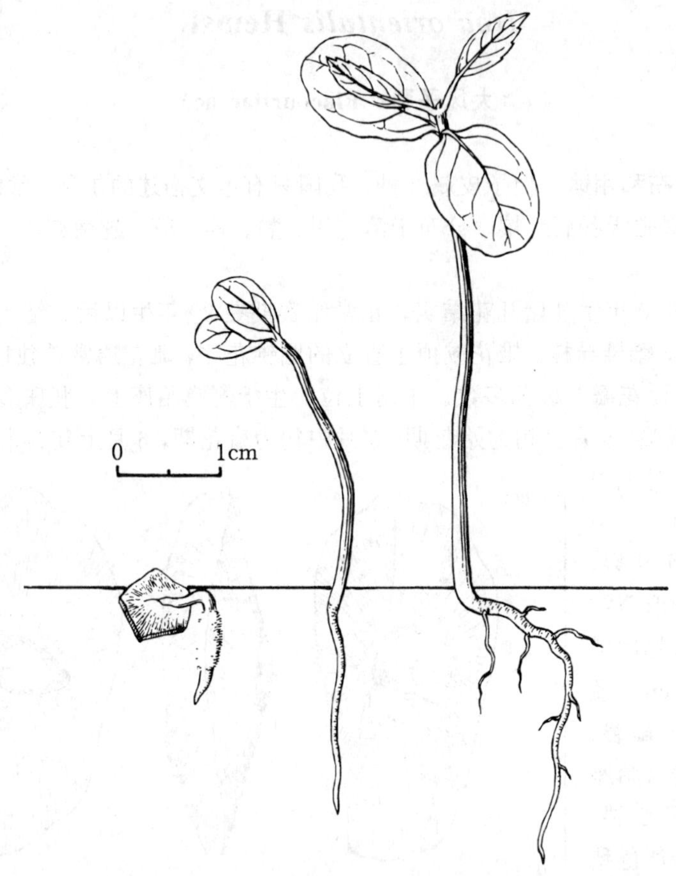

图 2　栀子皮种子萌发后第 4、12、25 天幼苗的生长情况

（黄应钦绘）

撒播。苗床经细致整地后，淋足水分，播下种子。每平方米播种 3～5g。幼苗木质化时移至容器继续培育 2 个月，或移至大田培育 1 年生苗出圃。

（林榕庚）

山拐枣

Poliothyrsis sinensis Oliv.

（大风子科　Flacourtiaceae）

生长习性、分布和用途　山拐枣属只有本文描述的 1 种，为我国特产。落叶乔木，高达 15m，分布于我国西南部至东部黔、滇、川、陕、豫、鄂、湘、粤、苏、浙等地石灰质的山坡杂木林中。除木材可以利用外还有固土效用。花具蜜腺，为蜜源树种。

开花结实　花单性，雌雄同株。圆锥花序顶生，直立疏松，长10～20cm。花淡绿色，渐变黄色。雌花多生于分枝顶端。萼片5，无花瓣。雄花具离生雄蕊20～25枚。雌花具退化雄蕊多数。子房上位，1室，侧膜胎座3，胚珠多数。花柱3，柱头2裂。1978年在南京地区观察，6月上旬～7月上旬开花，9～11月果实成熟。蒴果椭圆形，长约2cm，顶端急尖，被灰黄色绒毛，成熟时3(4)瓣裂。外果皮革质，具毡状毛。内果皮木质。种子多数，周围具翅，带翅长约1cm。去翅种子长卵形，长约5mm，径约2mm。种皮膜质，胚乳丰富，白色，富含淀粉。胚长约4mm，子叶顶端深凹成2裂形，在种子内摺成椭圆形。果实及种子剖面结构见图1。

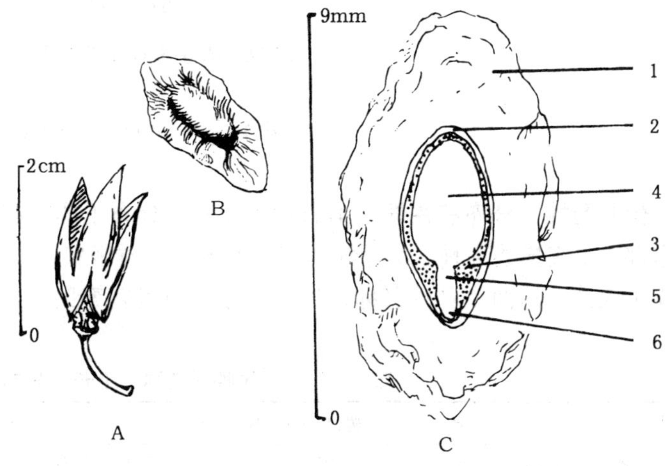

图1　山拐枣的蒴果外形（A）和种子外形（B）及其纵切面（C）
1. 种翅　2. 种皮　3. 胚乳　4. 子叶　5. 胚轴　6. 胚根
（史渭清绘）

果实的采收调制和种子贮藏　当蒴果呈灰棕色，顶端即将开裂或有小部分开裂时，整穗采下晾晒，蒴果充分开裂后，敲击脱粒，扬除瘪粒。干藏或湿沙层积供翌春播种。带翅种子千粒重约4.4g。

发芽和播种　种子有短期休眠习性。目前缺乏大量育苗的经验。1988年2月中旬江苏省植物研究所将上一年秋季采收干藏的种子在室外自然变温条件下层积，4月中旬～6月下旬陆续发芽，5天中发芽率为27.5%。在冰箱中连续低温（5～10℃）条件下层积的同一批种子，发芽过程35天，开始萌发的时间稍迟，但6月上旬已结束发芽，36天内发芽率为52.5%。出土萌发。初生叶对生（图2）。

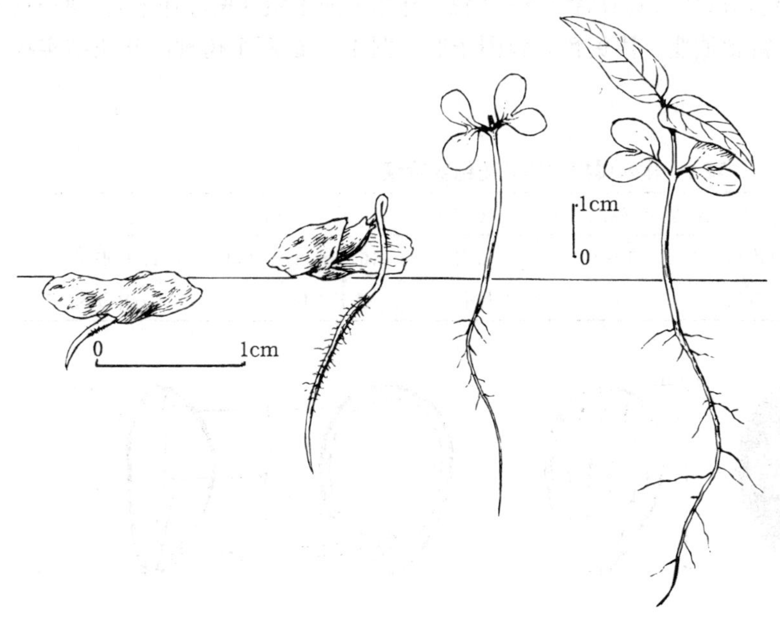

图2　山拐枣种子萌发后第1、2、7、20天的幼苗生长情况
（史渭清绘）

（何泽瑛）

柞木属
Xylosma G. Forst.

（大风子科 Flacourtiaceae）

生长习性、分布和用途 本属约100种，分布于热带和亚热带，我国有4种，产西南部至东部。本文描述2种。常绿灌木或小乔木，材质坚实，枝干有刺，可种植为绿篱。柞木叶、树皮可入药。这两个种的名称、生长、分布和用途等见表1。

表1 柞木属树种的名称、生长、分布和用途

中 名	学 名	树高(m)	胸径(cm)	分 布	用 途	供 稿
柞 木	*X. congestum* (Lour.) Merr.	25	80	华东、中南、西南及陕。日本、朝鲜半岛	材用、高篱、药用	1001
长叶柞木	*X. longifolium* Clos	7	15	滇、黔、粤、桂、琼、闽。越南、印度	材用、高篱	612

开花结实 花单性，雌雄异株。总状花序腋生，长1～2（5）cm，有细柔毛，具短梗。萼片4～6。无花瓣。雄花具多数雄蕊，花盘由较多腺体组成，位于雄蕊外围。雌花花盘圆盘状，边缘稍呈浅波状。子房上位，1室，具2侧膜胎座，每胎座有2至数枚胚珠。花柱短，柱头2～3裂。浆果球形，成熟时黑色，每个果实有种子2～3粒（柞木）或多数（长叶柞木）。种子具两片白色子叶，胚乳丰富，富含油脂，浅黄色，包围子叶（图1）。这两个树种的开花结实物候列入表2。

表2 柞木属树种的开花结实物候

树 种	开始结实年龄	正常结实年龄	结实间隔期（年）	花 期	果实成熟期	果实脱落期
柞 木	10～15	30以后	1～2	5月	10月～11月上旬	11月中旬
长叶柞木	6	15以后	0	11月中下旬	翌年3月中下旬	4月中下旬

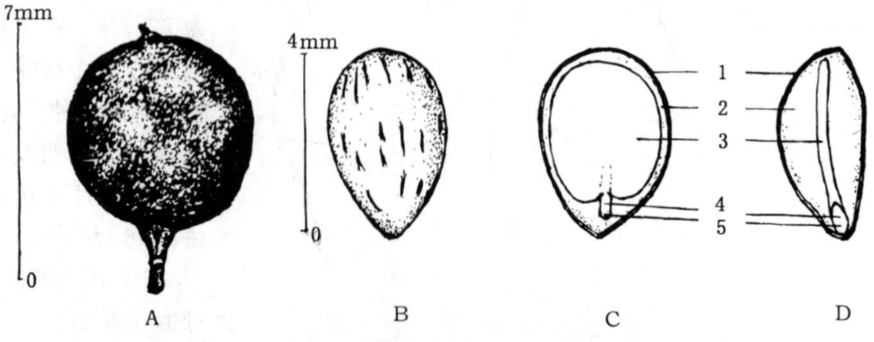

图1 柞木浆果的外形（A）、种子外形（B）及其纵切正面（C）和纵切侧面（D）

1. 种皮 2. 胚乳 3. 子叶 4. 胚轴 5. 胚根（田恒德绘）

果实采收调制和贮藏　浆果呈红黑色，果皮由脆硬变软即可采集果穗或直接摘取果实。果实采回后堆沤，淘洗，漂去杂质即得纯净种子。种子不宜曝晒，阴干后干藏。柞木和长叶柞木的种子性状见表 3。

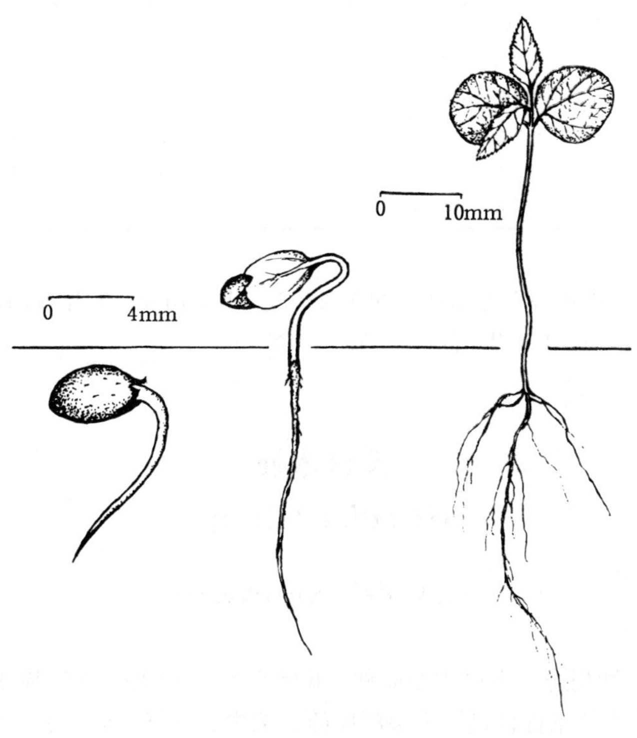

图 2　柞木种子萌发后第 2、6、23 天的幼苗生长情况
（田恒德绘）

表 3　柞木和长叶柞木的种子性状

树　种	果　径 (mm)	出种率 (%)	每果含种子数（粒）	种子的形状和颜色	种子的大小（mm）			净　度 (%)	千粒重 (g)	每千克种子数（万粒）
					长	宽	厚			
柞　木	3～5	50	2～3	倒卵形。浅棕色，表面具深褐色条纹	3～4	2～3	1～2	85～95	8.5～9	11～11.7
长叶柞木	4～6	25～35	3～6	椭圆形，具棱。深褐色	4～4.5	3.8～4.2	2.2～2.6	85～95	24～30	3.3～4.2

柞木种皮薄，结构致密，富有弹性。成熟的柞木果实含水量 25%～27%，气干种子含水量 8%～9%，萌动期种子含水量 40%～45%。

发芽和播种　种子无休眠习性，新鲜种粒在适宜条件下发芽整齐。据林业部南方林木种子检验中心的一次试验，柞木种子在 20℃、25℃的恒温或 20℃/30℃的变温条件下发芽，发芽率无显著差异，但在 35℃下发芽不良（表 4）。表 4 也列入了 1989 年广西林业研究所在南宁对长叶柞木所作的一次发芽测定结果。这两个树种也可用靛蓝或四唑溶液测定生活力。

表 4 柞木属树种种子的发芽能力及其测定条件

树 种	地点时间	测定温度（℃）	发芽势		发芽率	
			天 数	（%）	天 数	（%）
柞 木	江苏南京 1989	20，恒温	11	27	29	93
		25，恒温	12	40	30	94
		35，恒温	21	20	38	40
		20/30，变温	14	61	26	94
长叶柞木	广西南宁 1989	室外沙床，日均温 18 以上	7	80	11	84

春季播种。条播。柞木每平方米床面可播 2～3g。出土萌发。子叶椭圆形，幼苗健壮，20～35 天展现初生叶（图 2）。1 年生苗可以出圃。

（吴琼美）

天料木属

Homalium Jacq.

（天料木科 Samydaceae）

生长习性、分布和用途 本属约 200 种，我国有 10～12 种，本文描述 2 种。常绿或半落叶乔木，喜温暖，在水热条件优越、土层深厚的立地生长良好。适生于砖红壤和赤红壤地区。喜与多种树混交生长，纯林生长较差。材质优良，为热带珍贵用材树种。这 2 个树种的名称、生长、分布及用途见表 1。

表 1 天料木属树种生长情况、分布和用途

中 名	学 名	树高(m)	胸径(cm)	分 布	用 途	供 稿
光叶天料木①	*H. ceylanicum* (Gardn.) Benth.	30	80	藏、滇。斯里兰卡、印度、泰国、老挝、越南	材用	618
红花天料木（母生）	*H. hainanense* Gagnep.	40	100	琼、粤、桂。越南。滇、闽栽培	材用	601

① 光叶天料木原用学名 *H. laoticum* Gagnep. var. *glabratum* C. Y. Wu。在《中国植物红皮书》中列为渐危种

开花结实 约 10 年生开始开花结实，15 年生以后进入正常结实期。有大小年之分，丰年应多采收，贮藏备用。花两性，排成腋生或顶生的总状花序或圆锥花序，有数朵簇生在花序轴上，极少单生。花梗在中部上下有关节。萼管陀螺形，与子房的基部合生，裂片（4）5～8（12），宿存。花瓣与萼片同数，着生在花萼喉部，宿存。雄蕊与花瓣同数或多数，每一雄蕊或每一成束雄蕊与一花瓣对生，且有一腺体与每一萼片对生。子房 1 室，半下位。侧膜胎座 3～5（2～6），花柱与胎座同数：光叶天料木 3～4，红花天料木 5～6，仅生于子房分离部的侧壁上。每一胎座有胚珠 3 至多颗。据在广西南宁和云南西双版纳勐仑观察，红花天料木

1 年开花结实 2 次。这 2 个树种的开花结实物候期见表 2。

表 2　天料木属树种开花结实的物候期

树　种	观察地点	观察年份	开　花			果实成熟		果实脱落
			始　期	盛　期	末　期	始　期	盛　期	
光叶天料木	云南勐仑	1984～1987	3 月下旬	4 月	5 月中下旬	5 月中下旬	6 月上中旬	—
红花天料木	广西南宁	1978～1985	5 月中下旬	5 月下旬末	6 月中旬	8 月上旬	8 月上中旬	8 月中旬～8 月下旬末
			9 月上旬	9 月中下旬	10 月中旬	12 月中旬	翌年 1 月中旬	1 月中下旬
	海南	—	4 月	～	5 月	7 月	8 月上旬	—
			9 月	～	10 月	12 月下旬	翌年 1 月	—

蒴果，中部为宿存的萼片和花瓣所包围，纺锤形，顶端 4～6 瓣裂。红花天料木每果有种子 3～5 粒，光叶天料木 1 粒。种子有棱，种皮硬而脆。胚乳丰富。胚直，子叶扁平，叶状。果实及种子形态见表 3、图 1。

表 3　天料木属树种果实和种子的形态特征

树　种	果　实			种　子		
	形　状	大小（mm）	颜　色	形　状	大小（mm）	颜　色
光叶天料木	纺锤形	长 2～2.4 径 1.5～2	褐色	棱状长椭圆形	长 1.5	白色
红花天料木	纺锤形	长 2～2.5 径 1.5～2	灰褐色或褐色	棱状长椭圆形	极小，长约 1	白色

果实的采收调制和种子贮藏　红花天料木的种子品质以 8 月份成熟的为佳，但成熟期仍有差异，宜分批采集呈灰褐色或褐色的果实。可连果穗一起采下，晒干，待开裂后脱出种子。光叶天料木可于末花期在树下铺设塑料薄膜，收集凋落的花和果穗。由于萼片、花瓣和柱头宿存，种子也不脱落，因此可在晾干后除去果穗轴和其他杂质，其余部分即可作为播种材料，通称种子。种子含油脂，不宜曝晒，含水量约 10%～13%，宜随采随播。贮藏宜用瓦罐密封，内加相当于种子量 10%的氯化钙，贮藏期不宜超过半年。这 2 个树种果实的出种率和种子的净度、质量等数据见表 4。

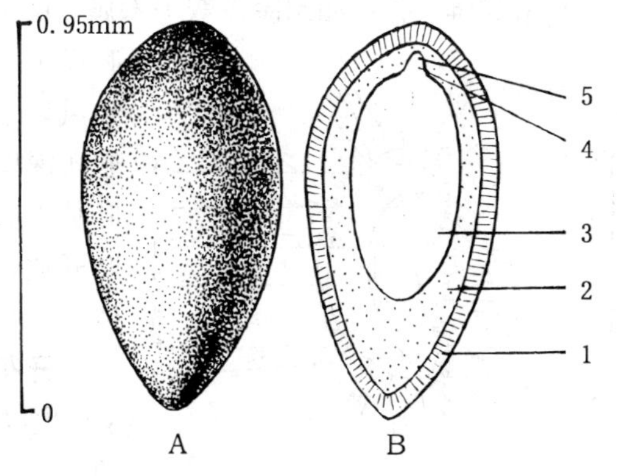

图 1　红花天料木种子外形（A）及其纵切面（B）
1. 种皮　2. 胚乳　3. 子叶　4. 胚轴　5. 胚根（黄应钦绘）

表 4　天料木属树种果实的出种率和种子净度、质量

树　种	出种率（%）	净　度（%）	千粒重（g）	每克纯净种子粒数（粒）
光叶天料木	1	40～60	0.7～0.8	1 300～1 400
红花天料木（广西）	1	40～60	0.6～0.8	1 200～1 700
（海南）	25	22	1.0	1 000

发芽和播种　种子无明显休眠现象。发芽时的日均温宜在 20℃以上。据广西经验，红花天料木种子容易丧失发芽能力，以在采种后 1 个月内播种为宜。此时发芽率约 30%，播后约 10 天发芽，20～30 天发芽完毕。中国林业科学院热带林业研究所对红花天料木种子测定前先浸种 24 小时，再用湿布包裹 4 天，室内置床后 5 天开始发芽，27 天时发芽率为 36%～46%；室外播种后 6 天开始发芽，18 天发芽率为 28%～36%（陈荷美，1978）。云南热带植物园在室内用培养皿对光叶天料木进行过发芽测定，基质为滤纸，置床后 7 天开始萌发，自发芽之日起算，13 天的发芽率为 22%～25%，最高为 30%。出土萌发。子叶出土后 8～10 天展出第一片初生叶。

撒播。红花天料木每平方米约播 10g，光叶天料木约播 12g。播后覆沙，遮荫。2～3 个月后苗高 12cm 时便可分床，再经 5～8 个月苗高 80～100cm，即可出圃造林。亦可扦插繁殖。

（王宏志）

弯　子　木

Cochlospermum vitifolium Spreng.

（弯子木科　Cochlospermaceae）

生长习性、分布和用途　弯子木属，我国只有本文描述的 1 种。落叶乔木，高 10m 以上，胸径 15～30cm。不耐寒，嫩叶易受冻枯黄。根系浅，不抗强风，在 9～10 级强风下，植株便会歪斜或倒伏。适生于肥沃深厚的沙壤土。产于热带亚洲。我国云南勐海常绿阔叶林中有分布，海南儋县、保亭及云南景洪有栽培。花先叶开放，颇美丽，观赏价值高。种子含丝状毛，可作填充体。

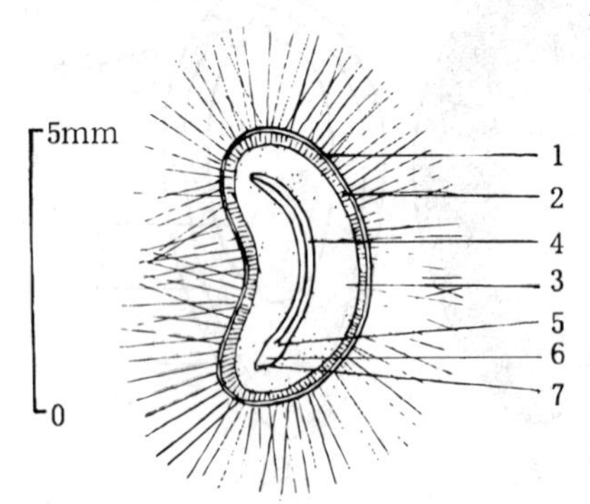

图 1　弯子木种子的纵切面
1. 外种皮　2. 内种皮　3. 胚乳　4. 子叶　5. 胚芽　6. 胚轴　7. 胚根
（黄应钦、黄光郁绘）

开花结实　3～4 年生开始开花结实，正常结实期在 8 年生以后，结实大小年现象不显著。花两性。圆锥花序顶生。花冠黄色，直径约 10cm。萼片 5，肉质，长 2cm，宽 1.2cm。花瓣 5（3 大、2 小），矩圆形，长 4.5cm，宽 2.2cm，中部凹陷。雄蕊多数。花药 2 室，条形，自顶端短的孔状裂缝开裂。子房上位，1 室，侧膜胎座，胚珠多数。花柱短，白色，长约 2～3mm，稍弯曲。花柄长 2～

3cm。据海南儋县的物候观察，2月中旬为始花期，2月下旬～3月中旬为盛花期，3月下旬为末花期；8月中旬果实开始成熟，8月下旬为成熟盛期。蒴果椭圆形。种子具毛，近肾形弯曲。胚乳丰富。胚镰刀状弯曲。种子长约5mm，径约3mm（图1）。

果实的采收调制和种子贮藏　8月中下旬，果皮由绿色转为褐黑色时应及时采收，否则果实开裂，种子会飘飞散失。通常是在树冠周围地面铺上塑料薄膜，用采种钩刀割断总果梗，果穗落下。收捡回的果实室内摊放几天，自然开裂后捡去果壳，从短纤维中选出种子。鲜果的出种率约33%。种子的净度一般为93%。千粒重为（30）33（35）g，每千克有纯净种子28 000～33 000粒。种子阴干后含水量降至接近10%时，放入塑料袋密封贮藏或运输，但贮藏期不能超过30天。

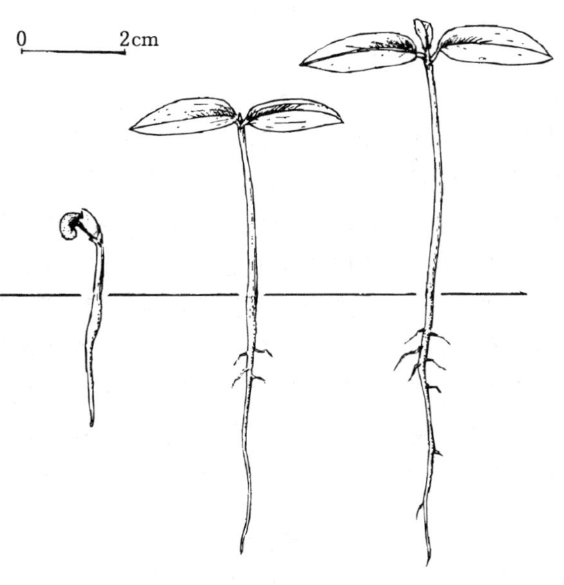

图2　弯子木种子萌发后第4、10、19天的幼苗生长情况
（黄光郁绘）

发芽和播种　种子无休眠习性。发芽时日均温宜在25℃以上。1987年8月20日，华南热带作物研究所用新采种子播入沙床作发芽测定，播后4天开始发芽，9月2日发芽终止，发芽率仅有13%。发芽率低可能是由于母树单生，自花授粉所致。出土萌发。种子萌发后3天子叶带壳出土，出土后5～6天子叶展开，再过10天左右初生叶展出。弯子木种子的萌发和幼苗初期生长情况见图2。

目前尚无大田育苗经验。种粒细小，发芽率又低，通常将种子密播于沙盘内，待初生叶发出，苗木半木质化时，移至容器继续培育。半年生苗出圃。用于庭院绿化的则需培育1年半生苗出圃。

（丁慎言）

土沉香（白木香）

Aquilaria sinensis（Lour.）Gilg

（沉香科　Aquilariaceae）

生长习性、分布和用途　沉香属15种，我国2种，本文描述1种。常绿乔木，高达25m，胸径达90cm。产琼、粤、桂、闽、台。越南、泰国等亦产。木材白色轻软，用作绝缘或漂浮材料。经甲酸刺激后，木材凝聚树脂，形成沉香，为名贵中药，品质与进口的*A. agallocha* Roxb. 所形成的沉香相同。树皮为优质纤维，种子富含油脂。为了采得沉香，分布比较集中的树木已被砍伐殆尽，在《中国植物红皮书》中被列为渐危种。

开花结实　10年生左右开始开花结实，正常结实年龄为20年生以后。花两性。伞形花序腋生或顶生。花黄绿色，有芳香。花萼钟状，檐5裂，喉部有10片连成一环的鳞片状花瓣，被毛。雄蕊10，1轮着生于萼管喉部。子房上位，卵状，2室，偶有1室或3室，每室1胚珠。据广西南宁1978～1984年观测，4月中旬末～下旬初花芽出现于当年生新枝的叶腋或枝梢，5月上旬末形成伞形花序；5月下旬末花始开，6月上旬为盛花期，中旬为末花期，全程约20天；6月下旬初幼果形成，7月上旬果实开始成熟，中旬为果实成熟盛期。蒴果，木质，成熟时由浅绿色转为淡黄色，倒卵形或扁椭圆形，长2.5～4cm，径1.4～2.6cm，密被绒毛，充分成熟后室背2裂。种子基部有一角状附属体，附属体的一端有2条丝线与果瓣内壁相连，开裂后将种子悬挂于果瓣上，约2日丝断，种子脱落。果瓣随种子落地或迟1～3日落地。种子1～2，黑褐色，表面有光泽，近卵形，两端尖，一侧中有棱纹，长13～18mm，宽6～8mm，厚5～6mm。附属体长为种子的1.5～2倍。胚乳白色略带黄色。种子形态见图1。

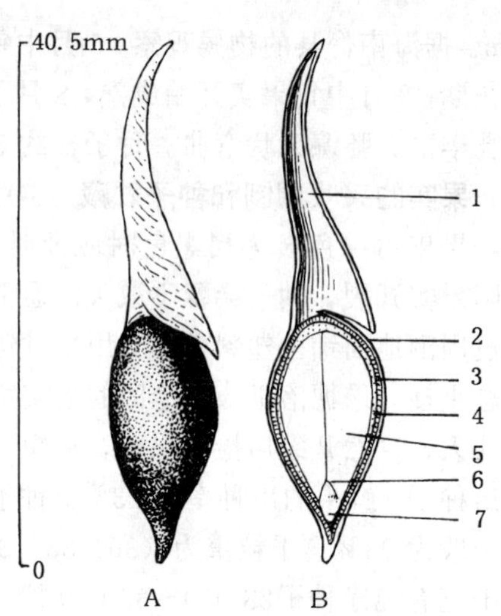

图1　土沉香种子外形（A）及其纵切面（B）
1. 角状附属体　2. 外种皮　3. 内种皮　4. 胚乳　5. 子叶　6. 胚芽　7. 胚根
（黄应钦绘）

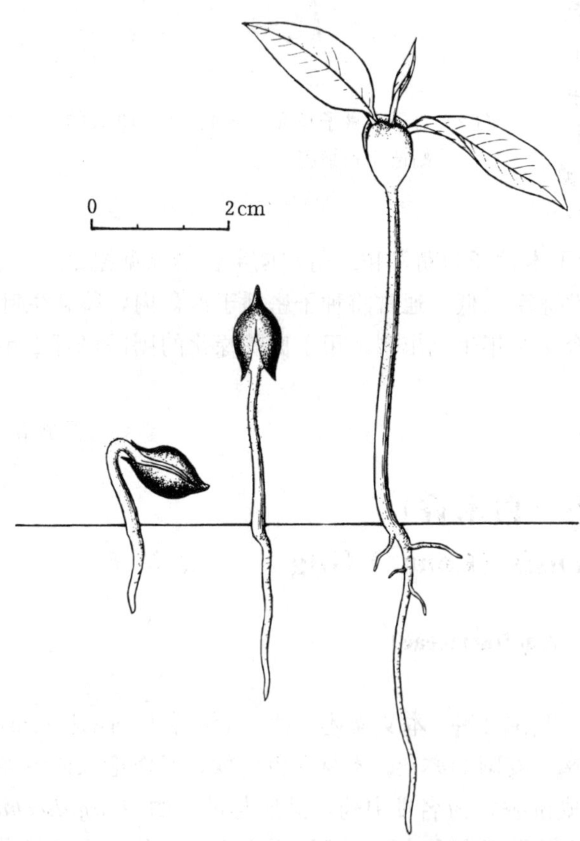

图2　土沉香种子萌发后第2、4、12天幼苗生长情况
（黄应钦仿《热带亚热带主要树种采种育苗技术》）

果实的采收调制和种子贮藏　果实成熟期比较整齐。当小部分果瓣开裂，大部分果实转呈黄绿色时，即可在树上采摘，或打落后在地面收捡。采得的果实置弱光下晒1天，或置室内通风处摊放2～4天，果瓣即可全部开裂。此时种子还连在丝上，用手搓动，脱去果瓣，即得纯净种子。鲜果出种率为15%～20%，净度约88%，种子(含附属体)千粒重190～260g，一般为230g，每千克有种子3 800～5 300粒。种子含水量30%～34%，忌脱水，不宜日晒，亦忌堆积。短期运输可混以湿沙，也可在调制前以果实的状态用竹筐包装，每筐以不超20kg为宜，以免堆积发热。常温条件下裸露存放的种子，40天即大部分失去发芽能力。混沙贮藏的种子，发芽能力可保持1～2个月。贮藏于5～10℃的低温条件下，大部分种子的发芽能力可保持

7～8个月。

发芽和播种　种子无休眠习性，宜随采随播。发芽时日均温宜在25℃以上。1982年7月上旬，广西林业科学研究所在室外播种床进行发芽测定，播种时日均温30℃上下。播后9天开始萌发，12～16天为发芽盛期，第19天发芽结束；从开始发芽至发芽高峰的7天中，发芽百分数为60%；从播种至发芽终了，以19天计，发芽率为75%。出土萌发。胚根萌动后约3天下胚轴直立，种壳脱落，8～9天子叶展开，初生叶出现。土沉香种子的萌发和幼苗早期生长情况见图2。

条播。新鲜种子播前不需作任何处理。存放过数天的种子，用清水浸泡2～3小时可促进发芽，每平方米约播25g（含附属体），覆土2～2.5cm。培育半年生或1年半生苗出圃。

（王宏志）

了哥王（南岭荛花）

Wikstroemia indica（L.）C. A. Mey.

（瑞香科　Thymelaeaceae）

生长习性、分布和用途　荛花属约50种。我国约38种，本文描述1种。常绿小灌木，高0.3～2m。萌芽力强。颇耐干旱贫瘠的酸性土。分布于长江以南的低山丘陵。越南至印度也有。树皮纤维为蜡纸、打字纸及人造棉的优质原料。叶消肿，根镇痛。种子富含油脂，可制皂。叶和种子均有毒。

开花结实　约3年生开始开花结实，5年生以后进入正常结实期，结实无大小年现象。花两性，黄绿色，数朵组成顶生短总状花序。花萼筒状，裂片4。无花瓣。雄蕊8，2轮排列于花萼筒的近顶部和中部，花丝短。花盘通常深裂成2或4鳞片。子房上位，倒卵形或长椭圆形，1室，1胚珠。柱头头状。据广西南宁1987～1989年观测，花期及结果期均很长：5月上旬～11月中旬陆续开花，花开后约40天果实成熟，从6月中旬～12月上旬均有果熟。花蕾和花、果常共存于一树。主要的花期为5月上旬开始，5月下旬为盛期，6月中旬盛期结束，6月中旬～7月中旬为果实成熟盛期。核果，成熟时由黄绿色转为鲜红色至暗紫黑色，长9～12mm，径5～7mm。外果皮薄膜质，有光泽。中果皮浆状。内果皮骨质，椭圆形。每果有核1粒。核倒卵形，褐色，长3～5mm，径2.5～3.5mm。种皮薄，无胚乳，胚大，子叶倒卵形（图1）。

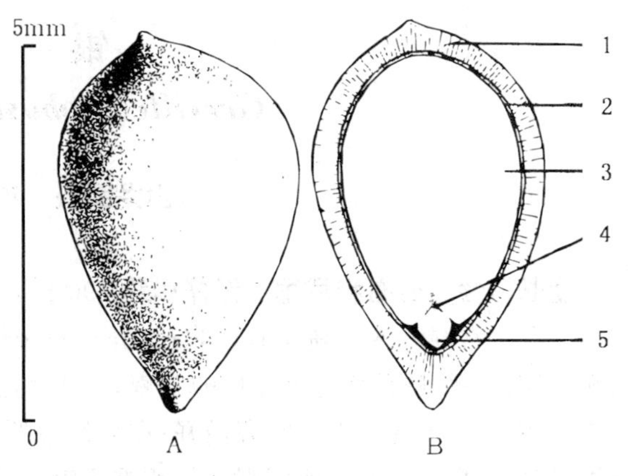

图1　了哥王果核外形（A）及其纵切面（B）

1. 内果皮　2. 种皮　3. 子叶　4. 胚芽　5. 胚根（黄应钦绘）

果实的采收调制和种子贮藏　果实成熟期较长，可以分期采摘，6～7月为主要采种期。采回的果实堆放1～2天，充分

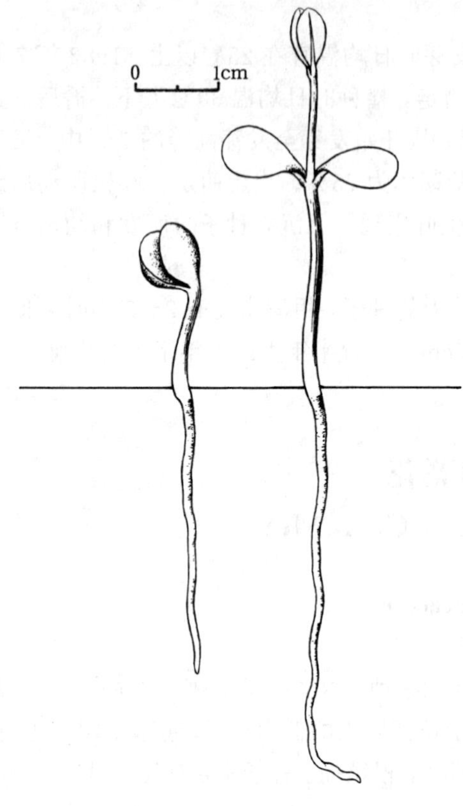

图 2 了哥王种子（核）萌发后第 5、9 天幼苗的生长情况

（黄应钦绘）

软熟后装入竹器或布袋，置水中搓揉，反复淘洗，清除果皮、浆肉等杂质，得到的果核用作播种材料，通称种子。鲜果的出籽率为 13%～16%。净度可以高达 98%～100%，含水量约 30%。千粒重约 33g，每千克有种子(核)3(2.8～3.3)万粒。种子忌脱水，不宜日晒。调制好的种子裸露存放的时间不宜超过 7 天，最好是随采随播。运输时需混以湿润木糠，用木箱或竹篓装运。贮藏时需混湿沙。常温下沙藏时间大约半年左右。一般 6～7 月采的种子，只宜沙藏至次年 2～3 月播种。5 月以后气温升高，种子发芽能力将明显下降。

发芽和播种 种子有休眠习性。发芽时气温宜在 20℃以上。1987 年 7 月中旬，广西林业科学研究所在室外沙床上对新采种子作过发芽测定：播前种子未作任何处理，播种时日均温 28～30℃，播后 77 天至 10 月上旬方始发芽。种子的发芽率较低，发芽盛期不明显。从开始发芽至发芽终结期，全程约半个月，发芽率为 36%。出土萌发。胚根萌发后约 4 天子叶出土，再过 2 天发出初生叶。了哥王种子（核）的萌发和幼苗初期生长情况见图 2。

撒播或条播。也可将种子撒于育苗沙盘内，每平方米约播 8～10g。发芽后约 50 天苗木基部木质化时，移至大田或容器内继续培育，半年或 3 个月出圃。

（王宏志）

银 桦

Grevillea robusta A. Cunn.

（山龙眼科 Proteaceae）

生长习性、分布和用途 银桦属约 190 种，主产马来西亚东部和大洋洲，本文只描述引入的 1 种。常绿乔木。高 40m，胸径 1m。喜暖热湿润气候，在我国也表现出一定的耐旱性。能耐轻霜。在引种栽培范围较广的云南，以年均温 15～18℃，极端最低温 -1℃，极端最高温为 36℃的气候条件下生长发育最好；在年均温低于 14℃或高于 20℃，极端最低温低于 -4℃，极端最高温高于 40℃的地区则生长严重受害。对土壤要求较严，在土层深厚、肥沃疏松、排水良好的微酸性土壤上发育良好。质地粘重、排水不良的地方生长差；钙质土不适宜生长。

银桦原产澳大利亚昆士兰州南部与新南威尔士北部沿海区域。我国主要引种地为川、滇、黔、桂、粤、琼、湘、赣、闽、台、浙。

树形挺拔秀丽，树干通直，花色艳丽，是优美的庭园观赏树和四旁绿化树。花富含蜜，是良好的蜜源树种。木材纹理直，色泽美观，切面光滑，宜作装饰材和胶合板材，以及家具、箱板等。

开花结实　开花结实年龄因树木生长环境而不同，有的 7～8 年，有的 10～11 年。大小年现象不明显。结实的多少受热量条件影响。例如云南年均温 17～19℃的地区开花结实较多；年均温在 15～16℃的地区能正常开花结实，但结实少；年均温在 21℃以上的地区开花结实少，甚至不结实。

银桦开花结实物候见表 1。

表 1　银桦的开花结实物候

观察地点	开花			果实成熟		种子散落		
	始期	盛期	末期	始期	盛期	始期	盛期	末期
云南勐海	3 月上旬	3 月中下旬	4 月上旬	6 月上旬	6 月中下旬	6 月中旬	6 月下旬	7 月上旬
云南昆明	4 月中旬	4 月下旬～5 月上旬	5 月中旬	8 月上旬	8 月中旬	8 月中旬	8 月下旬	9 月上旬
四川攀枝花	3 月下旬	4 月上旬～4 月中旬	4 月下旬	7 月上旬	7 月中下旬	7 月中旬	7 月下旬	8 月中旬
四川西昌	4 月下旬	5 月上中旬	5 月下旬	8 月中旬	8 月下旬～9 月上旬	8 月下旬	9 月上旬	9 月中旬

花两性。单被花，橙黄色，总状花序单个或数个聚生于无叶的短枝上，长 7～16cm，花序柄具蜜腺。花萼筒纤弱细长，稍弯曲，萼裂片 4，花瓣状，开后向外反卷。雄蕊 4，无花丝，花药附着于萼片的凹陷处。单心皮雌蕊，子房有柄，1 室，花柱细长且自萼筒裂缝中伸出，柱头常偏于一侧，宿存。胚珠 2，横生于子房的侧面。

蓇葖果，木质，卵状矩圆形，长 14～18mm，径 6～12mm，偏斜而扁，顶端具宿存花柱。果未熟时绿色，成熟后棕褐色，沿腹缝线开裂。每果内有种子 2 粒，倒卵形，扁平，棕褐色或深褐色，周边有膜质翅，连翅长 12～14mm，径 8～10mm，无胚乳（图 1）。

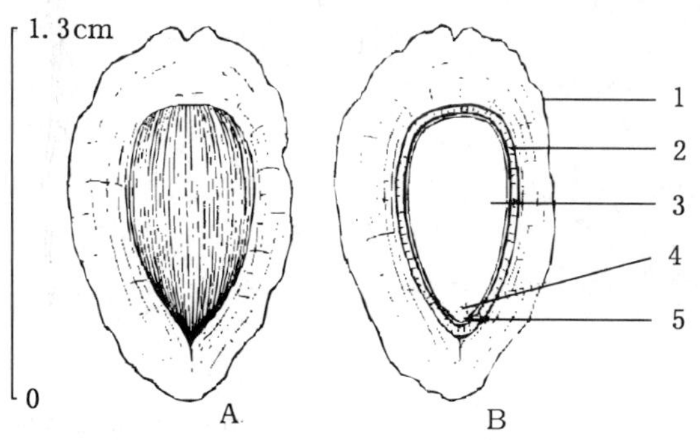

图 1　银桦种子外形（A）及其纵切面（B）
1. 种翅　2. 种皮　3. 子叶　4. 胚芽　5. 胚根
（黄应钦绘）

果实的采收调制和种子贮藏　果实成熟后如遇晴天果壳会加速开裂，种子易飞散。因此宜在果实变为棕褐色而果壳尚未裂开前及时将果枝剪下，或在地面铺放收种布直接收集种子。采回的果实经晾晒后脱出种子。生产上直接用带翅种子作为播种材料，通称种子。

干藏。将种子装入布袋置库房内或干燥

荫凉处越冬贮藏。如贮藏时间过长，种子生活力会显著下降。据检查，四川及云南所产银桦种子的净度为75%～90%，千粒重8.1～10.2g，每千克有种子9.8万～12.3万粒。

发芽测定　种子无休眠习性。新采收的种子发芽率在70%以上，1～3个月后发芽率在35%～65%，延至翌春，发芽率在20%左右。

发芽测定前常用始温45℃水浸种24小时，测定时多用25℃的恒温。

播种　在季风气候区，种子成熟正值雨季，宜随采随播：云南在7～8月，四川西南部在8～9月。非季风气候区可以春播，播种期在2月下旬～3月中旬。撒播或条播。播前用温水或冷水浸种24小时。撒播时每666m^{2}播种8～10kg，条播时每666m^{2}播种5～6kg。播后覆土5mm，轻压后盖草。随采随播的播后20～30天种子发芽出土，幼苗越冬要防霜冻。春播约50天发芽出土。幼苗期要搭棚遮荫。1年生苗高1m左右，2年生苗高可达3m。

出土萌发。子叶2，倒三角状卵形，长10～14mm，宽7～10mm，先端宽圆，基部箭形具两耳。银桦种子萌发后的幼苗生长情况见图2。

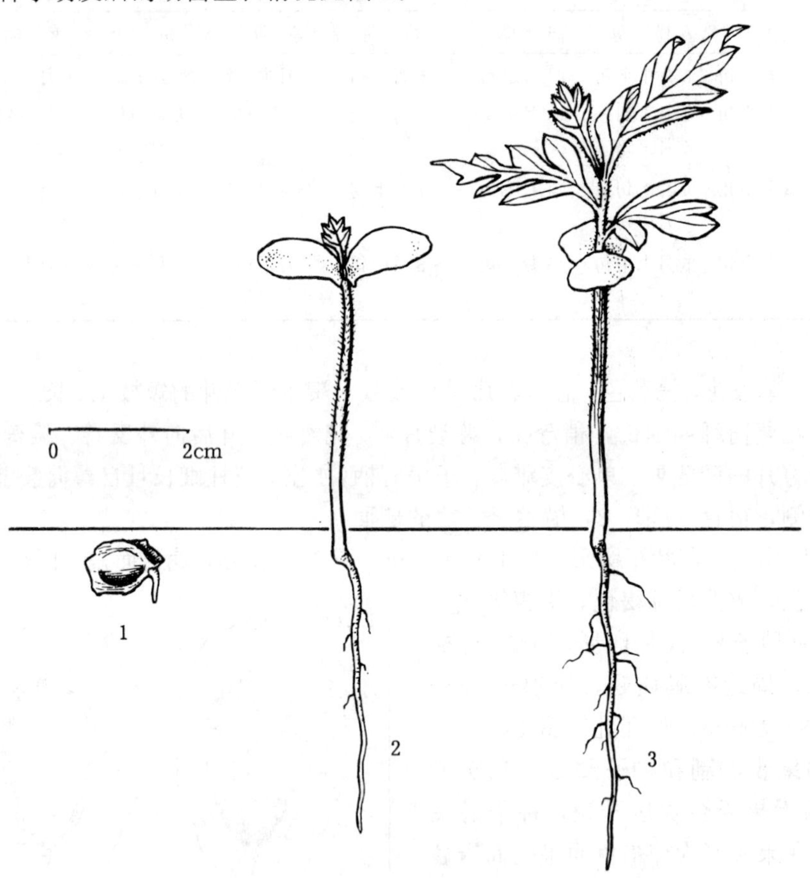

图2　银桦种子萌发后的幼苗生长情况

1. 幼根初露　2. 初生叶互生　3. 幼苗形成

（杨向晨、黄应钦仿《主要树木种苗图谱》）

（赵德铭）

山龙眼属

Helicia Lour.

（山龙眼科　Proteaceae）

生长习性、分布和用途　本属约 90 种，我国产 18 种，本文描述 3 种。常绿乔木。适生于质地疏松肥润的酸性土。它们的名称、生长、分布及用途见表 1。

表 1　山龙眼属树种的名称、生长、分布和用途

中　名	学　名	树高（m）	胸径（cm）	分　布	用　途
小果山龙眼（越南山龙眼）	*H. cochinchinensis* Lour.	20	40	川、滇、桂、粤、琼、湘、鄂、赣、闽、浙、台。越南、日本	水源、材用、油脂、制皂
倒卵叶山龙眼	*H. obovatifolia* Merr. et Chun	20	40	桂、粤、琼。越南	水源、材用、淀粉、食用
网脉山龙眼	*H. reticulata* W. T. Wang	10～15	20～30	滇、黔、桂、粤、湘、赣、闽	水源、蜜源、材用、淀粉、食用

开花结实　8～10 年生开始结实，正常结实期在 15 年生以后。结实大小年间隔期一般为 1 年，但有的种不明显。丘陵或低山的天然林中，单株结实量较少，但植株寿命长。引种至低丘、平原，开始结实期可提前 1～2 年且结实量多，但寿命短，约 20 年即现衰老。花两性，无花瓣，辐射对称。总状花序腋生。网脉山龙眼部分花序生于小枝叶痕处，花序长 8～15cm。萼筒细长，花蕾时直立，顶部棒状至近球形，开时裂片 4，条形，分离，外卷。雄蕊 4，着生于萼片檐部，无花丝。子房上位，无柄，1 室，胚珠 2，基生或侧生。花柱细长，顶部棒状，柱头小。广西南宁 1978～1984 年观察的物候期见表 2。

表 2　山龙眼属树种的开花结实物候期（广西南宁，1978～1984）

树　种	开花 始期	开花 盛期	开花 末期	果实成熟 始期	果实成熟 盛期	果实脱落期
小果山龙眼	5 月上旬	5 月中下旬	6 月中旬	10 月上旬	10 月中下旬	10 月下旬～12 月下旬
倒卵叶山龙眼	6 月上旬	6 月中旬	7 月上旬	11 月中旬	12 月上旬	12 月下旬～翌年 3 月中旬
网脉山龙眼	4 月上旬	4 月下旬	5 月中旬	10 月上旬	10 月下旬	11 月中旬～12 月上旬

坚果，圆形或椭圆形，不开裂，外层为肉质苞片所包裹，内有种子 1（稀 2）粒。种子近圆形或椭圆形，种皮膜质。无胚乳。子叶厚肉质，上部具皱纹。种子形态见表 3、图 1。

果实的采收调制和种子贮藏　果实成熟落地后有虫蛀现象，应在果实成熟盛期及时采摘，或用竹竿敲打，震落后在地面捡拾。由于果实的成熟期不甚一致，捡拾时应将已充分成熟同接近成熟的果实分别放置。充分成熟的果实装入竹筐，置清水中浸泡 2～3 天后，用手轻轻搓

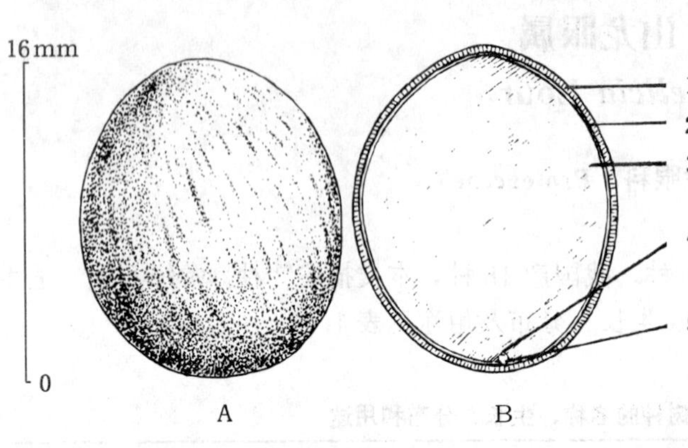

图 1　网脉山龙眼去除肉质苞片的坚果外形(A)及其纵切面(B)
1. 果皮　2. 种皮　3. 子叶　4. 胚芽　5. 胚根（黄应钦绘）

擦，不要弄伤果皮。漂去肉质苞片，所得纯净坚果用作播种材料，通称种子。接近成熟的果实应在室内堆放 2～3 天，苞片软熟后再按上述方法调制。数量较少时，也可不浸水而直接剥出种子（坚果)。种子忌失水，不能日晒。调制出来的种子裸露存放 1 周以上，发芽能力便明显下降。新调制的种子含水量为 40%～60%，不影响发芽的种子最低含水量在 30%以上。3 天以内的短期运输，可带苞片装运，运抵后再行调制。长途运输或贮藏，需先行调制，稍加晾干后混以湿沙。贮藏期一般为 4～6 个月。鲜果的出籽率及种子的净度、质量等数据见表 4。

表 3　山龙眼属树种果实和种子的形态特征

树　种	未熟果颜　色	成熟果实			种　子		
		形　状	大小（cm）	颜　色	形　状	大小（cm）	颜　色
小果山龙眼	绿色有光泽	近椭圆形，两端圆	长 1～1.8 径 0.8～1	蓝黑色	卵形或椭圆形	长 0.8～1.5 径 0.7～0.9	浅褐色
倒卵叶山龙眼	青　色	倒卵形、桃形或长圆形，两端尖	长 3～6 径 2～4.5	紫黑色	卵形，先端有突尖	长 2.7～5.5 径 1.7～4	褐色至棕褐色
网脉山龙眼	青绿色	椭圆状球形	长 1.5～3 径 1.5～2.5	黑色，有光泽	近球形	长 1～2.3 径 0.8～1.8	浅褐色

表 4　山龙眼属树种的出籽率和种子净度、质量

树　种	出籽率（%）	净度（%）	千 粒 重（g）	每千克纯净种子粒数（粒）
小果山龙眼	55～65	99	400～550	1 800～2 500
倒卵叶山龙眼	55～60	99	15 000～20 000	50～70
网脉山龙眼	50～60	99	1 000～2 000	500～1 000

发芽和播种　种子无明显休眠习性。发芽时的日均温宜在 18℃上下。1981 年春和 1987 年秋，广西林业科学研究所先后 2 次在室外沙床播种，对倒卵叶山龙眼和网脉山龙眼进行过发芽测定，结果见表 5。

留土萌发。倒卵叶山龙眼胚根萌发后约 41 天胚芽出土，出土后 10 天左右发出初生叶。网脉山龙眼胚根萌发后 12 天胚芽出土，出土后约 7 天发出初生叶。网脉山龙眼种子（坚果）的萌发和幼苗初期的生长情况见图 2。

表 5 山龙眼属树种的发芽能力及其测定条件（广西南宁，室外沙床）

树 种	温 度（℃）	发芽势（%）		发芽率（%）	
		计算天数	一般数值	计算天数	一般数值
倒卵叶山龙眼	20～25	—	—	44	65
网脉山龙眼	15～25	8	52	60	81

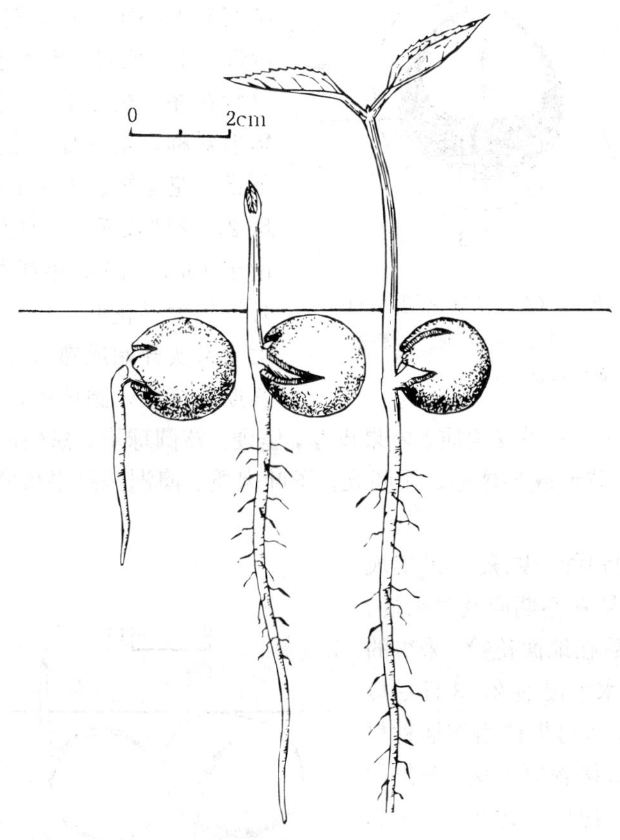

图 2 网脉山龙眼种子萌发后第 8、15、25 天幼苗的生长情况
（黄应钦绘）

条播。小果山龙眼、倒卵叶山龙眼和网脉山龙眼每平方米的播种量分别为 45～60g、1 500～2 000g 和 150～200g。培育 1 年生苗出圃。

（王宏志）

澳洲坚果

Macadamia tetraphylla L. Johnson

（山龙眼科 Proteaceae）

生长习性、分布和用途 澳洲坚果属约 10 种，我国引入本文描述的 1 种。常绿乔木，高达 12～20m，胸径 30cm。适生于排水良好，pH 值为 7～8 的土壤。忌海风。原产大洋洲东北部，我国粤、琼、桂、滇、浙等引种栽培。种子香甜可食。木材微红色，坚固美观，可用于细木工、农具和家具。多年来，澳洲坚果沿用 *M. ternifolia* F. v. Muell. 这个学名。据吴中伦考证，该种种子味苦，不能食。

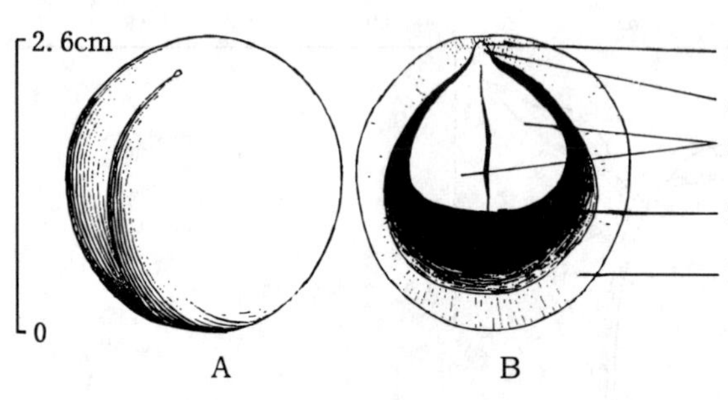

图 1 澳洲坚果果核外形（A）及其纵切面（B）
1. 内果皮 2. 种皮 3. 子叶 4. 胚轴 5. 胚根
（黄应钦绘）

开花结实 8～10 年生开始开花结实。结实有大小年，间隔期一般为 1～2 年。花两性，小，成对，橙黄色，组成顶生或腋生总状花序。苞片小，早落。花萼裂片 4，近辐射对称。无花瓣。雄蕊 4，生于萼裂片稍下部，花丝短。子房上位，1 室，无柄，胚珠 2。花柱长而直，柱头小，顶生。据广西南宁 1987～1988 年观察，3 月中旬始花，3 月下旬为盛花期，4 月上旬为末花期；8 月上旬果实开始成熟，8 月中下旬为果实成熟盛期。核果，圆球形，径 2.7～3.5cm，未熟时青绿色，熟时黄色。外果皮肉质。内果皮厚，坚硬。核圆球形，棕褐色，有光泽，径2.3～2.7cm，种子 1 或 2，球形或半球形。无胚乳。子叶肉质。澳洲坚果果核的外观和解剖构造见图 1。

果实的采收调制和种子贮藏 果实成熟落地后易受虫蛀。果熟盛期应及时采摘，或用竹竿敲打，震落后在地面捡拾。采得的果实装入竹筐，置清水中浸泡 2～3 日后搓擦，漂去皮肉和杂质，所得果核用作播种材料，通称种子。也可直接剥出果核。鲜果的出籽率约50%～70%。净度可达 99%。不影响种子发芽能力的含水量在 30%以上。千粒重 8 300（7 000～10 000）g，每千克有种子（核）100～140 粒。种子忌失水，不宜日晒。调制出来的种子宜混湿沙贮藏或运输，贮藏期不超过 6 个月。

发芽和播种 种子休眠现象不明显。发芽时日均温宜在 20℃以上。1987 年 8 月 10 日，广西林业科学研究所在室外沙床对未作处理的当年新采种子作过发芽测定：播种时日均温 24℃，10 月 1 日开始发芽，10 月 25 日发芽终止，发芽盛期不明显。从置床之日起算，77 天的发芽率为 70%。留土萌发。播后约 50 天萌发。胚轴生长约 7

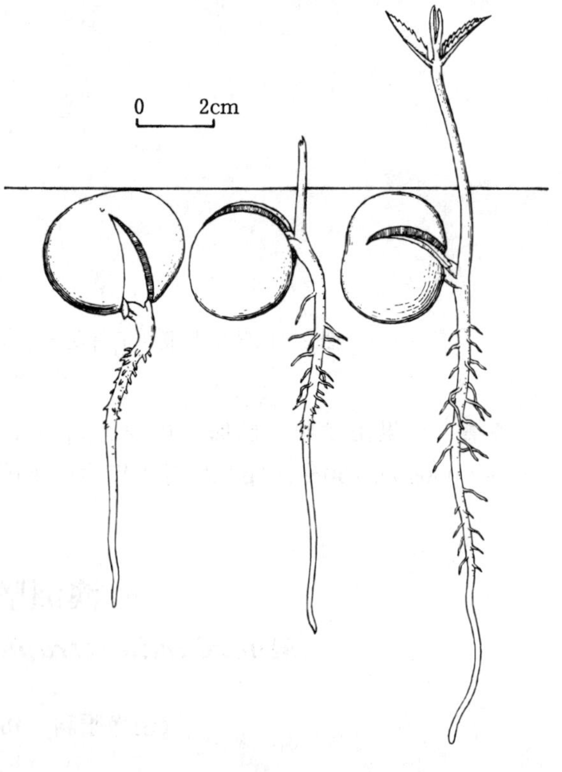

图 2 澳洲坚果种子萌发后第 5、7、15 天幼苗的生长情况
（黄应钦绘）

天左右上胚轴出土，再过 5 天发出初生叶。澳洲坚果果核的萌发和幼苗初期生长情况见图 2。

条播。混沙湿藏过的种子，播前浸水 2～3 小时，每平方米播种量为 380～500g，覆土约 2cm。培育 1 年生苗出圃。生产中多用 1 年生实生苗作砧木，嫁接育苗建立果园。

（王宏志）

海桐花属

Pittosporum Banks

（海桐花科　Pittosporaceae）

生长习性、分布和用途　海桐花属约 300 种，广泛分布于大洋洲、太平洋西南各岛屿、东南亚及东亚热带和亚热带。我国约 44 种，本文描述 2 种（表 1）。常绿小乔木或灌木。喜光亦能耐荫。适生于温暖湿润气候。有一定的抗旱、抗寒能力。海桐对土壤要求不严，在弱碱或中性壤土上生长较好，对多种有毒气体抗性强，是城市和工矿区绿化的重要树种。

表 1　海桐花属树种的名称、生长、分布和用途

中　名	学　名	树高（m）	分　布	用　途
崖花海桐（海金子）	*P. illicioides* Mak.	5	闽、台、浙、苏、皖、赣、鄂、湘、黔。日本	观赏、造纸、油脂
海　桐	*P. tobira* (Thunb.) Ait.	6	长江以南沿海各省；内地多为栽培。日本、朝鲜半岛	观赏、环保、药用、染料

开花结实　10～15 年生可以正常结实，无明显的大小年现象。花两性或杂性。伞形花序顶生或近顶生。海桐花白色，后变黄色；崖花海桐花黄色，均具芳香。萼片、花瓣和雄蕊均 5 数。海桐雄蕊 2 型，退化雄蕊的花丝短，约 2～3mm，花药不育；正常雄蕊的花丝长约 5～6mm。雌蕊由 3 心皮组成，侧膜胎座与心皮同数，胚珠多数。在同一地区，崖花海桐和海桐的花期和果熟期大致相同。在江苏花期 5 月，果熟期 10 月。浙江杭州花期为 4 月下旬～5 月上旬，果熟期 10 月下旬～11 月上旬，11 月中旬果实开裂，种子开始散落。

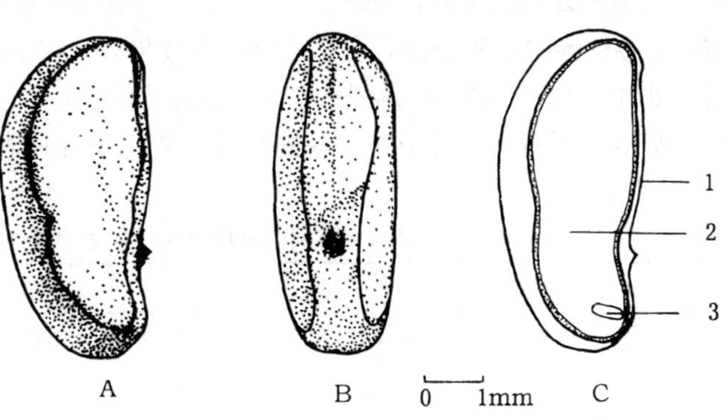

图 1　海桐种子外形的侧面（A）、腹面（B）及其纵切面（C）
1. 种皮　2. 胚乳　3. 胚
（童军平绘）

蒴果常 3～5 个着生，球形或卵形，果皮木质或薄木质，成熟时沿腹缝线 3 瓣裂。每果有种子 8～16 粒，多者达 23 粒，或 2 列着生于胎座中段，或生于子房内壁中部。种子暗红色或微暗红色，外被粘性油质物。胚乳丰富，胚小，见图 1。果实和种子的形态特征见表 2。

表 2 海桐花属树种果实和种子的形态特征

树 种	果 实			种 子		
	形 状	大小（cm）	颜 色	形 状	大小（mm）	颜 色
崖花海桐	椭圆形或三角状球形，有纵沟 3 条。果皮薄，厚约 1mm	长 0.9～1.6 径 0.7～1.2	橙黄色	肾形、半球形或三角状半球形。背面隆起，有 1 条凹槽。种脐内凹	长 4.0～6.0 宽 3.0～4.5	红色，干后变暗红色
海 桐	近球形，有 3 棱。果皮上有褐色瘤状体。果皮厚约 1.5mm	长 1.3～1.9 径 1.0～1.6	橙黄色	不规则多角形，但多为肾形。背面隆起，且布满凹点。有浅凹槽 1 条。种脐内凹	长 4.0～5.7 宽 2.0～3.5	鲜红色，干后变暗红色，有光泽

果实的采收调制和种子贮藏 果实由绿色转为橙黄色并微裂时，可直接从树上采摘。采回的果实置阳光下曝晒至开裂，再用草木灰拌擦脱粒。短期贮藏的可用布袋、塑料袋或厚纸袋盛装，也可置坛罐或磨口玻璃瓶内密封存放在阴凉处。贮藏 1 年以上的，宜在 3～5℃低温环境中密封保存。据笔者观察，1986 年 11 月中旬在浙江杭州采集的海桐种子在杭州室温条件下摊放，15 个月后全部丧失发芽能力。种子的净度和有关质量的数据详见表 3。

表 3 海桐花属树种的出种率和有关种子质量的数据

树 种	每千克果实数（个）	出种率（%）	净 度（%）	千粒重（g）		每千克纯净种子粒数（万粒）	
				一 般	变动范围	一 般	变动范围
崖花海桐	1 110～1 800	25～35	90～97	40	35～48	2.50	2.0～2.9
海 桐	600～950	20～30	90～97	30	24～44	3.33	2.2～4.2

发芽前的处理和发芽测定 这两个树种的种子均有休眠现象，崖花海桐的休眠程度比海桐深。层积处理能解除休眠。据笔者试验，先用浓硫酸处理 2 分钟，洗净种子外被的粘性油质物，再行湿沙层积处理，比单纯层积的效果好。在 15℃或 20℃的恒温条件下，这 2 个树种经过处理的种子都能顺利萌发。温度再高，例如 25℃恒温便明显地抑制海桐种子萌发，30℃时便基本上不萌发。种子的发芽能力及其测定条件见表 4。

表 4 海桐花属树种种子的发芽能力及其测定条件①

树 种	预处理	温 度（℃）	发芽势（%）			发芽率（%）	
			计算天数	一般数值	变动范围	计算天数	变动范围
崖花海桐	5～15℃变变层积 21 天	20	58	50	—	63	74～80
海 桐	2～8℃的室温层积 14 天	15	15	39	32～48	20	92～96
	5～15℃的变温层积 14 天	15	17	36	32～44	20	76～84

注：①测定前种子均经浓硫酸处理 2 分钟，洗净后再行层积处理。发芽基质为沙。测定中未加光照

播种　可随采随播，也可湿沙层积至翌春播种，但多行春播。每平方米播种 6～8g。条播，行距 20cm，覆土厚约 1cm，盖草。1988 年在浙江杭州室外盆播的种子发芽情况见表 5，出土萌发。种子萌发和幼苗生长情况见图 2。

海桐 1 年生苗高约 15cm。亦可扦插繁殖。

表 5　海桐花属树种种子萌发出土情况

树　种	预处理①	播种期	萌发出土			初生叶出现期	出苗率（%）
			始　期	盛　期	末　期		
崖花海桐	3.5～18℃室温层积 40 天	4 月中旬	9 月下旬	10 月上旬	10 月下旬	10 月下旬	81～96
海　桐	5～15℃变温层积 40 天	3 月中旬	4 月中旬	4 月下旬	5 月上旬	5 月中旬	78～92

注　①种子均经浓硫酸处理 2 分钟，洗净后再行层积处理

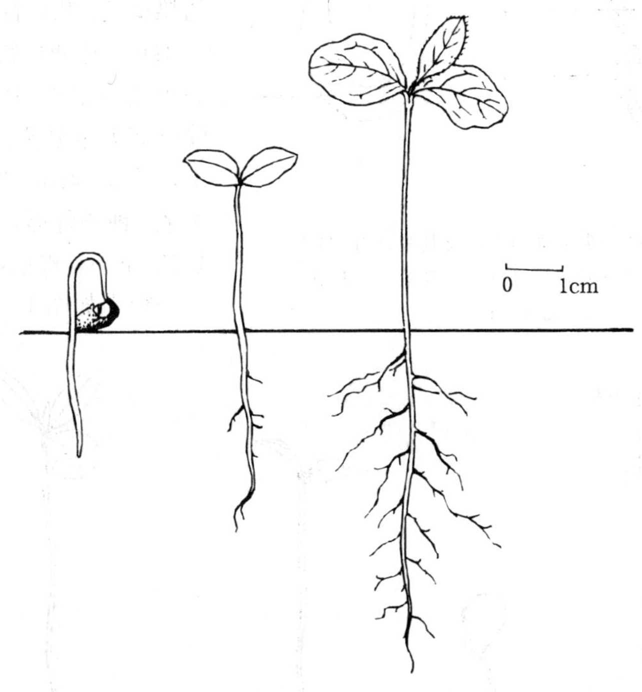

图 2　海桐种子萌发后第 1、11、50 天的幼苗生长情况
（童军平绘）

（史晓华）

刺山柑（老鼠瓜）

Capparis spinosa L.

（白花菜科　Capparidaceae）

生长习性、分布和用途　槌果藤属（山柑属）约 250 种，我国约 52 种。本文描述沙漠地

区产的刺山柑这一个种，也叫老鼠瓜，攀援状小灌木，枝条平卧呈辐射状。长可达4m，高0.5～1.0m。极耐干旱、瘠薄、沙埋，不耐盐碱。生于荒漠地带的戈壁、沙地、石质低山和山麓地带。寿命较长。产于新疆、甘肃西部和西藏。哈萨克斯坦、阿富汗、伊朗、土耳其至巴尔干半岛、欧洲南部、地中海地区亦产。用于防风固沙。种子含油率高达35%，可食用，花蕾可制酱，根皮入药。

开花结实 1年生即可开花结实，3年生开始大量结实，无大小年现象。花两性，直径2～4cm，花梗长2.5～4.0cm，单生叶腋，萼片4，其中1枚较大。花瓣4，白色、粉红色或紫红色，倒卵形，2片较大。雄蕊多数，长于花瓣。子房上位，心皮多数，通常1室，具长2～4(5)cm的雌蕊柄。吐鲁番地区始花期在5月上中旬，花期长约半年左右，从开花到果熟约需30天，从6月下旬开始陆续有大量果熟。浆果，椭圆形，无毛，长2～4cm，宽1.5～3cm，种子多数。种子肾形，种皮较硬，具光泽，长约3mm，褐色，胚弯曲，无胚乳。种子形态如图1。

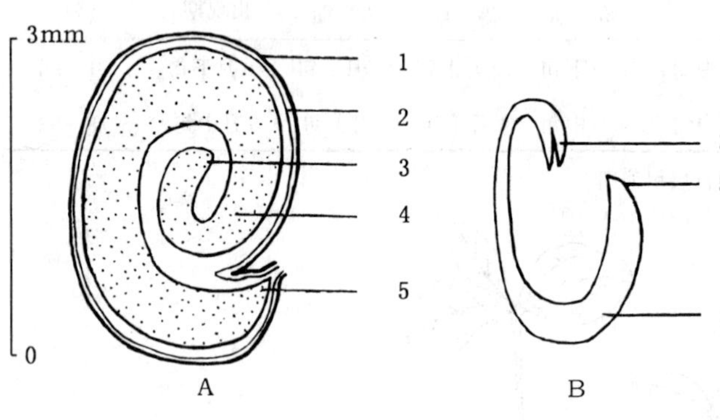

图1 刺山柑种子纵切面（A）及其中之胚（B）

1. 外种皮 2. 内种皮 3. 子叶 4. 胚轴 5. 胚根

（吴新安绘）

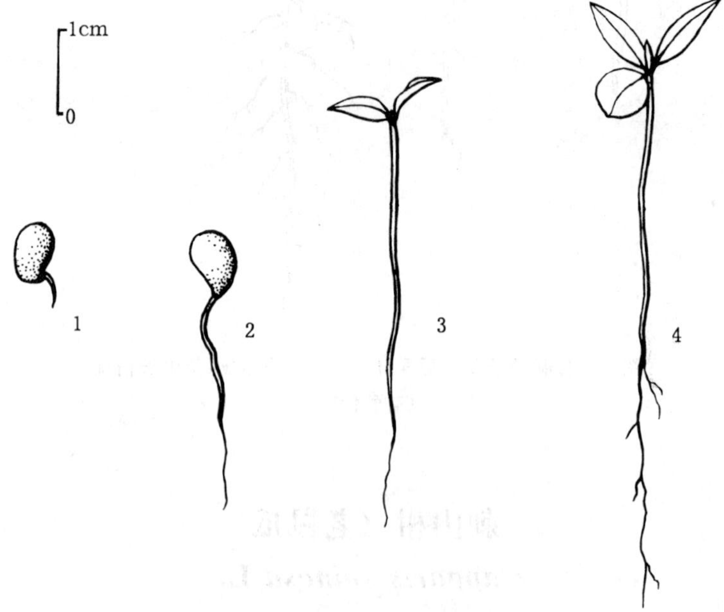

图2 刺山柑种子萌发及幼苗生长情况

1. 胚根伸出 2. 胚轴伸长 3. 子叶出土 4. 初生叶出现

（吴新安绘）

果实的采集调制和种子贮藏 在吐鲁番地区6月下旬开始大量果熟，即可采集。果实成熟后果皮干裂，由果肉胶结着的近椭圆形种子坠落在灌丛下，故在此期间可相隔10天左右卷起枝蔓在地上收集，亦可在枝蔓上直接摘取。采集的果实应及时浸泡、冲洗，除去果肉及杂质，即得纯净种子。出种率75%。种子千粒重3.2～3.7g，每千克纯净种子约30万粒。在干燥、通气和常温条件下贮藏3年发芽率为80%～90%，4～6年发芽率显著下降，但仍可保持在40%左右。

发芽和播种 将种子水浸2～3天，在10℃左右温度下层积1个月或在15～20℃下层积半个月。经层积的种子在20～30℃条件下发芽率最高。春播秋播均可，以秋播为好。条播，每平方米播种6～7g。秋播覆土厚1～2cm，春播3～4cm，播后不宜压实。出土前不宜浇水，以防土壤板结，影响出苗。子叶2枚，出土萌发。幼苗基本形成时可以间苗。种子萌发和幼苗形态如图2。

（周君英）

辣 木

Moringa oleifera Lam.

（辣木科 Moringaceae）

生长习性、分布和用途 辣木属约12种，原产北非和西亚，我国引入本文描述的1种。落叶乔木，高10～30m，胸径15～30cm。耐高温，适生的最高温为43～47℃，能够忍受的最低温为0℃左右。喜肥沃而排水良好的酸性沙壤土。世界热带各地有引种。我国滇、桂、粤、琼、闽有栽培。种子榨油，清彻透明，为钟表和精密仪器的润滑油。油味香，是香水的优良凝香剂。根有辛辣味，可作调味品。幼果和叶亦可食。木材可制作家具。花和皮供药用。

开花结实 8年生左右开始开花结实，正常结实年龄在15年生以后。结实大小年现象不明显。花两性，两侧对称。腋生圆锥花序。萼筒杯状，5裂。花瓣5，不相等，近轴2较小，外弯，侧方2向上，远轴的较大，直立。雄蕊2轮，发育的5枚着生于花盘边缘，与退化的5枚互生。花药背着，1室纵裂。子房上位，圆柱形，1室，具3侧膜胎座，胚珠多数，2列。花柱顶生，筒状，顶端平。花期较长，全年大半年均有花开，有果实成熟。据海南尖峰岭1986年观测，主花期自11月中旬开始，12月中旬为盛花期，翌年5月中旬为末花期，花期长达半年；6月下旬果实开始成熟，7～9月为果实成熟盛期。果为一长蒴果，剑状，下垂，先端尖。果长20～50cm，径约2.5cm，断面三角状，3

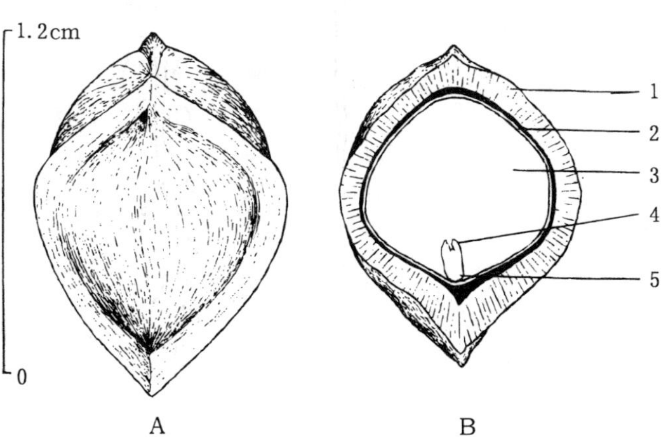

图1 辣木去翅种子外形（A）及其纵切面（B）

1. 外种皮 2. 内种皮 3. 子叶 4. 胚芽 5. 胚根

（黄应钦绘）

瓣，每瓣有肋纹 3 条。成熟时果皮由青色转为灰褐色。充分成熟后蒴果成 3 瓣开裂，种子散落。每果有种子 70～120 粒。种子 3 棱形，黑褐色，具 3 膜质翅，翅灰色。种壳木质，较脆。种子带翅长 3～4cm，宽 2～2.5cm。不带翅的种子长 9～12mm，径 8.5～11mm。无胚乳，子叶肉质。辣木种子的形态见图 1。

果实的采收调制和种子贮藏　果实成熟而尚未开裂时，用手或采种刀等器具将蒴果采下，置室内通风干燥处 2～3 天，果瓣开裂，种子脱出。除去果瓣等杂质即得带薄膜翅的种子，用作播种材料。鲜果出种率约 21%。种子净度约为 90%（主要杂质为种翅），含水量 15%～25%。带翅种子千粒重 220（200～250）g，每千克有纯净种子 4 000～5 000 粒。种子忌失水，不宜日晒，亦不能久藏。夏季采集的种子，裸露存放 1 周即失去发芽能力。运输时宜选择尚未开裂的果实，带果包装运输。8 月以前采集的种子，贮藏期宜在半个月以内。9 月以后采集的种子可以沙藏越冬，至次年春暖后播种。

发芽和播种　种子无休眠习性。日均温在 20℃以上时随采随播。1987 年 8 月 10 日（日均温 30℃），广西林业科学研究所在室外沙床进行过发芽测定，采用刚脱果瓣的新鲜种子，播前未作任何处理：播后第 5 天开始萌发，第 6 天即进入发芽盛期，第 9 天发芽结束；从发芽开始至发芽高峰的 3 天，发芽百分数为 60%；从播种之日起算，以 14 天计，发芽率为 65%。发芽比较快速整齐。留土萌发。种子萌动后约经 2 天，上胚轴出土，再过 2～3 天展出初生叶。辣木种子的萌发和幼苗初期生长情况见图 2。

条播。种子播前不作处理。沙藏或稍存放过的种子，需浸水 2～3 小时再行播种。每平方米播种 30～38g。也可将种子密播于沙床内，发芽后移至容器培育。培育百日苗或半年生苗，至次年春出圃。

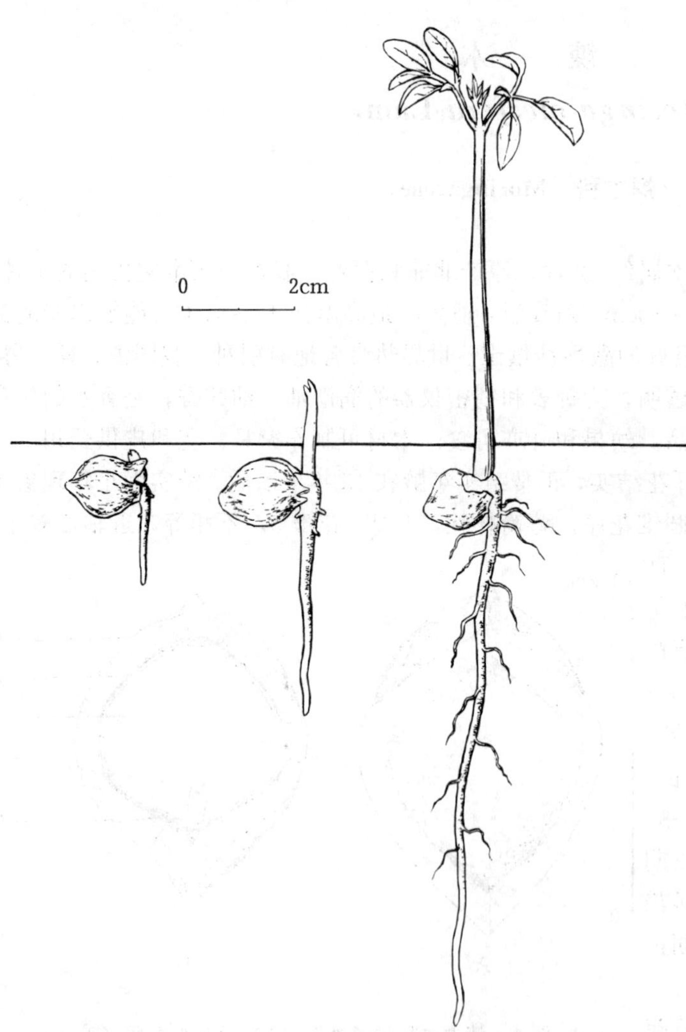

图 2　辣木种子萌发后第 2、3、7 天幼苗的生长情况

（黄应钦绘）

（王宏志）

红砂（枇杷柴）属

Reaumuria Hasselq. ex L.①

（柽柳科　Tamaricaceae）

生长习性、分布和用途　本属约 22 种，分布于欧、亚、非洲，我国 4 种，主要分布于西北及内蒙古的荒漠、半荒漠和干旱草原区域内。沙漠地区有 3 种，本文描述 2 种。落叶灌木或半灌木。叶革质或肉质，常具泌盐腺体。习生于低矮沙丘、覆沙地、沙砾质戈壁及山前砾石冲积扇上。喜光。根系发达，分枝密集，极耐旱。抗寒，耐盐碱。可药用，又是牲畜喜食的饲料。它们的名称、生长和分布情况见表 1。

表 1　红砂属树种的名称、生长和分布情况

中　名	学　名	株高（cm）	分　布
红　砂	*R. soongorica* (Pall.) Maxim.	15～25	西北、内蒙古。蒙古，中亚
黄花红砂（黄花枇杷柴）	*R. trigyna* Maxim.	15～30	内蒙古、甘、宁

开花结实　2～3 年生开始开花结实，3～4 年生大量结实，丰年间隔期不明显。花两性，单生叶腋，少数生枝端。萼片 5，分离，或萼钟状 5 裂（如红砂）。花瓣 5，白色、黄色或粉红色，先端钝或凹入，基部渐窄，内侧有 2 附属物。雄蕊 5～15，离生或基部结合成 5 束，与花瓣对生，宿存。子房上位，球形或卵形，1 室，具不完全隔膜。胚珠 2～5，生于基生的侧膜胎座上，花柱 3～5。开花结实物候见表 2。

表 2　红砂属树种的开花结实物候

树　种	观察地点	观察年份	开　花			果实成熟	
			始　期	盛　期	末　期	始　期	盛　期
红砂	甘肃民勤	1975～1981	7 月下旬	8 月上旬	8 月下旬	10 月上旬	10 月下旬
黄花红砂	内蒙古乌海	1988	7 月	8 月上旬	8 月下旬	8 月	9 月

蒴果，3～5 瓣裂，内含 1～3 枚种子。种子椭圆形，无芒，全部被毛。果实和种子的形态特征见表 3。种子形态见图 1。

果实的采收调制和种子贮藏　果实采集后揉去果皮，除去杂质。但种子不易收拾干净，红砂常见的净度只有 37%。在干燥阴凉处可贮存较长时间。据甘肃治沙研究所的一次测定，室内发芽率 53%。种子净度及质量见表 4。

① 有些学者另立红砂属（*Hololachna* Ehrenb.），将红砂的学名定为 *H. soongorica* (Pall.) Ebrenb.，本属（*Reaumuria* Hasselq. ex L.）则称枇杷柴属

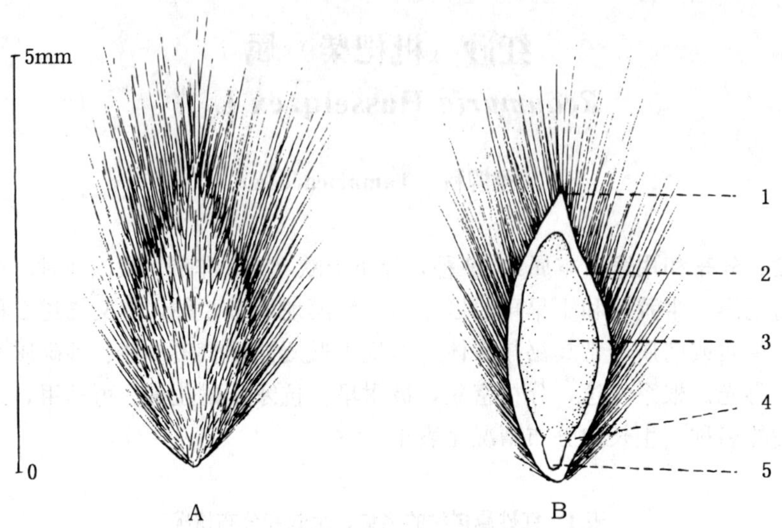

图 1 红砂种子外形（A）及其纵切面（B）

1. 种皮 2. 胚乳 3. 子叶 4. 胚轴 5. 胚根

（姚桂英绘）

表 3 红砂属树种果实和种子的形态特征

树 种	未熟果颜色	成熟果实			种 子		
		形 状	大小（mm）	颜 色	形 状	大小（mm）	颜 色
红砂	红绿色	纺锤形	长 5～6，径约 2，3 瓣裂	浅黄色	椭圆形	长 3～4	全被淡褐色毛
黄花红砂	—	椭圆形	长 8～10，光滑，3 瓣裂	黄色	长椭圆形	长 3.5～4 宽 1～1.5	全被浅褐色毛

表 4 红砂属树种的种子净度及质量

树 种	净度（%）	千粒重（g）	每克纯净种子粒数（粒）
红砂	15～37	1.1～1.5	600～900
黄花红砂	70～80①	1.2	800

① 指果实的净度

发芽和播种 种子容易发芽。红砂室内发芽率约为 50%，场圃发芽率 21%左右。曾经测定过一份黄花红砂种子，在白天给以 8 小时光照的条件下，8 天的发芽率为 56%，3 天的发芽势为 47%。播前用始温 50～55℃的水浸 36 小时即可播种。条播，行距 30～35cm，沟深 2cm，覆土厚度 0.5～1cm。播后 6～7 天即开始出苗。出土萌发。每平方米播种 0.6g。1 年生苗出圃。幼苗形态如图 2。

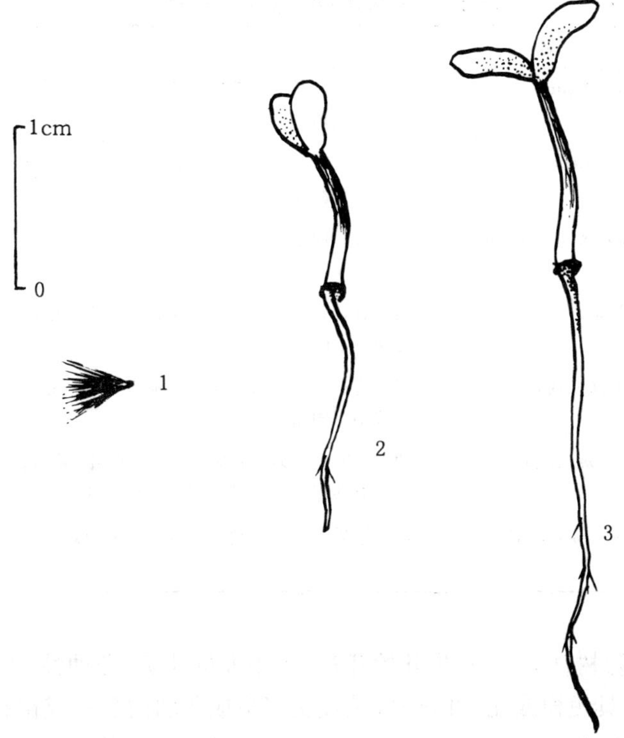

图 2　红砂种子萌发及幼苗生长情况
1. 种子　2. 子叶出土　3. 子叶展开
(姚桂英绘)

(李荣晨)

柽　柳　属
Tamarix L.

(柽柳科　Tamaricaceae)

生长习性、分布和用途　本属约 90 种，产欧洲、亚洲及非洲沙地、盐碱地和荒漠。我国有 18 种，主要分布于西北、华北，其中新疆有 15 种，占全国柽柳总数的 83%。本文描述 7 种，其中沙生柽柳为渐危种，已列入《中国植物红皮书》。落叶灌木或小乔木，高 2～5（7～8）m。叶细小，常呈鳞片状。喜光，耐寒，耐热，耐大气干旱及盐碱，也耐水湿。

本属树种生长迅速，萌生力强，是优良的固沙造林树种。树形美观，适应性强，花期长，适于庭园栽培。枝条细长，柔软而坚硬，用于各种工具小材、编织材料及薪炭材。嫩叶可做饲料，枝叶可入药。沙生柽柳根部常寄生管花苁蓉（*Cistanche tubulosa*），是著名中药。本属树种的名称、分布和用途见表 1。

表 1 柽柳属树种的名称、分布和用途

中名	学名	分布	用途
柽柳	*T. chinensis* Lour.	东北南部、华北、华东、中南。新、甘、陕、川、滇栽培	观赏、药用、燃料
长穗柽柳	*T. elongata* Ledeb.	新、青、甘、宁、内蒙古。蒙古、哈萨克斯坦、西伯利亚	固沙、饲料
甘肃柽柳	*T. gansuensis* H. Z. Zhang	新、甘、内蒙古	荒漠绿化、固沙
短穗柽柳	*T. laxa* Willd.	新、甘、青、宁、陕、内蒙古。蒙古、哈萨克斯坦、阿富汗、伊朗	固沙、饲料、绿化
细穗柽柳	*T. leptostachys* Bge.	新、青、甘、宁及内蒙古。蒙古、哈萨克斯坦等中亚国家	绿化、固沙
多枝柽柳（红柳）	*T. ramosissima* Ledeb.	西北、华北，以新疆沙漠地区最普遍。蒙古、哈萨克斯坦、阿富汗、伊朗、土耳其	固沙、饲料、药用
沙生柽柳（塔克拉玛干柽柳）	*T. taklamakanensis* M. T. Liu	新疆塔里木盆地流动沙丘区、甘肃敦煌	固沙、薪柴、饲料、管花苁蓉寄主

开花结实 多数树种 2～3 年生开始开花，年年大量结实。花两性，小，径 1.5～3(5)mm，极少数可达 7mm，具短梗或无。15～25 朵或更多集成柔弱的总状或穗状花序，侧生或顶生，或再组成圆锥状复花序。萼片 5 或 4，与花瓣同数。花粉红色、桃红色或紫红色，少数近白色。雄蕊 4～5，或 8～10，花丝分离，着生于花盘裂片间。子房上位，1 室，花柱 2～5，顶端扩大或呈棍棒状，胚珠多数着生于基生胎座上。许多种每年开花 2～3 次。花果物候期见表 2。

表 2 柽柳属树种开花结实物候

树种	观察地点	开花期	果实成熟期	种子散落期
柽柳	内蒙古巴彦淖尔盟	5～9 月	10 月	10 月
	新疆吐鲁番	5～8 月	6～9 月	6～9 月
长穗柽柳	内蒙古巴彦淖尔盟、伊克昭盟	5 月	6 月	6 月
	新疆吐鲁番	4～5 月	5 月下旬	5 月
甘肃柽柳	甘肃河西走廊	4 月下旬～5 月上旬	5～8 月	5～8 月
短穗柽柳	内蒙古阿拉善盟、巴彦淖尔盟	4 月下旬～5 月上旬	5～6 月	5～6 月
	新疆吐鲁番	3 月下旬～4 月上旬	4 月下旬～5 月上旬	4～5 月
细穗柽柳	新疆吐鲁番	6～7 月	7～8 月	7～8 月
多枝柽柳	新疆	5 月～10 月上旬	6～10 月	6～10 月
沙生柽柳	新疆吐鲁番	7～9 月	8～10 月	8～10 月

蒴果，3 瓣裂。种子多数，细小，顶端具芒柱，芒柱自基部被柔毛，长 0.5～0.7mm，个别种达 2～3mm。无胚乳。果实形态特征见表 3。种子形态见图 1。

表 3　柽柳属树种果实形态特征

树　种	形　状	大小（mm）	颜　色
柽柳	圆锥形	长 3～3.5，熟时 3 裂	—
长穗柽柳	卵状披针形	4～6	橙红色
甘肃柽柳	圆锥形	—	—
短穗柽柳	狭细圆锥形	3～6	—
细穗柽柳	长圆锥形	5～7	黄绿色或棕黄色
多枝柽柳	三角状圆锥形	3～4	—
沙生柽柳	圆锥形	长 5～7，宽 2.5	土黄色或黄灰色

果实的采收调制和种子贮藏　柽柳属各树种果熟期差异很大，有些在 5～6 月，有些在秋季，有些则从 4～9 月陆续开花结实。同一果穗上的果实成熟期亦极不一致。必须准确掌握不同种的果熟期，边熟边采。蒴果由绿变黄即表示种子成熟。过迟则果皮开裂吐絮，带簇生绒毛的种子极易随风飞散。采回后摊开晾晒，勤翻动。果实开裂，种子脱出后清除杂质。纯净种子放在通风干燥的室内晾干，种子含水量降到 10%时即可装袋干藏。新采种子发芽率可达 90%以上。一般情况下种子生活力可保持半年左右，半年后发芽率逐渐下降，下降程度与空气干湿程度有关。在干燥条件下，有些柽柳种子生命力丧失较快，不能隔年贮藏。例如长穗柽柳，采集 20 天后发芽率从 70%降至 25%，2 个月左右完全丧失发芽能力。但有些种如多枝柽柳、细穗柽柳等种子不易丧失发芽能力，贮藏至第 2 年 5～6 月间，发芽率仍能保持在 80%左右（表 4）。种子的质量在树种之间差异很大（表 5）。

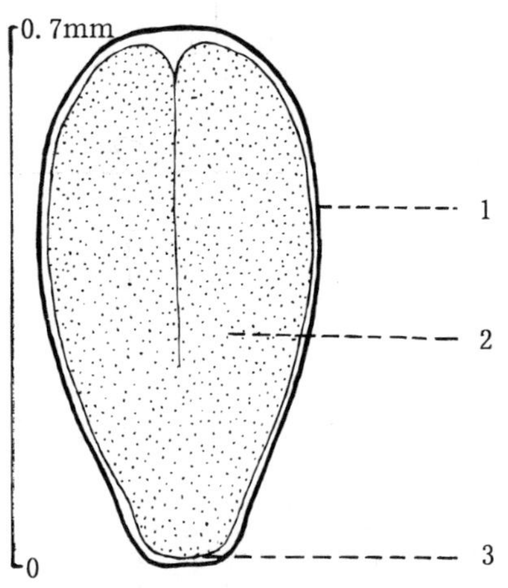

图 1　短穗柽柳种子的纵切面
1. 种皮　2. 子叶　3. 胚根
（吴新安绘）

发芽与播种　柽柳种子没有休眠习性，极易发芽。在湿润的沙地上播种，一般 24 小时即可发芽。在实验室 25～30℃条件下，3 天的发芽率可达 85%～90%。7 月以前采收的种子，最好随采随播。8 月以后采收的种子，可贮藏到第 2 年早春播种。播种可以采用水面落种法，24 小时即可发芽，2～3 天可出全苗。多枝柽柳每平方米播种 0.1～0.2g。出土萌发。沙生柽柳幼苗形态如图 2。柽柳属树种亦可插条育苗。

表 4　几种柽柳种子发芽率的变化

树　种	采集日期（年-月-日）	贮藏截止日期（年-月-日）	室内发芽率（%）
多枝柽柳	1964-08-24	1964-09-10	88
		1965-05-06	81
细穗柽柳	1964-08-24	1964-09-10	78
		1965-06-06	76
沙生柽柳	1963-10	1964-03-05	94
		1964-07-06	81

表 5　柽柳属树种的种子质量

树　种	千粒重（g）	种子长度（mm）	1g 纯净种子的粒数（万粒）
长穗柽柳	0.02	—	5
短穗柽柳	0.037	0.8	2.7
细穗柽柳	0.02	2.8	5
多枝柽柳	0.015	0.5	6.7
沙生柽柳	0.41	2～3	0.24

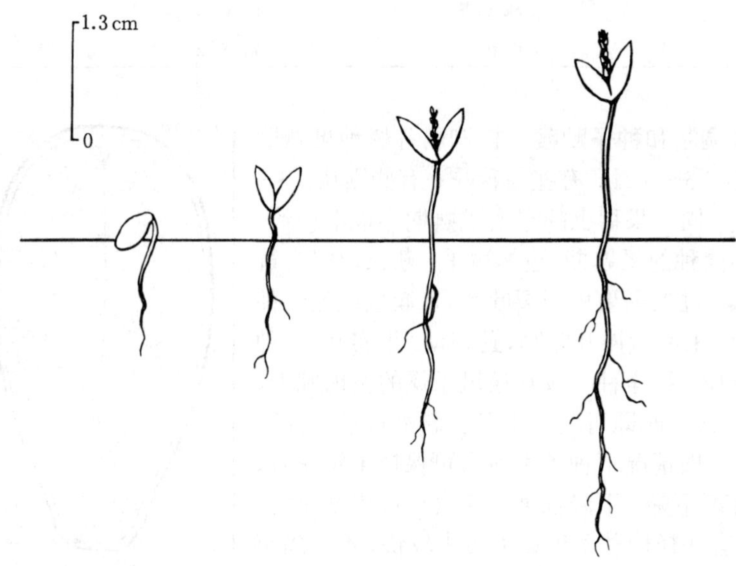

图 2　沙生柽柳萌发后第 3、8、20、30 天幼苗生长情况

（吴新安绘）

（周君英）

四　数　木

Tetrameles nudiflora R. Br.

（四数木科　Datiscaceae）

生长习性、分布和用途　四数木属只有本文描述的 1 种。落叶大乔木，高 25～45m，直径 60～120cm，具明显的大板根。喜高温湿润气候。幼树在通风且阳光水肥充足的条件下，生长迅速。产云南南部和西藏东南部，多生于海拔 500～700（1 000）m 的石灰岩山地雨林或沟谷雨林中。印度、斯里兰卡、中南半岛、马来半岛、印度尼西亚也有分布。为东南亚热带雨林典型的上层落叶树种。木材径级大，但材质较次，适于作室内用材、轻型箱板材和舢板等。由于过度采伐，且繁殖能力极差，数量日渐减少，被《中国植物红皮书》列为稀有种，亟需加强保护。

开花结实　10～12 年生开始开花结实，正常结实期在 20 年生以后，大小年现象不明显。花单性，雌雄异株，4 基数，无花瓣，开于叶前。雄花为顶生圆锥花序，有灰色短毛，无梗，花多而小；萼筒极短；雄蕊 4，与花萼裂齿对生，着生花盘周围。雌花排成下垂的穗状花序，花密而大，无梗；萼筒贴生于子房上；子房下位，1 室，具 4 个侧膜胎座，胚珠多数，顶端凹入；花柱 4，与萼裂片对生，柱头椭圆形。据西双版纳地区观测，花芽于 2 月底或 3 月初开放，始花期 3 月中旬，盛花期 3 月下旬～4 月上旬，末花期 4 月中旬，花期历时 1 个月左右；果实成熟始于 4 月下旬，盛期 5 月上旬，末期 5 月中旬。蒴果，卵状、坛状，长 4～5mm。成熟时果皮棕黄色，膜质。果自顶端花柱间开裂，散出细小种子。种子多且极小，椭圆形，微扁，长 0.5mm 以下，宽约 0.2mm，周围有褐色薄翅。有胚乳。胚直，圆柱状。四数木种子的形态和解剖结构见图 1。

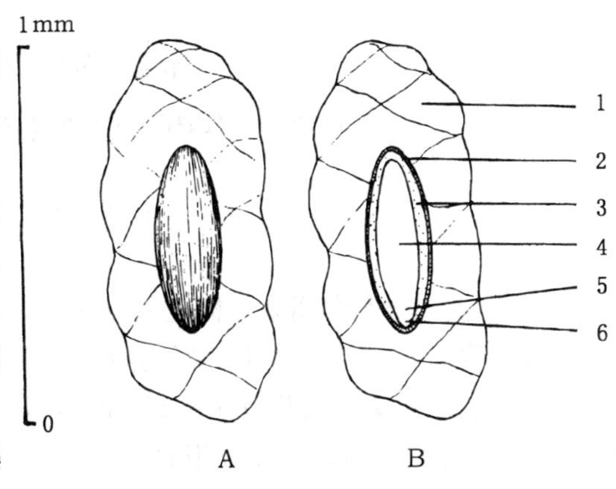

图 1　四数木种子外形（A）及其纵切面（B）
1. 种翅（外种皮）　2. 种皮　3. 胚乳　4. 子叶
5. 胚轴　6. 胚根
（马信祥、黄应钦绘）

果实的采收调制和种子贮藏　四数木树形高大，采摘困难，且果实成熟后蒴果裂开，种子随风飘散，待果穗自然脱落时果内的多数种子已经飘失。因此果实刚成熟时如遇大风，应及时捡拾吹落的果实，或在果实尚未开裂时用高枝剪带果穗采下。采集的果实放在纸箱或瓷盘中，果皮干燥开裂后稍加抖动，带翅的种子即可脱出。种子的净度在 80%左右。调制出来的种子可放室内无风处阴干。千粒重约 0.023g，每克有种子约 4 万粒。可以随采随播。室内干藏可以贮藏 1 年。在 8～9℃的低温下贮藏 2 年，发芽率仍在 70%左右。

发芽和播种　种子无休眠习性。发芽时的日均温宜在 20℃以上，气温高，发芽相应地也快。播前种子可不作任何处理。由于种粒细小，宜在装有细土的花盘内播种，播后不盖土。1989 年西双版纳热带植物园在室内用培养皿作过发芽测定：出土萌发，播后 4～5 天胚根开始萌发，发芽后 2 天子叶展开，5～10 天长出第 1 对初生叶，20 天左右第 3 片叶出现，约 22 天第 4 片叶出现。发芽的全程约为 12 天；从开始发芽至发芽的高峰日，7 天的发芽百分数为 66%；从播种之日起算，以 17 天计，发芽率为 76%。四数木种子的萌发和幼苗早期生长情况见图 2。

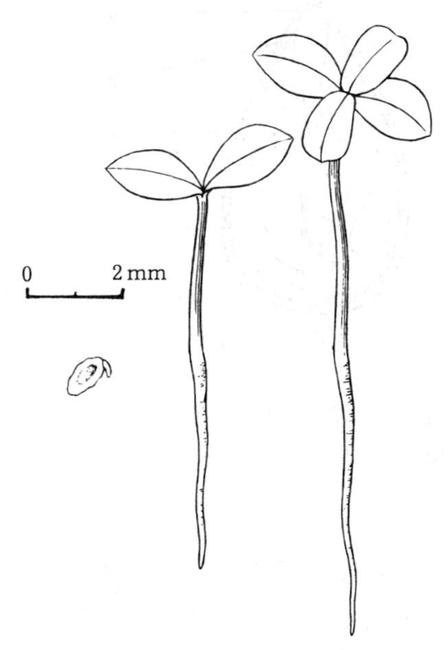

图 2　四数木种子萌发后第 1、3、10 天幼苗生长情况（马信祥、黄应钦绘）

将种子密播于播种盘内或设有塑料防雨棚的圃地上，发出 4 片初生叶后将幼苗移入容器或圃地继续培育。容器苗 3～4 个月出圃，大田苗 1 年生出圃。

（马信祥）

柄翅果（心叶蚬木）

Burretiodendron esquirolii（Lévl.）Rehd.

（椴树科 Tiliaceae）

生长习性、分布和用途 柄翅果属 4 种，我国 2 种，本文描述 1 种。落叶大乔木，高约 20m，胸径达 1m。适生于排水良好的酸性土，石灰岩发育的土壤上亦能生长。热带性树种，能耐－1℃左右的极端低温及短期霜冻，但忌冰雪，喜静风。分布于滇东南、黔南及桂西北，为我国特产。材质坚实，为珍贵用材。由于乱砍滥伐，数量已急剧减少，在《中国植物红皮书》中被列为渐危种。

开花结实 8 年生左右开始结实，20 年生以后进入正常结实年龄。结实丰年间隔期一般为 1 年。花单性，雌雄异株，数朵排列为聚伞花序。雄花苞片 2～3，萼片 5；花瓣 5；雄蕊约 30，花丝基部连生成 5 束；药 2 室，纵裂，雄蕊丛中有退化雌蕊。雌花子房具柄，5 室，每室有胚珠 2，生于中轴胎座上，花柱短。据广西南宁 1978～1984 年观察，5 月中旬始花，5 月下旬为盛花期，6 月上旬为末花期；8 月中旬果实开始成熟，8 月下旬为成熟盛期。果实成熟较为整齐。蒴果，长椭圆形或椭圆形，具 5 纵翅，长 3.5～4.4cm，径 2～2.6cm，黄褐色。果熟后 5 瓣裂，每果有种子 2～5 粒。种子三角形或倒卵形，先端尖，长 8～13mm，径 2.5～4mm，褐色或黄褐色，有光泽。子叶皱，回旋。柄翅果的果实和种子形态见图 1。

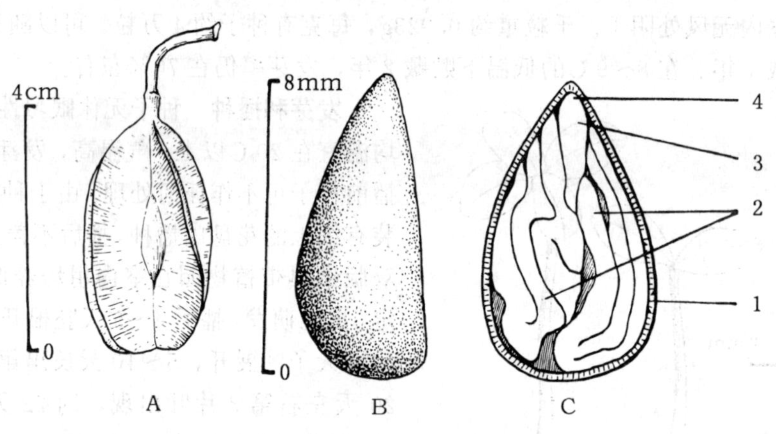

图 1 柄翅果蒴果外形（A）、种子外形（B）及其纵切面（C）

1. 种皮 2. 子叶 3. 胚轴 4. 胚根

（黄应钦绘）

果实的采收调制和种子贮藏 当果实由青色转呈黄色或黄褐色时，及时用采种刀带总果柄采下，一般不采用震落法，以免种子散失。采得的果实忌日晒，应置室内荫凉处，将果搓裂为 5 瓣，但不剥出种子，带瓣播种。鲜果的出籽率（带瓣）为 28％～35％。带瓣种子的含水量约为 11％。千粒重（带瓣）270（250～290）g，每千克有带着果瓣的种子 3 450～4 000 粒。去瓣种子千粒重约 142（138～148）g，每千克有去瓣种子 6 800～7 300 粒。常温条件下裸露存放的带瓣种子，发芽能力的保存期约20～30天。混沙贮藏的种子，越冬后发芽率略有降低，但尚可作生产用种。

发芽和播种　种子无休眠习性。发芽时日均温宜在20℃左右。发芽能力因种子贮藏时间的长短而差异颇大。随采随播的种子，发芽率可达60%，发芽也比较快而整齐。1981年广西林业科学研究所用沙藏越冬的种子在室外沙床作过发芽测定：3月上旬日均温16℃时播种，因气温低，播后33天才开始萌发，发芽盛期不明显，至5月初发芽结束，发芽率为32%。出土萌发。发芽后第3天子叶出土，约30天发出初生叶。柄翅果带瓣种子的萌发和幼苗早期生长情况见图2。

条播。播前浸种4～5小时。每平方米播带瓣种子约60g，覆土1～1.5cm。1年生苗出圃。

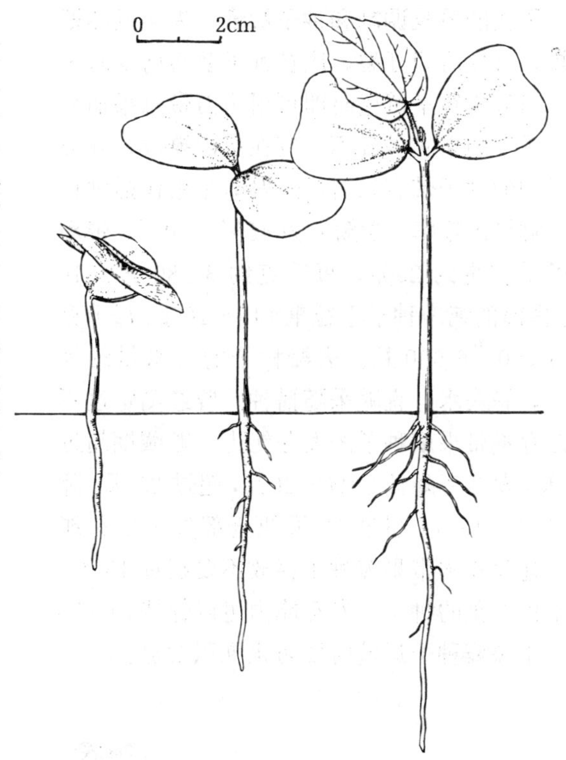

图2　柄翅果种子萌发后第5、10、35天幼苗的生长情况
（黄应钦仿《热带亚热带主要树种采种育苗技术》）

（王宏志）

蚬　木

Excentrodendron hsienmu（Chun et How）Chang et Miau

（椴树科　Tiliaceae）

生长习性、分布和用途　蚬木属5种，我国4种，本文描述1种。常绿大乔木，高达30m以上，胸径1m以上。喜肥沃湿润的石灰性土。分布于广西南部和云南东部。越南北部也有分布。为著名珍贵硬木，耐纵切，宜作砧板及高级家具。在《中国植物红皮书》中列为渐危种。

开花结实　约15年生开始开花结实，正常结实期在30年生以后，结实丰年间隔期一般为1年。花两性，稀单性，腋生圆锥花序。萼片5。花瓣5（3～9）。雄蕊24～40，基部连成5束。药2室，纵裂。子房无柄，5室，每室有胚珠2，着生于中轴胎座上。花柱5，极短。据1978～1984年广西南宁观测，3月下旬始花，4月上旬为盛花期；4月中旬为末花期，6月上旬果实开始成熟，6月中旬为果熟盛期。蒴果，未熟时绿色，熟后黄色，被短柔毛，椭圆形，具5纵翅。果熟后室间开裂，每室1种子，种子近似三角形或倒卵形，先端尖，长13～17mm，径4～7mm，棕褐色，有光泽。子叶回旋或褶叠。蚬木种子（带果瓣）的形态见图1。

果实的采收调制和种子贮藏　果实成熟期较短，且熟后即开裂，应在种子散落前及时采集。当果皮转呈黄色时即可用钩刀带总果柄采下。采回的果实忌日晒，应在室内摊放，开裂后得到的种子带有果瓣，一般即可用作播种材料。鲜果出籽率（含瓣）为 45%～50%，带瓣种子千粒重约 290g，每千克约 3 380～3 510 粒。去瓣的纯净种子千粒重 145～165g，每千克约 6 000～6 900 粒。去瓣种子的含水量约为 13%，忌失水，宜随采随播种。曾经测定过不同贮存期带果瓣种子的发芽能力：贮藏期超过 15 天，发芽率降低 10%～20%，超过 25 天，降低35%～40%；超过 50 天便全部失去发芽能力。建议蚬木带果瓣种子存放不要超过 15 天。混干沙贮藏的种子，发芽能力可以保持40～50 天。带果瓣种子运输时要防止堆积发热。

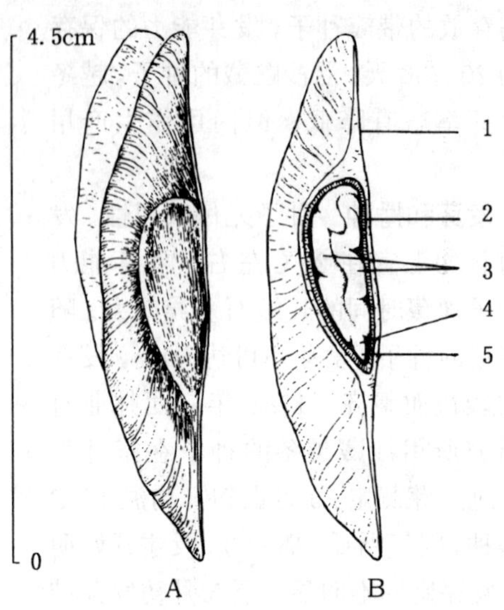

图 1　蚬木带果瓣种子的外形（A）及其纵切面（B）

1. 果瓣　2. 种皮　3. 子叶　4. 胚轴　5. 胚根

（黄应钦绘）

发芽和播种　种子无休眠习性。发芽时日均温宜在 25℃以上。1981 年和 1983 年 6 月下旬，广西林业科学研究所 3 次在室外沙床用新采种子作过发芽测定：播种时日均温约 28℃，播后第 2 天萌发并立即进入发芽盛期；3 天的发芽百分数为 75%；从播种之日起算，以 7～8 天计，发芽率为 90%。出土萌发。种子萌芽后第 3 天子叶出土，第 25～28 天发出初生叶。蚬木种子的萌发和幼苗初期生长情况见图 2。

条播。新鲜种子播前可不作处理。经过运输或存放数日的种子，播前应浸水 2～3 小时，捞出稍晾干播种。生产上一般带瓣播种，覆土1.5～2cm。每平方米播种 20～25g。幼苗期需遮荫，1～2 年生苗出圃。

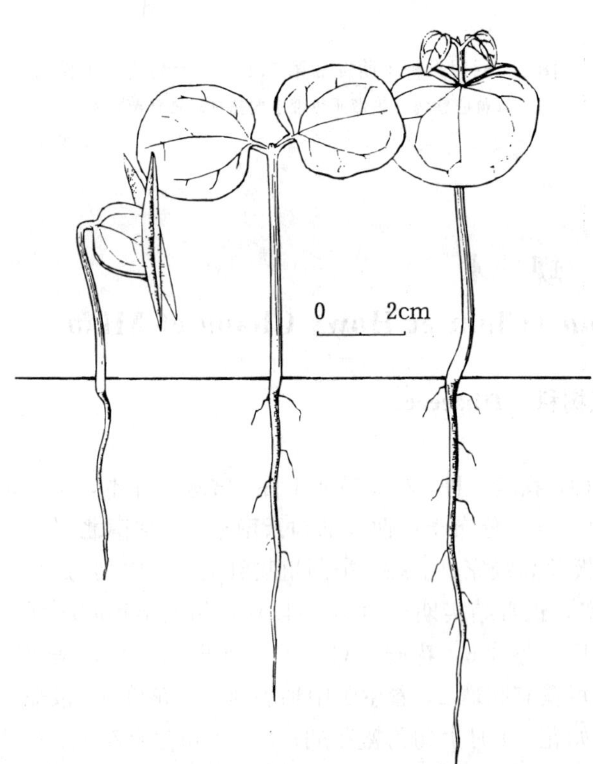

图 2　蚬木种子萌发后第 5、10、30 天幼苗生长情况

（黄应钦仿《热带亚热带主要树种采种育苗技术》）

（王宏志）

小花扁担杆（孩儿拳）

Grewia biloba G. Don var. *parviflora*（Bunge）Hand. -Mazz.

（椴树科　Tiliaceae）

生长习性、分布和用途　扁担杆属约 90 余种，我国 26 种，本文仅描述小花扁担杆这个变种。落叶灌木或小乔木，高 1～4m，最高可达 8m。喜生于丘陵、平原或低山阳坡灌丛中。耐寒、耐旱，也耐土壤瘠薄。产于东北南部、华北、华东及西南地区。朝鲜半岛也有分布。成熟果实红色，鲜丽，久留枝上，引人注目，可作观果树种。茎皮纤维可供纺织，根、枝、叶入药。

开花结实　每年开花结实，未见大小年现象。花两性。聚伞花序近伞形，与叶对生，具花 10 余朵或 3～4 朵。萼片 5，分离，绿色，条状披针形，长 5～6mm，有腺。花瓣 5，细小，淡黄色，长约 1.2mm。雄蕊多数，离生。子房上位，2～4 室，每室具 2 至数个倒生胚珠。花柱单一，柱头浅裂。据在北京地区观察，开花期自 6 月中旬～7 月上旬，6 月下旬最盛；果实 8 月下旬开始成熟，9 月下旬进入盛期，10 月中旬开始落果，10 月下旬落果最多，最迟至 11 月上旬。核果，中部稍缢缩，呈2（4）裂状，红色，径 8～12mm，长 8～9mm，每裂有 2 小核。果核倒宽卵形，或宽肾形，一侧边直，浅黄色，长 5.1～7.0mm，宽 4.1～5.4mm，厚3.1～3.6mm，表面有小凹坑。胚乳富含油脂。胚直，子叶宽而扁，胚根短，稍偏斜（图 1）。

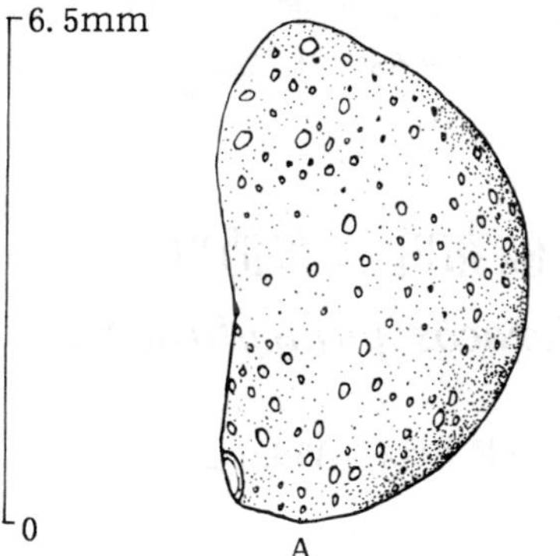

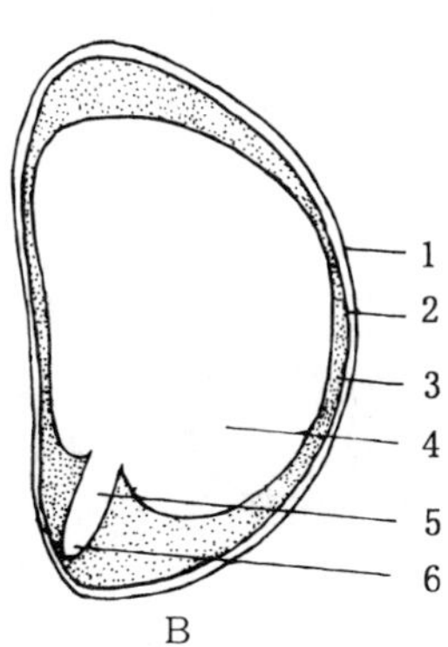

图 1　小花扁担杆果核外形（A）及其纵切面（B）
1. 内果皮　2. 种皮　3. 胚乳　4. 子叶　5. 胚轴　6. 胚根
（胡冬梅、林平绘）

果实的采收调制和种子贮藏　果实成熟时由绿色变为红色或橙色，果肉变软，果核变硬。为避免鸟类啄食，宜随熟随采。采后揉烂果肉，并用水漂除杂物，洗净表面。所得果核（分核）即为播种材料，通称种子。未洗净的果核在贮藏期间易生霉菌。出籽率 50%～60%。净度 80%～90%。发芽率 20%～30%。千粒重 41～46g。每千克有种子 2.1 万～2.4 万粒。种子晾干后装入袋内，存放于冷凉处即可。在北京地区一般室内条件下发芽能力可以保存 1 年。

发芽和播种　种子无休眠习性。发芽测定时可将种子排放在发芽皿内的滤纸床上。在20～25℃的温度条件下第 9 天开始发芽，约 1 个月发芽结束。出土萌发（图 2）。春季播种。床面条播，覆土 1.5cm。也可用分株方法繁殖。

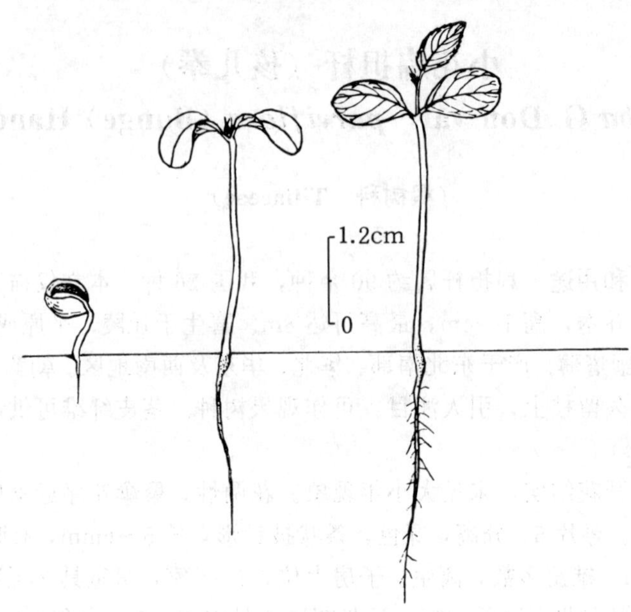

图 2　小花扁担杆种子萌发出土后第 1、6、13 天的幼苗生长情况

（胡冬梅、林平绘）

（刘长江）

破布叶（布渣叶）

Microcos paniculata L.

（椴树科　Tiliaceae）

生长习性、分布和用途　破布叶（布渣叶）属约 60 种，我国有 3 种，本文描述 1 种。半落叶乔木，高约 12m，胸径 40～50cm。能耐 0℃左右的极端低温。分布于滇、桂、粤、琼。越南、印度、印度尼西亚亦产。树皮纤维可作人造棉、编绳索。叶药用。果可食，种子榨油。

开花结实　10～12 年生开始结实，15 年生以后进入正常结实期，结实无大小年现象。花两性，具柄，排成 3 花的小聚伞花序，再复排成圆锥花序式，顶生或生于上部叶腋。花淡黄色。萼片 5，长于花瓣。花瓣 5，内侧具腺体。雄蕊多数。子房上位，球形，3 室，每室有胚珠 4～6。花柱单生，柱头钻形。据 1985～1987 年云南西双版纳勐仑的观察，4 月上旬花芽始现，4 月中旬～下旬花序形成，5 月上旬始花，5 月中下旬盛花，6 月上旬花期结束，幼果形成；9 月下旬果实开始成熟，10～11 月为果实成熟盛期。

核果，近球形或倒卵形。成熟时外果皮由绿色转变为黑褐色，表面有光泽。果长 8～10mm，径 6～7mm，成熟后 3～5 天脱落。中果皮除有少量的果肉外，多数为编织有序的纤维层。内果皮骨质，有棱。果核卵形或球形，长 6～7mm，径 3～4（7）mm。每核有种子 1 粒。种子具胚乳。子叶矩圆形。破布叶果核的形态见图 1。

果实的采收调制和种子贮藏　果实成熟盛期用竹竿敲打，在树下拾集熟落的果实。置水中搓揉淘洗，去掉果皮、果肉等杂质，所得附有纤维状毛(中果皮的残余物)的果核即为播种材料，通称种子。鲜果出籽率为45%～50%。种子(含纤维毛)千粒重为120～140g，每千克有种子7 100～8 500粒。调制后晾干的种子含水量为20%～25%。种子裸露存放1个月左右发芽能力便会明显下降。一般混干沙贮藏，发芽能力可保持半年。用干燥器贮藏效果亦佳。

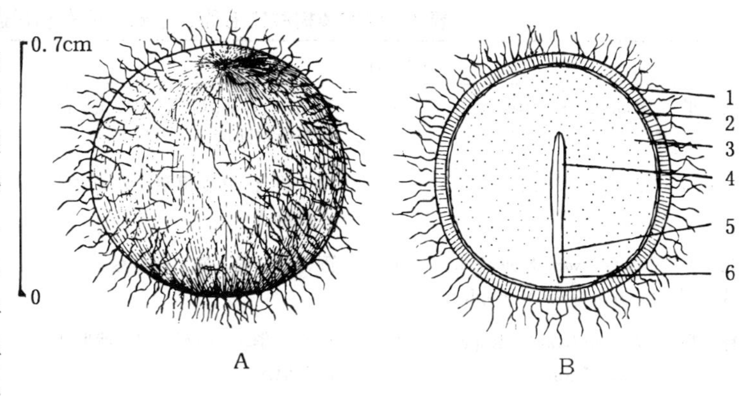

图1　破布叶果核外形（A）及其纵切面（B）
1. 内果皮　2. 种皮　3. 胚乳　4. 子叶　5. 胚轴　6. 胚根
（黄应钦绘）

发芽和播种　种子有休眠现象。发芽时日均温宜在22～25℃。1984年10月20日，西双版纳热带植物园用新采种子作过发芽测定，基质为细沙：翌年4月4日开始出土，4月20日发芽结束，没有明显的发芽盛期，发芽率为25%。1987年10月底采种，用干燥器贮藏至翌年7月上旬播种，播后15天开始发芽出土，也没有明显的发芽盛期，发芽全程21天，发芽率为30%。出土萌发。子叶出土后6天左右长出第1片初生叶，10天后长出第2片初生叶。

条播。播前用始温60℃水浸种至自然冷却，捞出后晾干即可播种。每平方米约播35g，覆土0.5～1.0cm。1年生苗出圃。

（马信祥）

椴　树　属

Tilia L.

（椴树科　Tiliaceae）

生长习性、分布和用途　本属约80种，主产北温带，我国有32种，本文描述4种1变种。落叶乔木。它们的名称、分布、生长和用途见表1。

开花结实　椴树属树种10年生以前生长较迟缓，10年以后加快，年高生长可达0.5m或更多。表2列出了紫椴等3个树种的结实年龄及丰年间隔期。

椴树属花两性，整齐，黄色、黄白色或白色，有香味，多成下垂的聚伞花序，花序梗下半部或基部有1枚带状苞片连生。萼片5。花瓣5。雄蕊多数，离生或基部连合成5束，有时具花瓣状退化雄蕊，且与花瓣对生。子房无柄，5室，每室具2胚珠。花柱细长，柱头5裂。果实核果状，稀浆果状，球形，卵形或倒卵形。各树种的物候期已汇入表3。果实内含1～3粒种子，一般仅1个发育。具胚乳，子叶掌状5裂(图1)。果实和种子的形态特征见表4。

表 1　椴树属树种的名称、生长、分布和用途

中　名	学　名	树高（m）	分　布	用　途	供　稿
紫　椴	*T. amurensis* Rupr.	30	东北、鲁、冀、晋。朝鲜半岛、俄罗斯远东地区	材用、蜜源、观赏、纤维	101、107
心叶椴	*T. cordata* Mill.	28	原产欧洲。新疆伊犁、塔城地区引种栽培	材用、蜜源、油脂	204
光　叶糯米椴	*T. henryana* Szysz. var. *subglabra* V. Engl.	15	苏、浙、皖、赣、鄂、豫	材用、蜜源、纤维	1003
糠　椴	*T. mandshurica* Rupr. et Maxim.	20	东北、华北。朝鲜半岛、俄罗斯远东地区	材用、药用、蜜源、观赏	101、107
南京椴	*T. miqueliana* Maxim.	15	苏、浙、皖、赣。日本	材用、纤维	1003

表 2　椴树属树种结实年龄及丰年间隔期

树　种	开始结实年龄	正常结实年龄（年）	丰年间隔期
紫　椴	15 年左右	15～50	0～1
心叶椴	8～10 年	—	0
糠　椴	15 年左右	15～30	0～1

表 3　椴树属树种开花结实物候期

树　种	观察年份和地点	花　期			果实成熟		果实散落期
		始　期	盛　期	末　期	初　熟	末　熟	
紫　椴	1963～1980 哈尔滨	7 月 11 日	7 月 13 日	7 月 17 日	8 月 16 日	10 月 6 日	晚秋、冬季或翌春脱落，初熟至脱落间隔期为 20～28 天
	1991 黑龙江五常县	7 月 13 日	7 月 18 日	7 月 25 日	8 月 30 日	10 月 7 日	10 月 10 日开始脱落
心叶椴	—	6 月中旬	～	7 月底	9 月	9 月	11 月脱落，老树可至翌春
光叶糯米椴	—	—	7 月	—	—	9 月下旬	—
糠　椴	1963～1980 哈尔滨	6 月下旬	～	7 月下旬	8 月 19 日	9 月 21 日	初熟至脱落间隔期为 10～18 天，晚秋落尽
	1991 黑龙江五常县	7 月 7 日	7 月 12 日	7 月 20 日	8 月 26 日	10 月 1 日	10 月 5 日开始脱落
南京椴	—	—	6 月	—	—	9 月下旬	—

果实的采收调制和种子贮藏　每年 9～10 月份果实呈现成熟颜色时采收。可以直接从树上采摘或用钩剪采摘，或用长杆打落后地面收集。采后经日晒或阴干，去杂，所得纯净的果实即为播种材料，通称种子。它们的出籽率、净度和有关质量的数据请见表 5。紫椴与糠椴种子含水量 10%～12%，密封瓶中置于低温室内，可贮 2～3 年。心叶椴种子含水量不低于 12%～15%，装入木箱或袋内置低温室内，也可存放 2～3 年。

表 4　椴树属树种果实的形态特征

树　种	成熟果实		
	形　状	大小（cm）	色　泽
紫　椴	球形或椭圆形，密被褐色绒毛	长 0.5～0.7 径 0.5～0.6	褐黄色
心叶椴	卵圆形或扁球形，有轻度棱角	长 0.5～0.7 径 0.4～0.5	灰褐色
光叶糯米椴	近球形	径 0.5～0.6	—
糠　椴	球形或椭圆形，密被黄褐色绒毛	长 0.8～1.0 径 0.7～0.9	黄褐色
南京椴	卵球形，基部有 5 棱，表面有星状毛	长 1.1 左右 径 0.7～0.9	棕灰色

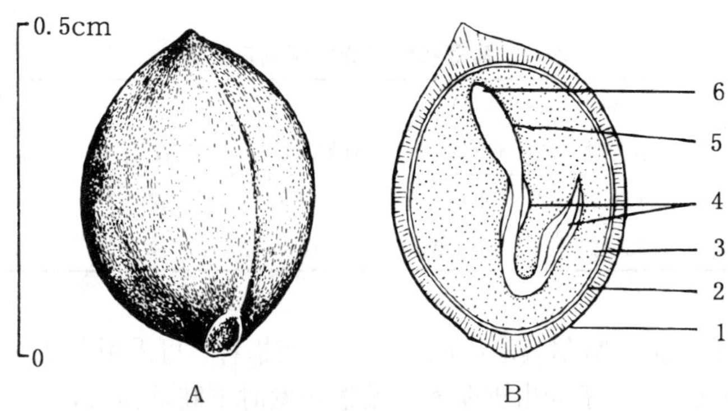

图 1　紫椴果实的外形（A）及其纵切面（B）

1. 果皮　2. 种皮　3. 胚乳　4. 子叶　5. 胚轴　6. 胚根

（梁鸣、黄应钦绘）

表 5　椴树属播种材料的有关数据

树　种	出籽率（%）	净度（%）	千粒重（g）	每千克纯净播种材料粒数（万粒）
紫　椴	75～90	85～95	35～38	2.6～2.9
心叶椴	80	95	24～27	3.7～4.2
光叶糯米椴	—	—	70～80	1.2～1.4
糠　椴	60～80	85～95	114～125	0.8～0.9
南京椴	—	—	165～268	0.37～0.61

发芽和播种　椴树属的果实就是播种材料，因果皮结构致密，透水性差，加上胚有生理性深休眠，播前需要变温层积催芽。未经层积处理的播后当年很少发芽，甚至不发芽。紫椴、心叶椴种子层积处理的关键是必须有 0～5℃的低温阶段，可以在低温层积后再用高温（10～20℃）层积或先高温层积再用低温层积。这样处理后发芽率一般可达 50%～80%。破伤种皮和浓硫酸伤蚀等方法的效果均不及低温层积。另外，还发现心叶椴 9 月上旬采摘带绿色的果实并立即秋播，翌春出苗整齐。糠椴和糯米椴采用过隔年埋藏法。表 6 汇编了 3 种椴树的层

积处理条件、层积期及其效果。表 7 列出了紫椴和糠椴发芽测定的实例。

表 6　椴树属树种果实层积处理条件、层积期及其效果

树　种	层积条件	层积效果
紫　椴	(1) 10～20℃下层积 30～45 天后转入 0～5℃下层积 45～60 天	种子裂口率达 30%，播后发芽率可达 40%～90%。层积前种子存放时间长的效果较差
	(2) 雪藏后混沙置于 20℃左右 20～30 天	
心叶椴	(1) 0～6℃层积 90 天后在 20℃左右层积 10 天	发芽率达 50%～80%
	(2) 20℃左右层积 30 天后，在 0～6℃下层积 60 天，再用 20℃左右层积 10 天	
糠　椴	埋藏 240～270 天	种子裂口率达 50%，播后发芽率可达 50%

表 7　紫椴和糠椴的发芽测定实例

树　种	预处理	发芽测定温度（℃）		发芽势（%）		发芽率（%）	
		有光时	无光时	计算天数	一般数值	计算天数	一般数值
紫　椴	见表 6（1）	15～25	8～12	10	42	15	63
糠　椴	见表 6	15～25	8～12	10	37	15	54

春播（部分亦可秋播），哈尔滨地区应于 4 月下旬播种。可采用垄播或苗床条播。椴树属 3 个树种的播种数据见表 8。子叶出土萌发。紫椴萌发时下胚轴伸长，子叶带壳出土，5～10 天后种壳脱落，子叶张开，10～15 天出现初生叶（图 2）。

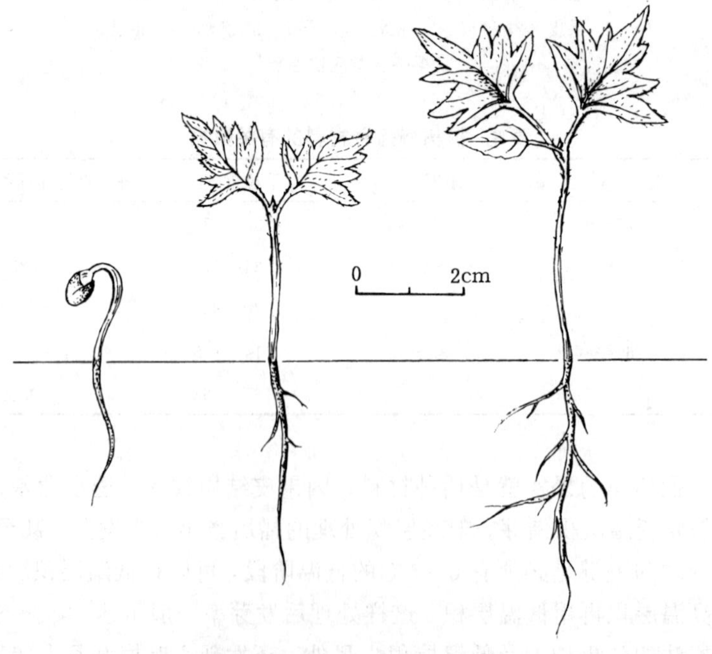

图 2　紫椴种子萌发后第 3、9、16 天的幼苗生长情况（曹雅范绘）

表 8　椴树属树种的播种育苗数据

树　种	播种方式	播种量 (kg/666m²)	覆土厚 (cm)	当年苗高 (cm)	出圃年龄 (年)
紫　椴	垄播	13	1.5～2.0	40～50	1～2
	床播	20～25	1.5～2.0	40～50	1～2
心叶椴	床播	12～15	1.5～2.0	—	2
糠　椴	垄播	20	2.5～3.0	15～25	1～2

（周陛勋、周德本）

杜　英　属
Elaeocarpus L.

（杜英科　Elaeocarpaceae）

生长习性、分布和用途　本属约 200 种，我国有 38 种，本文描述 8 种。常绿小乔木或乔木。喜湿热气候，不耐严寒。适生于肥沃湿润的酸性土。为培育香菇的优质材。它们的名称、生长、分布及用途见表 1。

表 1　杜英属树种的名称、生长、分布和用途

中　名	学　名	树　高 (m)	胸　径 (cm)	分　布	用　途	供　稿
尖叶杜英	*E. apiculatus* Mast.	25	60	琼、粤、桂、滇。中南半岛、马来西亚	材用、水源、香菇、蜜源	601
圆果杜英	*E. ganitrus* Roxb.	20	40	粤、琼、桂。南亚至东南亚	材用、香菇、水源、绿化	601
水石榕（海南杜英）	*E. hainanensis* Oliv.	8	15	粤、琼、桂、滇。越南、泰国	绿化、蜜源	601
薯豆杜英（胆八树）	*E. japonicus* Sieb. et Zucc.	20	40	滇、川、黔、桂、粤、琼、湘、赣、闽、皖、浙。日本、越南	材用、香菇、蜜源、绿化	601
绢毛杜英（绢光杜英）	*E. nitentifolius* Merr. et Chun	20	40	滇、桂、粤、琼	材用、水源	1002
锡兰杜英（皱皮橄榄）	*E. serratus* L.	20	40	原产印度东部至印度尼西亚。粤、琼、桂、滇栽培	果食用、材用、绿化	601
山杜英	*E. sylvestris* (Lour.) Poir.	25	50	川、滇、黔、桂、粤、琼、湘、赣、闽、台、浙、皖。越南、老挝、泰国	材用、栲胶、纤维、香菇	601
粗壮杜英	*E. tectorius* (Lour.) Poir.	20	70	滇、桂。越南	食用、材用、药用、蜜源	616

开花结实　水石榕和锡兰杜英 5～6 年生开始结实，其余 6 种约在 8 年生开始结实。正常

结实年龄在15年生以后，丰年间隔期1～2年，但不甚显著。花两性，稀杂性(绢毛杜英)，白色，下垂。总状花序腋生或着生于去年枝条上。萼片4～6。花瓣通常4～6，与萼片等长，分离，着生于花盘基部。花瓣顶端有撕裂，以水石榕最为显著，裂片多达30条，稀近全缘或有数浅齿。雄蕊多数(10～50)，着生于花盘内，集生成束。花药2室，线形，顶孔开裂，药隔有时突出成芒刺状或毛丛。子房上位，2～5室，每室有胚珠2～6，通常仅有1室发育，花柱钻状。水石榕子房无毛，其余7种均被毛。广西南宁和云南普文观察的开花结实物候期见表2。

表2 杜英属树种的开花结实物候期

树种	观察地点	观察年份	开花			果实成熟		果实脱落
			始期	盛期	末期	始期	盛期	
尖叶杜英	南宁	1987～1988	5月中旬	5月下旬	6月中旬	9月上旬	9月下旬	10月上旬～下旬
圆果杜英	南宁	1987～1989	4月中旬	4月下旬	5月中下旬	11月上旬	11月中下旬	11月下旬～12月中旬
水石榕	南宁	1978～1984	6月上旬	6月中旬	6月下旬	9月上旬	9月中旬	9月下旬～10月下旬
薯豆杜英	南宁	1978～1984	4月上旬	4月中旬	5月上旬	8月上旬	8月下旬	9月上旬～下旬
绢毛杜英	南宁	—	4月	～	5月	10月	～	11月
锡兰杜英	南宁	1978～1984	6月上旬	6月中旬	6月下旬	10月中旬	10月下旬	11月上旬～下旬
山杜英	南宁	1978～1984	5月中旬	6月上旬	6月中下旬	9月中旬	9月下旬	10月上旬～12月下旬
粗壮杜英	普文	1986～1987	6月中旬	6月下旬	7月上旬	9月下旬	10月中旬	11～12月

核果。外果皮薄膜状，有光泽。中果皮肉质。内果皮骨质，坚硬，表面具沟纹，1～5室，每室1种子。胚乳肉质。子叶薄而平坦。果实及种子形态见表3、图1。

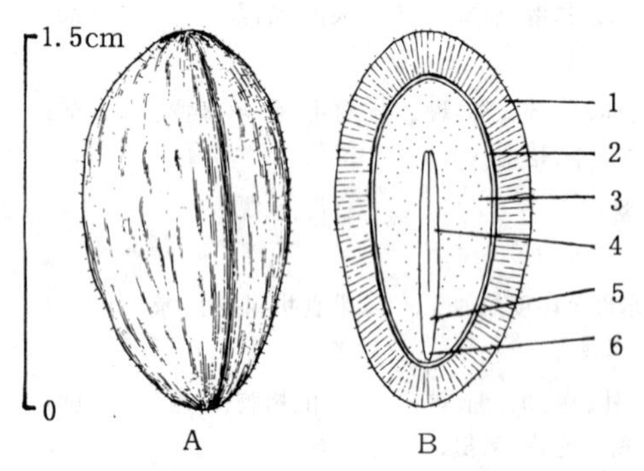

图1 绢毛杜英果核外形(A)及其纵切面(B)
1. 内果皮 2. 种皮 3. 胚乳 4. 子叶
5. 胚轴 6. 胚根
(黄应钦绘)

果实的采收调制和种子贮藏 果实成熟盛期用竹竿敲打或摇动树枝，震落后在地面捡拾。采得的果实，除锡兰杜英可以加工作凉果或剥出果肉腌渍外，一般是堆沤或置水中浸泡数天，软熟后装入竹筐，反复搓洗漂淘，清除皮肉等杂质。得出的果核用作播种材料，通称种子。种子(果核)摊放于室内通风荫凉处1～2天，将表层所附水晾干即可，不能曝晒，以免失去发芽能力。种子的含水量宜保持在25%上下。运输可在种子稍晾干后用麻袋或竹筐包装。贮藏时需混湿沙。在避光荫凉处贮藏期一般为半年左右。杜英属的出籽率和种子质量等数据见表4。

表 3　杜英属树种果实及种子(果核)的形态特征

树　种	未熟果颜　色	成熟果实			种　子(果　核)		
		形　状	大小(cm)	颜　色	形　状	大小(mm)	颜　色
尖叶杜英	青色	纺锤形或椭圆形	长 2.2～3.6 径 1.6～2.3	黄褐色	扁纺锤形,表面有不规则突起的尖刺	长 20～32 宽 14～20	深褐色
圆果杜英	青绿色	圆球形	径 1.5～2	青紫色	圆形,表面有不规则突起	径 12～18	深褐色
水石榕	淡绿色	长卵形	长 2～4 径 0.8～1.2	淡黄色或紫黄色	棒状,两端尖,表面具纤维	长 18～36 宽 6～8	浅褐色
薯豆杜英	青色	椭圆形	长 0.9～1.5 宽 0.7～0.9	紫蓝色	椭圆形,表面有 6 条棱纹,具纤维	长 7～13 宽 5～7	褐色
绢毛杜英	青色	椭圆形或长卵形	长 1.1～2.0 径 0.8～1.1	蓝黑色	长椭圆形,表面有不规则棱状突起	长 10～15 径 6～8	褐色
锡兰杜英	青色	卵形或椭圆形	长 1.8～2.5 径 1.4～1.7	淡黄色	椭圆形,具 3 条纵槽,表面粗糙	长 15～23 径 11～14	浅褐色
山杜英	青色	椭圆形	长 1～1.3 径 0.7～0.9	青紫色	椭圆形,表面有 6 条棱纹,被绒毛	长 8～11 径 5～7	褐色
粗壮杜英	青色	椭圆形	长 3～3.5 径 2～2.5	黄绿色	椭圆形,顶端尖,表面粗糙	长 25～32 径 17～22	褐色

表 4　杜英属树种果实的出籽率和种子净度、质量

树　种	出种率(%)	净度(%)	千粒重(g)	每千克种子粒数(粒)
尖叶杜英	35	97	1 500～1 900	600～670
圆果杜英	30～35	98	1 200～1 500	670～830
水石榕	60～70	90	500～900	1 100～2 000
薯豆杜英	36	92	180～230	4 300～5 600
绢毛杜英	—	—	300～350	2 800～3 300
锡兰杜英	22	97	1 200～1 600	630～830
山杜英	30	92	150～170	5 800～6 700
粗壮杜英	25～30	90	2 700～2 900	340～370

表 5　杜英属树种的发芽能力及其测定条件(室外沙床)

树　种	播种日期	发芽始期	发芽终期	发芽势(%)		发芽率(%)	
				计算天数	一般数值	计算天数	一般数值
尖叶杜英	1980-10-14	1980-10-28	1981-07-02	—	不明显	261	51
水石榕	1978-09-20	1978-11-15	1978-12-09	20	72	79	90
薯豆杜英	1987-09-01	1987-12-31	1988-04-30	—	不明显	242	50
锡兰杜英	1981-03-17	1981-04-22	1981-06-18	—	不明显	93	48
粗壮杜英	1988-03-05	1988-04-05	1988-06-05	25	45	92	60

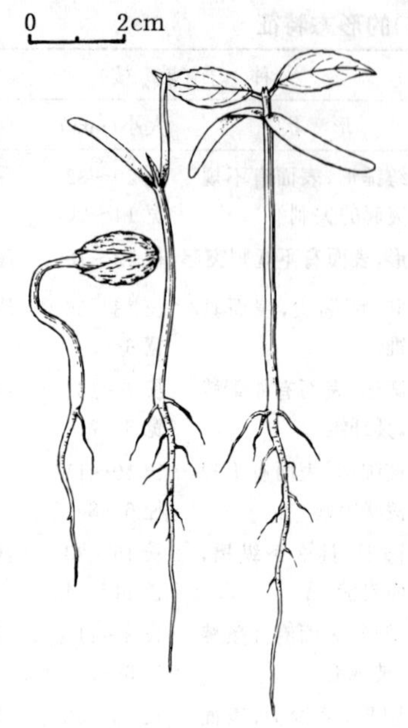

图 2 绢毛杜英种子萌发后第 5、7、15 天幼苗的生长情况

(黄应钦仿《热带亚热带主要树种采种育苗技术》)

发芽和播种 多数树种的种子无休眠习性，部分种有中等程度的休眠。除水石榕等少数种外，多数种发芽不整齐。发芽时日均温宜在 15℃以上。1978、1980 和 1987 年，广西林业科学研究所和云南林业科学研究所在南宁和普文对尖叶杜英等 5 个种作过发芽测试，情况见表 5。据某次在广西南宁测定，绢毛杜英种子的发芽率约为 60%。

除水石榕等少数树种外，一般不宜随采随播，以层积越冬后翌年 3 月播种较为适宜。经过层积催芽的种子，发芽过程可以缩短 2～3 个月。出土萌发。绢毛杜英的胚根萌发后 5 天子叶出土，7～8 天初生叶展出。其余数种，胚根萌发后 9～16 天子叶出土，20～30天初生叶展出。绢毛杜英种子的萌发和幼苗早期生长情况见图 2。

条播。每平方米尖叶杜英、锡兰杜英和圆果杜英播种 200～300g；水石榕播种 75～100g；山杜英播种 20～30g；绢毛杜英播种 40～50g；粗壮杜英播种 350～450g。培育 1 年生苗出圃。

(王宏志)

文 定 果

Muntingia colabura L.

(杜英科 **Elaeocarpaceae**)①

生长习性、分布和用途 文定果属 3 种，我国引入本文描述的 1 种。常绿小乔木，高达 10m。对土壤要求不严，抗风性能强。温度降至 0℃时，植株受冻害。原产热带美洲和西印度群岛，现世界热带地区已广泛栽培。我国琼、粤、桂、滇、闽、台零星栽植于庭园，花期长，供观赏。台湾嘉义、台南、高雄等地用作行道树。果味香甜可口，富含钙、磷、铁、胡萝卜素、维生素 C 以及维生素 B_2 和碱烟酸等。叶可代茶。

开花结实 2 年生即可开花结实，正常结实期在 5 年生以后，结实大小年现象不明显。花两性，单生或成对着生于上部小枝的叶腋。花萼合生，深 5 裂，很少 6～7 裂。花瓣白色，具瓣柄，全缘。花盘杯状。雄蕊多数，着生于花盘外面，花药纵裂。子房具柄，5～6 室，每室有胚珠多数。花柱极短，柱头 5～6 裂。据海南儋县的物候观察，全年有花开放，盛花期为 3～4 月，一年四季有果成熟，6～8 月为果熟盛期。果为多汁浆果，圆形，径 4～6mm，成熟时红色，内含多粒种子。种子椭圆形，极细小，长约 1mm。有胚乳。种子的形态见图 1。

① 也有学者将本属归于椴树科 Tiliaceae——编者

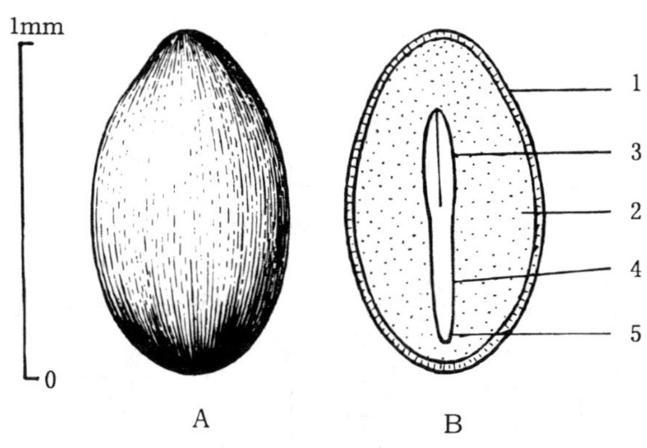

图 1　文定果种子外形(A)及其纵切面(B)
1. 种皮　2. 胚乳　3. 子叶　4. 胚轴　5. 胚根
(黄光郁、黄应钦绘)

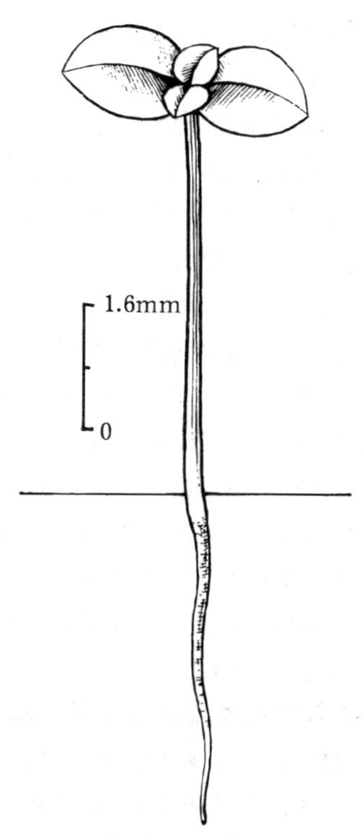

图 2　文定果种子萌发后第 9 天幼苗的生长情况
(黄光郁、黄应钦绘)

果实的采收调制和种子贮藏　果实由青色转为黄色或鲜红色时即可采收。果实成熟时用手采摘,或打落后在地面捡拾。采回的果实或食去果肉,或堆放 1～2 天,软熟后装入密布袋内,置水中将果浆挤出。文定果种子所附的果浆胶质较浓,要经过多次搓擦,反复漂洗过滤,方可得出播种材料。鲜果的出种率约为 0.8%。种粒极其细小,难以计算出种率及千粒重等数据。调制出来的种子忌日晒,贮藏期一般不宜超过半月。

发芽和播种　种子无休眠习性。发芽时日均温宜在 25℃以上。1989 年 6 月 6 日,华南热带作物研究所在室内以花盆的沙壤土作发芽基质,用新鲜种子作过发芽测定:播后半个月(6 月 20 日)开始发芽,6 月 29 日发芽结束。从开始发芽之日起算的 6 天中,发芽百分数为 40%;从播种之日起算,以 24 天计,发芽率为 50%。出土萌发。子叶出土后 7～9 天展出 1 对初生叶(图 2)。

目前尚无大田育苗经验,多用沙盘育苗。每平方米播种约 1g。幼苗稍呈木质化时移至容器内继续培育到次年春季出圃。亦可用高空压条繁殖。

(丁慎言)

猴　欢　喜

Sloanea sinensis (Hance) Hemsl.

(杜英科　Elaeocarpaceae)

生长习性、分布和用途　猴欢喜属约 120 种,我国 13 种,本文描述 1 种。常绿乔木,高达 20m,胸径 20～30cm。适生于中亚热带至北热带北缘的广大地区,对土壤水肥条件要求中等。分布于粤、琼、桂、黔、湘、赣、闽、台、浙、皖。越南亦产。木材有材用价值,树形美观,可供庭园观赏。

开花结实　约10年生开始开花结实，正常结实期在15年生以后。结实大小年现象不甚明显，但每年结实均不多。花两性，数朵生于小枝顶端或小枝上部叶腋，绿白色，下垂。花梗长2.5～5cm，具微柔毛。萼片4，阔卵形，长5～8mm，覆瓦状排列。花瓣4，长7～9mm，比萼片稍长，上部浅裂。雄蕊多数，有微柔毛，分离，着生于肥厚的花盘上。子房上位，密生短毛，3～7室，每室有胚珠多颗。花柱连合，钻形。本种分布广，花果期各地相差较大。据广西南宁1987～1989年观察，5月中旬为开花始期，5月下旬为开花盛期，6月上旬为末花期；9月上中旬果实开始成熟，9月下旬为果熟盛期。蒴果，木质，卵圆形，具刺毛，带刺长5.2～6.0cm，径4～5cm，未熟时绿色，熟时黄褐色。果实成熟后1个月内室背开裂为3～7果瓣。果瓣长2～3.5cm，厚3～5mm，针刺长1～1.5cm，内果皮紫红色。每果有种子约4粒。种子椭圆形或近卵形，黑色，长10～13mm，径6～8mm，具长约5mm的黄色假种皮，胚乳丰富，子叶平直，见图1。

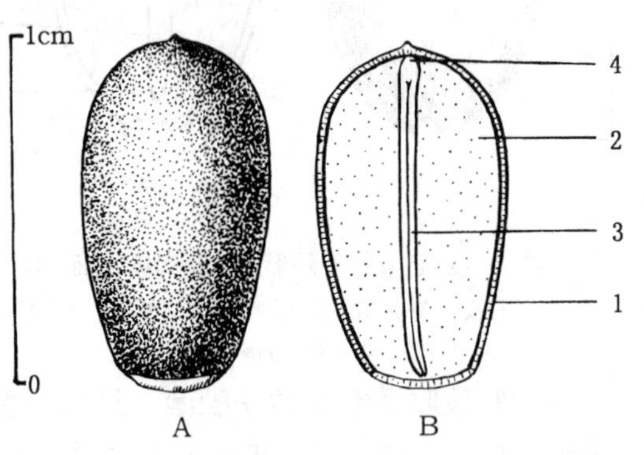

图1　猴欢喜种子（已去假种皮）外形（A）及其纵切面（B）
1. 种皮　2. 胚乳　3. 子叶　4. 胚根
（黄应钦绘）

果实的采收调制和种子贮藏　果实成熟盛期用采种刀或高枝剪采集，采时需戴手套，以防刺伤。采回的果实摊放在室内通风处。阴干，开裂后筛出种子。用水洗去假种皮，晾干，即得纯净种子。鲜果的出种率约1%～2%。种子净度可达98%。千粒重约310g，每千克纯净种子在2 900～3 700粒。种子忌失水，不能曝晒，宜随采随播。贮藏或运输时均应混以湿沙，贮藏期以不超过半年为宜。

图2　猴欢喜种子萌发后第5、15、30天幼苗的生长情况
（黄应钦绘）

发芽和播种　种子无明显休眠现象。发芽时气温宜在15℃以上。1982年11月2日，广西林业科学研究所用当年未作处理的新采种子，在室外沙床上作过发芽测定：当时气温为16℃，播后38天（12月10日）方始发芽，发芽盛期不明显，翌年1月7日发芽终止，发芽率为26%。出土萌发。胚根萌发后约10天子叶出土，再过20天长出初生叶。种子萌发和幼苗初期生长情况见图2。

点播。每平方米播种75～100g，覆土厚1～1.5cm。因发芽不甚整齐，宜先将种子密播于沙床，发芽后再分批移至苗圃。遮荫。一般1年生苗出圃。

（王宏志）

火绳树属
Eriolaena DC.

（梧桐科　Sterculiaceae）

生长习性、分布和用途　本属有17种，我国5种，本文描述2种。落叶小乔木。萌芽力强，耐修剪。喜质地疏松的酸性土壤。颇耐干热，忌严重霜冻。这2个种的名称、生长、分布及用途见表1。

表1　火绳树属树种的名称、生长、分布和用途

中　名	学　名	树高(m)	胸径(cm)	分　布	用　途
桂火绳（广西芒木）	*E. kwangsiensis* Hand.-Mazz.	7～12	10～18	桂、滇、粤	优质纤维胶虫寄主，瓶塞
火绳树（泡火绳）	*E. spectabilis* (DC.) Planch. ex Mast.	7～10	10～15	滇、黔、桂；四川有栽培。印度	胶虫寄主，纤维，瓶塞

开花结实　4～5年生开始开花结实。8年生以后进入正常结实期。结实大小年现象不明显，花期多雨或持续干旱会影响结实。正常年份结实量均较多。花两性。火绳树为聚伞花序，桂火绳为聚伞状总状花序，腋生。花白色。萼片5(4)。花瓣5(4)，与萼片近等长。雄蕊多数，花丝连合成筒，顶端分离，花药多数。子房上位，卵形，无柄，5～10室，每室有胚珠多颗，柱头5～10裂。开花结实的物候期见表2。

表2　火绳树属树种的开花结实物候期

树　种	观察地点和年份	开　花			果实成熟		果实脱落
		始　期	盛　期	末　期	始　期	盛　期	
桂火绳	南宁 1987～1988	5月中旬	5月下旬	6月中旬	8月下旬	9月上旬	9月下旬～10月中旬
火绳树	海南　1987	5月中下旬	6月上旬	6月下旬	9月上旬	9～10月	—
	云南　—	4月中旬	4月下旬	5月上旬	11月下旬	12～1月	—

表3　火绳树属树种果实和种子的形态特征

树　种	未熟果颜　色	成熟果实			种　子(去翅)		
		形　状	大小(cm)	色　泽	形　状	大小(mm)	颜　色
桂火绳	黄绿色	长椭圆状披针形或卵形，6～8条纵槽，无瘤状凸起	长3.5～5 径1.5～2	浅褐色，被淡黄色星状毛	不规则圆形或半圆形	长6～6.5 宽4～4.5 厚3～4	棕褐色
火绳树	黄绿色	卵状椭圆形或卵状锥形，果瓣连合处有沟槽，具瘤状凸起	长3～5 径约2.5	密被黄色星状毛	棱状椭圆形	长5～7 宽4～5 厚2～3	深褐色

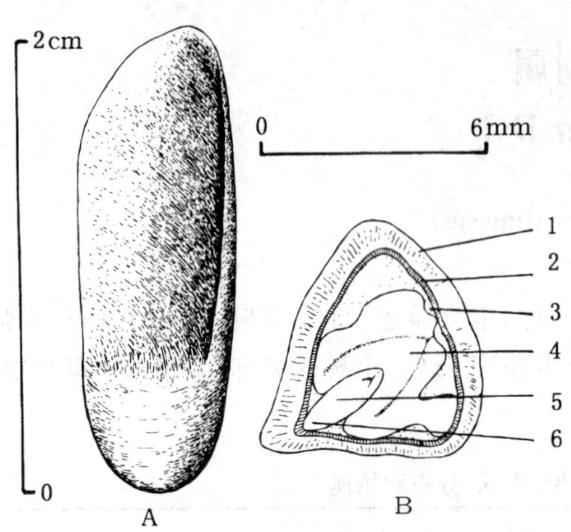

图 1　桂火绳带翅种子外形(A)及其去翅后纵切面(B)
1. 外种皮　2. 内种皮　3. 胚乳　4. 子叶
5. 胚轴　6. 胚根(黄应钦绘)

蒴果，成熟时由肉质变为木质，成熟后约1周成6～8瓣室背开裂。每果有种子40～50粒。种子具翅，果裂后飘散，果瓣在树上存留3～4个月脱落。种子具肉质胚乳。胚弯曲。果实及种子形态见表3、图1。

果实的采收调制和种子贮藏　采种宜选壮龄母树。果实成熟期较长。果熟盛期中选择形体较大的果实，用采种钩刀或高枝剪带果梗采下。采集的果实可堆放于室内3～5天，有利于部分果实的种子后熟，然后置日光下曝晒。果瓣开裂后用木棒轻加敲打，将脱出的种子用手搓揉，清除果瓣，扬去种翅等杂质即得纯净种子。果实的出种率和种子的净度、质量等数据见表4。

表 4　火绳树属树种果实出种率和种子净度、质量

树　种	出种率(%)	净度(%)	千粒重(g)	每千克纯净种子粒数(万粒)
桂火绳	12～20	90～98	17～23	4.3～5.9
火绳树	15～22	90～98	10～20	5～10

种子可以晒干贮藏，干种子的含水量一般为10%左右。贮藏时用布袋或坛罐均可，忌用塑料袋。种子可以随采随播，也可贮藏越冬至次年春播。常温条件下贮藏期约为1年。

发芽和播种　种子无休眠习性。发芽时的日均温在18℃左右。1987年10月下旬，广西林业科学研究所在室外沙床上用当年采集的新鲜种子作过发芽测定，播前种子未作任何处理，播后7天胚根开始萌动。2个种的发芽盛期均不明显。桂火绳从播种至发芽终了约13天，发芽率35%。火绳树从播种至发芽终了约20天，发芽率为40%。出土萌发。发芽后4～5天子叶出土，再过28～30天发出初生叶。桂火

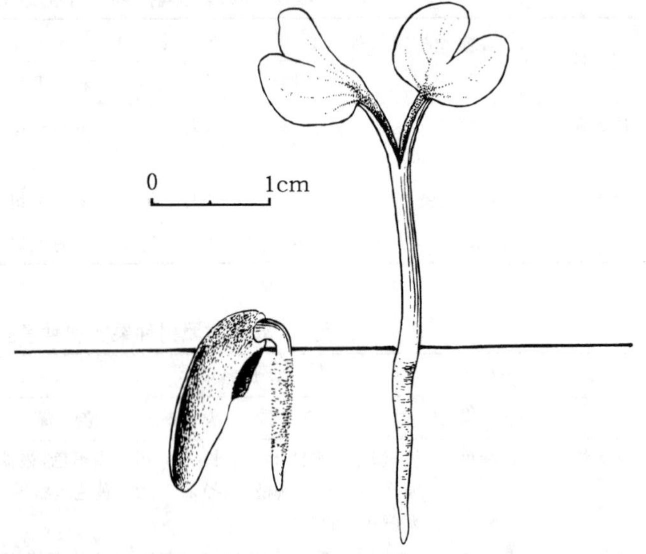

图 2　桂火绳种子萌发后第4、10天幼苗的生长情况
(黄应钦绘)

绳种子的萌发和幼苗早期生长情况见图 2。

据《云南主要树种造林技术》记载，火绳树种皮有蜡质，播种前曾用以下 4 种方法进行过预处理：用 96%左右的浓硫酸拌种 6 分钟，然后冲洗去掉蜡质，发芽率可达 66%～73%；5%草木灰水浸种 24 小时，发芽率 60%左右；沸水浸种 1 次，发芽率 36%～48%；沸水浸种至冷却后再用同样方法浸种 2 次，发芽率可达 70%以上。

播种期以 3 月下旬～4 月中旬为宜。条播。每平方米播种 4～5g，覆土 0.5～1cm。播后圃地应经常保持湿润。1 年生苗出圃。

（张声燕）

梧　　桐

Firmiana simplex (L.) W. F. Wight

（梧桐科　Sterculiaceae）

生长习性、分布和用途　梧桐属约 15 种，分布在亚洲和非洲东部，我国有 3 种，本文描述 1 种。落叶乔木，高达 16m。深根性。生长迅速，适应性强。梧桐分布广泛，从海南到华北各省都有。日本也有分布。

木材白色，轻软，强度中等，适于做箱匣、乐器和民房建筑用材。种子可焙食或榨油。茎、叶、花、果和种子均可药用。树皮纤维可造纸编绳。树干端直，翠叶浓荫，冠形优美，是著名的行道树和庭园观赏树。

开花结实　4 年生开始结实，以后结实量逐年增加。梧桐的开花结实物候情况见表 1。花两性，因心皮和雄蕊常只有一种发育，在功能上属于单性花。单性或杂性同株，顶生圆锥花序。花小，黄绿色，花瓣缺如。花萼 5 裂，披针形，向外反卷，外面密生黄色星状毛，内面仅在基部被柔毛。花药 15 个，不规则地聚集在雌雄蕊柄的顶端，呈球形。雌花的子房上位，球形，被毛，5 室，基部有退化雄蕊。子房具柄。蓇葖果 5，果瓣膜质，成熟前沿腹缝线 5 裂成叶状，匙形，长6～11cm，宽 1.5～2.5cm。成熟时果瓣黄褐色，外面被短茸毛或无毛。每个蓇葖有种子 2～4 粒，着生在果瓣边缘。种子圆球形，直径约 7mm，初为绿色，成熟时黄褐色至棕色，表面有网状皱纹。胚乳丰富。子叶两片极薄，子叶先端有小空腔(图 1)。

表 1　梧桐的开花结实物候

地　点	花　期			果实成熟期	种子脱落期	
	始　期	盛　期	末　期		始　期	末　期
北京	6 月中旬～6 月下旬	6 月中旬～7 月上旬	7 月上旬～7 月下旬	8 月上旬～9 月中旬	8 月上旬～9 月下旬	9 月中旬～11 月上旬
桂林	6 月上旬～6 月中旬	6 月上旬～6 月下旬	6 月下旬～7 月上旬	8 月中旬	8 月中旬～9 月中旬	—
成都	6 月中旬	6 月下旬	7 月上旬	8 月上旬（初熟期）	—	8 月下旬

果实的采收调制和种子贮藏　果实成熟失水后很快开裂，应及时采种，以免种子连同果瓣随风散落。多用人工采摘果穗，也可用竹竿等工具击断果梗。果实应摊开晾干，裂开后剥取种子。种子容易干燥失水而丧失发芽能力，以混沙湿藏为好。短期贮存的也可盛于布袋或干藏在密闭的容器内。据作者研究，室温下干藏 11 个月的种子发芽率仍高达 90%。种子气干含水量约 8%，千粒重 120～150g，每千克纯净种子 6 600～8 300 粒。

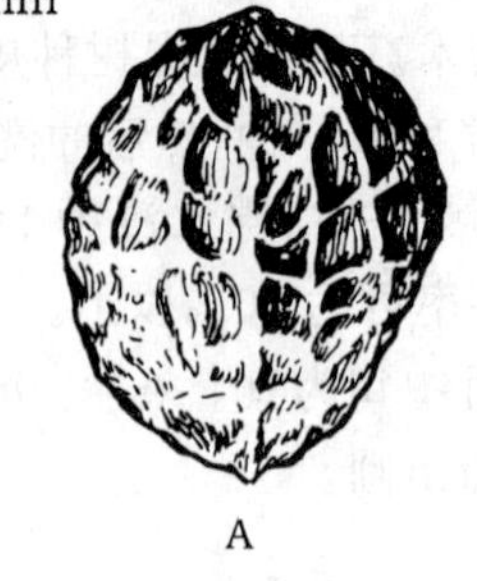

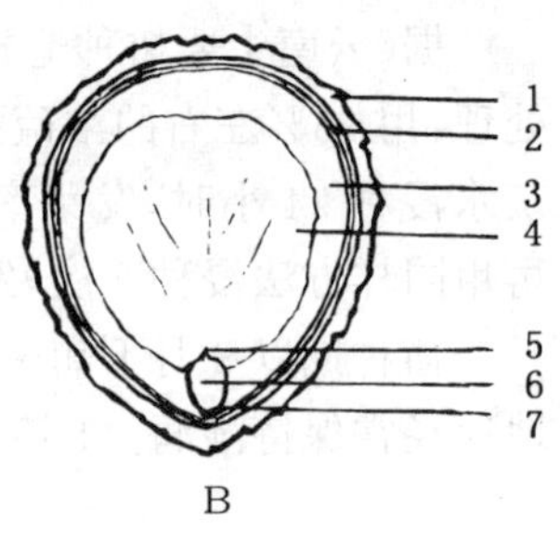

图 1　梧桐种子的外形(A)及其纵切面(B)
1. 外种皮　2. 内种皮　3. 胚乳　4. 子叶　5. 胚芽
6. 胚轴　7. 胚根(童军平、田恒德绘)

发芽前的预处理　梧桐种皮透水性和透气性都差，属强迫性休眠。未经催芽处理的种子在 30℃条件下培养 11 天，含水量只能升到 30%，仅及发芽所需最低含水量的 1/3，且呼吸强度仅为经过催芽处理的种子 1/3～1/6。据作者(王成霖，1989)研究，较好的催芽处理方法是用浓硫酸浸蚀：层积过的种子或浸泡过 1～2 天的种子，浓硫酸浸蚀以 30 分钟为好。机械破损种皮也有促进发芽的作用。另据李近雨和穆贵荣(1985)研究，干种子的酸蚀时间以 2.5 小时为宜。

发芽测定　种子发芽温度以 30℃为好，变温能加速发芽。光照对发芽无显著影响。酸蚀后的种子培养 4～5 天，发芽势为 33%～64%，约 4 周时发芽率达 90%～94%。

播种　冬播或春播，但多采用春播。条播，条距 30cm。每平方米播种约 38g，覆土厚约 1～1.5cm。干藏的种子播后发芽极不整齐，常有一部分种子延至下一年出土。层积 30 天的种子浓硫酸处理后播种，发芽期 60 天。1 年生苗高约 40cm。通常用 1m 以上的 2 年生移植苗供四旁栽植。

出土萌发。子叶两枚，倒卵形，先端钝圆或平截，基部圆或心形，长 28～ 35mm，宽 27～32mm，掌状脉 5 出，柄长 12～15mm。初生叶互生，第 1 叶小，宽卵形，第 2 叶较大，三角形，全缘。具主根。幼苗形态见图 2。

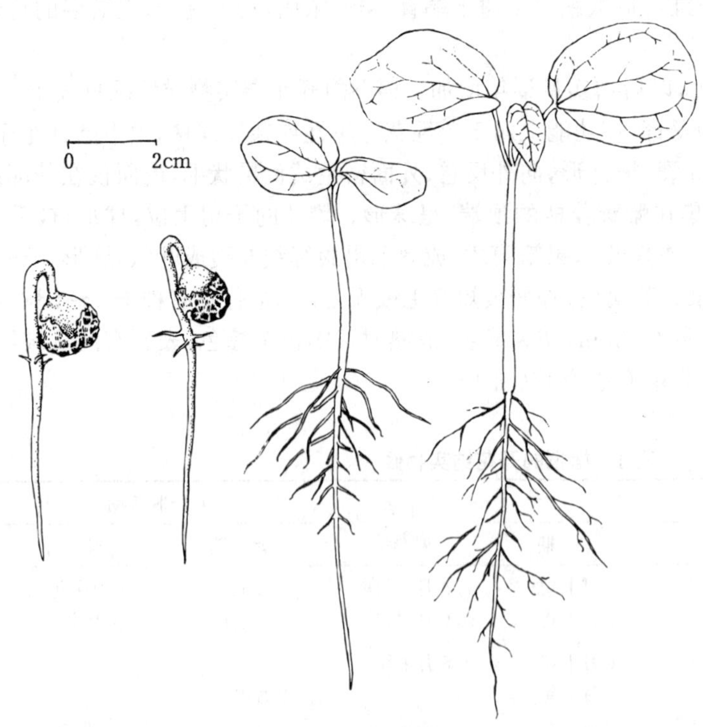

图 2　梧桐种子萌发后第 1、2、9、28 天幼苗的生长情况
(童军平绘)

(王成霖)

扭蒴山芝麻（火索麻）

Helicteres isora L.

（梧桐科　Sterculiaceae）

生长习性、分布和用途　山芝麻属约60种，我国9种，本文描述1种。小乔木，高4～5m。耐干旱，生于丘陵或灌木丛中。产于滇、桂、粤、琼。南亚、东南亚及大洋洲热带地区均有分布。树皮富含纤维，可织袋、编绳、造纸或作人造纤维。根入药，治慢性胃炎和胃溃疡。

开花结实　1～2年生开始开花结实，正常结实期在5年生以后，结实无大小年现象。每年采伐抚育，萌芽条当年便可开花结实。花两性。聚伞花序腋生，有花6～9朵，常2～3朵成簇。花萼长1.7cm，漏斗形，外面密被星状毛，4～5浅裂。花瓣5，红至紫红色，前面2片较大，爪具耳状附属物。雄蕊10，位于伸长的雌雄蕊柄顶端，花丝多少合生，内有退化雄蕊5。子房上位，具5棱，5室，每室胚珠多个。花柱5，线形。据广西南宁观察，开花始期为7月上旬，盛期为8月初，末期为10月中旬，11月上旬仍有零星开花；果实成熟始期为翌年3月上旬，成熟盛期在3月下旬～4月初，5月上旬为果实脱落始期，6月中旬果实全部脱落。蒴果，圆柱形，螺旋状扭曲，顶端具一喙尖，长4～5cm，径5～9mm。幼果青色，被星状毛，成熟时转为黑褐色，脱落时各室不开裂。每果有种子50～100粒。种子斜方形、棱状三角形或不规则形，褐色，有光泽，长2～2.8mm，宽与厚略等，约1～1.4mm。有肉质胚乳。子叶叶状而筒卷。种子形态见图1。

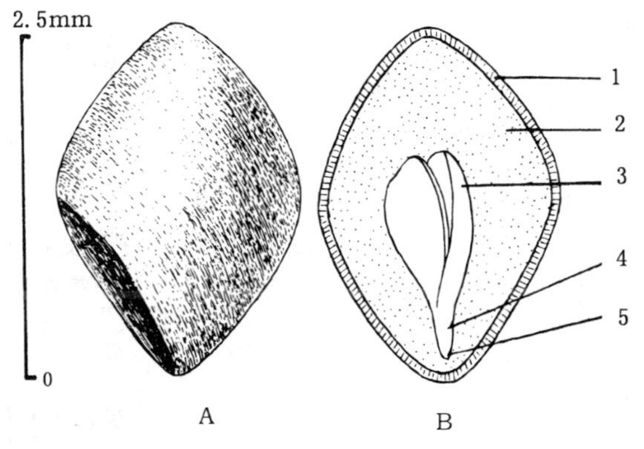

图1　扭蒴山芝麻种子外形(A)及其纵切面(B)

1. 种皮　2. 胚乳　3. 子叶　4. 胚轴　5. 胚根

（黄应钦绘）

果实的采收调制和种子贮藏　蒴果部分干枯或全部干枯时，用枝剪剪下。采得的果实置日光下晒3～5天。果壳干脆时用木棒敲打或搓揉，并筛出种子。鲜果出种率约为20%。种子净度一般为90%。千粒重约1.9g，每千克有纯净种子40万～67万粒。种子可以干藏，贮藏期为半年至1年。运输时用布袋包装即可。

发芽和播种　种子无休眠习性，发芽时日均温宜在20℃以上。1988年5月21日，广西林业科学研究所在室内作过发芽测定：室温约28℃，发芽基质为滤纸；置床后6天胚根伸出，发芽盛期不明显，发芽终于6月14日；从播种之日起算，以25天计，发芽率20%左右。出土萌发。子叶出土后3天展开，5天后初生叶出现(图2)。

条播。每平方米用种6～8g，覆土0.5cm。培育1年生苗出圃。

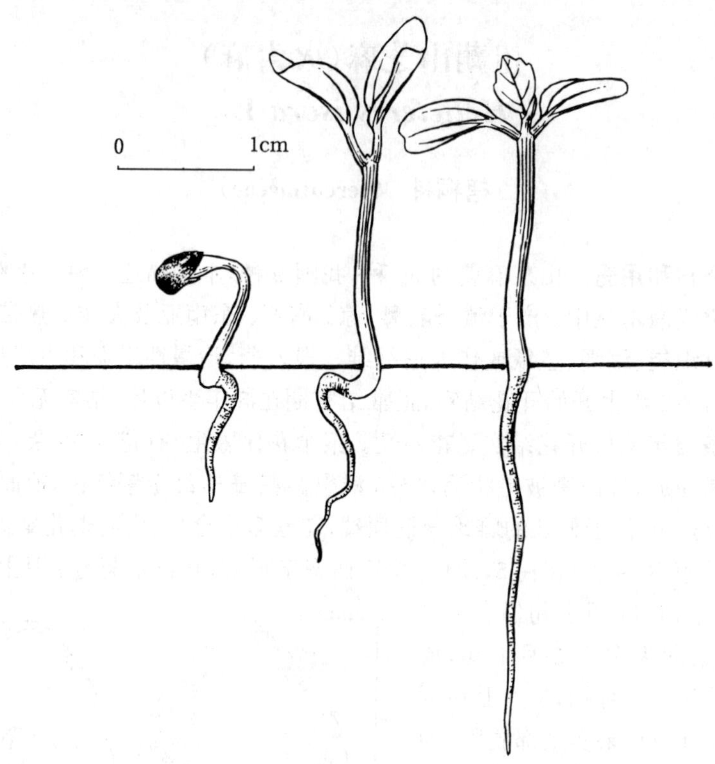

图 2　扭蒴山芝麻种子萌发后第 2、4、6 天幼苗的生长情况
（黄应钦绘）

（曾　玲）

银叶树属

Heritiera Dryand.

（梧桐科　Sterculiaceae）

生长习性、分布和用途　本属约 35 种，我国 3 种，本文描述 2 种。常绿乔木。喜高温高湿气候，不耐寒冷，忌霜冻。喜肥沃湿润的酸性土壤。材质优，是优良的建筑、造船和家具用材。这 2 个树种的名称、生长、分布及用途见表 1。蝴蝶树为海南特产，分布区狭小，又遭过度采伐，资源日益稀少，在《中国植物红皮书》中被列为渐危种。

表 1　银叶树属树种的名称、生长、分布和用途

中　名	学　名	树　高 (m)	胸　径 (cm)	分　布	用　途
长柄银叶树	*H. angustata* Pierre	15	50	琼、滇、桂、粤。柬埔寨	材用、纤维、淀粉、油脂
蝴蝶树	*H. parvifolia* Merr.	40	100	琼特产。粤、桂栽培	优质用材、纤维、淀粉

开花结实　约13年生开始结实，正常结实期在20年生以后。结实有大小年现象，间隔期为1年。花单性。聚伞花序作圆锥花序式排列，腋生或顶生。花小，红色（长柄银叶树）或白色（蝴蝶树）。花萼坛状或管状4～6浅裂。无花瓣，具花盘。雄花的雄蕊柄短，花药8～12成环状，聚集于蕊柄顶部；花盘厚，围绕在蕊柄基部。雌花子房上位，心皮3～5(6)，几分离；退化的雄蕊位于子房基部；花柱短，柱头与心皮同数，每心皮1胚珠。广西、海南观测的开花结实物候期见表2。

表2　银叶树属树种的开花结实物候期

树　种	观察地点和年份	开花			果实成熟		果实脱落
		始　期	盛　期	末　期	始　期	盛　期	
长柄银叶树	广西南宁 1987～1989	4月上旬	4月中旬	4月下旬	7月中旬	8月上旬	8月中下旬
蝴蝶树	海南　1988	4月上旬	4月中下旬	5月上旬	8月中旬	9月上旬	9月下旬～10月上旬

另据海南的观测资料，长柄银叶树10～11月第2次开花，翌年3～4月果熟，为小造果。果干燥。长柄银叶树果木质，坚硬，有长约1cm短翅。蝴蝶树果革质，有鱼尾状长翅，翅长2～4cm。两个树种的果实均不开裂。种子无胚乳，子叶特厚。果实及种子的形态见表3、图1。

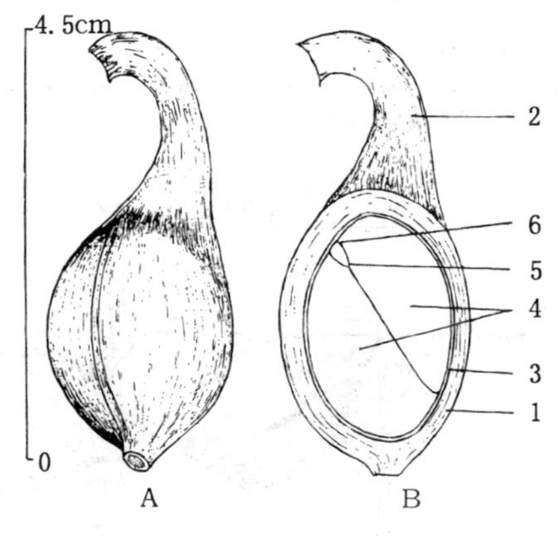

图1　长柄银叶树翅果外形(A)及其纵切面(B)
1. 果皮　2. 果翅　3. 种皮　4. 子叶　5. 胚芽　6. 胚根(黄应钦、田恒德绘)

果实的采收调制和贮藏　果实大部分成熟时，应及时摇动树枝震落或用竹竿打落，在地上收集。种子易遭虫蛀，不宜堆放，也忌脱水。采得的果实不宜在日光下曝晒，宜在通风荫凉处摊放1～2天，稍加晾干即可。生产上一般不剥除果壳，也不去翅，而将带翅果实直接用作播种材料，也通称种子。果实的质量见表4。

表3　银叶树属树种果实和种子的形态特征

树　种	果实			种子		
	形　状	大小(cm)	色　泽	形　状	大小(cm)	颜　色
长柄银叶树	近椭圆形，基部稍隆起，顶部有1扁短翅	长2～3.5 径1.5～2.2 连翅长4～4.5	黄褐色至黑褐色	椭圆形，种皮薄	长1.7～2.7 径1.4～2	棕褐色
蝴蝶树	椭圆形，顶部有鱼尾状长翅	长2.3～3 径2～2.5 连翅长6～7	锈褐色，被银色秕鳞	椭圆形或卵形	长1.7～2.5 径1.3～1.8	浅褐色

表 4 银叶树属树种果实的质量

树 种	千粒重(g)		每千克果实粒数(粒)	
	一 般	变动范围	一 般	变动范围
长柄银叶树	3 940	3 500～4 200	250	230～290
蝴 蝶 树	3 000	2 700～3 500	330	280～400

为防虫蛀,采得的果实可用始温 40℃水浸泡 2～3 昼夜,使幼虫在孵化过程中闷死,捞出稍加晾干即可播种或贮藏。宜随采随播,不耐久藏。浸泡后,常温条件下果实裸露可存放 20 天左右，过期发芽能力会逐渐降低。运输时应在浸泡稍晾干后用竹筐或麻袋包装，每筐（袋）不超过 20kg。如需贮藏越冬播种，则需分层混沙湿藏，时间不宜超过次年 3 月。

表 5 银叶树属树种的发芽能力及其测定条件①

树 种	温度(℃)	发芽势(%)	发芽率(%)	
			计算天数	一般数值
长柄银叶树	30～32	无明显发芽盛期	57	60
蝴 蝶 树	20～24	无明显发芽盛期	45	55

① 广西南宁,室外沙床

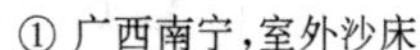

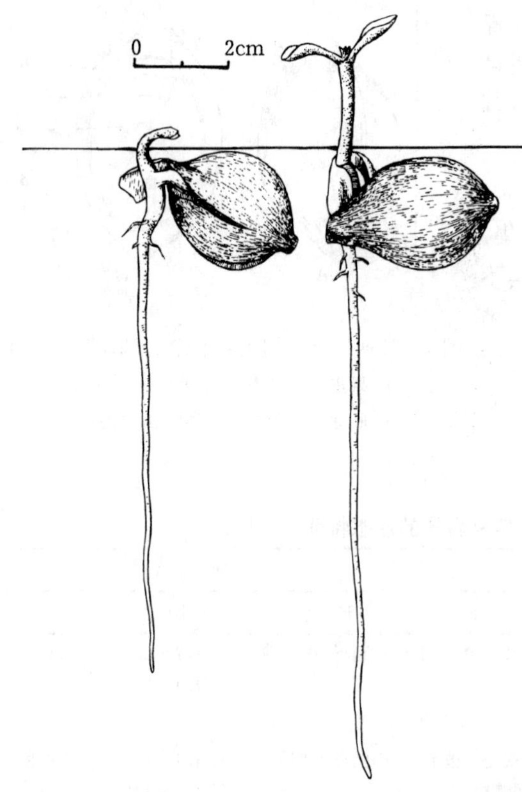

图 2 长柄银叶树种子萌发后第 5、8 天幼苗的生长情况（黄应钦绘）

发芽和播种 种子无明显休眠习性。发芽时日均温需在 20℃以上。1977 年 7 月 20 日（日平均气温 30℃）和 9 月下旬（日平均气温 22℃），广西林业科学研究所曾在室外沙床上用新采翅果对长柄银叶树和蝴蝶树作过发芽测定，播种后约 1 个月开始发芽，发芽情况见表 5。

长柄银叶树留土萌发。萌发后 4～5 天上胚轴伸出地面，7～8 天展出初生叶。蝴蝶树出土萌发，发芽约 3 天子叶出土，再过 4 天发出初生叶。长柄银叶树种子的萌发和幼苗初期生长情况见图 2。

条播。播种前如果皮已干，应置水中浸泡 2～3 昼夜。播时应将果实平放或将基部斜向下。每平方米播 380～500g，覆土约 3cm。幼苗宜遮荫，培育 1～2 年生苗出圃。

（张声燕）

翅子树属
Pterospermum Schreber

（梧桐科　Sterculiaceae）

生长习性、分布和用途　本属约40种，我国9种，本文描述2种。常绿或半落叶乔木，高20～30m，胸径40～80cm。生长较快速，喜光。颇耐干热。喜肥沃湿润的酸性土或钙质土。能耐轻霜，忌冰雪。为优质轻软木及优质纤维原料树种，根浸酒为著名的“枫荷桂”酒。这两个树种的名称、生长、分布和用途见表1。

表1　翅子树属树种的名称、生长、分布和用途

中　名	学名	树高（m）	胸径（cm）	分　布	用　途
翻白叶树（半枫荷、异叶翅子树）	*P. heterophyllum* Hance	20～30	40～60	琼、粤、桂、闽、台。越南	板材、纤维、胶虫寄主、油料、药用（浸酒治风湿）
窄叶半枫荷（剑叶翅子树）	*P. lenceaefolium* Roxb.	25～30	50～80	琼、粤、桂、滇。印度、缅甸、越南	家具用材，纤维、油料

开花结实　约10年生始开花结实，正常结实期在20年生以后。结实大小年间隔期一般为1年，不甚显著。花两性，腋生，单生或数朵聚生成聚伞花序；萼片5裂，花瓣5，白色，与萼裂片等长或稍短。雄蕊柱短，雄蕊15，每3枚集合成群，有退化雄蕊5，舌状，与雄蕊群互生。子房上位，5室，每室有倒生胚珠多颗，中轴胎座。花柱棒状，柱头有5纵沟。广西南宁1987年观测的开花结实物候期见表2。

表2　翅子树属树种的开花结实物候期①

树　种	开　花			果实成熟		果实脱落
	始　期	盛　期	末　期	始　期	盛　期	
翻白叶树	5月中旬	5月下旬	6月上旬	10月上旬	10月中旬	10月下旬～11月中旬
窄叶半枫荷	5月中旬	5月下旬	6月上旬	10月上旬	10月中旬	10月下旬～11月中旬

① 广西南宁，1987

另据海南观测，翻白叶树在海南为6月开花，10月果熟；窄叶半枫荷在海南5～6月开花，10月果实成熟。

蒴果，果实密被绒毛，成熟时由肉质转为木质，5室。翻白叶树每室有种子15～25粒，窄叶半枫荷每室有种子20～30粒。果实充分成熟后室背5瓣开裂，种子飘落，果瓣存留约3～4个月后方始脱落。种子具膜质长翅，有薄层胚乳，胚直立，子叶叶状。果实及种子形态见表3、图1。

表 3 翅子树属树种果实和种子的形态特征

树种	未熟果颜色	成熟果实			种子		
		形状	大小（cm）	颜色	形状	大小（mm）	颜色
翻白叶树	黄绿色	矩圆状卵形	长 5.8～7.0 径 2～2.5	黄褐色，被锈色短毛	扁椭圆形，具翅	长 8～12 宽 5～8 厚 2～3 带翅长 25～45	棕褐色
窄叶半枫荷	黄绿色	长椭圆状卵形	长 4.5～6 径 1.8～2.2	黄褐色，被锈色短毛	扁椭圆形，具翅	长 6～11 宽 4～7 厚 1～2 带翅长 20～35	棕褐色

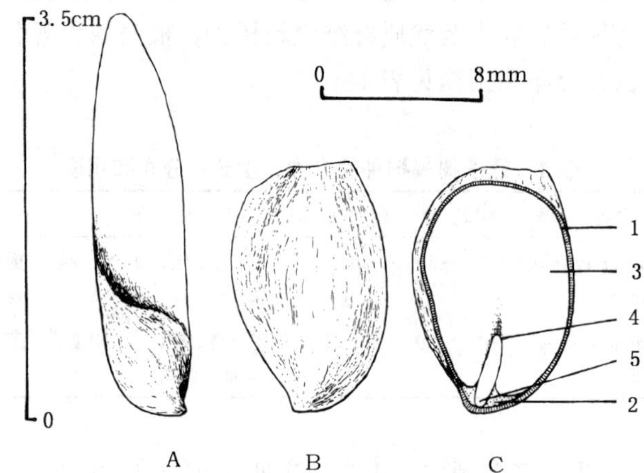

图 1 翻白叶树种子外形（A）、去翅种子外形（B）及其纵切面（C）
1. 种皮 2. 胚乳 3. 子叶 4. 胚芽 5. 胚根
（黄应钦绘）

果实的采收调制和种子贮藏 成熟的果与不成熟的果较难区别，当果实接近成熟时，应注意观察，发现有少数果开裂时，即及时采集，以免果裂种子飘失。通常用采种钩刀将果柄割断，在地面收捡。采集的果实，宜堆放在室内 3～5 天，然后摊开晾干或置日光下曝晒，果瓣开裂后，稍加翻动，即得种子。已脱出果瓣的种子不宜在烈日下曝晒，以免过分失水，降低种子发芽能力。果实出种率及带翅的种子质量等见表 4。

表 4 翅子树属树种果实出种率和种子净度、质量

树种	出种率（%）	净度（%）	千粒重（g）	每千克纯净种子粒数
翻白叶树	10～15	85	60～70	14 300～16 700
窄叶半枫荷	10～15	85	58～68	14 700～17 200

种子可以带翅贮藏或播种，也可以搓揉扬去种翅后供播种。干种子可装于布袋或坛罐内，置于室内通风干爽荫凉处贮藏。常温条件下贮藏期一般为半年左右，次年 5 月以后，发芽能力便明显下降。置 0～5℃的低温条件下，种子的生活力可保持 1 年以上。

表 5　翅子树属树种的发芽能力及其测定条件

树　种	基　质	室外温度（℃）	发芽势（%）		发芽率（%）	
			计算天数	一般数值	计算天数	一般数值
翻白叶树	沙	19～21	9	50	62	90
窄叶半枫荷	沙	19～21	8	42	53	84

发芽和播种　种子无休眠习性或有短期休眠。发芽时的日均温需在 19℃以上。1978 年和 1980 年广西林业科学研究所对 2 个树种在室外沙床进行过发芽测定，播种期分别为 3 月上旬和 3 月中旬，播前未作任何处理，两个树种都在播后 31～32 天开始萌出胚根，发芽时日均温为 19～21℃。翻白叶树的发芽时间较长，全程为 30 天，窄叶半枫荷发芽的全程为 22 天。发芽情况见表 5。

播在苗床上越冬的种子，2 月间如持续 1 周日均温达到 19℃，少数种子可以发芽，但发芽不整齐。出土萌发。胚根萌发后约 4 天，子叶出土，再过 4～5 天发出初生叶，见图 2。

条播或撒播。整地要求细致，每平方米播种 5～6g，覆土 0.5～1cm，经常保持圃地湿润，培育 1 年生苗出圃。

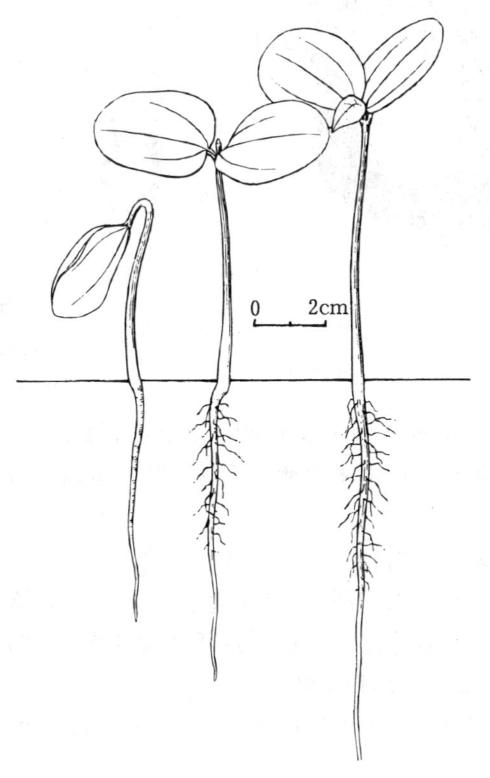

图 2　翻白叶树种子萌发后第 5、9、12 天幼苗生长情况

（黄应钦仿《热带亚热带主要树种采种育苗技术》）

（王宏志）

翅　苹　婆

Pterygota alata（Roxb.）R. Br.

（梧桐科　Sterculiaceae）

生长习性、分布和用途　翅苹婆属约 20 种，我国仅有本文描述的 1 种。常绿大乔木，高达 30m，胸径 40cm。喜光。喜湿热气候。适生于湿润的酸性土壤。为热带雨林上层树，分布于海南南部。越南、印度、菲律宾也有分布。为轻软用材，树形美，可供绿化观赏。

开花结实　8～12 年生始开花结实，正常结实期为 12～40 年生以后，结实有大小年现象，间隔期一般为 1 年。花单性，排成腋生的总状花序或圆锥花序，萼钟状，5 深裂。无花瓣。雄花的雄蕊柄圆柱状锥形，顶端扩展成杯状，花药约 20 集成 5 束；有退化雌蕊。雌花的雌蕊柄短，有 5 束不发育雄蕊，子房圆球形而被短柔毛，心皮 5，近分离，每心皮有胚珠多个；花柱 5，柱头膨大，辐射状。据海南尖峰岭 1988～1989 年的物候观察，花期较短：3 月中旬花芽出现，3 月下旬为始花期，4 月上旬为盛花期，4 月中旬开花结束，幼果形成；11 月果实开始成熟，12 月至翌年 1 月为果实成熟盛期，果实成熟后陆续脱落。蓇葖果木质，扁球形，棕褐色，被粉状短柔毛，径 6～12cm，每果有种子 25～30 多粒，种子褐色或浅褐色，扁圆形，长约 2.5cm，顶端有长而阔的翅，连翅长约 7cm（图 1）。

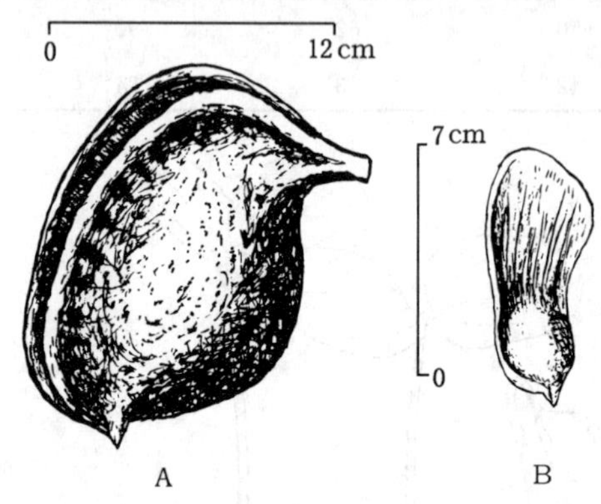

图 1　翅苹婆果实外形（A）和种子外形（B）
（田恒德绘）

果实的采收调制和种子贮藏　果实成熟期中用采种钩刀钩采，收集后放置 1～2 天，果实自然开裂或用力压裂，取出种子。鲜果按质量计算的出种率约为 50%，种子千粒重 1 500～2 100g，每千克种子粒数 500～730 粒。种子忌失水，不能日晒，亦忌堆放，应随采随播，不耐贮藏，运输可带果，运抵后再行调制。贮藏需混以润沙，贮藏期 1 个月以内。

发芽和播种　种子无休眠习性。发芽时日均温宜在 20℃以上。据中国林业科学研究院热带林业试验站的测定，鲜种子播种后 8 天开始萌动，从发芽之日起算，20 天的发芽百分数为 60%，从置床之日起算，38 天的发芽率为 85%。出土萌发。种子萌发后 12 天子叶出土，再过 2～3 天发出初生叶。

点播。每平方米播种 100～150g，覆土 2cm。培育百日苗或 1 年生苗出圃。

（符史深）

梭罗树属

Reevesia Lindl.

（梧桐科　Sterculiaceae）

生长习性、分布和用途　本属约 18 种，我国 14 种，本文描述 4 种。常绿乔木。对立地的要求较高，贫瘠地生长不良。分布于我国西南至东南部，喜马拉雅山东部亦产。木材结构致密，但耐腐性差，可作室内用具，树皮为优质纤维及造纸原料，种子可榨油。它们的名称、生长及分布情况见表 1。

开花结实　8～12 年生开始开花结实，正常结实期在 20 年生以后，结实大小年现象较明显，间隔期为 1 年。特寒年份或遭虫害也影响结实。花两性，排成聚伞状的伞房花序或圆锥花序，顶生。花萼钟状，5 浅裂。花瓣白色，5 片，具爪。花丝合生成管状，并与雌蕊柄贴生而形成雌雄蕊柄。花药 15 聚成头状。子房上位，5 室，具 5 条纵沟，每室有倒生胚珠 2 颗，柱头5裂。南宁、贵阳等地的开花结实物候期见表2。

表1　梭罗树属树种的名称、生长情况和分布

中　名	学　名	树　高（m）	胸　径（cm）	分　布	供　稿
长柄梭罗	*R. longipetiolata* Merr. et Chun	30	60	琼、粤、桂	601
梭罗树	*R. pubescens* Mast.	26	30～40	滇、川、黔、桂、粤、琼。南京栽培。南亚	620
上思梭罗	*R. shangszeensis* Hsue	18～20	30～40	桂、粤。越南	601
两广梭罗	*R. thyrsoidea* Lindl.	12～15	20～35	粤、桂、滇、黔、川。东南亚各国至印度	601

表2　梭罗树属树种的开花结实物候期

树　种	观察地点和年份	开花			果实成熟		果实脱落
		始　期	盛　期	末　期	始　期	盛　期	
长柄梭罗	南　宁 1987～1988	5月中旬	5月下旬	6月中旬	10月上旬	10月中旬	11月上旬
梭罗树	贵阳	3月	～	4月	—	10月	11月上旬
上思梭罗	南　宁 1983～1989	2月中旬	2月下旬	3月上旬	10月上旬	10月下旬	11月上旬～12月上旬
两广梭罗	南　宁 1984～1989	4月上旬	4月中旬	4月下旬	9月下旬	10月上旬	10月下旬～11月下旬

蒴果木质，密被绒毛，成熟后不脱落，约10天左右成5瓣室背开裂，每室1～2种子。种子具膜质翅，易飘失。果瓣宿存于树梢。种子具胚乳。子叶长圆形，平展，叶脉可见。果实及种子形态见表3、图1。

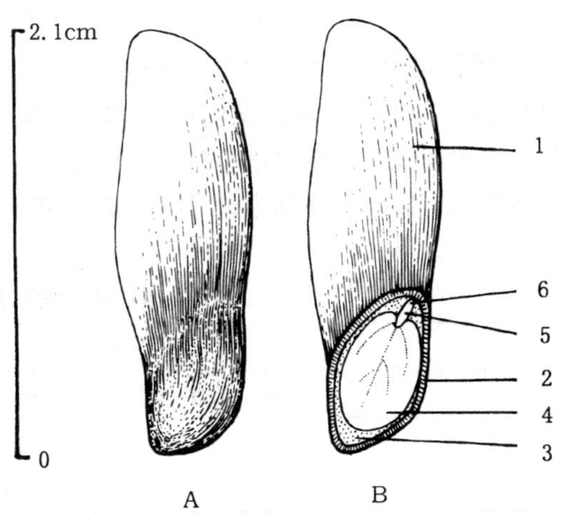

图1　两广梭罗种子外形（A）及其纵切面（B）
1. 种翅　2. 种皮　3. 胚乳　4. 子叶　5. 胚轴　6. 胚根
（黄应钦绘）

果实的采收调制和种子贮藏　成熟果实与未成熟果实较难区别，不同年份果实的成熟期略有差异，当果实接近成熟期时，应注意观察，以免果实开裂种子飘失。用高枝剪带果穗采下，堆放于室内，使种子在果内完成后熟。经过5天左右再将果实摊放于荫凉通风处，果瓣开裂后翻动或稍加敲打，种子脱出。清除果壳及未发育的种子，所得带翅种子通常用作播种材料。果实出种率和种子质量等见表4。

表 3 梭罗树属树种果实和种子的形态特征

树种	未熟果颜色	成熟果实			种子		
		形状	大小（cm）	色泽	形状	大小（mm）	颜色
长柄梭罗	青色	倒卵形至矩圆形	长 5～7 径 3～4	黄褐色	压扁状，卵形，下端带矩形膜质翅	长约 9，宽约 5，厚约 2.2，带翅长约 25	棕褐色
梭罗树	青黄色	近梨形	长 2.5～3.5	黄褐色	扁平，椭圆形，有翅	长 7～9，宽 4～6，厚 2，带翅长 15～25	棕褐色
上思梭罗	黄绿色	矩圆状倒卵形，具 5 棱	长 3～4.2 径 1.8～2.7	黄绿至黄褐色	扁平，椭圆形，带膜质翅	长 7～9.5，宽4.5～6，厚 2～2.6，带翅长 15～25	棕褐色
两广梭罗	青绿色	矩圆状倒卵形，具 5 棱	长 2.7～4 径 1.7～2.7	黄绿色	扁平，椭圆形，带膜质翅	长 7～9.5，宽4.5～5.5，厚 1.8～2.3，带翅长 20～26	棕褐色

表 4 梭罗树属树种果实出种率和种子净度、质量

树种	出种率（%）	净度（%）	千粒重（g）	每千克纯净种子粒数（粒）
长柄梭罗	3～5	95	45～54	18 500～22 200
梭罗树	3～5	95	35～45	22 000～28 600
上思梭罗	3～5	95	35～45	22 000～28 600
两广梭罗	3～5	95	35～42	23 800～28 600

种子可以干藏，但忌烈日曝晒，一般宜在室内通风处晾干，袋藏或用木箱等容器贮藏于室内干爽荫凉处。常温条件下，冬季贮藏期为 5～6 个月，夏季为 1～2 个月。一般秋冬季采种，次年春暖后播种。贮藏于 0～5℃的低温条件下，种子的生活力可保存 1 年以上。

发芽和播种 种子无明显休眠习性。日均温 16℃以上即可发芽，但发芽的持续期长，日均温 20℃以上时，发芽较整齐。1982 年和 1984 年，广西林业科学研究所在室外沙床上先后对上思梭罗和两广梭罗进行过发芽测定：11 月播种，次年 1 月即陆续发芽，持续期为 70 天，发芽势不明显，发芽率为 60%；3 月下旬末播种的，4 月中旬开始发芽。测定情况见表 5。

表 5 梭罗树属树种的发芽能力及其测定条件①

树种	发芽势（%）		发芽率（%）	
	计算天数	一般数值	计算天数	一般数值
上思梭罗	9	46	42	79
两广梭罗	6	34	40	65

① 室外沙床，20～24℃

出土萌发。胚根萌发后约经 4 天子叶出土，再经 6 天左右发出初生叶。两广梭罗种子的萌发和幼苗初期生长情况见图 2。

条播，每平方米用种 7～8g。也可密播于精细整地的播种床内，幼苗转木质化后移至容器或大田培育。容器育苗 3～4 个月出圃，大田育苗一般 1 年生苗出圃。

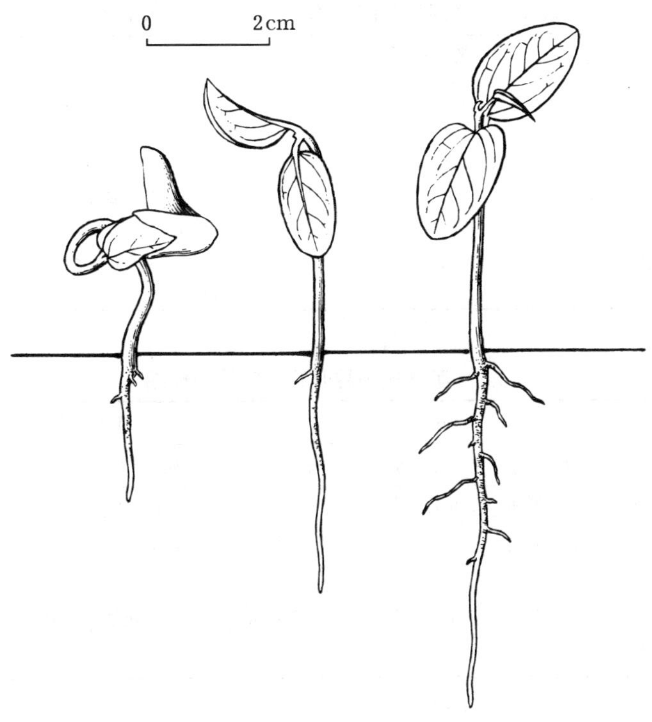

图 2　两广梭罗种子萌发后第 4、6、11 天幼苗生长情况
（黄应钦仿《热带亚热带主要树种采种育苗技术》）

（王宏志）

苹　婆　属
Sterculia L.

（梧桐科　Sterculiaceae）

生长习性、分布和用途　本属约 300 种，我国 22 种，又引入 3 种，本文描述 3 种。常绿乔木。适生于肥沃湿润的酸性土或石灰性土。种子可以炒食，树皮纤维质地优良，树形美观，可供庭院绿化，可以用于涵养水源、保持水土。其名称、生长、分布和用途见表 1。

开花结实　香苹婆 8～10 年开始结实，假苹婆与苹婆 10～12 年开始结实，结实丰年间隔期一般为 1 年。人工栽培而水肥条件好的地方，大小年的差异较小，野生条件下的差异较显著。花单性稀杂性。腋生圆锥花序。萼管状，5 浅裂或深裂，淡红色至紫红色。无花瓣。雄花的花药 10～15 聚生于雌雄蕊柄的顶端，包围着退化雌蕊。雌花的雌雄蕊柄很短，顶端有轮生的不育花药和发育的雌蕊；雌蕊由 5 个心皮粘合而成，每心皮有胚珠 2～15；花柱基部合生，柱头与心皮同数而分离。据广西及海南观测，花芽于 2 月上旬～3 月中旬形成。开花结实的物候期见表 2。

表 1 苹婆属树种的名称、生长、分布和用途

中 名	学 名	树高(m)	胸径(cm)	分 布	用 途
香苹婆	*S. foetida* L.	25	40	南亚、东南亚、大洋洲、非洲。琼、粤、桂、滇栽培	材用、纤维、观赏、食用
假苹婆	*S. lanceolata* Cav.	15	25	琼、粤、桂、黔、滇、川。中南半岛	材用、纤维、食用、观赏
苹 婆	*S. nobilis* Smith	30	60	闽、台、琼、粤、桂、黔、滇。东南亚	材用、纤维、食用、观赏

表 2 苹婆属树种开花结实的物候期

树 种	观察地点和年份	开花			果实成熟		果实脱落
		始 期	盛 期	末 期	始 期	盛 期	
香苹婆	海南 1979	5 月下旬	6 月上旬	6 月中下旬	9 月上旬	9 月中下旬	—
假苹婆	南宁 1981～1984	4 月上旬	4 月中旬	5 月上旬	6 月下旬初	6 月下旬末	6 月下旬～7 月上旬
苹 婆	南宁 1987～1989	3 月下旬初	4 月上旬	4 月下旬	7 月上旬初	7 月上旬末	7 月上旬～7 月中旬

蓇葖果多革质，稀木质，2～3 个，间有 1 或 4～5 个共生于 1 果柄上。果实成熟后腹缝开裂，露出种子但不立即脱落，能在树上维持 4～6 天。落地以后种子陆续脱出。如遇连续雨天或湿润环境，有些种子便在蓇葖果内发芽。香苹婆每蓇葖有种子 10～15 粒，假苹婆有种子2～7 粒，苹婆有种子 1～4 粒。种子有胚乳。胚伸直，子叶大。果实种子的形态特征见表 3、图 1。

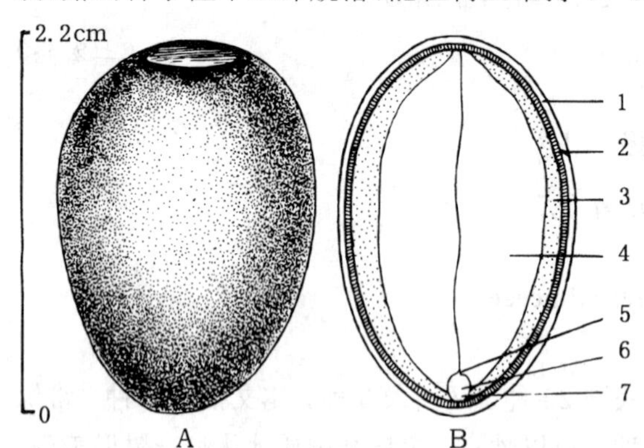

图 1 苹婆种子外形(A)及其纵切面(B)
1. 外种皮 2. 内种皮 3. 胚乳 4. 子叶 5. 胚芽 6. 胚轴 7. 胚根(黄应钦绘)

果实的采收调制和种子贮藏 大部分果实呈红色或黄色，少数蓇葖已开裂时，用采种刀带果柄采摘，不宜敲打或摇动。采收后，将已开裂且呈红黄色的同青色的蓇葖分别堆放，分期处理。蓇葖开裂后脱出种子。种子在蓇葖上着生较牢，需从蓇葖的背部敲打使其脱落，少数敲打不出的种子，再用手或工具挑出。除去蓇葖即得纯净种子。种子忌脱水，不能日晒，亦不耐久藏。常温条件下，种子的裸露贮藏期为 1 周以内。贮藏或运输均需混以湿沙。假苹婆和苹婆混沙贮藏的种子，贮藏期不宜超过 1 个月，香苹婆的种子可以沙藏至次年 3 月。种子的净度、质量等数据见表 4。

表 3　苹婆属树种果实和种子的形态特征

树　种	未熟果颜　色	成熟果实			种　子		
		形　状	大小（cm）	颜　色	形　状	大小（cm）	颜　色
香苹婆	青绿色	卵形，先端具喙，木质	长 8～12 径 5～8	黄色，内壁红色	椭圆形	长 1.8～2.0 径 0.8～1.0	黑色，有光泽
假苹婆	黄绿色	长椭圆形或长卵形，基部渐狭，先端具喙，革质	长 5～7 径 2～2.5	鲜红色	椭圆形或圆形	长 1.2～1.6 径 0.75～1.0	黑褐色
苹婆	黄绿色	矩圆状卵形，舟状，先端具喙，厚革质	长 5～9 宽 2～3.5	鲜红色	椭圆形、卵形或近圆形	长 1.8～2.4 径 1.2～1.6	黑褐色或棕褐色，有光泽

表 4　苹婆属树种的出种率和种子净度、质量

树　种	出种率（%）	净　度（%）	千 粒 重（g）		每千克纯净种子粒数		新鲜种子含水量（%）
			一　般	变动范围	一　般	变动范围	
香苹婆	30～50	98	1 450	1 300～1 600	690	620～770	65
假苹婆	35～45	99	800	750～850	1 250	1 180～1 300	56
苹　婆	40～50	99	4 500	3 500～7 000	220	140～300	70

发芽和播种　种子无休眠习性。发芽时的日均温需在 22℃以上。假苹婆和苹婆的种子成熟期正值盛夏高温，宜随采随播。香苹婆种子的成熟期较晚，在海南尚宜随采随播，大陆南部则宜沙藏越冬，春暖后再播。假苹婆和苹婆留土萌发，香苹婆出土萌发。1978 年 7 月，广西林业科学研究所用新鲜种子在室外沙床播种，对假苹婆和苹婆 2 个树种进行过发芽测定（播种时日均温在 30℃上下）：播后假苹婆第 4 天萌发，第 9～10 天上胚轴出土，第 14～15 天展出初生叶。苹婆播种后第 5 天萌发，上胚轴出土及展叶期均较假苹婆迟 3～4 日。1979 年 9 月下旬，曾将海南采集的香苹婆种子在南宁沙藏至 1980 年 1 月 20 日播种，播后 6 天开始萌发，发芽后 4～5 天子叶出土，再经 14～16 天发出初生叶，发芽进程较夏播的种子明显缓慢。发芽情况见表 5、图 2。

点播。每平方米香苹婆播种 75～85g，假苹婆 38～50g，苹婆 300～380g。覆土 2～3cm。香苹婆 1 年生苗出圃，其余 2 种育 1 年半生苗出圃。如果用于庭院绿化，需培育 2 年半～3 年生苗出圃。

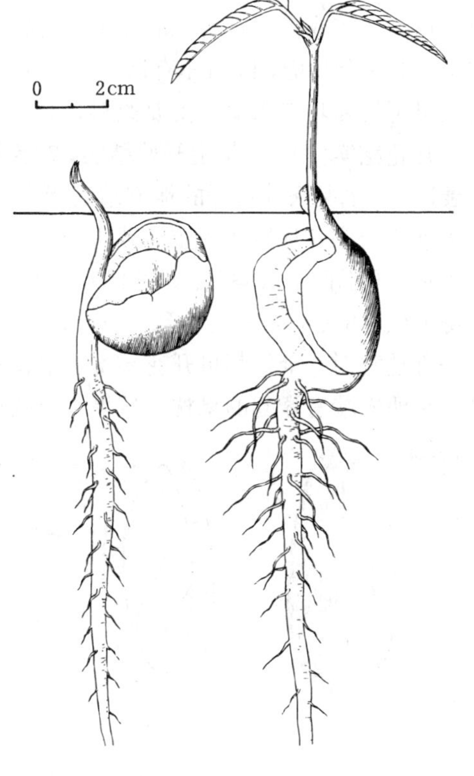

图 2　苹婆种子萌发后第 12、18 天幼苗生长情况
（黄应钦绘）

表 5　苹婆属树种的发芽能力及其测定条件（广西南宁，室外沙床）

树　种	温度（℃）	发芽势（%）		发芽率（%）	
		计算天数	一般数值	计算天数	一般数值
香苹婆	18～24	12	38	30	73
假苹婆	30	3	67	9	72
苹　婆	30	4	50	13	65

（王宏志）

可　　可

Theobroma cacao L.

（**梧桐科　Sterculiaceae**）

生长习性、分布和用途　可可属约 30 种，我国引入本文描述的 1 种。常绿小乔木，高6～10m。原产美洲中部和南部，现已广泛种植于世界热带地区。喜湿热环境，要求年平均气温在24～28℃。日均温低于 20℃时生长速度减慢，低于 15℃时停止生长，极端低温降至 10℃时嫩叶开始受害。温度高于 37℃时，生长受到抑制。喜年降雨 2 000mm 左右，且分布均匀。旱季开花不结实或极少结实。抗风力差。我国于 1922 年引入台湾，1954 年引入海南，近年滇南、桂南、粤南有栽培，闽南亦有试种。世界三大饮料之一。种子营养丰富，有兴奋滋补作用，是制造可可粉和巧克力糖的主要原料，又可榨油，供药用。果壳作饲料和肥料。

开花结实　4～5 年生开始结实，7 年生以后进入正常结实期。一般立地条件下 20 年后生长衰退，但水肥条件良好的地方结实能力可维持 50～60 年。结实大小年现象不明显。花两性，单生或为聚伞花序簇生于主干或主枝上。花小，浅黄色。萼 5 深裂。花瓣 5，上部匙形，中部窄条形，下部凹陷成盔状。雄蕊的花丝基部合成筒状。雄蕊 1～3 聚成一组，与退化雄蕊互生。子房上位，5 室，每室有胚珠 14～16，2 列，花柱圆柱形，柱头 5 裂。据海南儋县观察，在温湿条件适宜时，全年都可开花结实。主花期 5～11 月，果期 8 月～翌年 5 月，盛熟期为 2～4 月。果柄末端有隆起的果枕。核果，长圆形至纺锤形，长 15～20（30）cm，直径 8cm 左右，外有 10 条纵沟，呈瘿瘤状，深黄色或近红色，干后褐色。果皮厚，肉质，干后木质，厚 4～8mm，每室种子 12～14。种子白色，卵形或纺锤形，长 2～3.4cm，宽 1.2～1.5cm。每果有种子 33～38 粒。种子无胚乳，有一对大而折叠的含油子叶。可可种子的形态见图 1。

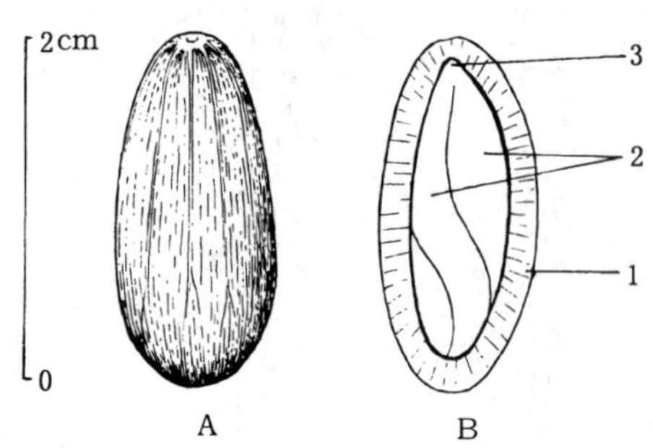

图 1　可可种子外形（A）及其纵切面（B）
1. 种皮　2. 子叶　3. 胚根
（黄应钦绘）

果实的采收调制和种子贮藏　当果实由绿色变成黄色或橙黄色，或由红色变为淡红色，外表具光泽，敲击便有响声时即可采收。采时用利刀切断果柄，切忌伤及果枕，以免影响次年结果。采集的果实用刀切开，取出种子，用木屑或谷壳擦洗种子附着的果肉，但不可伤及发芽孔。鲜果要保持湿润，忌曝晒。鲜果的出种率约25%。种子的净度可达98%。千粒重2 380(2 000～2 500)g，每千克有纯净种子420(400～500)粒。可可种子属于顽拗型，不耐干藏。据程必强和马信祥(1987)报道，用湿沙保存20～30天的可可种子发芽率仍为100%，未用湿沙的则已降为零。晾干表皮的种子含水量接近40%，易于发芽，调制后宜随即播种。短期贮藏需以果实的状态贮藏。果实在常温下可保存7～10天，超过14天果内的种子便丧失发芽能力。用蜡涂着果实放于荫凉处，可保存30天。用蜡涂着果实并用塑料袋包装，可保存40天。

发芽和播种　种子无休眠习性，应随采随播。发芽时日均温宜在25℃以上。1988年8月3日，华南热带作物研究所用沙床作过发芽测定：播后8天（8月10日）开始发芽，12天即发芽结束，有明显的发芽盛期。从开始发芽之日起算，3天中的发芽百分数为87.5%；从播种之日起算，12天的发芽率为99%。出土萌发。子叶出土后同时展开初生叶。可可种子的萌发和幼苗初期生长情况见图2。

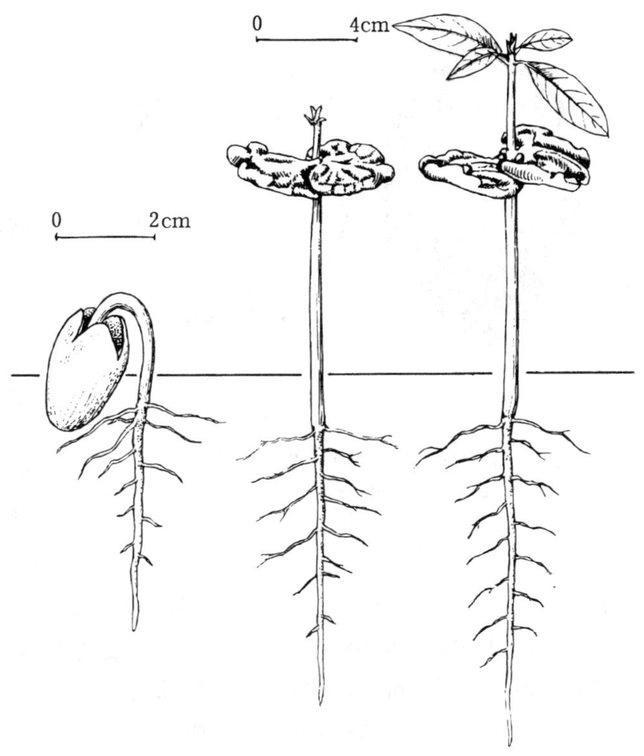

图2　可可种子萌发后第3、7、12天幼苗的生长情况

（黄光郁绘）

条播。每平方米播种360～500g，覆土1～2cm。半年生苗高30cm左右时即可定植。

（丁慎言）

木棉（攀枝花、英雄树）

Bombax ceiba L.

（木棉科　Bombacaceae）

生长习性、分布和用途　木棉属约50种，本文描述我国2种中的1种。落叶大乔木，高达30余米，胸径达1.5m。喜深厚肥沃的酸性土或钙质土。分布于川、滇、黔、桂、粤、琼、台、闽、赣。南亚、东南亚及澳大利亚北部等热带地区亦有。木材浸水后可用于建筑及制作各种家具。花色红艳壮丽，有英雄树及攀枝花之称，供观赏及药用。果皮绵毛纤维直，无绞合性，可作垫褥及枕头填充物，种子含油脂，供工业用。

开花结实　5～6年生开始结实，正常结实在12年生以后，结实大小年间隔期为1年。花

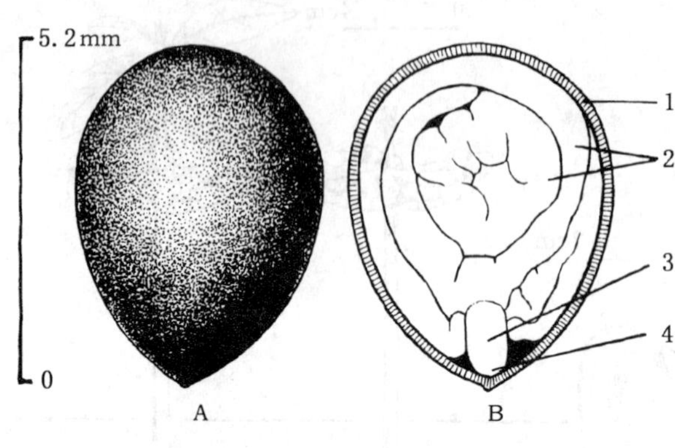

图1　木棉种子外形（A）及其纵切面（B）
1. 种皮　2. 子叶　3. 胚轴　4. 胚根
（黄应钦绘）

两性，单生枝顶叶腋。花大，径约10cm，先叶开放。萼杯状，长2～3cm，内面密被淡黄色绢毛，萼齿3～5，半圆形，高1.5cm，宽2.3cm。花瓣5，红色或橙红色，肉质，倒卵状矩圆形，长8～10cm，宽3～4cm，两面被星状柔毛。雄蕊多数，合生成管。花丝排成多轮，最外轮集生为5束，与花瓣对生，花丝较粗。花药1室，肾形，盾状着生，内轮部分花丝上部分2叉。子房上位，5室，每室有胚珠多数。花柱长于雄蕊，细棒状，柱头星状5裂。据广西南宁1978～1987年观测，2月中旬花蕾形成，着生于头年生的枝梢；3月中旬花始开，3月下旬～4月上旬为盛花期，4月中旬为末花期；5月下旬末果实开始成熟，6月上旬为果实成熟盛期。在云南勐仑，果实成熟于4～5月（程必强、马信祥，1987）。果实成熟时由黄绿色或暗绿色转变为棕黄色或暗棕色，肉质转木质，密被灰白色星状绒毛及长柔毛。蒴果，长椭圆形，基部渐小，长10～15cm，径3.5～5cm。充分成熟的果实室背5瓣开裂。果瓣革质，内有丝状绵毛，种子带绵毛飞落，果瓣在树上存留2～5天脱落。种子黑褐色，圆形、卵形或倒卵形，部分种子的一端喙尖，种壳较薄，长4～6mm，径3.5～4.5mm。无胚乳，子叶大而多折叠。种子形态见图1。

果实的采收调制和种子贮藏　大部分果实呈现棕黄色或浅棕色时，种子已发育饱满。果瓣尚未开裂前用采种刀钩下或上树打落果实。采回的果实室内堆放3～5天，使部分颜色较青的果实继续得到成熟，然后在阳光下摊晒1天或继续留在室内摊放，使果瓣开裂，清除果壳，即得带有绵毛的种子。用手搓揉，收捡绵毛后过筛，即得纯净种子。鲜果的出种率为15%～20%，每果有种子约130粒。过筛的种子净度可达92%～98%。种子千粒重35～50g，每千克有纯净种子2万～2.8万粒，阴干的种子3万～3.5万粒。种子的含水量18%～25%，不宜脱水。种子脱出绵毛后忌日晒，亦不耐久藏，宜随采随播。常温条件下

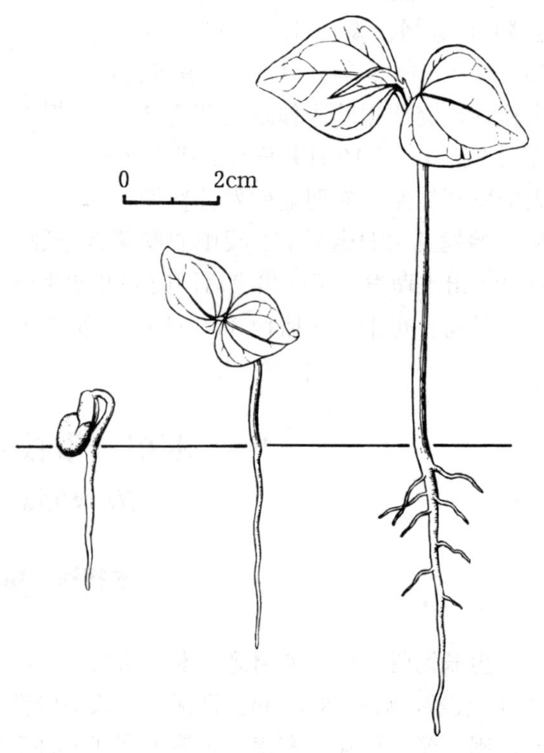

图2　木棉种子萌发后第5、7、9天幼苗的生长情况
（黄应钦仿《热带亚热带主要树种采种育苗技术》）

种子裸露存放半个月便会降低发芽能力。混沙贮藏的种子，发芽能力可以保持40～60天。一般认为种子应在当年7月以前播完，不宜贮藏越冬。但据程必强和马信祥（1987）报道，木棉种子耐干藏，密封瓶藏7～8个月后，发芽率有的仍达95%。

发芽和播种　种子无休眠习性。发芽时日均温宜在25℃以上。1981年7月，广西林业科学研究所用室外沙床对木棉的新鲜种子作过发芽测定：播后第3天开始发芽，第4天进入发芽盛期，从开始发芽至发芽高峰日的4天中，发芽百分数为53%；从播种之日起算，以10天计，发芽率为68%。出土萌发。播后5～6天子叶出土，7～8天初生叶出现。种子的萌发和幼苗早期生长情况见图2。

条播。随采随播的种子播前不需作任何处理。裸露存放5天以上或经过沙藏的种子，播前用清水浸种3～4小时，捞出后稍晾片刻即可播种。每平方米播种10～12g，覆土约1cm。半年生苗可以出圃。用于四旁绿化则需培育2～3年生苗。

木棉亦可扦插繁殖。

（林榕庚）

爪哇木棉（吉贝）

Ceiba pentandra（L.）Gaertn.

（木棉科　Bombacaceae）

生长习性、分布和用途　爪哇木棉属约10种，我国只引入本文描述的1种。落叶大乔木，高达30m，胸径80cm，侧枝大而轮生。忌6℃以下的连续低温和霜冻。喜肥沃湿润的酸性沙壤土。原产热带美洲，现亚洲、非洲热带地区广泛引种，以印度尼西亚的栽培量最大。我国琼、粤、桂、滇有栽培。丝状绵毛可作枕芯、床垫的填充物，又可作防热、隔音的绝缘材料。种子榨油，供工业用。可供观赏。

开花结实　3～4年生开始开花结实，正常结实年龄在12年生以后。结实大小年间隔期为1年。水肥条件充足的母树常不出现大小年现象。花两性，多数簇生于上部叶腋，稀单生。花萼钟状，不规则5裂，无毛，宿存。花瓣5，白色或玫瑰色，外部密被白色茸毛。雄蕊管短，花丝10，不等高分离或分为5束，花药肾形。子房上位，5室，无毛，每室有胚珠多数。花柱线形，柱头棒状，5浅裂。据海南的物候资料，1月中旬新叶尚未萌发时花蕾出现，单生或簇生于头年生枝上部的叶腋，2月中旬为始花期，下旬为盛花期，3月上旬为末花期；4月下旬果实开始成熟，5月中旬～下旬为果实成熟盛期。蒴果，长圆形或长椭圆形，向上渐窄，长7～16cm，径3～5cm，成熟时由青绿色转变为黄绿色或浅棕色，果瓣由肉质转为木质。果实充分成熟后室背5瓣裂，内面密生丝状绵毛，种子裹于其中，随绵毛飞散，果瓣亦相继

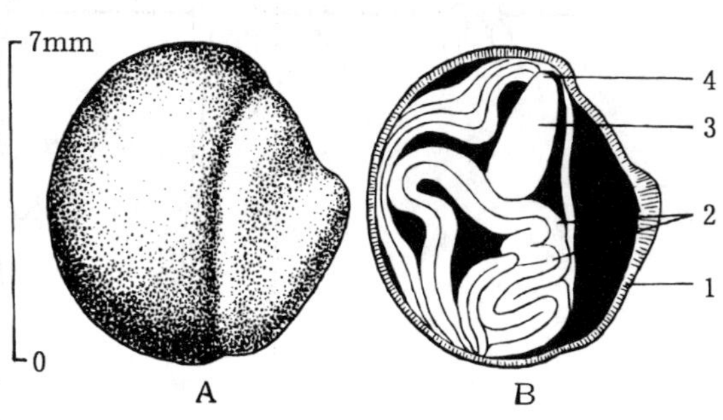

图1　爪哇木棉种子外形（A）及其纵切面（B）
1. 种皮　2. 子叶　3. 胚轴　4. 胚根
（黄应钦绘）

脱落。种子暗褐色，扁圆形或不规则圆形，种皮革质，较薄，有小半部具窿起状的不规则环状痕。种子径 5～7mm，少有胚乳，子叶回旋折叠。种子的形态见图 1。

果实的采收调制和种子贮藏 果实的成熟时期不及木棉整齐，可分期采收。或在少数果实开裂，大部分果实由青绿色转变为黄绿色，种仁发育饱满时，用采种刀钩采或用竹竿将果实打落。爪哇木棉枝条脆嫩，上树时应有安全防护措施。果实采回后按成熟程度分别堆放。已接近开裂的果实可置日光下曝晒或摊放于室内，其余尚呈黄绿色的继续堆放一处，青绿色的另行堆放，使其后熟。3～6 天后，将接近开裂的果实陆续选出曝晒或摊放，分批进行处理。种子脱出果瓣后便忌日晒，应将带有绵毛的种子收至室内搓揉，使种子脱出绵毛，筛去杂物，即得纯净种子。鲜果的出种率 20%～25%，每果出种子约 100 粒。种子净度可达 95%～99%，千粒重 58～65g。每千克有纯净种子 15 000～17 000 粒。种子含水量 17%～25%。据报道，海南尖峰岭的一份种子含水量约 15%，千粒重 73.3g，每千克有纯净种子 13 600 粒（陈荷美，1978）。种子忌脱水，不耐贮藏，宜随采随播。常温条件下裸露贮存的种子，发芽能力只能保持 10～15 天，超过半个月，发芽能力便明显下降。据程必强和马信祥（1987）报道，瓶装的种子 2～3 个月后发芽率为 20%～40%。短期贮藏或运输，宜混以潮润的锯木屑或少量绵毛。贮藏 1～2 个月的也需混以湿沙。在 5～10℃的低温条件下，种子可以保存 6～7 个月。

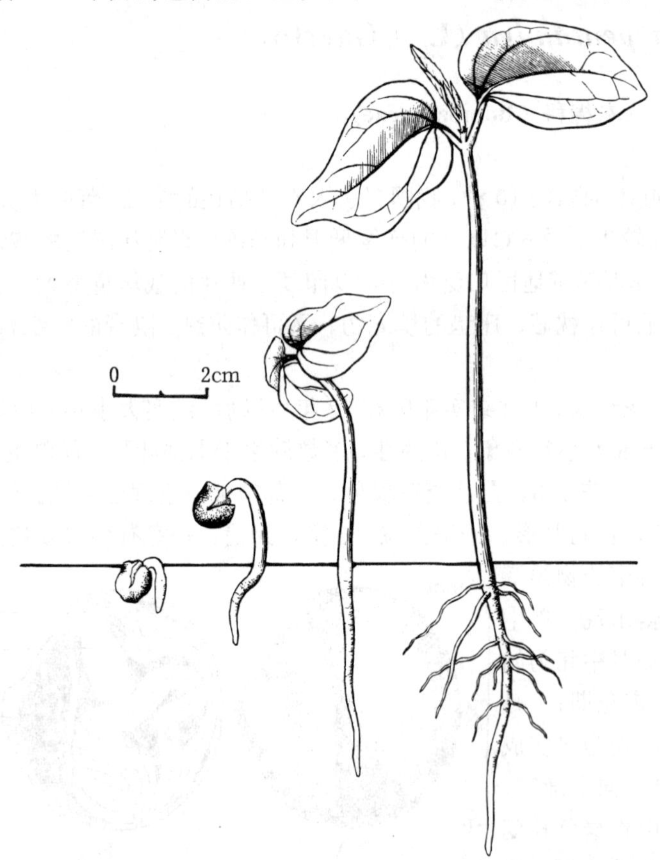

图 2 爪哇木棉种子萌发后第 2、4、8、13 天幼苗生长情况

（黄应钦绘）

发芽和播种 种子无休眠习性。发芽时日均温宜在 25℃以上。1987 年 6 月（日均温 29℃左右），广西林业科学研究所在室外沙床作过发芽测定（播前种子浸水 2～3 小时），播后 2～3 天开始萌发，并随即进入发芽盛期。开始萌发后的 4 天中，发芽百分数为 55%；第 8 天发芽结束，发芽率为 68%。海南尖峰岭场圃发芽率可达 87%～93%。出土萌发。发芽后 3～4 天子叶出土，12～14 天发出初生叶。种子的萌发和幼苗初期生长情况见图 2。

条播。每平方米播种 15～20g。播前将种子浸水 2～3 小时，捞出稍晾干后播种。覆土 1.5～2cm。半年生苗出圃。爪哇木棉也可扦插繁殖。

（林榕庚）

轻　木

Ochroma pyramidale（Cav. ex Lam.）Urban

（木棉科　Bombacaceae）

生长习性、分布和用途　轻木属只有本文描述的 1 种，我国引入栽培。常绿乔木，高 20m，胸径 50～80cm。寿命较短，根浅，抗风力差。抗寒力低，遇 4～6℃低温便受冻死亡。适生于潮湿深厚的森林土或冲积土，沼泽地不能生长。原产热带美洲及西印度群岛。台、滇、琼引种栽培。材质轻，强度较大，隔音隔热性好，浮力大，用于航空、航海、防震、冷冻装置等。

开花结实　2～3 年生开始开花结实，4 年生以后进入正常结实期，大小年现象不明显。花大型，两性，单生于上部叶腋。花萼漏斗状，5 裂，厚革质。花瓣 5，匙形，长 8～8.5cm，宽 1.3～1.8cm，白色或黄白色。雄蕊多数，雄蕊管长 9cm，上部扭转，无分离的花丝，花药 5～10，花药 1 室。子房上位，5 室，每室有胚珠多颗。花柱圆柱形，长 4.5cm，藏于雄蕊管内。柱头 5，相互扭转成螺旋状纺锤形，长 2.5cm，粗 5mm，伸出于花药之上。据海南乐东1988～1989 年的物候观察，1 月中旬花芽形成，1 月下旬～2 月上旬花蕾形成；2 月中旬为始花期，2 月下旬为盛花期，2 月下旬末～3 月上旬初为末花期；4 月中旬果实开始成熟，5 月～6 月中旬为果熟盛期。果实熟后脱落。蒴果，圆柱形或长圆形，长 13～22cm，革质，果皮内被浅褐色绵状簇毛，室背 5 瓣裂，裂后种子飘落。种子多数，倒卵形，褐色，状似芝麻而略大，裹于绵毛中。种皮近革质，长约 3mm，径约 1.5mm。具肉质胚乳。胚直立，子叶扁平。种子的形态见图 1。

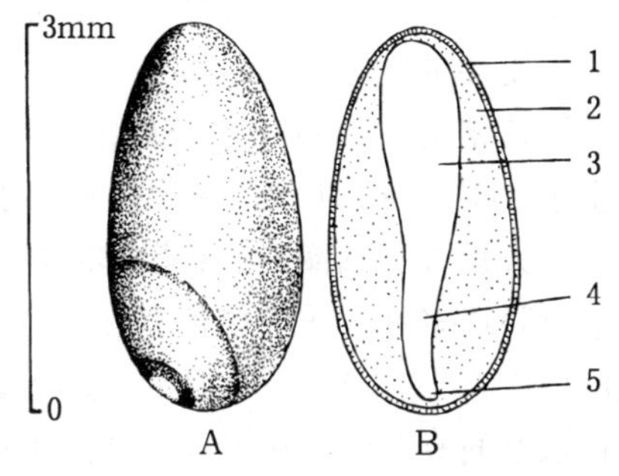

图 1　轻木种子外形（A）及其纵切面（B）
1. 种皮　2. 胚乳　3. 子叶　4. 胚轴　5. 胚根
（黄应钦绘）

果实的采收调制和种子贮藏　在果实成熟期内，分批用采种钩刀或高枝剪截断果梗，注意少伤树枝。采得的果实摊放于室内，开裂后置日光下曝晒，除去果瓣和绵毛等杂质，即得种子。净度可达 95%以上。种子可以晒干贮藏，含水量宜在 8%左右。纯净种子的千粒重 6.8（6.3～7.4）g，每千克有种子 15.8（14～16.5）万粒。发芽能力的保存期较长。常温下室内一般贮存的种子，发芽能力可以保持 1 年以上，晒干后置密封玻璃瓶或坛罐内贮藏，可以保持 2 年以上。

发芽和播种　种子无休眠习性。发芽时日均温宜在 20℃以上。经过贮藏的种子，播前用始温 60℃水浸种并在自然冷却的过程中浸泡 24 小时。据中国林业科学研究院海南热带林业试验站的发芽测定，室内置床后 4 天开始发芽，从发芽之日起算的 5 天，发芽百分数为 38%；从播种之日起算，以 13 天计，发芽率为 50%。另据《中国主要树种造林技术》一书记载，发芽率为 75%～84%。出土萌发。种子萌发后约 3 天子叶出土，6 天左右发出初生叶。

条播。每平方米播种 2～4g。苗木半木质化时间苗移苗，或将已萌发的幼芽移至容器内继续培育。圃地苗一般为半年至1年生出圃，容器苗2～3个月可出圃定植。

（符史深）

瓜 栗

Pachira aquatica Aubl.

（木棉科 Bombacaceae）

生长习性、分布和用途 瓜栗属2种，我国皆引入，本文描述1种。另1种大果瓜栗 *P. macrocarpa* (Cham. et Schlecht.) Walp. 在云南、海南栽培。常绿乔木，高达15m，胸径15～20cm。喜高温高湿气候，能耐较长期5～6℃低温及轻霜，忌风。适生于肥沃湿润的酸性沙壤土。原产热带美洲，现世界热带地区均有引种。我国闽、琼、粤、桂有栽培。材质轻软，可作漂浮物和隔热物。枝轮生，花大而艳丽，可供庭园观赏。种仁炒食味如花生，可榨油，供食用。皮纤维质优。

开花结实 4～6年生开始结实，正常结实期在8年生以后。大小年现象不明显。花两性，单生于1年生枝的叶腋。花萼杯状，边缘成波纹状。花瓣5枚，淡黄绿色，带状，长17～18cm，开后向外反卷成圆圈形。雄蕊基部合生成筒状，花丝多数，白色，花药1室，黄色。子房5室，每室有胚珠多颗，花柱白色，柱头5浅裂。由于各年冬春的积温不同，物候期也有差异。例如广西南宁1987年1～3月的月均温较常年高2～3℃，物候期也比一般年份提早10天：4月上旬瓜栗便有花蕾陆续形成，但花期不甚整齐；主花期在5月上旬开始，中旬为盛花期，下旬结束，6～8月间尚有少量花陆续零星开放；7月中旬果实开始成熟，7月下旬为盛果期，8～10月尚陆续有果实成熟。

蒴果，卵状椭圆形，具5条纵槽，基部中心下凹，长8～15cm，径6～10cm，成熟时由青绿色转变微黄色，室背5裂。果瓣内面具长绵毛，随种子脱落。种子近半圆形或不规则形。种粒的形态及大小差异较大，种子径1～2cm。种皮浅褐色，具多条不规则明显条纹环绕种子一周，腹部具脐状斑痕，与条纹基部相连。种子多胚，子叶回旋折叠。种子形态见图1。

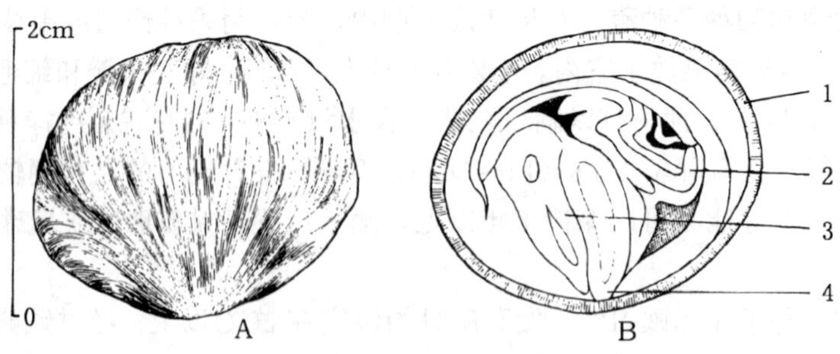

图1 瓜栗种子外形（A）及其纵切面，示多胚（B）

1. 种皮 2、3、4. 胚

（黄应钦绘）

果实的采收调制和种子贮藏 大部分果皮由青绿色转变为黄绿色，种仁饱满，挤压时有油脂外溢，即可采收。通常用手采摘或用采种钩刀钩落地面收集。采回堆放 2～4 天后置日光下曝晒 1 天，或摊放于通风处晾干，俟果瓣开裂后脱出种子。用于食用或榨油的种子置日光下晒干备用。用于育苗的种子则不能日晒。鲜果的出种率为 15%～20%，每果有种子 10～20 粒。种子的含水量 40%～52%，千粒重 2 600（1 900～3 500）g，每千克有纯净种子 380（280～520）粒。种子的净度可达 98%以上。种子宜随采随播种，不耐久藏。常温条件下裸露存放的种子发芽能力约可保持 20 天。10 月以后采集的种子，由于气温转低，可混湿沙贮藏至次年 3 月中旬播种。

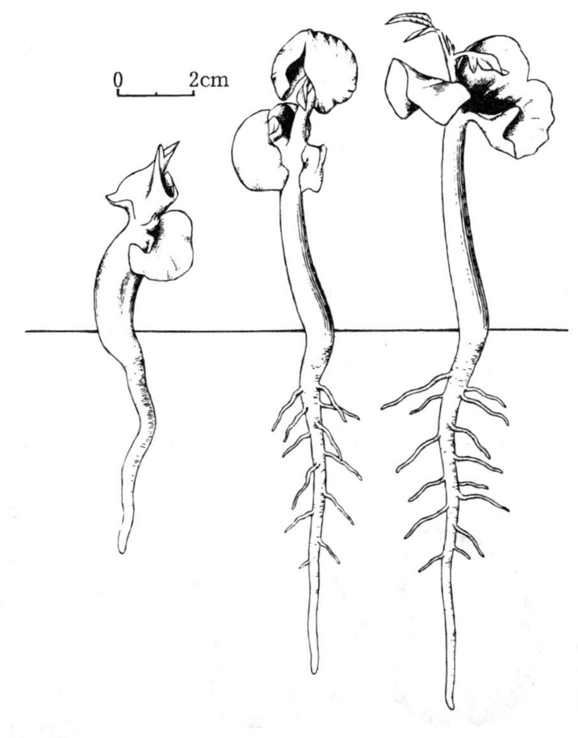

图 2　大果瓜栗种子萌发后第 4、6、9 天幼苗的生长情况
（黄应钦仿《热带亚热带主要树种采种育苗技术》）

发芽和播种　种子无休眠习性。发芽时日均温宜在 25℃以上。随采随播的种子发芽比较迅速整齐。1981 年 8 月，广西林业科学研究所在室外沙床作过发芽测定，播种时日均温 30℃，播前鲜种未作处理。播后 3～4 天萌发，第 5 天进入发芽盛期。从开始发芽至发芽高峰的 4 天，发芽百分数已达 65%；从播种至发芽终止，以 12 天计，发芽率为 80%。图 2 是大果瓜栗种子的萌发和幼苗初期生长情况。出土萌发。发芽后第 3～4 天子叶出土，第 7～8 天发出初生叶。

点播。播前将种子浸水 3～4 小时，捞出稍晾干即可播种。每平方米播种 300～500g，覆土 2～3cm。培育半年生苗出圃。

（林榕庚）

木　槿　属

Hibiscus L.

（锦葵科　Malvaceae）

生长习性、分布和用途　本属约 200 余种，我国约 24 种（包括引入种），本文描述 4 种。灌木或乔木。喜疏松肥沃的土壤，稍耐水湿，多数种能生长于酸性土和钙质土，少数种能耐盐碱。木芙蓉、木槿人工栽培品种颇多。主要分布于热带至亚热带地区。为重要的观赏和纤维树种，少数种的花萼和花可以食用。它们的名称、生长、分布及用途见表 1。

表 1　木槿属树种的名称、生长、分布和用途

中　名	学　名	生长情况	分　布	用　途	供　稿
木芙蓉	*H. mutabilis* L.	落叶大灌木，高 3～4m	湘原产，现广布于华北、华东、中南、西南、辽、陕	观赏、药用、纤维	605
玫瑰茄	*H. sabdariffa* L.	落叶亚灌木，高 2～4m	原产东半球热带地区。滇、桂、粤、琼、闽、台栽培	色素、果酱、纤维、油料、绿化	605
木槿	*H. syriacus* L.	落叶灌木，高 2～4m	原产华中，现华北、华东、中南、西南广为栽培	观赏、围篱、纤维、药用、蔬菜	403
黄槿	*H. tiliaceus* L.	常绿乔木，高约 10m	桂、粤、琼、闽、浙、台。南亚、东南亚	防风、固沙、观赏、纤维、材用、食用	605

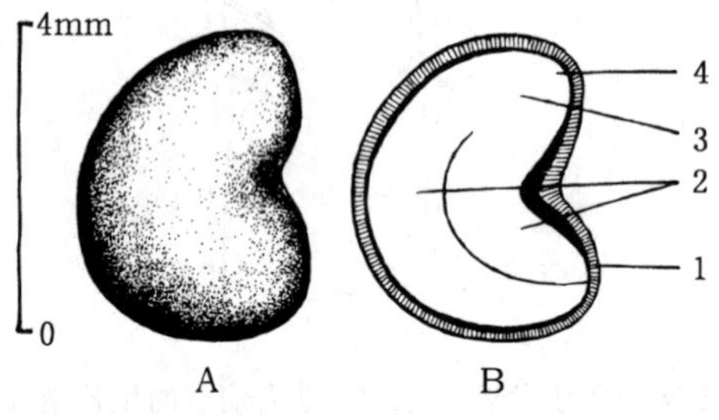

图 1　黄槿种子外形（A）及其纵切面（B）
1. 种皮　2. 子叶　3. 胚轴　4. 胚根
（黄应钦绘）

开花结实　木槿和木芙蓉 2～3 年生开始开花结实，正常结实期在 4 年生以后。玫瑰茄 1 年生即可开花结实。黄槿 6～8 年生开始开花结实，正常结实龄在 15 年生以后。结实均无大小年现象，特寒年份可影响结实，但不甚明显。花两性，腋生或顶生。木芙蓉、玫瑰茄和木槿花单生于枝顶叶腋。黄槿常数花排成聚伞花序，腋生或顶生。花冠钟状，黄色、白色、蓝色、淡紫色或由白变红色，无香味。总苞状小苞片 6～8（12）枚，分离或于基部合生。花萼杯状或钟形，5 裂。花瓣 5，基部与雄蕊柱合生。雄蕊柱顶端截平或 5 齿裂，花药多数，生于柱顶。子房上位，5 室，每室有胚珠 3 至多颗。花柱 5 裂，柱头头状。据南宁和北京观测，开花结实物候期如表 2。

蒴果，扁球形、卵圆形至椭圆形，成熟时由肉质变为革质，成熟后约 10 天左右成室背 5 瓣开裂。种子存留在瓣内 2～15 天后逐渐脱落。种子肾形，无胚乳，胚稍弯曲。果实及种子的形态见表 3、图 1。

表 2　木槿属树种的开花结实物候期

树　种	观察地点和年份	开　花			果实成熟		种子脱落
		始　期	盛　期	末　期	始　期	盛　期	
木芙蓉	南宁 1987	10 月上旬	10 月中下旬	11 月上旬	12 月下旬	翌年 1 月上旬	2 月上旬～3 月中旬
玫瑰茄	南宁 1987	7 月下旬	8 月上中旬	9 月上旬	9 月中旬	10 月中下旬	11 月下旬～12 月下旬
木槿	北京	6 月下旬	7～8 月	10 月上旬	9 月中旬	11 月中旬	10 月上旬～12 月上旬
黄槿	南宁 1987～1988	7 月上旬	7 月中下旬	11 月中旬	9 月中旬	9 月下旬	9 月上旬～12 月下旬

表 3　木槿属树种果实和种子的形态特征

树　种	未熟果颜色	成熟果实			种　子			每果种粒数
		形　状	大小（cm）	颜　色	形　状	大小（mm）	颜　色	
木芙蓉	黄绿色	扁球形，密被淡黄色毛	高 1.6～2.2 径 1.8～2.5	浅褐色	近肾形，背部被长毛	长 2～2.5，宽、厚各1.4～1.8	褐色	175～225
玫瑰茄	紫色	卵球形，密被粗毛	高 1.8～2.8 径 1.5～2.7	紫红色	近肾形，无毛	长 5～6 宽 4～4.5 厚 2.5～2.8	暗褐色	20～30
木槿	黄绿色	矩圆形至卵形，被绒毛	高 2～3 径 1.2～1.5	土黄色	近肾形，背部被棕色长毛	长 3～5 宽 3～4 厚 1.8～2.5	黄褐至黑褐色	—
黄槿	绿色	卵圆形或倒卵圆形，具短喙，被绒毛	高 1.3～2.5 径 1.2～1.8	浅黄褐色	近肾形或棱状卵形，无毛	长 3.5～4 宽 2.5～3 厚 2～2.5	暗褐色	13～16

果实的采收调制和种子贮藏　外果皮已转变为革质而尚未开裂时，即可采集。木芙蓉和玫瑰茄等灌木类的果实可以用手摘；黄槿多用高枝剪带总果梗剪下。采得的果实堆放室内，1～3 天后摊放于通风处或置日光下曝晒，待果瓣开裂，用木棒敲打，脱出种子。木槿属不饱满种粒较多，捡去果瓣后，玫瑰茄可用筛筛去瘪粒等杂质；木芙蓉和黄槿可用簸箕扬去杂质和瘪粒，即得纯净种子。种子的净度、质量等数据见表 4。

表 4　木槿属树种果实的出种率和种子净度、质量

树　种	出种率（%）	净度（%）	千粒重（g）	每千克纯净种子粒数（万粒）	含水量（%）
木芙蓉	25～40	85	3.5～4.5	22～29	8～11
玫瑰茄	1.5～2.5	99	40～48	2～2.5	10～13
木槿	20～30	85	15～17	5.8～6.6	—
黄槿	8～15	99	10～15	6.5～10	8～11

表 5　木槿属树种的发芽能力的测定①

树　种	预处理	发芽势（%）		发芽率（%）	
		计算天数	一般数值	计算天数	一般数值
木芙蓉	始温 50℃水浸种	7	32	13	40
玫瑰茄	—	—	盛期不明显	12	40
木槿	—	—	—	20	65
黄槿	—	—	盛期不明显	21	48

①　北京和广西南宁，室内，26～28℃

种子可以干藏，但不宜在烈日下曝晒过久，以免过分失水。干燥的种子装入袋或坛罐内，贮藏在干爽避光处。玫瑰茄的种子一般贮藏越冬后春播，贮藏期约为半年左右。木芙蓉和黄槿的种子可以随采随播，也可贮藏越冬，贮藏期约为7～8 个月。

发芽和播种　种子无休眠习性，播后 3～5 天即可发芽，发芽时日均温需在 20℃以上。1987 年 5 月和 9 月，广西林业科学研究所在室内自然光照条件下，在发芽皿内用当年新采的

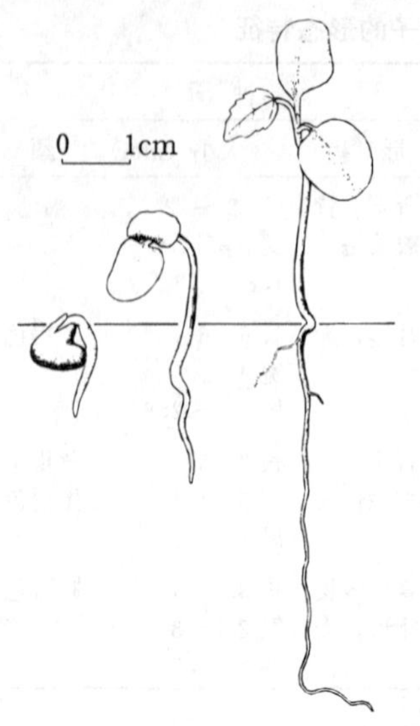

图 2　黄槿种子萌发后第 4、8、25 天幼苗的生长情况
（黄应钦绘）

种子对木芙蓉、玫瑰茄和黄槿 3 个种作过发芽测定；另外，北京植物园在室内对木槿种子作过测定：播前木芙蓉种子曾用温水浸种，其余 3 个种均未作过任何处理，测定时日均温为26～28℃。出土萌发。播后 4～5 天胚根开始萌发，发芽后约 3 天子叶伸出。玫瑰茄的子叶伸出后约 8 天发出初生叶。黄槿的子叶伸出后约需 13 天方出现初生叶。种子发芽及幼苗生长情况见表 5、图 2。

条播。使用当年新采种子，播前可不作任何处理。如果使用头一年采集而经过干藏的种子，播前可用始温 50℃水浸种半天，捞出稍晾干后播种。每平方米播种量：木芙蓉 1～1.5g，玫瑰茄 6～8g，木槿和黄槿 2.5g。整地要求细致，播后覆土 0.5～1cm。培育半年至 1 年生苗出圃。木芙蓉、木槿和黄槿多用扦插育苗。（曾　玲）

古　柯　属

Erythroxylum P. Br.

（古柯科　Erythroxylaceae）

生长习性、分布和用途　本属约 250 种，产热带亚热带地区，主产美洲及马达加斯加，我国 1 种，又引入 2 种。本文描述 2 种（其中引入 1 种）。灌木至小乔木，高 1～6m。它们的名称、生长、分布和用途见表 1。

表 1　古柯属树种的名称、生长、分布和用途

中　名	学　名	树高（m）	分　布	用　途	供　稿
东方古柯	*E. kunthianum*（Wall.）Kurz	1～6	滇、黔、桂、粤、琼、闽、湘、赣、浙。缅甸、印度、孟加拉	药用	605
爪哇古柯	*E. nova-granatense* Hierno	1～2	南美。亚洲热带栽培。台、琼、滇、粤、桂引种	药用	611

开花结实　2 年生开始开花结实，正常结实期在 3 年生以后。结实间隔期为 1 年。花两性，1～3 朵簇生于叶腋。花小，花冠白色或粉红色。萼钟状，5 深裂。花瓣 5，具爪；花瓣内面有 2 舌状附属体。雄蕊 10，花丝合生成筒状。子房上位，矩圆形，3 室，仅 1 室发育，有胚珠1～

2。花柱 3，基部多少连合，柱头头状。据 1987 年在广西南宁对东方古柯的观察，始花期在 5 月中旬，盛花期 5 月下旬，末花期 6 月中旬，花期不整齐，果熟时仍偶有少量花开；9 月中旬为果熟始期，10 月上旬果盛熟，11 月中旬为果熟末期。另据海南儋县观察，爪哇古柯 4 月和 8～10 月两次开花，7 月和 11 月两次果熟。浆果状核果，成熟时鲜红色，花萼宿存。东方古柯果实为长椭圆形略弯，长 1～1.2cm，径 0.5～0.6cm，果核长 0.6～0.8cm，径 0.3～0.4cm。爪哇古柯果实为椭圆形，长约 0.7cm，果核椭圆形，具有 6～8 条沟纹。有胚乳，胚直（图 1）。

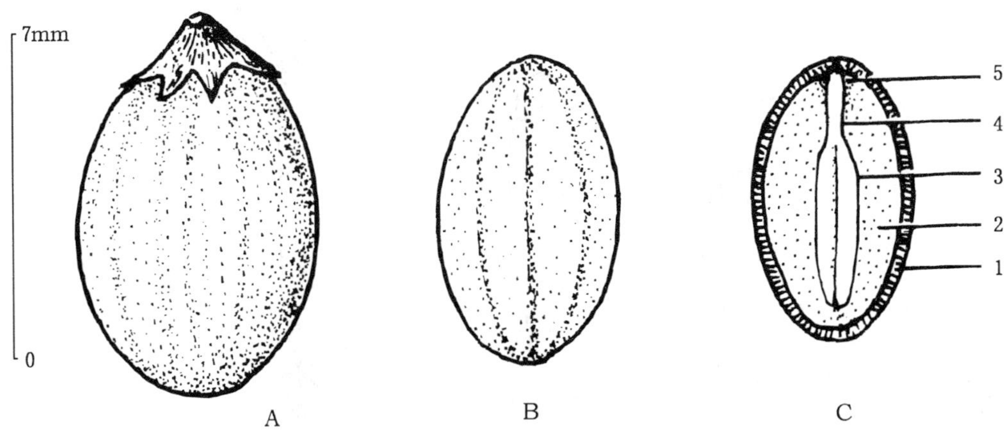

图 1　爪哇古柯果实外形（A）和果核外形（B）及其纵切面（C）

1. 内果皮和种皮　2. 胚乳　3. 子叶　4. 胚轴　5. 胚根

（黄光郁绘）

果实的采收调制和种子贮藏　果熟时用击落法从地面拾集。采回的果实置室内堆沤 1～3 天，待果皮软熟后装入箩筐内搓揉并置水中冲洗，淘出果核用作播种材料，通称种子。果实出籽率和质量等见表 2。

表 2　古柯属树种果实出籽率和净度、质量数据

树　种	出籽率（%）	净　度（%）	千粒重（g）		每千克种子粒数（万粒）	
			一　般	变动范围	一　般	变动范围
东方古柯	50～60	95～99	45	35～55	2.2	1.8～2.9
爪哇古柯	60	95～99	34	30～48	2.9	2.1～3.3

种子忌失水，不宜曝晒或裸露存放。种子含水量宜在 50%左右。运输或贮藏时需混以湿沙。东方古柯的混沙贮藏期为 6 个月，爪哇古柯为 1 个月以内。

发芽和播种　种子休眠现象不明显。发芽时的日均温应在 18℃以上。广西林业科学研究所 1987 年 10 月 24 日在室外沙床播种东方古柯，播种时日均温 23℃左右，11 月 30 日开始发芽，12 月 25 日发芽终止。从开始发芽到发芽终止共 26 天，从播种到发芽终止历时 63 天，发芽率为 42%。华南热带作物研究所 1988 年 11 月 15 日在室外沙床播种爪哇古柯，12 月 14 日开始发芽，翌年1月19日发芽结束。从开始发芽到发芽结束共37天，从播种之日到发芽结束历时67天，发芽率为62%。

条播。每平方米东方古柯播种2～4g，爪哇古柯2～3g，覆土0.5～1cm。1年生苗可以出圃。

出土萌发。胚根萌发后5天左右子叶带壳出土，12～16天子叶展开，真叶展现，生长情况见图2。

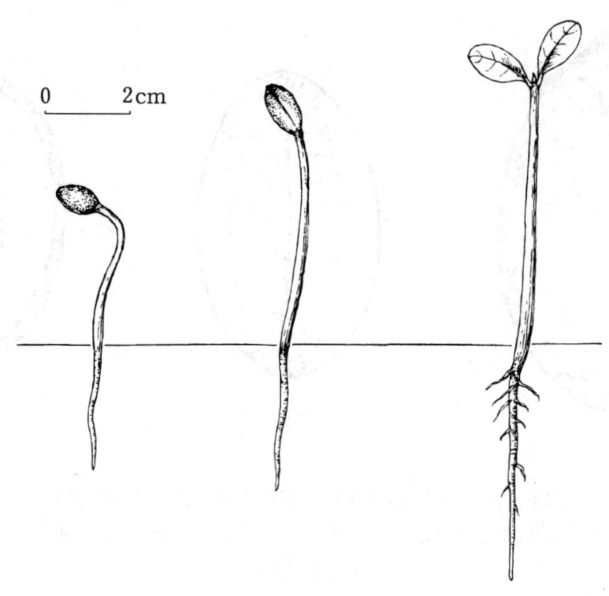

图2 爪哇古柯种子萌发后第2、10、12天幼苗生长情况
（黄光郁绘）

（曾　玲）

白　刺　属

Nitraria L.

（蒺藜科　Zygophyllaceae）

生长习性、分布和用途　本属8种，主要分布在北温带。我国有本文描述的4种。落叶灌木或小灌木。植株矮小，高0.2～2m。丛生，枝端针刺状，叶肉质。喜光，耐寒，耐盐碱，耐贫瘠干旱，抗风沙，根系发达，能耐沙压。旱生或强旱生植物。这4个种的名称、分布和用途见表1。

表 1 白刺属树种的名称、分布和用途

中 名	学 名	分 布	用 途
大白刺	*N. roborowskii* Kom.	内蒙古、宁、甘、新、青。蒙古	果可食用。薪材，固沙，饲料
小果白刺（盐生白刺）	*N. sibirica* Pall.	西北、华北及东北沿海盐渍化沙地。蒙古、西伯利亚、中亚	饲料，薪材，固沙
泡泡刺（球果白刺）	*N. sphaerocarpa* Maxim.	内蒙古西部、宁夏西部、甘肃西北部和新疆东部。蒙古	饲料，薪材，固沙
白刺（酸胖）	*N. tangutorum* Bobr.	西北、内蒙古、西藏东北	果可食用，薪材，固沙，饲料

开花结实 3～5 年生开始结实，5 年以后进入正常结实，结实未见有大小年现象。花两性，白色，形小而繁多。顶生蝎尾状聚伞花序。萼片 5，肉质，宿存。花瓣 5，雄蕊 10～15。子房上位，3 室，每室 1 胚珠。白刺属开花结实物候见表 2。

表 2 白刺属树种开花结实物候

树 种	观察地点和年份	开花			果实成熟		种子散落期
		始 期	盛 期	末 期	始 期	盛 期	
大白刺	—	—	6 月	—	7 月	8 月	8～10 月
小果白刺	甘肃民勤 1981～1984	5 月下旬～6 月上旬	6 月上旬	6 月中旬～6 月下旬	7 月上旬～7 月中旬	8 月上旬～8 月下旬	8～9 月
泡泡刺	甘肃	5 月	～	6 月	6 月	7 月	7～8 月
	内蒙古巴彦淖尔盟西部	5 月		6 月	6 月	7 月	6～7 月
白刺	甘肃民勤 1983～1987	5 月下旬～6 月上旬	～	6 月中旬	7 月上旬～7 月下旬	8 月上旬～8 月下旬	8～10 月

浆果状核果，含 1 种子。白刺属果实和种子的形态特征见表 3。种子形态见图 1。

表 3 白刺属树种果实和种子（果核）的形态特征

树 种	未熟果颜色	成熟果实			种子（果核）		
		形 状	大小（cm）	颜 色	形 状	大小（mm）	颜 色
大白刺	绿色	卵形	长 1.2～1.8 径 0.8～1.5	深红色，果汁紫黑色	窄卵形	长 8～10 宽 3～4	黄褐色
小果白刺	绿色	球形或椭圆形	长 0.6～0.8	暗红色，果汁暗蓝紫色	卵形，先端尖	长约 4～5	黄褐色

（续）

树 种	未熟果颜色	成熟果实			种子（果核）		
		形 状	大小（cm）	颜 色	形 状	大小（mm）	颜 色
泡泡刺	未熟果披针形，密被黄褐色柔毛	球形	径1.0	膨胀成球形，淡黄色，干膜质，无果汁	纺锤形，先端渐尖	8～9	黄褐色，表面具蜂窝状小孔
白刺	绿色	卵形或椭圆形	长0.8～1.2 径0.6～0.9	深红色，果汁玫瑰色	窄卵形	长5～6	黄褐色

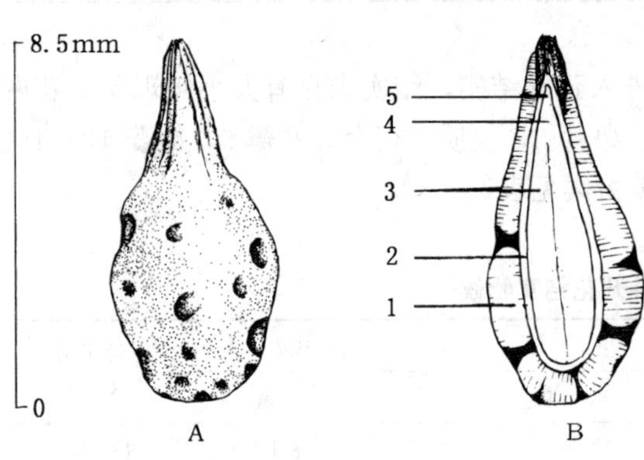

图1 大白刺种子（果核）外形（A）及其纵切面（B）
1. 内果皮 2. 种皮 3. 子叶 4. 胚轴 5. 胚根
（姚桂英绘）

果实的采收调制和种子贮藏 泡泡刺果实6月开始成熟，成熟的果实极易被风吹散，必须及时采收。用手直接采取，运至晒场摊晒碾碾，扬去果膜和杂质，取得纯净果核即为播种材料，通称种子，放在通风干燥处贮藏。适于贮藏的种子含水量6.5%～7.2%。其它3种白刺通常混生，采种时不易区分。果实成熟后易遭鼠害，应及时采摘。将采到的果实放在筛子里加水揉搓，滤去果汁和外果皮后晒干。白刺属种子的净度及质量见表4。种子含水量7.3%～8.3%时装入麻袋，放入种子库贮藏。白刺种皮坚硬，耐贮藏。甘肃民勤综合试验站1962年贮存在种子瓶内的白刺种子，到1978年发芽率仍有22%。

表4 白刺属树种的出籽率及种子净度、质量

树 种	出籽率（%）	净度（%）	千粒重（g）	每千克纯净种子粒数（万粒）
大白刺	20	90～95	36～40	2.5～2.8
小果白刺	20	90～95	12～14	7.1～8.3
泡泡刺	—	79～85	11～13	7.6～9.9
白刺	20	90～98	34～40	2.5～2.9

发芽和播种 白刺属核壳坚硬，属深休眠，播种前必须层积处理。层积前用始温60℃左右的温水浸泡24小时后，置0～15℃的条件下混沙层积，层积温度白天为15℃，夜间0℃，持续35～40天，发芽率为22%～42%。

春播。条播，行距20～25cm，覆土厚度2cm为宜。每平方米播种量为5～15g。在甘肃民勤，4月上旬播种，4月下旬出苗。出土萌发。白刺的幼苗形态如图2。

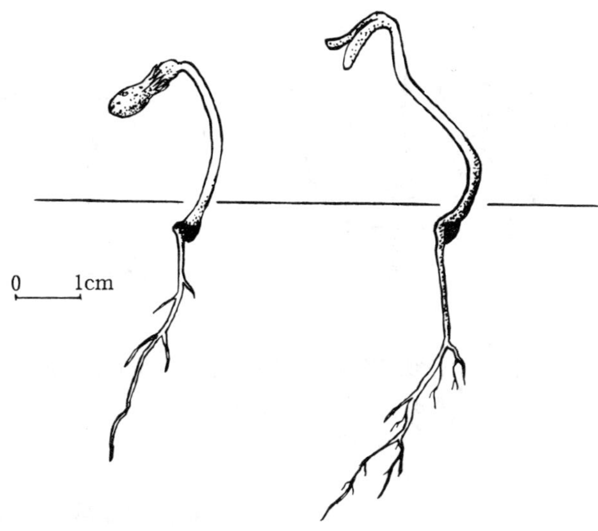

图 2　白刺种子播后第 20、35 天幼苗生长情况
（姚桂英绘）

（郭志中）

四　合　木

Tetraena mongolica Maxim.

（蒺藜科　Zygophyllaceae）

生长习性、分布和用途　四合木属只有 1 种，是我国特有的单种属植物。落叶灌木，高可达 90cm。基部分枝，弯曲。喜光，耐干旱瘠薄，强旱生植物。根系非常发达，是草原化荒漠的建群种之一。产于内蒙古鄂托克旗西北部黄河东岸，由乌海市到石嘴山市一带，为该地特有种，当地群众称为油柴。由于该植物湿时也极易燃烧，常被大量樵采用作燃料，资源破坏严重，在《中国植物红皮书》中被列为稀有种。

开花结实　7～8 年生开始结实，10 年左右进入正常结实。花两性，单生于叶腋。萼片 4，长圆形，长约 3mm，宽 2.5mm，被不规则丁字毛，宿存。花瓣 4，淡黄色或白色，椭圆形，基部稍狭具爪，长约 4mm，宽 2mm。雄蕊 8，排成 2 轮，外轮 4 个较短，内轮 4 个较长，花丝近基部有白色薄膜状附属物，具花盘。子房上位，4 深裂，被毛，4 室，花柱单一，丝状，着生于 4 深裂子房的基部。花期 5～6 月，果期 7～8 月。蒴果 4 深裂，室间开裂。果瓣长卵形或新月形，两侧扁，长 5～7mm，宽 3～4mm，内具 1 种子。种子矩圆状卵形，长 4mm，宽约 1mm，棕褐色，表面被小瘤状突起，无胚乳。种子形态如图 1。

果实的采集调制和种子贮藏　果实由绿色变为黄色标志成熟，应及时采集。可从树上直接摘取，装入布袋贮藏。种子较小，从果实取种时可直接碾取。果实净度 90%，种子净度 95%。种子千粒重 1.6～1.73g，每千克纯净种子 58.8 万～62.5 万粒。

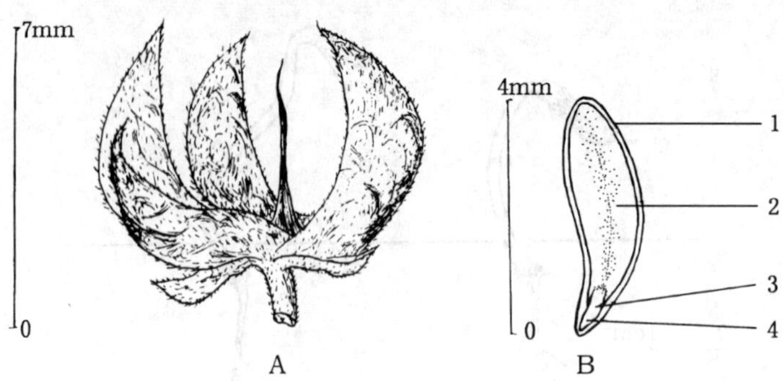

图 1　四合木开裂蒴果外形（A）及其种子纵切面（B）

1. 种皮　2. 子叶　3. 胚轴　4. 胚根

（姚桂英绘）

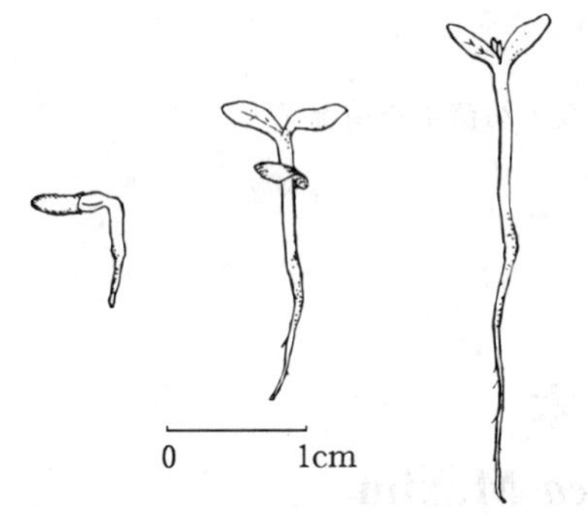

图 2　四合木种子萌发后第 2、4、8 天幼苗生长情况

（姚桂英绘）

发芽和播种　种子休眠期较短，极易解除。用凉水或始温 50℃左右温水浸泡 24 小时后进行发芽测定。每天 8 小时高温，16 小时的低温，测定 18 天，发芽率 30%～55%；置床第 2 天便有 32%的种粒发芽。播种育苗前用温水浸泡一昼夜，或浸种后混细沙催芽半月左右后播种。春季条播，播后覆细沙，以不见种子为度。每平方米播种量 0.6～1.2g。播后一周左右幼苗可以出土。子叶 2 枚。出土萌发。幼苗形态如图 2。

（赵美华）

霸　王

Zygophyllum xanthoxylum（Bge.）Maxim.

（蒺藜科　Zygophyllaceae）

生长习性、分布和用途　霸王属 100 余种，我国约 20 种，多为草本，本文描述 1 种。落叶灌木，高 70～150cm。强旱生植物。生于半荒漠和沙砾质河流阶地、低山山坡、碎石低丘和山前平原。产内蒙古、宁、甘、新、青。蒙古也有分布。固沙植物，叶肉质，可做饲料，根入药。

开花结实　5 年生开始开花结实，丰年间隔期 1～3 年。花两性，1～2 朵生于叶腋，萼片 4，绿色。花瓣 4，黄色，具爪。雄蕊 8，长于花瓣，花丝基部具鳞片状附属物。子房上位，3 室，花柱生于子房顶端，柱头不分裂。霸王的开花结实物候见表 1。

表 1　霸王的开花结实物候

观察地点（和年份）	开花			果实成熟			种子散落	
	始　期	盛　期	末　期	始　期	盛　期	末　期	始　期	盛　期
内蒙古阿拉善盟和巴彦淖尔盟	5 月上旬	5 月中下旬	6 月	6 月	～	7 月	6 月	7 月
甘肃民勤 1975～1981	4 月下旬	5 月上旬	5 月下旬	5 月下旬	7 月上旬	7 月下旬～8 月上旬	7 月	8 月

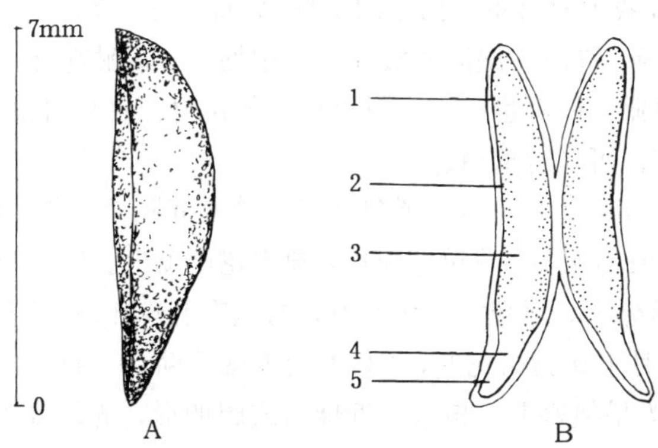

图 1　霸王种子外形（A）及其纵切面（B）
1. 种皮　2. 胚乳　3. 子叶　4. 胚轴　5. 胚根
（姚桂英绘）

蒴果具 3 宽翅（偶有 4 或 5 翅），宽椭圆形或近圆形，不开裂，长 1.8～3.5cm，宽 1.7～3.2cm。通常 3 室，每室 1 种子。种子肾形，黑褐色，长 6～7mm，宽约 2.5mm，具胚乳。种子形态如图 1。

果实的采收调制和种子贮藏　蒴果成熟后随即脱落，可以摘取或用木棒击落，收集后置于通风干燥处晒干。晒干后的果实不开裂，可搓去果翅果皮后置于室内或地下室贮藏。出种率 66%。也可带果翅贮藏。适于贮藏的种子含水量为 8%。种子净度 95%，千粒重 18.6g。每千克纯净种子 5.3 万粒。

发芽前处理和发芽测定　霸王种皮较薄，极易吸水膨胀。1988 年 4 月曾在室内试验，用药棉和滤纸做发芽基质，将浸过水的种子混以湿锯末置于 0～3℃的冰箱中 15 天。取出后在白天 30℃、夜间 20℃的变温条件下，7 天的发芽率为 92%，3 天的发芽势为 24%，发芽迅速而整齐。未经低温层积处理的种子，在同样的变温条件下，19 天的发芽率为 66%，9 天发芽势为 48%。同一份种子用靛蓝测定生活力为 81%。

播种　用始温 45℃水浸种 24 小时后播种。或浸种后混以相当于种子 3 倍的湿润基质（锯末、沙子均可），置 0～5℃条件下层积半月后播种。条播，行距 20～25cm，覆土厚度 1.5～2.5cm。每平方米播种量约 5g，播后 3～4 天即可发芽。出土萌发。子叶 2，肥厚，随后肉质多汁的初生叶陆续展现。1 年

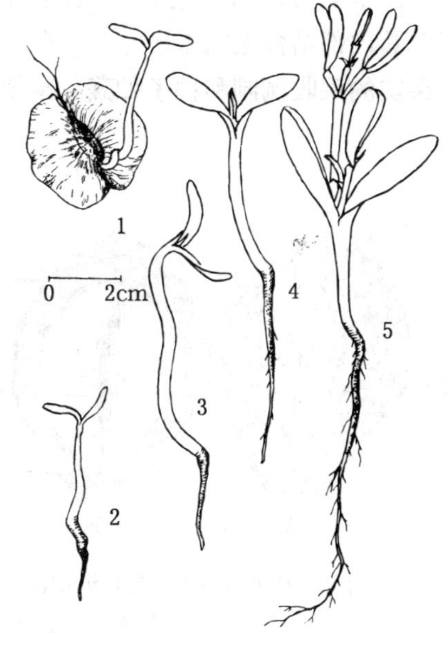

图 2　霸王种子的萌发和幼苗生长情况
1. 种子从果实中萌发　2、3. 子叶出土
4. 初生叶出现　5. 幼苗形成
（姚桂英绘）

即可出圃。幼苗形态如图 2。

（赵美华）

石 栗

Aleurites moluccana（L.）Willd.

（大戟科 Euphorbiaceae）

生长习性、分布和用途 石栗属共 2 种，我国只有本文描述的 1 种。常绿乔木，高达 25m，胸径 60cm。能耐 0℃左右低温，但苗期忌霜冻。喜肥沃湿润的酸性土至中性土，不耐贫瘠。分布于粤、桂、琼、滇。越南、泰国、马来西亚、印度也有。木材可作一般用材。种仁含油达 65%，为油漆、涂料原料；树皮可提制栲胶；可作行道树。

开花结实 约 8 年生开始结实，12～15 年生以后进入正常结实期。结实间隔期 1 年，但在光照及水肥条件好的地方，结实大小年现象不明显。花单性同株，圆锥花序顶生。花小，萼 2 深裂。花瓣 5。雄花：雄蕊 8～20 枚，花丝分离。雌花：子房 2 室，每室有胚珠 1 颗；花柱 2 裂。据 1987～1989 年在广西南宁观测，5 月下旬为始花期，6 月上旬为盛花期，6 月中旬为末花期；7 月下旬果实开始成熟，8 月中旬为果熟盛期。核果，近球形或阔卵形，先端喙尖，密被绒毛，未成熟时青灰色，成熟后暗绿色，长 4～6cm，径 3.4～5.5cm。核果成熟后 10～15 天脱落。每果有果核 2，稀 1～3 粒。果核黑褐色，扁椭圆形，表面有不规则突起，壳骨质，坚硬，长 2.4～2.9cm，径 2.1～2.6cm。胚乳丰富，胚大而伸直，子叶平卧于胚乳中。果核的外形及其内部结构见图 1。

果实的采收调制和种子贮藏 果实成熟盛期击落后从地面拾集，或待果实自然落地后捡

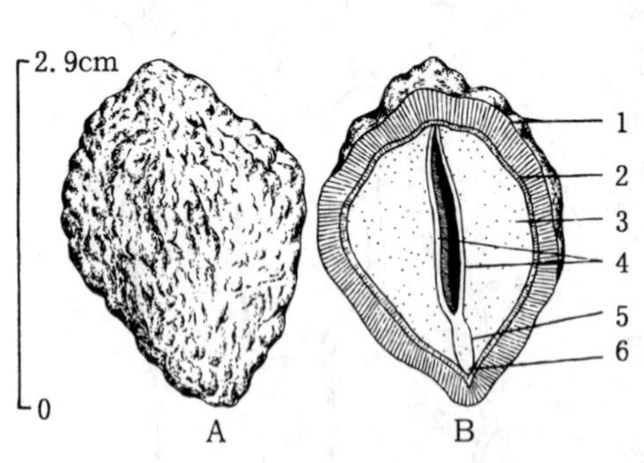

图 1 石栗果核外形（A）及其纵切面（B）
1. 内果皮 2. 种皮 3. 胚乳 4. 子叶
5. 胚轴 6. 胚根
（黄应钦绘）

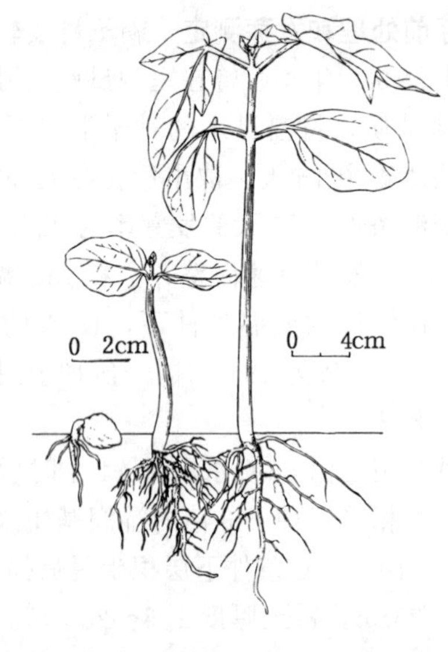

图 2 石栗种子萌发后第 5、12、25 天的幼苗生长情况
（黄应钦仿《热带亚热带主要树种采种育苗技术》）

拾。采集的果实堆沤于湿润处，果皮腐烂后装入竹筐，置水中冲捣漂淘数次。所得果核用作播种材料，通称种子。果实出籽率为20%～30%，千粒重约7 200g，每千克有种子约140粒。种子可以摊晒。摊晒后种子含水量降至10%～12%时，装入袋中干藏。干藏期一般为4～5个月。也可混沙贮藏于荫凉处，贮藏期可达1～1.5年。

发芽和播种　种子无明显休眠习性，但因核壳坚厚，发芽较难。未经摊晒的新调制出的种子，立即播于湿沙床内，80～100天开始发芽。但此时发芽的幼苗越冬易受冻害。一般以沙藏越冬，翌年春播。沙藏的种子，翌年3月下旬日均温在20℃左右时播种，播后50～60天开始发芽。干藏的种子在日均温20℃以上时播种，播后60～70天开始发芽。常见的发芽率约55%。

通常将种子密播于沙床内催芽，发芽后移于大田培育。1年生苗可出圃。

出土萌发。胚根萌发后7～8天子叶带壳出土，11～13天发出初生叶。种子萌发和幼苗生长情况见图2。

（王宏志）

木　奶　果

Baccaurea ramiflora Lour.

（大戟科　Euphorbiaceae）

生长习性、分布和用途　木奶果属约70种，我国有3种，本文描述1种。常绿乔木，高8～12m，胸径约60cm。能耐0.5℃左右的极端低温。适生于肥沃湿润的微酸性至酸性土壤。产云南、海南。越南也有分布。野生果树，老茎生花，可供观赏。

开花结实　10～13年生开始开花结实，15年生以后进入正常结实期，有大小年现象，间隔期1～2年。花单性，雌雄异株。花序狭圆锥状，生于老枝上。花小，无花瓣。雄花萼片4～5，不等大，覆瓦状排列。雄蕊4～8，花丝短，分离。雌花萼片4～6，较雄花大；子房2～5室，每室有胚珠2颗。花柱2～5，极短。雌花序逐渐延长，果穗长可达30cm。据1981年和1983年在云南西双版纳勐仑观察：1月下旬花芽始现，2月花序形成，3月上旬始花，3月中下旬进入开花盛期，4月中旬开花结束，幼果形成；6月底果实开始成熟，7月为果实成熟盛期。蒴果，卵形或近球形。成熟时外果皮由肉质变为木质，由淡黄色转变为淡红色至红色。果长2～3cm，径1.5～2.5cm，不开裂。假种皮肉质胶状，与种皮粘合。每果有种子1～3粒。种子卵形或扁椭圆形，具白色胚乳，子叶淡绿色，大而扁平，倒卵形。种子的外形及其内部结构见图1。

果实的采收调制和种子贮藏　果实成熟时用采种刀采摘。采回的果实用木棒敲破外果皮，取出被有假种皮的种子，在金属筛上搓擦或拌细沙搓擦，在清水中冲洗，取得种子。果实出种率为5%～7%，种子净度为96%，千粒重400g左右。每千克纯净种子约2 500粒。调制后稍晾干的种子含水量为51%～53%，忌失水，不宜日晒。贮藏时需混湿沙，贮藏期不超过半年。

发芽和播种　种子无休眠现象。发芽适宜的日均温为25℃左右。云南西双版纳热带植物

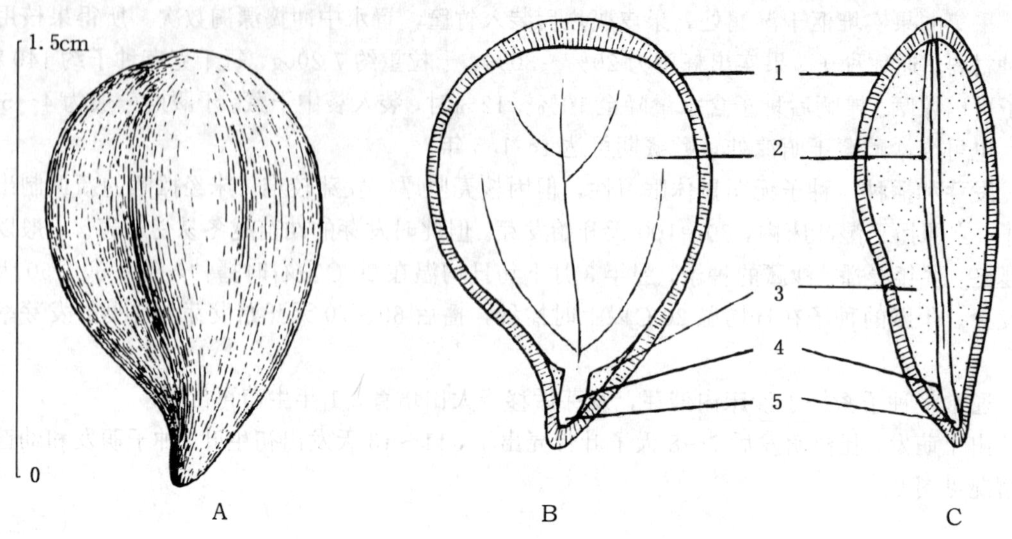

图 1 木奶果种子外形（A）及其纵切面（B、C）

1. 种皮 2. 子叶 3. 胚乳 4. 胚轴 5. 胚根

（黄应钦绘）

园于 1988 年 7 月 6 日用刚调制的种子在室外沙床进行发芽测定，7 月 16 日种子发芽出土，7 月 26 日发芽结束，无明显的发芽盛期。从播种之日起算，21 天的发芽率为 100%。

出土萌发。子叶出土后 1 个月左右出现第 1 片初生叶，45 天出现第 2 片初生叶。

点播。种子生命力维持的时间很短。调制后的种子置通风处存放半个月，含水量下降到 28%时就完全丧失发芽能力。因此必须随采随播。新采种子播前不必作任何处理。每平方米播种 35～60g，覆土 0.5～1.0cm。1 年生苗可出圃。

（马信祥）

重阳木属

Bischofia Bl.

（大戟科 Euphorbiaceae）

生长习性、分布和用途 本文描述该属仅有的 2 种，它们的名称、习性、分布和用途见表 1。秋枫为常绿或半常绿大乔木，树高 40m，胸径 130cm，有的达 230cm。重阳木为落叶乔木，树高 15m，胸径 50cm，稀达 100cm。喜温暖湿润气候。适生于微酸性及中性的湿润肥沃的沙质壤土。干旱瘠薄土壤上生长不良。

分布于我国秦岭、淮河流域以南。秋枫还分布在东亚、东南亚、南亚及澳大利亚。木材结构略粗，纹理直，较坚硬，耐用，赤色，有光泽，可代紫檀，供建筑、车船、家具等用。果肉可酿酒。种子含油率较高，可榨油供工业用。

表 1　重阳木属树种的名称、习性、分布和用途

中　名	学　名	习　性	分　布	用　途	供　稿
秋枫	*B. javanica* Bl.	常绿或半常绿	西南、华南、华中南部、闽、台。印度、越南、菲律宾、印度尼西亚、澳大利亚	材用	601
重阳木	*B. polycarpa* (Levl.) Airy-Shaw	落叶	西南、中南、华东，豫、鲁栽培	材用	502

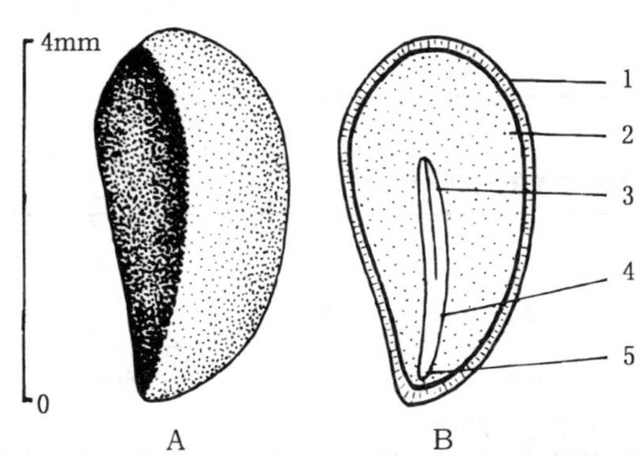

图 1　重阳木种子外形（A）及其纵切面（B）
1. 种皮　2. 胚乳　3. 子叶　4. 胚轴　5. 胚根
（黄应钦绘）

开花结实　6～8 年生开始开花结实。正常结实期在 20 年生以后。结实间隔期一般为 1 年。花单性，雌雄异株，无花瓣及花盘。秋枫圆锥花序腋生。重阳木总状花序下垂。雄花萼片 5，镊合状排列，雄蕊 5，分离，与萼片对生，花药大，退化雌蕊短而宽，具短柄。雌花萼片覆瓦状排列，子房上位，3（4）室，每室 2 胚珠，花柱2～4，长而肥厚，下垂，顶不裂。秋枫开花期 3 月下旬～4 月中旬。重阳木 4～5 月花与叶同时开放，花淡绿色。

重阳木属树种的开花结实物候见表 2。

果实浆果状，未熟时青绿色，成熟时淡红色至棕褐色或红褐色，球形或略扁，径 0.5～1.5cm。外果皮肉质，内果皮坚纸质。种子 3～6 粒，棕褐色或紫黑色，有光泽，肾形、扁圆形或矩圆形，长4～5mm（图 1），无种阜。种皮脆，胚乳肉质，白色，胚直立，子叶宽而扁平。这 2 个树种果实和种子的形态特征见表 3。

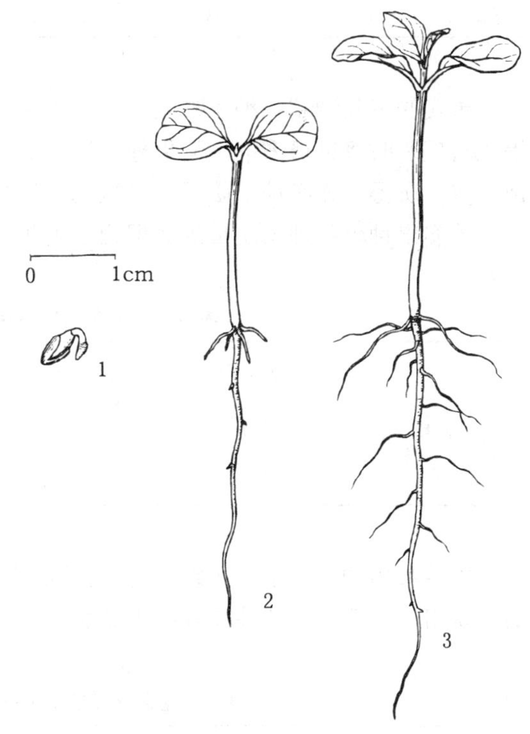

图 2　重阳木种子萌发后幼苗的生长情况
1. 幼根初露　2. 子叶展开　3. 初生叶互生
（黄应钦、杨向晨仿《主要树木种苗图谱》）

表 2 重阳木属树种的开花结实物候

树 种	观察地点	观察年份	开花 始期	开花 盛期	开花 末期	果实成熟 始期	果实成熟 盛期	果实散落 始期	果实散落 盛期	果实散落 末期
秋枫	广西南宁	1987	3月下旬	4月上旬	4月中旬	11月中旬	11月下旬～12月上旬	12月中下旬	～	翌年3月上旬
重阳木	四川荣县	1987～1988	4月上旬	4月中下旬	4月底	10月中旬	10月下旬～11月上旬	11月中旬	12月上中旬	翌年1月

表 3 重阳木属树种果实和种子的形态特征

树 种	果实 颜色	果实 形状	果实 径（cm）	种子 颜色	种子 形状	种子 长（mm）
秋枫	淡红色至棕褐色	球形	1～1.5	棕褐色	扁椭圆形	4.5～5
重阳木	红褐色	球形	0.5～0.7	棕红色或紫黑色	肾形	4

果实的采收调制和种子贮藏 成熟的果实易遭雀鸟啄食。当果实呈淡红色、棕褐色或红褐色时应及时将果穗采下。从果穗上摘下的果实堆沤数日，果皮腐烂后置水中搓揉，淘去果肉果皮，所得的纯净种子阴干后干藏或混沙贮藏。种子一般贮藏一个冬季，翌春播种。

果实出种率和种子质量的数据列入表 4。

表 4 重阳木属树种果实的出种率和种子质量

树 种	出种率（%）	净度（%）	千粒重（g）	每千克纯净种子数（万粒）
秋枫	10	93～98	16～25	4～6.2
重阳木	8～12	90～96	6.2～7.1	14～16

发芽和播种 种子无休眠习性。发芽测定前有用始温 45℃水浸种 24 小时的，也有不浸种的。常用的发芽测定条件及其结果见表 5。

表 5 重阳木属树种种子的发芽能力及其测定条件

树 种	发芽测定温度（℃）	发芽势（%） 计算天数	发芽势（%） 一般数值	发芽势（%） 变动数值	发芽率（%） 计算天数	发芽率（%） 一般数值	发芽率（%） 变动数值
秋枫	28	4	65	—	7	80	70～90
重阳木	25	6	42	36～45	12	65	60～70

春季播种。播种期 2 月中旬～3 月中旬。条播时条距 20cm，条幅 3～5cm，沟深 3cm，每平方米播种 5～8g。撒播时每平方米播种 10～15g。播后覆细土 0.5cm，轻压后洒水、盖草。重阳木播种后约 30 天种子发芽出土。1 年生苗高约 50cm。供行道树或在庭院栽植的，应在幼苗期进行移栽，在苗圃继续培育，待苗木高达 3m 许时出圃定植。出土萌发，子叶 2，椭圆状卵形，或阔卵形。秋枫胚根萌发后约 9 天子叶出土，约 50 天真叶出现。初生叶单叶互生，至第 8 或第 9 片始为 3 小叶复叶。重阳木种子萌发后的幼苗生长情况见图 2。

（阙再点）

蝴　蝶　果

Cleidiocarpon cavaleriei (Lévl.) Airy-Shaw

（大戟科　Euphorbiaceae）

生长习性、分布和用途　蝴蝶果属仅 2 种，我国只有本文描述的 1 种，在《中国植物红皮书》中列为稀有植物。常绿乔木，高达 30m，胸径 80cm。喜肥沃湿润的石灰质土壤，酸性土亦能生长，忌积水地。能耐短期－1℃低温及轻霜。分布于云南东南部、广西西南部及贵州南部。越南也有。种仁含脂肪达 35%，淀粉含量也较高，可供食用及榨油。材质中等，作一般用材。树形美观，常作为庭园树和行道树栽培。

开花结实　10 年生左右开始开花结实，正常结实期在 20 年生以后。结实间隔期 1～2 年。花单性同株或同序。由多个长穗状花序集成顶生的圆锥花序。雄花多数，着生于花序上部，雌花 1～6 朵在下部。无花瓣。雄花的花萼 3～5 深裂。雄蕊 3～5，花丝分离，花药背着，4 室，退化雌蕊短圆柱状。雌花的花萼 5～8 裂。子房具短柄，2 室，常仅 1 室发育，每室 1 胚珠。花柱 3，基部合生。柱头 3 深裂，每裂再 2～3 叉裂。据 1981～1985 年在广西南宁观测，4 月上旬花序形成，着生于头年生枝条的叶腋；4 月中旬花始开，下旬为盛花期，5 月上旬为末花期；5 月上旬～6 月中旬幼果开始形成，8 月中旬～9 月上旬果实成熟；8 月下旬～10 月上旬果实陆续脱落。果，核果状，不规则椭圆形或圆形，具长柄，基部细小，常因子房仅 1 室发育而偏肿，有时 2 室

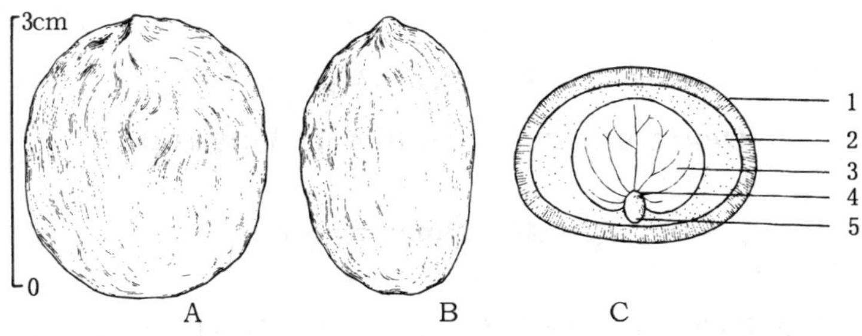

图 1　蝴蝶果种子外形（A、B）及其纵切面（C）

1. 种皮　2. 胚乳　3. 子叶　4. 胚芽　5. 胚根

（黄应钦绘）

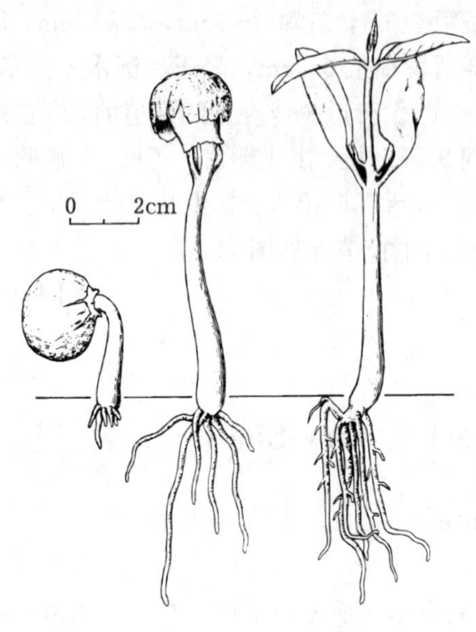

图 2 蝴蝶果种子萌发后第 3、6、20 天幼苗的生长情况
（黄应钦仿《热带亚热带主要树种采种育苗技术》）

发育则成双球形。外果皮薄，近壳质，成熟时由暗绿色或黄绿色转变为灰黄色，表面粗糙，密被绒毛。果长 3～4cm，径 2.5～3.5cm，略开裂或不开裂，花萼宿存。每果有种子 1 粒，稀 2 粒。种子扁球形，长 2.2～3cm，扁径 2.1～2.5cm。种壳木质，暗褐色或灰褐色，外表不平，呈凹凸状。胚乳丰富，胚横生于种子中部，子叶大而扁平，近圆形，见图 1。

果实的采收调制和种子贮藏 果实变为灰黄色时在地面拾集，或上树敲打摇动果枝，震落果实，在地面拾集。采回的果实按成熟程度分别堆放 1～2 天，剥出种子。种子忌脱水，不能日晒，亦忌堆沤发热。半个月以内的短期贮藏或运输宜带外果皮。运输宜用竹筐包装，每筐不超过 20kg，但播种前则需剥去外果皮。果实出种率约 60%。净度为 98%，千粒重 6 750g，每千克约有纯净种子 150 粒。种子含水量约为 30%。8 月以前采集的种子可随采随播或混湿沙贮藏约 1 月。9 月中旬以后采集的种子只宜随采随播，不宜越冬贮藏。

发芽和播种 种子无休眠现象。发芽时日均温宜在 22℃以上。1987 年 8 月 19 日，广西林业科学研究所在室外沙床作过发芽测定，当时日均温 30℃左右：播种后第 3 天即开始萌发，萌发后第 2 天便进入发芽盛期；从开始发芽至发芽高峰的 5 天中，发芽百分数为 70%；以 12 天计，发芽率为 90%。

条播。新采种子播前不必作任何处理。贮存过 1～3 个月的种子则需浸水 2～3 小时再行播种。每平方米约播 350g。因子叶较肥大，不易出土，播后覆土宜浅，一般为 1cm 左右。苗木多用于城镇绿化，通常用 2 年半生苗。

出土萌发。发芽后子叶 8～10 天出土，第 15～16 天发出初生叶（见图 2）。

（王宏志）

东 京 桐

Deutzianthus tonkinensis Gagnep.

（大戟科 Euphorbiaceae）

生长习性、分布和用途 东京桐属只有本文描述的 1 种，属稀有植物，已列入《中国植物红皮书》。常绿乔木，高达 18m，胸径 30cm。能耐 0℃左右低温，但苗期忌霜冻。喜肥沃湿润的酸性土和钙质土。分布于滇、桂。越南北部也有。为优质速生用材树，种子含油率 49%，供制皂及工业用。

开花结实 约 8 年生开始结实，正常结实期在 20 年生以后，结实间隔期一般为 1 年。花单性异株，伞房花序式的圆锥花序顶生。雄花萼钟状，短 5 裂，花瓣 5，白色。雄蕊 7，外 5 枚与花瓣对生，中央 2 枚粘合达中部。雌花萼裂片三角状延长，花瓣 5。花盘杯状，5 裂。子房 3 室，每室有下垂的胚珠 1 颗。花柱 3，极叉开，2 裂达中部。据 1987 年在广西南宁观察，5 月上旬为始花期，中旬为盛花期，下旬为末花期；8 月中旬果实开始成熟，8 月下旬～9 月上旬为果熟盛期，9 月下旬～11 月中旬果实陆续脱落。核果。外果皮未成熟时黄绿色，成熟后浅褐色被灰白色绒毛。果阔卵形或矩圆形，具 3 棱，顶端钝尖，长 2.8～3.5cm，径 3～3.8cm。每果有果核 3 粒，稀 2 或 4 粒。果核扁椭圆形或卵形，栗色而有光泽，长 1.6～2.1cm，宽1.5～1.9cm，厚 1.3～1.5cm。胚乳丰富，胚伸直，子叶大，椭圆形，见图 1。

果实的采收调制和种子贮藏　果实变为浅褐色时用高枝剪或采种刀采摘，或敲打果枝震落果实后在地面捡拾。果实的中果皮与内果皮间充满胶质，新采的果实需堆沤 10～15 天，并经常保持湿润，待果皮变软，胶质腐熟，易于洗脱时，用手剥去外果皮和中果皮，置水中搓洗，所得果核用作播种材料，通称种子。果实出籽率约为 54%，净度可达 98%。千粒重约 3 100g，每千克有种子 320 粒。种子可以稍加摊晾，但忌日晒或过分干燥。晾干的种子含水量为 15%～20%。一个月内的短期贮藏可用袋装。供翌年播种的应混湿沙层积。

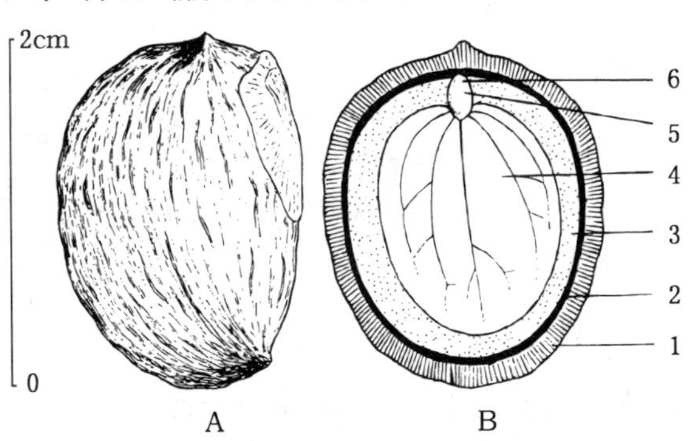

图 1　东京桐果核外形（A）及其纵切面（B）
1. 内果皮　2. 种皮　3. 胚乳
4. 子叶　5. 胚轴　6. 胚根
（黄应钦绘）

发芽和播种　种子有休眠现象，需湿沙层积越冬至翌年春末夏初播种。1987 年 8 月下旬，广西林业科学研究所用当年采集未经层积的种子作发芽测定，播后 10 天有 4%的种子发芽，到 9 月 4 日种子发芽过程中断，至 1988 年 5 月 4 日起又继续发芽，5 月 30 日结束，总发芽率为 85% 。发芽时的日均温需在 20℃以上。出土萌发。胚根萌发后约 10 天子叶出土，15 天展出初生叶。

春季播种。点播，每平方米播种 150～200g，覆土 2cm。1 年生苗可以出圃栽植。

（王宏志）

黄　桐

Endospermum chinense Benth.

（大戟科　Euphorbiaceae）

生长习性、分布和用途　黄桐属约 13 种，我国只有本文描述的 1 种。常绿大乔木，高达 35m，胸径达 100cm。能耐 0℃左右低温，但幼苗忌霜冻。适生于湿润凉爽的气候和质地疏松的酸性土，干旱贫瘠地生长不良。分布于琼、粤、桂、滇。越南也有。木材纹理直，结构细致，属优质轻软木，宜作室内器具及胶合板等。

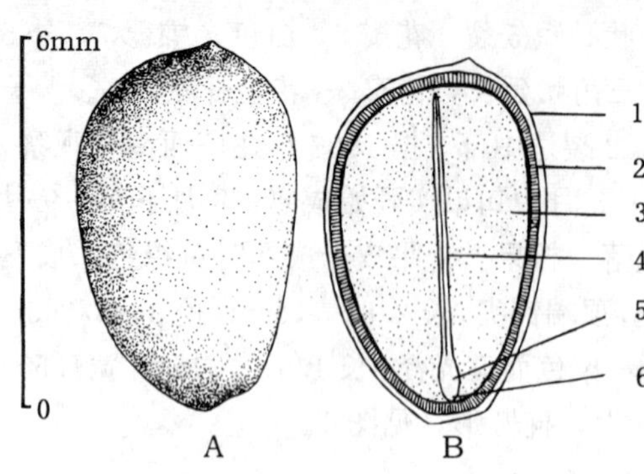

图 1　黄桐果核外形（A）及其纵切面（B）
1. 内果皮　2. 种皮　3. 胚乳　4. 子叶
5. 胚轴　6. 胚根
（黄应钦绘）

开花结实　6～8 年生开始结实，15 年生以后进入正常结实期。结实丰年间隔期为 1 年。花单性异株，无花瓣。雄花簇生成圆锥花序，长 16～20cm，雌花排成总状花序长 6～10cm，均腋生。雄花萼3～4 浅裂。雄蕊 5～8，着生于圆锥状凸起的花托上。雌花萼 5 齿裂。子房圆球形，2～3(6)室，每室 1 胚珠。花柱短，合生成一扁平 2～6 浅裂的盘状体。据在海南观测，花期5～6 月，8 月中旬果实开始成熟，下旬为果熟盛期，8 月下旬至 9 月中旬果实脱落。核果。外果皮和中果皮稍肉质，内果皮木质而坚硬。果实近球形，长约 1cm，径约 0.7cm，未成熟时青色，成熟时黄色，密被黄色星状毛，分离成 2 个不开裂的分果爿，无中轴。每果有果核 2。果核为略具三棱的长椭圆形，长 4.5～6mm，径2.5～3.5mm，具网状皱纹，无种阜。胚乳丰富，子叶扁平，见图 1。

果实的采收调制和种子贮藏　果实变为黄色时用高枝剪或采种刀截断果梗采集，也可以击落后从地面拾集。采集的果实室内堆放 1～2 天后置水中浸泡，搓擦漂洗，淘去果肉等杂质即得果核，用作播种材料，通称种子。果实出籽率为 11%。千粒重约 18g，每千克有种子 5.5 万粒。种子不宜脱水，忌日晒。混湿沙贮藏，贮藏期 5～6 个月。

发芽和播种　种子无休眠习性。发芽时要求日均温 20℃以上。1987 年 8 月 28 日，广西林业科学研究所用当年采集的种子在室外沙床播种，9 月 20 日开始发芽，10 月 6 日发芽终止，未见明显的发芽盛期。从播种之日起算，40 天的发芽率为 48%。

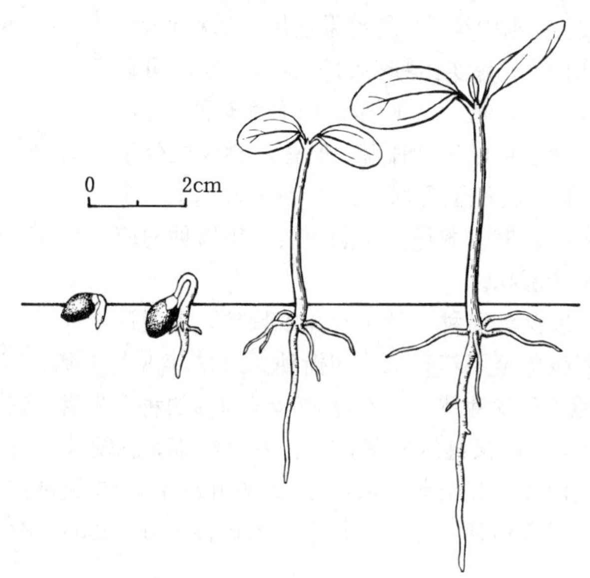

图 2　黄桐种子萌发后第 1、3、7、17 天的幼苗生长情况
（黄应钦绘）

出土萌发，胚根萌发后约 4 天子叶出土，17 天初生叶出现。种子萌发及幼苗生长情况见图 2。

撒播或条播。也可将种子密播于沙盘或经过细致整地的圃地，每平方米播种 40～ 50g，苗木转木质化时移植至大田继续培育。气候凉爽的山区可将种子条播至圃地，每平方米播种 5～8g，覆土厚 0.2cm。1 年生苗可出圃栽植。

（王宏志）

白　饭　树

Flueggea virosa（Willd.）Baill.

（大戟科　Euphorbiaceae）

生长习性、分布和用途　白饭树属约6种，我国2种，本文描述1种。落叶灌木，高2～4m。适生于酸性土。分布于滇、黔、桂、粤、琼、闽、台。印度、菲律宾也有。全株供药用，能清热解毒，治跌打损伤，外用能治脓疮湿毒。

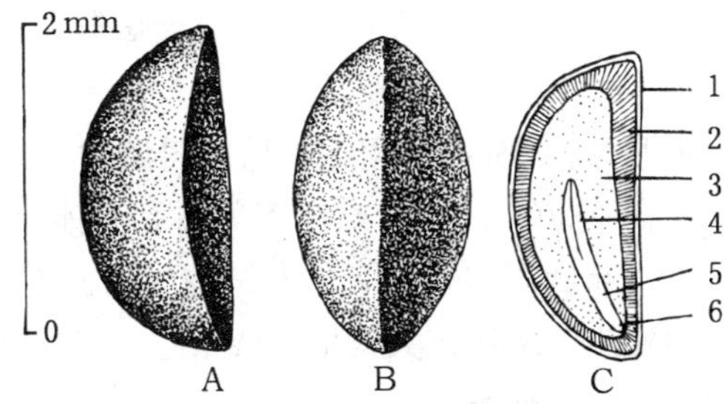

图1　白饭树种子外形（A、B）及其纵切面（C）
1. 外种皮　2. 内种皮　3. 胚乳　4. 子叶　5. 胚轴　6. 胚根
（黄应钦绘）

开花结实　2～3年生开始开花结实，正常结实在5年生以后，结实间隔期为1年。花单性，雌雄异株。单生或簇生于叶腋，或组成密伞花序。花小，无花瓣，淡黄色。雄花多数簇生，萼片5，近花瓣状，雄蕊5；花盘腺体5或花盘杯状，全缘，退化雌蕊3，基部连合，顶端分裂或弯曲。雌花萼片花瓣状，花盘环状有齿缺。子房3室，每室有胚珠2颗。据1988年在广西南宁观察：4月中旬开始开花，5月上旬为花盛期，5月中旬为花末期，但雄花物候期比雌花迟3～5天；7月上旬果始熟，9月上旬为果熟盛期，9月下旬为果熟末期。蒴果浆果状，近球形，径5～6mm。幼果绿色，成熟时转为白色。外果皮薄，中果皮肉质，有2裂的分果爿。每果有种子4～5粒，浅褐色，长1.8～2mm，径约1mm。种皮脆壳质，胚乳丰富，胚直，子叶扁平，见图1。

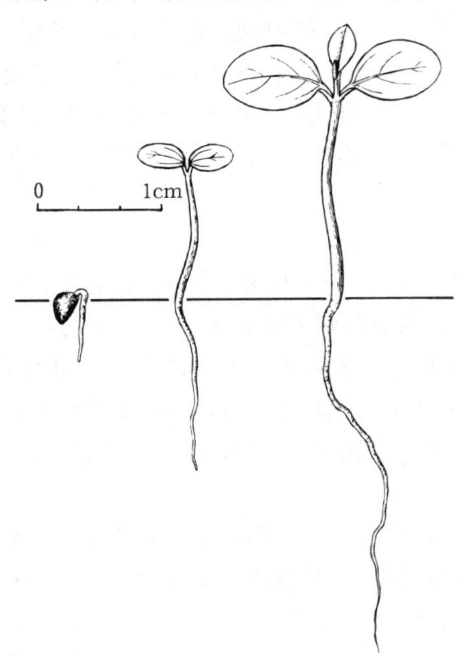

图2　白饭树种子萌发后第2、5、9天幼苗的生长情况
（黄应钦绘）

果实的采收调制和种子贮藏　9月果熟时即可采摘。采回的果实搓揉后置水中漂淘，洗去果皮等杂质得到种子。果实出种率约13%。种子净度95%，千粒重约1.4g，每克有纯净种子约710粒。种子晾干后用布袋盛装置于通风荫凉处贮藏，贮藏期约1年。

发芽和播种　种子无休眠习性。发芽时要求日均温在20℃以上。1987年广西林业科学研究所曾在室内所作过发芽测定，测定时室温约25℃。9月11日置床，16日开始发芽，10月21日结束，41天的发芽率为40%。

出土萌发，胚根伸出4天后子叶出土，再过4天初生叶出现。幼苗生长情况见图2。

条播。可随采随播，也可将种子贮藏至翌年春播，每平方米播种 2～2.5g，覆土 0.2cm。

（曾 玲）

橡 胶 树

Hevea brasiliensis（H. B. K.）Muell. -Arg.

（大戟科 Euphorbiaceae）

生长习性、分布和用途 橡胶树属约 20 种，我国引入本文描述的 1 种。常绿乔木，高 20m 以上，胸径 30～40cm。原产南美洲亚马孙河流域的巴西、秘鲁、哥伦比亚、委内瑞拉和圭亚那。1904 年我国引入云南德宏试种成功。1905 年引入台湾恒春种植。1906 年引入海南，现已在我南亚热带地区扩大栽培。浅根性，抗风力较弱。喜深厚肥沃，pH 值为 4.5～5.5 的酸性土壤。橡胶是重要的工业原料。种子含油，可作涂料、肥皂的原料。木材经化学处理后可制造高级家具。

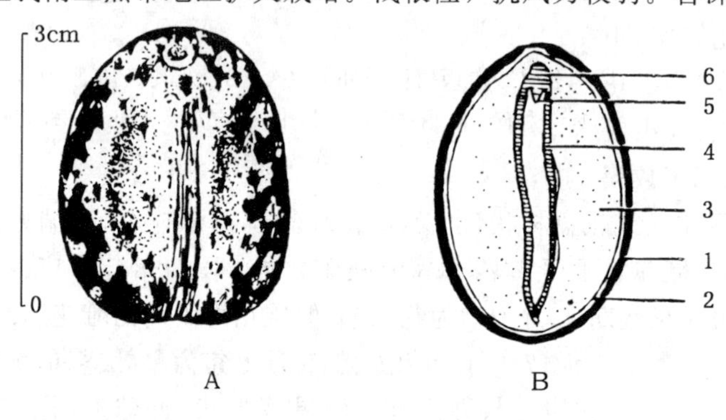

图 1 橡胶种子外形（A）及其纵切面（B）

1. 外种皮 2. 内种皮 3. 胚乳 4. 子叶 5. 胚芽 6. 胚根

（黄光郁绘）

开花结实 实生树约 8～10 年生开始结实，芽接树 3～4 年开花结实。花单性，雌雄同株并同序，组成圆锥花序状聚伞花序，腋生。聚伞花序中央为雌花，其余为雄花，无花瓣。萼片 5 裂。雄花：雄蕊 10，花丝合生成柱状。雌花具绿色花盘，比雄花大，生于花序的顶端；子房 3 室，每室 1 胚珠；无花柱，柱头粗壮。据在海南儋县观测：一般每年 2 次开花，3～4 月为主花期，称春花，5～6 月第 2 次开花，称夏花，也有 8～9 月第 3 次开花的；从受精到果实成熟需 4～5 个月；一般 8～9 月果实成熟，10～11 月为第 2 次果熟期，翌年 1 月为第 3 次果熟期。未熟时外果皮绿色，成熟时转为淡黄色，最后变为灰褐色。蒴果 3 裂，具 3 个 2 裂的分果爿，各含 1 粒种子。种子椭圆形，褐色，有斑纹，长约 3cm，宽约 1.5cm。外种皮比内种皮稍厚。胚乳丰富，胚直，子叶大而扁平，见图 1。

果实的采收调制和种子贮藏 果实转呈淡黄色时从树上直接采摘，或用采种刀钩落果实后在地面捡拾。也可待果实在树上自然干燥，内果皮弯扭爆裂，从林地拾集爆落的种子。采集的果实摊放于室内，1～2 天后种子自行弹出，即得播种材料。出种率约 42%，净度 98%，千粒重约 4 500g，每千克有纯净种子约 220 粒。新鲜种子含水量为 45%～50%，如果保存不当，失水很快，容易丧失发芽能力。在种子保管、包装、运输过程中，要保持荫凉、通风，忌曝晒，忌堆积，也忌水浸。短期贮存可摊放于荫凉处。时间在 2～3 天以内的短途运输，可用麻袋或草包装运；长途运输用竹箩包装，并用洗净的谷壳、锯木屑等作填充物，将种子隔层铺放，中间设置通气草把。在室内常温条件下裸露存放半个月，发芽率便降至60%以下。需要越冬的种子，宜混沙贮藏。方法是在空气流通干燥的室内，地面铺一层15～20cm的细沙，将种子同湿沙混合并堆至50cm高，再覆盖和圈围15～20cm的湿沙。贮藏期间注意保持湿润，但要防止过干过湿，避免发芽变坏。用此法贮藏3个月，种子发芽率可达70%～80%。少量种子可在灭菌处理后装于塑料袋中，每袋1～3kg，密封袋口，并用大头针穿2～5个小孔，贮放于室内阴凉处。这样贮藏4个月后，种子发芽率仍可保持在80%以上。

发芽和播种　种子无休眠习性。发芽时日均温宜在22℃以上。1988年9月23日，华南热带作物研究所在室外沙床上用新鲜种子作发芽测定，播后11天开始发芽，10月17日发芽结束。从播种之日起算，25天的发芽率为90%。

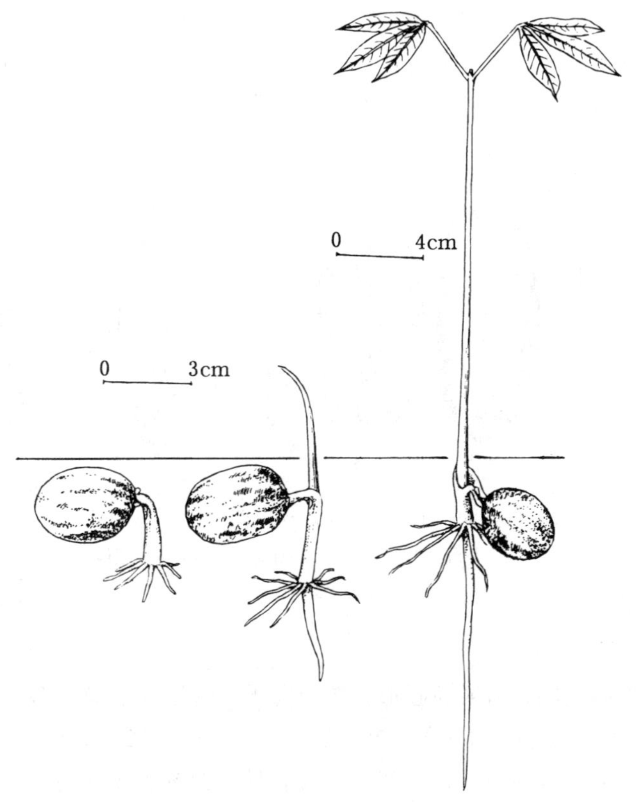

图2　橡胶树种子萌发后第2、12、19天的幼苗生长情况

（黄光郁绘）

留土萌发。种子发芽时，乳白色的胚根穿过发芽孔向下生长，胚芽随即伸出种壳外。此时古铜色幼芽露出地面向上生长，并不断加粗和伸长，4天后，可见到一对下垂的小初生叶，6天后由古铜色转为绿色，并由下垂状态展开为水平状。幼苗生长情况见图2。

育苗时通常将种子播于沙床，催芽后再行移床。播种时，种子平放，腹沟向下，覆土以微露种背为度，淋水保湿。待幼苗出土挺直，高约10cm时移至圃地，每666m^{2}育苗2 400株或3 300株。目前已有人工选育的优良品种。生产中多用无性苗，实生苗一般用作砧木。

（丁慎言）

麻　疯　树

Jatropha curcas L.

（大戟科　Euphorbiaceae）

生长习性、分布和用途　麻疯树属约175种，分布于热带、亚热带地区。我国约4种，栽培或半野生。本文描述1种。灌木或小乔木，高2～5m，多分枝。速生。喜肥沃土壤，酸性土或钙质土均宜生长。耐0℃左右的低温及轻霜。抗风力强。原产美洲热带，现广布于全世界热带地区。我国南海诸岛有分布。台、闽、粤、琼、桂、滇、黔、川等地栽培或半野生。多栽培作围篱及供观赏。种仁含油脂约50%，可作润滑油、制皂等用，并可催吐下泻，但有毒，忌食。

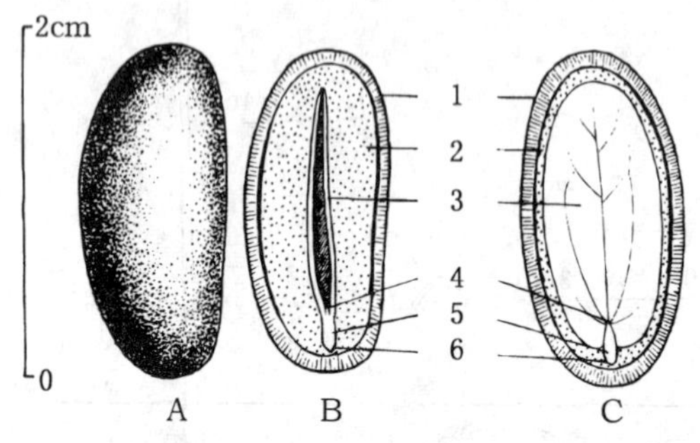

图1 麻疯树种子外形（A）及其纵切面（B、C）
1. 种皮 2. 胚乳 3. 子叶 4. 胚芽 5. 胚轴 6. 胚根
（黄应钦绘）

开花结实 3年生开始结实，正常结实期6～20年。结实间隔期不明显，立地条件好的母树，连年结果均较多。花单性，雌雄同株。二歧聚伞花序腋生。雌花顶生于花序中央或二歧分叉处。雄花萼片和花瓣各5，花瓣长于花萼。雄蕊10，2轮，内轮花丝合生，花盘腺体5。雌花花瓣短于花萼。子房上位，3室，每室1胚珠。花柱3，箭形，基部合生，顶端2裂。据在南宁观测，3月下旬～4月中旬花序形成，4月下旬为始花期，5月上旬为盛花期，中旬为末花期；6月下旬果实开始成熟，7月上旬为盛熟期。蒴果椭圆形，初为肉质，成熟后干燥，不脱落，在树上可维持10天左右，开裂为3个2瓣裂的分果瓣。果实长2.5～4cm，径2～3cm，每果内有种子3粒，间有2粒，种子黑褐色至黑色，扁椭圆形，长1.6～2cm，宽1～1.3cm，有肉质胚乳，胚伸直，子叶宽扁，长椭圆形（图1）。

果实的采收调制和种子贮藏 当大部分蒴果呈干燥状，少数已开裂时，用采种刀从树上带果柄采摘。采集的果实不宜在日光下曝晒，应按果实成熟程度，分别摊放于通风干燥的室内，待果实全部开裂后，分批抖出种子，再进行风筛，除去果壳和杂质，即得种子。干燥果实的出种率为75%，种子的净度为98%，含水量约20%，千粒重约670g，每千克纯净种子1 500粒左右。用于榨油供工业用的种子，宜置日光下曝晒，将种子的含水量降至13%以下。用于播种育苗的种子不能脱水，忌晒。裸露种子存放期不超过半个月。贮藏或运输，需混湿沙或稍湿的锯木屑，贮藏期不超过2个月。一般不宜越冬贮藏。

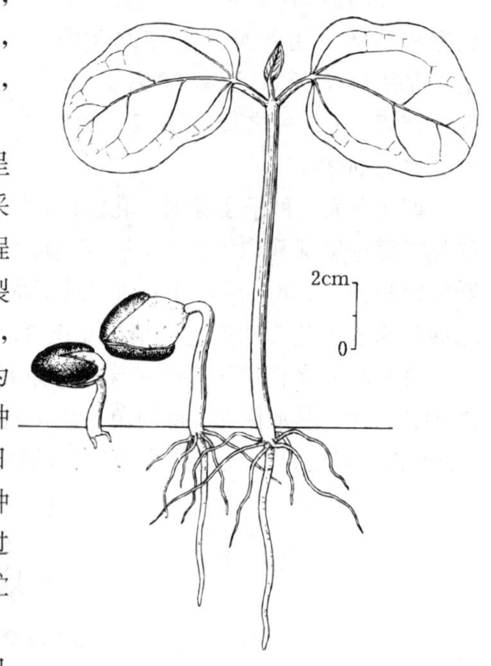

图2 麻疯树种子萌发后第3、5、18天幼苗的生长情况
（黄应钦绘）

发芽和播种 种子无休眠习性。发芽时要求日均温在22℃左右。1987年8月10日，广西林业科学研究所在室外沙床作过发芽测定：播种时日均温为30℃，播后第3天开始萌发，第5天子叶出土，第7天发出初生叶。9天的发芽率为68%。未经贮藏的新

鲜种子，发芽率可达80%～85%，场圃发芽率一般为60%左右。幼苗生长情况见图2。

条播。每平方米苗床播种45～55g。种子宜平放，覆土1.5～2cm，1年生苗可出圃。

（王宏志）

余　甘　子

Phyllanthus emblica L.

（大戟科　Euphorbiaceae）

生长习性、分布和用途　叶下珠属约600种，我国约35种，本文描述的这1种在云南亦称滇橄榄。落叶小乔木或灌木状，高1～3m，罕达14m。根系发达，穿透能力强，有很强的萌蘖能力。能耐干旱瘠薄。忌霜冻和风害。分布于川、黔、滇、桂、粤、琼、赣、闽、台。余甘子在福建的栽培历史悠久，但云南的资源居全国之首。马来西亚、中南半岛、印度也有分布。根据果实的形状、大小、成熟期和果实品质等性状，可以分出许多品种。姚小华、叶金好等(1993)和李昆、陈玉德等(1994)作过余甘子自然类型的调查并提出过优良类型的选择标准。果实可生食或作蜜饯，并可入药，古时即被认为具有医疗保健作用。据研究，余甘子的果汁能阻断强致癌物质N-亚硝基化合物在人体内的合成（侯开卫和刘凤书等，1989；刘凤书和侯开卫等，1988）。王炳三和叶金好等(1992)分析过余甘子的营养成分。据他们报道，各品种平均每百克果肉含408mg维生素C，且因受到抗氧化物质的保护而具有相当高的稳定性。树皮及叶可提制栲胶。根可药用。这个种经济价值很高，有开发前景。

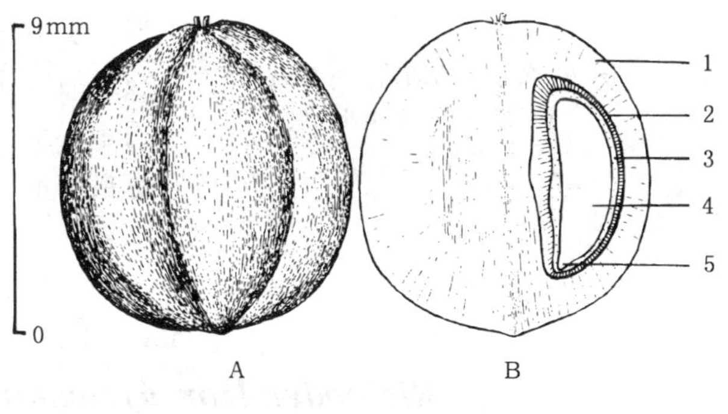

图1　余甘子除去肉质果皮后的果核外形（A）及其纵切面（B）

1. 内果皮　2. 种皮　3. 胚乳　4. 子叶　5. 胚根

（黄应钦绘）

开花结实　3～4年生开始结实，6年生后进入正常结实期。结实有大小年现象，但不明显。花单性，雌雄同株。3～6朵簇生于叶腋。萼片6，无花瓣。雄花花盘腺体6枚，分离，三角形，与萼片互生。雄蕊3～5，花丝合生。雌花花盘杯状，边缘撕裂状，包围子房达一半以上。子房3室，每室胚珠2。在闽南1年开花结实3次。4～5月开的春花，8～9月果熟；7～8月开的夏花，11～12月果熟；9～10月开的秋花，12月～翌年1月果熟。果肉质，核果状，扁球形或球形，淡绿色或淡黄色，间有不同程度的锈斑。果横径1.2～2.7cm，纵径1.0～2.4cm。单果重1.2～10.4g。中果皮肉质。内果皮骨质，近似果核，径约1cm。每果内有3（4）室，每室有种子2粒。种子褐色，肾形，长5～6mm，宽2～3mm。有薄层胚乳。子叶2枚，含油脂。种子形态见图1。

果实的采收调制和种子贮藏　果实由绿色变为淡绿色或淡黄色而呈半透明时，即已成熟。

春花所结的果实，种子饱满，发芽率高，一般留待 12 月充分成熟后采收。手工采摘，采得的果实令其腐烂，装入箩筐内，置水中冲捣后漂去果皮果肉，将洗净的果核捞起摊开曝晒，上盖塑料薄膜，以防核壳爆裂弹掉种子。核壳开裂后过筛，得到种子。根据李昆和陈玉德（1994）的资料推算的出籽率为 7%～20%。净度 85%～95%。千粒重约 13.8g，变动范围12.5～14.5g，每千克有种子 7.2 万粒，变动范围 6.9 万～8 万粒，每升约有种子 590g。种子可以干藏，但在常温条件下只能短期贮藏，即冬季采种，贮藏至翌年春夏季播种。含水量在 11% 以下的种子曾在 0～5℃的低温环境中密封贮藏 1 年，发芽能力未受影响。

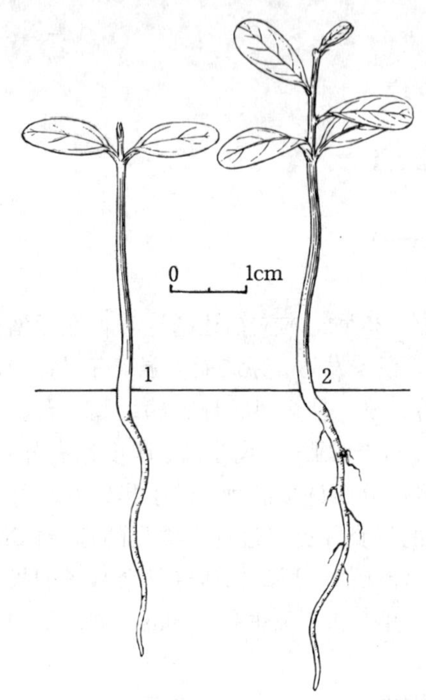

图 2　余甘子的幼苗生长情况
1. 子叶展开　2. 初生叶互生
（黄应钦绘）

发芽和播种　种子无休眠习性。发芽时的温度以 25℃左右为宜。播后第 4 天开始发芽，第 7 天的发芽百分数为 37%，第 29 天的发芽率为 58%（51%～67%）。

3～4 月播种。条播或撒播，每平方米播种 3～4g，覆土 0.5cm。播后 4～6 天发芽。出土萌发。子叶长椭圆形，长 15mm，宽 5mm。初生叶互生，长椭圆形，长 10～13mm，宽 3～4mm。以后长出分枝，分枝上的叶互生并排成两列。幼苗的初期形态见图 2。用半年至 1 年生苗进行移植，径粗 1.5～3cm 时可作嫁接砧木。也可以分蘖繁殖。

（李玉科）

拟　蓖　麻

Ricinodendron africanum Muell.-Arg.

（大戟科　Euphorbiaceae）

生长习性、分布和用途　在拟蓖麻属中，我国引入本文描述的 1 种。落叶小乔木，高 4～5m。原产热带非洲西部。1962 年华南热带作物研究所从加纳引入海南试种，生长、发育一般。喜光。适宜生长的年平均温度为 23～25℃，年降雨量 1 000～1 800mm，空气湿度70%～80%。当气温下降到 2.9℃时，叶片干枯脱落。在海南，冬季常有轻度冻害。喜肥沃深厚的沙质壤土。木材松软，不抗强风，遇 7～9 级大风，树冠便可能折断。木材比重轻，可作轻木的代用品，用作木筏、木器及纸浆原料。种子含油 45%～47%，为淡黄色干性油，供工业和医药用。

开花结实　8～10 年生开始开花结实，每年结实不多。花单性，雌雄异株。聚伞圆锥花序，顶生。花冠黄白色，坛状。萼片 5。雄蕊 15，基部有 5 个浅黄色腺体。苞片披针形，长 1.7cm。据海南儋县的观察，花期 4～5 月，果熟期 12 月～翌年 2 月。蒴果，近球形。种子近

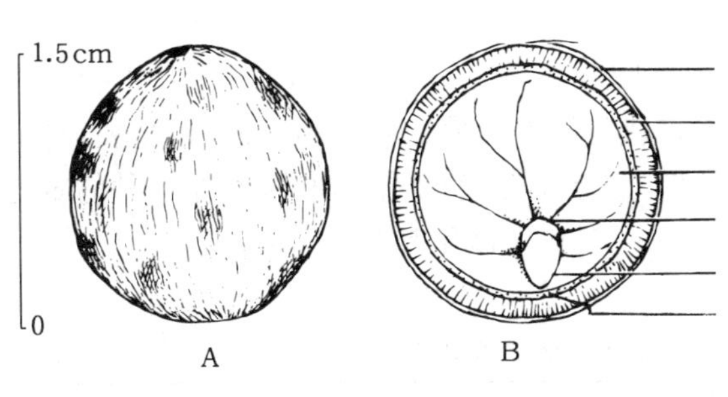

图 1　拟蓖麻种子外形（A）及其纵切面（B）
1. 外种皮　2. 内种皮　3. 胚乳　4. 子叶　5. 胚芽　6. 胚根
（黄光郁绘）

圆形，外种皮薄，内种皮厚，有少量胚乳，子叶近圆形（图 1）。

果实的采收调制和种子贮藏　果实转为深褐色，果壳变硬发亮时即可用采种刀将果实钩下。采集后立即将果实切开，从子房内的绒毛纤维物中取出种子，稍加晾干，即为播种材料。果实的出种率为 44%。种子净度可达 97%，千粒重约 1 250g，每千克约有种子 800 粒。新鲜种子含水量约 20%，忌失水，宜随采随播。贮藏时需混湿沙，贮藏期一般 1～2 个月。

发芽和播种　种子无休眠现象或仅有程度很浅的休眠现象。种子发芽时日均温宜在 20℃以上。1988 年 2 月 5 日，华南热带作物研究所在室外沙床上用新采种子作过发芽测定。可能因气温偏低，播后历时 2 个多月才于 4 月 15 日开始发芽，4 月 19 日发芽结束，且无明显的发芽盛期；发芽率为 60%。

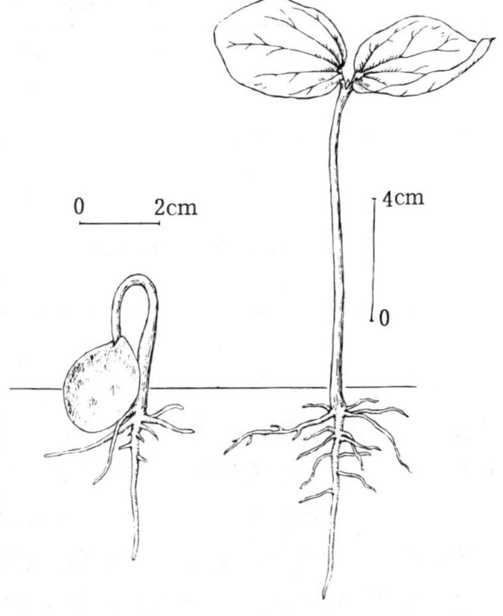

图 2　拟蓖麻种子萌发后第 3、17 天幼苗生长情况
（黄光郁绘）

出土萌发。种子萌发后 17 天初生叶开始展现，幼苗生长情况见图 2。

条播。播种前宜用 40～50℃温水浸种 1 昼夜，每平方米播种 100～130g，覆土 2cm。有 3～5 片初生叶时移植。1 年生苗可以出圃。

（丁慎言）

乌　桕　属

Sapium P. Br.

（大戟科　Euphorbiaceae）

生长习性、分布和用途　本属约 120 种，我国约 10 种，本文描述 2 种。浆果乌桕为常绿乔木，乌桕为落叶乔木。喜温暖湿润气候，对土壤要求不严，pH4.5～8.5 的沙壤和黏土都能生长，但在深厚、肥力高的壤质土上生长较好。浆果乌桕分布于滇南。乌桕分布于山东及长江流域以南各地，已有 1 000 多年的栽培历史，形成了许多农家品种。田荆祥和黎章矩等（1988）分析过浙江省乌桕优良无性系种子的理化性质。乌桕是著名的工业木本油料树种，种子外被白色蜡层，可提取桕脂（皮油）。钱领元和施拱生（1986）观察过蜡被的形成过程。桕脂广泛用于制造肥皂、蜡纸、金属涂擦剂并提取硬脂酸等。近年来还利用纯桕脂制取类可可脂生产巧克力。乌桕种仁榨出的桕油（青油）可以代替桐油，是油漆、油墨工业的重要原料。乌桕叶可作染料，根可入药，花为蜜源；秋叶变红，是点染秋色的重要树种。2 个树种的木材都可用于家具、农具、箱板。它们的名称、生长、分布和用途见表 1。

表 1　乌桕属树种的名称、生长、分布和用途

中　名	学　名	树　高（m）	胸　径（cm）	分　布	用　途	供　稿
浆果乌桕	*S. baccatum* Roxb.	30	100	滇南的西双版纳、思茅、临沧、德宏等地。缅甸、印度、马来西亚、印度尼西亚	材用、造纸	616
乌　桕	*S. sebiferum* Roxb.	15～20	60～100	华东、华中，川、滇、黔、桂、粤、陕、甘。日本、印度	油脂、材用、药用、观赏	803

开花结实　开花结实年龄因繁殖方法不同而有差异。浆果乌桕实生苗 8～10 年生开始开花结实，正常结实年龄在 15 年生以后。乌桕实生苗开始开花结实为 4～8 年，正常结实年龄也在 15 年生以后；嫁接苗开始开花结实为 2～3 年，正常结实年龄在 5 年生以后。

花单性，雌雄同株，常同序。无花瓣和花盘。雄花 6～15 朵组成小聚伞花序，再集生为长穗状复花序。雄花小，萼 2～3 浅裂或具细齿，雄蕊 2（3）。雌花序常位于雄花序下部，雌花单生，萼 2～3 裂。子房 2～3 室，每室 1 胚珠。花柱 2～3，分离或基部连合，柱头外卷。浆果乌桕的果为浆果，近球形。乌桕为蒴果，木质，三角状圆形或椭圆状球形。未熟时青绿色，成熟时黑褐色或紫黑色。浆果乌桕从幼果出现到种子成熟，一般为 120～130 天；果实成熟后不易脱落，挂果可至 12 月。乌桕从幼果出现到种子成熟一般需 170 天（早熟的 150 天，迟熟的 190 天）。乌桕种子在果爿上附着牢固，果壳开裂后 3 个月仍不脱落，直至春天树液开始流动时才逐渐掉落（黎章矩，1965）。表 2 列出了这两个树种开花结实物候的一般情况。乌桕的花期和果实成熟期还因品种而有较大的差异。

表 2　乌桕属两个树种的开花结实物候

树　种	观察地点和年份	开　花			果实成熟		种子散落期
		始　期	盛　期	末　期	始　期	盛　期	
浆果乌桕	云南普文 1988～1989	3 月下旬	4 月上旬～中旬	5 月上旬	8 月下旬	9 月上旬	12 月以后
乌桕	浙江临安 1962～1963	6 月上中旬	6 月中下旬	6 月下旬～7 月上旬	11 月上旬	11 月中旬	翌年 2 月以后

浆果乌桕每果含种子 1～2。乌桕每果含种子 3 粒，个别的含 4～5 粒。种子近球形，一侧扁平，黑色，外被白色蜡层。胚乳丰富。乌桕种子的形态及其内部结构见图 1。2 个树种果实和种子的形态特征见表 3。

表 3 乌桕属树种果实和种子的形态特征

树 种	果 实			种 子		
	形 状	大小（cm）	颜 色	形 状	大小（mm）	颜 色
浆果乌桕	近球形	1.0～1.3	紫黑色	近球形	5	—
乌桕	扁球形至近圆形	1.1～1.6	黑褐色	近球形	6～8	黑色（外被白蜡）

乌桕有两个变种。一是葡萄桕，它仅生一种雌雄同序的穗状花序，因形成的果穗状如葡萄而得名。另一种是鸡爪桕，结果枝上着生两种花序，开两次花，第1次在春梢顶上着生雄花序，雄花开放后脱落。第2次开的穗状花序，着生在雄花序的基部或在其基部抽生的几个二次梢梢顶。由于一个春梢可抽生几个雌雄同序的穗状花序，开花后组成复果序的果穗，状如鸡爪，故称鸡爪桕。

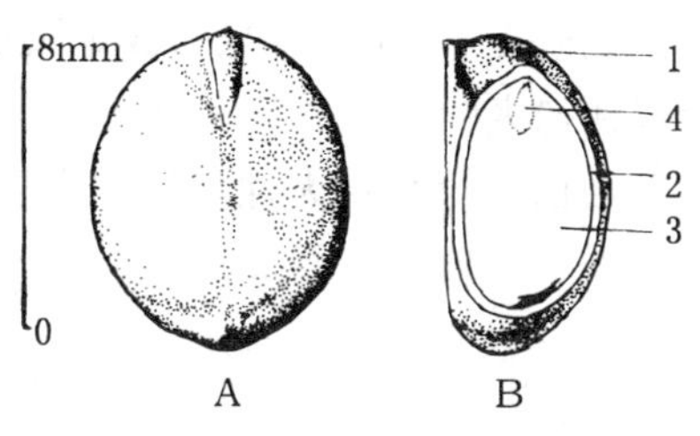

图 1 乌桕种子外形（A）及其纵切面（B）
1. 蜡被 2. 种壳 3. 胚乳 4. 胚
（童军平绘）

果实的采收调制和种子贮藏 果皮变为紫黑色时表明种子已经成熟。浆果乌桕采摘果穗，摘下的果实用水浸泡，果肉腐烂后搓去果皮，洗净，晾干，使种子含水量在10%～20%，用湿沙层积供翌春播种。乌桕蒴果开裂，果壳脱落露出洁白种子时，用采种刀连同果穗采下。脱粒后经风选去杂，晒1～2天，放入袋中或桶中置通风干燥处贮藏，翌春播种。常温下隔年贮藏会严重丧失发芽能力。2个树种的出种率和种子质量的数据见表4。

表 4 乌桕属树种的出种率、种子净度和质量等数据

树 种	出种率（%）	净度（%）	千粒重（g）		每千克纯净种子粒数	
			一 般	变动范围	一 般	变动范围
浆果乌桕	25～32	75～90	46	40～50	21 000	20 000～25 000
乌桕	—	96～99	200	160～360	5 000	2 800～6 300

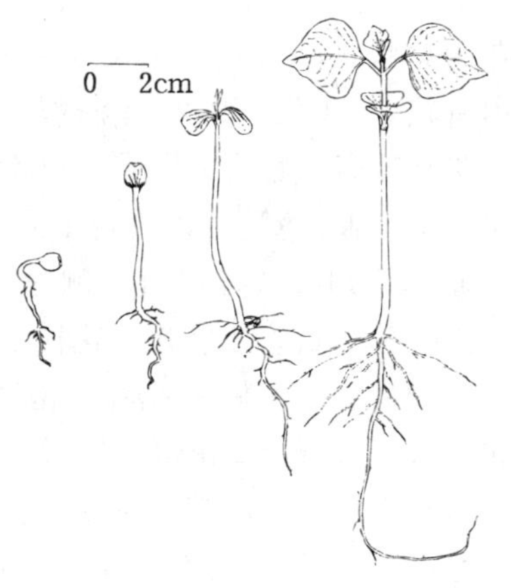

图 2 乌桕种子萌发后第4、10、15、55天幼苗的生长情况
（童军平绘）

发芽和播种 种子有休眠习性，实验室发芽率很低。一般认为是乌桕种子外层的蜡质影响种子吸水萌发，播前应当去蜡。通常是将种子放在桶缸中，用始温80℃水浸泡并立即搅拌，在冷却的过程中继续浸泡24小时，捞出放石臼中舂去蜡皮，再筛除蜡粉。钱领元和施拱生（1986）报道说除去蜡被的乌桕种子，在长达1个月的时间里发芽率也只有5%。他们推测过乌桕种子休眠的原因。据李晓洁（1987）报道，去蜡的乌桕种子用500μg/g赤霉素水溶液浸种48小时，在0～5℃条件下层积1个月，发芽率可达65%以上。

乌桕可冬播或春播，一般为春播。冬播在11～12月采种后播种。春播在2～3月，以早春播种为好。浆果乌桕以即采即播为好。条播，条距25～30cm。每666m^2的播种量，大粒乌桕种子为8kg，小粒种子为5kg，浆果乌桕约3kg。覆土约1.5cm，洒水，盖草，并经常保持土壤湿润。乌桕冬播出苗期约120天，春播约50天。在云南普文，浆果乌桕1月中旬播种后约15天发芽出土，开始出土后50余天发芽结束。

出土萌发。乌桕子叶矩形或矩圆形，稀椭圆状卵形。浆果乌桕子叶呈倒卵形。子叶出土10天后展现初生叶。

乌桕已广泛采用无性繁殖，主要是嫁接繁殖。乌桕种子的萌发和幼苗初期生长情况见图2。

（翁尧富）

滑　桃　树

Trewia nudiflora L.

（大戟科　Euphorbiaceae）

生长习性、分布和用途　滑桃树属只有本文描述的1种。落叶大乔木，高达35m，胸径40～80cm。喜酸性肥沃湿润的沙质土壤，不耐干旱瘠薄。分布于云南、广西、广东、海南。印度、斯里兰卡、中南半岛、菲律宾也有分布。材质稍轻软，可作室内家具及胶合板等用材，叶捣烂可敷治疥疮。

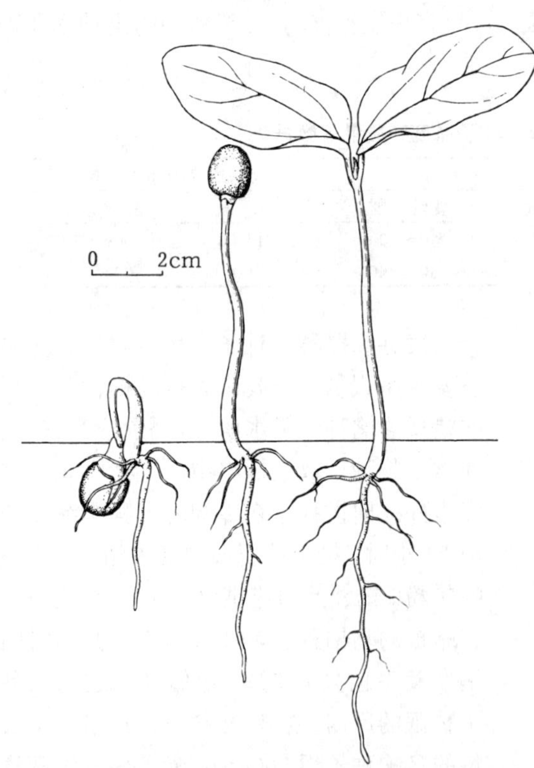

图1　滑桃树种子萌发后第7、15、46天幼苗的生长情况
（黄应钦仿《热带亚热带主要树种采种育苗技术》）

开花结实　10年生左右开始开花结实，正常结实年龄在15年生以后，结实间隔期一般为1年。花单性，雌雄异株。无花瓣和花盘。雄花为总状花序腋生，序长6～18cm，每苞片内有花2～3朵，萼3～4深裂。雄蕊75～95，花丝分离。雌花单生或2～4朵组成总状花序生于叶腋，每苞片内有花1朵，萼佛焰苞状，通常不规则2～4裂且一侧深裂。子房上位，2～4室，每室1胚珠。花柱2～4，基部稍连合，柱头长约2cm。据在海南观测，10月上旬～11月中下旬，花蕾、花序陆续形成，着生于当年生嫩枝的叶腋，花期较长；当年12月为始花期，翌年1～2月花陆续开放，3月开花结束；果实的成熟期也较长，8月上旬果实开始成熟，8月中下旬为成熟盛期，9～10月仍有少量果熟。核果，通常不开裂，扁球形，

成熟后，在树上着生约 1 周后脱落。未成熟时青色，成熟时转变为褐色，基部稍扁。果径2.5～3cm。外果皮略带肉质，内果皮薄壳质。每果有果核 2～4 粒。核深褐色，稍坚硬，扁圆形，长 7～9mm，宽 6.5～8.5mm。种子无种皁，胚乳肉质，子叶叶状。

果实的采收调制和种子贮藏　用击落法在地面捡拾果实。采回的果实堆沤 3 天左右，置水中浸泡搓洗，淘去外果皮，晾干后用小铁锤或木棒敲破果壳（中果皮）取出的果核即为播种材料，通称种子。也可不经浸水搓洗，采种后摊放于室内或置弱光下，将外果皮晾晒干，然后敲破果壳，取出种子。果实出籽率为 18%～25%。净度可达 98%。千粒重约 250g，每千克种子约 4 000 粒。种子含水量为 20%～30%。已脱壳的种子忌失水，不能日晒。短期运输的种子可以带壳。贮藏时需剥去果壳，并混湿沙。贮藏期一般不超过半年。

发芽和播种　种子休眠现象不明显。发芽时日均温宜在 20℃以上，气温低则发芽不整齐。9 月以后采集的种子宜沙藏越冬，至翌年 2～3 月播种。广西林业科学研究所 1979 年 11 月 27 日播种的滑桃树种子，至 1980 年 1 月 14 日才有少量开始发芽，4 月 12 日发芽结束，发芽全程近 3 个月，发芽率约为 50%。1982 年将采集的种子沙藏至翌年 3 月上旬播种，3 月中旬末开始发芽，4 月中旬发芽结束，发芽率约为 60%。

出土萌发。胚根萌发后 6～8 天子叶出土，再过 30～40 天，初生叶出现（图 1）。

条播。每平方米播种 12～16g。或将种子密播于沙床，待初生叶长出后移至圃地培育。每 666m^2 产苗约 1 万株。1 年生苗高约 80cm，可出圃。

（王宏志）

油　桐　属

Vernicia Lour.

（大戟科　Euphorbiaceae）

生长习性、分布和用途　油桐属有 3 种，分布于东亚。本文描述我国所产的 2 种（表 1）。落叶乔木、小乔木或灌木状。胸径 12～60cm，树高 2～15m。分布于北纬 19°～33°15′，东经 98°～122°的广大亚热带地区，包括黄河以南的 17 个省（自治区）。喜光。浅根性。喜温暖湿润环境，千年桐较油桐能适应较高的气温与较多的降水。适生于微酸性或中性富含腐殖质的土壤。种子中榨出的油称桐油，主要成分为桐酸，属于不饱和酸，是性质优良的涂料，在工业中用途极广。

油桐是我国特产，栽培和利用历史已逾千年。种内变异多。经过长期的自然选择和人工选择，形成了许多各具特点的地方品种，例如小米桐、大米桐、对年桐、柴桐等。

表 1　油桐属树种的名称、树高、分布和用途

中　名	学　名	树高（m）	分　布	用　途	供　稿
油桐（三年桐，光桐）	*V. fordii* (Hemsl.) Airy-Shaw	1.9～12	华东、华中、华南、西南、陕、鲁、甘、台	油料、材用、药用	501、507
千年桐（皱桐）	*V. montana* Lour.	5～15	桂、粤、琼、闽、赣、湘、皖、鄂、滇、川、黔、台	油料、材用、药用	501、507、612

开花结实 开花结实年龄因种和品种，也因生长环境和繁殖方法而异。油桐开花结实早于千年桐；对年桐早于小米桐、大米桐；大米桐一般晚于小米桐。同一品种，南缘产区比北缘产区早（表 2）。嫁接繁殖的比实生苗栽植的开花早。

表 2 油桐属树种的开始结实年龄和盛果期年龄

树种	开始结实年龄（年）			盛果期年龄（年）		
	中心产区	南缘产区	北缘产区	中心产区	南缘产区	北缘产区
油桐						
对年桐	2	2	2～3	3	—	5
小米桐	3	2～3	3～4	6～15	5～10	6～15
大米桐	4～5	3～5	4～5	8～25	6～15	8～25
柴　桐	3～4	2～3	3～4	6～15	5～10	6～15
千年桐	5～7	（2）4～5	—	10	30	—

花单性，着生在去年生枝条的顶端。油桐雌雄同株，其花在枝条抽出一段时间后开放。聚伞花序或多歧聚伞花序。油桐的花序构造有 3 种类型：单花、少花花序和多花花序。单花的一个花枝上常只着生 1 朵雌花，坐果后为单生果；少花花序具小花数朵至二三十朵，其主轴与花序基部生出的侧轴的顶端着生雌花，主轴及侧轴上二级轴的顶端也可能为雌花，坐果后为丛生果。多花花序的小花数十朵至二三百朵或更多，但雌花甚少，或全为雄花，被称为雄株或雄性化极强的植株，在选种上属淘汰对象。一些品种在幼龄期，如果立地条件较好，管理精细，花序上雌花比例较多，老树和栽培管理粗放的则雌花比例小。例如四川小米桐，5～7 年生时雌花与雄花的比例常为 1∶7，10～12 年生时为 1∶11，20 年生的植株为 1∶37，24 年生的达到 1∶65。

千年桐雌雄异株，偶有同株。雌花序常呈总状排列和圆锥花丛排列，每花序有小花十数朵至五六十朵。雄花序的小花数常达 300 朵。有的雄树雄花序上着生雌花若干朵，或是雌树雌花序上着生一些雄花，形成了雌雄同花序。

油桐属雌花的花萼 2～3 裂，绿色或紫红色，呈镊合状接合。花瓣 5～7，白色。油桐花瓣基部有黄红色或深红色脉纹，少数品种为黄色或紫蓝色。千年桐花白色或有红色脉纹。子房上位，3 室或 4～5 室，有的品种可达 8～12 室。每室 1 胚珠。花柱 3 或 4，2 裂。雄花的花萼 2 裂，花瓣白色，具 7～10 条浅红色条纹。雄蕊 8～20，花丝基部合生，上部离生，并排成 2 轮。

花芽是在当年生枝条的顶端发育过程中形成的。其分化期因产地不同而有先后。在福建福州，油桐花芽分化期在 7～11 月，千年桐在 7 月下旬～翌年 3 月中旬。在四川，有的油桐品种花芽分化期从 5 月上旬～11 月上旬，有的从 6～11 月。开花与温度关系密切。油桐开花的旬平均温度约 17℃，盛花期的旬平均温度约 19℃。千年桐开花的旬平均温度约 20℃，盛花期的旬平均温度约 24℃。因此，油桐属开花期通常在 3～5 月，果实成熟期在 9 月中旬～11 月上旬（表 3）。陈炳章和杨乾洪等（1990）连续 5 年在四川万县观察过海拔高度对油桐开花结实的影响。他们发现，海拔 900m 处的油桐，无论是始花期还是盛花期和终花期，都要相应地比海拔 400m 处的推迟 46 天。桐油的主要成分是桐酸。陈炳章（1988）1980 年测定过桐酸合成和累积的进程。

表 3　油桐属主要栽培品种的开花结实物候

树　种	观察地点	开　花			果实成熟期	
		始　期	盛　期	末　期	始　期	盛　期
油桐						
四川小米桐	四川万县	4 月上旬	4 月中旬	4 月下旬	10 月上旬	10 月中旬
贵州窄冠桐	贵州正安	4 月中旬	～	4 月下旬	11 月上旬	
广西隆林矮脚桐	广西隆林	4 月上旬	4 月	中 旬	9 月中旬	9 月下旬
江苏球桐	江苏高淳	4 月下旬	～	5 月中旬	10 月中旬	10 月下旬
千年桐						
福建软枝千年桐	福建漳浦	4 月下旬	～	5 月中旬	10 月中旬	10 月下旬

未成熟果实青绿色，成熟果实呈黄色至深红色或红黄色。核果。果皮脆壳质，不开裂，球形、扁圆形或三角状卵圆形。千年桐核果表面具 3 条纵棱和不规则网状皱纹。果长 4～8cm，径 4.7～7.2cm。每果含果核 3～5（12）粒。核壳厚，木质，坚硬，无种阜，褐色或深褐色，长 2～2.8cm，径 1.7～2.3cm。胚乳丰富，胚直，子叶宽而扁；少数种粒仅胚乳发育而无胚。油桐属树种的核果和果核的颜色、形状和大小见表 4。油桐果核的形态及其纵切面见图 1。

表 4　油桐属树种核果和果核的颜色、形状与大小

树　种	成熟核果				成熟果核				
	颜　色	形　状	长（cm）	径（cm）	颜　色	形　状	长（cm）	径（cm）	厚（cm）
油桐							2.2～2.8	1.7～2.2	1.3～1.7
小米桐类 大米桐类 对年桐类 柿饼桐类 柴桐类	黄色，黄褐色，深红色或暗红色	球形，近球形或扁圆形	4～6.7 5.9～6.8 4.5～7.4 5～5.9 6～8	4.7～6.5 5.5～6.7 4.9～6.8 4.9～7.1 5～5.7	褐色或深褐色	三角状卵形，表面有疣状突起			
千年桐	黄色，黄褐色或红黄色	三角状卵形，表面有 3 条纵棱和不规则隆起	—	4～6	浅褐色或深褐色	扁圆形	2～2.4	1.7～2.3	1.1～1.5

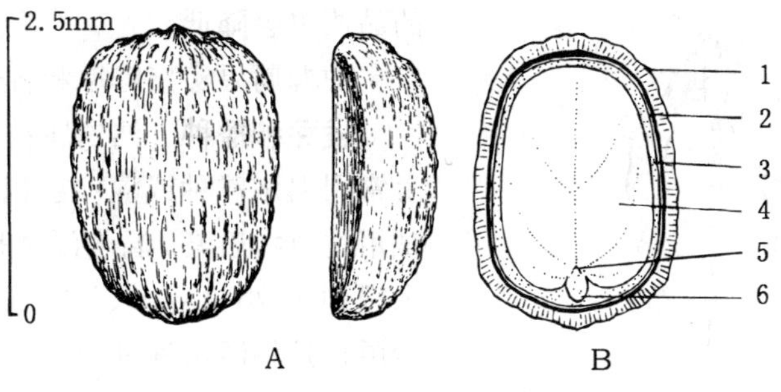

图 1　油桐果核外形（A）及其纵切面（B）
1. 内果皮　2. 种皮　3. 胚乳　4. 子叶　5. 胚芽　6. 胚根
（黄应钦、杨向晨绘）

油桐结实有大小年现象或歇年现象，但有的品种明显，有的品种则不明显。例如湖南葡萄桐、四川小米桐、湖北九子桐、河南股爪青、浙江少花丛生球桐等结实大小年现象显著；四川大米桐、广西龙胜大蟠桐、浙江座桐等大米桐类的结实大小年则不明显。

油桐果实成熟后自然脱落，持续约 15 天。千年桐果实成熟后不立即脱落，相当多的果实挂在果枝上，落果持续约 40 天。

表 5 油桐属树种果实出籽率和种子（果核）质量

树 种	品 种	出籽率（%）	千粒重（g）	每千克纯净种子数（粒）
油桐	四川小米桐	58.5	2 300～2 500	400～430
	贵州大米桐	44.6	2 240	450
	陕西小米桐	64.0	2 740	360
	湖北郧阳桐	58.5	5 700	180
	湖北五子桐	25.8	2 900	340
	安徽大扁球	68.0	3 500	280
	江苏小米桐	56.0	1 990	500
千年桐	福建软枝千年桐	45.8	3000	330

果实的采收调制和种子贮藏 果皮呈黄色、深红色或红黄色时，分别品种和优良单株组织采收。先熟先采收，后熟后采收。可上树采摘，或用高枝剪连同果柄剪下，或用竹竿敲击或在地面拾取自然脱落的果实。采回的桐果置荫凉通风处，上盖稻草，喷洒适量水分，堆沤15～20 天，待果皮变软后用人工或剥壳机剥取果核，摊于通风处晒干，去杂后将纯净果核包装入库。果核即为播种材料，通称种子。

通常采用干藏。将含水量不超过 10%的种子装入容器置荫凉通风处或库房内，大批量种子可散装于库房内。用于繁殖的种子一般是越冬贮藏。即使榨油用的桐籽，贮藏时间太长，出油率和油的品质也会降低。

油桐属果实出籽率和种子(果核)重量见表 5。

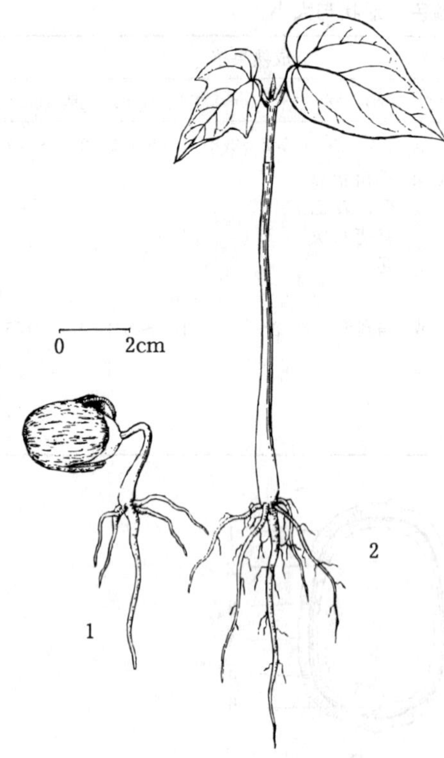

图 2 油桐萌发后的幼苗生长情况
1. 下胚轴延伸 2. 幼苗形成
（黄应钦、杨向晨绘）

发芽和播种 油桐属种子有休眠习性。生产上常用温水浸种或混沙层积来解除休眠，有人还试用过植物激素。温水浸种是用始温 50～70℃的水浸种 24 小时后，每天换水连续浸种 5～7 天。混沙层积是将温水浸过的种子与其厚度相等的湿沙分层铺放 4 周左右后播种。使用 50～100μg/g 的萘乙酸溶液，或 150～500μg/g 的赤霉素溶液处理种子，据说也能起到促进种子萌发的作用。

有即采即播的，但多采用春播。春播时间为 2 月下旬～3 月。条播，条距 20～30cm，条幅 3～5cm，每米长的播种行播 6～8 粒种子，大约每 666m^2 播种30～50kg。播后覆细土 4～5cm，轻压后洒水、盖草。油桐种子发芽的适宜温度为 25℃左右。经处理过的种子播后约 30 天开始发芽出土，50 天左右达发芽盛期。种子在发芽过程中应分批揭除盖草。田间发芽率一般为 75%，最高可达 90%以上。

出土萌发，子叶 2，较薄，胚乳丰富，油质，包着子叶。出土时下胚轴伸长，将种子带出土面，以后种皮与胚乳脱落，露出子叶。有时子叶伸展不出而与包于种皮内的胚乳一同脱落。子叶矩圆形，先端和基部圆，薄，柄无毛。初生叶第 1、2 片近对生，长卵形，先端渐尖，基部心形，全缘或有缺裂。下胚轴有突起的圆形或椭圆形皮孔，向上渐细。根颈向下骤细。主根不明显，根颈处着生侧根 4 条，与主根等粗（图 2）。油桐 1 年生苗高 50～80cm，千年桐60～90cm，可以出圃造林。

油桐还可嫁接繁殖，方法有芽接、枝接、顶芽劈接、腹接和切腹接等，其中板状芽接是油桐嫁接的主要方法。嫁接时间春秋皆可，春季在 3 月中旬～4 月上旬，秋季在 9～10 月。

（赵德铭）

茶　属

Camellia L.

（茶科　Theaceae）

生长习性、分布和用途　茶属约 220 种，分布亚洲热带至亚热带，我国约 190 种，本文描述 20 种和 1 变种。其中金花茶为稀有种，红皮糙果茶、长瓣短柱茶为渐危种，皆已列入《中国植物红皮书》。常绿灌木或小乔木，高 2～15m。生长速度稍慢。萌芽性强。大部分树种幼龄期耐荫，成龄树喜光。部分灌木类的树种长期耐荫。喜温暖湿润气候。适生于排水良好的酸性土壤，在钙质土上生长不良，部分树种较耐干旱贫瘠土。少数种能耐－8℃左右的低温。分布于我国南方各省。越南、缅甸、日本、印度、尼泊尔、不丹等也有分布。茶的嫩叶和芽焙制后是优良的饮料。种子富含油脂，可提取食用油或工业用油，其中油茶是我国重要的木本油料树种。开红花和黄花的树种是重要的观赏花木，红花类的园艺品种达 1 000 种以上，为世界著名花木。开黄花的称金花茶，为园林珍品。木材坚实，结构致密，可作柄具和工艺材。茶属 21 个树种（含变种）的名称、生长、分布和用途见表 1。

表 1　茶属树种的名称、生长、分布和用途

中　名	学　名	树高（m）	分　布	用　途	供　稿
尾叶山茶	*C. caudata* Wall.	2～3	桂、粤、滇、川、湘、台。越南、缅甸、不丹、印度	油料、材用	604
红皮糙果茶	*C. crapnelliana* Tutcher	7～10	桂、浙、闽、香港、九龙	观赏、油料、材用	604

（续）

中　名	学　名	树高（m）	分　布	用　途	供稿
厚叶红山茶	*C. crassissima* Chang et Shi.	2.5～5	赣	观赏、油料、材用	701
窄叶短柱茶	*C. fluviatilis* Hand. -Mazz.	2～5	桂、粤。缅甸、印度	油料、材用	604
瑶山糙果茶	*C. furfuracea* var. *yaoshanica* S. Y. Liang et Y. C. Zhong	2～4	桂	油料、材用	604
长瓣短柱茶	*C. grijsii* Hance	2～3	闽、鄂、湘、赣、桂	油料、观赏、材用	604
凹脉金花茶	*C. impressinervis* Chang et S. Y. Liang	5～7	桂	观赏、油料、材用	604
石果红山茶	*C. lapidea* Wu	5～7	桂、粤	观赏、油料、材用	604
长尾红山茶	*C. longicaudata* Chang et S. Y. Liang	5～7	桂、粤	油料、材用	604
大果红山茶	*C. magnocarpa* (Hu et Huang) Chang	5～8	桂、粤	观赏、油料、材用	604
金花茶	*C. nitidissima* Chi	2～4	桂。越南	观赏、油料、材用	604
钝叶短柱茶（小果油茶）	*C. obtusifolia* Chang（*C. meiocarpa* Hu，Ms.）	2～3	桂、粤、湘、鄂、赣	油料、薪柴	604
油茶	*C. oleifera* Abel	2～4	长江流域以南各省。日本	油料、材用、栲胶	604
宛田红花油茶	*C. polyodonta* How ex. Hu	4～8	桂、粤、湘、赣、川	观赏、油料、材用	604
云南山茶	*C. reticulata* Lindl.	4～15	滇	观赏、油料、材用	604
皱果茶	*C. rhytidocarpa* Chang et S. Y. Liang	2～3	桂、黔、湘	油料、材用	604
广宁红花油茶	*C. semiserrata* Chi	4～8	粤、桂	观赏、油料、材用	604
茶	*C. sinensis*（L）. O. Ktze.	3～8	长江流域及以南各地	饮料、油料、材用、药用	801
全缘红山茶	*C. subintegra* P. C. Huang	3～8	赣、湘	油料、育种材料、观赏	1002
越南油茶	*C. vietnamensis* T. C. Huang ex Hu	5～8	桂。越南	油料、材用	604
长毛红山茶	*C. villosa* Chang et S. Y. Liang	5～8	桂、湘、黔	油料、材用	604

开花结实　多数树种8～10年生开始开花结实，人工栽培的3～4年生即可开花结实。结

实间隔期通常为1年，加强培肥管理可使年年结实。花两性，单生或2～4朵簇生于枝梢或叶腋。苞片2～6或更多，萼片5～16，有时不分化为苞片及萼片，称苞被。花有单瓣、半重瓣和重瓣，单轮或数轮覆瓦状排列，白色、粉红色、红色、黄色。雄蕊多数，2～5列，花药丁字着生。子房上位，3～5室，每室4～6胚珠，花柱3～5，基部连合或分离。据广西、云南、江西等地观察，茶属多数树种的花芽于6月前后形成，着生于当年生春梢的叶腋，11月至翌年2月为主要开花期，9～11月果实成熟（表2）。油茶的落花落果现象比较严重，不同品种之间落花落果的数量差异极大。陈育松和胡玉琴（1963）报道过他们的观察结果。油茶果实还常在成熟开裂，尤海量和伍子和（1963）研究过这种现象的形成原因及其防治途径。

表2　茶属树种的开花结实物候

树种	观察地点	观察年份	开花			果实成熟		种子脱落期
			始期	盛期	末期	始期	盛期	
尾叶山茶	广西南宁	1986～1988	11月上旬	12月中旬	12月下旬	翌年10月中旬	10月下旬	10月下旬～11月中旬
红皮糙果茶	广西南宁	1986～1988	10月下旬	11月中旬	12月下旬	翌年11月上旬	11月中旬	11月中旬～12月上旬
厚叶红山茶	江西吉安	—	11月上旬	翌年1月下旬	4月中旬	9月上旬	9月下旬	10月上旬～11月上旬
窄叶短柱茶	广西南宁	1986～1988	11月中旬	12月上旬	12月下旬	翌年10月下旬	11月上旬	11月上旬～11月下旬
瑶山糙果茶	广西南宁	1986～1988	11月中旬	12月上旬	12月下旬	翌年10月下旬	11月上旬	11月中旬～12月下旬
长瓣短柱茶	广西南宁	1986～1988	11月下旬	12月中下旬	翌年1月上旬	10月下旬	11月上旬	11月上旬～11月下旬
凹脉金花茶	广西龙州	1984～1986	2月上旬	3月上中旬	4月上旬	12月上旬	12月中下旬	12月下旬～翌年1月
石果红山茶	广西南宁	1986～1988	1月中旬	2月上旬	2月中下旬	11月上旬	11月中旬	11月中旬～12月上旬
长尾红山茶	广西南宁	1986～1988	1月下旬	2月中旬	3月中旬	10月下旬	11月上旬	11月上旬～11月下旬
大果红山茶	广西南宁	1986～1988	12月下旬	翌年1月下旬	2月下旬	11月上旬	11月中旬	11月中旬～12月上旬
金花茶	广西南宁	1984～1985	11月下旬	12月中旬	翌年3月中旬	9月下旬	10月中旬	10月中旬～11月中旬
钝叶短柱茶	广西南宁	1978～1984	10月下旬	11月中旬	12月下旬	翌年10月下旬	11月上旬	11月中旬～12月上旬

（续）

树 种	观察地点	观察年份	开花			果实成熟		种子脱落期
			始 期	盛 期	末 期	始 期	盛 期	
油茶	广西南宁	1978～1984	10月下旬	11月中旬	12月下旬	翌年10月下旬	11月上旬	11月上旬～12月上旬
宛田红花油茶	广西南宁	1978～1984	12月下旬	翌年2月上旬	3月上中旬	11月上旬	11月中旬	11月中旬～12月上旬
云南山茶	云南	—	12月中下旬	翌年2月	3月下旬	9月上旬	9月中下旬	9月下旬～10月中旬
皱果茶	广西南宁	1986～1988	11月中旬	12月上旬	12月下旬	翌年11月上旬	11月中旬	11月中旬～12月上旬
广宁红花油茶	广西南宁	1984～1988	1月下旬	2月中旬	3月中旬	10月下旬	11月上旬	11月上旬～11月下旬
茶	广西南宁	1986～1988	10月上旬	11月上旬	翌年1月中旬	8月中旬	9月上旬	9月上旬～11月中旬
全缘红山茶	江西安福	1978～1980	11月上旬	翌年2月～3月	3月上旬	8月	9月	10月
越南油茶	广西南宁	1986～1988	12月中旬	翌年1月上旬	1月下旬	11月上旬	11月中旬	11月中旬～12月上旬
长毛红山茶	广西南宁	1986～1988	1月上旬	1月中下旬	2月中下旬	11月上旬	11月中旬	11月中旬～12月上旬

蒴果木质，室背开裂，3～5室，每室有种子1～6粒，或仅1室能育，每果有种子1～12粒。果实成熟后7～15天从顶端开裂，种子散落，果实中轴宿存。种子无胚乳，种皮革质坚硬。子叶较肥大。白花和红花类的子叶2枚，黄花类的子叶2～4枚。21个树种果实及种子的特征见表3、凹脉金花茶种子形态见图1。

果实的采收调制和种子贮藏 果实已经成熟，接近开裂时采集。采回的果实根据种子的使用目的分别调制。作榨油用的果实置阳光下曝晒，待果皮开裂脱出种子，晒干后供榨油。用于育苗的果实不宜曝晒雨淋，亦忌堆沤，应在干燥荫凉通风处阴干脱壳，茶籽含水量保持在30%左右。出种率、种子净度、质量等见表4。

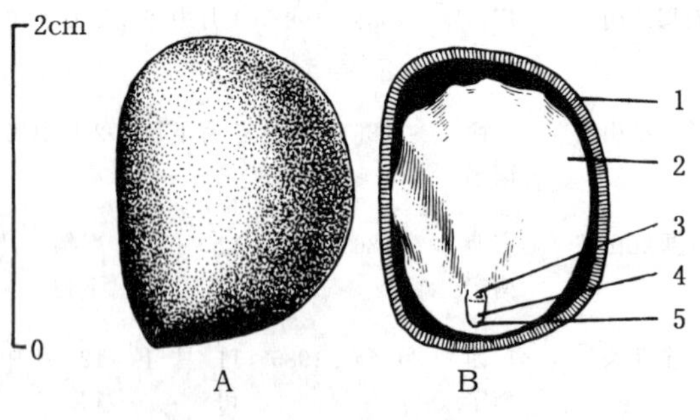

图1 凹脉金花茶种子外形（A）及其纵切面（B）
1. 种皮 2. 子叶 3. 胚芽 4. 胚轴 5. 胚根
（黄应钦绘）

育苗用种子忌失水，不宜日晒，亦忌裸露堆放。运输或1个月以内的短期贮存可暂不作调制，将采集的果实稍加晾干后，带果皮装

表 3　茶属树种果实和种子的形态特征

树　种	未熟果颜色	成熟果实			种　子		
		形　状	大小（cm）	颜　色	形　状	大小（mm）	颜　色
尾叶山茶	青绿色	椭圆形	长 1.2～1.9 径 1～1.5	褐色	椭圆形、半椭圆形、三棱状椭圆形	长 10～17 径 8～14	深褐色
红皮糙果茶	棕色	扁圆形或圆形，表面粗糙状隆起	径 7～12	棕褐色或浅褐色	棱状椭圆形、三角形、不规则形	径 18～32	深褐色
厚叶红山茶	青色	梨形，球形或橄榄形	长 5～8 径 5～7	棕色	三角形或不规则形	长 12～22 径 10～18	黑褐色或棕褐色
窄叶短柱茶	黄绿色	椭圆形	长 1.4～2.6 径 1.2～2	黄褐色	椭圆形、卵形或半椭圆形	长 12～18 径 8～15	深褐色
瑶山糙果茶	黄绿色	扁球形或球形，表面粗糙，小瘤状隆起	长 2.8～3.6 径 3.2～4.8	黄褐色	三角状椭圆形、三角形或不规则形	长 15～25 径 10～23	深褐色
长瓣短柱茶	黄绿色	卵圆形	径 2～2.5	黄褐色或淡褐色	卵形、椭圆形、半圆形、三棱状椭圆形或不规则形	径 11～20	深褐色
凹脉金花茶	青绿色	近三棱状、五棱状或不规则扁球形	长 2.3～3 径 4～5	褐色	半圆形，三棱状圆形或不规则形	长 16～22 径 15～20	深褐色
石果红山茶	棕色	阔圆形或圆形，或卵圆形	长 4.8～6.7 径 4.5～7	棕褐色	棱状椭圆形、半椭圆形、三角形或不规则形	长 18～29 径 13～29	深褐色
长尾红山茶	棕色	球形或卵圆形	长 4～5.6 径 3.5～5	棕褐色	三角状椭圆形或不规则形	长 18～38 径 15～26	深褐色
大果红山茶	棕色	扁圆形、圆形或卵圆形表面蜂窝状皱纹	长 6～8 径 8～10	棕褐色	三角状圆形、半椭圆形、棱状椭圆形或不规则形，表面具毛	长 20～27 径 15～26	深褐色
金花茶	青绿色	三棱或四棱状扁球形	径 3～6.5	黄绿色或带淡紫色	近球形或多样，具角棱	长 15～25 径 12～22	深褐色
钝叶短柱茶	绿色	圆形或倒卵圆形	长 2～2.7 径 1.6～2.5	浅赭色或浅黄色	圆球形或半圆形	径 11～21	深褐色
油茶	绿色	扁球形或球形，顶端有长柔毛	长 2.8～3.5 径 3～4	浅赭色或浅黄色	三角状圆形、棱状椭圆形、不规则形	长 15～22 径 10～20	深褐色
宛田红花油茶	棕色	扁球形或球形	长 5.5～11 径 6～12	棕褐色	三角状圆形、半椭圆形、棱状椭圆形，不规则形	长 15～28 径 14～23	深褐色
云南山茶	青绿色	近球形，顶端具十字形槽纹	长 3.2～4 径 3.8～4.5	黄褐色	三角状或不规则状	长 25～30 径 15～20	深褐色

（续）

树　种	未熟果颜色	成熟果实			种　子		
		形　状	大小（cm）	颜　色	形　状	大小（mm）	颜　色
皱果茶	黄绿色	扁圆形或近圆形，表面不规则隆起	长 2.5～3.5 径 3～4.5	黄褐色	近球形、半圆形，表面具绒毛	长 11～17 径 9～17	锈褐色
广宁红花油茶	棕色	扁圆形或圆形，表面稍粗糙	长 7～10 径 8～11	棕褐色	三角状圆形、半椭圆形、棱状椭圆形、不规则形，表面具毛	长 20～27 径 12～26	深褐色
茶	青绿色	肾形或三棱状扁圆形	长 1.5～2.2 径 1.4～2.6	浅黄至黄褐色	近圆形、半圆形、不规则形	长 11～22 径 10～19	深褐色或褐色
全缘红山茶	黄绿色	圆球形或卵球形	长 3.8～6 径 4～6	黄褐色	短圆柱形、不规则形	长 15～20 径 11～16	黑褐色
越南油茶	青绿色	圆形或扁圆形，表面稍粗糙	长 4～6 径 4.2～6.5	浅黄色	棱状椭圆形，不规则形	长 15～30 径 12～23	深褐色
长毛红山茶	绿色	扁圆形或不规则形	长 4～6 径 5～7	黄褐色	棱状椭圆形、半椭圆形、三角形、不规则形	长 18～29 径 13～29	深褐色

表 4　茶属树种果实出种率和种子净度、质量

树　种	鲜果出种率（%）	净度（%）	千粒重（g）	每千克纯净种子粒数
尾叶山茶	50	98	290	3 450
红皮糙果茶	20	98	8 000	130
厚叶红山茶	17～21	98	1 100	910
窄叶短柱茶	47	98	500	2 000
瑶山糙果茶	28	98	1 800	560
长瓣短柱茶	60	98	1 600	630
凹脉金花茶	70	98	1 970	510
石果红山茶	18	98	4 200	240
长尾红山茶	29	98	4 800	210
大果红山茶	8	98	3 800	260
金花茶	—	98	3 650	270
钝叶短柱茶	68	98	1 800	560
油茶	28	98	1 500	670
宛田红花油茶	11	98	2 600	380
云南山茶	20～30	98	4 200	240
皱果茶	44	98	1 000	1 000
广宁红花油茶	7.5	98	3 200	310
茶	51	98	1 300	770
越南油茶	35	98	2 500	400
长毛红山茶	18	98	4 200	240

入麻袋或竹篓内，播种时再剥出种子。如需越冬贮藏，可将调制好的种子混以湿沙层积贮藏。贮藏期一般为 2～4 个月：秋末或冬季采集的种子，贮藏至翌年 2 月中旬～3 月上旬播种。据研究（韩宁林，1984；韩宁林等，1991），在保湿条件下，油茶种子在 0～2℃的低温中可以保存 3 年以上。

发芽和播种　种子发芽时所需的日均温为 15℃左右。1987 年冬，广西林业科学研究所用当年新采的 18 个树种的种子进行发芽测定，测定时日均温为 14～22℃，其结果见表 5。

表 5　茶属树种的发芽能力及其测定条件

树　种	播种期（年-月-日）	测定温度（℃）	发　芽		发芽率（%）	
			始　期（年-月-日）	终　期（年-月-日）	计算天数	一般数值
尾叶山茶	1987-11-10	14～16	1987-12-22	1988-01-10	61	76
红皮糙果茶	1987-11-18	14～16	1987-12-18	1988-01-20	63	94
窄叶短柱茶	1987-11-23	14～20	1987-12-25	1988-03-01	99	81
瑶山糙果茶	1987-11-23	14～18	1987-12-23	1988-02-15	84	96
长瓣短柱茶	1987-11-23	14～20	1987-12-23	1988-02-28	97	86
凹脉金花茶	1987-11-15	14～16	1987-12-26	1988-01-12	59	74
石果红山茶	1987-11-23	14～20	1988-01-05	1988-02-18	87	90
长尾红山茶	1987-11-23	14～22	1987-12-24	1988-03-04	100	76
大果红山茶	1987-11-23	14～21	1988-01-06	1988-03-03	99	71
金花茶	1987-11-10	14～16	1987-12-18	1988-01-07	58	74
钝叶短柱茶	1987-11-23	14～22	1987-12-24	1988-03-04	100	82
油茶	1987-11-23	14～22	1988-01-10	1988-03-02	98	84
宛田红花油茶	1987-11-23	14～22	1988-01-04	1988-03-04	100	80
皱果茶	1987-11-10	14～20	1987-12-21	1988-02-24	106	87
广宁红花油茶	1987-11-23	14～22	1988-01-05	1988-03-02	98	72
茶	1987-09-01	18～22	1987-09-28	1987-11-12	72	74
越南油茶	1987-11-10	14～16	1987-12-18	1988-01-24	76	83
长毛红山茶	1987-11-23	14～20	1987-12-23	1988-02-28	97	72

留土萌发。萌发出土的早迟同温度有关。日均温在 20℃左右时，胚根萌发后约 6 天上胚轴出土，14 天初生叶出现。日均温在 15℃时，胚根萌发后约 15 天上胚轴出土，20～25 天初生叶出现。种子萌发和幼苗生长情况见图 2。金花茶类种子子叶 4 或 3，幼苗形态见图 3。

秋冬播种或春播。条播或点播，通常多采用条播。条距 20cm，条幅 10cm。点播时行距 20cm，穴距 15cm。冬播穴深 3～4cm，春播穴深 2～3cm，每穴播种 3～5 粒。播种量随树种而异。条播时，油茶每 666m^2 约播 100kg，茶约 60kg。播后覆土 2～3cm，并薄薄盖层稻草。盖草在 4 月下旬揭除。或将种子密播于细沙床内，发芽后移芽至容器，每容器 1 粒，在容器内培育约 3 个月出圃造林。近年来，本属人工栽培面积最大的茶和油茶，以及供观赏用的金花茶及红花油茶等，已选用优良无性系采用组织培养、扦插、压条、嫁接等方法繁殖，它们的杂交育种工作也在广泛开展。红花类已育出为数众多的新品种，金花茶近年也育出了新品种。

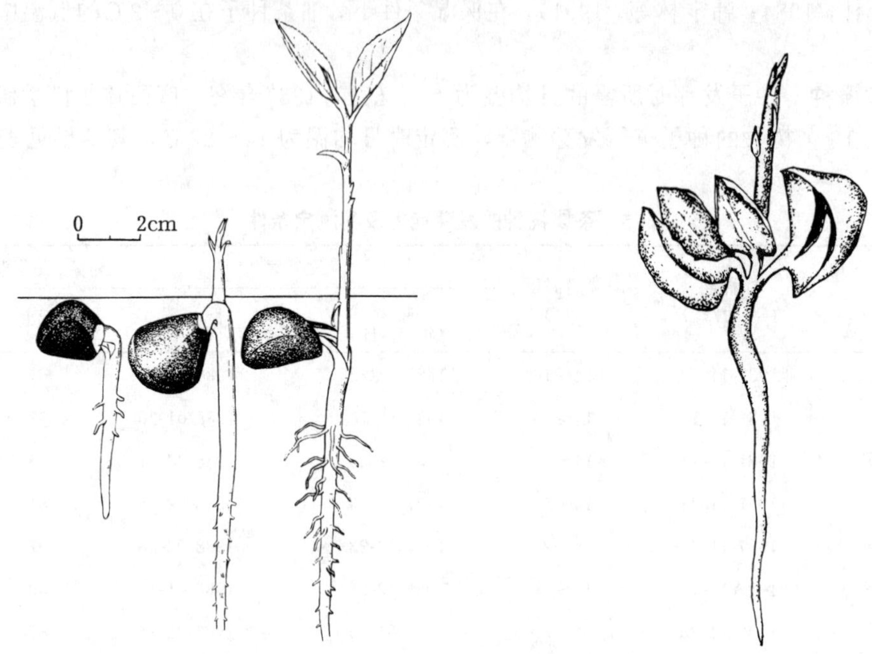

图 2 广宁红花油茶种子萌发后 5、15、30 天幼苗的生长情况
（黄应钦绘）

图 3 金花茶幼苗，示子叶和初生不育叶
（黄鹏成供稿、田恒德绘）

（林榕庚）

大头茶属
Gordonia Ellis

（茶科 Theaceae）

生长习性、分布和用途 大头茶属 40 种，我国 6 种，本文描述 2 种。常绿。多生于马尾松等针阔叶林内，颇耐荫。喜温暖湿润气候。适生于肥沃湿润的酸性土。这两个种的名称、生长情况、分布和用途见表 1。

表 1 大头茶属树种的名称、生长情况、分布和用途

中 名	学 名	树高（m）	胸径（cm）	分 布	用 途	供 稿
大头茶	*G. axillaris*（Roxb. ex Ker.）Dietr.	10～20	15～30	桂、粤、台	材用、栲胶	601
四川大头茶	*G. szechuanensis* Chang	4～8	6～15	桂、滇、川	材用	501

开花结实　5～10年生开始开花结实，正常结实期在20年生以后。寿命较长，100年老树仍能正常结实。结实间隔期一般为1年。花两性，白色，径7～12cm，单生或簇生于小枝顶端。小苞片与萼片覆瓦状排列，宿存。花瓣5片稀6片。雄蕊多数，花丝仅基部合生，花药丁字着生。子房上位，3～5室，每室有胚珠4～8。花柱顶端3～5裂。它们的开花结实物候见表2。

表2　大头茶属树种的开花结实物候

树种	观察地点 观察年份	开花			果实成熟		种子 散落期
		始期	盛期	末期	始期	盛期	
大头茶	广西南宁 1987～1989	8月中旬	8月下旬	9月上中旬	翌年9月下旬	10月上旬	10月下旬～11月下旬果开裂，种子散落
四川大头茶	四川荣县 1988～1989	10月上旬	10月中旬～11月上旬	11月中旬	翌年7月下旬	8月中旬	8月下旬～10月果开裂，种子散落

蒴果矩圆形或倒卵形，木质。成熟时果实黄褐色，室背3～5裂。每果有种子25～35粒。种子一端有膜质翅。种壳较厚硬，无胚乳，胚充满种子（图1）。种子形态特征见表3。

表3　大头茶属树种果实和种子的形态特征

树种	果实			种子		
	形状	大小（cm）	色泽	形状	大小（mm）	色泽
大头茶	矩圆形或倒卵形	长3～3.5 径2～2.5	褐色	三角状或扁长椭圆形	长7～10 径3～7	棕褐色
四川大头茶	椭圆形或矩形	长2.5 径1.5	黄褐色	斜倒三角形或棱形，背面微拱	长5～7 径3～4	黄褐色

果实的采收调制和种子贮藏　果实成熟盛期时用采种工具钩落果实或震动枝条，从地面捡拾果实。采回的果实摊放在干燥通风处，果壳开裂后敲打取出种子，搓揉并筛去种翅，即得净种。出种率为20%～25%。四川大头茶去翅种子千粒重约20g，每千克有纯净种子5万粒。大头茶去翅种子千粒重为21～25g，每千克有纯净种子4万～4.8万粒。种子忌烈日曝晒。运输时用布袋包装。贮藏于荫凉干燥处，或混湿沙贮藏。

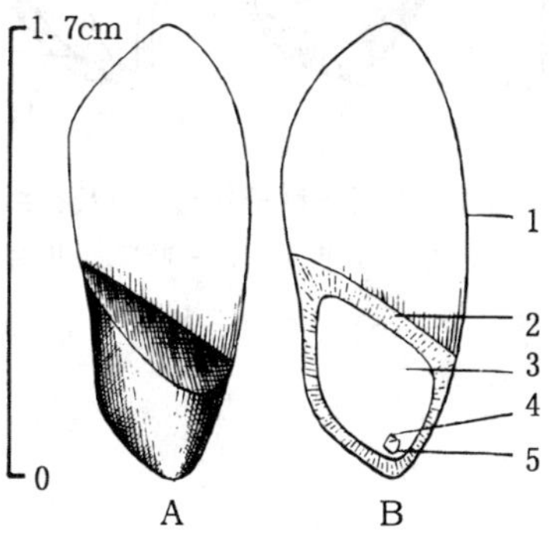

图1　大头茶带翅种子外形（A）及其纵切面（B）
1. 种翅　2. 种皮　3. 子叶　4. 胚芽　5. 胚根
（黄应钦绘）

发芽和播种　种壳较厚且硬，通常多用染色法测定种子生活力。四川大头茶在做生活力测定前，用始温45℃水浸种72小时（隔24小时换水1次）后，用0.5%的四氮唑溶液染色，常见的生活力约85%。1991

年1月19日，广西林业科学研究所用1990年冬采集的大头茶种子在恒温箱内作发芽测定，温度为28℃，基质为滤纸，播种后至2月18日开始发芽，5月10日发芽结束，发芽率为50%。

出土萌发。胚根萌发后7天子叶出土，约10天子叶展开，再过40天初生叶出现。

条播。每平方米约播5g。苗木木质化时间苗移栽，夏季需遮荫。1～2年生苗出圃。

（王宏志）

折柄茶（舟柄茶）

Hartia sinensis Dunn

（茶科 Theaceae）

生长习性、分布和用途 折柄茶属12种，产我国南部，其中3种分布至中南半岛北部。本文描述广义的1种。常绿小乔木，高6～9m，胸径约20cm。生长较缓慢，耐庇荫。适生于湿润凉爽的山地，低丘平原地生长不良。能耐短期－10℃低温。喜排水良好、土质疏松的酸性土。分布于滇、川、黔、桂、粤、湘、赣。材质坚实致密，为优质用材。也是涵养水源的树种。

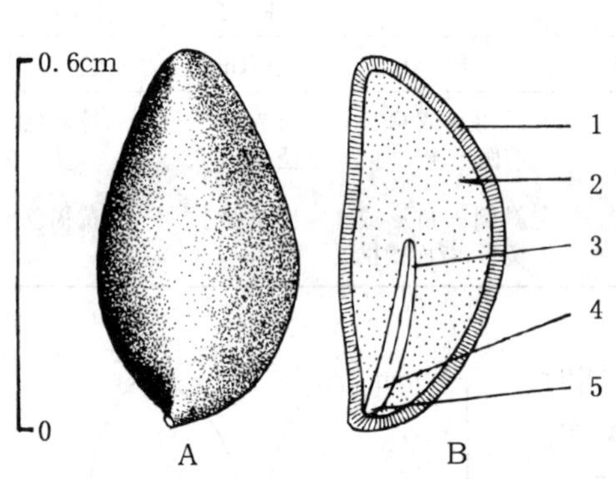

图1 折柄茶种子外形(A)及其纵切面(B)
1. 种皮 2. 胚乳 3. 子叶 4. 胚轴 5. 胚根
（黄应钦绘）

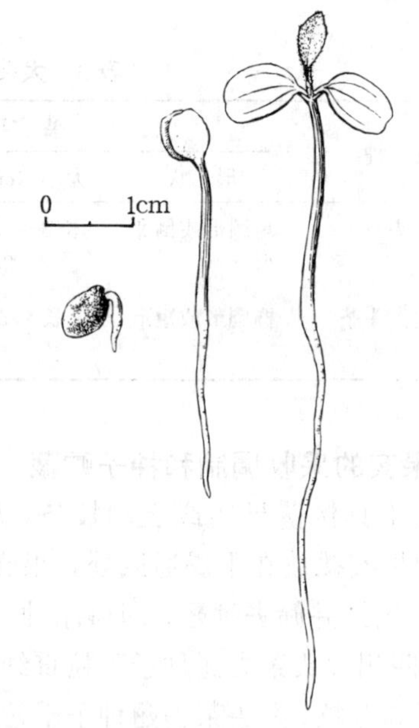

图2 折柄茶种子萌发后第5、10、25天幼苗的生长情况
（黄应钦绘）

开花结实 8～10年生开始开花结实，正常结实以后丰年间隔期为1年。花两性，单生于叶腋，白色。苞片2，有丝状毛。萼片5，宿存。花瓣5，卵圆形，基部合生成短管。雄蕊多数，花丝1/2合生成杯状。子房上位，5室，每室5～6胚珠。花柱5，合生，顶端5裂。据1975～1976年在广西大明山观测，10月下旬为始花期，11月上旬为盛花期，11月下旬为末花期；翌

年 10 月中旬果实开始成熟，10 月下旬为果熟盛期。蒴果，木质，为宿存花萼所包裹，卵形至椭圆形，先端渐尖，有 5 槽；未成熟时青色，成熟时黄色；长 1.3～1.9cm，径 1.2～1.5cm。10 月下旬～11 月中旬果实开裂，种子散落。每果有种子多粒。种子扁椭圆形或双凸镜状，棕褐色，长 4.5～6.5mm，径 3.5～5mm。种皮革质坚硬，胚乳丰富，胚伸直，子叶扁平（图 1）。

果实的采收调制和种子贮藏　果实盛熟时用采种刀钩落蒴果，在地面捡拾。采回的果实在室内摊开晾干。蒴果开裂后用棒轻敲，脱出种子，筛去果壳杂质，即得纯净种子。出种率约 5.8%。种子净度可达 96%，千粒重约 6.5g，每千克纯净种子有 15 万粒。种子可以稍耐脱水，但忌日晒。种子装入袋内置干燥荫凉通风处干藏。也可将种子混湿沙贮藏于 5℃的场所。

发芽和播种　种子无休眠习性。发芽时日均温宜在 18℃以上。1987 年广西林业科学研究所用当年新采种子在室内作过发芽测定。测定前种子未作处理。基质为滤纸。自然光。10 月 24 日播种，到 11 月 5 日日均温 19℃时种子开始发芽，11 月 8～17 日为发芽盛期，22 日发芽结束。从开始发芽之日起算，12 天发芽 36%；从播种之日起算，28 天的发芽率为 44%。出土萌发。胚根萌发后约 8 天子叶出土，22 天初生叶出现。种子萌发及幼苗生长情况见图 2。

撒播。将种子密播于沙盘或经过细致整地的苗床上，每平方米播种 8～10g，稍加覆土。幼苗木质化时移至大田培育。苗期需遮荫。1～2 年生苗出圃。

（王宏志）

木　荷　属

Schima Reinw. ex Bl.

（茶科　Theaceae）

生长习性、分布和用途　本属约 30 种，我国 19 种，本文描述 2 种。常绿乔木，高达 30m，胸径可达 1m。速生。喜酸性土壤，对肥力的要求不甚苛刻。能耐轻霜及短期－3℃低温。适于热带及亚热带地区与松、栲等混交，或用于营造防火林带。大片纯林生长较差。这两个树种的名称、生长情况、分布及用途见表 1。

表 1　木荷属树种名称、生长、分布和用途

中　名	学　名	树高（m）	胸径（cm）	分　布	用　途
木荷	*S. crenata* Korth.	30	120	苏、皖、浙、台、闽、赣、湘、粤、桂、黔、滇、川	材用、水源林、防火林
西南木荷（红荷木）	*S. wallichii* (DC.) Korth.	30	100	滇、川、黔、桂、湘、赣。老挝	材用、水源林、荒山先锋树种，防火树

开花结实　约在 8～10 年生开始开花结实，正常结实期在 12 年生以后。大小年间隔期一般为 1 年，但不甚明显。花两性，白色。西南木荷花簇生于枝端叶腋，梗长 1～1.5cm。木荷花单独腋生或顶生成短总状花序，梗长 1.2～4cm，常直立。小苞片 2。萼片 5，宿存。花瓣 5。

雄蕊多数，花丝贴生于花冠基部，花药丁字着生。子房通常 5 室，每室具胚珠 2～6。花柱 1，顶端 5 裂。在广西南宁观察的物候期见表 2。

蒴果，木质，基部萼片宿存，熟时室背 5 裂。每果木荷有种子 20 余粒，西南木荷约 15 粒。种子无胚乳，子叶扁平，长圆状。果实及种子形态见表 3、图 1。黄长辉（1993）报道过提高木荷母树林种子产量的措施及其效果。

表 2 木荷属树种的开花结实物候期①

树 种	开 花			果实成熟		种子
	始 期	盛 期	末 期	始 期	盛 期	散落期
木荷	5 月上中旬	5 月中下旬	5 月下旬	10 月下旬	11 月中旬	11 月下旬～12 月上中旬
西南木荷	5 月上旬	5 月中旬	5 月下旬	翌年 3 月中旬	3 月中下旬	4 月上旬～4 月中下旬

① 广西南宁，1978～1985

表 3 木荷属树种果实和种子的形态特征

树 种	未熟果颜色	果 实			种 子		
		形 状	大小（cm）	颜 色	形 状	大小（mm）	颜 色
木荷	绿色	扁球形	长 1～1.3 径 1.5～1.7	灰黑色或黑褐色	肾形，薄而扁平，有膜质翅	带翅 长 8～11 宽 4～6 厚 0.5～0.8	黄褐色
西南木荷	黄绿色	扁球形	长 1.4～1.8 径 1.7～2.0	黄褐色或紫黑色	肾形，薄而扁平，有膜质翅	带翅 长 8～11 宽 4～6 厚 0.5～0.8	黄褐色

果实的采收调制和种子贮藏 果实呈现黄褐色尚未开裂时采回果实，在荫凉处堆放 2～3 天后置阳光下曝晒至开裂，敲出种子，筛选，去除杂质，晾干，即得纯净种子。种子忌烈日曝晒，可在弱光下晒干或晾干。不影响种子发芽能力的含水量约 10%。普通干藏法即可，贮藏时间木荷不宜超过 5 个月，西南木荷不宜超过 3 个月。种子的净度、质量等数据见表 4。

表 4 木荷属树种鲜果出种率和种子净度、质量

树 种	出种率（%）	净 度（%）	千 粒 重（g）	每千克纯净种子粒数（万粒）
木荷	4～6	70～90	4～6	16.7～25
西南木荷	3～4	70～90	3.5～5.5	18.2～28.6

发芽和播种 种子无休眠习性或有短暂休眠现象。发芽时日均温需在 18℃以上。据王成霖和邵蓓蓓（1987）报道，在 0～5℃下层积 0 周、1 周、2 周和 4 周后的木荷种子，发芽率分别为 3%、37%、54%和 63%。广西林业科学研究所用室外沙床对这 2 个树种进行过发芽测定，测定材料为当年新采种子，测定前未作处理。木荷于 1981 年 1 月 11 日播种，1 月 25 日

开始发芽，2 月 25 日发芽终止。西南木荷于 1982 年 3 月 22 日播种，4 月 10 日开始发芽，5 月 8 日发芽终止。测定结果见表 5。

表 5　木荷属树种的发芽能力及其测定条件（室外沙床）

树　种	温度（℃）	发芽势（%）		发芽率（%）	
		计算天数	一般数值	计算天数	一般数值
木荷	14～16	13	55	44	78
西南木荷	22～24	12	50	46	60

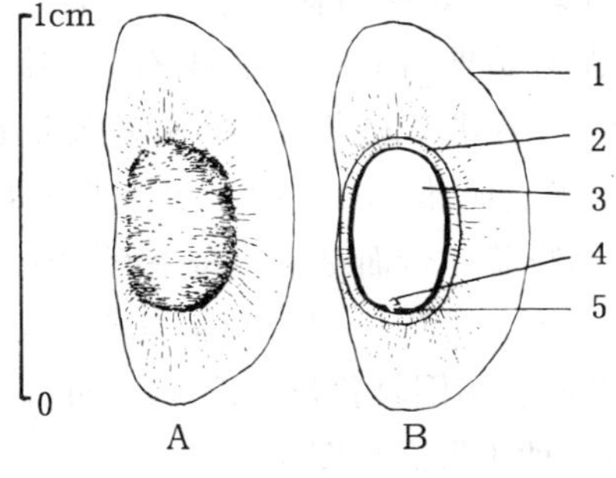

图 1　西南木荷种子外形（A）及其纵切面（B）
1. 种翅　2. 种皮　3. 子叶　4. 胚芽　5. 胚根
（黄应钦绘）

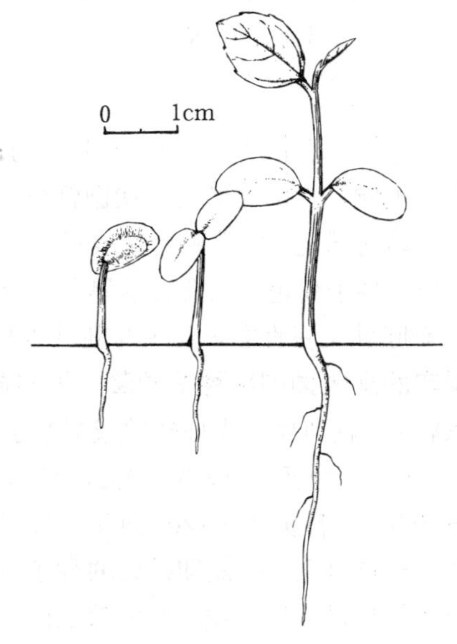

图 2　西南木荷种子萌发后第 6、9、20 天幼苗生长情况
（黄应钦仿《热带亚热带主要树种采种育苗技术》）

出土萌发。播种后 10～18 天开始萌发，5 天后子叶出土，再过 10 天初生叶展出。种子萌发和幼苗生长情况见图 2。

撒播或条播。覆土只宜稍盖种子。撒播时每平方米播种木荷 3～5g，西南木荷 4～6g。发芽后 1 月左右移植，每 666m^2 移苗 1.3 万～1.8 万株。条播时每平方米用种量约 4～5g。覆土 0.5cm。1 年生苗高 80cm，地径 0.8cm，可以出圃造林。也可培育 3 月生容器苗，在稀疏林下混交造林。

（王宏志）

紫　茎

Stewartia sinensis Rehd. et Wils.

（茶科　Theaceae）

生长习性、分布和用途　紫茎属约 13 种，我国约 6 种，本文描述 1 种。落叶灌木或小乔木，高 6～15m。要求凉爽湿润气候。适生于土层深厚、疏松肥沃的酸性黄壤。分布于豫、皖、

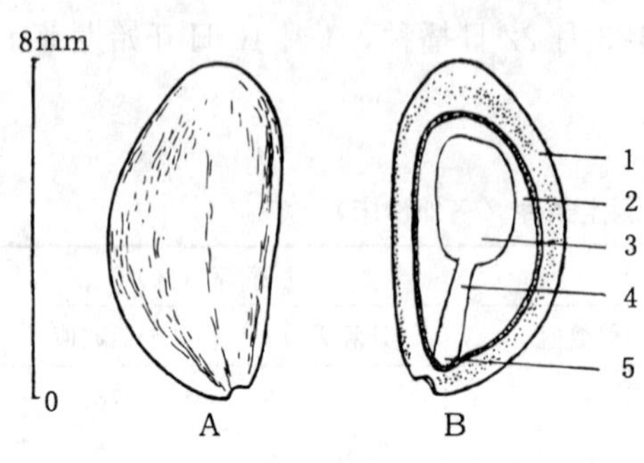

图 1　紫茎种子的外形（A）及其纵切面（B）
1. 种皮　2. 胚乳　3. 子叶　4. 胚轴　5. 胚根
（童军平、田恒德绘）

浙、闽、赣、鄂、湘、桂、黔、川、滇。木材坚硬致密，可作细木工用；种子油可食用或工业用；根、茎皮入药。本种亦可作观赏。紫茎是渐危种，已列入《中国植物红皮书》。

开花结实　结实大小年现象不明显。花两性，白色，直径约 4.5cm，单生叶腋。苞片 2，萼片 5，均叶状，宿存。花瓣 5，宽倒卵形，边缘啮蚀状。雄蕊多数，花丝向内弯曲。子房 5 室，每室有胚珠 2 枚。柱头 5 裂。紫茎的物候期因分布区的纬度、海拔高度而有差异，开花期有 5 月，有 6 月也有 7 月的，果实成熟始期有在 9 月，也有 10 月的。据 1989～1990 年在浙江杭州观察，开花始期在 5 月上旬，盛期在 5 月中旬，末期在 5 月下旬；10 月上中旬果实成熟，10 月中下旬种子开始散落。

蒴果木质，尖圆锥形或卵状圆锥形，有 5 棱，顶端喙状，全面被毛。每果有种子 10 粒。果实未成熟时绿色，成熟时褐色或黑褐色，长 1.4～1.8cm，宽 0.9～1.2cm，室背开裂，每室种子 2 粒。种子褐色，上有许多黑色小点，扁压，平凸状，周围具狭翅，长 7～9mm，宽 3～4mm，有胚乳。胚直立，几乎与胚乳等长，子叶扁平。种子形态见图 1。

果实的采收调制和种子贮藏　果实成熟时从树上采摘，或用竹竿击落于采种布上收集。果实采收后，置通风干燥处摊放或曝晒至开裂，脱出种子，再用风选或筛选清除杂物。出种率为 4%～6%。净度一般为 60%，变动在 45%～80%。千粒重 9～12g。每千克纯净种子 8.3 万～11.1 万粒。短期贮藏的种子可用布袋或塑料袋盛装后干藏。时间较长的，宜在 0～5℃环境中密封贮存。供翌春播种用的种子要混湿沙层积。

发芽前的处理　紫茎胚具有休眠特性，需在较低温度和湿润条件下完成生理后熟。作者曾试验过几种发芽前的处理方法，其中比较有效的处理是：用清水浸种 24 小时后，在 6～14℃条件下层积 45～60 天，或在 3～5℃条件下层积 60～75 天。层积时间过长，许多种子会在低温湿润的条件下萌发。

发芽测定　经过低温层积的种子，用沙作基质，置于 15～25℃变温条件下便会很快萌发。据试验，种子置床后第 2 天就开始萌发，5 天的发芽百分数一般为 31%，变动范围 19%～38%，24 天的发芽率为 76%～81%。发芽时不需要光照。

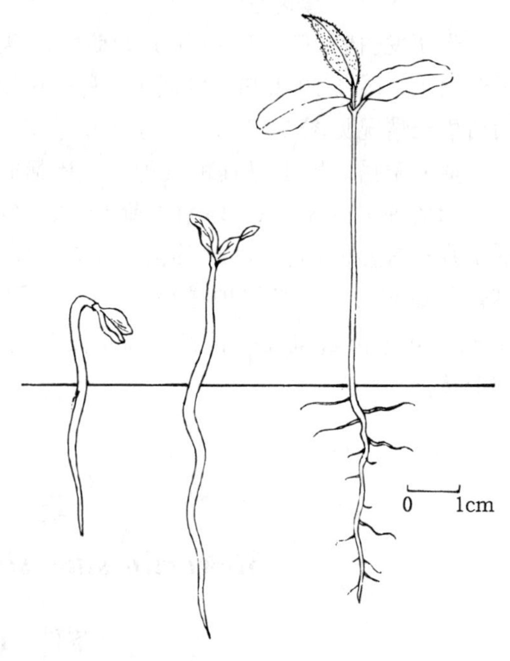

图 2　紫茎种子萌发后 3、5、11 天幼苗的生长情况
（童军平绘）

紫茎种子中的空粒较多，常常占到 1/3～1/2，可在层积前用清水浸种 24 小时，捞去浮在水面的空粒。

播种　秋播或春播，通常用经过层积的种子春播。未经层积的种子，春播后要延至第 2 年春才能陆续萌发出土。作者曾在浙江杭州对紫茎种子进行过室外发芽试验：1988 年 4 月 6 日曾将未经任何处理的干种子播于盆内，1989 年 3 月 24 日种子才发芽出土；1989 年 4 月 10 日将经过低温层积的种子播于盆内，播后 14 天开始发芽出土，23 天便发芽结束。条播，条距 20cm，播后覆土 1～2cm，上盖草。发芽出土后揭草，并作短期庇荫，充分浇水 1 次。留床1～2 年，用 2～3 年生苗造林。出土萌发，子叶 2 枚，出土后 8～10 天展出初生叶。幼苗生长情况见图 2。

（史晓华）

厚　皮　香

Ternstroemia gymnanthera (Wight et Arn.) Sprag.

（茶科　Theaceae）

生长习性、分布和用途　厚皮香属约 100 种，我国约 16 种，本文描述 1 种。常绿小乔木，高 8m，胸径约 20cm，生长较缓慢，耐庇荫。能耐－10℃左右的低温。适生于疏松湿润的酸性土，干旱裸露地生长不良。分布于皖、赣、鄂、湘、川、黔、滇、桂、粤、琼。日本、印度亦产。材质坚重细致，材色鲜艳，为优质用材，种子含油 18%～29%，可制作润滑油、油漆、肥皂。树皮含鞣质 21%～30%，可提制栲胶。

开花结实　约 10 年生开始结实，正常结实期在 20 年生以后。结实大小年间隔期为 1 年。花两性，单生叶腋。花冠开展，直径 1.5～2cm。萼片与花瓣各 5。雄蕊多数，2 轮。花丝短，基部合生并贴生于花瓣基部。子房上位，2～3 室，每室有胚珠 2，稀有 3～5。花柱 1，柱头顶端 3 裂。据 1987 年在海南观察，花期 5～6 月，9～11 月果实成熟。蒴果，未成熟时绿色，成熟时浅褐色，球形或卵形，径 1.5～2cm，为萼片及小苞片所支持，顶端有宿存花柱。果实

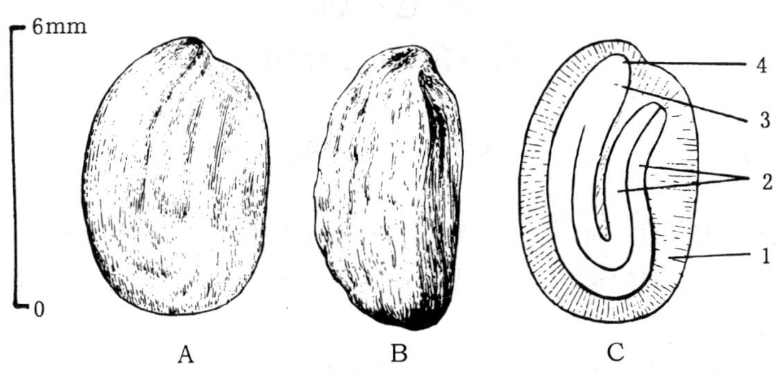

图 1　厚皮香种子正面（A）、侧面（B）及其纵切面（C）

1. 种皮　2. 子叶　3. 胚轴　4. 胚根

（黄应钦绘）

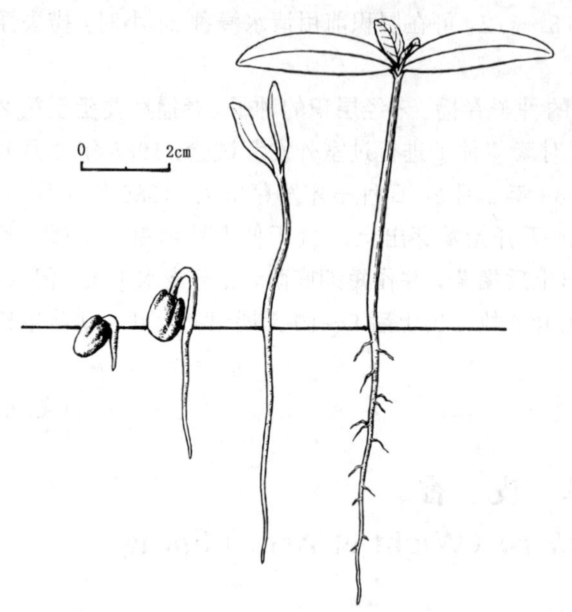

图 2　厚皮香种子萌发后第 2、6、12、30 天幼苗的生长情况
（黄应钦绘）

成熟后约 10 天开裂。种子初为鲜红色，存放后转为浅褐色或白色。每果有种子 1～4 粒。种子扁椭圆形，长 6～8.5mm，宽 5～6mm，厚 3～4.5mm。种皮坚硬，厚。无胚乳，胚弯曲，子叶长（图 1）。

果实的采收调制和种子贮藏　果盛熟而尚未开裂时上树用手采摘，或用钩刀截断果梗在树下捡拾。采得的果实可在室内堆放 2～3 天后摊放于通风处。当果皮充分裂开，种子已大部分外露时翻动或用木棒轻敲，即可脱出种子，筛去果皮等杂质即得纯净种子。果实出种率约为 23%。种子的净度可达 97%。千粒重约 100g，每千克有纯净种子 0.9 万～1.1 万粒。种子富含油脂，忌失水，不能日晒。可以果实的状态运输或短期贮藏，播种前再行调制。宜随采随播，也可混沙贮藏越冬，至翌年 2～3 月春播，贮藏期约为 4～6 个月。

发芽和播种　种子休眠现象不明显。发芽时日均温宜在 18℃以上。1987 年 11 月 12 日广西林业科学研究所在室外沙床上对当年新采种子作过发芽测定：播种后 46 天（12 月 28 日）开始发芽，无明显的发芽盛期，1988 年 1 月 8 日发芽过程终止。从播种之日起算，58 天的发芽率为 60% 。出土萌发。胚根萌发后 7 天子叶出土，27 天左右初生叶出现（图 2）。

条播。每平方米播种 12～18g，覆土 0.2cm。1 年生苗出圃。

（王宏志）

石笔木属
Tutcheria Dunn

（茶科　Theaceae）

生长习性、分布和用途　本属约 26 种，我国约 23 种，本文描述 4 种。常绿乔木。生长速度中等。忌严寒，但能耐短期－5℃低温。适生于较肥沃的酸性土。这 4 个树种的名称、生长、分布及用途见表 1。

开花结实　8～10 年生开始结实，大小年间隔期一般为 1 年，但不甚明显。花两性，单生于近枝顶叶腋。花冠浅黄色至白色。苞片 2。萼片 10（9～11），半宿存。花瓣 5，外面常有绢毛。雄蕊多数，花丝分离，5～8 轮，基部与花瓣连生。子房上位，3～5（6）室，每室有胚珠 2～5 颗，花柱几乎全部合生。广西南宁观察到的花果物候期见表 2。

表 1　石笔木属树种的名称、生长、分布和用途

中　名	学　名	树高（m）	胸径（cm）	分　布	用　途
六爿石笔木	*T. hexalocularia* Hu et S. Y. Liang	15	30	桂、粤	材用、水源、绿化、油脂
小果石笔木	*T. microcarpa* Dunn	9～12	20～25	闽、赣、湘、粤、桂	材用、水源
石笔木	*T. spectabilis*（Champ.）Dunn	10～15	20～30	闽、湘、粤、桂、滇	材用、水源、绿化、油脂
长毛石笔木	*T. wuiana* Chang	9～12	30	桂、粤	材用、水源、绿化

表 2　石笔木属树种的开花结实物候期（广西南宁）

树　种	观察年份	开　花			果实成熟		种子散落期
		始　期	盛　期	末　期	始　期	盛　期	
六爿石笔木	1978～1984	5 月中旬	5 月下旬初	5 月下旬末	11 月上旬	11 月中旬	11 月中旬～下旬
小果石笔木	1978～1984	6 月上旬	6 月中旬	6 月中下旬	11 月上旬	11 月中旬	11 月中旬～12 月中旬
石笔木	1987～1989	5 月下旬	6 月上旬	6 月中旬	11 月上旬	11 月中旬	11 月中旬～下旬
长毛石笔木	1987～1989	6 月上旬	6 月中旬	7 月上旬	11 月上旬	11 月中旬	11 月中旬～12 月中旬

表 3　石笔木属树种果实和种子的形态特征

树　种	未熟果颜色	果　实			种　子		
		形　状	大小（cm）	颜　色	形　状	大小（mm）	颜　色
六爿石笔木	黄绿色	扁圆形、长圆形以至桃形，5～6 裂	长 5.1～6.7 径 5～7.8	浅黄或黄红色	棱状梭形或不规则形	长 18～36 宽 8～16 厚 5.5～9	棕褐至黑褐色，有光泽
小果石笔木	青绿色	钝三棱状卵圆形，3 裂	长 2.5～3.3 径 2.1～2.6	黄绿或黄色	棱状椭圆或不规则形	长 10～17 宽 5.5～7.5 厚 3～4	棕褐至黑褐色，有光泽
石笔木	黄绿色	近球形至桃形，3～5 裂	长 2.8～3.5 径 2.8～3.8	浅黄色	棱状梭形或不规则形	长 16～34 宽 7～15 厚 3～4	棕褐至黑褐色，有光泽
长毛石笔木	青绿色	长卵形，3～4 裂	长 4～4.8 径 1.5～2	黄绿至黄色	棱状不规则形	长 10～23 宽 5～8 厚 3～5	棕褐至黑褐色，有光泽

蒴果，木质，密被绒毛，花柱及萼片宿存。果实成熟后 10～15 天从基部向顶端成 3～6 瓣裂，中轴不脱落，种子逐渐散落。种子有棱角。种壳骨质，形态不规则。外种皮厚，内种皮薄。无胚乳，子叶大，折叠。果实和种子的形态特征见表 3、图 1。

果实的采收调制和种子贮藏　果实成熟盛期采摘，或用采种钩刀将果实钩落后在地面捡拾。采得的果实堆放 3～5 天后置日光下曝晒，待果瓣开裂，种子脱出，经簸扬筛选即得纯净种子。种粒的大小差异较大，颗粒过小的种子不宜用于播种。种子可以晾干，但不宜在日光下

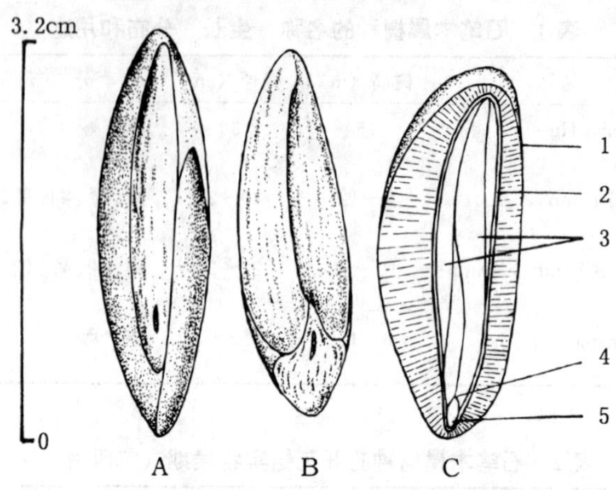

图1　石笔木种子外形（A、B）及其纵切面（C）
1. 外种皮　2. 内种皮　3. 子叶　4. 胚芽　5. 胚根
（黄应钦绘）

曝晒。种子的含水量宜在10%左右。种子的净度、质量等数据见表4。

表4　石笔木属树种的出种率、净度和质量

树　种	出种率（%）	净度（%）	千粒重（g）		每千克纯净种子粒数（粒）	
			一　般	变动范围	一　般	变动范围
六爿石笔木	18～22	99	1 330	1 100～1 500	750	660～900
小果石笔木	20～24	99	160	120～200	6 200	5 000～8 300
石笔木	18～22	99	600	550～650	1 600	1 500～1 800
长毛石笔木	14～16	99	230	180～280	4 300	3 600～5 600

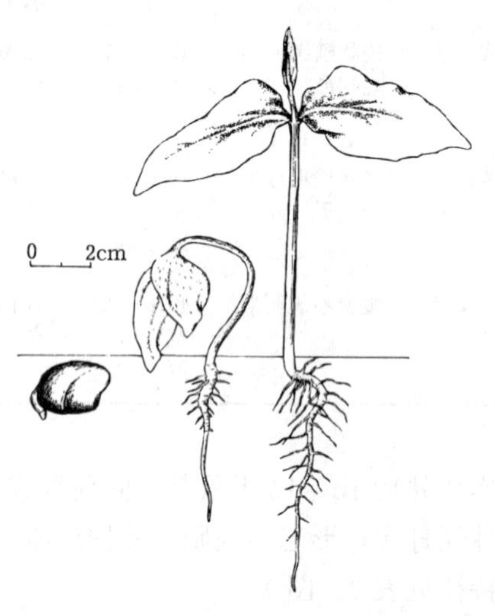

图2　六爿石笔木种子萌发后第2、8、25天幼苗的生长情况
（黄应钦仿《热带亚热带主要树种采种育苗技术》）

冬末春初的4个月，种子可以裸露贮藏于室内荫凉处；超过4个月的贮藏宜混以干沙，贮藏期一般为6～8个月。

发芽和播种　种子无明显休眠习性，播后30～45天可以发芽。发芽所需温度在13～15℃以上。1979年1～3月和1987年11月中下旬，广西林业科学研究所先后对这4个树种用室外沙床进行过发芽测定，测定前种子未作处理。测定六爿石笔木时气温20～22℃，测定其它3个树种时气温未超过15℃。4个树种的发芽历程都接近或超过2个月，没有明显的发芽盛期，发芽率均未超过40%。

出土萌发。胚根萌发后4～6天子叶出土。小果石笔木子叶出土后14天发出初生叶，其余 3 种需 20 天左右发出初生叶。种子萌发及幼苗生长情况见图 2。

条播。播前可以浸种 1 昼夜。每平方米播种量为：六爿石笔木 300～380g，石笔木 130～150g，小果石笔木和长毛石笔木 50～60g。覆土 2cm。1 年生苗出圃。

（王宏志）

水　东　哥

Saurauia tristyla DC.

（水东哥科　Saurauiaceae）

生长习性、分布和用途　水东哥属约 300 种，我国 13 种，本文描述 1 种。常绿灌木或小乔木，高 2～6m，径 6～10cm。生长速度慢，能耐荫蔽，分布于海拔 300～1 200m 的热带季雨林中，常见于沟谷溪流边。分布于滇、黔、桂、粤、琼。印度、马来西亚亦产。叶入药，鲜叶清热生肌，干叶粉调油外敷，治烫伤，果供鸟兽食。

开花结实　约 6～8 年生开始开花结实，正常结实期在 15 年生以后，结实大小年现象不明显。花两性，红色略带白色。聚伞花序具总梗，腋生于嫩枝或着生于老枝叶痕处，具茎花特征。萼片 5，不等大。花瓣 5，覆瓦状排列，基部合生。雄蕊多数（25～35），花丝短。子房上位，卵形，2～4（5）室，每室有胚珠多颗。花柱 2～4（5），下部合生，柱头头状。据在广西南宁及海南观察，开花期较长。南宁 4 月上旬为始花期，陆续开放至 7 月中旬；海南的花期为 3～7 月；果实成熟始于 6 月中旬，7 月中旬～8 月上旬为主要成熟期，9 月～10 月仍陆续有果成熟。浆果近球形，成熟时由青色转为灰白色，径 0.6～1.1cm，成熟后 1 周脱落。每果有种子数百粒。种子褐色，扁椭圆形或三角状扁椭圆形，长约 0.8mm，宽 0.6mm，有凹孔。种子藏于果肉内。胚乳稍丰富，胚伸直或稍弯，子叶短。

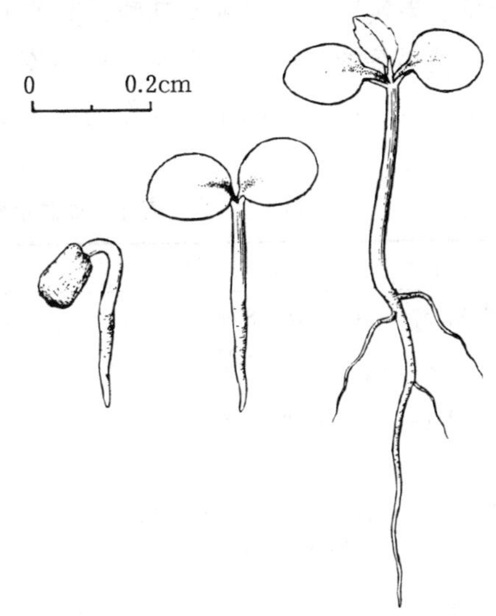

图 1　水东哥种子萌发后第 1、5、30 天幼苗的生长情况

（黄应钦绘）

果实的采收调制和种子贮藏　果实的成熟期长约 6 个月，采种的时期也较长。用于育苗的种子，宜在盛熟期选择颗粒大、发育饱满的果实采摘，或用竿杖击落后在地面捡拾。果实采回后在室内堆沤 1～3 天，果肉充分软熟时装入布袋，置水中反复搓揉，淘去果皮、果肉等杂质，即得种子。鲜果出种率约 2%。种子的净度一般为 93%～98%。种粒小，千粒重0.08～0.12g，每克纯净种子有 8 000～12 000 粒。种子忌日晒，不宜过分脱水，稍晾干即可贮藏或播种。如需运输，可采集接近成熟的果实，用小木箱或竹篓等包装，运抵后再行调制。也可将种子裹以湿润吸水纸保湿包

装运输。种子宜随采随播，一般不作贮藏。如需贮藏越冬，可用湿润吸水纸分层夹放，保湿贮藏，或混以细沙贮藏，以后混沙播种。

发芽和播种 种子无休眠习性。发芽时的日均温需在20℃以上。1987年8月下旬，广西林业科学研究所在室内自然光照条件下，用发芽皿垫吸水纸，对当年新采种子作过发芽测定（播种及发芽时日均温29℃，播前种子未作任何处理）。播后6天开始发芽，第4天进入盛期，第13天发芽终止。从发芽之日起算，8天中发芽百分数为60%；从播种之日起算，19天的发芽率为79%。出土萌发。胚根萌发后3天子叶伸出，再过22～25天初生叶出现（图1）。

撒播。种子混沙播下，每平方米播种0.1～0.2g。播后不覆土，仅用喷雾器淋水。幼苗转木质化后移至容器或大田继续培育。容器苗3个月出圃，大田培育1年生苗出圃。

（钟梅芳）

猕猴桃属
Actinidia Lindl.

（猕猴桃科 Actinidiaceae）

生长习性、分布和用途 猕猴桃属约54种，分布在亚洲马来西亚至西伯利亚东部的辽阔地带。我国是主产区，约52种和许多变种。本文描述4种，它们的名称、分布和用途见表1。大型落叶藤本。幼苗喜荫凉，忌强光直射。喜湿润，较耐寒，适应性强。生于较荫湿的林中或山坡、溪边，常缠绕其它乔木，或伏卧地上。中华猕猴桃的寿命可达百年以上。在我国史籍中早有记载，近三四十年才作为果树栽培。果实中维生素C含量特别丰富，比一般水果高十至数十倍。果实还含有多种氨基酸，营养丰富，甜酸适度，适于鲜食，也可以加工。作为果品，经济价值最高的是中华猕猴桃，已在不少国家引种栽培，它的鲜果及其加工品已成为国际商品。根、茎、叶、花药用。叶作饲料，茎皮造纸，花为蜜源。种子含干性油22%～24%。

表1 猕猴桃属树种的名称、分布和用途

中名	学名	分布	用途	供稿
软枣猕猴桃	*A. arguta* (Sieb. et Zucc.) Planch. ex Miq.	东北、华北、西北、华东、中南各省。朝鲜半岛、日本	食用、酿酒、药用	701
中华猕猴桃	*A. chinensis* Planch.	华东、中南、陕。欧洲、非洲、美洲、大洋洲及亚洲其它国家先后引种，新西兰栽培面积最大	食用、酿酒、药用	701
狗枣猕猴桃	*A. kolomikta* (Maxim. ex Rupr.) Maxim.	东北、冀、川、滇。俄罗斯远东、朝鲜半岛、日本	食用、酿酒、观赏	107
葛枣猕猴桃	*A. polygama* (Sieb. et Zucc.) Maxim.	东北、华中、西南、陕、甘、冀、鲁。俄罗斯远东、朝鲜半岛、日本	食用、酿酒	107

开花结实 中华猕猴桃实生苗3～6年开始结果，7～8年后进入盛果期；嫁接苗2～3年

开花结实，4～5 年后进入盛果期。盛果期可维持 40 年左右。中华猕猴桃雌花花蕾期 33～37 天，花期 3～5 天；雄花花蕾期 27～30 天，花期 3～6 天；单花寿命 2～4 天；雌花从受精到果实成熟约需 110～120 天。据观察，海拔每升高 100m，物候期推迟 0.56～3.60 天（李顺望等，1987）。猕猴桃属树种的开花结实物候见表 2。

表 2　猕猴桃属树种的开花结实物候期

树　种	观察地点	观察年份	花　性	开　花			果实成熟期
				始　期	盛　期	末　期	
软枣猕猴桃	黑龙江五常	1991	雄花 雌花	6 月 20 日 6 月 22 日	6 月 23 日 6 月 25 日	6 月 26 日 6 月 27 日	10 月初
中华猕猴桃	浙江杭州 福建厦门	1979 1983～1986	雌花 雌花 雄花	4 月 24 日 4 月 14 日～ 5 月 2 日 4 月 5 日～ 4 月 28 日	4 月 26 日 4 月 23 日～ 5 月 7 日 4 月 30 日～ 5 月 7 日	5 月 1 日 4 月 28 日～ 5 月 9 日 5 月 8 日～ 5 月 16 日	10 月上旬 9 月初～9 月底
狗枣猕猴桃	黑龙江五常	1991	雄花 雌花	6 月 2 日 6 月 4 日	6 月 5 日 6 月 7 日	6 月 7 日 6 月 10 日	10 月初
葛枣猕猴桃	黑龙江五常	1991	雄花 雌花	7 月 1 日 7 月 3 日	7 月 3 日 7 月 5 日	7 月 6 日 7 月 8 日	10 月初

花杂性或雌雄异株。聚伞花序或单花。萼片、花瓣 3～7，通常 5。雄蕊多数。软枣猕猴桃花药暗紫色，其余 3 种花药黄色。子房上位，瓶状、球形或圆柱状，多室。花柱离生。浆果，形状和大小随种而异。狗枣猕猴桃和葛枣猕猴桃的单果重一般在 5g 以下。软枣猕猴桃的果实稍大，单果重在 3～10g。中华猕猴桃果实最大，最重的可达 200g。据林承惠（1976）报道，安徽休宁县海拔 1 150m 处有一株 25 年生的中华猕猴桃，单株产果 1 700 余个，重约 72 千克。杨汉仕、张玉凤（1991）报道，在中华猕猴桃的花期和 15 天后的幼果期，用 20μg/g 的 2，4-D 液各喷施 1 次，同对照相比，平均单果重可以增加 62.3%，单株产量可以增加 118%。种子多数，细小，扁卵形，棕色，有光泽，表面有多角形网纹。种脐扁平呈嘴状。胚乳丰富，肉质，富含脂肪。胚直立，圆柱状，长约为种子一半以上，位于胚乳中央。子叶短，胚根靠近种脐。果实和种子的特征见表 3。图 1 绘出的是中华猕猴桃种子的外观和内部结构。

表 3　猕猴桃属树种果实种子特征

树　种	果　实			种　子		
	形　状	颜　色	大　小	形　状	颜　色	大小（mm）
软枣猕猴桃	球形至柱状长椭圆形。无宿存萼片	熟时绿色或深绿色	长 2～3cm 重约 3～10g	长椭圆形	褐色或棕褐色，有光泽	长约 2.5 宽约 1.8

（续）

树种	果实			种子		
	形状	颜色	大小	形状	颜色	大小（mm）
中华猕猴桃	近球形、圆柱形、倒卵形或椭圆形。宿存萼片反折	黄褐色，具小而多的淡褐色斑点	长 4～7cm 重 30～100g （最重 201g）	长椭圆形、似芝麻	金黄或棕色，有光泽	长 2.1～3.0 径 1.6～2.0
狗枣猕猴桃	柱状长圆形、卵形或球形，多为扁体长圆形。果熟时萼片脱落	未熟时淡绿色，熟时暗绿色并有深色纵纹	长 1.5～3.0cm 重约 1.5～5g	扁圆形	金黄色或褐色，有光泽	长约 2.0 宽约 1.2
葛枣猕猴桃	长圆形或卵圆形，顶端有喙，具宿存萼片	未熟时淡绿色，熟时黄色、淡橘红色	长 2.0～4.5cm 重约 2.0～7.0g	长圆形、扁	黄色，带白色，表面网状，略凹	长约 2.0 宽约 1.2

果实的采收调制和种子贮藏　采摘充分成熟的果实，在 20～25℃下堆放几天，装入袋中搓揉，用清水淘净果皮果肉等杂质，放在通风处阴干即得纯净种子。在 5℃左右的低温下混沙湿藏或干藏。果实和种子的质量见表 4。

表 4　猕猴桃属树种果实和种子质量的有关数据

树种	单株产果（kg）		单果重（g）		每果种子粒数		千粒重（g）	每克种子数（粒）	鲜果出种率（%）
	一般	可达	一般	最重	一般	变幅			
软枣猕猴桃	10～20	200	3～10	13	150～250	100～300	1.5	620～710	2.2～2.6
中华猕猴桃	15～20	500	30～100	200	200～500	100～1 000	1.2	450～1 100	—
狗枣猕猴桃	10～20	100	1.5～5	7.2	150～250	100～300	0.9	1 000～1250	2.2～2.6
葛枣猕猴桃	10～20	50	2～5	7.3	100～200	100～300	0.9	1 000～1 250	3.4～3.9

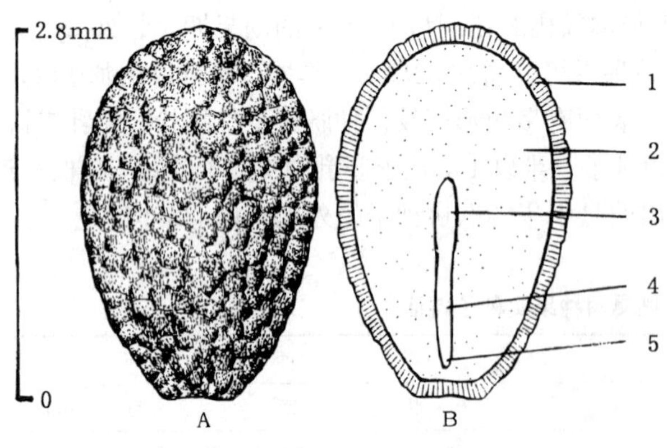

图 1　中华猕猴桃种子的外形（A）及其纵切面（B）
1. 种皮　2. 胚乳　3. 子叶　4. 胚轴　5. 胚根
（黄应钦绘）

发芽和播种　中华猕猴桃种子有休眠习性，播前需要层积（陈辉，1980）。据李顺望（1987）报道，低温（6℃）下贮存的中华猕猴桃种子用 500μg/g 的赤霉素浸泡 5 分钟，可以提高发芽率。赵隆钰（1987）播前将中华猕猴桃种子用沙搓擦几分钟，60 天的田间发芽率为 32%，未经沙擦的为 16%。梁根桃和严逸俊等人（1989）报道说，室外自然变温层积 64 天后播种的中华猕猴桃，平均出苗 37 株/10m^2，室内干藏后播种的则无一发芽。软枣猕猴桃和狗枣猕猴桃的种子无需预处理便能顺利萌发。在昼夜变温（低温时10℃，无光；高温时15～25℃，有光）的条件下盆播的这两个树种，10天的发芽率可以超过60%。在哈尔滨4月下旬～5月初播种，每平方米约播5g。只要床面保持潮润，狗枣猕猴桃和葛枣猕猴桃可以在播后约20天出苗。出土萌发。当年苗高20～30cm，2年生苗可以出圃。软枣猕猴桃种子的萌发和幼苗生长情况见图2。

徐本美（1984）对中华猕猴桃采用的室内发芽温度是15～20℃的变温或25℃的恒温。中华猕猴桃可以随采随播，也可以层积后春播。春播的季节在分布区的南部为2月底～3月初，北部为3月底～4月初。条播。每平方米大约播种2.5g。播后30～40天出苗，场圃发芽率约30%。生产上常在幼苗具3～5片初生叶时按4cm×12cm的株行距移栽。1或2年生苗出圃。

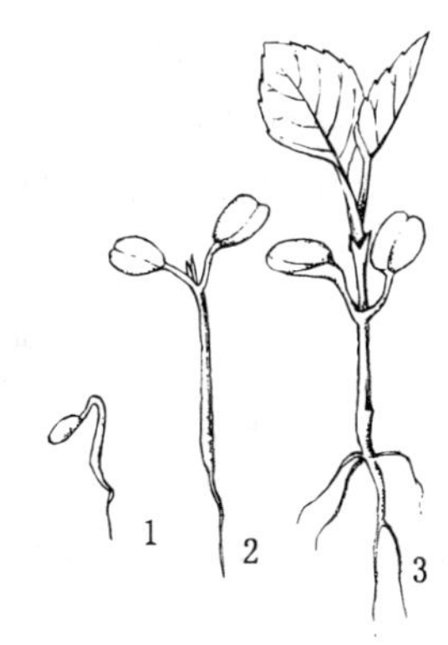

图2　软枣猕猴桃种子的萌发和幼苗生长情况

1. 下胚轴延伸　2. 子叶展开，初生叶出现　3. 幼苗形成

（田恒德仿《东北园林树木栽培》）

猕猴桃属的树种多用嫁接、根插和枝插繁殖。定植时雌雄株的比例为8∶1。

（杨国华）

龙脑香属

Dipterocarpus Gaertn. f.

（龙脑香科　Dipterocarpaceae）

生长习性、分布和用途　本属约76种，我国3种，引入约5种，本文描述2种。常绿大乔木，早期生长较慢。能耐－0.6℃的极端低温，苗期忌霜冻。适生于肥沃湿润疏松的酸性土壤，干旱贫瘠或黏重土生长不良。木材通直，坚硬，具香气，为珍贵用材。树干产树脂，根、干可提芳香油。这2个树种的名称、生长、分布和用途见表1。

表1　龙脑香属树种的名称、生长、分布和用途

中　名	学　名	树高（m）	胸径（cm）	分　布	用　途
云南龙脑香	*D. retusus* Bl.	25～50	48～130	滇南、藏东南。南亚至东南亚	材用、香料
龙脑香（羯布罗香）	*D. turbinatus* Gaertn. f.	25～60	50～150	滇南、藏东南。南亚至东南亚	材用、香料、药用

开花结实　10～15年生开始结实，20年生以后为正常结实期。结实有大小年现象，间隔期1～2年，尤以云南龙脑香最为明显。云南龙脑香5月上旬花芽始现，5月中旬～下旬花序形成，花蕾膨大。龙脑香2月下旬花芽始现，3月上中旬花序形成，花蕾膨大。花两性，排成总状花序，生于上部叶腋。每序有花3～7朵。花大，花冠粉红色，有芳香。花萼基部合生成

罐状或杯状，与子房离生，5裂，裂片2长3短。花瓣5，被柔毛。雄蕊多数（30～35)。花药2室，药隔附属体芒状或丝状伸出。子房上位，3室，每室2胚珠。在同一地区，云南龙脑香和龙脑香开花结实的物候期有明显差别。在西双版纳勐仑观察的物候期见表2。

表2 龙脑香属树种的开花结实物候期（西双版纳勐仑）

树种	观察年份	开花期			果实成熟期		果实脱落
		始期	盛期	末期	始期	盛期	
云南龙脑香	1987～1989	5月下旬	6月中旬	6月下旬	翌年1月下旬	2月下旬	2月下旬～3月上旬果实飘散
龙脑香	1986～1988	3月中旬	3月下旬	4月中旬	6月上旬	6月中旬	6月中旬果实飘散

坚果，包藏在花后膨大的萼筒内。2枚萼裂片增大为长翅，条状长圆形或披针形，具3～5条纵脉。每果具1种子。种子与果皮基部连合，无胚乳，胚充满种子。子叶大而厚，扭曲，包围胚根。果实和种子的形态分别见表3、图1。

表3 龙脑香属树种果实的形态特征

树种	未熟果颜色	形状	大小（cm）	色泽
云南龙脑香	淡绿色	陀螺形，表面有浮凸的脉纹，顶部裸露，具2枚萼片增大的长翅	长4.3～5.2 宽2.4～3.5 翅长10～22 基出5纵脉	茶褐色
龙脑香	淡绿色	陀螺形或椭圆形，表面有浮凸的脉纹，顶部裸露，具2枚萼片增大的长翅	长3～4.5 宽1.6～2.3 翅长12～15 基出3纵脉	茶褐色

果实的采收调制和贮藏 果实成熟时在树下周围拾集，或摇动树枝震落后收集。选取大而饱满的果实，剪去果翅即为播种材料，通称种子。采收当日的播种材料含水量，云南龙脑香为35%～40%，龙脑香为40%～45%。播种材料切忌日晒或久放。净度及质量等数据见表4。

表4 龙脑香属树种播种材料的净度和质量

树种	净度（%）	千粒重（去翅）(g)	每千克去翅果实粒数（粒）
云南龙脑香	96	2 200	450
龙脑香	96	3 800	260

发芽和播种 种子无休眠习性，成熟时在树上即开始萌动，胚突出，尤以龙脑香更为明显，宜随采随播。1986和1987年，西双版纳热带植物园在室外沙床进行过发芽测定：云南龙

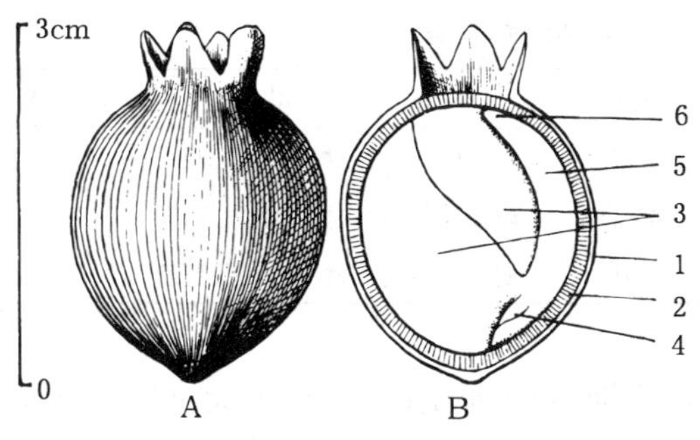

图 1　龙脑香去翅坚果外形（A）及其纵切面（B）

1. 萼筒　2. 果皮及种皮　3. 子叶　4. 胚芽　5. 胚轴　6. 胚根

（黄应钦绘）

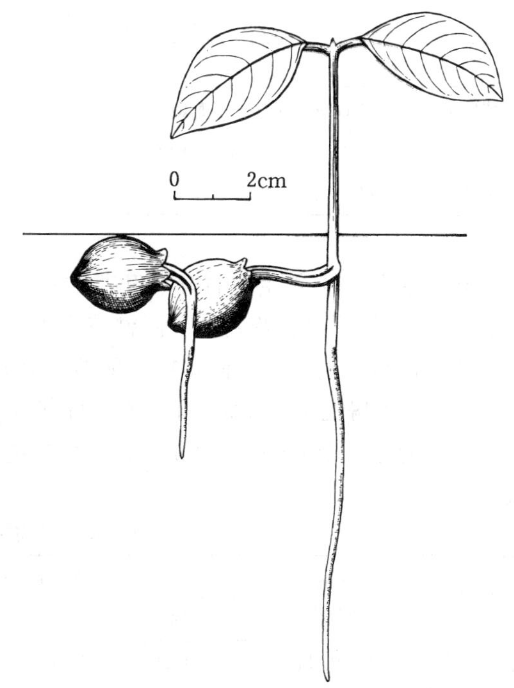

图 2　龙脑香种子萌发后第 6、15 天的幼苗生长情况

（黄应钦绘）

脑香发芽时的日均温为 20～23℃，2 月下旬播种后 15 天，即 3 月中旬开始发芽，第 30～40 天为发芽盛期，发芽全程约 65 天，发芽率为 85%。龙脑香发芽时的日均温为 25℃左右，6 月下旬播种后 5 天开始发芽，10～15 天为发芽盛期，发芽全程约 30 天，从发芽开始至高峰日的 10 天中发芽百分数为 70%，从播种至发芽终止，30 天的发芽率为 85%。

留土萌发。上胚轴萌出后 10 天左右展现初生叶（图 2）。

条播或点播。每平方米播去翅果 70～150g。播时应使胚芽一端向上或横置，覆土以不外露即可，不宜过厚。幼苗具 1～3 片初生叶时移苗补苗。1 年生苗出圃。

（程必强）

坡　垒　属

Hopea Roxb.

（龙脑香科　Dipterocarpaceae）

生长习性、分布和用途　本属约 104 种，我国 5 种。本文描述的 3 种都是我国特有的珍贵用材树种，由于过度采伐，残存大树极少，都被《中国植物红皮书》列为濒危种。常绿乔木，苗期及幼树生长缓慢，5 年生以后生长转快。苗期忌霜冻，成年后能耐 0℃极端最低气温及轻霜。喜肥沃深厚排水良好的沙壤土。材质坚实，极耐腐，有万年木之称，为著名珍贵用材，适用于高级家具、军工、造船和雕刻。树脂供药用。是热带雨林和季雨林的重要组成成分，能够涵养水源，保持水土。它们的名称、生长情况、分布和用途见表 1。

表 1　坡垒属树种的名称、生长、分布和用途

中　名	学　名	树高(m)	胸径(cm)	分　布	用　途	供　稿
狭叶坡垒	*H. chinensis* Hand.-Mazz.	22	65	桂	材用、水源、药用	604
坡垒（海梅）	*H. hainanensis* Merr. et Chun	25～30	60～85	琼。桂栽培	材用、水源、药用	604
毛叶坡垒	*H. mollissima* C. Y. Wu	35	70	滇南。越南	材用、水源、药用	616

开花结实　12 年生左右开始开花结实，正常结实期在 20 年生以后。结实大小年的间隔期为 1～2 年，较明显。花两性。圆锥花序顶生或生于上部叶腋。萼管极短，裂片 5，覆瓦状排列，花后宿存。花瓣 5。雄蕊 15，稀 10，排成 2 轮，花丝下部宽而扁平，花药卵形。药隔延伸成丝状。子房上位，近卵形，3 室，每室有胚珠 2。花柱短，锥状，具明显的花柱基。坡垒、毛叶坡垒的花序于 8～9 月形成；狭叶坡垒的花序于 4～5 月形成。它们开花结实的物候期见表 2。

表 2　坡垒属树种的开花结实物候期

树　种	观察地点和年份	开花期			果实成熟期		果实脱落期
		始　期	盛　期	末　期	始　期	盛　期	
狭叶坡垒	广西防城 1987～1989	5 月中旬	5 月下旬	6 月上旬	10 月上旬	10 月中下旬	10 月下旬～11 月中旬
坡垒	广西南宁 1987～1989	9 月中旬	9 月下旬	10 月上旬	翌年 4 月下旬	5 月上旬	5 月中旬～5 月下旬
毛叶坡垒	云南景洪 1988	7 月中旬	7 月下旬	8 月上旬	翌年 4 月中旬	4 月下旬	5 月上旬～5 月中旬

坚果，卵形，包裹于增长的萼裂片的基部，其中 2 裂片扩大成条状长圆形或倒披针形长翅。成熟后果实在树上可维持 4～6 天，随即带翅飘落，但遇大风天会提早飞散。每果有种子 1 粒。种子无胚乳，胚充满种子，子叶常扭曲。果实的形态特征见表 3、图 1。

表 3　坡垒属树种果实的形态特征

树　种	萼翅			去翅果实		
	形　状	大小（cm）	颜　色	形　状	大小（cm）	颜　色
狭叶坡垒	条状长圆形，有 11 条纵脉纹	长 8～11 宽 2～3	红色	卵状圆锥形，基部圆形，先端锐尖	长 1.8～2 径 0.9～1.2	棕褐色
坡垒	长倒卵形，有 7～9 条纵脉纹	长 4.5～7 宽 1～1.5	赭红	卵状圆锥形，基部圆形，先端锐尖	长 1.3～1.6 径 0.7～0.9	棕褐色
毛叶坡垒	倒披针形至椭圆形，变异较大，有 11～13 条纵脉纹	长 9～12 宽 1～3.7	鲜红	卵圆形，先端具短突尖	径 1～2	栗褐色具光泽

表 4　坡垒属树种播种材料的净度和质量

树　种	净度（%）	千粒重（去翅）（g）		每千克去翅果实粒数（粒）	
		一　般	变动范围	一　般	变动范围
狭叶坡垒	95	1 050	900～1 100	950	910～1 100
坡垒	95	440	350～500	2 250	2 000～2 860
毛叶坡垒	95	650	400～900	1 540	1 100～2 500

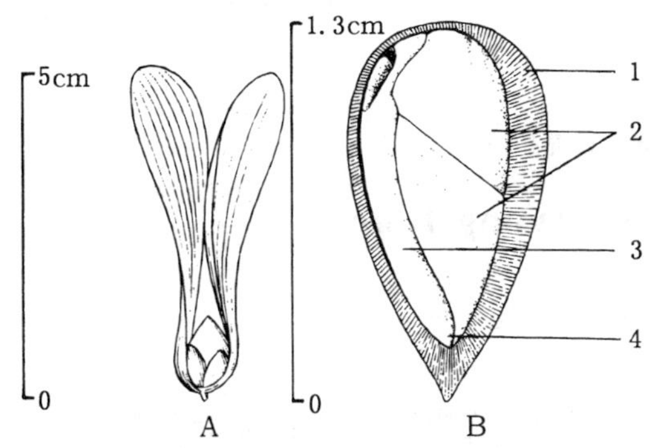

图 1　坡垒坚果外形（A）及其去翅后的纵切面（B）
1. 果皮及种皮　2. 子叶　3. 胚轴　4. 胚根
（黄应钦绘）

果实的采收调制和贮藏　成熟的果实容易飞散。当萼翅由青色开始转呈浅红色或桃红色而尚未变成鲜红色或赭红色时，即应选择无风天或午前静风时，用竹竿敲打或上树摇动震落后在地面捡拾。饱满的果实外被树脂油层，瘪粒或虫蛀粒则不具油层，应注意区分。果实采集后置室内阴晾处，剥去 2 片特长的萼翅即成播种材料，通称种子。果实此时的含水量多为 35%～40%。运输时应混以润沙，用竹筐或木箱包装，每筐（箱）不宜超过 15kg。忌日晒，不能裸露贮存。宜随采随播或运抵后即播，不耐久藏。坡垒和毛叶坡垒的种子成熟后，气温逐渐升高，混沙贮藏的时期不宜超过 1 个月。狭叶坡垒种子成熟以后气温逐渐降低，除即采即播外，种子可以混沙贮藏越冬，至次年 2～3 月播种。沙藏越冬的种子发芽率可保持在 60%左右。表 4 汇集的是这 3 个树种播种材料常见的质量数据。据陈荷美（1978）在海南尖峰岭测定，除去翅状萼片后，坡垒千粒重 614g。

发芽和播种　种子无休眠习性。充分成熟的种子在树上尚未散落时即有部分开始萌发。这 3 个树种多用去翅果实作播种材料，发芽特性基本相同。日均温 20℃以上时，采回的新鲜果实播后 3～5 天便能开始萌发。不同年份和不同贮藏期的果实，发芽的情况也会略有差异。贮藏期较长的，播种后发芽较慢。1978～1981 年的 4 年中，广西林业科学研究所利用室外沙床对经过半个月左右贮运的坡垒种子作过测定：播后 10～15 天开始萌发，在气温适宜的条件下，1 周内发芽结束；从发芽之日起至发芽高峰日的

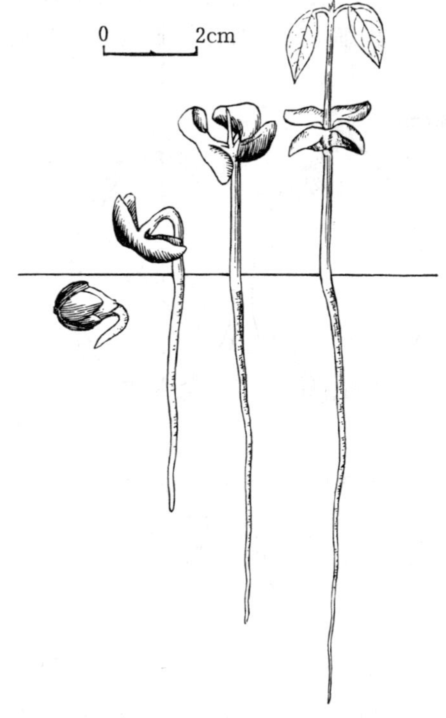

图 2　坡垒种子萌发后第 1、2、4、9 天幼苗的生长情况
（黄应钦绘）

5～7 天中，发芽百分数为 50%～70%；从播种之日起算的 17～22 天中，发芽率为65%～90%。在海南，坡垒的场圃发芽率曾达 97.5%（陈荷美，1978）。出土萌发。种子萌发后 2～3 天子叶出土，再过 4～6 天发出初生叶（图 2）。

条播。生产上大批量播种时一般不剥去果翅。带翅播种操作虽较不便，播种的深度也不易掌握一致，但对种子的发芽率和成苗率影响不大。每平方米苗床播种材料用量为：狭叶坡垒去翅果实 85～110g，带翅果 150～200g；坡垒去翅果 45～55g，带翅果 70～100g；毛叶坡垒去翅果 50～80g，带翅果 100～140g。覆土 2～3cm。苗期需遮荫。2～3 年生苗出圃。

（林榕庚）

望天树（擎天树）

Parashorea chinensis Wang Hsie

（龙脑香科　Dipterocarpaceae）

生长习性、分布和用途　柳安属约 11 种，我国只有本文描述的 1 种。常绿至半落叶大乔木，高达 40～65m，最高可达 80m，胸径 60～150cm，最大达 300cm。生长速度中等。适生于土层深厚、肥沃湿润的弱酸性土至钙质土。产于云南西南部和东南部、广西西南部，数量极少，在《中国植物红皮书》中列为稀有种。木材节少，纹理直，适用于胶合板、家具、建筑、造船及多种细木工，加工性能良好，是热带优良的用材树种。

开花结实　12～20 年生开始开花结实，30 年生以后进入正常结实期。有大小年现象，大小年间隔1～2 年，但每年都有植株开花结实。花两性，总状花序或圆锥花序腋生和顶生。顶生花序长 5～12cm，宽 3～5cm；腋生花序长 1.9～5.2cm。每个花序有花3～8 朵。花萼裂片 5，覆瓦状排列，外面 3 片较大，内面 2 片较小，果时增大。花瓣 5，黄白色，被毛。雄蕊12～15 枚，两轮排列。花药线状披针形，药隔附属体锥状。子房上位，圆锥状卵形，3 室，每室具 2 个倒悬胚珠，无明显花柱基，柱头微 3 裂。据 1988 年在云南勐腊观察，4 月中旬现蕾，4 月下旬始花，5 月上旬盛花，5 月中旬花期结束；7 月底果实开始成熟，8 月上中旬为果实盛熟期。坚果，卵状椭圆形，先端尖，果长 1.8～2.8cm，直径 1～1.5cm。果实表面密被灰色毡毛，外具萼翅 5 枚，3 长 2 短。长翅长6～7.5（9）cm，宽 1.2～1.8cm；小翅长3.5～5（6）cm，宽 0.6～0.8cm。成熟时果翅由黄绿色转变为褐色。果皮薄，果皮和种皮粘在一起。每果有 1 种子。每种子有 1～5 个胚，子叶厚，黄绿色，无胚乳。果实和种子的形态见图 1。

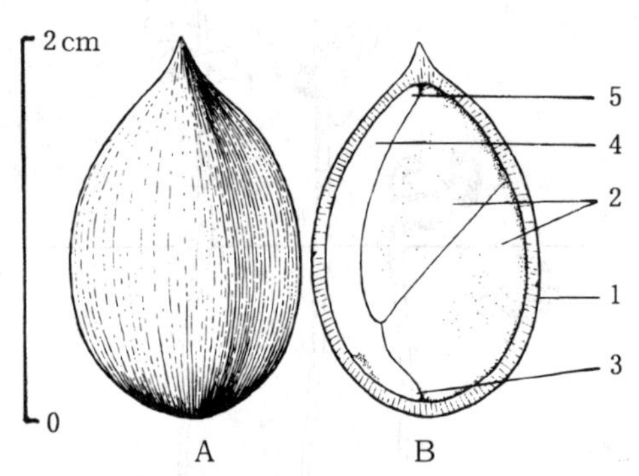

图 1　望天树去翅坚果外形（A）及其纵切面（B）
1. 果皮及种皮　2. 子叶　3. 胚芽
4. 胚轴　5. 胚根
（黄应钦绘）

果实的采收调制和贮藏　果实成熟时应及时在树下拾集，并剔除已发芽的种子、虫害种子和腐坏种子，去掉果翅即为播种材料，通称种子。据西双版纳热带植物园对勐腊地区望天树落地种子的统计，除已发芽的种子外，无虫的种子占 10%～30%，而广西采到的种子则虫害少见。去翅坚果质量相当于带翅坚果质量的 40%～50%。勐腊地区千粒重约 1 560g，每千克去翅坚果约 620～660 粒。广西的千粒重约 2 000g，每千克去翅坚果约 500 粒。鲜果含水量 45%～50%，忌失水，不宜日晒。短期运输可用湿润苔藓或拌湿润锯木屑，运抵后即播种，一般不作贮藏。

发芽和播种　种子极易发芽，在树下拾集到的种子大多数都已萌动或发芽，只宜随采随播。1988 年西双版纳热带植物园将当地新采种子播入室外沙盆，7 月 29 日播种，8 月 1 日出土，8 月 6 日发芽结束。从播种之日起算，9 天的发芽率为 65%。出土萌发。子叶出土后10～12 天张开，初生叶出现。

条播。用粗沙或纯净的细沙混拌催芽播种。每平方米播去翅果实 250～350g，覆细土0.5～1cm。也可密播于沙床内，发出初生叶后移入容器继续培育。1 年生苗可以出圃。

（马信祥）

粗壮娑罗双

Shorea robusta Gaertn.

（龙脑香科　Dipterocarpaceae）

生长习性、分布和用途　娑罗双属 190 余种，我国 2 种，本文描述 1 种。常绿或半落叶大乔木，高达 45m，胸径 120～150cm。喜高温高湿，忌10℃以下低温。不耐干旱贫瘠土，忌水淹。分布于西藏东南部。印度东北部也有分布。海南 1970 年引入栽培。材质坚硬致密，为建筑和器具珍贵用材。果可食，树脂入药。

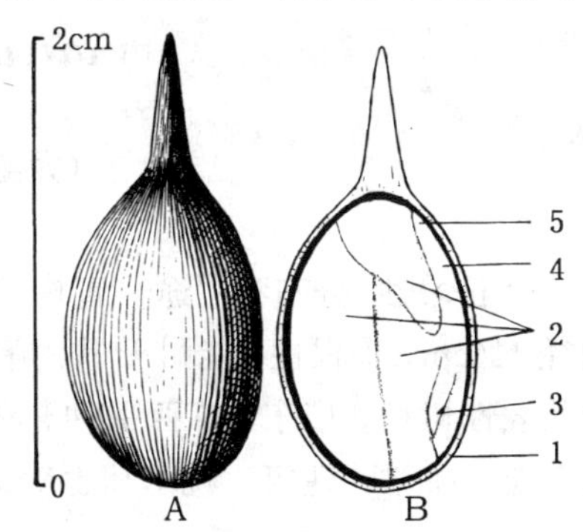

图 1　粗壮娑罗双坚果外形（A）及其纵切面（B）
1. 果皮及种皮　2. 子叶　3. 胚芽　4. 胚轴　5. 胚根
（黄应钦绘）

开花结实　6 年生开始开花结实，15 年生以后进入正常结实期，大小年现象不明显。花两性。圆锥花序腋生或顶生。花冠淡黄色，有短梗，每朵花下有 2 小苞片。萼管短，裂片 5，覆瓦状排列，常 3 片在外，2 片在内，宿存。花瓣 5，具毛。雄蕊多数，花药卵形，药隔附属体芒状或丝状。子房上位，卵形，3 室，每室具胚珠 2。据海南尖峰岭观察，2 月上旬花芽始现，2 月中旬花序形成，花蕾出现；2 月下旬为始花期，3 月中旬为盛花期，4 月上旬开花结束，幼果形成；5 月上旬果实开始成熟，5 月中旬末为果实成熟盛期。果实成熟后陆续脱落。坚果椭圆形，先端具长喙，成熟时果皮由淡黄色转变为褐色，萼翅由紫红色变为褐色，成熟后 3～4 天脱落。每果有种子 1 粒。果长 1.1cm，宽 0.9cm。种子无胚乳，胚充满种子。子叶厚，扭曲。粗壮娑罗双果实的形态见图 1。

果实的采收调制和贮藏　果实成熟期中在树下拾集，或摇动树枝震落后收集。除去果翅

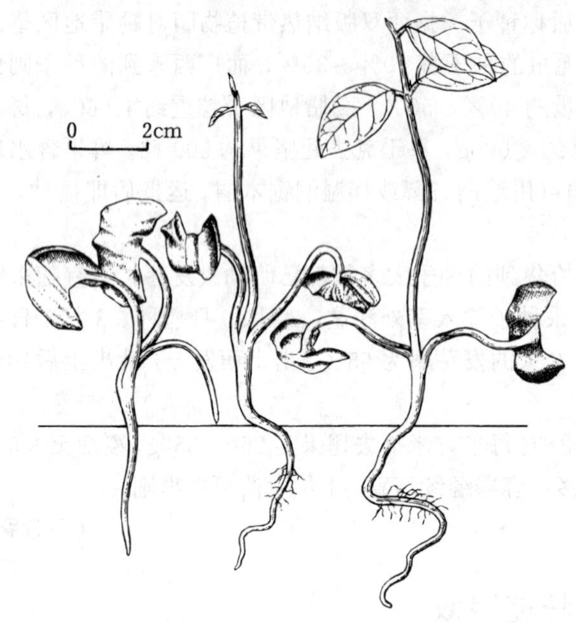

图 2　粗壮娑罗双种子萌发后第 1、2、5 天幼苗的生长情况
（黄应钦绘）

即为播种材料，通称种子。新鲜去翅坚果质量大约相当于带翅坚果质量的 70%。千粒重 725～1 040g，每千克去翅果实 960～1 400 粒。含水量 40%～45%，忌失水，不宜日晒或裸露存放。短期运输可用塑料袋拌少量湿沙或椰糠，运抵后即行播种，不作贮藏。

发芽和播种　种子无休眠习性，发芽时日均温宜在 25℃以上。据中国林业科学研究院热带林业试验站测定，播后 3～4 天开始发芽，从发芽开始至发芽高峰的 6 天，发芽百分数为 60%；从播种至发芽终了以 13 天计，发芽率为 85%。出土萌发。子叶出土后约 1 天发出初生叶（图 2）。

条播。每平方米播去翅果实120～200g，覆土 1.5～2cm。1 年生苗出圃。

（符史深）

青梅（青皮）

Vatica mangachapoi Blanco

（龙脑香科　Dipterocarpaceae）

生长习性、分布和用途　青梅属约 70 种，我国 3 种，本文描述 1 种。常绿大乔木，高 30m，胸径 120cm。深根性。能耐干旱、瘠薄和盐碱。分布于海南，为海南天然林中分布最普遍的热带代表性树种。广东、广西引种。泰国、马来西亚、印度尼西亚、菲律宾也有分布。材质优，为造船、桥梁、建筑、枕木和上等家具材。数量日渐减少，在《中国植物红皮书》中列为渐危种。

开花结实　5～8 年生开始结实，15 年生以上为正常结实期，大小年现象不明显。花两性，圆锥花序腋生或顶生。萼管短，裂片 5，镊合状排列。花瓣 5，白色，长于萼片 2～3 倍，旋转排列。雄蕊 15，两轮排列，外轮 10，内轮 5，花丝不等长，花药 4 室，药隔短凸尖。子房上位，球形，3 室，每室有胚珠 2，密生柔毛。花柱短，柱头头状。据海南尖峰岭 1985～1988 年观察，每年开花 2 次：第 1 次 4 月下旬为花蕾期，5 月下旬～6 月上旬为始花期，6 月中旬为盛花期，8 月中旬为末花期；8 月下旬～9 月初果实开始成熟，9 月中旬果实脱落。第 2 次开花在 7～9 月，翌年 2～3 月果熟，为小造。蒴果近球形，不裂。果径 6～8mm，下托增大的宿萼，2 长 3 短，其中最长的两片长 3～4cm，宽 1.1～1.6cm，基部合生成平盘状。成熟时果皮由青绿色变为赭色。种子无胚乳，胚充满种子。子叶肉质，绿色，2～3 裂。青梅果实的形态特征见图1。

果实的采收调制和贮藏　果实成熟盛期用高枝剪采下或用竹竿击落果实，在地面收集。除去宿萼即为播种材料，通称种子。新鲜带翅果实每千克约2 360粒。去翅后的千粒重240～360g，每千克有去翅果实2 800～4 200粒。据刘文明和宋学之（1989，1993）研究，新采收的青梅种子含水量约为40%，发芽率在95%以上，但不耐失水；在30℃、相对湿度70%的室内存放70小时，含水量便降至20%，种子死亡。他们建议的运输贮藏的安全措施是用含水量为27%～29%的椰糠（椰子壳加工后的粉碎物）作为介质，可以使青梅种子的含水量保持在30%～36%。

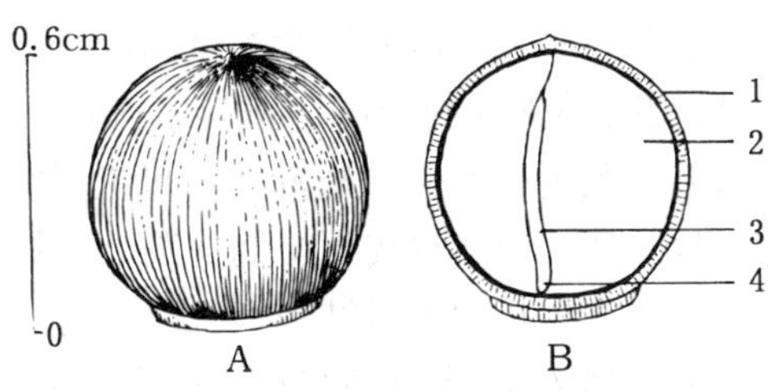

图1　青梅果实外形（A）及其纵切面（B）
1. 果皮及种皮　2. 子叶　3. 胚轴　4. 胚根
（黄应钦绘）

发芽和播种　种子无休眠习性，熟后即发芽。据试验，用新鲜果实播种，发芽率可达90%以上，存放5天后发芽率降至80%，7天后降至26%，10天后完全丧失发芽能力。3天以内无法播完的种子，应混湿沙存放，使其在沙中萌发，以后带芽播种。从播种之日起算，3天的发芽百分数约为60%～70%，5天的发芽率约为80%～90%。

条播。带翅新鲜果实每平方米播种50g；去翅鲜果每平方米播种35g。覆土2cm，苗期宜遮荫。培育1年生苗出圃。

（符史深）

杜鹃花属
Rhododendron L.

（杜鹃花科　Ericaceae）

生长习性、分布和用途　杜鹃花属约有900余种，广布于欧、亚和北美，澳大利亚的昆士兰仅产1种。亚洲种类最多，约有850种，我国约有650种，本文描述我国产的46种。其中濒危种大树杜鹃，渐危种蓝果杜鹃、似血杜鹃和大王杜鹃均已列入《中国植物红皮书》。常绿、半常绿或落叶乔木、小乔木、灌木或小灌木，也有附生或匍匐成垫状的。高0.15～25m。寒温带、暖温带、亚热带、热带均有生长。

我国仅新疆和宁夏未发现野生杜鹃花属的树种。集中分布在川、滇、藏的高山地区，共有400余种，粤、桂共有120余种，湘、黔、鄂约110种，闽、浙、台共有80余种，陕、甘、青共有40余种，赣、皖共有30余种，吉林有5种，苏、辽、黑各4种，内蒙古有3种，鲁、豫各2种，晋、冀各1种。一些分布区还陆续有新种发现。

垂直分布从海拔300～4 500m。从亚属的分类概念来看，常绿杜鹃亚属和有鳞杜鹃亚属主要分布于海拔1 000～4 500m的高山地带，马银花亚属主要分布于海拔1 000m以下地带（高原地区可至2 500m以上），羊踯躅亚属和落叶杜鹃亚属主要分布于海拔1 000m以下的地区（高原地区可至2 000m以上）。川、滇两省海拔（2 500）3 000～4 500m的地带杜鹃花种类最为丰富，集中了几乎所有原始类型和进化类型，是公认的现代杜鹃花的最大分布中心。

杜鹃花属树种种类丰富，花形多样，花色艳丽，是我国十大名花之一，加以树形奇雅，是世界著名的观赏植物。一些树种叶和花有香味，可以提炼芳香油，有的树种花可食用，花和果等可以入药，叶和树皮可提制栲胶。植株矮小的种类，枝条密集，根系盘结，有保持水土的作用。乔木树种的木材结构细，纹理直，可镟作木器及手工艺品。

本文描述的46个种的名称、树高、分布和用途见表1。

表1 杜鹃花属树种的名称、树高、分布和用途

中 名	学 名	树高（m）	分 布	用 途	供 稿
腺房杜鹃	*Rh. adenogynum* Diels	1～3	滇西北、川西北，海拔3 400～4 300m	观赏、保土	501
雪山杜鹃	*Rh. aganniphum* Balf. f. et Ward	0.3～0.5	滇西北、川西南、藏东南、青南、青东南，海拔（2 700）3 300～4 500m	观赏、保土	501
团花杜鹃	*Rh. amesisa* Rehd. et Wils.	1.5～9	滇西北、川西南、藏东南，海拔2 000～3 500m。缅甸	观赏、保土	501
粗枝杜鹃	*Rh. basilicum* Balf. f. et Forr.	3～9	滇西、滇西北，海拔2 400～3 900m。缅甸东北	观赏、保土、（花）食用	501
宽钟杜鹃	*Rh. beesianum* Diels	2～9	川西、川西南、滇西南、藏东南，海拔3 200～4 200m。缅甸	观赏、保土	501
锈红毛杜鹃	*Rh. bureavii* Fr.	1～4	川西南、川西北、滇东北、滇西，海拔2 800～4 200m	观赏、保土	501
毛喉杜鹃	*Rh. cephalanthum* Fr.	0.3～0.6	滇西北、川西南、川西，海拔3 600～4 200m	观赏、香料、保土	501
矮小杜鹃	*Rh. chamae-thomsonii* (Tagg et Forr.) Cowan et Davidian	0.15～0.25	滇西北、藏东南，海拔3 300～4 500m	观赏、保土	501
刚毛杜鹃	*Rh. crinigerum* Fr.	2～4	滇西北、藏东南，海拔2 400～3 800m	观赏、保土	501
蓝果杜鹃	*Rh. cyanocarpum* (Fr.) W. W. Sm.	1.2～6	滇西，海拔3 000～4 050m	观赏、保土	501
兴安杜鹃	*Rh. dauricum* Linn.	1～2	吉、黑、内蒙古，海拔500～1 600m。朝鲜半岛、日本、俄罗斯远东	观赏、药用、香料、保土	107
大白杜鹃	*Rh. decorum* Fr.	1～5	川、滇、黔西北，海拔2 000～4 000m	观赏、保土、（花）食用	501
马缨杜鹃	*Rh. delavayi* Fr.	2～15	滇南、滇西南、滇西北、桂西北、黔西北，海拔1 200～3 000m。缅甸	观赏、材用、药用、保土	501
泡泡叶杜鹃	*Rh. edgeworthii* Hook. f.	0.3～0.5	滇西北、川西南、藏东南，海拔2 400～3 400m。不丹、锡金、印度、缅甸	观赏、保土	501

（续）

中　名	学　名	树高（m）	分　布	用　途	供　稿
文雅杜鹃	*Rh. facetum* Balf. f. et Ward	2～6	滇西北，海拔 2 400～2 800（3 300）m。缅甸	观赏、保土	501
密枝杜鹃	*Rh. fastigiatum* Fr.	0.3～2.5	川西南、滇东北，海拔 3 000m	观赏、香料、保土	501
云锦杜鹃	*Rh. fortunei* Lindl.	3～4	粤、闽、浙、华中、西南，海拔 1 100～1 900m	观赏、香料、保土	704
镰果杜鹃	*Rh. fulvum* Forr. et W. W. Sm.	3～8	滇西北、藏东南，海拔 2 800～3 600m	观赏、保土	501
大树杜鹃	*Rh. giganteum* Forr. et Tagg	20～25	滇西南，海拔 2 400～2 600m	观赏、材用、香料、保土	501
似血杜鹃	*Rh. haematodes* Fr.	1～3	滇西，海拔 3 250～3 800m	观赏、保土	501
亮鳞杜鹃	*Rh. heliolepis* Fr.	1～5	滇西北、藏东南，海拔 3 000～4 000m。缅甸	观赏、保土	501
灰背杜鹃	*Rh. hippophaeoides* Balf. f. et W. W. Sm.	1～1.5	川西、川西南、滇西北，海拔 2 200～3 800m	观赏、保土、香料	501
隐蕊杜鹃	*Rh. intricatum* Fr.	0.2～1.5（5）	川西、川西南、滇西北，海拔 3 500～4 500m	观赏、保土	501
露珠杜鹃	*Rh. irroratum* Fr.	1～8	滇中、滇西北、川西南、藏东南，海拔 2 000～3 200m	观赏、保土	501
拉卜楞杜鹃	*Rh. labolergense* Ching et H. P. Yang	0.5	甘西南，海拔 3 500～3 900m	观赏、保土	501
满山红	*Rh. mariesii* Hemsl. et Wils.	1～3	川、鄂、皖、苏、浙、台、闽、赣、湘、粤、桂、黔，海拔 800～1 400m	观赏、药用	501
亮毛杜鹃	*Rh. microphyton* Fr.	0.3～2	滇、川西南、黔西南、桂东北，海拔 1 000～2 300（3 000）m	观赏、保土	501
羊踯躅	*Rh. molle* G. Don	1～2	川、黔、湘、鄂、皖、苏、赣、粤、闽、豫，滇有栽培，海拔 200～2 000m	药用、观赏	704
迎红杜鹃	*Rh. mucronulatum* Turcz.	1.5	东北、晋、冀、鲁、苏北，海拔 400m。日本、朝鲜半岛、俄罗斯远东	观赏	501
山育杜鹃	*Rh. oreotrephes* W. W. Sm.	1～5	川西南、滇东北、滇西北、藏东南，海拔 2 200～3 200m。缅甸	观赏、保土	501
马银花	*Rh. ovatum* (Lindl) Pl.	1～4	华东、湘、粤、川，海拔 500～1 400m	观赏	704
褐黄杜鹃	*Rh. phaeochrysum* Balf. f. et W. W. Sm.	1～1.4	川西南、川西北、川西、滇西北、藏东南，海拔 3 300～4 100m	观赏、保土	501

(续)

中 名	学 名	树高 (m)	分 布	用 途	供 稿
樱草杜鹃	*Rh. primuliflorum* Bur. et Fr.	0.5～1 (2)	川西、川西南、滇西北、藏东南，海拔 4 000～4 800m	保土、观赏、香料	501
光蕊杜鹃	*Rh. pronum* Tagg et Forr.	0.2～0.4	滇西北、藏东南，海拔 3 800～4 400m	观赏、保土	501
腋花杜鹃	*Rh. racemosum* Fr.	0.2～1	川西、川西南、滇中、滇西北、黔东南，海拔 1 800～3 000 (3 400) m	观赏、保土、香料	501
大王杜鹃	*Rh. rex* Levl.	5～10	川西南、滇中、滇东北，海拔 2 500～4 000m	观赏、保土、材用	501
红棕杜鹃	*Rh. rubiginosum* Fr.	1～3 (10)	滇、川、藏东南，海拔 (2 100) 2 800～4 200m。缅甸东北	观赏、保土、	501
血红杜鹃	*Rh. sanguineum* Fr.	约 1	滇西北、藏东南，海拔 2 800～4 200m	观赏、保土	501
锈叶杜鹃	*Rh. siderophyllum* Fr.	1.2～3	滇中、滇西北、川西南、黔西南，海拔 1 900～3 300m	观赏、保土、(花) 食用	501
杜鹃 (映山红)	*Rh. simsii* Pl.	1～2.5	西南、华东、华中、台、陕南、豫南，海拔 500～2 600m。泰国、越南、马来西亚	观赏、药用、保土	704
凸尖杜鹃	*Rh. sinogrande* Balf. f. et W. W. Sm.	5～10	滇西北、藏东南，海拔 2 000～2 900m。缅甸	观赏、保土、材用	501
草原杜鹃	*Rh. telmateium* Balf. f. et W. W. Sm.	0.3～1	川西、滇西北，海拔 (2 700) 3 800～4 500m	保土、观赏、香料	501
千里香杜鹃	*Rh. thymifolium* Maxim.	0.3～1.5	川西、青东、甘南、滇西北，海拔 2 800～3 000m	观赏、药用、香料、保土	501
川滇杜鹃	*Rh. traillianum* Forr. et W. W. Sm.	3～10	滇西北、川西、川西南，海拔 3 400～4 100m	观赏、保土	501
亮叶杜鹃	*Rh. vernicosum* Fr.	1～5 (8)	川西南、滇西、滇西北、藏东南，海拔 2 650～4 300m	观赏、保土	501
黄杯杜鹃	*Rh. wardii* W. W. Sm.	3～6	川西南、滇西北、藏东南，海拔 3 400～4 000m	保土、观赏	501

开花结实 5 年生左右开花。花两性，总状伞形花序、伞房花序或头状花序，通常顶生，稀单生，少数腋生或假顶生。花萼 5 裂或 5 齿，有时 6～10 裂，宿存。花冠轮状、漏斗状或钟状，稀管状或高脚碟状，略两侧对称，通常 5 裂，少有 7～8 裂，覆瓦状排列。雄蕊 8～10，稀 5 或 12～20，通常前倾，花丝无毛或向基部有柔毛或髯毛。花药无芒，顶孔开裂。花盘厚，圆齿状。子房上位，通常 5～10 室，偶有 12～16 室，每室有极多胚珠，密集于中轴胎座上。花柱细长劲直或粗短而弯曲。开花期和果实成熟期因树种及生境不同而有差异。蒴果未熟时

绿色，成熟时一般呈黄褐色、褐色、深褐色，也有呈淡蓝紫色（蓝果杜鹃）、黑色（褐黄杜鹃）、黑褐色（兴安杜鹃、马银花）、棕黑色（镰果杜鹃）、淡紫褐色（露珠杜鹃）的。蒴果由数心皮合成，木质、圆柱形、长圆柱形、狭圆柱形、椭圆状卵形、卵形或卵锥形，直，或弯，或微弯，或有棱，长 3～70mm（表 2）。未成熟种子黄色或淡黄色，成熟时多为深褐色（如腺房杜鹃、镰果杜鹃、云锦杜鹃），或黄褐色（如羊踯躅、马银花、杜鹃）。种子细小如锯屑状或粉末状，椭圆形、长圆形、卵圆形、矩形或披针形，有狭翅，偶有尾状附属物（如泡泡叶杜鹃、满山红、云锦杜鹃）。胚乳和胚极小。有的树种种子长 0.5～3mm。杜鹃花属树种的开花期、果实成熟期等列入表 2。

表 2　杜鹃花属树种的开花期、果实成熟期及形状和大小

树　种	开花期	蒴果成熟期	形　状	长（cm）	径（mm）
腺房杜鹃	5～7 月	11 月	长圆状椭圆形，有棱	1～1.5	8
雪山杜鹃	6～7 月	9 月	圆柱形，直	1.5～2.5	5～7
团花杜鹃	4～5 月	10～11 月	狭圆柱形，有棱	2.5～3.2	—
粗枝杜鹃	5～6 月	10～11 月	长圆柱形	2.5～4	8～10
宽钟杜鹃	4～6 月	9～10 月	长圆柱形，稍弯曲	2～4	—
锈红毛杜鹃	6～7 月	8～10 月	卵形	2～2.5	约 10
毛喉杜鹃	6～7 月	9～10 月	卵状	0.4	—
矮小杜鹃	5～6 月	10～11 月	圆柱形，有 5 棱	1.2	5
刚毛杜鹃	5～6 月	10～11 月	长圆柱形，微弯曲	1.5	4～5
蓝果杜鹃	4～5 月	10 月	圆柱形	1.5～2.5	6～10
兴安杜鹃	4 月下旬～5 月下旬	8 月中旬～10 月中旬	短圆柱形	1～2	—
大白杜鹃	4～6 月	11～12 月	长圆形	3～4	—
马缨杜鹃	3～5 月	9～11 月	长圆柱形	2	8
泡泡叶杜鹃	4～5 月	11 月	长圆状卵形，近球形	1～2	—
文雅杜鹃	5～7 月	11～12 月	短圆柱形	1.5	7
密枝杜鹃	5 月	9～10 月	卵状	0.3～0.4	2
云锦杜鹃	5～6 月	10 月中旬～11 月上旬	长椭圆形，有棱	2～3	1～1.5
镰果杜鹃	4～5 月	10～11 月	圆柱形，弯弓似镰	3～4	—
大树杜鹃	2～3 月	10～11 月	长圆柱形	4	15
似血杜鹃	3～5 月	10～11 月	圆柱形	1.5	—
亮鳞杜鹃	6～8 月	11 月	长圆形，圆柱形	1～1.5	—
灰背杜鹃	5～6 月	10（11）月	卵状	0.5～0.6	—
隐蕊杜鹃	5～6 月	9～10 月	卵状	0.3	1.5
露珠杜鹃	3～5 月	10～11 月	圆柱形	1.5～3.5	8

（续）

树　种	开花期	蒴果成熟期	形　状	长（cm）	径（mm）
拉卜楞杜鹃	7月	8～9月	卵圆形	0.3～0.4	—
满山红	4～5月	6～11月	椭圆状卵形	0.6～0.9	—
亮毛杜鹃	第1次3～5月 第2次10～11月	11月	卵圆形	0.4～0.9	—
羊踯躅	3月下旬～4月中旬	9～10月	圆柱状矩圆形	2～2.5	—
迎红杜鹃	4～5月	10月	圆柱形	1.3	—
山育杜鹃	5～6月	9～10月	圆柱形	1～1.2	5
马银花	5～6月	10～11月	卵圆形	0.5～0.8	—
褐黄杜鹃	5～6月	9～11月	长圆柱形，顶部微弯	1.5～3	4～6
樱草杜鹃	5月	9～10月	卵形	0.4～0.5	—
光蕊杜鹃	5～6月	11月	卵形	1	5
腋花杜鹃	3～4月	8～9月	椭圆形	0.7	—
大王杜鹃	4～5月	9～10月	圆柱形、椭圆形	4～4.5	8
红棕杜鹃	4～6月	10月	柱状	1～1.5	5～8
血红杜鹃	6～7月	10～11月	圆柱形	1～2	—
锈叶杜鹃	3～5月	11月	长椭圆形	1～1.3	—
杜鹃	3月底～4月	9～10月	卵圆形	0.8～1.2	—
凸尖杜鹃	4～5月	10～11月	圆柱形	4～7	约1.5
草原杜鹃	4～6月	10～11月	圆锥形	0.3～0.4	—
千里香杜鹃	5～6月	9～10月	卵状	0.3～0.4	2～2.5
川滇杜鹃	4～6月	10～11月	圆柱形	2	—
亮叶杜鹃	4～6月	8～10月	长圆柱形，稍弯曲	3～3.5	15
黄杯杜鹃	5～7月	8～9月	圆柱状，微弯曲	1.5～2.5	—

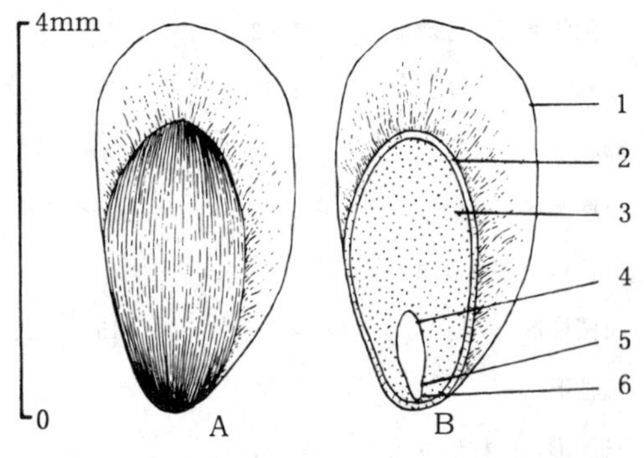

图1　云锦杜鹃带翅种子外形(A)及其纵切面(B)
1. 种翅　2. 种皮　3. 胚乳　4. 子叶
5. 胚轴　6. 胚根
(黄应钦绘)

云锦杜鹃种子外形及纵切面见图1。

果实的采集调制和种子贮藏　蒴果呈现成熟颜色，果瓣微裂时即可采集。采回的蒴果置阳光下摊晒，并经常翻动，种子从开裂的果瓣中脱出。种子有狭翅，偶有尾状附属物。脱出的种子不宜久晒，应及时收集。通常以带翅种子作为播种材料。蒴果出种率约1%。种子净度约70%。千粒重因树种不同而差异悬殊，有的在1g以上，有的只有0.05g，每千克纯净种子由几百万粒至几千万粒，见表3（张长芹等，1992；张敖罗等，1966；门玉岑，1986）。兴安杜鹃蒴果初熟至脱落约20天。

表 3　几种杜鹃花属树种种子的大小和质量

树　种	种子长（mm）	千粒重（g）	每克纯净种子粒数
团花杜鹃	—	1.2	800
宽钟杜鹃	2.2	0.2	5 000
兴安杜鹃	—	0.16～0.20	5 000～6 500
大喇叭杜鹃	—	1.4	700
皱叶杜鹃	2.2	0.11	9 000
油叶杜鹃	—	0.05	20 000
厚叶杜鹃	4.0	0.36	2 800
樱草杜鹃	—	0.08	12 000
凸尖杜鹃	—	1.2	800
亮叶杜鹃	3.2	0.25	4 000

杜鹃花属种子细小，且不耐贮藏，贮藏方法不当或时间过长，种子生命力会显著下降。据张敖罗等（1966）报道，宽钟杜鹃、皱叶杜鹃、厚叶杜鹃、亮叶杜鹃的种子在19℃常温下密封贮藏18个月，发芽率均在5%以下。不同种类的杜鹃种子耐贮性也有显著差异。例如同样是室温下瓶装贮藏后，海拔4 000m产的宽钟杜鹃种子发芽率已不及2%，采自海拔2 000m处的露珠杜鹃种子则发芽率仍为84%。生产上一般是越冬贮藏：将调制出的种子用容器盛装，置干燥凉爽处，或置5℃的低温库或冰箱内，贮藏1个冬季，翌春播种。杜鹃花属种类多，生境复杂，它们的种子贮藏问题需要进一步探索。

发芽和播种　张敖罗等(1966)观察过不同温度下红花杜鹃种子的发芽过程，结果是以在20℃条件下发芽最为顺利：开始发芽所需日数为14天，置床到发芽结束的时间为27天。在18℃、15℃和13℃的温度下，这2个时间则分别为20天和39天、24天和47天、38天和75天。

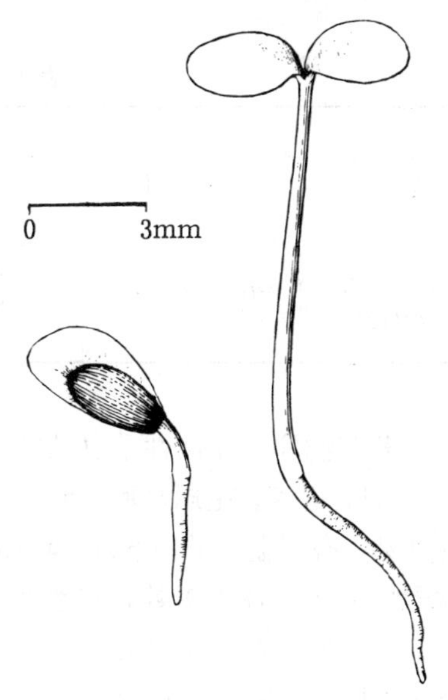

图 2　云锦杜鹃种子萌发后第3、8天的生长情况

（黄应钦绘）

任祝三(1988)的研究认为，杜鹃花属的种子是需光性种子，发芽必须有光照。在温度为20℃，每天12小时给以强度为1000lx光照的条件下，他所试验的腺房杜鹃等22个树种的种子，发芽率平均为72%；在没有光照的条件下，则平均不足1%。

常绿类杜鹃花用播种繁殖。有的地区主张即采即播，但多采用春播。长江流域以南播种期在3月或4月上旬。海拔升高或纬度北移，播种期推迟。哈尔滨地区的播种期约在5月上旬。

杜鹃花种子细小，多采用已作灭菌处理并配制好的培养土作室内盆播。播种量根据种子质量而定，一般每平方米播种2g左右。张长芹等（1992）认为，每0.5m^2的播种量应不少于 1 万粒。播后薄薄地铺盖一层剪碎的苔藓，并用喷雾器喷洒水雾，使苔藓充分湿润，上盖玻璃板，置通风良好的培育室或荫棚内，经常喷洒雾滴以保持苔藓湿润。播种 2 周后开始发芽，逐渐将盆移至阳光下并将玻璃板揭去。培养过程中应及时间苗，并分盆移植。她们播种的云锦杜鹃、羊踯躅和杜鹃，盆播的发芽率分别为 54%、30%和 70%。

出土萌发，子叶 2 片，近椭圆形。1 年生苗高 10cm 左右。一般培育 3～4 年后出圃。

落叶类杜鹃花也可采用种子繁殖，但多采用无性繁殖，方法有扦插、压条、嫁接和分蘖。扦插是常用的繁殖手段，春、夏、秋皆可进行，以夏季梅雨季节的效果最好。通常用瓦盆、木箱作插床，其内充填培养土。扦插后应经常喷洒水雾，保持土壤和空气的湿润。经 2～3 个月即可生根。生根较困难的可用生长素处理插穗。

云锦杜鹃种子萌发后的生长情况见图 2。

（赵德铭）

乌饭树（越橘）属

Vaccinium L.

（越橘科 Vacciniaceae）

生长习性、分布和用途 本属 300 余种，我国 80 余种，本文介绍 2 种。乌饭树为常绿灌木，高 0.5～3m，酸性土的指示植物，在 pH 为 3.8～4.5 的土壤上也能生长良好。笃斯越橘为落叶小灌木，高 15～80cm。耐水湿，喜光，性极耐寒，土壤解冻层不足 38cm 时便能开花。它们的名称、分布和用途见表 1。

表 1 乌饭树属树种的名称、分布和用途

中 名	学 名	分 布	用 途	供 稿
乌 饭 树	*V. bracteatum* Thunb.	长江流域以南至台、粤、滇。中南半岛、朝鲜半岛、日本	叶、果、根、药用，叶捣汁染米可做乌饭	1003
笃斯越橘（笃斯）	*V. uliginosum* L.	黑、吉、内蒙古东部。朝鲜半岛、蒙古、日本、俄罗斯远东地区	饮料、果酒、果酱，可提取食品色素	107

开花结实 花两性。总状花序腋生或顶生，或仅 3 朵花，稀单生。萼管与子房贴生，顶 4～5 裂，宿存。花冠壶形、钟形或筒状，裂片 4～5。雄蕊 8～10，花药背部有时有芒刺 2，顶孔开裂。有花盘。子房下位，4～5 室，稀 8～10 室，每室有胚珠数枚至多枚。浆果，多汁，顶端冠以宿萼。种子多数，胚乳丰富（图 1）。据在黑龙江省红星林业局（北纬 48°21′）观察（王洪学和王砚革等，1993），笃斯越橘 5 月下旬～6 月上旬始花，6 月中旬为盛花期，6 月下旬～7 月上旬果实迅速膨大并开始变色，7 月下旬～8 月上旬果实成熟，8 月中旬落果。笃斯越橘一般 2～3 年生开始结实，5 年生以前为盛果期，之后枝条逐渐枯死，结果部位外移。有大小年现象，间隔期一般为 1 年。王洪学和王砚革等（1993）报道过使笃斯越橘浆果高产的技术措施。这 2 个树种的开花结实物候见表 2。

表 2　乌饭树属树种的开花结实物候

树　种	观察年限和地点	花期			果期			果实成熟特征	果实散落期
		始　期	盛　期	末　期	始　期	盛　期	末　期		
乌饭树	—	6月	～	7月	10月	～	11月	紫黑色，微被白粉	—
笃斯越橘	1986～1987 黑龙江抚远县	6月上旬	6月上中旬	6月下旬	7月下旬	8月上中旬	9月上旬	紫黑色，被白粉	9月上中旬

表 3　乌饭树属果实与种子的形态特征

树　种	果实			种子		
	形　状	径（mm）	含种子粒数	形　状	长（mm）	宽（mm）
乌饭树	扁球形	4～6	8～27	呈带棱角的圆或椭圆形	2	—
笃斯越橘	球形或椭圆形	8～10	10～25	近橘瓣形	1～1.5	1.0

表 4　乌饭树属树种的出种率及种子质量

树　种	出种率（%）	净度（%）	千粒重（g）	每克纯净种子粒数
乌饭树	—	—	0.52	1 900
笃斯越橘	0.1	70～90	0.28～0.32	3 100～3 600

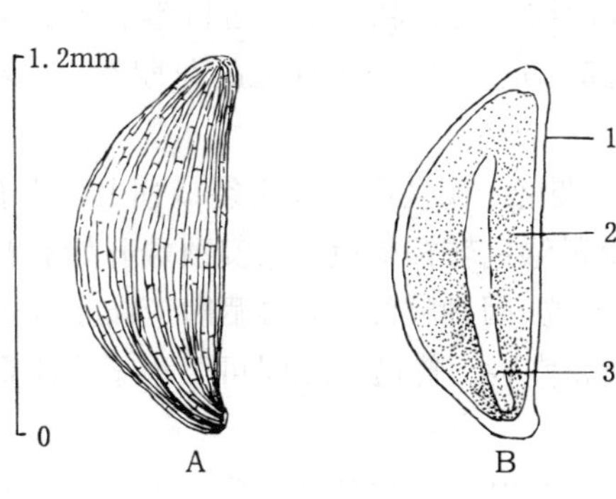

图 1　笃斯越橘种子的外形（A）及其纵切面（B）

1. 种皮　2. 胚乳　3. 胚

（梁鸣绘）

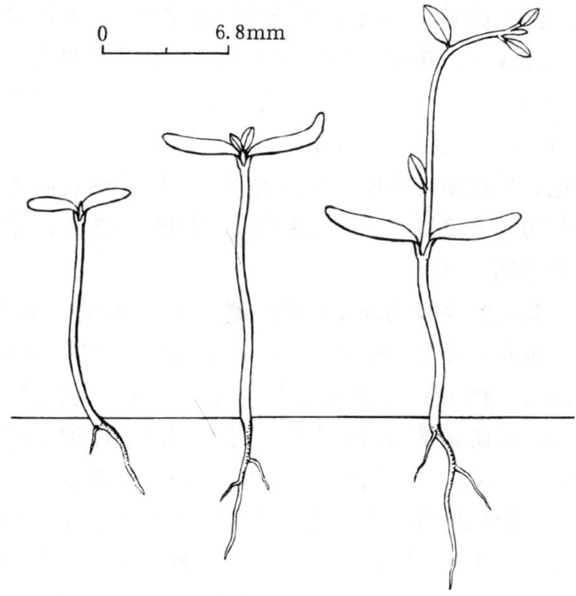

图 2　笃斯越橘萌发后第 9、16、28 天幼苗的生长情况

（许芝源绘）

果实的采收调制和种子贮藏 笃斯越橘7月末即可采摘，至9月中旬大部落尽。乌饭树果实10～11月采摘。可直接从树上采摘或摇落承接。采回后搓揉果实，用水漂去果皮、果柄与果肉，晾干后即得纯净种子，亦可从提取果汁后的废渣中获得种子。果实与种子的形态特征见表3，种子质量等数据见表4。

笃斯越橘种子在0～5℃下贮藏1年，发芽率未见明显降低。

发芽与播种 1988年3月，南京中山植物园将混沙湿藏的乌饭树种子播入圃地，40天后63%发芽，60天后发芽率达84%。笃斯越橘置于10～25℃温度下，60～70天仅10%发芽，余者不发芽；如先置于0～5℃下2个月，后转入5～20℃，发芽率可达60%以上。2种均为出土萌发。幼苗生长缓慢（图2）。笃斯越橘在哈尔滨市4月末至5月初播种，苗床条播，每平方米床面播种5g，覆土厚5mm，播后20天内保持床面湿润。这2个种野生资源丰富，少有栽培。

（周德本）

金 丝 桃

Hypericum monogynum L.

（金丝桃科 Hypericaceae ）

生长习性、分布和用途 金丝桃属约400余种，分布于北半球的温带和亚热带地区，我国有50余种，广布全国。本文描述金丝桃1种。半常绿小灌木，高达1m，分布于冀、晋、豫、陕、华中、华东、西南。日本也有。庭园栽培供观赏。果、根可入药。

开花结实 2～3年生开始开花结实，没有明显的结实大小年现象。花两性，黄色，顶生、单生或成聚伞花序，直径3～5cm。小苞片披针形。萼片5。花瓣5。雄蕊极多数，分离或基部合生为5束，长约2cm。子房上位，1室，胚珠多数，花柱细长，顶端5裂。蒴果卵圆形，成熟时室间开裂，花柱和萼片宿存。中轴胎座。种子细小，极多，梭形或香蕉形，长仅1mm，黑褐色。种皮薄，纸质，在放大镜下可以看到表面的整齐网纹。无胚乳。子叶近圆形。下胚轴圆柱形。果实和种子形态见图1。花期5～7月，果熟期8～9月。同一植株上的花期不一，果实成熟期也不一。

果实的采收调制和种子贮藏 由于花期长，蒴果成熟期不一，且树上果实裂口朝上，种子不易散落，所以采种期可以一直延续到12月。当蒴果由黄绿色转棕褐色，果实顶端开始裂开时即可分批采摘。摊晾至蒴果干燥完全开裂后轻轻击打脱粒。因种粒细小，摊晒时应防止种子漏失。用细眼网筛清除杂质。经调制的干燥种子可用纸袋盛装，普通干藏即可。种子千粒重0.04～0.05g，每克含种子约2万～2.5万粒。

发芽和播种 金丝桃种子无休眠习性，发芽不困难，室内发芽时对温度要求不严。一般在4月上中旬，气温10～15℃时播种，很快发芽，15～20天内发芽率可达80%～90%。由于种粒细小，场圃育苗宜用细土盆播后移栽。播后轻压，不必覆土，但需保持土壤湿润。出土萌发（图2）。可以扦插繁殖。

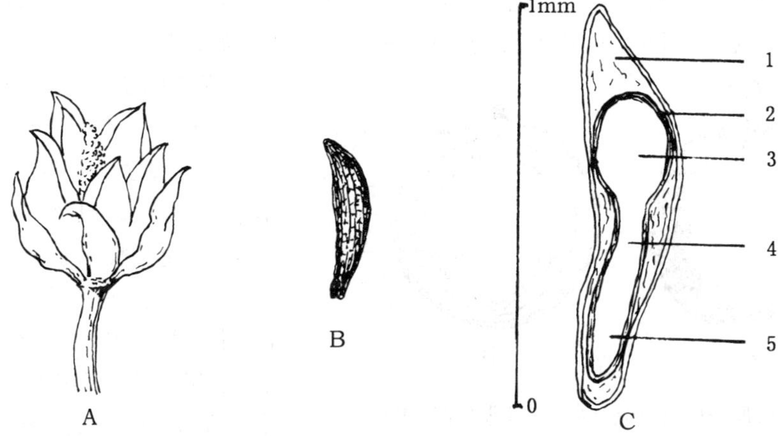

图 1　金丝桃的蒴果外形（A）、种子外形（B）及其纵切面（C）

1. 外种皮　2. 内种皮　3. 子叶　4. 胚轴　5. 胚根

（史渭清绘）

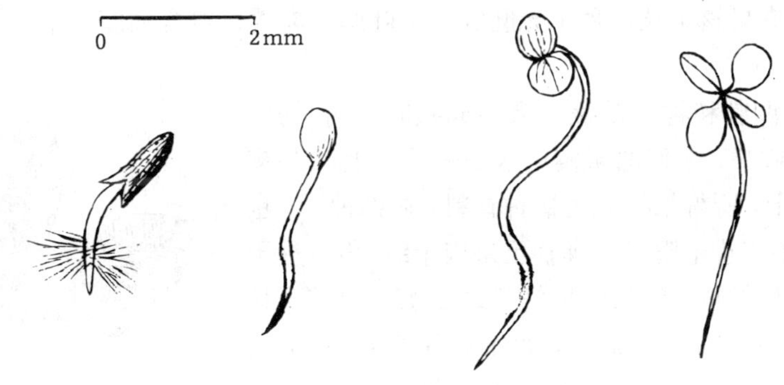

图 2　金丝桃种子萌发后第 1、2、3、5 天的幼苗生长情况

（史渭清绘）

（何泽瑛）

红厚壳（琼崖海棠）

Calophyllum inophyllum L.

（山竹子科　Clusiaceae）

生长习性、分布和用途　红厚壳（胡桐）属约 80 种，我国 4 种，本文描述 1 种。常绿乔木，树高 15～20m，胸径约 90cm，生长迅速。耐高温干旱，对土壤要求不严，红壤、滨海冲积沙土或盐碱土均能正常生长。分布于台、琼、粤沿海地区。南亚、东南亚、大洋洲、非洲、热带美洲亦有分布。木材坚实，为优质用材。种子榨油供工业用和医药用，精炼可食用。树脂可作漆油的防脆剂。可以用作海岸防护林、庭园观赏和蜜源树种。

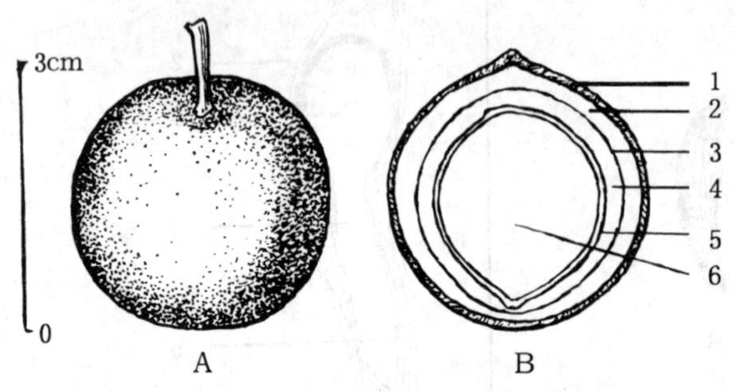

图1　红厚壳核果外形（A）及其纵切面（B）

1. 外果皮　2. 中果皮　3. 内果皮　4. 外种皮　5. 内种皮　6. 胚

（田恒德绘）

开花结实　3～5 年生可以开花结实，20～40 年生为结实盛期。结实有大小年现象，间隔期一般为 1 年。本种花两性，白色，芳香，径 2～2.5cm。花梗长 1.5～4cm。近顶生总状花序具花 7～11 朵。萼片 4。花瓣 4。雄蕊极多数，花丝基部合生成 4 束。子房上位，近球形，1 室，1 胚珠。花柱细长，蜿蜒状，柱头盾状。据海南乐东1987～1988 年观察，每年可开花 2 次：第 1 次 5 月底出现花蕾，6 月初始花，6 月中旬～8 月上旬盛花，8 月底第 1 次花期结束，10～11 月果熟；第 2 次 11 月～翌年 1 月开花，3～4 月果熟。核果球形，肉质，成熟时外果皮由青转黄褐色，径 2.5～3cm。果实成熟后陆续脱落。外果皮薄，中果皮肉质。每果有果核 1 粒。种子无胚乳。子叶厚，肉质，富含油脂。果实的形态特征见图 1。

果实的采收调制和种子贮藏　果实成熟时上树采摘，或用竹竿击落后收集。采回的果实浸水 2～3 天，擦去果皮果肉，洗净后晒干，所得果核即为播种材料，通称种子。也可不作调制，把果实置于阳光下曝晒至果肉干缩，将这种干缩的果实用作播种材料。鲜果的出籽率约 27%。果核千粒重 1 250g，变动在 1 000～1 600g，每千克种子（果核）约 800 粒，变动在600～1 000 粒。种子含水量约 38%。干果千粒重 4 300～5 500g，每千克干果 180～240 粒。可用布袋或麻袋运输。采集的果实晒干外果皮后堆放于凉爽干燥处，发芽能力可保持 1 年左右，有时可以贮藏至 2 年。

发芽和播种　种子无休眠习性，发芽时日均温宜在 22℃以上。据海南经验，播种前将果核浸水 2～4 小时，播种后 10～25 天发芽率 80%左右。用带有外果皮的果实浸清水（或尿水）2 昼夜播种，在高温多雨季节约需 2 个月方能发芽，发芽率可达 70%～80%。敲裂内果皮（果核），浸入 50%人尿 12 小时，稍阴干后播种，可提早半月左右发芽。留土萌发。具初生不育叶，对生（图 2）。

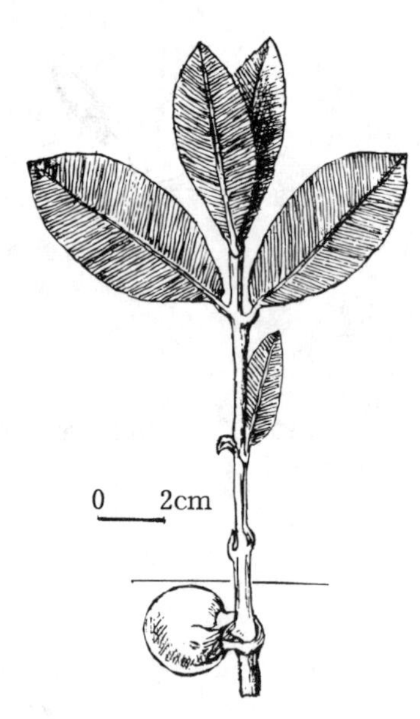

图 2　红厚壳幼苗（部分）

（田恒德绘）

条播或点播。每平方米播种 70～130g，覆土 2cm。一般培育 1 年生苗出圃。

（符史深）

山竹子（藤黄）属

Garcinia L.

（山竹子科　Clusiaceae）

生长习性、分布和用途　本属约 400 种，我国 21 种，又引入多种，本文描述 5 种。常绿乔木，热带性树种，喜高温湿润的气候，忌冰雪，少数种能耐－4℃左右的低温。多数种适生于土层深厚肥沃的酸性土壤，但有些种喜钙质土壤。这 5 个树种的名称、生长、分布和用途见表 1。其中金丝李分布范围狭窄，天然更新不良，已陷入濒危境地，在《中国植物红皮书》中列为渐危种。

表 1　山竹子属树种的名称、生长、分布和用途

中　名	学　名	树高（m）	胸径（cm）	分　布	用　途	供　稿
大苞藤黄	*G. bracteata* C. Y. Wu ex Y. H. Li	20	60	桂、滇。越南	材用、油脂、栲胶	603
多花山竹子	*G. multiflora* Champ.	20	80	滇、黔、桂、粤、琼、湘、赣、闽、台。越南	材用、油脂、食用、黄色素、栲胶	603
岭南山竹子	*G. oblongifolia* Champ.	15	40	粤、琼、桂。越南	材用、油脂、食用、栲胶、黄色素	603
金丝李（少脉山竹子）	*G. paucinervis* Chun et How	30	100	桂、滇。越南	材用、食用、栲胶、油脂、观赏	603
油山竹子	*G. tonkinensis* Vesque	20	40～60	琼。东南亚	材用、油脂、观赏	610

开花结实　大苞藤黄、多花山竹子、岭南山竹子、油山竹子 8～10 年生开始结实，金丝李 13～15 年生开始结实。前 4 个种在 15 年生以后进入正常结实年龄，大小年间隔期约 1 年，但不甚明显。金丝李在 25 年生以后进入正常结实年龄，隔 2～3 年才有 1 次结实大年。花单性，异株，稀杂性，单生或组成聚伞花序或圆锥花序。花冠橙黄色或淡黄色。萼片及花瓣各 4。雄花中雄蕊多数。大苞藤黄和岭南山竹子的花丝合生成一肉质体。多花山竹子和金丝李的花丝合生成 4 束围绕退化雌蕊。雌花较雄花稍大，有退化雄蕊；子房上位，卵圆形，1～ 12 室，每室胚珠 1 颗；花柱短或缺，柱头盾

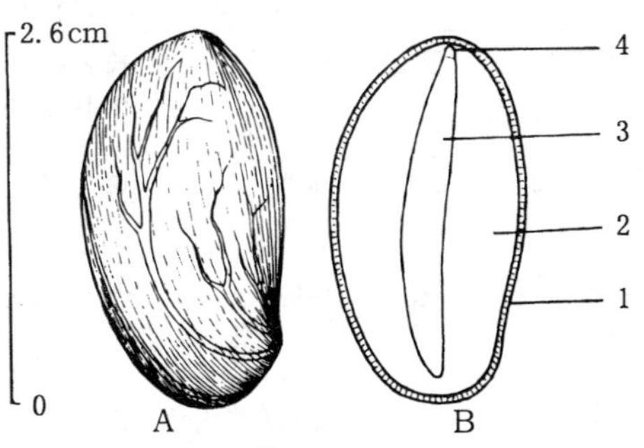

图 1　多花山竹子种子外形（A）及其纵切面（B）
1. 种皮　2. 子叶　3. 胚轴　4. 胚根
（黄应钦绘）

状。在广西南宁和海南儋县观察的开花结实物候期见表 2。

浆果，具革质外果皮，顶端具宿存柱头。种子有肉瓤状假种皮。大苞藤黄和金丝李的果单生，其余 3 种每果穗有果 4～10 个。每果有种子 2～7 粒。种子大，无胚乳，完全被伸直的胚所充满。果实及种子的形态见表 3、图 1。

表 2 山竹子属树种的开花结实物候期

树种	观察地点	观察年份	开花			果实成熟		果实脱落期
			始期	盛期	末期	始期	盛期	
大苞藤黄	南宁	1987～1988	雄花 4 月下旬	5 月上中旬	5 月下旬	11 月中旬	11 月下旬	11 月下旬～翌年 2 月上旬
			雌花 5 月上旬	5 月中旬	5 月下旬			
多花山竹子	南宁	1978～1982	雄花 4 月下旬	5 月上中旬	5 月下旬	10 月上中旬	11 月上旬	11 月上旬～12 月中旬
			雌花 4 月下旬末	5 月中旬	5 月下旬			
岭南山竹子	南宁	1987～1988	雄花 5 月上旬	5 月下旬	6 月中旬	8 月下旬	9 月中旬	9 月中旬～10 月中旬
			雌花 5 月中旬	5 月下旬	6 月下旬			
金丝李①	南宁	1978～1982	雄花 3 月下旬	4 月上旬	4 月下旬	9 月下旬	11 月上旬	11 月上旬～12 月下旬
			雌花 3 月下旬	4 月上旬末	4 月中下旬			
油山竹子	海南	1987	雄花 4 月上旬	4 月中旬	4 月下旬	11 月下旬	12 月中旬	12 月下旬
			雌花 4 月中旬初	4 月中旬	4 月下旬			

① 金丝李每年开花结实 2 次，表中所列为主要花果期。另一次花期在 5～8 月，翌年 3～7 月果熟

表 3 山竹子属树种果实和种子的形态特征

树种	未熟果颜色	成熟果实			成熟种子		
		形状	大小（cm）	色泽	形状	大小（cm）	色泽
大苞藤黄	青色	椭圆形	长 2.4～3.8 径 1.9～2.5	浅黄色，有褐色斑点	卵形或椭圆形	长 2～3.3 径 1.2～1.9	棕红色
多花山竹子	浅绿色	球形	径 3～4.5	黄绿色或黄色	近卵形，种皮具花纹	长 2.4～3.5 径 1.2	黄褐色至深褐色
岭南山竹子	浅绿色	球形	径 2.5～3.5	黄绿色或黄色	近肾形，表面粗糙	长 1.5～2 径 0.9～1.2	棕红色
金丝李	青色	椭圆形先端渐尖且稍弯	长 3～4.5 径 1.8～2.3	浅黄色至黄色	椭圆形，种皮具花纹	长 2.5～3 径 1.4～1.7	棕褐色
油山竹子	青色	椭圆形	长 3.5～4 径 2.8～3	淡黄色	椭圆形	长 2.5～3 径 1.5～2	棕色

果实的采收调制和种子贮藏 猴、狸等兽类喜食这几个树种的果实。当果实即将成熟时，

应掌握时机上树采摘或用采种钩刀割落后在地面收集。采回的果实可视成熟程度堆沤 2～3 天。软熟后有些种可食去果肉，洗净种子。一般情况是堆沤后将果置水中冲淘搓洗，漂去果皮等杂质即得种子。出种率和种子质量等数据见表 4。

表 4　山竹子属树种的出种率和种子净度、质量

树　种	出种率（%）	净度（%）	千粒重（g）		每千克纯净种子粒数（粒）	
			一　般	变动范围	一　般	变动范围
大包藤黄	56	96～100	6 700	5 000～8 000	150	125～200
多花山竹子	22	96～100	3 600	3 000～4 000	280	250～330
岭南山竹子	8	96～100	270	240～300	3 700	3 300～4 200
金丝子	55	96～100	4 500	4 000～5 000	220	200～250
油山竹子	27	97～99	4 500	4 000～5 000	220	200～250

种子忌失水，不能日晒。调制后的种子，稍晾干后即可混以湿沙供贮藏或运输。贮藏期一般 4～8 个月。

发芽和播种　种子有休眠现象。大苞藤黄、油山竹子种子的休眠期约 2 个月，多花山竹子和岭南山竹子的休眠期约为 6 个月。金丝李种子的休眠期颇不一致，多数种粒的休眠期约为 7～8 个月，部分种粒的休眠期为 3～4 个月，还有一部分种粒的休眠期可达 1 年以上。种子发芽所需的日均温在 18℃以上。1974、1981 和 1987 年，广西林业科学研究所和华南热作研究所先后用经过层积的种子在室外沙床上进行过发芽测定，结果见表 5。

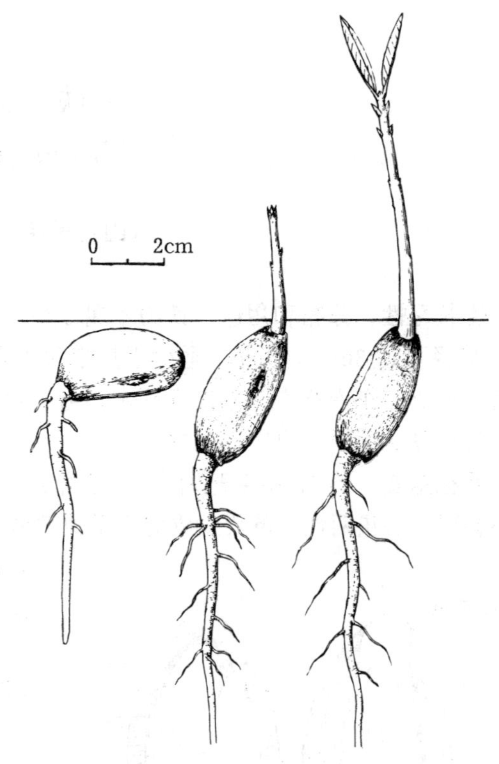

图 2　多花山竹子种子萌发后第 50、65、80 天幼苗的生长情况
（黄应钦仿《热带亚热带主要树种采种育苗技术》）

留土萌发。大苞藤黄的胚根萌发后约 45 天胚芽出土，具不育叶 2～4 对，70 天左右方展出初生叶。多花山竹子的胚根萌发后约 58 天左右胚芽出土，具不育叶 3 对，70 天左右方展出初生叶。金丝李胚根萌发后约 8～10 天胚芽出土，25～32 天初生叶展出。油山竹子萌发后 35 天初生叶展出。胚根与胚芽分别在种子的两端萌发，发芽后种壳中的子叶不脱落，不腐烂，苗木生长后逐渐变为苗根的基部或主侧根。多花山竹子的种子发芽及幼苗生长情况见图 2。

表 5 山竹子属树种的发芽能力及其测定条件（室外沙床）

树 种	温度（℃）	发芽势（%）		发芽率（%）	
		计算天数	一般数值	计算天数	一般数值
大苞藤黄	20～22	16	75	38	85
多花山竹子	22～24	—	不明显	180	85
岭南山竹子	22～24	15	72	34	90
金丝李	18～22	—	不明显	360	65
油山竹子	22～24	30	40	53	50

点播或条播。每平方米的播种量大约是大苞藤黄播种 500～700g，多花山竹子 250g，岭南山竹子 50～65g，金丝李和油山竹子 380～500g。覆土 1.5～2cm。大苞藤黄和金丝李的幼苗生长缓慢，需搭棚遮荫，培育 2～3 年方能出圃。多花山竹子、岭南山竹子及油山竹子 1 年生苗可以出圃。

（张声燕）

铁 力 木

Mesua ferrea L.

（山竹子科 Clusiaceae）

生长习性、分布和用途 铁力木属 3 种，我国只有本文描述的 1 种。常绿大乔木，高达 30m，胸径可达 3m。有板根。适生于热带和南亚热带肥沃湿润的酸性土。产滇、桂。粤、琼有栽培。南亚、东南亚亦产。珍贵硬木，为工业特用材。种仁含油约 78%，用于工业。嫩叶红色，树枝婆娑美观，可作风景树。

开花结实 约 15 年生开始结实，正常结实期在 30 年生以后。结实大小年间隔期为 1 年，但不甚明显。花两性，单朵或成对生于顶部枝条叶腋。花冠金黄色，径约 12cm。萼片及花瓣 4，覆瓦状排列。雄蕊多数，花丝纤细，短于花柱。子房上位，2 室，每室胚珠 2。花柱长，丝状，柱头盾状。年度之间开花结实的物候期差异很大。据广西南宁1983～1985 年观测，9 月下旬～10 月上旬为始花期，11 月中旬为盛花期，翌年 2 月下旬～3 月中旬果实成熟。但 1987～1988 年，同一株树的始花期在 5 月下旬，盛花期在 6 月中下旬，7 月下旬花期结束；9 月下旬果实开始成熟，10 月中下旬为成熟盛期。另据云南资料，4～7 月为开花期，9～12 月为果熟期。

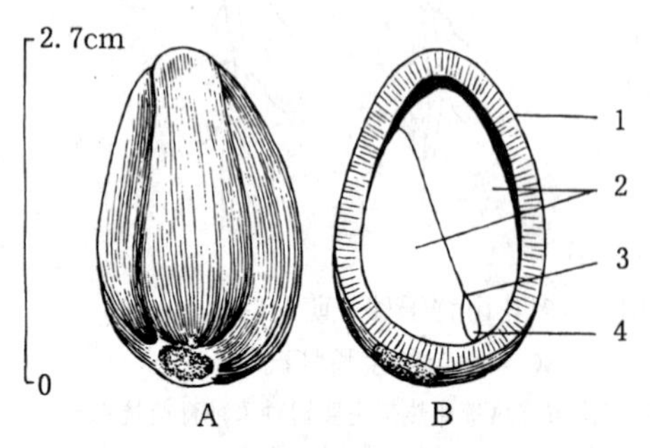

图 1 铁力木种子外形（A）及其纵切面（B）
1. 种皮 2. 子叶 3. 胚芽 4. 胚根
（黄应钦绘）

蒴果，未熟时黄绿色，成熟时深褐色；卵状球形，先端尖，基部为干枯的花被所包裹；长 2.6～

3.5cm，径2.5～3.5cm。果皮坚硬，革质，成熟后1～2个月成2～4瓣裂。种子1～4（5）粒。种子三棱状椭圆形、半圆形或不规则形，棕色；长1.9～2.9cm，宽1.5～2.3cm，厚0.7～1.6cm。种皮厚而坚硬。无胚乳，子叶大，富含油脂（图1）。

果实的采收调制和种子贮藏　果实成熟期较长，应分期分批采摘，或用采种钩刀将果实钩落地面捡拾。采得的果实可堆放2～3天，充分成熟后摊放于阴凉通风处或稍加曝晒，果壳开裂即得种子。出种率约30%。种子净度可达100%。含水量约28%。千粒重1 200～2 000g，每千克有纯净种子500～830粒。种子含油脂，不宜日晒，亦忌裸露贮藏，短期运输可在调制前装运。长途运输或贮藏则应在调制后混以湿沙。宜随采随播，贮藏期不宜超过半年。

发芽和播种　种子无休眠习性。发芽时日均温宜在20℃以上。经过贮藏的种子播后发芽不整齐，发芽期可持续2～3个月。新鲜种子随采随播则发芽快速整齐。1987年9月24日，广西林业科学研究所在室外沙床上用新鲜种子作过发芽测定：播种时日均温26℃上下，播后10天于10月3日开始发芽，10月10日进入盛期，23日发芽终止。从发芽之日起至高峰日的12天发芽百分数为60%；从播种之日起算，31天的发芽率为83%。留土萌发。胚根萌发后2～3天胚芽出土，10～12天展出初生叶。种子萌发及幼苗生长情况见图2。

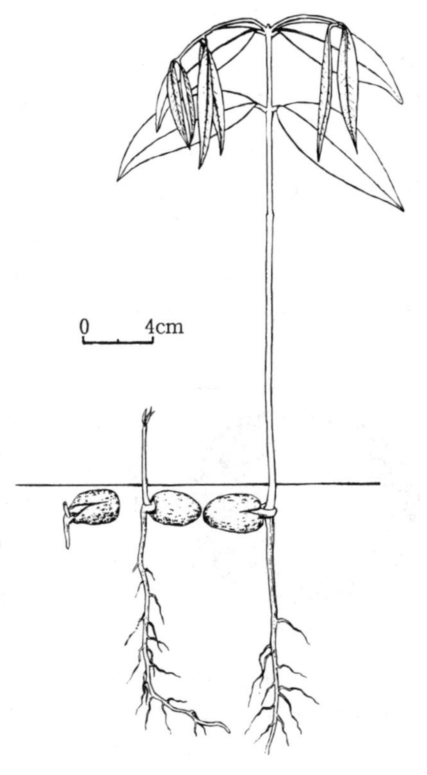

图2　铁力木种子萌发后第2、7、21天幼苗的生长情况

（黄应钦仿《热带亚热带主要树种采种育苗技术》）

点播。每平方米播种120～150g，覆土2cm。苗期需搭棚遮荫。2年生苗出圃。

（林榕庚）

肖　蒲　桃

Acmena acuminatissima（Blume）Merr. et Perry

（桃金娘科　Myrtaceae）

生长习性、分布和用途　肖蒲桃属约11种，我国只有本文描述的1种。常绿乔木，高20m，胸径可达60cm。对水肥条件要求不甚苛刻，肥力中等的酸性土壤生长良好。能耐0℃的极端低温及轻霜，但忌霜雪。分布于琼、粤、桂。印度、中南半岛、马来西亚、印度尼西亚、菲律宾亦有分布。材质优，作高级家具及细木工用。果可食。

开花结实　约7年生开始开花结实，正常结实期在15年生以后。结实大小年间隔期为1年，但不甚明显，正常年份每年均有结实。花两性。复聚伞花序，顶生，长4～8cm。花3朵集生。萼筒倒圆锥形，与子房合生。萼齿 4～5，不明显。花瓣 5，细小，白色。雄蕊多数，花丝极短，花药分叉，顶孔开裂。子房下位，2 室，每室有胚珠数颗。花柱短，基部较粗大。据南宁 1984～1987 年观测，花期及果成熟期均较长：5 月下旬始花，6 月上旬盛花，7 月上旬花期结束；9 月下旬早期开花的果实开始成熟，10 月上旬末～中旬为果熟盛期。各地果实的成熟期颇不一致，海南岛可迟至 12 月。

浆果，近球形，成熟时由青色转变为暗红色以至紫黑色；顶部有萼痕；果径 1.5～2.5cm，高 1.3～2.2cm。每果有种子 1 粒。种子近球形，青色，径 0.7～2.3cm，高 0.7～1.7cm。种皮极薄，与果皮粘合。有丰富胚乳。胚呈棒状，不分化。子叶藏于胚轴内，里面多裂（图 1）。

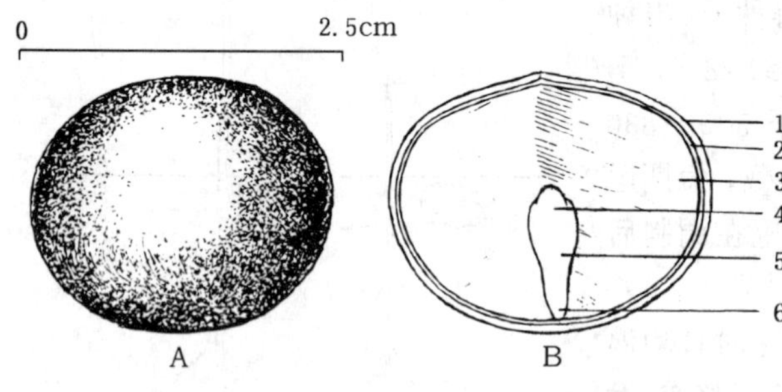

图 1 肖蒲桃浆果外形（A）及其纵切面（B）
1. 果皮 2. 种皮 3. 胚乳 4. 胚芽 5. 胚轴 6. 胚根
（黄应钦绘）

果实的采收调制和种子贮藏 果实成熟盛期用高枝剪带果穗剪下，也可用竹竿敲打或摇动树枝震落，在地面收集。果实易受虫蛀，采得的果实可浸水 2～3 天，杀虫并使果实充分软熟，搓去果皮，得出纯净种子。鲜果的出种率为 60%～69%。净度可以高达 99%。不影响种子发芽的含水量为 40%。千粒重 1 200（700～1 800）g，每千克纯净种子 800（600～1 400）粒。也可不洗去果皮，直接用果实播种或贮藏。3～5 天的短途运输可不洗去果皮，用竹筐装运，每筐以不超过 20kg 为宜。调制出来的种子运输时需混少量湿沙。种子不耐久藏，可以随采随播。常用的方法是混沙贮藏越冬，次年春暖后播种。贮藏期半年左右。超过次年 6 月，气温升高，贮藏的种子便会失去发芽能力。

发芽和播种 种子有短期休眠现象。发芽时日均温需在 20℃以上。1982 年 4 月上旬，广西林业科学研究所用沙藏越冬的种子，在室外沙床播种，播后约 1 个月开始发芽，5 月中旬至下旬为发芽盛期，6 月 1 日发芽终止。从发芽之日起算的 12 天，发芽百分数为 60%～70%；从播种之日起算的 55 天中，发芽率为 73%。留土萌发。胚根萌发后 2～3 天上胚轴出土，再过 4～5 天初生叶出现。

条播或点播。每平方米播种 60～75g，覆土 1～1.5cm。1 年生苗出圃。

（林榕庚）

红千层属

Callistemon R. Br.

（桃金娘科 Myrtaceae）

生长习性、分布和用途 本属约 20 种，我国引入 4 种，本文描述 3 种。常绿大灌木或小乔木。能耐轻霜，但忌霜雪或长时间 5℃的低温。对土壤肥力要求不严，能耐干旱贫瘠。原产澳大利亚，现全世界热带地区多有引种。花多为红色，形如瓶刷，颇美观，叶含芳香油。它们的名称、引种至我国的生长、栽培区和用途等情况见表 1。

表 1　红千层属树种的名称、生长、栽培区和用途

中　名	学　名	树高（m）	胸径（cm）	栽　培　区	用　途
红千层	*C. rigidus* R. Br.	4～6	10～15	粤、桂、滇、琼、闽	观赏、芳香油
柳叶红千层	*C. salignus* DC.	6～8	15～20	粤、桂、滇、琼、闽	观赏、芳香油
柳枝红千层	*C. viminalis* DC.	6～8	15～20	桂、粤	观赏、芳香油

开花结实　4～5 年生开始开花结实，8 年生以后进入正常结实期。结实大小年现象不明显，只要冬季无特殊寒害，次年均能正常结实。花两性，单生于苞片腋内，顶生于当年嫩枝梢，排列成穗状花序。开花后花序轴继续抽出新枝，苞片脱落。萼筒卵形，萼齿 5，脱落。花瓣 5，圆形。雄蕊多数，花丝长，本文所写 3 种均为红色，本属尚有少数种为黄色或白色。花药背着，药室平行，纵裂。子房下位，与萼筒合生，3～4 室，胚珠多数。花柱线形，柱头不扩大。在广西南宁观察到的开花结实物候见表 2。柳叶红千层在 9 月下旬～12 月下旬也有少量花开。

表 2　红千层属树种的开花结实物候期

树　种	观察地点	观察年份	开花期			果实成熟期	
			始　期	盛　期	末　期	始　期	盛　期
红千层	南宁	1978～1983	4 月中旬	5 月上旬	5 月中旬	8 月下旬	9 月中旬
柳叶红千层	南宁	1983～1987	3 月下旬	4 月上旬	4 月中旬	8 月中旬	9 月上旬
柳枝红千层	南宁	1983～1987	10 月上旬	10 月中旬	11 月上旬	4 月中旬	5 月上旬

蒴果，全部藏于萼管内，半球形，顶部截平或隆起，成熟时由红褐色转为褐色。果形较饱满。果实成熟后长时间不脱落，亦不开裂，宿存于树枝上。柳枝红千层和柳叶红千层的果实可存留 2～3 年，红千层可存留 10 年以上。每果有种子多颗。种子条状或四棱形，极小。种皮薄，无胚乳，胚直。发育饱满的种子与未授精的瘪粒不易区别。果实及种子的形态见表 3、图 1。

表 3　红千层属树种果实和种子的形态特征

树　种	未熟果颜色	果实			种子		
		形　状	大小（mm）	颜　色	形　状	大小（mm）	颜　色
红千层	灰绿色，被白毛	矩圆形，果缘稍平，果瓣微凹	高 5～6 径 7～8	灰褐色，被白毛	三棱状针形，先端尖	长 1～1.3 径约 0.2	棕褐色
柳叶红千层	绿色，被白毛	半球形，果缘稍平，有宿存萼	高 4～4.6 径 4.3～5	暗褐色或棕褐色	三棱状针形，先端尖	长 1～1.3 径约 0.15	黄褐色至棕褐色
柳枝红千层	浅赤褐色	半球形，果缘隆起	高 4～5.4 径 4.5～5.5	棕褐色至灰褐色	三棱状针形，先端钝	长 0.8～1.2 径 0.2～0.3	黄褐色至棕褐色

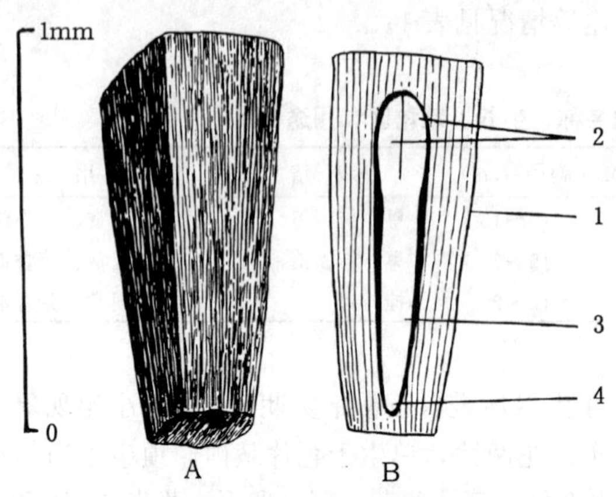

图1 红千层种子外形(A)及其纵切面(B)
1. 种皮 2. 子叶 3. 胚轴 4. 胚根
(黄应钦绘)

果实的采收调制和种子贮藏 一年四季均可采收。冬季可选用1～2年生的果枝,其余季节以2～3年生果枝的种子品质较优。着生4年以上的果实,种子仍有发芽能力,但发芽率较低。果无柄。柳叶红千层和柳枝红千层的枝条较稀疏,采集时宜用手或钩刀将果实从枝上剥下或刮下,不宜带枝条剪下。但有时果枝较密,例如红千层,也可剪下果枝。采集的果实垫上薄膜于无风天曝晒,果瓣开裂,经翻动或轻敲打后,种子即自行脱出,种子极轻细,瘪粒与发育饱满的种粒不易区分,播种材料中常含相当多的瘪粒。出种率和种子的质量数据见表4。

表4 红千层属树种果实的出种率和种子净度、质量

树 种	出种率(%)	千粒重(g)	每克种子粒数(万粒)
红千层	15.3	0.016～0.02	5～6.3
柳叶红千层	15.2	0.014～0.016	6.2～7.2
柳枝红千层	15.2	0.04～0.06	1.6～2.5

由于果实长期宿存树上不开裂,一般不作长期的种子贮藏,而在需要时随采随播。也可以将调制出来的种子晒干后装袋贮藏。常温条件下,种子的生活力可保持1年左右。贮存于5℃左右的低温条件下,保存期可达3年以上。

发芽和播种 种子无休眠习性。发芽时的日均温宜在20℃以上。播前种子可不作任何处理。1982和1987年,广西林业科学研究所用当年采后经过精选的种子在室内对3种红千层进行过发芽测定,发芽情况见表5。

表5 红千层属树种的发芽能力及其测定条件

树 种	发芽基质	室内温度(℃)	发芽势(%)		发芽率(%)	
			计算天数	一般数值	计算天数	一般数值
红千层	纸	26～28	5	14	13	20
柳叶红千层	纸	27～29	7	28	17	33
柳枝红千层	纸	27～29	5	10	15	16

出土萌发。种子萌发后约3天子叶出土。柳叶红千层和柳枝红千层子叶出土后23～28天发出初生叶，红千层则需经30～40天方始发出初生叶。如遇较低气温，子叶出土及初生叶发生期可延迟5～8天。种子萌发和幼苗生长情况见图2。

撒播。红千层和柳叶红千层每平方米育苗用种0.2～0.5g，柳枝红千层0.7～1.2g。播种床的整地要求特别细致，土壤粉碎过筛后耙平，稍加镇压。淋足水分后将种子均匀撒下，稀薄覆土，加盖薄膜，保温保湿并防止雨水冲击。生产上一般是将种子密播于透水的托盆或瓦盆内，待小苗稍木质化后移至圃地或容器内，培育1～2年生苗出圃。用于庭院绿化的红千层需培育3～4年生苗出圃。

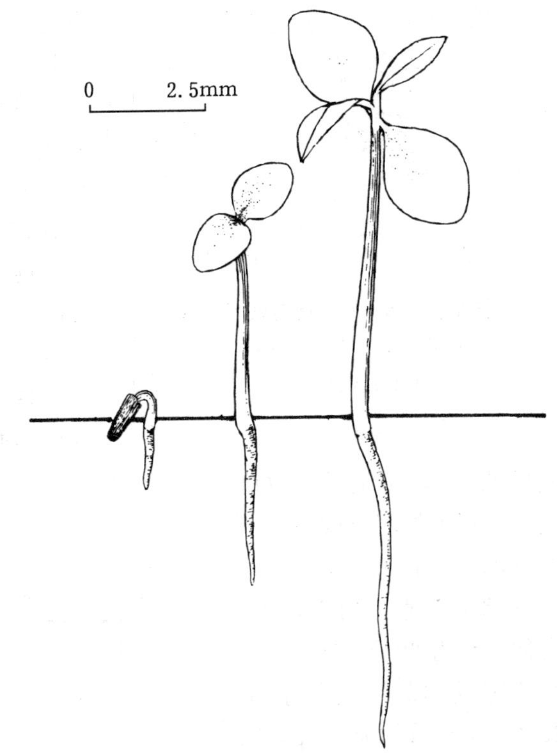

图2　红千层种子萌发后第1、8、40天幼苗的生长情况

（黄应钦仿《热带亚热带主要树种采种育苗技术》）

（林榕庚）

水翁（水榕）

Cleistocalyx operculatus（Roxb.）Merr. et Perry

（桃金娘科　Myrtaceae）

生长习性、分布和用途　水翁属约20余种，我国2种，本文描述1种。常绿乔木，高20～30m，胸径30～60cm。耐水湿，喜肥沃湿润酸性土。喜高温湿热环境，不耐0℃以下低温。分布于粤、琼、桂、滇。印度、中南半岛、印度尼西亚、大洋洲也有分布。为护堤防洪优良树种。材质优，花及叶可治感冒，根治黄疸性肝炎，嫩叶治乳腺炎。

开花结实　7～8年生开始结实，大小年间隔期一般为1年，但不甚明显。花两性，数朵组成聚伞花序，再排成圆锥花序生于无叶的老枝上，长6～12cm。萼筒半球形，长3mm。萼片连合成帽状体，长2～3mm，先端有短喙，花开放时整块脱落。花瓣4～5，分离，常附于帽状萼上一并脱落。雄蕊多数，长5～8mm，花药背着，纵裂。子房下位，2室，胚珠少数，花柱长3～5mm。据广西南宁1978～1984年观察，6月中旬为始花期，6月下旬为盛花期，7月

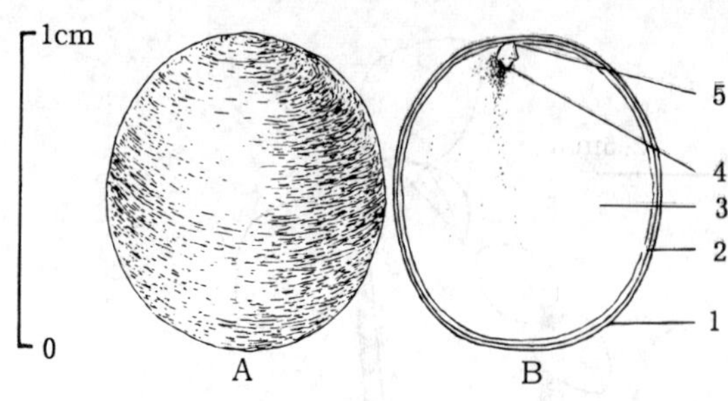

图1　水翁种子外形（A）及其纵切面（B）
1. 种皮　2. 胚乳　3. 子叶　4. 胚芽　5. 胚根
（黄应钦绘）

下旬为末花期，8月下旬果实开始成熟，9月上旬为果实成熟盛期。

浆果，宽卵圆形，顶端有残存的环状萼檐，未成熟时浅黄绿色，成熟时紫红色至紫黑色，高0.8～1.2cm，径1～1.4cm。成熟果在树上约存留7天，于9月上旬～10月中旬陆续脱落。每果有种子1粒。种子圆卵形，青色，径0.6～1.1cm。种皮薄，具胚乳，胚直，子叶肉质（图1）。

果实的采收调制和种子贮藏　果实成熟盛期，用竹竿敲打或摇动树枝，在地面捡拾。采得的果实忌堆沤，应及时装入竹筐置水中浸泡，用木棒轻捣或用手搓擦，淘去皮肉残渣即得纯净种子。生产上也常将果实直接用于播种。鲜果的出种率约为60%～70%。种子的净度可达100%。含水量50%～60%。千粒重约为310g，每千克纯净种子有2 900～4 000粒。种子忌失水，不耐贮藏。运输时宜将果实装入竹筐，每筐以不超过20kg为宜。已调制出来的种子运输时需混以湿沙。种子宜随采随播，不作贮藏。如当年不能播种，则需混以等量湿沙，贮藏于荫凉避光处，贮藏期一般为4个月。

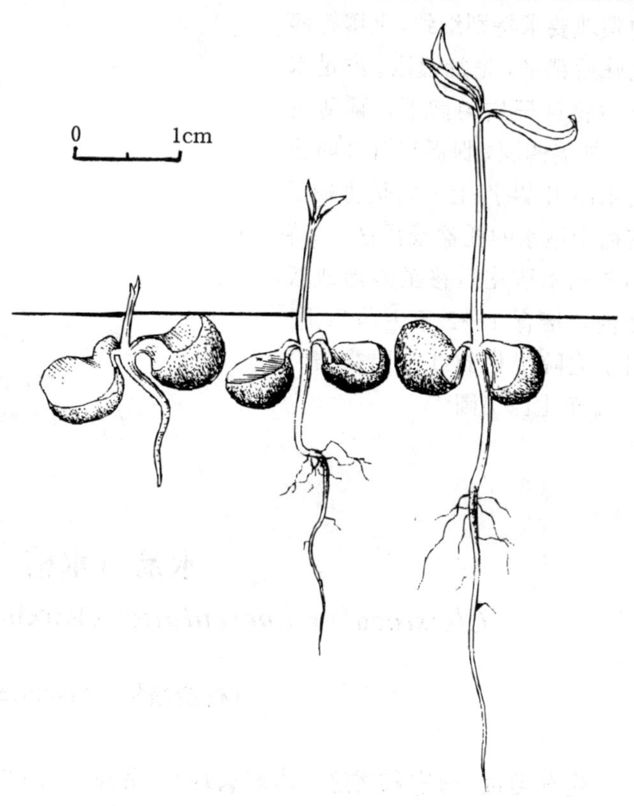

图2　水翁种子萌发后第3、7、10天幼苗的生长情况
（黄应钦绘）

发芽和播种　种子无休眠习性。发芽时日均温宜在20℃以上。1988年9月9日（气温27℃），广西林业科学研究所用当年新采而未作处理的种子在室外沙床作发芽测定：播后第2天开始发芽，随即进入萌发盛期，9月14日发芽终止。从开始发芽至发芽高峰日的3天，发芽百分数为90%，从播种至发芽终止的6天中发芽率为96%。留土萌发。播种后2天胚根萌发，第3天胚芽出土，7天左右初生叶出现（图2）。

9月播种。条播时，每平方米播种20～30g，次年5～6月出圃造林。也可以容器育苗，每容器播种1粒。

（林榕庚）

桉　属
Eucalyptus L' Herit

（桃金娘科　Myrtaceae）

生长习性、分布和用途　桉属约600种，广泛栽培的有40余种。我国引入200多种，本文描述20种。热带、亚热带常绿乔木，速生，喜光，深根，抗性强，病虫害少。适生于土层深厚疏松的酸性土。多数种要求年平均气温在18℃以上，最冷月平均气温不低于10℃。

桉属多数种产澳大利亚，少数种产菲律宾和新几内亚等岛屿。在全世界热带亚热带地方广为栽培。我国引种桉树始于1890年，早期引种地点为广东广州和广西龙州。目前桉树在我国最北缘已达北纬33°（甘肃省文县），最高海拔达2 000m（云贵高原）。全国已有16个省（自治区）栽培桉树，主要栽培区为华南地区。我国桉树造林面积仅次于巴西，居世界桉树人工林造林的第二位。

桉材主要用于造纸、建筑、矿柱、家具及薪柴等。副产品有精油、栲胶、黄酮、芦丁、蜂蜜等。桉树还广泛用于行道树及防护林带。桉属树种的名称、成年时的树高和胸径、我国栽培区和用途见表1。

表1　桉属树种的名称、成年时树高和胸径、我国栽培区和用途

中　名	学　名	树高（m）	胸径（cm）	栽培区	用　途	供　稿
黑桉	*E. aggregata* Decne. et Maid.	20～30	30～50	桂	材用、观赏	607
葡萄桉	*E. botryoides* Sm.	20～40	80～100	粤、桂、闽、川、滇、浙、赣、陕	材用、观赏	607
赤桉	*E. camaldulensis* Dehnh.	25～50	90～200	粤、桂、滇、川、湘、闽、鄂、浙、苏、陕、沪	材用、造纸、观赏	607
柠檬桉	*E. citriodora* Hook.	30～40	60～120	闽、粤、琼、桂、滇、川、浙、苏	材用、造纸、防风、精油、观赏	607
大花序桉	*E. cloeziana* F. v. Muell.	35～45	180～200	桂、粤、琼	材用、造纸	607
薄皮大叶桉	*E. crawfordi* Maid. et Blakely	25～30	40～60	粤、桂、川、闽、赣、湘	材用、造纸	608
窿缘桉	*E. exserta* F. v. Muell.	15～25	30～50	桂、粤、琼、滇、闽、赣、浙、川	材用、造纸、防风、药用	607

（续）

中 名	学 名	树高（m）	胸径（cm）	栽培区	用 途	供 稿
蓝桉	*E. globulus* Labill.	45～55	80～100	滇、川、黔、桂、赣	材用、精油、观赏、药用	607
巨桉	*E. grandis* Hill. ex Maid.	45～55	120～180	桂、粤、琼、闽	材用、造纸	607
斜脉胶桉	*E. kirtoniana* F. v. Muell.	25～30	40～50	粤、琼、桂、浙、湘、川、闽	材用、观赏	608
纤脉桉	*E. leptophleba* F. v. Muell.	12～18	20～30	粤、琼、桂、浙、湘、川	材用、观赏	608
斑皮桉	*E. maculata* Hook.	35～45	60～120	粤、琼、桂、闽、川、赣、苏	材用、造纸、观赏	608
直杆桉	*E. maideni* F. v. Muell.	45～60	70～100	滇、川、桂、粤	材用、造纸、精油、药用	607
粗皮桉	*E. pellita* F. v. Muell.	30～45	60～150	粤、琼、桂、滇、闽	材用	608
大叶桉	*E. robusta* Sm.	25～30	40～60	粤、桂、滇、黔、川、陕、湘、鄂、苏、闽、浙	材用、药用	608
柳桉	*E. saligna* Sm.	30～35	120～180	桂、粤、川、苏、闽	造纸、材用	607
谷桉	*E. smithii* R. T. Bak.	30～40	60～120	桂、粤、琼、闽	材用	607
细叶桉	*E. tereticornis* Sm.	30～45	90～150	粤、桂、赣、湘、川、浙、闽、滇、黔、陕	材用	608
托里桉	*E. torelliana* F. v. Muell.	20～30	20～35	桂、粤、闽	材用、观赏	607
尾叶桉	*E. urophylla* S. T. Blake	20～35	100～160	桂、粤、琼、闽	材用、造纸	608

开花结实 在华南通常 3～5 年生开始开花结实，比西南、华中等地稍早 1～2 年。8 年生以后进入正常结实年龄。一年四季都有不同的桉树开花，一般花期长 1～2 个月，蓝桉花期长达 3 个月。大小年现象不明显。开花结实情况大致有以下 5 类：

（1）当年孕蕾、开花、果熟的有窿缘桉、柳桉、巨桉、黑桉等。

（2）当年孕蕾、开花，次年果熟的有大叶桉、蓝桉、葡萄桉等。

（3）当年孕蕾，次年开花果熟的有细叶桉等。

（4）当年孕蕾，次年开花，第三年果熟的有直杆桉。

（5）1 年开花 2 次。其中一次为当年开花，当年果熟，另一次为当年开花，次年果熟，如赤桉、柠檬桉等。

花两性，淡黄色，通常腋生。伞形花序或圆锥花序，多为 3～10 朵，如窿缘桉、柠檬桉、赤桉、巨桉、柳桉、细叶桉、粗皮桉、谷桉、斜脉胶桉、黑桉、大叶桉、葡萄桉等。少数树种花单生，如蓝桉。萼筒钟形、倒圆锥形或半球形，基部与子房合生。帽状体外层为萼片，内层为花瓣，通常外层小而早落。雄蕊多数，呈数轮排列，多数分离。花药小，近基部着生或丁字着生于花丝之顶。药室 2 个平行或分离，顶端连合，纵裂或很少顶孔开裂。子房下位，3～6 室，每室有胚珠多数。广西南宁、广东雷州半岛、云南昆明等地 1968～1986 年间桉属树种开花结实的物候期见表 2。

表 2　桉属树种的开花结实物候期

树　种	观察地点	开　花			果实成熟	
		始　期	盛　期	末　期	始　期	盛　期
黑桉	广西南宁	7 月中旬	7 月下旬	8 月上旬	11 月中旬	12 月
葡萄桉	四川彭县	6 月下旬	7 月中旬	7 月下旬	翌年 6～7 月	
赤桉	广西南宁	第 1 次 3 月上旬	3 月中下旬	4 月上旬	10～11 月	
		第 2 次 10 月下旬	11 月中旬	11 月下旬	翌年 8～10 月	
柠檬桉	广西南宁	第 1 次 3 月中旬	4 月中旬	5 月上旬	9 月中旬	10 月
		第 2 次 10 月下旬～11 月			翌年 6～7 月	
大花序桉	广西南宁	5 月中旬	7 月	8 月中旬	7 月	9 月
薄皮大叶桉	广东梅县	5 月中旬	6 月	—	9 月	10 月
窿缘桉	广西南宁	6 月上旬	6 月下旬	7 月中旬	11 月中旬	12 月
蓝桉	云南昆明	9 月中旬	10 月上旬	10 月下旬	翌年 2 月上旬～3 月上旬	
巨桉	广西南宁	8 月上旬	8 月中旬	9 月上旬	11 月	12 月
斜脉胶桉	广东雷州	5 月	6 月	7 月	10 月	11 月
纤脉桉	广东雷州	—	9～11 月	—	翌年 5～6 月	
斑皮桉	广东雷州	—	12～1 月	—	翌年 6～8 月	
直杆桉	云南昆明	7 月上旬	7 月下旬	9 月中旬	翌年 2～3 月	
粗皮桉	广东雷州	—	8～9 月	—	翌年 5～6 月	
大叶桉	广东雷州	8 月上旬	8 月中旬	9 月中旬	翌年 6～7 月	
柳桉	广西南宁	7 月上旬	7 月中旬	8 月	11 月	12 月
谷桉	广西南宁	6 月上旬	6 月下旬	7 月中旬	11 月中旬	12 月
细叶桉	广东广州	—	1～4 月	—	6～7 月	
托里桉	广西南宁	3 月	4 月	—	7 月	8 月
尾叶桉	广东雷州	—	8～9 月	—	翌年 5～6 月	

蒴果，硬木质，半球形、卵形、坛形、钟形、圆锥形、圆筒形、梨形等。果实直径 0.6～2.5cm，顶端开裂为 3～6 个果瓣，果瓣内藏，或与萼筒平齐，或突出萼筒之外，果瓣位于果缘（果盘）的顶端。蒴果，成熟期依树种和分布地区不同而异，成熟的特征是果皮颜色由绿变

褐。少数受精的胚珠发育成种子，多数为“瘪粒”，混杂在种子中，比种子小而轻，色浅。蒴果成熟时的形态特征见表3。

表3 桉属树种的蒴果成熟特征

树种	长×宽（mm）	果形	果瓣	果缘
黑桉	10～12×8～9	钟形	4，略伸出	内陷
葡萄桉	7～9×7～9	圆筒形或枪筒形	4或5，下陷或与果盘平	窄且下凹或平
赤桉	6～8×5～6	半球形至球状截形	4，内曲伸出	宽，凸出
柠檬桉	10×10	坛形至卵形	3～4，下陷	宽，下凹
大花序桉	9×10	半球形	3～4，伸出或与果盘平	窄且平
薄皮大叶桉	9～11×6～9	钟形	稍突出	—
窿缘桉	5～9×6～11	半球形至阔圆锥形	3～5，伸出，弯曲	宽，凸起
蓝桉	10～15×15～25	球形至阔圆锥形，具有4条棱脊和瘤突	3～5，与果盘平齐	宽，凸起或平
巨桉	7～8×6～8	梨形至圆锥形、球形	5，顶端内曲并相应变钝，伸出或平	窄，下凹或略平
斜脉胶桉	7～10×10～14	钟形至梨形	4～5，伸出	窄至中等宽，平
纤脉桉	9～10×9～10	卵形，截头形至梨形	4，稍突出	小而不明显
斑皮桉	14～18×10～14	坛形至截卵形	3，深陷	宽，下凹
直杆桉	8～10×10～12	卵形至钟形，或圆锥形，具白霜	3～5，伸出或与果盘平	宽，平或微凸
粗皮桉	15～18×16～20	半球形至圆锥形，有2条棱脊	3～4，伸出	中，凸起或平
大叶桉	10～15×10～12	圆筒形至壶形	3～4，下陷或与果盘平	窄，下凹或平
柳桉	5～6×5～6	钟形或长椭圆形	3～4，顶端直或延伸并呈尖头，伸出或与果盘平	窄，下凹或略平
谷桉	6×5～6	截卵形，半球形或阔圆锥形	3或4，伸出	宽，略凸
细叶桉	6～9×8～10	半球形	4或5，伸出	宽，凸出
托里桉	10×15	壶形	3，下陷	宽，下凹
尾叶桉	6×8	杯形	—	—

多数种的种粒极细小。赤桉、窿缘桉、大叶桉、柳桉、巨桉、斜脉胶桉、尾叶桉、细叶桉等种子的千粒重仅0.122～0.460g，种子直径一般在1mm以下；少数种的种粒稍大，如蓝桉的千粒重2.8～3.4g，直径1.8mm；柠檬桉的千粒重4.5～4.7g，长约6mm，直径2～3mm。种皮薄而光滑，有光泽，黑色或暗棕色，有纹孔或形态各异的雕纹。无胚乳。胚具有2深裂、2浅裂或2全缘的子叶，折叠或扭曲在直生的胚轴上。种子形态见图1。

果实的采收调制和种子贮藏 蒴果成熟后在短期内一般不开裂，亦不脱落。除赤桉外，多数桉树种子成熟后可以存留在树上几个月甚至更长时间。大叶桉和蓝桉的种子能在树上存留1年以上也不完全散落。桉树的采种期因此并不严格要求与蒴果成熟期完全一致，但也不宜采收存留2年以上的蒴果。一般认为蒴果存留期长，种子会因养分损耗而质量降低。据广西的试验，大叶桉蒴果存留 3 年，经过精选，种子的发芽率也只有 65%，存留 2 年的为 85%，存留 1 年的则能达到 96.5%。窿缘桉蒴果存留 2 年后，种子发芽能力也明显降低。

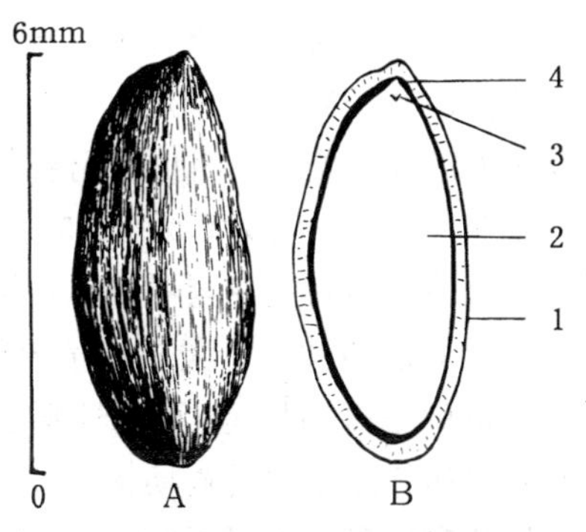

图 1　柠檬桉种子外形(A)及其纵切面(B)
1. 种皮　2. 子叶　3. 胚芽　4. 胚根
(黄应钦绘)

应当在无亲缘关系的优良林分中选择高大通直的优良单株采种。按母树编号分别采收，分别脱粒。同一个亚属的种间易于杂交，例如现在得到的柳窿桉（*E. saligna* × *E. exserta*）、巨尾桉（*E. grandis* × *E. urophylla*）等杂交种，表现出明显优势，已广泛应用于生产。通常用高枝剪或钩刀将果枝采下。由于干高枝脆，爬树采种时应注意安全并保护母树。同一枝条上常有不同成熟阶段的蒴果，也有种子早已散落的蒴果空壳，还有正在开放的花，要注意区分，采收发育良好的蒴果。

采回的蒴果铺放在竹垫或薄膜上，曝晒，厚约 3～5cm。有的蒴果当天就会脱粒，应立即收藏，以免连续曝晒而影响生活力。果瓣隆起或突出的蒴果，如窿缘桉、赤桉、斜脉胶桉、粗皮桉、细叶桉、谷桉等，在阳光下曝晒 6 小时，果瓣即大部分开裂，用木条轻敲，60%左右可以脱粒。果缘下陷的蒴果裂开较慢，有生命力的种子通常着生在蒴果底部，如柠檬桉、斑皮桉、托里桉、大叶桉等，则需晒 20 个小时以上，并反复敲打才能全部脱粒。除了有生命的种子以外，从蒴果中脱出来的还有没有受精的胚珠或中途败育的胚珠，可以统称“粉渣”。据分析，窿缘桉、大叶桉、葡萄桉和蓝桉的种子中，粉渣的含量分别为 91.4%、96.5%、64%和 40%。有些树种，例如赤桉、柠檬桉、蓝桉、大叶桉和柳桉，有生命力的种子同粉渣在颜色上有差异，如果确有需要，可以仔细地把它们区分出来。但是桉树种子颗粒一般都十分细小，很难分离。所以通常都用混有粉渣的种子作为播种材料。少数树种种子颗粒较大，可以扬去粉渣，得出纯净种子。部分树种每克播种材料中的可育性种子粒数见表 4。

表 4　桉属树种每克播种材料中可育性种子粒数

树　种	可育性种子粒数	树　种	可育性种子粒数
黑桉	810	纤脉桉	120
葡萄桉	507	斑皮桉	110
赤桉	770	直杆桉	110
柠檬桉	118	粗皮桉	70
大花序桉	140	大叶桉	560
薄皮大叶桉	220	柳桉	560
窿缘桉	360	谷桉	250
蓝桉	70	细叶桉	700
巨桉	815	托里桉	360
斜脉胶桉	550	尾叶桉	500

种子短期贮藏用布袋包装，挂放在室内通风干燥处。不同树种种子发芽能力的保存期各异。柠檬桉种子袋装于干燥处，6年还有10%以上的发芽率；同样条件贮藏的窿缘桉种子则完全丧失了发芽能力。四川省用袋装的大叶桉种子，存放3年发芽率仍有50%。据记载，赤桉种子存放7年，发芽率为28.5%；细叶桉种子贮存10年，也能保持同样高的发芽率。桉树种子不宜使用塑料袋密封贮藏。长期贮藏桉树种子，应当保持0～2℃的低温。据国外资料介绍，0～5℃的条件下，大多数桉树种子能贮藏10年。

少量种子短期邮运，可用塑料袋装，置于木盒内。大量种子短途调运宜用布袋，外套草包或麻袋。运输途中要防雨防潮。

发芽和播种 种子无休眠习性。播前一般不需处理。1986～1987年，广西林业科学研究所和广东雷州林业局曾对20种桉树作过室内发芽测定。测定时使用发芽皿，室温15～25℃，室内自然光照，基质是双层湿纸。用质量发芽率表示的测定结果见表5。

表5 桉属树种的发芽能力

树 种	每克种子的发芽粒数		开始发芽天数	终止发芽天数	温度（℃）
	一般数值	变动范围			
黑桉	657	—	7	10	25
葡萄桉	406	157～655	10	21	25
赤桉	670	205～1 135	5	10	30
柠檬桉	106	56～156	5	14	25～30
大花序桉	129	29～229	7	28	25
薄皮大叶桉	219	—	3	14	25～30
窿缘桉	267	94～440	5	21	25
蓝桉	73	35～111	5	14	20
巨桉	652	308～996	5	14	25
斜脉胶桉	440	—	5	14	25
纤脉桉	107	—	5	21	25
斑皮桉	109	70～148	5	14	25
直杆桉	85	—	5	14	20
粗皮桉	111	25～197	5	21	25
大叶桉	454	184～724	7	14	15～20
柳桉	538	225～851	5	14	25
谷桉	204	81～327	5	21	20～25
细叶桉	600	168～1 032	5	14	25～30
托里桉	323	213～433	5	14	25
尾叶桉	458	173～743	5	14	25

播种期在琼、粤、桂一般为1～2月，也有的在2～4月播种。滇、川、湘、赣的播种期多

在 2～4 月。育苗方式有苗床育苗和容器育苗 2 类。

出土萌发。桉属种子的萌发和幼苗生长过程大致可以分为以下几个时期：①发芽出土期，从种子发芽至子叶变绿。在冬末春初，这段时期长约 10～20 天，在夏季约 4～8 天。这段时期种子吸水发芽，组织幼嫩，要细致管护。②初生叶对生期，从着生 1 对初生叶至 5～6 对初生叶（苗高约 5～6cm）。在冬末春初这段时期长约 20 天，夏季约 10 天。③初生叶互生期，由初生叶对生转为互生（苗高约 6～11cm），在春季约 20～30 天。④速生期，由苗木枝叶在床面相接至苗高 18～20cm，每天可增高约 0.5～1.0cm，在春末夏初这段时期历时约 15 天。窿缘桉种子的萌发和幼苗生长情况见图 2。

苗木出圃规格：容器苗，苗高 20～25cm，苗龄 2～3 个月；裸根苗，苗高 70～150cm，苗龄 6～10 个月。

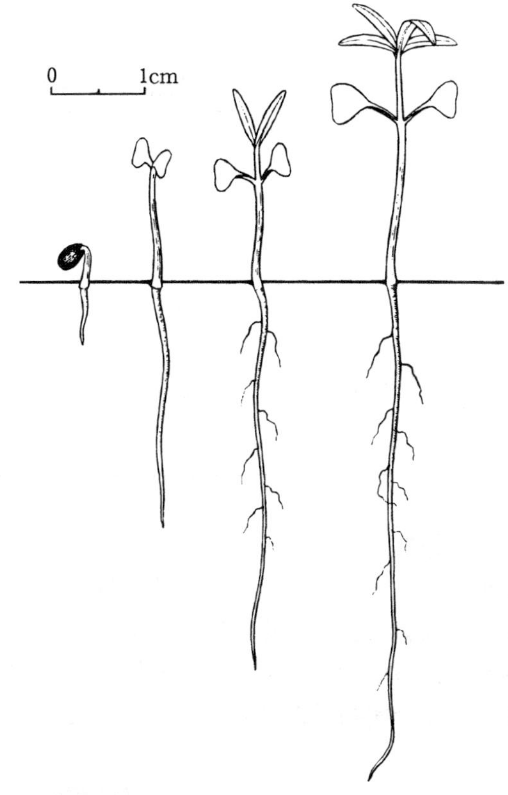

图 2　窿缘桉种子萌发后第 2、4、7、10 天幼苗生长情况
（黄应钦仿《热带亚热带主要树种采种育苗技术》）

（伍春魁）

番樱桃属
Eugenia L.

（桃金娘科　Myrtaceae）

生长习性、分布和用途　本属约 1 000 种，我国引入 4～5 种，本文描述 3 种。常绿灌木至小乔木，高 2～8m。原产热带美洲及印度尼西亚马鲁古群岛，世界热带各地广为引种。喜高温潮湿气候。气温降至 7℃以下，嫩叶和顶芽便开始受害。土壤以土层深厚、肥沃、排水良好、pH5～6 的沙壤土最为适宜。积水的沼泽地会引起根腐。这 3 个树种的名称、生长、分布及用途见表 1。

番樱桃所产的“丁香”为世界著名芳香材料，其花蕾、枝、叶和果实可直接利用或提取丁香油。油的主要成分为丁香酚（Eugenol，$C_{10}H_{12}O_2$），含量高达 84％～95％，可以提制香兰素，广泛用作食品、烟草的配香剂，医药上用作镇痛剂。中医以干燥花蕾入药称“ 公丁香”，近成熟的果实为“母丁香”。我国引种已有 30 多年的历史，目前在海南的儋县、屯昌、兴隆、七甲等地有较大面积的人工栽培。大叶番樱桃味淡，生产上不采用。

表 1 番樱桃属树种的名称、生长、分布和用途

中 名	学 名	树高（m）	胸径（cm）	分 布	用 途	供 稿
番樱桃（丁香）	*E. aromatica* Baill.	4～6	10～20	印度尼西亚。琼、滇引种	香料、药用治胃痛	610
大叶番樱桃（大叶丁香）	*E. caryophyllus* Merr. et Perry	4～6	10～20	印度尼西亚。琼、滇引种	香料、观赏、药用治霍乱、吐泻	610
红果仔	*E. uniflora* L.	2～6	10～20	巴西。琼、滇、粤、桂引种	食用、观赏	604

开花结实 番樱桃 4～6 年生开花结果，20 年生进入盛果期，经济寿命可达七八十年，每隔 4 年为一丰产年。红果仔 2～3 年生开始结实，无大小年习性。花两性。番樱桃为伞形花序腋生，花冠紫红色。红果仔花单生或数朵簇生于叶腋，花白色。萼筒短，萼齿 4。花瓣 4。雄蕊多数，药室平行，纵裂。子房下位，2～3 室，每室有多数横列胚珠。海南儋县及广西南宁观测的开花结实物候期见表 2。

表 2 番樱桃属树种的开花结实物候期

树 种	观察地点	观察年份	开花			果实成熟		果实脱落
			始 期	盛 期	末 期	始 期	盛 期	
番樱桃	儋县	—	2 月下旬	3 月中旬	4 月上旬	6 月	7 月上中旬	7 月下旬
大叶番樱桃	儋县	—	6 月下旬	7 月中旬	8 月中旬	9 月中旬	9～10 月	10 月下旬
红果仔	南宁	1984～1988	3 月上旬	3 月中旬～4 月上旬	4 月下旬	5 月下旬	6～7 月	8 月上旬

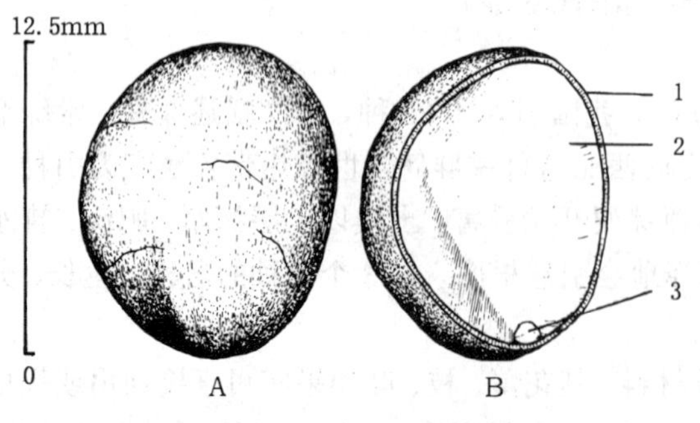

图 1 红果仔种子外形（A）及其纵切面（B）
1. 种皮 2. 胚乳 3. 胚（黄应钦绘）

浆果，近球形。番樱桃通常仅有 1 枚发育的种子。种子椭圆形，绿紫色，表面多皱纹，有 1 条明显的纵沟。红果仔果实成熟时由青色转为鲜红色，表面有光泽，每果有种子 1～3 粒；种子扁圆形或近三棱形，乳白色，径 0.7～1.4cm，厚 0.5～1cm；种皮薄，胚乳丰富，胚伸直，不分化，子叶藏于胚轴内。红果仔种子的外观和解剖结构形态见图 1。

果实的采收调制和种子贮藏　商品“丁香”收获花蕾和果实，采收期因产品不同和各地气候不同而异，一般花芽出现后6个月，花蕾含苞待放时即可采收，在海南儋县为3～4月，屯昌为5～7月。鲜花蕾晒4～6天至干脆即为商品的“公丁香”，成熟的果实干燥后为商品“母丁香”。供育苗用的应在果熟盛期摇动树枝震落，在地面捡拾，红果仔用手采摘。采收的果实装入布袋或竹箩，置水中搓擦，淘去果皮、果肉等杂质即得种子。出种率、净度、质量等见表3。

表3　番樱桃属树种的出种率和种子净度、质量

树　种	出种率（%）	净度（%）	千粒重（g）		每千克纯净种子粒数（粒）	
			一　般	变动范围	一　般	变动范围
番樱桃	24.4	92	400	300～700	2 450	1 430～3 330
大叶番樱桃	25.7	90	1 450	1 400～1 500	690	680～720
红果仔	15～20	92	630	400～800	1 580	1 200～2 500

种子的含水量约为40%，忌脱水，不宜日晒。贮藏时需混湿沙，贮藏期一般在1个月以内。据程必强和马信祥（1987）报道，从马来西亚空运来的番樱桃种子，10天以内的发芽率在90%以上，大约30天即降为8.6%。

发芽和播种　种子无休眠习性。发芽时日均温宜在25℃以上。1987年6月11日和1988年10月14日，华南热带作物研究所在室外沙床对番樱桃和大叶番樱桃作过发芽测定；广西林业科学研究所1987年6月8日在室外沙床上对红果仔作过发芽测定，结果见表4。

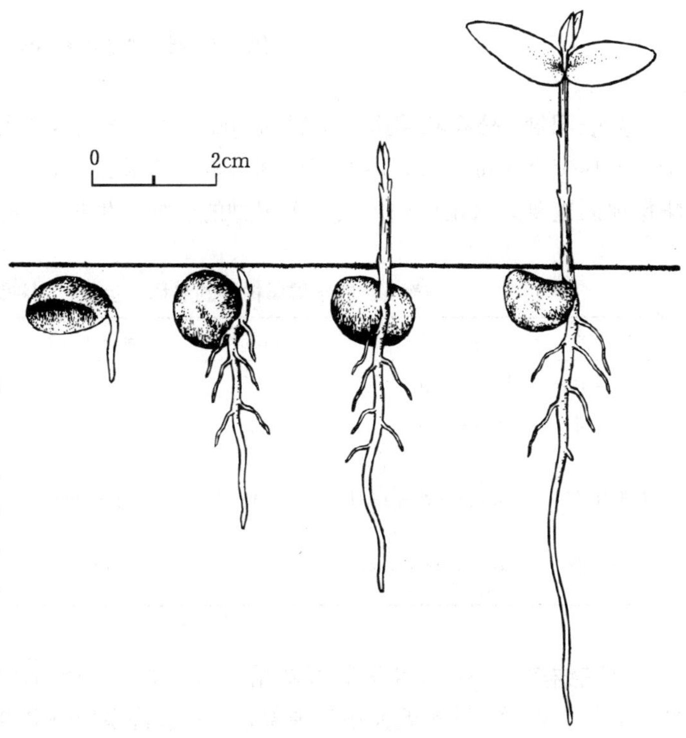

图2　红果仔种子萌发后第1、3、5、8天幼苗的生长情况

（黄应钦绘）

番樱桃和大叶番樱桃出土萌发，红果仔留土萌发。番樱桃种子萌动后14天子叶出土，17天发出初生叶，23～25天初生叶展开。大叶番樱桃种子萌动后8天发出初生叶，21天初生叶展开。红果仔种子萌动后3天上胚轴出土，7～8天发出初生叶，幼苗具不育叶6～8片。红果仔种子的萌发和幼苗早期生长情况见图2。

表 4 番樱桃属树种的发芽能力及其测定条件（室外沙床）

树 种	温度（℃）	发芽势（%）		发芽率（%）	
		计算天数	一般数值	计算天数	一般数值
番樱桃	28～30	12	66	38	88
大叶番樱桃	28～30	—	—	42	60
红果仔	28～30	15	50	37	70

点播，株行距 10cm×10cm。种子平放，使胚根向下。每平方米的播种量：番樱桃和红果仔 38～50g，大叶番樱桃 75～120g。覆土 0.3～0.5cm。培育半年或 1 年生苗出圃。

（丁慎言）

白千层属

Melaleuca L.

（桃金娘科 Myrtaceae）

生长习性、分布和用途 本属约 100 种，我国引入 6 种，本文描述 4 种。常绿乔木。适生于南亚热带至热带地区，原产大洋洲各地。喜酸性土，能耐干旱贫瘠的土壤及渍水地。我国华南地区重要的绿化树种。这 4 个树种的名称、生长、我国的引种栽培区及用途见表 1。

表 1 白千层属树种的名称、生长、引种栽培区和用途

中 名	学 名	树高（m）	胸径（cm）	引种栽培区	用 途
垂枝白千层	*M. armillaris*	18	25	广西	观赏、防风
白千层	*M. leucadendra* L.	25	40～50	广西、广东、海南、福建、台湾	观赏、香料、药用、防风
细花白千层	*M. parviflora* Lindl.	16	20～30	广西、广东	观赏
白树油树	*M. quinquenervia* S. T. Blake	25	35～40	广西、广东、福建	药用、香料

开花结实 4～6 年生开始开花结实，5～7 年生母树所结的种子即可育苗造林。正常结实期在 7 年生以后。结实无大小年现象，除冬季特寒影响开花外，其余年份均能正常开花结实。花两性，无柄，着生于当年生的嫩枝上，成穗状排列。花序轴无限生长，开花后继续长出新的枝叶。萼筒近球形或钟形。萼片 5。花瓣 5。雄蕊多数，绿白色。花丝基部合生成 5 束，并与花瓣对生。花药背部着生，药室平行，纵裂。子房下位，与萼筒合生，3 室，每室胚珠多数。花柱线形，柱头多少扩大。花期较长。白千层和白树油树每年开花结实 2 次。垂枝白千层和细花白千层开花期虽较长，但花期较整齐，每年只开花 1 次。广西南宁观测的开花结实物候期见表 2。

表 2　白千层属树种的开花结实物候期

树　种	观察地点	观察年份	开　花			果实成熟		果实开裂习性
			始　期	盛　期	末　期	始　期	盛　期	
垂枝白千层	南宁	1986～1987	12 月下旬	翌年 1 月下旬	2 月上旬	6 月下旬	7 月上旬	熟后 1 个月内顶端开裂
白千层	南宁	1978～1984	3 月中旬	4 月中旬	4 月下旬	9 月上旬	9 月下旬	数年不开裂不脱落
细花白千层	南宁	1983～1985	4 月上旬	4 月中旬	5 月上旬	9 月中旬	9 月下旬	1 年内不开裂
白树油树	南宁	1978～1983	7 月中旬	9 月中旬	11 月下旬	翌年 3 月下旬	4 月上旬	熟后 1 个月左右开裂

除了表 2 所列的主花期外，白树油树小造果为 4 月开花，9 月成熟；白千层小造果为 9 月下旬～12 月上旬开花，翌年 4 月上旬～5 月中旬成熟。12 月中旬～翌年 2 月上旬白千层还会有少数植株开花。

蒴果，包藏于宿存的萼筒内，半球形或近球形，成熟时由暗红色转为灰褐色。垂枝白千层和白树油树成熟后 1 个月左右顶端开裂，果实脱落。其余 2 个种成熟后长期不脱落，亦不开裂。种子细小，近三角形。种皮薄，无胚乳，胚直。果实和种子的形态见表 3。

表 3　白千层属树种果实和种子的形态特征

树　种	果　实			种　子		
	形　状	大小（mm）	色　泽	形　状	大小（mm）	色　泽
垂枝白千层	半球形，果缘隆起	高 2.8～3.5 径 3～3.8	灰褐色	刺状楔形	长约 1 径约 0.1	黄褐色至棕褐色
白千层	半球形，果缘突起	高 4～4.5 径 5～5.4	深灰色或灰褐色	刺状楔形	长约 1 径约 0.1	黄褐色至棕褐色
细花白千层	杯状，果缘突起	高 2.6～3.2 径 2.6～3.2	浅褐色或灰褐色	刺状楔形	长 0.6～1 径约 0.1	黄褐色至棕褐色
白树油树	半球形，果缘隆起	高 3.5～4.2 径 3.6～4.4	灰褐色	刺状楔形	长约 1 径约 0.1	黄褐色至棕褐色

果实的采收调制和种子贮藏　采收时宜用高枝剪剪下顶端果枝，也可以上树用枝剪剪取果枝，不能采割较粗的枝条。应注意选择剪取树冠中部东南向的果枝，因为树冠内部以及下垂纤弱的果枝，果实中瘪粒较多。采回的果枝剪去顶部枝叶，带果轴置日光下曝晒，俟顶端开裂后翻动敲打，脱出种子，除去果壳等粗杂物即得播种材料。饱满种子与瘪粒不易区分，生产上通常不再精选而直接使用。不同年份种子质量的差异较大。授粉较差的年份，种子的质量及发芽率均明显降低。正常年份果实的出种率（含瘪粒）及种子质量等数据见表 4。

表 4　白千层属树种果实出种率和种子质量（含瘪粒）

树　种	出种率（%）	千粒重（g）	每克纯净种子数（万粒）
垂枝白千层	12.2	0.015～0.03	3.3～7
白千层	10.6	0.03～0.04	2.5～3.3
细花白千层	11	0.015～0.03	3.3～7
白树油树	10	0.025～0.035	2.8～4

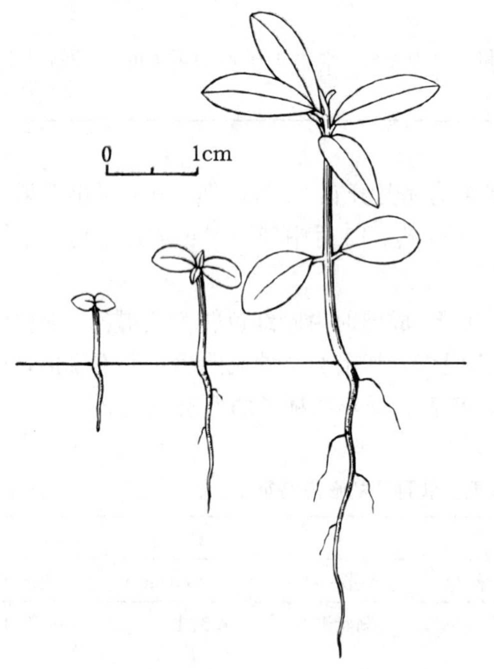

图 1　白千层种子萌发后第 3、30、45 天幼苗的生长情况
（黄应钦绘）

种子晒干后贮藏，也可随采随播。在常温条件下，用袋或其它容器盛装后置于室内通风干燥的阴凉处，生活力可保持 1 年。在 5℃左右的低温条件下贮藏，生活力可保持 3 年以上。

发芽和播种　种子无休眠习性。发芽时的日均温宜在 16℃以上。10 年生左右的母树种子发芽率比 20 年生以上母树的高 5%～10%。不同年份所采种子的发芽能力略有差异。播种前可不作任何处理。1987 年 9 月中旬，广西林业科学研究所用当年新采且经过精选的种子，在室内自然光照，日均温 28℃上下的条件下，对 3 种白千层作过发芽测定。垂枝白千层为 5 年生母树，白千层与白树油树的种子采自 20 年生母树。播种后垂枝白千层 3 天开始萌发，10 天发芽结束，其余 2 个种 4～5 天开始萌发，12～13 天发芽结束，发芽进程较垂枝白千层明显慢（表 5）。出土萌发。发芽后 2～3 天子叶出土，再过 30～34 天发出初生叶（图 1）。

表 5　白千层属树种的发芽能力及其测定条件

树　种	基　质	温度（℃）	发芽势（%）		发芽率（%）	
			计算天数	一般数值	计算天数	一般数值
垂枝白千层	滤纸	28	5	53	13	62
白千层	滤纸	28	—	不明显	16	28
白树油树	滤纸	28	7	43	18	48

撒播。播后覆盖薄膜防雨。也可将种子播于透水的浅口瓦盆或塑料盆内，出土后 20 天幼

苗开始转木质化时移植至场圃。1年生苗出圃。用于城镇绿化的需培育2～3年生苗出圃。

（林榕庚）

番　石　榴

Psidium guajava L.

（桃金娘科　Myrtaceae）

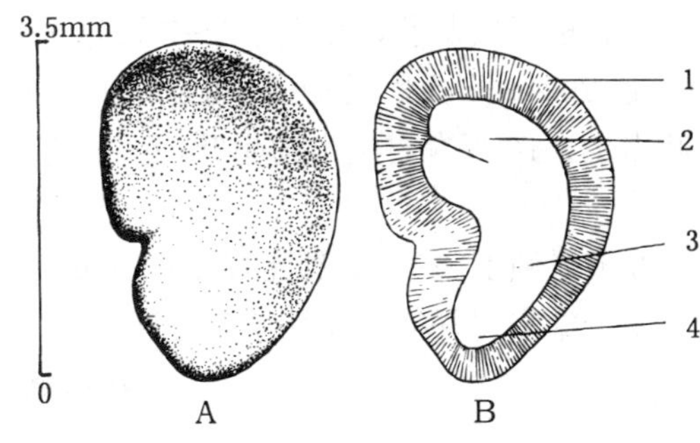

图1　番石榴种子外形（A）及其纵切面（B）
1. 种皮　2. 子叶　3. 胚轴　4. 胚根
（黄应钦绘）

生长习性、分布和用途　番石榴属约150种，我国引入2种，本文描述1种，已逸为野生种。常绿灌木或乔木，高可达13m，分枝低，地径可达20cm。能耐0℃左右低温及轻霜。原产南美洲。广东、广西、云南、四川、福建、台湾北回归线以南地区栽培。广西右江干热河谷出现野生纯林。栽培品种较多，引种历史久。喜肥沃湿润的酸性土。果供食用，嫩叶治腹泻，叶可代茶。

开花结实　实生树6年生左右开始开花结实，8年生以后进入正常结实年龄。无性繁殖的植株2年生开花结实。结实无大小年现象，水肥条件较好的立地可以连年丰产。花两性，通常1～3朵腋生。萼筒钟形，萼帽不规则4～5裂。花瓣4～5，白色。雄蕊多数，分离，排成多列，着生于花盘上。药椭圆形，近基部着生，药室平行，纵裂。子房下位，与萼筒合生，4～5室，胚珠多数。花柱线形，与雄蕊同长，柱头扩大。花果期均较长。据南宁1984～1987年观测，3月下旬～4月下旬花蕾陆续形成；5月上旬为开花始期，5月中旬至6月上旬为开花盛期，6月下旬为末期；7月上旬起果实陆续成熟，8月上中旬为盛果期；8月下旬仍有少数果实成熟；9月中旬又有少数花开，但所结果实不丰满。浆果，多肉，近球形或梨形，径3～5cm，长3～8cm，顶有宿存萼片；成熟时由青色转变为黄绿色或浅红色，果肉白色或黄色。成熟果在树上保持3～4天即脱落。每果有种子多颗。种子近肾形，稍弯曲，白色至黄色或胭脂红色，长3.2～3.8mm，宽2～3mm，厚1.5～2mm。种皮厚，坚硬。无胚乳。胚弯曲，肾形，胚轴长，子叶短（图1）。

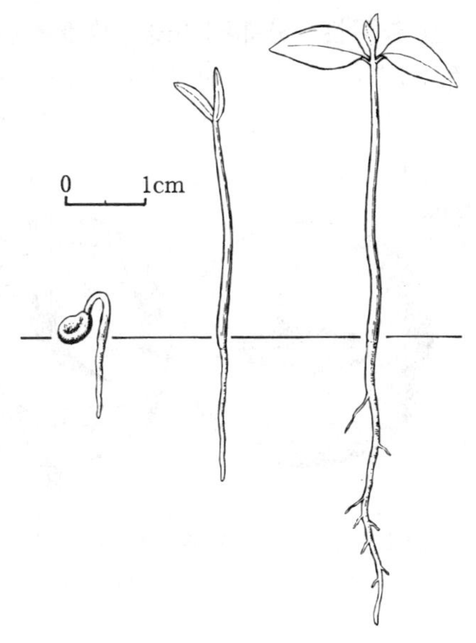

图2　番石榴种子萌发后第3、6、10天幼苗的生长情况
（黄应钦绘）

果实的采收调制和种子贮藏　果实成熟或接近

成熟时用手采摘，食去果肉，将果实中心部分的肉质胎座和种子堆沤 1～2 天，装入竹筐或布袋内置水中反复搓洗，淘去果瓤等杂质即得纯净种子。也可用过熟不堪食用的果实置水中搓洗，淘去果肉，得出种子。鲜果按质量计算的出种率约为 4%。种子净度可达 98%。千粒重约为 9.8g，每千克纯净种子粒数约9.5 万～10.5 万粒。种子的含水量 15%～20%，忌日晒，不宜失水。种子裸露存放 1 周左右即失去发芽能力。短期运输可以果实装运。已经调制洗净的种子，贮藏或运输时需混湿沙。混沙贮藏的时间一般不超过 3 个月。

发芽和播种 种子无休眠习性。可随采随播或混沙贮藏 1～2 个月播种。1987 年 9 月上旬，广西林业科学研究所用室外沙床对 8 月上旬采集经湿沙贮藏的种子作过发芽测定：播前用始温 50℃水浸种，播时日均温 28℃；播种后第 11 天胚根开始萌发，萌发后 2～7 天为发芽盛期；从开始发芽至高峰日的 10 天中，发芽百分数已达 70%；从播种至发芽终止的 28 天中，发芽率为 78%。出土萌发。种子萌芽后 4～5 天子叶出土，10 天左右发出初生叶，生长情况见图 2。

撒播。播前用始温 50℃水浸种 2～3 小时。每平方米播种 1.2～2g，覆土厚 1～2mm。出苗后移至容器或苗床培育。3 个月或 1 年生苗出圃。生产中多用扦插等无性繁殖方法育苗。

（林榕庚）

玫瑰木（三脉木）

Rhodamnia dumetorum（Poir.）Merr. et Perry

（桃金娘科 Myrtaceae）

生长习性、分布和用途 玫瑰木属约 20 种，我国有 1 种和 1 变种，本文描述 1 种。常绿小乔木，高 8～10m，胸径 10～15cm。生长较慢，颇耐荫蔽。适生于无霜冻的热带地区。对土壤肥力要求不苛，肥力中等的酸性土均能生长。产海南。广西有栽培。中南半岛、马来西亚亦产。材质坚重耐腐，为工业用材。枝叶繁茂，涵养水源能力强。

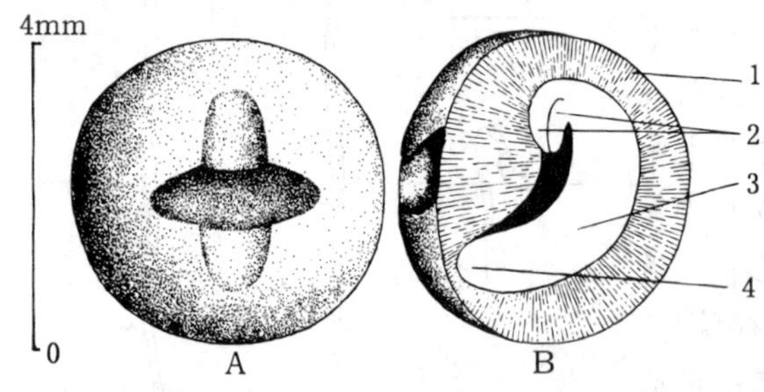

图 1 玫瑰木种子外形（A）及其纵切面（B）
1. 种皮 2. 子叶 3. 胚轴 4. 胚根
（黄应钦绘）

开花结实 6～8 年生开始开花结实，正常结实期在 15 年生以后。结实大小年现象不明显，一般年份均能正常结实。花两性，白色，常 3 朵排成腋生聚伞花序。萼筒卵形，萼片 4，宿存。花瓣 4。雄蕊多数，分离，排成多列。花药背部着生，药室纵裂。子房下位，与萼筒合生，1 室，有 2 个侧膜胎座。花柱线形，柱头盾状，胚珠多数。花果期均较长。据海南 1989 年观测，5 月中旬始花，5 月下旬～6 月中旬为盛花期，7 月上旬仍有少量花开，8 月中旬果实开始成熟。浆果，卵球形，冠以宿存的萼片。果成熟时由青色转变为暗褐色，径 6～8mm。种子棱状圆形、半圆形或不规则形，高 3～5mm，径 2～3mm。种皮厚而坚硬。无胚乳。胚弯曲，肾形。胚轴长，子叶短，胚根粗。种子的外形和解剖结构见图 1。

果实的采收调制和种子贮藏　果实成熟后易遭鸟啄食。果实大部分成熟后应及时用采种钩刀钩取，或直接采摘果穗。采回后摘下果实置室内堆沤 2～3 天，果皮充分软熟后置竹筐内，在水中反复搓擦，淘净果皮果肉等杂质，即得纯净种子。种子不宜脱水，忌日晒。处理出的种子在室内摊放晾干后，即可播种或贮藏。种子的净度可达 96%～98%。含水量约为 17%。千粒重 10～13g，每千克有纯净种子 7.7 万～10 万粒。种子因授粉不足而有部分空粒，与正常种子较难区别。宜随采随播，一般不作贮藏。如越冬后播种则需混湿沙贮藏于室内荫凉处，时间以半年左右为限，超过 8 个月，发芽能力便明显下降。沙藏的种子次年开春后即应播种，不宜迟至 5 月以后。

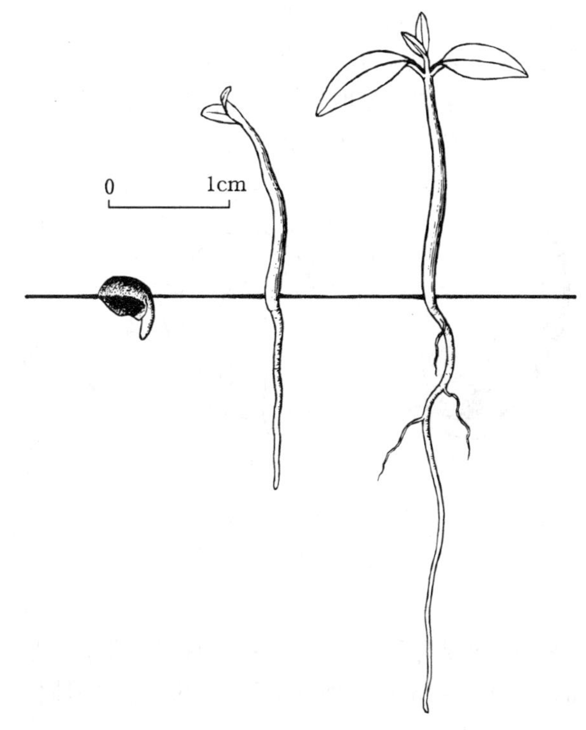

图 2　玫瑰木种子萌发后第 1、3、12 天幼苗的生长情况
（黄应钦绘）

发芽和播种　种子无休眠习性。1987 年 9 月 15 日，广西林业科学研究所在室内自然光照条件下，对当年新采的种子用培养皿作过发芽测定。置床时日均温 25℃，置床后 15 天胚根开始萌发，发芽后 13 天进入盛期，发芽全程为 23 天。从发芽之日起算的 18 天中，发芽百分数为 45%；从播种之日起算，以 38 天计，发芽率为 58%。出土萌发。发芽后 3 天子叶出土，再过 12 天发出初生叶，生长情况见图 2。

撒播。每平方米约播 10g，覆土 0.1cm。混沙贮藏的种子也可带沙播种。长出初生叶后移入容器或大田培育。也可条播，每平方米播种 2.5～4g。容器苗约 3 个月出圃。

（林榕庚）

桃　金　娘

Rhodomyrtus tomentosa（Ait.）Hassk.

（桃金娘科　Myrtaceae）

生长习性、分布和用途　桃金娘属约 20 种，我国只有本文描述的 1 种。常绿灌木，高约 2m。萌芽力强。耐贫瘠，喜生于低丘缓坡地，为酸性土指示植物。分布于台湾、福建、广东、海南、广西、云南、贵州及湖南最南部。南亚和东南亚亦产。花大而美丽。果可食，也可酿酒。根治风湿、肝炎，并可降低血脂。

开花结实　实生树3～4年生开始开花结实，正常结实期在6年生以后，结实无大小年现象。砍伐后的萌生树1年即可结实，2年后正常结实。花两性，玫瑰红色，径2～4cm，1～3朵簇生于叶腋。萼筒倒卵形，萼裂片5，革质，宿存。花瓣5。雄蕊多数，红色，分离，排成多列。花药背部及近基部着生，纵裂。子房下位，与萼筒合生，3室，每室有胚珠2列。花柱线形，长1cm，柱头头状。据广西南宁1988～1990年观测，5月上旬为始花期，5月中旬为盛花期，6月下旬为末花期；7月下旬果实开始成熟，8月上旬为果熟盛期。浆果，卵状壶形，外表被短绒毛，成熟时由青色转为红色，充分成熟时为紫黑色，成熟后在树上可存留4～6天，随后自行脱落。少数果在树上不脱落，逐渐干枯。果长1.5～2cm，径1～1.5cm，每室有种子2列，每果有种子60～80粒。种子扁平，近圆形或椭圆形，灰褐色至深褐色，径2～2.7mm，厚0.7～1mm。种皮坚硬。无胚乳。胚轴长，胚马蹄形弯曲，子叶短。种子的外形和解剖结构见图1。

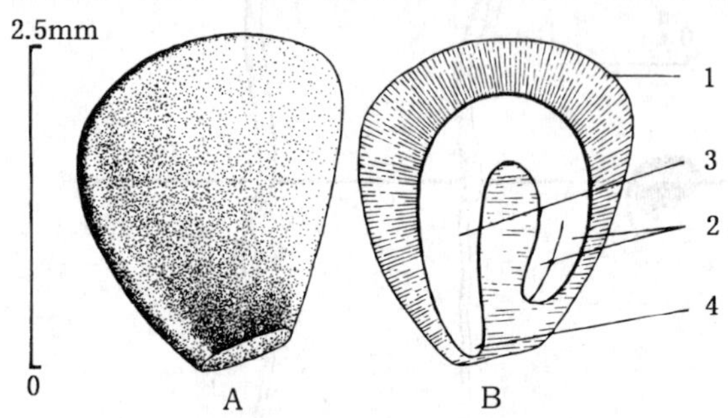

图1　桃金娘种子外形（A）及其纵切面（B）
1. 种皮　2. 子叶　3. 胚轴　4. 胚根
（黄应钦绘）

果实的采收调制和种子贮藏　果皮转为红色或紫色时即可采集。已经软熟的果实可以随采随调制。尚未软熟的果实采回后堆放1～2天，使果肉软化。处理方法可以有2种：一是吃去果肉，留下种子，浸水洗净后供播种。另一种方法是将果实洗净后用布袋包裹，挤出果汁供饮用，然后淘去杂质，得到纯净种子。也可以带果肉一起捣烂，置于水中搓洗，淘出种子。鲜果的出种率约为10%。种子的净度可达97%～99%。千粒重9～11g，每千克有纯净种子9万～11万粒。种子忌日晒，不能脱水，宜在采种的当年夏秋季，气温尚未转低时即播种，不做长期贮藏。如当年不能播种，贮藏时需混以湿沙，存放于干爽荫凉处。混沙贮藏的种子生活力约可保持半年，至次年春暖后播种，不宜

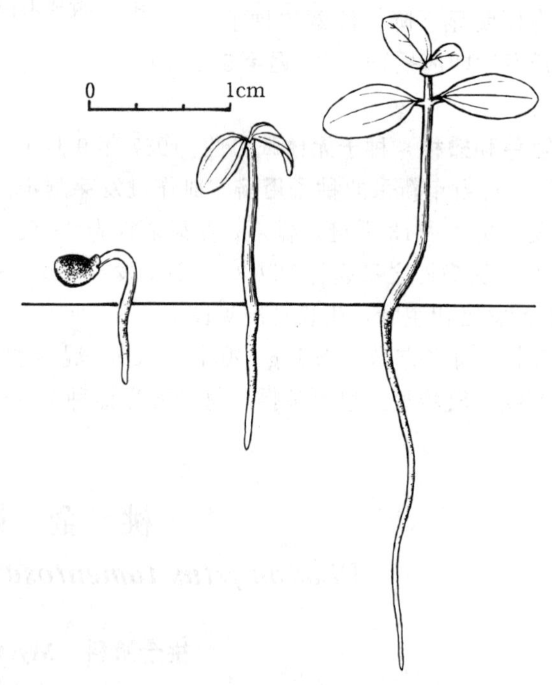

图2　桃金娘种子萌发后第3、15、45天幼苗的生长情况
（黄应钦绘）

贮藏越夏。袋装干藏或其它容器干藏的种子，约1月后即失去发芽能力。

发芽和播种　种子有短暂的休眠现象。发芽时日均温宜在20℃以上。1987年8月中旬，广西林业科学研究所在室内自然光照条件下，用发芽皿作过发芽测定：置床时日均温26～28℃，置床后47天胚根才开始萌发，无明显的发芽盛期；从发芽之日起算，26天中的发芽百分数为35%；发芽过程一直延续至11月上旬结束，以82天计，发芽率为49%。出土萌发。发芽后约3天子叶伸出，再过42天发出初生叶，生长情况见图2。

撒播。由于种子有短暂休眠，一般不宜随采随播。播前种子需作催芽处理，用湿润细沙或吸水纸层积催芽均可，至种子开始萌动时撒播。每平方米播种5～10g。幼苗转木质化后移至圃地或容器继续培育。3个月或1年生苗出圃。

（林榕庚）

蒲　桃　属
Syzygium Gaertn.

（桃金娘科　Myrtaceae）

生长习性、分布和用途　本属约500余种，我国70余种，本文描述7种。常绿灌木或乔木。对土壤要求不甚苛刻，但贫瘠地生长不良，耐水湿。适生于疏松湿润的酸性土，少数种也适于钙质土。主要分布在亚洲热带，少数种在大洋洲，非洲也有分布。我国主要分布区为粤、桂、滇、琼、闽，少数耐寒种也可分布到长江以南。为重要的绿化、水土保持及用材树种。它们的名称、生长、分布和用途见表1。

表1　蒲桃属树种的名称、生长、分布和用途

中　名	学　名	树高(m)	胸径(cm)	分　布	用　途
线枝蒲桃	*S. araiocladum* Merr. et Perry	10	15～20	琼、粤、桂。越南	绿化、水源、用材
短药蒲桃	*S. brachyantherum* Merr. et Perry	16～20	20～30	琼、粤、桂、滇	水源、保土、用材
赤楠（假黄杨）	*S. buxifolium* Hook. et Arn.	3～8	10～15	皖、浙、台、闽、赣、湘、粤、桂、黔。越南、日本	绿化、保土、薪柴
乌墨（乌楣、海南蒲桃）	*S. cumini* (L.) Skeels	20～30	40～80	台、闽、粤、琼、桂、滇。东南亚、澳大利亚	用材，皮含褐色染料和红色树脂
红鳞蒲桃（车辕木）	*S. hancei* Merr. et Perry	15～20	20～30	闽、粤、琼、桂	防风、固沙、用材、皮含鞣质
蒲桃（水蒲桃）	*S. jambos* (L.) Alston	10～15	30～40	台、闽、粤、桂、滇、黔。东南亚	绿化、果树、薪柴及小材、防风、固沙
竹叶蒲桃（杨柳蒲桃）	*S. myrsinifolium* (Hance) Merr. et Perry	5	10	琼、桂、滇	绿化、防风、固沙、薪柴

开花结实　灌木类树种 6～8 年生开始开花结实，乔木类树种 10～12 年生开始开花结实。灌木类正常结实期在 10 年生以后，乔木类在 20 年生以后。结实大小年间隔期为 1 年，特寒年份也影响结实。花两性，白色，芳香，3 至多朵排成聚伞花序式再组成圆锥花序，顶生或腋生。萼筒倒圆锥形，有时棒状。萼片 4～5。花瓣 4～5，分离或连合成帽状。雄蕊多数。花药细小，丁字着生，2 室，纵裂，顶端常有腺体。子房下位，2 或 3 室，每室有胚珠多数，花柱线形。据广西南宁观察，这 7 种蒲桃的开花结实物候情况见表 2。

表 2　蒲桃属树种的开花结实物候期（广西南宁）

树　种	观察年份	开　花			果实成熟		果实脱落
		始　期	盛　期	末　期	始　期	盛　期	
线枝蒲桃①	1986～1988	5 月中旬	6 月上旬	6 月下旬	10 月下旬	11 月上中旬	11 月上旬～12 月中旬
短药蒲桃	1986～1988	5 月中旬	6～7 月	8 月上旬	11 月中旬	12 月上旬	12 月中旬～翌年 1 月中旬
赤楠	1987～1989	5 月上旬	5 月下旬	6 月中旬	9 月下旬	10 月中旬	10 月中旬～11 月上中旬
乌墨	1978～1983	4 月中旬	4 月下旬	5 月上旬	7 月上旬	7 月下旬	8 月上旬～8 月下旬
红鳞蒲桃	1978～1983	7 月上旬	7 月中旬	8 月中旬	12 月下旬	翌年 1 月中旬	2 月上旬～3 月下旬
蒲桃②	1978～1983	3 月下旬	4 月中旬	5 月上旬	6 月中旬	7 月上旬	7 月上旬～8 月中旬
竹叶蒲桃③	1987～1988	5 月上旬	5 月下旬	6 月下旬	8 月下旬	9 月下旬	9 月下旬～12 月上旬

①　线枝蒲桃 10 月上旬有少量花开；②　蒲桃 12 月有少量花开；③　竹叶蒲桃 2 月上旬～3 月中旬有少量花开

浆果或核果状，果顶部有残存的环状萼檐。蒲桃果有种子 1～2 粒，多胚。其余 6 种每果有种子 1 粒。果实成熟后可在树上存留 4～6 天，随即脱落。种皮薄，种皮多少与果皮粘合。具胚乳，胚一般横生于种子中部。子叶肥厚，粘合成一块状体，有时多胚，子叶不粘合。果实及种子形态见表 3、图 1。

表 3　蒲桃属树种果实和种子的形态特征

树　种	未熟果颜　色	果　实			种　子		
		形　状	大小（mm）	颜　色	形　状	大小（mm）	颜　色
线枝蒲桃	黄绿色	近球形，上部有宿存萼檐	长 5～9 径 4～7	浅白色	圆球形	径 3.5～6 厚 3～5	褐色
短药蒲桃	黄绿色	圆球形或近球形	径 18～26	浅黄色	圆球形	径 7～12	乳白色
赤楠	绿色	卵圆形，顶冠以宿存萼檐	径 6～10	紫黑色	近圆球形	径 5～9 厚 5～8	紫红色

（续）

树　种	未熟果颜　色	果　实			种　子		
		形　状	大小（mm）	颜　色	形　状	大小（mm）	颜　色
乌墨	青色	椭圆形或壶形，上部有宿存萼筒	长 10～20 径 8～13	红至紫黑色	长椭圆形	长 9～15 径 5.5～9	淡紫红色
红鳞蒲桃	青色	椭圆形或近球形	长 10～17 径 6～13	紫黑色	卵形或椭圆形	长 6～9 径 5～7	浅褐色
蒲桃	青色或黄绿色	圆球形或扁球形	长 23～46 径 25～50	浅黄色	近圆形或椭圆形	径 15～20	浅棕色
竹叶蒲桃	绿色	近球形，上部有浅杯状萼檐	长 11～17 径 10～15	紫黑色	椭圆形或卵形	长 9～15 径 5～9	淡紫红色

果实的采收调制和种子贮藏　乔木类的蒲桃和短药蒲桃果形较大，可在果熟时上树采摘，或用采种钩刀等工具选择成熟的果实分期分批采摘。乌墨可用竹竿敲打或摇动树枝震落，在地面捡拾。赤楠和竹叶蒲桃等灌木类树种多用手采摘。采集的果实，蒲桃可食去果肉后得出种子；其余 6 种可堆沤 1～3 天，果肉充分软熟后装入布袋或竹筐，置水中冲捣搓擦，淘去果皮果肉等杂质，即得种子。果实的出种率、净度及质量等数据见表 4。

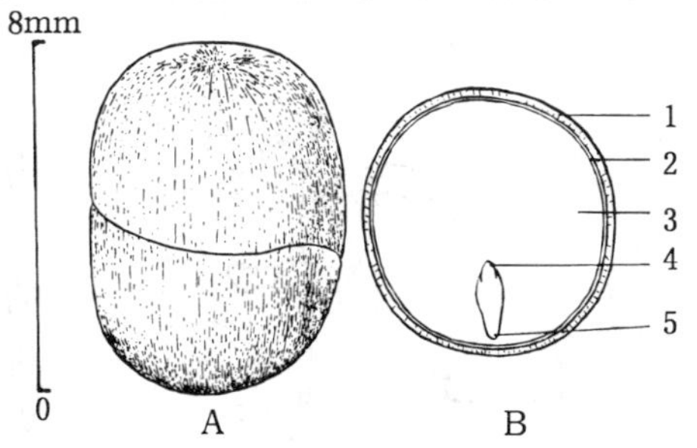

图 1　红鳞蒲桃种子外形（A）及其纵切面（B）
1. 种皮　2. 胚乳　3. 子叶　4. 胚芽　5. 胚根
（黄应钦绘）

表 4　蒲桃属树种果实的出种率和种子净度、质量

树　种	出种率（%）	净　度（%）	千粒重（g）	每千克纯净种子粒数（粒）
线枝蒲桃	40～50	99	45～55	18 000～22 000
短药蒲桃	11	99	300～500	2 000～3 300
赤楠	50	99	150～250	4 000～6 500
乌墨	25～30	98	300～500	2 000～3 300
红鳞蒲桃	45～55	99	160～250	4 000～6 300
蒲桃	19	99	4 000～4 600	220～250
竹叶蒲桃	50～60	98	290～340	2 900～3 500

种子不能脱水，忌日晒，亦不耐贮藏。夏、秋季成熟的种子宜随采随播。需作短期贮藏或运输的，可暂不调制脱粒或脱粒后混以湿沙。秋后成熟的种子经调制稍晾干后，混湿沙层积堆藏，贮藏期不宜超过半年。

表 5　蒲桃属树种的发芽能力及其测定条件（广西南宁）

树　种	基　质	温度（℃）		发芽势（%）		发芽率（%）	
		发芽箱	室　外	计算天数	一般数值	计算天数	一般数值
线枝蒲桃	滤纸	28	—	8	46	21	88
赤楠	滤纸	28	—	—	不显著	32	82
乌墨	细沙	—	26～28	—	不显著	23	60
红鳞蒲桃	细沙	—	26～28	—	不显著	35	60
蒲桃	细沙	—	26～28	7	70	17	96
竹叶蒲桃	滤纸	28	—	—	不显著	20	82

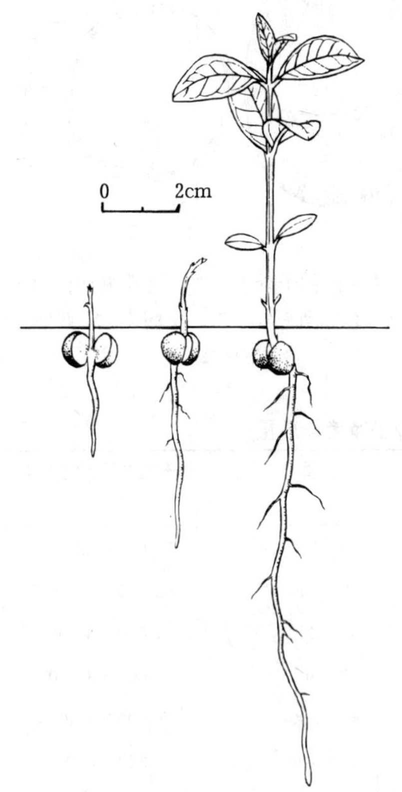

图 2　乌墨种子萌发后第 9、12、25 天幼苗的生长情况
（黄应钦绘）

发芽和播种　种子无休眠习性。发芽时日均温需在 20℃以上。1977～1987 年，广西林业科学研究所先后对乌墨、红鳞蒲桃、蒲桃等 3 种夏、秋季成熟的种子，在室外沙床作过发芽测定；对赤楠、线枝蒲桃、竹叶蒲桃等 3 种冬季成熟的种子，在室内发芽箱内作过发芽测定。测定结果见表 5。线枝蒲桃出土萌发，红鳞蒲桃和短药蒲桃半出土萌发，其余 4 种均为留土萌发。播种后线枝蒲桃、乌墨、蒲桃和竹叶蒲桃 3～6 天胚根萌发，发根后 7～10 天胚芽或子叶出土，再过 4～5 天初生叶出现。赤楠播种后 10 天胚根萌发，发根后 5～8 天胚芽出土，再过 10 天左右初生叶出现。红鳞蒲桃的发芽过程较长，播种后约 20 天胚根才开始萌发，发根后约 10 天胚芽出土，再过 12 天左右初生叶出现。蒲桃的种子多胚，每颗种子可发芽成苗 4～8 株。乌墨种子的萌发和幼苗生长情况见图 2。

点播。乌墨、短药蒲桃和竹叶蒲桃每平方米播种 50～60g；赤楠、红鳞蒲桃每平方米播种25～30g。线枝蒲桃种子较小，每平方米播种量一般为 60～75g。蒲桃每粒种子可发苗数株，宜先将种子密播，发芽后再行分株移植。一般培育半年至 1 年生苗出圃。园林绿化用苗需培育2～3 年出圃。

（林榕庚）

红　胶　木

Tristania conferta R. Br.

（桃金娘科　Myrtaceae）

生长习性、分布和用途　红胶木属约 22 种，我国引入本文描述的 1 种。常绿大乔木，高达 42m，胸径可达 180cm。喜深厚疏松的肥沃土壤，肥力较低的土壤只能长成中小径材。能耐 0℃左右的极端低温，但忌霜冻。原产澳大利亚及印度尼西亚。我国引种已有近百年的历史，粤、桂、闽南、浙南（平阳）有栽培。木材抗白蚁，耐水下使用。树冠广阔，绿荫效果好。

开花结实　6～8 年生开始开花结实，正常结实期在 15 年生以后。结实大小年间隔期不明显，除特寒年份影响当年结实外，一般年份均可正常开花结实。花两性。聚伞花序腋生，有花 3～7 朵。萼筒倒圆锥形，与子房的基部合生，萼裂片 5。花瓣 5，浅黄色，倒卵形或卵状圆形。雄蕊多数，花丝基部合生成 5 束，白色，与花瓣对生。花药背部着生，药室平行，纵裂。子房下位，3 室，每室有胚珠多颗。花柱比雄蕊短，柱头稍扩大。据广西南宁 1981～1987 年观测，5 月上旬为始花期，中旬为盛花期，下旬为末花期；11 月中旬果实开始成熟，11 月下旬至翌年 1 月上旬为果熟盛期。蒴果，半球形，顶部截平，果缘突起，果瓣内藏。果实成熟时由青色转为灰褐色，宿存萼筒木质，果瓣革质，高 9～13mm，径 8～13mm，果瓣微 3 裂。种子一般可留在果内 1 个月左右不散失。成熟的果实在树上约可存留半年左右，以后逐渐脱落。每果有种子 400～600 粒（含不发育的瘪粒）。种子极纤细，线状或稍弯曲，黄褐色，长 2～3.8mm，径约 0.3mm，无胚乳。

果实的采收调制和种子贮藏　当果实大部分成熟，少数果瓣微裂时，用高枝剪或采种刀割断总梗，在地面捡拾。如采种时果瓣已大部分开裂，则需在树上用手采摘，装入袋内，以免种子散失。采集的果实垫上薄膜，摊放于室内，或于无风天在阳光下曝晒。果瓣充分开裂后，两手分别持果，相互敲打，种子即可全部脱出。鲜果出种率为 12%～16%，每果出种约 0.05g（均含不发育瘪粒，以下同）。千粒重 0.1～0.15g，每克有种子 6 500～10 000 粒。种子可晒干后袋装贮藏于室内通风干爽处，或装入坛罐内密封贮藏。袋装种子贮藏期约为半年左右，密封的贮藏期为 1 年左右。

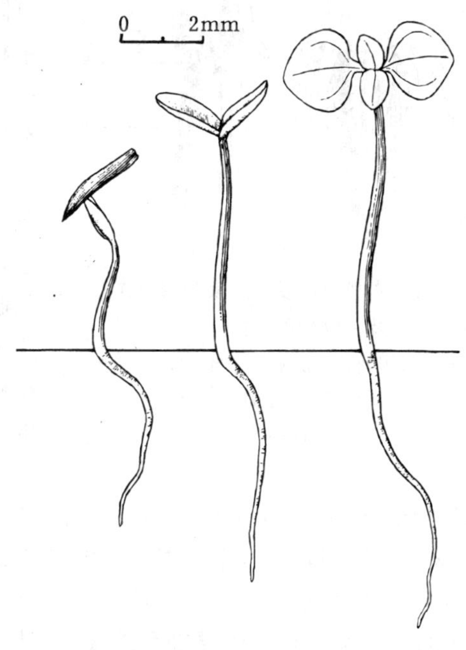

图 1　红胶木种子萌发后第 5、8、40 天幼苗的生长情况
（黄应钦绘）

发芽和播种　种子无休眠现象。发芽时的日均温需在 20℃以上。1987 年 11 月下旬，广

西林业科学研究所用当年成熟的新鲜种子，在发芽箱内用发芽皿进行过发芽测定，箱内保持28℃。播种后5天胚根开始萌动，发芽盛期不明显，12月中旬末发芽结束。从播种之日起算，27天的发芽率为11%。出土萌发。萌发后约4～5天子叶伸出，再过38天展出初生叶（图1）。

撒播。将种子密播于播种沙盘内，每平方米播种0.2～0.4g。苗木半木质化时移入容器或大田培育。容器苗百日左右出圃，大田苗一般1年生出圃。用于庭院绿化的苗木需培育2年。

（林榕庚）

木　榄

Bruguiera gymnorrhiza（L.）Savigny

（红树科　Rhizophoraceae）

生长习性、分布和用途　木榄属约7种，我国3种和1变种，本文描述1种。常绿乔木或小乔木，高达10m以上（分布在纬度较高的广西沿海，高度一般为6～8m）。适生于热带海洋淤泥深厚肥沃的浅海盐滩。可单独组成群落，也可与其它种类混生，是构成红树林的重要树种。不具支柱根，但常有密集的曲膝状呼吸根露出滩面。耐盐。分布于琼、粤、桂、闽、台及其沿海岛屿。南亚、东南亚、大洋洲和非洲也有分布。木榄林可防浪护堤，供鱼虾栖息并供观赏，木材可作薪柴、用材，树皮可提取栲胶。

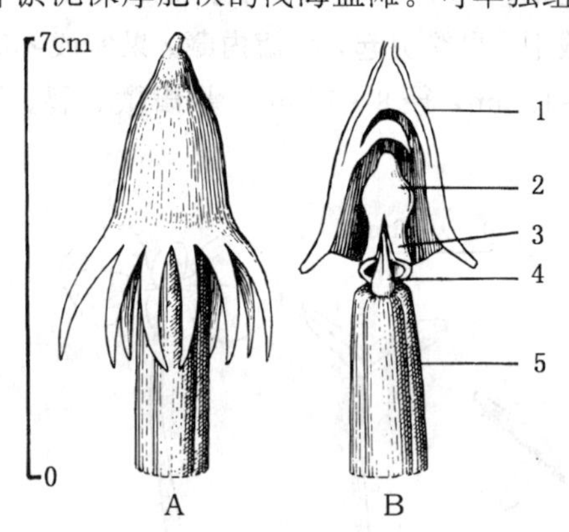

图1　木榄已萌发的果实外形（A）及其纵切面（B）

1. 果皮及种皮　2. 帽状子叶　3. 子叶鞘　4. 胚芽　5. 胚轴（部分）

（黄应钦绘）

开花结实　3～4年生开始开花结实，10年生左右进入正常结实期。无大小年现象，每年结实均较丰。花两性，单生于叶腋内。下弯花梗长0.8～2cm。花淡红白色或淡红黄色，直径2.5～3cm。萼筒近钟形，裂片11～13，条形，约与萼筒等长，长1.8～2cm。花瓣与萼裂片同数，较小，2深裂，裂隙有1刺毛。雄蕊为花瓣倍数，略短。子房下位，3室，每室2胚珠。花柱3棱柱状，柱头3裂。果包藏于萼筒内，两者合生，1室1种子，离开母体前萌发，俗称“胎生”。据广西北部湾沿海观察，木榄几乎全年开花，每年有两次果实成熟盛期，第1次在6～7月，第2次在9～10月。果实成熟后便在树上陆续发芽 。胚轴纺锤形，幼时暗红色，稍有棱角，熟时黑绿色，长15～25cm。胚轴可在树上悬挂数月之久。帽状子叶被内胚乳所包，子叶鞘长筒形。木榄果实的形态见图1。

果实（胚轴）**的采收调制和贮藏**　红树科树种（除生长在内陆的以外）的育苗材料不是种子，也不是普通意义上的果实，而是脱离母树前就已经萌发而突出果实之外的棒状体——胚轴。胚轴在树上发育成熟后要及时采收，以免大批脱落发根。也可捡收落下而尚未发根的胚轴。胚轴的净度可达100%，每千克有胚轴45～50条。采得的胚轴不需作任何处理，即可用于育苗或直接造林。胚轴忌失水，不能日晒，亦忌堆沤发热。如需短期存放，可堆成小堆放在室外荫凉处，并覆以湿草，贮存期一般在1周以内。运输时每200条左右为1捆，外裹湿草，再用竹筐装载。运输途中应防止失水或堆沤发热，运输时间以不超过3天为宜。

发芽和播种　木榄已经萌发的胚轴最好随采随播。一般按2m×3m的距离，把胚轴下端插入淤泥，深及1/3，1～2天内即可发根。为了避免海浪冲击新插的胚轴，提高造林成活率，近年的做法是在海湾静风处集中育苗，大约半年后，选在退潮时起苗，栽植造林。

（何祖家）

竹节树属
Carallia Roxb.

（红树科　Rhizophoraceae）

生长习性、分布和用途　竹节树属约10种，我国4种，本文描述2种。常绿乔木，为红树科植物生长在内陆的乔木树种。对土壤肥力的要求不甚苛刻，但要求湿润，干旱地生长不良。这2个种的名称、生长、分布及用途见表1。

表1　竹节树属树种的名称、生长、分布和用途

中　名	学　名	树高（m）	胸径（cm）	分　布	用　途	供　稿
竹节树	*C. brachiata* (Lour.) Merr.	30	60	粤、琼、桂。非洲东南、马达加斯加、南亚、东南亚、澳大利亚	材用、水源林	604
旁杞木	*C. longives* Chun ex Ko	15	30	滇、桂、粤	材用、水源林	618

开花结实　约10～12年生开始结实，结实大小年间隔期为1～2年。花两性。聚伞花序腋生，具短梗，每一分枝有花2～5朵。花小，无柄。萼钟形，6～7裂。竹节树花瓣白色，旁杞木白色或淡红色，同数。雄蕊数为花萼裂片数之2倍，同着生于波状花盘边缘，通常一半与花瓣对生，一半与萼片对生。花药4室，纵裂。子房下位，3～5室。据广西南宁市1987～1988年的物候观察，竹节树5月下旬始花，6月上旬为盛花期，10月下旬果实开始成熟，11月中旬为盛果期。另据1977～1978年在海南采种实地调查，10月上旬果实开始成熟，果期延续至11月中下旬。旁杞木在云南勐腊2月中下旬始花，3月上旬盛花，3月中下旬花期结束，5月中旬果始熟，下旬为盛熟期。浆果，圆

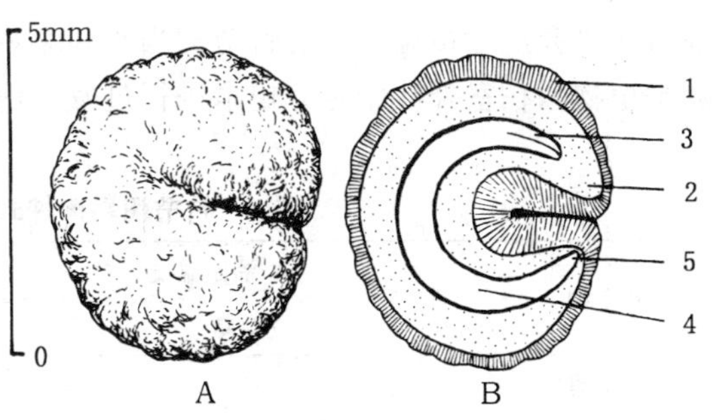

图1　竹节树种子外形（A）及其纵切面（B）
1. 种皮　2. 胚乳　3. 子叶　4. 胚轴　5. 胚根
（黄应钦绘）

球形，顶端冠以短三角形的萼齿，成熟后红色，在树上可留7天左右。每果有种子1～3粒。种皮粗糙。种子一侧内凹，有胚乳，胚弯曲。果实和种子形态见表2、图1。

表2 竹节树属树种果实和种子的形态特征

树种	未熟果颜色	果实			种子		
		形状	大小（mm）	颜色	形状	大小（mm）	颜色
竹节树	青绿色	圆球形	径6～8	浅红色	弯曲或扁圆形，具麻点	宽4.5～6.5 厚3～3.5	红棕色或紫红色
旁杞木	绿色	圆球形	径16～17	红色至紫红色	圆柱形或近肾形	长15～16 径3～4	深褐色

果实的采收调制和种子贮藏 种子成熟的持续时间长，一般不宜带果穗采下。可在果实成熟盛期清除地面灌木杂草，摇动树枝震落果实，在地面捡拾。采集的果实可堆放2～3天，充分软熟后装入布袋或竹筐，置水中搓洗，淘去果皮等杂质，即得纯净种子。鲜果的出种率、净度和质量见表3。

表3 竹节树属树种的出种率和种子净度、质量

树种	出种率（%）	净度（%）	千粒重（g）	每千克纯净种子粒数（万粒）
竹节树	28	95～100	30～50	2～3.3
旁杞木	40～50	95～100	100～125	0.8～1

种子的含水量为25%～35%，忌失水，不能日晒或裸露贮存。运输或贮藏均需混以湿沙，贮藏期一般在半年以内。

发芽和播种 种子无休眠习性。发芽时日均温宜在22℃以上。1987年11月下旬，广西林业科学研究所在恒温箱中对竹节树当年新采种子进行过发芽测定。同年5月23日，西双版纳热带植物园在室外沙床对旁杞木进行过发芽测定。测定结果见表4。

表4 竹节树属树种的发芽能力及其测定条件

树种	基质	测定温度（℃）		发芽势（%）		发芽率（%）	
		发芽箱	室外	计算天数	一般数值	计算天数	一般数值
竹节树	滤纸	28	—	16	63	34	77
旁杞木	沙	—	22～35	7	70	25	90

出土萌发。竹节树置床后8天开始萌动，胚根萌发后8天子叶带壳出土。子叶出土后40天左右种壳脱落。但此时子叶仍未开展，大约再经30天，2片子叶展开，再过12天，初生叶展出。旁杞木萌发后5～10天子叶带壳出土，并陆续脱壳，子叶展开，再过10～20天初生叶展现。竹节树种子发芽及幼苗生长情况见图2。

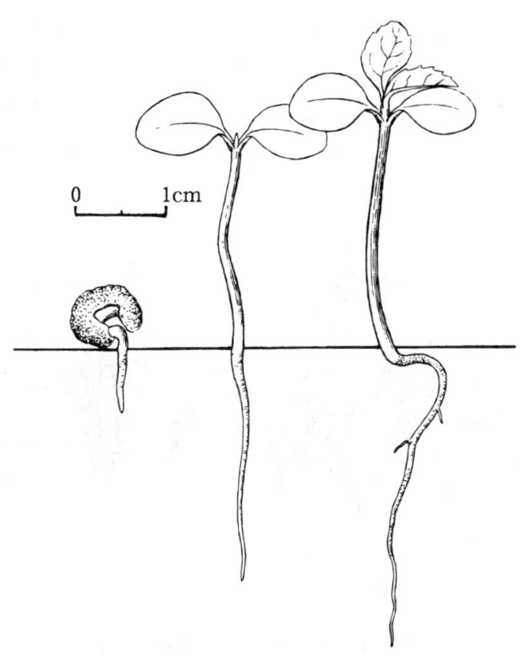

图 2　竹节树种子萌发后第 8、70、85 天幼苗的生长情况

（黄应钦绘）

条播。3 月中旬至下旬为适宜播种期。竹节树每平方米播种 7～10g，旁杞木 25～32g，覆土 0.5～1cm。苗期宜遮荫。1 年生苗出圃。

（林榕庚）

秋　茄　树

Kandelia candel（L.）Druce

（红树科　Rhizophoraceae）

生长习性、分布和用途　秋茄树属只有本文描述的 1 种。常绿灌木或乔木，高 3～10m，茎基部粗大，具板状根或密集支柱根。生于海涂内滩及有淤泥分布的外滩，与其它红树植物混生或单独形成群落。适应性强，耐海水浸渍又耐淡水泛滥。分布于桂、粤、琼、闽、台。浙江有引种。东南亚及日本亦有分布。秋茄树林防浪护堤，供鱼虾栖息。皮含鞣质 17%～26%。胚轴含淀粉，浸泡后可食用。干枝作材用或薪柴。树形优美，可供观赏。

开花结实　人工林 5 年生左右可以开花结实，10 年生进入结实盛期。大小年不明显，每年结实颇丰。据广西北部湾观察，7～8 月为开花盛期，次年 4～5 月为果熟盛期。另据海南观察，几乎全年开花结实。花两性。二歧聚伞花序腋生，有花 4（9）朵。萼 5～6 裂，裂片条形。花瓣与之同数，2 裂，每 1 裂片再分裂为数条丝状裂片，早落。雄蕊 20～25，分离或基部多少合生，花药 4 室，纵裂。子房下位，幼时 3 室，每室有 2 胚珠，结果时变为 1 室，仅 1 胚珠发育。花柱丝状，柱头 3 裂。果长椭圆形或圆锥形，中部为外反、宿存的花萼裂皮及一环状小苞片所围绕。果长约 5cm，径约 3cm，有种子 1 颗，偶有 2 颗。子叶帽状，子叶鞘筒状，包裹胚芽。下胚轴的尖端在顶部突出。种子脱离母体前发芽（胎生），将胚芽连下胚轴推出。胚轴瘦长，棒状无棱，先端尖锐，长达 20～30cm，径约 1cm，可悬挂在果实上 2～4 个月。果实及胚轴形态见图 1。

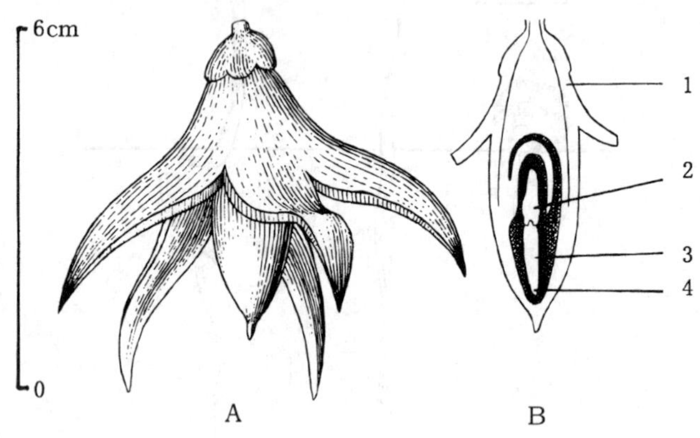

图 1　秋茄树果实外形（A）及其纵切面（B）

1. 果皮及种皮　2. 帽状子叶　3. 胚轴　4. 胚根

（黄应钦绘）

果实（胚轴）的采收调制和贮藏　胚轴充分发育后脱离母体插入淤泥中，1 天左右便迅速发根，亦有不少随波逐流飘泊至另一海湾淤泥处发根生长。人工造林多在胚轴成熟后用手采摘。胚轴的净度可达 98%以上，不需作任何处理，即可作育苗材料。每千克有胚轴 45（35～55）条。脱离母体的胚轴忌失水。3～4 天的短期存放宜以小堆堆放于荫凉处，覆盖湿草。运输时以 100 条左右为 1 捆，外裹湿草，运抵后即行插入已整好的淤泥中育苗。脱离母体的胚轴贮藏加运输以 1 周左右为限。

发芽和播种　造林方法有 2 种：一是直播造林，二是采用半年至 1 年生野生小苗造林。目前人工育苗造林尚不多见。直播造林的作法是将胚轴插入泥中，深及 1/3 左右，1～2 天便约有 95%的胚轴迅速发根，成为新的植株。

（何祖家）

山　红　树

Pellacalyx yunnanensis Hu

（红树科　Rhizophoraceae）

生长习性、分布和用途　山红树属约 8 种，我国只有本文描述的 1 种。常绿乔木，高15～20m，胸径 25～35cm。分布区属热带气候，但能耐短期 0.5℃左右低温。喜肥沃湿润的酸性至微酸性土壤。产云南南部，零星分布，《中国植物红皮书》列为稀有种，亟需保护。

开花结实　10 年生左右开始结实，15 年生以后为正常结实期。结实大小年现象不明显。花两性，单生或 2～5（6）丛生叶腋。花萼筒状，裂片 6（7），宿存。花瓣同数，白色，椭圆

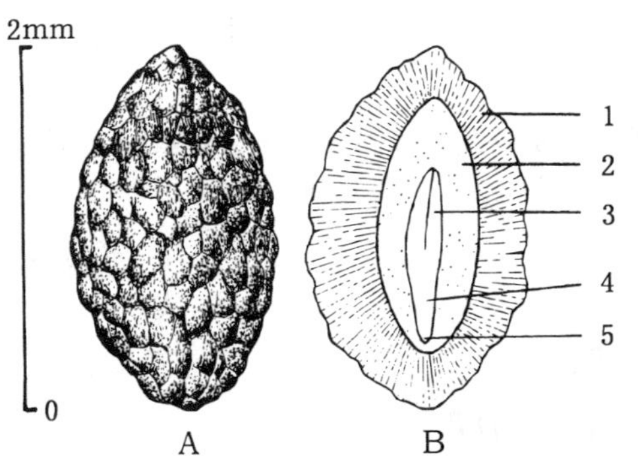

图1　山红树种子外形(A)及其纵切面(B)
1. 种皮　2. 胚乳　3. 子叶
4. 胚轴　5. 胚根
(黄应钦、马信祥绘)

形，边缘撕裂状，与萼裂片互生。雄蕊12(14)，生于萼筒喉部。花药近圆形，4室，纵裂。子房下位，球形，12室，每室有成束胚珠多数。花柱柱状，柱头盘状，6深裂，每一裂片再浅裂，成12裂。据1986年和1988年在云南西双版纳勐仑的观察，4月上旬花芽始现，4月下旬～6月上旬花蕾膨大，6月中旬始花，6月下旬～7月上旬为盛花期，7月中旬花期结束，幼果形成；10月上旬果实开始成熟，10月中旬～11月中旬为成熟盛期。浆果，单生，陀螺状，通常着生于落叶的老枝上。成熟时果实由绿色变为暗黄色，表面有光泽，长2.5～3.5cm，径1～1.5cm。果实顶端有6(7)宿存的花萼裂片，裂片披针形，长1～1.2cm，并宿存有花柱和柱头。果熟后3～5天落地，逐渐变成褐色。果皮肉质浆状。种子多数，椭圆形，长2～2.5mm，径0.8～1.2mm，褐色至深褐色。新鲜种子种皮光亮，半透明状，干后皱缩成凸凹窝孔状，一侧有种脊。胚乳白色，子叶较小。种子形态见图1。

果实的采收调制和种子贮藏　果实成熟期用竹竿敲打或摇动震落，在树下捡拾。采回的果实中有的果肉尚未软化，这部分果实可堆沤3～5天，待软化后再行调制。已经充分成熟软化的果实装入袋内，置水中反复搓揉，漂去果皮果肉等杂质，放入孔径0.3cm的金属筛里用水冲洗，将种子冲入筛下浅盘中，再用水漂去极细的杂质即得纯净种子。鲜果出种率0.5%～1%。种子净度为70%～90%。千粒重约1g，每千克有纯净种子95万～100万粒。含水量约24%，忌失水，不宜日晒。调制出来的种子混湿沙贮藏，保持湿润，室内贮藏可达3年以上。

发芽和播种　种子无休眠习性。发芽时日均温需在25℃以上，对光不敏感。1985年11月15日，西双版纳热带植物园用30℃的恒温对刚调制出的湿种子作过发芽测定，基质为滤纸。置床后9天(11月23日)开始发芽，12月12日发芽结束，没有明显的发芽盛期，发芽率为80%。另外将晾干的种子置床，置床后13天开始发芽，从开始发芽至发芽结束历时41天，发芽盛期不明显，发芽率为56%。用经湿沙层积贮藏到翌年5月进行发芽测定，播种后9天开始发芽，从开始发芽到发芽结束只需4天，发芽率高达90%。

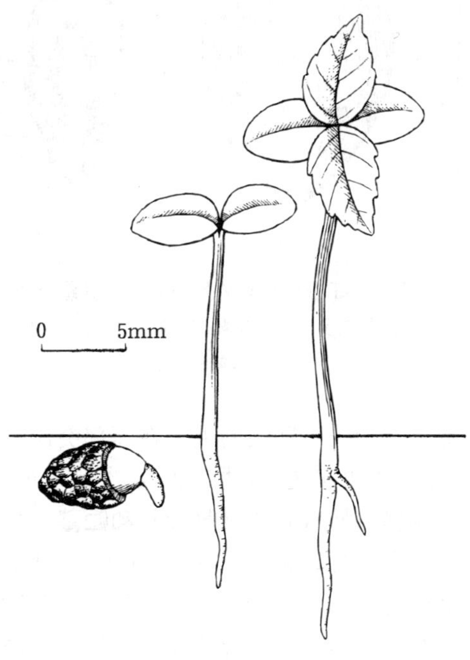

图2　山红树种子萌发后第2、10、25天幼苗的生长情况
(黄应钦、马信祥绘)

出土萌发。在日平均温度25℃左右时，子叶出土后约23天开始出现第1对初生叶，卵形，同

时长出 1 条很短的侧根。气温低时生长极缓慢，40～50 天才出现第 1 对初生叶。种子萌发和幼苗生长情况见图 2。

撒播。在播种盘内用细沙和过筛的火烧土作基质，每平方米播种 2～3g。淋水时可混以极稀薄的氮肥。幼苗转木质化时移入大田或容器，培育 1 年或 3 个月出圃。

（马信祥）

红 海 榄

Rhizophora stylosa Griff.

（红树科 Rhizophoraceae）

生长习性、分布和用途 红树属约 7 种，我国 3 种，本文描述 1 种。常绿乔木或大灌木，高 7～8m。常有密集高大的支柱根，适生于热带沿海淤泥深厚的潮淹滩涂，耐盐。分布于桂、粤、琼、闽、台。东南亚和大洋洲北部也有分布。可以防浪护堤，并为海生鱼虾提供栖息场所，还可作薪柴、用材、蜜源，富含鞣质，又供观赏。

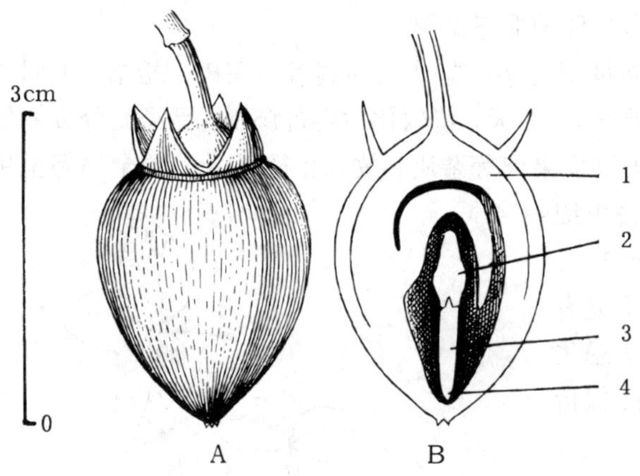

图 1 红海榄果实外形（A）及其纵切面（B）
1. 果皮及种皮 2. 帽状子叶 3. 胚轴 4. 胚根
（黄应钦绘）

开花结实 4 年生左右开始开花结实，8 年以后进入正常结实期，结实大小年不明显。较暖的年份几乎终年开花结实。据广西北部湾沿海的观察，大造花始期为 3 月中旬前后，4 月为盛花期，5～6 月为幼果期，9～10 月为果熟盛期。花两性。二歧聚伞花序有花 2～7 朵。总花梗生于当年生的叶腋或已落叶的腋部。萼 4 裂，宿存，裂片三角形。花瓣 4。雄蕊 8，4 枚萼上着生，4 枚瓣上着生。花丝极短，花药多室，瓣裂。子房半下位，2 室，每室有胚珠 2 颗。花柱长 4～6mm，柱头不明显 2 裂。果梨形至卵形，下垂，顶端收窄，绿色或褐色，平滑，具宿存萼，长 2～3cm。种子 1，帽状子叶被内胚乳所包。下胚轴的尖端在顶部突出，脱离母体前发芽，胚轴突出于果外而成棒状，下端稍尖，悬挂于果实之下，幼时红色，老呈绿色，长 20～ 40cm，粗 0.8～1.5cm。果实的形态见图 1。

果实（胚轴）的采收调制和贮藏 当大部分胚轴已成熟，棒状体开始脱落时，即可用手采摘。采集的胚轴不加任何处置即可用作育苗材料。每千克红海榄约有胚轴 40 条。已脱母体的胚轴忌失水，忌堆沤过久，不宜曝晒或裸露存放。短期存放也应在室外荫凉处，每堆数量不宜过多，上面用湿草覆盖或淋水保湿，堆放时间一般在 10 天左右，不宜久藏。运输时每 100 条左右捆成 1 包，外用湿草包裹，避免运输过程中失去水分或损伤胚轴。

发芽和播种 离开母体之前红海榄萌出的胚轴悬挂在果实下面可达数月之久（胎生）。胚轴发育成熟坠下插入淤泥后，1 天内就开始发根并固定下来。由于红海榄胚轴较长，海生动物及风浪海潮对它影响不很大，一般采用胚轴直播造林。胚轴插入淤泥后2～3天即可迅速发根，新鲜胚轴的发根率可达100%。每666m^{2}可插200株，需用胚轴5kg左右，近年也有先育苗后造林的，也可采集半年至1年生的野生苗造林。胚轴插入泥中的深度约占总长度的1/3左右，以免被浪冲走或被淤泥淹没。

（何祖家）

八　宝　树

Duabanga grandiflora（Roxb. ex DC.）Walp.

（海桑科　Sonneratiaceae）

生长习性、分布和用途　八宝树属有3种，我国1种，又引入1种。本文描述我国产的1种。常绿乔木，高40m，胸径1.5m。喜高温，耐0℃左右极端低温及轻霜，小苗忌霜冻。适生于肥沃湿润的酸性土。产滇南、桂西南。南亚、东南亚亦产。材质中等，可用于一般建筑及家具。枝轮生，花大而艳丽，供观赏。

开花结实　约8年生开始开花结实，正常结实年龄在15年生以后。结实大小年间隔期一般为1年，但不甚明显。花两性，直径2～4cm。伞房花序，顶生。花（4）5～6（8）基数。萼筒阔杯形，裂片长约2cm。花瓣近卵形，连柄长2.5～3cm。雄蕊多数，2轮排列。花丝长4～5cm。花药长圆形，长1～1.2cm，丁字着生。子房半下位，5～6室，有胚珠多颗。花柱长3～4cm，柱头微裂。据南宁1981～1984年的物候观测，4月下旬始花，5月上旬盛花，5月中旬为末花期；6月上旬果实开始成熟，6月中旬至下旬为果实盛熟期。蒴果，近椭圆形，有6条棱状条纹，萼片及弯长的花柱宿存，成熟时由淡黄色或青色转变为黄褐色至黑褐色。果高3～3.5（4）cm，径2.5～3（3.5）cm。果壳薄革质，充分成熟后从室背顶端向下开裂，呈6瓣，种子飞失。约1周左右，果瓣脱落，宿存的萼片最后脱落。种子极纤细，针棒状，弯曲，灰褐色，种皮向两端延伸成尖尾状，长4～7mm，径0.2～0.3mm。无胚乳，胚直立，子叶扁平（图1）。

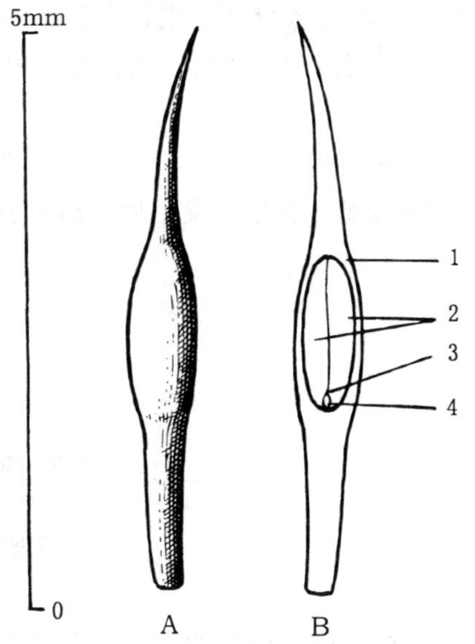

图1　八宝树种子的外形（A）及其纵切面（B）

1. 种皮　2. 子叶　3. 胚芽　4. 胚根

（黄应钦绘）

果实的采收调制和种子贮藏　应在果实成熟尚未开裂时，选择健壮母树的中部南向枝条将果穗采下，就地剪取果穗中部下部较大的果实。采回后置室内避风处，并垫上白纸或塑料薄膜，2～3天后果壳开裂，稍加敲打，种子便可脱出。鲜果出种率为15%～23%。每果内有种子万粒以上。种子净度可

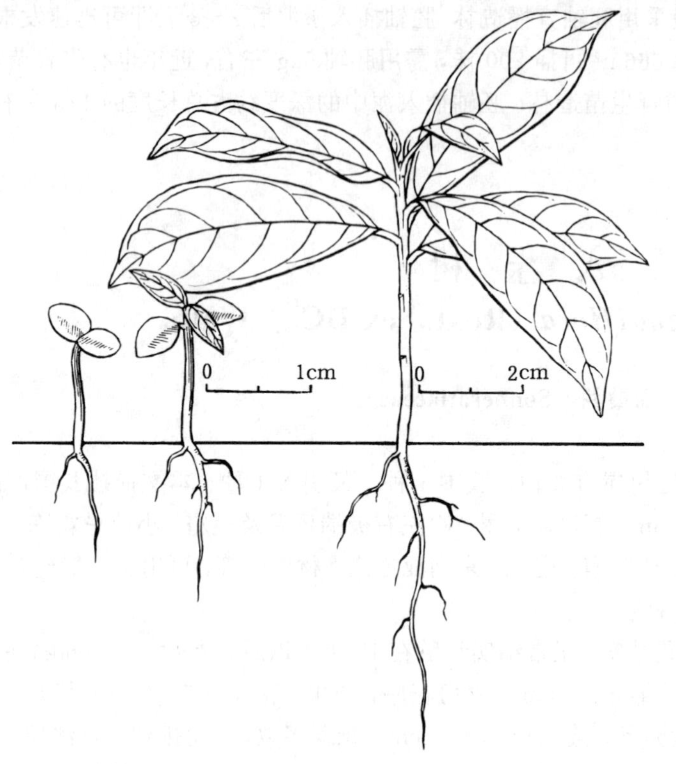

图 2　八宝树种子萌发后第 3、15、35 天幼苗的生长情况

（黄应钦绘）

达 90%，千粒重 0.04～0.06g，每克纯净种子有 1.7 万～2.5 万粒。种子轻细，忌日晒风吹。调制出的种子在室内稍晾干后，宜用瓶罐密封贮存，但不耐久藏。用布袋盛装贮存于室内的，半个月左右发芽能力明显降低；瓶装密封贮藏的，半年后发芽率尚有 35%左右；少量种子可附装在信封内寄送。如运输时间较长，可选择尚未开裂的成熟果实装木箱内寄运，以缩短裸露种子贮运的时间。

发芽和播种　种子无休眠习性。发芽时日均温应在 20℃以上。1981 年 7 月 9 日，日均温 30℃时，广西林业科学研究所用滤纸作基质，在透光良好的室内对鲜种作过发芽测定：置床后第 3 天开始萌动，并随即进入发芽盛期，盛期的 3 天中发芽百分数为 50%；第 11 天发芽结束，发芽率为 63%。出土萌发。种子萌发后第 2～3 天子叶伸出，再过 12～ 15 天发出初生叶。播种至圃地，一般需 10 日左右子叶出土。种子萌发和幼苗早期生长情况见图 2。

生产上常用播种盘盛装经过消毒的细沙，将种子密播于沙盘内，发芽后每日淋水，后期可掺入极少量氮肥。初生叶发出后移苗至容器或圃地继续培育，3 个月或 1 年生苗出圃。

（林榕庚）

石　榴

Punica granatum L.

（石榴科　Punicaceae）

生长习性、分布和用途　石榴科仅有石榴属 1 属 2 种，分布地中海地区至喜马拉雅及索科特拉岛。本文描述的这个种原产中亚的伊朗、阿富汗等地，约在公元前 2 世纪传入我国。落叶小乔木或灌木，通常高 3～5m，稀达 10m。枝顶常成尖锐长刺。喜光。在温暖地带生长良好，冬季休眠期能耐短期低温。对土壤酸碱度的适应性较强，在 pH 值 4.5～8.2 范围内均能生长，尤以富含石灰质的疏松肥沃的壤土和沙壤土最为适宜。

石榴种子外种皮肉质多汁，甜酸适度，可食用。果皮、根皮、花瓣均可入药。果皮、树皮和根皮含鞣质可提制栲胶。叶色翠绿，花形美而色泽鲜艳，常栽培供观赏。我国石榴栽培历史悠久，已沿着观花和取果食用两个方向培育出许多优良品种。

开花结实　实生苗5年生可以结实，分株苗和扦插苗3年生可以开花结实，10年左右进入盛果期。盛果期延续70～80年，寿命可达120年。结实丰年间隔期为0～1年。云南南部生长的石榴有3月下旬开始开花，7月中旬果熟的。据观察，北京1976年5月22日花蕾始现，26日花初开，6月2日花盛开，23日花期结束，幼果形成；8月18日果实成熟，9月2日开始落果。某些供观花的品种花期长达4个月。

花两性，顶生或近顶生，单生或几朵聚生或组成聚伞花序。萼筒钟状或管状，长2～3cm，裂片5～9，略外展，卵状三角形，长8～13mm，红色或淡黄色，革质，宿存。花瓣5～9，红色、黄色或白色，多皱褶，倒卵形，长1.5～3cm，宽1～2cm。雄蕊多数。花药背着。花丝无毛，长达1.3cm，花柱长超过雄蕊。子房下位，多室，呈上下叠生，上部5～7室，为侧膜胎座，下部3～7室，中轴胎座，胚珠多数。浆果，近球形，直径5～12cm，淡黄褐色或淡黄绿色、白色及红色。外果皮厚，革质，顶端具宿萼。每果有种子多数。种子钝角形，具棱角，红色至乳白色。外种皮肉质，晶莹多汁，味酸甜；内种皮木质，厚2～3mm。无胚乳，胚直立。子叶2枚，近白色，相互重叠旋卷呈长柱形（图1）。

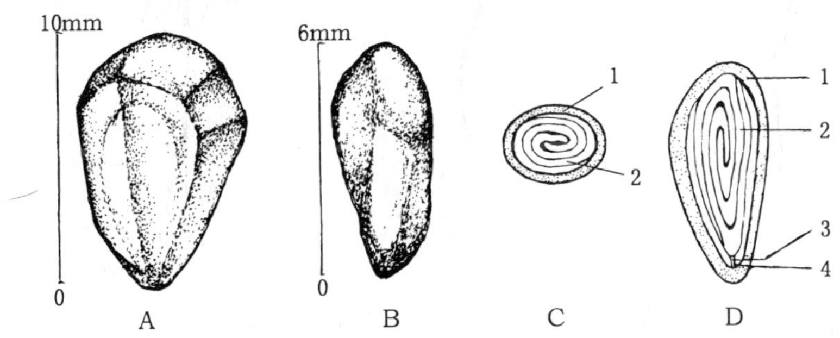

图1　石榴种子外形（A）、去掉外种皮的种子外形（B）及其横切面（C）和纵切面（D）
1. 内种皮　2. 子叶　3. 胚轴　4. 胚根
（A、B. 田恒德，C、D. 张世经绘）

果实的采收调制和种子贮藏　石榴花期较长，果实和成熟期不一致，通常先开放的花形成的果实大，成熟早。充分成熟的果实颜色变化不大，一般以果皮微裂为成熟标志。果实成熟后并不立即脱落。入秋后可分批采摘出现成熟特征的果实。采收时要防止伤害结果母枝，避免影响来年结实。剥去果皮即得种子，用细眼筐篓置水中轻揉淘洗，清除肉质外种皮，阴干后收藏。这种仅有内种皮的部分即为播种材料，通种种子。也可在生食后洗净得到播种材料。南京栽培的红壳石榴，果实出种率（含外种皮）为43%～49%，高的可达65%；千粒重一般为220g，每千克纯净种子约4 500粒。具外种皮的种子含水量达80%，除去外种皮含水量为55%～60%。清除外种皮的种子净度可达95%以上，含水量可晾干至13%～15%。石榴种子易于保存，阴干后即可混沙贮藏，也可采用普通干藏。含水量降至10%的石榴种子在低温（1～5℃）下干藏可以保存2年。

发芽和播种 石榴的内种皮木质，萌发较慢，但种子无生理休眠习性。在25℃恒温条件下28天的发芽率可达90%。3月下旬播种。条播。经过短期层积处理的种子播后发芽整齐。干藏越冬的种子可在播前浸种4小时。覆土1.5cm左右。播后15天左右即可发芽。出土萌发。子叶绿色，初期仍旋卷，经3～7天2片子叶才伸展，大约再经7天萌发初生叶（图2）。可以1年出圃，也可培育大苗出圃。除常用种子繁殖外，石榴还可以分株、压条、嫁接和扦插繁殖，以扦插应用最广。

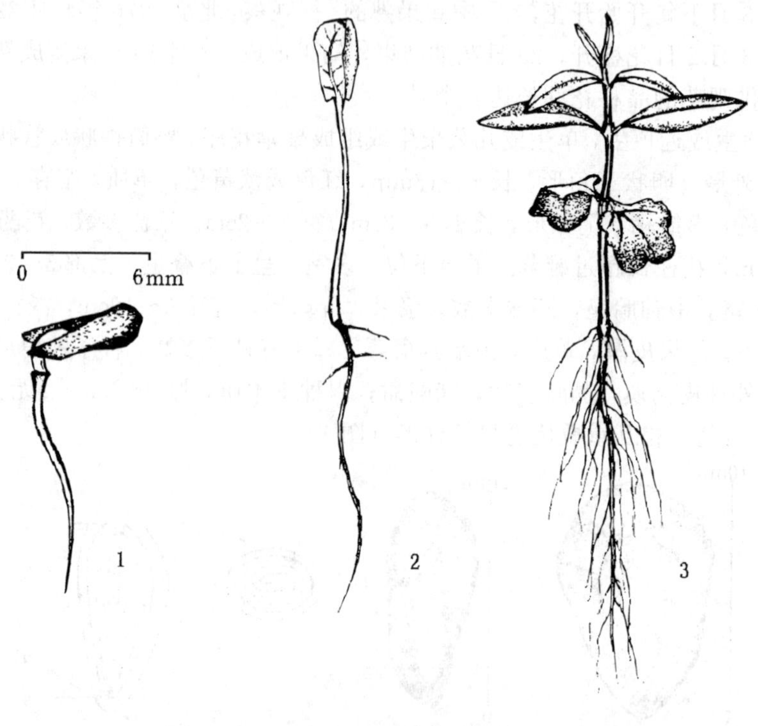

图2 石榴种子的萌发和幼苗生长情况

1. 幼根延伸 2. 子叶出土 3. 初生叶对生

（田恒德绘）

（杨余康）

使 君 子

Quisqualis indica L.

（使君子科 Combretaceae）

生长习性、分布和用途 使君子属约17种，我国产2种，本文描述1种。落叶攀援状灌木，长10m以上。适生于肥沃湿润疏松的酸性土。分布于湘、赣、闽、台、琼、粤、桂、滇、川、黔。印度、缅甸、菲律宾也有。花色鲜红，供棚架观赏。果为驱蛔虫特效药。

开花结实 4～6年生开始开花结实，正常结实期在10年生以后，结实无大小年现象。特

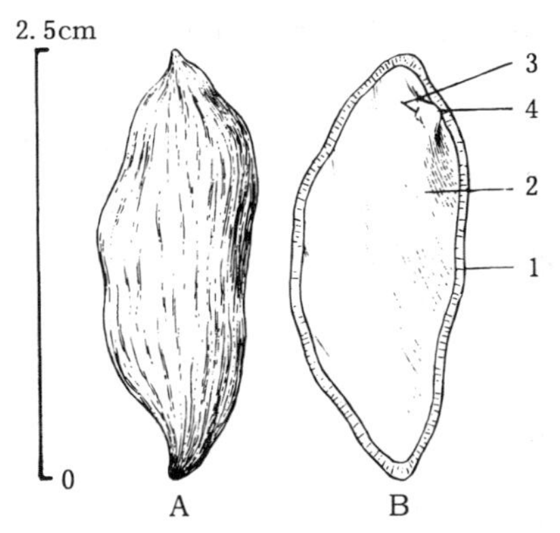

图 1　使君子种子外形(A)及其纵切面(B)
1. 种皮　2. 子叶　3. 胚芽　4. 胚根
(黄应钦绘)

寒年份结实量明显减少,正常年份每年结实量均较多。花两性,顶生或腋生。穗状花序似伞房花序状。苞片卵形至线状披针形,被毛。萼管长 5～9cm,萼齿 5 枚,小形。花瓣 5,长1.2～2.4cm,先端钝圆,初白色,后转红色。雄蕊 10,2 轮,外轮着生于花冠基部,内轮着生于萼管中部。子房下位,1 室,胚珠 3,垂悬于室壁顶端。花柱丝状,大部分与萼管靠合。据广西南宁1985～1988 年观察,花期较长,每年有 2 次明显的主花期和果实成熟期:第 1 次主花期始于 5 月中旬,5 月下旬为盛期,6 月下旬为末期,8 月上旬果实开始成熟,下旬为果熟盛期。第 2 次主花期始于 7 月中旬,8 月上旬为盛花期,9 月下旬花期结束,10 月下旬果实开始成熟,11 月上旬为果熟盛期。果革质,干燥,卵状纺锤形或椭圆形,具明显 5 棱,成熟时由青色转为褐色或青黑色;长 3～4.5cm,径 1.4～2.3cm,成熟后约 1 周脱落。每果有种子 1 粒。种子圆柱纺锤形或椭圆形,黄褐色,长 2.5～3cm,径 7～9mm。无胚乳,子叶大,表面不平。种子形态见图 1。

果实的采收调制和贮藏　部分果实转黑色时便可直接采摘或敲落后地上捡拾。生产上一般不剥去果皮而用果实直接播种。用于育苗的,宜选用外形发育饱满,颗粒较大,无虫蛀的果实,置室内摊放。果实可以稍稍晾干,但不能过度干燥,忌烈日下曝晒。稍晾干的果实千粒重约 1 600g,每千克约 620(580～670)粒。含水量约为 35%～45%。夏季和冬季成熟的果实均可用于播种。夏季成熟的果实可以不经贮藏而随采随播。10～ 11 月成熟的果实,因气温转低,不宜播种,需贮藏至次年春暖后播种。果实在冬季可裸露贮藏 30～40 天。贮藏期如超过 2 个月,需混湿沙,混沙贮藏的果实生活力可保持半年左右。

发芽和播种　种子无休眠习性。发芽时的日均温需在 20℃以上。经过贮藏的果实,播前宜浸水 8～10 小时。新采的果实播种前可不作任何预处理。1987 年 9 月中旬,广西林业科学研究所用新鲜种子在室外沙床作过发芽测定,播种时日均温 24℃,播后 9 天胚根开始萌发,发芽进程较慢,发芽盛期不明显,10 月下旬发芽终止,发芽率为 52%。留土萌发。发芽后 4 天上胚轴出土,具不育叶 2～4 枚,再过 2～3 天初生叶展开(图 2)。

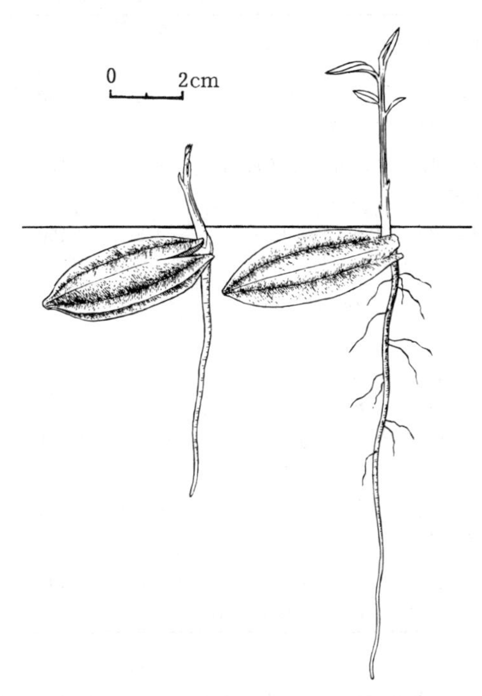

图 2　使君子种子(果实)萌发后第 4、10 天幼苗的生长情况
(黄应钦绘)

条播。也可在沙床密播，待胚根萌发后将芽苗移至苗床。每平方米播种量150～200g，覆土2～3cm。1年生苗出圃。

（林裕庚）

榄仁树（诃子）属

Terminalia L.

（使君子科 Combretaceae）

生长习性、分布和用途 本属约200种，我国产8种，又引入约8种，本文描述8种。多为落叶大乔木，少数为半落叶或季节性落叶大乔木。喜肥沃疏松的酸性土，贫瘠地生长不良。适生于无霜冻的热带地区，少数的耐寒种可分布至亚热带地区。主产于非洲热带及亚洲热带。这8个树种的名称、生长、分布及用途见表1。其中千果榄仁在《中国植物红皮书》中被列为渐危种。

表1 榄仁树属树种的名称、生长、分布和用途

中 名	学 名	树 高（m）	胸 径（cm）	分 布	用 途	供 稿
阿江榄仁（安心树）	*T. arjuna* Wight et Arn.	25	250	印度、斯里兰卡、马来西亚。滇、桂、粤、琼栽培	材用、油料、果治心病	604
毗黎勒（油榄仁）	*T. bellirica* (Gaertn.) Roxb.	18～35	100	滇南。东南亚至南亚	材用、鞣质、药用	618
榄仁树	*T. catappa* L.	25	100	粤、琼、台。滇栽培。南亚、大洋洲、南美	绿化、固沙、材用	604
诃子	*T. chebula* Retz.	15～30	100	滇南。粤、桂、琼栽培。南亚至东南亚	材用、果治喉痛，并含鞣质及黑、黄色素	604
海南榄仁（鸡尖）	*T. hainanensis* Exell.	20～25	50～100	琼特产。滇、桂、粤、闽、浙栽培	优质硬材，皮含鞣质	604
大翅榄仁	*T. macroptera* Guill. et Perr.	20～30	60～100	热带非洲。粤、琼、桂栽培	材质优，树皮可提制芳香原料	604
卵果榄仁（木勒榄仁）	*T. muelleri* Benth.	15～20	40～50	琼、滇、粤、桂、闽有栽培	绿化、用材、鞣质	604
千果榄仁（多果榄仁）	*T. myriocarpa* Huerck et M. -A.	40	200	桂、滇、藏。粤、闽栽培。南亚至东南亚	优质用材，鞣质	617

开花结实 千果榄仁10～12年生开始开花结实，其余7个树种8～10年生开始开花结实。始花后第3年所结的果实便可供育苗，但千果榄仁正常结实年龄在20年生以后，其余7种一般在15年生以后。结实大小年现象不明显。一般年份结实量均较多，特寒或特旱年份结实量才明显降低。千果榄仁为大型圆锥花序，海南榄仁多数为穗状花序成圆锥状排列，毗黎勒为穗状花序，其余5个种为穗状花序或呈圆锥状排列，腋生，间有顶生。毗黎勒和榄仁树具雄性及两性花，其余6种为两性花，偶有雄性花。花无梗，萼筒杯状，延伸于子房之上，萼

齿 5 或 4。无花瓣。雄蕊 10 或 8，2 轮，着生于萼筒上，花药背着。子房下位，1 室，花柱长，单一，伸出。胚珠 2，稀 3～4，悬垂。开花结实的物候期见表 2。

表 2　榄仁树属树种的开花结实物候期

树　种	观察地点	观察年份	开　花			果实成熟		果实脱落
			始　期	盛　期	末　期	始　期	盛　期	
阿江榄仁	南宁	1982～1984	6 月上旬	6 月上中旬	6 月下旬	翌年 3 月上旬	3 月下旬	5～6 月
毗黎勒	勐腊	1987	4 月上中旬	4 月中旬	4 月末	10 月中旬	11 月	11～12 月下旬
榄仁树	海南	—	4 月中旬	5 月中旬	6 月中旬	7 月上旬	9 月上旬	8 月中旬～10 月上旬
诃子	南宁	1986～1988	5 月上旬	5 月中旬	5 月下旬	10 月下旬	12 月中下旬	11 月下旬～翌年 3 月中旬
海南榄仁	南宁	1987～1989	7 月中旬	7 月下旬	8 月中旬	翌年 1 月下旬	2 月中旬	2 月下旬～3 月下旬
大翅榄仁	南宁	1983～1986	6 月上旬	6 月上中旬	6 月下旬	翌年 3 月上旬	3 月下旬	5～6 月
卵果榄仁	南宁	1987～1989	5 月中旬	5 月下旬	6 月上旬	10 月下旬	11 月中旬	12 月上旬～翌年 1 月中旬
千果榄仁	勐腊	1982	9 月中旬	9 月下旬	10 月中旬	11 月下旬	12 月中旬	翌年 1 月～2 月

果实成熟时转为黄色或红色，充分成熟时为黑褐色。果实成熟 1 周后开始脱落，但海南榄仁和大翅榄仁等果实可在树上悬挂 1～2 个月。果实成熟持续的时间较长。多数树种同一植株果实的成熟期先后长达 1 个月以上，诃子果实的成熟期先后长达半年。果为核果或假核果，肉质有时革质或木栓质，具棱或 2～5 翅。内果皮具厚壁组织。种子 1，无胚乳，子叶旋卷。果实及种子的形态特征见表 3、图 1。

表 3　榄仁树属树种果实和种子的形态特征

树　种	未熟果颜　色	成熟果实			种子（核）		
		形　状	大小（cm）	颜　色	形　状	大小（cm）	颜　色
阿江榄仁	黄绿色	椭圆形，具五棱	长 3.6～4.2 径 3.2～3.6	黄褐色	棱形	长 1.5～2.3 径 0.25～0.35	浅黄色
毗黎勒	绿色	卵形	长 4 径 3	灰绿色	卵形，具五棱	长 2～2.7 径 1.7～2	灰白色
榄仁树	青色	倒卵状椭圆形，两侧压扁，有棱或翅	长 3.5～5 宽 2.5～3.2 厚 2	青黑色	扁椭圆形，表面具网状粗纤维	长 3～4.5 宽 1.7～2.9 厚 1.3～2	浅褐色
诃子	青绿色	倒卵形至椭圆形，具 5～6 棱	长 2.7～4.5 径 2～2.3	绿带紫红色	椭圆形或倒卵形，先端突尖	长 1.7～2.1 径 0.9～1.3	浅黄色
海南榄仁	青色	三棱状椭圆形	长 2.5～3.5 径 1.5～2	黄色或青紫色	棱形	长 1～1.5 径 0.4	—
大翅榄仁	黄绿色	五棱状椭圆形	长 3～3.5 径 2.5～3	浅黄至黄褐色	棱形	长 1.8～2.4 径 0.3～0.37	淡黄色
卵果榄仁	浅黄绿色	卵状	长 1.6～1.9 径 1～1.5	暗紫红色	卵形，扁椭圆形，先端具短尖	长 1.4～1.7 宽 0.9～1.2 厚 0.8～0.95	灰褐色或浅褐色
千果榄仁	红色	卵形，具 3 翅，2 大 1 小	长 0.3 宽 1.2（带翅）	浅褐色至苍黄色	卵形	长 0.2 径 0.1	乳白色

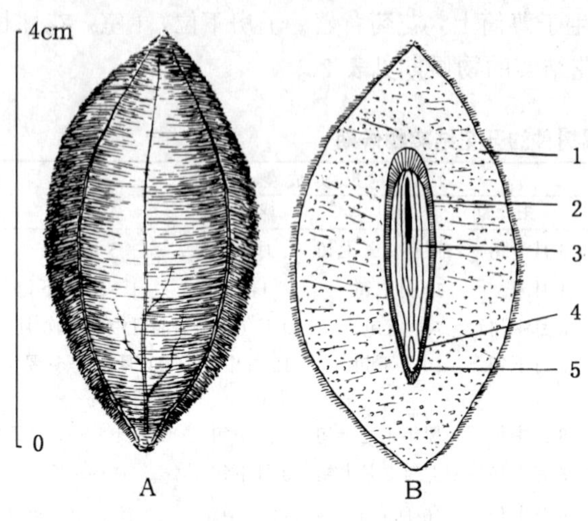

图 1　榄仁树果核外形（A）及其纵切面（B）
1. 内果皮　2. 种皮　3. 子叶　4. 胚轴　5. 胚根
（黄应钦绘）

果实的采收调制和种子贮藏　果实成熟盛期用采种钩刀将果穗采下，也可以摇动树枝或用竹竿敲打，震落后在地面收集。调制的方法大致分为 2 类。一类是具革质果皮的果实如阿江榄仁、海南榄仁和千果榄仁，可以带果皮贮藏而不作任何处理，播种时可连果播下；也可以在播前浸水 2～3 天，俟外层革质果皮软化后剥出果核。果实可以晾干在室内贮藏，不宜在日光下曝晒。贮藏期一般为 1～2 个月，冬季贮藏期为 3～4 个月；湿沙贮藏，生活力可延长 2～4 个月。据程必强和马信祥（1987）报道，千果榄仁耐干藏，普通室内干藏 9 个月后有的仍能保持 76%的发芽率，但有的发芽率也降到 31%。他们没有对这样巨大的差异作出解释。另一类调制方法是毗黎勒、诃子、卵果榄仁和榄仁树果皮肉质的果实，采集后堆沤数日至数十日，果皮果肉软化后置水中搓擦，淘出果核，用作播种材料，通称种子。种子不宜脱水，在室内稍晾干后即可混沙贮藏，贮藏期一般为半年左右。鲜果出籽（核）率及质量等见表 4。

表 4　榄仁树属树种果实的出籽率和种子①的质量

树　种	出籽率（%）	千粒重（g）	每千克粒数（粒）
阿江榄仁	带翅果	4 000～5 000	200～250
毗黎勒	40～50	3 000～3 800	260～340
榄仁树	20	4 500～5 500	180～220
诃　子	50	1 200～1 800	550～830
海南榄仁	带翅果	150～180	5 500～6 700
大翅榄仁	带翅果	4 200～5 200	190～240
卵果榄仁	70	650～850	1 170～1 540
千果榄仁	带翅果	3.2～3.6	280 000～310 000

①　阿江榄仁、海南榄仁、大翅榄仁和千果榄仁 4 种为带翅果，毗黎勒为假核果，其余 3 种为果核

发芽和播种　种子无休眠习性。发芽时日均温需在 20℃以上。未经处理的果实用于播种时，播前需用清水浸泡 1 昼夜。经过调制的种子，播前浸泡 1～2 小时或不作任何处理即可播种。1981～1987 年，广西林业科学研究所和西双版纳热带植物园在室外沙床和室内发芽皿中作过发芽测定（播种时日均温20～30℃）：千果榄仁播后 5～6 天开始发芽，再经 6～10 天发芽结束，4 天的发芽百分数为 25.5%，15 天的发芽率为 27%。其余 7 种，播后 3～4 周胚根开始萌发，且发芽进程较慢，从始期至终期需要 16～34 天，发芽盛期不明显。其中诃子的发芽较困难，不同年份种子的发芽率差异颇大：1960 年广西林业科学研究所湿沙层积后播种，发芽率曾达 20% 左右，是发芽率较高的年份；1986～1987 年曾用新鲜种子和沙藏种子播种 2

次，发芽率仅为5%～9%。如何促进这个树种的种子发芽，尚需进一步研究。另6个树种种子的发芽情况见表5。

表5　榄仁树属树种的发芽能力[①]

树　种	发芽率（%）	
	计算天数	一般数值
阿江榄仁	45	45
毗黎勒	52	75
榄仁树	54	72
海南榄仁	64	50
大翅榄仁	48	46
卵果榄仁	52	54

①　广西南宁，室外沙床，气温22～28℃

出土萌发。发芽后5～8天子叶出土，再过6～9天展出初生叶。榄仁树种子的萌发和幼苗初期生长情况见图2。

条播。带果播种的，覆土0.5～1cm，不宜过厚，以免影响子叶出土。每平方米播种量：阿江榄仁和大翅榄仁为380～500g（带翅果）；海南榄仁为25g；千果榄仁为4g；卵果榄仁为75～100g；毗黎勒为150～180g；榄仁树为380～500g。播种后苗床需经常保持湿润。1年生苗出圃。

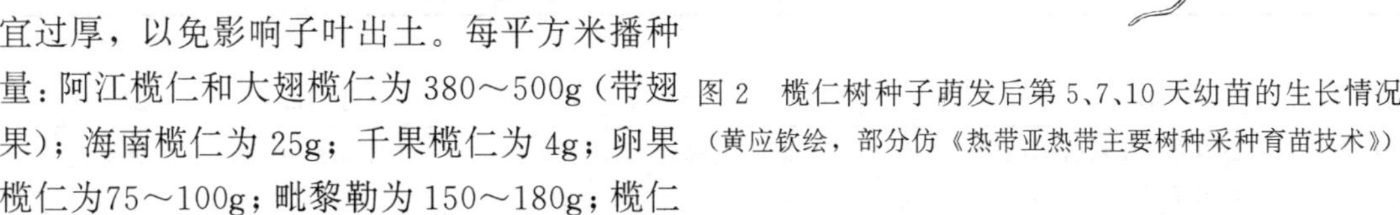

图2　榄仁树种子萌发后第5、7、10天幼苗的生长情况
（黄应钦绘，部分仿《热带亚热带主要树种采种育苗技术》）

（林榕庚）

野牡丹属
Melastoma L.

（野牡丹科　Melastomataceae）

生长习性、分布和用途　野牡丹属约100种，我国9种，本文描述2种。常绿灌木，高1～3m。适生于丘陵空旷山坡及山坡中下部山谷或水边。耐干旱瘠薄，是酸性土指示植物。分布于亚洲热带地区。这2个种都可入药。它们的名称、生长、分布和用途见表1。

表1　野牡丹属树种的名称、生长、分布和用途

中　名	学　名	树高（m）	分　布	用　途
野牡丹	*M. candidum* D. Don	1～2	滇、桂、粤、琼、闽、台。中南半岛	根叶入药，解毒消肿，收敛止血，观赏
毛　稔	*M. sanguineum* Sims	2～3	桂、粤、琼。印度、马来西亚、印度尼西亚	果可食，根叶入药

开花结实　2～3年生开始开花结实，正常结实期在5年生以后。野牡丹结实大小年现象不明显，毛稔间隔期为1～2年。花两性，单生或3～5朵组成伞房花序，顶生，基部具总苞2，披针形。花5数，萼坛状球形，被粗毛，檐5裂。野牡丹花冠玫瑰红色或粉红色，花瓣倒卵形。毛稔花冠紫红色，花瓣广倒卵形，上部略偏斜。雄蕊10，5长5短，长者药隔基部伸长，弯曲，末端2裂；短者药隔不伸长，基部具1对小瘤。花药顶孔开裂。子房半下位，5室，胚珠多数，密被刺毛。这2个种的开花结实物候期见表2。

表2　野牡丹属树种的开花结实物候期

树　种	观察地点和年份	开　花			果实成熟		果实脱落
		始　期	盛　期	末　期	始　期	盛　期	
野牡丹	广西南宁1987～1989	4月上旬	5月中旬	9月中旬	6月中旬	8月上中旬	8月中旬～11月下旬
毛稔	广西上思1987	3月下旬	6月上旬	10月上旬	7月上旬	10月上旬	翌年1月

蒴果，坛状球形，包于宿萼中。成熟时果呈环裂、顶孔开裂或不规则裂，种子露出。种子镶于肉质胎座上，长肾形。每室有种子多数。发育不良种子为浅紫红色或乳白色，极小。种皮表面凹凸不平。外种皮厚，内种皮薄。无胚乳，胚稍弯。果实和种子的形态特征见表3、图1。

表3　野牡丹属树种果实和种子的形态特征

树　种	果　实			种　子		
	形　状	大小（cm）	颜　色	形　状	大小（mm）	颜　色
野牡丹	坛状球形，密被黄褐色鳞片状糙毛	长1.0～1.6 径0.8～1.4	黄褐色	扁卵形	径0.5～0.6	紫黑色
毛稔	杯状球形，被红色长硬毛	长1.8～2.1 径1.8～2.2	黄绿色	三角状卵形	径0.4～0.5	黄褐色

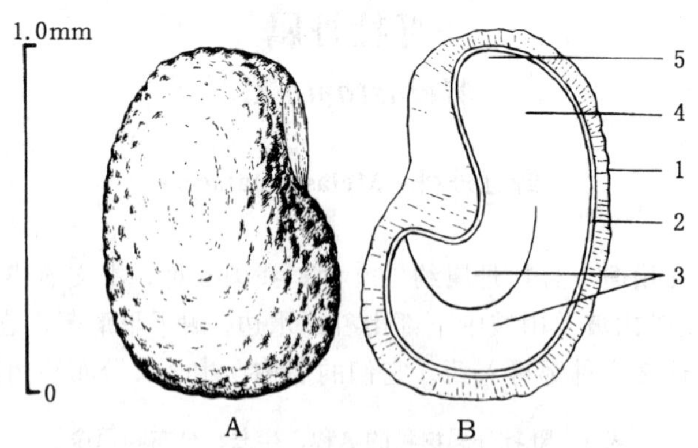

图1　野牡丹种子外形（A）及其纵切面（B）

1. 外种皮　2. 内种皮　3. 子叶　4. 胚轴　5. 胚根

（黄应钦绘）

果实的采收调制和种子贮藏　果实成熟盛期用手采摘。采得的果实在室内晾干1～2天，剥去果皮，放到光滑的小容器中，加入少量的水，用玻棒搅动使种子脱落，取出肉质胎座后用滤纸在漏斗上过滤，得到的种子连滤纸一起在室内无风处晾干。种子可以干藏，贮藏期约为 6 个月。种子的净度、质量等见表 4。

表 4　野牡丹属树种的出种率和种子净度、质量

树　种	出种率（%）	净　度（%）	千粒重（g）	每克纯净种子粒数（粒）
野牡丹	—	90	0.05	20 000
毛稔	6.4	90	0.08	12 500

发芽和播种　种子无明显的休眠习性。1987～1988 年，广西林业科学研究所在室内对这 2 个树种作过发芽测定，结果见表 5。

表 5　野牡丹属树种的发芽能力及其测定条件

树种	基质	室内温度(℃)	发芽势（%）		发芽率（%）	
			计算天数	一般数值	计算天数	一般数值
野牡丹	滤纸	28	6	15	16	24
毛稔	滤纸	18	—	不明显	20	8

出土萌发。胚根萌发后 3 天子叶出土，20 天后初生叶出现，生长情况见图 2。种子发芽率低，适于室内盆中密播后移植。播种盆的表层土应过细孔筛，淋足水分后再覆一层薄细土，用板压平，然后将种子均匀撒播。每平方米约播 1～2g。苗期用喷雾器淋水。约 1 年生苗出圃。

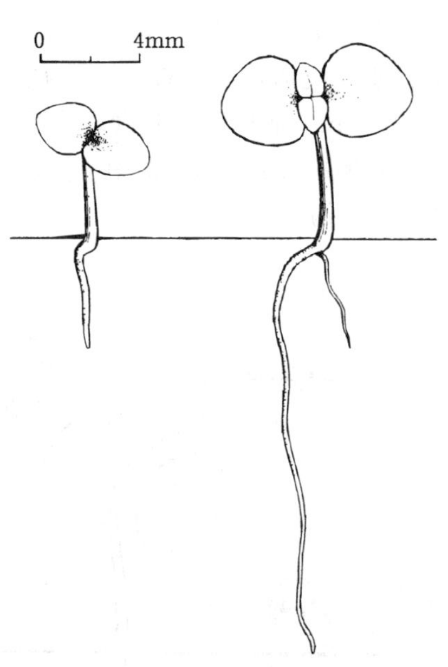

图 2　野牡丹种子萌发后第 5、20 天幼苗的生长情况
（黄应钦绘）

（曾　玲）

细叶谷木

Memecylon scutellatum (Lour.) Hook. et Arn.

（野牡丹科　Melastomataceae）

生长习性、分布和用途　谷木属约 130 种，我国约 11 种，本文描述 1 种。常绿灌木，高 2～4m。分布于琼、粤、桂。缅甸、越南至马来西亚也有。枝叶茂盛，树形经修剪可供观赏。

开花结实　2～3 年生开始开花结实，大小年现象不明显。花两性。聚伞花序腋生，长约 0.8～1cm。花萼杯状，檐部平截，具 4 短尖头。花瓣 4，蓝紫色，近椭圆形，长约 2.5cm。雄蕊 8，长约 3mm。花药短，纵裂。药隔膨大，伸长成圆锥形，脊上具 1 环状体。子房下位，1

室，胚珠 6～12 颗，生于特立中央胎座上。花柱丝状，单一。据海南屯昌所作的物候观察，5 月中旬为始花期，6 月中旬为盛花期，7 月上旬为末花期；翌年 2 月上旬果实开始成熟，2 月下旬为果熟盛期，3 月中旬为果熟末期。浆果状核果，近球形，直径6～7mm。果皮密布小瘤状突起。果实顶端具环状宿存萼檐。外果皮肉质，内果皮骨质，浅褐色。果核球形，径 4.5～5.5mm，种子 1。种皮薄，无胚乳，子叶折叠。细叶谷木果核的形态及其内部结构见图 1。

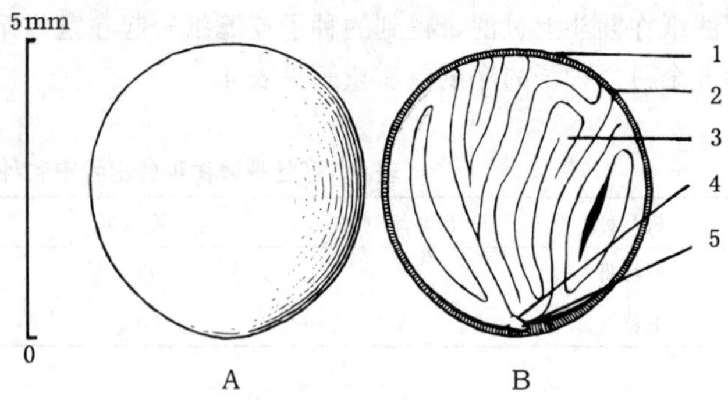

图 1　细叶谷木果核外形（A）及其纵切面（B）
1. 内果皮　2. 种皮　3. 子叶　4. 胚轴　5. 胚根
（黄应钦绘）

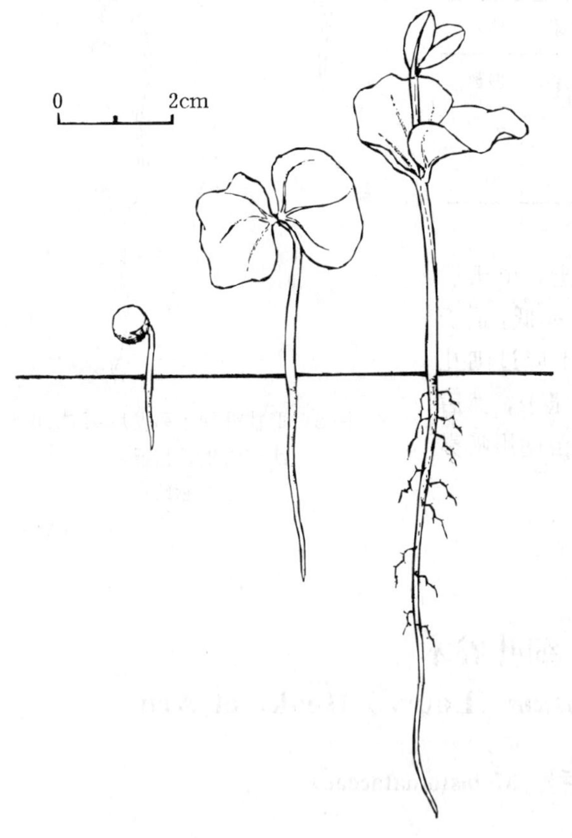

图 2　细叶谷木种子（核）萌发后第 10、15、28 天幼苗的生长情况
（黄应钦绘）

果实的采收调制和种子贮藏　果实成熟盛期用手采摘。采回的果实装入布袋搓揉后，置水中淘洗。所得果核即为播种材料，通称种子。鲜果的出籽率为 35%～40%。种子净度可达 95%～98%。千粒重约 105g，每千克有纯净种子约 7 100～12 500 粒。种子晾干后干藏，贮藏期半年至 1 年。

发芽和播种　种子休眠现象不明显。1988 年 4 月 8 日，广西林业科学研究所在室外沙床作过发芽测定，播种时日均温约为 26℃。4 月 20 日胚根伸出，第 3 天便进入发芽盛期，7 天中的发芽百分数为 35%；5 月 6 日发芽终止；以 28 天计，发芽率为 66%。出土萌发。子叶在胚根伸出后 10 天出土，26 天初生叶初现。细叶谷木的萌发和幼苗生长情况见图 2。

条播。每平方米播种 12～14g，覆土 1cm。半年生苗可以出圃，或移植修剪造形，2 年后供庭院绿化。

（曾　玲）

冬　青　属

Ilex L.

（冬青科　Aquifoliaceae）

生长习性、分布和用途　冬青属约400种，分布在亚洲、美洲的热带和温带地区，有些种产欧洲和大洋洲。我国约140种，产秦岭和长江以南各地。常绿或落叶乔木或灌木。枝叶浓密，可为小型野生动物和雀鸟提供隐蔽场所。核果入冬成熟，红色且经久不落，可在冬季为鸟类提供食物，也可供观赏。材质坚硬，可作细木工料。树皮可提取栲胶。叶、果、皮、根具药用价值。种子含油。本文描述6个种，它们的名称、生长习性、成年时树高以及分布和用途请见表1。

表1　冬青属树种的名称、成年时树高、分布和用途

中　名	学　名	习　性	树高（m）	分　布	用　途	供　稿
枸　骨	*I. cornuta* Lindl.	常绿灌木，稀小乔木	2～5	长江流域中下游各省。朝鲜半岛	观赏、药用	1001
榕叶冬青	*I. ficoidea* Hemsl.	常绿乔木	12	长江流域以南各省	材用	601
大叶冬青	*I. latifolia* Thunb.	常绿乔木	20	及台、华东。日本	材用、栲胶、药用、观赏	1001
大柄冬青	*I. macropoda* Miq.	落叶乔木	13	皖、浙、赣、鄂、湘。朝鲜半岛和日本	材用	1008
冬　青	*I. purpurea* Hassk.	常绿乔木	13	陕西及长江流域以南各省	材用、观赏、药用、栲胶	1001
铁冬青	*I. rotunda* Thunb.	常绿乔木或灌木	5～15	长江流域以南及台	染料、油脂、药用、栲胶、材用	601

开花结实　冬青属树种开始结实的年龄到来得较迟。本文描述的6个种中，榕叶冬青、大叶冬青、大柄冬青和铁冬青通常每年都能结实，没有明显的大小年现象。冬青常常隔1年大量结实1次，枸骨则每2～3年才有一次较多的结实。花单性异株，稀杂性。聚伞花序或伞形花序，多为簇生，很少单生。花小，白色、淡黄色或紫红色。花萼4～5，分离或基部合生，宿存。花瓣4～5。雄蕊与花瓣同数，着生于花瓣的基部，花丝短，花药长椭圆状卵形。子房上位，卵圆形，3室至多室。每室1～2胚珠生于中轴胎座上，下垂。花柱短，柱头通常4～6裂。浆果状核果，圆球形，内有2至多颗分核。核壳骨质，常有纵向槽或网纹槽，每个分核有1枚种子。胚小，胚乳丰富。果实和果核的外形以及种子的解剖构造见图1。这6个树种的开花结实习性见表2。

表 2 冬青属树种的开花结实习性

树 种	开始结实年 龄	观测地点和时间	花 期	花 色	果形和分核数	果熟期	结 实间隔期
枸 骨	15	南 京 1989～1990	4～5 月	黄绿色	球形，4 分核	10 月下旬～11 月中旬，鲜红色	2～3
榕叶冬青	7	—	3 月中旬～4 月下旬	白色或绿黄色	球形，4～6 分核	10 月中旬～11 月上旬，红色	0
大叶冬青	15	南 京 1989～1990	4 月上旬～5 月中旬	淡黄色	球形，4 分核	10 月中旬～11 月中旬，红色至褐色	0
大柄冬青	—	南 京 1989～1990	4 月下旬～5 月中旬	白色	球形，7～9 分核	9 月下旬～10 月上旬，红色	0
冬 青	15	南 京 1988～1989	5～6 月	紫红或淡紫色	椭圆形至近球形，4～5 分核	10 月中旬～11 月下旬，深红色	1
		南 宁 1984～1987	—			10 月下旬～11 月上旬，红色	0
铁 冬 青	7	南 宁 1984～1987	4 月上旬～4 月下旬	白色	球形至椭圆形，4～6分核	10 月下旬～11 月上旬，红色	0

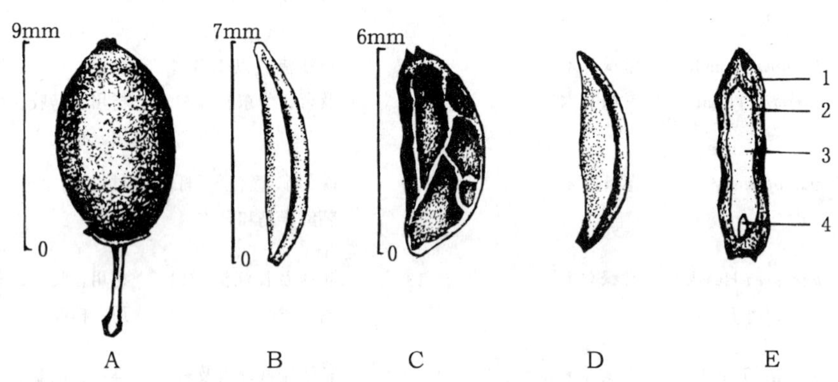

图 1 冬青核果外形（A）和分核外形（B）、枸骨分核外形（C）、铁冬青分核外形（D）和冬青分核的纵切面（E）

1. 核壳 2. 种皮 3. 胚乳 4. 胚

（田恒德绘）

果实的采收调制和种子贮藏 果实成熟后并不立即脱落，常在树上悬挂 1～2 个月甚至更长时间，有的种成熟果实呈红色或紫色，雀鸟喜食，如为获取种子仍需及时采收。枸骨叶缘有刺，雀鸟难以停留，受害较少，可以在冬季果实脱落前采集。采集时可摘取果实，也可敲击树干或果枝，在地面收集。核果用水浸渍 2～3 天，待果皮和果肉充分软熟后搓擦，淘洗，除去杂质后得到的分核即为播种材料，通称种子。种子忌曝晒，晾干即可收藏待用。这 6 个树种果实的出籽率、种子的净度和种子的质量数据已列入表 3。供来年春播的种子宜低温层积处理。气干种子密封贮藏于低温条件下，生活力可保持 1 年以上。

表 3　冬青属树种果实的出籽率、净度和种子质量

树　种	出籽率（%）	净度（%）	千粒重（g）	每千克种子粒数（万粒）
枸　　骨	30	85～96	18～21	4.7～5.6
榕叶冬青	16	80～95	0.8～1.2	83～125
大叶冬青	15	80～95	10～13.5	7.4 ～10
大柄冬青	18	95～100	15～17	5.8～6.7
冬　　青	20	80～95	8～11	9～12.5
铁 冬 青	17	85～95	4～6	17～25

发芽和播种　冬青属种子具有休眠习性，休眠期长短因种而异，有的较短，如榕叶冬青；有的长达 2 年，如枸骨。目前用切开法估测优良度，或用四唑染色法估测生活力。林业部南方林木种子检验中心曾对一份大柄冬青的种子用 1%四唑溶液浸泡 4 小时，测得生活力为 85%，但很难得到真实的发芽率来检验这个数值的可靠程度。广西林业科学研究所在室内测定榕叶冬青种子，基质为滤纸，恒温 28℃，置床后 28 天发芽百分数为 28%，42 天的发芽率为 42%。

本文所列冬青属 6 个种都可用种子繁殖。春播或秋播，条播或撒播均可。由于具有休眠习性，冬青属播前都需要催芽处理。目前已有报道的几种处理方法如下：①用夜间 20℃，白天 30℃的变温层积处理 60 天，而后在 5℃下处理 60 天。②大柄冬青和冬青可于播前 3～4 个月用湿沙层积至种粒"露白"播种。③我国南方冬季气温较高，果实采收脱粒后可以秋播。广西林业科学研究所 1988 年 10 月 18 日用当年新采的铁冬青种子在室外沙床上播种，播后 130 天胚根萌发，再 7 天后萌出子叶，1 个月后展出初生叶，1989 年 4 月发芽终止。

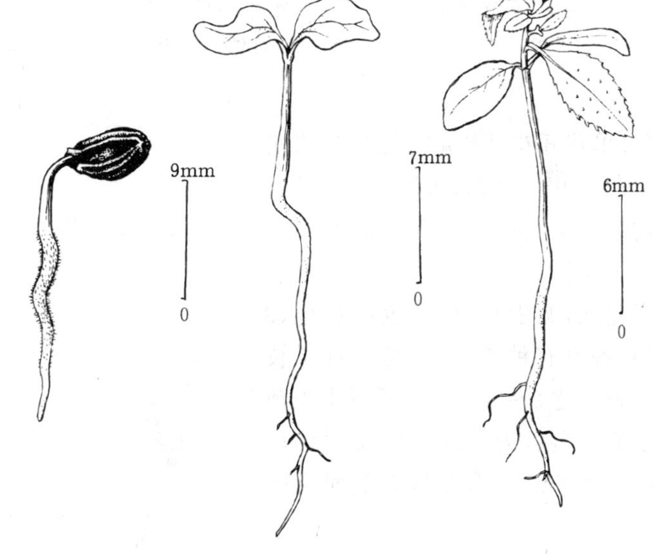

图 2　冬青属之一种的种子萌发后第 3、10、20 天的幼苗生长情况
（张世经绘）

冬青属的种子属于子叶出土萌发（图 2），发芽持续时间长。例如上述铁冬青发芽过程长达半年，枸骨播后第 3 年仍有种子萌发。一般用 2～3 年生苗出圃。如作为观赏栽植，可截干培育 3～4 年后出圃。有的地方常在枸骨树下挖掘天然下种苗移栽培育。

（吴琼美）

南　蛇　藤

Celastrus orbiculatus Thunb.

（卫矛科　Celastraceae）

生长习性、分布及用途　南蛇藤属约50种，我国20余种，本文仅描述南蛇藤1种。落叶，藤状灌木，长达12m，丛生。生于荒山坡、阔叶林边或灌丛内。分布于东北、华北、西北、华东及华南各省。朝鲜半岛、西伯利亚、日本也有。

庭园垂直绿化树种，入秋叶色变红，绽开的蒴果露出红色假种皮及白色种子。根、茎、叶及果壳入药。种子含油率达50%，可供工业用。

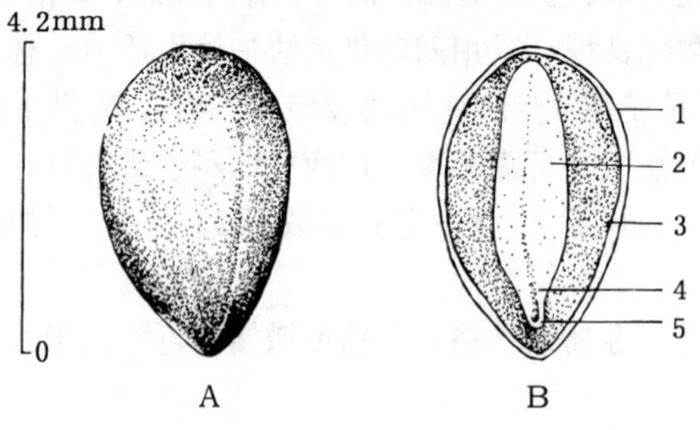

图1　南蛇藤种子外形（A）及其纵剖面（B）
1. 种皮　2. 胚乳　3. 子叶　4. 胚轴　5. 胚根
（梁鸣绘）

开花结实　3年生以上开始开花结实，5年生以后为正常结实年龄，几无间隔期。雌雄异株。聚伞花序腋生，3～7花。花小，淡绿色。雄花萼片5，花瓣5，雄蕊5，着生于杯状花盘边缘，退化雌蕊柱状。雌花子房上位，3室，基部包被于杯状花盘中，但不与之合生，柱头3裂，雄蕊不育。6～7月开花，9～10月果实成熟。据哈尔滨1963～1980年观察，平均始花期在6月9日左右，末期在7月10日左右。单花开花日数13天左右。果实初熟期在9月上旬，末期在10月上中旬。果实初熟至脱落间隔期22～30天。蒴果球形，径7～9mm，顶部有宿存花柱，刺尖状，橘黄色或黄色，3瓣裂，每室有种子1～2粒。种子长4～5mm，宽2～3mm。种子着生于蒴果基部，白色，为深红色肉质假种皮包被。种子的外形及其纵剖面见图1。

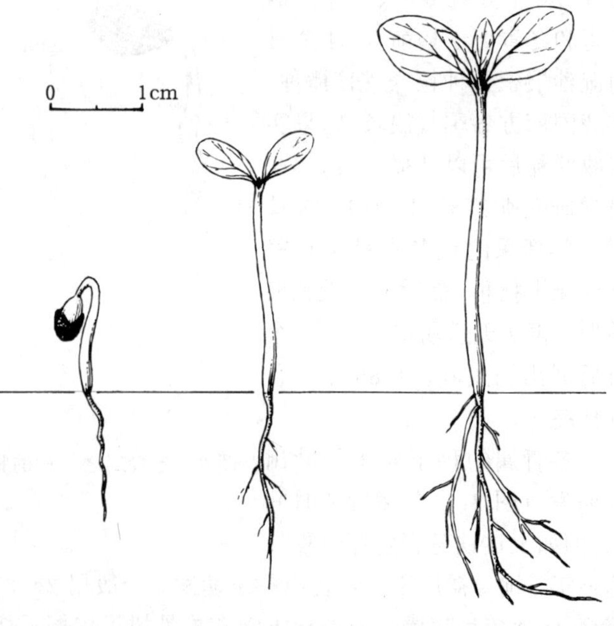

图2　南蛇藤种子萌发后第3、10、22天幼苗生长情况
（曹雅范绘）

果实的采收调制和种子贮藏　9～10月份，当蒴果含水量减少，种粒饱满，果皮稍裂时采种。从藤上摘取或捋下蒴果，晒干经敲打后去杂质，可得纯净种子。出种率44%。种子千粒重8～10g，发芽率60%，每千克种子有10万～12.5万粒。低温

(5℃左右)干藏至1.5年的种子发芽率未见显著降低。

发芽和播种 种子属于浅休眠类型,播前15～20天应在10～20℃下层积催芽。春季高床条播,每平方米播种4g,覆土厚1.5～2cm。出土萌发(见图2)。1年生苗高20～30cm,2年后出圃栽植。

(周德本、罗丽芬)

卫 矛 属

Euonymus L.

(卫矛科 Celastraceae)

生长习性、分布和用途 本属约200余种,产于欧洲、亚洲、美洲和大洋洲,我国约100余种,本文描述5种。落叶或常绿灌木或小乔木。喜光,能耐荫,喜温暖气候及肥沃湿润土壤。各树种的名称、生长、分布及用途见表1。

表1 卫矛属树种的名称、树高、分布和用途

中 名	学 名	树高(m)	分 布	用 途	供 稿
卫 矛	*E. alatus* (Thunb.)Sieb.	2～3	南北各省均有分布	观赏、药用、油料	1003
丝棉木(白杜、桃叶卫矛)	*E. bungeanus* Maxim.	4～8	东北南部、华北、西北东部、西南东部、华中、华东。朝鲜半岛	观赏、材用、药用、油料、橡胶	1004
大花卫矛	*E. grandiflorus* Wall.	10	西北东部、西南、鄂、湘	观赏、油料、橡胶	1003
冬青卫矛(大叶黄杨)	*E. japonicus* L.	5	我国南北各省均有栽培。原产日本	观赏、药用、橡胶、环保	1003
华北卫矛	*E. maackii* Rpur.	1～3	东北、华北。朝鲜半岛、日本、俄罗斯	观赏、材用、橡胶	403

开花结实 结实大小年现象不明显。花两性,稀杂性。腋生聚伞花序具多花,总花梗长。萼片和花瓣4～5数,淡绿色或淡黄绿色。花盘肥厚,肉质。雄蕊4～5,着生于花盘边缘,花丝短。子房上位,与花盘合生,3～5室,每室1～2胚珠。花柱短,柱头3～5裂。开花结实习性详见表2。蒴果,近球形或倒圆锥形,平滑,常4裂,具棱或翅,径约1cm,3～5室,每室1～2粒种子。果皮革质。种子卵形或椭圆形,常具一条明显种脊,种皮薄,包藏于红色或橘红色肉质假种皮内。胚直,呈匙形,浅黄色或绿色。子叶扁而宽。胚乳丰富,白色(详见表3和图1)。

表2 卫矛属树种的开花结实物候

树 种	开 始 结实年龄	丰年间隔期(年)	观察地点 年 限	花 期	果实成熟期	种子散落期
卫矛	—	0	—	4月上旬～5月上旬	10月中旬～11月中旬	—
			哈尔滨 1963～1980	5月20日～6月15日	9月5日～10月12日	初熟至脱落历时15～20天

（续）

树　种	开　始 结实年龄	丰年间隔 期（年）	观察地点 年　限	花　期	果实成熟期	种子散落期
丝棉木	5～8	1	—	5月中旬～5月下旬	10月下旬	—
			哈尔滨 1963～1980	6月10日～7月2日	8月30日～10月8日	初熟至脱落历时25～30天
大花卫矛	—	—	—	5月下旬～6月下旬	10月下旬	—
冬青卫矛	—	每年只结少量果实	哈尔滨 1963～1980	6月～7月	9月～11月	—
华北卫矛	—	0	哈尔滨 1963～1980	6月9日～7月3日	8月30日～10月8日	初熟至脱落历时25～30天

表3　卫矛属树种的花、果实和种子形态特征

树　种	花　色	果实			种子					
		形　状	颜　色	大小(mm)	形　状	颜　色	大小(mm)	假种皮	胚	含油率(%)
卫矛	黄绿色	4深裂，近球形	棕色带紫色	长8～10 宽12～14	球形或卵球形	褐色	长3.8～5.5 宽3.1～3.4 厚2.3～3.0	橘红色	浅黄色	44.4
丝棉木	黄绿色	倒圆锥形4棱，4裂	粉红色	长6.5～7 宽4～5	倒圆锥形或长三角形	淡黄或粉红色	长4～5.7 宽3.1～3.7 厚2.3～3	橘红色	淡黄色	—
大花卫矛	淡黄色	近球形，4棱，4深裂	浅红色	长10～13 宽8～10	长椭圆形、三角状倒圆锥形	亮黑色	长5～6.8 宽2.6～3.7 厚2.2～2.9	橘红色	绿色	52
冬青卫矛	淡绿色	扁球形，4浅裂	粉红色	长6～8 宽8～10	椭圆形或卵形，有棱或无棱	棕色	长4.7～6 宽3.4～4 厚3～3.8	橘红色	绿色	—
华北卫矛	黄色	4裂	粉红色	长8～10	椭圆形	红色	长3.9～6 宽2.5～3.6 厚1.8～2.7	橘红色	浅黄色	49

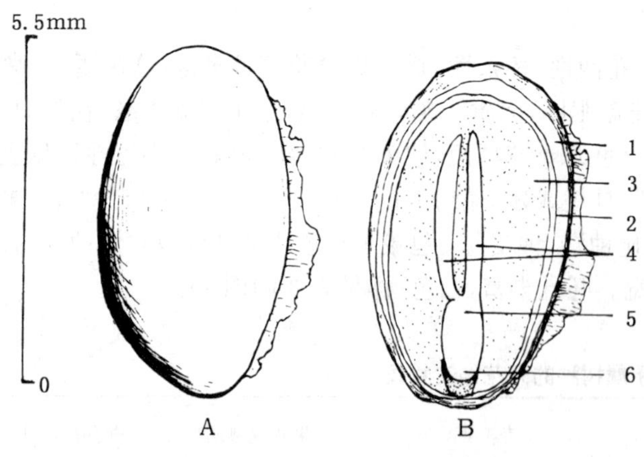

图1　卫矛种子外形（A）及其纵切面（B）
1. 假种皮　2. 种皮　3. 胚乳　4. 子叶　5. 胚轴　6. 胚根
（胡冬梅、林平绘）

果实的采收调制和种子贮藏　卫矛属的多数树种每年结实良好。果实充分成熟、果皮变色、部分蒴果开裂时即可采摘，或震落到铺有塑料布的地面上拾取。采得的鲜果摊开晾干，使果实充分开裂，揉搓脱出种子。经过风选和筛选除去杂物，即得到带有假种皮的种子。再在水中浸沤，搓去肉质假种皮后再行晾干。各树种的出种率及有关种子质量见表4。如不除去富含油脂的假种皮，贮藏期间容易招致虫害或滋生霉菌。但在揉搓假种皮时，切勿用力过度损伤种皮。晾干后的纯净种子可装袋干藏，或春播前低温层积保存种子；也可在1～3℃条件下密封低温干藏。在一般室内贮藏条件下，生活力约可保持2年。

表4　卫矛属树种的出种率和有关种子质量的数据

树　种	出种率（%）	净度（%）	千粒重（g）	每千克纯净种子粒数（万粒）
卫　矛	—	—	15～25	4～6.6
	（哈尔滨）60	—	21.8～24.8	4～4.6
丝棉木	5	—	26～30	3.3～3.8
	（哈尔滨）40	—	17.4～20	5.0～5.7
大花卫矛	—	—	23～25	4～4.3
冬青卫矛	—	—	6.5	15.3
华北卫矛	—	95～98	18～22	4.5～5.5
	（哈尔滨）40	—	20	5.0

发芽和播种　本属许多种的种子有休眠。供发芽测定的种子需先经0～10℃低温层积2～3个月。层积过的种子置发芽皿滤纸床上，在20～25℃条件下发芽期2个月左右。或用四唑法测定种子生活力。生产上可在采种后随即播种，在地里经受冬季低温，翌春即能发芽。卫矛发芽率可达70%～80%。未经低温层积处理的种子，春播后当年不能发芽，要到下一个春季萌发，且发芽率低。有的地方是在春季浸种2天后再混湿沙（湿度60%），在10～20℃温度下层积，40～60天即可萌动。出土萌发（见图2）。条播。每平方米播种20g，覆土1～1.5cm，

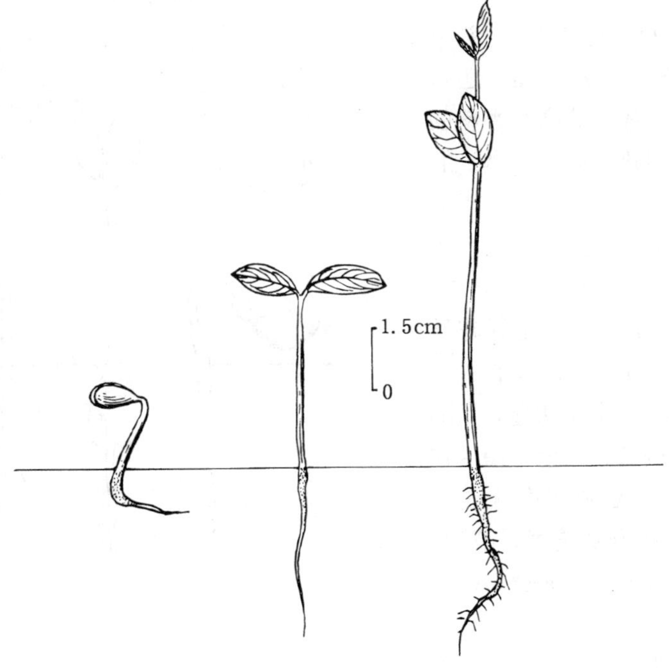

图2　卫矛萌发出土后第2、7、14天的幼苗生长情况

（胡冬梅、林平绘）

再用稻草或木屑保持床面湿润。当年生苗高可达 30～100cm。亦可用扦插方法繁殖。

（刘长江）

蒜头果（马兰后）

Malania oleifera Chun et S. Lee

（铁青树科　Olacaceae）

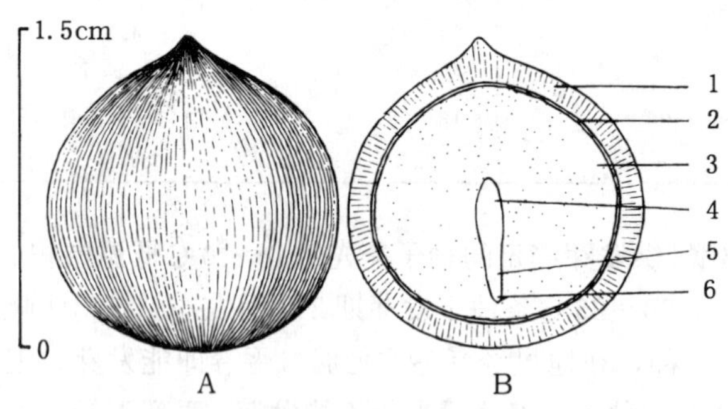

图 1　蒜头果果核外形（A）及其纵切面（B）
1. 内果皮　2. 种皮　3. 胚乳　4. 子叶　5. 胚轴　6. 胚根
（黄应钦绘）

生长习性、分布和用途　蒜头果属为我国特有属，只有本文描述的 1 种。常绿乔木，树高 15～25m，胸径 30～50cm。萌芽力强。耐－5℃低温，也耐 39℃高温，适生于石灰岩低山地区，砂岩、页岩发育的山地红壤、黄红壤也能生长。分布在滇东南和桂西，近年桂北也有发现。垂直分布在广西为海拔 300～800m，云南分布至 1 640m。种子出油率 31%，油脂中廿四（碳）烯-［15］-酸含量 67%，是合成麝香酮的主要原料。每 50kg 蒜头果油可以合成麝香酮 0.3～0.5kg，还可合成十五内酯、十五环酮等高级香料的定香剂。木材结构细致，为家具、船舶、木模、雕刻良材。大树残存不多，天然更新不良。《中国植物红皮书》将它列为稀有种，亟需保护。

开花结实　6～7 年生开始开花结实，正常结实期在 15 年生以后，有的百年老树结实仍较多。大小年明显，间隔期一般为 1 年。花两性，绿色，10～16 朵排成伞形花序状、复伞花序状或短总状花序状的蝎尾状聚伞花序，腋生。花萼筒小，上端 4（5）裂。花瓣 4（5），卵形，镊合状排列。雄蕊 8（10），2 轮。子房上位，长圆锥形，上部 1 室，下部 2 室，每室 1 胚珠，花柱 1 。据云南林业科学研究所观测，初花期在 4 月上旬，盛花期在 4 月中旬，末花期 4 月下

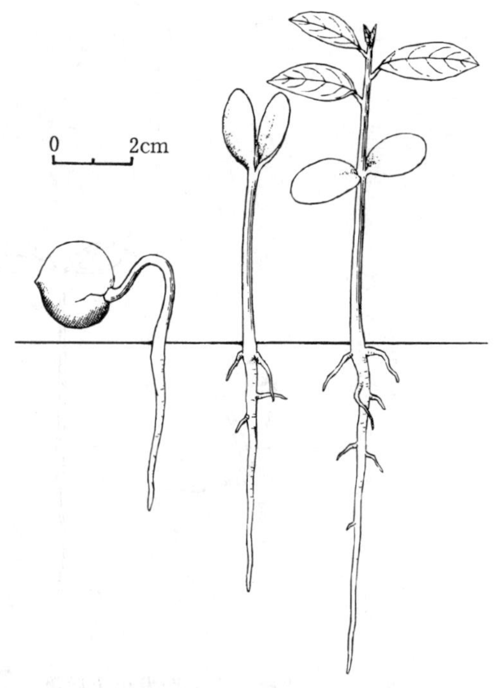

图 2　蒜头果萌发后第 3、6、20 天幼苗的生长情况
（黄应钦绘）

旬；果实发育期 5～9 月，10 月上旬至中旬成熟。果实未熟时绿色，熟时灰绿色或灰褐色。核果，具肉质果皮，扁球形，蒜头状，径 3.5～4.5cm，高 3～4cm。每千克有鲜果约 27 个。每果有种子 1 粒。果核球形、扁球形、蒜头形。核壳坚硬，木质，淡黄褐色。核径1.5～1.8cm。种皮膜质，有绒毛，黄褐色。种仁柔软，胚乳丰富，白色，胚直伸。蒜头果果核的外形及其解剖结构见图 1。

果实的采收调制和种子贮藏　果实成熟期中用竹竿敲打，或上树摇动树枝震落后在地面捡拾。果实收集后就地踩去果肉，运回洗净，所得果核用作播种材料，通称种子。鲜果的出籽率约 35%～38%，净度92%～100%。种子（果核）千粒重约 11kg，每千克约 70～110 粒。种子稍晾干时含水量为 25%～30%，忌失水，不能日晒或裸露贮存。贮藏或运输时均需混以湿沙，贮藏期 4 个月左右。

发芽和播种　种子无休眠习性。发芽时日均温宜在 22℃以上。宜随采随播，也可沙藏至翌年春播。据云南大田育苗的记载，随采随播的种子出苗率 74.5%；湿沙贮藏至次年 3 月播种，出苗率 57.4%。播后 21 天出苗，13 天的发芽百分数为 25%～30%，40～45 天发芽结束。出土萌发。子叶出土后 12 天露出初生叶。蒜头果的萌发和幼苗生长情况见图 2。

条播。播种沟深 5cm，宽 5cm，沟距 15cm。每平方米播种约 1kg，播后覆土 4～5cm。苗期需遮荫。1～2 年生苗出圃。

（张茂钦）

檀　　香

Santalum album L.

（檀香科　Santalaceae）

生长习性、分布和用途　檀香属约 20 种，我国引入 2 种，本文描述 1 种。常绿乔木，高 10～18m，胸径 18～25cm。生长稍慢。寄生性树种，已知寄主有 200 多种，以长春花（*Catharanthus roseus*）、铁刀木（*Cassia siamea*）、南洋楹（*Albizia falcataria*）等寄主较好。颇耐干热，原产地为干旱季节长的热带，能耐轻霜。喜质地疏松而排水良好的酸性土壤。原产太平洋岛屿，印度栽培最多，迈索尔邦有大面积森林，产量占全世界的 90%。我国引种有30～40 年的历史，目前台、粤、琼、闽、浙、桂、黔、川等省、自治区均有引种，已开花结实。世界著名的珍贵树种，材质极优，供雕刻、扇柄、工艺等用，油为名贵香料，木屑作焚香。

开花结实　8～10 年生开始开花结实，正常结实期在 20 年生以后。结实大小年间隔期为 1～3 年。花两性。三歧聚伞式圆锥花序，腋生或顶生。单被花。花被筒钟状，裂片 4（5），淡绿色。雄蕊 4（5），外伸，与花被片对生。花盘 4（5）裂，裂片鳞状，肉质，与雄蕊互生。子房半下位，1 室，有胚珠 1 颗。花柱红色，柱头浅 3（4）裂。据南宁 1987 年观测，檀香的始花期在 6 月下旬，7 月上旬为盛花期，7 月中旬花期结束；8 月上旬果实开始成熟，8 月中下旬为果熟盛期。此物候期与印度南部基本接近。核果，未熟时淡黄色，熟时转为紫色至黑紫色，近球形，高 0.8～1.2cm，径 0.9～1.3cm。外果皮膜质，有光泽。中果皮肉质多汁，顶部有灰白色环状花被痕。果实成熟后可在树上存留 5～7 天，然后脱落。每果有果核 1 粒。果核

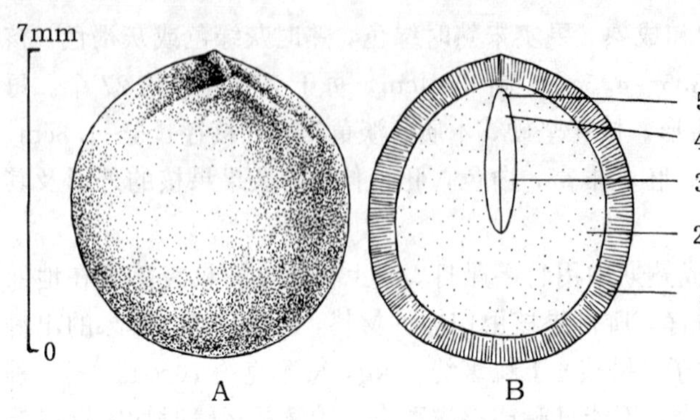

图 1 檀香果核外形（A）及其纵切面（B）

1. 内果皮及种皮 2. 胚乳 3. 子叶 4. 胚轴 5. 胚根

（黄应钦绘）

近圆形，具三棱线，顶部喙尖，棕褐色，高 6.5～8mm，径 6～7.5mm，内果皮厚。无外种皮。胚乳肉质，丰富，其内埋有一胚根朝上而伸直的胚。子叶圆柱状。檀香果核的外形及其解剖结构见图 1。

果实的采收调制和种子贮藏 果实的成熟期不甚整齐。采种宜选早期及盛期成熟的果实，直接采摘或用竹竿敲打震落后在地上捡拾。采回的果实室内堆沤 2～3 天，俟果肉充分软熟后装入竹筐或布袋，置水中反复搓揉，淘去果皮、果肉等杂质即得果核，用作播种材料，通称种子。鲜果的出籽率为 30%～40%。净度可达 99%。含水量约为 44%，忌脱水，不能日晒。种子（核）的千粒重约 190g，每千克有种子（核）5 000～5 600 粒。种子贮藏中需要保持湿润。已调制的种子裸露存放期不宜超过 1 周。运输或贮藏均需混以湿沙。可以随采随播，也可以混沙贮藏至次年春播。混沙的贮藏期一般为半年左右，不宜将种子贮藏度夏。

发芽和播种 种子无休眠习性。发芽时日均温宜在 20℃以上。1976 年和 1987 年，广西林业科学研究所在室外沙床作过发芽测定。1976 年采用的是引进的种子，3 月 20 日播种，4 月 13 日开始发芽，4 月末发芽终止，发芽率约为 80%。1987 年采用当地 14 年生幼龄母树所产的种子，采后在 9 月 1 日播种，10 月 2 日发芽，11 月 9 日发芽终止，发芽盛期不显著，发芽率为 60%。出土萌发。胚根萌发后约 7 天子叶出土，再过 3 天左右初生叶出现。檀香果核的萌发和幼苗初期生长情况见图 2。

目前国内外多将种子密播于沙床，苗高 10cm 左右时移至较大容器继续培育。移苗时应同时移入铁刀木、长春花等寄主树苗。培育的 1 年生苗随寄主树苗一同出圃。

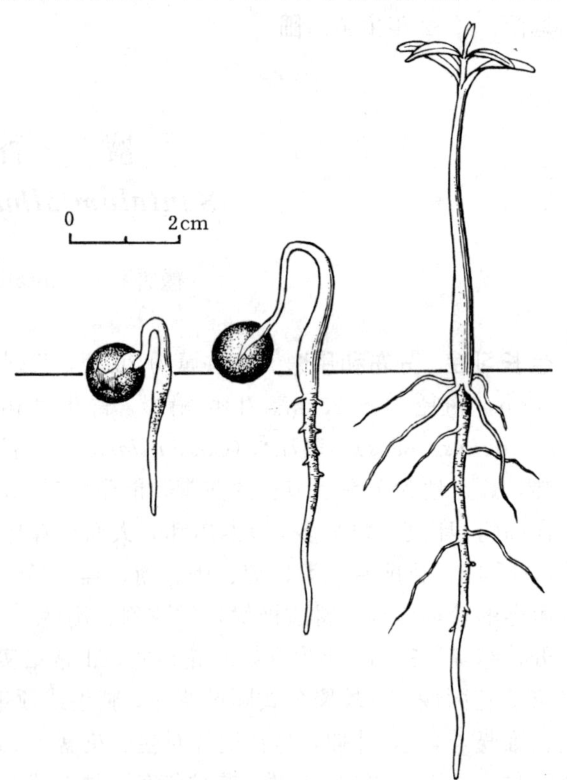

图 2 檀香种子（核）萌发后第 5、8、12 天幼苗的生长情况

（黄应钦绘）

（林榕庚）

胡颓子属

Elaeagnus L.

（胡颓子科　Elaeagnaceae）

生长习性、分布和用途　本属约 80 种，产亚洲、欧洲、北美。我国 55 种，各地均产，长江流域以南尤为普遍。本文描述 6 种。常绿或落叶灌木或小乔木，直立或攀援，通常具刺，稀无刺，全体被银白色或褐色鳞片或星状绒毛。速生，喜光，生活力强。本属经济价值颇高，不少种类的果实富含维生素、糖及有机酸，可食用。翅果油树种子含油脂，可食用、药用，花为蜜源。有的种类也是固沙造林树种。胡颓子属树种的名称、生长、分布及用途见表 1。

表 1　胡颓子属树种的名称、习性、树高、分布及用途

中　名	学　名	习　性	树高（m）	分　布	用　途	供　稿
沙枣	*E. angustifolia* L.	落叶乔木或小乔木	5～10	西北，华北、黑、辽、豫。皖、鲁、苏栽培。中亚、西亚至欧洲	食用，药用，蜜源，防风固沙	201
密花胡颓子（羊奶头）	*E. conferta* Roxb.	常绿大形木质藤本	长过 10	滇、桂。中南半岛、印度尼西亚、尼泊尔、印度	果食用，果汁制饮料	601
翅果油树（软毛胡颓子）	*E. mollis* Diels	落叶乔木或灌木	2～10	晋、陕	种子可榨油，出油达 30%～35%，食用，工业用，药用，水土保持	310
大果沙枣	*E. moorcroftii* Wall.	落叶乔木	10	藏、新、甘	食用，药用，蜜源，防风固沙	201
胡颓子	*E. pungens* Thunb.	常绿灌木	3～4	华东、中南、川、黔、陕。日本	药用，蜜源，果食用，观赏	1003
牛奶子	*E. umbellata* Thunb.	落叶灌木	1～4	西南、西北东南部、华北、华东、辽、鄂	药用，食用，观赏	1003

开花结实　花两性，稀杂性，单生或 1～7 花簇生于叶腋或叶腋短小枝上，成伞形总状花序。花部常被银白色或灰白色盾状鳞片或星状绒毛。花萼连合成筒，钟状或管状，在子房之上收缩，先端通常 4 裂。无花瓣。雄蕊 4，花丝短，不外露，着生于萼筒喉部，与裂片互生。子房上位，包被于花萼筒内，1 心皮，1 室，1 胚珠。花柱单一，细长，柱头偏向一边膨大或棒状。蜜花胡颓子每隔 1～2 年有 1 次结实丰年。本属开始开花结实年龄和开花结实物候见表 2。

表 2　胡颓子属树种开始开花结实年龄和开花结实物候

树　种	开始结实年龄（年）	观察地点	观测时间	开花期			果实成熟期	
				始　期	盛　期	末　期	始　期	盛　期
沙枣	4～5	内蒙古巴彦淖尔盟	—	5 月下旬	～	6 月下旬	9 月	10 月上旬

（续）

树种	开始结实年龄（年）	观察地点	观测时间	开花期			果实成熟期	
				始期	盛期	末期	始期	盛期
	—	宁夏中卫沙波头	—	5月上旬	5月中旬	5月下旬	10月上旬	10月下旬
	—	甘肃河西走廊	—	5月下旬	～	6月下旬	9月中旬	9月下旬
密花胡颓子	6～7	广西南宁	1987～1988	11月下旬	12月上中旬	12月下旬	翌年4月上旬	翌年4月中旬
翅果油树	3～4（萌生植株）	山西翼城	1973	4月下旬	5月上旬	5月中旬	9月上旬	～
大果沙枣	4～5	新疆	—	5月	～	—	9月	10月
胡颓子	—	江苏	—	9月	～	12月	翌年4月	翌年6月
牛奶子	6	江苏	—	5月	～	6月	8月	10月

瘦果为肉质化宿存的萼筒所包围，呈核果状，椭圆形或矩圆形，稀近球形，红色或黄红色。果核椭圆形或纺锤形，核壳为宿存萼筒去除外面肉质层后留下的骨质层，具8肋，内壁具白色丝状棉毛。内为瘦果，具薄层果皮，含1种子，具较硬的种皮。种子无胚乳，具2枚肉质子叶。本属果实和种子的形态特征见表3。沙枣果核的外形及其解剖结构见图1。

表3 胡颓子属树种果实和种子的形态特征

树种	果实类型	未熟果颜色	成熟果实			种子（果核）		
			形状	大小（cm）	颜色	形状	大小（mm）	颜色
沙枣	瘦果核果状	绿色	椭圆形	长 0.8～1.4 径 0.6～0.8	黄红色或粉红色，密被银白色鳞片	椭圆形	长7～14 径4	浅棕褐色
密花胡颓子	瘦果	青绿色密被白色斑点	椭圆形	长3.2～4 径2～2.5	深黄色或淡红色	纺锤形	长15～20 径7～9	棕褐色
翅果油树	瘦果	绿色	长果型（果椭圆形） 大宫灯（大圆） 小宫灯（小圆）	长2.86 径1.57 长1.8 径2 长1 径1.3	土黄色	纺锤状圆锥形或倒卵形	长12～20	栗褐色
大果沙枣	瘦果核果状	绿色	卵状矩圆形	长1.5～2 径1～1.5	黄褐色	窄椭圆形，先端钝，基部尖	长13～18	浅棕褐色
胡颓子	瘦果核果状	—	椭圆形	长 1.2～1.4 径 0.6～0.8	被褐色鳞片，熟时红色	长椭圆形	—	浅棕褐色
牛奶子	瘦果核果状	绿色	球形或卵圆形	长 0.6～0.9	被褐色鳞片，熟时红色	椭圆形	—	棕色

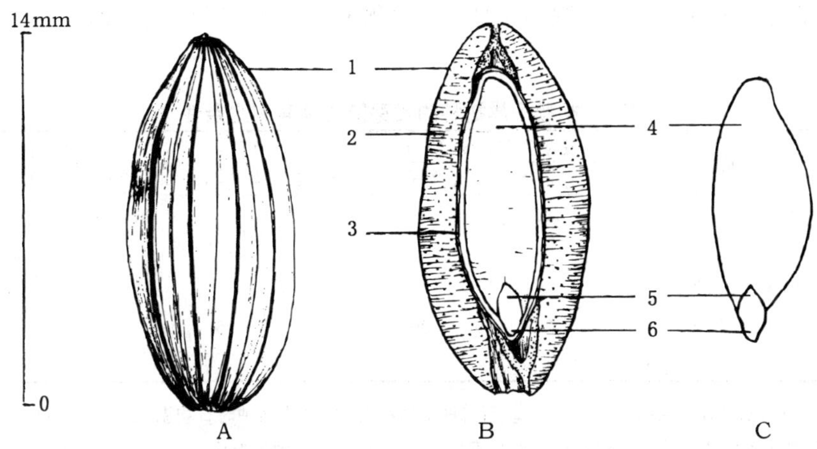

图1　沙枣果核外形（A）及其纵切面（B）和胚（C）

1. 核壳（肉质萼筒的坚硬内层）　2. 果皮　3. 种皮　4. 子叶　5. 胚轴　6. 胚根

（姚桂英绘）

果实的采收调制和种子贮藏　果实成熟后从树上直接采摘，或敲击枝条使其脱落。果实摊开干燥，或置水中搓洗淘净，漂去或晾干后筛去果肉后所得果核即为播种材料，通称种子。新鲜饱满的沙枣种子发芽率在90%以上，含水量在18%以下，在通风干燥处贮藏，5～6年以后发芽率尚在60%～70%。9月初翅果油树果皮变土黄色并与果核易分离时即可采收。采后摊在通风、干燥的室内或棚内阴干，碾压去杂即得作为播种材料的果核，用湿沙贮藏。密花胡颓子及胡颓子的种子无休眠或仅有暂短休眠，不耐贮藏，可随采随播。种子忌失水，不能日晒，运输和贮藏均应混以湿沙，贮藏期一般不宜超过2个月。据江苏植物研究所1979年、1982年、1988年试验，胡颓子在5～10℃条件下低温密封贮藏，贮藏期失水量不超过20%～30%，2年后其生活力仍能保持60%～90%，但贮藏超过3年（1980年5月～1983年3月）则全部丧失发芽能力。沙枣和翅果油树可以干藏。胡颓子属树种的出籽率、净度及质量见表4。

表4　胡颓子属树种的出籽率、净度及质量

树　种	出籽率（%）	净度（%）	千粒重（g）	每千克纯净种子粒数（万粒）
沙　枣	20～40	95	67～82	1.2～1.5
密花胡颓子	3	97	350～550	0.18～0.28
翅果油树	长果型 32	—	550	0.18
	大宫灯 24	—	770	0.13
	小宫灯 35	—	260	0.4
大果沙枣	30～40	99	295	0.34
胡颓子	—	—	67～99	1～1.5
牛奶子	5～10	—	22	4.5

发芽和播种　发芽前处理：胡颓子属有的树种种子有休眠习性，常用的解除方法是在0～－5℃下层积30～60天。沙枣果核有时表现出有硬粒，有人主张发芽前在硫酸中浸蚀半小时至1小时。发芽测定：胡颓子属种子休眠期有的较长，如沙枣、翅果油树；也有的树种

无休眠习性或仅有半月左右的休眠期（如密花胡颓子、胡颓子等）。它们的发芽能力及测定条件见表 5。

表 5　胡颓子属树种的发芽能力及其测定条件

树　种	基　质	发芽测定温度（℃）		发芽势（%）		发芽率（%）		
		有光时	无光时	天　数	数　值	持续天数	数　值	变动范围
沙枣	沙①	25	25	—	—	25～30	81	85～90
密花胡颓子	沙床②（室外）	—	—	—	—	18	78	70～90
翅果油树	沙床③（室外）	—	—	—	—	—	78	75～80
大果沙枣	沙床④（室内）	20	20	8	26	24	33	30～70
胡颓子	沙⑤	—	—	—	—	19	60	—

① 种子在 1～5℃条件下层积 30～60 天；② 新采种子当年播种，沙床播种时日均温 25～27℃；③ 经过层积的种子；④ 浓硫酸浸种 3 小时，再用温水浸 33 小时后室内盆播；⑤ 据少量种子推算

条播或点播。密花胡颓子点播，或沙床催芽后芽苗移栽至容器中。沙枣、翅果油树可按 50cm 左右的行距开沟点播。每平方米的播种量为：沙枣、大果沙枣 75～150g，蜜花胡颓子50～75g，翅果油树 38g。经过层积的种子播后 10 天可出芽。胡颓子属种子出土萌发者多，如沙枣、密花胡颓子和牛奶子。胡颓子则是留土萌发。初生叶第 1、2 对对生或近对生，以后互生。沙枣种子的萌发和幼苗的生长情况如图 2。

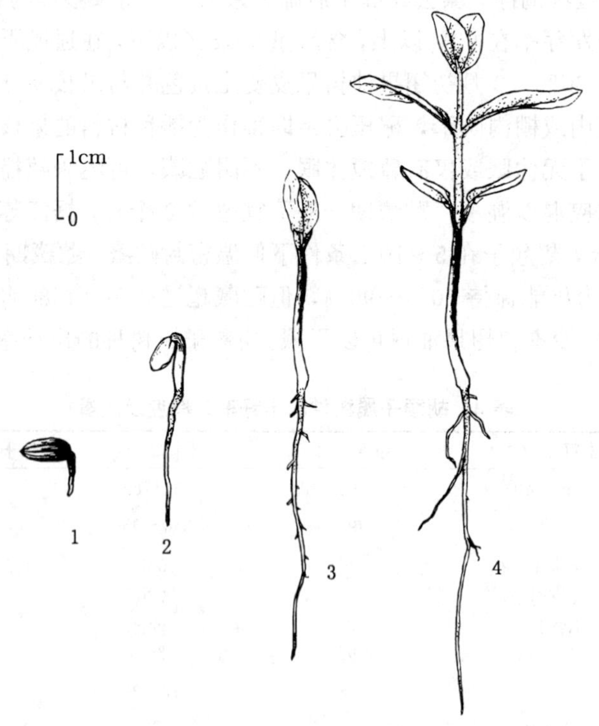

图 2　沙枣种子萌发后幼苗生长情况

1. 幼根延伸　2、3. 子叶出土　4. 初生叶对生

（姚桂英仿《主要树木种苗图谱》）

（李荣晨）

沙　棘　属

Hippophae L.

（胡颓子科　Elaeagnaceae）

生长习性、分布和用途　本属4种9亚种，分布于亚洲和欧洲的温带地区。我国有4种5亚种，产于北部、西部和西南地区，本文描述3种1亚种。落叶直立灌木或小乔木，具刺。幼枝密被鳞片或星状绒毛，老枝灰黑色。通常高1～5m，高山沟谷可达18m。喜光。适应性强，抗严寒风沙，耐大气干旱和高温，也耐水湿和盐碱，能在pH9.5的重碱性土以及含盐量1.1%的盐碱地上生长，但土壤不能过于粘重。果实可药用、酿酒、制糕点饮料。冯瑞英、陈碧珠等人（1993）测定过沙棘属3个种4个亚种采自26个地区的果肉和种子的含油量及其生理活性成分。根系发达，根蘖性强，生长迅速，为防风、固沙、水土保持的优良树种。沙棘属树种的名称、生长和分布情况见表1。黄铨和佟金权（1993，1992）、赵汉章和朱长进（1991）研究过中国沙棘表型性状的种群变异。康永善和陈学林（1992）研究过沙棘属植物的性状演化。

表1　沙棘属树种的名称、生长和分布

中　名	学　名	树高（m）	分　布
肋果沙棘	*H. neurocarpa* S. W. Liu et T. N. He	0.6～5	藏、青、川、甘
中国沙棘（醋柳）	*H. rhamnoides* L. subsp. *sinensis* Rousi	5（18）	内蒙古、冀、晋、陕、甘、青、川
柳叶沙棘	*H. salicifolia* D. Don	0.4～0.6 稀达1	藏南。尼泊尔、不丹、印度
西藏沙棘	*H. thibetana* Schlechtend	1～5	甘、青、川、藏

开花结实　4～5年生开始开花结实，无明显的结实周期性。花单性，雌雄异株。据寇纪烈和罗晶等（1988）在山西右玉县调查，在104m²的样方中，有雌株186株，雄株170株，雌雄比接近1∶1。陈炳浩（1988）转述前苏联的经验是，雌雄比应为1∶4，每公顷应有雄株70～80株。

沙棘属树种无花瓣。短总状花序生于枝腋。雌花序轴发育成小枝或棘刺，雄花序轴花后脱落。雄花先开放，生于早落苞片腋内，无花梗，花萼2裂，雄蕊4，花丝短。雌花单生叶腋，具短柄，花萼囊状，顶端2齿裂。子房上位，1心皮，1室，1胚珠，花柱丝状，柱头圆筒状，微伸出花外，急尖。果实为瘦果，外被肉质化的萼筒包围，呈核果状，近圆形或矩圆形，长5～12mm。果核即为瘦果，倒卵形或椭圆形，果皮革质，内含种子1粒，有胚乳。沙棘属开花结实物候见表2。果实和种子的形态特征见表3。沙棘果实的解剖结构见图1。

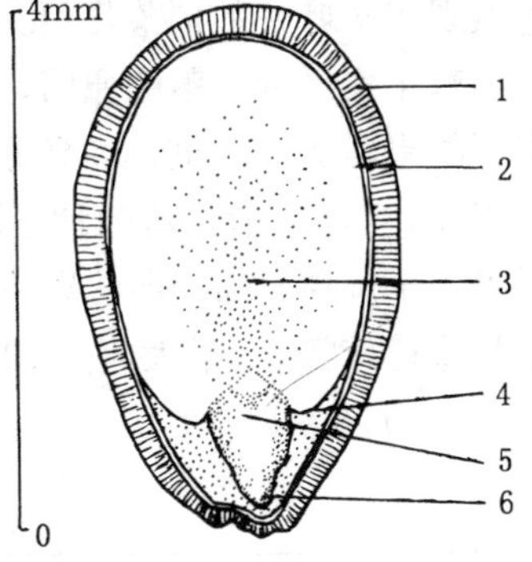

图1　中国沙棘果实的纵切面
1. 果皮　2. 种皮　3. 子叶
4. 胚乳　5. 胚轴　6. 胚根
（姚桂英仿《美国木本植物种子手册》）

表 2 沙棘属树种开花结实的物候

树 种	观察地点	花期（月）	果实成熟期（月）	果实散落期
肋果沙棘	青海	5～6	9～10	10 月～翌春
中国沙棘	山西	4～5	9～10	9 月～翌春
	内蒙古呼和浩特	5	9～10	9 月～翌春
柳叶沙棘	西藏	6	10	10 月～翌春
西藏沙棘	西藏	5～6	9	9 月～翌春

表 3 沙棘属树种果实和种子的形态特征

树 种	成熟果实			种 子（瘦果）		
	形 状	大小（mm）	颜 色	形 状	大小（mm）	颜 色
肋果沙棘	圆柱形，弯曲，具 5～7 纵肋。通常 6 纵肋	长 6～8（9）径 3～4	褐色、密被银白色鳞片	圆柱形	长 4～6	黄褐色
中国沙棘	球形或卵圆形	径 4～6	橙黄色或橘红色	阔椭圆形至卵形，稍扁	长 3～4.2	黑色或紫黑色具光泽
柳叶沙棘	圆形（通常 6 纵肋）或近圆形，多汁	长 8，径 6	橙黄色	阔椭圆形	长 5.5 径 3.2	黑色具光泽
西藏沙棘	阔椭圆形或近圆形，顶端具 6 条放射状黑色条纹，多汁	长 8～12 径 6～10	黄褐色	椭圆形或近橄榄形	长 5～6，径 2.8～3.2	深褐色

果实的采收调制和种子贮藏 沙棘果实成熟后长期不落，在树上能悬挂到翌年 3 月末。沙棘枝上有刺，摘取果实很不方便，一般在冬季果实冻结后，把果实直接敲打到塑料布上或容器内。这种方法采种效率高，果实纯净，便于调制，但可能影响长远的经济效益，破坏沙棘资源。朱长进（1991）在果实着色期即将结束的 8 月下旬，将果枝基部纵切 1～2cm，深及木质部，用浸有 250、500 和 1 000μg/g 脱落酸溶液的脱脂棉敷贴切口；2 天内 3 种浓度处理的都出现落果高峰，1 个月的累积脱果率均在 50%左右，未处理的则只有 14%。不过，外源脱落酸处理也表现出对叶片的催落作用，可能会影响树体的生长发育。张徵和张锐等人（1992）也探讨过沙棘果实经久不落的原因，但他们所用的各种试验处理都未能促使果柄形成离层。

采下的果实放在盆里，在半冻半化时用木棒捣碎，加水稀释，充分搅拌，然后装入布袋把果汁压净，并将带有皮渣的种子晾干，筛去果皮等杂质，所得的果核（瘦果）即为播种材料，通称种子。一般用普通干藏法即可。据缪礼科和孔抒清(1987)测定，中国沙棘 7 个产地之间种子千粒重比较接近，最大值为 10.61g，最小值为 7.89g。沙棘属树种的出籽率及种子质量见表 4。

表 4 沙棘属树种的种子净度和质量

树 种	出籽率（%）	净度（%）	千粒重（g）	每千克纯净种子粒数（万粒）
肋果沙棘	—	80～90	15～18	5.5～6.6
中国沙棘	7～10	85～95	7.8～10.6	9.4～12.8
西藏沙棘	—	95～98	15～16	6.2～6.6

发芽和播种　肋果沙棘种皮较厚较硬，种皮外被蜡层，妨碍吸水发芽，可在浸种后用1%的洗衣粉溶液搓洗去蜡。有的地方先用始温60～80℃的水浸种24小时，再将种子与3倍湿沙混合，置于0～5℃条件下层积约60天，再移至较温暖的地方催芽，至1/3的种子露白时播种。缪礼科和孔抒清（1987）报道，始温50℃水浸种48小时后，沙棘种子检验时适宜的发芽温度条件是23～25℃的恒温，每天光照8～10小时。正常情况下，肋果沙棘的发芽率可以达到90%，中国沙棘和西藏沙棘的发芽率可以达到70%。同缪礼科和孔抒清（1987）一样，肖东玉和盛芳（1988）也认为中国沙棘种子属于强制性休眠。他们的试验结果显示，在25℃的光照发芽器中，新鲜的完整种子30天的发芽率为62%，裸胚则两天便有98%发芽。他们建议室内测定前的最佳催芽方法是用4%的H_2O_2溶液浸种4小时。

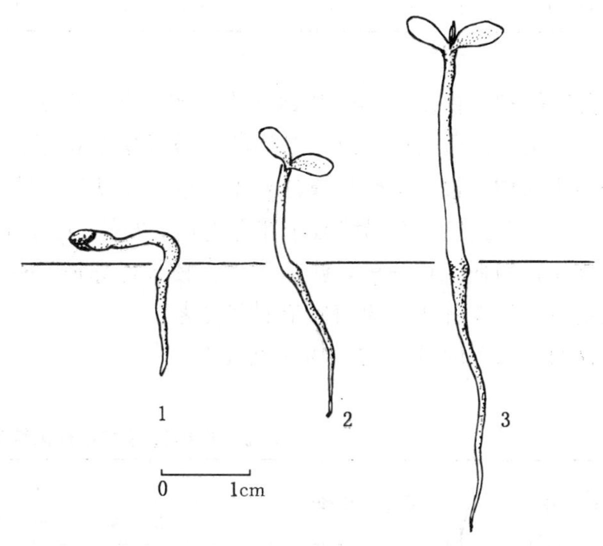

图2　中国沙棘种子的萌发和幼苗生长情况
1. 下胚轴伸长　2. 子叶出土　3. 初生叶出现
（姚桂英绘）

地表5cm温度达10℃时种子即可发芽，以15℃最为适宜。每平方米播种量为12～15g。条播，行距20～25cm，覆土厚0.5～1cm。出土萌发。7～10天幼苗出土。中国沙棘种子的萌发和幼苗早期生长情况如图2。

沙棘可用扦插繁殖育苗。据罗京和张桂龙等人（1992）报道，硬枝扦插的生根率接近100%。

（李荣晨）

枳　椇　属

Hovenia Thunb.

（鼠李科　Rhamnaceae）

生长习性、分布和用途　本属有3种，2变种，分布于中国、朝鲜半岛、日本和印度。本文描述的2种以长江流域以南为主要分布区（表1）。落叶乔木，胸径23～30cm，树高10～25m。喜温暖湿润气候和疏松深厚的肥沃土壤。

果实为清凉利尿药。肥大的肉质果梗含糖，霜后变为紫褐色，可以食用，民间常用以浸制“拐枣酒”，能治风湿。木材坚硬细致，可供建筑及细木工用。

表 1 枳椇属树种的名称、分布和用途

中 名	学 名	分 布	用 途	供 稿
枳椇	*H. acerba* Lindl.	华东、中南、西南及陕、甘。印度、尼泊尔、不丹和缅甸	材用、药用	902
北枳椇（拐枣）	*H. dulcis* Thunb.	华北、华东、华中及川、甘、陕。日本、朝鲜半岛	材用、药用	502

开花结实 枳椇 5 年生开始开花结实。北枳椇 7～8 年生开始开花结实。无大小年现象。花两性。花序排列成对称或不对称的二歧式聚伞花序，顶生或兼腋生。花小，黄绿色。萼片、花瓣和雄蕊均 5 枚。花瓣与萼片互生，生于花盘下。雄蕊与花瓣对生，花丝基部与瓣爪离生，背着药。子房上位，下部埋藏于花盘之内，3 室，每室 1 胚珠，花柱 3 裂。果梗分枝逐渐膨大变为肉质，扭曲，红褐色，味甜可食。核果近球形，径 5～8mm，外果皮革质，内果皮膜质。外果皮与内果皮分离。果每室有种子 1 粒。

这两个树种的开花结实物候见表 2。

表 2 枳椇属树种的开花结实物候

树 种	观察年份和地点	开花			果实成熟	
		始 期	盛 期	末 期	始 期	盛 期
枳椇	1986～1987 福建南平	6 月中旬	7 月上旬	7 月中旬	10 月中旬	10 月下旬
北枳椇	1987～1988 四川雅安	6 月上旬	6 月中旬	6 月下旬	9 月上旬	9 月中下旬

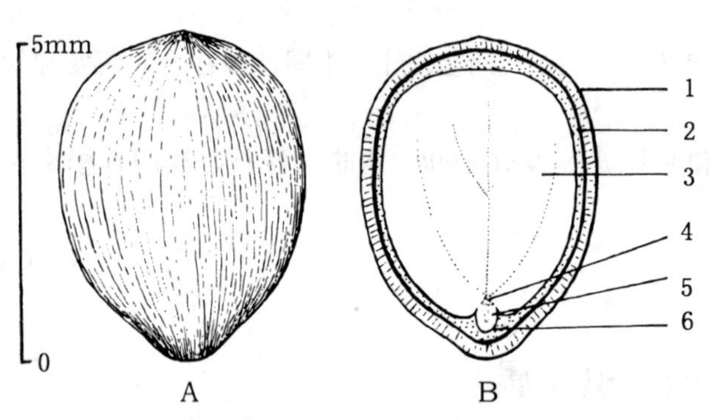

图 1 北枳椇种子外形（A）及其纵切面（B）
1. 种皮 2. 胚乳 3. 子叶 4. 胚芽 5. 胚轴 6. 胚根
（孟玲、黄应钦绘）

果实未熟时浅褐色，成熟时灰褐色、紫褐色、黄褐色或棕褐色，近球形。种子圆形、扁平，红褐色、深褐色或紫褐色，有光泽，背面凸起，腹面平而微凹或中部具棱，基部内凹，常具灰白色乳头状突起，长 3～5mm。种皮坚硬，胚乳丰富，胚小，直，子叶 2 枚（图 1）。枳椇属两个树种果实和种子的形态特征见表 3。

表 3 枳椇属树种果实和种子的形态特征

树 种	果实			种子（果核）		
	形 状	大小(mm)	颜 色	形 状	长 (mm)	颜 色
枳椇	近球形	5～6.5	黄褐色或棕褐色，有光泽	圆形，扁平	3～4	暗褐色，有光泽
北枳椇	近球形	6～8	灰褐色或紫褐色，有光泽	圆形，扁平	5	红褐色或深褐色，有光泽

果实的采集调制和种子贮藏　外果皮呈紫褐色或黄褐色或棕褐色并有光泽时，上树或在地上用高枝剪采下果枝。果梗肉质多汁，富含糖分，易于腐烂，不能长期堆积，应从果梗上剥下果实或连果梗晒干，搓揉果皮，去杂后即得种子。种子摊晾荫干。

通常种子只越冬贮藏，少有贮藏1年以上的。干藏。将含水量已降到8%的种子装入塑料袋或麻袋，置温度为5℃左右的库房内，或置通风、干燥、低温的地方。也可以混沙湿藏，沙的含水量约5%。

果实出种率和种子质量等数据见表4。

表4　枳椇属树种的出种率和种子质量

树　种	出种率（%）	净度（%）	千粒重（g）	每千克纯净种子数（万粒）
枳椇	72～80	98	11～15	6.7～9.1
北枳椇	85	95	16～18	5.5～6.3

表5　枳椇属树种种子的发芽能力及其测定条件

树　种	预　处　理	基　质	发芽测定	发　芽　率（%）		
			温度（℃）（无光）	计算天数	一般数值	变动范围
枳椇	始温60℃水浸种24～48小时	滤纸	25	30	77	66～85
北枳椇	始温60℃水浸种24～48小时	纱布	25	30	60	50～70

发芽和播种　种子无明显休眠习性。紫黑色或暗灰色种子多系空粒。

在室内作发芽测定，先将种子在始温60℃水中浸泡24～48小时，捞出后置纱布或滤纸上在25℃的恒温中发芽。发芽无明显峰值。两树种种子的发芽能力及其测定条件见表5。

春播。播期在2月下旬至3月上旬。条播时条距20～25cm，条幅5～10cm，播种沟深1～2cm。枳椇种子可以用0.15%福尔马林液灭菌，冲洗后浸种2天再播种。每平方米播种8～10g。干藏的北枳椇种子播前用始温60℃水浸种，自然冷却后再用冷水浸种48小时，每24小时换水1～2次，用下沉饱满种子播种。条播，每平方米播5～6g，播后覆细土1cm。这2个树种也可以撒播。干藏后经浸种处理的，播种20天后发芽出土。混沙湿藏的，播种15天开始发芽出土，再经30天发芽基本结束。

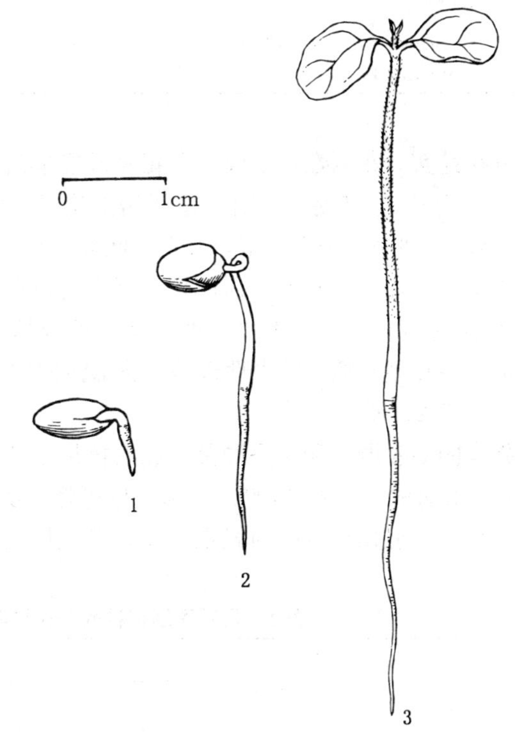

图2　北枳椇种子的萌发及其生长情况
1. 下胚轴延伸　2. 幼芽出土　3. 子叶展现
（黄应钦、孟玲绘）

出土萌发。子叶2，卵圆形，对生，

全缘，长 11～13mm，宽 10～ 11mm。初生叶开始时对生，尖端渐尖，叶缘浅锯齿，以后叶片呈互生。1 年生苗高 1m 以上，可出圃栽植。

枳椇和北枳椇还可用分根、分蘖、扦插等方法繁殖。

北枳椇种子萌发后的生长情况见图 2。

（阙再忠）

马甲子属

Paliurus Tourn. ex Mill.

（鼠李科　Rhamnaceae）

生长习性、分布和用途　本属有 6 种，产东亚与南欧，我国有 4 种，又引入 1 种 ，本文介绍 2 种（见表 1）。落叶乔木或灌木，在山地和平原杂林中野生，庭园也有栽培。

表 1　马甲子属树种的名称、生长习性、分布及用途

中　名	学　名	习　性	树高(m)	分　布	用　途
铜钱树	*P. hemsleyanus* Rehd.	落叶小乔木	13～15	陕、甘、豫、西南、中南、华东	鞣质、材用、枣树砧木
马甲子	*P. ramosissimus* (Lour.) Poir.	落叶多枝灌木	2～3(6)	陕、西南、中南、华东、台。越南、朝鲜半岛、日本	材用、药用、绿篱

开花结实　花两性，组成聚伞花序或聚伞圆锥花序，腋生或顶生。花小，黄绿色 ，直径约 5mm。萼片 5。花瓣 5。雄蕊 5。子房上位，大部分藏于花盘内。花盘肉质 ，五边形或圆形，边缘 5 或 10 浅裂。子房 3 室，每室 1 胚珠。花柱深 3 裂。核果杯状或草帽状，黄褐色或红褐色，周围具木栓质厚翅或革质薄翅，基部有宿存萼筒。内果皮发育成坚硬的核，具 3 室，每室含发育或不发育种子 1 枚。种子紫红色或棕褐色，扁圆形，表面有光泽，直径约 3.0～3.5mm。种皮硬革质，透水性极差。脐部浅凹穴略突起。胚乳量中等，包围胚。子叶近圆形，大而扁（表 2、图 1）。

果实的采收调制和种子贮藏　成熟核果不开裂，经霜后带果柄掉落，可以拾集，也可在果实变色、果柄枯干时采摘，晒干脱粒。也可将果实干藏到播种前脱粒。一般用小锤敲打或用石碾压碎核壳，但操作时必须轻重合度，以免伤及种子。

表 2　马甲子属树种的开花结实物候及果实、种子特征

树　种	开花期	果熟期	果实				种子	
			果径 (mm)	果梗长 (mm)	千粒重 (g)	出种率 (%)	千粒重 (g)	每千克数 (万粒)
铜钱树	5 月中下旬	7 月下旬～10 月上旬	28～35	10～17	70～80	49～66	10.5	9.5
马甲子	7 月上旬	10 月上旬	10～18	6～10	130	—	—	—

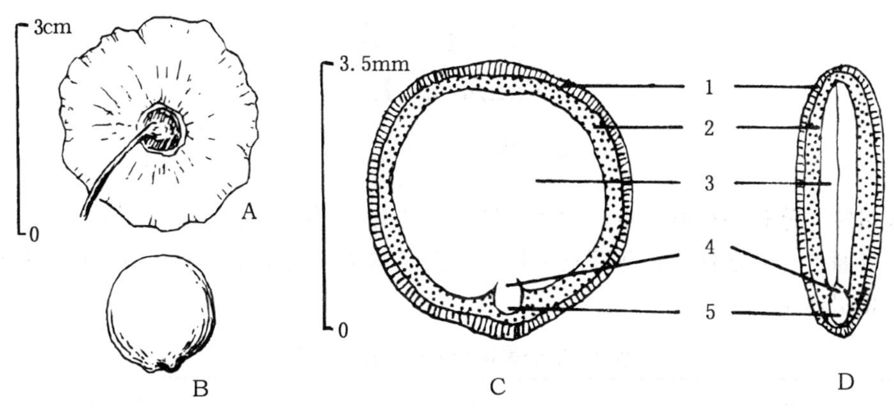

图 1　铜钱树核果外形（A）与种子外形（B）及其纵切正面（C）和侧面（D）
1. 种皮　2. 胚乳　3. 子叶　4. 胚轴　5. 胚根
（史渭清绘）

发芽和播种　未经处理的铜钱树种子发芽率低，发芽极不整齐。1988 年 3 月 9 日，本文作者曾取 1987 年秋季采收干藏的铜钱树果实脱粒取种，随即混以湿沙，分别放在室内和室外观察发芽情况。4～5 月间室内组有 6％的种粒萌发，室外组有 3.2％的种粒萌发，以后便停止发芽。1989 年 4 月中旬以后，二组又有少量种粒萌发。2 年合计的发芽率在室内组为 22％，在室外组为 16.7％。剩下的未发芽粒中，绝大部分种粒尚未吸胀。由此推测，铜钱树种子发芽困难的原因可能是种皮透水性差。

江苏省植物研究所曾经培育铜钱树作为枣树的砧木。据介绍，不加处理的种子播后发芽率仅 15％～20％，温水浸种可使发芽率提高到 60％。他们的做法是，春播前脱粒取种，浸于 50～60℃水，在保温条件下浸泡 4 小时，晾干，混沙撒播或条播，覆土 1.5cm，盖草保温，约 20～30 天出苗。子叶出土（图 2）。6 月上旬幼苗具 5～6 片初生叶时即可移栽。移栽时株距 10cm，行距为阔窄行相间，阔行 50cm，窄行 30cm。幼苗生长较快，1 年生苗粗可达 1cm，翌春 3 月中下旬～4 月即可作为砧木供嫁接用。

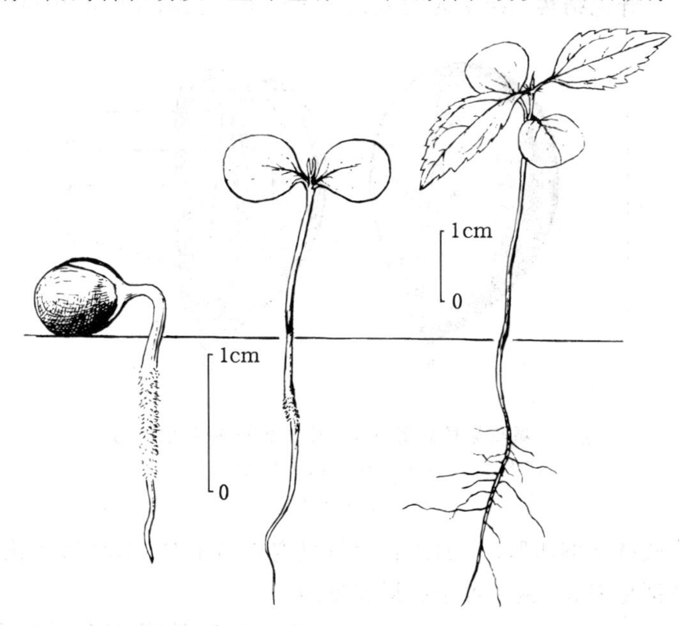

图 2　铜钱树种子萌发后第 1、5、25 天的幼苗生长情况
（史渭清绘）

（何泽瑛）

鼠　李　属

Rhamnus L.

（鼠李科　Rhamnaceae）

生长习性、分布和用途　本属约 200 种，我国有 57 种和 14 变种，南北均产。落叶或常绿，乔木或灌木，通常具刺，冬芽具鳞片或裸露。本文仅描述 3 种落叶小乔木或灌木，其名称、分布和用途见表 1。

表 1　鼠李属树种名称、分布和用途

中　名	学　名	分　布	用　途	供　稿
鼠　李	*Rh. davurica* Pall.	东北、内蒙古、冀、晋。朝鲜半岛、蒙古、西伯利亚	果、根皮药用，材雕刻、观赏	106
小叶鼠李	*Rh. parvifolia* Bunge	东北、内蒙古、冀、晋、鲁、豫、陕、甘。朝鲜半岛、蒙古、西伯利亚	果实药用，水土保持	109
乌苏里鼠李	*Rh. ussuriensis* J. Vass.	东北、内蒙古、鲁。朝鲜半岛、西伯利亚、日本	果、根皮药用，材雕刻、观赏	109

开花结实　结实无明显大小年现象。花两性或单性异株，稀为杂性。花小，淡绿或淡黄色，单生或数个簇生或排成腋生的聚伞花序或聚伞总状花序或聚伞圆锥花序。花萼钟形或漏斗状钟形，4～5 浅裂。花瓣 4～5，短于萼片，稀无花瓣。雄蕊 4～5，为花瓣所包裹。花盘薄，杯状，子房着生其上。子房上位，球形，2～4 室。每室 1 胚珠，花柱 2～4 裂。果为浆果状核果，基部为萼筒所包，2～4 分核（本文 3 种皆为 2 分核）。分核骨质或软骨质，开裂或不开裂，各有 1 种子。种子倒卵形或长圆状倒卵形，背面或背侧具纵沟或稀无沟。鼠李属开花结实物候见表 2，果实和种子形态特征见表 3。鼠李种子的形态见图 1。

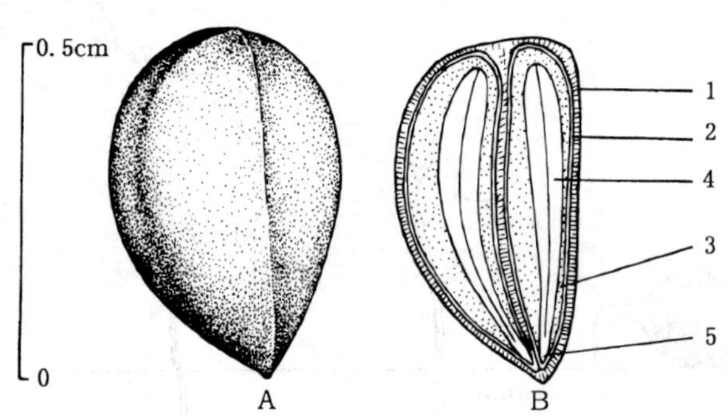

图 1　鼠李果核外形（A）及其 2 分核纵切面（B）
1. 内果皮　2. 种皮　3. 胚乳　4. 子叶　5. 胚根
（梁鸣、黄应钦绘）

表 2　鼠李属树种开花结实物候

树　种	观察年限地点	花期 始　期	花期 末　期	果期 始　期	果期 末　期	种子散落期
鼠　李	—	5 月	6 月	8 月	9 月	—
乌苏里鼠李	1963～1980 哈尔滨	6 月 4 日	6 月 20 日	8 月 30 日	10 月 6 日	初熟至脱落间隔 20～30 天
小叶鼠李	1963～1980 哈尔滨	5 月 18 日	6 月 4 日	8 月 26 日	10 月 14 日	初熟至脱落间隔 20～35 天

表 3　鼠李属树种果实和种子形态特征

树　种	未熟果颜　色	果　实			种　子（果核）	
		形　状	颜　色	大　小（cm）	形　状	颜　色
鼠　　李	绿色	近球形	黑色	长 0.6～0.7 径 0.5～0.6	长圆状倒卵形，侧扁，背侧有狭纵沟，沟与种子等长	黄褐色
乌苏里鼠李	绿色	球形或倒卵圆形	黑色	长 0.6～0.8 径 0.5～0.6	长圆状倒卵形，侧扁，背侧基部有短沟，上部有沟缝	黑褐色
小叶鼠李	绿色	球形或倒卵形	黑色	长 0.5～0.6 径 0.4～0.5	长圆状倒卵形，侧扁，背侧有长为种子4/5的纵沟	栗褐色

果实的采收调制和种子贮藏　果实充分变黑色即为成熟，树上摘取。堆沤熟透变软后捣碎，用水漂洗，除去果皮与果肉等杂物，晾干去杂后所得果核即为播种材料，通称种子。种子含水量应降到 10%以下，0～5℃下贮存 1 年后发芽能力无明显变化。出籽率、净度及种子质量见表 4。

表 4　鼠李属树种出籽率及种子（果核）质量

树　种	出籽率（%）	净　度（%）	千粒重（g）	每千克纯净种子粒数（万粒）
鼠　　李	40	95～98	25～30	3.8～4.0
乌苏里鼠李	40	95～98	20～25	4.5～5.0
小叶鼠李	10	75～85	17～20	5.0～5.9

发芽和播种　将种子浸水 1～2 天，种与湿沙按 1∶3 比例混合，在 10～25℃下层积，40～50 天可以发芽。如先在 0～5℃下处理 1 个月然后置于 15～25℃下，10 天即可发芽。秋播翌春出苗。苗床条播，每平方米播种量约 15g，覆土 1.5～2.0cm。出土萌发（见图 2）。幼苗期需防日灼，2 年生苗出圃栽植。

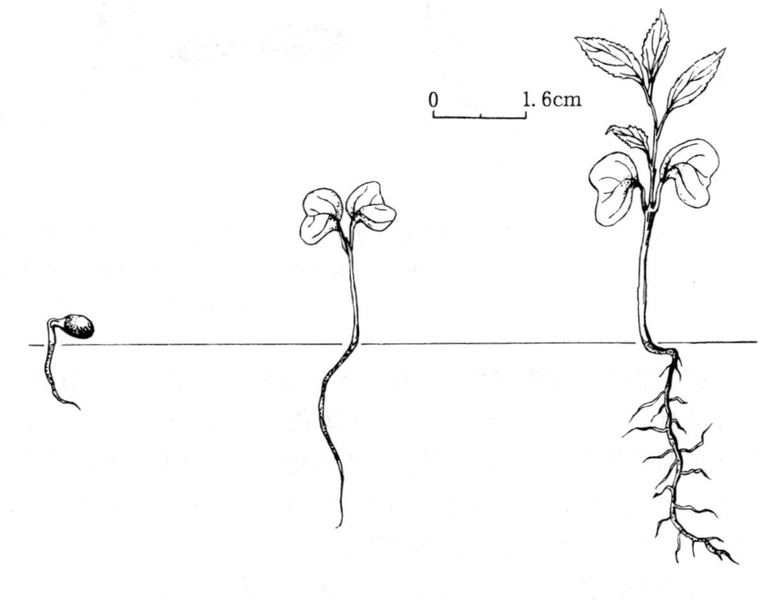

图 2　鼠李种子萌发后第 7、11、24 天幼苗生长情况

（梁鸣绘）

（周德本）

雀梅藤

Sageretia thea（Osbeck）Johnst.

（鼠李科　Rhamnaceae）

生长习性、分布和用途　雀梅藤属约 34 种，我国 16 种。本文描述的雀梅藤为此属的模式种。常生于丘陵山地林下或灌木丛中。分布于苏、浙、皖、赣、湘、鄂、川、滇、桂、粤、闽、台。印度、越南、朝鲜半岛、日本也有分布。果可食，嫩叶能代茶。常利用其有刺攀援特性栽培作为绿篱。树桩可作盆景、手杖。

开花结实　花两性，小，绿黄色，无柄，通常多数簇生排成顶生或腋生的疏散穗状或圆锥状穗状花序。花序长 2～5cm，密被短绒毛。萼片 5，三角状卵形，长 1mm。花瓣 5，匙形。雄蕊 5。花盘厚，杯状，5 裂。子房上位，大部分被花盘包围，基部合生。3 室，每室 1 胚珠。花柱极短，柱头 3 浅裂。浆果状核果近球形，直径 4～7mm，紫黑色或黑色，具 1～3 分核。核壳脆壳质。种子扁圆形，背凹腹凸，长 4mm，宽 3mm，厚 0.8mm。种脐部有凹穴，略突起。种皮浅褐色，革质，有光泽。内种皮膜质。胚乳层薄。胚绿色，子叶宽心形，大而薄，胚根短小（图 1）。雀梅藤秋花春实，10～11 月开花，幼果直到翌春树叶萌发后才发育，4～6 月陆续长大成熟。

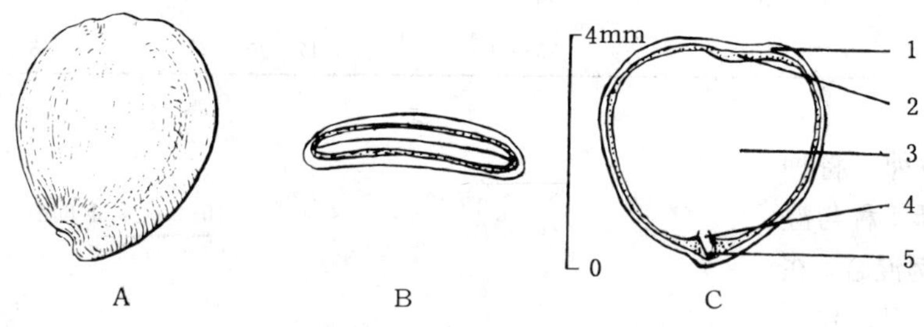

图 1　雀梅藤种子外形（A）及其横切面（B）和纵切面（C）

1. 种皮　2. 胚乳　3. 子叶　4. 胚轴　5. 胚根

（史渭清绘）

果实的采收调制和种子贮藏　果实成熟期极不一致而且成熟后随即脱落，应当根据色泽变化分批采摘。成熟果实的特征是由绿色经橘红、枣红最后变为黑色或紫黑色，质地由硬变软直至成为浆果状，种子则由白色变成浅褐色。大量采种比较困难。果实采收后搓揉漂洗，清除果肉、核壳后可得纯净种子。出种率约 15%。种子千粒重约 12.8g，每千克约 7.8 万粒。种子遇有湿润条件便可萌发，不耐贮藏。

发芽和播种　种子无休眠习性，采回调制结束后在常温下即能发芽，宜在夏季采后即播。本文作者 1989 年 5 月 18 日在南京采收种子，随即混拌湿沙置于常温条件下，7 天即开始发芽，11 天发芽结束。充分成熟的褐色种子发芽率可达 97%。种皮虽已变硬但仍呈白色的种子发芽率为 86%。出土萌发。子叶长心形。初生叶对生，以后真叶近对生。幼苗生长发育健壮，主根发达而侧根短小（图 2）。

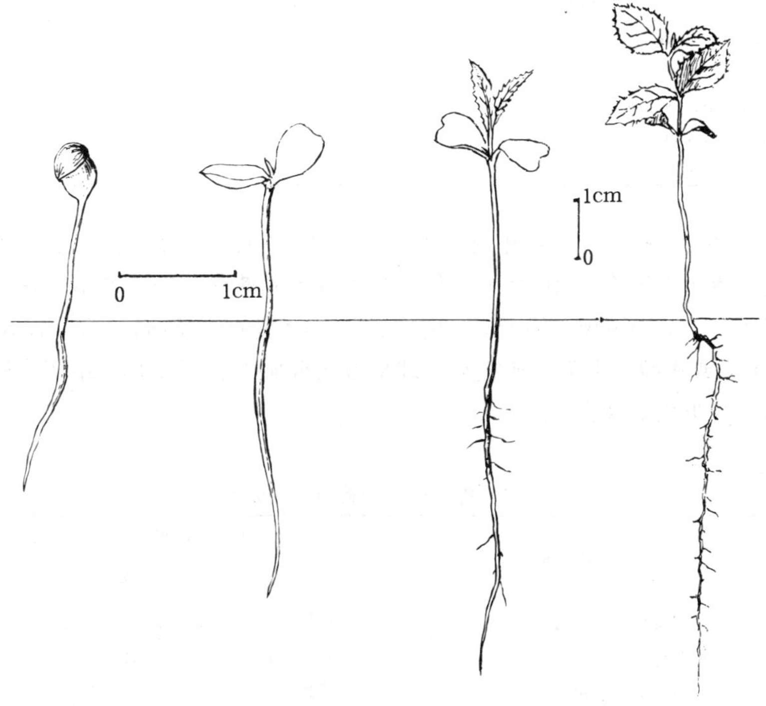

图 2　雀梅藤种子萌发后第 6、10、14、62 天的幼苗生长情况

（史渭清绘）

（何泽瑛）

枣　属

Ziziphus Mill.

（鼠李科　Rhamnaceae）

生长习性、分布和用途　本属约 100 种，我国 12 种 3 变种，本文描述 2 种 1 变种。灌木或乔木，落叶稀常绿，高 1～15m，胸径 30～70cm。寿命长。萌芽能力强，易发生根蘖长成树林。水平根系发达，深度较浅。适应性强，能耐寒、耐热，耐旱、耐涝。枣树对土壤要求不严，除沼泽和盐碱土外，山地、丘陵、高原、平原、河滩均有枣树生长，最宜在土层深厚肥沃、排水良好的土壤上栽培。花芳香为蜜源。核果肉质，营养丰富，供生食或作食品加工。据陈贻金和陈春雷等（1991）分析，3 个品种的红枣所含 16 种氨基酸中，有 8 种在人体内不能合成，另有 2 种在幼儿体内不能合成。枣仁入药，为滋补强壮剂。我国引进了优良的紫胶虫种信德胶虫（*Kerria sindica* Mahd.），滇刺枣是这个紫胶虫种的主要寄主（汪云，1994）。这 3 个树种的名称、树高、分布及用途见表 1。

表 1　枣属树种的名称、分布及用途

中　名	学　名	树高（m）	分　布	用　途	供　稿
枣	*Z. zizyphus* (L.) Meikle	10	东北南部、华北南部、西北、西南、中南、华东	食用	311
酸枣	*Z. zizyphus* var. *spinosa* (Bge.) Y. L. Chen	1～3	东北南部、华北、西北、豫、皖、苏	药用	311
滇刺枣	*Z. mauritiana* Lam.	15	川、滇、桂、粤、琼。闽、台栽培	用材，紫胶虫寄主	604

开花结实　约 2～6 年生开始开花结实，正常结实在 15 年生以后。结实大小年不明显。花两性，黄绿色，单生或数个至 10 余个密集成腋生的聚伞花序。花 5 基数，萼片卵状三角形。雄蕊与花瓣对生。子房半下位，陷于厚、肉质、5 或 10 裂的花盘内，2 室，稀 3 室，每室 1 胚珠。花柱 2，半裂或浅裂。果为肉质核果，球形或长椭圆形。枣属开花结实物候期见表 2。果实及种子形态特征见表 3 及图 1。

表 2　枣属树种开花结实物候期

树　种	观察地点	花　期	果实成熟期		采收期
			始　期	盛　期	
枣	稷山	5 月中旬～7 月下旬	8 月上旬	9 月下旬	10 月上旬
酸枣	北京	5 月上旬～8 月上旬	8 月下旬	9 月中下旬	9 月下旬
滇刺枣	南宁	第一次：5 月上旬～5 月下旬	9 月上旬	9 月下旬	10 月上旬
		第二次：8 月下旬～10 月下旬	翌年 1 月中下旬	翌年 3 月上旬	翌年 4 月下旬

表 3　枣属树种果实及种子（果核）形态特征

树　种	果　实			果实颜色		种　子（果 核）			
	形　状	径（cm）	长（cm）	未熟	成熟	形　状	径（cm）	长（cm）	颜　色
枣	球形或椭圆形	1.3～1.5	1.7～4.3	淡绿	紫红色	椭圆形或梭形	0.4～0.8	1.1～2.6	褐色
酸枣	球形或椭圆形	1.2～3.0	1.3～2.2	淡绿	紫红色	球形或椭圆形	0.7～0.9	0.9～1.5	褐色
滇刺枣	椭圆形或梨形	0.9～1.1	1.0～1.3	淡绿	紫黑色	圆球形或卵形	0.5～0.6	0.6～0.8	红褐色

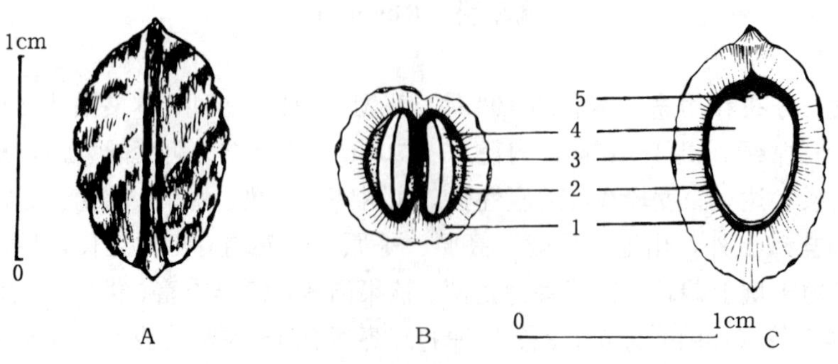

图 1　酸枣果核外形（A）及其横切面（B）和纵切面（C）

1. 内果皮　2. 种皮　3. 胚乳　4. 子叶　5. 胚根

（孟玲绘）

果实的采收调制和种子贮藏　果实成熟时人工采摘，或用竹竿击落果实或者震动枝条摇落果实。人工采摘的果实杂质少。在采果过程中可以随时弃去劣质果。沤制或用剥果机剥离果肉，所得的果核即为播种材料，通称种子。种子经过晾晒干燥后盛于布袋贮藏。果实出籽率及千粒重见表 4。

表 4　枣属树种果实出籽率及种子（果核）千粒重

树　种	出籽率（%）	千粒重（g）		每千克纯净种子粒数（粒）	
		一　般	变动范围	一　般	变动范围
枣	4～10	230	200～470	4 300	2 100～5 000
酸枣	14～26	200	180～400	5 000	2 500～5 600
滇刺枣	45～55	500	350～710	2 000	1 400～2 900

发芽和播种　种子放在缸中，用始温 80℃水烫种 10 分钟，捞出再用凉水泡 2 周，中间经常换水，然后层积催芽，待有 30%的种子萌动，并微露出胚根时即可播种。播种用宽窄行穴播：宽行间距 80cm，窄行间距 40cm，穴距 20cm，每穴播 5～6 粒种子，覆土 2cm。每平方米播种量约 25g。出土萌发。播后 20～30 天开始萌发，初生叶开始出现。当幼苗长到 3～4 片真叶时，进行间苗补栽，每穴留健壮苗 1～2 株。酸枣种子萌发和幼苗初期生长情况见图 2。

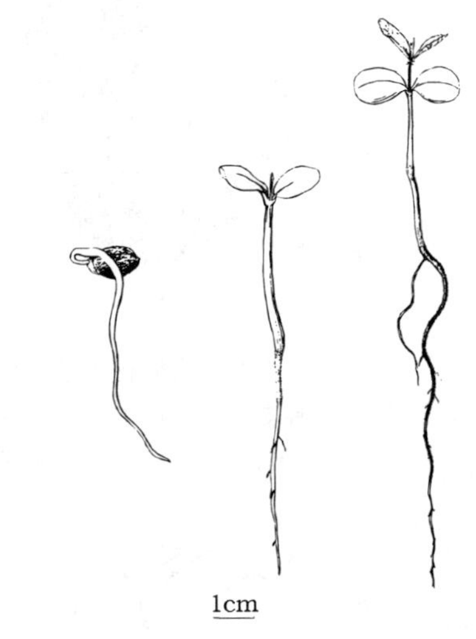

图 2　酸枣种子萌发出土后第 3、9、18 天幼苗生长情况

（孟玲绘）

（邓明全）

火　筒　树

Leea indica（Burm. f.）Merr.

（葡萄科　Vitaceae）

生长习性、分布和用途　火筒树属约 70 种，我国约产 8 种，本文描述 1 种。常绿至半落叶小乔木，高 4～5m，树干丛生，上部倾斜，干呈直立状。萌芽力强。颇耐荫蔽，适生于沟谷、疏林下和灌丛中。喜湿润的疏松土壤，酸性土或钙质土均宜。能耐短期 0℃左右低温，忌霜雪。分布于滇、黔、桂、粤、琼、台。中南半岛、印度尼西亚、日本也有分布。涵养水源作用强。叶及嫩枝为优质绿肥，干作薪柴。

开花结实　约 4～5 年生开始开花结实，正常结实期在 8 年生以后。结实无大小年现象。冬季 4～6℃低温的时间较长，或者花期暴雨较多的年份，结实量明显降低，其余年份均可正常结实。花两性。伞房状聚伞花序，长 7～13cm，与叶对生。萼 5 裂。花瓣同数，淡绿白色，基部合生，花冠径约 7cm。雄蕊 5，基部合生成一个 5 齿裂的柱，花药即生于裂齿间。子房上位，5 室，每室有胚珠 1 颗。据南宁 1987～1988 年观测，7 月下旬为始花期，8 月中旬为盛花期，9 月上旬为末花期，9 月中旬果实开始成熟，10 月中旬为果熟盛期。花期和果实成熟期均较长。浆果，成熟时由黄绿色转变为黑色，扁球形，高 6～8mm，径 8～10mm。成熟的果实在树上约可存留 1 周，从 9 月上旬～11 月中旬陆续有果实成熟并脱落。每果有种子 3～5 粒。种子三角状圆形，褐色，高 3.5～4.0mm，径 2.5～3.5mm。外种皮薄，内种皮厚。胚乳丰富，胚弯曲。种子的形态及其内部结构见图 1。

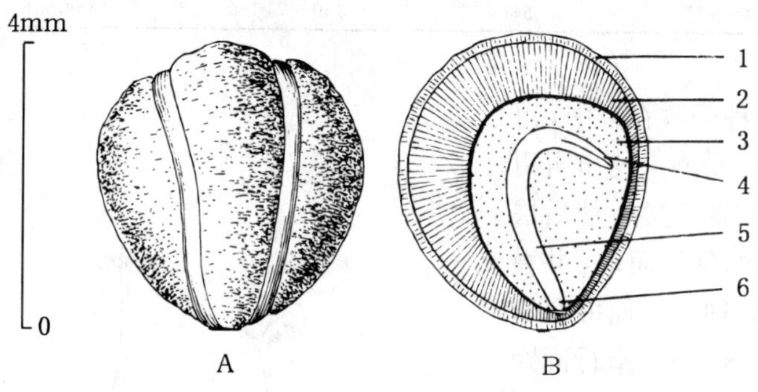

图 1　火筒树种子外形（A）及其纵切面（B）

1. 外种皮　2. 内种皮　3. 胚乳　4. 子叶　5. 胚轴　6. 胚根

（黄应钦绘）

果实的采收调制和种子贮藏　果实成熟盛期，用高枝剪或采种钩刀截断总梗，将颗粒较大的果实摘下堆沤 2～3 天。浆果充分软熟后装入竹箩或布袋，置水中搓擦，淘去果皮果肉及果柄等杂质，取出下沉的纯净种子。鲜果的出种率约 16%，种子的净度可达 96%～98%。千粒重约 25g，每千克有纯净种子 3.5 万～5 万粒。种子含水量约 25%，忌脱水，不宜日晒，运输时需混以湿润锯木屑。贮藏前将混有木屑的种子浸水，浮去木屑，取出下沉的种子，稍晾干后混以过筛的细沙，贮藏期约为半年左右。

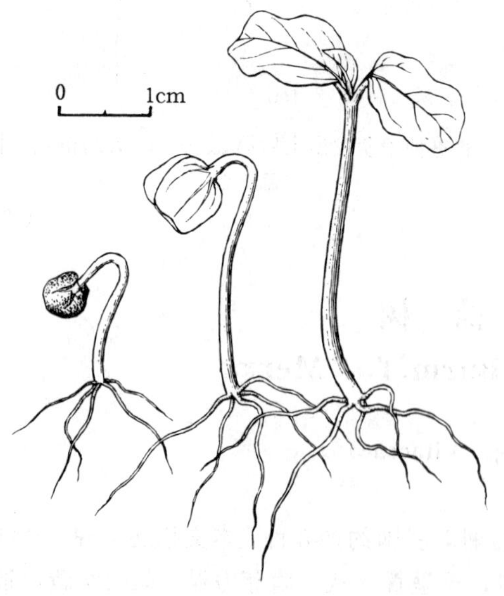

图 2　火筒树种子萌发后第 7、10、30 天幼苗的生长情况

（黄应钦绘）

发芽和播种　种子无明显休眠现象。发芽时日均温需在 20℃以上。1987 年广西林业科学研究所在室内用发芽箱作过发芽测定，箱内温度保持 28℃，基质为滤纸。11 月上旬末置床，12 月中旬末方始发芽，发芽

盛期不明显。1988年1月上旬初发芽终止，发芽率为50%。出土萌发。胚根萌发后约经8天子叶伸出，再过7天初生叶展现。火筒树种子的发芽和幼苗早期形态见图2。

撒播。种子播在盆内或经过细致整地的苗床上。每平方米播种8～10g。发芽后约1个月，苗木半木质化时移入容器或圃地，培育3个月或1年生苗出圃。

（林榕庚）

扁　担　藤

Tetrastigma planicaule（Hook. f.）Gagnep.

（葡萄科　Vitaceae）

生长习性、分布和用途　崖爬藤属约90种，我国约产30种，本文描述1种。常绿大型木质藤本，茎扁，长达30m以上，基部宽达40cm，分枝圆柱状，具卷须。能耐荫蔽。生于山谷密林中，喜湿润肥沃的森林土壤，酸性土和钙质土均宜。粘土和干旱贫瘠地生长不良。能耐0℃左右的低温，但忌冰冻。分布于闽、粤、琼、桂、滇、黔。印度、越南也有。藤茎供药用，能祛风去湿。枝叶涵养水源，可作荫棚植物。

开花结实　约6年生开始开花结实，正常结实期在12年生以后。结实大小年现象不明显，水肥条件充足的地方每年结实均多。花杂性。复伞形聚伞花序，腋生。花小，绿色，4基数，花萼全缘。花瓣宽卵状三角形。雄蕊4，较子房短。花盘不显著。子房上位，2室，宽圆锥形，每室有胚珠2。柱头4裂，辐射状。据广西南宁1987～1988年对攀援于荫棚架上12年生植株的观测，7月中下旬为始花期，8月中旬为盛花期，9月上旬为末花期；9月中旬果实开始成熟，10月上旬至中旬为果熟盛期。浆果，椭圆形或倒卵形，顶端有一褐灰色环痕，成熟时由青色转为橙红色，熟后可在树上存留5天左右并逐渐脱落。果长1.6～2.6cm，径1.3～2.3cm。每果有种子1～4粒。种子深褐色，扁椭圆形，长1.2～1.5cm，宽6～7.5 mm，厚4～5.5mm，两侧各有纵棱1条。种皮厚度不均。胚乳丰富，胚伸直。扁担藤种子的形态及其解剖构造见图1。

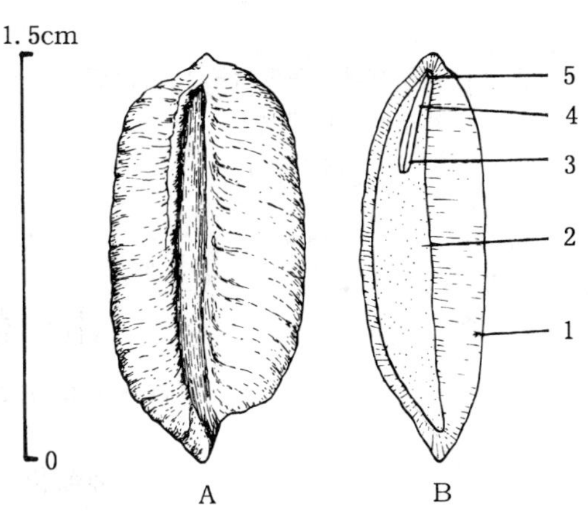

图1　扁担藤种子外形（A）及其纵切面（B）
1. 种皮　2. 胚乳　3. 子叶　4. 胚轴　5. 胚根
（黄应钦绘）

果实的采收调制和种子贮藏　果实成熟盛期上树采摘，或用采种钩刀从总果梗处割断，在地面收集果实。密林中采种较难，一般只能在果熟盛期摇动攀藤，震落果实后在地面捡拾。采得的果实堆沤2～4天，全部软熟后装入竹筐或布袋，置水中反复搓擦，淘去皮肉等杂质即得纯净种子。浆果出种率约13%。种子的净度可达96%～98%。千粒重约350g，每千克有纯净种子2 500～3 400粒。种子的含水量约30%，忌失水，不宜日晒，宜随采随播。如需运输，

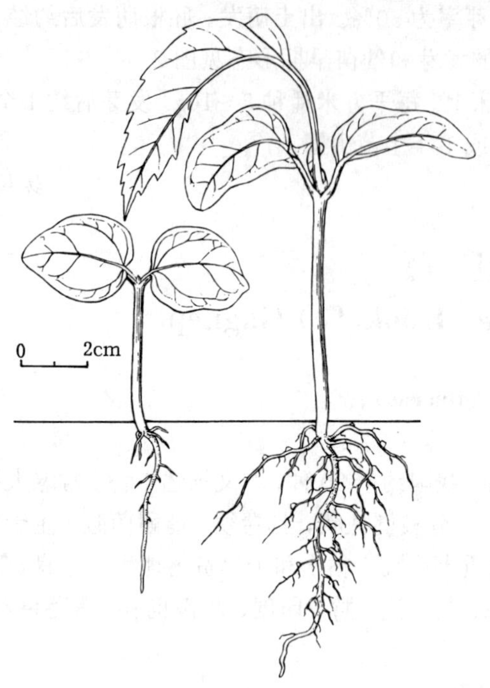

图 2　扁担藤种子萌发后第 15、40 天幼苗的生长情况
（黄应钦绘）

宜混以湿润锯木屑，用袋、箱或竹箩等包装。贮藏时则需混以湿沙，贮藏期一般为半年左右。10 月采集的种子宜在翌年 3～4 月播种。

发芽和播种　种子无明显休眠现象。发芽时日均温宜在 18℃以上。1987 年 10 月上旬末，广西林业科学研究所在室外沙床用当年新采种子作发芽测定，播后 36 天至 11 月中旬方始发芽，发芽时日均温 18～20℃，发芽后 2～4 天为萌发盛期，约 1 周发芽结束。以高峰期的 4 天计，发芽百分数为 50%；从播种日起以 43 天计，发芽率为 80%。出土萌发。胚根萌发后约经 6 天子叶出土，再过 30 天左右初生叶出现。种子萌发和幼苗早期生长情况见图 2。

目前只有少量育苗，供立体绿化用。撒播。每平方米播种 20～30g。以后将幼苗移至容器或圃地继续培育，3 个月或 1 年生苗出圃。也可用压条法育苗。

（林榕庚）

山　葡　萄

Vitis amurensis Rupr.

（葡萄科　Vitaceae）

生长习性，分布和用途　葡萄属约 60 余种，我国产 27 种，本文仅介绍山葡萄 1 种。落叶藤本，通常以与叶对生的卷须攀援上升。枝条粗壮，野生条件下有长达 15m、主干直径10～20cm 的植株。常攀援于乔灌木上。生长快，耐寒，能忍受－40～－50℃严寒。分布于东北、华北和华东。朝鲜北部，俄罗斯东西伯利亚亦有分布。山葡萄是一种经济价值很高的野生果树：山葡萄酒在葡萄酒中独树一帜，深受国内外欢迎；果汁可做饮料，果可食。东北各地已广泛引种栽培，并已选育出一些优良品种。

开花结实　野生山葡萄一般 3 年生以上开始结实，大小年间隔一般为 1 年。栽培的山葡萄 3 年生便开始有稳定的产量，其经济栽培年龄预计最少可达 30 年。花单性，雌雄异株，花小，黄绿色。圆锥花序与叶对生，雌花序长 9～15cm。雌花：萼片小，5 裂，花瓣 5，顶部合生呈帽状，早落，具 5 退化雄蕊，子房上位，2 室，每室 2 胚珠。花柱短。雄花序长 7～12cm，雄花：雄蕊 5，长 2.5～3mm，雌蕊退化。据哈尔滨 1963～1980 年观测，开花最早为 6 月 10

日，末期最迟为6月26日，单花开花日数7天。果实初熟为9月13日，末熟期为10月6日，初熟至脱落间隔25天左右。据吉林左家观察，植株萌芽期4月26～30日，开花始期至末期为6月1～13日，果实初熟至末熟期为8月30日～9月16日。浆果，肉质，球形，黑色，有蓝白色粉霜，直径7～13mm，最大达15mm。每果含种子1～4粒，最多达6粒。种子梨形（卵圆形），褐色带红色，长5～6mm，宽4～5mm；有背、腹面之分，背面中部有合点，基部突起处为喙。种皮坚硬，胚乳白色，丰富，胚位于喙附近（见图1）。

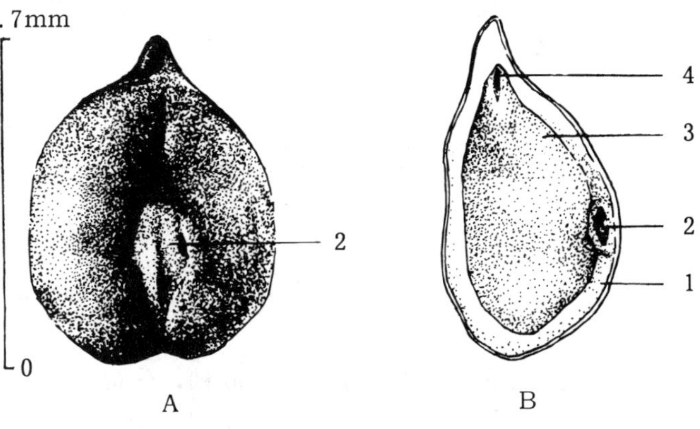

图1　山葡萄种子外形（A）及其纵切面（B）

1. 种皮　2. 合点　3. 胚乳　4. 胚

（梁鸣绘）

果实的采收调制和种子贮藏　成熟浆果体积有较明显的增长并迅速着色，果粉随之出现，种子亦由绿色逐渐变为棕褐色，种皮变硬。采回的果穗放在室内3～5天，压碎果肉，洗净。或将新鲜果穗去杂后榨出果汁酿酒，将果渣漂洗，晾晒，即得纯净种子。经过高温发酵的种子发芽率低，出苗生长不齐。鲜果出种率一般为12%，变动范围9%～15%。种子净度85%～95%。千粒重一般为30（23～49）g，每千克有种子33 000（25 000～43 000）粒。种子装麻袋、木箱等容器在种子库贮藏可保存2年，4年后发芽力全部丧失。

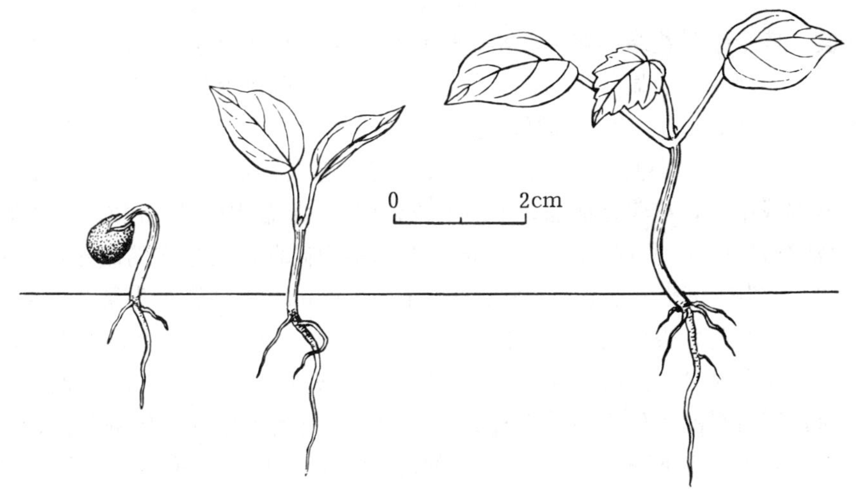

图2　山葡萄种子萌发后第7、19、25天幼苗的生长情况

（梁鸣、黄应钦绘）

发芽和播种　种子调制后混沙室外层积，翌春解冻后随气温升高而发芽。如发芽不充分，可移至20～25℃条件下，催芽至约有1/3种子萌发时播种。亦可采用先高温（20～25℃），后低温（1～5℃），再高温（20～25℃）；或先低温（冷冻）后高温（20～35℃），再低温（1～5℃）

这 2 种变温处理。哈尔滨地区 5 月上旬播种（塑料大棚中 3 月下旬～4 月上旬播种）。苗床条播或垄播（塑料大棚内点播）。苗床条播时每平方米播种量 25g 左右，覆土 1.5cm 左右。出土萌发。种子萌发时从喙部裂开，胚根伸出，随后胚轴迅速生长，将子叶伸出地面，经 20 多天后第 1 片初生叶出现（见图 2）。当年苗高 30～40cm，可出圃。塑料大棚中的当年生苗可做砧木苗，用于嫁接。

（周德本、穆福忠）

紫金牛属
Ardisia Swart
（紫金牛科 Myrsinaceae）

生长习性、分布和用途 本属约 400 种，分布于热带美洲、大洋洲、太平洋诸岛及亚洲东部和南部。我国约 69 种，产长江流域以南各地。常绿灌木，稀乔木。本文描述 2 种，它们的名称、生长习性、树高、分布和用途见表 1。

表 1 紫金牛属树种的名称、生长习性、树高、分布和用途

中 名	学 名	生长习性和树高	分 布	用 途	供 稿
密鳞紫金牛	*A. densilepidotula* Merr.	6～8（15）m	海南	材用、药用	605
紫金牛	*A. japonica*（Thunb.）Bl.	近蔓生，具匍匐根状茎，直立茎高可达 30cm	陕西及长江流域以南各地，但海南未发现。朝鲜半岛、日本	观赏、药用	1007

开花结实 紫金牛开花结实年龄较早，密鳞紫金牛的结实始于 8～10 年。这 2 个种每年都能结实，大小年现象不明显。花两性。紫金牛花序近伞形，腋生。密鳞紫金牛为近伞形花序构成圆锥花序，顶生或近顶生。花萼、花冠 5 裂，基部合生。紫金牛花冠粉红色或白色，具腺点。密鳞紫金牛花冠粉红色至紫红色，无腺点或稀具腺点。雄蕊与花冠裂片同数且对生，并着生于花冠基部或中部，花药无横隔。子房上位，1 室，紫金牛胚珠 15 枚，密鳞紫金牛胚珠约 14 枚，通常都只有 1 枚发育。这 2 个种的开花结实物候期以及果实的成熟特征见表 2。核果状浆果，球形。花柱和花萼常宿存。紫金牛果径 5～6mm，密鳞紫金牛果径 6～8mm。外果皮微肉质，内果皮坚脆，表面具明显纵向线纹，内含种子 1 粒。种子为胎座的膜质残余物所盖，球形，基部稍有内凹，胚乳丰富。紫金牛胚横生（图 1），且常有多胚现象。1992 年本文作者曾对南京地区所采的 1 份样品拍摄软 X 射线照片，并对实物作了解剖分析，发现该样品中单胚的种子占 31%，双胚的占 34%，三胚的占 24%，四胚的占 9%，五胚的占 2%。

表 2 紫金牛属树种的开花结实物候期和果实特征

树 种	观察地点	花 期	果 期	果实成熟特征
密鳞紫金牛	海南	5～6 月	8～10 月	黄绿色变为紫红色至紫黑色，表面无腺点
紫金牛	南京	6～7 月	11～12 月	鲜红色后转黑色，表面多少有腺点

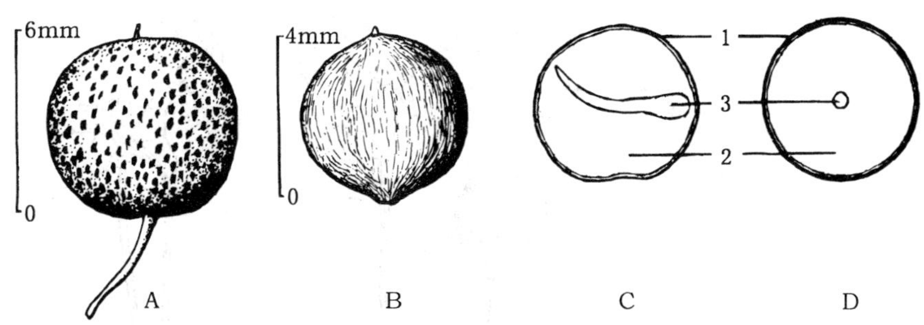

图 1　紫金牛果实（A）和果核（B）的外形及其纵切面（C）、横切面（D）
1. 种皮　2. 胚乳　3. 胚
（田恒德绘）

果实的采收调制和种子贮藏　果期较长，例如紫金牛果实成熟于 11～12 月，但迟至 5～6 月仍可见果实悬挂，且植株矮小，采摘相当方便。密鳞紫金牛果实丛集枝条顶端，采收也很容易。果实易为鸟雀啄食，常有鸟播现象，应在果实出现成熟特征时及时采收。选充分成熟的果实水浸 1 昼夜，搓揉淘洗，除净果皮、果肉等杂质，所得果核即为播种材料，通称种子。这 2 个树种的出籽率、净度和种子的质量数据已列入表 3。关于它们的贮藏寿命和适宜的贮藏条件，目前还无确切的研究资料。

表 3　紫金牛属树种的出籽率、净度和种子质量

树　种	出籽率（%）	净度（%）	千粒重（g）	每千克种子粒数（万粒）
密鳞紫金牛	55	97～99	80～110	0.9～1.3
紫金牛	38	97～99	80～90	1.1～1.3

发芽和播种　这 2 个树种的发芽习性和播种育苗方法，目前还知之不多。从 1987～1988 年在江苏南京和广西南宁的少量试验来看，紫金牛的种子无休眠习性，在 25℃条件下，35 天的发芽率为 95%，发芽高峰期在第 26 天。密鳞紫金牛则有休眠现象。1987 年 10 月 15 日广西林业科学研究所将当年新采的密鳞紫金牛播入沙床，播前种子未加处理，到 1988 年 4 月 25 日，即播后 192 天，日均温升至 18℃时才开始萌发；从发芽之日起算的 30 天内发芽率为 25%～38%，且无明显发芽高峰。

本属树种有些是子叶留土萌发，有些是子叶出土萌发。A.B. 伦德勒在《有花植物分类学》（中译本）第 2 册中提到，紫金牛的“子叶一直留于种子内”。但据我们观察，本文描写的密鳞紫金牛和紫金牛这 2 个种都是出土萌发。密鳞紫金牛胚根伸出后约 10 天子叶带壳出土，再过 10 天种壳脱落，再经 12 天初生叶展现。紫金牛的种壳有的脱落较迟，且 2 片子叶有的双双萎缩脱落，有的其中之一萎缩脱落，并在幼茎上留下脱落痕迹；留下的子叶变成绿色，成为光合作用器官，并逐渐增大。紫金牛初生叶近对生，宽卵形，长 7～10mm，边缘有细锯齿，齿缝间无边缘腺点，羽状脉，叶两面有突起腺点；下胚轴长 1.2～2cm，有短毛，主根发达，侧根短而细。紫金牛种子的萌发及其多胚现象见图 2。

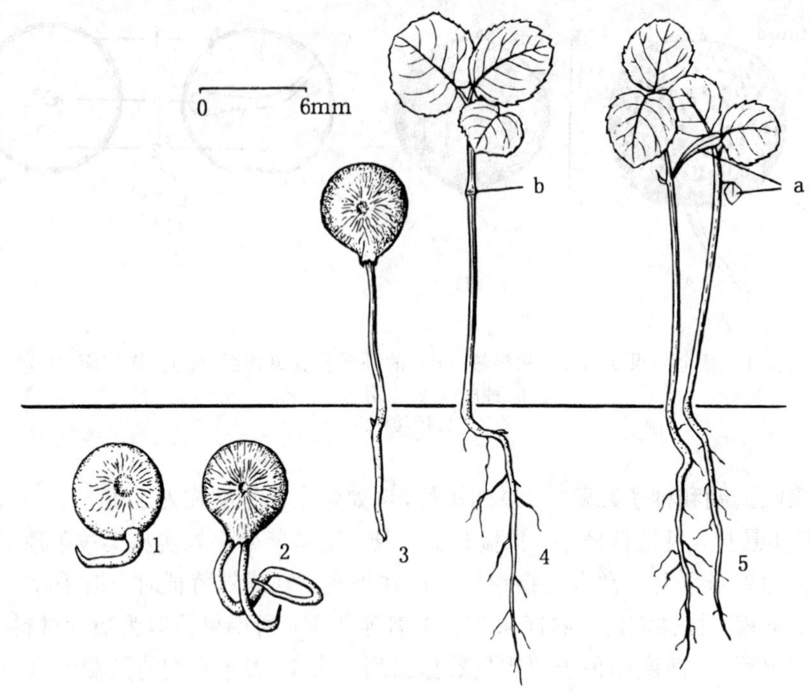

图 2 紫金牛种子的萌发和幼苗生长情况

1. 胚根延伸 2. 双胚萌发 3. 核壳出土 4、5. 初生叶近对生

a. 子叶 b. 子叶痕

(田恒德绘)

(沈永宝)

杜 茎 山

Maesa japonica (Thunb.) Moritzi ex Zoll.

(紫金牛科 Myrsinaceae)

生长习性、分布和用途 杜茎山属约 200 种，我国 29 种，本文描述 1 种。常绿灌木，高 1～3m，稀至 5m。生长缓慢，耐庇荫。喜肥沃湿润的森林土壤，干旱贫瘠地生长不良。分布于长江流域以南各地。日本、越南也有。可与乔木混种，作水源涵养林。嫩叶可代茶。全株入药，有生肌、解毒及祛风消肿之效。

开花结实 6～8 年生开始开花结实，结实大小年现象不明显。花两性。总状花序或圆锥花序，腋生，通常单生，有时 2～3 个聚生。花萼漏斗形，5 裂。花冠白色，筒状，5 裂，长 3～4mm。雄蕊 5，纤弱，不伸出花冠外。子房半下位，1 室，冠以宿存萼片及花柱，有多数胚珠生于特立中央胎座上。据广西南宁观测，2 月上中旬始花，3 月中下旬为盛花期，9 月中下旬果实开始成熟，10～11 月为成熟盛期，成熟期较长。另据海南记载，花期 1～3 月，果实成熟期 7～10 月。浆果，球形或卵形，具明显的腺状条纹，径 4～4.5mm。每果有种子数百粒，镶于空心的胎座内。种子细小，带棱状圆球形，黑褐色，径 0.3～0.5mm，具肉质胚乳。

果实的采收调制和种子贮藏　植株矮小，成熟后直接采摘。采得的果实堆沤 4～7 天，浆果软熟后装入紧密布袋，置水中反复搓擦，淘去果皮等杂质，清洗，即得纯净种子。鲜果出种率约 5%。种子的净度可达 95%。千粒重约 0.05g，每克有纯净种子 1.5 万～2.5 万粒。种子不宜失水，忌日晒。短期运输可直接装运果实，运抵后再行调制。已调制的种子宜随即播种，如需贮藏应混细沙，贮藏期一般在半年以内。

发芽和播种　种子无休眠现象。发芽时要求日均温度在 20℃以上。1987 年 12 月 18 日，广西林业科学研究所在发芽箱内用当年新采种子作过发芽测定，置床前种子未经任何处理，基质为滤纸，温度保持 28℃。置床后 9 天（12 月 26 日）开始发芽，次年 1 月 2 日为发芽盛期，1 月 12 日发芽终止。从开始发芽至发芽高峰的 8 天中发芽百分数为 14%。从置床之日起算，26 天的发芽率为 18%。出土萌发。胚根萌发后约 4 天子叶出土，再过 50 天左右新叶展出。种子萌发和幼苗早期生长情况见图 1。

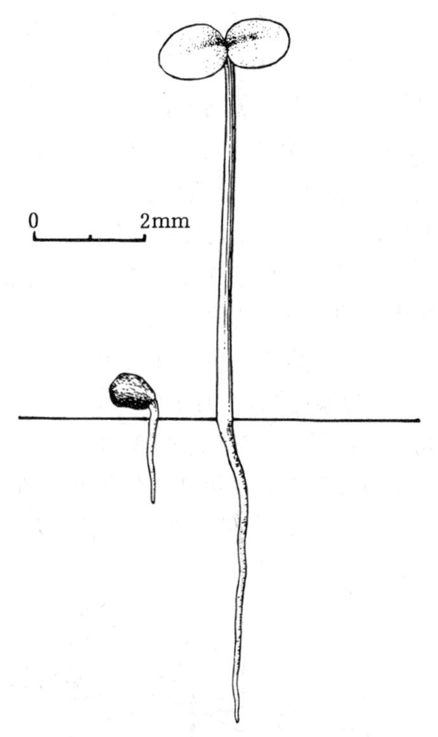

图 1　杜茎山种子萌发后第 4、15 天幼苗的生长情况

（黄应钦绘）

混细沙撒播于播种盆或细致整地的播种床。每平方米播种 0.3～0.5g。初生叶出现后约 2 个月，苗木半木质化时，移至大田继续培育。幼苗需搭棚遮荫。一般需培育 2 年方宜出圃。

（林榕庚）

密　花　树

Rapanea neriifolia (Sieb. et Zucc.) Mez

（紫金牛科　Myrsinaceae）

生长习性、分布和用途　密花树属约 200 种，我国有 7 种，本文描述 1 种。常绿乔木，高 20m 以上，胸径 35cm。速生。极耐荫蔽。适生于空气凉爽、土壤肥沃湿润的山谷或山腰下部的稀疏林下，裸露荒山生长不良。分布于华南、华东各省、自治区。日本，缅甸、越南也有。木材坚硬，可制作高级家具和车轴。重要的水源林树种。树皮可提制栲胶，叶捣烂后可医治外伤。

开花结实　约 5 年生开始结实，结实大小年间隔期常为 1 年，但不甚明显。花单性异株。伞形花序或花簇生，有花 3～10 朵。苞片广卵形，具疏缘毛。花梗粗壮，长 2～4mm。花萼仅基部连合，裂片 5，卵形。花冠白色或淡绿色，稀紫红色，5 深裂。雄蕊在雌花中退化，在雄花中着生于花冠中部，与花瓣对生。花丝极短，花药略小于花瓣。子房上位，卵形或椭圆形，无毛，1室，有胚珠数颗着生于特立中央胎座上。花柱极短，柱头伸长，顶端平扁，基部圆柱形，长约为子房的2倍。据广西南宁1978～1984年观察，1年中有2次开花结实现象：第1次的始花期在3月上旬，3月中旬为盛花期，3月下旬为末花期，8月下旬果实开始成熟，10月上旬为果熟盛期；第2次的始花期在7月上旬，7月中下旬为盛花期，8月上旬为末花期，11月下旬为果熟始期，12月中下旬为果熟盛期。两造果实均可用于播种。浆果状，核果，球形，有较长的果柄及4～5裂的萼片，未熟时青色，成熟时灰白色或黑色，径4～8mm。果肉薄，具1核。内果皮较坚脆，外表有皱纹，内层光滑。核近圆形或不规则形，浅绿色，径4mm，高5mm。种子1，基部有明显凹陷或顶部有小圆形凹陷。胚乳丰富，胚横置，稍弧弯。密花树的果核外形及其解剖结构见图1。

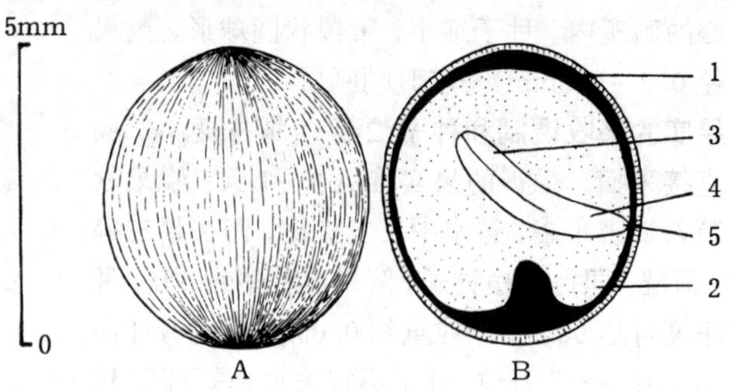

图1　密花树果核外形（A）及其纵切面（B）
1. 内果皮及种皮　2. 胚乳　3. 子叶　4. 胚轴　5. 胚根
（黄应钦绘）

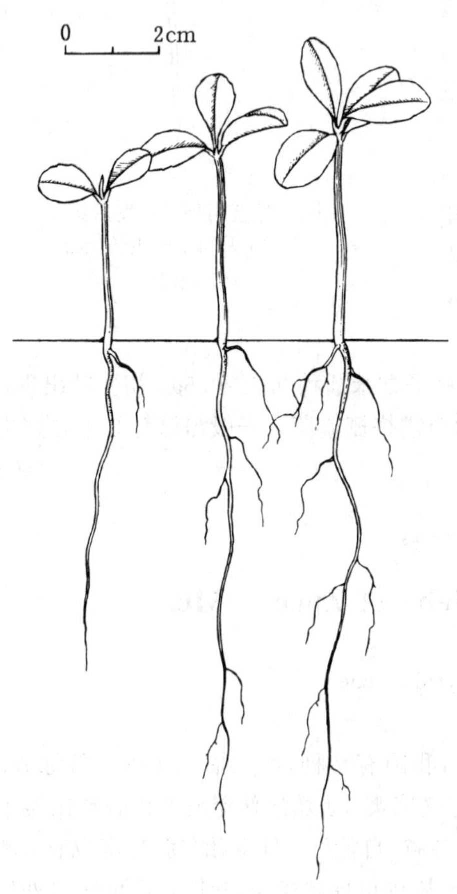

图2　密花树种子萌发后第35、45、56天幼苗的生长情况
（黄应钦仿《热带亚热带主要树种采种育苗技术》）

果实的采收调制和种子贮藏　果实密集于树干上，可上树采摘或用采种钩刀采集。采得的浆果置阴凉处摊放1～2天，不需处理即可用于播种，或分层置湿润沙床内催芽。也可搓洗并淘去果肉，得出果核作为播种材料，通称种子。鲜果出籽率为40%～50%。种子（核）含水量约20%。千粒重约55g，每千克有种子（核）18 000（16 000～20 000）粒。种子不宜脱水，忌曝晒，不宜久藏。运输时需混沙，用木箱包装，短期贮藏亦应混以等量湿沙。

发芽和播种　少数种子无明显休眠习性，大部分种子呈深休眠。发芽时的日均温13～19℃即可。1981年1月5日，广西林业科学研究所用未作处理的新采种子，在室外沙床作发芽测定。播后种子分2次发芽，第1次发芽始于1981年3月10日，结束于

5 月 20 日，发芽率约 8%。第 2 次发芽始于 1982 年 1 月 2 日，终于 3 月 10 日，发芽率为 32%。2 次总发芽率为 40%。出土萌发。胚根萌发后 18 天下胚轴延伸，部分拱出地面，过 12 天子叶出土，再过 7 天展出初生叶。密花树幼苗早期生长情况见图 2。

条播。种子发芽极不整齐，发芽全程可持续至春播后的第 2 年，一般不宜将种子直接播至圃地。播前应当层积催芽，而将已经萌动的种子移至苗床。也可将种子密播于苗床上任其陆续出苗，并搭棚遮荫，至第 2 年春再移植培育。2～3 年生苗出圃。

（张声燕）

蜡烛果（桐花树）

Aegiceras corniculatum（L.）Blanco

（蜡烛果科　Aegicerataceae）

生长习性、分布和用途　蜡烛果属 2 种，我国只有本文描述的 1 种。常绿灌木或小乔木，高 1～4.5 m。热带海岸滩涂红树林内最常见的树种之一。多单独组成群落，也常与其他树种混生。大部分生于河口咸水淡水交界处以至离海边数千米的河岸。对盐度的适应性很广，既适应高盐度，又适应盐度较低的水域。分布于粤、桂、闽、琼、台沿海及南海诸岛。印度、中南半岛至菲律宾及澳大利亚南部均有。蜡烛果林可防浪护堤，是海滨鱼虾理想的生活场所。优质薪材，树皮含鞣质，可制栲胶。花为蜜源。

开花结实　4 年生左右开始结实，大小年不明显。花两性。伞形花序顶生或腋生，有花10～24。萼片 5，覆瓦状排列，宿存。花冠白色，直径 1cm，钟形，裂片 5，卵形，花后脱落。雄蕊 5，长约 8mm，花丝基部合生成管。药室内具横隔膜，分为若干小室。子房上位，1 室，卵形，与花柱无明显界限，连成一圆锥体，有胚珠多颗。据在广西北部湾沿海观察，花期 3～4 月，幼果期 4～5 月，果熟期 8～9 月。另据《中国植物志》记述，花期 12 月～翌年 1～2 月，果期 10～12 月，有时花期 4 月，果期翌年 2 月。蒴果，圆柱形，新月状弯曲，顶端稍尖，革质，长 6～8cm，直径 6mm，宿存萼紧包基部。每果有种子 1 粒，与果同形。种子无胚乳，胚长而弯曲，子叶合生成圆筒状。种子在母体上发芽后仍留在果皮内（“隐胎生”现象）。果实脱落吸水，种子胀破果皮后，胚轴伸出并插入泥中，很快生根固定。

果实的采收调制和种子贮藏　果皮转入黄褐色即可采摘。采得的果实不需作任何处理即可作为育苗材料。每千克有果实约 2 300 粒。果实忌失水，不宜日晒，亦忌堆沤。短期存放宜选阴凉处并覆盖湿草。运输时用袋或竹箩包装，途中应保持湿润。贮存或运输期一般应在 1 周以内。

发芽和播种　1983 年 8 月下旬在广西钦州沿海作过育苗试验。播前将果实放在竹箩内，置于有流水的小湾，每昼夜浸泡 5～6 小时，捞起后置荫湿处，稍加翻动，不使发热。一般情况下 3 天后胚轴开始胀大，第 4～5 天为萌发盛期，6～7 天发芽终止。萌发盛期即可播于静水淤泥深厚的平坦滩涂上，逐粒插播。插入淤泥深达果长的一半左右。株行距 30cm×30cm，每 666m^2 用种 3～3.5kg，产苗 6 000～8 000 株。1 年生苗高 30～40cm，可以出圃造林。

（何祖家）

柿　属
Diospyros L.

（柿树科　Ebenaceae）

生长习性、分布和用途　本属约500种，主产热带，我国58种7变种，本文描写9种1变种。乔木或灌木，落叶或常绿。树皮及未熟果实富含鞣质，可以提制柿漆（柿油）及栲胶原料。柿是我国著名的果树，栽培历史悠久，果型大，味甜汁多，营养丰富，既可生食，又可制成各种加工食品，干制后可代粮食。柿的树冠庞大，入秋果色鲜红，富有观赏价值。柿属其它树种的果实多数都可食用，或作砧木嫁接柿树。木材材质坚硬，结构细，供建筑、家具、器具及细木工用。它们的名称、习性、生长和分布见表1。

表1　柿属树种名称、习性、生长和分布

中　名	学　名	习　性	树高(m)	胸径(cm)	分　布	供　稿
柬埔寨黑柿（黑柿）	*D. embryopteris* Pers.	常绿或半常绿小乔木	5～6	8	原产东南亚。琼、粤、桂栽培	604
乌材	*D. eriantha* Champ.	常绿乔木	16	50	桂、粤、琼、台。越南、老挝、马来西亚、印度尼西亚	604
粉叶柿（浙江柿）	*D. glaucifolia* Metc.	落叶乔木	17	50	华东	1003
柿	*D. kaki* L. f.	落叶大乔木	10～14 (27)	65	原产长江流域。现自辽西、长城经甘南折入川、滇，此线以南，东至台，多有栽培	1003
野柿	*D. kaki* var. *silvestris* Makino	落叶乔木	8～12	—	西南、中南、华东	1003
君迁子	*D. lotus* L.	落叶乔木	10～20 (30)	25～35 (130)	东北南部、华北、西北东部、西南、中南、华东。西亚、南欧	302
罗浮柿	*D. morrisiana* Hance	常绿或半常绿乔木	20	30	西南东部、中南、东南、台。越南	604
油柿	*D. oleifera* Cheng	落叶乔木	14	40	浙南、皖南、闽、赣、粤北、桂、湘	1003
异色柿（毛柿）	*D. philippensis* (Desr.) Gurke	常绿大乔木	—	—	台恒春半岛、兰屿。琼、粤、桂、滇、闽栽培。菲律宾、马来西亚	611
老鸦柿	*D. rhombifolia* Hemsl.	落叶小乔木或灌木	2～8	—	华东	1003

开花结实　花单性，雌雄异株或杂性。雄花成短聚伞花序，花部4～5基数。萼4（3～7）深裂，绿色。雌花单生叶腋，花萼与果实同时增大。花冠壶形或钟形，黄白色，4～5（3～7）浅裂或深裂。雄蕊4至多数，通常16。子房2～16室，每室1～2胚珠。浆果，肉质，基部

有宿存萼片，果实由子房壁发育而成。种子扁平，大。胚乳丰富，均一或嚼烂状；子叶叶状(见图 1)。花期 5～6 月，果实成熟于 9～10 月（详见表 2)。在海南岛，乌材每年开花 2 次，一次在 2 月，一次在 7～8 月。柿属果实和种子的形态特征已汇入表 3。

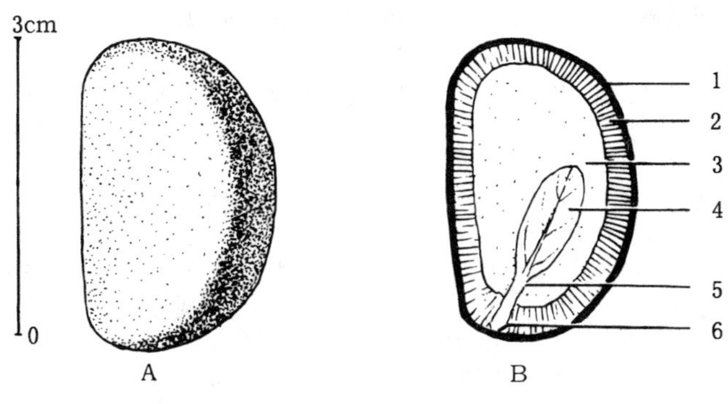

图 1　异色柿种子外形（A）及其纵切面（B）

1. 外种皮　2. 内种皮　3. 胚乳　4. 子叶　5. 胚轴　6. 胚根

（孟玲绘）

表 2　柿属树种的开花结实物候期

树　种	观察地点和年份	开　花			果实成熟		果实脱落期
		始　期	盛　期	末　期	始　期	盛　期	
柬埔寨黑柿	南宁 1987～1988	5 月上旬	5 月下旬	6 月上中旬	翌年 2 月中旬	3 月上旬	3 月中旬～6 月上旬
乌材	南宁 1987～1988	5 月下旬	6 月上旬	6 月中旬	12 月中旬	12 月下旬	翌年 1 月上旬～4 月上旬
粉叶柿	南京	5 月上旬	5 月下旬	6 月上旬	10 月中旬	11 月上旬	—
柿	南京	5 月上旬	5 月中旬	5 月下旬	9 月上中旬	10 月中旬	—
君迁子	北京 1956～1987	5 月下旬	5 月下旬～6 月上旬	6 月上中旬	9 月下旬～10 月上旬	10 月中旬～11 月上旬	—
罗浮柿	南宁 1987～1988	4 月下旬	5 月上旬	5 月中旬	10 月中旬	10 月下旬	11 月上旬～12 月上中旬

表 3　柿属树种果实和种子的形态特征

树　种	未熟果颜色	成熟果实			种　子		
		形　状	大小（cm）	色　泽	形　状	大小（cm）	色　泽
柬埔寨黑柿	浅黄色	近球形	径 1.1～1.3	黑紫色	扁椭圆形	长 1～1.2 宽 0.4～0.6 厚 0.3～0.55	棕褐色有光泽
乌材	青绿色	卵形先端渐尖	长 1.2～1.9 径 0.8～1.2	棕红色至黑紫色	椭圆形，扁圆形，三棱形	长 1.1～1.4 宽 0.6～0.9 厚 0.37～0.75	棕褐色至黑褐色

（续）

树 种	未熟果颜色	成熟果实			种 子		
		形 状	大小（cm）	色 泽	形 状	大小（cm）	色 泽
粉叶柿	绿色	球形较高	径 1.5～3	橘红色外被白霜	扁长圆形	长约 1.2 宽约 0.8	淡褐色 略有光泽
柿	绿色	扁球形或长圆形	长 4～9 径 3.5～8.5	橙红色或橙黄色	扁椭圆形	因品种而异 长约 2.0 宽约 1.0	棕色
野柿	绿色	近球形	长 3～5 径 2～5	—	扁平，半圆形	长约 1.5 宽 0.8～0.9 厚 0.25～0.3	褐色
君迁子	浅棕色	近球形	径 1.5 左右	黑色	扁椭圆形	长 1.0～1.5 宽 0.4～0.8 厚 0.3～0.4	棕色有光泽，表面有不规则胶丝状纹理
罗浮柿	青绿色	近球形	径 1.3～1.8	浅黄色 有光泽	扁椭圆形	长 1.1～1.4 宽 0.6～0.8 厚 0.3～0.5	栗色有光泽
油柿	绿色	扁球形或卵圆形	长（3）4.5～7（8） 径约 5（8）	暗黄色 呈油污状	扁椭圆形	长 2～2.3 宽 1.2～1.3 厚 0.3～0.5	棕色
异色柿	绿色	扁球形	径约 8.0	新鲜时黄色，成熟时近黑色，密被锈色毛	扁椭圆形	长 2～3 宽 1.0～1.8	淡褐色
老鸦柿	绿黄色无斑或橘红色带不均匀黑斑	卵球形或长卵形，顶端突尖，有长柔毛	长 1.7～3.0 径 1.2～2	红色有蜡质有光泽	半球形，近三棱形	长约 1 宽约 0.6	褐色

果实采收调制和种子贮藏 充分成熟的果实采回后除去果肉，取出种子，用水洗净后稍稍阴干。11 月下旬前后用湿沙层积，待来春播种。柿属的出种率、种子的净度和质量的数据见表 4。

表 4 柿属树种鲜果出种率、种子净度和质量

树 种	出种率（%）	净度（%）	千粒重（g）	每千克种子粒数（万粒）
柬埔寨黑柿	7.0	97～100	90～130	7 700～11 000
乌材	40.0	97～100	200～300	3 300～5 000
粉叶柿	—	—	350～400	2 500～2 900
柿	—	—	950～1 100	900～1 100
野柿	—	—	200～300	3 300～5 000
君迁子	13.6	97～100	85～100	10 000～12 000
罗浮柿	22.0	97～100	85～105	9 500～12 000
油柿	—	—	800～1 100	900～1 250
异色柿	20	96～99	3 500～3 800	260～290
老鸦柿	—	—	175～220	4 500～5 700

发芽和播种 柿属种子无休眠习性，发芽时日均温要求在 20℃以上。1987 年 7 月底，华

南热带作物学院在海南儋县室外沙床上测定过异色柿新鲜种子的发芽能力：播后约20天开始发芽，从开始萌发之日起算，12天的发芽率为80%～85%。1988年在广西南宁测定过当年采收的柬埔寨黑柿、罗浮柿、乌材种子的发芽能力，播种时种子均未做任何处理。7月12日在室外沙床播种柬埔寨黑柿，当时日均温为29℃，播后12天开始发芽，8月25日发芽终止，从发芽之日起算，20天的发芽百分数为63%，31天的发芽率为75%～90%。11月25日和12月20日先后在发芽箱内对罗浮柿和乌材进行发芽测试，基质为滤纸，温度保持28℃，发芽高峰不明显，从发芽之日起算，罗浮柿28天的发芽率为80%～95%；乌材30天发芽率80%～95%。据孙秀琴和田树霞(1993)报道，采自北京和河北邢台的君迁子，以1%琼脂为基质，在30℃/20℃的昼夜变温和黑暗条件下，10天发芽率可达90%以上。他们的试验材料还显示，无论测定前是否浸种，无论浸种水的始温是20℃、45℃还是60℃，君迁子种子开始萌发的天数、平均发芽日数以及最终发芽率都无显著差异。

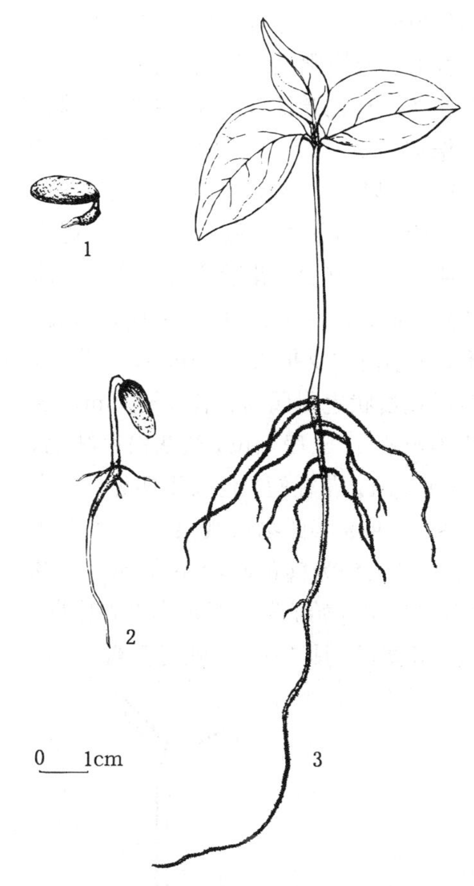

图2　君迁子种子萌发后幼苗生长情况
1. 幼根伸出　2. 子叶出土　3. 初生叶互生
(孟玲绘)

播种后出土萌发。胚根萌发所需时间在柬埔寨黑柿为12天，罗浮柿为9天，乌材为11天；再过8、22和10天子叶带壳出土。柬埔寨黑柿子叶出土后约3天即展开并发出新叶，罗浮柿和乌材出土后8～12天种壳脱落，子叶与初生叶同时展现。君迁子、柿一般在3～4月上旬春播，也可以在11～12月上旬秋播。君迁子种子萌发和幼苗生长的情况见图2。

(钱耀明)

血胶树（锈毛梭子果）

Eberhardtia aurata（Pierre ex Dubard）Lec.

（山榄科　Sapotaceae）

生长习性、分布和用途　梭子果属约3种，我国2种，本文描述1种。常绿乔木，高15～

20m，胸径 30～40cm。喜高温高湿的气候，成龄树能耐 0℃左右的极端低温，苗期忌霜冻，不耐冰雪。适生于疏松肥沃湿润的酸性土，粘重土或干旱贫瘠土生长不良。分布于滇、桂、粤。越南北部亦产。种仁含油 55%，供食用和制皂。木材为优质用材。

开花结实　5～7 年生开始结实，正常结实期在 15 年生以后。结实大小年间隔期为 1 年，但不甚明显。花两性，簇生于叶腋。花萼具短管，裂片 2～3 (4)。花冠合瓣，管近圆筒形，乳白色，无毛，裂片 5；每一裂片 3 深裂，中间条形，先端内向反折长 2～3mm，为花瓣，两侧为膜质花瓣状之附属物，长 4～5mm。发育及退化雄蕊各 5。子房上位，被灰白色绒毛，5 室，每室胚珠 1。花柱短，柱头不明显。据广西南宁 1987～1988 年的物候观察，3 月中下旬为始花期，4 月上旬为盛花期，4 月下旬为末花期；9 月中旬果实开始成熟，下旬为果熟盛期。果核果状，近球形，具 5 棱，成熟时黄褐色，密被锈毛；长 2.5～3.5cm，径 2.5～3cm；顶端花柱遗迹形成突尖，具宿存萼。果实成熟后 1 个月内不脱落。9 月下旬～10 月中旬，果棱隆起处（室背）纵裂，种子弹落地面。每果有种子 5 粒，少有 6 或 4 粒。种子扁椭圆形，栗色，具光泽，长 2～3cm，宽 1.2～1.8cm，厚 0.8～1.1cm。种脐长圆形，从腹面延伸至种子的一端。种皮坚脆，具油质胚乳。子叶椭圆形，胚根向下。种子的外观形态和内部结构见图 1。

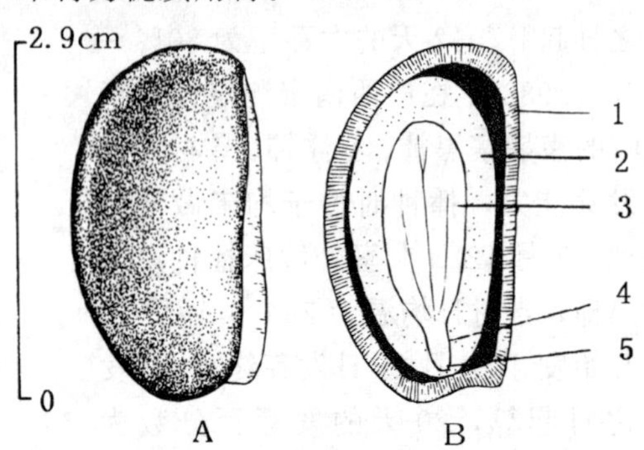

图 1　血胶树种子外形（A）及其纵切面（B）
1. 种皮　2. 胚乳　3. 子叶　4. 胚轴　5. 胚根
（黄应钦绘）

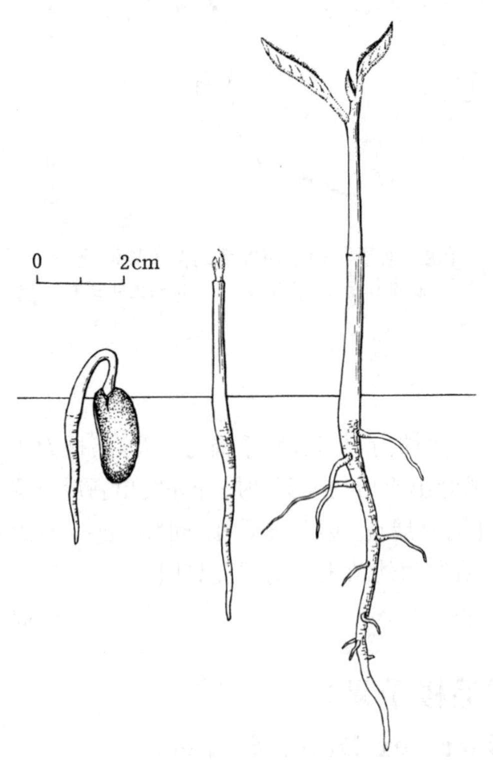

图 2　血胶树种子发芽后的幼苗生长情况
1. 幼根伸长　2. 子叶脱落、幼芽生长　3. 初生叶互生
（黄应钦仿《热带亚热带主要树种采种育苗技术》）

果实的采收调制和种子贮藏　果实成熟盛期用采种钩刀采下，也可在果实开裂，种子落地后的 1 周内在地面捡拾。落地较久的种子发芽率明显降低，可供榨油，不宜用于育苗。采得的果实可摊放于室内，俟果皮干缩纵裂，种子自行弹出。鲜果出种率为 50%～60%。净度可达 100%。千粒重约 1 600（1 300～2 000）g，每千克有纯净种子 620（500～800）粒。种子的含水量约 20%，忌日晒，不宜脱水。种子裸露存放于室内 20 天即全部失去发芽能力。贮藏期为 4～6 个月。

发芽和播种　种子无明显休眠现象。发芽时日均温宜在22℃以上。1981年，广西林业科学研究所在室外沙床上用当年新采种子作过发芽测定：9月中旬播种，10月上旬开始发芽，发芽盛期不明显；10月下旬以后由于气温降低，停止发芽，至次年4月上旬又继续发芽，5月中旬发芽结束；发芽过程持续半年以上，发芽率为82%。出土萌发。下胚轴萌发后约10天，上胚轴从子叶基部发出，壳与种脱离。上胚轴出土后子叶即脱落，约过7天发出初生叶。子叶脱落处具环状痕。血胶树种子的萌发及幼苗初期生长情况见图2。

条播或撒播。宜将种子密播于沙床催芽，发芽后将芽苗移至大田继续培育。每平方米育苗用种子75～120g。1年生苗出圃。

（林榕庚）

蛋　黄　果

Lucuma nervosa A. DC.

（山榄科　Sapotaceae）

生长习性、分布和用途　蛋黄果属约100种，我国引入2种，本文描述1种。小乔木，高7～12m。原产热带美洲和西印度群岛。现世界热带地区广为栽培。我国滇、桂、琼、粤、闽引种栽培。为著名果品，供食用、制果酱、果汁等。

开花结实　4～5年生开始开花结实，6年生以后正常结实，大小年现象不明显。花两性，1～2（3～4）朵生于叶腋，花梗长1.2～1.7cm。萼裂片5（6～7），卵形或阔卵形，长7mm。花冠长约1cm，裂片（4）6，狭卵形，长约5mm。能育雄蕊5，花丝钻形，长约2mm。花药心状椭圆形，长约1.5mm。退化雄蕊5，狭披针形至钻形，长3mm。子房上位，圆锥形，长3～4mm，5室，每室1胚珠。花柱圆柱形，长4～5mm，柱头头状。花的各部分除花柱、花萼内侧无毛外，均被黄白色绒毛。据广西南宁1987～1988年的物候观察，每年开花结实2次：第1次为4月下旬花蕾初现，花期始于5月下旬，盛期在6月中旬，末期6月下旬；10月中旬果实开始成熟，果熟盛期在11月上旬，果熟末期在12月末。另一次开花始期为8月20日左右，翌年3月为果实成熟盛期。

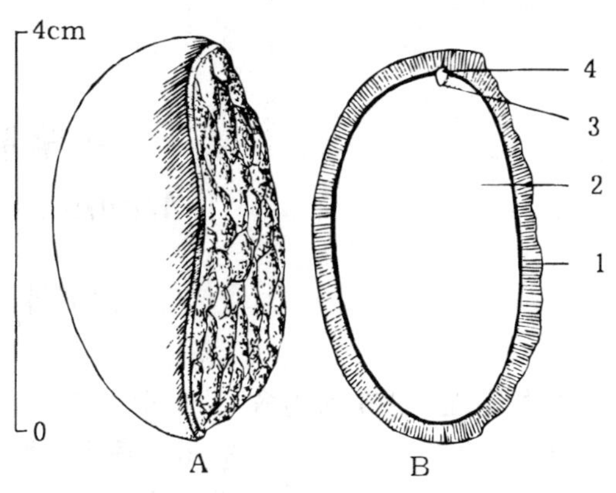

图1　蛋黄果种子外形（A）及其纵切面（B）
1. 种皮　2. 子叶　3. 胚芽　4. 胚根
（黄应钦绘）

浆果，倒卵形或圆形，径约8cm，幼时被黄白色细绒毛。成熟时脱落至光滑或被薄蜡质。外果皮薄，中果皮肉质，肥厚，蛋黄色。每果有种子2～4枚。种子长肾形或椭圆形，压扁，长4～5cm，径2～2.5cm，黄褐色，具光泽，有隐约可见的褐、黄相间纵向条纹。种脐侧生，长圆形或不规则，表面凹凸不平，几与种子等长。外种皮厚壳质，内种皮膜质。无胚乳，子叶肥厚。蛋黄果种子的外形和内部结构见图1。

果实的采收调制和种子贮藏　根据果实

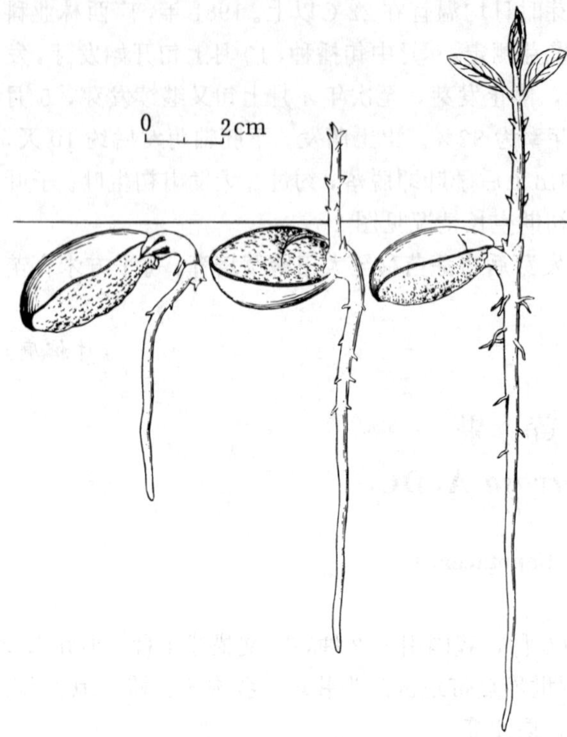

图 2　蛋黄果种子萌发后第 50、65、80 天幼苗的生长情况
（黄应钦绘）

成熟特征分批采摘，采下后小心轻放，以免碰伤果皮流脂，影响果的品质。采收后放置室内 5～8 天，或用薄膜密封 2～3 天，或喷洒适量催熟剂催熟。催熟后果实用手压有弹性感，外果皮和中果皮易分离，剥开果皮有熟蛋黄香味。中果皮利用后洗净得出种子。鲜果的出种率为15%～20%。种子的净度可达 99% 。千粒重为 4 000（3 500～5 000）g，每千克有纯净种子200～300 粒。种子不宜日晒或裸露存放，以免失水后内外种皮分离或紧贴，子叶悬空而失去发芽能力。贮藏及运输均需混以湿沙或湿润锯木屑。贮藏期为半年以内。

发芽和播种　种子无休眠习性，发芽时日均温需在 20℃左右。1987 年 12 月 1 日，广西林业科学研究所在室外沙床作过发芽测定，当时日均温 14℃左右。1988 年 2 月 20 日日均温升至 20℃时发芽，3 月 4 日进入发芽盛期，3 月 20 日发芽结束。从开始发芽至发芽高峰，14 天的发芽百分数为 73%。从播种至发芽终止，以 82 天计，发芽率为 87%。留土萌发。胚根伸出 60 天以后，上胚轴出土，具不育叶 6 枚左右，10 天以后初生叶展现。蛋黄果种子的萌发和幼苗初期生长情况见图 2。

点播。每平方米播种量约为 150～200g，覆土 3～4cm。苗期需遮荫。2 年生苗出圃。也可用压条法育苗。

（曾　玲）

紫荆木属
Madhuca J. F. Gmel.

（山榄科　Sapotaceae）

生长习性、分布和用途　本属约 85 种，我国产 2 种，又引入 1 种，本文描述 3 种。常绿大乔木，高 25～30m，胸径 80cm。适生于肥沃湿润的酸性土或钙质土。热带珍贵硬木和油料树种。种仁含油 50%左右，可食用。它们的名称、生长、分布及用途见表 1。其中海南紫荆木是海南特有的树种，木材为特种用材，资源已急剧下降。紫荆木是我国南部珍贵用材树种之一，资源已趋枯竭。这 2 个树种都亟需保护，在《中国植物红皮书》中被分别列为渐危种和稀有种。

表 1　紫荆木属树种的名称、生长情况、分布和用途

中　名	学　名	树高（m）	胸径（cm）	分　布	用　途	供　稿
海南紫荆木	*M. hainanensis* Chun et How	25～30	80	海南特产	油料、材用	613
长叶紫荆木（长叶马府油树）	*M. longifolia* (L.) Gmel.	25	80	印度、缅甸。海南栽培	油料、造船、材用、药用	611
紫荆木（子京木、滇木花生）	*M. pasquieri* (Dubard) Lam.	30	80	滇、桂、粤。越南	油料、材用	605

开花结实　长叶紫荆木6～8年生开始开花结实，正常结实期在15年生以后。海南紫荆木和紫荆木10～12年生开始开花结实，正常结实期在20年生以后。结实大小年间隔期均为1年，但不甚明显。花两性，多朵簇生或单生于叶腋。萼片4，排列成互生的2轮，常被毛，宿存。花冠无毛，白色或淡黄色，6～18裂。海南紫荆木雄蕊28～30，长叶紫荆木16，紫荆木（16）18～22（24），2～3裂。子房上位，6～10室，每室1胚珠。花柱突出，宿存。紫荆木每年开花结实2次，主花期始于8月下旬，翌年3～4月果实成熟；少部分枝条于2～3月开花，8～9月果熟。这3个树种的开花结实物候期见表2。

表 2　紫荆木属树种的开花结实物候期

树　种	观察地点	观察年份	开花			果实成熟		果实脱落
			始　期	盛　期	末　期	始　期	盛　期	
海南紫荆木	海南乐东	1985～1989	4月下旬	5月中旬～6月中旬	9月中旬	翌年3月上旬	3月中旬	3月下旬～4月上旬
长叶紫荆木	海南儋县	1988	—	5月	—	—	8月	9月
紫荆木	广西南宁	1978～1984	8月下旬	9月中旬	12月中旬	翌年3月下旬	4月中旬	4月中旬下旬

浆果，椭圆形或卵形，萼宿存。种子1～5，种脐长椭圆形或条形。无胚乳，子叶扁平，肥厚，富含油脂。果实及种子的形态见表3、图1。

表 3　紫荆木属树种果实和种子的形态特征

树　种	果实			种子		
	形　状	大小（cm）	色　泽	形　状	大小（cm）	色　泽
海南紫荆木	阔卵形，近球形	长2.5～3 宽2～2.8	黄绿色，被短柔毛	扁椭圆形	长2～2.5 宽0.8～1.2	褐色，具光泽
长叶紫荆木	椭圆形	—	淡绿色	弯月形	长2.5～3 宽0.7～1	黄褐色，具光泽

（续）

树种	果实			种子		
	形状	大小（cm）	色泽	形状	大小（cm）	色泽
紫荆木	椭圆形或卵形	长 2.8～3.8 宽 1.9～2.1	红色转黑色，被锈色毛	扁椭圆形	长 2.1～3.3 宽 1.2～1.7	黄褐色或浅栗色，具光泽，有长圆形疤痕

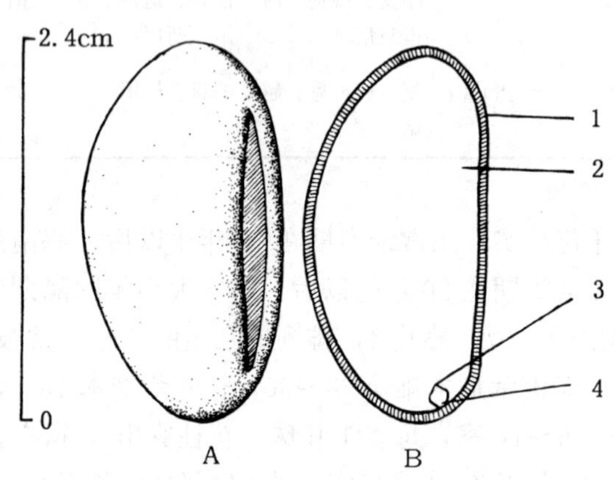

图 1 紫荆木种子外形（A）及其纵切面（B）
1. 种皮 2. 子叶 3. 胚芽 4. 胚根
（黄应钦绘）

果实的采收调制和种子贮藏 果实鸟兽喜食，应趁成熟盛期及时采收。紫荆木以采 4 月成熟的大造果为宜。直接采摘或摇动树枝震落后在地面捡拾。采得的果实可堆放 2～3 天，果皮软熟时装入竹箩，置水中搓洗，淘去果皮等杂质，稍晾干即得纯净种子。种子富含油脂，忌脱水，不宜曝晒。种子的含水量应保持在 30%以上。运输和短期存放均需混以湿沙。宜随采随播，一般不作贮藏。果实的出种率及种子的净度、质量等见表 4。

表 4 紫荆木属树种果实出种率和种子净度、质量

树种	出种率（%）	净度（%）	千粒重（g）	每千克纯净种子粒数
海南紫荆木	20	96～100	1 000～1 330	750～1 000
长叶紫荆木	34	95～99	1 300～1 700	600～770
紫荆木	21	96～100	1 400～2 000	500～700

发芽和播种 种子无休眠习性。发芽时日均温需在 20℃以上。1987 年 8 月 23 日，华南热带作物研究所用未作处理的当年新采长叶紫荆木种子，在室外沙床作过发芽测定，播后 19 天开始发芽。1982 年 4 月 30 日，广西林业科学研究所用当年新采而未作处理的紫荆木种子，也在室外沙床作过发芽测定，播后 37 天开始发芽。测定情况见表 5，其中海南紫荆木由中国林业科学研究院热带林业试验站测定。

表 5 紫荆木属树种的发芽能力及其测定条件

树种	测定条件		发芽势（%）		发芽率（%）	
	基质	室外温度（℃）	计算天数	一般数值	计算天数	一般数值
海南紫荆木	沙壤土	24～29	7	50	29	93
长叶紫荆木	沙	24～29	8	80	32	95
紫荆木	沙	24～27	10	40	56	70

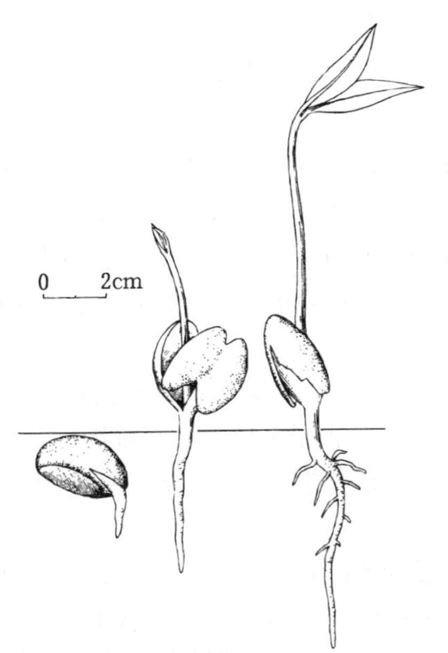

图 2 紫荆木种子萌发后第 4、20、30 天幼苗生长情况
（黄应钦仿《热带亚热带主要树种采种育苗技术》）

紫荆木属种子萌发出来的幼苗，下胚轴短，生长慢。长叶紫荆木为留土萌发，发芽后 11 天初生叶展现。海南紫荆木和紫荆木为出土萌发，胚根萌出后 16～20 天子叶出土，初生叶同时出现。由于这 2 个树种的胚轴短，发芽出土后，肉质的子叶接近地表，也称地面萌发。紫荆木的种子萌发及幼苗初期生长情况见图 2。

条播或点播。长叶紫荆木条播，每平方米约播 75～85g，培育 1 年生苗出圃。海南紫荆木点播每平方米约播 60～75g。紫荆木点播，每平方米约播 100g，覆土 1.5～2cm。后面这 2 个树种苗期需强度遮荫，培育 1～3 年生苗出圃。

（曾 玲）

人 心 果

Manilkara zapota（L.）Van Royen

（山榄科 Sapotaceae）

生长习性、分布和用途 铁线子属约 70 种，我国仅产 1 种，又引入栽培 1 种，本文描述引入的 1 种。常绿乔木，高 15～20m。能耐 0℃左右的极端低温，忌霜冻。喜肥沃湿润酸性土。原产热带美洲，琼、粤、桂、闽、滇引种栽培。为热带著名水果，乳汁为制口香糖胶原料，树冠卵形浓绿，为优良庭院观赏树种。皮入药。

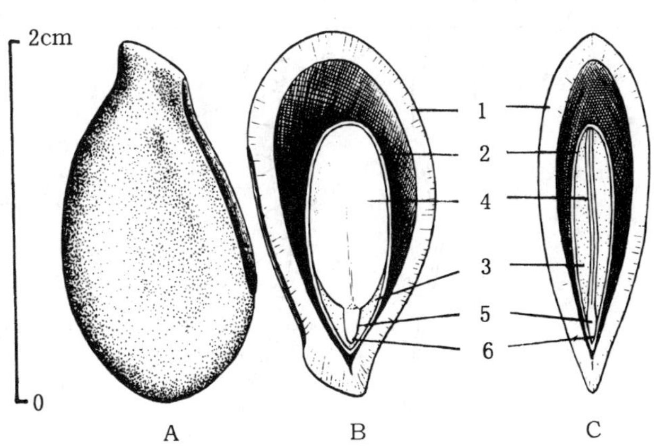

图 1 人心果种子外形（A）及其纵切面（B、C）
1. 外种皮 2. 内种皮 3. 胚乳 4. 子叶 5. 胚轴 6. 胚根
（黄应钦绘）

开花结实 5～6 年生开始开花结实，高枝压条苗 2 年生即可开花结实，正常结实期在 10 年生以后。结实无大小年现象，冬季

特寒会影响开花结实。花两性，1～2 朵生于枝顶叶腋，间有 10～13 朵集生于当年新枝上部，每枝结出的果实只有 3～5 个。花蕾各部分被锈色毛。花萼 6 裂，2 轮。花冠白色，6 裂，每裂片背部有 2 等大的附属物。能育雄蕊 6，生于花冠的喉部。退化雄蕊 6，花瓣状，与花冠裂片互生。子房上位，6～8（14）室，每室 1 胚珠。花柱圆柱形，基部略粗。据广西南宁观察，每年开花结实 2 次：第 1 次为主花期，5 月中旬始花，盛花期在 5 月下旬，末花期 6 月中旬；果实成熟始期 7 月上旬，果熟盛期 8 月上旬，果熟末期 9 月上旬，果实脱落期为 9 月下旬。第 2 次花期在 9 月上旬至下旬，果熟期在翌年 4～5 月，花期不整齐。除了这 2 次开花外，常年还有零星花开。

浆果，卵圆形、纺锤形或球形，长 4～8cm，径 5～7cm。外果皮棕褐色，粗糙，小片状脱落。中果皮厚肉质，富含糖分。每果有种子 1～7 粒。种子近梭形，压扁状，褐色至黑褐色，有光泽，外种皮脆壳质，长 18～23mm，宽 7～11mm，厚 5～6.5mm。种脐侧生条状。具胚乳，子叶薄叶状。种子形态见图 1。

果实的采收调制和种子贮藏 在树上选择已经成熟，果径 6～7cm 或更大的果实，用手采摘，轻放，以免损伤而降低品质。果实采回后堆放 3～7 天，果肉便软化变甜，味道可口。用手掰开果实便可取得种子。鲜果出种率为 2%～4%。种子净度可达 99%。千粒重约 510(400～600)g，每千克有纯净种子 1 960（1 670～2 500）粒。种子忌失水，不宜日晒或裸露存放。短途运输及贮藏均需混以湿沙，贮藏期为3～6 个月。

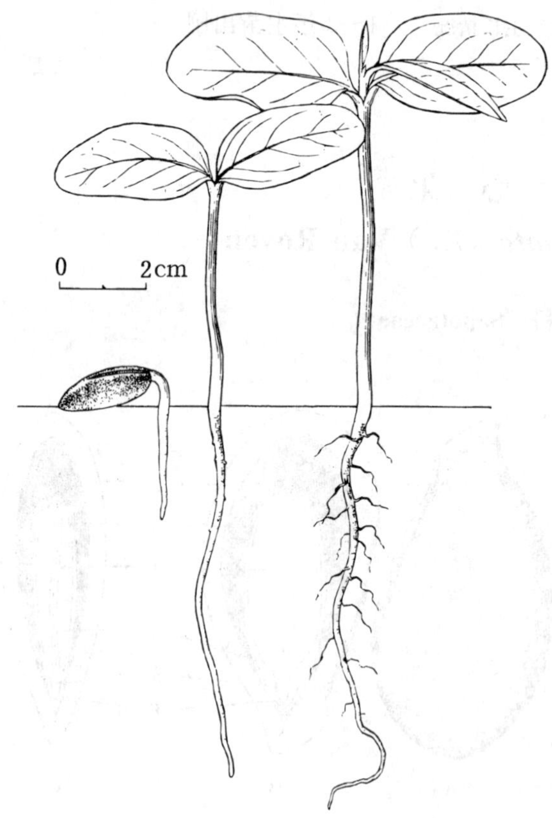

图 2 人心果种子萌发后第 12、20、35 天幼苗的生长情况

（黄应钦绘）

发芽和播种 种子无休眠习性。发芽时日均温宜在 22℃以上。1988 年 4 月 14 日，广西林业科学研究所在室外沙床上用新鲜种子作过发芽测定，当时日均温 26℃。播后 10 天开始发芽，5 月 18 日发芽进入盛期。从开始发芽至发芽高峰的 25 天，发芽 58%。5 月 28 日发芽结束。从播种至发芽终止，44 天的发芽率为 70%。出土萌发。下胚轴伸出 12 天后子叶出土，再过 18 天初生叶展现。人心果种子的萌发和幼苗早期生长情况见图 2。

点播。每平方米播种 100～160g，覆土 1.5～2cm。培育 2 年生苗出圃。作果树栽培时常用高枝压条苗。

（曾 玲）

桃　榄

Pouteria annamensis (Pierre) Baehni

（山榄科　Sapotaceae）

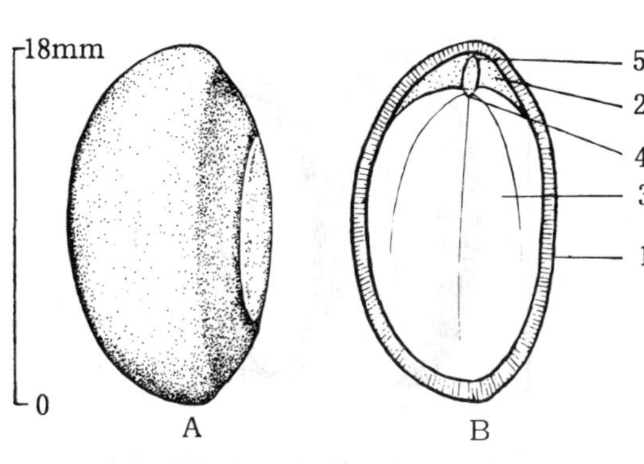

图 1　桃榄种子外形（A）及其纵切面（B）
1. 种皮　2. 胚乳　3. 子叶　4. 胚芽　5. 胚根
（黄光郁、黄应钦绘）

生长习性、分布和用途　桃榄属约 50 种，我国产 2 种，本文描述 1 种。常绿乔木，高10～20m，胸径 15～30cm。适生于中海拔林中，村边宅旁有时偶见。分布于海南、广东、广西。越南也有。果味香甜，可食用。民间用树皮治疗毒蛇咬伤。木材可制小型家具、农具。

开花结实　花两性，数朵成簇状腋生。萼 5 裂。花冠白色，阔圆筒状，裂片 5。能育雄蕊 5，着生于花冠喉部，具退化雄蕊。子房上位，圆锥形，5 室，密被锈色柔毛，每室有胚珠 1 颗。据海南儋县观察，每年开花 2 次：一次为 5 月开花，9 月果熟；另一次为 9 月开花，次年 4 月下旬果熟。浆果，无柄，苹果状，顶端钝，长宽约 2.5～4cm，成熟时由绿色转紫红色，内含 2～5 粒种子。种子侧向压扁，褐色，光亮，长 1.2～1.8cm。种皮脆壳质，淡黄色。种脐位于腹面，长圆形。具胚乳，子叶叶状。种子的形态见图 1。

果实的采收调制和种子贮藏　在每年 2 次结实中，宜选采 4 月下旬的大造果。当果实由青色转为黄绿色而带暗红色时，即可用竹竿打落树上的果实，或者摇动树枝震落后在地面收集。采集的果实摊放于室内数天，待果皮果肉软化后，食去果肉，或装入布袋置水中搓擦，淘去果皮等杂质，即得播种材料。鲜果出种率约 17%。种子的净度可达 98%。千粒重 1 100～1 200g，每千克有纯净种子 800～900 粒。种子忌失水，不能日晒，稍晾干的种子含水量约为 27%。已调制好的种子可以混湿沙贮藏，贮藏期不宜超过 4 个月。

发芽和播种　种子无休眠现象。发芽时日均温宜在 20℃以上。1989 年 5 月 2 日，华南热带作物研究所在室外沙床用新鲜种子作发芽测定，6 月 17 日开始发芽，7 月 6 日发芽结束，没有明显的发芽盛期，发芽率为 30%。出土萌发。发芽后 9～12 天展出初生叶。桃榄种子的萌发和幼苗初期的生长情况见图 2。

点播。每平方米播种 100～125g，覆土 0.5cm 左右。苗期需遮荫。1～2 年生苗出圃。

（邓春媛）

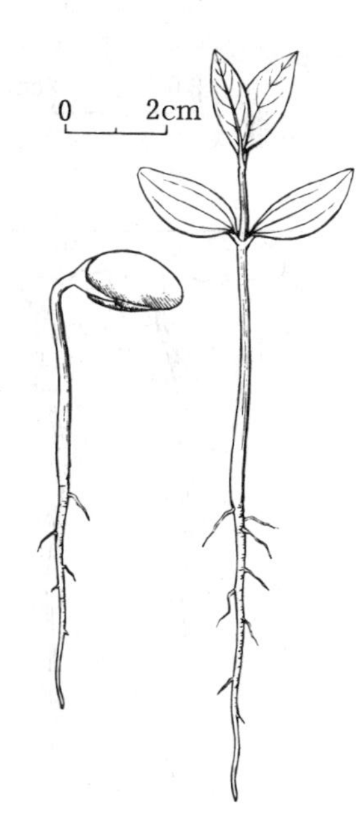

图 2　桃榄种子萌发后第 9、14 天幼苗的生长情况
（黄光郁、黄应钦绘）

神 秘 果

Synsepalum dulcificum Daniell

（山榄科 Sapotaceae）

生长习性、分布和用途 神秘果属约10种，我国只引入本文描述的1种。常绿灌木，高约1m。生长缓慢。忌霜冻，不耐持续1周以上的5℃的低温。喜肥沃酸性土。原产西非热带地区。海南及云南西双版纳在露地引种成功，在广州和南宁正常年份可安全越冬，特寒年份仍有冻害。树冠圆整，颇美观。果可食，吃后再吃酸味食物，可转酸味为甜味，故有“神秘果”之称。

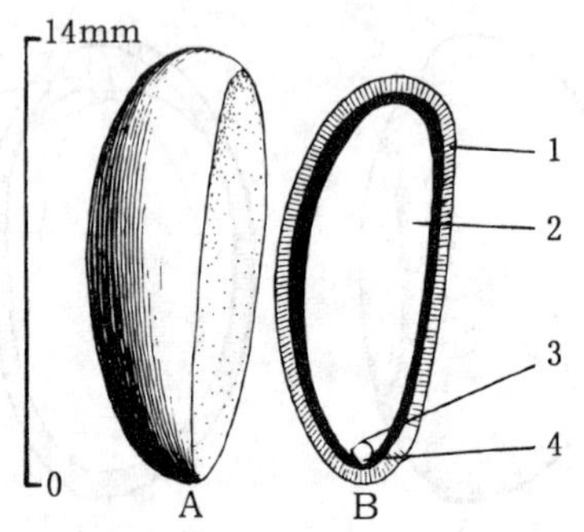

图1 神秘果种子外形（A）及其纵切面（B）
1. 种皮 2. 子叶 3. 胚芽 4. 胚根
（黄应钦绘）

开花结实 5～6年生开始结实，正常结实期在10年生以后，大小年现象不明显。在我国，影响结实的主要原因是冬春低温。受冻的嫩枝一般需1年后方能正常结实。花两性，1或3朵簇生于半年至1年生枝的叶腋或叶痕附近。花柄极短或近于无。萼浅褐色，被绒毛。萼筒约包裹花冠的3/5，径2mm，高4～5mm，萼片5裂。花冠黄白色，5裂，裂片倒卵形，长6～8mm，宽2～3mm。雄蕊多数，白色，长6～7mm。子房上位，花柱白色，突出于花冠顶部，长0.8～1cm，柱头钝尖。据广西南宁1986～1988年观察，花期特长：6月下旬始花，7月下旬～8月中旬开花较多，翌年3月中下旬为末花期；8月中旬至翌年5月下旬陆续有果成熟。在热量较高的海南南部，一年四季陆续有花开放，有果实成熟。

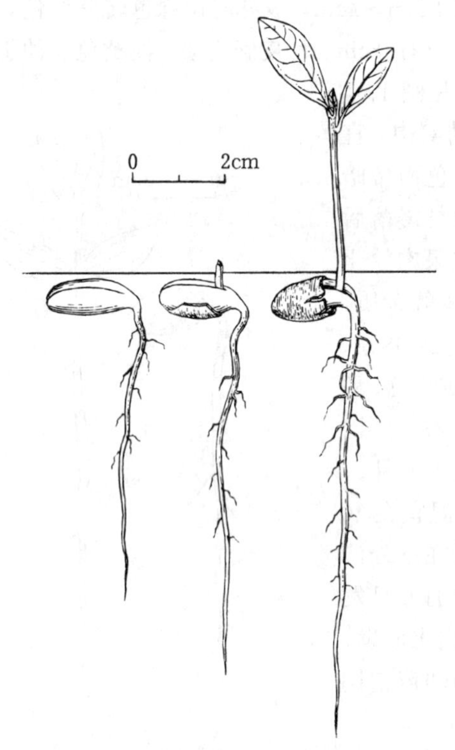

图2 神秘果种子萌发后第10、25、50天幼苗的生长情况
（黄应钦绘）

浆果，未成熟时青绿色，熟时红色，长椭圆形，长2.1～2.4cm，径0.9～1.2cm。每果有种子1粒。种子椭圆形，长1.1～1.5cm，径6～7.5mm，大半边褐色，有光泽。种脐在另侧，阔大，白色，粗糙，几与种子等长。无胚乳，子叶长椭圆形。种子的形态和内部结构见图1。

果实的采收调制和种子贮藏 采收时选摘红色果实。食去果肉，或堆沤后置水中搓擦洗净，即得种子。鲜果出种率约31%。净度可达98%。千粒重一般为335g，每千克有纯净种子2 700～3 300粒。种子的含水量宜保持在30%上下，忌失水，不能日晒或裸露存放。运输或贮藏均需混以湿沙，贮藏期一般为半年左右。

发芽和播种　种子有休眠习性。发芽时日均温宜在20℃以上。1988年11月10日，广西林业科学研究所用当年新采的种子，在室外沙床作发芽测定，当时日均温在20℃以上。翌年4月30日（日均温23℃）开始发芽，5月30日发芽终止。无明显的发芽盛期。发芽率为50%。留土萌发。胚根萌发后25天胚芽出土，再过9天左右展出初生叶。神秘果种子的萌发和幼苗初期生长情况见图2。

目前尚无批量播种育苗经验。一般是将种子点播于沙盘内，发芽后移至花盆或容器继续培育。2～3年生苗出圃定植。

（林榕庚）

柑　橘　属
Citrus L.

（芸香科　Rutaceae）

生长习性、分布和用途　柑橘属植物是柑橘类果树中最重要的植物，世界上的商品柑橘几乎全是本属树种。目前，世界公认的有17种，我国连引入栽培的共约15种。本文描述7种及45个品种（表1）。常绿小乔木或灌木，高1.5～7（10）m。喜温暖湿润环境，适生于微酸性红壤或黄壤，以在土层深厚、疏松肥沃、富含有机质的微酸性沙质壤土上生长发育最好。

柑橘属分布在长江流域以南，年平均温度在15℃以上，极端最低温度不低于－3（－9)℃，年降水量在1 000mm及其以上的广大丘陵地区（见表1）。

表1　柑橘属树种的名称、分布和用途

中　名	学　名	分　布	用　途	供　稿
酸　橙	*C. aurantium* L.	浙、闽、苏、湘、川、滇、赣。越南、缅甸、印度、日本	药用、砧木、香料、食用	501、507
柚（抛、栾、文旦）	*C. grandis* (L.) Osbeck	原产东南亚。浙、鄂、湘、川、滇、黔、赣、粤、桂、闽、台有栽培	食用、药用	501、507
宜昌橙	*C. ichangensis* Swingle	鄂西、湘西、川、滇、黔、桂北、陕南、甘南	药用、砧木	501、507
柠　檬	*C. limon* (L.) Burm. f.	发现于马来西亚，后传入欧、美。桂、粤、台、川、滇等有栽培	食用、药用、香料	501、507
檸　檬	*C. limonia* (L.) Osbeck	粤、桂、闽、台、琼、湘、黔、滇、川	砧木、食用	501、507
宽皮柑橘	*C. reticulata* Blanco	粤、闽、浙、皖、赣、湘、鄂、川、滇、黔、桂	食用、药用	501、507
甜橙（广柑、黄果）	*C. sinensis* (L.) Osbeck	粤、川、闽、湘、浙、桂、鄂、赣、滇、黔	食用、药用	501、507

果实大多供生食，也可制蜜饯果汁。花和果皮制作香料或入药，种子亦可作药用。

本属品种、品系复杂，国内各产区对于种类、品种命名多不一致，有同种异名的，也有同名异种的。本文对树种用拉丁学名命名，对于品种则采用公认的名称。

以下是表1各树种的主要品种及其主要产地：

酸橙：朱栾（浙）、枸头橙（浙）、代代（闽、苏、浙）。柚：沙田柚（桂）、金兰柚（粤）、文旦柚（闽）、绾溪密柚（闽）、四季柚（浙）、安江香柚（湘）、斋婆柚（赣）、红密柚（鄂）、梁平柚（川）、垫江柚（川）、夔柚（川）、五步红心柚（川）、长寿沙田柚（川）、台湾晚白柚（台）、台湾麻豆文柚（台）。柠檬的主要品种有尤力克、里斯本、维拉弗兰卡、费米耐劳，均为四川引进栽培的外来品种。檬檬：红檬檬和白檬檬。宽皮柑橘：南丰蜜橘、椪柑（滇、川、湘、赣、浙、粤、闽、台）、温州蜜橘（川、桂、湘、赣、浙）、四会柑（粤）、本地早橘（川、鄂、湘、粤、浙）、年橘（粤、湘）、红橘（陕、川、黔、湘、鄂、浙、闽）、朱橘（鄂、湘、赣、苏、浙）、茶枝柑（粤）、焦柑（川、粤、闽、台）。甜橙：桐子柑（川）、冰糖柑（川）、柳橙（川、湘、粤）、新会橙（粤）、锦橙（川、鄂）、雪柑（粤、闽、台）、佛灵夏橙（四川栽培）、五月红甜橙（川）、脐橙（川、湘）、血橙（川、湘）。

开花结实　实生苗约8年生开始开花结实。嫁接苗2～3年生即开花结实。结实有大小年，通常间隔1～2年出现1次丰年。

花两性，单生或簇生于当年生枝顶端或叶腋，或排列成总状花序。萼杯状，2～5裂，宿存，结果时常增大。花瓣4～8片，白色为主，间有黄白色。雄蕊为花瓣数的4～6倍，插生花瓣的四周，花丝常合生成束。子房上位，6～18室，每室有胚珠数颗。

除柠檬1年多次开花外，其他种和品种都是1年开花1次。开花期多在4月（表2）。

表2　柑橘属树种（品种）的开花结实物候

名　称	产　地	开花盛期	果实成熟期
沙田柚	广西柳州	—	11月下旬
南丰蜜橘	江西南丰	4月中旬	11月上旬
红　橘	四川江津	4月上中旬	11月下旬
	陕西城固	—	11月中下旬
温州蜜橘	湖南衡山	4月下旬	11月中旬
本地早橘	浙江黄岩	5月上旬	10月中下旬
	广东潮汕	3月下旬	11月中旬～12月中旬
甜　橙	四川江津	4月上中旬	12月上中旬
	福建龙溪	—	11月中旬
	湖北宜昌	—	11月中旬
蕉　柑	广东潮汕	—	12月上旬～翌年1月
	福建龙溪	—	12月中旬
椪　柑	福建龙溪	—	11月下旬

柑橘属树种的花芽分化期与品种以及当年的气候、树势和结果量有关。大多数柑橘属树种的花芽分化是在10月或11月～翌年春季萌芽前完成的。1年多次开花的柠檬及其品种，花

芽分化对外界条件要求不严，其分化进程也较快，冬季枝条抽生 1 个月即能完成花芽分化，并陆续开花结实。甜橙、椪柑与雪柑的花芽分化期列入表 3。

表 3　柑橘属几个品种的花芽分化时期

名　称	观察地点	分化前期	开始分化期	花萼形成期	花瓣形成期	雄蕊形成期	雌蕊形成期
甜橙	四川北碚	11 月上中旬	12 月下旬～翌年 1 月上旬	2 月上旬	2 月下旬	2 月下旬～3 月上旬	3 月上中旬
雪柑	福建福州	12 月下旬～翌年 1 月上旬	—	1 月上旬至 2 月上旬	2 月上中旬	2 月中下旬	2 月下旬
椪柑	福建福州	1 月中旬～2 月上旬	—	2 月中旬	2 月下旬	2 月下旬～3 月上旬	3 月上中旬

柑橘属各品种的成熟期受立地条件、品种特性和管理水平等的影响而有差异，大多数在 10～12 月成熟（见表 2），少数品种在第 2 年 5 月中旬成熟，如佛灵夏橙、五月红甜橙等。通常将果实在 10 月份成熟的视为早熟种，11～12 月成熟的为中熟种，第 2 年 1 月以后成熟的为晚熟种。判断果实成熟的主要依据是果皮着色情况及果实风味。风味成熟的标志包括果汁中酸的减少，含糖量的增加，可溶性固体物的增加，果汁的增加，组织软化和叶绿素的消失等。

柑橘属果实未熟时青绿色或浅橙色，成熟时果皮橙黄色、淡黄色、橙红色或红色。果为柑果（或称橙果），是芸香科植物特有的果实类型。外果皮薄，由小而排列紧密的厚角细胞组成，内含色体并有挥发油分泌腔，使外果皮具有特殊的香味。中果皮白色，同外果皮分界线不甚明显，由疏松的无色细胞组成。内果皮分成若干室，向内产生许多肉质多汁的囊毛（也有的简称"汁囊"），呈柄状或纺锤状。囊毛包围着生于胎座上的种子。果皮及表面情况因种而异。有果皮甚薄的（檬檬）、有果皮较厚，不易剥离的（柠檬、甜橙）、有果皮与果瓣易剥离的（宽皮柑橘）、有果皮甚厚的（柚类）、有果皮平滑的（宜昌橙）、有果皮较平滑的（甜橙）、有果皮较粗糙的（酸橙）。柑果球形、扁圆形、卵状椭圆形、长椭圆形或梨形。种子黄白色、绿色或淡绿色，卵形或倒卵形。子叶及胚为乳白色的有柚、柠檬等；子叶及胚为绿色的有檬檬，子叶及胚为深绿色间有淡绿色的有宽皮柑橘。

宽皮柑橘果实及种子形态见图 1。几种柑橘的果实和种子的颜色、形状、大小见表 4。

表 4　柑橘属树种成熟果实和种子形态特征

树　种（含品种）	果　实			种　子		
	颜　色	形　状	大小（cm）	颜　色	形　状	大小（cm）
酸橙	橙黄色或橙红色	球形或近球形	6～8	黄白色	卵形	0.8～1.2
柚	淡黄色或橙黄色	球形或梨形	10～20	黄白色	楔形	1.2～1.8
宜昌橙	橙黄色或淡黄色	椭圆形	4.5～6	黄白色	卵形或卵状楔形	1.2～1.7
柠檬	淡黄色	长椭圆形或卵圆形	5～10	黄白色	卵形	0.7～1.2

（续）

树种（含品种）	果实			种子		
	颜色	形状	大小（cm）	颜色	形状	大小（cm）
檸檬	淡黄色或橘红色	圆形或扁圆形	4～6	黄白色	卵形	0.6～1.0
宽皮柑橘	淡橙黄色至深橙黄色，橙红色至深红色	圆球形或扁圆形	4～8	淡黄白色、绿色或淡绿色	倒卵形	0.5～0.8
甜橙	橙黄色、橙红色至深橙红色	亚球形、卵状椭圆形、长椭圆形	5～12	黄白色	楔形或卵形	0.6～1.0

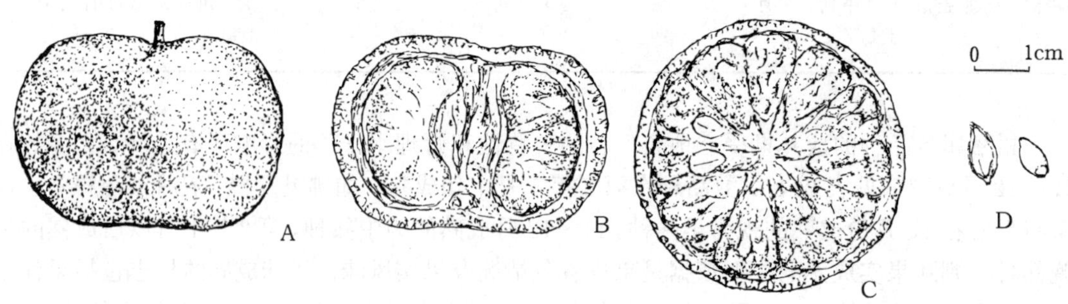

图1 宽皮柑橘果实外形（A）及其纵切面（B）和横切面（C）及种子外形（D）

（田恒德绘）

果实的采收调制和种子贮藏 柑橘属树种的采果期因用途不同而略有差异。目的在于取得种子时应让果实在树上充分成熟，使种子达到最饱满程度。用于生食的果实，须达到该品种固有的色泽、风味和香气，肉质变软时采收。贮藏用的果实，应当在果皮大部分转变为成熟时的颜色，肉质尚坚实而未变软时采摘。目的在于加工制作果汁、果酒、果酱的，宜在充分成熟时采收；制作罐头、蜜饯的常在八成熟时采收。柠檬类果实只要充分成长，果皮尚未转黄时便可采收。制作药材用的果实，应在幼果时采收。阴雨天、降雾时及中午烈日下均不宜采果。常用采果剪剪取果实：平口剪用来剪取宽皮柑橘类果蒂凸起的果实；剪口部分弯曲的对口剪用来剪取甜橙类果蒂凹下的果实。采果时还应准备盛果容器，使用采果凳或采果梯。在树上采果一般是从下而上，由外而内进行，并注意不要扯破蒂部的果皮。

作种用的果实采回后，用人工或机械剥取种子。取出的种子用温水洗净种子上的粘液后阴干，然后混湿沙贮藏于阴凉处。种子与湿沙的比例为1：4，沙的含水量约10%。种子与湿沙均匀混合后堆积的高度不超过45cm，表面盖10cm的河沙，再盖1层薄膜，每隔7天左右翻动1次，如发现沙已变干可适当喷水以保持湿润，但不宜过湿，以免种子霉变。

柑橘属树种每个果实所含种子数因种、品种不同而异。除无核柑橘外，甜橙类的锦橙、新会橙、佛灵夏橙等每果含几粒种子；宽皮柑橘的本地早橘、年橘、朱橘每果有种子约10粒；茶枝柑、红橘10～20余粒；四会柑有20余粒；四季柚平均有10余粒；文旦柚、安江香柚、梁平柚每果的种粒数则稍多。

柑橘属树种果实的出种率和种子质量列入表 5。

表 5　柑橘属树种果实出种率和种子质量

树　种	出种率（%）	千粒重（g）	每千克纯净种子数（粒）
酸　橙	2.5～3.0	130～170	5 800～7 700
柚	3.9～4.5	190～240	4 100～5 300
宜昌橙	3.0～3.5	180～230	4 300～5 600
柠　檬	1.5～2.0	130～180	5 600～7 700
檬　檬	0.5～1.2	80～130	7 700～12 500
宽皮柑橘	1.0～1.5	90～130	7 700～11 100
甜　橙	2.0～2.5	120～150	6 600～8 300

发芽和播种　在室内作发芽测定，先用始温 45℃水浸种 24 小时，捞出后置于垫有纱布或滤纸的培养皿内，在 25～28℃的恒温箱内发芽，新鲜种子 1 周后种子破壳萌发，发芽率在 80%以上。如需快速测定，可用 1%的四唑溶液染色。

生产上春季播种较普遍，也可在采种后立即播种。春播在 1～3 月，南亚热带柑橘区可在 1 月播种，中亚热带柑橘区一般在 2 月播种，北亚热带柑橘区在 3 月播种。条播或撒播。条距 10cm，条幅 5cm，或根据圃地情况和需要而定。播前用始温 40～45℃水浸种 24 小时，或浸种后将种子置于 25℃室内催芽，待胚根初露时播种。培育砧木的播种量列入表 6。播种要均匀。播种后用细土覆盖，覆土厚度视种子大小而定，大粒种子覆土厚度是种子直径的 2～3 倍，小粒种子是种子直径的 3～5 倍，覆土后轻压，浇水，再用稻草覆盖。播种后约 20 天开始发芽，30 天左右达发芽盛期。幼苗过密时可以间苗移植。

留土萌发。1 年生苗高 18cm 左右。宽皮柑橘的幼苗形态见图 2。

表 6　柑橘属几种砧木品种的播种量（g/m²）

名　称	撒　播	条　播
酸　橙	180～220	150～180
酸　柚	220～250	170～220
红檬檬	150～180	75～100
红　橘	150～170	120～150
甜　橙	250～280	210～250

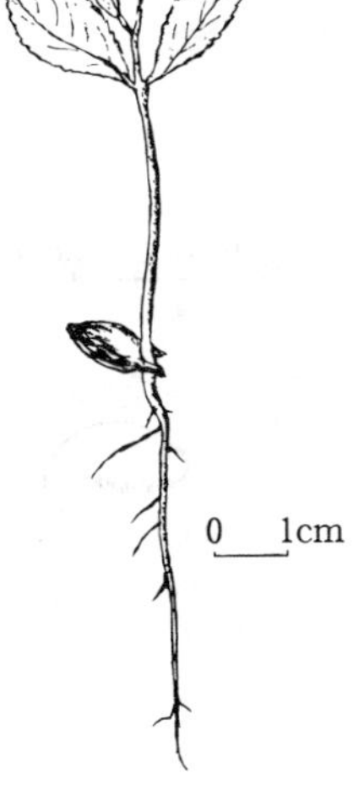

图 2　宽皮柑橘幼苗形态
（田恒德绘）

柑橘属树种还有嫁接、扦插、压条等繁殖方法，用得最多的是嫁接。

（赵德铭）

黄 皮

Clausena lansium（Lour.）Skeels

（芸香科 Rutaceae）

生长习性、分布和用途 黄皮属有25种，我国约9种，本文描述黄皮这1种。常绿小乔木，高5～6m。适生于肥沃湿润地，酸性土或钙质土均宜。分布于滇、川、黔、桂、粤、琼、闽、台。东南亚等热带地区亦广为栽培。南方水果之一，材作小农具及柄材，果皮去疳积，通小便。

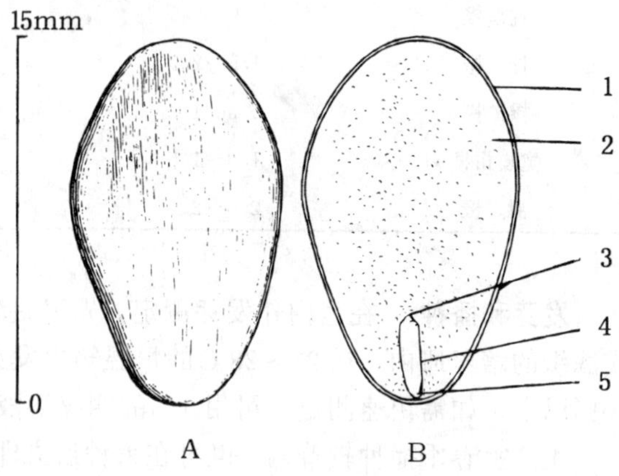

图1 黄皮种子外形（A）及其纵切面（B）
1. 种皮 2. 子叶 3. 胚芽 4. 胚轴 5. 胚根
（黄应钦绘）

开花结实 实生树5～6年生开始开花结实，无性繁殖苗一般2年生便可结实，结实大小年间隔期一般为1年。花两性。聚伞状圆锥花序腋生或顶生，长15～25cm。花小，淡黄色，萼片5裂。花瓣5。花盘短，内生。雄蕊8～10，生于花盘基部四周。子房4～5室，每室2胚珠。据广西南宁1978～1982年观察，3月下旬为始花期，4月上旬为盛花期，4月中旬为末花期；7月上旬果实开始成熟，7月中旬为果实盛熟期。人工培育的早熟或晚熟品种，果熟期可提早或推迟1个月左右。肉质浆果，未成熟时绿色，成熟后黄色或暗黄色，球形、卵形或倒卵形，长1～2cm，径1～1.5cm，被毛。种子2～5粒，稀6粒。种子扁椭圆形，青绿色，长1.2～1.6cm，宽0.6～1cm，厚3.5～5.5mm。种皮薄，无胚乳，胚伸直，子叶厚，近椭圆形（图1）。

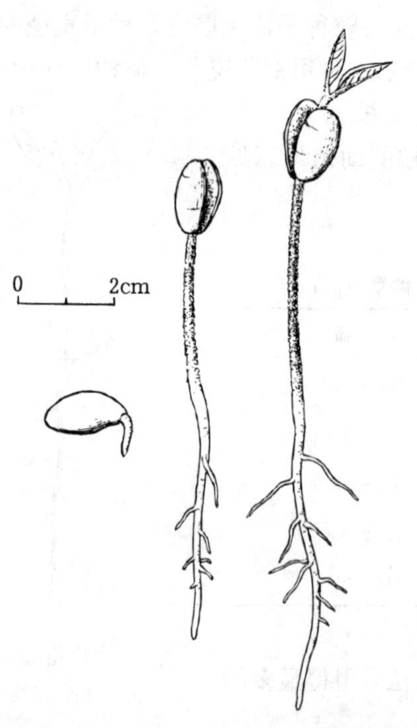

图2 黄皮种子萌发后第2、5、7天幼苗生长情况
（黄应钦绘）

果实的采收调制和种子贮藏 用枝剪截断总果梗，将果采下，即为商品黄皮。果为热带名果，一般食去果肉后，将种子收集加以洗净，即可供育苗用。出种率为20%～30%，种子净度一般为98%。种子忌失水，不能日晒或裸露存放。不影响种子发芽的最低含水量在

30% 以上。千粒重 240～300g，每千克纯净种子 3 300～4 200粒。种子对脱水和低温敏感，是典型的顽拗型种子。宋松泉和傅家瑞（1993）推测黄皮种子很可能在采收前就开始了萌发过程。生产上通常为随采随播，不作贮藏。运输时可混以湿沙或锯木屑，用竹篓或木箱包装。湿沙贮藏的贮藏期在 1 个月以内。

发芽和播种　种子无休眠习性。发芽时日均温宜在 20℃以上。1988 年 7 月 16 日广西林业科学研究所用当年新采且未作任何处理的种子，在室内滤纸上作过发芽测定，播种时日均温为 29℃，播后第 3 天开始发芽，随即进入盛期。从发芽开始至发芽高峰日的 4 天发芽 79%；从播种至发芽终止，9 天的发芽率为 98%。出土萌发。胚根萌发后约 3 天子叶带壳出土，出土后的第 3 天子叶展开，第 5 天发出初生叶（图 2）。

条播。每平方米播种 20～30g。1 年生苗可作嫁接砧木。用实生苗造林一般需培育 1 年半生苗方宜出圃。生产中多无性繁殖育苗。

（林榕庚）

楝叶吴茱萸

Evodia meliaefolia（Hance）Benth.

（芸香科　Rutaceae）

生长习性、分布和用途　吴茱萸属约 45 种，我国约 22 种，本文描述 1 种。常绿乔木，高达 20m。产于亚洲中亚热带南部至北热带地区，对水肥条件要求高。可耐轻霜及短期－6℃左右的低温。分布于闽、台、琼、粤、湘、桂、滇。越南也有。种子含油，可制肥皂和润滑剂，亦可入药，有行气消积之效。材质较优，易加工，宜于作家具、农具，也用于建筑。叶为优质绿肥。

开花结实　4～5 年生开始开花结实，正常结实期在 8 年生以后，结实大小年现象不明显。花单性异株。聚伞状圆锥花序，顶生。花极小，白色。雄花序长12～15cm；萼 5 深裂；花瓣 5，椭圆形；雄蕊 5 枚，白色，花药黄色，花丝下部被毛，有退化子房。雌花序长 8～11cm；花瓣较大；子房上位，球形深 4 裂，径 2～3mm，4 室，每室 2 胚珠；花柱短，柱头头状。据广西南宁 1978～1985 年观察，7 月上旬为开花始期，7 月中下旬为盛期，8 月上旬为末期；9 月中旬果实开始成熟，9 月下旬末果实盛熟。果扁圆形，长 4～5mm，径 7～8mm，未熟时绿色，成熟时紫红色至灰红色，果皮具网状皱纹。成熟后 1 个月内果实开裂，成 4 分果瓣，种子脱落。每果瓣含种子 1，稀 2。种子卵形或半圆形，黑色有光泽，长2.8～3.3mm，径 2～2.6mm。胚乳丰富，胚伸

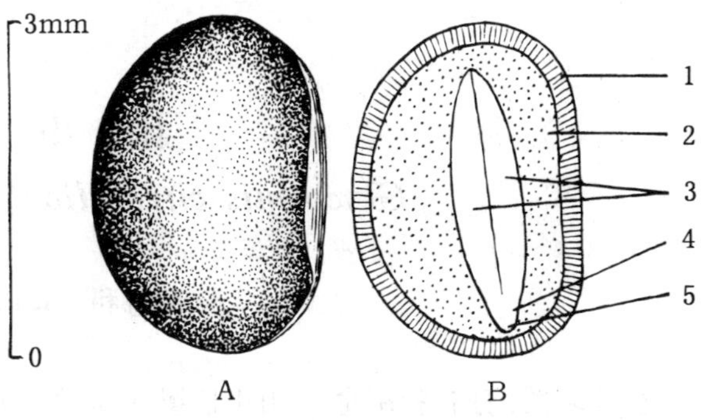

图 1　楝叶吴茱萸种子外形（A）及其纵切面（B）
1. 种皮　2. 胚乳　3. 子叶　4. 胚轴　5. 胚根
（黄应钦绘）

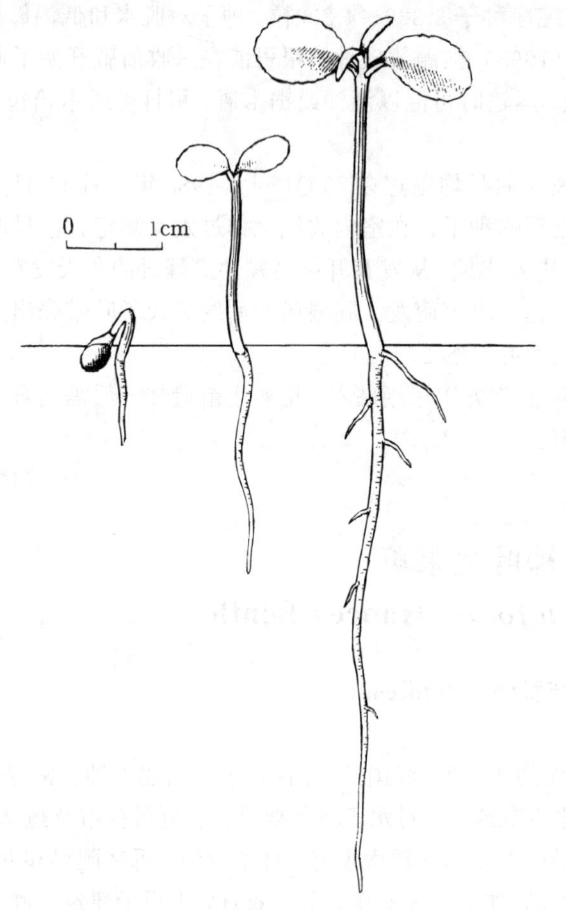

图 2　楝叶吴茱萸种子萌发后第 5、7、10 天幼苗的生长情况
（黄应钦绘）

直，子叶大。楝叶吴茱萸种子的外形和解剖结构见图 1。

果实的采收调制和种子贮藏　果熟后开裂，应及时采收。采得的果实摊于室内阴干，开裂后搓去果皮等杂质即得纯净种子。鲜果的出种率为 35%～40%。种子净度为 85%～95%。千粒重 (9) 10 (12) g，每千克有纯净种子 (8) 10 (11) 万粒。种子含油脂，忌失水，不能日晒或裸露存放，含水量宜保持在 13%左右，宜随采随播。运输时需混细沙，用布袋或木箱包装。贮藏亦需混湿沙，贮藏期为 6 个月以内。

发芽和播种　种子无休眠习性。发芽时日均温宜在 15℃以上。1989 年 10 月 14 日，广西林业科学研究所用当年新采且未经任何处理的种子，在室外沙床上作过发芽测定。11 月 1 日开始发芽，发芽盛期不明显，12 月 10 日发芽终止，发芽率为 50%。出土萌发。胚根萌发后约 5 天子叶出土，再过 4 天初生叶展现。楝叶吴茱萸种子的萌发和幼苗初期生长情况见图 2。

撒播。播前需精细整地。每平方米约播 2.5～3.5g，覆土 0.5cm。培育 1 年生苗出圃。

（林榕庚）

山　小　橘

Glycosmis citrifolia（Willd.）Lindl.

（芸香科　Rutaceae）

生长习性、分布和用途　山小橘属约 60 种，我国约 10 种，本文描述 1 种。常绿灌木，高 2～3m。适于湿润疏松的酸性土或钙质土，多生长在立地条件较好的疏林下或灌丛中。能耐短期 0℃左右的低温，但忌冰雪。分布于闽、台、琼、粤、桂、黔、滇。越南亦产。果可食。根、叶入药，治风湿、跌打、感冒及蛇咬伤。

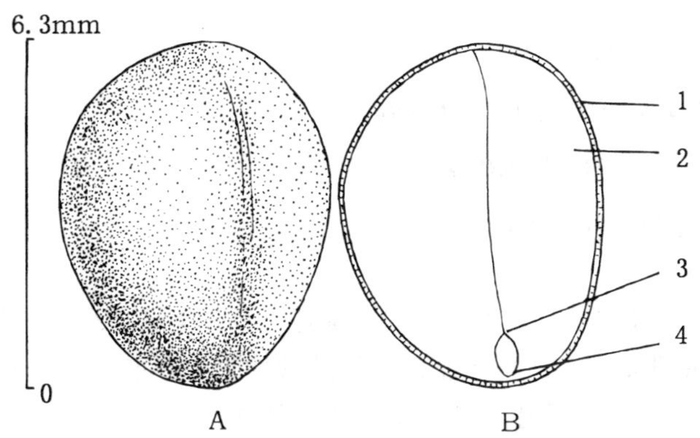

图 1　山小橘种子外形（A）及其纵切面（B）
1. 种皮　2. 子叶　3. 胚芽　4. 胚根
（黄应钦绘）

开花结实　约 3～4 年生开始开花结实，正常结实期在 8 年生以后。结实大小年间隔期不明显。低温或长期干旱会影响结实。花两性。聚伞花序多为 5 数，腋生或顶生。萼 4～5 齿裂。花瓣白色，4～5，覆瓦状排列。雄蕊8～10。子房扁圆形，平滑，2～3 室，每室 1 胚珠。花柱短而宽，宿存。花期较长。据广州及海南的物候资料，每年 6～9 月均陆续有花开放，8 月中旬果实开始成熟，9～10 月下旬陆续成熟，无明显的成熟盛期。浆果，近球形，扁而偏斜，径 0.7～1cm，未熟时青绿色，成熟时淡红色或朱红色，半透明。成熟后 5～ 7 天脱落。每果有种子 1 粒。种子圆形或半圆形，淡紫红色，径 5～7mm。种皮薄，无胚乳，子叶肥厚。种子的外形和解剖构造见图 1。

果实的采收调制和种子贮藏　当果实呈红色时分批采摘。采回的果实堆沤 2～3 天，充分软熟后装入筐内，置水中反复搓擦，淘去果皮等杂质，得出种子。种子的净度可达 96%。鲜果的出种率约为 34%。洗净后表面晾干的种子含水量为 32%～52%。种子千粒重约 133g，每千克有纯净种子 6 800～8 000 粒。种子忌失水，不宜日晒，亦忌裸露存放，宜于随采随播，一般不作贮藏。种子运输时需混以种子体积一半左右的湿沙。未经调制的种子不宜运输。如需贮藏越冬，也需混以种子体积一半以上的湿沙浅层堆积，贮藏期一般在 5 个月以内。当年秋后所采的种子，翌年 3 月前播种。

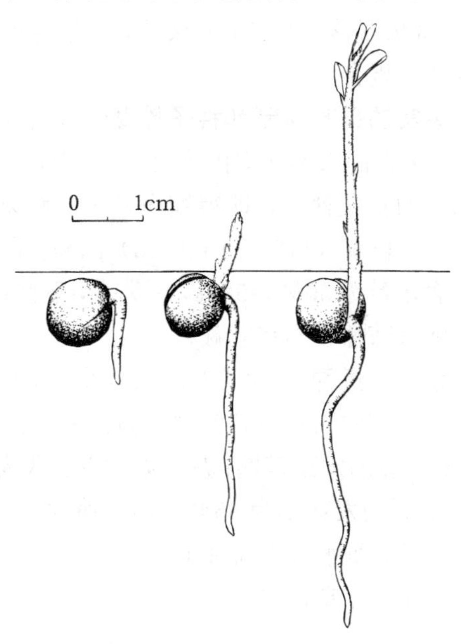

图 2　山小橘种子萌发后第 3、8、14 天幼苗的生长情况
（黄应钦绘）

发芽和播种　种子无明显休眠习性。发芽时日均温宜在 20℃以上。1987 年 9 月 5 日，广西林业科学研究所在室内用当年新采种子作发芽测定，基质为滤纸，室内自然光，发芽期的日均温为 26～27℃。9 月 16 日开始萌发，发芽盛期不明显，9 月 19 日发芽终止，发芽率为 88%。留土萌发。胚根萌发后约 4 天上胚轴伸出，具不育叶 3～5 枚，再过 4～6 天初生叶出现。山小橘种子的萌发和幼苗初期生长情况见图 2。

撒播。播前种子可不作处理，但整地要求细致。每平方米播种 10～20g，播后稍加覆土。也可将种子浅播于沙床内，发芽后2～3 个月移至大田培育。1 年生苗出圃。

（张声燕）

小 芸 木

Micromelum integerrimum (Buch. -Ham.) Roem.

（芸香科 Rutaceae）

生长习性、分布和用途 小芸木属约 11 种，我国 2 种，本文描述 1 种。常绿小乔木，高 5～8m，胸径 5～14cm。能耐 0～2℃的短期低温。适生于肥沃湿润的酸性土。分布于滇、桂、粤。老挝、越南亦产。木材为一般用材。皮、叶药用。花、果供提芳香油。

开花结实 4～5 年生开始结实，8 年生进入正常结实期，无大小年现象。花两性。伞房花序式圆锥花序，长 10～20cm，簇生。萼杯状，3～5 裂。花冠淡黄色，花瓣 5，分离。雄蕊 10，着生于花盘的周围。子房上位，2～4 室，每室 2 胚珠。据 1985～1987 年西双版纳勐仑观察，8 月下旬花序始现，着生于枝顶端，9 月上旬～10 月上旬花蕾膨大，翌年 2 月中下旬花盛开，3 月上旬花期结束，幼果形成；5 月中旬果实开始成熟，6 月上旬为果实成熟盛期。浆果，矩圆形或椭圆形，成熟时由黄绿色转变为橘红色，果皮有腺点。果长 1.2cm，径 9.5mm，熟后 4～6 天脱落。果皮肉质浆状，每果有种子 2～4 粒。无胚乳。胚似肾形，绿色，长 7～9mm，子叶薄，折合。

果实的采收调制和种子贮藏 果实成熟时用手或采种刀成串采集，也可击落后在树下捡拾。采得的果实在室内堆沤 3～5 天，待果肉软化后装入缸中搓擦，加水淘去果肉等杂质以及漂浮的瘪粒种子，即得纯净种子。鲜果的出种率为 25%～30%。种子的净度可达 95%～98%。千粒重（110）115（125）g，每千克有纯净种子 8 000～9 000 粒。调制后稍加晾干的种子含水量为 50%～58%，忌失水，不宜日晒或裸露存放。运输时需混湿润锯木屑或椰糠。产地宜随采随播，不宜贮藏。

发芽和播种 种子无休眠习性。发芽时日均温宜在 22℃左右。1988 年 5 月 19 日，西双版纳热带植物园在室外沙盆用新鲜种子作过发芽测定：5 月 24 日开始发芽，6 月 9 日发芽结束，有明显的发芽盛期；从开始发芽至发芽高峰，8 天的发芽百分数为 70%；从播种至发芽终止，22 天的发芽率为 95%。出土萌发。子叶出土后 7～10 天展出初生叶。

条播。每平方米播种 15～20g，覆土 1～2cm。幼苗具 2～5 片初生叶时可以移植。培育半年生或 1 年生苗出圃。

（程必强）

九 里 香

Murraya paniculata (L.) Jacks.

（芸香科 Rutaceae）

生长习性、分布和用途 九里香属约 12 种，我国约产 8 种，本文描述 1 种。常绿灌木至小乔木，高 4～8m。萌芽力强，极耐修剪。庇荫下及强阳光下均能生长良好。喜空气湿润、土

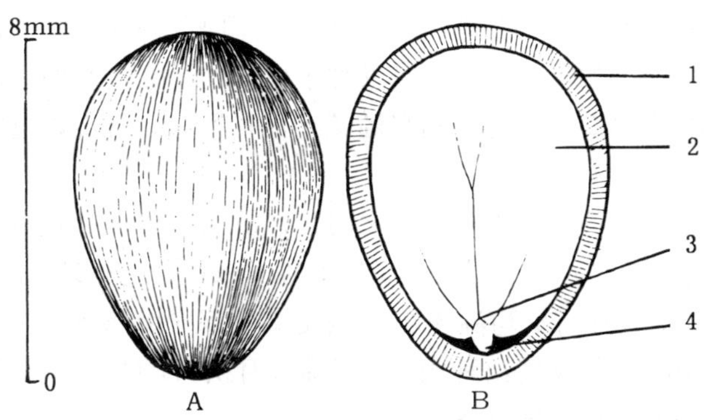

图 1　九里香种子外形（A）及其纵切面（B）
1. 种皮　2. 子叶　3. 胚芽　4. 胚根
（黄应钦绘）

质肥沃的生境，但干燥贫瘠地也能缓慢生长。酸性土或钙质土均宜。能耐短期－2℃左右低温，忌冰雪。分布于台、闽、琼、粤、桂、湘、黔、滇。中南半岛、马来半岛也有。宜作庭院绿篱或盆景。花芳香，可提精油。木材可作雕刻及工艺品。根叶入药，止痛、止血，治跌打及蚊虫咬伤。

开花结实　3～5 年生开始开花结实，正常结实期在 10 年生以后。结实无大小年现象，但庇荫下及干旱贫瘠地结实量较少。花两性，单生或 3 至数朵组成聚伞花序，腋生或顶生。萼极小，5 裂。花冠白色，花瓣 5。雄蕊 10，长短相间。子房上位，2 室，每室有胚珠 1～2 颗。花柱棒状，粗于子房，柱头极增广。据南宁 1987～1989 年观察，花期较长，每年 4 月～12 月，不同的植株均陆续有花开放：主花期 5 月下旬开始，6 月中旬为盛期，7 月上旬花期终止；6 月～翌年 3 月均有果实零星成熟，主要果熟期为 9～10 月。

浆果，纺锤状或椭圆形，先端锐尖，未熟时青色，熟时红色，长 1～1.4cm，径 0.7～1.1cm，成熟后约 5 天脱落。每果有种子 1～3 粒。种子浅褐色，心形或桃形，长 6～8mm，宽 4～6mm，厚 3～4mm，一侧呈龟背状，一侧扁平。种皮被棉质毛。种子无胚乳。子叶肥厚，倒卵形。种子的外形和内部结构见图 1。

果实的采收调制和种子贮藏　从 6 月～翌年 3 月都有果实陆续成熟，可以分批采摘。采得的果实堆沤 1～3 天，俟浆果充分软化后置水中搓擦，淘去果皮等杂质，即得纯净种子。鲜果的出种率约为 45%。种子的净度可达 98%，含水量为 25%～30%，千粒重 55～

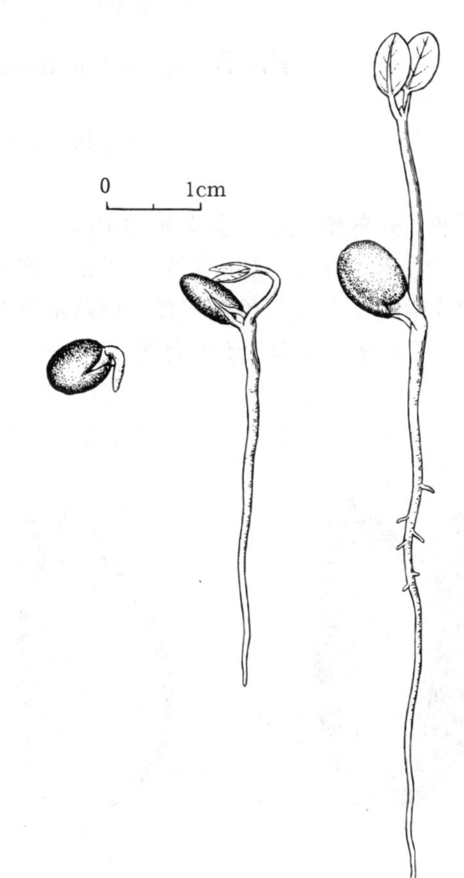

图 2　九里香种子萌发后第 1、5、9 天幼苗的生长情况
（黄应钦绘）

75g，每千克有纯净种子13 000～18 000粒。调制出来的种子不宜失水，忌日晒或裸露贮存。调制前浆果亦不宜长期存放。种子调制出来以后如需运输，应混以种子体积约1/3过筛的细湿沙，用密编竹筐或厚尼龙袋包装。种子的成熟期长，可以随采随播，如需贮藏，宜混以过筛的湿沙。冬季贮藏期一般为2～4个月。

发芽和播种　种子无休眠习性。发芽时的日均温需在20℃以上。1987年，广西林业科学研究所曾在室内用滤纸作基质，对新采的种子进行过发芽测定，播前种子未作任何处理。9月7日置床，11日开始发芽，16日发芽终止，发芽时日均温为26～28℃，发芽率为90%。出土地面萌发。胚根萌发后4～5天子叶带壳出土，但不脱出种壳，上胚轴同时拱出，再过2～3天初生叶展现。九里香种子的萌发和幼苗初期生长情况见图2。

撒播。将种子密播于沙床或经过细致整地的播种床，播前淋足水分，播后覆盖稀薄细土。每平方米约播30g。发芽后约半年移至大田培育1年出圃。整形供绿化用的大苗，需培育2～3年方可出圃。也可以扦插育苗，扦插成活率可达60%以上。

（张声燕）

黄檗（黄波罗）

Phellodendron amurense Rupr.

（芸香科　Rutaceae）

生长习性、分布和用途　黄檗属约10种，主要分布于亚洲东部，我国有2种，本文仅描述1种。落叶乔木，高10～15m（最大22m），胸径50cm（最大1m）。深根性，侧根发达，抗风，耐火，较耐寒，适生于冷湿气候，但苗木栽植初期易受冻害。湿润肥沃的腐殖土上生长良好，不耐干旱瘠薄及水湿，萌生力强。分布于东北、华北，最北到北纬52°，最南达39°。朝鲜半岛、日本、俄罗斯远东亦产。木材黄色至黄褐色，材质坚韧，纹理美观，耐湿，耐腐，富有弹性，加工容易，为上等家具、造船、飞机、枪托、胶合板等珍贵用材；重要的木栓树种，内皮是贵重药材（黄柏），主要成分为黄连素；可作染料；果实可提芳香油，花为蜜源。

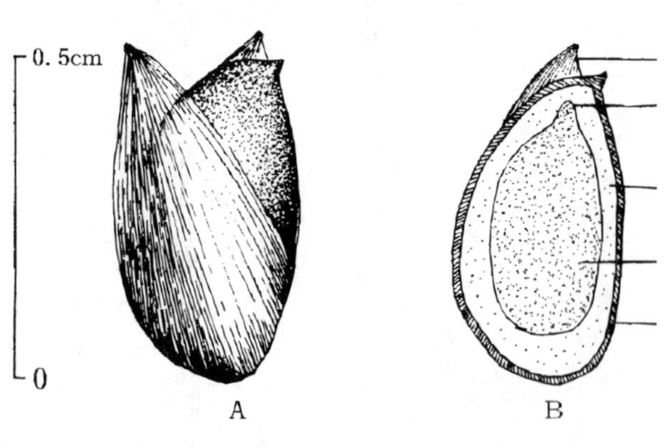

图1　黄檗果核和种子外形（A）及其纵切面（B）
1. 核膜（内果皮）　2. 种皮　3. 胚乳　4. 子叶　5. 胚根
（梁鸣绘）

开花结实　15～20年生开始结实。花单性，雌雄异株。聚伞或伞房圆锥花序。花黄绿色，萼片5。花瓣5。雄花的雄蕊5，有退化雌蕊。雌花的退化雄蕊鳞片状。子房5室，每室1胚珠，柱头5裂。平均每穗有果21个。浆果状核果，近球形，径约0.8～1cm，内有粘质；初时绿色，成熟时蓝黑色，有光泽，具特殊香气及苦味。果实有5室，每果含果核3～5粒。果核长约

5mm，宽约 3mm，厚约 2mm，半椭圆形，有灰黑色膜质外皮，即内果皮，有小腺点或皱纹。种皮石质，较硬。见图 1。种子丰年间隔期 2～3 年。花期 5～6 月，果熟期 9～10 月。据哈尔滨 1963～ 1976 年观察，花蕾花序出现期为 5 月下旬，始花期在 6 月上旬，盛期 6 月中旬，末期 6 月下旬；果实成熟期 9 月中旬，果落始期 9 月中旬，果落末期 11 月上旬；初熟至脱落间隔期 25～40 天。罗丽芬和陈华豪等(1986)1985 年在哈尔滨观察过黄檗的开花结实物候进程。

果实的采集调制和种子贮藏　9 月下旬～10 月上旬果实形态成熟后不立即脱落，但易为鸟类啄食，应及时从树上采摘。采后堆放（100～150kg 一堆），使果皮变软，捣碎后淘去果皮等杂物，所得果核即为播种材料，通称种子。种子应置通风良好处阴干，使含水量降至 9%～11%。鲜果出种率约 5%～12%，净度 85%～95%。据罗丽芬和陈华豪等人（1986）对 50 份样品的测定，平均千粒重 14.26g。每千克有纯净种子 5.9 万～7.7 万粒。空气干燥时，应薄薄地摊放，以防发热发霉；长期存放则应低温密封干藏。在 0～5℃，空气相对湿度 40%～70%，种子含水量 8%～10%的条件下，寿命可保持 4～5 年。

发芽和播种　种子有轻度休眠，发芽率 60%～70%。新种子播前需经 1 个月低温（0～5℃）层积处理，陈种子则需经 60～70 天低温层积。混雪埋藏效果甚佳，在 12 月末积雪不化时进行，翌春播种前 15～20 天取出，并在 15～25℃温度下保持种子湿润，至 1/3 的种粒裂嘴时播种。春播。多为垄播，播种量每平方米 7～8g，或床播每平方米 15g。覆土 1.5cm 左右。出土萌发，见图 2。1 年生苗高 20～30cm，可出圃。

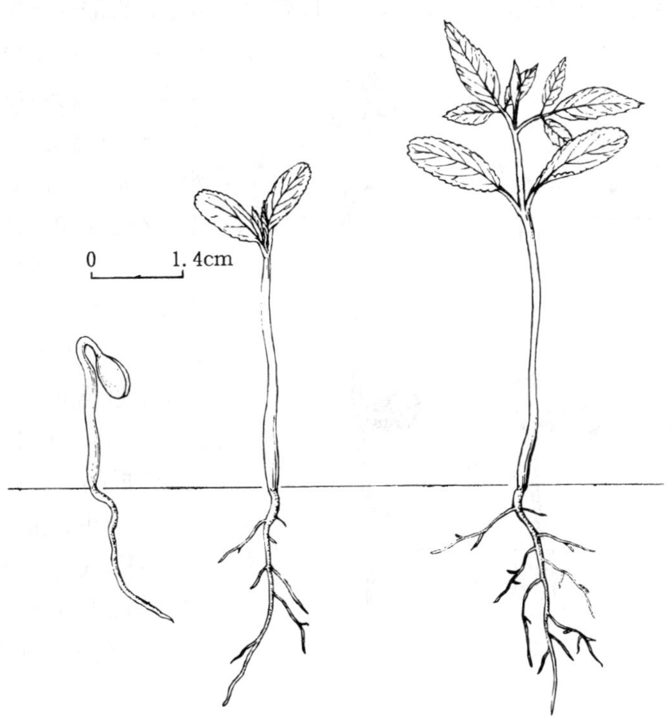

图 2　黄檗种子发芽出土后第 4、10、29 天幼苗生长情况

（曹雅范绘）

（周德本、邹　铨）

枳（枸橘）

Poncirus trifoliata（L.）Raf.

（芸香科 Rutaceae）

生长习性、分布与用途 枳属仅有本文描述的1种，原产长江流域各地，目前北起鲁、豫，

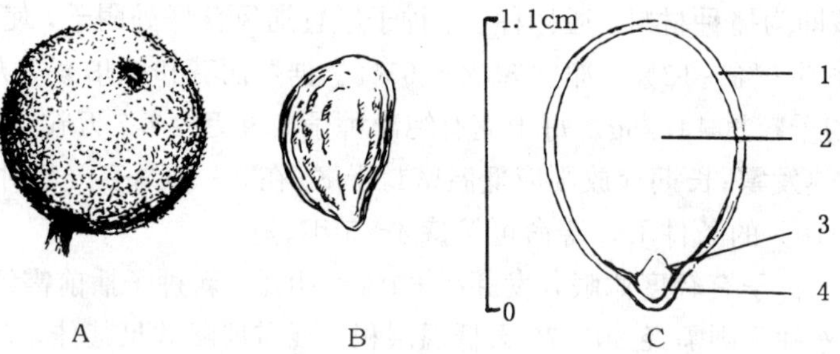

图1 枳的柑果外形（A）、种子外形（B）及其纵切面（C）
1. 种皮 2. 子叶 3. 胚轴 4. 胚根
（史渭清绘）

南至粤、桂均有栽培。落叶灌木或小乔木，生长健壮，耐寒，喜酸性土壤，不耐碱性土。小枝丛密，具粗硬扁刺，多栽培作绿篱，常用作柑橘类的砧木，果入药。

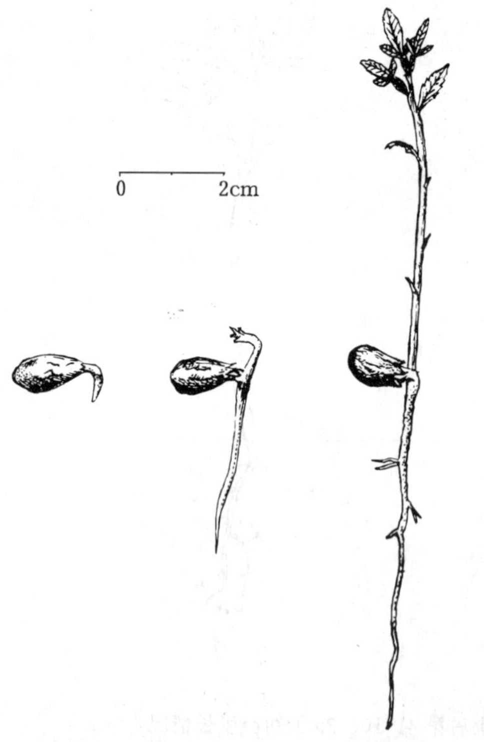

图2 枳的种子萌发后第2、14、35天的幼苗生长情况
（1、2. 史渭清绘；3. 田恒德仿《主要树木种苗图谱》）

开花结实 开始开花结实树龄为4～8年，正常结实寿命可长达几十年至上百年。花两性，白色，近无柄，单生或成对腋生于老枝上，先叶开放。花瓣、萼片均为5。雄蕊8～10（20），长短不等。子房上位，6～8室，每室有胚珠4～8。花柱粗短，柱头增大呈头状。柑果圆球形，径3.5～5cm，成熟时黄色，密被茸毛，6～8瓣，果肉少，富含胶质，味酸苦，不能食用。每果含种子（30）40～50（60）粒。种子斜卵形，长1～1.4cm，外种皮革质，内种皮膜质。无胚乳。子叶椭圆形，肥厚，胚根部短小（图1）。花期4～5月。果熟期9～10月。幼果通常有1～2次生理落果。

果实的采收调制与种子贮藏 软熟的黄色果实采摘后摊放在阴凉

处，待到开始腐烂时剥果取种，用清水淘洗至种皮上无粘液，漂除杂质、瘪粒，阴干，忌曝晒。不需立即播种的种子应混湿沙贮于阴凉处，经常翻动检查，使沙的含水量保持在10%左右。过度失水的种子种皮皱缩，会降低甚至丧失发芽能力，过分潮湿或通气不良则会发热腐烂。因此以带果摊放贮存，播前剥取种子较为妥当。据广西柑橘研究所报道，花谢后90天左右的青熟果内的种子已具有较高的发芽能力（90%以上），但不耐贮藏，必须立即播种。成熟新鲜种子的千粒重为160～225g，每千克含4 400～6 300粒。净度可达100%。

发芽和播种　1988年9～10月，本文作者曾在南京采摘黄熟果实，剥出种子，12月中旬混湿沙在室温条件下层积。据观察，该份种子1989年4月初开始萌发，5月中旬萌发结束，有明显的发芽盛期，发芽率为91%。萌发期间室内气温约为15～20℃。经过层积的种子可在早春2～3月播种，4～5月出苗。从青熟果实中取出的种子应在7月下旬～8月播种。园艺栽培实践中促进种子萌发和减少苗期病害的方法是温汤淋种：35～40℃温水浸种1小时后浸入冷水中半天，取出种子放在垫有秸草的箩筐中，盖草，每天继续用35～40℃温水均匀淋浇3～4次并加翻动，5～9天后胚根微露时即可播种。条播，行距15～20cm，每米长的播种沟内约播20～30粒。覆土1～1.5cm，盖草，每平方米用种50～100g。播后25～40天出苗。留土萌发。种子有多胚现象，常从一粒种子中萌出两株幼苗。幼苗具初生不育叶1～4片，以后为1片或2片披针形或倒卵形的单叶，再以后才出现3小叶复叶。主根发达，侧根疏生（图2）。

（何泽瑛）

花　椒　属

Zanthoxylum L.

（芸香科　Rutaceae）

生长习性、分布和用途　本属约250种，我国45种13变种，本文描述4种。落叶或常绿、半常绿，灌木或小乔木。在深厚的沙壤和石灰性土壤上生长最好。可作油料、调味、香料树及庭园绿化栽培。花椒种子含油量达25%～30%，出油率22%～25%；青花椒种子含油量30%～40%，都可供食用和工业用。种子和果皮均可入药。这4个树种的名称、树高、分布见表1。

表1　花椒属树种的名称、树高和分布

中　名	学　名	树高（m）	分　布	供　稿
竹叶椒	*Z. armatum* DC.	1～2	晋、鲁、豫、华东、华中、西南	1003
花　椒	*Z. bungeanum* Maxim.	3～7	除东北、内蒙古外均有分布，主产鲁、冀、晋、甘	310
青花椒（崖椒）	*Z. schinifolium* Sieb. et Zucc.	1～3	辽东、华北、西北、华中、华南、西南	1003
野花椒	*Z. simulans* Hance	1～2	华北、华东、华中	1003

开花结实　通常3～5年生开始开花结实，7～8年生或10年生进入盛果期。河北涉县12年生的花椒树每株每年可产鲜果6.4kg（常剑文和田玉堂，1988）。花椒植株20～25年生逐渐

衰老，寿命最高可达 40～45 年。花单性或杂性，同株或异株，常簇生或组成聚伞状圆锥花序，顶生或腋生。萼片 3～8，花瓣与萼片同数或无花瓣。雄花有雄蕊 4～8，退化雌蕊短柱状，2～4 裂。雌花无退化雄蕊或极小呈鳞片状，雌蕊由 2～5（7）离生心皮组成，2～5 室，每室有 2 并生胚珠。花期因树种及产地而异，多在 3～8 月。花椒分布广，又有许多栽培品种，果实的成熟很不一致。聚合蓇葖果，熟后蓇葖瓣裂，具两层果皮：外果皮革质，浅绿色、紫红色或灰褐色，常密生突起的油点；内果皮纸质，淡黄色。每蓇葖果含 1 球形种子，或稀为 2 半球形种子。竹叶椒中圆球形种子与半球形种子数量的比例大约是 2.5∶1。成熟果皮纵裂后，可以看到种子同蓇葖果子房柄中心有细丝状“种柄”连接。花椒属种皮硬骨质，黑色有光泽。胚乳肉质，含油脂较多。这 4 个树种的开花结实物候见表 2，果实和种子特征见表 3。种子形态见图 1。常剑文和田玉堂（1988）报道过对花椒花芽分化的观察结果。

表 2 花椒属树种的开花结实物候期

树 种	观察地点	观察年份	花 期	果实成熟期
竹叶椒	河南	—	3～5 月	6～8 月
花椒	河北涉县	1983～1984	4 月下旬～5 月中旬	7 月中旬～8 月中旬
青花椒	辽宁丹东	—	6～7 月	10～11 月
野花椒	河南	—	4～5 月	6～8 月

表 3 花椒属树种果实和种子的形态特征

树 种	成熟果实			种 子	
	形 状	大小（mm）	颜 色	形 状	大小（mm）
竹叶椒	球形	—	红棕色至暗棕色	卵球形或半球形	径 3.5～4.0
花椒	球形，密生瘤状油点	径 4～6	褐红色或紫红色	圆卵状	径约 3.5
青花椒	球形，先端常有 1 短喙状尖头	—	灰绿色至棕绿色	卵状球形	径约 4.0
野花椒	倒卵状球形	—	黄棕色或紫红色	近球形或半球形	径约 4.0

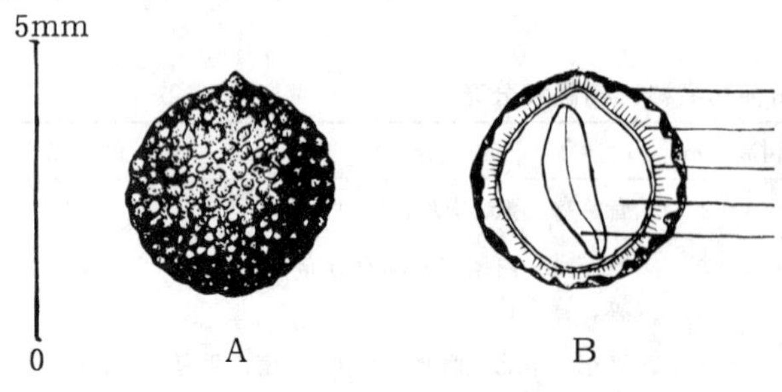

图 1 花椒果实外形（A）及其横切面（B）
1. 外果皮 2. 内果皮 3. 种皮 4. 胚乳 5. 子叶
（孟玲绘）

果实的采收调制和种子贮藏 果实成熟时要及时采摘。花椒中有些栽培品种成熟的果皮很快开裂，种子容易散落，更要求在几天之内采摘完毕。采收脱粒后曝晒会使种子丧失发芽能力，应在通风干燥处阴干。果皮裂口，种子脱出后，除去杂物便得纯净种子。出种率及种子质量见表 4。竹叶椒圆球形种子的千粒重约 18.6g，半球形种子的千粒重约为 14g。野花椒球形种子千粒重 17.8～21.4g，半球形种子千粒重约 14.7g。种子可放在罐中加盖置于干燥、阴凉的室内。竹叶椒的种子在5℃下可以密封贮藏2年。

表4　花椒属出种率和种子质量

树　种	出种率（%）	千粒重（g）	每千克纯种粒数（万粒）
竹叶椒	—	14.0～18.6	5.4～7.1
花椒	28.0	12.5～22.0	4.5～8.0
青花椒	—	10.2	9.8
野花椒	—	14.7～21.4	4.7～6.8

发芽和播种　花椒属种子有休眠习性，生产上常用秋播，或用湿沙（或牛粪，或黄土和牛粪的混合物）使种子在层积状态下越冬，供翌年春播。《中国主要树种造林技术》中列举了花椒种子的许多越冬处理方法，读者可以参阅。1988年3月7日，何泽瑛（未刊稿）将秋季采收冬季用纸袋干藏的野花椒种子混湿沙装入塑料袋，在南京的室外条件下层积，4月26日起开始萌发，发芽高峰出现在5月初，5月底发芽终止，全部发芽率仅17%。未发芽的种子中大部分空瘪，或胚乳已经发黄变质，一部分种粒则仍呈休眠状态。这表明，野花椒种子发育不良，空瘪粒多，也表明春播前的冬季干藏更不能解除野花椒种子的休眠。1978年11月丹东市林业科学研究所将青花椒种子在室外深坑内埋藏越冬，翌年4月取出种子用5%的洗衣粉、5%的纯碱液和60℃的始温水淘洗，再在常温清水中浸泡1小时，搓洗后在清水中继续浸泡3天，再混湿沙待播。经过这样的处理，场圃发芽率达到了58%，比室内干藏种子的播种效果好。竹叶椒种子需要在5℃下层积120天。

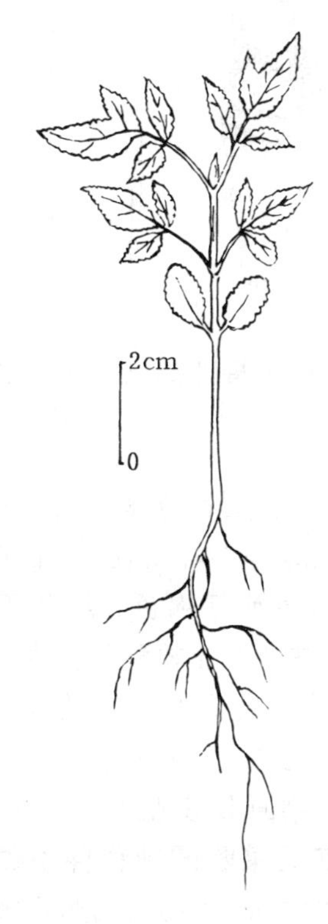

图2　野花椒幼苗形态
（田恒德仿《中国东北主要树木幼苗图说》）

花椒属种子发芽测定的报道不多。竹叶椒种子的发芽以20～30℃的昼夜变温为好，不过发芽进程也需要45～60天。秋季播种可以使种子在圃地里接受低温层积。但秋播前，以及层积处理过的种子在春播之前，生产上还常用碱性溶液浸泡，这是因为人们认为它们的种壳坚硬，油质多，不易透水，不过这方面的对比试验并不多。开沟条播，行距20cm。每平方米播种10～18g。

出土萌发。野花椒有多胚现象，1粒种子内常有大小不等的胚2～4枚，且都可以萌发。竹叶椒的少数种子也有多胚。文献中还未见到关于花椒属多胚现象的报道。野花椒幼苗的形态见图2。

（张毅萍）

臭　椿

Ailanthus altissima Swingle

（苦木科　Simaroubaceae）

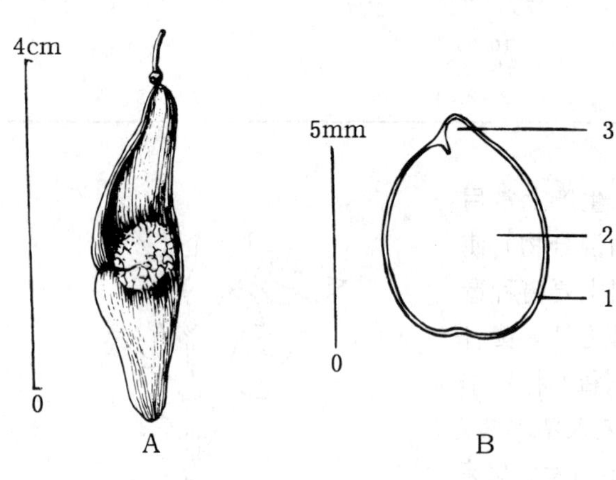

图 1　臭椿翅果外形（A）及其种子纵切面（B）
1. 种皮　2. 子叶　3. 胚根
（孟玲绘）

生长习性、分布和用途　臭椿属约10种，我国5种，本文描述1种。落叶乔木。树高30m，胸径可达1m以上。原产我国北部和中部，现广布全国各地。生长较快，适应性强，繁殖容易，病虫害少，木材具有多种用途。树姿美观，抗烟尘，耐盐碱，是城市工矿区绿化及盐碱地造林的重要树种。

开花结实　花单性或杂性异株。圆锥花序顶生。花小，淡黄绿色。萼5裂。花瓣5。雄蕊10，着生于花盘基部，花盘10浅裂。子房上位，心皮5，每室1胚珠。仅花柱合生，柱头5裂。聚合翅果1～5簇生。翅果长圆状纺锤形，长3～5cm，宽0.8～1.2cm。花期5～7月。据侯惠宗（1956）1953、1954年在北京妙峰山观察，初花期5月30日，末花期6月8日。翅果9～10月成熟。翅果熟时呈褐黄色或红褐色，经冬不落。翅果中部含1枚种子。种子扁平，圆形或倒卵形，径6～8mm，无胚乳。臭椿翅果的外形和种子的剖面形态见图1。

果实的采收调制和种子贮藏　8月中旬～10月，成熟的翅果变为褐色时将果穗采下，翻晒4～5天，去掉总梗和杂物。翅果即为播种材料，通称种子。经过净种和干燥后干藏。1987年秋北京采收的一批种子，1988年3月测定，种子净度92.5%，含水量8.15%，优良度66%，千粒重28～32g，每千克带翅种子3.0万～3.5万粒。种子发芽率60%～85%。曾有一批含水量为

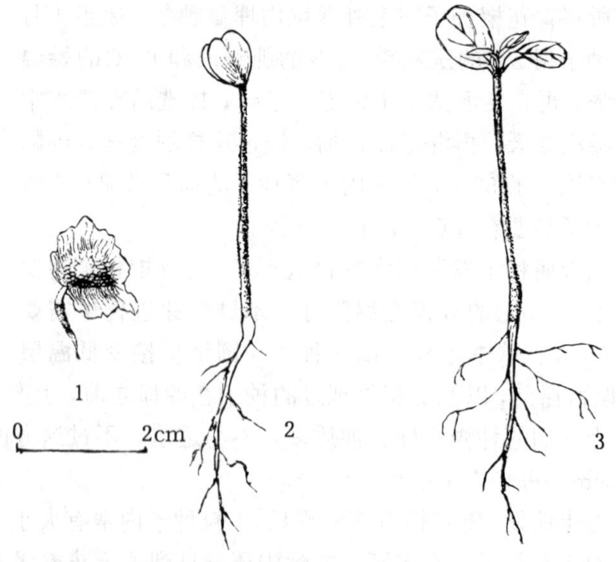

图 2　臭椿种子萌发出土后第1、7、10天幼苗的生长情况
（孟玲绘）

8.8%，发芽率为58%的种子，在普通地下室（最高温度20℃）用布袋贮藏，2年后发芽率降到36%；密封贮藏两年后发芽率为50%。

发芽和播种　曾经比较过30℃、36℃和40℃3种温度对臭椿发芽的影响。结果是30℃以下发芽率最高，40℃发芽率明显下降。苗圃春季播种前可用始温40℃水浸泡，冷却后再以室温水浸种1昼夜，捞出后在向阳处盖草帘催芽，每天用水冲洗1～2次，一般催芽10天左右。条播。行距40cm，播幅5cm，沟深3cm，每米长播种约50粒，覆土1～1.5cm，轻压，洒水。室外发芽适宜温度为9～15℃。播种后4～6天幼芽开始出土，10～15天出齐。出土萌发。子叶2，近圆形或倒卵形，先端圆或平截，基部楔形，近革质，淡绿色，具短柄。种子萌发和幼苗早期形态见图2。每666m^2产苗8 000株，其中成苗（苗高1m，地径1cm）7 000株。

根蘖性强，也可分蘖、分根繁殖。

（陶章安）

鸦　胆　子

Brucea javanica（L.）Merr.

（苦木科　Simaroubaceae）

生长习性、分布和用途　鸦胆子属6种，我国2种，本文描述1种。常绿灌木或小乔木，高3～5m。适生于湿润的酸性土或钙质土。分布于滇、桂、粤、琼、台。印度至大洋洲也有分布。种仁含油率约55%，入药，有杀虫、治疟、止痢之功效，是国产治痢药的主要成分之一。

开花结实　2～3年生开始开花结实，正常结实期在8年生以后，大小年现象不明显，冬春低温持续期长会影响当年结实。花单性异株，少有两性。排成腋生或顶生的圆锥花序。花小，暗紫色。萼4裂，裂片卵形。花瓣4，长椭圆状披针形。雄花有雄蕊4，着生于花盘外缘裂片间，并具不孕性雌蕊。雌花具心皮4，分离，并具半不孕性雄蕊4，心皮各有1悬垂胚珠。据广西南宁1987年观察，5月下旬为现蕾期，6月初为始花期，6月下旬进入盛花期，7月中旬为末花期；果熟期始于10月下旬，果熟盛期在11月中旬；果实脱落期始于11月上旬，12月下旬果实全部脱落。另据观察，云南景洪的物候期比南宁约早10～15天。

核果，椭圆形，长8～10mm，径6～7mm，成熟时由青紫色变为紫黑色，具突起的网纹。每果有果核1颗，椭圆形略扁，长7～8mm，径5～6mm。核壳（内果皮）骨质，有龟壳状花纹凸起，灰褐色。种子卵形，具薄种皮，无胚乳。子叶肥大，腹平，背凸，富含油脂。鸦胆子果核的形态见图1。

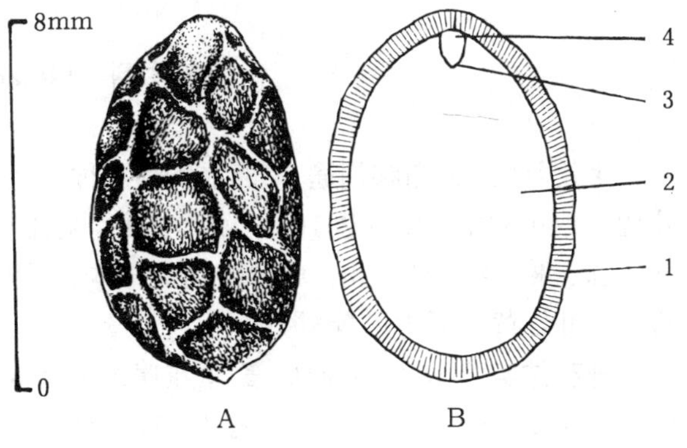

图1　鸦胆子果核外形（A）及其纵切面（B）

1. 内果皮及种皮　2. 子叶　3. 胚芽　4. 胚根

（黄应钦绘）

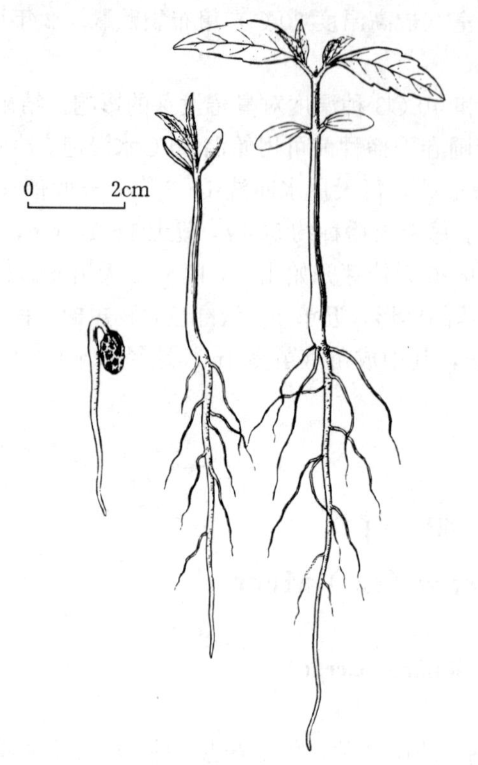

图 2 鸦胆子种子（果核）萌发后第 10、13、20 天幼苗的生长情况
（黄应钦绘）

果实的采收调制和种子贮藏 果实成熟盛期用手采摘。采得的果实堆放 3～5 天，装入布袋或竹篓，加适量细沙，搓揉后置水中淘出果核，用作播种材料，通称种子。鲜果出籽率约 32%。净度可达 98%。千粒重 60～70g，每千克有种子（核）14 200～16 600 粒。种子不宜曝晒，应在晾干后干藏，贮藏期约为半年。混以湿沙，贮藏期为 1 年。

发芽和播种 种子有中度休眠习性。当年所采种子宜贮藏到翌年春天播种。广西林业科学研究所 1987 年 11 月 18 日用室内沙盘作过发芽测定（当时温度约 25℃），1988 年 4 月 20 日方始发芽，发芽盛期不明显，5 月 15 日发芽终止，发芽率为 68%。出土萌发。胚根伸出后 8 天子叶出土，又 2 天后初生叶展现。鸦胆子种子（核）的萌发和幼苗初期生长情况见图 2。

点播。每平方米约播 6～8g，覆土 1cm。1 年生苗木可以出圃。

（曾 玲）

苦木（苦树）

Picrasma quassioides（D. Don）Benn.

（苦木科 Simaroubaceae）

生长习性、分布和用途 苦木属约 8 种，产热带和亚热带，我国 2 种，本文描述 1 种。落叶灌木或小乔木，高可达 10m。耐旱抗寒，对土壤要求不严。分布于陕、晋、冀、豫、苏、皖、赣、鄂、湘、桂、滇、川。朝鲜半岛、日本、尼泊尔、不丹、印度也有。树皮苦，药用或为农药。秋叶红黄，可为风景树。

开花结实 花杂性异株。聚伞花序组成圆锥花序腋生。花小，黄绿色。萼片 4～5，卵形，宿存。花瓣 4～5，倒卵形。雄蕊 4～5，着生于花盘基部。花盘 4～5 浅裂。子房上位，离生心皮 4～5，各有 1 枚胚珠。心皮在花盘上面发育长大，果常由 2～4 个肉质小核果组成，外具宿萼。成熟期不一致。小核果，椭圆形，长约 7～8mm，成熟时深蓝色，外果皮厚膜质，内果皮形成核壳，硬脆而不厚，呈偏斜的卵圆形，长约 6～7mm。表面平滑，米灰色，上有不规则的褐色花斑。白色凸起的长形种脐位于种子腹面一端。种皮膜质。无胚乳，子叶阔大肥厚，胚根短小，不显露（图 1）。花期 4～5 月，果熟期 8～10 月。

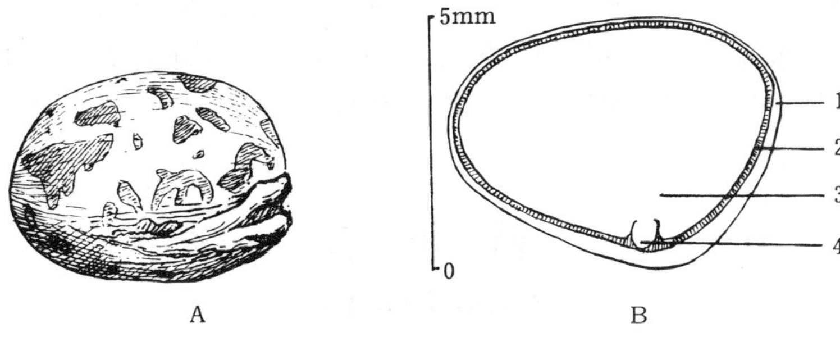

图 1　苦木的果核外形（A）及其纵切面（B）

1. 内果皮　2. 种皮　3. 子叶　4. 胚根

（史渭清绘）

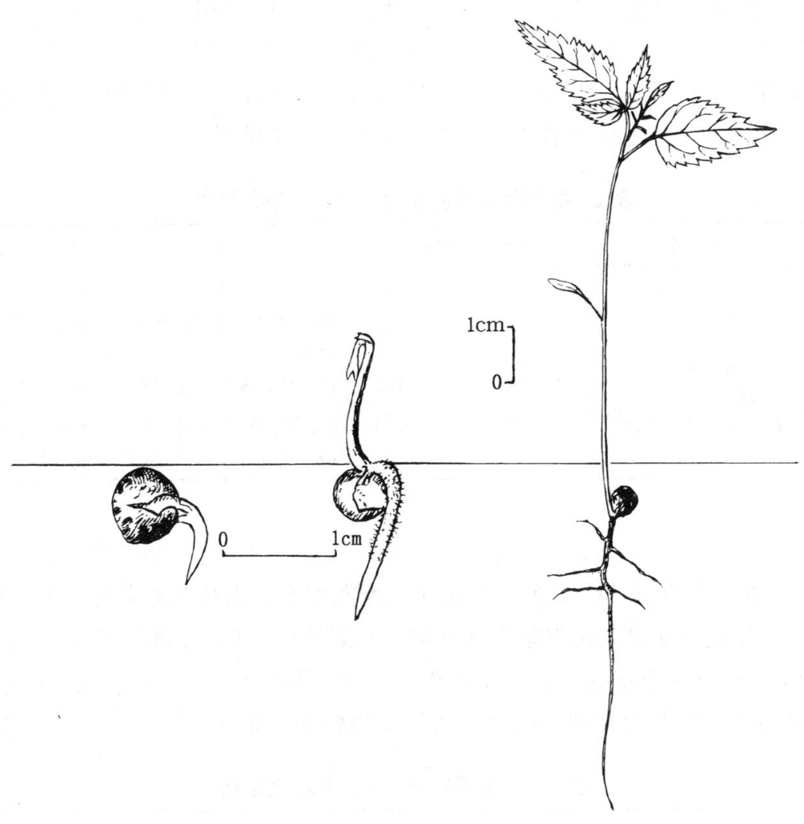

图 2　苦木种子萌发后第 1、27、165 天的幼苗生长情况

（史渭清绘）

果实的采收调制和种子贮藏　苦木为杂性异株，结实率低，果实成熟期极不一致，且熟后即落，采收比较困难。当果实由绿色转为红色至深蓝色时即标志成熟，宜及时分批采收。采收时动作要轻，以免震落成熟了的果实。洗去外果皮、果肉等杂质，晾干。所得果核即为播种材料，通称种子。种子需保持较高含水量，通常混沙湿藏，不宜干藏。千粒重 46～52g，每

千克含种子1.9万～2.2万粒。

发芽与播种　种子有休眠习性，宜随采随播或在自然变温下层积至翌春3月播种。解除休眠后发芽整齐，播后1个月发芽结束，场圃发芽率约35%。留土萌发，初生叶互生，第1～2叶为单叶，后为3小叶复叶。苗期生长不快（图2）。

（何泽瑛）

橄　榄　属

Canarium L.

（橄榄科　Burseraceae）

生长习性、分布和用途　本属约100种，我国产7种，又引入栽培1种。本文描述3种。常绿乔木。能耐0℃左右的极端低温，但苗期忌霜冻。喜肥沃湿润的酸性沙壤土，村边宅旁多栽植，干旱贫瘠土生长不良。中果皮肉质，供食用。木材轻软，材质中等，易受虫蛀，经处理可作板材、家具。这3个种的名称、生长、分布及用途见表1。

表1　橄榄属树种的名称、生长、分布和用途

中　名	学　名	树高（m）	胸径（cm）	分　布	用　途
橄榄	*C. album*（Lour.）Rauesch.	20～30	50～150	台、闽、琼、粤、桂、滇、川。越南、日本、马来西亚	食用、材用、油脂、核供雕刻并治鱼骨梗喉
三角榄（方榄）	*C. bengalense* Roxb.	20～25	50～120	滇、桂。南亚至东南亚	食用、材用、绿化
乌榄	*C. pimela* Leenh.	20～30	50～150	台、闽、琼、粤、桂、滇。中南半岛	油脂、食用、材用、绿化、根入药治风湿

开花结实　8～10年生开始开花结实，正常结实期在20年生以后，结实大小年间隔期为1年。花单性异株，少有杂性。聚伞圆锥花序（雌花序有时退化为总状花序），腋生。花萼杯状，3～5裂（在雄花中明显，在雌花中常近截平）。花瓣3～5，长圆状倒卵形，肉质较厚，乳白色。雄蕊6，1轮，生于花盘外侧，在雌花中不育。子房上位，3室，每室有并生的上转胚珠2颗。花柱顶生。广西南宁观测的开花及结实物候期见表2。

表2　橄榄属树种的开花结实物候期

树　种	观察年份	开花			果实成熟		果实脱落
		始　期	盛　期	末　期	始　期	盛　期	
橄榄	1981～1985	4月下旬	5月上旬	5月中旬	9月中旬	9月下旬	9月下旬～10月下旬
三角榄	1983～1985	5月下旬	6月上旬	6月中旬	10月中旬	10月下旬	10月下旬～11月下旬
乌榄	1981～1985	4月中旬	4月下旬	5月上旬	9月中旬	9月下旬	9月下旬～10月下旬

花开后3～5天凋谢，15天左右幼果形成，130天左右果实开始成熟。核果，外果皮薄膜

状，中果皮肉质。内果皮（果核）骨质，3 室，每室常有种子 1 粒，其中 1～2 室常不育。种皮褐色，无胚乳，种仁富含油脂。子叶掌状分裂为 3 小叶，卷叠。成熟后的果实可在树上存留 5～7 天然后脱落。成熟果实的特征及果核的形态见表 3、图 1。

表 3　橄榄属树种果实和果核的形态特征

树　种	未熟果颜　色	成熟果实			果　核		
		形　状	大小（cm）	颜　色	形　状	大小（cm）	颜　色
橄榄	浅绿色	椭圆形	长 2.6～4 径 1.7～2.3	淡黄绿色	梭状，两端锐尖，骨质，表面有浅沟槽	长 2.5～3.5 径 1～1.5	褐色
三角榄	浅绿色	钝三棱状椭圆形	长 5～6.5 径 2.5～3.5	黄绿色	三棱状梭形，骨质	长 3.6～5 径 1.5～2.5	褐色
乌榄	青色	椭圆形	长 3.5～5 径 1.7～2.5	紫黑色	纺锤状，两端钝尖，骨质	长 2.5～4 径 1～2	淡黄褐色

果实的采收调制和种子贮藏　果实成熟盛期上树采摘，也可用竹竿敲打或摇动树枝，将果实震落后在地面收集。采得的果实应按树种分别调制。乌榄的果肉与果核容易分离，调制的方法是，将已熟的果实用沸水浸烫 1～3 分钟，果肉稍软时取出摊开，用小刀或棉线从果实中部割断 1 圈，将果肉分为两半脱开，即可脱出果核。三角榄和橄榄的果肉与果核不易分离，又因外果皮和中果皮可以生食，多带核加工食用。这两个树种用于育苗的种子，可以食去果皮果肉，脱出果核；也可以将果实淋湿堆沤，俟果皮呈腐烂状时装入竹筐置水中冲捣，洗净后所得果核用作播种材料，通称种子。种子的质量等数据见表 4。

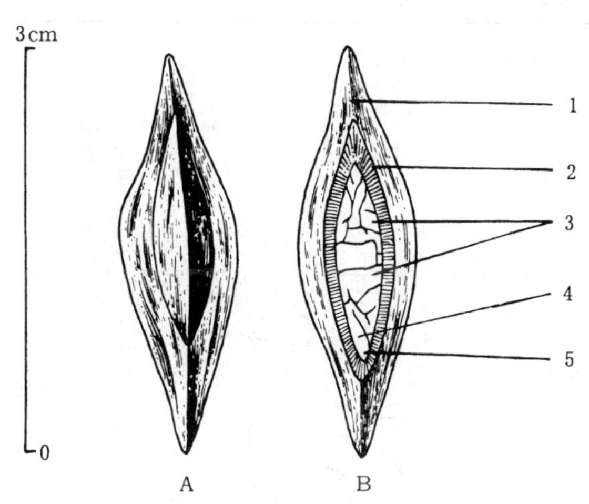

图 1　橄榄果核外形（A）及其纵切面（B）
1. 内果皮　2. 种皮　3. 子叶　4. 胚轴　5. 胚根
（黄应钦绘）

表 4　橄榄属树种的出籽率和种子净度、质量

树　种	出籽率（%）	千粒重（g）		每千克种子（核）粒数（粒）	
		一　般	变动范围	一　般	变动范围
橄榄	30～40	1 690	1 090～2 300	590	430～920
三角榄	35～45	4 550	4 000～5 100	220	200～250
乌榄	35～45	3 100	1 950～4 250	320	230～510

果核坚实，运输或贮藏均较易。运输时可用麻袋或竹筐包装。1 个月以内的短期贮藏可以不作任何处理，堆放于室内室外均可。较长期的贮藏需混以湿沙或用湿沙同种子分层堆积贮藏。贮藏期为 4～6 个月。

发芽和播种　种子休眠现象不明显。发芽时的日均温宜在 18℃以上。1982 年 10 月上旬，广西林业科学研究所在室外沙床上对 3 种橄榄当年新采的种子作过发芽测定，播种时日均温 24℃，播后约 1 个月即 11 月上旬陆续发芽，12 月中旬发芽终止，发芽期的日均温为 16～20℃。出土萌发。胚根萌发后 3～4 天子叶出土，再过 4～5 天初生叶展现，发芽情况见表 5、图 2。

表 5　橄榄属树种的发芽能力及其测定条件①

树　种	温　度（℃）	发芽势（%）		发芽率（%）	
		计算天数	一般数值	计算天数	一般数值
橄榄	16～20	20	66	65	85
三角榄	18～20	22	60	66	78
乌榄	18～20	25	62	68	80

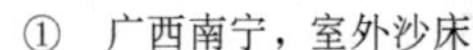

①　广西南宁，室外沙床

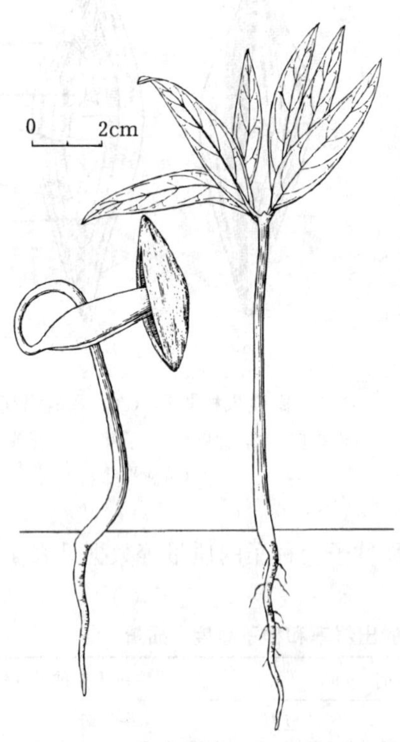

图 2　乌榄种子萌发后第 5、9 天幼苗的生长情况
（黄应钦仿《热带亚热带主要树种采种育苗技术》）

常用的催芽方法有以下 2 种：一种为果核调制后，置烈日下的水泥或石板地面曝晒，并趁热置冷水中浸泡，使核外表发生裂纹，播后易于吸水。另一种方法是调制后即置湿沙床中，种子萌发后再移至圃地培育。多数果核每核发苗 1 株，少数发苗 2～3 株，可分开移植。每平方米育苗用种量在橄榄为 100～125g，三角榄和乌榄为300～370g。除催芽移苗外，也可点播，覆土 2～3cm。育 1 年生苗出圃。用于庭院绿化的需培育 2～3 年生大苗。集约经营的榄园多用高压或嫁接苗造林。

（张声燕）

多花嘉榄（多花白头树）

Garuga floribunda Decne. var. *gamblei* (King ex Smith) Kalkm.

（橄榄科　Burseraceae）

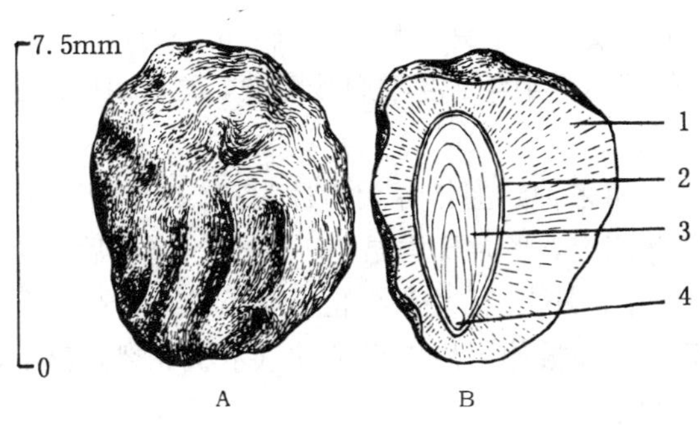

图 1　嘉榄果核外形（A）及其纵切面（B）
1. 内果皮　2. 种皮　3. 子叶　4. 胚根
（黄应钦绘）

生长习性、分布和用途　嘉榄（白头树）属有 4～5 种，我国产 4 种，包括本文描述的这个变种。落叶乔木，高达 35m，胸径 80cm。喜肥。适生于潮湿的山谷、坡地或山腰下部的杂木林中，酸性土和钙质土均宜，干旱贫瘠土生长不良。分布于滇、桂、粤、琼。印度、孟加拉也有。其正种则分布于菲律宾、马来西亚、印度尼西亚、大洋洲东北部至西太平洋群岛。木材可作农具、家具。落叶前羽叶鲜红色，可作热带至南亚热带的观叶树种。

开花结实　6～7 年生开始开花结实，10 年生以后为正常结实期，大小年现象不明显。花杂性。圆锥花序，6～8 序集生于枝顶。每序有花 120～150 朵。花小，黄色，径 3～4mm。萼钟状，5 裂，三角形。花瓣 5，矩圆形，镊合状排列，长 2～3mm，宽 1～1.5mm。花盘杯状。雄蕊 10，着生于花盘边缘。子房上位，球形，5 室，每室 2 胚珠。花柱柱状，柱头 5 浅裂。据广西南宁 1988～1990 年观察，2 月末为现蕾期，4 月下旬为始花期，5 月中旬进入盛花期，5 月 20 日左右为末花期；果实成熟始于 8 月下旬末，9 月中旬进入盛期，10 月中旬为果实脱落末期。核果，不规则球形或扁圆形，成熟时

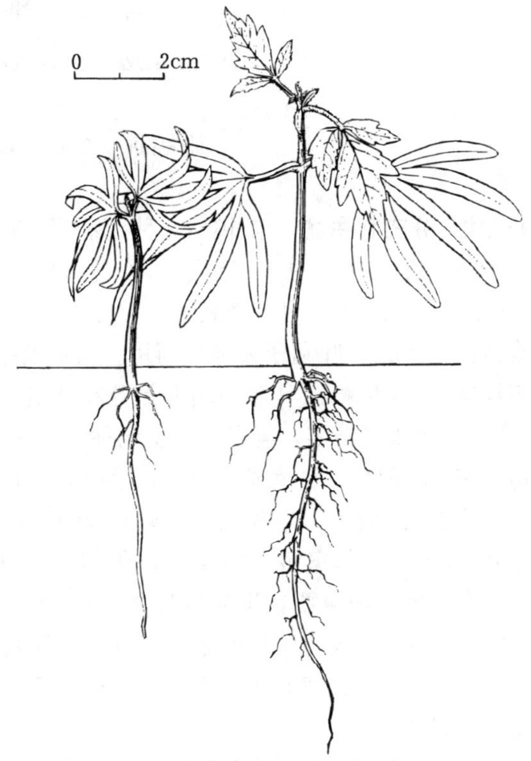

图 2　嘉榄种子萌发后第 6、9 天幼苗的生长情况
（黄应钦绘）

由绿色转为绿黄色。中果皮肉质。果径 1.5cm，高 1～1.2cm。每果有核 1～4，骨质，表面凹凸不平，黄褐色，径 6～8mm。种皮薄，无胚乳。子叶掌状分裂，复合，卷折。图 1 是嘉榄（羽叶白头树）*C. pinnata* Roxb. 之果核形态。

果实的采收调制和种子贮藏 采种前先清理地上落果。趁果熟盛期用高枝剪或采种钩刀将果穗剪下或钩下，在地面收集果实。采回的果实在室内堆放 1 个月左右，待外果皮部分或大部分变为黑褐色后，搓擦并淘去果皮、果肉等杂质，所得果核用作播种材料，通称种子。鲜果的出籽率约 40%。千粒重 190（140～240）g，每千克有种子（核）5 260（4 100～7 100）粒。种子含水量15%，可以晾干干藏，贮藏期约为半年。混干润沙贮藏，贮藏期可达 1 年。运输时可以用袋或竹篓包装。

发芽和播种 种子有中等程度休眠，当年采种到翌年春天播种为宜。1987 年 9 月 26 日，广西林业科学研究所曾在室外沙床作过发芽测定，当时日均温 25～28℃。播种后 182 天，即翌年 3 月下旬末方始发芽，发芽盛期不明显，5 月 10 日发芽终止，发芽率约 20%。发芽率低的原因可能是采种母树年龄较小。出土萌发。胚根伸出后 5 天子叶（掌状 5 裂）出土，再过 2 天初生叶出现。嘉榄幼苗初期生长情况见图 2。

点播。每平方米播种 50～60g，覆土 1～1.5cm。1 年生苗出圃。

（曾 玲）

阳 桃

Averrhoa carambola L.

（阳桃科 Averrhoaceae）

生长习性、分布和用途 阳桃属 2 种，我国只有本文描述的 1 种。常绿乔木，高 10～15m，胸径 20～40cm。生长速度中等。能耐－1℃的极端低温及轻霜。喜肥沃湿润的酸性土。原产华南，滇、桂、粤、琼、闽、台广为栽培，历史悠久。马来西亚亦产，世界热带各地有栽培。果可生食或盐渍、蜜饯。叶利尿去毒供药用。树冠及果形美，为南方重要果树和观赏树。

开花结实 实生树 10 年生左右开始开花结实，正常结实期在 15 年生以后，无性繁殖的园艺品种 2～3 年生可结实，但多数果实无种子。育苗用种多采自实生树。无大小年现象，每年结实均较多。花两性，白色或淡紫色，近钟形，组成聚伞花序，着生于 1～3 年生的枝干上或叶腋。萼片 5，红紫色。花瓣 5，倒卵形，旋转排列。雄蕊 10，2 轮，外轮 5 个较短且无花药。子房上位，5 室，每室多数胚珠，中轴胎座，花柱 5。据广西南宁 1978～1984 年观察，1 年中多季开花结实，但主要花果期是两次：一次为 5 月上旬始花，5 月下旬～6 月上旬盛花，6 月下旬为末花期。这次花所结的果实 8 月上旬开始成熟，8 月中下旬为盛果期。另一次为 7 月下旬始花，8 月中下旬盛花，9 月中旬为末花期。这次花的果实翌年 1 月中旬开始成熟，2 月中下旬为盛果期。

浆果，成熟时由青绿色转变为淡黄色以至黄色，5 棱状椭圆形，长 5～8cm，径 4～4.5cm。果实成熟后 4～6 天脱落。每果有种子 1～10 颗。种子扁椭圆形，两端尖，浅褐色，长 1.1～1.7cm，宽 5～7mm，厚 2～2.5mm，种皮薄，有胚乳。胚直伸，子叶长椭圆形。阳桃种子的

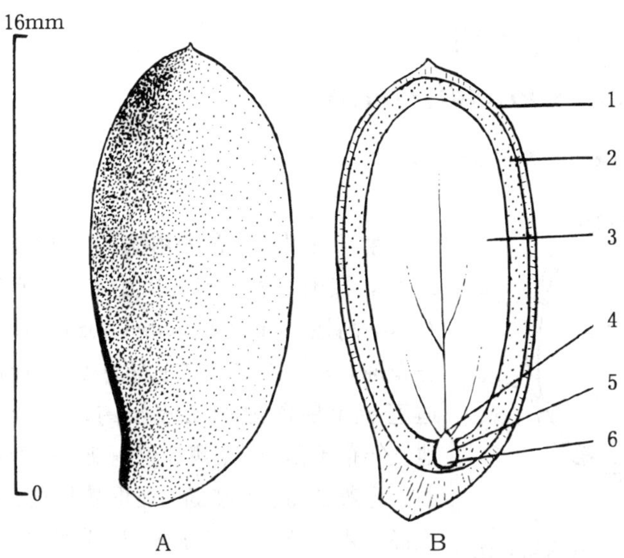

图 1　阳桃种子外形（A）及其纵切面（B）

1. 种皮　2. 胚乳　3. 子叶　4. 胚芽　5. 胚轴　6. 胚根

（黄应钦绘）

外形和解剖构造见图 1。

果实的采收调制和种子贮藏

供盐渍和蜜饯用的果实多在尚未充分成熟时即已采集。供育苗用的，需待果实转为黄色后方可采集；也可在大熟季节在树下捡拾成熟后自然脱落的果实。采得的果实可堆沤数日，果肉软化后置水中搓洗，淘去果皮、果肉等杂质，得出种子。鲜果的出种率为0.4%～0.5%。种子的净度可达96%。千粒重约 92g，每千克有纯净种子 0.9 万～1.2 万粒。种子的含水量为 33%，不宜失水，忌日晒，裸露存放的时间不宜超过 3 天。运输或贮藏均需混以湿沙，混沙贮藏时间一般为 3 个月。

发芽和播种　种子无明显休眠现象。发芽时日均温应在 20℃左右。1981 年 3 月上旬，日均温 15～17℃时，广西林业科学研究所用新鲜种子在场圃播种，播后 40 天发芽，开始发芽后 5 天进入盛期，发芽率为 65%。1987 年 4 月 3 日，日均温 18～26℃时在室外沙床作发芽测定，播后 35 天开始萌发，5 月 20 日发芽结束。从开始发芽至发芽高峰，以 7 天计，发芽 61%；从播种之日起算，以 47 天计，发芽率为 70%。出土萌发。胚根萌发后 4 天子叶出土，再过3～4 天初生叶展出。阳桃种子的萌发和幼苗初期生长情况见图 2。

条播。每平方米播种 10～12g，覆土 1.5～2cm。1 年生苗可作砧木，2～3 年生大苗供庭院栽培。园艺上多用压条苗或嫁接苗。

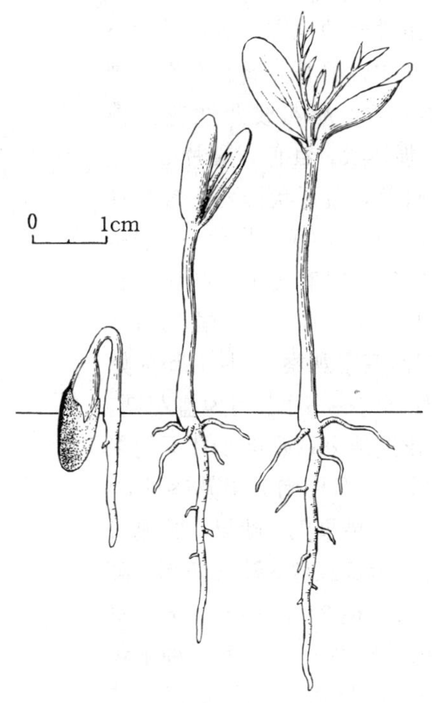

图 2　阳桃种子萌发后第 4、7、10 天幼苗的生长情况

（黄应钦仿《热带亚热带主要树种采种育苗技术》）

（张声燕）

大叶山楝

Aphanamixis grandifolia Bl.

（楝科　Meliaceae）

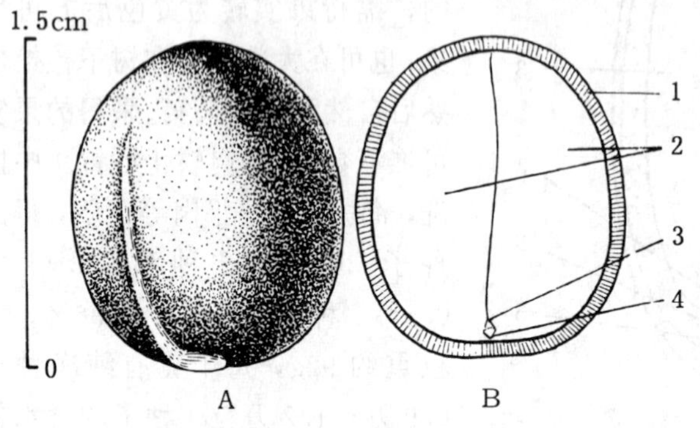

图 1　大叶山楝去除假种皮的种子外形(A)及其纵切面(B)
1. 种皮　2. 子叶　3. 胚芽　4. 胚根
(黄应钦绘)

生长习性、分布和用途　山楝属约 25 种，我国 3 种，本文描述 1 种。常绿乔木，高达 25m，胸径 50～60cm。幼苗忌霜冻，大树能耐 0℃左右的极端低温。适宜于土层深厚肥沃疏松的疏林或平地。分布于滇、桂、粤。东南亚、中南半岛、马来半岛、爪哇岛、加里曼丹岛、帝汶岛、伊里安岛也有分布。材质优，宜于造船、建筑及军工等用。种仁含油率 54.6%，供工业用。

开花结实　10 年生左右开始开花结实，正常结实年龄在 18 年生以后。水肥条件好的地方，结实无大小年现象。花单性异株，偶有杂性。雄花为圆锥花序。雌花或两性花为穗状花序。萼杯状，红色，裂片 5。花瓣 3，绿白色，覆瓦状排列。雄蕊管近球形，花药 6，内藏。花盘小。子房上位，3 室，每室有叠生 2 胚珠。花柱极短，柱头大，具 3 棱。据 1981～1985 年广西南宁观察，6 月上旬为始花期，6 月下旬～7 月上旬为盛花期，7 月下旬花期结束；翌年 2 月下旬果实开始成熟，3 月中旬为果熟盛期，4 月中旬仍有零星果熟。蒴果，成熟时由青色转变为黄红色，钝三角形，高 1.9～3.2cm，径 2～3.1 cm，干后室背 3 裂。每果有种子 2～3 粒。种子包含于已开裂的果瓣内，果瓣可维持 3～5 天不脱落。种子外层为橙红色假种皮所包裹。种子近卵形，长1.4～2.4cm，径 1～1.7cm。种皮黑褐色，表面有光泽，沿基部至内侧有一明显

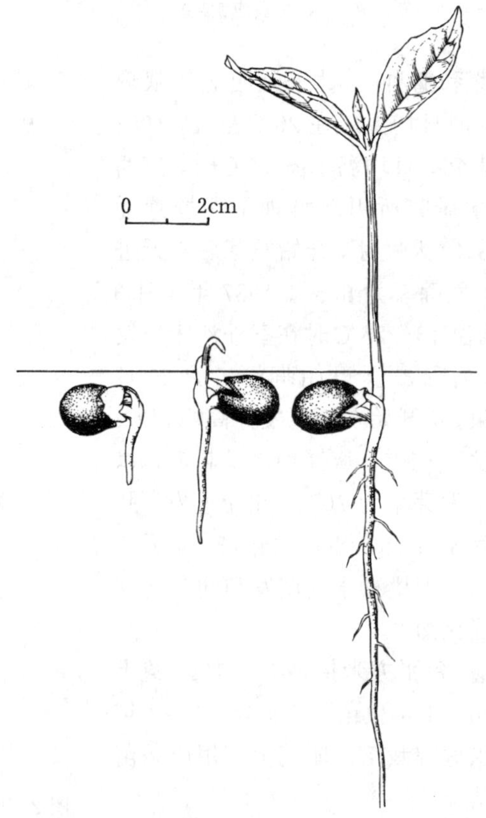

图 2　大叶山楝种子萌发后第 2、3、10 天幼苗的生长情况
(黄应钦仿《热带亚热带主要树种采种育苗技术》)

白色条形纵纹的种脐。无胚乳，子叶肥大。大叶山楝种子的外观和内部结构见图 1。

果实的采收调制和种子贮藏　果实已部分开裂时带果穗采下，不宜用竹竿敲打，以免种子散失。从采得的果穗上摘下果实并按成熟程度分别堆放。从开裂的果实中剥出种子。堆放 1 周左右仍不开裂的果实，即不宜作为播种材料。种子剥出后置水中浸泡、搓洗，并淘净假种皮，得出种子。鲜果的出种率约 33%。种子净度可达 98%。千粒重 1 000（700～1 400）g，每千克有纯净种子约 700～1 400 粒。种子含水量 35%～45%，忌失水，不宜日晒。短期运输可带果壳包装。已调制出来的种子需混湿沙方能贮藏或运输。贮藏以不超过 2 个月为宜。

发芽和播种　种子无休眠现象。1981 年 3 月 12 日，日均温 16℃时，广西林业科学研究所在室外沙床上用新鲜种子作过发芽测定：4 月 10 日（日均温约 20℃）开始发芽，开始发芽后 5 天左右进入盛期，5 月 3 日发芽结束；从开始发芽之日至发芽高峰，15 天发芽 60%；从播种之日起算，以 52 天计，发芽率为 80%。留土萌发。子叶留于种壳内。胚根萌发后第 3 天上胚轴出土，约 1 周以后发出初生叶。大叶山楝种子的萌发和幼苗早期生长情况见图 2。

条播。每平方米约播 60～75g，覆土约 2cm。育 1 年生苗出圃。

（张声燕）

墨西哥椿

Cedrela mexicana Roem.

（楝科　Meliaceae）

生长习性、分布和用途　洋椿属 9 种，我国引入 2 种，本文描述 1 种。落叶乔木，具板根。速生，15 年生高 14m，胸径 21.6cm。原产中美洲和南美洲。1964 年我国从古巴引进种子，在海南儋县试种，生长发育正常。对土壤要求不甚苛刻，适生于石灰岩发育的钙质土壤，积水的沼泽地不宜种植。根系浅，不抗台风。1983 年 7 月 17 日在 10 级台风侵袭下，小枝折断。木材近似桃花心木，质地松软，心材粉棕色至红棕色，比重 0.37～0.75，适用于高级家具和室内装饰，是南美洲出口的商品用材。

开花结实　10 年生左右开始开花结实，正常结实期在 15 年生以后。花两性。圆锥花序顶生。萼筒状，5 裂。花瓣 5，紫罗兰色。雄蕊 8～10，着生在瓶状的管口上。子房上位，瓶状，5 室，具长于子房的子房柄。据海南儋县观察，花期在 6 月，果熟期为翌年 4～5 月。蒴果，卵圆形，长 3～3.8cm，宽 1.5～1.8cm，成熟时褐黑色，上有许多白色皮孔，成熟后 5 瓣裂，每室有种子 6～9 粒。种子轻，具翅，无胚乳。胚直立，子叶长椭圆形。墨西哥椿种子的外形及

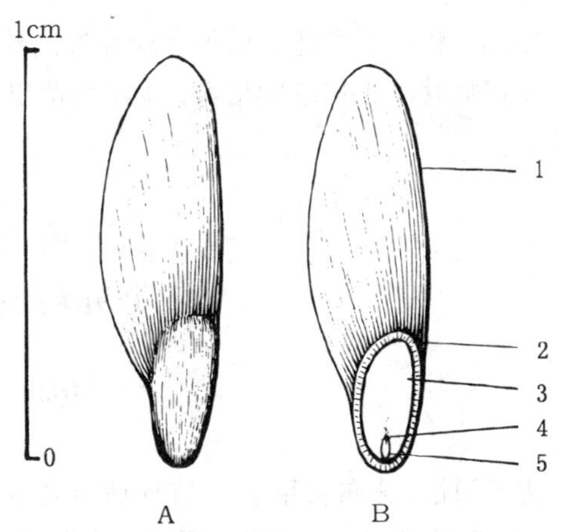

图 1　墨西哥椿种子外形（A）及其纵切面（B）
1. 种翅　2. 种皮　3. 子叶　4. 胚芽　5. 胚根
（黄应钦绘）

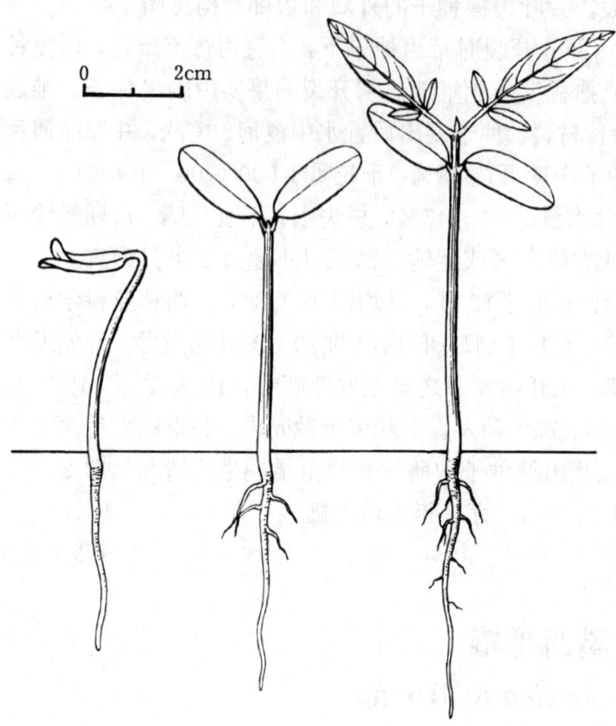

图 2　墨西哥椿种子萌发后第 1、3、22 天幼苗生长情况
（黄光郁、黄应钦绘）

其解剖构造见图 1。

果实的采收调制和种子贮藏　4 月底～5 月上旬果实由青绿色转变成褐色时即可采收。种子较轻，成熟的果实落地爆裂，种子随之散失，采收时要避免种子散失。用手采摘或用高枝剪、采种钩刀等将果实钩落，在地面捡拾。采回的果实摊放室内数天后自行开裂，种子脱出，去除杂质即为播种材料。可以去翅，但常不去翅。鲜果的出种率约为 20%。种子的净度一般在 90%左右。千粒重约 19g（原产地千粒重 22.9g），每千克有带翅种子 5 万～5.5 万粒。种子含水量约 27%。已调制好的种子混湿沙贮藏于坛罐等容器内。室内常温条件下可以贮藏 2～4 个月。

发芽和播种　种子无休眠习性。发芽时日均温宜在 20℃以上。1981 年 4 月和 1989 年 4 月，华南热带作物研究所先后 2 次在室外沙床用新鲜种子作过发芽测定：播后 8 天左右开始发芽，10～11 天进入发芽盛期，16 天后发芽结束；从开始发芽至发芽高峰，3 天中发芽 70%；从播种之日起算，16 天的发芽率为 85%。出土萌发。发芽后 6 天展出初生叶。种子萌发和幼苗初期生长情况见图 2。

撒播。将种子密播于播种盘或经过细致整地的播种床上。每平方米播种 8～10g，稀薄覆土。出苗后移至大田继续培育。1 年生苗出圃。

（邓春媛）

麻　楝　属

Chukrasia A. Juss.

（楝科　Meliaceae）

生长习性、分布和用途　麻楝属只有本文描述的 1 种和 1 变种。常绿至半落叶大乔木。能耐 0℃的极端低温，但苗期忌霜冻。适生于肥沃湿润的疏松土壤，宜村边宅旁种植。对酸性土或微碱性土均能适应，但干旱贫瘠或黏重土生长不良。木材坚硬适中，具香气，为速生珍贵用材。树皮可退热。种子含油达 50%，供工业用。这个种和变种的名称、生长及分布见表 1。

表 1　麻楝属树种的名称、生长和分布

中　名	学　名	树高（m）	胸径（cm）	分　布
麻楝	*C. tabularis* A. Juss.	30	50～70	琼、粤、桂、滇、藏。南亚、东南亚
毛麻楝	*C. tabularis* A. Juss. var. *velutina* (Wall.) King	50	150	滇、黔、桂、粤。印度、斯里兰卡

开花结实　8～10 年生开始结实，16 年生以后进入正常结实期。结实有大小年现象，但不甚明显。花两性，4～5 数。圆锥花序，顶生，少数生于上部叶腋，花序短于叶。花萼杯状，萼齿 4～5。花瓣长椭圆形，旋转排列。雄蕊管圆柱形，10 钝齿裂，花药 10，着生于齿的内面。花盘退废。子房具柄，甑状，3 室，少有 4～5 室，每室有 2 列的胚珠多粒。花柱粗壮，柱头头状，约与花药等高。在同一地区，麻楝与毛麻楝开花结实的时间大体相近，但麻楝的开花期稍晚，果熟期稍早。在云南，毛麻楝 4～5 月开花，11～12 月果实成熟。广西南宁观察的物候期见表 2。

表 2　麻楝属树种的物候期（广西南宁）

树　种	观察年份	开　花			果实成熟		种子散落
		始　期	盛　期	末　期	始　期	盛　期	
麻楝	1978～1984	5 月中旬	5 月下旬	6 月上旬	10 月下旬	11 月中旬	11 月中旬～翌年 3 月下旬果实陆续开裂，种子飘散
毛麻楝	1981	5 月上旬	5 月中旬	5 月下旬	11 月上旬	11 月下旬	12 月上旬～翌年 3 月下旬种子陆续飘散

蒴果，木质，成熟后 1～3 月逐渐从顶端室间开裂为 3～5 果瓣，果瓣 2 层，从具 3～5 翅的中轴上分离。每果有种子 150～180 粒，成 2 行，覆瓦状排列于中轴上。种子下端具膜质翅，无胚乳。子叶叶状，倒卵形，胚根依附于子叶边缘。果实和种子的形态特征见表 3、图 1。

表 3　麻楝属树种果实和种子的形态特征

树　种	未熟果颜　色	成熟果实			种　子		
		形　状	大小（cm）	颜　色	形　状	大小（cm）	颜　色
麻楝	黄绿色	近球形或椭圆形，表面粗糙，有淡褐色瘤状体，先端凸尖	长 3.8～4.3 径 2.9～3.3	深褐色	扁平，椭圆，具翅	宽 0.4～0.5，带翅长 1.2～2	黄褐色
毛麻楝	黄褐色	椭圆形，表面粗糙，有黑色小瘤体，先端凸尖	长 4～5 径 3～3.5	深褐或黑褐色	扁平，椭圆，具翅	宽 0.5，带翅长 1.2～2	黄褐色

果实的采收调制和种子贮藏 当果实成熟盛期,少数果瓣呈现开裂时,可用采种刀带果穗采下。选摘果穗中下部的果实,置通风干燥处摊放或曝晒至开裂,脱出种子。种子宜在弱光下晒干或晾干,含水量低于9%方可贮藏。短期存放的种子可用布袋或塑料袋盛装,置坛罐内密封存放于阴凉处,发芽能力可以保存1年。贮藏1年以上的宜在5℃左右的低温环境中保存。种子的净度、质量等数据见表4。

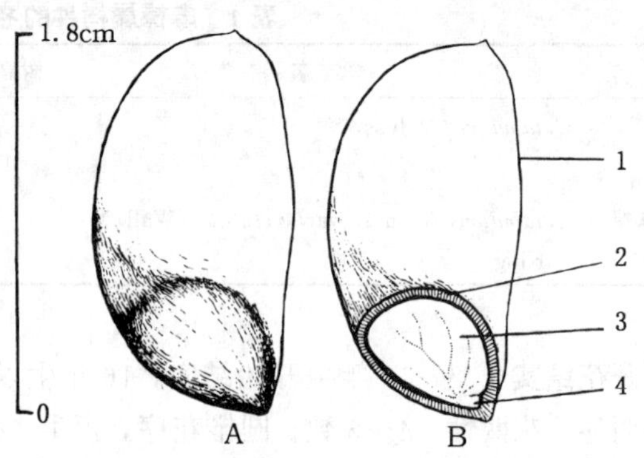

图1 麻楝种子外形(A)及其纵切面(B)
1.种翅 2.种皮 3.子叶 4.胚根
(黄应钦绘)

表4 麻楝属树种的出种率和种子净度、质量

树种	出种率(%)	净度(%)		千粒重(g)		每千克纯净种子粒数(万粒)	
		一般	变动范围	一般	变动范围	一般	变动范围
麻楝	21①	80	75～90	12	10～15	8.3	6.7～10
毛麻楝	22	80	75～90	13	10～17	7.7	5.9～10

注 另据记载,广东12月中旬～1月采果,出种率10%～17.9%;海南1～2月采果,出种率7.9%

发芽和播种 种子无明显休眠现象。发芽时的日均温宜在20℃左右。1981年3月中旬和1983年4月中旬,广西林业科学研究所先后2次在室外沙床上对麻楝进行过发芽测定:4月中旬播种后7天开始萌发,9～10天为发芽盛期;3月中旬播种后12天开始萌发,14～15天为发芽盛期;发芽全程约为5天;从开始发芽至发芽高峰日,3天中发芽约70%;从播种至发芽终止,14～18天的发芽率约80%。另据云南资料,毛麻楝的发芽率为60%～80%。出土萌发。子叶出土后4～5天发出初生叶。麻楝种子的萌发和幼苗初期的生长情况见图2。

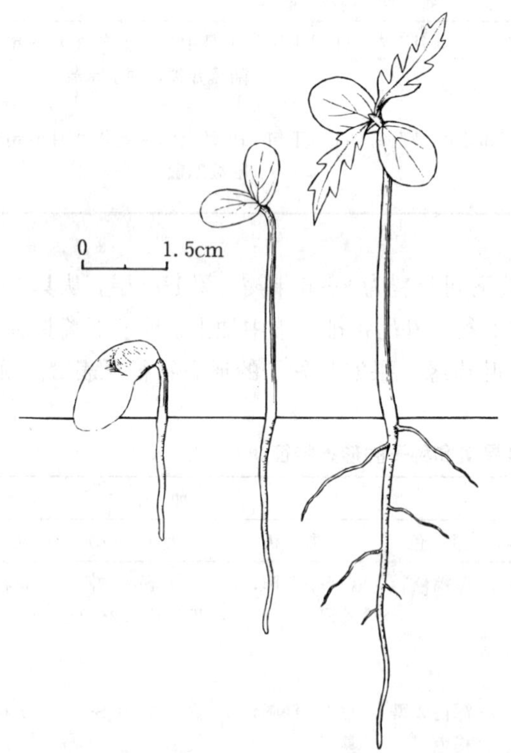

图2 麻楝种子萌发后第2、3、6天幼苗的生长情况
(黄应钦仿《热带亚热带主要树种采种育苗技术》)

播种前浸种2～3小时,捞出晾干后即可播种。条播。每平方米播种2～2.5g,覆土0.5 cm。1年生苗出圃。

(王宏志)

灰毛浆果楝

Cipadessa cinerascens（Pell.）Hand.-Mazz.

（楝科　Meliaceae）

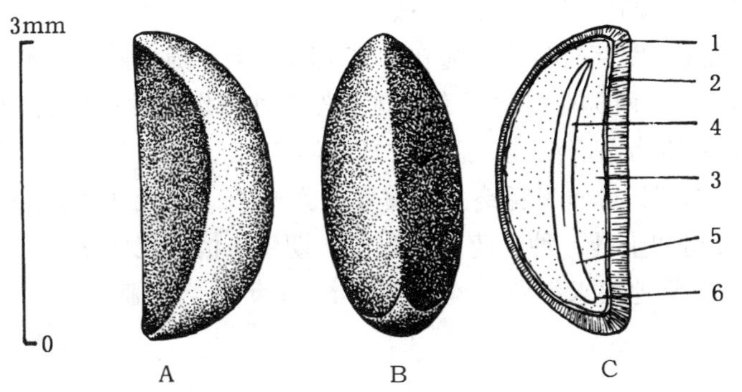

图 1　灰毛浆果楝果核外形（A、B）及其纵切面（C）
1. 内果皮　2. 种皮　3. 胚乳　4. 子叶　5. 胚轴　6. 胚根
（黄应钦绘）

生长习性、分布和用途　浆果楝属 4 或 5 种，我国产 2 种，本文描述 1 种。常绿灌木或小乔木，高 4～10m。适生于湿润而较肥沃的酸性土。忌冰雪。分布于桂、黔、川、滇。越南亦有。种子含油约 11%，可制皂。全株或根、叶入药，治感冒、腹泻、风湿、跌打、蛇咬等。

开花结实　4～6 年生开始结实，正常结实期在 10 年生以后，结实大小年现象不明显。花两性。圆锥花序腋生，长 5～15cm。萼浅杯状，5 齿裂。花瓣 5，白色、淡黄色至黄色，径3～4mm。雄蕊 10，花丝线形，基部合生成杯状体，顶端分裂为 2 尖齿，花药生于尖齿间。花盘短杯状。子房球形，无毛，5 室，每室有胚珠 2 颗。花柱短，直立，柱头半球形。据广西南宁 1987 年观测，1 年中开花结实 2 次：第一次 6 月上旬始花，6 月中旬为盛花期，7 月中旬为末花期，8 月上旬果实开始成熟，下旬为果熟盛期；第 2 次 8 月中旬始花，9 月上旬结束，10 月中旬～下旬果实成熟。

核果，浆果状，未熟时黄绿色，熟时赭红色至黑色，圆球形，径 4.6～6mm。果熟后 7～10 天脱落。果具 5 棱，内含 4～5 核。核软骨质，近半月形，灰白色，长 3～4mm，宽 1.5～2mm，厚 1.5～1.8mm，内有种子 1～2。种子具棱，微作肾形。有胚乳，胚近弧形。灰毛浆果楝果核的外形和内

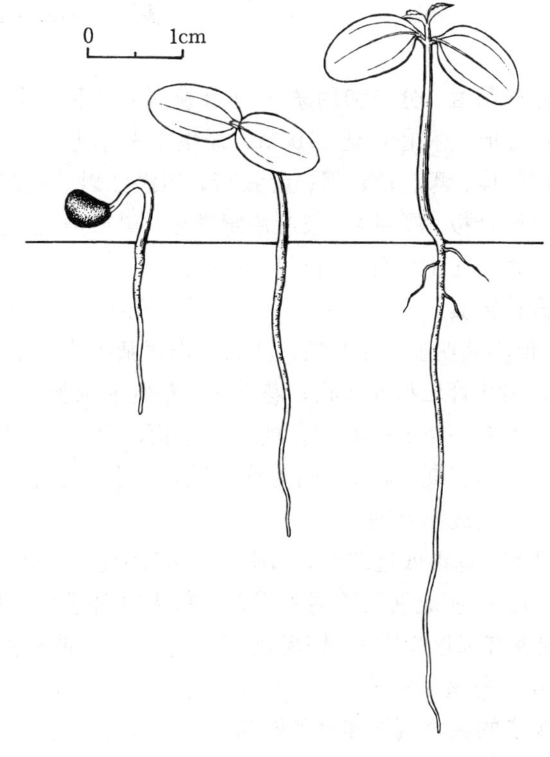

图 2　灰毛浆果楝种子萌发后第 3、4、8 天幼苗的生长情况
（黄应钦绘）

部结构见图 1。

果实的采收调制和种子贮藏　果熟盛期时用高枝剪剪断总果梗，在地面选摘饱满的果实。采回的果实堆沤 1～3 天，待核果肉质外层软熟后置水中搓擦洗净，得出果核即为播种材料，通称种子。鲜果的出籽率约 15%。种子（果核）的含水量 13%～16%，净度可达 96%～99%，千粒重为 4～5g，每千克有纯净种子（果核）20 万～25 万粒。种子（核）可以适当晾干，但不能日晒。贮藏时需混湿沙，贮藏期一般为半年左右。

发芽和播种　种子无明显休眠现象。发芽时日均温需在 20℃以上。当年随采随播或贮藏至翌年 3 月播种均可。1987 年 9 月 4 日，广西林业科学研究所在室外播种床作过发芽测定，种子为当年新采，播前未作任何处理，基质为湿润细沙，播种时日均温 24℃。播后 27 天（9 月 30 日）开始发芽，发芽盛期不明显，10 月 26 日发芽终止，发芽率为 52%。出土萌发。胚根萌发后约 3 天子叶出土，再过 4 天展出初生叶。种子萌发及幼苗初期生长情况见图 2。

条播。每平方米播种 1.5～2.5g。整地要求细致，覆土 0.5cm。1 年生苗出圃。

（张声燕）

非洲楝（非洲桃花心木、塞楝）
Khaya senegalensis (Desr.) A. Juss.

（楝科　Meliaceae）

生长习性、分布和用途　非洲楝属有 8 种，我国引入本文描述的 1 种。常绿大乔木，树高 20～30m，胸径可达 270cm。适生于深厚肥沃、湿润而排水良好的土壤。原产非洲热带地区。我国桂、粤、琼、闽、台栽培，但湛江以北便难以正常开花结实。优质用材，用于家具、建筑及胶合板。冠幅大，枝叶浓绿婆娑，树形美观，已成为我国华南南部四旁绿化和荒山造林的一个优良速生树种。可以作为桃花心木（*Swietenia mahagoni* (L.) Jacq.）的代用材。

开花结实　约 10～12 年生开始开花结实，结实盛期在 20 年生以后。大小年间隔期常为 1 年，但不甚明显。花两性。复聚伞花序腋上生。花萼 4 裂，裂片几达基部。花瓣 4，黄白色，分离。雄蕊管坛状或杯状，花药 8，着生于雄蕊管内近顶端。花盘杯状，红色。子房上位，4 室，每室有胚珠 8～16 颗。柱头圆盘状，上面具 4 槽。据海南尖峰岭 1978～1987 年观察，4 月中旬为始花期，4 月下旬为盛花期，5 月下旬为末花期，翌年 3 月上旬果实开始成熟，3 月下旬为果实成熟盛期。

蒴果，球形或近球形，木质，未熟时绿色，熟时褐色带灰白色，表面粗糙，长 5～7cm，径 4～6cm。果实成熟时顶端 4 瓣裂。每果含种子数十粒。种子宽，扁平，横生，椭圆形至近圆形，周围有圆形膜质翅，棕褐色。种子浅褐色。带翅种子长 13～20mm，宽 17～25mm，厚1.5～2.5mm。去翅后种子长 7～9mm，宽 12～18mm，厚 1.2～1.6mm。

果实的采收调制和种子贮藏　果熟盛期采集。采下的果实不宜在强光下曝晒，可摊放于室内通风处，任其自然开裂，去掉杂质，取出种子，晾干，即可播种或贮藏。鲜果的出种率约为 6%～10%。种子净度 85%～95%，含水量约为 15%。千粒重 178（160～200）g，每千克有纯净种子 5 600（5 000～6 250）粒。种子用普通方法贮存容易丧失发芽能力。据中国林业科学研究院热带林业站在海南的试验，用放有无水氯化钙的干燥器密封贮藏，28 个月后，种子发芽率仍在 90%左右；用瓦坛拌鲜草木灰密封贮藏，种子发芽能力亦可保持 10 个月。

发芽和播种　种子无休眠习性。发芽时日均温宜在 20℃以上。1987 年 6 月 19 日，广西林业科学研究所曾在室外沙床上用当年新采而未作处理的种子作过发芽测定（播种时日均温 29℃）：6 月 28 日开始发芽，6 月 30 日为发芽盛期，7 月 6 日发芽终止；自发芽之日起算，5 天发芽 51%；从播种之日起算，17 天的发芽率为 88%。留土萌发。播种后 9 天胚根萌发，3 天后胚芽出土，再过 5 天长出初生叶。

播前用始温 25～40℃水浸种，并在自然冷却过程中继续浸泡 24 小时，能提高发芽率 6%～9%，提早 3 天发芽。条播或撒播。每平方米播 20g，覆土 0.5cm 并盖薄草。1 年生苗高可达 1m，可以出圃造林。也可以将芽苗移至容器，3 个月苗高 35cm 以上时可供造林。

（张声燕）

楝　属

Melia L.

（楝科　Meliaceae）

生长习性、分布和用途　本属约 20 种，分布于东半球的热带和亚热带，我国有 3 种，本文描述 2 种。落叶乔木，胸径 25～35cm，有达 1m 的，树高 8～15m，可达 20～35m。速生。喜温暖湿润气候。不耐寒，不耐旱，忌积水，但对土壤适应性强，在酸性以至轻盐碱土上均能生长。楝树在含盐量 0.25%以下的盐碱土上生长良好，川楝则在厚层紫色土上生长良好，但在土层深厚肥沃、疏松湿润、排水良好的土壤上生长更快。

楝树分布于黄河流域以南各省。川楝主产我国西南，湘、鄂、甘也有分布（表 1），华东、华南引种栽培。多用于四旁栽植。

表 1　楝属树种的名称、分布和用途

中　名	学　名	分　布	用　途
楝树（苦楝）	*M. azedarach* L.	豫、陕、晋南、冀南、鲁南、甘南、川、鄂、湘、闽、滇、黔、粤、桂、台	材用、药用、栲胶
川楝（金铃子）	*M. toosendan* Sieb. et Zucc.	川、滇、黔、甘、豫、鄂、湘、桂。浙、苏、闽、赣、粤引入栽培。日本、越南、泰国、印度、印度尼西亚、斯里兰卡	材用、药用、栲胶

楝树材质轻软坚韧，纹理粗但较均匀，有光泽，易加工。川楝较硬，材性较脆，纹理细致美观，耐腐性强。这 2 个树种都是建筑、家具、农具、船舶、乐器和板壁等的用材。果实可入药。种子可榨油，供制肥皂、润滑剂等。树皮及叶含鞣质，可提制栲胶。树皮纤维可用于纺织和造纸。

开花结实　开始开花结实年龄因种源及母树生长立地条件而不同。一般 4～5 年生开花结实，也有 3 年生开花结实的，7～8 年生开始进入盛果期。在盛果期内大小年现象不明显。

花两性，白色或青紫色，圆锥花序腋生于当年新生枝的下段，形成多个花序，多花，多少被粉末状星毛，开放时有香气。楝树花序长 12～24cm，川楝花序长 9～17cm，多为 10～13cm。萼片 5～6。花瓣 5～6，分离，远长于萼，倒卵状长圆形、倒披针形至线状匙形。雄蕊合生成 1 管，圆筒形，稍短于花瓣，边缘 10～12 齿裂，裂齿再 2～3 细裂，管部有线条 10～12，花药 10～12，着生于雄蕊管内侧上缘的裂齿间，内藏或部分突出。花盘环状。子房近球形。楝树子房 4～6 室，川楝子房 6～8 室，每室有胚珠 2 颗。花柱较子房长，近圆柱形，3～6 裂。花期从 3 月上旬～5 月下旬，果实成熟期多在 11 月，也有在 8 月下旬、甚至有在翌年 1 月的(表 2)。

表 2 楝属树种的开花结实物候期

树 种	观察地点	开 花			果实成熟期	果实脱落	
		始 期	盛 期	末 期		始 期	末 期
楝树	山东济南	5 月中旬	5 月中旬	5 月下旬	11 月上旬	—	—
	陕西西安	5 月中旬	5 月中旬	5 月下旬	8 月下旬	9 月上旬	—
	江苏南京	5 月上旬	5 月上旬	5 月中旬	10 月下旬	11 月中旬	—
	浙江杭州	4 月下旬	4 月下旬	5 月上旬	11 月上旬	11 月中旬	—
	福建厦门	4 月上旬	4 月中旬	4 月下旬	9 月上旬	9 月下旬	翌年 1 月下旬
	广西桂林	4 月中旬	4 月中旬	4 月下旬	12 月上旬	12 月下旬	—
	广东广州	3 月下旬	3 月下旬	4 月中旬	10 月上旬	10 月下旬	—
	湖北武汉	4 月下旬	4 月下旬	5 月上旬	11 月中旬	12 月中旬	—
	四川成都	4 月中旬	4 月中旬	4 月下旬	10 月上旬	10 月中旬	翌年 1 月下旬
川楝	云南昆明	4 月初	4 月上旬	4 月下旬	翌年 1 月中旬	翌年 2 月上旬	翌年 2 月下旬
	四川西昌	3 月上旬	3 月下旬	4 月上旬	11 月下旬	12 月中旬	—
	四川雅安	4 月下旬	5 月上旬	5 月中旬	11 月下旬	翌年 1 月下旬	翌年 3 月上旬

果实未熟时青绿色，成熟时淡黄色、黄白色或栗棕色。核果，球形至椭圆形或长圆状球形。楝树核果小于川楝，前者一般长 1.5～2cm，径 1.5cm，后者一般长约 2.5～3cm，径1.5～3.2cm。据本文作者在四川雅安观察，楝树的结果枝平均果穗数为 4～5，每穗的成果数为1～7 个；川楝结果枝平均果穗数为 6～9，每穗成果数为 1～6 个。核果的外果皮薄，表面散生橙黄色疣点，中果皮肉质，内果皮木质坚硬。果核淡黄白色，表面粗糙，有棱，呈龙骨状隆起，棱的中央有凹槽，先端尖，中部膨大，基部深陷。楝树果核椭圆形，(4) 5～6 室，每室有种子 1 粒。川楝果核椭圆状球形或近球形，6～8 室，每室 1 粒种子。果核内的种子紫黑色，椭圆形，微扁，有光泽。胚乳薄，肉质，围绕着胚。胚根和子叶均呈蜡质状，白色。每果平均含种子数及其大小见表 3。

表 3 楝属树种果核的棱数、平均含种子粒数及其大小

树 种	观察地点	果核类型	果核棱数	果 核 (cm)		每果核平均含种子数 (粒)	种 子 (mm)	
				长	径		长	径
楝树	四川	大型果	5～6	1.2 以上	1.2～1.25	4.4	10.6～11	3.5～3.8
	雅安	中型果	5～6	1～1.9	1.0～1.1	2.8	10～10.5	3.3～3.5
		小型果	5～6	0.9～1.0	0.9～0.95	1.4	8.6～9.9	3.3～3.5
川楝	四川	大型果	7～8	1.5 以上	1.5～1.6	4.2	11.4～11.6	4.1～4.3
	雅安	中型果	6～8	1.3～2.0	1.3～1.4	3.0	10～11	3.8～4
		小型果	6～7	1～1.29	1～1.2	1.2	9.3～9.5	3.8～4

健壮的散生楝属树种结实量一般都多。混生的，密度大光照不足的则结实量少。剥割树皮或天气干旱或土壤肥力差皆能影响结实量。

川楝果核的外形及纵切面见图 1。

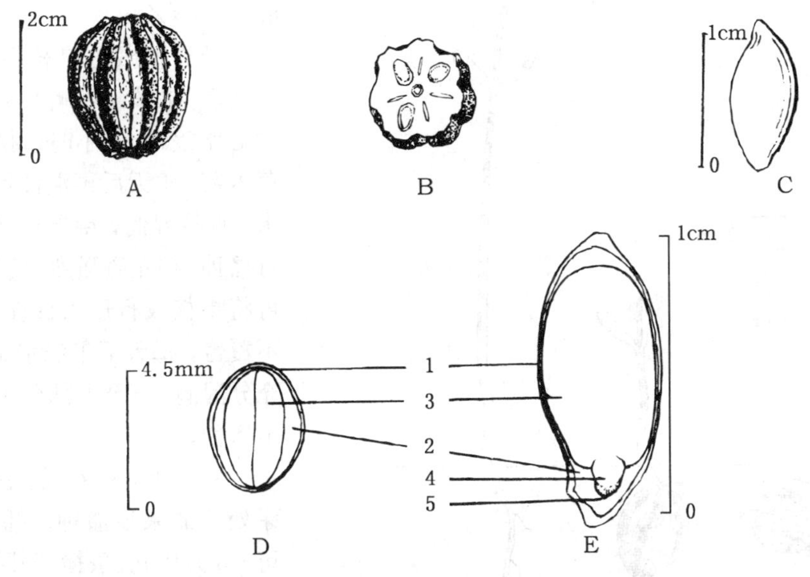

图 1 川楝果核外形 (A) 及其横切面 (B) 和种子外形 (C) 及其横切面 (D) 和纵切面 (E)

1. 种皮 2. 胚乳 3. 子叶 4. 胚轴 5. 胚根

(孟玲绘)

果实的采集调制和种子贮藏 选择健壮树木作母树。核果变为黄白色，果皮微皱时采果。用竹竿敲击果枝，使核果落地收集。果熟后悬挂树上经久不落，一直到翌年树木发芽时才大量脱落，因此采种期较长。采集的果实堆沤数天，或在堆沤处每天用温水淋洒 1～2 次，待果肉软化时放入筐内搓揉，漂去果肉杂质，洗净阴干即得果核。生产上以果核作为播种材料，通称种子。也有不除去果皮果肉而将核果晒干作为播种材料的。采后即播的可以不去果肉。调制出的果核通常采用干藏，散装或袋装置库房内。也有混沙贮存的。核果出籽率及其质量见表 4。

表 4　楝属树种的核果出籽率、种子质量

树　种	出籽率（%）	净度（%）	千粒重（g）		每千克纯净果核数（粒）	
			一　般	变动范围	一　般	变动范围
楝树	38.6	98～99	680～700	—	1 430～1 470	—
	25～35	—	700	550～830	1430	1 200～1 800
	25～45	—	—	700～800	—	1 250～1 430
川楝	25.5	98～99	—	1 100～1 560	—	640～900
	20～25	98～99	1 690	1 250～1 880	590	530～800

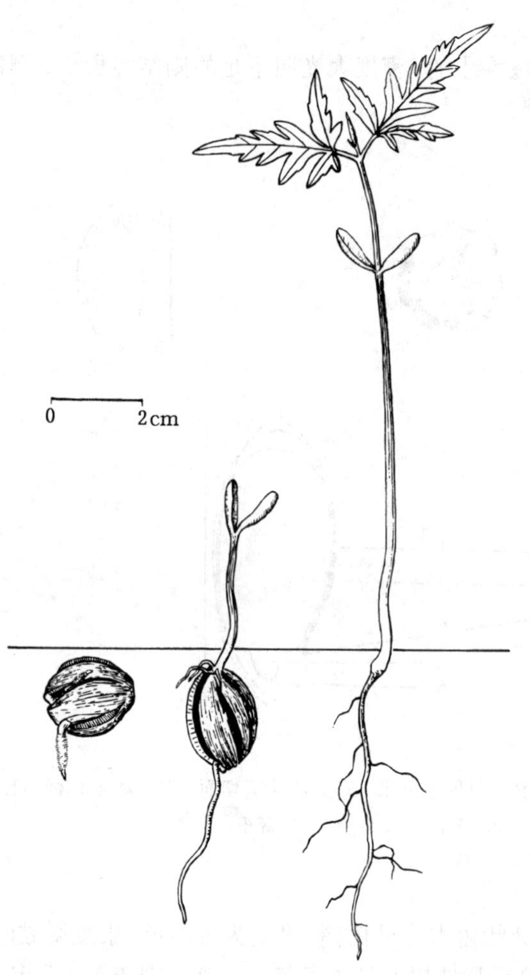

图 2　川楝种子萌发后的幼苗生长情况
1. 幼根初露　2. 子叶展现　3. 初生叶对生
（孟玲、黄应钦绘）

发芽和播种　楝属树种果核坚硬，通常敲碎果核取出种子，用 0.5%的四唑溶液染色测定生活力。种子生活力在 80%～95%，多为 82%～88%。

经过贮藏的果核播种前宜作催芽处理。果核先曝晒 2～3 天，用始温 80℃水浸种 24 小时，捞出果核混拌草木灰，1 天后再混沙堆放，经常淋水，保持湿润，待果核开始裂嘴时即行播种。在光照强烈，气温高的地区，可将果核成行压入沙床 1/2 或 1/3，不覆盖，每天下午浇水 1 次，使沙床充分湿润，直至大部分果核开裂时取出播种。

采后即播或春播。春播用经过催芽处理的果核播种。播种期 2 月下旬～3 月中旬。条播，条距 20～30cm，条幅 3～5cm，播种沟深 2cm，每米长播种沟播 15～20 粒。每 666m^2 用种量楝树为 20～25kg；川楝为 30～50kg，播后覆细土 2cm，盖草。当年采果后即播的，翌年 3 月可发芽出土。春播的播后 1 月左右发芽出土。每个果核能发出数株幼苗，苗高 8cm 左右时，在阴雨天分床移植，每簇留 1 株健壮幼苗。以 1 个果核生 1 株幼苗计算，楝树场圃发芽率为 40%～65%，川楝平均为 50%。

出土萌发，子叶 2，线状长椭圆形，肉质，微厚。初生叶对生，以后互生，叶为 3 深裂单叶或 3 小叶复叶，顶端裂片或顶生小叶菱形，侧裂或侧生小叶卵状披针形，边缘有不规则缺裂，第 5（或第 4）叶为 5 小叶羽状复叶，最后为二回羽状复叶。川楝种子萌发后的幼苗生长情况见图 2。1 年生苗高 1m 许，可出圃造林。

萌芽力极强，可分根繁殖。张瑞军和林万钊等人（1989）用 1 年生良种苗的根段扦插，待幼苗长到 15～20cm 时采作插穗，用 NAA 等处理后再行扦插，成活率可达 90%，当年苗高可达 2m 以上，可以出圃。

（阙再旦）

桃花心木属

Swietenia Jacq.

（楝科　Meliaceae）

生长习性、分布和用途　本属有 7～8 种，本文描述我国引入的 2 种。常绿大乔木。喜温暖湿润气候和土层较深厚肥力中等以上的土壤。为世界著名的珍贵用材树种。它们的名称、生长、分布及用途等见表 1。

表 1　桃花心木属树种的名称、生长、分布和用途

中　名	学　名	生长情况	分　布	用　途
大叶桃花心木	*S. macrophylla* King.	高 40m，胸径 200cm，速生，耐寒性较差	原产热带美洲，现世界热带地区均有引种。我国闽、桂、粤、琼栽培	高级家具和装饰品的优良用材
桃花心木	*S. mahagoni* (L.) Jacq.	高 30m，胸径 150cm，生长较慢，较耐低温	原产中美北部，现世界热带地区均有引种。我国滇、桂、粤、琼栽培	世界著名的红木之一，高级家具和装饰品优良用材

开花结实　8～10 年生开始结实，结实盛期在 15 年生以后。结实大小年间隔期一般为 1 年，但不甚明显。花两性。圆锥花序顶生或腋生。花小，白色。萼 5 齿裂，小。花瓣 5。雄蕊 10，花丝合生成壶状，顶部 10 齿裂，花药内藏。花盘环状。子房上位，5 室，每室有胚珠多数，排成 2 列。花柱小，柱状，柱头宽，盘状。据广西南宁和海南尖峰岭观察，这 2 个种的开花结实物候期见表 2。

表 2　桃花心木属树种的开花结实物候期

树　种	观察地点和年份	开花			果实成熟		种子散落
		始　期	盛　期	末　期	始　期	盛　期	
大叶桃花心木	海南尖峰岭 1978～1987	3 月下旬	4 月上中旬	4 月下旬	翌年 2 月上中旬	2 月下旬	3 月下旬

（续）

树 种	观察地点和年份	开花 始期	开花 盛期	开花 末期	果实成熟 始期	果实成熟 盛期	种子散落
桃花心木	广西南宁 1978～1984	5月下旬	6月上旬	6月中旬	翌年2月中下旬	3月上旬	3月中旬
	海南尖峰岭 1978～1987	4月下旬	5月上旬	5月中旬	翌年3月下旬	4月上旬	4月上中旬

蒴果，卵形，木质，熟时5瓣裂，顶端室间开裂后从具5棱宿存的中轴上分离。每室有种子10粒以上。种子着生于中轴上，上端具长而阔的翅，翅长2.5～4cm。有胚乳。胚侧生于种子中部。果实及种子的形态见表3、图1。

表3 桃花心木属树种果实和种子的形态特征

树 种	果实 形 状	果实 大小（cm）	果实 颜 色	种子 形 状	种子 大小（mm）	种子 颜 色
大叶桃花心木	卵形	径5～7 长7～10	棕褐色，粗糙	翅状	长15～28 宽10～12 厚5～6	种翅棕色，种子浅褐色
桃花心木	卵形	径4.5～5.7 长5.8～7	棕褐色，粗糙	翅状	长10～16 宽5～8 厚1～2	种翅棕褐色，种子浅褐色

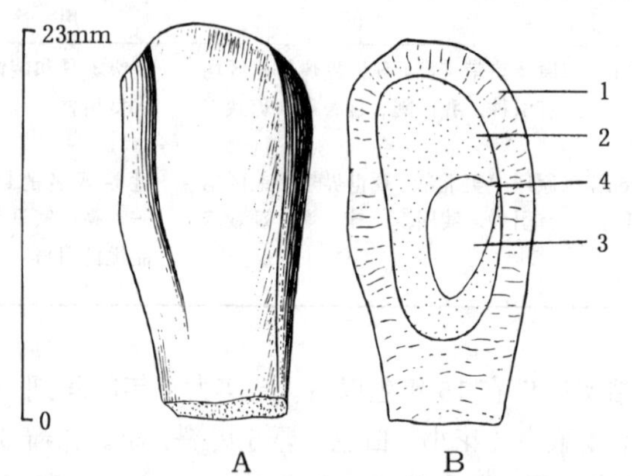

图1 大叶桃花心木剪去部分种翅的种子外形（A）及其纵切面（B）
1. 种皮 2. 胚乳 3. 子叶 4. 胚根
（黄应钦绘）

果实的采收调制和种子贮藏

果实成熟盛期用采种钩刀或高枝剪采集。采得的果实不宜在强光下曝晒，可摊放于室内通风处任其自然开裂，晾干，取出种子，即可用于播种或贮藏。种子发芽能力保存期短，宜随采随播。在常温条件下贮藏3个月以后，发芽率便降至20%以下，贮藏5个月以后全部丧失发芽能力。1个月以上的贮存，应置于5℃左右的低温环境。鲜果出种率及种子的净度、质量等见表4。

表4 桃花心木属树种的出种率和种子净度、质量

树 种	出种率（%）	净度（%）	千粒重（g） 一 般	千粒重（g） 变动范围	每千克纯净种子粒数（粒） 一 般	每千克纯净种子粒数（粒） 变动范围
大叶桃花心木	9.5	85～95	365	340～490	2 740	2 000～2 950
桃花心木	5.7	85～95	77	60～90	13 000	11 000～16 700

发芽和播种 种子无休眠习性。发芽时的日均温宜在 20℃以上。1987 年 6 月和 1982 年 3 月，广西林业科学研究所曾在室外沙床上用当年新采种子，对大叶桃花心木和桃花心木进行过发芽测定，播前种子未作任何处理，发芽时气温分别为 29℃和 21℃。测定结果见表 5。

表 5 桃花心木属树种的发芽能力①

树 种	温度（℃）	发 芽 势（%）		发芽率（%）	
		计算天数	一般数值	计算天数	一般数值
大叶桃花心木	29～30	4	48	19	75
桃花心木	21～22	8	50	52	60

① 广西南宁，室外沙床

留土萌发。大叶桃花心木播种后 11 天胚根萌发，过 2 天胚芽出土，出土后 4 天左右发出初生叶。桃花心木播种后 32 天胚根萌发，2～3 天后胚芽出土，出土后 3～5 天发出初生叶。桃花心木种子的萌发和幼苗初期的生长情况见图 2。

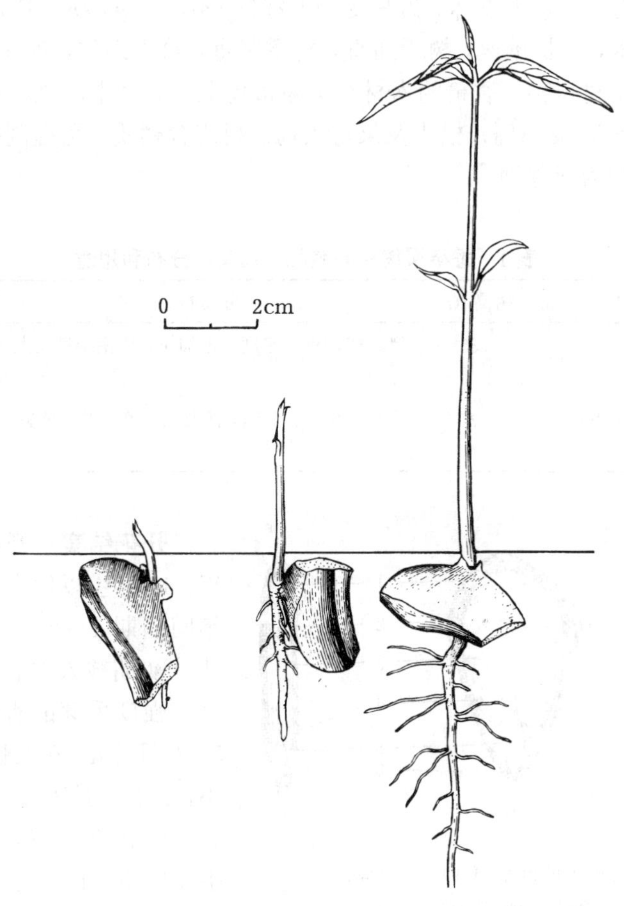

图 2 桃花心木种子萌发后第 3、5、10 天幼苗的生长情况
（黄应钦绘）

条播。大叶桃花心木和桃花心木每平方米的播种量分别为50～75g和8～12g。培育1年生苗出圃。

（张声燕）

香椿属

Toona Roem.

（楝科　Meliaceae）

生长习性、分布和用途　本属约15种，分布于亚洲和大洋洲。我国有3种5变种，本文描述香椿和红椿这两个种（表1）。香椿的分布范围比红椿广而偏北，中心区域在黄河流域和长江流域之间，北缘在辽宁南部和内蒙古南部，但至今已不多见；南抵粤、桂北部，西及川、滇、黔。香椿多为零星栽培，对钙质土、中性土、酸性土均能适应。红椿分布在滇、桂、粤的亚热带地区，皖南有少量天然林，为其变种毛红椿 var. *pubescens* (Fr.) Hand.-Mazz.，适于酸性土。落叶乔木，生长迅速，树干通直，材质坚重，纹理细致，色泽红润，略具芳香，易加工，为速生珍贵用材树种。香椿的木材在国际市场上素有"中国桃花心木"之称。香椿嫩芽可食；种子含油量达38.5%；根皮及果可入药。树皮含鞣质，可提制栲胶。在《中国植物红皮书》中红椿被列为渐危种。

表1　香椿属树种的名称、树高、分布和用途

中　名	学　名	树高（m）	分　布	用　途
红椿	*T. ciliata* Roem.	35	滇、桂、粤。印度、孟加拉、中南半岛及大洋洲	材用、栲胶
香椿	*T. sinensis* (A. Juss.) Roem.	25	华北、华中、华南，西及川、滇、黔。朝鲜	材用、食用、药用

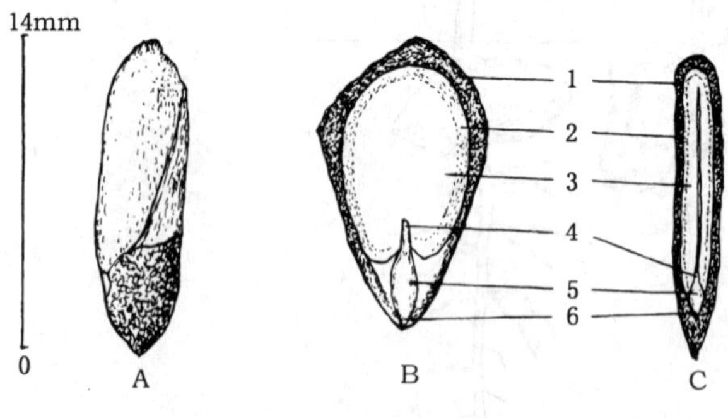

图1　香椿带翅种子的外形（A）、种子纵切正面（B）和纵切侧面（C）

1. 种皮　2. 胚乳　3. 子叶　4. 胚芽　5. 胚轴　6. 胚根

（田恒德绘）

开花结实　开始结实年龄在7～12年，15年生以后进入正常结实年龄。丰年间隔期1～2年。香椿嫩芽常在春季吐叶初期被人取食，对结实影响极大。每年强度采芽的香椿树难以开花结实。5～6月开花。花两性。圆锥花序顶生。花小，白色，具芳香。萼短小，管状，5齿裂，稀全缘。花瓣5。雄蕊5，分离，着生于肉质5棱的花盘上。子房柄厚，成5棱形的短柱，子房5室，每室有胚珠8～12枚。蒴果，木质，椭圆状卵形，长2.5～3.5cm，9～10月成熟，成熟时纵向5

裂。种子多数，三角形或略呈菱形，一面扁平，另一面稍隆起，长5～7mm。香椿种子上端具长翅，翅长10～12mm。红椿种子两端具翅。种翅膜质。子叶叶状。胚乳白色。但在发育过程中，各粒种子胚乳被胚吸收的程度很不一致：有的种粒具薄层胚乳；有的种粒胚乳完全消失；有的种粒一片子叶的外面仍有胚乳存在，另一片子叶的外面已无胚乳。图1选绘的是一粒具有完整胚乳的种子。

表2　香椿属树种果实的成熟特征、种子性状及发芽能力

树　种	果实成熟期和成熟特征	种子性状			
		外　形	千粒重（g）	每千克种子粒数（万粒）	发芽率（%）
红椿	6～8月，青绿色变为深褐色，外被显著的苍白色皮孔	种子上端具长翅，下端具短翅，种子连翅长1.5～2.2mm。亮褐色	3～3.5	28.5～33.3	70～95
香椿	10～11月，果实由绿色转为黄褐色	种子略扁平，上端具矩圆形薄翅，翅长1～1.2cm。种子一端成弧形，另一端渐尖，长5～7mm。红褐色	10～15	6.6～10	70～90

果实的采收调制和种子贮藏　红椿和香椿的蒴果幼时绿色，成熟后褐色，开裂，带翅的种子随即飞散飘落，难以收集。应在果实表现出成熟特征（请参见表2）而尚未开裂前连同果梗摘下，摊晾，待果壳张开后抖动果梗脱粒，筛去果壳、果轴后，搓揉去翅，扬去杂质，但通常并不去翅。鲜果出种率4%～6%，净度可达90%以上，精细筛选可达100%。这2个树种的种子形态特征和种子质量数据已列于表2。种子含水量8%～13%。种子脱粒、净种、晾干后一般用袋装贮于通风干燥处，在常温下作越冬贮藏。在干燥、密闭、低温（0～5℃）条件下，发芽能力可保持2年。据张德纯（1993）试验，香椿种子充分干燥后密封在塑料瓶中，保存在－17℃条件下，1年后种子发芽率仍可达80%。

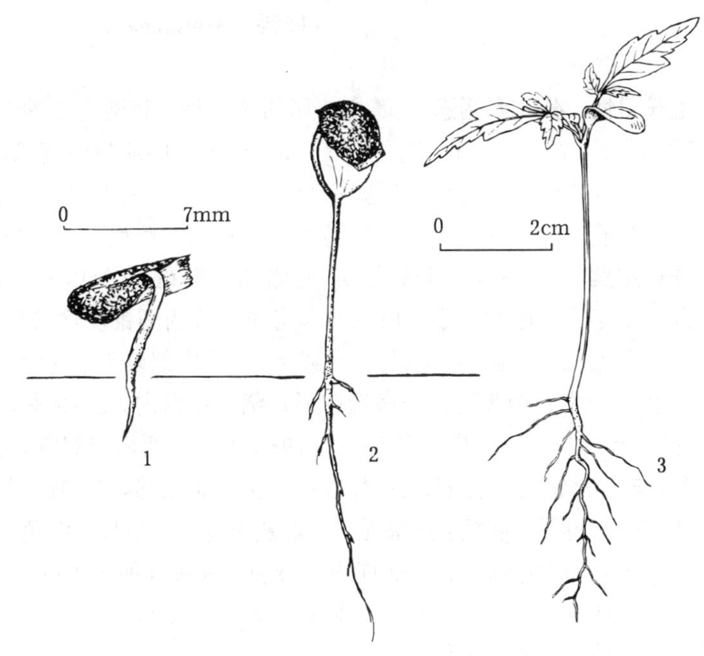

图2　香椿种子的萌发和幼苗生长情况

1. 下胚轴延伸，种壳出土　2. 子叶露出　3. 初生叶为对生的三小叶复叶

（1、2. 田恒德绘，3. 张世经仿《主要树木种苗图谱》）

发芽和播种　种子无休眠习性，发芽迅速。据陈幼生（1985）测定，在22℃、25℃和28℃的恒温下，香椿种子的绝对发芽率都可以达到100%；其中25℃

为最适温度，平均发芽日数为 9.7 天。22℃和 28℃下的发芽进程明显延长，平均发芽日数分别为 12.5 天和 14.1 天；在 18℃的温度条件下绝对发芽率下降到 56%，平均发芽日数延长到 18.3 天；在 15℃的温度下则难以发芽；但如置于 15℃或 15℃以下一段时间再转入 25℃，则发芽会相当迅速、整齐；在 25℃/15℃的昼夜变温和每天 10 小时光照条件下发芽更快，置床 3 天开始发芽，5 天达到高峰，10 天发芽基本结束。张德纯（1993）认为香椿萌发的最适温度为 17～22℃。香椿种子发芽率通常为 70%～90%，精选种子发芽率可以接近 100%。新采收的红椿种子发芽率可达 98%。

春播。播前种子无需处理。条播或撒播。条距 25～30cm。撒播时，红椿每平方米约播 6g。条播时香椿每平方米约播 10g。播后覆土，盖草，并经常保持土壤湿润。当地面温度白天达到 20℃时，播后 10 天开始发芽出土，再经 10 天大多数种壳脱落，子叶张开。1 年生红椿苗高约 60cm，香椿 80～120cm，可以出圃。出土萌发。子叶 2，长卵形（红椿），宽椭圆形（香椿）。初生叶为对生的 3 小叶复叶（图 2），以后互生，再以后为 5 小叶羽状复叶。红椿幼苗小叶边缘锯齿比香椿少。香椿也可根插繁殖或用根蘖苗繁殖，起苗后留在圃地的残根也可培育新苗。

（吴琼美）

鹧鸪花（海木）

Trichilia connaroides（W. et A.）Bentv. f. *glabra* Bentv.

（楝科　Meliaceae）

生长习性、分布和用途　鹧鸪花属约 300 种，我国有 2 种 1 变种 1 变型，本文描述 1 变型。落叶乔木，高 5～10m，胸径约 12cm。适生于中亚热带南缘以至北热带的低山丘陵地区，对水肥条件要求较高。可耐一般霜冻及短期－3℃左右低温。分布于桂、滇、黔。南亚、东南亚亦产。材质轻软适中，纹理直，易加工，抗虫，耐腐，宜作家具及胶合板等。

开花结实　6～8 年生开始结实，正常结实期在 12 年以后。结实大小年现象较明显，间隔期通常为 1 年。花两性，小，白色。伞房花序式的圆锥花序腋生。萼钟状，4～5 裂，覆瓦状排列。花瓣 4～5，矩圆形。雄蕊管 8 或 10 裂至中部以下。花丝长短相间，顶端有 2 齿。花药 8 或 10 枚，着生 2 裂齿间。子房近球形，藏于花盘内，2～3 室，每室有并生 2 胚珠。花柱与子房近等长，柱头 2～3 齿。据 1978～1983 年在广西南宁观察，4 月上旬始花，4 月中下旬为盛花期，5 月上旬为末花期；10 月中旬果实开始成熟，10 月下旬为果熟盛期。

蒴果，近球形，有柄，光滑无毛，未熟时青色，熟时黄红色或粉红色，长约 1.2～1.8cm，径 1～1.6cm。果实熟后约 1～2 周内 2 瓣裂。每果有种子 1 粒。种子扁圆形，无翅，具白色假种皮，种皮栗褐色，径 8～14mm，高 6～9mm。无胚乳。子叶厚，并生，平卧，胚根向上，藏于子叶间。种子的外形和解剖构造见图 1。

果实的采收调制和种子贮藏　果实成熟盛期用采种钩刀钩断果穗，或上树用手采摘。采回的果实在室内摊放，开裂后抖出种子，晾干，去除杂质，即得纯净种子。鲜果的出种率为 30%～40%。净度一般为 95%，变动范围为 92%～98%。种子含水量约 30%，千粒重约 500g，每千克有纯净种子 1 600～2 500粒。种子忌失水，不能曝晒，宜随采随播。贮藏或运输均需混以湿沙或湿润锯屑，贮藏期以不超过 5 个月为宜。

发芽和播种　种子有休眠习性。发芽时日均温需在 15℃以上。1978 年，广西林业科学研究所用当年新采种子在室外沙床作过发芽测定：11 月 7 日播种，当时气温为 18℃左右：播后 65 天（翌年 1 月 10 日）方始发芽，1 月 22 日为发芽盛期，2 月 22 日发芽终止；从发芽开始至发芽高峰日的 12 天内，发芽百分数为 40%；从播种至发芽终止，107 天的发芽率为 60%。留土萌发。胚根萌发后约 10 天胚芽出土，出土后 4 天左右发出初生叶。

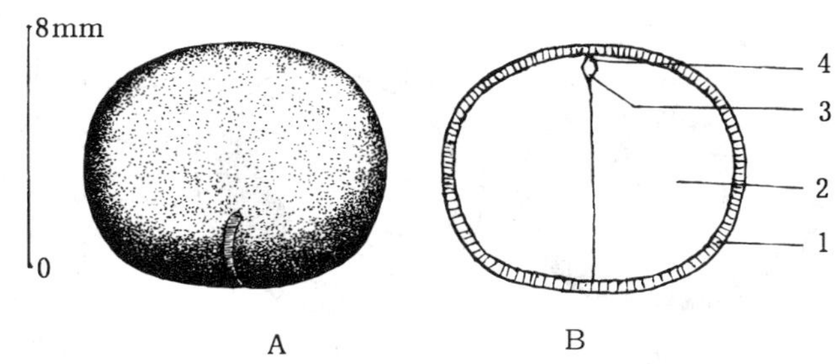

图 1　鹧鸪花种子外形（A）及其纵切面（B）
1. 种皮　2. 子叶　3. 胚芽　4. 胚根
（黄应钦 、田恒德绘）

点播。每平方米播种 40～50g，覆土 1.5～2cm。由于发芽延续时间长，可将种子密播于沙床，发芽后 1 月左右，按 30cm×10cm 的行株距将芽苗移至圃地继续培育。1 年生苗出圃。每 $666m^2$ 产苗约 1 万株。

（张声燕）

田林细子龙

Amesiodendron tienlinense H. S. Lo

（无患子科　Sapindaceae）

生长习性、分布和用途　细子龙属 3 种，我国均有，本文描述 1 种。常绿乔木，高 12～20m，胸径 50cm。喜生于高温高湿的静风低谷，忌冰雪。适生于石灰岩地区，也能在酸性黄壤或红壤上生长。根系发达，在石缝中也能扎根。分布于桂、黔。分布区域狭窄而又砍伐过度，残存种群极少，《中国植物红皮书》列为渐危种。材质坚重，不受虫蛀，为优良用材，用于建筑及制作高级家具。种仁含淀粉，可作饲料并酿酒；又含油 43%，供工业用。

开花结实　约 12 年生开始开花结实，正常结实年龄在 20 年生以后。结实大小年间隔期为 1 年，较明显。花单性，同株。聚伞圆锥花序腋生于枝上部或丛生于小枝顶端。萼浅杯状，5 深裂。花瓣 5，稀 6 或 7，白色，基部内侧有鳞片，鳞片顶端 2 裂，反折，密被长毛。雄花的雄蕊 8，花丝密被硬毛，花药肥大，被毛。雌花的子房上位，陀螺形，密被硬毛，3 室，每室 1 胚珠。据广西南宁 1987～1989 年观察，7 月上旬为始花期，7 月中旬为盛花期，7 月下旬为末花期；9 月中旬果实开始成熟，10 月上旬为成熟盛期，10 月中下旬仍有少量果熟。

蒴果，深裂为 3 果爿，仅 1 个稀 2 个发育。蒴果的发育果瓣近球形，成熟时由青色转变为褐色。果壳木质，较薄，无瘤状凸起。果径 1.3～1.8cm。熟时果实暂不脱落，沿室背开裂，2～3 天后种子脱落。空果壳在树上存留 1～2 个月后陆续脱落。每果有种子 1 粒。种子黄褐色，

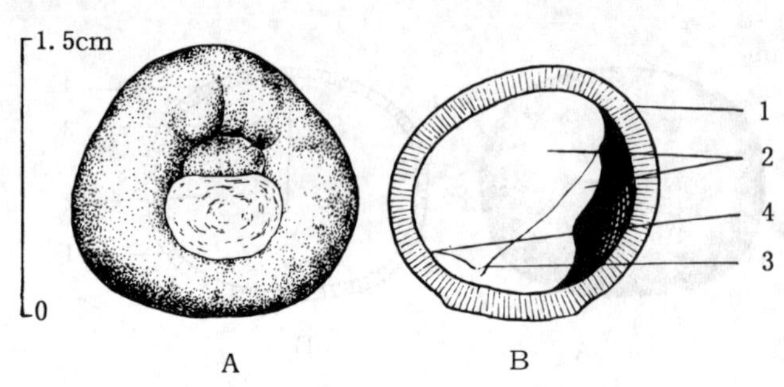

图1　田林细子龙种子外形（A）及其纵切面（B）
1. 种皮　2. 子叶　3. 胚芽　4. 胚根
（黄应钦绘）

扁球形，径1.2～1.6cm，高1.2～1.5cm。种皮有光泽，基部有横椭圆形大种脐。无胚乳。胚弯拱，侧生，子叶肥大，叠生（图1）。

果实的采收调制和种子贮藏　成熟果实所含养分丰富，易遭鼠类盗食。种子落地后又易受虫蛀。因此应在果实部分开裂或将开裂而尚开未落地时，用采种刀带果穗采下，或敲打枝条果穗，果实脱落后在地面捡拾。采回的果实按成熟程度分开，果壳尚带青色的可堆放数日。已开裂和接近开裂的，可摊放于室内，稍阴干后果壳开裂，即可取得种子。种子的净度可达98%～100%，千粒重1 880（1 600～2 000）g，每千克有纯净种子530（500～630）粒。调制出来的种子含水量40%～50%，忌脱水，不能日晒。裸露种子存放1周便会失去发芽能力。如需运输，可以在采收之后直接运输果实。如果需要运输调制出来的种子，则应将种子混以湿沙。贮藏时宜混湿沙分层堆积，贮藏期以半年左右为限。

发芽和播种　种子无休眠习性或有短期休眠。发芽时日均温宜在18℃以上。1979年11月28日，广西林业科学研究所在室外沙床作过发芽测定：由于播后气温转低，直至次年3月12日种子方始发芽；3月下旬进入发芽盛期，4月27日发芽结束；从开始发芽至发芽高峰的24天中，发芽百分数为52%；从播种至发芽终止，以150天计，发芽率为81%。留土萌发。种子萌发后约1周上胚轴出土，出土后5天左右发出初生叶。

条播。每平方米播种量150～200g。播前浸种3～5小时，播后覆土2～3cm。1年生苗出圃。

（张声燕）

滨　木　患

Arytera littoralis Bl.

（无患子科　Sapindaceae）

生长习性、分布和用途　滨木患属约25种，我国只有本文描述的1种。常绿小乔木或灌木，高5～13（20）m，胸径约20cm。分布于琼、粤、桂、滇。东南亚至新几内亚岛广有分布。材质坚韧，宜制农具、家具等。种子可作油料。

开花结实　5～6年生开始开花结实，正常结实期在15年生以后，大小年现象不明显。花单性，同株或异株。聚伞圆锥花序腋生。萼杯状，5裂。花瓣5，与萼近等长，内侧基部有小鳞片2。雄花的雄蕊8。花丝线状，常被毛，在雌花中较短。雌花子房上位，倒卵形，2（3）室，每室1胚珠。花柱线形，先端2浅裂。据广西南宁1986～1988年观察，6月下旬花始开，

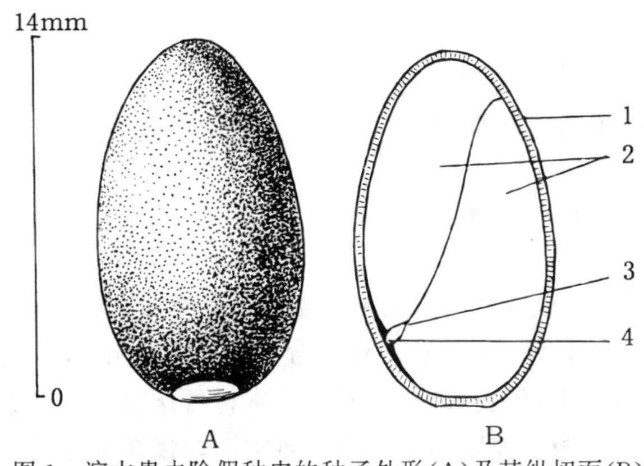

图 1　滨木患去除假种皮的种子外形(A)及其纵切面(B)
1. 种皮　2. 子叶　3. 胚芽　4. 胚根
(黄应钦绘)

7 月上旬为盛花期，7 月下旬为末花期；11 月上旬果实开始成熟，11 月下旬为果熟盛期，12 月上旬为果熟末期，翌年 3 月果实脱落殆尽。

蒴果，深裂为 2（稀 3）果瓣，通常仅有 1（或 2）个发育。发育果瓣椭圆形，成熟时室背开裂。果皮革质，橙红色至黑色。果长 1.2～1.7cm，宽 7～9mm。种子卵状椭圆形，长1～1.4cm。种皮薄壳质，枣红色，为透明的假种皮包裹。无胚乳。子叶肥大，叠生，胚根位于种子 1 侧。种子的外观和内部构造见图 1。

果实的采收调制和种子贮藏　果熟盛期用手或高枝剪带果穗采下。在室内摊放3～5 天后果实自行开裂。脱落出来的种子置水中洗净假种皮，即得纯净种子。鲜果的出种率为20%～30%。种子忌失水，不能曝晒或干藏。种子的净度一般为 95%。千粒重 1 000（900～1 100）g，每千克有纯净种子 1 000（900～1 100)粒。运输或贮藏均需混以湿沙，混沙贮藏期 3 个月左右。

发芽和播种　种子无休眠习性。发芽时日均温宜在 14℃以上。1987 年 12 月 28 日，广西林业科学研究所在室外沙床作过发芽测定：播后 19 天（1988 年 1 月 15 日）开始发芽，1 月 28 日发芽终止；从发芽之日起算的 8 天中，发芽百分数为 53%；从播种至发芽终止，以 31 天计，发芽率为 80%。留土萌发。胚根萌发后约 6 天胚芽出土，出土后约 7 天初生叶展现。滨木患种子的萌发和幼苗初期生长情况见图 2。

条播。每平方米播种 100～150g，覆土 1cm。1 年生苗出圃。

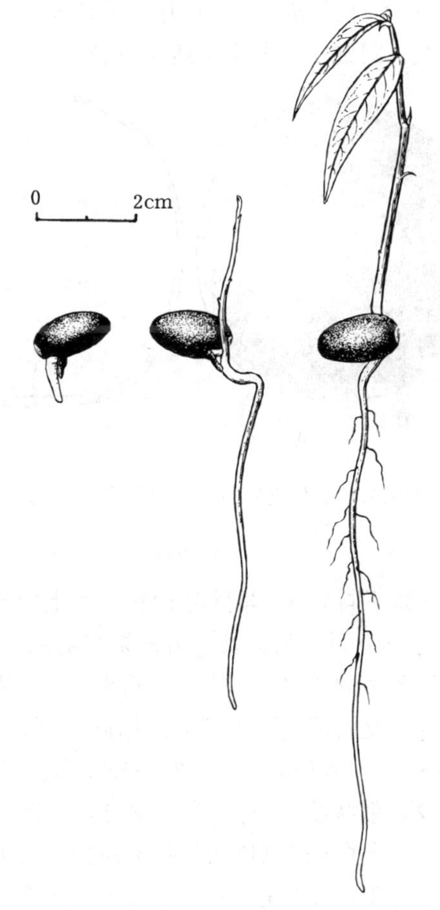

图 2　滨木患种子萌发后第 1、8、15 天幼苗的生长情况
(黄应钦绘)

（张声燕）

龙 眼

Dimocarpus longan Lour.

（无患子科 Sapindaceae）

生长习性、分布和用途 龙眼属约20种，我国4种，本文描述1种。常绿乔木，高5～20m。不耐寒。根系发达，有菌根菌共生。分布于闽、粤、琼、桂、台、黔、滇、川。印度、马来西亚、越南和菲律宾也有。栽培品种繁多。果可生食或制桂圆干；木材可作家具，供雕刻；树皮和根含鞣酸，可染渔网；种子可酿酒。野生龙眼是珍贵的种质资源，仅存于海南西南部低山丘陵季雨林中，现存数量日渐稀少，已被《中国植物红皮书》列为渐危种。

开花结实 实生树8～10年生开始结实，正常结实年龄在20年生以后。无性繁殖的栽培品种3～5年生开始结实。结实有大小年现象，丰年间隔期1～2年。花杂性，同株，多为单性，稀两性。圆锥状聚伞花序顶生或腋生。花萼5，深裂。花瓣5，直径4～5mm。雄花黄白色，雄蕊7～10枚，花丝长6～8mm，药黄色，纵裂，雌蕊退化，仅留一红色的小突起。雌花深黄色，子房2～3室，花柱合并，柱头叉裂，具退化雄蕊7～8枚。据在福建南部观测：4月中旬～6月上旬开花，但花期随着气候、树势、品种等许多因素而不同。较暖的地区及年份开花早；树势壮，抽穗早的植株开花也早；品种不同，花期更不一致。果实成熟期亦随品种、气候不同而异：早熟品种在8月上中旬，中熟品种在8月下旬～9月上旬，晚熟品种在9月中旬～10月中旬。果实为核果状荔果。关于龙眼果实的类型，学者间说法不一。有的认为是特殊类型的浆果，并特称为荔果，有的则写作浆果状。龙眼的果实纯由心皮发育而成，成熟后果皮完全石质化而不开裂。据此，严楚江（1964）认为龙眼的果实是坚果。果扁圆形或球形，高2.1～2.9cm，径2.3～3cm。果皮黄褐色，幼时有瘤状突起，熟时近平滑，具不同的龟状纹及放射状纹。种子1粒，扁圆形至圆形，径1.1～1.5cm，高1～1.2cm，棕黑色至棕红色，光滑。种子为白色透明状的肉质假种皮所包围。但Van der Pijl（1957）认为，这是一层肉质化的种皮，而不是真正来源于珠柄的假种皮。种脐白色，稍突起。无胚乳，子叶2枚，含淀粉和皂苷。龙眼种子的形态见图1。

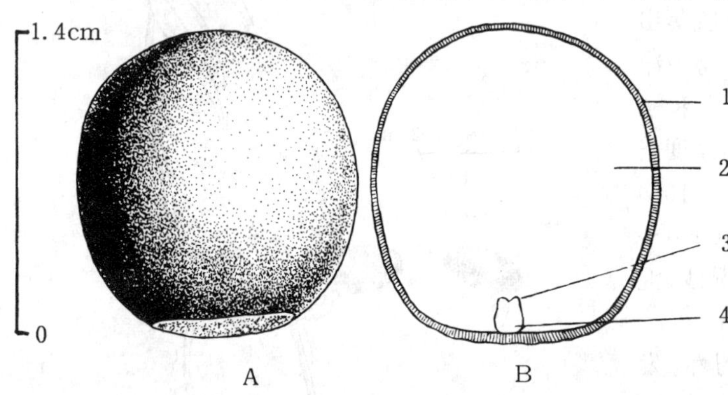

图1 龙眼去除假种皮后的种子的外形（A）及其纵切面（B）
1. 种皮 2. 子叶 3. 胚芽 4. 胚根
（黄应钦绘）

果实的采收调制和种子贮藏 当果壳变薄而光滑，果肉白色透明而味甜，种皮转呈棕黑色至棕红色时，用枝剪采收。采摘时应在果穗基部折断，断口要整齐，防止撕裂而影响以后抽梢。果实食用后得到种子或剥离取出种子，或利用罐头厂加工废弃的种子，洗净即为播种材料。100kg鲜果可出种子20～25kg，含水量40%以上。千粒重1 190（1 000～1 400）g，每千克有种子840（710～1 000）粒。种子寿命短，失水会丧失发芽能力，属于顽拗型，不能久藏。在荫凉处混沙湿藏可保存15～30天。彭业芳和傅家瑞（1994）讨论过发育过程中龙眼种子的脱水忍耐力。

发芽和播种　种子无休眠习性。播后第四天开始发芽，第十天发芽结束。在20～30℃的室温下发芽率90%～100%。一般在8～9月份随采随播。条播，每666m²播种75～140kg，覆土1cm并盖草。留土萌发。胚根和下胚轴伸出后上胚轴出土。初生叶具2小叶，对生，紫红色，以后逐渐变为绿色。龙眼种子的萌发和幼苗初期形态见图2。秋播的幼苗，翌年春梢萌发前移植。移植后培育1～2年可作砧木供嫁接，或培育3年生实生苗供园林绿化。生产上多用嫁接苗。

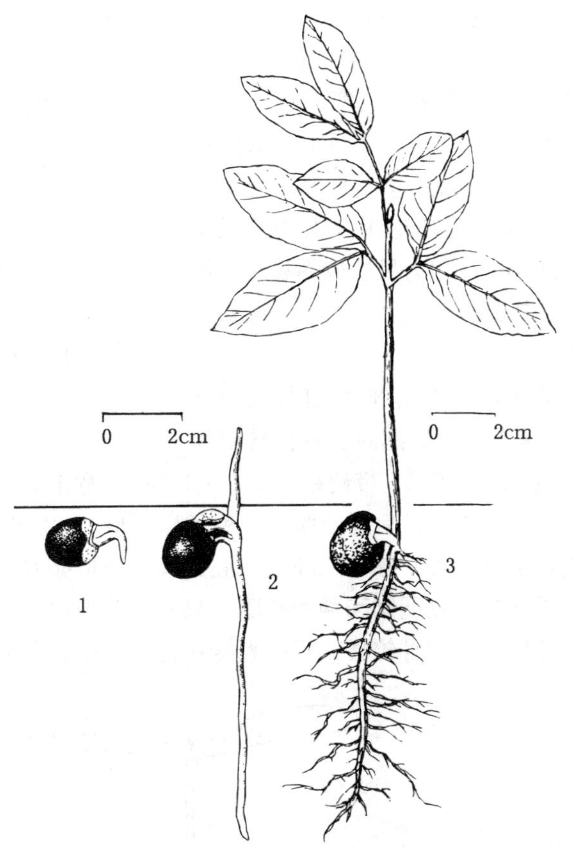

图2　龙眼种子萌发和幼苗生长情况
1. 胚根伸出　2. 上胚轴延伸，子叶留土　3. 初生叶出现
（黄应钦、田恒德绘）

（李玉科）

车桑子（坡柳）

Dodonaea viscosa（L.）Jacq.

（无患子科　Sapindaceae）

生长习性、分布和用途　车桑子（坡柳）属约60种，我国只有本文描述的1种。常绿灌木，高1～3m。适应性较强，能耐庇荫，裸露荒坡亦能生长，喜酸性土。热带树种，产于我国东南、华南、以及西南金沙江河谷。全世界热带、亚热带地区广泛分布。覆盖荒山和保持水土的重要灌木。种子油可制皂，根及叶供药用。

开花结实　4～5年生开始开花结实，正常结实年龄在8年生以后。结实无大小年现象，连年结实量甚多。花单性，雌雄异株。圆锥花序、总状花序或伞房花序，顶生或腋生。花小，绿黄色，约3mm，辐射对称。萼片4，无花瓣。雄花的雄蕊7或8。花丝短，长不及1mm。花药长圆形，长2.5mm，药隔突出。雌花子房上位，椭圆形，外有胶状粘液，2或3室，每室

2 胚珠。花柱长 4～6mm，比子房长 2～3 倍，柱头 2～3 裂。据 1985～1987 年在广西南宁观察，每年开花结实 2 次：第 1 次在 4 月上旬形成花序，5 月上旬始花，5 月中旬为盛花期，5 月下旬为末花期；7 月中旬果实开始成熟，下旬为果实成熟盛期。第 2 次花期始于 8 月上旬，11 月果实成熟。

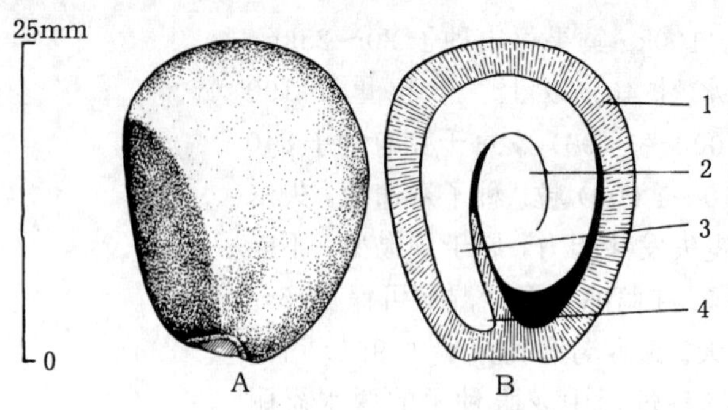

图 1 车桑子种子外形（A）及其纵切面（B）
1. 种皮 2. 子叶 3. 胚轴 4. 胚根
（黄应钦绘）

蒴果，果瓣 2～3，背延伸成一膜质翅，具短柄，成熟时由青色转变为黄褐色，近似蝶形。带翅果高 1.2～2.2cm，宽 1.5～2.5cm，成熟后暂不脱落，在树上约可维持 1 个月左右，以后逐渐脱落。种子扁圆形，宽2.5～3.5mm，厚 1.4～1.7mm。种皮膜质或纸质，黑色。种脐厚。无胚乳，胚旋卷，子叶条形。种子的外形和内部结构见图 1。

果实的采收调制和种子贮藏 果实充分成熟后即宜采收，不宜在树上挂果过久，以免降低发芽能力。树身较矮，可以直接在树上采摘。带翅果易受风吹失，采后应装入袋中。运回后放入室内摊放，不宜在室外摊放或曝晒。摊放时要保持干爽，防止水湿，免遭霉烂。果瓣易开裂，反复搓揉，果瓣即可全部开裂，脱出种子。扬去果瓣等杂质即得种子。果实的出种率为 35%～15%，净度可达 98%或更高。千粒重 28（25～30）g，每千克有纯净种子 3.5（3.3～4）万粒。种子含水量约 12%。调制出来的种子不宜曝晒，亦不耐久藏。7 月成熟的种子贮藏期不宜超过 1 个月。11 月成熟的种子可以贮藏越冬，至次年 3 月播种。贮藏时宜带果瓣装袋挂于室内通风处，不宜裸露存放。

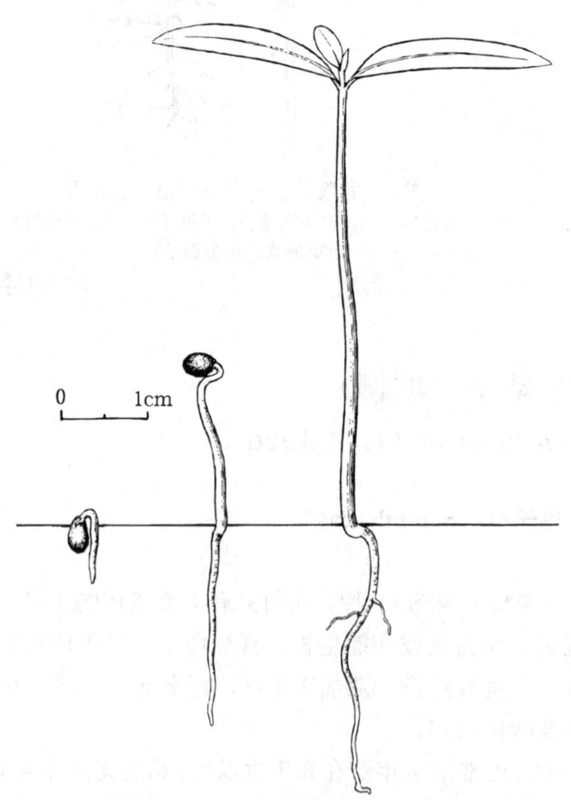

图 2 车桑子种子萌发后第 1、3、13 天的幼苗生长情况
（黄应钦绘）

发芽和播种 种子无休眠习性。发芽时日均温宜在 20℃以上。1978 年，广西林业科学研究所在室内培养皿内用滤纸作基质，进行过发芽测

定，置床时日均温为 30℃：8 月 7 日置床，11 日开始萌发，9 月 4 日发芽结束；从种子萌发至发芽高峰日的 14 天中，发芽百分数为 45%；从播种至发芽终止的 28 天中，发芽率为 56%。场圃出苗率一般在 40%左右。出土萌发。胚根萌发后约 1～2 天子叶伸出，再过 12 天初生叶展现。车桑子种子的萌发和幼苗初期生长情况见图 2。

条播或撒播。每平方米播种 5～8g。生产上多用带翅果作为播种材料，趁无风天播种，每平方米约播带翅果 15～20g。覆土约 0.5cm。1 年生苗出圃。

（张声燕）

伞　花　木

Eurycorymbus cavaleriei（Lévl.）Rehd. et Hand.-Mazz.

（无患子科　Sapindaceae）

生长习性、分布和用途　伞花木属只有本文描述的 1 种，我国特有。落叶小乔木或乔木，高 6～20m。能耐−8℃的极端低温。适生于肥沃湿润的中性至酸性土。星散分布于黔、滇、桂、粤、湘、鄂、赣、闽、台。涵养水源效果好。种子可榨油。木材可制家具。资源已日渐稀少，《中国植物红皮书》列为稀有种。

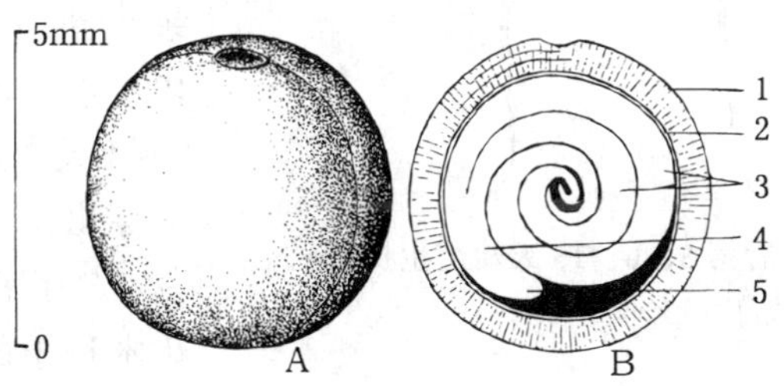

图 1　伞花木种子外形（A）及其纵切面（B）
1. 外种皮　2. 内种皮　3. 子叶　4. 胚轴　5. 胚根
（黄应钦绘）

开花结实　5～6 年生开始开花结实，8 年生以后进入正常结实。花单性异株。伞房花序状的圆锥花序顶生，长 10～30cm，宽 15～18cm，密被灰黄色细柔毛。苞片线状披针形，被微柔毛。花小，白色，芳香。雄花柄极纤细，长约 2mm，雌花柄稍粗壮，长约 4mm，均被灰色小茸毛。萼片 5。花瓣 5，长椭圆状匙形，长约 2mm，外被长柔毛。花盘环状，无毛。雄蕊 8（7），花丝细长，长约 4mm。雌花子房上位，3（4）浅裂，3（4）室，每室有胚珠 2 颗。花柱着生于子房裂片间，线形，直立。据贵州贵阳观察，3 月下旬叶芽开展，5 月上旬花蕾出现；5 月下旬始花，6 月上旬为盛花期，6 月中旬为末花期；6 月中旬幼果形成，10 月上旬果实成熟；熟时室背开裂，10 月中旬种子散落。

蒴果，椭圆形或球形，深 3 裂，通常仅 1 个，稀 2 个果瓣发育。果长 7～8mm，径 5～6mm，密被灰褐色短毛。每一发育果瓣中 1 粒种子。种子球形，种皮黑色有光亮，径 4～5mm。种子中缝有微突起白痕。种脐长椭圆形，锈红色。种皮厚。无胚乳。胚旋卷。伞花木种子的外观及其内部结构见图1。

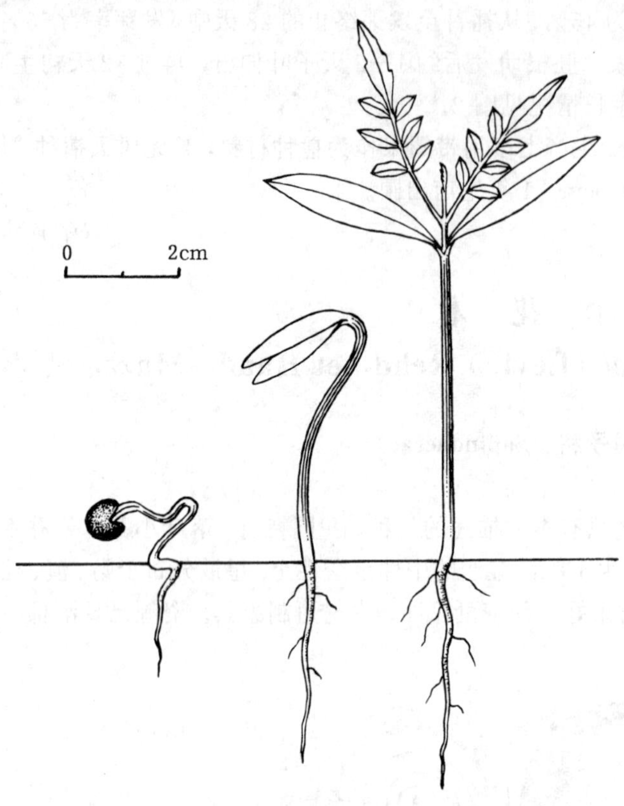

图2　伞花木种子萌发后第3、6、15天幼苗生长情况
（黄应钦绘）

果实的采收调制和种子贮藏　成熟的果实开裂后，种子极易散落，应及时采集。当种子变黑，少数果实呈现开裂时，用手或用高枝剪将总果梗截断，采下果穗，置于筛中晾干。待果瓣完全开裂后搓揉，除去果壳及杂物即得种子。鲜果出种率约65%。种子净度约96%。千粒重约61g，每千克有纯净种子16 400粒。种子可以干藏，但忌在裂日下曝晒。短期存放可装入布袋，悬挂在室内阴凉通风处。超过1月的需混干沙贮藏。超过半年以上则需置0～5℃低温下贮藏。

发芽播种　种子无休眠现象。贵州林业科学研究所的育苗经验是，经过沙藏的种子播后20天左右即可发芽出土。干藏的种子播前用始温40℃水浸种，自然冷却后继续浸种2～3天，这样的种子播后约1个月发芽出土。发芽期可持续1个月左右，发芽率80%左右。出土萌发。伞花木种子的发芽及幼苗初期生长情况见图2。

条播。每平方米约播7～10g，覆土0.5～1cm。幼苗期用喷雾器淋水。1年生苗出圃。

（钟梅芳）

栾　树　属

Koelreuteria Laxm.

（无患子科　Sapindaceae）

生长习性、分布和用途　本属有4种，我国3种，本文描述1种1变种。落叶乔木，高可达20m。全缘叶栾树在深厚肥沃、湿润疏松的土壤上生长迅速。栾树耐干旱、瘠薄，也耐盐渍性土。深根性，萌芽力强。它们的名称、分布和用途见表1。

开花结实　花杂性同株稀异株，黄色，大型聚伞圆锥花序顶生。花两侧对称。萼片5稀4。花瓣4稀5，略不等大。花盘厚，偏于一边，上端有圆裂齿。雄蕊8，生于花盘之内，花丝分离，被长柔毛。子房上位，3室，每室2胚珠。花柱短或长，柱头3裂或近全缘。蒴果中空，膨胀如囊，室背开裂为3果瓣。果瓣膜质，有网状脉纹。全缘叶栾树花序长30～45cm，花期8～9月。果10～11月成熟，椭圆形或近球形，顶端钝圆，长4～7cm，径3.5～5cm，淡紫红色，老熟时褐色。每室种子2或1，黑褐色，球形，径4～6mm。栾树花序长25～40cm，花色淡黄，中心带紫色，花期6～8月，果（8）9～10月成熟。果圆锥形或卵形，顶端渐尖，长4～6cm，径2～3.5cm，幼果红色，熟时黄色。每室1种子，黑色，球形，径6～8mm。种子无胚乳，胚旋卷，胚根稍长。开花结实物候见表2。种子形态见图1。

表1　栾树属树种的名称、分布和用途

中　名	学　名	分　布	用　途	供　稿
全缘叶栾树（黄山栾树）	*K. bipinnata* Franch var. *integrifoliola* (Merr.) T. Chen	华东、中南	观赏、材用	801
栾　树	*K. paniculata* Laxm.	东北、华北、华东、西南及陕、甘。朝鲜半岛、日本	观赏、材用、染料、油脂	301

表2　栾树属树种的开花结实物候期

树　种	观察地点	观察年份（年）	开花		果实成熟期	种子散落期
			始　期	末　期		
全缘叶栾树	南京	1981，1982	9月上旬	9月下旬	10～11月	冬季
栾树	北京妙峰山	1953，1954	6月上旬	6月中旬	8～10月	10～11月

果实的采集调制和种子贮藏　全缘叶栾树10～11月果实逐渐转变为红褐色时即可采摘。采后日晒1～2天，置于袋内揉搓脱粒。栾树9～10月间逐渐成熟开裂（8月间有少数蒴果变色或开裂，系虫害所致，应注意区分），在成熟蒴果开裂前采摘，脱粒净种，水选除去空粒。种子可以密封干燥贮藏。一般情况罐藏的全缘叶栾树种子，发芽能力可以保存10年左右。出种率和种子质量的数据见表3。

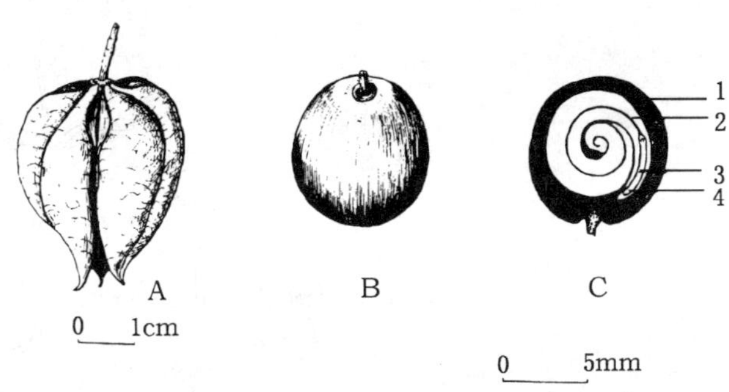

图1　栾树蒴果外形（A）、种子外形（B）及其纵切面（C）
1. 种皮　2. 子叶　3. 胚轴　4. 胚根
（孟玲绘）

表3　栾树属树种的出种率和种子质量

树　种	出种率（%）	净度（%）	千粒重（g）	每千克纯净种子粒数（万粒）
全缘叶栾树	20	—	57～100	1～1.8
栾树	66	97	148	0.57～0.77

发芽和播种 全缘叶栾树有的种批有休眠，需要低温层积才能充分发芽；有的种批种子无休眠现象，以白天30℃(8小时)，夜间20℃(16小时）变温发芽较好。栾树种皮有由石细胞构成的角质层，阻碍吸收水分，且种皮含有脱落酸（ABA)，一般要求浓硫酸处理1～3小时，用清水浸泡2天，再层积3～4个月才能较好地发芽（李近雨，1987)。在露天深1m地下(0～5℃）混沙层积132天后播种，场圃发芽率达93%；对照（温水浸种）发芽率只有12%。某次室内发芽测定结果见表4。

表4 栾树属树种种子室内发芽测定结果举例

树 种	预处理	温 度（℃）		发芽率（%）
		有光时	无光时	
全缘叶栾树	—	30	20	90
栾树	硫酸浸种1小时后层积90天	30	20	76

全缘叶栾树可以秋播或低温层积5～6个月后春播。每平方播种10～30g。条播，行距20cm，上盖木屑或稻草。出土萌发。子叶2。初生叶为羽状复叶，近对生。播种后要防鼠害。1年生苗高1.0～1.5m。栾树秋播或春播。春季播种的需层积100～130天。条播，覆土厚1～1.5cm。每平方米播种20～38g。出土萌发。子叶2。1年生苗高40cm，地际直径0.7cm。栾树种子的萌发和幼苗初期生长情况见图2。

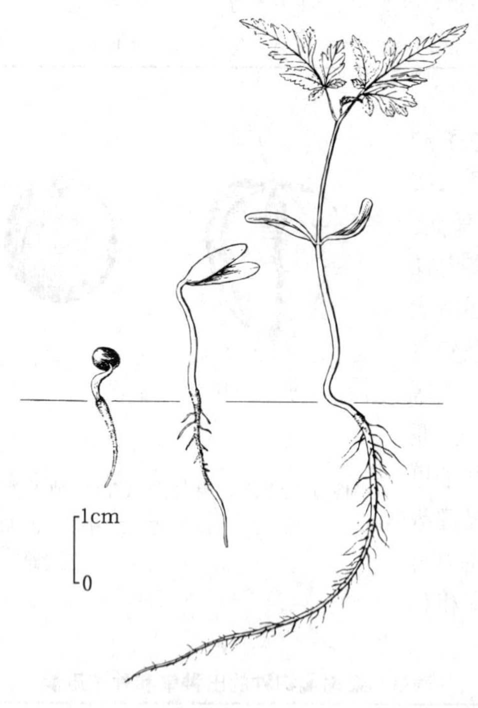

图2 栾树种子萌发出土后第1、3、7天的幼苗生长情况

（孟玲绘）

（陶章安）

荔　　枝
Litchi chinensis Sonn.

(无患子科　Sapindaceae)

生长习性、分布和用途　荔枝属有2种，菲律宾和我国各1种，本文描述我国的这一种。我国的荔枝为常绿大乔木，胸径可达1.3m以上。适生于温暖湿润的气候环境和肥沃深厚的酸性至石灰性土壤。根系发达，有菌根菌共生。抗风力强。分布于闽、台、粤、琼、桂、黔、滇、川。东南亚及太平洋岛屿、澳大利亚、南非和美洲有引种。栽培品种有160多个。果为著名佳果，可食，也可制成罐头和果干。种子含淀粉37%，可酿酒、入药。木材坚韧致密，可用于船舶、农具和家具。木材和根含鞣酸，可染渔网。花为优良蜜源。产于海南岛的野生荔枝(*L. chinensis* var. *euspontanea* Hsue)木材名贵，也是重要的种质资源，因数量日减稀少，在《中国植物红皮书》中列为渐危种。

开花结实　8～10年生开始结实，正常结实年龄在20年生以后。无性繁殖的植株2～4年生开始结实。丰年间隔期1～2年。花杂性，同株。聚伞圆锥花序顶生。花小，绿白色。萼杯状，4或5浅裂。无花瓣。花盘肉质。雄蕊通常8。子房倒心形，2裂稀3裂，2～3室，每室1胚珠。花柱着生子房裂片间，柱头2或3裂。

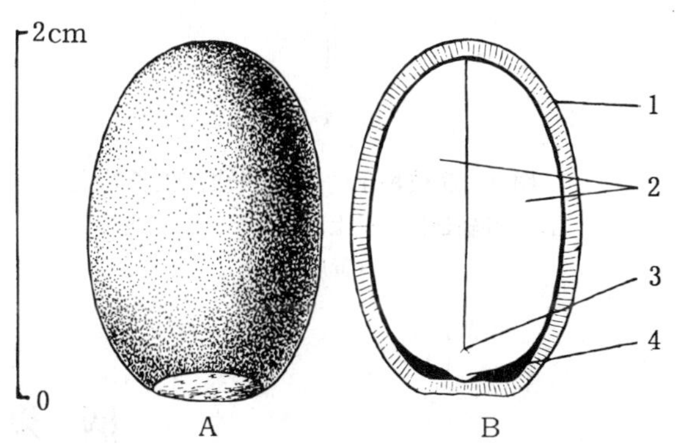

图1　荔枝去除假种皮后的种子外形(A)及其纵切面(B)
1. 种皮　2. 子叶　3. 胚芽　4. 胚根
(黄应钦绘)

花果期因纬度、品种不同而异。

"乌叶"品种在福建诏安比较北的漳州早1周开花，更北的泉州、莆田又比漳州迟1周开花。4～5月开花，6～7月果熟。早熟品种"三月红"5月上旬成熟，晚熟品种"下番枝"迟至8月上旬成熟。果为坚果(严楚江，1964；并请参见本书第848页龙眼属一文)。果深裂为2或3果瓣，通常仅1个稀2个发育，卵形或近球形，直径2～3.5cm。果皮未成熟时青绿色，成熟时暗红色或红色，具凸起的瘤点。每果有种子1粒，具肉质白色半透明的假种皮(请参见本书龙眼属一文)。种子长椭圆形，长1.6～2.4cm，宽1～1.7cm，厚0.7～1.2cm，棕褐色，具种柄，无胚乳，子叶2枚(图1)。

果实的采收调制和种子贮藏　充分成熟后采集。采收时从果枝与果穗交界的膨大处折断。果实食用后或食品加工后得到种子，洗净，稍晾干即可播种。种子忌失水，不能曝晒或裸露存放，属于顽拗型种子。彭业芳和傅家瑞(1994)研究过种子发育过程中脱水忍耐力的变化。夏季常温条件下种子很快干燥变色，丧失发芽能力，不宜久藏。远途运输需混苔藓或湿沙，用木箱或箩筐包装，可保持1～2周。将种子放在勤加更换的清水中可保持2～3周。俞旭平和方坚等人(1994)将从新鲜果实中剥出的发芽率为100%的荔枝种子，用海藻酸钙制成胶丸，

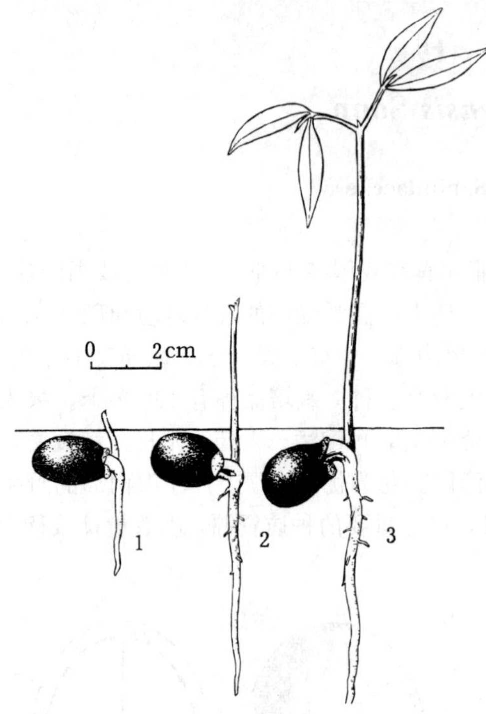

图 2　荔枝种子的萌发和幼苗生长情况
1. 下胚轴延伸　2. 上胚轴出土　3. 初生叶出现
（黄应钦绘）

在 10℃的环境中存放，80 天后，胶丸保存的种子发芽率可达 53.3%，对照种子的发芽率则在 40 天时已降为零。每百千克鲜果出种 8～12kg，较差的品种可出种 20kg 以上。种子含水量约为 48%。净度 96%～100%。千粒重 2 000～2 500g，每千克有种子 400～500 粒。每升种子 690g。

发芽和播种　种子无休眠习性。播后第 3 天开始发芽，第 10 天发芽结束，发芽率 82%，变幅为 75%～90%。宜随采随播。条播。每 666m^2 播种 150～200kg，覆土 1cm，并盖草保持土壤湿润。留土萌发。胚根和下胚轴伸出后 7～10 天上胚轴出土。不育的初生叶苞片状，早落。初生叶具 2 小叶，对生，紫红色，以后逐渐变为绿色。主根发达（图 2）。翌年春梢萌发前移植。实生苗的树形高大，多用于“四旁”绿化栽植。生产上通常培育高压苗定植。

（李玉科）

柄　果　木

Mischocarpus sundaicus Bl.

（无患子科　Sapindaceae）

生长习性、分布和用途　柄果木属约 12 种，我国 3 种，本文描述 1 种。常绿乔木，高达 17m，胸径 40cm。喜肥沃湿润的酸性土。通常生于滨海地区林中，内陆少见。分布于琼、桂。印度、马来西亚、菲律宾、新几内亚岛西至澳大利亚亦产。木材结构细，耐腐，色泽美，颇珍贵，供高级家具、装饰、工艺等用。

开花结实　约 10 年生开始开花结实，正常结实期在 20 年生以后。结实颇多，大小年现象不明显。花小，单性同株或异株，辐射对称。聚伞圆锥花序，近基部分枝，或不分枝而呈总状，腋生、侧生或近顶生。花萼杯状，5 裂。花瓣全部退化。花盘杯状，无毛。雄蕊 8，长 2～3mm，近无毛，着生花盘之内。子房上位，具柄，3 室，最后多退化为 1 室，每室 1 胚珠。柱头 3，外弯。据广西南宁 1987 年观测，10 月下旬为始花期，11 月上旬为盛花期，11 月下旬为末花期；翌年 3 月下旬果实开始成熟，4 月上旬为果熟盛期。蒴果，梨形，具子房柄，着生种子部分直径 7～10mm；未熟时黄红色，成熟后深红色，成熟后约 10 天果实开裂。每果有种子1粒，少有2粒。种子倒卵形、近圆形或半圆形，具紫色假种皮。种子长5～8mm，径4.5～6.5mm，种皮薄，壳质，枣红色。无胚乳。胚弯拱，充满种子，子叶叠生。

果实的采收调制和种子贮藏　果实成熟盛期直接采摘或用高枝剪带果穗采下，室内摊放1～3天。果穗上的果实大部分成熟时稍加敲打，震落果实，少数不脱落的果实可弃而不用。将果实摊放于室内通风处，开裂后搓去果壳，得出的种子装入布袋，置水中搓擦，洗去假种皮，即得纯净种子。以下关于出种率、净度和种子质量的数据是根据褐叶柄果木（*M. pentapetalus* (Roxb.) Radlk.）种子测得的。鲜果出种率约为30%，种子的净度一般90%～95%。千粒重180（150～200）g，每千克有纯净种子5 500（5 000～6 700）粒。种子含水量约20%，忌失水，不能日晒，运输或贮藏均需混以湿沙。宜随采随播。贮藏期一般为3～6个月。混沙贮藏至次年春播的种子，发芽率下降至10%左右。

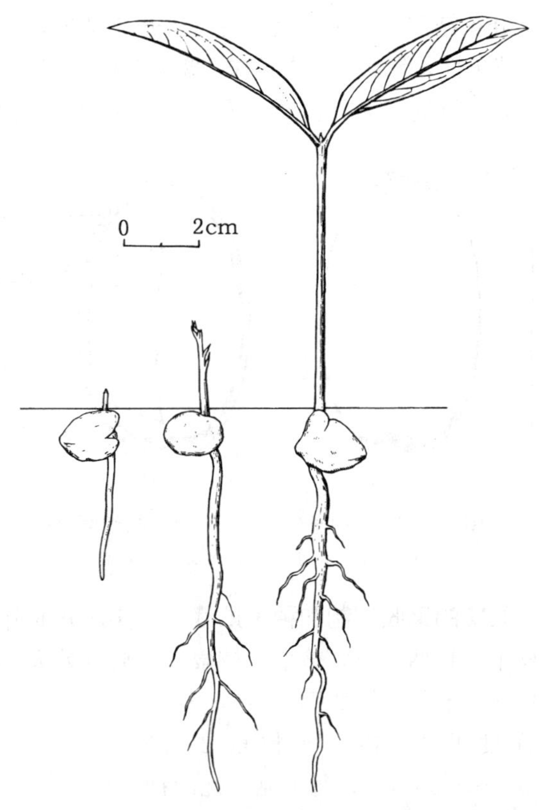

图1　褐叶柄果木种子萌发后第4、7、12天幼苗的生长情况
（黄应钦仿《热带亚热带主要树种采种育苗技术》）

发芽和播种　种子无休眠现象。发芽时日均温需在20℃以上。1980年1月，广西林业科学研究所用头年沙藏的种子，在室外沙床上作过发芽测定。播后3个半月即4月上旬开始发芽，5月下旬发芽终止，发芽率仅10%。1988年4月中旬用当年新采种子在室外沙床播种，播后20天开始发芽，发芽率约为70%。留土萌发，胚根萌发后约4天上胚轴呈直线状伸出，出土后7～8天展出初生叶。褐叶柄果木种子的萌发及幼苗初期生长情况见图1。

条播。整地要求细致。每平方米播种25～35g，覆土0.5cm。幼苗需适当遮荫。1年生苗可以出圃定植。

（张声燕）

红　毛　丹

Nephelium lappaceum L.

（无患子科　Sapindaceae）

生长习性、分布和用途　韶子属约38种，我国有3种，本文描述引入的1种。常绿乔木，高10～15m。抗风力强。土壤以深厚肥沃、排水良好的冲积土为宜。原产亚洲热带，现世界热带地区已广为引种。我国海南 20 世纪 50 年代引种栽培，生长发育良好，广东和台湾以及云南南部也有栽培。热带抗风树种，并可作材用，供观赏，果供食用。

开花结实 高压苗栽植后 2 年开花结果，实生树 10 年生左右开始结实，大小年现象不明显。花单性，雌雄同株或异株。聚伞圆锥花序，顶生或腋生。花辐射对称。萼杯状，5 或 6 裂。无花瓣，开时有芳香。雄花雄蕊 6～8，着生于肉质环状花盘内，花丝被长毛。雌花子房倒心形，2 裂，少有 3 裂；2 室，少有 3 室，每室 1 胚珠，通常仅 1 室发育，密被瘤状体。据海南儋县观察，4 月上旬花始开，中旬为盛花期，下旬为末花期；8 月中下旬果实开始成熟，9 月上中旬为果熟盛期。果为核果状。果皮革质，有软刺，果爿阔椭圆形，连刺长约 5cm，宽约 4.5cm，刺长约 1cm。种子与果爿近同形。假种皮肉质，与种皮粘连，包裹种子全部。无胚乳，胚稍直，子叶肥厚。种子的外观和内部结构见图 1。

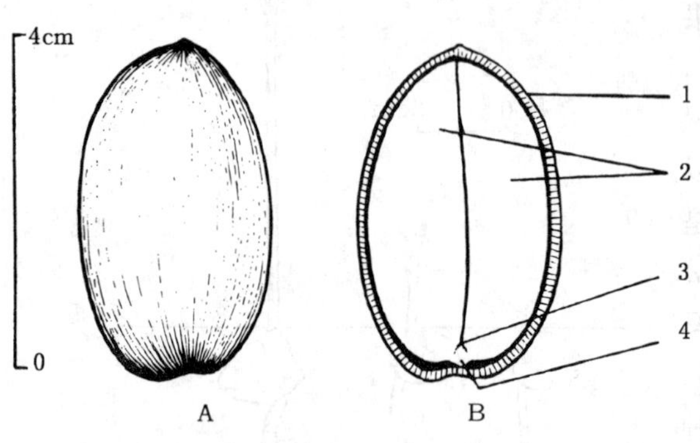

图 1 红毛丹种子外形（A）及其纵切面（B）
1. 种皮 2. 子叶 3. 胚芽 4. 胚根
（黄应钦绘）

果实的采收调制和种子贮藏 果实成熟时外果皮由青色变成红色。这种成熟标志出现时，在树下用长柄钩刀将整个果枝截下，放置数天。待果皮变软，生食后得出种子，洗净即为播种材料。鲜果的出种率约 18%。种子的净度可达 98%。千粒重 2 500（2 000～3 000）g，每千克有纯净种子 400（330～500）粒。生食后稍加阴干的种子含水量约为 38%。已调制出来的种子忌失水，不宜日晒或裸露存放。稍晾干后的种子混适量的湿沙，用木箱包装进行短期运输或贮藏。贮藏期为半个月左右。

发芽和播种 种子无休眠习性，采后即播。发芽时日均温宜在 25℃以上。1988 年 8 月 8 日，华南热带作物研究所在室外沙床用新鲜种子作过发芽测定，播后 12 天开始发芽。从开始萌发至发芽高峰，3 天中发芽 90%；从播种至发芽终止，17 天的发芽率为 95%。留土萌发。发芽后 7 天初生叶展开。种子的萌发和幼苗初期生长情况见图 2。

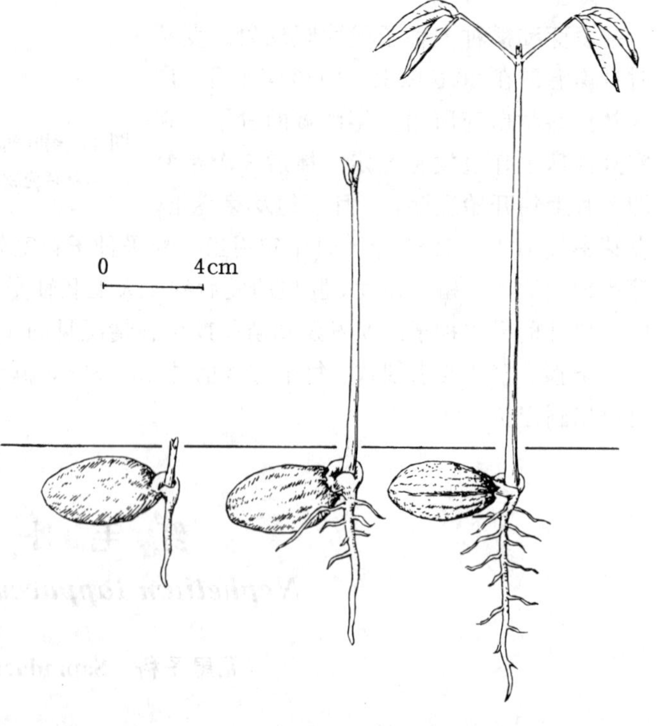

图 2 红毛丹种子萌发后第 2、6、9 天幼苗的生长情况
（黄光郁绘）

条播。每平方米播种 100～

125g，覆土约1cm。1年半生苗出圃定植或作嫁接砧木。生产中多用高枝压条育苗。

（邓春媛）

番龙眼属

Pometia J. R. et G. Forst.

（无患子科　Sapindaceae）

生长习性、分布和用途　番龙眼属约8种，我国只有本文描述的2种。常绿大乔木，具大板根。能耐2℃左右的极端低温。适生于肥沃湿润的酸性和微酸性土。木材坚实而重，是优良的工业和建筑用材。绒毛番龙眼是沟谷雨林中高大的上层优势树种之一，植株数量已迅速减少，在《中国植物红皮书》中列为渐危种。它们的名称、生长、分布和用途见表1。

表1　番龙眼属树种的名称、生长、分布和用途

中　名	学　名	树　高（m）	胸　径（cm）	分　布	用　途	供　稿
番龙眼	*P. pinnata* J. R. et G. Forst.	30	100	台。滇栽培。菲律宾至萨摩亚群岛	材用、水源林	617
绒毛番龙眼	*P. tomentosa* (Bl.) Teysm. et Binn.	40～45	100	滇。南亚、东南亚	材用、水源林	618

开花结实　约10年生开始结实，正常结实期在20年生以后。番龙眼大小年现象不甚明显；绒毛番龙眼大小年现象明显，丰年间隔期3～5年。花单性或杂性，同株。聚伞圆锥花序顶生或腋生，长25～40cm。萼杯状，深5裂，镊合状排列。花瓣5，黄色。花盘杯状。雄蕊5，花药小。子房上位，倒心形，2裂，2室，每室1胚珠。花柱长，丝状，顶部旋扭。这2个树种的开花结实物候期见表2。

表2　番龙眼属树种的开花结实物候期

树　种	观察地点年份	开花 始期	开花 盛期	开花 末期	果实成熟 始期	果实成熟 盛期	果实脱落
番龙眼	云南景洪 1988	5月下旬	6月上旬	6月中下旬	7月下旬	8月上旬	8月上中旬
绒毛番龙眼	云南勐腊 1985～1988	6月上中旬	6月中下旬	6月末	7月下旬	8月上中旬	8月下旬

果实核果状，深裂为2果爿，通常仅有1室发育。果皮厚，中果皮海绵质，内面平滑。种子与果爿同形。种皮革质，除顶部外，覆有胶质黄色假种皮。无胚乳。胚弯拱，子叶横叠。果实及种子的形态见表3、图1。

表 3 番龙眼属树种果实和种子的形态特征

树 种	果实			种子		
	形 状	大小（cm）	颜 色	形 状	大小（cm）	颜 色
番龙眼	椭圆形	长 3～3.5 径 2.3～2.6	紫褐色	扁椭圆形	长 2.3～2.5 径 1.8～2	基部透明胶质
绒毛番龙眼	狭椭圆形	长 2.5～3.5 径 1.6～2.5	深红色	扁椭圆形	长 2～3 径 1.2～2	基部透明胶质

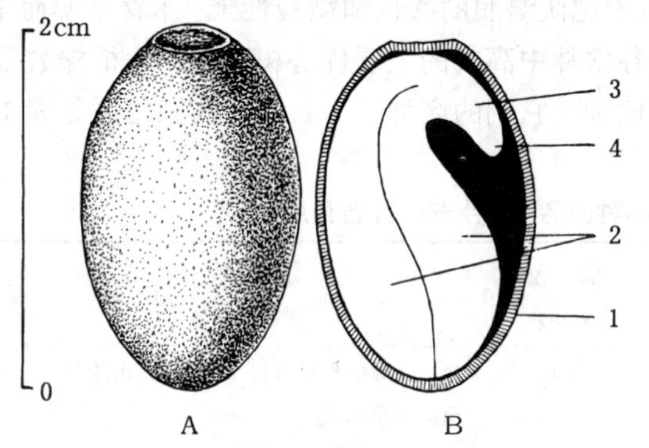

图 1 绒毛番龙眼洗去假种皮后的种子外形（A）及其纵切面（B）
1. 种皮 2. 子叶 3. 胚轴 4. 胚根
（黄应钦绘）

果实的采收调制和种子贮藏 果实成熟时在地上拾集，堆放 3～5 天后挤压，使果皮破裂。将覆有假种皮的种子在筛上搓擦或拌细沙搓擦，淘净杂质，洗出纯净种子。此时种子的含水量约为 25%～39%，忌失水，不宜日晒或裸露存放。运输或贮藏均需混以湿沙，贮藏期为 3 个月。鲜果的出种率、种子的净度和质量见表 4。

表 4 番龙眼属树种的出种率和种净度、质量

树 种	出种率（%）	净 度（%）	千粒重（g）		每千克纯净种子粒数	
			一 般	变动范围	一 般	变动范围
番龙眼	25～35	80～95	2 600	2 200～3 200	380	310～460
绒毛番龙眼	35～40	85～96	2 500	2 000～3 000	400	300～500

发芽和播种 种子无休眠习性。发芽时日均温宜在 25℃左右。1988 年 8 月，云南普文林场和西双版纳热带植物园在室外分别对番龙眼和绒毛番龙眼进行过发芽测定，结果见表 5。

表 5 番龙眼属树种的发芽能力及其测定条件

树 种	基 质	温 度（℃）	发芽势（%）		发芽率（%）	
			计算天数	一般数值	计算天数	一般数值
番龙眼	圃地	26～30	6	70	21	90
绒毛番龙眼	沙	26～30	5	80	15	95

出土萌发。子叶出土后 4～5 天展开，同时展出初生叶。绒毛番龙眼种子的萌发和幼苗初

期生长情况见图 2。

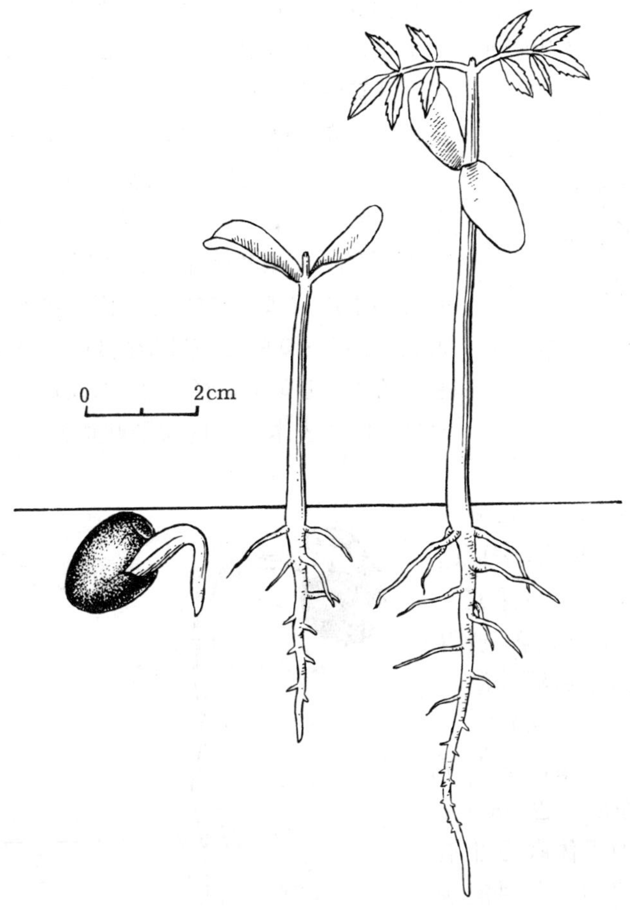

图 2　绒毛番龙眼种子萌发后第 1、3、6 天幼苗的生长情况

（黄应钦绘）

点播。宜随采随播。每平方米播种 500～680g，覆土 1～2cm。出现第一对初生叶时可以移植到容器内继续培育 3 个月出圃。大田育苗一般需 10 个月左右方可出圃。

（龙素珍）

无　患　子

Sapindus mukorossi Gaertn.

（无患子科　Sapindaceae）

生长习性、分布和用途　无患子属约 13 种，分布在美洲、大洋洲和亚洲热带地区，我国有 4 种，本文描述 1 种。落叶乔木，高达 20m。喜温暖湿润气候。稍能耐寒。酸性土、钙质土均能生长。萌芽力弱，不耐修剪。枝、干质地较脆，当风处常易风折。分布于我国东部、南部至西南部。日本、韩国、缅甸、老挝、越南、印度也有。木材结构细，质硬而脆，不耐腐，易加工，供作箱板、器具、木梳等。果肉含皂素可代肥皂洗涤纺织品。根、果入药。种仁可榨油制肥皂和润滑油。枝叶广展，冠大荫浓，秋色黄艳，为行道树和庭园蔽荫的优良树种。

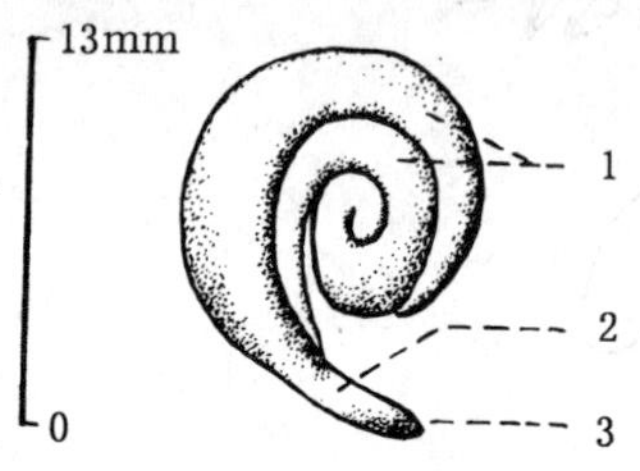

图 1 无患子种子的弯胚
1. 子叶 2. 胚轴 3. 胚根
（张世经绘）

开花结实 花通常两性。圆锥花序顶生，长 15～30cm。花小，整齐。萼片、花瓣各 5。花瓣具长爪，内面基部有 2 个耳状小鳞片。雄蕊 8，花丝下部被长毛。子房上位，3 室，每室 1 胚珠，仅 1 室（稀 2 室）发育。花期 5～6 月。果9～10 月成熟。核果状浆果，近球形，径1.6～2.3cm，果皮褐黄色，肉质，熟时半透明，表面起皱纹，果实基部具有盘状疤痕，为不发育的心皮残留的痕迹。果实内含 1 粒种子，黑色，光亮，近球形，径 1.2～1.5cm，无假种皮。外种皮硬骨质，内种皮木质，内藏子叶重合拳卷的弯胚（图 1）。无胚乳。

果实的采收调制和种子贮藏 当果实呈褐黄色而略带透明，表面起皱纹时即可击落拾集。采回的果实浸水沤烂，搓掉果肉并冲洗干净。所得纯净种子置通风处阴干后混沙湿藏或干藏。果实出种率 50%～60%，种子千粒重 1 180～1 590g，每千克 850～630 粒。种子净度可达 100%。

发芽和播种 种子休眠习性不明显。即采即播或春季 3 月播种混沙湿藏过的种子。越冬干藏的种子需浸种催芽。条播，行距 20～25cm，每平方米播种 120～150g。发芽率 60%～80%。北方 1 年生苗高 30～50cm，南方 80～100cm。出土萌发，子叶肉质肥厚，其中 1 片较大，1 片较小，略不对称（图 2）。初生叶为 3～4 对小叶的奇数羽状复叶，近对生，以后互生。

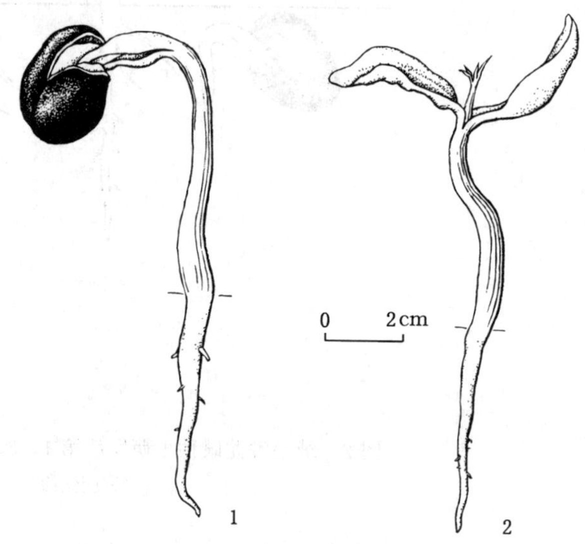

图 2 无患子种子的萌发和幼苗初期生长情况
1. 子叶即将脱出种皮 2. 初生叶近对生
（张世经仿《主要树木种苗图谱》）

（黄鹏成）

文冠果（木瓜）

Xanthoceras sorbifolia Bge.

（无患子科 Sapindaceae）

生长习性、分布和用途 文冠果为单种属，产于辽、内蒙古、冀、晋、豫、陕、甘、宁。

黑、吉、鲁、皖、赣有栽培。朝鲜半岛亦产。落叶灌木或小乔木，高 2～5（8）m，胸径 90cm。喜光，耐干旱，抗寒，在黄土丘陵、冲积平原、固定沙地和石质山区都能生长，在土层深厚、肥沃、通气良好的土壤上生长较好，低湿地及盐碱地则生长不良。我国北方重要的油料树种，种子含油率高，油质好，可供食用、药用和化工用。

开花结实　3～4 年生开始结实，20 年左右进入结实盛期，种子产量比较稳定。由开花到种子成熟约 90～100 天。花杂性，雄花和两性花同株，辐射对称，总状花序。顶生花序为两性花，可授粉结实，侧生花序为雄花。萼片 5。花瓣 5，白色，内侧基部有紫红色斑点。花盘 5 裂，裂片与花瓣互生，背面顶端具角状附属体，橙黄色。雄蕊 8，内藏，花药药隔的顶端和药室的基部均有 1 球状腺体。子房上位，矩圆形，3 室，各具胚珠 7～8，2 列。花柱顶生，直立，柱头乳头状。文冠果在不同地区的开花结实物候见表 1。

表 1　文冠果开花结实物候期

观察地点	观察年份	开花			果实成熟		种子散落	
		始期	盛期	末期	始期	盛期	始期	盛期
内蒙古	1971	5 月上旬	5 月中旬	5 月下旬	7 月上旬	8 月中旬	8 月中旬	8 月中旬
呼和浩特	1972	5 月上旬	5 月中旬	6 月上旬	7 月上旬	8 月中旬	8 月中旬	8 月下旬
内蒙古	1961	5 月上旬	5 月中旬	5 月下旬	—	8 月中旬	8 月中旬	8 月中旬
伊克昭盟	1962	5 月中旬	5 月下旬	5 月下旬	—	8 月上旬	—	—
	1963	5 月中旬	5 月中旬	5 月下旬	—	8 月上旬	—	—
	1965	5 月下旬	5 月下旬	5 月下旬	7 月中旬	8 月中旬	—	—
吉林吉林市	—	5 月下旬～6 月上旬	5 月下旬～6 月上旬	6 月上旬	8 月上旬	8 月下旬	8 月下旬	9 月下旬
江西景德镇	1975	3 月下旬～4 月上旬	4 月上中旬	4 月上中旬	4 月中下旬	～	—	—
青海西宁市	—	5 月下旬	6 月上旬	6 月中旬	—	9 月下旬	—	—

蒴果，椭圆形或球形，黑褐色，直径 3～6cm，熟时室背 3 瓣裂，3 室，每室具种子（1）4～6（8）粒。果皮厚，木栓质或革质。种子近球形，直径 1～1.8cm，未成熟种子白色，成熟后为黑色，有光泽。种皮厚，革质，种子无胚乳。异形双子叶，其中 1 枚肥大，1 枚瘦小，均向一面卷曲，大子叶包着小子叶。种子形态如图 1。

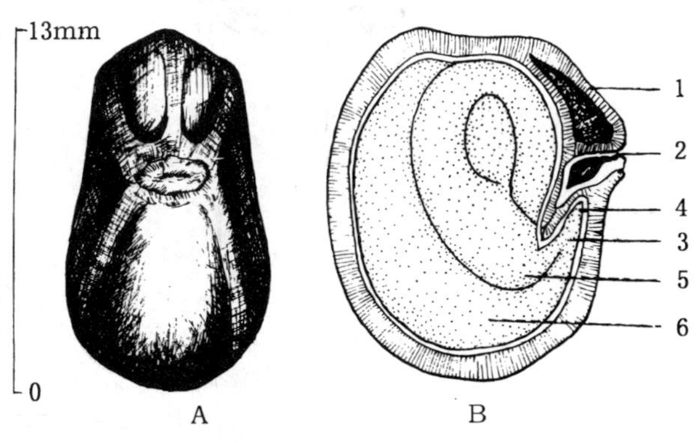

图 1　文冠果种子外形（A）及侧面纵切面（B）
1. 种皮　2. 珠孔　3. 胚轴　4. 胚根　5. 小子叶　6. 大子叶
（姚桂英绘）

果实的采集调制和种子贮藏　当果皮由绿变黄，由光滑变粗糙，先端开裂，种子由红褐色变为黑褐色且有光泽时，即可采种。一般人工直接摘果，高大树木可用高枝剪或采种钩刀。采下的果实放在阴凉通风处。数日内果皮开裂，除去果皮，将种子放在通风良好的室内阴干，当种子含水量降至 10%左右时即可贮藏。种子净度在 95%以上，出种率 50% 左右。千粒重 600～1 250g，最大可达 2 850g，平均约为 1 000g。每千克有种子 800～1 700 粒，平均约为 1 000 粒。

发芽和播种 文冠果的种皮较厚，种子吸水困难，播前必须经过处理。种子处理方法有 2 种，一是秋末冬初低温层积，翌春播前半个月取出，再置于 20～25℃的条件下高温催芽，约 1/4 的种子露白即可播种。层积与未层积的种子场圃发芽率差异很大。据试验，水浸 3 天，场圃发芽率 33%；层积 1 个月，场圃发芽率 58%；层积 5 个月，场圃发芽率可达 68%。另一种方法是高温层积催芽，即播前 20 天左右，将种子放入始温 50℃左右的水中浸 3 昼夜（或用浓硫酸浸 3 小时并彻底冲洗后再浸入温水中），每天换水，捞出后混 3 倍湿沙置于 20～25℃条件下高温催芽，大约 1/3 的种粒露白即可播种。据对一份样品测定，这样处理的种子发芽率 73%，场圃发芽率为 69%。

10cm 处地温达 10℃时即可播种。每平方米播种 50～60g，行距 15cm，点播，株距 10～15cm。播种时种粒宜横放土中，沟深 4～6cm，覆土 3～4cm，10～15 天开始出苗，15～20 天基本出齐。留土萌发。种子萌发和早期幼苗形态如图 2。

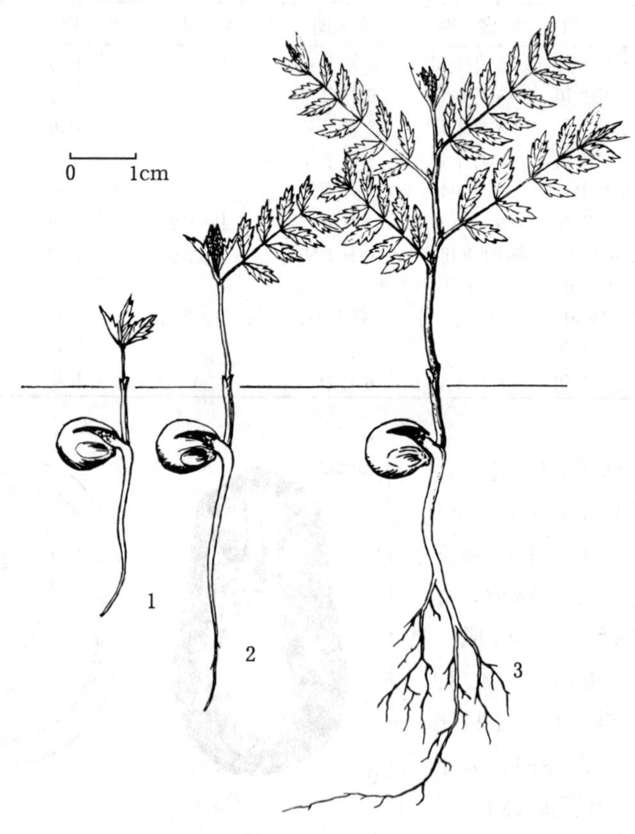

图 2 文冠果种子的萌发和幼苗生长情况

1. 子叶留土萌发 2. 初生叶展开 3. 幼苗形成

（姚桂英绘）

文冠果还可插根育苗。

（李荣晨）

伯乐树（钟萼木）

Bretschneidera sinensis Hemsl.

（伯乐树科　Bretschneideraceae）

生长习性、分布和用途　伯乐树属只有本文描述的1种，是我国特有的单种科残遗种，在《中国植物红皮书》中列为稀有种。落叶乔木，高达20m，胸径60cm。喜生于肥沃湿润之地。具萌芽力。零星分布在川、黔、滇、桂、湘、鄂、赣、粤、闽、浙、台。越南也有分布。是优良的用材树种和观花、观果树种。

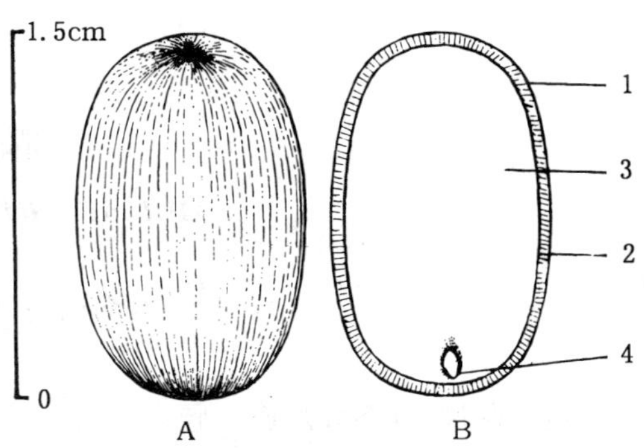

图1　伯乐树种子洗去肉质外种皮后的外形（A）及其纵切面（B）
1. 中种皮　2. 内种皮　3. 子叶　4. 胚根
（黄应钦绘）

开花结实　15年生开始结实，20年生后为正常结实期，结实常有大小年之分。花两性。总状花序顶生，长20～40cm。花萼阔钟形，5浅裂。花瓣5，不相等，粉红色。雄蕊8（5～9），紫红色，短于花瓣。子房上位，3～5室，每室有胚珠2枚，几并列，自中轴下垂。蒴果，木质，桃形，红褐色，长3.0～4.5cm，径2.0～3.2cm，熟时3～5裂，稀2裂，内有种子2～3（1～6）粒。种子大，椭圆形，长1.1～1.8cm，径0.7～1.1cm。外种皮（也有学者认为是假种皮）肉质，金黄色或橘黄色，中种皮白色骨质，内种皮膜质。无胚乳，子叶肥大，胚根短。盛花期在5～6月，果实10月中旬到下旬成熟。

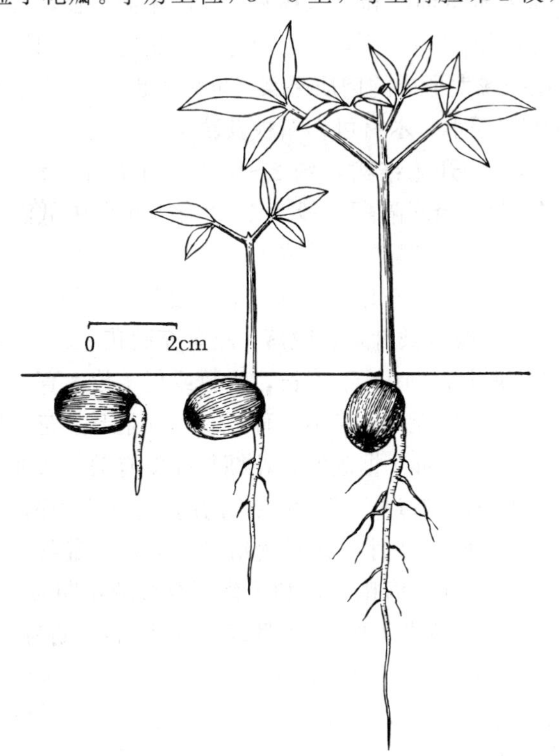

图2　伯乐树种子萌发后第4、20、36天的幼苗生长情况
（黄应钦绘）

果实的采收调制和种子贮藏　果实由青绿色转为红褐色时即可采收。采回置通风处晾干，开裂后剥出种子，浸水中洗去肉质外种皮，阴干1～2天即为播种材料，通称种子。鲜果出种率约20%，洗去外种皮后的纯净播种材料约占完整种子鲜重的65%。千粒重715～780g，每千克有纯净种子1280～1400粒。湿沙层积越冬。伯乐树种子的形态特征见图1。

发芽和播种　种子有休眠现象。经过层积解除休眠的种子，室内发芽率可达

80%～100%，场圃发芽率约85%。留土萌发。上胚轴圆柱形，长7.5cm，径2～3mm，绿色，无毛。初生叶为3小叶羽状复叶，对生。第3叶以上互生，开始仍为3小叶。主根直伸，侧根发达，黄白色。伯乐树种子的萌发和幼苗生长情况见图2。

春播或随采随播。条播，条距25cm，播幅15cm，播种沟深5cm。每条播种沟播2行，株距3～5cm，覆土厚3cm，盖草。每平方米播种30～38g，每666m^{2}产苗2.5万～3.0万株。1年生苗高18cm，2年生苗出圃。用1年生苗根扦插，成活率可达70%～80%(陈辉等，1986)。

（杨国华）

腰 果

Anacardium occidentale L.

（漆树科 Anacardiaceae）

生长习性、分布和用途 腰果属约15种，我国只引入本文描述的1种。常绿乔木，高12～15m，胸径20～30cm。适生于排水良好的肥沃土壤，也耐干旱、贫瘠，在沙地、石砾地和砖红壤地也能生长，但在排水不良及石灰性土壤上则生长不良。5～6℃的气温或轻霜均会造成死亡。原产西印度群岛和美洲热带，现广植于全球热带地区。我国滇、桂、粤、琼、台、闽有栽培，但在广州、南宁露地不能越冬。热带重要木本油料和干果树种。种仁味美可食，含油44%，为高级食用油。腰果壳味涩有毒，含壳油40%～50%，是合成橡胶、调制高级油漆和彩胶作色剂的原料，亦可用于治疗麻疯病。花托膨大形成的肉质果梨（也称假果），可作水果或制成饮料。树皮所产树脂可除虫或制不退色的墨水。木材可作箱板及薪材。

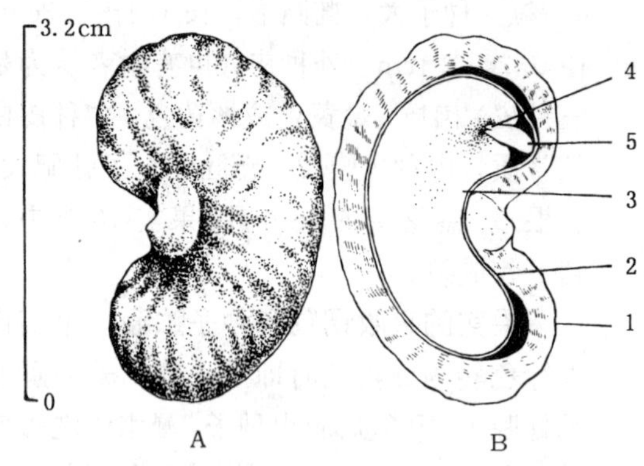

图1 腰果果核外形（A）及其纵切面（B）
1.内果皮 2.种皮 3.子叶 4.胚芽 5.胚根
（黄应钦绘）

开花结实 约2～4年生开始开花结实，大小年间隔期一般为1～2年。花期中低温多雨对结实影响大。花小，杂性，排成顶生、分枝的圆锥花序。萼深5裂，覆瓦状排列。花瓣5，线状披针形，脱落。花盘充满花萼的基部。雄蕊7～10，不等长，少数或仅1枚发育。子房上位无柄，略不对称，1室，1胚珠。花柱侧生，线形。据观察，花期与纬度有关，纬度高的地区有迟开花、迟结果的趋势。在海南南部，花期从12月一直延续到翌年4月，盛花期为2～3月。海南北部和云南西双版纳花期为1～5月，盛花期为3～4月。湛江地区盛花期为4～5月，果期为4～7月。从幼果形成至果实成熟约需40天。果实下部为花托膨大形成的果梨（肉质果托），长3～7cm，最宽处4～5cm，成熟时呈橙黄色或紫红色。核果，肾形，两侧压扁，熟时灰褐色，长2.5～3.4cm，宽约1.8cm，厚1.3～1.8cm，位于梨形肉质的花托上，果皮硬。种子无胚乳。子叶与果同形，富含油质。腰果果核的外形和内部结构见图1。

果实的采收调制和种子贮藏　果实成熟期长。成熟期中每 3～5 天收集一次落地果实。如要利用果梨，则需每天采果，晾干后取出核果，去除外果皮和中果皮及杂质，所得果核即为播种材料，通称种子。净度可达 99%。千粒重5 360（4 950～5 700）g，每千克有种子（核）180（170～200）粒。种子富含油脂，忌失水，不耐贮藏，宜随采随播。贮藏或运输均应混以湿沙，贮藏期以不超过 1～2 个月为宜。

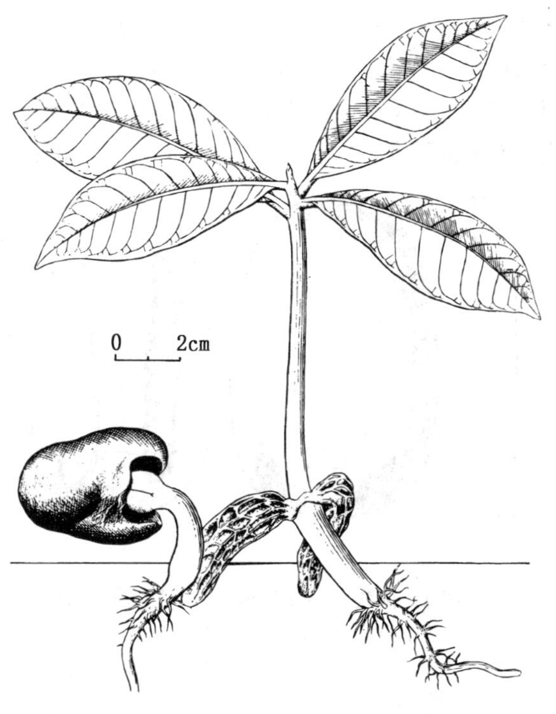

图 2　腰果种子(核)萌发后第 5、16 天幼苗的生长情况
(黄应钦绘)

发芽和播种　种子无休眠习性。发芽时日均温宜在 20℃以上。1989 年 6 月 5 日，广西林业科学研究所在室外沙床上，用当年新采的种子（播前未作处理）作过发芽测定：当时日均温约 28℃，播后第 5 天开始发芽，并随即进入发芽盛期，6 月 13 日发芽终止。自发芽之日起算，2 天中发芽 75%；自播种之日起算，8 天的发芽率为 95%。留土萌发或地面萌发。播后 5 天胚根萌出，约 8 天下胚轴拱出，再过 3 天子叶留于地下，如脱出核壳则露出土面，同时出现初生叶。腰果种子（果核）的萌发及幼苗初期的生长情况见图 2。

条播。随采随播，播种期不应迟于 9 月。播前浸种 1～2 天。播时核脐向上 ，种子倾斜，这样发芽早，出苗齐。播种深度 2～6cm。每平方米播种 300～380g。腰果裸根苗造林成活率较低，生长亦较差，现多采用容器育苗，每个容器播种 1 粒，次春出圃定植。生产中还广泛采用嫁接或扦插繁殖。

（张声燕）

南酸枣（五眼果）

Choerospondias axillaris（Roxb.）Burtt et Hill

（漆树科　Anacardiaceae）

生长习性、分布和用途　南酸枣属只有本文描述的 1 种。落叶乔木，高达 25m，胸径 60～80cm。在肥沃湿润的酸性或微碱性土壤上能生长成材。分布于长江以南。尼泊尔、印度、泰国、中南半岛及日本南部也有。木材为优质材。果可食或酿酒。树皮可作纤维并可入药，主治烫伤。果核可制活性炭。

开花结实　8～10 年生开始结实，15 年生以后进入正常结实期，大小年现象不明显。花单性或杂性异株。花冠淡紫红色。雄花和假两性花排成聚伞状圆锥花序，长 4～10cm。花萼浅杯状，5 裂。花瓣 5，开花时外卷。雄蕊 10，与花盘裂片互生。雄花无不育雌蕊。雌花单生于上部叶腋，子房上位，卵圆形，5 室，每室 1 胚珠。花柱 5，柱头头状。据 1979～1982 年广西南宁观察，3 月上旬花芽始现，着生于当年新枝叶腋，3 月中旬至下旬雄花和假两性花序形成，雌花蕾膨大；4 月上旬花盛开，4 月中旬花期结束，幼果形成；7 月上旬果实开始成熟，中下旬为果实成熟盛期。核果，椭圆形或卵形，熟时外果皮由青色转变为浅黄色，表面有光泽，长 2.2～3.5cm，径 1.6～2.5cm，成熟后 3～4 天自落。中果皮肉质浆状。由骨质内果皮包被的果核与果同形，长 1.8～2.5cm，径 1.2～1.5cm。果核顶端有 5 个（少数为 4 或 6 个）较大的椭圆形眼孔，孔具膜质盖，基部有 5 个圆形凹点，成辐射状排列。每果核有种子 3～5 粒。无胚乳，胚伸直，子叶长。果核的外形和内部解剖构造见图 1。

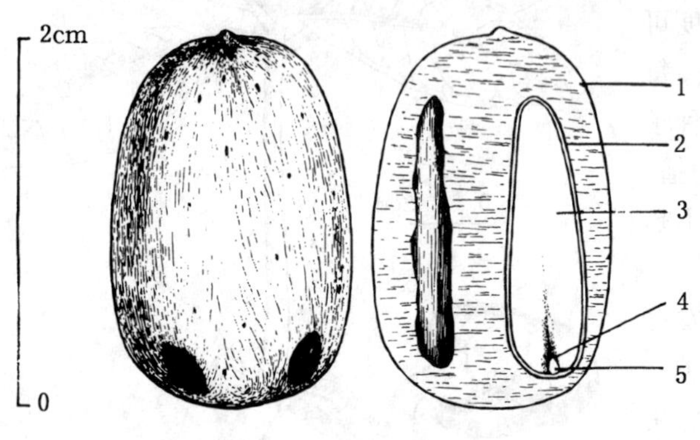

图 1　南酸枣果核外形（A）及其纵切面（B）
1. 内果皮　2. 种皮　3. 子叶　4. 胚芽　5. 胚根
（黄应钦绘）

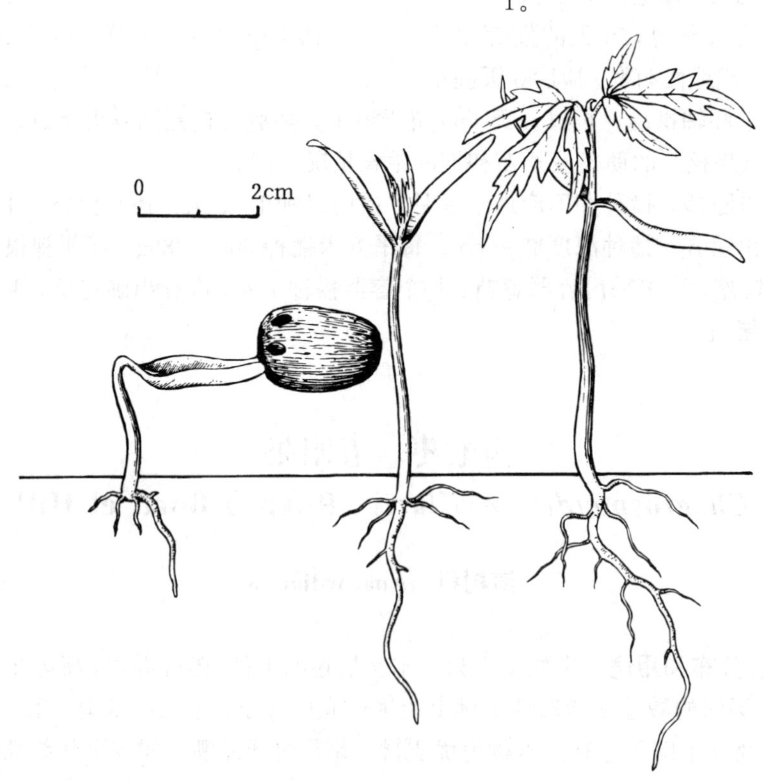

图 2　南酸枣果核萌发后第 5、7、11 天幼苗的生长情况
（黄应钦仿《热带亚热带主要树种采种育苗技术》）

果实的采收调制和种子贮藏　果实成熟期在树下拾集或摇动树枝震落后收集。果实可以在加工果汁或生食后得出果核；也可将果实堆沤数日，果皮和果肉软化后置水中捣烂，淘洗，所得果核即为播种材料，通称种子。鲜果的出籽率为28%～35%。净度可达100%。千粒重1 750（1 600～2 000）g。每千克有纯净种子（果核）570（500～650）粒。调制后稍加晾干的果核含水量为25%～30%，忌失水，不宜日晒。短期运输可用麻袋或竹筐直接装运果实，运抵后再行调制；也可调制后拌少量湿润锯木屑运输。贮藏时需混湿沙，贮藏期以不超过8个月为宜。

发芽和播种　种子有程度较深的休眠。调制出来的果核，以湿沙层积至翌年2月播种为好。1981年7月10日，广西林业科学研究所在室外沙床用新鲜果核作过发芽测定：播后历时半年多，至翌年1月23日才开始发芽，3月16日发芽结束，且无明显的发芽盛期，发芽率为60%。经过层积催芽至翌年2月中旬播种，则播后15天左右即开始萌发，从发芽开始至发芽高峰期，15天发芽50%；从播种至发芽终止，45天的发芽率为60%～70%。出土萌发。萌发后约5天子叶带果核出土，随即展出初生叶。南酸枣的萌发和幼苗初期生长情况见图2。

点播。每平方米播种100～125g，覆土2～3cm。部分果核发苗2～3株或4～5株，可以分株移植。1年生苗出圃。

（王宏志）

红叶黄栌（红叶）

Cotinus coggygria Scop. var. *cinerea* Engl.

（漆树科　Anacardiaceae）

生长习性、分布和用途　黄栌属约5种，我国有3种3变种，本文描述红叶黄栌（又名灰毛黄栌）这个变种。落叶灌木或小乔木，高3～6m。生于背阴山坡林中或灌丛中。耐寒抗旱，能在瘠薄山地或盐碱不甚重的土地生长，忌水涝。产冀、鲁、豫、鄂、川。间断分布于东南欧。叶片秋季经霜鲜红，北京称为“西山红叶”。多数不孕花花梗在花后伸长，被长毛，羽毛状，粉红色，久留树上，宛如炊烟，国外又称“烟树”，为优美的观赏树。又可作荒山造林先锋树种。木材可提取黄色染料。树皮和叶可提栲胶。叶含芳香油。枝、叶入药，有消炎、清湿热之功效。

开花结实　结实大小年现象不明显。花杂性。圆锥花序顶生，长约20cm。花小，黄绿色，仅少数发育。花萼5裂，宿存。花瓣5，长为花萼的2倍。雄蕊5，着生于环状花盘下部。子房上位，偏斜，压扁，1室，1胚珠。花柱3，侧生而短，柱头小而不显。据北京地区观察，开花始期在4月下旬或5月初，3～5天后进入盛花期，5月10日左右为末花期，有时延至5月中旬；果实成熟始期为6月初，6月上旬进入盛熟期，成熟后2～3天即脱落，6月中旬大部分脱落。核果，小，歪肾形，略扁，暗红褐色，长3.5～5.5mm，宽2.7～3.7mm，厚1.5～1.7mm。外果皮和中果皮干燥后皱缩，外果皮表面有网状纵肋条。内果皮骨质。种子小，肾形。种皮膜质。无胚乳。胚大，子叶薄片状，下胚轴弯曲，含油脂（见图1）。

果实的采收调制和种子贮藏　嫩果暗绿色，成熟时暗红褐色。果壳干硬时剪下或从果穗

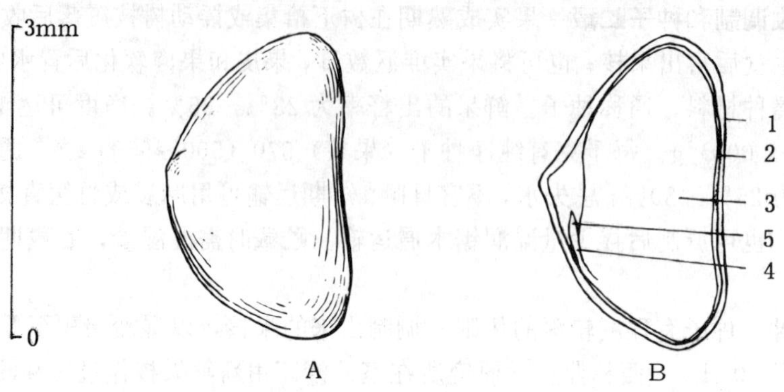

图 1　红叶黄栌果核外形（A）及其纵切面（B）

1. 内果皮　2. 种皮　3. 子叶　4. 胚轴　5. 胚根

（胡冬梅、林平绘）

上捋下果实。有风时果穗会被吹落，也可在地面收集。果穗装入袋内揉搓，使果实与果梗分开，风选去除杂物和秕粒，晾干，有的地方还搓去果皮、果肉；但通常是直接用干燥的核果作为播种材料，通称种子。核果的出籽率（果核）为 30%。净度 95%。千粒重 13.8（11～14.7）g，每千克 7.2（6.8～9）万粒。发芽率约 65%。干燥的核果装袋，置冷凉处干藏。在北京地区一般室内贮藏可保持发芽能力 1～2 年。

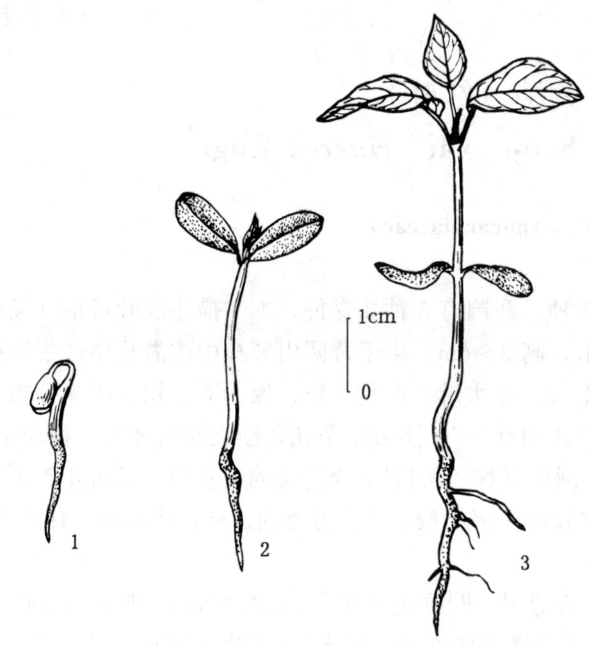

图 2　红叶黄栌种子萌发和幼苗生长情况

1. 子叶出土　2. 初生叶初露　3. 第 1、2 片初生叶对生，以后互生

（胡冬梅、林平仿《主要树木种苗图谱》）

发芽和播种　果核具不透性木质果壳和休眠胚。发芽测定时可将层积过的核果（即播种体）置于发芽盒中的滤纸床上，在恒温 20℃的条件下第 7 天开始发芽。或用四唑法测定生活力。在北京地区育苗，可在 1 月上旬浸种 24 小时后放在背阴处低温层积，2 月中旬移至背风向阳处催芽，3 月下旬种子开始萌动后播种。低床条播或垄播。每平方米床面播种 2.5g。覆土 1.5～2cm。播后 2～3 周可出苗。1 年苗高可达 1m 左右。也可将新采收的种子层积 2 个月后在 8～9 月间播种，每平方米播种 2～2.5g。越冬时覆盖保护。出土萌发。第 1、2 片初生叶对生（图 2）。也可用压条或分蘖繁殖。

（刘长江）

人面子属

Dracontomelon Bl.

（漆树科　Anacardiaceae）

生长习性、分布和用途　本属约 8 种，我国只有本文描述的 2 种。常绿大乔木，高达35～40m，胸径 100～160cm。具板根。喜肥沃湿润的深厚土壤，酸性土或钙质土均宜，但贫瘠土上生长不良。人面子果肉可食并可入药，木材用于家具或建筑。这 2 个种的种子含油，可以制皂或用于润滑。这 2 个种的名称、生长、分布及用途见表 1。

表 1　人面子属树种的名称、生长、分布和用途

中　名	学　名	树　高（m）	胸　径（cm）	分　布	用　途	供　稿
人面子	*D. duperreanum* Pierre	40	160	琼、粤、桂、滇。越南	材用、食用、药用、绿化	603
大果人面子	*D. macrocarpum* H. L. Li	35	100～130	滇	材用、油料、绿化	617

开花结实　10～15 年生开始开花结实，正常结实期在 20 年生以后。结实大小年间隔期一般为 1 年。水肥条件差的地方结实甚少或不结实。花两性。圆锥花序顶生或腋生。花萼 5 裂。花瓣 5，绿白色。雄蕊 10，与花瓣等长，花丝线状钻形。花药长圆形，丁字着生，内侧向纵裂。花盘碟状，不明显浅裂。心皮 5，合生。子房 5 室，每室 1 胚珠。花柱 5，下部分离，上部合生。柱头近尖塔形，5 角。据广西南宁及云南勐腊观察，这 2 个种的开花结实物候期见表 2。

表 2　人面子属树种的开花结实物候期

树　种	观察地点和年份	开花			果实成熟		果实脱落
		始　期	盛　期	末　期	始　期	盛　期	
人面子	南宁 1981～1984	5 月中旬	5 月下旬	6 月中旬	10 月中旬	10 月下旬	10 月下旬～11 月
大果人面子	勐腊　—	4 月中旬	4 月下旬	5 月上旬	9 月下旬	10 月	11 月

核果，球形或扁球形，上半部具小瘤体，未熟时青色，熟时黄色。中果皮肉质。核骨质，顶部有孔数个，表面 1 侧有裂状凹纹，状如人面。种子无胚乳。子叶厚，平凸状。这 2 个种果实和果核的形态见表 3、图 1。

表 3　人面子属树种果实和果核的形态特征

树　种	果实			果核		
	形　状	大小（cm）	颜　色	形　状	大小（cm）	颜　色
人面子	球形或扁圆形	1.7～2.6	黄绿色	扁圆形	宽 1.1～1.4 厚 1～1.3	深褐色
大果人面子	扁圆球形	4～5.5	黄绿色	三棱状椭圆形	宽 1.8～2.5 厚 1.5～2.2	深褐色

果实的采收调制和种子贮藏 用于食品加工的人面子果实，可在未成熟时即行采集。用于育苗的人面子和大果人面子，则应清除树下灌丛杂草，在成熟盛期用竹竿敲打或摇动树枝，将果实震落后在地面捡拾。人面子果肉的食用价值高，调制时用刀剥去果肉后将果核堆沤，并洒上草木灰，促使残留的果肉腐坏，再用水淘洗干净。大果人面子果肉无食用价值，可在采集后将完整的果实堆沤数日，俟果皮果肉软化，装入竹筐，置水中捣冲，淘去皮肉等杂质。这两个树种的果核便是播种材料，通称种子。果核忌失水，不能日晒或干藏。运输或贮藏均需混以湿沙，贮藏期为半年以内。鲜果的出籽率以及净度、质量等见表 4。

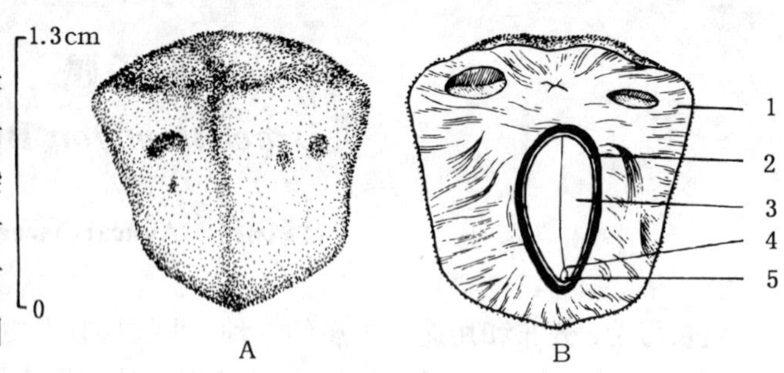

图 1 人面子果核外形（A）及其纵切面（B）
1. 内果皮 2. 种皮 3. 子叶 4. 胚芽 5. 胚根
（黄应钦绘）

表 4 人面子属树种出籽率和果核的净度、质量

树 种	出籽率（%）	净度（%）	千粒重（g）	每千克果核粒数（粒）
人面子	16	85～95	2 000～2 700	370～500
大果人面子	25～30	80～90	14 000～15 000	65～72

发芽和播种 人面子的种子有休眠习性，但种粒间的休眠深度有差异。发芽时的日均温应在 15℃以上。1979 年广西林业科学研究所用当年新采果核，在南宁室外沙床作过发芽测定：10 月上旬播种，当时日均温 26℃左右，至翌年 2 月上旬开始发芽，发芽盛期不明显，发芽期持续约 4 个月，至 6 月中旬发芽终止，发芽率 65%。大果人面子果核的吸水困难，发芽极不整齐。为缩短发芽时间，可将果核反复浸晒 7～8 天，经过这种处理，播后 14 天便开始发芽，发芽全程为 40～50 天，发芽率约 40%。出土萌发。胚根萌发后 8～10 天子叶出土，再过 4～7 天发出初生叶。人面子的萌发和幼苗初期生长情况见图 2。

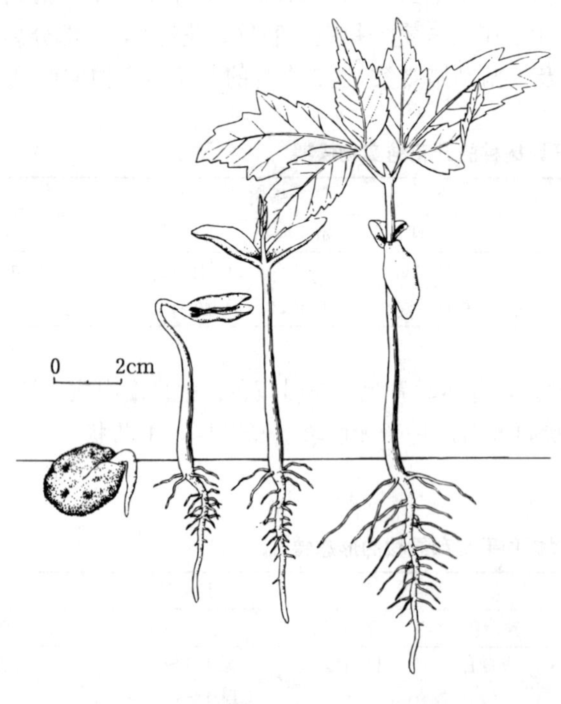

图 2 人面子种子(核)萌发后第 2、9、13、20 天幼苗的生长情况
（黄应钦仿《热带亚热带主要树种采种育苗技术》）

由于发芽的持续期长，一般不直接播于圃地，而是将种子密播于沙床

进行催芽，覆土 1.2～2cm，萌发后再移至圃地。人面子育苗每平方米用果核 100～150g。大果人面子育苗每平方米用果核 750～1 000g。1～3 年生苗出圃。

（张声燕）

厚　皮　树

Lannea coromandelica（Houtt.）Merr.

（漆树科　Anacardiaceae）

生长习性、分布和用途　厚皮树属约 70 种，我国只有本文描述的 1 种。落叶乔木，高 5～10m，胸径约 20cm。土壤以深厚肥沃的沙质壤土为宜。抗风性能强，耐干旱。分布于滇、桂、琼、粤。中南半岛、印度至印度尼西亚也有。木材纹理通直，易加工，心材较耐腐，宜作箱板和家具，并可用作造纸原料。树皮含 10％鞣质，内含红色染料，渔民常用浸出液涂染鱼网。茎皮纤维强韧，可织粗布或制绳索。种子可榨油。

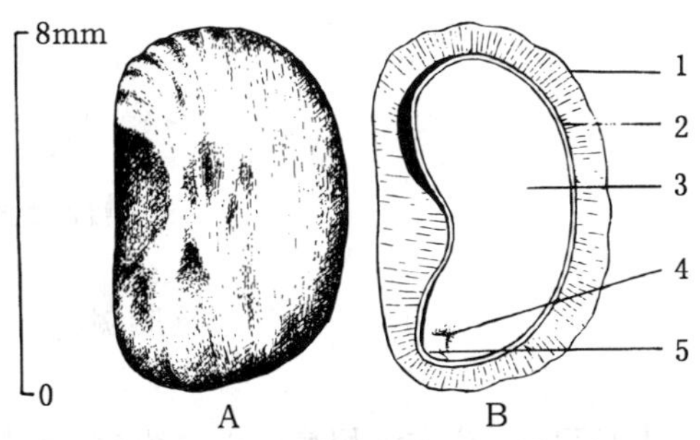

图 1　厚皮树果核外形(A)及其纵切面(B)
1. 内果皮　2. 种皮　3. 子叶　4. 胚芽　5. 胚根
（黄应钦绘）

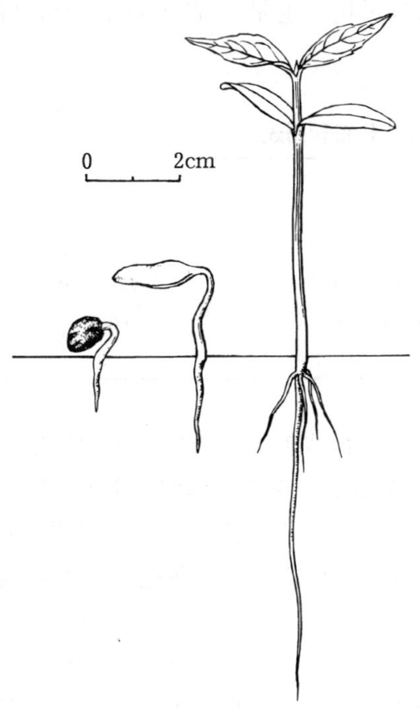

图 2　厚皮树种子萌发后第 3、5、10 天幼苗的生长情况
（黄应钦绘）

开花结实　5～8 年生开始开花结实，正常结实期在 15 年生以后，大小年现象不明显。花单性。同株或异株，排成顶生分枝或不分枝的总状花序，雄花序长 15～30cm，分枝；雌花序较短，簇生枝顶。花小，黄色而带紫色。花萼 4 裂。花瓣 4。雄蕊 8，与花瓣等长，生于花盘边缘。花药卵形或箭形，在雌花中极短，不育。花盘环状。子房上位，4 室，每室 1 胚珠，常退化为 1 室。花柱 3 或 4，侧生。柱头盾状。据海南儋县观察，3 月中旬花始开，3 月下旬～4 月中旬为盛花期，4 月下旬为末花期；5 月下旬果实开始成熟，6 月上中旬为果熟盛期。核果，近肾形，压扁，4 室，仅 1 室发育，先端具残存花柱，紫红色，长 8～10mm，宽 5mm。外果皮和中果皮薄。内果皮厚，坚硬。果核扁肾形，白色。种子无胚乳，胚稍弯。果核的外观和内部结构见图 1。

果实的采收调制和种子贮藏　当果实成熟盛期，外果皮由青色变成红色时即可采收。通常用采种刀钩下果穗摘采果实。采得的果实摊放于室内

1～2 天，待果皮软化后装入袋内，置水中搓擦冲洗。所得果核用作播种材料，通称种子。鲜果的出籽率约 24%。净度可达 95%。千粒重 90（80～100）g，每千克有种子（核）1.1（1～1.25）万粒。种子含水量约 41%，忌失水，不宜日晒或干藏，短期贮藏也需混以湿沙。

发芽和播种 种子无休眠习性。发芽时日均温宜在 22℃以上。1989 年 5 月 5 日，华南热作研究院在室外沙床作过发芽测定：播后 7 天（5 月 11 日）开始发芽，5 月 18 日发芽结束，没有明显的发芽盛期，发芽率为 30%。出土萌发。子叶出土后 2 天初生叶展开。厚皮树果核的萌发和幼苗初期生长情况见图 2。

条播。秋播。每平方米约播 7～10g，覆土 0.5～1cm。培育约 8 个月，至翌年早春出圃。亦可用枝条扦插繁殖。

（邓春媛）

杧 果 属

Mangifera L.

（漆树科 Anacardiaceae）

生长习性、分布和用途 本属约 50 余种，我国 5 种，人工栽培品种较多，本文描述 2 种。常绿大乔木。杧果在印度树高达 25m，胸径可达 2.5m。无性繁殖的良种果园，树高一般为4～6m。幼龄期稍耐庇荫，成龄树较喜光。喜高温，能耐 0℃左右的极端低温，苗期忌霜冻，花期忌低温阴雨天气。喜疏松肥沃的酸性土。为热带名贵果树，也是重要的庭园和行道绿化树种、材用树种。它们的名称、分布及用途见表 1。

表 1 杧果属树种的名称、生长、分布和用途

中 名	学 名	树 高 (m)	胸 径 (cm)	分 布	用 途
杧果	*M. indica* L.	15～20	60～100	滇、桂、粤、琼、闽、台。南亚 东南亚	食用、绿化、材用、止咳、染料
扁桃	*M. persiciformis* C. Y. Wu et T. L. Ming	20～25	80～120	桂、滇、黔	绿化、材用、食用、止咳

开花结实 实生杧果 7～8 年生、扁桃 10～12 年生开始结实，寿命长，正常结实期分别在 15 年生和 30 年生以后。结实大小年间隔期为 1～5 年不等。花期喜高温、干燥。影响结实的主要原因是我国多数地区春季气温偏低且有阴雨天气，每年开花虽多，但只有较少的年份能大量结实。无性繁育的果园，多选用适生品种或辅以人工授粉和防寒等措施，每年结实较正常。花小，杂性，黄色或黄白色。圆锥花序顶生或着生于上部叶腋，花序长 10～20cm。萼片 4～5。花瓣 4～5。雄蕊 5，仅 1 个发育，不育雄蕊（1）2～4。子房上位，斜卵形和球形，1 室，1 胚珠。花柱近顶生和顶生。杧果不同年份的花期颇不一致。据广西南宁观察，多数年份 1～2 月虽有花开，但在自然条件下一般不能发育成果，3 月下旬以后所开的花，座果率较高。这两个树种的物候期见表 2。在印度和越南的热带地区，杧果多季开花结实。

表 2　杧果属树种的物候期

树　种	观察地点	观察年份	开　花			果实成熟	
			始　期	盛　期	末　期	始　期	盛　期
杧果	广西南宁	1978～1987	1月下旬～3月下旬	2月上旬～4月中旬	4月中旬	6月下旬	7月中旬
扁桃	广西南宁	1978～1987	3月上旬～中旬	3月中旬～4月上旬	4月中旬	6月中旬	6月下旬

核果，肾形或卵形，顶端稍弯，未成熟时青色，成熟后黄绿色至黄色。中果皮肉质，鲜黄色或淡黄色，味香甜。果核压扁，木质坚硬，表面具纤维。种皮薄，无胚乳，胚略有弯曲，胚根侧向，子叶肥厚，内面扁平。这 2 个树种果实和果核的形态特征见表 3；扁桃果核的外形及其剖面见图 1。

表 3　杧果属树种果实和果核的形态特征

树　种	未熟果颜色	成熟果实			种　子（果核）		
		形　状	大　小（cm）	颜　色	形　状	大　小（cm）	颜　色
杧果	青绿色	肾形、扁椭圆形，顶端微弯	长 5～10（16） 宽 3～4.5（8）	淡黄至黄色	扁椭圆形或近肾形，表面具软纤维	长 3～12 宽 2.5～6 厚 1.5～2.5	淡黄色
扁桃	青色	卵形或桃形，顶部稍尖	长 3.6～5 径 3～4	黄绿色	扁椭圆形，表面具粗纤维	长 3～4.5 宽 2～3.5 厚 1.5～2.5	淡黄色

果实的采收调制和种子贮藏　通常上树直接采摘，也可用高枝剪或采种钩刀截断果梗，在离地 0.5～1m 高处垫细网或布单承接，以免果实落地破损。杧果为名贵果品，扁桃的市场价值亦较高，采集的果实一般供食用。食用或加工后的果核置水中浸泡 1～2 天，搓擦去纤维上残余的果肉即为播种材料，通称种子。种子属于顽拗型，忌失水。据王晓峰和傅家瑞（1990，1991）报道，室温下晾干 8 天，含水量便从新采收时的 70%左右下降到 39%，相应的发芽率由 100%下降为零。他们还研究过利用人工培养基保存杧果离体胚的技术（王晓峰和傅家瑞，1990）；研究过氧气浓度对杧果种子贮藏寿命的影响（王晓峰和傅家瑞，1994）。杧果和扁桃都适于随采随播。运输时可混以湿润锯木屑，用麻袋

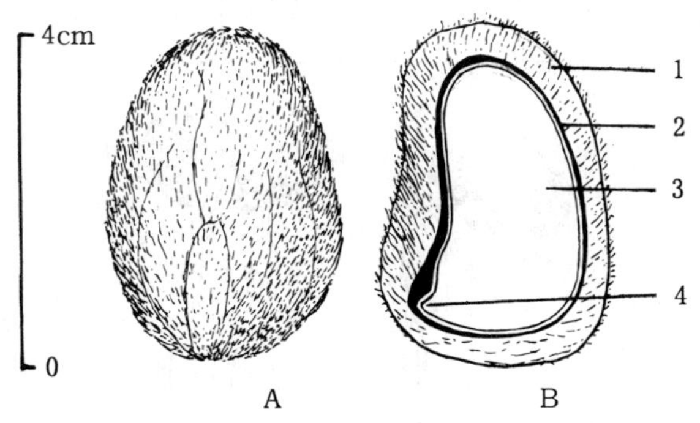

图 1　扁桃果核外形（A）及其纵切面（B）
1. 内果皮　2. 种皮　3. 子叶　4. 胚根
（黄应钦绘）

或木箱包装。贮藏需混湿沙，贮藏期一般 1～2 个月。果实出籽率及种子质量等见表 4。

表 4　杧果属树种鲜果出籽率和种子(核)质量

树　种	出籽率（%）	千粒重（g）		每千克种子粒数	
		一　般	变动范围	一　般	变动范围
杧果	10～15	25 000	15 000～50 000	40	20～70
扁桃	27～30	8 500	7 000～11 000	118	90～150

发芽和播种　种子无休眠习性。发芽时日均温需在 20℃以上。据法国学者研究，新鲜的杧果种子在 5℃和 10℃下即能发芽，发率分别为 80%和 99%，但幼苗会迅速死亡；在15℃、20℃、25℃、30℃和 35℃下发芽率均为 100%（见王晓峰和陈润政译文，1990）。1981 年 7 月 7 日和 1988 年 8 月 16 日，广西林业科学研究所采用当年新鲜种子，不作处理，分别对扁桃和杧果进行室外发芽测定，播种后扁桃 11 天、杧果 5 天开始发芽，发芽情况见表 5。

表 5　杧果属树种的发芽能力及其测定条件

树　种	温　度（℃）	发芽势（%）		发芽率（%）	
		计算天数	一般数值	计算天数	一般数值
杧果	28～30	8	50	20	70
扁桃	29～30	8	74	27	86

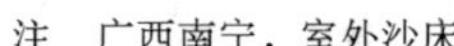

注　广西南宁，室外沙床

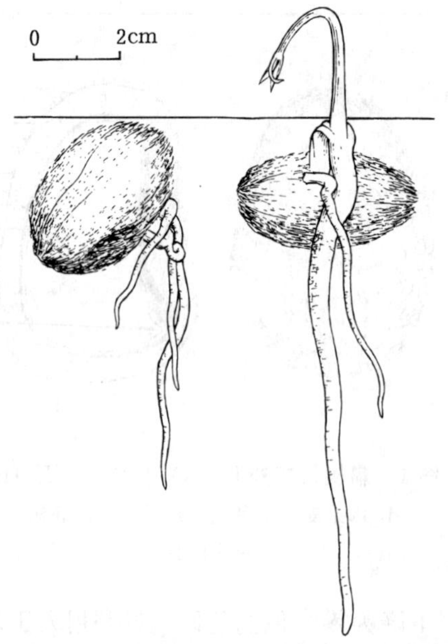

图 2　扁桃种子（多胚）萌发后第 5、9 天的幼苗生长情况
（黄应钦仿《热带亚热带主要树种采种育苗技术》）

留土萌发。杧果胚根萌发后 7～8 天胚芽出土，出土后 2～3 天发出新叶。扁桃胚根萌发后 8～10 天胚芽出土，出土约 4 天发出初生叶。少数扁桃的种子，1 颗可发芽 2～3 株，可分株移植。扁桃幼苗的生长情况见图2。

条播，或将种子密播于催芽苗床，发芽后再行移植。育苗用种每 666m² 杧果为 400～600kg，扁桃为 200～300kg。培育 1 年半生苗出圃。用于城镇绿化，需培育 3 年生以上大苗出圃。杧果多用无性繁殖。田金文和曾有才等（1991）将 ABT 生根粉用于杧果扦插，成活率近 90%。嫁接时实生苗用作砧木。

（张声燕）

黄连木属
Pistacia L.

（漆树科　Anacardiaceae）

生长习性、分布和用途　本属约 10 种，产地中海区域、亚洲和美洲中南部。我国 2 种，本文描述 3 种。黄连木、清香木原产我国，阿月浑子相传在唐朝或唐朝前 1～2 个世纪由波斯引种到我国。落叶或常绿，乔木或灌木。生长缓慢，寿命长。抗性强，但喜生于肥沃、湿润、排水良好的壤土中。木材坚固致密，可用于细木工工艺。黄连木属这 3 个树种的名称、习性、分布及用途见表 1。

表 1　黄连木属树种的名称、习性、分布及用途

中　名	学　名	习性和树高 (m)	分　布	用　途	供　稿
黄连木	*P. chinensis* Bunge	落叶 20～30	华北(除内蒙古)、华东、中南、西南，陕、甘。菲律宾	材用。种子含油率 42%，出油率 20%～30%，工业及食用。根、枝、叶、皮做农药，枝等可提芳香油	310
阿月浑子	*P. vera* L.	落叶 5～7(10)	新疆塔里木盆地栽培。中亚和西亚	干果，重要木本油料，工业及医药广泛应用，荒山、干旱地区绿化	203
清香木	*P. weinmannifolia* Poiss. ex Franch.	常绿 2～8 (10～15)	西南。缅甸	材用、药用	619

开花结实　8～15 年生开始结实，20 年生以后进入正常结实期，结实大小年现象特别明显。雌雄异株。圆锥或总状花序，长 4～25cm，腋生。花小，密生，无花瓣。雄花花被片 2～8，雄蕊 3～6（7）。雌花花被片 3～10。子房上位，1 室，1 胚珠。花柱短，柱头 3 裂。核果，近球形或长圆形，外果皮革质，内果皮骨质。果实含 1 粒种子，种皮膜质，无胚乳，子叶肉质或较薄。黄连木属树种开花结实习性见表 2，果实和种子（果核）形态特征见表 3，阿月浑子种子（ 果核）的解剖结构见图 1。

表 2　黄连木属树种开花结实物候

树　种	开始结实年龄	观察地点	花　期	果实成熟期	果实散落期
黄连木	8～10	河北涉县	3～4 月中旬	9～10 月	9～11 月
阿月浑子	10～15	新疆喀什疏附县	4 月中旬	7～9 月	7～9 月
清香木	10～15	云南西双版纳勐仑	2～3 月	5～6 月	5～6 月

表 3　黄连木属树种果实和种子形态特征

树　种	未熟果颜色	成熟果实			种子（果核）		
		形　状	大小（cm）	颜　色	形　状	大小（cm）	颜　色
黄连木	黄白色	倒卵状扁球形	径 0.5～0.8，端具小尖头	铜绿色	—	—	—
阿月浑子	黄绿色	圆形或长椭圆形	长 2～2.5 径 1～1.3	红褐色	球形或椭圆形	长 0.6～2.0 宽 0.5～1.0	白色
清香木	黄绿色	近球形	径 0.6～0.9	紫红色	矩圆形或椭圆形	长 0.6～0.8 径 0.5	—

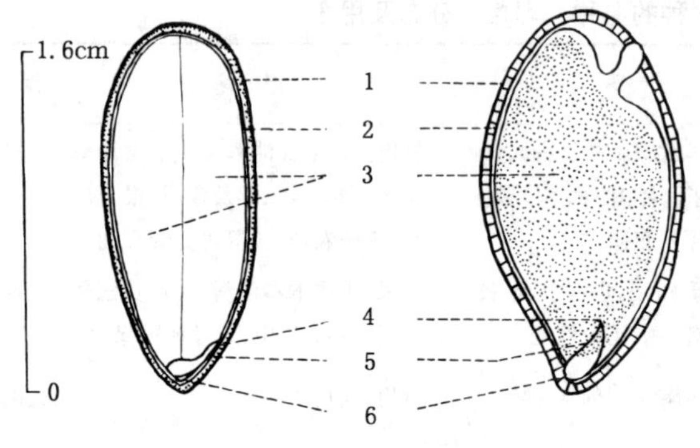

图 1　阿月浑子果核的两个纵切面

1. 内果皮　2. 种皮　3. 子叶　4. 胚芽　5. 胚轴　6. 胚根

（吴新安绘）

果实的采收调制和种子贮藏　果实夏末秋初成熟，熟后 8～10 天开始散落。果实成熟盛期将果穗剪下或用手摘取，亦可用木棒轻轻击落收集。黄连木正常果实呈铜绿色；被黄连木种子小蜂（*Eurytoma plotnikovi* Nikolskaya）蛀食一空的果实大小同正常果实区别不大，但呈红色，应当集中销毁。将正常核果放在缸中浸泡或堆沤数日，揉搓淘洗，除去果皮蜡质后所得的果核即为播种材料，通称种子，晾干后播种或贮藏。阿月浑子干缩后的果皮很难清除。但采回的鲜果当天用水浸湿，轻加搓揉即可脱去外皮，洗净风干数日，所得果核装入布袋，放在阴凉干燥室内贮藏。黄连木、清香木种子不耐久藏，忌失水，春播或短途运输要混沙或木屑进行湿藏。每千克纯净种子粒数在树种间差异很大，其出籽率、净度及质量见表 4。

表 4　黄连木属树种出籽率、种子净度及质量

树　　种	出籽率（%）	净　度（%）	千粒重（g）		每千克纯净种子(果核)粒数(万粒)	
			一　般	变动范围	一　般	变动范围
黄连木	—	90～95	92	—	1.1	—
阿月浑子	35～48	85～100	590	485～800	0.17	0.13～0.21
清香木	60～75	75～90	95	80～100	1.05	1～1.3

发芽和播种　据邵蓓蓓（1989）试验，黄连木种子的萌发障碍在于种皮，层积可以克服

这种障碍：未层积的种子在各种发芽温度下发芽率均小于 10%，在 0～5℃下层积 60 天后发芽率可以达到 89.5%。据本文作者试验，在 10～25℃下培养 10～15 天，未经处理的阿月浑子发芽率可达 80%～90%；清香木发芽率较低，仅 20%～30%。黄连木属这 3 个树种春播及秋播均可。阿月浑子因种皮厚，吸水慢，春播前也需短期层积催芽。

条播或点播：黄连木行距 30cm，每平方米播种 15～20g，覆土 2～3cm。阿月浑子每平方米播种 120g 覆土 3～5cm。播后 7 天开始出土，30 天幼苗可出齐。黄连木属子叶 2 枚。萌发方式多样：阿月浑子为留土萌发；清香木、黄连木为出土萌发，但黄连木的下胚轴仅将子叶顶出土面，可称地面萌发。图 2 是阿月浑子种子萌发和幼苗的生长情况。

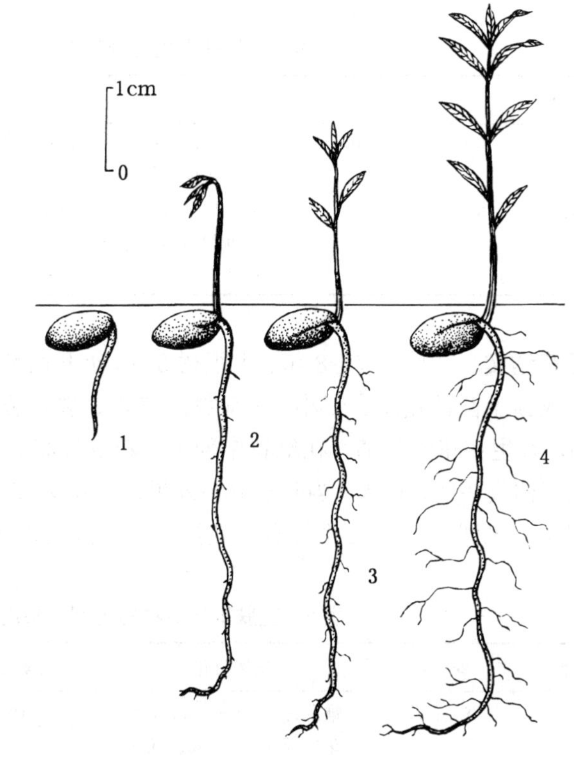

图 2　阿月浑子种子(核)的萌发和幼苗生长情况

1. 幼根生长　2. 幼芽出土　3. 初生叶对生，第 1、2 片偶为互生　4. 幼苗形成

(吴新安绘)

(周君英)

盐肤木属

Rhus L.

(漆树科　Anacardiaceae)

生长习性、分布和用途　盐肤木属约 250 种，分布于亚热带和暖温带。我国 6 种，除黑龙江、内蒙古、新疆、青海外均有分布，皆可作五倍子蚜虫寄主植物，但以盐肤木上的虫瘿较好，称“角倍”，其余称“肚倍”，质量较次。本文描述盐肤木和引入的火炬树两种。喜光，对土壤要求不严，耐干旱、瘠薄，后者更耐寒、耐盐碱，喜生河谷沙滩、堤岸及沼泽地缘，亦能生砂砾质土上，垂直根较浅，水平根发达，根萌芽性极强。落叶灌木或小乔木，它们的名称、分布和用途见表 1。

表 1 盐肤木属树种的名称、分布和用途

中 名	学 名	分 布	用 途	供 稿
盐肤木	*Rh. chinensis* Mill.	除内蒙古、黑、新、青外，各地均有分布。东亚、南亚	鞣料、药用、塑料、油料、材用、观赏	1002
火炬树	*Rh. typhina* L.	原产加拿大东南部和美国中部，本世纪50年代引入我国华北、西北	水土保持、鞣料、薪炭、油料、药用、蜜源、观赏	302

开花结实 盐肤木7～8年、火炬树3～5年开始开花结实。花单性或杂性，同株或异株。多花组成顶生圆锥花序。花小，5基数。花萼5裂，裂片覆瓦状排列。花瓣5，覆瓦状排列。雄蕊5，着生在花盘基部，在雄花中伸出。花盘环状。子房上位，1室1胚珠。花柱3，基部略合生，柱头头状。核果，当年成熟。外果皮与中果皮连合，内果皮分离，中果皮无蜡质。每果1果核（见图1)。这2个树种的花期、果期及其形态特征见表2。

表 2 盐肤木属树种的花期、果期及其形态特征

树 种	花 期	花序及花	果实成熟期	果实和果核特征
盐肤木	8～9月	雄花序长30～40cm，雌花序长15～20cm。花黄白色。子房密被白色微柔毛	10月	核果扁球形，径3.8～5.6mm，橙红色或紫褐色，表面被盐粒状结晶及灰白色短毛。萼片和花柱宿存。果核扁圆形，长2.6～2.9mm，宽2.9～3.1mm，厚约1.8mm。核壳（内果皮）灰绿色，有光泽，骨质
火炬树	5月下旬～7月	雌雄异株，花序长10～20cm，密被绒毛。花淡绿白色。子房、花柱被红色短刺毛	9月	果及果核均较盐肤木小。核果扁球形，被红色短刺毛，聚生为火炬状果穗。果核扁肾形，长1.8～2mm，宽2.4～2.6mm，厚约1.5mm。核壳（内果皮）黑褐色，坚硬，骨质

果实的采收调制和种子贮藏 盐肤木和火炬树果实成熟后，果穗在枝上能滞留到次年果梗腐烂才脱落，采种期可长达半年。剪取果穗，晾3～5天，捋下果实，用石臼捣碎或石磙碾碎外果皮和中果皮，漂洗干净，晾干，所得果核即为播种材料，通称种子。火炬树的果实也可以在3%的碱水中浸泡2小时后搓去果皮，晒干，簸去杂质。果实出籽率达70%。盐肤木果核千粒重7.7～9g，每千克纯净果核11万～13万粒。火炬树果核千粒重7.5～10g，每千克有纯净果核10万～13.3万粒。这2个树种的种子用普通的方法便可贮藏越冬。在0～5℃下密封贮藏的火炬树种子可以存放2年。

发芽和播种 同盐肤木属的许多树种一样，本文描述的这2个种的种子（果核）外壳坚硬，透性较差，不经处理很难发芽。李晓洁和徐元（1989）曾用室温水和始温45℃的水浸泡火炬树种子，没有收到效果；用浓硫酸浸种也只能部分地消除硬粒。在她们的报道中，处理火炬树最有效的办法是用100℃的水浸烫并在自然冷却过程中保持24小时，发芽率可以达到

88%，供试种子中只有1%仍保持硬粒状态。据她们研究，经过沸水浸烫的火炬树种子，在30℃/25℃、30℃/20℃、25℃/20℃的昼夜变温下，或者是在25℃和20℃的恒温下，无论有无光照都能正常发芽。国外也有在室温下用酸腐蚀的办法处理火炬树的，据说发芽进程比沸水浸烫的短。盐肤木和火炬树使用过的播前处理方法还有：①将果核用凉水浸泡一昼夜，捞出与湿沙混合（比例为1：2），放在温度约20℃的室内，每隔1天翻动1次，并始终保持湿润状态，直至果核裂嘴时播种。②用85℃左右的水浸烫果核，在不停地搅拌中令其自然冷却并继续浸泡2昼夜，每天换水1次，然后捞出装在麻袋内，置于15℃条件下，每天翻动1～2次，待约50%的果核裂嘴时播种。

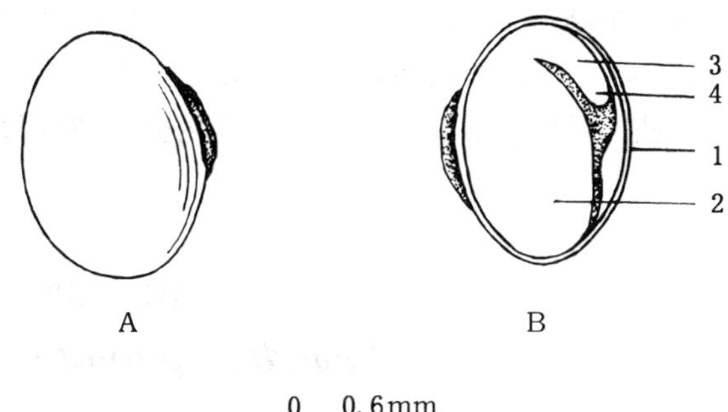

图1　火炬树果核外形（A）及其纵切面（B）
1. 内果皮及种皮　2. 子叶　3. 胚轴　4. 胚根
（孟玲绘）

可以秋播，也可以用经过处理的种子春播。条播。春旱地区，苗床开沟后应在沟内浇水，待水渗透后播种。盐肤木行距25cm，每平方米约播种2g。净度70%以上、发芽率约50%的火炬树种子，如果行距30cm，每平方米约播1.5g。在这样的技术条件下，每平方米可以育出高1m、地径1cm的苗木25株（刘培华和薛守辉，1985）。播后覆土约1cm，然后盖草。1周后发芽出土。出苗后逐渐揭除盖草。出土萌发。火炬树种子萌发及幼苗初期生长情况见图2。

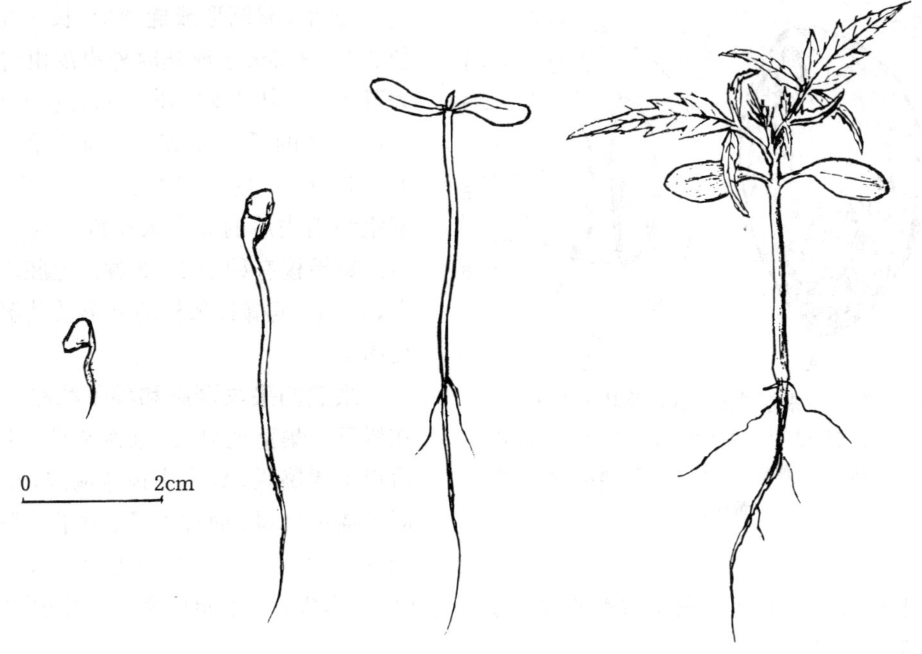

图2　火炬树种子萌发后第4、6、8、28天的幼苗生长情况
（孟玲、田恒德绘）

子叶匙状矩圆形，平展；下胚轴长12～30mm，圆柱形，深红色，无毛。初生叶为3小叶羽状复叶，第一、二两片对生，第三叶以上互生。

盐肤木可以扦插繁殖。火炬树多用分蘖或埋根繁殖。

（黄鹏成）

槟　榔　青

Spondias pinnata（L. f.）Kurz

（漆树科　Anacardiaceae）

生长习性、分布和用途　槟榔青属约12种，我国产3种，本文描述1种。落叶乔木，高10～15m，胸径约50cm。能耐－1℃左右的极端低温。适于土层较厚、较为湿润的酸性至中性土。产于云南、广西和海南。广东引种。印度、斯里兰卡、缅甸、泰国、柬埔寨、越南、马来西亚、菲律宾、印度尼西亚亦产。果及幼叶可食，树皮可提栲胶。

开花结实　8～12年生开始结实，15年生以后为正常结实期。大小年现象不明显，但偶有不开花的年份。花先叶开放，杂性。聚伞圆锥花序顶生，长25～35cm。花小，白色。萼短杯状，4～5裂。花瓣4～5，芽中镊合状排列。雄蕊10，花丝线形，具乳突体。花盘大，10裂。子房上位，5室，每室1胚珠，花柱顶生。据1983～1988年在云南西双版纳勐仑观察，1月底或2月初花芽始现，2月中下旬花序形成，3月上中旬始花，3月下旬盛花，4月上中旬花期结束；10月中旬果实开始成熟，11月至翌年1月为果实成熟盛期。

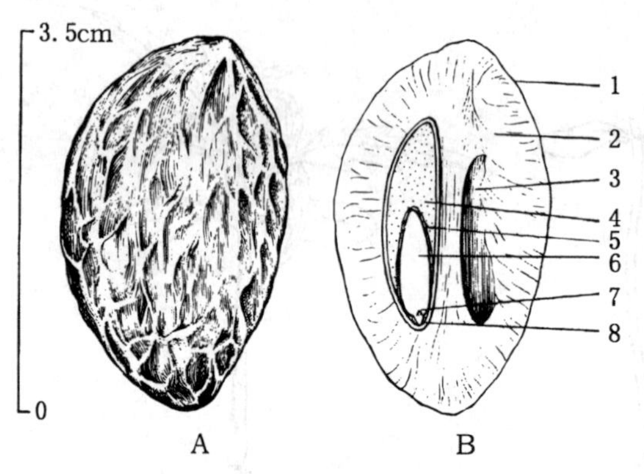

图1　槟榔青果核外形（A）及其纵切面（B）

1. 内果皮外层　2. 内果皮内层　3. 不育子房室　4. 大空腔　5. 种皮　6. 子叶　7. 胚芽　8. 胚根

（黄应钦绘）

核果，卵圆形或椭圆形，长3.5～5cm，径2.5～3.5cm。成熟时外果皮由青色转变为黄褐色。中果皮肉质。内果皮外层为密集纵向排列的纤维质和少量疏松的软组织，无刺状突起；内果皮内层木质，坚硬，有薄壁组织消失后的5个大空腔，与子房室互生。每果核有种子1～3颗。无胚乳。胚稍大，肉质。槟榔青果核的外形及其解剖结构见图1。

果实的采收调制和种子贮藏　盛熟期在树下拾集落地果实，堆沤数日。果皮软化后用木棒捣烂，置水中搓揉淘洗，取出果核即是播种材料，通称种子。鲜果出籽率约为28%。净度可达95%。千粒重约7 600g，每千克有果核120～140粒。调制后果核含水量约为40%，忌失水，不能日晒。贮藏时需混湿沙，贮藏期6～8个月。

发芽和播种　种子有中等程度的休眠。发芽时日均温以23℃左右为宜。调制所得的果核以湿沙层积至翌年2～3月播种为好。1988年11月8日，西双版纳热带植物园在室外沙盆用新鲜果核作过发芽测定：播后历时 4 个月才开始萌发，没有明显的发芽盛期，发芽全程为 53 天，发芽率为 70%。经湿沙层积至翌年 2 月中旬播种，则播后 28 天左右便开始萌发，发芽全程约 40 天，仍无发芽盛期，发芽率为 80%。出土萌发。子叶出土后 5 天左右展开，同时出现初生叶。槟榔青果核的萌发和幼苗初期生长情况见图 2。图 2 所绘的初生叶 1、2 为单叶，对生。据记载（张若蕙等，1993），亦有第 1、2 初生叶为 3 或 5 小叶的复叶，对生。

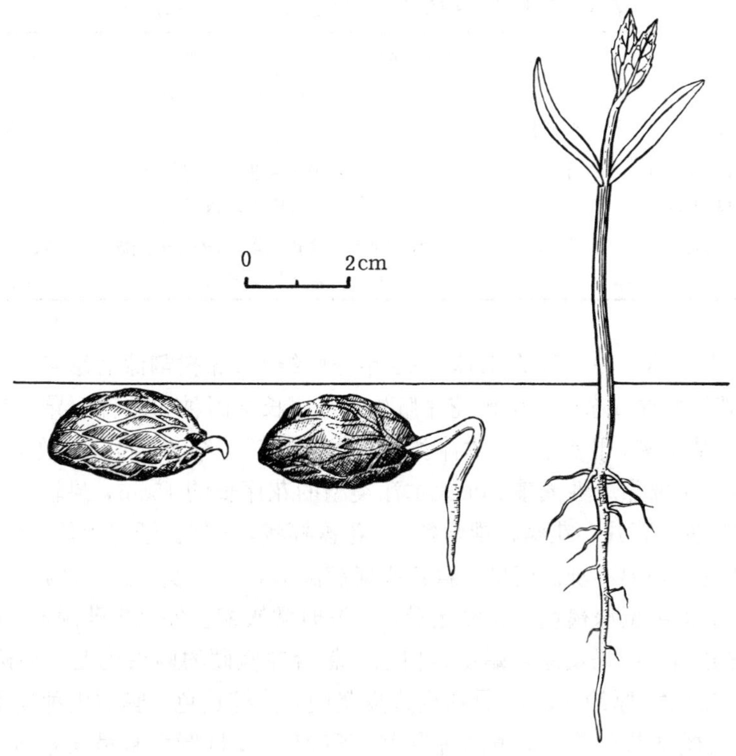

图 2　槟榔青种子萌发后第 1、3、10 天幼苗的生长情况

（黄应钦绘）

点播。每平方米播种 750～1 000g，覆土 1～2cm。1 年生苗出圃。

（马信祥）

漆　树　属

Toxicodendron（Tourn.）Mill.

（漆树科　Anacardiaceae）

生长习性、分布和用途　本属约 20 余种，分布于亚洲东部和北美至中美，我国有 15 种。本文描述 3 种，它们的名称、生长、分布和用途见表 1。落叶乔木或灌木，多生于海拔 1 800（2 500）m 以下的阳坡、沟旁疏林中，是提供漆、蜡等工业原料的重要经济树种。分布区域广。其中漆树为我国特产，19 世纪 70 年代传入欧美。从漆树韧皮部割取的生漆，是优良的防腐、防锈涂料，用途广泛；从果皮提取的蜡和从种子榨取的油，都是重要的工业原料。我国漆树栽培历史悠久，形成了许多农家品种，一般分为3个形态类型：大木漆类型为大乔木，树高可达15m以上；小木漆类型则为小乔木，树高4～8m；介于这两者之间的小大木漆类型树高8m左右。

表1　漆树属树种的名称、生长习性、分布和用途

中　名	学　名	生长习性	树高（m）	分　布	用　途	供　稿
野漆	*T. succedaneum* (L.) O. Kuntze	小乔木或灌木	10	华南、华东、西南、河北	栲胶、蜡、油、药用	1003
木蜡树	*T. sylvestre* (Sieb et Zucc.) O. Kuntze	乔木	10	华北南部至华南、西南，以长江中下游最多	蜡、油、漆	1003
漆树	*T. vernicifluum* (Stokes) F. A. Barkl.	乔木	4～15，因品种不同而差异较大	除新、黑、吉外几遍全国，栽培历史悠久	漆、蜡、油、栲胶、药用	511

开花结实　天然林8～9年、人工林5～6年开始结实。漆树割漆后结实有大小年现象，间隔期1～2年。雌雄异株或杂性。圆锥花序腋生，花序长度因种和品种而异。大木漆类型的花序长30cm以上，结实多，果实小。小木漆类型基本上不开花结实，偶有少数花序，长仅10cm，且种子多为空粒，主要靠无性繁殖。小大木漆类型的花序长约15cm，果较大。花小，直径仅约1～2mm，黄绿色。萼片、花瓣、雄蕊各5。花盘杯状，5裂。子房上位，1室，花柱3，或花柱1而柱头3裂。果穗下垂。核果，扁圆或扁椭圆形，有尖或无尖，偏斜，一般宽大于高，成熟后淡黄色、灰褐色至深褐色，表面光滑，干燥时常皱缩。外果皮薄膜质，脆，常具光泽，成熟时与中果皮分离。中果皮厚，蜡质，白色，粘合在坚硬的内果皮上。果核扁肾形，宽约6～7mm，长4～5mm，厚约2mm。骨质内果皮坚硬，淡棕黄色。膜质内种皮不明显。种子具胚乳或近于无，或微量贴在子叶上侧的种皮上。子叶扁平，椭圆而略呈长方形，下胚轴及胚根较长，横卧于胚的下端。木蜡树果核的外形和横切面与纵切面见图1。花期5～6月，果熟期9～11月（表2）。果熟后短期不落。

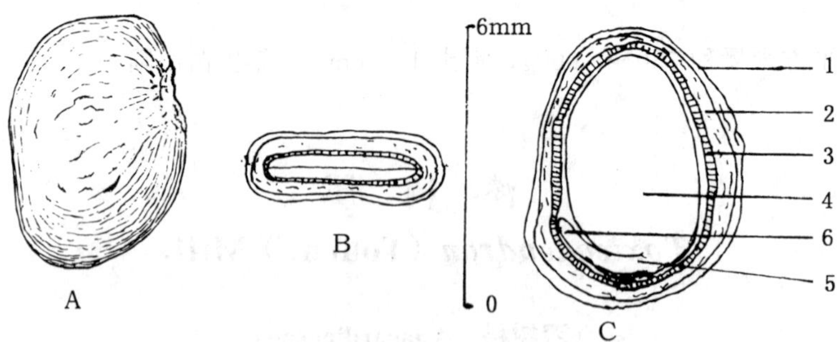

图1　木蜡树果核外形（A）及其横切面（B）和纵切面（C）

1. 内果皮　2. 外种皮　3. 内种皮　4. 子叶　5. 胚轴　6. 胚根

（史渭清绘）

表 2　漆树属树种花果特性及开花结实物候期

树　种	圆锥花序长（cm）	花直径（mm）	果实直径（mm）	开花期（月）	果熟期（月）
野漆	5～11	2	6～8	6	9～10
木蜡树	6～10	—	8～10	5～6	10～11
漆树	11～35	1	6～8	5～6	9～11

漆树的开花结实物候期还因品种类型而有差异。表 3 是王道植和徐玉蓉（1988）在贵阳地区对漆树的 6 个品种所作观察的部分资料。据他们观察，这 6 个品种在贵阳地区的开花从始期到末期大约经历 20 天，每朵小花从花开到花谢大约经历 4～5 天。整个花期中雄花大量散粉期同雌花最适受粉期重合的小花占总花量的比例常仅 50%～66%，且为时只有 6～7 天。这段时间放蜂能促进结实。他们还发现，这 6 个品种自交的座果率不高，最低的为零，最高的也只有 23%，表明漆树具有防止自交衰退的某种机制。但 6 个品种自由授粉座果率最高只有 57%，低的只有 35%，仅占雌花序小花数的 1/3～1/2。他们推测，可能的原因是部分小花的花期不遇、花期中传粉昆虫数量不足以及部分小花败育。

表 3　贵阳地区漆树 6 个品种的开花结实物候期

品　种	花的性别	开　花			果实形态成熟期
		始　期	盛　期	末　期	
官大木	雌花	5 月上旬	5 月中旬	5 月中旬	8 月中旬
	雄花	5 月上旬	5 月上旬	5 月中旬	
红光大木	雌花	5 月中旬	5 月中旬	5 月中旬	8 月中旬
	雄花	5 月上旬	5 月中旬	5 月中旬	
红漆大木	雌花	5 月上旬	5 月中旬	5 月中旬	9 月上旬
	雄花	5 月上旬	5 月中旬	5 月中旬	
肤胭皮	雌花	5 月上旬	5 月中旬	5 月中旬	9 月上旬
	雄花	5 月上旬	5 月上旬	5 月中旬	
小大木	雌花	5 月上旬	5 月中旬	5 月中旬	9 月上旬
	雄花	5 月上旬	5 月上旬	5 月中旬	
粉红皮	雌花	5 月上旬	5 月中旬	5 月中旬	9 月中旬
	雄花	4 月下旬	5 月上旬	5 月中旬	

果实的采收调制和种子贮藏　漆树的品种多，应分别品种采收调制。漆树属的核果常有雀鸟啄食，有益于生态平衡。但为了取得播种材料，仍应在核果出现成熟特征之时选择成熟果穗，剪下摊晾 3～5 天后除去果梗等杂质，再作进一步调制。漆树属果实调制的重要内容是除去外果皮特别是中果皮，俗称脱蜡。所得的果核即为播种材料，通称种子。脱蜡有干法和

湿法2种。干法脱蜡是采用臼、磨、碾或碾米机之类的简单工具粉碎外果皮和中果皮，强使它们同果核分离，筛去粉碎物即得果核。有的地方在干法脱蜡时拌入谷壳等物，增加摩擦力，又能吸附部分蜡质。干法脱蜡所得的果核无需晾晒，便于运输贮藏。湿法脱蜡有烫种和退蜡2个步骤：向盛放果实的容器注入沸水并不断搅拌，直至水温降到30～40℃时撇去浮在水面的空瘪粒。倒出容器中烫种的水，加入按100：50：1（鲜果：水：碱）的质量比配制的碱液（可以用洗衣粉代替碱），浸泡约10分钟后加入适量细沙搓揉，直至果核变为黄白色且有粗糙感，倒掉碱液并用清水多加冲洗，所得果核晾干即为播种材料。湿法脱蜡比较彻底，但贮藏或运输前必须晾干。安志文（1982）认为湿法脱蜡还具有促进发芽的作用。

漆树属树种的出籽率约为55%～65%，净度可达90%以上。果核的质量数据已列入表4。含水量8%左右的漆树种子低温密封贮藏，发芽能力可以保持3～5年。短期存放的可以干藏，供来年春播的也可以混沙湿藏。

表4 木蜡树及漆树的出籽率、千粒重和发芽率

树 种	出籽率（%）		千粒重（g）	每千克种子（核）数（万粒）	发芽率（%）
木蜡树	55		60	1.6	70～80
漆树	65	大木漆类型：	23～30	3.3～4.3	40～90
		小大木漆类型：	28～34	2.9～3.6	20～50 （空粒较多）

发芽和播种 漆树属种子发芽困难，发芽迟缓且极不整齐，不少种粒甚至要到播后的第二个生长季节才会萌发。有人推测是坚硬的核壳妨碍吸水；也有人认为是胚或胚乳中存在着发芽抑制物质。据李近雨（1988）观察，漆树内果皮（核壳）的外层是一层栓化细胞，第2层是细胞狭长、排列紧密的栅状组织，再向内是5～6层石细胞。他认为，正是这3类壁厚而坚硬的细胞迫使漆树种子处于休眠状态。他观察过漆树种子的酸蚀过程：内果皮表层的栓化细胞容易受到酸蚀而被破坏；栅状组织则耐酸力强，因此漆树种子酸蚀处理0.5、1.0和2.0小时的效果差异不大。他还推测，硫酸腐蚀漆树种子的作用似乎不在于破坏栓化细胞层，而在于消除发芽孔处的某种堵塞物。

有一种处理漆树种子的方法是，用沸水浸烫后转入80℃碱水中搓洗，称为高温脱蜡；再在始温80℃水中浸种36小时，称为高温浸种；最后在28～38℃的环境中催芽，称为高温催芽。据说经过这种“三高”法处理的漆树种子，播后40天的出苗率能达到96%（邓中美，1986）。李近雨（1988）则发现，在60～80℃的范围内，随着浸种水温的升高，漆树种子的腐烂粒越来越多，他因此认为漆树不宜烫种。徐玉蓉和王道植等人（1989）以肤胭皮和官大木这2个品种为材料研究过漆树种子的预处理，认为最佳方法是浓硫酸浸种1～1.5小时，彻底冲洗后在2～5℃下层积20～30天。李近雨（1988）认为3～5℃下层积超过15天就可以解除漆树种子的休眠。

为了促进漆树种子发芽，《中国主要树种造林技术》列举的方法是浸种催芽：经过脱蜡的漆树种子水浸 10～20 天，经常换水，种子开始膨胀时装入筐篓，每天用 25℃左右的温水淋浇 2～3 次，每隔 2～3 天用温水淘洗 1 次，约 10 天后少量种子裂嘴露白时取出播种。徐玉蓉和王道植等（1989）建议，经过酸蚀和层积的漆树种子发芽测定时的最适温度是 15～20℃。李近雨报道（1988）说，层积后的漆树种子应在 17℃的温度下发芽，25℃的发芽环境会引起二次休眠。徐本美（1988）也得到过类似的结果：经过酸蚀和层积的漆树种子发芽的最适温度是 15～20℃，发芽率可达 90％；在这个温度范围之外，发芽率显著降低，10℃时降到 30％，30℃时只有 9％。

漆树属种子应在采收、脱蜡、催芽后及早冬播，或层积后春播。条播，条距 50cm。漆树的大木漆每 666m² 播种 6～10kg，小大木漆每 666m² 播种 15～20kg。覆土厚 1～1.5cm，盖草。经过层积的种子春播后约 20～30 天发芽出土，冬前播种的出苗期可提早 20～25 天。出土萌发。本属有的种，例如本文未收的小漆树（*T. delavayi*（Franch.）F. A. Barkl.），为留土萌发。子叶质地较厚。初生叶对生，有的种为单叶，有的种为 3 小叶复叶。木蜡树的初生叶为单叶（图 2），稀为三小叶复叶。产区南部 1 年生苗可以出圃，长江以北多用 2 年生苗出圃。每 666m² 产苗约 1 万株。漆树可以分根繁殖。

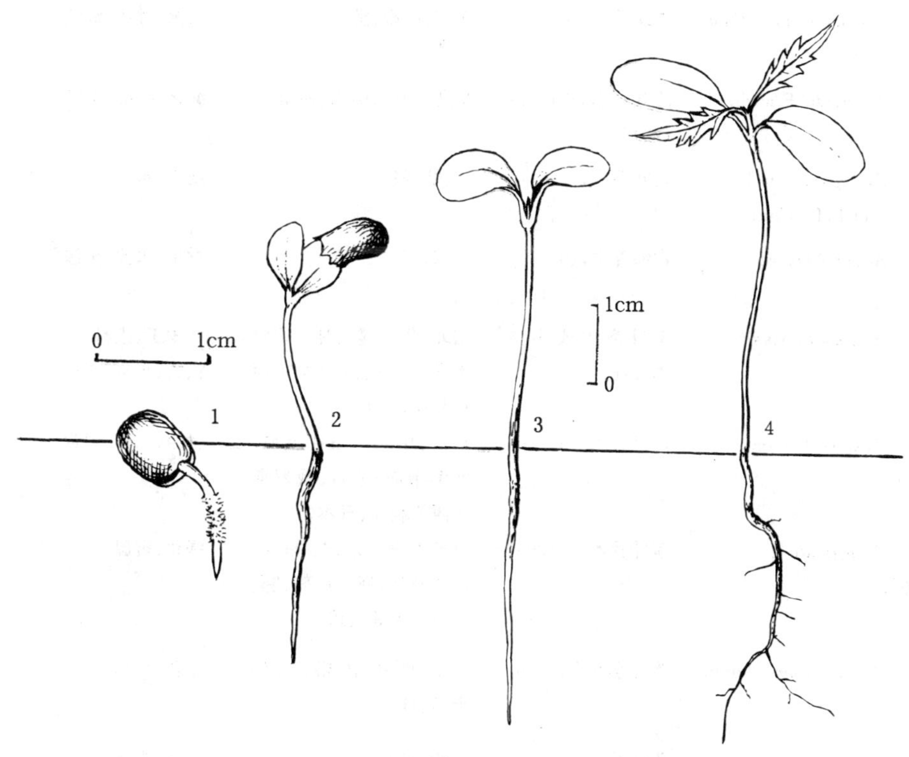

图 2　木蜡树种子（核）的萌发和幼苗生长情况

1. 胚根延伸　2. 子叶带壳出土　3. 子叶平展　4. 初生叶对生

（史渭清绘）

（何泽英）

槭　　属

Acer L.

（槭树科　Aceraceae）

生长习性、分布和用途　本属约200余种，分布于北温带，我国140余种，本文介绍12种。木材可以用于细木工。花为蜜源。有的种子含油脂。叶形秀丽，入秋多呈红色，供观赏。它们的名称、生长、分布和用途见表1。

表1　槭属树种的名称、生长、分布和用途

中　名	学　名	生长(m)	分　布	用　途	供　稿
髯毛脉槭（簇毛槭）	*A. barbinerve* Maxim.	落叶乔木，5～15	东北。朝鲜半岛，东西伯利亚	细木工、观赏	104
樟叶槭	*A. cinnamomi-folium* Hay.	常绿乔木，10～20	东南、中南、黔	水源、材用、蜜源	603
青榨槭	*A. davidii* Franch.	落叶乔木，10～15(20)	华北、华东、中南、西南	绿化、纤维、单宁	801
秀丽槭	*A. elegantulum* Fang et P. L. Chiu	落叶乔木，9～15	浙、皖、赣	观赏、水源	801
罗浮槭（红翅槭）	*A. fabri* Hance	常绿乔木，10	中南、川	材用、水源、蜜源	603
茶条槭	*A. ginnala* Maxim.	落叶灌木或小乔木，2(6)	东北、华北、陕、甘。朝鲜半岛、蒙古、俄罗斯东西伯利亚、日本	细木工、染料、饮料、观赏、蜜源	104
色木槭（五角槭、地锦槭）	*A. mono* Maxim.	落叶乔木，15～20	东北、华北至长江流域。朝鲜半岛、蒙古、俄罗斯东西伯利亚、日本	材用、油料、观赏	102
梣叶槭（糖槭、复叶槭）	*A. negundo* L.	落叶乔木，15(20)	原产北美。东北、华北、西北及江、浙、皖、赣、鄂、湘、川、黔等栽培	绿化、蜜源	104
鸡爪槭	*A. palmatum* Thunb.	落叶小乔木，3～5	华东、华中、滇、黔。朝鲜半岛、日本	绿化、药用	801
天目槭	*A. sinopurpurascens* Cheng	落叶乔木，10	浙西、赣北	绿化、观赏	1003
元宝槭	*A. truncatum* Bge.	落叶乔木，8～10	东北南部、华北、陕西	油料、鞣料、绿化、观赏、蜜源	401
花楷槭	*A. ukurunduense* Trautv. et Mey	落叶乔木，8～10(15)	东北。朝鲜半岛、俄罗斯东西伯利亚、日本	材用、观赏	104

开花结实　10 年生左右开始结实，20 年生以后进入正常结实。元宝槭的丰年间隔期为1～2 年，其余树种的结实周期不甚明显。这 12 个树种的开花结实物候期见表 2，果实和种子的形态特征见表 3。

表 2　槭属树种的开花结实物候

树　种	观察年份和地点	花　期			果　期		成熟特征	果实散落期
		始　期	盛　期	末　期	始　期	末　期		
髯毛脉槭	—	5 月	～	6 月	9 月	9 月	黄褐色	10 月上旬
樟叶槭	1987～1988 广西南宁	3 月中旬	3 月下旬	4 月上旬	10 月上中旬	10 月下旬	浅褐色至深褐色	11 月上旬～12 月中旬
青榨槭	—	4 月	～	4 月	9 月	9 月	黄褐色	—
秀丽槭	—	5 月	～	5 月	9 月	9 月	淡黄色	—
罗浮槭	1987～1988 广西南宁	3 月下旬	4 月上旬	4 月中旬	10 月中旬	10 月下旬	黄褐色至棕褐色	11 月上旬～12 月中旬
茶条槭	1963～1980 哈尔滨	5 月	～	6 月	8 月上旬	9 月下旬	深褐色	10 月中旬～深冬
	1991 黑龙江五常	5 月 16 日	5 月 31 日	6 月 7 日	9 月 1 日	9 月 29 日	淡黄色或淡黄褐色	10 月上旬
色木槭	1991 黑龙江五常	5 月 8 日	5 月 12 日	5 月 20 日	9 月 4 日	10 月 2 日	—	10 月上旬
桦叶槭	1963～1980 哈尔滨	4 月下旬	～	5 月	8 月下旬	10 月中下旬	淡黄褐色	9 月下旬～翌春
鸡爪槭	1963 杭州	4 月 2 日	4 月 8 日	4 月 16 日	（盛期 9 月 28 日）		淡棕黄色	10 月 20 日
天目槭	—	4 月	～	4 月	9 月	9 月	—	—
元宝槭	（20 年观察）北京	4 月中旬	4 月中旬	4 月下旬	9 月下旬	10 月下旬	黄褐色	—
花楷槭	—	5 月	～	6 月	9 月	9 月	黄褐色	10 月上旬

表 3　槭属树种果实和种子形态特征

树　种	未熟果颜　色	成熟果实			小　坚　果		
		翅展角度	大　小（连翅，cm）	色　泽	形　状	大　小（mm）	颜　色
髯毛脉槭	—	果翅成钝角，约 120°	长 3（3.5）～4.0	黄褐色	卵圆形	—	—
樟叶槭	黄绿色	翅果张成锐角或近直角	长 2.8～3.2 宽 0.8～1.2 厚 0.32～0.38	淡黄褐色至深褐色	椭圆形，种子心脏形或麦粒状	长 3～5 径 2～4	棕褐色
青榨槭	淡绿色	张成钝角或几乎成水平	长 2.5～3.0	黄褐色	卵圆形	—	—
秀丽槭	淡紫色	开张近水平	长 2.0～2.3	淡黄色	凸起近于球形	径 6	—
罗浮槭	紫红色	张开成钝角	长 2.4～3.0 宽 0.7～1.2 厚 0.36～0.44	黄褐色至棕褐色	椭圆形，种子心脏形或锥形	长 4～5.5 径 3～4	棕褐色

（续）

树　种	未熟果颜　色	成熟果实			小　坚　果		
		形　状	大　小（连翅，cm）	颜　色	形　状	大　小（mm）	颜　色
茶条槭	黄绿色	张开近于直立或锐角	长 2.5～3.0 宽 0.8～1.0	果翅褐色带紫红色	扁平或长圆形	长 8 宽 5	褐色
色木槭	紫褐色	张开成锐角或近于钝角	长 2.0～2.5 翅宽 0.5～0.8，果梗长 2	淡黄色或淡黄褐色	压扁状或稍凸出	长 10～13 宽 5～8	—
梣叶槭	黄绿色	张开成锐角约 70°	长 3.0～3.5	淡黄褐色	长圆形，两面略突起，中央部凹入	—	—
鸡爪槭	紫红色	张开成钝角	长 2.0～2.5 宽 1.0	淡棕黄色	球形	径 7	—
天目槭	—	交角约 50°	长 3.5。翅宽 1，梗长 0.8～1	黄褐色	小坚果特别凸起，脉纹显著	径 1	
元宝槭	青绿色	张成直角或钝角，似元宝	长 2.5～3.0 翅宽 0.7～0.9	黄褐色	压扁状	长 13～18 宽 10～12	—
花楷槭	淡红色	张开成直角	长 1.5～2.0 宽 0.6	黄褐色	卵圆形	径 6	—

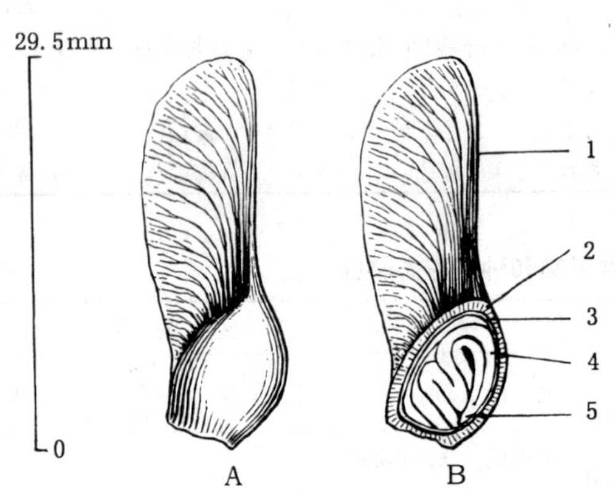

图 1　元宝槭果实外形（A）及其纵切面（B）
1. 果翅　2. 果皮　3. 种皮　4. 子叶　5. 胚根
（黄应钦绘）

花小，整齐。雄花与两性花同株或异株，稀单性异株。总状花序、圆锥花序或伞房花序。花 5 数，罕 4 数。萼片有时合生，有时无花瓣。花盘常大而为环形，稀分裂或无花盘。雄蕊 4～10，常为 8。子房上位，2 室，每室 2 胚珠。花柱 2 裂，柱头通常反卷。果为 2 枚相连的小坚果，侧面有长翅也称翅果。小坚果含种子 1 枚，无胚乳，胚卷折（图 1）。

果实的采收调制和种子贮藏　翅果成熟后脱落期较长，逐渐随风飘落，早落者质量较好，故应及时采集。上树用钩镰从立木上采摘或敲击后在地上收集。采后晾晒 3～5 天，去杂（生产上多不去翅）后所得纯净翅果即为播种材料，通称种子。含水量 9%～11%，在低温下密封干藏可保存 3 年，在 15℃下可以保存 2 年，普通干藏不可超过 1 年。槭属的出籽率和种子质量见表 4。

发芽和播种　樟叶槭和罗浮槭的种子无明显休眠习性。除它们以外，本属的种子大都需要低温层积催芽。王成霖和邵蓓蓓（1987）将青榨槭种子放在 0～5℃下分别层积 0 天、10 天、

表 4　槭属树种的出籽率和种子质量

树　种	出籽率（%）	净　度（%）	千粒重（g）		每千克纯净种子粒数（万粒）	
			一　般	变动范围	一　般	变动范围
髯毛脉槭	—	—	—	35～40	—	2.5～2.9
樟叶槭	28～32	25～40	带翅 55	40～70	1.8	1.4～2.5
			去翅 14	10～20	7.1	5.0～10.0
青榨槭	—	—	31	29～33	3.2	3.0～3.4
秀丽槭	—	—	29	28～31	3.4	3.2～3.6
罗浮槭	35～42	25～40	带翅 68	50～85	1.5	1.2～2.0
			去翅 29	25～32	3.5	3.1～4.0
茶条槭	40	70～80	12	—	8.3	—
色木槭	50	90～95	95	—	1.0	—
梣叶槭	65	90～95	—	32～40	—	2.5～3.2
鸡爪槭	—	—	12	11～13	8.1	7.5～8.8
天目槭	—	—	200	—	0.5	—
元宝槭	—	90～95	—	带翅 136～186	—	0.54～0.74
				去翅 125～175	—	0.57～0.80
花楷槭	50	80～90	20	—	5.0	—

20 天和 34 天后，得到的发芽率分别是 6%、36%、88%和 91%。在同样的温度下他们将秀丽槭分别层积 0 周、2 周、6 周和 9 周后，得到的发芽率分别是 2%、17%、80%和 88%。1988 年，广西林业科学研究所曾在恒温 28℃的发芽箱中测试过樟叶槭 1 份种子样品的发芽能力，基质为滤纸，置床前种子未作任何处理，置床后第 8 天开始发芽，到第 13 天止发芽粒数已达 47%。鸡爪槭种皮和胚均含发芽抑制物质，阻碍萌发。据张化疆（1987）报道，梣叶槭种子含有发芽抑制物质，在 15～18℃下层积 40 天后抑制物质的活性显著减弱。青榨槭、秀丽槭、花楷槭和鸡爪槭的种子属于低温发芽型，在高于 10℃或 15℃的环境中发芽受到抑制。孙秀琴和田树霞（1991）认为，元宝槭的果皮和种皮并不阻碍种子吸水，发芽比较困难的主要原因是果皮和种皮限制了对胚的氧气供应，可能同果皮特别是种皮中含有单宁物质有关。据她们研究，在 2～5℃下层积 40～50 天，元宝槭便可顺利萌发；在室内检验时，剥去果皮和种皮的元宝槭，在恒温 25℃、每昼夜光照 8 小时的条件下 1.5 天即可萌发，5 天的发芽率可以高达 98%。据邵蓓蓓（1989）试验，保留果皮、剥去果皮以及剥去种皮的青榨槭，在 20℃下培育 2 周的发芽率分别为 0%、30%和 98%。她因此认为青榨槭的胚并无生理休眠习性，难于发芽的原因在于果皮和种皮的阻抑作用。据她报道，在 0～5℃下层积 30 天便可解除青榨槭的发芽障碍，在 30℃/15℃的昼夜变温条件下培育 15 天，发芽率可达 84%。她还发现秀丽槭也有类似情况，在 0～5℃下层积 30 天后，利用 15℃的恒温测得的室内发芽率为 87%；未经层积的则发芽率仅为 2%。槭属部分树种的发芽能力及其测定条件见表 5。槭属部分树种育苗时的播前处理、播种期和有关技术数据见表 6。

表 5 槭属树种的发芽能力及其测定条件

树 种	预处理	温度（℃）		发芽势（%）		发芽率（%）	
		有光时	无光时	计算天数	变动范围	计算天数	一般数值
樟叶槭	发芽箱内恒温 28℃，8 天	—	28	5	60～85	—	—
		—		10	73	—	—
青榨槭	室温浸种 24 小时	—	10	19	54～65	31～38	70
	0～5℃层积 30 天	—	15	8	22～36	24	89
		—	15～30	7	28～42	15	84
秀丽槭	室温浸种 24 小时，0～5℃层积 30 天	—	15	14	16～32	30	87
鸡爪槭	室温浸种 24 小时，0～5℃层积 30 天	—	5 左右	—	—	92～93	90
元宝槭	2～5℃层积 40～50 天或剥去果皮种皮	25	25	—	—	5	98

表 6 槭属树种种子处理及育苗一览表

树 种	播前处理	播种期	播种量（g/m^2）	出苗期（天）	出苗率（%）	1 年生苗高（cm）
髯毛脉槭	低温层积 6 个月以上（10℃以上层积则不发芽或很少发芽）	4 月下旬～5 月上旬	15	20～30	30	30～40
樟叶槭	发芽箱内保持 28℃	—	—	8～12	60～85	—
青榨槭	浸种 24 小时，0～5℃层积 30 天	3 月	10～12	36	62	—
秀丽槭	浸种 24 小时，0～5℃层积 70 天	2 月	12～14	22	80	—
茶条槭	0～10℃或 5～10℃层积 40～50 天	4 月下旬～5 月上旬	15～20	15～25	70	70～80
色木槭	秋播或室外层积 6 个月	4 月下旬～5 月上旬	13～20	20～30	50	30～40
梣叶槭	播前 5～10℃层积 30～50 天	4 月下旬～5 月上旬	12～15	15～30	80	90～130
鸡爪槭	浸种 24 小时，0～5℃层积 42 天	2～3 月	3～4.5	37	67	35～65
元宝槭	浸种 24 小时，层积 40～50 天	3 月中旬～4 月中旬	23～38	15	—	—
花楷槭	层积催芽 6 个月，10℃以上连续催芽则不发芽或很少发芽	4 月下旬～5 月上旬	15	20～30	30	30～40

大部分苗木可当年出圃。樟叶槭、青榨槭、茶条槭、梣叶槭、鸡爪槭为出土萌发（见图 2），元宝槭、色木槭接近于地面萌发。

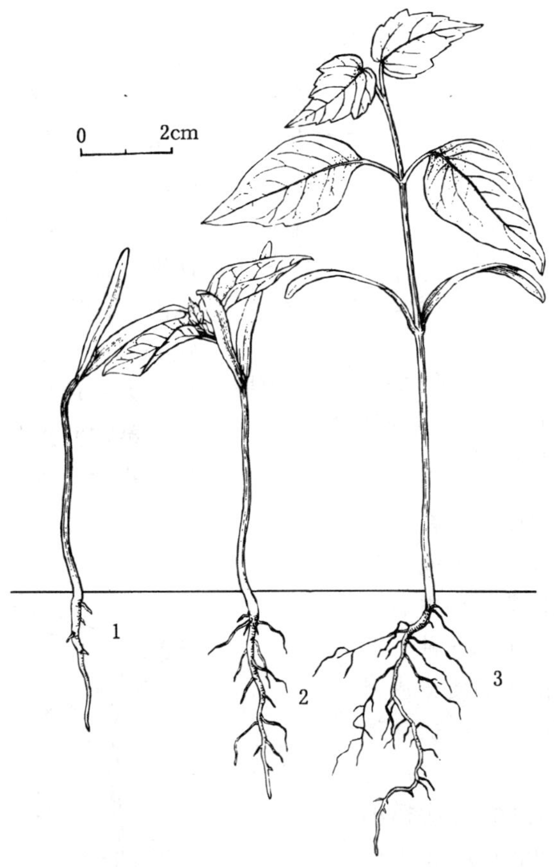

图 2　梣叶槭种子萌发和幼苗生长情况
1. 子叶出土　2. 初生叶单叶对生　3. 幼苗形成
（黄应钦仿《主要树木种苗图谱》）

（周德本、邹铨）

云南金钱槭
Dipteronia dyerana Henry

（槭树科　Aceraceae）

生长习性、分布和用途　金钱槭属 2 种，均为我国特产，本文描述 1 种。落叶乔木，高 7～13m。喜温凉湿润，也较耐干旱，适生于深厚湿润的山地黄棕壤。产云南东南部和贵州西南部。果实特异，形如金钱，观赏价值高。花为蜜源，材质优。本种资源已渐稀少，有濒危的可能，《中国植物红皮书》列为稀有种，亟需保护。

开花结实　5～10 年生开始开花结实，15 年生以后为正常结实期，结实量大，无大小年现象。花杂性，雄花与两性花同株。圆锥花序顶生或腋生，长 20～35cm。萼片 5，卵形或椭圆形，长于花瓣。花瓣 5，白色，宽卵形。雄蕊 8。子房上位，扁形，2 室，柱头 2。据云南文

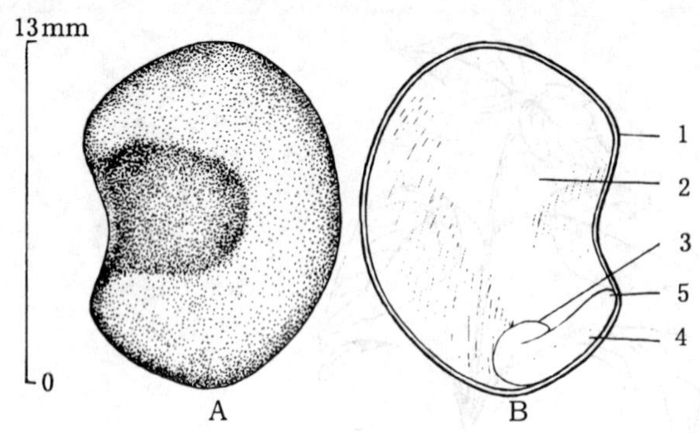

图 1　云南金钱槭小坚果(去翅)外形(A)及其纵切面(B)
1. 果皮和种皮　2. 子叶　3. 胚芽　4. 胚轴　5. 胚根
(黄应钦绘)

山的物候观察，4 月开花，花后即形成幼果，9 月下旬为果实成熟始期，10 月中旬为果实成熟盛期，10 月下旬为果实成熟末期。小坚果 2，基部连合，为阔翅所围绕。翅薄革质或膜质，近圆形，径 3.5～6cm，嫩时绿色，后转黄褐色。果梗细，长 2.5～3cm。小坚果扁卵形，一侧内凹，褐色，长 13mm，宽 12mm，厚 3mm。种子无胚乳，胚弯，倒生。小坚果（去翅）的外观和解剖结构见图 1。

果实的采收调制和种子贮藏　果实成熟后短期内不脱落。当果实达到形态完全成熟时，可用竹竿敲打或直接采摘。采得的果实在室内摊放晾干或在弱光下晒干。干燥的带翅果即可作为播种材料。为了便于操作，也为了加速种子发芽，生产上多将果实周围的膜质阔翅剥去，取出小坚果，用作播种材料，通称种子。每千克带翅果约 1 800～2 000 粒。去翅小坚果约为鲜果原重的 50%。去翅小坚果千粒重 270（220～330）g，每千克有去翅小坚果 3 700（3 000～4 500）粒。运输和贮藏都可带翅装袋。干藏，用袋挂于室内通风处，也可去翅后混干沙贮藏，贮藏期可达 2 年。

发芽和播种　种子无休眠或有短期休眠。发芽时日均温宜在 19℃上下。1989 年 2 月 21 日，云南林业科学研究所用经过层积催芽的种子在室外遮荫苗床进行过发芽测定：播后 32 天种子开始发芽，发芽出土时日均温为 19℃，最高温为 25.3℃，最低温为 9.3℃；从开始发芽之日起的 15 天中，发芽百分数为 65%；从播种之日起算，以 61 天计，发芽率为 75%。出土萌发。种子发芽时主根迅速向下伸展，并形成侧根，子叶多数带壳出土，1～2 天后子叶伸开，壳脱落，3～5 天初生叶出现。云南金钱槭的萌发和幼苗初期生长情况见图 2。

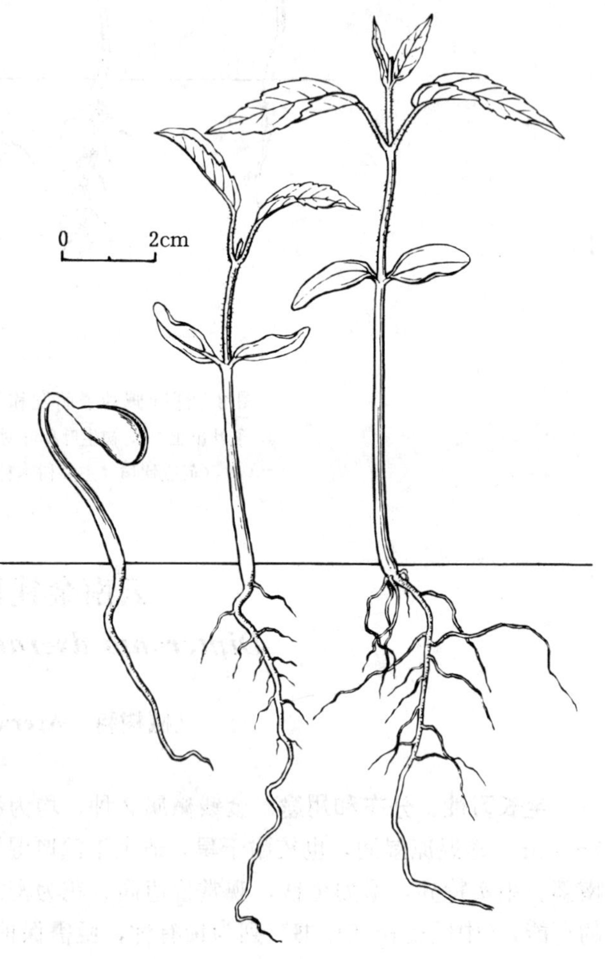

图 2　云南金钱槭种子萌发后第 2、7、10 天幼苗的生长情况
(黄应钦绘)

条播。每平方米播带翅果 100～150g 或去翅小坚果 50～75g。1 年生苗出圃。

（李达孝）

七叶树属

Aesculus L.

（七叶树科　Hippocastanaceae）

生长习性、分布和用途　本属 30 余种，分布于东南亚、欧洲东南部和北美。我国有 10 余种，本文描述 3 种 1 变种。落叶乔木，以西南部的亚热带地区为分布中心，北达黄河流域，东到江、浙，南抵粤北。树荫浓密，花序长而形态独特，颇具观赏价值，常被栽培为庭园遮荫树和行道树。这 3 个种的名称、成年树高、分布和用途见表 1，其中云南七叶树在《中国植物红皮书》中列为渐危种。

表 1　七叶树属树种的名称、成年树高和胸径、分布及用途

中　名	学　名	树　高 (m)	胸　径 (cm)	分　布	用　途	供　稿
七叶树	*A. chinensis* Bunge	25	50～60	冀南、晋南、豫北、陕南。	观赏、材用、药用	1001
云南七叶树	*A. wangii* Hu ex Fang	28	50～80	滇东南	材用、药用、饲料、观赏	615
天师栗（猴板栗）	*A. wilsonii* Rehd.	15～20	40～50	豫西南、鄂西、湘西、赣西、粤北、川、黔、滇东	观赏、药用	1001

开花结实　10～20 年生开始结实。七叶树和云南七叶树每年都能结实。花杂性，雄花与两性花同株，不整齐。聚伞圆锥花序呈圆筒形，顶生。花萼钟形或管状，上段 4～5 裂，大小不等，排列成镊合状。花瓣 4～5，倒卵形、倒披针形或匙形，基部爪状，大小不等。花盘全部发育成杯状或仅一部分发育，微分裂或不裂。雄蕊 5～8，着生于花盘。子房上位，无柄，3 室。花柱细长，不分枝，柱头扁圆形。胚珠每室 2 枚，重叠。蒴果 1～3 室，种子仅 1～2 枚发育良好。春季开花，当年秋季果实成熟。这 3 个树种开始结实的年龄、花期、果实成熟期列入表 2。

表 2　七叶树属树种的开花结实习性

树　种	开始结实年龄	花　期	果实成熟期
七叶树	15～20	4～5 月	9～10 月
云南七叶树	10～15	—	9～10 月
天师栗	10～15	4～5 月	9～10 月

同一树种不同年份的开花结实物候期进程常有明显的变动。1983、1987 和 1988 年南京林业大学校园内七叶树的变种浙江七叶树 *A. chinensis* Bge. var. *chekiangensis* (Hu et Fang)

Fang 的开花结实物候观察资料如表 3。

表 3　浙江七叶树的开花结实物候期①

观察年份	开花			果实成熟脱落		
	始　期	盛　期	末　期	始　期	盛　期	末　期
1983	5 月 3 日	5 月 12～20 日	6 月 3 日	9 月 5 日	9 月 27 日～10 月 1 日	10 月 23 日
1987	5 月 1 日	5 月 10～20 日	6 月 1 日	9 月 11 日	9 月 23 日～10 月 2 日	10 月 27 日
1988	4 月 20 日	5 月 1～10 日	6 月 3 日	9 月 1 日	9 月 24 日～10 月 10 日	10 月 21 日

①　观察地点：南京林业大学校园

七叶树属的蒴果较大，果皮薄革质，成熟时室背 3 裂。种子大而重。种皮栗色至栗褐色，表面光滑，坚硬，基部具明显种脐。种子的外形和种脐的大小可以作为识别树种的参考。七叶树属这 3 个树种果实和种子的形态以及种子质量的有关数据见表 4。无胚乳。子叶肥厚，胚根扁平，卧伏于种子背面（图 1）。

表 4　七叶树属树种成熟果实和种子的形态及种子质量

树　种	蒴　果	种　子	种子重（克/粒）	每千克种子粒数
七叶树	球形或倒卵形，黄褐色，有密集的斑点，径 3～5cm	近球形，栗褐色，径 2～3.5cm。种脐淡黄褐色，约占种皮表面的 1/2	10～35	28～100
云南七叶树	扁球形，先端有短尖头，黄褐色，有黄色斑点，有瘤状突起，径 6～8cm	近球形，暗栗褐色，径 6cm。种脐白色，约占种皮表面 1/2 以上	78～157	6～13
天师栗	卵圆形或近于梨形，黄褐色，长 5～6cm，顶端有短尖头，有斑点，成熟时常 3 裂	近球形，直径 3～3.5cm，栗褐色，种脐淡白色，近于圆形，约占种皮表面 1/3 以下	20～55	18～50

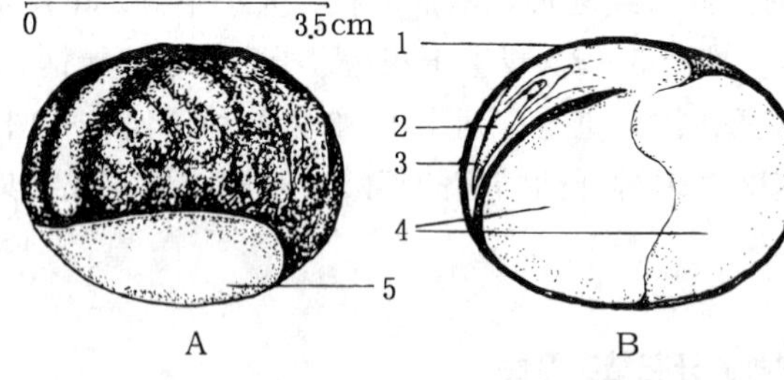

图 1　浙江七叶树种子的外形（A）及其纵切面（B）
1. 种皮　2. 胚轴　3. 胚根　4. 子叶　5. 种脐
（田恒德绘）

果实的采收调制和种子贮藏　蒴果微黄或开始开裂时即可从树上采摘，也可以震落或待自然脱落后从地面拾集。室内阴干，待蒴果开裂后剥出种子。蒴果出种率可达 70%～80%。净度可达 100%。种子忌过分干燥。七叶树新鲜种子含水量达 65%～70%。在平均气温 14℃、相对湿度 80%的室内裸藏 10 天的浙江七叶树种子，含水量降至 8%～10%，这时种皮皱缩，已经丧失生活力。七叶树属的种子也不能密封贮藏，否则会窒息致死。发芽率可保持5～6 个月。七叶树属种子含水量需保持在 50%～60%，常与湿沙混藏。

发芽和播种　种子无休眠习性。在 15～25℃范围内均能发芽。室内测定时一般在置床后 10 天开始萌发。据云南林业科学研究所测定，9 月 20 日室内沙床播种的云南七叶树，10 月 27

日开始发芽，11 月 4 日发芽结束，没有明显的发芽盛期；发芽率一般为 80%，变动范围为 65%～85%。通常随采随播，也可混沙湿藏后精选优良种子春播。点播，行距 25cm，株距 5cm，覆土 2～5cm。圃地不宜灌溉过度，否则种子容易腐坏。圃地春季播种 3～4 周开始发芽。留土萌发。发芽时上胚轴自子叶柄间萌发出土，有或无初生不育叶，初生叶为对生的 5 小叶掌状复叶（图 2）。场圃发芽率可达 65%～80%。通常 1 年生苗即可出圃。作为行道树或庭园绿化的可移植培育 3～5 年后定植。

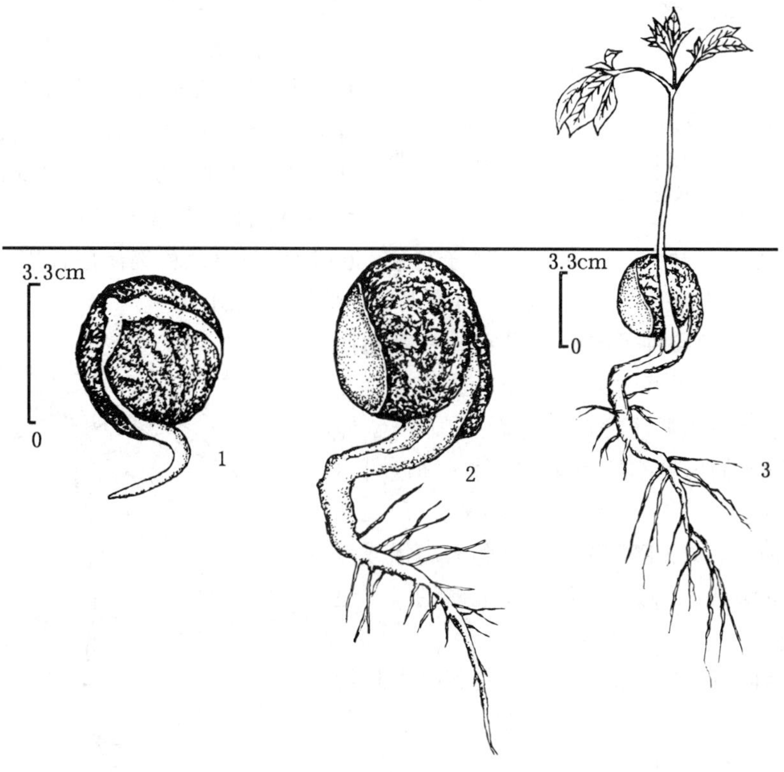

图 2　浙江七叶树种子萌发后第 1、8、24 天的幼苗生长情况

（田恒德绘）

（吴琼美）

野　鸦　椿

Euscaphis japonica（Thunb.）Kanitz

（省沽油科　Staphyleaceae）

生长习性、分布和用途　野鸦椿属 3 种，分布于中国、日本。我国 2 种，本文描述 1 种。落叶灌木或小乔木，高 3～8m。适生于温暖湿润之地，较耐瘠薄。分布于陕、晋、豫、皖、江苏及以南各省，东南达台湾。朝鲜半岛亦产。种子含油脂，可制肥皂。树皮提栲胶。根及干果入药，有祛风除湿之效。

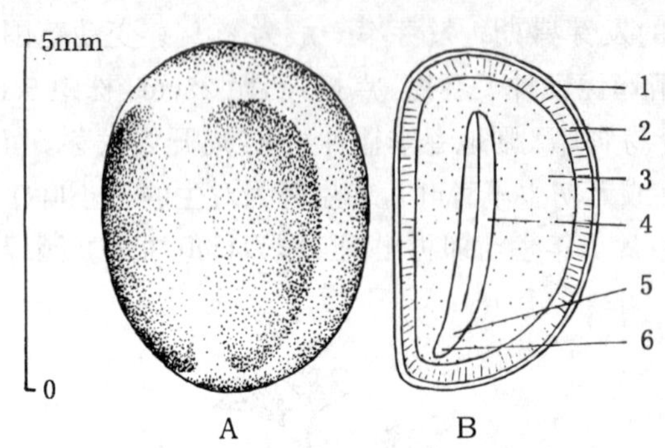

图 1　野鸦椿去除假种皮的种子外形(A)及其纵切面(B)
1. 假种皮　2. 种皮　3. 胚乳　4. 子叶　5. 胚轴　6. 胚根
(黄应钦绘)

开花结实　3～4 年生开始开花结实，正常结实期在 5 年生以后。结实大小年现象不甚明显。花两性。圆锥花序顶生。花黄白色，径约 5mm。萼片、花瓣和雄蕊均为 5。萼片宿存。雄蕊着生于花盘外缘。花盘盘状。子房上位，心皮 3，分离。据广西南宁 1978～1981 年观察，4 月中下旬为始花期，5 月上旬为盛花期，5 月下旬为末花期；8 月中旬果实开始成熟，9 月上旬为果熟盛期。果由 3 个分离的蓇葖构成，长圆形或近球形，具不规则脉纹，未熟时绿色，成熟果红绿色或褐色。果皮软革质。每个蓇葖长 8～20mm，径 7～12mm，含种子 1～3 粒。种子褐色，近球形，具蓝黑色肉质假种皮，径 4～6mm。有胚乳。胚直，子叶扁平，圆形。种子的外观和内部结构见图 1。

果实的采收调制和种子贮藏　果实成熟盛期摘取呈红绿色的果实。采回的果实摊开，让其自行开裂，敲出种子，搓去肉质假种皮，洗净后稍晾干，即得纯净种子。鲜果的出种率约为 50%。种子净度可达 98%。千粒重 40～50g，每千克有纯净种子 2 万～2.5 万粒。种子含油脂，忌日晒和失水，宜随采随播或沙藏。运输时需混湿沙。贮藏亦需混沙，贮藏期以不超过 5 个月为宜。

发芽和播种　种子有休眠习性。发芽时的日均温需在 15℃以上。1989 年 11 月 28 日，贵州林业科学研究所用当年新采的种子，经短期沙藏后，在室外沙床作过发芽测定，当时日均温为 18～20℃。播后 102 天（翌年 3 月 10 日）方始发芽，3 月 14～16 日为发芽盛期，4 月 8 日发芽终止。发芽势为 32%，发芽率为 50%。出土萌发。胚根萌发后 5 天子叶出土，再过 7 天初生叶出现。野鸦椿的幼苗早期形态见图 2。

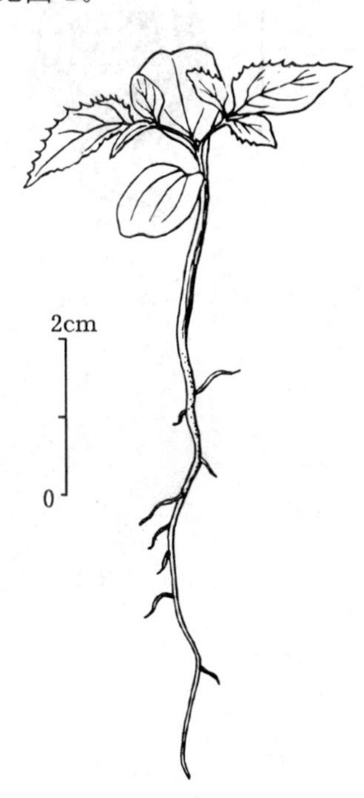

图 2　野鸦椿的幼苗形态
(黄鹏成供稿，张世经临)

条播，每平方米约播 12～15g。也可将种子密播于苗床上，发芽生长至高 10cm 左右时再移植培育。1 年生苗出圃。

（钟梅芳）

银　鹊　树（瘿椒树）

Tapiscia sinensis Oliv.

（省沽油科　Staphyleaceae）

生长习性、分布和用途　银鹊树属有3种，均产我国，本文只描述1种。我国特有的稀有、古老树种，第四纪冰川的孑遗植物，已列入《中国植物红皮书》。落叶乔木，高可达34m（胡知定等，1989），胸径达1.3m（余茂林，1983）。速生。喜气候温暖、雨量充沛、冬无严寒、夏无酷暑之地。适生于湿润而排水良好的酸性山地黄壤或黄棕壤。分布于浙、皖、闽、赣、鄂、湘、桂、滇、黔、川、陕。树干通直，木材轻软，结构细，纹理直，易加工，不翘不裂，是建筑、胶合板、造纸用材。花黄色，有香气。秋叶黄灿，果色紫黑，是优良的观赏树种。叶可入药，树皮可作木纤维原料。

开花结实　7年生开始结实，9年生时即大量结果（陶金川，1990）。开花结实物候期见表1。

表1　银鹊树的开花结实物候期

观察地点和年份	花芽出现期	开花			果熟期	果实脱落	
		始期	盛期	末期		始期	末期
浙江杭州1978	5月2日	6月10日	6月13日	6月17日	8月22日	8月25日	8月28日
江苏南京	5月上旬	6月上旬	6月中旬	6月下旬	7月下旬	—	—

花杂性，两性花与雄花异株。圆锥花序腋生。雄花序由长而纤弱的穗状花序构成，花密集。两性花序粗短，花单生于苞腋。萼筒状钟形，5裂。花瓣5。雄蕊5，突出。子房上位，1室，基生，倒生，1胚珠。花柱长过雄蕊。果穗长达10cm。核果，近球形，长7～8mm，熟时由黄绿色转黄红色，最后呈紫黑色。果柄肥大，长7～8mm。外果皮肉质。果核圆形，径5mm，熟时黑色。种阜三角形，有角质胚乳。

果实的采收调制和种子贮藏　成熟果实喜被鸟类啄食并传播种子。当果皮由黄绿色转为黄红色时将果穗梗截下，薄摊阴凉通风处2～3天，搓掉果肉，洗净，阴干，所得的果核即为播种材料，通称种子。千粒重40～55g，每千克18 000～24 000粒。混沙湿藏或流水贮藏。

发芽和播种　种子有休眠现象，可以带果肉冬播，也可以去除果肉层积越冬后春播。条播，条距25cm，沟深3cm。每米长播种沟播种30粒，每666m^{2}约播5kg。复土厚2cm。出土萌发。发芽率50%～60%。春播40天后发芽。子叶2，绿色或黄绿色，基部三出脉。子叶柄淡紫色。初生叶2，单叶对生，有托叶2，叶成熟后托叶立即脱落。第3叶以上互生。第3～4叶为3小叶羽状复叶，以后小叶数渐次增多，多数由5～7（少数9～11）小叶构成羽状复叶。下胚轴上部红色而带绿色，下部红色。根淡黑色。每米长播种行产苗15株，每666m^{2}产苗1万株。1年生苗高可达1m。萌芽力强。扦插繁殖以7月中旬最好，生根率90%以上。银鹊树种子萌发及幼苗生长情况见图1。

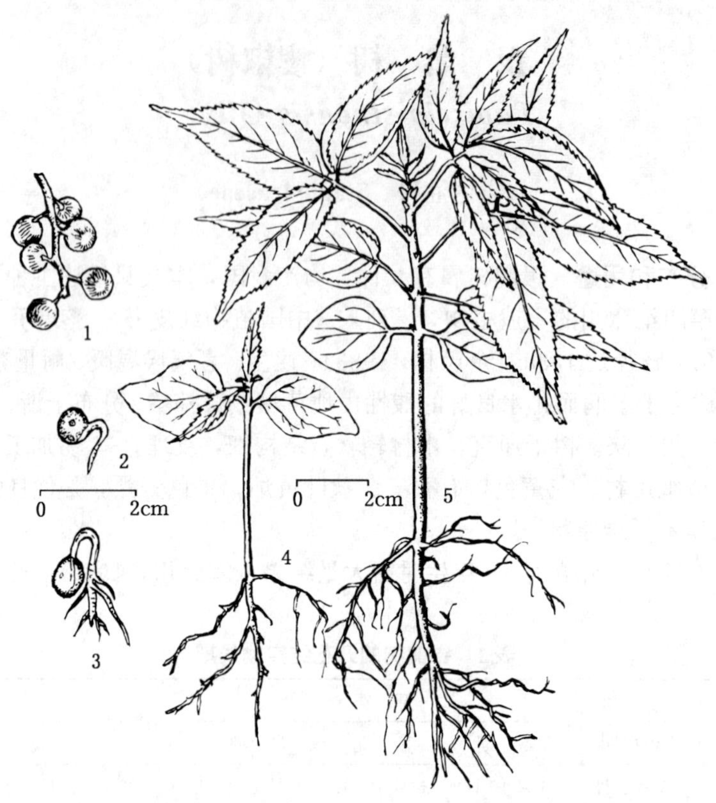

图 1　银鹊树的果穗外形、种子萌发及幼苗生长情况

1. 果穗（部分）　2. 胚根初生　3. 下胚轴延伸，侧根生出

4. 第 1 对初生叶出现　5. 3～5 小叶的初生叶互生

（黄鹏成供稿，田恒德临）

（杨国华）

大果山香圆

Turpinia pomifera（Roxb.）DC.

（省沽油科　Staphyleaceae）

生长习性、分布和用途　山香圆属 30～40 种，热带亚洲和热带美洲间断分布。我国有 13 种，本文描述 1 种。落叶乔木 ，高 8～15m。分布于滇、桂。印度、越南也有。树干通直，用作板材或制作家具等。

开花结实　5～7 年生开始开花结实，结实大小年现象不明显。花两性。圆锥花序顶生，花序短于叶，长至 21cm，较粗壮。萼片 5，宿存。花瓣 5，覆瓦状排列。雄蕊 5，着生于花盘裂齿外。花丝扁平，花药长圆状披针形。子房上位，3 裂，3 室，每室有胚珠多颗。花柱 3，柱头头状。据广西南宁 1987 年观察，显蕾期 12 月下旬，始花期 2 月上旬，盛花期 4 月初，末花期 4 月下旬；果实成熟始于 6 月中旬，果熟盛期在 8 月上旬，果熟末期在 8 月下旬，果脱落末期为 11 月上旬。浆果，近球形，稍不规则，直径 1.5～2.5cm，表面粗糙，成熟时由青带紫色转为浅黄色。每果有种子 5～10 粒。种子椭圆形或不规则圆形，长 6～8mm，径 4～6mm。外种皮硬膜质，棕褐色至深褐色。胚乳少量。胚伸直，子叶扁平。大果山香圆种子的外形和解剖构造见图 1。

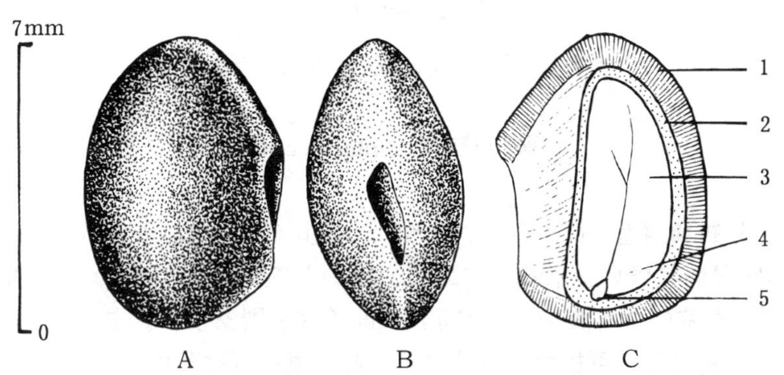

图 1　大果山香圆种子外形（A、B）及其纵切面（C）

1. 种皮　2. 胚乳　3. 子叶　4. 胚芽　5. 胚根

（黄应钦绘）

果实的采收调制和种子贮藏　果熟盛期摇动树枝，震落果实，在地面收集。果实采回后可以在室内堆沤 2 周左右，如已充分成熟，也可以立即调制。用手挤压果实即可得到种子。鲜果的出种率为 5%～10%。种子净度可达 97%。千粒重 120（100～150）g，每千克有纯净种子 8 300（6 700～10 000）粒。种子可以适当干燥，但不宜曝晒。晾干后用布袋包装存放在通风处，也可以混沙贮藏，贮藏期约为半年。

发芽和播种　种子无休眠习性。采后即可播种，也可以贮藏至翌年春天播种。1987 年 10 月 17 日，广西林业科学研究所在室外沙床作发芽测定，播种时日均温约 22℃：11 月 17 日开始发芽，11 月 19 日～21 日发芽进入盛期，11 月 23 日发芽终止；从开始发芽至发芽高峰日的 5 天发芽 60%；从播种之日起算，以 38 天计，发芽率为 90%。出土萌发。胚根露出 3 天后子叶出土，再过 3 天初生叶展现。这个树种种子的萌发和幼苗初期生长情况见图 2。

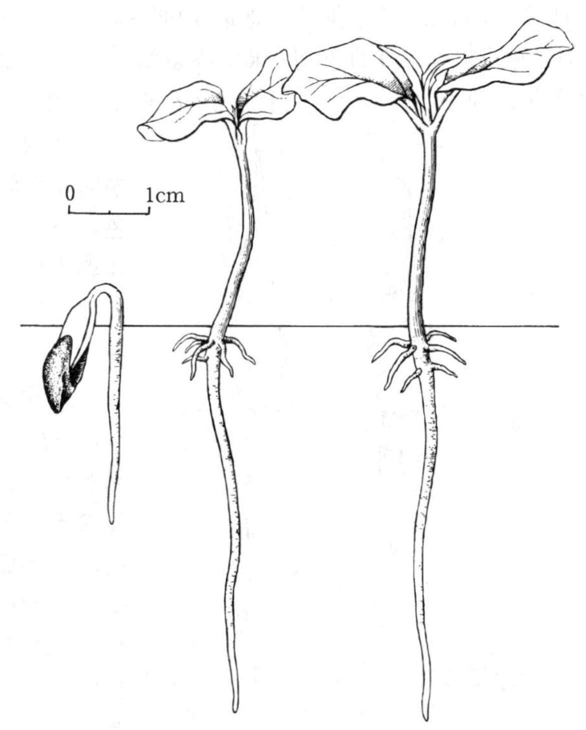

图 2　大果山香圆种子萌发后第 3、6、8 天幼苗的生长情况

（黄应钦绘）

点播。每平方米约播 6～8g，覆土约 1cm。1 年生苗出圃。幼苗生长较粗壮，每 666m² 可产苗 7 500 株。

（曾 玲）

大叶醉鱼草

Buddleja davidii Franch.

（醉鱼草科 Buddlejaceae）

生长习性、分布和用途 醉鱼草属约 100 种，我国约 40 种，本文描述 1 种。落叶灌木，高达 5m。喜温暖湿润气候，忌水涝，较耐寒、耐旱。产于西南、西北东部、华中、晋、豫、皖、苏、浙，国内庭院中常有栽培。日本亦产。欧美许多国家引种。本种花色鲜艳，花序大，花期长，为庭园观赏树种，可提取芳香油。花、叶、根皮可供药用。

开花结实 实生苗 3 年生即开花结实。结实大小年现象不明显。花两性。多数小聚伞花序集成穗状的圆锥花序，长 15～25cm，顶生，直立或下垂。花萼钟状，4 裂。花冠淡紫色，4 裂，喉部橙黄色，花冠筒部长约 1cm。雄蕊 4，着生于花冠筒中部，几无花丝。子房上位，2 室。花柱单一，柱头 2 裂。据北京地区 1986～1988 年观察，开花始期在 6 月上旬，开花盛期在7～8 月，开花末期在 9 月下旬；果实成熟始于 10 月中旬，成熟盛期在 11 月上旬，成熟末期在 11 月中旬；种子散落始于 11 月上旬，盛期 11 月中旬。

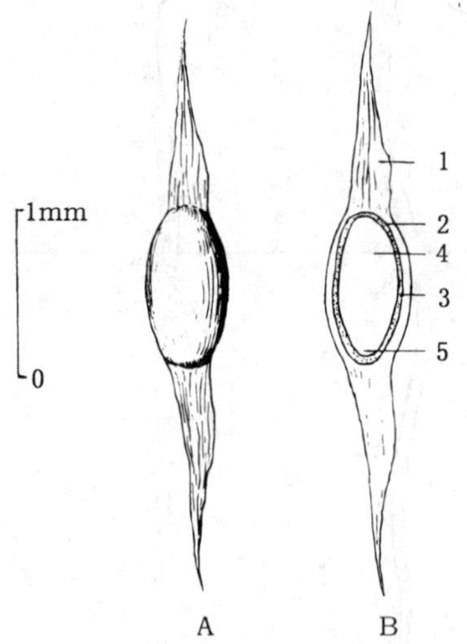

图 1 大叶醉鱼草种子外形(A)及其纵切面(B)
1. 种翅 2. 种皮 3. 胚乳 4. 子叶 5. 胚根
(胡冬梅、林平绘)

蒴果，条状矩圆形，黄褐色，长 6～8 (10) mm，熟时从室间 2 瓣裂。种子多数，长纺锤形至条形，黄色，长 3～4.5mm，宽 0.3～0.4mm，两端有白色膜质渐尖翅，种子本体仅占全长的 1/5。胚直立，线型，几乎充满种腔，子叶 2 片，有少量胚乳（见图 1）。种粒极小，易于风力传播。

果实的采收调制和种子贮藏 花期长达 4 个月，果实陆续成熟。当果实由暗褐色变为黄褐色，蒴果始见开裂时，便可剪下果穗或捋下蒴果（不要花序顶端未成熟果实），摊开晾晒。果瓣开裂后种子即可脱出，经细孔筛筛选，去除杂质，即得比较纯净的种子。出种率0.5%～1%。净度 30%～50%。发芽率 40%～60%。千粒重 0.009～0.014g。每克种子 7 万～11 万粒。将干燥种子装入袋内，置于冷凉处贮藏。在北京地区一般室内保存，发芽力可保持 1 年。

发芽和播种 发芽测定可以用滤纸床，在恒温 25℃的条件下第 3 天开始发芽，7 天发芽结束。出土萌发（见图 2）。因种子细小，播种覆土要浅，幼苗期要适当遮荫，防曝晒。亦可先播在室内盆中，再移入露地。也可用分株、嫩枝扦插方法繁殖。

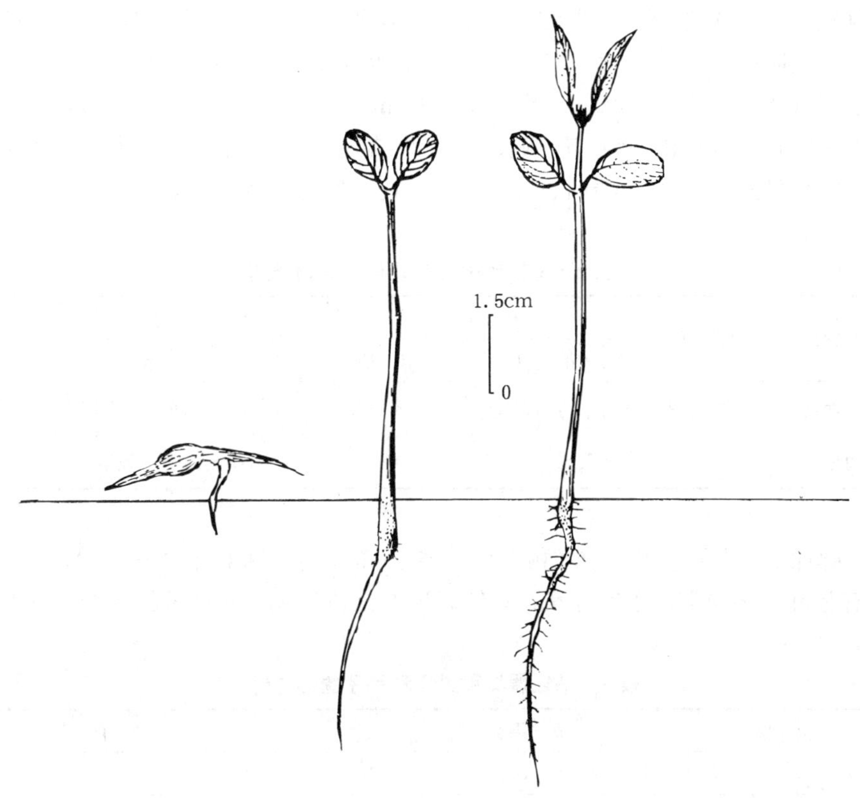

图 2　大叶醉鱼草萌发出土后第 1、5、13 天的幼苗生长情况
（胡冬梅、林平绘）

（刘长江）

马　钱　属
Strychnos L.

（马钱科　Strychnaceae）

生长习性、分布和用途　本属约 200 种，我国 9 种，本文描述 2 种。小枝常变态成为螺旋状曲钩，老枝有时变态为硬刺。种子有剧毒，供药用。它们的名称、生长习性、分布和用途见表 1。

表 1　马钱属树种的名称、生长、分布和用途

中　名	学　名	生长习性	分　布	用　途	供　稿
牛眼马钱	*S. angustiflora* Benth.	木质藤本，长 2～3m	琼、桂、粤	药用、灭鼠、酿酒	603
马钱	*S. nux-vomica* L.	小乔木，高 5～6m	印度、柬埔寨、缅甸、老挝。琼、粤、桂引种	麻醉剂、兴奋剂	611

开花结实 5～8年生开始开花结实，正常结实期在10年生以后。结实大小年间隔期一般为1年，但不甚明显。花两性。聚伞花序，顶生于短侧枝上，直径3～4cm。花4～5基数，长8～12mm，具短梗。萼裂片卵状三角形，长约1mm，急尖。花冠白色或淡黄色，有香味。花冠筒长约3～4mm，裂片狭，约与筒等长，外弯。雄蕊4～5，生于花冠喉部。子房上位，无毛，2室，每室有胚珠数颗。花柱长约8mm。开花结实的物候期见表2。

表2 马钱属树种的开花结实物候期

树 种	观察地点	观察年份	开 花			果实成熟		果实脱落
			始 期	盛 期	末 期	始 期	盛 期	
牛眼马钱	广西南宁	1987～1988	4月下旬	5月中旬	5月下旬	8月中旬	9月上中旬	翌年3月下旬
马钱	海南儋县	—	4月	5月	—	11月	12月～翌年3月	4月

浆果，圆球形。成熟果脱落前在树上可维持4～5个月。每果有种子1～2，少有3～6。种子盘状。有胚乳，胚伸直，子叶叶状。马钱属这2个树种果实和种子的形态见表3、图1。

表3 马钱属树种果实和种子的形态特征

树 种	未熟果颜色	成熟果实			种 子		
		形 状	大小(cm)	颜 色	形 状	大小(mm)	颜 色
牛眼马钱	绿色	圆球形	2.5～4	黄红或浅红色	扁圆形	径10～18 厚4～6.5	浅黄色
马钱	黄绿色	圆球形	2.5～4	棕红色	扁圆形	径12～20	浅黄色

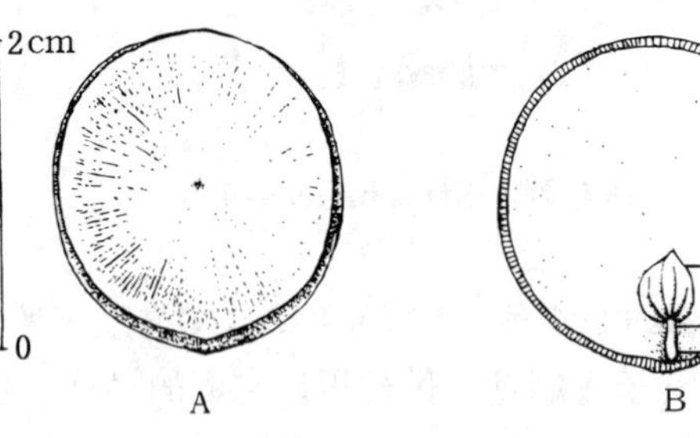

图1 马钱种子外形（A）及其纵切面（B）
1.种皮 2.胚乳 3.子叶 4.胚轴 5.胚根
（黄光郁绘）

果实的采收调制和种子贮藏 用手采摘。采回的果实堆沤数日，待果皮充分软熟后装入竹筐或布袋，置水中搓擦，淘去果皮、果肉等杂质，得出种子。种子含水量约为20%～30%，忌失水，不能日晒或干藏。运输或贮藏均需混以湿沙，贮藏期为4～6个月。出种率以及净度、质量等数据见表4。

表 4　马钱属树种果实的出种率和种子净度、质量

树　种	出种率（%）	净　度（%）	千粒重（g）		每千克纯净种子粒数	
			一　般	变动范围	一　般	变动范围
牛眼马钱	20～30	98	700	600～800	1 400	1 200～1 700
马钱	17～20	98	800	700～1 000	1 200	1 000～1 400

发芽和播种　牛眼马钱的种子有程度较深的休眠，马钱种子无休眠现象，发芽时日均温需在 23℃以上。1987 年 10 月上旬，广西林业科学研究所在室外沙床用当年新采的牛眼马钱种子作发芽测定：10 月 5 日播种，翌年 5 月 12 日方始发芽，6 月 10 日发芽终止。1988 年 6 月 23 日，华南热带作物研究所用当年所采的马钱种子，在室外沙床作发芽测定，播后于 6 月 30 日开始发芽，7 月 18 日发芽终止。测定情况见表 5。

表 5　马钱属树种的发芽能力及其测定条件

树　种	基　质	室外温度（℃）	发芽势（%）		发芽率（%）	
			计算天数	一般数值	计算天数	一般数值
牛眼马钱	沙	25～28	—	不明显	250	55
马钱	圃地	28～30	12	50	26	75

出土萌发。胚根萌发后约 15 天子叶出土，出土后 14～16 天展出初生叶。马钱种子的萌发和幼苗初期生长情况见图 2。

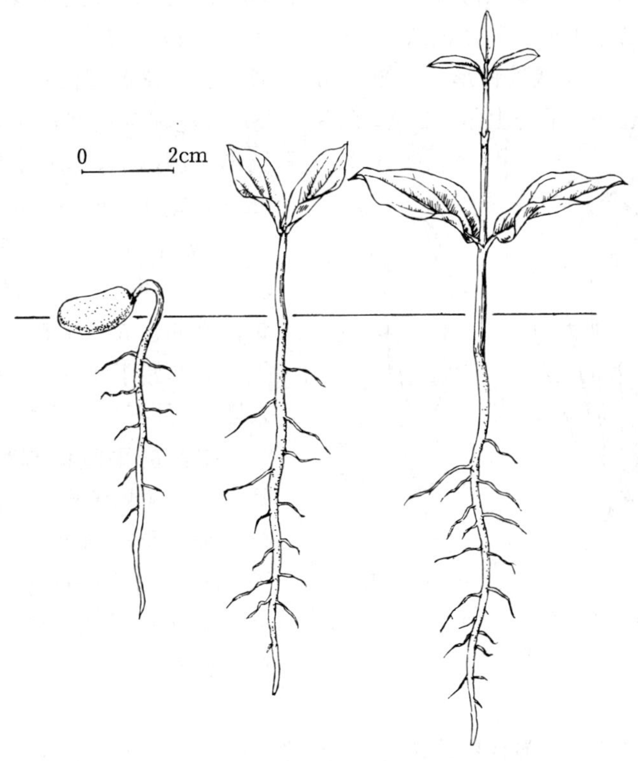

图 2　马钱种子萌发后第 8、15、29 天幼苗的生长情况

（黄光郁绘）

条播。牛眼马钱的种子播前需用湿沙层积催芽半年，马钱则视采种时间，湿沙层积催芽2～4个月。4月下旬～6月下旬播种。每平方米播种75～125g，覆土2cm。育苗至翌年或第3年春出圃。苗期需遮荫。

（张声燕）

雪　柳

Fontanesia fortunei Carr.

（木犀科　Oleaceae）

生长习性、分布和用途　雪柳属有2种，本文描述我国产的1种。落叶灌木或小乔木，高达5～9m，栽培的多呈灌木状。暖温带树种，分布于赣、浙、苏、皖、鲁、豫、陕、晋、冀和东北等地。喜光好湿，亦耐干旱，平原山野都能适应，对土壤要求不严，除盐碱土外均能生长，在湿润肥沃之处生长迅速，一二年即能起到绿化效果。萌芽力强，极耐修剪。枝柔软，易剪扎整形，宜作绿篱、绿屏，以及在河边、池畔、草坪边缘配植。作为树丛林缘，防风和防尘固沙林的中下层树种也很相宜。对二氧化硫等有害气体有一定的抗性。茎皮纤维可制人造棉，茎枝编筐箩，嫩叶晒干可代茶饮。

开花结实　4～6年生开始结实，10年生以后为正常结实期，大小年现象不明显。花两性。圆锥花序，着生于当年生枝，顶生的长2～6cm，腋生的较短。花淡红白色。花萼微小，量杯状，4裂。花冠裂片4，卵状披针形，长2～3mm，顶端钝，花冠筒极短。雄蕊2，着生于花冠基部，花丝伸出花冠外。子房上位，2室，每室2胚珠。花柱圆柱状，柱头2叉。翅果，宽椭圆形或倒卵形，扁平，周围有窄翅，长7～9mm，宽4～5mm，顶端微凹，花柱宿存，10月成熟。每果有种子1～2粒。种子具胚乳，子叶薄。北京地区开花初期在5月上旬，盛期在5月中旬，末期在5月中下旬；果实成熟始于10月上旬，盛期10月中旬，末期10月下旬。果实和种子形态见图1。

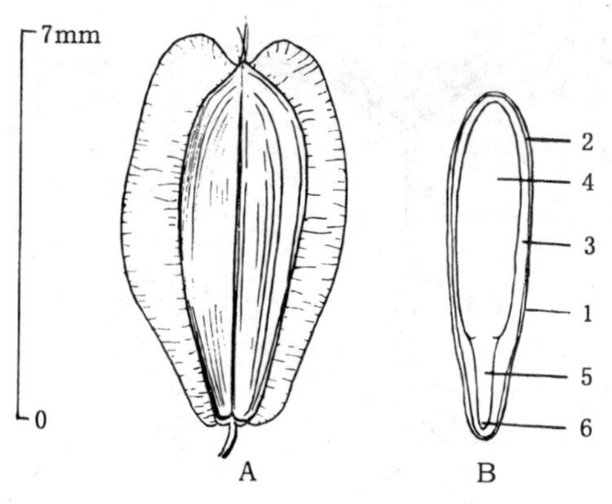

图1　雪柳翅果外形（A）及其纵切面（B）
1. 果皮　2. 种皮　3. 胚乳　4. 子叶　5. 胚轴　6. 胚根
（胡冬梅、林平绘）

果实的采收调制和种子贮藏　9～10月成熟时翅果黄褐色，树上采收。晾干，去杂净种。所得翅果即为播种材料，通称种子。净度95%～98%，千粒重5.6～8.2g，每千克有纯净种子12万～18万粒。适于干藏或密封贮藏，但含水量不能高于6%～8%。

发芽和播种　种子无明显休眠现象。发芽时的日均温宜在20～25℃，发芽率一般为70%，变动范围为60%～80%。出土萌发。子叶2片。初生叶对生（图2）。采种翌年3月上旬播种，4月中旬前后出土。条播，播种量每平方米约2.5g。通常采用扦插繁殖。

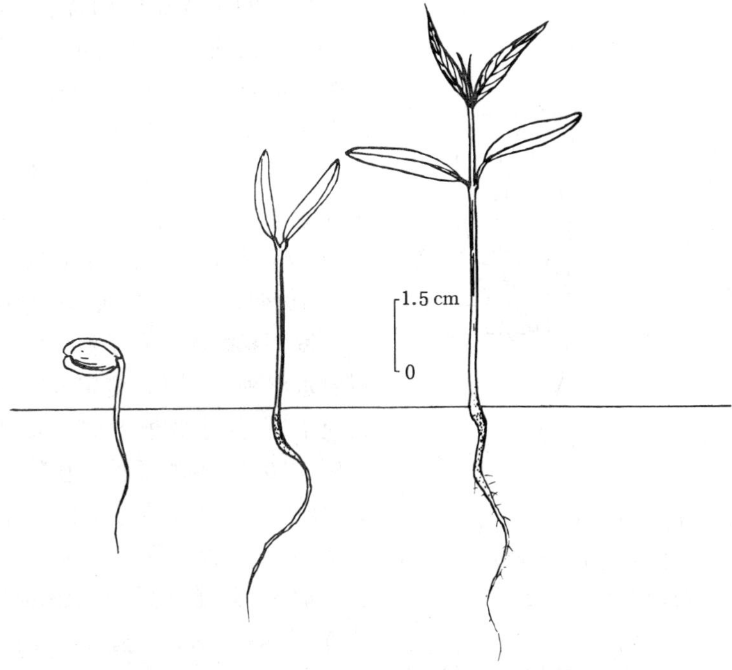

图 2　雪柳种子发芽出土后第 1、6、16 天的幼苗生长情况
（胡冬梅、林平绘）

（印佩文）

连　翘

Forsythia suspensa（Thunb.）Vahl

（木犀科　Oleaceae）

生长习性、分布和用途　连翘属 8 种，地中海、西亚和东亚间断分布。我国 4 种，本文描述 1 种。落叶灌木，高 2～4m。耐旱、怕涝、耐寒，对土壤要求不严。原产华北、西北，华中、华东均有栽培。花黄色，先叶开放，是早春重要观赏树种。果壳有清热、散风、解表之功效，是重要的中药材，种子可提取油脂。

开花结实　2～3 年生开始结实。结实没有大小年现象，但在荫蔽和积水处不易成花，或开花后结果极少。花两性，1～3（6）朵生于叶腋。花萼 4 深裂，宿存。花冠钟状，裂片 4，倒卵状椭圆形。雄蕊 2，着生于花冠筒基部。子房上位，2 室，每室多枚胚珠。柱头 2 裂。据北京地区多年观察，花芽于夏季形成，一般情况是翌年 3 月中下旬始花，4 月上旬盛花，4 月底末花；果实通常在 9 月上旬开始成熟，9 月中旬盛熟，9 月下旬为成熟末期。蒴果，黑褐色，卵圆形，先端有短喙，具瘤点。果长 1.5～2cm，径约 1cm，室背开裂为 2 片木质或革质的果瓣。每果 30～40 粒种子。成熟种子红褐色，狭椭圆形，扁平，具膜质翅，长 5.5～6mm，宽 1.6～2.4mm。种子有胚乳。连翘种子形态见图 1。

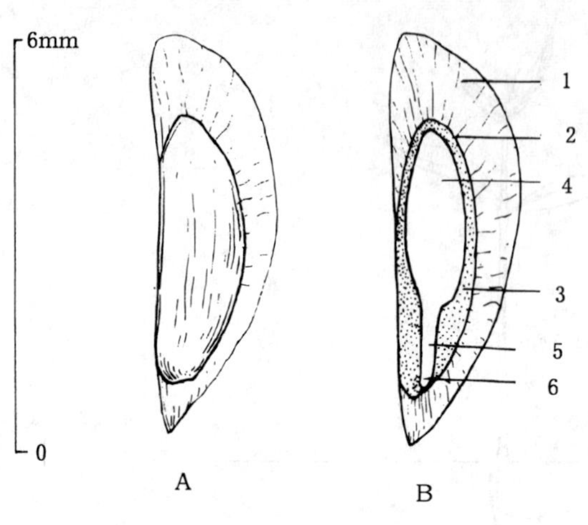

图1　连翘种子的外形（A）及其纵切面（B）
1. 种翅　2. 种皮　3. 胚乳　4. 子叶　5. 胚轴　6. 胚根
（胡冬梅、林平绘）

果实的采收调制和种子贮藏　连翘树体矮小，可在蒴果由黄褐色变为黑褐色尚未开裂时摘取蒴果。采回后置阳光下曝晒，敲打，脱出种子。经风选、筛选，去除杂质，分离出纯净种子。通常不去翅。种子净度95%～98%，千粒重4.4～5.3g，每千克有种子19万～23万粒。将调制所得的种子晾晒干燥，使含水量达到10%以下，置普通库或冷库干藏。

发芽和播种　种子浅休眠。发芽测定前常用温水浸种12小时。测定温度为20～28℃，每天应有8小时给予光照。30天发芽率可达40%～60%。春季播种。通常于播前温水浸种8～12小时，捞出混沙2倍，置于背风向阳处催芽10～15天，待有20%～30%种子裂嘴时立即播种。条播。每平方米床面播种5～8g，播种深度0.8～1cm。播种后10～15天幼芽出土。出土萌发。连翘幼苗形态见图2。当年苗高60～90cm。每公顷可产苗6.8万～9.0万株。亦可用扦插、压条、分株繁殖。生产上多用扦插繁殖，硬枝或嫩枝扦插均可。

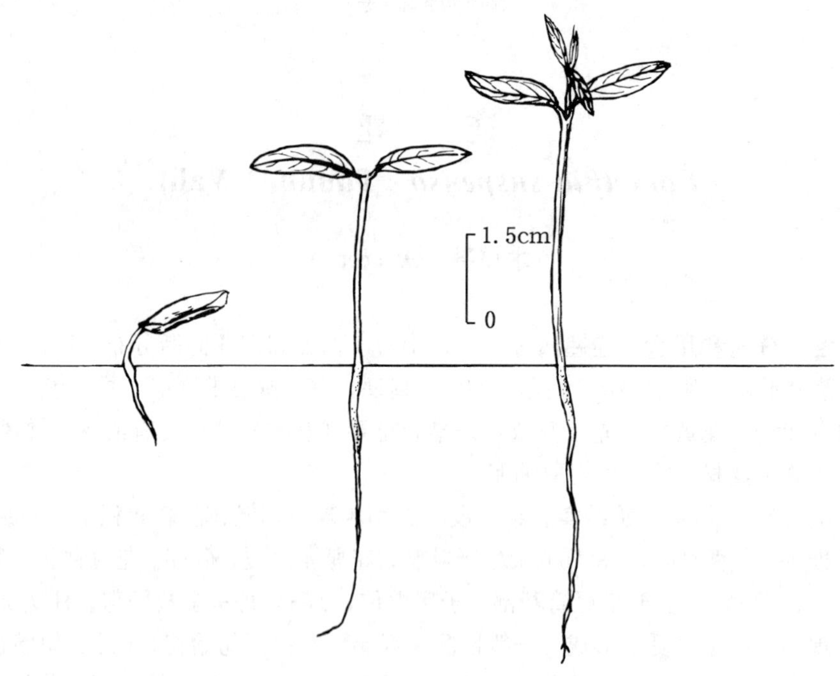

图2　连翘种子萌发后第1、7、15天的幼苗生长情况
（胡冬梅、林平绘）

（宋廷茂）

白蜡树属
Fraxinus L.

（木犀科　Oleaceae）

生长习性、分布和用途　本属约70种，北温带分布。我国约20种，又引入约8种。本文描述6种（表1），除光蜡树为常绿乔木外，均为落叶乔木，是优良用材及园林观赏树种，或放养蜡虫，生产白蜡。多数种类喜光或稍耐庇荫，喜湿润气候及潮湿肥沃土壤，少数种类稍耐瘠薄及轻度盐碱，耐寒，适应性较强，生长较快，萌芽力强。

表1　白蜡树属树种名称、树高、分布及用途

中　名	学　名	树高（m）	分　布	用　途	供　稿
白蜡树	*F. chinensis* Roxb.	15	东北南部、华北、华东、中南至西南	用材、编条、虫蜡、入药	305
光蜡树	*F. griffithii* C. B. Clarke	6～15	中南、滇、藏、陕南、台。日本南部、菲律宾、印度尼西亚、印度	用材	1006
水曲柳	*F. mandshurica* Rupr.	35	东北、华北、陕、甘。朝鲜半岛、俄罗斯、远东、日本	用材	305
美国红梣	*F. pennsylvanica* Marsh.	—	原产加拿大东南至美国东部。我国北京至长江中下游多有引栽	用材、行道树、观赏	203
新疆小叶白蜡	*F. sogdiana* Bunge	25	新疆伊犁河谷。南北疆、甘、青及东北引栽。欧洲、俄罗斯、中亚	用材、绿化、防护林	203
绒毛白蜡	*F. velutina* Torr.	—	原产美国西南。冀、晋、鲁、京、津引栽	用材、绿化、行道树、防护林	305

开花结实　本属树木常在6～8年生开始开花结实，正常结实年龄可以延续70～100多年，结实每隔1～3年有1个丰年，或没有明显的大小年现象。花两性、单性或杂性异株。圆锥花序或呈总状或近簇生，花序顶生或腋生于当年生枝上，或侧生于前1年生枝上。花小，双被、单被或无被，4基数。花萼钟状或杯状。花瓣白色，4或2花瓣，稀6花瓣，或无花瓣。雄蕊2，稀较多。子房上位，2（3）室，每室2胚珠。翅果，顶端具条形或匙形扁平翅，长2.5～5cm，宽0.4～0.8cm。果凸出或稍扁，具1稀2种子，新疆小叶白蜡有10%～40%的翅果含2粒种子。种子长圆柱形，两端略尖，长0.8～1.5cm，种皮栗褐色，具胚乳，子叶扁平（图1）。翅果幼时绿色，成熟时黄绿色，散

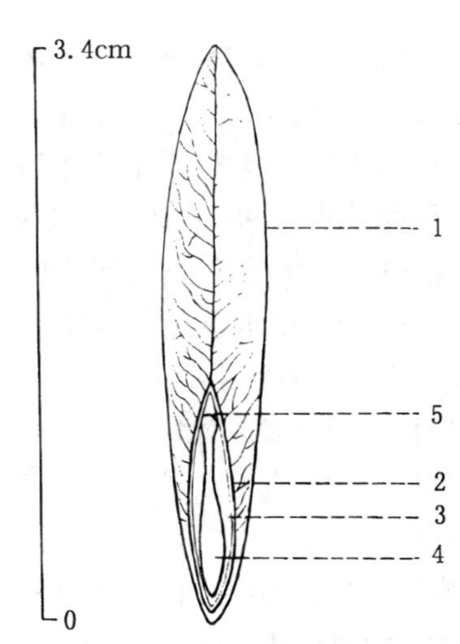

图1　新疆小叶白蜡翅果纵切面
1. 果翅及果皮　2. 种皮　3. 胚乳　4. 子叶　5. 胚根
（吴新安绘）

落时呈土黄色。开花结实物候见表 2。

表 2 白蜡树属树种开花结实物候

树 种	观察地点	开花 始期	开花 盛期	开花 末期	果实成熟 始期	果实成熟 盛期	果实散落期
白蜡树	北京	4 月中旬	4 月中下旬	4 月下旬	9 月下旬	10 月	11～12 月
光蜡树	海南	5 月	～	7 月	7 月	8 月	8 月
	南京	5 月下旬	～	6 月	11 月	11 月下旬	11 月下旬
水曲柳	东北	5 月	5 月中下旬	6 月	9 月上旬	10 月中旬	10 月中旬
美国红梣	新疆	4 月上旬（雄花早 4～5 天）	4 月中下旬	4 月下旬	9 月中旬	9 月下旬	11 月～翌年 3 月
新疆小叶白蜡	新疆	4 月上中旬（雄花早 5～8 天）	4 月中旬	4 月下旬	9 月下旬	10 月	11～12 月
绒毛白蜡	北京	4 月中旬（雄花早 4～5 天）	～	4 月下旬	9 月	10 月	11～12 月

果实的采集调制和种子贮藏 翅果由绿变黄时即开始成熟，剪下果穗或果枝，放在通风处晾干，通常不去翅，除去夹杂物，所得的纯净翅果即为播种材料，通称种子，净度可达 90%以上。新鲜果实含水量 30%左右，待含水量降至 9%～13%时，装入麻袋置通风室内或种子库贮藏，也可以随采随播。这 6 个树种的出籽率、种子净度、千粒重见表 3。

表 3 白蜡树属树种播种材料的出籽率、净度及千粒重

树 种	出籽率（%）	千粒重（g）	每千克纯净种子粒数（万粒）
白蜡树	50	21.3～29.7	3.4～4.7
光蜡树	—	22～24.4	4.1～4.5
水曲柳	80	64.5	1.6
美国红梣	75	21.7～27.4	3.6～4.6
新疆小叶白蜡	—	42～51.3	1.9～2.4
绒毛白蜡	—	18.6～22.0	4.5～5.4

发芽和播种 白蜡树属种子有休眠习性。白蜡树、水曲柳和新疆小叶白蜡属于深休眠。没有破除休眠的水曲柳种子当年不能出苗。赵海珍（1983）研究过水曲柳种子各个部位内源激素的变化。解除白蜡树属种子休眠的常用方法是层积和秋季播种。据凌世瑜、董愚得（1983）和凌世瑜（1986）研究，水曲柳的休眠有两方面的原因：一是胚发育不全，20℃下的暖层积比 5℃下的冷层积更有利于胚的发育；二是果皮、种皮和胚乳等包被组织对萌发的障碍，需要一定时间的低温湿润条件才能克服这种障碍；暖层积加冷层积的几种组合（3 个月＋5 个月、4 个月＋4 个月或 5 个月＋3 个月）能使水曲柳的发芽率达到 90%；GA 能促进水曲柳胚的发育，因此 GA 处理后暖层积 3 个月加冷层积 3 个月也能使发芽率达到 90%；但 GA 不能取代种子对低温的要求：如果仅仅是暖层积，无论是层积前还是层积后用 GA 处理，或是层积前和层积 1～3 个月后 2 次用 GA 处理，且无论是处理果实还是处理种子，如果没有冷层积，则发芽率都为零。白蜡树属种子常见的催芽处理方法见表 4。

条播，稀用撒播，行距约 60cm，播幅约 10cm。开沟播种，覆土约 3～4cm。这几个树种常用的播种量见表 5。种子不经处理可在 10～11 月播种，灌好冬水，翌年 4 月地温 10℃以上时即萌芽出土。经过层积处理的种子 4 月中下旬～5 月上旬播种，覆土 3cm，10 天左右可出苗。

表 4　白蜡树属树种常用的催芽处理方法

树　种	催芽方法
白蜡树	①2～5℃低温层积 2～3 个月 ②秋播
光蜡树	1988 年在南京低温层积 3 个月，发芽率达 78%
水曲柳	暖（25℃）层积 3～4 个月，再低温（2℃）层积 3～4 个月
美国红梣	①放入土坑中混沙层积约 3 个月 ②用 60℃水浸 10 分钟，冷水浸 24 小时，再混沙层积约 20 天 ③10 月下旬播种
新疆小叶白蜡	①10 月中下旬播种 ②用始温 60℃水浸种或冷水泡 3～5 天，再置入 15～20℃室内保温层积约 1 个月
绒毛白蜡	①秋播 ②用始温 40～50℃水浸种 24 小时，再保持 25℃层积催芽

出土萌发。当年苗高可达 50～150cm，每 666m^2 产苗 1.5 万～3.0 万株。1 年生苗可以出圃造林或移栽后继续培育大苗。白蜡树属也可以用萌蘖、埋条或插条繁殖。幼苗形态图见图 2。

表 5　白蜡树属树种常见的播种量和产苗量

树　种	发芽率（%）	播种量（kg/666m^2）	产苗量（万株/666m^2）
白蜡树	60	3～5	2～3
光蜡树	78	—	—
水曲柳	50～60	15	—
美国红梣	80～85	4～5	1.5～2.0
新疆小叶白蜡	90	4～5	4.0
绒毛白蜡	75～85	3～5	1.0

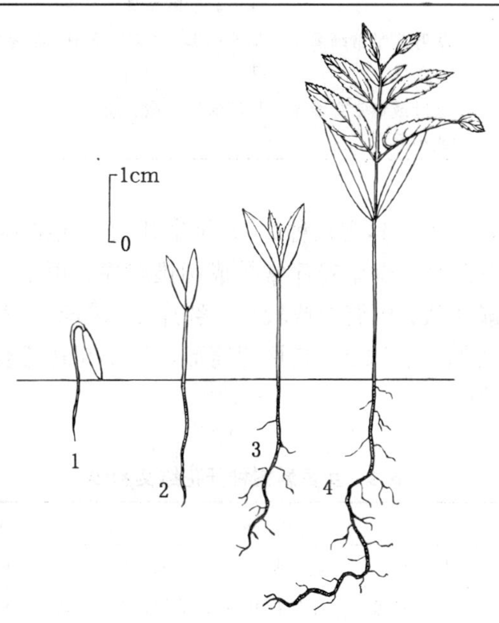

图 2　新疆小叶白蜡种子萌发和幼苗生长情况
1. 下胚轴延伸　2. 子叶开展　3. 初生叶初现　4. 幼苗形成
（吴新安绘）

（王木林）

女　贞　属
Ligustrum L.

（木犀科　Oleaceae）

生长习性、分布和用途　本属50余种，主产东亚。我国约38种，分布秦岭以南各地，本文描述4种。喜温暖湿润气候，适宜于深厚肥沃而湿润的微酸性或微碱性土壤，干燥瘠薄之地生长不良。耐修剪，萌发力强。水蜡树和小叶女贞较耐寒。女贞凌冬青翠，为温带地区不可多得的常绿乔木，对大气中二氧化硫、氯气、氟化氢及铅蒸气等有较强抗性，并有滞尘抗烟功能，可选为行道树，尤可栽培作为工矿企业有害气体、粉尘源的隔离带。几种女贞皆为常用的绿篱树种。几种女贞可以放养白蜡虫，以收取白蜡。女贞木材可用于细木工。根、树皮、叶、果入药。果又可酿酒，种子油可制肥皂。几种灌木女贞叶可代茶。几个树种的苗木可作嫁接桂花、丁香的砧木。这几个种的名称、习性、分布和用途见表1。

表1　女贞属树种的名称、习性、分布和用途

中　名	学　名	习　性	分　布	用　途
女　贞	*L. lucidum* Ait.	常绿乔木	华东、中南、西南、陕、晋。冀栽培。日本	材用、绿化、药用、放养白蜡虫，果酿酒
水蜡树	*L. obtusifolium* Sieb. et Zucc.	落叶灌木	华东、华中　陕、甘；辽、冀栽培。朝鲜半岛、日本	绿篱、叶代茶
小叶女贞	*L. quihoui* Carr.	落叶或半常绿灌木	华北南部、华东、华中、西南、陕	绿篱、放养白蜡虫、叶代茶、种子入药、花可提芳香油
小蜡树	*L. sinense* Lour.	落叶或半常绿灌木或小乔木	长江流域以南、陕、鲁	绿篱、放养白蜡虫、叶代茶、果酿酒

开花结实　女贞8～12年生，其它几种3～5年生开始开花结实，并分别在15年生或7～8年生后进入正常结实。花两性。聚伞花序常排成圆锥花序，顶生。花小。花萼钟形，4齿裂或近全缘。花冠白色，近漏斗状，冠筒短或较长。裂片4。雄蕊2，生于花冠筒上。子房上位，球形，2室，每室有悬垂的倒生胚珠2。花柱圆筒形，长不超过雄蕊，柱头稍肥厚，近2裂。4个种的开花结实物候见表2。

表2　女贞属树种开花结实物候

树　种	观察地点 观察年份	开花 始　期	开花 盛　期	开花 末　期	果实 成熟期	果实脱落 始　期	果实脱落 末　期
女　贞	洛阳 1975～1979	6月上中旬	6月中下旬	6月下旬、7月上旬	10月上中旬	—	—
	西安 1975～1979	6月中下旬	6月中下旬	6月下旬、7月中旬	10月下旬～11月上旬	—	—
	扬州 1975～1979	6月中下旬	6月下旬	7月上中旬	10月中旬	—	—

（续）

树　种	观察地点 观察年份	开　花			果实 成熟期	果实脱落	
		始　期	盛　期	末　期		始　期	末　期
女贞	杭州 1976、1978	6 月中下旬	6 月下旬	6 月下旬、7 月上旬	—	—	—
	成都 1975、1976	6 月上旬	6 月中旬	6 月下旬、7 月上旬	11 月上旬	翌年 1 月下旬～2 月中旬	
	宜宾 1975～1978	5 月中下旬	5 月下旬～6 月下旬	6 月中旬～7 月上旬	11 月中旬	12 月上旬	
	赣州 1978、1979	5 月中旬	5 月下旬～6 月上旬	6 月中旬	—	—	—
	昆明 1978		6 月下旬	7 月中旬	—	—	—
水蜡树	沈阳 1975～1978	6 月上中旬	6 月上中旬	6 月中下旬	9 月上旬～下旬	9 月下旬	11 月下旬
小叶女贞	盐城 1975～1980	5 月中下旬	5 月下旬	6 月上中旬	—	—	—
	贵阳 1977	5 月上旬	5 月中旬	5 月下旬	—	—	—
	重庆 1977	4 月下旬	5 月上旬	5 月上旬	6 月中旬	—	—
小蜡树	杭州 1978、1979	5 月下旬	5 月下旬	5 月下旬	11 月中旬	11 月下旬	12 月中旬

核果浆果状。果皮黑色、蓝黑色或蓝紫色，常被蜡质白粉。子房 2 子室均匀发育则形成 2 分核，果成倒卵状椭圆形、宽椭圆形或近球形；仅 1 子室发育则形成 1 果核，另 1 不发育之子室则成 1 卵状椭圆形之薄片，贴生于发育果核弯曲的腹面，因而果实常呈肾形，且花柱斜生于一侧。内果皮（核壳）膜质、纸质或带木质，常具多条纵沟。一般 1 核仅具 1 种子，偶有 2 种子则内果皮常于室背开裂，因此每果有种子 1～2，稀 3 或 4。种皮薄，亦具纵沟纹。胚乳丰富，子叶扁平，卵形，胚根短，见图 1。这 4 种女贞成熟果实的形状、颜色和大小见表 3。

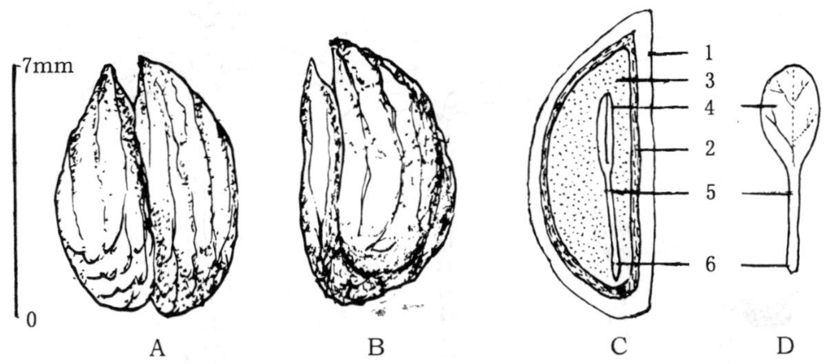

图 1　女贞种子 2 分核发育（A）、1 分核发育（B）及其纵切面（C）和胚（D）

1. 内果皮　2. 种皮　3. 胚乳　4. 子叶　5. 胚轴　6. 胚根

（田恒德绘）

表 3　女贞属树种果实形状、颜色和大小

树　种	形　状	颜　色	大　小	
			长（mm）	径（mm）
女　　贞	长圆形或肾形	蓝黑色或蓝紫色	9～11	6～7
水 蜡 树	宽椭圆形或近球形	黑色	7～8	6.5
小叶女贞	宽椭圆形或近球形	蓝黑色	8～9	5
小蜡树	近球形	黑色	—	4～5

果实的采收调制和种子贮藏 女贞属各树种的果实成熟期不一致，核果成熟后也不会立即脱落，女贞核果能延至翌年2～3月才脱落。因此采收时间长，在果实呈现蓝紫色、蓝黑色或黑色时可用高枝剪或枝剪剪取果穗。女贞应选生长健壮、主干通直的20～40年生母树，于11～12月时采收。剪下果穗，将果实捋下，浸于水中5～7天，搓去果皮、果肉及不发育之分核，洗净晾干，所得果核用作播种材料，通称种子。女贞果实出籽率25%，种子（核）千粒重36g，每千克有2.8万粒。小叶女贞果实出籽率60%，种子（核）千粒重25g，每千克约4万粒。种子净度在95%以上。种子调制后装袋干藏或混沙贮藏。

发芽和播种 女贞属种子发芽率在50%～70%。播种季节常影响种子的发芽率：在南京地区，凡是12月～翌年1月播种的女贞，4月中旬即开始发芽，场圃发芽率和成苗率均较高。干藏而延至3月中旬播种的，到6月才开始出土；4～5月播种的，其场圃发芽率和成苗率都极低，故以随采随播或冬播为好。王成霖和邵蓓蓓（1987）报道说，未经层积的女贞种子，发芽率只有14%，在0～5℃下层积10周后发芽率也只有22%，而在室温下层积5周后得到的发芽率为73%。北方地区将小叶女贞调制贮藏后，常于2月下旬用凉水浸种2～3小时，捞出加沙2～3倍，混合均匀，藏于窖内或冷库中；也可堆置于背阴处，厚度以30cm为宜，至3月初解冻后将种子置于背风向阳处，增温催芽，并经常检查、翻动和浇水，3月底前后种子即可萌动，俟有30%左右种子萌发时即可取出播种。

条播。南方用高床，北方常用低床。每$10m^2$女贞播种200～300g，小叶女贞150～200g。水蜡树介于两者之间，小蜡树则少于150g。沟深1～1.5cm，覆土后盖草。出土萌发。女贞种子的萌发和幼苗的早期生长情况见图2。多数种类可以扦插或压条繁殖。

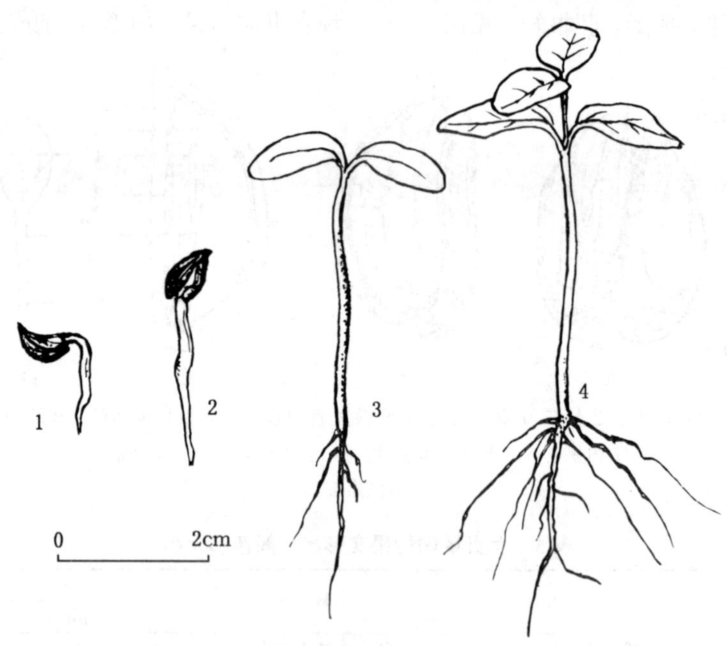

图2 女贞种子萌发及幼苗生长情况

1、2. 胚轴延伸 3. 子叶出土 4. 初生叶对生

（田恒德仿《主要树木种苗图谱》）

（黄鹏成）

木犀榄属
Olea L.

（木犀科　Oleaceae）

生长习性、分布和用途　本属约40种，我国产14种，又引入1种，本文描述2种（表1）。常绿乔木或灌木。喜光，速生。具有发达和伸展的根系。最适宜在冬湿夏旱型的气候区生长，能耐－7～－11℃的短时低温。喜质地疏松排水良好的石灰质土壤。油橄榄分布于地中海沿岸地区，我国引入栽植40多个品种和类型，并选育出10多个优良品种。陈宪初（1988）概述过我国引种油橄榄的主要限制性生态因子。为促进油橄榄结实，章树文和李龙山等（1988）在汉中地区作过施肥试验。滨木犀榄分布于亚洲热带，我国海南多作庭园绿化树。果实富含油脂及维生素A、D、E、K。橄榄油是世界著名的食用油，又是医药等方面的特用油。薛益民和王笑山等人（1988，1989）对引种到陕西汉中地区的油橄榄主要品种果实的经济性状作过分析，并对油用品种和餐用品种分别提出了选择标准。木材坚硬致密，是制造农具、家具、雕刻的优良用材。

表1　木犀榄属树种名称、树高、分布和用途

中　名	学　名	树　高 (m)	分　布	用　途	供　稿
滨木犀榄	*O. brachiata* (Lour.) Merr. ex Groff.	2～5	琼。亚洲热带地区	观赏	605
油橄榄	*O. europaea* L.	10～20	原产地中海沿岸地区。陕西以南各省引种栽培	食用，药用	311

开花结实　3～7年生开始结实。10年生以后进入正常结实年龄。结果盛期可延续40～60年。有隔年结实现象。

花两性或单性，或杂性异株。圆锥花序腋生。花瓣及萼片4裂，花冠辐射对称。花淡黄色、白色或淡绿色。雄蕊2。子房上位，2（稀3）室，每室具2胚珠。花柱较短，柱头2裂。核果，核壳（内果皮）质地坚硬，表面具沟棱，缝合线不明显，厚1～3mm，通常1粒，稀2粒种子。开花结实物候见表2。种子（果核）形态特征见表3及图1。

表2　木犀榄属开花结实物候

树　种	观察地点	观察年份	开　花			果实成熟		采收期
			始　期	盛　期	末　期	初　熟	盛　熟	
滨木犀榄	海南屯昌	1986	2月下旬	3月上旬	3月中旬	6月下旬	7月下旬	8月中旬
油橄榄	陕西城固	1980	5月25日	5月28日	5月31日	9月下旬	11月下旬	11月下旬
	湖北武昌	1985	5月12日	5月15日	5月19日	—	—	—
	（品种佛奥）	1987	5月13日	5月16日	5月20日	—	—	—

表 3 木犀榄属树种果实和种子(果核)的形态特征

树 种	果 形	果 色		果实大小(cm)		种子特征			
		未熟果	成熟果	长	径	形 状	长(cm)	径(mm)	颜 色
滨木犀榄	椭圆形或球形	粉绿	紫红	1～1.3	0.7～1.0	椭圆形	0.6～0.9	5～6	褐色
油橄榄	椭圆形或卵圆形	淡绿	紫黑	2.2～4.1	1.9～2.2	椭圆或长卵形	1.6～3.1	8～9	浅褐色

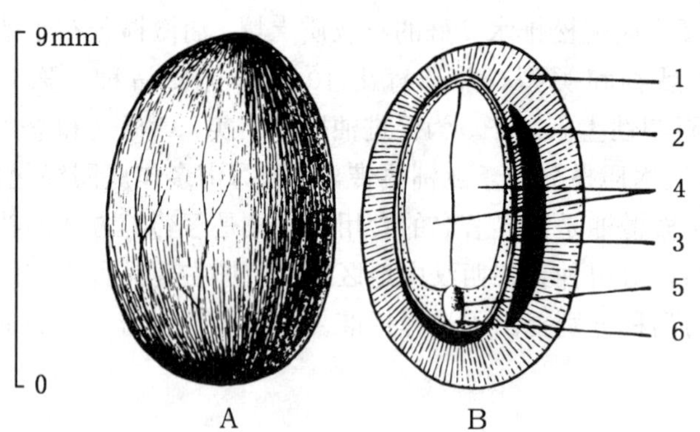

图 1 滨木犀榄果核外形（A）及其纵切面（B）
1. 内果皮 2. 种皮 3. 胚乳 4. 子叶 5. 胚轴 6. 胚根
（黄应钦绘）

果实的采收调制和种子贮藏 通常用手工采收。油橄榄果实成熟期长且很不一致，果型不大，每个工作日只能采收 30kg 以上。陈坤荣和方文德（1987）以昆明地区两株佛奥品种为材料，在果实完全变褐时喷施 1 200μg/g 的乙烯利，5 天后摇动果枝，落果率达到 100%。采回的果实用手工或果肉剥离机剥离果肉；或堆放于室内 3～4 天，待软熟后置水中搓揉，漂净皮肉杂质，所得果核即为播种材料，通称种子。摊放在通风良好的室内阴干，待含水量在 10%以下时装于麻袋或布袋内，存放在通风、阴凉、干燥的室内架上贮藏。果实出籽率及千粒重见表 4。

表 4 木犀榄属树种的出籽率和种子质量

树 种	出籽率 (%)	净 度 (%)	千粒重（g）		每千克纯净种子粒数	
			一 般	变动范围	一 般	变动范围
滨木犀榄	40	97～99	110	80～140	9 000	7 100～12 500
油橄榄	20	92～97	420	350～1 200	2 400	800～2 900

发芽和播种 种子有休眠习性。用清水浸泡种子 1～2 天，换水 2～4 次。浸过的种子置于 0.02%～0.05%的赤霉素溶液中浸种 24 小时，清水洗净，阴干后用洗净的中粒河沙层积催芽，当种子有 30%以上裂口时即可播种。适于种子发芽的日平均温度在 16～27℃。高床条播或点播。油橄榄每平方米床面播种 0.2～0.6kg，覆土厚 0.5～2.0cm。播后塑料膜覆盖。油橄榄出土萌发。播后 15 天左右胚根伸长，侧根出现，并把种壳和子叶顶出地面。10 天左右种壳脱落，子叶张开并由淡黄色转为绿色。30 天左右主根伸长，侧根增多并扩展，同时出现 2 片初生叶。当幼苗具有 4～5 对真叶时移植，株距 15～20cm，行距 30～40cm。滨木犀榄则是留土萌发。油橄榄种子萌发情况见图 2。

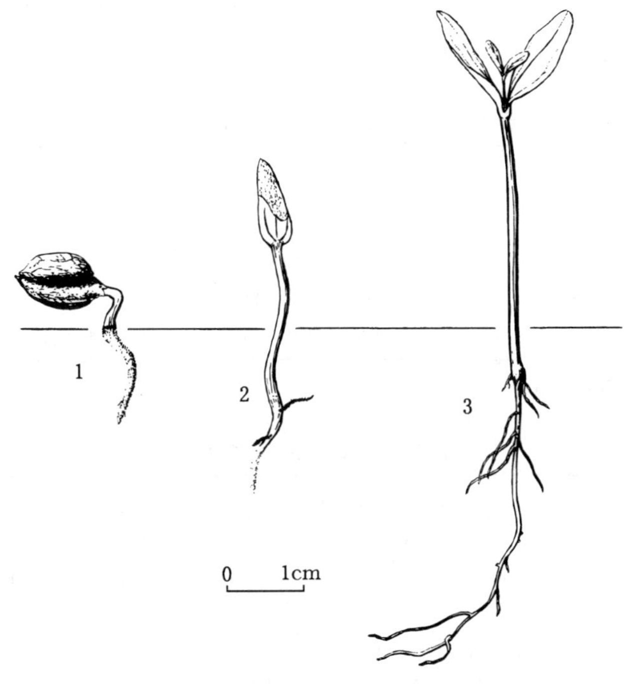

图 2　油橄榄种子萌发及幼苗生长情况
1. 下胚轴延伸出土　2. 子叶初露　3. 初生叶对生
（孟玲绘）

（邓明全）

桂　　花

Osmanthus fragrans Lour.

（木犀科　Oleaceae）

生长习性、分布和用途　木犀属约 40 种，东亚—北美间断分布。我国约 27 种，本文描述 1 种。常绿小乔木或乔木，树高 12～30m，胸径 10～20cm，最粗可以超过 2m。生长较慢，寿命长。陕西勉县武侯墓前 2 株丹桂—“护墓双桂”，据记载已经经历 1 700 余年，至今开花仍较繁茂。喜温暖，较耐荫。适生于肥沃而排水良好的沙质壤土。原产我国西南部。印度、尼泊尔、柬埔寨也有分布。黄河下游以南地区普遍栽培。名贵观赏花木之一，香飘数里。花可作香料，也可食用。在江苏吴县，8～10 年生的单株产花 0.25～0.5kg，30～50 年生产 15～20kg，百年大树可产 80kg。木材具光泽，纹理美丽如犀，故有木犀之称，为雕刻良材。花、果、叶、皮、根均可药用。

桂花有许多变种和栽培品种，分类的依据主要是花期和花色。主要变种的特征见表 1。有些栽培品种能开花但不能结实。

开花结实　实生苗始花年龄较晚，嫁接苗当年可以开花。花两性。聚伞花序簇生于叶腋。

花萼盘状，长1mm，4裂，边缘啮蚀状。花冠4裂，橙红色至白色，极香。雄蕊2，花丝极短，着生于花冠筒近顶部。子房2室，每室有胚珠2枚。开花结实的物候见表1、表2。

表1 桂花主要变种的形态特征和花期①

变 种	学 名	形态特征	花 期
金 桂	var. *thunbergii* Mak.	花金黄色，易脱落，香气浓，产花量高	9月下旬
银 桂	var. *latifolius* Mak.	花奶黄至淡黄色，香气颇浓	比金桂略晚
丹 桂	var. *aurantiacus* Mak.	花橙红色或橙黄色，香气较淡	9月下旬
四季桂	var. *semperflorens* Hort.	花黄色或淡黄色，1年中数次开花，香气淡	春、夏、秋

① 据南京中山植物园《花卉园艺》

表2 桂花开花结实的物候

观察地点	观察年份	开花			果实成熟期	落果	
		始 期	盛 期	末 期		始 期	末 期
江苏南京	1963～1964	9月下旬	10月上旬	10月上旬～中旬	4月下旬～5月下旬	4月下旬～5月下旬	—
安徽芜湖	1964	9月22日	9月26日	10月8日	4月10日	4月28日	5月8日
广西桂林	1965～1979	9月中旬～10月下旬	9月下旬～10月下旬	10月中旬～下旬	2月下旬～4月中旬	3月中旬～5月上旬	4月上旬～6月上旬
广西雁山	—	9月上旬～10月下旬	9月上旬～10月下旬	10月中旬～11月上旬	2月中旬～4月下旬	3月上旬～5月上旬	3月下旬～5月下旬
广西南宁	1978～1984	9月中旬～下旬	10月上旬	11月上旬	4月上旬～中旬	—	—
湖南常德	1975～1980	9月上旬～下旬	9月中旬～下旬	9月下旬～10月上旬	4月上旬～5月上旬	4月中旬～5月上旬	4月下旬～5月下旬

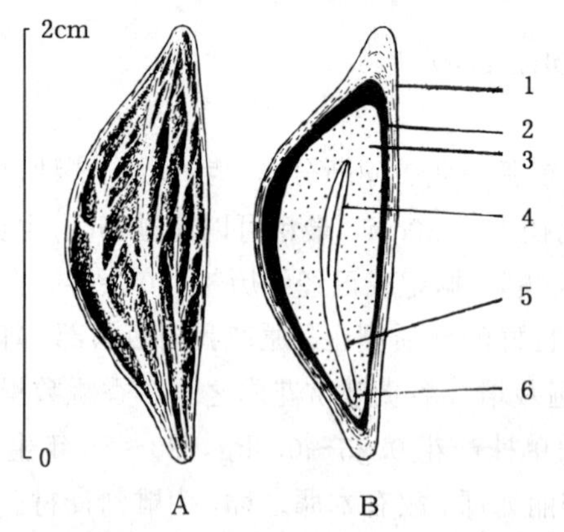

图1 桂花果核的外形（A）及其纵切面（B）
1. 内果皮 2. 种皮 3. 胚乳 4. 子叶 5. 胚轴 6. 胚根
（黄应钦绘）

核果椭圆形，长1.5～2.3cm，径0.8～1.2cm，熟时果皮由绿色转为紫红色或紫黑色，熟后自落。外果皮膜质，中果皮肉质，内果皮骨质。果核中有种子1枚。果核长1.2～2.1 cm，径5.5～6.5 mm，纺锤形，黄色或米黄色，表面有纵棱脉纹7～9条，由基部直达先端，脉纹上分出细小分叉，弯曲不正，脉纹间形成不规则的深沟。种脐圆形，周围隆起，中心成不明显的乳头状突起。有胚乳。胚直立。果核的外形及其剖面形态见图1。

果实的采收调制和种子贮藏 采摘或摇落果实，堆沤三数日，待果皮软化后浸水搓去果皮果肉，淘去空粒，忌曝晒，晾干1～2天后所得果核即为播种材料，通称种子。

鲜果出籽率25%～30%，千粒重约260（190～310）g。每千克种子约3 700（3 200～5 500）粒。晾干后的种子用0.05%高锰酸钾浸4分钟，晾干后混沙湿藏。

发芽和播种　种子有休眠习性，生产上常用湿沙层积催芽后春播，也可以当年秋播。种子发芽率为50%左右，高的可达70%～80%。每平方米约播50g。出土萌发。子叶2片，黄色转绿色，宽披针形。子叶出土后3～4天发出对生的初生叶2片。主侧根均较粗壮，分布较均匀。发芽和幼苗生长情况见图2。每666m^{2}产苗2.5万～3万株。1年生苗高15～20cm。一般用嫁接、压条或扦插繁殖。

（杨国华）

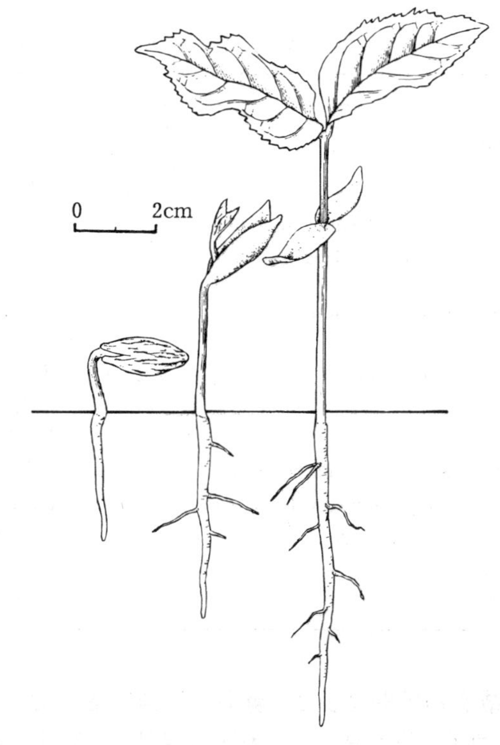

图2　桂花种子萌发后第4、10、30天的幼苗生长情况
（黄应钦绘）

丁　香　属
Syringa L.

（木犀科　Oleaceae）

生长习性、分布和用途　本属约30种，分布于亚洲、欧洲的温带及亚热带。我国24种，本文描述5种和2变种，其中1种从欧洲引入（表1）。落叶灌木或小乔木。喜光，稍能耐荫；喜肥沃湿润土壤，忌水涝，也不耐潮湿。本属花期早，花序大，具香气，为我国著名的优良园林观赏树种。花枝做切花，叶和果皮入药。

开花结实　3～5年生开始结实。结实盛期在10年生以后。不同种类的结实率有较大差异。北京丁香和暴马丁香结实率最高，几乎每一花序都能大量结实。华北紫丁香、白丁香和欧洲丁香只有花序中的部分花结实。花叶丁香不结实或仅少量花序上的极少数花朵结实，且多为秕籽。花两性，组成大圆锥花序，顶生或腋生于上1年的小枝上。萼钟状，4齿裂或截形，宿存。花冠长漏斗状，具深浅不同的4裂片，蓝紫色、白色或红色及紫红色，有香气。雄蕊2，着生于花冠筒上，内藏或突出于花冠外。子房上位，2室，每室具2倒生胚珠。本属各树

表 1 丁香属树种的名称、成年树高、分布和用途

中 名	学 名	树高（m）	分 布	用 途	供 稿
华北紫丁香	*S. oblata* Lindl.	4～5	东北南部、华北、西北东部。朝鲜半岛	观赏，药用	403
白丁香	*S. oblata* Lindl. var. *affinis* (Henry) Lingdelsh.	4～5	豫。华北地区普遍栽培	观赏	403
北京丁香	*S. pekinensis* Rupr.	5～13	华北南部、西北东部	观赏，蜜源	403
花叶丁香	*S. persica* L.	2～3	西北、西南。阿富汗、伊朗、地中海地区及中欧。京、晋、鲁、豫有栽培	观赏，香料	403
暴马丁香	*S. reticulata* (Bl.) Hara var. *mandshurica* (Maxim.) Hara	10	东北、华北、西北、华中。俄罗斯、朝鲜半岛、日本	材用、蜜源、芳香、观赏	108
红丁香	*S. villosa* Vahl.	3	东北、华北、陕。朝鲜半岛	观赏	404
欧洲丁香	*S. vulgaris* L.	5～7	原产欧洲。京、晋、鲁有栽培	观赏，香料	403

种开花结实物候期见表 2。蒴果，长椭圆形，尖端喙状，果皮革质，2 室，沿室背开裂，每室 1～2 枚种子。种子扁片状，表面有纵棱。1 室具 1 粒种子时，种子为椭圆形，有 1～1.5mm 的边翅围绕；1 室具 2 粒种子时，种子为长椭圆形，两侧边无翅，仅顶端有窄翅，一侧边缘略呈压扁状。不同树种种子的形态特征稍有差异（表 3）。种皮木质，脆，易折断。有胚乳。胚大，匙形（图 1）。

表 2 丁香属树种开花结实物候期

树 种	观察年份和地点	花期 始期	花期 盛期	花期 末期	果期 始期	果期 盛期	种子散落期 始期	种子散落期 盛期	种子散落期 末期
华北紫丁香	1986～1988 北京	4 月中下旬	4 月下旬	5 月上旬	8 月中旬	8 月下旬	8 月下旬	9 月下旬	10 月上旬
白丁香	1986～1988 北京	4 月中下旬	4 月下旬	5 月上旬	8 月中旬	8 月下旬	8 月下旬	9 月下旬	10 月上旬
北京丁香	1986～1988 北京	5 月下旬	6 月上旬	6 月上旬	10 月中旬	10 月下旬	10 月下旬	11 月上旬	11 月上旬
花叶丁香	1986～1988 北京	4 月中下旬	4 月下旬	5 月上旬	8 月中旬	8 月下旬	8 月下旬	9 月下旬	10 月上旬
暴马丁香	1963～1980 哈尔滨	6 月上旬	6 月下旬	7 月上旬	8 月下旬	9 月中下旬	—	11 月下旬	—
红丁香	1963～1980 哈尔滨	5 月下旬	6 月中旬	6 月下旬	8 月下旬	10 月上旬	9 月中旬	～	10 月中旬
欧洲丁香	1986～1988 北京	4 月中下旬	4 月下旬	5 月上旬	8 月中旬	8 月下旬	8 月下旬	9 月下旬	10 月上旬

表 3 丁香属树种的种子形态特征

树 种	形 状	大 小（mm）	颜 色
华北紫丁香	椭圆形	长 10～11，宽 4～4.5	黑褐色，翅黄褐色
白丁香	椭圆形	长 9～10，宽 4～5	黑褐色，翅褐色
北京丁香	长椭圆形	长 10～12.5，宽 3～5	棕褐色，翅棕褐色
花叶丁香	长椭圆形	长 10～12，宽 3～3.5	棕褐色，翅棕褐色
暴马丁香	长椭圆形	长 9～15，宽 4～7	棕褐色，翅色略浅
红丁香	长椭圆形	长 9～10，宽 3～4	红褐色，翅淡红褐色
欧洲丁香	椭圆形至长椭圆形	长 9～11，宽 3～4.5	红褐色，翅色略浅

果实的采收调制和种子贮藏 蒴果由绿色变为黄褐色或黑褐色并有少数果实开裂时，即可剪取整个果穗，在通风处摊开晾干。蒴果充分开裂后，翻动敲打脱粒，或用手倒提果穗，震落种子。暴马丁香和北京丁香种子不易脱落，在自然状态下甚至经冬不落。风选去杂时勿使风力过猛而把种子吹掉。调制中不要折断种子。不同树种的出种率、净度、千粒重、每千克粒数详见表 4。

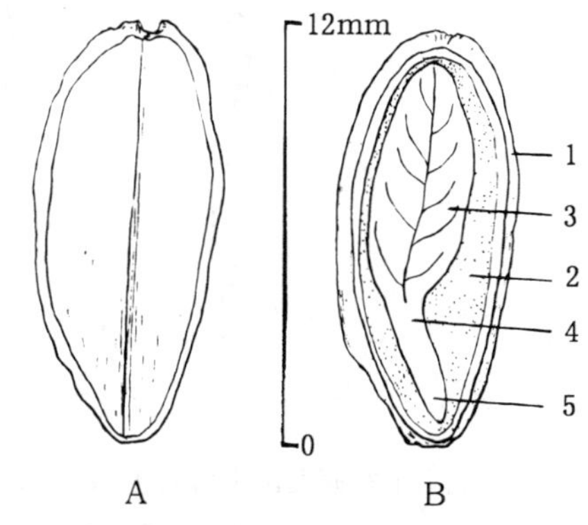

图 1 北京丁香种子外形(A)及其纵切面(B)
1. 种皮 2. 胚乳 3. 子叶 4. 胚轴 5. 胚根
（胡冬梅、林平绘）

种子晾干后，如短期贮藏，可装入纸袋、布袋或麻袋中，置于干燥、通风、温度平稳的室内存放。在这样的条件下贮存 2 年，生活力不会明显下降。长期贮藏则需将种子充分干燥（含水量 13%以下），装入密闭容器或塑料袋内，放在 0～5℃的低温下贮藏。曾经在一般室内存放过北京丁香和欧洲丁香种子，1 年后北京丁香发芽率为 56%～80%、欧洲丁香的发芽率为 50%～78%。

表 4 丁香属树种的出籽率及种子质量

树 种	地 点	出籽率（%）	净度（%）	千粒重（g）	每千克纯净种子粒数（万粒）
华北紫丁香	北 京	3～5	—	8～14	7.1～12.5
	北 京	15	—	8	12.5
白丁香	哈尔滨	10	—	9	11
北京丁香	北 京	10～15	—	11～12.5	8～9
	北 京	9	—	14.2	7
花叶丁香	北 京	0～1	—	9.4～10.6	9.4～10.6
暴马丁香	哈尔滨	25	—	18～24	4.1～5.5
红丁香	哈尔滨	2.5	70～80	6.7	15
欧洲丁香	北 京	3～5	60～70	5～6.7	14.9～20

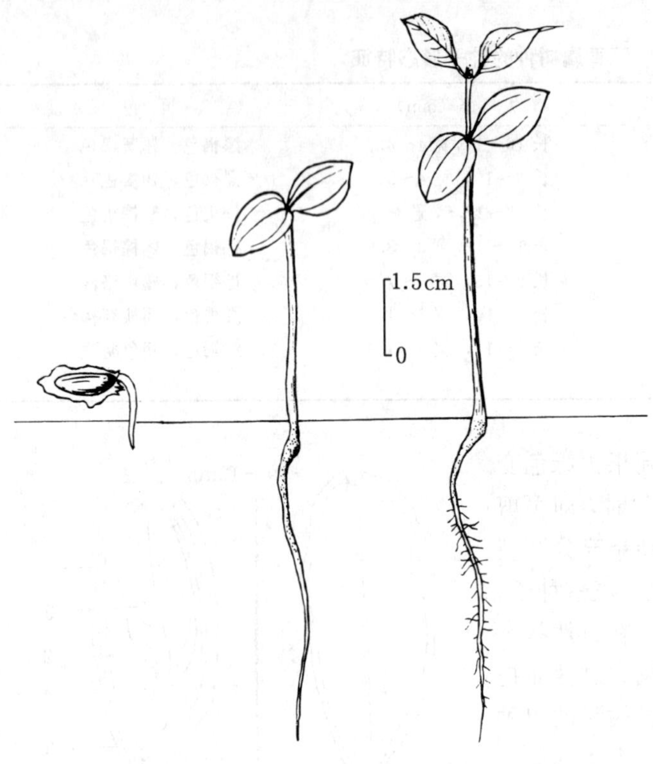

图 2 北京丁香萌发出土后第 1、8、18 天幼苗生长情况
(胡冬梅、林平绘)

发芽和播种 发芽测定可在发芽皿内的滤纸床上，在 20～25℃的温度条件下进行。北京丁香第 6～9 天开始发芽，29～32 天结束，发芽率约 72%；欧洲丁香第 7 天开始发芽，22 天发芽结束，发芽率约 61%。出土萌发（图 2）。

可以秋播。如果春播，所用种子需层积 30～90 天。据孙锦等(1982)报道，3 月中旬播种前浸种 1～2 小时后，将种子混以 2～3 倍的湿沙，放在背风向阳处催芽，5～10 天后部分种子萌动即可取出播种。如种子稀少，也可在室内盆播，出苗至长出 4～5 对初生叶时移栽。暴马丁香发芽比较困难，通常需在 1～5℃下层积 60～90 天后再转入 10～20℃下催芽 7～10 天。条播。每平方米床面播种 5g，留苗约 30 株。当年苗高 0.5～0.8m。也可用压条、扦插、分株或嫁接法繁殖。

(刘长江)

糖 胶 树

Alstonia scholaris (L.) R. Br.

(夹竹桃科 Apocynaceae)

生长习性、分布和用途 鸡骨常山属约 50 种，我国 6 种，本文描述 1 种。常绿大乔木，高可达 40m，胸径 1.25m。适生于肥沃湿润土壤。幼龄期忌霜冻，成年树可耐 0℃左右的极端低温。分布于滇、桂。湘、粤、琼、台有栽培。南亚、东南亚和澳大利亚也有分布。根、皮、叶均含生物碱，供药用，全株为疟疾、发汗、健胃药。根、皮可治慢性支气管炎，外用止血。乳汁可提制口香糖原料。树形美观，可以用作行道树。

开花结实 约 10 年生开始开花结实，正常结实期在 20 年生以后。影响结实的主要因素是低温和虫害。结实有大小年现象，间隔期 1～3 年不等。花两性。稠密的聚伞花序顶生，被柔毛。花萼短，5 裂。花冠白色，高脚碟状，筒中部以上膨大，内被柔毛，裂片 5，短，向左覆盖。雄蕊 5，着生于花冠上，内藏。子房上位，为 2 枚离生心皮组成，被柔毛，每心皮有胚

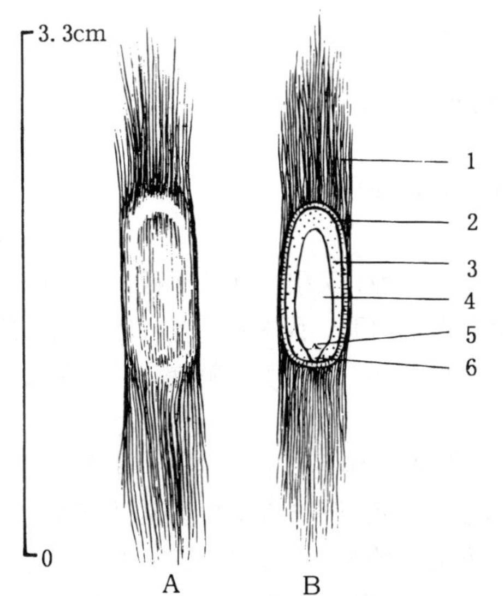

图 1　糖胶树种子外形（A）及其纵切面（B）
1. 长缘毛　2. 种皮　3. 胚乳　4. 子叶　5. 胚芽　6. 胚根
（黄应钦绘）

珠多颗。花柱长，丝状，柱头棍棒状，顶端 2 深裂。花盘环状。据广西南宁 1987～1988 年观察，4 月下旬为开花始期，5 月中旬为盛花期，5 月下旬为末花期；6 月中旬果实开始成熟，7 月中旬为果熟盛期，8～11 月成熟果实陆续开裂，种子飘落。蓇葖双生，长线状披针形，微弯扭，细长如豆角，长 20～28（57）cm，径 3～5mm，未成熟时青色，成熟时黑褐色。每果具种子多粒。种子扁长圆形，两端具长缘毛，红棕色至深褐色，长 5～7mm，宽约 1.5mm，厚约 0.6mm，带毛长 3～4cm。胚直，有大而平展的子叶，被包于丰富的胚乳中（图 1）。

果实的采收调制和种子贮藏　果实成熟后要及时采收，以防开裂后种子飞散。用采种钩刀或高枝剪连同总果梗一起采下。采得的果实置避风处曝晒，待开裂后轻轻敲打或搓擦，去除果壳杂质即得纯净种子。果实出种率为 14%～16%，种子净度为 80%，千粒重 4.3g，每千克有纯净种子 23（20～29）万粒。种子可干藏，含水量应在 10%以下。用布袋包装，悬挂于室内通风处，贮藏期 6～8 个月。

发芽和播种　种子无休眠习性。发芽时气温需在 20℃以上。1988 年 6 月 17 日，广西林业科学研究所对当年新采且未处理的种子，用滤纸作基质，在室内作过发芽测定。温度为 28℃，光照为室内自然光。置床后 3 天开始发芽，第 6 天进入发芽盛期，第 9 天发芽终止。自置床之日起计算，6 天的发芽百分数为 53%，9 天发芽率为 70%。出土萌发。胚根萌发后 3 天子叶出土，再过 6 天发出初生叶，生长情况见图 2。

撒播。选无风天播种，每平方米约播 1～2g。初生叶出土后移至圃地培育。1 年生苗出圃。

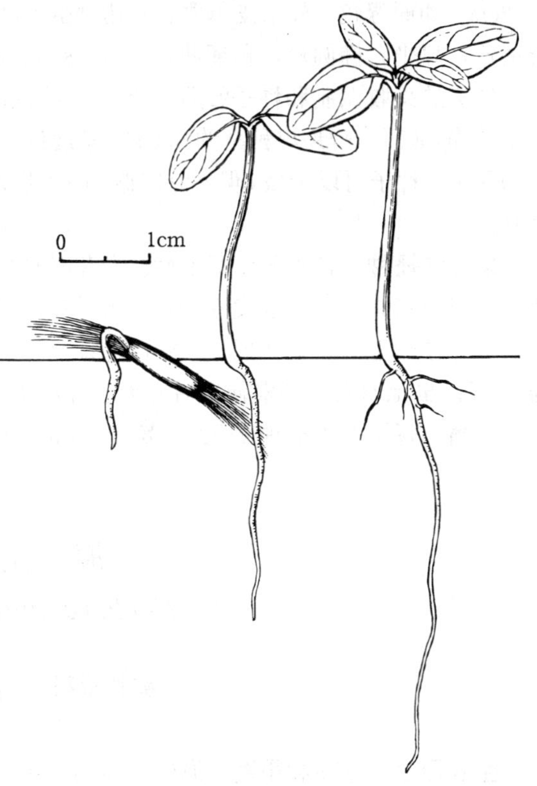

图 2　糖胶树种子萌发后第 3、5、9 天幼苗的生长情况
（黄应钦绘）

（张声燕）

刺 黄 果

Carissa carandas L.

（夹竹桃科 Apocynaceae）

生长习性、分布和用途 假虎刺属约 36 种，我国 2 种，又引入栽培 3 种，本文描述引入的 1 种。常绿灌木，高 2～4（6）m，原产印度至印度尼西亚。我国台、粤、琼、桂、黔引种栽培。果可作糕饼果馅及果酱。树枝刺长而锐利，枝叶浓密，花果色泽鲜艳，可作绿篱并供观赏。

开花结实 2～3 年生开始开花结实，正常结实期在 5 年生以后。结实大小年间隔期为 1 年。花两性。聚伞花序顶生，稀腋生。花白色稍带玫瑰红色。萼 5 裂，内面基部有腺体。花冠高脚碟状。花冠筒长 2cm，裂片 5，向右覆盖，比花冠筒略短。雄蕊 5，着生于花冠筒上，离生，内藏。子房上位，2 室，每室有胚珠 12～15 颗。花柱圆柱状，柱头长圆形，顶端 2 裂。据广西南宁观察，始花期在 3 月下旬，盛花期在 4 月中旬末，末花期在 5 月中旬；果实成熟始于 9 月中旬，盛期在 10 月中旬，末期为 11 月初。浆果，球形至椭圆形，长 1.5～2.5mm，径 1～2cm，幼时绿色。外果皮革质，成熟时由鲜红色转为紫黑色。每果有种子 12～30 粒。种子浅黄色，扁平，椭圆形，稍翘曲，长 6～8.5mm，宽 4～6mm，厚 1～2mm，具肉质内胚乳。

果实的采收调制和种子贮藏 果实成熟盛期采摘紫黑色果实。掰开果实，剥出种子，置水中洗净，晾干。鲜果出种率约 5.8%。净度约 97%。千粒重 25g，每千克有纯净种子 4（3.6～4.4）万粒。种子可以干藏，但不宜曝晒。晾干后用布袋包装挂于室内通风阴凉处，贮藏期3～6 个月。

发芽和播种 种子无休眠习性。发芽时日均温宜在 20℃以上。1987 年 9 月 14 日广西林业科学研究所在室内作过发芽测定，当时室内日均温约 27℃，基质为吸水纸。置床后 3 天开始发芽并立即进入发芽盛期。从播种之日起算的 9 天中，发芽百分数为 85%，15 天的发芽率为 95%。出土萌发。胚根伸出后 4 天子叶出土，26 天初生叶初现。

条播。每平方米播种 2～3g，覆土 5mm。1 年生苗出圃。

（曾 玲）

海 杧 果

Cerbera manghas L.

（夹竹桃科 Apocynaceae）

生长习性、分布和用途 海杧果属约 9 种，我国只有本文描述的 1 种。常绿小乔木，高 8m，胸径约 6～20cm。适生于热带海边湿润地，引种至内陆酸性沙壤土上亦能正常生长发育。苗期忌霜冻。分布于桂、粤、琼、台。亚洲、大洋洲热带地区也有分布。富含乳汁，果实有剧毒，以核仁最毒。种子榨油作燃料，木材作小径材用。可作海岸防潮树种，也可供观赏。

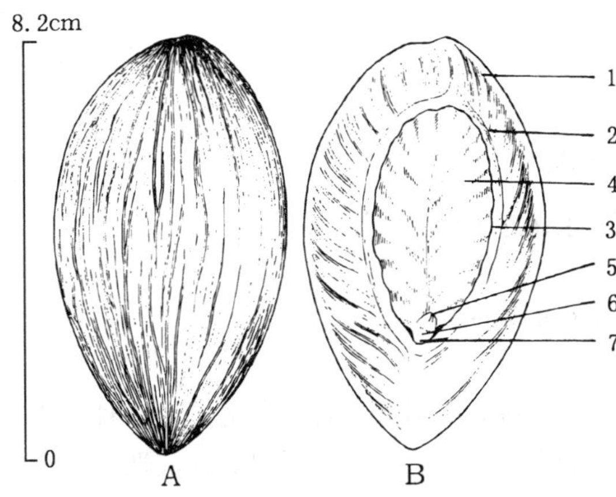

图 1　海杧果去外果皮果实外形(A)及其纵切面(B)
1. 中果皮　2. 内果皮　3. 种皮　4. 子叶
5. 胚芽　6. 胚轴　7. 胚根
(黄应钦绘)

开花结实　5～6 年生开始结实，正常结实期在 10 年生以后，结实大小年现象不明显。花两性。聚伞花序顶生，总花梗长 5～25cm。花萼 5 深裂，内无腺体。花冠白色，高脚碟状，直径约 5cm。冠筒长 2～3cm，喉部红色，被柔毛，裂片 5，向左覆盖。雄蕊 5，花药内藏。无花盘。子房由 2 离生心皮组成，每心皮有胚珠 4 颗。花柱丝状，柱头球形，基部环状，顶端浑圆，2 裂。据广西南宁 1987～1988 年观察，5 月上旬为始花期，5 月末～7 月为盛花期，9 月上旬为末花期；9 月下旬果实开始成熟，10～11 月为果熟盛期。另据海南观察，花期为 3～10 月，果期为 9 月～翌年 2 月。核果双生或单生，椭圆形或卵形，未成熟时青绿色，成熟后红色或紫红色，长 5～ 8.5cm，径 3.2～6cm。外果皮肉质，薄而平滑，中果皮纤维质，内果皮木质。核卵形或椭圆形，长 5～7cm，径 3～5cm，浅褐色至暗褐色。无胚乳。胚大，伸直，子叶厚，肉质，边缘波状（图 1）。

果实的采收调制和种子贮藏　果实成熟期长。最好选择成熟盛期的果实，直接采摘或打落后在地面捡拾。采得的果实堆沤数日，使肉质外果皮腐烂脱落，装入竹筐置水中，用木棒冲捣，除去纤维状中果皮。不宜用手搓洗，以防中毒。洗净果肉纤维杂质即得果核，用作播种材料，通称种子。鲜果出籽率约 55%。千粒重 27（18～34）kg，每千克有果核 37（30～55）粒。种子忌失水，不能日晒或裸露存放，含水量宜保持在 40%以上。宜随采随播。贮藏时需混以湿沙，贮藏期一般为 3～5 个月。

发芽和播种　种子有休眠习性。发芽时日均温宜在 14℃以上。1987 年 10 月 20 日，广西林业科学研究所用新采种子在室外沙床作过发芽测定：播后 102 天至 1988 年 1 月 31 日开始发芽，发芽盛期不明显，2 月 25 日发芽终止，发芽率为 83%。留土萌发。胚根萌发后约 15 天胚芽出土，出土后 10 天左右发出初生叶（图 2）。

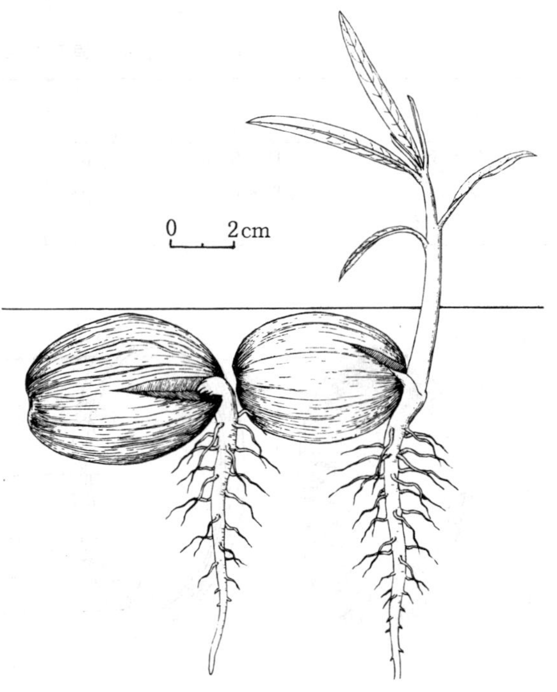

图 2　海杧果种子萌发后第 10、30 天幼苗的生长情况
(黄应钦绘)

点播或催芽移植。目前生产上尚无大批量育苗。苗期应注意防霜。一般培育 1 年生苗出

圃。如用于庭园绿化，需培育2年生苗。

（张声燕）

云南蕊木

Kopsia officinalis Tsiang et P. T. Li

（夹竹桃科　Apocynaceae）

生长习性、分布和用途　蕊木属约30种，我国3种，又引入1种，本文描述1种。常绿乔木，高12～15m，胸径约20cm。能耐轻霜及较长期5℃低温。喜肥沃湿润的酸性土。产于云南南部。广西南宁等地栽培。果和叶治咽喉炎，皮治水肿。花果较美，可供观赏。

开花结实　5～6年生开始结实，正常结实年龄在10年生以后。每年结实均较丰，无大小年现象。花两性。聚伞花序复总状，顶生。萼5深裂，覆瓦状排列。花冠白色，高脚碟状，冠筒细长，5裂片向右覆盖。雄蕊5，着生于花冠筒喉部，花丝短，药圆卵形。花盘为2枚线状披针形的舌状片组成，与心皮互生。子房上位，心皮2，分离，每心皮有胚珠2颗。花柱细长，柱头增厚，顶端短2裂。据广西南宁观察，1年中开花结实多次，除冬季外，其余3季均有结实。主要花果期每年有3次，见表1。

表1　云南蕊木主要开花结实的物候（广西南宁）

次　数	开　花			果实成熟	
	始　期	盛　期	末　期	始　期	盛　期
第1次	3月下旬	4月下旬	5月中旬	5月中旬	6月中旬
第2次	6月下旬	7月上旬	7月中旬	8月中旬	8月下旬
第3次	8月下旬	9月上旬	—	10月上旬	10月中旬

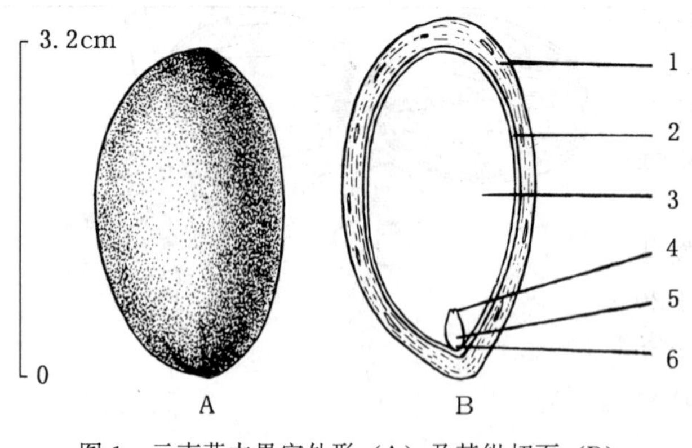

图1　云南蕊木果实外形（A）及其纵切面（B）
1. 果皮　2. 种皮　3. 子叶　4. 胚芽　5. 胚轴　6. 胚根
（黄应钦绘）

核果，双生。果实成熟时由青色转变为紫色至黑色，扁椭圆形，一侧略弯，中部稍隆起。另一侧半圆形，基部下凹。果长3.2～4cm，径1.7～2.3cm。外果皮薄，中果皮纤维状木质，内果皮木质。每果有种子1粒，稀2粒。种子椭圆形，长2～2.7cm，径1～1.3cm。种皮纸质。无胚乳。胚直，子叶肥大，白色或粉红色。成熟果实在树上可维持20天，以后脱落。如遇多雨天气，6～8月成熟的部分果实在树上发芽后方脱落。果实形态见图1。

果实的采收调制和贮藏　果实充分成熟时直接在树上采摘，不宜敲打或摇动震落，以免损伤树上的花及未成熟果。采收后不宜堆沤或日晒，而应摊放于室内。由于果肉的乳汁极多，操作不便，加以种皮纸质，去除果皮剥取种子时容易损坏种仁，目前都是将果实直接用作播种材料，通称种子。果实千粒重 5 900（5 500～6 300）g，每千克有果实 170（160～180）粒。含水量约 60%，宜随采随播，不耐久藏。常温条件下室内存放 1 周，果皮便会皱缩但尚不影响发芽能力。如需作 1 个月以上的贮藏，需混湿沙。混沙贮藏以 1 个半月为限。

发芽和播种　种子无休眠习性。发芽时日均温宜在 20℃以上。随采随播的，播前不需作任何处理，选出大粒饱满的果实作播种用。如因存放后果皮呈现皱纹，可在播前浸水 3～5 小时，俟吸足水分后再播种。1987 年 5 月 19 日，广西林业科学研究所用当年新采果实在室外沙床作发芽测定，当时日均温 25～27℃。5 月 26 日开始萌发，6 月 3 日进入发芽盛期，6 月 20 日发芽结束。从发芽开始至发芽高峰的 16 天中，发芽百分数为 60%；从播种至发芽终了，以 32 天计，发芽率为 85%。由于种子成熟后在树上存留的时间不同，播种后的发芽快慢也有差异。成熟后在树上存留超过 10 天的，播后 2～4 天即开始萌发，一般为播后 7 天左右开始萌发。留土萌发。胚根萌发后 4～5 天上胚轴出土，再过 3～4 天展出初生叶（见图 2）。

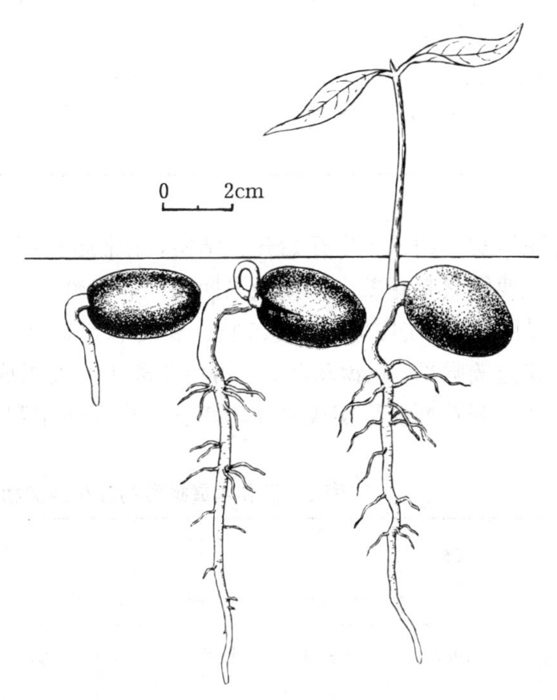

图 2　云南蕊木种子萌发后第 1、6、12 天幼苗的生长情况
（黄应钦绘）

条播。播时应使果实基部向下或平放。每平方米播果实 370～500g，覆土约 2cm。1～2 年生苗出圃。

（张声燕）

萝芙木属
Rauvolfia L.

（夹竹桃科　Apocynaceae）

生长习性、分布和用途　本属约 135 种，我国 7 种，又引入 5 种，本文描述 2 种。常绿灌木，高 2～4m，分枝多。喜肥沃湿润的森林土壤，裸露地或干旱贫瘠地生长不良。它们的名称、生长、分布及用途见表1。

表1　萝芙木属树种的名称、生长、分布和用途

中　名	学　名	树高(m)	分　布	用　途	供　稿
萝芙木	*R. verticillata* (Lour.) Baill.	3	滇、黔、桂、粤、琼、台。越南	降血压、退热、治胆囊炎、癫痫、疟疾及蛇咬并杀虫	603
催吐萝芙木	*R. vomitoria* Afzel. ex Spreng.	4	原产热带非洲。滇、粤、桂栽培	根降血压，治呕吐、下泻、茎皮治高热、疟疾并助消化，乳汁治腹痛	619

开花结实　约3年生开始开花结实，正常结实期在6年生以后。结实无大小年现象。影响结实的主要因素是花期中长期低温阴雨。花两性。伞形或聚伞花序顶生。萼5裂，花冠高脚碟状，裂片5，向左覆盖。雄蕊5枚。萝芙木雄蕊着生白色花冠筒中部，催吐萝芙木雄蕊着生粉红色花冠筒喉部。花盘环状。子房由2离生心皮组成，2室，每室有胚珠1。花柱圆柱状，柱头棍棒状。开花期及果实成熟期均较长。广西南宁观测的开花结实物候期见表2。

表2　萝芙木属树种的开花结实物候期（广西南宁）

树　种	观察年份	开　花			果实成熟		果实脱落
		始　期	盛　期	末　期	始　期	盛　期	
萝芙木	1987	3月下旬	4月下旬	6月中下旬	6月下旬	8月中下旬	7月上旬～9月中旬
催吐萝芙木	1986	3月下旬	4月中旬	6月上旬	7月中旬	8月下旬	8月上旬～9月下旬

核果，双生。外果皮秃净有光泽，中果皮浆状，内果皮木质。果核内含1种子。果实成熟后可在树上着生10天左右。种子有胚乳，胚伸直，子叶扁平。果实及果核的形态见表3、图1。

表3　萝芙木属树种果实和果核的形态特征

树　种	未熟果颜色	果　实			果　核		
		形　状	大小(mm)	色　泽	形　状	大小(mm)	色　泽
萝芙木	绿色转暗红色	椭圆形或卵圆形	长8～10 径5～7	紫黑色	扁椭圆形，表面凹凸不平	长7～9 宽4.5～5.5 厚2～2.5	浅褐色
催吐萝芙木	绿色转浅红色	圆球形	径7～10	红色	桃形，表面呈凸凹状	长5～8 宽4～5.8 厚2.5～3.8	乳白色

果实的采收调制和种子贮藏　成熟盛期直接采摘。果实采回后堆放1～3天，充分软熟后装入竹筐或布袋置水中搓擦，淘去果皮等杂质。所得果核用作播种材料，通称种子。稍晾干的种子含水量为15%～30%，忌失水，不能日晒。短期运输可直接运输果实，运抵后再行调制。5天以上的运输或贮藏需在调制后混以湿沙，贮藏期一般为半年左右。种子（核）的质量等数据见表4。

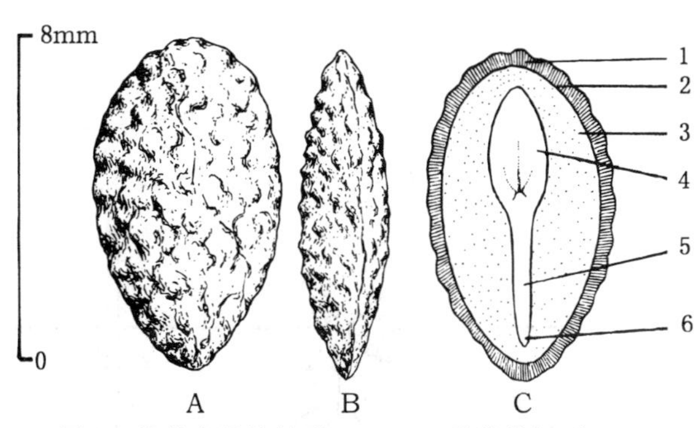

图1　萝芙木果核外形（A、B）及其纵切面（C）
1. 内果皮　2. 种皮　3. 胚乳　4. 子叶　5. 胚轴　6. 胚根
（黄应钦绘）

表4　萝芙木属树种的出籽（核）率和种子（核）质量

树　种	出籽率（%）	千粒重（g）	每千克种子（核）粒数
萝芙木	30～35	25～30	33 000～40 000
催吐萝芙木	13～15	35～42	24 000～29 000

发芽和播种　种子无休眠习性。发芽时日均温需在23℃以上。1987年7月中旬和1987年9月下旬，广西林业科学研究所先后对萝芙木和催吐萝芙木进行过发芽测定，发芽情况见表5。催吐萝芙木10月中旬发芽以后由于气温转低，大部分种子延迟至翌年4月中旬继续发芽，4月下旬终止。

表5　萝芙木属树种的发芽能力及其测定条件（广西南宁，室外沙床）

树　种	温度（℃）	发芽势（%）		发芽率（%）	
		计算天数	一般数值	计算天数	一般数值
萝芙木	28～30	6	35	20	65
催吐萝芙木	—	—	不明显	192	65

出土萌发。萝芙木宜随采随播，每平方米播种约5g。催吐萝芙木以沙藏至翌年4月上旬播种较宜，每平方米播种约7g。覆土1.5cm。半年或1年生苗出圃。萝芙木种子的萌发和幼苗生长情况见图2。

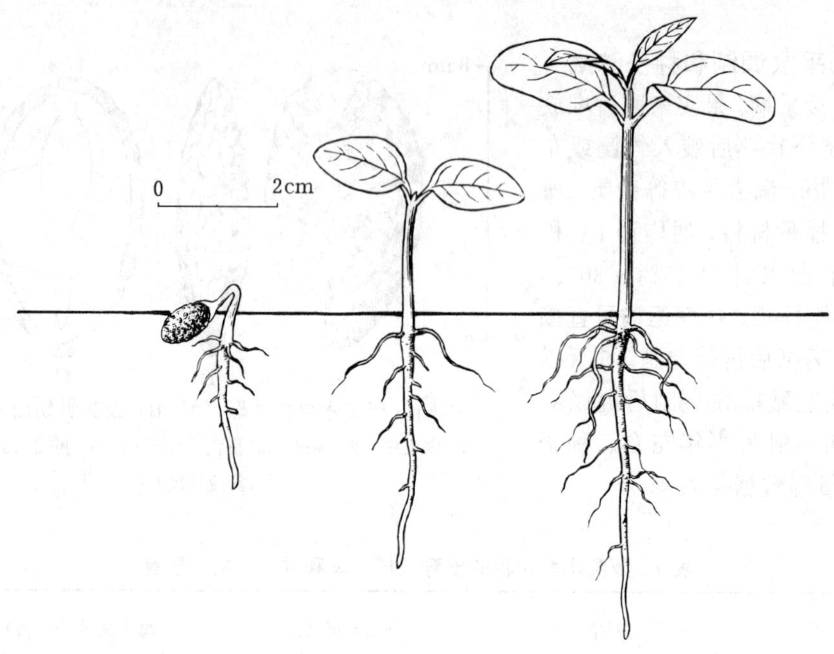

图 2　萝芙木种子萌发后第 2、10、20 天幼苗的生长情况

（黄应钦绘）

（张声燕）

黄花夹竹桃

Thevetia peruviana（Pers.）K. Schum.

（夹竹桃科　Apocynaceae）

生长习性、分布和用途　黄花夹竹桃属约 15 种，均产热带非洲和热带美洲。我国引入栽培 2 种，本文描述 1 种。常绿小乔木，高达 6m，胸径约 12cm。喜肥沃湿润土壤，较耐水湿，忌干旱贫瘠，忌霜冻。原产热带美洲。我国台、闽、粤、琼、桂、滇庭院多有栽培。长江流域城市温室栽培。花期长，颇美观。种子的乳状汁液有剧毒。种子可榨油，作肥皂原料或杀虫剂。果有强心作用。

开花结实　约 2 年生开始结实，正常结实期在 5 年生以后，结实无大小年现象。遇特寒年份，嫩枝受冻害，会影响当年结实。花两性。聚伞花序顶生或腋生于嫩枝上部，有花 2～6 朵，或单花。花萼 5 深裂，内面基部具腺体。花冠黄色，漏斗状，花冠喉部具 5 枚被毛鳞片。花冠裂片 5，向左覆盖，比冠筒长。雄蕊 5 枚，着生于花冠喉部。无花盘。子房上位，2 室，2 深裂，每室有胚珠 2 颗。据广西南宁 1986～1988 年观测，4 月下旬为始花期，5～7 月为盛花期，10 月中旬为末花期；10 月上旬果实开始成熟，10 月下旬为成熟盛期，11 月上旬～翌年 1 月上旬果实脱落。核果，未成熟时绿色，成熟时转黄绿色至黑色，扁三角状球形，顶端具 2 渐尖，高 2.5～3.5cm，径 3～4cm。内果皮木质，坚硬，浅褐色，椭圆形或扁椭圆形，2

室，中间有两条痕纹交叉成十字状，径2.7～3.4cm，高1.4～2cm。每室有2稀1种子。种子无胚乳，子叶肥大（图1）。

果实的采收调制和种子贮藏　果实成熟盛期选摘饱满大型的果实。果实采回后可堆沤至外果皮和中果皮呈腐烂状时装入竹筐，置水中搓洗，淘净果皮等杂质，得出果核用作播种材料，通称种子。鲜果出核率为50%～60%。含水量25%～40%。千粒重约2 900g，每千克有果核340（280～400）粒。种子忌失水，不能日晒。运输时可以果实的状态装运，运到目的地后再行调制。调制出来的种子应混以湿沙，贮藏期一般为半年左右。

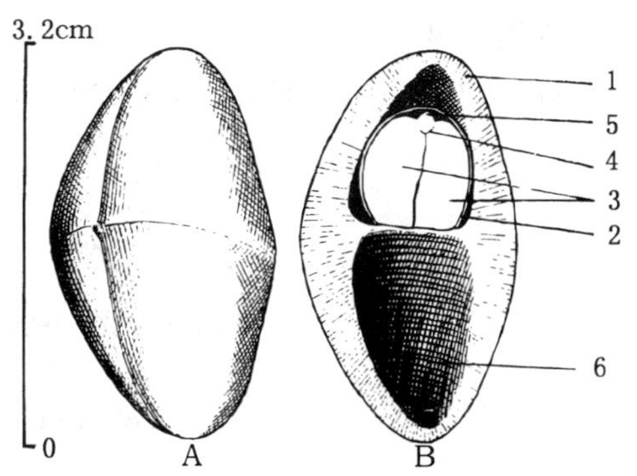

图1　黄花夹竹桃果核外形（A）及其纵切面（B）
1. 内果皮　2. 种皮　3. 子叶　4. 胚芽
5. 胚根　6. 另一种子空腔
（黄应钦绘）

发芽和播种　种子无休眠习性。发芽时日均温需在20℃以上。1987年11月10日，广西林业科学研究所用当年新采种子在室外沙床作发芽测定。播种时日均温21℃左右，但播后气温转低，种子不发芽，至翌年3月20日才开始发芽。发芽无明显盛期，4月15日发芽终止，发芽率为62%。出土萌发。胚根萌发后约25天子叶出土，再过2天初生叶展现。种子发芽及幼苗生长情况见图2。

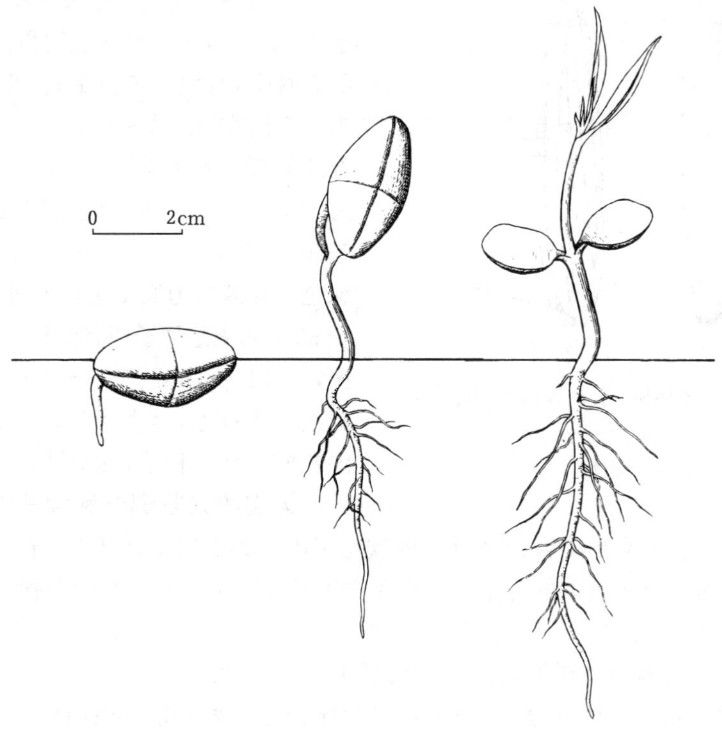

图2　黄花夹竹桃种子萌发后第20、27、30天幼苗的生长情况
（黄应钦绘）

种子密播于湿沙床内或经过细致整地的苗床上。种子萌发后，将芽苗移至大田。每平方米用种约 100～200g。1 年生苗出圃。扦插成活率可达 60%以上，生产中常用扦插育苗。

（张声燕）

倒吊笔

Wrightia pubescens R. Br.

（夹竹桃科　Apocynaceae）

生长习性、分布和用途　倒吊笔属约 23 种 8 亚种，我国产 6 种，本文描述 1 种。落叶乔木，高 8～20m，胸径达 60cm。适生于山麓疏林土壤深厚肥沃的谷地或平坦地。分布于粤、琼、桂、黔、滇。印度、泰国、越南、柬埔寨、马来西亚、印度尼西亚、菲律宾和澳大利亚也有分布。木材纹理通直，结构细致，可作上等家具、雕刻、乐器用材，商品材称银木，皮富含纤维，茎皮药用。树形美观，可供庭园观赏。

开花结实　5～10 年生开始开花结实，大小年不明显。花两性。聚伞花序顶生，长约 5cm。萼片 5，内面基部具鳞状腺体。花冠漏斗状，白色、浅黄色或粉红色，裂片 5，长圆形，向左覆盖。副花冠分裂成 10 鳞片，呈流苏状。无花盘。雄蕊 5，突出，花药箭头形，伸出花喉之外。子房由 2 粘生心皮组成，胚珠多数。花柱丝状，向上逐渐增大。柱头头状卵形。据广西南宁 1987～1989 年观察，花期始于 4 月上旬，盛于 5 月下旬，终于 6 月上旬；果熟期始于 12 月末，盛于翌年 2 月～3 月下旬；4 月上旬以后果逐渐开裂，种子飞散，4 月末果全部开裂。果为蓇葖，2 个粘生，线状倒披针形，长 15～25cm，直径 1～2cm。果实成熟时由青色转黄绿色，开裂后为黑褐色，表面粗糙。每果有种子 60～90 粒。种子线状纺锤形，乳黄色至浅褐色，长 11～15mm，宽 1～2mm。种子顶端具淡黄色绢质种毛，毛长 2～3.5cm。无胚乳，子叶内卷成筒状。种子形态见图 1。

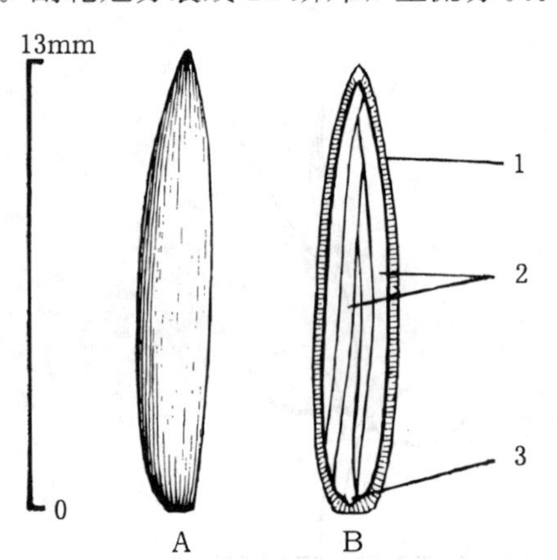

图 1　倒吊笔种子外形（A）及其纵切面（B）
1. 种皮　2. 子叶　3. 胚根
（黄应钦绘）

果实的采收调制和种子贮藏　果实盛熟时用高枝剪剪下，于无风处日晒 1～2 天，或放在室内让其自然干燥开裂。开裂后震动果瓣，脱出种子。从不开裂的果实剥取的种子，播种后子叶不易展开成苗。果实出种率为 20%～23%。净度约 95%。千粒重约 10g，每千克有纯净种子 10（8.3～12.5）万粒。短途运输以装运果实为宜。贮藏时用小容器分装或混以细沙。贮藏期 4～8 个月。

发芽和播种　种子无休眠习性，可以随采随播，或贮藏至翌年初播种。1987 年 2 月 19 日广西林业科学研究所在室内作过发芽测定，室温控制在 25～28℃。播后 4 天胚根伸出，2 月 25 日发芽进入盛期。从发芽之日起算，3 天的发芽百分数为 66%；发芽终期为 2 月 28 日，9

天的发芽率达95%左右。出土萌发。子叶出土后4～5天种壳脱落，子叶展开，约6～8天初生叶出现（图2）。

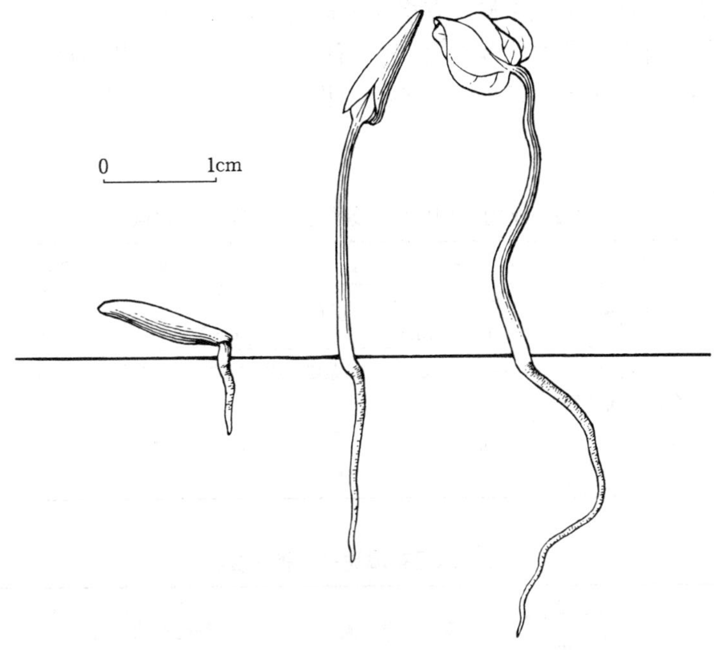

图2　倒吊笔种子萌发后第2、4、6天幼苗的生长情况
（黄应钦绘）

条播或撒播。每平方米用种量分别为1.5g、2g左右。1年生苗可以出圃。

（曾　玲）

水团花属

Adina Salisb.

（茜草科．Rubiaceae）

生长习性、分布和用途　本属约20种，分布于亚洲和非洲的热带、亚热带地区。我国产8种，其中常见的是本文描述的水团花和鸡仔木2个种。性喜水湿，多生于溪旁、山坡湿地及山地阴湿疏林中。它们的名称、生长习性、成年时的树高、分布和用途见表1。

表1　水团花属树种的名称、生长习性、树高、分布与用途

中　名	学　名	生长习性	树　高(m)	分　布	用　途
水团花	*A. pilulifera* (Lam.) Franch. ex Drake	落叶灌木至小乔木	5	长江以南各省。越南、日本	材用、纤维、药用、固堤
鸡仔木（水冬瓜）	*A. racemosa* (Sieb. et Zucc.) Miq.	半常绿或落叶乔木	6～14	西南、湘、粤、台、浙、皖、苏	材用、纤维

开花结实　花两性，小，极多数，通常5数，密集于花序托上，成一单生的球形头状花序或再排成总状花序式。萼筒有棱。花冠长漏斗状，5裂。雄蕊5。花盘杯状。子房下位，2室，胚珠多数。蒴果室间开裂成2瓣，有宿存中轴。每果含种子2～3或8～10粒。这两个种的花、果形态特征和开花结实的物候期见表2。种子极微小，长椭圆形略扁，或具棱，两端有翅。胚乳量中等，胚发育完全，长及种子的4/5（图1）。种子的形态特征和种子质量的数据见表3。

表2　水团花属树种花果特征与开花结实物候

树　种	头状花序		花　冠			蒴　果		每果种子粒数	开花期	果熟期
	直　径（cm）	排列式	长（mm）	直　径（mm）	颜色	长（mm）	形状			
水团花	1.5～2	单1或2～3个顶生或腋生	5～7	2～3	紫红	2～3	楔形	2～3	6～7月	9～10月
鸡仔木	2～2.3	10多个排成总状	8～9	2	淡黄	4～5	楔形	6～8	7～9月	10～11月

表3　水团花属树种的种子性状

树　种	形　状	长　度（mm）	颜　色	表　面	种翅形状	千粒重（g）	每克种子粒数
水团花	长卵形，具纵棱	1.5～2	褐色	被短绒毛	翅较短，较不规则	0.12	8 000
鸡仔木	扁椭圆形，无纵棱	3.5～4	浅棕	无毛	上端翅尖长，下端翅燕尾形	0.05	20 000

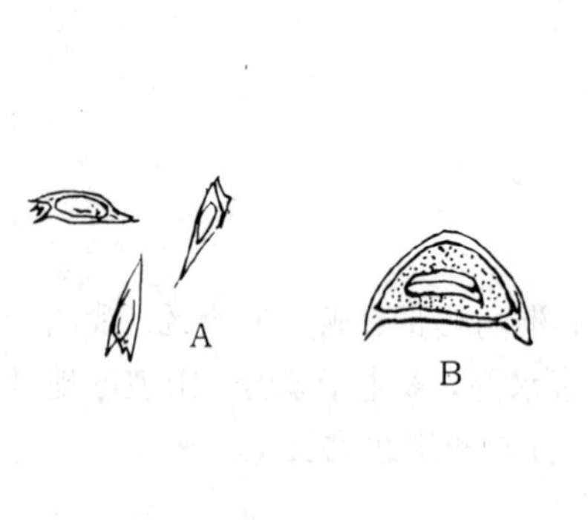

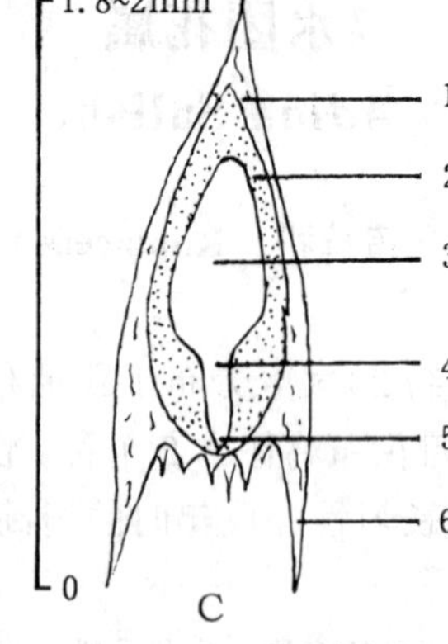

图1　水团花种子外形（A）及其横切面（B）和纵切面（C）
1. 种皮　2. 胚乳　3. 子叶　4. 胚轴　5. 胚根　6. 翅膜
（史渭清绘）

果实的采收调制和种子贮藏

种子成熟的形态标志是球状果穗变为褐色，总梗、果柄干枯，果球由硬而紧变为松软，此时便应及时采摘，以免蒴果开裂，种子散落。采下的果球晒干后稍加搓揉即可散开脱出种子。种子极其细小且具翅，最好用细眼网筛剔除果瓣、中轴、干花药等杂质。一般净度较低。

发芽和播种　1988年春和1989年春夏期间曾在南京室内条件下摸索过水团花和鸡仔木的发芽习性，可以初步看出以下几点：①发芽需要十分充足的水分，甚至可以在被水淹没的沙上发芽；②发芽适温为32℃上下；③以滤纸作为发芽基质为好；④置床后约20天开始萌发，60天内发芽率为27%～30%；⑤未萌发的种子并无异常表现，但不能萌发的原因尚不清楚。水团花的播种期应在 4 月底～5 月初，鸡仔木应在 7 月初。撒播，不必覆土，但应保持床面有充足的水分。出土萌发。出苗后幼苗纤小，早期发育缓慢(图 2)，需要精细管理。

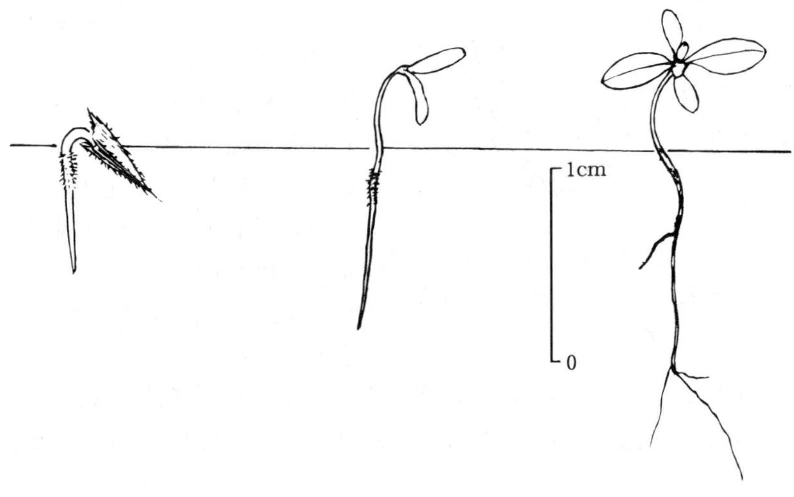

图 2　水团花种子萌发后 1、2、35 天的幼苗情况
（史渭清绘）

（何泽瑛）

团花（黄梁木）

Anthocephalus chinensis（Lam.）Rich. ex Walp.

（茜草科　Rubiaceae）

生长习性、分布和用途　团花属 3 种，我国只有本文描述的 1 种。常绿大乔木，树高 30m 以上，胸径可达 1m 以上。速生，有“奇迹树”之称。喜光。幼苗忌霜冻，大树能耐 0℃左右极端低温及轻霜。喜肥沃湿润的酸性沙壤土。分布于滇、桂。琼、粤、闽有栽培。越南、马来西亚、缅甸、印度等亚洲热带地区亦有分布。材质轻软，不裂不变形，宜作板材、牙签、火柴杆及胶合板。树形雄伟，供观赏。

开花结实　8 年生左右开始开花结实，正常结实年龄在 15 年生以后。结实大小年间隔期为 1 年，不甚明显。花两性。圆球形头状花序单生枝顶。花黄色，5 数。萼筒光滑，裂片矩圆形。花冠漏斗状，裂片披针形。雄蕊生于花冠筒喉部，花丝短，药顶部削尖。子房下位，上部 4 室，下部 2 室。花柱突出，柱头纺锤形。胚珠多颗 。据广西南宁 1980～1984 年的物候观测，4～6 月头状花序形成，着生于头年生枝条的叶腋，7 月上旬为始花期，中旬为盛花期，下旬为末花期；次年 1 月果实开始成熟，1 月下旬～2 月中旬为主要果熟期。果熟后 1 周左右脱落。在西双版纳，果实成熟期为 10～12 月（程必强、马信祥，1987）。聚合果成球形头状体，成熟时由青色转变为黄色，由硬转化为柔软状，由多数革质小坚果融合而成。小坚果细小，每聚合果有 1 万～1.2 万粒。小坚果棕褐色，棱状圆形、近三角形或不规则形，果皮粗糙，长0.5～

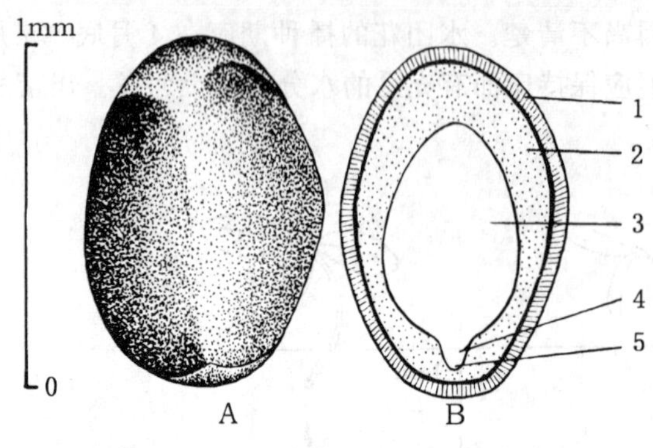

图1　团花小坚果外形(A)及其纵切面(B)(浸水膨胀后)
1. 果皮和种皮　2. 胚乳　3. 子叶　4. 胚轴　5. 胚根
(黄应钦绘)

0.6mm，径0.3～0.5mm。种皮极薄，有肉质胚乳，胚直，子叶卵形（图1）。

果实的采收调制和种子贮藏　果实成熟后，鼠、鸟喜食，应及时用竹竿敲打或用钩刀采下聚合果，堆沤于室内，腐熟变黑后装入紧密布袋，置水中搓揉淘洗。小坚果脱离果肉下沉后，漂去果皮、果肉等杂质，反复多次，即得纯净小坚果。小坚果便是播种材料，通称种子。每个聚合果可得种子0.5g。鲜果的出籽率约为2%，净度可达95%以上。千粒重约0.05g，每克有纯净种子（小坚果）约20 000粒。可以晒干贮藏。常温条件下，用布袋或木箱等包装贮藏，发芽能力可保持1年不降低。密封或5℃低温贮藏，发芽力可保持2年以上。程必强和马信祥（1987）用瓶干燥贮藏12个月的团花种子，发芽率有的仍达95%。

发芽和播种　种子无明显休眠习性。发芽时日均温宜在20℃以上。1981年4月16日，日均温20℃左右时，广西林业科学研究所用当年所采种子，在室内用滤纸作基质，用发芽皿进行过发芽测定：置床后22天即5月7日开始萌发，发芽盛期不明显，至6月9日发芽结束，发芽率为50%。另外一份种子经过0～5℃低温处理2个月，用同一方法测定，发芽率一般为80%，最高可达90%。出土萌发。胚根萌发后约5天子叶伸出，再过21天发出初生叶，生长情况见图2。

撒播。幼苗易染立枯病，需用不带病菌的土壤或细沙作成播种盘，将种子拌以过筛的火烧土撒播。幼苗发出4片初生叶时移至容器或圃地继续培育。早期需遮荫，用喷雾器淋水。3个月或1年生苗出圃。

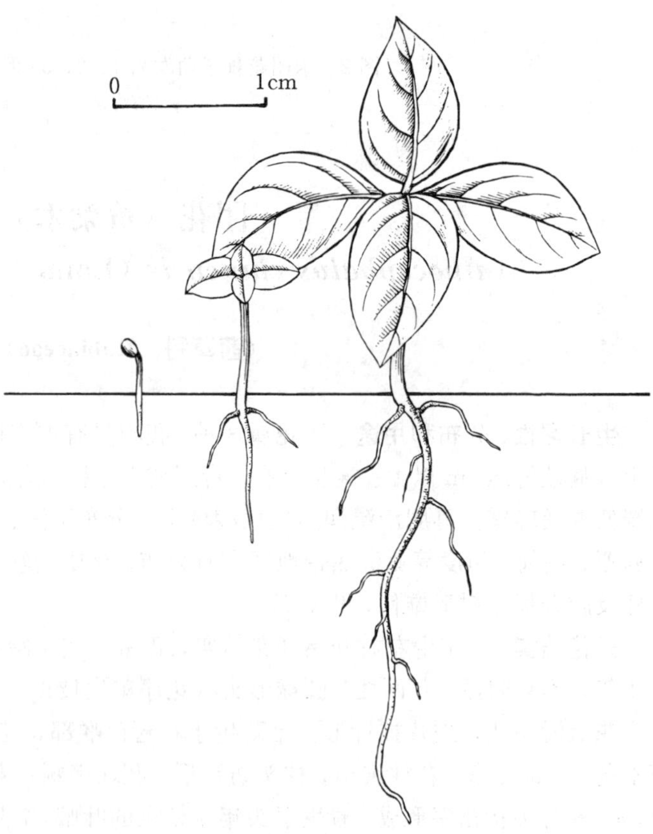

图2　团花种子萌发后第5、26、32天幼苗的生长情况
(黄应钦仿《热带亚热带主要树种采种育苗技术》)

（张声燕）

铁　屎　米

Canthium parvifolium Roxb.

（茜草科　Rubiaceae）

生长习性、分布和用途　鱼骨木属约200种，我国产4种，本文描述1种。半落叶具刺小乔木，高5～6m，胸径15～20cm。能耐0℃左右极端低温。适生于肥沃湿润的酸性至微酸性土壤。产云南南部。亚洲和非洲的热带地区亦产。果可食。

开花结实　8～10年生开始开花结实，15年生以后进入正常结实期，没有大小年现象。花杂性，单朵至数朵簇生于叶腋。萼筒短，花冠筒管状，裂片4～5，外反，绿白色至浅黄色。雄蕊4～5，生于花冠喉部，花丝短，花药近基部背着。花盘杯状。子房下位，2室，每室有胚珠1颗。花柱伸出。据1986～1988年云南西双版纳勐仑的物候观察，3月上中旬花芽出现，3月下旬～4月上旬花蕾出现，4月中旬花始开，以后便陆续开放，没有盛花期；6月下旬第1次花期结束，幼果形成，同时又陆续出现花蕾；9～10月为第2次花期；7月中下旬第1次果开始成熟，8月为果实盛熟期；12月～翌年3月第2次果陆续成熟，数量比第1次少。

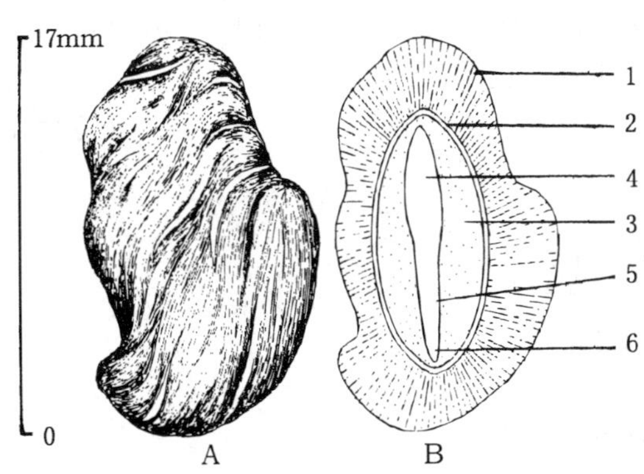

图1　铁屎米果核外形（A）及其纵切面（B）
1. 内果皮　2. 种皮　3. 胚乳　4. 子叶　5. 胚轴　6. 胚根
（黄应钦绘）

核果，卵圆形或椭圆形，长3.5～4.5cm，径约3～4cm，成熟时由绿色转变为黄色，表面光滑，成熟后2～3天脱落。中果皮肉质浆状，内果皮骨质。果核具棱，一侧凹入，长约1.5～1.8cm，径约6～8mm。每果有核1～2粒。种子具胚乳，胚条形，位于种子中部直达两端。果核形态及其解剖构造见图1。

果实的采收调制和种子贮藏　果实成熟盛期在树下拾集，或摇动树枝震落后收集。采得的果实堆沤1～3天，捣烂果皮，挤出果核，在水中搓洗后即为播种材料，通称种子。鲜果出籽率

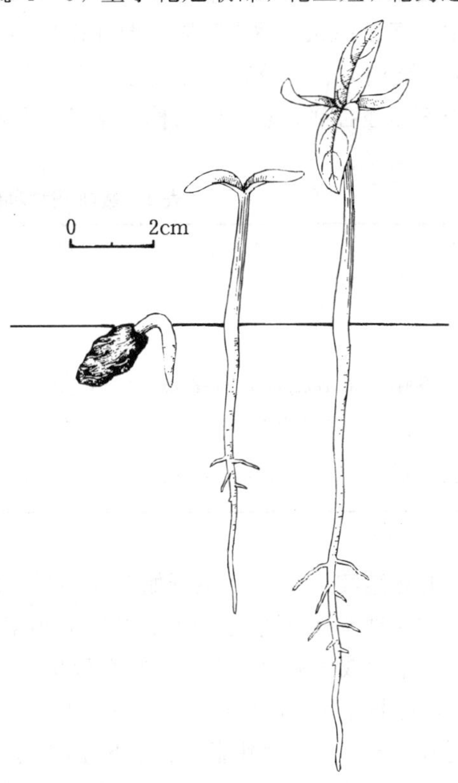

图2　铁屎米种子萌发后第3、15、22天幼苗的生长情况
（黄应钦绘）

36%～38%，果核的净度可达 98%。千粒重约 520g，每千克有果核 1 800～2 000 粒。调制后稍晾干时含水量 29%。贮藏时宜混湿沙，贮藏期半年左右。

发芽和播种 种子有短期休眠现象，播前可用热水浸泡。发芽时的日均温宜在 20～25℃。1987 年 12 月，西双版纳热带植物园用新鲜种子作过发芽测定，基质为细沙，播后 115 天开始出土，发芽全过程历时 20 天，没有明显的发芽盛期，发芽率 72%。出土萌发。子叶出土后19～20 天展开，同时有 1 对初生叶出现。种子萌发和幼苗的生长情况见图 2。

点播。播前用始温 60℃水浸泡，并在自然冷却的过程中继续浸泡 24 小时，捞出后稍晾干即可播种。每平方米播 75～85g，覆土 1～2cm。1 年生苗出圃。

（马信祥）

咖 啡 属
Coffea L.

（茜草科 Rubiaceae）

生长习性、分布和用途 咖啡属约 40 种，我国引入 5 种，本文描述 3 种。常绿小乔木或灌木，高 2～5m。适生于肥沃湿润微酸性的沙壤土。咖啡为世界三大饮料之一。咖啡豆含有脂肪、蛋白质、淀粉等多种成分。除饮用外，还可提取咖啡碱，医药上用作麻醉剂、镇痉剂、兴奋剂和强心剂。它们的名称、生长、分布及用途见表 1。

表 1 咖啡属树种的名称、生长、分布和用途

中 名	学 名	树高（m）	分 布	用 途
小果咖啡	*C. arabica* L.	1～2	埃塞俄比亚。华南、西南、台引种栽培	饮料、观赏、药用
中果咖啡	*C. canephora* Pierre ex Froehn.	2～4	扎伊尔。海南引种栽培	饮料、观赏、药用
大果咖啡	*C. liberica* Bull. ex Hiern	2～5	利比里利。琼、台引种栽培	饮料、观赏、药用

开花结实 2～3 年生开始开花结果，6～8 年生进入盛产期，可连续收获 20～30 年，在良好的管理条件下可达 50 年。结实大小年现象较明显。具有多次开花和花期集中的特性。花两性，单生或组成聚伞花序，数个簇生于叶腋。花白色，芳香。萼管短，萼檐截平或 4～5 齿裂。花冠漏斗状或高脚碟状，裂片（4）5～8（11）。雄蕊 4～8，着生于花冠的喉部或之下。花药近基部背着。花盘肿胀。子房下位，2 室，少有 1 或 3 室。每室 1 胚珠。花柱线形或略粗大，柱头 2 裂，线形或钻形。海南观测的物候期见表 2。

表 2　咖啡属树种的开花结实物候期（海南儋县）

树　种	始花期	盛花期	果实成熟期
小果咖啡	2 月	3～5 月	9～11 月
中果咖啡	11 月	2～3 月	2～4 月
大果咖啡	3 月	4～5 月	6～7 月

核果，幼果绿色，成熟时呈红色或紫红色，椭圆形。小果咖啡长 9～16mm，大果咖啡长 19～21mm，中果咖啡近球形，径 10～12mm。每果有 2 核，各有 1 种子。核半椭圆形，中间有一条纵沟。咖啡豆为除去薄种皮的种仁。种子胚乳丰富，胚在胚乳的基部一侧（图 1）。

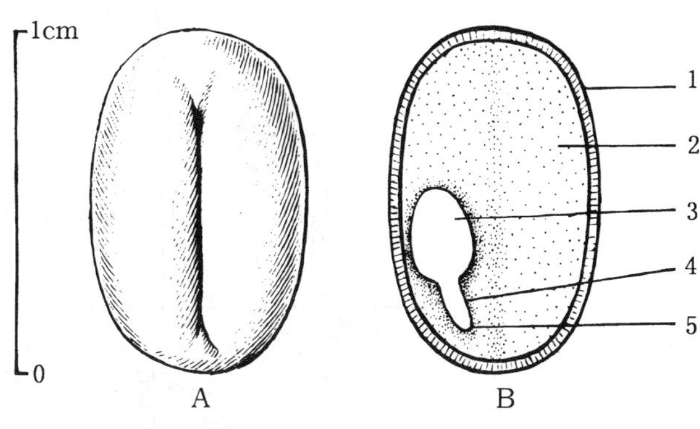

图 1　小果咖啡果核外形（A）及其纵切面（B）
1. 内果皮和种皮　2. 胚乳　3. 子叶　4. 胚轴　5. 胚根
（黄光郁、黄应钦绘）

果实的采收调制和种子贮藏　果皮由青色变红色时即为成熟，直接采摘。果肉与果核较难分离。一般是在果实盛熟期在优良母树上选择充分饱满、大小一致、具两粒种子的果实直接作为播种材料。也可以除去外果皮和中果皮，用果核作播种材料。方法是将果实放入箩筐内，加适量的草木灰和少量沙，搓揉洗净即得果核，通称种子。调制时勿伤及发芽孔。调制出来的果核置通风处阴干 3～4 天，忌日晒，待内果皮（核壳）变白色时即可播种。鲜果出籽率、种子净度和质量见表 3。

表 3　咖啡属树种的出籽率和种子净度、质量

树　种	出籽率（%）	净度（%）	千粒重（g）		每千克种子（核）粒数	
			一　般	变动范围	一　般	变动范围
小果咖啡	48	98	260	200～300	3 850	3 300～5 000
中果咖啡	48	98	270	210～330	3 700	3 000～4 760
大果咖啡	49	98	790	750～830	1 250	1 200～1 300

已调制的种子可用 1kg 装塑料袋密封包装，置于干燥处贮存，或混湿沙贮藏于室内阴凉处。贮藏期一般不超过 3 个月。

发芽和播种　种子无休眠现象，宜随采随播。发芽时的日均温宜在 25℃左右。3 种咖啡的发芽情况基本接近。1988 年 2 月 28 日，华南热带作物研究所在室外沙床对中果咖啡进行过发芽测定：播后 42 天（4 月 11 日）开始发芽，4 月 15～24 日为发芽盛期，4 月 28 日发芽结束，具明显的发芽盛期；从发芽之日起算，13 天中发芽百分数为 80%；从播种之日起算，59 天的发芽率为 90%。出土萌发。发芽后约 10 天子叶出土，20 天左右展出初生叶，生长情况见图 2。

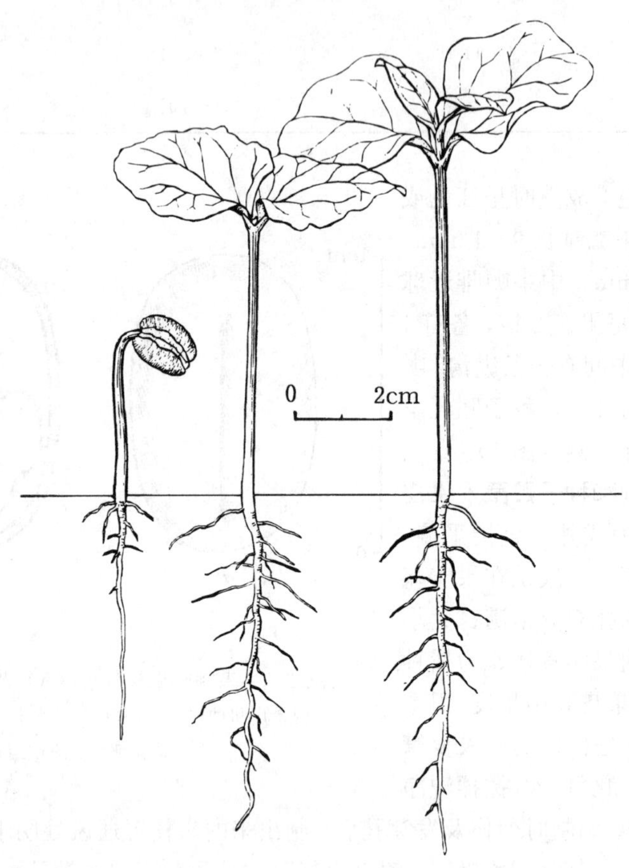

图 2 小果咖啡种子萌发后第 40、55、85 天幼苗生长情况
（黄光郁、黄应钦绘）

播种前最好用始温 40℃水浸种半小时，捞出晾干后撒播于沙床上，覆沙 1cm。发芽后将幼苗移至大田培育。小果咖啡和中果咖啡每平方米播种 25～30g，大果咖啡 60～75g。苗期宜遮荫。1 年生苗出圃。

（邓春媛）

香 果 树

Emmenopterys henryi Oliv.

（茜草科 Rubiaceae）

生长习性、分布和用途 本属仅有本文描述的 1 种，我国特产，属稀有种，已列入《中国植物红皮书》。落叶乔木，寿命长，树高 25（30）m，胸径 50（300）cm。浅根性。适生于肥沃湿润、土层深厚的酸性山地黄壤或黄棕壤。分布于华东、华中、西南及陕、甘。

木材洁白，色纹美观，易加工，干后不裂，是雕刻、家具和建筑用的良材，树皮纤维可制蜡纸和人造棉。树姿优美，花大而艳丽，是优良的观赏树种。

开花结实　30 年生左右的壮龄树才能开花结实，结实间隔期 2～4 年。座果率很低。结果枝往往大量枯死，次年抽梢细弱，不能形成花芽，需要经过两三年才能恢复树势（汪祖潭、杨逢春等，1988）。花两性，硕大，聚伞花序排成顶生的圆锥花序状。花萼近陀螺形，5 裂，花后脱落，但有些花的萼裂片中有 1 片扩大成叶状，白色，宿存于果实上。花冠漏斗状，被柔毛，顶端 5 裂，淡黄色。雄蕊 5，与花冠裂片互生。子房下位，2 室，胚珠多数。4～5 月花序开始分化，开花始期在 7 月中下旬，盛期在 8 月上旬，末期在 8 月中下旬。9 月上中旬幼果形成，10 月底～11 月成熟。果实未成熟时青绿色，成熟时红色。蒴果，近纺锤状，长 3～5cm，径 0.5～1.5cm，成熟后室间开裂为 2 果瓣。种子多数，淡黄色，小而扁平，长椭圆形，周围有阔翅，表面具网状纹（图 1）。

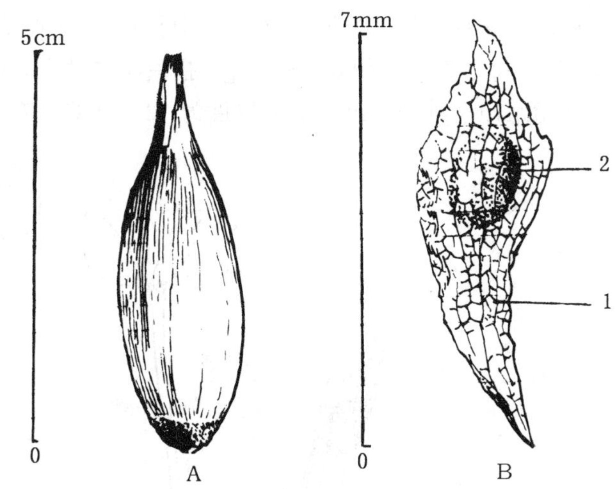

图 1　香果树蒴果外形（A）及其种子外形（B）
1. 种翅　2. 种子（童军平绘）

果实的采收调制和种子贮藏　果实变为红色时应及时采摘。采回的果实摊于通风处晾干，蒴果开裂后抖出种子，经筛选、风选去除杂质，即得纯净种子。每果约有种子 200 粒或更多。果实出种率 20%～30%。千粒重 0.3～0.6g，每克有纯净种子 1 600～3 300 粒。纯净种子晾干后装入布袋，置通风干燥的阴凉处贮放。

据作者试验，在常温下香果树种子生活力可保持 1 年。精选过的、含水量约为 10%的种子在 0～5℃条件下贮藏 1 年后，在 20～30℃变温和光照下的发芽率相当于贮藏前的 94%；贮藏 2 年后，经低温层积催芽 1、2、3、4 周后，在 30℃条件下发芽，其发芽率分别相当于贮藏前的 39%、45%、59%和 63%。

发芽和播种　种子有休眠习性，据推测是因为含有发芽抑制物质。据王成霖和邵蓓蓓（1986）及管康林（1985）的试验研究，种子在 15～30℃恒温下发芽率极低，仅为 2%～5%。打破休眠的最佳发芽条件是 20～30℃变温加 8 小时光照，其次是变温，再次是红光照射 15 天以上。红光照射少于 3 天的种子仍不能萌发，照射 5 天后，种子萌发反应已不能被远红光逆转。低温层积 3～4 周也可以解除休眠。

表 1 列出了几份香果树种子的发芽能力及其测定条件。

表 1 香果树种子的发芽能力及其测定条件举例

种子来源	温度（℃）		每天光照	发 芽 势（%）			发芽率（%）	
	低 温	高 温	小时数	计算天数	一般数值	变动范围	计算天数	数值
湖北九宫山	20	30	8	11～14	50	46～53	16～19	92
湖北九宫山	20	30	—	14～19	43	38～48	24～31	64
浙江天目山	—	30	红光照射 10 天以上，光照 1 000lx	—	—	—	20	65
湖北九宫山①	25	—	—	8	26	—	18	63

① 0～5℃下层积 3 周

播种期在 3 月上旬～中旬。撒播，亦可条播，条距 20～25cm。种子小，播种时宜混 3～8 倍的细沙。每平方米播种 0.3～0.5g。播后稍加镇压，盖土宜薄，以不见种子为度。床面搭棚，以防雨水冲动种子。出苗最低温度为 13～18℃。在浙江，3 月份播种的出苗期约 60 天。

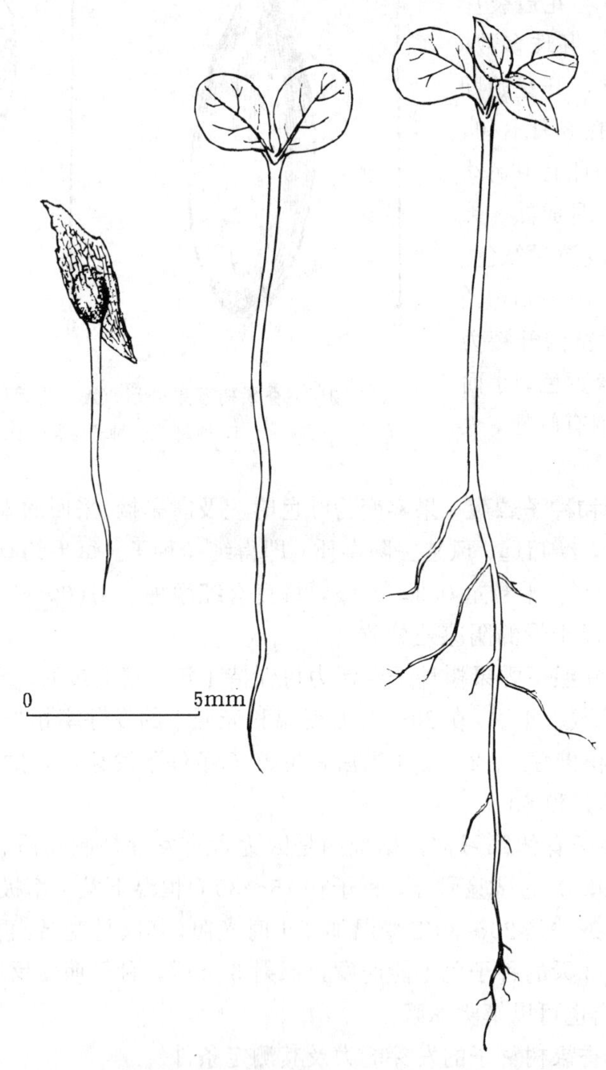

图 2 香果树种子萌发后第 7、13、36 天幼苗生长情况

（童军平绘）

出苗率一般为20%～50%。幼苗出土时要遮荫。幼苗期生长缓慢，第二年后生长转快。1年生苗留圃培育，用2年生苗造林。

出土萌发，子叶2枚，平展，卵形。子叶出土后40天左右长出初生叶，对生，广卵形。主根明显。种子萌发和幼苗生长情况见图2。

可以扦插繁殖，插穗以半木质化萌条和萌蘖条为好。

（王成霖）

栀　子　属
Gardenia Ellis

（茜草科　Rubiaceae）

生长习性、分布和用途　本属约250种，分布于亚、欧、非热带、亚热带地区。我国有4种，产西南至东部。本文描述3种，它们的名称、生长习性、分布和用途见表1。庭园常见栽培的多为栀子的变种栀子花（var. *grandiflora* Nakai），花形较大，重瓣，但通常不结实，本文未收。

表1　栀子属树种的名称、生长习性、分布和用途

中　名	学　名	生长习性	分　布	用　途	供　稿
栀子	*G. jasminoides* Ellis	常绿灌木，高约1m	长江流域以南至川、滇、黔。日本、越南	药用、色素、染料、香料、观赏	1003
（大）黄栀子	*G. sootepensis* Hutch.	半落叶乔木，高达15m	滇南至西双版纳	色素、香料	619
狭叶栀子	*G. stenophylla* Merr.	常绿灌木，高可达3m	桂、粤、琼。越南	色素、香料、观赏	605

开花结实　花两性，大，芳香，白色或淡黄色，单生枝端或叶腋。萼筒与子房合生，卵形或倒圆锥形，5～7裂，裂片长1～2cm，宿存。花冠高脚碟状，5～11裂。雄蕊5～11，生于花冠喉部。子房下位，1室，胚珠多数生于2～6个侧膜胎座上。浆果（对栀子属果实的类型各种文献说法不一，本文暂取浆果说——作者），革质或肉质，圆柱状或卵形，有纵棱，顶端有宿存萼裂片。种子多数，常与肉质胎座粘结成一体。种子形状大小不很一致，呈不规则扁圆形或扁椭圆形。外种皮角质，橙黄色。胚乳半透明，略硬。子叶薄。果实和种子的外形及解剖结构见图1。3个种的花果形态及开花结实物候等见表2。

果实的采收调制和种子贮藏　果实由绿转黄色或橙色时即表示成熟。成熟的果实2～3个月仍不脱落，有时有鸟雀啄食。采集的果实先置室内堆沤3～7天，变软后搓揉漂洗，除去杂质，将沉在水下的种子取出晾干即可播种。忌日晒。可混湿沙贮藏供春播，也可带果晾干贮存到播前剥取种子。3个树种种子的有关数据见表3。

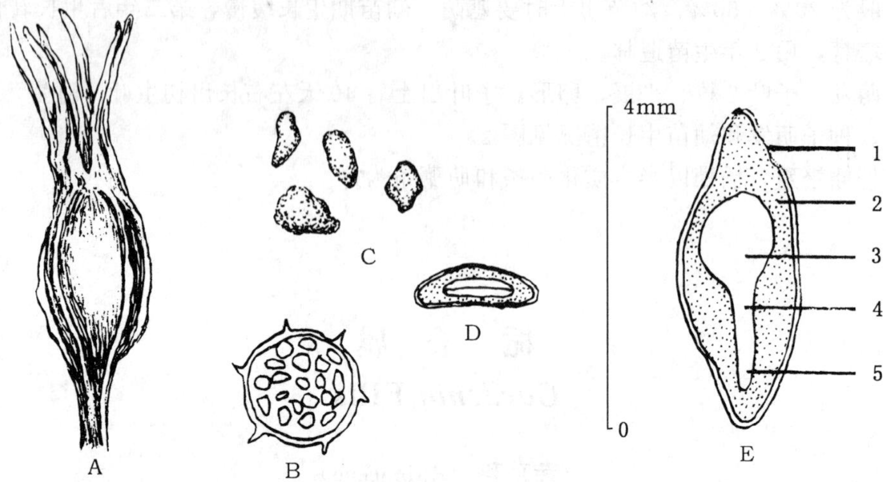

图 1　栀子果实外形（A）及其横切面（B），种子外形（C）及其横切面（D）和纵切面（E）

1. 种皮　2. 胚乳　3. 子叶　4. 胚轴　5. 胚根

（史渭清绘）

表 2　栀子属树种的花果形态及开花结实物候

树　种	观察地点	开始结实年龄	花　期	花冠筒长（cm）	花冠直径（cm）	果　长（cm）	果　径（cm）	果熟期
栀子	江苏南京	—	5 月中旬～6 月下旬	2～3	4～5	2.5～3	1.5～2	11 月下旬～12 月上旬
黄栀子	云南勐仑	8～10	3 月下旬～5 月下旬	7～8	9～10	4～6	2.6～3.6	翌年 3 月中旬～5 月上旬
狭叶栀子	广西上思	3～5	5 月上旬～6 月上旬	4～6	3～4	1～2.5	—	10 月中旬～11 月上旬

表 3　栀子属树种种子的有关数据

树　种	出种率（%）	每果含种子数（粒）	净　度（%）	长（mm）	宽（mm）	厚（mm）	千粒重（g）	每千克纯净种子数（万粒）
栀子	—	80～120	—	3～4	2～3	0.5～0.6	3.2～3.6	27.7～31.2
黄栀子	6～9	＞10	86	—	—	—	7.5～9	11～13
狭叶栀子	35	＞100	95	3～4	2～3	0.5～0.8	3.8～4.7	21～26

发芽和播种　种粒大小和成熟程度差异较大。在南京对栀子种子的一次解剖观察发现，空粒和发育不饱满粒占 45%以上。种子充分水选可以剔除空瘪粒。发芽适宜的温度为 20℃左右。发芽较快且有明显的发芽盛期。剔除空瘪粒后的发芽率较高。3 个树种的一次发芽测定结果列入表 4。

表 4　栀子属树种发芽测试结果

树　种	测试地点	播种时间	发芽条件	开始萌发至萌发结束的时间	发芽率（%）
栀子	江苏南京	1988 年 2 月 3 日	室外，沙	5 月 3～20 日，共 18 天	96
黄栀子	云南勐仑	1988 年 4 月 9 日	室外，沙	4 月 18～28 日，共 11 天	88
狭叶栀子	广西南宁	1987 年 10 月 24 日	室内，滤纸	11 月 5 日～12 月 2 日，共 28 天	62

狭叶栀子可以在 10～11 月采后即播，也可同其它 2 个种一样春播。黄栀子条播时每平方米用种量 10～14g。狭叶栀子撒播时每平方米播 10～15g。覆土约 1cm。出土萌发。从种壳出土到子叶展开以及到初生叶（对生）出现的过程较长（图 2），因此可以集中播种，待幼苗具有 1～3 对初生叶时移栽。生产上常用扦插繁殖，优点是成活率高，生长快，开花早。

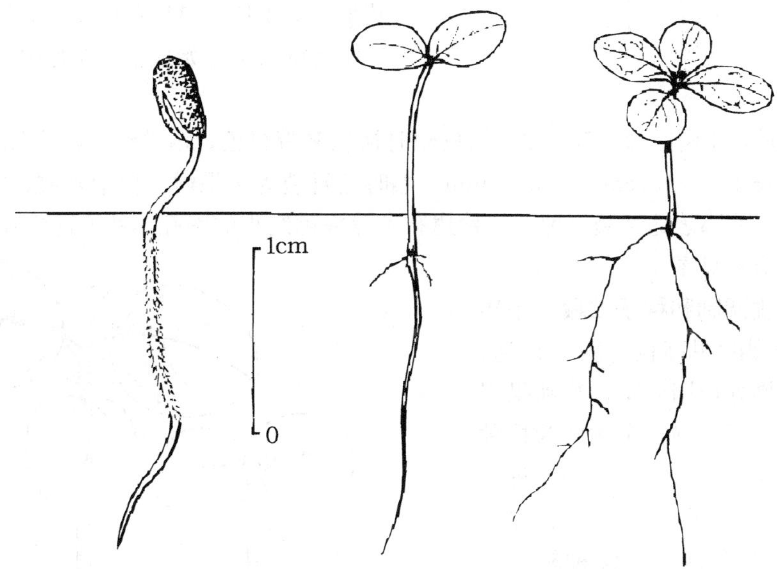

图 2　栀子种子萌发后第 2、30、60 天的幼苗生长情况
（史渭清绘）

（何泽瑛）

云南土连翘

Hymenodictyon excelsum（Roxb.）Wall.

（茜草科　Rubiaceae）

生长习性、分布和用途　土连翘属约 20 种，我国 2 种，本文描述 1 种。落叶乔木，高10～15m，胸径 50～80cm。产云南南部，适生于肥沃湿润的酸性至中性土。树皮药用，材作网模木。

开花结实　10～15 年生开始开花结实，18 年生以后为正常结实期。大小年现象不明显，

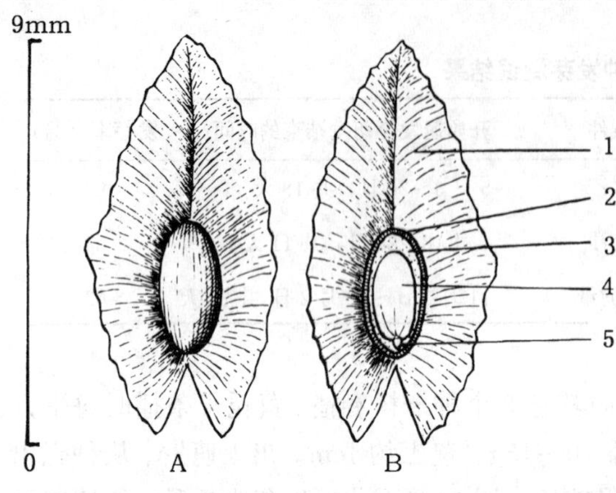

图 1　云南土连翘种子外形（A）及其纵切面（B）
1. 种翅　2. 种皮　3. 胚乳　4. 子叶　5. 胚根
（黄应钦、马信祥绘）

但多年中也有开花结实极少的年份，例如在1979～1988 年的 10 年之间，1984 年就罕见开花结实。花两性。花小而密，紫红色。圆锥花序腋生，长 10～20cm，弯垂。总花梗上端有两片具柄的叶状苞片，有网纹。萼管卵形，5～6 裂，脱落。花冠漏斗状，裂片 5。雄蕊 5，着生冠筒喉部之下。花丝短，压扁，花药基着，药隔膨大，内藏。花盘杯状。子房下位，2 室，每室有胚珠极多数。花柱长，突出，柱头棒状。据 1980～1988 年在云南西双版纳勐仑观察，6 月上旬花芽始现，着生在头年生枝条的叶腋处；6 月中下旬花序形成，7 月上旬始花，7 月中下旬花盛开，8 月上中旬花期结束；翌年 2 月底或 3 月上旬果实开始成熟，3 月中下旬为果实成熟盛期。

蒴果，倒垂，长椭圆形，有 2 槽。果皮幼时褐色略带绿色，熟时褐色，无光泽，表面有灰白色斑点。果长 1.2～1.7cm，径 5～8mm。熟时室背裂为 2 果瓣。种子多数，薄。种皮延伸成膜质的宽翅，基部之翅 2 裂。种子和种翅布满深褐色细小的网孔。种子近椭圆形，有胚乳，胚伸直。种子形态见图 1。

果实的采收调制和种子贮藏　果实成熟而尚未开裂时用高枝剪剪下果穗，摊放在通风干燥处或阳光下，待蒴果完全裂开后抖出种子，筛去杂物即为播种材料。果实出种率为 25%～28%。种子净度一般为 80%，千粒重 2.7（2.5～2.8）g，每千克有种子（含种翅）37（35～40）万粒。调制后种子含水量约 11.8%。密封干藏或低温干藏，贮藏期 3 个月。

发芽和播种　种子无休眠现象。发芽时日均温宜在 23～25℃。1989 年，西双版纳热带植物园在室内培养皿中作过发芽测定，基质为过滤纸：3 月 28 日播种，4 月 1 日开始萌发，发芽全程为 10 天；从开始发芽之日起算，7 天的发芽百分数为 90%；从播种之日起算，14 天的发芽率为 95%。出土萌发。子叶出土后 15～20 天长出初生叶，生长情况见图 2。

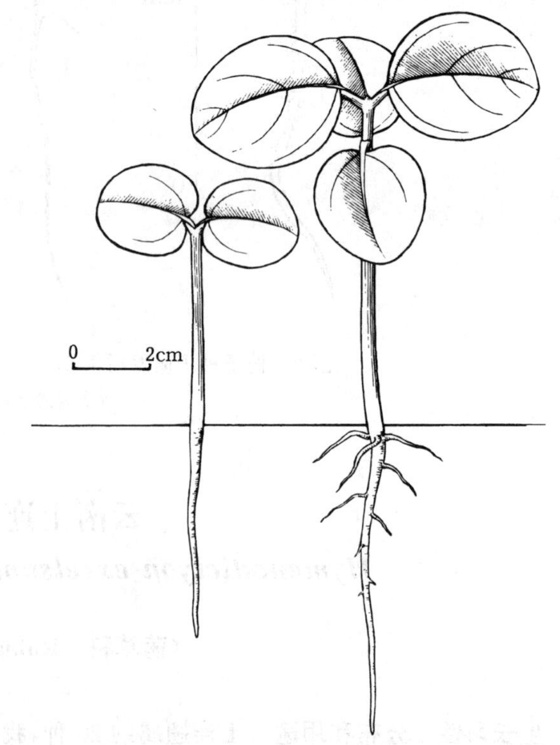

图 2　云南土连翘种子萌发后第 10、34 天幼苗的生长情况
（黄应钦、马信祥绘）

条播。每平方米播种 1.2～2g，覆土以不见种子为度。1 年生苗可以出圃定植。

（马信祥）

龙　船　花

Ixora chinensis Lam.

（茜草科　Rubiaceae）

生长习性、分布和用途　龙船花属约 400 种，我国约产 11 种，本文描述 1 种。常绿灌木，高 0.5～1.5m，颇耐荫，裸露地亦生长良好。喜湿润酸性土。能耐轻霜，但忌冰雪。分布于台、闽、粤、琼、桂。马来西亚、印度尼西亚亦产。华南重要观赏花木，全株入药，能散瘀止血，又能催产。

开花结实　2～3 年生即开花结实，正常结实期在 5 年生以后。结实无大小年现象，每年结实颇多。花两性，具短梗，排成顶生伞房花序式、三歧分枝的聚伞花序。萼筒卵形，萼檐短，4 浅裂，裂片宿存。花冠红色或黄红色，花冠筒长 3～3.5cm，裂片 4，倒卵形或近圆形。雄蕊 4，着生于花冠筒喉部，花药背着。花盘肉质。子房下位，2 室，胚珠单生。花柱突出，柱头 2。花期特长，据广西南宁 1986～1988 年的物候观察：3 月中旬始花，4 月上旬～9 月下旬为盛花期，11 月上中旬为末花期，6 月中旬～12 月中旬陆续有果实成熟。成熟果在树上可存留 1 个月左右，以后陆续脱落。

核果状浆果，未熟时红色，成熟时紫色，表面有光泽，近球形，径 6～9mm。每果有种子 1～2 粒。种子灰白色或浅褐色，圆形或扁圆形，腹面下陷，宽 4～6mm，厚 4～5.5mm。胚横生，胚乳丰富（图 1）。

果实的采收调制和种子贮藏　选摘成熟盛期的饱满果实。采回的果实堆沤 2～3 天，待果皮充分软熟后，装入布袋置水中搓洗，淘去果皮等杂质，即得纯净种子。鲜果出种率约 31%。种子的含水量约为 27%。千粒重 45g，每千克有纯净种子 2.2（2～2.5）万粒。种子忌失水，不能日晒，亦忌裸露存放。需要运往其它地方时，宜采集接近成熟初呈红紫色的果实，用布袋或竹筐包装运输，运抵后再行调制。种子不宜随采随播。贮藏时需混以湿沙，贮藏期为 6～8 个月。

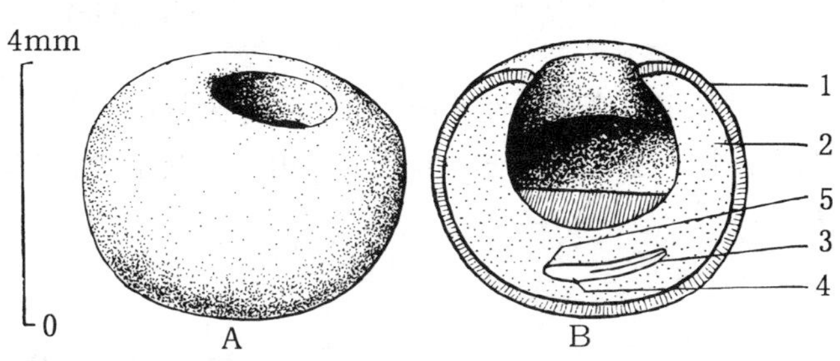

图 1　龙船花种子外形（A）及其纵切面（B）

1. 种皮　2. 胚乳　3. 子叶　4. 胚轴　5. 胚根

（黄应钦绘）

发芽和播种　种子个体间的休眠程度有很大差异，有的无休眠习性，有的则有休眠。1987 年 9 月 9 日，广西林业科学研究所用当年采集的新鲜种子在室外沙床作发芽测定，播前未作任何处理。播种时日均温在 28℃左右。10 月 14 日开始发芽，12 月 2 日前后发芽较多，翌年 1～2 月发芽中断，3 月以后又陆续发芽，至 6 月 2 日终止。发芽盛期不明显，发芽率为 75%。

出土萌发。胚根萌发后约 8 天子叶带壳出土，出土后 20～25 天种壳脱落，子叶展开，再过30～60 天初生叶展出。种子发芽及幼苗生长情况见图 2。

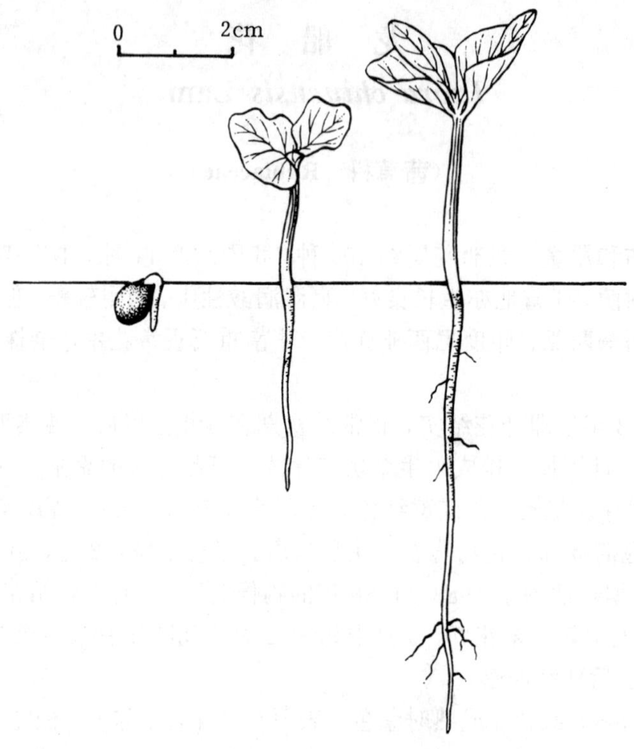

图 2 龙船花种子萌发后第 5、25、70 天幼苗的生长情况

（黄应钦绘）

种子可以密播于发芽床，萌发后将芽苗移至大田或容器内培育。每平方米用种约 8g。扦插成活率可达 80%左右。园艺上多用无性繁殖法育苗，1 年生无性苗可开花结实，上盆供观赏。

（林榕庚）

南 山 花

Prismatomeris tetrandra（Roxb.）K. Schum.

（茜草科 Rubiaceae）

生长习性、分布和用途 三角瓣花属约 25 种，我国 3 种，本文描述 1 种。灌木或小乔木，高 1～5m，耐庇荫，为下层林组成树种。常见于山溪两旁之湿润地。分布于滇、桂、粤。印度、马来西亚、菲律宾亦有分布。重要的药用植物，根入药，治风湿性关节炎、肝炎、矽肺等。

开花结实 3～4 年生开始开花结实。正常结实期在 8 年生以后。结实大小年间隔期为 1～

2 年。花单性，雌雄同株。伞形花序近顶腋生，有花数朵至多朵，雄花多数，而雌花较少。雄花萼筒小，陀螺形。雌花萼倒卵形，萼筒杯状，顶端截平。花冠高脚碟状，筒圆柱形，长 2～2.5cm，裂片 5（4）。雄蕊 4～5，着生于花冠筒内，花丝短，花药近基部背着，内藏。花盘垫状。子房下位，2 室，每室 1 胚珠。据广西南宁 1988 年的物候观察，花期始于 4 月下旬，花盛期 5 月上旬，花期终于 5 月下旬；果实成熟期始于 9 月下旬，盛期在 10 月末。同一植株果实成熟不整齐。成熟的果实在树上存留 7～15 天后脱落。浆果状核果，肉质，近球形，径0.8～1cm，红紫色至黑紫色。每果有种子 2 粒，常有 1 粒不发育。种子近球形，径 6～8mm，腹部深凹。种皮膜质，黑褐色。胚乳肉质或角质。

果实的采收调制和种子贮藏　果熟盛期直接采摘。采得的果实置水中搓揉，淘去果肉等杂质，除去 空瘪粒，所得果核即为播种材料，通称种子。鲜果出籽（核）率约 51%。种子含水量 35%～50%，净度 99%。千粒重 306（250～350）g，每千克有纯净种子（果核 ）3 270（2 860～4 000）粒。种子忌失水，运输时需混以湿沙，或以果实形态运输。贮藏需混湿沙，贮藏期约为 6 个月。

发芽和播种　种子有休眠习性。广西林业科学研究所 1987 年 10 月 17 日用当年新采种子在室外沙床播种。播种后 77 天开始发芽，无明显发芽盛期，翌年 1 月 30 日发芽终止，发芽率为 50%。出土萌发。子叶带壳出土，出土后约 50 天初生叶展现。

条播。播前种子应层积越冬，春暖时播种。每平方米播种 50～60g，覆土 1～1.5cm。苗期需遮荫。1 或 2 年生苗出圃。亦可用扦插法繁殖。

（曾　玲）

凌　霄　属

Campsis Lour.

（紫葳科　Bignoniaceae）

生长习性、分布和用途　本属仅 2 种，1 种产我国和日本，另一种产北美，但我国已广为引种栽培（见表 1）。落叶木质藤本，以气根沿树干或墙壁攀登。生于山谷、小河边及疏林中，可护堤保土，也可药用，用于垂直绿化时可作墙壁、棚架、假山的荫蔽材料，供观赏。2 个种的习性和用途极相似。南京中山植物园已培育出大量杂交苗。

表 1　凌霄属树种的名称及分布

中　名	学　名	分　布
凌霄	*C. grandiflora* (Thunb.) Loisel.	黄河流域至长江流域中下游以南。日本
美国凌霄	*C. radicans* (L.) Seem.	原产北美。苏、浙一带普遍引种栽植

开花结实　栽植后 3～5 年开始开花结实，无明显的大小年。花两性。圆锥花序顶生。完全花，大型，橙黄色至橙红色。花萼筒钟状。花冠漏斗状，中部以上扩大，5 裂片，边缘略歪

斜，顶端圆。雄蕊 4，2 强，不外露。子房上位，2 室，基部为花盘围绕。蒴果，长荚状，2 室，室背开裂为 2 果瓣。每果含种子约 250～380 粒。种子略扁，两侧有透明翅。种皮褐色，膜质。胚乳不明显。子叶薄，扁椭圆形，顶部深凹，下胚轴和胚根短小（图 1）。南京中山植物园 1979～1987 年观察的开花结实物候和花果形态见表 2。

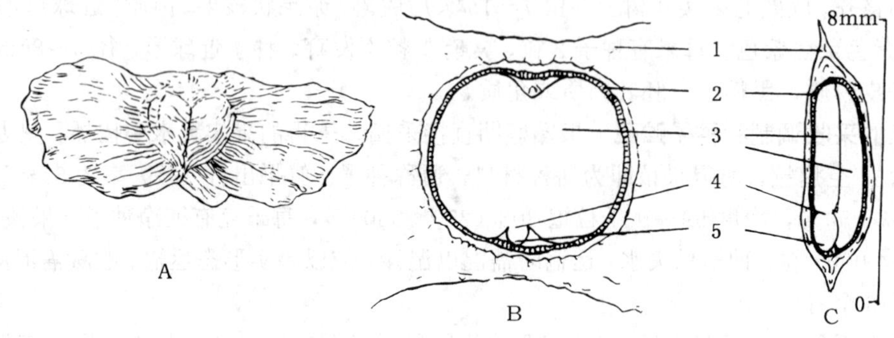

图 1　凌霄种子的外形（A）及其纵切正面（B）和纵切侧面（C）
1. 种翅　2. 种皮　3. 子叶　4. 胚轴　5. 胚根
（史渭清绘）

表 2　凌霄属树种在南京地区的开花结实物候和花果的形态

树　种	开花期	花　色	花冠径（cm）	萼　齿	果熟期	果长（cm）	果径（cm）
凌霄	6～8 月	橙黄	7	披针形	10～11 月	16～17（20）	1.5～2
美国凌霄	7～10 月	橙红	4	三角形	11～12 月	12～16	2～2.5

果实的采收调制和种子贮藏　蒴果成熟开裂后，带翅的种子随风飘散。当蒴果由绿色转灰褐色而尚未裂开时，应及时采下，摊晾干燥后剥开取种，或敲击脱粒，去杂净种。调制后净度可达 98%以上，但空瘪粒较难清除。种子发育情况较差，一般饱满种子仅占 30%～50%。种子普通干藏可保存 1 年。凌霄种子千粒重 8.2g，每千克有种子约 12 万粒。美国凌霄千粒重 7.1g，每千克约 14 万粒。

发芽和播种　种子有某种程度休眠。1989 年 2 月中旬，本文作者将头年采后干藏的美国凌霄种子用湿沙层积，分置于室内、室外和 5～10℃的冰箱 3 种条件之下。结果是，无论是室内还是室外自然温度层积，种子都是在 4 月初开始萌发，4 月 18～20 日发芽达到高峰，5 月初发芽结束。一直层积在冰箱中的种子也能发芽，只不过时间略有推迟：5 月 6 日开始萌发，5 月下旬发芽达到高峰，6 月 13 日发芽结束。3 种条件下的发芽率差异不大。这 2 个种的空瘪种粒较多，一般发芽率只有 30%～50%。如能经过精选，发芽率可达 90%以上。干藏种子可在 2～3 月播种，薄覆土，4～5 月出苗。出土萌发。初生叶 2 或 4，对生，单叶，以上为 3 小叶复叶（图 2），再以后小叶数逐渐增多。苗高 15～20cm 后需立支柱引附枝条。可以扦插繁殖。

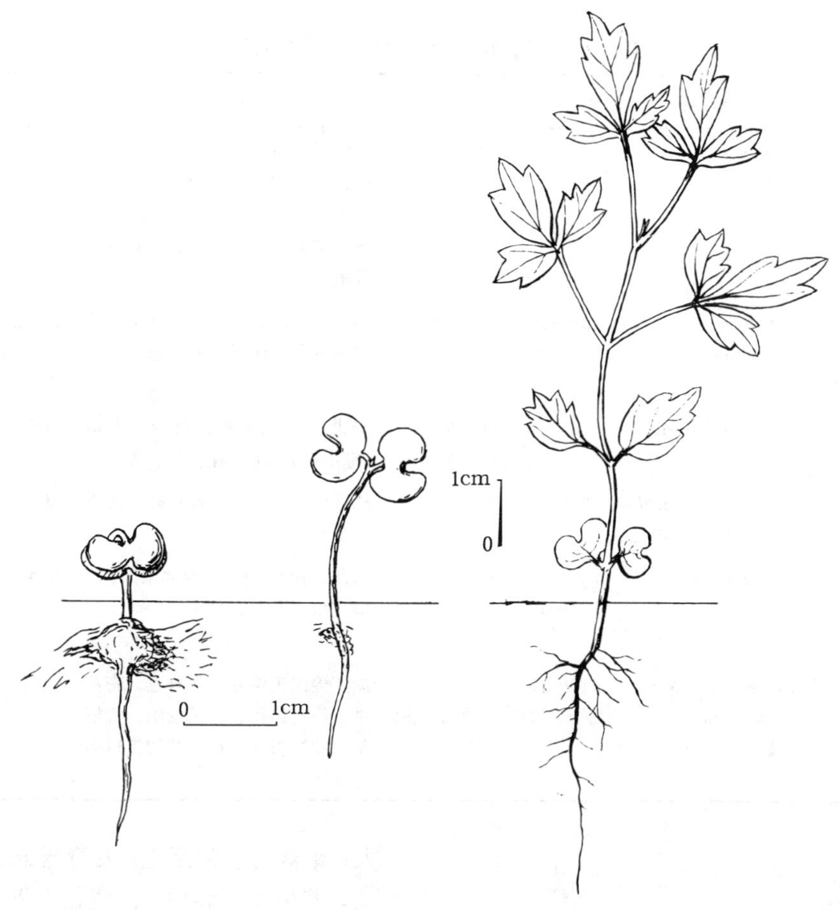

图 2　凌霄种子萌发后第 1、10、70 天的幼苗生长情况
（史渭清绘）

（何泽瑛）

梓　树　属
Catalpa Scop.

（紫葳科　Bignoniaceae）

生长习性、分布和用途　梓树属约 13 种，分布北美、西印度群岛和东亚。我国 5 种，数变种、数变型，并引入 3 种。本文描述 5 种 1 变型，其中 2 种从北美引入（表 1）。落叶乔木。较喜光，喜深厚肥沃土壤，生长迅速。

开花结实　梓树属树种 5～8 年生或 12 年生开始开花，20～30 年后大量开花，可持续 100～200 年或更长。花两性，组成总状花序、伞房状总状花序或聚伞状圆锥花序，顶生。花蕾圆球形，萼 2 深裂，裂片半圆形，镊合状排列。花冠 2 唇形，上唇 2 裂较小，下唇 3 裂较

表 1 梓树属树种名称、生长、分布和用途

中 名	学 名	树高（m）	胸径（cm）	分 布	用 途	供 稿
紫葳楸（美国梓、南方楸）	*C. bignoniodes* Wall.	15～20	50，原产地可达 100	原产美国路易斯安那、密西西比、亚拉巴马、佐治亚、佛罗里达。我国上海、南京、杭州、安庆、青岛、沈阳、石河子、丽江有零星栽植	观赏	1008
楸树（金丝楸、梓桐、线楸）	*C. bungei* C. A. Mey.	20～30	70～190	华北、陕、豫、皖、苏、浙，以黄河流域和长江以北较普遍	材用，观赏，蜜源，药用	302
灰楸（山楸、银楸）	*C. fargesii* Bureau	18～21	70～90（178）	华北、西北东部、西南、中南（琼不产），皖。南京有栽培	材用，观赏	1002
滇楸（紫花楸、光灰楸）	*C. fargesii* f. *duclouxii*（Dode）Gilmour	25～32	60～84	滇、桂、黔、川、鄂、陕。南京有栽培	材用，观赏	504
梓树（黄花楸、水桐、河楸）	*C. ovata* Don	8～15（20）	60	东北、华北、华东、西北东部、西南、华南北部。日本	材用，观赏	302
黄金树（美国楸、北方楸）	*C. speciosa*（Ward. ex Barney）Engelmann	10～12（原产地 38）	85（原产地 135）	原产美国田纳西、阿肯色、密苏里、伊利诺伊、印第安纳、肯塔基。我国长江流域和黄河流域多有栽培	观赏	1003

大。雄蕊与裂片互生，发育雄蕊 2 枚，花丝长，着生于下唇内，内藏，花药 2 室。退化雄蕊 2 或 3，着生上唇内，花药不育或缺。子房上位，2 心皮合生，2 室，中轴胎座，胚珠多数，倒生。花柱细长，柱头 2 裂，舌状。楸树、灰楸、滇楸结果不正常，有的只开花不结果，自花不孕。蒴果，长而细，略似豇豆荚，当年成熟，两瓣裂，果皮栗褐色或黑褐色，略有皮孔。种子多而轻，2～4 行着生于质厚的隔膜上。种子横椭圆形或条形，扁平，背腹面色泽相似，种皮薄，有横纹。种子两端翅状，有白色纤维质长丝状毛。种子无胚乳。子叶扁平，由左右两片长椭圆形的分叶组成，胚根位于两片子叶对合的中央紧缩处。梓树属的花序、花的数目和开花结实物候期已汇入表 2，果实和种子特征汇入表 3。种子形态见图 1。

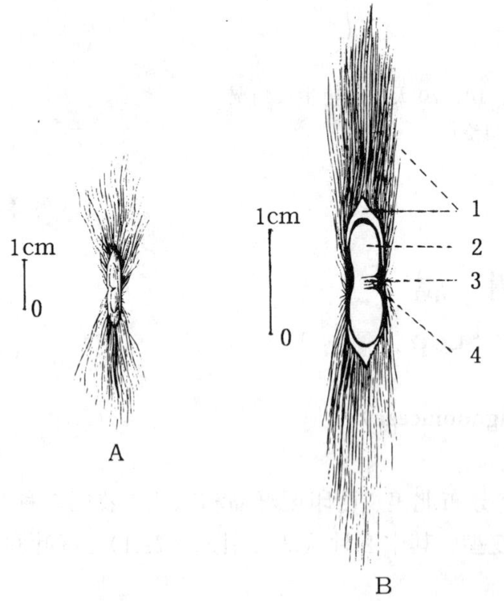

图 1 楸树种子外形（A）及其纵切面（B）
1. 种皮及纤维质长毛 2. 子叶 3. 胚轴 4. 胚根
（孟玲绘）

表 2　梓树属树种花序、花数和物候期

树　种	花　序	花　数（朵）	观察地点	开　花			果实成熟		种子散落期
				始　期	盛　期	末　期	始　期	盛　期	
紫葳楸	聚伞状圆锥	20＋	南京	5月上旬	5月中旬	6月上旬	10月		—
楸树	总状或伞房状总状	5～20	山东	4～5月			8～9月		—
			陕西	4月下旬～5月上旬			8～9月		—
			北京	5月上旬	5月中旬	5月下旬	10月上旬	10月下旬	10月下旬～11月上旬
灰楸	聚伞状圆锥或伞房状总状	7～30	山东	5月			9～10月		—
			陕西	5月下旬～6月上旬			9月		—
			湖南	4～5月			9～10月		—
滇楸	聚伞状圆锥	6～30	昆明	3月下旬	4月上旬	4月下旬	10月下旬	11月上旬	11月下旬～12月上旬
			南京	5月上旬～下旬			9月中旬～下旬		—
			陕西	5月			9～10月		—
梓树	聚伞状圆锥	100～130	北京	5月中旬	5月中旬	5月下旬	9月下旬	10月上旬	10月下旬～11月上旬
			南京	5月下旬～6月上旬			9～10月		—
			陕西	5月下旬～6月上旬			9月下旬～10月		—
			山东	6～7月			8～10月		—
黄金树	聚伞状圆锥	10＋	江苏	5月		—	9月		—
			山东	5月		—	10月		—

表 3　梓树属树种蒴果和种子的形态特征

树　种	成熟果实			种　子		
	形　状	大　小（cm）	色　泽	形　状（去毛）	大　小 连毛宽、高（mm）	色　泽
紫葳楸	细圆柱形	长 20～35 径 0.8～1.5	褐色	椭圆形	宽 40～60 高 6～8	灰褐色
楸树	细长圆柱形	长 25～50 径 0.4～0.5	黄褐色	条形	宽 35～48 高 2～2.5	紫褐色
灰楸	细长圆柱形	长 50～80 径 0.3～0.5	紫褐色	条形	宽(50)70～75(80) 高 2～2.5	灰褐色
滇楸	细长圆柱形	长 54～104 径 0.3～0.5	淡紫褐色	条形	宽 50～56 高 2～2.5	灰黄色
梓树	长筷状	长 22～25（35） 径 0.5～0.7	深褐色	椭圆形	宽 22～26（30） 高 2.5～3	淡褐色
黄金树	圆柱形	长 25～55 径 1.3～1.8	棕褐色	矩圆形	宽 38～45 高 6～10	灰褐色

果实采收调制和种子贮藏　选择 15～30 年生健壮母树采种。当果实由黄绿色变成灰褐色或黄褐色，顶端微裂时，即可采收。蒴果采回后摊晒，用木枷轻击，促使开裂，脱出种子，去杂净种后干藏。滇楸失去果瓣后种子容易失水，丧失发芽力，常在采收后曝晒 2～3 日，将干蒴果成束捆扎，吊挂通风干燥处，等来春播种时再脱去果瓣。这几个树种的出种率和种子净度、质量见表 4。

表 4　梓树属树种出种率和种子净度、质量

树　种	出种率（%）	净度（%）	千粒重（g）	每千克纯净种子粒数（万粒）
紫葳楸	40	85～95	9～12	8.3～11.1
楸树	50	65～85	3.8～5.5	18～26
灰楸	50	70～90	3.3～4.8	20.8～30
滇楸	50	70～90	4.0～5.0	20～25
梓树	50	80～90	3.7～4.5	22～27
黄金树	10～35	75～90	12～32	3.1～8.3

发芽和播种　种子纸质而薄，透性好，无明显的休眠习性。室内检验时，在每天光照 8 小时和 30～20℃的昼夜变温条件下，梓树种子置床 2 天即开始萌发，7 天后发芽率可达 94%（孙秀琴和田树霞，1990）。育苗时一般在 3 月下旬～4 月上旬播种。播前用 20～30℃温水浸

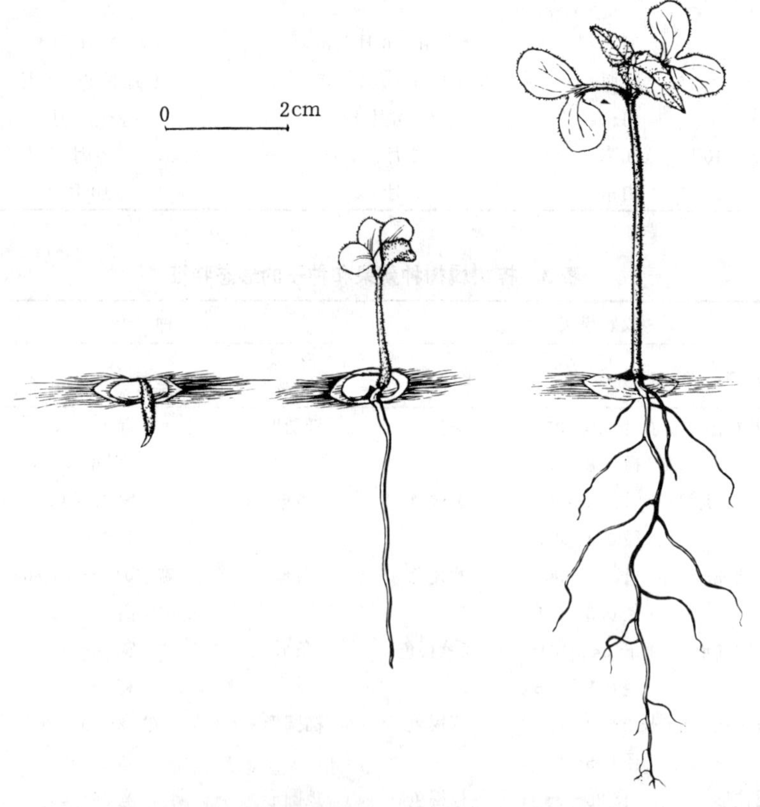

图 2　楸树种子萌发出土后第 1、8、23 天幼苗生长情况

（孟玲绘）

泡 2～4 小时（亦可不浸种）后条播或撒播。或浸泡后捞出，混以 3～4 倍湿沙，堆在室内或盆内进行催芽。在催芽期间要注意及时喷水，保持湿度，每天翻动 1～2 次，使上下湿度、温度保持均匀，约经 10 天，有 30%出芽，此时即可均匀播下。筛细土或细沙覆盖，约 0.5cm 厚，10～20 天发芽出土。梓树属种子有时秕粒占 40%～60%，空瘪粒很难剔除，所以发芽率变幅较大，在 35%～90%。如果开花结实情况良好，梓树的发芽率也可达 96%。因此播种量应视不同树种及种子质量而定，每平方米在 2.5～7.5g。出土萌发。种子发芽情况见图 2。

梓树属的树种，特别是楸树，多用根段扦插繁殖。

（钱耀明、黄鹏成）

蓝 花 楹

Jacaranda acutifolia Humb. et Bonpl.

（紫葳科　Bignoniaceae）

生长习性、分布和用途　蓝花楹属约 50 种，我国引入栽培 2 种，本文描述 1 种。落叶乔木，高约 15m。喜光。浅根性。适生于肥沃、湿润的土壤。速生，不耐寒。原产巴西。川、桂、粤、琼、闽栽培。供庭园观赏。材质轻软，可作家具。

开花结实　3～5 年生开始结实，8 年以上进入正常结实期。有大小年现象，但不明显。花两性。圆锥花序顶生。萼筒顶端 5 齿裂。花冠蓝色，冠筒细长，下部微弯，上部膨大，长约 5cm，呈 2 唇状。发育雄蕊 4，2 强，退化雄蕊 1，棒状。花盘厚，垫状。子房上位，2 室，胚珠多数。据 1987 年和 1988 年在福建福州观察，每年开花 2 次。第 1 次开花始于 5 月上旬，盛花期 6 月中旬，终花期 7 月上旬。第 2 次开花始于 8 月下旬，花量少，到 10 月中旬花期结束。第 1 次开的花翌年 2 月上旬果实成熟，3 月下旬开始开裂。第 2 次开的花，翌年 4 月上旬果实成熟，5 月上旬开始开裂。另据广西南宁观察，4 月中旬始花，下旬为盛花期，8 月下旬果始熟，9 月上旬为果熟盛期。

蒴果，木质，阔卵圆形，扁平，长 4～6.6cm，宽 3～5.5cm。未熟果青绿色，成熟后黄褐色，不脱落。每果有种子 40～100 粒，一般 70 粒。种子灰褐色，扁平，心形至扁圆形，宽0.5～0.7cm，周围带有蝶形薄翅，连翅宽 1.5～2.2cm。种子无胚乳，子叶 2 枚，含有油脂。种子的形态见图 1。

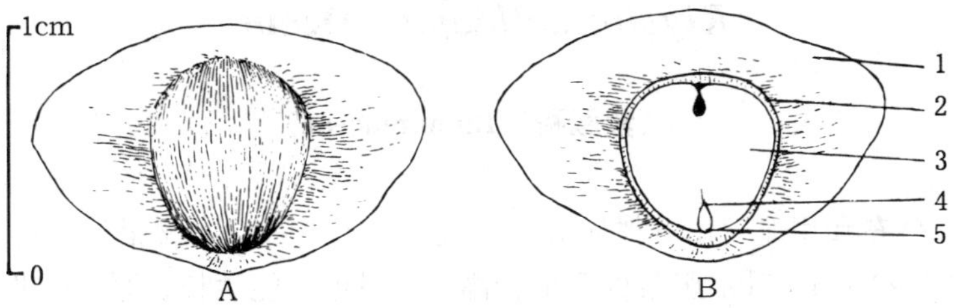

图 1　蓝花楹种子的外形（A）及其纵切面（B）

1. 翅　2. 种皮　3. 子叶　4. 胚芽　5. 胚根

（黄应钦绘）

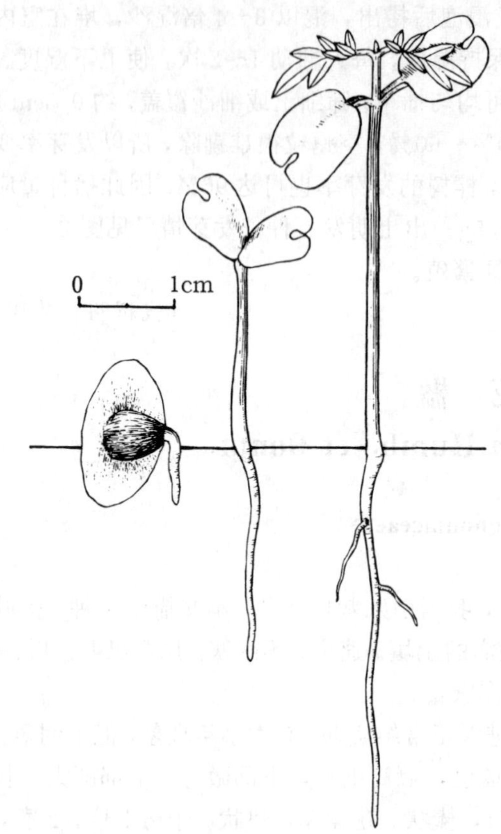

图 2　蓝花楹种子的萌发和幼苗生长情况
1. 胚根下延　2. 子叶展现　3. 初生叶出现
（黄应钦绘）

果实的采收调制和种子贮藏　果熟期长，应在蒴果由青绿色变黄褐色后分批采收。可用竹竿敲落果实后在地面收集，采回后晒干或阴干。果实仅有室背微裂而不完全开裂，须用螺丝刀撬开果壳，收集种子，经晒干即可贮藏。出种率 6%～8%，成品种子要求净度 95% 以上。含水量应在 8%以下。千粒重为 8.2～10.8g，每千克有种子 9.2 万～12.2 万粒。福建省林木种苗总站曾在 1988 年 4 月进行过贮藏试验。贮藏前种子发芽率 84%，将种子含水量降到 3.2%，用塑料薄膜袋密封，分别置于 0～5℃和室温条件下贮藏。当年 12 月测定，种子发芽率分别为 68%和 19%。这表明，即使含水量很低，蓝花楹的种子也不能在常温条件下贮藏。

发芽和播种　种子无休眠习性。室内发芽时以 15～25℃并加光照的效果较好。温度超过 30℃对发芽不利。

春末夏初随采随播。播后 4～8 天发芽。一般发芽率为 73%，发芽势为 34%。无发芽能力的种子通常为腐烂粒和空粒。条播或撒播，每平方米播种 5～8g，播后覆土 0.3cm。出土萌发。子叶圆形至长圆形，上有一缺口，径约 5～7mm，初为黄绿色，后转为绿色。初生叶为一回羽状复叶，半年以后长出二回羽状复叶。幼苗形态见图 2。1 年生苗可以出圃，庭院绿化需培育 2 年生大苗出圃。

（李玉科）

吊　瓜　树

Kigelia aethiopica Decne.

（紫葳科　Bignoniaceae）

生长习性、分布和用途　吊灯树属 3 种，我国引入 2 种，本文描述 1 种。常绿乔木。高约 15m，胸径 25～30cm。对水肥条件要求较高，干旱地不能生长。原产西非热带，现已广泛种植于全球热带的许多地区。我国于 20 世纪 50 年代从东南亚引入试种成功。琼、粤、桂、滇有栽培。花下垂形似宫灯，果形多样，如瓜吊树上，观赏价值高。果肉可作缓泻剂。枝条具韧性，抗风性能强。

开花结实　6～7 年生开始开花结实，10 年生以后为结实盛期。每年结实颇多，无明显的大小年现象。影响结实的主要因素，除水肥条件外，主要为温度。花两性。大型圆锥花序，长约 100～120cm，生于老茎，下垂。花萼钟状，不规则开裂。花冠紫红色，筒圆柱状，裂片 2 唇形，上唇 2 裂，下唇外弯，3 裂。雄蕊 4，2 强。子房上位，1 室，胚珠多数。花期、果期均较长。据海南儋县观察，主要花期在 4～5 月，主要果熟期在翌年 3～4 月，其它季节也有果实成熟。果不开裂。果皮木质，粗糙，坚硬。果实长圆柱形，长 30～47cm，直径 12～15cm，重 4.5～6.5kg，内含种子多粒。种子扁平，近椭圆形，黑褐色，无胚乳（图 1）。

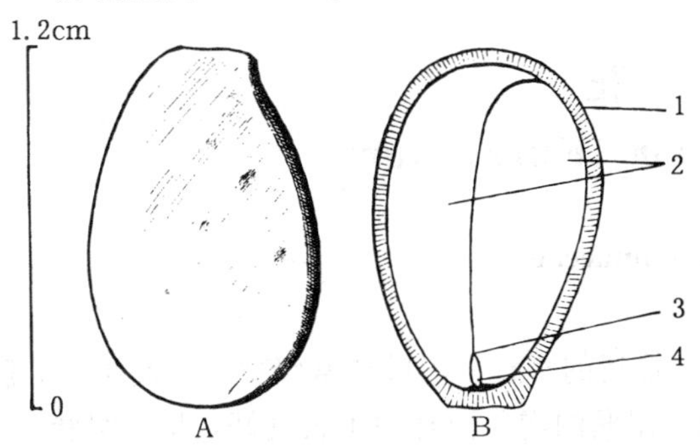

图 1　吊瓜树种子外形（A）及其纵切面（B）
1. 种皮　2. 子叶　3. 胚芽　4. 胚根
（黄应钦、黄光郁绘）

果实的采收调制和种子贮藏　果实由青灰色变成暗银灰色时就可采收，用采种钩刀切断果梗。果皮坚硬，较难取出种子，需用锋利的柴刀将果实劈开，再用小尖刀将种子逐个挖出，洗净紧贴种子上的果肉后得到纯净种子。鲜果出种率约 2.4%。净度可达 96%，千粒重 140～235g，每千克有纯净种子 4 250～7 140 粒。种子忌曝晒或干藏，含水量宜保持在 40%左右，运输时以果实的形态装运，运抵后再调制。贮藏时可将整个果实置于通风处，或将调制出来的种子混以湿沙贮藏，贮藏期一般不宜超过 1 个月。

发芽和播种　种子无休眠习性，宜随采随播。发芽时日均温宜在 20℃以上。1988 年 10 月 21 日，华南热带作物研究所在室外沙床上用新鲜种子作过发芽测定，播后 1 个多月（12 月 8 日）开始发芽，2 月 23 日发芽结束。没有明显的发芽盛期。发芽率为 30%。出土萌发。萌发后 6 天展出初生叶（图 2）。

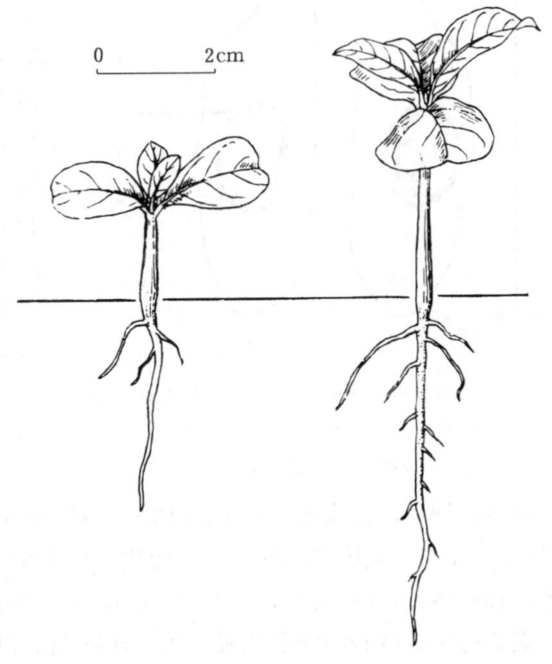

图 2　吊瓜树种子萌发后第 7、23 天幼苗生长情况
（黄光郁绘）

条播。每平方米播种 38～60g，覆土 5～8mm。培育 1 年生苗出圃。

（邓春媛）

火　烧　花

Mayodendron igneum（Kurz）Kurz

（紫葳科　Bignoniaceae）

生长习性、分布和用途　火烧花属只有本文描述的1种。落叶或常绿乔木，高15m，胸径30～50cm。能耐0℃左右的极端低温。在肥沃湿润的沟谷雨林中生长良好，呈常绿状。干热河谷、比较湿润的沟谷低地也适宜生长。瘠薄的干坡虽能生长，但生长缓慢，长势差，旱季全落叶。适于酸性至中性土壤。产滇、桂、粤、台。越南、老挝、缅甸亦产。可作庭园观赏树及行道树。花可作野菜，木材材用。

开花结实　10～12年生开始开花结实，15年生以后进入正常结实。有大小年现象，丰年间隔期1～2年。花两性。短总状花序5～13朵着生于短侧枝顶或小枝干上。总花梗长约2.5cm。花萼佛焰苞状，宿存。花冠橙黄色至金黄色，坛状。冠筒长约6cm，裂片5，半圆形，长约5mm。雄蕊4，两两成对，近等长，着生于花冠筒近基部。花药个字形着生。子房上位，2室，长圆柱形。花柱细长，柱头2裂，舌状扁平。据1985～1988年云南西双版纳勐仑观察，1月上中旬花芽始现，1月下旬花序形成，2月上旬始花，2月中旬～3月上旬盛花，3月下旬第1次开花结束；以后至10月前还有多次开花，有的年份可以结实2次，第2次果少。

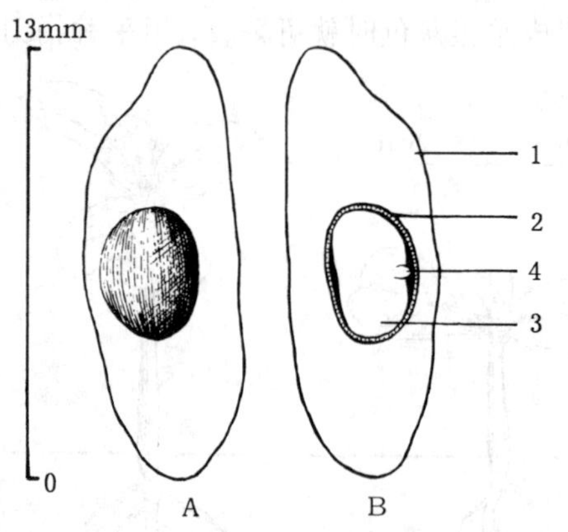

图1　火烧花种子外形（A）及其纵切面（B）
1. 种翅　2. 种皮　3. 子叶　4. 胚根
（黄应钦、马信祥绘）

蒴果，细长，长线形，长15～45cm，径3～7mm。成熟时果皮由青色转变为黑褐色。果皮革质，2瓣裂，隔膜细圆柱形，木栓质。种子在胎座两侧2列，极多数，褐色，极薄，横矩圆形，两侧具银白色膜质翅，长边连翅宽1.3～1.5cm，短边2～4mm。无胚乳，子叶宽(图1)。

果实的采收调制和种子贮藏　果实成熟期上树摘取，或用钩刀带总梗采下。采得的果实置日光下或通风处晾晒，待果实裂开后抖出种子，筛去杂质即为播种材料。果实出种率8%～9%。种子净度一般为85%。千粒重约0.85g，每克有种子1 200（1 100～1 250）粒。密封干藏可贮藏1年左右。

发芽和播种　种子无休眠习性。发芽时的日均温宜在25℃左右。1986年西双版纳热带植物园在室内用培养皿进行发芽测定，基质为滤纸，播后6天开始发芽。从发芽之日起算，6天发芽78%；从播种之日至发芽终了，22天的发芽率为84%。出土萌发。子叶出土后15天左右长出初生叶，种子萌发和幼苗生长情况见图2。

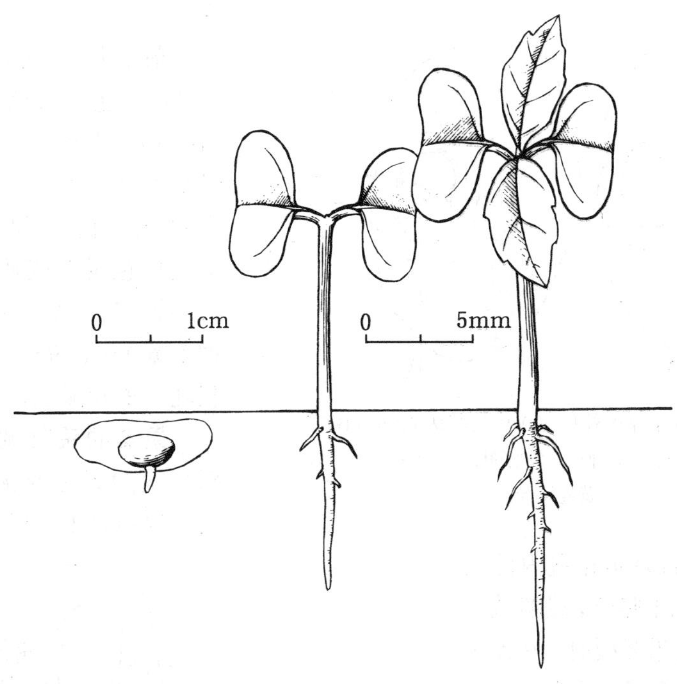

图 2　火烧花种子萌发后第 1、10、25 天幼苗的生长情况
（黄应钦、马信祥绘）

条播。每平方米播种 0.3～0.5g，覆土以不见种子为度。1 年生苗可以出圃。

（马信祥）

木蝴蝶（千张纸）

Oroxylum indicum（L.）Vent.

（紫葳科　Bignoniaceae）

生长习性、分布和用途　木蝴蝶属 1 或 2 种，我国只有本文描述的 1 种。落叶乔木，高 7～12m，胸径约 18cm。适生于温暖湿润的气候环境，能耐短期 0℃左右的低温，幼苗忌霜冻。喜肥沃湿润的酸性土。分布于川、滇、黔、桂、粤、琼、闽、台。印度、马来西亚等南亚和东南亚国家亦产。木材可作火柴杆、造纸。种子入药，消炎镇痛。叶作绿肥。

开花结实　5～6 年生开始开花结实，正常结实期在 10 生年以后。结实大小年间隔期为 1 年，但不甚明显。花两性。总状花序顶生。总花梗粗壮，长约 30cm。花萼肉质，紫色，阔钟形，顶部近平截，2 裂。花冠筒肉质，长 9cm，钟状或圆柱形，5 裂，微 2 唇形，橙红色或紫色，少数为白色带紫色斑条。雄蕊 5，4 枚较长，花丝基部被绵毛。子房上位，2 室。花柱长约 6cm，柱头 2 片裂，舌状，宽约 1cm。据南宁 1987 年观察，6 月上旬为始花期，8 月上中旬为盛花期，9 月上旬为末花期；9 月下旬，早期开花所结的果实开始成熟，10 月上中旬为果熟

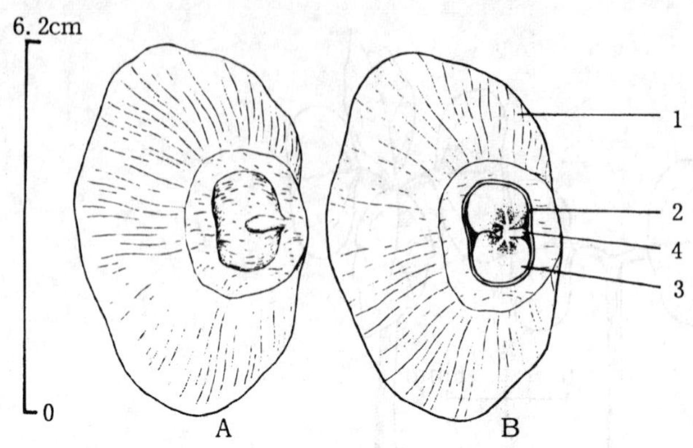

图 1　木蝴蝶种子外形（A）及其纵切面（B）
1. 种翅　2. 种皮　3. 子叶　4. 胚根
（黄应钦绘）

盛期。

蒴果大，长而扁平，未熟时绿色，成熟时紫色或褐色，木质，长 30～90（100）cm，宽 5～ 8.5cm。成熟后约经半个月果实开裂，种子飘落。每果有种子数百粒。种子扁平，除基部外全被白色透明膜质翅包围，近蝴蝶状。种子带翅宽 4～7cm，长 3～4cm。种子宽 1～1.5cm，长 0.8～1cm。无胚乳，子叶深裂，形态见图 1。

果实的采收调制和种子贮藏　蒴果变为紫色或褐色时，用高枝剪或采种钩刀于果梗处截断，选出早期和盛期开花所结的果实在室内陈放 2～3 天，再置日光下曝晒，或摊放于通风处晾干。当蒴果开裂时，在无风处剥开果瓣，脱出种子，通常不作去翅。鲜果出种率约 40%。种子可带翅置弱光下晒干或晾干，含水量在 9% 以下时方可贮藏。种子的净度约 92%。带翅千粒重 80g，每千克有带翅种子 1.2（1～1.4）万粒。可装入袋内悬挂于室内通风阴凉处，也可装入坛罐内贮藏，贮藏期一般为半年左右。

发芽和播种　种子无休眠习性。发芽时日均温宜在 18℃以上。1987 年广西林业科学研究所用当年新采种子在室内作发芽测定：11 月 4 日置床，至 11 月 15 日（日均温 19℃）开始发芽，19～25 日为发芽盛期，30 日发芽终止；从发芽之日起算，10 天发芽 70%；从播种日至发芽终了，27 天的发芽率为 84%。出土萌发。胚根萌发后 5 天子叶出土，约 35 天发出初生叶。本蝴蝶种子的萌发及幼苗生长情况见图 2。

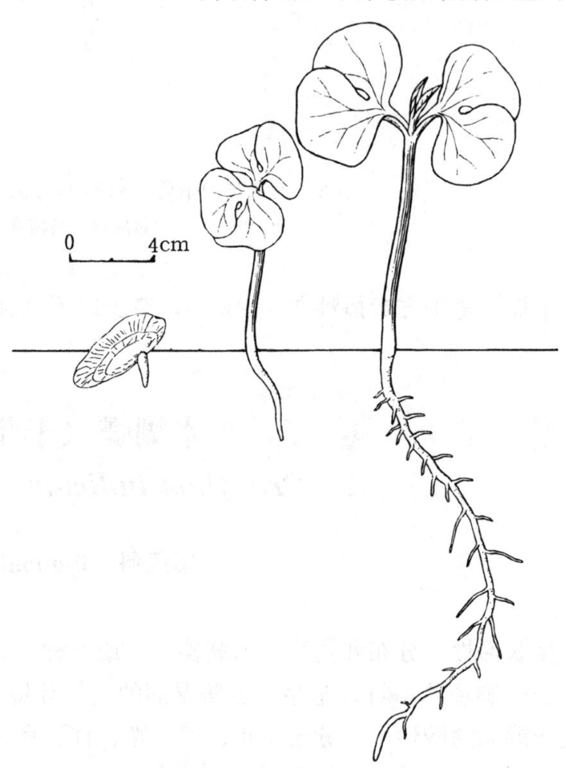

图 2　木蝴蝶种子萌发后第 4、15、35 天幼苗的生长情况
（黄应钦绘）

撒播或条播。整地要求细致，播种时注意将种翅摆平。每平方米播种 6～ 8g，覆土 0.5cm。培育 1 年生苗出圃。

（林榕庚）

菜豆树属
Radermachera Zoll. et Moritzi

（紫葳科　Bignoniaceae）

生长习性、分布和用途　本属40余种，我国约产7种，本文描述2种。落叶乔木。萌芽力极强。能耐短期－4℃左右低温，但不耐冰冻。喜疏松湿润土壤，适生于石灰岩山区，酸性土也有少量分布，粘重干旱贫瘠土生长不良。它们的名称、生长、分布及用途见表1。

表1　菜豆树属树种的名称、生长、分布和用途

中　名	学　名	树高(m)	胸径(cm)	分　布	用　途
海南菜豆树	*R. hainanensis* Merr.	20～25	40	桂、琼。粤引种	材用、叶作绿肥
菜豆树	*R. sinica* (Hance) Hemsl.	15～20	30～40	滇、桂、粤、湘、台。越南	材用、果入药、叶作绿肥，观赏

开花结实　7～8年生开始开花结实，正常结实期在15年生以后。结实丰年间隔期为1年。花两性。总状花序或圆锥花序。海南菜豆树花序腋生或着生于老枝上，菜豆树花序顶生。花萼钟状，5齿裂。花冠漏斗状钟形，冠筒长6～7cm，橙红色或白色至淡黄色，裂片5，微呈2唇形。发育雄蕊4，微2强，内藏，另1退化。花盘杯状，肉质。子房上位，圆柱形。胚珠极多数，排成多列。花柱纤细，柱头舌状扁平。广西南宁和海南观察的开花结实物候期见表2。

表2　菜豆树属树种的开花结实物候

树　种	观察地点和年份	开花			果实成熟		种子散落
		始　期	盛　期	末　期	始　期	盛　期	
海南菜豆树①	海南1982～1984	6月上旬	7月上旬	8月下旬	9月上旬	10月上旬	9月中旬～10月中旬
菜豆树	南宁1987～1989	6月中旬	6月下旬	7月上旬	9月上旬	9月中下旬	10月上旬～下旬

① 另据《海南主要经济树木》记载，海南菜豆树在海南1年开花结实2次：第1次6月开花，8～9月果实成熟；第2次11月至次年1月开花，2～3月果实成熟

蒴果，细长圆柱状，常旋扭。果皮薄革质，成熟后1个月内室背裂开。隔膜厚而贯连，近圆柱状。每果有种子千粒以上。种子扁平，两侧有膜质翅。无胚乳，子叶扁平，先端稍凹。果实及种子的形态见表3、图1。

表 3 菜豆树属树种果实和种子的形态特征

树 种	果实			种子		
	形 状	大小（cm）	颜 色	形 状	大小（mm）	颜 色
海南菜豆树	细长圆柱状，微扭曲	长 20～40 径 0.5～0.7	灰黄色	长椭圆形扁平，稍卷曲	带翅 宽 9～11 长 3～4	翅白色，种子褐色
菜豆树	细长圆柱状，旋扭	长 42～75 径 1～1.3	黄褐色	长椭圆形扁平，微卷曲	带翅 宽 14～20 长 4～6	翅白色，种子褐色

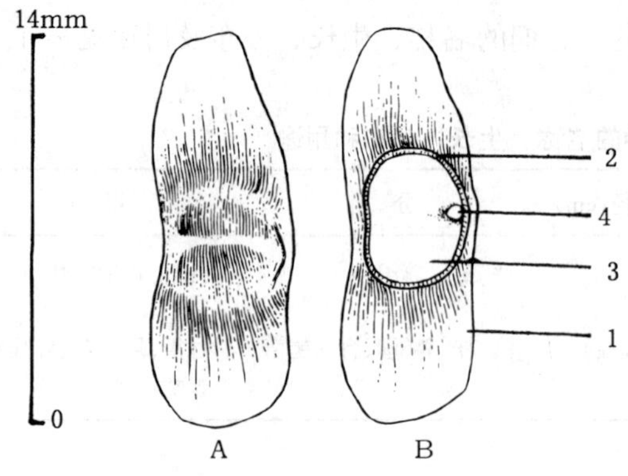

图 1 菜豆树种子外形（A）及其纵切面（B）
1. 种翅 2. 种皮 3. 子叶 4. 胚根
（黄应钦绘）

果实的采收调制和种子贮藏

果实成熟盛期，选择向阳的外部果枝，用高枝剪或钩刀将果穗采下。采集的果实可摊放于室内通风处，数天后果皮自行开裂，用手抖动或轻轻敲打即可脱出种子，常不去翅。除去果皮及隔膜等杂质即得纯净种子。种子宜置于无风的弱光下晒干或摊放在室内晾干，含水量低于 9%方可贮藏。装入布袋悬挂于室内通风凉爽处，冬季贮藏期一般为半年左右。也可随采随播。出种率和种子的净度、质量等见表 4。

表 4 菜豆树属树种的出种率和种子净度、质量

树 种	出种率（%）	净度（%）	千 粒 重（g）	每克纯净种子粒数
海南菜豆树	20～22	90	0.7～1	1 000～1 400
菜豆树	22～24	92	2～3	330～500

发芽和播种 种子无休眠习性。发芽时的日均温宜在 20℃以上。1987 年，广西林业科学研究所在室内用当年新采种子进行了发芽测定：海南菜豆树于 9 月 8 日置床，9 月 15 日开始发芽，16～18 日为发芽盛期，20 日发芽终止；菜豆树于 10 月 7 日置床，12 日开始发芽，14～18 日为发芽盛期，25 日发芽终止。发芽情况见表 5。

表 5 菜豆树属树种的发芽能力及其测定条件①

树 种	温度（℃）	发芽势（%）		发芽率（%）	
		计算天数	一般数值	计算天数	一般数值
海南菜豆树	26～28	4	46	13	54
菜豆树	23～26	6	67	18	89

① 广西南宁，室内滤纸床，自然光

出土萌发。发芽后 2～4 天子叶出土，12～16 天发出初生叶（图 2）。

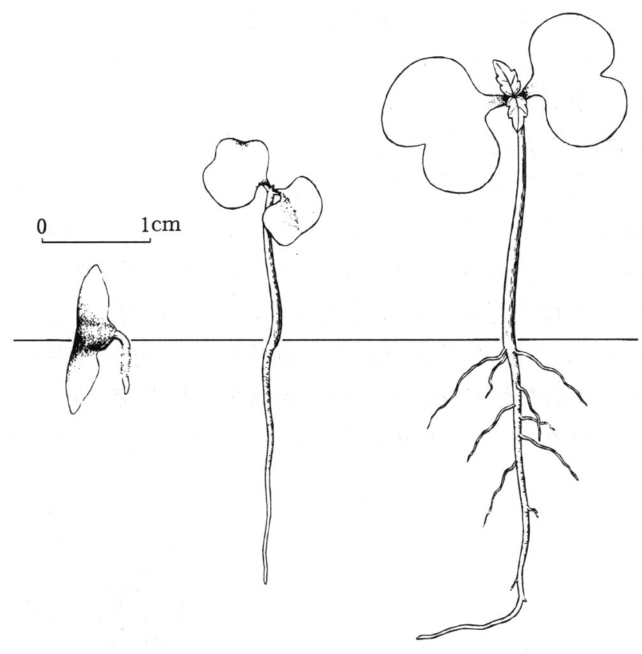

图 2　菜豆树种子萌发后第 2、7、14 天幼苗的生长情况

（黄应钦绘）

撒播。每平方米播种 0.3～0.5g。10 月播种，次年春苗木转木质化后移至大田培育。也可早春播种，春末夏初移植。1 年生苗出圃。

（朱积余）

羽　叶　楸

Stereospermum tetragonum（Wall.）DC.

（紫葳科　Bignoniaceae）

生长习性、分布和用途　羽叶楸属约 24 种，我国 3 种，本文描述 1 种。乔木，高达 10～18m，胸径 15～25cm。生于石灰岩山的次生阔叶林中。分布于滇、黔、桂、粤、台。越南、柬埔寨、缅甸、马来西亚、印度尼西亚和斯里兰卡也有分布。木材可作梁、柱和坑木等。花序大，蒴果长，可作园林观赏树种。

开花结实　约 7～8 年生开始开花结实，正常结实期在 15 年生以后，结实丰年间隔期为 1 年。花两性。聚伞状圆锥花序顶生，长 25cm。花萼钟状，萼齿 3～5。花冠筒状，一侧肿胀，内白色，外黄带微红色，裂片 5，近等大，2 唇形。雄蕊 4，2 强，内藏。花盘垫状。子房上位，2 室，胚珠多数。花柱丝状，柱头丝状，柱头 2 裂。据广西龙州 1987 年观察，始花期在 5 月上旬，盛花期 6 月上旬，末花期在 7 月中旬；果实成熟始于 9 月中旬，果熟盛期在 10 月中旬末，果熟末期为 11 月中旬。果实成熟后 20 天左右开裂，种子脱落。

蒴果，长方柱形，有明显 4 纵棱，稍弯曲，绿褐色，长 30～70cm，宽 6～7.5（10）mm。果皮厚，木质。每果有种子 40～60 粒。隔膜厚，近圆柱形，木栓质，种子脱落后有下陷的穴。种子两侧具膜质翅，每侧翅宽 10mm，高约 5mm，白色。种子扁圆形，连翅宽约 28mm，高 4～ 5mm，厚约 2mm，乳黄色，无胚乳。

果实的采收调制和种子贮藏　果实成熟盛期，在蒴果尚未开裂时，用高枝剪将蒴果剪下。采收的蒴果在无风处置弱光下或室内摊放。蒴果开裂后轻轻抖动，种子脱落。鲜果出种率约 10%。种子净度一般为 90%。千粒重 20（18～22）g，每千克有带翅种子 50 000（45 000～55 000)粒。种子不宜在强光下曝晒，但可晾干后干藏，贮藏期约为 1 年。

发芽和播种　种子无休眠习性，可以随采随播，亦可以贮藏至翌年春播。发芽时日均温宜在 20℃以上。1987 年 10 月 7 日，广西林业科学研究所在室内作发芽测定，当时室温约26℃，基质为吸水纸：置床 4 天后开始发芽，次日便进入盛期；从发芽之日起算，3 天中发芽 40%；10 月 28 日发芽终止，以 21 天计，发芽率为 58%。出土萌发。子叶出土 8 天后初生叶出现（图 1）。

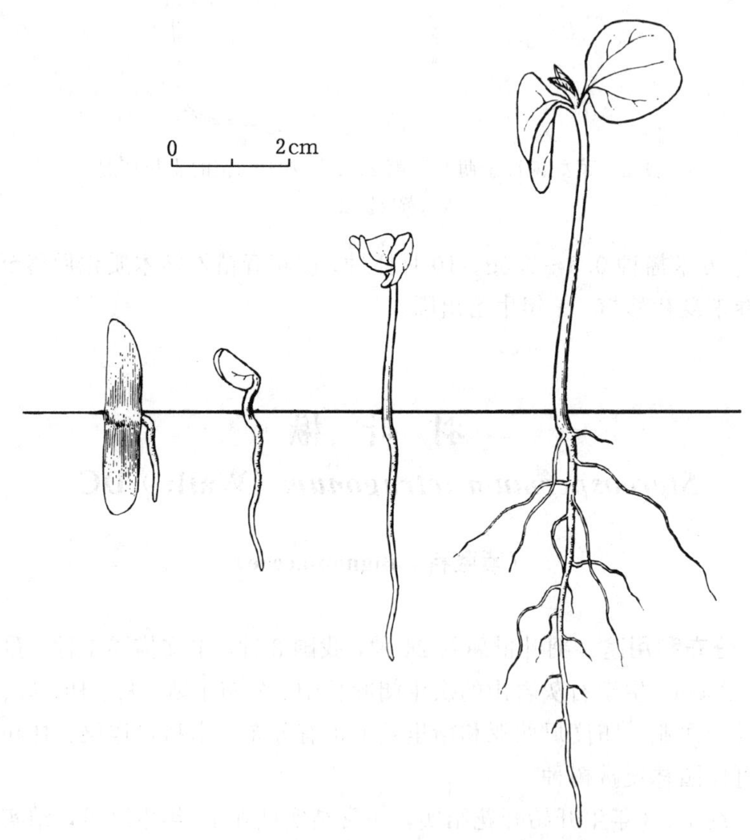

图 1　羽叶楸种子萌发后第 1、3、5、10 天幼苗的生长情况

（黄应钦绘）

条播。每平方米播种 2～3g，覆土 5mm。随采随播的种子，当年幼苗需防霜冻。育半年至 1 年生苗出圃。

（曾　玲）

破　布　木

Cordia dichotoma Forst. f.

（厚壳树科　Ehretiaceae）

生长习性、分布和用途　破布木属约 250 种，我国产 5 种，本文描述 1 种。常绿乔木，高 20m，胸径 40～60cm。适生于肥沃湿润的酸性土。分布于滇、桂、粤、琼、闽、台。印度、越南、澳大利亚东北部、新喀里多尼亚岛等也有分布。木材供建筑、造纸，果可食并作药用。

开花结实　6～7 年生开始开花结实，10 年生以后进入正常结实期，大小年间隔不明显。花杂性，两性花和雄花异株。聚伞花序顶生。两性花萼长 5mm，5 裂，裂片三角形。花冠白色，钟状，长约 8mm，5 裂，裂片比筒部长。雄蕊 5，花丝短。子房上位，4 室，花柱基部合生，2 裂，柱头 4。雄花似两性花，花丝较长，退化雌蕊近球形。据广西南宁市郊 1987～1989 年观察，2 月中旬末花蕾出现，3 月上旬为始花期，4 月上旬为盛花期，4 月下旬初为末花期；9 月初为果实成熟始期，9 月上旬末为果熟盛期，9 月中旬末为果熟末期；10 月上旬果实脱落。核果，近球形，直径 1～1.5cm，花萼宿存。果熟时由青黄色转变为橙红色。每果 4 室，果核 4 粒。核扁，近圆形，表面凹凸不平，浅褐色，长 0.7～1.0cm，厚 5～7mm。无胚乳。

果实的采收调制和种子贮藏　果实成熟盛期用采种钩摇动果枝，震落后在地面捡拾。采集的果实装入筐内置水中搓擦，淘出果核，用作播种材料，通称种子。鲜果出籽率约 37%。千粒重 175（165～185）g。每千克有种子（核）5 700（5 400～6 000）粒。种子可以晾干 3～5 天，不影响发芽的含水量约为 15%。短期贮藏可用布袋包装挂于室内通风阴凉处。2 个月以上的贮藏需混湿沙，贮藏期约为 1 年左右。

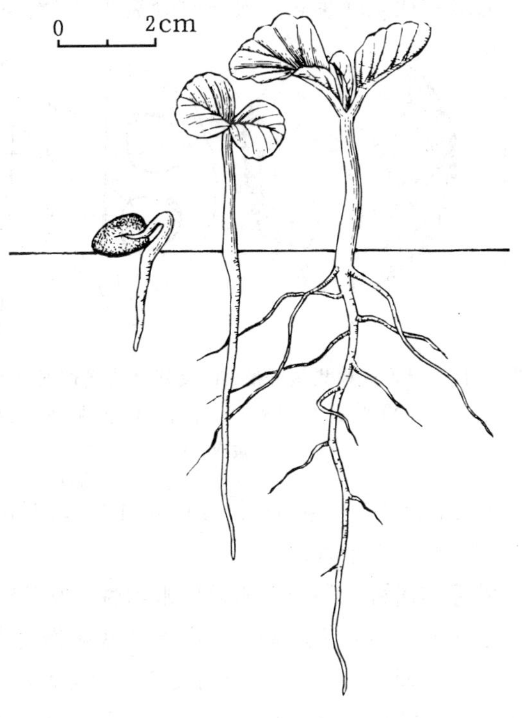

图 1　破布木种子萌发后第 11、16、24 天幼苗的生长情况

（黄应钦绘）

发芽和播种　种子有中等程度的休眠习性。1987 年 9 月 21 日，广西林业科学研究所用室外沙床作过发芽测定：播后 103 天（翌年 1 月 2 日）开始萌发，发芽盛期不明显，1988 年 3 月 30 日发芽终止，发芽率为 50%。出土萌发。发芽后 11～13 天子叶出土，24 天后初生叶出现（图 1）。

条播。每平方米约播 25～30g，覆土1～1.5cm。半年生苗或 1 年生苗出圃。

（曾　玲）

倒卵叶厚壳树

Ehretia acuminata B. Br. var. *obovata* (Lindl.) Johnst. [*E. thyrsiflora* (Sieb. et Zucc.) Nakai]

（厚壳树科　Ehretiaceae）

生长习性、分布和用途　厚壳树属约50种，我国约产11种3变种，本文描述1变种。落叶乔木，高15～18m，胸径30～40cm。能耐－10℃左右低温。要求土质肥沃，肥力较差的土地仅能生长成高10m以下的小乔木。分布北自陕南、豫南、鲁南，南至华东、华南、西南。印度、越南、菲律宾、日本、朝鲜半岛也有分布。木材作家具和建筑，嫩芽可作蔬菜，树皮作染料。

开花结实　7～8年生开始开花结实，正常结实期在15年生以后。结实丰年间隔期为1年。花两性，芳香，无柄。圆锥花序顶生或近顶部腋生。萼5裂。花冠白色，筒短，裂片5，长2～4mm。雄蕊5，着生于花冠筒上，花丝细长，花药外露。子房上位，2室，每室2胚珠，花柱2，合生至中部以上，柱头2枚。据广西南宁1987～1989年观察，4月上中旬花始开，下旬为盛花期，5月上旬为末花期；7月中旬果实开始成熟，下旬为果熟盛期，8月上旬～9月上旬果实脱落。核果，球形，径3～5mm。未熟时果绿色，成熟后外果皮橘红色或黄色。果核表面有皱纹或呈蜂巢状，成熟时分裂为2个各具2种子的分核。分核半圆形，壳坚硬，表面粗糙，黄褐色，长2.5～2.8mm，厚1.4～1.6mm。种皮薄，具少量胚乳，子叶扁平（图1）。

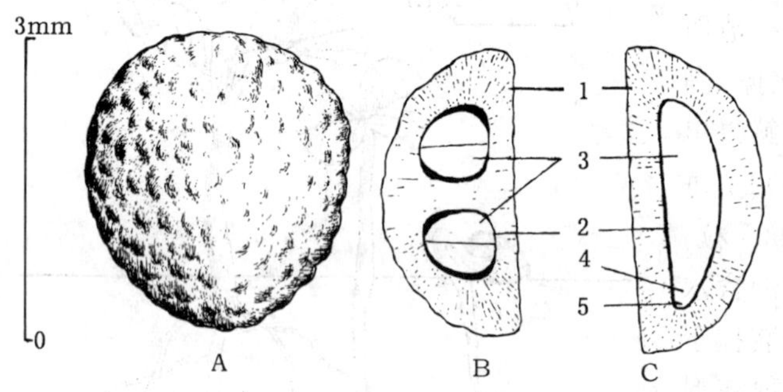

图1　倒卵叶厚壳树果核外形(A)及其分核横切面(B)、纵切面(C)
1. 内果皮　2. 种皮　3. 子叶　4. 胚轴　5. 胚根
（黄应钦绘）

果实的采收调制和种子贮藏

果熟盛期用高枝剪或采种钩刀将果穗采下。摘下果实堆沤1～2天后装入布袋，置水中搓洗，淘去果皮等杂质。所得分核用作播种材料，通称种子。鲜果出籽率约14%。种子的含水量15%～18%，净度可达96%。分核千粒重7.5(7～9) g，每千克有种子(分核)13(11～14)万粒。种子可以晾干，但不宜日晒。贮藏时需混湿沙，贮藏期6～8个月。

发芽和播种　种子无明显休眠习性。发芽时日均温宜在20℃以上。1987年7月，广西林业科学研究所用当年新采种子在室外沙床作过发芽测定：7月20日播种，日均温28～30℃，9月24日（日均温27℃）开始发芽，发芽盛期不明显，11月1日发芽终止。发芽率为40%。出土萌发。胚根萌发后约6天子叶出土，11天初生叶出现。

条播。整地要求极细致，播前用始温40～50℃水浸种，并在室内催芽，部分种子萌动后播种。每平方米播种3～4g，覆土0.2cm。1年生苗出圃。

（朱积余）

海　榄　雌

Avicennia marina (Forsk.) Vierh.

（马鞭草科　Verbenaceae）

生长习性、分布和用途　海榄雌属约10种，我国只有本文描述的1种。灌木，高1.5～6m。单独成群落或与其它种类红树混生。指状呼吸根群密集露出滩面。广布于多潮带，常为滩涂半泥半沙或半淤泥深厚处的先锋树种，适应于在高盐度地段生长。分布于桂、粤、琼、台、闽沿海地区。热带亚洲、大洋洲和非洲均有分布。成林地段为鱼虾提供栖息场所并防浪护堤。果实浸泡去涩后可炒食，并治痢疾。

开花结实　5年生左右开花结实，正常结实期在8年生以后，大小年现象不明显，一般年份结实量均较多。花两性。聚伞花序密集成头状。花萼杯状，顶端5裂，外面有茸毛。花冠黄褐色，钟状，顶端4裂。雄蕊4，花丝极短，着生于冠筒喉部。子房上位，不完全的4室，每室有胚珠1，着生于有4翼的中轴胎座上。柱头2裂。据广西北部湾的观察，初花期3月，盛花期5月，幼果期4～6月，果熟期8～9月。果为一稍呈肉质的蒴果，成熟时淡灰黄色，扁球形，直径1.2～1.5cm。每果有种子1粒，离母体前发芽。胚轴不突出于果实外。

果实的采收调制和种子贮藏　果实发育成熟已经发芽而尚未坠落时及时采摘。已萌发的果实，每千克约400粒。采得的果实即为育苗材料，忌失水或堆沤发热，不宜日晒和大堆存放。短期存放宜放成小堆或摊放在荫凉处，上面加盖湿草。运输时宜用竹筐包装，每筐以20kg为宜。贮存或运输时间以不超过1周为宜。

发芽和播种　在母体上已经萌芽的果实，采后可不作任何处理，随采随播。部分尚未萌动的果实，可装于竹箩内置水体流动的海湾内浸种催芽：每天浸泡5～6小时，其余时间取出放在荫湿处，经常翻动，不使发热沤坏。1～2天后胚轴吸水膨胀萌发，即可选择风平浪静的天气在肥沃平坦的滩涂上播种。播时胚根向下，埋入土中1/3即可。过深种子会腐烂，过浅会随海水飘走。1年生苗高30cm左右即可出圃造林。也可不经过育苗，直接用萌动的果实造林。

（何祖家）

海州常山

Clerodendrum trichotomum Thunb.

（马鞭草科　Verbenaceae）

生长习性、分布和用途　赪桐属约400种，主产东半球热带和亚热带，少数至温带，我国34种6变种，多数分布于西南、华南。本文描述1种。落叶灌木或小乔木，高1.5～10m，喜凉爽湿润气候及向阳环境，一般土壤均可生长，耐旱和耐盐碱性较强。分布于陕、甘、辽、华北、华东、中南、西南及台湾。朝鲜半岛、日本、菲律宾北部也有分布。深秋季节，宿存的红色萼似红花盛开，颇为美观，是优良的庭园绿化树种。根、叶、花入药，有镇痛、利尿和降低血压之效。枝叶作农药可杀红蜘蛛、棉蚜虫和地下害虫。

开花结实　花两性，组成顶生或腋生的伞房状聚伞花序，通常二歧分枝。花蕾时花萼绿白色，后变紫红色，基部合生，中部略膨大，有5棱脊，顶端5深裂，裂片三角状披针形或卵形，急尖，花后增大，宿存。花冠白色或带粉红色。花冠筒细，长约1.2～2cm，顶端5裂，裂片长圆形，长5～10mm，宽3～5mm。雄蕊4，长而弯曲，伸出花冠外。子房上位，不完全4室，每室1胚珠。花柱略短于雄蕊，柱头2裂。花期长，6～9月，果熟期9～11月。

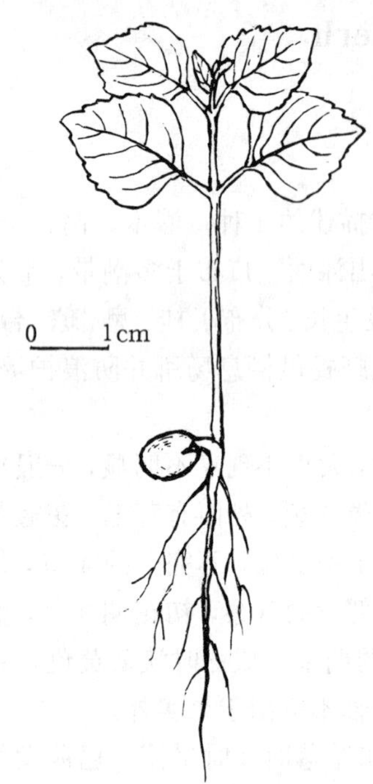

图1　海州常山幼苗形态
（田恒德绘）

浆果状核果，常为宿存增大的紫红色花萼所包被。果近球形，径6～8mm。外果皮肉质，蓝色或蓝紫色。果外面常有4浅槽或成熟后分为4分核，或因发育不全而为1～3分核。种子长圆形，无胚乳。

果实的采收调制和种子贮藏　10～11月果熟呈蓝紫色时剪取果穗，捋下果实，浸于水中。搓去宿萼及肉质外皮后取得的果核，即为播种材料，通称种子。调制后沙藏于冷凉高燥处，2～3月间勤加翻动，以防种子腐烂变质，并可促其发芽一致。千粒重约19g，每千克约5.3万粒。

发芽和播种　据北京园林部门经验，3月下旬～4月中旬取出经过沙藏催芽的种子，高床撒播，每$10m^2$用种80～100g，可产苗300～400株。留土萌发。幼苗形态见图1。也可扦插繁殖。

（黄鹏成）

假　连　翘

Duranta repens L.

（马鞭草科　Verbenaceae）

生长习性、分布和用途　假连翘属约36种，我国引入本文描述的1种。常绿灌木，主干不直立，枝有刺，高达3m。萌芽力强，颇耐修剪。喜湿润的酸性土，能耐水湿，耐轻霜，忌冰雪。原产美洲热带。我国华南地区多栽培，常逸为野生。花果期长，颇美观，为重要的绿篱植物。果治疟疾，叶捣烂可治痈肿。

开花结实　2～3年生开始结实，5年生以后为正常结实期。结实无大小年现象，除特寒年份嫩枝受冻外，每年结实颇多。花两性。总状花序顶生或腋生，排成圆锥状。花萼顶端有5齿，宿存，果时增大。花冠蓝色或紫蓝色，高脚碟状，不相等的5裂片向外开展。雄蕊4，2长2短，着生于冠筒中部，内藏。子房上位，由4个2室的心皮组成8室，每室胚珠1。花柱短，内藏。据广西南宁1986～1988年观察，从4月中旬～12月中旬均连续有花开，其中5～9月

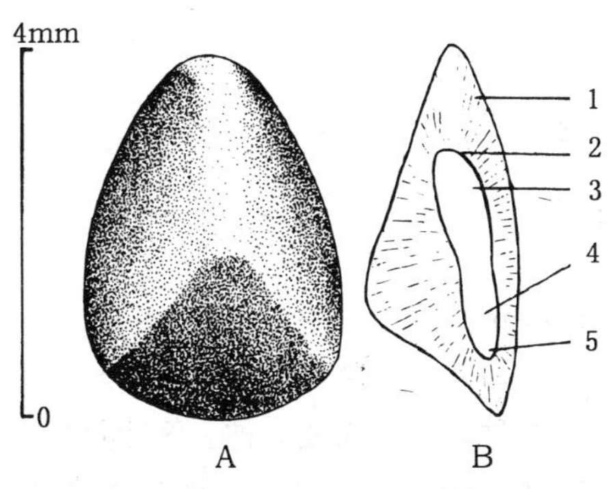

图 1　假连翘小核外形（A）及其纵切面（B）
1. 内果皮　2. 种皮　3. 子叶　4. 胚轴　5. 胚根
（黄应钦绘）

为盛花期。全年均有果实成熟，6～11 月为成熟盛期。

核果，未熟时青色，成熟后橘红色或金黄色，有光泽，近球形或倒卵形，顶端喙尖，高 1～1.2cm，径 0.6～1cm。果实完全包藏在扩大的花萼内，成熟后在树上存留约 2 个月。每果有小核 4（5）。小核近扁三角状，浅褐色，长 2.8～4mm，宽 2.4～2.8mm，厚 2～2.2mm。内果皮厚而坚硬。每核 2 室，每室 1 种子。种子无胚乳，胚伸直（图 1）。

果实的采收调制和种子贮藏　全年都有果实可采，但最好是在果实成熟盛期采摘。采得的果实可堆沤 4～6 天，软熟后装入竹筐内，置水中捣烂搓擦，淘去果肉等杂质。所得小核用作播种材料，通称种子。鲜果出籽率约 16%。净度 92%～98%。含水量 18%～25%。千粒重 7.5（6～9）g，每千克有种子 13（11～16）万粒。种子不宜过分干燥，忌日晒或半月以上的裸露陈放。运输宜以果实形态装运，运抵后再行调制。果实的采收期特长，一般都是随采随播，不作贮藏。如需贮藏种子，宜混以湿沙，贮藏期为半年左右。

发芽和播种　种子无休眠现象。发芽时日均温需在 20℃以上。1987 年 8 月 19 日，广西林业科学研究所在室内培养皿中用当年新采种子作发芽测定。当时日均温 27℃。基质为滤纸，光照为室内自然光。播后 11 天（8 月 30 日）开始发芽，至 10 月 6 日终止，无明显的发芽盛期，发芽率为 45%。出土萌发。胚根萌发后约 4 天子叶出土，再过 6 天发出初生叶。种子发芽及幼苗生长情况见图 2。

在过筛的细土上均匀撒播，每平方米播种 3～5g，用细孔筛稍加覆土。幼苗木质化后移至大田培育。也可扦插育苗。一般培育 1 年生苗出圃。

（朱积余）

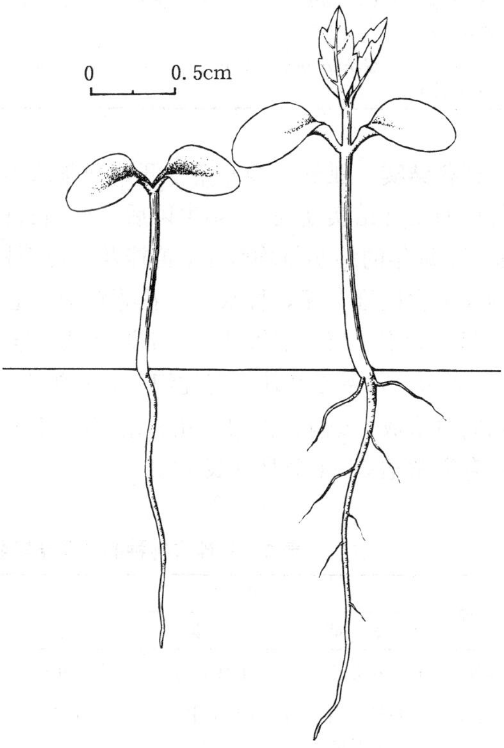

0　0.5cm

图 2　假连翘种子萌发后第 6、12 天幼苗生长情况
（黄应钦绘）

石　梓　属
Gmelina L.

（马鞭草科　Verbenaceae）

生长习性、分布和用途　石梓属约35种，我国有7种，本文描述3种。落叶乔木。喜肥沃深厚湿润的酸性土。材质优良。云南石梓的材质与世界驰名的柚木近似。这3个种的名称、分布及用途见表1。在《中国植物红皮书》中，云南石梓被列为稀有种，苦梓被列为渐危种。

表1　石梓属树种的名称、生长、分布和用途

中　名	学　名	树高(m)	胸径(cm)	分　布	用　途
云南石梓（石梓）	*G. arborea* Roxb.	35	100	滇、藏南部。华南栽培。东南亚及印度	珍贵用材
亚洲石梓（刺石梓）	*G. asiatica* L.	6～8	10～15	粤、桂、黔。南亚、东南亚	围篱、小径材
苦梓（海南石梓）	*G. hainanensis* Oliv.	15	50	琼、粤、桂、赣	珍贵用材

开花结实　亚洲石梓6年生开始开花结实，其余2种7～8年生开始开花结实。云南石梓和苦梓的正常结实期在15年生以后，亚洲石梓在10年生以后。云南石梓的结实大小年现象不显著，每年的结实量均较丰，苦梓及亚洲石梓结实的丰年间隔期为1年。花两性。聚伞花序排列成顶生圆锥花序，稀腋生。花萼钟形，宿存，5裂，具腺点，或近平截（亚洲石梓）。花冠略呈二唇形，筒下部管状，上部漏斗状。雄蕊4，2长2短。花药2室，2分叉。子房上位，球形，4室，每室1胚珠。花柱细弱，柱头不相等2裂。据广西南宁观察，花芽与叶芽均于4月初同时形成。云南石梓的开花和结实均较整齐。苦梓和亚洲苦梓的开花结实期很不整齐，花果共存期常达3～4个月（表2）。

表2　石梓属树种的开花结实物候（广西南宁，1983～1987）

树　种	开花			果实成熟		果实脱落
	始　期	盛　期	末　期	始　期	盛　期	
云南石梓	4月中旬末	4月下旬	4月下旬末	5月中旬	5月下旬	5月下旬～6月下旬
亚洲石梓	主花期4月下旬（6～9月陆续有花开）	5月上中旬	5月下旬	6月上旬	7月中旬(8～10月尚有零星果熟)	6月中旬～10月上旬(6月下旬为主要落果期)
苦梓	主花期5月中旬（7～10月陆续有花开）	6月上旬	6月下旬	8月上旬	8月下旬～9月下旬（10月～翌年2月尚有零星果熟）	8月上旬～11月下旬

核果，肉质，倒卵形。外果皮光滑，中果皮肉质，内果皮骨质，内含种子1～4粒。果核基部的一端有一马耳形的长斜缺口。种子无胚乳，子叶椭圆形。成熟果实和果核的形态特征见表3、图1。

表3　石梓属树种果实和果核的形态特征

树　种	未熟果颜色	成熟果实			果　核	
		形　状	大小（cm）	颜　色	大小（cm）	颜色
云南石梓	青绿色	倒卵状圆形，顶端有1小浅凹	长3～3.5 径2.3～3	黄色，光滑无毛	长1.3～2.5 径0.7～1.3	浅褐色
亚洲石梓	黄绿色	倒卵形，顶端有1黑点，基部花萼高5～6mm	长2.5～3.2 径2～2.5	淡黄至黄色，有光泽	长1.2～1.8 径0.6～1.0	浅褐色
苦梓	浅绿色	椭圆形，顶端有1凹环，基部花萼高1.6～2cm，黑褐色	长2.5～3.2 径1.9～2.3	灰黄绿色，表面被短白毛	长1.0～1.6 径0.5～0.8	褐色

果实的采收调制和种子贮藏　云南石梓的成熟期颇整齐，当果实大熟时，用竹竿敲打或摇动树枝，震落果实，在树下捡拾。亚洲石梓及苦梓的果实成熟期很不一致，通常在果实大熟时轻摇树枝，使已经成熟的果实落地，尚未成熟的果实和花则不被震落。采集的果实堆沤数天，肉质果充分软化后置水中用木棒冲击并加搓擦，淘去皮肉等杂质，得出果核，用作播种材料，通称种子。果核的含水量约11%。出籽率和果核的净度、质量等见表4。据魏素梅和李炎香（1989）报道：常温条件下袋装贮存的云南石梓种子，半年后发芽率从72%降至12%，12个月后全部失去发芽能力；在盛有硅胶的干燥器中贮存12个月，以及在5℃的冰箱中贮存6个月的云南石梓则保存得相当完好。5～8月采集的种子，生产上随采随播，一般不作贮藏。9月以后采集的种子，因气温转低，可混湿沙贮藏，至次年3月中旬播种。

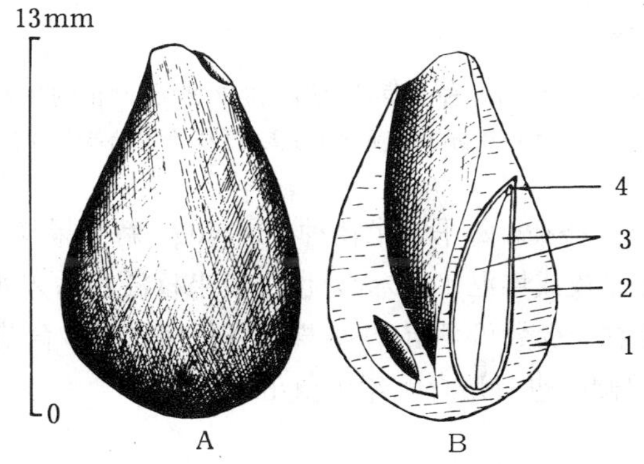

图1　云南石梓果核外形（A）及其纵切面（B）
1. 内果皮　2. 种皮　3. 子叶　4. 胚根
（黄应钦绘）

表4　石梓属树种果核出籽率及其净度和质量

树　种	出籽率（%）	净度（%）	千粒重（g）		每千克纯净果核粒数	
			一般	变动范围	一般	变动范围
云南石梓	5～6	98	500	400～700	2 000	1 430～2 500
亚洲石梓	5～7	98	420	350～500	2 380	2 000～2 850
苦梓	5.5～8	97	400	350～450	2 500	2 200～2 850

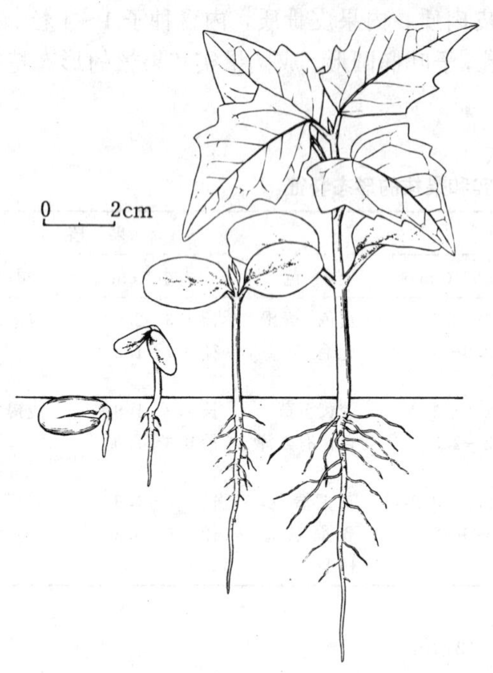

图 2 云南石梓种子萌发后第1、3、8、20天幼苗的生长情况
（黄应钦仿《热带亚热带主要树种采种育苗技术》）

发芽和播种 种子无休眠习性。发芽时日均温宜在20℃以上。夏季随采随播的种子，播后8～10天开始萌动。1989年8月广西林业科学研究所用3种石梓的新鲜果核，在相同的气温条件下，在室外沙床作发芽测定，播种时日均温为29℃，3个种的发芽情况很接近：播后第9天开始萌发，第12～18天为发芽盛期；从发芽之日起算，9天中发芽60%；从播种至发芽终期以22天计，发芽率为72%，场圃发芽率一般为65%。魏素梅和李炎香(1989)、李炎香和谭天泳等(1990)报道过云南石梓播前的催芽技术。据他们介绍，果核在石灰浆中浸沤7天，或者反复浸种摊晒7天，播后7天发芽率可达80%以上；未经处理的则发芽缓慢，发芽率也不高：17天的发芽率只有74%。出土萌发。萌发后约3天子叶出土，再过4～5天发出初生叶。云南石梓的萌发和幼苗早期的生长情况见图2。

条播。每平方米播种38～50g。新鲜果核播前可不作处理。经过贮藏的果核，播前可浸水1昼夜或堆沤1～2天，也有将果核浸水后置水泥地曝晒，促进种子发芽的。每果核一般发苗1株，少数可发苗2～3株，可分株移植。育苗期一般为半年左右，头年夏季播种，次年3月出圃，也有培育1年生苗出圃的。

（朱积余）

黄毛豆腐柴

Premna fulva Craib

（马鞭草科 Verbenaceae）

生长习性、分布和用途 豆腐柴属约200种，我国约产44种，本文描述1种。常绿灌木至乔木，高6～12m。在肥沃湿润的酸性土或石灰岩山麓生长较快，干旱贫瘠土壤生长不良。分布于黔、桂、滇。泰国、越南、老挝也有分布。能改良土壤。根、茎药用，治跌打刀伤和风湿骨痛，俗有“战骨”之称。

开花结实 3～4年生开始开花结实，8年生以后为正常结实期，大小年现象不明显。花两性，聚伞花序伞房状，顶生，长2.5～6（10）cm，宽4～9（17）cm。花萼钟状，近2唇形，5裂，花后略增大，宿存。花冠绿白色，4裂，近2唇形，上唇1裂，下唇3裂。雄蕊4，2强，花药褐色。子房上位，4 室，每室 1 胚珠，柱头 2 裂。据 1987～1988 年广西南宁观察，花蕾期 3 月下旬，始花期 4 月中旬，盛花期 4 月下旬，5 月上旬～6 月上旬尚有少量开花；6 月下旬果实开始成熟。同一果穗中的果实成熟程度参差不齐，即使在果熟盛期，果穗上尚有少量幼果。果实成熟后约 15 天逐渐脱落，落果末期为 8 月上旬。

浆果状核果，倒梨形，萼宿存，成熟时由青紫色转为紫黑色，果长 6～7mm，径 4～5mm。每果有果核 1 枚，梨形。长 4～5mm，径约 3mm。核壳骨质，表面凹凸不平。种子长圆形，种皮薄，无胚乳，子叶扁平。

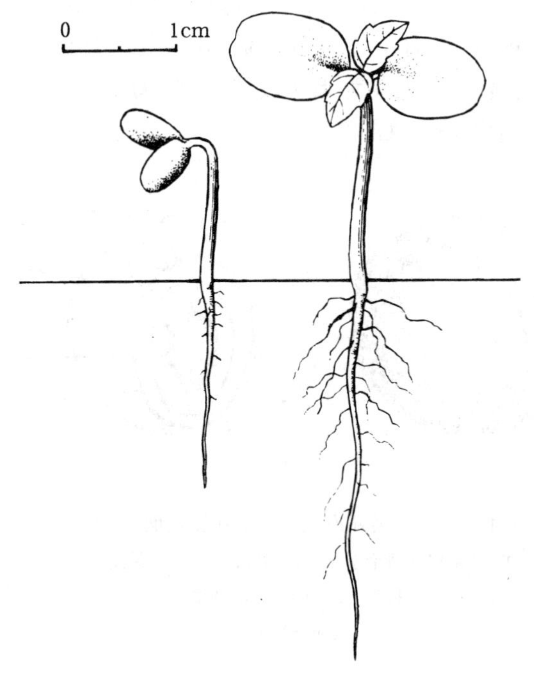

图 1　黄毛豆腐柴种子萌发后第 4、10 天幼苗的生长情况
（黄应钦绘）

果实的采收调制和种子贮藏　果熟盛期用枝剪或直接采摘果穗。摘下的果实堆放 3～7 天，如果成熟度高，也可不经堆放而直接置水中搓擦，淘洗后得出果核，用作播种材料，通称种子。鲜果出籽率为 25%～30%。净度为 92%～98%。千粒重 31(25～35)g，每千克有种子（核)32 200 (28 600～40 000) 粒。调制晾干时的含水量为 20%～25%。种子不宜日晒，宜随采随播。贮藏时需混以湿沙，贮藏期约为 5～7 个月。

发芽和播种　种子无明显的休眠习性。1987 年 7 月 17 日，广西林业科学研究所在室外沙床上用新采种子作发芽测定，播种时日均温约为 28℃。播种后 16 天开始发芽，8 月 26 日发芽结束，发芽率为 45%，无明显的发芽盛期。出土萌发，子叶出土后 3～4 天初生叶出现(图 1)。

条播。每平方米播种 3.5～5g，覆土约 5mm。也可用撒播，苗木转木质化时移植。通常培育 1 年生苗出圃。

（曾　玲）

柚　木

Tectona grandis L. f.

（马鞭草科　Verbenaceae）

生长习性、分布和用途　柚木属 3 种，本文描述我国引入栽培的 1 种。落叶大乔木，高可达 50m 以上，胸径 2.5m。对水肥条件要求高，喜肥沃湿润的沙壤土，粘重土及干旱瘠瘦地均不宜。能耐轻霜及短期 0℃左右的低温，但忌严重霜冻。原产印度、泰国、缅甸、马来西亚、印度尼西亚，现已广泛引种至亚洲、非洲、美洲各国热带地区。我国滇、桂、琼、桂、粤、闽、台引种栽培。柚木的木材结构细致，花纹美观，轻重适中，坚软适度，富弹性，耐磨损，耐腐蚀，耐水湿，抗虫、抗火等性能均极好，为世界著名的最优用材之一，用于船舶、枪托、车辆、建筑及高级家具，木屑浸水可治皮肤病或煎水治咳嗽。

开花结实 6～8年生开始开花结实，正常结实期在20年生以后。每年大量开花，但花期对天气条件敏感，结实大小年现象明显。花两性。二歧聚伞花序组成圆锥花序，顶生，长25～40cm。花萼钟状，5～6短裂，具白色星状绒毛，宿存，果时增大。花冠白色，有芳香，上部5～6裂。雄蕊5～6，着生于冠筒上，伸出花冠外。子房上位，4室，每室1胚珠。花柱线形，柱头顶端2浅裂。据广西南宁1978～1984年观察，7月下旬为始花期，8月上旬为盛花期，8月中下旬为末花期，翌年3月下旬果实开始成熟，4月上旬为果实成熟盛期。曾庆波和丁美华等（1989）报道过柚木在海南尖峰岭的开花结实物候进程。

核果，包藏于扩大的花萼内，球形或近球形，径1.2～1.8cm。外果皮密被毡状细毛，未成熟时黄绿色，成熟时茶褐色。内果皮骨质。成熟果在树上约保留1个月，4月中下旬～5月上旬陆续脱落。每果有1个由内果皮形成的含4室之核，常有2～3室的种子不发育。核扁椭圆形，浅褐色，被毡状绒毛，长4.5～6.5mm，宽3～4mm，厚1.5～2mm。种子矩圆形，种皮薄，无胚乳，胚直。柚木核果的形态见图1。

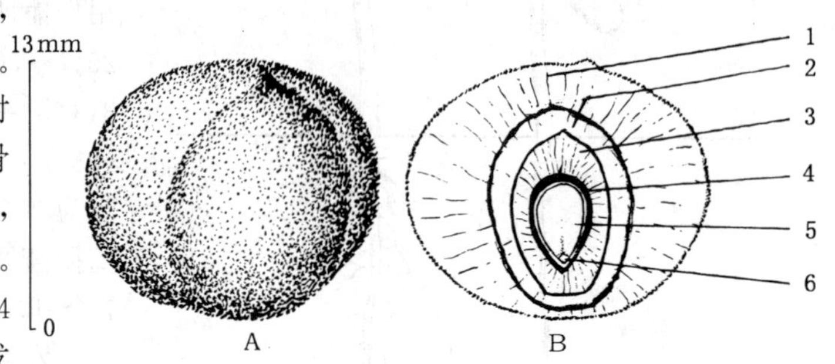

图1 柚木核果外形（A）及其纵切面（B）

1. 外果皮（毡毛层） 2. 中果皮 3. 内果皮

4. 种皮 5. 子叶 6. 胚根

（黄应钦绘）

果实的采收调制和种子贮藏 果实成熟盛期用竹竿敲打或用高枝剪采集。采得的果实仍包藏在扩大的宿存花萼囊中，反复曝晒后搓去花萼囊，所得的核果即为播种材料，通称种子。净度一般为70%。干时千粒重为910（700～1 125）g。每千克有果1 100（900～1 400）粒。运输时可用布袋包装。贮藏时将含水量降到10%左右，用干燥器或瓦罐密封贮藏，发芽力可保持1年左右。

发芽和播种 果核骨质坚硬，被有木栓状绒毛，发芽率低而且发芽迟缓。广西林业科学研究所1982年用前1年未经处理的种子进行发芽试验，发芽基质为细沙，8月29日播种，播后8天开始发芽，发芽盛期不明显，发芽率为28%。另据广西夏石林业试验站和海南尖峰岭热带林业研究所试验，播种前用石灰沤种可显著提高发芽率。他们的方法是，把生石灰放入水中溶解，搅成糊状。将种子放入灰浆内，搅拌混匀并覆盖稻草防止干燥。一星期后绒毛变黑，取出种子洗去石灰，放入臼内轻轻捣去果上绒毛，洗净后置沙床催芽。用此法处理的种子，发芽率可达80%。湿沙层积催芽30～50天，使果核吸足水分，也可提高发芽率。

柚木发芽困难，一般认为是因为果皮具有不透水性，胚具有生理后熟，或者是存在发芽抑制物质。宋学之和刘文明等（1991）通过研究否定了这3种推测，认为主要原因是中果皮和内果皮具有巨大的机械束缚力。据他们观察，柚木果实有绒毛型和光皮型2种类型。只要发芽条件适宜，绒毛型的柚木不经处理也能得到80%以上的发芽率。光皮型的则需要利用预处理措施或播后的管理手段消除果皮对萌发的束缚。可以采取的办法有：反复若干天（例如7天）的浸种和摊晒交替处理，直至内果皮显出裂纹；用石灰浆液浸沤，待中果皮变软后利用机械力量除去中果皮；也可以在沙床上"露播"，即果蒂一端朝下，轻压入沙，使另一端微微露出沙面，沙面干燥时少量淋水，但应防止水分过多影响通气。国外有报道说，柚木播后不覆土的发芽率为51.2%，覆土1.5cm和3.0cm的发芽率分别为16.2%和5.8%，覆土4.5cm则发芽率极低。对于室内发芽，宋学之和刘文明等人（1991）建议的条件是40℃/30℃的昼夜变温，每天给以10～15小时5 000lx以上的光照。

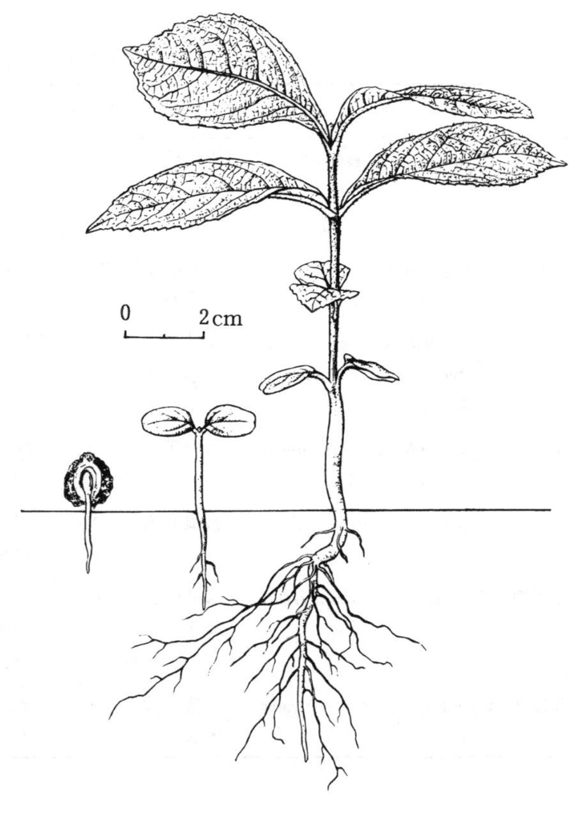

图2　柚木种子萌发后第5、10、35天幼苗生长情况
（黄应钦仿《主要树木种苗图谱》）

播种期为3～5月，每平方米密播500g。不盖草。出土萌发。子叶出土后10天生出初生叶（图2）。2～3个月后幼苗长出5对初生叶时移至圃地培育。至翌年春，苗高0.7m，地径1.5cm以上时出圃定植。

（朱积余）

牡　荆　属
Vitex L.

（马鞭草科　Verbenaceae）

生长习性、分布和用途　本属约250种，我国有14种7变种及3变型，本文描述4种3变种，它们的名称、生长习性、成年时树高和用途见表1。荆条较耐旱、耐瘠薄、耐寒。灰毛牡荆为山地雨林树种，单叶蔓荆为海边沙滩常见植物。

开花结实　灰毛牡荆8～10年生、荆条3年生、山牡荆7～8年生、越南牡荆5～6年生开始开花结实，未见大小年现象。花两性，白色、浅蓝色、浅蓝紫色或淡黄色。聚伞花序组成圆锥状、伞房状或近穗状花序，顶生或腋生。萼钟状，宿存。花冠略长于萼，二唇形，上唇2裂，下唇3裂，中裂较大。雄蕊4。子房2～4室，每室1～2胚珠；花柱丝状，柱头2裂。

表 1 牡荆属树种名称、生长、分布和用途

中 名	学 名	生长习性	树高(m)	分 布	用 途	供 稿
灰毛牡荆	*V. canescens* Kurz	常绿乔木	8～20	赣、鄂、湘、粤、琼、桂、西南。印度、缅甸、泰国、老挝、越南、马来西亚	材用、药用	616
黄荆	*V. negundo* L.	落叶灌木或小乔木	2～4	长江以南，北达秦岭淮河。东南亚、非洲东部、玻利维亚	药用、蜜源、香料	1003
牡荆	*V. negundo* var. *cannabifolia* (Sieb. et Zucc.) Hand.-Mazz.	落叶灌木或小乔木	2～4	华东、中南、西南、冀。日本	药用、香料	1003
荆条	*V. negundo* var. *heterophylla* (Franch.) Rehd.	落叶灌木	4	陕、甘、辽、冀、晋、豫、鲁、苏、皖、赣、湘、川、黔。日本	药用、蜜源、香料	403
山牡荆	*V. quinata* (Lour.) Will.	常绿乔木	10	浙、赣、闽、湘、粤、琼、桂、台。日本、印度、马来西亚、菲律宾	材用	605
单叶蔓荆	*V. trifolia* var. *simplicifolia* Cham.	落叶匍匐灌木	1.5	辽、冀、鲁、华东、粤、台。日本、印度、缅甸、泰国、越南、马来西亚、澳大利亚、新西兰	药用、香料、防沙、固堤	1003
越南牡荆	*V. tripinnata* (Lour.)Merr	半落叶乔木或灌木	4～8	琼。缅甸、越南、柬埔寨、马来西亚	材用	605

核果球形、卵形或倒卵形，棕褐色、紫黑色或黑色，基部包有宿萼或果柄。果实成熟时不裂。越南牡荆外果皮与中果皮贴合，肉质。由内果皮发育成的核壳木质或骨质。荆条和灰毛牡荆的核壳顶端有一凹陷孔并具膜质盖；越南牡荆的顶端开裂如鸟喙状。果核内含种子 1～4 枚。黄荆的种子长倒卵形，长 1.8～2.2mm，宽 1.1mm，种皮厚膜质，棕灰色；胚乳层薄或近于无；子叶厚。牡荆属具无限花序，从下到上陆续开花结实，花期果期长，果实成熟期不一致。牡荆核果及种子的外形和内部结构见图 1。牡荆属树种的开花结实物候见表 2。

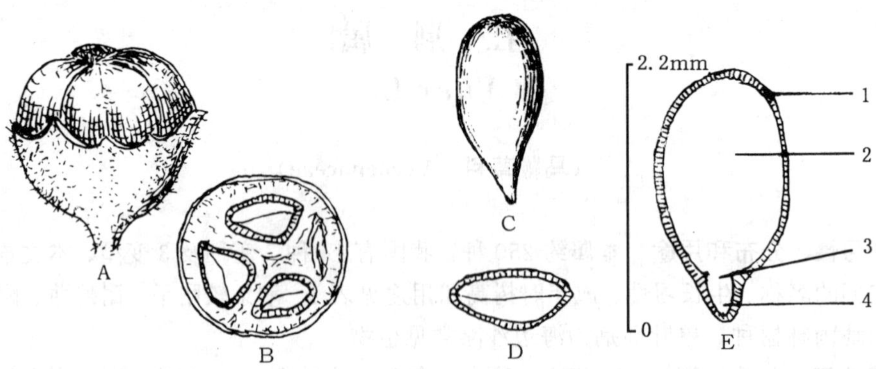

图 1 牡荆的核果外形（A）及其横切面（B），种子外形（C）及其横切面（D）和纵切面（E）

1. 种皮 2. 子叶 3. 胚轴 4. 胚根

（史渭清绘）

表 2　牡荆属树种的开花结实物候

树　种	开始结实树龄（年）	观察地点	观察年份（年）	开花期			果实成熟期		果实散落期
				始　期	盛　期	末　期	始　期	盛　期	
灰毛牡荆	8～10	云南南岛河	1987、1989	5 月	5 月中旬	6 月中下旬	9 月	10 月	—
黄荆	—	江苏南京	1979、1988	6 月上旬	7 月中旬	7 月下旬	8～9 月	9 月下旬	10 月下旬～11 月
牡荆	—	江苏南京	1988	6 月	7 月中旬	8 月	8～9 月	9 月下旬	10～11 月
荆条	3	北京	1966、1968	6 月上旬	6 月下旬	8 月上旬	9 月中旬	10 月	10 月下旬～11 月上旬
山牡荆	7～8	广西南宁	1987、1988	5 月中旬	5 月下旬	6 月上旬	9 月上旬	9 月中旬	9 月下旬～11 月中旬
单叶蔓荆	—	江苏南京	1979	7 月	8 月上旬	—	9 月	10 月	10 月下旬～11 月
越南牡荆	5～6	广西南宁	1987、1988	5 月上旬	5 月中旬	5 月下旬	7 月中下旬	8 月上旬	8 月下旬～10 月上旬

果实的采收调制和种子贮藏　同一植株的果实成熟时间也很不一致，应根据果实变色、核壳变硬等标志分期采收，或在大部分果实成熟时整穗剪下或捋下晒干。有时就以这种带有宿萼和短柄的干燥核果作为播种材料，或搓去果皮果肉，洗净晾干，以取得的果核作为播种材料，通常都称种子。可能是由于授粉不良，一般每个果实都有 1～3 个空室或全为空室，因此必须加大采种量以满足需要，并尽可能筛除没有充分发育的小粒果实。充分干燥的核果或果核用纸袋、布袋贮存在通风、干燥、阴凉处待播。北京地区在全年温度波动不大（10～24℃）的地下室用纸袋或布袋盛装的荆条种子，存放 1 年后发芽率仍有 30%。本文描述的 7 个树种的种子特征见表 3。

表 3　牡荆属树种核果和果核的形态及其质量

树　种	核果		果核		出籽率（%）	千粒重（g）	每千克粒数（万粒）
	形　态	大小(mm)	形　态	大小(mm)			
灰毛牡荆	近球形或长椭圆状倒卵形，紫红色，有光泽	长 10 径 7	—	—	55～65	（核） 90～220	（核） 0.4～1.1
黄荆	近球形，棕褐色，有光泽。宿萼与果近等长	长 2～2.5 径 2～2.5	—	—	—	（果） 6.6～8.4	（果） 12～15

（续）

树　种	核　果		果　核		出籽率	千粒重	每千克粒数
	形　态	大小(mm)	形　态	大小(mm)	(%)	(g)	(万粒)
牡荆	近球形，棕褐色或近黑色	长 2～2.8 径 2.3～2.8	—	—	—	(果) 6.2～6.7	(果) 15～16
荆条	卵球形或近球形，黑褐色	长 3～3.5 径 1.7～2.4	—	—	—	(果) 9～11.2	(果) 8.9～11
山牡荆	球形或倒卵形，顶端近截形，黑色。宿萼圆盘状	长 7～10 径 6～9	卵形，黄褐色，表面有不规则孔或槽沟，顶端有小缺口	长 5～6.5 径 4.5～6.5	32	(核) 65～85	(核) 1.2～1.5
单叶蔓荆	近球形，黑色，果皮具腺点，有芳香。宿萼被灰白色绒毛	径 5	—	—	—	(果)26.4	(果)3.8
越南牡荆	近球形，浅黄色，干后黑色	长 12～15 径 12～16	扁椭圆形或倒卵形，浅褐色，顶端有鸟喙状缺口	长 10～13 径 7～8.5	20	(核) 250～320	(核) 0.3～0.4

发芽和播种　牡荆属种子无明显的休眠习性，但空粒多，发芽率较低。在室内作发芽测定时，黄荆种子的适宜温度为 30℃，12 天的发芽率约 40%；温度低到 10℃以下发芽就会受到抑制。单叶蔓荆发芽较慢，有 1 份种子在 15～20℃的温度下历时 100 天发芽率为 62%。据孙秀琴和田树霞（1988）报道，荆条种子室内发芽测定时无需事先浸种，特别是不能用高温水烫种；发芽的适宜条件是恒温 25℃，每天应有 8 小时给予 1 000～1 200lx 的光照，在这样的条件下 3 天便可开始萌发。

秋播，或湿沙层积后春播。据云南林业科学研究所试验，1988 年秋采集调制后的灰毛牡荆种子于 12 月 11 日播种，翌年 4 月 11 日开始发芽出土，5 月 30 日发芽结束，发芽率为 20%，没有明显的发芽盛期；经过层积的种子 1989 年 3 月 10 日播种后 20 天开始萌发，至 4 月底发芽率约为 40%。据广西林业科学研究所试验，混沙贮藏约半年的山牡荆种子于 4 月播于室外沙床，当时日均温为 22～24℃，播后 13 天开始发芽，又 8 天为发芽盛期，52 天的发芽率为 50%。1987 年 9 月，广西林业科学研究所将当年采集未经处理的越南牡荆种子播于室外沙床，播后 16 天开始发芽，再经 29 天发芽率为 55%，无明显的发芽高峰。灰毛牡荆每平方米播种 28～33g。荆条每平方米播种 15g。覆土约 1cm。

出土萌发。第 1、第 2 对初生叶为单叶。牡荆种子的萌发及幼苗初期生长情况见图 2。

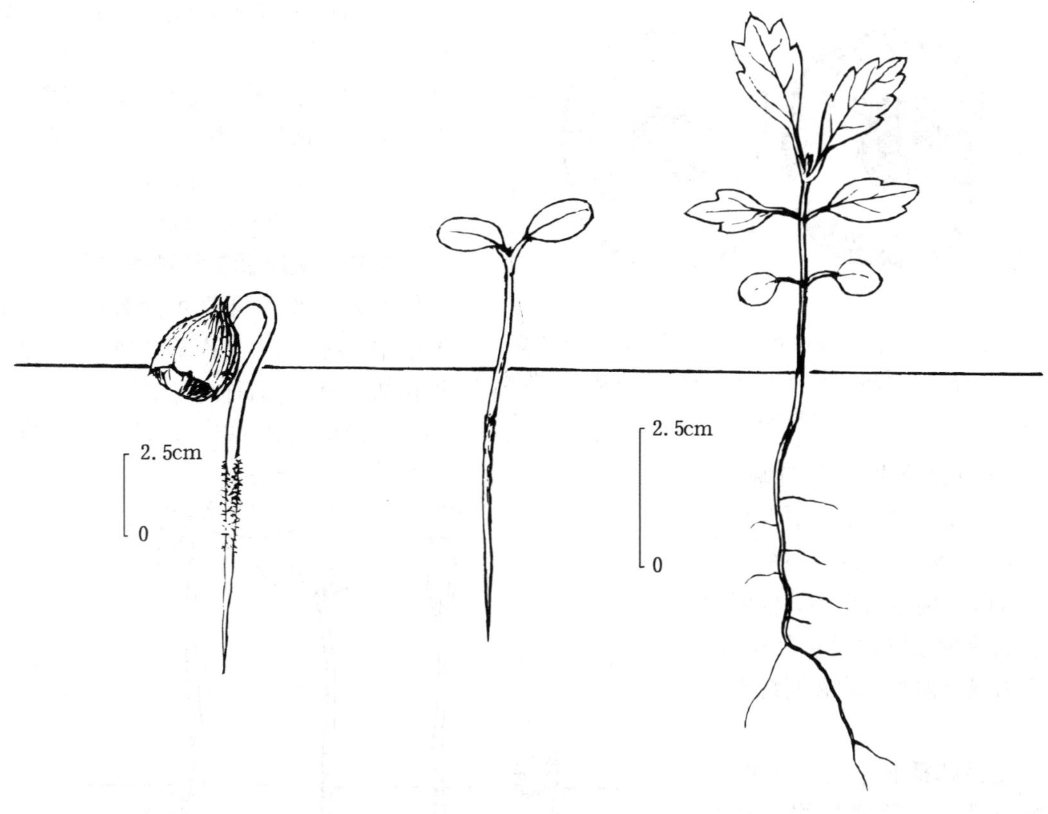

图 2　牡荆种子萌发后第 3、18、66 天的幼苗生长情况
（史渭清绘）

（何泽瑛）

樟叶木防己

Cocculus laurifolius DC.

（防己科　**Menispermaceae**）

生长习性、分布和用途　木防己属约 10 种，我国约产 3 种，本文描述 1 种。常绿直立灌木，高达 3m，部分枝条下垂，呈攀援状。喜温暖湿润的气候环境，能耐－10℃左右的低温，适生于湿润疏松的酸性土或钙质土，干旱地或裸露地生长不良。分布于闽、台、粤、桂、湘、黔、滇。印度、东南亚、日本也有分布。根供药用，散寒健胃，理气止痛。

开花结实　5～7 年生开始结实，12 年生以后进入正常结实期。花单性，雌雄异株。聚伞状圆锥花序腋生，少单生。雄花萼片、花瓣 6。雄蕊 6，长约 1mm。雌花萼片、花瓣与雄花相似，退化雄蕊 6，雌蕊 3，分离。据广西南宁 1987 年观察，雌雄花同期开放，5 月中旬为始花期，6 月上旬为盛花期，6 月下旬为末花期；9 月上旬果实开始成熟，9 月下旬为果熟盛期。核

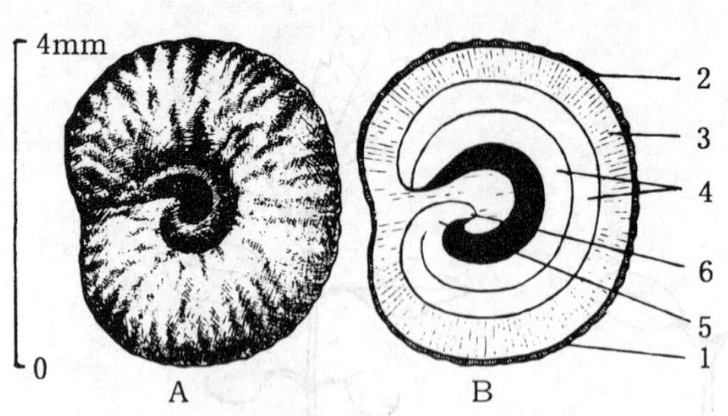

图 1　樟叶木防己果核外形（A）及其纵切面（B）
1. 内果皮　2. 种皮　3. 胚乳　4. 子叶　5. 胚轴　6. 胚根
（黄应钦绘）

果，近两侧压扁圆形，未熟时青色，成熟时蓝褐色，长约 5mm。内果皮两侧压扁开口，中间具穿孔的隔膜。每果有果核 1 粒，卷缩成球形，中间凹入，灰色，径 3.5～4mm，厚 2.5～3mm。有胚乳，胚弯曲，子叶半圆柱状（图1）。

果实的采收调制和种子贮藏　果实成熟期人工采摘。采回堆沤数日，软熟后置水中搓擦，淘去果肉等杂质，所得果核用作播种材料，通称种子。鲜果出籽率约 19%，净度可达 97% 。种子含水量约 12%，不宜过分失水，忌日晒。千粒重 15（12～18）g，每千克有种子（核）67 000（55 000～83 000）粒。可以直接运输果实，也可调制后混拌少量湿沙后包装运输。贮藏需混以湿沙，贮藏期约为半年。

发芽和播种　种子无明显休眠习性。发芽时日均温宜在 18℃以上。1987 年 10 月 15 日，广西林业科学研究所用当年采集的新鲜种子，在室外沙床播种作过发芽测定，播前对种子未作任何处理：播后 19 天（11 月 3 日）开始发芽，11 月 8～15 日为发芽盛期，21 日发芽终止；从发芽之日起算的 12 天中发芽 62%；从播种之日至发芽终期，37 天的发芽率为 75%。出土萌发。胚根萌发后约 6 天子叶出土，21 天发出初生叶。种子发芽及幼苗生长情况见图 2。

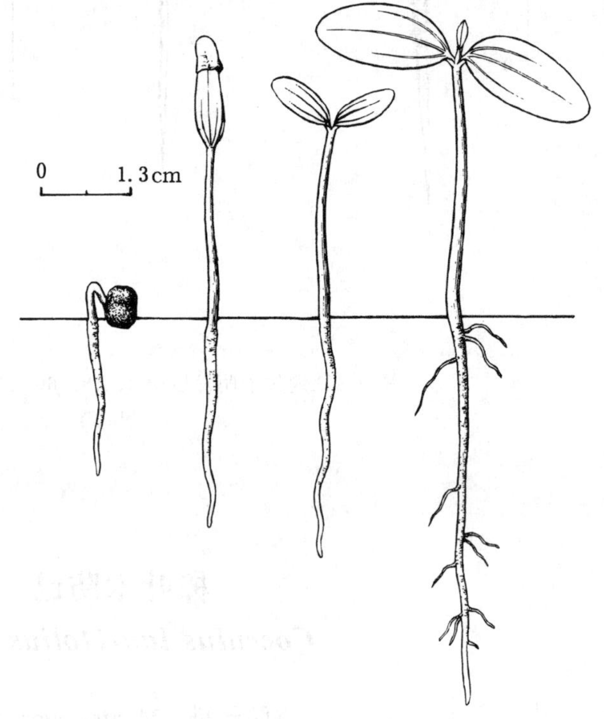

图 2　樟叶木防己种子萌发后第 6、10、15、21 天幼苗的生长情况
（黄应钦绘）

撒播。将种子密播于细致整地的苗床或沙盘内。每平方米播种 2～3g。苗木转木质化时移植。也可条播。覆土 2mm。1 年生苗出圃。

（朱积余）

南　天　竹

Nandina domestica Thunb.

（南天竹科　Nandinaceae）

生长习性、分布和用途　本属仅本文描述的 1 种，产我国和日本。常绿灌木，高 1～2m，生于山地疏林下和灌木丛中，为钙质土的指示植物。性喜温暖多湿和通风良好的环境。分布于苏、浙、皖、赣、鄂、川、陕、桂。全株供药用，各地庭园多栽培用于观赏。

开花结实　1～2 年生的小苗移栽后 4～5 年开始开花，5～6 年开始结实，树高 1m 以上即可大量结实。两性花，白色。圆锥花序顶生，长 20～25cm。萼片与花瓣相似，多轮，每轮 3 片。雄蕊 6，离生。子房 1 室，有胚珠 2。浆果，球形，果径 5～6mm，鲜红色或偶有黄白色的品种。果内含种子 1～2 枚。种子浅钵状或盘状。种皮薄，与胚乳紧贴，浅棕色，干燥后呈灰褐色。胚乳丰富，灰白色，半透明，质地坚韧。果实成熟时胚尚未发育完全，长仅约 1mm（图 1）。据在南京地区观察，5 月下旬～6 月初开始开花，6 月上中旬为盛花期，6 月下旬花期结束，幼果形成；9～11 月果实转为红色，种子变硬，预示成熟。

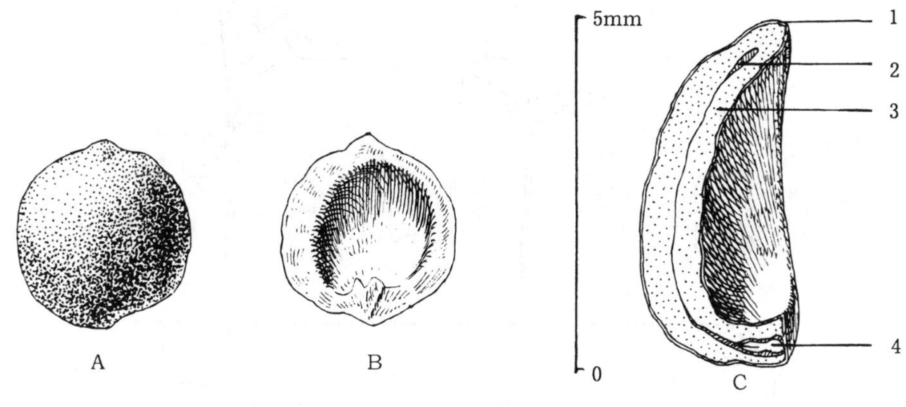

图 1　南天竹种子背面（A）、腹面（B）及其纵切面（C）

1. 种皮　2. 胚腔　3. 胚乳　4. 胚

（陈荣道绘）

果实的采收调制和种子贮藏　果熟后短期不落。如无鸟雀啄食，密集鲜艳的浆果可在树上保留至翌年 1 月，成为冬季庭园点缀。剪插室内观赏，可到 2 月仍不变色。每一果穗含果多至 200～400 粒。一般在 11～12 月初采摘。整果存放，也可搓去果皮果肉取得种子，淘净晾干后在 5～10℃下混湿沙层积，供春季播种。出种率 23％～36％，千粒重 25～43g，每千克含种子 2.3 万～4.0 万粒。生活力可以保存 1 年以上。

发芽和播种　种子有深休眠习性。据本文作者在南京观察，秋季采收的种子到 1 月底胚的长度为 0.6～1.0mm，可以称为胚的形态发育早期。播种后 6 月底胚长 0.8～3.0mm，可以视为形态发育中期。9～10 月胚的长度为 4～5mm，几与种子等长，可以认为此时的胚已经到了形态成熟末期，接下来便可大量萌发。这表明，南天竹种子休眠的主要原因是，胚需要有将近 10 个月的时间在自然温度下完成形态发育，当然，还可能同种子内部激素等物质的转化缓慢有关。目前发芽测定的报道很少。表 1 是 1987～1988 年在南京室外和室内自然温度条件所作的一次发芽测定结果。

表 1 南天竹种子在南京自然温度条件下的发芽测定结果①

播前处理	置床或播种日期（年-月-日）	集中发芽时间（年-月-日）	发芽率（%）
室外层积	1988-02-13	1988-10-10～1988-10-18	97.8
室内带果干藏	1988-04-09	1988-10-31～1988-11-08	92.0

① 采种期：1987 年 12 月 24 日。湿沙，自然变温

可以随采随播，也可层积后春播或室内带果干藏后春播。干藏越冬的种子播前需要浸种。出土萌发（图 2）。下胚轴基部白色而微带绿色，中上部黄绿色。1～2 年生苗出圃。生产上多用分株繁殖。

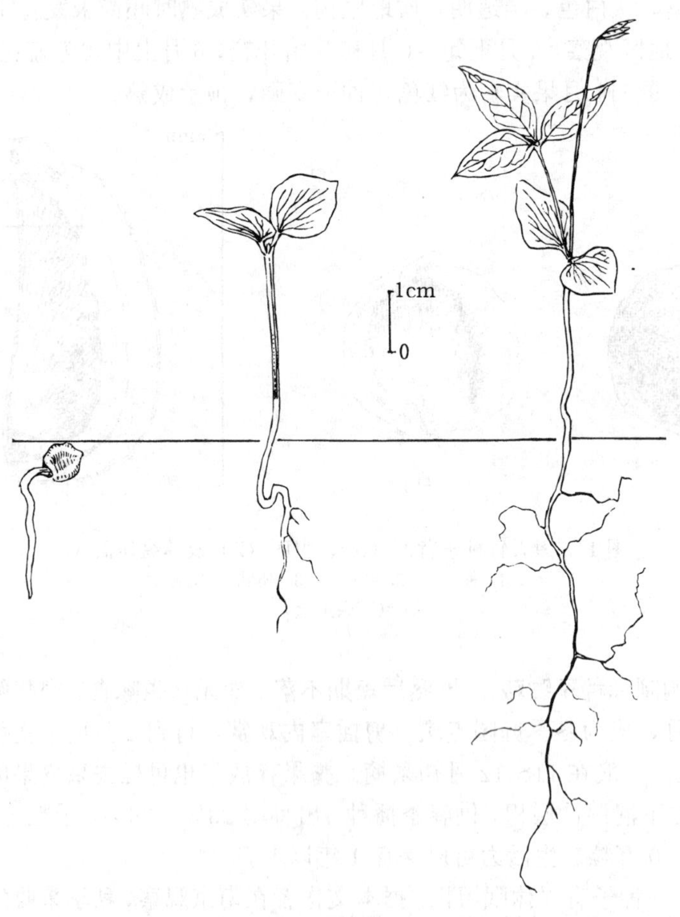

图 2 南天竹种子萌发后 2、10、45 天的幼苗生长情况
（陈荣道绘）

（何泽瑛）

小　檗　属

Berberis L.

（小檗科　Berberidaceae）

生长习性、分布和用途　小檗属约500种，广布于亚洲、欧洲、美洲和非洲。我国约200种。本文描述的4种（表1）都是落叶灌木。枝常具单一或分叉针状刺，为叶之变态。高2～3m。稍喜光，耐寒，耐旱，对土壤要求不严。富含小檗碱，根、茎入药，有消炎效果。果变红后酸甜可食。叶秋季变紫红色，可供观赏。

表1　小檗属树种的名称、生长、分布和用途

中　名	学　名	树高(m)	分　布	用　途	供　稿
大叶小檗	*B. amurensis* Rupr.	1～3	东北、华北、西北东部。俄罗斯、朝鲜半岛、日本	药用,观赏,刺篱	107
细叶小檗	*B. poiretii* Schneid.	2	东北南部、华北。蒙古、俄罗斯、朝鲜半岛	药用	403
刺叶小檗	*B. sibirica* Pall.	1	黑龙江与内蒙古大兴安岭山区。蒙古、俄罗斯西伯利亚	观赏,药用	404
日本小檗	*B. thunbergii* DC.	3	华北。日本也有。我国北方普遍栽培	观赏,刺篱,药用	401

开花结实　3～4年生开始结实。未见明显的结实间隔期。花两性。总状或伞形花序,簇生或单生于叶刺腋部的短枝顶端。花黄色,后叶开放。萼片6,花瓣状。花瓣6,较小。雄蕊6,花药瓣裂。子房上位,1室,内有1至数枚胚珠。浆果,成熟后变为红色或黑色,在树上宿存一段时间,果柄干枯后果实散落(表2)。大叶小檗和日本小檗每个果实有种子2粒,细叶小檗果实中仅1粒种子。种子红褐色,宽倒卵形或椭圆形,长0.4～0.6cm,径0.2～0.3cm。胚乳丰富(图1)。

表2　小檗属树种的开花结实物候

树　种	观察时间和地点	开花		果实成熟			果实散落	
		始　期	末　期	颜　色	始　期	末　期	始　期	末　期
大叶小檗	1963～1980 哈尔滨	5月中旬	6月上旬	红色	8月末	10月中下旬	9月中旬	11月上旬
细叶小檗	1963～1980 河北省	6月	7月	红色	7月	8月	8月	10月

（续）

树 种	观察时间和地点	开花		果实成熟			果实散落	
		始 期	末 期	颜 色	始 期	末 期	始 期	末 期
刺叶小檗	1963～1980 哈尔滨	5月中下旬	6月上旬	—	8月末	10月中旬	9月中旬	10月下旬
	1963～1980 哈尔滨	5月下旬	6月中旬	暗红色	8月末	10月上旬	9月中旬	10月下旬
日本小檗	北京植物园	5月	—	红色	9月中旬	11月上旬	10月上旬	11月下旬

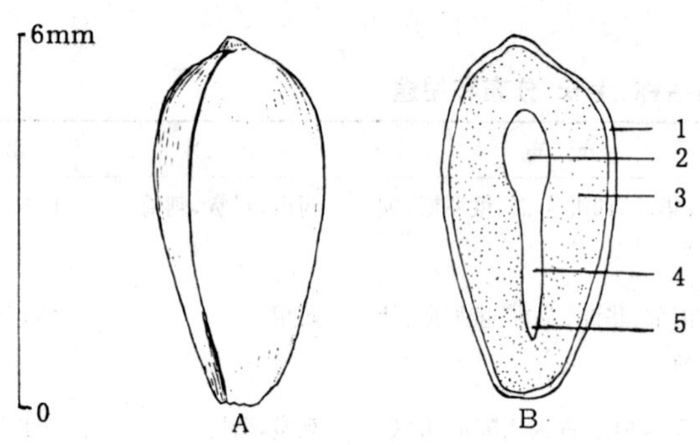

图1 细叶小檗种子外形(A)及其纵切面(B)
1. 种皮 2. 子叶 3. 胚乳 4. 胚轴 5. 胚根
(胡冬梅、林平绘)

果实的采收调制和种子贮藏 8～9月份果实由杏黄色转红时开始采摘。浆果采后堆放，促其腐熟变软，捣碎后搓洗，漂去果皮、果肉，阴干去杂即得纯净种子。出种率40%左右。根据一份样品测得的小檗属各个树种种子品质状况见表3。种子可以干藏，在常温下贮藏1年后发芽能力未见明显降低。

发芽和播种 种子有浅休眠习性，播前温水浸种2～3天后湿沙层积20～50天，待有30%的种子裂嘴即可播种。春季播种，高垅或床面条播，播种量每平方米5～8g，播深1～1.5cm。出土萌发。经过催芽的种子播后7～10天出土，20天左右出现初生叶(图2)。当年苗高15～20cm，2年出圃。7～9月嫩枝扦插成活率很高，亦可压条或分株繁殖。

表3 小檗属树种的出种率、净度、种子质量及发芽率

树 种	出种率(%)	净度(%)	千粒重(g)	每千克纯净种子粒数(万粒)	发芽率(%)
大叶小檗	40	80～90	18～19	5.2～6.3	90
细叶小檗	40	95～98	15.6	6.4	40
刺叶小檗	33	90～95	16～19	5.3～6.3	80
日本小檗	40	95～98	12.3	8.1	34

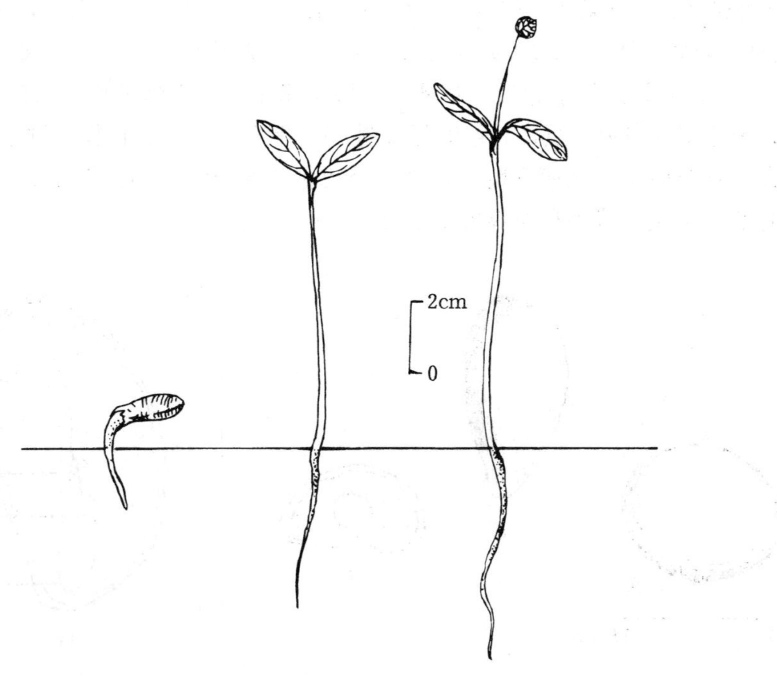

图 2　细叶小檗萌发出土后第 2、6、15 天的幼苗生长情况
（胡冬梅、林平绘）

（宋廷茂）

十大功劳属
Mahonia Nutt.

（小檗科　Berberidaceae）

生长习性、分布和用途　本属约 100 种，分布于亚洲、美洲。我国约有 50 种，主要分布在西南部和南部。本文描述常见栽培的 2 个种(表 1)。常绿灌木，喜阴湿。药圃、庭园多栽培。

表 1　十大功劳属树种的名称、树高、分布和用途

中　名	学　名	树高（m）	分　布	用　途
阔叶十大功劳	*M. bealei* (Fort.) Carr.	4	陕、豫、皖、浙、闽、赣、鄂、湘、桂、黔、川	药用、观赏
十大功劳	*M. fortunei* (Lindl.) Fedde	2	川、桂、鄂、浙	药用、观赏

开花结实　花两性。总状花序直立，数个花序簇生枝顶。花黄色，具小苞片 1。萼片花

瓣状，9 枚排成 3 轮。花瓣 6，较内轮萼片小。雄蕊 6，分离。子房上位，单生，有胚珠 4～5。柱头无柄，盾状。浆果，球形或长圆形，成熟后暗蓝色或蓝黑色，表面光滑，有白粉。果内含种子 1～2（4）粒。种子卵形，基部尖而腹部略平。种皮薄纸质，浅棕褐色，不耐碰擦。胚乳丰富，半透明乳白色，质嫩，包围胚（图 1）。果实成熟时胚长仅及种子的 1/2～1/3，以后逐渐长大，直至萌发。在南京地区，这 2 个树种的开花结实物候期完全不同，阔叶十大功劳为春花夏实，十大功劳为秋花春实（表 2）。

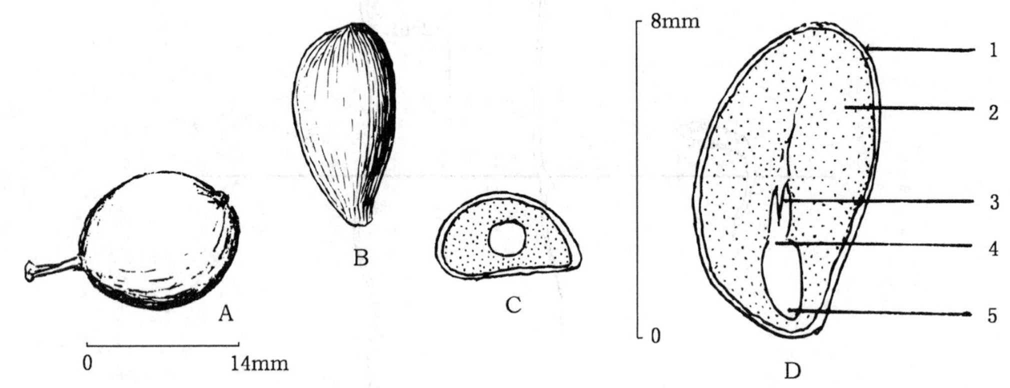

图 1 阔叶十大功劳的浆果外形（A）、种子外形（B）及其横切面（C）和纵切面（D）
1. 种皮 2. 胚乳 3. 子叶 4. 胚轴 5. 胚根
（史渭清绘）

表 2 十大功劳属树种的开花结实特性（南京地区）

树 种	花序		花		开花期	果熟期	果落期
	长（cm）	簇生数	颜 色	梗 长（mm）			
阔叶十大功劳	5～10	6～9	黄褐色	4～6	3 月～4 月上旬	5 月下旬～6 月中旬	6 月下旬
十大功劳	3～5	4～8	浅黄色	1～4	9 月～10 月上旬	翌年 3 月中旬	3 月下旬

果实的采收调制和种子贮藏 阔叶十大功劳结实多，每 1 果穗常具几十、上百甚至 200 枚果实，采集容易。十大功劳则不能普遍结实，果穗较小。不同果实的成熟期不完全一致，且熟后易落，最好分期分批采收。采后先带果在室内摊晾，等采收结束再集中轻搓、漂洗、去杂取种。洗净的种子不沾水，稍晾即干，忌曝晒，也要防止过分干燥。随采随播或摊晾后短期存放。这 2 个树种的种子都不耐贮藏：过湿堆放容易引起霉烂，也容易失水而降低发芽能力。种子又无休眠习性，胚可以在湿润条件下继续长大并发芽。因此混沙湿藏虽然可以保持发芽能力，但容易造成过早萌发。贮藏时要严格控制种子含水量和贮藏温度。一般带果保潮贮藏比剥出种子干藏为好。有关果实和种子的数据见表 3。

表 3　十大功劳属树种果实和种子的数据

树　种	每穗含果实数（粒）	果　长（mm）	果　径（mm）	新鲜种子			
				长（mm）	径（mm）	千粒重（g）	每千克种子粒数（万粒）
阔叶十大功劳	40～100	12～16	7～12	7～9	4～6	35～70	1.4～2.9
十大功劳	25～40	7～9	6～7	5～6	4	37～39	2.5～2.7

发芽与播种　这 2 个树种的种子无休眠习性，极易发芽。江苏省植物研究所在南京地区所作的一次发芽测定结果见表 4。

表 4　十大功劳属树种发芽测定条件及其结果

树　种	采种日期（年-月-日）	置床日期（年-月-日）	室　温（℃）	萌发日期（月-日）	发芽势		发芽率（%）
					（%）	天　数	
阔叶十大功劳	1988-05-24	1988-06-01	27±	06-11～07-16	不明显	—	91
十大功劳	1989-03-25	1989-03-30	15～25	04-05～05-25	95.4	5	99.5

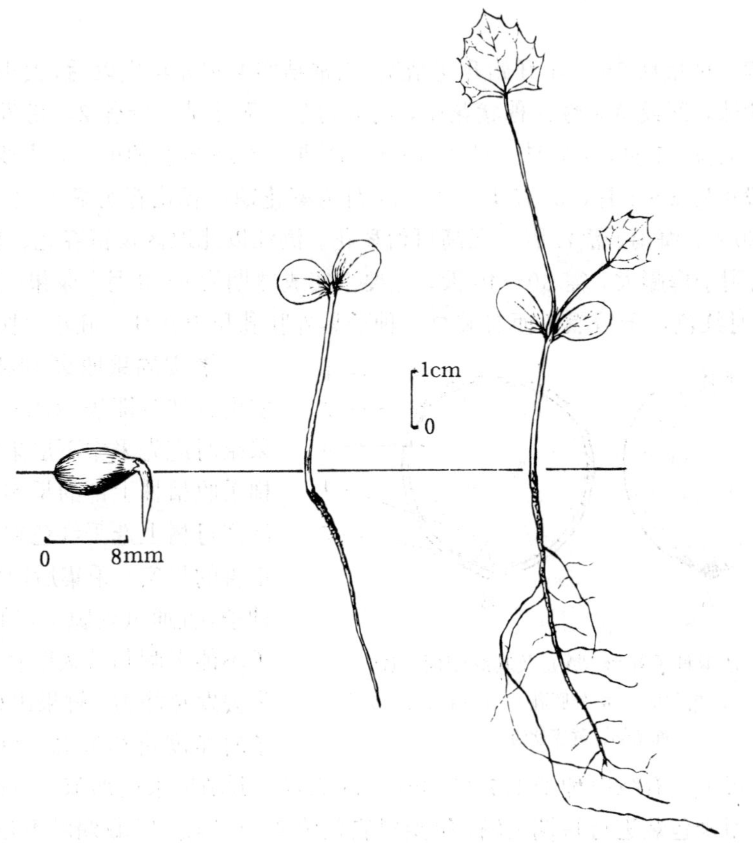

图 2　阔叶十大功劳种子萌发后第 2、10～60 天和 300 天的幼苗生长情况
（史渭清绘）

苗圃播种宜在种子成熟采收后随即进行。播后约1个月出苗。出土萌发。幼苗生长缓慢（图2），在高温高湿季节容易死亡，宜遮荫。

（何泽瑛）

胡　椒

Piper nigrum L.

（胡椒科 Piperaceae）

生长习性、分布和用途 胡椒属约2 000种，我国约产60种，本文描述引入的1种。常绿木质攀援藤本。喜高温潮湿的静风环境，年均温21℃以上的无霜地区能正常生长发育，不抗强风。要求深厚疏松、排水良好、pH值5.5～7、富含有机质的壤土。原产印度西南海岸西高止山脉的热带雨林，现已广泛种植于热带地区。1951年我国从马来西亚引入海南琼海试种成功，现滇、桂、琼、粤、闽、台有栽培。胡椒是世界重要的香辛植物之一，供调味并作防腐香料，又入药，健肠胃。未熟果实干后果皮皱缩而黑，称黑胡椒；成熟果实脱皮后色白，称白胡椒。

开花结实 种植后2～3年开始开花结实，正常结实期在5年生以后，大小年现象不明显。花通常单性同株，间或有杂性。穗状花序，与叶对生。无花被。雄蕊2，花药肾形，2室，花丝粗短。子房上位，1室，1胚珠。柱头3～4（5）裂。在我国各种植区，主要花期在春、夏、秋3季，一般年份3～5月，5～7月及8～11月为盛花期，春花着实率40%～50%，秋花着实率60%～70%。海南、湛江因气温高可留秋花。湛江以北地区仅留春花、夏花。从花穗初露、开花授粉到子房膨大，需30～40天。主要果实成熟期为6～8月。浆果，球形，无柄，径3～4mm，熟时红色，干后黑色而有皱纹。种子具外胚乳和内胚乳，胚小（图1）。

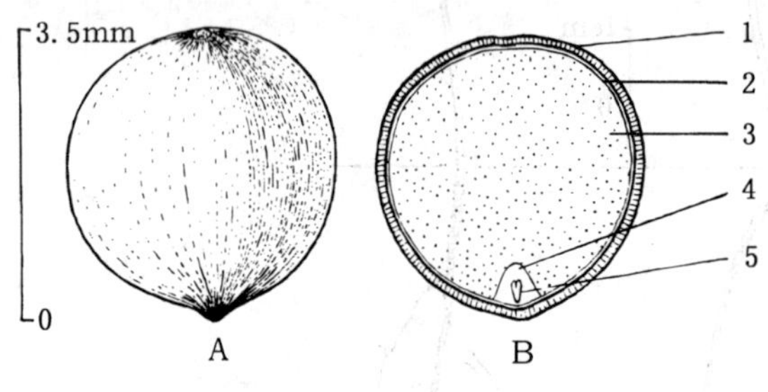

图1 胡椒种子外形（A）及其纵切面（B）

1. 外种皮 2. 内种皮 3. 外胚乳 4. 内胚乳 5. 胚

（黄光郁、黄应钦绘）

果实的采收调制和种子贮藏 果实由青转黄即表示成熟，应及时采摘。采果时应先采中下层果实，然后用三脚梯采收植株上部的果实。育苗用种应在高产母树上选采红色粒大、饱满、无病虫害的果实。采集后即脱去果皮，洗净种子，置通风处晾干，随即催芽播种。种子不能曝晒和长久贮存，超过1个月即失去发芽能力。鲜果出种率约57%。种子的净度可达93%～99%。千粒重60（40～75）g，每千克有纯净种子1.7（1.3～2.5）万粒。调制出来稍加晾干后的种子含水量约17.5%，可以用来包装进行短期运输，每塑料袋包装2.5～5kg。贮藏期以不超过1个月为宜。

发芽和播种 种子无休眠习性，可随采随播。发芽时日均温要求在22℃以上。为加速种子发芽，提高发芽率，播前宜浸种12～24小时，取出放入湿润细沙中催芽。1988年8月1日，华南热作研究所在室外沙床上用新鲜种子作发芽测定：先催芽，胚根露出白点时取出播种。播后 28 天（8 月 29 日）开始发芽，9 月 21 日发芽结束，没有明显的发芽盛期，发芽率为 65%。出土萌发。下胚轴成弓状出土，约 1 周左右种壳脱落（图 2）。

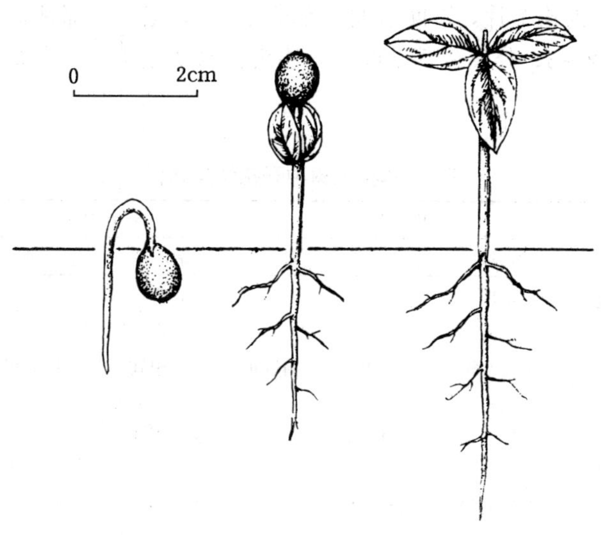

图 2　胡椒种子萌发后第 4、6、25 天幼苗的生长情况
（黄光郁绘）

点播。每平方米播种 3.8～5g，覆土 0.5～1cm。当幼苗抽出 2～3 片初生叶时，及时移入容器继续培育。苗高达 30cm 左右时可以定植。也可用压条、扦插等无性繁殖方法育苗。

（丁慎言）

木蓼（针枝蓼）属
Atraphaxis L.

（蓼科　Polygonaceae）

生长习性、分布和用途　本属约 25 种，我国 11 种 1 变种，分布于草原、半荒漠和荒漠带。本文描述 2 种。落叶灌木或小灌木。喜光，根系发达，耐旱，耐寒，耐高温，耐瘠薄，适应性强。沙埋后发生不定根，地上部分生长更旺盛。花期长，是荒漠地区的蜜源树种，又是良好的饲料及优良固沙造林的先锋树种。木蓼属这 2 个种的名称、分布和用途见表 1。

表 1　木蓼属树种的名称、分布及用途

中　名	学　名	分　布	用　途
沙木蓼	*A. bracteata* A. Los.	内蒙古、宁、甘（内蒙古鄂尔多斯和阿拉善地区特有种）	固沙、饲料，各种家畜都喜食
东北木蓼（东北针枝蓼）	*A. manshurica* Kitag.	内蒙古、冀、辽西	固沙、饲料，骆驼喜食

开花结实　2～3年生开始结实，3～4年便进入结实盛期。花小，两性，白色或粉红色，1至数朵簇生于节部托叶鞘状的苞腋内，成顶生或侧生的总状花序。花被5或4，2轮，外轮2片较小，常外反卷，内轮3片，果期增大。雄蕊（6）8，花丝基部常扩大。子房上位，1室，具3棱或扁平。花柱2～3，分离或基部联合，柱头粗棒状或头状。木蓼属开花结实物候见表2。

表2　木蓼属树种开花结实物候

树　种	观察地点和年份	开　花			果实成熟		果实散落	
		始　期	盛　期	末　期	始　期	盛　期	始　期	末　期
沙木蓼	内蒙古西部地区 —	6月	9月	—	6月	9月	7月	9月
	甘肃民勤 1975～1981	5月上旬	5～8月	霜冻前	7月中旬	7月下旬	7月下旬	霜冻
	宁夏中卫沙坡头 1956～1980	5月上旬	5月中旬	—	5月下旬	6月上旬～10月上旬	—	—
东北木蓼	甘肃民勤 1975～1981	5月上旬	5～8月	—	7月中旬	7月下旬	—	—
	内蒙古西部地区 —	7月	9月	—	7月	9月	8月	9月

木蓼属果实为瘦果，双凸或卵状三棱形，顶端尖，基部楔形，包藏于扩大的花被片内。胚稍弯，偏于一侧。种子有丰富的胚乳，形态特征见表3。木蓼果实的形态如图1。

表3　木蓼属树种果实形态特征

树　种	大小（mm）	颜　色
沙木蓼	长4.8～6.0 宽2.2～3.0	深褐色，有光泽
东北木蓼	长3.0～5.0 宽2.0～3.0	暗褐色，略有光泽

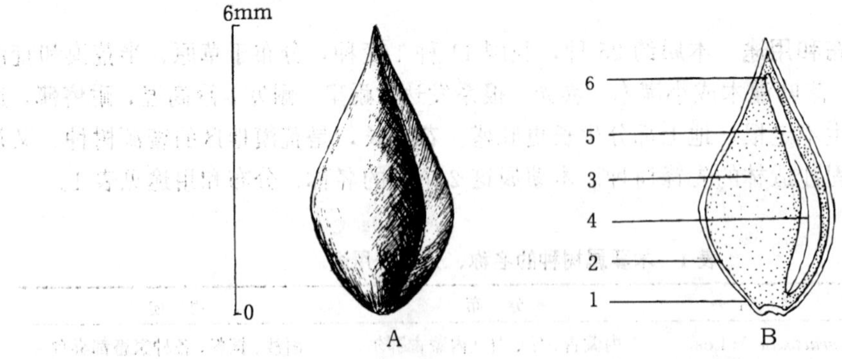

图1　沙木蓼果实外形（A）及其纵切面（B）

1. 果皮　2. 种皮　3. 胚乳　4. 子叶　5. 胚轴　6. 胚根

（姚桂英绘）

果实的采收调制和种子贮藏　据甘肃民勤观察，花期从5月中旬开始，到霜冻时还继续开花，果实在8月～9月下旬陆续成熟。果实成熟期不一致，不易采收。果实小而轻，易随风吹走，应熟一批采一批。采得的果实清除杂质后即为播种材料，通称种子。在通风凉爽的5℃的条件下可贮藏较长时间。出籽率、净度和种子质量见表4。

表4　木蓼属树种的出籽率、净度和种子质量

树　种	出籽率（%）	净　度（%）	千粒重（g）	每千克纯净种子（果实）粒数（万粒）
沙木蓼	33	80～90	5	20
东北木蓼	—	80～90	6.5	15.3

发芽和播种　木蓼属种子发芽迅速。在呼和浩特市所作的一次室内发芽测定结果如表5。

表5　木蓼属树种的发芽能力

树　种	温度（℃）		发芽势（%）			发芽率（%）		
	有　光	无　光	计算天数	一般数值	变动范围	计算天数	一般数值	变动范围
沙木蓼	30	20	3	7	5～10	12	25	20～30
东北木蓼	30	20	3	5	—	12	20	—

低床条播。每平方米约播3g。气温达16℃以上时播种发芽率高，出苗快，播种后15天左右出苗。出土萌发。幼苗形态如图2。

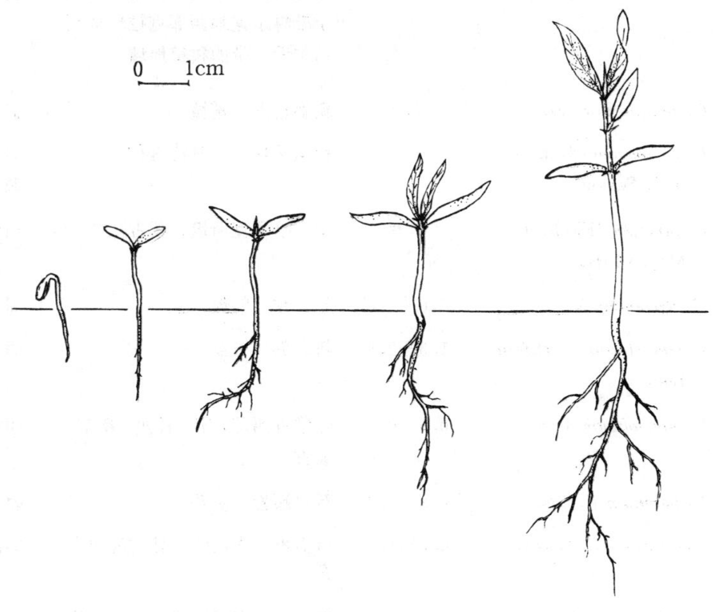

图2　沙木蓼种子萌发后第1、16、30、35、45天幼苗生长情况

（姚桂英绘）

（李荣晨）

沙拐枣属

Calligonum L.

（蓼科 Polygonaceae）

生长习性、分布和用途 本属约80种，产北非、西亚及南欧荒漠和半荒漠地带。我国约20种，全部生于西北和内蒙古沙漠地区。灌木，多分枝。枝常曲折，有关节。叶退化成鳞片状。旱生，喜光，极耐干旱，并有抗高温、耐盐碱、耐风蚀、抗沙埋的能力，生长迅速，生活力强，易繁殖，寿命在20年以上。是优良的固沙造林树种，可作饲料。本文描述13种，其名称、分布及用途见表1。

表1 沙拐枣属树种的名称、生长、分布及用途

中 名	学 名	树高（m）	分 布	用 途
无叶沙拐枣	*C. aphyllum* (Pall.) Gurke	1～2	新疆伊犁地区。中亚、欧洲东南	固沙、饲料、薪炭
乔木状沙拐枣	*C. arborescens* Litv.	3～5	原产宁、甘，新引种栽培	固沙、饲料、薪炭、蜜源
头状沙拐枣	*C. caput-medusae* Schrenk	2～5	原产中亚。内蒙古、甘，新引种栽培	固沙、饲料、薪炭，蜜源
心形沙拐枣	*C. cordatum* Eug. Kor.	0.5～1.0	新疆精河、霍城。其群系仅见于准噶尔盆地西部克拉玛依至乌尔禾一带的狭窄地域	固沙、薪材
密刺沙拐枣	*C. densum* Borszcz.	1.0	新疆精河、霍城	固沙、饲料、薪炭
精河沙拐枣（艾比湖沙拐枣）	*C. ebi-nuricum* Ivan ex Y. D. Saskov	1.0	新疆艾比湖一带特有种	固沙、饲料、薪炭、蜜源
泡果沙拐枣	*C. junceum* (Fisch. et Mey.) Litv.	1.0	新。哈萨克斯坦、蒙古	固沙、饲料、薪炭
新疆沙拐枣	*C. klementzii* A. Los.	0.6～1.0	新、甘肃敦煌	固沙、饲料、薪炭
白皮沙拐枣（白茎沙拐枣）	*C. leucocladum* (Schrenk) Bge.	1.5～2.0	新北部。中亚	固沙、饲料、薪炭、
沙拐枣（蒙古沙拐枣）	*C. mongolicum* Turcz.	0.3～0.6	内蒙古西部，宁、甘西、新东。蒙古	固沙、饲料、薪炭
小果沙拐枣	*C. pumilum* A. Los.	0.3～0.4	新疆鄯善、伊吾	固沙、饲料、薪炭
昆仑沙拐枣（塔里木沙拐枣）	*C. roborowskii* A. Los.	0.5～1.0	新疆塔里木盆地、甘肃河西走廊	固沙、饲料、薪炭
红皮沙拐枣（红果沙拐枣）	*C. rubicundum* Bge.	0.5～1.0	新疆额尔齐斯河流域和布尔津沙地。中亚	固沙、饲料、薪炭

开花结实　花两性，单生或数朵排成疏散的花束。单被花，花被片 5，淡红色，果期不增大，膜质，反折。雄蕊（8）12～18，基部结合。子房上位，具 4 棱，1 室 1 胚珠，花柱 4，柱头头状。1～2 年生开始结实，2～3 年生即可正常结实。每年结实量都很高。在新疆北部沙漠地区 5 月底开花，6 月底～7 月初果熟。不少树种在新疆吐鲁番地区 1 年 2 次结实：第 1 次 5 月初开花，5 月底 6 月初果熟；第 2 次 9～10 月开花，10 月底至 11 月初果熟。部分树种的个别植株甚至 1 年 3 次开花结实。沙拐枣属树种开花结实的物候汇入表 2。

表 2　沙拐枣属树种的开花结实物候

树　种	观察地点	开　花		果实成熟	
		始　期	末　期	始　期	末　期
无叶沙拐枣	新疆伊犁	4 月下旬	6 月	5 月下旬	6 月中下旬
乔木状沙拐枣	内蒙古阿拉善盟	5 月	6 月	6 月	—
	宁夏中卫沙坡头	5 月下旬	—	7 月上旬	—
	新疆吐鲁番	4 月中旬	5 月中旬	4 月下旬	6 月上旬
		9 月下旬	11 月上旬	10 月中旬	11 月上旬
头状沙拐枣	甘肃民勤	5 月	6 月	6 月	9 月
	新疆吐鲁番	4 月中旬	5 月中旬	4 月下旬	6 月上旬
		10 月上旬	11 月上旬	10 月中旬	11 月上旬
心形沙拐枣	新疆吐鲁番	4 月中旬	5 月上旬	4 月下旬	5 月下旬
		9 月下旬	10 月下旬	10 月上旬	10 月下旬
密刺沙拐枣	新疆吐鲁番	4 月中旬	5 月上旬	4 月下旬	5 月中旬
		9 月上旬	10 月下旬	9 月上旬	10 月下旬
精河沙拐枣	新疆艾比湖	6 月	—	6 月	—
	新疆吐鲁番	4 月中旬	5 月中旬	4 月下旬	5 月中旬
		6 月中旬	11 月上旬	6 月下旬	11 月上旬
泡果沙拐枣	新疆吐鲁番	4 月上旬	4 月下旬	4 月下旬	5 月下旬
		10 月上旬	10 月下旬	10 月中旬	10 月下旬
新疆沙拐枣	甘肃民勤	5 月	7 月	5 月	7 月
白皮沙拐枣	新疆北部	5 月	7 月	5 月	7 月
	新疆吐鲁番	4 月上旬	4 月下旬	4 月下旬	5 月下旬
		9 月下旬	10 月下旬	10 月上旬	10 月下旬
沙拐枣	内蒙古西部	5 月	7 月	5 月	7 月
	甘肃民勤	5 月上旬	6 月中旬	6 月下旬	7 月中旬
小果沙拐枣	新疆东部	5 月	6 月	5 月	6 月
昆仑沙拐枣	新疆塔里木	5 月	6 月	6 月	7 月
红皮沙拐枣	甘肃民勤	6 月	7 月	6 月	7 月
	新疆吐鲁番	4 月上旬	4 月下旬	4 月下旬	5 月下旬

果为瘦果，直或弯曲，具4棱，沿棱肋着生膜质或革质的刺毛或翅，或在刺毛顶罩有膜质囊包于整个瘦果外部。种子长椭圆形、圆柱状或四棱形。胚直，有胚乳。沙拐枣属果实的形态描述见表3。沙拐枣的种子外形及其纵切面。见图1。

表3 沙拐枣属树种果实（播种材料）的形态特征

树种	未熟果颜色	成熟果实		
		形状	大小（mm）	颜色
无叶沙拐枣	—	圆形	长16～20，径15～18	红色
乔木状沙拐枣	鲜绿色	宽卵形	长20～30，径20～25	黄色或棕红色
头状沙拐枣	鲜绿色	球形	径20～25	红色
心形沙拐枣	淡绿色	近圆形或宽椭圆形	长16～21，径13～19	褐色
密刺沙拐枣	淡绿色	近球形	径20	浅黄色
精河沙拐枣	黄色	尖卵形	长约15或较小	红色
泡果沙拐枣	—	球形或近球形	径8～10	黄色或红色
新疆沙拐枣	黄色	圆卵形	长15～20	红色或褐色
白皮沙拐枣	—	卵形	长13～15，径11～13	红色
沙拐枣	绿色	宽椭圆形	长8～12	—
小果沙拐枣	黄色	卵圆形	长7～10	黄色或浅红色
昆仑沙拐枣	黄色	宽卵形或近圆形	长13～15（18）	浅红色或红褐色
红皮沙拐枣	红色	卵形或圆卵形	长10～16，径9～14	紫红色或红褐色

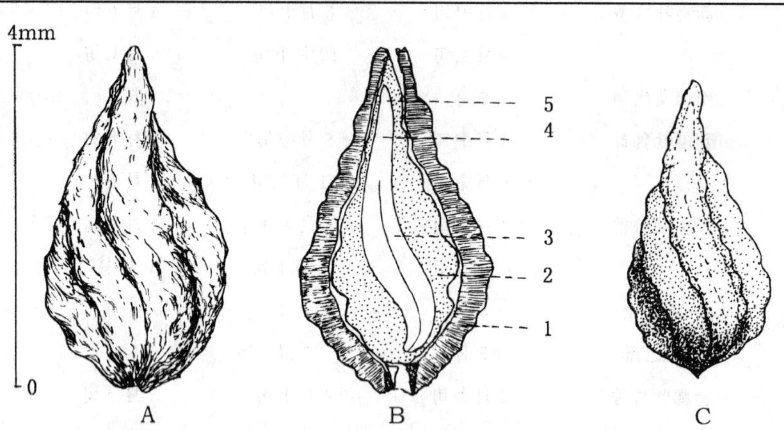

图1 沙拐枣种子外形（A）及其纵切面（B）和胚体（C）

1. 种皮 2. 胚乳 3. 子叶 4. 胚轴 5. 胚根

（姚桂英绘）

果实的采集调制和种子贮藏 沙拐枣属树种有些是边开花，边结果，边成熟，成熟后大部分果实立即脱落。这类树种应分期分批采种。有的树种果实成熟后隔一段时间才脱落，它们成熟的标志是果翅或刺毛干燥，果皮坚硬成木质，一旦出现这些特征便应及时采集。可以直接摘取果实，也可以击落后在地面收集，筛去杂质后的果实即为播种材料，通称种子，而无需从果实中取出种子。沙拐枣属播种材料的质量数据见表 4。

表 4　沙拐枣属播种材料的净度和质量①

树　种	净度（%）	千粒重（g）	每千克纯净播种材料粒数（万粒）
无叶沙拐枣	85～95	45	2.2
乔木状沙拐枣	70～85	90～95	1.1
头状沙拐枣	78～95	108～154	0.65～0.93
心形沙拐枣	70～85	88①	1.1①
密刺沙拐枣	90～98	48～114	0.88～2.1
精河沙拐枣	85～95	50	2.0
泡果沙拐枣	75～85	25.4～49.0	2.0～3.9
新疆沙拐枣	85～95	50～54	1.8～2.0
白皮沙拐枣	95～98	35～45	2.2～2.8
沙拐枣	73～85	58～66	1.5～1.7
小果沙拐枣	82～90	16～22	4.5～6.3
昆仑沙拐枣	75～85	21～24	4.2～4.8
红皮沙拐枣	95～98	106～128	0.78～0.94

① 据 600 粒种子质量推算

发芽和播种　沙拐枣属播种材料的外壳坚硬，吸水困难，播前需催芽处理。低温层积 1～2 个月，翌春种子露白时即可播种。也可以用浓硫酸浸种 3～4 个小时后用水冲洗，再放入温水中浸泡 3 昼夜，置于 20～25℃条件下催芽，种子露白时即可播种。对无叶沙拐枣、精河沙拐枣、白皮沙拐枣、沙拐枣和小果沙拐枣，邱国玉（1989）比较过秋播、无处理春播和 36%盐酸溶液浸种后春播的效果，认为秋播的发芽率明显地高于其它两种方法。张鹤年和张希明等（1988）曾将乔木状沙拐枣和头状沙拐枣用流水浸泡 7 天，待大部分种子吐白时夏播，据报道，5 天即可出苗，7 天苗齐。对于沙拐枣属的发芽测定方法和测定条件，目前经验还不多。一般是用滤纸作发芽床，下垫脱脂棉；发芽温度多用变温，每 24 小时中 8 小时给予 30℃并加光照，16 小时在无光条件下维持 20℃。表 5 所列 11 个树种发芽能力的数值就是在这种条件下测得的。

表 5　沙拐枣属树种的发芽能力

树　种	发芽势（%）		发芽率（%）		
	计算天数	一般数值	计算天数	一般数值	变动范围
乔木状沙拐枣	6	23	15	—	60～80
头状沙拐枣	11	18	13	—	50～80
心形沙拐枣	6	11	12	33	—
密刺沙拐枣	2	81	19	—	90～100

（续）

树 种	发芽势（%）		发芽率（%）		
	计算天数	一般数值	计算天数	一般数值	变动范围
精河沙拐枣	9	2	13	32	—
泡果沙拐枣	4	21	13	—	75～90
新疆沙拐枣	4	64	9	—	75～90
白皮沙拐枣	9	5	13	21	—
沙拐枣	3	21	13	—	50～70
小果沙拐枣	3	22	—	48	—
昆仑沙拐枣	2	18	9	21	—

条播，行距 30cm。乔木状沙拐枣、头状沙拐枣、心形沙拐枣、密刺沙拐枣和红皮沙拐枣，它们的千粒质量接近或超过 100g，每平方米播种 25～38g；千粒重在 50g 左右的树种，例如无叶沙拐枣、精河沙拐枣、泡果沙拐枣、新疆沙拐枣等，每平方米约播种 15g。覆土厚 5～8cm，播后 7～10 天可出土。出土萌发。子叶 2 枚，肥厚肉质。沙拐枣幼苗形态如图 2。

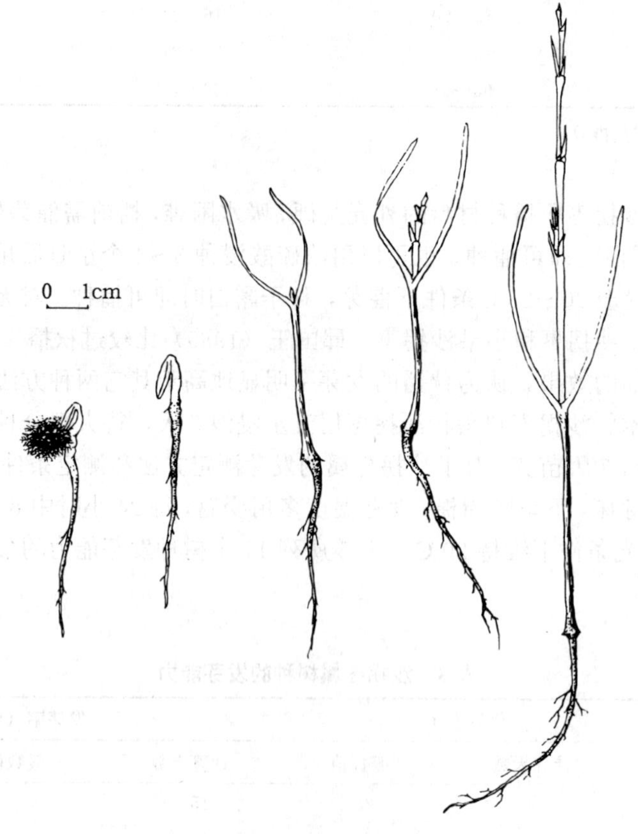

图 2 沙拐枣种子（果实）萌发后第 7、11、17、37、44 天的幼苗生长情况

（姚桂英绘）

据张孝仁（1986）报道，曾经在甘肃武威的流动沙丘上，用乔木状沙拐枣、头状沙拐枣、网状沙拐枣和红皮沙拐枣扦插造林，据认为，只要扦插后2个月内沙丘水分含量在2%～3%以上且技术措施得当，成活率可以超过80%。

（李荣晨、赵美华）

梭　梭　属
Haloxylon Bge.

（藜科　Chenopodiaceae）

生长习性、分布和用途　本属约11种，分布于地中海及中亚。我国2种，均为渐危种，已列入《中国植物红皮书》。本文描述2种（表1）。落叶小乔木或灌木，我国西北地区优良的固沙造林树种。梭梭还是中药肉苁蓉［*Cistanche ambigua*（Bge.）G. Bech.］的主要寄主。

表1　梭梭属树种的名称、分布和用途

中　名	学　名	分　布	用　途
梭梭（盐木）	*H. ammodendron*（C. A. Mey.）Bunge	内蒙古、甘、宁、青、新。中亚、蒙古、俄罗斯西伯利亚	固沙、燃料、饲料
白梭梭	*H. persicum* Bunge ex Boiss. et Buhse	新疆北部。甘、宁、内蒙古栽培。伊朗、阿富汗、哈萨克斯坦	固沙、燃料、饲料

梭梭植株高3～8m，最高达10m；白梭梭植株矮小，一般为2～7m。它们的叶退化成极小的鳞片状，而用当年绿色嫩枝进行光合作用。细胞液浓度大，渗透压很高，抗脱水力强，能在年降水量仅有10mm而蒸发量高达3 000mm的沙地及戈壁上生长。耐干旱、风蚀、沙埋，极耐沙割。梭梭抗盐能力极强，茎枝内盐分高达14%～17%，故又称“盐木”。喜光，不耐庇荫。耐酷热严寒。气温高达43℃，沙面地温达60～70℃甚至80℃情况下，仍能正常生长；也能忍耐－40℃的低温。发热力强，是极好的燃料。也是良好的饲料。

开花结实　3～5年生开始结实，10～15年大量结实，20年后进入衰老期。4月下旬～5月下旬开花，花小而数量繁多，花期梭梭为5～8天（在新疆艾比湖盆地的甘家湖地区，1985～1986年2年观察，梭梭花期可延续22天）。白梭梭花期为10～15天。梭梭属具有2次休眠的特征，花后子房暂不发育而进入夏眠，直至9月中旬气候凉爽后才开始发育成果实。果实成熟后陆续飘落，11月底进入冬眠。开花结实物候期见表2。

表2　梭梭属树种的开花结实物候期

树　种	观察地点和年份	开　花			果实成熟		果实散落	
		始　期	盛　期	末　期	始　期	盛　期	始　期	末　期
梭梭	甘肃民勤 1983～1987	4月下旬～5月中旬	5月上旬～5月中旬	5月中旬～6月上旬	10月中旬	10月下旬	10月下旬	10月下旬～12月上旬

（续）

树　种	观察地点和年份	开花 始期	开花 盛期	开花 末期	果实成熟 始期	果实成熟 盛期	种子散落 始期	种子散落 末期
梭梭	内蒙古西部	4月	～	5月	10月下旬	11月上旬	10月下旬	11月上旬
	新疆吐鲁番	4月下旬	～	5月上旬	10月下旬		10月下旬以后	11月上旬
白梭梭	甘肃民勤 1978～1979	4月下旬～5月中旬	5月上旬	5月上旬～5月中下旬	10月上旬	10月下旬	10月下旬	10月中下旬～11月下旬
	新疆吐鲁番	4月下旬	～	5月上旬	10月下旬		10月下旬以后	～

花两性。单被花，黄色，对生于2年生枝条侧生短枝上。具2小苞片，舟状，与花被等长。花被片5，绿色，膜质，果时于背部上方生翅状附属物。雄蕊5，与花被片对生，着生于杯状花盘上。子房上位，1室，基部陷入花盘内，花柱极短，柱头2～5，胚珠1，弯生。胞果，扁球形或半球形，顶面微凹，果皮肉质，与种子贴伏。种子横生，扁圆形，种皮膜质，无胚乳。胚绿色，螺旋状卷曲。梭梭属果实和种子形态特征见表3和图1。

表3　梭梭属果实和种子形态特征

树　种	成熟果实 形状	成熟果实 大小（cm）	成熟果实 颜色	种子 形状	种子 大小（mm）	种子 颜色
梭梭	果翅半圆形，干膜质，基部心形	果翅宽5～8	黄褐色	螺旋状	2.5	黑褐色
白梭梭	果翅扇形或近圆形，基部楔形或圆形，干膜质	果翅宽4～7	淡黄褐色	螺旋状	2.5	黑褐色

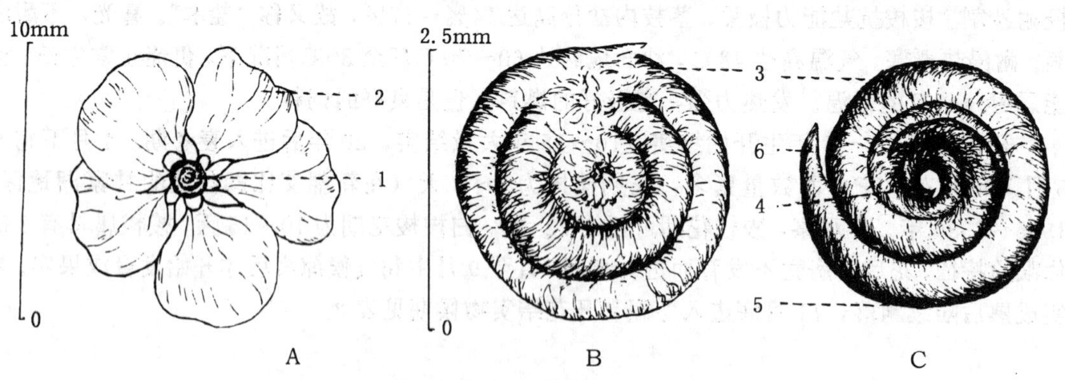

图1　梭梭胞果外形（A）、去翅后胞果外形（B）及其横切面（C）
1. 胞果　2. 果翅　3. 果皮及种皮　4. 子叶　5. 胚轴　6. 胚根
（姚桂英绘）

果实的采集调制和种子贮藏 成熟的胞果遇风极易脱落飞散，应在果实由绿色转呈黄褐色时及时摇动或击打树枝，使果实落在下方承接的采种布上。新采集的果实含水量较大，应及时摊晒。调制的主要内容是清除果翅。这是因为带翅果实的体积比纯净果实大 6～10 倍，不便运输贮藏，播种时还容易随风飞散，不便操作，播后与土壤也接合不良。果翅的吸湿能力强，常常影响种子贮藏效果。摊晒后轻加碾压，果翅即可碎落。清除果翅和其它杂质后所得的纯净胞果即为播种材料，通称种子。净度可达 80%以上。适于贮藏的含水量为 6.2%～6.4%。梭梭属树种的出籽率和种子质量见表 4。

表 4 梭梭属树种的出籽率和种子质量

树 种	出籽率（%）	净度（%）	千粒重（g）	每克纯净种子（胞果）粒数
梭梭	77	80～95	3～3.5	280～330
白梭梭	73	80～95	3.5～4.7	210～280

梭梭种子生命力持续时间较短，贮藏时间不宜过长。带有膜质果翅的胞果，用麻袋贮藏半年后发芽率由 95%降低到 40%左右，7 个月后全部丧失发芽能力。去果翅后贮藏半年发芽率仍能保持在 90%左右，1 年之后发芽率下降到 40%～50%，2 年后发芽率仅 10%左右。白梭梭种子一般条件下贮藏半年，发芽率 80%左右，保存 9 个月便完全丧失发芽能力，若无贮藏条件则不能过夏。

发芽和播种 梭梭种子没有休眠习性，发芽迅速，在适当的温度条件下，2 小时后即开始发芽。白梭梭 1 天内发芽可达 30%，2 天为 60%，3 天为 80%，1 周内发芽结束，发芽率达 90% 以上。梭梭属树种的发芽能力及其测定条件见表 5。

表 5 梭梭属树种室内发芽能力及其测定

树 种	发芽测定温度（℃）		发芽势（%）		发芽率（%）	
	昼	夜	计算天数	一般数值	计算天数	一般数值
梭梭	20	10	2	92.0	9	93
白梭梭	20	10	2	82.0	8	95

表 6 梭梭种子的室内发芽率与场圃发芽率

室内发芽率（%）	20～30	30～40	50～60	60～70	70～80	80～90	90～100
场圃发芽率（%）	0	4	11	24	30	35	40

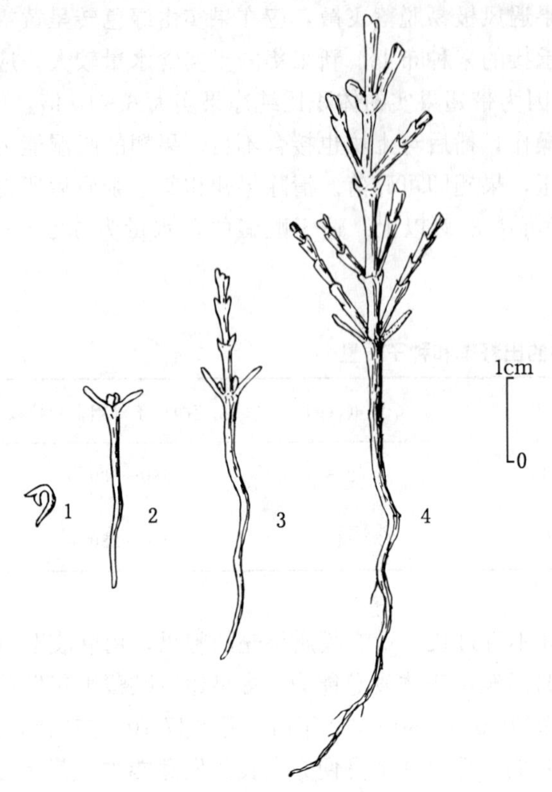

图 2　梭梭种子萌发后第 2、18、28、48 天的幼苗生长情况
1. 子叶出土　2. 初生叶出现　3. 上胚轴伸长　4. 幼苗形成
（姚桂英绘）

梭梭在含盐量 1%～2%时生长良好，种子的发芽能力在基质含盐量达 5%时还不受影响，只有达到 8%时才丧失发芽能力。白梭梭耐盐力不及梭梭，基质含盐量达到 3%时即丧失发芽力。梭梭种皮黑褐色，极薄，极易吸水膨胀。子叶 2 枚，条状绿色，紧卷曲于螺旋状胚的中心。种子内已有发育完全的胚根和子叶，只要种皮能够吸胀破裂，胚根就突破种皮而萌发。

在沙区，白天地表化冻 5cm 左右时即可播种梭梭。梭梭发芽对地温要求低，在白天化冻，夜间又结冰的早春天气里，它能顺利发芽出土。据测定，晚间温度降至－9℃时，仅 5% 的幼苗受冻致死。为预防白粉病及根腐病，播前可用 0.1%～0.2%的高锰酸钾或硫酸铜水溶液浸种，20～30 分钟后捞出，晾干拌沙播种。梭梭室内发芽率高，但场圃发芽率常仅为 35%左右。表 6 是一般情况下，梭梭同室内发芽率相对应的场圃发芽率。

每平方米的播种量，梭梭为 5g，白梭梭为 5～10g。条播，行距 25～30cm，播幅 5cm。覆土不宜过厚，沙壤土约覆土 1cm，不要超过 1.5cm，沙土可以稍厚。播后要镇压。2 种梭梭幼苗期就很耐旱。出土萌发。梭梭的幼苗形态如图 2。

（郭志中）

紫　薇　属

Lagerstroemia L.

（千屈菜科　Lythraceae）

生长习性、分布和用途　本属约 55 种，我国 16 种，又引进栽培 3 种，本文描述 5 种。落叶或常绿，乔木或灌木。喜温暖湿润的气候，适生于排水良好、土层深厚肥沃的中性及微酸性或微碱性土壤。花期长，花朵繁茂，丛生于枝顶，色泽艳丽，是园林、庭园及游览区的观赏树种。木材坚硬，可作小型用具和细木工。栽培品种甚多。紫薇属的名称、生长和分布情况见表 1。

表 1　紫薇属树种的名称、生长和分布

中　名	学　名	树　高 (m)	胸　径 (cm)	分　布	供　稿
大萼紫薇	*L. flos-reginae* Retz.	15～20	10～15	原产东南亚。1979 年引种在海南儋州	610
紫薇	*L. indica* L.	8	8	华北、华东、华中、华南及西北	309
福建紫薇	*L. limii* Merr.	4～6	4	浙、闽、鄂。南京引栽	1001
大花紫薇	*L. speciosa* (L.) Pers.	15～20	30～40	原产印度等国，粤、琼、桂、闽引种	603
绒毛紫薇	*L. tomentosa* Presl.	20～30	60～100	滇西南。缅甸、泰国、越南	618

开花结实　4～6 年生开始开花结实，正常开花结实在 10 年生以后。结实有大小年现象，但不甚明显。花两性，圆锥花序顶生或腋生，花萼陀螺形或半球形，5～9 裂。花瓣通常 6，具长爪，瓣片波状皱缩，花瓣浅红色、紫红色或白色。雄蕊 6 至多数，花丝细长。子房 3～6 室，每室胚珠多数，花柱长，柱头头状。开花期及果实成熟期详见表 2。

表 2　紫薇属树种的开花结实物候期

树　种	观察地点	观察年份	开花 始　期	开花 盛　期	开花 末　期	果实成熟 始　期	果实成熟 盛　期	种　子散落期
大萼紫薇	海南儋州	1988（春花） （秋花）	4 月中旬 9 月	5 月中旬 10 月	6 月上旬 11 月	7 月 11 月	8 月 12 月	9 月 12 月下旬
紫薇	北京	1987	7 月中旬	8 月中旬	9 月中旬	9 月下旬	10 月中旬	11 月上旬～中旬
福建紫薇	南京	1988	5 月下旬	6 月中旬～7 月中旬	7 月下旬	10 月中旬	11 月中旬	12 月上旬～翌年 1 月上旬
大花紫薇	广西南宁	1984	6 月上旬	7 月上旬	8 月下旬	9 月中旬	9 月下旬	10～11 月
绒毛紫薇	云南勐腊	1988	5 月上旬	5 月中旬	5 月下旬	11 月下旬	12 月～翌年 2～3 月	翌年 3～4 月

蒴果，木质，广椭圆形或近球形，基部为宿存的花萼包围，室背开裂，3～6 果瓣。种子多数，顶端有翅，种皮薄，无胚乳（图 1）。果实及种子特征见表 3。

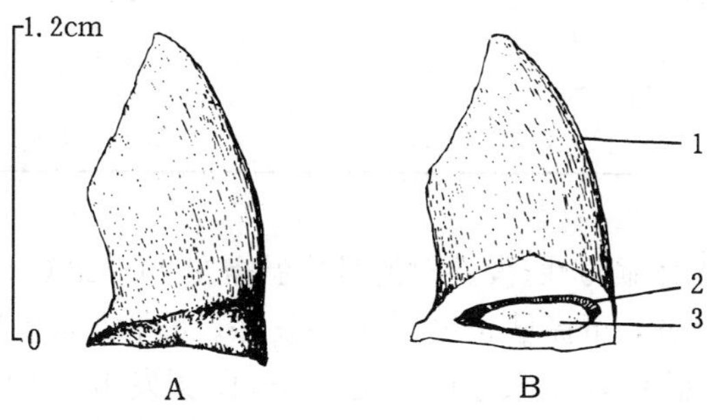

图 1　大萼紫薇种子外形（A）及其纵切面（B）

1. 种翅　2. 种皮　3. 胚

（黄光郁绘）

表 3 紫薇属树种果实和种子的形态特征

树种	未熟果颜色	成熟果实			每果种子粒数	种子		
		形状	大小（cm）	颜色		形状	大小（mm）	颜色
大萼紫薇	—	卵圆形	长 1.2 宽 0.7～0.8	—	14～16	—	—	—
紫薇	浅绿色	广椭圆形至近球形	1.0～1.4	绿褐色至暗褐色	4～6	三角状扁卵圆形	宽 2～3，带翅长 6～8	暗褐色
福建紫薇	青绿色	球形或卵状椭圆形	0.7～0.8	褐色	多数	不规则三角形	2～3	黄褐色
大花紫薇	绿色	倒卵形或圆球形	长 1.8～2.5 宽 1.1～1.8	黄赤色或浅褐色	近 100	扁三角形	宽 4～6，带翅长 11～14	浅褐色至褐色
绒毛紫薇	黄绿色	卵形至椭圆形	长 1.2～1.5 宽 0.8～1.0	黑色	20～30	—	—	深褐色

果实的采收调制和种子贮藏 果实成熟季节摘取或用采种钩将果枝钩下，也可摇动树木使果实脱落。果实收集后置通风干燥处摊放或在日光下摊晒，经常翻动直至果实开裂，脱出种子。经筛选得到的纯净种子含水量降至 10%以下时方可贮藏。短期贮藏可装入布袋或塑料袋。种子置坛罐内密封存入低温干燥阴凉处，发芽能力可保存 1 年。贮存 1 年以上，宜置 5℃以下的低温环境。种子的净度、质量等见表 4。

表 4 紫薇属树种的鲜果出种率和种子净度、质量

树种	出种率（%）	净度（%）	千粒重（g）	每克纯净种粒数
大萼紫薇	28	92～96	4.5～5.5	180～220
紫薇	13～15	85～95	1.8～2.6	380～560
福建紫薇	26～30	—	1.2～1.9	520～840
大花紫薇	21	92～98	7～11	90～140
绒毛紫薇	28	20～30	2.5～3.5	280～400

发芽和播种 种子无休眠习性。发芽时的日均温需在 20～28℃。如果层积催芽处理 20 天左右，发芽会更加迅速整齐，可提早 6～7 天结束发芽。每平方米播种量 2～3g。采用撒播或宽条状带播。出土萌发。播后 4～5 天子叶出土，10～15 天发出初生叶（图 2）。当年苗高40～80cm，可出圃定植或移床培养大苗。据孙秀琴和田树霞（1990）报道，紫薇室内发芽测定时最好是用白天 30℃、夜间 20℃的变温，且在白天的 8 小时中给以 1 000～1 200lx 的光照。

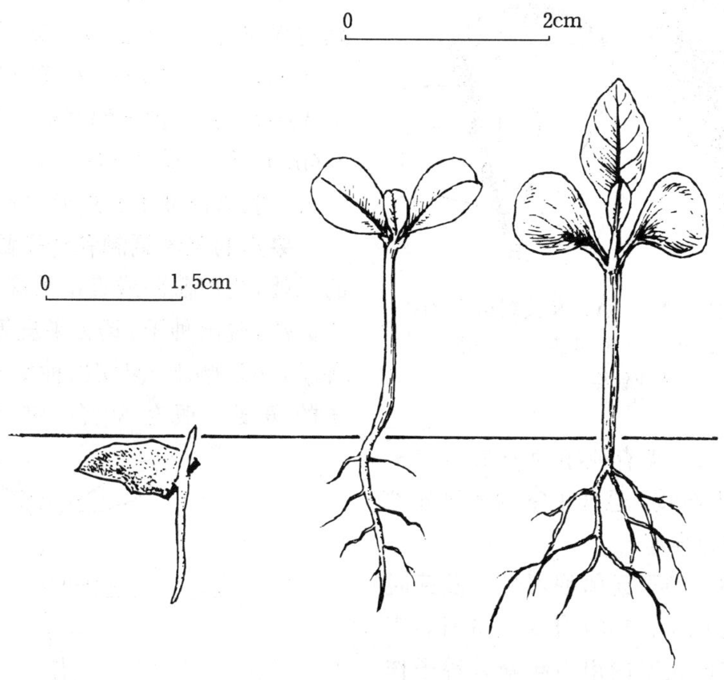

图 2　大萼紫薇种子萌发后第 2、12、19 天的幼苗生长情况
（黄光郁绘）

（宋朝枢）

散沫花（指甲花）

Lawsonia inermis L.

（千屈菜科　Lythraceae）

生长习性、分布和用途　散沫花属只有本文描述的 1 种。常绿大灌木至小乔木，高 3～6m，多分枝，小枝呈刺状。较耐水湿。原产地可能在东非和东南亚，我国滇、桂、粤、琼、闽、浙庭院多有栽培，长江流域及以北地区多为盆栽，冬季放入温室。可以修剪成各种形状供观赏或作围篱。花极芳香。叶捣烂可产生 1 种红色染料，可染指甲。

开花结实　3～4 年生开始结实，正常结实期在 8 年生以后。结实无大小年现象，每年结实颇多。花两性，白色、玫瑰色或朱红色。圆锥花序顶生，长 7～20cm。萼筒极短，4 裂。花瓣 4，着生于萼筒之顶，与裂片互生。雄蕊通常 8 枚，成对着生于花瓣间。子房上位，4 室，花柱丝状，高出于雄蕊。柱头钻状。据广西南宁 1987～1988 年观察，开花期及果实成熟期均长，从 5 月下旬～9 月下旬陆续有花，其中 6～8 月为盛花期；9 月上旬～12 月上旬陆续有果实成熟。

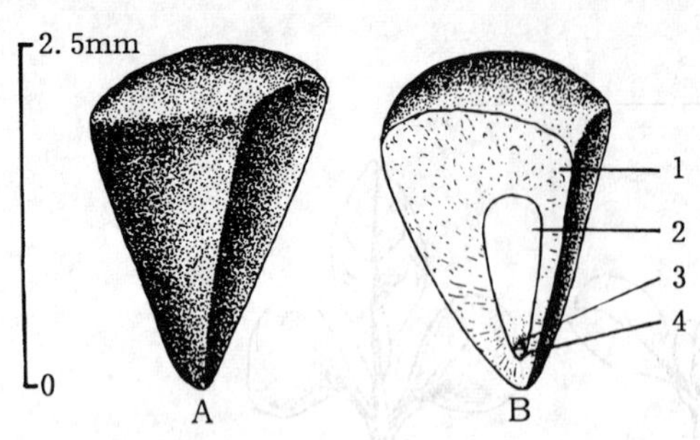

图 1　散沫花种子外形（A）及其纵切面（B）
1. 种皮　2. 子叶　3. 胚芽　4. 胚根
（黄应钦绘）

蒴果，球形或扁球形，未熟时青绿色，成熟时黄褐色。果皮薄，革质。果高 4.5～6.5mm，径 5～7mm，不规则开裂。每果有种子 40～50 粒。种子倒锥形，黄褐色，长 2～2.6mm。种皮厚海绵质。无胚乳，胚伸直，子叶大。散沫花种子的外形和解剖结构见图 1。

果实的采收调制和种子贮藏　选择成熟的果穗，用剪截断或直接采摘。采得的果实用手搓揉，脱出种子，扬去果皮等杂质即得纯净种子，可以晒干。果实出种率 40%～60%，种子的净度一般在 80%～95%。千粒重 1.9（1.7～2.1）g，每千克有纯净种子 53（47～59）万粒。可装入布袋，悬挂于室内通风处贮藏，贮藏期一般为半年。

发芽和播种　种子无休眠习性。发芽时日均温需在 20℃以上。1987 年 9 月 5 日，广西林业科学研究所在室内用当年新采种子作过发芽测定。置床前种子未作任何处理，培养基质为滤纸，光照为室内自然光，日均温在 27℃上下：9 月 10 日开始发芽，至 9 月 14 日为发芽盛期，24 日发芽终止；从发芽之日至发芽高峰日，6 天发芽 35%；从播种至发芽终了，19 天的发芽率为 60%。出土萌发。胚根萌发 3 天左右子叶出土，幼苗细弱，子叶出土后约 23 天方发出初生叶（图 2）。

撒播。将种子密播于播种盘或经过精细整地的圃地。每平方米约播 1g。幼苗转木质化后移至大田继续培育。1 年生苗出圃。扦插成活较易，生产中常用扦插育苗。

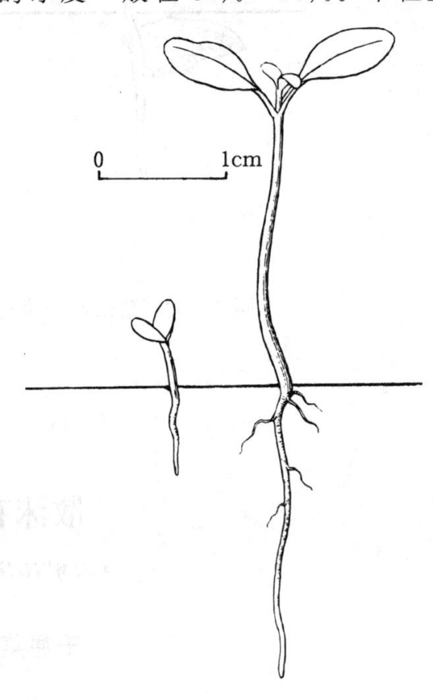

图 2　散沫花种子萌发后第 5、25 天幼苗的生长情况
（黄应钦绘）

（朱积余）

虾　子　花

Woodfordia fruticosa（L.）Kurz

（千屈菜科　Lythraceae）

生长习性、分布和用途　虾子花属 2 种，我国只有本文描述的 1 种。常绿或半落叶灌木，高 2～3（5）m，枝条横展下垂。喜肥力较高的湿润酸性土。热带树种，耐干热，也耐短时间

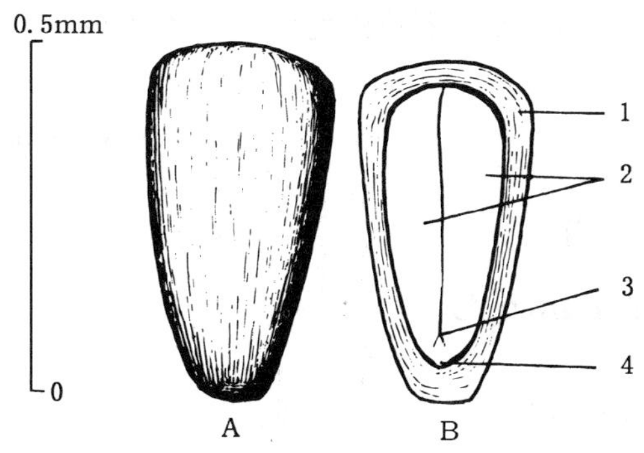

图 1　虾子花种子外形（A）及其纵切面（B）
1. 种皮　2. 子叶　3. 胚芽　4. 胚根
（黄应钦绘）

小。据广西南宁 1987～1989 年观察，2 月上旬～中旬花芽形成于 1～2 年生枝的叶腋，3 月上旬花始开，花期较长，3 月中旬～4 月中旬为盛花期，5 月上旬为末花期；5 月中旬果实开始成熟，5 月下旬～6 月上旬为果熟盛期。蒴果，线状长椭圆形，包藏于宿存萼筒内，膜质，长 10～15mm，径 4～6mm；成熟时由青色转变为灰褐色，5 月下旬～6 月下旬陆续室背开裂为 2 果瓣，种子散落。种子扁椭圆形或倒卵形，红褐色，长 0.5～0.6mm，宽 0.3～0.4mm，厚约 0.1mm。无胚乳，胚伸直，充满种子（图 1）。

果实的采收调制和种子贮藏　产果量甚丰。选择成熟盛期的果实，带果穗采摘。采得的果实置室内无风处阴干，果瓣开裂即得种子。鲜果出种率为 29%左右，每果有种子500～800 粒。种子的净度约 85%。种子极轻细，千粒重仅约 0.01g，每克有种子约 10 万粒。种子不宜在日光下曝晒，但可以干藏。短期贮藏可用布袋包装。贮藏 1 年以上需用坛罐密封，置避光荫凉处。贮藏 2 年以上需置 5℃低温环境。

0℃左右的极端低温。分布于粤、桂、滇。印度、斯里兰卡、缅甸、越南、印度尼西亚及马达加斯加也有分布。花多于叶，颇艳丽，供观赏。全株含鞣质，可提制栲胶。花干燥后可治痢疾、肝病、烫伤等。

开花结实　2 年生即开花结实，正常结实年龄在 4 年生以后。每年结实甚丰，无大小年现象。花两性，生于腋生的短聚伞状圆锥花序上，长约 3cm。萼筒花瓶状，橘红色，6 齿裂，长 9～15mm。花瓣小而薄，淡黄色，着生于萼齿间。雄蕊 12，突出萼外。子房上位，矩圆形，2 室，胚珠多数。花柱线形，超过雄蕊，柱头

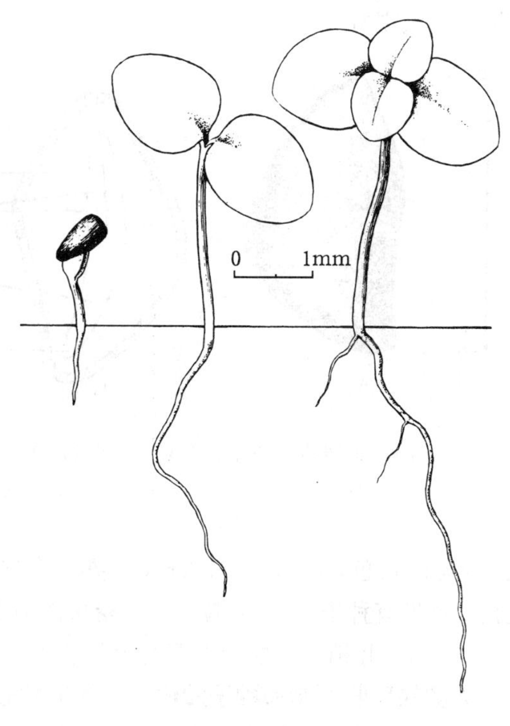

图 2　虾子花种子萌发后第 3、8、14 天幼苗的生长情况
（黄应钦绘）

发芽和播种　种子无休眠习性。发芽时日均温宜在 20℃以上。1987 年 7 月 29 日，日均温 29℃时，广西林业科学研究所在室内用滤纸作基质进行过发芽测定：8 月 1 日即开始萌发，发芽后第 3 天进入盛期，至 8 月 18 日发芽结束；从发芽之日起算，8 天发芽 42%；从播种之日起算，以 21 天计，发芽率为 52%。出土萌发。胚根萌发后第 3 天子叶伸出，再过 12 天发出初生叶。种子萌发和幼苗生长情况见图 2。

撒播。一般将种子密播于发芽盘内或经过细致整理、过筛平整的播种床上，播后加盖防

雨薄膜,精细管理。待幼苗发出初生叶约1个月左右,再移至容器或圃地培育。8个月或1年生苗出圃。生产上多用压条繁殖。

(朱积余)

番 茉 莉

Brunfelsia americana L.

(茄科 Solanaceae)

生长习性、分布和用途 番茉莉属有25～30种,我国引入栽培3种,本文描述1种。常绿灌木,高1.2～1.6m。喜高温湿润的气候环境,能耐0℃左右极端低温,苗期忌霜冻。适生于湿润的酸性土,干旱贫瘠土生长不良。原产美洲安的列斯群岛。我国滇、桂引种,用于庭院绿化,供观赏。

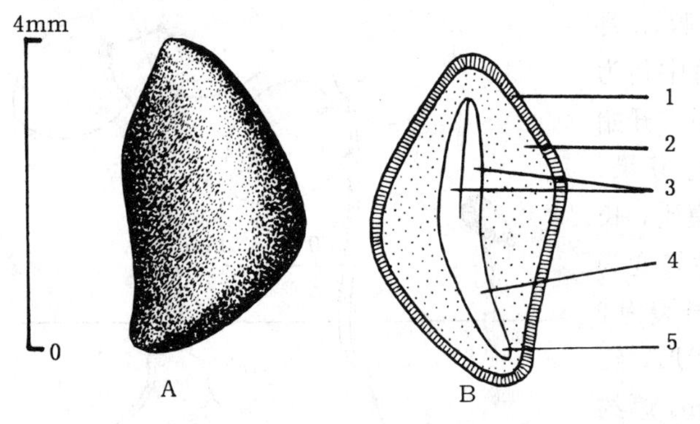

图1 番茉莉种子外形(A)及其纵切面(B)

1.种皮 2.胚乳 3.子叶 4.胚轴 5.胚根

(黄应钦绘)

开花结实 3～4年生开始结实,正常结实期在8年生以后。结实无大小年现象,但越冬嫩枝受寒害,会影响次年结实。花两性,单生、顶生或腋生。萼钟状,5裂。花冠高脚碟状,筒长,黄色,檐部5裂。雄蕊4,着生于花冠上,内藏。子房上位,2室,胚珠多数。据广西南宁1987年观察,5月下旬为始花期,6月中旬为盛花期,7月下旬为末花期;9月上旬果实开始成熟,下旬为成熟盛期。浆果,近球形,有两条纵槽,未熟时青色,成熟时黄色,径1.3～1.6cm,成熟后在树上可存留20～30天,9月下旬～10月中旬陆续脱落。每果有种子22～24粒。种子菱状椭圆形,棕褐色,长3～4.3mm,宽1.5～2.3mm,厚1.5～2mm。胚稍弯,埋于丰富的内胚乳中(图1)。

果实的采收调制和种子贮藏 选摘颗粒较大、发育饱满的成熟果实。采得的果实堆放数日,软熟后装入布袋置水中搓洗,淘去果皮等杂质即得纯净种子。鲜果出种率约17%,种子的净度可达95%～99%。种子可以适当阴干,但忌日晒。含水量12%～20%的种子千粒重7.5(6～9)g,每千克有纯净种子13(11～17)万粒。直接运输果实,到达目的地后再行调制。供贮藏的种子宜在晾至半干时混以湿沙,贮藏期一般为半年。

发芽和播种 种子无休眠习性。发芽时日均温宜在20℃以上。1987年广西林业科学研究所用当年新采种子,在室内作发芽测定,基质为过滤纸,光照为室内自然光,发芽时日均温24～26℃:9月26日播种,10月9日开始发芽,9～11日为发芽盛期,17日发芽终止;从开始萌发至发芽高峰日,3天发芽80%;从播种至发芽终止,20天的发芽率为90%。出土萌发。胚根萌发后8天子叶出土,18天发出初生叶。发芽及幼苗生长情况见图2。

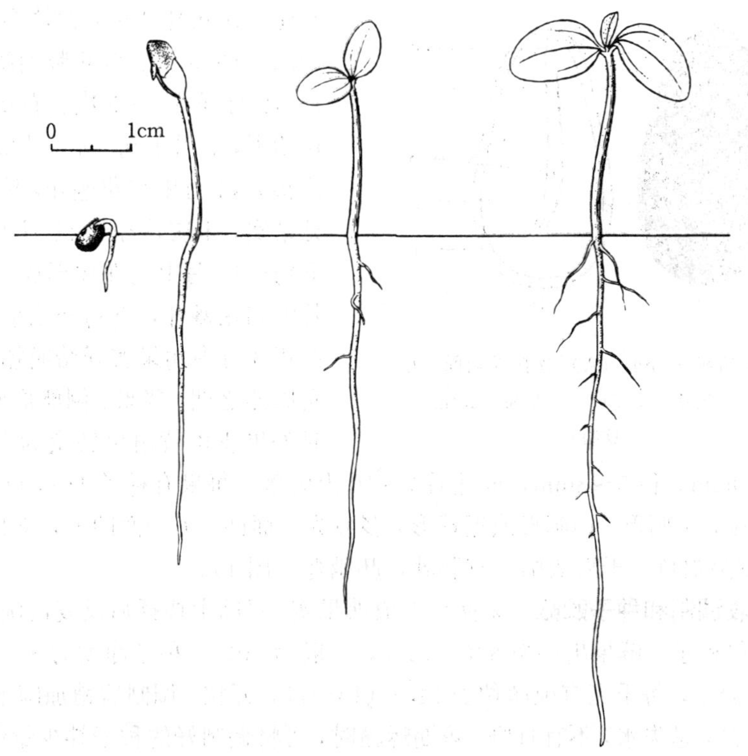

图 2 番茉莉种子萌发后第 7、12、14、18 天幼苗生长情况
（黄应钦绘）

撒播。将种子密播于经过细致整理的播种床或沙盘内，每平方米约播 4g。幼苗转木质化时移至大田培育。1 年生苗出圃。亦可用扦插、高空压条法繁殖。

（朱积余）

夜 香 树

Cestrum nocturnum L.

（茄科 Solanaceae）

生长习性、分布和用途 夜香树属约 160 种，我国引入栽培 3 种，本文描述 1 种。直立或半攀援状常绿灌木，高 1～3m。萌芽力强。能耐－3℃的短期低温，忌霜冻。适生于肥沃湿润、排水良好的酸性土。原产美洲西印度群岛，现广泛种植于热带地区。滇、桂、粤、琼、台、闽有栽培。花芳香味浓，入夜尤香，故名。供提浸膏或精油，作天然调香原料。通常用于庭园绿化，供观赏。

开花结实 2～3 年生开始结实，4 年生以后为正常结实期。大小年现象不明显，但因败育而结实率很低。花两性。伞房式聚伞花序，顶生或腋生。花萼钟状，萼齿 5。花冠白色至黄绿色，高脚碟状，裂片 5。雄蕊 5，着生于花冠筒中部。花药极短，褐色，内藏。子房上位，

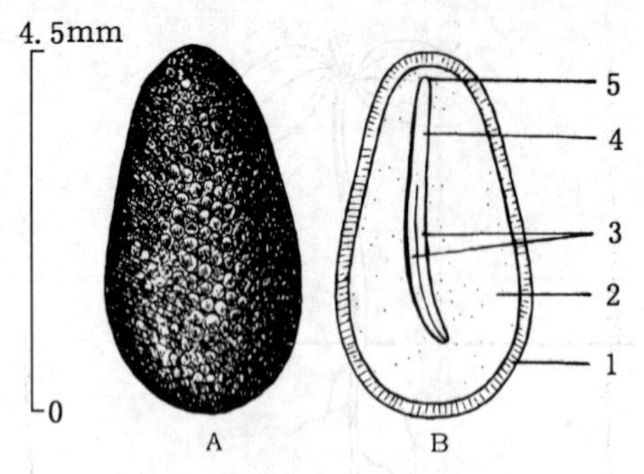

图 1　夜香树种子外形（A）及其纵切面（B）
1. 种皮　2. 胚乳　3. 子叶　4. 胚轴　5. 胚根
（黄应钦绘）

2 室，有胚珠 1～4 枚，多数为 2 枚。据 1986～1989 年在西双版纳勐仑观察，每年 4～12 月开花 4～5 次：首次花芽在 3 月中旬始现，3 月下旬～4 月上旬花序形成，花蕾膨大，4 月中旬花盛开，下旬开花结束，少见结实；末次花芽于 11 月中旬始现，11 月下旬～12 月上旬花序形成，花蕾膨大，12 月中旬花盛开，下旬开花结束，幼果形成，翌年 3 月中旬果实开始成熟，4 月上旬为果实成熟盛期。浆果，倒卵形或卵圆形，成熟时外果皮由黄白色转变为白色，表面无光泽。果长 6～10mm，径 5～8mm，成熟后 7～10 天脱落。每果有种子 1～4 粒，多数为 2 粒。种子形状不规则，半圆形、椭圆形或近球形，多数为一面凸，另一面凹入，长 2～4.5 mm，径 1～2mm。种皮黑褐色，干后皱缩。有胚乳，胚稍弯（图 1）。

果实的采收调制和种子贮藏　采摘树上成熟果实，置筛上搓揉后反复淘洗，漂去果皮、果肉及果柄，即得种子。鲜果出种率 8%～12%，一般为 10%。种子净度为 85%～95%。千粒重 6.8（6～7.5）g，每千克有纯净种子 14.7（13～17）万粒。调制后稍加晾干的种子含水量约为 40%～53%，忌失水，不宜日晒。短期运输时，可将调制好的种子拌少量湿润锯木屑，运抵后及时播种。贮藏时需混湿沙，贮藏期以不超过 1 个月为宜。

发芽和播种　种子无休眠习性，宜随采随播。发芽时日均温宜在 20℃以上。1989 年 4 月 6 日，西双版纳热带植物园在室内用培养皿，对新鲜种子作过发芽测定，当时日均温为 23～25℃：置床后 5 天（4 月 10 日）开始发芽，4 月 11～12 日为发芽盛期，4 月 14 日发芽结束；从开始发芽之日至发芽高峰，5 天发芽 76%；从置床之日起算，以 9 天计，发芽率为 90%。出土萌发。子叶带种壳出土。

点播或撒播。每平方米播种 1.2～2.5g，覆土以不见种子为限。幼苗具 2～3 片初生叶时移植，继续培育 10～12 个月即可出圃定植。生产上多用扦插育苗，扦插的成活率一般可达 80%。也可分株繁殖。

（程必强）

枸　杞　属

Lycium L.

（茄科　Solanaceae）

生长习性、分布和用途　本属约 80 种，产西半球温带及亚热带，我国有 7 种，分布于我国南北各省，北起陕、甘、冀、豫，南至闽、粤、川、滇，主产西北、华北。本文描述 2 种。灌木，小枝通常具刺。喜光，耐干旱瘠薄，宜冷凉气候，耐寒，耐盐碱，忌高温及水涝。对土壤质地要求不严，适应性强，但以排水良好、土质肥沃的中性或微碱性沙壤土为好。果实及根皮为名贵药材，嫩叶可食用，又是保土灌木。其名称、分布和用途见表 1。

表 1　枸杞属树种的名称、分布及用途

中　名	学　名	分　布	用　途
宁夏枸杞	*L. barbarum* L.	华北、西北。以宁夏中宁栽培历史最久，最著名	根、果、叶药用，果做滋补强壮剂，沙地造林
枸杞	*L. chinense* Mill.	辽宁以南，云南以北，东自台，西至甘，分布极广	根、果入药，嫩叶作蔬菜，种子榨油，水土保持

开花结实　3～4 年生开始结实，5 年生以后进入盛果期，40 年左右结实量下降。宁夏枸杞在宁夏、甘肃一带 1 年 2 次开花结果：第 1 次花期在 5 月上旬，果熟期 6 月中旬～8 月上旬；第 2 次 8 月开花，9～10 月果实成熟。山东、河北等地 4 月下旬～10 月开花结果陆续不断。花两性，单生或簇生于叶腋。花萼钟状，3～5 齿裂，宿存。花冠漏斗状，4～5 裂，紫红色或淡紫色。雄蕊 5～4，着生于花冠筒部。子房上位，2 室，柱头 2 浅裂，胚珠多数或少数。枸杞属开花结实物候见表 2。

表 2　枸杞属树种的开花结实物候

树　种	观察地点	开　花			果实成熟		果实散落期
		始　期	盛　期	末　期	始　期	盛　期	
宁夏枸杞①	宁夏中宁	5 月上中旬	～	10 月上旬	6 月中旬	11 月中旬	分批脱落
	北京	5 月	～	9 月	6 月	10 月	10 月以后
枸杞	内蒙古呼和浩特	6 月上旬	—	—	7 月下旬	—	7 月以后
	辽宁、黑龙江	7 月	～	8 月	—	10 月	10 月以后
	甘肃民勤	4 月下旬	～	6 月中旬	6 月下旬	7 月下旬	7 月以后
	北京	6 月	～	9 月	9 月	10 月	10 月以后

① 开花和果实成熟集中在 5～6 月和 8～9 月两季

果为浆果，具肉质果皮，球形或长圆形，有宿萼，熟时红色。种子扁平，少数至多枚。种皮骨质，密布网纹状凹穴。胚乳丰富，胚弯曲，子叶半圆棒状。枸杞属果实和种子形态特征见表 3。枸杞果实和种子形态见图 1。

表 3　枸杞属树种果实和种子的形态特征

树　种	果　实			种　子		
	形　状	大小（cm）	颜　色	形　状	大小（mm）	颜　色
宁夏枸杞	圆形或长圆形	长 0.5～3，径 0.5～1.2 种子 20～50 粒	黄棕色	扁肾形	长 2	黄白色
枸杞	圆形或长圆形	长 0.5～3，径 0.5～1.2	橘红色	扁肾形	长 2～3，宽 1.5～2	黄色

果实的采集调制和贮藏　从6月中旬～11月上旬，选择呈现红色或橘红色、柔软有浆的果实分批采收。要轻采轻放，防止压伤。采下的鲜果放在席子上摊晒，厚约1.5cm，切忌翻动，以免果实变黑，约10天左右即可干燥，去果柄，含水量为10%～20%时，放在通风干燥处贮藏。也可将采下果实置始温30～50℃的温水中浸泡24小时，轻加揉搓后淘出种子，干燥后密封贮藏。种子净度及质量见表4。

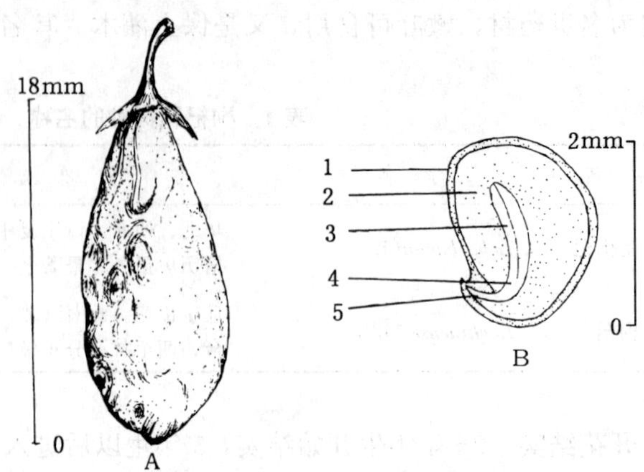

图1　枸杞浆果外形（A）及其种子纵切面（B）
1. 种皮　2. 胚乳　3. 子叶　4. 胚轴　5. 胚根
（姚桂英绘）

表4　枸杞属树种出种率和种子的净度、质量

树　种	鲜果出种率（%）	净　度（%）	千粒重（g）	每克纯净种子粒数
宁夏枸杞	20～25	85～90	0.8	1 250
枸杞	35～40	85～90	1.9～2.4	410～530

发芽前处理　生产上常用3倍湿沙同种子混拌，放在20℃左右的室内催芽，保持湿润，待种子有30%露白时再拌湿润土灰播种。有的场圃用始温30～50℃水浸种1昼夜，种子吸胀后播种。

发芽测定　曾经在室温20～25℃的条件下测定过宁夏枸杞的发芽能力，发芽基质为脱脂棉及滤纸，8天发芽率为99%；3天计算发芽势为64%。枸杞发芽率为20%～30%。用靛蓝或四唑测定生活力时，先用30℃水浸泡2天，沿边缘剪破种皮，挤出种胚，进行染色。

播种　条播，行距15～20cm，每平方米播种0.4～0.6g。覆土厚度0.5～1.0cm。播后6～7天即可出苗。出土萌发。子叶2枚，肉质较厚。幼苗形态如图2。

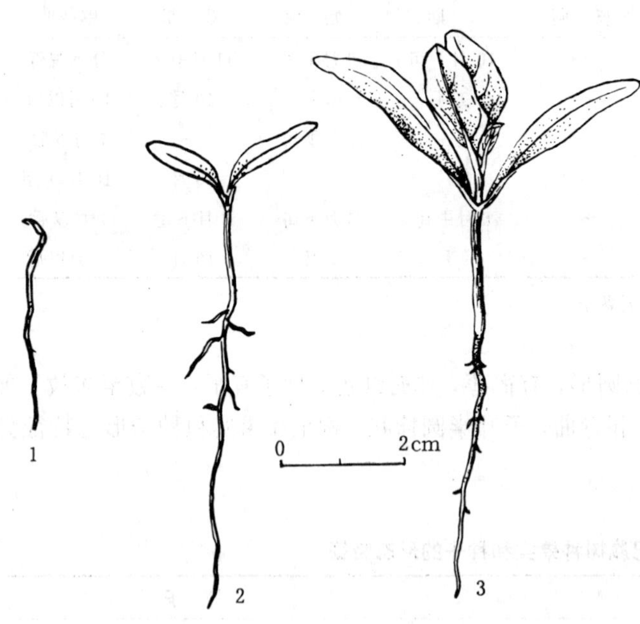

图2　枸杞幼苗生长情况
1. 子叶出土　2. 子叶张开（一子叶较小）　3. 初生叶互生
（姚桂英仿《主要树木种苗图谱》）

枸杞还可用分株、扦插、压条等方法繁殖。

（李荣晨）

大　花　茄

Solanum wrightii Benth.

（茄科　Solanaceae）

生长习性、分布和用途　茄属有 1 500～2 000 种，我国约产 39 种，多为草本，本文描述引入的 1 种。常绿或半落叶小乔木，高达 5～7m，胸径 7～10cm。能耐短期 5～6℃低温，忌霜冻。喜肥沃湿润的酸性土壤，干旱贫瘠地生长不良。原产南美洲玻利维亚和巴西。滇、桂、粤、琼引种栽培。花期特长，目前多栽培供观赏。

开花结实　3～4 年生开始结实，正常结实期在 6 年生以后。结实无大小年现象，影响结实的主要原因是水肥条件差和冬季的低温寒害。花两性。聚伞花序或总状花序，顶生或着生于嫩枝干上，少有腋生。萼 4～5 齿裂。花冠紫蓝色，5 裂。雄蕊 5，着生于花冠筒喉部，花丝短，药长于花丝。子房上位，2 室，胚珠多数。据广西南宁 1986～1987 年观察，5 月中旬为始花期，5 月下旬～8 月上旬陆续有花开放，盛期不明显，8 月下旬花期基本结束；8 月上旬果实开始成熟，8 月中旬～11 月下旬陆续有果成熟，9～10 月为果实成熟盛期。浆果，圆球形，未熟时青色，成熟时黄褐色，径3.5～5cm。果实成熟后约 4 天脱落。每果有种子数百粒，种子扁椭圆形，棕灰色，长 2.5～3mm，宽 2～2.8mm，厚约 1.2 mm。胚弓曲，埋于内胚乳中（图 1）。

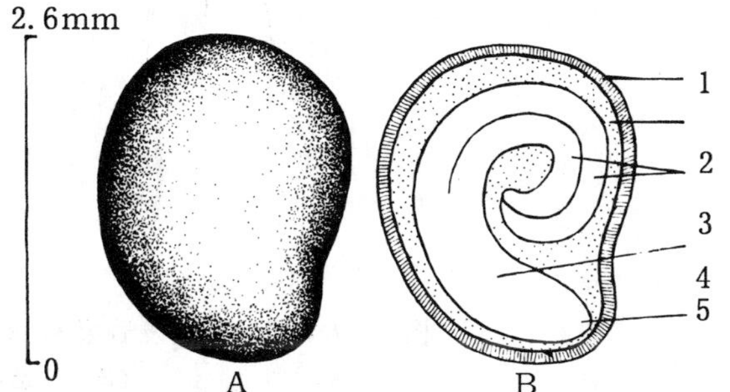

图 1　大花茄种子外形（A）及其纵切面（B）
1. 种皮　2. 胚乳　3. 子叶　4. 胚轴　5. 胚根
（黄应钦绘）

果实的采收调制和种子贮藏　当果实由青色转黄色时选择果形圆整、发育良好的采摘。采集后室内堆放 2～3 天，浆果充分软熟后装入紧密布袋置水中搓擦，淘去果皮等杂质即得纯净种子，摊放于室内阴干。鲜果出种率 6%～10%。种子净度可达 92%～99%。千粒重 4～5g，每克有纯净种子 200～250 粒。种子可以晾干，但不宜过分曝晒。适于贮藏的含水量在 10%左右。贮藏时可用袋或坛罐盛装，置室内干爽阴凉处，贮藏期一般约为 6 个月。

发芽和播种　种子无明显的休眠现象，但发芽进程稍缓慢。发芽时日均温宜在 20℃以上。1987 年 9～10 月，广西林业科学研究所用当年新采种子在室内作过发芽测定，基质为滤纸，利用室内自然光，当时日均温在 22℃上下。9 月 9 日置床，10 月 17 日开始发芽，发芽盛期不明显，11 月 18 日发芽终止。发芽率为 35%。出土萌发。胚根萌发后约 5 天子叶出土，11 天发出初生叶。种子萌发及幼苗生长情况见图 2。

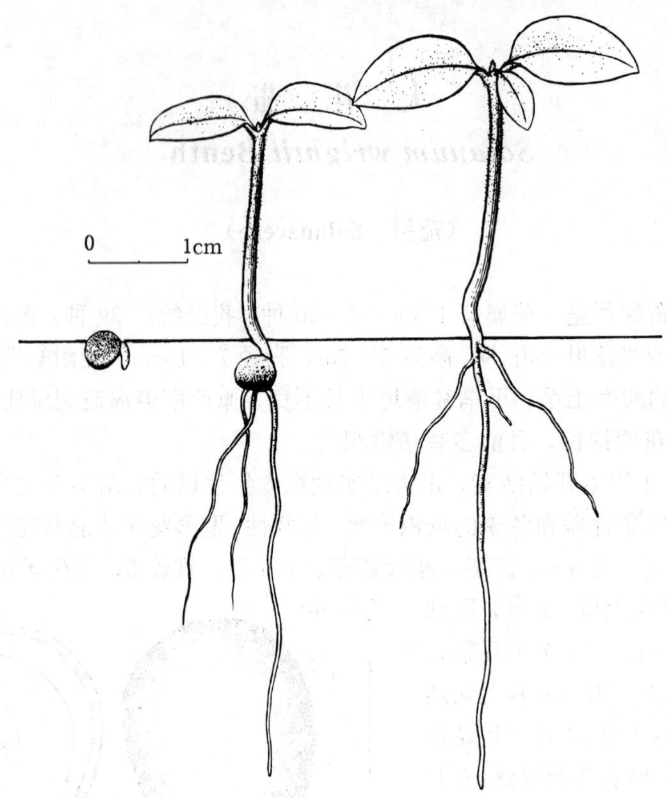

图 2　大花茄种子萌发后第 1、8、12 天幼苗的生长情况

（黄应钦绘）

撒播。密播于经过细致整地的播种床或沙盘内。每平方米的播种量约 5～6g。幼苗半木质化时移至圃地继续培育。1 年生苗出圃。

（朱积余）

泡　桐　属
Paulownia Sieb. et Zucc.

（玄参科　Scrophulariaceae）

生长习性、分布和用途　泡桐属共 7 种，我国全产，除东北北部、内蒙古、新疆北部、西藏等地外，全国均有分布，栽培或野生，有些地区正在引种。本文描述 6 种。高大落叶乔木，但在热带为常绿。白花泡桐在越南、老挝，毛泡桐在朝鲜半岛、日本也有分布。适应性较强，忌积水，在较瘠薄的低山丘陵或平原地也能生长。材质优良，轻而韧，具有很强的防潮隔热性能，导音性好。不翘不裂，易于加工。叶、花、果均可入药。这 6 个树种的名称、生长、分布和用途见表 1。

表 1　泡桐属树种的名称、生长、分布和用途

中　名	学　名	生　长			分　布	用　途	供　稿
		树龄	树高(m)	胸径(cm)			
楸叶泡桐	*P. catalpifolia* Gong Tong	10	—	41	鲁、冀、晋、豫、陕，通常栽培，太行山有野生	材用	301
兰考泡桐	*P. elongata* S. Y. Hu	19	12	104	冀、豫、晋、陕、鲁、鄂、皖、苏，多为栽培，豫、川有野生	材用、农桐间作	303
川泡桐	*P. forgesii* Franch.	18	—	55	鄂西、湘西、川、黔、滇，野生或栽培，引种到皖、豫、粤	材用	502
白花泡桐	*P. fortunei* (Seem.) Hemsl.	25	31	100	华东、中南、西南，野生或栽培。近年晋、冀、豫、陕引种	材用、行道、庭院树	301
台湾泡桐	*P. kawakamii* Ito	—	—	—	鄂、湘、皖、浙、闽、台、粤、桂，多为野生，四川等有引种	材用	301
毛泡桐	*P. tomentosa* (Thunb.) Steud.	5	8	30.6	辽南、冀、豫、晋、陕、鲁、皖、苏、鄂、赣、甘，通常栽培，西部省区有野生。日本、朝鲜半岛、欧洲、北美有栽培	材用、绿化	303

开花结实　花两性。顶生圆锥花序由多数具总梗或无总梗（川泡桐）的聚伞花序组成，花序成圆锥形、金字塔形或圆柱形。萼 5 深裂，裂片稍不等，后方 1 枚较大，宿存。花冠大，裂片 5，2 唇形，紫色或白色，漏斗状钟形或管状漏斗形。雄蕊 4，2 强，着生于花冠筒基部。子房 2 室，每室胚珠多数，中轴胎座，花柱细长，柱头 2 裂。开始开花年龄因树种和立地条件而差异较大。一般情况下毛泡桐、台湾泡桐 2～4 年生，楸叶泡桐 6～8 年生，川泡桐 7～8 年生开始开花结实。开花时间和花期长短也因种类、自然条件不同而有差异。大体上纬度每升高 1°，花期延迟一天半。例如广西桂林（北纬约 25°）3 月中旬初花，郑州（北纬约 35°）4 月初初花，北京（北纬约 40°）初花期则在 4 月中下旬（蒋建平，1990）。花期延续时间 15～ 30 天，以台湾泡桐延续的时间最长，楸叶泡桐最短。这 6 个树种在不同地点的开花结实物候期见表 2。在同一地点，不同树种的初花期相差很大。例如同在郑州市区，1974 年楸叶泡桐、兰考泡桐和毛泡桐的初花期分别为 4 月 1 日、6 日和 13 日，盛花期分别为 4 月 10 日、13 日和 5 月 17 日，末花期分别为 4 月 20 日、5 月 2 日和 5 月 10 日（蒋建平，1990）。

表 2　泡桐属树种的开花结实物候期

树　种	观察地点	观察年份	开　花				果实成熟期
			始花期	盛花期	末花期	天　数	
楸叶泡桐	陕西武功	1975	4 月 5 日	4 月 9 日	4 月 20 日	15	7～8 月
兰考泡桐	河南郑州	1976	4 月 16 日	4 月 22 日	5 月 5 日	19	9 月下旬～10 月中旬
川泡桐	四川雅安	1987	4 月上旬	4 月中旬	5 月初	—	10 月中下旬
白花泡桐	浙江长兴	1978	4 月 1 日	4 月 10 日	4 月 25 日	24	10 月～11 月上旬
台湾泡桐	浙江长兴	1975	4 月 17 日	4 月 23 日	5 月 15 日	29	8～9 月
毛泡桐	北京	—	4 月 25 日	4 月 28 日	5 月 4 日	9	9 月 18 日～11 月 3 日

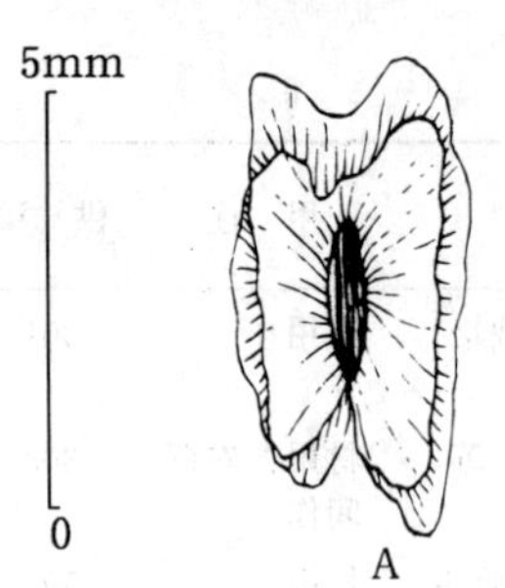

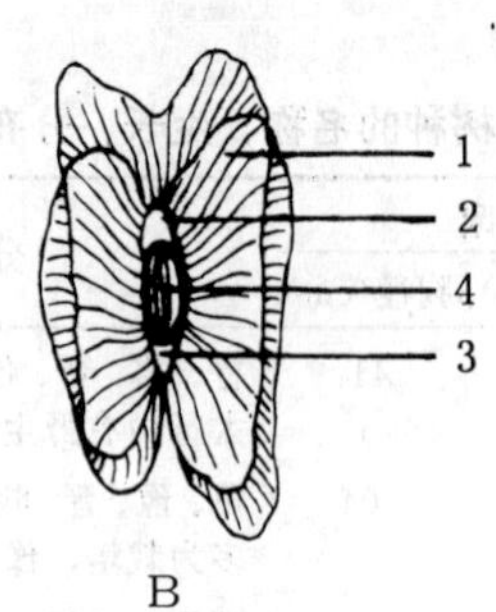

图1 川泡桐种子外形（A）及其纵切面（B）
1. 膜质翅 2. 种皮 3. 胚乳 4. 胚
（孟玲绘）

泡桐授粉后，子房迅速膨大形成果实。7月份果实外形停止生长，颜色由绿变褐，果期约为180～200天。泡桐结实能力差异很大，毛泡桐、川泡桐、台湾泡桐和部分白花泡桐结实力较强，兰考泡桐、楸叶泡桐结实较少。蒴果，木质，2瓣裂或不完全4瓣裂。种子小而多，两侧有乳白色膜质翅，具少量胚乳（图1）。果实和种子形态特征见表3。每个蒴果内含种子数量和种子质量见表4。在正常的天然授粉情况下，毛泡桐、川泡桐、台湾泡桐座果率在30%～50%以上，白花泡桐为20%～30%，兰考泡桐、楸叶泡桐在0.1%以下。

表3 泡桐属树种果实和种子形态特征

树种	蒴果			种子	
	形状	长(cm)	果皮厚(mm)	形状	连翅长(mm)
楸叶泡桐	椭圆形	4.5～5.5	2～2.8	椭圆形扁平，翅灰白色	2.5～5
兰考泡桐	卵形，稀卵状椭圆形	3.5～5	1～2.5	—	4～5（6）
川泡桐	卵圆形或椭圆状卵形	3.5	1	长圆形稍扁平，暗灰色	4.5～5（6）
白花泡桐	长圆形或长圆状椭圆形	6～8	3～6	扁平，矩圆形	6～8（9）
台湾泡桐	卵圆形，果皮薄	2.5～4	1	长圆形，种翅褐色	2.5～3.5（4）
毛泡桐	卵圆形	3～4.5	1	—	2.5～4（5）

表4 泡桐属树种种子粒数和千粒重

树种	每果种子粒数	千粒重（mg）	每克种子粒数（粒）
楸叶泡桐	300～500	150	6 700
兰考泡桐	500～800	230	4 300
川泡桐	1 000以上	220	4 500
白花泡桐	1 000以上	290	3 600
台湾泡桐	—	400	2 500
毛泡桐	1 000以上	260	3 800

果实的采收调制和种子贮藏 秋季果壳由绿色变成黄褐色，个别果实先端稍开裂时即可采种。同一个树种在不同的地区果实成熟期也不一致。例如白花泡桐在杭州地区成熟期为10月下旬～11月上旬，而桂林地区则是11月中旬。果实采下后晾晒5～7天，使果壳开裂脱出种子。除去杂质，并使种子充分干燥。

种子贮藏环境条件要求阴凉、干燥而又通风，防止受热受潮。据饭三男（1978）研究，毛泡桐种子在2～5℃种子库内，或在常温室内贮藏于放有硅胶的玻璃干燥器内，5年仍可保持70%以上的发芽率；同一批种子在－20℃冷藏库内或在常温室内纸袋贮藏，5年后发芽率均为0。据在四川雅安研究，川泡桐在0～5℃冰箱内双层塑料袋密封贮藏，2年半后仍保持59%的发芽率。泡桐属种子贮藏条件及贮藏效果见表 5。

表 5　泡桐属树种种子贮藏条件及贮藏效果

树　种	贮藏地点	贮藏条件	贮藏期限	发芽率（%）
楸叶泡桐	北京	常温室内纸袋	3 年	0
兰考泡桐	北京	常温室内纸袋	2 年半	52
川泡桐	四川雅安	常温室内双层塑料袋	1 年 3 个月	1
	四川雅安	0～5℃冰箱内双层塑料袋	2 年半	59
白花泡桐	北京	常温室内纸袋	2 年半	94
毛泡桐①	日本东京	A. 常温室内纸袋	5 年	0
		B. 室内玻璃干燥器（内放硅胶）	5 年	72～88
		C. 2℃冷藏库	5 年	77～93
		D. －20℃冷藏库	5 年	0
		E. 5℃冷藏库	5 年	76～89

①　材料引自（饭三男，1978）

发芽和播种　对这 6 种泡桐种子发芽测定的结果（管康林等，1982）表明，每日光照 8 或 12 小时，20℃/30℃或 25℃/30℃变温条件能明显提高发芽率；而黑暗的恒温条件下发芽率几乎接近 0。周佑勋（1988）的研究也证实，白花泡桐有较强的休眠习性，在黑暗中完全不能发芽；一次性给光 10 分钟便能出现发芽反应，但发芽率仍极低；一次性给光的时间延长，发芽率也随着增加，但以每天连续给光 8 小时为最好。她的研究还证明，白花泡桐室内的发芽最适温度是 30℃/20℃的昼夜变温，其次是 35℃的恒温，在 15℃或 40℃下便不能萌发。她建议田间播种后用地膜覆盖，为发芽创造良好的光、温和水分条件。泡桐种子的发芽测定条件和发芽能力见表 6。

表 6　泡桐种子发芽测定条件和发芽能力举例

树　种	产　地	发芽条件		发芽率	测定时间
		光　照	温度（℃）	（%）	（年-月）
楸叶泡桐	北京	每日 8 小时	20/30	56	1988
兰考泡桐	河南	每日 8 小时	25/30	26	1986
	河南	每日 8 小时	25	22	1986
	河南	黑暗	25/30	1	1986
	河南	黑暗	25	0	1986
川泡桐	四川沐川	黑暗 10 天后再每日 8 小时	20～22/30～32	51	—
	四川沐川	每日 8 小时	20～22/30～32	77	—
白花泡桐	贵州安龙	每日 8 小时	20/30	21	1988-02
	重庆黔江	每日 8 小时	20/30	90	1988-03
	浙江浦江	每天 10 分钟	20/28	37	1982（发表）
	浙江浦江	每日 12 小时	20/28	70	1982（发表）
台湾泡桐	四川沐川	每日 8 小时	25/30	93	1988-03
毛泡桐	河南开封	每日 8 小时	25	84	1986
	河南开封	每日 8 小时	25	44	1986
	河南开封	黑暗	25/30	3	1986
	河南开封	黑暗	25	1	1986

河南地区一般播种期为3月下旬～4月中旬；南方2月上中旬即可播种。每平方米播种量0.6～1.2g。播种前以始温40℃水浸种10分钟，再以冷水浸种24小时，后混沙催芽6～8天。出土萌发。子叶2（图2）。

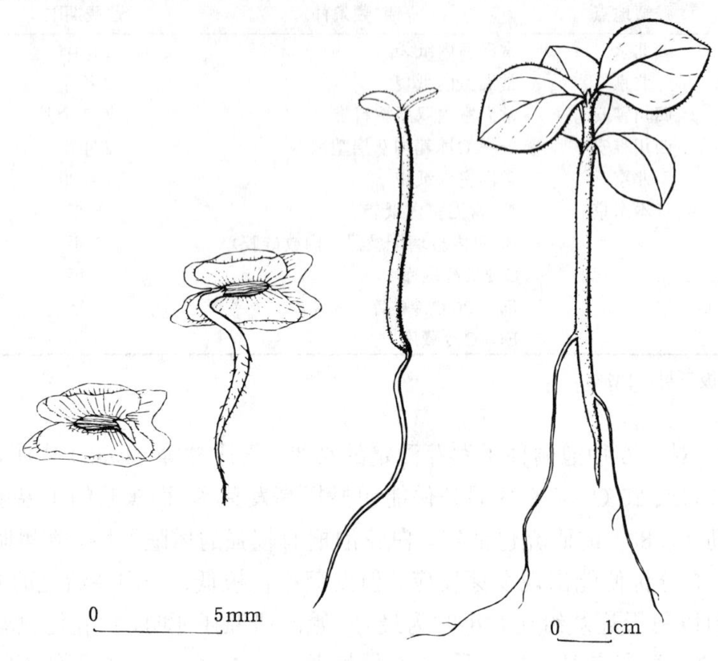

图2　川泡桐种子萌发第1、3、7、35天幼苗生长情况
（孟玲绘）

用营养钵进行芽苗移栽，可以节省种子，效果好。先在光照发芽器内用20/30℃的变温条件发芽，10天左右出苗，苗长1cm左右子叶展开时移栽。每个营养钵内栽苗1～2株。另一种营养钵育苗法是在温室内电热床（30～35℃）上育苗（张维栋，1986）。播种育苗当年苗高可达3m。

泡桐除播种育苗以外，应用较多的是埋根育苗。

（陶章安）

毛叶丁香罗勒

Ocimum gratissimum L. var. *suave*（Willd.）Hook. f.

（唇形科　Labiatae）

生长习性、分布和用途　罗勒属约100～150种，我国连引入栽培的有5种，本文描述引入的1变种。常绿直立灌木，高达2.5m。能耐－3℃左右的短期低温。在肥沃湿润的酸性或中性土壤上均能正常生长。原产热带非洲（马达加斯加），热带各地广为栽培。我国滇、桂、粤、琼、闽、台、浙、苏有栽培。茎叶果穗极香，可提芳香油，又可入药，有祛风湿、健胃、镇痛之效。种子含油脂。

开花结实　栽植后 1 年生即开花结实，2～6 年生为正常结实期，无大小年现象。花两性，黄白色。总状花序顶生或腋生，常组成复总状花序，长 11～20cm，径 6～21cm。萼钟状，5 齿，上部中齿宽大，边缘下延，下 2 齿极小，完全合生成 2 刺芒。花冠长于萼，上唇 4 浅裂，近相等，下唇矩圆形，全缘。雄蕊 4，花药卵圆形。子房上位，1 室，1 胚珠。据 1983～1987 年云南西双版纳观察，7 月中旬花芽始现，7 月下旬～9 月中旬花序形成，花蕾膨大，10 月中旬花盛开，12 月中旬开花结束；12 月上旬果实开始成熟，翌年 1～2 月为果实盛熟期。成熟时果穗由绿色转变为褐色，10～15 天脱落。小坚果，近球形。种子 1 粒，褐色，表面有光泽，径约 1mm。无胚乳，子叶薄。

果实的采收调制和种子贮藏　果熟盛期在树上采摘，采集的果穗垫以塑料薄膜置日光下晒干、翻动、敲打或搓揉，并用筛清除杂物，得出坚果，用作播种材料，通称种子。鲜果的出籽率为 15%。种子的净度为 90%～96%。千粒重 0.55（0.45～0.8）g，每克约 1 200～2 200 粒。调制后的种子含水量约为 10%。运输时可用木箱或塑料袋包装。贮藏时可用一般容器干藏。常温条件下室内干燥瓶藏 12 个月后发芽率为 76%（程必强和马信祥，1987）。

发芽和播种　种子无休眠习性。发芽时日均温宜在 20℃以上。1981 年 3 月 31 日，西双版纳热带植物园在室外播种盆内用新采的种子作发芽测定，播后 9 天（4 月 8 日）开始发芽，4 月 25 日发芽结束，有明显的发芽盛期。从发芽之日起算，15 天发芽 50%；从播种至发芽终了，26 天的发芽率为 76%。出土萌发。子叶出土后 5～7 天展出初生叶。

撒播。每平方米播种 1.5～2g，不覆土。60～90 天，幼苗具 3～5 片初生叶时定植。生产中也有不经移植，直接将种子条播至大田培育。通常就地栽培8～10 个月即采割收获。采种母树可留至第 3 年。

（程必强）

黑龙江百里香

Thymus amurensis Klok.

（唇形科　Labiatae）

生长习性、分布和用途　百里香属约 400 种，我国 12 种，产西藏、青海及黄河以北地区，本文仅介绍黑龙江百里香 1 种。矮小落叶灌木，茎纤细，常在基部分枝，径约1～1.5mm，匍匐。不育枝生自茎末端或基部，直立或上升，高 1.5～8cm。花枝直立，高 6～10cm。喜光，耐寒，耐干旱瘠薄，可生于砾石坡地。产于大兴安岭北部。俄罗斯远东地区也有分布。全株可提芳香油，为优良的羊饲料，可供观赏。

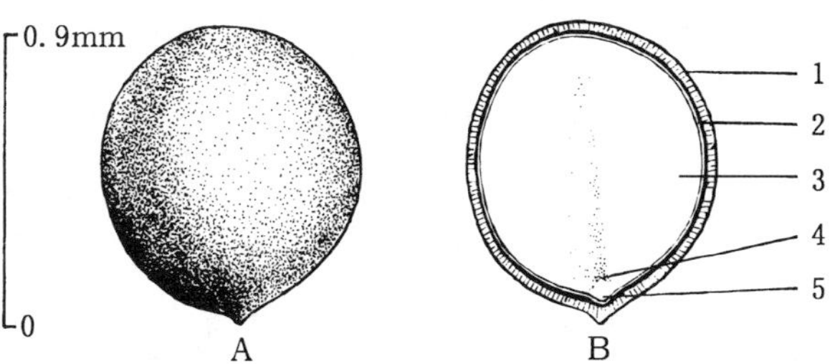

图 1　黑龙江百里香小坚果的外形（A）及其纵切面（B）
1. 果皮　2. 种皮　3. 子叶　4. 胚芽　5. 胚根
（梁鸣、黄应钦绘）

开花结实　约 2～3 年生开始结实。

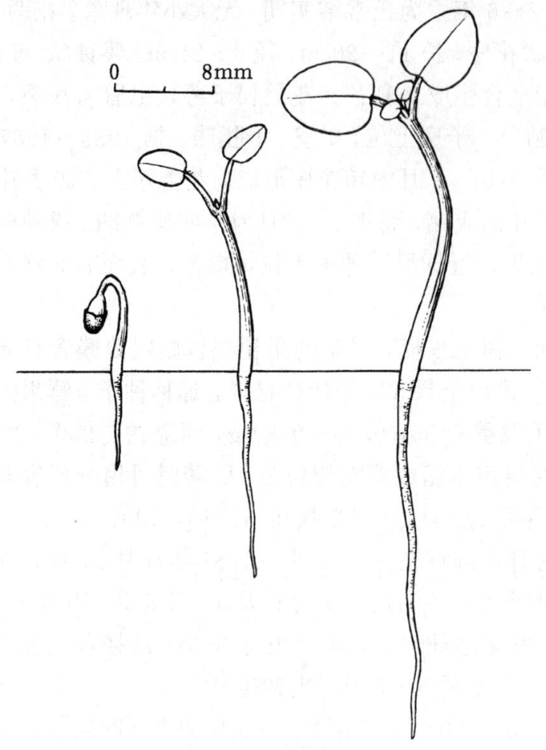

图 2　黑龙江百里香种子萌发后第 4、9、17 天幼苗生长情况
（曹雅范、黄应钦绘）

花两性，2 唇形，轮伞花序聚成头状，生于花枝顶端，下部常有疏轮伞花序。花萼管状钟形，长 4.5～5mm。花冠长约为花萼的 2 倍，淡红紫色，被白长柔毛。雄蕊 4，2 强。子房上位，2 室，每室 2 胚珠，花柱单一，柱头 2 裂。花期 6 月上旬～9 月上旬，开花期长达 80～90 天。果常由 4 个小坚果组成。小坚果近球形，径 0.7～1.0mm，黑色，光滑，种子无胚乳（图 1）。果期 7 月中旬～10 月上旬，边开花边成熟，成熟后不易脱落。

果实的采收调制和种子贮藏　7 月中旬果实陆续成熟，可陆续采集，采后晾干搓揉，风选后所得纯净小坚果即为播种材料，通称种子。净度 85%～95%，千粒重 0.25～0.30g，每克有 3 000～4 000 粒。曾经将未脱粒的小坚果在低温条件下（0～5℃）贮藏 2 年，发芽率未见明显降低。

发芽和播种　种子极易发芽，7 月中旬～8 月上旬采收的种子可随采随播，5～7 天出苗；当年苗高 10～15cm，可安全越冬。苗床条播，每平方米播种 0.5g，覆土厚 0.5cm 左右，发芽率可达 80%～90%，出土萌发（图 2）。8 月中旬～10 月采的种子翌春播种。

（周德本）

单子叶植物 MONOCOTYLEDONEAE

文　竹

Asparagus setaceus（Kunth）Jessop

（假叶树科　Ruscaceae）

生长习性、分布和用途　天门冬属约 300 种，我国约产 24 种，本文描述 1 种。常绿木质藤本，攀援或缠绕，长达 2～4m。能耐－2℃左右的短期低温。在肥沃湿润的酸性或中性土上能正常生长。原产非洲南部。滇、桂、粤、琼、闽、台可以露地栽培，间有野生，其它地区室内盆栽。植株清秀文雅，可作盆景，供室内观赏，也是插花重要的配叶。

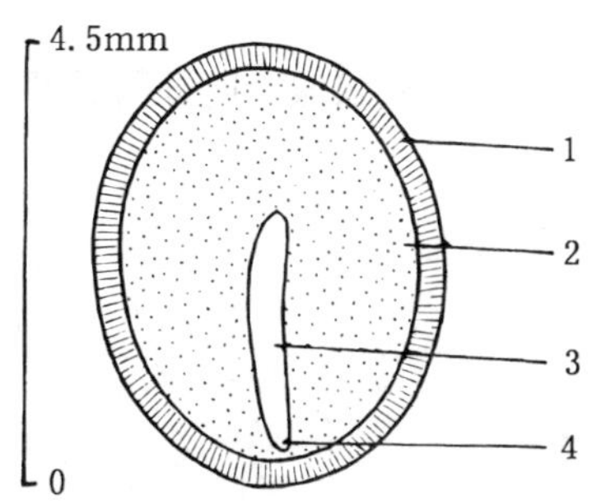

图 1　文竹种子的纵切面
1. 种皮　2. 胚乳　3. 圆筒状胚　4. 胚根
（黄应钦、马信祥绘）

开花结实　3～5 年生开始结实，8 年生以后为正常结实期，大小年现象不明显。花两性，小，花冠白色，单生，径 4mm 左右。花被片 6，2 轮。雄蕊 6 枚，着生于裂片的基部。子房上位，3 室，每室有 2 胚珠。据 1986～1988 年云南西双版纳观察，10 月上旬花芽始现，着生于当年或 1 年生新枝退化叶腋的背面，10 月中旬花蕾膨大，11 月上旬花盛开，11 月中旬开花结束，幼果形成；翌年 2 月上旬果实开始成熟，2 月下旬～3 月上中旬为果熟盛期。浆果，小，近球形，成熟时果皮由黄绿色转变为紫黑色，径 5.5～10mm，成熟后 5～8 天脱落。每果有种子 1～2 粒，球形或半球形，具光泽，径 3.3～4.5mm。种子胚乳丰富，胚小，直。

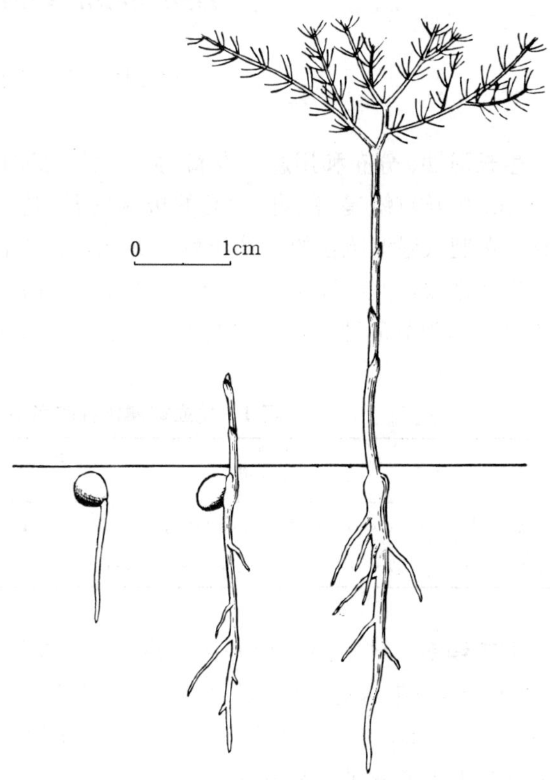

图 2　文竹种子萌发后第 3、10、17 天幼苗的生长情况
（黄应钦、马信祥绘）

文竹种子的解剖结构见图 1。

果实的采收调制和种子贮藏 果实成熟盛期在藤上采摘，或摇动植株震落后捡拾。采得的果实堆沤 2～4 天，待果肉软化后装入布袋搓揉，淘去果皮等杂质以及飘浮的瘪粒种子，取出沉水的种粒晾干即为播种材料。鲜果出种率约 40%。种子的净度可达 90%～98%。千粒重为 32～40g，每千克有纯净种子 25 000～33 000 粒，调制后稍晾干，不影响发芽能力的含水量为 35%～38%。贮藏时需混以湿沙，贮藏期以不超过 6 个月为宜。

发芽和播种 种子有休眠习性，种粒间的休眠深度有差异。发芽时的日均温宜在 22℃以上。1988 年 2 月 18 日，西双版纳热带植物园在室外沙盆用鲜种作过发芽测定：播后历时 75 天，至 5 月 4 日开始发芽，9 月 20 日结束，没有明显的发芽盛期，发芽率为 55%。留土萌发。胚根萌发后3～5 天胚芽鞘出土，将鞘叶或鳞叶 3～5 枚带出土面，约 7～12 天展出枝状的假叶(见图 2)。

撒播。每平方米播种 6～10g，覆土约 1cm。幼苗高 3～5cm，具枝状假叶时即可移至盆内，作盆景培育。

（程必强）

龙血树属

Dracaena Vand. ex L.

（龙舌兰科 Agavaceae）

生长习性、分布和用途 本属约 150 种，我国 6 种，本文描述 2 种。常绿灌木或乔木，高 3～15m。生长较缓慢，能耐－2℃的极端低温。适生于石灰岩山区的紫色土或红色土壤，耐旱，嗜钙，在肥沃湿润疏松的酸性土壤上亦能正常生长。干含树脂，可提制“血竭”，有止血、活血、化瘀之效，为伤科要药，是分布区狭窄的稀有树种，小花龙血树已濒于灭绝，《中国植物红皮书》列为渐危种。它们的名称、生长、分布和用途见表 1。

表 1 龙血树属树种的名称、生长、分布和用途

中 名	学 名	树 高 (m)	胸 径 (cm)	分 布	用 途
小花龙血树	*D. cambodiana* Pierre ex Gagnep.	3～15	10～140	滇、琼、桂。亚洲热带	观赏、药用
剑叶龙血树	*D. cochinchinensis* (Lour.) S. C. Chen	3～12	14～20	滇、桂。亚洲热带	观赏、药用

开花结实 15～20 年生开始开花结实，25 年生以后进入正常结实期。结实大小年现象明显，3～7 年开花结实 1 次。花两性。圆锥花序顶生，长 30～200cm。花被漏斗状，裂片 6。雄蕊 6，花丝线形，花药长圆形，丁字着生。子房上位，无柄，3 室，每室有胚珠 1 枚。在云南西双版纳勐仑观察的物候期见表 2。

表 2　龙血树属树种的开花结实物候期（云南勐仑）

树　种	观察年份	开　花			果实成熟		果实脱落
		始　期	盛　期	末　期	始　期	盛　期	
小花龙血树	1983～1988	3 月中旬	3 月下旬	4 月上旬	7 月中旬	8 月上旬	7 月下旬～8 月下旬
剑叶龙血树	1980～1987	8 月上旬	8 月下旬	9 月中旬	翌年 1 月下旬	2 月中旬	2 月下旬～3 月下旬

浆果。每果有种子 1～3 粒。种皮黄白色。有胚乳。胚体剑形。果实和种子的形态特征见表 3、图 1。

表 3　龙血树属树种果实和种子的形态特征

树　种	未熟果颜色	成熟果实			种　子		
		形　状	大小（cm）	颜　色	形　状	大小（mm）	颜　色
小花龙血树	褐色	近球形或扁圆形，表面光滑，有 3 条纵向条纹	高 1.1～1.5 径 1～1.12	橘红色	近球形	径 5.5～7.5	黄白色
剑叶龙血树	褐色	近球形或扁圆形，表面光滑，有 3 条纵向条纹	高 1.3～1.8 径 1.2～1.3	橙红色	近球形	径 6.4～9.6	黄白色

果实的采收调制和种子贮藏　果实盛熟期用采种刀带果穗采下，或摇动结果枝震落后拾集。采得的果实中有的已经软熟，可当日调制。其余的可在室内堆沤 2～4 天，待果肉软化后装入缸内搓揉，置水中淘洗，清除果肉等杂质以及飘浮的瘪粒种子，取出沉水的种子，即为播种材料。调制后稍加晾干的种子含水量为 50%左右，忌日晒。短期运输可用木箱包装，及时运抵目的地。贮藏时需混等量湿沙，贮藏时间以不超过 6 个月为宜。种子的净度、质量等见表 4。

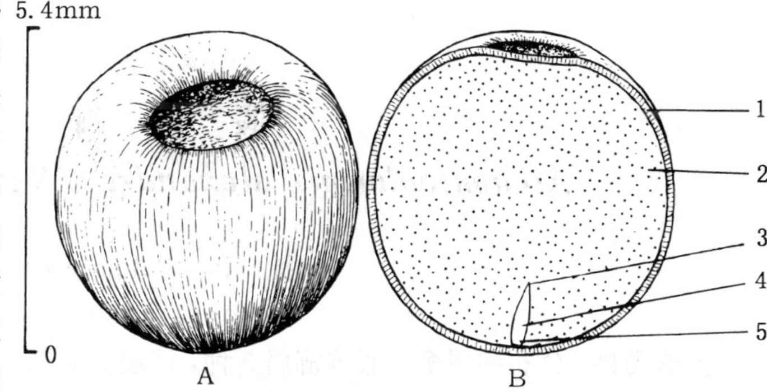

图 1　小花龙血树种子外形（A）及其纵切面（B）
1. 种皮　2. 胚乳　3. 子叶吸器　4. 胚轴　5. 胚根
（黄应钦绘）

表 4　龙血树属树种出种率和种子净度、质量

树　种	出种率（%）	净度（%）	千粒重（g）	每千克纯净种子粒数
小花龙血树	33	90～98	180	5 500
剑叶龙血树	35	90～99	250	4 000

发芽和播种　小花龙血树种子无休眠现象，剑叶龙血树种子有休眠现象。发芽时的日均

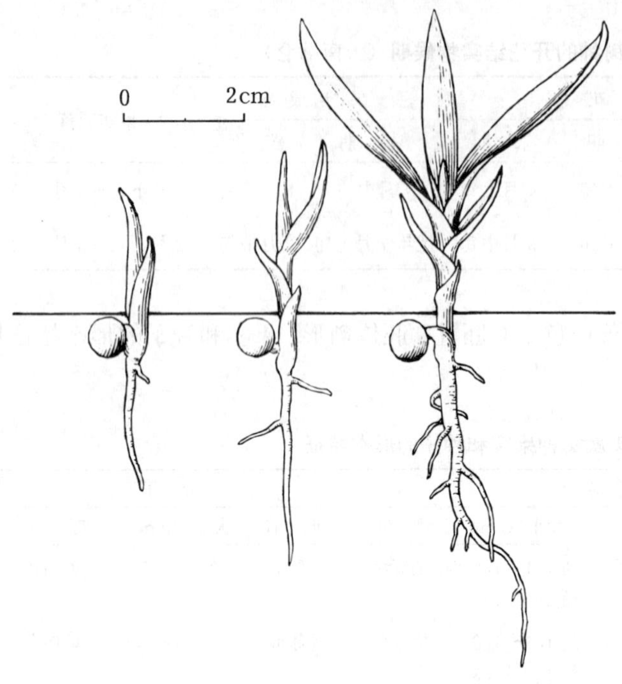

图 2　小花龙血树种子萌发后第 25、50、90 天幼苗的生长情况
(黄应钦绘)

温宜在 20℃以上。1980 年和 1983 年,西双版纳热带植物园在室外沙盆作过一次发芽测定:小花龙血树种子 8 月 3 日播种,播后约 45 天(9 月 18 日)开始发芽,10 月 10 日结束,发芽全程约 70 天,没有明显的发芽盛期,发芽率为 66%。剑叶龙血树种子 1 月 23 日播种,播后约 65 天开始发芽;3 月下旬播种后约 50 天开始发芽,发芽持续期约为 235 天,发芽率为 85%。留土萌发。胚根萌发后 8～10 天先出叶出土,25 天展出初生叶(图 2)。

条播。每平方米可播种 38～62g。播前用清水浸种 10～12 小时。覆土 1～2cm。幼苗具2～4片叶时移植,苗期需遮荫。2 年生苗出圃。用于庭园绿化的需培育 3～4 年生苗出圃。

(程必强)

假　槟　榔

Archontophoenix alexandrae Wendl. et Drude

(棕榈科　Palmaceae)

生长习性、分布和用途　假槟榔属 3 种,产澳大利亚。我国引入 2 种,本文描述 1 种。常绿乔木,茎单生,高达 20m。苗期忌霜冻,适生于较肥沃的酸性土壤。华南地区城镇广为栽培,是重要的庭院观赏树和街道、广场风景树。

开花结实　12～15 年生开始结实,正常结实期在 20 年生以后,大小年间隔期为 1 年。肉穗花序着生于叶丛下,多分枝,佛焰苞 2 枚。花单性,雌雄同株。雄花萼片 3,三角状圆形,覆瓦状排列;花瓣 3,镊合状排列;雄蕊通常 9～10,花丝极短。雌花萼片与花瓣卵圆形,各 3;子房 1 室,柱头 3;胚珠生于侧壁上。据广西南宁 1987～1988 年观察,7 月中旬为始花期,7 月下旬～8 月中旬为盛花期,8 月下旬为末花期,10 月上旬果实开始成熟,11 月中下旬为成熟盛期;另有少数植株开花结实无定

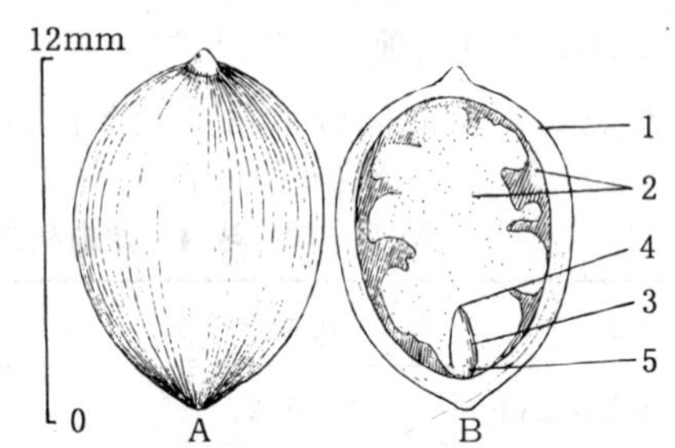

图 1　假槟榔果核外形(A)及其纵切面(B)
1. 内果皮　2. 种皮内层及胚乳　3. 胚芽鞘
4. 吸器　5. 胚根鞘
(黄应钦绘)

期。核果，椭圆形，未熟时青绿色，成熟后鲜红色，长1～1.5cm，径0.8～1.3cm。外果皮薄，中果皮纤维质。果核卵形，长0.9～1.2cm，径0.6～1.0cm，浅褐色或黄褐色。每核内有种子1粒。种皮内层与胚乳犬牙交错，使胚乳成嚼烂状。假槟榔果核的外形和解剖结构见图1。

果实的采收调制和种子贮藏　果实成熟盛期用采种钩刀截断总果梗，或用竹竿敲打，果实落地后在地面捡拾。采集的果实堆沤或浸水5～7天，俟果皮呈腐烂状时置水中搓擦漂洗，淘净果皮等杂质，所得果核用作播种材料，通称种子。鲜果出籽率为30%～40%。净度可达95%～99%。千粒重490（450～530）g，每千克有种子（核）2 000（1 800～2 200）粒。忌失水，不能日晒，含水量应保持在20%以上。1周左右的贮藏可以裸露存放。较长时间的贮藏需混以湿沙，贮藏期一般为半年以内。

发芽和播种　种子无明显休眠习性。发芽时日均温宜在18℃以上。1988年11月24日，广西林业科学研究所在室外沙床上用当年新采种子作发芽测定（当时日均温20℃，播后气温转低）：1989年3月12日（日均温约18℃）开始发芽，4月3日为发芽盛期，4月21日发芽终止；从开始发芽至发芽高峰，20天发芽64%；从播种至发芽终了，148天的发芽率为85%。留土萌发。胚根萌发后14天鳞叶出土，27天左右发出初生叶，40～50天初生叶展开。假槟榔种子（果核）的萌发和幼苗生长情况见图2。

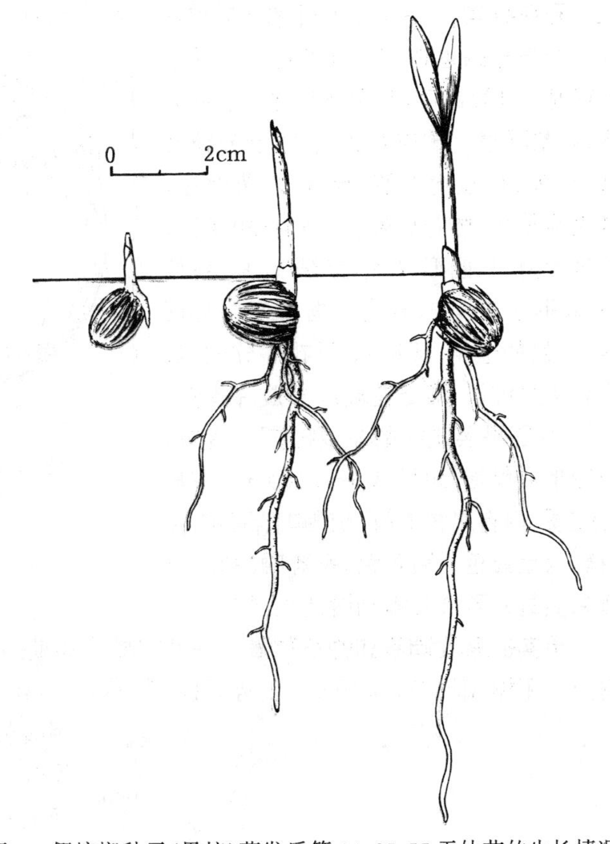

图2　假槟榔种子（果核）萌发后第14、27、55天幼苗的生长情况
（黄应钦仿《热带亚热带主要树种采种育苗技术》）

种子沙藏越冬，翌年3月初播种。条播，每平方米约播40g。1年生以后移植，培育3～4年生苗出圃。

（朱积余）

槟　榔

Areca cathecu L.

（棕榈科　Palmaceae）

生长习性、分布和用途　槟榔属约54种，我国引入本文描述的1种。常绿乔木，高达20m。

喜高温潮湿环境，适生于年均温 23℃以上的地区，气温 25～28℃时生长迅速，16℃时落叶，5～6℃时植株受冻害。宜深厚肥沃的酸性土。原产马来西亚，现广泛种植于世界热带各地。琼、台、闽、粤、桂、滇有栽培。果实药用，有驱虫、消滞、消肿、固齿、通便、防痢等功用。树形秀丽，可供观赏。

开花结实　7～8 年生开始开花结实，部分品种 4～5 年开花结实，20 年生以后为结实盛期，经济寿命达 60 年，结实大小年现象不明显。肉穗花序，着生叶丛之下。花单性，雌雄同序。雄花多数，着生于分枝上部，花萼、花冠均 3 裂；雄蕊 6，花丝短，退化雌蕊 3。雌花数朵生于分枝的基部；子房上位，长圆形，1 室，胚珠 1，基生；柱头 3，退化雄蕊 6，合生。据海南儋县观察，花果期全年；每年 10 月花芽分化；随着花穗的伸长，雌雄花陆续开放并授粉；4～6 月是果实成熟的主要时期。核果，长圆形或卵形。果长 3.5～6(8)cm。果皮纤维质，内有果核 1 枚。成熟时果实由青色转为金黄色或橙黄色。胚乳嚼烂状。槟榔果实的外形及其解剖构造见图 1。

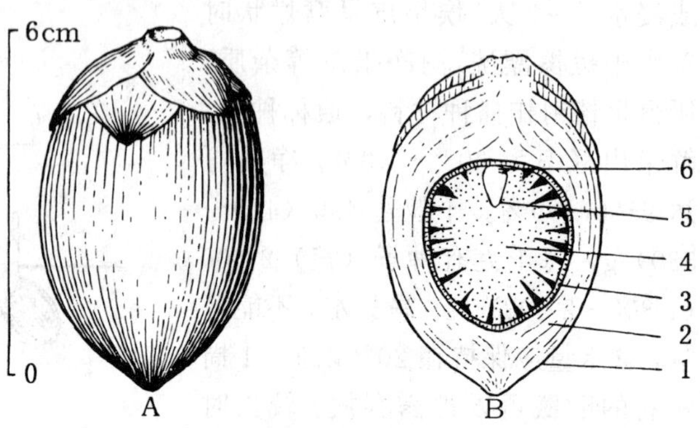

图 1　槟榔核果外形（A）及其纵切面（B）
1. 外果皮　2. 中果皮　3. 内果皮及种皮
4. 胚乳　5. 吸器　6. 胚根鞘
（黄光郁绘）

果实的采收调制和种子贮藏　在果实成熟盛期的 4～6 月采种，采种母树的选择标准是，树身上下粗细均匀，节间短，长势旺盛，树龄在 20 年生左右。当果穗上的果实大部分成熟时，用高枝剪或采种钩刀带果穗采下。摘取果实，堆放于室内 2～3 天，任其充分成熟，不需作任何调制，即可作为播种材料，仍通称种子。果实千粒重约 2（1.8 万～2.2 万）万克，每千克有果实 50（40～60）粒。忌曝晒，应置阴凉处摊开存放，贮藏期不宜超过 2 个月。运输时可用麻袋或竹筐包装。

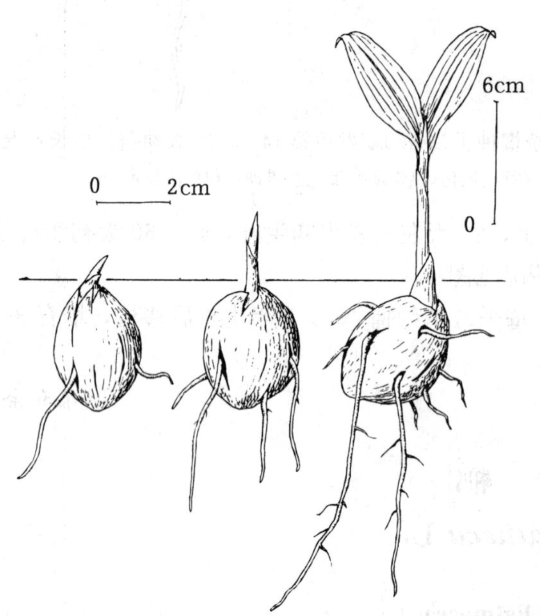

图 2　槟榔核果萌发后第 40、55、70 天幼苗的生长情况
（黄光郁绘）

发芽和播种　种子无休眠习性。播前先将果实置阴凉处堆积催芽，每堆以不超过 1 500 个为度，堆高 15～20cm，上盖 5～7cm 厚的湿稻草，经常淋水。如此处理约 2 个月，芽长3～5cm，发根 4 条以上时按 30cm×30cm 的株行距移至苗圃栽植。也可以用沙床催芽。1988 年 4 月 20 日，华南热带作物研究所在室外沙床用新鲜果实作发芽测定：播后 58 天开始萌发；从发芽之日起算，30 天发芽 50%；从播种至发芽终止，103 天的发芽率为 68%。留土萌发。果

实下部先发根 2～4 条后从四周发根多条。发根后约 1 个月鳞叶出土，70 天展出 2 片初生叶。槟榔核果的萌发和幼苗生长情况见图 2。

每平方米育苗用果量 750～850g。通常 1 年生苗出圃。用于庭院绿化的需培育 2～3 年生苗出圃。

（邓春媛）

山葵（金山葵、皇后葵）

Arecastrum romanzoffianum Becc. var. *australe* Becc.

（棕榈科　Palmaceae）

生长习性、分布和用途　山葵属有 1 种和数变种，我国引入本文描述的这个变种。常绿乔木，高达 14m 或更高。适生于酸性沙壤。较耐干热。原产巴西至阿根廷北部，现广植于热带、亚热带地区。闽、粤、琼、桂、滇有栽培。通常作行道树，或供庭园绿化。

开花结实　约 10 年生开始开花结实，正常结实期在 15 年生以后。结实大小年间隔期一般为 1 年，但不甚明显。肉穗花序生于下部叶腋中，多分枝，排成圆锥花序式。佛焰苞 1，木质，舟状，长达 1.5m，背面有多数纵沟纹。花单性，雌雄同株。雄花生于花序上部，雌花生于下部。花被片 6。雄花卵状披针形，雄蕊 6。雌花阔卵圆形，子房上位，3 室，柱头 3。据 1987 年广西南宁观察，5 月上中旬为始花期，6 月上旬为盛花期，7 月中下旬为末花期；每一花序开花的时间为 15～20 天；9 月下旬为果实成熟始期，10 月中旬为果熟盛期；果实成熟后 1～3 个月开始脱落。核果，倒卵形或卵形，长 2～2.5cm，径 1.5～2.2cm，未熟时黄绿色，成熟时黄红色。花被片宿存。中果皮纤维质。内果皮骨质，坚硬，近基部有 3 个萌发孔，内无空腔和液汁。每果含果核 1 粒。果核卵形或圆锥形，表面附纤维，灰褐色，长 1.6～2cm，径1.2～1.6cm，胚小，胚乳均匀（图 1）。

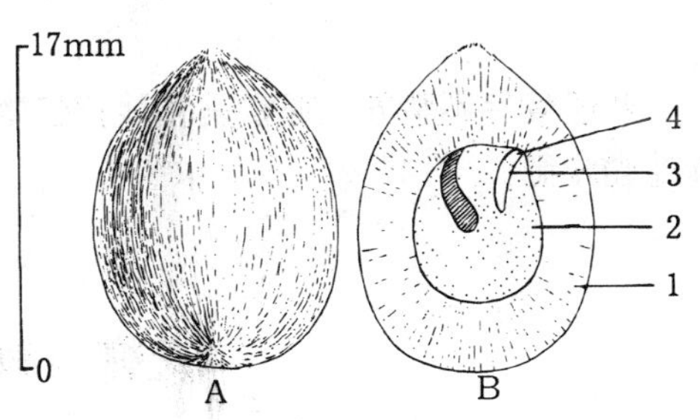

图 1　山葵果核外形（A）及其纵切面（B）
1. 内果皮及种皮　2. 胚乳　3. 胚芽鞘　4. 胚根鞘
（黄应钦绘）

果实的采收调制和种子贮藏　同一果穗的果实大部分呈黄红色时，用钩刀将果穗割下，打脱果实。将果实堆沤数日，待中果皮腐烂时置水中用木棍冲捣，切勿用手搓擦，以免皮肤过敏。淘去果皮杂质，所得果核用作播种材料，通称种子。果实出籽（核）率为 38% 左右。千粒重 2 650（2 300～3 000）g；每千克有果核 380（330～430）粒。种子忌失水，不宜曝晒。运输时宜混以湿沙或湿锯木屑，用木箱包装。也可以果实的状态运输，运抵后再按上法调制。贮藏时应混湿沙，贮藏期为半年左右。

发芽和播种　种子有休眠习性。发芽时气温需在 20℃以上。广西林业科学研究所在室外沙床用当年新采而未作任何处理的种子进行过发芽测定：1987 年 10 月 8 日播种，1988 年 5 月 4 日开始发芽，发芽盛期不明显，8 月 8 日发芽终止，发芽率为 60%。留土萌发。胚根萌发 15 天后鳞叶从子叶鞘中抽出，过 10 天鳞叶伸长出土面并发出第二片鳞叶，再过 15 天初生叶抽出并展开。山葵种子（果核）的萌发和幼苗生长情况见图 2。

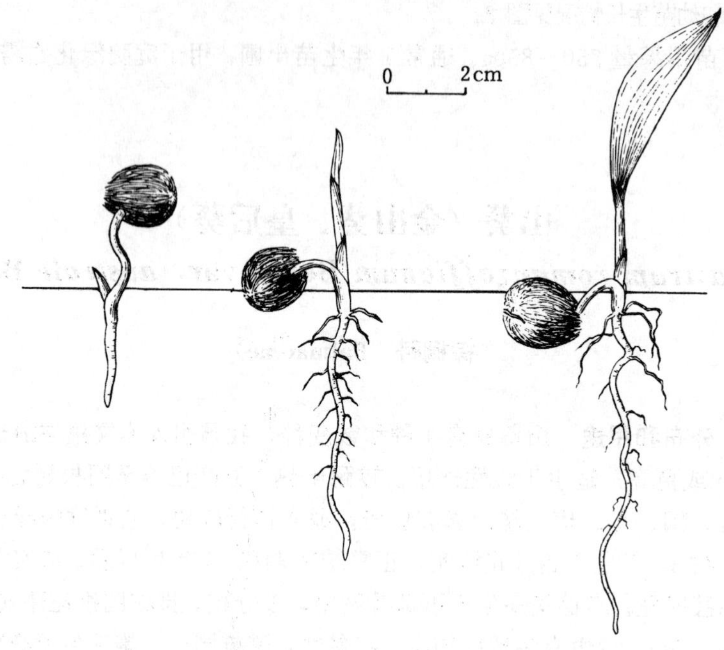

图 2　山葵种子（果核）萌发后第 15、30、45 天幼苗的生长情况

（黄应钦绘）

种子发芽不整齐。通常是将种子密播在沙床中催芽，发芽后陆续移至苗圃地培育。2～3 年生苗出圃。

（朱积余）

桄榔（砂糖椰子、糖棕）

Arenga pinnata（Wurmb.）Merr.

（棕榈科　Palmaceae）

生长习性、分布和用途　桄榔属约 11 种，我国 2 种，本文描述 1 种。乔木，高达 12m，茎粗壮。分布于粤、琼、桂、滇。印度、斯里兰卡、马来西亚、菲律宾、澳大利亚也有分布。供观赏。花序液汁制糖。髓心可提取淀粉。叶鞘纤维可制绳索和刷子。

开花结实　约 12 年生开始开花结实，正常结实期在 20 年生以后。结实大小年现象明显，间隔期常为 3 年。肉穗花序腋生，总花梗粗壮，下垂，分枝多，长达 1.5m。佛焰苞 5～6 枚，披针形。花单性，雌雄花通常单生于不同花序轴上，有时 3 朵聚生，则雌花生于 2 朵雄花之间。雄花：萼片 3，近圆形；花瓣 3，革质，长圆形，长 1.5～2cm，宽 0.4cm，镊合状排列；雄蕊 70～80 枚，花丝短；无退化雌蕊。雌花：近球形，萼片 3，较宽；花瓣 3，阔卵状三角形，长 1.3cm，镊合状排列；退化雄蕊多数或缺；子房上位，近球形，3 室；每室 1 胚珠，直立，基生，柱头圆锥状。据广西南宁 1986 和 1987 年观察，始花期 5 月下旬，盛花期 6 月上旬，末花期 6 月中旬；果实成熟始期在翌年 8 月上旬，盛期 8 月中下旬。浆果，倒卵状球形，径 3.5～5cm，未熟时青色，成熟时暗绿色。每果含种子 3 粒。种子黑褐色，三角状椭圆形，长 2.3～3.1cm，宽 1.7～2cm，厚 1.1～1.4cm。胚乳均匀。桄榔种子的形态和内部结构见图 1。

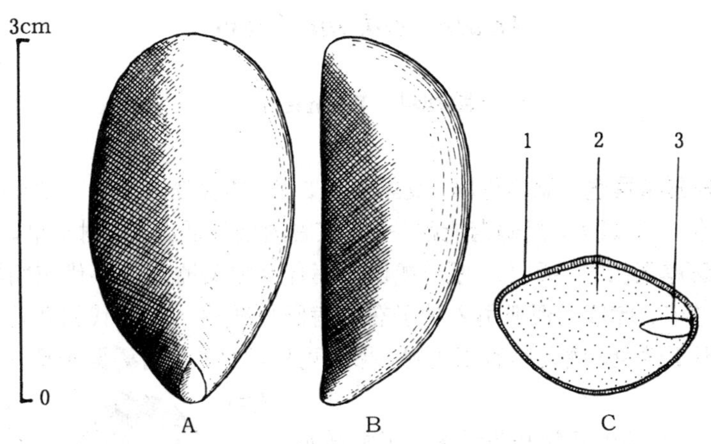

图 1　桄榔种子外形正面（A）、侧面（B）及其横切面（C）

1. 种皮　2. 胚乳　3. 胚

（黄应钦绘）

果实的采收调制和种子贮藏　果实成熟盛期用采种刀割断总果梗，摘取果实。采得的果实堆沤至果肉软化后，置竹箩浸于水中，用木棒捣去果皮果肉。搓洗时切勿用手，以免引起皮肤过敏。洗净种子，稍稍晾干即可。果实出种率为 40%～60%。种子净度可达 99%。千粒重 4 300（3 800～4 500）g，每千克有纯净种子 230（200～260）粒。种子忌失水，不宜日晒。运输时可以果实的状态装运，运抵后再行调制。也可以将调制出来的种子混湿锯屑包装运输。贮藏时需混湿沙，贮藏期半年左右。

发芽和播种　种子无明显休眠现象，宜随采随播，发芽时气温宜在 20℃左右。1987 年，广西林业科学研究所在室外沙床上对新采种子作过发芽测定，播前种子未作任何处理：8 月 10 日播种，9 月 30 日开始发芽，发芽盛期不明显，12 月 10 日发芽终止，发芽率为 50%。留土萌发。由于子叶联结的伸长，种壳被带出土面。发芽后 30 天，鳞叶出现，过 125 天发出初生叶。桄榔种子的萌发和幼苗生长情况见图 2。

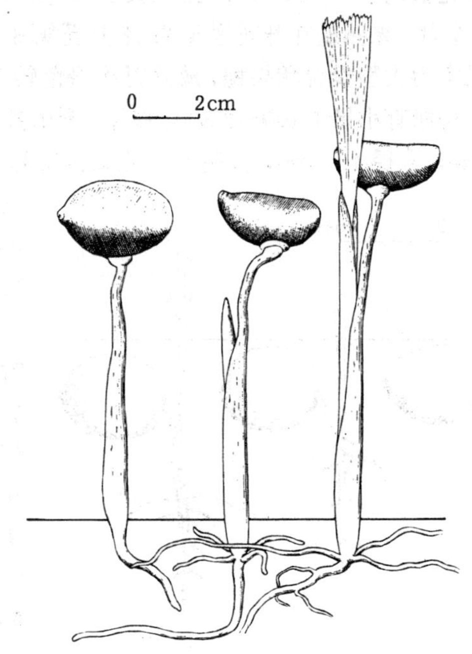

图 2　桄榔种子萌发后第 20、30、130 天幼苗的生长情况

（黄应钦绘）

通常是将种子密播于沙床，发芽后陆续移至苗圃培育。每年再分床移植1次。苗期需遮荫。一般培育3年生苗出圃。（朱积余）

亚塔棕（科洪棕、巴西榈）

Attalea cohune Mart.

（棕榈科　Palmaceae）

生长习性、分布和用途　亚塔棕属，我国引入2种，本文描述1种。常绿乔木，高达20m。适生区的年均温在22℃以上，但幼苗能耐－2℃左右的极端低温。较耐干旱，抗风力强，适生于微酸到中性的深厚肥沃的冲积土。原产墨西哥南部至中美洲。60年代初期华南热带作物科学研究所从斯里兰卡、印度尼西亚引入种子在海南儋县试种，80年代初开花结实。种仁含油60%～70%，为不干性油，供食用和工业用，果壳可作活性炭。树姿挺拔雄伟，是优良的园林绿化树种。

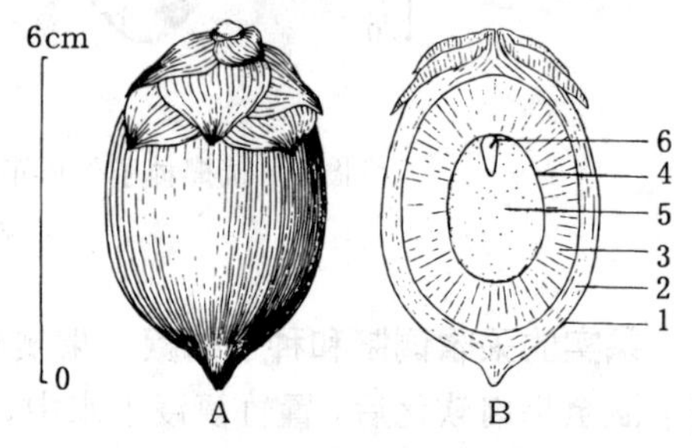

图1　亚塔棕核果外形(A)及其纵切面(B)
1. 外果皮　2. 中果皮　3. 内果皮　4. 种皮
5. 胚乳　6. 胚
（黄应钦绘）

开花结实　约15年生开始开花结实，正常结实期在25年生以后，大小年现象不明显。肉穗花序腋生。花单性，雌雄异株或同株异序，雄花多于雌花。雌花序比雄花序长1倍以上。花冠黄色，萼片、花瓣3裂。据海南儋县观察，每年5～11月均有雄花开放。雌花大约在8～11月开放。从授粉至果实成熟约需14个月，果熟期在开花翌年的12月至第3年的1月。果实为大型聚合状果穗，成熟时由绿色转变为黄褐色。每穗有小果1 400～2 300。核果，形似椰子，长5～7cm，直径4～5cm，顶端有一乳头状突起。每果核有种子1～3粒，多为1粒。内果皮厚壳质。胚乳丰富，胚较小。亚塔棕果实的形态和解剖结构见图1。

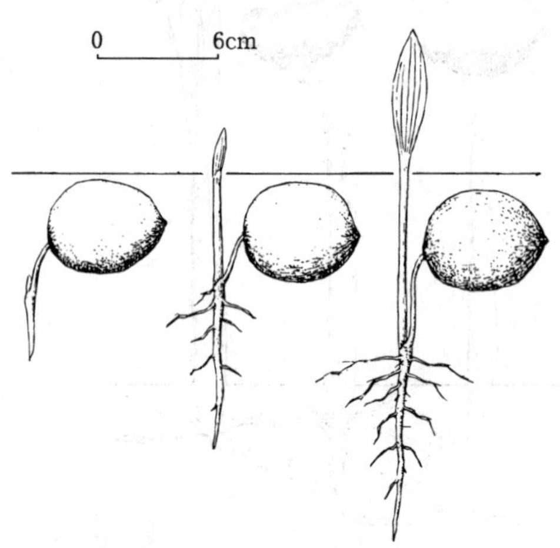

图2　亚塔棕果实萌发后第60、80、100天幼苗的生长情况
（黄光郁绘）

果实的采收调制和种子贮藏　果实充分成熟时用利刀砍下果穗（穗长125～135cm，穗幅50～55cm，重80～140kg。）从穗上摘下小核果，不作任何处理即为播种材料，通称种子。鲜果含水量在30%以上，采回后的果实在室内通风处摊开晾干，忌晒，待含水量下降到15%左右时放在10℃左右的冷库中贮存，或直接播入沙床中催芽。每个鲜果重25～71g，每千克有鲜果约20个，晾干后的果实每千克约26～30个。运输时可用麻袋或木箱包装，运抵后即播入沙床催芽。

发芽和播种　种子无休眠习性或有短期休眠

现象。发芽时日均温宜在25℃以上。内果皮坚硬，透水性差，发芽慢，播前需湿沙催芽。1985年2月，华南热带作物研究所用沙床为600个果实催芽，当时日均温约20℃，因气温低，延续到5月下旬才发芽出土，且不整齐。115天共发芽70株，发芽率为11.6%。同年5月用低温贮存的400粒果实播于沙床，播种时日均温25℃以上，60天发芽6株，发芽率为1.5%。种子发芽率很低，可能是因授粉不良或果壳坚硬等原因，尚在研究中。常温下堆放于室内的果实，3个月即全部丧失发芽能力。留土萌发，5月中旬发芽的植株，6月3日苗高4cm，地下部分长8cm。亚塔棕的萌发和幼苗生长情况见图2。苗期需遮荫。1～3年生苗出圃。

（邓春媛）

白藤（鸡藤）

Calamus tetradactylus Hance

（棕榈科 Palmaceae）

生长习性、分布和用途 省藤属约375种，我国产20余种，本文描述1种。有刺藤本，簇生。茎纤弱，径约1cm，长约3～5m，常高攀。耐荫。喜高温多湿、阴凉肥沃的森林环境，常生于林中。分布于滇、桂、粤、琼、闽。茎可编织各式藤器。全株药用，解毒。

开花结实 约10～13年生开始开花结实，结实大小年现象不明显。肉穗花序鞭状，分枝4～7。佛焰苞管状，宿存。花单性异株。雄花花萼杯状，长约1.5mm，浅3裂；花瓣矩圆形，长约2mm；雄蕊6枚，具退化雌蕊。雌花子房上位，被鳞片，为不完全的3室，每室1胚珠；柱头3，近无柄；退化雄蕊连合成杯状。据广西南宁1982和1987年观察，10月中旬为开花始期，10月中下旬为盛花期，11月上中旬为末花期；翌年5月下旬果实开始成熟，6月上中旬为果熟盛期。浆果，圆球形，径6～7.5（10）mm，未熟时青绿色，熟时绿白色。外果皮薄壳质，被以紧贴的覆瓦状排列的鳞片。种子通常1粒，圆球形，表面有麻点及雕纹，坚硬，黑褐色，径约5～6mm。胚乳质硬。种子的形态及其解剖结构见图1。

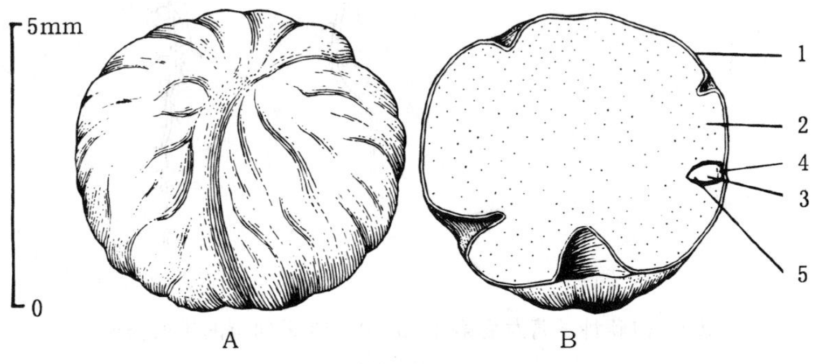

图1 白藤种子外形（A）及其纵切面（B）

1. 种皮 2. 胚乳 3. 胚芽鞘 4. 胚根鞘 5. 吸器

（黄应钦绘）

果实的采收调制和种子贮藏 果熟盛期用枝剪或采种钩刀截断总果梗，敲打果穗，脱下或摘下果实。果实堆沤数日后置水中搓洗，淘去果肉即得纯净种子。果实出种率为75%～

85%。种子净度 97%。千粒重 136（125～140）g，每千克有纯净种子 7 350（7 140～8 000）粒。种子忌失水，不宜裸露存放，包装运输时需拌湿沙或锯屑。贮藏需混湿沙，贮藏期 7～8 个月。

发芽和播种　种子有中度休眠。发芽时日均温需在 20℃以上。1987 年 6 月 8 日，广西林业科学研究所用当年新采种子，不作处理，在室外沙床作过发芽测定：播后 117 天（10 月 3 日）开始发芽（当时气温 25℃），10 月 9 日进入发芽盛期，10 月 12 日发芽终止；从开始发芽至发芽高峰，6 天发芽 40%；从播种之日起算，126 天的发芽率为 55%。留土萌发。胚根萌发后约 3 天鳞叶出土，过 12 天发出初生叶。白藤种子的萌发和幼苗生长情况见图 2。

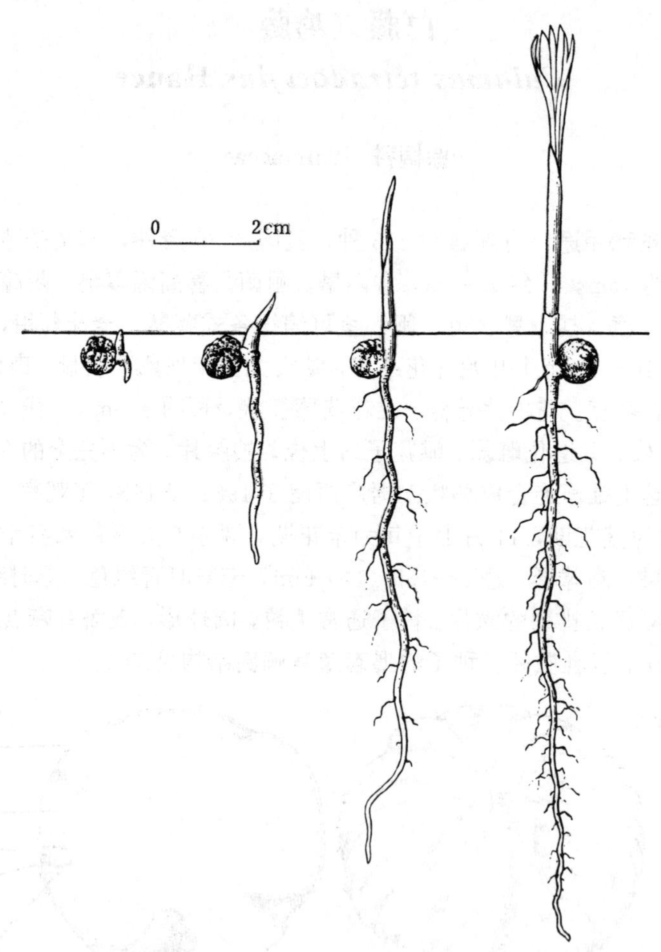

图 2　白藤种子萌发后第 3、5、8、13 天幼苗的生长情况

（黄应钦绘）

条播或芽苗移植。种子休眠期较长，宜密播于沙床催芽，萌发后再移至大田继续培育。每平方米播种 40g 左右，覆沙厚 5mm。苗期需遮荫。一般培育 1 年生苗出圃。

（朱积余）

鱼尾葵属

Caryota L.

（棕榈科　Palmaceae）

生长习性、分布和用途　本属 12 种，我国有 4 种，本文描述 2 种。适生于南亚热带至北热带的丘陵及平原地区，多生长于石灰岩山区，酸性土也能正常生长，对土壤水肥条件的要求中等。它们的名称、生长、分布及用途见表 1。

表 1　鱼尾葵属树种的名称、生长、分布和用途

中　名	学　名	树　高 (m)	胸　径 (cm)	分　布	用　途
短穗鱼尾葵	*C. mitis* Lour.	丛生 5～8	8～10	粤、琼、滇、桂。亚洲热带地区	庭园绿化、茎含淀粉可食
鱼尾葵	*C. ochlandra* Hance	单生 20	20	粤、琼、滇、桂、闽。亚洲热带地区	庭园绿化、茎含淀粉可食、材作工艺材、根药用

开花结实　约 10～12 年生开始结实，正常结实期在 15 年生以后。结实有大小年现象，间隔期一般为 1 年，但不甚明显。肉穗花序腋生。佛焰苞 3～5 枚，管状。花单性，3 朵聚生，中间为雌花。花部 3 数。雄花：萼片圆形，分离，覆瓦状排列；花瓣线状长圆形；雄蕊 9 至多数。雌花：萼片圆形；花瓣卵状三角形；子房上位，3 室，柱头 3 裂。广西南宁观察的开花结实物候期见表 2。

表 2　鱼尾葵属树种的开花结实物候期（广西南宁）

树　种	观察年份	开　花			果实成熟		果实脱落
		始　期	盛　期	末　期	始　期	盛　期	
短穗鱼尾葵	1987～1989	5 月中旬	5 月下旬末	12 月下旬	6 月下旬	8 月中旬	8 月中旬～翌年 4 月中旬
鱼尾葵	1978～1984	4 月上旬	4 月上旬～6 月上旬	10 月	9 月下旬	10 月中旬	9～10 月均有果熟，熟后逐渐脱落

鱼尾葵的开花期较长，4～10 月均有花开，主花期在 4 月下旬～6 月上旬，每穗开花时间为 3～5 天。

浆果状核果，球形。果皮肉质。短穗鱼尾葵每果有果核 1 颗，鱼尾葵有果核 1～2 颗。胚乳嚼烂状，胚位于种子的一侧。果实及果核的形态见表 3、图 1。

表 3 鱼尾葵属树种果实和果核的形态特征

树 种	未熟果颜色	果实			果核		
		形 状	大小（cm）	颜 色	形 状	大小（cm）	颜 色
短穗鱼尾葵	青绿色	球形	径 1.5～2.6	赭红色或紫红色	圆球形至扁球形	高 1.2～1.45 径 1.2～1.7	紫黑色
鱼尾葵	黄绿色	球形	径 2.3～2.9	红色或紫红色	半圆形或圆形，一侧内凹	高 1.2～1.5 宽 1.5～2.5 厚 1～1.5	黑色或暗褐色

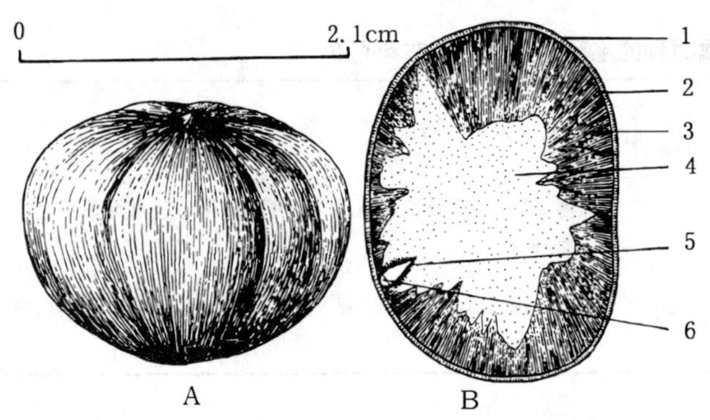

图 1 鱼尾葵果核外形（A）及其纵切面（B）
1. 内果皮 2. 种皮 3. 发育的内珠被（种皮内层）
4. 胚乳 5. 吸器 6. 胚根鞘
（黄应钦绘）

果实采收调制和种子贮藏 果实成熟不甚整齐。当同一果穗的果实大部分呈红色时，可用高枝剪或采种钩刀带果穗采下。采得的果实可用以下两种方法处理。一是堆沤数日，俟外果皮腐烂时置水中搅拌，洗净果核即为播种材料，通称种子。另一方法是将果实摊放于通风处，果皮干缩起皱后剥出果核。果皮及果穗含有毒质，可刺激皮肤发痒，处理时应注意防护。种子忌失水，不宜曝晒。洗净的种子可混沙贮藏运输，贮藏期一般为半年以内。鲜果的出籽（核）率及果核的净度、质量等见表 4。

表 4 鱼尾葵属树种果实出籽（核）率和果核的净度、质量

树 种	出籽率（%）	净度（%）	千粒重（g）		每千克果核粒数	
			一 般	变动范围	一 般	变动范围
短穗鱼尾葵	30～50	99	2 030	1 900～2 200	490	450～530
鱼尾葵	25～30	99	2 350	2 100～2 650	420	380～480

发芽和播种 这 2 个树种的种子都有中度或中度以上休眠习性，短穗鱼尾葵的休眠程度比鱼尾葵深。它们都宜层积至翌年春播。发芽时的日均温需在 20℃左右。1987 年 10 月和 1988 年 11 月，广西林业科学研究所在室外沙床上对短穗鱼尾葵和鱼尾葵作过发芽测定，都是当年新采种子，播前未作任何处理。测定结果见表 5。

表 5　鱼尾葵属树种的发芽能力[①]

树　种	发芽势（%）	发芽率（%）	
		计算天数	一般数值
短穗鱼尾葵	不明显	265	72
鱼尾葵	不明显	167	70

①　广西南宁，室外沙床

留土萌发。在上节所述的条件下，短穗鱼尾葵种子播后 224 天方始萌发，萌发后 25 天抽出鳞叶，约 60 天初生叶展出。鱼尾葵种子播后 106 天开始萌发，萌发后 60 天抽出鳞叶，97 天初生叶展现（图 2）。

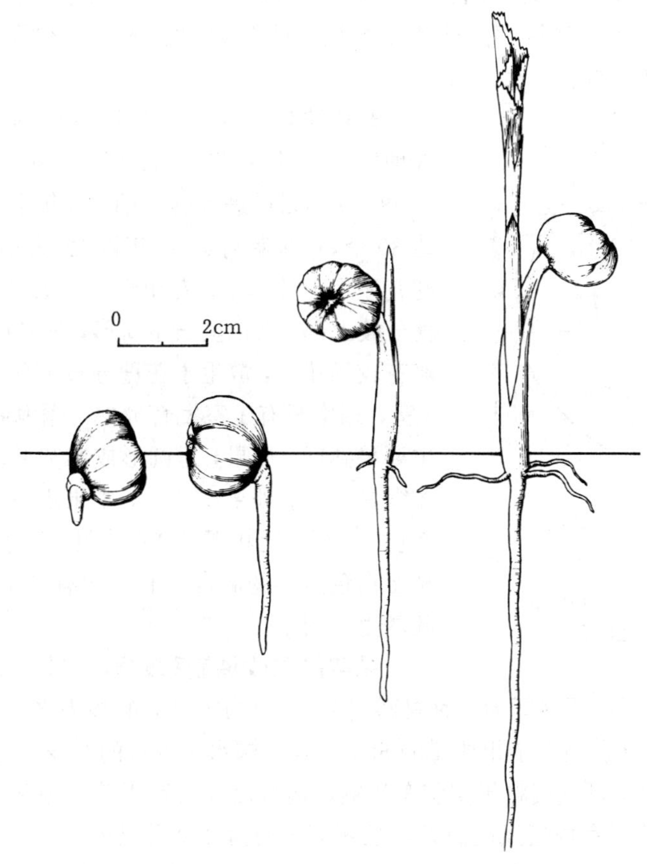

图 2　鱼尾葵种子萌发后第 5、20、65、100 天幼苗的生长情况

（黄应钦仿《热带亚热带主要树种采种育苗技术》）

通常将种子密播于沙床或圃地上催芽，每平方米播种量：短穗鱼尾葵为 1～1.3kg，鱼尾葵为 1～1.6kg。种子发出鳞叶后 3 个月内移植至圃地培育。一般育 3～4 年生苗供园林绿化。

（朱积余）

椰　子

Cocos nucifera L.

（棕榈科　Palmaceae）

生长习性、分布和用途　椰子属只有本文描述的 1 种。常绿乔木，高 30m 以上。适生于高温高湿、阳光充足的热带地区，年均温 24℃以上方能正常开花结实；最适宜的生长区，年均温应在 26℃以上，年降水量应有 1 500～2 000mm 且分布均匀。土壤以海滨和河岸冲积土最为适宜。主要分布在南北纬 20°以内，尤以赤道滨海地区最多。主产国依次为菲律宾、印度尼西亚、印度和斯里兰卡。我国琼、粤南、桂南、滇南、台南、闽南有栽培，以海南栽培最多。椰子是重要木本油料作物。椰肉含脂肪 33%，蛋白质 4%左右，可生食或制成椰蓉、椰奶、椰糖等。椰麸可作精饲料。椰纤维可作绳索、扫把、刷子和地毯，椰壳可供雕刻手工艺品。干材可作梁柱、板材。树形雄伟，供观赏。

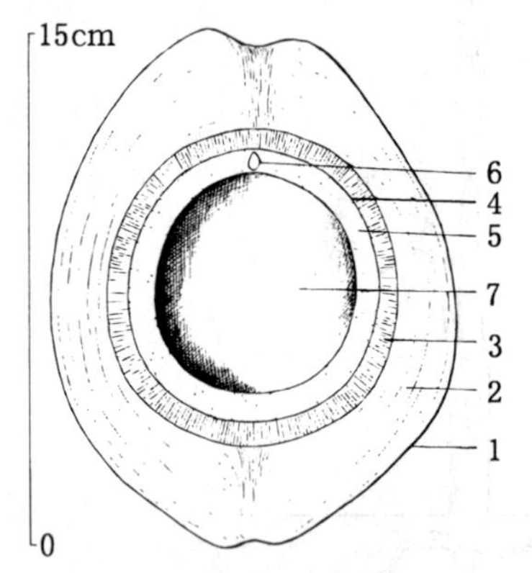

图 1　椰子核果纵切面
1. 外果皮　2. 中果皮　3. 内果皮　4. 种皮
5. 胚乳　6. 胚　7. 内腔
（黄应钦、黄光郁绘）

开花结实　不同的品种开花结实有先后，高干品种栽种后 6～7 年开花结实；短干品种栽种后 3～3.5 年结实。8～9 年以后进入盛果期，20 年生时单株年产椰果可达 80 个，产果龄可达 80 年以上。大小年现象不显著。肉穗花序生于叶丛中，花单性，雌雄同穗。雄花聚生于花穗分枝上部。雄花呈三角筒形，花被 6，2 轮，雄蕊 6 枚。雌花较雄花大，散生于花穗分枝下部，球形。子房上位，3 室，通常只有 1 室发育成熟。据海南观察，7～9 月为主要果熟期。核果，呈圆形或椭圆形，径 12～25cm，顶端微具 3 棱。外果皮革质，中果皮厚，纤维质。内果皮骨质，坚硬，近基部有 3 萌发孔。种子 1 颗，种皮薄，紧贴着白色坚实的胚乳。胚乳内有 1 富含液汁的大空腔。胚基生（图 1）。

果实的采收调制和贮藏　果实成熟后即可上树采摘。采得的果实不经任何调制即为播种材料。椰子为植物中最大的核果之一，每个果重1 500～2 000g。新鲜果核和胚乳肉占全果质量的 50%，其中椰水又占核的 50%。忌失水，不宜日晒。运输时不需任何包装，可直接将果实装车运输，运抵后及时置于苗床催芽。贮藏的方法较简单，通常是将椰果置于室内通风阴凉处，贮藏期一般为 2 个月以内。

发芽和播种　椰果有短期休眠现象。在自然条件下，其发芽过程是从椰果成熟后约 60 天开始，发芽快慢和发芽率的高低取决于果实成熟的程度和天气条件。用作育苗的椰果要及时催芽，以免降低发芽能力。1988 年 8 月中旬，华南热带作物研究所用当地新采的椰果作过发芽测定：10 月中旬（即播后 60 天左右）开始发芽，90 天后进入发芽盛期，翌年 1 月中旬发芽结束。从开始发芽至发芽高峰日，发芽率为 59%；从播种之日至发芽终了，发芽率为 65%。留土萌发。胚芽萌发后 48 天长出燕尾状初生叶，胚根尚包含在果内（图 2）。

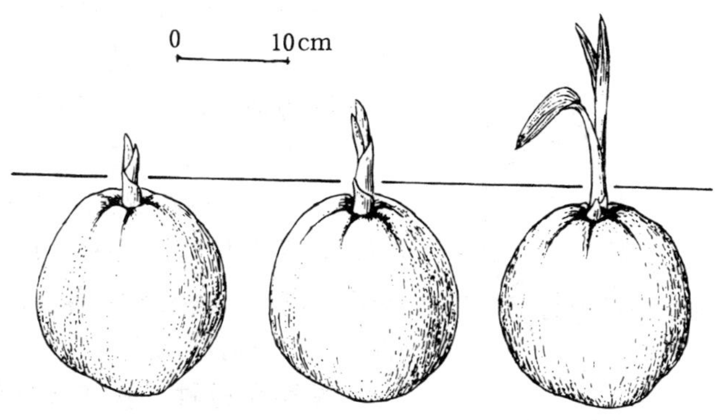

图 2　椰子核果萌发后第 4、24、48 天幼苗的生长情况
（黄光郁绘）

播种前先行催芽：选择平坦通风的树荫下，挖沟深 20cm，长度依地形而定，将椰果倾斜 45°排放沟内，覆土深及果实高度的 2/3，当芽高 20cm 左右时移至圃地继续培育。株行距约 40cm×40cm，正方形或三角形排列，覆土深及椰果高度的 2/3 或将全果掩埋。幼苗期需适度荫蔽，及时淋水。1 年生苗高约 80～100cm，可以出圃定植。

（邓春媛）

油棕（油椰子）
Elaeis guineensis Jacq.

（棕榈科　Palmaceae）

生长习性、分布和用途　油棕属 2 种，本文描述引入的 1 种。常绿大乔木，高达 20m。年均温 22℃以上方能正常开花结实。气温低于 17.5℃时生长显著缓慢，果实发育不良；低于 15℃时，几乎停止生长；低于 10℃时，果穗严重冻坏；低于 4℃时，成龄树 98%严重冻害，幼苗 35%冻死。年降雨量要求 1 800～3 000mm，且分布均匀。抗风力较差。原产西非热带地区。我国琼、台、粤、桂和云南西双版纳有栽培。果肉和果仁均可榨油，供食用或制人造奶油，工业上用以制皂或作润滑油。在东南亚产油量高，有“世界油王”之称。

开花结实　种后 2～3 年生开始开花结实，6～7 年生进入盛果期，经济寿命 25～30 年，自然寿命长达 80 年以上。佛焰花序短而厚，生于叶丛中。花单性，雌雄同株。雄花小，为稠密的穗状花序，雄蕊 6，花丝合生成 1 管。雌花较大，雌花序密集成圆头状，每花下托 1 具长刺的苞片，子房上位，3 室。据海南观察，花期全年，果熟期颇不一致，一年四季有果实成熟。核果，卵形或倒卵形，长 4～5cm，宽 3cm。果聚合成稠密的果束，熟时橙红色。外果皮海绵质，含油分。中果皮纤维质。内果皮骨质，坚硬，顶端有 3 萌发孔。每果有种子 1 粒，稀 2～3 粒。果核的形态见图 1。

果实的采收调制和种子贮藏　外果皮由黑色变为橙红色时及时采收。油用的果实可以一次将果穗全部割下。用于育苗的，须待果实充分成熟后从果穗上逐粒摘取。采得的果实浸水约

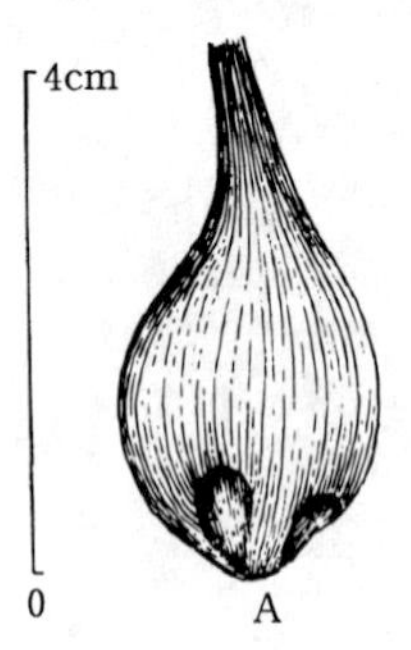

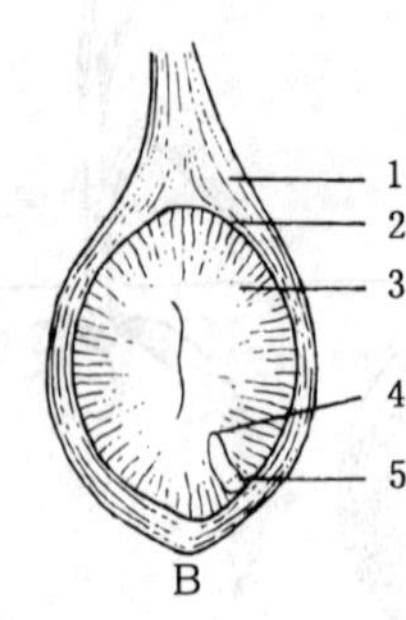

图 1　油棕果核外形（A）及其纵切面（B）
1. 内果皮　2. 种皮　3. 胚乳　4. 吸器　5. 胚根鞘
（黄光郁绘）

7 天，捣去果肉并将果核浸在 30%的草木灰水中，经过 6～7 天的脱脂处理，置清水中冲洗。洗净的果核即为播种材料，通称种子。鲜果的出籽率约 40%。千粒重在 1 350～2 310g。每千克有种子（核）430～740 粒。调制后稍晾干的种子含水量约 17%，不宜日晒。运输时以 500 粒为一组装入 38cm×45cm 的塑料薄膜袋，并加封扎。贮藏宜混湿沙，贮藏时间以不超过半年为宜。

发芽和播种　种子无休眠习性或有短期休眠。发芽前宜用 37～40℃的高温保湿催芽。1988 年 9 月 6 日，华南热带作物研究所对种子进行湿热处理：每 500 粒种子装入一只 30cm×25cm 塑料袋，密封，放在 38～40℃恒温箱中催芽。催芽过程中种子含水量过低时，应立即补充水分，保持种子潮湿。10 月 26 日，即催芽后 50 天开始发芽，66 天进入发芽盛期，12 月 14 日发芽终止。有明显的发芽盛期。从开始发芽至发芽高峰日，22 天，发芽 70%。从播种至发芽终了，以 99 天计，发芽率为 80%。留土萌发。发芽后 15 天鳞叶出土，50 天可见到半开的初生叶，80 天后初生叶展开，生长情况见图 2。

容器育苗。每 666m² 用种 100～140kg。将湿热处理催芽中已发芽的种子及时取出，点播至小容器或沙床内。出现初生叶时移至 50cm×40cm 的大容器内继续培育。1 年生苗出圃。

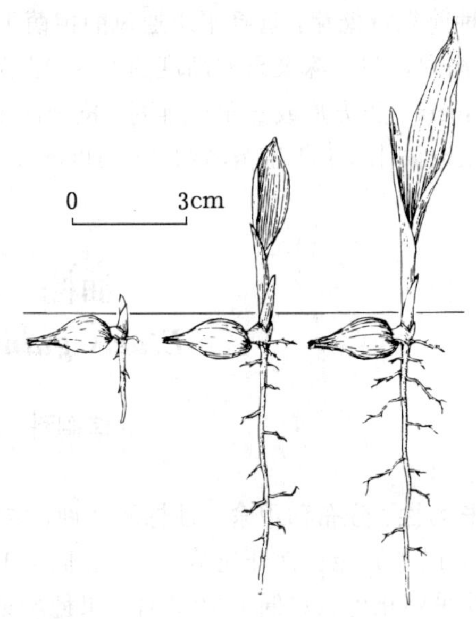

图 2　油棕种子萌发后第 15、50、80 天幼苗的生长情况
（黄光郁绘）

（邓春媛）

蒲　葵

Livistona chinensis（Jacq.）R. Br.

（棕榈科　Palmaceae）

生长习性、分布和用途　蒲葵属约 30 种，我国 4 种，又引入 1 种，本文描述 1 种。常绿乔木，高可达 20m，单干直立，叶集生于顶部。能耐 0℃左右低温。分布华南，以广东新会栽培最多。南京以北盆栽温室越冬。中南半岛也有分布。供绿化观赏，可制葵扇、牙签，药用。

开花结实　约15年生开始开花结实，正常结实期在20年生以后。结实大小年现象明显，间隔期常为1年。肉穗花序排成圆锥花序式，长达1m，腋生，分枝疏散。佛焰苞棕色，筒状，革质，2裂。花两性，黄绿色，小，无柄，长2mm，花萼、花冠均3裂。雄蕊6，花丝合成1环。子房上位，由3个近分离的心皮组成，3室，每室1胚珠，花柱短。据广西南宁1984～1989年观察，始花期3月中下旬，盛花期4月中旬，末花期5月上中旬；果实成熟始期在10月上旬，盛期在11月上旬，果实在翌年1月下旬～5月下旬散落。核果，椭圆形，状如橄榄，长1.8～2.3cm，径1.3～1.5cm，未熟时青色，熟时外果皮黑色或蓝黑色。中果皮肉质，内果皮骨质。每果有果核1粒，长椭圆形，浅褐色，长1.5～2.1cm，径0.9～1.2cm。胚乳均匀，胚侧生（图1）。

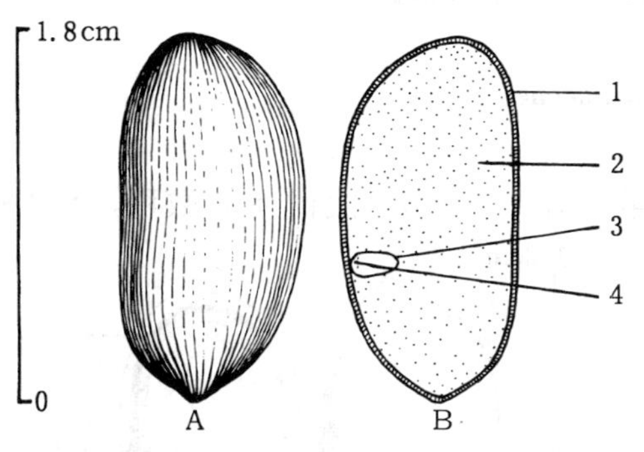

图1　蒲葵果核外形（A）及其纵切面（B）
1. 内果皮及种皮　2. 胚乳　3. 吸器　4. 胚根鞘
（黄应钦绘）

果实的采收调制和种子贮藏　果实成熟后应及时采收，以防虫蛀。通常用钩刀带总梗割下，在地下捡拾或摘取。采回的果实堆沤数日，果皮软化后置水中搓去外果皮和果肉。所得的果核稍加晾干即可用作播种材料，通称种子。果实出籽率60%～70%。千粒重1 060（800～1 300）g，每千克有种子（核）940（770～1 250）粒。种子忌失水，不宜日晒。可以以果实的状态运输，运抵后再行调制。也可将调制出来的种子混沙装入木箱或麻袋运输。贮藏需混湿沙，贮藏期一般为1年。

发芽和播种　种子无休眠习性。可随采随播，发芽时日均温宜在20℃以上。1988年，广西林业科学研究所在室外沙床上用当年新采的种子作过发芽测定：播前种子未作任何处理，10月7日播种，第5天开始发芽，第8天进入发芽盛期，第11天发芽终止。自发芽之日起算，4天发芽54%，从播种之日起算，以12天计，发芽率为94%。留土萌发。发芽后6天鳞叶从子叶鞘中抽出，过38天初生叶从鳞叶中抽出，再过8～12天初生叶展开。生长情况见图2。

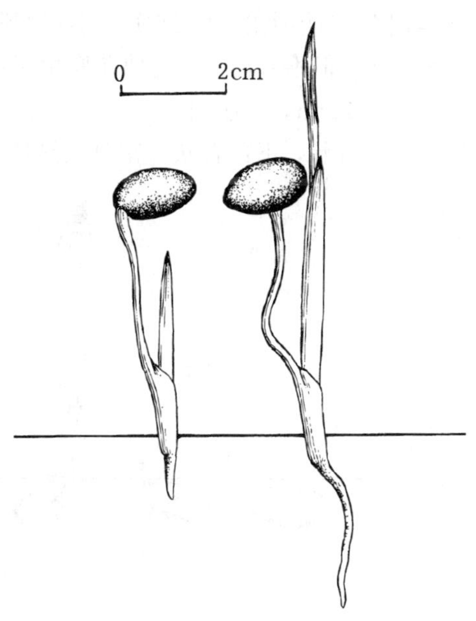

图2　蒲葵种子萌发后第15、43天幼苗的生长情况
（黄应钦绘）

条播。培育3～4年生苗出圃。

（朱积余）

软叶刺葵

Phoenix roebelenii O. Brien.

（棕榈科　Palmaceae）

生长习性、分布和用途　刺葵属约17种，我国有2～3种，并引入栽培数种，本文描述1种。灌木，茎单生，高2～3m，生长较慢。适生于温暖湿润的环境，能耐轻霜，忌冰雪，极耐荫。对土壤水肥条件的要求中等。产云南澜沧江边。中南半岛也有分布。我国粤、琼、桂露地栽培，其它城市多有室内盆栽。南亚热带和北热带庭园绿化的重要观赏树种，也是重要盆栽荫生植物。

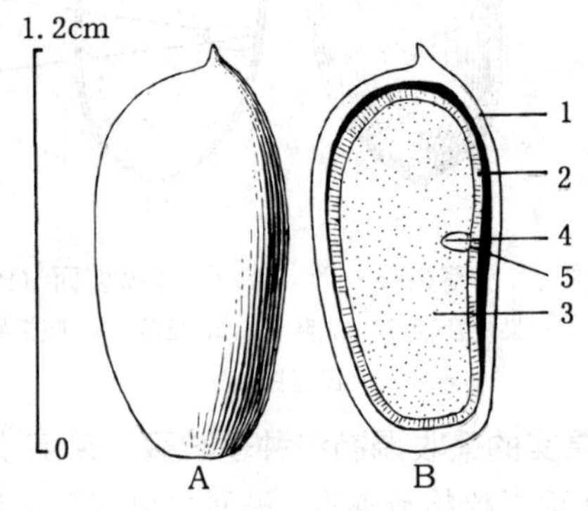

图1　软叶刺葵浆果外形(A)及其纵切面(B)
1. 果皮　2. 种皮　3. 胚乳　4. 吸器　5. 胚根鞘
（黄应钦绘）

开花结实　约10年生开始开花结实，正常结实期在15年生以后。结实大小年间隔期一般为1年，但不甚明显。肉穗花序生于叶丛中，长30～50cm，花序轴扁平。佛焰苞1枚，革质，上部舟状，下部管状。花单性异株。雄花穗的佛焰苞约与花穗等长，雌花穗的佛焰苞短于花穗。雄花花萼杯状，3齿裂，花瓣3，披针形，镊合状排列；雄蕊6，花丝极短。雌花卵圆形，长约4mm；心皮3，分离，每室1胚珠；无花柱，柱头钩状。据广西南宁1987～1989年观察，始花期为5月上旬，盛花期为5月中旬，末花期为5月下旬；7月下旬果实开始成熟，8月中旬为果熟盛期，果实成熟后1～2个月脱落。浆果，矩圆形或长椭圆形，未熟时青色，熟时枣红色，长1.2～1.5cm，径5～7mm，具尖头。果肉薄，有枣味。每果有果核1粒。核长椭圆形，背部有一沟槽，褐色，长10～12mm，宽约5mm，厚4～4.4mm。胚乳角质。果实及种子形态见图1。

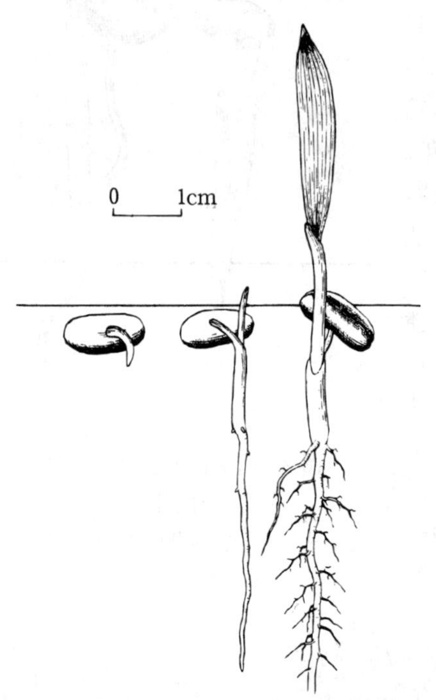

图2　软叶刺葵种子萌发后第7、35、60天幼苗的生长情况
（黄应钦绘）

果实采收调制和种子贮藏　当同一果穗的果实大部分呈红色时，可带果穗

采下。采摘下来的果实可堆沤数日，待果皮腐烂，置水中搅拌，淘去果皮杂质，即得种子。鲜果出种率为25%左右。种子净度可达98%。千粒重170（100～240）g，每千克有种子5 880（4 160～10 000)粒。种子忌失水，不宜曝晒。洗净的种子稍晾干后即可播种。贮藏或运输均需混以湿沙，贮藏期6～8个月。

发芽和播种　种子有中等程度休眠习性。发芽时气温需在18℃以上。1988年9月6日，广西林业科学研究所用当年新采而未作任何处理的种子，在室外沙床进行发芽测定，当时气温为28℃。播种后186天（1989年3月10日）方始发芽，发芽盛期不明显，发芽率为62%。留土萌发。种子萌发35～40天后抽出鳞叶，过20天发出初生叶（图2)。

通常将种子密播于沙床催芽，每平方米约播40～60g。待种子发出初生叶后3个月内，移至圃地培育。一般培育3年生苗出圃。

（朱积余）

钩　叶　藤

Plectocomia microstachys Burret

（棕榈科　Palmaceae）

生长习性、分布和用途　钩叶藤属约20种，我国产2种，本文描述1种。常绿有刺攀援灌木，长约1～1.5m，茎粗约为2cm。耐庇荫，适生于土质肥沃的热带密林中。基部有丛生的叶鞘和叶柄，密生针状刺。有文献说植株开花1次即死亡。分布于海南。栽培作刺篱，也可供编织。

开花结实　10～15年生开花结实。肉穗花序生于上部叶腋内。花单性，雌雄异株。雄花穗长约70cm。小佛焰苞长圆形，长2.2～2.5cm，内藏1小穗状花序，有花8～12朵。雄花花萼小，3裂，花瓣3，其中1片舟状，雄蕊6。雌花较大，花被于花后增大，退化雄蕊6，子房3室。据海南观察，花期为9～10月，果实成熟期为翌年6月下旬，成熟盛期为7月中下旬。浆果，未熟时青绿色，熟后黑褐色，圆形至椭圆形，长2～3cm，径2～2.5cm。每果有种子1颗，稀3颗。种子棕褐色至黑褐色，扁椭圆形，长1.5～2cm，宽1.5～1.7cm，厚1.2～1.4cm。胚乳充满种子，胚位于一侧。种子的形态及其解剖构造见图1。

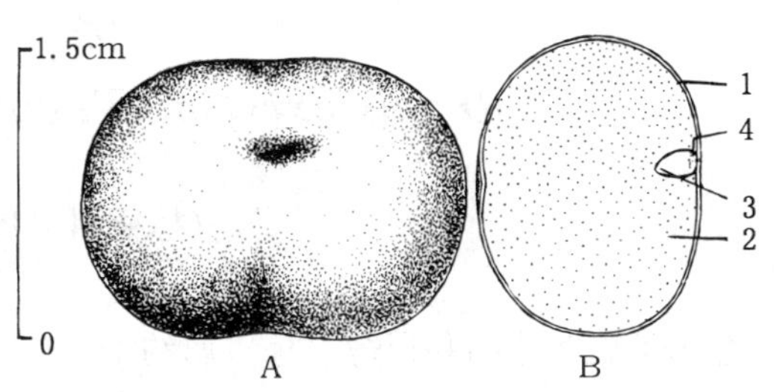

图1　钩叶藤种子外形（A）及其纵切面（B）
1. 种皮　2. 胚乳　3. 吸器　4. 胚根鞘
（黄应钦绘）

果实的采收调制和种子贮藏　用枝剪截断总果梗，将果穗置地面打落果实。采得的果实堆沤或浸水约1周，俟果皮软熟后装入竹筐置水中用木棒反复冲捣，不宜用手搓擦。淘去果皮

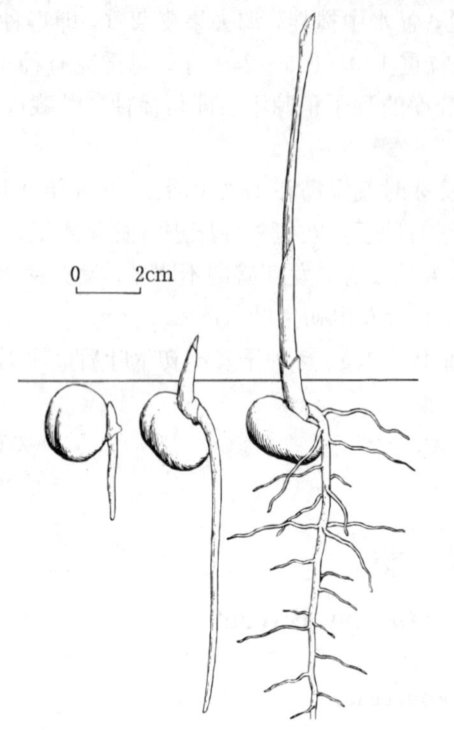

图 2 钩叶藤种子萌发后第 4、6、15 天幼苗的生长情况
（黄应钦绘）

等杂质即得种子。鲜果出种率为 30%左右。种子的净度可达 97%～100%，含水量约 20%。千粒重 3 200（2 900～3 500）g，每千克有纯净种子 300（280～350）粒。种子忌失水，不能日晒。可以直接运输果实，运抵后再行调制，也可将已调制的种子混以湿锯木屑装袋运输。一般宜随采随播。如需贮藏应混以湿沙，贮藏期约为 8 个月。

发芽和播种 种子无明显休眠现象。发芽时日均温宜在 25℃以上。1987 年 8 月 10 日，广西林业科学研究所在室外沙床上用当年从海南采集的种子作发芽测定，当时日均温 28℃：播后 39 天（即 9 月 17 日）开始发芽并进入盛期，10 月 18 日发芽终止；从开始发芽之日起算，5 天发芽 78%；从播种至发芽终了，70 天的发芽率为 88%。留土萌发。胚根萌发后 4 天鳞叶出土，12 天发出初生叶（见图 2）。

目前尚无批量播种育苗经验。条播或催芽后点播。苗期宜遮荫。培育 1 年半或 2 年生苗出圃。

（朱积余）

棕 竹

Rhapis excelsa（Thunb.）Henry ex Rehd.

（棕榈科 Palmaceae）

生长习性、分布和用途 棕竹属约 15 种，我国产 7 种，本文描述 1 种。丛生灌木，茎高 2～3m，直径 2～3cm。钙质土或酸性土均能生长。分布于我国东南部至西南部。日本也有。可供绿化观赏，茎作手杖、伞柄，根、叶入药。

开花结实 约 5～6 年生开始开花结实，正常结实期在 10 年生以后。结实大小年间隔期 1～2 年。肉穗花序 长 30cm，多分枝，佛焰苞 2～3 枚，管状，被棕色弯卷绒毛。花单性，雌雄异株，无梗。雄花较小，淡黄色，无柄；花萼杯状，3 齿裂，长 1.5mm；花冠倒卵形或棒状，浅 3 裂，长 1mm；雄蕊 6，无退化雌蕊。雌花较大，花萼与雄花相似，花冠则较短；心皮 3 枚，分离；花柱短；每心皮胚珠 1，基生；退化雄蕊 6 枚。据广西南宁 1978 年和 1979 年观察，5 月上旬为始花期，5 月中旬为盛花期，5 月下旬为末花期；11 月下旬果实开始成熟，12 月上旬为果熟盛期，12 月中下旬～翌年 2 月上中旬果实脱落。

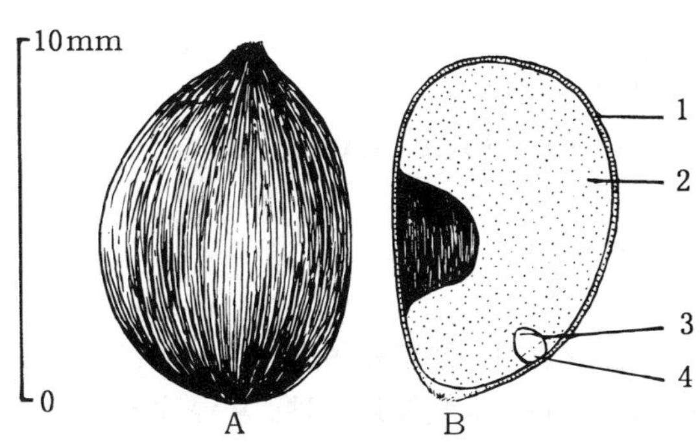

图1　棕竹种子外形（A）及其纵切面（B）
1. 种皮　2. 胚乳　3. 吸器　4. 胚根鞘
（黄应钦绘）

浆果，通常由1心皮发育而成，球状倒卵形，宿存的花冠管不变成实心的柱体；未熟时绿色，熟时黄褐色至黑褐色，径0.8～1cm。种子单生，球形或近球形，浅褐色，表面有网状粗纤维，径6～9mm，有胚乳，胚生于一侧（图1）。

果实的采收调制和种子贮藏　果实成熟盛期采摘。采回的果实洒水堆沤数日，果皮变软后置竹箩内浸泡，搓洗，淘去果皮果肉和杂质，即得纯净种子。种子忌脱水，不宜日晒。鲜果出种率约70%。种子净度可达97%；千粒重210（190～230）g，每千克有纯净种子4 760(4 350～5 260)粒。一般以果实的状态包装运输，运抵后再行调制。需要时调制后的种子混湿沙贮藏，贮藏期一般为1年。

发芽和播种　种子无休眠习性或有短期休眠。发芽时气温宜在20℃以上。1987年广西林业科学研究所曾在室外沙床上用当年新采的种子作发芽测定：播前种子未作任何处理，11月23日播种。因气温低，种子迟至1988年4月26日才开始发芽，发芽盛期为5月3日，5月6日发芽终止。从发芽开始期至发芽高峰，8天，发芽52%；从播种至发芽终止，以166天计，发芽率为60%。留土萌发。胚根萌发后15天第1片鳞叶出土，再过17天初生叶从第2片鳞叶中抽出。棕竹种子的萌发和幼苗生长情况见图2。

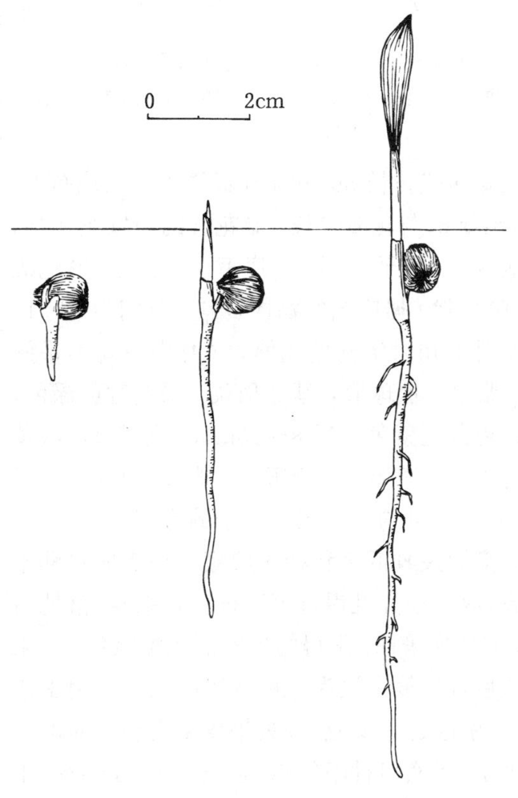

图2　棕竹种子萌发后第10、20、32天幼苗的生长情况
（黄应钦绘）

通常将种子密播于沙床催芽，每平方米播0.3kg，发芽后移至圃地继续培育。早期需遮荫。培育2～3年生苗出圃或上盆栽植。亦可分株繁殖。

（朱积余）

王棕（大王椰子）

Roystonea regia（H. B. K. ）O. F. Cook

（棕榈科 Palmaceae）

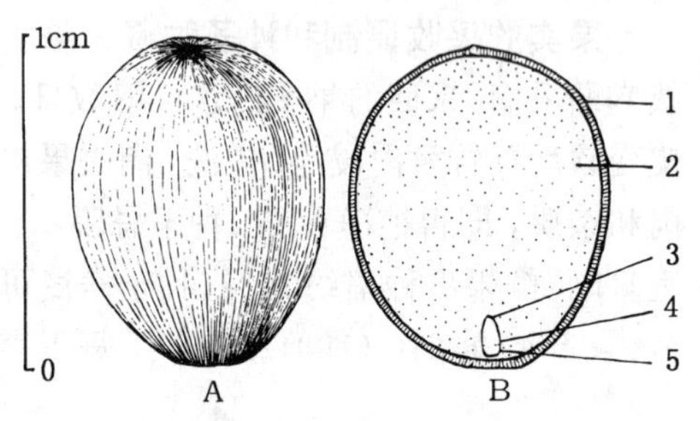

图 1 王棕种子外形（A）及其纵切面（B）
1. 种皮 2. 胚乳 3. 吸器 4. 子叶鞘 5. 胚根鞘
（黄应钦绘）

生长习性、分布和用途 王棕属约 17 种，我国引入 2 种，本文描述 1 种。常绿乔木，干单生，无分枝，常于中部膨大，高 10～20m。抗风力强，适生于肥沃湿润的酸性沙壤土。原产古巴。滇、桂、粤、琼、闽、台引种栽培。为热带名贵观赏树，适于作为行道树和庭园观赏栽培。种子可为鸽的饲料。

开花结实 约 15 年生开始开花结实，正常结实期在 20 年生以后，每年结实较多，大小年现象不明显。肉穗花序生于叶鞘束下，多分枝，排成圆锥花序式，长 50～60cm 或更长。佛焰苞 2，早落。花小，白色。单性同株。雄花长 6～7mm，雄蕊 6，与花瓣等长。雌花长约为雄花之半，花冠壶状，3 裂至中部；子房上位，3 室；退化雄蕊 6，鳞片状。据广州的物候资料，花期为 10 月，次年 7 月下旬果实开始成熟，8 月中下旬为果熟盛期。浆果，近球形，基部稍狭，未熟时黄绿色，熟时红褐色至紫色，径 8～13mm。种子 1，淡黄色，椭圆形至卵形，一侧压扁，长 8～10mm，径 6～8mm。胚乳均匀。王棕种子的形态见图 1。

果实的采收调制和种子贮藏 果实成熟期不很整齐，应分期分批用竹竿敲打已熟果实，落地后捡拾；或登梯采摘。采得的果实洒水堆沤数日，果皮变软后置竹箩中浸泡，用木棒冲捣，勿用手搓洗，以免引起过敏反应，淘去果皮果肉和杂质即得纯净种子。果实出种率约 50%。净度可达 98%。千粒重 198（180～220）g，每千克纯净种子 5 000（4 500～5 500）粒。种子忌脱水，不宜日晒，运输或贮藏均需混以湿沙。贮藏期一般为 1 年。

发芽和播种 种子个体间休眠程度有差异。发芽时日均温需在 20℃以上，采种时正值高温季节，宜随采随播。1981 年，广西林业科学研究所在室外沙床用当年新采种子作过发芽测定，播前未

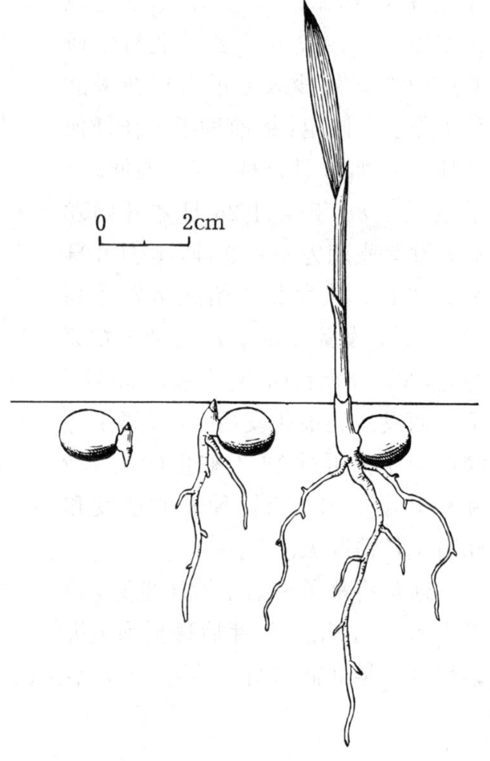

图 2 王棕种子萌发后第 7、15、42 天幼苗的生长情况
（黄应钦绘）

作任何处理。9 月 12 日播种，当时日均温为 27℃。播后 22 天少数种子胚根萌发，同时发出第 1 片鳞叶。发芽盛期不明显，迟至 1982 年 7 月 14 日发芽终止。发芽率为 60%。留土萌发。第 1 片鳞叶长出后 10 天长出第 2 片鳞叶，过 15 天长出第 3 片鳞叶，再过 10 天发出初生叶。王棕种子的萌发和幼苗初期生长情况见图 2。

撒播。通常是将种子密播于沙床，每平方米约播 0.3kg，发芽后陆续移至苗圃培育。通常培育 3 年生以上的大苗出圃。（朱积余）

棕　　榈

Trachycarpus fortunei Wendl.

（棕榈科　Palmaceae）

生长习性、分布和用途　棕榈属约 8 种，我国约 5 种，本文只描述 1 种。常绿，乔木状，高可达 15m。耐荫，幼树尤为突出。生长缓慢，寿命长。四川都江堰市一株“古山棕”已有 1 000 余年的高龄。喜温暖、肥沃、湿润环境。原产我国。淮河、秦岭以南温暖湿润的多雨地区广泛栽培。棕片（叶鞘）纤维拉力强，抗腐蚀，耐水湿，可制绳、缆、毯、垫。木材用于小型建筑和制作工艺品。嫩花苞可食。花、果入药。种子可提蜡，又富含淀粉和蛋白质，出粉率 88%～90%，可作饲料。叶形别致，可供观赏。对工业废气有较强的抵抗能力。从 70 年代初期开始，浙江等地流行棕榈腐烂病，李文虎和林栎等（1988）对危害程度作过调查并对防治措施提出过建议。

种内有变异。根据棕片性状、干节长度、叶形和结实习性等差异，四川雅安、峨眉等地分为：大木棕（结实迟且有间隔期，种子大，肾形）、小木棕（结实早而多，无间隔期，种子小，近圆形）和竹棕。

开花结实　8～10 年生开始结实，15 年生可正常结实。肉穗花序生于叶丛中，有明显的大花苞。花单性，雌雄异株。花小，萼和花瓣 3 片。雄花粟米状，黄白色，雄蕊 6 枚。雌花花萼绿色，花瓣黄色，子房 3 室，心皮基部合生，柱头反曲。开花结实的物候期见表 1。

表 1　棕榈的开花结实物候

观察地点	观察年份	现蕾期	开花			果实	
			始期	盛期	末期	成熟期	脱落期
陕西西安	1977	4 月 22 日	4 月 28 日	4 月 30 日	5 月 8 日	10 月 27 日	—
	1979	4 月 16 日	4 月 28 日	5 月 2 日	5 月 15 日	11 月 27 日	—
	1980	4 月 11 日	4 月 28 日	5 月 2 日	5 月 10 日	12 月 1 日	—
浙江杭州	1978	4 月 24 日	5 月 5 日	5 月 12 日	5 月 17 日	11 月 15 日	11 月 25 日
湖南常德	1975	3 月 23 日	4 月 19 日	4 月 24 日	4 月 27 日	10 月 24 日	11 月 15 日
	1976	4 月 10 日	4 月 25 日	4 月 30 日	5 月 2 日	11 月 2 日	—
	1979	3 月 12 日	4 月 19 日	4 月 22 日	4 月 28 日	11 月 8 日	12 月 29 日
广西桂林	1980	4 月 5 日	4 月 11 日	5 月 8 日	6 月 9 日	10 月 10 日	—

核果，球形、扁圆形或略呈肾形，熟时由青绿色转为蓝黑色，表面光滑，微被白粉或淡灰蓝色粉。果径10～13mm。外果皮肉质，内果皮骨质。果核长8～10mm，宽6～7mm，黑褐色，肾形，腹面中间有宽而浅的凹槽。有胚乳，胚生于一侧。果核的形态见图1。

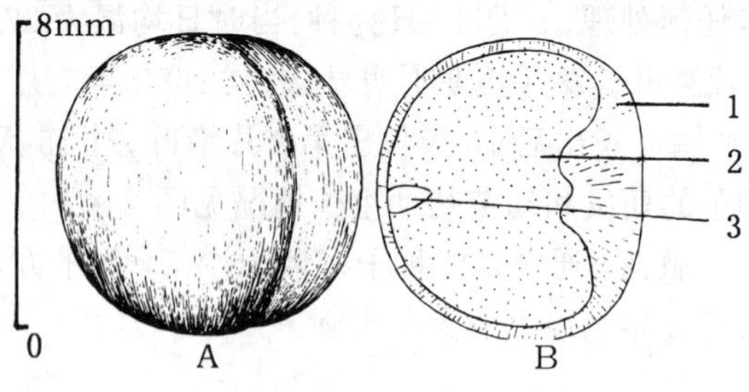

图1 棕榈果核的外形（A）及其纵切面（B）

1. 内果皮 2. 胚乳 3. 胚

（黄应钦绘）

果实的采收调制和种子贮藏 霜降前后，果皮由青灰褐色变成黄褐色时即已成熟。小雪后果皮变为黑褐色时采集的种子常常会表现出“硬化”，发芽迟缓。因此仍应在果熟后及时用刀将果穗割下，摊晾阴干后脱出果核即为播种材料，通称种子。出籽率65%～85%，净度95%～99%。千粒重350～480g，每千克有种子（核）2 100～2 900粒。幼龄母树的种子较小。

发芽和播种 种子有休眠习性。一般用切开法测定种子优良度。播前将种子浸草木灰水3～5天，或堆沤3～4天后搓去果实上的蜡质。干藏越冬的用始温60～70℃水浸种1天，以后每天早晚仍各用温水浸种1次，保持种子湿润，经10～15天种子萌动时播种。或在5～10℃下层积30天，去蜡后即可播种。过分干燥的种子播后有时会迟至次年才发芽。气温为25℃时，21天的发芽率为40（50～70）%。条播。每666m^{2}播种50～70kg，产苗4万～6万株。1年生苗高仅3cm左右，一般2～3年生后出圃。

留土萌发。单子叶筒状萌发。子叶分为3部分：下部粗圆筒状为子叶鞘，中部细长如柄为子叶联结，联结子叶上部留于种子中的吸器部分和子叶下部的子叶鞘部分。《主要树木种苗图谱》（1978）对棕榈的萌发和幼苗形态作过详细描述。

棕榈种子（核）的萌发和幼苗生长情况见图2。

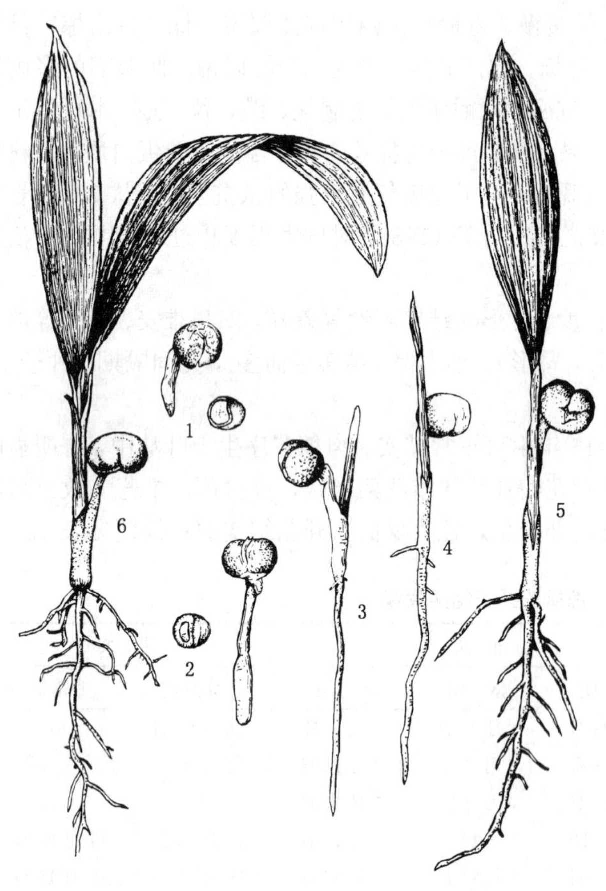

图2 棕榈种子的萌发及幼苗生长情况

1、2. 幼根及子叶鞘伸出果核 3. 初生叶（鳞叶）从子叶鞘抽出 4、5. 第2叶从鳞叶抽出 6. 幼苗形成

（田恒德仿《主要树木种苗图谱》）

（杨国华）

巴山木竹属

Bashania Keng f. et Yi

（禾本科　Gramineae）

生长习性、分布和用途　巴山木竹属已知3种，本文描述2种。灌木状或乔木状，秆高1～7m。喜温寒湿润气候和肥沃的微酸性土壤。产于川、甘、陕、鄂、滇、黔的山区。可以制作农具器物、造纸，也是大熊猫的食料。其名称、秆高、用途和分布见表1。

表1　巴山木竹属竹种的名称、秆高、分布和用途

中　名	学　名	秆高（m）	分　布	用　途
冷箭竹	*B. fangiana*（A. Camus）Keng f. et Yi	1～2.5	川西、滇、黔	材用、饲料
巴山木竹	*B. fargesii*（E. G. Camus）Keng f. et Yi	5～7(10)	川北、鄂西、陕南、甘东南	材用、饲料

开花结实　为一次性开花结实竹类，开花后秆枯死，鞭变黑。花两性。总状花序或圆锥状花序，小穗长2.1～5.4cm，每小穗含花5～10。颖片2。外稃7脉，内稃具2脊。鳞被3，不相等。雄蕊3，花丝长8～20mm。子房上位，卵圆形，胚珠1，柱头2，羽毛状。

巴山木竹开花期在4月，偶有在3月下旬开花的。田星群（1986）在陕西佛坪自然保护区观察过巴山木竹林的开花结实。他发现，在同一鞭系上的立竹中，老壮龄竹先开花；进入开花盛期后则不分竹龄大小均同时开花；从小花开放到花药枯萎历时4～14小时。他还观察到，1984年陕西洋县的竹林开花时，由于连绵阴雨，授粉受精率只有9.7%，成种率仅占授粉受精率的5%；而天气候条较好时，授粉受精率为85.3%，成种率为受精率的18.4%，完全发育的胚占受精率的85.6%。巴山木竹开花4天后子房膨大，由乳白色变为淡绿色，16天左右果实灌浆，40天左右果实成熟。未熟时绿黄色，成熟时褐色，有光泽。颖果，长卵形，长7～12mm，径3.5mm，深锈色或近黑色，有内外稃各1枚，芒不明显，腹沟明显（见图1）。颖果成熟期为5～6月。

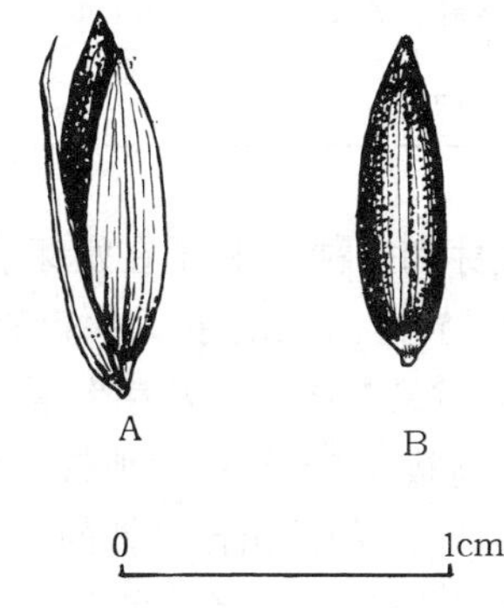

图1　巴山木竹具内稃、外稃的颖果苞(A)及颖果外形(B)
（叶可佳绘）

冷箭竹开花期在4～5月，果实成熟期为8～9月。未熟时黄绿色，成熟时褐色，有光泽。颖果长9mm，径1.6mm，全被外壳紧包，芒长3mm，具腹沟。胚乳粉状，胚小。

巴山木竹属竹种的开花结实物候见表2。

表 2 巴山木竹属竹种的开花结实物候

竹 种	观察地点和年份		开 花			果实成熟		果实散落期
			始 期	盛 期	末 期	始 期	盛 期	
冷箭竹	四川卧龙自然保护区 1985～1986			4～5 月		8 月中旬	9 月上旬	9 月下旬
巴山木竹	陕西佛坪自然	1981	4 月上旬	4 月中旬	4 月下旬	5 月下旬	6 月上旬	—
	保护区(岳坝)	1982	4 月中旬	4 月中旬	4 月下旬	5 月中旬	6 月上旬	6 月下旬
		1983	4 月上旬	4 月上旬	4 月中旬	5 月中旬	5 月下旬	—
		1984	3 月下旬	4 月上旬	4 月中旬	5 月中旬	5 月下旬	—

果实的采收调制和贮藏 巴山木竹果实 5 月下旬～6 月初完成生理成熟，6 月中下旬达到完全成熟即可采收。冷箭竹果实 8 月底 9 月初完成生理成熟，9 月中下旬采收较好。判断成熟的形态标志是果实变为褐色。剪摘果穗或人工击落。采回的果实摊晾阴干，搓揉后再风选或筛选去杂，清除稃片后的颖果即为播种材料，通称种子。

巴山木竹属颖果目前只作越冬贮藏。将含水量已下降到大约 10％的颖果，与颖果质量 5％的生石灰混合装入瓦罐内密封贮藏于 5℃以下的环境里，翌年 5 月播种。巴山木竹属的出籽率及净度、质量见表 3。

表 3 巴山木竹属竹种的出籽率及颖果的净度和质量

竹 种	出籽率（％）	净度（％）	千粒重（g）	每千克粒数（万粒）
冷箭竹	74.6	75	6.7	14.9
巴山木竹	92.7	93	42.2	2.4

发芽和播种 种子在发芽前用苔藓、湿沙在低温通气条件下层积，或用始温 40℃水浸种 1 小时，置蛭石上发芽。冷箭竹种子在 12℃条件下可以发芽，随温度升高发芽速度加快，但至 28℃时又趋缓慢，发芽率常为 14％～20％。

在海拔 1 500m 以上地带，以 5～6 月播种为宜。据四川卧龙自然保护区试验，冷箭竹春播（5～6 月）的幼苗比冬播的高 21.8cm。通常采用穴播或条播。穴播的穴距 20cm，穴深 1cm，每穴播种 4～6 粒。条播的条距 20cm，条幅 3～4cm，沟深 1cm。播种量以种子的发芽率及单位重量粒数来确定。播后覆细土 1cm，轻压后洒水，盖草。播种 2 周后开始发芽出土，再经 1 周后达发芽高峰阶段，其后逐渐转慢，经 1 个月基本停止发芽。小苗高约 8cm 时分床移植，一般 3 年生出圃造林。

留土萌发。冷箭竹种子萌发初期，基部背面露出一点白色胚根鞘，主根伸出胚根鞘，胚芽鞘即逐渐从种壳露出，胚根比胚芽生长快。主根长到约 1cm 时，从胚芽鞘顶端长出第 1 片鞘状叶（苗叶），随着根系的生长，抽出第 2 片鞘状叶后才开始发出正常叶（图 2）。

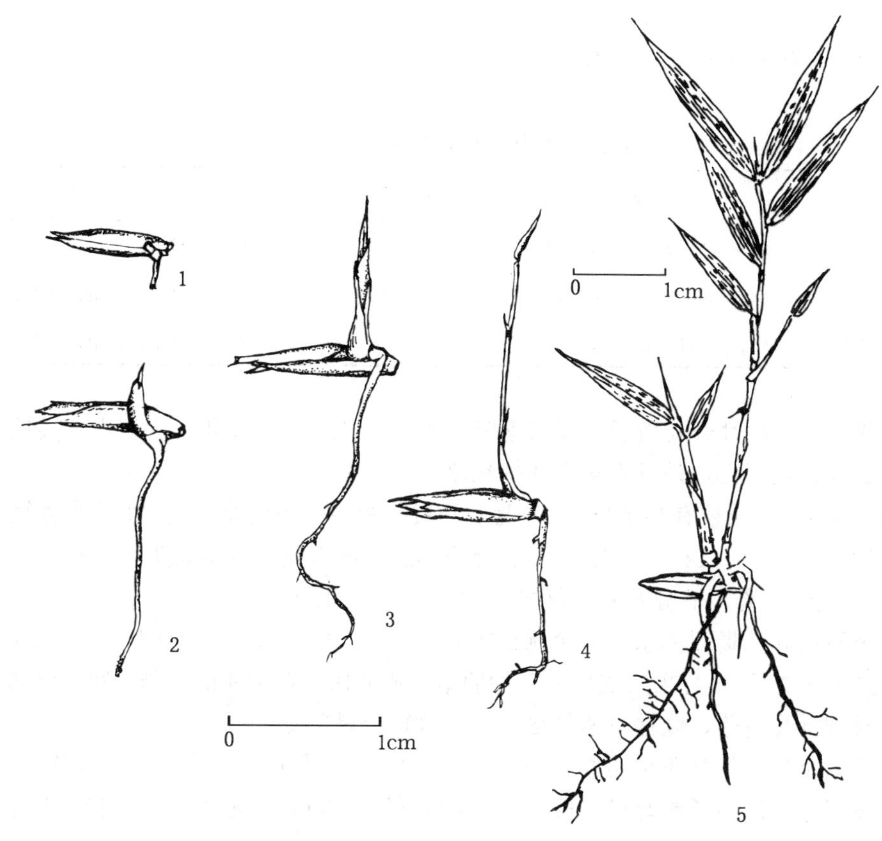

图2　冷箭竹种子的萌发过程及幼苗生长情况

1. 胚根鞘突出　2. 种子根及胚芽鞘形成　3. 苗叶形成　4. 叶片形成　5. 幼苗形态

（叶可佳、田恒德绘）

巴山木竹属竹种都可以用母竹分蔸、埋鞭等无性方法繁殖。

（向性明）

方　竹　属

Chimonobambusa Makino

（禾本科　Gramineae）

生长习性、分布和用途　方竹属已发表20种，4变种，分布在我国、日本及东南亚。我国约10种，本文描述3种。灌木状或小乔木状，地下茎复轴混生，秆圆筒形或微呈四方形，高2～7（10）m。基部数节生有气根状之刺瘤，秆箨宿存而易于在秆上腐坏，箨片很小。喜较温凉气候，适生于微酸性山地黄棕壤或山地黄壤。主产西南地区和秦岭南坡，少数种产东南沿海各省。

竹笋味鲜美，供食用。秆供造纸、制作家具、农具、伞柄、竹杖等用；叶和笋也是大熊猫的食料。

方竹属 3 个竹种的名称、分布和用途见表 1。

表 1 方竹属竹种的名称、分布和用途

中 名	学 名	秆高（m）	分 布	用 途
箐竹	*C. monophylla* Hsueh et Yi	3～7	川、黔	材用、食用、饲料、观赏
八月竹	*C. szechuanensis* (Rendle) Keng f.	2～6	川	材用、食用、饲料、观赏
金佛山方竹	*C. utilis* (Keng) Keng f.	4～10	川、黔	材用、食用、饲料、观赏

开花结实 一次性开花竹类。据张家贤（1985）报道，1984 年，贵州赤水县境内的金佛山方竹大面积开花死亡，是近百年未遇的现象。

方竹属花两性。花枝有些延长，末端为 1 开展的圆锥花序或总状花序，假小穗簇生于枝腋内。每小穗含花 3～6（11）。颖片 1～3。外稃膜质或厚纸质，顶端锐尖，内稃具 2 脊，鳞被 3。雄蕊 3，花丝分离。子房无毛，花柱 2，柱头 2，羽毛状。

据张家贤记载：贵州赤水县的金佛山方竹 1984 年 4 月 1～10 日开花，10～20 日为扬花期；5 月 25 日～6 月 15 日果实成熟并开始脱落；凡全株开花结实的竹株，种子成熟后便大量失水，小枝和叶片脱落，全株迅速枯萎死亡，竹鞭变黑腐烂。

方竹属 3 个竹种的开花结实物候期见表 2。果实未成熟时青绿色，成熟时紫绿色或褐色。坚果，果皮坚厚，椭圆形或肾形，长 8～18mm，径 3～8mm。无胚乳（图 1）。方竹属果实的形态特征见表 3。

表 2 方竹属竹种的开花结实物候

竹 种	开花期	果实成熟期	果实脱落期
箐竹	4 月	5 月上中旬	6 月下旬
八月竹	4 月	5～6 月	6 月下旬
金佛山方竹	4～5 月	6 月上旬	6 月下旬

表 3 方竹属竹种的果实形态特征

竹 种	颜 色	形 态	大小（mm）	
			长	径
箐竹	紫绿色	肾形	8～18	4～8
八月竹	褐色	椭圆形	10～12	3～4
金佛山方竹	褐色	椭圆形	10～15	6～8

果实的采收调制和种子贮藏 果实颜色变为紫绿色或褐色时即可采集。据张家贤（1985）观察，金佛山方竹有少数种子成熟后在竹枝上就已萌发。通常用人工敲击竹秆将果实

击落收集，经轻打搓揉除去杂物，装袋运回。运回的果实应立即调制，可在筛上搓揉，再用清水冲洗，清除了内稃外稃后置通风处阴干。通常以坚果作为播种材料，通称种子。贮藏用种子含水量不超过11%。将种子与种子质量0.6%的甲基托布津或某些杀虫粉剂拌合，装入瓦罐内，置5℃以下贮藏；或在罐内加入种子质量5%的生石灰密封贮藏于海拔2 000m左右的地带。

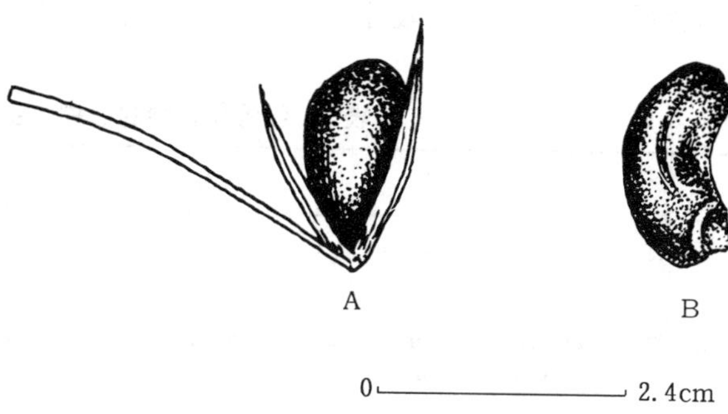

图1　箐竹带有内稃、外稃的坚果（A）及其外形（B）
（叶可佳绘）

方竹属竹种的种子净度和质量见表4。

表4　方竹属竹种的种子净度和质量

竹　种	净度（%）	千粒重（g）	每千克种子粒数（粒）
箐竹	96	113	8 800
八月竹	96	102	9 800
金佛山方竹	95	167①	5 900

①　张家贤（1985）记载为113g

发芽和播种　种子无休眠习性。箐竹种子发芽时种子含水量为42%，适于发芽的温度为25℃，发芽率可达95%（张家贤，1985）。1976年6月在四川马边县采集的八月竹种子，虽经层积处理，发芽率仍然只有14%。

播种期5～6月，以随采随播为好。穴播或条播。条距20cm，条宽3～4cm，沟深1cm。金佛山方竹条播时每平方米用种25g。播后覆土1cm，轻压后洒水、盖草，保持湿润。播种1周后开始发芽。留土萌发。经30～50天幼苗可长出叶片3～7枚。幼苗高约7cm时分蔸移植。2年生苗高20cm以上，可出圃造林。

方竹属竹种均可无性繁殖。其方法有母竹分蔸、鞭根诱导、埋鞭等。

（向性明）

牡　竹　属

Dendrocalamus Nees

（禾本科　Gramineae）

生长习性、分布和用途　牡竹属约40多种，我国有近30种，本文描述2种1变种。地下茎合轴型乔木状竹。秆丛生，高6～25m，直立或外倾，梢常下垂呈吊丝状。节间圆筒形，每节具多数分枝，主枝常发达。箨耳常不明显或缺如，箨舌明显。适宜于土层深厚、湿润肥沃的钙质土、微酸性至酸性土壤。它们的名称、生长、分布及用途见表1。

表1 牡竹属竹种的名称、生长、分布和用途

中 名	学 名	秆高(m)	胸径(cm)	分 布	用 途
麻 竹	*D. latiflorus* Munro	20～25	15～25	滇、川、黔、桂、粤、台、闽	笋用、建筑、纤维、篾用、观赏
吊丝竹	*D. minor*(McClure)Chia et H. L. Fung	8～12	6～8	桂、粤	笋用、篾用、纤维、观赏
花吊丝竹	*D. minor* var. *amoenus* Dai et Huang	10～16	8～14	桂	笋用、篾用、纤维、观赏

开花结实 牡竹属是多年生一次开花植物，竹丛开花结实后竹秆枯萎死亡。同一竹林中的少数竹丛开花，并不蔓延及其它竹丛。开花竹丛一般在开花前1年出笋量明显减少，枝梢出现缩小的变型叶片。花两性。花枝无叶或有时具叶。小穗无柄，数枚至多数簇生于花枝各节上，甚至密集成球状。小穗具1至多数小花，小穗轴不具关节。颖1至多数。外稃似颖，向小穗顶端依次变狭长。下部小花的内稃具2脊，最上的完全小花或仅具1花的内稃无脊。鳞被通常缺。雄蕊6枚，花丝分离。子房上位，卵形，顶端被毛，具短柄。花柱长而被毛，柱头单一，羽状。

一年四季均可开花，盛花期在2～6月。同一批花芽的生长发育同步，开花期集中在2～3天内，前一批花凋落时，另一批花芽又处在孕育中。每批开花的间歇时间随气温递增而有变化：2～4月间约间隔1～1.5个月开一批花；4～5月约隔1个月开一批花；5～6月每半个月左右就开一批花。一批花开后20天左右果实成熟。据广西南宁的观察，牡竹属的开花结实物候期见表2。颖果长圆形。果皮薄脆，易与种子分离。牡竹属果实的形态特征见表3。

表2 牡竹属竹种的开花结实物候（广西南宁）

竹 种	开 花			果实成熟		成熟果实颜色	果实散落期
	始 期	盛 期	末 期	始 期	盛 期		
麻 竹	1月	2～6月	7～12月	3月	4～7月	稃黄色，果皮淡褐色	4～7月
吊丝竹	1月	2～6月	7～12月	3月	4～7月	稃黄色，果皮灰褐色	4～7月
花吊丝竹	1月	2～6月	7～12月	3月	4～7月	稃黄色，果皮灰褐色	4～7月

表3 牡竹属竹种颖果的形态特征

竹 种	形 状	大小（mm）	颜 色
麻 竹	卵球形，腹部有长沟	长8～12，径4～6	淡褐色
吊丝竹	椭圆形，腹部有长沟	长4～7，径2～3	灰褐色
花吊丝竹	椭圆形，腹部有长沟	长5～8，径2～3	灰褐色

果实的采收调制和贮藏　果实成熟后极易自然脱落，应及时采收。零星开花的竹子，自然结实率极低，采种较难。常用的方法是将开花竹砍去中上部，仅留基部1.3m左右的竹秆，既方便管理和观察，又方便采种。牡竹属结实期喜光，林缘和空旷地的开花竹比林中的结实率高，若开花竹处于荫蔽中，需将周围遮荫植株砍去。花期加施磷钾肥可提高结实率。采得的颖果可直接用作播种材料，也可在清除稃壳后用作播种材料，都通称种子。种子宜随采随播，如需贮藏至翌年春播，可置于4℃左右的低温条件下贮藏。种子的质量见表4。

表4　牡竹属竹种的种子（颖果）质量

树　种	带内外稃的千粒重（g）	去内外稃的千粒重（g）	每千克纯净种子(去稃)粒数(万粒)
麻　竹	51～57	45～50	2.0～2.2
吊丝竹	15～20	10～16	6.2～10.0
花吊丝竹	24～26	19～23	4.3～5.3

发芽和播种　种子无休眠习性，可随采随播，也可在4℃左右的低温条件下贮藏至翌年3月播种。1978年6月，广西林业科学研究所在室外沙床作过发芽测定，结果见表5。

表5　牡竹属竹种的发芽能力①

竹　种	发芽势（%）		发芽率（%）	
	计算天数	一般数值	计算天数	一般数值
麻　竹	18	60	35	70
吊丝竹	18	45	35	60
花吊丝竹	18	50	35	65

①　广西南宁室外沙床，气温25～30℃

条播。麻竹每平方米约播15～18g，吊丝竹和花吊丝竹每平方米约播7～10g。覆土1～1.5cm。苗期宜稀疏遮荫。1～2年生苗出圃。生产中多用次生枝扦插育苗或移母竹造林。

（戴启惠）

箭　竹　属

Fargesia Franch.

（禾本科　Gramineae ）

生长习性、分布和用途　箭竹属约有76种，除总花箭竹产于尼泊尔东部及印度外，其余竹种均产我国。本文描述2种（表1）。灌木状竹类，丛生或近散生，地下茎合轴型，秆柄粗短，两端不等粗，实心。秆高2～5m，可达7m，直径1～3cm，浅根性。耐荫。喜温暖湿润气候。适生于排水良好的酸性或微酸性山地棕壤、山地灰棕壤和山地褐色土。

主产川、甘、陕等省中山和亚高山地带，海拔 1 500～3 800m，常生于亚高山针叶林或针阔叶混交林下。是大熊猫全年的主要食料。竹秆供编织和农作物支架，也可造纸。对保持水土也有重要作用。

表 1 箭竹属竹种的名称、分布和用途

中 名	学 名	分 布	用 途
缺苞箭竹	*F. denudata* Yi	川、甘、陕	材用、饲料
紫箭竹	*F. nitida* (Mitf.) Keng f.	川、甘、陕	材用、饲料

开花结实 多年生一次开花植物。开花不受地上部分年龄限制，而是地下茎达到成熟阶段就可开花。因此老竹和当年生竹常常同时开花，先是个别竹株开花，逐渐发展到整个竹丛，大约 3～5 年便会全林开花。花两性，总状花序，着生于顶生或侧生的小枝顶端，花穗下具一组由叶鞘扩大的佛焰苞，最上的一佛焰苞与花穗等长或长过花穗。花穗初开时由佛焰苞开口之一侧露出。小穗柄细长，含数花。颖 2 枚。外稃先端小尖头状，具数脉。内稃等长或短于外稃。鳞被 3，具缘毛。雄蕊 3，花丝分离，花药黄色。子房椭圆形，花柱 1 或 2，柱头 2～3。

竹林将开花时发笋量减少，幼竹和笋的地径变小。7 月开始孕穗，翌年 4 月开花。开花时花药抽出颖外，柱头包在颖内，花丝长，花柱短。5～6 月为开花盛期，此时的花易授粉结实，9 月（末期）开的花一般不能形成果实。9～10 月果实成熟。每丛竹子全部开花结实后，竹株养分耗尽，竹叶脱落，不能进行光合作用，直至枯黄死亡。

箭竹属 2 个竹种的开花结实物候见表 2。

表 2 箭竹属竹种的开花结实物候

竹 种	观察地点	观察年份	开 花			果实成熟	
			始 期	盛 期	末 期	始 期	盛 期
缺苞箭竹	四川平武	1985～1987	4 月	5～6 月	9 月	9 月下旬	10 月上旬
紫箭竹	四川南坪	1985	4 月	5～6 月	9 月	9 月下旬	10 月上旬

果实成熟前绿色，成熟时褐色。颖果。外果皮与种子连生。颖果纺锤形，顶端具芒，腹沟窄而浅，具外颖、内颖。缺苞箭竹果实棕褐色，长 8.4～9mm，径 1.9mm。紫箭竹果实黄褐色，长 8～8.6mm，径 1.8mm。胚乳粉状，胚小。

果实的采收调制和种子贮藏 10 月上中旬，当果实变为褐色并有甜味，含水量减少到 30%～40%时即应采集。敲击竹秆，使果实落在事先铺放的收种布上，除去粗枝杂物，收集入袋运回调制。也可以用枝剪剪下果枝后摘取果穗，或伐倒竹秆，剪取果穗，再经搓揉获取果实。采回的果实放在通风的室内或室外晾干后搓揉，除去杂质，所得颖果即为播种材料，通称种子。它们的净度、质量和数量见表 3。

表 3　箭竹属竹种种子（颖果）的净度和质量

竹　种	净度（%）	千粒重（g）	每千克纯净种子数（万粒）
缺苞箭竹	95	9.1	11.0
紫箭竹	95	7.4	13.5

种子宜于干藏。据四川林业科学研究所试验，含水量6%～11%的种子在5℃以下的环境里，或在海拔2 500m以上年均温为3℃的干燥处，瓦罐密封可以贮藏1年。

发芽和播种　箭竹属种子无明显休眠习性。四川林业科学研究所对缺苞箭竹和紫箭竹种子用几种预处理方法，在20℃恒温或12℃/20℃昼夜变温下，20天左右得到的发芽率都在70%左右。

箭竹属竹种种子发芽的适宜温度为14～24℃。海拔1 500m以上地带5月中下旬的气温能够符合这个要求，是这个地带适于播种的季节。穴播时，穴距20cm，穴深约1.5cm，每穴播种3～4粒，每平方米床面约播10g。条播时，条距20cm，条幅3～4cm，沟深1cm，每平方米床面约播15～17g。播后覆细土，以覆盖种子为度，并喷水盖草。留土萌发。播种后20天左右发芽出土。出土小苗应适时分蘖移栽。高山地区生长期短，1年生苗高一般约15cm。2年生苗可出圃栽植。

（向性明）

毛　竹　属

Phyllostachys Sieb. et Zucc.

（禾本科　Gramineae）

生长习性、分布和用途　毛竹属约50种，我国40余种，本文描述3种。常绿乔木状，高6～15m。地下茎单轴散生，顶芽通常不出土，在土壤中延伸成竹鞭。秆散生，节间圆筒形，但分枝的节间在出枝一侧具沟槽。枝条在各节上通常2枚。主产亚洲东部。我国黄河流域以南，南岭以北为分布中心。适生于土层深厚、湿润肥沃，中性偏酸性土壤。竹材在建筑、家具、农具和日常生活中用途极广。笋食用。它们的名称、生长、分布和用途见表1。

表 1　毛竹属竹种的名称、生长、分布和用途

中　名	学　名	秆高(m)	胸径(cm)	分　布	用　途
桂　竹	*Ph. bambusoides* Sieb. et Zucc.	6～10 (15)	6～12 (16)	黄河流域及以南各省(自治区)	材用、篾用、食用
毛　竹	*Ph. edulis* (Carr.) H. de Leh.	10～15 (20)	7～12 (20)	长江流域及以南各省(自治区)。日本有栽培	材用、篾用、食用
假毛竹	*Ph. kwangsiensis* Hsiung et Dai	6～16	4～10	桂、湘	材用、篾用、食用

开花结实 多年生一次开花植物，随着年龄的增长，进入发育成熟阶段的竹林，遇到适宜的环境条件就会开花结实，结实后竹秆枯萎死亡。据20世纪60年代在广西昭平观察，毛竹在开花前应当是发笋的一个大年，不发笋或发笋很少；应当换叶时不发新叶，老叶逐渐由绿变黄，以至枯萎凋落；在花穗轴抽出前后，绝大部分叶片脱落。开花与否与竹株年龄无关，初期多为零星竹株开花，随即蔓延至全林。完成全林开花的全部过程一般要3～5年。开花的竹株，有的全株开花，有的只是竹秆一侧的枝条开花。竹株年龄在2年以内的嫩竹结实率低，甚至只开花不结实；3年以上的老竹既开花也结实；5年和6年生的竹株结实量最多。

花两性，花序为圆锥状、复穗状或头状，由多数小穗组成，生于枝顶或小枝上部叶丛间。小穗外被叶状或苞片状佛焰苞。小穗轴具关节。颖1～3，有时缺。外稃顶端锐尖，内稃具2脊，先端具2尖头。鳞被3。雄蕊3，花丝分离而细长。子房上位，具柄，花柱3，细长，柱头3裂。颖果果皮与种皮紧密相连，难以分离。广西昭平、南宁、融水和浙江安吉观察的物候期见表2。

表2 毛竹属竹种开花结实的物候期

竹 种	观察地点	开花 始 期	开花 盛 期	开花 末 期	果实成熟 始 期	果实成熟 盛 期	果实散落期
桂 竹	广西融水	4月	5～6月	7～9月	7月	8～10月	8～10月
	浙江安吉	—	4～5月；9～11月	—	—	—	—
毛 竹	广西昭平	4月	5月	6月	8月	9～10月	9～10月
假毛竹	广西南宁	5月	6月	7月	7月	8月	8～9月

张文燕和马乃训(1989)观察过桂竹的开花过程。在她们测定的13个竹种中，桂竹花粉在人工培养基上的萌发率最低，仅17.1%。张文燕和马乃训(1990)在花粉散放完毕时检查过桂竹柱头上的花粉粒数，发现55.6%的柱头没有接受花粉或仅有一粒花粉，只有20%的柱头接受了5粒或5粒以上的花粉。乔士义(1984)观察过毛竹的胚胎发育过程。颖果成熟后为黄色的内稃和外稃包被。除去内稃和外稃的颖果为深褐色，具光泽。形态特征见表3、图1。

表3 毛竹属竹种果实的形态特征

竹 种	未熟果颜 色	成熟果实（种子）形 状	成熟果实（种子）大小(mm)	成熟果实（种子）颜 色
桂 竹	浅褐色	椭圆形，腹部稍扁平，具长沟	长5～5.5，径2～2.2	褐色到深褐色
毛 竹	浅褐色	长椭圆形，腹部稍平，具纵沟	长5～8，径1.5～2	褐色到深褐色
假毛竹	浅褐色	椭圆形，腹部稍平，具纵沟	长4～4.5，径1.8～2.0	褐色至深褐色

果实的采收调制和种子贮藏 果实成熟后容易散落，应及时从伐倒竹上剪下果枝，晒干后打落果实，用风选法除去空粒和杂物，所得颖果用作播种材料，通称种子。置日光下晒干，含水量一般要求在10%以下。运输或贮藏可用麻袋包装，装袋前按质量掺入0.3%的粉状杀虫剂，与种子拌匀。种子不宜久藏，随着贮藏时间的延长，发芽率会很快下降。在常温条件下

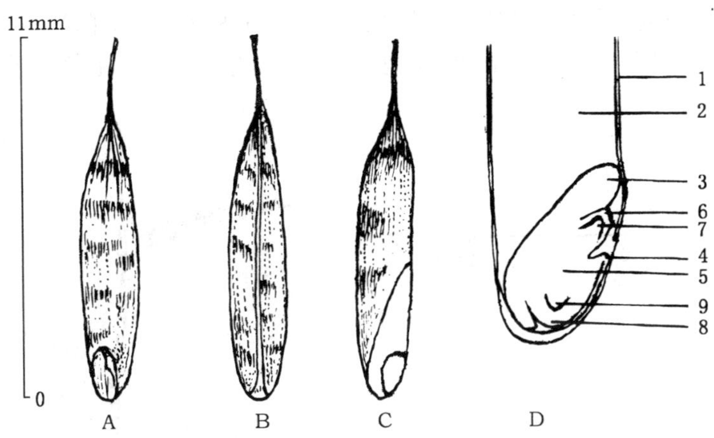

图 1　毛竹颖果外形背面(A)、腹面(B)、侧面(C)及其纵切面部分放大(D)
1. 果皮及种皮　2. 胚乳　3. 盾片（内子叶）　4. 外胚叶（外子叶）　5. 胚轴
6. 胚芽鞘　7. 胚芽　8. 胚根鞘　9. 胚根
（田恒德仿《中国主要树木幼苗形态》）

贮藏 7 个月以上的种子，常常就会完全丧失发芽能力。在 0～8℃的低温条件下，发芽能力可以保持 2 年。种子的质量见表 4。

表 4　毛竹属竹种的种子（颖果）质量

树　种	带内外稃千粒重（g）	去内外稃千粒重（g）	每千克纯净种子(去内外稃)粒数(万粒)
桂　竹	37.5	30.7	3.3
毛　竹	17.5	11.5	8.7
假毛竹	21.2	14.8	6.8

发芽和播种　种子无休眠习性。可以随采随播。一般发芽率为 10%～20%，较好的可达 20%～30%，经过精选的种子可达 40%以上。影响种子发芽可能的原因，一是采收时种子尚未成熟；二是授粉受精不良，半饱满粒和瘪粒多，四唑染色显示有些种粒虽有胚乳，但胚并没有发育起来；三是贮存时霉变，丧失了发芽能力。1974 和 1978 年，广西林业科学研究所用精选过的种子做发芽测定，播后日均温为 10～15℃时，开始发芽时间约 15 天，日均温上升到 20℃以上时，开始发芽时间仅需 9 天。发芽情况见表 5。

表 5　毛竹属竹种的发芽能力及其测定条件

竹　种	预处理	基　质	温度（℃）		发芽势（%）		发芽率（%）	
			室　内	室　外	计算天数	一般数值	计算天数	一般数值
桂　竹	浸种	河沙	15～20	—	20	40	50	60
毛　竹	浸种	河沙	15～20	—	20	30	50	40
假毛竹	—	壤土	—	10～15	30	20	60	25

留土萌发。萌发时胚根鞘突破种皮，胚根（种子根）从中穿出，随后胚芽鞘也破皮而出，胚芽向上出土，是为第一鞘叶，长出3、4鞘叶后，才出生正常叶。毛竹种子的萌发和幼苗早期生长情况见图2。

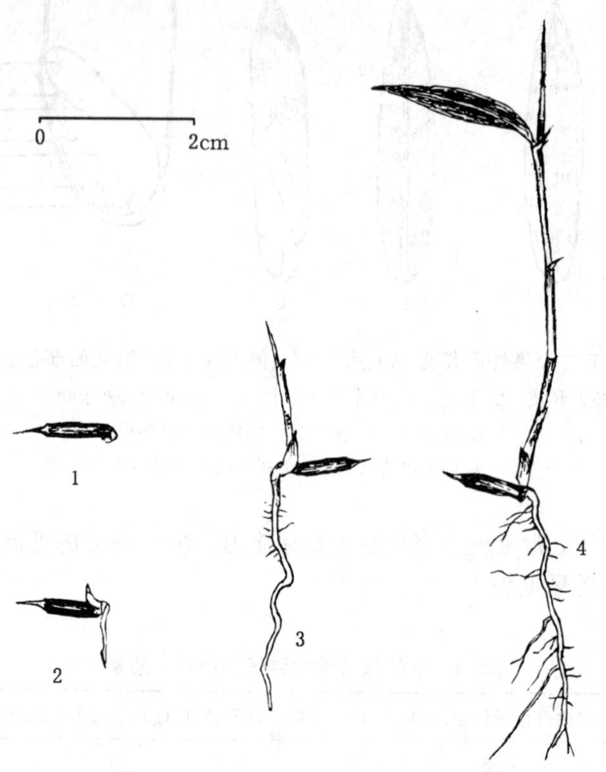

图2 毛竹种子萌发及幼苗初期生长情况

1. 胚根鞘初露 2. 胚芽鞘出现 3. 生出鞘叶 4. 出现正常叶

（田恒德仿《主要树木种苗图谱》）

条播。毛竹和假毛竹每平方米播种8～10g，桂竹12～15g，覆土约1.5cm。苗期需遮荫。育苗1～2年，幼苗丛生，分苗出圃时适当留苗，以后的3～5年便可继续分苗出圃。生产中多用移母竹造林。

（戴启惠）

筇 竹 属

Qiongzhuea Hsueh et Yi

（禾本科 Gramineae）

生长习性、分布和用途 筇竹属有7种，分布于我国西南部，为特有属。本文描述3种，其中筇竹为稀有植物，已列入《中国植物红皮书》。灌木状竹类，高2～7m。地下茎是复轴型。秆圆筒形或基部数节略呈方形，秆环不隆起至极度隆起而呈一圆脊，秆基部数节不具根刺。常成片生长于常绿阔叶林下，耐荫湿。喜气候温暖，雨量充沛，空气湿度大，日照少的环境。适生于微酸性的山地黄壤或棕色森林土。主产于川、滇海拔 1 500～ 2 500m 的山区。鲜笋白色，是著名笋用竹种。秆用于造纸和编织，是四川大凉山山系大熊猫的主要食用竹种。

筇竹属 3 个竹种的名称、生长、分布和用途见表 1。

表 1　筇竹属竹种的名称、生长、分布和用途

中　名	学　名	秆高（m）	分　布	用　途
平　竹	*Q. communis* Hsueh et Yi	3～7	川、黔、鄂	材用
实竹子	*Q. rigidula* Hsueh et Yi	2～5	川南、川西南	材用、饲料
筇　竹	*Q. tumidinoda* Hsueh et Yi	2.5～6	川、滇	材用、食用、观赏、饲料

开花结实　多年生一次开花结实植物。花两性。花序主轴各节具 1 大型苞片，着生 1 至数枚分枝，其顶具 1 小穗，下部为 1 组小苞片包被。小穗的颖片和基部小花的外稃常呈苞片状。花枝有时混杂有具叶小枝，无毛，小穗轴扁平，无毛，基部微被白粉。小穗含 3～8 花，绿色或暗绿色。颖 2 或 3 枚，内稃短于外稃。鳞被 3，后方 1 片披针形，两侧 2 片卵形。雄蕊 3，花药紫色或黄色。子房呈倒卵形或椭圆形，花柱 1，顶端生羽毛状柱头 2 枚。

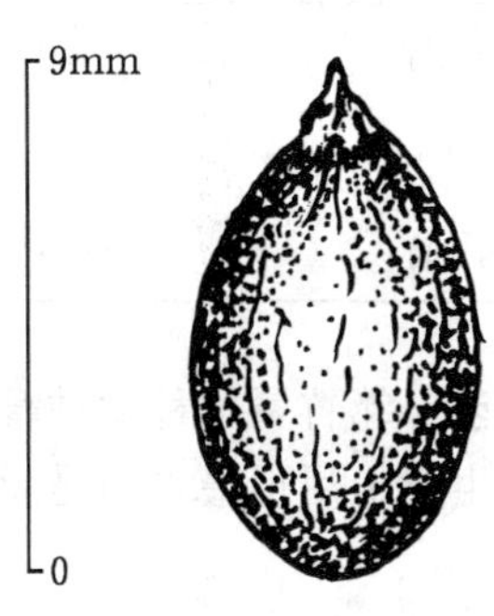

图 1　实竹子坚果外形
（叶可佳绘）

开花期 3～5 月。早开的花，果实早成熟，后开的花，果实晚成熟。因此在开花期中，花与果实同时存在。果实成熟期 5～6 月（表 2）。坚果，厚皮质，紫绿色或暗绿色，倒卵状长椭圆形或椭圆形，光滑无毛，长 8～13mm，径 4～7mm，顶端具宿存花柱。实竹子坚果外形见图 1。筇竹属 3 个竹种的果实形态特征见表 3。

表 2　筇竹属竹种的开花结实物候

竹　种	观察地点	开花期	果实成熟期	果实散落期
平　竹	四川丰都	3 月	5 月	6 月
实竹子	四川屏山	3～4 月	5～6 月	7 月
筇　竹	四川雷波	4～5 月	5～6 月	7 月

表 3　筇竹属竹种成熟果实的形态特征

竹　种	颜　色	形　状	长（mm）	径（mm）
平　竹	紫绿色	椭圆形	9～13	4～7
实竹子	紫绿色	倒卵状椭圆形	8～11	4～5
筇　竹	紫绿色	倒卵状椭圆形	10～12	4～5

果实的采收调制和种子贮藏 当果实颜色变为暗绿色或紫绿色时即可采集。通常在竹下铺放收种布，将竹秆上的果实击落或摇落在收种布上，除去粗枝杂物后装袋运回。果皮厚1～2mm，多为肉质。可以在浸泡后搓揉，用水冲去果皮，再置通风阴凉处晾干。也可以不去果皮，直接将果实摊放晾干。生产上常以坚果作为播种材料，通称种子。

适于贮藏的种子含水量为6%～11%，以7%为最好，贮藏地以高海拔（1 500m以上）地带最理想。贮藏前加入种子总质量0.6%的甲基托布津或某些杀虫粉剂，置瓦罐内密封贮藏。还可将采回的新鲜果实，放入湿沙或蛭石或苔藓层积，基质为种子的3倍。

筇竹属3个竹种的出籽率及种子净度、质量和数量见表4。

表4 筇竹属竹种的出籽率及种子净度、质量和数量

竹 种	出籽率（%）	净度（%）	千粒重（g）	每千克种子粒数（万粒）
平 竹	89.0	95	88.2	1.1
实竹子	88.7	96	67.6	1.5
筇 竹	88.9	95	90.1	1.1

育苗 筇竹属竹种播种育苗的实际经验不多，可以参照本书前面所述各竹类介绍的育苗方法。生产上用得最多的是无性繁殖。

（向性明）

玉山竹属

Yushania Keng f.

（禾本科 Gramineae）

生长习性、分布和用途 玉山竹属有47种，多分布于亚洲，模式种原产台湾至菲律宾，非洲刚果有3种，印度1种，我国有43种。本文描述国产的2种。竹秆较高大，秆柄细长，两端粗细近一致，散生。地下茎合轴型。耐荫，多生于林下。是亚热带高山地带特有竹类。喜温凉气候和湿润土壤。主产我国西南部。垂直分布在1 300～3 500m。可作造纸、农具、编织、箫、笛原材料，也是大熊猫食料。

玉山竹属2个竹种的名称、高度、分布和用途见表1。

表1 玉山竹属竹种的名称、高度、分布和用途

中 名	学 名	秆高（m）	分 布	用 途
石棉玉山竹	*Y. levigata* Yi	3～5	川	材用、饲料
光亮玉山竹	*Y. lineolata* Yi	2～5	滇	材用

开花结实　多年生一次性开花植物。开花前常出现竹叶变黄或叶片缩小脱落等现象。花两性。总状花序或圆锥花序，其下方有苞片附托，小穗露出。叶鞘不扩大成佛焰苞。花枝长6～30cm。小穗含花4～9。颖片2，鳞片2，鳞被3，雄蕊3，花丝细长，花柱短，柱头2枚。风媒花。

开花期5～6月。开花时花药抽出颖外，柱头包在颖内。授粉1个月后，果实陆续成熟。因开花期长，果实成熟期不整齐，大部分果实9～10月成熟。果实未熟时绿色，成熟时紫褐色。颖果，长椭圆形，长7～8mm，径1～1.5mm，有宿存花柱。退化的种皮与果皮相愈合，不易分离。果实数与种子数同。腹沟明显，具脐。胚乳粉质白色，胚小，子叶1。

玉山竹属竹种的开花结实物候见表2。

表2　玉山竹属竹种的开花结实物候

竹　种	观察地点	开花期	果实成熟期	果实脱落期
石棉玉山竹	四川石棉	5～6月	9～10月	11月
光亮玉山竹	云南临沧	5～8月	9～10月	11月

果实的采收调制和种子贮藏　果实呈紫褐色且变硬时敲击或震动竹秆，使果实落于铺在地面的收种布上。也可以用枝剪从伐倒的竹秆上剪摘果穗。从果穗上搓下果实，其中常常混有大量杂质，应进行风选、筛选，除去杂质而得纯净颖果，阴干后即为播种材料，通称种子。玉山竹属2个竹种播种材料的净度和质量见表3。

表3　玉山竹属竹种播种材料的净度和质量

竹　种	净度（%）	千粒重（g）	每千克粒数（万粒）
石棉玉山竹	95	5.8	17.2
光亮玉山竹	94	5.6	17.8

颖果贮藏的适宜含水量为6%～10%，而以7%较好。置瓦罐内加入种子质量0.5%的生石灰密封贮藏。

育苗　玉山竹属播种育苗的实际经验不多，可以参照本书所写的禾本科其它竹种。生产中常用的繁殖方法仍然是无性繁殖，例如母竹分蔸移栽、鞭根诱导等。

（向性明）

树种中名索引

（按汉语拼音顺序排列）

树种中名首字笔画索引

四画

五画

六画

七画

八画

九画

十画

十一画

十二画

十三画

十四画

十五画

十六画

十七画

十八画

十九画

二十画

二十一画

树种拉丁学名索引

（按拉丁字母顺序排列）

B

C

D

E

F

G

H

I

J

K

L

M

N

Q

R

T

U

V

W

X

Y

Z

参考文献

（按作者姓名汉语拼音顺序排列）

一、专著

《安徽木本植物》编写组．安徽木本植物．合肥：安徽科学技术出版社，1983

常自力等．木本药用植物栽培与加工．北京：科学出版社，1990

陈机主编．植物发育解剖学．济南：山东大学出版社，1996

陈俊愉，程绪珂．中国花经．上海：上海文化出版社，1990

陈嵘．造林学各论．中华农学会，1933

陈永胜，谷凤坤．林木种子与苗木活力．哈尔滨：黑龙江科学技术出版社，1994

陈植．观赏树木学．北京：中国林业出版社，1984

方文培主编．中国四川杜鹃花．北京：科学出版社，1986

冯国楣主编．云南杜鹃花．昆明：云南人民出版社，1983

冯国楣主编．中国杜鹃花（第1册）．北京：科学出版社，1986

冯自诚，徐梦龙．甘南树木图志．兰州：甘肃科学技术出版社，1994

福建省科学技术委员会．福建植物志．福州：福建科学技术出版社，1982

傅立国等．中国植物红皮书·稀有濒危植物（第1册）．北京：科学出版社

干铎主编．中国林业技术史料初步研究．北京：农业出版社，1964

A. 耿德逊著．马骥译．双子叶植物科志．北京：科学出版社，1959

广西植物研究所．广西植物志（1卷）．南宁：广西出版社，1991

广西林业局，广西林学会．阔叶树种造林技术．南宁：广西人民出版社，1980

管中天．四川松杉植物地理．成都：四川人民出版社，1982

《贵州植物志》编委会．贵州植物志（第1卷）．贵阳：贵州人民出版社，1982

J. 哈钦松．中国科学院植物研究所译．有花植物科志（Ⅰ）双子叶植物．上海：商务印书馆，1954

J. 哈钦松．中国科学院植物研究所译．有花植物科志（Ⅱ）单子叶植物．上海：商务印书馆，1955

黑龙江祖国医药研究所．中国刺五加研究．哈尔滨：黑龙江科技出版社，1981

侯宽昭．广州植物志．北京：科学出版社，1956

侯宽昭．中国种子植物科属词典（修订本）．北京：科学出版社，1982

《华北树木志》编写组．华北树木志．北京：中国林业出版社，1984

华东师范大学，上海师范学院．种子植物属种检索表（上册、下册）．北京：人民教育出版社

《华南主要经济树木》编写组．华南主要经济树木．北京：农业出版社，1976

黄普华，卓丽环．中国东北主要树木幼苗图说．哈尔滨：东北林业大学出版社，1992

火树华主编．树木学（第2版）．北京：中国林业出版社，1992

胡适宜．被子植物胚胎学．北京：人民教育出版社，1982

江苏植物研究所．江苏植物志（上册）．南京：江苏科学技术出版社，1977

江苏植物研究所．江苏植物志（下册）．南京：江苏科学技术出版社，1982

凌麓山等．油桐栽培．北京：中国林业出版社，1983

李璠．中国栽培植物发展史．北京：科学出版社，1984

刘小媛，黄应钦，王惠英等．热带亚热带主要树木果实图谱．南宁：广西科学出版社，1989

刘业经．台湾植物志．台北：国立中央大学农学院出版委员会．1981

刘瑛心．中国沙漠植物志第1卷（1985年）、第2卷（1987年）．北京：科学出版社
李跃阶，杨生福等．青海木本植物志．西宁：青海人民出版社，1987
A. B. 伦德勒．钟补球译．有花植物分类学（第1册）．北京：科学出版社，1958
马常耕主编．华山松种源选择研究．北京：北京农业大学出版社，1992
马常耕主编．落叶松种源选择．北京：北京农业大学出版社，1992
马常耕主编．白榆种源选择研究．西安：陕西科学技术出版社，1993
缪勉之，张仲卿，方荫才等．湖南主要经济树种．长沙：湖南科学技术出版社，1982
南京林产工业学院《主要树木种苗图谱》编写小组．主要树木种苗图谱．北京：农业出版社，1978
南京林学院树木学教研组主编．树木学．北京：农业出版社，1964
《内蒙古植物志》编写组．内蒙古植物志（2～6卷）．呼和浩特：内蒙古人民出版社，1978～1980
牛春山主编．陕西树木志．北京：中国林业出版社，1990
裴鑑等．江苏南部种子植物手册．北京：科学出版社，1959
彭德纯，林睦就．植物活化石——银杉．长沙：湖南出版社，1984
祁承经，孙希儒，林仕榕．湖南植物名录．长沙：湖南科学技术出版社，1987
祁树雄等．中国桉树．北京：中国林业出版社，1989
全国油桐科研协作组．中国油桐主要栽培品种志．长沙：湖南科学技术出版社，1983
《山东树木志》编写组．山东树木志．济南：山东科学技术出版社，1984
《树木学》（南方本）编写委员会．树木学（南方本）．北京：中国林业出版社，1994
《四川植物志》编辑委员会．四川植物志第1卷（1981年）、第2卷（1983年）．成都：四川人民出版社
孙锦等．园林苗圃．北京：中国建筑工业出版社，1982
孙可群等．花卉及观赏树木栽培手册．北京：中国林业出版社，1985
王发祥，梁惠波主编．中国苏铁．广州：广东科技出版社，1996
王名金，刘克辉等．树木引种驯化概论．南京：江苏科学技术出版社，1990
王宏志主编．热带亚热带主要树种采种育苗技术．南宁：广西人民出版社，1985
王宏志主编．热带亚热带主要树种物候图谱．南宁：广西人民出版社，1988
王挺良．秃杉．北京：中国林业出版社，1995
吴征镒主编．中国植被．北京：科学出版社，1980
吴征镒主编．西藏植物志．第3卷．北京：科学出版社，1986
吴征镒主编．云南植物志．第1～4卷．北京：科学出版社，1976～1986
吴中伦等．国外树种引种概论．北京：科学出版社，1982
吴中伦主编．杉木．北京：中国林业出版社，1984
西南林学院，云南省林业厅．云南树木图志．昆明：云南科学技术出版社，1988
徐化成主编．林木种子区划．北京：中国林业出版社，1990
徐化成主编．油松地理变异和种源区划．北京：中国林业出版社，1992
严楚江．花果形态学．福州：福建人民出版社，1964
云南省林业科学研究所．云南主要树种造林技术．北京：中国林业出版社，1985
俞新妥．杉木．福州：福建科学技术出版社，1983
张若惠等．浙江珍稀濒危植物．杭州：浙江科学技术出版社，1994
张若惠，刘洪锷，汪祖潭．中国主要树木幼苗形态．北京：科学出版社，1993
郑勉．中国种子植物分类学，上册（1959年）、中册，第1分册（1960年）、中册．第2分册（1960年）．上海：上海科学技术出版社
中国科学院地理研究所．中国动植物物候观测年报，第1号（1963年）．北京：科学出版社，1965；第2号

（1964～1965年），北京：科学出版社，1977；第3号（1966～1972年）．北京：科学出版社，1977；第4号（1973～1974年）．北京：科学出版社，1982；第5号（1975～1976年）．北京：科学出版社，1986；第6号（1977～1978年）．北京：科学出版社，1986；第7号（1979～1980年）．北京：地质出版社，1988；第8号（1981～1982年）．北京：地质出版社，1988；第9号(1983～1984年)，北京：地质出版社，1989；第10号(1985～1986年)．北京：测绘出版社，1989；第11号(1987～1988年)，北京：科技出版社，1990

中国科学院华南植物研究所，广东省植物研究所．海南植物志．北京：科学出版社，1977

中国科学院昆明植物研究所．云南种子植物名录（上册、下册）．昆明：云南人民出版社，1984

中国科学院植物研究所．中国主要植物图说（豆科）．北京：科学出版社，1955

中国科学院植物研究所．中国高等植物图鉴，第1册（1980年）、第2册（1972年）、第3册（1974年）、第4册（1975年）、第5册（1976年）、补编第1册（1981年）、补编第2册（1983）．北京：科学出版社

中国科学院《中国植物志》编写委员会．中国植物志．第7卷（1978）、第20卷第1分册（1982年）、第20卷第2分册（1984年）、第21卷（1979年）、第25卷第2分册（1979年）、第27卷（1979）、第31卷（1984年）、第34卷第1分册（1984年）、第35卷第2分册（1979年）、第36卷（1974年）、第37卷（1985年）、第38卷（1986年）、第39卷（1988年）、第45卷第1分册（1980年）、第46卷（1981年）、第47卷第1分册（1985年）、第48卷第1分册（1982年）、第49卷第1分册（1989年）、第49卷第2分册（1984年）、第52卷第2分册（1983年）、第53卷第1分册（1984年）、第54卷（1978年）、第58卷（1979年）、第60卷第1分册（1987年）、第63卷（1977年）、第65卷第1分册（1982年）、第67卷第2分册（1979年）．北京：科学出版社

中国林木种子公司主编．林木种实病虫害防治手册．北京：中国林业出版社，1988

中国林业科学研究院，河南商丘林业局．泡桐研究．北京：中国林业出版社，1982

中国农业科学院果树研究所主编．中国果树栽培学（第3卷）．北京：农业出版社，1959

《中国树木志》编辑委员会．中国树木志，第1卷（1983年）、第2卷（1985年）、第3卷（1998年），北京：中国林业出版社

《中国树木志》编委会．中国主要树种造林技术．北京：中国林业出版社，1981

周德本．东北园林树木栽培．哈尔滨：黑龙江科学技术出版社，1986

周家骏，高林主编．优良阔叶树种造林技术．杭州：浙江科学技术出版社，1985

周以良等．黑龙江树木志．哈尔滨：黑龙江科学技术出版社，1986

周政贤等．杜仲．贵阳：贵州人民出版社，1979

浅川澄彦，胜田柾，横山敏孝．日本の树木种子．日本林木育种协会，1981

Forest Service, U. S. Department of Agriculture. Seeds of Woody Plants in the United States, 1974. Washington D. C., .U. S. A.

二、期刊文献

安志文．漆树飞机播种造林．中国生漆，1982. 1（3）：28～31

敖复，高永丽，杜坤．气调贮藏油松催芽种子的效果．林业科学，1988. 24（2）：209～215

白志华译．利用植物的药效反应鉴定植物的雌雄性．林业科技通讯，1986（6）：32～33

白仲奎．板栗空篷率发生及防治试验报告．林业科技通讯，1988（7）：5～9

蔡克孝，孙鸿有．金钱松种子的贮藏研究．种子，1988（5）：13～15

曹泽猷．杜松种子萌发及育苗技术的研究．林业科技通讯，1993（1）：6～9

常剑文，田玉堂．花椒花芽分化的初步观察．林业科技通讯，1988（4）：22～24

常培英，刘曼玲．华北落叶松种子发芽促进和发芽指标的研究．林业科学，1989. 25（2）：157～161

陈炳浩．沙棘种植园建园技术的探讨．林业科技通讯，1988（1）：9～11

陈炳章，杨乾洪，郭纯福．海拔高度对油桐生长结实影响的研究．林业科学研究，1990. 3（2）：195～199
程必强，马信祥．五十种热带植物种子的贮藏与寿命．种子，1987（3）：35～38
陈荷美．海南岛热带林木种子品质检验．热带林业科技，1978（3）：34
陈辉．层积、变温对提高中华猕猴桃种子发芽的试验（庐山植物园）．植物研究资料汇编，1980（2）：17
陈辉，王正刚．伯乐树的根插育苗（庐山植物园）．植物研究资料汇编，1986（1）：17
陈开秀，周君英等．利用太阳能干燥云杉球果试验初报．新疆林业，1988（1）：9
陈坤荣，方文德．乙烯利促进油橄榄落果试验初报．林业科技通讯，1987（6）：13～14
陈坤荣，方文德，李卓杰等．珙桐种子休眠生理研究（摘要）．种子，1990（4）：70
陈胜．球果害虫侵害对火炬松种实发育的影响．林业科学研究，1993. 6（2）：231～233
陈宪初．油橄榄的适生环境条件．林业科技通讯，1988（9）：15～16
陈晓阳，陈振炳，卢云凤．侧柏种苗性状在群体各层次上的变异．林业科学研究，1992（6）：16～17
陈耀华，阎万祥．北京地区毛白杨实生苗培育的研究．林业科技通讯，1991（3）：22～23
陈贻金，陈春雷，陈必芳等．红枣营养成分定量分析．林业科技通讯，1991（3）：25～27
陈益泰，何贵平，李恭学．杉木种子发芽和苗木高生长的近交效应．林业科学研究，1989. 2（5）：420～426
陈幼生．香椿 11 个种源种子发芽习性的研究．Forest Ecology and Management. 1985（10）：269～281
陈幼生，方升佐．杉木种子园结实量近期预测方法的研究．南京林业大学学报，1990，14（3）：1～5
陈幼生，吴琼美，陈智建等．湿地松种子衬比射线检验方法的研究．中南林学院学报，1992，12（2）：112～115
陈幼生，吴琼美，陈智建等．软 x 射线衬比法测定马尾松种子发芽能力的研究．林业科学研究，1993，6（5）：583～587
陈幼生，吴琼美，陈智建等．油松种子软 x 射线检验方法的研究．林业科技开发，1994，29（1）：2～3
陈幼生，吴琼美，陈智建等．杉木种子软 x 射线衬比检验方法的研究．南京林业大学学报，1993，17（1）：15～20
陈幼生，喻方圆．利用开花结实信息段预测中幼龄杉木球果产量的研究．林业科学，1993，29（5）：443～448
陈育松，胡玉琴，欧阳适．油茶落花落果的初步观察．林业科学，1963. 8（3）：267～270
陈运大．红松种子质量与苗木质量的关系．东北林学院学报，1981（4）：79
迟健．初论杉木种子园的园址选择．林业科学研究，1988（1）：57～65
迟健，邵蓓蓓．杉木球果和种子发育节律的研究初报．林业科学研究，1988. 1（4）：445～449
迟健，胡德治，谢正成等．浙粤两省 6 个杉木种子园施肥技术和效应研究．林业科学研究，1993. 6（1）：19～26
邓中美．用“三高”法处理漆树种子育苗经验．林业科技通讯，1986（2）：25～26
邓荫伟，黄连桂．提高银杏产量的技术经验——人工辅助授粉．林业科技通讯，1986（1）：17～20
董鸿运，班青，乔玉玲．杉木马尾松种子人工老化过程中某些生理生化变化规律的研究．种子，1987（2）：14～18
董丽芬等．白皮松种子休眠原因的初步研究．陕西林业科技，1987（1）：15
杜宏彬．杉木种子品质及涩粒的初步研究．亚林科技，1980（4）：42～48
杜宏彬等．檫树种子生活力的简易测定法．浙江林业科技，1982. 2（3）：22
杜宏彬．檫树种子贮藏试验．亚林科技，1982（3）：31～33
方升佐，许献文，裴忠诚等．关于用球果比容确定杉木种子园种子采收期的探讨．种子，1989（5）：27～30
范前炎．日本柳杉引种及栽培技术的研究．湖北林业科技，1982（1）：24～26
樊汝汶，叶建国，尹增芳．鹅掌楸种子和胚胎发育的研究．植物学报，1992. 34（6）：437～442
樊汝汶，尹增芳，尤录祥．中国鹅掌楸花芽分化的细胞形态学观察．南京林业大学学报，1990. 14（2）：26～32
范亚奇．樟子松球果烘裂取种生产工艺的试验研究．东北林业大学学报，1983. 11（3）：108～114

范忠仁．樟子松震动采种的探讨．林业机械，1985（5）：11
饭冢三男．5年保存レちキソ种子の发芽率．89回日林论，1978：203～204
冯瑞芝，陈碧珠，连文瑛等．不同种沙棘油的生理活性成分比较．林业科技通讯，1993（1）：20～22
傅德志．裸子植物一新科——竹柏科．植物分类学报，1992．30（6）：515～528
傅辉恩，刘建勋．华北落叶松引种试验研究．林业科学，1987．23（4）：406～413
符梅忠，傅志远，蒋国兴．杉木籽粒败育的研究．浙江林业科技，1989．9（2）：15～19
高捍东．用软x射线摄影术测定马尾松种子发芽能力的研究．南京林业大学学报，1998，22（3）：16～20
高捍东．用IDX法测定马尾松和樟子松种子发芽能力的研究．南京林业大学学报，1998，22（4）：28～30
高捍东，沈永宝，苑兆和．用过氧化物同功酶分析鉴定松属种子的真实性．中南林学院学报，1995(1)：30～32
顾姻，王传水，赵昌民等．悬钩子属种质的评价．植物资源与环境，1996．5（3）：6～13
管康林．香果树种子的光萌发特性的初步研究．浙江林学院学报，1985．2（2）：43～46
管康林．山茱萸种子苗床快速发芽技术．种子，1990（5）：5～8
管康林等．泡桐种子萌发的需光性研究．浙江林学院研究报告，1982（7）：3
管康林，陶银周．濒危树种——天目铁木的现状．繁殖．浙江林学院学报，1988．5（1）：90～92
韩宁林．油茶种子冷藏技术的研究．林业科技通讯，1984（12）：7～9
韩宁林，高继银，吴继武等．油茶无性系早实丰产配套技术的研究．林业科学研究，1991．4（5）：479～485
何方，姚文华等．中国油桐品种数量分类的研究．见：中国油桐科技论文选．中国林业出版社，1988
何福基．杉木种子涩粒成因的初步研究．种子，1985（6）：4～6
李滨，黄志中，吴琼美等．云南松采种期与种子品质关系的研究．南京林业大学学报，1989，13（1）：80～88
贺善安，郝日明，汤思杰．鹅掌楸致濒的生态因素研究．植物资源与环境，1996．5（1）：1～8
侯惠宗．北京妙峰山树木花期的调查．林业科学，1956．2：56
侯开卫，刘凤书，杨臣武等．余甘子对强致癌物质N-亚硝基化合物在人体内合成的阻断作用．林业科学研究，1989．2（1）：55～57
胡之定等．国家重点保护树种．湖北林业科技，1989（增刊）
黄鹏成．中国油茶属一新种．南京林产工业学院学报，1981（2）：107～111
黄启强．马尾松种子园花粉产量的研究．林业科技通讯，1988（5）：10～12
黄铨，佟金权．中国沙棘主要经济性状的变异．种子，1992．5（2）：18～21
黄铨，佟金权．中国沙棘的表型结构与种群变异．林业科学研究，1993．6（2）：175～182
黄少甫，王雅琴，赵治芳等．香榧性别的早期鉴定．林业科学研究，1990．3（2）：127～132
黄玉国．刺楸种子胚休眠的研究．东北林业大学学报，1986．14（1）：39～44
黄玉国．刺楸种子层积后熟过程中两种磷酸脂酶的作用．种子，1987（4）：12～14
胡宝信．冷冻贮藏可以延长秃杉种子发芽力．林业科技通讯，1993（6）封2
胡云楚，周培疆，宋昭华等．刺槐种子发芽过程的微量热法研究．种子，1994（6）：16～17
姜国武，杨锡章．提高华东黄杉种子有胚率的初步研究（杭州植物园）．植物引种驯化集刊（第4集），1985
蒋恕．杉木开花结籽的解剖学观察．南京林产工业学院学报，1980（1）：109～115
江西油桐种源资源调查科研协作组．江西省油桐种源资源调查研究．见：中国油桐科技论文选．中国林业出版社，1988
金万昌，李海臣，周荫祥．红松种子秋埋覆膜催芽试验．林业科技通讯，1989（7）：11～12
居翔汉，徐正法．日本柳杉、日本扁柏生长情况的调查．林业科技通讯，1986（12）：18～21
康永善，陈学林．沙棘属植物的性状演化及其意义．沙棘，1992（2）：16～20
孔凡彬．马尾松开花结实规律及种子产量预测预报研究．南京林业大学硕士学位研究论文，1994

寇纪烈，罗晶，郭振新等．改造野生沙棘林为沙棘园的方法及效益．林业科技通讯，1988（5）：17～18
梁根桃，严逸伦，钱金城等．猕猴桃种子贮藏处理与发芽试验．浙江林业科技，1989. 9（3）：34～36
梁玉堂，龙庄如，徐明广．银杏种子的后熟处理．林业科技通讯，1987（2）：28
李碧涛．杜仲种子发芽测定方法的探讨．林业科技通讯，1991（9）：20～21
李淑娴，陈幼生，吴琼美．湿地松种子活力测定方法的研究．南京林业大学学报，1996，20（3）：8～13
李财德．小北湖红松结实量的调查．林业勘察设计，1983（2）：39
李长喜，徐化成．巴山松和油松地理分界和分类关系的数值分析．林业科学，1989. 25（1）：14～21
李德文等．小米桐雌雄花比例变化的初步研究．四川林业科学研究所1964年科学文集，第50页
李风华，于曙明．秃杉在我国的分布与生长．亚热带林业科技，1987. 15（3）：215～220
李华．50种经济作物种子和果核的鉴别．江西庐山植物园，1981.11
李华．日本扁柏引种驯化的研究（庐山植物园）．植物研究资料汇编，1984（1）：14
李家玉，齐之尧，叶大华．断根对银杏幼苗根系的促进作用．林业科技通讯，1993（7）：23～24
李锦清，韩宁林．杉木1.5代种子园喷施满果粉效果分析．林业科技通讯，1991（8）：24～26
李锦文，崔红．油松、华北落叶松种子贮藏试验研究．种子，1988（4）：18～20
李近雨．漆树种子的休眠生理及催芽技术的初步研究．种子，1988（3）：17～19，11
李近雨．核桃种子烂种原因及发芽条件的研究．林业科学，1994. 30（1）：18～24
李近雨，穆贵荣．青桐种子休眠和萌发生理的初步研究．河北林业科技，1985（2）：8～12
李昆，陈玉德，谷勇等．云南野生余甘子果实类群及其分布特点研究．林业科学研究，1994，7（6）：606～611
林承惠．安徽休宁发现猕猴桃优良单株．林业科技通讯，1976（2）：27
凌世瑜．赤霉素对水曲柳种子解除休眠的作用．林业科学，1986. 22（1）：78～85
凌世瑜，董德愚．水曲柳种子休眠生理的研究．林业科学，1983. 19（4）：349～359
林坚，郑光华，张庆昌．杜仲种子休眠原因及发芽特性的研究．种子，1989（2）：8～10
林坚，郑光华，张庆昌．杜仲种子贮藏方法和活力鉴测的研究．种子，1990（1）：9～11
林平，张卓文，管圣全等．姥山林场杉木初级种子园球果类型及其种子分布特征．浙江林学院学报，1990. 7（4）：361～364
林夏馨．大力推广杉木摇落采种法．林业勘察设计，1982（4）：43
林祖思等．杉木种子涩粒的空间分布．福建林学院学报，1990，10（3）：230～236
刘橙，宿仁敬．红松种子长期贮藏试验阶段总结．林业科技通讯，1983（7）：4
刘国凡，邓廷秀．不同紫色土几种树苗结瘤及其对植株生长的影响．生态学报，1983. 3（4）：350～353
刘家德，路永祺．珙桐育苗试验．湖南林业科学，1981（4）：22～24
刘继国．草河口地区红松人工林结实规律的初步研究．林业科学，1963. 8（1）：41
刘培华，王小纪．山茱萸低产树改造试验研究．林业科学研究，1991（2）：19～22
刘培华，薛守辉．火炬树育苗技术．林业科技通讯，1985（3）：1～2
刘仕俊．油桐花芽分化的初步观察．植物学报，1966. 14（2）：180～182
刘文明，宋学之．青皮种子主要贮藏条件的研究（Ⅰ）．种子含水量与控制．林业科学研究，1989. 2（3）：214～220
刘文明等．海南红豆种子生理特性及萌发影响因素的研究．林业科学研究，1990. 3（4）：351～357
刘文明等．失水对青皮种子劣变的影响．林业科学研究，1993. 6（2）：162～166
刘益康，王冰．落叶松球果花蝇生物学特性及防治研究．林业科技通讯，1993（1）：5～7
刘玉壶，周仁章．广东，云南木兰科一新种．植物研究，1986. 6（2）：139～141
刘玉华．日本柳杉的引种与种源选择．赣州林业科技通讯，1982（5）：28

刘振陆，王洪魁．落叶松实小卷蛾防治的初步研究．林业科技通讯，1985（1）：24～26
李卓然，陈润政，傅家瑞等．珙桐种子休眠和萌发中酸性磷酸酶同工酶的研究．西南林学院学报，1989．9（1）：11～12
李文彪，叶栎，周宁一．棕榈腐烂病的研究．浙江林学院学报，1988．5（1）：8～15
李文荣，齐力旺，韩有志．山西华北落叶松天然林的地理分布和种群变异规律的研究．林业科学，1992．28（6）：493～501
李文钿，朱彤．胡杨花粉和胚囊的发育．林业科学研究，1988．1（2）：132～137
李文钿等．胡杨的受精作用和胚胎发育，1989．2（1）：1～8
李晓储，黄利斌．高锰酸钾液处理对杉木种子发芽率的影响．种子，1990（5）：38～41
李晓洁．赤霉素对乌桕种子发芽及幼苗生长的影响．种子，1987（6）：20～22
李晓洁．华山松种子及其发芽的地理变异研究．植物生态学与地植物学学报，1991．15（1）：27
李晓洁．黑松种子实验室发芽条件研究．种子，1990（1）：22～23
李晓洁，徐化成．白皮松种子发芽习性及其种源变异的研究．林业科学，1989．25（2）：97～105
李晓洁，徐元．火炬树种子休眠解除与发芽条件研究．种子，1989（3）：23～24
李炎香，谭天泳，魏素梅等．石梓栽培技术的研究．林业科学研究，1990．3（5）：446～453
李永多等．红花尔基樟子松生长状况及结实规律的调查．林业科学，1981．17（3）：306
李玉俊，朱勇．旱生植物半日花的水分生理研究．林业科技通讯，1991（3）：7～8
黎章矩．乌桕生长发育年周期．林业科学，1965．10（3）
黎章矩，曹燕如．山茱萸花芽分化的研究．浙江林学院学报，1986．3（2）：37～40
黎章矩，钱莲芳，钱光林．山核桃保花保果技术研究．林业科学，1993．29（4）：360～365
龙耀华等．水杉一新变种．植物研究，1984（1）：146～148
陆康年等．樟子松球果烘裂取种工艺的研究．东北林学院学报，1982（2）：104
罗成荣．柏木种子活力探讨．四川林学院硕士学位研究论文，1989
罗京，张桂龙，张小林．沙棘硬枝扦插不同处理方法试验研究．林业科技通讯，1992（11）：5～10
罗丽芬，陈华豪，王凤霞等．黄波罗人工林结实调查研究．东北林业大学学报，1986．14（增刊）：15～19
罗丽芬，王彩荣，林淑云．红松种子贮藏效果及生活力测定效果比较．东北林业大学学报，1983．11（2）：44～49
卢小根．苏铁播种育苗试验．林业科学研究，1988（2）：24～25
马常耕．杉木种子园种子生物学特性的地区及年度变异．林业科学研究，1989．2（1）：59～66
马常耕，王建华．云杉、落叶松等树种种子发芽对水分胁迫的反应．林业科学研究，1994．7（1）：121～124
马常耕，张云跃，安定国．华山松抗寒性的地理变异．林业科学研究，1990.3（2）：113～118
马正山等．香榧生物学特性的初步研究．浙江林业科技，1983．3（1）：31～35
门玉岭．哈尔滨地区栽培灌木开花历．自然资源研究，1980（2）：8～10
门玉岭．哈尔滨地区绿化树种采种历．森林植物研究，1986：42～79
缪礼科，孔抒清．沙棘属种子检验分析．陕西林业科技，1987（2）：13～14
莫钊志，喻方圆等．杉木种子产量近期预测方法的研究．广西林业科学，1996（2）：72～77
南方十四省区杉木种源试验协作组．杉木种源地理变异的研究．全国杉木地雷种源试验会议，1984
南京林产工业学院球果干燥科研组．杉木球果人工干燥工艺的研究．林业科技通讯，1979（6）：8～10
欧阳准，余文彪．福建省油桐农家品种类型．见：中国油桐科技论文选．中国林业出版社，1988
潘盛荣．如何采收秃杉树种．黔东南林业信息，1987（9）：4
潘志刚，游应天．厚荚相思的引种及种源试验．林业科学研究，1994．7（5）：498～505
彭长根，楚国忠，郭晶华．江西大岗山年珠林场白颈长尾雉和白鹇的秋季食物组成．林业科学研究，1994，7

(5)：574～578
彭金贵，萧国钦．墨西哥柏引种试验初报．四川林业科技，1984．5（1)：39～40
彭金贵等．墨西哥柏扦插育苗试验．四川林业科技，1984．5（2)：60～61
彭业芳，傅家瑞．荔枝和龙眼种子发育过程，发芽率与脱水忍耐力变化．种子，1994（3）：1～5
钱莲芳．柳杉花芽分化的观察．浙江林学院学报，1988．5（2)：173～179
钱莲芳等．金钱松嫁接技术．浙江林学院学报，1990．7（3)：281
钱领元，施拱生．乌桕籽“蜡被”形成过程的研究．浙江林学院学报，1986．3（1)：1～5
钱万杰．山楂种子休眠与萌发生理的初步研究．植物生理学通讯，1984（5)：17～20
乔士义等．毛竹的胚胎发育观察．竹类研究，1984（1)：7～17
秦国峰，汪昌铭．马尾松开花结实规律的初步研究. 林业科学研究，1991．4（3)：328～331
秦慧贞，李碧媛．鹅掌楸雌配子体败育对生殖的影响．植物资源与环境，1996．5（3)：1～5
裘国宝等．浙江紫茎属植物小志．云南植物研究，1988．10（1)：55～59
邱国玉．沙拐枣属植物最佳育苗方法．林业科技通讯，1989（12)：3～7
齐之尧，李家玉．银杏绿枝单芽扦插．林业科技通讯，1991（3)：29～31
全国杉木种源试验协作组．杉木种源区划分的研究．林业科学研究，1994．7（增刊)：130 ～144
全国杉木种源试验协作组．杉木种源变异的研究．林业科学研究，1994．7．（增刊)：117～129
任立中等．板栗空篷率的研究．安徽农学院学报，1988（3)：26
任钦良等．低丘红壤引种香榧初获成果．浙江林业科技，1981（4)：168
任钦良等．香榧育苗技术．亚林科技，1983（3)：34～37
任祝三．光与激素对于杜鹃花种子发芽的研究．云南植物研究，1988（1)：73～75，77～78
任祝三等．水青树种子的发芽．全国植物园与植物资源保护利用学术讨论会资料，1990
陕西省柞水县核桃板栗研究所．板栗石窖贮藏．林业科技通讯，1985（2)：11～13
邵蓓蓓．榔榆种子发芽条件的研究．亚热带林业科技，1987．15（2)：113～117
邵蓓蓓．层积和变温对10种林木种子萌发的影响．林业科技通讯，1989（2)：4～7
沈永宝．IDX法测定湿地松种子生活力的研究．南京林业大学学报，1996，20（3)：35～38
沈永宝，陈幼生，吴琼美等．杉木种子园中无性系的开花结实规律．东北林业大学学报，1996，24(4)：113～116
盛炜彤等．杉木生长区气候区划的初步研究．林业科学，1981（1)：50～57
佘祥威等．低温层积对日本五针松种子萌发及生理生化的初步研究．种子，1987（6)：16
歙县林科所经济林调查组．歙县香榧的初步调查．徽州林业科技，1982．(1)：1～2、14～16
史忠礼等．香榧种子休眠的研究．植物学报，1973．15（12)
史忠礼，许月明，高智慧等．杉木种子及胚胎发育生化变化的研究．种子，1990（1)：77～78
史晓华．玉兰种子的采收和贮藏．植物杂志，1982（6)：27
史晓华．木兰属种子的调制和贮藏．江西林业科技，1986（6)：41
史晓华．玉兰种子休眠和萌发生理的研究．林业科学，1987（营林专刊)：77～82
史晓华，史忠礼．浙江楠种子休眠生理初探．浙江林学院学报，1990．7（4)：377～382
史晓华，田丽浩．紫楠种子休眠生理的研究．种子，1988（1)：32～33
宋纯清．喜树用途多．植物杂志，1981（6)：2
宋松泉，傅家瑞．黄皮种子萌发过程中的一些生理生化变化．种子，1993（4)：1～3
宋廷茂，白兆瑞，王晓梅．华北落叶松育苗播种量研究初报．林业科技通讯，1985（1)：2～5
宋学之，刘文明．红椎种实主要贮藏条件的研究．林业科学研究，1992．5（2)：134～141
宋学之，刘文明，邱坚峰．柚木种实处理及催芽技术的研究．林业科学研究，1991．4（6)：616～621
宋学之，刘文明，邱坚峰．柚木种实萌发生理的研究．林业科学研究，1991．4（5)：471～478

苏冬梅．翅荚木种子活力的测定．中南林学院学报，1991．11（1）：103～106
苏冬梅．短期高温高湿处理可提高翅荚木种子活力．林业科技通讯，1992（10）：12～13
苏梦云．香榧性别的生化鉴定研究．经济林研究，1987．5（2）：1～7
孙昌高，潘晓飙，方坚等．药用植物种子生理的研究（Ⅳ）——山茱萸种子休眠机制的初步研究．种子，1988（1）：30～32
孙昌高，徐秀英，梅捍卫等．玉兰种子在层积处理期间的形态变化．种子，1987（1）：22～25
孙慧明．树木种子发芽条件试验初报．江西林业科技，1980（4）：26～33
孙秀琴，田树霞．荆条种子萌发生理条件的研究．林业科学研究，1988．1（6）：688～690
孙秀琴，田树霞．4种乔灌木种子发芽特性研究初报．林业科技通讯，1990（4）：15～16
孙秀琴，田树霞．元宝枫种子休眠生理的研究．林业科学研究，1991．4（2）：185～190
孙秀琴，田树霞．两个产地君迁子种子萌发的研究．林业科学研究，1993．6（1）：88～91
谭一凡．南方红豆杉种子后熟生理的研究．中南林学院学报，1991．11（2）：200～206
谭志一．红松种子休眠与脱落酸及外种皮的关系．中国科学，1983（9）：899～906
唐光楚，江兆般等．天山云杉物候生态观测及其与林业生产关系的初步研究．新疆林业科技文集，1981（3）：41～47
陶金川．银鹊树的地理分布与引种．南京林业大学学报，1990（2）：34～39
陶金川，宗世贤，杨志斌．珙桐的地理分布与引种．浙江林学院学报，1986．3（1）：25～33
陶金川等．天女花的地理分布与气候的关系．浙江林学院学报，1988．5（1）：104～107
陶章安．白皮松等13个树种层积催芽的初步观察．林业科学，1956（1）：48
田荆祥，黎章矩，吴美春．浙江省乌桕优良无性系种子理化性质及脂肪酸分析．浙江林学院学报，1988．5（1）：1～7
田金文，曾有才，陈道俊．杧果扦插育苗试验初报．林业科技通讯，1991（7）：20～21
田星群．巴山木竹开花习性的观察．竹类研究，1986（2）：56～57
万才淦．光、温和化学处理对水青树种子发芽的影响．武汉植物学研究，1986．4（3）：257～260
万才淦．光、温和化学处理对领春木种子发芽的影响．种子，1988（1）：24～26
万才淦，王诗云，白明旭．3种珍稀濒危植物种子的休眠习性．种子，1990（5）：65
万才淦，张炳坤．水红木种子的发芽特性初探．种子，1994（2）：31～32
王白坡，戴文圣，钱银才．樱桃的特性及早结果早丰产技术的研究．浙江林学院学报，1990．7（2）：111～115
王白坡，仰新民，包根潮．中国樱桃授粉结实和果实发育的研究．浙江林学院学报，1990．7（1）：15～21
王必农，钱尤德等．银杉种子繁殖试验成功．植物杂志，1980（2）：13
王炳三，叶金好，郑荣章等．余甘子营养（化学）成分研究．林业科学研究，1992．5（2）：170～176
王成霖．梧桐种子休眠类型和促进发芽的研究．安徽农学院学报，1985．16（2）：128～131
王成霖，邵蓓蓓．我国亚热带30个树种种子发芽条件的研究．亚林科技，1986（3）：3～12
王成霖，邵蓓蓓．我国亚热带30个树种种子发芽条件的研究（续）．亚林科技，1986（4）：38～49
王成霖，邵蓓蓓．林木种子的低温层积．亚热带林业科技，1987．15（3）：222～227
王定跃，苏铁科形态结构、系统分类与演化研究，南京林业大学博士论文，2000.6
王道植，徐玉蓉．6个漆树品种开花习性研究．种子，1988（5）：24～29
王东馥等．油松种子贮藏前后活力与内源激素的变化．种子，1985（1）：11
王洪学，王砚革，叶林等．笃斯越橘集约经营技术的研究．林业科技通讯，1993（1）：14～16
王华緘．杉木种子的选种．中国林业，1956（7）：6～7
王继贵．川北古驿道行道古柏考．四川林业科技，1991．12（1）：97，99

王景章，王有才．日本落叶松种子园各无性系花期的观察研究．林业科技通讯，1985（3）：3～5
王九龄．昆明地区云南松结实特性的初步观察．林业科学，1979．15（4)：303～307
王丽娟，衣俊鹏．樟子松种子园种子败育原因初探．林业科技通讯，1990（9)：12～15
王林．桑树硬枝扦插新技术．林业科技通讯，1993（12)：21～22
王嫩良，方炳发，童修耀等．姥山杉木种子园高产稳产经营技术的研究．林业科技通讯，1991（9)：1～4
王培蒂．马尾松种子贮藏方法的试验．亚热带林业科技，1987．15（4)：294
王培蒂，秦国峰．马尾松胚胎发育的观察研究．林业科学研究，1990．3（5)：441～445
王培蒂，秦国峰．长期贮藏对马尾松种子品质的影响．林业科学研究，1994．7（2)：193～198
汪企明等．日本五针松种子萌发形态和催芽方法的初步研究．江苏林业科技，1986（2)：6
王文章．红松种子抑制物质的初步研究．东北林学院学报，1978（1)：91
王文章等．红松种子休眠与种皮的关系．东北林业大学学报，1986．14（2)：83
王文章，陈杰，刘恩举．红松种子抑制物质的提取、分离和鉴定．中国科学，1980（9）：899～906
汪五星．天目木姜子播种育苗技术．林业科技通讯，1986（7)：23
王晓峰．板栗保鲜贮藏概况．种子，1993（3)：30～32
王晓峰，陈润政译．4种热带树木顽拗型种子的贮藏．种子，1990（5)：73～75
王晓峰，傅家瑞．杧果种子的脱水与贮藏研究（摘要）．种子，1990（1)：79
王晓峰，傅家瑞．氧气浓度对杧果种子贮藏寿命的影响．种子，1994（1)：1～3
王晓茹，沈熙环．油松种子园花粉飞散规律的研究．林业科学，1987．23（1)：1～10
王锡林，田乃祥，赵仁学．胡杨播种育苗技术．林业科技通讯，1985（1)：1～3
汪云．滇刺枣地理种源初探．林业科学研究，1994．7（3)：334～336
王赵民，张健忠，陈弈良等．疏伐促进杉木种子园开花结实的研究．林业科学研究，1994．7（6)：624～628
王赵民，吴隆高，王嫩良等．GA_3等三种植物激素生长调节剂对杉木结实和种子品质的影响．林业科技通讯，1993（9)：27～29
王忠信等．泡桐种子育苗新技术——芽苗移栽．泡桐，1985（2)：34～35
汪祖谭．香果树的繁育技术及木材物理学性质．浙江林业科技，1982．2（3)：1～3
汪祖谭．香果树的育苗方法．林业科技通讯，1982（4)：3～5
汪祖谭，杨逢春，陶银周．香果树芽的生长发育与花芽分化．浙江林学院学报，1988．5（3）306～311
宛志沪等．湿地松结实量与气候因子的关系及在其安徽省布局的探讨．农业气象，1986（1）
魏素梅，李炎香．石梓种子的发芽试验．林业科学研究，1989．2（2)：185～189
翁尧富等．金钱松种子贮藏条件的研究．林业科技开发，1991（4）
温远光，刘世荣．杉木物候期地理变化规律及其与生产力关系的研究．林业科学，1994．30（4)：313～319
吴楚材．不同贮藏方法对杜仲种子发芽的影响．林业科技通讯，1990（4)：17～18
武春生，曹诚一，刘友樵．云南油杉种子小卷叶蛾新属新种记述及其生物学特性的研究．林业科学，1987．23（2)：150～161
吴洪广．柏木丽松叶蜂研究初报．四川林业科技，1984．5（3)：40～45
吴美春，施拱生，田荆祥．山苍子的栽培与利用．浙江林学院学报，1988．5（1)：36～42
吴天林，朱德俊，荣文琛．马尾松花粉高产种源的初步选择．林业科学研究，1993．6（3）：327～331
吴侠中，胡素梅．水杉芽苗移栽育苗成功．林业科技通讯，1993（6)：10～11
吴振多，王昌杰，张绪卿等．草袋包装橡种室内贮藏效果好．林业科技通讯，1986（11)：30～31
吴中伦．杉木地理分布的研究．地理学报，1955，21（3)：285
萧东玉，盛芳．沙棘种子快速催芽方法的探讨．种子，1988（2)：33～34
谢善高，莫金莲．1年生肉桂育苗新方法．林业科技通讯，1993（11)：24～25

谢早荣．银钟花的繁殖与栽培．南岳树木园通讯，1986（8)：20～21
邢章美．北美红杉的扦插育苗．浙江林学院学报，1988．5（1）108～110
徐本美．樱桃种子的发芽初探．种子，1988（2)：35～37
徐本美．八个漆树品种种子的萌发试验．中国生漆，1988（1)
徐本美，白克智．杜仲苗的生理矮化现象．种子，1995（6)：54～55
徐本美，闪崇辉，董靖和等．山楂田间播种当年出苗实验．种子，1984（1)：5，22
徐本美，闪崇辉，邢北任等．裂口处理对桧柏种子的萌发作用．种子，1989（6)：16～18
徐本美，张治明，张会金．蔷薇种子的萌发与休眠的研究．种子，1993（1)：5～10
薛益民，王笑山，淡克德等．油橄榄不同品种经济形状的研究（Ⅰ）．油用品种果实的研究．林业科学研究，1988．1（5)：499～507
薛益民，王笑山，淡克德等．油橄榄不同品种果实经济形状的研究（Ⅱ）．餐用品种果实．林业科学研究，1989．2（2)：142～148
徐光田．秃杉扦插试验．四川林业科技，1990．1（1)：41
徐化成，唐季林．油松种子发芽的生态学及其与种源的关系．林业科学，1989．25（6)：493～501
徐明柏．马尾松球果脱粒设备生产试验初报．四川林业科技，1986．7（4)：42
许绍惠，韩忠环，刘财富．东北地区刺楸种子休眠原因及解除休眠的研究．林业科学研究，1991（2)：1～4
许绍远．金钱松生长特性与林分结构的研究．浙江林学院学报，1990．7（4)：297～306
徐有源，刘露．贵州省道真县沙河林区银杉的调查研究．植物生态学与地植物学丛刊，1983．7（1)：52，56
徐玉蓉，王道植，徐本美．不同预处理对漆树种子萌发的影响．种子，1989（1)：20～25
严圭．林木种子贮藏寿命的研究．南京林业大学硕士学位研究论文，1982
颜松河．南靖县杉木采种工作的体会．中国林业，1954（9)：10～11
杨国华等．马尾松种子贮藏条件研究．江西林业科技，1988（3)：9
杨汉仕，张玉凤．2，4-D对中华猕猴桃果实效应初报．林业科技通讯，1991（9)：22
阳含熙等．杉木生态特征研究．载《林业科学研究报告》1958．北京：中国林业出版社
杨思平．秃杉育苗最佳播种量的研究．林业科学研究，1993（12)：11
杨业勤．珙桐种子育苗．林业科技通讯，1982（9)：6～9
杨镇，王志彦．板栗嫩枝扦插育苗研究初报．林业科技通讯，1990（4)：封2～封3
姚庆渭，黄鹏成．江苏省珍贵用材树种的研究——楸树属．热带林业科技，1978（3)：22～33
姚庆渭，黄鹏成．江苏省珍贵用材树种的研究（2）——红豆树、美国肥皂荚．南林科技，1978（3)：44～53
姚庆渭，黄鹏成．松科各属种子的研究．南京林产工业学院学报，1980（1)：28～41
姚庆渭，黄鹏成．东亚和北美楸树属和肥皂荚属的种类及其系统．南京林产工业学院学报，1980（1)：122～125
姚显明，刘天斌．油松飞播造林应用HL粉剂拌种防止鼠害的研究．林业科技通讯，1991（7)：25～28
姚小华，叶金好，盛荣能等．余甘子优良类型选择．林业科学研究，1993．6（3)：299～305
姚秀玲，杨德基，王小平．小叶杨种子贮藏试验研究．林业科学，1993．29（2)：172～175
叶培忠，陈岳武，蒋恕．杉木种子生活力变异的研究．南京林产工业学院学报，1981（3)：22～32
叶培忠等．杉木自然类型的研究．林业科学，1964，9（4)：297～310
尤海量，吴志和．油茶裂果的形成及其防治的研究．林业科学，1963（1)：81～85
喻方圆等．马尾松种子发芽标准的探讨．种子，1996（1)：7～9
喻方圆，陈幼生．可见半面树冠球果估测法的探讨．林业科技通讯，1992（7)：6～8
喻方圆，陈幼生等．福建省杉木种子园种子产量预测方法的研究．南京林业大学学报，1996，20（2)：9～14
喻方圆，萧石海，余荣卓等．用球果切开法预测杉木种子的产量和质量．种子，1992（4)：19～21

庾府诗．连香树和领春木繁殖技术初报．衡阳林业科技，1983（1）：45～47
庾府诗．珍贵树种引种驯化及繁殖技术的研究．南岳树木园通讯，1984（6）：8～12
余茂林．珍贵优良树种香果树、银鹊树的种子育苗．安徽林业科技，1980（4）：34～35
余茂林．乡土速生树种——银鹊树．皖西林业科技，1983（1）
余茂林，李成智．香果树造林学性质的观察．皖西林业科技，1982（1）：21～26
余清珠，吴越．刺楸繁育途径的试验研究．林业科技通讯，1986（8）：20～22
于淑兰．解除白皮松种子休眠和促进发芽的试验研究．种子，1983（1）：24～28
于淑兰．油松、侧柏种子贮藏试验研究．种子，1988（6）：23～28
于淑兰，孙秀琴，胡春姿等．落叶松种子活力测定和田间育苗对比试验研究．种子，1993（3）：18～22
余象煜等．杉木败育种子及其涩粒物质的研究．西北植物学报，1989，9（4）：252～256
俞新妥．杉木种子瘪粒形成观察初报．福建林学院学报，1960（1）：1～7
俞旭平，方坚，盛来军等．贮藏在海藻酸钙胶丸中的荔枝种子活力．种子，1994（5）：34～35
俞志林，许仲明．板栗保鲜贮藏．种子，1987（4）：59
于卓，孙祥．羊柴种子发芽率低的原因分析．种子，1993（3）：12～15
曾垂惠．柏木丽松叶蜂生物学特性及防治试验．林业科学，1984．20（3）：332～335
曾广文，傅远志，符梅忠．杉木种子吸胀期间亚细胞结构的发育．林业科学研究，1989．2（2）：124～126
曾庆波，丁美华，邱坚锐．海南省尖峰岭十八种热带乔木物候谱．林业科学研究，1989.2（3）：291～295
张敖罗，冯桂华．杜鹃的有性繁殖．见：植物引种驯化集刊第二册，中国科学院植物园工作委员会编辑．北京：科学出版社，1966
张德纯．香椿种子．种子，1993（6）：67
张海廷，黄泽昌，黄质彬．红松采种期、脱粒时间对种子成熟度的影响．林业科技，1990（5）：19
张鹤年，张希明，买买提依提等．流沙地沙拐枣夏季直播造林的初步研究．林业科技通讯，1988（3）：16～18
张宏达．山茶属植物的系统研究．中山大学学报（自然科学）论丛，1981（1）
张化疆．椤叶槭种子中抑制物质研究简报．植物生理学通讯，1987（1）：22～23
张建国．喷洒乙烯利和萘乙酸混合液促进核桃果皮开裂的研究．林业科技通讯，1990（6）：23～25
张家贤．金佛山方竹开花结实及更新途径的调查研究．林业科技通讯，1985（3）：2～3
张家贤．箐竹开花结实和更新调查情况．竹子研究汇刊，1985．4（1）：88
张纪卯，陈巧女，赖文胜．长序榆育苗密度试验．林业科技通讯，1993（11）：17～19，15
张良诚，郭维明，陈永盛．红松种子后熟生理的实验研究．东北林学院学报，1981（4）：43
张瑞军，林万剑，王怀金等．楝树嫩芽扦插育苗技术研究．林业科技通讯，1989（4）：11～13
张若蕙，张金谈，邹达明等．天目铁木的花及花粉形态．浙江林学院学报，1988．5（1）：93～96
张若蕙，沈锡康，杨逢春．天目铁木生长节律的观察．浙江林学院学报，1990．7（1）：58～62
章树文，李龙山，吕平会．油橄榄施肥效果的初步研究．林业科技通讯，1988（8）：3～5，9
张涛．沙冬青生理结构特性的研究．林业科学，1988．24（4）：508～509
张万雄，刘君．兴安落叶松优良采种林分划分方法的研究．林业科学研究，1992（5）：13～15
张维栋．泡桐电热床营养钵温室育苗试验．泡桐，1986（1）：37～38
张微，张锐，张慧．沙棘不落果原因探讨．林业科学，1992．28（1）：76～79
张文燕，马乃训．竹类植物花期的生物学特性．林业科学研究，1989．2（6）：596～600
张文燕，马乃训．竹类植物花粉的生活力和自然授粉．林业科学研究，1990．3（3）：250～255
张文燕，马乃训，陈红星．竹类花粉形态及萌发试验．林业科学研究，1989．2（1）：67～70
张孝仁．流动沙丘沙拐枣扦插造林试验．林业科技通讯，1986（8）：24～25

张卓文．杉木、马尾松球花形态特征变异性研究初报．中南林学院学报，1987. 7 (2)：199～204

张卓文，林平．杉木花粉生态学特性研究．林业科学，1990. 26 (5)：410～418

张卓文，许大明，朱红东等．湿地松球花特性及其种子园配子比例研究．种子，1994 (5) ：20～22

赵海珍．激素对水曲柳种子休眠萌发的影响．东北林业大学学报，1983 (11)：7～11

赵汉章，朱长进．中国沙棘果实性状的地理变异及果用种源的选择．沙棘，1991 (4)：15～18

赵锦年，陈胜，黄辉．马尾松种子园松实小卷蛾的研究．林业科学研究，1991. 4 (6)：662～667

赵锦年，陈胜，黄辉等．果梢斑螟对马尾松球果和雄花序枝生长发育的影响．林业科学研究，1989. 2 (3)：300～303

赵锦年，陈胜，周世水．马尾松林油松球果小卷叶蛾发生及防治．林业科学研究，1993. 6 (6)：666～671

赵锦年，陈晓，陆增发．球果害虫对湿地松球果发育的影响．林业科学研究，1988. 1 (4) ：453～455

赵君以．杨树种子隔年贮藏试验．林业科学，1958 (1)：37～43

赵书喜．杂交马褂木嫁接试验．林业科技通讯，1991 (8)：22～24

赵自富．云南油桐品种及分布．见：中国油桐科技论文集．北京：中国林业出版社，1988

郑振鸿．铅笔柏种子休眠及快速催芽条件的研究．种子，1994 (4)：15～18

中国科学院四川分院林研所园林绿化室．杉木开花结实及产量预测的研究．林业科学，1960 (1)：2～13

中国科学院林业土壤研究所．杉木人工林结实规律．载《林业科技资料》第二辑，1973

钟淑英，李基平．云南松、思茅松的种子寿命．云南林业科技，1986 (2)：23

周坚，樊汝汶．鹅掌楸属两种植物花粉品质和花粉管生长的研究．林业科学，1994. 30 (5)：405～411

周武忠．常绿油麻藤．植物，1986 (5)：27

周学泉．落叶松种子的鉴别．种子，1989 (1)：59～62

周佑勋．光皮桦种子休眠和萌发特性的研究．林业科技通讯，1984 (4)：5～8

周佑勋．火炬松种子的休眠与萌发特性．植物生理学通讯，1987. (5)：22

周佑勋．白花泡桐种子萌发特性的研究．林业科技通讯，1988 (1)：22～24

周佑勋．喜树种子休眠和萌发特性的研究．林业科技通讯，1989 (8)：22～25

周佑勋，段小平．华南五针松种子休眠生理的研究．中南林学院学报，1993. 13 (2) ：123～126

周佑勋，胡春姿．阔瓣含笑种子休眠生理的初步研究．林业科技通讯，1990

周政贤．贵州遵义杜仲生物学特性及遵义杜仲林场经营问题．林业科学，1958 (2) ：129～147

周兆祥，陈经梧，傅深渊．山茱萸果实的化学成分．浙江林学院学报，1988. 5 (1)：63～70

庄茂长．秤锤树．见：中国花经．上海文化出版社，1990. 476

朱长进．脱落酸与沙棘果实脱落关系研究．林业科学研究，1991 (4)：333～336

朱长进，刘庆香，赵丽华等．生长调节剂与板栗生长、成花及结果的研究．林业科学研究，1992. 5 (3)：311～316

◎苏铁
Cycas revoluta Thunb.
大孢子叶聚成松散"球状"，
露出将成熟的种子(向其柏摄)

◎鹅掌楸
Liriodendron chinense (Hemsl.) Sarg. 聚合果
(高捍东、马有基摄)

◎南方红豆杉
Taxus chinensis (Pilg.) Rehd.var. *mairei* (Lemee et Lévl.) Cheng et L.K.Fu
带肉质假种皮种子(高捍东、马有基摄)

◎柳杉
Cryptomeria fortunei Hooibrenk.
当年未熟球果和去年种子已经散落的球果
(高捍东、马有基摄)

◎圆柏
Sabina chinensis (L.) Ant. 雌球花、雄球花和球果
(高捍东、马有基摄)

◎观光木
Tsoongiodendron odorum Chun.
聚合果（邹惠渝摄）

◎醉香含笑
Michelia macclurei Dandy. 花、果（邹惠渝摄）

◎木瓜
Chaenomeles sinensis (Thouin) Koehne. 梨果
（高捍东、马有基摄）

◎闽楠
Phoebe bournei. (Hemsl.)Yang 浆果状核果
（黄鹏成摄）

◎西南栒子
Cotoneaster franchetii Boiss. 梨果（高捍东、马有基摄）

◎桢楠
Phoebe zhennan S.Lee et F.N.Wei 浆果状核果
（高捍东、马有基摄）

◎石楠
Photinia serrulata Lindl. 梨果
（高捍东、马有基摄）

◎无忧花
Saraca asoca (Roxb.) De Wilde. 伞房花序（陈幼生摄）

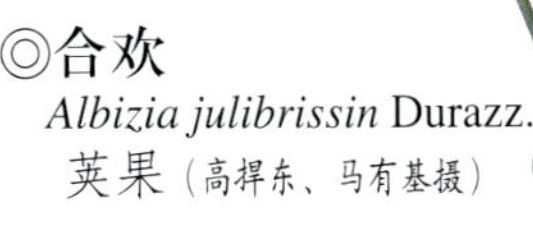

◎槐树
Sophora japonica L. 荚果（高捍东、马有基摄）

◎合欢
Albizia julibrissin Durazz.
荚果（高捍东、马有基摄）

◎黄檀
Dalbergia hupeana Hance 荚果
（高捍东、马有基摄）

◎毛梾

Cornus walteri Wanger. 核果（高捍东、马有基摄）

◎喜树

Camptotheca acuminata Decne 坚果成头状果穗（高捍东、马有基摄）

◎秤锤树

Sinojackia xylocarpa Hu 木质果（高捍东、马有基摄）

◎白簕

Acanthopanax trifoliatus (L.) Merr.

伞形花序组成复伞形或圆锥状果穗（高捍东、马有基摄）

◎金银忍冬

Lonicera maackii (Rupr.) Maxim. 浆果（高捍东、马有基摄）

◎榔榆
Ulmus parvifolia Jacq. 翅果（高捍东、马有基摄）

◎天目琼花
Viburnum sargentii Koehne var. *calvescens* Rehd. 核果
（高捍东、马有基摄）

◎薄壳山核桃
Carya illinoensis K.Koch 核果
（高捍东、马有基摄）

◎二球悬铃木
Platanus hispanica Muench.
坚果组成头状果穗
（高捍东、马有基摄）

◎青檀
Pteroceltis tatarinowii Maxim. 具翅坚果

◎海桐
Pittosporum tobira (Thunb.) Ait. 蒴果
（高捍东、马有基摄）

◎梧桐
Firmiana simplex (L.)W.F.Wight 蓇葖果（高捍东、马有基摄）

◎乌桕
Sapium sebiferum (L.) Roxb. 蒴果（高捍东、马有基摄）

◎南京椴
Tilia miqueliana Maxim . 坚果（高捍东、马有基摄）

◎油桐
Vernicia fordii (Hemsl.) Airy-Shaw 核果（高捍东、马有基摄）

◎铁冬青
Ilex rotunda Thunb. 核果（高捍东、马有基摄）

◎柿
Diospyros kaki L.f.
浆果（高捍东、马有基摄）

◎枳
Poncirus trifoliata (L.) Raf. 柑果
（高捍东、马有基摄）

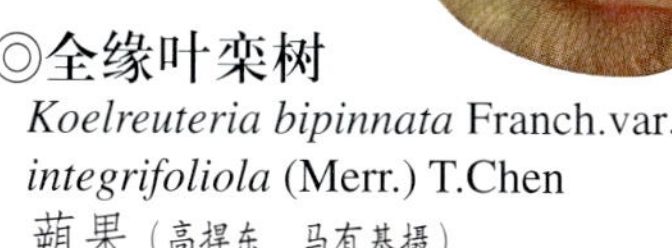

◎全缘叶栾树
Koelreuteria bipinnata Franch.var. *integrifoliola* (Merr.) T.Chen
蒴果（高捍东、马有基摄）

◎黄连木
Pistacia chinensis Bunge. 核果，感染虫害的呈红色
（高捍东、马有基摄）

◎苦楝
Melia azedarach L. 核果（高捍东、马有基摄）

◎羽叶槭

Acer negundo L. 小坚果具翅或成双翅果（高捍东、马有基摄）

◎紫薇

Lagerstroemia indica L. 蒴果（高捍东、马有基摄）

◎浙江七叶树

Aesculus chinensis Bunge var. *chekiangensis* (Hu et Fang) Fang 蒴果（高捍东、马有基摄）

◎光蜡树

Fraxinus griffithii C.B.Clarke 翅果（高捍东、马有基摄）

◎棕榈

Trachycarpus fortunei (Hook.f.)H.Wendl. 核果（高捍东、马有基摄）